Methods for General and Molecular Bacteriology

Methods for General and Molecular Bacteriology

Philipp Gerhardt, *Editor-in-Chief*
Editor, II. Growth; Editor, III. Molecular Genetics; Editor, VI. General Methods
Department of Microbiology,
Michigan State University, East Lansing, Michigan 48824

R.G.E. Murray
Editor, I. Morphology
Department of Microbiology and Immunology,
University of Western Ontario, London, Ontario, Canada N6A 5C1

Willis A. Wood
Editor, IV. Metabolism
The Agouron Institute, 505 South Coast Boulevard, La Jolla, California 92037

Noel R. Krieg
Editor, V. Systematics
Department of Biology,
Virginia Polytechnic Institute and State University, Blacksburg, Virginia 24061

American Society for Microbiology
Washington, D.C.

Cover:

Escherichia coli and its DNA. (Micrograph © K.G. Murti, courtesy Visuals Unlimited.)
Gram-stained *Desulfovibrio gigas*. (Micrograph © F. Widdel, courtesy Visuals Unlimited.)
Proteus mirabilis on deoxycholate agar. (Photograph by Noel R. Krieg.)
Autoradiogram of a sequencing gel. (See p. 442, this volume.)

1325 Massachusetts Ave., N.W.
Washington, DC 20005

Library of Congress Cataloging-in-Publication Data

Methods for general and molecular bacteriology/
Philipp Gerhardt, editor-in-chief; R.G.E. Murray,
Willis A. Wood, Noel R. Krieg, editor.
p. cm.
Rev. ed. of: Manual of methods for general
bacteriology.
Includes index
ISBN 1-55581-048-9
1. Bacteriology—Laboratory manuals.
I. Gerhardt, Philipp. 1921– . II. Manual of
methods for general bacteriology.
QR65.M26 1994
589.9'0078—dc20 93-38049
CIP

Printed in the United States of America

CONTENTS

AUTHORS

Luis A. Actis (Chapter 16. Plasmids)
Department of Microbiology and Immunology, L220, Oregon Health Sciences University, 3181 S.W. Sam Jackson Park Road, Portland, Oregon 97201; (503) 494-7599

Ronald M. Atlas (Chapter 19. Polymerase Chain Reaction)
Department of Biology, University of Louisville, Louisville, Kentucky 40292; (502) 588-6721

Michael Bagdasarian (Chapter 18. Gene Cloning and Expression)
Department of Microbiology, S110 Plant Biology Laboratory, Michigan State University, East Lansing, Michigan 48824; (517) 353-8619

Mira M. Bagdasarian (Chapter 18. Gene Cloning and Expression)
Department of Microbiology, S110 Plant Biology Laboratory, Michigan State University, East Lansing, Michigan 48824; (517) 353-8619

W. Emmett Barkley (Chapter 29. Laboratory Safety)
Office of Laboratory Safety, Howard Hughes Medical Institute, 4000 Jones Bridge Road, Chevy Chase, Maryland 20815-6789; (301) 215-8825

Asim K. Bej (Chapter 19. Polymerase Chain Reaction)
Department of Biology, University of Alabama at Birmingham, Birmingham, Alabama 35294-1170; (205) 934-8308

Terry J. Beveridge (Chapter 3. Electron Microscopy)
Department of Microbiology, College of Biological Science, University of Guelph, Guelph, Ontario, Canada N1G 2W1; (519) 824-4120, ext. 3366

John A. Breznak (Chapter 6. Physicochemical Factors in Growth)
Department of Microbiology, Michigan State University, East Lansing, Michigan 48824-1101; (517) 355-6536

Barbara J. Brown (Chapter 13. Gene Mutation)
Ecogen, Inc., Langhorne, Pennsylvania 19047-1852; (215) 757-1590
(Formerly at Department of Genetics, University of Georgia, Athens, Georgia)

Bruce C. Carlton (Chapter 13. Gene Mutation)
Ecogen, Inc., Langhorne, Pennsylvania 19047-1852; (215) 757-1590
(Formerly at Department of Genetics, University of Georgia, Athens, Georgia)

Roger M. Cole (Chapter 3. Electron Microscopy)
6200 Maiden Lane, Bethesda, Maryland 20817-6602; (301) 320-3734
(Formerly at National Institute of Allergy and Infectious Diseases, Bethesda, Maryland)

Ralph N. Costilow (Chapter 6. Physicochemical Factors in Growth)
14406 Cameo Way, Sun City, Arizona 85351; (602) 972-5050
(Formerly at Department of Microbiology, Michigan State University, East Lansing, Michigan)

Rosalie J. Cote (Chapter 7. Nutrition and Media)
American Type Culture Collection, 12301 Parklawn Drive, Rockville, Maryland 20852; (301) 231-5542, ext. 544

Jorge H. Crosa (Chapter 16. Plasmids)
Department of Microbiology and Immunology, L220, Oregon Health Sciences University, 3181 S.W. Sam Jackson Park Road, Portland, Oregon 97201; (503) 494-7583

Roy Curtiss III (Chapter 14. Gene Transfer in Gram-Negative Bacteria)
Department of Biology, Box 1137, Washington University, St. Louis, Missouri 63130; (314) 935-6819

Simon M. Cutting (Chapter 15. Gene Transfer in Gram-Positive Bacteria)
Department of Microbiology, University of Pennsylvania School of Medicine, Philadelphia, Pennsylvania 19104; (215) 573-3513

Lacy Daniels (Chapter 22. Chemical Analysis)
Department of Microbiology, University of Iowa, Iowa City, Iowa 52242; (319) 335-7780

Frans J. de Bruijn (Chapter 17. Transposon Mutagenesis)
Center for Microbial Ecology and Plant Research Laboratory, A306 Plant Biology Building, Michigan State University, East Lansing, Michigan 48824; (517) 353-2229

Raymond N. Doetsch (Chapter 2. Determinative and Cytological Light Microscopy)
Department of Microbiology, University of Maryland, College Park, Maryland 20742; (301) 454-2848

Stephen W. Drew (Chapter 10. Liquid Culture)
Merck & Co., Inc., WS3DE-25, P.O. Box 100, Whitehouse Station, New Jersey 08889-0100; (908) 423-3112

Eric Eisenstadt (Chapter 13. Gene Mutation)
Office of Naval Research, Biological Sciences Division, 800 N. Quincy Street, Arlington, Virginia 22217-5000; (703) 696-4596

Joseph O. Falkinham III (Chapter 28. Nucleic Acid Probes)
Department of Biology, Virginia Polytechnic Institute and State University, Blacksburg, Virginia 22217-5000; (703) 696-4596

Stanley Falkow (Chapter 16. Plasmids)
Department of Microbiology and Immunology, Stanford University, Stanford, California 94305; (415) 723-9187

Philipp Gerhardt (Methodology for General and Molecular Bacteriology; Introduction to Growth; Chapter 9. Solid, Liquid/Solid, and Semisolid Culture; Chapter 10. Liquid Culture; Introduction to Molecular Genetics; Introduction to General Methods; Chapter 31. Records and Reports)
Department of Microbiology, Michigan State University, East Lansing, Michigan 48824-1101; (517) 355-7530

Robert L. Gherna (Chapter 7. Nutrition and Media; Chapter 12. Culture Preservation)
American Type Culture Collection, 12301 Parklawn Drive, Rockville, Maryland 20852; (301) 231-5542

Richard S. Hanson (Chapter 22. Chemical Analysis)
Gray Freshwater Biological Research Institute, University of Minnesota, P.O. Box 100, Navarre, Minnesota 55392-0100; (612) 471-8476

William G. Hendrickson (Chapter 20. Nucleic Acid Analysis)
Department of Microbiology, MC 790, University of Illinois at Chicago, P.O. Box 6998, Chicago, Illinois 60680; (213) 996-5600

John G. Holt (Chapter 8. Enrichment and Isolation)
Department of Microbiology, Michigan State University, East Lansing, Michigan 48824-1101; (517) 336-2459

John L. Johnson (Chapter 26. Similarity Analysis of DNAs; Chapter 27. Similarity Analysis of rRNAs)
Department of Anaerobic Microbiology, Virginia Polytechnic Institute and State University, Blacksburg, Virginia 24061-0305; (703) 231-7127

Arthur L. Koch (Chapter 11. Growth Measurement)
Department of Biology, Indiana University, Bloomington, Indiana 47401; (812) 855-5036; FAX (812) 855-6705

Susan F. Koval (Chapter 4. Cell Fractionation)
Department of Microbiology and Immunology, Health Sciences Centre, University of Western Ontario, London, Ontario, Canada N6A 5C1; (519) 661-3439

Noel R. Krieg (Chapter 8. Enrichment and Isolation; Chapter 9. Solid, Liquid/Solid, and Semisolid Culture; Introduction to Systematics; Chapter 25. Phenotypic Characterization)
Department of Biology, Virginia Polytechnic Institute and State University, Blacksburg, Virginia 24061-0406; (703) 231-5912

Joseph S. Lam (Chapter 5. Antigen-Antibody Reactions)
Department of Microbiology, University of Guelph, Guelph, Ontario, Canada N1G 2W1; (519) 824-4120

Robert E. Marquis (Chapter 24. Permeability and Transport)
Department of Microbiology and Immunology, School of Medicine and Dentistry, University of Rochester, Rochester, New York 14642; (716) 275-1674

Tapan K. Misra (Chapter 20. Nucleic Acid Analysis)
Department of Microbiology, MC 790, University of Illinois at Chicago, P.O. Box 6998, Chicago, Illinois 60680; (312) 996-9609

R. G. E. Murray (Introduction to Morphology; Chapter 1. Light Microscopy; Chapter 2. Determinative and Cytological Light Microscopy)
Department of Microbiology and Immunology, Health Sciences Centre, University of Western Ontario, London, Ontario, Canada N6A 5C1; (519) 661-3455

Lucy M. Mutharia (Chapter 5. Antigen-Antibody Reactions)
Department of Microbiology, University of Guelph, Guelph, Ontario, Canada N1G 2W1; (519) 824-4120

J. R. Paterek (Chapter 21. Physical Analysis)
Institute of Gas Technology, 3424 South State Street, Chicago, Illinois 60616-3896; (312) 567-3896
(Formerly at Salk Institute Biotechnology/Industrial Associates, San Diego, California)

Allen T. Phillips (Chapter 23. Enzymatic Activity)
Department of Molecular and Cell Biology, Paul M. Althouse Laboratory, Pennsylvania State University, University Park, Pennsylvania 16802; (814) 865-1247

Jane A. Phillips (Chapter 22. Chemical Analysis)
College of Biological Sciences, University of Minnesota, St. Paul, Minnesota 55108; (612) 624-2789

Terry J. Popkin (Chapter 3. Electron Microscopy; Chapter 30. Photography)
1809 Snowdrop Lane, Silver Spring, Maryland 20906-3361; (301) 949-2464
(Formerly at National Institute of Allergy and Infectious Diseases, Bethesda, Maryland)

David L. Provence (Chapter 14. Gene Transfer in Gram-Negative Bacteria)
Department of Biology, Box 1137, Washington University, St. Louis, Missouri 63130; (314) 935-4662

John H. Richardson (Chapter 29. Laboratory Safety)
2982 Darbyshire Court, Atlanta, Georgia 30345; (404) 938-5137
(Formerly at Environmental Safety and Health Office, Emory University School of Medicine, Atlanta, Georgia)

Carl F. Robinow (Chapter 1. Light Microscopy; Chapter 2. Determinative and Cytological Light Microscopy)
Department of Microbiology and Immunology, Health Sciences Centre, University of Western Ontario, London, Ontario, Canada N6A 5C1; (519) 679-3751

Silvia Rossbach (Chapter 17. Transposon Mutagenesis)
Center for Microbial Ecology and Plant Research Laboratory, A306 Plant Biology Building, Michigan State University, East Lansing, Michigan 48824; (517) 353-2009

Carl A. Schnaitman (Chapter 4. Cell Fractionation)
5401 E. Rockridge Road, Phoenix, Arizona 85018-1921; (602) 840-1356
(Formerly at Department of Microbiology, Arizona State University, Tempe, Arizona)

Robert M. Smibert (Chapter 25. Phenotypic Characterization)
1201 Glen Cove Lane, Blacksburg, Virginia 24060; (703) 552-2295
(Formerly at Department of Anaerobic Microbiology, Virginia Polytechnic Institute and State University, Blacksburg, Virginia 24061-0305)

G. Dennis Sprott (Chapter 4. Cell Fractionation)
Institute of Biological Sciences, National Research Council of Canada, 100 Sussex Drive, Ottawa, Ontario, Canada K1A 0R6; (613) 993-5291

Marcelo E. Tolmasky (Chapter 16. Plasmids)
Department of Microbiology and Immunology, L220, Oregon Health Sciences University, 3181 S.W. Sam Jackson Park Road, Portland, Oregon 97201; (503) 494-7599

W. A. Wood (Introduction to Metabolism; Chapter 21. Physical Analysis)
The Agouron Institute, 505 South Coast Boulevard, La Jolla, California 92037; (619) 456-5300

Philip Youngman (Chapter 15. Gene Transfer in Gram-Positive Bacteria)
Department of Genetics, The University of Georgia, Athens, Georgia 30602-7223; (706) 542-1417

REVIEWERS

Richard L. Anderson
Michael Bagdasarian
Albert Balows
Brian H. Belliveau
Robert A. Bender
Claire M. Berg
Douglas E. Berg
David R. Boone
Zachary Burton
David M. Carlberg
Lester E. Casida, Jr.
A. M. Chakrabarty
Thomas R. Corner
Ronald L. Crawford
Cecil S. Cummins
Pierre-Marc Daggett
Frans J. de Bruijn
Alma Dietz
Gary M. Dunny
Eric Eisenstadt
George L. Evans
Joseph O. Falkinham III
Beverly M. Guirard
Gary L. Gustafson
Franklin M. Harold
George D. Hegeman
Donald R. Helinski
William G. Hendrickson
Stanley C. Holt
Harry T. Horner
Karin Ippen-Ihler
Mahendra K. Jain
Vivian Jonas-Taggart
Clarence I. Kado
Jordan Konisky
Donn J. Kishner
Noel R. Krieg
Lee R. Kroos
Terry A. Krulwich
Donald J. Le Blanc
Richard E. Lenski
James R. Lupski
Frederick J. Marsik
J. Justin McCormick
Michael J. McInerney
Robert W. McKinney
Edward A. Meighen
Leonard E. Mortenson
R. G. E. Murray
Kent R. Myers
David P. Nagle, Jr.
Walter G. Niehaus
John R. Norris
Patrick J. Oriel
S. John Pirt
David A. Power
Ian Poxton
David L. Provence
Louis B. Quesnel
William S. Reznikoff
Charles E. Richardson
Ivan L. Roth
William F. Seip
Joseph Shiloach
Simon D. Silver
Peter H. A. Sneath
Esmond E. Snell
John R. Sokatch
James T. Staley
Robert W. Stieber
Michele van de Walle
Donald Vesley
Stefan Wagener
George M. Weinstock
Friedrich W. Widdel
Philip J. Youngman

PREFACE

The objective of *Methods for General and Molecular Bacteriology* remains the same as for its predecessor, *Manual of Methods for General Bacteriology*: "to meet the need for a compact, moderately priced handbook of reliable, basic methods for practicing general bacteriology in the laboratory." However, the field of general bacteriology has widened in the decade since the first edition was published and now encompasses the methods of molecular biology, hence the broadening in the title and contents of this new edition. The book is intended to serve as a first resource for undertaking a laboratory method and then for seeking further information or alternative methods in general and molecular bacteriology.

The scope of *Methods for General and Molecular Bacteriology* covers all kinds of bacteria, archaeobacteria as well as eubacteria, without particular emphasis on any one kind and indeed with emphasis against relying heavily on *Escherichia coli* as a prototype. All of the bacteria described in *Bergey's Manual of Systematic Bacteriology* and *The Prokaryotes*, plus new bacteria yet to be discovered, are sought to be covered by the methods in this book, which is intended as a laboratory manual of methods to complement these systematics treatises and the various textbooks of general microbiology. Microbes other than bacteria (viruses, fungi, algae, protozoa, and mammalian cells) are not covered. The approach is generalized for both pure and applied bacteriology, omitting the methods that are specialized for a particular application such as in medicine or industry, but including the basic methods that are widely applicable.

The audience for this new edition again embraces serious students beyond the elementary level, as well as researchers and teachers in microbiology and allied disciplines. Although published in the United States by the American Society for Microbiology, the first edition became widely sought and used throughout the world, as exemplified by an authorized Russian translation and its publication in the former USSR and by several unauthorized reprintings in other countries. The audience for the present edition is foreseen to be even wider.

As stated in the Preface to the earlier *Manual*, "this is a 'how to' manual. Principles are introduced sufficient to understand why and how the method is conducted. Procedures are described step-by-step with enough detail so that the user can carry out the method without further reference to another source. Common problems, pitfalls, and precautions are pointed out. Examples, illustrations, tabulations, and schematics are used wherever possible. Selections are made of the methods used most frequently and reliably; for less important methods (and for sophisticated, high-technology methods), references are given to guide the user to original literature (or commercial sources) for complete descriptions. As much as possible, a method is generalized so that it might be applied for use with an unknown, newly isolated bacterium. However, a method often is exemplified with a specific bacterium familiar to the author or widely used. The authors frequently suggest commercial sources for equipment and materials, but these suggestions are not meant either to endorse or to exclude a particular product, and any omissions are unintentional." Information on further commercial sources is obtainable from annual buyers' guide issues of journals such as *American Laboratory*, *Chemical Engineering*, and *Science*, and from *Biotech Buyers' Guide*, published annually by the American Chemical Society.

As for the earlier *Manual*, the present book "is organized into sections corresponding to main subjects of general bacteriology and then into chapters corresponding to relevant areas of methodology. Each chapter is independent and self-contained, with a table of contents to help the reader see the organization in outline and find a specific method. A decimal numbering system is used to facilitate identification, cross-referencing, and indexing (special effort was taken to make the cross-referencing and indexing comprehensive). Headings are topical and brief so as to enable easy identification of a method. Literature citations are listed at the end of each chapter: first, general references with full citations and annotations; next, specific articles."

The present edition was prompted not only by the time interval since publication of the first version in 1981 but also by the emergence of new methods, particularly those in molecular biology which were limited to a few chapters in the first version because of the then rapidly changing methodology. These important methods are expanded in the present edition, notably as eight chapters in the section on Molecular Genetics and three chapters in the section on Systematics. Other notable additions are the chapters on Antigen-Antibody Reactions, Photography, and Records and Reports. Throughout the book, each chapter was either revised or completely rewritten. New authors provided new perspectives in many chapters, while the contributions of the original authors are recognized with coauthorship if substantial portions of their writing are retained. The external refereeing and the internal editing processes were again multiple and rigorous, essentially the same as for an ASM journal article. Suggestions for improvements from users of the first edition were incorporated into this one, and further constructive criticism is welcomed.

Philipp Gerhardt
Editor-in-Chief

ACKNOWLEDGMENTS

Sincere and appreciative acknowledgment is made to the following people and institutions who have contributed to producing this book:

- The authors and editors, who served without remuneration or royalty, for their splendid cooperation and knowledgeability in writing and editing, and their institutions, which subsidized the considerable time, staff support, and often direct costs for manuscript preparation
- The reviewers, for diligent and constructive criticism which greatly contributed to the quality of the book
- The staff of the ASM Books Division, particularly Patrick J. Fitzgerald, Director, and Eleanor S. Tupper, Senior Production Editor, and to Yvonne Strong for meticulous copy editing and Barbara Littlewood for comprehensive indexing
- The ASM Publications Board, particularly Barbara H. Iglewski, Chair, and the late Helen R. Whiteley, for policy direction
- The Biological Sciences Program of the U.S. Army Research Office, which provided partial financial support for the efforts of the Editor-in-Chief through Grant DAAL03-90-G-0146
- Vera M. Gerhardt, for support in every way and sustenance at every turn for the efforts of the Editor-in-Chief

All of us are privileged to share in this contribution to the advancement of bacteriology.

Philipp Gerhardt
Editor-in-Chief

Methodology for General and Molecular Bacteriology

PHILIPP GERHARDT

Bacteriology became a science only after unique methods were developed, and they are responsible for the continuing influence of bacteriology on subsequently developed fields such as virology, immunology, molecular biology, and biotechnology. Leeuwenhoek's microscope opened an entirely new perspective of life. Koch's introduction of the pure-culture technique and Pasteur's uses of immunological response and chemical analysis remain as influential now as then. Enzymatic analysis, electron microscopy, and the more recent molecular biology techniques have also powerfully influenced modern research. "The solution to a major problem in science often begins with the introduction of a new method . . . which then opens a pathway of derivative ideas and suggests experiments that lead to the problem's ultimate solution" (15).

The ways in which techniques and technology (methods and methodology) have influenced science generally and in historical perspective are the subjects of a pair of trenchant essays by Hall (7) and Crease (4) which are worth quoting here. The developments of gene cloning, DNA sequencing, and scanning tunneling microscopy are among those cited to examplify the enormous transforming power of techniques. "In the absence of an essential technique, a researcher or a field flounders, developing elegant theories that cannot be decisively accepted or rejected—no matter how many intriguing circumstantial observations are available. But with a key technique in hand, the individual and field move ahead at almost terrifying speed, finding the right conditions to test one hypothesis after another. Conversely, new techniques often uncover new phenomena that demand new theories to explain them. Almost all great techniques were invented by people in a hurry to get someplace else scientifically, and a researcher's ability to answer important scientific questions successfully often depends on how closely theory is tethered to technique" (7). "Techniques play a key role in everything from Nobel-class discoveries to routine measurements carried out by technicians. Although technique is a word that has meaning in a wide variety of contexts, in science it generally refers to a practice that can be repeated to produce measurements or prepare objects for measurement or manipulation. A technique, one might say, is a 'knowledge-producing tool'" (4).

The methodology for general and molecular bacteriology is defined in this book with a sectioning of content reflecting that in several exemplary generalized textbooks in the field. Thus, the content includes sections on morphology, growth, molecular genetics, metabolism, systematics, and general methods. These sections then are divided into chapters, and they are further divided into main groups of methods. Often the divisions are arbitrary, so some overlap occurs. This is managed with cross-referencing in the text and indexing at the end, facilitated by a decimal number identification system for each method.

The main general resource for the reader seeking further review information on methods is the multivolume work started in 1969, *Methods in Microbiology* (11), and a new series started in 1988, *Modern Microbiological Methods* (6). Original papers and short reviews on methods are contained in periodicals such as the *Journal of Microbiological Methods* (Elsevier, Amsterdam), *BioTechniques* (Eaton Publishing Co., Natick, Mass.), and *Applied and Environmental Microbiology* (American Society for Microbiology [ASM], Washington, D.C.). The Meynells' *Theory and Practise in Experimental Bacteriology* (10) provides an excellent book resource for many older procedures, and ASM's *Manual of Microbiological Methods* (13) also continues to be useful. Videotape programs provide useful teaching aids for certain basic techniques that need demonstration (1). In addition to these general reference resources, there are a number of methodological books and journals that deal with a specific area of bacteriology, and these are referenced in the appropriate chapter.

Methods for applied fields are obtainable in a number of references, including ASM's companion books *Manual of Clinical Microbiology* (2), *Manual of Clinical Laboratory Immunology* (12), and *Manual of Industrial Microbiology and Biotechnology* (5). The American Public Health Association, Washington, D.C., has a long history of publishing manuals for its relevant fields, notably *Standard Methods for the Examination of Water and Wastewater* (3) and *Standard Methods for the Examination of Dairy Products* (9), but also other manuals for the examination of bacterial infections, foods, etc. Methods for specific genera and species are included in the volumes of *Bergey's Manual of Systematic Bacteriology* (8) and of *The Prokaryotes* (2a, 14).

REFERENCES

1. **American Society for Microbiology, Board of Education and Training.** 1990. Videotape programs in microbiology. American Society for Microbiology, Washington, D.C.
This informative collection of videotapes depicts selected laboratory procedures. Also available is an eight-part laboratory manual, Basic Microbiology Techniques. The pro-

cedures include microscopy, aseptic technique, pure culture techniques, staining techniques, and pipetting.

2. **Balows, A., W. J. Hausler, Jr., K. L. Herrmann, H. D. Isenberg, and H. J. Shadomy (ed.).** 1991. *Manual of Clinical Microbiology,* 5th ed. American Society for Microbiology, Washington, D.C.

2a. **Balows, A., H. G. Truper, M. Dworkin, W. Harder, and K.-H. Schleifer (ed.).** 1992. *The Prokaryotes. A Handbook on the Biology of Bacteria: Ecophysiology, Isolation, Identification, Applications,* 2nd ed. Springer-Verlag, New York.
The methodology in the initial edition of this comprehensive treatise (see annotation for reference 14) is continued in this second edition, which is greatly expanded in size and content (four volumes, 237 chapters, 4,126 pages). It differs fundamentally in that the systematics are organized on the basis of phylogenetic relationships, which are drawn mainly from 16S rRNA analyses. Altogether this treatise on organismic bacteriology is even more valuable than its predecessor.

3. **Clesceri, L. S., A. E. Greenberg, and R. R. Trussell.** 1989. *Standard Methods for the Examination of Water and Wastewater,* 17th ed. American Public Health Association, Washington, D.C.

4. **Crease, R. P.** 1992. The trajectory of techniques: lessons from the past. *Science* **257**:350–353.

5. **Demain, A. L., and N. A. Solomon (ed.).** 1986. *Manual of Industrial Microbiology and Biotechnology.* American Society for Microbiology, Washington, D.C.

6. **Goodfellow (ed.).** 1988–1990. *Modern Microbiological Methods.* John Wiley & Sons, New York.
This series of books is directed primarily toward active research workers but can also introduce novice readers to the methods within a specialty. Specific step-by-step protocols are provided after each chapter. To date the following books are published: Methods in Aquatic Bacteriology, by B. Austin (1988); Bacterial Cell Surface Techniques, by I. Hancock and I. Poxton (1988); and Molecular Biological Methods for Bacillus, by C. R. Harwood and S. M. Cutting (1990).

7. **Hall, S. S.** 1992. How technique is changing science. *Science* **257**:344–349.

8. **Holt, J. G. (ed.).** 1984–1989. *Bergey's Manual of Systematic Bacteriology.* The Williams & Wilkins Co., Baltimore.
The four volumes of this treatise cover the gram-negative bacteria of general, medical, or industrial importance (vol. 1), the gram-positive bacteria other than actinomycetes (vol. 2), the archaeobacteria, cyanobacteria, and remaining gram-negative bacteria (vol. 3), and the actinomycetes (vol. 4). Preceded by eight editions of Bergey's Manual of Determinative Bacteriology, this version was substantially expanded in scope to include, importantly here, methods for enrichment and isolation, culture maintenance, and testing of special characteristics.

9. **Marshall, R. T. (ed.).** 1993. *Standard Methods for the Examination of Dairy Products,* 16th ed. American Public Health Association, Washington, D.C.

10. **Meynell, G. G., and E. Meynell.** 1970. *Theory and Practise in Experimental Bacteriology,* 2nd ed. Cambridge University Press, Cambridge.
An excellent, compact (347 pages) resource book containing the rationale and procedure for many methods in general bacteriology. There are sections on growth; culture media; oxygen, carbon dioxide, and anaerobiosis; sterilization; light microscopy; quantitation; and genetic technique.

11. **Norris, J. R., and D. W. Ribbons (ed.).** 1969–1990. *Methods in Microbiology,* vol. 1–22. Academic Press, Inc., New York.
This is perhaps the primary resource for further information or alternative methods in general and molecular bacteriology. The style of presentation varies with the many authors and articles, usually that of a narrative review with extensive references to original papers, but often without stepwise directions like those in this book. Volumes 1 to 3, 5 to 9, and 18 cover various methods for general bacteriology. Volumes 17 and 21 dwell on plasmids; volume 19 addresses systematics; volume 20 addresses electron microscopy; and volume 22 addresses microbial ecology. Volume 4 on mycology and volumes 10 to 16 on typing methods for pathogenic bacteria are less relevant.

12. **Rose, N. R., E. Conway de Macario, J. L. Fahey, H. Friedman, and G. M. Penn (ed.).** 1992. *Manual of Clinical Laboratory Immunology,* 4th ed. American Society for Microbiology, Washington, D.C.

13. **Society of American Bacteriologists, Committee on Bacteriological Technique.** 1957. *Manual of Microbiological Methods.* McGraw-Hill Book Co., Inc., New York.
This 315-page book continues to be useful, particularly because of its chapters on staining methods, preparation of media, and routine tests for the identification of bacteria.

14. **Starr, M. P., H. Stolp, H. G. Truper, A. Balows, and H. G. Schlegel (ed.).** 1981. *The Prokaryotes. A Handbook on Habitats, Isolation, and Identification of Bacteria.* Springer-Verlag, New York.
This two-volume, 2,284-page "handbook" is a major authoritative resource for 22 groups and hundreds of genera of eubacteria and archaeobacteria. "The presentation of methods was considered to be of primary importance, especially those methods that have been found by experts to be reliable and satisfactory. Therefore, a part of each chapter dealing with a specific organism or a group of organisms has been designed as a 'cookbook,' offering detailed recipes for enrichment, isolation, cultivation and conservation. Stolp and Starr have a 40-page introductory essay on principles of isolation, cultivation and conservation." Altogether, this is an extraordinarily valuable treatise of organismic bacteriology.

15. **Stossel, T. P.** 1985. An essay on biomedical communications. *N. Engl. J. Med.* **313**:123–126.

Section I

MORPHOLOGY

Introduction to Morphology

R. G. E. MURRAY

Microscopy provides the primary approach to the study of bacteria and is essential to the recognition of their associations in nature. Bacteriology has had to take full advantage of simple light microscopy as well as sophisticated approaches to providing both the highest possible resolution and effective contrast. These ubiquitous procaryotic organisms are, indeed, small cells. The basic descriptions needed to classify the diverse forms start with shape, growth habit, microscopic observations of staining reactions, and motility. Now that the components of bacterial cells can be resolved to the level of macromolecules and subcellular structures, the methods applying light and electron microscopy provide essential additions to descriptions as well as the means for monitoring fractionation procedures. We would be hard put to know what we are studying in our fractions without information from microscopy and the specificity provided by antigen-antibody reactions to add to the results of biochemical analysis.

The light microscope is available to and needed by all who study microorganisms. Because effective use is so desirable but so seldom attained, some detailed advice is offered in Chapter 1. Modern microscopes conceal in elaborate housings all the same features as those that are out in the open and adjustable on the more old-fashioned instruments. The descriptions of the operations required to attain good images are therefore based on the simplest of microscopes, which can give excellent results. The procedures known to work are described as clearly as possible, and attention is drawn to pitfalls so that the user may learn to avoid them. The main concern is to make all types of microscopy more useful, effective, and rewarding. The need for illustration is minimal since the basic characteristics and design of microscopes are depicted in detail in most texts and in the references given with Chapter 1.

The methods described in Chapter 2 are concerned with approaches to the study of the properties and behavior of bacteria in life, the determination of morphological and staining characteristics to assist description and identification, and the use of special preparations to attain the best resolution for cytological studies by light microscopy. It is recognized that bacteria can be few and far between in many environments, so that methods of sampling and attaining a reasonably prompt assessment of the nature and diversity of the biota in a sample can be crucial. Therefore, attention is given at the outset to sampling methods.

The electron microscope is a more specialized instrument and is usually operated and guarded by professional electronmicroscopists who instruct, assist, and advise those who need the detail and the resolution available in electron micrographs. As is true for light microscopy, the preparation of the specimen for electron microscopy is as important to the results obtained as are the microscope and the operator. Therefore, much attention must be paid to the selection of the preparative method to be used for a specific task. Chapter 3 describes electron microscopy methods and contains carefully chosen references that exemplify appropriate application and interpretation of the images obtained by these techniques.

Most of the components of bacterial cells can be isolated for structural, biochemical, and genetic characterization, with great advantage to the study of structure and function and in parallel to molecular biology studies. Cell fractionation methods involve breaking open cells and then separating fractions in a manner appropriate to the special requirements of the cell components and the peculiarities of many kinds of bacteria. These methods are described in Chapter 4. It is important to emphasize that both light and electron microscopes, particularly the latter, are essential to effective monitoring of the identity and cleanliness of subcellular fractions. The well-chosen application of appropriate methods for both kinds of microscopy is of great assistance in the control of fractionation techniques and has often prevented egregious errors.

A limited number of cytochemical methods for light microscopy (some are indicated in Chapters 2 and 3) assist in the recognition and identification of subcellular structures, and even fewer are used in electron microscopy. Fortunately, the ability to produce polyvalent antibodies in animal sera with specificity for specific structures and definitive components or to produce monoclonal antibodies to specific epitopes of macromolecules allows the development of reagents for the accurate recognition and localization of specific markers in a cell structure or macromolecule. Thus, the application of antigen-antibody reactions is an essential skill in morphological and cytological studies, as it is also in the identification of specific bacteria. Chapter 5 provides guidance for those unused to raising antibodies and describes varied serological methods that can be applied to cells, cell fractions, and soluble extracts with appropriate techniques for detecting reactions. The applications range beyond fluorescence labeling of antibody for light microscopy or colloidal-gold-labeled antibody for electron microscopy, because classical serological techniques (agglutination, precipitation, immunodiffusion, immunoblotting, cross-adsorption, etc.) can all be applied to the analysis of cells and their fractions.

Chapter 1

Light Microscopy

R. G. E. MURRAY and CARL F. ROBINOW

Light microscopy is made easy, interesting, and useful for bacteriological purposes if at least four different kinds of instruments are readily available, permanently set up for work, and maintained in good order. These are (i) a rugged, well-equipped, but uncomplicated microscope for "encounters of the first kind," (ii) a phase-contrast microscope, (iii) an optimally equipped and adjusted microscope for high resolution of detail and the best of photomicrography, and (iv) an instrument equipped for fluorescence microscopy. Many modern microscopes are equipped to combine two or more of these functions and are very satisfactory if the transitions are easy to accomplish. An additional instrument, a stereoscopic "dissecting" microscope, is valuable for undertaking the isolation of bacteria from nature on agar plates when colonies early in growth are small and close together.

A variety of excellent microscopes are available. Those from first-rank manufacturers are all of comparable quality, and offer a more than adequate range of accessories. The differences involve mostly the support structures, and the choice is likely to be based on convenience factors and personal preferences. Other choices may involve technical requirements, and it is always advisable to make direct comparisons of the possible instruments and the alternative accessories for a given material under given conditions of work. The really sophisticated instruments are expensive. With care, good work can be done with first-class optics fitted to a general-purpose or even a student microscope stand. However, it is hard to equal the modern instruments from the best manufacturers. It is not necessary to be specific about manufacturers. What matters is an awareness that a good microscope (whatever its age), a discerning choice of components, and a knowledge of how to use them enhance the results and the pleasure derived from the work, save time, and improve accuracy.

The descriptions that follow are based on an old-fashioned basic microscope with separate light source, mirror, and adjustable and centerable components. The optical principles are no different for the enclosed modern instruments with substage illumination, which seem simpler to use but can suffer from similar problems. The use of poorly adjusted and poorly illuminated microscopes is so widespread that most scientists are no longer aware of what microscopes can attain and are astounded when given a demonstration of the best resolution obtainable with even a basic microscope when it is used in accord with the principles outlined below (see sections 1.1 and 1.2).

1.1. BASIC MICROSCOPY

The microscopes assigned for general use in the classroom and in the research or routine laboratory can be of variable vintage, quality, and condition. The applications to the basic requirements of bacteriology are simple, and the questions answered with its help are as follows: Are there objects of bacterial size and staining properties? Are they consistent in size and shape? Are the bacteria gram positive or gram negative? Is there more than one kind of organism present? . . . and so on to more sophisticated questions. These primary questions can usually be answered without attaining the highest resolution, so a simplified procedure suffices as long as the user remembers that higher resolution demands improvement by adjustments at almost every step.

The beginner would be well advised to have at hand a fully illustrated manual for microscopy, of which there are many available (4, 6, 9, 18).

1.1.1. Instrumentation

The basic microscope (Fig. 1) should be equipped with the following.

1. Oculars (eyepieces) of at least 10× magnification. Prolonged observation is more comfortable with the use of a binocular system. The oculars fit into the microscope tube, which may be fixed or adjustable in length (to 180 mm, or whatever tube length the manufacturer recommends). Observations on bacterial structure at high resolution are assisted by higher-magnification (12× or 15×) compensating oculars (see section 1.2.2). Another circumstance in which high-power (15×) oculars assist vision is when scanning growth on agar plates with a 10× objective.

2. Low-power, dry objectives (e.g., 10× and 40×). These are essential for looking at growing cultures and for identifying rewarding areas for study at high magnification, according to the nature of the preparation. Ideally, the objectives (including the oil-immersion objective) should all be parfocal, i.e., close to being in focus when exchanged by the nosepiece carrier.

3. An achromatic oil-immersion objective (90× to 100×). This is essential to bacteriologists for observing the finest details in specimens.

4. A stage with a mechanical slide-holding device or with simple spring clips. If a mechanical stage is provided, it should be easily removable so that petri plates may be examined at low power.

5. A substage condenser of the "improved Abbe type." This is essential for use with oil-immersion objectives, because it provides the quality and quantity of light they require. It should have an iris diaphragm (aperture diaphragm) and a movable filter carrier incorporated below the lower lens. Some condensers have a movable top element (essential for work with high-power objectives) that tilts out of the way for purposes of low-power examination. A condenser that has three lenses, is aplanatic, and, when under oil, has a numerical aperture of 1.3 to 1.4 is preferable when high-resolution as well as general use is expected (see section 1.2.1).

6. A light source, either built-in and of limited adjustability or (preferably) external and movable. In the latter case an adjustable two-sided mirror (flat on one face and concave on the other) is fixed under the condenser and directs the light up the optical axis of the microscope by tilting. There are many lamps from which to choose. The simplest form, a frosted bulb behind a "bull's-eye" lens, is adequate for much work. However, the use of a lamp with a "collector" lens, capable of projecting a sharp image of the light source and provided with a filter carrier as well as an iris diaphragm (field diaphragm), results in greatly improved images of details in stained preparations, especially when Koehler illumination (see section 1.2.7) is used. A frosted-glass diffusing screen held in the filter carrier of the condenser can be of value for low-power microscopy, especially when a three-element condenser is used.

1.1.2. Operating Steps

Before operating the microscope, realize that two basic kinds of bacteriological preparations are used for simple microscopy: a specimen under a coverslip, usually a wet mount (see Chapter 2.2.1.1), or a stained smear on the slide surface (see Chapters 2.2.1 and 2.2.4). The former preparation is suitable for examination with all objectives. Start with a dry objective, either low power (10×) or "high-dry" (40×) to see whether the specimen is rewarding and to select an area for study or to detect motility. Then use an oil-immersion objective (90× to 100×) for more definitive observation. A stained smear is usually not suitable for study with dry objectives (although they may be used to locate the bacteria when the smear is very thin) because the smear must be covered with immersion oil to provide a clear and translucent image of the bacteria. The steps described below assume that a progression in levels of magnification will be used, although one can go directly to using an oil-immersion objective.

Supposing that a slide preparation is to be examined by use of the simple illumination source described above, the steps in operating the basic microscope are as follows.

1. With the eyepiece and 10× objective in place, the condenser lens fully up and almost level with the stage, and the aperture diaphragm of the condenser fully open, tilt the mirror to provide illumination up the optical axis. Adjust the objective by using the coarse control wheel to focus the light field, and then center it with the mirror. Use the *flat* mirror for most microscopy; the concave mirror is used only for low-power microscopy when a condenser is not being used.

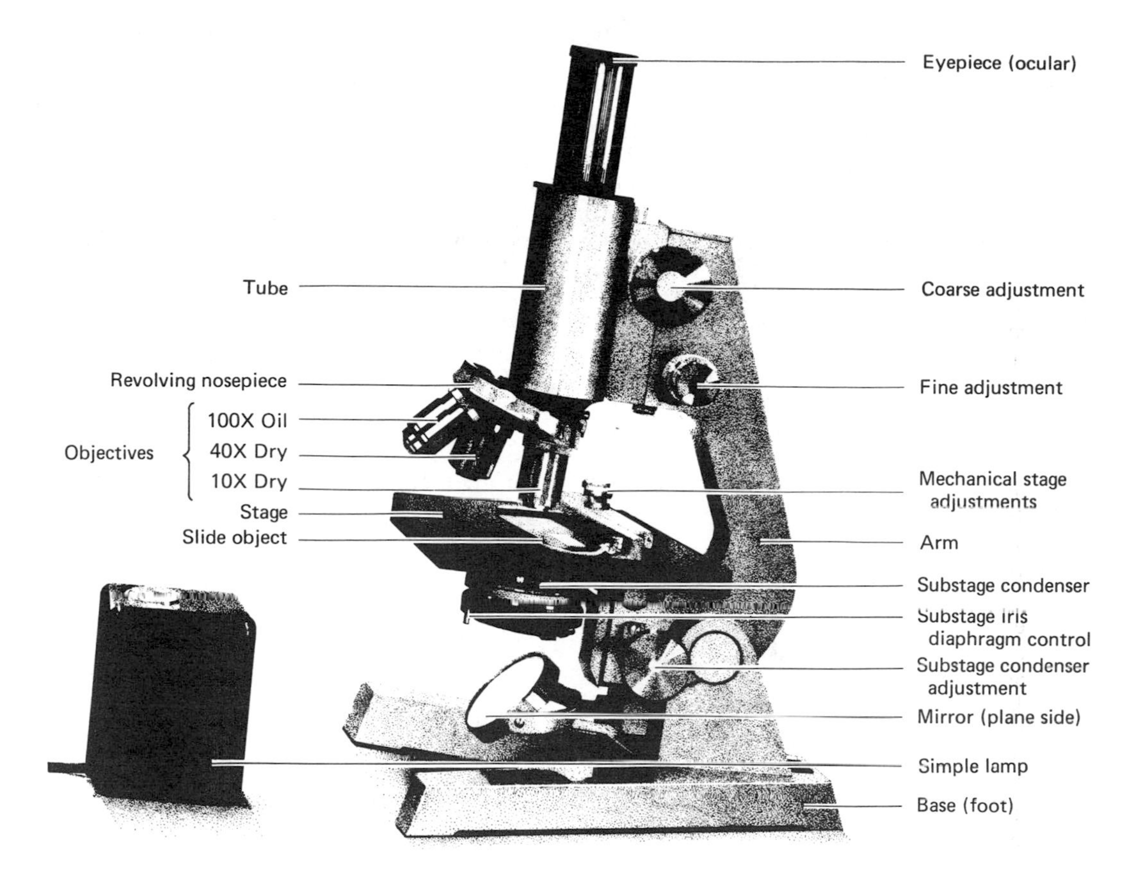

FIG. 1. Basic microscope and its parts.

2. Place a suitable slide on the stage, and focus the objective to get a sharp (not necessarily well-illuminated) image of the specimen.

3. Improve the illumination by adjusting the position of the condenser and the aperture diaphragm to get the best image and appropriate contrast. The simplest way to determine the optimum condenser position is to remove the eyepiece and look down the tube at the visible illumination of the back of the objective lens. While looking, adjust the position of the condenser to give the fullest and most homogeneous illumination of the objective. Any lack of concentration can be recognized and adjusted by tilting the mirror. Reduce glare (light scattering from lens mounts) by closing the aperture diaphragm of the condenser (if available) just enough to impinge on the bright disk. This will approximate "critical illumination"; i.e., the light source is centered and focused in the plane of the specimen. This whole procedure applies to *all* objective lenses as long as they are first focused on the specimen. Alternatively, or to check that the critical illumination is properly adjusted, place a wire loop against the lamp or against the frosted glass of other illuminators (the effective light source) while the objective is in focus on the specimen. Then adjust the condenser to get the sharpest image of the wire loop on the specimen.

4. Replace the eyepiece to examine the specimen, and make any final adjustments for the best illumination and focus.

5. Turn the nosepiece to any other objective desired. Many microscope makers provide parfocal objectives (i.e., the focal position for one is nearly identical to that for all the others). Clearance is sufficient for the dry lenses. If the lenses are from different manufacturers or otherwise mismatched, it is wise to raise the tube a little before bringing into place and focusing the oil-immersion objective, which has a small focal distance (often 0.2 mm). The clearance will be even smaller when the specimen is mounted under a coverslip (which should be no. 1 thickness).

6. If going direct to oil immersion, apply a drop of oil over the specimen on the slide before placing it on the stage. If the specimen has been scanned for a rewarding area with a lower power, it is convenient to apply the oil drop to the slide in place with the nosepiece turned halfway to the correct objective and before swinging the oil-immersion objective into position.

7. Using and focusing the oil-immersion objective takes a little practice and care even if the objective has a spring-loaded front end to protect both the lens and slide from the ham-fisted operator. While looking from the side, lower the objective slowly onto the oil drop and down just a little more so that the lens almost touches the slide. Using the coarse focusing wheel and watching in the eyepiece, look for the specimen while raising the objective *gently*. When the specimen is found, use the fine focus wheel. Adjust the illumination as above, if needed. Focus with care; contact between the objective

and the slide can crack the slide and damage the lens mounting. Adjust for critical illumination as described in step 3 above.

8. On a binocular microscope, adjust the eyepiece tube that has a knurled ring for adjusting tube length so that both eyes are in exact focus.

9. If the image is hazy, moves, or is only dim and not sharp, there is something wrong. It may be necessary to clean the front lens of the objective (with lens paper *only*). A common cause of hazy images is bubbles entrapped in the immersion oil (take out the eyepiece and look down the tube; the bubbles are then easily identified). To remove the bubbles, raise the objective, turn the nosepiece, wipe off the oil with lens tissue (the bubbles usually come up onto the objective), and then turn back the objective and refocus. A poor image may also be caused by having accidentally closed the diaphragm, moved the lamp, or moved the mirror. See section 1.1.7 for other troubleshooting suggestions.

1.1.3. Operating Rules

Remember the following simple rules of microscopy and practice them.

Rule 1. The focused image requires the best possible quality and quantity of light.

Rule 2. If there is too much light, it is best to reduce it with a neutral-density filter or with a voltage controller. The light can also be moved farther away, but then the condenser must be refocused. With oil-immersion (high-aperture) objectives, avoid reducing the illuminating aperture (by lowering the condenser or closing the aperture diaphragm more than 10% of the opening), although this stratagem is helpful for low-aperture dry objectives and for looking at living, unstained cells.

Rule 3. If there is too little light for the objective used, either there is something in the way, the alignment has shifted, or a more appropriate illumination system (lamp and condenser) is needed.

Rule 4. The specimen must be part of a (nearly) homogeneous optical system and therefore must be mounted in oil, water, or another optically appropriate medium. No details can be discerned in dried, stained films of bacteria viewed with dry objectives.

Rule 5. Keep all lenses clean and free from dust, fingerprints, noseprints, and the grime from oily eyelashes (with lens paper *only*).

1.1.4. Comfortable Microscopy

Long hours at a microscope are needlessly tiring if the microscopist cannot sit comfortably and upright while looking into the instrument. Inclined eyepiece tubes are no help if the table or chair height is inappropriate and causes stooping and straining. Binocular viewing also contributes to restful study.

Chairs (including stools in teaching laboratories) should be adjustable. For the very tall, a block of wood can be cut and used on a table of fixed height to raise the microscope to just the right position. The whole object of these adjustments is to allow the microscopist to look down the tube with minimal deflection of the neck or back. A comfortable position minimizes strain and the misery that can result.

1.1.5. Immersion Oils

Immersion oils are designed for use with oil-immersion lenses and provide an environment of the correct refractive index for the front lens of the objective. They also generate a homogeneous system, approximating the refractive index of glass, and they "clarify" the dried cells of bacteriological slide preparations by permeating and embedding the stained cells.

The commercially formulated oils recommended by microscope manufacturers have been tested to be sure that they do not have acidic or solvent effects on lenses or their mountings; substitutes are not recommended and may be a false economy. Certain properties can be chosen for special purposes (5), such as "very high viscosity" oil to fill wide gaps or to stay in place on horizontal or inverted microscopes, and "low-fluorescence" oil for fluorescence microscopy. For general purposes use immersion oil of a fairly high viscosity (e.g., type B, high-viscosity oil of 1,250 cP from Cargille Laboratories Inc., Cedar Grove, NJ 07009), which stays more or less where it is placed and creeps less than low-viscosity oils. The only problem with this level of viscosity is that bubbles are easily entrained, but this is avoidable.

Oil-immersion objectives should always be cleaned after use or, at least, at the end of a day's work. Use lens paper moistened with xylol or another appropriate solvent as recommended by the lens manufacturer. Ethanol and methanol are recommended by Wild-Leitz for its lenses, but xylol is also safe to use. The modern nonoxidizing oils are less of a problem than the traditional cedarwood oil (which hardens, oxidizes, and leaves annoying residues), but they still slowly alter and may also creep where not wanted. *CAUTION:* Modern immersion oils are safe, but before 1972 nondrying oils were often formulated with polychlorinated biphenyl compounds, which are now believed to be carcinogenic and toxic, so look out for old bottles!

Never use more oil than necessary; usually 1 drop will do. A good "oil bottle" will help to reduce mess; e.g., use a glass bottle with a broad base and an applicator (glass rod or wire loop) attached to the cap. Among the best is a double-chambered bottle in which the stopper is formed into a vessel for oil and the base is made to hold the solvent for cleaning. Squeeze-bottles tend to generate bubbles unless used carefully.

Oiling the condenser to the slide, which is necessary for highest resolution with the oil-immersion objective (see section 1.2.1), can be a messy procedure. It demands a high-viscosity oil and some simple precautions both when mounting and when demounting a preparation, as follows:

1. Place a drop of high-viscosity oil on the top lens of the condenser, and then lower it below the stage level. Avoid bubbles in the oil drop.

2. Place a drop of oil on the underside of the slide below the area you wish to examine.

3. Set the slide down on the microscope stage with the drop central in the hole, and hold the slide there while adjusting the mechanical stage or clips to keep the slide in place.

4. Raise the condenser so that drop meets drop and until the top condenser lens is fully covered with oil.

5. Place a drop of oil over the specimen, and focus with the objective. Then adjust the condenser to give Koehler illumination (see section 1.2.7).

6. When demounting the preparation, lower the condenser before picking up the slide. Lateral movement or wide scanning may lead to oil being deposited or creeping under the stage. Remember this, and clean the stage underside when cleaning the top lens (with lens tissue!) after use.

7. Only condensers with integral or screw-on top elements are suitable for oil immersion. Movable "swing-out" top elements on some "universal" condensers (i.e., those that can be slid aside with a lever for low-power work) must be oiled with great care or not at all because subsequent cleaning is difficult.

8. Cleaning condensers after use is the same as for objectives. Remove the condenser from its substage mount to avoid (or clean up) oil that has flowed around the condenser housing. Remove excess oil with lens tissue, and then clean the lenses with fresh lens tissue moistened with xylol or the appropriate solvent recommended by the microscope maker.

1.1.6. Care and Cleaning

Microscopes are remarkably tough and durable. With reasonable care and protection from the elements (particularly the hostile acid-laden air of some laboratories), a microscope will last a lifetime or more. Take care of the microscope as follows.

1. Protect the microscope from dust and grit. Put a dust cover over the microscope when it is not in use; the best is a rigid, transparent, and all-enclosing glass or plastic "bell." Do not allow dust to accumulate anywhere; it drifts into the lenses and mechanisms.

2. The moving parts, especially the rackwork and gears, should be cleaned and treated with new grease at long intervals (the grease for model-train gears from a hobby shop works well). Do not use thin oil on gears or bearing surfaces; the tube or condenser may then sink by its own weight.

3. Clean the stage regularly, and mop up any spills.

4. Keep the lenses clean, and, especially, clean up after a session of oil-immersion work. *Never* use the finger instead of an appropriate lens tissue (see below).

5. Do not attempt to repair an objective lens or the fine adjustments. These are best left to professionals. Good microscopes deserve a professional cleaning and adjustment every 20 years or so! Microscope dealers usually have or can recommend a repair facility.

6. Keep the tube closed at all times with an eyepiece, and keep all objective mounts filled or plugged, to minimize dust in the tube.

The "old soldiers" among microscopes can be cleaned, repaired if necessary, and put back to good use. The brass bodies on ancient ones (How elegant they are! How fine their lenses may be!) can be rubbed down with a lightly oiled cloth or one lightly moistened with tarnish remover and then a dry cloth. Modern black finishes should just be kept clean.

Clean lenses make for a dramatic improvement in image quality. Optical surfaces need special care in cleaning, and some general rules apply.

1. Keep a good supply of lens tissue (lens paper), a soft brush (artist's watercolor brush, 0.6 cm), and a nebulizer bulb with a short narrow rubber tube attached near the microscopy area. Keep them scrupulously clean and in a dust-free box.

2. Nonadherent dust can be easily removed by using the brush, puffing with the rubber bulb, or wiping with lens tissue after breathing on the optical surface.

3. Finger marks, other adherent grease and dirt must be polished off with lens tissue doubled over a finger and barely moistened with xylol or the recommended solvent (see section 1.1.5). Do not flood lenses with solvents, and *never* use (without specific recommendation) alcohol, ether, or acetone for fear of penetrating the cement between lenses. If the lens has a raised mount, clean the edges with moistened lens tissue wrapped around an applicator stick.

4. If xylol is not effective, try using tissue moistened with distilled water.

5. The usual way of cleaning oil from objective and condenser lenses is to wipe most of it away with lens tissue and finish with tissue barely moistened with xylol. Oil and grease can also be efficiently removed with a freshly broken piece of polystyrene foam (common packing material) by pressing it against the objective front lens and rotating it (H. Pabst quoted by James [9]). The foam has lipophilic properties. Xylol dissolves the foam, so do not use both.

6. Never take an objective apart to clean the components, because the interlens distances are critical. Dust can be puffed from the back lens or lifted with a superclean brush.

7. Dust on oculars is a constant problem and produces dark spots which move around as the lens is rotated. The eyepieces are quite simple in construction (top and bottom lenses with a fixed diaphragm between), and both lenses unscrew. If cleaning the external top and bottom surfaces does not remove a spot, there may be particles on the inner surfaces. Be careful of "bloomed" lens surfaces (antireflective coating which appears iridescent blue), and use a brush or tissue with care.

8. *Keep fingers off optical surfaces!*

1.1.7. Troubleshooting

The list of technical problems, causes, and remedies shown in Table 1 is taken (with permission) from the excellent book by James (9). For additional advice, consult other books (see section 1.9), an experienced microscopist, or an instrument technician to identify the problem. If the problem is optical, work along the light path in systematic order from the lamp and then through the condenser, specimen, objective, and ocular. Adjust at each step, and determine the effect of adjustment on the problem; this is where Table 1 is useful.

1.2. HIGH-RESOLUTION MICROSCOPY

The "research microscope" comes into its own for the resolution of fine detail at high magnification. Its use is not called for in determining the outcome of a Gram test, but it will help to settle questions about flagella and matters requiring the perception of fine detail; also, high resolution allows the sharpest photomicrographs. For such purposes it is essential to pay close attention to the quality of the optical components. Achieving high resolution as well as freedom from chromatic and spherical aberration requires attention to basic principles.

Table 1. Common problems in light microscopy, and their causes and remedies[a]

Problem	Possible causes	Remedies
Coarse adjustment is too stiff	Mechanism adjustment was faulty	With many stands, adjust simply by moving the two coarse-control knobs in opposite directions
	Dirt in rackwork	Clean and apply new grease
Tube or stage sinks spontaneously under its own weight (image drifts out of focus)	Incorrect adjustment of rackwork	As with first remedy
	Lubrication with too-thin oil	As with second remedy
	Faulty adjustment of focus control	As with first remedy
Micrometer movement is blocked to one side	Fine adjustment at the end of its travel	Bring a 10× objective into position with the revolving nosepiece, set the fine focus control at the middle of its range, and then focus with the coarse adjustment
Drift of focus with the slightest movement of the fine adjustment (especially with oil-immersion objectives)	Objective insufficiently screwed into the revolving nosepiece	Seat the objective fully into the nosepiece
	Surface of the coverslip stuck to the objective by the layer of oil	Use less viscous immersion oil; clip specimen firmly
Veiled, spotty image	Dirt or grease (i) on the eyepiece (spots move when the eyepiece is rotated in the tube), (ii) on the objective, (iii) on the coverslip (spots move when the specimen is shifted), or (iv) on any surface of the illumination apparatus	Clean where necessary
Sharply focused spots or specks in the image which change and disappear when the condenser is moved up and down	Dirt (i) on the light source, or on the diffusing screen in front of it with critical illumination, (ii) on the cover plate of a built-in lamp, or (iii) on a filter near the cover plate with Koehler illumination	Clean where necessary; when the contaminated surface cannot be reached, change the focusing of the condenser slightly or tolerate the problem for the sake of resolution
Hazy image, which cannot be brought sharply into focus	Wrong immersion medium (oil instead of air, air instead of oil, air bubble in oil)	Use correct immersion medium
	Transparent contamination on objective front lens	Clean where necessary
	Coverslip too thick, or too thick a layer of mounting medium	Use a correct coverslip or an appropriate objective
	Irregularly distributed remnants of immersion oil on the coverslip when using high-power dry objective	Clean the coverslip with dry cloth or paper tissue; beware of solvents, as some may weaken or dissolve mounting medium
	Slide upside down on the stage (only with high-power objectives)	Invert the slide; make sure a label is not stuck to the wrong side of the slide
Object field partially illuminated	Filter holder partially in the light path	Move the filter holder
	Objective not clicked into position	Reposition the objective
	Condenser (or swing-out lens) not in the optical axis	Realign the condenser
Object drifts diagonally when focusing	A lens is not centered in the optical axis	Check concentration of the lenses at all points in the optical path
Object field unevenly illuminated	Mirror not correctly in position	Reposition the mirror
	Condenser not centered (with critical illumination)	Realign the condenser
	Irregularity in light source and/or diffusing screen (with critical illumination)	Move the condenser slightly up and down; use ground glass in front of the light source
Drift of cloud across the field; after this, image out of focus (oil-immersion)	Air bubble in the immersion oil; oil in image space with a dry objective	Wipe off the oil from the specimen, and set up anew; clean the slide and objective carefully
Sharply delineated bright spots in the image	Transverse reflections in the interior of the microscope (often sickle or ring shaped)	Try another eyepiece; use correct Koehler illumination
	Longitudinal reflections in the tube, causing round light spots	Use lenses with antireflective coatings; change the combination objective-eyepiece

(Continued)

Table 1. *(Continued)*

Problem	Possible causes	Remedies
Unsharp bright spots in the image	Contamination at a lens surface, on upper or under side of the object; or air in immersion oil of the condenser (differentiate as explained before)	When localization and removal of the contamination are not possible, reduce the effect by opening the condenser diaphragm somewhat more

[a]Reprinted from reference 9 with permission.

1.2.1. Resolution Requirements

Resolution (R) is the shortest distance between points of detail which will still appear as a distinct gap in the images, visual or photographic. $R = \lambda/2NA$, where λ is the wavelength of the light used and NA is the numerical aperture, a measure of the light-gathering power of the objective. The purpose of a good research microscope is to help make R as small as possible. The smallest values of R require not only an objective of high NA but also a condenser of equivalent quality (see section 1.2.3) and the use of short-wavelengths light). Unfortunately, light of the most effective wavelength (namely, UV light of about 365 nm) is not perceived by the eye and does not pass through glass. Indirect methods of microscopy, requiring a microscope with lenses (lamp, condenser, and objective) made of quartz (permeable in various degrees to UV light), take advantage of the short wavelength of UV light, as is the case for fluorescence microscopy. The best results with visible light are obtained in the green/yellow region of the spectrum because it is the wavelength to which the eyes are most sensitive (close to the mercury green spectral line) and for which microscope objectives are designed to transmit with a minimum of aberrations. So much for the λ part of the expression for resolution.

The NA of oil-immersion objectives is usually a value of 1.30 or 1.32, and the best ones may be 1.40. The numerical aperture represents the sine of one-half of the angle described by the cone of light admitted by the objective lens, multiplied by the refractive index of the medium through which light passes on its way to the lens. The maximum value of the sine function cannot exceed 1.0, but the refractive index of the medium through which light passes on the way from the condenser to the objective can be raised to that of optical glass (i.e., 1.5150) by the use of immersion oil. It is this fact that allows lenses of NA of greater than 1.0 to be filled with light. Provided that certain other conditions are met, the resolving power of the costly objectives with NA of 1.30 to 1.40 engraved on their glittering housings can be utilized to the fullest.

1.2.2. Objectives and Oculars

High-aperture achromatic or apochromatic objectives are spherically correct for one or three colors, respectively, in the center of the field of view (where most of the work is done at high magnification). The two kinds of objectives work almost equally well, price notwithstanding, especially when narrow-band-pass (interference) filters are used.

The objective best suited to the task in hand should be complemented by an ocular of commensurate quality. What is needed for apochromatic oil-immersion objectives are compensating eyepieces. "They are designed to correct the lateral chromatic error of magnification inherent in objective lenses of high corrections and high apertures"; experience amply proves the soundness of the additional advice of Shillaber (15) that "the compensating ocular should be of high power, preferably 15× or 20×. The ability of apochromatic objectives to take a high-power eyepiece should be fully utilized; otherwise, the fine detail that they are capable of bringing out in a photomicrograph is likely to be lost."

For those who wear glasses with major corrections, it is convenient and helpful to buy high-eyepoint oculars which are made so that glasses may be worn while observing.

1.2.3. Condensers

Another important station in the path of the light from lamp to object to eye is the substage condenser. Objective NA values of 1.30 to 1.40 will be fully utilized only if the condenser NA is at least as high. Such condensers are usually also achromatic or aplanatic and should match the optical properties of the objectives described above.

Remember the role of the refractive index in the expression for NA: the full resolving power of an objective of high NA is utilized only when the medium through which light passes between the condenser and the lens system of the objective is homogeneous and glasslike all the way. Therefore, the maximum resolution of an oil immersion lens requires oil between the condenser and the object slide as well as between the slide and the objective lens. The condenser iris diaphragm (aperture diaphragm) restricts the beam and thus affects the NA attained (see section 1.2.7).

1.2.4. Mirrors

A mirror is used to reflect the beam of light along the axis of the condenser. The adjustable mirror issued with old-style microscopes has a plane side for work with a condenser and a concave side for work at low powers without a condenser. The plane mirror is made by depositing metal on the underside of an optically flat disc, does well in everyday work, and is durable. However, it is not the best for optimum illumination, because the front and back of the glass that covers the reflecting metal produce separate and overlapping reflections of the light source or of the opening of the field diaphragm when the Koehler procedure is followed, thus adding undesirable scattered light to the final image. Instead, use a "first-surface mirror," which lacks protective coverings and gives but a single reflection.

1.2.5. Filters

The light used for microscopy is modified in intensity or wavelength by the use of filters. Neutral-density filters (aluminized optical flats) allow the brightness to be adjusted for comfortable vision. Color filters restrict the wavelength used. Two kinds of color filters are available: expensive "interference filters" of narrow band pass consist of two half-silvered glass plates facing each other across a precisely measured gap; Wratten filters consist of colored gelatin mounted between two 5-cm^2 flats of optically polished glass. Wratten filters deteriorate with time because air creeps in from the sides, but they come in a wide range of colors and densities and are valuable accessories. A red filter (Wratten no. 29) or an orange filter (Wratten no. 22) is helpful in making dense structures transparent or enhancing structures stained with blue dyes. A green filter (Wratten no. 11 or no. 58B) is the most generally useful because it is the kind of light to which the human retina is maximally sensitive and because many commonly used histological stains are red. Their contrast is greatly enhanced when the light is green, the complementary color.

1.2.6. Light Sources

Optimum performance of a highly resolving objective depends on optimum illumination. The field of view must be evenly filled with light of a brightness sufficient for photography. This requires a more complex lamp than that effective for ordinary observation. The light from the usual coiled filaments is unevenly distributed over their surface and therefore over their image unless diffused by ground glass. The best structureless, uniformly bright sources of light are small mercury or zirconium arcs and the less expensive conventional projection lamps with tungsten bead or ribbon filaments that can be imaged to fill the condenser. The Koehler system of illumination below is designed to provide the appropriate quality of light and requires that the source be focused in the plane of the object being viewed. Therefore, the microscope lamp should be equipped with a lens system to project a sharp image of the light source into the entrance plane of the substage condenser. In front of this lens, there should be an iris diaphragm, the so-called field diaphragm (see below), which can restrict the area of illumination without interfering with the quality of the light. There should also be a carrier for filters.

1.2.7. Optimum (Koehler) Illumination

The Koehler system of illumination is useful because it provides appropriate illumination of the field for high-resolution microscopy and makes good use of high-quality optics. It is achieved by focusing an image of the filament or light source at the level of the condenser diaphragm (i.e., the lower focal plane of the condenser), when the condenser is in the correct position relative to the specimen. Under these conditions, an objective lens in focus on the specimen will be fully illuminated whatever the size of the source. Effectively, this applies to every part of a source, so that there is even illumination across the field of view even when a coil filament lamp is used. The principle can be applied usefully to both high-power dry and oil-immersion objectives.

With eyepiece, objective, and condenser of matching optical quality and with a first-surface mirror receiving filtered green light from a bright and homogeneous source, the research microscope is ready to be set up in the best way for work and needs only a rewarding specimen preparation to show its worth.

The steps allowing achievement of Koehler (nearly optimum) illumination for an oil-immersion objective are as follows.

1. With a specimen slide in place, switch on the lamp and adjust the condenser to give the smallest bright spot on the specimen. This approximates the focal position of the condenser.

2. Hold a small white card against the underside of the condenser, and, observing in the mirror, focus the image of the lamp filament on the card and move the lamp so that the image covers the opening of the condenser. Alternatively, focus the image on the closed condenser (aperture) diaphragm. If the physical arrangement is awkward, use a small mirror for observation or approximate the focal position on a card placed over the mirror.

3. Insert appropriate filters in the light path, and fully open all diaphragms.

4. Oil the condenser with 1 drop of high-viscosity oil on the top lens, and raise it to meet a similar drop on the underside of the slide (see section 1.1.5).

5. If necessary, scan the preparation with a low-power objective to identify a rewarding area, using reasonable centration and illumination (section 1.1.1), and mark it by the coordinates provided on many mechanical stages.

6. Apply oil to the top of the slide (see section 1.1.5 for precautions), change to the oil-immersion objective, and focus on the specimen. A fuzzy or moving image may indicate bubbles in the oil; check by looking down the tube after removing the eyepiece.

7. Adjust the field (iris) diaphragm of the lamp to a minimum opening, and bring it into view with the mirror. Focus the image margin with the substage condenser in the plane of the specimen (i.e., so that the objective and condenser are both in focus). Readjust the field diaphragm so that its image just impinges on the field of view and centers exactly with the mirror. The optical alignment is then correct.

Note: If the margins of the field diaphragm give an uneven color fringe when using white light, the condenser needs centration. To center it, turn the two adjustment knobs to give movement toward the red side, centering the aperture with the mirror, and stop when the color fringes are symmetrical.

8. Open the field diaphragm until it just impinges on the margin of the field of view. This prevents illumination of the specimen outside the field of view, which scatters light into the image plane. The field diaphragm does not affect resolution in the area illuminated and can be left in view.

9. Removing the ocular (which should be 12× or 15×), look down the tube and restrict the condenser aperture diaphragm to abolish glare or reflections. This restriction should not be more than 1/10 the diameter of the fully illuminated back lens; excess restriction lowers resolution.

10. Replace the eyepiece, and insert filters appropriate to observation or photography (section 1.2.5).

The same principles apply to adjustment of the illumination provided by modern integral substage systems or to incident-light systems (3). However, the centration of the light for these systems is attained by adjusting the position of the bulb in its mount with centering screws. The filament image is focused by positioning the lamp forward or backward in its sleeve, because the diaphragm and the lenses are fixed in place. The point of reference for alignment is usually an integral field diaphragm.

The same principles apply also to dry lenses, but oil immersion of the condenser is then unnecessary as long as the condenser is of adequate quality.

1.3. DARK-FIELD MICROSCOPY

Dark-field microscopy provides a useful means of looking at wet mounts of unstained specimens and detecting very small structures by reflected and diffracted light, revealed like motes of dust in a beam of sunlight or like planets and moons in a night sky. It is performed by illuminating the specimen with a hollow cone of light such that only the light diffracted by objects in the field of view is transmitted up the microscope tube to the eye or camera; the beams forming the cone of light focused on the specimen are at too low an angle to be captured by the objective. The result is a field of bright objects (spirochetes, bacteria, particles, organelles, etc.) against a dark background. It is an appropriate technique for all powers of objectives, but it has some requirements that impose the need for specialized condensers for use with oil-immersion objectives.

Low-power (10×) and "high dry" (45×) objectives can be used in a dark-field mode by using an ordinary condenser equipped with a central patch stop in its filter holder. This patch stop allows light paths only in the periphery of the condenser and so attains a hollow cone of light effective at a particular condenser position. This means that the NA of the condenser must be considerably greater than that of the objective being used in order to attain an appropriate illumination angle, and the condenser diaphragm must be completely open. The cone of light accepted by an objective limits the usefulness of patch stops, so they do not work for objectives with an NA greater than 0.60 to 0.65. A simple version of low-power dark-field microscopy with 10× and 45× objectives can also be attained by using a phase-contrast condenser (see section 1.4); it is useful but does not provide the highest-quality image. Few problems are encountered when dry objectives are used with a condenser particularly designed for dark-field microscopy, and an excellent image can be expected.

Oil-immersion objectives (which collect light at an angle proportional to the NA) require specialized condensers of high NA, which must be oiled to the slide so that an effective cone of light can be produced. These reflecting condensers consist of either a paraboloid mirror with a central stop, or, more effectively, a central spherical reflecting surface which reflects much of the light entering the condenser to a cardioid (curved) peripheral mirror on the perimeter, reflecting a very low-angled beam in the form of a cone (refer to texts for optical diagrams).

However, when these condensers are used, most oil-immersion objectives do not provide an adequately dark ground without the presence of devices to restrict scatter from the margins and mountings of their lenses. Excellence requires an objective with an adjustable diaphragm in the back focal plane or restriction in the back focal plane by use of a "funnel stop" (a metal tube with a hole in the lower end) that fits inside the objective to block transmitted rays scattered from the periphery of the objective.

Intensive light is necessary to show very thin or very small objects. Sunlight from a heliostat has been used to show bacterial flagella in action.

It is sometimes difficult to get a good dark background when there is poor centration of the condenser relative to the objective. This requires some trial and error to gain symmetrical illumination. It is helpful to find the best setting for the condenser by using tongue or cheek scrapings mounted in water under a coverslip; these large epithelial cells are easy to find. All the precautions about oiling the condenser and slide (section 1.1.5) must be followed. But given a well-aligned microscope, all that matters then is the intensity, geometry, and quality of the cone of light with an appropriate objective.

The focus of the condenser is crucial in dark-field microscopy. The apex of the cone of light *must* be at the level of the specimen, which is also the plane of focus of the objective. Above this focal point is a dark conical space, in which the front lens of the objective is placed, symmetrical to the dark conical space formed below the specimen above the condenser. When the dark-field condenser is brought up to the slide, with immersion oil making a complete contact, a bright ring of light is projected on the specimen. As the condenser is raised further, the light comes to a bright intense spot, which must be close to the best focus. Because the high-aperture condensers for oil-immersion microscopy have a very short focal length, it is important to use a thin slide (ca. 0.8 mm thick) so that the condenser can focus through it on the specimen. Standard slides (ca. 1.2 mm) may be too thick, thus making high-resolution dark-field microscopy impossible.

For many years, dark-field examination of exudates and media was the accepted method for demonstrating spirochetes. It is impressive to be able to see *Leptospira* spp. in low-power dark ground by using 10× objective and 15× oculars, but it is even more when they appear in all their glory under oil immersion at an effective NA of 1.3. It is a pity that these beautiful images are seldom seen nowadays because of the convenience of phase-contrast microscopy.

A procedure for oil-immersion dark-field microscopy is as follows.

1. See that the microscope and illumination system are fully aligned, as for bright-field microscopy.

2. Replace the condenser with the dark-field condenser, and apply 1 drop of high-viscosity oil to the top lens.

3. Make a wet mount of the specimen on a clean, thin slide. If it is very clean, it is helpful to make a small wax pencil mark under the coverslip for preliminary focus.

4. Raise the condenser to meet the specimen slide, and observe the ring of light. Focus this to give the

smallest spot. An asymmetrical ring of light above or below focus indicates that the condenser is not properly illuminated.

5. Focus the oil-immersion objective on the specimen. The objective should be equipped in the body with a funnel stop or with an iris diaphragm controlled by a knurled ring. In the latter case, start with the diaphragm half closed and make final adjustments later for maximum darkness of the field.

6. Move the specimen slowly to a useful field, and make final adjustments in the condenser focus for maximum brightness of the reflections from the specimen.

1.4. PHASE-CONTRAST MICROSCOPY

Phase-contrast microscopy is a system for gaining contrast in a translucent specimen without the help of stains (9, 14, 19) and has the advantage of using high-resolution optical components. Stained specimens or biological material with opaque areas form images in a microscope because the various components transmit different amounts of light, so that amplitude and wavelength are modified by absorption and scattering in the specimen. Most cells and their components do not cause enough amplitude modification to give useful contrast in the image. However, their materials are translucent and have different refractive indices from neighboring structures and from the mounting medium, and this causes different degrees of phase retardation in the light beam passing through a structure, compared with beams that have passed through other material or only through the mounting medium. The purpose of the phase microscope is to take advantage of changes in phase, converting these into amplitude differences to form an image with enhanced contrast.

An instrument that is permanently set up for phase-contrast microscopy is preferable to one that provides alternating service with ordinary optics on the same stand. Phase-contrast objectives are not suitable for the best bright-field work, despite their high quality. It is better to have objectives dedicated to each purpose. Phase-contrast microscopy demands much more light than is adequate for the examination of stained specimens with ordinary optics. A green filter should be used to reduce unavoidable chromatic aberration.

One of the advantages of phase-contrast over the less-available dark-field microscopy is that the former uses the full aperture of the objective lens. Therefore, the best of phase-contrast objectives should be used. This in turn implies the use of a substage condenser with a suitably high aperture and of 12× to 15× compensating eyepieces, which fully utilize the fine resolution that the objectives are capable of giving. Medium-power (40× or 60×) oil immersion phase-contrast objectives are available and provide crisp, rewarding images for work with large cells. For the very best results, the condenser and slide must be connected by oil when high-power (90× or 100×) objectives with numerical apertures higher than 1.0 are used. However, an oiled condenser is not required for all applications. A dry condenser is adequate for observing the shape and form of bacteria and their spores and for checking motility.

Phase contrast works because a phase annulus is inserted at the lower focal plane of the condenser (to generate a cone of light) and because a phase plate is incorporated in the back focal plane of the objective. An image of the annulus will thus be formed on the phase plate. Some light beams are diffracted by the specimen (e.g., a living cell) to go through all of the objective field, and some go directly through the specimen and the phase plate ring; both take part in forming an image. The phase ring in the objective is a flat ring-form groove in an optical flat plate; its size fits the cone of the direct light beams generated by the condenser annulus. The groove is formed so that the phase ring is less thick glass, causing at least a 0.25 wavelength difference in phase retardation compared with a direct (nondiffracted) beam. The phase difference of the diffracted and nondiffracted beams, some of which are also retarded by passing through a structure in the specimen, causes either destructive interference (dark contrast) or constructive interference (bright contrast) when the beams form the image. Therefore, it is essential to center the annulus plate in the condenser with respect to the analyzer plate in the back focal plane of the objective. This is achieved by imaging both with a focusing eyepiece (or "telescope") and by centering the visible rings by using the two adjusting knobs on the sides of the condenser. The bright image of the annulus should be completely enclosed by the grey annulus of the phase plate when centration and condenser positions are correct; complete filling of the annulus is not essential.

1.4.1. System Assembly

1. Set up the microscope as for Koehler illumination (section 1.2.7) without the annulus in place (use the "bright-field" position for the usual rotatable carrier in the condenser of a phase microscope), and adjust the phase objective to focus on a specimen.

2. Revolve the condenser carrier to bring the appropriate annulus into register below the condenser for the objective in use.

3. Insert a focusing telescope instead of the eyepiece and focus on the grey ring image of the phase plate, which has distinct inner and outer edges.

4. Look in the telescope for all or part of a very bright ring of light, the image of the annulus, which has to be brought into register with the objective phase ring, seen as faint rings for each side of the groove in the phase plate. Achieve centration either by manipulating the condenser centration knobs or by nudging the annulus with a knurled ring on the carrier (systems vary according to manufacturer).

5. Replace the telescope with the eyepiece, adjust the fine focus, and regulate the light intensity (by using a transformer) for comfortable viewing. The image should not shift asymmetrically on focusing; if it does, recheck step 4.

6. If the image is unsatisfactory, check that all surfaces are clean, that the specimen is not too thick, that there are no bubbles in the oil if you are using immersion, and that the correct annulus is in place. The image quality deteriorates if any part of the direct beam falls outside of the annulus in the objective phase plate or is not fully concentric. Accurate focusing of the condenser is essential.

1.4.2. Specimen Preparation

The specimens for phase microscopy should be as thin as possible and must be mounted in a fluid or gel and under a coverslip to give a homogeneous background to the images. For general work, a clean standard slide and coverslip are satisfactory.

When intracellular detail of living organisms is wanted, attention must be paid to the refractive index of the medium in which the cells are mounted. This is most easily provided by dissolving gelatin or bovine serum albumin (15 to 30%) in the medium. According to the concentration used, this equalizes or brings closer together the refractive indices of the contents and medium, since phase retardation is proportional to the refractive index multiplied by the light path. The disturbing bright halos that usually surround dark-contrast images are then reduced or abolished, and the cells are more grey than black. As an additional benefit, the cells then show internal structures that are not easily discerned in life, e.g., nucleoids and granules other than the obvious lipid droplets, or developing endospores. When the mounting fluid and an adjacent structure are identical in refractive index, the structure will disappear; this technique of immersion refractometry can be used to measure the density and mass of structures (13).

1.4.3. Phase Condenser for Low-Power Dark Field

The condenser phase plate is, essentially, a form of patch stop, and so, given suitable geometry, it should produce dark-field images with ordinary low-power objectives. The annulus for oil-immersion phase commonly gives a reasonable semblance of dark-field optics (section 1.3) with an ordinary 10× objective and sometimes with a 45× objective, after a little fiddling with centration and the position of the condenser.

1.5. INTERFERENCE MICROSCOPY

Interference microscopes have definite applications in bacteriology for discerning the structure of cells. Such microscopes are expensive, which makes them less available, and their operation and the interpretation of the images obtained is best learned by practice with experts. Like phase-contrast microscopes, they attain contrast in images of translucent specimens by detecting phase changes induced in light that traverses cell components with different masses and refractive indices. The two systems differ in that the interference microscope develops separate object and reference beams, rather than forming an image from the direct and diffracted elements of a single beam from a condenser annulus as in the phase-contrast microscope (9, 14, 17, 19).

Interference microscopes provide superior images of the internal morphological details of cells without the halos that surround cells when viewed by phase-contrast microscopy. A great advantage of the former is that phase changes may be measured to provide quantitative cytological data. Furthermore, the lack of an annulus avoids deterioration of the image by optical effects from cell structures. The lenses are of high quality, are used to their effective numerical aperture, and will operate in either transmitted- or epi-illumination modes. The quality of light and the adjustment of the microscope to give critical or Koehler illumination at the outset are important.

The Nomarski-type interference microscope uses polarized light (white or monochromatic) from a filter at the source to fill the condenser through a Wollaston (birefringent) prism. The prism generates two polarized beams at right angles, each filling the lenses. The resulting cone of light traversing the specimen and illuminating the objective is a complex of these two beams; in effect, they are very close together, in parallel, and uniform. They act as object and reference beams according to what they traverse in the specimen. The beams are recombined above the objective by another Wollaston prism, allowing interference (constructive or destructive) to take place. The plane-polarized beam is recaptured by an "analyzer" polarizing filter, which is usually in a fixed orientation (this means that the polarizing filter at the source has to be rotated to give the appropriate orientation). If white light is used, with appropriate manipulation parts of the image will contain interference colors which are related to the amount and sign of the phase change. This information and photometric measurements will allow generation of mass data.

In some interference microscope systems (including the Nomarski type), the image has a pseudo-three-dimensional appearance, which is a consequence of an angle of shear between the beams.

1.6. FLUORESCENCE MICROSCOPY

Fluorescence microscopy employs all the principles of optics described for the research microscope (section 1.2). The differences in practice and design relate to generation and transmission of wavelengths of light suitable to the excitation of fluorochrome stains and of natural fluorescence in the specimen. The secondary emitted wavelengths are detected as an image of a fluorescing object. Because the excitation process usually requires short wavelengths in the near-UV or blue range, the lamp (a high-pressure mercury vapor arc lamp) and any lens between the lamp and object must be made of material (quartz) appropriate for passage of that range of wavelengths. A first-surface mirror is essential (to avoid an interfering glass layer). The immersion oil for the objective and condenser must be nonfluorescent (e.g., a special synthetic formulation or sandalwood oil). Quartz slides and coverslips must be used. Good advice on details comes from specific (10, 20) and general texts.

Most important are the light source and the arrangement of filters in the light path, which vary in mechanical arrangement among manufacturers (who provide appropriate sets of filters for a specific fluorochrome), but all meet the following demands.

1. The lamp must be able to excite the fluorochrome in use, and two types, of differing utilities, are available: (i) high-pressure mercury-vapor arc lamps produce a large proportion of UV rays and the widest spectrum of wavelengths, so that any fluorochrome will respond to the illumination; and (ii) quartz-halogen-tungsten filament lamps provide light rich in the blue end of the visible spectrum, so that there is a more limited response of fluorochromes, requiring other excitation wavelengths.

2. There must be filters between the lamp and the specimen. The first, when using UV light, is a (long-

wave) heat-absorbing filter. Then there is one filter or a combination of filters designed to pass the wavelength required for excitation and to absorb most other wavelengths. The combination may be narrow-band interference filters passing the blue (490-nm) rays capable of exciting fluorescein isothiocyanate.

3. There must be blocking or barrier filters between the objective and the eye. These have to absorb the exciting radiation, which may be a major component, except in the case of dark-field or epi-illumination systems. UV rays are damaging to the eyes, and a "stop" or barrier filter is an important protection. Also, longer (colored) wavelengths may be emitted by the specimen apart from those due specifically to the fluorochrome. Consequently, a suppression filter will assist in selecting the wavelength for the intended image.

1.6.1. Fluorescence Systems

The illumination systems for fluorescence microscopy take several forms: (i) transmitted light through a substage condenser; (ii) dark-field illumination through a specialized substage condenser; and (iii) incident illumination (epi-illumination) through a specialized objective, in which case the objective acts as both condenser and objective. The second and third systems have the advantage of an effective dark-field system, because the optical path is such that a direct beam does not go to the eye and there is minimal interference with the detection of the weaker fluorescent light from the specimen. In the third system, which is the best, the objective acts as the condenser and suffers no centration problem; therefore, the weak fluorescence suffers minimum attenuation from specimen thickness. The exciter beam is reflected down to the objective from a side or rear port in the microscope tube by a beam-splitting "mirror" or prism that reflects the exciting wavelength but transmits visible light back from the objective to the eye.

The cells or materials to be examined react with fluorochrome stains (e.g., xanthenes, acridines, and quinolines [see Chapter 2.1.4]), with substrates that release fluorochromes inside a cell (e.g., fluorescein diacetate), and with antibody conjugated to a fluorochrome (see Chapters 2.4.4 and 5.5.6) to allow localization or recognition.

The manner of using the microscope and its filters may be unique to the instrument, so consult the manual provided by the microscope manufacturer.

There is only a small light response (less than 1%) in fluorescent emission. Also, only a small proportion of the specimen usually fluoresces, and the responses of most fluorochromes tend to fade with time. In brief, the light level emitted is low. Therefore, it is more effective to use incident illumination (epi-illumination), if available, rather than transmitted light. Therefore, heed the following advice.

1. Work in a dark or well-screened room.
2. Do not use any unnecessary lenses or filters (no ground glass).
3. If there is an auxiliary lens, check the position that gives the most light. Open the condenser (aperture) diaphragm completely, but close the field diaphragm to the best position, as for bright-field microscopy (section 1.2.7).
4. Use eyepieces of low magnification (8×), because the intensity of fluorescence perceived decreases exponentially with the total magnification.
5. Use high-speed and high-contrast films for photomicrography, either color or black-and-white (see Chapter 28.3.3.1).

1.6.2. Confocal Scanning Microscopy

The elaborate equipment (2,16) required for confocal scanning microscopy, utilizing an intense beam of light from a laser, is designed to scan the sample by illuminating and imaging one very small area at a time in a single focal plane of the specimen. It is confocal because the scanning and the image are both attained through the objective. An epi-illumination system focuses a small spot of light at a plane in the specimen. This illuminated spot is imaged through a conjugate aperture, which accepts only the direct beams (but not the diffracted beams) for forming the image. The specimen is scanned through the objective by a moving beam producing a series of spots of light or aperture images. A raster scan allows the synthesis of a complete image in a detector system, and it is displayed on a monitor. Only structure that is in focus will form an image. This imaging system will operate either with direct imaging by visible light or with fluorescence imaging by UV light.

There are two major forms of the scanning equipment. One form involves tandem multiple-aperture arrays, and the other involves a regulated movement of a very fine laser beam. The principle is straightforward, but the equipment is very specialized and requires its own instruction manuals and expert users.

The advantages of this method of microscopy are that it can be used at high magnification with the best of epi-illumination objectives (including oil-immersion) to study large objects that scatter a lot of light in other modes of microscopy. It can focus on the structure of a surface and, particularly, can allow examination of structure within large cells. As far as bacteria are concerned, a major use is the study of the interactions of bacteria with or within eucaryotic cells.

Information is increased by using confocal scanning microscopy as a form of fluorescence microscopy; photomultiplier imaging reduces the problems with dim signals, and the confocal system evades fluorescent interference from features above and below the plane of focus. Because images can be generated at known depths in the specimen, a computer correlation of a stacked series of images generates three-dimensional information about the specimen.

1.7. CELL MEASUREMENT

It is important to determine the magnification imposed on bacterial cells either in projections for drawings or on photomicrographs. Equally important, a range of dimensions may be needed as part of experiments or for the description of cells for purposes of classification. Although a number of measuring techniques may be applied, they all must be based on a measurement standard provided by a stage micrometer; the use of the stage micrometer is clearly described in texts (11, 15).

1.7.1. Stage Micrometer

A stage micrometer is a slide on which a number of lines have been ruled by a grating engine or deposited photographically to show a precise scale. Usually, there are 10 parallel lines at 0.1 mm (100 μm) spacing and 10 parallel lines at 0.01 mm (10 μm) spacing. Image the micrometer slide with the same care as given to a specimen. Project the image on the ground glass of a photomicrographic camera for direct measurement, or record the image on film under the same conditions as for photomicrographs. A number of measurements of the appropriate intervals are then used to generate either the magnification on the film, the length required for a bar scale to apply to micrographs, or a set of measurements of the cells under study.

1.7.2. Eyepiece Graticule Micrometer

An eyepiece graticule micrometer is an optically flat glass disc, about 12 mm in diameter, with an arbitrary but accurately proportioned scale engraved on it. The scale is usually in numbered "units," each with 10 divisions. The graticule disc is placed in the plane of the intermediate image inside a Huygenian eyepiece, which has a diaphragm at the plane of the intermediate image where the graticule disc can be rested (access is obtained by unscrewing the top element of the eyepiece). In more complex eyepieces the diaphragm is below the lower field lens, and so there are fewer problems with parallax.

Each graticule must be calibrated for each eyepiece-objective combination to be used. With a stage micrometer and an appropriate set of ruled lines in focus, rotate the eyepiece so that the graticule scale is normal to the rulings and note a number of coincidence points. Then, it is easy to derive a statistical measurement of the graticule unit in micrometers to two decimal places. Final direct measurements against the real specimen should not be considered more precise than one decimal place.

There are graticules ruled in squares for counting cells (see Chapter 2.1.4) or in circles for rough sizing of cells. When counting, relate the squares to the specimen area and count the cells lying within the squares and touching two of the sides. In this application, it is advantageous to have a special eyepiece allowing focus on the grid rulings by rotation of the top lens.

1.7.3. Eyepiece Screw Micrometer

The eyepiece screw micrometer is a specialized eyepiece (sometimes called a filar micrometer) for direct measurements after calibration against the rulings of a stage micrometer. The units are on a wheel (with or without a vernier reading scale) to be read against a fixed mark, and the wheel drives a hairline across the field allowing a measurement in calibrated units by difference. As with an eyepiece graticule micrometer, the cursor line and the object must be in exact focus to minimize distortion by parallax.

1.8. PHOTOMICROGRAPHY

All types of light microscopy can make exact records on film, whatever the wavelength of light being used. The technical requirements are important, in all cases, so that the most faithful record may be obtained. There are excellent books available that amplify and explain the requirements (3, 4, 6, 8, 13, 15, 18). Chapter 28 deals with all aspects of photography in bacteriology, including photomicrography (28.3.3).

Three types of photomicrography apparatus are in together with general use, as follows.

1. A roll-film camera back and shutter (usually for 35-mm film), together with an integral device that includes the ocular and a beam-splitting prism with a side viewing arm to allow focusing, forms the working unit. The whole device rests on the microscope tube.

2. Integral cameras have been developed by a number of microscope makers and come complete with a built-in light, an exposure meter, and timing controls. These cameras are usually adequate and produce effective routine photomicrographs.

3. An old-fashioned light-tight bellows is sometimes used. It is located on a stand on which a microscope can also be placed. A light-excluding sleeve is fitted at the bottom of the bellows and meshes with its mate, which slips onto the top of the microscope tube. A plate carrier is fitted at the top of the bellows and can be made for either cut film or instant (Polaroid) negative film.

In all cases the success of photomicrography and the resulting photographic print depends on:

1. A first-class specimen preparation, appropriately mounted and exactly focused.

2. Good optics, properly aligned and optimally (Koehler) illuminated (section 1.2.7) to give the best possible image, and a suitable choice of color filters (section 1.2.5).

3. A camera-microscope assembly that is free from vibration.

4. Appropriate photographic materials and processing technique appropriate to giving optimum grain size with adequate contrast and grey scale (see Chapter 28.2.2 and 28.3.3).

5. An appropriate final magnification, attained by printing with an enlarger.

The camera-microscope assembly can present some problems, because fixing photographic devices on the microscope tube tends to allow the transmission of vibrations. Partly, this is because pressing the shutter release removes the beam splitter before activating the shutter, and each action can generate persisting vibrations. When a shutter integral to the camera is used, it should be of the iris diaphragm type, and a focal-plane shutter must be avoided. Always use a cable release to actuate the shutter (see Chapter 28.3.3). Shutters for the timing of exposures are best put in the light path between lamp and microscope. Vibration is no problem with the bellows cameras as long as the light-tight collar around the eyepiece allows the camera and the microscope to be independent and not to touch each other. Vibrations must be kept to a minimum; a sturdy, heavy table that is not attached to a wall is a help.

Film for black-and-white photomicrography should be a panchromatic film of ASA 60 to 100 for general purposes (to keep the exposures at a manageable level). For phase-contrast photomicrography, a faster film (ASA 300 to 400) is needed because the light levels are much lower. An exposure meter allows repetition of values established by test (see Chapter 28.2.1).

If an old bellows-type photomicrographic camera is available, acquire it because it is the best apparatus for the highest quality work, even though it is not the most convenient to use.

When exposing negatives and printing micrographs, think about the final magnification and the detail to be shown. Use film processing that enhances contrast with reasonably fine grain (see Chapter 28.2.2.4), as specified by the manufacturer. Grain in a developed negative dictates that you should not enlarge the film more than 2 to 2.5 times in printing. Good microscopy of most ordinary bacteria allows a total magnification of ×1,500, which is needed for fine detail and can be attained on the film (see section 1.6.1. for the determination of magnification). The larger images taken on a bigger area (4 by 5 in. [ca. 10 by 13 cm]) using cut film in an old-fashioned camera, for which enlargements of 1.25 to 1.5 times are usually sufficient, are better than higher enlargements of much smaller images on 35-mm film. If the micrographs are not as sharp and as good as they appeared in the properly adjusted microscope, try again. Focusing is critical, whatever the camera used. The trick with bellows cameras is to use a plain glass insert and to focus on the image obtained with a focusing magnifier adjusted to focus on a mark on the inside of the glass. The traditional ground glass and black hood can still be used effectively.

Prints should be enlarged to a format that allows easy visibility of the important detail, with the object occupying most of the frame.

1.9. REFERENCES

Books on light microscopy and photomicrography, ranging from encyclopedias to paperbacks, are readily available. However, books can give only basic principles. Satisfying microscopy is best learned through direct instruction and practical experience, not a small part of which is the art of preparing worthwhile specimens. No single book is directed toward microscopy for bacteriologists, but the principles are the same for all users. The list presented below is representative of the books available.

1. **Barer, R.** 1968. *Lecture Notes on the Use of the Microscope.* Blackwell Scientific Publications, Oxford.
   The advice of a master microscopist.
2. **Boyde, A.** 1990. Confocal optical microscopy, p. 185–204. In P. J. Duke and A. G. Michette (ed.), *Modern Microscopies.* Plenum Press, New York.
   A description of equipment and applications.
3. **Bradbury, S.** 1976. *The Optical Microscope in Biology.* Edward Arnold, London.
   A short paperback book that deals clearly with matters of resolution and modern forms of microscopy.
4. **Bradbury, S.** 1984. *An Introduction to the Optical Microscope.* Oxford University Press, Oxford.
   A short handbook with simple explanations of lens systems.
5. **Cargille, J. J.** 1975. *Immersion Oil and the Microscope.* Technical reprint 10-1051. R. P. Cargille Laboratories, Cedar Grove, N.J.
   Another excellent booklet produced by a manufacturer.
6. **Culling, C. F. A.** 1974. *Modern Microscopy—Elementary Theory and Practice.* Butterworth & Co., London.
   Another short paperback book.
7. **Eastman Kodak Co.** n.d. *Photography through the Microscope.* Eastman Kodak Co., Rochester, N.Y.
   An example of the booklets produced by manufacturers involved in aspects of microscopy. These booklets are obtainable from such firms and their agents in updated versions and are generally excellent.
8. **Engle, C. E. (ed.).** 1968. *Photography for the Scientist.* Academic Press, Inc., New York.
   General aspects of scientific photography, including photomicrography.
9. **James, J.** 1976. *Light Microscopic Techniques in Biology and Medicine.* Martinus Nijhoff Medical Division, Amsterdam.
   A fine book on theory and practice, with emphasis on the latter. It provides good advice on special and advanced techniques, including phase-contrast, interference, dark-field, polarization, and fluorescence microscopy.
10. **McKinney, R. M., and W. B. Cherry.** 1985. Immunofluorescence microscopy, p. 891–897. *In* E. H. Lennette, A. Balows, W. J. Hausler, and H. J. Shadomy (ed.), *Manual of Clinical Microbiology,* 4th ed. American Society for Microbiology, Washington, D.C.
11. **Mollring, F. K.** 1981. *Microscopy from the Very Beginning.* Carl Zeiss, Oberkochen, Germany.
    Another excellent booklet produced by a manufacturer.
12. **Quesnel, L. B.** 1971. Microscopy and micrometry, p. 1–103. *In* J. R. Norris and D. W. Ribbons (ed.), *Methods in Microbiology,* vol. 5A. Academic Press Ltd., London.
13. **Quesnel, L. B.** 1972. Photomicrography and macrophotography, p. 276–358. *In* J. R. Norris and D. W. Ribbons (ed.), *Methods in Microbiology,* vol. 7B. Academic Press Ltd., London.
    Quesnel's two articles are useful resources for optical details and practical advice on microscopy and photomicrography for microbiology in particular.
14. **Ross, K. F. A.** 1967. *Phase Contrast and Interference Microscopy for Cell Biologists.* Edward Arnold, London.
    An excellent explanation of the theory and practice of phase-contrast and interference microscopy. Also practical discussion of photographic techniques applied to microscopy.
15. **Shillaber, C. P.** 1944. *Photomicrography in Theory and Practice.* John Wiley & Sons, Inc., New York.
    Nothing is likely to replace this classic text, which deals exhaustively but readably with the properties of objective lenses, oculars, and condensers. It sets out the practice of good illumination, weighs the advantages of different mounting media, and deals with both theoretical and practical bench microscopy; however, it antedates phase microscopy.
16. **Shuman, H., J. M. Murray, and C. Di Lullo.** 1989. Confocal microscopy: an overview. BioTechniques **7:**154–613.
    This review includes an example of three-dimensional reconstruction.
17. **Slayter, E. M.** 1970. *Optical Methods in Biology.* John Wiley & Sons, Inc., New York.
    A source book for the theoretical bases of most forms of microscopy and for analytical processes including diffraction, spectroscopy, and related optical techniques. It is concerned with principles and not practice.
18. **Smith, R. F.** 1990. *Microscopy and Photomicrography—a Working Manual.* CRC Press, Inc., Boca Raton, Fla.
    A professionally illustrated procedural manual with minimal theory. Useful for a beginner with no experience.
19. **Spencer, M.** 1982. *Fundamentals of Light Microscopy.* Cambridge University Press, Cambridge.
    A useful general survey.
20. **Wang, Y.-L., and D. L. Taylor (ed.).** 1989. *Fluorescence Microscopy of Living Cells in Culture. Part A. Fluorescent Analogs, Labelling Cells, and Basic Microscopy.* Academic Press, Inc., New York.
    A volume with helpful technical advice.

Chapter 2

Determinative and Cytological Light Microscopy

R. G. E. MURRAY, RAYMOND N. DOETSCH, and C. F. ROBINOW

Specific morphological details are required to characterize a bacterium; these are usually determined by means of light microscopy, but some require electron microscopy (see Chapter 3). Some of the light microscopy methods employed are time honored and trace their genesis to the early days of bacteriological science.

The first approach to the study of natural populations is morphological classification, an assessment of relative numbers, and an idea of the complexity of the bacterial mix in advance of any attempt at cultivation. More specialized methods estimate the proportions of growing and nongrowing cells and the productivity of the populations (9, 12). Despite the problems in choice of method and accuracy of results (2), light microscopy provides most of the basic biological information about an ecosystem, e.g., in the extreme thermal gradients around hot springs (25).

The establishment and maintenance of pure cultures (and even the struggle to maintain impure cultures of difficult bacteria) require control by microscopy. Frequently, contamination plagues the cultures maintained by those less practiced in bacteriology, so that regular microscopy and cultural control must be encouraged. The identification or description of bacteria inevitably requires the application of determinative staining methods, including those suitable for recognition of shape, behavior, or life cycles, as well as for the measurement of cells.

The information gleaned from these classical techniques may be useful, but it is important to understand the methodological limitations because any laboratory manipulation may introduce some alteration of form and structure, albeit small in most cases. Despite shortcomings, microscopic observations of bacteria are necessary, but usually not sufficient, factors for identifying them (see Chapter 24). Nevertheless, errors in identification are often traceable to mistakes in judging the shape, Gram reaction, and motility of a new isolate.

2.1. SAMPLING

It is difficult to observe living bacteria directly in natural habitats, and information so obtained may be meagre. There are several reasons for this.

1. Not all environments contain large populations of bacteria per unit mass. Organisms in marine and lake waters, for example, usually must be concentrated by retention on membrane filter discs (cellulose or polycarbonate) or by centrifugation to obtain sufficient numbers for study. Waters are often low in nutrients, and the organisms in them have maintained life by use of their own components, so that division often leads to extremely small cells. The best sites for productive growth may be provided by interfaces on suspended solids, surfaces of rocks, sand particles or soil grains in sediments, and the surfaces of water plants; these sites adsorb and provide nutrients, making adhesion to surfaces advantageous. The bacteria that populate the glass-water interface in laboratory distilled-water bottles are witness to this principle.

2. Certain environments may support too many bacteria per unit mass and must be diluted for examination or cultivation. Sewage sludge and feces contain an astonishing array of bacteria and other microorganisms intermixed with particulate materials. Certain soils and marine muds contain only a few bacteria mixed with much opaque colloidal matter, some of which may be hard to distinguish from bacterial forms.

3. Most bacteria constituting the natural flora of an environment do not reveal particularly distinctive morphological features, such as the star-shaped *Prosthecomicrobium* species, the sheaths of *Sphaerotilus* species, and the trichomes of *Caryophanon* species.

2.1.1. Air

Air can be sampled for the presence of suspended bacteria, although the numbers in a unit volume may be quite small, by depositing them on or in nutrient medium and characterizing the resultant growth by microscopy and other determinative methods. The most common situations for air sampling involve aerosol formation in industry or in hospitals.

The simplest approach to air sampling is the disposition of open petri dishes containing a suitable nutrient agar near a suspect source and is applicable when the concentration of organisms is relatively large. When large volumes of air have to be sampled and the organisms have to be characterized, bubble a defined volume (use a flowmeter in the system) through a flask of fluid medium, and then subject it to plate counting (8).

A degree of quantitation is also possible by using a slit-sampler device that impinges air on an agar surface

(1). The slit-sampler is usually a portable apparatus and has a mechanical pump evacuating a chamber in which a petri dish of a nutrient agar is mechanically rotated under a slit (adjustable width) whose length is exactly the radius of the plate. This slit permits a measured volume of air to contact the surface as the plate turns, and the colonies that develop on incubation can be related to volume by calculation with data obtained from a flow meter.

Filter sampling methods for bacteria (1) are derived from those used to sample particulates of industrial or public health importance. Filter membranes and suitable filter holders for aerosol assay are provided by all the producers. The pore size used for bacteria is usually 0.45 or 0.8 μm. Bacterial spores and vegetative cells are both counted (section 2.1.4), given that the cell content/volume assayed is appropriate, either by performing direct microscopic count or by depositing the plate-sized (90-mm) membrane on a nutrient surface for a count of CFU after incubation and identification. However, many vegetative bacterial cells die rapidly on an air filter because of excessive desiccation, retention on that surface is uncertain, and viable counting is inaccurate. However, the cells can be collected in a fluid medium, and a defined volume is then passed through a fluid-passing filter (usually 0.2 to 0.45 μm) for viable counting as described above. Replicate filters are required for multiple differential media and to allow for variation in the number of accumulated cells.

It is also possible to do direct counting of bacteria on membrane filters by using epifluorescence microscopy, as is done for waters (section 2.1.4).

2.1.2. Water

Since bacteria in aqueous suspension have optical properties similar to those of water and give minimal contrast in ordinary transmission light microscopy, examine specimens by positive or negative phase-contrast or dark-field microscopy. If these are not available, lower the condenser and reduce the illumination to increase the contrast obtainable by ordinary microscopy.

Direct microscopic examination of marine water samples generally reveals few bacteria. Use filtration techniques to concentrate organisms for estimation of total number per unit volume, as well as for direct visualization of different morphological types. Centrifuging water samples from most lakes, rivers, and estuaries leads to a remarkably large gelatinous pellet, which includes polymers of diverse origins, entrapped particulates, and a diversity of bacteria and other microorganisms. Wet mounts of the pellets examined by phase-contrast microscopy or preparations stained with simple basic dyes reveal the biota for preliminary morphological classification (section 2.2.1).

Membrane filters provide a rapid and effective means of sampling from fluids for direct microscopic counting of bacteria. They can also be used for viable counting by laying a duplicate filter membrane on top of a suitable nutrient agar plate and allowing colonies to form. In both approaches, the bacterial content of the volume filtered has to be appropriate. Combining the approaches of direct total counts and viable counts allows assessment of the proportion of viable organisms in a great variety of aqueous samples, from natural waters to foods and oil or solvent emulsions. Each type of sample requires a selection of appropriate filters and methods. A comprehensive technical review of membrane filtration methods is provided by Brock (1), and other articles give a good account of particular methods (4–6, 12).

A recommended filtration procedure for direct counting (36) is as follows.

1. Prestain polycarbonate filters (diameter, 25 mm; pore diameter, 0.2 μm) for 5 min in a 0.2% (wt/vol) solution of Irgalan black (color index acid black no. 107; Union Carbide Corp., New York, N.Y.) in 2% (vol/vol) acetic acid.
2. Rinse the filter in cell-free distilled water and place it wet on a cell-free glass filter apparatus (Millipore Corp., Bedford, Mass.). Prestained filters, dried after rinsing, can be stored for future use.
3. Fix the cells in the water sample by adding neutralized (with $BaCO_3$) glutaraldehyde to a concentration of 0.1% (vol/vol).
4. Add the fixed sample to the filter funnel, and then add acridine orange (40% dye content) to the sample at a concentration of 0.1% (wt/vol) in 0.02 M Tris (pH 7.2) (at 20°C) to make a final concentration of 0.02% (wt/vol).
5. After staining for 3 min, draw the water sample-acridine orange through the filter membrane by suction (122 mmHg).
6. Remove the membrane from the filter apparatus and place it over a drop of immersion oil (type I, F, or A; Cargille Laboratories, Inc., Cedar Grove, N.J.) on a glass microscope slide. Place another drop of oil on top of the membrane, followed by a glass coverslip. The steps must be performed rapidly to prevent the filter membrane from becoming dry. Examine the preparation by epifluorescence microscopy with a 100- or 200-W halogen lamp, a BG-12 excitation filter, an LP-510 barrier filter, and an FT-510 beam splitter. Bacteria on this filter will fluoresce green, and individual morphological types may be observed.

Another, time-honored technique for scanning the biota that attach to surfaces, popularized by A. T. Henrici decades ago, involves suspending slides or coverslips for hours or days in the water being sampled and retrieving them for staining and microscopy. Many organisms from fresh and salt water are able to attach to surfaces.

2.1.3. Soil

Nonselective techniques are available for direct visualization of soil bacteria (19, 49). Special glass capillaries of rectangular cross-section are available as "Microslides" in various widths (Carlson Scientific Inc., Matteson, Ill.). These have been developed with one wall quite thin (0.17 mm). The "peloscope" consists of a bundle of several (say five) rectangular capillary tubes 1 to 2 cm long. Fill the tubes by capillarity with sterile distilled water. Attach the bundle to a holder (with an appropriate depth mark on it) by means of a rubber band, and place it vertically in the soil. After suitable periods, remove a sample capillary, wipe the outside with moist lens paper, and observe through the thin wall the "microbial landscapes" that grow into the capillaries. Pelscopic capillaries may be made into permanent preparations by use of special techniques (49).

Another simple means of obtaining a selection (but by no means a complete one) of soil organisms for microscopic examination consists of placing the soil sample in the bottom of a watch glass or small dish, adding water *below* the sample by use of a Pasteur pipette until a water surface is formed above the sample, and floating coverslips or electron microscope grids on the surface to sample the organisms in the surface film for examination.

2.1.4. Direct Total-Cell Counting

Direct counts can be made on bacterium-retaining filters either by use of epifluorescence microscopy after appropriate staining of the cells on a portion of the filter membrane or by use of phase microscopy on a clarified portion of membrane (1). The counting is done for large numbers by using an appropriate eyepiece graticule ruled in squares (counting the cells within a square or touching two of the sides, counting a statistically sufficient number of squares, and relating the counts to the total area counted, the filtration area, and the volume filtered). For smaller numbers, use a statistically appropriate number of total fields of the objective in use (which has to be calibrated as to the area observed for subsequent relation of the counts to the filtration area and volume filtered). Use filters of fine porosity (0.2 μm), because many bacteria in waters are small. Filters used for fluorescence microscopy should have minimum reflectance and be stained black (as purchased or, before use, with Irgalan black [section 2.1.2] for 5 min and then rinsed). The filters should also be tested to be sure they do not fluoresce, which can be a problem with some filter membranes. Other than this, there is no clear choice between membrane materials. The size of filter, e.g., 25 or 90 mm in diameter, must be appropriate for the volumes and cell concentrations involved and for the available support apparatus (e.g., syringe attachment or vacuum system).

The filters must be made transparent for counting by phase or ordinary microscopy. Dry the filters after staining them with a basic dye (0.01% thionine). Clear a cut-out portion of the filter with a drop of immersion oil or xylene on the slide and on the membrane before covering with a coverslip. When doing filtrations in the field, dry the filter immediately after filtration; later, wet a cut-out portion for appropriate staining and then dry for microscopy (as above).

The most usual method used for a total count is the acridine orange direct count (often referred to as AODC) by fluorescence microscopy (1, 3, 4, 7). Stain the bacteria either by adding acridine orange to the sample to give an acridine orange concentration of 0.01% before filtration or by holding a small volume of 0.01% acridine orange on the filter for 1 min after filtering the sample. Other fluorochromes, such as 0.01% fluorescein isothiocyanate, can be used. Diluent or rinsing waters should be of an appropriate quality and free of cells; e.g., samples from the sea require filter-sterilized artificial seawater medium. Formaldehyde (2%) may be used to prevent growth in samples. Published applications of the methods should be consulted, and consideration should be given to the sampling and counting strategies (1, 9, 14, 36). Also see Chapter 1.6.1.

2.1.5. Direct Viable-Cell Counting

Viable-cell counting from a natural sample (*without* added formaldehyde) requires filtration of a set of appropriate dilutions and placing of the filters on a suitable nutrient agar for counting the colonies formed after incubation. Viable counts can also be an estimate based on the observed frequency of dividing cells, although pitfalls abound (1, 46), or an estimate based on recognition and counting of cells on a membrane filter that are synthetically active and incorporate a given substrate (1, 7, 58, 59). Relating the figures obtained by the latter methods to the total direct count allows determination of the proportions of live to dead bacteria. The accuracies of these methods are not great and can be open to question (2, 4, 9, 14). Cells in water samples do not necessarily revive easily for plate counting, and the sampling and dilution errors make for variability in results.

The **nalidixic acid method** is representative of the approaches to viable counts made by estimating synthetic activity in bacteria (9). As with all viable counting methods, sterility and aseptic technique are important. In this method the intention is to let the cells grow bigger and more stainable but not divide. The procedure compares a fixed (2% formaldehyde) aliquot of the sample with an unfixed aliquot to which is added 0.025% yeast extract as a growth substrate plus 0.002% nalidixic acid as an inhibitor of DNA synthesis. Incubate the amended aliquot for a period of 6 h. Filter the samples, and stain with 0.01% acridine orange or fluorescein isothiocyanate for epifluorescence microscopy. Viable cells elongate when exposed to yeast extract and nalidixic acid, inducing nonseptate filaments in bacillary forms. Elongated cells are enumerated as viable cells. Gram-positive cells are relatively insensitive to nalidixic acid, so in some environmental samples the nalidixic acid method may underestimate their numbers and one of the following methods may be more appropriate.

Vital fluorogenic dyes, such as 0.02 to 0.05% fluorescein diacetate, in a diluent appropriate to the sample, are nonfluorescent when taken up by cells but are hydrolyzed by the nonspecific esterases in cells that are probably alive but remain unchanged in cells that are probably dead (20, 39). The released fluorescein is easily detected by fluorescence microscopy (with either transmitted illumination or epi-illumination), and the method may be applied to samples on filters as described above for quantitation by epifluorescence microscopy.

Microradioautography can be applied to detection of the uptake and incorporation into active cells of ^{3}H-labeled glucose, acetate, amino acids, or thymidine (59). Specific activities of the labeled substrates in the samples after treatment should be in the region of 0.2 to 0.4 μCi ml^{-1} for substrate concentrations of 1 to 2 μg liter^{-1}. Treat the samples for 2.5 h, and then fix them in 2% formaldehyde. Filter the cells onto 0.2-μm porosity filters for application in a darkroom to an appropriate autoradiographic emulsion-coated slide (NTB-2; Eastman Kodak, Rochester, N.Y.), which is held in a light-tight cassette for a 3-day exposure. Remove the filter in a darkroom, leaving the organisms behind in the emulsion, which is then developed and fixed (see Chapter 28.5) to give clusters of silver grains around labeled cells. Then stain the preparation with acridine orange. After this, the emulsion must be destained in citrate buffers, covered by spraying a thin layer of gelatin (2% containing 0.05% chrome alum) to obscure irregularities impressed by the Nuclepore filter, and topped with a coverglass for epifluorescence microscopy. The method is complex, and the original papers should be consulted (58, 59).

2.1.6. Detection of Specific Bacteria

Recognition of specific bacteria in the environment by microscopy requires the application of either immunofluorescence techniques (see Chapter 5.5.6), direct serological procedures (see Chapter 5.3.10), or molecular methods (12) by using species-specific RNA or DNA probes (see Chapters 26 through 28).

2.2. PREPARATION

Microscopy is effective and important in the primary characterization of organisms either in the sample or growing in liquid media, but it is most often applied to preparations from colonies on agar plates. A variety of staining methods used to determine the characters useful for description and identification of bacteria are compiled in more specialized reference works (3, 7, 10, 12, 13, 17). These microscopic observations may determine the procedures to be followed for isolation and further characterization.

Descriptions of colony form on agar media are based on appearance to the naked eye when colonies on petri plates are large enough, after 1 to several days of growth. Survey of the large variety of organisms from nature that grow on solid nutrient media, rich in peptones, etc., is made difficult by the overgrowth of the rapid growers (or, worse, fungi) obscuring the low-frequency and slow-growing bacteria. The better policy at the outset is to use an appropriate medium for common bacteria but to dilute the nutrient concentration ca. 1:10 or to use a medium containing 0.1 to 0.3% each of peptone and yeast extract, plus salts and trace elements. Restrictive nutrition and temperatures of incubation (15, 22, or 30°C for mesophiles) are helpful, because growth is relatively slow and the colonies are smaller and separate, even if close together. It is then possible to recognize various colony forms by using a binocular (stereo) dissecting microscope with the help of a moveable light or tilting stage. With this setup, very small colonies can be picked up by using a steady hand and a straight wire, small loop, or fine capillary drawn at the tip of a Pasteur pipette for making a transfer to fresh media or to a preparation for high-power microscopy. Effective procedures have been developed for the cultivation of specific bacteria from specific environments or for the selective culture of particular kinds of bacteria (see Chapters 7 and 8).

Determinative observations of form and growth habit are made on living cells from nature or from cultures, and more definitive cellular features are made on fixed and stained preparations. Equally important is the checking of cultures (especially cultures in liquid media) for morphological consistency with what is believed to be growing in them; the detection of contaminants requires constant vigilance.

Nature provides niches of an extraordinary variety, with and without a wide range of extraneous materials, containing all possible variations in numbers and species of biota. The best ways of sampling and studying the bacteria in these niches, both in situ and in vitro, involve ingenious application and modification of the approaches outlined in this manual and of the methods applied in specialized publications and journals.

CAUTION is required in the handling of specimens, loops, slides with droplets on them, and coverslips, as well as in many of the procedures in making and manipulating preparations for microscopy, because they offer manifold opportunities for the contamination of fingers, benches, clothing, and equipment. Appropriate aseptic precautions and attention to technical details and hazards are essential, especially when working with pathogens (Chapter 29), but many organisms are opportunists and must be respected. Remember that even fixed preparations of organisms containing endospores may retain a level of viability and are therefore a hazard.

Tools for handling bacterial specimens include a large loop (diameter, 3 mm), a small loop (diameter, 1 mm), and a straight inoculating needle. Fashion them from platinum or nichrome wire (the latter is preferable for the needle) and of such a length and thickness that, when mounted in a handle of glass or aluminum, they are easily controlled for purposes of streaking plates, sampling colonies, and inoculating tubes of media. Use platinum alloyed with 5 to 10% iridium to harden it and make it springy and controllable. In addition, a supply of sterile Pasteur pipettes is almost essential.

2.2.1. Living-Cell Suspensions

There are several approaches to the examination of living bacterial cells, either in a natural specimen or in culture. The initial preparation is usually a "smear" or film made by spreading a droplet by use of a wire loop over a few square centimeters of a microscope slide and then drying and fixing it (section 2.2.4) before simple staining (section 2.2.5). Cultures grown in liquid media can be used with the proviso that dilution with sterile medium (e.g., by adding a loopful to a drop of medium on a slide) is generally required for reducing the population for microscopic study to a useful concentration. Cultures grown on solid media can be suspended on a microscopic slide with a sterile loop or needle in a drop of the sterile liquid medium or a diluent (*not tap water*) to a faint turbidity. Collect cells that are in very low concentration either by centrifugation or on a membrane filter (e.g., a 0.2-μm membrane filter in a syringe adapter), which is gently pressed on a slide to transfer cells for simple staining (60). Remember that the age and conditions of a culture may affect the size and shape of cells, their surface components (flagella, pili, and capsules), their inclusions, and the selection of mutants. The culture conditions must be clearly defined when describing bacterial form and structure. Examination of living cells under conditions appropriate for growth is important to defining growth habit (e.g., mycelial or differentiated forms) and behavior (e.g., motility), as described for the following methods.

2.2.1.1. Wet mounts

In preparing "wet mounts" of specimens, place a loopful or 1 drop (ca. 0.05 to 0.1 ml) of sample on a clean and "degreased" (by prior heating for 20 min at 400°C) microscope slide, 0.8 to 1.0 mm thick, and cover it with a glass coverslip. The latter should be 18 to 22 mm^2 and from 0.13 to 0.16 mm thick (no. 1 thickness) to allow an oil-immersion objective to focus through it. To prevent convection currents, "drifting," and drying, seal the edges of the coverslip to the slide by applying Vaspar (a mixture of equal parts of petrolatum and paraffin wax), petrolatum, birthday-candle wax (using the wick as a brush after melting the wax, not lighting it, in a pilot flame), or clear nail polish.

The motility of *strictly* aerobic bacteria can be observed only for a brief time in these preparations, because the bacteria cease moving once the oxygen is depleted. The inclusion of small air bubbles prolongs activity. Heat from the light source may interfere with motility, and it may be advisable to interpose a heat filter (a plane-sided vessel of water) if observations are to be made for a long period. A green filter is easy on the observer's eyes and it utilizes the best corrections of the lenses.

Phase-contrast microscopy is recommended for examining wet mounts, and the thinnest possible film should be used for best results. Placing a piece of blotting paper (remember hazards) over the coverslip before sealing assists in drawing off sufficient fluid to give a satisfactory thin film for viewing and dries the exposed surfaces. The best view of the cells may be obtained if a thin film of 1.5% agar or 20% gelatin is dried on the surface of the slide to absorb the drop of culture, as it is covered immediately with a coverslip.

2.2.1.2. Hanging-block mounts

Cut a block from a nutrient agar plate showing early growth of bacteria. Suspend the block from a coverslip over a hole in a thick slide made of clear plastic. The live bacteria are still surrounded by the growth medium, and they lie flat against the coverslip, preserving the natural arrangement, which is not possible in the usual smear preparations. Because of the thick block and optical problems, the resolution is not of a high order. Despite these shortcomings, the technique is readily adapted to determining growth habits or to monitoring the effects of toxic chemicals and antibiotics on growth and form.

2.2.1.3. Hanging-drop mounts

Make a hanging-drop mount by placing a drop of specimen at the center of a clean but not degreased coverslip. Invert it over the well of a depression slide in such a manner that the drop does not move or contact the sides of the well. Apply a small drop of water at the outside edge of the coverslip to provide a seal, and maintain the coverslip in place.

Hanging-drop preparations have disadvantages in that concavity of the well, curvature of the drop, and increased thickness of the depression slide introduce optical aberrations. It is difficult to focus on a single organism, especially if it is motile, because of the depth of the medium. It is best to focus initially on the edge of the drop. Some of these problems are avoided by using slides with flat-bottomed wells.

A useful device is a thick (2 to 3 mm) slide made of clear plastic with a hole (1 to 1.5 cm in diameter) drilled through its center. Place a coverslip with a hanging drop over the hole, and hold it in place by a touch of immersion oil or water to the edge of the coverslip. The chamber may be closed, if desired, by attaching a coverslip to the underside.

2.2.2. Slide Cultures

2.2.2.1. Agar slide cultures

For longer observation under better conditions for microscopy, slide cultures are superior to hanging blocks and can be made thin enough to be studied under effective optical conditions. It is best to use phase microscopy, although much can be learned by using ordinary optics. Dip clean, sterile slides twice (or more times for thicker layers) into molten agar medium in a petri dish. Wipe the agar from the underside with dampened tissue. Inoculate a selected area of the upper surface under a dissecting (stereoscopic) microscope, using a finely drawn-out glass fiber at the end of a Pasteur pipette or glass rod. Wet the tip of the fiber with peptone water, lightly touch it to a colony or culture of the organism, immediately move it to a drop of peptone water on the fresh agar surface, and spread. Trim away the agar around the selected area, place a coverslip on top, and seal the edges with wax. Practice rigorous aseptic technique. If the square of nutrient agar is smaller than the coverslip, there is a supply of air and minimal drying. If oxygen is crucial to growth, flood a thin agar layer on top of a thin polyethylene or polycarbonate film stretched over a hole in a plastic slide (section 2.2.1.2) to allow access of oxygen from below through the gas-permeable film (27). Appropriate modifications and circumstances can be contrived for anaerobes. These slide cultures can give adequate information on shape, size, and growth habit.

2.2.2.2. Gelatin-agar slide cultures

Changes in the internal detail of bacterial cells are made visible in phase microscopy by using a solid nutrient agar medium of high refractility (21), which can be provided by adding 14 to 18% gelatin (14). The higher concentrations are suitable for gram-positive bacteria, whereas gram-negative bacteria may need lower concentrations. This technique allowed Mason and Powelson (41) to monitor and photograph the patterns of nuclear division in growing bacteria. Polyvinylpyrrolidine (up to 30%) can be used instead of the gelatin (56).

If desired, slide cultures can be fixed for staining procedures. The coverslip, with undisturbed culture and medium attached, must be removed from the slide for appropriate cytological fixation and staining (sections 2.2.7.1 and 2.2.7.2). This can be facilitated by fashioning a "handle" of a strip of Mylar tape or a self-stick label to allow the coverslip to be lifted up with the medium. It is also possible to fix and embed these preparations for electron microscopy, but such flat embeddings (31, 48) need considerable skill in handling and sectioning.

2.2.2.3. Petri dish cultures

Living bacteria, on or in agar, are clearly seen through an inverted petri dish by using a 10× or lower-power objective, and the nature and condition of what is being examined are established without further manipulation. Two points about work with petri dishes are worth making: (i) the best illumination for observations made through the bottom of a dish is provided by the concave mirror, after removing the condenser; and (ii) the distortion of images because of scratches in the glass or unevenness in shape of the bottom of the dish can be cancelled out by a drop of immersion oil topped with a coverslip to give a homogeneous light path and an unflawed surface.

2.2.3. Negative Staining

Negative staining provides the simplest and often the quickest means of gaining information about cell shape, cell breakage, and refractile inclusions in cells such as sulfur and poly-β-hydroxybutyrate granules, and about spores (15).

Procedure

1. Place a droplet of 7% (wt/vol) nigrosin on a coverslip (no. 1 thickness).
2. Mix a small sample of bacterial culture or natural specimen into the droplet.
3. Place another coverslip, rotated at 45°, over the droplet, and slide the two coverslips apart to form a thin film on each. Alternatively, spread the droplet over most of the coverslip with a loop until the fluid starts to dry and there are thicker and thinner parts of the film visible.
4. Let the film dry *completely.*
5. Place the coverslip on a slide, *film side down,* and fix it in place with several spots of wax along two edges (use the wick of a birthday candle as a brush after a quick melting of the wax near a flame, without lighting the wick).
6. Observe appropriately thin areas of the film under oil immersion. Only experience teaches how much of a culture to mix into how large a droplet and the appearance of an effectively spread film.

These "air-mounted" preparations reveal bacteria unstained and standing out brightly against a sepia background. They can be used to monitor the progress and extent of cell disruption and disintegration (see Chapter 4). Negatively stained preparations should not be used for making cell length and width measurements (see below), because the capsule or slime layer outside the cell wall may exclude the nigrosin and because they display dried, unfixed, and partially collapsed cells. Furthermore, the negatively charged particles of colloidal nigrosin do not react with the bacterial surface because, at physiological pH, it also is negatively charged. Consequently, the dark film dries so that the bacterium appears to be somewhat larger than in life, even if no capsule is present.

Nigrosin solutions can acquire contaminating organisms. A little formaldehyde (0.5%) helps prevent growth and does not harm the solution. Flame-sterilize loops, because the presence of dead bacteria leads to confusion.

India ink also forms a sort of negative stain when added to a wet mount and can display large capsules effectively if the coverslip is pressed down tightly.

2.2.4. Fixation of Smears and Suspensions

A smear of bacterial cells (section 2.2.1) merely dried on a microscope slide will wash off, partially or completely, during staining unless some stratagem is used to ensure that the cells are made to stick to the glass. Furthermore, some important structural components of cells change shape or position or are digested during the process of making a preparation. Preservation of structure usually requires a treatment causing inactivation of enzymes and attaining the cross-linking of macromolecules. Both of these stratagems are termed "fixation," but the latter process should be called chemical fixation. Whatever the fixation procedure, it is important not to assume that all cells are killed. Remember that spores and some vegetative cells are remarkably resistant to heat and to chemical killing. Also, remember that aerosol droplets can be formed and can spread to the surroundings by any manipulation of fluids with loop or pipette and that fingers are easily contaminated.

2.2.4.1. Heat fixation

Heat fixation is the most common procedure used for stabilizing bacterial smears. Allow the smear to dry completely. Pass the underside of the slide several times over the flame of a Bunsen burner to induce adherence. The smear is now ready for simple staining (section 2.2.5), Gram staining (section 2.3.5), or other procedures. Remember that when considering morphological interpretations, the drying and heating kill most vegetative cells and cause some shrinkage and distortion.

2.2.4.2. Chemical fixation

Chemical fixation provides more accurate preservation of shape and structure, although it is more time-consuming than heat fixation. Useful procedures are as follows.

a. **Fixation with an aldehyde.** Add Formalin to a suspension to give a 5% solution (vol/vol, i.e., ca. 1.7% formaldehyde), and hold for a few minutes before making and drying a smear. This is a simple and adequate preparation for Gram staining. Alternatively, use 3% (vol/vol) glutaraldehyde in the same fashion.
b. **Osmium tetroxide fixation of wet films.** Prior to drying a smear, expose it to the fumes from 1% aqueous OsO_4 for 2 min in a closed vessel (see section 2.2.7.2 for precautions). Cell arrangements will be better preserved.

c. Fixation on an agar surface

(i) Spread a suspension on an agar surface to give a suitable distribution of cells, and let it soak into the agar; the bacteria may be allowed to grow for a short time or not, as requirements dictate. Cut out a block of about 1.0 by 0.5 cm, and then lift and invert it, cells down, onto a no. 1 coverslip. Immerse the coverslip and block in Bouin's fixative (section 2.2.7.1), or 3% glutaraldehyde for fixation through the agar for 45 min or more. While the coverslip and block are under the fixative, flick off the block with a needle; most of the cells should remain attached to the coverslip. The coverslip can then be washed for 10 to 20 min in a vessel with a gentle stream of water or washed in 70% ethanol with several changes before simple staining. The procedure is given in detail in section 2.2.7.1.

(ii) Alternatively, spread a loopful of suspension on an agar block (ca. 1 cm^2) supported on a slide and allow the fluid to soak into the agar. Place the slide in a closed vessel with OsO_4 vapor for 2 min, as above. Make an impression transfer of the cells onto either a slide or a coverslip for further processing, either as with a smear or as for a cytological preparation (section 2.2.7.2). However, OsO_4-fixed bacteria do not stain well with simple stains unless subsequently treated with 70% ethanol or another chemical fixative.

Detailed instructions and suggestions for the fixation, handling, staining, and mounting of coverslip preparations for cytological purposes are given in section 2.2.7. There is considerable advantage to microscopy on chemically fixed and lightly stained organisms on coverslips, which are *mounted in water and not allowed to dry at any time in the processing.*

2.2.5. Simple Staining

Morphological studies of bacteria in culture or in natural specimens are generally done on heat-fixed smears that are stained with basic dyes. When a single dye is used, the process is referred to as "simple staining." Simple staining of smears on slides is accomplished conveniently on a rack made of glass rods linked by short lengths of rubber tubing or of linked brass rods supporting slides over a sink or suitable vessel for disposal of stains and wash water. Coverslip preparations are best handled in Columbia staining dishes, as described in section 2.2.7.

a. Procedure for slides

1. Make a smear (section 2.2.1) of either living or chemically fixed cells (section 2.2.4.2).
2. Let the smear dry completely.
3. Heat fix (section 2.2.4) to gain adherence to the glass.
4. Flood the heat-fixed or chemically fixed smear briefly with a solution of a basic dye such as crystal violet (for 10 s) or methylene blue (for 30 s), and then gently rinse with tap or distilled water and blot dry with absorbent paper.

b. Preparation of staining solutions

Crystal violet (Hucker formula)

Crystal violet	2 g
Ethanol, 95% (vol/vol)	20 ml
Ammonium oxalate, 1% (wt/vol) aqueous	80 ml

Dissolve the crystal violet in the ethanol. Then add the ammonium oxalate solution, and allow to stand for 48 h before use.

Methylene blue (Loeffler formula)

Methylene blue chloride	1.6 g
Ethanol, 95% (vol/vol)	100 ml
Potassium hydroxide, 0.01% (wt/vol) aqueous	100 ml

Prepare a saturated solution of methylene blue by adding the dye to the ethanol. Then add 30 ml of the supernatant solution to the potassium hydroxide.

If the cells stain lightly or not at all with one of the dyes, it may be necessary to stain them with carbolfuchsin (section 2.3.6.1), and they may even be acid fast (section 2.3.6). A few seconds of exposure to this dye is usually enough for most bacteria; it is a very intense dye and overdoes the staining for most bacteria.

2.2.6. Permanent Mounts

For extended preservation, the stained and dried film of bacteria on a coverslip should be cleared with xylene and then inverted on a drop of Canada balsam on a slide. Wipe excess balsam away with a tissue moistened with xylene. The balsam hardens in a day. Neutral mounting media consisting of a plastic (polystyrene) dissolved in solvent (toluene or xylene) and containing a plasticizer (tricresyl phosphate) are marketed under such names as Permount (Fisher Scientific, Pittsburgh, Pa). Some of these media may be painted onto a stained film on a glass microscope slide without the need for a coverslip. Permount is neutral and does not become acid or discolor with age, nor does it tend to trap bubbles under the coverslip. Stained films of bacteria on coverslips, mounted in water and sealed with petrolatum, clear nail polish, or candle wax, can be kept if evaporation is prevented. They remain useful for only a few days kept in a moist chamber, but they provide excellent material for photomicrography.

2.2.7. Cytological Preparations

The advent of electron-microscopic methods has diminished in part the drive to gain the maximum cytological information from light-microscopic methods. However, these methods are very useful in the general study of bacteria and in bridging the gaps in the level of resolution between the detection of components in the living cell on one hand and in electron microscope images on the other. The preparation of specimens for cytological light microscopy is not difficult, and the results are rewarding both aesthetically and scientifically (15, 44, 45, 52).

Useful items are as follows.

1. Coverslip forceps: Forceps with flat, nonserrated gripping surfaces bent at an angle of 30°, as used by stamp collectors, are convenient for picking up coverslips. Curved, pointed, and nonserrated jeweller's forceps are also useful for this purpose.

2. A knife for cutting, lifting, and transferring agar blocks: All sorts of tools will do this, but real knives or

scalpels are ruined by sterilizing in a flame. Therefore, obtain some thick Nichrome wire and hammer a narrow leaf shape at the tip on an anvil. Hold this wire knife, cut fairly short, in a loop holder, and sterilize it in a flame for use.

3. Coverslips: Use no. 1 thickness coverslips, because they must be thin enough for the oil-immersion objective to attain focus on the cells through the glass sheet.

4. Columbia staining jars: These make for easy handling of coverslip preparations during storage, rinsing, and staining, and they are made to take 22-mm² coverslips (A. H. Thomas Co., Philadelphia, Pa.).

2.2.7.1. Bouin fixation

The most lifelike preparations are obtained by fixing cells in situ, and these preparations are the most useful for the description of shape and growth habits of individual cells and the effects of treatments on these cells; it is important to consider observing living cells as a control for fixation artifact (section 2.2.1). Grow the bacteria on a solid medium in a petri dish that has been inoculated to provide, after only a few cell divisions, either a lawn of cells in a single layer or separate microcolonies in which the cells have not piled upon each other. Check the appropriate state of the growth by low-power (10× objective) microscopy on the plate during growth. If preparations must be made from fluid media, transfer an appropriate amount of growth to an area on a well-dried agar medium and allow the fluid to sink into the agar. In each case, cut blocks of agar from a region showing appropriate distribution of cells. For the best results, cut a block about 1 by 0.5 cm and invert it, *cells down,* onto a 22-mm² no. 1 coverslip without sliding or pushing it on the glass. The cells are then in contact with the glass and retain their arrangement. Pick up the coverslip with the block attached, and slip it gently into a petri dish containing enough Bouin fixative to cover the block completely.

Bouin fixative

Saturated aqueous picric acid	75 ml
Formalin	25 ml
Glacial acetic acid	5 ml

Mix and store in a stoppered bottle. The solubility of picric acid is 1.4 g/100 ml at 20°C, and the saturated solution (i.e., with undissolved crystals in the bottom) should be kept in stock for preparing the fixative when needed. *CAUTION:* Keep picric acid crystals soaked in water, because the dry crystals may be an explosive hazard.

Several blocks can be fixed in a dish. Leave the block undisturbed for enough time for the fixative to penetrate completely (generally, 45 min to 4 h) and to fix the cells to the coverslip. The block may come off by itself, but if not, hold the coverslip down with forceps and flick the block upward with the point of the wire knife, avoiding a sliding movement. Then rinse the coverslip in water and transfer it to 70% ethanol, where it may be kept until staining is undertaken. All these steps are facilitated by transfer of the coverslips to a series of Columbia staining jars.

Stain the Bouin-fixed preparation with dilute solutions of basic dyes. Appropriate stains are 0.01% solutions of crystal violet, thionine, or methylene blue, applied for 30 to 90 s, as determined by trials. Other stains for cytoplasmic inclusions may be useful (section 2.3.12). The basic dyes stain the ribosome-rich cytoplasm preferentially on these better-preserved cells, and the nucleoplasms are unstained so that they are revealed, negatively, in conformations that are remarkably similar to those seen in phase-contrast microscopy of living cells. Unfortunately, Bouin fixation is not appropriate for staining the nucleoplasms with Giemsa stain, even after RNase treatment or acid hydrolysis.

The cell walls are unstained (most convincingly shown in filamentous *Bacillus* spp.), but they can be displayed effectively as follows. Treat the Bouin-fixed cells with 5% (wt/vol in water) tannic acid for 20 min, wash, and then stain with 0.01% crystal violet for a sufficient time to give intense staining of the wall. Alternatively, treatment of the fixed preparation with a saturated solution of mercuric chloride for 5 min also allows staining of the surface with basic dyes such as crystal violet, thionine, and Victoria blue (52). However, no method is entirely reliable, and gram-negative bacteria are the most difficult to treat for display of walls.

Mount the preparations in water or diluted stain (which is helpful if the intensity of staining is low for photomicrography), blot gently to remove excess water from the top and edges of the coverslip, and seal to the slide with wax, Vaspar, or nail varnish. The preparations can be dehydrated and made permanent by using a resin mountant, but the result is usually disappointing. Permanence is best achieved by photomicrography, for which these preparations are ideal.

Some cells do not adhere strongly to the coverslip after fixation through the agar or in the impression film technique (section 2.2.7.2). This problem can be reduced by first coating the coverslip surface intended for the cells with a thin film of serum, egg albumin, or 8% filtered egg white in distilled water and letting it dry. This film may provide faint background staining in the final preparation, but the result is very satisfactory. A similar coating of polylysine, which provides a positively charged interface, is as effective or more so (42).

2.2.7.2. Osmium tetroxide vapor fixation

Appropriate distributions of cells on the surface of agar blocks, whether grown there or transferred from other media, are fixed conveniently for light microscopy purposes by using vapors of osmium tetroxide or of aldehydes such as formaldehyde or glutaraldehyde. This is not effective in destroying all enzymes and therefore is not necessarily sufficient fixation for cytochemical purposes. The procedure for OsO_4 fixation is described because it is the most generally useful. The critical steps are as follows.

1. Prepare a 1 or 2% (wt/vol) solution of osmium tetroxide in distilled water. The chemical comes in weighed, crystalline form in a sealed ampoule, which is broken and dropped into the appropriate volume of water. The chemical is slow to dissolve, so one must make it up 1 or 2 days ahead of need. *CAUTION*: OsO_4 is irritating, and the vapor is harmful to eyes and mucous membranes; therefore, use a fume hood. Store the chemical in a cool, dark place, preferably in brown bottles, to prevent formation of a

black, inactive product. (If there is a lot of black precipitate, it can be reactivated by adding H_2O_2.)
2. Place agar blocks, cells upward and supported on a slide or a portion thereof, in a closed, wide-mouthed, screw-top jar containing a few milliliters of the OsO_4 solution. A bed of glass beads in the bottom supports the slide above the fixative.
3. Allow the vapor to be absorbed for 1.5 to 2 min.
4. Remove the slide, and place a coverslip on the block. Hold the edge of the coverslip with coverslip forceps, and lift the coverslip quickly and vertically to separate it from the block. Alternatively, invert the block onto a coverslip and then flick it off with a knife point so that some of the cells are imprinted on the coverslip.
5. Immediately put the coverslip into 70% ethanol in a Columbia jar, in which it can be stored until ready for staining.

Staining procedures can be those suggested for Bouin-fixed preparations (section 2.2.7.1). The cytoplasm stains directly with basic dyes, such as thionine, and the nucleoplasms do not stain. The latter, however, show as narrower clefts in the stained cell than is the case in the living cell in phase-contrast microscopy. The OsO_4-fixed preparation is suitable for the staining of nucleoplasms with Giemsa stain (section 2.4.3.1) after either digestion with RNase (100 μg/ml in 1 mM $MgSO_4$ for 30 min) or hydrolysis with acid. The latter procedure is as follows.

1. Put a coverslip bearing fixed cells in a 1 N solution of HCl at 60°C for 6 to 8 min.
2. Wash the coverslip in water.
3. Stain in Giemsa solution (10 drops of stock stain per 10 ml of 0.067 M phosphate buffer [pH 6.8]) in a Columbia jar for 10 min or more for most gram-negative species. A superior Giemsa stain is Gurr's "R66," sold as a stock solution (section 2.5).
4. Rinse with buffer, and examine to see whether the cells are adequately stained. (A water-immersion objective is helpful but not essential for this step.)
5. Mount in water or diluted stain. The latter is useful for gram-negative bacteria, because the cytoplasm can be very pale. Seal with wax or nail polish.

The nucleoplasms stain reddish-purple, and the cytoplasm is somewhat pink. Step 1 of this procedure is the basis for the Feulgen test for cytochemical detection of DNA, so that nucleoplasms stain red after hydrolysis and exposure to the Schiff reagent (50). In essence, the same procedure is used to reveal polysaccharide structures (section 2.3.12.3).

Impression preparations, vapor fixed with OsO_4, can be very helpful in the study of changes in cells treated in manifold ways, whether in agar or fluid cultures, e.g., cell changes following phage infection (44). Note, however, that the anionic nucleoplasms respond to the cationic environment (i.e., salts in the medium) at the time of fixation and that this response, if excessive, leads to chromatin condensation (45).

2.3. CHARACTERIZATION

The methods described below have been found satisfactory for revealing characteristics of determinative value for a large number of different bacteria. These techniques, however, are amenable to improvements and modifications; indeed, the literature abounds with statements of minor alterations found necessary to obtain the best results in specific situations. Cytologically rewarding preparations require great care in fixation, staining, handling, and microscopy, as described above (section 2.2.7). The following are sound routine methods, but some of the procedures, especially those for cytoplasmic inclusions, are most effective if applied to chemically fixed preparations with drying avoided at any stage.

2.3.1. Size

Bacterial cell lengths and widths cannot be measured with the degree of precision one might desire, because of certain unavoidable technical difficulties. The boundaries of living bacteria do not appear sharp when examined by phase-contrast or bright-field microscopy, and in phase microscopy they and internal details are obscured by halos. The halo problem can be reduced by mounting the organisms in a medium of a refractive index close to that of their cells, which is done simply by using solutions of gelatin from 15 to 30% (wt/vol) (16), the best results being determined by experiment for each kind of organism and the detail to be observed. A less precise method to reduce the halos involves drying, ahead of time, a thin film of 15 to 20% (wt/vol) gelatin on slides and using these to make wet mounts. The fluid swells the gelatin, and the cells become embedded in it; this immobilizes them under the coverslip and allows better phase relations between the organism and its environment. Similar preparations made on a thin layer of agar are also very useful.

Cells in fixed and stained preparations all suffer some degree of distortion, depending on the technique used; hence, one obtains only an approximation of the true dimensions. Dried, stained films, even if prepared from Formalin-fixed material, do not reveal the cell wall. Tannic acid-crystal violet staining (section 2.2.7.1) does reveal the cell wall and thus makes bacteria appear appreciably wider than they look when live in a wet medium or stain after drying. This staining procedure does not work well with gram-negative bacteria.

Microscopic measurements are made with either a calibrated ocular micrometer or a filar micrometer eyepiece. Calibration is attained by using a stage micrometer, which provides a set of engraved lines at defined intervals (Chapter 1.7). After calibration, the stage micrometer is no longer required, but new calibrations must be made for each objective used.

2.3.2. Shape

For determinative and descriptive purposes (Chapter 25), individual bacterial shapes are designated as straight, curved, spiral, coccobacillary, branching, pleomorphic, square ended, round ended, tapered, clubbed, or fusiform. Perhaps star shaped, stalked, and lobular should be added to these classical terms. Arrangements of individual bacteria are described as single, pairs, short chains (fewer than five bacteria), long chains (five or more bacteria), packets, tetrads, octets, clumps, filaments, and branching or mycelial forms.

When assigning the descriptive terms, it is assumed that natural groupings are not disturbed. For example, excessive shaking may break long chains or cause clumping of single cells. To minimize the breakup of long chains or other fragile associations, add formaldehyde to the culture to a final concentration of 1 to 2% (vol/vol). Centrifuge after 15 min, and resuspend the culture in water for making films or negative stains. This is highly recommended for streptococci. The influence of culture age and medium composition also must be taken into account. When there are doubts, there is no substitute for examining early growth in the undisturbed state (sections 2.2.1 and 2.2.7).

2.3.3. Division Mode

The mode of division used by a given bacterium is ordinarily not seen at first glance, and continuous observation of the living organism in a suitable medium is required (26, 47). Simple staining procedures may not adequately reveal whether division is binary, ternary, by budding, or by fragmentation. It may be necessary to use cell wall staining (section 2.2.7.1) and even electron microscopy (Chapter 3) to determine this behavior.

2.3.4. Motility

Translational movement of bacteria by flagellar propulsion may be observed in wet mounts (section 2.2.1) of specimens by use of, in most cases, the low-power or high-dry objectives. Bacteria vary in their translational velocity, and slow organisms must be differentiated from those showing only Brownian motion. Some flagellated nonmotile ($Fla^+ Mot^-$) mutants have been described. To ascertain the presence of flagella in doubtful cases, as well as to determine flagellar distribution (polar, peritrichous, lateral), staining procedures (section 2.3.11) and electron microscopy (Chapter 3.5) may be required.

When a standard condenser and an oil-immersion objective are used, the light must be reduced by the aperture diaphragm and the condenser must be lowered to improve contrast. Much better resolution and contrast are attained with phase microscopy and (most dramatically) with dark-field microscopy. If the specialized condensers needed for this latter technique are not available, an adequate dark field is possible by the stratagem of using a phase microscope condenser, with the oil-immersion phase plate in place, for illuminating the specimen observed with standard (*not* phase) 10× or 45× objectives (Chapter 1.4.3) and high-power eyepieces (12× to 16×). The condenser is not immersed, but its position for attaining the best dark field is discovered by trial. This trick makes it possible to see even spirochetes under low power!

Observation of living cells sometimes detects another motion effectively described as **twitching** (34), which is a sudden movement or change of position that is believed to be associated with fimbriae (common pili) on the cells.

Some procaryotic protists exhibit a peculiar type of translational movement known as **gliding** when in contact with a solid surface. Gliding is a comparatively slow, stately, intermittent progression parallel to the longitudinal axis of the organism. It is characterized by frequent directional changes and the absence of external locomotory organelles (40). Gliding is often not obvious on the primary medium of isolation but is usually facilitated on solid media containing very small amounts of the nutrients appropriate to the group of bacteria involved. It is best observed on an agar slide culture or directly on an agar plate with a high-dry objective or with a low-power objective and high-power eyepiece. Single bacteria away from the margin of the colony and tracks or trails are often considered indicative of gliding motility. Young, active cultures of filamentous gliders form wavy, curly, or lacelike patterns. Translocation by gliding movement can occur at velocities of 10 to 15 μm/s.

Gliding should be differentiated from **swarming**, which is an expression of flagellar motility on agar under special conditions. Swarming is the active spreading of growth on relatively dry agar surfaces in which the bacteria move as groups or microcolonies (34) or, in some instances (e.g., *Proteus* species), as associations of filamentous swarmer cells. A continuously shifting pattern of organisms develops on the agar surface, ranging from short, tongue-like extensions and isolated comet-shaped groups cruising away from the mass to interlacing bands interspersed with empty areas.

2.3.5. Gram Staining

Gram staining is the most important differential technique applied to bacteria. In theory, it should be possible to divide bacteria into two groups, gram positive and gram negative; in practice, there are instances when a given bacterium is gram variable. Numerous modifications of Gram staining have been published since the method was first developed by Christian Gram in 1884. In essence, the cells are stained with crystal violet and then treated with an iodine solution as a mordant. The intensely stained cells are then washed with ethanol. Gram-positive cells retain the stain, whereas gram-negative cells do not. To make the contrast between the two results obvious, the preparation is counterstained with a contrasting red dye (e.g., safranin) so that gram-negative cells are easily seen.

The Gram reaction is determined by the interaction of crystal violet and iodine and by the integrity and the structure of the cell wall, most particularly the molecular architecture of the peptidoglycan (murein) layer of the eubacteria (24, 28, 54). Cell walls of gram-positive cells do not allow the extraction of the crystal violet-iodine complex from the cytoplasm by a solvent. If the walls are broken or their structure is compromised by autolysis, exposure to lysozyme, or exposure to a wall-targeted antibiotic (e.g., penicillin), the complex is easily extracted and the gram-positive bacteria become gram-negative. The crystal violet dye is small enough to penetrate the interstices of the wall, but the dye-iodine complex is too big to exit (28).

The description of novel bacteria requires recording the Gram reaction, but this description should also include the structural profile of the cell wall observed in sections by electron microscopy (Chapter 3).

The smear of bacteria on the slide can be made directly for routine Gram-staining purposes from a liquid or solid culture, as for simple staining (section 2.2.1) but is best made from a Formalin-fixed, washed sample. The few extra minutes spent suspending the bacteria in 5% (vol/vol) Formalin and concentrating and washing them by centrifugation pay off in good preservation of size and shape, good staining, and the absence of messy precipitates from liquid culture media. In either case, a thin smear should be used for air drying and heat fixation (thick clumps are hard to decolorize).

Gram-negative bacteria may seem to be gram positive if the film is too thick and the decolorization is not completed. Gram-positive organisms, on the other hand, may seem to be gram negative if the film is over-decolorized; this happens particularly if the culture is in the maximum stationary phase of growth. Some *Bacillus* species are gram positive for only a few divisions after spore germination. Furthermore, gram-positive organisms will seem to be gram negative if the integrity of their cell walls is breached by autolysis, external enzyme action (e.g., with muramidase), or direct mechanical damage. It is advisable to prepare light films (faint turbidity) of young, actively growing cultures for best results, since older cultures tend to give variable reactions. A wise precautionary measure is to use known gram-positive and gram-negative organisms as controls.

It is important to standardize the Gram staining procedure, and two methods that give equivalent results are those of Hucker and of Burke (22, 23), as follows.

2.3.5.1. Hucker staining method (7, 23)

Solution A

Crystal violet (certified 90% dry content)	2.0	g
Ethanol, 95% (vol/vol)	20	ml

Solution B

Ammonium oxalate	0.8	g
Distilled water	80	ml

Mix A and B to obtain the crystal violet staining reagent. Store it for 24 h, and filter it through paper before use.

Mordant

Iodine	1.0	g
Potassium iodide	2.0	g
Distilled water	300	ml

Grind the iodine and potassium iodide in a mortar, and add water slowly with continuous grinding until the iodine is dissolved. Store the mordant in amber bottles.

Decolorizing solvent

Ethanol, 95% (vol/vol)

Counterstain

Safranin 0, as a 2.5% (wt/vol) alcoholic solution in 95% (vol/vol) ethanol, 10 ml

Distilled water, 100 ml

Procedure

1. Place the slide on a staining rack, and flood the thin, air-dried, heat-fixed smear with the crystal violet staining reagent for 1 min.
2. Wash the smear in a gentle and indirect stream of tap water for 2 s.
3. Flood the smear with the iodine mordant for 1 min.
4. Wash the smear in a gentle and indirect stream of tap water for 2 s, and then blot the film dry with absorbent paper.
5. Flood the smear with 95% ethanol for 30 s with agitation, and then blot the film dry with absorbent paper.
6. Flood the smear with the safranin counterstain for 10 s.
7. Wash the smear with a gentle and indirect stream of tap water until no color appears in the effluent, and then blot the film dry with absorbent paper.

2.3.5.2. Burke staining method (3)

Stain solution A

Crystal violet (certified 90% dye content)	1.0	g
Distilled water	100	ml

Bicarbonate solution B

Sodium bicarbonate	1.0	g
Distilled water	100	ml

Iodine mordant

Iodine	1.0	g
Potassium iodide	2.0	g
Distilled water	100	ml

Prepare this mordant as for Hucker's method.

Decolorizing solvent (*CAUTION*: highly flammable)

Ethyl ether	1 volume
Acetone	3 volumes

Counterstain

Safranin O (85% dry content)	2.0	g
Distilled water	100	ml

Procedure

1. Flood the thin, air-dried, heat-fixed smear with solution A, add 2 or 3 drops of bicarbonate solution B, and let stand for 2 min.
2. Rinse off the stain with the iodine mordant, and then cover the smear with fresh iodine mordant for 2 min.
3. Wash off the mordant with a gentle and indirect stream of tap water for 2 s; blot *around* the stained area with absorbent paper, but do not allow the smear to dry.
4. Add the decolorizing solvent dropwise to the slanted slide until no color appears in the drippings (less than 10 s), and allow the smear to air dry.
5. Flood the smear for 5 to 10 s with the counterstain.
6. Wash the smear in a gentle and indirect stream of tap water until no color appears in the effluent, and then blot the smear dry with absorbent paper.

2.3.6. Acid-Fast Staining

A second important tinctorial property of certain bacteria is that of not being readily decolorized with acid-alcohol after staining with hot solutions of carbolfuchsin. Acid-fast bacteria include the actinomycetes, mycobacteria, and some relatives which, it is believed, react in this way because of limited permeability of the waxy components of the cell wall. Dormant endospores are also acid fast, but the cause is unknown.

Two methods are recommended for determining the acid-fast reaction, as follows.

2.3.6.1. Ziehl-Neelsen staining method (7)

Carbolfuchsin stain

Basic fuchsin	0.3	g
Ethanol, 95% (vol/vol)	10	ml
Phenol, heat-melted crystals	5	ml
Distilled water	95	ml

Dissolve the basic fuchsin in the ethanol, and then dissolve the phenol in the water. Mix, and let stand for several days. Filter through paper before use.

Decolorizing solvent

Ethanol, 95% (vol/vol)	97	ml
Hydrochloric acid (concentrated)	3	ml

Counterstain

Methylene blue chloride	0.3	g
Distilled water	100	ml

Procedure

1. Place a slide with an air-dried and heat-fixed smear on a slide carrier over a trough. Cut a piece of absorbent paper to fit the slide, and saturate the paper with the carbolfuchsin stain. Carefully heat the underside of the slide by passing a flame under the rack or by placing the slide on a hot plate until steam rises (without boiling!). Keep the preparation moist with stain and steaming for 5 min, repeating the heating as needed (*CAUTION:* overheating causes spattering of the stain and may crack the slide). Wash the film with a gentle and indirect stream of tap water until no color appears in the effluent.
2. Holding the slide with forceps, wash the slide with the decolorizing solvent. Immediately wash with tap water, as above. Repeat the decolorizing and washing until the stained smear appears faintly pink.
3. Flood the smear with the methylene blue counterstain for 20 to 30 s, and wash with tap water as above.
4. Examine under oil immersion. Acid-fast bacteria appear red, and non-acid-fast bacteria (and other organisms) appear blue.

2.3.6.2. Truant staining method (61)

It is also possible to determine the acid fastness of an organism by fluorescence techniques. A good example is the Truant technique. An advantage of this procedure is that slides of clinical specimens suspected of containing mycobacteria, for example, may be screened by using a 60× rather than a 100× objective; hence, an entire slide may be screened in a short time.

Fluorescent staining reagent

Auramine O, CI 41000	1.50	g
Rhodamine B, CI 749	0.75	g
Glycerol	75	ml
Phenol (heat-melted crystals)	10	ml
Distilled water	50	ml

Mix the two dyes with 25 ml of the water and the phenol. Add the remaining water and the glycerol, and mix again. Filter the resulting fluorescent staining reagent through glass wool.

Decolorizing solvent

Ethanol, 70% (vol/vol)	99.5	ml
Hydrochloric acid (concentrated)	0.5	ml

Counterstain

Potassium permanganate	0.5	g
Distilled water	99.5	g

Procedure

1. Flood a lightly heat-fixed smear with the fluorescent staining reagent for 15 min.
2. Wash the slide with a gentle and indirect stream of distilled water until no color appears in the effluent.
3. Flood the smear with the decolorizing solvent for 2 to 3 min, and then wash with distilled water as above.
4. Flood the smear with the permanganate counterstain for 2 to 4 min.
5. Wash the slide with distilled water as above, blot with absorbent paper, and dry.

Examine with a fluorescence microscope equipped with a BG-12 exciter filter and an OG-1 barrier filter. Acid-fast bacteria appear as brightly fluorescent, yellow-orange cells in a dark field; non-acid-fast cells are dark.

2.3.7. Endospores

Mature, dormant endospores of bacteria, when viewed unstained, are sharp edged, even sized, and strongly refractile, shining brightly in a plane slightly above true focus. It is the core (protoplast) of the spore that is refractile; the surrounding cortex, coat, and exosporium appear dark and often difficult to discern. Electron microscopy is needed for determining the details of these peripheral structures. Do not assume that any highly refractile body within a bacterium is an endospore, particularly if the inclusion body is irregularly sized and if information concerning heat resistance is lacking (see Chapter 25); large poly-β-hydroxybutyrate granules often appear in the cytoplasm of *Bacillus* spp. before sporulation and can be confusing.

2.3.7.1. Popping test

A direct test is provided by a visible phenomenon (15) occurring in dormant endospores after immersion in an acid oxidizer (e.g., 0.1% $KMnO_4$ in 0.3 N HNO_3). The spore cortex ruptures, and a portion of the spore protoplast including all of the nucleoplasm, now readily stainable with basic dyes, is herniated through the aperture. The dramatic suddenness of the event (after about 5 min in the solution) makes the term "popping test" appropriate. The test can be conveniently performed by mounting a dried film of spores on a coverslip in the reagent for 10 to 20 min; the consequent popping is visible with the oil-immersion objective without staining, although staining does make the popping more easily visible.

2.3.7.2. Negative staining

There are several methods for staining endospores in bacteria, although most of them stain bacteriologists better than they stain spores. The simplest method is a negative stain (section 2.2.3), which yields lifelike preparations (15). In this method, mix a small loopful of 7% (wt/vol) aqueous nigrosin on a coverslip with an appropriate amount of culture, spread it into a thin film, and air dry. Invert the coverslip, and place it film side down on a microscope slide (so the film remains in air when the coverslip is examined under oil) and maintained in place with several spots of candle wax. Endospores appear as highly refractile spherical or ellipsoidal bodies both within the bacterial soma and free.

Endospores strongly resist staining by simple dyes until they germinate. Dormant endospores, once stained, are quite resistant to decolorization and hence are acid fast (section 2.3.6). A useful positive-staining method for bacterial endospores is that of Dorner, as follows.

2.3.7.3. Dorner staining method (29)

1. Heat fix the organism on a glass slide, and cover with a square of absorbent paper cut to fit the slide.
2. Saturate the absorbent paper with carbolfuchsin stain, and steam for 5 to 10 min, as described above (section 2.3.6).
3. Remove the blotting paper, and decolorize the film with acid-alcohol (3 ml of concentrated HCl in 97 ml 95% ethanol) for 1 min; rinse with tap water, and blot dry. Use forceps to handle the stain-covered slides.
4. Place a drop of 7% nigrosin (section 2.2.3) on the slide, cover with another slide, and draw the two apart to form a thin film of the nigrosin over the stained smear.
5. Examine under oil immersion. Vegetative cells appear colorless, the endospores appear red, and the background appears black to sepia.

A variation of Dorner's technique is to mix in a test tube an aqueous suspension of bacteria with an equal volume of carbolfuchsin. Immerse the tubes in a boiling-water bath for 10 min. Then mix a loopful of 7% (wt/vol) aqueous nigrosin on a glass slide with one loopful of the boiled carbolfuchsin-organism suspension, and air dry in a thin film. Keep spreading with the loop until the smear begins to dry. The results are as indicated above, but the procedure is less messy.

2.3.7.4. Schaeffer-Fulton staining method (57)

In the Schaeffer-Fulton technique, 0.5% (wt/vol) aqueous malachite green is used instead of carbolfuchsin. Air dry the specimen on a glass slide, heat fix, cover with absorbent paper saturated with the dye, and place over a boiling-water bath for 5 min. Wash the slide in tap water, and counterstain the film with safranin for 30 s as in Gram staining (section 2.3.5). Then wash the slide and blot dry. Endospores appear bright green, and vegetative cells appear brownish red.

2.3.8. Cysts

Bacterial cysts (e.g., as in *Azotobacter* spp.) stain weakly with simple stains and generally appear as spherical bodies surrounded by thick but poorly staining walls. The following method (62) is useful for demonstrating *Azotobacter* cysts, including forms developed prior to the appearance of a mature cyst.

Reagent

Glacial acetic acid	8.5	ml
Sodium sulfate (anhydrous)	3.25	g
Neutral red	200	mg
Light green S.F. yellowish	200	mg
Ethanol, 95% (wt/vol)	50	ml
Distilled water	100	ml

Add the chemicals and dyes to the water with continuous stirring for 15 min. Filter through a membrane filter (pore diameter, 0.5 μm).

Procedure

Immerse the specimen in the reagent, and examine the wet preparation. Vegetative azotobacters are yellowish green; early encystment stages appear with a darker green cytoplasm somewhat receded from the outer cell wall, from which it is divided by a brownish-red layer. In mature cysts the central body appears dark green and is separated by the unstained intine from the outer, brownish red exine.

2.3.9. Cell Walls

Visualization of cell walls by light microscopy can assist in decisions about cell division and cell shape. Cell walls are not stained by the usual simple stains unless the walls are treated with tannic acid as a mordant, as in the method described for Bouin-fixed cells (section 2.2.7.1). Such preparations (52) require careful high-resolution microscopy (Chapter 1.2.7). Electron microscopy is needed for determination of the structural profile of the cell wall and details of septation (Chapter 3).

2.3.10. Capsules and Slime Layers

Capsules and slime layers are produced by bacteria capable of forming them under specific culture conditions and are best demonstrated in wet preparations because the highly hydrated polymers constituting them are distorted and shrunk by drying and fixation.

The capsules of specific pathogens can be displayed effectively (e.g., on pneumococci, *Haemophilus influenzae,* and meningococci) by use of antisera specific for the capsule type, and this can provide a presumptive identification. Suspend the clinical exudate or the organism in a drop of antiserum containing 0.01% methylene blue under a coverslip. The cell is stained and the surrounding capsule becomes outlined by a dark-blue line in a short time. For determinative purposes, this is known as the **Neufeld quellung (or capsule-swelling) test.**

Duguid's method is the best and simplest of the three capsule stains that follow.

2.3.10.1. Duguid staining method (30)

1. Place a large loopful of India ink on a clean slide, and mix in a loopful of culture; then place a glass coverslip over this in such a way that only part of the mixture is covered.

2. Press firmly down on the coverslip, using several thicknesses of absorbent paper, until the ink is a sepia color beneath the cover glass.
3. Examine with high-dry and oil-immersion lens systems. Capsules appear as clear zones around the refractile organism and against the brownish black background full of dancing particles of India ink.

2.3.10.2. Hiss staining method (35)

1. Mix a loopful of culture with a drop of normal horse serum or skim milk on a glass slide.
2. Air dry and *gently* heat fix the film.
3. Cover the film with crystal violet stain (crystal violet, 0.1 g; distilled water, 100 ml), and heat until steam rises (section 2.3.6.1).
4. Wash off the crystal violet with a 20% (wt/vol) aqueous solution of copper sulfate ($CuSO_4{\cdot}5H_2O$). Blot dry.
5. Examine under oil immersion. Capsules appear faint blue, and organisms appear dark purple.

2.3.10.3. Anthony staining method (18)

1. Stain an air-dried film for 2 min with a 1% (wt/vol) aqueous solution of crystal violet.
2. Wash off the dye with a 20% (wt/vol) aqueous solution of copper sulfate, drain, and blot dry.
3. Examine under oil immersion. Capsules appear light blue, and organisms appear dark purple.

2.3.11. Flagella

Although Koch devised a technique for staining flagella over a century ago, no easy or constantly reliable method is yet available. Since the width of individual bacterial flagella lies below the limits of resolution for transmission light microscopy (flagella are 10 to 30 min in diameter), it is necessary to "tar and feather" them, so to speak, to make them visible. In general, nonmotile bacteria do not possess flagella, but there are cases of organisms having "paralyzed" flagella (Fla^+ Mot^-).

Best results are obtained when using superclean slides; this reduces drying artifacts and precipitation of stains. It is important to expose the cells, in handling and in making the smear, to a minimum of agitation. Flagella are sheared off easily by drying and mechanical forces.

More information about the form and disposition of flagella is obtained by electron microscopy (Chapter 3.5). As in preparations for light microscopy, the drying forces involved in making preparations for electron microscopy can separate flagella from cells and can make it difficult to distinguish between polar and lateral flagella when they are sparse.

Some methods for staining flagella are labeled as simple (53), but few are simple in practice. Some combine the elements of several methods, as does the method of Heimbrook et al. (33), which works in most hands and can be useful when others fail. The following methods have enjoyed good success for many years.

2.3.11.1. Gray staining method (32)

Solution A

Tannic acid (20% [wt/vol] aqueous)....... 2 ml
Potassium alum [$KAl(SO_4)_2{\cdot}12H_2O$], saturated aqueous .... 5 ml
Mercuric chloride, saturated aqueous ... 2 ml
Basic fuchsin (3% [wt/vol] in 95% [wt/vol] ethanol)............................. 0.4 ml

Add the fuchsin dye to the other ingredients, preferably immediately before use, and filter the solution through paper.

Solution B

Ziehl's carbolfuchsin (section 2.4.6)

Procedure

1. Ordinary organisms may be grown on an agar slant of appropriate nutrient properties; 0.5 h before sampling, add a small volume of 1% peptone broth, into which the organisms swim. Decant this fluid, and free it from medium constituents by centrifugation. Resuspend the pellet *gently* (to prevent loss of flagella) with 10% (vol/vol) aqueous Formalin to produce a light, faint turbidity.
2. Allow a loopful of suspension to run down a tilted (45° angle) glass slide, and air dry the film so formed.
3. Flood the slide with solution A, and allow it to remain for 6 min. The exact time will have to be determined experimentally.
4. Wash solution A off the slide with distilled water.
5. Place a piece of blotting paper over the film, and flood the slide with solution B for 3 min.
6. Remove the blotting paper, wash the slide with distilled water, and gently blot dry.
7. Examine under oil immersion. Flagella and bacterial bodies stain red. A heavy coating around cells may be indicative of large numbers of fimbriae, requiring confirmation by electron microscopy.

An important part in the success of this technique depends upon the use of *absolutely clean* and grease-free slides. Either heat a slide by holding it in a Bunsen flame until it shows yellow around the edges and allow it to cool, or degrease a dish of loosely packed slides by heating at 400°C for 20 min in a muffle furnace. After cooling, use the slides immediately or store them in closed containers. Alternatively, a slurry of Bon Ami cleanser, allowed to dry on slides in a thin film and removed with a tissue, leaves slides very clean. Solutions or solvents, such as chromic acid or 50% (vol/vol) alcohol, are not recommended for cleaning slides to be used for staining flagella.

2.3.11.2. Leifson staining method (38)

Solution A

Sodium chloride..................... 1.5 g
Distilled water .... 100 ml

Solution B

Tannic acid.......................... 3.0 g
Distilled water .... 100 ml

Solution C

Pararosaniline acetate .... 0.9 g
Pararosaniline hydrochloride......... 0.3 g
Ethanol, 95% (vol/vol)................ 100 ml

Mix equal volumes of solutions A and B; then add 2

volumes of this mixture to 1 volume of solution C. The resulting dye solution may be kept under refrigeration for 1 to 2 months.

Procedure

1. Prepare an air-dried film on a slide as in steps 1 and 2 of Gray's method (section 2.3.11.1).
2. Using a wax glass-marking pencil, draw a rectangle around the film.
3. Flood dye solution on the slide within the confines of the wax lines. Leave for 7 to 15 min, the best time to be determined experimentally.
4. As soon as a golden film develops on the dye surface and a precipitate appears throughout the film, as determined by illumination under the slide, remove the stain by floating off the film with gently flowing tap water. Air dry.
5. Examine under oil immersion. Bacterial bodies and flagella stain red.

2.3.12. Cytoplasmic Inclusions

Many bacteria grown under certain culture conditions produce, as a result of metabolic reactions, deposits within the cytoplasm which are termed inclusions. Among these are deposits of fat, poly-β-hydroxybutyrate, polyphosphate, starch-like polysaccharides, sulfur, and various crystals. Some of these are polymerized waste products, and others are food reserves. Several staining procedures can be used to reveal some of these inclusions, which are also evident by both ordinary, phase-contrast, and interference microscopy because of their refractility.

2.3.12.1. Staining of poly-β-hydroxybutyrate

A convenient method for demonstrating these inclusions in bacteria is as follows.

1. Prepare a heat-fixed film of the specimen on a slide or use a Bouin-fixed preparation on a coverslip (section 2.2.7.1), and immerse in a filtered solution of 0.3% (wt/vol) Sudan black B made up in ethylene glycol. Stain for 5 to 15 min (determine the time experimentally).
2. Drain and air dry the slide.
3. Immerse and withdraw the slide several times in xylene, and blot dry with absorbent paper.
4. Counterstain for 5 to 10 s with 0.5% (wt/vol) aqueous safranin.
5. Rinse the slide with tap water and blot dry.
6. Examine under oil immersion. Poly-β-hydroxybutyrate inclusions appear as blue-black droplets, and cytoplasmic parts of the organism appear pink.

2.3.12.2. Staining of polyphosphate

Polyphosphate inclusions consisting of unbranched chains (often called metachromatic granules) may be detected as follows.

1. Prepare a heat-fixed film of the specimen on a glass slide.
2. Stain for 10 to 30 s in a solution of Loeffler methylene blue (section 2.2.5) or 1% (wt/vol) toluidine blue.
3. Rinse the slide with tap water, and blot dry.
4. Examine under oil immersion. With methylene blue, polyphosphate granules appear as deep-blue to violet spheres and the remaining cytoplasm appears light blue. Toluidine blue stains metachromatically, and the granules appear red in a blue cytoplasm. A more selective stain is attained by using acidified toluidine blue (0.3% in 0.5% acetic acid [vol/vol]) and mounting the stained cells in 0.5% acetic acid.

2.3.12.3. Periodate-Schiff staining of glycogen-like polysaccharides (37)

Glycogen-like inclusions can be revealed by using the following technique.

1. Prepare a heat-fixed smear of the specimen on a glass slide.
2. Flood the slide for 5 min with periodate solution (20 ml of 4% [wt/vol] aqueous periodic acid, 10 ml of 0.2 M aqueous sodium acetate, 70 ml of 95% ethanol; protect the solution from light!).
3. Wash the slide with 70% (vol/vol) ethanol.
4. Flood the slide for 5 min with a reducing solution containing 300 ml of ethanol, 5 ml of 2 N hydrochloric acid, 10 g of potassium iodide, 5 ml of sodium thiosulfate pentahydrate, and 200 ml of distilled water. Add the ethanol and then the hydrochloric acid to the solution of potassium iodide and sodium thiosulfate in distilled water. Stir, allow the sulfur precipitate to settle, and then decant the supernatant for use.
5. Wash the slide with 70% (wt/vol) ethanol.
6. Stain for 15 to 45 min (the time to be determined experimentally) with the following solution (Schiff reagent). Dissolve 2 g of basic fuchsin in 400 ml of boiling distilled water, cool to 50°C, and filter through paper. Add 10 ml of 2 N hydrochloric acid and 4 g of potassium metabisulfite to the filtrate. Stopper tightly, and allow to stand for 12 h in a cool, dark place. Add about 10 ml of 2 N hydrochloric acid until the reagent, when dried on a glass slide, does not show a pink tint. Store the reagent in the dark.
7. Wash the smear several times in a solution consisting of 2 g of potassium metabisulfite and 5 ml of concentrated hydrochloric acid in 500 ml of distilled water.
8. Wash the slide with tap water, and counterstain the smear with a 0.002% (wt/vol) aqueous solution of malachite green for 2 to 5 s.
9. Wash the slide in tap water and blot dry.
10. Examine under oil immersion. Polysaccharides appear red, and other cytoplasmic components appear green.

2.3.12.4. Alcian blue staining of acidic polysaccharides (10)

Alcian blue 8GX (43) is one of a family of copper pathalocyanin dyes which can be used to reveal acidic polysaccharides often found in capsules and polysaccharide inclusions. Use 1% (wt/vol) solution of Alcian blue in 95% (vol/vol) ethanol, with carbolfuchsin as the counterstain. The procedure is as follows.

1. Heat-fix a film of the specimen on a glass slide.

2. Stain for 1 min with a 1:9 dilution (in water) of the alcoholic Alcian blue solution prepared as above.
3. Wash the film in tap water, and dry.
4. Counterstain (take care not to overstain!) for about 5 s with carbolfuchsin (section 2.3.6), wash immediately with tap water, and air dry.
5. Examine under oil immersion. Polysaccharides appear blue, and other cytoplasmic constituents appear red.

2.3.13. Nuclear Bodies

The display of bacterial nucleoids requires special care in fixation and in mounting the preparation for high-resolution microscopy. They can be seen in living cells (section 2.2.1) by phase-contrast microscopy as long as the refractive index of the surround is raised to a sufficient level (14 to 18% gelatin in a nutrient agar or 15 to 30% gelatin in a fluid mount; the effective concentration requires experiment). Interference microscopy (Chapter 1.5) may be used if it is available.

Bacterial nuclei are not easily stained directly with basic dyes. Consequently, after Bouin fixation (section 2.2.7.1) of actively growing cells, staining with 0.01% thionine, and mounting in water, the cytoplasm is stained intensely because of the ribosomal proteins and nucleic acids. The nuclear areas appear as unstained clefts and "butterfly" patterns in the stained cell and represent the general distribution of the DNA plasms.

Nuclei are best displayed by fixation with osmium vapor followed by hydrolysis in 1 N HCl at 60°C to remove RNA and form aldehyde groups in the DNA plasms (the procedures are described in section 2.2.7.2). The cells stained with Giemsa stain and mounted in water or diluted stain show nuclei that appear an intense reddish purple. The microscopic contrast is greatest when a green light filter is used. The exact timing for hydrolysis may require some experiment but is usually 5 to 8 min.

2.4. SPECIAL TECHNIQUES

A great number of techniques based on light microscopy are special to particular areas of bacteriology. Medical bacteriology, in particular, has had to develop many methods to assist decisions about procedure, to attain an early presumptive diagnosis, to detect organisms that are difficult to stain or that occur in small numbers, and to confirm the identity of a specific organism. These uses and attendant precautions are detailed in organism-oriented chapters of the *Manual of Clinical Microbiology,* in which the staining methods are separately summarized (7), as well as in general reference works (section 2.6.1). Here, we give examples of some special methods not routinely used.

2.4.1. Spirochetes

Spirochetes do not stain well with ordinary stains, but free-living spirochetes show up well negatively stained in nigrosin films (section 2.2.3). A number of spirochetes have a width near the limits of light microscope resolution; consequently, these bacteria are best observed in life by direct dark-field or phase-contrast microscopy. Spirochetes stained with Giemsa (a Romanowsky-type stain used for haematology [10]) fluoresce bright golden yellow when observed by dark-field microscopy.

The following positive stain generally gives good results.

Fontana silver staining

Solution A (fixative)

Acetic acid	1 ml
Formalin	2 ml
Distilled water	100 ml

Solution B

Ethanol, absolute

Solution C (mordant)

Phenol	1 g
Tannic acid	5 g
Distilled water	100 ml

Solution D

Add a solution of 10% (wt/vol) aqueous ammonium hydroxide dropwise to a 0.5% (wt/vol) aqueous solution of silver nitrate until the precipitate, which initially forms, just redissolves.

Procedure

1. Using slides cleaned in the manner described for making suitable flagellar stains (section 2.3.11), air dry a film of the specimen on a slide.
2. Fix the film in solution A for 1 to 2 min.
3. Rinse the film in solution B for 3 min.
4. Cover the film with solution C, and heat until steam rises for 30 s.
5. Wash the film with distilled water, and air dry.
6. Cover the film with solution D, and heat until steam rises and the film appears brown. Wash the film and air dry.
7. Examine under oil immersion. Spirochetes appear brownish black.

2.4.2. Mycoplasmas

Mycoplasmas are probably the smallest bacteria observable with light-microscopic techniques and are hard to recognize individually because of extreme polymorphism. They are parasitic in animal and plant tissues, and their growth and study require specialized technical skill (51). The inexperienced would be well advised to seek expert help when faced with the need to recognize, cultivate, and identify mycoplasmas.

The mycoplasmas lack cell walls, and their cells are bounded only by the cytoplasmic membrane; consequently, special media (Chapter 8.2.5.2) are needed to maintain osmotic balance for cell integrity as well as to satisfy complex nutritional requirements. The individual cells tend to be irregular, but a number of species and genera each have definite overall shapes (star shaped, round, goblet shaped, filamentous, and helical) which can be recognized by phase-contrast or dark-field microscopy (Chapter 1.3 and 1.4), usually assisted by electron microscopy of both negatively stained and sectioned preparations (Chapter 3). Some species show a gliding motility on the surfaces of cells in the tissues they infect. However, the mycoplasmas growing in a complex medium, in tissue culture, or in plant or animal tissues can be very hard to distinguish among the very large variety of confusing artifacts and tissue components.

The organisms grow slowly on appropriate nutrient agar plates, and the colonies are microscopic, often less than 100 μm in diameter. Plates must be examined with a stereoscopic dissecting microscope (magnification ×20 to ×60), using oblique illumination to recognize the characteristic "fried-egg" colonies, which have a central plug penetrating the agar and a superficial spreading periphery. The colony margins may also be examined in situ at high magnification by placing a droplet of methylene blue on the colony and covering it with a coverslip for examination by transmitted light with high-power objectives.

The standard stained smears made by using heat-fixed films and Gram stained or stained with basic dyes are useless and uninformative. Stained preparations of mycoplasmas require careful chemical fixation with whatever staining procedure is used, and staining with Giemsa (section 2.4.3.1) or Wright stain (7) is preferable for cytological testing. The osmotic requirements for integrity of shape and form must be maintained until the chemical fixation is complete. The best preparations for light microscopy are made by fixing the organisms with Bouin fixative through a block of mycoplasma agar medium onto a coverslip for staining either with basic dyes such as thionine or with Giemsa stain (section 2.2.7.1).

Similar methods and precautions are needed for microscopic study of the wall-less bacteria derived from a variety of gram-positive bacteria and known as pleuropneumonia-like organisms.

2.4.3. Rickettsiae

Rickettsiae are obligately intracellular parasites (except for *Rickettsia quintana*) that are generally rod shaped, occur in pairs, and show marked pleomorphism. Infected tissue samples may be examined directly by fluorescence microscopy with specific antisera labeled with fluorescein isothiocyanate. With the Gram stain, rickettsiae appear gram negative.

Rickettsiae may be seen in films of infected tissue by direct microscopic examination after staining by the Giemsa or Gimenez procedure (see below). Most rickettsiae appear as pleomorphic coccobacillary organisms varying in length from 0.25 to 2.0 μm. Pairs and short chains are most frequently observed. *Coxiella burnetii* is the smallest (0.25 by 1.0 μm), and it appears as a bipolarly staining rod in the cytoplasm of infected cells. The rickettsiae that cause typhus and spotted fever are generally larger (0.3 to 0.6 μm by 1.2 μm). Spotted fever rickettsiae are found in the nuclei and cytoplasm of infected cells and often appear to be surrounded by a halo, sometimes mistaken for a capsule.

2.4.3.1. Giemsa staining method

Stock solution

Giemsa powder	0.5	g
Glycerol	33	ml
Absolute methanol (acetone free)	33	ml

Dissolve the Giemsa powder in the glycerol at 55 to 60°C for 1.5 to 2 h. Add the methanol, mix thoroughly, allow to stand, and decant. Store the sediment-free stock solution at room temperature. For use, dilute 1 part of the stock solution with 40 to 50 parts of distilled water or M/15 phosphate buffer at pH 6.8.

Procedure

1. Air dry the specimen on a glass slide, fix in absolute methanol for 5 min, and again air dry.
2. Cover the slide with freshly diluted Giemsa stain for 1 h.
3. Rinse the slide in 95% (vol/vol) ethanol to remove any excess dye, and air dry.
4. Examine under oil immersion for the presence of basophilic intracytoplasmic organisms, which will be small, pleomorphic, and purple.

2.4.3.2. Gimenez staining method

Stock solution

Basic fuchsin	10	g
Ethanol, 95% (vol/vol)	100	ml
Phenol, 4% (wt/vol) aqueous	250	ml
Distilled water	650	ml

Dissolve the dye in the ethanol, and then add the other ingredients. Allow the stock solution to stand at 37°C for 48 h. For use, dilute the stock solution 1:2.5 with phosphate buffer (pH 7.45) prepared by mixing 3.5 ml of 0.2 M NaH_2PO_4, 15.5 ml of 0.2 M Na_2HPO_4, and 19 ml of distilled water. Then filter the solution, which will keep for 3 to 4 days but should be filtered before each use.

Procedure

1. Heat fix the smear, and cover it with the staining solution for 1 to 2 min.
2. Wash the smear with tap water, and counterstain for 5 to 10 s with a 0.8% (wt/vol) aqueous solution of malachite green.
3. Wash the smear again with tap water, and make a second application of the malachite green.
4. Rinse the slide thoroughly in tap water and blot it dry.
5. Examine under oil immersion. Rickettsiae appear as reddish-staining bacteria against a green background.

2.4.4. Legionellae

The members of the family *Legionellaceae* cause problems of recognition and detection because several species in the genera *Legionella*, *Fluoribacter*, and *Tatlockia* cause life-threatening disease, yet they and other species are common in fresh waters (including tap water) and inhabit some species of amebae. Because they grow only on very special media, a degree of suspicion of their presence is needed. The appearance of simple to pleomorphic bacilli in tissues and exudates or cultures from likely sources revealed by the nonspecific Gimenez (section 2.5.3.1) or Gram stain may suggest the use of the **direct fluorescent antibody test** (Chapter 5.5.6) as a means of presumptive determination. This test involves the use of several group-specific polyvalent pools of fluorescent antibody. Once a group is identified, a species-specific fluorescent antibody can be used. False-positive reactions are rare and are usually laboratory induced (e.g., because of incorrect reagent or contaminations) when they occur. The commercial antibody is usually conjugated with fluorescein isothiocyanate. The procedure (53), which is an example of similar direct fluorescent antibody tests for other organisms, is as follows.

Preparation

Make smears, impressions of the cut surface of tissues, or spreads of tissue scrapings on alcohol-cleaned slides (preferably those with a 1-cm Teflon circle to retain the reagents). Distilled water and aqueous reagents must be filter sterilized, and the use of vessels and moist chambers (petri dish with moist filter paper) must be regulated to avoid any bacterial contamination of the slides from either other specimens or the environment, which may harbor members of the *Legionellaceae*.

Procedure

For a smear, air dry and gently heat fix on a slide. For a spread of tissue cells and impressions, fix in filter-sterilized 10% neutral formalin (add $CaCO_3$ to the bottle, and let it settle). Add enough of the appropriate conjugated serum to spread over the entire surface within the Teflon circle. Use a moist chamber to prevent evaporative drying during the 20- to 30-min reaction at room temperature. Immerse the slide, first in buffered saline and then in distilled water. Drain, mount the smear in a drop of buffered glycerol (9 volumes of glycerol, 1 volume of 0.5 M $NaHCO_3$-Na_2CO_3 buffer [pH 9.0]), add a coverslip, and then seal with clear nail polish. Examine in a fluorescence microscope, preferably by epi-illumination (Chapter 1.6.1), with appropriate exciting and barrier filters for the fluorochrome in use.

Useful practical hints and a more extensive discussion of components are given by McKinney and Cherry (11).

2.5. SOURCES OF STAINING REAGENTS

Most of the laboratory requirements are available from the supply companies, but some stains are not widely stocked. E. Merck, Darmstadt, Germany, is the major source of biological stains, and this multinational company has acquired many of the suppliers of former times: BDH Chemicals, Harleco, Hopkin & Williams, J. T. and E. Gurr Co., and Raymond A. Lamb Co. A wide range of stains and other reagents are available through the E. Merck subsidiaries, and some of them are as follows.

EM Science, Gibbstown, N.J.
BDH Chemicals, Toronto, Ontario, Canada
BDH Chemicals, Poole, Dorset, United Kingdom
Merck Japan Ltd., Tokyo, Japan

2.6. REFERENCES

2.6.1. General References

1. **Brock, T. D.** 1983. *Membrane Filtration: A Users Guide and Manual.* Science Tech Inc., Madison, Wis.
A very complete guide and instruction in the use and interpretation of membrane filter techniques.
2. **Brock, T. D.** 1987. The study of microorganisms *in situ:* progress and problems, p. 1–17. *In* M. Fletcher, T. R. G. Gray, and J. G. Jones (ed.), *Ecology of Microbial Communities.* Cambridge University Press, Cambridge.
An assessment of the applications of counting techniques in microbial ecology.
3. **Clark, G.** 1973. *Staining Procedures Used by the Biological Stain Commission,* 3rd ed. The Williams & Wilkins Co., Baltimore.
A resource for methods used in association with "H. J. Conn's Biological Stains" (10).
4. **Daley, R. J.** 1979. Direct epifluorescence enumeration of native aquatic bacteria: uses, limitations, and comparative accuracy, p. 29–45. *In* J. W. Costerton and R. R. Colwell, (ed.), *Native Aquatic Bacteria: Enumeration, Activity, and Ecology.* American Society for Testing and Materials, Philadelphia.
A detailed overview of acridine orange direct-counting techniques.
5. **Francisco, D. E., R. A. Mah, and A. C. Rabin.** 1973. Acridine orange epifluorescent technique for counting bacteria in natural waters. *Trans. Am. Microsc. Soc.* **92:**416–421.
A pioneer paper in epifluorescent counting.
6. **Hall, G. H., J. G. Jones, R. W. Pickup, and B. M. Simon.** 1990. Methods to study the bacterial ecology of freshwater environments. *Methods Microbiol.* **23:**181–210.
A modern compilation of methods for bacteriological assessment of freshwaters.
7. **Hendrickson, D. A., and M. M. Krenz.** 1991. Reagents and stains, p. 1289–1314. *In* A. Balows, W. J. Hausler, Jr., K. L. Herrmann, H. D. Isenberg, and H. J. Shadomy (ed.), *Manual of Clinical Microbiology,* 5th ed. American Society for Microbiology, Washington, D.C.
A part of this chapter gives details of staining methods useful in clinical microbiology.
8. **Herbert, R. A.** 1990. Methods for enumerating microorganisms and determining biomass in natural environments. *Methods Microbiol.* **22:**1–39.
A useful review and assessment of approaches to estimating biomass in environments.
9. **Karl, D. M.** 1986. Determination of *in situ* biomass, viability, metabolism, and growth, p. 85–176. *In* J. S. Poindexter and E. R. Leadbetter (ed.), *Bacteria in Nature,* vol. 2. Plenum Press, New York.
Descriptions and assessments of the methods applied to determination of biomass and viable counts in the field.
10. **Lillie, R. D.** 1977. *H. J. Conn's Biological Stains,* 9th ed. The Williams & Wilkins Co., Baltimore.
An extensive compendium of stains and dyes with discussion of their chemical structure and applications. Also see reference 3.
11. **McKinney, R. M., and W. B. Cherry.** 1985. Immunofluorescence microscopy, p. 891–897. *In* E. H. Lennette, A. Balows, W. J. Hausler, Jr., and H. J. Shadomy (ed.), *Manual of Clinical Microbiology,* 4th ed. American Society for Microbiology, Washington, D.C.
A useful source of advice on the practice of immunofluorescence microscopy.
12. **Newell, S. Y., R. D. Fallon, and P. S. Tabor.** 1986. Direct microscopy of natural assemblages, p. 1–48. *In* J. S. Poindexter and E. R. Leadbetter (ed.), *Bacteria in Nature,* vol. 2. Plenum Press, New York.
A survey with useful references to studies in the field.
13. **Norris, J. R., and D. W. Ribbons (ed.).** 1971. *Methods in Microbiology,* vol. 5A. Academic Press, Inc., New York.
The material presented in the first part of this volume is particularly useful as a source of information supplementary to that presented here. The relevant chapters are as follows:
I. Microscopy and micrometry, by L. B. Quesnel, p. 1–103.

II. Staining bacteria, by J. R. Norris and H. Swain, p. 105–134.
III. Techniques involving optical brightening agents, by A. M. Paton and S. M. Jones, p. 135–144.
IV. Motility, by T. Iino and M. Enomoto, p. 145–163.

14. **Pickup, R. W.** 1991. Development of molecular methods for the detection of specific bacteria in the environment. *J. Gen. Microbiol.* **137:**1009–1019.
A discussion of the strategies for sampling and detecting bacteria in the environment with special reference to genetically modified bacteria.
15. **Robinow, C. F.** 1960. Morphology of bacterial spores, their development and germination, p. 207–248. *In* I. C. Gunsalus and R. Y. Stanier (ed.), *The bacteria,* vol. 1, Academic Press, Inc., New York.
A classic account of bacterial spores and their study using light microscopy. The same volume contains other illustrated articles on bacterial structure.
16. **Robinow, C. F.** 1975. The preparation of yeasts for light microscopy. *Methods Cell Biol.* **11:**1–22.
Practical advice on using gelatin-agar slide cultures.
17. **Rosebrook, J. A.** 1991. Labeled-antibody techniques: fluorescent, radioisotopic, and immunochemical, p. 79–86. *In* A. Balows, W. J. Hausler, J., K. L. Herrmann, H. D. Isenberg, and H. J. Shadomy (ed.), *Manual of Clinical Microbiology,* 5th ed. American Society for Microbiology, Washington, D.C.
The applications of immunomicroscopy in clinical microbiology.

2.6.2. Specific References

18. **Anthony, E. E., Jr.** 1931. A note on capsule staining. *Science* **73:**319.
19. **Aristovskaya, T. V.** 1973. The use of capillary techniques in ecological studies of microorganisms, p. 47–52. *In* T. Rosswall (Ed.), *Modern Methods in the Study of Microbial Ecology.* Swedish National Science Research Council, Stockholm.
20. **Babiuk, L. A., and E. A. Paul.** 1970. The use of fluorescein isothiocyanate in the determination of the bacterial biomass of grassland soil. *Can. J. Microbiol.* **16:**57–62.
21. **Barer, R., R. F. A. Ross, and S. Thczk.** 1953. Refractometry of living cells. *Nature* (London) **171:**720–724.
22. **Bartholomew, J. W., and T. Mittwer.** 1952. The Gram stain. *Bacteriol. Rev.* **16:**1–29.
23. **Bartholomew, J. W.** 1962. Variables influencing results, and the precise definition of steps in gram staining as a means of standardizing the results obtained. *Stain Technol.* **37:**139–155.
24. **Beveridge, T. J., and J. A. Davies.** 1983. Cellular responses of *Bacillus subtilis* and *Escherichia coli* to the Gram stain. *J. Bacteriol.* **156:**846–858.
25. **Brock, T. D.** 1978. *Thermophilic Microorganisms and Life at High Temperatures.* Springer-Verlag, New York.
26. **Casida, L.** 1972. Interval scanning photomicrography of microbial cell populations. *Appl. Microbiol.* **23:**190–193.
27. **Casida, L. E., Jr.** 1982. *Ensifer adhaerens* gen. nov., sp. nov.: a bacterial predator of bacteria in soil. *Int. J. Syst. Bacteriol.* **32:**339–345.
28. **Davies, J. A., G. K. Anderson, T. J. Beveridge, and H. C. Clark.** 1983. Chemical mechanism of the Gram stain and synthesis of a new electron-opaque marker for electron microscopy which replaces the iodine mordant of the stain. *J. Bacteriol.* **156:**837–845.
29. **Dorner, W.** 1926. Un procédé simple pour la coloration des spores. *Le Lait* **6:**8–12.
30. **Duguid, J. P.** 1951. The demonstration of bacterial capsules and slime. *J. Pathol. Bacteriol.* **63:**673.
31. **Girbardt, M.** 1965. Eine Zielschnittmethode für Pilzzellen. *Mikroscopie* **20:**254–264.
32. **Gray, P. H. H.** 1926. A method of staining bacterial flagella. *J. Bacteriol.* **12:**273–274.
33. **Heimbrook, M. E., W. L. L. Wang, and G. Campbell.** 1989. Staining bacterial flagella easily. *J. Clin. Microbiol.* **27:**2612–2615.
34. **Henrichsen, J.** 1972. Bacterial surface translocation: a survey and a classification. *Bacteriol. Rev.* **36:**478–503.
35. **Hiss, P. H., Jr.** 1905. A contribution to the physiological differentiation of Pneumococcus and Streptococcus. *J. Exp. Med.* **6:**317–345.
36. **Hobbie, J. E., R. J. Daley, and S. Jasper.** 1977. Use of Nucle-pore filters for counting bacteria by fluorescence microscopy. *Appl. Environ. Microbiol.* **33:**1225–1228.
37. **Hotchkiss, R. D.** 1948. A microchemical reaction resulting in the staining of polysaccharide structures in fixed tissue preparations. *Arch. Biochem.* **16:**131–141.
38. **Leifson, E.** 1951. Staining, shape, and arrangement of bacterial flagella. *J. Bacteriol.* **62:**377–389.
39. **Lundgren, B.** 1981. Fluorescein diacetate as a stain of metabolically active bacteria in soil. *Oikos* **36:**17–22.
40. **MacRae, T. H., and H. D. McCurdy.** 1976. The isolation and characterization of gliding motility mutants of *Myxococcus xanthus. Can. J. Microbiol.* **22:**1282–1292.
41. **Mason, D. J., and D. M. Powelson.** 1956. Nuclear division as observed in live bacteria by a new technique. *J. Bacteriol.* **71:**474–479.
42. **Mazia, D., G. Schatten, and W. Sale.** 1975. Adhesion of cells to surfaces coated with polylysine. *J. Cell Biol.* **66:**198–200.
43. **Mowry, R. W.** 1956. Alcian blue techniques for the histochemical study of acidic carbohydrates. *J. Histochem. Cytochem.* **4:**407.
44. **Murray, R. G. E., and J. F. Whitfield.** 1953. Cytological effects of infection with T5 and some related phages. *J. Bacteriol.* **65:**715–726.
45. **Murray, R. G. E., and J. F. Whitfield.** 1956. The effects of the ionic environment on the chromatin structures of bacteria. *Can. J. Microbiol.* **2:**245–260.
46. **Newell, S.Y., and R. R. Christian.** 1981. Frequency of dividing cells as an estimator of bacterial productivity. *Appl. Environ. Microbiol.* **42:**23–31.
47. **Noller, E. C., and N. N. Durham.** 1968. Sealed aerobic slide culture for photomicrography. *Appl. Microbiol.* **16:**439–440.
48. **Patton, A. M., and R. Marchant.** 1978. An ultrastructural study of septal development in hyphae of *Polyporus biennis. Arch. Microbiol.* **118:**271–277.
49. **Perfil'ev, B. V., and D. R. Gabe.** 1969. *Capillary Methods of Investigating Micro-Organisms.* Oliver and Boyd, Edinburgh.
50. **Piekarski, G.** 1937. Zytologische Untersuchungen an Bakterien mit Hilfe der Feulgenschen Nuclealreaktion. *Arch. Mikrobiol.* **8:**428–439.
51. **Razin, S., and J. G. Tulley (ed.).** 1983. *Methods in Mycoplasmology,* vol. 1. Academic Press, Inc., New York.
52. **Robinow, C. F., and R. G. E. Murray.** 1953. The differentiation of cell wall, cytoplasmic membrane and cytoplasm of gram-positive bacteria by selective staining. *Exp. Cell Res.* **4:**390–407.
53. **Rogers, F. G., and A. W. Pasculle.** 1991. *Legionella,* p. 442–453. *In* A. Balows, W. J. Hausler, Jr., K. L. Herrmann, H. D. Isenberg, and H. J. Shadomy (ed.), *Manual of Clinical Microbiology,* 5th ed. American Society for Microbiology, Washington, D.C.
54. **Ryu, E.** 1937. A simple method for staining bacterial flagella. *Kitasato Arch. Exp. Med.* **14:**218–219.
55. **Salton, M. R. J.** 1963. The relationship between the nature of the cell wall and the Gram stain. *J. Gen. Microbiol.* **30:**223–235.
56. **Schaechter, M., J. P. Williamson, J. R. Hood, and A. L. Koch.** 1962. Growth, cell and nuclear divisions in some bacteria. *J. Gen. Microbiol.* **29:**421–434.
57. **Shaeffer, A. B., and M. Fulton.** 1933. A simplified method of staining endospores. *Science* **77:**194.
58. **Tabor, P. S., and R. A. Neihof.** 1982. Improved method for determination of respiring individual microorganisms in natural waters. *Appl. Environ. Microbiol.* **43:**1249–

1255.
59. **Tabor, P. S., and R. A. Neihof.** 1984. Direct determination of activities for microorganisms of Chesapeake Bay populations. *Appl. Environ. Microbiol.* **48:**1012–1019.
60. **Traxler, R. W., and J. L. Arceneaux.** 1962. Method for staining cells from small inocula. *J. Bacteriol.* **84:**380.
61. **Truant, J. P., W. A. Brett, and W. Thomas.** 1962. Fluorescence microscopy of tubercle bacilli stained with auramine and rhodamine. *Henry Ford Hosp. Med. Bull.* **10:**287–296.
62. **Vela, G. R., and O. Wyss.** 1964. Improved stain for visualization of *Azotobacter* encystment. *J. Bacteriol.* **87:**476–477.

Chapter 3

Electron Microscopy

TERRY J. BEVERIDGE, TERRY J. POPKIN, and ROGER M. COLE

Bacteriologists are dedicated to the study of very small cells. For this reason, microscopy in one form or another is inescapable. Chapters 1 and 2 have explained the uses and advantages of light microscopy, and this chapter deals with electron microscopy (EM). Of all the research techniques available to bacteriologists, EM is the only one that can give shape, form, and position to ultrastructural components of bacterial cells (such as ribosomes, bilayers, and plasmids) and to the viruses that afflict them. There is a renaissance of interest in EM since new preparatory procedures and equipment have made possible extreme resolution in well-preserved material. The chromosome and ribosomes are found to be dispersed throughout most of the cytoplasm, and enveloping layers are more complicated than was first suspected. The accurate identification and determination of macromolecular distributions have provided a better understanding of the integrative steps required in cellular metabolic processes and of their alteration during changes in growth or during division.

These are exciting times for the microscopy of microorganisms, and it is important to recognize that microscopy holds an esteemed and necessary position in almost all aspects of bacteriology. EM assists the formation of a conceptual bridge between cellular functions, the chemistry of the parts, and the behavior of the macromolecules that make up structure. Students of biochemistry, molecular biology, biotechnology and industrial processes, and environmental studies all really do require a microscopic view at one time or another. It is to these casual users, as well as to neophytes of EM, that this chapter is aimed. Simple, routine types of EM are described together with hints based on experience of how best to go about preparing samples and maintaining equipment. The chapter provides a guide to the kind of information that can be obtained from basic techniques and suggests combinations of techniques for more complete information.

The first level of ultrastructural information is provided by transmission electron microscopy (TEM) and scanning electron microscopy (SEM). These are tried and true techniques which have, over the last 30 years, deciphered the shape, size, constituent arrangements, and cellular interactions of a great many bacteria (125, 137). Of the two microscopies, SEM is better for determining the surface appearance or topography of cells but usually does not have enough resolving power to elucidate more than general features. SEM is a good technique for deciphering cell-cell associations, growth habits, and the topography of cell aggregates such as biofilms. TEM offers greater resolving power in a variety of preparations and is capable, with suitable material, of detecting the positions of cellular macromolecules. Many procedures are suitable for TEM of bacteria, including negative stains, thin sections, and shadowed replicas; most of this chapter concentrates on them. It is important to know what sort of information can be expected from these major techniques and their limitations.

There are a number of EM books available and recommended (1–10), but nothing can replace the experience of the regular operators of an EM unit. The instruction from and the friendship of these highly qualified personnel should be nurtured. No chapter or textbook on EM can make an expert microscopist; hands-on experience with guidance from someone who knows is essential.

Many new and complex derivatives of EM are available, all with impressive and complicated names, e.g., scanning transmission electron microscopical elemental point mapping using energy-dispersive X-ray spectroscopy (STEM-EDS). Some cannot even be considered to involve typical electron microscopes, but they are still a type of EM since they depend on electrons or weak atomic forces for image generation, e.g., scanning tunneling microscopy (STM) and atomic force microscopy (AFM). To add to the confusion, new specimen-processing methods abound, e.g., cryofixation by freeze-"plunging" or freeze-"slamming," freeze-substitution, and thin frozen films. Preparative techniques allowing the localization of component macromolecules have also become crucial in varied areas of biology; these are extremely important techniques in modern bacteriology and are mainly responsible for the renaissance of interest in ultrastructure. These techniques and the operation of their equipment require expert, dedicated personnel; they are available to most of us through collaboration with more specialized EM laboratories.

3.1. INSTRUMENTATION

3.1.1. Microscopy Procedures

An electron microscope is an extremely complicated and expensive instrument. It is looked after by specialists and maintained, because of the expense of parts and labor, by service contracts with the manufacturer. The instrument should not be abused. Instruments for TEM are frequently encountered in microbiology departments, since they can resolve the fine detail of small cells and viruses. Most instruments have multiple users to justify the capital expense and the complex operating problems. Such a laboratory often provides access to facilities for section cutting, shadow casting, freeze fracturing, freeze-etching, and perhaps other kinds of EM. Those in charge of units serving other users establish rules and standards of operation, which protect the integrity of the equipment and the work that comes out of it, and are strict in enforcing them. Experience says that casual users of EM fit the description given by a museum guard of its young patrons: "Lady, they will do *anything*." To this one must add, "and not tell you what it was." To assist troubleshooting and maintenance, the user must be considerate of the instrument (some parts are delicate, and it *does* matter which knobs you twist) and of the next user (a "golden rule") and must be scrupulously honest about anything done or not done.

Some procedural points to remember are as follows.

1. Get checked out on every instrument and procedure that is needed not only to know what to do but also so that limitations at each step in training are known.

2. If equipment doesn't work, behaves strangely, or requires unusual force, *stop* in a safe position, think, and seek help.

3. Do not be tempted to undertake repairs. Own up and confess to the problem.

4. Keep records, even if that is not the habit of the unit, and see that operating times are recorded and every malfunction is reported.

5. Make notes of the steps in procedures, have them at hand when at work, and use a checklist, which should include the following: positions of switches and settings at start-up; start-up procedures; readiness requirements; alignment procedures for everyday operation for condenser adjustment, centering of objective aperture, and astigmatism compensation; setting for wide-field scanning; operating procedures and camera operation; warning signs and responses to them; standby and safety actions; turn-off procedure; and position of switches and settings after turn-off.

Good preparations, procedures, patience, and operational skills will be rewarded by good micrographs and will be appreciated by both the supervisor and the microscope.

3.1.2. Transmission Electron Microscope and Resolving Power

The value of EM lies in its capacity to resolve objects that cannot be resolved by any form of light microscopy. The wavelength of electrons, which can be shortened as the accelerating voltage is increased, permits the resolution of objects as small as 2 Å (0.2 nm or 0.0002 μm), whereas the limit of light microscopy lies close to 0.2 μm. In practice, the best modern transmission electron microscopes probably achieve a resolution between 0.5 and 1.0 nm with an optimum specimen under the best conditions of operation. However, biological specimens are not infinitely thin and their substance contributes to "noise," so the practical limit of real resolution, (e.g., in thin biological membranes such as bacterial S-layers), is about 1.0 to 1.5 nm. Even then it takes real skill and a well-tuned instrument to do better than 2.0 to 2.5 nm. Thick specimens and support films increase electron scattering, and the resolution deteriorates accordingly.

It is useful to recall some of the limitations of EM. Because specimens must be examined in a high vacuum to permit electron flow, the cells obviously will be dried and killed (unless they are held frozen in a special cryospecimen holder at extremely low temperatures). In addition, typical accelerating voltages (60 to 100 kV) subject the specimen to so much energy that many chemical bonds (including covalent linkages) are broken, molecular conformations are altered, and substantial organic mass is lost. For best results, specimen exposure to the electron beam should be kept to a minimum. The nature of electron penetration through biological specimens is such that the degree of specimen contrast is poor and, in contrast to resolution, decreases with an increase in accelerating voltage. Consequently, the specimen must be as thin as possible and selectively stained or contrasted by salts of heavy metals, which scatter electrons so that a useful image can be projected onto the viewing screen or film.

Because of the increased resolution, useful magnification can also be greatly increased over that obtainable by light microscopy. Instruments in current use achieve enormous direct magnifications, but, unless there is a need to resolve macromolecules, most applications are well served by magnifications of ×5,000 to ×15,000 (allowing subsequent photographic enlargements of 5 to 10×). Higher magnifications require more skill and a very well-tuned instrument.

The value of the increased resolution of EM in bacteriology is the ability to elucidate details of bacterial cell structure, provided that the preparative methods and the choice of EM techniques are suitable to the structure involved. For example, the use of heavy-metal salts to surround the bacterium and to penetrate into surface irregularities supplies contrast by differential impedance of electrons, thus producing the effect known as negative staining (section 3.2.2), analogous to the use of nigrosin in light microscopy. However, unless there are breaks in the cell wall or membrane to allow penetration of the contrasting metal salts, no intracytoplasmic features can be seen by this method. Macromolecular arrangements and cellular components may be studied by this technique if cell breakage and fractionation (Chapter 4) are accomplished before negative staining.

Bacterial cells are too thick for adequate resolution of internal structures, but sectioning (into slices 50 to 100 nm thick) of chemically fixed and stained cells embedded in plastic allows their cytoplasmic arrangements to be discerned. Heavy-metal "stains" of the sections are needed to provide differential contrast, because biological elements (C, H, O, N, etc.) do not scatter electrons proficiently. Most of the electrons of the electron beam pass cleanly through the specimen, and these unscattered electrons do not carry structural information. However, some electrons do have their trajectories altered as they go through the specimen (remember, this is *transmission* EM) by elastic or inelastic scattering; this scattering by structures in the specimen gives structural information about the specimen. The elastic scattering is responsible for the high fidelity of TEM images.

Electron beams are focused by powerful and symmetrical electromagnetic lenses stacked on top of one another. Essentially, the column of a TEM consists of condenser lenses (which focus the electron beam on a small area on the specimen) and magnifying lenses (functioning as objective, diffraction, intermediate, and projector lenses) immediately below. Electrons cannot be seen, so reliance is placed on a phosphorescent screen, which is excited by the incident electrons to emit visible light. The emitted light intensity is low, and a darkened room is required to clearly see the images. Electron-sensitive photographic film is necessary to record the images.

3.1.3. Scanning Electron Microscope

In SEM, the electron beam rapidly scans the surface of the specimen to induce radiations (low-energy secondary electrons or backscattered primary electrons). These form an image on a cathode ray tube by synchronizing the movement of the electron beam in the microscope (and the information being collected) with the information displayed (as a raster) on the tube; the process is similar to the formation of a television picture. The specimen must first be chemically fixed, dried (best by critical-point drying or freeze-drying to avoid distortion), and given a thin metal coating to make it conductive (sputtered in vacuo with gold, platinum, or a 60:40 platinum-palladium mixture; the last of these gives a less granular metal layer). The resolution achieved by SEM has been markedly improved since its introduction (at best about 3.0 nm in commercially available models) but is still lower than that routinely available by TEM. Newer low-voltage SEM instruments are beginning to approach TEM resolution. SEM can display only the images of surfaces, but it does so directly in a three-dimensional mode that avoids the tedious preparation of replicas of intact, fragmented, or frozen and fractured bacteria used in TEM for similar purposes. In bacteriology, SEM is most useful for discerning surface appendages and other surface structure and for defining shape and topographic relationships as in colonies or on surfaces of infected tissues (109–117).

Preparation of specimens for SEM involves fixation on a conductive surface such as graphite that can be affixed to the specimen holder. Vacuum coating with a heavy metal prior to examination is usually required to avoid specimen charging and to give strong surface signals to form the image. Because of the time and expertise needed for these procedures, the poorer resolution of SEM, and the unavailability of the instrument in many microbiological laboratories, SEM is not likely to be used for a "first look" by students of bacterial structure. TEM and negative staining (section 3.2.2) can more easily and rapidly supply such information. However, SEM is a useful method for looking at microorganisms in situ (e.g., to examine surfaces from natural environments for adhering microorganisms).

3.1.4. Specialized Electron Microscopes

This section is intended to identify and give a general explanation of highly sophisticated EM equipment and techniques but not procedures or methods for their use. They have been chosen because they are important to research in bacteriology. Special equipment and expert operators are required, so suitable problems must be taken to them. However, it is necessary to have a broad familiarity with them so as to know what they may be able to contribute to a research project.

3.1.4.1. Scanning transmission electron microscope

Microscopes for scanning transmission electron microscopy (STEM) combine some of the attributes of SEM and TEM; like SEM, a narrow electron beam is scanned back and forth over the specimen; like TEM, a transmitted signal is recorded. The great advantage of STEM is that a highly concentrated and narrow electron beam can be directed onto and through a thicker than normal specimen. Because STEM uses a scanning mode, secondary and backscatter electron detectors can be placed above the specimen so that, at a flick of a switch, STEM becomes SEM. Secondary electrons produce topographical images of the specimen, whereas the backscattered electrons provide density information. An energy-dispersive X-ray spectrometer can be added so that the X-ray emission spectrum can be sampled and compositional analysis of the elements in the specimen can be obtained. By selecting the energy band of a single element (e.g., Ca or Mg) and using a computer to combine the data with raster coordinates obtained in the STEM mode, a "point" map of the distribution of that element can be obtained (82, 90a, 91, 92, 95). Most bacteria have overwhelming C, H, O, and N concentrations, but fluctuating concentrations of P, S, and metal ions not commonly associated with cell structures (e.g., toxic heavy metals) can also be monitored by STEM. Magnetosomes (Fe), polyphosphate granules (P), sulfur granules (S), and bacterium associated minerals such as clays (Si) are also readily identified. The discrimination of the elemental composition of bacteria and the detection of toxic heavy metals from the environment complexed to bacteria (129, 130) have provided some exciting recent examples of structural analysis by STEM.

3.1.4.2. High-voltage electron microscope

Although STEM can be used for thick specimens, there is a practical limit (0.5 to 1.0 μm) to beam penetration. The higher energies provided by high-voltage electron microscopy (HVEM) must be used for thicker specimens. Accelerating voltages of 1 to 5 MeV are currently possible and can penetrate biological specimens up to 10 μm thick. The very expensive instruments for HVEM require two or three floors of a building for their housing, and the few in operation are often national facilities. The Electron Microscopy Society of America, P.O. Box EMSA, Woods Hole, MA 02543, has a list of HVEM facilities, and the *EMSA Bulletin* has descriptions of each one in special articles during 1984 and 1985.

HVEM has had little use in bacteriology (125, 126, 137) since bacteria are so small and powerful penetrating power is rarely required. The tremendous depth of focus of HVEM allows stereoimaging of thicker specimens so that the precise location and depth of organelles and infecting pathogens can be obtained. For the accurate study of pathogenic mechanisms, such as the phagocytosis and discrimination of phagosomes versus primary or secondary lysosomes within the confines of tissues, HVEM has advantages. Whole tissue cells can be visualized (instead of tedious serial sections) and cell-cell and cell-bacterium relationships can be more readily appreciated by HVEM.

3.1.4.3. Specialized configurations for spectroscopy

There are three basic ways to obtain elemental analyses by electron microscopy: **energy-dispersive X-ray spectroscopy** (EDS), **X-ray wavelength-dispersive spectroscopy** (WDS), and **electron energy loss spectroscopy** (EELS). The application of EDS has already been mentioned (section 3.1.4.1). This can be done by SEM but is better done by STEM because higher-resolution point elemental mapping can be included for single elements. TEM with a detector will serve if no elemental imaging is required and if a full-spectrum analysis is satisfactory. As a specimen is irradiated by the electron beam, its constituent atoms emit "signature" X-rays, and the X-ray emission spectrum is read by a detector placed above and to the side of the specimen. The X-ray detector must be separated from the high vacuum of the EM column, usually by a beryllium foil which limits access to X rays with energies above those of beryllium. For this reason, low-atomic-number elements (e.g., C, H, N, and O) are not easily resolved, although recent advances with other separating foils and "windowless" detectors are making the detection of such elements feasible.

WDS is another technique for obtaining elemental analysis; it is restricted to SEM and relies on physical crystals (e.g., those of lithium fluoride or lead stearate) to separate and expand the X-ray emission spectrum focused on the detector. Analysis of the spectrum provides compositional analysis. Compared with EDS, WDS requires long counting times and can be used only if the bacteriological material is very stable under the electron bombardment of the beam.

EELS requires a detector underneath the specimen (usually below the viewing screen) which is aligned to the optical axis and is, therefore, restricted to TEM and STEM since their configuration provides the transmitted signal that has passed through the specimen for analysis. Only the inelastically scattered electrons are used for EELS. These electrons have lost energy after atomic interaction with the specimen and have lower energies and longer wavelengths; both of these traits give them an elemental signature of the atom(s) with which they interacted in the specimen. Although these inelastic electrons lessen the quality of a TEM image by producing chromatic aberration, they are very useful for compositional analysis. Since the EELS detector operates in the vacuum of the EM column without metal foil filters (unlike EDS), the technique can resolve a full spectrum of elements including those of low atomic number. Compared with EDS, long counting times are required and highly beam-sensitive bacteriological samples cannot be accurately analyzed unless low-temperature (cryo-) stages are used to stabilize the specimen. However, so far, EELS is the best technique for low atomic number analyses.

One of the most exciting advances in EM uses a modified version of EELS and is called **electron spectroscopic imaging** (ESI) (90). This configuration of EM separates the energy lines in the inelastic signal and allows the operator to choose a single elemental signal as the sole information signal to use in the formation of a TEM image. As a consequence, an elemental image at TEM resolution is formed. This technique is so new that it has been only rarely used in bacteriology; however, because of the need for very fine compositional analyses, ESI holds tremendous promise.

A comprehensive list of articles describing these methods of elemental analysis and their use in bacteriology can be found in references 82 to 92 and 129.

3.1.4.4. Scanning tunneling microscope and atomic force microscope

Molecular or even atomic detail was the promise of scanning tunneling microscopy (STM) and atomic force microscopy (AFM). However, experience tells us that hard, current-conducting surfaces such as those possessed by graphites, silicates, and metal foils are easier to image by STM and AFM than are nonconducting, pliable material like bacteria. For STM an extremely sharp, charged probe (usually made of tungsten) is brought toward the specimen surface until a tunneling current is established, usually when the probe tip is within 0.5 nm of the surface. The probe tip is then moved over the specimen surface in the form of a raster at very close intervals while the tunneling current is kept constant. Since the magnitude of the current depends entirely on the distance between the probe tip and the specimen, the tip must constantly be raised and lowered as it traces its path over the surface. Amplification of this precise movement of the tip along the x, y, and z axes and detection of the exact position of the tip during the rastering provides a high-resolution topographical map of the specimen surface, so much so that the atomic arrangement of graphite surfaces has been visualized (122).

Although several biological structures (e.g., protein-DNA complexes) have been revealed by STM, it is apparent that these soft structures are deformable and poorly conducting unless a vaporized metal layer coats their surfaces. Special precautions and equipment are necessary to visualize uncoated bacteriological specimens (118–120). Furthermore, although commercial instruments for STM and AFM are available, a good background in physics is required for their operation, maintenance, and repair.

The instruments for AFM are quite similar to those for STM in appearance but do not rely on tunneling currents for their operation. Instead, the probe tip is brought so close to the specimen that only the weak repulsive atomic forces of the surface keep it buoyant; the tip force is on the order of 1 nN. The tip is rastered over the surface, and the x, y, and z movements are monitored by highly complicated detectors. Since electrical currents are not involved between the tip and the surface, the topography of hydrated specimens is possible. So far, the fidelity of AFM has not been as good as that of STM, but, for biological specimens, AFM may be the better choice (121, 122a).

3.1.5. Computer-Enhanced Imaging (93–108)

Diffraction is another method for structural analysis applicable to crystals and highly ordered macromolecular arrays. **X-ray diffraction**, which uses a beam of X rays, has been influential in deriving molecular structure at the atomic level and in initially understanding the molecular conformation of proteins and nucleic acids. However, like X rays, electrons also diffract and a TEM instrument can focus the incident electrons on extremely small areas. In fact, if the usual magnifying lenses are turned off in TEM while viewing a periodic object, an electron diffraction pattern will be obtained. Since the electron beam is concentrated on a small region of the specimen, this is called **selected-area electron diffraction** (SAED). High-level periodicity of most bacteriological specimens is quite sensitive to electron bombardment, and SAED is restricted to organic structures with a high proportion of covalent bonding (105), to inorganic mineralized material such as magnetosomes (125), and to external mineral deposits (129, 130).

Bacteriologists avoid the difficulty of beam damage due to SAED by using another form of diffraction on micrographs of paracrystalline structures, **optical diffraction** (OD). High-resolution, low-intensity electron images are taken by TEM, and the resulting photographic negatives are subjected to a beam of coherent light from a laser. Periodic, regular, and repeating structures on the photographic negative act as a diffraction grating, and the resulting information (as diffracted light) focused by appropriate lenses along an optical bench forms a diffraction pattern or transform. The reflections seen in the transform correspond to the spacing and symmetrical arrangement of the particles seen in the photographic negative.

OD is performed because the EM photographic negative suffers from noise which detracts from the overall quality of the image. The granularity of the heavy-metal stains (especially in negative stains), unequal stain distribution, small distortions in crystal lattices, and the grain of the silver emulsion of the EM negative all contribute to the inherent noise, thus masking detail. The noise is randomly distributed throughout the photographic image, will be so distributed in the optical transform, and will be seen as low-intensity background illumination after diffraction. Since the regular structures within the specimen have the same periodicity, these will be displayed as intense reflections (spots) separated from and arranged with respect to one another according to the natural specimen spacings and arrangements. Because of the inverse relationship between spacings on the negative and those seen in the transform, reflections which are widely separated describe small spacings in the original specimen.

The optical transform makes more obvious the high-order structure within the specimen. In fact, a physical mask, placed at the position of the transform so that only those reflections derived from the periodic structure can pass through, will allow the reconstruction of a virtually noiseless image. This approach to image enhancement was instrumental in clarifying the perception of the subunit arrangement of such diverse paracrystalline structures as flagella, pili, S layers, spinae, and parasporal bodies. The advent of computers (and, more recently, microcomputers) has caused a major improvement in image processing. As a result, OD is used only for the survey of negatives to assess their suitability for processing. Selected negatives are digitized by either a high-resolution television camera or a narrow-beam densitometer, and the information is then fed into a computer. Two basic computer programs are used: **Fourier analysis**, which deals with highly periodic objects, and **correlation averaging**, which is used for more irregular shaped objects (such as ribosomes) or periodic objects with lattice defects. References 93 to 108 are good sources of additional information, some outlining the general principles as well as presenting specific examples (95, 102, 103).

3.2. TECHNIQUES FOR TRANSMISSION ELECTRON MICROSCOPY

This section describes a limited number of TEM methods that, from the authors' experience, are considered useful to the worker who has need of basic anatomical information about bacteria. Emphasis is placed on the use of negatively stained preparations and thin-sectioned materials for examination by TEM. No attempt is made to describe the operation and maintenance of the microscope (2, 6, 8) or of other associated instruments. Materials that are required are mentioned (section 3.7), and their sources are given (section 3.8). It is assumed that instruments are functional and experienced operators are available for assistance and instruction. The aim here is to supply the neophyte with helpful information on techniques and interpretation. Although methods and instrumentations other than those emphasized are mentioned, details will be found only by reference to selected texts, other manuals, and articles in journals. General reading is provided in references 1 to 10. More specialized sources of information are given in references 11 to 148.

3.2.1. Preparation

3.2.1.1. Grids (11–14)

Specimens in the form of thin sections, intact organisms, and cell components must be supported in the electron beam in accurate relation to the objective pole pieces of the microscope. This support has to be relatively transparent to the electron beam and is provided by a film (section 3.2.1.2). To accomplish this, thin metal discs with a regular pattern of holes (grids) are usually covered with thin plastic or carbon films. Standard grids, of a diameter (2.3 or 3.0 mm) appropriate to the microscope used, are commercially available (see section 3.8, *k, o, p, r, u, x, z,* and *cc*). Grids are made of several materials and are available in several mesh sizes, but the most commonly used grids are made of copper and have a mesh size of 200 (mesh size indicates the lines per inch). The grids have a dull (matt) and shiny (bright) side. Before support films are applied, the grids must be clean and therefore must be treated with acetone and/or other solvents. However, supply houses can furnish precleaned grids.

3.2.1.2. Support films (11–14)

Although uncovered grids may be used to support sections cut from epoxy resins (see 3.2.3.6), most sections as well as suspensions of bacterial cells for metal shadowing or for negative staining must be supported on thin films covering the holes in the grid. The film materials commonly used are nitrocellulose or collodion (Parlodion), polyvinyl formol (Formvar), or carbon. Films of the first two substances, which are plastics, are frequently stabilized by evaporative deposition of a thin additional layer of carbon. Formvar is the most satisfactory plastic-film material and is used as follows.

1. Have available the materials listed in section 3.7.1.
2. Prepare a working solution of 0.2% Formvar in ethylene dichloride (chloroform can also be used) from a 0.5% Formvar stock solution and the solvent. Mix thoroughly.
3. Degrease microscope slides by washing with soap (or detergent) and water, and rinse thoroughly in distilled or deionized water. *Blot* dry with lint-free tissue (lens tissue), which minimizes static charge buildup; *never* wipe dry. Cleaned slides may be stored indefinitely by being wrapped in folds of lint-free tissue.
4. Fill a Coplin jar with the working solution of 0.2% Formvar. Add 4 or 5 drops of ethylene dichloride to another jar and cover it; this "drying jar" thus has a solvent atmosphere to facilitate even drying of the Formvar-coated slides.
5. Place two or three slides in the Coplin jar containing the Formvar solution, immersed to three-quarters of their length. Allow to stand for 3 to 5 min. Withdraw one slide at a time, and immediately place it in the drying jar. When a slide is dry, withdraw it and place it on lint-free tissue. With an alcohol-cleaned razor edge, cut the Formvar film on the slide around the entire perimeter of both sides. Be careful that small glass chips do not land on the Formvar film!
6. Place a 6-in. (ca. 15-cm)-high, flat-bottomed, glass "crystallizing" dish on black paper in a tray to catch overflow. Completely fill the dish with deionized water to form a convex meniscus at the top. Keeping a clean glass rod in contact on all sides with the top edge of the dish, slide the rod across the top from edge to edge to skim off the meniscus and remove all surface dust and contaminating film.
7. While holding the coated slide vertically over the dish, *carefully* drop 10% hydrofluoric acid from a Pasteur pipette onto the cut marks on the Formvar film. (Discard the pipette after each use. The use of hydrofluoric acid etches the Formvar from the glass and replaces, in a more reproducible fashion, the common procedure of breathing on Formvar coatings to free them from the glass.) *Immediately* but *slowly*, immerse the slide vertically into the water. The Formvar films on both sides will slowly release from the glass surface and float off onto the meniscus of the water surface. Inspect the films visually, preferably under a fluorescent light source, to judge uniformity and thickness. They should appear gray to silver-gray, representing thicknesses of ca. 60 nm according to interference color charts. The theoretically ideal thicknesses of 10 to 20 nm are probably rarely achieved. If uniformity or thickness of the films is unsatisfactory, repeat the above steps with appropriate modifications, as follows. If films are not uniform, the coated slides must be dried for longer or immersed in the water more slowly. If the films are too thick, dilute the Formvar solution with ethylene dichloride. If films are too thin, add a little of the 0.5% stock solution to the working solution of Formvar and repeat the process.
8. Once satisfactory films are obtained on the water surface, use forceps to carefully align the grids (dull side down) onto the films. To maintain a degree of consistency, avoid using film areas with obvious irregularities. Usually a film will accommodate 25 to 50 individual grids.
9. To retrieve the film-covered grids, cut a 9.0-cm diameter circle of filter paper (Whatman no. 1) in half, hold the cut edge over the Formvar-covered grids, and lower it gently onto the grid surfaces until the paper has uniformly absorbed water. (Some filter papers are so "hairy" that the fibers damage the film; avoid these.) Push the paper under the meniscus, and then bring the paper back up through the water surface so that the paper supports the undersides of the grids.
10. Place the semicircle of filter paper supporting the grids onto another piece of filter paper (grid side up) in a petri dish, and partially cover the dish with its lid while the contents dry.
11. After drying, lift the grids by the edges with finely pointed forceps (Dumont no. 5). The plastic film, if suitable, will break cleanly around the perimeter of the grid and will be retained over the grid. Films that lift away from the filter paper as the grid rises or that do not break cleanly and evenly are too thick; some areas of the film may show this property while others are satisfactory. Films that are too thin will tear readily in the electron beam. Films that have large holes (which may result from use of the exhaling technique instead of the use of hydrofluoric acid during the stripping procedure of step 7) may fail to hold liquid and will be obvious on examination in the microscope. Holes also result from too high an ambient humidity during drying.

Because of ionization and consequent changes in electrostatic charge, plastic films expand and contract to cause specimen drift as the electron beam is intensified and decreased. Any resulting movement will degrade the photographic image in proportion to the magnification used. Often, adjustments of beam intensity and a brief wait will establish a level of stability. Formvar films prepared as described above are quite satisfactory for routine purposes, with magnifications up to $\times 50{,}000$. For fine detail and higher magnifications, however, the stability of plastic films is improved by the deposition of a thin (2.0 to 3.0 nm) layer of carbon. Thin films of pure carbon avoid drift and improve resolution even better because of a more even and finer substructure.

Formvar-covered grids are usually placed within a carbon evaporator and are carbon coated by the heating of carbon electrodes in a 1.33 to 0.13 nPa vacuum. Evaporators are essential equipment of all EM laboratories; make sure that you have an operation manual and have been "checked out" on their use. Grids without Formvar can also be carbon coated. Evaporate carbon onto a freshly cleaved mica surface under the same conditions; float the carbon film from the mica and place grids on this film in the same way that Formvar grids are prepared. Grids with only a carbon coating are *extremely* fragile and must be handled with care. They are especially useful for high resolution of subcellular components and viruses.

No matter how good the specimen preparation, its accurate visualization will clearly depend on the quality of the support film on the grid. For most routine investigations, Formvar- and carbon-coated 200-mesh copper grids are recommended. The appropriate thickness of the coat can only be gauged by experience and from the light-grey discoloration of the supporting filter paper for the raft of grids.

3.2.2. Negative Staining

3.2.2.1. Stain choice

Negative staining is the highest-resolution technique that is easily available. It can be thought of as a thin embedding since the sample is immersed in a thin film of a dilute heavy-metal salt, which is then dried down to a glassy consistency surrounding the object. Negative staining is especially good for visualizing subcellular particles such as ribosomes, cell wall or membrane fragments, viruses, and (in its most refined form) protein macromolecules. Most material does not have to be chemically fixed before a negative stain is applied, but some fragile specimens may require prefixation (see section 3.2.2.2, step 5). When objects are negatively stained, they appear white against the dark background of the stain. Since the heavy-metal solution penetrates in and around the contours of subunit arrangements, fine infrastructure can often be seen.

Several different heavy-metal salts, usually in the range of 0.5 to 4.0% (dry wt/vol), have been used as negative stains. In the authors' experience, ammonium molybdate is the most useful because it spreads well, does not react with proteins or most other components in a suspension, and can be readily adjusted in tonicity as required. However, solutions of sodium or potassium phosphotungstate are perhaps the most widely used and are generally satisfactory at a neutral pH. Contrary to popular belief, phosphotungstic acid is almost never used, because of its very low pH; it is usually brought to neutral pH with sodium or potassium hydroxide to become a phosphotungstate salt. Uranium salts are also widely used because at pH below 4.5 they dissociate into small uranyl (UO_2^{2+}) ions, which penetrate into surface irregularities and therefore improve the definition of fine detail. At higher pH values, the uranyl ions polymerize into a wide range of larger ions and can even precipitate from solution. Uranyl acetate, which is generally available in the EM laboratory because it is also used for staining of thin sections, is useful in concentrations of 1 to 2%, but its pH is very low (about 2.5 to 4.0). When uranyl acetate is used as a negative stain, both positive and negative staining may be seen on different areas of the same grid. In addition, uranyl acetate acts as a fixative for proteins and nucleic acids, contracts and rounds membrane-bounded cells (e.g., mycoplasmas), disrupts some protein structures, and should be used only with well-washed preparations. Uranyl oxalate is more useful then uranyl acetate because the oxalate anion is more tightly complexed to the uranyl cation and resists precipitation at more neutral pH. Uranyl formate, sodium silicotungstate, and various other salts are also sometimes used. Of all negative stains, uranium (uranyl) has the highest atomic number (Z_U = 92) and provides the highest specimen contrast. (*CAUTION*: Salts of uranium are slightly radioactive and toxic. Care in both use and disposal is advised.)

Some workers recommend the addition of very small amounts of wetting agents (such as peptone, glycerol, serum albumin, starch, bacitracin, or sucrose) to some stains (but not uranyl salts) to improve spreading (11–14). Not all investigators find this useful or necessary, especially if high magnifications are to be used and uncluttered backgrounds are desired.

3.2.2.2. Staining procedures

There are many variations among the negative stains used and the procedures for applying them (11–14). A standard procedure is a two-step method of first applying the specimen to the grid and then staining with 2% aqueous ammonium molybdate, as follows.

1. Press a small piece of Parafilm M (American National Can. Co. [see section 3.8 for addresses of suppliers]) onto the laboratory bench surface, and then remove the tissue covering to provide a clean surface. Since the Parafilm surface is hydrophobic, a small drop of the sample suspension can easily be applied with a Pasteur pipette; the diameter of the drop should approximate the diameter of the grid. Also place a small drop of 2% ammonium molybdate on the Parafilm (for step 3).
2. Place a coated grid, with the film side down, on the drop of sample suspension. The surface tension of the meniscus spreads the suspended solids and allows them to adsorb to the grid. Large debris falls to the bottom of the drop and will not obscure the grid. Usually 15 to 60 s is enough time for particle adsorption. Small concentrations of detergents, organic solvents, or other surface-active agents can reduce adsorption and affect stainability. Trial and error will determine the optimum time for adsorption, which is also affected by the nature and charge (hydrophilic properties) of the film and the size and nature of the particles. To improve both the adherence and distribution of particles, the hydrophilicity of the support film can be improved by the use of coated grids that have been stored in a refrigerator or exposed to UV irradiation, to "glow discharge" in a high-vacuum evaporator, or to discharge from a Tesla coil. The exposures minimize electrostatic charges.
3. With thin-pointed tweezers, remove the grid. By touching the edge of the grid to a piece of filter paper (Whatman no. 1), remove the small drop still adhering to the film side. Quickly touch this side to the small drop of 2% ammonium molybdate on the Parafilm.

4. After 15 to 60 s, withdraw the drop of stain in the fashion described above. Staining times may require some experimentation, especially with different sorts of specimens.
5. Allow the sample to dry and examine it in the electron microscope. For very delicate samples (such as bacterial S layers), macromolecular arrangements can be perturbed by the high surface tensions during drying. This effect can be generally overcome by prefixation with 0.1 to 1.0% (vol/vol) glutaraldehyde in water or a suitable buffer. This is easily done by incorporating the fixative into the floating-grid/drop method (step 2 above) and washing (section 3.2.2.3). (*CAUTION*: Glutaraldehyde is volatile and toxic, so prefixation should be done in a suitably vented fume cabinet or hood.)

There are several alternative procedures for negative staining of bacterial suspensions. One involves mixing equal volumes (or other ratios) of stain and suspension, placing the grid onto a drop of the mixture, and then withdrawing excess fluid with filter paper, either immediately or after an appropriate interval of time (e.g., 5 to 10 s). It is usually best to do this immediately after mixing, since interactions of stain and particles can, with time, produce confusing results. Sometimes it is advantageous to pick up a thin film of the mixture in a platinum-iridium loop just bigger than the grid (e.g., 3.2 mm) and to break this flat film on a grid resting on blotting paper, to supply a more uniform particle distribution. The grid is then allowed to dry and is examined.

Occasionally, stain-suspension mixtures are sprayed onto grids, but this requires special apparatus. This procedure is most useful for quantitation of viruses or other small particles (13) and should be done in a special chamber for safety.

A particularly useful technique for bacteria is as follows. Place the coated grid surface film side down on very early growth of a bacterial lawn or on a colony on nutrient agar. Remove the grid, and proceed immediately with negative staining. This technique permits examination of the natural state of the bacteria and their topographic relationships, undisturbed by pipetting or other shearing procedures that may remove flagella, pili, or other delicate structures. Be careful not to take up too many cells, since the electron beam will not penetrate them. The procedure is also useful for a quick determination of bacteriophage type (e.g., T_4 versus λ) produced by a given plaque on an indicator lawn.

3.2.2.3. Washing

A useful accessory procedure after application of a suspension but before negative staining (in the two-step method) is to wash the grid to eliminate excessive debris, salt, and protein that can be precipitated by some stains and that can obscure detail. Washing can be done by touching the coated grid surface rapidly and repeatedly in succession to the surfaces of several fresh drops of distilled water or other fluid on Parafilm. Another washing procedure is to place the grid, film side down, on the wet surface of a thin strip of Whatman no. 1 filter paper bent and placed in a small beaker filled with washing fluid (usually distilled water or dilute buffer); the strip withdraws the washing fluid by capillary action over the edge and down from the completely filled beaker. In either example, the grid is stained immediately after removal of excess fluid.

In addition to water, various solutions may be tried as washes, but these can be risky. Mineral salts, organic compounds, pH, and other conditions are sometimes essential for the maintenance of structure, but they can also produce an obscuring film as the specimen dries. In fact, during drying, tremendous concentration gradients can be produced, and these can denature proteins. Salts have a tendency to form large precipitates on the grid. Of all washing fluids other than water, 1% (wt/vol) aqueous ammonium acetate seems best since it is nonreactive with bacteria, mycoplasmas, membranes, bacteriophages, and most intracellular components and because any excess is volatile in the vacuum of the electron microscope.

3.2.2.4. Pitfalls

For examination of negatively stained specimens, it is particularly important to balance contrast and resolution by use of an appropriate accelerating voltage (usually 60 to 80 kV). Contamination of the microscope column, apertures, and specimen by evaporated stain or other substances is minimized by the use of a liquid-nitrogen-cooled anticontamination device (or cold trap) and of self-cleaning gold objective apertures in the microscope. Specimen drift is avoided by patience, experimentation with beam intensity, and use of carbon-coated plastic or pure carbon support films. Beam damage to, or alteration of, fine structure, as well as "puddling" of stain about or on the specimen, can be minimized by learning to achieve focus as quickly as possible with the lowest possible beam intensity and making the shortest possible photographic exposure that produces a good, accurate photographic image. A short exposure at a relatively low magnification usually produces the best results: by trial and error, each microscopist will determine the magnifications most easily focused and producing the most consistent results.

Solutions of stains and wash solutions (buffers, salts, and even deionized, distilled water) may become contaminated with bacteria over time, and a quick examination of nigrosin films by light microscopy will soon confirm the condition of the solutions (Chapter 2.2.3). Reduce the problem by keeping working stains in the refrigerator in a rack of small vials, which are filled as needed (with sterile precautions) from stock bottles. Contaminated solutions may be filtered but are best made up anew. A convenient method for storage and filtered delivery is the use of a plastic syringe with an attached 0.45- or 0.20-μm Swinney filter. Solutions of uranyl acetate require protection from light; this is done by covering the container with aluminum foil.

The distribution of negative stain and specimens on the support film may be uneven or unsatisfactory; it is therefore efficient to prepare two or more grids of the same material. A grid should be scanned thoroughly at low magnification. A satisfactory grid prepared by the two-step method will show a gradient of specimen and stain distributed from one side to the other (remember, fluid was withdrawn by touching one *edge* of the grid to the filter paper). If the particles are in low concentration in the original suspension or if drop withdrawal techniques are unsatisfactory, specimens and stain may be found only on one small area of the grid.

Ammonium molybdate produces less contrast than does phosphotungstate or uranyl acetate, and its use may require some practice. However, the photographic negatives (with appropriate development) will have more contrast than anticipated and present a great range of printable tones. Even molybdate tends to heavily surround most whole bacteria with an almost impenetrable mass of stain; shorter staining times and less concentrated (0.1 to 0.5%) solutions are often required for good results with these relatively large objects.

3.2.2.5. Expectations

Negative staining is the simplest and fastest EM method that can be used, but to obtain very good results it can take many different attempts with several combinations of staining and specimen concentrations. This can be frustrating, but do not give up; the results are well worth the effort!

What sort of structural information on bacteria can be obtained by negative-staining techniques? The size and shape of the cell can be ascertained. The presence of flagella (perhaps already suggested by motility studies) can be affirmed, and their distribution and fine structure can be determined. Other surface appendages such as pili (fimbriae) or the presence of S layers or capsular materials can be revealed. Even the differences between walled and wall-less procaryotes (mycoplasmas or bacterial L forms) become apparent.

Sometimes negative stains penetrate through the surface layers of bacteria so that membrane invaginations (e.g., mesosomes [which are considered to be artifacts]) and the internal membranes of photosynthetic, methylotrophic, and nitrifying bacteria can be seen. Indeed, bacteria can be lysed and their component parts can be isolated (see Chapter 4) so that they can be negatively stained, and this is one of the most powerful methods of ultrastructural determination. The subunit arrangements of periodic structures, such as S layers, flagella, pili, spinae, and porins (after delipidation of outer membranes) are also revealed. Even minute structural differentiations within some appendages can be seen; e.g., filament, hook, and basal body rings of a flagellum. With experience, gram-positive and gram-negative bacteria can be readily differentiated by their negatively stained envelope profiles.

Examples of the applications of negative staining may be found in references 124–128, 132, 136, 137, and 141.

3.2.3. Thin Sectioning (15–18)

For conventional thin sectioning, bacteria must be chemically fixed, washed, dehydrated, and embedded in a liquid plastic that then solidifies with appropriate properties for the cutting of very thin sections. The sections are stained with solutions of heavy-metal salts to give sufficient contrast for examination by TEM. The chemicals used in this technique (fixatives, stains, dehydration solvents, and plastics) can be very damaging to the specimen, and structural artifacts can be introduced. An experienced microscopist should help with the assessment of the initial preparations. Although many different combinations can be used, it is generally adequate to use sequential fixation in glutaraldehyde and osmium tetroxide, or in osmium tetroxide alone, followed by embedding in Epon 812, LR White, or Spurr's medium (section 3.7.4). There are other choices of embedding plastics, but these three work reasonably well for routine purposes and are a good starting point. Once experience is gained with a sample and the way it reacts to fixation and embedding, modifications can be tried.

CAUTION: Chemical fixatives, heavy-metal stains, and plastic resins are chosen because of their high reactivity with and penetration into biological specimens. Remember, *bacteriologists are made out of biological stuff too. These are highly toxic chemicals and should be treated with care!* Surgical gloves and a well-vented fume cabinet (hood) with the glass door pulled down to protect the face are recommended.

3.2.3.1. Fixation with glutaraldehyde and osmium tetroxide

1. To the cell suspension or liquid culture, add an equal volume of 5% glutaraldehyde in aqueous 100 mM HEPES (*N*-2-hydroxythylpiperazine-*N*′-2-ethanesulfonic acid) buffer (pH 6.8) containing 2 mM $MgCl_2$ (final glutaraldehyde concentration, 2.5%). Establish the pH of the growing culture immediately before fixation, and alter the pH of the fixative solution accordingly. (0.2 M sodium cacodylate is another effective buffer often used for the fixative, but it contains arsenate, which vaporizes and is toxic, so be careful!) See sections 3.7 and 3.8 for sources and chemical formulations. Alternatively, centrifuge the cells and suspend the pellet in 2.5% glutaraldehyde in HEPES buffer. A wash in fresh medium or in an appropriate buffer may precede this step. For some procaryotes, especially some mycoplasmas, it may be preferable to make up the glutaraldehyde (perhaps in a different concentration) in fresh culture medium to balance osmolality or pH.
2. Allow the cells to fix for 2 to 4 h at room temperature. Actually, the time is flexible; any time beyond 20 min is usually adequate, and bacteria held in glutaraldehyde for several days may be shipped across the country if appropriately packaged to meet transport regulations for dangerous chemicals.
3. Wash the pellet twice in HEPES buffer, centrifuging between washes. (Some laboratories prefer to change the type of buffer here to Kellenberger's Veronal-acetate buffer [see section 3.7]).
4. Suspend the washed pellet in 1% osmium tetroxide for 2 to 4 h at room temperature. (The glutaraldehyde fixed pellet may be enrobed in agar prior to this step; see steps 6 to 8.) *CAUTION*: If the odor of osmium tetroxide is detectable, eyes may be subject to dangerous vapors. For safety, a commercially available eyewash station (Nalge Co.) should be available, properly labeled, and strategically located.
5. Centrifuge and wash the pellet (now black) once by suspension in the buffer. Recentrifuge.
6. Make up and melt 2% Noble agar in the buffer, and cool to 45°C. Add 2 or 3 drops to the pellet, and mix quickly with a warm Pasteur pipette.
7. With the pipette, quickly dispense the suspension onto an alcohol-cleaned glass slide, and allow it to solidify. The bacteria should be in good concentra-

tion and evenly distributed. With a clean razor blade, dice into 1-mm cubes.

8. Suspend the cubes for 15 min at room temperature first in deionized water to equilibrate them and then in 2% aqueous uranyl acetate for 2 h at room temperature. Wash with two to five changes of deionized water. The reason for replacing the buffer with water is that UO_2^{2+} will polymerize from solution at pH above 4.5, and it is best to have an unbuffered system. Make sure that a phosphate-buffered system has not been used (unless the cells are well washed of it), because uranyl and phosphate will rapidly complex together and precipitate from solution.

3.2.3.2. Fixation with osmium tetroxide only

Since the osmium tetroxide procedure does not include the glutaraldehyde step in 3.2.3.1, it is faster than the previous method and (usually) gives adequate results.

1. Centrifuge the cell suspension or culture.
2. Wash the pellet of cells by suspension in fresh culture medium or Veronal-acetate buffer, centrifuge, and suspend in tryptone medium or tryptic soy broth. (The use of tryptone or tryptic soy broth is optional, but the use of Veronal-acetate buffer usually improves definition of ribosomes and nucleoids [44].)
3. Centrifuge and suspend the pellet in 0.1% osmium tetroxide in Veronal-acetate buffer. Fix for 2 to 4 h at room temperature.
4. Proceed as in steps 3 through 8 in section 3.2.3.1.

3.2.3.3. Dehydration and embedment with plastic medium

1. Wash the uranyl acetate-treated agar cubes twice in 70% ethanol for 15 min each. (Use 2- to 3-ml volumes in 4.5-ml [1-dram] screw-cap glass vials.)
2. Transfer to 80 and 95% ethanol, successively, for 15 min each.
3. Transfer to 100% ethanol for 30 min.
4. Replace with 1.0 ml of fresh 100% ethanol. Add 1.0 ml of complete plastic medium, and mix thoroughly but gently. Allow to stand, with occasional mixing, for 30 min or until cubes sink to the bottom of the vial. (The medium is most easily dispensed by use of disposable plastic syringes.)
5. Add another 1.0 ml of plastic medium, and repeat the process.
6. Replace the mixture with 2.0 ml of plastic medium, and allow to stand for 1 to 4 h until the cubes settle to the bottom. (Some laboratories have automatic processors, such as the LKB or RMC Ultraprocessors, which perform steps 1 to 6.)
7. Insert typed labels into size 00 Beem capsules (in an appropriate holder), and add to each an equal volume of complete plastic medium. (Use 0.55 ml, so that the final block fits precisely into the microtome chuck.) Embeddings can also be done in gelatin capsules (available at a local drug store) or in flat molds.
8. With a sharpened wooden applicator stick, place one infiltrated cube (from step 6) into each filled Beem capsule. The cube should sink to the bottom of the capsule within 30 min.
9. With caps *off* (to allow volatilization of any residual solvent), place the capsules in a 60 to 79°C oven for polymerization for the specified time (Epon 812, 48 h; LR White, 18 h; and Spurr, 16 h; some plastics (e.g., LR White) cure better in the absence of oxygen). Immediately on removal, the blocks in the capsules may feel slightly spongy, but they will harden to optimal cutting quality after 2 to 4 h at room temperature.

3.2.3.4. Other fixatives and embedding media

Many different fixatives have been devised for special purposes. These include permanganates, which have applications for thick-walled microorganisms such as yeasts and fungi, and other aldehydes (e.g., acrolein, formaldehyde, crotonaldehyde). Principles of their use (including effects of pH, buffers, tonicity, concentration, temperature, and duration of fixation) are discussed in numerous books (15–18). Similarly, numerous embedding media are used (15–18), and their selection depends on their infiltration into cells, polymerization and cutting character, miscibility with water, and availability.

Other embedding media in common use include Durcupan, Araldite, Maraglas, Vestopal, various methacrylates for special purposes, and numerous epoxy resins. The reader is referred to the literature and to local experts for advice on the use and properties of these media. A new variety of methacrylates, the Lowicryls, are particularly useful for low-temperature embeddings and the preservation of antigenic sites for colloidal-gold antibody labeling (section 3.6.2) (19–22). *CAUTION*: Lowicryls can be highly allergenic, and surgical gloves must be worn.

3.2.3.5. Block trimming

Before sectioning, remove the hardened blocks from Beem or other capsules and trim them to proper size and shape. Various methods of trimming have been described (2, 5, 6, 8). Newer models of ultramicrotomes often incorporate trimming devices, and separate specimen trimmers are also available. The preparatory procedure is as follows.

1. With pliers or a Beem Capsule Press (E. F. Fullam), press firmly over the entire capsule to loosen the enclosed block. Now press from the pyramidal end, and proceed upward to force out the cured block. Alternatively, release blocks by cutting the capsule lengthwise on two sides with a razor blade. Mount the block in a holder, pyramidal end up, and place the holder in a microtome chuck.
2. Under a stereo-binocular microscope, use an alcohol-cleaned razor blade to carefully trim the top of the pyramid (which will be the front or cutting face when mounted in the ultramicrotome) until the cell mass has been reached. Cut down and away from this surface on all four sides at an angle of about 30° (the angle of the pyramid formed by the block in the capsule is about 60°) to leave a small truncated pyramid (having a face about 1.0 mm^2) that encompasses the specimen to be sectioned. With care and practice, this may be adequate trimming—especially if the face is cut in the shape of a trapezoid. However, grooves inevitably left by the razor blade are rather large and run at right angles to the face: as a result,

water may run onto the face during sectioning and interfere with good results. Better to further fine trim the face and pyramidal sides in the microtome with a glass knife, so that any grooves on the sides run parallel to the edges of the block face.

3.2.3.6. Thin sectioning

Because procedures and equipment differ in each laboratory, no attempt is made here to describe the sectioning process. Descriptions of pitfalls are thoroughly described in EM texts as well as in operating manuals available from the instrument makers (2, 5, 6, 8).

3.2.3.7. Knives

Both glass and diamond knives are used for cutting ultrathin sections. Glass knives, which are less expensive, are usually prepared by the use of a knife-making apparatus, which is available in all EM laboratories. The making of glass knives is described in texts (2, 5, 6, 8) and in manuals accompanying the instruments. For sectioning, the choice of knife depends on financial limitations, operator choice and experience, and the polymer used for embedding. In some instances, glass knives are preferable and produce superior results. A diamond knife produces more consistent results as the operator becomes familiar with its edge and other qualities and when a standard embedding medium is consistently used. The angle of the edge of the diamond as the diamond sits in its holder must be specified when purchasing a diamond knife; an angle of 42° to 45° is satisfactory for most conventional embedding plastics.

3.2.3.8. Staining

Even though most embedding procedures have a uranyl acetate staining step, additional staining of the thin section is usually required for added contrast. Uranyl acetate followed by lead citrate is the most common staining procedure. For uranyl acetate staining, use the following procedure.

1. Fill an appropriate number of caps of size 00 Beem capsules to a positive meniscus with 1% uranyl acetate, made up in either water or 70% ethanol. For aqueous solutions, drops may be placed on a sheet of clean dental wax or Parafilm M.
2. Float each grid that bears a section, section surface down, on the surface of the stain, and withdraw a little fluid with a Pasteur pipette to center the grid in the cap.
3. Cover with a petri dish lid to maintain humidity and protect from dust, and allow to stain for 5 min. Longer staining times (up to 30 min) may be required for hard-to-stain structures such as mycobacterial cell walls.
4. Remove the grids with grid tweezers, immerse three times in distilled water to wash, and dry on filter paper.

Some workers use uranyl acetate staining only, but the degree of contrast may be relatively feeble. Lead citrate (section 3.7.4) is commonly used to supply added contrast. For unknown reasons, uranyl acetate staining followed by lead citrate staining gives a much more intense stain than either alone. Lead citrate is conveniently stored in and dispensed from a plastic syringe, thus excluding air, since lead precipitates on exposure to CO_2.

For staining with lead citrate, place a strip of Parafilm M in a large (6-in. [ca. 15-cm]) petri dish in which is placed a small (2-in.) dish containing a few pellets of NaOH. Add the stain as a single drop on the Parafilm, and cover the larger dish with its top. Soon the internal atmosphere becomes alkaline because of the NaOH pellets and removes residual CO_2, which might affect staining. To stain the sections, float grids section side down on the drop of stain, cover the dish, and stain for 3 to 7 min. Remove the grid, and, while holding it with tweezers, wash in 0.02 N NaOH and rinse several times in freshly boiled cooled water. Touch the grid edge to filter paper to drain, and place the grid on filter paper to dry.

3.2.4. Cryofixation and Freeze-Substitution (31–51)

Cryofixation has improved the preservation of the ultrastructure of procaryotes. Cryofixation relies on the very rapid freezing of cells so that they are fixed physically without the aid of toxic chemicals. The freezing requires specialized devices which either plunge the specimen into an ultracold liquid cryogen (usually propane or ethane held at liquid-nitrogen temperature of −196°C) or slam the specimen onto an ultracold polished metal block (either copper or gold at −196°C or colder). These devices do not have to be complicated and can be manufactured for a reasonable price by university machine shops (39, 40). Ice crystals are not formed by the technique; rather, vitreous ice is formed. The resulting preservation of cells is so good they can be thawed back to life. Detailed accounts of freezing theory are available (31, 36, 37, 45–50).

Highly specialized cryo-EM laboratories may possess a cryostage on one of their microscopes, but they are rare and will not be considered further. Freeze-substitution, however, is a technique available to any EM laboratory which can cryofix samples. Once the samples are frozen, chemical fixatives are substituted into them at low temperature so that the samples are chemically fixed without thawing. At the same time, the water (ice) is substituted with an organic solvent. Once substitution is complete the specimen is ready for plastic resin infiltration and can, eventually, be thin sectioned like a conventional plastic block.

A simple "plunging"/freeze-substitution method (39, 40) is as follows.

1. Wash cells free of culture medium in 50 mM HEPES buffer at pH 6.8 (other buffers and pHs can be used) by centrifugation (an Eppendorf centrifuge is satisfactory). Resuspend the pellet in a volume of molten 2% (wt/vol) Noble agar equal to the volume of the pellet. Rapidly spread this suspension into a 20 to 30-μm-thick film over a cellulose-ester filter (Gelman Sciences, Ann Arbor, Mich.) by using the edge of a sterile glass microscope slide (the film is so thin that it can hardly can be seen and the spreading requires a few tries to gain experience). Be sure cellulose-ester filters are used, since other composites may

dissolve during the substitution. While spreading the film or while waiting for the plunger to be available, ensure that a humid atmosphere is maintained over the agar film; drying artifacts are not wanted! Humidification can easily be accomplished by rigging a simple humidity chamber consisting of the bottom of a large glass petri dish and a smaller-diameter glass beaker containing a large, damp (but not dripping) sponge. Force the sponge into the bottom of the beaker, and turn the beaker upside down over the agar film in the petri dish. Make sure that this humidity chamber has had time to equilibrate before putting the agar film inside.

2. Fill the cryogen reservoir with propane (or ethane) after it is cooled to liquid-nitrogen temperature. A 1-liter Dewar flask with a brass cylinder in the middle to hold the propane can be easily manufactured. The top of the brass cylinder should hold 50 to 25 ml of propane for good heat exchange. If it can be stirred by a magnetic stirrer to reduce convection, so much the better. Propane is liquid at these low temperatures and is not volatile. *CAUTION*: Once the freeze-plunging is finished, the propane will warm to room temperature and hence become volatile and *EXPLOSIVE!* Make sure that the freeze-plunging is done in a well-vented fume cabinet and that the fan in the cabinet is of a safe, sparkless variety. Turn the magnetic stirrer off and unplug it once the propane begins to warm up. All workers in the laboratory must be forewarned to turn off Bunsen burners, stoves, heaters, etc., and not operate electrical switches.
3. Immediately before plunging, cut small wedges from the agar film-cellulose ester filter composite. These must be pointed and resemble arrowheads ca. 5 mm long and 2 to 3 mm at their base. They are inserted into the plunger so that the pointed end hits the cryogen first. This will allow the specimen to be oriented for thin sectioning and ensure that the best frozen (pointed) region is sectioned.
4. For the freeze-substitution, use 7-ml glass scintillation vials filled with 2 ml of substitution medium. Consistent results are obtained with 2% (wt/vol) osmium tetroxide and 2% (wt/vol) uranyl acetate in anhydrous acetone, although other media can be used (40). "Anhydrous" is an important criterion since all water (ice) must be completely replaced. Store the acetone with a hygroscopic molecular sieve in it (sodium aluminosilicate with pore size of 0.4 nm [Sigma Chemical Co.]), and the sieve should also be added to the substitution media in the vials. Freeze the medium in each vial at −196°C before use.
5. The time elapsed between spreading the bacteria as an agar film on the cellulose-ester filter, plunging, and transferring the frozen cells to the surface of the frozen substitution medium must be short, not exceeding 2 min. Seal vials by their screw caps, and hold at −80°C in an ultracold freezer for 72 h. During this period the specimen remains frozen, the substitution medium melts, the ice is replaced by acetone, and the cells become chemically fixed, stained, and dehydrated. The molecular sieve takes up the released water.
6. Allow the vials to come to room temperature, wash the specimen free of fixative with anhydrous acetone, and infiltrate plastic resin into the specimen as described above (section 3.2.3.3). Section and stain the plastic embeddings as usual (sections 3.2.3.6 and 3.2.3.8).

3.2.5. Direct Microscopy

Bacterial preparations on Formvar or collodion support films may be examined directly by TEM without the benefits of contrast enhancement by negative staining. The bacteria may be grown on or can be placed on the coated grids and then examined at any stage of growth. However, information obtained by such means is limited and of no advantage unless one is interested in the inherent electron opacity of the specimen. The minimal increase in time involved in negative staining (section 3.2.2) will yield much more structural information.

Other methods of direct examination include high-resolution dark-field electron microscopy and electron-optical phase microscopy. Because of special instrument requirements and limitation of their applications to particular types of specimens, these techniques are of little importance to investigators beginning the study of bacterial structure. Similarly, HVEM, which with anticipated refinements may permit examination of thicker specimens without excessive beam damage, is a specialized procedure (section 3.1.4.2).

3.2.6. Metal Shadowing (1, 5, 8, 23–25)

Because of the growing popularity and availability of SEM, the use of direct metal shadowing of TEM preparations has declined. However, like SEM, metal shadowing is a method for eliciting the surface topography and dimensions of bacteria. Furthermore, because of the greater resolving power of TEM, these preparations provide greater resolution than those by SEM.

Shadowing was among the first techniques used to visualize bacteria by EM. It is especially useful for determining the shapes of bacteria and viruses, the surfaces that coat them (e.g., S layers and capsids), and the existence and disposition of appendages (e.g., flagella, pili, and phage tail fibers). The vaporized metal hits the structures at an angle (10° to 45°), congeals on them, and mimics fine detail. Where surface elevations on the specimen limit exposure to the shadowing substance, little metal accumulates. Consequently, a gradation of metal shadow coats the specimen according to the elevations in its topography, resulting in a high-contrast, three-dimensional representation which can be seen by TEM. There are several methods of using shadowing (e.g., shadow-casting, freeze-etching, and rotary shadowing); each has its particular advantage, and outlines of these techniques follow.

3.2.6.1. Shadow-casting of whole mounts (1, 5, 8)

Shadow-casting involves the deposition of an electron-dense metal (e.g., platinum) at a specific angle onto the specimen while it is in a vacuum evaporator (2, 6). Shadow-casting is especially useful for determining the topography of bacterial surfaces. Unlike negative stains, which can permeate through the various surface layers to the cytoplasmic membrane (and therefore produce moirés of superimposed structure), metal shadowing reveals only the upper surface of the specimen. In addition to delineating surface irregularities or periodic structures such as S layers, the technique can be used to measure the vertical height of structures in the specimen by measuring shadow length and calculating geometrically from the known shadow angle. This process can be aided by inclusion on the grid of a known standard such as latex spheres (obtained from EM suppliers) of a given size for purposes of calibration.

Shadowing is conducted as follows.

1. Wash the specimen to be examined free from extraneous debris. It may or may not be fixed first.
2. Place the material on a Formvar-coated grid, and allow to dry. Attach the grid to double-sided sticky tape on a glass microscope slide, together with a small piece of white paper to serve as a visual indicator of the amount of metal deposited during the evaporation.
3. Place the slide on the specimen support head of the vacuum evaporator.
4. Wrap Pt wire around the tungsten V-filament held between the electrodes. The V is formed from standard 30-mil (ca. 0.2-cm) tungsten wire, and 7 to 10 mm of 10-mil (ca. 0.05-cm) Pt wire is wrapped on it. Before doing this, it is essential to heat the tungsten filament to white heat in vacuo and then reopen the evaporator and affix the Pt wire. (This procedure degasses the filament, improving the vacuum during subsequent metal evaporation and relieving strains produced in the wire during its bending, thus prolonging filament life.) Evaporators, such as Balzers freeze-etchers, are equipped with carbon and platinum evaporation sources. Simple but reliable sources consist of two graphite rods (one sharpened like a pencil) which touch one another, so that carbon is evaporated when a voltage of ca. 30 V is passed through them. For platinum evaporation, a small length of Pt wire is wrapped around one of the rods and a voltage is applied. More expensive and elaborate systems use electron beam guns as evaporation sources.
5. With the Pt wire wound around the apex of the tungsten V-filament, place this metal source approximately 12 cm above the grid at the desired angle. Establish a good vacuum (10^{-6} mmHg) in the evaporator, turn up the controls to heat the metal quickly to its melting point, and hold it there until all is evaporated or until the shadow on the visual filter paper indicator appears correct. This judgment requires some practice and experience. Some evaporators have quartz monitors which measure the metal thickness by increased electrical conductance.
6. Sometimes it is useful to omit the Pt and to shadow with tungsten oxide so as to produce a minimal background and finer granularity than that obtained with Pt. To do this, heat the tungsten V-filament *in air* until it is coated with yellow tungsten oxide. Scrape off the oxide with a scalpel from all the wire except the apical few millimeters, and proceed with the shadowing as described above.

3.2.6.2. Freeze fracturing and freeze-etching (26–30)

Freeze fracturing is a difficult technique because it requires rapid freezing of a paste of cells to below liquid N_2 temperature, and the frozen pellet must be maintained at low temperature (−100 to −196°C) to preserve the structure accurately. At these temperatures, the paste will fracture if hit with a knife, and the cleavage line will run through or around cells, thus exposing their constituents and surface structure. Membranes are frequently cleaved through their hydrophobic domain, so that *intramembrane* structure is revealed. The fracturing is done in vacuo, and the water in the exposed ice surface immediately begins to sublime, revealing structural components in the cells, their membranes, or the external milieu; this is the etching process. Freeze-etching is one of the few techniques besides thin sectioning that exposes the inside structure of bacteria. However, the exposed structure in the frozen-cleaved surface of the pellet then has to be replicated as a carbon-supported metal replica of that surface. This means that a complex apparatus for freeze-cleaving, etching, shadow-casting, and carbon deposition is required, all at controlled low temperature. Equipment allowing this series of critical steps in vacuo is provided by Balzers A/S and by Denton Vacuum Inc. The replica is examined by TEM and looks like a shadowed preparation.

Freeze fracturing (with or without etching) is a form of cryofixation, which was discussed in section 3.2.4. Both are carried out in sequence, starting with fracturing and followed by etching in a specialized apparatus, a "freeze-etcher." Instead of plunging a thin bacterial film into a cryogen held at liquid-nitrogen temperature (as is the case for freeze-substitution [section 3.2.4]), the paste of cells is formed into a 1-mm-diameter drop and placed on a brass, copper, or (better because it is acid resistant) gold planchet before plunging. Although propane or ethane can be used as a cryogen, Freon 22 (Du Pont) held at liquid-nitrogen temperature has been the traditional cryosolvent even though it has a lower freezing rate than the first two. (These cryosolvents are chosen because they do not "boil" and thus inhibit rapid freezing, as would liquid N_2 alone). Freon 22 is safer, being nonflammable, but it is environmentally unfriendly and should be replaced.

The frozen paste is placed on an ultracold stage in the freeze-etcher, the stage temperature is set to −100°C, and the chamber is evacuated to ca. 10^{-6} mmHg. Once the vacuum and stage temperature are stable, the frozen drop of cell paste is cleaved with a microtome knife (usually a razor blade mounted on a rotary arm designed to strike the specimen). The fracture follows the lines of least bond energy, so that the fracture travels over cells and through cells and frequently splits membranes (e.g., the plasma membrane).

Initially, the fracture surface through the cell paste is smooth and flat. As the ice sublimes, the etching exposes ribosomes and DNA where the fracture has cut through cytoplasm, and, of course, it also exposes the edges and surfaces of walls and membranes. Etching increases the topography of the entire fracture plane and provides more informative interfaces; if deep enough, it will also show the external wall surface of a whole cell that was ice embedded. Generally, etching times of 20 to 60 s are used, after which the surface is shadowed with Pt as explained in section 3.2.6.1. Etching times can be decreased and "deep-etching" can be attained if the microtome arm is cooled to −196°C and placed over (but not touching) the fracture surface as a cold trap to speed sublimation.

The surfaces and interfaces now revealed in the ice can be visualized only by making a *replica*. This is done by shadow casting (section 3.2.6.1) a thin metal layer (Pt) at an angle to the fractured surface, which gives relief to show structure, and by supporting this thin layer with a continuous layer of carbon (i.e., a Pt-carbon replica now sits on top of the etched fracture face and precisely mimics its structure). Carbon coating is done in the same way that EM grids are coated (section 3.2.1.2), except that it is performed in the freeze-etcher. Finally, the ice and organic constituents of the frozen cell mass (which are too thick for EM) must be removed to provide the final replica for TEM, as follows.

Let air into the chamber, and allow the temperature to rise so that the paste begins to melt. Then cleanse the replica of adhering cellular debris so that the replica can be mounted on an EM grid. That is, immerse the paste/replica into a small petri dish containing concentrated sulfuric acid. (*CAUTION*: This is a strong acid!) Float the replica and the adhering debris on the acid surface for ca. 60 min, allowing the organic cellular constituents to char. Use a loop 4 to 5 mm in diameter, made out of 10-mil Pt wire and mounted in a suitable holder (a bacteriological inoculation loop holder works fine), and carefully bring the loop under the replica to pick it up. Carefully transfer the replica to a dish of distilled water. The acid in the loop and the water in the dish mix violently and boil, but, surprisingly, the replica remains intact and floats on the surface if the Pt-to-C ratio of the replica is correct. Using the loop, transfer the replica to a 5% sodium hypochlorite solution (diluted from commercial bleach solution) for 20 min to oxidize the remaining organics and remove the charred debris from the replica.

Wash the replica free of salts and hypochlorite by transferring two to five times into fresh distilled water by using the loop. Surprisingly, this is the most tricky step because the cleaned replica tends to roll up on itself. If this happens, a quick transfer to concentrated sulfuric acid and then back to water will often generate a violent straightening. Some EM suppliers sell fluids which can be used to straighten replicas; these are organic solvents with high vaporization points which, when dropped on the floating rolled-up replica in water, violently alter the surface tension and flatten the replica.

Mount the replica on a 200- or 400-mesh uncoated copper grid; the 200-mesh grid is best because there is more viewing space. Mounting can be difficult because the replica will always prefer to stay at the surface of a water drop and not at the grid surface. Using the loop, pick up the replica in a thin film of water, ensuring that the replica is in the center of the film. Now, taking the grid (held by EM tweezers at the edge and bent at a right angle to the edge held by tweezers), slowly move it through the center of the loop so that the film of water breaks, leaving the replica attached to the grid. Some beads of water will remain on the grid, and the replica may be floating on them. With pointed wedges of filter paper, carefully absorb the water and center the replica on the grid.

The technique of freeze-etching and the production of good replicas can be difficult, but they are well worth the trouble. A good replica can be beautiful!

3.2.6.3. Rotary shadowing of DNA and RNA

The spreading of nucleic acids on fluid interfaces to make preparations for electron microscopy after rotary metal shadowing has come into widespread use. It is an essential method in many aspects of molecular biology and is well described in appropriate texts (23–25). Single- and double-stranded nucleic acids are too thin to be seen clearly by usual EM methods, but once spread and dried, the polymers have enough height to be shadowed with Pt. In fact, the metal adds to their bulk and makes them easier to see. They are even more apparent when a rotary stage is used in the evaporator, which spins the specimen during shadowing, adding even more Pt to their mass and ensuring that all sides of the polymer are covered. Other than this, the process is similar to that outlined in section 3.2.6.1.

3.3. PHOTOGRAPHY

3.3.1. Film

The working and reference material of the electron microscopist consists of electron micrographs, so the proper taking and processing of photographs is therefore essential. Most microscopes are designed for sheet or roll film or both. Roll film (35 mm) is cheap, effective, and easily stored. A single roll usually provides 40 to 50 exposures. Sheet film is larger (usually 10 by 8 cm) and consists of a photographic emulsion on a plastic (Estar) base. Since sheet film is larger, not as many can be stored in the microscope (e.g., only 20 to 40 exposures).

The major problem with all of these films is that the emulsion and the backing absorb and retain water vapor from the air. Therefore, it is wise to keep film in a series of aluminum desiccators with solid lightproof lids, which are tapped to take a vacuum needle valve. Cut the 35-mm film into appropriate lengths to fit the camera; separate and stack sheet film to allow quicker drying. Fill a tray in the bottom of the desiccator with phosphorus pentoxide. (*CAUTION*: This reacts violently with moisture, especially that on fingers!) The pentoxide has to be stirred at each opening (in the darkroom) and replaced as needed. A modest vacuum and the pentoxide are effective in drying the film after a few days of storage. This prevents long pumpdown times in the electron microscope as a result of outgassing.

35-mm Fine Grain Positive film (Kodak) is satisfactory for routine purposes when the electron microscope is equipped to handle roll film. For plate cameras, use Kodak Electron Image film no. 4489 or Ilford Technical EM film. It is important to first review the recommendations and data supplied by the manufacturer of any film (or paper) considered for use. Experience will then indicate an appropriate routine exposure, development, and degree of contrast desired in the negative. Negatives can be easily scratched, and it is best to keep them in transparent sleeves for protection, handling, and easy scanning. Film characteristics and the darkroom requirements are fully discussed in Chapter 30.2.

3.3.2. Printing

Skill in focusing to achieve sharp negatives is obtained by experience, but it is helpful to recall that, at high magnifications, the exact focus is not necessarily the best for printing the negative. In-focus images often do not have enough contrast for easy viewing; slightly under-focus images may make better prints. For publication, micrographs should have slightly more contrast than normal since the published print loses contrast. It is therefore a good habit to take three to five "through-focus" pictures of each field desired, trying to have exact focus near the middle step. Each step ranges from 0.05 μm (at ×30,000 on the film) to 0.4 μm (at ×4,000 on the film). If you have greatly over- or under-focused images, they become blurred because of Fresnel fringes which occur around the objects being viewed. These are most evident at high magnifications.

The printing of electron micrographs requires more than ordinary attention to detail. A high-quality enlarger is essential, and many choices are available (Chapter 30.2.1.4). Remember that point-source enlargers, although the most critical, will show every scratch in the film! Automated photosensor exposure devices are helpful for reproducibility, once calibrated for each paper used, but nothing can replace an experienced worker. The choice of paper grades should be reasonably wide because of some unavoidable variation among negatives and the emphasis desired. This can be addressed by use of a multigrade paper system with appropriate filters, but a collection of bromide papers with grades of "hardness" from 2 to 6 gives the best results. Similarly, although out of vogue, cellulose-backed papers give the best tone and clarity. These can be difficult to obtain unless purchased in bulk (ca. 1,000 sheets), but they store well in a refrigerator. Plastic-based (resin-coated) papers are more common and require different processing. Large EM units may use automated processors and contrast-controlled enlargers. Further information is given in 30.3.4.

It is not always necessary to print all negatives; this takes time and can be expensive. Negatives can be used in a slide projector and displayed on a screen at appropriate magnification. For this purpose, less sophisticated early-model projectors are often better since they are more easily adapted to strips or rolls of film. In fact, projection of negatives is a very good way to perform accurate measurements and statistical counts. A slide or negative of a measuring ruler will give a magnification factor for the projector and can be used for calibration each time the projector is used.

Because microscopes, exposure arrangements, films, and processing differ among EM laboratories, no attempt is made here to present detailed procedures. Consult with experienced investigators within their own units. Also, see Chapter 30.

3.4. INTERPRETATION

Interpretation is not just seeing what is there; the EM image must also be related to what is already known about bacterial anatomy. Therefore, it is useful to have some idea of the field, and it is assumed that the student will have been exposed, through courses or some reading, to the general features of bacterial ultrastructure. Unfortunately, pictorial atlases in which one can readily compare different genera are rare or hard to find. Specific references on interpretation, including some special reviews (125–128, 137–142, 145), are listed in section 3.9.3.

In a more precise sense, "interpretation" may be construed as getting the most information out of an electron micrograph, e.g., the details of fine periodic structure or macromolecular arrangements such as in flagella or accessory protein layers of the cell envelope. For such advanced work, acquisition of a satisfactory micrograph is only the first step, and then methods of image enhancement and reconstruction can be used (section 3.1.5) (93–108). However, remember that it is rarely necessary to go this far; general information (about size, cell shape, mode of division, enveloping layers, flagellation, and internal membranes) is also useful and important to recognize and record (125).

3.5. MICROSCOPY TECHNIQUE SELECTION

The proper selection of EM technique is usually taken for granted, but it requires experience and is a source of confusion to students. Inexperience usually dictates a "shotgun" approach that includes the most sophisticated techniques. Don't "aim for the stars" right from the start. Best start with light microscopy (Chapter 2) to gain preliminary information; e.g., is the culture pure, what is the shape of the bacterium, is it gram negative or gram positive, is it motile? Light microscopy with stains and phase-contrast microscopy will answer these simple questions. *All* electron microscopy takes a great deal of time and effort and uses complicated, expensive equipment. Simple procedures are usually best, at least initially. Figure 1 describes a good stepwise approach to visualizing bacteria and their structures by EM. It is a flow diagram which starts at a simple level and works toward more exotic techniques. Obviously, not all structures and techniques can be included, but there is enough information to assess the utility of methods and the times required to accomplish each individual step.

3.6. MACROMOLECULE IDENTIFICATION TECHNIQUES

The identities and positions of bacterial macromolecules, either within the cytoplasm or on the surfaces of cells, are often matters of importance to research and require appropriate EM methods; e.g., it can be essential to know the cellular site of enzymes, selected molecular species, and virulence factors or to monitor the synthesis and incorporation of cellular components within the cell. This requires the use of specialized electron-dense probes that can both search out and identify (with great precision) the molecules of interest, which can be in low copy number. Several different techniques are used, and many of them require the production of specific antibody (Chapter 5.3 and 5.4); most are lengthy, and some are both technically complicated and difficult.

3.6.1. Polycationized Ferritin (52–56)

Polycationized ferritin (PCF) is a large (ca. 11-nm) protein multimer with iron atoms in its core, which is

GENERAL CELL SHAPE

and

INTERCELLULAR ASSOCIATIONS

light microscopy for preliminary observations
followed by negative stain EM (5-30 min) (3.2.2)*

CELL SURFACE

Shadow-casting of whole cells (60 min) (3.2.6.1)

NUMBER AND ALIGNMENT OF ENVELOPING LAYERS

conventional embedding of whole cells and thin-sections (3-6 days) (3.2.3.1)

CONFIRMATION OF THE EXISTENCE OF ENVELOPING LAYERS AND THEIR MORE ACCURATE PRESERVATION

freeze-etching (1 day) or freeze-substitution of whole cells (3-6 days) (3.2.6.2 and 3.2.4)

INTERNAL STRUCTURES

Conventional embedding of whole cells and thin-sections (capsules are best seen if stabilizing agents such as ruthenium red or antibody are added: 3-6 days) (3.2.3.1)

CONFIRMATION OF THE EXISTENCE OF AND MORE ACCURATE PRESERVATION OF INTERNAL STRUCTURES

freeze-etching (1 day) or freeze substitution of whole cells (3-6 days) (3.2.6.2 and 3.2.4). These techniques are especially good for revealing the internal membranes of phototrophs, methylotrophs, nitrifiers, etc.

MORE ACCURATE RESOLUTION

cell breakage, isolation and purification of the various components (several days to weeks) (Chapter 4)

S LAYERS, FLAGELLA AND FIMBRIAE

negative stain (5-30 min) (3.2.2), shadow-casting (60 min) (3.2.6.1), rotary shadowing (60 min) (3.2.6.3) or freeze-etching (1 day) (3.2.6.2)

Optical (OD) or electron diffraction (SAED) (3.1.5) to determine the quality of the preparations and macromolecular spacings and arrangements; 5-30 min, but the production of preparations which are good enough can take days or weeks.

Computer enhancement of image (usually Fourier analysis is used; 1 to several days (3.1.5)

OUTER AND PLASMA MEMBRANES, MUREIN SACCULI, AND GRAM POSITIVE WALLS

negative stain (5-30 min)(3.2.2) shadow-casting (50 min)(3.2.6.1) or rotary shadowing (60 min) (3.2.6.3)

freeze-etching (to reveal particle distributions within membranes; 1 day) (3.2.6.2)

RIBOSOMES

negative stain (5-30) (3.2.2) or shadow-casting (a low angle approaching 10° is best; 60 min) (3.2.6.1)

computer enhancement of image (correlation averaging to bring out fine detail; 1 to several days) (3.1.5)

DNA AND RNA

Kleinschmidt spreading of the polymers (5-60 min) (3.2.6.3)

rotary shadowing (3.2.6.3) (adds bulk to the nucleic acid strands to make them easier to see; 60 min)

INTERNAL GRANULES

the choice of method depends on the composition of the granule E.G., MAGNETOSOMES, CARBOXYSOMES and GAS VESICLES - negative stain (5-30 min) (3.2.2) POLYHYDROXYBUTYRATE (PHB), SULPHUR, GLUCOSE and POLYPHOSPHATE GRANULES- these are best seen in intact cells by thin section (3.2.3). They can be extracted by conventional fixations so be careful. PHB granules can be recognized in freeze-etching (3.2.6.2) by their tendency to melt and "string-out" (124); 1 to 6 days)

Compositional Analysis - EDS, EELS or ESI (good material takes only 1-5 min to analyse) (3.1.5) (Remember, conventional processing for EM and isolation/purification procedures can extract native constituents. It is best to look at intact cells or their isolated structures using a cryo-technique).

readily seen by EM. It is a useful probe to determine electronegative sites on bacteria. Ferritin is purified from mammalian spleens and is chemically modified so that its overall charge character is electropositive. Its size does not allow PCF to penetrate into bacteria, but it attaches to surfaces to reveal electronegative sites. PCF can be used for unstained whole mounts (55), in negative stains (53, 54), in freeze-etchings (54), or in thin sections of prelabeled cells (56). Since most bacterial surfaces are anionic (124–127), extremely low, nonsaturating concentrations of PCF can be used to probe for surface sites with the most electronegativity (55). Furthermore, because most bacterial capsules are also anionic, PCF can be used to stabilize and stain this hard-to-preserve material (52). PCF can also be used to probe for electronegative sites on isolated material such as gram-positive walls (55, 56). Other smaller electropositive probes (e.g., cytochrome *c* [127]) and electronegative probes (e.g., polyglutamic acid) are available, but they are not electron dense and require sophisticated methods of detection (127). PCF is available through most commercial EM supply houses (section 3.8).

3.6.2. Colloidal-Gold Antibody Probes (57– 81)

Colloidal gold can be made as gold sols (section 3.6.2.1) to specific particle sizes (2 to 1,000 nm) in the laboratory and is also available from commercial sources (section 3.8). The gold is reduced from chlorauric acid to its colloidal metallic form, which is extremely electron scattering. There is still debate over the surface properties of the gold particles; some authorities believe the surface to have a net electrostatic charge, whereas others believe it to be hydrophobic. If chemical reduction is complete and metallic Au^0 is formed, it should be charge neutral and therefore hydrophobic. Whatever the mechanism of union, there is good attachment of colloidal gold to highly specific recognition molecules such as antibody, protein A, protein G, avidin, and lectins. Once coated with these organic molecules, the gold particle becomes a highly specific marker for electron microscopy. In fact, gold of different sizes and coated with different recognition molecules can be used to probe the same sample; the size of the gold particle is used to distinguish the different macromolecules. Because bacteria are so small, gold particles of 5 to 50 nm are usually used.

3.6.2.1. Colloidal-gold production

Colloidal gold is chemically reduced from an aqueous solution of chlorauric acid ($HAuCl_4$) by using reducing agents such as white phosphorus, sodium citrate, or tannic acid. The reaction produces a gold sol in which the particles grow with time, which allows selection of a size range. A single absorption peak (λ_{max}) for small particle sols is seen at ca. 510 nm, and this peak shifts toward 550 nm as the particles grow. Visually, the color of the sol changes from pale orange to purple to brown as the particle size increases. This colorimetric change allows choice of the approximate particle size.

The "white phosphorus" method (63, 79) is frequently used for the production of extremely small colloidal gold particles (2.0 to 50 nm). This is an exothermic reaction, and care must be taken. (*CAUTION*: Use a fume cabinet and keep a solution of $CuSO_4$ on hand to extinguish burning phosphorus.) Other methods of colloidal-gold production are available (63, 66, 79), and usually each EM laboratory has a preferred method. All glassware must be scrupulously clean.

Colorimetric analysis of the gold sol can give only approximate particle sizes, and accurate measurements with unstained suspensions on EM grids must be made. If the gold particles do not stick to the grid, try polylysine-coated grids (0.01% [wt/vol] aqueous polylysine dried on to the support film). TEM will show a range of particle sizes, which tend to aggregate because of surface interaction, but most will be of a similar size. To store the sol for more than 1 or 2 weeks, it is best to stabilize the particles with Carbowax 20M (a type of polyethylene glycol), bovine serum albumin (BSA), or diaminoethane-derivatized dextran to inhibit aggregation (67, 69). The dextran stabilization is best because it resists aggregation in electrolyte solutions such as phosphate-buffered saline (PBS).

The production of colloidal gold is relatively easy, takes little time, and is inexpensive, yet many microbiological laboratories have little desire to immerse themselves in inorganic-chemical reactions and the difficulty of macromolecular adsorption phenomena once the colloidal gold is produced. Most EM suppliers have a range of gold sol sizes that can be purchased. They also supply sols that have been precoated with antibody (e.g., anti-mouse or anti-rabbit antibody), protein A, protein G, lectin, or avidin. For some reason, sols obtained from commercial sources, whether they are precoated or not, can be quite expensive. They are not always of an exact size or of high specificity (no matter what the advertisements say!) and their size should be checked by TEM on unstained mounts.

3.6.2.2. Probe production

Colloidal gold by itself has no specificity and is therefore not a complete probe. It must be coated with highly specific recognition molecules, most often antibody (Chapter 5.3 and 5.4). In general, strong adsorption of proteins to gold surfaces occurs at pH values slightly above the isoelectric points of the proteins (i.e., their zwitterion is preferred, and interfacial tension is maximal). However, immunoglobulin G (IgG) molecules (even when affinity purified) can be a problem since they display a spectrum of isoelectric points. An alkaline pH of ca. 9.0 is best. Because IgG can aggregate when subjected to an excess of salts or when concentrated, dilute buffers should be used (e.g., 2 mM Borax buffer [pH 9.0]) and the antibody concentration should never exceed 1 mg ml^{-1}. Centrifuge the IgG solution at 100,000 x *g* for 1 h before use to remove aggregates. Before adsorbing the antibody to the gold, some workers determine the "minimal protecting concentration" of the antibody as a guide (63). This is the amount

FIG. 1. Flow diagram for the structural analysis of bacteria. From the top down the techniques go from simple to complex, and each has the section reference. The times given for each procedure suppose that the procedure is set up and ready for use. The techniques described are for TEM, but SEM can also be appropriate for assessing general shape, growth habit, and intercellular associations.

of antibody that will protect a given volume of colloidal gold against flocculation when optimal pH and electrolyte concentration are used.

For adsorption of the antibodies to colloidal-gold particles, follow DeMey's method (63), which is useful for both IgG and IgM antibody classes. Of the two immunoglobulin classes, IgG always seems to bind more efficiently to the gold. Monoclonal antibodies have a narrower isoelectric point profile than do polyclonal antibodies. Carry out all reactions at room temperature, as follows.

1. Suspend the stabilized gold particles in 20 mM Tris-HCl buffer containing 1% BSA, 150 mM NaCl, and 20 mM NaN_3 (to discourage bacterial contamination) and adjust to pH 8.2 to 9.0. (For the higher pH levels, use a Borax buffer or, for more physiological conditions, use PBS at pH 6.8 to 7.6.) The colloidal-gold concentration is usually adjusted to 20 to 80 $\mu g\ ml^{-1}$. Since the reaction area of the gold particle depends on particle size and since there may be a range of sizes in the gold sol, trial and error will optimize the reaction conditions. Without suitable analytical equipment, it is difficult to accurately measure gold concentration by light absorption since the gold particles both absorb and scatter light at 520 nm. To help estimate gold concentration, 70 mg ml^{-1} of 15-nm gold particles gives an optical density at 520 nm of 0.25. If the gold particle size is known from TEM, an optical density determination will help estimate the gold concentration. Usually, small gold particles require more protein than larger particles do.
2. Centrifuge at low speed to remove gold aggregates (for 5-nm gold, use 4,800 x *g* for 20 min; for 15-nm gold, use 250 x *g* for 20 min).
3. Stir the gold sol rapidly, and slowly add the antibody. The antibody concentration is usually between 20 and 80 μg per ml of sol and must be above the minimal protecting concentration (63). After 2 min, add enough of a 10% BSA–20 mM Tris-HCl (pH 8.2 to 9.0) solution to bring the reaction mixture back to 1% BSA. This depends on the volume of antibody added. The BSA blocks the remaining adsorption sites on the gold particles that are not occupied by antibody and will also stop aggregation of the gold particles.
4. Centrifuge to pellet the antibody-coated particles (uncoated gold and unadsorbed antibody will stay in the supernatant fluid). Use 60,000 x *g* for 1 h for 5-nm gold and 15,000 x *g* for 1 h for 15-nm gold.
5. Carefully remove the supernatant, resuspend the pellet in a few drops of the 1% BSA buffer, and store overnight at 4°C. This period of storage ensures that the antibody-gold linkage becomes stable. Resuspend and centrifuge the following morning.
6. Resuspend 5-nm gold to an A_{520} of 0.250, and resuspend 15-nm gold to A_{520} of 0.50. As long as NaN_3 is in the suspension, it can be stored at 4°C for several weeks. The addition of glycerol to 50% (vol/vol) allows it to be frozen. Not all preparations freeze well. To be safe, freeze a small portion for 1 to 2 days and check its activity after thawing.

Because there is never absolute control over gold particle size (section 3.6.2.1), there will always be a small range of sizes. If there is a strict size requirement, the gold can be centrifuged through a 10 to 30% isopycnic sucrose gradient (use a Beckman SW41 centrifuge rotor for 45 min at 41,000 rpm) to separate the particles into specific sizes. Make sure to wash or dialyze the sucrose from the gold before use.

There are also probe molecules other than antibody that can be used to target the gold (60, 61, 63, 65, 67, 76, 78, 79). These are attached to the gold particles as outlined above for antibody, and include the following.

1. Protein A, specific for IgG and to a lesser extent IgM antibodies; and protein G, specific for IgG but also has good affinity for IgM. Because of their high specificity to antibodies, protein A and protein G are frequently used in the "indirect-labeling" technique (section 3.6.2.4; see also Chapter 5.3.4.2 and 5.3.4.3) since they are also used in affinity chromatography (Chapter 5.3.4). Protein A is a constituent of *Staphylococcus aureus* cell walls which has high affinity for IgGs, and it has been used extensively over the last decade in immunogold labeling (73). Protein G, which is extracted from group G streptococcal cell walls and is much newer (61), also has high affinity for IgG but seems to react more strongly to IgM than protein A does (60). Since monoclonal antibodies are frequently of the IgM class, protein G is a powerful addition to the immunolabeling arsenal.

2. Lectins. Lectins are specific for a range of carbohydrates (63, 67, 68, 79); e.g., concanavalin A is specific for α-D-mannoside and α-D-glucoside, and wheat germ agglutinin is specific for *N*-acetyl-β-D-glucosamine.

3. Avidin. Avidin is found in hen egg white and is highly specific for *d*-biotin (vitamin H) (63, 68, 79).

3.6.2.3. Direct antigen labeling

Colloidal gold coated with antibody is a very specific electron-dense probe used to find a specific antigen. Bacterial surface components contain important antigens (e.g., lipopolysaccharides, outer membrane proteins, capsule polymers), and these can be located on intact cells. Because antibodies have different affinities for their antigens and because they can exhibit nonspecific adsorption, each time a new colloidal-gold–antibody probe is produced from different serum, the labeling conditions may have to be modified. A general recipe follows.

1. Float Formvar carbon-coated grids on a bacterial suspension (section 3.2.2.2). Process them one at a time.
2. Blot dry and float on a 1 to 3% BSA solution in PBS (pH 7.4) for 15 min. Nonspecific adsorption of many proteins occurs, and the BSA acts as a non-electron-dense blocking agent. Alternatively, use a 1 to 3% solution of skim milk powder, which contains many proteins. Use the same droplet technique as in section 3.2.2.2 for negative staining.
3. Wash the grid by serial transfer on 3 to 5 drops of 0.1% BSA (or skim milk) in PBS.
4. Blot dry, and float the grid on the colloidal gold-antibody suspension in PBS for 10 to 30 min.
5. Wash as in step 3. The final wash should be on droplets of distilled water to reduce salting-out of the PBS in the TEM instrument. Too much washing in water can reduce the labeling, and too little washing produces severe salt effects and forms precipitates during drying. It is best to follow the washing by looking at a grid after each wash in the TEM instru-

ment. This means that several grids should be ready to process.
6. Whole mounts of the cells without further staining can be suitable to observe the surface-labeling pattern, but a negative stain can often help. Ammonium molybdate (1%) at pH 7.0 to 7.4 (section 3.2.2.1) is best because it has low contrast and the gold can easily be seen. Do the staining carefully and quickly so as not to disturb the labeling pattern. Do not use uranyl acetate, since its low pH can dissociate the gold probe from its binding site.

The most difficult aspect of using a gold probe is finding the best probe concentration for accurate labeling of the antigen. Even with blocking agents, there can be high background labeling, and this makes the accuracy of the probe questionable. Finding the best probe concentration is difficult since every antibody preparation has a different avidity to an antigen. Trial and error is the only method, and this means that steps 1 through 5 must be repeated until the best concentration is found. For this reason, controls are very important and must be processed through all steps 1 to 5. Uncoated (without antibody) gold and gold coated with nonspecific antibody should not label the specimen. Specific antibody (without the gold) should compete for the antigen sites with the gold probe (coated with specific antibody). Steps 1 to 5 can also be used for lectin-coated gold.

It is possible to label more than one surface antigen and to distinguish one from another by using two different-sized colloidal gold particles (e.g., 5- and 15-nm gold). Each particle size is coated with a separate antibody. The two probes can be mixed for step 4 or they can be used one after the other. The more probes used, the greater is the difficulty in obtaining optimal labeling.

3.6.2.4. Indirect antigen labeling

For convenience it is often best to use indirect gold labeling since a single gold probe can be used against a wide battery of antigens. For example, protein A-colloidal gold is specific for IgG and will recognize the IgG antibody that has specifically adsorbed to the cell. Care must be taken to wash unbound IgG away before treatment with the gold probe. This is an indirect labeling method because the gold probe is not directly linked to the antigen but is only indirectly linked by the underlying IgG. Therefore, the actual gold particle is distanced from the antigen site by the span of an IgG and a protein A molecule, but these distances are usually insignificant for most labeling studies. If the specific antibody is a monoclonal antibody, there is a good chance it will be an IgM. Protein A has a low binding efficiency for IgM, and so it is preferable to use protein G, which has a higher affinity for this class of antibody (61). If it is possible to link biotinyl residues to the desired molecule, then avidin (usually streptavidin)-gold can be used as the probe (63).

For indirect labeling, the same general approach outlined in section 3.6.2.3 is used but for step 4 the specific antibody (not linked to gold) is used alone. After being washed with 3 to 5 drops of 0.1% BSA in PBS, the grid is floated on protein A-gold (or protein G-gold) for 10 to 30 min and then washed as before. If appropriate, the grid can be negatively stained.

3.6.2.5. Thin-section antigen labeling

Sometimes mounts of whole cells will not provide enough information about how labeled surface antigen components are arranged with respect to the underlying enveloping layers (e.g., labeled outer membrane proteins compared with the peptidoglycan layer). Label these cells as in section 3.6.2.3 or 3.6.2.4, and fix/embed them as in section 3.2.3.1. Since the cells are labeled before fixation, the gold probe is chemically fixed to the cell surface (e.g., the outer membrane), and this positioning survives the trauma of embedding and curing. Thin sections show the position of the label with respect to the underlying envelope layers.

The difficulties will be greater if the intent is to label antigen constituents inside cells or to monitor the location of a component over time as it migrates through cellular space. The cells cannot be lysed to show their internal constituents, since this destroys their location and cellular associations. The solution is to gold-probe label thin sections of fixed cells after they have been cut and mounted on grids, by using either the direct or indirect methods. This can be very difficult; controls, to ensure labeling specificity, are essential.

Chemical fixation denatures bacterial constituents by altering their molecular folding and by increasing their covalent bonding. Dehydration and embedding in plastics remove all free cellular water and, presumably, most of the hydration shells which surround bacterial macromolecules; thus, folding is again altered. All of these changes will affect the retention of antigenicity. This is not so important when surfaces are prelabeled with immunogold before fixation (as above), but care must be taken for postlabeling of thin sections. The labeling of thin sections is a compromise. There must be some degree of chemical fixation, and the embedding plastic must not be too denaturing; also, both steps must be done without losing too much antigenicity. Once again trial and error is the best approach.

Osmium tetroxide is a strong oxidizing agent and should not be used in fixation. Formaldehyde seems to be the preferred fixative, but it is not a bifunctional agent and will not cross-link. Glutaraldehyde is bifunctional and is not as damaging as osmium tetroxide, but in high concentration it reduces antigenicity. Uranyl acetate, because of its low pH and binding character, also denatures macromolecules and reduces antigenicity. For dehydration, ethanol is not as damaging as acetone or propylene oxide. Most of the usual plastics (e.g., Epon 812, Vestopal, Araldite) do not preserve antigenicity very well; LR White or Lowricryl K4M is better. It seems like a "no-win" situation, but there are reasonable chances of success. Use the following procedure.

Fixation. Use a mixture of 2% formaldehyde (made fresh from paraformaldehyde; see section 3.7.4) and 0.1% glutaraldehyde in 50 mM HEPES buffer (pH 7.0) for 60 min. Wash with buffer a few times to remove unbound fixative. (Proceed as in section 3.2.3.1, but do not fix with osmium tetroxide or perform an en bloc stain with uranyl acetate.)

Dehydration. Use a graded ethanol series and repeat twice with 100% ethanol, as in section 3.2.3.3.

Embedding. Infiltrate by using an ethanol : LR White graded series (50 to 100% plastic medium as in section

3.2.3.3). Cure the block at 65°C for 18 h. Cut thin sections as usual, but do not stain with uranyl acetate or lead citrate. Lowicryl K4M (21, 22) can also be used for embedding and has the advantage that embedding and curing with UV can be carried out at −80°C to help retain antigenicity. *CAUTION*: Lowicryl can induce severe allergic responses, so wear surgical gloves!

Labeling sections with gold probe. Cut enough sections to make a number of grids for trials and controls. Float the grids section side down on 1 to 3% BSA (or skim milk) in PBS for 15 min to block nonspecific reactions. Wash with 2 to 5 drops of 0.1% BSA in PBS. Float the grid on the gold probe suspension for 30 to 60 min, and wash as before. Check by TEM to see how well the section is labeled (there has been no heavy-metal staining so far, and only the gold will be seen). Repeat the labeling on a fresh grid until a satisfactory labeling pattern is seen and the background is low. Controls with uncoated gold and nonspecific antibody-coated gold should be negative. Remember that only the antigens exposed on the surface of the section will be labeled and then only if they are not denatured; unless the antigen is present in very high copy number, expect only low labeling counts, which may be only slightly above background and control counts. Statistical analysis of the counts can be helpful (63).

Staining of labeled sections can be tricky, and it is best to work out optimal gold labeling on unstained sections first. Preserving gold-labeling patterns after staining with lead citrate is unsatisfactory; instead use only uranyl acetate to stain the sections. It can be used at 1 to 2% (wt/vol) either before or after labeling with the gold probe. Either way, the labeling efficiency of the gold may be reduced (and this is another reason for knowing how unstained sections label). If stained before, some antigen sites may be blocked by UO_2^{2+}; if stained after, some gold probe may be desorbed by UO_2^{2+} and low pH. Keep the period of uranyl acetate staining as short as possible for best results.

Sections can also be indirectly labeled with protein A- or protein G-gold. For this, the specific antibody is first adsorbed to the grid, the grid is washed, and the indirect gold probe is adsorbed onto the first antibody. Sometimes, because gold counts are so low on thin sections, it is useful to try to amplify the gold labeling. This is also an indirect technique and uses a second intermediate antibody between the first, specific, adsorbed antibody and the gold probe; e.g., if the first specific antibody is a rabbit IgG, once it is adsorbed to the section, antibodies directed against rabbit IgG (maybe goat antibodies) would be used to decorate the first antibody at several places along its length. Next, anti-goat–colloidal gold (or protein A-gold) would be used to decorate all of the adsorbed goat antibodies. This three-step procedure "amplifies" the initial single antibody to become labeled with several gold particles and makes the antigenic site easier to see.

3.6.3. Enzyme Identification

The precise localization of enzymatic activities as they occur within a cell is a long-standing objective of histochemistry and now of cytochemistry at the level of EM. Methods differ with the type of enzyme and, therefore, with the possibilities for devising an artificial or labeled substrate that is normally transported and allowed to react to give a visible reaction product with either light microscopy or EM. The subject area is a large one and has engendered a good deal of research that cannot be even summarized in this limited space. The reader is referred to available texts and to the several journals dealing exclusively with histochemistry and cytochemistry (2, 6, 10). Remember, colloidal-gold–antibody probes may be an easier or more precise way of labeling enzyme antigens than the use of a reaction product procedure.

3.7. MATERIALS

This section provides a catalog of useful items for electron microscopy. Commercial sources for most of the materials are indicated by the letters in parentheses, which refer to the list in section 3.8.

3.7.1. Plastic Support Films

Formvar powder (*a, k, o, p, r, u, x, z, cc*)
Formvar solutions (*o, r, cc*)
Ethylene dichloride (*b, e, m, n, y, aa*)
Clean glass microscope slides
Lint-free photowipe tissues (*a, k, o, p, r, u, x, t, cc*)
Large metal (or plastic) pan or tray
Large glass crystallizing dish with straight sides and flat bottom. The best crystallizing dish is one measuring 200 by 80 mm (culture dish; Vitro) (*w, dd*). The top edge must be even and smooth. The bottom and sides must be permanently covered with *black electrical tape* to permit visualization of the floating films against a black background.
Glass rod longer than the diameter of the crystallizing dish
Coplin glass staining jars for microscopic slides (*f, m, w, dd*)
Pasteur pipettes and rubber bulbs
Hydrofluoric acid, 10% (in Teflon container) (*b, e, m, n, y, aa*)
Razor blades (alcohol washed)
Filter paper, Whatman no. 1, 9.0 cm diameter (*f, m*)
Petri dishes, standard, 9.0-cm diameter
Scissors
Grid tweezers, fine-pointed no. 5 Dumont or equivalent (*a, k, o, p, r, u, x, z, cc*). (A loop or two of Nichrome wire placed around the base can be pushed forward to clamp the tips on the grid, thus preventing its accidental dropping as well as the cost of special expensive tweezers designed for this purpose.)
Copper grids, precleaned (*a, k, o, p, r, u, x, z, cc*)
Light source, fluorescent, adjustable

3.7.2. Carbon Films

High-vacuum evaporator and accessories (*d, g, q*)

3.7.3. Negative Staining

Coated grids, prepared as described in section 3.2.1.2 or from commercial sources (*a, p, r, u, x, t, cc*)

Sticky tape (Time Tape) (*f, m, w*)
Glass microscope slides
Zerostat gun (*i, w*) to discharge static charges
Pasteur pipettes
Ammonium acetate (*b, e, m, n, q, y, aa*)
Ammonium molybdate (*b, e, m, n, q, y, aa*)
Parafilm (*c, m*)
Potassium or sodium phosphotungstate (*b, e, m, n, q, y, aa*)
Uranyl acetate (*b, e, k, m, n, p, q, u, y, aa, cc*)

3.7.4. Fixing, Embedding, Sectioning, and Staining

Glutaraldehyde (*a, k, o, p, u, x, z, cc*)

Paraformaldehyde. To generate formaldehyde, add 2 g of paraformaldehyde (*b, e, m, n, q, y, aa*) to 25 ml of water and heat to 60 to 70°C. A white suspension is formed. With rapid stirring, add 1 to 3 drops of 1 M NaOH until the suspension clears. Cool to room temperature, and use immediately. Holding life is about 4 to 8 h.

Sodium cacodylate buffer. This may be purchased as buffer or as a kit with premeasured ingredients (*cc*). Otherwise, make up sodium cacodylate (*a, k, o, p, u, x, z, cc*) to 0.2 M in water and adjust the pH to 7.2 with HCl. (*CAUTION*: Cacodylate is toxic; avoid contact and inhalation of arsenical vapor; use a fume hood).

Veronal-acetate buffer. This may be purchased as the buffer (*r, x*) or made up from sodium Veronal (*x*) according to the following formula.

Stock solution
Sodium Veronal (barbitone sodium), 2.94 g
Sodium acetate, hydrated, 1.94 g
Sodium chloride, 3.40 g
Distilled water, 100.00 ml

Working buffer. Prepare daily as follows.
Stock solution, 25.0 ml
Distilled water, 96.5 ml
HCl, 1 N, 3.5 ml
$CaCl_2$, 1 N (110.98 g/liter), 1.25 ml

Dispense the 25 ml of stock solution into a 125-ml bottle. In a 100-ml graduated cylinder, add 90 ml of water and 3.5 ml of 1 N HCl, and bring to 100 ml with water. Add to the stock solution in the bottle, add 1.25 ml of 1 N $CaCl_2$, and stir. (*Note*: If cells to be washed in this buffer were previously exposed to phosphates, as in culture media or a phosphate buffer, it is necessary to have available working buffer made *without* $CaCl_2$ for use as initial washes to avoid precipitates of calcium phosphate. Complete buffer is then used as a final wash before proceeding.)

HEPES buffer. This may be purchased from a number of chemical supply houses as HEPES (*b, e, m, n, q, y, aa*). It should be made up in distilled water to a stock solution of 1 M (pH 7.2) and diluted as needed.

Safety goggles
Osmium tetroxide (*a, k, o, p, u, x, z, cc*)
Noble agar (*h*)
Uranyl acetate (*a, k, o, p, u, x, z, cc*)
Tryptone (*h*)
Tryptic soy broth (*h*)
Glass vials, screw cap, 1 dram (*w, dd*)
Respirator mask, Gasfoe (*v*)
Disposable plastic containers, 4.5 oz (ca. 130 ml) (*f, l, m*)

Epon 812 embedding resin. Epon 812 (*a, k, o, p, u, x, z, cc*) is an epoxy-based resin and is one of the most commonly used embedding plastics. It relies on the polymer's strained epoxide ring, which can be broken at 60°C to reanneal with those of adjacent polymers to form a solid plastic block. The reaction is controlled by the addition of nadic methyl anhydride (NMA; a chemical "hardener" which increases bonding) and dodecylsuccinic anhydride (DDSA; a chemical "softener" which decreases bonding). The reaction is catalyzed by tridimethylaminoethyl phenol (DMP-30). Epon 812 is quite viscous and can be difficult to infiltrate into bacteria, but it has good thin-sectioning characteristics and has a fine grain when examined by TEM. Lot numbers from chemical suppliers may have subtly different curing characteristics and may require some fine adjustments to the basic recipe. The major manufacturer of Epon 812 no longer makes this plastic resin, but stocks are still available. Some resins are very similar, e.g., Polybed 812 (*z*). Chemicals are usually stored at 4°C for better shelf life; remember to warm them to room temperature before opening so that water vapor does not condense in them.

A procedure for preparing Epon 812 is as follows. In separate disposable 250-ml beakers make:

Mixture A
Epon 812, 40 g
DDSA, 46.6 g

Mixture B
Epon 812, 42 g
NMA, 33.5 g

Mix these thoroughly for ca. 15 min. Weigh out 6 g of mixture A into each of 13 or 14 screw-cap bottles (20-ml scintillation vials work well), and then add 4 g of mixture B to each (some of mixture B will probably be left over). Mix thoroughly for ca. 15 min, and then freeze at −20°C until ready for use. When required, take a bottle from the freezer, warm to room temperature, and add 0.3 ml of DMP-30. Stir for ca. 30 min before use of this complete mixture.

Spurr medium. This is supplied as a kit or as ingredients (*a, k, o, p, u, x, z, cc*) to be used as follows to make the stock mixture.

Stock mixture
ERL-4206 (vinylcyclohexene dioxide), 30 g or 27.0 ml
DER-736 (diglycidyl ether of propylene glycol epoxy resin), 24 g or 22.8 ml
NSA (nonenylsuccinic anhydride), 78 g or 76.5 ml

In a chemical hood or wearing an appropriate mask, add the three components in the above order to a disposable container, mixing as added on a magnetic stirrer. Mix for 5 min at ambient temperature. This stock mixture, which lacks accelerator, can be stored tightly covered at −4°C for 3 to 5 months. If so stored, allow ample time to equilibrate to room temperature on removal and before opening, to avoid water of condensation that may destroy the mixture.

Complete mixture
To make the complete mixture, put 22 ml of the stock mixture in a disposable container, add 0.24 ml of the accelerator S-1 (dimethylaminoethanol), and stir on a magnetic stirrer for several minutes at room temperature. (For each bacterial pellet or other sample to be embedded, but *not* each agar cube resulting from the sample, allow at least 6 ml of complete mixture. This volume will take care of mixtures in steps 4 to 6 of the dehydration and embedment procedure and will provide for the filling of three Beem capsules. Obviously, if more than three blocks from each specimen are desired, more mixture will be needed. Also, if more than three pellets or specimens are to be handled at a time, the initial volume of the complete mixture will have to be greater than 22 ml; but make up in the same proportions.) The useful life of the complete mixture is only 2 days.

LR White Resin. LR White resin (*t*, but also available through North American distributors [*o, p, r, u, cc*]) is an acrylic-based, hydrophilic plastic which comes as a fluid. It is cured at 60°C for 18 to 24 h and does not require the addition of chemical hardeners or accelerators, which makes it very simple to use. After curing, the surface of the plastic block may be sticky because of a thin layer of partially cured resin at the surface which was exposed to air (oxygen). If this is a problem, a vacuum oven can be used. The advantages of LR white resin are its low viscosity (much lower than epoxy-based resins) for good infiltration, its low toxicity, and its simplicity of use. This resin works well for immunogold labeling (section 3.6.2.5). It comes in hard, medium, and soft grades.

Beem capsules (size 00) (*a, k, o, p, u, x, z, cc*).
Drying oven
Centrifuge, Brinkmann 3200 (Eppendorf), and appropriate conical tubes with caps
Pliers
Razor blades, single edge
Ultramicrotome
Plate glass, special for glass knives, cut in appropriate strips for knife maker
Glass-knife maker (*s*)
Diamond knife (*j*), an expensive alternative to the glass knife
Dental wax
Glass weighing bottles
Plastic tissue culture dish, 35 by 100 mm
NaOH pellets
Petri dishes, glass or plastic
Lead citrate. This is available ready-made from commercial sources (*a, k, o, p, u, x, z, cc*). It can also be prepared as follows. To a 50-ml volumetric flask, add:

Lead nitrate, 1.33 g
Sodium citrate, 1.76 g
Distilled water, 30.00 g

Shake for 1 min, and allow to stand at ambient temperature for 30 min. Add 8 ml of freshly prepared 1 N NaOH, and dilute to 50 ml with water.

3.8. COMMERCIAL SOURCES

a. Agar Aids Ltd., 66a Cambridge Road, Stansted CM24 8DA, England
b. Aldrich Chemical Co. Inc., 1001 West Saint Paul Ave., Milwaukee, WI 53233
c. American National Can Co., Greenwich, CT 06836 (This can be obtained through major scientific suppliers such as Fisher Scientific Co. [*m*].)
d. Balzers A/S, 8 Sagamore Park Rd., Hudson, NH 03051
e. Boehringer Mannheim Canada Ltd., 200 Micro Boulevard, Laval, Quebec H7V 3Z9, Canada
f. Curtin-Matheson Scientific, Inc., 10727 Tucker St., Beltsville, MD 20705
g. Denton Vacuum Inc., Cherry Hill Industrial Center, Cherry Hill, NJ 08003
h. Difco Laboratories, P.O. Box 1958A, Detroit, MI 48232
i. Dishwasher, Inc., 1407 North Providence Road, Columbia, MO 65201 (or local photo stores)
j. Dupont Instruments, Biomedical Division, Newtown, CT 06470
k. Electron Microscopy Sciences, P.O. Box 251, Fort Washington, PA 19034
l. Falcon Labware Division, 1950 Williams Drive, Oxnard, CA 93030
m. Fisher Scientific Co., 1 Reagent Lane, P.O. Box 375, Fairlawn, NJ 07410
n. Fluka Chemika-Biochemika, Fluka Chemie Ag, Industriestrasse 25, CH-9470 Buchs, Switzerland
o. Ernest F. Fullam, Inc., P.O. Box 444, Schenectady, NY 12301
p. J. B. EM Services Inc., P.O. Box 693, Pointe-Claire, Dorval, Quebec H9R 4S8, Canada
q. J. T. Baker Inc., 22 Red School Lane, Phillipsburg, NJ 08865
r. Ladd Research Industries, Inc., P.O. Box 901, Burlington, VT 05401
s. LKB Instruments, Inc., 12221 Parklawn Drive, Rockville, MD 20852
t. London Resin Co., Ltd., P.O. Box 34, Basingstoke, Hampshire RG25 2EX, England
u. Marivac Ltd., 5821 Russell Street, Halifax, Nova Scotia B3K 1X5, Canada
v. Mine Safety Appliances Co., Pittsburgh, PA 15208
w. Nalge Co., Division of Sybion Corp., Rochester, NY 14602
x. Ted Pella Co., P.O. Box 510, Tustin, CA 92680
y. Pierce, P.O. Box 117, Rockford, IL 61105
z. Polysciences, Inc., Paul Valley Industrial Park, Warrington, PA 18976
aa. Sigma Chemical Co., P.O. Box 14508, St. Louis, MO 63178
bb. Sorg Paper Co., Middletown, Ohio 45042
cc. SPI Supplies, P.O. Box 342, West Chester, PA 19380
dd. Tousimis Research Corp., P.O. Box 2189, Rockville, MD 20852
ee. Wheaton Scientific (Vitro), 1000 North 10th Street, Millville, NJ 08332

3.9. REFERENCES

3.9.1. General References

1. **Aldrich, H. C., and W. J. Todd (ed.).** 1986. *Ultrastructure Techniques for Microorganisms.* Plenum Press, New York.
One of the few books dedicated to EM techniques used on microorganisms, including chapters on conventional techniques (such as chemical fixation/embedding and freeze-etching) and also on more recent techniques (such as cryopreparation, image analysis, and immunolabeling). This book deserves a place in every microbiology EM laboratory.
2. **Glauert, A. M. (ed.)** 1972–1991. *Practical Methods in Electron Microscopy*, vol. 1 to 15. American Elsevier Publishing Co., Inc., New York.
Glauert, a distinguished electron microscopist with an interest in bacterial structure, has edited this remarkable series on state-of-the-art EM techniques. Most university

libraries have the entire set, and it is well worth while to look through them. Some of the volumes are available as paperbacks.

3. **Glauert, A. M.** 1980. Fixation, dehydration and embedding of biological specimens, part 1 of vol. 3 of A. M. Glauert (ed.), *Practical Methods in Electron Microscopy*, North-Holland Publishing Co., Amsterdam.
This is a handy 200-page paperback book which describes a full range of fixatives, buffers, stains, and plastics. Very good to have for laboratory recipes.
4. **Hayat, M. A.** 1975. *Positive Staining for Electron Microscopy*. Van Nostrand Reinhold Co., New York.
An in-depth treatment of the common and not-so-common stains used in biology.
5. **Hayat, M. A.** 1986. *Basic Techniques for Transmission Electron Microscopy*. Academic Press, Inc., New York.
This 400-page paperback book is filled with helpful hints and general recipes for processing biological samples for TEM. It deserves a place in every EM laboratory.
6. **Hayat, M. A. (ed.)** 1990. *Principles and Techniques of Electron Microscopy*, vol. 1 to 9. Van Nostrand Reinhold Co., New York.
You cannot do electron microscopy in biology without running into Hayat's name. This is a very successful multivolume set, which deals with a full range of important techniques. Like Glauert's series, this set is available in most university libraries and should be examined. There is a single book that goes by the same name, is 469 pages long, and is put out by CRC Press, Inc. Boca Raton, Fla., which you also may want to look at.
7. **Lewis, P. R., and D. P. Knight.** 1977. Staining methods for sectioned material. Part 1 of vol. 5 of A. M. Glauert (ed.), *Practical Methods in Electron Microscopy*. North-Holland Publishing Co., Amsterdam.
A comprehensive guide to the staining of sectioned material which includes histochemical techniques.
8. **Meek, G. A.** 1976. *Practical Electron Microscopy for Biologists*. John Wiley & Sons, Inc., New York.
Simple descriptions and explanations make this an easy reference for most aspects of EM. Don't let the publication date be discouraging; modern EMs still operate by the same principles. Available in paperback.
9. **Moses, N., P. S. Handley, H. J. Busscher, and P. G. Rouxhet (ed.).** 1991. *Structural and Physico-Chemical Methods for Microbial Cell Surface Analysis*. VCH Publishing, New York.
Although techniques other than EM are described, this book is dedicated to the elucidation of bacterial surfaces and surveys the most up-to-date methodology.
10. **Sommerville, J., and U. Scheer (ed.).** 1987. *Electron Microscopy in Molecular Biology: A Practical Approach*. IRL Press, Oxford.
This book does not deal with microorganisms but offers several methods for visualizing nucleic acids and proteins once they have been extracted from cells.

3.9.2. Specific References by Subject

3.9.2.1. Grids, support films, and negative staining

11. **Baumeister, W., and M. Hahn.** 1978. Specimen supports, p. 1–112. *In* M. A. Hayat (ed.), *Principles and Techniques of Electron Microscopy. Biological Applications*, vol. 8. Van Nostrand Reinhold Co., New York.
Good information on grids and support films.
12. **Dawes, C. J.** 1971. *Biological Techniques in Electron Microscopy*, Barnes and Noble, Inc., New York.
An old reference book, but p. 107 to 114 contain several useful tips on making support films, and p. 146 to 148 on negative stains.
13. **Horne, R. W.** 1965. Negative staining methods, p. 328–355. *In* D. Kay (ed.), *Techniques for Electron Microscopy*. F. A. Davis Co., Philadelphia.
Useful hints on negative staining from one of the pioneers of the technique.
14. **Kay, D.** 1976. Electron microscopy of small particles, macromolecular structures, and nucleic acids, p. 177–214. *In* J. R. Norris and U. W. Ribbons (ed.), *Methods in Microbiology*, vol. 9. Academic Press, Inc., New York.
Negative staining which deals entirely with microorganisms.

3.9.2.2. Sectioned preparations

15. **Dawes, C. J.** 1971. *Biological Techniques in Electron Microscopy*, p. 17–105. Barnes and Noble, Inc., New York.
An older reference, but it is still worthwhile.
16. **Glauert, A. M.** 1975. Fixation, dehydration and embedding of biological specimens, p. 1–216. *Practical Methods in Electron Microscopy*, vol. 3. American Elsevier Publishing Co., New York.
Still one of the best, and treats all general aspects of tissue preparation for thin sectioning.
17. **Hayat, M. A.** 1986. *Basic Techniques for Transmission Electron Microscopy*, Academic Press, Inc., New York.
An in-depth treatment of processing from fixation all the way to thin sectioning.
18. **Hobot, J. A.** 1990. New aspects of bacterial ultrastructure as revealed by modern acrylics for electron microscopy. *J. Struct. Biol.* **104:**169–177.
A discourse on the newer acrylic plastics that can be cured from −80 to +60° C with an emphasis on bacteria.

3.9.2.3. Low-temperature embedding resins

19. **Acetarin, J.-D., E. Carlemalm, and W. Villiger.** 1986. Developments of new Lowicryl resins for embedding biological specimens at even lower temperatures. *J. Microsc.* **143:**81–88.
20. **Armbruster, B. L., E. Carlemalm, R. Chiovetti, R. M. Caravito, J. A. Hobot, E. Kellenberger, and W. Villiger.** 1982. Specimen preparation for electron microscopy using low temperature embedding resins. *J. Microsc.* **126:**77–85.
21. **Carlemalm, E., M. Garavito, and W. Villiger.** 1982. Resin development for electron microscopy and an analysis of embedding at low temperature. *J. Microsc.* **126:**123–143.
22. **Carlemalm, E., W. Villiger, J. A. Hobot, J.-D. Acetarin, and E. Kellenberger.** 1985. Low temperature embedding with Lowicryl resins: two new formulations and some applications. *J. Microsc.* **140:**55–63.

3.9.2.4. Visualizing macromolecules using metal shadowing

23. **Coggins, L. W.** 1987. Preparation of nucleic acids for electron microscopy, p. 1–29. *In* J. Sommerville and U. Scheer (ed.), *Electron Microscopy in Molecular Biology: A Practical Approach*. IRL Press, Oxford.
24. **Glenney, J. R., Jr.** 1987. Rotary metal shadowing for visualizing rod-shaped proteins, p. 167–178. *In* J. Sommerville and U. Scheer (ed.), *Electron Microscopy in Molecular Biology: A Practical Approach*. IRL Press, Oxford.
25. **Zentgraf, H., C. T. Bock, and M. Schrenk.** 1987. Chromatin spreading, p. 81–100. *In* J. Sommerville and U. Scheer (ed.), *Electron Microscopy in Molecular Biology: A Practical Approach*. IRL Press, Oxford.

3.9.2.5. Freeze fracturing and freeze-etching

26. **Benedetti, E. L., and P. Favard (ed.).** 1973. *Freeze-Etching: Techniques and Applications*. Société Française de Microscopie Electronique, Paris.
By one of the masters of the technique.
27. **Bullivant, S.** 1973. Freeze-etching and freeze-fracturing, p. 67–112. *In* J. K. Koehler (ed.), *Advanced Techniques in*

Biological Electron Microscopy. Springer-Verlag, New York.
By one of the masters of the technique.

28. **Chapman, R. L., and L. A. Staehelin.** 1986. Freeze-fracture (-etch) electron microscopy, p. 213–240. *In* H. C. Aldrich and W. J. Todd (ed.), *Ultrastructure Techniques for Microorganisms.* Plenum Press, New York.
By one of the masters of the technique. This up-to-date chapter is devoted to microorganisms.
29. **Hui, S. W. (ed.).** 1989. *Freeze-Fracture Studies of Membranes.* CRC Press, Inc., Boca Raton, Fla.
One of the most recent books on freeze-etching which is dedicated to the preparation of membranes for freeze-fracture and the interpretation of their fracture planes.
30. **Nanninga, N.** 1973. Freeze-fracturing of microorganisms. Physical and chemical fixation of *Bacillus subtilis,* p. 151–179. *In* E. L. Benedetti and P. Favard (ed.), *Freeze Etching: Techniques and Applications.* Société Française de Microscopie Electronique, Paris.
Devoted to the freeze-etching of microorganisms.

3.9.2.6. Cryotechniques

Since one of the most recent advances in the preparation of biological specimens for EM is low-temperature ultrarapid freezing, a more extensive bibliography is given. The references below range from general books on the principles involved in rapid freezing and the formation of vitreous ice to specific journal papers which deal specifically with bacteria. Freeze-substitution and thin frozen films are covered.

31. **Adrian, M., J. Dubochet, J. Lepault, and A. W. McDowall.** 1984. Cryo-electron microscopy of viruses. *Nature* (London) **308:**32–36.
Thin, frozen films of virus particles exemplify exquisite detail against a background of vitreous ice.
32. **Amako, K., Y. Meno, and A. Takade.** 1988. Fine structures of the capsules of *Klebsiella pneumoniae* and *Escherichia coli* K-1. *J. Bacteriol.* **170:**4960–4962.
33. **Amako, K., K. Murata, and A. Umeda.** 1983. Structure of the envelope of *Escherichia coli* observed by the rapid-freezing and substitution fixation method. *Microbiol. Immunol.* **27:**95–99.
34. **Amako, K., K. Okada, and S. Miake.** 1984. Evidence for the presence of a capsule in *Vibrio vulnificus. J. Gen. Microbiol.* **130:**2741–2743.
35. **Amako, K., and A. Takade.** 1985. The fine structure of *Bacillus subtilis* revealed by the rapid-freezing and substitution-fixation method. *J. Electron Microsc.* **34:**13–17.
References 32 to 35 are some of the earliest and best freeze-substitution studies on bacteria. Amako was one of the first to apply the technique to bacteria.
36. **Beckett, A., and N. D. Read.** 1986. Low-temperature scanning electron microscopy, p. 45–86. *In* H. C. Aldrich and W. J. Todd (ed.), *Ultrastructure Techniques for Microorganisms.* Plenum Press, New York.
A helpful chapter on the advantages of low temperatures for SEM.
37. **Carlemalm, E., W. Villiger, J.-D. Acetarin, and E. Kellenberger.** 1985. Low temperature embedding, p. 147–164. *In* M. Müller, R. P. Becker, A. Boyde, and J. J. Wolosewick (ed.), *The Science of Biological Specimen Preparation for Microscopy and Microanalysis 1985.* Scanning Electron Microscopy, Inc., AMF O'Hare, Ill.
This Biozentrum group in Basel, Switzerland, developed the concept of low-temperature embedding and the use of Lowicryl resins.
38. **Dubochet, J., A. W. McDowall, B. Menge, E. N. Schmid, and K. G. Lickfeld.** 1983. Electron microscopy of frozen-hydrated bacteria. *J. Bacteriol.* **155:**381–390.
Shows some of the first and best frozen thin sections of bacteria.
39. **Graham, L. L., and T. J. Beveridge.** 1990. Evaluation of freeze-substitution and conventional embedding protocols for routine electron-microscopic processing of eubacteria. *J. Bacteriol.* **172:**2141–2149.
40. **Graham, L. L., and T. J. Beveridge.** 1990. Effect of chemical fixatives on accurate preservation of *Escherichia coli* and *Bacillus subtilis* structure in cells prepared by freeze-substitution. *J. Bacteriol.* **172:**2150–2159.
41. **Graham, L. L., R. Harris, W. Villiger, and T. J. Beveridge.** 1991. Freeze-substitution of gram-negative eubacteria: general cell morphology and envelope profiles. *J. Bacteriol.* **173:**1623–1633.
References 39 and 40 detail the biochemical preservation of bacteria and relate it to ultrastructural detail by comparing conventional methods with freeze-substitution. Reference 41 applies the freeze-substitution technique to a wide range of gram-negative bacteria. These articles will assist any microbiologist contemplating the use of freeze-substitution.
42. **Hobot, J. A.** 1991. Low temperature embedding techniques for studying microbial cell surfaces, p. 127–150. *In* N. Mozes, P. S. Handley, H. J. Busscher, and P. G. Rouxhet (ed.), *Structural and Physico-Chemical Methods for Microbial Cell Surface Analysis.* VCH Publishing, New York.
A good review of the current techniques that use low-temperature methacrylates to better preserve antigenicity and structure.
43. **Hobot, J. A., E. Carlemalm, W. Villiger, and E. Kellenberger.** 1984. Periplasmic gel: new concept resulting from the reinvestigation of bacterial cell envelope ultrastructure by new methods. *J. Bacteriol.* **160:**143–152.
44. **Hobot, J. A., W. Villiger, J. Escaig, M. Maeder, A. Ryter, and E. Kellenberger.** 1985. Shape and fine structure of nucleoids observed on sections of ultrarapidly frozen and cryosubstituted bacteria. *J. Bacteriol.* **162:**960–971.
References 43 and 44 profoundly altered our perception of gram-negative periplasmic spaces (i.e., the concept of a "periplasmic gel") and the cytoplasmic distribution of the bacterial chromosome.
45. **Kellenberger, E., E. Carlemalm, and W. Villiger.** 1985. Physics of the preparation and observation of specimens that involve cryoprocedures, p. 1–20. *In* M. Müller, R. P. Becker, A. Boyde, and J. J. Wolosewick (ed.), *The Science of Biological Specimen Preparation for Microscopy and Microanalysis 1985.* Scanning Electron Microscopy, Inc., AMF O'Hare, Ill.
46. **Robards, A. W., and U. B. Sleytr.** 1985. Low temperature methods in biological electron microscopy. Vol. 10 of A. M. Glauert (ed.), *Practical Methods in Electron Microscopy.* Elsevier Science Publishing, Amsterdam.
47. **Roos, N., and A. J. Morgan.** 1990. *Cryopreparation of Thin Biological Specimens for Electron Microscopy: Methods and Applications.* Royal Microscopy Society handbook 21. Oxford University Press, Oxford.
48. **Sitte, H., K. Neumann, and L. Edelmann.** 1985. Cryofixation and cryosubstitution for routine work in transmission electron microscopy, p. 103–118. *In* M. Müller, R. P. Becker, A. Boyde, and J. J. Wolosewick (ed.), *The Science of Biological Specimen Preparation for Microscopy and Microanalysis 1985.* Scanning Electron Microscopy, Inc., AMF O'Hare, Ill.
49. **Steinbrecht, R. A., and K. Zierold (ed.).** 1987. *Cryotechniques in Biological Electron Microscopy.* Springer-Verlag KG, Berlin.
50. **Stewart, M., and G. Vigers.** 1986. Electron microscopy of frozen hydrated biological material. *Nature* (London) **319:**631–636.
References 45 to 50 are books and chapters which describe the principles involved in rapid freezing, the properties of different ice forms, freeze-substitution, frozen thin sections, and thin frozen films.
51. **Umeda, A., Y. Ueki, and K. Amako.** 1987. Structure of the *Staphylococcus aureus* cell wall determined by the freeze-substitution method. *J. Bacteriol.* **169:**2482–2487.
Other than Bacillus subtilis, few gram-positive vegetative cells have been freeze-substituted. Here we have Staphylococcus aureus.

3.9.2.7. Polycationized ferritin

52. **Jacques, M., M. Gottschalk, B. Foiry, and R. Higgins.** 1990. Ultrastructural study of surface components of *Streptococcus suis. J. Bacteriol.* **172:**2833–2838.
One of the most recent PCF studies on capsules.
53. **Luckevich, M. D., and T. J. Beveridge.** 1989. Characterization of a dynamic S layer on *Bacillus thuringiensis. J. Bacteriol.* **171:**6656–6667.
Shows the use of PCF and optical diffraction to characterize an elusive S layer.
54. **Sára, M., and U. B. Sleytr.** 1987. Charge distribution of the S layer of *Bacillus stearothermophilus* NRS 1536/3c and importance of charged groups for morphogenesis and function. *J. Bacteriol.* **169:**2804–2809.
Shows both negative stains and freeze-etching on PCF-labeled cell surfaces.
55. **Sonnenfeld, E. M., T. J. Beveridge, and R. J. Doyle.** 1985. Discontinuity of charge on cell wall poles of *Bacillus subtilis. Can. J. Microbiol.* **31:**875–877.
Limiting concentrations of PCF are used to locate the most anionic sites on the cell wall.
56. **Sonnenfeld, E. M., T. J. Beveridge, A. Koch, and R. J. Doyle.** 1985. Asymmetric distribution of charge on the cell wall of *Bacillus subtilis. J. Bacteriol.* **163:**1167–1171.
Shows the use of PCF for labeling electronegative sites on B. subtilis walls in thin section, before and after neutralization of charges. Eventually carboxyl and phosphoryl groups are shown to be the major wall anions.

3.9.2.8. Immunogold labeling

The ability to precisely label distinct antigens in bacteria has had a profound influence on the elucidation of microbial ultrastructure; for the first time in microbiology, the exact location of macromolecules and even distinct domains (epitopes) of macromolecules is possible. Because of the importance of the technique, this extensive bibliography includes recipes, reviews, and its use in bacteriology. See also the references in Chapter 5.

57. **Acker, G., D. Bitter-Suermann, U. Meier-Dieter, H. Peters, and H. Mayer.** 1986. Immunocytochemical localization of enterobacterial common antigen in *Escherichia coli* and *Yersinia enterocolitica. J. Bacteriol.* **168:**348–356.
58. **Acker, G., and C. Kammerer.** 1990. Localization of enterobacterial common antigen immunoreactivity in the ribosomal cytoplasm of *Escherichia coli* cells cryosubstituted and embedded at low temperature. *J. Bacteriol.* **172:**1106–1113.
Two of the more recent immunogold studies on a distinct Escherichia coli antigen.
59. **Behnke, O., T. Ammitzboll, J. Jessen, M. Klokker, K. Nilausen, J. Tranum-Jensen, and L. Olsson.** 1986. Non-specific binding of protein-stabilized gold sols as a source of error in immunocytochemistry. *Eur. J. Cell Biol.* **41:**326–338.
60. **Bendayan, M.** 1987. Introduction of the protein-G-gold complex for high-resolution immunocytochemistry. *J. Electron Microsc. Techniques* **6:**7–13.
One of the first studies to use protein G-gold.
61. **Bendayan, M., and Garzon, S.** 1988. Protein-G-gold complex: comparative evaluation with protein-A-gold for high-resolution immunocytochemistry. *J. Histochem. Cytochem.* **36:**597–607.
Shows that protein G is better for labeling IgM antibodies which make up the widest spectrum of monoclonal antibodies.
62. **Birrel, B. G., K. K. Hedberg, and O. H. Griffith.** 1987. Pitfalls of immunogold labeling: analysis by light microscopy, transmission electron microscopy and photoelectronmicroscopy. *J. Histochem. Cytochem.* **35:**843–853.
63. **DeMey, J.** 1984. Colloidal gold as a marker and tracer in light and electron microscopy. *EMSA Bull.* **14:**54–66.
A useful review including some of the most frequently used recipes.
64. **Dürrenberger, M., M.-A. Bjornsti, T. Uetz, J. A. Hobot, and E. Kellenberger.** 1988. Intracellular localization of histone-like protein HU in *Escherichia coli. J. Bacteriol.* **170:**4757–4768.
One of the most carefully controlled studies of thin-section labeling dealing with bacteria. See also reference 70.
65. **Dürrenberger, M. B.** 1989. Removal of background label in immunocytochemistry with the apolar Lowicryls by using washed protein A-gold-precoupled antibodies in a one-step procedure. *J. Electron Microsc. Techniques* **11:**109–116.
References 59, 62, 63, and 65 describe some of the problems to be aware of with colloidal-gold labeling.
66. **Frens, G.** 1973. Controlled nucleation for the regulation of the particle size in monodisperse gold suspensions. *Nature* (London) **241:**20–22.
One of the original papers describing the production of colloidal gold and the control of its particle size.
67. **Geohagen, W. D., and G. A. Ackerman.** 1977. Adsorption of horseradish peroxidase, ovomucoid and anti-immunoglobulin to colloidal gold for the indirect detection of concanavalin A, wheatgerm agglutinin and goat anti-human immunoglobulin G on cell surfaces at the electron microscopic level: a new method, theory and application. *J. Histochem. Cytochem.* **25:**1187–1200.
A good paper for recipes; it is especially good for describing conditioning of gold particles before adsorption of specific antibody.
68. **Hayat, M. A. (ed.)** 1991. *Colloidal Gold: Principles, Methods and Applications,* vol. 3. Academic Press, Inc., New York.
69. **Hicks, D., and R. S. Molday.** 1984. Analysis of cell labeling for scanning and transmission electron microscopy, p. 203–219. *In* J. P. Revel, T. Barnard, and G. H. Haggis (ed.), *The Science of Biological Specimen Preparation for Microscopy and Microanalysis.* Scanning Electron Microscopy, Inc., AMF O'Hare, Ill.
References 68 and 69 review the methodology used in immunolabeling. As an encompassing book on the topic, reference 68 is one of the most up-to-date.
70. **Hobot, J. A., M.-A. Bjornsti, and E. Kellenberger.** 1987. Use of on-section immunolabeling and cryosubstitution for studies of bacterial DNA distribution. *J. Bacteriol.* **169:**2055–2064.
This goes together with reference 64.
71. **Humbel, B. M., and W. Schwarz.** 1989. Freeze-substitution for immunochemistry, p. 115–134. *In* A. J. Verkleij and J. L. M. Leunissen (ed.), *Immuno-Gold Labeling in Cell Biology.* CRC Press, Inc., Boca Raton, Fla.
A good review.
72. **Kellenberger, E., M. Dürrenberger, W. Villiger, E. Carlemalm, and M. Wurtz.** 1987. The efficiency of immunolabel on Lowicryl sections compared to theoretical predictions. *J. Histochem. Cytochem.* **35:**959–965.
A good explanation why Lowicryl resins are useful for immunolabeling.
73. **Romano, E. L., and M. Romano.** 1977. Staphylococcal protein A bound to colloidal gold: a useful reagent to label antigen-antibody sites in electron microscopy. *Immunochemistry* **14:**711–715.
One of the first articles to describe the use of protein A-colloidal gold for indirect labeling.
74. **Roth, J.** 1982. The protein A-gold (pAg) technique—a qualitative and quantitative approach for antigen localization on thin sections, p. 107–133. *In* G. R. Bullock and P. Petrusz (ed.), *Techniques in Immunocytochemistry,* vol. 1. Academic Press, Ltd., London.
75. **Roth, J.** 1986. Post embedding cytochemistry with gold-labeled reagents: a review. *J. Microsc.* **143:**125–137.
76. **Roth, J., M. Bendayan, and L. Orci.** 1978. Ultrastructural localization of intracellular antigens by the use of protein A-gold complex. *J. Histochem. Cytochem.* **26:**1074–1081.

Roth and Bendayan are two of the pioneers in immunogold labeling. References 74 and 75 are good overviews, and reference 76 describes the use of protein A-colloidal gold on thin sections.

77. **Slot, J. W., and H. J. Geuze.** 1984. A new method of preparing gold probes for multiple-labeling cytochemistry. *Eur. J. Cell Biol.* **38:**87–93.
78. **Slot, J. W., and H. J. Geuze.** 1991. Sizing of protein A-colloidal gold probes for immunoelectron microscopy. *J. Cell Biol.* **90:**533–536.
Recipes for making and sizing colloidal gold.
79. **Smit, J., and W. J. Todd.** 1986. Colloidal gold labels for immunocytochemical analysis of microbes, p. 469–517. *In* H. C. Aldrich and W. J. Todd (ed.), *Ultrastructure Techniques for Microorganisms.* Plenum Press, New York.
One of the few review chapters which deals with the immunolabeling of microorganisms.
80. **Walker, P. D., I. Batty, and R. O. Thomas.** 1971. The localization of bacterial antigens by use of the fluorescent and ferritin labeled antibody techniques, p. 219–254. *In* J. R. Norris and U. W. Ribbons (ed.), *Methods in Microbiology,* vol. 5A. Academic Press, Inc., New York.
81. **Williams, M. A.** 1977. Autoradiography and immunochemistry, vol. 6 of A. M. Glauert (ed.), *Practical Methods in Electron Microscopy.* Elsevier/North-Holland Publishing Co., New York.
References 80 and 81 are two good review chapters filled with recipes and techniques.

3.9.2.9. Compositional analysis

Accurate compositional analysis requires exquisitely preserved structure, and cryopreservation is becoming the preferred preparatory method.

82. **Aldrich, H. C.** 1986. X-ray microanalysis, p. 517–525. *In* H. C. Aldrich and W. J. Todd (ed.), *Ultrastructure Techniques for Microorganisms.* Plenum Press, New York.
A chapter dedicated to EDS applied to microorganisms.
83. **Chandler, J. A.** 1977. X-ray microanalysis in the electron microscope. Part II of vol. 5 of A. M. Glauert (ed.), *Practical Methods in Electron Microscopy.* North-Holland Publishing Co., Amsterdam.
84. **Chang, C.-F, H. Shuman, and A. P. Somlyo.** 1986. Electron probe analysis, X-ray mapping and electron energy-loss spectroscopy of calcium, magnesium, and monovalent ions in log-phase and in dividing *Escherichia coli* B cells. *J. Bacteriol.* **167:**935–939.
One of the few compositional studies of the natural electrolyte concentrations in Escherichia coli. It applies EDS and EELS with a field-emission electron source.
85. **Goldstein, J. I., D. E. Newbury, P. Echlin, C. Fiori, and E. Lifshin.** 1981. *Scanning Electron Microscopy and X-Ray Microanalysis: a Text for Biologists, Materials Scientists, and Geologists.* Plenum Press, New York.
86. **Hren, J. J., J. I. Goldstein, and D. C. Joy.** 1979. *Introduction to Analytical Electron Microscopy.* Plenum Press, New York.
87. **Hutchinson, T. E., and A. P. Somlyo.** 1981. *Microprobe Analysis of Biological Systems.* Academic Press. Inc. New York.
References 84 to 87 are good overviews of the topic. The physics and theory are emphasized and reference 86 may be better to start with. See also reference 89 for a good short review.
88. **Johnstone, K., D. J. Ellar, and T. C. Appleton.** 1980. Location of metal ions in *Bacillus megaterium* spores by high-resolution electron probe X-ray microanalysis. *FEMS Microbiol. Lett.* **7:**97–101.
One of the first EDS studies applied to bacteria. In this case, endospores are analyzed.
89. **Morgan, J. A.** 1985. *X-Ray Microanalysis in Electron Microscopy for Biologists.* Royal Microscopical Society handbook 5. Oxford University Press, Oxford.
A good, short, simple overview of EDS which is highly recommended. It comes in paperback.
90. **Ottensmeyer, F. P.** 1984. Electron energy loss analysis and imaging in biology, p. 340–343. *In* G. W. Bailey (ed.), *Proceedings of the 42nd Annual Meeting of the Electron Microscopy Society of America.* San Francisco Press, Inc., San Francisco.
A short synopsis of ESI by its pioneering spirit.
90a. **Sigee, D. C., J. Morgan, A. T. Sumner, and A. Warley.** 1993. *X-Ray Microanalysis in Biology.* Cambridge University Press, New York.
Another good overview of the topic.
91. **Stewart, M., A. P. Somlyo, A. V. Somlyo, H. Shuman, J. A. Lindsay, and W. G. Murrell.** 1980. Distribution of calcium and other elements in cryosectioned *Bacillus cereus* T spores, determined by high-resolution scanning electron probe X-ray microanalysis. *J. Bacteriol.* **143:**481–491.
92. **Stewart, M., A. P. Somlyo, A. V. Somlyo, H. Shuman, J. A. Lindsay, and W. G. Murrell.** 1981. Scanning electron probe x-ray microanalysis of elemental distribution in freeze-dried cryosections of *Bacillus coagulans* spores. *J. Bacteriol.* **147:**670–674.
References 91 and 92 are two meticulous compositional studies on bacterial endospores. They use frozen sections and point-map the elements within the sectioned spores.

3.9.2.10. Image processing

Computer processing has been used to clarify several bacterial structures such as ribosomes, pili, flagella, spinae, and S layers. The latter have been the most studied because of their paracrystallinity and planar symmetry, and three-dimensional structure is available for several arrays. Only research articles which use image processing in rather novel ways are listed below, and the review articles will point toward more extensive bibliographies.

93. **Baumeister, W., and H. Engelhardt.** 1987. Three-dimensional structure of bacterial surface layers, p. 109–154. *In* R. Harris and R. W. Horne (ed.), *Electron Microscopy of Proteins,* vol. 6. Academic Press Ltd., London.
Baumeister's group has published extensively on the three-dimensional aspects of bacterial S layers, and this is a good review chapter.
94. **Beeston, B. E. P., R. W. Horne, and R. Markham.** 1973. Electron diffraction and optical diffraction techniques. Part II of vol. 1 of A. M. Glauert (ed.), *Practical Methods in Electron Microscopy.* North-Holland Publishing Co., Amsterdam.
An early overview of diffraction techniques which are used to analyze periodic structure to high-order resolution.
95. **Beveridge, T. J., D. G. Sprott, and P. Whippey.** 1991. Ultrastructure, inferred porosity, and Gram-staining character of *Methanospirillum hungatei* filament termini describe a unique cell permeability for this archaeobacterium. *J. Bacteriol.* **173:**130–140.
One of the most recent uses of image processing on bacterial structure. It is intimately linked with reference 105, and image processing is used to help define the permeability characteristics of this unusual bacterium.
96. **Deatherage, J. F., K. A. Taylor, and L. A. Amor.** 1983. Three-dimensional arrangement of cell wall protein of *Sulfolobus acidocaldarius. J. Mol. Biol.* **167:**823–852.
One of the three-dimensional studies of a bacterial S layer.
97. **Engel, A., A. Massalski, H. Schindler, D. L. Dorset, and J. P. Rosenbusch.** 1985. Porin channel triplets merge into single outlets in *Escherichia coli* outer membranes. *Nature* (London) **317:**643–645.
One of the first articles using computer processing to show the complexity of the pores in gram-negative outer membranes.

98. **Messner, P., D. Pum, and U. B. Sleytr.** 1987. Characterization of the ultrastructure and the self-assembly of the surface layer of *Bacillus stearothermophilus* strain NRS 2000/3a. *J. Ultrastruct. Mol. Struct. Res.* **97:**73–88.
Shows how image processing can be used to unravel complex moiré patterns in S layer assembly products.
99. **Misell, D. L.** 1978. Image analysis, enhancement and interpretation. vol. 7 of A. M. Glauert (ed.), *Practical Methods in Electron Microscopy.* North-Holland Publishing Co., Amsterdam.
A fine overview of the entire topic.
100. **Shaw, P. J., G. Hills, J. A. Henwood, J. E. Harris, and D. B. Archer.** 1985. Three-dimensional architecture of the cell sheath and septa of *Methanospirillum hungatei. J. Bacteriol.* **161:**750–757.
A three-dimensional structure of a very complicated surface.
101. **Sleytr, U. B., P. Messner, D. Pum, and M. Sàra (ed.).** 1988. *Crystalline Bacterial Cell Surface Layers.* Springer-Verlag, Berlin.
Almost 200 pages of up-to-date information on bacterial S layers including several articles containing state-of-the-art image processing.
102. **Stewart, M.** 1986. Computer analysis of ordered microbiological objects, p. 333–364. *In* H. C. Aldrich and W. J. Todd (ed.), *Ultrastructure Techniques for Microorganisms.* Plenum Press, New York.
References 102 and 103 are good reviews on image processing by one of the best researchers in the field. Reference 102 deals exclusively with microbial structure.
103. **Stewart, M.** 1988. An introduction to computer image processing of two-dimensionally ordered biological structures. *J. Electron Microsc. Techniques* **9:**325–358.
104. **Stewart, M., and T. J. Beveridge.** 1980. Structure of the regular surface layer of *Sporosarcina ureae. J. Bacteriol.* **142:**302–309.
A novel computer manipulation in which the two faces of an S layer are separated from one another and then realigned on top of each other to prove that pores pass through a p4 S layer.
105. **Stewart, M., T. J. Beveridge, and D. G. Sprott.** 1985. Crystalline order to high resolution in the sheath of *Methanospirillum hungatei*: a cross-beta structure. *J. Mol. Biol.* **183:**509–515.
This article takes the S layer with the smallest spacing (2.8 nm) down to a resolution of 0.47 nm to reveal protein folding. See also reference 95.
106. **Stewart, M., T. J. Beveridge, and T. J. Trust.** 1986. Two patterns in the *Aeromonas salmonicida* A-layer may reflect a structural transformation that alters permeability. *J. Bacteriol.* **166:**120–127.
Because this was a distorted lattice, both correlation averaging and Fourier analysis had to be used. Two separate p4 lattices were found on the same S layer and are correlated with the permeability and virulence of this fish pathogen.
107. **Stewart, M., and R. G. E. Murray.** 1982. Structure of the regular surface layer of *Aquaspirillum serpens* MW5. *J. Bacteriol.* **105:**348–357.
One of the first to decipher and separate the S layers of a moiré pattern.
108. **Trachtenberg, S., and D. J. DeRosier.** 1987. Three dimensional structure of the frozen-hydrated flagellar filament. The left-handed filament of *Salmonella typhimurium. J. Mol. Biol.* **195:**581–601.
A recent article using frozen thin films of bacterial flagella and image enhancement to produce three-dimensional structure. Thin helical objects such as flagella are among the most difficult structures to analyze, and DeRosier's group is one of the best at it.

3.9.2.11. Scanning electron microscopy

109. **Hayat, M. A. (ed.).** 1974. *Principles and Techniques of Scanning Electron Microscopy. Biological Applications, vol. 1.* Van Nostrand Reinhold Co., New York.
110. **Hayes, T. L.** 1973. Scanning electron microscope techniques in biology, p. 153–214. *In* J. K. Koehler (ed.), *Advanced Techniques in Biological Electron Microscopy.* Springer-Verlag, New York.
111. **Joy, D. C.** 1984. Resolution in the low-voltage SEM, p. 444–449. *In* G. W. Bailey (ed.), *Proceedings of the 42nd Annual Meeting of the Electron Microscopy Society of America.* San Francisco Press, Inc., San Francisco.
References 109 to 111 are overviews of the techniques used for SEM, and references 111 and 116 describe the advantages of using low-voltage SEM.
112. **Kondo, I.** 1978. Scanning electron microscopy of bacterial viruses, p. 306–316. *In* M. A. Hayat (ed.), *Principles and Techniques of Scanning Electron Microscopy. Biological Applications,* vol. 6. Van Nostrand Reinhold Co., New York.
113. **Kormendy, A. C.** 1975. Microorganisms, p. 82–108. *In* M. A. Hayat (ed.), *Principles and Techniques of Scanning Electron Microscopy. Biological Applications,* vol. 3. Van Nostrand Reinhold Co., New York.
114. **Nickerson, A. W., L. A. Bulla, Jr., and C. P. Kurtzman.** 1974. Spores, p. 159–180. *In* M. A. Hayat (ed.), *Principles and Techniques of Scanning Electron Microscopy. Biological Applications,* vol. 1. Van Nostrand Reinhold Co., New York.
115. **Passmore, S. M., and B. Bole.** 1976. Scanning electron microscopy of microbial colonies, p. 19–29. *In* R. Fuller and D. W. Lovelock (ed.), *Microbial Ultrastructure. The Use of the Electron Microscope.* Academic Press, Inc., New York.
References 112 to 115 show how conventional SEM can be applied to microorganisms.
116. **Pawley, J.** 1984. SEM at low voltage, p. 440–443. *In* G. W. Bailey (ed.), *Proceedings of the 42nd Annual Meeting of the Electron Microscopic Society of America.* San Francisco Press, Inc., San Francisco.
Along with reference 111, this reference shows the advantages of low-voltage SEM.
117. **Yoshii, A., J. Tokumaga, and J. Tawara.** 1975. *Atlas of Scanning Electron Microscopy in Microbiology.* Igaku Shoin, Ltd., Tokyo (The Williams & Wilkins Co., Baltimore).

3.9.2.12. Scanning tunneling and atomic force microscopies

These are the newest types of microscopies and have revealed startling atomic detail of surfaces and molecular detail of certain organic films. So far, they have not been used extensively in microbiology.

118. **Beveridge, T. J., G. Southam, M. H. Jericho, and B. L. Blackford.** 1990. High resolution topography of the S-layer sheath of the archaeobacterium *Methanospirillum hungatei* provided by scanning tunneling microscopy. *J. Bacteriol.* **172:**6589–6595.
The highest-resolution topographical analysis of a bacterial surface by STM. The article outlines the difficulties involved in the STM of bacterial surfaces and the various techniques used to overcome them. The "sewing-machine" STM technique may be the best way to obtain molecular topography on bacterial surfaces.
119. **Blackford, B. L., M. O. Watanabe, D. C. Dahn, M. H. Jericho, G. Southam, and T. J. Beveridge.** 1989. The imaging of a complete biological structure with the scanning tunneling microscope. *Ultramicroscopy* **27:**427–432.
120. **Blackford, B. L., M. O. Watanabe, M. J. Jericho, and D. C. Dahn.** 1988. STM imaging of the complete bacterial cell sheath of *Methanospirillum hungatei. J. Microsc.* **152:**237–243.
References 119 and 120 outline some of the initial attempts in using STM on bacterial surfaces.
121. **Egger, M., F. Ohnesorge, A. L. Weisenhorn, S. P. Heyn, B. Drake, C. B. Prater, S. A. Could, P. K. Hansma, and H. E.**

Gaub. 1990. West lipid-protein membranes imaged at submolecular resolution by atomic force microscopy. *J. Struct. Biol.* **103:**89–94.
A recent article which shows what AFM can do.

122. **Golovinchenko, J. A.** 1986. The tunneling microscope: a new look at the atomic world. *Science* **232:**48–53.
A good, readable article on the operation of the STM.

122a. **Southern, G., M. Firtel, B. L. Blackford, M. H. Jericho, W. Xu, P. J. Mulhern, and T. J. Beveridge.** 1993. Transmission electron microscopy, scanning tunneling microscopy, and atomic force microscopy on the cell envelope layers of the archaebacterium *Methanospirillum hungatei* GP1. *J. Bacteriol.* **175:**1946–1955.
The best correlation to this date between TEM, STM, and AFM on bacterial surfaces.

3.9.3. Specific References on Interpretation

It is all very well to read about and learn the many techniques used in electron microscopy and to process bacteria for ultrastructural analysis, but it is an additional problem to interpret the images produced by the techniques. It is best to consult with local EM experts, but remember that most biological electron microscopists deal with eucaryotic cells! Procaryotic structuralists around the world are glad to help, so send good micrographs to them and seek advice. In the meantime, here is a list of articles to help in interpretation.

123. **Avakyan, A. A., P. N. Katz, and I. B. Pavlova.** 1972. *Atlas of Bacteria Pathogenic for Man and Animals.* Meditsyna, Moscow.
This is a difficult book to obtain and is in Russian.

124. **Beveridge, T. J.** 1981. Ultrastructure, chemistry, and function of the bacterial wall. *Int. Rev. Cytol.* **72:**229–317.
A review which correlates bacterial wall chemistry in structural perspective and discusses what it means in a functional context.

125. **Beveridge, T. J.** 1989. The structure of bacteria, p. 1–65. *In* E. R. Leadbetter and J. S. Poindexter (ed.), *Bacteria in Nature: A Treatise on the Interaction of Bacteria and Their Habitats,* vol. 3. Plenum Press, New York.
One of the most comprehensive compilations of bacterial structure, it should be useful reading before interpreting initial EM images.

126. **Beveridge, T. J., and J. W. Costerton (ed.).** 1988. Shape and form [of bacteria]. *Can. J. Microbiol.* **34:**363–420.
A compilation of short reviews by experts in the field. The topics range from structural design strategies to growth and taxis.

127. **Beveridge, T. J., and L. L. Graham.** 1991. Surface layers of bacteria. *Microbiol. Rev.* **55:**684–705.
An up-to-date review on modern perception of the structure of bacterial surfaces. Cryotechniques, diffraction, image processing, immunogold labeling, and probes for charge are all covered.

128. **Costerton, J. W.** 1979. The role of electron microscopy in the elucidation of bacterial structure and function. *Annu. Rev. Microbiol.* **33:**459–479.
A good overview which has stood the test of time.

129. **Ferris, F. G., W. S. Fyfe, and T. J. Beveridge.** 1987. Bacteria as nucleation sites for authigenic minerals in a metal-contaminated lake sediment. *Chem. Geol.* **63:**225–232.

130. **Ferris, F. G., W. S. Fyfe, and T. J. Beveridge.** 1988. Metallic ion binding by *Bacillus subtilis*: implications for the fossilization of microorganisms. *Geology* **16:**149–152.
Both references 129 and 130 reveal the remarkable capacity of bacteria to sorb metals from solution and to develop fine-grained minerals. EDS and SAED are used to identify minerals and the entire microbiogeochemical process is correlated with the development of microfossil structures.

131. **Ghosh, B. K. (ed.).** 1981. Organization of prokaryotic cell membranes, vol. 1 and 2. CRC Press, Inc. Boca Raton, Fla.
A two volume set which covers various aspects of plasma membranes (including that of mycoplasmas), mesosomes, photosynthetic membranes, and gas vesicles.

132. **Glauert, A. M., M. J. Thornley, K. J. I. Thorne, and U. B. Sleytr.** 1976. The surface structure of bacteria, p. 31–47. *In* R. Fuller and D. W. Lovelock (ed.), *Microbial Ultrastructure. The Use of the Electron Microscope.* Academic Press, Inc., New York.
Emphasis is on the use of TEM to decipher bacterial surfaces.

133. **Graham, L. L., T. J. Beveridge, and N. Nanninga.** 1991. Periplasmic space and the concept of the periplasm. *Trends Biochem. Sci.* **16:**328–329.

134. **Graham, L. L., R. Harris, W. Villiger, and T. J. Beveridge.** 1991. Freeze-substitution of gram-negative eubacteria: general cell morphology and envelope profiles. *J. Bacteriol.* **173:**1623–1633.
References 133 and 134 redefine the concept of the periplasm and the periplasmic space. Also see reference 135.

135. **Hobot, J. A., E. Carlemalm, W. Villiger, and E. Kellenberger.** 1984. Periplasmic gel: a new concept resulting from the reinvestigation of bacterial cell envelope ultrastructure by new methods. *J. Bacteriol.* **160:**143–152.

136. **Hodgkiss, W., J. A. Short, and P. D. Walker.** 1976. Bacterial surface structures, *In* R. Fuller and W. Lovelock (ed.), *Microbial Ultrastructure. The Use of the Electron Microscope,* p. 49–71. Academic Press, Inc., New York.
An earlier overview which is still worthwhile. It is found in the same book and expands reference 132.

137. **Holt, S. C., and T. J. Beveridge.** 1982. Electron microscopy: its development and application to microbiology. *Can. J. Microbiol.* **28:**11–53.
Describes the beginnings of electron microscopy and its value to microbiology as it has advanced up to modern times.

138. **Kellenberger, E.** 1989. Bacterial chromatin (a critical review of structure function relationships), p. 3–25. *In* K. W. Adolph (ed.), *Chromosomes: Eukaryotic, Prokaryotic and Viral,* vol. III. CRC Press, Inc., Boca Raton, Fla.
Both references 135 and 138 discuss some of the hallmark freeze-substitution studies on bacteria, as well as explain how they have altered our previous knowledge and should be correlated with references 133 and 134. See also ref. 39 to 41.

139. **Koch, A. L.** 1983. The surface stress theory of microbial morphogenesis. *Adv. Microb. Physiol.* **24:**301–367.
This review does not emphasize EM, but it is essential reading for students who are attempting to understand bacterial shape.

140. **Krell, P. J., and T. J. Beveridge.** 1987. The structure of bacteria and molecular biology of viruses. *In. Rev. Cytol. Suppl.* **17:**15–88.
An overview of procaryotic and viral structure which was aimed at advanced university undergraduate studies.

141. **Murray, R. G. E.** 1978. Form and function I: bacteria, p. 2/1–2/31. *In* J. R. Norris and M. H. Richmond (ed.), *Essays in Microbiology.* John Wiley & Sons, Inc., New York.
Essential reading for all students, and written by one of the true experts.

142. **Rogers, H. J.** 1983. Bacterial cell structure. *Aspects of Microbiology 6.* American Society for Microbiology, Washington, D.C.
A paperback book that is very readable and was written by an expert dedicated to the chemical aspects of structure.

143. **Sleytr, U. B.** 1981. Morphopoietic and functional aspects of regular protein membranes present on prokaryotic cell walls. *Cell Biol. Ser. Monogr.* **8:**1–26.

144. **Sleytr, U. B., and A. M. Glauert.** 1982. Bacterial cell walls and membranes, p. 41–76. *In* J. R. Harris (ed.), *Electron Microscopy of Proteins,* vol. 3. Academic Press, Inc., New York.
Both references 143 and 144 deal with the surfaces of bacteria, especially with how S layers encapsulate some of these cells.

145. **Smith, D. G.** 1982. Bacterial appendages, p. 105–153. *In* J. R. Harris (ed.), *Electron Microscopy of Proteins,* vol. 2. Academic Press, Inc., New York.
A good review on bacterial flagella, pili, spinae, and endospore appendages which is still relevant.

146. **Stoltz, J. F. (ed.).** 1991. *Structure of Phototrophic Prokaryotes.* CRC Press, Inc., Boca Raton, Fla.
The most up-to-date review of the structure of cyanobacteria and the purple and green bacteria and their internal, photosynthetic membranes.

147. **van Iterson, W. (ed.).** 1984. Inner structures of bacteria. *Benchmark Papers in Microbiology,* vol. 17. Van Nostrand Reinhold Co., Inc., New York.

148. **van Iterson, W. (ed.).** 1984. Outer structures of bacteria. *Benchmark Papers in Microbiology,* vol. 18. Van Nostrand Reinhold Co., Inc., New York.
References 147 and 148 are two wonderful volumes in which van Iterson has compiled the significant papers showing progressive advance of bacterial ultrastructure. Only the significant features of each paper are printed and are followed by editorial comments by an experienced microscopist.

Chapter 4

Cell Fractionation

G. DENNIS SPROTT, SUSAN F. KOVAL, and CARL A. SCHNAITMAN

This chapter presents techniques for the fractionation of cellular components, including organelles and appendages, beginning at the external surface and ending with those internally located. The structural and physiological diversity of bacteria makes it impossible to present a universal method for cell fractionation or to summarize all of the varied approaches developed to fit the peculiarities of each bacterium. Nevertheless, an attempt is made to give a fair account of methods applicable to a wide range of bacteria. These include representatives of both the eubacterial and archaeobacterial lineages, which are phylogenetically far apart.

Large variations in genomic content, often leading to changes in structure, exist even within the subgroupings of bacteria such as the methanogens; these structural changes affect cell fractionation, even among different strains of the same species. Indeed, a rational fractionation scheme can be developed only on the basis of solid knowledge of a particular organism. Since the information needed is often lacking for newly isolated or poorly studied bacteria, some preliminary exploration may be essential. It is important, for example, to determine at what stage in the growth cycle and in which growth medium the component of interest is produced optimally. Such factors can influence not only the yield but also the chemical structure of cell components.

It is strongly recommended that the isolation of any cellular component be monitored by electron microscopy at all stages of the procedure. This often provides the only adequate means of monitoring the fractionation and state of intactness of the cell component. This is especially important during the preliminary exploration of a bacterium.

Glassware must be carefully cleaned and sometimes sterilized to avoid contamination from stray microbial cells and enzymes, especially if the fraction is to be stored. Also, it is preferable to store samples at −20°C or below, provided that the component resists freeze-thaw damage. Samples are usually stored for short periods at 4°C, with the growth of microbial contaminants inhibited temporarily with poisons such as 0.05% sodium azide.

Containment facilities may be needed to comply with safety regulations if potentially pathogenic bacteria are used (Chapter 29), as well as to protect against oxidative inactivation of specific components of anaerobic bacteria (Chapter 6).

The fractionation of cytoplasmic proteins is not discussed, since gel electrophoresis and the chromatography of soluble proteins are described in Chapter 21. These basic methods would apply to the electrophoretic analysis of membrane proteins.

4.1. CELL BREAKAGE

Because of differences in the architecture of bacterial cell surfaces, there is perhaps no greater area of variability, or importance, than in the initial choice of breakage method. This choice depends on the lytic susceptibility of the organism and on the effect of the lysis method on the structural intactness and functional activity to be preserved in the isolated component(s). The factors germane to this chapter on cell fractionation, and often dictating the approach needed, include the differences in wall structures and bonding forces responsible in maintaining cell shape.

The Gram staining reaction subdivides eubacteria into gram-negative, gram-positive, and gram-variable species on the basis of wall structure (Chapter 2.3.5) and may be useful as a first step in assigning a method for breakage. The Gram reaction has limited value in this context for the archaeobacteria because they, in general, lack the outer membrane type of wall characteristic of gram-negative eubacteria. Archaeobacteria presently consist of the methanogenic, extremely halophilic, and sulfur-dependent groupings and have at least five different envelope structures (67). Methanogens having only an S-layer are relatively fragile, whereas the sheathed methanogens and those with lysozyme-insensitive pseudomurein (with or without an accompanying S-layer) may be more difficult to break. *Thermoplasma* and *Methanoplasma* species resemble the eubacterial mycoplasmas with respect to the complete absence of a wall structure.

The extent of breakage can be assessed by comparing the bacterial count of the suspension before and after treatment. For routine assessments it is adequate to view wet mounts by phase-contrast microscopy or by negative staining with nigrosin (Chapter 2.2.3). Microscopy, providing direct cells counts (Chapter 2.1.4), has an advantage over techniques which monitor loss of cell viability, since viability may be lost without a concomitant release of internal contents. Alternatively, it is possible to make measurements which reflect the release of cytoplasmic contents into the supernatant following standardized centrifugation (e.g., 8,000 × *g* for 15 min). Markers for cytoplasm include specific cytoplasmic enzyme activities, ions such as K^+, or UV-absorbing cofactors or nucleic acids.

4.1.1. Mechanical Breakage

Mechanical methods for cell breakage are the most universally successful. Each method, however, must be carefully assessed in the light of the species studied and the cell fraction to be isolated.

4.1.1.1. Pressure shearing

Pressure shearing with the **French pressure cell** is probably the most widely used and useful method of cell breakage, particularly for gram-negative bacteria (including the cyanobacteria) and some gram-positive eubacteria (particularly the gram-positive bacilli). Gram-positive cocci are not broken readily by this device, and ballistic disintegration or enzymatic digestion of the cell wall is more effective for breaking them. The French pressure cell is effective for many of the archaeobacteria, although *Methanobrevibacter* and *Thermoplasma* spp. are particularly resistant.

The French cell consists of a steel cylinder, piston, and pressure relief valve fitted with a replaceable nylon ball. Additional equipment includes a loading stand and a 10-ton (89,000-N) motor-driven hydraulic press which can deliver a constant force over a range of piston speeds. The Aminco pressure cell, accessories, and a suitable hydraulic pump can be obtained from SLM Instruments, Inc. (Urbana, Ill.). Alternatively, the pump, which is the most expensive part of the apparatus, can be purchased through a local supplier of hydraulic equipment, although commercial units may require some modification by a machine shop to fit the cell unit safely and satisfactorily. Aminco cells are available with maximum capacities of 3.7, 35, or 40 ml of sample. Both the 35- and 40-ml cells can be fitted with a one-way return valve for continuous operation with larger volumes of sample or for repeated processing. The larger cells are recommended for most applications, since they are more easily controlled than the smaller ones.

To use a French pressure cell, load the cell suspension (up to 30% cells by volume) into the pressure cell, taking care to avoid a gas pocket, and place the pressure cell assembly in the hydraulic press. Close the relief valve, and place the entire assembly in the hydraulic press. Lower the hydraulic press piston onto the cell piston, and allow it to reach the maximum force desired. Then slowly open the relief valve to the point where the piston moves downward while maintaining pressure. The cells are broken by the high shear forces generated as the suspension passes through the small orifice of the relief valve.

This method has several advantages. Heating can be minimized by precooling the press cylinder and piston; heat is generated only when the cells pass through the relief valve. The effluent temperature may rise by 5 to 10°C, but the effluent can be rapidly cooled if it is collected in a metal tube or beaker in an ice bath. Anaerobiosis may be maintained by flushing the chamber with argon or N_2 while loading the cell suspension with a syringe preflushed with N_2 and collecting the effluent in a serum vial continuously flushed with N_2. Breakage is virtually instantaneous, and the broken suspension is not subjected to additional shear forces which could damage subcellular particles. If a satisfactory hydraulic press is used, very reproducible conditions can be obtained. The extent of breakage is not influenced by the density of the cell suspension, the growth phase at which the cells are harvested, or the breakage medium in which the cells are suspended. Multiple passes are done as needed.

a. Breakage procedure for *Escherichia coli* cells

Harvest cultures, wash them once, and resuspend them at 0.01 to 0.1 times the original culture volume in 0.01 M HEPES (*N*-2-hydroxyethylpiperazine-*N'*-2-ethanesulfonic acid) pH 7.4 at 0°C. Other buffers such as phosphate may be used, but Tris buffer or chelating agents should be avoided because they cause outer membrane damage. Add a small amount of pancreatic RNase and DNase I (approximately 0.1 mg/ml) (omit Mg^{2+}, to encourage ribosome dissociation), and break the suspension with the French pressure cell operated at a cell pressure of 20,000 lb/in^2. After breakage, add $MgCl_2$ to give a final concentration of 1 mM, and remove unbroken cells by centrifugation at $5,000 \times g$ for 5 min. The purpose of the Mg^{2+} is to allow DNase to act. RNase acts in the absence of Mg^{2+}. Omission of the DNase allows the isolation of short strands of DNA by this breakage method; viscosity is generally not a problem because of DNA shearing.

The **Ribi cell fractionator** is similar in principle to the French pressure cell, except that the Ribi device provides higher pressure (which is more effective with endospores and some of the more resistant cells), a cooling system, and containment suitable for pathogens. The Ribi fractionator is no longer obtainable commercially.

4.1.1.2. Ultrasonic disintegration

A variety of ultrasonic probe devices are available to break cells. The rapid vibration of the probe tip produces high-intensity sound waves, which generate microscopic gas bubbles. Creation of these transient cavities (cavitation) is thought to create high-shear gradients by microstreaming (53). This breakage method is tempting to use, since the probe devices are relatively inexpensive and effective. However, there are inherent disadvantages.

The major disadvantage limiting the usefulness of this method is that breakage is not instantaneous; a cell suspension must be treated for 30 s to several minutes before a reasonable proportion of the cells are broken. During this time, subcellular particles released from broken cells are subject to the same high shear forces as unbroken cells are. Membrane vesicles are degraded to small lipoprotein fragments, which can no longer be sedimented in the ultracentrifuge, and extensive redistribution of membrane proteins between various membranes occurs (as, for example, between the inner and outer membranes of gram-negative bacteria).

In addition, it is difficult to control the temperature of the sample during breakage. Foaming can cause protein denaturation, and the cavitation phenomenon promotes oxidation of oxygen-sensitive enzymes and unsaturated lipids. Problems of foaming and temperature control can be minimized by subjecting the sample to several short bursts (15 to 30 s) rather than one continuous treatment, with cooling periods between the bursts. Oxidation problems can be minimized by covering the sample with argon gas during treatment. It is very difficult to obtain reproducible breakage, since the effectiveness of the probe depends on sample viscosity and on the size, shape, and composition of the sample vessel (for example, a glass beaker is far more effective than a plastic one, which will absorb some of the ultrasonic energy). *CAUTION:* Wear ear protection, unless the probe is enclosed in a sound-absorbing cabinet. Avoid holding the vessel in the fingers, because ultrasound is transmitted through glass.

Although ultrasonic treatment is not a good method for primary cell breakage, it is useful for lysis of spheroplasts (as in the procedure of Osborn and Munson [100] for separation of inner and outer membranes of gram-negative bacteria), for dispersing clumps of subcellular organelles, and for suspending centrifuge pellets. It is useful also for small-scale preparation of walls or envelopes, to be used for the initial identification of S-layers (section 4.4.3.4). In these procedures, very short bursts of ultrasonic energy are required so that subcellular organelles are not damaged.

4.1.1.3. Ballistic disintegration

Ballistic disintegration describes a variety of methods in which bacteria are broken by the shear forces developed when a suspension of cells together with small glass or plastic beads is shaken or agitated violently. A variety of commercial devices have been used for this procedure, and they are described in considerable detail in reviews by Salton (110) and by Hughes et al. (53).

Two devices used frequently in early studies were the Mickle shaker and a shaker fitted to the shaft of an International centrifuge (145). Neither of these devices had any provision for adequately cooling the sample during shaking, and it was necessary to interrupt the shaking frequently to cool the sample container. These devices are no longer available and have been replaced by the **Braun disintegrator** (B. Braun Melsunger Apparatebau, Melsungen, Germany), which is available from many laboratory supply houses in the United States. The Braun disintegrator has a 65-ml sample container which is shaken horizontally at 2,000 to 4,000 oscillations per min. Cooling is provided by a stream of liquid CO_2 delivered to the sample container. Most samples are disintegrated in 3 to 5 min at a temperature of less than 4°C.

The Braun disintegrator is often the method of choice for isolation of cell walls and particulate enzymes from gram-positive bacteria. The following procedure, described by Work (145), can be used with most organisms.

Mix a cell suspension (30 ml, containing 20 to 50 mg [dry weight]/ml) in a suitable buffer with 20 ml of glass beads (Ballotini no. 12; diameter, 100 to 200 μm) in the sample container (an air space of at least 20% of the container volume is essential for breakage), and pass liquid CO_2 through the apparatus for 0.5 min to cool the container. Shake the container at 3,000 strokes per min for 3.5 to 5.0 min, depending on the species. The beads may be removed by low-speed centrifugation or by passage through coarse-grade sintered glass. Because glass beads may liberate alkali when shaken violently, with potential damage to alkali-sensitive components, they must be acid washed, usually with nitric acid, before use. Suitable glass beads are supplied by Sigma Chemical Co. (St. Louis, Mo.). Some workers prefer to use plastic beads (108) to minimize enzyme denaturation. When the procedure described above is used for the isolation of cell walls, the sample should be treated promptly as described below to inactivate autolytic enzymes (145).

Bacterial endospores may be broken with the Braun disintegrator by shaking dry spores with dry, acid-washed glass beads (5 g of dry spores plus 15 g of beads) as described above and then suspending the sample in buffer (126). Beads of 470 μm diameter are appropriate; acid-washed beads of 425 to 600 μm are available from Sigma.

The **Bead-Beater** is sold by Biospec Products (Bartlesville, Okla.) as an apparatus for homogenizing and disrupting microorganisms. To use this apparatus, mix glass beads with up to 20 g (dry weight) of cells (0.5 mm in diameter for yeasts, fungi, and algae; 0.1 to 0.15 mm in diameter for bacteria) in a 30- or 60-ml chamber. A Teflon rotor within the chamber is driven by a conventional blender motor, providing breakage by bombardment with the beads. An external jacket may be filled with ice-water to minimize heating during the 2 to 3 min normally needed to achieve breakage. For small samples (approximately 1 ml), a mini-Bead-Beater is available.

4.1.1.4. Solid shearing

Cells may be broken by grinding cell paste or lyophilized cells with abrasives such as alumina by using an agate mortar and pestle; they may also be broken by forcing a frozen suspension, or paste, through a small hole or slit. Hughes et al. (53) have provided a complete description of these methods.

4.1.1.5. Freezing and thawing

Freezing and thawing has long been used as a method to release high-molecular-weight DNA from a wide variety of cells ranging from plants to bacteria. The efficiency of the process depends on the organism and the operator. A washed cell pellet is frozen in liquid N_2 and ruptured by grinding the frozen cells with a pestle in a precooled mortar. This procedure of lysis has been used recently in the isolation of DNA from *Methanothermus fervidus* (141) and several other methanogens (56) and involves a further treatment with sodium dodecyl sulfate (SDS).

4.1.1.6. Microwave disruption

Some microorganisms are lysed by microwaves; for *E. coli* this lysis is attributed primarily to thermal effects (34). This phenomenon has specific application in molecular biology, since colonies may be blotted onto nylon filters, soaked in 5% SDS solution, and microwaved for 2.5 min (650 W) to obtain lysis. This procedure denatures and fixes released DNA to the filters for subsequent hybridization (17).

4.1.2. Chemical Lysis

4.1.2.1. Detergent uses and actions

Detergents have three uses in cell fractionation. When nonmembranous organelles such as ribosomes and nucleoids are sought, detergents provide a gentle means of lysing the cells once the integrity of the peptidoglycan (gram-positive bacteria) or outer membrane (gram-negative bacteria) has been damaged. Detergents are used to selectively solubilize the cytoplasmic membrane of gram-negative eubacteria while leaving the outer membrane intact (section 4.4.2). Detergents can also be used to remove membrane contamination from ribosomes, polysomes, gram-positive cell walls, or other cell fractions resistant to a specific detergent.

Detergents are amphipathic molecules, meaning that the molecules have both hydrophilic and hydrophobic regions, and they are sparingly soluble in water. At very low concentrations, detergents will form true solutions in water. As the concentration is increased, additional molecules of detergent will aggregate to form micelles, in which the hydrophilic regions are exposed to water and the hydrophobic regions are shielded from water on the inside of the micelle. The concentration at which micelles begin to form, as the amount of detergent added to water is increased, is designated as the critical micelle concentration (CMC). The CMC and the size and shape of the detergent micelle are characteristic of each detergent. Membranes undergo various degrees of disintegration at different detergent concentrations. At low concentrations lysis or rupture is seen first, at intermediate detergent-to-membrane ratios (0.1 to 1 mg/mg of membrane lipid) membrane proteins may be selectively extracted, and at higher ratios (2 mg/mg of lipid) the membrane may be solubilized by formation of soluble micelles of lipid-detergent, protein-detergent, and lipid-protein-detergent (48). Excellent reviews by Helenius et al. (48) and Zulauf et al. (148) summarize the properties of many of the detergents used in cell fractionation.

Detergents may be grouped into three classes, which differ in micelle properties, protein binding, and response to other solutes. These classes are ionic detergents (cationic, anionic, and zwitterionic), nonionic detergents, and bile salts.

4.1.2.2. Ionic detergents

The most commonly used ionic detergents are SDS (also known as sodium lauryl sulfate), sodium *N*-lauryl sarcosinate (Sarkosyl), alkyl benzene sulfonates (common household detergents), and quaternary amine salts such as cetyltrimethylammonium bromide (CTAB). Ionic detergents tend to form small micelles (molecular weights of about 10,000) and exhibit a rather high CMC (the CMC for SDS is about 0.2% at room temperature in dilute buffers). The CMC and the micelle solubility of ionic detergents are strongly influenced by the ionic strength of the solution and the nature of the counterions present. For example, a 10% solution of SDS is soluble down to about 17°C, whereas a similar solution of Tris dodecyl sulfate is soluble at 0°C. Potassium dodecyl sulfate is soluble only at elevated temperatures, and K^+ must be excluded from all buffers when this detergent is used. The guanidinium salt of SDS is also insoluble.

Detergents such as SDS, which have a hydrophilic group that is strongly ionized, are not affected by pH and are not precipitated by 5% trichloroacetic acid. Ionic detergents bind strongly to proteins, and, for SDS, this usually results in unfolding and irreversible denaturation of proteins. This is the basis of their use in polyacrylamide gel electrophoresis (Chapter 21.5.1). Although ionic detergents can pass through dialysis membranes, they bind strongly to proteins and cannot be removed completely by dialysis.

4.1.2.3. Nonionic detergents

Commercial nonionic detergents include the polyoxyethylene detergents such as Triton X-100, Nonidet P-40, Brij 58, and Tween 80 (48). These detergents are not pure but are made up of a mixture of related compounds. In general, the detergents have a high micelle molecular weight (50,000 or greater) and a low CMC (0.1% or less), which limit their usefulness in procedures such as gel filtration or electrophoresis and make them difficult to remove by dialysis. Although their properties are not strongly affected by pH or ionic strength, they may be precipitated by 5% trichloroacetic acid. The advantage of these detergents is that they bind primarily to hydrophobic sites on the surface of proteins and thus do not cause extensive denaturation or loss of biological activity. Since nonionic detergents are uncharged, they do not interfere with separations which are based on the charge character of the cellular component. As noted in section 4.4.2, the resistance of the outer membrane of enteric bacteria and many other gram-negative bacteria to dissolution in nonionic (114) or weakly ionic (33) detergents allows the application of these detergents to the isolation of cell surface components.

There are a number of new nonionic detergents consisting of acyl derivatives of sugars, which have been synthesized for scientific rather than commercial applications. Examples include octyl-β-D-glucopyranoside, octyl-β-D-thioglycopyranoside, and dodecyl-β-D-maltoside. These are available from biochemical supply houses such as Calbiochem Inc. (San Diego, Calif.) or Sigma. These detergents are similar to Triton X-100 in their membrane protein-solubilizing properties, but they have several advantages. They are available as pure compounds of defined structure, and they generally have a much higher CMC, which makes them more useful in chromatography and facilitates removal by dialysis or gel filtration. They are available in different acyl chain lengths, which in turn results in a range of properties such as CMC, ease of dialysis, and micelle molecular weight. High cost is a major disadvantage.

4.1.2.4. Bile salts

Bile salts are salts of sterol derivatives, such as sodium cholate, deoxycholate, or taurocholate. Because of the poor packing ability of the bulky sterol nucleus, these detergents form small micelles (often just a few molecules), and, unlike other detergents, the micelle molecular weight is a function of the detergent concentration. Since these detergents are salts of very insoluble weak acids with pK_as in the range of 6.5 to 7.5, they must be used at alkaline pH. To avoid solubility problems, stock solutions are often prepared by dissolving the free acid in excess NaOH. Ionic composition, pH, and total detergent concentration affect their usefulness and must all be maintained constant when these detergents are used.

For release of genomic DNA from rod-shaped halobacteria, sodium taurocholate may be preferred to sodium deoxycholate, since it dissolves more readily in the presence of high salt concentrations. The lysis of cells by bile salts is effective for only select members of the archaeobacteria and eubacteria (66); also see Chapter 25.1.10.

4.1.2.5. Problems with detergents

In excess, detergents act by forming mixed micelles with lipids or by binding to proteins to form soluble protein-detergent complexes. Since most detergents are used at concentrations well above the CMC, the ratio of detergent to protein or lipid is much more important than the actual detergent concentration. As a general rule of thumb, use at least 2 to 4 mg of detergent per mg of sample protein to ensure a satisfactory excess of detergent. For example, if 2% Triton X-100 is used for membrane solubilization, the maximum protein content of the sample should be less than 5 to 10 mg/ml.

Triton X-100, one of the most useful nonionic detergents, presents problems in protein assays because it has an aromatic residue which prevents measurement of A_{280} and because it (and other nonionic or cationic detergents) forms a cloudy precipitate in chemical protein assays. Proteins can be assayed in the presence of these detergents by the modification of the procedure of Lowry et al. described by Peterson (104), in which an excess of SDS is added to the sample. This forms stable mixed micelles of the detergents which do not interfere with the assay.

Triton X-100 and other similar nonionic detergents are soluble in an ethanol-water mixture, and proteins dissolved in these detergents may be freed from detergent by ethanol precipitation as follows. Place the sample on ice, and add 2 volumes of ice-cold absolute ethanol with stirring. Allow the sample to stand overnight in a freezer, and collect the protein precipitate by centrifugation. Efficient precipitation requires a protein concentration of at least 0.2 mg/ml. Dilute samples may be concentrated by using an ultrafiltration apparatus (Amicon Corp., Beverly, Mass.) with a filter of appropriate pore size, although this is limited by the possibility that detergent micelles will also be concentrated. Triton detergents can also be removed by an SM-2 resin (Bio-Rad Laboratories, Richmond, Calif.) (52).

SDS can be removed from samples by acetone precipitation as follows. Stir 6 volumes of anhydrous acetone into the sample at room temperature, and recover the precipitate by centrifugation. Wash the precipitate several times with acetone-water (6:1, vol/vol). Since the precipitate is often waxy and difficult to work with, it may be dispersed in water with a Potter-Elvchjem homogenizer and then lyophilized.

An alternative method for SDS removal, used often in DNA purification, is precipitation of the potassium salt of SDS made by the dropwise addition of potassium acetate to a final concentration of 0.3 to 0.5 M.

4.1.2.6. Dithiothreitol

Dithiothreitol will induce lysis of the sheathed methanogens in the absence of an osmotic protectant, provided that the pH is moderately alkaline (122). The procedure can be scaled up to allow the isolation of cellular organelles or enzymes from *Methanospirillum hungatei* or *Methanosaeta concilii*, but it is generally ineffective with most other methanogens (94, 122).

4.1.3. Osmotic Lysis

Osmotic lysis in hypotonic solution may be achieved for some bacteria without the need of first weakening the cell wall. Extreme halophiles such as *Halobacterium* species, as well as many moderate halophiles, require salts to maintain their structural integrity, and they lyse in deionized water. The lytic susceptibility of a new culture should therefore be tested by resuspension in low-ionic-strength solutions. Because of its simplicity, this is often recommended as a first option for lysis, provided that the component to be isolated is stable under these conditions. Gas vesicles have been isolated from *Halobacterium salinarium* following lysis by salt dilution (119).

The nature of the cations to which bacterial cells were previously exposed can affect their susceptibility to lysis in water (81, 89). Eighteen gram-negative marine bacteria and two gram-negative terrestrial bacteria were found to give a range of responses in lytic susceptibility in distilled water ranging from those that lysed even after exposure to Mg^{2+}-containing solutions to those that were sensitized by washing in 0.5 M NaCl (e.g., *Pseudomonas aeruginosa*). Others resisted lysis in distilled water even after being washed with NaCl (e.g., *E. coli*). Many of the methanogenic bacteria which have S-layers as the sole wall component (e.g., *Methanococcus* species such as *Methanococcus voltae* and *Methanococcus jannaschii*), and others such as the moderate halophile *Methanosarcina mazei*, lyse in water.

A general procedure for osmotic lysis in low-ionic-strength solutions, from Laddaga and MacLeod (81) is as follows.

1. Harvest cells (exponential growth phase normally) at 8000 x *g* for 10 to 20 min, and wash them in an equal volume of 0.5 M NaCl solution up to three times (the outer membrane of gram-negative bacteria can sometimes be removed by further washes in 0.5 M sucrose).
2. Resuspend the cell pellet in deionized water.

Extraordinary resistance to lysis may require an individualized procedure. The sulfur-dependent, wall-less thermophile *Thermoplasma acidophilum*, which grows optimally at pH 2, can be lysed by raising the pH to 6 or 7. The organism otherwise resists mechanical breakage, nonionic detergents, pronase, and trypsin and is osmotically stable (82).

A procedure combining chemical and osmotic techniques is effective with a wide variety of methanogens, as follows (94).

1. Centrifuge a 10-ml sample of cell culture (0.3 to 0.5 mg [dry weight]/ml), freeze overnight at $-20°C$ as cell paste, and thaw at room temperature.
2. Resuspend frozen-thawed cells in 10 ml of ABE buffer (50 mM ammonium bicarbonate, 50 mM disodium EDTA [pH 8.0]). Chelators of divalent cations (e.g., EDTA) potentiate lysis in cases when the cell wall has cationic cross-bridging.
3. Add pronase to give 40 μg/ml, and incubate for 1 h at 37°C. If DNA is to be isolated, nucleases in the pronase (Boehringer Mannheim Biochemicals, Indianapolis, Ind.) must be inactivated by autodigestion in 50 mM Tris-HCl (pH 8.0) for 1 h at 37°C. *Note:* proteinase K cannot be substituted for pronase with some strains.
4. Add dithiothreitol (200 μg/ml) for 15 min at 23°C.
5. Finally, add SDS (1 mg/ml) and heat the mixture at 55°C for 15 min to promote lysis.

Note: not all steps are necessary for all strains; the pronase digestion step is likely to be inappropriate for most applications, excluding nucleic acid isolations, and can often be omitted.

4.1.4. Enzymatic Lysis

Osmotic lysis may be achieved in the many cases when osmotically sensitive cell forms can be generated by enzymatic activity, as described in section 4.2.

4.1.5. Lysis by Boiling

Lysis by boiling (111) has application in the isolation of closed-circular plasmid DNA (Chapter 16). To obtain lysis of *E. coli* by boiling, harvest the cells from 500 ml of medium and wash in a solution of ice-cold 0.1 M NaCl–10 mM Tris · HCl (pH 8.0)–1 mM EDTA (pH 8.0). Resuspend the cell pellet in 10 ml of the above solution containing 5% Triton X-100, and add 1 ml of freshly prepared lysozyme solution (10 mg/ml in 10 mM Tris · HCl [pH 8.0]). Incubate for 30 min at 23°C to allow the lysozyme time to act on the peptidoglycan exposed by the action of EDTA on the outer membrane. Heat the cells in an Erlenmeyer flask with an open flame for 40 s beyond the boil. Immediately cool the flask in ice-cold water.

4.1.6. Alkali Lysis

Alkali lysis (111) combines the effects of EDTA, NaOH, and the detergent SDS at 0°C. Obtain a washed cell pellet as described above, and resuspend it in 50 mM glucose–10 mM EDTA–25 mM Tris · HCl (pH 8.0). Dilute the suspension with an equal volume of cold 0.2 N NaOH and 1% SDS solution.

4.2. GENERATION OF PROTOPLASTS AND SPHEROPLASTS

The strength of most bacterial walls resides in the murein component, which may be damaged or destroyed by specific enzymes (e.g., lysozyme), leading to osmotically sensitive cells. The surrounding medium is generally hypotonic (or can be made so by dilution), which results in lysis whose extent can be monitored by phase-contrast microscopy. If lysis is prevented by adding 0.3 to 0.8 M sucrose and 0.5 to 10 mM Mg^{2+} to the reaction mixture, the round and osmotically sensitive cells remain intact as protoplasts (from gram-positive cell types with the plasma membrane remaining) or spheroplasts (from gram-negative cell types with both outer membrane and plasma membrane remaining). Other osmotic stabilizers may be used (e.g., polyethylene glycol, sodium succinate, and certain salts from 0.2 to 0.8 M). However, structural variations in the murein substrate leads to differences in sensitivity to murein-lytic enzymes. For example, the extent of O acetylation of murein directly affects its susceptibility to egg white lysozyme (28). Since the extent of O acetylation may increase during the stationary phase (59), it is generally recommended that exponential-phase cultures be used. Archaeobacteria either contain a lysozyme-insensitive pseudomurein or lack the murein-pseudomurein type of structure altogether.

The enzymes that degrade murein can be divided into three classes: carbohydrases, acetylmuramyl-L-alanine amidases, and peptidases (61). Those most commonly used in forming protoplasts and spheroplasts are the muramidase subgroup of the carbohydrases, notably lysozyme.

Osmotically sensitive forms may also be generated by interactions with antibiotics. Penicillin G causes specific inhibition of murein biosynthesis and leads to unbalanced growth and lysis of penicillinase-negative bacteria in a hypotonic environment. In the presence of hypertonic sucrose, osmotically sensitive spheres are formed from many rod-shaped gram-negative eubacteria (spheroplasts) and gram-positive eubacteria (protoplasts), such as *Bacillus subtilis* and *B. megaterium.* For the technique to work effectively, rapidly growing mid-logarithmic-phase cells are used. The method, with pros and cons of its use, has been described in detail by Kaback (63). *Staphylococcus aureus* forms protoplasts during growth in media containing sublethal doses of D-cycloserine and 1 M sucrose (134).

4.2.1. Protoplasts from Gram-Positive Bacteria

A procedure for generating protoplasts from gram-positive bacteria that are sensitive to lysozyme is as follows. Wash the bacteria in a buffer of pH 6 to 8, and resuspend (1 to 10 mg [dry weight]/ml) in the same buffer containing hypertonic sucrose (usually 0.2 to 1.0 M) and lysozyme at 0.1 to 1.0 mg/ml (Sigma). Treatment for 30 min at room temperature should result in complete protoplast formation. The enzyme may be used over a broad pH range, with optimum pH of 6 to 7, and over a wide temperature range from 4 to 60°C (27, 109). Lysozyme may be inhibited by the presence of salts in excess of 0.2 M (121).

Many gram-positive bacteria are not sensitive to egg white lysozyme but may have murein that is sensitive to muramidases of broader substrate range (Table 1). These include lysostaphin from *"Staphylococcus staphylolyticus,"* mutanolysin from *Streptomyces globisporus* (Sigma), and a fungal muramidase from *Chalaropsis* sp. (Miles Laboratories, Inc., Elkhart, Ind.). The latter enzyme, unlike hen egg white lysozyme, is active against murein acetylated at C-6 of muramic acid (46). Caution should be used when working with these enzymes since, unlike egg white lysozyme, they are often contaminated with proteases, lipases, or nucleases.

Methods can sometimes be found to cause the murein of insensitive strains to become sensitive to lysozyme. For example, *Mycobacterium smegmatis* becomes sensitive to lysozyme after prolonged exposure to DL-methionine (146). Protoplasts of *B. megaterium* can be formed following incubation for 2 h at 30°C in nutrient broth containing 100 U of penicillin per ml. The treated cells are washed once in 0.85% saline and resuspended for about 1 h in 0.3 M sucrose containing 500 μg of lysozyme per ml (90). Partially lysed and weakened cell forms of cyanobacteria (two *Oscillatoria* spp.) can be made by pretreatment in the presence of ampicillin followed by treatment with lysozyme (135).

Clinical isolates of *Staphylococcus* spp. can be made more sensitive to lysostaphin by growth in P medium lacking glucose to limit slime production (39).

4.2.2. Spheroplasts from Gram-Negative Bacteria

Virtually all gram-negative eubacteria have a murein structure that is sensitive to lysozyme. However, since

Table 1. Some commonly used methods to lyse bacteria and to form osmotically sensitive cell forms in hypertonic solutions

Method	Sensitive representative bacteria	Reference
Lysozyme	*Bacillus megaterium*	121
	Bacillus subtilis	121
	Micrococcus lysodeikticus	121
Lysozyme-EDTA-Tris (pH 8)	*Salmonella typhimurum*	63
	Escherichia coli	63
	Enterobacteria	63
	Pseudomonas aeruginosa	63
	Micrococcus denitrificans	63
	Azotobacter vinelandii	63
	Bacillus subtilis	63
	Bacillus megaterium	63
	Clostridium thermoaceticum	63
	Aquaspirillum serpens	21
	411 gram-positive/gram-negative eubacteria	133
	Thiobacillus strain A2	86
NaCl/sucrose-lysozyme	5 *Alteromonas* spp.	81
	7 *Vibrio* spp.	81
	2 *Cytophaga* spp.	81
	4 *Pseudomonas* spp.	81
	3 *Alcaligenes* spp.	81
EDTA-Tris (pH 8.6)	*Pseudomonas aeruginosa*	37
	Alcaligenes faecalis	38
	Escherichia coli	38
Mutanolysin ±1 M NaCl	*Streptococcus mutans*	118
	Streptococcus sanguis	118
	Streptococcus salivarius	118
Lysoamidase	*Staphylococcus aureus*	105
Penicillin	Various gram-positive and gram-negative eubacteria	63
Lysozyme-penicillin[a]	*Fusobacterium varium*	19
	Enterococcus faecium	19
Serine protease	*Methanobacterium formicicum*	18
Pronase	*Methanosarcina barkeri*	62
Dithiothreitol (pH > 8)	*Methanospirillum* spp.	11
	Methanosaeta concilii	11
Pseudomurein endopeptidase	*Methanobacterium thermoautotrophicum*	70

[a]Lysozyme susceptibility follows growth in medium containing penicillin G.

the outer membrane of gram negative bacteria is impermeable to lysozyme, the integrity of the outer membrane must be compromised before lysozyme will act. This can be done in various ways (Table 1). The basic lysozyme-EDTA procedure for forming spheroplasts and lysis of a wide variety of gram-negative eubacteria (63) is as follows.

1. Harvest bacteria in mid- to late exponential growth phase and wash twice with 10 mM Tris · HCl (pH 8.0) at 4°C. Resuspend the cells (1 g [wet weight]/80 ml), and stir magnetically at room temperature in 30 mM Tris · HCl (pH 8.0) containing 20% sucrose.
2. Add 10 mM potassium EDTA (pH 7.0) and 0.5 mg of lysozyme per ml (final concentrations), and incubate for 30 min at room temperature. Add lysozyme slowly under the surface of the liquid to decrease aggregation of the spheroplasts.
3. If lysis is required, dilute the sucrose either directly by dialysis or by centrifugation and resuspension of the pellet.

Some modifications to this method are described by Osborn and Munson (100). Factors influencing the efficiency of spheroplast formation include the growth medium (cells grown in a rich medium are more suitable than cells from defined media), the temperature at which the cells are harvested (0°C versus room temperature), factors which might destroy the permeability barrier to sucrose, and the concentration of lysozyme.

4.2.3. Osmotically Sensitive Forms from Archaeobacteria

The formation of osmotically sensitive forms in the archaeobacteria is less well understood. Spheroplasts of *Halobacterium* spp. may be formed by resuspending the cells in 0.1 M 2(*N*-morpholino)ethanesulfonic acid (MES) buffer (pH 7.0) containing 0.5 M sucrose, 0.25 M NaCl, and 0.01 M $MgCl_2$ (57). Also, *Methanospirillum* spp. and *Methanosaeta concilii* form spheroplasts when exposed to dithiothreitol at alkaline pH (section 4.1.2.6).

Strains of *Methanosarcina barkeri*, which have cell walls composed largely of protein, form metabolically active protoplasts when incubated at 37°C with pronase (2 mg/ml) and 0.5 M sucrose (62). During growth some bacteria enter a lytic phase, often as a result of nutrient depletion. In these cases, lysis may be prevented by incorporating the osmotic protectant directly into the medium. This concept may be used to form protoplasts during growth limitation of *Meth-*

anobacterium bryantii in a medium deficient in nickel and ammonium but containing 20 mM $MgCl_2$ to prevent lysis (55), and to form protoplasts from *Methanosarcina barkeri* cultures stabilized by 0.3 M sucrose or glucose (24). Finally, a pseudomurein endopeptidase has been discovered in *Methanobacterium wolfei* and found to be effective in preparing protoplasts of *Methanobacterium thermoautotrophicum* stabilized with 0.8 M sucrose (70). The usefulness of this enzyme is limited by its O_2 sensitivity, but it is the first enzyme known to lyse the pseudomurein sacculus. It is possible to purify this pseudomurein-degrading enzyme aerobically and to recover activity by incubation under reducing conditions with $Na_2S \cdot 9\ H_2O$, or dithiothreitol (98).

4.3. CENTRIFUGATION

The techniques of differential centrifugation and equilibrium density gradient centrifugation are widely used in the purification of components from cell lysates. Detailed discussions of both the theoretical and practical aspects of centrifugation are provided in comprehensive reviews (64, 107, 129).

The sedimentation rate of particles subjected to a centrifugal force is proportional to the force applied, and also to the properties of the suspension fluid, as shown below for a spherical particle:

$$v = \frac{d^2 (p_p - p_l) \times g}{18n}$$

where v is the sedimentation rate, d is the particle diameter, p_p is the particle density, p_l is the liquid density, g is the centrifugal force, and n is the viscosity of the liquid. At constant centrifugal force and viscosity, the sedimentation rate is proportional to the size of the particle and to the difference between the density of the particle and the liquid. Density gradient centrifugation can be based on either the size difference among particles (rate-zonal centrifugation) or the density difference among particles (equilibrium or isopycnic centrifugations).

4.3.1. Differential Centrifugation

In differential centrifugation, samples are centrifuged for a given time at a given speed, resulting in a supernatant and a pellet fraction. This technique is useful for separation of particles with very different sedimentation velocities. For example, centrifugation for 5 to 10 min at 3,000 to 5,000 × *g* will pellet intact bacterial cells while leaving most cell fragments (walls, membranes, appendages) in the supernatant. Cell wall fragments and large membrane structures can be pelleted by centrifugation at 20,000 to 50,000 × *g* for 20 min, whereas centrifugation at 200,000 × *g* for 1 h is required to pellet small membrane vesicles or ribosomes.

Low-speed centrifugations may be performed anaerobically by using standard anaerobic methods and glass centrifuge tubes modified to accept serum bottle closures (123).

4.3.2. Rate-Zonal Centrifugation

Zonal centrifugation is effective for separating subcellular structures having a similar buoyant density but differing in shape or particle size, e.g., ribosomal subunits, classes of polysomes, and various forms of DNA molecules. Centrifugation is carried out in either swinging-bucket rotors or specially designed zonal rotors, and a shallow **linear gradient** (usually sucrose) is used to prevent convection in the tubes or the chamber of the zonal rotor during centrifugation. Preparation of linear gradients is described in section 4.3.5.1. A density range is chosen such that during the separation the particle densities are always higher than the liquid. The sample is applied as a zone or narrow band at the top of the gradient, and the run is terminated before the separating particles reach the bottom of the tube. Times and speeds must be determined empirically. For subcellular particles, a 15 to 40% (wt/vol) sucrose gradient is commonly used, and centrifugation at 100,000 × *g* for 1 to 4 h is sufficient for separation of most subcellular particles.

The use of conventional sucrose or glycerol gradients results in an increase in the concentration of the gradient material and therefore of the liquid viscosity as the particle moves down the gradient. Consequently, the rate of movement can decrease as the particle moves downward, even though the centrifugal force increases, and can ultimately result in less resolution between separating zones. When a cesium chloride gradient is used (section 4.3.3), the viscosity decreases with increasing concentration, resulting in an accelerating gradient and better separations.

4.3.3. Equilibrium and Isopycnic Density Gradient Centrifugation

Equilibrium and isopycnic density gradient centrifugations are often used to separate particles and membrane fractions on the basis of buoyant density instead of sedimentation velocity. Membrane fragments derived from the same subcellular membrane may differ greatly in size and hence in sedimentation velocity, but they should have the same buoyant density. The sample is centrifuged in a solute density gradient (sucrose gradients are commonly used for membranes and organelles with a density of less than 1.3 g/ml; cesium chloride is commonly used for denser structures such as viruses) until an equilibrium state is reached at which each particle has migrated to a point in the gradient where the particle has the same density as the surrounding solution (Table 2). Since sucrose solutions are relatively viscous, preformed gradients are generally used in equilibrium centrifugation.

Cesium chloride solutions have a low viscosity, so that preformed gradients of this solute are difficult to prepare, and the technique of isopycnic (isodensity) gradient centrifugation is used. In this latter technique, the sample is mixed with enough cesium chloride to provide a density equal to the average density of the subcellular particles. This homogeneous suspension is placed in the centrifuge, and the cesium chloride gradient is formed during centrifugation as a result of the sedimentation of the cesium chloride in the centrifugal field.

Since the sedimentation velocity of a particle becomes progressively lower as the particle approaches the region in the gradient where it has the same density as the solution, very long centrifugation times may be

Table 2. Buoyant densities of fractions isolated from bacterial cells

Fraction	Buoyant density (g/ml)	Medium	Source	Reference
Plasma membrane	1.14–1.16	Sucrose	Gram-negative enterics	100
	1.17	Sucrose	*M. hungatei*	124
Outer membrane	1.22	Sucrose	Gram-negative enterics	100
Sheath	1.35	Sucrose	*M. hungatei*	124
Fimbriae	1.117–1.137	Sucrose	*E. coli*	68
	1.3	CsCl	*P. aeruginosa*	101
Spinae	1.37	CsCl	Marine pseudomonad	30
Flagella[a]	1.30	CsCl	*E. coli*	25
Spores	1.27	Percoll	*C. perfringens*	132
	1.21	Percoll	*B. subtilis*	132

[a]Intact flagella

required to approach equilibrium. This is particularly true for small membrane vesicles such as chromatophores or plasma membrane fragments from cells broken with a French press, since the sedimentation velocity of these vesicles is low even in the absence of the gradient. If centrifugal forces on the order of 100,000 to 200,000 × *g* are used, at least 18 h is required for a reasonable separation and periods as long as 72 h are needed for critical separations (section 4.4.5.3).

4.3.4. Rotor Selection

There are basically four types of rotors used for centrifugal separations, (i) swinging (swing-out) bucket, (ii) fixed angle, (iii) vertical tube, and (iv) zonal. It is important to choose the correct rotor to achieve optimum separation results. This topic is discussed in references 64 and 107.

Swinging-bucket rotors are recommended for most preparative rate-zonal separation methods because of the relatively long path length. Vertical-tube rotors are also useful for rate-zonal separations, especially for large particles such as cells or subcellular organelles. The main advantage of vertical-tube rotors is the shorter time needed to achieve separation compared with that for other rotor types. Fixed-angle rotors are not normally used for rate-zonal separation.

For isopycnic separations, sample particles will cease sedimenting when they reach their buoyant density, and wall effects are much less important than for rate-zonal separation. Swinging-bucket, fixed-angle, and vertical-tube rotors may all be used for isopycnic centrifugation; however, the density range and slope of an equilibrium gradient will depend on the type of rotor used for isopycnic centrifugation. Fixed-angle and vertical-tube rotors have some advantages over swinging-bucket rotors for these separations. Because the effective path length is shorter for these rotors than for swinging-bucket rotors, the period of centrifugation is considerably shorter for equilibrium banding of particles. The increased number of tubes in fixed-angle and vertical-tube rotors means that sample capacity per rotor is much greater.

A wide variety of materials have been used for the manufacture of centrifuge tubes and bottles. It is important to choose a material that is compatible with the sample and the gradient medium. Centrifuge tubes should be filled (to within ca. 3 mm of the tube rim) to prevent collapse of the tubes during centrifugation. An exception is thick-walled tubes, which are available for most fixed-angle and swinging-bucket rotors. Tubes should be capped (with an "O"-ring sealing system) if the sample is radioactive or biohazardous and in fixed-angle rotors to support thin-walled tubes during centrifugation. This tedious aspect of ultracentrifugation is avoided by using swinging-bucket rotors, in which it is unnecessary to cap the individual tubes. Fortunately, Beckman (Fullerton, Calif.) has introduced narrow-necked tubes (Quick-Seal tubes) that can be heat sealed and do not need a cap. Quick-Seal tubes can be used in all swinging-bucket and most fixed-angle rotors and must be used in vertical-tube rotors. Tube size must be compatible with the rotor and the sample volume. Many Beckman rotors can accommodate smaller tubes by using adapters or spacers. The Beckman *g*-Max system uses a combination of short polyallomer Quick-Seal tubes and floating spacers to permit shorter run times at maximum *g* force.

4.3.5. Gradient Media

4.3.5.1. Sucrose and Ficoll

Sucrose concentrations in gradients are reported in the literature in several ways: density, molar concentration of sucrose, percent sucrose by weight, and percent sucrose per unit volume (expressed as % [wt/vol] or grams/100 ml). Standard chemistry handbooks contain tables (8) listing the concentrative properties of aqueous sucrose solutions which relate to these various units. The most useful unit is percent sucrose by weight. When preparing sucrose solutions, assume that water or dilute buffers have a density of 1 g/ml. Therefore, to prepare a 54% (wt/wt) sucrose solution, for example, dissolve 54 g of sucrose in 46 ml of water. This will give 100 g of solution, and since the density of a 54% (wt/wt) sucrose solution is 1.2451 g/ml, the final volume will be 80.3 ml. Preparing solutions in this manner avoids the necessity of having to make viscous solutions up to fixed volumes. Sucrose allows a density range up to 1.35 g/ml.

Ficoll 400, supplied by Pharmacia (Uppsala, Sweden), is a 400,000-molecular-weight hydrophilic polymer of sucrose. Solutions of up to 50% (wt/vol) Ficoll can be made, which allows a density range of up to 1.2 g/ml. Its advantages over sucrose for density gradient centrifugation lie in the lower osmotic pressure, which can be important in preserving the morphology/activity of subcellular fractions.

a. Preparation of linear (continuous) gradients

A simple device for preparing linear gradients is covered in Chapter 21.3.1.1.

Satisfactory gradients can be prepared by hand-layering a series of sucrose solutions (e.g., in 2% steps) in a centrifuge tube and allowing the gradient to "age" overnight in order for diffusion to produce a linear gradient. It is not necessary to age gradients that will be used for centrifuge runs of 24 h or longer, since diffusion during centrifugation will result in a linear gradient. Sucrose gradients are often prepared as a series of five to seven layers. When wettable tubes such as cellulose nitrate or polycarbonate are used, each layer can be added by allowing it to run down the side of the tube from a pipette held at an angle against the side of the tube. When nonwettable tubes such as polyallomer or polypropylene are used, the pipette tip must be kept in contact with the meniscus during addition of the solution to avoid mixing. A variety of commercial gradient-forming devices are available, but these are unnecessary if the molecular weight of the gradient medium is less than 1,000. Diffusion of higher-molecular-weight molecules such as Ficoll 400 is too slow, requiring a gradient maker for preparation of linear gradients.

Step (discontinuous) gradients are prepared by layering solutions of different density in the centrifuge tube. The sample is layered on top, and centrifugation is started immediately.

b. Methods for fractionating sucrose gradients

The gradients may be pumped out through a stainless-steel cannula inserted into the bottom of the tube by using a peristaltic pump. Alternatively, the bottom of the tube, or the side just below the band, may be pierced with a needle, and samples can then be collected by gravity flow. Although simple to perform, these methods in which samples are collected starting from the bottom of the gradient can lead to cross-contamination of upper bands with those appearing lower in the tube. In cases when the component of interest is the upper band, the gradient should be fractionated by upward displacement. The centrifuge tube is fitted with a tight-fitting rubber bung pierced by a stainless-steel cannula sufficiently long to extend to the bottom of the tube. A sucrose solution about 5% greater in density than the bottom of the gradient is pumped through the cannula, thus displacing the gradient through a short exit tube leading to a fraction collector.

4.3.5.2. Cesium chloride and potassium bromide

Cesium chloride and potassium bromide are discussed in section 4.3.3 with isopycnic techniques. CsCl allows a density range up to 1.81 g/ml. KBr can also be used and is less expensive than CsCl (65).

4.3.5.3. Percoll

Percoll is a density gradient medium containing colloidal silica particles that are rendered nontoxic by a coating of polyvinylpyrrolidine (103). Percoll is recommended for the separation of cells because it has low osmolarity, low viscosity, and large particle size (30 nm) so that it does not permeate bacterial walls or membranes. It is used also to separate viruses and subcellular organelles. It can be purchased from Pharmacia as a sterile suspension which may be diluted as desired. Gradients may be either preformed or spontaneously formed during the centrifugation run, to produce a density range up to 1.3 g/ml. Since Percoll cannot penetrate walls or membranes, intact cells band at the buoyant densities of the intact cell (40). In contrast, gradient media of smaller particle size (e.g., Metrizamide and Nycodenz [see below]) permeate integumental structure and thus indicate the buoyant densities of the protoplasts, as exemplified by lysozyme-sensitive spores (85). Density marker beads of differing densities and colors can be purchased (Pharmacia) for calibration. This is especially important because the rotor angle and tube size affect the geometry (shape) of the gradient formed.

4.3.5.4. Metrizamide and Nycodenz

Metrizamide, or 2-(3-acetamido-5-*N*-methylacetamido-2,4,6-triiodobenzamido)-2-deoxy-D-glucose (Nyegaard and Co., Oslo, Norway; Sigma) is a density gradient medium that has a molecular weight of 789, a density of 2.17 g/ml, and very low viscosity. This last property is particularly useful in applications in which it is desirable to isolate membrane fractions by floatation to assess the association of the membrane with mutant proteins or products of genes cloned into multicopy vectors. Spheroplast lysates of *E. coli* are brought to a density of 1.29 g/ml with Metrizamide and placed in a centrifuge tube. The sample is layered over with a 1.27 g/ml Metrizamide solution. Rapid floatation of membranes occurs while the gradient is being formed during centrifugation, leaving cytoplasmic proteins near the bottom of the tube. The authors of the above method (130) use to advantage the short sedimentation distance and high force obtainable in the Beckman TL100 benchtop ultracentrifuge.

Nycodenz (Nyegaard and Co.) is an iodinated compound improved, compared with Metrizamide, by increased heat stability, absence of toxicity, and resistance to microbial degradation (84). Excellent booklets on the use of Nycodenz and Metrizamide in gradient centrifugation are available from the manufacturer.

4.3.5.5. Renografin

Renografin (E. R. Squibb & Sons, Princeton, N.J.; also Urografin), a pharmaceutical for image enhancement in radiology, is sometimes still used as a gradient medium for cell fractionation despite the following adverse properties: a mixture of two ionic iodine-diazotrizoate compounds plus salts and certain additives, resulting in high osmolarity, high water activity, and intermediate viscosity (132).

4.4. ISOLATION OF CELL COMPONENTS

4.4.1. Appendages

4.4.1.1. Flagella

Intact flagella of most bacteria consist of three major components: filament, hook, and basal body. The first two components are external to the cell surface and are relatively easy to isolate from other cell components. Flagellar filaments can be removed mechanically from cells by shear forces in a blender or Potter-Elvehjem homogenizer. This technique removes the filament component of flagella without damaging the cell walls of most organisms. Experience has shown that the hook is usually not attached to the sheared filament. Indeed, filaments can be sheared from cells, the cells can be placed in fresh medium, and regrowth of the filament can be monitored. When flagella are released spontaneously from autolyzed cells, the hook often remains attached to the filament.

Note: check the culture supernatant by electron microscopy (use negative staining) after harvesting for any filaments which have been removed either during growth or centrifugation. Often concentration and ultracentrifugation of the culture supernatant will produce a good yield.

a. Procedure for the isolation of flagellar filaments

1. Grow 1 to 3 liters of bacteria.
2. Harvest cells by centrifugation.
3. Resuspend cells in approximately 10% of the culture volume in 0.5 M HEPES or Tris buffer (pH 7.5). The density of the cell suspension is important for the removal of filaments.
4. Homogenize cells in a rotating-blade blender (Waring, Sorvall Omni-mixer) at maximum speed for about 60 s. Whatever brand of blender is used, the volume of cell suspension must at least cover the blades. For some bacteria, flagella can also be sheared by forcing the cells through a 22-gauge needle and subsequently homogenizing the cells in a Potter-Elvehjem Teflon-glass tissue grinder.
5. Sediment the deflagellated cells by centrifugation (6,000 to 10,000 × *g* for 15 min)
6. Centrifuge the cell-free supernatant at 100,000 x *g* for 90 min to pellet the filaments.
7. The sheared filaments can be further purified by equilibrium density gradient centrifugation with KBr (65). Resuspend the pellet of filaments from step 6 into 10 ml of buffer (HEPES or Tris) overnight at 4°C. Add 5 g of KBr (Sigma; ca. 99%), and dissolve the KBr with gentle stirring. Centrifuge at 210,000 × *g* for 24 h at 5°C (Beckman SW41Ti or SW50.1 rotor).

A single diffuse band of filaments usually occurs in the bottom third of the centrifuge tube. Remove the band from the top with a Pasteur pipette. KBr can be removed by dialysis against distilled water or buffer, and the filaments can be collected by ultracentrifugation at 100,000 × *g* for 90 min.

b. Procedure for isolation of intact flagella

The DePamphilis and Adler (25) method can be used successfully with *E. coli* and *B. subtilis*. Refer to the original paper for further comments on each step of the procedure, which may be of use for work with other bacteria. In all steps, care must be taken not to break off the basal bodies from the filaments.

1. Grow 1 to 3 liters of bacteria.
2. Harvest cells by centrifugation.
3. Gently resuspend cells in 50 ml of 20% (wt/wt) sucrose (final volume, ca. 70 ml).
4. To form protoplasts and spheroplasts, add in the following order 7 ml of 1 M Tris · HCl (pH 7.8), 2 ml of 0.25% lysozyme in 0.1 M Tris · HCl (pH 7.8) containing 0.2 M NaCl, and 6 ml of 0.1 M EDTA in 0.1 M Tris · HCl (pH 7.8). Mix the suspension after addition of each reagent, and add the EDTA within 30 s after the lysozyme. Incubate at 30°C for 1.5 h with gentle shaking. A good yield of intact flagella depends on formation of at least 80% protoplasts or spheroplasts.
5. Add 7 ml of 20% (wt/wt) Triton X-100. This should produce a clear, viscous lysate.
6. To free the flagellum preparation of DNA, add 0.8 ml of 1 M $MgCl_2$ followed by 1.5 mg of DNase I. Incubate the mixture at 30°C for 20 min. Do not add Mg^{2+} before the Triton X-100, since this prevents solubilization of the outer membrane of gram-negative bacteria.
7. Dilute the lysate immediately to 260 ml with cold 0.1 M Tris · HCl (pH 7.8) containing 0.5 mM EDTA (Tris-EDTA buffer). Rapidly pour 87 ml of cold saturated $(NH_4)_2SO_4$ (in Tris-EDTA buffer) into the diluted lysate to give 25% saturation. Stir the suspension slowly at 5°C for 2 h.
8. Centrifuge the suspension at 12,000 × *g* for 25 min in a fixed angle rotor. Collect the viscous white material floating at the meniscus and adhering to the side of the tube. Discard the fluid.
9. Rinse the tube with Tris-EDTA buffer, and combine with recovered material to give a final volume of 40 ml.
10. Add 1 ml of 20% Triton X-100.
11. Dialyze the suspension at 5°C against 1 liter of Tris-EDTA buffer. The initially turbid suspension should clear completely on dialysis.
12. Dilute the dialyzed preparation to 50 ml with Tris-EDTA buffer. To separate the intact flagella from other cellular debris, layer in duplicate tubes 25 ml of the diluted preparation on a gradient consisting of 2 ml of 20% (wt/wt) sucrose and 3 ml of 60% sucrose (in Tris-EDTA). After centrifugation for 1 h at 40,000 × *g* in a swinging-bucket rotor (Beckman SW28), remove and discard the liquid above the sucrose layers.
13. Add 5 ml of Tris-EDTA buffer to the sucrose layers. Mix gently. Remove sucrose by dialysis overnight against Tris-EDTA buffer. Dilute the suspension to 25 ml with buffer, and add 0.15 ml of 20% Triton X-100. Centrifuge at 4,000 × *g* for 10 min to remove unlysed cells and large aggregates. Retain the supernatant.
14. To further purify the intact flagella, use equilibrium density gradient centrifugation in CsCl. Dilute the supernatant material from step 13 to 27 ml with Tris-EDTA buffer. Add 12.1 g of CsCl (all at once), and dissolve rapidly. Centrifuge at 50,000 × *g* for 50 h (Beckman SW28 rotor). A band of flagella should be present near the center of the gradient. Collect this band, and remove the CsCl by dialysis against Tris-EDTA buffer. The appearance and number of bands in CsCl gradients and the purity and quality of intact flagella recovered are markedly influenced by the rate of addition of CsCl and the age of the cells at harvesting.

c. Isolation of intact flagella from archaeobacteria

Kalmokoff et al. (65) devised a procedure appropriate to *Methanococcus voltae*, whose cell envelope consists of only the plasma membrane and a protein surface array (section 4.4.3.4). This procedure is based on the temperature-induced phase separation of the nonionic detergent Triton X-114. At temperatures above 20°C, Triton X-114 separates into two phases: hydrophilic proteins separate into the upper aqueous phase, and hydrophobic integral membrane proteins separate into the lower detergent-rich phase. Since the envelopes of archaeobacteria (with the profile described above) contain few hydrophilic proteins, this technique allows a selective enrichment of the wall protein and flagella. For eubacteria it is still essential to prepare spheroplasts first, as with the DePamphilis and Adler technique (25). For osmotically fragile methanogens, cells can be lysed by dilution into distilled water. Crude flagellar preparations are purified as described by Kalmokoff et al. (65). The procedure is not suitable for organisms such as *Methanospirillum hungatei*, whose flagellar filaments are sensitive to Triton.

Procedure

1. Prepare spheroplasts (section 4.2), if possible.
2. Lyse spheroplasts or osmotically sensitive cells (from 1 liter of medium) in distilled water with gentle mixing. Add DNase (1.5 mg) with 1 mM $MgCl_2$ to reduce viscosity.
3. Centrifuge the lysate at 22,000 × *g* for 30 min at 4°C.
4. Resuspend envelopes in 20 ml of cold 10 mM Tris · HCl (pH 7.5), and treat with 1% (vol/vol) (final concentration) Triton X-114 for 30 min at 4°C with occasional mixing. For the corresponding step with eubacteria, add 5 mM EDTA with the detergent.
5. Incubate samples at 37°C to induce phase separation. Centrifuge at 300 × *g* for 3 min. Do not centrifuge at higher speeds, since the flagella will be peleted in the lower, detergent phase.
6. Remove the upper, aqueous phase (approximately 17 ml), and concentrate to 1 ml with an ultrafiltration apparatus (e.g., model 8010 PM30 filter; Amicon Corp.). Check the sample by electron microscopy for the presence of flagella.

4.4.1.2. Fimbriae (pili)

Fimbriae (also called pili) are proteinaceous, filamentous surface structures composed of identical subunits. They are narrower in diameter (less than 10 nm) than flagella (eubacterial flagella, 20 nm; methanogenic flagella, 11 nm) and have no insertion organelle or basal structure. They are often found on flagellated cells, in which case isolation and purification will involve separation of these two kinds of appendages. They are very stable protein assemblies and are firmly attached, making them more resistant to shear forces than are flagella.

a. Procedure for removal of fimbriae from cells

1. Grow 1 liter of cells in appropriate medium.
2. Harvest the cells by centrifugation.
3. Resuspend the cells in 50 ml of 10 mM Tris · HCl (pH 7.2).
4. Remove fimbriae by blending for 2 min at maximum speed in a rotating-blade blender (e.g., Waring, Sorvall Omni-Mixer).
5. Remove cells by differential centrifugation (4.3.1.).
6. Retain the supernatant fluid ("shear supernatant") which contains the fimbriae.

b. Procedures for purification of fimbriae

Procedure 1. Fimbriae can be separated from other contaminating materials in the shear supernatant by differential centrifugation. A detailed description of the purification of gonococcal fimbriae is given by Heckels and Virji (47). Brinton et al. (16) successfully precipitated the common fimbriae of a nonflagellated *E. coli* strain from the shear supernatant by addition of 0.1 M $MgCl_2$ (final concentration). The fimbriae are collected by centrifuging, resuspending in buffer, and reprecipitating twice with 0.1 M $MgCl_2$.

Procedure 2. The shear supernatant (containing 0.5 M NaCl) of sheared *P. aeruginosa* cells contains flagella and fimbriae, which are both precipitated by the addition of polyethylene glycol 6000 (101). After 18 h at 4°C, collect the precipitate by centrifugation and resuspend the pellet in 10% (wt/wt) $(NH_4)_2SO_4$ (pH 4). After 2 h at 4°C, the fimbriae form a precipitate while the flagella remain in suspension. Remove the flagella by repeating the $(NH_4)_2SO_4$ precipitation step. Dissolve the final pellet in water, dialyze to remove the $(NH_4)_2SO_4$, and subject the suspension to CsCl density gradient centrifugation.

Procedure 3. The flagella and fimbriae of an *E. coli* strain precipitate on addition of ammonium sulfate to give 20% saturation (68). Resuspend the pellet in buffer, and layer on top of a linear sucrose gradient (15 to 50% [wt/wt] sucrose). Remove contaminating flagella by incubating the fimbriae for 1 h at 37°C in 0.4% SDS. Separate the dissociated flagella from the intact fimbriae by gel filtration on Sepharose CL-4B by using a buffer containing 0.4% SDS and 0.1% EDTA. Only fimbriae elute in the void volume.

When appropriate, isolation is assisted by taking advantage of the buoyant density for fimbriae versus other structures (Table 2).

4.4.1.3. Spinae

Spinae are nonprosthecate, rigid, helical protein structures that extend outward from the cell surface of some eubacteria (29); they may coexist with flagella and require appropriate culture conditions for expression (31). They are readily detached from the cell surface by shear forces, but the base may remain on the cell. Spinae are not efficiently detached from the cell by homogenization in a Potter-Elvehjem homogenizer. Their large size and characteristic high density (Table 2) allow purification by differential and density gradient centrifugation. Easterbrook and Coombs (30) described the following procedure for isolation of spinae from a marine pseudomonad grown under conditions that repressed flagellum synthesis.

1. Harvest cells from 1 liter of medium, and wash by centrifugation with resuspension in 100 ml of 0.15 M NaCl, or an appropriate buffer, according to the origin of the culture.
2. Homogenize the cells in a rotating-blade (Waring type) blender for 1 to 2 min, and sediment sheared cells by centrifugation.
3. Centrifuge the supernatant at 65,000 × *g* for 20 min to pellet the spinae. Spinae detached during growth may be recovered from the culture supernatant by retention on membrane filters (pore size, 0.45 μm; Millipore Corp., Bedford, Mass.).

4. Resuspend the spinae in a small volume of 0.15 M NaCl or buffer, and layer on a CsCl gradient (density, 1.3 to 1.4). Centrifuge at 60,000 × *g* for 90 min. Collect the band, and remove the CsCl by dialysis against 0.15 M NaCl or buffer.

4.4.2. Walls and Envelopes

The term "cell wall" is used to define all structural components external to the plasma membrane and should not be used in reference to the murein component alone. Bacterial walls have proved to be sufficiently hardy to withstand cell disruption procedures (section 4.1.1). Historically, one of the first successful bacterial fractionation techniques was the isolation of the rather homogeneous cell walls from gram-positive bacteria (110). The gram-negative bacteria have a complex cell wall consisting of murein, periplasmic components, and outer membrane with or without surface embellishments. It is obvious that the cell wall of a gram-negative bacterium cannot be isolated as the physiologically active wall, because the periplasmic components, which form part of the structure, are released during mechanical breakage. In addition, it has proved difficult to remove all of the plasma membrane from the murein and outer membrane of many gram-negative bacteria, probably because of adhesion zones between the plasma membrane and the outer membrane. Therefore, this fraction as isolated from gram-negative bacteria is more accurately referred to as an envelope fraction. The cell wall of many archaeobacteria consists solely of a paracrystalline protein surface array, and simple mechanical or osmotic lysis techniques usually result in the preparation of an envelope fraction.

Although many wall components can be isolated from whole cells, it is often advantageous to start with an envelope or wall fraction to reduce contamination of the desired wall components with cytosol constituents.

a. Procedure for obtaining crude cell wall fraction

Best separations of envelope or wall fractions from the cell contents are achieved following mechanical or osmotic breakage (section 4.1) followed by differential centrifugation (section 4.3.1). Perform procedures at 0 to 4°C and as rapidly as possible to minimize enzymatic degradation of murein by autolysins, which is a definite problem with many *Bacillus* species. It may be necessary to heat the wall preparations at ca. 80°C after the first centrifugation or even to inactivate the enzymes by heating at 100°C for 5 to 10 min. RNase and DNase may be included during disruption of cells, since disruption via the French press does not shear nucleic acids as effectively as shearing with glass beads. Use an initial low-speed spin to remove unbroken cells, followed by centrifugation at 30,000 to 45,000 × *g* for 20 min to pellet the envelope or wall fraction. Inevitably, some larger envelope fragments will sediment with the intact cells, but the low-speed spin is necessary to ensure purity of the final fraction. The plasma membrane is disrupted by shear forces and usually reseals to form small vesicles which require much higher centrifugal forces (ca. 100,000 × *g*) to sediment. Thus, the crude wall or envelope fraction obtained at 30,000 to 45,000 × *g* does not usually contain much plasma membrane other than that present via adhesion zones to the gram-negative outer membrane. The choice of subsequent purification methods depends on the purpose of the preparation and is considered separately for the wall fraction of gram-positive and envelope fraction of gram-negative bacteria.

b. Procedure for isolation of gram-positive walls (Fig. 1)

1. Wash the crude cell walls with saline (0.9% NaCl) or buffers (50 mM Tris-HCl [pH 8] or 50 mM HEPES [pH 7.8]) to remove soluble proteins. If necessary, use 1 M NaCl to displace cytoplasmic contaminants (110) and subsequently wash the walls with distilled water to remove the salt. Avoid proteolytic enzymes to digest soluble proteins, because important wall proteins can be covalently bound to the peptidoglycan.
2. Suspend the washed walls in a minimum volume of cold water, and pour this into 4% SDS held in a boiling-water bath to give a concentration of 30 mg (dry wt)/ml. Stir occasionally for 15 min or until all pigmentation is lost.
3. Recover the walls by centrifugation (45,000 × *g* for 15 min at 20°C), and wash the wall pellet repeatedly by suspension and centrifugation, first with 0.9% NaCl and then with water at 20°C to remove traces of SDS. The purified walls of gram-positive bacteria are white and devoid of any cellular pigments. The walls can be stored in suspension at −20 to −70°C or lyophilized. Verify the purity of the product by electron microscopy.

c. Procedure for isolation of gram-negative envelopes (Fig. 2)

1. Wash the crude envelope fraction several times with saline (0.9% NaCl) or buffers (50 mM Tris-HCl [pH 8] or 50 mM HEPES [pH 7.8]). It is advisable to include 1 mM $MgCl_2$ and/or $CaCl_2$. Mg^{2+} serves to maintain the integrity of the plasma membrane, and Ca^{2+} maintains the integrity of walls and S-layers; use will vary with intention.
2. Treat the preparation with RNase (100 μg/ml) for 2 to 3 h at 37°C, and wash again. For many gram-negative bacteria this final preparation is suitably devoid of plasma membrane and cytosol contaminants, which should be verified by electron microscopy.

Selective solubilization of the plasma membrane by detergents provides a method for examining outer membrane proteins without degradation of the murein layer. The basis for selective solubilization is that the outer membrane of *E. coli* is insoluble in 2% Triton X-100 in the presence of Mg^{2+} (e.g., 10 mM) (25, 115) or in Sarkosyl (33), whereas the plasma membrane is soluble. The fraction obtained should be referred to as the "detergent-insoluble fraction" rather than the "wall" since, especially for other gram-negative bacteria, minor polypeptides, lipopolysaccharide (LPS), or lipids of the outer membrane may also be solubilized and thus a native outer membrane may not be obtained. However, quite the opposite effect is obtained with *Treponema pallidum* (102), whose outer membrane has unusual properties and is solubilized by Triton X-114 while the plasma membrane remains intact (22).

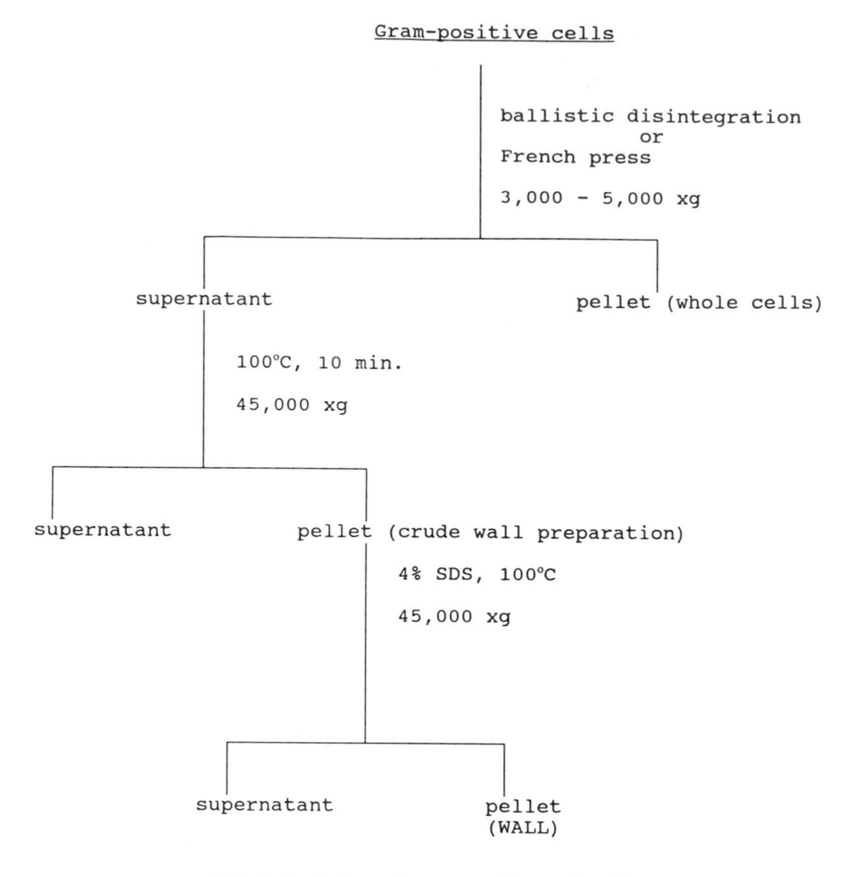

FIG. 1. Isolation of gram-positive cell walls.

d. Purification of murein-outer membrane complex

Purification of the murein-outer membrane complex from the crude envelope fraction by density gradient separation in sucrose is described by Schnaitman (114). This method separates inner (plasma) and outer membranes, but it must be noted that the murein is still associated with the outer membrane and has to be removed by lysozyme if purified outer membrane is required.

A total outer and inner membrane fraction can be obtained by lysis of EDTA-lysozyme spheroplasts (96). However, EDTA is known to solubilize some outer membrane components (especially in *P. aeruginosa* [43]) or to result in reorganization of the outer membrane structure (87), and thus a native outer membrane is not obtained. The French press method avoids the use of EDTA and is useable with large amounts of cells of a variety of species, for which spheroplasting methods may not be known.

The following technique for isolation of plasma and outer membranes depends on successful spheroplast formation and lysis, as described by Osborn and Munson (100). Subsequent separation by sucrose density gradient centrifugation (section 4.3.3) takes advantage of the fact that the outer membrane has a higher buoyant density than the plasma membrane (Table 2).

e. Procedure for isolation of membranes of enteric bacteria

1. Grow 1 to 2 liters of bacteria to mid-exponential phase.
2. Prepare spheroplasts (section 4.2).
3. Lyse the spheroplasts by dilution with water or brief sonication. (If the spheroplast preparation contains a significant number of intact rods, lysis by osmotic shock is advantageous since the unlysed cell can be removed by centrifugation at 1,200 × *g* for 15 min.)
4. Recover envelopes from the supernatant by centrifugation at approximately 100,000 × *g* for 2 h in a fixed-angle rotor.
5. Resuspend the pellets in 3 to 5 ml of 10 mM HEPES buffer (pH 7.4) with the aid of a glass tissue homogenizer. Avoid Mg^{2+} salts in the buffer, because this prevents subsequent separation of inner and outer membranes by isopycnic density gradient centrifugation.
6. Wash the envelope fraction twice, and resuspend in 0.5 to 2 ml of 10 mM HEPES buffer (pH 7.4).
7. Layer the envelopes onto a discontinuous sucrose gradient (section 4.3.5.1) to separate outer membranes from plasma membranes. For a large-scale preparation (from 1 to 2 liters of culture), use a Beckman SW28 rotor. Layer 6.3 ml each of 50, 45, 40, 35, and 30% sucrose over a cushion (1.5 ml) of 55% sucrose. Make all sucrose solutions in 10 mM HEPES (pH 7.4). As much as 5 mg of membrane protein (in 2 to 3 ml) can be added to one tube. Centrifuge at 110,000 × g (R_{max}) for 18 to 40 h. Equilibrium is attained at 30 to 36 h, but adequate separation is obtained by 18 h.

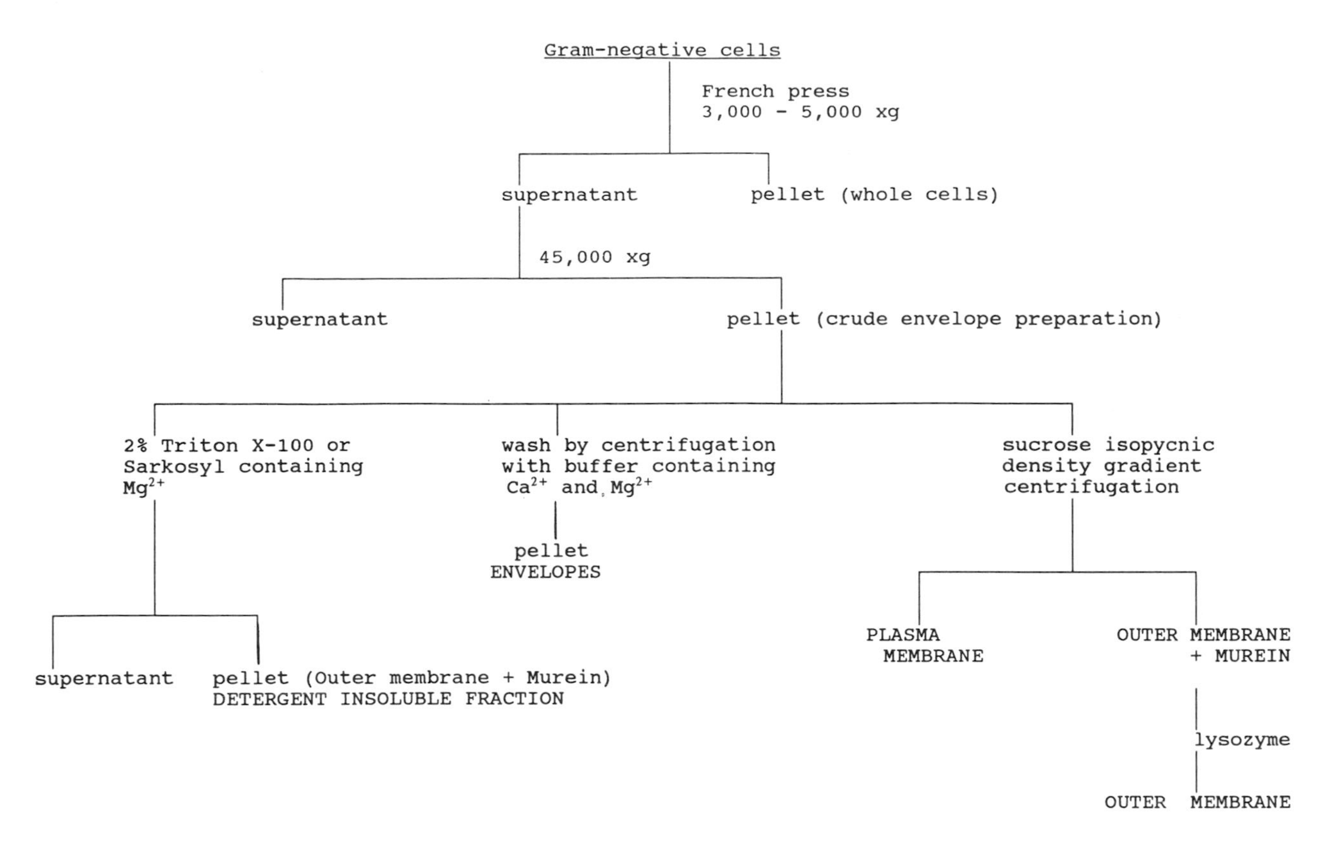

FIG. 2. Isolation of gram-negative envelopes and related fractions.

8. Collect fractions by piercing the bottom of the tube with an 18-gauge syringe needle and collecting counted numbers of drops in a series of test tubes for readings of A_{280}. Pool peak fractions, dilute approximately fourfold with HEPES buffer (to a final sucrose concentration of 10% or less), and centrifuge at 100,000 × *g* for 2 h. Resuspend the washed pellets in a small volume of distilled water, again with the aid of a glass tissue homogenizer.
9. The identification of peaks as outer or plasma membrane is based on well-characterized markers: e.g., buoyant density (Table 2), enzyme activities, outer membrane 3-deoxy-D-mannooctulosonic acid (KDO), plasma membrane cytochromes.

4.4.3. Wall Macromolecules

4.4.3.1. Murein and pseudomurein

Most procaryotic cells have a rigid layer (R-layer) in the cell wall that is covalently linked and resistant to dissociation by ionic detergents such as SDS. The R-layer of eubacteria consists of murein (peptidoglycan), whereas archaeobacteria either possess pseudomurein (modified murein) or have a completely different wall construction (67). For example, the archaeobacterial genera *Methanosarcina* and *Halococcus* possess complex anionic heteropolysaccharides. Many archaeobacteria have cell walls that consist solely of noncovalently-linked protein or glycoprotein subunits (S-layers) (section 4.4.3.4).

Murein can be prepared from eubacteria either with or without prior cell disruption. On a small scale, the intact murein sacculus can be isolated from both gram-positive and gram-negative bacteria based on its insolubility in boiling SDS solutions. A general method outlined by Hancock and Poxton (42) follows.

a. Procedure for isolation of murein sacculi without prior cell disruption

1. Suspend a wet paste of lyophilized bacteria in ice-cold water (2 to 4 mg [dry weight]/ml). Mix mechanically or with a hand homogenizer to obtain a homogeneous suspension, and then add it dropwise (immediately) to an equal volume of boiling 8% (wt/vol) SDS. Bring the mixture to boiling again as quickly as possible (to help in the inactivation of autolysins), and boil for 30 min with continuous stirring.
2. Cool the sample, and leave it overnight at room temperature. Collect the insoluble wall material by centrifugation at 30,000 × *g* for 15 min (for gram-positive bacteria) or 130,000 × *g* for 60 min (for gram-negative bacteria). To avoid precipitation of SDS, do not refrigerate the samples.
3. Reextract the insoluble residue twice more with boiling 4% SDS, until the murein sacculus appears free of other cellular constituents as judged by electron microscopy and chemical analysis.
4. Wash this murein sacculus preparation by resuspension and centrifugation, four times with water, twice with 2 M NaCl, and four more times with water. The

final residue may be frozen or lyophilized for storage.

A better separation of murein from the cytosol and membranes is often achieved after mechanical disruption of cells and extraction of the envelope fragments. This method was described in section 4.4.2. for gram-positive eubacteria. The same principles apply to gram-negative eubacteria, except that a larger starting culture volume is necessary, to compensate for the smaller proportion of murein in these cells.

As a final cautionary note, remember that in cultures of enteric bacteria and possibly other gram-negative bacteria allowed to enter the stationary phase, the outer membrane proteins may become cross-linked in a way that prevents their solubilization by boiling SDS. Care must be taken to use actively growing exponential-phase cultures.

4.4.3.2. Teichoic acids

Complex polysaccharides, including murein, make up the major portion of the cell walls of the gram-positive eubacteria. Often, accessory (secondary) polymers of glycerol or ribitol teichoic acids (wall teichoic acids), acidic teichuronic acid polysaccharides, or other polysaccharides are covalently linked to the murein strands. Lipoteichoic acids (membrane teichoic acids), also found in most gram-positive eubacteria, are anchored through the lipid moiety to the outer surface of the plasma membrane with the hydrophilic chains extending toward, and in some strains through, the cell wall (139). Isolation methods for these various accessory polymers have been reviewed critically (42a), and will not be addressed in detail here.

Teichoic acids may be released from the murein by acidic or basic hydrolysis of sensitive phosphodiester linkages between teichoic acid and the C-6 of muramic acid in the murein strand. Degradation of the accessory polymer can be a problem depending on the exact structure of the teichoic acid. If extensive damage occurs, use less harsh methods such as lysozyme or neuraminidase digestion of murein from a purified wall fraction (section 4.4.2). However, the enzymatic digestion may leave murein moieties close to the point of linkage to the teichoic acid polymer. Protocols for extraction with trichloroacetic acid, acidic buffers, NaOH, or *N,N'*-dimethylhydrazine and by enzymatic hydrolysis are given by Hancock and Poxton (42a).

4.4.3.3. Lipopolysaccharide

LPS is ubiquitous, with very few exceptions (e.g., members of the family *Deinococcaceae*), as a major surface component of gram-negative bacteria. However, although there are common structural features in almost all kinds of LPS, the structure varies so widely among different strains and species of bacteria that the choice of a suitable extraction method may be quite difficult. In addition, most individual species can produce more than one kind of LPS. A single strain of bacteria may produce multiple forms of LPS continuously during balanced growth, discontinuously as a result of genetic changes such as phase variation, or in response to physiological signals such as temperature, culture density, and nutrition. Some of the extraction procedures (35, 142) were developed before this heterogeneity was fully appreciated, and thus they may not yield a representative preparation. Because LPS is so frequently the subject of research and presents so many problems, an extensive discussion follows.

Most enteric bacteria and many *Pseudomonas* strains produce a large form of LPS termed the S (or smooth) form, which has three parts: (i) a lipid termed lipid A, which is a disaccharide of glucosamine containing both O-linked and N-linked fatty acids; (ii) a complex core oligosaccharide of about 10 sugars, which is attached at its reducing end to the lipid via the eight-carbon sugar, KDO; and (iii) a polysaccharide called the O antigen, which is attached to the nonreducing end of the core oligosaccharide. The O antigen consists of a three-, four-, or five-sugar monomer which is repeated about 15 to 20 times to form a long unbranched polymer. Since the O-antigen polymers are not all of the same length, the S form of LPS will often appear on gels as a ladder of discrete bands, each differing from the next by the size of one monomer. Strains which produce an S-form LPS also produce a smaller LPS species termed the R (or rough) form, which is very heterogeneous and contains only two parts, a lipid A and a varied population of core molecules that do not carry O antigens. In enteric bacteria, the S and R forms each represent about 50% of the total number of LPS molecules. The terms S and R refer to smooth and rough colony designations given to strains which did or did not produce an O antigen, respectively. This LPS distinction is somewhat less applicable since it was discovered that the strains which exhibit a smooth colony produce a mixture of S and R forms of LPS.

Many gram-negative bacteria (including genera such as *Haemophilus* and *Neisseria*) never produce an S form of LPS but instead produce molecules which consist only of lipid A and an antigenically complex, often heterogeneous oligosaccharide somewhat similar to the core present in R-form LPS from enteric bacteria. These molecules are sometimes termed lipooligosaccharides (LOS) to distinguish their properties from those of S-form LPS molecules which bear O antigens. These molecules are similar in extraction properties to the R form of LPS from enteric bacteria.

The choice of methods for the isolation of LPSs in gram-negative bacteria depends mostly on the presence or absence of O-antigenic side chains. The classical method for isolation of smooth LPS from enteric and other gram-negative bacteria was described by Westphal and Jann (142) and utilizes hot aqueous phenol. This method is not successful in extracting R-form LPS from whole cells. Galanos et al. (36) developed another method involving aqueous phenol, chloroform, and petroleum ether (PCP) for extraction of R-form LPS. Another procedure described by Darveau and Hancock (23) for the simultaneous extraction of both S- and R-form LPS has broader application. Although originally described for *P. aeruginosa*, it has been used successfully for other organisms. LPS isolated by any of the above procedures is sufficiently pure for subsequent chemical analyses and SDS-polyacrylamide gel electrophoresis (SDS-PAGE) (described below). LPS isolated by the phenol-water method preserves the reactivity of pseudomonad LPS with antibodies in Western immunoblots and is better in this regard than LPS isolated by the Darveau-Hancock method (23). Isolated LPS can be converted to a soluble uniform salt form (Ca^{2+}, Na^{+}, triethylammonium) by electrodialysis to the free-acid form (deionized) and neutralization with various bases (35).

a. Proteinase K digestion of bacterial lysates and analysis by SDS-PAGE

A breakthrough in the field of LPS analysis occurred when Hitchcock and Brown (50) devised the following simple procedure for chemotype analysis of LPS by SDS-PAGE in crude lysates of whole cells or outer membrane preparations.

1. Remove cells from an agar plate with a sterile swab, and resuspend them in 10 ml of phosphate-buffered saline (PBS; pH 7.2) to a turbidity of 200 Klett units (blue filter, Klett Summerson colorimeter). Since some bacteria produce a different LPS when grown on agar rather than in broth, and since some also produce different LPSs at different stages of growth, it is recommended to standardize by use of late-log-phase cells from a broth culture. Centrifuge cells from 10 ml of an overnight culture, and resuspend in PBS to an A_{525} of 0.5 to 0.6. The volume of culture used may vary according to the organism under investigation. Use a portion (1.5 ml) of the cell suspension for LPS extraction. An equal amount (1 to 10 mg) of lyophilized cells may also be used.
2. Sediment bacteria from 1.5 ml of suspension by centrifugation in a microcentrifuge for 3 min.
3. Resuspend the pellet in 50 μl of SDS-PAGE lysis buffer (2% SDS, 4% 2-mercaptoethanol, 10% glycerol, and 0.002% bromophenol blue in 1 M Tris-HCl buffer [pH 6.8]), and heat at 100°C for 10 min.
4. Add 25 μg of proteinase K (Boehringer Mannheim Biochemicals, Indianapolis, Ind.), and incubate at 60°C for 1 h.

If multiple samples are to be prepared, enough proteinase K solution can be prepared for all. However, it cannot be stored after it is dissolved in lysing buffer. The digested samples may be frozen at this point. Prior to running the gel, mix 5 μl of digested sample with 10 μl lysing buffer and load 3 to 6 μl of this dilution per 5- by 0.75-mm well (16- by 18-cm gels). Proteinase K is not stable during prolonged storage, and it should be replaced if a background of protein bands is visible on the gels.

A cleaner picture is obtained if proteinase K-digested samples (50 μl) are extracted with an equal volume of 90% phenol at 65°C for 15 min (vortex every 5 min) (91). Centrifuge samples in a microcentrifuge for 10 min. Transfer the aqueous phase to a new microcentrifuge tube, and extract once with 10 volumes of ethyl ether to remove traces of phenol. Aspirate the upper ether phase, and mix the lower phase 1:1 with sample buffer as above. Cleaner lanes are also obtained if the Hitchcock and Brown procedure is applied to envelope or outer membrane preparations (5).

b. Procedure for Tricine–SDS-PAGE

Successful electrophoresis and visualization of the LPS by silver staining of the aldehydes, resulting from periodate oxidation of LPS core sugars, requires three things. First, the LPS must contain enough lipid to give the resulting LPS-SDS complex a negative charge so that it will migrate during electrophoresis. Second, the LPS must be precipitated by the isopropanol-acetic acid-water solution used to fix the gels. Third, the LPS core must contain at least one pair of adjacent unsubstituted hydroxyl groups which can be cleaved by periodate oxidation to generate aldehydes that can react with silver. Most LPS molecules meet these criteria. The Tricine-SDS gel system is modified from that of Lesse et al. (84) and gives vastly increased resolution (especially of R-form LPS) over that obtained by the older glycine-SDS system used by Hitchcock and Brown (50).

1. The bottom buffer is 0.2 M Tris base (24.2 g/liter) adjusted to pH 8.9 with concentrated HCl. The top buffer is 0.1 M Tris base (12.1 g/liter), 0.1 M Tricine (17.9 g/liter), and 0.1% SDS. A 4× gel buffer stock solution consists of 4 M Tris base (48.5 g/100 ml, heat to dissolve), adjusted to pH 8.45 with concentrated HCl, and 0.4% SDS. For the best results, add dry SDS to the top buffer and gel buffer stock just before use.

2. The running gel (per 20 ml) consists of 9 ml of 40% acrylamide (final concentration, 18%), 3.6 ml of 2% bisacrylamide (final concentration, 0.36%), 5 ml of 4× gel buffer, 2.1 ml of glycerol, and 0.3 ml water. Predissolved acrylamide and bisacrylamide stock solutions can be obtained from AMRESCO (Solon, Ohio). Degas the gel solution under vacuum for 20 min, and add 4 μl of *N,N,N′,N′*-tetramethylethylenediamine (TEMED) and 40 μl of 10% ammonium persulfate. After casting the gel, overlay it with water-saturated butanol. When polymerization is complete, wash out the butanol with 1× gel buffer. The gel can be stored overnight with an overlay of 1X gel buffer. For convenience, a stock solution of 10% ammonium persulfate can be prepared in advance, dispensed into small single-use vials, and stored frozen for up to 6 months.

3. The stacking gel (per 10 ml, 0.74 M Tris [final concentration]) contains 1.125 ml of 40% acrylamide (final concentration, 4.5%), 0.4 ml of 2% bisacrylamide, 1.86 ml of gel buffer, and 6.62 ml of water. Degas the gel solution under vacuum for 10 min, and add 12 μl of 10% ammonium persulfate. Wells should be rinsed well with top buffer before samples are added.

4. For a 16-cm by 0.75-mm gel with 20.5-mm wells, perform electrophoresis at a constant current of 35 to 38 mA for about 4.5 h. The actual running time will depend on the size of the LPS and the separation desired. Cool the electrophoresis apparatus to room temperature with recirculating water.

c. Procedure for silver staining of LPS

Wear gloves for handling the gel and staining solution, both for safety and to prevent contamination. To prevent formation of a silver carbonate precipitate, it is important that the NaOH used be carbonate free and that the NH_4OH be fresh. It is convenient and inexpensive to purchase NaOH as a predissolved carbonate-free 1 N solution, so that bottles can be discarded after they have been opened a few times. Use deionized water, and agitate the solution throughout all procedures. Because of the frequent decanting of washes, it is convenient to fix, wash, and stain gels in small baking dishes on a rocker platform adjacent to a vacuum aspiration apparatus.

1. Fix the gel overnight in 200 ml of freshly prepared 25% (vol/vol) isopropanol in 7% (vol/vol) acetic acid. Decant the solution, and oxidize the gel for 5 min in a freshly prepared solution consisting of 150 ml of water, 1.05 g of periodic acid, and 4 ml of the 25% isopropanol–7% acetic acid fixing solution.
2. Wash the gel eight times, for 15 to 20 min each time, with 200 ml of water. This extensive washing is to remove small aldehydes resulting from periodate oxidation.
3. Stain for 10 min in a freshly prepared solution consisting of 28 ml of 0.1 N NaOH, 1.25 ml of concentrated (29.4%) NH_4OH, 5 ml of freshly prepared 20% silver nitrate, and 115 ml of water. If a dark-brown precipitate forms in the staining solution, add concentrated NH_4OH dropwise until it just disappears. After staining, wash the gel four times for 10 min each time, with 200 ml of water.
4. Develop in 250 ml of a solution which contains (per liter) 50 mg of citric acid and 0.5 ml of 37% formaldehyde. Developing usually requires 5 to 10 min and can be stopped when the desired band intensity is reached by soaking the gel for 60 min in 200 ml of water containing 10 ml of 7% acetic acid. If the gel is to be stored for a long period, repeat the fixation step to minimize band fading.
5. Soak the gel overnight in water containing 5% glycerol prior to drying on filter paper. Because of the high acrylamide content, dry the gel on a heated dryer with a high vacuum and a refrigerated or dry-ice trap to minimize cracking.

The Hitchcock-Brown (50) method has several advantages which make it particularly suitable both for characterizing the LPS present in a bacterial culture to aid in the choice of a suitable extraction method and for monitoring other LPS extraction and purification procedures. Because it does not involve extraction and is independent of the size or complexity of the LPS, it gives an accurate picture of the distribution of various forms of LPS in a given culture or preparation. The procedure is very useful for screening large numbers of clinical or environmental isolates. Depending on the branching pattern of the core carbohydrate, most LPS molecules yield either one or two aldehyde groups which are attached to the lipid and remain fixed to the gel after prolonged washing. Therefore the intensity of the silver stain reflects the number of molecules and not the number of sugar residues per molecule prior to cleavage. This means that the silver stain pattern provides both qualitative and semiquantitative information on the distribution of different size classes of LPS in a sample. With minor modifications in gel strength and running time, the Hitchcock-Brown procedure can be used to characterize the polymerization number and size distribution of ladders of S form LPS, and relative amounts of S- and R-form LPSs, the chemotype of LPS produced by various mutants, and the size and the properties of LOS. In addition to its visualization by silver staining, LPS prepared by the Hitchcock-Brown procedure and separated by SDS-PAGE is ideal for detection by Western immunoblotting, lectin binding, or autoradiography.

d. Procedure for phenol-water extraction for S-form LPS (142)

Dry redistilled phenol can be purchased (molecular biology grade; Boehringer Mannheim) and stored at 4°C for several months. This avoids the hazards associated with redistilling crystalline phenol and the instability of liquefied phenol. To prepare 90% phenol, melt by heating a 500-g bottle (cap loosened) of the dry phenol to approximately 70°C in a microwave oven (or water bath) and add 61 ml of water while hot. *CAUTION:* Do not attempt to weigh by removing crystals of phenol from the bottle. If lesser amounts are required, weigh 90 g of the liquefied phenol and add 11 ml of water.

1. Thoroughly resuspend lyophilized bacteria to a concentration of 5% (wt/vol) in distilled water. For large amounts of LPS for carbohydrate analysis, resuspend 20 g of cells in 350 ml of water. For smaller amounts of LPS for column chromatography, SDS-PAGE, and Western immunoblotting, 1 to 2 g of lyophilized cells (in 30 to 60 ml of distilled water) is sufficient. However, comparison of LPS extraction with phenol-water from an equivalent starting dry weight of cells for spheroplast lysates, whole cells, and cells broken by grinding revealed that the spheroplast lysates gave the highest yields (60). As an alternative to using lyophilized cells, sensitive strains may be treated with EDTA-lysozyme and the extract may be digested with RNase and DNase (plus 20 mM $MgCl_2$) and then with trypsin. Purified LPS can also be extracted from RNase-treated envelopes (section 4.4.2) preextracted with chloroform-methanol to remove lipid (143) (use 50 ml of water/g of lipid-depleted lyophilized walls). This LPS preparation is usually free of contaminating nucleic acids.
2. Heat the aqueous suspension at 70°C in a water bath for 10 min.
3. Heat an equal volume of 90% (vol/vol or wt/vol) phenol to 70°C, and add it to the bacterial suspension. A glass Erlenmeyer flask is a suitable vessel for the phenol mixture.
4. Stir vigorously with a mechanical stirrer in a water bath at 70°C for 15 min.
5. Transfer the mixture to 50-ml centrifuge tubes (polycarbonate tubes must *not* be used), and cool on ice to about 10°C.
6. Centrifuge the emulsion at 10,000 × *g* for 30 min. Three layers are formed: an upper aqueous phase (containing the LPS), a lower phenol layer, and an insoluble residue (containing murein) at the interface of the two layers. Note that exceptions exist (e.g., *Actinobacillus* [*Haemophilus*] *pleuropneumoniae* serotype 4) in which significant amounts of LPS are soluble in the phenol phase (2). This possibility must be excluded before the phenol phase is discarded.
7. Remove the upper aqueous layer with a Pasteur pipette, and retain it.
8. Extract the phenol phase and any insoluble residue again at 70°C with a volume of water equivalent to that removed in the aqueous phase. Combine the aqueous layers.
9. Dialyze the aqueous extract against distilled water for 2 to 3 days to remove the phenol.

10. Lyophilize and dissolve in 0.9% saline. Avoid rotary evaporation when unbuffered (acidic) phenol is present, since structural damage such as loss of *O*-acetyl groups may occur. Use saline rather than distilled water to dissolve the LPS as an aid in protein removal.
11. Centrifuge at 27,000 × *g* for 30 min to remove any insoluble residue.
12. If the phenol extraction was performed on whole bacterial cells, about half of the organic material in the crude extract will be RNA. This contamination can be avoided by extraction of cleaned walls (envelopes). LPS can be separated from the RNA by ultracentrifugation of the extract at 100,000 × *g* for 10 h (shorter times may result in partial recovery of LPS). The LPS should sediment to form a clear, gelatinous pellet. Experience has shown that phenol-extracted LPS of some bacteria does not sediment efficiently by ultracentrifugation. If the pellet has an opaque base, it is not free of contaminating material and must be washed with water by ultracentrifugation. Alternatively, before ultracentrifugation treat the dialyzed suspension with RNase and DNase (1 μg/ml) and $MgCl_2$ (final concentration, 10 mM) at 37°C for 1 h. Resuspend the final pellets in a small volume of distilled water, and lyophilize.

If a low yield is obtained, LPS should be extracted from envelopes as mentioned above. The advantage of this approach is that the RNA has been digested prior to phenol extraction and the dialyzed aqueous extract can be lyophilized directly.

e. Procedure for PCP extraction for R-form LPS

The procedure described below (5) is modified from that of Galanos et al. (36) and gives a good yield of LPS from R-form strains such as *E. coli* K-12 and from a variety of mutants that produce different chemotypes of LPS including deep-rough and heptose-deficient strains. The physical properties of purified R-form LPS are very much determined by the size of the carbohydrate core. For example, lyophilized LPS from wild-type *E. coli* K-12 which contains a complete core is a fluffy powder that is readily dispersed in water, whereas a similar preparation of LPS from a deep rough *rfaG* mutant which lacks the hexose outer core sugars and has only the inner heptose-KDO region is a dense waxy solid that can be dispersed in water only by heating and vigorous homogenization.

The efficiency of extraction of R-form LPS by PCP and the purity of the final product are determined by the way the bacteria are treated prior to extraction. Unlike other extraction procedures, PCP extraction gives a poor yield with isolated outer membrane preparations and is most effective if the bacteria are intact. It is important that the bacteria are harvested in log phase, free of broken or damaged cells, completely dry, and free of extraneous salt from the medium and that they are in the form of a loose, granular powder which is easily dispersed in the extraction mixture. The following procedure is designed for isolation of LPS from two 6-liter batches of *E. coli* K-12 in late logarithmic growth phase and yields about 120 mg of purified LPS or about 3 to 5% of the weight of the dry cells. It can be easily scaled down.

1. After harvesting, process cultures promptly through the drying procedure without overnight storage or freezing of cell pellets. Chill a 6-liter culture, and harvest the cells by centrifugation. Resuspend the pellets with a Potter homogenizer to a total volume of 150 ml in cold deionized water. Distribute the suspension into six 40-ml centrifuge tubes, and pellet cells by centrifugation in a Sorvall SS-34 rotor for 10 min at 12,000 × *g*. Suspend the cells as a thick paste in a minimum amount of cold water (no more than 1 ml per tube plus 1 to 2 ml for rinsing the tubes) by using a test tube as a homogenizer.
2. Transfer the cell paste to the bowl of a blender or Sorvall Omni-mixer, add 90 ml of cold methanol, and homogenize for 30 s at maximum speed. The suspension should be smooth and free of clumps. Transfer to a Corex bottle, and centrifuge for 10 min in a Sorvall GSA rotor at 8,000 × *g*. Discard the supernatant, and resuspend the pellet in 90 ml of cold acetone; homogenize with a blender as above, and centrifuge as above. Repeat the acetone washing step once more. Drain off all of the acetone, cover the mouth of the Corex bottle with a sheet of tissue or lens paper secured with a rubber band, place in a large lyophilizer jar, and dry overnight on a lyophilizer. The pellet will "pop" during drying, and the resulting chunks of dried bacteria can easily be crushed to a fine powder with a spatula. The dried material can be stored in a vacuum desiccator at room temperature until PCP extraction.
3. *CAUTION:* Prepare PCP in a hood, and use gloves and extreme care for all steps. PCP (300 ml) is prepared by combining with vigorous stirring 40 ml of liquified phenol (see above for preparation), 100 ml of chloroform, and 160 ml of petroleum ether (ACS grade; boiling range, 37 to 52°C). This generally forms a hazy emulsion. Add about 1 ml of melted dry phenol to break the emulsion, which will result in a few small drops of a separate phenol-rich phase which tends to adhere to the walls of the container. This can be removed by pouring the PCP through a piece of fluted Whatman filter paper in a glass funnel. The filtrate is a clear and stable single phase and can be stored for up to 1 month in a stoppered container at room temperature.
4. To extract the bacteria, combine the dried cells from two 6-liter batches (about 4 g [dry weight] total). Place in the bowl of a 100-ml Sorvall Omni-mixer or other microblender (a beaker and a rotary probe homogenizer of the Braun type can also be used), and add 50 ml of PCP. Homogenize at maximum speed for about 30 s. The purpose is to suspend the bacteria completely without breaking the cells, and the suspension should have a smooth consistency and resemble a milk shake. Transfer to a Corex bottle, and centrifuge in the GSA rotor for 5 min at 9,200 × *g*. Decant, and save the supernatant, which contains the bulk of the LPS. To increase the yield, remove the very rubbery pellet, break it into pieces with a spatula, and homogenize it as above with a second 50-ml portion of PCP. Centrifuge as above, combine the supernatants, and filter through Whatman no. 1 paper in a small Buchner funnel.
5. It is necessary to concentrate the sample. A rotary evaporator is not recommended, since the solution may foam strongly. Transfer portions of the sample

to a 100-ml beaker, place on a hot plate or heating block which is just hot to the touch (about 50°C), and evaporate under a stream of N_2 gas until the entire sample has been concentrated to about 15 ml. Small flakes or a skin may form on the sides of the beaker. Allow to cool, and add a small amount of water (no more than 2 ml) dropwise while stirring with a Teflon-coated stirring rod until no more waxy precipitate forms. Transfer to a Corex tube (some precipitate will remain on the sides of the beaker), and centrifuge at 12,000 × *g* for 10 min in a Sorvall SS-34 rotor. Depending on the type of LPS and the extent to which the chloroform has been removed, it will either form a firm white pellet or float on top as a waxy cake. Remove the clear phenol liquid (aspirate with a hypodermic syringe with a long needle if necessary), and test to be sure that precipitation of the LPS is complete. Precipitation is complete if no more precipitate forms when a few drops of water or several volumes of methanol are added. Add about 10 ml of methanol to the pellet or waxy cake in the Corex tube, and use an additional 10 ml to wash any precipitate from the sides of the precipitation beaker into the tube. Mix the methanol suspension in the Corex tube vigorously, and centrifuge at 12,000 x *g* for 15 min. Carefully aspirate the methanol from the fragile pellet, and wash the precipitate twice more by suspending it in 20 ml of methanol with a Teflon-coated stirring rod and centrifuging.

6. Drain off the methanol, and cap the tube with a piece of tissue or lens paper secured with a rubber band. Evaporate the solvent for about 2 h on a lyophilizer, and suspend the dry pellet as completely as possible in 24 ml of water containing 0.1 mM $MgCl_2$ by using a Teflon Potter homogenizer. Depending upon the nature of the LPS, it may appear as an opalescent waxy suspension or as a clear, very viscous solution. Centrifuge in an ultracentrifuge at 200,000 × *g* for 4 h. The pellets are sometimes transparent and invisible until the supernatant is poured off. These are suspended by thorough homogenization in about 25 ml of water, frozen, and lyophilized. It is important that the samples do not thaw during lyophilization and that they are completely dry before being removed from the lyophilizer.

With some samples, the LPS cannot be removed from an aqueous solution by ultracentrifugation. If this occurs, the LPS can be recovered from the aqueous solution by precipitation with 5 volumes of ethanol or acetone. This is somewhat less satisfactory, since it also precipitates any nucleic acids or proteins in the preparation.

To test for protein and nucleic acid content, dissolve the LPS at a concentration of 1 mg/ml in 0.1% SDS solution. It may be necessary to boil the sample to dissolve it in the SDS. Protein can be determined by a standard Lowry assay (Chapter 22.7.6), and nucleic acid can be determined by measuring the A_{260} and A_{280} versus a blank of SDS solution.

f. Darveau-Hancock (23) procedure for S- and R-form LPS

Day 1

1. Resuspend 500 mg of lyophilized bacterial cells in 15 ml of 10 mM Tris · HCl buffer (pH 8.0) containing 2 mM $MgCl_2$, 100 μg of pancreatic DNase I per ml (add 1.5 mg directly), and 25 μg of pancreatic RNase per ml (add 750 μl of 500 μg/ml stock). In this and all subsequent steps, "Tris · HCl buffer" refers to 10 mM Tris · HCl buffer (pH 8.0).
2. Pass this suspension through a French press (section 4.1.1) at 16,000 lb/in^2. Sonicate for two 30-s intervals at maximum output. Add an additional 100 μg of DNase per ml and 25 μg of RNase per ml. Incubate cell lysate at 37°C for 2 h with gentle shaking.
3. Add the following to the cell lysate: 5 ml of 0.5 M tetrasodium EDTA dissolved in Tris · HCl buffer, 2.5 ml of 20% (wt/vol) SDS in Tris · HCl buffer, and 2.5 ml of Tris · HCl buffer. Adjust the pH to 9.5. Vortex carefully, and incubate the lysate for 1 h at 37°C with gentle shaking. This incubation period is necessary for bacteria whose outer membrane may not dissociate as readily as that of *P. aeruginosa.*
4. Ultracentrifuge the suspension at 50,000 × *g* for 30 min at 15°C to sediment intact murein (23,500 rpm; Beckman type 40 rotor).
5. Remove the supernatant, and add pronase (protease from *Streptomyces griseus* [Sigma]) at 5 mg/25 ml to give a final concentration of 200 μg/ml. Incubate overnight at 37°C with gentle shaking.

Day 2

6. If a precipitate forms after overnight incubation, centrifuge the mixture at 120 × *g* for 10 min. Discard the precipitate.
7. Precipitate LPS by adding 2 volumes of 0.375 M $MgCl_2$ in 95% ethanol and cooling to 0°C in a −20°C freezer or an ethanol–dry-ice bath. Do not allow the temperature to go below −5°C. Centrifuge the mixture at 12,000 × *g* for 15 min at 0°C. (Precool the rotor and centrifuge tubes.)
8. Resuspend the pellet in 17.5 ml of Tris · HCl buffer. Add 2.5 ml of 20% SDS and 5 ml of 0.5 M tetrasodium EDTA (both in 10 mM Tris · HCl). Sonicate for two 30-s intervals as in step 2.
9. Adjust the pH to 8 by dropwise addition of 2 N HCl.
10. Incubate the sample at 85°C for 30 min in a water bath, and then cool the sample to room temperature.
11. Adjust the pH to 9.5 with 2 N NaOH. Add pronase to 25 μg/ml (625 μg/25 ml), and incubate overnight at 37°C with gentle shaking.

Day 3

12. Precipitate LPS as in step 7 with 2 volumes of 0.375 M $MgCl_2$ in ethanol. Centrifuge at 12,000 × *g* for 15 min at 0°C. If the supernatant remains turbid, add a further 1 volume of 0.375 M $MgCl_2$ in 95% ethanol and repeat the precipitation and centrifugation steps.
13. Resuspend the pellet in 15 ml of Tris · HCl buffer, sonicate as described above, and centrifuge at 1,000 rpm (150 × *g*) for 5 min in a clinical centrifuge. For some bacteria, this pellet may contain a large amount of LPS. Resuspend the pellet in Tris · HCl buffer, dialyze against distilled water, and lyophilize.
14. Make up the supernatant from the low-speed centrifugation in step 13 to 25 ml with Tris · HCl buffer, and add $MgCl_2$ to 25 mM (125 mg/25 ml). Ultra-

centrifuge at 200,000 × *g* for 4 h at 15°C (50,000 rpm; Beckman type 70.1 Ti rotor). Resuspend the pellet in distilled water, and lyophilize.

4.4.3.4. S-layer proteins

Many bacteria in nature possess paracrystalline arrays of protein, forming surface layers (S-layers) on their cell walls (74, 120). Electron microscopy has shown that S-layers have hexagonal, tetragonal, or oblique symmetry. S-layers are widely distributed in nearly every major taxonomic group of both eubacteria and archaeobacteria. A notable exception is the absence of S-layers in the members of *Enterobacteriaceae*. S-layers, when present, are usually the outermost component of the cell wall.

Most S-layers are composed of a single homogeneous protein or glycoprotein species. The energy and information for assembly are contained within the individual protein monomers, and thus some S-layer proteins will self-assemble into the paracrystalline array. S-layers are noncovalently associated with the underlying cell wall (eubacteria) or plasma membrane (most archaeobacteria) via a combination of ionic and hydrogen bonds or, in some cases, hydrophobic interactions. The same forces also maintain the integrity of the S-layer itself. The stability and conformation of the protein are often dependent on divalent cations (most often Ca^{2+} and occasionally Mg^{2+}), whereas the ability to integrate with the wall depends on ionic strength. A valuable feature of S-layers with regard to their isolation is that they usually remain associated with the underlying envelope components after mechanical or osmotic disruption of cells. Most techniques for the isolation and purification of S-layers thus involve the initial disruption of cells and subsequent differential centrifugation to separate cell envelope fragments. S-layers cover the entire cell surface, and their presence is not dependent on the growth stage. Cells should therefore be harvested at maximum cell density.

Reagents that are effective in removing S-layers from the cell surface often disrupt the subunit interactions of the array. Therefore, it should be specified whether the intact paracrystalline S-layer or the component polypeptide is to be isolated. These are considered separately below.

a. Isolation of intact S-layers

The isolation of intact S-layers is most successful with gram-positive bacteria and archaeobacteria. In gram-negative bacteria, the close association with the outer membrane makes this isolation nearly impossible (76, 95). Most S-layers are resistant to dissociation by the nonionic detergent Triton X-100, but the plasma membrane dissolves during Triton X-100 treatment of cell walls of gram-positive bacteria (section 4.4.2). The contaminating plasma membrane fragments are then removed by washing. The wall remaining may be treated with lysozyme to remove the peptidoglycan underlying the S-layer. For the archaeobacteria that have an S-layer as the sole wall component, with notable exceptions (e.g., *Methanoculleus marisnigri*), treatment of envelopes with 0.5% Triton X-100 at 65°C for 30 min solubilizes the plasma membrane (75). The crude, intact S-layer is collected by centrifugation (72).

Procedure (10)

1. Grow 3 liters of cells in complex medium to the mid-exponential phase.
2. Prepare cell walls as described for gram-positive bacteria in section 4.4.2.
3. Treat the walls with 100 ml of 1% (vol/vol) Triton X-100 for 30 min at room temperature.
4. Wash the walls by centrifugation three times with 250 ml of 0.05 M sodium phosphate buffer containing 1 mM $MgCl_2$.
5. Resuspend the final pellet in 250 ml of buffer containing 950 U of egg white lysozyme per ml, and incubate at room temperature for 6 h.
6. Wash the residual intact S-layer with buffer by centrifugation at 45,000 × *g* for 30 min, and then dialyze against distilled water.
7. The final S-layer preparation can be stored frozen at −20°C as a suspension, or lyophilized.

Some S-layers are remarkably stable in the presence of chemical perturbants. The intact hexagonal S-layer of *Deinococcus radiodurans* can be isolated following treatment of cells or outer membrane vesicles with 2% SDS at 20 or 60°C for 2 to 12 h (131), which dissolves the outer membrane away from a stable protein array. The inner perforate S-layer of *Lampropedia hyalina* is resistant to SDS at 24°C (6). The structured sheath of *Methanospirillum hungatei* is very resistant to heat or chemical solubilization (12). The S-layer of the extremely thermophilic archaeobacterium *Thermoproteus tenax* is so resistant to chemical denaturants (72) that most of its constituent subunits appear to be covalently linked.

b. Isolation of S-layer proteins

Methods for the isolation of S-layer proteins have been described by Koval and Murray (76) and Messner and Sleytr (95). No single isolation procedure will apply to all organisms, because S-layers vary considerably in mechanical stability and ease of dispersion into subunits. The various procedures that have been used for the specific extraction of S-layer proteins are described below. Detergents can be used to solubilize S-layer proteins, but they usually result in nonspecific extraction of other wall-associated proteins.

General procedure for extraction of S-layer proteins with guanidine hydrochloride and urea

The chaotropic agents guanidine hydrochloride and urea have been used successfully with both eubacteria and archaeobacteria. Most S-layer proteins are readily solubilized from the underlying wall components by moderate concentrations of guanidine hydrochloride (1 to 2 M) or urea (4 to 6 M). The effective concentration must be determined for each individual S-layer by using SDS-PAGE and also electron microscopy to monitor by negative staining the removal of wall and the presence of subunits in extracts (Chapter 3.2.2). The choice of chaotropic agent will require a trial of procedures, e.g., the S-layer proteins of *Aquaspirillum serpens* strains are more susceptible to proteolysis during extraction with guanidine hydrochloride than during extraction with urea (76). However, the use of urea is not compatible with some subsequent biochemical analyses of the polypeptide, such as N-terminal amino acid sequencing.

1. Grow 1 to 3 liters of cells in the appropriate medium to mid-exponential phase.
2. Prepare cell envelopes (section 4.4.2.).
3. Treat the cell envelopes with 1% (vol/vol) Triton X-100 to remove contaminating plasma membrane. This is an optional step but reduces the total number of proteins monitored by SDS-PAGE.
4. Wash the walls by centrifugation with 10 mM HEPES buffer (pH 7.4).
5. Resuspend the walls in the appropriate concentration of guanidine hydrochloride or urea, and incubate at 37°C for 2 h. If necessary, protease inhibitors (e.g., 1 mM phenylmethylsulfonyl fluoride) may be included at this and subsequent stages (9).
6. Recover the envelopes by centrifugation, and reextract with the chaotropic agent.
7. Combine the extracts (i.e., supernatants), and dialyze at 4°C against several changes of distilled water or the appropriate cation.
8. Concentrate the solubilized protein in a lyophilizer or an ultrafiltration apparatus with a PM 10 filter (Amicon Corp.). Store at −20°C as a concentrated solution or in a lyophilized state.
9. In the event that fragments of the S-layer form assembly products on removal of the chaotropic agent, a precipitate will appear in the dialysis bag. Collect the precipitate by centrifugation at 6,000 × *g* for 15 min, and clean from soluble contaminants by washing and centrifugation. If a specific cation (e.g., Mg^{2+} or Ca^{2+}) is required for assembly and not included, amorphous aggregates will form on removal of the chaotropic agent.

Procedure for extraction of S-layer proteins with metal-chelating agents

Because many S-layers, particularly those from gram-negative eubacteria, are dependent on cations (Ca^{2+} or Mg^{2+}) for attachment and assembly, it is important to provide for the necessary cations during growth and for the preparation and storage of cell envelopes. For some species, extraction of cell envelopes with 10 mM EDTA or ethylene glycol-bis(β-aminoethyl ether)-*N*,*N*,*N′*,*N′*-tetraacetic acid (EGTA) will solubilize the S-layer protein. Follow the procedure above, substituting metal-chelating agents for the chaotropic agents in steps 5 and 6. For S-layer proteins that strongly bind cations, dialysis against the chelating agent may be necessary (76). Cation-dependent S-layer proteins from gram-negative bacteria are usually not capable of self-assembly in the absence of a suitable template (e.g., outer membrane vesicles or S-layer$^-$ cell surfaces).

Other approaches

The synthesis of S-layer proteins is usually closely regulated and not overproduced, but there are two examples to date in which S-layer proteins have been detected in the supernatant of early stationary-phase cultures. *Bacillus brevis* 47 secretes large amounts of its two S-layer proteins (147). *Campylobacter fetus* also secretes its S-layer protein (antigen a) into the culture medium in a soluble form, along with only two or three additional antigens. This S-layer protein has also been isolated by treatment of cells with 0.2 M glycine hydrochloride at pH 2.2 (26). The S-layer protein of *Azotobacter vinelandii* is unusual in that it is removed from the cell surface by distilled-water washes (14). The S-layer of *A. vinelandii* is stabilized by Ca^{2+} and Mg^{2+}, but the stability is not as great as for other S-layers such as those of *Aquaspirillum* spp., which require the use of a metal-chelating agent for dissociation.

4.4.3.5. Exopolysaccharides

Acidic exopolysaccharides cover the surface of many gram-negative and gram-positive bacteria. They may form a capsule composed of high-molecular-weight polysaccharides attached to the cell surface and seen by microscopy (Chapter 2.3.10), or they may produce slime either loosely attached to the cell surface or released to the culture fluid. Capsules are sometimes referred to as **glycocalyx, extracellular polymeric substance,** or **exopolymers.** Growth conditions including carbon-to-nitrogen ratio, phase of growth, temperature, and growth in laboratory media can markedly affect the production of exopolysaccharide. Properties and structures of exopolysaccharides were reviewed in detail recently (54, 58).

Various methods are described to isolate the high- and low-molecular-weight exopolysaccharides (42b). Bacteria may be grown and harvested from solid media in cases in which nondialyzable medium ingredients contaminate the final preparation. The following method of Altman et al. (1) produces pure exopolysaccharides from *A. pleuropneumoniae* strains and is generally applicable for the isolation of capsules or slime. During isolation of exopolysaccharides, it is important to ensure that extracts are clarified by centrifugation. Since these polymers are very water soluble, haziness indicates impurities.

Procedure

1. Harvest bacteria, usually in late exponential to stationary phase, as a centrifuged pellet. Attempt to recover exopolysaccharides from the culture medium supernatant, unless their absence is established. Dialyze this supernatant against water, and concentrate by lyophilizing and dissolving in 0.9% saline. Recover exopolysaccharide by precipitating with 5 volumes of 95% ethanol. Dissolve the precipitate in distilled water.
2. Wash the bacterial pellet from step 1 with 0.9% saline (the supernatant may be combined with the medium supernatant and processed in step 1), and extract with "phenol-water" as described for LPS (section 4.4.3.3, steps 1 to 12). *CAUTION:* In rare cases, exopolysaccharides may be in the phenol phase (13).
3. Digest exopolysaccharide preparation from steps 1 and 2 (ultracentrifugation supernatant plus 20 mM $MgCl_2$) with RNase and DNase followed by trypsin.
4. Dialyze against water, and recover by precipitating with 5 volumes of 95% ethanol.
5. Precipitate with 1% cetyltrimethylammonium bromide. Store overnight at 4°C, and recover by centrifugation. Dissolve in 10% NaCl. Precipitate with ethanol, dialyze extensively against water, and lyophilize.
6. Obtain pure exopolysaccharide as the void-volume fraction following gel filtration on a column of Sephadex G-50 or as a homogeneous peak emerging in the salt gradient of a DEAE-Sephacel ion-exchange column (Pharmacia).

4.4.4. Periplasm

All sorts of bacteria have a gelled region (periplasm or periplasmic space) on the external surface of the plasma membrane and filling the space between the plasma membrane and wall. In the gram-negative bacteria (51) the periplasmic components appear to fill any space between the plasma and outer membranes, so that the murein strands run through it. Specific proteins in the periplasm account for 10 to 15% of the cell protein (49, 51), and a class of membrane-derived oligosaccharides (116) is present also. Periplasmic contents are released to the medium when the outer membrane of gram-negative bacteria is damaged or removed by the various means described above.

4.4.4.1. Chloroform release of periplasm

Chloroform release of periplasmic proteins (3) maintains activity of the periplasmic amino acid-binding proteins and appears to be little influenced by strain variation, thus providing a rapid procedure with which to compare these proteins among various strains. The mechanism of the chloroform effect is unknown, but it seems not to be the expected result of preferential loss of the outer membrane.

4.4.4.2. Osmotic release of periplasm

Periplasmic proteins also may be released by the osmotic shock procedure described by Nossal and Heppel for exponential-phase cells of *E. coli* (99). This method can result in the necessity of centrifuging large volumes during scale-up and recovering periplasmic contents from large volumes of shock fluid. The procedure is as follows.

1. Harvest cells, and wash twice with about 40 volumes of cold 10 mM Tris · HCl (pH 7.1) containing 30 mM NaCl.
2. Resuspend 1 g (wet weight) of cells in 40 ml of 33 mM Tris · HCl (pH 7.1) at 24°C.
3. In stage 1, dilute the cells, with rapid stirring, with 40 ml of 40% sucrose in 33 mM Tris · HCl (pH 7.1). Then add sufficient 0.1 M disodium EDTA (pH 7.1) to give 0.1 mM EDTA.
4. Place the suspension on a rotary shaker at 180 rpm for 10 min at 24°C.
5. Centrifuge the cells for 10 min at 13,000 × *g* in a refrigerated centrifuge.
6. In stage 2, rapidly disperse the well-drained pellet in 80 ml of ice-cold 0.5 mM $MgCl_2$ solution (water has also been used) by using a rubber spatula. Gently stir the suspension in an ice bath for 10 min, and centrifuge.
7. Carefully remove the supernatant containing the osmotic shock proteins.

4.4.4.3. Calcium release of periplasm

High concentrations of Ca^{2+} at low temperature (7) provide a third method to render the outer membrane of gram-negative eubacteria permeable to macromolecules and so to release periplasmic proteins. Increased permeability of the outer membrane results in enhanced DNA uptake in transformation assays and accounts for the ability to reconstitute active transport by adding osmotic shock proteins to a binding protein-deficient mutant of *E. coli* (7). The procedure is as follows.

1. Harvest about 2 × 10^9 cells in the exponential growth phase.
2. Wash the cells at room temperature with 5-ml portions of 50 mM Tris · HCl (pH 7.2) followed by 1 ml of 100 mM potassium phosphate buffer (pH 7.2).
3. Wash the cells at 0°C with 1 ml of 50 mM Tris · HCl (pH 7.2) containing 300 mM $CaCl_2$.
4. Resuspend the cells in 50 μl of the above Tris-Ca^{2+} buffer for reconstitution and transformation assays.

4.4.5. Plasma Membranes

The plasma (cytoplasmic) membrane was first isolated to obtain compositional information on its lipid (ca. 25 to 35%) and protein components and to study electron transport, but membrane vesicles were later shown to have advantages over bacterial cells for studies of active transport (63).

In general, the murein portion of the cell wall is first removed to form protoplasts or spheroplasts (section 4.2), which are then lysed to provide either the plasma membrane or the plasma membrane plus outer membrane. When this is done osmotically, vesicles are produced with membranes oriented primarily as in the cell. When lysis is done by French pressure cell treatment (section 4.1.1.1), a high proportion of inside-out vesicles are generated (63, 71). Orientation can be ascertained by electron-microscopic observations of freeze-fractured preparations (Chapter 3.2.6.2). DNase and RNase are added to remove DNA and RNA, which tend to adhere to the membranes.

Osmotic lysis may not be the method of choice when special precautions are needed to preserve membrane structure and activity. For example, in an extreme halophile, *Halobacterium halobium*, functional membrane vesicles were isolated following brief sonication of the cells suspended in 4 M NaCl–50 mM Tris · HCl (pH 7) (88). Also, to prevent membrane disaggregation which occurs at pH > 6 in *Thermoplasma acidophilum*, plasma membranes may be prepared by sonicating the cells in 0.05 M acetate buffer at pH 5 (82). As pointed out above (section 4.1.1.2), the use of sonication can have serious disadvantages. Repeated freezing and thawing of osmotically sensitive cells or cell forms can be used as an alternative breakage method for membrane isolation (93).

It may not be convenient or possible to remove the cell wall prior to breakage. Membranes may be isolated from French pressure cell lysates by sucrose density-gradient centrifugation (section 4.3.5.1), although lower yields and lower purity may result for gram-negative eubacteria. Good separation of plasma membrane from outer membrane requires the absence of Mg^{2+} and maintenance of a low ionic strength at all points in isolation, assisted by the incorporation of 5 mM EDTA in the sucrose gradient (100) (section 4.4.2).

4.4.5.1. Gram-negative eubacteria

EDTA not only helps to remove the outer membrane of gram-negative eubacteria (83) but also helps to remove bound RNA from the cytoplasmic membranes. If EDTA is used, Mg^{2+} is re-added at a later step, since it is needed for DNase activity. In Kaback's procedure (63) described here, lysis is followed by differential centrifugations, with extensive homogenizations designed to aid in the fractionation of residual or partially lysed cells. It should be noted that this procedure was developed for *E. coli* ML, a strain in which the spheroplast outer membrane is particularly sensitive to disruption and removal by EDTA. With *E. coli* K-12 or smooth strains of *E. coli* or *Salmonella* spp., a large amount of outer membrane remains attached to the cytoplasmic membrane vesicles. If outer membrane contamination is a problem, its removal should be monitored by SDS-PAGE examination of the membrane preparations for the presence of outer membrane proteins.

Procedure

1. Prepare protoplasts or spheroplasts from washed cells by a method which solubilizes all or much of the cell wall murein (e.g., EDTA-lysozyme).
2. Centrifuge the osmotically fragile cell forms at 16,000 × *g* for about 10 min (or until the supernatant is clear).
3. Resuspend the pellet in the smallest possible volume of 0.1 M potassium phosphate (pH 6.6) containing 20% sucrose and 20 mM $MgSO_4$. To obtain a uniform suspension, add DNase and RNase to give 10 μg of each per ml after dilution in step 4 and homogenize in a Teflon and glass homogenizer.
4. Lyse by dilution of the protoplast-spheroplast suspension into 300 to 500 volumes of 50 mM potassium phosphate (pH 6.6) preequilibrated to 37°C. Incubate for 15 min at 37°C with vigorous shaking.
5. Incubate for another 15 min after adding potassium EDTA (pH 7.0) to give 10 mM.
6. Add $MgSO_4$ to give 15 mM, and continue incubation for 15 min.
7. Collect membranes by centrifuging the lysate at 16,000 × *g* for 30 min (alter the time, if necessary, to obtain a clear supernatant).
8. Resuspend by vigorous homogenization in 0.1 M potassium phosphate (pH 6.6) containing 10 mM EDTA at 0°C. Add $MgSO_4$, DNase, and RNase to final concentrations of 20 mM, 100 μg/ml, and 100 μg/ml, respectively, and incubate for 30 min at 37°C. Centrifuge at 45,000 × *g* for 30 min or until the supernatant is clear. Homogenize the pellet in 0.1 M potassium phosphate (pH 6.6) containing 10 mM EDTA.
9. Subject the membrane suspension to low-speed centrifugation (800 × *g* for 30 min) to remove any partially lysed or whole cells. The pellet can be rehomogenized and low-speed centrifugation can be repeated to increase the yield of membranes. Carefully remove membranes in the supernatant and repeat low-speed centrifugation as needed to remove cells from the preparation; monitor by phase-contrast microscopy.
10. Wash the membranes four to six times with 0.1 M potassium phosphate–10 mM EDTA, resuspending between steps by homogenizing.

4.4.5.2. Sheathed methanogens

In cases when a sheath or cell wall is not solubilized during the spheroplasting procedure, it is necessary to purify the membranes from the outer layers of the cell either by differential centrifugation or by density gradient centrifugation. *Methanospirillum hungatei* is such an example, in which the individual cells are encased in a very resilient sheath. Centrifugation through 65% (wt/vol) sucrose removes the large sheath and is followed by equilibrium density gradient centrifugation to complete the purification of the plasma membrane (124).

Procedure

1. Harvest cells of *Methanospirillum hungatei,* and wash in water to prevent Mg^{2+} inhibition of spheroplast formation by dithiothreitol.
2. Resuspend the cells in 100 ml of spheroplast solution (50 mM NH_4HCO_3 [pH 9.5], 0.5 M sucrose, 50 mM dithiothreitol), and incubate at room temperature for 45 min. When the sucrose is 0.5 M, spheroplasts do not emerge from the sheath tubes.
3. Centrifuge the cells at 10,000 × *g* for 10 min, and resuspend the pellet in 100 ml of 50 mM HEPES buffer (pH 7) containing 0.1 M sucrose, 10 mM $MgCl_2$, 20 μg of RNase per ml, and 20 μg of DNase per ml. This concentration of sucrose allows the spheroplasts to emerge from the sheath tubes prior to lysis and prevents entrapping of the plasma membrane fragments within the sheath.
4. Promote lysis of the spheroplasts by passing them once through a French pressure cell at 55,000 kPa. Osmotic lysis may also be used.
5. Centrifuge the mixture at 177,000 × *g* for 2 h, and homogenize the pellets in 10 to 20 ml of water with a loose-fitting hand-held Potter-Elvehjem-type homogenizer.
6. Separate plasma membranes (which are pink) from the sheath and any unbroken cells on a discontinuous sucrose density gradient (section 4.3.5.1) consisting of 65 and 70% (wt/vol) steps. Use a Beckman SW28 rotor with 30 ml of sucrose per tube, and load up to 4 ml of sample. Centrifuge at 110,000 × *g* for 6 h at about 4°C.
7. Collect the top membrane fraction, dilute it 1:1 with water, and load it onto a second discontinuous gradient consisting of 40, 45, and 50% steps (110,000 × *g* for 48 h). This allows separation of the plasma membranes (appearing close to the top of the 40% step) from a brown band of hydrogenase at the top of the 45% step. Membranes are recovered as above, giving about 8% of the starting cell dry weight.

4.4.5.3. Photosynthetic membranes

The isolation of bacterial photosynthetic membranes has been one of the more demanding exercises in cell fractionation, especially with regard to defining the structural relationship of these intracytoplasmic membranes to the plasma membrane.

The culture conditions under which photosynthetic bacteria are grown have a profound effect on the composition and extent of intracytoplasmic membrane structure. Slight differences in light intensity and variations in temperature can influence both growth rate and pigmentation. For any organism, the conditions of culture must be defined precisely.

Many of the procedures described for isolation of photosynthetic membranes have as their goal the preparation of a photochemically active unit containing the bulk of the photopigments. Mechanical procedures for cell disintegration can produce a photochemically active preparation consisting of regular small vesicles. In view of the structural variety of intracytoplasmic membranes found in photosynthetic bacteria, some change in structure must have occurred. A structurally intact intracytoplasmic membrane array may best be isolated by lysis of osmotically sensitive spheroplasts (section 4.2.2).

Sykes (129) has compiled an extensive description of procedures for isolation of photosynthetic membranes from the main groups of photosynthetic bacteria. The diversity and complexity of these membranous structures rule out the possibility of providing a single procedure for their preparation.

As an example, the isolation of chlorosomes (chlorobium vesicles) from *Chlorobium limicola* 6230 is described here. The method of Schmidt (113) with slight modifications is used and results in structurally intact chlorosomes which are morphologically distinct from plasma membrane vesicles.

Procedure

1. Grow 2 to 5 liters of bacteria.
2. Harvest cells by centrifugation, wash, and resuspend to 1/10 the original culture volume in 50 mM Tris · HCl buffer (pH 8).
3. Pass this suspension through a precooled French press (section 4.1.1) at 16,000 lb/in^2 in the presence of pancreatic DNase (100 μg/ml).
4. Remove whole cells and large particles by centrifugation at 12,000 × *g* for 15 min.
5. Treat the supernatant with lysozyme (50 μg/ml) and EDTA (1 mM) for 30 to 45 min at 37°C.
6. Collect the chlorosome and plasma membrane vesicles by centrifugation at 100,000 × *g* for 2 h.
7. Resuspend the pellet in Tris · HCl buffer to a final protein concentration of 5 to 7 mg/ml.
8. Briefly homogenize the suspension (using a Potter-Elvehjem Teflon-glass tissue grinder).
9. Layer 1 ml of the membrane homogenate onto a discontinuous sucrose gradient (25 to 55% [wt/wt]), and centrifuge at 100,000 × *g* (R_{max}) for 18 h (24,000 rpm; Beckman SW28 rotor).
10. Remove bands from the top of the gradient with a Pasteur pipette. The chlorosomes are concentrated in the 25% (wt/wt) sucrose layer. A pure plasma membrane fraction is found in the 35% (wt/wt) sucrose layer. To obtain these purified fractions, it may be necessary to repeat the density gradient centrifugation two or three times.

4.4.5.4. Purple and red membranes

Pigmented extreme halophiles, including *H. halobium* and *H. cutirubrum*, have plasma membranes with differentiated areas which can be isolated as purple membrane sheets (80, 127). During growth under O_2-*limited* conditions in the light, the plasma membrane is largely purple as a result of the retinal protein chromophore known as **bacteriorhodopsin**. Biochemically the isolated purple membrane consists of one protein band of molecular weight 20,000 on SDS-PAGE gels and ether lipids including sulfated lipids (79).

Cells grown *aerobically* produce mostly a red plasma membrane containing the red C_{50} carotenoids called **bacterioruberins.** On SDS-PAGE gels, the red membrane yields at least six bands ranging from 10,000 to 60,000 in molecular weight (the 20,000 molecular-weight band is absent) and ether lipids which are nonsulfated (79).

Methods to isolate the purple and red membranes have been summarized by Kates et al. (69) and are based on separation on sucrose density gradients following removal of stabilizing salts.

Procedure

1. Select growth conditions carefully to enrich for purple or red membranes as desired (69, 79).
2. Harvest cells from six 1.5-liter cell batches (10,000 × *g* for 20 min), and wash twice in salt solution (250 g of NaCl per liter, 9.8 g of $MgSO_4$ per liter, 2 g of KCl per liter [pH 6.5]).
3. Resuspend in 20 ml of salt solution containing 10 mg of DNase, and stir for 20 min at room temperature.
4. Dialyze at 4°C against 6 liters of distilled water for 6 h with two changes of water (one every 2 h).
5. Centrifuge the dialysate at 10,000 × *g* for 20 min to remove cell debris. Centrifuge the supernatant at 50,000 × *g* for 1.5 h.
6. Suspend the pellet in 200 ml of distilled water, and centrifuge at 10,000 × *g* for 20 min. Discard the pellet, and recentrifuge the supernatant at 50,000 × *g* for 1.5 h.
7. Resuspend the membrane pellet in 18 ml of water, and layer on discontinuous sucrose density gradients consisting of 1.5 and 1.3 M steps. Centrifuge in a swinging-bucket rotor at 260,000 × *g* for 18 h. The red membrane appears at the top of the 1.3 M step, and the purple membrane appears at the top of the 1.5 M step.
8. Remove the membrane bands, and dialyze against water to remove sucrose prior to a second gradient of 1.5 and 1.3 M steps for purple-membrane purification or 1.3 and 1.0 M sucrose steps for red-membrane purification.

4.4.6. Nucleoids

The DNA-containing structure of procaryotic cells is generally referred to as the nucleoid (meaning nucleus-like). Other terms are **nucleus, chromosome, nuclear body, nuclear region,** or **chromatin body.** Nucleoids are visible in living cells by phase-contrast microscopy (Chapter 1.4) or by bright-field microscopy after fixation and staining (Chapter 2.3.13). The nucleoid has a lobular or coralline shape with clefts and is highly dispersed throughout the ribosome-containing cytoplasm (15). The structural organization of the in vivo nucleoid is dependent on the constraints that fold and compact the DNA. As the DNA unfolds, the sedimentation and viscometric properties of the nucleoid change. Methods for the isolation of genomic DNA (Chapter 26.2) do not retain the tertiary structure of the nucleoid. Investigations into the chemical basis of the structural organization of the nucleoid, as well as membrane attachment sites, require procedures for the isolation of intact nucleoids. Virtually all research on isolated nucleoids has been carried out with *E. coli*.

Procedure for isolation of radiolabeled nucleoids (128; with modifications [144])

1. Grow *E. coli* to the exponential phase in M9 medium supplemented with 0.2% Casamino Acids.
2. Label the cells (approximately 2×10^8 cells per ml) for 30 min at 37°C with 50 to 100 μCi of [^{3}H]thymidine (10 to 50 Ci/mmol).
3. Chill the culture quickly to 0°C, and harvest by centrifugation.
4. Resuspend cells in 0.2 ml of a cold solution (4°C) containing 10 mM Tris (pH 8.1), 10 mM sodium azide, 20% (wt/vol) sucrose, and 0.1 M NaCl.
5. Add 0.05 ml of a freshly made solution containing 50 mM EDTA, 4 mg of lysozyme per ml, and 0.12 M Tris (pH 8.1). Mix carefully and quickly.
6. After 30 s at 4°C, remove from ice and add 0.25 ml of a solution containing 1% (wt/vol) Brij 58 plus 0.4% (wt/vol) deoxycholate, 2 M NaCl, and 10 mM EDTA to complete the lysis. The Na^+ is important as a counterion to maintain the folded conformation. Incubate at room temperature until the mixture has cleared (10 to 30 min). Mix the components slowly and continuously by rotary motion. Brij 58 is a polyoxyethylene ether detergent (Sigma).
7. Centrifuge the lysate for 5 min at 4,000 × *g* and 4°C. Some of the cellular DNA will precipitate.
8. Layer the supernatant fraction onto a 5-ml sucrose gradient (10 to 30% [wt/vol]) containing 0.01 M Tris · HCl (pH 8.1), 1.0 M NaCl, 1 mM EDTA, and 1 mM mercaptoethanol. Centrifuge at 30,000 × *g* for 30 min in a Beckman SW 50.1 rotor. The nucleoid sediments as a distinct band and, when removed from the gradient, remains compact. It is approximately 80% DNA by weight and includes a small amount of protein and nascent RNA.

The isolated nucleoid is either relatively free of cell envelopes or associated with them depending on the conditions for cell lysis. Lysis in molar NaCl at 20 to 25°C results in nucleoids with few envelope components. Lysis at 4°C results in envelope-associated nucleoids. Isolation of completely envelope-free nucleoids requires a longer lysozyme treatment and use of an ionic detergent such as Sarkosyl (83, 144). Other macromolecules may bind to the nucleoid during cell lysis and must be recognized as an artifactual association. The effect of lysis and physiological conditions on the nucleoid is discussed by Korch et al. (73).

4.4.7. Ribosomes

The cytoplasm of bacterial cells as observed by electron microscopy in thin sections is filled with an abundance of ribosomes. During ribosome isolation, all steps are conducted on ice or at 4°C to retard nuclease action, and it is also prudent to sterilize all glassware and tubes. The RNase I-deficient *E. coli* D_{10} is especially suitable for ribosome isolation (125). In some cases, for example in ribosome isolation from *Bacillus subtilis*, the protease inhibitor phenylmethylsulfonyl fluoride may be added as an ethanolic solution to give a final concentration of 1.0 mM prior to cell harvesting (117). The growth phase may be important as well, since *E. coli* ribosomes prepared for in vitro protein synthesis are most active for tRNA binding and elongation if prepared from cells early in mid-log growth phase (106). Frozen-thawed cells are generally broken by grinding with alumina (125), although breakage by sonication or French pressure cell treatment may be used also. Crude ribosomes may be purified by the ammonium sulfate procedure of Kurland (78) or on gradients of sucrose (106) or CsCl (41). Readers are referred to these publications to choose the method most appropriate for their application.

Procedure for ribosome isolation from frozen-thawed cells (106, 125)

1. Chill the cells in mid-exponential phase rapidly, and harvest them to obtain wet cell pellets. Freeze the wet pellets rapidly, and store them at −80°C.
2. Thaw the frozen cells (usually 25 g to several hundred grams wet weight) in a few milliliters of 10 mM Tris · HCl (pH 7.5)–6 mM $MgCl_2$–30 mM NH_4Cl–4 mM 2-mercaptoethanol or 0.5 mM dithiothreitol.
3. Break the cells by grinding with a mortar and pestle for about 10 min with 2 to 2.5 g of levigated alumina (Fisher Scientific Co., Pittsburgh, Pa.) per g of wet cell paste. While grinding is continued, slowly add the above buffer solution (containing DNase) to a final volume of about 2 ml/g of cell paste and 5 μg of DNase per ml.
4. Centrifuge the lysate to remove the alumina (16,000 × *g* for 10 min) and then to remove cell debris (27,000 × *g* for 45 min). A wash of the alumina pellet with the buffer solution may increase the ribosome yield by up to 60% (97). Combine these supernatants prior to the second low-speed centrifugation.
5. Harvest the crude ribosomes from the supernatant at 47,000 × *g* for 18 h. The use of higher centrifugation speeds (e.g., 150,000 × *g* for 3 h or 200,000 × *g* for 1 h) is reserved for cases when pressure-induced dissociation of the 70S ribosome is not a concern (106).
6. Rinse the ribosome pellets in the buffer solution, and resuspend them in the same buffer with gentle stirring over about 2 h. Clarify the suspension by centrifuging at 3,100 × *g* for 5 min. Yields are measured by monitoring A_{260}, with about 300 U expected per g of wet cells.
7. Gradient centrifugation or $(NH_4)_2SO_4$ fractionation may be used to purify these crude ribosomes. Archaeobacteria have 70S ribosomes, which can be isolated from methanogens (32, 92) and *H. cutirubrum* (92) by methods similar to those used for other bacteria.

4.4.8. Inclusion Bodies

The isolation and analysis of glycogen (Chapter 22.3.4) and poly-β-hydroxybutyrate granules (Chapter 22.4.9) are described elsewhere.

Inorganic polyphosphates (linear polymers of phosphate in anhydrous linkage) accumulate in a wide variety of bacteria including *Methanosarcina* spp. (112) and have been observed on addition of P_i to cells previously phosphate starved (44, 77). The terms **metachromatic, volutin,** and **Babes-Ernst granules** refer to deposits of polyphosphates that can be stained for light microscopy (Chapter 2.3.12.2).

Procedure for isolation of polyphosphate inclusion bodies (77)

1. Extract the cells (5 ml, 10^9 cells per ml) twice with 5-ml aliquots of 0.5 N cold perchloric acid for 5 min at 20°C, centrifuging to obtain acid-soluble polyphosphates of 2 to 10 residues of phosphoric acid.
2. Extract the residue with 5 ml of ethanol for 30 min at 20°C, and centrifuge.
3. Add 5 ml of ethanol-ether (3:1 vol/vol) to the cell residue from step 2, and heat in a water bath (fume cupboard) to approach boiling for 1 min. Centrifuge the cooled mixture to obtain lipid-depleted residue.
4. Extract the residue from step 3 twice with 5-ml aliquots of hot 0.5 N perchloric acid for 15 min at 70°C, and centrifuge to remove particulates. Separate nucleic acids from the longer polyphosphate chains found in the hot perchloric acid extract by adsorption of the nucleic acids on charcoal (Norit A; BDH Chemicals, Poole, England). The polymers may be sized by PAGE (20).

4.4.9. Endospores

Procedures for the production of high yields of free, dormant spores by spore-forming bacteria generally involve growth in a moderately rich medium until the nutrients are depleted and autolysis occurs. Supplementation of the medium with minerals, especially calcium (9b), selection of a smooth-colony variant (9b), and growth at above-optimal temperatures (9a) may be necessary to produce maximally resistant spores. High aeration rates (on an agar surface, in a shaker flask, or in a sparged fermentor) are necessary for the aerobic sporeformers and prevent the accumulation of contaminating poly-β-hydroxybutyrate. The release of endospores by mechanical means and centrifugation at high gravities should be avoided because highly resistant spores are fragile to these procedures. Dormant spores can be stored by lyophilization or by suspension in absolute alcohol or distilled water. Specific conditions are prescribed for producing spores from a variety of sporeformers (9a, 9b, 37a, 45, 140).

Methods for isolating highly purified spores (85, 132) are based on their high density (Table 2). A given spore population may be heterogeneous with respect to density, resulting in more than one density band on density gradient centrifugation. For example, spores of *Bacillus stearothermophilus* 7953 separate in a gradient of a low-molecular-weight medium (such as Metrizamide or Nycodenz) into bands of germinated spores, lysozyme-resistant dormant spores, and lysozyme-susceptible dormant spores. The banding patterns are attributed to differences in the penetration of the gradient medium into the spore integument. In a gradient of Percoll, whose colloidal particles do not penetrate the spore integument, only one density band is obtained. Thus, Percoll provides a suitable medium to determine the wet density of the entire spore, with values comparing favorably with direct mass measurements, whereas Metrizamide or Nycodenz are suitable to determine the wet density of the spore protoplast in lysozyme-susceptible spores and to completely separate dormant spores from germinated spores, vegetative cells, parasporal crystals, and debris.

4.4.10. Gas Vesicles

Gas vesicles appear as refractile areas in light microscopy, and these areas are full of vesicles in electron microscopy. Reviews are available for the vacuolate cyanobacteria (138) and other vacuolate bacteria (137). It is diagnostic of gas vesicles that they collapse on the application of pressure, generally of about 500 kPa. The cylindrical structure of the gas vesicles consists of a monolayer of protein molecules wound around the cylinderlike ribs. This vesicle is permeable to gases but not to water. Gas vacuolate bacteria are found in aquatic species ranging from phototrophs to the extremely halophilic archaeobacteria and *Methanosarcina barkeri* FR1 (4).

Gas vesicles are isolated from osmotic lysates of osmotically sensitive cells or cell forms by the flotation method of Walsby (136). A general method follows, taken from details given for *H. salinarium* (119), cyanobacteria (135), and *Methanosarcina barkeri* (4).

Procedure for isolation of gas vesicles

1. Prepare spheroplasts or protoplasts for bacteria which do not lyse readily in hypotonic solutions.
2. Break the cells gently by osmotic means to prevent collapse of the gas vesicles. Perform lysis in 1 mM $MgSO_4$ containing 10 to 20 μg of DNase per ml to reduce viscosity for the subsequent floatation step.
3. Centrifuge the lysate for 16 h in a swinging-bucket rotor at 60 to 300 × *g*. Minimize the pressure by using a maximum depth of 5.0 cm of liquid. Gas vesicles float to the surface (138a).
4. Add 1% Triton X-100 for 30 min at 37°C to aid in removal of plasma membrane, and repeat floatation. Other detergents have been used, e.g., 0.1% SDS or 1% Tween 20; however, 1% SDS or Sarkosyl can weaken the vesicle.
5. Wash the surface gas-vesicle layer several times in 20 mM Tris buffer (pH 7.5) by flotation.

4.5. REFERENCES

1. **Altman, E., J.-R. Brisson, and M. B. Perry.** 1986. Structural studies of the capsular polysaccharide from *Haemophilus pleuropneumoniae* serotype 1. *Biochem. Cell Biol.* **64:**707–716.
2. **Altman, E., M. B. Perry, and J.-R. Brisson.** 1989. Structure of the lipopolysaccharide antigenic O-chain produced by *Actinobacillus pleuropneumoniae* serotype 4 (ATCC 33378). *Carbohydr. Res.* **191:**295–303.
3. **Ames, G. F.-L., C. Prody, and S. Kustu.** 1984. Simple, rapid, and quantitative release of periplasmic proteins by chloroform. *J. Bacteriol.* **160:**1181–1183.
4. **Archer, D. B., and N. R. King.** 1984. Isolation of gas vesicles from *Methanosarcina barkeri. J. Gen. Microbiol.* **130:**167–172.
5. **Austin, E. A., J. F. Graves, L. A. Hite, C. T. Parker, and C. A. Schnaitman.** 1990. Genetic analysis of lipopolysaccharide core biosynthesis by *Escherichia coli* K-12. *J. Bacteriol.* **172:**5312–5325.
6. **Austin, J. W., and R. G. E. Murray.** 1987. The perforate component of the regularly-structured (RS) layer of *Lampropedia hyalina. Can. J. Microbiol.* **33:**1039–1045.
7. **Bakau, B., J. M. Brass, and W. Boos.** 1985. Ca^{2+}-induced permeabilization of the *Escherichia coli* outer membrane: comparison of transformation and reconstitution of binding-protein-dependent transport. *J. Bacteriol.* **163:**61–68.

8. **Barber, E. J.** 1976. Viscosity and density tables, p. 415–418. *In* G. D. Fasman (ed.), *Handbook of Biochemistry and Molecular Biology,* 3rd ed., vol. 1. *Physical and Chemical Data.* CRC Press, Inc., Cleveland.
9. **Barrett, A. J.** 1980. Introduction: the classification of proteinases. *CIBA Found. Symp.* **75:**1–13.
9a. **Beaman, T. C., and P. Gerhardt.** 1986. Heat resistance of bacterial spores correlated with protoplast dehydration, mineralization, and thermal adaptation. *Appl. Environ. Microbiol.* **52:**1242–1246.
9b. **Beaman, T. C., J. T. Greenamyre, T. R. Corner, H. S. Pankratz, and P. Gerhardt.** 1982. Bacterial spore heat resistance correlated with water content, wet density, and protoplast/sporoplast volume ratio. *J. Bacteriol.* **150:**870–877.
10. **Beveridge, T. J.** 1979. Surface arrays on the wall of *Sporosarcina urea. J. Bacteriol.* **139:**1039–1048.
11. **Beveridge, T. J., G. B. Patel, B. J. Harris, and G. D. Sprott.** 1986. The ultrastructure of *Methanothrix concilii,* a mesophilic aceticlastic methanogen. *Can. J. Microbiol.* **32:**703–710.
12. **Beveridge, T. J., M. Stewart, R. J. Doyle, and G. D. Sprott.** 1985. Unusual stability of the *Methanospirillum hungatei* sheath. *J. Bacteriol.* **162:**728–737.
13. **Beynon, L. M., M. Moreau, J. C. Richards, and M. B. Perry.** 1991. Structure of the O-antigen of *Actinobacillus pleuropneumoniae* serotype 7 lipopolysaccharide. *Carbohydr. Res.* **209:**225–238.
14. **Bingle, W. H., J. L. Doran, and W. J. Page.** 1984. Regular surface layer of *Azotobacter vinelandii. J. Bacteriol.* **159:**251–259.
15. **Bohrmann, B., W. Villiger, R. Johansen, and E. Kellenberger.** 1991. Coralline shape of the bacterial nucleoid after cryofixation. *J. Bacteriol.* **173:**3149–3158.
16. **Brinton, C. C., A. Buzzell, and M. A. Lauffer.** 1954. Electrophoresis and phage susceptibility on a filament-producing variant of the *E. coli* B. bacterium. *Biochim. Biophys. Acta* **15:**533–542.
17. **Buluwela, L., A. Forster, T. Boehm, and T. H. Rabbitts.** 1989. A rapid procedure for colony screening using nylon filters. *Nucleic Acids Res.* **17:**452.
18. **Bush, J. W.** 1985. Enzymatic lysis of the pseudomurein-containing methanogen *Methanobacterium formicicum. J. Bacteriol.* **163:**27–36.
19. **Chen, W., K. Ohmiya, and S. Shimizu.** 1986. Protoplast formation and regeneration of dehydrodivanillin-degrading strains of *Fusobacterium varium* and *Enterococcus faecium. Appl. Environ. Microbiol.* **52:**612–616.
20. **Clark, J. E., H. Beegen, and H. G. Wood.** 1985. Isolation of polyphosphate from *Propionibacterium shermanii. Fed. Proc.* **44:**1079.
21. **Coulton, J. W., and R. G. E. Murray.** 1977. Membrane-associated components of the bacterial flagellar apparatus. *Biochim. Biophys. Acta* **465:**290–310.
22. **Cunningham, T. M., E. M. Walker, J. N. Miller, and M. A. Lovett.** 1988. Selective release of the *Treponema pallidum* outer membrane and associated polypeptides with Triton X-114. *J. Bacteriol.* **170:**5789–5796.
23. **Darveau, R. P., and R. E. W. Hancock.** 1983. Procedure for isolation of bacterial lipopolysaccharides from both smooth and rough *Pseudomonas aeruginosa* and *Salmonella typhimurium* strains. *J. Bacteriol.* **155:**831–838.
24. **Davis, R. P., and J. E. Harris.** 1985. Spontaneous protoplast formation by *Methanosarcina barkeri. J. Gen. Microbiol.* **131:**1481–1486.
25. **De Pamphilis, M. L., and J. Adler.** 1971. Purification of intact flagella from *Escherichia coli* and *Bacillus subtilis. J. Bacteriol.* **105:**376–383.
26. **Dubreuil, J. D., S. M. Logan, S. Cubbage, D. NiEidhin, W. D. McCubbin, C. M. Kay, T. J. Beveridge, F. G. Ferris, and T. J. Trust.** 1988. Structural and biochemical analysis of a surface array protein of *Campylobacter fetus. J. Bacteriol.* **170:**4165–4173.
27. **Dunn, R. M., M. J. Munster, R. J. Sharp, and B. N. Dancer.** 1987. A novel method for regenerating the protoplasts of thermophilic bacilli. *Arch. Microbiol.* **146:**323–326.
28. **Dupont, C., and A. J. Clarke.** 1991. Dependence of lysozyme-catalysed solubilization of *Proteus mirabilis* peptidoglycan on the extent of O-acetylation. *Eur. J. Biochem.* **195:**763–769.
29. **Easterbrook, K. B.** 1989. Spinate bacteria, p. 1991–1993. *In* J. G. Holt (ed.), *Bergey's Manual of Systematic Bacteriology,* vol. 3. The Williams & Wilkins Co., Baltimore.
30. **Easterbrook, K. B., and R. W. Coombs.** 1976. Spinin: the subunit protein of bacterial spinae. *Can. J. Microbiol.* **22:**438–440.
31. **Easterbrook, K. B., and S. Sperker.** 1982. Physiological controls of bacterial spinae production in complex medium and their value as indicators of spina function. *Can. J. Microbiol.* **28:**130–136.
32. **Elhardt, D., and A. Böck.** 1982. An in vitro polypeptide synthesizing system from methanogenic bacteria: sensitivity to antibiotics. *Mol. Gen. Genet.* **188:**128–134.
33. **Filip, C., G. Fletcher, J. L. Wulff, and C. F. Earhart.** 1977. Solubilization of the cytoplasmic membrane of *Escherichia coli* by the ionic detergent sodium lauryl sarcosinate. *J. Bacteriol.* **115:**717–722.
34. **Fujikawa, H., H. Ushioda, and Y. Kudo.** 1992. Kinetics of *Escherichia coli* destruction by microwave irradiation. *Appl. Environ. Microbiol.* **58:**920–924.
35. **Galanos, C., and O. Lüderitz.** 1975. Electrodialysis of lipopolysaccharides and their conversion to uniform salt form. *Eur. J. Biochem.* **54:**603–610.
36. **Galanos, C., O. Lüderitz, and O. Westphal.** 1969. A new method for the extraction of R lipopolysaccharides. *Eur. J. Biochem.* **9:**245–249.
37. **Gilleland, H. E., Jr., J. D. Stinnett, I. L. Roth, and R. G. Eagon.** 1973. Freeze-etch study of *Pseudomonas aeruginosa:* localization within the cell wall of an ethylenediaminetetraacetate-extractable component. *J. Bacteriol.* **113:**417–432.
37a. **Gould, G. W.** 1971. Methods for studying bacterial spores. *Methods Microbiol.* **6A:**361–381.
38. **Gray, G. W., and S. G. Wilkinson.** 1965. The effect of ethylenediaminetetraacetic acid on the cell walls of some gram-negative bacteria. *J. Gen. Microbiol.* **39:**385–399.
39. **Gruter, L., and R. Laufs.** 1991. Protoplast transformation of *Staphylococcus epidermidis. J. Microbiol. Methods* **13:**299–304.
40. **Guerrero, R., C. Pedros-Alio, T. M. Schmidt, and J. Mas.** 1985. A survey of buoyant density of microorganisms in pure cultures and natural samples. *Microbiologia* (Madrid) **1:**53–65.
41. **Hamilton, M. G.** 1971. Isodensity equilibrium centrifugation of ribosomal particles: the calculation of the protein content of ribosomes and other ribonucleoproteins from buoyant density measurements. *Methods Enzymol.* **20:**512–521.
42. **Hancock, I. C., and I. R. Poxton.** 1988. Isolation and purification of cell walls, p. 55–65. *In* I. C. Hancock and I. R. Poxton (ed.), *Bacterial Cell Surface Techniques.* John Wiley & Sons Ltd., Toronto.
42a. **Hancock, I. C., and I. R. Poxton.** 1988. Teichoic acids and other accessory carbohydrates from Gram-positive bacteria, p. 79–88. *In* I. C. Hancock and I. R. Poxton (ed.), *Bacterial Cell Surface Techniques.* John Wiley & Sons Ltd., Toronto.
42b. **Hancock, I. C., and I. R. Poxton.** 1988. Isolation of exopolysaccharides, p. 121–125. *In* I. C. Hancock and I. R. Poxton (ed.), *Bacterial Cell Surface Techniques.* John Wiley & Sons Ltd., Toronto.
43. **Hancock, R. E. W., and H. Nikaido.** 1978. Outer membranes of Gram-negative bacteria. XIX. Isolation from *Pseudomonas aeruginosa* PAO1 and use in reconstitution and definition of the permeability barrier. *J. Bacteriol.* **136:**381–390.
44. **Harold, F. M.** 1963. Accumulation of inorganic polyphosphate in *Aerobacter aerogenes.* I. Relationship to growth and nucleic acid synthesis. *J. Bacteriol.* **86:**216–221.
45. **Harwood, C. R., and S. M. Cutting.** 1990. *Molecular Biological Methods for Bacillus.* John Wiley & Sons, Inc., New York.

46. **Hash, J. H., and M. V. Rothlauf.** 1967. The N,O-diacetylmuramidase of *Chalaropsis* species. *J. Biol. Chem.* **242:**5586–5590.
47. **Heckels, J. E., and M. Virji.** 1988. Detection and preparation of surface appendages, p. 67–72. *In* I. C. Hancock and I. R. Poxton (ed.), *Bacterial Cell Surface Techniques.* John Wiley & Sons Ltd., Toronto.
48. **Helenius, A., D. R. McCaslin, E. Fries, and C. Tanford.** 1979. Properties of detergents. *Methods Enzymol.* **56:**734–749.
49. **Heppel, L. A.** 1971. The concept of periplasmic enzymes, p. 223–247. *In* L. I. Rothfield (ed.), *Structure and Function of Biological Membranes.* Academic Press, Inc., New York.
50. **Hitchcock, P. J., and T. M. Brown.** 1983. Morphological heterogeneity among *Salmonella* lipopolysaccharide chemotypes in silver-stained polyacrylamide gels. *J. Bacteriol.* **154:**269–277.
51. **Hobot, J. A., E. Carlemalm, W. Villiger, and E. Kellenberger.** 1984. Periplasmic gel: new concept resulting from the reinvestigation of bacterial cell envelope ultrastructure by new methods. *J. Bacteriol.* **160:**143–152.
52. **Holloway, P. W.** 1973. A simple procedure for the removal of Triton X-100 from protein samples. *Anal. Biochem.* **53:**304–308.
53. **Hughes, D. E., J. W. T. Wimpenny, and D. Lloyd.** 1971. The disintegration of microorganisms, p. 1–54. *In* J. R. Norris and D. W. Ribbons (ed.), *Methods in Microbiology,* vol. 5B. Academic Press, Inc., New York.
54. **Jann, B., and K. Jann.** 1990. Structure and biosynthesis of the capsular antigens of *Escherichia coli. Curr. Top. Microbiol. Immunol.* **150:**19–42.
55. **Jarrell, K. F., J. R. Colvin, and G. D. Sprott.** 1982. Spontaneous protoplast formation in *Methanobacterium bryantii. J. Bacteriol.* **149:**346–353.
56. **Jarrell, K. F., D. Faguy, A. M. Hebert, and M. L. Kalmokoff.** 1992. A general method of isolating high molecular weight DNA from methanogenic archaea (archaebacteria). *Can. J. Microbiol.* **38:**65–68.
57. **Jarrell, K. F., and G. D. Sprott.** 1984. Formation and regeneration of *Halobacterium* protoplasts. *Curr. Microbiol.* **10:**147–152.
58. **Jennings, H. J.** 1990. Capsular polysaccharides as vaccine candidates. *Curr. Top. Microbiol. Immunol.* **150:**97–127.
59. **Johannsen, L., H. Labischinski, B. Reinicke, and P. Giesbrecht.** 1983. Changes in the chemical structure of walls of *Staphylococcus aureus* grown in the presence of chloramphenicol. *FEMS Microbiol. Lett.* **16:**313–316.
60. **Johnson, K. G., and M. B. Perry.** 1976. Improved techniques for the preparation of bacterial lipopolysaccharides. *Can. J. Microbiol.* **22:**29–34.
61. **Jollès, P.** 1969. Lysozymes: a chapter of molecular biology. *Angew. Chem. Int. Ed.* **8:**227–294.
62. **Jussofie, A., F. Mayer, and G. Gottschalk.** 1986. Methane formation from methanol and molecular hydrogen by protoplasts of new methanogen isolates and inhibition by dicyclohexylcarbodiimide. *Arch Microbiol.* **146:**245–249.
63. **Kaback, H. R.** 1971. Bacterial membranes. *Methods Enzymol.* **22:**99–120.
64. **Kaempfer, R., and M. Meselson.** 1971. Sedimentation velocity analysis in accelerating gradients. *Methods Enzymol.* **20:**521–528.
65. **Kalmokoff, M. L., K. F. Jarrell, and S. F. Koval.** 1988. Isolation of flagella from the archaebacterium *Methanococcus voltae* by phase separation with Triton X-114. *J. Bacteriol.* **170:**1752–1758.
66. **Kamekura, M., D. Oesterhelt, R. Wallace, P. Anderson, and D. J. Kushner.** 1988. Lysis of halobacteria in Bacto-Peptone by bile acids. *Appl. Environ. Microbiol.* **54:**990–995.
67. **Kandler, O., and H. König.** 1985. Cell envelopes of archaebacteria. *Bacteria* **8:**413-457.
68. **Karch, H., H. Leying, K.-H. Buscher, H.-P. Kroll, and W. Opferkuch.** 1985. Isolation and separation of physiochemically distinct fimbrial types expressed on a single culture of *Escherichia coli* O7:K1:H6. *Infect. Immun.* **47:**549–554.
69. **Kates, M., S. C. Kushwaha, and G. D. Sprott.** 1982. Lipids of purple membrane from extreme halophiles and of methanogenic bacteria. *Methods Enzymol.* **88:**98–111.
70. **Kiener, A., H. König, J. Winter, and T. Leisinger.** 1987. Purification and use of *Methanobacterium wolfei* pseudomurein endopeptidase for lysis of *Methanobacterium thermoautotrophicum. J. Bacteriol.* **169:**1010–1016.
71. **Kobayashi, H., J. van Brunt, and F. M. Harold.** 1978. ATP-linked calcium transport in cells and membrane vesicles of *Streptococcus faecalis. J. Biol. Chem.* **253:**2085–2092.
72. **Konig, H., and K. O. Stetter.** 1986. Studies on archaebacterial S-layers. *Syst. Appl. Microbiol.* **7:**300–309.
73. **Korch, C., S. Ovrebø, and K. Kleppe.** 1976. Envelope-associated folded chromosomes from *Escherichia coli:* variations under different physiological conditions. *J. Bacteriol.* **127:**904–916.
74. **Koval, S. F.** 1988. Paracrystalline protein surface arrays on bacteria. *Can. J. Microbiol.* **34:**407–414.
75. **Koval, S. F., and K. F. Jarrell.** 1987. Ultrastructure and biochemistry of the cell wall of *Methanococcus voltae. J. Bacteriol.* **169:**1298–1306.
76. **Koval, S. F., and R. G. E. Murray.** 1984. The isolation of surface array proteins from bacteria. *Can. J. Biochem. Cell Biol.* **62:**1181–1189.
77. **Kulaev, I. S.** 1975. Biochemistry of inorganic polyphosphates. *Rev. Physiol. Biochem. Pharmacol.* **73:**131–157.
78. **Kurland, C. G.** 1971. Purification of ribosomes from *Escherichia coli. Methods Enzymol.* **20:**379–381.
79. **Kushwaha, S. C., M. Kates, and W. G. Martin.** 1975. Characterization and composition of the purple and red membrane from *Halobacterium cutirubrum. Can. J. Biochem.* **53:**284–292.
80. **Kushwaha, S. C., M. Kates, and W. Stoeckenius.** 1976. Comparison of purple membrane from *Halobacterium cutirubrum* and *Halobacterium halobium. Biochim. Biophys. Acta* **426:**703–710.
81. **Laddaga, R. A., and R. A. MacLeod.** 1982. Effects of wash treatments on the ultrastructure and lysozyme penetrability of the outer membrane of various marine and two terrestrial gram-negative bacteria. *Can. J. Microbiol.* **28:**318–324.
82. **Langworthy, T. A.** 1978. Membranes and lipids of extremely thermoacidophilic microorganisms, p. 11–30. *In* S. M. Friedman (ed.), *Biochemistry of Thermophily.* Academic Press, Inc., New York.
83. **Leive, L.** 1965. Release of lipopolysaccharide by EDTA treatment of *Escherichia coli. Biochem. Biophys. Res. Commun.* **21:**290–296.
84. **Lesse, A. J., A. A. Campagnari, W. E. Bittner, and M. A. Apicella.** 1990. Increased resolution of lipopolysaccharides and lipooligosaccharides utilizing tricine-sodium dodecyl sulfate-polyacrylamide gel electrophoresis. *J. Immunol. Methods* **126:**109–117.
85. **Lindsay, J. A., T. C. Beaman, and P. Gerhardt.** 1985. Protoplast water content of bacterial spores determined by buoyant density sedimentation. *J. Bacteriol.* **163:**735–737.
86. **Lu, W., A. P. Wood, and D. P. Kelly.** 1983. An enzymatic lysis procedure for the assay of enzymes in *Thiobacillus* A2. *Microbios* **38:**171–176.
87. **Lugtenberg, B., and L. van Alphen.** 1983. Molecular architecture and functioning of the outer membrane of *Escherichia coli* and other Gram-negative bacteria. *Biochim. Biophys. Acta* **737:**51–115.
88. **MacDonald, R. E., and J. K. Lanyi.** 1975. Light-induced leucine transport in *Halobacterium halobium* envelope vesicles: a chemiosmotic study. *Biochemistry* **14:**2882–2889.
89. **MacLeod, R. A.** 1985. Marine microbiology far from the sea. *Annu. Rev. Microbiol.* **39:**1–20.
90. **Manoharan, R., E. Ghiamati, R. A. Dalterio, K. A. Britton, W. H. Nelson, and J. F. Sperry.** 1990. UV resonance Raman

spectra of bacteria, bacterial spores, protoplasts and calcium dipicolinate. *J. Microbiol. Methods* **11**:1–15.

91. **Marolda, C. L., J. Welsh, L. Dafoe, and M. A. Valvano.** 1990. Genetic analysis of the O7-polysaccharide biosynthesis region from the *Escherichia coli* O7:K1 strain VW 187. *J. Bacteriol.* **172**:3590–3599.
92. **Matheson, A. T., M. Yaguchi, W. E. Balch, and R. S. Wolfe.** 1980. Sequence homologies in the N-terminal region of the ribosomal 'A' proteins from *Methanobacterium thermoautotrophicum* and *Halobacterium cutirubrum. Biochim. Biophys. Acta* **626**:162–169.
93. **Mayer, F., A. Jussofie, M. Salzmann, M. Lübben, M. Rohde, and G. Gottschalk.** 1987. Immunoelectron microscopic demonstration of ATPase on the cytoplasmic membrane of the methanogenic bacterium strain GÖ1. *J. Bacteriol.* **169**:2307–2309.
94. **Meakin, S. A., J. H. E. Nash, W. D. Murray, K. J. Kennedy, and G. D. Sprott.** 1991. A generally applicable technique for the extraction of restrictable DNA from methanogenic bacteria. *J. Microbiol. Methods* **14**:119–126.
95. **Messner, P., and U. B. Sleytr.** 1988. Separation and purification of S-layers from Gram-positive and Gram-negative bacteria, p. 97–104. *In* I. Hancock and I. Poxton (ed.), *Bacterial Cell Surface Techniques.* John Wiley & Sons, Toronto.
96. **Miura, T., and S. Mizushima.** 1968. Separation by density gradient centrifugation of two types of membranes from spheroplast membranes of *Escherichia coli* K12. *Biochim. Biophys. Acta* **150**:159–161.
97. **Moore, P. B.** 1979. The preparation of deuterated ribosomal materials for neutron scattering. *Methods Enzymol.* **59**:639–655.
98. **Morii, H., and Y. Koga.** 1992. An improved assay for a pseudomurein-degrading enzyme of *Methanobacterium wolfei* and the protoplast formation of *Methanobacterium thermoautotrophicum* by the enzyme. *J. Ferment. Bioeng.* **73**:6–10.
99. **Nossal, N. G., and L. A. Heppel.** 1966. The release of enzymes by osmotic shock from *Escherichia coli* in exponential phase. *J. Biol. Chem.* **241**:3055–3062.
100. **Osborn, M. J., and R. Munson.** 1974. Separation of the inner (cytoplasmic) and outer membranes of gram negative bacteria. *Methods Enzymol.* **31A**:642–653.
101. **Paranchych, W., P. A. Sastry, L. S. Frost, M. Carpenter, G. D. Armstrong, and T. H. Watts.** 1979. Biochemical studies on pili isolated from *Pseudomonas aeruginosa* strain PAO. *Can. J. Microbiol.* **25**:1175–1181.
102. **Penn, C. W., A. Cockayne, and M. J. Bailey.** 1985. The outer membrane of *Treponema pallidum:* biological significance and biochemical properties. *J. Gen. Microbiol.* **131**:2349–2357.
103. **Pertoft, H., T. C. Laurent, R. Seljelid, G. Akerstrom, L. Kagedal, and M. Hirtenstein.** 1979. The use of density gradients of Percoll[R] for the separation of biological particles, p. 67–72. *In* H. Peeters (ed.), *Separation of Cells and Subcellular Elements.* Pergamon Press, Toronto.
104. **Peterson, G. L.** 1977. A simplification of the protein assay method of Lowry *et al.* which is more generally applicable. *Anal. Biochem.* **83**:346–356.
105. **Petrov, V. V., V. Y. Artzatbanov, E. N. Ratner, A. I. Severin, and I. S. Kulaev.** 1991. Isolation, structural and functional charcterization of *Staphylococcus aureus* protoplasts obtained using lysoamidase. *Arch. Microbiol.* **155**:549–553.
106. **Rheinberger, H.-J., U. Geigenmüller, M. Wedde, and K. H. Nierhaus.** 1988. Parameters for the preparation of *Escherichia coli* ribosomes and ribosomal subunits active in tRNA binding. *Methods Enzymol.* **164**:658–670.
107. **Rickwood, D. (ed.).** 1984. *Centrifugation: A Practical Approach,* 2nd ed. IRL Press, Oxford.
108. **Ross, J. W.** 1963. Continuous-flow mechanical cell disintegrator. *Appl. Microbiol.* **11**:33–35.
109. **Salton, M. R. J.** 1957. The properties of lysozyme and its action on microorganisms. *Bacteriol. Rev.* **21**:82–99.
110. **Salton, M. R. J.** 1974. Isolation of cell walls from Gram-positive bacteria. *Methods Enzymol.* **31**:653–667.
111. **Sambrook, J., E. F. Fritsch, and T. Maniatis.** 1989. *Molecular Cloning: A Laboratory Manual,* 2nd ed. Cold Spring Harbor Laboratory, Cold Spring Harbor, N.Y.
112. **Scherer, P. A., and H.-P. Bochem.** 1983. Ultrastructural investigation of 12 *Methanosarcinae* and related species grown on methanol for occurrence of polyphosphatelike inclusion. *Can. J. Microbiol.* **29**:1190–1199.
113. **Schmidt, K.** 1980. A comparative study on the composition of chlorosomes (chlorobium vesicles) and cytoplasmic membranes from *Chloroflexus aurantiacus* strain OK-70-fl and *Chlorobium limocola f. thiosulfatophilum* strain 6230. *Arch. Microbiol.* **124**:21–31.
114. **Schnaitman, C. A.** 1970. Protein composition of the cell wall and cytoplasmic membrane of *Escherichia coli. J. Bacteriol.* **104**:890–901.
115. **Schnaitman, C. A.** 1971. Solubilization of the cytoplasmic membrane of *Escherichia coli* by Triton X-100. *J. Bacteriol.* **108**:545–552.
116. **Schulman, H., and E. P. Kennedy.** 1979. Localization of membrane-derived oligosaccharides in the outer envelope of *Escherichia coli* and their occurrence in other gram-negative bacteria. *J. Bacteriol.* **137**:686–688.
117. **Sharrock, W. J., and J. C. Rabinowitz.** 1979. Fractionation of ribosomal particles from *Bacillus subtilis. Methods Enzymol.* **59**:371–382.
118. **Siegel, J. L., S. F. Hurst, E. S. Liberman, S. E. Coleman, and A. S. Bleiweis.** 1981. Mutanolysin-induced spheroplasts of *Streptococcus mutans* are true protoplasts. *Infect. Immun.* **31**:808–815.
119. **Simon, R. D.** 1981. Morphology and protein composition of gas vesicles from wild type and gas vacuole defective strains of *Halobacterium salinarium* strain 5. *J. Gen. Microbiol.* **125**:103–111.
120. **Sleytr, W. B., and P. Messner.** 1988. Crystalline surface layers in procaryotes. *J. Bacteriol.* **170**:2891–2897.
121. **Spizizen, J.** 1962. Preparation and use of protoplasts. *Methods Enzymol.* **5**:122–134.
122. **Sprott, G. D., J. R. Colvin, and R. C. McKellar.** 1979. Spheroplasts of *Methanospirillum hungatei* formed upon treatment with dithiothreitol. *Can J. Microbiol.* **25**:730–738.
123. **Sprott, G. D., and K. F. Jarrell.** 1981. K^+, Na^+, and Mg^{2+} content and permeability of *Methanospirillum hungatei* and *Methanobacterium thermoautotrophicum. Can. J. Microbiol.* **27**:444–451.
124. **Sprott, G. D., K. M. Shaw, and K. F. Jarrell.** 1983. Isolation and chemical composition of the cytoplasmic membrane of the archaebacterium *Methanospirillum hungatei. J. Biol. Chem.* **258**:4026–4031.
125. **Staechelin, T., and D. R. Maglott.** 1971. Preparation of *Escherichia coli* ribosomal subunits active in polypeptide synthesis. *Methods Enzymol.* **20**:449–456.
126. **Steinberg, W.** 1974. Properties and developmental roles of the lysyl- and tryptophanyl-transfer ribonucleic acid synthetase of *Bacillus subtilis:* common genetic origin of the corresponding spore and vegetative enzymes. *J. Bacteriol.* **118**:70–82.
127. **Stoeckenius, W., and R. Rowen.** 1967. A morphological study of *Halobacterium halobium* and its lysis in media of low salt concentration. *J. Cell Biol.* **34**:365–393.
128. **Stonington, O. G., and D. E. Pettijohn.** 1971. The folded genome of *Escherichia coli* isolated in a protein-DNA-RNA complex. *Proc. Natl. Acad. Sci. USA* **68**:6–9.
129. **Sykes, J.** 1971. Centrifugation techniques for the isolation and characterization of sub-cellular components from bacteria, p. 55–207. *In* J. R. Norris and D. W. Ribbons (ed.), *Methods in Microbiology,* vol. 5B. Academic Press, Inc., New York.
130. **Thom, J. R., and L. L. Randall.** 1988. Role of the leader peptide of maltose-binding protein in two steps of the export process. *J. Bacteriol.* **170**:5654–5661.
131. **Thompson, B. G., R. G. E. Murray, and J. F. Boyce.** 1982. The association of the surface array and the outer mem-

brane of *Deinococcus radiodurans. Can. J. Microbiol.* **28:**1081–1088.

132. **Tisa, L. S., T. Koshikawa, and P. Gerhardt.** 1982. Wet and dry bacterial spore densities determined by buoyant sedimentation. *Appl. Environ. Microbiol.* **43:**1307–1310.
133. **Vandenbergh, P. A., R. E. Bawdon, and R. S. Berk.** 1979. Rapid test for determining the intracellular rhodanese activity of various bacteria. *Int. J. Syst. Bacteriol.* **29:**339–344.
134. **Virgilio, R., C. Gonzalez, N. Muñoz, T. Cabezon, and S. Mendoza.** 1970. *Staphylococcus aureus* protoplasting induced by D-cycloserine. *J. Bacteriol.* **104:**1386–1387.
135. **Walker, J. E., P. K. Hayes, and A. E. Walsby.** 1984. Homology of gas vesicle proteins in *Cyanobacteria* and *Halobacteria. J.Gen. Microbiol.* **130:**2709–2715.
136. **Walsby, A. E.** 1974. The isolation of gas vesicles from blue-green algae. *Methods Enzymol.* **31:**678–686.
137. **Walsby, A. E.** 1981. Gas-vacuolate bacteria (apart from cyanobacteria), p. 441–447. *In* M. Starr, H. Stolp, H. Truper, A. Balows, and H. G. Schlegel (cd.), *The Prokaryotes.* Springer-Verlag, New York.
138. **Walsby, A. E.** 1981. Cyanobacteria: planktonic gas-vacuolate forms, p. 224–235. *In* M. Starr, H. Stolp, H. Truper, A. Balows, and H. G. Schlegel (ed.), *The Prokaryotes.* Springer-Verlag, New York.

138a.**Walsby, A. E., and B. Buckland.** 1969. Isolation and purification of intact gas vesicles from a blue-green alga. *Nature* (London) **224:**716–717.

139. **Ward, J. B.** 1981. Teichoic and teichuronic acids: biosynthesis, assembly, and location. *Microbiol. Rev.* **45:**211–243.
140. **Warth, A.D.** 1978. Relationship between the heat resistance of spores and the optimum and maximum growth temperature of *Bacillus* species. *J. Bacteriol.* **134:**699–705.
141. **Weil, C. F., D. S. Cram, B. A. Sherf, and J. N. Reeve.** 1988. Structure and comparative analysis of the genes encoding component C of methyl coenzyme M reductase in the extremely thermophilic archaebacterium *Methanothermus fervidus. J. Bacteriol.* **170:**4718–4726.
142. **Westphal, O., and K. Jann.** 1965. Bacterial lipopolysaccharides. *Methods Carbohydr. Chem.* **5:**83–91.
143. **Wilkinson, S. G., L. Galbraith, and (in part) G. A. Lightfoot.** 1973. Cell walls, lipids, and lipopolysaccharides of *Pseudomonas* species. *Eur. J. Biochem.* **33:**158–174.
144. **Worcel, A. and E. Burgi.** 1972. On the structure of the folded chromosome of *Escherichia coli. J. Mol. Biol.* **71:**127–147.
145. **Work, E.** 1971. Cell walls, p. 361–418. *In* J. R. Norris and D. W. Ribbons (ed.), *Methods in Microbiology,* vol. 5A. Academic Press, Inc., New York.
146. **Yabu, K., and S. Takahashi.** 1977. Protoplast formation of selected *Mycobacterium smegmatis* mutants by lysozyme in combination with methionine. *J. Bacteriol.* **129:**1628–1631.
147. **Yamada, H., N. Tsukagoshi, and S. Udaka.** 1981. Morphological alterations of cell wall concomitant with protein release in a protein-producing bacterium, *Bacillus brevis* 47. *J. Bacteriol.* **148:**322–332.
148. **Zulauf, M., U. Fürstenberger, M. Grabo, P. Jäggi, M. Regenass, and J. P. Rosenbusch.** 1989. Critical micellar concentrations of detergents. *Methods Enzymol.* **172:**528–538.

Chapter 5

Antigen-Antibody Reactions

JOSEPH S. LAM and LUCY M. MUTHARIA

Reactions between antigen and antibody are usefully exploited in many areas of life science research (1–4), certainly including bacteriology. Antibodies have many virtues as biological reagents, including specificity for an antigen, availability, and the usually visible secondary reactions due to the divalent nature of the antibody molecule. The monoclonal antibody (MAb) technology developed by Köhler and Milstein (49) allows production of unlimited quantities of antibodies against virtually any molecule. The MAbs have unique epitope specificity, and the reaction is reproducible. Although MAbs have obvious advantages over conventional polyclonal antibodies, one should also consider their disadvantages, which include cost and the involvement of a great deal of labor. In most cases, the generation of polyclonal antibodies requires nothing more than the antigen, a rabbit, and a syringe. The resulting polyclonal antisera are adequate for most needs in bacteriology at a fraction of the cost of MAb production. Since there is an immense volume of material concerning all aspects of antigen-antibody reactions in the literature, the objective of this chapter is to provide readers with simple and useful protocols and an introduction to some of the more novel techniques.

5.1. ANIMAL HANDLING

The success of antibody production is dependent on many factors including the quality of the immunizing antigen, the dosage of the immunogen, the route of immunization, the species of animal used, and the state of health of the animals. With few exceptions, animals are used to generate the antibodies of choice; therefore knowledge of animal handling, knowledge of the guidelines of animal usage, and awareness of animal welfare are of primary concern for anyone who needs to produce antibodies.

5.1.1. Care Guidelines

Institutional animal facilities and programs should be operated in accordance with the governmental requirements and recommendations outlined in the *Guide for the Care and Use of Laboratory Animals*, the Animal Welfare Act, and other applicable federal regulations and policies. In most cases, the use of animals for immunization and the production of antibodies can be classified in the acute-care and low-pain category. However, it is necessary for the users to be aware of the established guidelines. In brief, all animals used should be cared for, transported, and handled in accordance with the Animal Welfare Act or its equivalent; the procedures used should be designed and performed with due consideration of their relevance to human and animal health; the appropriate use of animals is imperative, and, when necessary, the minimum and appropriate numbers of animals should be considered; sedation and anesthesia are to be used to minimize distress and pain; animals that would otherwise suffer chronic pain or distress that cannot be relieved should be painlessly killed after or during the procedure; appropriate living conditions should be provided for each species; and investigators and other personnel should be appropriately trained and qualified. A copy of the *Guide for the Care and Use of Laboratory Animals* can easily be obtained from the Animal Care Committee of an institution or directly from the Animal Welfare Division of the U.S. National Institutes of Health.

5.1.2. Species

Rabbits, mice, and rats are the species of laboratory animals most commonly used for antibody production. The preference for any particular species depends on the volumes of serum required and may also be influenced by the specificity of the generated serum. In general, rabbits are the animals of choice for polyclonal antibody production because of their size and, more importantly, their gentle nature and the ease of handling and the commercial availability of anti-rabbit immunoglobulin antibodies (second antibodies) that are already conjugated to either enzymes, biotin, fluorescence reagents, or electron-dense colloidal-gold particles. For MAb productions, one would naturally choose either mice or rats because most of the myeloma cell lines available are of murine origin. Taking into consideration the points discussed in section 5.1.1 above, no individual should handle animals without some form of training and demonstration of the procedures by qualified personnel. The details of how to handle each individual species of animals will not be described here but should be learned from experienced and qualified personnel. For detailed descriptions of techniques for handling animals, consult reviews on the subject (5–8).

5.1.3. Age and Size

5.1.3.1. Rabbits

Many breeds of rabbits, weighing from 2 to 7 kg, are readily available from commercial sources. For antibody production, and when large quantities are required, the larger animals should be considered. Lop-eared rabbits, which have a body weight of up to 7 kg and large ears and large veins to allow easy bleeding, are suitable. However, the most frequently used rabbit is the New Zealand White rabbit. The adult weighs between 5 and 6 kg and has the added advantage of a white skin and easily visible veins. Generally, 8- to 10-week-old animals weighing approximately 2 kg are purchased; by the time that they have been immunized and are ready to be bled, they would have been kept for another 4 to 6 weeks. Female rabbits are usually more tame and easy to handle. It is not uncommon to find researchers keeping rabbits for more than 6 months for continuous immunization and bleeding. However, be warned that, by that stage, the rabbits may have outgrown the standard-sized cages used by most institutions, and long-term care is expensive. As an alternative, consider using several smaller rabbits such as the Dutch Belted or the San Juan rabbit strains and pooling the sera to obtain a larger volume. This ensures that the standard-sized cages will be adequate to meet the animal-holding requirements.

5.1.3.2. Mice

Many laboratory strains are available from commercial suppliers, and some institutions may even breed some of the more common strains such as the Swiss White, which is often used for microbiological work. Because of the small size of the animals, one cannot expect to obtain more than 0.2 to 0.3 ml of blood unless one is ready to sacrifice the animal by cardiac puncture. BALB/c mice are the source of myeloma cell lines, and, for obvious reasons, these mice are used for immunization to obtain primed splenocytes for plasmacytoma fusion experiments for MAb production. Animals are usually purchased at 5 to 8 weeks postweaning, weighing approximately 20 g. After being housed for the duration of the immunization procedures, they are large enough to yield sufficient amounts of splenocytes.

5.1.3.3. Rats

The commonly available strains are the Hooded Lister and the two albinos, the Wistar and the Sprague-Dawley. Rats are stronger and more agile than mice, so they must be handled with caution. Do not attempt to handle rats without some proper training. The size of the animals ordered depends on how long they are to be kept. Animals that are 5 to 8 weeks postweaning can weigh between 75 to 150 g. They are slightly small, but, after they have been kept for the duration of the immunization and are ready to be bled, they should have grown to sufficient size to give a good yield of blood.

A frequently asked question is: How many animals are needed for generation of sufficient antibodies? If the amount of antigen is not limiting, more than one animal should be used, because, even in genetically identical animals, each one will respond to the same antigen differently (3). Therefore, immunize more than one animal with the same antigen, and, if they provide adequate titers of antibody, pool the sera collected. For rabbits, use a minimum of two; for mice use at least groups of five; and for rats use a minimum of three animals. *Note:* Antisera should not be pooled when the objective of the study is to evaluate the response of different animals to an antigen or to monitor the antibody response during the course of immunization.

5.1.4. Injection Routes and Dosages

Many factors determine the "immunogenicity" of an antigen. Among them, the dose of the antigen used and the route of injection can significantly influence the outcome. The routes of injections commonly used on the three species of animals described above include intradermal (i.d.), subcutaneous (s.c.), intramuscular (i.m.), intraperitoneal (i.p.), and intravenous (i.v.). Other routes of immunization (such as ingestion, inhalation, and skin application) are also possible, but the concern is with the more common laboratory practices. The i.d., s.c., and i.m. routes are normally chosen because of the slow release of the injected antigens, which can enhance an effective humoral response. The i.p. route of injection allows an antigen to come in contact with the lymphatics sooner than in the above-described methods. Also, i.p. injection is easy in rodents. Mice are often immunized by this method to prime their spleen cells for MAb production; however, this route is not recommended for rabbits because of their large size and the difficulty one would experience in practice. The i.v. route is more often used for administering booster dose for a quick response to the antigen, such as when one wishes to produce agglutinating antibodies against microbial surface antigens. When this route is chosen, one should be cautious to exclude air bubbles or adjuvants from injecting material. The practice of injecting into the footpad of rabbits to get the desired slow release of antigen should never be used. The footpad is a very sensitive site, and the method causes unnecessary severe pain and discomfort to the animal. For all injections, the materials should be suspended in isotonic saline/buffer to avoid burning sensations to the animal from the materials. Very low doses of antigen may cause either nonresponsiveness or tolerance, whereas very high doses can also cause tolerance. In general, between 50 and 500 μg of foreign antigen should be injected to elicit a good response. Table 1 summarizes the maximum volumes of materials that can be injected into animals via the various routes of injection.

Table 1. Volume of antigen-adjuvant mixtures for injections

Animal	Maximum vol (ml) for route of injection:				
	i.d.	s.c.	i.m.	i.p.	i.v.[a]
Rabbit	0.1	0.4	2.0	5.0	1.0
Mouse	0.1	0.1	0.2	1.0	0.2
Rat	0.1	0.2	0.5	2.5	0.5

[a]Freund's adjuvants should not be used with i.v. injections.

5.1.5. Immunization Protocols

Circulating antibodies against a specific antigen do not appear in significant amounts until at least 7 days after an immunization. Most of the antibodies of an early (or primary) response are of the immunoglobulin M (IgM) class, whereas antibodies from later (secondary) response are mostly IgG. However, if the antigen used is pure carbohydrate, the secondary response will be mainly IgM. The amount of antibody formed after a second or booster injection of any given antigen is usually much greater than that formed after the first injection; therefore, for the production of high-titer antibodies the following schedule can be used (for rabbits):

1. *Test bleed.* Collect preimmune serum (before the first injection) for use as a control to detect cross-reactive antibodies.
2. *Day 1.* Inject 0.1 ml (1:1 emulsified mixture of protein antigens with Freund's complete adjuvant; see section 5.2.4) at each of four sites s.c., usually at the dorsal part of the body near the shoulders and the hips). *Note:* Freund's complete adjuvant should not be used more than once, because repeat injections can cause granuloma formation and ulceration of the tissue at the site of injection (33). The reaction to only one injection with this adjuvant is usually mild.
3. *Day 4.* Repeat the day 1 injection steps but with Freund's incomplete adjuvant (see section 5.2.4). After this step, allow the animals to rest for 14 days so that the primary response subsides to a "baseline level" (otherwise a secondary response may not be achieved).
4. *Day 18.* Inject 1.0 ml (1:1 mixture of protein antigens with Freund's incomplete adjuvant) i.m. into the thigh muscles of one leg of the animal.
5. *Day 22.* Bleed, and collect serum.
6. *Day 25.* Repeat the above step of intramuscular injection on another leg of the animal.
7. *Day 29.* Bleed and collect serum, and pool the serum with the previous sample.

The i.m. injections can be kept up once a month or once every 2 weeks, and serum can be collected 3 to 4 days later. To avoid batch-to-batch variability, the immune sera collected at different times should be pooled unless one is attempting to monitor immune response throughout a time course. The above general immunization schedule has been used successfully on a wide range of antigens including soluble proteins, enzymes, whole bacterial cell suspensions, lipopolysaccharide antigens, and polysaccharide-protein conjugates. Variations can be made at some of the steps; for instance, i.v. injection, *without adjuvant,* may be given as a booster (on days 18 and 25) to elicit a more rapid response before bleeding to collect the antibodies.

5.1.6. Blood Collection

Blood should be collected in a dry, sterile container without any coagulant.

5.1.6.1. Rabbits

Restrain the animal by wrapping a large towel around it so that only the head and the ears are exposed. This is more efficient than the use of commercially built rabbit-holding boxes, since animals will get nervous about them; they may kick or bounce violently and could sustain severe injury such as a dislocated spine. Routinely, blood is withdrawn from the marginal vein of the ears of the animal. The part of the vein closer to the head of the animal can be occluded by using a paper clip or by holding firmly with the fingers to allow for a better flow of blood at the more proximal part of the vein to be punctured. Xylene can be applied to the base of the ear (not the tip) to dilate the veins. It is also essential to remove xylene by wiping the ear with alcohol at the end of the procedure to avoid further irritation to the animal. A small cut (superficial venesection) to a dilated vein with the tip of a fresh scalpel blade will open it up. The droplets of blood that emerge are collected in a suitable container, usually a screw-cap glass bottle. With sufficient practice, 20 to 30 ml of blood can easily be obtained by this old-fashioned but well-exploited method. If the cut is small enough, the bleeding will stop rapidly when the procedure is completed and gentle pressure is applied by hand with cotton wool on the cut. The cut will heal within a short time. Alternatively, the rabbit is lightly anesthetized and a 22-gauge needle can be inserted into the vein, bevel up; blood is allowed to drip from the needle into a collection container. If sterile blood is required, it can be withdrawn directly from the vein into a syringe and needle or into a Vacutainer (Becton Dickinson; for supplier addresses, see section 5.6). The latter will require proper instruction and practice to avoid causing hematoma to the veins of the animal.

Larger volumes of blood can be obtained by bleeding from the central ear artery or by cardiac puncture on anesthetized animals. The latter is not recommended for routine and repeated use on an animal but is useful if the animal is to be exsanguinated. An inexperienced animal handler should definitely be coached by qualified individuals if the cardiac puncture procedure is to be attempted. The following steps can be followed for bleeding from the central ear artery.

1. At 15 min prior to bleeding, administer s.c. or i.m. a dose of anesthetic (either 0.125 ml/kg of rabbit weight of Innovar-Vet or 0.4 ml/kg of a mixture containing 1 mg of oxymorphone and 0.5 mg of acepromazine). Innovar-Vet contains 20 mg of droperidol per ml and 0.4 mg of fentanyl per ml (9, 5.6.n.). The drugs described will both tranquilize the animal and induce vasodilation. The effect of the drugs will last 45 to 60 min.

2. Dilate the central ear artery by warming the ear with a lamp or gentle friction.

3. Insert a 20- or 21-gauge needle into the central ear artery, bevel up, and allow blood to drip into container. Again, for collection of sterile blood, aspirate the blood into a sterile syringe. To prevent hematoma formation, apply pressure over the puncture site until bleeding has ceased; it will take a few minutes since an artery of the animal is involved.

5.1.6.2. Mice

Only a small volume of blood can be obtained from a mouse because of the small size of the animal. The easiest and most humane way is to put the animal under light anesthesia, dip the tail into a beaker of warm water to dilate the tail veins, and then nick a tail vein (usually at about the one o'clock position relative to the dorsal part of the tail) with a sharp scalpel. The tail can be massaged lightly while blood is withdrawn into a Pasteur pipette. Only approximately 0.2 ml of blood can be obtained this way. Alternatively, blood can be withdrawn from the venous plexus, located in the orbit behind the eyeball. Naturally, the latter is a more delicate operation and should not be attempted without instruction. From 1 to 3 ml of blood can be obtained from the heart or from the heart cavity immediately after the animal is sacrificed.

5.1.6.3. Rats

Blood can usually be collected from the tail vein by the procedure described above for the mouse. Again, once the animal is sacrificed, larger amounts can be obtained from the heart.

5.1.7. Serum Collection

Blood withdrawn from animals should be allowed to clot at room temperature for 30 min; it can then be kept at 37°C for another 30 min. Serum can then be separated and centrifuged to get rid of all blood cells. Alternatively, some laboratories prefer to keep clotted blood at 4°C overnight for complete shrinkage of the clot before removing the serum. The latter procedure will yield more serum per milliliter of blood and will preserve antibodies from hydrolysis by the action of naturally occurring IgG proteases.

5.2. ANTIGEN PREPARATION

5.2.1. Particulate Antigens

5.2.1.1. Bacterial cells

Whole bacterial cultures are naturally immunogenic because of the "foreignness" and the complex nature of the bacterial cells. Therefore, the use of whole-cell suspensions to immunize animals is common, provided that a protective or a polyvalent antibody response against all components of the organism is desired (38). An effective dose of whole-cell suspension (section 5.1.5) ranges from approximately 5×10^8 to 1×10^{10} cells per ml. The estimation of cell counts can be achieved by one of the many standard methods including serial dilutions and plate counts, or measurements of optical density at 660 nm (OD_{660}) (Chapter 11). An OD_{660} of ≥ 0.6, given in comparison with *Escherichia coli*, usually indicates a cell concentration of 10^9. Alternatively, compare the cell suspension to McFarland standards (11), which are made up of various concentrations of $BaSO_4$. These $BaSO_4$ suspensions can be used to give a rough estimate of bacterial cell concentration from 10^8 to 10^{10} cells per ml. Comparison of the test and reference suspensions is made easier if they are held in front of a printed page.

In some cases, live attenuated vaccines are necessary; however, "infecting" animal hosts with live organisms may have some unexpected effects. Apart from the fact that live organisms may be more toxic and can cause distress to the animals, live bacterial cells may undergo antigenic and phase variations due to the pressure of host defense mechanisms. Bacterial cells can be killed either by heating at 100°C for 30 min or by suspending the cells in formalinized saline overnight, at a final Formalin concentration of 0.3%. The cells should be centrifuged and washed twice in isotonic saline to remove the free Formalin. Killing of the cells should be ascertained before being used for injections.

As a common practice, antigens should be divided into at least 10 doses and stored frozen with or without emulsifying with adjuvants. This way, the consistency of each dose can be ensured. Repeat thawing and freezing of emulsified antigen should be avoided.

5.2.1.2. Sheep erythrocytes as carriers

Both carbohydrate and protein antigens can easily be adsorbed onto sheep erythrocytes (SRBC) to render them agglutinable with antiserum specific for the "add-on" antigens. Thus, hemagglutination is a very easy and convenient method to assess antigen-antibody reactions. For the same reasons, SRBC can easily be used as a carrier for purified carbohydrate antigens, which by themselves are normally not very immunogenic in rabbits (12). By treating the SRBC with 0.005% tannic acid, low-molecular-weight protein antigens can also be adsorbed.

Attachment of crude LPS antigens onto SRBC

1. Boil an overnight culture of gram-negative bacteria for 1 h; centrifuge at $2{,}000 \times g$ for 10 min to sediment cell debris. The supernatant contains crude lipopolysaccharide (LPS) antigen extract.
2. Mix 4.5 ml of bacterial supernatant extract to 6 ml of 2.5% SRBC at a bacterial extract-to-SRBC ratio of

3:4 (vol/vol). (*Note*: commercially available SRBC usually are prepared in 2.5% solutions in citrate buffers and are usually stable at 4°C for at least 1 month.) Isotonic solutions such as phosphate-buffered saline (PBS) or Alsever's solution (see recipe below) should be used for all manipulations of the SRBC suspension. Incubate at 37°C for 30 min with occasional shaking.
3. Sediment the SRBC by centrifugation at 200 × *g* for 10 min, and wash the cells twice with 10 ml of saline or isotonic buffers.
4. Resuspend the pellet with 6 ml of PBS. This suspension is now ready for injection (Table 1) or for use in hemagglutination assays.

The attachment of LPS onto SRBC has the advantage of presenting the antigens of choice on highly immunogenic particulate carriers. In addition, the amount of LPS presented to the animals will be much smaller than a dose of pure LPS; therefore, this type of immunogen will be relatively nontoxic to the animals while eliciting a response to the LPS antigen.

Alsever's solution (citrate saline solution)

Alsever's solution is an isotonic, anticoaggulant blood preservative that permits the storage of whole blood at refrigerated temperatures for 10 weeks or more.

Dextrose	20.50	g
Sodium citrate (dihydrate)	8.00	g
Citric acid (monohydrate)	0.55	g
Sodium chloride	4.20	g
Distilled water	to 1	liter

Tanning of SRBC for coating with protein antigens

1. Add 3 ml of 0.005% tannic acid into a centrifuge tube containing 3 ml of 2.5% SRBC. Incubate at 37°C for 10 min.
2. Centrifuge the cells at 2,000 × *g* for 5 min, and wash once in 5 ml of PBS. Centrifuge as before, and resuspend the pellet in 3 ml of PBS.
3. To each centrifuge tube containing 3 ml of tanned SRBC, add 3 ml of 0.3 mg/ml soluble protein, mix gently, and incubate at 37°C for 15 min.
4. Centrifuge, and wash twice as above. Resuspend each pellet in 3 ml of PBS diluent. These cells are now ready for injection or for passive hemagglutination.

5.2.1.3. Bacterial cells as carriers

Bacterial smooth LPS containing O-antigen sugars is extremely immunogenic, while it is often difficult to raise antibodies to the LPS-core oligosaccharide epitopes of gram-negative organisms. To elicit a response to epitopes of the core region, antigens such as pure oligosaccharides, lipid A, or rough LPS can be attached to bacterial cells of a rough strain, e.g., *Escherichia coli* J5, by the method of Galanos et al. (13) with modifications by Bogard et al. (10), as follows.

1. Prepare 5 x 10^9 heat-killed cells of the rough mutant per ml of 1% (vol/vol) acetic acid.
2. Heat the cell suspension to 100°C for 1 h.
3. Wash three times in distilled water.
4. Lyophilize.
5. Dissolve lipid A or rough LPS in 0.5% (vol/vol) triethylamide at a concentration of 1 mg/ml; then add the lyophilized acid-treated bacteria to a final concentration of 1 mg/ml.
6. Stir slowly for 30 min at room temperature.
7. Dehydrate the mixture in vacuo with a Speed-Vac centrifuge (Savant Instruments Inc.).

5.2.2. Soluble Antigens

Complete antigens (immunogens) usually have a high molecular weight, although some naturally occurring immunogens such as insulin (molecular weight 6,000) and RNase (molecular weight 14,000) have a fairly low molecular weight. Molecules with a molecular weight of 3,000 to 5,000 are not good immunogens. As a rule of thumb, polypeptides larger than molecular weight 20,000 should be reasonably immunogenic and should induce a T-dependent response with an increased level of specific IgG. Smaller peptides can be conjugated onto larger protein carriers, adsorbed onto particulate carriers (such as SRBC, as described in section 5.2.1), or polymerized by treatment with a cross-linking fixative such as 0.5% (final concentration) glutaraldehyde for 1 h at room temperature. The material is dialyzed against buffer to remove any excess cross-linking agent. The disadvantage of glutaraldehyde treatment is the irreversible modification of certain "native" epitopes.

The purity of soluble antigens from bacteria depends on the extraction methods used. Soluble cell products such as toxins and enzymes are usually purified by a combination of column chromatography including ion-exchange, gel filtration, chromatofocusing, and affinity methods; these methods are further enhanced by more recently developed high-pressure liquid chromatography (HPLC) and fast protein liquid chromatography (FPLC) systems (Chapter 21). Chapter 4 specifically deals with fractionation of antigens from cells. The higher the purity, the easier it will be to raise specific antibodies against the antigen of interest. The following protocols represent some of the more frequently used "quick methods" for purifying antigens from polyacrylamide gels, provided that the protein band of interest has been identified.

5.2.2.1. Purification

Protein antigens are usually separated by the discontinuous sodium dodecyl sulfate-polyacrylamide gel electrophoresis (SDS-PAGE) methods of Laemmli (16) or, if native protein is important for eliciting antibody response to conformational epitopes, by omitting the ionic detergent SDS. The more commonly used methods for identifying the location of the band of interest in the gel are either staining a vertical strip of the gel to locate the antigen band or staining the entire gel lightly. The former method has the advantage of avoiding chemical fixation of the entire gel and the antigen of interest. Normally, 0.05% Coomassie brilliant blue R-250 is used. The sensitivity is roughly 1 to 2 μg per band in a lane of standard width. If the more sensitive silver staining method has to be used to see the band, the quantity is insufficient for immunization.

Direct use of antigens in gel slices

1. Excise the band of interest with a razor or scalpel blade. Before throwing away the rest of the gel, stain it to check the accuracy of the excision.

2. Wash the gel slices in PBS or distilled water for a few minutes. Chop the gel into thin slices with a razor blade or repeatedly squeeze the gel pieces through the barrel of a standard 5-ml syringe with a small amount of buffer. If the latter method is used, perform a final squeezing step through the syringe with a 21-gauge needle attached.
3. For immunization, the homogenate could be used directly or lyophilized and ground to a fine powder. However, it is more effective to electroelute the proteins from these chopped-up gel slices before injection, since acrylamide may not be easily degradable in the animal host.

5.2.2.2. Electroelution

Antigens can be electroeluted from gel slices under conditions described by Leppard et al. (17). Briefly, place gel slices into a small dialysis tube containing 1 ml of 0.2 M Tris acetate (pH 7.4), 1.0% SDS, and 100 mM dithiothreitol per 0.1 g of wet polyacrylamide gel; then place the dialysis tube in a horizontal electrophoresis chamber. Alternatively, place gel slices into an ISCO sample cup (Canberra Packard) with a dialysis membrane attached to the bottom of both chambers. Use running buffer containing 50 mM Tris acetate (pH 7.4), 0.1% SDS, and 0.5 mM sodium thioglycolate. Run for 3 h at 100 V. Remove the gel slices, and stain them with Coomassie blue to verify removal of the antigens. Then dialyze the protein solutions against distilled water, and lyophilize. The lyophilized sample can now be incorporated into an emulsion with adjuvant for immunization.

5.2.2.3. Extraction

As an alternative to the above procedures, antigen bands can be transferred onto nitrocellulose membranes, and the band of interest can be identified and finely ground for injection (15). This method has the advantage of the resolution of protein separation of the usual electrophoresis method as well as easy location of the band of interest.

1. Perform the usual electrophoresis (SDS-PAGE) technique (Chapter 21.5.2).
2. Transfer the protein by electrophoretic blotting as described in section 5.5.5.
3. Stain the nitrocellulose with either India ink (14) or Ponceau S (3). The latter stain is reversible and thus should not affect the antigenicity of the antigen of interest. If the protein bands are barely visible with these two stains, it means that there may not be a sufficient quantity for immunization.
4. Identify the band of interest, and excise with a scalpel blade.
5. Cut the nitrocellulose band into very small pieces, add 200 μl of buffer or PBS, and sonicate for 10 bursts of 10 s each. Cool between each burst.
6. The resultant fine suspension can now be used for injection with or without adjuvant.

5.2.3. Haptens

Antigens of molecular weight below 10,000 are usually weak immunogens. Therefore, to raise specific antibody it is necessary to conjugate low-molecular-weight molecules with little or no immunogenicity (haptens) onto carriers before immunizing animals. The carriers can be protein molecules, such as bovine serum albumin (BSA) or keyhole limpet hemacyanin (KLH), or particulate antigens (section 5.2.1). For more details on the choice of peptide sequences and the strategy on coupling of peptides to carriers, consult the appropriate chapters in the volume by Harlow and Lane (3).

A common procedure for conjugating haptens onto protein carriers involves glutaraldehyde, which cross-links molecules through their amino groups. The following procedure has been used successfully to conjugate cadmium-binding peptide (18) and can be used for conjugating other very small peptides with 9 to 25 amino acids onto BSA.

1. Add 2 ml of 0.2% (vol/vol) glutaraldehyde to a mixture containing 3 mg of a peptide in a 1-ml volume and 1 mg of BSA in a second 1-ml volume. To decrease the rate of coupling, the buffer used in the reaction mixture should be above the pK_a of the amino group (if known) so that NH_2 and not NH_3 will be targeted.
2. Incubate the mixture at room temperature for 2 h with gentle mixing.
3. Dialyze for 24 h against several changes of 5 mM Tris-HCl (pH 8.0).
4. Concentrate the contents of the dialysis bag by using either a Speed Vac (Savant Instruments, Inc.) or by sprinkling dry flakes of polyethylene glycol (PEG; molecular weight 6,000 or higher) on the surface of the dialysis bag placed in a Pyrex dish. This process can be accomplished in a matter of minutes if sufficient PEG flakes are used to totally embed the dialysis bag.

Alternatively, a higher concentration of glutaraldehyde, 1 to 2% solution, can be used to cross-link amino groups of the peptide of interest such that polymerization of the small peptide will occur. Polymerization will increase the size of the antigen, but it may also interfere with the epitopes of interest.

5.2.4. Adjuvants

With few exceptions, adjuvants are always used to enhance antibody response to an antigen. Reviews by Edelman (25) and Warren et al. (41) give more specific details on the comparison of adjuvants. Table 2 summarizes commonly used adjuvants and their characteristics.

5.2.4.1. Freund's adjuvants

Both the complete and the incomplete forms of Freund's adjuvant are readily available commercially (Sigma Chemical Co.).

1. To antigens in aqueous form (usually suspended in isotonic saline), add an equal volume of Freund's adjuvant (either the complete or the incomplete form). Mix the two ingredients vigorously until a thick emulsion is formed.

   To generate the emulsion, take the mixture into a glass syringe fitted with an 18-gauge needle and then

Table 2. Adjuvants and their characteristics

Adjuvant	Characteristics	References(s)
Freund's complete adjuvant (FCA)	The most commonly used adjuvant. Mix 1:1 (vol/vol) with antigen until thick emulsion is formed. Serves as a depot for slow release of antigens. An emulsion of oil (Bayol F), detergent (mannide mono-oleate), and extract from mycobacteria. Must never be used for i.v. injections. Disadvantages: (i) elicits granuloma at the site of injection if used repeatedly; (ii) the emulsion is difficult to prepare.	29
Freund's incomplete adjuvant (FIA)	Contains the oil and detergent only and no bacterial extract. Mix 1:1 (vol/vol) with antigen as above. Do not use in i.v. injections. Normally used to replace FCA in subsequent injections.	29
Alum	An aluminum hydroxide salt (gel-like) which also facilitates slow release of antigens. Attracts immunocompetent lymphocytes to the area of injection, thereby able to induce an improved antibody response. Elicits IgG_1 and IgE response in mice. Is less toxic than FCA.	36, 40
Liposomes	Could be made as lecithin-cholesterol-dicetyl phosphate in molar ratio of 7:2:1. Is useful to entrap antigens for slow release. Advantages: (i) functions as a depot of antigen and provides a better presentation of incorporated antigens; (ii) helps antigens to retain T-cell dependency. Disadvantage: entrapped antigens that are not exposed on the outer lipid layer may not be seen by the immune system.	19, 31, 39
Lipopolysaccharide (LPS)	A B-cell mitogen Stimulates natural (nonspecific) antibody if administered without antigen, but stimulates specific antibody response when injected with an antigen. Usually, the lipid A region is the immunopotentiating component; however, mannan O side chain sugars from *Klebsiella* strain 03 and *Escherichia coli* 09 were shown to have adjuvant effects as well. Toxicity is its major disadvantage.	30, 37
Bordetella pertussis	A human pathogen that causes whooping cough. Lipooligosaccharide (LOS) and pertussis toxin (PT) are its principal adjuvant components. LOS enhances antibody response while PT potentiates cell-mediated immunity. Toxicity is its major disadvantage.	41
Muramyl dipeptide	The active component of mycobacterial extract used in FCA. Is prepared in many forms and derivatives. Stimulates specific antibody response to a wide variety of natural antigens such as bacteria, viruses, and fungi. Is nontoxic	27
Quil A	Is composed of saponin, a mixture of water-soluble triterpene glycosides from the South American tree *Quillaia saponaria*. Advantages: is highly surface active and forms stable complexes known as ISCOMS (immunostimulating complexes) with viral envelopes. Has also been reported to be superior to FCA in eliciting antibody response to human serum albumin, but the effectiveness of this adjuvant with bacterial antigen is unknown. Is toxic to small animals.	32, 34, 35
Syntex adjuvant formulation 1 (SFA-1, Syntex)	A synthetic MDP. Was shown to stimulate IgG_2 to human serum albumin in mice.	32
Ribi adjuvant system (Ribi)	Is composed of monophosphoryl lipid A (MPL). From 5 to 50 μg of MPL was shown to augment humoral response to polysaccharide antigens. Also augments response to protein antigens. Is nontoxic and has low viscosity; thus antigen-adjuvant mixtures are easy to prepare.	20, 21, 28
Poly (A-U)	Has enhanced immunogenicity to inactivated bacterial and viral vaccines.	22

force the material repeatedly in and out of the syringe until it is almost impossible to move the syringe plunger. Alternatively, mix the antigen and adjuvant vigorously in a vortex until a thick emulsion develops. The emulsion is deemed ready when a droplet of it does not disperse immediately when added to a saline solution.
2. Take up the volume needed for injection into a fresh syringe, and add a needle of an appropriate size. The immunogen preparation is then complete.

5.2.4.2. Alum

The procedure of Chase (23) can be used, as follows.

1. Add 50 ml of 10% (wt/vol) aluminum potassium sulfate (dissolved in distilled water) to a 50-ml conical centrifuge tube. Then add 22.8 ml of 0.25 N NaOH dropwise to the tube while vortexing.
2. Allow the mixture to settle for 10 min at room temperature.
3. Sediment the $Al(OH)_3$ by centrifugation at 1,000 × *g* for 10 min. Wash with 50 ml of distilled water, and repeat the centrifugation. Discard the supernatant.
4. The pellet contains the $Al(OH)_3$, or alum adjuvant, which has a binding capacity of 50 to 200 μg of protein antigen.
5. Add an aliquot of antigen to the alum, and allow the two components to mix at room temperature for 10 min. Centrifuge at 10,000 × *g* for 10 min. Check the supernatant for the amount of unbound antigen. The sediment is now ready for injection.

5.2.4.3. Liposomes

Liposomes are micelles that take the shape of concentric spheres consisting of phospolipid bilayers. Proteins and other antigens can be trapped inside the liposomes. Furthermore, depending on the nature of the interaction between antigen molecules and the liposomes, hydrophilic parts of antigens can be well exposed on the liposomes. A commonly used mixture of lipids (composed of egg lecithin, cholesterol, and stearylamine in molar ratios of 7:2:1) forms positively charged liposomes. If negatively charged liposomes are desired, phosphatidic acid or dicetylphosphate replaces stearylamine. In a study by Allison and Gregoriadis (19), negatively charged liposomes are reported to be superior to positively charged ones as adjuvants to elicit antibody response to diphtheria toxin. Although there are obviously many possible combinations of phospholipids, the general protocol of Claassen et al. (24) can be used to generate liposomes as follows.

1. Dissolve 75 mg of phosphatidylcholine and 11 mg of cholesterol in choloroform in a round-bottom flask. A thin film will develop after low-vacuum rotary evaporation at 37°C for a short time or under an atmosphere of nitrogen.
2. Prepare BSA-liposomes by adding 10 mg of BSA in 10 ml of PBS to the flask with the thin film of lipid and incubating at 37°C for 1 h with gentle rotation. Alternatively, sonicate the mixture for 30 min.
3. Wash the BSA-liposomes three times with PBS (100,000 × *g* for 30 min) to remove unattached BSA. Resuspend the pellet in a small volume of saline or PBS. The amount of bound BSA can be assessed by standard protein assay procedures. The antigen-incorporated liposomes are ready to be used for injections.

5.2.4.4. Ribi adjuvant systems

Very effective adjuvant systems are available commercially (Ribi ImmunoChem Research, Inc.). These systems contain a combination of mycobacterial cell wall skeletons, trehalose dimycolate, and a nontoxic lipid A derivative, monophosphoryl lipid A. The adjuvant effect appears to be based on its ability to inactivate suppressor T cells while stimulating a polyclonal B-cell response (21, 26, 28). These systems are proven nontoxic in human as well as animal use.

5.3. CONVENTIONAL ANTIBODY PREPARATION

For conventional antibody preparation, the immune serum is expressed from the clot by centrifugation and approximately 52% of the blood volume can be collected as serum (section 5.1.7). The two principal types of proteins in serum are globulins and albumin, which can be fractionated by various means. The most common combination of purification methods involves ammonium sulfate precipitation, followed by dialysis to get rid of the salts and anion-exchange chromatography. This combination is tedious but inexpensive and can effectively isolate all the IgG molecules from serum. Gel filtration may be used instead of anion exchange when IgM purification is desired. More recently, the use of affinity chromatography can dramatically speed the process, but, more importantly, it can help to target the specific antibody that one really wants if purified antigen is incorporated in affinity column matrices. Affi-Gel Blue (Bio-Rad Laboratories) can easily replace the salt precipitation step because the Cibacron Blue dye in this affinity gel matrix (1.9 mg/ml) has a strong affinity for binding albumin at 11 mg/ml; therefore it can be used as a one-step method to yield a relatively clean IgG fraction. HPLC and FPLC methods involving a high-pressure pump and special columns both provide excellent speed and resolution for the separation of antibodies from other serum proteins. However, the equipment is expensive and may not be readily available. For additional information on (enzyme) protein purification, see Chapter 21.7.

5.3.1. Ammonium Sulfate Precipitation

Ammonium sulfate is the most widely used salt for the precipitation of proteins. It has the advantage of a high solubility that is only minimally dependent on temperature, varying only about 3% between 0 and 25°C. In contrast, Na_2SO_4, which is called for in some methods, is five times as soluble at 25 as at 0°C. Proteins are polyvalent ions, with surface charges that interact with water molecules through hydrogen bonding. Since SO_4^{2-} ions also attract water molecules, they compete with the proteins and, at increased concentration, will strip away the solvation layer. The protein molecules will then have an increased tendency to interact with each other and precipitate out of solution. The gamma globulin fraction, containing antibodies, is obtained on repeated precipitations with $(NH_4)_2SO_4$ added to 33.3% saturation solution. Unwanted proteins including albumin will remain in solution. A higher yield can be achieved by a single 50% saturation, although a small amount of albumin may be coprecipitated.

Procedure

1. To prepare a saturated stock solution of $(NH_4)_2SO_4$, add 100 g of the crystals per 100 ml of distilled water and leave it stirring for 1 to 2 days on a magnetic stir plate at room temperature. There should always be undissolved crystals in the bottom, otherwise the solution is not saturated. Check and adjust the pH to 7.0. Store at 4°C since the antibody precipitation is usually performed at this temperature.
2. To 20 ml of serum (already cooled to 4°C), slowly add (dropwise) 10 ml of saturated $(NH_4)_2SO_4$ from the stock solution while the solution is stirred on a magnetic stir plate. Continue the stirring for at least 2 h at 4°C.
3. Sediment the pellet by centrifugation at 10,000 × *g* for 10 min.
4. Resuspend the pellet with a small volume of buffer (5 ml), e.g., PBS or TBS (50 mM Tris, 1.5 M NaCl) at pH 7.0, and then add 2.5 ml of saturated $(NH_4)_2SO_4$ as in step 2.
5. Sediment the pellet by centrifugation as in step 3.
6. Resuspend the pellet with 10 ml of buffer, and transfer it into a dialysis bag.
7. Dialyze overnight at 4°C against two or three changes of 2-liter volumes of buffer. The dialyzed material can now be stored frozen or can be lyophilized prior to further steps such as anion-exchange chromatography.

5.3.2. Ion-Exchange Chromatography

For the purification of IgG, ion-exchange chromatography is a natural second step after ammonium sulfate precipitation to get rid of albumin. The most commonly used matrix for antibody purification is a matrix to which an ionizable (DEAE) group is attached. DEAE-Sephacel (cellulose) and DEAE-Sepharose are conveniently supplied (Pharmacia) preswollen and ready to use. These matrices have a very high capacity, and a 10-ml column can be used to bind 100 to 200 mg of protein. At pH 8.0, the matrix has a strong positive charge for binding antibodies. Immunoglobulins have isoelectric points in the range of 6 to 8; therefore they will be negatively charged (anionic) at pH 8.0 and will bind to DEAE at low salt concentration. Antibodies can easily be eluted either by increasing the concentration of competing ions (e.g., Cl^-) in the column buffer or by lowering the pH. The recovery of antibodies should be almost 100%.

Procedure

1. Pack and equilibrate a DEAE-Sephacel or DEAE-Sepharose column (1.5 by 10 cm) with 10 mM Tris buffer (pH 8.0) at room temperature. To avoid formation of air bubbles which will diminish the efficiency of the column, the buffer and gel slurry should be allowed to equilibrate to room temperature before starting. The size of the gel bed will be determined by the amount of the antibodies to be bound. As described above, a 10-ml column can be packed for every 100 mg of protein to be bound. If fancy columns are not available, a plastic disposable syringe will suffice as long as some fiberglass is used as a screen to prevent leakage of the slurry.
2. Wash the column with at least 5 column volumes of the 10 mM Tris buffer. Gravity feed with a head pressure of 100 to 200 mm is sufficient (48) if a peristaltic pump is not available.
3. Apply the ammonium sulfate-precipitated and dialyzed antibody solutions onto the column. If the dialyzed antibodies have been lyophilized, reconstitute them in the gel buffer (10 mM Tris).
4. Wash the column with 2 to 5 bed volumes of Tris buffer. Collect these fractions to ensure that antibodies are not lost.
5. Develop the column with a linear gradient of increasing concentrations of NaCl made from 100 ml of 10 mM Tris buffer in the first chamber and 100 ml of 0.3 M NaCl in 10 mM Tris buffer in the second chamber of the gradient maker. The usual total volume of the gradient is 10 to 20 times the bed volume. Allow the gradient to run for 4 h, and collect 5-ml fractions. Gradient makers can be purchased from commercial sources; as long as both chambers are of identical size, linear gradients can easily be achieved. The solutions in the two chambers can be mixed by magnetic stirrers. Mouse IgG2a usually is eluted earlier, and IgG1 is eluted later. Mouse IgG3 may not be stable at low ionic strength; therefore the concentration of the Tris buffer used for purification of IgG3 should be increased to 50 or 100 mM.

5a. Alternatively, the column can be eluted with a stepwise salt gradient. Use a series of 50-ml volumes of 10 mM Tris (pH 8.0) containing no salt, and then use increasing salt concentrations of 50, 100, 150, 200, 250, and 300 mM NaCl, respectively. Most of the antibodies should be eluted by the 200 mM NaCl step.

6. Monitor the fractions by using a spectrophotometer fitted with a UV monitor or by assaying for proteins. The purity of the antibody eluted can be assessed by standard agarose gel electrophoresis or SDS-PAGE.

5.3.3. Gel Filtration

Gel filtration chromatography separates proteins by their molecular size. Since IgM has a molecular weight of approximately 900,000, it can be separated easily from other antibodies with average molecular weight of approximately 150,000. Many commercial gel matrices are suitable for this purpose, including Sepharose 6B (Pharmacia), Sephacryl S-500 (Pharmacia), Sephadex G-200 (Pharmacia), Ultrogel AcA 22 (LKB), and Bio-Gel P-300 (Bio-Rad). Sephacryl (a mixture of dextran and acrylamide) and Ultrogel (a mixture of agarose and acrylamide) are more resistant to compression, are easier to pack, and can be run at higher pressures and flow rates than gels such as Sephadex (48). The columns used are typically long and thin (15 to 30 mm in diameter). When packed and properly degassed, the gel generally occupies approximately 70% of the column, and the space between the gel beads occupies the remaining 30%. Either Tris or PBS can be used at pH 7 to 8 and a salt concentration of at least 100 mM to prevent adsorption effects. Sodium azide at 10 mM may be added to running buffers to avoid microbial growth. For best resolution, the flow rate of buffers should be low, which means that it will take 1 to 3 days to perform a proper run. An obvious disadvantage besides the time this takes to perform is the dilution of the antibodies being eluted. However, the fractions are easily concentrated by lyophilization or polyethylene glycol (section 5.2.3).

5.3.4. Affinity Chromatography

Antigen covalently bound onto a chromatographic gel matrix provides specific binding sites to isolate the antibody of choice from serum. Some of the commonly known gels are CNBr-activated Sepharose 4B and Sepharose 6B (Pharmacia), which are ready for direct coupling of antigen through their primary amino groups. For obvious reasons, direct coupling of high-molecular-weight microbial polysaccharides onto CNBr-activated gels is not possible until polysaccharides are conjugated onto protein carriers. More appropriately, the microbial polysaccharides can be coupled through binding of their carboxyl groups to one of the following gels: AH-Sepharose 4B (Pharmacia), Affi-Gel 102 (Bio-Rad), or Aminoethyl Bio-Gel (Bio-Rad). Detailed descriptions of the principles and procedures of coupling ligands to these affinity gel matrices can be found in a monograph entitled *Affinity Chromatography—Principles and Methods*, which can be obtained free of charge from Pharmacia.

5.3.4.1. Affi-Gel Blue

In addition to the aforementioned activated gel matrices for preparing affinity columns, a product from Bio-Rad known as Affi-Gel Blue is worth mentioning (44). Affi-Gel Blue (50- to 100-mesh) is a beaded, cross-linked agarose with covalently attached Cibacron Blue F3GA dye. There is approximately 1.9 mg of dye per ml of gel, and the dye has a very high affinity for albumin. It can bind ≥ 11 mg of albumin per ml of the Affi-Gel Blue. Thus, a single passage of serum through a minicolumn of this gel can yield a relatively pure immunoglobulin fraction with the albumin virtually all bound. This method is very rapid when compared with the steps involved in ammonium precipitation, dialysis, and concentration to isolate immunoglobulins. Minicolumns of 5 to 10 ml can be prepared in Econo-columns (Bio-Rad) or in syringes. To reconstitute a column, albumin can be removed by using 3.0 M potassium thiocyanate or 8 M urea (48) and then the column can be reequilibrated with running buffer of near neutral pH.

5.3.4.2. Protein A-agarose

Protein A is a cell wall component of *Staphylococcus aureus* which was found to have a strong affinity for the Fc region of IgG. It has a molecular weight of 42,000 and consists of four globular immunoglobulin-binding sites at the N-terminal end. It interacts with immunoglobulins of 65 mammalian species (52). Purified protein A can be covalently attached to gel beads such as agarose to be used as affinity gels for the purification of IgG. The binding of protein A to IgG is generally at least fivefold stronger than binding to IgM. However, protein A-agarose columns can be used to purify IgM if buffers of high ionic strength and high pH are used (45). Protein A-agarose matrix can easily be obtained from Pharmacia or Bio-Rad; the latter is known as the MAPS (monoclonal antibody purification system), which can be purchased as a kit with the column and all the necessary buffers. In general, IgG2a, IgG2b, and IgG3 can be easily purified by this method and IgG1 and IgM can also be purified when high pH and high ionic strength are used (46). The antibodies are usually eluted with acidic buffers such as citrate or glycine-HCl buffer (pH 5.0 or lower). The buffer conditions described below for HPLC purification of MAbs can be used. Naturally, normal columns should be run at a much lower flow rate, such as 2 to 5 ml/h, compared with 5 to 10 ml/min in the HPLC systems. To avoid clogging up of the rather expensive protein A-agarose columns, the methods described for pretreatment of ascitic fluid could be used. Samples should at least be passed through a 0.45- to 0.22-μm-pore-size filter before being applied to the column. Acid-eluted antibodies should be neutralized with 1 M Tris (pH 8.0) (48) immediately after elution from the column to avoid any loss of antibody activity.

5.3.4.3. Protein G

Protein G is a cell wall component of ß-hemolytic group C or G streptococci (43). Native protein G has a strong affinity for the Fc region of all IgGs, including IgG classes that have low affinity for protein A; however, this protein also binds albumin (42). Recombinant protein G in which the albumin site has been eliminated is now available for immunoglobulin purification and immunodetection procedures (42).

5.4. MONOCLONAL ANTIBODY PREPARATION

Since the first report by Köhler and Milstein (49) of cell fusion between a hypoxanthine-aminopterin-thymidine (HAT)-sensitive variant of the MOPC-21 myeloma cells and spleen cells immunized with SRBCs, the techniques of generating hybridomas have been well exploited by scientists in all disciplines of life sciences. The main advantages of MAbs over conventional antibodies include reproducibility, specificity, improved sensitivity, and the ability to produce unlimited amounts. These advantages usually outweigh the expense and the fact that people involved in the production of MAbs require a fair amount of training. However, one should consider the "real need" for these MAbs before embarking on the process. For most intended purposes, there is nothing wrong with rabbit antibodies, which will cost a fraction of the cost of MAb production. For consideration of legal aspects in the use of animals for MAb production, consult the review by Kuhlmann et al. (50). The following protocol is for the production of murine MAbs by a cell fusion technique with a special mixture of PEG and dimethyl sulfoxide (DMSO).

5.4.1. Production Protocol

Most protocols used for this purpose are based on the protocol of Galfré and Milstein (47) with PEG as a fusogen instead of Sendai virus as used in earlier studies. The following is a brief outline of the procedure described by Lam et al. (51) for the production of MAbs against lipopolysaccharide antigens; all the reagents are from Sigma.

1. In a 50-ml sterile centrifuge tube, mix 10^8 spleen cells in 10 ml of serum-free medium (either RPMI 1640 or Dulbecco modified Eagle medium [DMEM]) from an immunized animal (BALB/c mouse) with 10^7 myeloma cells in 10 ml of the above medium. This ratio of 10:1 of the two cell types may be varied depending on personal experience.
2. Centrifuge at 400 x *g* for 5 min.
3. Resuspend the pellet with 20 ml of serum-free medium, and split the cell suspension into four aliquots of 5 ml each, and put into 15-ml sterile centrifuge tubes. The purpose of splitting the cells is to obtain four pellets, which will provide better exposure to the fusogen step below. Repeat step 2.
4. To one of the four cell pellets, dispense 0.5 ml of fusogen (40% PEG [molecular weight, 1,050], 4% DMSO) in small dropwise quantities *in exactly 2 min.* (Keep the other three tubes with the cell pellets on ice until ready for use.)
5. Add 20 ml of serum-free medium over 3 to 5 min.
6. Centrifuge at 400 × *g* for 10 min.
7. Resuspend the cell pellet gently with 12.5 ml of medium containing 20% fetal calf serum. It is acceptable to leave some clumps. Repeat steps 4 through 7 for each of the remaining cell pellets.
8. Combine the four fused cell suspensions, each 12.5 ml, and transfer to a large (75-mm²) tissue flask. Incubate overnight at 37°C under 5% CO_2. This allows the cells to recover from the PEG shock before subjecting them to the selective medium containing HAT (in medium containing 20% fetal calf serum).
9. On the next day, add 50 ml of medium containing 2 × HAT, which can be obtained as a 50 × stock from Sigma, to yield a cell density of 5 x 10^4 to 5 x 10^5 myeloma cells per ml. This is the concentration in which myeloma cells grow best.
10. Add feeder cells, either normal spleen cells or thymocytes, at a concentration of 10^5 cells per ml.
11. Transfer appropriate aliquots of the cell suspensions to 24-well or 96-well tissue culture plates.
12. Leave the fused cells undisturbed for 10 to 14 days in an incubator set at 37°C with 5% CO_2 and 95% humidity. Hybrid clones that can survive the HAT selection will appear as plaques on the bottom of the wells. These plaques or colonies of clones can be confirmed by viewing on an inverted microscope.
13. Assay for specific antibody activities against the immunizing antigen (e.g., by enzyme-linked immunosorbent assay [ELISA; section 5.5.4]), and identify the wells with positive activities.
14. Clone cells in the positive wells by the limiting dilutions procedure of Oi and Herzenberg (54). The medium used should contain 20% fetal calf serum, hypoxanthine, and thymidine, which can be obtained as a 50× stock from Sigma.
15. Reclone cells as in step 14.
16. Culture supernatant from the twice-cloned cell lines usually contains microgram quantities of MAbs, which are ready to be used for most antigen-antibody reactions. To produce larger quantities (such as milligram-per-milliliter amounts) of MAbs, the method of ascites production described below can be used.

5.4.2. Isotype Determination

Before an MAb can be further characterized or exploited, it is necessary to identify the antibody isotype including class, subclass, and whether it has a kappa or a lambda light chain. Commercially produced anti-mouse antibody reagents are readily available such that there no longer is a need to produce antibodies against each of the immunoglobulin isotypes to be identified. Depending on which commercial mouse antibody-typing kit one purchases, the technique used may be the simple double immunodiffusion test, the more sensitive passive agglutination test, a dot-blotting method with precoated nitrocellulose strips, or the highly sensitive ELISA method. Table 3 summarizes the various characteristic commercial kits.

Table 3. Commercial source of immunoglogulin isotype determination reagents

Commercial reagents or kits	Source and catalog no.	Technique used	Remarks
Mouse Typer Isotyping Kit	Bio-Rad no. 172-2055	ELISA	Complete kit includes rabbit anti-mouse sera, goat anti-rabbit antibody-peroxidase conjugate, substrate solutions, and anti-κ and anti-λ light chain reagents.
Monoclonal Typing Kit—Mouse, immunodiffusion	ICN-Flow no. 64-690	Gel diffusion	Simple to do, but one has to wait 24 to 48 h before reading the results.
Monoclonal Typing Kit—Mouse, hemmagglutination	ICN-Flow no. 69-216	Hemagglutination	More sensitive than gel immunodiffusion; can detect 1 to 20 μg of MAbs.
Mouse monoclonal antibody isotyping reagents	Sigma no. n-ISO-2	Choice of users	Six reagents provided to identify all the class and subclasses, but no anti-light-chain reagents. The reagents can be used in either ELISA or immunodiffusion methods.
Sigma Immuno Type Kit	Sigma no. n-ISO-1	Dot blot	Precoated strips and reagents for color development are provided.

5.4.3. Ascites Production

The term "ascites" refers to an accumulation of fluid in the abdominal cavity. In human medicine, it is usually associated with circulation congestion due to disorders of the heart, lungs, kidneys, or liver. For research purposes, ascites is induced in mice as a source of highly concentrated antibody production because these fluids contain the same level of antibody as do the rest of the body fluids, i.e., 10 to 20 mg of antibody per ml. This level is at least 100 to 1,000 times higher than that in tissue culture supernatant of hybridoma cell lines. Both Freund's incomplete adjuvant and Pristane, a defined mineral oil (Aldrich Chemical Co.) (55, 56) can induce ascites in mice. Pristane has been particularly useful for priming BALB/c mice for the injection of myeloma cells to induce antibody production. An average of 5 to 10 ml of ascites fluid can be generated by this procedure. However, before this route of antibody production is chosen, it is necessary to realize that both Freund's adjuvant and Pristane cause immunosuppression yet induce tumor formation in the peritoneal cavity of the injected animals. Without doubt, such a process will cause stress in the animals; therefore, prolonged injection of the animals and more than one harvest of ascites from the animals should be avoided to minimize the stress inflicted. The procedure is outlined below.

1. For Pristane priming, inject 0.1 ml of Pristane into a BALB/c mouse and, 1 week later, inject 0.2 ml of Pristane i.p. (Most published protocols recommend one or two doses of 0.5 ml of Pristane for priming; however, increasing the second dose to 0.5 ml or higher does not increase the volume of ascites fluid recovered. By using two smaller doses of Pristane as described, animals are subjected to less stress and the survival time is longer than when 0.5 ml is used.)
2. Pristane priming, optimally, should not occur more than 60 days prior to injection of the hybridoma cells.
3. One week after the second dose, inject 10^6 hybridoma cells (in 0.1 ml) i.p. into the animal. *Note:* A solid tumor sometimes occurs, either as a result of the Pristane-priming conditions or as a result of a "low-secreting" characteristic of a certain cell line (48).
4. Once the abdominal area of the animal begins to enlarge, watch the animal closely to avoid overextension of the peritoneal cavity and skin. When the bulging of the abdomen is deemed large enough, drain the animal aseptically with a syringe and a 20-gauge needle. (At this point, the animal cannot keep itself clean and will appear stressed and scruffy; therefore it should be killed by euthanasia.)
5. Remove the needle and dispense the fluid into sterile 15-ml conical centrifuge tubes with caps. Sediment the cells by centrifugation at 1,000 × *g* for 10 min.
6. Separate the fluid (which contains the antibodies) from the pellet. *Note:* The pellet contains perhaps the most healthy hybridoma cells that one can get, thus they can be resuspended in HT- or HAT-containing DMEM to be propagated for freezing and storage. Cell lines that are suspected to contain contaminating microbes can actually be cleaned up by passage through an animal by following this protocol. The normal leukocytes from the ascites will eventually die and should not pose any problem to the purity of the hybridomas.
7. The ascites fluid is ready to be used for antigen-antibody reactions, with or without further purification.

Many undesirable components such as lipid, adjuvants, and fibrins are included in ascites collected from the animals. These impurities can clog affinity columns and the more expensive HPLC columns and must therefore be removed before purification steps are performed. The following examples can be used separately or in combination to achieve this goal.

1. Add $CaCl_2$ to a final concentration of approximately 1 mM (as per Bio-Rad instructions for Affi-Prep Protein A column). Allow the mixture to sit for 2 h at 4°C. Use a wooden applicator stick or other suitable device to remove the fibrin clot and any lipid. Centrifuge at 10,000 × *g* for 10 min at 4°C, collect the clear supernatant, and filter it through a 0.22- or 0.45-μm-pore-size sterile filter. The supernatant is now ready for injection into HPLC columns.
2. Ascites can also be poured through a small Sephadex G-25 column to adsorb clogging materials. The amount of Sephadex to use is 1 ml/ml of ascites. The column should be preequilibrated with 50 mM Tris adjusted to pH 8.6 with HCl. Again, filter the eluant with a sterile filter as described above before subjecting it to further purification (see protein A-Sepharose protocols [Pharmacia]).
3. Glass wool can also be used to adsorb clogging agents. Dilute ascites samples 1:1 with buffer (50 mM Tris [pH 8.6]), and apply it to a small column made of either syringes or Pasteur pipettes. A small volume may be retained by the glass wool.

5.4.4. Antibody Purification

Since any of the methods described in section 5.3 can be used to purify MAbs, this section will be devoted to describe two quick methods, an HPLC one and a less expensive affinity protocol, that have been used successfully to purify MAbs of both IgG and IgM isotypes (Fig. 1).

5.4.4.1. Protein A

Although protein A has been used effectively to purify IgG antibodies, it generally binds poorly to IgM and murine IgG1. When protein A-Sepharose affinity chromatography was used to purify IgM, the range of recovery was reported at 14 to 29% (46, 53). In recent years, many companies (Pharmacia, Bio-Rad, and Pierce) have demonstrated in their publications that efficient binding between IgM and protein can be accomplished by using buffers of high pH and high ionic strength. Pharmacia reported the use of a 1.5 M glycine buffer containing 3 M NaCl (pH 8.9), while other companies preferred to protect the recipes of their buffers (e.g., the MAPS [monoclonal antibody purification systems] buffers of Bio-Rad). The following procedure can be used for puri-

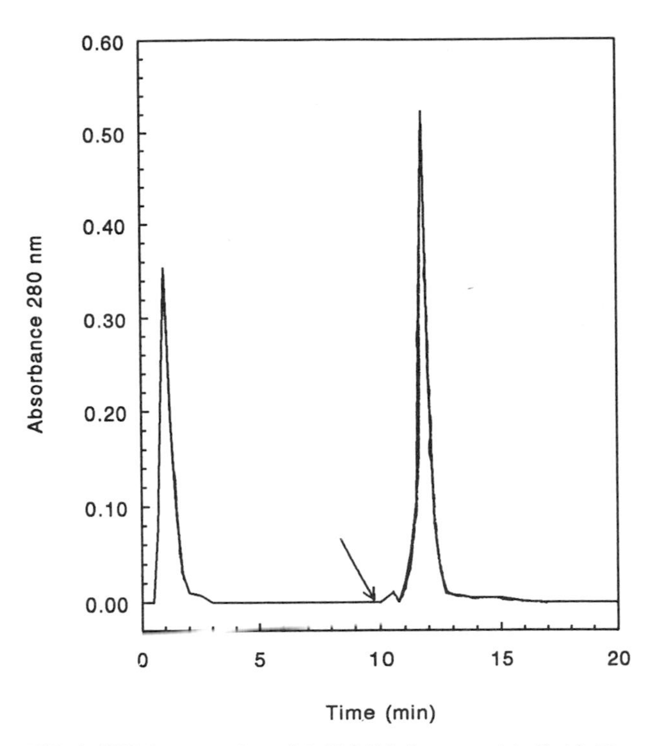

FIG. 1. HPLC separation of IgM MAb from ascitic fluid. The first peak is unbound material, and the second peak is pure IgM antibodies eluted after addition of elution buffer 1. The arrow indicates the time when elution buffer is added.

fying either IgM or murine IgG1. A variation for purifying other isotypes is summarized in Table 4.

Procedure

1. Equilibrate a column (Bio-Rad Affi-Prep Protein A preparative cartridge, binding capacity at approximately 35 mg of immunoglobulin) with binding buffer (BB; see Table 4 for the recipe) for approximately 12 min at 5 to 10 ml/min (15 column volumes). Alternatively, a protein A Superose column (Pharmacia) can be used. The MAPS buffer from Bio-Rad could also be used; however, these premade buffers are more expensive.
2. Dilute samples (1:5) with BB, and filter through a 0.45- or 0.22-μm-pore-size filter if the ascites has not been pretreated as described above.
3. Load 0.2 to 1.0 ml of the diluted sample, and run the sample through the column at 5 to 10 ml/min. The flow rate may be reduced if the column used cannot withstand the pressure of a high flow rate. All the proteins that do not bind to the matrix should elute within the first 5 min. To remove the unbound proteins, run the column for 5 to 10 min such that the OD_{280} spectrum remains at the base line for at least 5 min.
4. Elute the immunoglobulin fraction with the first elution buffer (EB1; 0.1 M citric acid buffer [pH 4] containing 0.15 M NaCl) for at least 5 min at the same flow rate as before.
5. Repeat the above with the second elution buffer (EB2; 0.1 M citric acid buffer [pH 3] containing 0.15 M NaCl) at 5 to 10 ml/min for 5 min.
6. Wash the column with PBS-containing azide, and store and seal the column in the same buffer. *Note:* The eluted immunoglobulin should not be stored in the elution buffer, since the acidity may be deleterious to the IgM if the antibody is exposed to it for too long.
7. Dialyze the eluted IgM against PBS (two 4-liter volumes) overnight at 4°C.
8. Concentrate to the desired volume by placing the IgM solution into dialysis bags and sprinkling PEG flakes (molecular weight 20,000) over the tubing. This process should not take any longer than 30 min to 1 h.
9. Assay for the total protein content, and analyze the sample by SDS-PAGE to ensure that only the heavy-chain (approximately 55 to 60 kDa) and light-chain (approximately 28 kDa) bands are seen (Fig. 2).

Figure 1 shows the purification of IgM Mab from ascitic fluid by HPLC by using this method.

5.4.4.2. DEAE Affi-Gel Blue

The DEAE Affi-Gel Blue method has been described by Bruck et al. (44) as a one-step procedure for the isolation of IgG free of protease, nuclease, and albumin. Recovery was reported to be as high as 77 to 80%. Note that DEAE Affi-Gel Blue is not the same as the Bio-Rad Affi-Gel Blue described above (5.4.4.1, step 1). IgG will be bound to the DEAE functional groups and therefore must be eluted.

Procedure

1. Preequilibrate a small column containing 5 to 10 ml of DEAE Affi-Gel Blue gel matrix with buffer (20 mM Tris-HCl [pH 7.2]).
2. Apply 1 to 5 ml of ascites (pretreated as described above). Wash the column with 2 bed volumes of the same buffer.
3. Wash the column with 1 bed volume of buffer containing 20 mM Tris-HCl and 25 mM NaCl (pH 7.2) to elute transferrin.
4. Elute the MAbs with 3 bed volumes of buffer containing 20 mM Tris-HCl (pH 7.2) plus 50 mM NaCl.

Table 4. Conditions for HPLC purification of various immunoglobulin isotypes

Isotypes of MAb	Binding buffer	Elution buffer I	Elution buffer II
IgM and murine IgG1	1.5 M glycine–3 M NaCl (pH 8.9)	0.1 M citric acid–0.15 M NaCl (pH 4.0)	0.1 M citric acid–0.15 M NaCl (pH 3.0)
Murine IgG2a and IgG3	50 mM Tris-HCl (pH 8.6)	0.1 M citric acid (pH 5.0)	0.1 M citric acid (pH 4.0)
Murine IgG2b	50 mM Tris-HCl (pH 8.6)	0.1 M citric acid (pH 4.0)	0.1 M citric acid (pH 3.0)

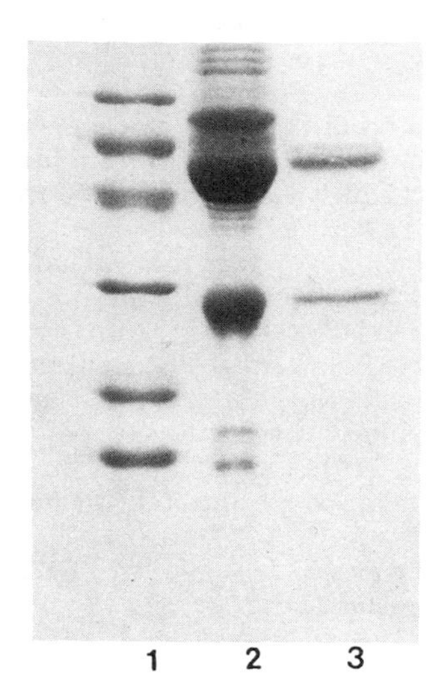

FIG. 2. SDS-PAGE profiles of IgM MAb purification by HPLC. Lanes: 1, molecular mass standards in descending order of 97.4, 66.2, 42.7. 31, 21.5, and 14.4 kDa; 2, ascitic fluid before purification (note the large numbers of bands that are normally found in serum proteins or body fluid of an animal); 3, purified IgM MAb eluted from column (second major peak of Fig. 1) (note the presence of only two major bands representing the heavy and the light chain at apparent molecular masses of 60 and 28 kDa, respectively).

5.5. DETECTION OF REACTIONS

Antibody-antigen reactions have a wide application in all areas of biological, medical, and biochemical research in which antibodies are used as probes for the presence, structure, and even function of a given antigen. In many instances, there is a limited amount of the antibody, antigens, or both reagents, so that the economical use of these reagents becomes paramount. Furthermore, in diagnostic laboratories, screening systems should be easy to perform routinely on a large number of samples yet should be rapid, sensitive, and specific.

The classical or traditional serological methods (agglutination, precipitation, immunodiffusion, and complement fixation) have long been used in diagnostic laboratories. These techniques have greatly contributed to our understanding of the antibody-antigen interactions and to the development of new assay systems. In recent years, these techniques have been replaced as routine screening procedures in many areas of research by the highly sensitive and rapid immunoassay techniques such as ELISA and Western immunoblotting. The reaction mechanisms and application of these techniques are discussed below.

5.5.1. Agglutination

Agglutination reactions are among the easiest of immunological tests to perform and evaluate and have long been used in bacterial identification and serological classification (57, 59, 62). In these techniques, the antigen, generally a suspension of cells or antigen-coated particles, is mixed with the test antibody; a positive reaction is observed as aggregation or clumping and is termed "agglutination." The titer of an antiserum is the highest dilution of the serum that gives a visible aggregation reaction with the serum. If the antigen is heterologous to the antibody or if no antigen is present, the suspensions remain unchanged.

For agglutination to occur, the antigen must have more than one antigenic epitope to allow formation of large aggregates of cross-linked antigen-antibody complexes. Bacterial antigens such as lipopolysaccharides (LPS; O antigens), carbohydrate-capsular antigens (K antigens), or flagella (H antigens) contain repeat antigenic epitopes, and whole cells have a large number of antigens; both are ideal candidates for agglutination reactions. The antigens used in agglutination reactions can be particulate, such as whole cells and antigen-coated particles (section 5.2.1).

Diagnostic polyclonal antiserum should be extensively adsorbed with a strain of the bacterium that is deficient in the target antigen, for example, a nonflagellated, unencapsulated, or O-antigen-deficient strain. Adsorption reduces nonspecific cross-reactive activity and increases the specificity of the serum. The adsorbed polyclonal antiserum contains a mixture of antibodies of different affinities to different antigenic determinants, and the combined activities allow detection of an organism bearing these antigens. The major problems encountered in the use of polyclonal antisera include batch-to-batch variation in antibody quality and residual nonspecific activity. In addition, although adsorption increases the specificity of the serum, there is considerable reduction of antibody titer. These problems are virtually eliminated by use of the highly specific MAbs in serotyping, given an appropriately exposed specific epitope. Despite the high affinity of MAbs, their application can be limited by the fixed affinity of these reagents. This problem can be overcome by using mixtures of specific MAbs. MAbs are invaluable in the serotyping of nontypable and polyagglutinable bacterial isolates and in the identification of organisms expressing a specific antigenic determinant. The choice of polyclonal versus monoclonal antisera depends on the problem under investigation.

Bacterial agglutination techniques are routinely used for detection of antibodies to bacterial surface antigens, notably to the LPS, cell wall antigens, capsular polysaccharide antigens, cell wall protein in streptoccocci, and flagellar antigens. Serological classification of bacteria is based on selective agglutination of the bacterial cells by antibodies to these antigens (57, 59, 62).

5.5.1.1. Qualitative slide agglutination

1. Suspend a loopful of bacterial cells (or other particulate antigens) in 1.5 ml of normal saline (0.85% [wt/vol] NaCl).
2. Mix 25 μl or a loopful of the bacterial suspension with an equal volume of the appropriately diluted test antiserum on a glass slide.
3. Use cell suspensions mixed with normal saline or nonimmune serum as negative controls to detect nonspecific agglutination of the cells.
4. Aggregation should be visible in about 1 min, by eye or at low-power microscopy.

Note: (i) Do not use bacterial cells that autoagglutinate in normal saline or water and in the absence of an antibody in these reactions. Minor autoagglutination may be overcome by using 0.1% (wt/vol) skim milk in the buffers to block the adherence or by suspending the cells in 1:100 dilution of normal serum. These reagents, when used, should also be used in the control assay.

(ii) LPS O-antigen-deficient strains may agglutinate with more than one serotype-specific serum.

(iii) Heat serum to be used in agglutination of fresh bacteria at 56°C for 30 min to inactivate complement.

(iv) Dilute the serum for use to avoid a prozone effect, i.e., failure to agglutinate at high serum concentrations.

5.5.1.2. Microtiter plate agglutination

1. Make doubling dilutions of the test serum in normal saline or in PBS (pH 7.4; see section 5.5.4) containing 150 mM NaCl. Add Safranin O dye (0.005% [wt/vol]) to the buffer to make it easier to read the results. (Agglutinated cells appear as a pink mat.)
2. Dispense 25-μl volumes of the serum in duplicate series in a microtiter plate. Use normal sera as a control.
3. Adjust the bacterial suspension to 10^9 cells per ml by absorbance (OD_{600} of 0.6) or by comparison with McFarland standards (11).
4. Add an equal volume of the bacterial suspensions to each well containing the serum. Cover the plate with Saran food wrap to prevent evaporation.
5. Incubate the plates at room temperature for 5 to 60 min. Incubation at lower temperatures may give indeterminate results and/or nonspecific clumping.
6. Record and compare the time of visible agglutination for different sera and antigens. Record agglutination as (++++) to (+) for complete to slight agglutination and (−) for no agglutination, seen when the cells remain as a smooth suspension. Express agglutination titers as the highest dilution of serum that gives a visible agglutination of the cells. Plate agglutination allows the testing of several antisera and antigens at the same time.

5.5.1.3. Passive agglutination

In the passive agglutination technique, specific antigens (or antibodies) are adsorbed or chemically immobilized on erythrocytes (section 5.2.1) or latex beads. These sensitized particles are then mixed with the test antisera (slide or microtiter assays), and agglutination is observed. The large size of the antigen (antibody)-coated particles enhances the visibility of the antigen-antibody reaction. This technique allows the use of extracted and soluble antigens rather than whole cells (which contain a complex of antigens) in detection of homologous antibodies by agglutination. Inhibition of passive agglutination by addition of exogenous purified antigen to the reaction mixture can be used to confirm the specificity of the reaction.

Latex beads have several advantages over erythrocytes in passive agglutination assays. The beads are immunologically and chemically inert, can be stored for long periods without damage, have a uniform size, and are available in different colors. A mixture of different-color beads, each color sensitized by a specific antigen (antibody), can be used to detect the presence of different antibodies (antigens) in the same sample in an economical and time-saving manner (59, 61). Antigens can be passively adsorbed or chemically coupled to surface-modified latex beads. Chemical-coupling procedures are described in detail by McIllmurray and Moody (62). The procedure described below is for passive agglutination of protein antigens to latex beads (Difco).

Procedure

1. Wash 1 ml of bead suspension (10%, wt/vol) in 40 ml of glycine-saline buffer. (Buffer: glycine, 14.0 g; sodium chloride, 17.0 g; sodium hydroxide, 0.7 g; sodium azide, 1.0 g. Dissolve in 800 ml of distilled water; adjust to pH 8.2 with 1 M NaOH; make up to 1 liter.) Centrifuge at 12,000 × *g* for 15 min. Repeat the wash.
2. Resuspend the beads in 20 ml of the buffer. Resuspend the proteins in the buffer, and add 300 to 500 μg of protein to the beads. Place the flask on a rotary shaker at low speed, and mix the suspension at room temperature for 30 min.
3. Wash three times in 20 ml of the buffer.
4. Resuspend in 20 ml of the buffer containing 0.1% skim milk or 1% (wt/vol) BSA (to block all remaining protein-binding sites), and incubate for 30 min. Centrifuge the beads, and resuspend in glycine-saline buffer containing 0.1% (wt/vol) sodium azide. At this point the suspension should show no spontaneous agglutination of the sensitized particles. If clumping occurs, it may indicate that excess antigen was used in the coupling reaction, which should be repeated. Store the suspension at 4°C in acid-clean glass or plastic containers which have minimal protein binding.
5. Mix 25 μl of the activated latex beads and the test sample. Rock the mixture gently, and observe agglutination against a dark background.

5.5.1.4. Coagglutination

In the coagglutination technique, developed by Kronvall (60), *S. aureus* cells are coated with test IgG antibodies and then are used to detect the presence of antigens in serum. *S. aureus* Cowan 1 (ATCC 12598) produces large amounts of cell wall-bound protein A, which has a high affinity for the Fc portion of IgG antibodies from a range of animals (52). The immobilized antibodies retain their antigen recognition properties. The IgG-coated cells are mixed with the serum or antigen preparation to detect the presence of the specific antigens.

5.5.2. Quellung Reaction

Encapsulated bacteria can be rapidly identified by the capsule-swelling Quellung reaction. In this assay, a drop of the bacterium is mixed with a drop of the antiserum or capsule-specific antibody and then mixed with an equal amount of methylene blue dye. The cells are observed under a light microscope. The binding of the antibody to the capsule appears to increase the capsular volume, possibly by stabilizing the structure. The capsules are then more visible and refractile and are easily identified because they show a sharply stained margin. This reaction is used in identification of encapsulated *Streptococcus pneumoniae, Haemophilus influenzae* type b, and *Meningococcus* strains (64).

5.5.3. Precipitation

The interaction of a multivalent *soluble* antigen with the homologous divalent antibody results in the formation of a cross-linked lattice of the two components. If the antigen has more than one antibody-binding site, and if there are sufficient quantities of both reagents to attain optimal proportions, the lattice is visible as a precipitate. Precipitation, however, occurs only at a narrow range of antibody and antigen concentration, called the **equivalence zone**; in an excess of either component, the precipitate dissolves. The rate at which precipitation occurs is dependent on the ratio of antibody to antigen in solution and on the avidity of the interaction. Although precipitation reactions can be performed in solution (ring test), the most commonly used procedure is precipitation in an agar gel matrix, called immunodiffusion, which eliminates the necessity of finding the optimal proportions of antigen and antibody.

5.5.3.1. Ring precipitation

1. Set up a series of small-bore glass tubing in a block of plasticine.
2. Using a capillary pipettor, add an equal amount of antiserum to each tube. If a pipettor is not available, attach a long capillary tube to a rubber tube containing a mouthpiece and use the device to add the reagents to the glass tubing. Fill the tubing from the bottom to avoid trapping air bubbles.
3. Prepare serial twofold dilutions of the antigen in normal saline as follows. Put 0.9 ml of saline into 10 tubes. Add 0.1 ml of the antigen to the first tube, and mix well. Take 0.1 ml from the first tube, put it into the second tube, mix well, and then take 0.1 ml and put it into the third tube. Repeat the procedure with all 10 tubes. This will give antigen dilutions of 1:2 to 1:1,024.
4. *Very* carefully layer the antigen solution on the antisera. A sharp interface must be obtained between the two solutions. Include a control tube containing saline instead of antigen.
5. Incubate the tubes at 37°C for 1 h and, if necessary, overnight at 4°C. Observe against a black background.

The antibody and antigen will diffuse in solution, forming immune complexes. The complexes are stable where the two components are at equivalence, i.e., where the ratios are optimal for precipitation. Rapid precipitation is observed when serum of high affinity and avidity is used. This method is extremely exacting and laborious for routine serological testing and uses a large amount of reagents. However, it is an excellent method to demonstrate precipitation reactions to serology classes.

5.5.3.2. Radial single and Ouchterlony double immunodiffusion

The detection of antigen-antibody reaction complexes was made much easier by the introduction of agar as a diffusion matrix. Agar and agarose (Chapter 9.1.1) have chemically inert matrices which do not influence the free diffusion of the molecules. At low concentrations, the agar or agarose forms a network of large pores which allows free movement of molecules within the matrix. The rate of diffusion of each molecule is a function of its molecular weight and structure. Molecules of similar structural properties will diffuse at similar rates and in this way will form a concentrated front.

In **radial single immunodiffusion,** antigens are allowed to diffuse from a well into an antibody-containing gel; precipitin rings are formed in the gel around the antigen well.

In the classical **Ouchterlony double immunodiffusion** (DID) method (68), antigen-containing wells are placed around a central antibody-containing well and the components are allowed to diffuse radially into the gel. Antigen-antibody complexes form in the agar zone between the wells, and precipitation occurs at the point at which concentrations of the reagents attain equivalence. The diffusion of complex antigens against the homologous polyclonal antiserum results in the formation of a complex pattern of precipitin lines in the gel. If two adjacent antigen wells containing a common antigen are tested against the homologous antibody, the precipitin line between the antibody and the antigen wells is continuous, representing a reaction of "identity" (Fig. 3B, wells 1 and 3). Reactions of "nonidentity" are seen as precipitin lines that cross each other and represent antibody reactions to distinct antigens (Fig. 3C, wells 1 and 2). A line of "partial identity" is formed between two antigen wells that contain a related (continuous precipitin) but not identical (spur) antigen (Fig. 3A and C, wells 2 and 3). The spur projects toward the well of the antigen possessing the distinct antigenic determinants. The position of the precipitin lines is determined by the local concentration of the reactants in the gel and is a function of the original concentration in the well and of the diffusion rates of the reactants.

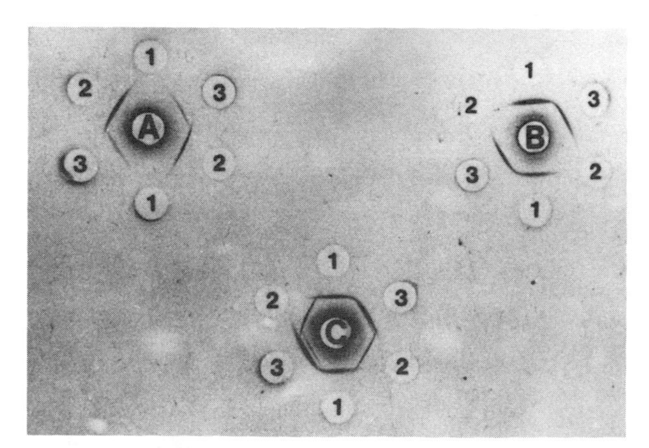

FIG. 3. Double immunodiffusion profiles of antisera to proteases of *Aeromonas hydrophila*. Each well was filled with 10 μl of either an antigen or an antiserum preparation. Wells: A, rabbit antiserum to protease A; B, rabbit antiserum to protease B; C, a 1:1 (vol/vol) mixture antiserum to protease A and protease B, respectively. The antigens are placed into the outer wells. Well 1, protease B from strain 1. Well 2, protease A from strain 2. Well 3, extracellular products of strain 3. (Courtesy of K. A. Leung and R. M. W. Stevenson.)

DID procedure

Barbitone buffer (pH 8.2; ionic strength, 0.08)

12 g of sodium barbital (5′,5-diethylbarbituric acid, Na salt [Fisher Scientific] in 800 ml of distilled water

4.40 g of barbital (5′,5-diethylbarbituric acid [Fisher Scientific] in 150 ml of distilled water at 95°C

Mix the two solutions, and adjust to pH 8.2 with 5 M NaOH; then bring to a final volume of 1 liter.

Gel preparation

a. Precoating gel: Make up 0.5% (w/v) agarose in distilled water. Boil to dissolve.
b. Resolving gel: Dissolve 2 g of electrophoresis-grade agarose (SeaKem Medium Electroendosmosis, $-M_r$ = 0.16 to 0.19 [Mandel Scientific]) in 50 ml of the Barbitone buffer and boil to dissolve the agarose. Add 50 ml of the buffer to make a 1% (wt/vol) buffered agarose solution. Store at 4°C for up to 4 weeks.

Making templates

Immunodiffusion assays can be performed on a glass plate (such as a microscope slide) or on gel bond film (Mandel Scientific).

1. Place the glass slide or gel bond film on a leveling table.
2. Pour a thin film of the precoating gel on glass plate or on the hydrophilic face of the gel bond film (a drop of water placed on this face will spread out).
3. Pour 0.18 ml/cm² of the buffered agarose on the coated plate or film to form the immunodiffusion template.
4. Punch the required pattern of wells (20 mm in diameter and 1 cm apart) in the agar with a sharp gel punch.
5. Pipette 5 to 10 μl of the sample into the wells. Do not overfill the wells.

Incubation

Develop the templates in a humid box for 12 to 48 h. Observe the precipitin lines against a black surface. The visibility of the precipitates can be enhanced by soaking the plates in 4% (wt/vol) tannic acid in saline.

Pressing, washing, and staining templates

1. Place the template on a filter paper. Fill the wells with distilled water.
2. Press for 15 min with several pieces of Whatman No. 1 filter paper, a glass pane, and several books.
3. Soak the template in saline for 20 min. Repeat the washing and pressing process twice.
4. Soak the template in distilled water for 20 min.
5. Air dry the template.
6. Stain by placing the template in Coomassie blue solution (see recipe below) for 15 min.
7. Destain with the destaining solution (as below without the Coomassie dye). The staining is reversible, and it is possible to completely destain the gel. If this occurs, restain the gel.
8. Air dry the stained gel.

Staining solution

Coomassie brilliant blue	0.5	g
95% Ethanol	45	ml
Glacial acetic acid	45	ml
Distilled water	10	ml

Place the solution on a magnetic stirrer and dissolve the Coomassie overnight before use. The staining solution is reusable.

5.5.3.3. Rocket immunoelectrophoresis and counterimmunoelectrophoresis

Immunoprecipitation reactions in gels are accelerated by the active movement of antigens and antibodies toward each other under the influence of an electrical field. The sensitivity of the technique is increased by eliminating the dilution of reactants that occurs in immunodiffusion assays. Rocket immunoelectrophoresis and counterimmunoelectrophoresis (CIE) are analogous to the single radial and double Ouchterlony immunodiffusion assays, respectively.

In rocket immunoelectrophoresis (66) the antigens are placed in wells at the cathode side of the gel and then electrophoresed into the antibody-containing agarose gel. Immunoprecipitates are observed as peaks radiating into the gel. The area below each precipitin peak is a function of the antigen concentration and is inversely proportional to the antibody concentration in the gel.

In counterimmunoelectrophoresis (65) antigens and antibodies are placed in paired wells (diameter 3 mm) cut 1 cm apart in agar coated slides. The antigens are placed in wells at the cathode side while the antibodies are at the anode side. The gels, antigens, and antibodies are all in Barbital-Tris (pH 8.6) buffer. At that pH, most bacterial antigens have an overall negative charge as a function of the carboxyl groups and will migrate to the anode under electrophoresis. The antibodies are neutral or very weakly charged and will drift (with the water molecules) toward the cathode by electroendosmosis. The antibodies and antigens therefore migrate toward each other in a concentrated fashion. Precipitation is detected in 20 to 60 min. CIE of complex antigens mixtures against the homologous polyclonal antisera results in a complex pattern of precipitin lines.

Procedure for CIE

1. *Equipment*: Flat-bed electrophoresis chamber (water cooled if available) and power supply.
2. *Buffer*: Barbituric-Tris (pH 8.6). This buffer is used to make the gels, to dissolve the antigens and antibodies, and for the cathode and anode wells.

Barbituric acid (0.02 M)	4.48	g/liter
Tris base (0.07 M)	8.86	g/liter
Calcium lactate (0.03 M)	0.108	g/liter

3. *Gel:* 1% (wt/vol) SeaKem Medium Electroendosmosis agarose ($-M_r$ = 0.16 to 0.19). Dissolve the agar in the buffer by boiling. Store at 4°C. Heat to redissolve, and keep at 56°C for use.
4. *Casting gels*: Cast the gels on the hydrophilic side of Gel Bond film (Mandel Scientific). Use 0.18 ml of the agarose per cm² of film. Punch paired antigen and antibody wells in the agar. Place the gel in the electrophoresis chamber, and load 2 to 4 μl of sample per well.
5. *Electrophoresis*: Perform electrophoresis at 10 V/cm. Use five layers of Whatman no. 1 paper or Kimwipe strips as wicks to connect the gel to the buffer during electrophoresis. Boil the strips in water before use to remove impurities.

5.5.3.4. Crossed immunoelectrophoresis

Crossed (two-dimensional) immunoelectrophoresis (XIE) is a powerful and highly sensitive analytical technique that combines the prior separation of complex antigens by electrophoresis with immunoelectrophoresis (70). XIE has been used to investigate the immunogenicity of bacterial cell lysates and membrane antigens. Figure 4 shows XIE of *Pseudomonas aeruginosa* cell lysates against the homologous polyclonal rabbit serum.

In the first dimension, antigens are electrophoretically separated into zones on agarose or polyacrylamide gels on the basis of molecular weight. The gel strips containing the separated antigens are then placed along the cathode side of a rectangular slab of agarose gel containing the antibodies. This is the second-dimension gel. An intermediate gel with or without additives is cast between the antigen- and antibody-containing gels. This gel can be used to stack the antigens or identify specific antigens. The antigens are then electrophoresed perpendicularly into the second-dimension gel, and, on interaction of the antigens with the homologous antibodies, immunoprecipitates are formed in the second dimension gel. A complex pattern of overlapping peaks is obtained when several antigens are electrophoresed into a polyclonal antibody-containing gel matrix. The area below each precipitin peak is a function of the antigen concentration and is inversely proportional to the antibody concentration in the second-dimension gel. The position of specific antigens can be defined by incorporation of a MAb or specific antibody in the intermediate gel. The antigen will react with the antibody and form a precipitin in the intermediate gel.

The direct application of XIE to investigation of antigen structure as well as immunogenicity are exemplified by crossed affino-immunoelectrophoresis (XAIE). In this form of XIE, a lectin such as concanavalin A is incorporated into the intermediate gel. The lectin will bind to and precipitate the respective glycosylated antigen as the antigens are electrophoresed from the first-dimension gel through the separating gel. The precipitin peak corresponding to that antigen will be identified by comparison of lectin-containing and lectin-deficient gels. The antigenic and structural properties of immunogens can therefore be investigated by XAIE (70).

XAIE procedure

Dissolve 1% SeaKem agarose in Barbituric-Tris buffer (pH 8.6), and keep the agar at 56°C. Warm all sera or antibody solutions to 56°C before use.

First dimension

1. Place a glass plate (10 by 10 cm) on a leveling table. Pour 18 ml of the 1% Barbituric-Tris-buffered agarose on the plate. Allow it to set.
2. Punch 10 sample wells (diameter 2 mm) in the gel, and remove the gel plugs. The wells should be 1 cm apart and 2 cm from the edge of the plate.
3. Place the gel in the flat-bed electrophoresis chamber. Connect the gel to the buffer in the electrode wells with wicks. Do not turn on the power source.
4. Pipette 4-μl aliquots of the antigens into nine wells. Place 2 μl of the tracking dye (0.05% bromophenol blue) and indicator protein (0.1% BSA) mixture in the last well. Do not overfill the wells. Close the lid, and apply voltage.
5. Electrophorese the samples until the tracking dye has moved 2.8 cm away from the well. Turn off the power supply.
6. Using a long sharp blade, cut the sample lanes and trim to 1- by 6-cm strips (from well to tracking dye front). This is the first-dimension gel.

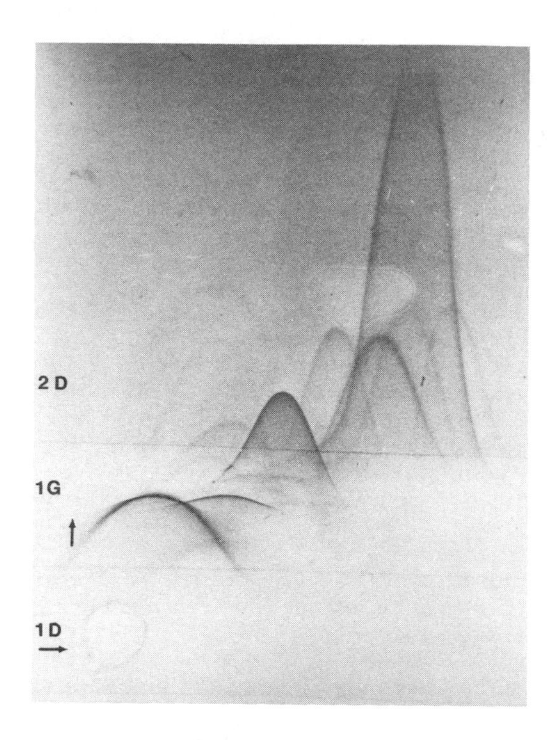

FIG. 4. XIE of *Pseudomonas aeruginosa* cell lysate and the homologous polyclonal antiserum. A 10-μl portion of the lysate was loaded in the first dimension gel (1D), the intermdediate gel (IG) contained NaCl, and the second dimension gel (2D) contained antiserum.

Intermediate gel

1. Lift the first-dimension gel strips with the blade, and place on one end (on the *hydrophilic* face) of a Gel Bond film (5 by 6 cm). The films should be on a leveling table.
2. Place a metal bar 1 cm away from the first-dimension agar gel strip. This is the space for the intermediate gel.
3. Add 0.1 ml of the prewarmed nonimmune serum into 1.5 ml of the 1% agarose at 56°C. Gently mix the gel, avoiding air bubbles.
4. Pour the gel into the intermediate gel space. Allow to set, and remove the metal bar. This gives a continuous first-dimension and intermediate gel strip.

The nonimmune serum will complex and absorb nonspecific activities thus improving the quality of the gel. Specific antibodies, antigens, lectins, or saline can be incorporated into this gel, depending on the aim of the experiment.

Second-dimension gel

1. Add 0.5 ml of the test serum to 3.5 ml of the 1% agarose at 56°C. Mix well by swirling the gel in the water bath. Avoid air bubbles.

2. Pour the gel onto the rest of the Gel Bond film, and allow to set.
3. Place the gel in the electrophoresis unit with the first-dimension gel on the cathode side. Connect the wicks. Close the lid. Electrophorese at 3 V/cm for 16 to 18 h.
4. Remove the plate and proceed to press, wash, and stain as described in 5.5.3.2.

5.5.4. Enzyme-Linked Immunosorbent Assay (ELISA)

The design and development of ELISA by Engvall and Perlmann (74) provided a major revolution in diagnostic immunology. The application and versatility of ELISA were greatly facilitated by the design of the microtiter plate. ELISA has become the most widely used immunoassay technique, with applications in routine diagnostic serology and in virtually all areas of biological sciences in which antibodies are used as research tools. In all immunoassays (including ELISA, and Western and dot immunoblots) the antigen-antibody reactions are monitored by enzyme activities rather than by direct observations of immune complexes. The popularity of the ELISA is due to its high sensitivity and specificity at low cost; amplification systems allowing detection of nanogram quantities; adaptability to automation; elimination of hazardous reagents; use of color reactions that can be evaluated visually; and simultaneous screening of many samples in the same microtiter plate with as little as 50-μl total volume.

The two common requirements in all immunoassay procedures are (i) the effective immobilization of the antigen or antibody onto a suitable solid matrix without loss of function, the adsorbed reagent then being used to capture the antibody or antigen from the test sample; and (ii) the availability of suitable reaction indicator systems for detection of the bound antigen-antibody complexes. Immunoassays depend on the fact that enzymes or chemically active groups can be covalently linked to immunoglobulins, under the appropriate conditions, and that the resultant complex retains the immunological and the chemical activities. The high stability, specificity, and activity of the enzyme complexes are crucial to the high sensitivity seen in the immunoassay techniques.

5.5.4.1. Requirements

The sensitivity, reproducibility, and specificity of the ELISA depend on three major factors: (i) the optimal concentration and stability of the immobilized or coating reagent, usually the antigen; (ii) the reduction of nonspecific activities by blocking reagents; and (iii) the choice of indicator system. These are discussed below because of the importance of these factors to the design and success of the assay. The reviews by Ekins (73), Nakamura et al. (85), and Tijseen (84) may also be consulted.

Optimum concentration of antigen required to coat the matrix. The actual amount of antigen adsorbed onto the plate is crucial to the sensitivity of the assay. Generally, the optimal concentration range of antigens in the coating solution is 0.2 to 10 μg of the purified antigens per ml, whereas complex mixtures and cell lysates require about 5 to 25 μg/ml. Too much antigen in the coating buffer results in complexes of the proteins in solution, which then bind to the matrix in "stacks" (rather than as layers) that are removed in subsequent washes, resulting in insufficient antigen coating. Too little antigen leads to selective binding of high-affinity antibodies so that low-affinity antibodies are excluded from the assay, resulting in highly inaccurate results when hyperimmune serum and early-infection serum are compared. Consequently, several methods have been described for the estimation of the optimal antigen concentrations required to obtain the highest sensitivity reproducibly (80). The aim is to achieve an antigen concentration that is slightly below the concentration needed to saturate all the antigen-binding sites on the matrix. It is assumed that this concentration will give a one-layer distribution of the antigen on the matrix and allow optimal interaction with the antibodies in subsequent steps.

Stability of antigen. Antigen adsorption is influenced by the binding capacity of the matrix and the stability of the bound antigen. The sensitivity of the ELISA is influenced by antigen concentrations, and therefore the matrix should have high adsorptive capacity without denaturing the antigen. There should also be minimal loss of antigen after absorption.

Binding of antigens to microtiter plates is due mainly to hydrophobic interactions rather than size, ionic, or charge interactions (71). Therefore, the buffer systems used in the adsorption step reflect the chemical and structural properties of the antigen that determine its solubility rather than the effect of the buffer in antigen binding. For example, bicarbonate buffers (pH 9.6), phosphate buffers (pH 7.2), and water have been used in the adsorption of soluble antigens. Inorganic solvents (78, 82), detergents (79), and glutaraldehyde (86) have been used for adsorption of lipidic, glycosylated, and integral membrane protein components. Detergents are reported to reduce binding of some antigens to ELISA plates and should be avoided whenever possible (76). Integral membrance components, however, require that the coating buffer contain detergents at concentrations above the critical micelle concentrations for solubility. Several protocols have been used to enhance the immobilization of insoluble, complex, or small antigens on the solid matrices. Some of these procedures include the use of polyvinyl chloride plates derivatized to contain aldehyde groups for glycoprotein adsorption (79), avidin-coated plates to bind biotinylated peptides (75), and pretreatment of polystyrene tubes with 0.2% glutaraldehyde for adsorption of small peptides (86). Improved antigen adsorption is obtained following antigen denaturation, a process that may expose hydrophobic moieties of the antigen (72). $MgCl_2$ in the coating buffer enhances adsorption of LPS (77), whereas preincubation of rough LPS with polymyxin enhances binding of the molecules to microtiter plates (2, 83). In a novel procedure, mercuric chloride is adsorbed onto plates precoated with BSA-glutathione complexes for titer determination of the HgCl-specific antibodies (87).

The most common type of solid matrices used in ELISA are modified polyvinyl chloride and polystyrene microtiter plates, beads, or tubes, obtained from commercial sources. Antigen binding can be improved by treatment of the matrix before or after antigen adsorption.

Blocking reagents. All immunoassays require blocking reagents to saturate the excess binding sites on the matrix, thereby preventing nonspecific adsorption of reactants and reducing the background values. The best blocking reagent should have no affinity for the test antigens, immunoglobulins, or indicator system. Examples of blocking reagents include 0.1% (vol/vol) Tween 80, 3 to 10% (vol/vol) skim milk, 2 to 10% (vol/vol) fetal bovine serum, 1 to 3% (wt/vol) gelatin, 1 to 5% (wt/vol) BSA, and various combinations of these reagents. Fetal bovine serum may increase the background when used as the blocking reagent because of nonspecific binding of some preparations of antibodies or antigens to these reagents. Gelatin above 0.5% (wt/vol) may solidify at the normal room temperatures and will block ELISA plate-washing apparatus; 5% (wt/vol) skim milk is recommended for blocking, and 2% (wt/vol) skim milk is recommended for all other steps. In laboratories that routinely use immunoassays, the cost of the blocking reagent could be an important consideration, and skim milk is cheap!

Indicator system. The sensitivity of immunoassays is vastly increased by use of labeled indicator systems that allow the detection of extremely low concentrations of immunoreactants. There are four main groups of indicator conjugates used in immunoassays, as follows.

(i) Immunoglobulin-enzyme conjugates. The success of immunoglobulin-conjugates depends on the bifunctional nature of the immunoglobulin molecule and on the high stability and activity of the enzymes following conjugation. The enzymes are covalently conjugated to the immunoglobulin molecule without affecting the antigen recognition and binding properties of the antibody molecule. The most commonly used conjugates are alkaline phosphatase (AP) and horseradish peroxidase (HRPO) conjugates, because of their stability and cost. Others include ß-galactosidase, glucose oxidase, glucoamylase, carbonic anhydrase, and acetylcholinesterase (81). Recently, fluorescent probes and fluorometers for ELISA have been described (81).

The most commonly used immunoglobulin conjugates are the anti-species antibodies (second antibodies). For example, goat anti-mouse IgG or IgM [or the respective $F(ab')_2$] enzyme-conjugated antibodies are used in all assays in which the test antibody is from a mouse. The use of anti-species antibodies eliminates the need to purify or label the test antibody for each assay. A large variety of highly purified, high-affinity, enzyme-labeled anti-species antibodies are available commercially, for example from Cedarlane Laboratories, BIO/CAN Scientific, and Boehringer-Mannheim.

(ii) Ligand-receptor complexes. Ligand-receptor complex systems make use of the specific and extremely high affinity of avidin (egg white glycoprotein) or streptavidin complexes (from *Streptomyces avidii*) for biotin, a small water-soluble vitamin. Each single molecule of avidin can bind four molecules of biotin with an affinity of 10^{-15} M^{-1}. The biotin-avidin complex is stable to high pH, proteolytic enzymes, and organic solvents.

The test or second antibody [or the respective $F(ab')_2$ fragments] is covalently labeled with biotin in a simple procedure (section 5.5.5). The streptavidin or streptavidin-biotin complexes conjugated to enzymes (HRPO or AP) are then used as the indicator. Streptavidin complexes are preferred over avidin because they have low backgrounds due to an isoelectric point close to neutral pH, and hence little nonspecific interaction with charged groups or glycoproteins. Also, streptavidin is not glycosylated and will not bind to tissue or cell lectins. In addition, the binding of one streptavidin-biotinylated-HRPO complex to a single biotin-labeled second antibody (or biotinylated antigen) greatly amplifies the resultant positive reaction. The amplification obtained by this detection system allows the detection of very low levels of antigen-antibody activity, thereby increasing the sensitivity of the assay several fold.

(iii) Protein A conjugates. Protein A is obtained by purification from the cell walls of *Staphylococcus aureus* and has a well-known affinity for the Fc domains of IgG classes and subclasses from a diverse group of animals (52). The main disadvantage of protein A in immunoassays is that the binding to antibodies is sensitive to changes in ionic strength and pH of buffers. Protein A may also show nonspecific binding to bacterial antigens.

(iv) Lectins. Lectins are proteins that recognize and bind to limited carbohydrate groups. Biotinylated or enzyme-conjugated lectins such as biotinylated concanavalin A (Con A) can be used to detect the binding of glycosylated antigen moieties to immobilized antibodies in antigen-capture ELISA. These lectin conjugates can also be used to detect glycosylated molecules on Western immunoblots directly.

5.5.4.2. System classification

The ELISA systems can be placed into two main classes based on (i) the immobilized or matrix-bound component and (ii) the antibody on which the indicator molecule is localized.

(i) The immobilized or matrix-bound component. If the *antigen* is immobilized or adsorbed onto the matrix, the ELISA procedure is then used to screen test sera for presence of antibodies that react with the immobilized antigen. This procedure is called the **antigen capture ELISA**, and it has wide application in, for example, detection of antibodies in serum from patients or animals that may have been immunized (vaccinated) with a specific antigen or are suspected of having been exposed to or infected by a specific pathogen. If an *antibody* of known specificity is adsorbed onto the matrix, the ELISA procedure is then used to screen test samples for the presence of antigens that react with the immobilized antibody. This procedure is also called antigen capture ELISA and has application in the diagnosis of ongoing infections in which the immobilized antibody is used to capture circulating antigens in serum. Another application is in detection of specific contaminants in test samples, e.g., microorganisms or metals in water.

(ii) The antibody on which the indicator molecule is localized. In the **direct ELISA** the indicator molecule (enzyme, biotin, or fluorochrome) is directly conjugated to the first or primary antibody; i.e., the antigen-specific antibody is labeled with the indicator molecule. In these assays the labeled primary antibody is added to plates coated with the test antigen. If the antibody recognizes antigenic determinants on the test antigen, the antibody will remain bound on the plates. The bound antibody is detected by addition of the specific substrate (enzyme-labeled antibody) or can be visualized by fluorescence microscopy (fluorochrome-labeled antibody). Direct ELISA is a simple one-step procedure. There are several disadvantages to using direct ELISA: first, every test antibody has to be labeled, and second, the sensitivity of the assay depends on how many antibody molecules are bound (or on the number of specific antigenic determinants) on the antigen.

In the **indirect ELISA** the primary antibody is not labeled. The detector molecules are conjugated onto a secondary (or second) antibody. This labeled second antibody is often an anti-species antibody; hence if the primary antibody is from a rabbit, the second antibody can be a goat or sheep anti-rabbit antibody. Indirect ELISA has several advantages. First, the second antibody can be directed toward the immunoglobulin Fc domain (anti-immunoglobulin class) or pooled immunoglobulins (anti-all classes). Second, a single labeled anti-species second antibody can be used for all assays in which the primary antibodies are of the same animal species; for example, an enzyme-labeled goat anti-rabbit immunoglobulin antibody can be used in all assays in which the primary antibody is generated in rabbits regardless of the immunoglobulin class and antigen specificity of the primary antibody. Third, labeled anti-species antibodies are available from many commercial suppliers (for example, Sigma, Amersham, and Pierce). Finally, the use of a second antibody increases the sensitivity of the assay; i.e., more than one second-antibody molecule can potentially bind to the primary antibody (which is bound onto the antigen), and hence there is an amplification step built into the antibody detection reactions.

5.5.4.3. Procedure

The following procedure describes an indirect "checkerboard" ELISA that can be used to determine the optimal working concentration of antigen and antibody. The optimal antigen concentration can then be used in assays to screen hybridoma supernatants or serum samples for specific antibody activity.

a. Reagents for ELISA

1. *Coating buffer:* 0.05 M carbonate-bicarbonate buffer (pH 9.6)
   Solution A: 21.2 g of Na_2CO_3 per liter
   Solution B: 16.8 g of $NaHCO_3$ per liter

To prepare the working buffer, mix 20 ml of solution A and 42.5 ml of solution B and adjust to 250 ml with distilled water. The pH should be 9.6. Add 0.05 g of NaN_3. Store at 4°C in a dark bottle for up to 4 weeks. *Note*: Detergents or salts may be added to this buffer depending on the properties of the antigen. In addition, other buffers and solutions may be used as coating buffers.

2. *PBS* (pH 7.4)
   Solution A: 31.2 g of $NaH_2PO_4 \cdot 2H_2O$ per liter
   Solution B: 28.39 g of Na_2HPO_4 or 71.7g of $Na_2HPO_4 \cdot H_2O$ per liter

The working buffer is prepared as follows: Mix 47.50 ml of solution A and 202.50 ml of solution B; adjust to 800 ml with distilled water; add 8.75 g of NaCl, and make up to 1 liter. The pH should be 7.4.

3. *Wash buffer*
   PBS containing 0.05% Tween 20.

4. *Blocking reagent*
   PBS containing 5% (wt/vol) skim milk (Difco). This should be made fresh every day.

5. *Reagents for alkaline phosphatase conjugates*
   Substrate buffer. Sodium carbonate (0.05 M; pH 9.8) containing 10 mM $MgCl_2$. To prepare the substrate buffer, use the buffer stocks made for the coating buffer (step 1). Mix 27.5 ml of solution A with 35 ml of solution B and make up to 250 ml with distilled water. Add 0.05 g of $MgCl_2 \cdot 6H_2O$. Check the pH before use. Store in a dark bottle at 4°C.
   Alternatively, use 1 M diethanolamine buffer (pH 9.8) containing 0.5 mM $MgCl_2$.
   AP substrate: Dissolve in the buffer 1 mg of *p*-nitrophenyl phosphate per ml. The substrate can be obtained from Sigma in the form of 5-mg tablets. Store substrates at −20°C.

Read the yellow color at 400 nm. Color development can be stopped by addition of 10 μl of 3 M NaOH to the wells.

6. *Reagents for horseradish peroxidase conjugates*
   Substrate buffer. Make citric acid buffer (pH 4.0) by dissolving 0.2 g of citric acid in 90 ml distilled water. Adjust to pH 4.0 with 1 M NaOH. Store at room temperature. *HRPO substrate.* Immediately before use, add to 10 ml of the citric acid buffer, 5 μl of 30% H_2O_2 and 75 μl of 10 mg/ml [2,2-azino-bis(ethylbenzthiazoline)-6-sulfonic acid] (Sigma).

Read the green color at 414 nm. Color development can be stopped by addition of 0.08 M NaF to the wells.

b. Protocol for ELISA

1. *Antigen coating or adsorption.* Make several dilutions of the antigen, in the coating buffer, at concentrations between 10 and 100 μg/ml for complex antigen mixtures or between 0.2 and 10 μg/ml for pure antigens. Mix very well. Distribute in triplicate in the vertical rows, with 50 to 100 μl of each antigen dilution per well. Cover the plates with plastic food wrap (such as Saran wrap or Stretch-and-Seal), and incubate at 4°C for 16 h. Adsorption of antigens onto the matrix increases with exposure time. Add the appropriate proteases inhibitors to the coating buffer when proteases are suspect in the antigen preparations. Alternatively, if the antibody reacts with the denatured antigens, boil the sample prior to adsorption.
2. *Washing.* Shake out the solution from the wells with a strong flick of the wrist. Wash the plates by directing a jet of wash buffer from a wash bottle into each well. Wash three times with PBS-Tween, incubating for 5 min each time. Rinse with PBS. Shake out all the buffer between each wash.
3. *Blocking.* Add 200 μl of 5% skim milk to each well. Incubate for 2 h at 37°C or 4 h at room temperature. Rinse three times with PBS without incubations.
4. *Test antibody.* Prepare dilutions of the test antibody in 2% (wt/vol) skim milk and add to the antigen-

coated wells (in a volume equal to that of the antigen). Generally, use primary immune sera at a dilution of 1:10 and hyperimmune sera at dilutions of 1:100 to 1:1,500 or higher. Aliquot the serum in horizontal rows such that each serum dilution is tested against all antigen dilutions, i.e., a checkerboard titration. Incubate as in step 2, and repeat the washes.

5. *2nd-antibody.* Make a dilution of the labeled (conjugated) second antibody in the 2% skim milk. Repeat the incubation and washing.
6. *Substrate.* Prepare the appropriate substrates for the conjugate used, and add to the wells in volumes equal to that of the coating antigen. Incubate the plates at 37°C, and take readings (at the appropriate wavelength for the substrate) at 30, 45, and 60 min.
7. *Data analysis.* Plot the percent absorbance (calculated as a percentage of the maximum reading) against the antigen concentrations. From the graph, calculate the lowest antigen concentration that will give 50% of the maximum reading for the time point; this is the antigen concentration to use for subsequent ELISA. The antibody titer at that antigen concentration can then be used in the positive controls.
8. *Notes.* It is important that each plate contain the following control wells: (i) an antigen blank, (ii) a first-antibody blank, and (iii) homologous and nonrelated antigen controls. The antigen-deficient control is used to zero the plate reader. Checkerboard ELISA can also be used to establish the optimum concentrations of conjugates and antibody. In each assay only one parameter is varied at a time. Incubations can also be carried out at 37°C for 2 h or at room temperature for 4 h.

5.5.5. Immunoblotting of Antigens

Immunoblotting (Western blotting, electroblotting, immunoelectroblotting, immunoblotting) (92) is a popular method that has largely replaced other immunoelectrophoretic procedures. In this technique the antigenic and electrophoretic properties of a molecule are characterized simultaneously. The technique can be applied to study any molecule that retains its functional antigenic structure following SDS-PAGE (16).

The initial step is separation of the antigenic mixtures by PAGE and then separation of the molecules on the basis of size (urea or SDS-PAGE), charge (Isoelectric focusing [IEF]), or both charge and size (IEF-SDS-PAGE) (see Chapter 21 for details). The separated components are then electrophoretically transferred onto a support matrix, often nitrocellulose. The immobilized antigens are allowed to react with the test antibodies and then with the enzyme- or ligand-conjugated antibodies and are visualized by addition of the specific chromogenic substrate in a process similar to the ELISA. The substrates used in immunoblotting differ from those used in ELISA in that the product of the enzymatic reaction is insoluble and precipitates at the site of formation. The reaction is observed as a labeled band.

Although immunoblotting was originally used to investigate protein antigens, the development of nitrocellulose matrices with improved macromolecule binding has made it possible to study glycosylated proteins and lipids, lipoproteins, proteoglycans, viral antigens, and LPSs. Some antigens show reduced antibody binding following SDS-PAGE and immunoblotting. This problem can be reduced considerably by renaturation of the antigenic sites in the blotted proteins (88). Conformational antigenic epitopes that may be subject to "alteration or dissociation" following SDS-PAGE cannot be investigated by this method. Such antigenic determinants may be indicated by ELISA, dot immunoblotting (section 5.5.5.3), or immunofluorescence using the organism. Because of the improved resolution of the antibody-binding molecules in the test samples, the immunoblotting technique has been used in clinical laboratories to identify false-positive ELISA results.

Assay parameters. The parameters that should be optimized for the immunoblotting technique are (i) the concentration of acrylamide that gives the best separation of the antigens and (ii) the binding efficiency of the antigens to the nitrocellulose.

Application. Immunoblotting as an assay has the same sensitivity as ELISA. However, it has several advantages over ELISA: the adsorption of proteins on the nitrocellulose is not affected by detergents; less sample is required; and information on structural properties of the antigenic molecule is obtained. Nitrocellulose-bound molecules can also be chemically modified, and the nature of the antigenic epitopes can be examined; for example, periodate oxidation of the blots can be used to differentiate periodate-sensitive and insensitive antigenic determinants (93). Furthermore, immobilized antigens of interest can be excised and used in immunization of animals to generate specific antibodies (section 5.2.2).

Nitrocellulose matrices. A wide variety of nitrocellulose matrices with improved antigen-binding capacities are available, including, Hybond (Pierce, Amersham), BioTrace HP (Gelman), and polyvinyl chloride (Millipore). Nitrocellulose is a brittle matrix and tears easily. When possible, nylon-backed membranes, which are less fragile and have very high molecule-binding capacity, should be used.

5.5.5.1. General procedure

Similar methods can be found described in detail in the Bio-Rad immunoblotting pamphlets.

a. Reagents

1. *Transfer buffer* (pH 8.1–8.4)

Tris base	12.12 g
Glycine	57.48 g
Methanol	800 ml

Dissolve the Tris in 3,200 ml of distilled water. Add the glycine and then the methanol. Store in a cold place.

2. *Incubation and wash buffers*
   10 mM PBS (pH 7.2). Add 8.76 g of NaCl, 0.2 g of KCl, 1.44 g of $Na_2HPO_4 \cdot 2H_2O$, and 0.2 g of KH_2PO_4 to 1 liter of water. Alternatively, 10 mM Tris-HCl buffer containing 150 mM NaCl (TBS) can be used.
3. *Wash buffer*
   PBS or TBS containing 0.2% (vol/vol) Tween 20.
4. *Blocking solutions*
   Either 3% gelatin, 5% skim milk, or 3% BSA dissolved in PBS or TBS may be used. For gelatin, heat

the buffer to dissolve the gelatin. Make stock solutions by autoclaving the gelatin and storing in 100-ml volumes at 4°C. Dissolve before use.

5. *Dilution buffers*
Dilute all antibodies and conjugates in PBS or TBS containing 0.8 to 1.0% gelatin, 2% BSA, or 2% skim milk.

6. *Developing reagents*
Several reagents are described in the literature for blot development, and premade kits are available commercially (e.g., from Bio-Rad). It is cheaper to make the reagents as described below.
(i) Horseradish peroxidase conjugates
Solution A: 50 mg of 3,3′-diaminobenzidine tetrahydrochloride (Sigma) in 50 ml of PBS
Solution B: 30 mg of cobalt chloride in 50 ml of PBS
Pour first solution A, then solution B, on the blot. Add 1 ml of 30% (stock) H_2O_2. Agitate during the staining. 4-Chloro-1-naphthol and aminoethylcarbazole are alternate reagents that are less sensitive than 3,3′-diaminobenzidine tetrahydrochloride but may show less background.
(ii) Alkaline phosphatase conjugates
Solution A: 300 mg of nitroblue tetrazolium (Sigma) in 10 ml of 70% dimethyl formamide (DMF; 3 ml of water and 7 ml of DMF) (Sigma).
Solution B: 150 mg of 5-bromo-4-chloro-3-indolyl phosphate-toluidine (Sigma) in 10 ml of DMF. Store solutions A and B at −20°C.
Bicarbonate buffer (pH 9.8): 8.4 g of $NaHCO_3$ and 0.203 g of $MgCl_2 \cdot 6H_2O$ in 1 liter of distilled water.
Before use, add 1 ml each of solutions A and B to 100 ml of the bicarbonate buffer.

b. Protocol
It is important to remember the following.

1. Use prestained or biotinylated molecular weight standards to facilitate accurate identification of the immunolabeled bands.
2. Do not touch the nitrocellulose paper with bare hands.
3. Be economical, as nitrocellulose is expensive.
4. Always place the nitrocellulose membrane on the anode (+) side of the chamber.
5. Remove all air bubbles from the sandwich, or the transfer will not be uniform.
6. Cool the buffer during high-voltage antigen transfer.
7. Silver stain the gel after blotting to confirm the efficiency of transfer.

Sandwich assembly
1. Cut a piece of nitrocellulose equal to the size of the gel to be blotted, float it carefully on distilled water or transfer buffer, and let it wet evenly. If parts of the nitrocellulose membrane do not wet after 15 min of soaking, discard it. Meanwhile, cut two pieces of Whatman filter paper. Soak the papers and the fiber pads in the transfer buffer.
2. Carefully remove one glass plate from the gel. Remove the stacking gel, and nick one corner of the gel for orientation.
3. Partially submerge one plate of the blot sandwich holder in a dish containing transfer buffer, and lay a soaked fiber pad and then a filter paper on the submerged plate. Next, place the gel on the filter paper and pour some buffer over the gel to reduce the chances of trapping air bubbles. Place the nitrocellulose membrane over the gel. Remove all air bubbles. Lastly, place the other filter paper and fiber pad on the sandwich.
4. Clamp the sandwich together. Place the sandwich in a transfer chamber with the gel side toward the cathode.
5. Connect the leads, and transfer at 100 V for 1 to 2 h with cooling or at 30 V for 16 h at room temperature or 4°C.
6. After transfer, remove the nitrocellulose membrane (Western blot or blot) and mark the corner corresponding to the nicked corner of the gel with a pencil. Place the blot in the blocking buffer, and block for 4 to 16 h at room temperature or 1 h at 37°C. At this point, rinse the blot in PBS, air dry, and seal in a plastic bag or wrap in plastic food wrap at −20°C till use. *Do not* fold the blot.

5.5.5.2. Staining

a. Biotin-streptavidin amplification method
All blocking and incubation steps for the biotin-streptavidin amplification method can be carried out for 1 h at 37°C, 2 to 4 h at room temperature, or 16 h at 4°C. The volume of reagents or buffers used should be sufficient to cover the blot completely. Small dishes should be used at the antibody and reagent steps to save these reagents. Agitate the nitrocellulose blot throughout the following procedure.

1. Incubate the blot in the blocking solution.
2. Wash the blot for 5 min in the wash buffer. Repeat twice more.
3. Place the blot in the test antibody, and incubate. As a general estimation, use a dilution of antibody that gives an absorbance of 50% of the highest absorbance as determined by ELISA. Hybridoma culture supernatants are often used at a 1:5 dilution or undiluted.
4. Wash for 15 min in a sufficient volume of wash buffer. Repeat twice more.
5. Add the biotin-conjugated second antibody, diluted as indicated for commercial preparations. Incubate.
6. Repeat step 4.
7. Add a dilution of streptavidin-HRPO or streptavidin-biotin-HRPO complex, and incubate for 30 min at room temperature.
8. Repeat step 4.
9. Place the immunoblots in the appropriate developing solution.
Wash the stained immunoblots extensively to stop color development. Dry between paper towels, and store out of direct light.

b. Detection of periodate-sensitive epitopes on antigens immobilized on nitrocellulose membranes (93)
1. For oxidation, incubate blots after transfer in 50 mM sodium acetate (pH 4.5) containing 10 mM periodic acid, for 1 h in the dark. Incubate control blots in the sodium acetate buffer without the periodate.
2. Rinse in the sodium acetate buffer.

3. For reduction, incubate blots in 50 mM sodium borohydride in PBS for 15 min at room temperature.
4. Wash extensively in PBS.
5. Proceed with step 1 of the blot immunostaining procedure above.

c. Dye staining of antigens

Proteins immobilized on nitrocellulose can be stained with amido black (91), the reversible Ponceau S stain (3), India ink stain (14), or fluorescein isothiocyanate (89). Transfer efficiency can be assessed by the following method.

1. Wash the blot four times for 10 min each in PBS containing 0.1% Tween 20. Rinse well with water between the washes. The washing process is necessary to remove SDS, which interferes with the stain.
2. Prepare 1 μl/ml Pelican India Ink solution in PBS–0.1% Tween 20. Mix for at least 1 h.
3. Stain the blot with the India ink for 1 to 18 h at room temperature, with agitation.
4. Destain in several changes of distilled water. India ink will stain both protein and nonprotein components.

5.5.5.3. Dot blotting

Dot immunoblotting is an indirect ELISA conducted on a nitrocellulose matrix. The assay exploits the known ability of nitrocellulose to bind and immobilize a large variety of molecules in comparison with the ELISA plate matrix. In this procedure, 2-μl volumes of the test sample are dotted onto the nitrocellulose and allowed to dry. Dilute samples (e.g., from chromatographic columns) can be concentrated on the paper by repeated applications, allowing the sample to dry after each application. The dot blot is then placed in blocking reagent for 4 to 16 h at room temperature or for 1 h at 37°C. Process the dot blot as described in section 5.5.5.2. Whole-cell lysates, bacterial colonies, culture supernatants, and serological samples can be tested by this method. The major advantage of dot blots over ELISA is the high sensitivity, easy execution, and rapid results. The antigen-immobilizing step of the ELISA is eliminated. The disadvantage is that the results are not quantitative.

5.5.5.4. Glycoprotein

Several biotinylating agents can be used to label cell surfaces and soluble and immobilized proteins. Biotinylation is rapid and easy to perform, and the reaction is easily terminated by washing the cells or by dialysis. The best results are obtained at high protein concentrations. The biotinylated molecules can be detected directly by addition of streptavidin-enzyme complexes. Biotinylation may be an added advantage in antigen capture assays, in which a second-antigen-binding antibody may not be available. Biotinylated molecules can also be immobilized on avidin-coated plates.

The succinamide esters of biotin such as Sulpho-NHS biotin (with and without a spacer arm) reacts with primary amine groups on the protein. The target amino acids are lysine residues. If the lysine residues are in the antigen recognition pocket of antibodies, there is potential for interference with the antigen-binding properties. The method described below uses the hydrazide derivative of biotin, which couples to aldehyde groups on carbohydrates (90). This label does not seem to affect antibody activity and shows increased sensitivity compared with the biotin-ester-labeled molecules.

Reagents

1. *Biotinylating reagent.* Dissolve biotin aminocaproyl hydrazide in DMF at 50 mg/ml.
2. *Protein.* Prepare antibody solution at 3 mg/ml in 0.1 M sodium acetate buffer (pH 5.5).
3. *Oxidizing reagent.* Dissolve 100 mM Sodium metaperiodate in the sodium acetate buffer (pH 5.5). The periodate oxides carbohydrates to generate aldehyde groups, the target of the hydrazide group. Use of excess periodate will result in production of carboxylic groups and reduce the efficiency of biotinylation.
4. Add an equal amount of 80 mM sodium sulfite to quench the periodate.

Biotinylation

1. Add 0.1 ml of the sodium metaperiodate reagent to 1 ml of the antibody solution. Incubate on ice for 30 min in the dark.
2. Add 0.25 ml of the sodium sulfite solution, and incubate at room temperature for 5 min to destroy all the periodate.
3. Add 20 μl of biotin-LC-hydrazide to give a final concentration of 1 mM. Incubate for 15 min at room temperature.
4. Dialyze the labeled antibody extensively against PBS with 0.02% sodium azide. Omit the azide if peroxidase-coupled reagents are to be used in the immunoassays.
5. Concentrate the antibody to 10 mg/ml, and store at 4°C in azide, or add 1% BSA and store at −20°C.

5.5.6. Immunofluorescence

Fluorescent or immunofluorescent antibody techniques (FAT) work on the same principle as the ELISA, except that in FAT the antibodies are labeled with chemical compounds which emit a fluorescent signal on activation by UV radiation of suitable wavelength. The emitted fluorescence is then observed by a fluorescence microscope (Chapter 1.6.1.). The visual impact of the fluorescent dye is convincing evidence of an antigen-antibody interaction.

In the basic or direct fluorescent assay, cells, tissues, or smears of the antigens are fixed on a glass slide and treated with the fluorochrome-labeled antibody, and the resulting fluorescence is observed. The assay can also be performed on unfixed cells in solution to study cell surface antigens. Great care must be taken to ensure the integrity of the cells during this procedure. The reader is referred to excellent reviews by Johnson and Holborow (94) and McKinney (95) on the technique, reagents, and application.

Application. Immunofluorescence is still the most useful technique, in terms of simplicity and speed in clinical diagnostic laboratories, for screening of intracellular and extracellular pathogens in tissues, blood smears, and respiratory exudates. Although direct fluorescent antibody assays with fluorochrome-labeled polyclonal antibodies can be used, most laboratories use the indirect fluorescent antibody technique (IFAT). The increased brightness of the labeled sample and the availability of commercial fluorochrome-labeled second antibody make the IFAT a very useful technique. The specificity of IFATs has been improved by the development of antigen-specific MAbs. One advantage of IFAT over ELISA is that the person reading the slides can often distinguish cross-reacting organisms in tissues by morphology of the labeled cells, for example, paired cocci as opposed to single coccal cells.

Another important area of application is the fluorescence-activated cell sorter technique. Cells are incubated with fluorescence-labeled immunoglobulin, and the labeled cells are fractionated on the basis of the intensity of the fluorescent signal or by the type of signal when antibodies conjugated to different fluorochromes are used. Fluorescence-labeled lectins have also been important in studies of cell surface glycosylation.

Limitations in using the technique. The best results are obtained by using an epifluorescence microscope to read the slides. This is expensive equipment for most laboratories. The method can be tedious for screening large numbers of samples and may not be sensitive enough. The diagnostic potential of the assay in early infections is also complicated by background noise due to low-level cross-reactivities and endogenous tissue fluorescence. Lastly, interpretation of the results requires skill and is largely subjective, particularly for semiquantitative judgments and borderline-positive cases.

The sensitivity of fluorescence assays has been greatly improved by the development of (i) fluorescent labels such as lanthanide chelates which emit powerful signals on activation, and (ii) highly sensitive fluorometers to quantitate these short-lived (microsecond) signals. The cost of the equipment is prohibitive to most laboratories. For these various reasons, FAT has been replaced in many areas of research and diagnostics by the highly sensitive and labor-saving ELISA as a screening procedure and by the more analytical immunogold-labeling procedures (Chapter 3.6.2.) in immunocytochemistry. The most common application of IFAT is in diagnostic laboratories where frozen tissue is examined for the presence of pathogens.

Fluorochromes. The most commonly used fluorochromes are fluorescein isothiocyanate (FITC) and tetramethyl rhodamine isothiocyanate (TRITC or rhodamine). Antibodies and anti-antibodies conjugated to these fluorochromes can be obtained commercially. The protocol for labeling of antibodies with FITC and TRITC is described by Johnson and Holborow (94). Because the fluorescence of FITC is green and that of rhodamine is red, it is possible to examine the distribution or presence of two different antigens on the same sample by using antibodies of two or more specificities, each labeled with the different fluorochromes.

Protocol

In the IFAT described below, the test antibody is unlabeled and a fluorochrome-labeled anti-species second antibody is used. Alternatively, the antibody can be labeled with fluorescein, followed by an anti-fluorescein antibody conjugated to FITC. In this method the fluorescein is used as a ligand. A more popular and highly sensitive process uses biotin-labeled anti-immunoglobulins followed by fluorochrome-labeled streptavidin or streptavidin-biotinylated fluorochrome complexes.

1. Prepare a suspension of the test cells, and apply 10 μl to the slide. Immediately aspirate all the suspension to leave a thin smear. Air dry the smear. Fix the smear by passing the slide over a frame, by heating in an oven at 60°C for 5 min, or by dipping for a few seconds in dry acetone or methanol at −20°C. Tissue extracts are smeared on the slides and fixed in the same manner. Acetone or methanol fixation will permeabilize membranes and allow cells in tissues to be detected more readily than heat fixation. This treatment also allows binding of antibodies to intracellular antigens that would otherwise not be accessible to antibodies in the intact cell. For best results, use commercially available slides that have been treated for improved cell binding.
2. Add a drop of the test antibody to the fixed sample, and incubate at room temperature for 5 min. Place the slides on a covered, moist chamber (petri dish, sandwich box containing a wet paper towel) to avoid drying out during the staining. Include positive and negative controls with specific and nonspecific antisera in the assay.
3. Rinse the slides with a gentle stream of PBS (pH 7.2), and place them in a bath of PBS for 5 min. Replace the PBS twice.
4. Add a drop of the fluorochrome-labeled anti-species antibody. Incubate and wash as above. Most fluorochrome-conjugated antibodies are used at dilutions of between 1:16 and 1:32.
5. Shake the slides to remove most of the buffer. Add the mounting fluid, consisting of 90 ml of glycerin in 0.5 M bicarbonate buffer (pH 9.6), since fluorescence is enhanced under alkaline conditions. Store the slides in the dark at 4°C.
6. To score the results, view the labeled cells under oil, using the fluorescent microscope in a darkened room. Use the right filters for the fluorochrome; FITC fluoresces apple-green at 515 nm. Score samples as follows: no fluorescence as 0; faint fluorescence as (+) to intense fluorescence as (+ + + +).

Problems and troubleshooting. When everything fluoresces (including the negative controls) there is nonspecific binding of the fluorochrome-labeled antibody, which is used at a very high concentration in these assays. Centrifugation of the antibody before use or preabsorption with unlabeled cells will remove aggregates and "sticky" proteins. Another problem is that some serum proteins will bind antibodies in a nonspecific manner, and cells grown in such media should be washed well before staining.

The test antibody should also be preabsorbed with unrelated cells to reduce nonspecific background. This is especially important when polyclonal antiserum is used. Repeated freeze-thawing of the conjugated antibody results in complex formation in the preparation. These complexes are seen as bright spots of fluorescence on the slide. The labeled preparations are stable at 4°C in presence of 0.02% (wt/vol) sodium azide to prevent bacterial growth.

If a low-level background persists, try preincubating the smear with a blocking reagent and diluting the antisera in PBS-blocker (section 5.5.4).

5.6. COMMERCIAL SOURCES

Aldrich Chemical Co., Inc., Milwaukee, Wis.
Amersham Corp., Arlington Heights, Ill.
Becton-Dickinson, Rutherford, N.J.
BIO/CAN Scientific, Mississauga, Ontario, Canada
Bio-Rad Laboratories, Richmond, Calif.
Boehringer-Mannheim GmbH Biochemica, Mannheim, Germany
Canberra-Packard (Canada, Ltd.), Mississauga, Ontario, Canada
Cedarlane Laboratories, Hornby, Ontario, Canada
Difco Laboratories, Detroit, Mich.
Gelman Sciences Inc., Ann Arbor, Mich.
Fisher Scientific, Unionville, Ontario, Canada
LKB Instruments Inc., Gaithersburg, Md.
Mandel Scientific Company, Guelph, Ontario, Canada
Millipore, Bedford, Mass.
Pharmacia LKB Biotechnology, Inc., Uppsala, Sweden
Pierce Chromatographic Specialities, Brockville, Ontario, Canada
Pitman-Moore, Davis, Calif.
Ribi ImmunoChem Research, Inc., Hamilton, Mont.
Savant Instrument Inc., Hicksville, N.Y.
Sigma Chemical Co., St. Louis, Mo.
Syntex Inc., Mississauga, Ontario, Canada

5.7. REFERENCES

5.7.1. General References

1. **Grange, J. N., A. Fox, and N. L. Morgan.** 1988. *Immunological Techniques in Microbiology.* Blackwell Scientific Publications, Oxford.
2. **Hancock, I. C., and I. R. Poxton.** 1988. *Bacterial Cell Surface Techniques. Modern Microbiological Methods.* Wiley-Interscience Publication.
   Provides specific protocols for studying immunochemistry of cell-surface antigens.
3. **Harlow, E., and D. Lane.** 1988. *Antibodies: A Laboratory Manual.* Cold Spring Harbor Laboratory Publications.
   An inexpensive book that every lab using antibodies should have at hand.
4. **Weir, D. M. (ed.)** 1986. *Handbook of Experimental Immunology,* 4th ed., Blackwell Scientific Publications, Oxford.
   Long recognized as a standard reference for laboratory immunological procedures.

5.7.2. Specific References

5.7.2.1. Animal handling

5. **Herbert, W. J., and F. Kristensen.** 1986. Laboratory animal techniques for immunology, p. 133.1–133.36. *In: Handbook of Experimental Immunology,* 4th ed. D. M. Weir, C. Blackwell, and L. A. Herzenberg (ed.). Blackwell Scientific Publications, Oxford.
6. **Kingham, W. H.** 1971. Techniques for handling animals, p. 281–299. *In* J. R. Norris and D. W. Ribbons (ed.), *Methods in Microbiology.* Academic Press Ltd., London.
7. **Smelser, J. F.** 1985. Rabbits: a practical guide for the veterinary technician. *Vet. Tech.* **6:**121–128.
8. **Stewart, K. L., E. L. Johnstone, and J. A. Vecera.** 1988. The laboratory mouse and rat. *Vet. Tech.* **9:**264–271.
9. **Tillman, P., and C. Norman.** 1983. Droperidol-fentanyl as an aid to blood collection in rabbits. *Lab. Anim. Sci.* **33:**181–182.

5.7.2.2. Antigen preparation

10. **Bogard, C. W., Jr., D. L. Dunn, K. Abernethy, C. Kilgarriff, and P. C. Kung.** 1987. Isolation and characterization of murine monoclonal antibodies specific for gram-negative bacterial lipopolysaccharide: association of cross-genus reactivity with lipid A specificity. *Infect. Immun.* **55:**899–908.
11. **Campbell, D. H., J. S. Garvey, N. E. Cremer, and D. H. Sussdorf.** 1970. *Methods in Immunology,* 2nd ed. W. A. Benjamin, Inc., New York.
12. **Diano, M., A. Le Bivic, and M. Hirn.** 1987. A method for the production of highly specific polyclonal antibodies. *Anal. Biochem.* **166:**224–229.
13. **Galanos, C., O. Luderitz, and O. Westphal.** 1971. Preparation and properties of antisera against the lipid-A component of bacterial lipopolysaccharide. *Eur. J. Biochem.* **24:**116–122.
14. **Hancock, K., and V. C. W. Tsang.** 1983. India ink staining of proteins on nitrocellulose paper. *Anal. Biochem.* **133:**157–162.
15. **Knudsen, K. A.** 1985. Proteins transferred to nitrocellulose for use as immunogen. *Anal. Biochem.* **147:**285–288.
16. **Laemmli, U. K.** 1970. Cleavage of structural proteins during the assembly of the head of bacteriophage T4. *Nature* (London) **227:**680–685.
17. **Leppard, K., N. Totty, M. Waterfield, E. Harlow, J. Jenkins, and L. Crawford.** 1981. Purification and partial amino acid sequence analysis of the cellular tumor antigen, p53, from mouse SV40-transformed cells. *EMBO J.* **2:**1993–1999.
18. **Rauser, W. E., A. A. Quesnel, J. S. Lam, and G. G. Southam.** 1988. An enzyme-linked immunosorbent assay for plant cadmium binding peptide. *Plant Sci.* **57:**37–43.

5.7.2.3. Adjuvants

19. **Allison, A. C., and G. Gregoriadis.** 1974. Liposomes as immunological adjuvants. *Nature* (London) **252:**252.
20. **Baker, P. J., J. R. Hiernaux, M. B. Fauntleroy, B. Prescott, J. Cantrell, and J. A. Rubach.** 1988. Inactivation of suppressor T-cell activity by nontoxic monophosphoryl lipid A. *Infect. Immun.* **56:**1076–1083.
21. **Baker, P. J., J. R. Hiernaux, M. B. Fauntleroy, P. W. Stashak, B. Prescott, J. L. Cantrell and J. A. Rudbach.** 1988. Ability of monophosphoryl lipid A to augment the antibody response of young mice. *Infect. Immun.* **56:**3064–3066.
22. **Branche, R., and G. Renoux.** 1972. Stimulation of rabies vaccine in mice by low doses of polyadenylic:polyuridylic complex. *Infect. Immun.* **6:**324–325.
23. **Chase, M. W.** 1967. Production of antiserum. *Methods Immunol. Immunochem.* **1:**197–209.
24. **Claassen, E., N. Kors, and N. van Rooijen.** 1987. Immunomodulation with liposome: the immune response elicited by liposomes with entrapped dichloromethyl-diphosphonate and surface associated antigen or hapten. *Immunology* **60:**509–515.
25. **Edelman, R.** 1980. Vaccine adjuvants. *Rev. Infect. Dis.* **2:**370–383.
26. **Ekwunife, F. S., C. E. Taylor, M. B. Fauntleroy, P. W. Stashak, and P. J. Baker.** 1991. Differential effects of monophosphoryl lipid A on expression of suppressor T cell activity in lipopolysaccharide-responsive and lipopolysaccharide-defective strains of C3H mice. *Infect. Immun.* **59:**2191–2194.
27. **Ellouz, F., A. Adam, R. Ciorbaru, and E. Lederer.** 1974. Minimal structural requirements for adjuvant activity of

bacterial peptidoglycan derivatives. *Biochem. Biophys. Res. Commun.* **59**:1317–1325.
28. **Fitzgerald, T. J.** 1991. Syphilis vaccine: up-regulation of immunogenicity by cyclophosphamide, Ribi adjuvant and indomethacin confers significant protection against challenge infection in rabbits. *Vaccine* **9**:266–272.
29. **Freund, J.** 1956. The mode of action of immunological adjuvants. *Adv. Tuberc. Res.* **7**:130–148.
30. **Gery, I., J. Kruger and S. Spiesel.** 1972. Stimulation of B lymphocytes by endotoxin reactions of thymus-deprived mice and karyotypic analysis of dividing cells in mice bearing T_6T_6 thymus graft. *J. Immunol.* **108**:1088–1091.
31. **Gregoriadis, G., D. Davis, and A. Davis.** 1987. Liposomes as immunological adjuvants: antigen incorporation studies. *Vaccine* **5**:145–151.
32. **Kenney, J. S., B. W. Hughes, M. P. Masada, and A. C. Allison.** 1989. Influence of adjuvants on the quantity, affinity, isotype and epitope specificity of murine antibodies. *J. Immunol. Methods* **121**:157–166.
33. **Kripke, M. L., and D. W. Weiss.** 1970. Studies on the immune responses of Balb/c mice during tumor induction by mineral oil. *Int. J. Cancer* **6**:422–430.
34. **Morein, B., K. Lövgren, S. Höglund, and B. Sundquist.** 1987. The ISCOM: an immunostimulating complex. *Immunol. Today* **8**:333–338.
35. **Morein, B., B. Sundquist, S. Hoglund, K. Dalsgaard, and A. Osterhaus.** 1984. ISCOM, a novel structure for antigenic presentation of membrane proteins from enveloped viruses. *Nature* (London) **308**:457–459.
36. **Nicholson, K. G., D. A. J. Tyrrell, P. Harrison, C. W. Potter, R. Jennings, A. Clark, G. C. Scheld, J. M. Wood, R. Yelts, V. Seagroatt, A. Higgins, and S. G. Anderson.** 1979. Clinical studies of mononvalent inactivated whole virus and subunit A/USSR/77(H_1H_1) vaccine serological and clinical reactions. *J. Biol. Stand.* **7**:123–136.
37. **Ohta, M., N. Kido, T. Hasegawa, H. Ito, Y. Fujii, T. Arakawa, T. Komatsu, and N. Kato.** 1987. Contribution of the mannan O side-chains to the adjuvant action of lipopolysacchride. *Immunology* **60**:503–507.
38. **Seppala, I. J. T., and O. Makela.** 1984. Adjuvant effect of bacterial LPS and/or alum precipitation in response to polysaccharide and protein antigens. *Immunology* **53**:827–836.
39. **Shek, P. N., and B. H. Sabiston.** 1981. Immune response mediated by liposome associated protein antigens. I. Potentiation of the plaque forming cell response. *Immunology* **45**:349–356.
40. **Taub, R. N., A. R. Krantz, and D. W. Dresser.** 1970. The effect of localized injection of adjuvant material on the draining lymph node. I. *Histol. Immunol.* **18**:171–186.
41. **Warren, H. S., F. R. Vogel, and L. A. Chedid.** 1986. Current status of immunological adjuvants. *Annu. Rev. Immunol.* **4**:369–388.

5.7.2.4. Conventional and monoclonal antibodies

42. **Bjorck, L., W. Kastern, G. Lindahl, and K. Wideback.** 1987. Streptococcal protein G, expressed by streptococci or *Escherichia coli,* has separate sites for human albumin and IgG. *Mol. Immunol.* **24**:1113–1122.
43. **Bjorck, L., and G. Kronvall.** 1984. Purification and some properties of streptococcal protein G, a novel IgG-binding reagent. *J. Immunol.* **133**:969–974.
44. **Bruck, C., D. Portetelle, C. Gilneur, and A. Bollen.** 1982. One step purification of mouse monoclonal antibodies from ascitic fluid by DEAE Affi-gel blue chromatography. *J. Immunol. Methods* **53**:313–319.
45. **Cassone, A., A. Torosantucci, M. Boccanera, G. Pellegrini, C. Palma, and F. Malavasi.** 1988. Production and characterisation of a monoclonal antibody to a cell surface, glucomannoprotein constituent of Candida albicans and other pathogenic Candida species. *J. Med. Microbiol.* **27**:233–238.
46. **Ey, P. L., S. J. Prouse, and C. R. Jenkin.** 1978. Isolation of pure IgG1, IgG2a and IgG2b immunoglobulins from mouse serum using protein A-Sepharose. *Immunochemistry* **15**:429–436.
47. **Galfré, G., and C. Milstein.** 1981. Preparation of monoclonal antibodies: strategies and procedure. *Methods Enzymol.* **73**:1–46.
48. **Goding, J. W.** 1986. *Monoclonal Antibodies: Principles and Practice,* 2nd ed. Academic Press Ltd., London.
49. **Köhler, G., and C. Milstein.** 1975. Continuous cultures of fused cells secreting antibody of predefined specificity. *Nature* (London) **256**:495–499.
50. **Kuhlmann, I., W. Kurth, and I. Ruhdel.** 1989. Monoclonal antibodies: *in vivo* and *in vitro* production on a laboratory scale, with consideration of the legal aspects of animal protection. *Altern. Lab. Anim.* **17**:73–82.
51. **Lam, J. S., L. A. MacDonald, M. Y. C. Lam, L. G. M. Duchesne, and G. G. Southam.** 1987. Production and characterization of monoclonal antibodies against serotype strains of *Pseudomonas aeruginosa. Infect. Immun.* **55**:1051–1057.
52. **Lindmark, R., K. Thoren-Tolling, and J. Sjoquist.** 1983. Binding of immunoglobulins to protein A and immunoglobulin levels in mammalian sera. *J. Immunol. Methods* **62**:1–13.
53. **Mariani, M., M. Cianfriglia, and A. Cassone.** 1989. Is mouse IgM purification on protein A possible? *Immunol. Today* **10**:115–116. (Letter.)
54. **Oi, V. T., and L. A. Herzenberg.** 1980. Immunoglobulin-producing hybrid cell lines, p. 351–372. *In* B. B. Mishell and S. M. Shiigi (ed.), *Selected Methods in Cellular Immunology.* W. H. Freeman, San Francisco.
55. **Tung, A. S., S.-T. Ju, S. Sato, and A. Nisonoff.** 1976. Production of large amounts of antibodies in individual mice. *J. Immunol.* **116**:676–681.
56. **Wilner, M. A. E., H. D. Troutman, F. W. Trader, and I. W. McLean.** 1963. Vaccine potentiation by emulsification with pure hydrocarbon compounds. *J. Immunol.* **91**:210–229.

5.7.2.5. Agglutination

57. **Edwards, P. R., and E. W. H. Ewing.** 1972. Identification of *Enterobacteriaceae,* 3rd ed. p. 48–67. Burgess Publishing Co., Minneapolis, Minn.
58. **Fung, J. C., and R. C. Tilton.** 1985. Detection of bacterial antigens by counterimmunoelectrophoresis, coagglutination, and latex agglutination, p. 883–890. *In* E. H. Lennette, A. Balows, W. J. Hausler, Jr., and H. J. Shadomy (ed.), *Manual of Clinical Microbiology,* 4th Ed. American Society for Microbiology, Washington, D.C.
59. **Handfield, S. G., A. Lane, and M. B. McIllmurray.** 1987. A novel coloured latex test for detection and identification of more than one antigen. *J. Immunol. Methods* **97**:153–158.
60. **Kronvall, G.** 1973. A rapid slide agglutination method for typing pneumococci by means of specific antibody adsorbed to protein A-containing staphylococci. *J. Med. Microbiol.* **6**:187–190.
61. **Lim, P. L., and K. H. Ko.** 1990. A tube latex test based on colour separation for detection of IgM antibodies to either of two different microorganisms. *J. Immunol. Methods* **135**:9–14.
62. **McIllmurray, M. B., and M. D. Moody.** 1986. Latex agglutination, p. 9–28. *In* R. B. Kohler (ed.), *Antigen Detection To Diagnose Bacterial Infections,* vol. 1. CRC Press, Inc., Boca Raton, Fl.
63. **Oakley, C. L.** 1971. Antigen-antibody reactions in microbiology p. 174–217. *In* J. R. Norris and D. W. Ribbons (ed.), *Methods in Microbiology,* vol. 5A. Academic Press, Ltd., London.

5.7.2.6. Capsule swelling (Quellung reaction)

64. **Merill, C. W., J. M. Gwaltney, J. O. Hendley, and M. A. Sande.** 1973. Rapid identification of pneumococci. *N. Engl. J. Med.* **288**:510–512.

5.7.2.7. Immunodiffusion

65. **Bjerrum, O. J., and Bøg-Hansen.** 1975. Immunochemical gel precipitation techniques in membrane studies, 378–426. *In* A. H. Maddy (ed.), *Biochemical Analysis of Membranes.* John Wiley & Sons, Inc., New York.
66. **Laurell, C. B.** 1966. Quantitative estimation of proteins by electrophoresis in agarose gel containing antibodies. *Anal. Biochem.* **15:**45–52.
67. **Leung, K. A., and R. M. W. Stevenson (University of Guelph).** Personal communication.
68. **Ouchterlony, O.** 1968. *Handbook of Immunodiffusion and Immunoelectrophoresis.* Ann Arbor Science Publishers, Ann Arbor, Mich.
69. **Tilton, R.** 1983. Procedures for the detection of microorganisms by counter immunoelectrophoresis, p. 87–96. *In* J. D. Coonrod, L. J. Kunz, and M. J. Ferraro (ed.), *The Direct Detection of Microorganisms in Clinical Samples.* Academic Press, Inc., Orlando, Fla.
70. **Weeke, B.** 1973. Crossed immunoelectrophoresis, p. 45–56. *In* N. H. Axelsen, J. Krøll, and B. Weeke (ed.), *A Manual of Quantitative Immunoelectrophoresis.* Blackwell Scientific Publications, Oxford.

5.7.2.8. ELISA

71. **Cantarero, L. A., J. E. Butler, and J. W. Osborne.** 1980. The adsorptive characteristic of proteins for polystyrene and their significance in solid phase immunoassays. *Anal. Biochem.* **105:**375–382.
72. **Conradie, J. D., M. Govender, and L. Visser.** 1983. ELISA solid phase: partial denaturation of coating antibody yields a more efficient solid phase. *J. Immunol. Methods* **59:**289–299.
73. **Ekins, R.** 1981. Merits and disadvantages of different labels and methods of immunoassay, p. 5–16. *In* A. Voller, A. Bartlett, and D. Bidwell (ed.), *Immunoassays for the 80's.* University Park Press, Baltimore.
74. **Engvall, E., and P. Perlmann.** 1971. Enzyme linked immunosorbent assay (ELISA): quantitative assay of immunoglobulin. *Immunochemistry* **8:**871–879.
75. **Fischer, P. M., and M. E. H. Howden.** 1990. Direct, enzyme-linked immunosorbent assay of anti-peptide antibodies using capture of biotinylated peptides by immobilised avidin. *J. Immunoassay* **11:**311–327.
76. **Gardas, A., and A. Lewartowska.** 1988. Coating of proteins to polystyrene ELISA plates in the presence of detergents. *J. Immunol. Methods* **106:**251–255.
77. **Ito, J. I., A. C. Wunderlich, J. Lyons, C. E. Davis, D. G. Gurney, and A. I. Braude.** 1980. Role of magnesium in enzyme-linked immunosorbent assay for lipopolysaccharide of rough *Escherichia coli* strain J5 and *Neisseria gonorrhoeae. J. Infect. Dis.* **142:**532–537.
78. **Lee, B. Y., D. Chatterjee, C. M. Bozic, P. J. Brennan, D. L. Cohn, J. D. Bales, S. M. Harrison, L. A. Androu, and I. M. Orme.** 1991. Prevalence of serum antibody to the type-specific glycopeptidolipid antigens of *Mycobacterium avium* in human immunodeficiency virus-positive and -negative individuals. *J. Clin. Microbiol.* **29:**1026–1029.
79. **Lutz, H. U., P. Stammler, and E. A. Fischer.** 1990. Covalent binding of detergent-solubilised membrane glycoproteins to "Chemobond" plates for ELISA. *J. Immunol. Methods* **129:**211–220.
80. **Munoz, C., A. Nieto, A. Gaya, J. Martinez, and J. Vives.** 1986. New experimental criteria for optimisation of solid phase antigen concentration and stability in ELISA. *J. Immunol. Methods* **94:**137–144.
81. **Pitzurra, L., E. Blast, A. Bartoli, P. Marconi, and F. Bistoni.** 1990. A rapid objective immunofluorescence microassay. Application for detection of surface and intracellular antigens. *J. Immunol. Methods* **135:**71–75.
82. **Reggiardo, Z., E. Vasquez, and L. Schnaper.** 1980. ELISA tests for antibodies against mycobacterial glycolipids. *J. Immunol. Methods* **34:**55–60.
83. **Scott, B. B., and G. R. Barclay.** 1987. Endotoxin-polymixin complexes in an improved enzyme-linked immunosorbent assay for IgA antibodies in blood-donor sera to gram-negative endotoxin core glycolipids. *Vox Sang.* **52:**272–280.
84. **Tijseen, P.** 1985. Practice and Theory of Enzyme Immunoassays, p. 9–327. *In* R. H. Burdon and P. H. Van Knippenberg (ed.), *Laboratory Techniques in Biochemistry and Molecular Biology.* Elsevier Science Publishing Co., Inc., New York.
85. **Voller, A., and D. E. Bidwell.** 1985. Enzyme immunoassays, p. 77–86. *In* W. P. Collins (ed.), *Alternate Immunoassays.* John Wiley & Sons, Inc., New York.
86. **Weigand, K., C. Birr, and M. Sutter.** 1981. The hexa- and pentapeptide extension of proalbumin II. Processing on specific antibodies against the synthetic hexapeptide. *Biochim. Biophys. Acta* **670:**424–427.
87. **Wylie, D. E., L. D. Carlson, R. Carlson, F. W. Wagner, and S. M. Schuster.** 1991. Detection of mercuric ions in water by ELISA with a mercury-specific antibody. *Anal. Biochem.* **194:**381–387.

5.7.2.9. Immunoblotting

88. **Birk, H. W., and H. Koepsell.** 1987. Reaction of monoclonal antibodies with plasma membrane proteins after binding to nitrocellulose: renaturation of antigenic sites and reduction of non-specific antibody binding. *Anal. Biochem.* **164:**12–22.
89. **Houston, B., and D. Peddie.** 1989. A method for detecting proteins immobilised on nitrocellulose membranes by *in situ* derivatization with fluorescein isothiocyanate. *Anal. Biochem.* **177:**263–267.
90. **O'Shannessy, D. J., P. J. Voorstad, and R. H. Quarles.** 1987. Quantitation of glycoproteins on electroblots using the biotin-streptavidin complex. *Anal. Biochem.* **163:**204–209.
91. **Schaffner, W., and C. Weissman.** 1973. A rapid, sensitive and specific method for the determination of protein in dilute solution. *Anal. Biochem.* **56:**502–514.
92. **Towbin, M., T. Staehlin, and J. Gordon.** 1979. Electrophoretic transfer of proteins from polyacrylamide gels to nitrocellulose sheets: procedure and some application. *Proc. Natl. Acad. Sci. USA* **76:**4350–4354.
93. **Woodward, M. P., W. W. Young, Jr., and R. A. Bloodgood.** 1985. Detection of monoclonal antibodies specific for carbohydrate epitopes using periodate oxidation. *J. Immunol. Methods* **78:**143–153.

5.7.2.10. Immunofluorescence

94. **Johnson, G. D., and E. J. Holborow.** 1986. Preparation and use of Fluorochrome conjugates, chap. 18.1–18.20. *In* D. M. Weir (ed.), *Handbook of Experimental Immunology,* vol. 1, 4th ed. Blackwell Scientific Publications, Oxford.
95. **McKinney, R. M.** 1986. Immunofluorescence microscopy: reagents and technique, p. 35–49. *In* R. B. Kohler (ed.), *Antigen Detection To Diagnose Bacterial Infections,* vol. 1. *Methods.* CRC Press Inc., Boca Raton, Fl.

Section II

GROWTH

Introduction to Growth

PHILIPP GERHARDT

"The study of the growth of bacterial cultures . . . is the basic method of Microbiology" (5). This opening assertion by Monod in his classic 1949 essay marked the start of a new era in studies of bacterial growth, in which the process became formulated in terms of a few basic parameters that could be expressed quantitatively. Prior to that landmark the methodology of bacterial growth had been essentially qualitative, starting with the historic work of Robert Koch. Today, both qualitative and quantitative aspects are essential to this "basic method of Microbiology," and details are presented in this section.

In Chapter 6, the methods to control physicochemical factors in bacterial growth are substantially revised and extended by Breznak from the prior version by Costilow (1). The principles and techniques for cultivating anaerobes are particularly updated, and the factors of light, diffusion gradients, magnetic fields, and viscosity are added.

Chapter 7 on nutrition and media by Cote and Gherna is completely reorganized and rewritten for practical usefulness. Defined and undefined components, design, and sterilization of media are described first and are followed by descriptions of nutritionally generalized media for 2 main and 18 special groups of bacteria. The prior and complementary version by Guirard and Snell (3), more biochemical in perspective, contained extensive tables and references of defined and undefined media for 19 systematic groups and about 250 species of bacteria.

The enrichment techniques in Chapter 8, succinctly described by Holt and Krieg, are organized into basic strategies. Each strategy is exemplified by one or a few types or species of bacteria but is applicable to many other known or new ones. Once it becomes predominant in a population, the desired organism is isolated by one of four basic procedures. With the ingenious new technique of optical trapping and manipulation, a single cell type can be isolated and grown from a mixed population without prior enrichment.

There follow in Chapter 9 updated methods for solid, liquid/solid, and semisolid culture and in Chapter 10 methods for liquid culture. Notable in the latter is a section on high-density culture strategies.

The quantification of bacterial growth is rigorously dealt with by Koch in Chapter 11. Various methods are described for directly and indirectly measuring numbers and mass in populations. This chapter also describes statistical and computer methods for growth measurement.

The foregoing culturing aspects of growth then give way to culture preservation in Chapter 12 by Gherna, which contains a new discussion of the use of culture collections.

These seven chapters on bacterial growth are quite diverse, and each contains general and specific reference lists for further information and alternative methods. For bacterial growth in general, however, there are unfortunately no methodology reference books or journals as there are, for example, in molecular genetics. Fortunately, however, there are several excellent books and review articles to which one may turn for background reading on the principles of bacterial growth. One certainly could well start with Monod's short but classic essay, *The Growth of Bacterial Cultures* (5). (Incidentally, in this same volume is Lederberg's equally classic essay, *Bacterial Variation*). A 1974 monograph edited by Dawson (2) contains a collection of articles that pioneered the study of bacterial growth. Pirt's 1975 monograph, *Principles of Microbe and Cell Cultivation* (6), is essential background reading. More recently there is the treatise by Ingraham, Maaloe, and Neidhardt, *Growth of the Bacterial Cell* (4). Currently and beyond these, the reader should turn to the introductory statements of principles of bacterial growth that are incorporated into the following chapters of this section, together with their reference lists.

REFERENCES

1. **Costilow, R. N.** 1981. Biophysical factors in growth, p. 66–78. *In* P. Gerhardt, R. G. E. Murray, R. W. Costilow, E. W. Nester, W. A. Wood, N. R. Krieg, and G. B. Phillips (ed.), *Manual of Methods for General Bacteriology.* American Society for Microbiology, Washington, D.C.
2. **Dawson, P. S. S. (ed.)** 1974. *Microbial Growth.* Halsted Press, New York.
3. **Guirard, B. M., and E. E. Snell.** 1981. Biochemical factors in growth, p. 79–111. *In* P. Gerhardt, R. G. E. Murray, R. N. Costilow, E. W. Nester, W. A. Wood, N. R. Krieg, and G. B. Phillips (ed.), *Manual of Methods for General Bacteriology.* American Society for Microbiology, Washington, D.C.
4. **Ingraham, J. L., O. Maaloe, and F. C. Neidhardt.** 1983. *Growth of the Bacterial Cell.* Sinauer Associates, Inc., Sunderland, Mass.
5. **Monod, J.** 1949. The growth of bacterial cultures. *Annu. Rev. Microbiol.* **3:**371–394.
6. **Pirt, S. J.** 1975. *Principles of Microbe and Cell Cultivation.* John Wiley & Sons, Inc., New York.

Chapter 6

Physicochemical Factors in Growth

JOHN A. BREZNAK and RALPH N. COSTILOW

Some physicochemical factors affecting bacterial growth are controlled primarily by the constituents of the culture medium (hydrogen ion activity, water activity, osmotic pressure, and viscosity). Others are controlled by the external environment (temperature, oxygen, light, hydrostatic pressure, and magnetic-field strength). The oxidation-reduction potential is controlled by both the medium and the environment. All of these factors can influence the growth rate, cell yield, metabolic pattern, and chemical composition of bacteria. The control of hydrogen ion activity, temperature, and oxygen supply is critical with every bacterial culture, whereas the control of oxidation-reduction potential is of major importance in culturing obligately anaerobic bacteria.

6.1. HYDROGEN ION ACTIVITY (pH)

pH is a measure of the mean hydrogen ion activity of a solution (a_{H^+}) and is defined as follows: pH = $-\log_{10} a_{H^+}$. The activity of the H^+ ion is the product of its molar concentration and its activity coefficient, which, in turn, is a function of the total ionic strength of the solution (μ). In dilute solutions, where μ ranges from 0 to 0.25, the activity coefficient of the H^+ ion (at 25°C) ranges from 1 to 0.85, respectively. Most routinely used bacteriological media can be considered ionically dilute solutions. For example, in a typical glucose-salts medium, μ is about 0.02. Therefore, the activity coefficient of the H^+ ion is close to unity, and hence a_{H^+} is essentially equal to the hydrogen ion concentration, $[H^+]$. Consequently, pH $\approx -\log [H^+]$ for most bacteriological media.

The pH scale is typically thought of as ranging from 0 ($[H^+] = 1$ M) to 14 ($[H^+] = 10^{-14}$ M), although values outside of this range are possible. Most known bacteria grow over a relatively narrow range of pH, usually best near neutrality (pH 7.0). However, an ever increasing number of **extremophiles** continue to be recognized and isolated from nature (59). Many of these grow optimally at very low pH **(acidophiles)** or very high pH **(alkaliphiles).** Some of these organisms also grow at high temperatures and salinities. Accordingly, for precise definition of the pH optimum for growth of such bacteria, it is important to keep in mind the factors that affect the activity coefficient of the hydrogen ion and hence the a_{H^+}. Among these factors are temperature, ionic strength, ion charge, dielectric constant, and the physical size of various ions in the solution. In practice, the influence of such factors on pH is taken into account by standardizing the pH meter with a pH reference solution whose composition and temperature are as close as possible to those of the solution being measured (e.g., the bacterial culture fluid).

6.1.1. pH Measurement

For accurate pH measurements of virtually all types of solutions, use a pH meter with a glass electrode. Two types of glass electrode systems may be used. One consists of a **pH-measuring electrode** paired with a separate **pH reference electrode** (Hg/Hg_2Cl_2 or Ag/AgCl), both of which are put into the solution being measured. Another system consists of a single electrode containing elements of both the pH-measuring and the reference electrode, referred to as a **pH combination electrode.** This type of electrode is popular because its generally compact size facilitates the measurement of small volumes contained in test tubes. Epoxy electrodes (pH-measuring, pH reference, and pH combination types) are also available and are generally more durable than glass electrodes. However, epoxy electrodes are not recommended for use with certain organic solvents; consult with the manufacturer for the solvent tolerance of such electrodes if this is a concern.

See Chapter 21.2.1 for a further description of hydrogen ion electrode systems, procedures used for calibration of pH meters, and proper care of electrodes. Also see references 14 and 22 for details of the measurement and control of pH.

When accurate pH measurements are desired, observe the following precautions.

1. Adjust the temperature of the buffer used for standardization of the pH meter to the same temperature as the sample. The pH of buffers changes with temperature; e.g., the pH of standard phosphate buffer is 6.98 at 0°C, 6.88 at 20°C, and 6.84 at 37°C. See the *Handbook of Chemistry and Physics* (21) for pH values of standard buffers at different temperatures. If pH measurements are to be made on numerous samples at different temperatures, a pH meter with a temperature compensation control is desirable, so calibration with standard buffer need be done at only one temperature. In any case, keep in mind that an average error of only 0.003 pH unit per degree Celsius exists at one pH unit from standardization; this is often less than the required accuracy for measurements.

2. Standardize the pH meter with a buffer that is near the expected pH of the sample and of similar ionic strength. If accurate readings are to be made across a fairly wide range of pH values (e.g., 3 pH units), a meter with a slope control is desirable and is used after two-point calibration. In this procedure, the pH meter is standardized with a buffer whose pH is at or just outside of one end of the expected range, the electrodes are placed in a second standard buffer representing the other end of the expected range, and then the pH meter readout is adjusted to the appropriate pH by using the slope control. A frequently encountered sample of relatively high ionic strength is seawater. The pH of such samples should be made after standardization with a buffer consisting of 0.4186 M NaCl, 0.0596 M $MgCl_2$, 0.02856 M Na_2SO_4, 0.005 M $CaCl_2$, 0.02 M Tris, and 0.02 M Tris-HCl. The pH of this buffer is 8.835 at 5°C, 8.517 at 15°C, 8.224 at 25°C, and 7.953 at 35°C (22).

3. If the sample must be stirred (e.g., with a magnetic stirrer) during measurement, stir the calibrating buffer in the same manner, since stirring usually alters the signal of the pH electrode.

4. Determine the pH of media after sterilization, not before. Autoclaving of media frequently results in a significant change in the pH owing to the escape of CO_2 or the precipitation of alkaline-earth phosphates. Even filtration may have a significant effect on the pH of media.

5. Just prior to use, check the pH of media that have been stored.

6. Some pH electrodes (those with linen fiber junctions) do not give accurate pH readings with Tris buffers. Be sure that the electrodes used with Tris buffers are recommended for such use by the manufacturer. In addition, reference electrodes may be especially sensitive toward chelators (e.g., EDTA) or sulfides. For such instances, probes with salt-bridged reference electrodes should be used.

In instances when accuracy is not required, such as in the preparation of routine media, the pH may be measured by use of pH indicator dye solutions or pH paper. By proper selection of either, the pH can be estimated within 0.2 pH unit. Indicator solutions and papers, and their corresponding color standards for various pH ranges, are available from most scientific supply companies. Common pH indicators and their useful pH ranges are listed in Table 1. Also see Chapter 25.4.15.

A number of pH indicators are frequently incorporated into culture media to demonstrate pH changes during growth of bacteria. Select the appropriate indicator for the pH range of interest, and make sure that it does not inhibit growth of the organism. All of these pH indicators are poorly soluble in water but are effective at very low concentrations (less than 0.01%) in media.

Table 1. pH indicators and their useful ranges of pH

Indicator[a]	pH range	pK_a	Color	
			Acid	Alkali
Thymol blue	8.0–9.6	8.9	Yellow	Blue
Metacresol purple	7.4–9.0	8.3	Yellow	Purple
Cresol red	7.2–8.8	8.3	Yellow	Red
Phenol red	6.8–8.4	7.9	Yellow	Red
Bromthymol blue	6.0–7.6	7.0	Yellow	Blue
Bromcresol purple	5.2–6.8	6.3	Yellow	Purple
Methyl red	4.4–6.0	5.2	Yellow	Red
Bromcresol green	3.8–5.4	4.7	Yellow	Blue
Bromphenol blue	3.0–4.9	4.0	Yellow	Blue
Thymol blue	1.2–2.8	1.5	Red	Yellow

[a]Prepare 0.04% solutions by solubilizing 0.1 g in the smallest possible volume (10 to 30 ml) of 0.01 N NaOH, and dilute to 250 ml.

6.1.2. pH Buffers

Many bacteria consume and/or produce significant amounts of acidic or basic ions during growth. This commonly occurs during anaerobic fermentation and during aerobic dissimilation of the salts of organic acids. Unless acid production and consumption are balanced, large changes in pH may occur during growth, and growth may be significantly inhibited. The pH of cultures may be controlled to some extent by incorporating pH buffers (substances that resist change in pH) into the medium. Continuous pH control may be achieved by the automatic addition of acid or base (section 6.1.3).

pH buffers usually are mixtures of weak acids and their conjugate bases. In complex media, the acidic and basic groups of organic molecules (such as proteins, peptides, and amino acids) provide some buffering capacity. Owing to the variety of compounds present in such media, there may be a fair degree of buffering action over a wide range of pH. However, the buffering capacity at any given pH varies widely with the types and concentrations of organic molecules present.

In cultivating many bacteria, it is necessary to include a buffer system in the medium. Consider the following in selecting a buffer for use: (i) the pH desired; (ii) possible inhibitory or toxic effects of the buffer; (iii) possible utilization of the buffer by the culture; and (iv) possible binding of di- and trivalent metal ions by the buffer.

Weak acids and bases buffer most effectively at the pH where they are 50% dissociated. This pH is equal to the pK_a (the negative logarithm of the dissociation constant) of the acid or base. The effective range of a buffer is ca. ± 1 pH unit from its pK_a. A number of compounds that have been used as buffers in growth media are listed in Table 2, along with their pK_a values. The **Good buffers,** developed by Good et al. (37), have greatly expanded the range of buffers useful in biological systems. The Good buffers have the advantages of a pK_a between 6.0 and 8.0, high aqueous solubility, impermeability by biological membranes, minimum interaction with mineral cations, enzymatic and hydrolytic stability, minimum participation in biochemical reactions, minimum absorbance between 240 and 700 nm, and usually no toxicity. In addition, they show minimum effects due to ionic composition of solutions, concentration, and temperature. A disadvantage is that they are slightly more expensive than some of the other buffers. Note that the salts of citric, phthalic, and succinic acids buffer over a wide range of pH because these acids have more than one dissociable hydrogen ion and hence more than one pK_a.

Buffers containing the compounds in Table 2 are prepared by one of the following two general procedures.

1. Prepare a solution of the acid or base at twice the desired concentration, and titrate with either NaOH or HCl to the desired pH by using a pH meter. Dilute to final volume with distilled or deionized water. For example, to prepare 1 liter of 0.1 M Tris-hydrochloride buffer at pH 7.6, make 500 ml of a 0.2 M solution of Tris, titrate with HCl to pH 7.6, and then dilute to 1 liter. Dilution has only a small effect on the final pH. If this effect is important, make a final correction before adding the last few milliliters of water during dilution.

2. Prepare equimolar concentrations of the acidic and basic components of the buffer system. With the aid of a magnetic stirrer and a pH meter, titrate one solution with the other until the desired pH is obtained. If the desired pH of the buffer is near the pK_a of the acid component, the ratio of the volume of acidic solution to basic solution is about 1:1; if it is 1 pH unit above the pK_a, the ratio is about 1:10; and if it is 1 pH unit below the pK_a, the ratio is about 10:1. For example, to prepare about 100 ml of 0.1 M sodium phosphate buffer at pH 7.0, mix solutions of 0.1 M NaH_2PO_4 (monobasic) and 0.1 M Na_2HPO_4 (dibasic) in approximately equal proportions and then adjust to pH 7.0 by adding more of the appropriate solution as determined with a pH meter. To prepare 100 ml of the phosphate buffer at pH 6.0, titrate

Table 2. pK_a values of chemical compounds used in buffers

Compound	pK_a at: 25°C	pK_a at: 20°C
Citric acid	3.13, 4.77, 6.40	
Phthalic acid	2.89, 5.51	
Barbituric acid	4.01	
Oxalic acid	4.19	
Succinic acid	4.16, 5.61	
Acetic acid	4.75	
Monobasic phosphate	7.21	
Tris	8.08	
Boric acid	9.24	
Glycine	9.87	
MES [2-(*N*-morpholino)-ethanesulfonic acid]		6.15
ADA (*N*-2-acetamidoiminodiacetic acid)		6.60
PIPES [piperazine-*N,N'*-bis(2-ethanesulfonic acid)]		6.80
ACES (*N*-2-acetamido-2-aminoethanesulfonic acid)		6.90
BES [*N,N-bis*-(2-hydroxyethyl)-2-aminoethanesulfonic acid]		7.15
MOPS [3-(*N*-morpholino)-propanesulfonic acid]		7.20
TES [3-((tris[hydroxymethyl]methyl)amino)ethanesulfonic acid]		7.50
HEPES [*N*-2-hydroxyethylpiperazine-*N'*-2-ethanesulfonic acid]		7.55
HEPPS [*N*-2-hydroxyethylpiperazine-*N'*-3-propanesulfonic acid]		8.00
Tricine [*N*-tris(hydroxymethyl)methylglycine]		8.15

95 ml of the monobasic solution with the dibasic solution.

Although the Good buffers are becoming increasingly popular, the most commonly used buffers in bacterial growth media are phosphate, Tris-hydrochloride, citrate, and acetate. These four systems will buffer over almost the entire range of pH permitting bacterial growth. Solutions used for their preparation and the appropriate volumes required to buffer at various pH values are given in Table 3. As noted above, the actual pH of any buffer is temperature dependent. Tables for preparing buffers of other types and at different temperatures are available elsewhere (4, 5, 14, 21).

When cultures are incubated in atmospheres enriched with CO_2, the CO_2-bicarbonate equilibrium is an important consideration and, in fact, can be exploited as a buffer system itself at a pH between 6.0 and 8.0. CO_2 dissolves in and reacts with water to form carbonic acid which readily ionizes to form a proton and a bicarbonate anion as follows.

$$CO_2\text{ (gas)} \overset{K_s}{\rightleftharpoons} CO_2\text{ (dissolved)} \underset{-H_2O}{\overset{+H_2O}{\rightleftharpoons}} H_2CO_3 \overset{K_{a1}}{\rightleftharpoons} H^+ + HCO_3^-$$

The formation of carbonic acid and hence the acidity created by it are dependent on the concentration of dissolved CO_2, which, in turn, is dependent on the concentration of CO_2 in the gas phase in accord with Henry's law. K_s is the solubility constant of CO_2 (which is temperature dependent). Inasmuch as the concentration of H_2CO_3 in solution is usually very low, the dissociation constant (K_{a1}) usually refers to the reaction,

$$CO_2\text{ (dissolved)} + H_2O \rightarrow H^+ + HCO_3^-$$

The Henderson-Hasselbach equation for this system is

$$\text{pH} = \text{p}K' + \log \frac{[HCO_3^-]}{[CO_2\text{ (dissolved)}]}$$

where pK' (the negative logarithm of K_{a1} for the reaction shown above) normally takes into account the activity of H_2O (which is 1 by definition and is close to 1 for dilute solutions), which can itself be considered a constant at a given temperature. The pK' value is about 6.4 at 25°C. Therefore, the molar ratio of dissolved CO_2 to HCO_3^- is about 1:1 at this pH. CO_2-bicarbonate buffers are effective over a pH range of about 5.4 to 7.4. If the pH of the medium is below 5.0, essentially no bicarbonate is present, but an increase in CO_2 concentration in the atmosphere will still result in a decrease in pH as a result of the formation of carbonic acid. In an atmosphere of 50 to 100% CO_2, the medium must contain a substantial amount of bicarbonate (e.g., $NaHCO_3$) to maintain a pH level near neutrality. The high salt concentration that results may be toxic to some bacteria.

Table 3. Formulas for buffers frequently used in bacteriological media[a]

pH	Value of x to be used in buffer formulas			
	Acetate buffer[b]	Citrate-phosphate buffer[c]	Phosphate buffer[d]	Tris-hydrochloride buffer[e]
4.0	41.0	30.7		
4.2	36.8	29.4		
4.4	30.5	27.8		
4.6	25.5	26.7		
4.8	20.0	25.2		
5.0	14.8	24.3		
5.2	10.5	23.3		
5.4	8.8	22.2		
5.6	4.8	21.0		
5.8		19.7	46.0	
6.0		17.9	43.85	
6.2		16.9	40.75	
6.4		15.4	36.75	
6.6		13.6	31.25	
6.8		9.1	25.5	
7.0		6.5	19.5	
7.2			14.0	44.2
7.4			9.5	41.4
7.6			6.5	38.4
7.8			4.25	32.5
8.0			2.65	26.8
8.2				21.9
8.4				16.5
8.6				12.2
8.8				8.1
9.0				5.0

[a]Data adapted from reference 5.
[b]Formula: x ml of 0.2 M acetic acid + (50 − x) ml of 0.2 M sodium acetate, diluted to 100 ml.
[c]Formula: x ml of 0.1 M citric acid + (50 − x) ml of 0.2 M Na_2HPO_4, diluted to 100 ml.
[d]Formula: x ml of 0.2 M NaH_2PO_4 + (50 − x) ml of 0.2 M Na_2HPO_4, diluted to 100 ml.
[e]Formula: 50 ml of 0.2 M Tris + x ml of 0.2 M HCl, diluted to 100 ml.

After appropriate substitution, the Henderson-Hasselbach equation can be solved to determine the amount of bicarbonate needed in a medium for buffering at a given pH when the atmosphere contains various percentages of CO_2. For this calculation, use the following equation.

$$\log [HCO_3^-] = pH - pK' + \log [P\alpha(\%CO_2)(5.87 \times 10^{-7})]$$

where P is the atmospheric pressure in millimeters of Hg; α (often referred to as the Bunsen absorption coefficient) is the volume of gas (normalized to the volume it would occupy at a 0°C and 760 mmHg) that is absorbed by a unit volume of water (at the temperature of the measurement) under a gas pressure of 760 mmHg; $\%CO_2$ is the volume of CO_2 per unit volume of gas phase $\times$ 100; and 5.87×10^{-7} is a factor that converts α to moles per liter at 760 mmHg (67). If atmospheric pressure varies between 720 and 760 mmHg (96.0 and 101.3 kPa), the variation has little effect on the calculated $[HCO_3^-]$. Therefore, in most instances the following simplified equation can be used (calculated for $P = 740$ mmHg).

$$\log [HCO_3^-] = pH - pK' + \log [\alpha(\%CO_2)(4.35 \times 10^{-4})]$$

When consulting published values for α, use tabulated values that take into account the contribution of solvent (in this case water) vapor pressure to the total gas pressure of 760 mmHg; otherwise a small error will be introduced in the calculation. This can be seen from Table 4, which lists values for α that do and do not consider the vapor pressure of water and which also lists values for pK' at various temperatures. However, values for α and pK' are also influenced by the presence of other ions in solution. Therefore, in practice slight adjustments may have to be made in the amount of HCO_3^- included in the liquid phase, or in the concentration of CO_2 used in the gas phase, to achieve the desired pH of the medium before inoculation. See reference 67 for a more detailed discussion of CO_2-bicarbonate buffers.

The pH of media used for growing bacteria that produce large amounts of acid (e.g., lactic acid bacteria) may be very difficult to control with soluble buffers. However, the acid can be neutralized as fast as it is formed by the inclusion of finely ground $CaCO_3$ (chalk) in the medium. In fact, a good way of visualizing acid production by colonies growing on agar plates is to include 0.3% finely ground chalk in the medium. Make sure that the chalk is well suspended in the molten agar when the plate is poured. Colonies that produce acid will be surrounded by a clear zone. The acid can also be withdrawn by use of a dialysis or other product removal culture system (Chapter 10.3.5 and 10.3.6) or can be neutralized by the addition of alkali with a system under continuous, automatic pH control (see below).

Table 4. Values of α (CO_2 solubility) and pK' at various temperatures

Temp (°C)	α[a]		pK'[b]
	A	B	
20	0.878	0.859	6.392
25	0.759	0.738	6.365
30	0.665	0.642	6.348
35	0.592	0.565	6.328
40	0.530	0.494	6.312

[a]Milliliters of CO_2 (reduced to 0°C and 760 mmHg) that will dissolve in 1 ml of pure water at a gas pressure of 1 atm (760 mmHg) at the stated temperature. The values for α in column A do not consider the contribution of water vapor pressure to the total gas pressure (35a, 51). The values in column B have been recalculated from q values in references 35a and 51 and do consider the contribution of water vapor pressure to the total gas pressure.
[b]$pK' = \log K_{a1}$ for the reaction CO_2 (dissolved) $+ H_2O \xrightarrow{K_{a1}} H^+ + HCO_3^-$. Data from reference 67.

6.1.3. Continuous Control of pH

In instances when the pH of a culture must be controlled within a narrow range, an automatic pH-control system should be used. Such a system should include a pH meter, steam-sterilizable pH electrodes, a controller and controls for setting the desired pH limit, and pumps for the addition of acid and/or base. Most such systems are linked to a recorder to provide a continuous record. Automatic systems are available from a number of sources (e.g., New Brunswick Scientific Co., New Brunswick, N.J.; The VirTis Co., Inc., Gardiner, N.Y.; and Cole-Parmer Instrument Co., Chicago, Ill.). Likewise, steam-sterilizable pH electrodes, both dual and combination, are available from a number of sources (e.g., Beckman Instruments, Inc., Fullerton, Calif.; Ingold Electrodes, Lexington, Mass.; and Leeds and Northrup, North Wales, Pa.).

Check pH control systems at frequent intervals for accuracy by using a separate pH meter. Problems that may be encountered are (i) improper or inadequate grounding of the system; (ii) establishment of pH control points (i.e., upper and lower limits) that are too narrow, resulting in almost constant addition of acid and base and hence in significant dilution of the culture; (iii) aging and deterioration of the pH glass electrode as a result of repeated steam sterilization; and (iv) contamination of the reference electrode diaphragm by continuous exposure to medium constituents and/or metabolic products of the bacteria. The most common contaminants of the diaphragms are proteins and silver sulfide precipitate. The latter is formed by reaction of the silver chloride in the reference electrode with sulfur-containing substances in the medium. Remove the proteins by soaking the electrode in a protease (e.g., pronase) solution, and remove the silver sulfide deposits with an acidic solution of thiourea.

For further discussion of automatic pH control and sterilization of electrodes, see Chapter 10.3.1 and reference 14.

6.2. WATER ACTIVITY AND OSMOTIC PRESSURE

Water must be available to cells for metabolism and growth. However, the mere presence of water in a medium does not ensure its availability, which is determined by the water activity (a_w) of the medium. The a_w represents the mole fraction of the total water molecules that are available and is equal to the ratio of the vapor pressure of the solution to that of pure water(p/p_0). This ratio is equal to the fractional relative humidity (%RH/100) of the atmosphere above the medium at equilibrium. Temperature variation within the range in which most bacteria grow has little effect on the a_w of the medium.

The osmotic pressure (π) of a solution, expressed in atmospheres, is related to a_w as follows.

$$\pi = \frac{RT}{V_w} \ln a_w$$

where R is the gas constant (0.0821 liter atm mol^{-1} K^{-1}), T is the absolute temperature (degrees Celsius + 273), and V_w is the volume of 1 mol of water. For a more detailed presentation of these relationships, see references 1, 15–17, 19, and 40.

The minimum a_w values at which bacteria grow vary widely, but the optimum values for most species are greater than 0.99. Some **halophilic bacteria** require sodium ions and grow best in the presence of high concentrations of sodium chloride at which $a_w \approx 0.80$. Variations in a_w may affect growth rates, cell composition, and metabolic activities of bacteria. Many solutes have very specific effects at concentrations below those required to limit the availability of water. Therefore, determinations of growth-limiting a_w values should be conducted by using media in which a_w has been adjusted with more than one solute. The a_w and osmotic pressure of bacteriologic media are best controlled by adding nonnutrient solutes such as sodium chloride, potassium chloride, sodium sulfate, or mixtures of such salts. Scott (60) has detailed methods for calculating the a_w of nutrient media and for calculating the required concentrations of salt(s) to attain a given a_w value.

For considerations of osmotic pressure relative to bacterial permeability and transport, see Chapter 24.3.2. See references 1, 17, 19, and 60 for reviews of the physiological effects of a_w on bacteria.

6.2.1. Measurement of Water Activity

A number of methods have been developed for the measurement of a_w. Special instruments available for this purpose are the Brady Array (Thunder Scientific Co., Albuquerque, N.M.), the Relative Humidity Indicator (General Eastern Corp., Watertown, Mass.), the Hygrometer (Hygrodynamics, Silver Spring, Md.), and the Sina-Scope (Sina Ltd., Zurich, Switzerland; marketed in the United States by Beckman Instruments, Inc., Cedar Grove, N.J.). The advantages and disadvantages of these instruments have been reviewed (16, 19).

The results of a study (50) of methods for determining a_w indicated that there is considerable variability among the values obtained by different procedures and that the significance of values reported beyond the second decimal place is doubtful. A statistical analysis (65) of measurements made with the electronic Sina-Scope instrument (over the range of 0.755 to 0.967 a_w) indicated a variation of less than $\pm$ 0.01.

6.2.2. Measurement of Osmotic Pressure

The most commonly used instruments for measuring osmotic pressure are based on the reduction of vapor pressure (e.g., model 833; Jupiter Instrument Co., Jupiter, Fla.) or freezing point depression (e.g., 3MO+ Microosmometer; Advanced Instruments, Norwood, Mass.). Other instruments measure the osmotic pressures developed across a semipermeable membrane (e.g., model 230; Jupiter Instrument Co.). Although some of these instruments are calibrated for direct readout, several standard solutions should be run simultaneously with the measured solutions.

6.3. TEMPERATURE

The temperature of incubation dramatically affects the growth rate of bacteria, because it affects the rates of all cellular reactions. In addition, temperature may affect the metabolic pattern, nutritional requirements, and composition of bacterial cells.

The growth rates of most bacteria respond to temperature in a manner similar to that shown in Fig. 1. The useful range of temperature at or near the optimum is usually quite narrow, and the maximum temperature for growth is only a few degrees (3 to 5°C) above the optimum. By contrast, the minimum temperature for growth may be 20 to 40°C below the optimum. It is frequently necessary to incubate a bacterial culture for a prolonged period to measure the growth rate near the

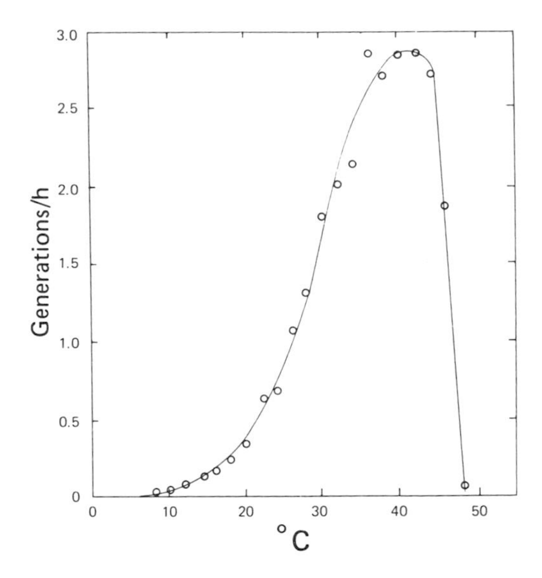

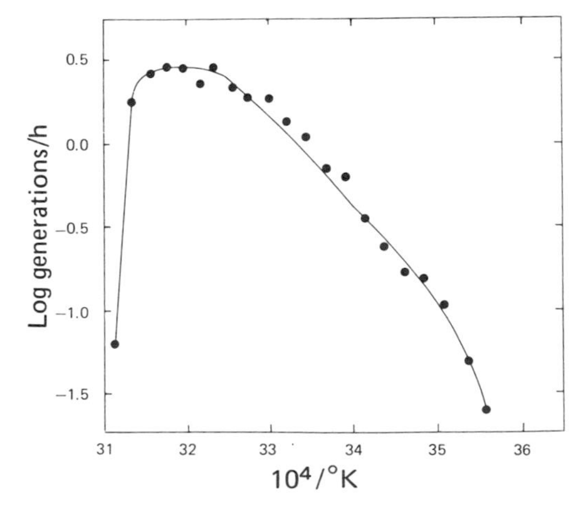

FIG. 1. Effect of temperature on the generation time of *E. coli*, as depicted by an arithmetic plot (top) and by a semilogarithmic Arrhenius plot (bottom). Data from reference 44.

minimum temperature; for example, one *Escherichia coli* culture required over 41 h for one generation at 8°C compared with 21 min at the optimum temperature plateau of 36 to 42°C (44).

Over a limited range of temperature below the optimum, the changes in growth rates of a bacterium are comparable to the responses of chemical reaction rates to temperature. Therefore, if the logarithm of the growth rate is plotted against the reciprocal of the absolute temperature (an **Arrhenius plot** [Fig. 1]), a linear slope is observed in a limited range (15, 44, 58). For most bacteria, the temperature coefficient (Q_{10}) for the growth rate in this limited range is about 2; i.e., the growth rate doubles with a 10°C increase in temperature. The slope of such an Arrhenius plot becomes essentially zero around the optimum growth temperatures and almost vertical near the maximum temperature for growth.

Generally, bacteria not only grow more slowly but also die more rapidly at temperatures above the maximum for growth. Consequently, incubation of a bacterial culture at temperatures above its optimum requires precise temperature control. To be safe, incubate a culture below its optimum temperature to an extent determined by the variability of the incubator.

The temperature within ordinary convection-type air incubators may vary by several degrees Celsius, even more if opened frequently. Incubators equipped with circulating fans or water jackets maintain a more constant temperature, but variations in temperature of 1°C are not uncommon. Nevertheless, for routine cultivation of bacteria, an array of air incubators, including models designed for low-temperature incubation, are commercially available from scientific supply companies.

For accurate measurements of the effect of temperature on bacterial growth rate, however, use a stirred water bath whose temperature is controlled by a thermostat with a sensing element that is immersed in the circulating water; such baths are available commercially and include models for high- and low-temperature incubation. For accurate measurements, also be sure to use an accurately calibrated thermometer. To assess the temperature response of a bacterium, measure growth rates at a number of carefully controlled temperatures. Plot these data as illustrated in Fig. 1. If the general shapes of the resulting curves vary greatly from the normal pattern, repeat the determinations while closely monitoring the stability of the temperature of the water bath.

Practical procedures and precautions relative to temperature control of bacterial growth are as follows.

1. Measure temperatures at various positions in air incubators by using thermometers inserted into flasks or stoppered tubes of water.

2. Air convection incubators may vary widely in temperature. Use circulation fans, and check to see that they are operating efficiently. Position cultures in the incubators so that "dead" air spaces are not created. Do not open incubators more frequently than absolutely necessary.

3. Surface cooling due to evaporation of media in air incubators must be controlled in critical studies of temperature effects. When possible, use tightly stoppered culture vessels. Where free air exchange is necessary, as for aerobic bacteria, use incubators with provision to maintain high relative humidities. This is exceedingly important at high incubation temperatures, at which loss in culture volume could present a serious problem, particularly with shaken cultures.

4. Use 50% polyethylene glycol antifreeze or light silicone oil in baths used for high-temperature or low-temperature incubation.

Temperature gradient incubators have been reviewed by Patching and Rose (58). These incubators provide temperatures that exceed the entire temperature range of growth for a bacterium. Different designs allow for incubation of organisms with various oxygen requirements. If large numbers of determinations are to be made, the construction and use of a gradient incubator may be desirable. A commercial source for a temperature gradient incubator is not known to the authors, but one can be constructed easily in a machine shop.

6.4. PRESSURE

High hydrostatic pressure may adversely affect the growth and survival of airborne, terrestrial, and surface water bacteria, although little or no effect on most species has been observed at pressures ranging from 1 to 100 atm (1 atm = 101,325 N/m² = 760 mmHg = 1.01325 bars = 101.325 kPa). However, with increased exploration of the deep sea, **barophilic bacteria,** which show optimum growth at hydrostatic pressures significantly greater than atmospheric, have been recognized (72). Some of these have proven to be obligate barophiles. For example, strain MT41 grows optimally at 690 bars and also grows at 1,035 bars, a pressure close to that at the depth of its origin (10 km). However, this strain does not grow at 346 bars or less, pressures found at average depths of the sea (73).

Studies of bacteria growing under high pressures are limited and currently represent a rather specialized domain of research, so details on the methods used will not be described here. However, Marquis (11) has described experimental procedures for applying high pressures to cultures, and Jannasch and colleagues (45, 63) have described equipment that allows sampling, manipulation, and even isolation of cultures under high hydrostatic pressure without decompression. The effects of hydrostatic pressure on the physiology and metabolism of bacteria are discussed in references 10, 11, and 12.

6.5. OXYGEN AND OTHER GASES

Gases frequently constitute substrates for bacteria, whether serving as an oxidizable energy source (e.g., H_2, CH_4, CO), a terminal electron acceptor of aerobic respiration (e.g., O_2), or a source of nitrogen (e.g., N_2). Consequently, the metabolism and growth rates of bacteria are often dependent on the concentration of gas in solution. Among the most prevalent concerns in this regard are aerobic and facultative bacteria, whose growth rate and yield may depend critically on the concentration of oxygen in solution. Hence, the following discussion will focus on oxygen as an example, although the principles are also true for virtually any gas that dissolves in water.

Unlike most nutrients, oxygen is relatively insoluble in water (< 10 mg/liter) and may quickly become limiting in liquid bacterial cultures unless special precautions are taken to ensure that it is supplied and dissolved continuously during growth. Principal factors to consider in supplying dissolved oxygen to liquid cultures (aeration) are as follows.

1. For availability of gaseous oxygen to the gas/liquid interface, the opening of the culture vessel must be sufficiently large and stoppered or covered with porous material so as to allow maximum exchange of the atmospheres inside and outside the vessel.

2. Large gaseous surface-to-liquid volume ratios in static cultures can be maintained only by use of very shallow liquid volumes in flat-bottom vessels such as Erlenmayer or Fernbach flasks. Either rotary or reciprocal shaking of flasks greatly increases the surface area of a given volume of medium and hence increases the efficiency of mass transfer. This effect can be further enhanced, when rotary shaking is used, by creation of additional turbulence in the flask either by making indentations ("dimples" or "baffles") in the side of the flask or by placing a stainless-steel coil spring in the flask. The oxygen-solution rate obtained in a shaken flask decreases rapidly as the volume of culture increases and as the density of cells increases.

3. With large volumes of cultures (>0.5 liter), it is usually necessary to force air through (sparge) the liquid. The efficiency of oxygen dissolution in sparged cultures depends primarily on the number and smallness of the air bubbles and on the time of contact between the bubbles and the liquid. The use of gas spargers with small pores results in small bubbles but may induce foaming. The time of contact can be maximized by using culture vessels with high height-to-diameter ratios, by vigorously stirring the cultures, and by using baffles in the vessel.

4. Finally it should be remembered that, according to Henry's law, the mass of gas that dissolves in a definite mass of liquid at a given temperature is directly proportional to the partial pressure of that gas. Therefore, if agitated liquid cultures are incubated under (or sparged with) a gas phase containing O_2 at concentrations greater than atmospheric (i.e., >21% [vol/vol]), the amount of dissolved oxygen that can be provided to growing cells is increased. Gas solubilities are often expressed in terms of an Ostwald coefficient, L, which is the ratio of the volume of gas absorbed to the volume of the absorbing liquid, both measured at the same temperature. The partial pressure of the gas must be designated, and tabulated values are usually reported for the pure gas at 1 atm. Multiplication of L by the decimal fraction of a particular gas in the gas phase (e.g., 0.21 for O_2 in normal air) yields the equilibrium concentration of that gas in water, as milliliters of gas per milliliter of H_2O, at the specified temperature T. Further multiplication by (273.15)(1,000)/(22.4)(absolute temperature in Kelvin) will express the equilibrium concentration as micromoles of gas per milliliter of water, or millimolar, at the specified temperature T.

Table 5 gives Ostwald coefficients for various gases at three different temperatures. Using this table and the information above, the concentration of O_2 in water in equilibrium with normal air at 1 atm and 25°C is calculated to be 0.0065 ml of O_2/ml of H_2O, or 0.267 mM. A more extensive tabulation of Ostwald coefficients and other constants for various gases is given by Wilhelm et al. (69).

See Chapter 10 for more details of aeration systems for liquid cultures. The principles of aeration are reviewed in references 15 and 20.

6.5.1. Dissolved-Oxygen Measurement

Dissolved oxygen is most conveniently measured polarographically with an oxygen electrode. The principles and operating procedures involved are described in Chapter 21.2.3. Procedures for using such an instrument to determine the rates of oxygen solution in cultures are described in detail in references 6, 15, and 20.

6.5.2. Measurement of Oxygen Absorption Rate

Oxygen absorption rates (OAR) are expressed in millimoles per liter per minute and are readily determined chemically by measuring the rate of oxidation of sulfite to sulfate by dissolved oxygen in the presence of a copper catalyst. This method provides information on the oxygenation efficiency of an aeration system. However, the actual OAR in broth cultures may be quite different from that observed in sodium sulfite solutions. Procedures for measuring OAR in media in the presence or absence of biomass are described elsewhere (2, 15, 20). The following is a simple method (34) for measuring OAR by sulfite oxidation.

1. Instead of a growth medium, use the appropriate volume of a sodium sulfite solution containing 0.001 M copper sulfate, and operate the system at the same temperature and under the same conditions of aeration to be used for cultures. The appropriate concentration of sodium sulfite to use depends on the aeration conditions under study (see below).

2. At various intervals after initiating aeration, pipette duplicate 5-ml samples of the sulfite solution into 1- by 10-in. (2.5- by 25-cm) test tubes each containing a small pellet of dry ice. The rising CO_2 stirs the sample during titration and blankets the sample to prevent further oxidation. If partial freezing occurs, warm the sample before completing the titration.

3. Add a few drops of a 10% starch solution, and titrate to a permanent blue end point with a freshly standardized solution of iodine. Standardize the iodine solution against a standard sodium thiosulfate solution. The normality of the iodine solution should be about one-fifth that of the sulfite solution.

4. Calculate the OAR as follows.

$$\text{OAR} = \frac{\text{(milliliters of titration difference} \times \text{normality of iodine)}}{4} \times \frac{1{,}000\ \text{ml}}{5\ \text{ml}} \times \frac{1}{\text{min}}$$

The concentrations of sulfite and iodine most appropriate for any given efficiency of aeration may be calculated by substitution in this equation. For example, if the OAR is about 3.0, the use of a 0.15 N iodine solution would result in a titration difference of about 4 ml for a

Table 5. Ostwald coefficients for various gases[a]

Temp (°C)	Ostwald coefficient for:							
	O_2	H_2	N_2	CO	CH_4	C_2H_2	C_2H_4	C_2H_6
20	0.0334	0.0194	0.0169	0.0249	0.0367	1.108	0.1275	0.0514
25	0.0311	0.0191	0.0159	0.0233	0.0340	1.013	0.1162	0.0453
30	0.0293	0.0190	0.0151	0.0221	0.0316	0.935	0.1068	0.0405
35	0.0277	0.0189	0.0145	0.0211	0.0296	0.870	0.0990	0.0367
40	0.0265	0.0189	0.0140	0.0202	0.0280	0.816	0.0925	0.0337

[a]Milliliters per milliliter of H_2O at a partial gas pressure of 1 atm and the specified temperature.

10-min interval of oxygen absorption by the solution. The initial sodium sulfite concentration should be about five times the iodine concentration, or 0.75 N.

6.6. ANAEROBIOSIS

There is often no sharp line of demarcation between oxic and anoxic conditions in environments. Accordingly, there are bacteria that grow either with or without oxygen **(facultative anaerobes)** and those that prefer or require low oxygen tensions for growth **(microaerophiles)**. *Spirillum volutans*, for example, is a true microaerophile in that it requires the presence of oxygen at low concentrations (1 to 12% O_2 in the gas phase) to initiate growth (49). See Chapter 9.4.1 for methods of cultivating microaerophilic bacteria.

Anaerobes are usually defined as bacteria that are unable to grow in the presence of oxygen. In practice, bacteria that are unable to grow on or near the surface of solid or semisolid media in air at atmospheric pressure are considered to be anaerobic. **Nonstringent (aeroduric, aerotolerant) anaerobes** are able to grow on the surface of agar plates with low but significant levels of oxygen in the atmosphere (38), whereas **stringent (obligate) anaerobes** die, or their growth is inhibited, almost immediately on exposure to such an environment. Driving most of the oxygen out of a liquid medium by boiling or by repeated evacuation and flushing with an oxygen-free gas may not be sufficient to permit the growth of stringent anaerobes, especially if growth is to be initiated from inocula of low cell density. This is because stringent anaerobes require not only the absence of oxygen but also a low oxidation-reduction potential of the medium.

See references 7–9, 10a, 13, and 18 for general principles and methods for growing anaerobic bacteria. Particular methods for anaerobic growth are also included in Chapters 7 and 10 of this volume.

6.6.1. Oxidation-Reduction Potential (E_h)

The **oxidation-reduction (redox) potential** (E_h) provides a useful scale for measuring the degree of anaerobiosis (8). Simply stated, the E_h is a measure of the tendency of a solution to donate or receive electrons (i.e., to become oxidized or reduced). Measurements of E_h are expressed in units of electrical potential relative to the potential of the hydrogen electrode, which is assigned a value of zero at a pH of zero at 25°C and under an H_2 pressure of 1 atm. Under these conditions, the E_h of the hydrogen electrode is equal to E_H (the potential of the standard hydrogen electrode [zero voltage]). E_0 is the **standard redox potential** of a 50% reduced substance, based on the standard hydrogen electrode. At pH 7.0, the E_h of the hydrogen electrode at 25°C under 1 atm of H_2 is −0.413 V, and it is designated E_0', the prime denoting pH 7.0. Likewise, E_0' is the standard redox potential of any 50% reduced substance at pH 7.0, based on the standard hydrogen electrode. The larger (i.e., the more positive) the E_0', the stronger the oxidizing properties of the respective redox couple, and vice versa. However, the E_h of a complex solution such as a culture medium, although a real and measurable quantity, actually represents the net E_h of a multitude of individual redox reactions, not all of which may be in equilibrium with the cell or freely reversible. As Morris (13) puts it, "the E_h of a complex solution is conceptually the equivalent of discussing the strength of a complex mixture of different acids, bases and buffers." Nevertheless, E_h measurement and control are of practical importance for the culture of anaerobes.

Oxygen is a strong oxidizing agent, and in normal laboratory media dissolved oxygen is the agent primarily responsible for raising the E_h. Positive E_h values resulting from dissolved oxygen inhibit the growth of most anaerobic bacteria. By contrast, positive E_h values created by the presence of other chemicals in a medium may not affect the growth of even stringent anaerobes (57, 68). Therefore, although no specific tolerances for E_h can be set for various anaerobic bacteria, most anaerobes are inhibited at E_h values higher than −100 mV. Some stringent anaerobes will not initiate growth at redox potentials higher than −330 mV.

The theoretical concentration of oxygen in pure water at 30°C under an atmosphere of air is 1.48×10^{20} molecules per liter; at an E_h of −330 mV the oxygen concentration is 1.48×10^{-55} molecules per liter (8). Viewed another way, if all but the last molecule of O_2 were removed from 1 liter of water, the E_h should still be about +480 mV based on theoretical calculations (8). Pirt (15) has found, however, on the basis of empirical observations, that oxygen behaves anomalously in solution. He suggests that the oxygen concentration of an aqueous solution at 25°C and an E_h of −140 mV will be about 10^8 molecules per ml, i.e., less than 1 molecule per bacterium in a dense culture. In any case, the above examples illustrate the difficulty of creating strongly reducing conditions merely by removing most of the oxygen from liquid media. The simplest way around this problem is to add a reducing agent to an anoxic medium to lower and poise the E_h in a range that permits growth of the more stringent anaerobic bacteria. In this respect, reducing agents act in a manner analogous to that of buffers, which poise the pH of a medium. This is discussed in Chapter 6.6.2.

6.6.1.1. Electrometric measurement of E_h

The E_h of a solution is most accurately measured electrometrically. Detailed procedures and precautions for such determinations are described in a review by Jacob (9). However, performance of such measurements is not practical during routine preparation of media and so is not discussed here.

6.6.1.2. Dyes sensitive to E_h

E_h-sensitive dyes are used widely to estimate the E_h of media and of cultures, especially of anaerobes. The most useful dyes are those that are reversibly oxidized and reduced, are colored in the oxidized state and colorless in the reduced state, and are nontoxic. Each dye becomes reduced at a different E_h, and the E_h at which it is 50% oxidized or reduced at pH 7.0 is its standard redox potential, E_0'.

Various useful dyes and their standard redox potentials are listed in Table 6. The total range of E_h covered (from completely oxidized to completely reduced) by a redox dye having a two-electron transition is about 120 mV at constant pH. For example, methylene blue (E_0' = 11 mV) is almost completely oxidized (full color) at an E_h of 71 mV and almost completely reduced (colorless) at an E_h of −49 mV (7).

The midpoint potential (E_0) of dyes varies with the pH. For example, the E_h values at which methylene blue is 50% reduced are 101, 11, and −50 mV at pH 5.0, 7.0, and 9.0, respectively (9). The exact change in E_0 with pH is variable among dyes. However, the E_0 for most dyes increases (become more electropositive) by 30 to 60 mV per unit of decrease in pH, and vice versa.

A number of redox dyes are toxic to certain bacteria even at very low concentrations. Unless a dye is known to be nontoxic, use it only in control tubes or flasks of medium handled in exactly the same manner as the medium used for cultures. In such instances, use the dyes at the lowest possible concentration, because the dye may alter the E_h of certain media in which the redox potential is weakly poised (9). Resazurin is a widely used redox dye, because it is generally nontoxic to bacteria and is effective at concentrations of 1 to 2 μg/ml. When incorporated into media, this indicator first undergoes an irreversible reduction step to resorufin, which is pink at pH values near neutrality. This first reduction step can occur when media are heated under an O_2-free atmosphere. The second reduction step to hydroresorufin (which is colorless) has an E_0' of −51 mV, so the resorufin/hydroresorufin redox couple becomes totally colorless at an E_h of about −110 mV (7). This usually requires the addition of a reducing agent to the medium (see below). When a very low E_h must be ascertained, phenosafranine (E_0' = −252 mV) may be incorporated into the medium, but it is often inhibitory. Titanium(III) citrate is both a reducing agent (E_0' = −480 mV) and E_h indicator (74); it becomes colorless when completely oxidized (its use will be described below along with that of other reducing agents). It should be kept in mind, however, that although redox dyes facilitate the preparation and use of anoxic media for anaerobes, they are poor indicators of the actual E_h of the medium. They only indicate that a *minimum* E_h has been achieved, i.e., resorufin in a medium at pH 7.0 will be colorless whether the E_h is −110, −200, or −300 mV.

Table 6. Standard redox potentials of various dyes at pH 7.0 and 30°C[a]

Dye	E_0' (mV)
Methylene blue	11
Toluidine blue	−11
Indigo tetrasulfonate	−46
Resorufin[b]	−51
Indigo trisulfonate	−81
Indigo disulfonate	−125
Indigo monosulfonate	−160
1,5-Anthraquinone sulfate	−200
Phenosafranine	−252
Benzyl viologen	−359

[a]Data from reference 9, where a more complete list is to be found.
[b]Formed from resazurin by reduction (section 6.6.1.2).

6.6.2. Reducing Agents

Reducing agents are added to most anaerobic media to depress and poise the redox potential at optimum levels. Such agents must be nontoxic at the concentration used and create an E_h low enough in the medium for the particular organism under study. The reducing agents most widely used in anaerobic cultures are listed in Table 7. All of these agents have an E_0' low enough to completely reduce resazurin, but only those with an E_0' less than −300 mV are likely to promote growth of the most stringent anaerobes. Most of these agents owe their reducing character to the presence of a reduced sulfur moiety (S^{2-}, HS^-, or H_2S). However, for organisms that are sensitive to such compounds, titanium(III) citrate or H_2 (+$PdCl_2$) may be tried. Amorphous FeS may also be suitable, since its solubility product is so low (3.7×10^{-19}) that little free sulfide will exist in the medium (30). O_2-consuming preparations of bacterial membranes may also be incorporated into liquid media as a type of reducing agent. Such preparations are available commercially (Oxyrase, Inc., Ashland, Ohio) and may be useful for growth of anaerobes that are sensitive to other reducing agents.

For maximum effectiveness, prepare stock solutions of reducing agents under O_2-free gas and with O_2-free water, sterilize them, and store them under O_2-free gas (see procedures for preparation of reduced media, below). Add appropriate concentrations of the reducing agent to the medium just prior to use.

Facultative bacteria may be used to reduce the E_h of media for anaerobic bacteria. Growth of a bacterium such as *E. coli* in the medium prior to inoculation will scavenge residual oxygen and reduce the E_h to low levels if the culture is completely protected from oxygen. The *E. coli* cells may then be heat killed prior to inoculation with the anaerobe. It is necessary to be sure that the facultative bacterium used does not interfere with the growth of the desired anaerobe, either by depleting essential nutrients or by producing toxic products. This procedure has been used to isolate methanogens, which are among the most stringent of all anaerobes (61). The principle is also evidenced in infectious disease, in which infection of a contaminated wound by staphylococci predisposes it to subsequent infection by clostridia.

Table 7. Chemical reducing agents for anoxic media

Agent[a]	E_0' (mV)	Reference	Concn in media
Sodium thioglycolate	< −100		0.05%
Cysteine · HCl	−210	33	0.025%
$Na_2S \cdot 9H_2O$[b]	−270	64	0.025%
FeS (amorphous, hydrated)	< −270	30	4 μg/ml
Dithiothreitol	−330	33	0.02%
H_2 (+$PdCl_2$)	−420	24	Variable[c]
Titanium(III) citrate[d]	−480	46, 73	1–4 mM

[a]Stock solutions may be autoclaved and stored under O_2-free gas.
[b]At pH 7, about half the added sulfide will exist as gaseous H_2S and half will exist as HS^-.
[c]Insoluble $PdCl_2$ is included in the medium at ca. 330 μg/ml and acts as a catalyst for reduction of the medium by H_2, which can be included in the gas phase. For H_2-consuming bacteria such as methanogens, which are usually grown under a gas phase of H_2-CO_2 (80/20 [vol/vol]), the H_2 thereby acts as a substrate and medium reducing agent.
[d]Prepared from commercial 20% solutions of $TiCl_3$ as described in reference 46. Purchase $TiCl_3$ in small volumes under N_2, and prepare stock solutions of the citrate salt as needed.

6.6.3. Techniques for Nonstringent Anaerobes

The procedures for preparing media and for cultivating and transferring nonstringent anaerobic bacteria (including most clostridia) are not difficult (18). All operations can be conducted in the normal atmosphere by taking a few precautions to prevent excessive exposure of media and bacterial cells to oxygen. Some useful procedures for such cultures are as follows.

1. Prepare and dispense media in containers that provide for a small surface-to-volume ratio for the medium (e.g., 16- by 150-mm to 18- by 150-mm test tubes, Florence flasks, round-bottom flasks).

2. Screw-cap bottles or vials completely filled with freshly boiled media are excellent for the prolonged incubation of anaerobes that produce little or no gas (e.g., photosynthetic bacteria).

3. Making liquid media semisolid, by incorporating 0.05 to 0.1% agar, reduces convection currents and is useful for test tube cultures.

4. A thick layer (ca. 1 cm) of sterile, molten Vaspar (50% petrolatum, 50% paraffin) may be poured on the surface of inoculated media in test tubes. When solidified, the Vaspar discourages diffusion of oxygen into the liquid. A disadvantage is that it is a nuisance to clean tubes that have contained Vaspar, since it is insoluble in water.

5. Autoclave media without a reducing agent present, if possible, and add the agent from a sterile stock solution after the medium has cooled to about 40 to 50°C.

6. Do not store media for prolonged periods at any temperature. Never store media in a refrigerator, because the solubility of oxygen in water increases as the temperature decreases. If possible, store media under an anoxic atmosphere, e.g., N_2.

7. Include resazurin (1 mg/liter) in the medium. If the top one-third of the medium is pink when ready for use, boil and cool the medium before inoculation.

8. When possible, use a fairly large inoculum (2 to 10%, vol/vol) of actively growing cells.

9. If cells are to be diluted before inoculation, use the growth medium or a freshly autoclaved diluent that contains a reducing agent.

10. Some nonstringent anaerobes (e.g., *Clostridium perfringens*) may be spread on the surface of a fresh agar medium containing a reducing agent in a petri dish, provided that the dish is subsequently incubated in a jar or chamber free from oxygen. However, cells of some common anaerobes will not tolerate even this brief exposure to air.

11. Petri plates with freshly prepared media that are inoculated with cells of nonstringent anaerobes or with spores of stringent anaerobes must be incubated in an anoxic environment. For this purpose, one of the following systems may be used.

(i) **Vacuum desiccator.** Use a vacuum desiccator only for reasonably aerotolerant anaerobes. Evacuate three times to a partial vacuum (about 500 mm Hg), refilling each time with the gas desired. Use nitrogen, argon, or hydrogen or mixtures of these gases. Also use 5 to 10% carbon dioxide, which some anaerobes require for growth.

(ii) **Anoxic jars.** One anoxic jar is the Torbal jar, model AJ-3 (Baxter Diagnostics, Inc., McGaw Park, Ill.). Evacuate and fill the jar with a gas mixture containing a significant percentage of H_2 (5 to 10%, vol/vol). Torbal jars contain a room temperature catalyst, on the surface of which O_2 is reduced to H_2O by the H_2 of the gas mixture. Although the jars are constructed almost entirely of stainless steel and therefore are opaque, they contain a redox indicator which is visible through a transparent holder and indicates when anoxic conditions have been attained. Catalysts should be reactivated periodically by heating to 160°C for 1 h.

(iii) **Pouches or jars with gas generators.** The gas generator systems most commonly used are the GasPak (BBL Microbiology Systems, Cockeysville, Md.) and the Anaerobe Culture System (Difco Laboratories, Detroit, Mich.); both use Plexiglas jars. The principle is the same as that of the Torbal jar, but evacuation is not absolutely required (although anoxic conditions are attained more rapidly in the jars if the system is purchased with a vented lid to allow preliminary flushing of the assembled jar with an O_2-free gas). Introduce an H_2 + CO_2 generator envelope into the jar just prior to sealing. Some of these generators require activation by the addition of water; others are entirely self-contained and include catalyst and redox indicator. After activation of the gas generation unit, close the bag or jar quickly and observe for water condensation within 30 min. The slow diffusion of oxygen through or out of the Plexiglas jar or plastic pouch is counteracted by the relatively large amount of reductant (H_2) generated by the gas generator envelope. If a redox indicator is not present (either in the medium or in the gas generation unit), methylene blue indicator strips may be inserted in the transparent containers before sealing and observed for decolorization. If such systems are used for the cultivation of anoxygenic phototrophs and the jars are placed in front of an incandescent light bulb(s), the temperature inside the jars may drift several degrees above ambient owing to the strong absorption of infrared light by CO_2. Therefore, temperature control may be necessary.

6.6.4. Hungate Technique for Stringent Anaerobes

Many anaerobes found in the gastrointestinal tract, sewage sludge, and other anoxic habitats require very low redox potentials to initiate growth. Therefore, special precautions must be taken to protect media and cells from even brief exposure to oxygen. The fundamentals of two general procedures for doing this are described below and in section 6.6.5.

A roll-tube technique was described in 1950 by Hungate (43), who pioneered it for the isolation and maintenance of pure cultures of stringent anaerobic bacteria. Although many modifications of the technique have evolved, the basic aspects of its execution have remained essentially unchanged. Its major advantages are that it requires little special apparatus and allows the use of defined, O_2-free atmospheres for cultivating specific groups such as the methanogens (23). Clear, well-illustrated descriptions of the various modifications have been published (8, 10a, 23, 31, 52, 55). The *Anaerobe Laboratory Manual* (7) published by the Virginia Polytechnic Institute and State University describes procedures for handling large numbers of anaerobic cultures rapidly. Essentially all the supplies and equipment required for the V.P.I. Anaerobic Culture System are commercially available (Bellco Glass Inc., Vineland, N.J.).

Even the most detailed descriptions of some modifications of the Hungate technique are frequently difficult to master without demonstration. It is best to visit a laboratory where the technique is in use. Therefore, only the basic steps and simple procedures used in the technique will be described here, as follows.

6.6.4.1. Removal of oxygen from gases

The gases that are used to replace air (generally N_2, Ar, CO_2, or H_2 or mixtures of these) must be treated to remove traces of oxygen. This treatment may be accomplished in one of the following ways.

1. Pass the gas through a heated (350 to 400°C) glass column containing copper filings, which provide a large surface area for "scrubbing" out traces of O_2 (8). Upon trapping O_2, the copper will begin to turn black because of formation of copper oxide ($2Cu^0 + O_2 \rightarrow 2CuO$). Cu^0 can be regenerated by purging the column with gas containing at least 3% H_2 until the characteristic brick-red color of reduced copper returns to the filings. (*CAUTION*: Do not use pure H_2 unless all O_2 is swept from the column first, or an explosion may result.) Since water vapor is formed during the regeneration procedure ($H_2 + CuO \rightarrow Cu^0 + H_2O$), divert the humid effluent from the column to a vented fume hood to avoid wetting the downstream gas lines. If copper filings are unavailable, cupric oxide "wire" (no. C-474; Fisher Scientific, Pittsburgh, Pa.) can be used after it is reduced to Cu^0 as indicated above.

2. Pass the gas through a room temperature catalyst cartridge (e.g., an Indicating Oxy-Trap, an Indicating Oxy-Purge, or an Oxy-Purge N cartridge [Alltech Associates, Inc., Deerfield, Ill.] positioned upstream from one of the indicating cartridges). Other such traps are available from various gas suppliers (e.g., Matheson Gas Products, Joliet, Ill.).

3. Bubble the gas through a solution of titanium(III) citrate, prepared as described in Table 7. Use a gas washing bottle. The solution will become colorless when it is completely oxidized, and it must then be replaced.

6.6.4.2. Preparation of prereduced media

Combine the heat-stable ingredients of the medium (omit the reducing agent) in a round-bottom flask, and add a glass boiling chip. Boil the solution gently while passing a stream of oxygen-free gas over the surface. Continue the gassing while the medium cools to about 50 to 60°C, at which time the medium can be dispensed or the flask can be stoppered. The choice of gas or gas mixture will depend on the medium being used (e.g., CO_2 must be included if the medium is to be buffered with a CO_2-$NaHCO_3$ buffering system) and the anaerobic species to be cultivated. For gassing, use a glass probe (such as that illustrated in Fig. 2) equipped with a bent needle long enough to end just above the liquid level in the vessel; this, in turn, should have a relatively long thin neck to discourage reentry of air. If the medium is to be used soon after autoclaving, a reducing agent may be added at this time provided that it is heat stable; otherwise, it should be added just before inoculation. Dispense the medium into tubes or bottles equipped with stoppers (section 6.4.4.4) while exercising the following precautions.

1. Maintain a constant flow of oxygen-free gas over the surface of the medium during transfer, and flush

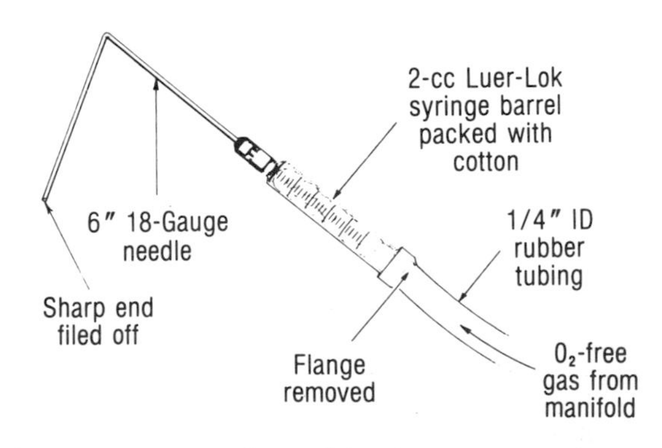

FIG. 2. Gassing cannula used for the Hungate technique. At least two are needed: one for the vessel to be inoculated or filled with medium, and one for the vessel containing the inoculum or the medium to be dispensed. After assembly, autoclave the cotton-filled glass syringe and needle, dry in a drying oven at 100°C, allow to cool, and connect to butyl rubber tubing. Thereafter, flame the needle to sterilize it before inserting it into a vessel. This procedure also permits a constant check that gas is flowing through the needle, since the issuing gas should make a visible dent in the flame. Tubes or flasks must be constantly gassed when open.

each tube or bottle with the gas before and during the transfer.

2. Fill the pipette used for transferring medium with the gas prior to drawing the medium into it. Use a pipette bulb or a pistol-shaped, diaphragm-type pipette pump (various types are available from Fisher Scientific) for transfer. Do not draw gas or liquid into the pipette faster than gas is entering the vessel.

3. After transfer, stopper the tube or bottle without allowing the entrance of air; this is a critical step and requires practice. Place the stopper alongside the needle in the mouth of the tube, and continue gassing for a few seconds. Withdraw the needle rapidly while pushing in the stopper. Seat the stopper with a twisting motion, but avoid excessive force, especially if tubes other than thick-walled Bellco-type anaerobe culture tubes are used (section 6.4.4.4).

Alternatively, the flask of oxygen-free medium may be stoppered as described above. The stopper is then clamped or wired in place, and the medium is autoclaved in bulk (32). However, it must then be dispensed as described above but with an aseptic technique. This procedure increases the risks of contamination and of traces of oxygen entering the sterile medium.

The gassing time required to remove air from a vessel (e.g., a tube) before it is stoppered will depend on the flow rate of gas and the vessel headspace volume. As a general rule, gas each tube for at least 10 s with a per-minute flow rate equal to 25 times the anticipated final headspace volume. For example, assume that 10 ml of medium is dispensed into a tube, leaving 10 ml of headspace, and that the tube is then gassed for 10 s with a probe delivering 250 ml of O_2-free gas per min. The fraction, x, of original atmospheric oxygen left in the tube will be e^{-kt}, where k is the dilution rate (25/min) and t is 0.17 min. Thus, $2.303 \log x = -kt = -(25)(0.17)$ or $x = 1.4 \times 10^{-2}$. Therefore, the amount of O_2 left in the tube after 10 s will be 21% × 0.014 = 0.29% (vol/vol), which is a total of about 1.2 μmol at room temperature (and probably much lower, since this calculation assumes complete mixing of the gas phase and does not consider the gassing of the tube that took place before and during medium addition). Inasmuch as reducing agents are usually incorporated into media at a final concentration of about 1 mM, the tube will contain 10 μmol total reducing agent, or at least a 10-fold molar excess over O_2. Flow rates of gases can be controlled and monitored with any of a variety of gas flow meters available from many gas products distributors.

For sterilization, the stoppers of tubes or bottles must be secured in position. If normal-shaped stoppers are used, they must be clamped in place. A special press for holding an entire rack of stoppered culture tubes is available commercially (Bellco). Clamps for single tubes may be constructed from wire coat hangers (23). Hungate-type anaerobic tubes equipped with a flanged butyl rubber stopper and a screw cap with a 9-mm opening to hold the stopper in place (no. 2047-16125; Bellco) do not require a clamp. Tubes or bottles with serum bottle necks and equipped with a butyl rubber septum stopper may also be used (55); the stopper may be held in place with an aluminum seal (Wheaton Scientific, Millville, N.J.). An aluminum-sealed tube assembly with a thick butyl rubber stopper was developed by Balch and Wolfe (26) and is also available commercially (no. 2048-18150; Bellco). Rubber materials (stoppers, septa, gas tubing, etc.) should be *butyl* rubber if at all possible, since butyl rubber is least permeable to oxygen.

6.6.4.3. Inoculation and transfer

When either Hungate tubes (8, 52) or tubes or bottles with serum bottle necks (26, 55) are used, inoculation and transfer of cultures may be accomplished with disposable syringes and hypodermic needles. Before use, flush the syringes several times with sterile, oxygen-free gas by drawing in and expelling gas that flows through a sterile tube or bottle or by inserting a 25-gauge syringe needle into the wide bore of an 18-gauge gassing needle. Inoculate tubes with conventional-shaped rubber stoppers by quickly inserting a sterile gassing cannula in beside the stopper as it is removed, adding inoculum, and restoppering as described above. The specimen or culture from which the inoculum is taken should also be under a stream of oxygen-free gas and should be adjacent to the tube of media to be inoculated.

If serological or Pasteur pipettes are used for culture transfer, be sure that they are made of glass and fill them with sterile O_2-free gas prior to transfer. Transfer can also be made with stainless-steel or platinum loops. The V.P.I. Anaerobic Culture System (Bellco) is very useful for rapid inoculation of different media with a single culture (7).

6.6.4.4. Roll tubes, shake tubes, and bottle plates

Instead of petri dishes, roll tubes or shake tubes are often used for the isolation of single colonies and for the estimation of viable populations of stringent anaerobes. Prepare agar medium as above, dispense it in anoxic tubes, autoclave it, and cool it to 45°C while the tube stoppers are still clamped in place. If the tubes of medium have been prepared in advance, be sure that stoppers are clamped before boiling. Add reducing agent if necessary. Transfer samples of specimens or dilutions made in prereduced medium or diluent to tubes as described above, and mix with the molten medium (but avoid frothing). To prepare a roll tube, roll the tubes horizontally under a cold-water tap until the molten agar solidifies as a shell on the inside wall of the tube. Try to coat the walls of the tube uniformly. A mechanical spinner that simplifies the procedure is available commercially (no. 7790-44125; Bellco).

Prepare shake tubes in a similar manner, except that after the inoculum is added to the molten agar, invert the tube several times and then allow the agar to solidify as an agar deep, with the tube held in an upright position. Each shake tube or roll tube is, in effect, a pour plate in a tube.

Roll tubes, prepared without inoculating the molten agar, may also be subsequently streaked with a specimen by starting at the bottom and, while simultaneously gassing, rotating the tube as the streaking loop is drawn straight up. A device for rotating a tube for streaking is also available commercially (no. 7790-33333; Bellco), and the technique is clearly depicted by Wolfe (23).

A rubber-stoppered bottle plate is the anoxic analog of a streak plate (42), and special bottles for this purpose are available commercially (no. 2535-50020; Bellco). These bottles discourage the water of syneresis of solidified agar from smearing the streaked agar surface. However, normal prescription bottles can also be used with care (66).

6.6.5. Anoxic Chambers for Stringent Anaerobes

A flexible plastic anoxic glove box chamber is as efficient as the roll-tube method in isolating anaerobic bacteria (24, 25). This type of chamber (often referred to as a Freter-type anaerobic glove box) has been tested and used in many laboratories. Even the extremely oxygen-sensitive methanogenic bacteria can be safely handled in such chambers, with only minor modifications (26, 36, 47). A number of glove box chambers are now available from different commercial sources. One widely used type is manufactured by Coy Laboratory Products, Inc. (Ann Arbor, Mich.).

The primary advantages of anoxic chambers are that they permit the use of standard bacteriological techniques, including spread plates, replica plating, and antibiotic sensitivity testing; they allow preparation of media in a conventional manner; and they require no special training to operate. However, a sizable initial investment is required, and a significant amount of laboratory space is occupied. In addition, anoxic chambers require constant supervision to ensure anaerobiosis. Also, there is some inconvenience in working with gloves, and it is necessary to anticipate the need for media well in advance of use. Nevertheless, the combined use of the Hungate techniques and a properly functioning anoxic chamber makes it possible to conduct almost any kind of experiment with the most stringent anaerobic bacteria.

Following are some precautions and considerations in the use of an anoxic chamber.

1. Chambers that can be heated to serve as incubators are available. At 30 or 37°C chamber temperature, hands become warm and perspiration makes it difficult to remove the gloves. Even with a chamber at room temperature, it is helpful to use cotton or nylon gloves under the rubber gloves when working for long periods. In many cases, it is preferable to have a small incubator inside the chamber. Alternatively, inoculated petri dishes or other cultures may be placed in jars, which can be sealed and removed to an incubator (26).

2. Gas mixtures used in the chamber must contain 5% to no more than 10% hydrogen. The most commonly used mixtures are 5% CO_2 and 10% H_2 in nitrogen or 10% H_2 in nitrogen. If the bacteria under study require CO_2, necessary precautions should be taken to ensure that the CO_2 does not become limiting. Gas mixtures may be purchased, but they are expensive. A gas-mixing apparatus is recommended (Matheson).

3. Open flames cannot be used in the chamber. Install an electric incinerator (e.g., Steri-Loop; Bacteriology Incinerator; Baxter Scientific Products, McGaw Park, Ill.) or a hot-wire incandescent flaming device operated by an external foot switch (Coy Laboratory Products). The latter device may be used to sterilize inoculating loops, the mouths of culture vessels, and other surfaces.

4. Installation of a forced-air filter such as those used in germfree hoods (Standard Safety Equipment Co., Palatine, Ill.) will help prevent contamination.

5. Change the palladium catalyst at frequent intervals. Once a week is usually adequate, but twice weekly or more may be necessary when there is heavy usage. Unless poisoned by H_2S, the catalyst can be regenerated indefinitely by heating at 160°C for 2 h. Cool and return the catalyst to the chamber promptly after regeneration. Check the catalyst for activity occasionally by directing a stream of H_2 over the cold catalyst, which should heat up quickly. *CAUTION*: Never expose hot catalyst to hydrogen because a violent reaction may occur.

6. H_2S irreversibly poisons the catalyst. Whenever possible, grow H_2S-producing cultures in closed containers and open them outside the glove box. To continually scrub H_2S from the chamber atmosphere, fill a tray with activated charcoal (8 to 12 mesh) on a layer of cheesecloth and place the tray under the palladium catalyst tray to trap any H_2S. This treatment reduces the H_2S level in a chamber from 100 to 1 μl/liter in 30 min (26, 54).

7. Keep the humidity at a low level. Place a relative-humidity indicator and a large tray of silica gel in the chamber. When the humidity exceeds 50%, change the silica gel. Regenerate the gel by heating at 160°C for 1 h.

8. Incubate plasticware in the chamber at least overnight before use, to allow dissolved oxygen to diffuse out. When possible, keep agar plates in plastic bags and keep media in screw-cap vials or bottles to minimize evaporation.

9. Keep the anaerobic chamber under positive pressure. There is some gas exchange through common plastics and even more through polycarbonate and silicone plastics or rubber. If the chamber is used frequently, the concentration of H_2 should remain high enough in the chamber as a result of opening and flushing the entry lock. However, it may be necessary to add the gas mixture about twice a month to maintain good positive pressure. Loss of pressure is readily observed in flexible plastic chambers, and a rapid loss is a sign of leaks (see below). The hydrogen level should be monitored. Open, empty serum vials held in the chamber may be periodically sealed and removed, and the gas therein may be analyzed for H_2 by gas chromatography. Alternatively, specific gas monitors are available and can be placed in the chamber to provide continuous measurement of hydrogen and oxygen levels (Coy Laboratory Products; Chemical Sensor Development Corp., Torrance, Calif.). The oxygen level should be kept below 5 ppm.

10. Watch for leaks in the system. Whenever the positive pressure is lost unusually quickly or when the humidity stays high, it is likely that there is a leak. A leak is also indicated when an E_h indicator such as resazurin or phenosafranine in reduced media fails to remain colorless. The most common place for a leak to develop is in a rubber glove. This can be checked by placing a large beaker of water inside the chamber, immersing each glove in it, and observing for bubbles. If gloves are replaced as directed by the manufacturer, significant gas exchange will not occur. Leaks in the plastic covering of the chamber may occur around seams and are difficult to find visually. A dilute solution of soapy water, when spread over parts of the chamber, may help locate leaks by bubble formation, but this is somewhat messy. An electronic leak detector (Coy Laboratory Products) is very useful in locating leaks in the covering.

11. Medium prepared in a conventional manner outside the chamber may be placed in the chamber while still warm, but the vessel should be tightly sealed (under sterile anoxic gas) during entry; otherwise, the vacuum phase of the entry lock may cause the medium to boil and erupt. However, reduction of such media to an E_h low enough to allow growth of stringent anaerobes (i.e., to a colorless resorufin/hydroresorufin end point) will occur only if the medium contains sufficient reducing agent. This may be furnished by including endogenous medium constituents (e.g., unspecified components of yeast extract), by adding specific reducing agents (e.g., cysteine) to the medium just before entry into the anoxic chamber, or by incorporating $PdCl_2$ powder (ca. 300 μg/ml) into media and allowing reduction to occur from the H_2 normally present as 5 to 10% of the anoxic chamber gas (24).

6.7. LIGHT

Light is of primary importance in the cultivation of **photosynthetic bacteria.** Their light-absorbing photopigments include bacteriochlorophylls and carotenoids in the anoxygenic phototrophs and chlorophylls, carotenoids, and phycobilins in the oxygenic phototrophs. As a group, the phototrophs absorb light in virtually all regions of the visible spectrum, as well as invisible (to the human eye) light in the near-infrared region. Absorption of near-infrared wavelengths is primarily by bacteriochlorophylls *a* and *b* of the purple bacteria (48). Another group of bacteria (the members of the family *Halobacteriaceae*) possess carotenoids and bacteriorhodopsin (a retinal-containing pigment) which absorb light strongly in the 500- to 650-nm range. The latter pigment functions as a light-driven proton pump, enabling cells to synthesize ATP by a nonclassical photophosphorylation reaction (62).

For photosynthetic growth of such organisms in the laboratory, selection of appropriate light sources and measurement of the quality and quantity of light used for illumination are critical. An excellent discussion of the production and measurement of photosynthetically useable light has been given by Carr (3), and so only the general considerations will be dealt with here.

6.7.1. Light Measurement

In most bacteriological literature, the wavelength of light is expressed as nanometers, where 1 nm = 10^{-9} m = 10 Å. Luminous flux, the emission of a light source that actually illuminates an object, is best expressed in terms of absolute energy such as ergs per square centimeter per second or watts per square meter, where 1 erg cm^{-2} s^{-1} = 10^{-3} W m^{-2}. Luminous flux is most accurately measured by means of a heat detector (thermopile, bolometer). A Kettering radiant power meter (Scientific Instruments, Inc., Lakeworth, Fla.) is one such instrument. It will also measure light in the infrared region that is invisible to the eye but used by photosynthetic microorganisms. There are also special photocells with sensitivity to infrared light (3), and these are different from conventional light meters, which are insensitive to light of wavelengths longer than 650 nm. In any case, for measurement of polychromatic light used in photosynthesis studies, use a photocell (with any necessary filters) with a relative spectral response that is similar to that of the in vivo spectrum of the organism. In reporting photosynthesis studies, be sure to state the light source and the type and spectral sensitivity of the photocell used for flux measurement.

6.7.2. Light Sources and Filters

Tungsten filament incandescent light bulbs have broad emission spectra, with considerable emission in the near-infrared region (800 to 1,500 nm) and a maximum at about 900 nm. They should be used to illuminate bacteriochlorophyll-containing (anoxygenic) phototrophs. However, the maximum emission of tungsten bulbs depends on the temperature of the bulb, which in turn depends on its wattage. At higher wattages, the emission spectrum is shifted 10 to 30 nm toward the UV. The luminous flux also decreases with time as the filament deteriorates, and hence it is important to check the illumination intensity periodically. Emission spectra for various bulbs can be obtained from the manufacturer of the bulb.

A disadvantage of tungsten bulbs is their heat emission, and caution must be exercised to control the incubation temperature of illuminated cultures. A water filter (e.g., a relatively thin, water-containing, transparent bottle) placed between the light source and the culture can accomplish this, but remember that water absorbs strongly in the near-infrared region. For monochromatic light $I = I_0 e^{-\alpha}$, where I_0 is the original intensity of radiation, I is the intensity after passing through 1 cm of water, and α is the absorption coefficient. Table 8 shows the values of α for water with respect to various wavelengths of light.

Fluorescent-lamp emissions range from about 400 to 700 nm, with maximum intensities between 550 and 650 nm. However, fluorescent lamps do not emit much light in the near-infrared region and so are mainly useful for chlorophyll-containing (oxygenic) phototrophs.

Absorption and interference filters can be used to provide a selected range of wavelength of illumination. Popular among these are the various Kodak "Wratten"-type filters, which consist of water-soluble dyes suspended in gelatin. These filters are sandwiched between glass plates for routine use. Such filters can, for example, be used to enrich anoxygenic phototrophs by absorbing the lower wavelengths of light absorbed by chlorophylls of oxygenic phototrophs and to enrich for the near-infrared-absorbing anoxygenic phototrophs containing bacteriochlorophyll *b*.

Regardless of the light source, light intensity can be most easily controlled by altering the number of lamps

Table 8. Absorption coefficients (α) for water at 20°C[a]

Wavelength (nm)	α (cm^{-1})
760	0.026
970	0.460
1,190	1.05
1,450	26.0
1,940	114

[a]Data from reference 35.

or by inserting neutral filters between the light source and the culture. Such neutral filters may consist of partially darkened photographic plates or layers of wire mesh or cheesecloth.

6.8. DIFFUSION GRADIENTS

Gel-stabilized, one- and two-dimensional diffusion gradient systems have been used to study the response of individual cultures and microbial communities to gradients of pH, salt concentration, and redox potential. Although their use is not yet widespread, they hold great potential for studying the behavior of bacteria in a system that more closely mimics natural habitats, in which diffusion gradients are pervasive. Diffusion gradient systems have been recently reviewed by Wimpenny et al. (70, 71).

6.9. MAGNETIC FIELDS

Some motile bacteria that possess magnetite-containing magnetosomes, e.g., *Aquaspirillum magnetotacticum* (53), align with the Earth's geomagnetic field, which has a strength of about 1 G. As a consequence, they show a biased swimming behavior; i.e., they are north seeking or south seeking. This behavior has been termed magnetotaxis and is thought to help these microaerophilic bacteria orient toward aquatic sediments, where dissolved-oxygen concentrations are lower than in surface water (29). By contrast, the growth of various bacteria is stimulated or inhibited, depending on the field strength (50 to 900 G) and frequency of a pulsed magnetic field (56). Equipment for exposing bacteria to magnetic fields is described in references 29 and 56. A microscope equipped with Helmholtz coils to observe magnetotactic behavior is described in reference 28.

6.10. VISCOSITY

Spirochetes are a group of coiled bacteria that possess endoflagella (periplasmic flagella, axial fibrils) and are capable of swimming through viscous solutions that readily immobilize other flagellated bacteria (27, 39). The possible advantages of such ability have been discussed by Harwood and Canale-Parola (41). The viscosity-inducing agents used for such studies include polyvinylpyrrolidone, methyl cellulose, Ficoll (ICN Biomedicals, Inc., Costa Mesa, Calif.), and agar.

6.11. REFERENCES

6.11.1. General References

1. **Brown, A. D.** 1990. *Microbial Water Stress Physiology.* John Wiley & Sons, Inc., New York.
2. **Brown, D. E.** 1970. Aeration in the submerged culture of microorganisms, p. 127–174. *In* J. R. Norris and D. W. Ribbons (ed.), *Methods in Microbiology,* vol. 2. Academic Press, Inc., New York.
3. **Carr, N. G.** 1970. Production and measurement of photosynthetically useable light, p. 205–212. *In* J. R. Norris and D. W. Ribbons (ed.), *Methods in Microbiology,* vol. 2. Academic Press, Inc., New York.
4. **Datta, S. P., and A. K. Grzybowski.** 1961. pH and acid-base equilibria, p. 19–58. *In* C. Long, E. J. King, and W. M. Speery (ed.), *Biochemists' Handbook.* D. Van Nostrand Co., Inc., New York.
Excellent treatment of pH and buffers, including ionic strength and buffer value. Also, extensive tables of buffers, dissociation constants of weak acids and bases, and pH indicators.
5. **Gomori, G.** 1955. Preparation of buffers for use in enzyme studies. *Methods Enzymol.* **1:**138–146.
6. **Hitchman, M. L.** 1978. *Measurement of Dissolved Oxygen.* (*Chemical Analysis,* vol. 49) John Wiley & Sons, Inc., New York.
7. **Holdeman, L. V., E. P. Cato, and W. E. C. Moore.** 1977. *Anaerobe Laboratory Manual,* 4th ed. Virginia Polytechnic Institute and State University, Blacksburg.
Excellent descriptions and illustrations of anaerobic techniques.
8. **Hungate, R. E.** 1969. A roll tube method for cultivation of strict anaerobes, p. 117–132. *In* J. R. Norris and D. W. Ribbons (ed.), *Methods in Microbiology,* vol. 3B. Academic Press, Inc., New York.
Good description of the Hungate technique.
9. **Jacob, H. E.** 1970. Redox potential, p. 92–123. *In* J. R. Norris and D. W. Ribbons (ed.), *Methods in Microbiology,* vol. 2. Academic Press, Inc., New York.
Describes electrometric measurement of redox potential and the effects of bacteria on the potential.
10. **Jannasch, H. W., and C. D. Taylor.** 1984. Deep-sea microbiology. *Annu. Rev. Microbiol.* **38:**487–514.

10a. **Ljungdahl, L. G., and J. Wiegel.** 1986. Working with anaerobic bacteria, p. 84–96. *In* A. L. Demain and N. A. Solomon (ed.), *Manual of Industrial Microbiology and Biotechnology.* American Society for Microbiology, Washington, D.C.

11. **Marquis, R. E.** 1976. High-pressure microbial physiology. *Adv. Microb. Physiol.* **14:**159–241.
A good review of how high pressures affect microorganisms and their activities.
12. **Marquis, R. E., and P. Matsamura.** 1978. Microbial life under pressure, p. 105–158. *In* D. J. Kushner (ed.), *Microbial Life in Extreme Environments.* Academic Press, Inc., New York.
13. **Morris, J. G.** 1975. The physiology of obligate anaerobiosis. *Adv. Microb. Physiol.* **12:**169–246.
14. **Munro, A. L. S.** 1970. Measurement and control of pH values, p. 39–89. *In* J. R. Norris and D. W. Ribbons (ed.), *Methods in Microbiology,* vol. 2. Academic Press, Inc., New York.
Contains several useful tables for preparing buffers at different ionic strengths.
15. **Pirt, S. J.** 1975. *Principles of Microbe and Cell Cultivation.* John Wiley & Sons, Inc., New York.
Excellent chapters on oxygen supply and demand, and on effects of temperature, hydrogen ion concentration, and water activity.
16. **Prior, B. A.** 1979. Measurement of water activity in foods: a review. *J. Food Prot.* **42:**668–674.
Reviews and evaluates methods for measuring water activity.
17. **Scott, W. J.** 1957. Water relations of food spoilage microorganisms. *Adv. Food Res.* **3:**84–123.
Principles of water activity, relationship to osmolality, methods of control of water activity, and effects on bacteria.
18. **Sutter, V. L., D. M. Citron, and S. M. Finegold.** 1980. *Wadsworth Anaerobic Bacteriology Manual,* 3rd ed. C. V. Mosby Co., St. Louis.
An excellent, how-to manual written mainly for clinical laboratories. A useful appendix includes a list of pitfalls in anaerobic work.
19. **Troller, J. A., and J. H. B. Christian.** 1978. *Water Activity and Food.* Academic Press, Inc., New York.
Excellent, authoritative, well-written monograph with pertinent chapters on basic concepts, methods, enzyme reactions, microbial growth, and microbial survival.

20. **Wang, D. I. C., C. L. Cooney, A. L. Demain, P. Dunnill, A. E. Humphrey, and M. D. Lilly.** 1979. *Fermentation and Enzyme Technology,* p. 157–193. John Wiley & Sons, Inc., New York.
Describes principles and procedures for measurement of oxygen transfer.
21. **Weast, R. C. (ed.).** 1989–1990. *CRC Handbook of Chemistry and Physics,* 70th ed., p. D-144–D-150 and D-161–D-165. CRC Press, Inc., Cleveland.
Tables of buffers, pH of standard buffers at different temperatures, acid-base indicators, and dissociation constants of acids and bases.
22. **Westcott, C. C.** 1978. *pH Measurements.* Academic Press, Inc., New York.
23. **Wolfe, R. S.** 1971. Microbial formation of methane. *Adv. Microb. Physiol.* **6:**107–146.
Illustrated description of the Hungate technique.

6.11.2. Specific References

24. **Aranki, A., and R. Freter.** 1972. Use of anaerobic glove boxes for the cultivation of strictly anaerobic bacteria. *Am. J. Clin. Nutr.* **25:**1329–1334.
25. **Aranki, A., S. A. Syed, E. B. Kenney, and R. Freter.** 1969. Isolation of anaerobic bacteria from human gingiva and mouse cecum by means of a simplified glove box procedure. *Appl. Microbiol.* **17:**568–576.
26. **Balch, W. E., and R. S. Wolfe.** 1976. New approach to the cultivation of methanogenic bacteria: 2-mercaptoethanesulfonic acid (HS-CoM)-dependent growth of *Methanobacterium ruminantium* in a pressurized atmosphere. *Appl. Environ. Microbiol.* **32:**781–791.
27. **Berg, H. C., and L. Turner.** 1979. Movement of microorganisms in viscous environments. *Nature* (London) **278:**349–351.
28. **Blakemore, R. P.** 1981. Magnetic navigation in bacteria. *Sci. Am.* **245:**58–65.
29. **Blakemore, R. P.** 1982. Magnetotactic bacteria. *Annu. Rev. Microbiol.* **36:**217–238.
30. **Brock, T. D., and K. O'Dea.** 1977. Amorphous ferrous sulfide as a reducing agent for culture of anaerobes. *Appl. Environ. Microbiol.* **33:**254–256.
31. **Bryant, M. P.** 1972. Commentary on the Hungate technique for culture of anaerobic bacteria. *Am. J. Clin. Nutr.* **25:**1324–1328.
32. **Bryant, M. P., and L. A. Burkey.** 1953. Cultural methods and some characteristics of some of the more numerous groups of bacteria in the bovine rumen. *J. Dairy Sci.* **36:**205–217.
33. **Cleland, W. W.** 1964. Dithiothreitol, a new protective reagent for SH groups. *Biochemistry* **3:**480–482.
34. **Corman, J., H. M. Tsuchiya, H. J. Koepsell, R. G. Benedict, S. E. Kelley, V. H. Feger, R. G. Dworschak, and R. W. Jackson.** 1957. Oxygen absorption rates in laboratory and pilot plant equipment. *Appl. Microbiol.* **5:**313–318.
35. **Curcio, J. A., and C. C. Petty.** 1951. The near infrared absorption spectrum of liquid water. *J. Opt. Soc. Am.* **41:**302–304.
35a. **Dean, J. A.** 1979. *Lange's Handbook of Chemistry,* 12th ed. McGraw-Hill Book Co., New York.
See reference 51 for an earlier edition.
36. **Edwards, T., and B. C. McBride.** 1975. New method for the isolation and identification of methanogenic bacteria. *Appl. Microbiol.* **29:**540–545.
37. **Good, N. E., G. D. Winget, W. Winter, T. N. Connolly, S. Izawa, and R. M. M. Singh.** 1966. Hydrogen ion buffers for biological research. *Biochemistry* **5:**467–477.
38. **Gordon, J., R. A. Holman, and J. W. McLeod.** 1953. Further observations on production of hydrogen peroxide by anaerobic bacteria. *J. Pathol. Bacteriol.* **66:**527–537.
39. **Greenberg, E. P., and E. Canale-Parola.** 1977. Relationship between cell coiling and motility of spirochetes in viscous environments. *J. Bacteriol.* **131:**960–969.
40. **Griffin, D. M.** 1981. Water and microbial stress. *Adv. Microb. Ecol.* **5:**91–136.
41. **Harwood, C. S., and E. Canale-Parola.** 1984. Ecology of spirochetes. *Annu. Rev. Microbiol.* **38:**161–192.
42. **Hermann, M., K. M. Noll, and R. S. Wolfe.** 1986. Improved agar bottle plate for isolation of methanogens or other anaerobes in a defined gas atmosphere. *Appl. Environ. Microbiol.* **51:**1124–1126.
43. **Hungate, R. E.** 1950. The anaerobic mesophilic cellulolytic bacteria. *Bacteriol. Rev.* **14:**1–49.
44. **Ingraham, J. L.** 1958. Growth of psychrophilic bacteria. *J. Bacteriol.* **76:**75–80.
45. **Jannasch, H. W., C. O. Wirsen, and C. D. Taylor.** 1982. Deep-sea bacteria: isolation in the absence of decompression. *Science* **216:**1315–1317.
46. **Jones, G. A., and M. D. Pickard.** 1980. Effect of titanium(III) citrate as a reducing agent on growth of rumen bacteria. *Appl. Environ. Microbiol.* **39:**1144–1147.
47. **Jones, W. J., W. B. Whitman, R. D. Fields, and R. S. Wolfe.** 1983. Growth and plating efficiency of methanococci on agar media. *Appl. Environ. Microbiol.* **46:**220–226.
48. **Kondratieva, E. N., N. Pfennig, and H. G. Trüper.** 1992. The phototrophic prokaryotes, p. 312–330. *In* A. Balows, H. G. Trüper, M. Dworkin, W. Harder, and K.-H. Schleifer (eds.), *The Prokaryotes,* vol. I, 2nd ed. Springer-Verlag, New York.
49. **Krieg, N. R.** 1984. The genus *Spirillum* Ehrenberg 1832, 38[AL], p. 90–93. *In* N. R. Krieg and J. G. Holt (ed.), *Bergey's Manual of Systematic Bacteriology,* vol. 1. The Williams & Wilkins Co., Baltimore.
50. **Labuza, T. P., K. Acott, S. R. Tatini, R. Y. Lee, J. Fink, and W. McCall.** 1976. Water activity determination: a collaborative study of different methods. *J. Food Sci.* **41:**910–917.
51. **Lange, N. A.** 1961. *Handbook of Chemistry,* 10th ed. McGraw-Hill Book Company, Inc., New York.
See reference 35a for a later edition.
52. **Macy, J. M., J. E. Snellen, and R. E. Hungate.** 1972. Use of syringe methods for anaerobiosis. *Am. J. Clin. Nutr.* **25:**1318–1323.
53. **Maratea, D., and R. P. Blakemore.** 1981. *Aquaspirillum magnetotacticum* sp. nov., a magnetic spirillum. *Int. J. Syst. Bacteriol.* **31:**452–455.
54. **Miguel, A. H., D. F. S. Natusch, and R. L. Tanner.** 1976. Adsorption and catalytic conversion of thiol vapors by activated carbon and manganese dioxide. *Atmos. Environ.* **10:**145–150.
55. **Miller, T. L., and M. J. Wolin.** 1974. A serum bottle modification of the Hungate technique for cultivating obligate anaerobes. *Appl. Microbiol.* **27:**985–987.
56. **Moore, R. L.** 1979. Biological effects of magnetic fields: studies with microorganisms. *Can. J. Microbiol.* **25:**1145–1151.
57. **Onderdonk, A. B., J. Johnston, J. W. Mayhew, and S. L. Gorbach.** 1976. Effect of dissolved oxygen and E_h on *Bacteroides fragilis* during continuous culture. *Appl. Environ. Microbiol.* **31:**168–172.
58. **Patching, J. W., and A. H. Rose.** 1970. The effects and control of temperature, p. 23–38. *In* J. R. Norris and D. W. Ribbons (ed.), *Methods in Microbiology,* vol. 2. Academic Press, Inc., New York.
59. **Schlegel, H. G., and H. W. Jannasch.** 1992. Prokaryotes and their habits, p. 75–125. *In* A. Balows, H. G. Trüper, M. Dworkin, W. Harder, and K.-H. Schleifer (ed.), *The Prokaryotes,* vol. I, 2nd ed. Springer-Verlag, New York.
60. **Scott, W. J.** 1953. Water relations of *Staphylococcus aureus* at 30°C. *Aust. J. Biol. Sci.* **6:**549–564.
61. **Smith, P. H., and R. E. Hungate.** 1958. Isolation and characterization of *Methanobacterium ruminatium* n. sp. *J. Bacteriol.* **75:**713–718.
62. **Stoeckenius, W., and R. A. Bogomolni.** 1982. Bacteriorhodopsin and related pigments of halobacteria. *Annu. Rev. Biochem.* **52:**587–616.
63. **Taylor, C. D., and H. W. Jannasch.** 1976. A subsampling technique for measuring growth of bacterial cultures under high hydrostatic pressure. *Appl. Environ. Microbiol.* **32:**355–359.

64. **Thauer, R. K., K. Jungermann, and K. Decker.** 1977. Energy conservation in chemotrophic anaerobic bacteria. *Bacteriol. Rev.* **41:**100–180.
65. **Troller, J. A.** 1977. Statistical analysis of a_w measurements obtained with the Sina scope. *J. Food Sci.* **42:**86–90.
66. **Uffen, R. L., and R. S. Wolfe.** 1970. Anaerobic growth of purple nonsulfur bacteria under dark conditions. *J. Bacteriol.* **104:**462–472.
67. **Umbreit, W. W., R. H. Burris, and J. F. Stauffer.** 1964. *Manometric Techniques,* 4th ed. Burgess Publishing Co., Minneapolis.
68. **Walden, W. C., and D. J. Hentges.** 1975. Differential effects of oxygen and oxidation-reduction potential on the multiplication of three species of anaerobic intestinal bacteria. *Appl. Microbiol.* **30:**781–785.
69. **Wilhelm, E., R. Battino, and R. J. Wilcock.** 1977. Low-pressure solubility of gases in liquid water. *Chem. Rev.* **77:**219–262.
70. **Wimpenny, J. W. T., and D. E. Jones.** 1988. One-dimensional gel-stabilized model systems, p. 1–30. *In* J. W. T. Wimpenny (ed.), *CRC Handbook of Laboratory Model Systems for Microbial Ecosystems,* vol. II. CRC Press, Inc., Boca Raton, Fla.
71. **Wimpenny, J. W. T., P. Waters, and A. Peters.** 1988. Gel-plate methods in microbiology, p. 229–251. *In* J. W. T. Wimpenny (ed.), *CRC Handbook of Laboratory Model Systems for Microbial Ecosystems,* vol. I. CRC Press, Inc., Boca Raton, Fla.
72. **Yayanos, A. A.** 1986. Evolutional and ecological implications of the properties of deep-sea barophilic bacteria. *Proc. Natl. Acad. Sci. USA* **83:**9542–9546.
73. **Yayanos, A. A., A. S. Dietz, and R. Van Boxtel.** 1981. Obligately barophilic bacterium from the Mariana Trench. *Proc. Natl. Acad. Sci. USA* **78:**5212–5215.
74. **Zehnder, A. J. B., and K. Wuhrmann.** 1976. Titanium (III) citrate as a nontoxic oxidation-reduction buffering system for the culture of obligate anaerobes. *Science* **194:**1165–1166.

Chapter 7

Nutrition and Media

ROSALIE J. COTE and ROBERT L. GHERNA

The ubiquitous presence of bacteria gives evidence of their remarkable ability to utilize almost any substance the Earth has to offer as a source of energy or for the essential elements required for growth. Although no one organism has the capacity to exploit all of the diverse metabolic biochemistry represented in the various modes of bacterial nutrition, there are numerous generalized bacteria such as *Escherichia coli* that can flourish over a wide range of nutritional conditions. Its metabolic complex of fully operational synthetic pathways also gives *E. coli* the **prototrophic** ability to thrive on a simple medium containing only a single compound as a source of carbon and energy and a few inorganic salts to supply the other essential elements required for growth.

However, other bacteria lack the ability in their metabolic machinery to synthesize one or more necessary nutrients and must be supplied with the missing compound in order to grow. The lactic acid bacteria, for example, require a growth medium supplemented with a variety of compounds (vitamins, amino acids, etc.) that they are unable to synthesize; such compounds are termed growth factors. Organisms that require the addition of a growth factor to their medium are **auxotrophic** for that compound. Bacteria with all gradations of complexity in their requirements for the recognized nutrients are known. Other bacteria have unidentified growth requirements which must be satisfied with complex natural materials such as peptone, ruminal fluid, serum, or yeast extract. In addition to this array of bacteria that can be grown in vitro, the obligately parasitic bacteria grow only in the presence of living animal, plant, or other bacterial cells which provide unidentified nutrients or substitute for metabolic deficiencies. A final group of bacteria is made up of the **syntrophs,** which can be grown in laboratory media only in cocultivation with other specific bacteria. For example, the obligate proton-reducing, H_2 transfer-dependent, anaerobic syntrophs "*Syntrophobacter*" spp. (3), *Syntrophomonas* spp. (40), and *Syntrophospora* spp. (64) are isolated and maintained in vitro only in cocultivation with H_2-utilizing anaerobes such as methanogens.

The following pages provide practical information about the essentials of bacterial nutrition, with particular emphasis devoted to medium composition, preparation, and use for the cultivation of about 200 representative bacteria. Additional information on bacterial nutrition may be found in texts by Koser (33) and Stanier et al. (57). The articles by Bridson and Brecker (5) and Bridson (4) contain detailed material about medium composition and design. LaPage et al. (37) and Guirard and Snell (22) provide medium formulations for maintenance and growth of numerous bacteria. The American Type Culture Collection *Catalogue of Bacteria and Bacteriophages* (18) includes formulations for defined and undefined media for the propagation of all strains available in the catalogue. A new handbook by Atlas (1) provides the formulations, preparation, and uses of more than 1,500 microbiological media. Media for particular purposes are described throughout the present book, notably enrichment media in Chapter 8.

7.1. NUTRITIONAL REQUIREMENTS

7.1.1. Energy Sources

All bacteria require an exogenous energy source for growth, and individual species vary tremendously in the array of compounds utilized for this purpose. **Autotrophic** bacteria derive energy from light, as with the photosynthetic bacteria and cyanobacteria, or by the oxidation of one or more inorganic elements or compounds. (The genus *Nitrobacter*, for example, derives energy by the oxidation of nitrites to nitrates.) With respect to photosynthesis, the kind of light provided under conditions of laboratory cultivation is important. Photosynthetic bacteria utilize light in the red region (>750 nm) of the spectrum and should be grown under tungsten bulbs, whereas the cyanobacteria require the blue wavelengths (<700 nm) emitted by fluorescent tubes. **Heterotrophic** bacteria obtain energy from the oxidation or dissimilation of reduced-carbon compounds. Some bacteria have the capacity to straddle both autotrophic and heterotrophic modes of nutrition depending on the available energy source. *Aquaspirillum autotrophicum* uses hydrogen for energy in an atmosphere enriched with hydrogen and carbon dioxide but uses organic compounds such as succinate for the same purpose when cultivated in air.

7.1.2. Carbon, Nitrogen, and Sulfur

Carbon, nitrogen, and sulfur are major elements required by all bacteria for cell syntheses. Autotrophs are capable of utilizing CO_2 as the sole carbon source; heterotrophs require organic carbon compounds. Although the nitrogen-fixing bacteria and some sulfur bacteria are capable of assimilating, respectively, nitrogen and sulfur in elemental form, most genera obtain their exogenous supply of these three essential nutrients from inorganic or organic compounds.

Organic carbon compounds generally serve a dual purpose for bacteria: as a source of energy and as a supply of carbon and other elements required for cell structure and function. Sole carbon sources for bacteria range from molecules as simple as carbon dioxide to those as complex or unusual as high-molecular-weight hydrocarbons and degradation-resistant pesticides. Because of its high frequency of utilization by many organisms, glucose is often used as the single carbon source in bacteriological media. However, certain species show little or no ability to assimilate glucose (or any other carbohydrate, organic acid, or alcohol) as the sole source of carbon: e.g., *Saprospira* species preferentially use mixtures of amino acids or peptones as sources of carbon and energy as well as for their supply of nitrogen and sulfur.

Although nitrogen-fixing bacteria (e.g., *Azotobacter* and *Rhizobium* species) can acquire nitrogen directly from the atmosphere, most bacteria normally obtain this element in its most defined forms as inorganic salts such as NH_4Cl, $(NH_4)_2SO_4$, $NaNO_3$, or KNO_3. The ability to utilize one or more of these inorganic salts is a characteristic of autotrophs; however, heterotrophs vary considerably, even within a genus, in this capacity. The use of peptones or similar hydrolysates in media will satisfy the nitrogen requirement for organisms incapable of assimilating the nutrient from inorganic salts.

In an energy-yielding reaction, sulfur bacteria such as *Thiobacillus thiooxidans* can oxidize elemental sulfur and further metabolize it for cellular requirements. Most other bacteria can synthesize essential sulfur compounds from sulfate supplied as various inorganic salts; for those that cannot, sulfur is added to medium formulations as hydrogen sulfide, as cystine or methionine, or as peptones high in these amino acids. The use of the sulfur compounds cysteine and hydrogen sulfide as reducing agents in anaerobic media is unrelated to their nutritional function and is discussed elsewhere (Chapter 6.6.2).

7.1.3. Mineral Ions

The inorganic ion requirements of bacteria have been extensively studied. This research has been complicated by (i) the ubiquitous occurrence of metal ions, (ii) the small amounts of many mineral elements required to permit growth, (iii) the difficulty in removing these ions from media to demonstrate a requirement, (iv) the sparing effects of one metal ion on the requirement for another, (v) the antagonistic effects between certain related metal ions, and (vi) the occurrence of metal ions in many biological products as chelates of various stabilities, thus affecting their availability to bacteria. For a detailed treatment of this topic, the extensive precautions necessary for its study, and the techniques involved, the reviews by Hutner et al. (27), Knight (32), and Snell (55) should be consulted. Hughes and Poole (26) have reviewed the theoretical and practical considerations of metal speciation with respect to microbial growth.

The nutritionally essential metal ions serve bacteria in a number of functions: (i) as activators or cofactors of a variety of enzymes (e.g., potassium, magnesium, and manganese), (ii) in membrane transport (e.g., potassium and sodium), and (iii) as components of molecules or structural complexes (e.g., calcium chelated within the spore protoplasts of gram-positive bacteria). For practical purposes, the mineral ion requirements for bacteria cultivated in defined media are satisfied by the addition of salts containing Ca^{2+}, K^+, Na^+, Mg^{2+}, Mn^{2+}, Fe^{2+} or Fe^{3+}, PO_4^{3-}, and SO_4^{2-} in milligram-per-liter concentrations. Trace amounts of several other ions (e.g., Zn^{2+}, Cu^{2+}, Co^{2+}, and MoO_4^{2-}) are oftentimes included in micro- or nanogram proportions. Results of investigations demonstrating the roles of nickel, selenium, and tungsten in enzyme functions in anaerobes (14, 28, 63) have led to the inclusion of these elements in complete trace-metal formulations.

When carefully constructed trace-element stock solutions are added to an appropriate chemically balanced medium (e.g., medium 11 or 18), no permanent precipitates will form with autoclaving or with subsequent incubation of the medium. Two trace-element solutions useful in bacteriological media are listed in Table 1.

With complex media, all these ions are generally present as contaminants in medium components such as peptones and extracts and no supplementation is required for routine growth. The use of tap water to supply trace minerals in the preparation of media is discouraged because of the deleterious precipitates formed with hard (alkaline) water as well as the potentially inhibitory effects of chemically polluted tap water supplies. Most metals (in particular the divalent and trivalent cations Ca^{2+}, Fe^{2+}, Fe^{3+}, and Mg^{2+}) tend to form insoluble hydroxides or phosphates at neutral to alkaline pH, making these elements unavailable for use to bacteria. The formation of precipitates of this type in nonchelated or poorly designed media is spontaneous and hastened by heat (autoclaving).

In preparing concentrated stock solutions of inorganic salts for dilution into synthetic media, it is fre-

Table 1. Trace-element solutions for use in bacteriological media

A. Trace-Element Solution Ho-Le (18)

H_3BO_3	2.85 g
$MnCl_2 \cdot 4H_2O$	1.80 g
$FeSO_4 \cdot 7H_2O$	1.36 g
Sodium tartrate	1.77 g
$CuCl_2 \cdot 2H_2O$	26.90 mg
$ZnCl_2$	20.80 mg
$CoCl_2 \cdot 6H_2O$	40.40 mg
$Na_2MoO_4 \cdot 2H_2O$	25.20 mg
Distilled water	1.0 liter

Dissolve chemicals one at a time, and adjust solution to pH 4.0 with H_2SO_4 to retard precipitation. Use 1.0 ml of trace-element solution per liter of medium.

B. Modified Balch's Trace-Element Solution (1a)

Nitrilotriacetic acid	1.5 g
$MgSO_4 \cdot 7H_2O$	3.0 g
$MnSO_4 \cdot H_2O$	0.5 g
NaCl	1.0 g
$FeSO_4 \cdot 7H_2O$	0.1 g
$CoCl_2 \cdot 6H_2O$	0.1 g
$CaCl_2$	0.1 g
$ZnSO_4 \cdot 7H_2O$	0.1 g
$CuSO_4 \cdot 5H_2O$	10.0 mg
$AlK(SO_4)_2 \cdot 12H_2O$	10.0 mg
H_3BO_3	10.0 mg
$Na_2MoO_4 \cdot 2H_2O$	10.0 mg
$NiSO_4 \cdot 6H_2O$	30.0 mg
Na_2SeO_3	20.0 mg
$Na_2WO_4 \cdot 2H_2O$	20.0 mg
Distilled water	to 1.0 liter

Dissolve NTA in about 500 ml of water, and adjust to pH 6.5 with KOH to dissolve. Add remaining salts one at a time. Bring final volume to 1 liter with distilled water. Use 10 ml of trace-element solution per liter of medium.

quently useful to adjust them to low pH with the addition of concentrated H_2SO_4 and to incorporate minimal amounts of nontoxic chelating agents (e.g., EDTA or nitrilotriacetic acid [NTA]) into the medium to prevent precipitation. The amount of chelating agent in the final medium is an important factor in determining the availability of metal ions, and the concentrations of these chelators must be carefully balanced with those of trace metals if abundant and sustained growth is to be attained (26). Just as the formation of precipitates diminishes the availability of trace metals, the overchelation of a medium also results in a metal-starved condition for bacteria. As a general guideline, EDTA is added to media in maximum concentrations of 0.5 to 2.0 mg/liter and the NTA concentration ranges from 15 to 150 mg/liter. The slight but potentially carcinogenic nature of NTA may be of concern in its use for certain applications.

In addition to the two powerful synthetic chelating agents listed above, the following medium components function as mild chelators. (i) The first are **individual amino acids,** particularly glycine and histidine, as well as the peptones and extracts composed of mixtures of amino acids. (ii) Second are **carboxylic acids** (acetate, citrate, succinate, and tartrate). The citrate iron salts ferric citrate and the more soluble ferric ammonium citrate are the most frequently encountered chelated iron supplements in media formulations; however, both ferric salts have indefinite chemical compositions and iron content, which might limit their use under certain rigorously defined media conditions. (iii) Third are **porphyrin-containing molecules** (hemin and hemoglobin). The incorporation of small amounts (5 mg/liter) of hemin, initially solubilized with NaOH, into complex media can often replace whole blood or hemoglobin in satisfying the soluble-iron requirement for a number of fastidious species including *Haemophilus* and *Neisseria* species. The factor X growth requirement for *Haemophilus influenzae* and other species is an iron-porphyrin that is replaceable by hemin (25). (iv) Fourth are **siderophores.** A number of compounds synthesized by bacteria and excreted can chelate iron and permit efficient assimilation of the element by an organism. The most widely recognized example is that of mycobactin produced by *Mycobacterium phlei* under conditions of iron deficit and required by *M. paratuberculosis* for sustained growth. The mycobactin requirement for this organism may be replaced by supplementation of the growth medium with a high (1%) concentration of ferric ammonium citrate (45). Siderophore requirements of other bacteria have also been shown to be bypassed by elevated amounts of mildly chelated iron compounds.

In one final note on trace metals, it is extremely important to evaluate the composition of the medium and the effect of the components on the bioavailability of the trace metals. This is critical when the medium is to be used in toxicity or resistance studies. Failure to consider the binding of metals by medium components can lead to a misinterpretation of the data. A reevaluation of bacterial resistance to selenium and silver (11) yielded smaller numbers than those reported earlier (29, 39). Tilton and Rosenberg (59) noted in their study on the reversal of silver inhibition that their test media, both agar and broth, absorbed as much as 400 μg of Ag^+ per ml. Deprivation of both nutritionally essential and toxic metal ions can occur in poorly constructed or manipulated media formulations.

An excellent discussion of the various factors affecting the speciation of metals and their bioavailability in media can be found in the review by Hughes and Poole (26).

7.1.4. Carbon Dioxide, Hydrogen, and Oxygen

The assimilation of carbon dioxide as the sole carbon source by autotrophic bacteria has been discussed in section 7.1.2; however, all heterotrophs are also able to fix this molecule, albeit to a much lesser degree. All growing cells require CO_2 for a number of metabolic functions such as the conversion of acetyl coenzyme A (acetyl-CoA) to malonyl-CoA in lipid synthesis. Although much of this CO_2 can be recycled within the cell, an exogenous supply of the molecule is necessary to initiate metabolic processes.

The extended lag phase characteristic of a small inoculum in a large volume of medium is considered to occur in part because of a suboptimal CO_2 tension present in the initial highly aerobic medium (42). A large inoculum from a rapidly growing culture can alter the O_2/CO_2 ratio of its medium much faster than an inoculum from a stationary-phase culture or one from a freeze-dried preparation does.

Some pathogenic organisms such as *Neisseria* and *Brucella* spp. require increased (5%) carbon dioxide levels, particularly at the time of isolation (although the need may eventually be lost with continuous subculture). A candle jar or an anaerobe jar used in conjunction with CO_2- or microaerophilic atmosphere-generating envelopes creates the carbon dioxide-enriched atmosphere necessary to support the growth of these **capnophilic bacteria** if CO_2 incubators are not available.

The use of molecular hydrogen as an energy source has been discussed in section 7.1.1. For this purpose, a culture is most easily cultivated in a butyl rubber-stoppered vessel containing a headspace atmosphere of 80% hydrogen and 20% carbon dioxide. Elevated concentrations of hydrogen can be explosive in the presence of air (oxygen), and suitable safety precautions should be followed when working with these compressed-gas mixtures. From 3 to 10% hydrogen is often added to gas mixtures used in the cultivation of strict anaerobes to reduce catalytically any residual molecular oxygen in the environment to water. The major pool of combined hydrogen used nutritionally by heterotrophic bacteria, however, comes from the cellular metabolism of organic substances such as amino acids, carbohydrates, and water.

The oxygen present in all cellular material originates from water and other nutritional substances assimilated by the cell (35). When all other essentials are present for adequate nutrition of a culture, the oxygen necessary for synthetic requirements will not be a limiting factor.

However, free molecular oxygen from the air functions as the terminal hydrogen and electron acceptor in the cytochrome pathway of aerobic and facultative bacteria. A molecular-oxygen deficiency in the bacterial environment decreases the energy-yielding capacity of the cytochrome system, thereby limiting cell growth and culture yields. Under conditions of oxygen deprivation, facultative organisms shift to anaerobic pathways and produce metabolic end products different from those found with adequate oxygenation. Lowered oxygen tensions, along with a concomitant rise in carbon dioxide levels, can occur in any incubator filled with actively respiring cultures. Standing broth cultures, especially those which form a pellicle of growth at the surface of the medium, give rise to distinct aerobic and anaerobic cell populations within a single vessel. See Chapters 6 and 10 for an elaboration of aeration needs.

7.1.5. Amino Acids, Peptides, and Proteins

Bacteria very frequently require one, several, or many of the naturally occurring L-amino acids. Although this requirement stems from loss of the ability to synthesize such amino acids, this deficiency may be conditional rather than absolute. For example, the alanine requirement of *Lactobacillus delbrueckii* subsp. *delbrueckii* (ATCC 9649) can be bypassed by the addition of vitamin B_6 to the medium (44).

Even in nonconditional auxotrophs, the addition of an excess of one amino acid may result in an increased requirement for another amino acid. This effect derives from the competition of the two amino acids for a single transport system. Such antagonistic interrelationships can be avoided by the addition of an appropriate peptide of the limiting amino acid, because the peptide is absorbed via an independent transport system. This phenomenon has led to occasional reports of the requirement for specific peptide growth factors. However, there is as yet no known instance of an obligatory requirement for a preformed peptide for bacterial growth, although peptides are usually readily utilized (and frequently desirable) sources of essential amino acids.

Amino acid auxotrophs grow well in complex media composed of peptones and other protein hydrolysates, with no additional supplementation of the limiting amino acid necessary. In chemically defined minimal media, the addition of 10 to 50 mg of a required amino acid per liter generally suffices to promote growth. Glutamic acid is added at the higher level listed since it is required by most organisms in greater concentration than other amino acids are. The reported requirement of certain bacteria for specific proteins deserves further study. The proteins may be acting solely as a source of amino acids, as detoxicants, or in other ways (sections 7.1.7 and 7.2.3).

Except for glutamine, which must be filter sterilized, amino acids are stable to autoclaving. Cysteine solutions should be prepared just prior to use since this compound is oxidized to cystine on exposure to air. All naturally occurring L-amino acids are sufficiently soluble in warm water in concentrations used for medium preparation, with the exception of cystine, which can be dissolved by warming in dilute hydrochloric acid. Tyrosine, having minimal solubility, becomes more soluble in the presence of additional amino acids or with the addition of a base such as KOH or NaOH.

7.1.6. Vitamins

Vitamins play catalytic roles within the cell, usually as components of coenzymes or as prosthetic groups of enzymes. As with the amino acids, the presence or absence of other growth factors in a medium may mask the need for a particular vitamin. One particular strain of *E. coli* (ATCC 10799; NCIB 8134) displays a vitamin B_{12} requirement in minimal medium; however, the requirement is bypassed with the addition of sufficient methionine. For growth in complex media, yeast extract supplementation at a concentration of 0.1 to 0.5% generally more than meets the minimal vitamin requirements of bacteria; for some strains, yeast extract stimulates growth at levels as low as 0.05%. A useful vitamin mix to supplement defined medium formulations is listed in Table 2. See section 7.3.5 for necessary precautions in conducting vitamin assays.

Vitamins that are required by one species or another and some of their interrelationships with other components of the growth medium are discussed briefly below.

7.1.6.1. *p*-Aminobenzoic acid

p-Aminobenzoic acid (PABA) serves as a precursor for the biosynthesis of folic acid and thereby also influences the metabolism of thymine, methionine, serine, the purine bases, and vitamin B_{12}. The presence of these compounds in the medium may decrease or eliminate

Table 2. Vitamin solution for use in bacteriological media

Balch's Vitamin Solution (1a)	
p-Aminobenzoic acid	5.0 mg
Folic acid	2.0 mg
Biotin	2.0 mg
Nicotinic acid	5.0 mg
Calcium pantothenate	5.0 mg
Riboflavin	5.0 mg
Thiamine HCl	5.0 mg
Pyridoxine HCl (vitamin B_6)	10.0 mg
Cyanocobalamin (vitamin B_{12})	100.0 μg
Thioctic acid (lipoic acid)	5.0 mg
Distilled water	1.0 liter

Dissolve vitamins one at a time in distilled water; adjust to pH 7 with NaOH, if necessary, to dissolve biotin and folic acid. Filter sterilize stock solution to retard fungal contamination during long-term storage, and store under refrigeration in the dark. Use 10 ml of vitamin solution per liter of medium.

the requirement of a given bacterium for PABA in the medium.

7.1.6.2. Folic acid group

Several different forms of folic acid occur naturally and differ in their availability for individual bacteria. Pteroyl-L-glutamic acid and its N^{10}-formyl derivative are fully active for *Lactobacillus casei* and *Enterococcus faecalis*, whereas pteroyltriglutamic acid, while fully active for *L. casei*, is only slightly active for *E. faecalis;* pteroylheptaglutamic acid does not support growth of either bacterium. N^5-Formyltetrahydropteroylglutamic acid (folinic acid; leucovorin) is active for both *L. casei* and *E. faecalis.*

Folic acid functions as a cofactor in synthesis of thymine, purine bases, serine, methionine, and pantothenic acid; however, the vitamin requirement is frequently eliminated if all of these compounds are added to the medium. Folic acid is only slightly soluble in water, but the ammonium salt is very soluble. The vitamin is stable to autoclaving.

7.1.6.3. Biotin

Biotin participates in several biosynthetic reactions that require CO_2 fixation, including synthesis of oxalacetate and of fatty acids. Biocytin (N^ϵ-biotinyl-L-lysine) also occurs naturally; it is as active as biotin for *L. casei.* Biotin is only slightly water soluble, but its salts are freely water soluble. Biotin is stable to autoclaving and to acids, but it is readily oxidized to the sulfoxide and sulfone. These oxidation products also promote growth of certain bacteria but only if they can convert them to biotin. Oxidation is a problem only in extremely dilute (<1 μg/ml) solutions and only if other reducing agents are absent.

7.1.6.4. Nicotinic acid and its derivatives

Most nicotinic acid auxotrophs utilize nicotinic acid and its amide interchangeably. The coenzyme forms of the vitamin, NAD and NADP, function in redox reactions in cellular metabolism. A few organisms (e.g., *Haemophilus influenzae*) cannot synthesize NAD from nicotinic acid; such organisms require the preformed coenzyme (the diagnostic factor V requirement for this genus). NAD is destroyed by autoclaving, acids, and alkalis; when required, it is filter sterilized and added aseptically to the separately autoclaved medium. Nicotinic acid and its amide are stable to autoclaving at neutral pH.

7.1.6.5. Pantothenic acid and related compounds

Pantothenic acid is a component of CoA and the acyl carrier protein (ACP). A few bacteria utilize the intact coenzyme as a growth factor, but many more require pantothenic acid or pantetheine for growth. The requirement of individual species for one or another of these related compounds derives from differences in their synthetic abilities or transport capabilities.

Free pantothenic acid and many of its salts are very hygroscopic; it is marketed as the calcium salt, which is only slightly hygroscopic, freely soluble in water, and stable to autoclaving. Like its more complex derivatives, pantothenic acid is readily hydrolyzed, and thus its growth-promoting activity is destroyed (for all bacteria requiring the intact vitamin) by acid or alkaline hydrolysis.

7.1.6.6. Riboflavin and its derivatives

Although relatively few bacterial species require preformed riboflavin, they all apparently contain the vitamin and some synthesize it in large enough quantities to serve as useful commercial sources.

Riboflavin 5′-phosphate (flavin mononucleotide) and flavin adenine dinucleotide are the major coenzyme forms of riboflavin; they function in redox reactions. All three compounds are equally active in supporting growth of a riboflavin auxotroph, *L. casei.*

Riboflavin is slightly soluble in cold water but is made easily soluble by warming. It is destroyed by visible light, especially at neutral pH or above; aqueous solutions should be stored in the dark. Riboflavin is stable to autoclaving.

7.1.6.7. Thiamine

A large number of bacteria are auxotrophic for thiamine, its precursors, or its coenzyme form, thiamine pyrophosphate (TPP_i; cocarboxylase). TPP_i functions in the decarboxylation of α-keto acids and in the *trans*-ketolase reaction. Since thiamine is cleaved into its component moieties when aqueous solutions at pH 5.0 or above are autoclaved, the solutions should be filter sterilized or autoclaved separately at pHs below 5.0 if the intact vitamin is required.

7.1.6.8. Vitamin B_6 group

Pyridoxine, pyridoxal, pyridoxamine, and their phosphorylated derivatives compose the vitamin B_6 group. There is much species variation in the growth response of vitamin B_6 auxotrophy to these various forms. Although pyridoxal is preferentially used by some bacteria (21, 44), adequate amounts of it are formed during autoclaving if sufficient pyridoxine is included in the medium. A few bacteria require pyridoxamine 5′-phosphate for growth; for most auxotrophs, however, the phosphorylated forms are inactive, apparently because such bacteria lack the necessary transport systems (35). Pyridoxal 5′-phosphate and pyridoxamine 5′-phosphate participate in a large number of reactions involving the synthesis and degradation of the naturally occurring α-amino acids.

Although all three forms of the vitamin are stable to heat, pyridoxal and pyridoxamine react with many naturally occurring compounds; if it is important that the vitamin forms remain unchanged, they should be sterilized separately from medium components. All forms of the vitamin are labile to light, especially at alkaline pH.

7.1.6.9. Vitamin B_{12}

Vitamin B_{12} (cyanocobalamin) appears to be synthesized in nature for the most part by microorganisms which serve as the commercial source of the vitamin. The coenzyme form of vitamin B_{12} (cyanocobalamin coupled to adenine nucleoside) is as active as the vitamin itself in promoting growth. The coenzyme functions in a number of isomerization reactions in metabolism and also (directly or indirectly) in the biosynthesis of deoxyribonucleosides, methionine, and perhaps other compounds.

Vitamin B_{12} coenzyme is unstable to light and should be filter sterilized. In contrast, vitamin B_{12} itself is relatively stable to both light and autoclaving.

7.1.6.10. Lipoic acid

Lipoic acid (thioctic acid) is an essential growth factor for several lactic acid bacteria. Its essential role in pyruvate and α-ketoglutarate oxidation can be bypassed by many lipoate auxotrophs in media that contain acetate. Lipoic acid, dihydrolipoic acid, and a variety of mixed disulfides all have about equal growth-promoting activities for *Enterococcus faecalis*. Lipoic acid is stable to acid and to autoclaving but is labile to oxidizing agents. Lability to oxidation does not usually present a problem because of the presence of reducing agents (e.g., glucose) in the growth medium.

7.1.6.11. Vitamin K

Strains of a few bacterial species (e.g., *Provotella melaninogenicus*) (19, 38) exhibit the need for vitamin K or related compounds. Vitamin K_3 (menadione; 2-methyl-1,4-naphthoquinone) is destroyed by light, alkalis, and reducing agents; vitamin K_1 (2-methyl-3-phytyl-1,4-naphthoquinone) is more heat stable and more biologically active for some bacteria. Both forms of the vitamin are fat soluble and best added to media from an ethanol stock solution. To retain activity of the compound, vitamin K stock solutions should be prepared fresh at monthly intervals and stored under refrigeration in the dark.

7.1.7. Other Organic Growth Factors

7.1.7.1. Choline

Choline has been shown to be required only by some pneumococci, apparently as a precursor of certain cellular lipids.

7.1.7.2. Purines and pyrimidines

The interrelationships of these compounds with folic acid and vitamin B_{12} were mentioned in section 7.1.6. Many bacteria are incapable of synthesizing purines and pyrimidines. For example, *Vibrio cholerae* ATCC 14033 has a purine requirement satisfied by the addition of 5 mg of either adenine or guanine per liter to a mineral-based medium (11). Guanine and xanthine are dissolved by warming with a minimal amount of hydrochloric acid. The other compounds are soluble in hot water. Purine salts (e.g., adenine sulfate) are more soluble than are the free bases.

7.1.7.3. Fatty acids and related compounds

Essential fatty acids, particularly the unsaturated fatty acids, are sometimes toxic, even when supplied in the low concentrations at which they promote growth (47). When used as a medium supplement, fatty acids are frequently added together with detoxifying absorbents such as albumin, proteins, or starch. Middlebrook and Cohn 7H10 Agar (13, 51), a standard medium for the cultivation of mycobacteria, includes oleic acid necessary for the growth of the tubercle bacillus as well as bovine serum albumin as a detoxifier. Lamanna et al. (35) suggest that the detoxifiers, although at first absorbing excess lipid material, may then slowly release the compounds back into the culture medium at levels tolerated by bacteria.

The requirement for fatty acids shows group specificity rather than an absolute requirement; that is, any of several active fatty acids are frequently incorporated unchanged into essential lipid material. However, mevalonic acid (a precursor of isoprenoid compounds) is a distinct growth factor required by some *Lactobacillus* species.

A mixture of volatile fatty acids can effectively replace the need for ruminal fluid in media for some ruminal bacteria. Table 3 lists the composition of a volatile fatty acid mix for the cultivation of these organisms.

Most members of the class *Mollicutes* (anaeroplasmas, mycoplasmas, spiroplasmas), unlike other bacteria, require sterols (e.g., cholesterol) in addition to fatty acids for growth. The sterols are supplied in the serum component of standard mycoplasma medium formulations.

Glycerol is often added as the carbon source in media for lipid-requiring bacteria. Although other carbon sources may also be metabolized by these strains, glycerol specifically seems to promote an increase in cellular lipid content.

Fatty acids and other lipids are generally stable to autoclaving. The high-molecular-weight fatty acids and other lipids should be autoclaved separately from other medium components to avoid the formation of soapy precipitates. High-molecular-weight fatty acids and other lipids are soluble in organic solvents such as ethanol; the low-molecular-weight volatile fatty acids are water soluble.

7.1.8. Miscellaneous Compounds

7.1.8.1. Polyamines

A number of organisms (e.g., some mycoplasmas, *Haemophilus parainfluenzae* ATCC 7901 [23, 25], *Neis-*

Table 3. Volatile fatty acid mix for bacteriological media (8)

Acetic acid	17.0 ml
Propionic acid	6.0 ml
N-Butyric acid	4.0 ml
N-Valeric acid	1.0 ml
Isovaleric acid	1.0 ml
Isobutyric acid	1.0 ml
DL-α-Methylbutyric acid	1.0 ml

Pipette the given volumes of fatty acids into a container that has a tight-fitting closure, and mix well. The objectionable odor of the free fatty acids can be reduced by neutralizing the pH of this stock solution with NaOH pellets. Use 3.1 ml of the fatty acid mix per liter of medium.

seria [43], *Veillonella atypica* [52]) are stimulated in growth by polyamines (putrescine, spermidine, spermine). The role of these compounds within the cell has not been fully clarified; however, they are implicated in DNA and RNA function. In vitro polyamine requirements have been shown to be replaceable by high levels of magnesium (12).

7.1.8.2. Diaminopimelic acid

A few strains of *E. coli* developed for the purposes of genetic manipulation are auxotrophic for diaminopimelic acid, a component by bacterial cell walls, to lessen the possibility that these attenuated organisms will escape the laboratory and proliferate in the environment. DL-α, ϵ-Diaminopimelic acid at a concentration of 5 mg/liter in defined media will support the growth of these strains.

7.1.8.3. Unidentified growth factors

Substances produced when glucose is autoclaved with phosphate or in the presence of amino acids (16, 17) and several undetermined factors present in yeast extract, serum, and other natural materials may be necessary for or stimulatory to the growth of some bacteria.

7.1.9. Determination of Nutritional Requirements

The ability to study bacteria fully necessitates the ability to cultivate and maintain these organisms in vitro. Unfortunately, there are too many examples of poorly constructed "growth media" formulations in which organisms survive despite the medium conditions, not because of them. The urge to define a medium chemically without taking the time and expense to do it well results only in an inevitable decrease in the viability of a culture after prolonged maintenance under such conditions. The appearance of bacterial growth in a test tube or plate is not necessarily indicative of the overall health of the culture in that particular medium. On microscopic examination, shriveled or deformed cells, numerous spheroplasts, or the presence of poly-β-hydroxybutyrate inclusions within the cells are all indications that the nutritional requirements of the organism have not been satisfactorily met.

A newly isolated, nonfastidious heterotrophic bacterium should be cultivated on a complex medium that supports good, sustained growth and retains a cellular morphology reflective of the organism in its natural habitat. Only after a stock culture has been properly maintained on a complex medium and/or preserved should additional studies be conducted to further elucidate the nutritional requirements of the organism. Methods for defining nutritional growth conditions are discussed in Section 7.3.

7.2. COMPONENTS OF UNDEFINED MEDIA

Most media used for routine laboratory growth of bacteria as well as those used for large-scale industrial fermentations are undefined and are constructed from a combination of protein hydrolysates, mixed carbon sources, and supplements. These formulations may also be amended with specific buffering agents to help prolong bacterial growth by stabilizing the pH shifts caused by the accumulation in the medium of acidic or basic waste products released by the actively metabolizing cultures. The most common undefined components used in bacteriological media are described below.

7.2.1. Peptones, Hydrolysates, and Extracts

Early bacteriologists grew their cultures on solid media composed of heated and coagulated proteins such as egg, serum, or gelatin or on bits of sterilized meat. Some media based on whole protein are still in use today, the most notable being the Lowenstein-Jensen formulation (51), which incorporates egg for the cultivation of mycobacteria. In these formulations, solid substrates of liquid proteins such as egg or blood (serum) are formed by simmering the tightly capped tubes containing liquid medium at a slant in an 85°C water bath for 45 min until the protein coagulates in a process called **inspissation.**

However, the success of whole protein as a complex nutrient depends on the strong proteolytic ability of a culture used as an inoculum; unfortunately, not all bacteria have the metabolic ability to degrade protein. Therefore, the evolution of media formulated with predigested proteins as a more assimilable source of essential nutrients gave rise to a marked increase in the isolation rates for the metabolically diverse heterotrophic bacteria. Formulations based on protein digests remain the backbone for complex media currently used in bacteriology.

Many proteinaceous materials can serve as raw matter for the digests used in media, but meat, milk, and plant (soybean) hydrolysates are the most widely used for laboratory purposes. The proteins from these raw materials are digested by acid, alkali, or enzymatic hydrolysis. The protein source and the method and degree of digestion are important in determining the suitability of a particular digest in a specific medium; although medium manufacturers rigidly evaluate the digestion products for generalized growth or reaction parameters, lot-to-lot variations are expected, particularly with respect to those nebulous undefined factors that promote growth of recalcitrant organisms.

The heat and low pH associated with acid hydrolysis destroy the amino acid tryptophan and, to a lesser degree, serine and threonine in acid-digested peptones.

Further treatment in the manufacture of an acid hydrolysate also limits vitamin content in these products. In addition, acid hydrolysates have high salt content generated during the neutralization stage of the process. Plant hydrolysates are high in fermentable carbohydrates, which limit their use in media for fermentation tests. Enzymatic digests best conserve the intact nutritional content of the original protein source.

The following are characteristics of those peptones and hydrolysates most familiar to bacteriologists. Bridson and Brecker (5) give excellent details of industrial methods as well as nutrient compositions for commercially available peptones. The technical manuals offered by media manufacturers (e.g., Becton Dickinson Microbiology Systems [BBL], Difco, Oxoid [13, 48, 51]) also provide compositional analyses and suggestions for use for each of their peptone products. Peptones and hydrolysates are generally used in media at 0.5 to 1.0% concentrations.

7.2.1.1. Acid-hydrolyzed casein

Acidicase (BBL; for addresses of suppliers, see section 7.6), Casamino Acids (Difco), Hy-Case (Sheffield), and other acid-hydrolyzed caseins consist predominantly of free amino acids but are deficient in tryptophan and cystine. They are used in vitamin assay media, which are also low in these growth factors, and are also used for toxin production. They have high salt content (30 to 40%), although some manufacturers also feature salt-free acid hydrolysates (e.g., Hy-Case SF [Sheffield]).

7.2.1.2. Enzymatic (pancreatic) digest of casein

Enzymatic digests of casein include Casitone (Difco), NZ Amine and NZ Case (Sheffield), Trypticase Peptone (BBL), tryptone (Difco and Oxoid); they are also from other manufacturers. They consist mostly of free amino acids and small peptides and are suitable in general growth media for many heterotrophs. High levels of tryptophan give good indole reactions. The digests are low in carbohydrates and are suitable for fermentation test media.

7.2.1.3. Enzymatic digest of meat protein

Enzymatic digests of meat protein include Myosate and Thiotone (BBL), Primatone (Sheffield), Bacto Peptone and Proteose Peptone (Difco), Bacteriological Peptone and Proteose Peptone (Oxoid). They consist of a general mixture of amino acids and peptides suitable for the cultivation of most heterotrophs. Peptones with the designation "proteose" are hydrolyzed to have a high peptide content and are used for the cultivation of fastidious strains. Thiotone (BBL) is high in the sulfur amino acids and can be used in a basal medium for the detection of hydrogen sulfide.

7.2.1.4. Enzymatic digest of plant (soy) protein

Enzymatic digests of plant protein include Phytone (BBL), Soytone (Difco), Soya Peptone (Oxoid), and NZ Soy Peptone (Sheffield). They are high in carbohydrates and vitamins. They are not useful in fermentation studies; they support rapid growth of many bacteria, but acid production from intrinsic carbohydrates may also cause a rapid culture decline.

7.2.1.5. Mixed hydrolysates

Mixed hydrolysates include Biosate (yeast extract and casein digest) (BBL), Polypeptone (casein and meat digests) (BBL), and Tryptose (mixed enzymatic sources) (Difco; Oxoid). They support good growth of many bacteria, sometimes performing better than peptones from single sources.

The aqueous, processed extracts of meat or yeast are frequently added to complex media to supplement the nutritional aspects of peptones. The extracts provide water-soluble amino acids, carbohydrates, nucleic acid fractions, organic acids, and some vitamins lost from hydrolysates during their manufacture. The extracts also supply trace minerals to a medium. Both meat and yeast extracts are normally added to media at a final concentration of 0.3 to 0.5%.

7.2.1.6. Meat extract

Meat extracts are mostly from beef (BBL and Difco) and Lab-Lemco (Oxoid). The Difco product is a paste, whereas both BBL and Oxoid extracts are powders. Meat extracts are free from fermentable carbohydrates but contain glycolic and lactic acids and creatinine, which can serve as carbon sources.

7.2.1.7. Yeast extract

Yeast extract (BBL, Difco, and Oxoid) consist of autolysates of baker's yeast and are high in amino acids, peptides, water-soluble vitamins, and carbohydrates. Yeast extract is not suitable for use in fermentation or sole-carbon-source test media. The powdered yeast extracts are not equivalent to the growth-promoting aspects of fresh yeast extract solutions (e.g., from GIBCO Laboratories) used in media for mycoplasmas and should not be substituted for them.

7.2.2. Carbon Sources

Carbohydrates are the carbon sources most frequently added to media for heterotrophic bacteria. When they are combined directly with the other medium components and autoclaved, a number of chemical reactions with other components (specifically the mono and disaccharides) may occur that can appreciably compromise the original nutritional composition of the medium. The **Maillard reaction** caused by an interaction of amino acids or peptones with the glycosidic hydroxyl group of sugars, results in the pronounced darkening of autoclaved media containing these reactants. A similar darkening reaction occurs between carbohydrates and phosphates (16), especially in neutral to alkaline media. Although there are reports of some organisms being stimulated by these breakdown products (15), browning reactions are generally regarded as growth limiting and undesirable in media (36).

Although the heat sterilization of concentrated stock solutions of carbohydrates with subsequent incorporation into the balance of a sterile basal medium does alleviate the browning reactions, not all sugars are stable to autoclaving in concentrated solutions. Except for starch and glycogen, which, because of their high molecular weights, are best autoclaved, filter sterilization through a 0.2-μm-pore-size membrane is the best method of sterilization for carbohydrates, especially when they are being used as test substrates in sole-carbon-source experiments.

Not all heterotrophs are able to utilize **organic acids** as sole carbon sources; however, the differential use of these compounds by metabolically active genera (e.g., *Pseudomonas* and *Vibrio* species) can serve as useful tests for characterization purposes. Most organic acids have limited solubility in water; raising the pH of an aqueous organic acid solution with sodium hydroxide (i.e., making a sodium salt of the acid) significantly increases their solubility.

As noted above (sections 7.1.2 and 7.1.9), **peptones** and **amino acids** can be utilized as carbon sources by a number of bacteria. Growth of an organism in a 1% peptone solution is indicative of this ability. Not all amino acids are utilized by a bacterium for this purpose, and, as above, the growth reactions with single amino acids as the carbon source can serve for diagnostic purposes.

Fatty acids can serve as carbon sources for certain organisms. *Leptospira* species require long-chain fatty acids as a major source of energy, which, for routine cultivation, is supplied by the addition of 5 to 10% rabbit serum to a peptone-based medium. Tween 80 or a defined mixture of esterified fatty acids (61) can substitute for serum for a number of *Leptospira* strains. *Desulfovibrio* strain DSM 2056 (34) can utilize butyrate and higher fatty acids up to stearate as carbon sources. These compounds are generally neutralized in aqueous solutions for incorporation into media.

7.2.3. Supplements

Many fastidious bacteria are best grown on complex media with additional undefined supplements. Although many of these organisms are cultivable without the supplements, growth and long-term viability can be superior when they are present. Whether necessary for growth or merely stimulatory, the undefined supplements provide a diverse combination of growth factors, detoxifiers, chelators, buffering capacity, or other support nutrients required for optimal cell metabolism.

Undefined supplements frequently reflect the milieu of a bacterial isolate's natural habitat: for example, isolation media for lactic acid bacteria from dairy products could well include **skim milk,** and those for strains isolated form vegetable matter could include **tomato** or **fruit juice.** In either case, the supplements provide numerous growth factors (amino acids, purines and pyrimidines, vitamins) known to be required by this group of organisms.

Bacteria pathogenic to animals are often grown in media supplemented with 5 to 15% **defibrinated blood** or **serum.** Hemolytic reactions differ between rabbit and sheep bloods, but for many organisms growth on either is usually equal. The inhibitory aspects of fresh sheep blood for some bacteria such as *Haemophilus* spp. (25) are diminished when **chocolatized blood** is used (i.e., when the medium containing the blood is heated to 70 to 80°C for 15 min). Sheep blood is used with greater frequency than is rabbit blood for routine cultivation and quality control procedures because of its lower cost and greater availability. The use of human blood or serum is to be avoided whenever possible because of the health risks associated with potentially contaminated material.

Defibrinated blood has a short shelf life and should be refrigerated (not frozen) immediately on receipt and incorporated into media within a week. Other components of a medium formulation help to stabilize the erythrocytes, and, with proper refrigerated storage, blood-containing media are usable for at least 1 month.

Serum (5 to 10%) is used as an alternative to blood, especially when the presence of erythrocytes interferes with examination or harvesting of the bacterial culture. Horse and fetal bovine sera are most widely used in bacterial formulations. All sera should be obtained from U.S. Department of Agriculture-inspected herds to decrease the possibility of transmitting bacterial or viral pathogens. Freshly collected and processed sera can develop precipitates upon thawing; these precipitates should be removed by filtration before use in media. Serum stored at −20°C for 1.5 years still supports the growth of fastidious bacteria. Sera, unless gamma globulin free, are normally heat inactivated (i.e., held at 56°C for 30 min) before use in bacteriological media to limit the potentially growth-inhibiting protein complement.

Soil extract has been used in media for the cultivation or sporulation of numerous soil organisms including *Azotobacter, Bacillus, Rhizobium, Streptomyces,* and *Vitreoscilla* species. In many instances for the *Vitreoscilla* species, 20 mM Ca^{2+} in the form of calcium chloride can replace the need for soil extract (46). For other organisms, vitamin supplementation can be an effective soil replacement. For most strains, however, the exact stimulatory factors of soil have not yet been determined. Fertile, chemically uncontaminated soil should be used for the preparation of extracts.

Complex media for the cultivation of ruminal bacteria are supplemented with 10 to 30% **ruminal fluid,** which nutritionally augments a basal medium with amino acids, carbohydrates, fatty acids, peptides, and vitamins in addition to a number of intermediate metabolites (49). The replacement of ruminal fluid with a defined volatile fatty acid mixture (Table 3) can satisfy the nutritional needs of some ruminal organisms; however, growth is always greatly stimulated when the fluid itself is included in a medium.

7.2.4. Buffers

Buffers are compound which, in solubilized dissociated states, resist fluctuations in pH. Within the narrow chemical confines of this definition (i.e., an aqueous solution in which the interacting components of the buffer pair and a single acid or alkali exerting hydrogen ion changes to the systems are the only elements in contention), all buffers function well.

However, when subject to the complex conditions of any bacterial growth medium, buffers are subject to many chemical interactions in addition to changes in hydrogen ion concentration. Most buffers are reactive compounds that (i) can bind with other substances in a medium formulation, (ii) can serve as a nutritional source for microorganisms, (iii) are susceptible to degradative microbial activities, or (iv) can respond variably to temperature modulations. The ideal buffer would be inert to all extraneous factors other than pH and would also be effective throughout the pH range of bacterial metabolic activity. To date there exists no universal buffer that meets all these criteria. Consequently, the choice of a buffer and its method of application in any medium must be made with caution, for, as Good et al. (20) have commented, "It is impossible even to guess how many exploratory experiments have failed, how many reaction rates have been depressed, and how many processes have been distorted because of imperfections of the buffers employed."

Buffers used in bacteriological media can be described as natural or synthetic. The naturally occurring buffer systems are most prone to the drawbacks listed above; however, most commonly used growth media are based, by intent or happenstance, on the innate buffering characteristics of their complex nutritional components. Since the 1960s a number of synthetic buffers have been developed, of which several have excellent potential for more widespread use in bacteriological media. This section briefly explains the advantages and disadvantages of some of the more frequently used buffers found in formulations of media. The chemistry of buffer action is discussed in Chapter 6.1.2.

7.2.4.1. Natural buffering agents

Carboxylic acids. Acetic, citric, and succinic acids and their salts are effective but metabolically degradable buffers covering the lower (less than pH 6) ranges of pK values.

Sodium bicarbonate-carbon dioxide. Sodium bicarbonate-carbon dioxide buffer is well suited for physiological or neutral conditions (pH 6.8 to 7.4). Both components can be metabolized by autotrophic organisms, but this buffer is commonly used in the cultivation of anaerobic bacteria. Tightly sealed containers or CO_2 incubators are required to maintain the gaseous buffer phase. Consult Chapter 6 for an elaboration of this gas-liquid buffer and anaerobic media techniques.

Potassium and sodium phosphates. In 1928 Clark (10) published a series of phosphate buffer tables spanning the pH range frequently encountered in biological work. The name "Sorensen's phosphate" is still applied to the disodium phosphate salt used by Sorensen during the early part of the century in his studies of hydrogen ions and buffering activity. Soluble phosphates such as potassium, sodium, and certain polyphosphates are all effective buffers but suffer the limitations of being microbial nutrients. Furthermore, without the intervention of chelating compounds, the solubilized phosphates can bind with the cationic alkaline earth metals (Ca^{2+}, Fe^{2+}, Fe^{3+}, and Mg^{2+}) to form insoluble and metabolically unavailable precipitates of these elements. Within the pH range of these buffers (pH 6 to 8), the process is irreversible and spontaneous. At room temperature, an unchelated medium containing trace-metal salts and buffered with phosphates will develop a haze on standing unless it is very carefully designed and prepared. The elevated temperatures associated with autoclaving increase the rate of precipitation in these media. When autoclaved together, phosphates also complex with glucose at neutral to alkaline pH values. Trace metals appear to catalyze the reaction but are not necessary for its occurrence.

Proteins, peptones, and amino acids. The zwitterionic nature of amino acids accounts for the buffering capacity of proteins and their derivatives. Because amino acids can function as either acids or bases, these individual compounds can be used to form buffers with either acidic or basic pK_a values. Glycine, for example, when titrated against hydrochloric acid, buffers most strongly at pH 2.3; when titrated against sodium hydroxide, the pK_a value is 9.6. The amino acid mixtures of peptones and hydrolysates yield buffered solutions between pH 6.8 to 7.0. These peptone-buffered systems are stable to autoclaving; however, they interact chemically at high temperatures with other medium components such as glucose and excess concentrations of phosphates. Amino acid-based buffers are metabolized by bacteria.

7.2.4.2. Synthetic buffers

Tris. Tris mixed with Tris HCl or various carboxylic acids yields buffers useful in the range of pH 7.0 to 9.0. Conversely, Tris HCl can be combined with sodium hydroxide or dibasic phosphates to form buffers effective for the same pH values. Tris buffers are routinely used at 0.05 M; increasing or decreasing the molarity, as with many buffers, causes a slight shift in pH from that noted at 0.05 M. The temperature effect of Tris buffers can be significant for some precise applications. For every degree Celsius above 25°C, Tris buffers will decrease approximately 0.025 pH unit; for every degree below 25°C, the pH increases approximately 0.03 unit (53). Tris is known to be inhibitory to some organisms (e.g., *Sporocytophaga myxococcoides*).

"Good" buffers. As an aid to biological research, Good et al. (20) in 1966 synthesized and studied a number of predominantly zwitterionic compounds that might satisfy the need for effective but inert buffers. Although the potential for bacterial degradation exists with these synthetic buffers, their overall advantages, including lack of precipitation with medium components, marginal concentration and temperature pH effects, and excellent solubilities, make the Good buffers worthwhile alternatives to any of the above-mentioned buffers. The following buffers have been satisfactorily used in bacteriological media.

ACES. N-(2-Acetamido)-2-aminoethanesulfonic acid (ACES) has a pK_a of 6.8 at 25°C and is useful for pH 6.1 to 7.5.

HEPES. N-2-Hydroxyethylpiperazine-*N'*-2-ethanesulfonic acid (HEPES) has a pK_a of 7.5 at 25°C and is useful for pH 6.8 to 8.2. Some inhibitory effects of HEPES that could be reversed by the addition of pyruvate have been reported for certain eukaryotic cell lines (56); however, this phenomenon has not been observed with bacteria.

MES. 2-(*N*-Morpholino)ethanesulfonic acid (MES) has a pK_a of 6.1 at 25°C and is useful for pH 5.5 to 6.7.

TES. N-Tris(hydroxymethyl)methyl-2-aminoethanesulfonic acid (TES) has a pK_a of 7.5 at 25°C and is useful for pH 6.8 to 8.2.

Since all these compounds are acids, the addition of sodium hydroxide or other alkaline ionic or zwitterionic compounds in a bacteriological medium will create the buffer pair. ACES, HEPES, MES, and TES are used at 0.01 to 0.05 M. Also see Chapter 6.1.2.

7.3. DESIGN OF MEDIA

The ultimate purpose of any bacterial growth medium is to provide the nutritional requirements and to establish the physicochemical conditions (pH, redox levels, etc.) necessary to maintain the viability of a culture. Numerous variations on this basic theme are used in bacteriology to isolate and enrich new cultures and to study the metabolic abilities of a given strain.

7.3.1. Undefined Growth Media

Most heterotrophic bacteria are routinely cultivated on complex media which support sustained growth, performance, and viability. This kind of medium is normally undefined in terms of its composition (i.e., the exact chemical composition of the nutrients are undetermined), but typically includes (i) an enzymatic hydrolysate of protein to provide nitrogen (as well as a potential supply of carbon and energy), (ii) a carbohydrate (usually glucose) to satisfy the general needs for a carbon and energy source, (iii) various salts to satisfy inorganic ion requirements, (iv) yeast extract to ensure that needs for vitamins are met, and (v) supplements to contribute any additional, or ratio of factors not met by the above ingredients.

As the nutritional requirements of the bacterium become better understood, a semidefined medium can be described in which the undefined extracts and hydrolysates are replaced by components of known chemical composition. It is best to work with one group of nutrients at a time rather than to attempt to manipulate all of the undefined additives at once. For example, one might first replace the undefined nitrogenous peptone with inorganic salts [NH_4Cl, $(NH_4)_2SO_4$, or KNO_3].

If good growth and cellular appearance persist after repeated transfers in the revised medium, one might determine the vitamin requirements (if any) by eliminating the yeast extract or by replacing it with a vitamin mixture of known composition. The exact requirements of a vitamin-requiring strain can be further defined by eliminating components from the vitamin mixture one by one, or in small groups, until the necessary growth factor is revealed (by a cessation in growth once the key vitamin is omitted).

Lack of growth with a defined inorganic nitrogen source may be indicative of amino acid or purine/pyrimidine requirements or may be suggestive of an organism, such as *Saprospira grandis* or *Halobacterium salinarium*, that prefers the organic nitrogen complement of peptones (mixed amino acids and peptides) rather than carbohydrates as suitable energy substrates. Again, the exact nutritional nature of the peptone requirement can be determined by replacing it with a defined and complete amino acid-purine/pyrimidine mix such as those used in assay media.

If growth is satisfactory with the defined amino acid mixture, the process of elimination may once again be used to identify the specific amino acid requirement. Unexpected problems may occur during this process since an amino acid imbalance (as discussed in section 7.1.5) as well as the omission of an essential amino acid can cause a cessation of growth.

Organisms that utilize a particular amino acid to satisfy an auxotrophic deficiency are stimulated by additions of the substance at as little as 5 to 50 mg/liter. Organisms which utilize amino acids for energy require much higher concentrations (0.2 to 0.5%) to initiate growth.

7.3.2. Defined Growth Media

Once all the undefined chemical characteristics of a medium have been replaced by known components, one may continue to refine the chemically defined formulation to create nutritional conditions under which the concentration of each component is adjusted to the minimal level at which it supports growth of the organism. Most formulations, however, are defined only in terms of their components and not with respect to the minimal required concentrations of each nutrient.

Failure to grow on the resulting defined medium may reflect the existence of an unidentified growth factor, the presence of an imbalance of known growth factors in the medium, or the requirement for known growth factors which were supplied or spared by the crude supplements but not by the mixtures of the synthetic compounds tested. Detailed discussions of the nutritional requirements of bacteria (e.g., reference 21) should be consulted in that event.

The final selection of a medium depends on the purpose of the experiment as well as the organism involved. *E. coli* prototrophs thrive with very simple formulations of defined media. However, complicated defined media such as those developed for fastidious bacteria are expensive in terms of the stockpile of chemicals used as well as the time needed for preparation. One chemically defined medium (9) for the cultivation of *Neisseria* spp. has 53 nutritional components. Unless one is actually involved in studies in which such defined conditions are mandatory, it may be more practical to work with an undefined or semidefined formulation.

7.3.3. Enrichment and Isolation Media

Enrichment media selectively encourage the growth of specific bacteria in mixed populations. Enrichment media may be designed to be inhibitory (or otherwise growth limiting) to organisms other than the desired species or to favor the proliferation of a particular species by incorporating unusual energy, nitrogen, or carbon sources.

Isolation media are used to select (i.e., purify) a given species from a mixed culture. These formulations may be inhibitory to the undesired species or, conversely, may support the growth of contaminants as well as the desired species to permit a visible distinction between isolated colonies on solid media and thus facilitate colony selection.

Chapter 8 provides detailed information on enrichment and isolation procedures.

7.3.4. Maintenance and Characterization Media

Maintenance media are nutrient formulations used for routine cultivation or expansion of a bacterial culture. These formulations may be either defined or undefined in terms of chemical composition; however, heterotrophic cultures that have been under long-term preservation often revive better on complex rather than defined media. Media that are excessively rich in nutrients (especially carbon and energy sources) can result in rapid growth but early decline of a culture.

Characterization media are formulations used to identify the morphological or biochemical attributes of a bacterial culture. Many observations of the former aspects (size, pigmentation, motility, luminescence, etc.) are done on maintenance media, since these formulations should reflect the appearance of an organism under field conditions.

For phenotypic characterization purposes, maintenance media supplemented with appropriate substrates can also serve as basal formulations to study an organism's degradative (enzymatic) capabilities, such as hydrolysis tests for esculin, gelatin, starch, Tween, and tyrosine. Maintenance formulations may also function as the basal medium for inhibition tests (e.g., crystal violet, tellurite).

However, the nutritive basis of maintenance media can obscure some metabolic activity or limit adaptive- or inducible-enzyme function. Therefore, for certain purposes, phenotypic characterization media must be tailored to enhance the detection of specific biochemical characteristics (e.g., methyl red/Voges-Proskauer, amino acid decarboxylases, or *O*-nitrophenyl-β-D-galactopyranoside tests). Chapter 25 of this manual and reference 41 describe the preparation methods, principles, and precautions for many biochemical tests used in bacteriology.

For phenotypic characterization tests of utilizable substrates (e.g., sole nitrogen or sole carbon sources), the basal medium must be free of any alternative substrates that will obscure results. Carbohydrate-free peptones can provide the nitrogen in media for carbon utilization tests only if the organism is not metabolically able to also utilize the amino acids in peptones for energy procurement. The best results for utilization tests are obtained with chemically defined mineral-based media.

7.3.5. Assay Media and Procedure

Assay media can be described as characterization media that have been refined to permit quantitative determinations regarding the growth response of a bacterium to a particular substrate, growth factor, or inhibitor. Bacteria have been widely used in quantitative assays for a wide variety of substances that promote or inhibit growth: vitamins, amino acids, trace elements, pyrimidine bases, polyamines, deoxyribonucleosides, deoxyribonucleotides, drugs, antibiotics, and other chemotherapeutic agents (21, 30). Some of these bacteriological methods have been replaced by instrumental methods (e.g., amino acid analyzers, high-performance liquid chromatography) but are still useful in many applications. Until a compound is isolated and characterized, its effect on the growth of an appropriate assay organism often is the only means of monitoring its purification from a natural material.

Quantitative growth-response assays are based in principle on the arithmetic (linear) response in the amount (not the rate) of growth as the quantity of a limiting nutrient is increased within a defined range. The reproducibility of microbiological assays is usually given as ±10%; however, under carefully controlled conditions it may be less than ±3%. In general, an assay for an essential nutrient can be performed with a higher degree of accuracy than can an assay for a stimulatory one.

For vitamin and amino acid assays, *Lactobacillus*, *Enterococcus*, and *Pediococcus* species have been particularly useful since they have multiple growth factor requirements; i.e., they require the addition of a large number of nutrients to the growth medium. One bacterium can be used for the assay of several vitamins or amino acids. For instance, an assay method employing *Pediococcus acidilactici* ATCC 8042 permits the determination of all amino acids except alanine (58); alanine can be determined by using the same medium plus folinic acid with *P. acidilactici* ATCC 8081 as the test bacterium.

Bacteriological assays for specific growth factors must always be conducted simultaneously with a standard (control) assay to assess the response of the test organism against known concentrations of a growth factor under the same conditions as the unknown. General guidelines for preparing and conducting an amino acid or vitamin standard assay with bacteria are outlined in the following paragraphs. Cautionary notes for amino acid or vitamin assays are indicated as necessary. Detailed instructions for a number of assays using biological (bacterial) methods are available in the *Difco Manual*, 10th ed. (13); *Official Methods of Analysis of the Association of Official Analytical Chemists* (62); and *The U.S. Pharmacopeia XXII/The National Formulary XVII* (60). Consult Kavanagh (30, 31), Hewitt and Vincent (24) and Snell (54) for extensive information about assay procedures.

7.3.5.1. General guidelines for assay procedures

Reagents for medium preparation. All chemicals should be reagent grade or better. Any peptones (e.g., Casamino Acids) used as a complex nitrogen source in vitamin assays should be previously checked to ensure that they are indeed vitamin free. The use of commercially available powdered basal media in vitamin assays is not recommended without lot-by-lot testing by the user since many of these preparations are contaminated at significant levels with the growth factor being investigated, making baseline determination difficult if not impossible. All water used in the preparation of media and stock solutions should be freshly distilled or deionized (type IIB or better) and not stored in carboys or containers that have held water supplies for extended periods. **Water storage** containers, unless emptied and flushed daily, are often contaminated with a bacterial population that contributes waste products and metabolites, in addition to the cells themselves, to a laboratory's "purified" water supply.

Glassware and equipment. Test tubes, flasks, beakers, closures, etc., must be carefully washed in a free-rinsing detergent, repeatedly rinsed in distilled or deionized water as specified above, and air dried. Any watery film or spots on the glassware indicate the need for more vigorous cleaning. This treatment is usually sufficient to clean the equipment used for amino acid or similar assays in which the test substrates are added in milligram or higher concentrations. However, soaking the material in an acid bath prior to final rinsing may be necessary for some vitamin assays (e.g., biotin or cyanocobalamin) in which activity is detectable at nanogram levels; baking the clean glassware at 250 to 300°C for 3 h will remove any trace levels of these vitamins. The use of new (not previously used) glassware in addition to disposable plastic pipettes and test tubes makes the cleaning effort easier but does not totally eliminate the process.

Basal medium preparation. Assay basal media are normally prepared at double strength to permit the addition of different amounts of an amino acid or vitamin standard from a single stock solution. The resulting nutrient solution is then diluted to its working (single-strength) concentration with distilled water. Basal medium formulations depend on the bacterial strain used, since the test organism as well as the growth factor is being assayed. The references cited at the beginning of this section include many formulations for various assay procedures.

Inoculum preparation. The bacterial test culture must be previously cultivated in a complex medium to establish good growth. The choice of this medium is dependent on a particular assay method. Just prior to the assay procedure, the inoculum must be washed free of residual nutrients from the complex medium. This step is critical, and failure to perform it properly causes most assay failures. For amino acid assays and also for vitamins such as thiamine or nicotinic acid that are utilized by bacteria in significant concentrations (nanograms per milliter), three to five centrifugations and resuspensions in 0.85% sodium chloride may be enough to deplete the growth factor from the inoculum. However, many published procedures for biotin and cyanocobalamin do not produce sufficiently vitamin-depleted inocula, since these vitamins elicit growth responses at very low concentrations (e.g., 0.01 ng/ml for demonstrable vitamin B_{12} response with *Lactobacillus* species) and standard growth curves derived from such conditions will be ill defined at low to middle ranges. The best method to produce an inoculum relatively free of cellular reserves of the required growth factor is to make rapid subcultures (every 18 to 30 h depending on the bacterial growth rate) of the test strain in vitamin assay media containing limited concentrations of the required growth factor. Final washing in physiological saline, as noted above, will complete the inoculum preparation.

7.3.5.2. Performing an assay (e.g., folic acid)

The following discussion demonstrates the step-by-step procedures for conducting a folic acid assay with *Lactobacillus casei* subsp. *rhamnosus* ATCC 7469.

a. Preparation of folic acid standard

1. Prepare a 200 μg/ml vitamin solution by dissolving 20 mg of dried folic acid in 80 ml of 20% ethanol.
2. Adjust to pH 10 with 0.1 N NaOH to dissolve the folic acid.
3. Readjust the solution to pH 7.0 with 0.05 N HCl.
4. Bring the total volume of solution to 100 ml with 20% ethanol.
5. Dilute 1 ml of this solution to 1 liter with distilled water in a volumetric flask. (The vitamin concentration at this point is 200 ng/ml.)
6. Dilute 1 ml of the 200 ng/ml vitamin solution to 1 liter with distilled water in a volumetric flask. (The vitamin concentration at this point is 0.2 ng/ml.)
7. Filter sterilize this folate standard solution, and store under refrigeration if not used immediately in preparing the assay medium dilutions outlined below.

b. Assay medium dilutions

1. Prepare the folic acid casei assay basal medium in double-strength (2×) concentration, and prepare the assay standards as described in Table 4. *Note:* Use commercial assay media if lot performance is satisfactory, or consult Kavanaugh (30) for alternate formulations for the 2× assay medium.

c. Assay method

1. Stab inoculate Bacto-Lactobacilli Agar AOAC (no. 0900; Difco), and incubate at 35 to 37°C for 24 to 48 h.
2. Subculture from the stab into 10 ml of Bacto-Lactobacilli Broth AOAC (no. 0901; Difco), and incubate at 35 to 37°C for 24 h.
3. Centrifuge the culture, and decant the supernatant.
4. Resuspend cells in 10 ml of 0.85% NaCl, and centrifuge. Decant the supernatant.
5. Perform step 4 two more times.
6. Resuspend the cell pellet in 10 ml of 0.85% NaCl.
7. Dilute the cell suspension 5:100 with 0.85% NaCl.
8. Use 1 drop of this suspension to inoculate each 10-ml assay tube of folic acid casei assay medium containing the various amounts of folate outlined under (b) above.
9. Incubate at 35 to 37°C for 24 to 48 h until there is no increase in turbidity in tube 7 over a 2-h period.
10. For the turbidimetric method, zero a spectrophotometer (at 540 nm) with an uninoculated assay medium tube 1. Then read the optical density of the inoculated blank (tube 1), and zero the spectrophotometer again against this tube. Read the remainder of the inoculated tubes (tubes 2 to 7), and plot increasing optical densities against the inoculated blank. Results should be similar to those shown in Fig. 1.

Table 4. Assay medium dilutions for folic acid assay[a]

Tube no.	Vol (ml) of:			ng of folate per tube
	2× assay medium	Distilled water	Folate standard	
1	5.0	5.0	0.0	0.0
2	5.0	4.5	0.5	0.1
3	5.0	4.0	1.0	0.2
4	5.0	3.0	2.0	0.4
5	5.0	2.0	3.0	0.6
6	5.0	1.0	4.0	0.8
7	5.0	0.0	5.0	1.0

[a]Sterilize completed medium at 121°C for 10 min, and cool rapidly.

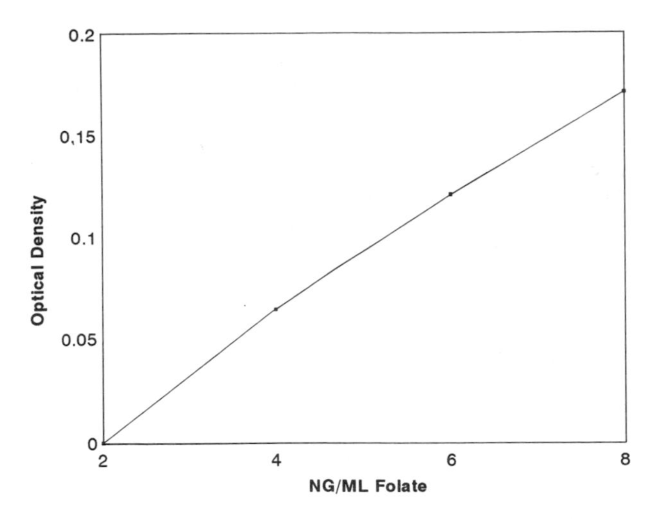

FIG. 1. Turbidimetric response of *Lactobacillus casei* subsp. *rhamnosus* to increasing concentrations of folate.

7.4. STERILIZATION OF MEDIA

All nutrient media as well as nonnutrient solutions (e.g., buffers used for dilutions) that are used in the cultivation of bacteria must be sterilized immediately after preparation to limit the microbial contamination inherent in this material. This section describes the various sterilization methods available to bacteriologists and highlights the advantages and disadvantages of each. Consult Chapter 29.8 for additional information regarding the topic of sterilization.

7.4.1. Autoclaving

Most media used in bacteriology are sterilized wholly or in part by steam sterilization or autoclaving, although this process is not the best method for maintaining the chemical or nutritional integrity of many formulations. The heat and high pressure associated with autoclaving contribute to an oxidatively destructive sterilization process, and all deleterious chemical reactions (e.g., precipitations, browning, substrate breakdown) mentioned in this chapter are attributable to or enhanced by autoclaving. Many of these adverse reactions can be avoided by physical separation of the reactants during the autoclaving process by following these guidelines.

(i) Autoclave (or, better still, filter sterilize) glucose separately from amino acids/peptones or phosphate components.

(ii) Autoclave phosphates separately from amino acids/peptones or other mineral salt components.

(iii) Autoclave mineral salt components separately from agar, since this gelling agent can contain appreciable amounts of calcium and magnesium salts that can interact with other medium components, especially other salts, to form insoluble precipitates.

(iv) Avoid autoclaving media at alkaline pH (>pH 7.5). The nutrient and mineral portions of media designed for alkalophilic organisms should be autoclaved at neutral pH and then adjusted to appropriate alkalinity with sterile stock solutions of alkaline buffer, Na_2CO_3, etc., after cooling.

(v) Avoid autoclaving agar solutions at less than pH 6.0, since the gelling properties of the polysaccharide will be destroyed by acid hydrolysis. Solid agar media for acidophiles can be constructed by preparing and autoclaving the acidic, heat-stable nutrient fraction of the formulation as a double-strength solution (e.g., 1 liter's worth of components prepared in 500 ml of distilled water) and the agar as a separate double-strength solution (e.g., 15 or 20 g of agar in 500 ml of distilled water). The two autoclaved solutions can then be aseptically combined at ~50°C and distributed as necessary into plates or tubes.

(vi) Filter sterilize heat-labile components. Consult a good chemical reference such as *The Merck Index* (7) for heat stabilities and other physical data regarding chemicals used in media formulations.

The standard operational definition of autoclaving ("sterilize at 121°C for 15 min") deserves clarification. The phrase refers to the temperature of the contents of the container being held at 121°C for 15 min, not to the fact that the autoclave itself has been set to run for that temperature and time. Therefore, large containers with large volumes of liquid will require longer sterilization times than will small containers with lesser volumes, and autoclaves filled to capacity with containers will require longer running times than will autoclaves holding only one or two small items. The exact time required to sterilize a given volume depends on each autoclave, but a 30-min autoclave cycle is not an unreasonable length of time at 121°C to sterilize the contents of a 2-liter Erlenmeyer flask containing 1 liter of medium.

The concept that sterility was achieved with 121°C for 15 min originated with early studies of the spores of *Bacillus stearothermophilus*, which, at that time, was regarded as the most thermally resistant organism. This organism is still used as the basis for a number of commercially available autoclave bioindicators (e.g., Kilit [BBL]), which should be used to spot check the efficacy of one's autoclave or sterilization practices. However, many other bacterial spores with thermal resistance greater than those of *B. stearothermophilus* have been isolated in recent years, rendering the original definition more practical than absolute. Since most laboratories work with mesophilic organisms, the probability of being hampered by gross contamination with an extreme thermophile (which grows only at temperatures above 70°C) is not very great; thermophile laboratories, where the bioburden of heat-resistant spores is much greater, are, of course, aware that the nature of their organisms requires significantly longer sterilization times and higher temperatures.

As a final note regarding autoclaving: steam penetration of the item is necessary to achieve its sterilization; therefore, autoclaving is not satisfactory for the sterilization of nonaqueous solutions such as oils or high-molecular-weight hydrocarbons. Perkins' book, *Principles and Methods of Sterilization in Health Sciences* (50), gives detailed treatment of autoclaving as well as other sterilization methods.

7.4.2. Filtration

Heat-labile or volatile liquids should be sterilized by filtration; any liquid medium in which the complete (soluble) chemical integrity must be retained can also be sterilized by this method. The commercial availability of presterilized filtration systems (e.g., Nalge, Corning) in small to medium volumes (50 to 1,000 ml) makes this sterilization technique almost foolproof for small-scale laboratories, compared with the equipment and preparation times required for earlier filtration methods.

Membrane filters are preferred over depth filters for sterilization. The choice of membrane composition depends on its chemical compatibility with the solution to be sterilized. Those composed of mixtures of cellulose acetate or nitrate are generally suitable for the sterilization of most media and their components, whereas nylon or Teflon membranes should be used for aggressive solvents.

Membrane filters with pore sizes of 0.2 μm are regarded as sterilization grade although some small organisms (e.g., *Pseudomonas diminuta*, mycoplasmas, rickettsia, viruses) can pass through these filters. In recent years, 0.1-μm membranes have been developed, but they are not readily available in commercial presterilized filtration systems.

Since membrane filters have limited loading capacity (i.e., they quickly clog), solutions with precipitates or other debris as well as viscous or proteinaceous solutions such as sera should be prefiltered through nonsterile fibrous depth filters and higher-porosity (0.45-μm) membranes before being passed through the final, sterile 0.2-μm filter. The technical information published by major filtration manufacturers (e.g., Gelman, Millipore) will be of much assistance to individuals unfamiliar with this technology. Brock's monograph (6) gives practical as well as theoretical treatment of membrane filtration methods.

7.4.3. Tyndallization

The process of heating a solution to 100°C for 30 min on three consecutive days is called **tyndallization.** It is not truly a sterilization mechanism but a method of reducing the mesophilic and psychrophilic bioburden of a solution not sterilizable by other means. In bacteriology today, this technique is used principally to prepare solutions of powdered elemental sulfur for thiobacilli and other organisms that use the substance as an energy source. Since elemental sulfur is insoluble and melts into solid globules at higher temperatures, tyndallization is the preferred treatment method.

7.5. MEDIA FOR REPRESENTATIVE BACTERIA GROUPS

There is no one "best medium" for the cultivation of any specific organism, and a quick scan of the literature reveals numerous variations on the themes. The following section presents selected maintenance media suitable for the cultivation of many bacterial genera. The formulations are chosen to exemplify the wide diversity of general nutritional requirements represented in the bacterial world and to show the nutritional similarities that exist between many taxonomically distinct species. When applicable, comments are added to show how a medium may be adjusted to provide conditions suitable for other nutritionally distinct yet ecologically similar species or to shift an organism from one nutritional mode to another (e.g., autotrophism to heterotrophism). Commercially available dehydrated preparations are cited for some standard medium formulations; however, the generic, classical formulation is also provided for individuals who prefer the flexibility of preparing (and modifying) a medium according to their own needs.

A significant change in format has been adopted for the bacterial key and medium section presented in this book. In the *Manual of Methods for General Bacteriology (MMGB)*, media were frequently taken from journal articles describing the initial isolation and preliminary growth studies of the organisms listed in the bacterial key. However, current knowledge and practice regarding many of these bacteria indicate that they grow perfectly well on standard bacteriological media available in most laboratories. Therefore, approximately 200 diverse bacteria are listed here and reference is made to their ability to grow on 1 or more of 20 selected media. Also included as footnotes are some newer organisms such as *Aquaspirillum magnetotacticum, Borrelia* spp., and *Legionella* spp., since these organisms have unique medium requirements. Although the number of medium formulations is limited, specific instructions for the successful preparation of each formulation have been significantly expanded in an effort to assist individuals with minimal experience in medium preparation methods. For additional medium formulations and bacterial genera and species, the reader should consult Tables 1 through 13 in Chapter 7 of *MMGB* (22). It should be noted, however, that more recent studies may have resulted in some alterations in those listed media and culture conditions.

The bacterial key and media in this edition are derived principally from the ATCC *Catalogue of Bacteria and Phages,* 18th Ed. (18). In some instances, medium formulations have been adjusted to expand their nutritional base for use with a wider spectrum of organisms.

As a cautionary note, these formulations are designed to promote growth; they are not constructed to maintain specific genotypes or other physiological characteristics of any strain, unless noted.

Table 5 lists species of bacteria representing the major groups in *Bergey's Manual of Systematic Bacteriology.* Also given are the optimal growth temperature and a number code designating one or more of the listed medium formulations capable of supporting the growth of each species. When genera alone are cited, the suggested medium will support the growth of many species; however, some strains may require specific growth conditions or unusual media. Since most laboratories cannot afford the time or expense of stocking a different medium for each culture maintained, the list is limited to 20 media formulations.

Generally, dissolve medium ingredients one at a time in the total volume of water; adjust the pH with HCl or NaOH. Heat the medium to boiling, with stirring, to dissolve agar or any other components that do not dissolve at room temperature. Sterilize at 121°C for 15 min. Check the medium after autoclaving to ensure that the pH has not drifted during the sterilization process. Other specific techniques or procedures are cited with each formulation as needed.

Table 5. Key to bacteria, incubation temperature, and growth media

Bacterium	Optimal temp (°C)	Medium (text section)
Acetobacter spp.	26	7.5.13
Acholeplasma granularum	37	7.5.19[a]
Acholeplasma laidlawii	37	7.5.19[a]
"*Achromobacter hartlebii*"	26	7.5.1
"*Achromobacter viscosus*"	26	7.5.4
Acinetobacter calcoaceticus	26–37	7.5.1
Actinobacillus lignieresii	37	7.5.14
Acinetobacter lwoffii	26–37	7.5.1, 7.5.14
Actinomadura spp.	26	7.5.17
Actinomyces pyogenes	37	7.5.14
Actinomyces viscosus	37	7.5.1
Actinoplanes spp.	26	7.5.17
Aerococcus viridans	37	7.5.14, 7.5.20
Aeromonas spp.	20–30	7.5.1
Agrobacterium spp. (marine)	26	7.5.4
Agrobacterium spp. (nonmarine)	26	7.5.1
Alcaligenes faecalis	26–35	7.5.1
Alteromonas spp.	15–26	7.5.4
Amycolata spp.	26	7.5.17
Amycolatopsis spp.	26	7.5.17
Anabaena spp.	26	7.5.3
Aquaspirillum magnetotacticum	26	7.5.1[b]
Aquaspirillum spp.	26	7.5.1
Arthrobacter spp.	26–30	7.5.1
Asticcacaulis excentricus	26	7.5.15
Aureobacterium spp.	26	7.5.1
Azomonas spp.	26	7.5.5
Azospirillum spp.	26	7.5.5
Azotobacter spp.	26	7.5.5
Bacillus acidocaldarius	55	7.5.1[c]
Bacillus circulans	26–37	7.5.1, 7.5.14
Bacillus marinus	20	7.5.4
Bacillus psychrosaccharolyticus	20	7.5.1
Bacillus spp.	26–37	7.5.1
Bacillus stearothermophilus	55	7.5.1
Beijerinckia indica	26	7.5.5
Bifidobacterium spp.	37	7.5.2
Blastococcus aggregatus	26	7.5.18
Bordetella bronchiseptica	37	7.5.1
Borrelia spp.	37	—[d]
Brevibacterium spp.	30	7.5.1
Brucella spp.	37	7.5.1
Budvicia aquatica	26	7.5.1
Calothrix sp.	26	7.5.3
Caryophanon spp.	26	7.5.15
Caulobacter spp. (marine)	26	7.5.18
Caulobacter spp. (nonmarine)	26	7.5.15
Cedecea spp.	26–35	7.5.1
Cellulomonas spp.	26–30	7.5.1
Chaemisiphon sp.	26	7.5.3
Chlorobium limicola	26–30	7.5.9
Chromatium buderi	26–30	7.5.9
Chromatium vinosum	26–30	7.5.9
Chromobacterium spp.	26	7.5.1
Chryseomonas luteola	30	7.5.1
Citrobacter sp.	37	7.5.1
Clavibacter spp.	26	7.5.1
Clostridium spp.	30–37	7.5.2
Clostridium spp. (thermophiles)	50–60	7.5.2
Colwellia psychroerythrus	10	7.5.4
Corynebacterium diphtheriae	37	7.5.1, 7.5.14
Corynebacterium flavescens	26	7.5.1
Corynebacterium glutamicum	30	7.5.1
"*Corynebacterium tuberculstearicum*"	37	7.5.13
Curtobacterium spp.	26	7.5.1
Cytophaga spp. (marine strains)	26	7.5.4, 7.5.18[e]
Cytophaga spp. (nonmarine strains)	26	7.5.1, 7.5.18
Deinococcus spp.	26–30	7.5.1
Deleya spp.	26	7.5.4
Dermocarpa sp.	26	7.5.3
Derxia gummosa	26	7.5.13
Desulfobacter salexigens	26–30	7.5.11
Desulfovibrio baarsii	37	7.5.11
Desulfovibrio desulfuricans	30–37	7.5.11
Edwardsiella tarda	37	7.5.1
Enterobacter cloacae	37	7.5.1
Enterococcus faecalis	37	7.5.1, 7.5.14
Erwinia spp.	26–30	7.5.1
Erysipelothrix spp.	37	7.5.14
Escherichia coli	37	7.5.1
Eubacterium cellulosolvens	37	7.5.2
Eubacterium combessi	37	7.5.2
Eubacterium moniliforme	37	7.5.2
Fischerella sp.	26–30	7.5.3
Flavobacterium spp. (marine strains)	26	7.5.4, 7.5.18[e]
Flavobacterium spp. (nonmarine strains)	26	7.5.1
Flexibacter spp.	26	7.5.4, 7.5.18[e]
Flexithrix dorotheae	26	7.5.4, 7.5.18[e]
Francisella novicida	37	7.5.14
Frateuria aurantia	26	7.5.13
Gemella haemolysans	37	7.5.14
"*Gloeobacter violaceus*"	26	7.5.3
Gloeocapsa sp.	26	7.5.3
Gloeothece sp.	26	7.5.3
Gluconobacter oxydans	26	7.5.13
Haemophilus influenzae	37	7.5.14
Hafnia alvei	37	7.5.1
Haloarcula vallismortis	37	7.5.6
Halobacterium cutirubrum	37	7.5.6
Halobacterium saccharovorum	37	7.5.6
Halobacterium salinarium	37	7.5.6
Halomonas halodurans	26	7.5.4
Heliobacterium chlorum	26	7.5.10
Herpetosiphon spp.	26	7.5.18
Hyphomicrobium indicum	26–30	7.5.4
Hyphomonas spp.	26–30	7.5.4
Instrasporangium calvum	26	7.5.17
Janthinobacterium livifum	26	7.5.1
Kingella denitrificans	37	7.5.14
Klebsiella spp.	30–37	7.5.1
Kurthia zopfii	26	7.5.1
Lactobacillus spp.	30–37	7.5.20
Lactococcus lactis	37	7.5.14, 7.5.20
Lampropedia hyalina	26	7.5.15
Leclercia adecarboxylata	37	7.5.1
Legionella spp.	37	—[a,f]
Leuconostoc mesenteroides	26	7.5.20
Leuconostoc dextranicum	26	7.5.20
Listeria monocytogenes	37	7.5.14
Lysobacter spp.	30	7.5.1
Methanococcus vannielii	30–37	7.5.12
Methanosarcina acetovorans	37	7.5.12
Methanosarcina barkeri	30	7.5.12

(Continued)

Table 5. *(Continued)*

Bacterium	Optimal temp (°C)	Medium (text section)	Bacterium	Optimal temp (°C)	Medium (text section)
Microbacterium spp.	30–37	7.5.1	*Rhodobacter* spp.	26	7.5.10[g]
Microbispora spp.	26–37	7.5.17	*Rhodococcus equi*	37	7.5.1
Micrococcus spp. (marine strains)	26–30	7.5.4	*Rhodococcus rhodochrous*	26–37	7.5.1, 7.5.17
Micrococcus spp. (nonmarine strains)	26–30	7.5.1, 7.5.14	*Rhodocyclus gelatinosus*	26	7.5.10
Micromonospora spp.	26	7.5.17	*Rhodomicrobium vannielii*	26	7.5.10
Microscilla spp.	26	7.5.4, 7.5.18[e]	*Rhodopseudomonas palustris*	26	7.5.10
Microtetraspora viridis	26	7.5.17	*Rhodospirillum rubrum*	26	7.5.10
Moraxella spp.	37	7.5.1, 7.5.14	*Saccarothrix aerocolonigenes*	26	7.5.17
Morganella morganii	37	7.5.1	*Salmonella* spp.	37	7.5.1
Mycobacterium phlei	37	7.5.1	*Saprospira grandis*	26	7.5.18
Mycobacterium spp.	37	7.5.8	*Serratia* spp.	26–30	7.5.1
Mycoplana spp.	30	7.5.1	*Shigella boydii*	37	7.5.1
Mycoplasma bovis	37	7.5.19[a]	*Shigella dysenteriae*	37	7.5.14
Mycoplasma felis	37	7.5.19[a]	*Sphingobacterium multivorum*	26	7.5.1
Mycoplasma hominis	37	7.5.19[a]	*Sporocytophaga myxococcoides*	26	7.5.18
Mycoplasma orale	37	7.5.19[a]	*Sporosarcina halophila*	30–35	7.5.4
Neisseria spp.	37	7.5.14	*Sporosarcina ureae*	26	7.5.1
Nitrobacter spp.	26	7.5.7	*Staphylococcus* spp.	37	7.5.1
Nitrococcus mobilis	26	7.5.7	*Streptococcus* spp.	37	7.5.14
Nitrosococcus oceanus	26	7.5.7	*Streptomyces* spp.	26	7.5.17
Nitrosolobus multiformis	26	7.5.7	*Streptoverticillium* spp.	26	7.5.17
Nocardia asteroides	26–30	7.5.1	*Synechococcus* sp.	26	7.5.3
Nocardia spp.	26–30	7.5.17	*Synechocystis* sp.	26	7.5.3
Nocardiopsis spp.	26	7.5.17	*Thermoactinomyces candidus*	40–50	7.5.1
Nostoc spp.	26	7.5.3	*Thermoactinomyces* spp.	40–50	7.5.17
Oceanospirillum spp.	26	7.5.4	*Thiobacillus neopolitanus*	30	7.5.16
Oerskovia turbata	26–37	7.5.1	*Thiobacillus thiooxidans*	26	7.5.16
Oscillatoria sp.	26	7.5.3	*Thiobacillus thioparus*	30	7.5.16
Pasteurella spp.	37	7.5.14	*Veillonella* spp.	37	7.5.2[h]
Pediococcus spp.	26–30	7.5.20	*Vibrio* spp.	15–30	7.5.4
Pedomicrobium ferrugineum	26	7.5.15	*Vitreoscilla stercoraria*	30	7.5.15
Photobacterium spp.	20–26	7.5.4	*Weeksella virosa*	37	7.5.1
Pirellula staleyi	26	7.5.4	*Xanthomonas campestris*	26	7.5.1
Proteus spp.	37	7.5.1	*Yersinia* spp.	26–30	7.5.1
Providencia spp.	37	7.5.1	*Zoogloea ramigera*	26	7.5.1
Pseudomonas spp. (marine strains)	20–30	7.5.4			
Pseudomonas spp. (nonmarine strains)	26–37	7.5.1			

[a]These strains require 5% CO_2 for optimal growth.

[b]With loss of magnetite, for retention of magnetic properties cultivate microaerophilically in containers filled to capacity with a mineral medium containing the following (per liter): $NaNO_3$, 5g; KH_2PO_4, 0.68 g; modified Balch's minerals (Table 1B), 10 ml; vitamins (Table 2), 10 ml; filter-sterilized sodium pyruvate, 1.0 g; sodium succinate, 1.0 g; sodium acetate, 0.05 g; sodium ascorbate, 35 mg; 0.01 M ferric quinate, 2 ml. Adjust medium to a final pH of 6.8.

[c]Adjust medium after sterilization to pH 4.

[d]Formulation per liter: CMRL 1066 (10×) (without glutamine, without bicarbonate) (no. 330–1540; GIBCO), 100 ml; Neopeptone (no. 0119; Difco), 5.0 g; bovine albumin (fraction V), 50.0 g; HEPES, 6.0 g; sodium citrate, 0.7 g; glucose, 5.0 g; sodium pyruvate, 0.8 g; *N*-acetylglucosamine, 0.4 g; sodium bicarbonate, 2.2 g; distilled water, 900.0 ml. Adjust to pH 7.6. Add 200 ml of 7% gelatin (aqueous), and filter sterilize. Add sterile rabbit serum to 6% (final concentration). Some lots of bovine albumin (fraction V) do not support optimal growth of *Borrelia* species. The medium has a short shelf life; use within 1 month of preparation.

[e]Freeze-dried marine cultures should be revived in medium from section 7.5.18; strongly growing cultures can be maintained in medium from section 7.5.4.

[f]Formulation per liter: ACES, 10.0 g; yeast extract, 10.0 g; charcoal (no. C5510; Sigma), 2.0 g; L-cysteine HCl · H_2O, 0.4 g; soluble ferric PP_i, 0.25 g; agar (for solid medium), 15.0 g; distilled water, 1.0 liter. Add all the ingredients except cysteine and ferric PP_i. Adjust to pH 6.9 with KOH, and heat to boiling to dissolve the agar. Sterilize at 121°C for 15 min, and cool to −50°C. Separately filter sterilize the cysteine and ferric PP_i in 10 ml of distilled water. Add cysteine and then ferric PP_i to the medium. Check pH, and readjust if necessary with sterile KOH. Swirl medium while dispensing to maintain charcoal in suspension.

[g]Add 3.0% NaCl to the medium for marine strains.

[h]Add 1.5% sodium lactate, 0.1% Tween 80, and 3 mg of putrescine per ml to the medium from section 7.5.2. Adjust to a final pH of 7.4 under gas phase with sodium bicarbonate.

7.5.1. Nutritionally Generalized Aerobic Heterotrophs

Commercial media. Tryptic Soy Agar/Broth (Difco); Trypticase Soy Agar/Broth (BBL); Tryptone Soya Agar/Broth (Oxoid).

Generic formulation (per liter). *For liquid medium:* pancreatic digest of casein, 17 g; papaic digest of soybean meal, 3.0 g; sodium chloride, 5.0 g; dipotassium phosphate, 2.5 g; glucose, 2.5 g. *For solid medium:* reduce casein digest to 15 g; omit glucose; add 5 g of agar. Final pH, 7.3 ± 0.2. Retain glucose in the solid medium for organisms requiring carbohydrates as energy sources (e.g., *Bacillus psychrosaccharolyticus*).

7.5.2. Nutritionally Generalized Anaerobic Heterotrophs

Commercial media. Reinforced Clostridial Agar (BBL; Oxoid); Reinforced Clostridial Medium/Agar (Difco).

Generic formulation (per liter). Yeast extract, 3 g; beef extract, 10 g; pancreatic digest of casein, 10 g; glucose, 5 g; sodium chloride, 5 g; sodium acetate, 3 g; soluble starch, 1 g; L-cysteine HCl, 0.5 g; agar, 15.0 g. For Reinforced Clostridial Medium (Difco), reduce agar to 0.5 g. Final pH, 6.8 ± 0.2.

If Reinforced Clostridial Medium (Difco) is prepared anaerobically by using Hungate methods (see Chapter 6) under an atmosphere of 80% N_2, 10% H_2, and 10% CO_2 and buffered with 2.75 g of sodium bicarbonate per liter to pH 7.0 under the same gas phase, the resulting preparation is satisfactory for a number of clinical anaerobes in addition to the clostridia.

7.5.3. Cyanobacteria

$NaNO_3$	1.5 g
K_2HPO_4	40.0 mg
$MgSO_4 \cdot 7H_2O$	75.0 mg
$CaCl_2 \cdot H_2O$	36.0 mg
Citric acid	6.0 mg
Ferric ammonium citrate	6.0 mg
EDTA (disodium salt)	1.0 mg
Na_2CO_3	20.0 mg
Trace metals*	1.0 ml
Purified agar (for solid medium)	10.0 g
Distilled water	1.0 liter

**Trace metals:* H_3BO_3, 2.86 g; $MnCl_2 \cdot 6H_2O$, 1.81 g; $ZnSO_4 \cdot 7H_2O$, 222 mg; $Na_2MoO_4 \cdot 2H_2O$, 39 mg; $CuSO_4 \cdot 5H_2O$, 79 mg; $Co(NO_3)_2 \cdot 6H_2O$, 49.4 mg; distilled water, 1.0 liter.

Adjust medium to a final pH of 7.1.

Cultivate cyanobacteria under fluorescent light of 2,000 to 3,000 lux.

For marine strains: add 1.0 μg of vitamin B_{12} per liter and 10.0 g of NaCl per liter; raise final pH to 8.0.

For nitrogen-fixing strains: omit $NaNO_3$.

7.5.4. Marine Bacteria

Commercial medium. Marine Agar/Broth 2216 (Difco).

Classical formulation (per liter). Bacto Peptone, 5 g; Bacto Yeast Extract, 1.0 g; ferric citrate, 0.1 g; sodium chloride, 19.45 g; magnesium chloride, 8.8 g; sodium sulfate, 3.24 g; calcium chloride, 1.8 g; potassium chloride, 55 mg; sodium bicarbonate, 16 mg; potassium bromide, 80 mg; strontium chloride, 34 mg; boric acid, 22 mg; sodium silicate, 4 mg; sodium fluoride, 2.4 mg; ammonium nitrate 1.6 mg; disodium phosphate, 8 mg; Bacto Agar (for solid medium), 15 g. Final pH, 7.6 ± 0.2.

For luminescent strains: add 1.0 g of calcium carbonate per liter and 3.0 ml of glycerol per liter.

7.5.5. N_2-Fixing Bacteria

KH_2PO_4	0.4 g
K_2HPO_4	0.1 g
$MgSO_4 \cdot 7H_2O$	0.2 g
NaCl	0.1 g
$FeCl_3$	10.0 mg
$Na_2MoO_4 \cdot 2H_2O$	2.0 mg
Sodium malate*	5.0 g
Yeast extract	50.0 mg
Distilled water	1.0 liter

*Or other utilized carbon source (see below).

Adjust medium to final pH of 7.2 ± 0.1.

Medium may be supplemented with 1 to 2 g of NH_4Cl per liter or 5 g of peptone per liter for purposes other than nitrogen fixation. Certain isolates may require the addition of a complete vitamin mixture (Table 2) or supplementation with soil extract (see below) for sustained growth.

Soil extract

Humus-enriched garden soil, air dried	80.0 g
Na_2CO_3	0.2 g
Distilled water	200.0 ml

Autoclave the soil extract at 121°C for 60 minutes. Allow the sediment to settle, and centrifuge the supernatant. Any excess extract can be sterilized by autoclaving and stored at room temperature for future use. Supplement media with 10 to 20% soil extract.

For Azomonas agilis: add 10% soil extract, replace malate with 2% glucose; adjust medium to a final pH of 6.0.

For Azomonas insignis or Azotobacter beijerinckii: replace malate with 2% sucrose.

For Azomonas macrocytogenes: add 10% soil extract; replace malate with 2% mannitol; adjust medium to a final pH of 6.0.

For Azotobacter chroococcum and A. vinelandii: add 10% soil extract; replace malate with 2% mannitol; adjust medium to a final pH of 7.6.

For Beijerinckia indica: replace malate with 2% glucose; adjust to a final pH of 6.0.

For Derxia gummosa: replace malate with 2% glucose.

For Azospirillum spp.: use medium without changes.

7.5.6. Extreme Halophiles

NaCl	250.0 g
$MgSO_4 \cdot 7H_2O$	10.0 g

KCl	5.0 g
$CaCl_2 \cdot 2H_2O$	13.0 mg
Yeast extract	10.0 g
Tryptone	2.5 g
Agar (for solid medium)	20.0 g
Distilled water	to 950.0 ml

To avoid precipitation of liquid medium during autoclaving, sterilize without pH adjustment. Set the final pH of medium aseptically *after* autoclaving with ~50 ml sterile 1 M HEPES–NaOH at pH 7.4 to approximately 7.4 ± 0.2.

For denitrifiers (e.g., Haloferax denitrificans, Haloarcola vallismortis): add 3.0 g of KNO_3 per liter to stimulate growth in large volumes of broth or for anaerobic conditions.
For alkalophiles (e.g., Halobacterium saccharovorum): aseptically adjust to a final pH of 7.8 after autoclaving.

Agar media with high salt concentrations gel very rapidly at normal pouring temperatures (~50°C). Best manipulate them on a hot plate or pour them soon after removal from the autoclave.

Most general-purpose combination pH electrodes do not give precisely accurate readings against elevated sodium chloride concentrations; electrodes specific for sodium ions are available. If using general-purpose electrodes for pH adjustments with halophilic media, allow 2 to 3 min for equilibration in the halophilic medium to establish a reading; allow the electrode to reequilibrate, and adjust the calibration, if required, in standard buffer solutions for at least 30 min prior to use for other pH measurements.

7.5.7. NH_4- and NO_2-Oxidizing Bacteria

$(NH_4)_2SO_4$	1.32 g
$MgSO \cdot 7H_2O$	0.38 g
$CaCl_2 \cdot 2H_2O$	20.00 mg
$MnCl_2 \cdot 4H_2O$	200.00 μg
$Na_2MoO_4 \cdot 2H_2O$	100.00 μg
$CoCl_2 \cdot 6H_2O$	2.00 μg
$ZnSO_4 \cdot 7H_2O$	100.00 μg
K_2HPO_4	87.00 mg
Phenol red, 0.5%	0.25 ml
EDTA, ferric salt	1.00 mg
Distilled water	1.00 liter

Adjust the medium with 0.5 M K_2CO_3 for a final pH of 7.5.

Maintain the pH during cultivation by countering any acid shift, noted by a change in the color of the indicator, with additional drops of sterile potassium bicarbonate to the medium.

For Nitrosolobus spp.: use above formulation.
For Nitrosococcus spp.: use 700 ml of seawater and 300 ml of distilled water in preparing the medium.
For Nitrosomonas spp.: replace phenol red with 125 μg of cresol red per liter, and maintain at pH 7.5 to 8.0.
For Nitrobacter spp. (freshwater strains): replace ammonium sulfate with 0.22 g of $NaNO_2$ per liter; omit phenol red.
For Nitrobacter spp. (marine strains): replace ammonium sulfate with 0.22 g of $NaNO_2$ per liter; omit phenol red; prepare with 700 ml of seawater and 300 ml of distilled water.

7.5.8. Mycobacteria

Commercial media. Middlebrook 7H9 Broth with ADC Enrichment (BBL; Difco) and Middlebrook 7H10 Agar with OADC Enrichment (BBL; Difco).

Generic formulation (per liter): *Basal medium:* ammonium sulfate, 0.5 g; glycerol (omit for glycerophobic strains), 2.0 ml; monopotassium phosphate, 1.0 g; dipotassium phosphate, 2.5 g; sodium citrate, 0.1 g; magnesium sulfate, 50 mg; calcium chloride, 0.5 mg; zinc sulfate, 1 mg; copper sulfate, 1 mg; L-glutamic acid, 0.5 g; ferric ammonium citrate, 40 mg; pyridoxine, 1 mg; biotin, 0.5 mg; malachite green (for agar only), 0.25 mg; agar (if needed), 15 g. pH 6.6 ± 0.2. *ADC enrichment:* Bovine albumin Fraction V, 5 g; glucose, 2 g; beef catalase, 3 mg; distilled water, 100 ml. *OADC enrichment:* ADC enrichment plus oleic acid, 50 mg; sodium chloride, 0.85 g. Filter sterilize enrichments, and add aseptically to sterile basal medium.

For Mycobacterium paratuberculosis: add 0.5 g of Tween 80 per liter and 2 mg of mycobactin J per liter.
Note: Newly isolated zoopathic strains may require egg-based media.

7.5.9. Anaerobic Phototrophic Sulfur Bacteria

Solution 1

$CaCl_2 \cdot 2H_2O$	1.06 g
Distilled water	1.0 liter

Distribute 80-ml quantities into 130-ml screw-cap bottles, and sterilize by autoclaving at 121°C for 15 min. Cool to room temperature.

Solution 2

$NaHCO_3$	4.5 g
Distilled water	900.0 ml

Solution 3

Heavy-metal solution (see below)	50.0 ml
Vitamin B_{12} (2 mg/100 ml of distilled water)	3.0 ml
Vitamin solution (see below)	15.0 ml
KH_2PO_4	1.0 g
KCl	1.0 g
NH_4Cl	0.8 g
$MgCl_2 \cdot 6H_2O$	0.8 g
Distilled water	32.0 ml

Sterilize solutions 2 and 3 through a 0.2-μm-pore-size membrane filter, and equilibrate by bubbling with 100% CO_2 gas for 30 min to establish the buffering system. Then add solution 4.

Solution 4

$Na_2S \cdot 9H_2O$	1.8 g
Distilled water	30.0 ml

Prepare solution 4 just prior to adding to solutions 2

and 3. Quickly neutralize solution 4 to approximately pH 7.8 with concentrated H_2SO_4, and sterilize through a 0.2-μm membrane filter. Immediately add to solutions 2 and 3 under CO_2 gas.

Heavy-metal solution

EDTA, sodium salt	1.5 g
Trace element solution (see below)	6.0 ml
$FeSO_4 \cdot 7H_2O$	0.2 g
$ZnSO_4 \cdot 7H_2O$	0.1 g
$MnCl_2 \cdot 4H_2O$	20.0 mg
Distilled water	1.0 liter

Trace-element solution

$AlCl_3$	1.0 g
KI	1.0 g
KBr	0.5 g
LiCl	0.5 g
$MnCl_2 \cdot 4H_2O$	7.0 g
H_3BO_3	11.0 g
$ZnCl_2$	1.0 g
$CuCl_2$	1.0 g
$NiCl_2$	1.0 g
$CoCl_2$	1.0 g
$SnCl_2 \cdot 2H_2O$	0.5 g
$BaCl_2$	0.5 g
$Na_2MoO_4 \cdot 2H_2O$	0.5 g
$NaVO_3 \cdot H_2O$	0.1 g
Sodium selenite	0.5 g
Distilled water	to 3.6 liters

Dissolve each salt separately in distilled water, and combine into one solution. Adjust to ~pH 3.5 with concentrated H_2SO_4, and bring to total volume with additional distilled water. Mix thoroughly before use.

Vitamin solution

Biotin	0.2 mg
Nicotinic acid	2.0 mg
Thiamine HCl	1.0 mg
p-Aminobenzoic acid	0.1 mg
Pantothenic acid, calcium salt	0.5 mg
Pyridoxamine 2HCl	5.0 mg
Distilled water	100.0 ml

Complete medium. Combine solutions 2, 3, and 4 under the 100% carbon dioxide gas phase. Dispense this solution into the bottles containing the sterile calcium chloride solution (solution 1), filling the bottles to full capacity. Hold the medium for 24 h before inoculation, to allow the fine precipitate to form and settle. Final pH, 6.8.

Cultivate strains anaerobically under tungsten light.

For marine strains: add sodium chloride to a final concentration of 2.5% in completed medium.

7.5.10. Anaerobic Phototrophic Nonsulfur Bacteria

Yeast extract	10.0 g
Ho-Le Trace Elements (Table 1A)	1.0 ml
Distilled water	1.0 liter

Adjust medium, after autoclaving, to pH 7.1 with sterile 5% K_2HPO_4.

Cultivate strains anaerobically under tungsten light.

7.5.11. Sulfate-Reducing Bacteria

Solution A

KH_2PO_4	0.2 g
NH_4Cl	0.3 g
NaCl	1.0 g
Na_2SO_4	3.0 g
$CaCl_2 \cdot 2H_2O$	150.0 mg
$MgCl_2 \cdot 6H_2O$	0.4 g
KCl	0.5 g
Resazurin	1.0 mg
Distilled water	870.0 ml

Solution B

Modified Balch's trace elements (Table 1B)	10.0 ml

Solution C

$NaHCO_3$	5.0 g
Distilled water	100.0 ml

Solution D

Sodium lactate	3.5 g
Distilled water	to 10.0 ml

Solution E

Vitamins (Table 2)	10.0 ml

Solution F

$Na_2S \cdot 9H_2O$	0.4 g
Distilled water	10.0 ml

Boil solution A for a few minutes to purge oxygen from the medium. Cool under an 80% N_2–20% CO_2 gas mixture until the pH drops below 6 to prevent any oxygen from reentering the medium and to establish the gaseous phase of the buffer. Autoclave anaerobically under the same gas phase and cool. Autoclave solutions B, D, E, and F separately under 100% nitrogen. Filter sterilize solution C, and purge under 100% carbon dioxide to remove dissolved oxygen. Add solutions B through F to sterile cooled solution A in the sequence listed. Dispense aseptically and anaerobically into sterile containers under 80% N_2–20% CO_2. Final pH, 7.1 to 7.4.

For marine strains (e.g., Desulfovibrio salexigens): add 25 g of sodium chloride per liter.

For fatty acid-requiring strains (e.g., Desulfovibrio baarsii): add 25 g of sodium chloride per liter; replace sodium lactate with 0.7 g of sodium butyrate per liter, 0.3 g of sodium caproate per liter, and 0.15 g of sodium octanoate per liter; adjust to a final pH of 7.7.

7.5.12. Methanogens (2)

NaOH	4.0 g
Yeast extract	2.0 g
Trypticase peptone (BBL)	2.0 g
Mercaptoethanesulfonic acid	0.5 g
NH_4Cl	1.0 g
$K_2HPO_4 \cdot 3H_2O$	0.4 g

$MgCL_2 \cdot 6H_2O$	1.0 g
$CaCl_2 \cdot 2H_2O$	0.4 g
Modified Balch's trace elements (Table 1B)	10.0 ml
Resazurin	1.0 mg
$Na_2S \cdot 9H_2O$	0.25 g
Energy substrates	as noted below*
Distilled water	to 1.0 liter

Dissolve NaOH, and equilibrate under 70% N_2–30% CO_2. Add all components except sodium sulfide and energy substrates. Using anaerobic methods, dispense medium into culture vessels containing the 70% N_2–30% CO_2 atmosphere. Add sulfide and energy substrates to cooled, autoclaved basal medium from anoxic stock solutions. Final pH, 7.2.

**For Methanosarcina barkeri:* add 40 mM trimethylamine HCl.
For Methanosarcina acetivorans: add 50 mM sodium acetate.
For Methanococcus vannielii: add 1.8% sodium chloride and 0.1% sodium acetate; cultivate with 80% H_2–20% CO_2 at 2 atm.

7.5.13. Acetobacteria

Yeast extract	5.0 g
Peptone	3.0 g
Mannitol	25.0 g
Agar	15.0 g
Distilled water	1.0 liter

7.5.14. Blood-Utilizing Bacteria

Trypticase Soy Agar (BBL)*	40.0 g
Distilled water	950.0 ml

*See section 7.5.1.

Heat to boiling with stirring to dissolve ingredients. Sterilize at 121°C for 15 min.
Cool to 50°C and add aseptically:
Sterile defibrinated sheep blood .... 50.0 ml

For Haemophilus and Neisseria species: "chocolatize" above medium by adding blood to sterile, liquefied medium and by heating to 70°C for 10 min before pouring; both genera may also require 10 ml of IsoVitaleX Enrichment (BBL) per ml added to medium after chocolatizing.

7.5.15. Budding and Appendaged Bacteria

Peptone	2.0 g
Yeast extract	1.0 g
Ho-Le Trace Elements (Table 1A)	1.0 ml
$CaCl_2 \cdot 2H_2O$	1.5 g
Sodium acetate	1.0 g
Agar	15.0 g
Distilled water	1.0 liter

Adjust medium to final pH of 7.4.
Trace elements and calcium chloride may be replaced by 50 ml of soil extract (as in section 7.5.5).

7.5.16. Sulfur Oxidizers

NH_4Cl	0.4 g
KH_2PO_4	4.0 g
K_2HPO_4	4.0 g
$MgSO_4 \cdot 7H_2O$	0.8 g
$CaCl_2$	0.03 g
$FeCl_3 \cdot 6H_2O$	0.02 g
$MnSO_4 \cdot H_2O$	0.02 g
$Na_2S_2O_3$	5.0 g
Distilled water	1.0 liter

Sterilize phosphates separately to retard precipitation.

For Thiobacillus thiooxidans: replace sodium thiosulfate with 1 g of elemental sulfur per 100-ml aliquot of medium; delete dibasic potassium phosphate and reduce monobasic potassium phosphate to 2 g; adjust to pH 4.5, and sterilize by tyndallization (section 7.4.3).

7.5.17. Actinomycetes

Yeast extract	4.0 g
Malt extract	10.0 g
Glucose	4.0 g
Agar	20.0 g
Distilled water	1.0 liter

Adjust to a final pH of 7.3.

7.5.18. Gliding Bacteria

Trypticase (BBL)	5.0 g
KNO_3	0.5 g
Glucose (filter sterilize)	2.0 g
Vitamins (Table 2), filter sterilized	10.0 ml
HEPES	4.0 g
Sodium α-glycerophosphate	0.1 g
$MgSO_4 \cdot 7H_2O$	0.1 g
$CaCl_2$	0.1 g
Ho-Le Trace Elements (Table 1A)	1.0 g
Agar (for solid medium)	15.0 g
Distilled water	to 1.0 liter

Adjust to final pH of 7.4.

For marine strains: omit magnesium and calcium salts and trace elements; prepare medium in seawater (artificial or natural) (see marine salt content in section 7.5.4 or seawater formulation in Chapter 24).
For cellulolytic strains: omit glucose; add strips of Whatman no. 1 filter paper to tubes of broth or overlay on solidified agar.

7.5.19. Mycoplasmas

Heart infusion broth (no. 0038; Difco)	700.0 ml
Heat-inactivated GG-Free horse serum (no. 210-6270; GIBCO)	200.0 ml
Fresh yeast extract solution (no. 360-8180; GIBCO)	100.0 ml

Sterilize the heart infusion broth by autoclaving at 121°C for 15 min. Filter sterilize the horse serum and yeast extract; add aseptically to cooled basal medium. For solid medium, add 10 g of Agar Noble (Difco) to autoclavable fraction. Final pH, 7.4.

7.5.20. Lactobacilli

Commercial medium. Lactobacilli MRS Broth (no. 0881 Difco).

Generic formulation (per liter). Bacto Proteose Peptone no. 3 (Difco), 10 g; beef extract, 10 g; yeast extract, 5 g; glucose, 20 g; Tween 80, 1 g; ammonium citrate, 2 g; sodium acetate, 5 g; magnesium sulfate, 0.1 g; manganese sulfate, 0.05 g; disodium phosphate, 2 g. Final pH, 6.5 ± 0.2.

7.6. COMMERCIAL SOURCES

7.6.1. Dehydrated Media Products

*BBL Products, Becton Dickinson Microbiology Systems, P.O. Box 243, Cockeysville, MD 21030
Telephone: 1-800-638-8663
*Difco Laboratories, P.O. Box 331058, Detroit, MI 48232
Telephone: 1-800-521-0851
Oxoid Products, 217 Colanade Road, Nepean, Ontario K2E 7K3, Canada
Telephone: (613) 226-1318
*Sheffield Products, Humko-Sheffield Chemical, P.O. Box 630, Norwich, NY 13815

7.6.2. Chemicals and Reagents

GIBCO/BRL Life Technologies, Inc., P.O. Box 68, Grand Island, NY
Telephone: 1-800-828-6686
Sigma Chemical Company, P.O. Box 14508, St. Louis, MO 63178
Telephone: 1-800-325-3010

7.6.3. Membrane Filtration Products

*Corning Incorporated, Science Products Division, Corning, NY 14831
Telephone: 1-800-222-7740
*Gelman Sciences, Inc., 600 South Wagner Road, Ann Arbor, MI 48106
Telephone: 1-800-521-1520
Millipore Corporation, 448 Grandview Drive, South San Francisco, CA 94080
Telephone: 1-800-632-2708
*Nalge Company, a subsidiary of Sybron Corporation, 75 Panorama Creek Drive, Box 20365, Rochester, NY 14602

*BBL, Difco, Corning, Gelman, and Nalge products are distributed through various scientific supply companies. Sheffield products are available from Sigma Chemical Co.

7.7. REFERENCES

1. **Atlas, R. M.** 1993. *Handbook of Microbiological Media.* CRC Press, Inc., Boca Raton, Fla.

1a. **Balch, W. E., G. E. Fox, L. J. Magrum, C. R. Woese, and R. S. Wolfe.** 1979. Methanogens: reevaluation of a unique biological group. *Microbiol. Rev.* **43:**260–296.

2. **Boone, D. R.** 1990. *Catalog of Strains of the Oregon Collection of Methanogens.* Oregon Graduate Institute of Science & Technology, Beaverton, Oreg.

3. **Boone, D. R., and M. P. Bryant.** 1980. Propionate-degrading bacterium, *Syntrophobacter wolinii* sp. nov. gen. nov., from methanogenic ecosystems. *Appl. Environ. Microbiol.* **40:**626–632.

4. **Bridson, E. Y.** 1978. Natural and synthetic culture media for bacteria, p. 91–281. *In* M. Rechcigl, Jr. (ed.), *CRC Handbook Series in Nutrition and Food, Section G. Diets, Culture Media, Food Supplements,* vol. III. CRC Press, Inc., Cleveland, Ohio.

5. **Bridson, E. Y., and A. Brecker.** 1970. Design and formulation of microbial culture media, p. 229–293. *In* J. R. Norris and D. W. Ribbons (ed.), *Methods in Microbiology,* vol. 3A. Academic Press, Inc., New York.

6. **Brock, T. D.** 1983. *Membrane Filtration: A User's Guide and Reference Manual.* Science Tech., Inc., Madison, Wis.

7. **Budavari, S., M. J. O'Neil, A. Smith, and P. E. Heckelman (ed.).** 1989. *The Merck Index.* Merck & Co., Inc., Rahway, N.J.

8. **Caldwell, D. R., and M. P. Bryant.** 1966. Medium without rumen fluid for nonselective enumeration and isolation of rumen bacteria. *Appl. Microbiol.* **14:**794–801.

9. **Catlin, W.** 1973. Nutritional profiles of *Neisseria gonorrhoeae, Neisseria meningitidis,* and *Neisseria lactamica,* in chemically defined media and the use of growth requirements for gonococcal typing. *J. Infect. Dis.* **128:**178–194.

10. **Clark, W. M.** 1928. *The Determination of Hydrogen Ions.* The Williams & Wilkins Co., Baltimore.

11. **Cote, R. J., and R. L. Gherna.** Unpublished data.

12. **Davis, B. D., R. Dulbecco, H. N. Eisen, H. S. Ginsberg, and W. B. Wood, Jr.** 1973. *Microbiology,* 2nd ed. Harper & Row, Publishers, Inc., Hagerstown, Md.

13. **Difco Laboratories, Inc.** 1984. *Difco Manual,* 10th ed. Difco Laboratories, Inc., Detroit.

14. **Ellefson, W. L., W. B. Whitman, and R. S. Wolfe.** 1982. Nickel-containing factor F_{430}: chromophore of the methylreductase of *Methanobacterium. Proc. Natl. Acad. Sci. USA* **79:**3707–3710.

15. **Field, M. F., and H. C. Lichstein.** 1957. Factors affecting the growth of propionibacteria. *J. Bacteriol.* **73:**96–99.

16. **Finkelstein, R. A., and C. E. Lankford.** 1957. A bacteriotoxic substance in autoclaved culture media containing glucose and phosphate. *Appl. Microbiol.* **5:**74–79.

17. **Fulmer, E. I., A. C. Williams, and C. H. Werkman.** 1931. The effect of sterilization of media upon their growth promoting properties toward bacteria. *J. Bacteriol.* **21:**299–303.

18. **Gherna, R., P. Pienta, and R. Cote (ed.).** 1992. *American Type Culture Collection Catalogue of Bacteria and Phages,* 18th ed. American Type Culture Collection, Rockville, Md.

19. **Gibbons, R. J., and J. B. MacDonald.** 1960. Hemin and vitamin K compounds as required factors for the cultivation of certain strains of *Bacteroides melanogenicus. J. Bacteriol.* **80:**164–170.

20. **Good, N. E., G. D. Winget, W. Winter, T. N. Connolly, S. Izawa, and R. M. M. Singh.** 1966. Hydrogen ion buffer for biological research. *Biochemistry* **5:**467–477.

21. **Guirard, B. M., and E. E. Snell.** 1962. Nutritional requirements of microorganisms, p. 33–93. *In* I. C. Gunsalus and R. Y. Stanier (ed.), *The Bacteria,* vol. 4. Academic Press, Inc., New York.

22. **Guirard, B. M., and E. E. Snell.** 1981. Biochemical factors in growth, p. 79–111. *In* P. Gerhardt, R. G. E. Murray, R. N. Costilow, E. W. Nester, W. A. Woods, N. R. Krieg, and G. B. Phillips (ed.), *Manual of Methods for General Bacteriology.* American Society for Microbiology, Washington, D.C.

23. **Herbst, E. J., and E. E. Snell.** 1948. Putrescine as a growth factor for *Haemophilus parainfluenzae. J. Biol. Chem.* **176:**989–990.
24. **Hewitt, W., and S. Vincent.** 1988. *Theory and Application of Microbiological Assay.* Academic Press, Inc., San Diego, Calif.
25. **Holt, L. B.** 1962. The growth-factor requirements of *Haemophilus influenzae. J. Gen. Microbiol.* **27:**317–322.
26. **Hughes, M. N., and R. K. Poole.** 1991. Metal speciation and microbial growth—the hard (and soft) facts. *J. Gen. Microbiol.* **137:**725–734.
27. **Hutner, S. H., L. Provasoli, A. Schatz, and C. P. Haskins.** 1950. Some approaches to the study of the role of metals in the metabolism of microorganisms. *Proc. Am. Philos. Soc.* **94:**152–170.
28. **Jones, J. B., and T. C. Stadtman.** 1977. *Methanococcus vannieliii:* culture and effects of selenium and tungsten on growth. *J. Bacteriol.* **130:**1404–1406.
29. **Kaur, P., and D. V. Vadehra.** 1986. Mechanism of resistance to silver ions in *Klebsiella pneumoniae. Antimicrob. Agents Chemother.* **29:**165–167.
30. **Kavanagh, F.** 1963. *Analytical Microbiology.* Academic Press, Inc., New York.
31. **Kavanagh, F.** 1972. *Analytical Microbiology,* vol. II. Academic Press, Inc., New York.
32. **Knight, S. G.** 1951. Mineral metabolism, p. 500–506. *In* C. H. Werkman and P. W. Wilson (ed.), *Bacterial Physiology.* Academic Press, Inc., New York.
33. **Koser, S.** 1968. *Vitamin Requirements of Yeast and Bacteria.* Charles C Thomas, Publisher, Springfield, Ill.
34. **Laanbroek, H. J., and N. Pfenning.** 1981. Oxidation of short-chain fatty acids by sulfate reducing bacteria and in marine sediments. *Arch. Microbiol.* **128:**330–335.
35. **Lamanna, C., M. F. Mallette, and L. Zimmerman.** 1973. *Basic Bacteriology: Its Biological and Chemical Background,* 4th ed. The Williams & Wilkins Co., Baltimore.
36. **Lankford, C. E., J. M. Ravel, and H. H. Ramsey.** 1957. The effect of glucose and heat sterilization on bacterial assimilation of cystine. *Appl. Microbiol.* **5:**65–73.
37. **LaPage, S. P., J. E. Shelton, and T. G. Mitchell.** 1970. Media for the maintenance and preservation of bacteria, p. 1–133. *In* J. R. Norris and D. W. Ribbons (ed.), *Methods in Microbiology,* vol 3A. Academic Press, Inc., New York.
38. **Lev, M.** 1959. The growth-promoting activity of compounds of the vitamin K group and analogues for a rumen strain of *Fusiformis nigrescens. J. Gen. Microbiol.* **20:**697–703.
39. **Lindblow-Kull, C., A. Shrift, and R. L. Gherna.** 1982. Aerobic, selenium-utilizing bacillus isolated from seeds of *Astragalus crotalariae. Appl. Environ. Microbiol.* **44:**737–743.
40. **Lorowitz, W. H., H. Zhao, and M. P. Bryant.** 1989. *Syntrophomonas wolfei* subsp. *saponavida* subsp. nov., a long-chain fatty-acid-degrading, anaerobic, syntrophic bacterium; *Syntrophomonas wolfei* subsp. *wolfei* subsp. nov.; and emended descriptions of the genus and species. *Int. J. Syst. Bacteriol.* **39:**122–126.
41. **MacFaddin, J. F.** 1976. *Biochemical Tests for Identification of Medical Bacteria.* The Williams & Wilkins Co., Baltimore.
42. **Mandelstam, J., K. McQuillen, and I. Dawes (ed.).** 1982. *Biochemistry of Bacterial Growth,* 3rd ed. Blackwell Scientific Publications, Boston.
43. **Martin, W. H. M., M. J. Pelczar, Jr., and P. A. Hansen.** 1952. Putrescine as a growth factor requirement for *Neisseria. Science* **116:**483–484.
44. **McNutt, W. S., and E. E. Snell.** 1950. Pyridoxal phosphate and pyridoxamine phosphate as growth factors for lactic acid bacteria. *J. Biol. Chem.* **182:**557–562.
45. **Merkal, R. S., and B. J. Curran.** 1974. Growth and metabolic characteristics of *Mycobacterium paratuberculosis. Appl. Microbiol.* **28:**276–279.
46. **Murray, R. G. E. (University of Western Ontario).** 1986. Personal communication.
47. **Nieman, C.** 1954. Influence of trace amounts of fatty acids on the growth of microorganisms. *Bacteriol. Rev.* **18:**147–163.
48. **Oxoid Limited.** 1976. *The Oxoid Manual.* Oxoid Limited, Basingstoke, England.
49. **Pelczar, M. J., Jr., R. D. Reid, and E. C. S. Chan.** 1977. *Microbiology,* 4th ed. McGraw-Hill Book Co., New York.
50. **Perkins, J. J.** 1976. *Principles and Methods of Sterilization in Health Sciences.* Charles C Thomas, Publisher, Springfield, Ill.
51. **Power, D. A. (ed.), and P. J. McCuen.** 1988. *Manual of BBL Products and Laboratory Procedures,* 6th ed. Becton Dickinson Microbiology Systems, Cockeysville, Md.
52. **Rogosa, M., and F. S. Bishop.** 1964. The genus *Veillonella.* II. Nutritional studies. *J. Bacteriol.* **87:**574–580.
53. **Sigma Chemical Co.** 1980. *Technical Bulletin no. 106B.* Sigma Chemical Co., St. Louis, Mo.
54. **Snell, E. E.** 1950. Microbiological methods in vitamin research, p. 327–505. *In* P. Gyorgy (ed.), *Vitamin Methods,* vol. 1. Academic Press, Inc., New York.
55. **Snell, E. E.** 1957. Microbiological techniques: inorganic ions, p. 547–574. *In* J. H. Yoe and H. J. Koch, Jr. (ed.), *Trace Analysis.* John Wiley & Sons, Inc., New York.
56. **Spierenburg, G. T., F. T. J. J. Oerlemans, J. P. R. M. van Laarhoven, and C. H. M. M. de Bruyn.** 1984. Phototoxicity of N-2-hydroxyethylpiperazine-N′-2-ethanesulfonic acid-buffered culture media for human leukemic cell lines. *Cancer Res.* **44:**2253–2254.
57. **Stanier, R. Y., M. Doudoroff, and E. A. Adelberg.** 1970. *The Microbial World,* 3rd ed. Prentice-Hall, Inc., Englewood Cliff, N.J.
58. **Steele, B. F., H. E. Sauberlich, M. S. Reynolds, and C. A. Baumann.** 1949. Media for *Leuconostoc mesenteroides* P-60 and *Leuconostoc citrovorum* 8081. *J. Biol. Chem.* **177:**533–551.
59. **Tilton, R. C., and B. Rosenberg.** 1978. Reversal of the silver inhibition of microorganisms by agar. *Appl. Environ. Microbiol.* **35:**1116–1120.
60. **U.S. Pharmacopeial Convention, Inc.** 1989. *The U.S. Pharmacopeia XXII/The National Formulary XVII.* U.S. Pharmacopeial Convention, Inc., Rockville, Md.
61. **Vogel, H.** 1961. Growth of leptospira in defined media. *J. Gen. Microbiol.* **26:**223–230.
62. **Williams, S. (ed.).** 1984. *Official Methods of Analysis of the Association of Official Analytical Chemists.* Association of Official Analytical Chemists, Inc., Arlington, Va.
63. **Yamamoto, I., T. Saiki, S.-M. Liu, and L. G. Ljungdahl.** 1983. Purification and properties of NADP-dependent formate dehydrogenase from *Clostridium thermoaceticum,* a tungsten-selenium-iron protein. *J. Biol. Chem.* **258:**1826–1832.
64. **Zhao, H., D. Yang, C. R. Woese, and M. P. Bryant.** 1990. Assignment of *Clostridium bryantii* to *Syntrophospora bryantii* gen. nov., comb. nov., on the basis of a 16S rRNA sequence analysis of its crotonate-grown pure culture. *Int. J. Syst. Bacteriol.* **40:**40–44.

Chapter 8

Enrichment and Isolation

JOHN G. HOLT and NOEL R. KRIEG

Pure cultures of bacteria are rarely encountered in natural habitats, yet much of the present knowledge of the properties of and interactions among bacteria is based on studies with pure cultures. Consequently, it is often necessary to isolate into pure culture the various kinds of bacteria that coexist in a habitat. Enrichment techniques are a major preliminary step in obtaining pure cultures. Enrichments also make it possible to assess the differential effects of environmental factors imposed on mixed microbial populations and permit the selection of organisms capable of attacking or degrading particular substrates or of thriving under unusual conditions.

Successful isolation of a given organism into pure culture requires a sufficiently high proportion of that organism in the mixed population. Isolation is easiest when the organism is the numerically dominant member of the population. Enrichment methods are designed to increase the relative numbers of a particular organism by favoring its growth, its survival, or its spatial separation from other members of the population. Biophysical enrichments make use of such conditions as growth temperature, heat treatment, sonic oscillation, or UV irradiation to kill or inhibit the rest of the population. Such methods may also take advantage of some physical property of the desired organism, such as its size or motility, which can allow the organism to be preferentially separated from the rest of the population. Biochemical enrichments employ toxic agents to kill or inhibit the rest of the population without affecting the desired organism; alternatively, they may provide nutrient sources that can be used preferentially by a particular component of the mixed population. Biological enrichments may make use of specific hosts for selective growth of a particular organism, or they may take advantage of some pathogenic property, such as invasiveness, which the rest of the population does not possess. In many enrichment procedures several physical, chemical, and/or biological methods may be used in combination to achieve a maximum effect.

Most enrichments are carried out in closed systems, such as batch cultures in a flask or tube, where the concentrations of nutrients and metabolic products in the culture vessel continually change during bacterial growth. Open systems have also been used for enrichment. For example, the use of a chemostat (Chapter 10.2.1) enables one to provide a constant environment for cultivation of bacterial cells by continuously supplying a growth-limiting nutrient and continually removing metabolic products. Alteration of the dilution rate controls the concentration of the growth-limiting nutrient, which differentially affects the growth rate of the various organisms in mixed culture, making it possible for one or another member of the mixed population to become predominant; for example, see section 8.2.11.2.

Bacteria are usually isolated from enrichment cultures by spatially separating the organisms in or on a solid medium and subsequently allowing them to grow into colonies. For organisms that cannot grow on solid media, dilution to extinction by allowing separation of cells into individual tubes of a liquid medium can be used. Because such ordinary isolation methods do not absolutely ensure purity, more difficult methods may be needed whereby an individual bacterial cell from a mixed population is spatially isolated under a microscope before being cultured into a clone.

This chapter is designed to demonstrate the multiplicity and in many instances the considerable ingenuity of enrichment and isolation methods for bacteria by presenting specific, selected examples. Although many different methods are described, the chapter is not intended to be a comprehensive compendium for all bacterial taxa, and the general references (1 to 10) given at the end of the chapter should be consulted for additional methods and organisms. For enrichment and isolation of mutant cells, see Chapter 13.

8.1. BIOPHYSICAL ENRICHMENT

8.1.1. Low-Temperature Incubation (87)

Low temperature retards the growth of many mesophilic bacteria but not psychrophilic and psychrotrophic bacteria. Therefore, incubation of enrichment cultures at 0 to 5°C can favor the growth of the latter organisms.

8.1.1.1. Psychrophiles

Inoue (59) isolated psychrophilic bacteria from Antarctic soil by spreading dilutions of the soil over plates of a glucose-yeast extract-peptone agar and incubating for 14 to 24 days, with the temperature maintained throughout at below 5°C. In this manner nine strains of obligately psychrophilic bacteria with maximum growth temperatures of ca. 20°C were obtained (60).

8.1.1.2. *Listeria monocytogenes*

It may be difficult to isolate *Listeria monocytogenes* from material heavily contaminated with other bacteria (e.g., feces, silage, sewage, and clinical specimens from the cervix, vagina, meconium, nasopharynx, or tissues). Such materials may be enriched by the following procedure (110). Add a 1- to 2-g sample of the suspected material to each of two flasks containing 100 ml of sterile enrichment broth. Examples of suitable enrichment broths are TN medium (section 8.5.68) and PTN medium (section 8.5.51). Incubate one flask at 37°C and the other at 4°C. Plate 0.1-ml samples from the flask incubated at 37°C onto Tryptose agar (Difco) daily for 7 days. Incubate the plates at 37°C for 24 h, and examine the plates under oblique lighting for the presence of distinctive blue-green colonies. Plate 0.1-ml samples from the flask incubated at 4°C onto Tryptose agar at 7-day intervals for a period of up to 2 months. Incubate and examine the plates as described for the broth held at 37°C. Although the incubation at 37°C may give results more quickly, the cold enrichment usually gives a higher proportion of positive cultures. The reason why cold incubation enriches for *L. monocytogenes* has been variously attributed to an ability to survive longer than many other bacteria at 4°C (138), an ability to multiply at 4°C (114), and release of the organisms from an intracellular location (38).

8.1.2. High-Temperature Incubation

High-temperature incubation favors the selection of thermophilic bacteria because the growth of other organisms is inhibited at high temperature. The definition of "thermophilic" varies depending on the particular organisms being studied. The term usually refers to organisms that are unable to grow at temperatures below ca 45 to 50°C; however, it has also been applied to other organisms, for example, *Campylobacter* species (e.g., *Campylobacter jejuni*, *C. coli*, and *C. lari*) that can be cultured at 42°C but not 25°C. Another term, "extreme thermophiles," is usually applied to organisms that can grow at temperatures above 70°C.

8.1.2.1. Thermophiles in milk

For milk thermophiles, plate serial dilutions of milk in standard methods agar (section 8.5.60) and incubate the cultures at 55°C (12).

8.1.2.2. *Thermus* species

Thermus aquaticus strains have an optimum growth temperature of 70°C, a maximum of 79°C, and a minimum of 40°C. *T. ruber* strains have an optimum temperature of 60°C, a maximum of 70°C, and a minimum of 37°C (137). For enrichment (26, 137), inoculate 0.5- to 1.0-ml samples of microbial mats, water from hot springs, thermally polluted water, or hot tap water into

10-ml portions of *Thermus* medium (basal salts medium [section 8.5.10] plus 0.1% pancreatic digest of casein and 0.1% yeast extract). Incubate in covered water bath shakers for 1 to 3 days at 70 to 75°C for *T. aquaticus* or 55 to 65°C for *T. ruber*. For isolation, streak turbid cultures onto the *Thermus* medium solidified with 2 to 3% agar. Wrap the petri dishes in household plastic wrap to avoid drying out during the incubation period (25), and incubate them aerobically at the appropriate optimum temperature. Colonies of *T. aquaticus* are yellow to pale or colorless. Colonies of *T. ruber* are red (137).

8.1.2.3. *Thermoplasma* species

Thermoplasma species are best enriched and isolated from coal refuse piles (21, 37). Filter a liquid sample through a membrane filter with a pore size of 0.45 μm and then through a second filter with a pore size of 0.22 μm. Incubate aerobically in *Thermoplasma* isolation medium (section 8.5.66) at 55°C for 4 to 6 weeks or until turbidity develops. The organism can be isolated either by performing dilution to extinction in liquid medium (section 8.4.3) or by obtaining colonies on medium solidified with 10% (wt/vol) hydrolyzed starch of the type used for gel electrophoresis (111). Adjust the pH to ca. 3 immediately before pouring the plates. Incubate the plates within a sealed chamber in a humid atmosphere consisting of ca. 60% air and 40% CO_2 (vol/vol). *Thermoplasma* colonies appear within 7 days (or more), are small and colorless to brownish, and have a characteristic "fried-egg" appearance when viewed under a dissecting microscope (111).

8.1.3. High-Temperature Treatment

Treatment of a sample at an appropriate high temperature may select heat-resistant bacteria from a mixture of microorganisms. The heat treatment may be relatively mild (as in pasteurization), which even certain vegetative cells can survive, or much more severe, which only sporeformers can survive.

8.1.3.1. Thermoduric organisms in milk (12)

Pasteurize a sample of raw milk by heating it to 62.8°C and holding it at this temperature for 30 min. Cool, add dilutions to standard methods agar (section 8.5.60), and incubate the plates at 32°C. For instance, *Microbacterium lacticum* can be isolated with the same heat treatment and subsequent plating on yeast extract milk agar (section 8.5.76) with incubation at 32°C for 7 days.

8.1.3.2. Mesophilic sporeformers

Heat water samples or soil suspensions at 80°C for 10 min, and then streak onto plating media. Isolation media for *Bacillus* species differ according to the special nutritional needs of the species (e.g., use of uric acid as substrate for *B. fastidiosus*), and the reader is referred to references 31 and 135 for details. For *Clostridium* species, streak a roll tube of prereduced chopped-meat agar (Chapter 25.3.11) or streak a plate of freshly prepared agar medium and incubate it in an anaerobe jar.

Another approach is to add some of the sample in which spores are suspected to a suitable growth medium, heat at 80°C for 10 min and then incubate the broth for about 1 day to permit growth to occur before streaking onto solid media. For *Bacillus* species, nutrient broth containing 0.5% yeast extract is usually a suitable growth medium; for *Clostridium* species, prereduced starch broth (PY broth [Chapter 25.3.30] + 1.0% soluble starch) is often satisfactory.

If small numbers of spores are present in the samples, the proportion may be increased by adding the sample to a suitable sporulation medium and incubating the culture for various periods (2 to 21 days) prior to the heat treatment. For most *Bacillus* species, a suitable sporulation medium is nutrient broth (nutrient agar [Chapter 25.3.27] without the agar component) plus 0.5% yeast extract, 7×10^{-4} M $CaCl_2$, 1×10^{-3} M $MgCl_2$, and 5×10^{-5} M $MnCl_2$. Also see Chapter 4.4.9. For most *Clostridium* species, no one medium is optimum for production of spores, but chopped-meat medium (Chapter 25.3.11) with and without glucose often supports sporulation (55).

The heat treatment employed may have to be modified for some types of sporeformers, since the endospores of some bacterial strains are not as heat resistant as others. Try a lower temperature (e.g., 75 or 70°C) if satisfactory results are not obtained at the usual temperature of 80°C.

8.1.3.3. Thermophilic sporeformers (12)

Thermophilic sporeformers may be present, for example, in various sweetening agents used in ice cream. Initially prepare a 20% (wt/vol or vol/vol) solution of the sweetening agent (beet or cane sugar, lactose, cerelose, invert syrup, corn syrup, maple syrup, liquid sugar, or honey), and heat to 100°C for 5 min to kill nonsporeformers. Dilute with sterile water to give a final concentration of 13.3%.

For thermophilic, aerobic flat-sour organisms, plate 2-ml portions of the heated, diluted sugar solution in glucose-tryptone agar (section 8.5.28). Incubate in a humid chamber at 55°C for 48 h. Characteristic surface colonies are round, are 2 to 5 mm in diameter, have a typical opaque central spot, and are surrounded by a yellow halo in the medium. Subsurface colonies are compact and may be pinpoint in size; these should be subcultured and streaked onto glucose-tryptone agar.

For thermophilic, anaerobic hydrogen sulfide producers, add samples of the heated, diluted sugar solution to deep tubes of sulfite agar (1% tryptone or Trypticase, 0.1% sodium sulfite, 2% agar). The tubes of medium should be melted and cooled to 55°C prior to inoculation. Allow the tube contents to solidify, and incubate at 55°C for 72 h. Sulfide producers will form blackened spherical areas in the medium.

8.1.4. Drying

A number of bacteria that normally reside in the soil and produce spores, cysts, or other dormant forms are resistant to dehydration. A soil sample is air dried to kill many of the vegetative cells and is then plated onto a suitable agar medium. This method is useful for members of the genera *Bacillus*, *Azotobacter*, and *Arthrobacter* (31).

8.1.5. Motility

Motility has been used as the basis for several ingenious methods for enrichment and isolation. The principle is that bacteria that swim in liquid media or swarm or glide across solid media may be able to outdistance other microorganisms.

8.1.5.1. Swarming motility of *Clostridium tetani* (116)

Inoculate the specimen (soil, animal feces, clinical material) onto a small area of a freshly prepared plate of blood agar (Chapter 25.3.8). Incubate the plate at 37°C in an anaerobic jar for 1 day, and examine the agar surface carefully for evidence of swarming (a thin layer of growth that has spread outward from the inoculated area). It may be helpful to scrape the surface of the medium with a needle to verify the occurrence of swarming. Suspend some growth from the edge of the swarming area in broth, and streak onto solid media containing 5% agar to obtain isolated colonies.

8.1.5.2. Swimming motility of treponemes (51–53, 103–105)

The following method can be used for enrichment of cultivable *Treponema* species, many of which occur as part of the normal flora of the oral cavity and intestinal tract of humans and animals. Cut a well into the center of a thick layer of suitable agar medium (section 8.5.52) contained in either petri dishes or beakers. The well should be at least 7 mm deep and 2 to 10 mm in diameter and should not be cut to the bottom of the dish. Inoculate the well with the specimen (sample from the oral cavity, intestinal contents, or feces). Inoculate large wells (diameter, 10 mm) with up to 0.2 ml of sample; inoculate small wells (diameter, 2 mm) by stabbing ca. 2 mm obliquely into one side of the well. Do not allow any of the sample to be deposited on the surface of the agar. Immediately place the plate in an anaerobic jar, and incubate at 37°C for 4 to 7 days. In contrast to most other bacteria, treponemes can migrate through agar media. Look for treponemal growth occurring as a "haze" in the medium at some distance from the well. Subculture a sample from the outermost portions of this hazy region into a suitable prereduced semisolid medium (e.g., broth containing 0.15% agar). Since more than one kind of treponeme may be present, streak the subculture onto a roll tube of prereduced medium to obtain isolated colonies. These appear as hazy, whitish, dense areas in the medium.

A similar approach has been successfully used for the isolation of anaerobic marine spirochetes (27).

8.1.5.3. Swimming motility of *Spirillum volutans* (58, 77, 102)

Obtain mixed cultures of the giant microaerophilic bacterium *Spirillum volutans* by preparing a hay infusion with stagnant pond water. After a surface scum develops, examine samples taken just beneath the scum. Look for very large spirilla (1.4 to 1.7 μm in diameter and up to 60 μm long) with bipolar flagellar fascicles that are clearly visible by dark-field microscopy. Enrich the culture by inoculating some of the hay infusion into Pringsheim soil medium (section 8.5.49) and incubating at room temperature. Even with this enrichment, *S. volutans* will be vastly outnumbered by other bacteria. However, it can be isolated by using a capillary tube (102), as follows. Soften the center of a short section of a sterile, cotton-plugged piece of 5-mm-diameter glass tubing in a flame. Pinch the tubing with square-ended forceps until it is almost closed. Reheat the flattened portion, and draw it out rapidly to form a long capillary tube 15 to 30 cm long and 0.1 to 0.3 mm wide, oval in cross section. Seal the ends of the capillary in a flame. Break the capillary near one end with sterile forceps, and draw up 10 to 20 cm of sterile Pringsheim soil medium (supernatant). Then dip the capillary into the enrichment culture, and draw up another 2 to 4 cm, making sure that no air space occurs between the sterile medium and the culture. Seal the tip of the capillary, leaving a small air space. Mount the capillary on the stage of a ×100 microscope. *S. volutans* is often able to swim faster than the other bacteria in the enrichment culture and thus can reach the distal end of the capillary first. As soon as some spirilla reach the distal end, break the capillary behind them, expel the spirilla into a tube of semisolid CHSS medium (section 8.5.21), and incubate at 30°C. Confirm the purity of the cultures by phase-contrast microscopy.

8.1.5.4. Gliding motility of cytophagas, flexibacters, and myxobacters

Spread dilutions of the specimen (soil, water, or animal dung which has been in contact with the soil) onto agar media that have a low nutrient concentration (e.g., 0.1% tryptone or Trypticase, or 1/10-strength nutrient agar). Another procedure is to smear terrestrial plant leaf material, algal fronds, or marine plants on nutrient-poor media. Cytophagas and flexibacters are able to migrate on the surface of solid media; their colonies can be recognized as thin, often nearly translucent colonies with fingerlike projections, which develop far beyond the streak of deposition line. Subculture from these colonies. The incorporation of penicillin G (15 U/ml) and chloramphenicol (5 μg/ml) into the agar media often helps to suppress the growth of other bacteria (131). Many of the fruiting myxobacters are bacteriolytic and can be enriched on bacterium-rich substrates to the point at which they produce characteristic fruiting bodies, which can then be transferred to agar media. Place natural materials such as dung from herbivorous animals, decaying plant material, or bark onto moistened paper contained in petri dishes or household plastic crisper boxes, and incubate for up to 3 weeks.

Myxococci can be enriched on sterilized, urine-free rabbit dung pellets (preferably from rabbits fed on nonantibiotic-containing feeds) placed on moistened soil in a petri dish (80). It is best to moisten the substrate with a solution of cycloheximide (30 to 50 mg per liter of distilled water) to inhibit fungi. Keep the preparations moist while incubating. Check periodically for the presence of characteristic fruiting bodies with the aid of a dissecting microscope. Transfer material from the fruiting body to an agar medium by using a sterile fine metal or glass needle (80). As with other gliding bacteria, pure cultures can be obtained by repeatedly removing cells from the leading edge of the colony and inoculating them onto the center of a fresh plate of agar medium. Bacteriolytic species can be inoculated onto an agar medium containing a suspension of heat-killed bacteria such as *Escherichia coli.*

8.1.5.5. Gliding motility of *Beggiatoa* species (29, 43, 109, 120)

Beggiatoa species form colorless filamentous chains of cells, which exhibit gliding motility on surfaces. The organisms occur in aerobic freshwater or marine environments rich in H_2S; the cells can oxidize the sulfide to elemental sulfur, which is deposited intracellularly.

For enrichment, first prepare extracted hay by the following procedure. Cut dried hay into small pieces, and extract by boiling the hay in a large volume of water. Change the water three times during the extraction. The final wash should have an amber color. Drain the hay, and place it on trays at 37°C to dry. The enrichment medium consists of 0.8% (wt/vol) of the dried, extracted hay in tap water, distributed in 70-ml volumes into 125-ml cotton-stoppered Erlenmeyer flasks. Sterilize by autoclaving. Inoculate the flasks with 5-ml portions of mud containing decaying plant materials, as from small ponds, lakes, or streams. Incubate the flasks at ca. 25°C for 10 days.

Effective enrichments are indicated by a strong odor of H_2S and the development of a white film on the surface of the medium and on the submerged upper walls of the flasks. Examine the film for the characteristic filaments of *Beggiatoa* organisms. To free the filaments of contaminants, remove a tuft of filaments with a loop needle or microforceps and place it into a vial of sterile basal salts solution (section 8.5.10) plus 5 mM neutralized sodium sulfide. Tease the tuft apart, swirl, and transfer to another wash bath. Repeat this process five times. Place the washed filaments on the dried surface of a 1.6% agar plate for about 1 min to absorb excess fluid, and then place them on the surface of plates containing a medium consisting of 0.1% yeast extract, 0.1% sodium acetate, and 2 mM sodium sulfide. View the plates after 24 h with a dissecting microscope, and use a fine sterile needle to transfer to a fresh plate those filaments growing from the mass that are free of contaminants.

8.1.6. Filterability

The ability of some bacteria to pass through a membrane filter has been used to advantage for enrichments. The initial filtration of samples containing *Thermoplasma* species has already been mentioned (section 8.1.2.3). The small size and plastic properties of these wall-less cells allow them to pass through the pores of a 0.45-μm membrane filter. During enrichment of bdellovibrios, the samples are filtered before being placed with potential prey cells (section 8.3.1). A somewhat different approach consists of placing a sterile filter on top of a suitable agar medium in a petri dish and inoculating the surface of the filter with the sample containing the desired organism. If the organism is small, motile and, preferably, flexible, it may be able to migrate through the filter pores into the agar below, where it can be isolated after growth by removing the filter and subculturing from the agar. The method is particularly useful for organisms that grow slowly and are easily overgrown by contaminants during isolation attempts. The principle is exemplified by enrichment procedures for spirochetes such as cultivable, anaerobic *Treponema* spp. and facultatively anaerobic *Spirochaeta* spp. and for thin, flagellated organisms such as *Aquaspirillum gracile* and *Serpens flexibilis.*

8.1.6.1. Spirochetes

The following method can be used for enrichment of cultivable *Treponema* species from the oral cavity and intestinal tract of humans and animals (55, 115). Prepare a petri dish of a suitable agar medium such as RGCA-SC medium (section 8.5.52). Place a membrane filter (pore size, 0.15 μm) on the surface of the agar. Place an O-ring (25 to 30 mm in diameter) that has been lightly coated with vacuum grease on top of the filter. Place several drops of the diluted specimen (sample from the oral cavity, intestinal contents, or feces) onto the center of the O-ring. Incubate the petri dish in an anaerobic jar at 37°C for 1 to 2 weeks. Treponemes are small enough to migrate through the filter pores and penetrate the underlying agar, where they grow as a haze in the medium. Remove the O-ring and membrane filter from the agar surface, and remove a plug of agar from the hazy region with a Pasteur pipette. Examine the plug by dark-field microscopy for treponemes and contaminants. Subculture and purify as described in section 8.1.5.2. The inclusion of polymyxin B (800 U/ml) and nalidixic acid (800 U/ml) in the agar medium often helps to suppress the growth of contaminants.

For enrichment of facultatively anaerobic *Spirochaeta* strains (27), place a membrane filter with a pore size of 0.3 or 0.45 μm on the surface of a suitable agar medium, such as GYPT medium (section 8.5.30). Place a drop of pond water or water-mud slurry on the disc near the center. Incubate the plate aerobically at 22 to 30°C for 12 to 18 h. Remove the filter from the medium, and continue incubating the plate. Within the agar medium, the spirochetes form a subsurface veil-like growth that expands toward the periphery of the plate and away from colonies of contaminating bacteria that may also have passed through the filter pores. Obtain pure cultures by serial dilution in deep tubes of the GYPT medium: *Spirochaeta* colonies resemble a cotton ball or veil-like growth with a denser center.

8.1.6.2. Thin, flagellated bacteria

Aquaspirillum gracile is the thinnest of the aerobic, freshwater spirilla, having a cell diameter of 0.2 to 0.3 μm. It can be isolated by the following procedure (28). Place a membrane filter with a pore size of 0.45 μm on the surface of the isolation agar (section 8.5.4), and deposit 0.5 ml of pond or stream water in the center of the filter. Incubate the plates for 1.5 to 2.0 h at room temperature. Remove the filter, and continue the incubation for 3 days or longer. *A. gracile* is thin enough to penetrate the pores of the filter into the underlying agar. Look for spreading, semitransparent areas of growth within the agar medium, and subculture from these areas. Other small bacteria (small vibrios, cocci, or short rods) may also pass through the filter to form small colonies on the surface of the isolation medium; these colonies can be distinguished easily from the subsurface, spreading, semitransparent growth typical of *A. gracile.*

Serpens flexibilis can be found in the sediment of eutrophic freshwater ponds. The cells are rod-shaped with a diameter of 0.3 to 0.4 μm and move through agar gels in a characteristic serpentine manner. For enrichment, place a sterile membrane filter with a pore size of 0.3 to 0.45 μm on the surface of the isolation agar (54). Deposit a small amount of pond water/mud slurry on the center of the filter. Incubate for 6 to 12 h at 30°C, remove the filter, and incubate the plate for 2 to 4 days. The organism grows as a subsurface veil; remove a sample from the edge of this veil, and streak onto a second plate of sterile medium to obtain a pure culture.

8.1.7. Visible Illumination for Phototrophs

Unlike chemotrophic bacteria, phototrophic bacteria can use light as a source of energy. This ability has been used for enrichment of these bacteria: by providing light as the sole energy source, growth of phototrophs will be favored over that of chemotrophs. However, chemotrophs may still be able to grow to some extent by using organic compounds produced by the phototrophs.

8.1.7.1. Unicellular cyanobacteria

For freshwater cyanobacteria (formerly called blue-green algae), inoculate samples from ponds, streams, or reservoirs into tubes of BG-11 medium (section 8.5.12). For cyanobacteria from marine environments, use MN medium (section 8.5.40); some grow poorly in this medium, so ASN-III medium (section 8.5.6) is sometimes preferable. Wash all glassware well, and then rinse successively in tap water, concentrated nitric acid, and deionized water. Incubate enrichment cultures in an illuminated water bath at 35°C; this temperature will inhibit the growth of most eucaryotic algae. Use a light intensity of 2,000 to 3,000 lux (186 to 279 ft-candles, or approximately the illumination provided by a 55-W incandescent desk lamp at a distance of 10 to 12 in. [25 to 30 cm]); some cyanobacteria may require lower intensities (500 lux or less). Examine the enrichment cultures periodically until there is evidence of development of cyanobacteria. Then streak onto media solidified with 2% bacteriological-grade agar. Incubate plates under illumination (<500 lux) at 25°C in air or in an atmosphere slightly enriched with CO_2; clear plastic boxes such as household vegetable crispers make good chambers to help prevent evaporation. Use a dissecting microscope to detect the compact, deeply pigmented colonies of nonmotile cyanobacteria. Restreaking several times may be necessary to eliminate nonphototrophic bacterial contaminants. For motile cyanobacteria, enrich by placing a small patch of culture material at one side of a petri dish of agar medium. Illuminate the dish from the opposite side. The cyanobacteria will respond phototactically by gliding across the agar toward the region of higher light intensity. When some of the organisms have reached the opposite side, subculture them to a new plate and repeat the procedure until nonphototrophic bacterial contaminants have been eliminated. This technique has also been used effectively with some filamentous cyanobacteria (117).

For slow-growing cyanobacteria such as members of the order *Pleurocapsales*, primary cultures are best established by direct isolation of colonies on solid media rather than by preliminary cultivation in a liquid medium (132). During transport of rock chips, mollusk shells, or macroalgae from intertidal zones to the laboratory, keep the samples in closed bottles or tubes containing a damp piece of filter paper; do not submerge the samples in seawater, since this promotes development of contaminants. In the laboratory, suspend material scraped from the natural substrates in sterile liquid medium, and streak several plates directly from the suspension in addition to preparing a liquid enrichment culture. If the suspension contains many contaminants, wash it repeatedly in sterile medium by low-speed centrifugation to reduce the level of contamination before streaking plates.

Confirm the purity of cultures of cyanobacteria by microscopic observation and also by inoculating into complex media and incubating in the dark at 30°C, where the cyanobacteria will not grow.

See section 8.1.12.1 for a further enrichment technique for cyanobacteria.

8.1.7.2. Purple nonsulfur bacteria (22)

Although termed purple nonsulfur bacteria, a number of these species can in fact use sulfide as an electron donor for growth, but they do so only when the sulfide is maintained at low, nontoxic concentrations. Consequently, the organisms are ordinarily cultured with organic electron donors. Rather specific enrichment for particular species is possible, depending on the carbon source provided. For enrichment, only substrates that cannot be fermented by nonphototrophic organisms are provided in the enrichment medium.

The purple nonsulfur bacteria can be readily isolated from freshwater and marine sediments and are less frequently isolated from field, lawn, or garden soils. Use a basal enrichment medium (section 8.5.8), and supplement it with a suitable carbon source. (The particular carbon source used depends on the genus and species desired; consult references 3 and 8 for specific compounds.) Place 0.1 g of the specimen into a screw-cap tube, and fill the tube completely with the enrichment medium, freshly made to minimize oxygenation. Tighten the cap so that there is no air space at the top, and incubate the culture at ca. 25°C. Illuminate the culture continuously with incandescent light (not fluorescent light); use a 50- or 75-W lamp at a distance of 40 to 60 cm. Make sure that the lamp does not heat the cultures. Look for development of turbidity with a brown, yellow, or pink tinge in 3 to 7 days. Transfer a drop of culture to a second tube of medium for a secondary enrichment. For purification, use an agar medium containing 1% yeast extract or peptone, 0.2% sodium malate, and 1.5% agar. Use the shake tube method (section 8.4.2), or use pour plates and incubate in an anaerobic jar. Illuminate the cultures during incubation. Many of the purple nonsulfur bacteria will also grow in the dark under an air atmosphere or under microaerobic conditions, although the colonies are less highly pigmented than when grown anaerobically in the light.

8.1.7.3. Green sulfur bacteria

The green sulfur phototrophs are not as easy to cultivate as the green nonsulfur forms, although massive developments ("blooms") of them are often readily visible to the eye, particularly in marine and brackish environments. Unless inocula from such blooms are available, a useful enrichment approach is to set up a **Winogradsky column.** This method provides anaerobic conditions and a long-lasting supply of H_2S, and successive blooms of the sulfur phototrophs (and many other microorganisms as well) usually result.

To prepare a Winogradsky column, obtain mud from freshwater, brackish, or marine environments (for example, mud from the edge of a freshwater pond or stream lake or from a salt marsh). Mix 3 parts of mud with 1 part of $CaSO_4 \cdot H_2O$. Add some insoluble organic material such as finely shredded filter paper or small pieces of roots from aquatic plants. If paper is used, also add a small amount of NH_4MgPO_4. Pour the mixture into a tall glass cylinder (at least 5 cm in diameter) to a height of at least 15 cm, stirring to avoid air pockets. Hold the cylinder in the dark for 2 to 3 days to minimize the development of oxygenic photosynthetic organisms. Expose the cylinder to incandescent light or to diffuse daylight at 18 to 25°C during subsequent incubation. Anaerobic decomposition of the organic material in the column (with concomitant production of CO_2, alcohols, fatty acids, hydroxy acids, organic acids, and amines) and the formation of H_2S from the $CaSO_4$ will provide an appropriate array of microhabitats in which the sulfur phototrophs can thrive and form distinctive purple, red, or green patches on the sides of the glass column or layers or bands in the water column above the sediment-water interface.

Use Pasteur pipettes to obtain organisms from the glass surfaces or from the distinctive layers for microscopic examination, isolation, or further enrichment in liquid cultures. Alternatively, sequentially remove portions of the sediment with a spoon or spatula to expose the various zones of growth.

Enrichment can also be accomplished in defined liquid media. Select for specific sulfur phototrophs by varying the wavelength and intensity of the illumination used, the temperature of incubation, and/or the type and concentration of the electron donor. For a clear, detailed exposition of the many intricacies of such enrichment and the subsequent isolation of sulfur phototrophs into pure culture, see the very useful essay by Van Niel (125).

8.1.8. Visible Illumination To Enhance Pigment Production

Some bacteria will respond to visible light by producing colored pigments, which can greatly aid the differentiation and isolation of these organisms. For example, *Brevibacterium* spp. produces a distinctive orange pigment only when exposed to visible light. Incubate plates in the light during the period of active growth, not after colonies have developed fully.

8.1.9. Sonic Oscillation Selection of *Sporocytophaga myxococcoides*

Enrichment of the microcyst-forming cellulolytic cytophaga *Sporocytophaga myxococcoides* takes advantage of a sonication-resistant form in the organism's life cycle (72). Place 100 ml of sporocytophaga medium (section 8.5.58) into a 500-ml Erlenmeyer flask, and inoculate with ca. 0.1 g of soil, mud, or plant material. After incubation at 30°C for 7 to 10 days with moderate agitation, remove 5 ml of culture, subject it to sonic oscillation for 15 to 30 s, and then use it to inoculate a secondary enrichment flask. After ca. 5 to 7 days, look for a distinctive yellow hue in the flask, and use phase-contrast microscopy to look for both microcysts and vegetative cells on and around the cellulose fibers. Subject a portion of this culture to sonic oscillation, and then use the pour plate method to obtain isolated colonies, as described in section 8.2.12.5.

8.1.10. Antiserum Agglutination

The use of antiserum may enrich for the minority organism in a two-membered population. In a study of a mixed culture in which two species of marine spirilla occurred were present—a large and a small organism—the smaller organism was found to be greatly predominant. This made it impossible to obtain isolated colonies of the larger organism by plating; moreover, none of a variety of selection methods was applicable. To enrich for the larger spirillum, the smaller organism was isolated and used to immunize a rabbit. When the resulting antiserum was added to the mixed culture, the smaller spirilla agglutinated and settled to the bottom of the tube. The supernatant, now containing a high proportion of the large spirilla, was used to obtain isolated colonies.

8.1.11. Cell Density

The ability of some bacteria to float or to sink in aqueous environments or to form a band during density gradient centrifugation has been used to separate these organisms from contaminants.

8.1.11.1. Flotation on water for *Lampropedia hyalina* (88)

The strictly aerobic, tablet-forming coccus *Lampropedia hyalina* forms a characteristic hydrophobic pellicle on the surface of liquid media. If complex media are inoculated with material from habitats rich in organic matter, the lampropedias often form contaminated pellicles containing their very distinctive cells. If such pellicles develop, transfer them with a Pasteur pipette to a plate containing soil seeded with starch or wheat grains covered with water. After incubation, pellicles containing the tablets of *L. hyalina* (as shown by microscopy) can be transferred to agar media for purification.

8.1.11.2. Swirling for *Achromatium oxaliferum* (71)

The relatively high density of cells of *Achromatium oxaliferum* is caused by the accumulation of intracellular $CaCO_3$, and this property can be exploited to concentrate the organism from natural samples. Place a small amount of sediment from a suspected source in the bottom of a large beaker, and cover it with ca. 1 cm of slightly alkaline water. Tilt the beaker, and gently swirl the contents to separate the achromatia as a thin, white deposit directly above the sediment. Transfer the cells to another beaker with a Pasteur pipette, and repeat the process until the achromatia are free of contaminants.

8.1.11.3. Density gradient centrifugation

The buoyant density of bacteria in pure culture and in samples from natural aquatic environments has been studied by density gradient centrifugation in Percoll gradients, and the average density of a representative bacterium is 1.080 pg μm^{-3} (49). There do not seem to be significant differences in density among bacteria; however, the density of a given bacterium can change by as much as 7% from the average as a result of the formation of inclusions such as sulfur and phosphorus storage materials or the formation of capsules and/or gas vesicles. Such inclusions may be responsible for the occurrence of two or more bands of organisms when samples from natural environments are subjected to density gradient centrifugation (49). Therefore the density gradient technique may be useful in enriching for organisms with different types of inclusions.

Another enrichment application of density gradient centrifugation is the use of Urografin gradients to completely separate endospores from vegetative cells in sporulating cultures of *Bacillus* spp. (122).

For further information on density gradient centrifugation, see Chapter 4.3.3 and 4.3.5.

8.1.12. Radiation Resistance

Some bacteria are resistant to high doses of radiation such as UV light, X-irradiation, and gamma-radiation, and this can be used for selection purposes by killing the less resistant contaminants that may be present in a mixed culture. The following examples illustrate this principle.

8.1.12.1. UV irradiation for cyanobacteria

It is often difficult to obtain cultures of cyanobacteria free from bacterial contaminants, which frequently penetrate and live in the gelatinous sheaths that surround the cells and filaments of cyanobacteria (section 8.1.7.1). However, by treating a suspension with UV light for an appropriate period, it is possible to kill the contaminating bacteria and yet recover viable cyanobacteria. Success has been reported with the following method (46). Place a dilute suspension of cyanobacteria in a quartz chamber and irradiate with 275-nm UV light from a quartz-jacketed mercury vapor lamp. Agitate the suspension by continuous stirring during the incubation. At periodic intervals during the irradiation, remove samples and prepare a large number of dilution cultures from each sample. In the dilution cultures that show growth of the cyanobacteria, test for bacterial contamination by performing microscopic examination and by inoculating a variety of bacteriological media. One disadvantage of this method is that the final pure culture may contain cyanobacteria in which mutations have occurred because of the UV light treatment.

8.1.12.2. X- and gamma-radiation resistance for deinococci (89, 90)

Members of the family *Deinococcaceae*, although not sporeformers, are highly resistant to X- and gamma-irradiation, and this property can be used to select for them. Samples suspected of containing deinococci (such as ground meat, fish, fecal samples, and sawdust) can be irradiated with 1.0 to 1.5 Mrad of gamma-radiation, killing most or all other vegetative cells. A few endospores of contaminating bacteria may survive, but if, before irradiation, the sample suspension is incubated for a time in a medium that allows endospores to germinate, this decreases the number of surviving sporeformers. After irradiation, plate the samples onto a suitable medium such as TGYM medium (section 8.5.63) and incubate at 25 to 30°C. Look for colonies that are pink, orange-red, or red.

Deinococci can also be selected from soil by preparing a slurry of the soil and exposing the supernatant in a petri dish to UV light at 600, 900, and 1,200 J m^{-2} (90).

8.1.13. Magnetic Field for Magnetotactic Bacteria (86)

A number of aquatic bacteria contain small enveloped crystals of the iron oxide magnetite, which enable the cells to become oriented in a magnetic field. The bacteria are able to move along the magnetic lines of force and, in natural bodies of water, will also move downward, since the Earth's geomagnetic field has a downward component. Magnetotactic bacteria can be enriched from natural samples, such as pond water and sewage, by the following procedure. Place the water and sediment in a 2-liter beaker wrapped in opaque material and covered with plastic wrap. Attach a stirring-bar magnet to the beaker, with the south pole (use the north pole in the Southern hemisphere) touching the side of the beaker and positioned about 5 to 8 cm above the sediment. The magnetotactic cells will congregate near the magnet and can be removed with a pipette.

A simple "racetrack" for selection of magnetotactic bacteria can be constructed (139). Briefly, prepare a capillary tube from a Pasteur pipette and seal the small end in a flame. Fill the sealed capillary with filter-sterilized water (preferably water collected from the surface layer of sediment from which the inoculum is to be taken) by means of a syringe and needle. The large, nonsealed end of the capillary forms a well whose bottom can be plugged with a tiny wad of cotton. Place a drop of sediment containing magnetotactic bacteria into the well, and position a stirring-bar magnet near the opposite, sealed end of the capillary. Magnetotactic bacteria migrate from the well through the cotton plug, travel along the capillary, and arrive at the sealed end in a few minutes. The migration can be followed by dark-field microscopy. Harvest the accumulated cells by aseptically breaking the capillary near the sealed end and removing the contents with a narrow sterile pipette.

At this writing, only very few magnetotactic bacteria have been isolated in pure culture (19, 23, 83, 107).

8.2. BIOCHEMICAL ENRICHMENT

8.2.1. Alkali Treatment for Mycobacteria (18)

Add a maximum of 10 ml of the suspected sputum sample to a sterile, disposable, plastic 50-ml conical centrifuge tube with a leakproof and aerosol-free plastic screw cap. Add an equal volume of *N*-acetyl-L-cysteine–sodium hydroxide solution (section 8.5.44). Tighten the screw cap, and mix well in a Vortex mixer for a maximum of 30 s. Allow to stand at room temperature for 15 min. The *N*-acetyl-L-cysteine functions as a mucolytic agent; it converts the thick sputum to a thin, watery consistency. An extra pinch of the powder may be needed to liquefy highly mucoid sputum samples. The function of the sodium hydroxide is to destroy many of the contaminants present in the sputum; the mycobacteria are relatively resistant to the alkaline treatment. Fill the centrifuge tube to within 1 cm of the top with sterile 0.067 M phosphate buffer to neutralize the action of the sodium hydroxide. Centrifuge at 3,600 × *g* for 15 min to concentrate the mycobacteria. Carefully decant the supernatant into a splash-proof container. Add 1 to 2 ml of sterile water or buffer to suspend the sediment, and use this suspension to inoculate suitable culture media (see, e.g., section 8.5.35).

CAUTION: Biosafety level 2 practices, containment equipment, and facilities are recommended for culturing sputum samples, provided that aerosol-generating manipulation of such specimens are conducted in a class I or II biological safety cabinet. Biosafety level 3 practices, containment equipment, and facilities are recommended for activities involving the propagation and manipulation of cultures that may contain *Mycobacterium tuberculosis* (100). See Chapter 29 for biosafety methods.

8.2.2. Incubation at Alkaline pH for Vibrios and *Sporosarcina ureae* (18, 32)

To select for *Vibrio cholerae* from a fluid stool sample or rectal swab, directly streak the specimen heavily onto a selective medium such as TCBS agar (section 8.5.62). If there are only small numbers of *V. cholerae* in the specimen, inoculate 20-ml portions of alkaline peptone water (section 8.5.3), incubate for 5 h at 35°C, and then streak this culture heavily onto the TCBS agar. The alkalinity of the peptone water (pH 8.4) and the TCBS (pH 8.6) inhibits the growth of most contaminants. On the TCBS agar, colonies of *V. cholerae* are yellow (sucrose fermenting) and oxidase positive.

Sporosarcinae urea in soil samples can also be isolated on alkaline agar media (pH 8.5). Streak the medium (tryptic soy-yeast extract agar with urea [section 8.5.69]) with a dilution of the soil, incubate at 30°C, and check colonies microscopically for the presence of packets or tetrads of cocci, which may be motile. To test for endospore formation, transfer suspected colonies to nutrient agar to which has been added 20 ml of a 10% (wt/vol) solution of urea (previously sterilized by filtration) and 50 mg of $MnSO_4 \cdot H_2O$ per liter. Incubate at a temperature below 25°C. Endospores are round and refractile by phase-contrast microscopy and are located centrally or laterally within the cocci.

8.2.3. Acid Treatment for *Legionella* Species

Isolation of *Legionella* species, particularly those from the environment, can sometimes be facilitated by acidification of samples to pH 2.2, which kills contaminants more quickly than it does the *Legionella* species. The following procedure is based on the isolation of *Legionella shakespearei* from water taken from an evaporative cooling tower (127). Pass 1 liter of water suspected of containing *Legionella* bacteria through a 0.45-μm membrane filter. Cut the filter into small pieces with sterile scissors, and suspend the pieces in a tube containing 20 ml of sterile distilled water. Cap the tube, and shake vigorously by hand for 1 min to suspend the bacteria. Centrifuge 10 ml of the suspension at 4,000 rpm for 30 min, and then remove and discard 9 ml of the supernatant. Acidify the remaining 1 ml with 9 ml of pH 2.2 KCl-HCl buffer (add 5.3 ml of 0.2 N HCl and 25 ml of 0.2 N KCl to 100 ml of distilled water, adjust the pH to 2.2, and sterilize by filtration). Allow to stand for 5 min. Spread 0.1-ml portions onto the surface of BCYE agar plates (section 8.5.11) supplemented with glycine (3 g/liter), polymyxin B sulfate (79,200 IU/liter), vancomycin (5 mg/liter), and cycloheximide (80 mg/liter). Incubate the plates at 35°C for 7 days in a vessel containing a highly humid air atmosphere. Look for colonies that exhibit a "ground-glass" appearance when viewed under obliquely transmitted light. Presumptive identification of an isolate as being a member of the genus *Legionella* is based on demonstration of small (diameter, 0.5 μm), aerobic, gram-negative rods that require iron and cysteine for growth.

8.2.4. Incubation at Acidic pH

Most bacteria are inhibited by highly acidic conditions, but some are able to thrive, e.g., certain bacteria that live on fruit, are used in cheese ripening, or oxidize reduced-sulfur compounds to sulfuric acid. Incubation at low pH values can be very useful for the selection and isolation of these organisms.

8.2.4.1. Lactobacilli (78)

For lactobacilli in cheddar cheese, streak dilutions of the cheese on modified Rogosa's medium (section 8.5.42). This medium has a pH of 5.35 because of an acetic acid/acetate buffer system. At this pH, lactobacilli such as *Lactobacillus casei* and *L. plantarum* will form colonies, but the common dairy organism *Lactococcus lactis* will not.

8.2.4.2. *Thiobacillus thiooxidans* (128)

Inoculate shallow layers of *Thiobacillus thiooxidans* medium (section 8.5.67) with samples from soil, mud, or water (marine mud is the most reliable source). Look for a drop in pH to 2.0 after 3 to 4 days or more, which virtually ensures the predominance of this acid-tolerant organism. Purify by streaking on the solidified medium.

8.2.4.3. *Frateuria* species (121)

Members of the genus *Frateuria* are commonly found on fruit and can be enriched on *Frateuria* isolation medium (section 8.5.25), which has an initial pH of 4.5. Inoculate the medium with fruit samples, and incubate at 30°C. Streak plates of GYC agar (section 8.5.29) with material from the enrichment. GYC agar contains insoluble $CaCO_3$, and clear zones will form around the *Frateuria* colonies.

8.2.5. Inhibition by Toxic Metals

The salts of heavy metals such as tellurium, thallium, and selenium can, at low concentrations, exert an inhibitory effect on many bacteria. The use of a particular metal salt at an appropriate concentration can allow some bacteria to grow while inhibiting the growth of others. Some examples follow.

8.2.5.1. Tellurium inhibition for corynebacteria and certain streptococci

Potassium tellurite inhibits gram-negative bacteria and most gram-positive bacteria when used at a suitable concentration. To select for corynebacteria such as *Corynebacterium diphtheriae*, use potassium tellurite at a concentration of 0.0375%, as in cystine tellurite blood agar (section 8.5.22). Colonies of corynebacteria are gray or black as a result of reduction of the tellurite. For selection of certain streptococci (*Streptococcus mitis, S. salivarius*) and enterococci, use potassium tellurite at a concentration of 0.001%, as in mitis-salivarius agar (section 8.5.39). *S. mitis* forms tiny blue colonies, *S. salivarius* forms larger blue "gumdrop" colonies, and enterococci form small blue-black colonies.

8.2.5.2. Thallium inhibition for mycoplasmas and enterococci (17, 67)

Use thallous acetate at a concentration of 0.023%, as in E agar (section 8.5.23) and E broth (section 8.5.24), to select for mycoplasmas such as *Mycoplasma pneumoniae* from the respiratory tract. Extract specimens collected on swabs into 2 ml of soybean-casein digest broth (section 8.5.56) containing 0.5% bovine serum albumin. Inoculate 0.1-ml amounts of the suspension into biphasic E medium (section 8.5.23) and also onto plates of E agar. Incubate the cultures at 37°C aerobically in sealed containers. Examine the plates at intervals for up to 30 days with a dissecting microscope (magnification, × 20 to × 60) for the appearance of minute colonies (diameter 10 to 100 μm) with a typical "fried-egg" appearance. Examine the biphasic cultures microscopically by looking through the side of the tube for "spherules" (fluid medium colonies); also observe the cultures for a decrease in pH (yellowing of the phenol red indicator). Inoculate E agar plates from the biphasic medium to obtain isolated colonies. For principles of biphasic media, see Chapter 9.3.

Thallous acetate has also been used at a concentration of 0.1%, as in thallous acetate agar (section 8.5.64) to select for enterococci.

8.2.5.3. Selenite inhibition for salmonellae

Use sodium hydrogen selenite at a concentration of 0.4%, as in selenite F broth (section 8.5.54), to temporarily suppress the growth of coliforms while allowing salmonellae to grow. In addition to directly streaking stool samples onto selective and nonselective agar media, inoculate selenite F broth heavily (ca. 1 g or 1 ml of stool specimen in 8 to 10 ml of broth). Incubate this enrichment at 35 to 37°C for 12 to 16 h, and then streak onto the plating media.

8.2.6. Phenylethanol Inhibition for Gram-Positive Cocci

Use phenylethanol in agar media at a concentration of 0.25%, as in phenylethyl alcohol agar (section 8.5.48), to inhibit the growth of gram-negative bacteria, particularly *Proteus* species, when these occur in mixed culture with gram-positive cocci. For example, one application is the isolation of coagulase-positive staphylococci from a stool specimen.

8.2.7. Dye Inhibition for Gram-Negative Bacteria, Mycobacteria, and Arthrobacters

Gram-positive bacteria are generally inhibited by lower concentrations of triphenylmethane dyes (such as crystal violet, basic fuchsin, brilliant green, and malachite green) than are gram-negative bacteria. For example, the presence of malachite green at a concentration of 1:4,000,000 in media will inhibit the growth of *Bacillus subtilis*, and 1:1,000,000 will inhibit the growth of staphylococci; however, concentrations of 1:30,000 to 1:40,000 are required to inhibit *Escherichia coli* or *Salmonella typhi* (45). Brilliant green is used in media for the confirmed test for coliforms (brilliant green lactose bile broth [section 8.5.17]) and in several selective media for members of the *Enterobacteriaceae*, such as salmonella-shigella agar (section 8.5.53) and brilliant green agar (section 8.5.16). Crystal violet is also used for selection of members of the *Enterobacteriaceae*, as in violet red bile agar (section 8.5.72) and MacConkey agar (section 8.5.36). Mycobacteria are very resistant to dyes; malachite green is often incorporated into media used for isolation of *Mycobacterium tuberculosis* to inhibit the growth of contaminants, as, for example, in Lowenstein-Jensen medium (section 8.5.35). Methyl red has been used to select for soil arthrobacters (section 8.5.5); the dye inhibits other gram-positive bacteria but not arthrobacters (50).

8.2.8. Salt Inhibition

The growth of some bacteria is not inhibited by high concentrations of NaCl, and some bacteria actually require a high NaCl concentration for growth (for instance, the red extreme halophiles such as *Halobacterium* species need at least 15% NaCl). Regardless of whether it is required or merely tolerated, NaCl has proven useful as a selective agent for certain groups of bacteria.

8.2.8.1. Staphylococci

Most *Staphylococcus* species grow in media containing 10% NaCl. For selection of staphylococci from foods and miscellaneous environments, mannitol salt agar (section 8.5.37), which contains 7.5% NaCl, is a useful medium. Growth of many other genera of bacteria is suppressed at this concentration of salt.

8.2.8.2. Halobacteria (47)

The archaeobacterial genera *Halobacterium* and *Halococcus* occur in heavily salted proteinaceous materials (such as salted fish), in salterns, and in the Dead Sea and other highly saline lakes. They can often be isolated from samples of solar salt. They require a high concentration of NaCl for good growth (ca. 25% NaCl) and cannot grow with less than ca. 15% NaCl. In fact, they are killed by even brief exposures to salt concentrations less than ca. 15%. The use of NaCl concentrations of 20% or greater makes their selection from natural sources a simple matter. For an example of a suitable enrichment medium, see section 8.5.31. Incubate inoculated flasks at 37°C with agitation, and subsequently obtain isolated colonies by streaking from the enrichment cultures onto plates of medium solidified with 2% agar. Incubate the plates at 37°C for 3 to 14 days in plastic bags to prevent excessive drying. Look for the development of pink or red colonies.

8.2.8.3. *Halomonas* species (130)

Halomonads are eubacteria that tolerate both high and low salt concentrations; all known species grow in NaCl concentrations from 0.2 to 25%, depending on the type of medium used (130). They have been isolated from a wide variety of saline environments, including solar salterns, the Dead Sea, manganese nodules from the ocean, underground salt formations, and the Antarctic. They can be isolated on high-salt-containing media such as CAS medium (section 8.5.19). They form white to yellow colonies, not pink or red as do the archaeobacterial halophiles.

8.2.9. Bile Inhibition for Enteric Bacteria

Bile or bile salts are often incorporated into culture media as selective agents for intestinal bacteria. There are some exceptions to this selectivity; for example, the plague bacillus *Yersinia pestis* can grow in pure bile even though it is not an intestinal organism (91). However, *Y. pestis* is closely related to *Y. pseudotuberculosis*, an intestinal organism, so perhaps the bile tolerance is not surprising. Nevertheless, the rule holds sufficiently well to make bile an important selective agent for gram-negative enteric rods and for enterococci. For selection of members of the family *Enterobacteriaceae*, bile or bile salts are incorporated into such media as MacConkey agar (section 8.5.36), salmonella-shigella agar (section 8.5.53), violet red bile agar (section 8.5.72), and many others. For enterococci, bile esculin agar (section 8.5.13) has proved to be an excellent selective and differential medium.

8.2.10. Antibiotic Inhibition

Antibiotics provide one of the easiest ways to devise a selective medium for strains of a particular species or genus. It makes no difference whether the organisms have medical importance or not; for instance, they may merely be harmless soil or water organisms. Obtain several reference strains of the bacterial taxon, and test them against a wide spectrum of individual antibiotics. Identify antibiotics to which all the strains are resistant, and incorporate several of these antibiotics into an appropriate sterile agar medium. When samples from nature are streaked onto the medium, the desired organisms, if present, will be able to grow, whereas many of the contaminant organisms will be suppressed by the antibiotics. A few specific examples follow.

(i) Because mycoplasmas lack cell walls, they are resistant to even very high concentrations of penicillin—concentrations that inhibit most other bacteria. For instance, the antibiotic is used at a concentration of 194 U/ml in mycoplasma isolation media such as E agar (section 8.5.23).

(ii) Penicillin is useful in the isolation of *Bordetella pertussis* from the nasopharynx because it helps to suppress the growth of members of the normal flora while permitting *B. pertussis* to form characteristic pearl-gray ("mercury-drop") colonies in ca. 4 days. The clinical specimen (obtained by means of a nasopharyngeal swab) is streaked onto plates of Bordet-Gengou agar (section 8.5.15) containing penicillin (0.5 U/ml).

(iii) Antibiotics are necessary when attempting to isolate *Campylobacter jejuni* or other *Campylobacter* species from stool specimens, in order to suppress the members of the normal flora that otherwise would rapidly outgrow the campylobacters. One example of a medium suitable for isolation of *C. jejuni* is Modified Campy BAP (section 8.5.41), which contains cephalothin, polymyxin B, trimethoprim, vancomycin, and amphotericin B.

(iv) Three antibiotics—vancomycin, colistin, and nystatin—aid in the selection of *Neisseria meningitidis* from nasopharyngeal samples (to detect healthy carriers of the organism) and of *Neisseria gonorrhoeae* from clinical specimens (urethral exudates, cervical swabs, etc.). These antibiotics are incorporated into Thayer-Martin agar (section 8.5.65) and help to suppress the growth of members of the normal flora. The vancomycin suppresses gram-positive contaminants, the colistin acts mainly against gram-negative contaminants, and the nystatin is an antifungal agent.

Table 1 lists some other examples of bacteria for which antibiotics serve as selective agents. Do not overlook the use of cycloheximide, nystatin, and other antifungal agents as general inhibitors of fungi in selective media, especially when the source of the inoculum is the soil.

8.2.11. Dilute Media

Many bacteria are residents of soil and aquatic habitats low in nutrients and have difficulty growing in rich media. Also, many potential contaminants cannot compete in dilute media, so that the shortage of nutrients becomes a selective factor. Below are some examples of the use of dilute media for isolation of various soil or water bacteria.

8.2.11.1. Caulobacters (97)

Caulobacters are prosthecate bacteria that can grow at levels of nutrients that do not support good growth of many contaminants. To samples of water from ponds, streams, or lakes or to samples of tap water add 0.01%

Table 1. Examples of the use of antibiotics in selective media

Bacterium	Antibiotic	Concn	Medium	Reference(s)
Agrobacterium biovar 1	Penicillin G Streptomycin Cycloheximide Tyrothricin Bacitracin	97,500 U/liter 30 mg/liter 250 mg/liter 1 mg/liter 6,500 U/liter	*Agrobacterium* biovar 1 isolation medium (85.1)	68
Agrobacterium biovar 2	Cycloheximide Bacitracin Tyrothricin	250 mg/liter 100 mg/liter 1 mg/liter	*Agrobacterium* biovar 2 isolation medium (85.2)	68
Brucella spp.	Bacitracin Polymyxin B Cycloheximide Vancomycin Nalidixic acid Nystatin	25 U/ml 5 U/ml 100 μg/ml 20 μg/ml 5 μg/ml 1 U/ml	Brilliant green lactose bile broth (8.5.17) plus antibiotics	34, 42
Micrococcus spp. (from skin)	Furazolidone	20 mg/liter	*Frateuria* enrichment medium (8.5.25)	36, 69, 129
Nocardia spp.	Demeclocycline[a] or methacycline	5 μg/ml 10 μg/ml	Diagnostic Sensitivity Test Agar (Oxoid) plus antibiotics	93
Pseudomonas spp.	Cycloheximide Nitrofurantoin Nalidixic acid	900 μg/ml 10 μg/ml 23 μg/ml	*Pseudomonas* isolation medium (8.5.50)	48
Spirochaeta spp.	Rifampin	2 μg/ml	*Sphaerotilus* medium (8.5.57)	27

[a]Formerly known as demethylchlortetracycline.

peptone and incubate aerobically at 20 to 25°C in bottles or flasks loosely covered with paper or aluminum foil. Examine the surface film daily by phase-contrast microscopy. When prosthecate bacteria occur in a relative proportion of ca. 1 in 10 or 20 cells (usually in about 4 days), streak the surface film onto plates of tap water agar containing 0.05% peptone and incubate at 10°C. The low peptone level allows the caulobacters to form tiny colonies but does not allow heavy overgrowth by other bacteria. After 4 days or more of incubation, select microcolonies of caulobacters under a dissecting microscope and transfer as patches to a richer medium (0.2% peptone, 0.1% yeast extract, 0.02% $MgSO_4 \cdot 7H_2O$, 1.0% agar). After 2 days of incubation, prepare wet mounts from the patches to detect caulobacters. Obtain isolated colonies by streaking growth from the patches onto fresh plates. For marine caulobacters, add 0.01% peptone to samples of stored seawater and incubate at 13°C for ca. 7 days. When microscopic observation indicates the development of a suitable proportion of prosthecate bacteria, streak the surface film onto plates of seawater agar containing 0.05% peptone and incubate at 25°C. Subculture microcolonies to fresh plates as patches, and later purify by streaking.

8.2.11.2. Aquatic spirilla (136)

In dilute media, aerobic chemoheterotrophic spirilla (the genera *Aquaspirillum* and *Oceanospirillum*) can often compete successfully with other bacteria for the nutrients present. For freshwater spirilla, add 1% peptone or yeast autolysate to samples of source water (e.g., water from stagnant ponds) and incubate at room temperature for ca. 7 days or until the spirilla become numerous. Then add part of this initial culture to an equal part of the source water, and sterilize by autoclaving. Inoculate this mixture from the unsterilized portion of the original culture. After incubation and further development of the spirilla, dilute a portion of the second culture with more source water, sterilize the mixture, and inoculate it from the unsterilized portion. Continue to deplete the nutrients in this manner until the spirilla predominate. Obtain colonies by streaking onto plates of MPSS agar (section 8.5.43).

Another approach depends on low levels of nitrogen sources (136). Supplement samples of source water with 1% calcium malate or lactate, and incubate at room temperature for ca. 1 week. Make a serial transfer into sterile source water containing 1% of the carbon source, and incubate. Continue in this manner (three or four serial transfers) until the spirilla predominate. In this method it is important not to add a nitrogen source such as NH_4Cl in order to prevent overgrowth of the spirilla by contaminants. For marine spirilla, mix the seawater sample with an equal volume of Giesberger base medium (section 8.5.27) supplemented with 1% calcium lactate (136). After incubation, remove a portion of the original culture and incubate. Continue to successively deplete the nitrogen content by repeating this procedure until the spirilla become the predominating organisms. Obtain isolated colonies by streaking onto MPSS agar (section 8.5.43).

Another way to enrich for spirilla is to use a continuous-culture system that provides low levels of nutrients. For instance, in chemostat experiments with a mixture of a marine spirillum and a pseudomonad (62), the growth rate of the spirillum exceeded that of the pseudomonad when the dilution rate of the chemostat was decreased to the point at which the limiting carbon and energy source (lactate) fell below 10 mg/liter. In other experiments the growth rate of the spirillum exceeded that of *Escherichia coli* when the lactate concentration fell below 5 mg/liter (63). In similar experiments with a freshwater spirillum and a pseudomonad, when the lactate concentration fell below ca. 0.09 mg/liter the pseudomonad was eliminated from the chemostat as a nongrowing population (82). The more efficient scavenging ability of the spirillum for lactate may be attributable to a lower K_m and a higher V_{max} of the transport system for lactate and also to a higher surface-to-volume ratio for the spirillum (82).

Under starvation conditions, spirilla appear to have a survival advantage (81). This may be related to their ability to form intracellular reserves of poly-ß-hydroxybutyrate under conditions of prior growth with limiting levels of carbon and energy sources. The role of poly-ß-hydroxybutyrate in bacterial survival has also been reported for other types of bacteria (39). Consequently, starvation conditions should be considered as a possible way to select for bacteria that form this polymer.

8.2.11.3. *Sphaerotilus* species

The sheathed bacteria of the genus *Sphaerotilus* occur in streams contaminated with sewage or organic matter and form slimy tassels attached to submerged surfaces. They can also be isolated from rivers, open drains, or ditches where there is no initial evidence of their presence. The enrichment procedure takes advantage of the ability of *Sphaerotilus* species to grow at very low nutrient concentrations. To 50-ml volumes of *Sphaerotilus* medium (section 8.5.57), add 25-ml volumes of the water sample or 1-, 5-, and 10-ml portions of settled sewage or the settled liquor from various stages of sewage treatment. Incubate at 22 to 25°C for 5 days. Examine microscopically for evidence of filamentous growth after day 2. Obtain pure cultures by selecting a filament from the enrichment broth and streaking it onto plates of a solid medium (0.05% meat extract, 1.5% agar). Incubate plates for 24 h at 25°C, and examine under a dissecting microscope for the typical curling filaments of *Sphaerotilus* species. Transfer isolates to a Trypticase-glycerol broth (Trypticase [BBL Microbiology Systems, Cockeysville, Md.], 5 g; glycerol, 5 g; distilled water, 1,000 ml [pH 7.0 to 7.2]), and incubate for up to 2 weeks. *Sphaerotilus* species form a heavy surface pellicle in 2 to 3 days, and the underlying broth remains clear. If turbidity develops, contamination has occurred. Even without development of turbidity, *Sphaerotilus* species should be reisolated from the surface pellicle by an additional streaking onto meat extract agar to ensure purity. To confirm that the isolated organisms form a sheath, place a small piece of slime growth on a slide in a drop of water, apply a coverslip, and press down on the coverslip with blotting paper. Place a very small drop of 1% crystal violet solution at the edge of the coverslip so that it flows into the preparation by capillary action. After 30 s, press again with blotting paper to remove excess dye and observe with a bright-phase oil-immersion lens. Both the cells and the sheath should be clearly visible.

8.2.12. Special Substrates

If a culture medium contains a sole carbon or nitrogen source that can be used only by a particular species or group of bacteria, the growth of that species or group will be favored over that of other organisms that may be present. However, be aware that other organisms may be able to grow to some extent by using products synthesized by the favored organisms; thus, this method is not absolutely selective but may merely increase the proportion of the desired organism in the population.

8.2.12.1. Tryptophan for pseudomonads (76)

To enrich for pseudomonads capable of using tryptophan as the sole carbon and nitrogen source, inoculate a 250-ml Erlenmeyer flask containing 40 ml of tryptophan medium (section 8.5.70) with ca. 0.1 g of soil. Incubate with shaking at 25°C for 5 to 7 days. Transfer 0.1 ml to a second flask of medium, and incubate for 2 to 3 days. After a further serial transfer, obtain pure cultures by streaking onto tryptophan medium solidified with 15 g of agar per liter. After 1 to 3 days of incubation, subculture isolated colonies to agar slants.

8.2.12.2. N_2 for *Azospirillum* and *Azotobacter* species

Azospirillum brasilense and *A. lipoferum* are microaerophilic nitrogen fixers associated with the roots of a variety of plants; they also occur in soil (40, 41). Place washed root pieces 5 to 8 mm long, macerated with a forceps, into nitrogen-free semisolid medium (Nfb medium [section 8.5.45]). Alternatively, inoculate the medium with a loopful of soil. Incubate with agitation for 40 h at 32°C; then test the enrichment culture for acetylene-reducing activity (Chapter 25.2.7). Be careful not to disturb the dense subsurface pellicle that forms in the medium, since this may stop nitrogenase activity. If the culture reduces acetylene, enrich further by a serial transfer to fresh Nfb medium. Examine by phase-contrast microscopy, and look for plump, curved, motile rods ca. 1 μm wide and filled with intracellular granules. Then streak onto plates of Nfb medium (solidified with 1.5% agar) containing 20 mg of yeast extract per liter. After 1 week, look for small, white, dense colonies and transfer to Nfb semisolid medium. For final purification, streak the Nfb culture onto BMS agar (section 8.5.14). Look for the development of typical pink, often wrinkled colonies.

Unlike *Azospirillum* spp., which are obligate microaerophiles under nitrogen-fixing conditions, *Azotobacter* spp. have mechanisms to protect oxygen-labile nitrogenase from oxygen and can fix nitrogen aerobically. Inoculate 0.1 g of soil into 100 ml of nitrogen-free *Azotobacter* medium (section 8.5.7) contained in a 1-liter flask. Incubate at 30°C on a shaking machine. Observe the culture microscopically at periodic intervals for the development of large, ovoid cells that are 2 μm or more in diameter. Prepare a secondary enrichment culture, and purify by obtaining isolated colonies on nitrogen-free agar medium (*Azotobacter* medium solidified with 1.5% agar).

See also section 8.2.12.14 for a discussion of the use of unusual carbon sources for isolation of nitrogen fixers.

8.2.12.3. H_2 for *Aquaspirillum autotrophicum* (13)

Collect bacteria on the surface of membrane filters (pore size, 0.45 μm) by filtering water samples taken at different depths (e.g., 3 and 7 m) from eutrophic lakes. Place the filters, bacterium side up, onto the surface of mineral agar plates (section 8.5.38). Incubate under an atmosphere of 60% H_2, 30% air, and 10% CO_2 at 30°C. Subculture growth that develops on the filter surface to mineral agar plates. Obtain pure cultures by repeated streaking on mineral agar. Microscopically, look for spirilla 0.6 to 0.8 μm wide with bipolar tufts of flagella. Confirm hydrogen autotrophy by demonstrating that both H_2 and CO_2 are required for growth on the mineral medium.

8.2.12.4. Methanol for hyphomicrobia (16)

Members of the genus *Hyphomicrobium* are able to use one-carbon compounds such as methanol or methylamine as sole carbon sources. Add the inoculum (5.0 ml of pond or ditch water, or 0.3 g of mud or soil—all preferably with a low organic content) to stoppered bottles (75- to 125-ml capacity), and add nonsterile *Hyphomicrobium* medium (section 8.5.32) through which nitrogen has been bubbled to provide anaerobic conditions. Fill the bottles completely with the medium. Incubate in the dark at 30°C. Be sure that the bottles remain completely filled by adding fresh medium if necessary. Hyphomicrobia generally develop in ca. 8 days. Monitor their development by phase-contrast microscopy: look for rod-shaped cells with pointed ends or oval, egg-shaped, or bean-shaped forms, which produce filamentous outgrowths (prosthecae) that vary in length and may show branching. Prepare a secondary enrichment, this time with sterile medium. Finally, streak onto solidified *Hyphomicrobium* medium and incubate aerobically to obtain isolated colonies.

8.2.12.5. Cellulose for cellulolytic cytophagas

Various methods and media have been devised for enrichment of cellulolytic cytophagas (99). A plate culture method for enrichment is as follows. Place autoclaved, round, filter paper discs, such as are used in chemistry laboratories, on the surface of plates of ST6CX agar (section 8.5.59). Inoculate the surface of the filter papers with soil, rotting plant material, or drops from water samples. For soil, inoculate the filter papers in a regular pattern at different places with a few grains of soil by tamping the soil down on the filter with a glass rod or distributing it with a swab. Incubate the plates at 25 to 30°C. After 4 to 5 days and periodically up to 1 to 2 weeks, look for glassy, translucent, yellow to orange spots on the paper. Make transfers as early as possible from the margins of the areas of cellulose decomposition to fresh plates of medium with several small pieces of filter paper. Look for slender, flexible cytophaga cells by phase-contrast microscopy in the areas of cellulolysis. Obtaining pure cultures from enrichments can be difficult; for the various methods and their limitations, see reference 99.

8.2.12.6. Agar for agarolytic cytophagas

Some facultatively anaerobic marine cytophagas are able to hydrolyze and ferment agar; this trait is a valuable selective feature that can be used for their enrichment. Fill glass-stoppered bottles or screw-cap tubes to the top with nonsterile Veldkamp medium (section 8.5.71). Inoculate with marine mud from areas with decaying algae, stopper the bottles, and incubate them in the dark at 30°C. Look for development of turbidity, gas formation, and a drop in pH in 3 to 7 days. Isolate colonies by the shake tube method (section 8.4.2) with sterile Veldkamp medium containing 2% agar. Look for the development of colonies which, on microscopic examination, are composed of cells that exhibit the flexing movements characteristic of cytophagas. Subculture the colonies to media containing 1% agar and 1% yeast extract to demonstrate softening or liquefaction of the agar.

For other methods of obtaining agarolytic cytophagas, see reference 99.

8.2.12.7. Lactate for propionibacteria

For most fermentative bacteria, lactate is an end product of fermentation, not a beginning substrate. The relatively unusual ability of propionibacteria to ferment lactate to propionate and CO_2 provides a basis for selection of these anaerobic organisms. Swiss-type cheeses are a good source of propionibacteria because these bacteria are the ripening agents. Fill a screw-cap culture tube (25-ml capacity) with a freshly boiled and cooled medium consisting of 4% sodium lactate and 1% yeast extract. Add ca. 0.2 g of $CaCO_3$ to the tube, and inoculate with a small piece of Swiss-type cheese. Tighten the screw cap, and incubate the tube at 30°C. Look for the development of reddish-brown turbidity in ca. 5 to 7 days. Obtain isolated colonies by streaking the culture onto sterile lactate-yeast extract agar and incubating anaerobically in a CO_2-enriched atmosphere, such as that provided by a GasPak (BBL).

8.2.12.8. Ammonium or nitrite for nitrifying bacteria (84, 98)

Use the basal nitrification medium described in section 8.5.46. For ammonium oxidizers, which oxidize ammonium to nitrite, supplement the medium with 0.5 g of $(NH_4)_2SO_4$ per liter. For nitrite oxidizers, which oxidize nitrite to nitrate, use 0.5 g of $NaNO_2$ per liter. Prepare a 1:10 dilution of soil, and inoculate each flask of the enrichment medium with 1.0 ml of the dilution. Incubate at 28°C. At weekly intervals, test for disappearance of the ammonium or nitrite by removing samples of the enrichment culture to a spot plate. For ammonium, test 3-drop samples with Nessler's solution (Chapter 25.4.11); an orange or yellow color indicates ammonium. For nitrite, add 3 drops of solution A and 3 drops of solution B (Chapter 25.4.13) to 0.5-ml samples; a red color indicates nitrite. Compare the intensity of the colors with those obtained with a set of dilutions from standard solutions of ammonium or nitrite, and replace the amount of substrate lost from the enrichment culture by adding fresh substrate from a sterile stock solution. If no nitrite is present in the enrichment culture, test for the formation of nitrate by adding a small amount of zinc dust after the nitrite test reagents; the zinc will reduce any nitrate present to nitrite, which will then yield a red color. Continue to incubate the enrichment culture, and replace ammonium or nitrite at weekly intervals until a population of nitrifiers is built up as indicated by microscopic observation.

The isolation of nitrifiers into pure culture is extremely difficult. Completely inorganic media must be used, and the colonies formed are tiny, ca. 100 μm in diameter. Moreover, nitrifiers grow very slowly and are therefore often overgrown by contaminants. Using sterile enrichment medium as a diluent, prepare a series of 10-fold dilutions from the enrichment. Place 1.0-ml samples of each dilution into petri dishes; then add silica gel medium prepared with double-strength enrichment medium as described in Chapter 9.1.3. Mix the culture dilution with the medium immediately, and allow the medium to solidify. Incubate the plates in a humid atmosphere at 28°C, and examine periodically under a microscope for the development of tiny colonies. With a Pasteur pipette freshly drawn out in a flame to a fine tip, subculture well-isolated colonies into sterile medium. Determine the ability of the subcultures to use ammonium or nitrite, and also test for contaminants by streaking samples onto organic media such as nutrient agar (no growth should occur). Purify any apparently pure cultures in silica gel medium several more times, each time selecting well-isolated colonies under the microscope, testing for the ability of subcultures to use ammonium or nitrite and for the presence of contaminants.

8.2.12.9. Nitrate plus organic acids for pseudomonads

Some members of the aerobic genus *Pseudomonas* can be enriched and isolated by taking advantage of their ability to use nitrate as a source of cellular nitrogen, together with their ability to use the salts of various organic acids as carbon and energy sources (118). Add 0.1 g of soil or mud or 0.1 ml of pond or river water to 3 ml of succinate-salts medium (section 8.5.61). Incubate at 30°C to enrich for members of the fluorescent group of pseudomonads, such as *Pseudomonas putida, P. fluorescens,* or *P. aeruginosa.* Incubate at 41°C to select for *P. aeruginosa* (*P. putida* and *P. fluorescens* cannot grow at this temperature). Enrich for *P. acidovorans* by substituting glycolate, muconate, or norleucine for succinate in the enrichment medium and incubating at 30°C. Some pseudomonads not only can use oxygen as a terminal electron acceptor but can also use nitrate, and these pseudomonads can often be selected by increasing the nitrate concentration of the enrichment medium to 1.0% and by filling screw-cap tubes completely with the medium to establish oxygen-limiting conditions, which favor nitrate respiration. After preparing secondary enrichment cultures of pseudomonads, obtain isolated colonies by streaking agar plates (use the appropriate enrichment media solidified with 15 g of agar per liter). Incubate the plates aerobically at 30°C.

8.2.12.10. Sulfate plus organic acids for sulfate reducers

Members of the anaerobic genus *Desulfovibrio* and most of the members of the genus *Desulfotomaculum* are able to oxidize lactate in the presence of sulfate, with the latter being reduced to H_2S. Inoculate screw-cap tubes or stoppered bottles with soil, mud, water, or fecal material, and fill the vessels completely with lactate medium (section 8.5.34). Incubate at 30°C. A blackening of the medium indicates sulfate reduction. Prepare a secondary enrichment before streaking onto solid media. To select for *Desulfotomaculum* species, heat enrichment cultures at 70°C for 10 min before streaking solid media. Obtain isolated colonies of the sulfate reducers by streaking plates of Iverson medium (section 8.5.33) and incubating them under a hydrogen atmosphere for 7 to 10 days at 30°C. To enrich for *Desulfotomaculum acetoxidans* (134), which cannot use lactate but can use acetate, use the enrichment medium of Widdel and Pfennig (section 8.5.73), which contains acetate as the oxidizable substrate. Use phase-contrast microscopy to look for development of motile straight or slightly curved rods that are 1.0 to 1.5 μm wide and 3.5 to 9.0 μm long and contain spores and also bright refractile areas. Use the shake tube method (section 8.4.2) to obtain isolated colonies. After 3 weeks of incubation, pick colonies and heat them at 70°C for 10 min; *D. acetoxidans* is a sporeformer and can be selected by this heat treatment.

8.2.12.11. Elemental sulfur for *Desulfuromonas acetoxidans*

Desulfuromonas acetoxidans obtains energy for growth by anaerobic sulfur respiration, with acetate, ethanol, or propanol serving as the carbon and energy source. Add samples of anaerobic, sulfide-containing water or mud from freshwater or marine sources to sterile screw-cap bottles. Fill the bottles with Pfennig and Biebl medium (section 8.5.47), in which flowers of sulfur are suspended; leave a small air bubble. Tighten the caps, and incubate the bottles at 28°C for 2 weeks with agitation (e.g., on a rotary shaker at 150 rpm). The action of the glass beads in the medium will gradually grind the sulfur to a very fine suspension. Positive enrichments can be recognized by a strong odor of H_2S. After two serial transfers in Pfennig and Biebl medium supplemented with a vitamin solution (section 8.5.73), allow the sulfur to settle and examine the supernatant for a faint turbidity of cells. Observe microscopically for small rods, 0.4 to 0.7 μm by 1 to 4 μm, some of which may be motile. Obtain isolated colonies by the shake tube procedure (section 8.4.2). To get a fine, homogeneous distribution of sulfur in the agar tubes, add 3 drops of an autoclaved polysulfide solution (10 g of $Na_2S \cdot 9H_2O$ and 3 g of flowers of sulfur dissolved in 15 ml of distilled water) per 50 ml of the agar medium. Look for development of pink to ochre colonies. Because the sulfide formed by *D. acetoxidans* eventually causes inhibition of growth (no more than 0.1% H_2S can be tolerated), an alternative enrichment method has been devised in which the sulfur medium is inoculated not only with the mud or water sample but also with a pure culture of a green sulfur bacterium (of the family *Chlorobiaceae*). The latter organism continuously consumes the H_2S formed by *D. acetoxidans* by reoxidizing it to elemental sulfur, allowing fast-growing and highly enriched cultures which can be directly used to isolate pure cultures (96).

8.2.12.12. Toluene for toluene oxidizers

To enrich for toluene-oxidizing bacteria, use cotton-stoppered flasks containing basal inorganic medium A or B (section 8.5.9). Inoculate with a small quantity of moist soil previously exposed to toluene vapor (i.e., incubated in a closed chamber containing a beaker of water saturated with toluene) for several days, or use fresh soil. Incubate the flasks at 25 to 30°C for 1 to 3 weeks in a closed chamber containing a beaker of water saturated with toluene. When growth occurs, obtain isolated colonies by streaking onto solidified medium and incubating the plates at 25 to 30°C in a toluene-containing atmosphere.

8.2.12.13. Bacterial cells as substrate for myxococci

Myxococcus species lyse the cells of other bacteria by means of bacteriolytic enzymes and use the compounds liberated from the bacteria for growth. Take advantage of this in isolating myxococci from samples of soil, water, or plant material (79, 95, 113). Obtain bacterial cells for use as substrate (*Enterobacter aerogenes* is suitable and convenient) by removing an entire 4-mm colony from the surface of an agar medium or by centrifuging the bacteria from a broth culture and washing them several times. Make a streak or smear of the bacteria ca. 1 cm wide and 4 cm long on the surface of a plate of water-agar medium (1.5% agar in distilled water). The smear of cells should be sufficiently thick to be barely visible to the eye. At one end of the smear, place two or three particles of soil or bits of plant material (e.g., bark or leaf). After 2 to 3 days and at daily intervals thereafter, examine the plates with a dissecting microscope for evidence of dissolution of the bacterial smear near the added particles. Also look for trails of gliding bacteria and development of fruiting bodies (usually yellow, orange, or pink) on and at the sides of the smear. Transfer the fruiting bodies found the greatest distance from the smear to an agar medium such as 0.2% tryptone or Casitone (Difco Laboratories, Detroit, Mich.) and 1.5% agar; crush each fruiting body in a drop of sterile water between two slides before streaking, in order to liberate the myxospores. The incorporation of cycloheximide (25 µg/ml) in both the nonnutritive agar and the subsequent plating medium will retard the growth of fungi, thereby aiding the isolation of the myxococci.

8.2.12.14. Unusual carbon sources for nitrogen fixers

The use of unusual carbon sources in nitrogen-free enrichment media can be exploited for the isolation of many of the N_2-fixing species (124). The use of mannitol in *Azotobacter* medium (section 8.5.7) is a case in point. Very few contaminants are able to metabolize mannitol, which gives azotobacters a selective advantage. Ethanol is also a good selective carbon source to use for azotobacters.

Table 2 lists other examples of N_2 fixers that can be enriched and isolated by using specific compounds as sole carbon sources in nitrogen-free enrichment media.

Table 2. Some unusual carbon sources for free-living nitrogen fixers[a]

Bacterium	Carbon source
Azotobacter vinelandii	L-Rhamnose Ethylene glycol Erythritol D-Arabitol
Azotobacter beijerinckii	L-Tartrate *o*-Hydroxybenzoate D-Glucuronate D-Galacturonate
Azotobacter armeniacus	Caprylate
Azomonas spp.	Benzoate
Beijerinckia indica	Citrate
Beijerinckia mobilis	Formate Benzoate
Derxia gummosa	Methane Methanol

[a]See references 20, 106, 123, and 124.

8.2.13. Chemotactic Attraction

Actinoplanetes are members of the actinomycete family *Actinoplanaceae* and produce sporangia, which release motile spores. They were first isolated by using baiting techniques (section 8.3.5), but better recovery was achieved by using their attraction to chloride ions (94). A special chamber can be constructed (94), or a plastic tissue culture tray can be used. If using the latter, cut a 3-mm-deep channel in the connecting bridge between two wells. Place a soil sample (0.5 g or less) in the bottom of both wells. Cover the soil with sterile water to a level of 2 mm above the bottom of the connecting channel, and let stand at 30°C for 1 h. Place a 1-µl capillary (Micro-caps; Drummond Scientific Co., Broomall, Pa.) filled with sterile phosphate buffer (5 to 10 mM [pH 6.8] containing 2 mM KCl), keeping the tip of the capillary 1 mm below the surface of the water. Let stand for 1 h, remove the capillary, wash with a few drops of sterile water, and blow the contents into 1 ml of sterile water or buffer. Spread a portion of the suspension on the surface of a plating medium such as casein-starch agar (section 8.5.20). After incubation, most of the colonies, especially the pigmented ones, should be actinoplanetes.

8.3. BIOLOGICAL ENRICHMENT

8.3.1. Bacterial Parasitism by Bdellovibrios (119)

The tiny vibrios called bdellovibrios are widely distributed in soil and water and are capable of attaching to a wide variety of gram-negative bacteria, penetrating the cell wall, and multiplying within the periplasmic space, with consequent lysis of the host bacteria. The method for isolating bdellovibrios resembles that used for bacteriophages in many respects, with the difference that most bdellovibrios have a broad host range. Suspend 50 g of a soil sample in 500 ml of tap water, and shake vigorously for 1 h. Centrifuge the suspension for 5 min at 2,000 × *g* to remove the larger particles. Pass the supernatant through membrane filters of decreasing pore size: 3.0, 1.2, 0.8, 0.65, and 0.45 µm. Mix 0.5 ml of the final filtrate with 0.5 ml of a suspension (ca. 5×10^{10} cells per ml) of the host bacterium (for example, *Enterobacter aerogenes* or *Pseudomonas fluorescens*). Add the mixture to 4 ml of molten semi-solid YP medium (section 8.5.77), mix, and pour over the surface of a plate of solid YP medium. After overnight incubation, examine the plates for plaques (areas of lysis). If plaques form within 24 h, they are attributable to bacteriophages rather than to bdellovibrios. Mark such plaques so that they will not be confused with plaques formed by bdellovibrios, which take at least 2 days to appear and grow larger with time. Cut out plaques suspected to be caused by bdellovibrios, suspend them in YP solution, and prepare a dilution series to be applied to lawns of host bacteria to obtain plaques that are well isolated. Examine one of the plaques by phase-contrast microscopy; look for tiny, highly motile vibrios ca. 0.3 μm wide. Suspend material from a plaque in YP broth, pass it through a membrane filter with a pore diameter of 0.45 μm, dilute the filtrate, and plate it onto lawns of the host bacteria. After three successive plaque isolations, the bdellovibrio strains will represent the descendants of a single bdellovibrio cell.

8.3.2. Plant Symbiosis by Rhizobia (11)

The nodules found on the roots of legumes represent a natural enrichment system for symbiotic nitrogen-fixing bacteria of the genera *Rhizobium* and *Bradyrhizobium*. Obtain nodulated roots of alfalfa, soybeans, or red clover, and wash the soil from them. Remove a nodule from the root, leaving a small portion of the root attached to the nodule. Use a camel-hair brush to remove any soil still adhering to the nodule while holding the nodule under running water. Submerge the nodule in a 1:1,000 solution of $HgCl_2$ for 3 to 6 min; move the nodule around occasionally with sterile forceps. Transfer the nodule to 75% ethanol, and agitate it in the solution for several minutes. Remove it to sterile water, and agitate it for several minutes. Add 1 ml of sterile water to each of six sterile petri dishes. Transfer the nodule to the first dish, and crush it with sterile forceps. Mix the exudate with the water. Transfer one or two loopfuls of the suspension to the sterile water in the second dish, and mix. Continue to serially dilute in this manner for the remaining dishes. To each of the dilutions, add molten yeast extract mannitol agar (section 8.5.75) at 45°C. Incubate the solidified plates at room temperature, and subculture from well-isolated colonies.

A baiting technique for enrichment of rhizobia, similar to the enrichment of plant-pathogenic bacteria (section 8.3.5), can also be used. Plant susceptible legume species in soil naturally infested with the rhizobia being sought, and isolate the bacteria from the nodules that form, in the manner described above.

8.3.3. Animal Parasitism (44)

Inoculation of a host animal with a mixed culture containing a pathogen can select for the latter. The pathogen will predominate in the infected animal, often occurring in pure culture in the blood and tissues. Nonpathogenic contaminants are inhibited or destroyed by the defense mechanisms of the animal. For instance, if 1 ml of emulsified sputum containing *Streptococcus pneumoniae* and other bacteria is injected intraperitoneally into a mouse, a pure culture of the pneumococci can be obtained 4 to 6 h later by inserting a sterile, sharp-tipped capillary pipette through the abdomen and collecting some of the peritoneal fluid for cultivation on blood agar plates. A number of other animal pathogens can be similarly enriched by inoculating a host animal. For instance, *Borrelia* spp. can be selected by injecting the blood or tissue suspensions from infected arthropods into young or suckling mice (66), and pathogenic *Leptospira* spp. can be selected by injecting contaminated soil, mud, or water into weanling hamsters or guinea pigs (64).

8.3.4. Plant Parasitism

Plant-pathogenic bacteria can be enriched from their locus of infection by planting a susceptible plant in the area thought to be infected by the pathogen. If the "bait" becomes infected, the pathogen can be isolated from the diseased portions of the plant by appropriate isolation techniques for that pathogen.

8.3.5. Baiting for Actinoplanetes

The chemotactic attraction to chloride has already been mentioned as an enrichment method for actinoplanetes (section 8.2.13). However, actinoplanetes were first isolated in enrichments for aquatic molds, in which baits of various types were floated on water covering soil samples (35). Most of the water molds and actinoplanetes produce motile sporangiospores which are released from the soil-borne mycelium and will colonize the bait. For this type of enrichment of actinoplanetes, place the soil in a petri dish and cover it with charcoal-treated water. Place baits such as pollen grains (e.g., from *Pinus* species), seeds, human hair, or pieces of leaves on the surface of, or partially submerged in, the water. Incubate the plates for a few days, and periodically check the bait with a dissecting microscope for the characteristic mycelium and stalked sporangia of the actinoplanetes. The colonized baits can then be transferred with forceps to the surface of a suitable medium such as casein-starch agar (section 8.5.20).

8.4. ISOLATION

Relatively few methods are available for the isolation of bacteria into pure cultures. Isolation is most commonly done by obtaining individual colonies in or on a solidified nutrient medium by using either a streak plate or pour plate method. However, obtaining a single colony does not always ensure purity, since colonies can arise from aggregates of cells as well as from individual cells. In slime producers, contaminants may be enmeshed in the chains of filaments formed by these organisms. Enrichment may be performed with selective media, but it is best to use nonselective media for purification because contaminants are more likely to grow and be detected on such media. Even with nonselective media, it is best not to pick (subculture) colonies too soon, because slow-growing contaminants may not yet have made their presence known. Also, do not assume that a single plating guarantees purity.

A pure culture should yield colonies that appear similar, and microscopic observation of the culture should reveal cells that are reasonably similar in appearance, particularly in regard to cell diameter and Gram reaction. There are, of course, some exceptions to these criteria; for example, colonies growing from a pure culture may exhibit smooth-rough variation; coccoid bodies, cysts, and spores may occur together in pure cultures of various organisms; and some organisms may show Gram stain variability. Nevertheless, the criteria are generally useful and apply in most cases.

8.4.1. Spatially Streaking or Spreading on Solid Medium

There are many methods for streaking plates of solid media (streak plates), but the one illustrated in Fig. 1 almost invariably yields well-isolated colonies, even when done by a novice. Alternatively, spread dilutions of a mixed culture onto the surface of plates of solid media (for details, see Chapter 11.3.2). For anaerobes, plates streaked or spread under an air atmosphere can subsequently be incubated in an anaerobic jar. Solid media for anaerobes should be freshly prepared and streaked within 4 h to avoid accumulating too much dissolved oxygen. Even so, it takes some time for an anaerobe jar to remove oxygen and establish anaerobic conditions; the use of roll tubes containing prereduced media eliminates this difficulty entirely (55–57). Such tubes are prepared by spinning sealed tubes of melted prereduced media so that the agar solidifies on the walls of the tubes as a thin layer. The method of streaking a roll tube is illustrated in Fig. 2. A roll tube can also

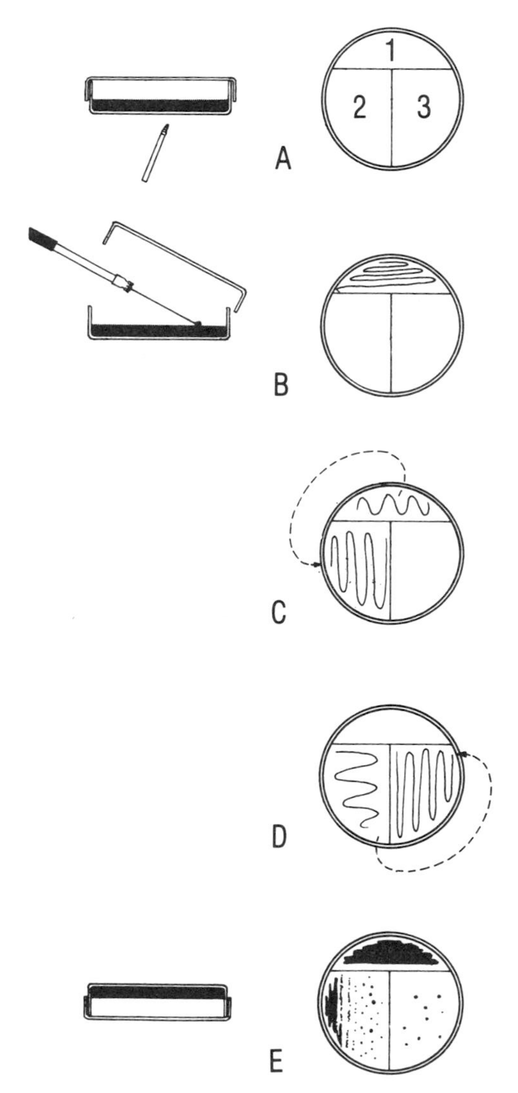

FIG. 1. Useful streak plate method for obtaining well-isolated colonies. (A) With a glass marker pencil, draw a "T" on the bottom of the petri dish to divide the plate into three sections. (B) Streak a loopful of culture lightly back and forth on the surface of the agar over section 1 as shown. Raise the lid of the dish just enough to allow the streaking to be done, then replace it. Flame sterilize the loop, and allow it to cool (15 s). (C) Draw the loop over section 1 as shown, and immediately streak back and forth over section 2. Flame the needle, and allow it to cool. (D) Draw the loop over section 2 as shown, and then streak back and forth over section 3. (E) Incubate the dish in an inverted position as shown, to prevent drops of condensed water on the lid from falling onto the agar surface. Section 1 will develop the heaviest amount of growth, and section 2 or 3 will usually have well-isolated colonies.

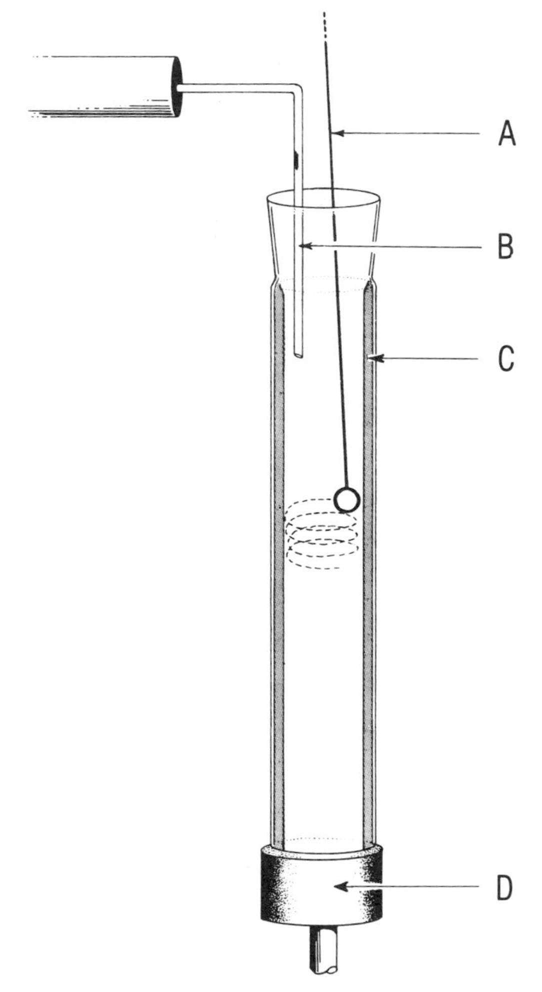

FIG. 2. Streaking an anaerobic roll tube (55). (A) Loop needle (platinum or stainless steel; nichrome will cause oxidation of the medium). (B) Gassing cannula for continuous purging of the tube with oxygen-free gas. (C) Prereduced agar medium coating the inner wall of the tube. (D) Motor-driven tube holder for rotating the roll tube during streaking. Insert the needle with a loopful of inoculum to the bottom of the tube, press the loop flat against the agar, and draw it upward. After streaking in this manner for one-fourth of the way up the tube, turn the loop so that it is perpendicular to the agar (as shown), and continue to streak upward to the top. Remove the gas cannula, replace the rubber stopper in the tube, and incubate the culture in a vertical position.

be inoculated by adding a dilute suspension of cells and then rotating the tube to spread the cells over the surface. Also see Chapter 6.6.4.4.

8.4.2. Serially Diluting in Solidified Medium

The simplest method for preparing a pour plate is to inoculate a tube of sterile, melted agar medium (cooled to 46 to 60°C) with a loopful of the sample, mix, pour the inoculated medium into a petri dish, and allow it to solidify. However, the sample often has to be diluted to produce well-isolated colonies, and the best approach is to use a series of 10-fold dilutions of the sample. Add 1.0-ml portions of each dilution to petri dishes, add 15 to 20 ml of the melted agar medium, mix by rotating the dishes several times, and allow the plates to solidify. For anaerobes, dilutions of the sample can be mixed with the melted, cooled (to 45°C) prereduced medium in roll tubes just before coating the walls of the tubes.

A disadvantage of diluting in agar media is that many of the isolated colonies are submerged in the agar and can be removed only by being dug out with a sterile instrument or punched out with a sterile Pasteur pipette. Another disadvantage is that the bacteria to be isolated must be able to withstand temporarily the 45 to 50°C temperature of the molten agar.

The **shake tube** method has often been used for the isolation of anaerobic bacteria, especially phototrophic and sulfate-reducing bacteria (125). For this method, prepare a series of sterile tubes held in a water bath at 45°C and half fill them with molten agar medium at the same temperature. Inoculate the first tube of the series with a few drops of the mixed culture, and mix gently with the medium. Then transfer 1/10 of the medium to the second tube of the series. Place the first tube in a vertical position, and immerse the bottom part in cold water to solidify the agar. Mix the contents of the second tube, transfer 1/10 of the medium to the third tube, and cool the second tube. Proceed in this manner for the remaining tubes of the series. Then overlay the solidified agar in each tube with a melted, sterile mixture (1:1) of paraffin and paraffin oil (mineral oil) to a depth of ca. 2 cm. This seals the medium from the air. During solidification of the wax, the plug may contract, resulting in an incomplete seal; in this case apply mild local heating and tap the tube to remove any air bubbles under the seal. After the bacteria have grown, select a tube that contains well-isolated colonies, and remove the agar from the tube by first melting and discarding the paraffin seal and then inserting a sterile capillary pipette between the glass wall of the tube and the agar. Push the tip of the pipette down to the bottom of the tube, and apply air pressure to push the column of agar out of the tube into a sterile dish. Dissect the agar to remove the desired colonies.

8.4.3. Serially Diluting to Extinction in Liquid Medium

Serial dilution to extinction in liquid medium is useful when the desired organism cannot grow on solid media. A prerequisite is that the organism desired must be the predominant member of the mixed population. Prepare a dilution of the mixed culture such that, when aliquots are added to a large number of tubes of growth medium, the mean number of bacteria inoculated per tube will be <0.05. In other words, if 100 ml of the dilution contained a total of 5 bacteria, and if 1.0-ml aliquots were inoculated into 100 tubes, the mean number of bacteria inoculated per tube would be 0.05. On incubation, most of the tubes would show no growth, but the few that did would be likely to have received only a single bacterium ($P = 0.975$). The smaller the mean number of bacteria inoculated per tube, the greater is the probability that the growth in the tube arose from a single bacterium. It is therefore imperative that most of the tubes inoculated exhibit no growth, so that the few tubes that do exhibit growth will have a high likelihood of having been inoculated with a single cell. For a discussion of the theoretical aspects of this method, see reference 85 and Chapter 11.4.

8.4.4. Isolating Single Cells

A review of the various methods used for single-cell isolation may be found in reference 65. When many isolations are required, the use of a mechanical micromanipulator is advisable. For occasional needs, the following procedure described by Lederberg (73) is useful. On the back of a clean microscope slide, draw a grid of 5-mm squares with India ink. Sterilize the face of the slide in a flame. After the slide cools, coat it with paraffin (mineral) oil to a depth of ca. 0.5 mm. (It is not necessary to sterilize the oil.) Heat 4-mm glass tubing in a flame, and draw it out to form a capillary with a terminal diameter of ca. 0.1 mm. Attach rubber tubing to the opposite end. Dilute the culture to a density of 10^6 to 10^7 bacteria per ml in medium, and draw up the suspension into the capillary by applying suction to the rubber tubing. Deposit a drop of suspension from the capillary at the center of each square, under the oil. The drops will adhere to the glass and will flatten out to a diameter of 0.1 to 0.2 mm. Scan the flattened drops by phase-contrast or dark-field microscopy, and determine which drops contain only a single cell of the desired type of organism. Frequently, such a single cell can be recovered by repeatedly flushing the drop in and out of a capillary pipette containing sterile medium.

Another procedure is to add a small amount of sterile medium to the drop, incubate the slide in a container of oil until a clone develops, and remove some of the cells of the clone with a capillary pipette. Since the growth conditions for the clone are semianaerobic, this procedure may not work for strictly aerobic organisms.

An ingenious new technique of optical trapping (aptly also called "optical tweezers") has been developed to isolate and manipulate microscopic particles, including bacteria, while they are being viewed through a high-resolution light microscope (15). The technique is based on the use of a single strongly focused laser beam at very low intensity so that a particle becomes trapped from a suspension into the focus spot in a horizontally movable chamber; when the chamber is moved, the trapped particle is dragged along to another position in the chamber. Although not yet perfected, a technique for isolating a single bacterial cell for cultivation involves the use of a low-intensity infrared laser to prevent killing and a completely enclosed shallow multicompartment manipulation chamber. After a single cell is optically trapped, it is dragged from the observation compartment containing the cell suspension into a culture compartment containing growth medium, which is then shut off or removed so that the single-cell culture can be separately observed and eventually subcultured. The technique thus enables a single cell type to be isolated from a predominant or mixed population without prior enrichment. Equipment for optical trapping is available commercially (Cell Robotics Inc., Albuquerque, N.M.), and a review of the technique and the scientific literature to date is available from this source (26a).

8.5. MEDIA AND REAGENTS

8.5.1. *Agrobacterium* Biovar 1 Isolation Medium (108)

Mannitol	10.0	g
$NaNO_3$	4.0	g
$MgCl_2$	2.0	g
Calcium propionate	1.2	g
$MgHPO_4 \cdot 3H_2O$	0.2	g
$MgSO_4 \cdot 7H_2O$	0.1	g
$NaHCO_3$	0.075	g
Magnesium carbonate	0.075	g
Agar	20.0	g
Distilled water	1,000	ml

Adjust pH to 7.1 with 1 N HCl, autoclave, and cool. Add the following compounds aseptically to give the indicated final concentrations (in milligrams per liter).

Berberine	275
Sodium selenite	100
Penicillin G (1,625 U/mg)	60
Streptomycin sulfate	30
Cycloheximide	250
Tyrothricin	1
Bacitracin (65 U/mg)	100

8.5.2. *Agrobacterium* Biovar 2 Isolation Medium (92)

m-Erythritol	5.0	g
$NaNO_3$	2.5	g
K_2HPO_4	0.1	g
$CaCl_2$	0.2	g
NaCl	0.2	g
$MgSO_4 \cdot 7H_2O$	0.2	g
Fe-EDTA solution (0.65%, wt/vol)	2.0	ml
Biotin	2.0	μg
Agar	18.0	g
Distilled water	1,000	ml

Adjust pH to 7.0 with 1 N NaOH, autoclave, and cool. Add the following compounds aseptically to give the indicated final concentrations (in milligrams per liter).

Cycloheximide	250
Bacitracin	100
Tyrothricin	1
Sodium selenite	100

8.5.3. Alkaline Peptone Water for *Vibrio cholerae*

Peptone	10.0	g
NaCl	5.0	g
Distilled water	1,000	ml

Adjust pH to 8.4 with 1 N NaOH and sterilize by autoclaving at 121°C for 15 min.

8.5.4. *Aquaspirillum gracile* Isolation Medium (28)

Peptone	5.0	g
Yeast extract	0.5	g
Tween 80 (sorbitan monooleate polyoxyethylene)	0.02	g
K_2HPO_4	0.1.	g
Agar	10.0	g
Distilled water	1,000	ml

Adjust to pH 7.2. Boil to dissolve the agar. Sterilize at 121°C for 15 min.

8.5.5. *Arthrobacter* Selective Agar (50)

Trypticase Soy Agar (BBL)	4.0	g
Yeast extract (Difco)	2.0.	g
NaCl	20.0	g
Cycloheximide	0.1	g
Methyl red	160.0	mg
Distilled water	1,000	ml

Sterilize a stock solution of the methyl red, and add it aseptically to the rest of the autoclaved medium to give the indicated final concentration. Adjust the pH to that of the soil being sampled.

8.5.6. ASN-III Medium (101)

NaCl	25.0	g
$MgCl_2 \cdot 6H_2O$	2.0	g
KCl	0.5	g
$NaNO_3$	0.75	g
$K_2HPO_4 \cdot 3H_2O$	0.02	g
$MgSO_4 \cdot 7H_2O$	3.5	g
$CaCl_2 \cdot 2H_2O$	0.5	g
Citric acid	0.003	g
Ferric ammonium citrate	0.003	g
EDTA, disodium magnesium salt	0.0005	g
Na_2CO_3	0.02	g
Trace-metal mix A5 (see BG-11 medium [section 8.5.8])	1.0	ml
Deionized water	1,000	ml

After autoclaving and cooling, the pH of the medium should be 7.5.

8.5.7. *Azotobacter* Medium

Mannitol	2.0	g
K_2HPO_4	0.5	g
$MgSO_4 \cdot 7H_2O$	0.2	g
$FeSO_4 \cdot 7H_2O$	0.1	g
Distilled water	1,000	ml

Adjust pH to 7.3–7.6. Sterilize at 121°C for 15 min.

8.5.8. Basal Enrichment Medium Members of the *Rhodospirillaceae* (125)

Solution A

$NaHCO_3$	2.0	g
Distilled water	25.0	ml

Sterilize by filtration (positive pressure).

Solution B

NH_4Cl	1.0 g
KH_2PO_4	0.5 g
$MgCl_2$	0.5 g
(NaCl, for organisms from brackish or marine environments)	(20–30 g)
Trace-metal solution (see below)	1.0 ml
Distilled water	975 ml

Sterilize at 121°C for 15 min.

Solution C

$Na_2S \cdot 9H_2O$	3.0 g
Distilled water	200 ml

Sterilize in a flask with a Teflon-covered magnetized stirring bar at 121°C for 15 min. When cool, add 1.5 ml of sterile 2 M H_2SO_4 with stirring.

To prepare medium, combine solutions A and B. Adjust the pH to 7 with sterile Na_2CO_3 and H_3PO_4 as required. At the time of inoculation of primary enrichment cultures, add 1 ml of solution C per 100 ml of medium. For subsequent transfers, when members of the *Rhodospirillaceae* have established themselves, omit the Na_2S and substitute $MgSO_4$ for the $MgCl_2$ in the medium.

Trace-element solution

EDTA, disodium salt	500 mg
$FeSO_4 \cdot 7H_2O$	200 mg
$ZnSO_4 \cdot 7H_2O$	10 mg
$MnCl_2 \cdot 4H_2O$	3 mg
H_3BO_3	30 mg
$CoCl_2 \cdot 6H_2O$	20 mg
$CuCl_2 \cdot 2H_2O$	1 mg
$NiCl_2 \cdot 6H_2O$	2 mg
$Na_2MoO_4 \cdot 2H_2O$	3 mg
Deionized water	1,000 ml

Dissolve the EDTA in a portion of the water. Separately dissolve the other ingredients in water and add them to the EDTA solution. Adjust the solution to ca pH 3, and bring to a final volume of 1,000 ml.

8.5.9. Basal Inorganic Media A and B (33)

Medium A

Solution 1

$(NH_4)_2SO_4$	1.2 g
$CaCl_2 \cdot 2H_2O$	0.1 g
$MgSO_4 \cdot 7H_2O$	0.1 g
Ferric citrate	0.002 g
Distilled water	1,000 ml

Sterilize at 121°C for 15 min.

Solution 2

K_2HPO_4	0.2. g
KH_2PO_4	0.1 g
Distilled water	200 ml

Sterilize at 121°C for 15 min.

To prepare medium, combine solutions 1 and 2 aseptically. For a solid medium, use 20 g of agar per liter.

Medium B

K_2HPO_4	0.8 g
KH_2PO_4	0.2 g
$CaSO_4 \cdot 2H_2O$	0.05 g
$MgSO_4 \cdot 7H_2O$	0.5 g
$FeSO_4 \cdot 7H_2O$	0.01 g
$(NH_4)_2SO_4$	1.0 g
Distilled water	1,000 ml

Sterilize at 121°C for 15 min. For a solid medium, use 20 g of agar per liter.

8.5.10. Basal Salts Medium for *Thermus* Species (25, 26)

Solution 1

Nitrilotriacetic acid	1.0. g
$CaSO_4 \cdot 2H_2O$	0.6 g
$MgSO_4 \cdot 7H_2O$	1.0. g
NaCl	0.08 g
KNO_3	1.03 g
$NaNO_3$	6.89 g
Na_2HPO_2	1.11 g
Distilled water	1,000 ml

Solution 2

$FeCl_3$ solution	0.028 g
Distilled water	1,000 ml

Solution 3

$MnSO_4 \cdot H_2O$	0.22 g
$ZnSO_4 \cdot 7H_2O$	0.05 g
H_3BO_3	0.05 g
$CuSO_4$	0.0016 g
$Na_2MoO_4 \cdot 2H_2O$	0.0025 g
$CoCl_2 \cdot 6H_2O$	0.0046 g
Distilled water containing 0.5 ml of H_2SO_4 per liter	1,000 ml

To prepare 1 liter of basal salts medium, combine 100 ml of solution A, 10 ml of solution B, and 10 ml of solution C, adjust the pH to 8.2, and bring to a final volume of 1,000 ml with distilled water. Sterilize by autoclaving. To make the complete medium for *Thermus* species, add 1.0 g of tryptone and 1.0 g of yeast extract per liter of basal salts medium before autoclaving. For a solid medium, add 15 g of agar per liter, boil to dissolve the agar, and then autoclave.

8.5.11. BCYE Agar for *Legionella* Species

Add 10.0 g of ACES buffer (*N*-2-acetamindo-2-aminoethanesulfonic acid) to 500 ml of distilled water, and dissolve by heating in a water bath at 45 to 60°C. Mix this solution with 440 ml of distilled water to which 40 ml of 1.0 N KOH has been added (see note below). Add the following ingredients: activated charcoal (Norit SG: no. C5510; Sigma Chemical Co., St. Louis, Mo.), 2.0 g; yeast extract, 10.0 g; and agar, 17.0 g. Dissolve by boiling, autoclave at 121°C for 15 min, and cool to 50°C. Aseptically add L-cysteine · HCI · H_2O solution (0.4 g in 10 ml of distilled water, sterilized by filtration), followed by ferric PP_i solution (0.25 g of soluble ferric PP_i in 10 ml of distilled water, sterilized by filtration). The ferric PP_i must be kept dry and stored in the dark until used; it is not usable if the color changes from green to yellow or brown. Do not use heat over 60°C to dissolve the ferric PP_i; a 50°C water bath is satisfactory. Check the pH, and adjust if necessary (see below). Dispense 20-ml portions of the complete medium into petri dishes; swirl the medium between pouring plates to keep the charcoal particles suspended. The pH of the final solid medium should be 6.9 ± 0.05 at room temperature. The pH is critical since legionellas do not grow below pH 6.8 or above pH 7.0

Note: When checking the pH, hold the bulk medium at 50°C while pouring one plate and checking its pH. When necessary, adjust the bulk medium with either 1.0 N KOH or 1.0 N HCl. Note that the pK_a of ACES buffer is influenced by temperature (0.02/°C); consequently, this must be considered with all pH determinations.

8.5.12. BG-11 Medium (101)

$NaNO_3$	1.5	g
$K_2HPO_4 \cdot 3H_2O$	0.04	g
$MgSO_4 \cdot 7H_2O$	0.075	g
$CaCl_2 \cdot 2H_2O$	0.036	g
Citric acid	0.006	g
Ferric ammonium citrate	0.006	g
EDTA, disodium magnesium salt	0.001	g
Na_2CO_3	0.02	ml
Trace-metal mix A5 (see below)	1.0	ml
Deionized water	1,000	ml

After autoclaving and cooling, the pH of the medium should be 7.4.

Trace-metal mix A5

H_3BO_3	2.86	mg/ml
$MnCl_2 \cdot 4H_2O$	1.81	mg/ml
$ZnSO_4 \cdot 7H_2O$	0.222	mg/ml
$Na_2MoO_4 \cdot 2H_2O$	0.439	mg/ml
$CuSO_4 \cdot 5H_2O$	0.079	mg/ml
$Co(NO_3)_2 \cdot 6H_2O$	0.0494	mg/ml

8.5.13. Bile Esculin Agar (BBL or Difco)

Solution A

Beef extract	3.0	g
Peptone	5.0	g
Agar	15.0	g
Distilled water	400	ml

Solution B

Oxgall	40.0	g
Distilled water	400	ml

Solution C

Ferric citrate	0.5	g
Distilled water	100	ml

Combine the three solutions, and heat to 100°C for 10 min. Sterilize at 121°C for 15 min. Cool to 50°C. Add aseptically 100 ml of a 1% solution of esculin (sterilized by filtration). Dispense into sterile tubes (for slants).

8.5.14. BMS Agar (Potato Agar) (40)

Washed, peeled, sliced potatoes	200	g
L-Malic acid	2.5	g
KOH	2.0	g
Raw cane sugar	2.5	g
Vitamin solution (see section 8.5.45)	1.0	ml
Bromothymol blue (0.5% alcoholic solution)	2	drops
Agar	15.0	g
Distilled water	1,000	ml

Place the potatoes in a gauze bag. Boil in 1,000 ml of water for 30 min; then filter through cotton, and save the filtrate. Dissolve the malic acid in 50 ml of water, and add 2 drops of bromothymol blue. Add KOH until the malic acid solution is green (pH 7.0). Add this solution together with the cane sugar, vitamins, and agar to the potato filtrate. Make up the volume to 1,000 ml with distilled water. Boil to dissolve the agar. Sterilize at 121°C for 15 min.

8.5.15. Bordet-Gengou Agar, Modified (75)

Place 125 g of washed, peeled, and sliced potatoes in a gauze bag. Submerge in a mixture of 10.0 ml of glycerol and 500 ml of water. Boil until the potatoes are soft; then strain through the gauze into the water-glycerol. Allow the fluid to stand in a tall cylinder until the supernatant is relatively clear. Decant the supernatant, and make up to 1,000 ml with distilled water. Add 5.6 g of NaCl. Heat. Add 22.5 g of agar, and dissolve by boiling with constant stirring. Sterilize at 121°C for 15 min, and cool to 45°C. Add 200 ml of defibrinated sheep blood (Chapter 25.3.8) aseptically, and mix. Dispense into petri dishes. (Bordet-Gengou Agar Base is available from BBL or Difco.)

8.5.16. Brilliant Green Agar (BBL or Difco)

Pancreatic digest of casein USP	5.0	g
Peptic digest of animal tissue USP	5.0	g
Yeast extract	3.0	g
NaCl	5.0	g
Lactose	10.0	g
Sucrose	10.0	g
Phenol red	0.08	g
Brilliant green	0.0125	g
Agar	20.0	g
Distilled water	1,000	ml

Adjust pH to 6.9. Boil to dissolve agar. Sterilize at 121°C for 15 min.

8.5.17. Brilliant Green Lactose Bile Broth (available from BBL or Difco as Brilliant Green Bile 2%)

Solution A

Peptone	10.0	g
Lactose	10.0	g
Distilled water	500	ml

Solution B

Oxgall	20.0	g
Distilled water	200	ml

Mix solutions A and B. Make up to 975 ml with distilled water. Adjust pH to 7.4. Add 13.3 ml of a 0.1% aqueous solution of brilliant green. Make up the final volume to 1,000 ml with distilled water. Dispense into tubes containing inverted gas vials, and sterilize at 121°C for 15 min.

8.5.18. Brucella Agar (BBL or Difco)

Pancreatic digest of casein	10.0 g
Peptic digest of animal tissue	10.0 g
Glucose	1.0 g
Yeast autolysate	2.0 g
NaCl	5.0 g
Sodium bisulfite	0.1 g
Agar	15.0 g
Distilled water	1,000 ml

Adjust pH to 7.0. Boil to dissolve agar. Sterilize at 121°C for 15 min.

8.5.19. CAS Medium (130)

Yeast extract	1.0 g
Casamino Acids (Difco, not "vitamin free")	7.5 g
Proteose Peptone no.3 (Difco)	5.0 g
Sodium citrate	3.0 g
$MgSO_4 \cdot 7H_2O$	20.0 g
$Fe(NH_4)_2(SO_4)_2$	5.0 mg
K_2HPO_4	7.5 g
NaCl	80.0 g
Distilled water	1,000 ml

Adjust pH to 8.0 with NaOH. Store only in the dark; discard if crystals form.

8.5.20. Casein-Starch Agar (94)

Soluble starch	10.0 g
Casein	1.0 g
K_2HPO_4	0.5 g
$MgSO_4$	5.0 g
Agar	15.0 g
Distilled water	1,000 ml

Adjust pH to 7.0 to 7.5. Boil to dissolve agar. Sterilize at 121°C for 15 min.

8.5.21. CHSS Medium for *Spirillum volutans*

Acid-hydrolyzed casein, vitamin free, salt free (ICN Nutritional Biochemicals, Cleveland, Ohio)	2.5 g
Succinic acid (free acid)	1.0 g
$(NH_4)_2SO_4$	1.0 g
$MgSO_4 \cdot 7H_2O$	1.0 g
NaCl	0.1 g
$FeCl_3 \cdot 6H_2O$	0.002 g
$MnSO_4 \cdot H_2O$	0.002 g
Distilled water	1,000 ml

Adjust pH to 7.0 with 2 N KOH. For semisolid medium, add 1.5 g of agar and boil to dissolve the agar. Sterilize at 121°C for 20 min.

8.5.22. Cystine Tellurite Blood Agar (75)

Heart infusion agar (blood agar base; Chapter 25.3.34)	500 ml
Agar	2.5 g

Adjust pH to 7.4. Boil to dissolve agar. Sterilize at 121°C for 15 min. Cool to 56°C. Add the following ingredients aseptically.

Defibrinated rabbit or sheep blood (Chapter 25.3.8)	25 ml
0.3% potassium tellurite solution (sterilized by autoclaving)	75 ml
L-Cystine, powdered	22 mg

Stir while dispensing medium into petri dishes to keep the cystine suspended.

8.5.23. E Agar (75)

Papaic digest of soy meal USP	20.0 g
NaCl	5.0 g
Agar	10.0 g
Deionized water	1,000 ml

Heat to dissolve ingredients. Cool, and adjust pH to 7.4 with NaOH. Dispense, and sterilize at 121°C for 15 min. Cool to 50°C. To 65 ml of solution, add aseptically 10 ml of yeast dialysate (see below), 25 ml of horse serum, 2 ml of penicillin (10,000 U/ml), and 1 ml of 3.3% thallous acetate (sterilized by filtration). (*CAUTION*: Thallium salts are poisonous!) Dispense 5-ml amounts into petri dishes (10 by 35 mm), and incubate overnight at room temperature before use.

Yeast dialysate.
Suspend 450 g of active dried yeast in 1,250 ml of distilled water at 40°C. Heat at 121°C for 5 min. Dialyze against 1 liter of distilled water at 4°C for 2 days. Discard the dialysis sac and its contents. Sterilize the dialysate at 121°C for 15 min. Store in a freezer.

Biphasic medium. For a biphasic medium, aseptically dispense 3-ml amounts of E agar into sterile screw-cap tubes (16 by 125 mm). After the medium solidifies, overlay it with 3 ml of E broth (section 8.5.24). Store at room temperature.

8.4.24. E Broth (75)

Papaic digest of soy meal USP	20.0 g
NaCl	5.0 g
Glucose	10.0 g
Phenol red, 2% aqueous solutions	2.0 ml
Deionized water	1,000 ml

Adjust pH to 7.6. Dispense and sterilize at 121°C for 15 min. After the broth cools, add the same supplements as for E agar (section 8.5.23).

8.5.25. *Frateuria* Enrichment Medium (121)

Glucose	10.0	g
Ethanol	5.0	g
Yeast extract	5.0	g
Peptone	3.0	g
Acetic acid	0.3	ml
10% potato extract	1,000	ml

8.5.26. Furazolidone (FTO) Agar (36,69,129)

Peptone	10.0	g
Yeast extract	5.0	g
NaCl	5.0	g
Glucose	1.0	g
Agar	12.0	g
Distilled water	1,000	ml

Adjust pH to 7.0, boil to dissolve agar and autoclave at 121°C for 15 min. Cool to 48°C, and add 100 ml of a 0.02% acetone solution of furazolidone with slow stirring. Leave flask open or loosely covered in a water bath for 3 to 5 min to allow the acetone to evaporate.

8.5.27. Giesberger Base Medium (136)

NH_4Cl	1.0	g
K_2HPO_4	0.5	g
$MgSO_4$	0.5	g
Distilled water (see below)	1,000	ml

Adjust to pH 7. Sterilize at 121°C for 15 min. (*Note:* For marine organisms, substitute seawater for the distilled water.)

8.5.28. Glucose-Tryptone Agar (12)

Tryptone (Difco) or Trypticase (BBL)	10.0	g
Glucose	5.0	g
Agar	15.0	g
Bromocresol purple	0.04	g
Distilled water	1,000	ml

Adjust pH to 6.7. Boil to dissolve agar. Sterilize at 121°C for 15 min.

8.5.29. GYC Agar (121)

Glucose	50.0	g
Yeast extract	10.0	g
$CaCO_3$	25.0	g
Distilled water	1,000	ml

Boil to dissolve agar. Sterilize by autoclaving at 121°C for 15 min.

8.5.30. GYPT Medium (27)

Cellobiose	0.2	g
Trypticase (BBL)	0.1	g
Yeast extract	0.1	g
L-Cysteine	0.05	g
Rifampin	0.2	mg
Agar	1.0	g
Tris-HCl buffer (1 M)	5.0	ml
Distilled water	20.0	ml
Seawater	75.0	ml

Replace the seawater with distilled water if attempting the isolation of freshwater strains. Sterilize the rifampin by filtration, and add aseptically to the autoclaved medium. Add 0.1 mg of resazurin to 100 ml of the medium if the medium is to be prereduced.

8.5.31. Halophile Medium (112)

Casamino Acids (Difco)	7.5	g
Yeast extract	10.0	g
Trisodium citrate	3.0	g
KCl	2.0	g
$MgSO_4 \cdot 7H_2O$	20.0	g
$FeCl_3$	0.023	g
NaCl	250	g
Distilled water	1,000	ml

Dissolve the solutes in 800 ml of the distilled water, and adjust the pH to 7.5 to 7.8 with 1 N KOH. Autoclave the medium at 120°C for 5 min, and filter to remove the precipitate. Adjust the pH to 7.4 with 1 N HCl, and make the medium up to 1,000 ml. Sterilize by autoclaving. For a solid medium, add 20 g of agar per liter, boil to dissolve the agar, and then autoclave. Dispense agar medium into petri dishes at 60 to 70°C to prevent premature solidification.

8.5.32. *Hyphomicrobium* Medium (16)

K_2HPO_4	1.74	g
$NaH_2PO_4 \cdot H_2$	1.38	g
$(NH_4)_2SO_4$	0.5	g
$MgSO_4 \cdot 7H_2O$	0.2	g
$CaCl \cdot 2H_2O$	0.025	mg
$FeCl_2 \cdot 4H_2O$	3.5	mg
Methanol	5.0	ml
KNO_3	5.0	g
Trace-element solution (see below)	0.5	ml
Deionized water	1,000	ml

Adjust pH to 7.0 with NaOH. Do not sterilize when using the medium for enrichment. For sterile medium for purification, sterilize at 121°C for 15 min prior to adding the methanol. For solid medium to be used for isolation, add 15 g of agar per liter and omit the KNO_3; sterilize at 121°C for 15 min prior to adding the methanol. Remove the oxygen from all liquid media by bubbling nitrogen through them just before use; the KNO_3 in these media serves as the electron acceptor under anaerobic conditions.

Trace-element solution

$ZnSO_4 \cdot 7H_2O$	50	mg
$MnCl \cdot 4H_2O$	400	mg
$CoCl_2 \cdot 6H_2O$	1	mg
$CaSO_4 \cdot 5H_2O$	0.4	mg
H_3BO_3	2,000	mg

$Na_2MoO_4 \cdot 2H_2O$	500	mg
Distilled water	1,000	ml

8.5.33. Iverson Medium (61)

Trypticase Soy Agar (dehydrated; BBL)	40.0	g
Agar	5.0	g
Sodium lactate, 0.4% solution	600	ml
$MgSO_4 \cdot 7H_2O$	2.0	g
Ferrous ammonium sulfate	0.5	g
Distilled water	400	ml

Adjust pH to 7.2 to 7.4. Boil to dissolve agar. Sterilize at 121°C for 15 min.

8.5.34. Lactate Medium

Yeast extract	1.0	g
Sodium lactate	4.0	g
NH_4Cl	0.5	g
K_2HPO_4	1.0	g
$MgSO_4 \cdot 7H_2O$	0.2	g
$CaCl_2 \cdot 2H_2O$	0.1	g
$FeSO_4 \cdot 7H_2O$	0.1	g
Na_2SO_4	0.5	g
(NaCl, for marine organisms)	(20–30	g)

Sterilize at 121°C for 15 min.

8.5.35. Lowenstein-Jensen Medium (BBL or Difco)

KH_2PO_4	2.4	g
$MgSO_4 \cdot 7H_2O$	0.24	g
Magnesium citrate	0.60	g
Asparagine	3.6	g
Potato flour	30.0	g
Glycerol	12.0	ml
Distilled water	600	ml
Homogenized whole eggs	1,000	ml
Malachite green, 2% aqueous solution	200	ml

Dissolve the salts and asparagine in the water. Add the glycerol and potato flour, and autoclave at 121°C for 30 min. Cleanse whole eggs, not more than 1 week old, by scrubbing with 5% soap solution. Allow to stand for 30 min in soap solution, and rinse thoroughly in cold running water. Immerse the eggs in 70% ethanol for 15 min, remove, and break into a sterile flask. Homogenize by shaking with sterile glass beads. Filter through four layers of sterile gauze. Add 1 liter of the homogenized eggs to the flask of cooled potato-salt mixture. Add the malachite green. Mix well, and dispense into sterile, screw-cap tubes (20 by 150 mm) 6 to 8 ml per tube. Slant the tubes, and inspissate them at 85°C for 50 min. Incubate for 48 h at 37°C to check sterility. Store the tubes in a refrigerator with their caps tightly sealed.

8.5.36. MacConkey Agar (BBL or Difco)

Peptone (Difco) or Gelysate (BBL)	17.0	g
Proteose Peptone (Difco) or Polypeptone (BBL)	3.0	g
Lactose	10.0	g
NaCl	5.0	g
Crystal violet	0.001	g
Agar	13.5	g
Distilled water	1,000	ml

Adjust pH to 7.1. Boil to dissolve agar. Sterilize at 121°C for 15 min.

8.5.37. Mannitol Salt Agar (BBL or Difco)

Beef extract	1.0	g
Proteose Peptone no. 3 (Difco) or Polypeptone (BBL)	10.0	g
NaCl	75.0	g
Mannitol	10.0	g
Phenol red	0.025	g
Agar	15.0	g
Distilled water	1,000	ml

Adjust pH to 7.4. Boil to dissolve agar. Sterilize at 121°C for 15 min. Cool to 55°C (45°C may not prevent premature solidification), and pour into petri dishes.

8.5.38. Mineral Agar (13)

$Na_2HPO_4 \cdot 12H_2O$	9.0	g
KH_2PO_4	1.5	g
$MgSO_4 \cdot 7H_2O$	0.2	g
NH_4Cl	1.0	g
Ferric ammonium citrate	0.005	g
$CaCl_2 \cdot 2H_2O$	0.010	g
Trace-elements solution (see below)	3.0	ml
Double-distilled water	1,000	ml

Adjust pH to 7.1. Sterilize at 121°C for 15 min. (For a solid medium, add 17 g of agar per liter, boil to dissolve the agar, and sterilize by autoclaving.) After the medium has cooled to 45 to 56°C, add sufficient $NaHCO_3$ solution (sterilized by filtration with positive pressure) to give a final concentration of 0.5 g of $NaHCO_3$ per liter.

Trace-element solution

$ZnSO_4 \cdot 7H_2O$	10	mg
$MnCl_2 \cdot 4H_2O$	3	mg
H_3BO_3	30	mg
$CoCl_2 \cdot 6H_2O$	20	mg
$CuCl_2 \cdot 6H_2O$	0.79	mg
$NiCl_2 \cdot 6H_2O$	2	mg
$Na_2MoO_4 \cdot 2H_2O$	3	mg
Double-distilled water	1,000	ml

8.5.39. Mitis-Salivarius Agar (30) (Difco)

Tryptone (Difco)	10.0	g
Proteose Peptone no. 3 (Difco)	5.0	g
Proteose Peptone (Difco)	5.0	g
Glucose	1.0	g
Sucrose	50.0	g
K_2HPO_4	4.0	g
Trypan blue	0.075	g
Crystal violet	0.0008	g

Agar	15.0	g
Distilled water	1,000	ml

Adjust pH to 7.0. Boil to dissolve agar. Sterilize at 121°C for 15 min. Cool to 50 to 55°C. Add 1.0 ml of 0.1% potassium tellurite (sterilized by filtration), mix, and dispense into petri dishes.

8.5.40. MN Medium (101)

$NaNO_3$	0.75	g
$K_2HPO_4 \cdot 3H_2O$	0.02	g
$MgSO_4 \cdot 7H_2O$	0.038	g
$CaCl_2 \cdot 2H_2O$	0.018	g
Citric acid	0.003	g
Ferric ammonium citrate	0.003	g
EDTA, disodium magnesium salt	0.0005	g
Na_2CO_3	0.02	g
Trace-metal mix A5 (see BG-11 medium; section 8.5.11)	1.0	ml
Seawater	750	ml
Deionized water	250	ml

After autoclaving and cooling, the pH of the medium should be 8.3.

8.5.41. Modified Campy BAP (24)

Prepare Brucella agar (section 8.5.17), sterilize by autoclaving, and cool to 50°C. Aseptically add defibrinated whole blood (10%, vol/vol; see Chapter 25.3.8). Add the following antimicrobial agents from filter-sterilized stock solutions to give the indicated final concentrations.

Cephalothin	15	μg/ml
Polymyxin B	2.5	IU/ml
Trimethoprim	5	μg/ml
Vancomycin	10	μg/ml
Amphotericin B	2	μg/ml

8.5.42. Modified Rogosa Medium (78)

Solution A

$MgSO_4 \cdot 7H_2O$	11.5	g
$MnSO_4 \cdot 4H_2O$	2.8	g
$FeSO_4 \cdot 4H_2O$	0.08	g
Distilled water	100	ml

Solution B

Yeast extract	6.0	g
Diammonium hydrogen citrate	2.4	g
KH_2PO_4	7.2	g
Glucose	24.0	g
Tween 80	1.2	g
Distilled water	100	ml

Solution C. Add 6.0 ml of solution A to 100 ml of solution B. Heat gently until the ingredients are dissolved. Then add 60 ml of 4 M sodium acetate–acetic acid buffer (pH 5.37), and make up to a final volume of 200 ml with distilled water. The final pH should be 5.0.

Solution D

Separated raw milk (adjusted to pH 8.5)	1,000	ml
Trypsin	5	g
Chloroform	10	ml

Incubate at 37°C for 24 h, steam for 20 min, filter while hot, and adjust the pH to 6.65 with glacial acetic acid (ca. 0.5 ml per liter).

To prepare complete medium, add 19 g of agar to 700 ml of solution D, and dissolve by autoclaving at 121°C for 20 min. While the mixture is hot, add 185 ml of solution C (previously warmed to 50°C). Make up to 1,000 ml with hot solution D. A sample diluted with 3 parts of warm water should have a pH of 5.35 ± 0.05 at 30°C. Dispense the medium in 10-ml quantities into tubes, and store in a refrigerator without sterilizing. When preparing plates, melt the medium with as little heating as possible to avoid darkening and the formation of a precipitate.

8.5.43. MPSS Agar

Peptone (Difco)	5.0	g
Succinic acid (free acid)	1.0	g
$(NH_4)_2SO_4$	1.0	g
$FeCl_3 \cdot 6H_2O$	0.002	g
$MnSO_4 \cdot H_2O$	0.002	g
Agar	15.0	g
Distilled water (see note)	1,000	ml

Adjust to pH 7.0 with KOH. Boil to dissolve agar. Sterilize at 121°C for 15 min. (*Note:* For marine organisms, substitute seawater for the distilled water.)

8.5.44. *N*-Acetyl-L-Cysteine–Sodium Hydroxide Reagent (44)

Combine equal volumes of 4% NaOH and 2.94% sodium citrate·$3H_2O$. Dissolve 0.5% *N*-acetyl-L-cysteine powder in this solution. Use within 24 h. *Note:* It is best to use distilled water for preparation of the solutions to minimize chances of adding acid-fast tap water contaminants to the specimens.

8.5.45. Nfb Medium (40)

L-Malic acid	5.0	g
K_2HPO_4	0.5	g
$MgSO_4 \cdot 7H_2O$	0.2	g
NaCl	0.1	g
$CaCl_2$	0.02	g
Trace-metal solution (see below)	2.0	ml
Bromothymol blue (5% alcoholic solution)	2.0	ml
Fe EDTA (1.64% solution)	4.0	ml
Vitamin solution (see below)	1.0	ml
KOH	4.0	g
Agar	1.75	g
Distilled water	1,000	ml

Adjust pH to 6.8. Boil to dissolve agar. Sterilize at 121°C for 15 min. Cool to 45 to 50°C, and dispense 4-ml amounts into 6-ml serum vials with rubber diaphragms.

Trace-metal solution

$Na_2MoO_4 \cdot 2H_2O$	0.2	g
$MnSO_4 \cdot H_2O$	0.235	g

H_3BO_3	0.28 g
$CuSO_4 \cdot 5H_2O$	0.008 g
$ZnSO_4 \cdot 7H_2O$	0.024 g
Distilled water	200 ml

Vitamin solution

Biotin	10 mg
Pyridoxine·HCl	20 mg
Distilled water	100 ml

Heating to nearly boiling is required to dissolve the biotin.

8.5.46. Nitrification Medium (98)

Na_2HPO_4	13.5 g
KH_2PO_4	0.7 g
$MgSO_4 \cdot 7H_2O$	0.1 g
$NaHCO_3$	0.5 g
$FeCl_3 \cdot 6H_2O$	0.014 g
$CaCl_2 \cdot 2H_2O$	0.18 g
Distilled water	1,000 ml

Place 75-ml amounts of medium into 250-ml Erlenmeyer flasks, and sterilize at 121°C for 15 min. For nitrite oxidizers, add 0.5 g of $NaNO_2$ per liter prior to sterilization. For ammonium oxidizers, sterilize a stock solution of $(NH_4)_2SO_4$ separately from the basal medium and add aseptically to give a final concentration of 0.5 g/liter.

8.5.47. Pfennig and Biebl Sulfur Medium (96)

KH_2PO_4	1.0 g
NH_4Cl	0.3 g
$MgSO_4 \cdot 7H_2O$	1.0 g
$MgCl_2 \cdot 6H_2O$	2.0 g
(NaCl, for marine organisms)	(20.0 g)
$CaCl_2 \cdot 2H_2O$	0.1 g
Trace-element solution (see below)	10.0 ml
2 M H_2SO_4 solution	2.0 ml
Distilled water	1,000 ml

Sterilize by autoclaving. When cool, add the following components from sterile stock solutions.

Sodium acetate	0.5 g
$NaHCO_3$ (sterilized by filtration under positive pressure)	4.0 g
$Na_2S \cdot 9H_2O$	0.3 g
Biotin	20 μg

Adjust the pH to 7.8. Grind highly purified flowers of sulfur in a mortar together with distilled water, and sterilize at 112 to 115°C for 30 min; decant the excess water. For every 50 ml of medium, add a pea-sized amount of the sulfur. Also add several glass beads to each bottle of medium.

Trace-element solution

EDTA disodium salt	500 mg
$FeSO_4 \cdot 7H_2O$	200 mg
$ZnSO_4 \cdot 7H_2O$	10 mg
$MnCl_2 \cdot 4H_2O$	3 mg
H_3BO_3	30 mg
$CoCl_2 \cdot 6H_2O$	20 mg
$CuCl_4 \cdot 2H_2O$	1 mg
$NiCl_2 \cdot 6H_2O$	2 mg
$Na_2MoO_4 \cdot 2H_2O$	3 mg
Deionized water	1,000 ml

8.5.48. Phenylethyl Alcohol Agar (BBL or Difco)

Pancreatic digest of casein USP	15.0 g
Papaic digest of soya meal USP	5.0 g
NaCl	5.0 g
Phenylethyl alcohol	2.5 g
Agar	15.0 g
Distilled water	1,000 ml

Adjust pH to 7.3. Boil to dissolve the agar, and sterilize at 118°C for 15 min.

8.5.49. Pringsheim Soil Medium (102)

Place one wheat or barley grain in a large test tube, and cover with 3 to 4 cm of garden soil. Fill the tube almost to the top with tap water. Sterilize the medium at 121°C for 30 min.

8.5.50. *Pseudomonas* Isolation Medium (48)

Trypticase Soy Agar (BBL) is supplemented with the following (per liter).

Basic fuchsin	9.0 mg
Cycloheximide	0.9 g
2,3,5-Triphenyl tetrazolium chloride	0.14 g
Nitrofurantoin (see below)	10.0 mg
Nalidixic acid (see below)	23.0 mg

Add appropriate amounts of stock solutions of nitrofurantoin and nalidixic acid, sterilized by filtration, to the autoclaved medium to give the indicated final concentrations.

8.5.51. PTN Medium for *Listeria* Species (133)

Potassium thiocyanate	3.75 g
Nutrient broth, hydrated (Oxoid no. 2)	q.s. to 1,000.00 ml
Nalidixic acid solution	19.0 ml

Autoclave the nutrient broth containing the potassium thiocyanate at 121°C for 15 min. Cool, and add the nalidixic acid solution aseptically to the sterile medium.

Nalidixic acid solution. Dissolve 0.1 g of crystalline nalidixic acid in 10 ml of 1 N NaOH; when dissolved, add 9 ml of sterile distilled water.

8.5.52. RGCA-SC Medium (55)

Glucose	0.0248 g
Cellobiose	0.0248 g
Soluble starch	0.05 g

$(NH_4)_2SO_4$	0.1	g
Resazurin solution (0.025%)	0.4	ml
Distilled water	20	ml
Salts solution (Chapter 25.3.30)	50	ml
Rumen fluid (Randolph Biologicals, Houston, Texas)	30	ml
Cysteine hydrochloride	0.05	g
Hemin solution (Chapter 25.3.30).	1.0	ml
Vitamin K_1 solution (Chapter 25.3.30)	0.02	ml

Prepare in a manner similar to that described for PY broth (Chapter 25.3.30). Dispense 10-ml amounts into tubes containing 0.2 g of agar, and autoclave. To prepare the final medium, melt the contents of two tubes and cool to 50°C. To each tube, aseptically add the following.

Sterile inactivated rabbit serum (heated at 60°C for 4 h)	1.5	ml
Cocarboxylase solution, 0.025% (sterilized by filtration)	0.2	ml
Sterile prereduced PY broth (Chapter 25.3.30)	1.0	ml

Mix, and pour the contents of both tubes into a petri dish. Allow to solidify. Use immediately, or store in an anaerobic jar until needed.

8.5.53. Salmonella-Shigella (SS) Agar (BBL or Difco)

Beef extract	5.0	g
Peptone	5.0	g
Lactose	10.0	g
Bile salts mixture	8.5	g
Sodium citrate	8.5	g
Sodium thiosulfate	8.5	g
Ferric citrate	1.0	g
Brilliant green	0.33	g
Neutral red	0.025	g
Agar	13.5	g
Distilled water	1,000	ml

Adjust pH to 7.0. Heat to boiling to dissolve agar. Do not sterilize by autoclaving. Cool to 42 to 45°C and dispense into petri dishes.

8.5.54. Selenite F Broth (74) (BBL or Difco)

Polypeptone (BBL) or Tryptone (Difco)	5.0	g
Lactose	4.0	g
Na_2HPO_4	10.0	g
Sodium hydrogen selenite	4.0	g
Distilled water	1,000	ml

Adjust pH to 7.0. Use immediately without sterilization, or place tubes in flowing steam for 30 min and store until needed.

8.5.55. *Serpens* Isolation Medium (54)

Yeast extract	0.2	g
Peptone	0.1	g
Hay extract (see below)	10.0	ml
Agar	1.0	g
Distilled water	90	ml

Prepare hay extract by boiling hay in 100 ml of water for 15 min and clarifying the mixture by centrifugation. Adjust pH to 7.0 with KOH before sterilization.

8.5.56. Soybean-Casein Digest Broth, also known as Trypticase Soy Broth (BBL) or Tryptic Soy Broth (Difco)

Pancreatic digest of casein USP	17.0	g
Papaic digest of soy meal USP	3.0	g
NaCl	5.0	g
K_2HPO_4	2.5	g
Glucose	2.5	g
Distilled water	1,000	ml

Adjust pH to 7.3. Sterilize at 118 to 121°C for 15 min.

8.5.57. *Sphaerotilus* Medium (14)

Sodium lactate	100	mg
NH_4Cl	1.7	mg
KH_2PO_4	8.5	mg
K_2HPO_4	21.5	mg
$Na_2HPO_4 \cdot 7H_2O$	34.4	mg
$MgSO_4 \cdot 7H_2O$	22.5	mg
$CaCl_2$	27.5	mg
$FeCl_3 \cdot 6H_2O$	0.25	mg
Distilled water	1,000	ml

Adjust pH, if necessary, to 7.1 to 7.2. Dispense 50 ml volumes into French square bottles. Sterilize at 116°C for 15 min.

8.5.58. *Sporocytophaga Medium*

Whatman Chromedia 11	10	g
KNO_3	0.5	g
$MgSO_4 \cdot 7H_2O$	0.2	g
$CaCl_2 \cdot 2H_2O$	0.1	g
$FeCl_3$	0.02	g
Distilled water	1,000	ml

8.5.59. ST6CX Medium for Cellulolytic Cytophagas (99)

$(NH_4)_2SO_4$	1.0	g
$MgSO_4 \cdot 7H_2O$	1.0	g
$CaCl_2 \cdot 2H_2o$	1.01	g
$MnSO_4 \cdot H_2O$	0.1	g
$FeCl_3 \cdot 6H_2O$	0.2	g
Yeast extract (Difco)	0.02	g
Agar	10.0	g
Distilled water	1,000	ml

Boil to dissolve agar. Sterilize at 121°C for 15 min. From a separately autoclaved stock solution, add sufficient K_2HPO_4 to give a final concentration of 0.1%.

From a filter-sterilized stock solution, add sufficient cycloheximide to give a final concentration of 25 μg/ml. Also add 1.0 ml of the following filter-sterilized trace-element solution per liter of medium.

Trace-element solution

$MnCl_2 \cdot 4H_2O$	100 mg
$CoCl_2$	20 mg
$CuSO_4$	10 mg
$Na_2MoO_4 \cdot 2H_2O$	10 mg
$ZnCl_2$	20 mg
$LiCl_2$	5 mg
$SnCl_2 \cdot 2H_2O$	5 mg
H_3BO_3	10 mg
KBr	20 mg
KI	20 mg
EDTA, Na-Fe^{3+} salt (trihydrate)	8 g

8.5.60. Standard Methods Agar (12) (BBL or Difco)

Pancreatic digest of casein USP	5.0 g
Yeast extract	2.5 g
Glucose	1.0 g
Agar	15.0 g
Distilled water	1,000 ml

Adjust pH to 7.0. Boil to dissolve agar. Sterilize at 121°C for 15 min.

8.5.61. Succinate-Salts Medium

Sodium succinate	4.0 g
KNO_3	0.5 g
K_2HPO_4	0.5 g
$MgSO_4 \cdot 7H_2O$	0.2 g
$CaCl_2 \cdot 2H_2O$	0.1 g
$FeSO_4 \cdot 7H_2O$	0.2 g

Adjust to pH 7.0. Sterilize at 121°C for 15 min.

8.5.62. TCBS Agar (2) (BBL)

Sodium thiosulfate	10.0 g
Sodium citrate	10.0 g
Oxgall	5.0 g
Sodium cholate	3.0 g
Sucrose	20.0 g
Pancreatic digest of casein USP	5.0 g
Peptic digest of animal tissue USP	5.0 g
Yeast extract	5.0 g
NaCl	10.0 g
Iron citrate	1.0 g
Thymol blue	0.04 g
Bromothymol blue	0.04 g
Agar	14.0 g
Distilled water	1,000 ml

The final pH should be 8.6. Heat with agitation, and boil for 1 min. Cool to 45 to 50°C, and dispense into petri dishes. Do not autoclave.

8.5.63. TGYM Medium (90)

Tryptone	5.0 g
Yeast extract	3.0 g
Glucose	1.0 g
DL-Methionine	1.0 g
Agar	15.0 g
Distilled water	1,000 ml

Adjust pH to 7.0. Boil to dissolve agar. Sterilize by autoclaving at 121°C for 15 min.

8.5.64. Thallous Acetate Agar

Thallous acetate (*CAUTION:* Thallium salts are poisonous!)	1.0 g
Peptone	10.0 g
Yeast extract	10.0 g
Glucose	10.0 g
Agar	13.0 g
Distilled water	1,000 ml

Adjust pH to 6.0. Boil to dissolve agar, and sterilize at 118°C for 15 min. Cool to 45 to 50°C, and add aseptically 10 ml of triphenyl tetrazolium chloride (1% aqueous solution, sterilized by filtration). Dispense into petri dishes.

8.5.65. Thayer-Martin Agar

This medium is made most conveniently by using commercial concentrates and solutions or by using prepared media (BBL or Difco).

Solution A

Pancreatic digest of casein USP	15.0 g
Peptic digest of animal tissue USP	15.0 g
Cornstarch	2.0 g
K_2HPO_4	8.0 g
KH_2PO_4	2.0 g
NaCl	10.0 g
Agar (see note)	20.0 g
Distilled water	1,000 ml

Boil to dissolve agar. Sterilize at 121°C for 15 min. Cool to 50°C. *Note:* Finegold et al. (44) recommend increasing the agar to 30 g.

Solution B

Hemoglobin, dry	20.0 g
Distilled water	1,000 ml

Add dry power gradually to a little water to make a smooth paste; then gradually add the rest of the water. Sterilize at 121°C for 15 min.

Solution C

Vitamin B_{12}	0.010 g
L-Glutamine	10.0 g
Adenine	1.0 g
Guanine hydrochloride	0.03 g
p-Aminobenzoic acid	0.013 g
L-Cystine	1.10 g
Glucose	100.0 g
NAD	0.250 g
Thiamine PP_i (cocarboxylase)	0.100 g
Ferric nitrate	0.020 g
Thiamine hydrochloride	0.003 g
Cysteine hydrochloride	25.9 g
Distilled water	1,000 ml

Sterilize by filtration.

Solution D

Vancomycin	30 mg
Colistin	75 mg
Nystatin	125,000 U
Distilled water	100 ml

Aseptically combine 1,000 ml of solution A, 1,000 ml of solution B, 20 ml of solution C, and 20 ml of solution D. Dispense into petri dishes.

8.5.66. *Thermoplasma* Isolation Medium (37)

KH_2PO_4	3.0 g
$MgSO_4 \cdot 7H_2O$	1.02 g
$CaCl_2 \cdot H_2O$	0.25 g
$(NH_4)_2SO_4$	0.2 g
Yeast extract	1.0 g
Deionized water	1,000 ml

Adjust to pH 2.0 with 10 N H_2SO_4. After autoclaving, add 25 ml of separately sterilized 40% glucose solution to give a final glucose concentration of 1.0%.

8.5.67. *Thiobacillus thiooxidans* Medium (128)

$Na_2S_2O_3 \cdot 5H_2O$	10.0 g
KH_2PO_4	4.0 g
K_2HPO_4	4.0 g
$MgSO_4 \cdot 7H_2O$	0.8 g
NH_4Cl	0.4 g
Trace-metal solution (see below)	10.0 ml
Deionized water	1,000 ml

Sterilize at 121°C for 15 min. Adjust to pH 3.5 to 4.0 with sterile H_2SO_4. For a solid medium, add 15 g of agar per liter before adjusting the pH. Boil to dissolve the agar, sterilize by autoclaving, and cool to 45 to 50°C. Adjust pH to 3.5 to 4.0 with sterile H_2SO_4.

Trace-metal solution

EDTA	50.0 g
$ZnSO_4 \cdot 7H_2O$	22.0 g
$CaCl_2$	5.54 g
$MnCl_2 \cdot 4H_2O$	5.06 g
$FeSO_4 \cdot 7H_2O$	4.99 g
$(NH_4)_6Mo_7O_{24} \cdot 4H_2O$	1.10 g
$CuSO_4 \cdot 5H_2O$	1.57 g
$CoCl_2 \cdot 6H_2O$	1.61 g
Distilled water	1,000 ml

Adjust to pH 6.0 with KOH.

8.5.68. TN Medium for *Listeria* Species (70)

Glucose	2.0 g
Thallous acetate	2.0 g
Nalidixic acid solution	7.6 ml
Nutrient broth, hydrated (Oxoid no. 2)	q.s. to 1,000 ml

Combine ingredients and sterilize by autoclaving at 121°C for 15 min.

Nalidixic acid solution

Dissolve 0.1 g of crystalline nalidixic acid in 10 ml of 1 N NaOH; when dissolved, add 9 ml of distilled water.

8.5.69. Tryptic Soy-Yeast Extract Agar with Urea

Tryptic Soy Broth, dehydrated (Difco)	27.5 g
Yeast extract (Difco)	5.0 g
Glucose	5.0 g
Agar	15.0 g
Distilled water	1,000 ml

Adjust pH to 8.5 with NaOH before autoclaving. After sterilization, add sufficient urea solution (sterilized by filtration) to give a final urea concentration of 1%.

8.5.70. Tryptophan Medium (76)

$MgSO_4 \cdot 7H_2O$	0.2 g
K_2HPO_4	1.0 g
$MnCl_2 \cdot 4H_2O$	0.002 g
$FeSO_4 \cdot 7H_2O$	0.05 g
$CaCl_2$	0.02 g
$Na_2MoO_4 \cdot 2H_2O$	0.001 g
L-Tryptophan	1.0 g
Distilled water	1,000 ml

8.5.71. Veldkamp Medium (126)

NaCl	30.0 g
KH_2PO_4	1.0 g
NH_4Cl	1.0 g
$MgCl_2 \cdot 6H_2O$	0.5 g
$CaCl_2$	0.04 g
$NaHCO_3$	5.0 g
$Na_2S \cdot 9H_2O$	0.1 g
Ferric citrate, 0.004 M solution	5.0 ml
Trace-element solution (see below)	2.0 ml
Powdered agar	5.0 g
Yeast extract	0.3 g
Distilled water	1,000 ml

Adjust to pH 7.0. Use without sterilization for primary enrichment. For a solid medium to be used for pour plates, prepare the medium at double strength with the omission of the powdered agar, and adjust the pH. Sterilize by filtration (positive pressure). Prepare a 4% solution of agar in freshly distilled boiling water, and sterilize by briefly autoclaving. Bring both the double-strength broth and the molten agar to 45 to 50°C, and combine equal volumes aseptically.

Trace-element solution

H_3BO_3	2.8 g
$MnSO_4 \cdot H_2O$	2.1 g
$Cu(NO_3)_2 \cdot 3H_2O$	0.2 g
$Na_2MoO_4 \cdot 2H_2O$	0.75 g
$CoCl_2 \cdot 6H_2O$	0.2 g
$Zn(NO_3)_2 \cdot 6H_2O$	0.25 g

Deionized water	1,000 ml

8.5.72. Violet Red Bile Agar (12) (BBL or Difco)

Yeast extract	3.0 g
Peptone (Difco) or Gelysate (BBL)	7.0 g
Bile salts mixture	1.5 g
Lactose	10.0 g
NaCl	5.0 g
Neutral red	0.03 g
Crystal violet	0.002 g
Agar	15.0 g
Distilled water	1,000 ml

Adjust to pH 7.4. Boil to dissolve agar. Cool to ca. 45°C, and use for pour plates. After the inoculated medium has solidified, overlay it with more medium to prevent surface growth and spreading of colonies.

8.5.73. Widdel and Pfennig Medium (134)

Sodium acetate	1.23 g
Na_2SO_4	2.84 g
KH_2PO_4	0.68 g
$MgCl_2$	0.19 g
NH_4Cl	0.32 g
$CaCl_2$	0.07 g
Trace-element solution (see below)	10 ml
Distilled water	1,000 ml

Sterilize at 121°C for 15 min. Then add the following ingredients from sterile stock solutions.

$FeCl_2$ (from acidified stock solution)	0.00025 g
$NaHCO_3$ (sterilized by positive pressure filtration)	1.68 g
Na_2S (sterilized by filtration)	0.117 g
Vitamin solution (sterilized by filtration; see below)	5.0 ml
Vitamin B_{12} (sterilized by filtration)	20 μg

Adjust the pH of the medium to 7.1 with sterile H_3PO_4. After inoculating media, add 0.0315 g of $Na_2S_2O_4 \cdot 2H_2O$ per liter from a freshly prepared stock solution sterilized by filtration. For marine organisms, and 20 g of NaCl and 1.14 g of $MgCl_2$ per liter of medium.

Trace-element solution

$ZnSO_4 \cdot 7H_2O$	10 mg
$MnCl_2 \cdot 4H_2O$	3 mg
H_3BO_3	30 mg
$CoCl_2 \cdot 6H_2O$	20 mg
$CuCl_2 \cdot 2H_2O$	1 mg
$NiCl_2 \cdot 6H_2O$	2 mg
$Na_2MoO_4 \cdot 2H_2O$	3 mg
Deionized water	1,000 ml

Vitamin solution

Biotin	0.2 mg
Niacin	2.0 mg
Thiamine	1.0 mg
p-Aminobenzoic acid	1.0 mg
Pantothenic acid	0.5 mg
Pyridoxine·HCl	5.0 mg
Distilled water	100 ml

Dissolve ingredients. Sterilize by filtration. Store at 4°C.

8.5.74. Wiley and Stokes Alkaline Medium

Yeast extract	20.0 g
$(NH_4)_2SO_4$	10.0 g
Tris buffer	15.7 g
Distilled water	1,000 ml

Prepare stock solutions of each ingredient separately and sterilize by autoclaving. (The pH of the Tris buffer solution should be 9.0.) Combine the ingredients aseptically from the sterile stock solutions. The final pH should be 8.7. For a solid medium, incorporate 15 g of agar per liter.

8.5.75. Yeast Extract Mannitol Agar

Mannitol	10.0 g
K_2HPO_4	0.5 g
$MgSO_4 \cdot 7H_2O$	0.2 g
NaCl	0.1 g
$CaCO_3$	3.0 g
Yeast extract	0.2 g
Agar	15.0 g
Distilled water	1,000 ml

Boil to dissolve agar. Sterilize at 121°C for 15 min.

8.5.76. Yeast Extract Milk Agar

Yeast extract	3.0 g
Peptone	5.0 g
Fresh whole or skim milk	10.0 ml
Agar	15.0 g
Distilled water	1,000 ml

Dissolve the yeast extract and peptone by steaming, and adjust the cooled medium to pH 7.4. Add agar and milk, and autoclave at 121°C for 20 min. While hot, filter through paper pulp and adjust pH to 7.0 at 50°C. Dispense, and autoclave at 121°C for 15 min. Final pH should be 7.2.

8.5.77. YP Medium (119)

Yeast extract	3.0 g
Peptone	0.6 g
Distilled water	1,000 ml

Adjust pH to 7.2. Sterilize at 121°C for 15 min. For semisolid YP medium to be used as an overlay, add 6.0 g of agar per liter; add 19.0 g of agar per liter for a solid medium. Boil to dissolve agar, and sterilize by autoclaving.

8.6 REFERENCES

8.6.1. General References

1. **Aaronson, S.** 1970. *Experimental Microbial Ecology.* Academic Press, Inc., New York.

Contains a wealth of detailed methods for enrichment and isolation of a great variety of bacteria.

2. **Balows, A., W. J. Hausler, Jr., K. L. Herrmann, H. D. Isenberg, and H. J. Shadomy (ed.).** 1991. *Manual of Clinical Microbiology,* 5th ed. American Society for Microbiology, Washington, D.C.
Principles and methods for enrichment and isolation of pathogenic bacteria, rickettsias, viruses, and fungi.
3. **Balows, A., H. G. Trüper, M. Dworkin, W. Harder, and K.-H. Schleifer (ed.)** 1992. *The Prokaryotes. A Handbook on the Biology of Bacteria: Ecophysiology, Isolation, Identification, Applications,* 2nd ed. Springer-Verlag, New York.
A comprehensive treatment of the procaryotes including isolation and enrichments of all types.
4. **Collins, V. G.** 1969. Isolation, cultivation and maintenance of autotrophs, p. 1–52. *In* J. R. Norris and D. W. Ribbons (ed.), *Methods in Microbiology,* vol. 3B. Academic Press, Inc., New York.
A comprehensive treatment of the principles and techniques for enrichment and isolation of photo- and chemoautotrophs.
5. **Krieg, N. R., and J. G. Holt (ed.)** 1984. *Bergey's Manual of Systematic Bacteriology,* vol. 1. The Williams & Wilkins Co., Baltimore.
6. **Lebeda, D. P.** 1990. *Isolation of Biotechnological Organisms from Nature.* McGraw-Hill Publishing Co., New York.
Covers both procaryotic and eucaryotic microorganisms and presents selected techniques for isolation of organisms of potential biotechnological importance.
7. **Sneath, P. H. A., N. S. Mair, M. E. Sharpe, and J. G. Holt (ed.).** 1986. *Bergey's Manual of Systematic Bacteriology,* vol. 2. The Williams & Wilkins Co., Baltimore.
8. **Staley, J. T., M. P. Bryant, N. Pfenning, and J. G. Holt (ed.).** 1989. *Bergey's Manual of Systematic Bacteriology,* vol. 3. The Williams & Wilkins Co., Baltimore.
9. **Veldkamp, H.** 1970. Enrichment cultures of prokaryotic organisms, p. 305–361. *In* J. R. Norris and D. W. Ribbons (ed.), *Methods in Microbiology,* vol. 3A. Academic Press, Inc., New York.
Emphasizes the theoretical aspects of enrichment cultures and presents methods for the enrichment of specific organisms.
10. **Williams, S. T., M. E. Sharpe, and J. G. Holt (ed.).** 1989. *Bergey's Manual of Systematic Bacteriology,* vol. 4. The Williams & Wilkins Co., Baltimore.
References 5, 7, 8, and 10 contains descriptions of all the genera of procaryotes with discussions of their enrichment and isolation.

8.6.2. Specific References

11. **Allen O. N.** 1957. *Experiments in Soil Bacteriology,* 3rd. ed. Burgess Publishing Co., Minneapolis.
12. **American Public Health Association** 1960. *Standard Methods for the Examination of Dairy Products,* 11th ed. American Public Health Association, New York.
13. **Aragno, M., and H. G. Schlegel.** 1978. *Aquaspirillum autotrophicum,* a new species of hydrogen-oxidizing, facultatively autotrophic bacteria. *Int. J. Syst. Bacteriol.* **28:**112–116.
14. **Armbruster, E. H.** 1969. Improved technique for isolation and identification of *Sphaerotilus. Appl. Microbiol.* **17:**320–321.
15. **Ashkin, A., J. M. Dziedzic, and Y. Yamane.** 1987. Optical trapping and manipulation of single cells using infrared laser beams. *Nature* (London) **330:**769–771.
16. **Attwood, M. M., and W. Harder.** 1962. A rapid and specific enrichment procedure for *Hyphomicrobium* spp. *Antonie van Leeuwenhoek J. Microbiol. Serol.* **38:**369–378.
17. **Barnes, E. M.** 1956. Methods for the isolation of faecal streptococci (Lancefield Group D) from bacon factories. *J. Appl Bacteriol.* **18:**193–203.
18. **Baron, E. J., and S. M. Finegold.** 1990. *Bailey & Scott's Diagnostic Microbiology,* 8th ed. C. V. Mosby Co., St. Louis.
19. **Bazylinski, D. A., R. B. Frankel, and H. W. Jannasch.** 1988. Anaerobic magnetite production by a marine, magnetotactic bacterium. *Nature* (London) **334:**518–519.
20. **Becking, J.-H.** 1984. Genus *Beijerinckia* Derx 1950, p. 315. *In* N. R. Krieg and J. G. Holt (ed.), *Bergey's Manual of Systematic Bacteriology,* vol. 1. The Williams & Wilkins Co., Baltimore.
21. **Belly, R. T., B. B. Bohlool, and T. D. Brock.** 1973. The genus *Themoplasma. Ann. N.Y. Acad. Sci.* **225:**94–107.
22. **Biebl, H., and N. Pfennig** 1981. Isolation of members of the family *Rhodospirillaceae,* p. 267–273. *In* M. P. Starr, H. Stolp, H. G. Trüper, A. Balows, and H. G. Schlegel (ed.), *The Prokaryotes. A. Handbook on Habitats, Isolation, and Identification of Bacteria.* Springer-Verlag, New York.
23. **Blakemore, R. P., D. Maratea, and R. S. Wolfe.** 1979. Isolation and pure culture of a freshwater magnetic spirillum in chemically defined medium. *J. Bacteriol.* **140:**720–729.
24. **Blaser, M. J., I. D. Berkowitz, F. M. LaForce, J. Cravens, L. B. Reller, And W.-L. L. Wang.** 1979. Campylobacter enteritis: clinical and epidemiologic features. *Ann. Intern Med.* **91:**179–184.
25. **Brock, T. D.** 1984. Genus *Thermus* Brock and Freeze 1969, p. 333–339. *In* N. R. Krieg and J. G. Holt (ed.), *Bergey's Manual of Systematic Bacteriology,* vol. 1. The Williams & Wilkins Co., Baltimore.
26. **Brock, T. D., and H. Freeze.** 1969. *Thermus aquaticus* gen. n. and sp. n., a nonsporulating extreme thermophile. *J. Bacteriol.* **98:**289–297.

26a. **Buican, T. N.** 1992. *Optical Trapping: Instrumentation and Biological Applications.* Cell Robotics Inc., Albuquerque, N.M.

27. **Canale-Parola, E.** 1984. Genus I. *Spirochaeta* Ehrenberg 1835, p. 41. *In* N. R. Krieg and J. G. Holt (ed.), *Bergey's Manual of Systematic Bacteriology,* vol. 1. The Williams & Wilkins Co., Baltimore.
28. **Canale-Parola, E., S. L. Rosenthal, and D. G. Kupfer.** 1966. Morphological and physiological characteristics of *Spirillum gracile* sp. n. *Antonie van Leeuwenhoek J. Microbiol. Serol.* **32:**113–124.
29. **Cataldi, M. S.** 1940. Aislamiento de *Beggiatoa alba* en cultivo puro. *Rev. Inst. Bacteriol. Dept. Nac. Hig.* (Buenos Aires) **9:**393–423.
30. **Chapman, G. H.** 1944. The isolation of streptococci from mixed cultures. *J. Bacteriol.* **48:**113–114.
31. **Claus, D., and R. C. W. Berkeley.** 1986. Genus *Bacillus* Cohn 1872, p. 1114–1120. *In* P. H. A. Sneath, N. S. Mair, M. E. Sharpe, and J. G. Holt (ed.), *Bergey's Manual of Systematic Bacteriology,* vol. 2. The Williams & Wilkins Co., Baltimore.
32. **Claus, D., and F. Fahmy.** 1984. Genus *Sporosarcina* Kluyver and Van Niel 1936, p. 1204. *In* P.H.A. Sneath, N. S. Mair, M.E. Sharpe, and J. G. Holt (ed.), *Bergey's Manual of Systematic Bacteriology,* vol. 2. The William & Wilkins Co., Baltimore.
33. **Claus, D., and N. Walker.** 1964. The decomposition of toluene by soil bacteria. *J. Gen. Microbiol.* **36:**107–122.
34. **Corbel, M. J., and W. J. Brinley-Morgan.** 1984. Genus *Brucella* Meyer and Shaw 1920, p. 377–382. *In* N. R. Krieg and J. G. Holt (ed.), *Bergey's Manual of Systematic Bacteriology,* vol. 1. The Williams & Wilkins Co., Baltimore.
35. **Couch, J. N.** 1949. A new group of organisms related to *Actinomyces. J. Elisha Mitchell Sci. Soc.* **65:**315–318.
36. **Curry, J. C., and G. E. Borovian.** 1976. Selective medium for distinguishing micrococci from staphylococci in the clinical laboratory. *J. Clin. Microbiol.* **4:**455-457.
37. **Darland, G., T. D. Brock, W. Samsonoff, and S. F. Conti.** 1970. A thermophilic, acidophilic mycoplasma isolated from a coal refuse pile. *Science* **170:**1416–1418.
38. **Davis, B. D., R. Dulbecco, H. N. Eisen, H. S. Ginsberg, W. B. Wood, and M. McCarty.** 1973. *Microbiology,* 2nd. ed. Harper & Row, Hagerstown, Md.

39. **Dawes, E. A., and P. J. Senior.** 1973. The role and regulation of energy reserve polymers in microorganisms. *Adv. Microb. Physiol.* **10:**135–266.
40. **Döbereiner, J., and V. L. D. Baldani.** 1979. Selective infection of maize roots by streptomycin-resistant *Azospirillum lipoferum* and other bacteria. *Can. J. Microbiol.* **25:**1264–1268.
41. **Döbereiner, J., I. E. Marriel, and M. Nery.** 1976. Ecological distribution of *Spirillum lipoferum* Beijerinck. *Can. J. Microbiol.* **22:**1461–1473.
42. **Farrell, I. D.** 1974. The development of a new selective medium for the isolation of *Brucella abortus* from contaminated sources. *Res. Vet. Sci.* **16:**280–286.
43. **Faust, L., and R. S. Wolfe.** 1961. Enrichment and cultivation of *Beggiatoa alba. J. Bacteriol.* **81:**88–106.
44. **Finegold, S. M., W. J. Martin, and E. G. Scott.** 1978. *Bailey and Scott's Diagnostic Microbiology,* 5th ed. The C. V. Mosby Co., St. Louis.
45. **Freeman, B. A.** 1977. *Burrows' Textbook of Microbiology,* 21st ed. The W. B. Saunders Co., Philadelphia.
46. **Gerloff, G. C., G. P. Fitzgerald, and F. Skoog.** 1950. The isolation, purification and culture of blue-green algae. *Am. J. Bot.* **37:**216–218.
47. **Gibbons, N. E.** 1969. Isolation, growth and requirements of halophilic bacteria, p. 169–183. *In* J. R. Norris and D. W. Ribbons (ed.), *Methods in Microbiology,* vol. 3B. Academic Press, Inc., New York.
48. **Grant, M. A., and J. G. Holt.** 1977. Medium for the selective isolation of members of the genus *Pseudomonas* from natural habitats. *Appl. Environ. Microbiol.* **33:**1222-1224.
49. **Guerro, R., C. Pedrós-Alió, T. N. Schmidt, and J. Mas.** 1985. A survey of buoyant density of microorganisms in pure cultures and natural samples. *Microbiologia* **1:**53–65.
50. **Hagedorn, C., and J. G. Holt.** 1975. A nutritional and taxonomic survey of *Arthrobacter* soil isolates. *Can. J. Microbiol.* **21:**353–361.
51. **Hampp, E. G.** 1957. Isolation and identification of spirochaetes obtained from unexposed canals of pulp-involved teeth. *Oral Surg. Oral Med. Oral Pathol.* **101:**1100-1104.
52. **Hanson, A. W.** 1970. Isolation of spirochaetes from primates and other mammalian species. *Br. J. Vener. Dis.* **46:**303–306.
53. **Hanson, A. W., and G. R. Cannefax.** 1964. Isolation of *Borrelia refringens* in pure culture from patients with condylomata acuminata. *J. Bacteriol.* **88:**111–113.
54. **Hespell, R. B.** 1984. Genus *Serpens* Hespell 1977, p. 373. *In* N. R. Krieg and J. G. Holt (ed.), *Bergey's Manual of Systematic Bacteriology,* vol. 1. The Williams & Wilkins Co., Baltimore.
55. **Holdeman, L. V., E. P. Cato, and W. E. C. Moore (ed.).** 1977. *Anaerobe Laboratory Manual,* 4th ed. Virginia Polytechnic Institute and State University, Blacksburg.
56. **Hungate, R. E.** 1950. The anaerobic cellulolytic bacteria. *Bacteriol. Rev.* **14:**1–49.
57. **Hungate, R. E.** 1969. A roll tube method for cultivation of strict anaerobes, p. 117–132. *In* J. R. Norris and D. W. Ribbons (ed.), *Methods in Microbiology,* vol. 3B. Academic Press, Inc., New York.
58. **Hylemon, P. B., J. S. Wells, Jr., J. H. Bowdre, T. O. MacAdoo, and N. R. Krieg.** 1973. Designation of *Spirillum volutans* Ehrenberg 1832 as type species of the genus *Spirillum* Ehrenberg 1832 and designation of the neotype strain of *S. volutans. Int. J. Syst. Bacteriol.* **23:**20–27.
59. **Inoue, K.** 1976. Quantitative ecology of microorganisms of Syowa Station in Antarctica and isolation of psychrophiles. *J. Gen. Appl. Microbiol.* **22:**143–150.
60. **Inoue, K., and K. Komagata** 1976. Taxonomy study on obligately psychrophilic bacteria isolated from Antarctica. *J. Gen. Appl. Microbiol.* **22:**165–176.
61. **Iverson, W. P.** 1966. Growth of *Desulfovibrio* on the surface of agar media. *Appl. Microbiol.* **14:**529–534.
62. **Jannasch, H.** 1967. Enrichments of aquatic spirilla in continuous culture. *Arch. Mikrobiol.* **69:**165–173.
63. **Jannasch, H.** 1968. Competitive elimination of *Enterobacteriaceae* from seawater. *Appl. Microbiol.* **16:**1616–1618.
64. **Johnson, R. C., and S. Faine.** 1984. Genus I. *Leptospira* Noguchi 1917, p. 64. *In* N. R. Krieg and J. G. Holt (ed.), *Bergey's Manual of Systematic Bacteriology,* vol. 1. The Williams & Wilkins Co., Baltimore.
65. **Johnstone, K. I.** 1969. The isolation and cultivation of single organisms, p. 455–471. *In* J. R. Norris and D. W. Ribbons (ed.), *Methods in Microbiology,* vol. 1. Academic Press, Inc., New York.
66. **Kelly, R. T.** 1984. Genus IV. *Borrelia* Swellengrebel 1907, p. 58. *In* N. R. Krieg and J. G. Holt (ed.), *Bergey's Manual of Systematic Bacteriology,* vol. 1. The Williams & Wilkins Co., Baltimore.
67. **Kenny G. E.** 1974. *Mycoplasma,* p. 333–337. *In* E. H. Lennette, E. H. Spaulding, and J. P. Truant (ed.), *Manual of Clinical Microbiology,* 2nd ed. American Society for Microbiology, Washington, D.C.
68. **Kersters, K., and J. De Ley.** 1984.Genus III. *Agrobacterium* Conn, p. 247. *In* N. R. Krieg and J. G. Holt (ed.), *Bergey's Manual of Systematic Bacteriology,* vol. 1. The Williams & Wilkins Co., Baltimore.
69. **Kocur, M.** 1986. Genus I. *Micrococcus* Cohn 1872, p. 1005. *In* P. H. A. Sneath, N. S. Mair, M. E. Sharpe, and J. G. Holt (ed.), *Bergey's Manual of Systematic Bacteriology,* vol. 2. The Williams & Wilkins Co., Baltimore.
70. **Kramer, P. A., and D. Jones.** 1969. Media selective for *Listeria monocytogenes. J. Appl. Bacteriol.* **32:**381–394.
71. **la Rivière, J. W. M., and K. Schmidt.** 1989. Genus *Achromatium* 1893, p. 2132. *In* J. T. Staley, M. P. Bryant, N. Pfennig, and J. G. Holt (ed.), *Bergey's Manual of Systematic Bacteriology,* vol. 3. The Williams & Wilkins Co., Baltimore.
72. **Leadbetter, E. R.** 1963. Growth and morphogenesis of *Sporocytophaga myxococcoides. Bacteriol. Proc.* 1963, p. 42.
73. **Lederberg, J.** 1954. A simple method for isolating individual microbes. *J. Bacteriol.* **88:**258–259.
74. **Leifson, E.** 1936. New selenite enrichment media for the isolation of typhoid and paratyphoid (*Salmonella)* bacilli. *Am. J. Hyg.* **24:**423–432.
75. **Lennette, E. H., A. Balows, W. J. Hausler, Jr., and J. P. Truant (ed.).** 1980. *Manual of Clinical Microbiology,* 3rd ed. American Society of Microbiology, Washington, D.C.
76. **Lichstein, H. C., and E. L. Oginsky.** 1965. *Experimental Microbial Physiology.* W. H. Freeman and Co., San Francisco.
77. **Linn, D. M., and N. R. Krieg.** 1971. Occurrence of two organisms in the type strain of *Spirillum lunatum:* rejection of the name *Spirillum lunatum* and characterization of *Oceanospirillum maris* subsp. *williamsiae* and an unclassified vibrioid bacterium. *Int. J. Syst. Bacteriol.* **28:**131–138.
78. **Mabbitt, L. A., and M. Zielinska.** 1956. The use of a selective medium for the enumeration of lactobacilli in cheddar cheese. *J. Appl. Bacteriol.* **18:**95–101.
79. **McCurdy, H. D.** 1963. A method for the isolation of myxobacteria in pure culture. *Can. J. Microbiol.* **8:**282–285.
80. **McCurdy, H. D.** 1989. Order *Myxococcales* Tchan, Pochon, and Prévot 1948, p. 2142. *In* J. T. Staley, M. P. Bryant, N. Pfenning, and J. G. Holt (ed.), *Bergey's Manual of Systematic Bacteriology,* vol. 3. The Williams & Wilkins Co., Baltimore.
81. **Matin, A., C. Veldhuis, V. Stegeman, and M. Veenhuis.** 1979. Selective advantage of a *Spirillum* sp. in a carbon-limited environment. Accumulation of poly-ß-hydroxybutyric acid and its role in starvation. *J. Gen. Microbiol.* **112:**349–355.
82. **Matin, A., and H. Veldkamp.** 1978. Physiological basis of the selective advantage of a *Spirillum* sp. in a carbon-limited environment. *J. Gen. Microbiol.* **106:**187–197.
83. **Matsunaga, T., T. Sakaguchi, and F. Tadokoro.** 1991. Magnetite formation by a magnetic bacterium capable of growing aerobically. *Appl. Microbiol. Biotechnol.* **35:**651–655.

84. **Meiklejohn, J.** 1950. The isolation of *Nitrosomonas europaea* in pure culture. *J. Gen. Microbiol.* **4:**185–191.
85. **Meynell, G. G., and E. Meynell.** 1965. *Theory and Practice in Experimental Bacteriology.* Cambridge University Press, London.
86. **Moench, T. T.** Genus *"Bilophococcus,"* p. 1889. *In* J. T. Staley, M. P. Bryant, N. Pfenning, and J. G. Holt (ed.), *Bergey's Manual of Systematic Bacteriology,* vol. 3. The Williams & Wilkins Co., Baltimore.
87. **Morita, R. Y.** 1975. Psychrophilic bacteria. *Bacteriol. Rev.* **39:**144–167.
88. **Murray, R. G. E.** 1984. Genus *Lampropedia* Schroeter 1886, p. 405. *In* N. R. Krieg and J. G. Holt (ed.), *Bergey's Manual of Systematic Bacteriology,* vol. 1. The Williams & Wilkins Co., Baltimore.
89. **Murray, R. G. E.** 1986. Genus I. *Deinococcus* Brooks and Murray 1981, p. 1039. *In* P. H. A. Sneath, N. S. Mair, M. E. Sharpe, and J. G. Holt (ed.), *Bergey's Manual of Systematic Bacteriology,* vol. 2. The Williams & Wilkins Co., Baltimore.
90. **Murray, R. G. E.** 1992. The family *Deinococcaceae,* pp. 3733–3744. *In* A. Balows, H. G. Trüper, M. Dworkin, W. Harder, and K.-H. Schleifer (ed.), *The Prokaryotes. A Handbook on the Biology of Bacteria: Ecophysiology, Isolation, Identification, Applications,* 2nd ed. Springer-Verlag, New York.
91. **Myrivk, Q. N., N. N. Pearsall, and R. S. Weiser.** 1974. *Fundamentals of Medical Bacteriology and Mycology.* Lea & Febiger, Philadelphia.
92. **New, P. B., and A. Keer.** 1972. Biological control of crown gall: field measurements and glass-house experiments. *J. Appl. Bacteriol.* **35:**279–287.
93. **Orchard, V. A., M. Goodfellow, and S. T. Williams.** 1977. Selective isolation and occurrence of nocardiae in soil. *Soil Biol. Biochem.* **9:**233–238.
94. **Palleroni, N. J.** 1980. A chemotactic method for the isolation of *Actinoplanaceae. Arch. Microbiol.* **128:**53–55.
95. **Peterson, J. E.** 1969. Isolation, cultivation and maintenance of the myxobacteria, p. 185–210. *In* J. R. Norris and D. W. Ribbons (ed.), *Methods in Microbiology,* vol. 3B. Academic Press, Inc., New York.
96. **Pfennig, N., and H. Biebl.** 1976. *Desulfuromonas acetoxidans* gen. nov. and sp. nov., a new anaerobic, sulfur-reducing, acetate-oxidizing bacterium. *Arch. Microbiol* **110:**3–12.
97. **Poindexter, J. S.** 1964. Biological properties and classification of the *Caulobacter group. Bacteriol. Rev.* **28:**231–295.
98. **Pramer, D. A., and E. L. Schmidt.** 1964. *Experimental Soil Microbiology.* Burgess Publishing Co., Minneapolis.
99. **Reichenbach, H.** 1989. Genus 1. *Cytophaga* Winogradsky 1929, p. 2015–2050. *In* J. T. Staley, M. P. Bryant, N. Pfennig, and J. G. Holt (ed.), *Bergey's Manual of Systematic Bacteriology,* vol. 3. The Williams & Wilkins Co., Baltimore.
100. **Richardson, J. H., and W. E. Barkley.** 1988. *Biosafety in Microbiological and Biomedical Laboratories,* 2nd. ed. U.S. Department of Health and Human Services, HHS publication no. (NIH) 88-8395, U.S. Government Printing Office, Washington, D.C.
101. **Rippka, R., J. Deruelles, J. B. Waterbury, M. Herdman, and R. Y. Stanier.** 1979. Generic assignments, strain histories and properties of pure cultures of cyanobacteria. *J. Gen. Microbiol.* **111:**1–61.
102. **Rittenberg, B. T., and S. C. Rittenberg.** 1962. The growth of *Spirillum volutans* in mixed and pure cultures. *Arch. Mikrobiol.* **42:**138–153.
103. **Rosebury, T.** 1962. *Microorganisms Indigenous to Man.* McGraw-Hill Book Co., New York.
104. **Rosebury, T., and G. Foley.** 1942. Isolation and pure cultivation of the smaller mouth spirochaetes by an improved method. *Proc. Soc. Exp. Biol. Med.* **47:**368–374.
105. **Rosebury, T., J. B. McDonald, S. A. Ellison, and S. G. Engel.** 1951. Media and methods for separation and cultivation of oral spirochaetes. *Oral Surg. Oral Med. Oral Pathol.* **4:**68–85.
106. **Sampaio, M.-J. A. M., E. M. R. da Silva, J. Döbereiner, M. G. Yates, and F. O. Pedrosa.** 1981. Autography and methylotrophy in *Derxia gummosa,* p.447. *In* A. H. Gibson and W. E. Newton (ed.), *Current Perspectives in Nitrogen Fixation,* Canberra, Australia, Dec. 1–5, 1980. Elsevier/North Holland Biomedical Press, Amsterdam.
107. **Schleifer, K.-H., D. Schüler, S. Spring, N. Weizenegger, R. Amann, W. Ludwig, and M. Köhler.** 1991. The genus *Magnetospirillum* gen. nov.: description of *Magnetospirillum gryphiswaldense* sp. nov. and transfer of *Aquaspirillum magnetotacticum* to *Magnetospirillum magnetotacticum* comb. nov. *Syst. Appl. Microbiol.* **14:**379–385.
108. **Schroth, M. N., J. P. Thompson, and D. C. Hildebrand.** 1965. Isolation of *Agrobacterium tumefaciens–A. radiobacter* group from soil. *Phytopathology* **55:**645–647.
109. **Scotten, H. L., and J. L. Stokes.** 1962. Isolation and properties of *Beggiatoa. Arch. Mikrobiol.* **42:**353–368.
110. **Seeliger, H. P. R., and D. Jones.** 1986. Genus *Listeria* Pirie 1940, p. 1235–1245. *In* P. H. A. Sneath, N. S. Mair, M. E. Sharpe, and J. G. Holt (ed.), *Bergey's Manual of Systematic Bacteriology,* vol. 2. The Williams & Wilkins Co., Baltimore.
111. **Segerer, A. H., and K. O. Stetter.** 1992. The genus *Thermoplasma,* p. 712–718. *In* A. Balows, H. G. Trüper, M. Dworkin, W. Harder, and K.-H. Schleifer (ed.), *The Prokaryotes. A Handbook on the Biology of Bacteria: Ecophysiology, Isolation, Identification, Applications,* 2nd ed. Springer-Verlag, New York.
112. **Sehgal, S. N., and N. E. Gibbons.** 1960. Effect of some metal ions on the growth of *Halobacterium cutirubrum. Can. J. Microbiol.* **6:**165–169.
113. **Singh, B.** 1947. Myxobacteria in soils and composts: their distribution, number and lytic action on bacteria. *J. Gen. Microbiol.* **1:**1–10.
114. **Slack, J. M., and I. S. Snyder.** 1978. *Bacteria and Human Disease.* Year Book Medical Publishers, Chicago.
115. **Smibert, R. M., and R. L. Claterbaugh.** 1972. A chemically-defined medium for *Treponema* strain PH-7 isolated from the intestine of a pig with swine dysentery. *Can. J. Microbiol.* **18:**1073–1078.
116. **Smith, L. D., and V. R. Dowell.** 1974. *Clostridium,* p.376–380. *In* E. H. Lennette, E. H. Spaulding, and J. P. Truant (ed.), *Manual of Clinical Microbiology,* 2nd ed. American Society for Microbiology, Washington, D.C.
117. **Stanier, R. Y., R. Kunisawa, M. Mandel, and G. Cohen-Bazire.** 1971. Purification and properties of unicellular blue-green algae (order *Chroococcales). Bacteriol. Rev.* **35:**171–205.
118. **Stanier, R. Y., N. J. Palleroni, and M. Doudoroff.** 1966. The aerobic pseudomonads: a taxonomic study. *J. Gen. Microbiol.* **43:**159–271.
119. **Stolp, H., and M. P. Starr.** 1963. *Bdellovibrio bacteriovorus* gen. et sp. n., a predatory, ectoparasitic, and bacteriolytic microorganism. *Antonie van Leeuwenhoek J. Microbiol. Serol.* **29:**217–248.
120. **Strohl, W. R.** 1989. Genus I. *Beggiatoa* Trevisan 1842, p. 2095–2096. *In* J. T. Staley, M. P. Bryant, N. Pfennig, and J. G. Holt (ed.), *Bergey's Manual of Systematic Bacteriology,* vol. 3. The Williams & Wilkins Co., Baltimore.
121. **Swings, J., J. De Ley, and M. Gillis.** 1984. Genus III. *Frateuria* Swings et al. 1980, p. 211. *In* N. R. Krieg and J. G. Holt (ed.), *Bergey's Manual of Systematic Bacteriology,* vol. 1. The Williams & Wilkins Co., Baltimore.
122. **Tamir, H., and C. Gilvarg.** 1966. Density gradient centrifugation for the separation of sporulating forms of bacteria. *J. Biol. Chem.* **241:**1085–1090.
123. **Tchan, Y.-T., and P. B. New.** 1984. Genus I. *Azotobacter* Beijerinck 1901, p. 220. *In* N. R. Krieg and J. G. Holt (ed.), *Bergey's Manual of Systematic Bacteriology,* vol. 1. The Williams & Wilkins Co., Baltimore.
124. **Thompson, J. P., and V. B. D. Skerman.** 1979. *Azotobacteraceae: The Taxonomy and Ecology of the Aerobic Nitrogen-Fixing Bacteria.* Academic Press Ltd., London.

125. **Van Niel, C. B.** 1971. Techniques for the enrichment, isolation, and maintenance of the photosynthetic bacteria. *Methods Enzymol.* **23:**3–28.
126. **Veldkamp, H.** 1961 A study of two marine agar-decomposing, facultatively anaerobic myxobacteria. *J. Gen. Microbiol.* **26:**331-342.
127. **Verma, U. K., D. J. Brenner, W. L. Thacker, R. F. Benson, G. Vesey, J. B. Kurtz, P. J. L. Dennis, A. G. Steigerwalt, J. S. Robinson, and C. W. Moss.** 1992. *Legionella shakespearei* sp. nov., isolated from cooling tower water. *Int. J. Syst. Bacteriol.* **42:**404–407.
128. **Vishniac, W., and M. Santer.** 1957. The thiobacilli. *Bacteriol Rev.* **21:**195–213.
129. **Von Rheinbaben, K. E., and R. M. Hodlak.** 1981. Rapid distinction between micrococci and staphylococci with furazolidone agars. *Antonie van Leeuwenhoek J. Microbiol. Serol* **47:**41–51.
130. **Vreeland, R. H.** 1992. The family *Halomonadaceae*, p. 3181–3188. *In* A. Balows, H. G. Trüper, M. Dworkin, W. Harder, and K.-H. Schleifer (ed.), *The Prokaryotes. A Handbook on the Biology of Bacteria: Ecophysiology, Isolation, Identification, Applications*, 2nd ed. Springer-Verlag, New York.
131. **Warke, G. M., and S. A. Dhala.** 1968. Use of inhibitors for selective isolation and enumeration of cytophagas from natural substrates. *J. Gen. Microbiol.* **51:**43–48.
132. **Waterbury, J. B., and R. Y. Stanier.** 1978 Patterns of growth and development in pleurocapsalean cyanobacteria. *Microbiol. Rev.* **42:**2–44.
133. **Watkins, J., and K. P. Sleath.** 1981. Isolation and enumeration of *Listeria monocytogenes* from sewage, sewage sludge and river water. *J. Appl. Bacteriol.* **50:**1–9.
134. **Widdel, F., and N. Pfennig.** 1977. A new anaerobic, sporing, acetate-oxidizing sulfate-reducing bacterium, *Desulfotomaculum* (emend.) *acetoxidans. Arch. Microbiol.* **112:**119–122.
135. **Wiley, W. R., and J. L. Stokes.** 1962. Requirement of an alkaline pH and ammonia for substrate oxidation by *Bacillus pasteurii. J. Bacteriol.* **84:**730–734.
136. **Williams, M. A., and S. C. Rittenberg.** 1957. A Taxonomic study of the genus *Spirillum* Ehrenberg. *Int. Bull. Bacteriol. Nomencl. Taxon.* **7:**49–111.
137. **Williams, R. A. D., and M. S. Da Costa.** 1992. The genus *Thermus* and related microorganisms, p. 3745–3753. *In* A. Balows, H. G. Trüper, M. Dworkin, W. Harder, and K.-H. Schleifer (ed.), *The Prokaryotes. A Handbook on the Biology of Bacteria: Ecophysiology, Isolation, Identification, Applications*, 2nd ed. Springer-Verlag, New York.
138. **Wilson, G. S., and A. A. Miles.** 1964. *Topley and Wilson's Principles of Bacteriology and Immunity*, 5th ed. The Williams & Wilkins Co., Baltimore.
139. **Wolfe, R. A., R. K. Thauer, and N. Pfennig.** 1987. A 'capillary racetrack' for isolation of magnetotactic bacteria. *FEMS Microbiol. Ecol.* **45:**31–35.

Chapter 9

Solid, Liquid/Solid, and Semisolid Culture

NOEL R. KRIEG and PHILIPP GERHARDT

Culture media prepared in the solid state, in the form of firm gels, have been used in bacteriology since adopted by Robert Koch. The most important uses of solidified media stem from their enabling separated colonies to arise from individual cells in a population diluted into or onto a solidified medium. Thus, the streak plate is a simple but effective technique for isolating pure cultures of bacteria (Chapter 8.4.1) and the pour plate, spread plate, and layered plate are similarly valuable for enumerating viable bacteria (Chapter 11.3). Other single-cell techniques that rely on solid culture are multiple-point inoculation with velveteen (Chapters 13.5.4 and 25.2.2.3) and the auxanographic method (Chapter 25.2.2.1). Solid media are also useful in mass culture, bioautography, and physiological studies of bacterial cells.

The purpose of this chapter is to describe the nature of the main solidifying agents and their uses in several solid, liquid/solid, and semisolid culture techniques. The general subject has been reviewed by Codner (1).

9.1. SOLIDIFYING AGENTS

9.1.1 Agar

Agar is extracted from certain red marine macroalgae and is the most commonly used solidifying agent for bacteriological media. Agar mainly consists of two polysaccharides, agarose (the gelling component) and agaropectin, with the former making up about 60% of the mixture. When first extracted, agar is contaminated by algal cell debris and various impurities, most of which must be removed before the product is suitable for bacteriological purposes. A summary of extraction and purification procedures can be found in reference 4. For information about the chemical composition and structure of agar, see references 4 and 24.

Agar is available in various commercial grades, but for most culturing purposes "bacteriological" grade is satisfactory. Lower grades of commercial agar may contain troublesome impurities, e.g., starches, fatty acids, Cu^{2+}, or bleaching agents that are toxic; elevated levels of Ca^{2+} and Mg^{2+} that can cause precipitates; and thermophilic spores that resist usual autoclaving procedures. "Special", "select", "Noble," and "purified" grades contain decreased levels of impurities and thus are suitable for electrophoretic, nutritional, enrichment and isolation, genetic, recombinant DNA, serological, and other special applications. Whenever an agar is used for a special purpose, pretest to ensure effectiveness, check with the manufacturer's technical service for suitability and analysis, and, if possible, use a single control lot number.

A laboratory procedure for washing bacteriological-grade agar free from impurities for special-purpose use is as follows (15). Soak granular agar in about 10 volumes of distilled water under refrigeration for several hours, and filter; do this 10 times over a period of 2 days. Soak the resulting agar with an equal volume of 95% ethanol for 12 h at room temperature, and filter; resoak the agar in fresh ethanol for 4 h, and filter again. Add the agar to boiling 95% ethanol, bring the alcohol to a boil again, and filter. Discard the filtrate, and spread out the washed agar to dry at room temperature.

The most important properties of agar for bacteriological work are as follows: agar is not enzymatically degraded by most bacterial species; agar gels are stable up to 65°C or higher, yet molten agar does not gel until cooled to ca. 40°C; and agar gels have a high degree of transparency.

To prepare an agar-solidified medium, first adjust the pH of the liquid medium to the desired value and then add the granular agar. Bacteriological or higher grades of agar will not alter the pH of the medium appreciably. For solid medium, use 15 to 20 g of agar per liter; for isolating highly motile organisms, even higher concentrations of agar may be required. For semisolid medium, use 1 to 4 g/liter, depending on the consistency required. Different brands and grades of agar may require different concentrations to achieve a particular degree of firmness, and the instructions of the manufacturer should always be consulted in this regard.

After adding the agar to a liquid medium, heat the mixture to boiling and dissolve the agar completely. If the medium is being heated over a flame or on a hot plate, stir it constantly during heating to prevent the agar from settling to the bottom of the container where it can caramelize and char; then bring the medium to a rolling boil for 1 min or so, being careful to avoid having it foam up and over the edge of the container. Alternatively, the agar can be dissolved by placing the mixture in a steamer or microwave oven for an appropriate period. After the agar is melted, mix to ensure uniformity in concentration throughout the medium. Dispense the molten medium into tubes, flasks, or bottles, and sterilize in an autoclave.

When an agar medium with a pH of 6.0 or less is required, initially prepare and sterilize the medium at a pH greater than 6.0; otherwise, the agar will be hydrolyzed during heating and will fail to solidify adequately when the medium is later cooled. Once the medium has been sterilized and cooled to 45 to 50°C, add sufficient sterile acid aseptically to achieve the final pH value desired. It is useful to prepare an extra portion of medium to experiment with in order to determine the correct amount of acid to add to the main batch.

For preparing tubes of slanted medium, place the hot tubes from the autoclave in a tilted position and allow them to cool and solidify. For preparing petri dishes of medium, first cool the molten medium to 45 to 50°C in a water bath and then dispense the medium aseptically into the bottom halves of the sterile dishes (usually 15 to 20 ml per dish), taking care to replace the tops after each dish is poured. Cover the dishes with newspaper or other insulation to prevent condensation of moisture underneath their tops as the agar solidifies.

After the agar solidifies, drops of moisture (water of syneresis) may form on the surface. This moisture should be evaporated before the plates are inoculated. Usually, storage of the dishes in an inverted position overnight at room temperature will result in sufficient drying of the agar. For more rapid drying, invert the dishes on the shelf of a 45°C gravity-convection incubator or in a sterile hood and adjust the agar-containing halves so that they are slightly ajar. Incubate in this manner until the water of syneresis disappears. For some procedures, such as in the use of velveteen replicators (Chapters 13.5.4 and 25.2.2.3), dishes that have been dried more extensively may be required.

For storage of agar dishes under conditions that prevent severe drying, place them inverted in closed polyethylene bags (the bags in which plastic petri dishes are packaged make excellent storage bags for media). Agar dishes can be stored in this way at room temperature for 4 to 5 weeks and under refrigeration for even longer periods. Media containing blood or other heat-labile components should always be refrigerated.

Store all culture media, liquid and solid, in the dark. Storage under illumination, especially sunlight (as near a window), may result in photochemical generation of hydrogen peroxide or other toxic forms of oxygen, which can render the media inhibitory for the growth of bacteria (23, 28). Microaerophiles appear to be particularly reluctant to grow on media that have been subjected to illumination (11).

9.1.2. Carrageenan

Carrageenan, long used in the food and dairy industry, has also been used as a solidifying agent for bacteriological media. Also known as Irish moss or vegetable gelatin, carrageenan is extracted from certain red marine macroalgae. There are several types of carrageenan: kappa, lambda, mu, and iota (24). The potassium salt of kappa carrageenan is capable of forming rigid transparent gels, which can be an effective substitute for agar in many bacteriological media (14, 29).

Carrageenan is considerably less expensive than agar. Like agar gels, carrageenan gels can be used for streaking or spreading inocula. Carrageenan is not degraded by most species of bacteria, and gels stable to temperatures of 60°C can be prepared. However, carrageenan does have certain limitations. One is the high temperature at which liquid carrageenan media must be dispensed into petri dishes (55 to 60°C), which probably precludes, for most species, the use of carrageenan for enumerating viable bacteria by the method of incorporating dilutions of bacterial suspensions into a molten medium before it solidifies and precludes its use in making media containing blood.

Variability may exist between lots of carrageenan (29). Some investigators have found carrageenan to be unsuitable for semisolid media, whereas others have found it to be satisfactory (14). In contrast to agar, carrageenan may cause alterations in the pH of some media during preparation and sterilization; therefore, determine whether and to what magnitude such pH changes occur for any given kind of medium. It may be necessary to compensate for such changes by preparing the medium at a different initial pH value or by increasing the buffering capacity of the medium. Media containing a phosphate buffering system seem to be particularly likely to exhibit a large decrease in pH (14).

The preparation of carrageenan media is similar to that of agar media. Carrageenan type I (Sigma Chemical Co., St. Louis, Mo.) and Gelcarin GP-812 (FMC/Marine Colloids, Rockland, Maine) are commercial grades that consist predominantly of kappa carrageenan and contain a lesser amount of lambda carrageenan; they are satisfactory for general purposes. For gels stable during incubation to 45°C, use 2.0% carrageenan; for gels stable to 60°C, use 2.4% carrageenan (14). After adding carrageenan to a liquid medium, boil the medium to dissolve the carrageenan completely and then sterilize by autoclaving. Cool the sterile medium to 55 to 60°C, dispense into petri dishes, and allow to solidify. Because of its high viscosity, it is difficult to dispense carrageenan media by pipetting; instead, dispense these media by pouring from a bottle or flask. Remove water of syneresis in the same manner as described above for solidified agar.

9.1.3. Silica Gel

Silica gel is an inorganic solidifying agent used in media for solid culture of autotrophic bacteria in the complete absence of organic substances. Such inorganic media can be supplemented with various organic compounds to study the ability of heterotrophic bacteria to use these compounds as sole carbon sources. Vitamin requirements can also be determined by the use of silica gel media.

A variety of methods are available for preparing silica gel media (4). The method of Funk and Krulwich (7) is simple and reliable and yields clear gels that are firm enough to be streaked lightly with an inoculating needle or to be used for spreading an inoculum with a glass rod. Proceed as follows.

Prepare the following solutions.

Double-strength liquid nutrient medium. Prepare as usual but in double strength, and sterilize by autoclaving.

Potassium silicate solution. Add 10 g of powdered silica gel (certified grade 923, 100 to 200 mesh; Fisher Scientific Co., Pittsburgh, Pa.) or silicic acid (reagent grade; J. T. Baker Chemical Co., Phillipsburg, N.J.) to 100 ml of 7% (wt/vol) aqueous KOH, and dissolve by heating. Dispense 20-ml amounts into flasks, and sterilize in an autoclave.

Phosphoric acid solution. Prepare a 20% solution of *o*-phosphoric acid (85%, certified grade; Fisher Scientific Co.).

Add 20 ml of the sterile double-strength liquid nutrient medium to 20 ml of the sterile potassium silicate solution. Rapidly add a measured amount of the phosphoric acid solution sufficient (as determined from prior test, approximately 4 ml) to provide a pH of 7.0. Mix the solutions, and immediately pour into two petri dishes. The medium will begin to solidify in 1 min and will become firm in 15 min. Remove water of syneresis in the same way as described above for solidified agar. After inoculating petri dishes, incubate in a moist atmosphere to prevent drying and cracking of the gels.

9.1.4. Gellan Gum

Gellan gum (GELRITE; Kelco Division of Merck & Co., Inc., San Diego, Calif.) is a purified heterosaccharide polymer from *Pseudomonas* species, which, with added Mg^{2+} or Ca^{2+}, forms a thermostable gel and so can be used as an agar substitute in selected bacteriological media (21). Gellan gum has comparably wide applicability and better clarity than bacteriological-grade agar and has proven especially advantageous for culturing and quantifying thermophilic bacteria at temperatures to 120°C and pressures to 265 atm (26,851 kPa) (6, 13). Prepare solid media solidified with gellan gum in accordance with the manufacturer's or other published directions or after appropriate experimentation.

9.1.5. Pluronic Polyol

Pluronic polyol, a block copolymer of polypropylene oxide and ethylene oxide, is a nontoxic solidifying agent which liquefies on cooling and gels into a semirigid clear jelly at temperatures of 11 to 32°C depending on the polyol concentration. This unusual solidifying agent has been used to isolate heat-sensitive bacteria and to enrich for denitrifying, sulfate-reducing, and methanogenic bacteria (8).

9.1.6. Gelatin

Gelatin was the first solidifying agent used (by Robert Koch) for bacteriological media, but it was soon replaced by agar because gelatin is liquefied by some species and melts at a relatively low temperature. At the 12% concentration effective for solidification, it melts at 28 to 30°C. Gelatin presently is used mainly as a specific hydrolysis test substrate in systematics (Chapter 25.1.29).

9.1.7. Starch

Starch of the type used for gel electrophoresis has been used as a solidifying agent in culture media for thermoacidophilic archaeobacteria such as *Thermoplasma*, *Acidianus*, and *Sulfolobus* species (19, 20). For *Thermoplasma* species, starch plates have been reported to be superior to agar and gellan gum plates (20). Media containing 10 to 12% (wt/vol) starch are boiled to dissolve the polymer, poured like agar into petri dishes, and allowed to solidify and to dry overnight at 4°C (19).

9.1.8. Solidifying Agents for Macromolecule Separations

The following solidifying agents are used for chromatographic and electrophoretic separation of macromolecules: the agarose component of agar, derivatives of cellulose and dextran, polyacrylamides, hydroxyapatite, and styrene-divinyl-benzene copolymer (Chapter 21.3 and 21.5).

9.2. SOLID-CULTURE TECHNIQUES

9.2.1. Gel Surface Culture

The maximum density of bacterial cells is obtained by growth as a colony on a solid surface. Only the interstitial (intercellular) space of the colony is occupied by liquid, although this may represent 20 to 30% of the total volume (27% for close-packed spheres, regardless of their size). If the inoculum is highly diluted, each cell develops into a distinct colony; such a procedure may be desirable even for mass culture if uniformity of colony characteristics is required before harvesting of the solid culture (for example, if one wants only a variant that produces "smooth" colonies). More frequently, the inoculum is not diluted and is spread evenly over the surface so that confluent solid growth results. The solid surface usually is that of an agar or otherwise solidified medium. However, a number of techniques have been described in which colonial growth is obtained on a membrane over a reservoir of liquid or solidified medium (section 9.2.2).

Mass culture of bacteria on a solid surface has certain advantages over that submerged in a liquid medium (Chapter 10), as follows.

(i) Solid cultures are already concentrated, so there is no need to use a centrifuge or other means for harvesting the cells. Instead, pipette a small amount of sterile saline or buffer onto the surface of the agar cultures and then suspend the growth in the fluid by using an appropriately bent sterile glass rod. Crops of cells can be obtained in this manner by harvesting growth from the surface of agar media contained in numbers of petri dishes, in flat culture bottles (e.g., Roux bottles), or in flat dishes (e.g., covered Pyrex household baking dishes). The avoidance or minimization of centrifugation (which produces aerosols) may be particularly useful for safely obtaining masses of pathogenic or otherwise harmful bacteria.

(ii) Solid cultures are relatively free from macromolecular components and completely free from particulate components of the nutrient medium, because these tend to be held within the agar gel. Furthermore, solid cultures are also relatively free from small molecular nutrients and their own metabolic products because these tend to be diluted into the much larger volume of the medium. Consequently, solid cultures may be particularly useful for preparing antigens or for other purposes for which cell purity is important.

(iii) Solid cultures may yield results that are otherwise unobtainable. For example, fruiting bodies of myxobacteria and endospores of certain *Bacillus* species are better produced from growth on solid medium (sometimes this is the only medium from which they can be obtained). The reasons for such results include the ability to retain physical associations among cells, the maximum oxygen supply from the atmosphere, and the diffusion and dilution away from the immediate cell environment of metabolic products (e.g., acids) that otherwise would inhibit growth or development.

(iv) Solid culture is particularly useful in the isolation and cultivation of single cells (12) (Chapter 8).

On the other hand, solid culture of bacteria has certain disadvantages compared with liquid culture, as follows.

(i) Solid culture is limited in the scale to which it can be increased (e.g., one can effectively produce gram amounts of cells, but dekagram amounts become difficult and hectogram or kilogram amounts are impossible in the laboratory).

(ii) Solid cultures are not homogeneous in physiological properties of the cells. For example, the cells at the top of an aerobic colony (which is likely to be 1,000 cells deep) will be nutrient starved but oxygen rich, whereas the opposite situation will prevail at the bottom of the colony.

(iii) Solid cultures often yield a small number of cells from a given amount of medium.

9.2.2. Membrane Surface Culture

Solid culture of bacteria can also be accomplished on the surface of a dialysis or microfiltration membrane in contact with an underlying reservoir of liquid or solidified medium. The principles and development of such colonial growth with membranes have been reviewed by Schultz and Gerhardt (18). This technique is particularly useful for viable counts of dilute concentrations of cells in air (Chapter 2.1.1) or water (Chapter 2.1.2) after membrane filtration, for studying conjugational transfer of plasmids (27), and for studying microbial interactions (16).

9.2.3. Bioautography

Bioautography is a version of paper chromatography in which the growth of bacteria is used as a highly sensitive indicator for locating the positions of certain compounds on a paper chromatogram. The method has the advantage of specifically detecting compounds with biological activity, which chemical or radioisotopic detection systems lack. The method is particularly applicable to locating the position of growth factors on chromatograms of spent culture media or cell extracts when the concentration of the growth factors is so low as to preclude the use of ordinary detection systems. For example, the location of as little as 5 to 10 ng of folic acid on a chromatogram can be determined by use of bioautography. For a good example of bioautography applied to the detection of growth factors in cell extracts, see reference 22. Bioautography is also widely used by the pharmaceutical industry for the detection of antibiotic agents on paper chromatograms (30).

For the detection of growth factors, prepare an agar medium containing all of the nutrients and growth factors required to support dense growth of the indicator strain, except for the particular growth factor to be tested. Inoculate the sterile molten medium at 45 to 50°C with a washed suspension of the indicator organism. Pour this "seeded" agar medium into a sterile, covered dish large enough to accommodate the paper chromatogram (a Pyrex household baking dish is often suitable). After the medium has solidified, place the dried paper chromatogram face down on the agar surface, taking care to avoid air spaces between the agar and the paper. Trace the outline of the chromatogram on the underside of the dish with a glass-marking pen, and also mark the location of the origin and the solvent front. (It is best to keep the dish horizontal while marking the underside to avoid dislodging the agar.) After incubation, remove the chromatogram from the agar surface and look for areas of dense growth in the agar that indicate the location of the growth factor. Flooding the agar surface with water often helps to increase the visibility of these areas.

For such bioautography to be successful, it is usually helpful to grow the inoculum for the agar in a broth containing a minimum concentration of the particular growth factor to be tested; this is to avoid accumulation of excess growth factor within the bacteria, which might allow them to grow nonspecifically throughout the seeded agar medium. For maximum sensitivity of detection, it is helpful to use a high concentration of washed cells to seed the agar, so that easily visible areas of dense growth will occur where the growth factor is located on the chromatogram. Perform preliminary experiments to determine the most suitable concentration of the inoculum. Contaminating bacteria that may be present on the paper chromatogram usually do not interfere with the experiments, since their growth occurs as discrete colonies, which are not easily confused with the more dense, diffuse growth of the indicator organisms within the agar. However, attempt to keep contamination to a minimum. For example, store the dried chromatograms in sterile envelopes until needed, and apply the chromatograms to seeded agar in a room or cabinet where there are no dust-carrying air currents.

For bioautography of antibiotic agents, the agar medium should be nutritionally complete to allow the indicator organism to grow throughout the dish except at regions of the chromatogram where the antibiotic substance is located. Because of the high potency of some antibiotic substances, it is sometimes advisable to leave the paper chromatogram on the agar surface for only a short time (e.g., 10 min) and then remove it before incubating the dish. If the chromatogram is left on the agar surface, the zones of growth inhibition may be too large to be clearly resolved.

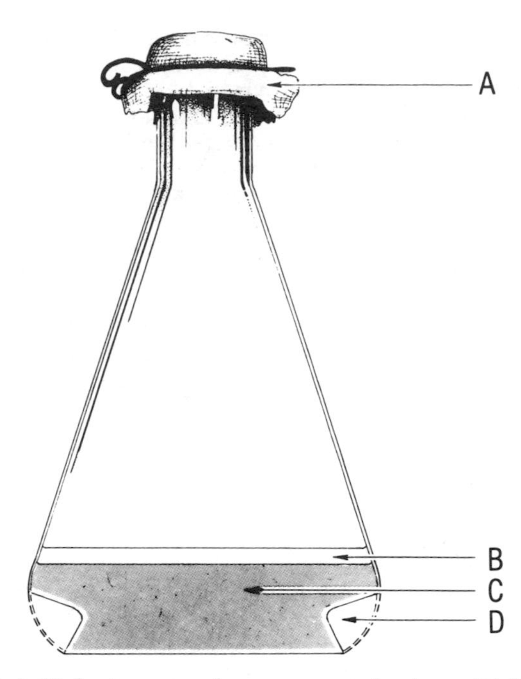

FIG. 1. Biphasic system for concentrated culture. (A) Cotton and gauze pad to provide an adequate supply of air (Chapter 10.1.2.1). (B) Liquid overlay of broth medium or water diffusate. (C) Solid base of agar medium. (D) Flask indentations to hold the agar base and prevent its breakup during incubation on a shaking machine.

9.3. LIQUID/SOLID (BIPHASIC) CULTURE TECHNIQUES

The advantages of solid-surface culture can be retained and the disadvantages can be obviated by using a liquid/solid culture technique called biphasic culture (26). Both techniques essentially represent a form of dialysis or diffusion culture in which an interface separates the culture from the nutrient and product reservoir. The principles and development of interfacial dialysis culture in the context of membrane dialysis culture have been reviewed by Schultz and Gerhardt (18), and membrane dialysis culture techniques are described in this volume in the chapter on liquid culture (Chapter 10.3.5). A biphasic system for concentrated bacterial culture was systematically studied by Tyrrell et al. (26).

The system used for the biphasic technique consists simply of a thick layer of solidified nutrient medium overlaid with a thin layer of nutrient broth (Fig. 1). To prepare such a system, partially fill the container with hot medium containing 2 to 3% agar. After the agar is solidified, overlay it aseptically with a small volume of broth, inoculate, and incubate. If the bacteria are aerobic, clamp the container on a mechanical shaker to provide aeration and agitation of the broth during incubation. The culture is confined to the liquid overlay but has diffusional access to the reservoir of nutrients in the solidified base, and consequently it becomes densely concentrated. Populations in excess of 10^{11} cells per ml can be obtained in this way, and the technique is apparently applicable to any type of bacterium.

Since the movement of nutrients is dependent on diffusion, the agar base should be limited to about 5 cm in depth. The ratio of solid to liquid should be at least 4:1 and less than 10:1; the lower ratio provides a better yield, and the higher ratio provides a greater concentration of cells. An Erlenmeyer flask is a convenient container, but indentations at its base are helpful to hold the agar in place if a mechanical shaker is used; rectangular containers also are useful in this way. Also, use a greater percentage of agar if breakup occurs during shaking.

If all of the medium components are incorporated into the agar base, the overlay can be distilled water. After an overnight equilibration period, the resulting clear diffusate in the overlay can be inoculated. For bacteria such as gonococci, which normally must be grown in a turbid medium enriched with blood and starch and which yield sparse populations, a relatively clean and dense population of cells can be harvested from such a diffusate overlay (9).

A variety of bacteria have been preserved in the short term by subculturing in a biphasic medium (Chapter 12.1.1).

Another example of a biphasic culture system is that devised by Castaneda (5) for the isolation of *Brucella* species from blood (Fig. 2). However, the principle of dialysis is involved to only a minor extent. To prepare such a system, first sterilize the 3% agar medium in a cotton-stoppered rectangular bottle; then place the bottle on its side so that the molten medium will solidify as a layer on the wall of the bottle. Stand the bottle upright, and add sterile liquid medium aseptically. Seal the bottle with a rubber diaphragm. For some bacteria, such as *Brucella abortus*, it may be necessary to replace the air in the bottle with an atmosphere containing 10% carbon dioxide. To use the bottle, inject a blood sample

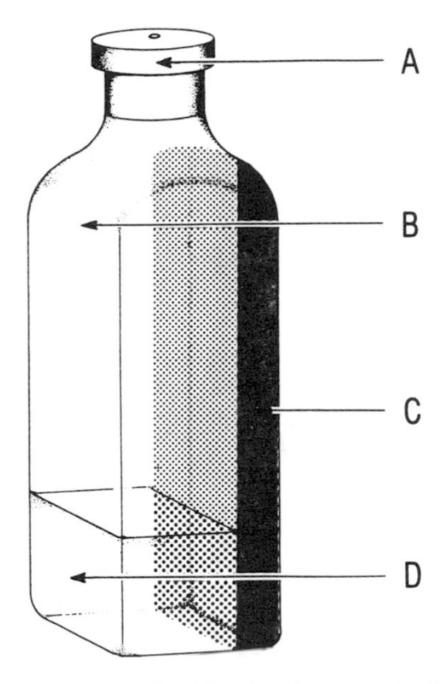

FIG. 2. Biphasic system for blood cultures on both solid and liquid media. (A) Serum bottle stopper through which the blood sample is injected. (B) Atmosphere containing air and 10% carbon dioxide. (C) Agar-solidified culture medium. (D) Liquid culture medium.

through the rubber diaphragm by means of a syringe. Incubate the bottle in an upright position. At 2-day intervals, tilt the bottle so that the liquid medium wets the entire surface of the agar and drains back down to the bottom. Bacteria present in the liquid medium will thus inoculate the agar surface and develop into visible colonies. The method has several advantages over the use of broth, as follows. (i) It is unnecessary to open the bottle repeatedly to withdraw samples from microscopic observation or subculturing in order to detect bacterial growth. This minimizes the chance of laboratory infections with highly infectious organisms such as *Brucella* species and also minimizes the chance of contaminating the culture system. (ii) Organisms that grow poorly in broth and develop only a slight turbidity may be able to grow better on the surface of the agar. (iii) Growth of bacteria in the presence of colored or turbid substances in broth can be easily detected by observing the development of colonies on the agar surface. A study by Hall et al. (10) indicates the usefulness of such a culture system for detecting the growth of a number of pathogenic bacteria.

9.4. SEMISOLID CULTURE TECHNIQUES

Semisolid media contain a low concentration of gelling agent (e.g., 0.1 to 0.4% agar) and have a soft, jelly-like consistency. Such media are useful for cultivating microaerophilic bacteria and for studying various aspects of motility and chemotaxis.

9.4.1. Microaerophiles

Microaerophilic bacteria have a strictly respiratory type of metabolism and exhibit oxygen-dependent growth when oxygen is provided in the absence of alternative terminal electron acceptors. However, they cannot grow under an atmosphere containing the level of oxygen present in air (21% oxygen) and must be grown under microaerobic conditions, i.e., under low levels of oxygen (Chapter 25.1.62). For example, *Campylobacter jejuni* generally grows best in a 6% oxygen atmosphere. Also, certain nitrogen-fixing bacteria that have a respiratory type of metabolism may grow as aerobes when supplied with a source of fixed nitrogen (such as ammonium sulfate) but can grow only as microaerophiles in nitrogen-deficient media. For example, *Azospirillum* species grow best in a 1% oxygen atmosphere under nitrogen-fixing conditions; the nitrogenase complex would be inactivated by excess oxygen. The general subject of microaerophily has been reviewed by Krieg and Hoffman (11).

Microaerophiles can be grown in liquid media or on solid media if a suitably low level of atmospheric oxygen is supplied, but this is unnecessary and an air atmosphere suffices if semisolid media containing 0.1 to 0.4% agar are used. The gelling agent stratifies the medium so that convection currents cannot mix the oxygen-rich upper layers of the medium with the underlying layers. The only way oxygen can penetrate into the depths of a tube of semisolid medium is by diffusion. Thus, a tube of semisolid medium provides an oxygen gradient, with the surface of the medium being the most highly oxygenated region. After the medium has been inoculated by stabbing deeply with a loop, microaerophiles begin to grow at a distance below the surface where the oxygen level is most suitable, i.e., at a point where the rate of oxygen uptake by the organisms is balanced by the rate at which oxygen is reaching the organisms by diffusion. Growth begins in the form of a thin disc visible several millimeters or centimeters below the surface; as the growth increases, with a consequent increase in oxygen uptake, the disk becomes denser and migrates closer to the surface. Eventually, the cells may form a very dense, diffuse layer just below the surface. In contrast, anaerobes begin to grow in the lower levels of a semisolid medium but may also eventually grow throughout as they develop their own anaerobic conditions.

If tubes of semisolid media are stored for long periods, oxygen may eventually diffuse deeply into the media. Place such tubes in a boiling-water bath for several minutes to drive off the dissolved oxygen, and then allow the tubes to cool and resolidify.

The microaerophile *C. jejuni* can be easily cultivated in tubes of semisolid Brucella medium (Brucella broth containing 0.15 to 0.3% agar [Chapter 8.5.14]). The nitrogen-fixing *Azospirillum* species grow well in a semisolid nitrogen-deficient malate medium (Chapters 8.5.37 and 25.3.4), and large crops of these cells for physiological studies can be obtained by using 1-liter Roux bottles containing 150 ml of culture medium stratified with 0.05% agar (17). This medium forms an 8-mm-deep layer on the wide, flat bottom of the Roux bottle. The agar concentration is high enough to allow the microaerophile to grow under aerobic conditions but low enough to permit the culture to be poured or pipetted from the Roux bottle after incubation into a sterile vessel. Sterile phosphate buffer is then added to the vessel, and the cells are collected by centrifugation and washed.

Some microaerophiles such as *C. jejuni* are not only microaerophilic but also **capnophilic**; i.e., they require increased levels of carbon dioxide for growth. However, *C. jejuni* cultures grown in semisolid medium do not require incubation in an atmosphere enriched in carbon dioxide, whereas cultures grown in liquid media or on solidified media under microaerobic conditions do require addition of at least 3% (vol/vol) carbon dioxide to the gaseous atmosphere in the culture. In semisolid media the organisms apparently provide their own carbon dioxide as a product of their respiration, and this carbon dioxide is retained in the microenvironment of the cells because of stratification of the medium.

9.4.2. Motility

Semisolid media can also be used to determine whether a bacterial strain is motile. Inoculate a tube of motility medium (broth containing 0.4% agar) with a straight needle to one-half the depth of the tube. During growth, motile bacteria will migrate from the line of inoculation to form diffuse turbidity in the surrounding medium; nonmotile bacteria will grow only along the line of inoculation.

A modification of this method is to place a piece of open-ended glass tubing into the semisolid medium prior to sterilization (Fig. 3). Inoculate only the upper part of the medium within the glass tubing. If the organism is motile, it will migrate downward, emerge from the bottom of the tubing, and then migrate upward to the surface of the medium, where it will grow extensively. This method is also helpful for selecting highly motile organisms for use in preparing H antigens for immunization. However, the method is suitable only for facultative organisms capable of metabolism under anaerobic conditions, such as members of the family *Enterobacteriaceae*. Strictly aerobic organisms such as *Pseudomonas* species will not migrate downward to the bottom of the glass tubing.

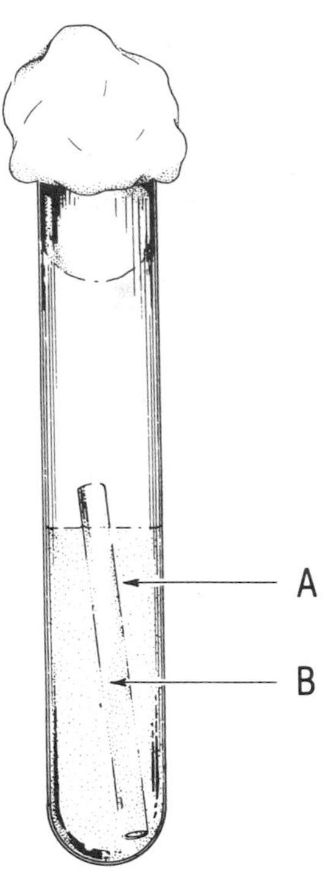

FIG. 3. Semisolid medium (A) into which a piece of glass tubing has been placed prior to autoclaving. The medium inside the glass tubing is inoculated. Motile facultative bacteria will migrate downward inside the tubing, emerge from the bottom of the tubing, and then migrate upward to the surface of the medium, where they will grow extensively.

9.4.3. Chemotaxis

Semisolid media are also useful in chemotaxis studies. For example, a semisolid medium containing an oxidizable carbon and energy source can be used to investigate positive chemotaxis in *Escherichia coli* (2). When a petri dish containing the medium is inoculated heavily at its center, the bacteria migrate outward as an expanding ring of cells that consumes all of the oxidizable substrate as the cells move. When two oxidizable carbon sources are provided in the medium, two rings which migrate at different rates are formed. In contrast to capillary-tube methods for studying chemotaxis, the oxygen supply on the agar plate is never exhausted; thus, the migrating rings form only in response to self-created substance gradients and not in response to self-created oxygen gradients.

Negative chemotaxis also has been studied by using semisolid media (25). *E. coli* cells are suspended in the medium at a concentration sufficient to give visible turbidity, and the inoculated medium is dispensed into a petri dish. After the medium gells, a plug of hard agar (2% agar) containing the suspected repellent compound is inserted into the medium. If the compound is a repellent, the surrounding bacteria migrate away from the plug and leave a clear zone that becomes visible in ca. 30 min.

The use of plates of semisolid media also makes it possible to select nonchemotactic mutants (3, 25). In the case of positive chemotaxis, nonchemotactic mutants (and also nonmotile mutants) fail to respond to a self-created substrate gradient and remain near the center of the petri plate, where they can be removed for subculturing and purification. In the case of negative chemotaxis, the nonchemotactic mutants can be isolated from the clear zone surrounding the hard agar plug.

9.5. REFERENCES

9.5.1. General Reference

1. **Codner, R. C.** 1969. Solid and solidified growth media in microbiology, p. 427–454. *In* J. R. Norris and D. W. Ribbons (ed.), *Methods in Microbiology*, vol. 1. Academic Press, Inc., New York.

9.5.2. Specific References

2. **Adler, J.** 1966. Chemotaxis in bacteria. *Science* **153:**708–716.
3. **Armstrong, J. B., J. Adler, and M. M. Dahl.** 1967. Non-chemotactic mutants of *Escherichia coli. J. Bacteriol.* **93:**390–398.
4. **Bridson, E. Y., and A. Brecker.** 1970. Design and formulation of culture media, p. 229–295. *In* J. R. Norris and D. W. Ribbons (ed.), *Methods in Microbiology*, vol. 3A. Academic Press, Inc., New York.

5. **Castaneda, M. R.** 1947. A practical method for routine blood cultures in brucellosis. *Proc. Soc. Exp. Biol. Med.* **64:**114–115.
6. **Deming, J. W., and J. A. Baross.** 1986. Solid medium for culturing black smoker bacteria at temperatures to 120°C. *Appl. Environ. Microbiol.* **51:**238–243.
7. **Funk, H. B., and T. A. Krulwich.** 1964. Preparation of clear silica gels that can be streaked. *J. Bacteriol.* **88:**1200–1201.
8. **Gardner, S., and J. G. Jones.** 1984. A new solidifying agent for culture media which liquefies on cooling. *J. Gen. Microbiol.* **130:**731–733.
9. **Gerhardt, P., and C. G. Hedén.** 1960. Concentrated culture of gonococci in clear liquid medium. *Proc. Soc. Exp. Biol. Med.* **105:**49–51.
10. **Hall, M. M., C. A. Mueske, D. M. Iltrup, and J. A. Washington II.** 1979. Evaluation of a biphasic medium for blood cultures. *J. Clin. Microbiol.* **10:**673–676.
11. **Krieg, N. R., and P. S. Hoffman.** 1986. Microaerophily and oxygen toxicity. *Annu. Rev. Microbiol.* **40:**107–130.
12. **Johnstone, K. I.** 1969. The isolation and cultivation of single organisms, p. 455–471. *In* J. R. Norris and D. W. Ribbons (ed.), *Methods in Microbiology,* vol. 1. Academic Press, Inc., New York.
13. **Lin, C. C., and L. E. Casida, Jr.** 1984. GELRITE as a gelling agent in media for growth of thermophilic microorganisms. *Appl. Environ. Microbiol.* **47:**427–429.
14. **Lines, A. D.** 1977. Value of the K^+ salt of carrageenan as an agar substitute in routine bacteriological media. *Appl. Environ. Microbiol.* **34:**637–639.
15. **Meynell, G. G., and E. Meynell.** 1965. *Theory and Practice in Experimental Bacteriology.* Cambridge University Press, London.
16. **Nordbring-Hertz, B., M. Veenhuis, and W. Harder.** 1984. Dialysis membrane technique for ultrastructural studies of microbial interactions. *Appl. Environ. Microbiol.* **47:**195–197.
17. **Okon, Y., S. L. Albrecht, and R. H. Burris.** 1976. Carbon and ammonia metabolism of *Spirillum lipoferum. J. Bacteriol.* **128:**592–597.
18. **Schultz, J. S., and P. Gerhardt.** 1969. Dialysis culture of microorganisms: design, theory, and results. *Bacteriol. Rev.* **33:**1–47.
19. **Segerer, A. H., and K. O. Stetter.** 1992. The order Sulfolobales, p. 684–701. *In* A. Balows, H. G. Trüper, M. Dworkin, W. Harder, and K.-H. Schleifer (ed.), *The Prokaryotes. A Handbook on the Biology of Bacteria: Ecophysiology, Isolation, Identification, Applications,* 2nd ed. Springer-Verlag KG, Berlin.
20. **Segerer, A. H., and K. O. Stetter.** 1992. The genus *Thermoplasma,* pp. 712–718. *In* A. Balows, H. G. Trüper, M. Dworkin, W. Harder, and K.-H. Schleifer (ed.), *The Prokaryotes. A Handbook on the Biology of Bacteria: Ecophysiology, Isolation, Identification, Applications,* 2nd ed. Springer-Verlag KG, Berlin.
21. **Shengu, D., M. Valiant, V. Tutlane, E. Weinberg, B. Weissberger, L. Koupal, H. Gadebusch, and E. Stapley.** 1983. GELRITE as an agar substitute in bacteriological media. *Appl. Environ. Microbiol.* **46:**840–845.
22. **Sirotnak, F. M., G. J. Donati, and D. J. Hutchison.** 1963. Folic acid derivatives synthesized during growth of *Diplococcus pneumoniae. J. Bacteriol.* **85:**658–665.
23. **Sneath, P. H. A.** 1955. Failure of *Chromobacterium violaceum* to grow on nutrient agar, attributed to hydrogen peroxide. *J Gen. Microbiol.* **13:**1.
24. **Towle, G. A., and R. L. Whistler.** 1973. Hemicellulose and gums, p. 198–248. *In* L. P. Miller (ed.), *Phytochemistry,* vol. 1. Van Nostrand Reinhold Co., New York.
25. **Tso, W., and J. Adler.** 1974 Negative chemotaxis in *Escherichia coli. J. Bacteriol.* **118:**560–576.
26. **Tyrrell, E. A., R. E. MacDonald, and P. Gerhardt.** 1958. Biphasic system for growing bacteria in concentrated culture. *J. Bacteriol.* **75:**1–4.
27. **VanElsas, J. D., J. M. Govaert, and J. A. van Veen.** 1987. Transfer of plasmid pFT30 between bacilli in soil as influenced by bacterial population dynamics and soil conditions. *Soil Biol. Biochem.* **19:**639–647.
28. **Waterworth, P. M.** 1969. The action of light on culture media. *J. Clin. Pathol.* **22:**273–277.
29. **Watson, N., and D. Apiron.** 1976. Substitute for agar in solid media for common usages in microbiology. *Appl. Environ. Microbiol.* **31:**509–513.
30. **Weinstein, M. J., and G. H. Wagman (ed.).** 1978. *Antibiotics: Isolation, Separation and Purification.* Elsevier Scientific Publishing Co., New York.

Chapter 10

Liquid Culture

PHILIPP GERHARDT and STEPHEN W. DREW

In this chapter are described various systems for the cultivation of bacterial cells in liquid culture ranging in volume from 10 ml to 100 liters and for harvesting and purifying the cells once growth is complete. The culture systems provide for the cultivation of a suspension of a single type of bacterium in aqueous medium with various degrees of control over the physical conditions. The systems are applicable with some modification to the mixed cultivation of more than one type of bacterium, but a discussion of this aspect of bacterial cell culture (58) is outside the scope of this book.

The principles and measurement of bacterial cell growth (see Chapter 11) are basic not only to the practice of general and molecular bacteriology but also to applications in traditional industrial fermentations and in new biotechnology processes. Bioengineering principles and mathematical models have increasingly become integral in methods of bacterial cell culture, as evidenced in this chapter and in the literature. Background reading on liquid culture of bacteria might well start with Pirt's classic monograph (5) and then proceed to recent texts (1–3, 6).

The techniques used to measure growth obscure the fact that all bacterial cultures are grossly heterogeneous and really monitor average values, which describe the growth of a population rather than of individual cells. Although all members of a particular population may be genetically identical, the individual members vary with respect to doubling time, age, composition, metabolic characteristics, and size. The magnitudes of these variations are influencd by the environment and can often be minimized by careful design and development of the system for cultivation. Most of the systems described here are designed to minimize heterogeneity within the population of a pure culture.

In such systems, the growth behavior of a bacterial population can be predicted by the simple relationship

$$\frac{dX}{dt} = \mu X \qquad (1)$$

where dX is the increase in amount of cell mass (biomass), dt is the time interval, X is the amount of biomass, and μ is the specific growth rate, representing the rate of growth per unit amount of biomass and having dimensions of reciprocal time (l/t).

If μ is constant, integration of equation 1 shows that X will increase exponentially with time as follows:

$$\ln \frac{X}{X_0} = \mu t \qquad (2)$$

where ln is the natural logarithm (to the base $e = 2.303$) and X_0 is the amount of biomass when $t = 0$.

It follows by rearrangement of equation 2 that the final biomass concentration X, is

$$X = X_0 e^{\mu t} \qquad (3)$$

Bacterial growth that follows this relationship is called **exponential** or **logarithmic growth.**

The relationship between the biomass doubling time (t_d) and the specific growth rate is found by letting $X = 2X_0$ at $t = t_d$ in equation 2 and solving the equation as follows:

$$t_d = \frac{\ln 2}{\mu} = \frac{0.693}{\mu} \qquad (4)$$

Equations 1 to 4 predict the growth of bacteria in simple systems in which the factors influencing growth are constant but do not allow prediction of deviation from constant-growth **(steady-state)** conditions. The techniques for continuous cultivation (section 10.2) very nearly establish steady-state conditions of theoretically infinite duration, whereas the techniques for batch cultivation allow significant changes in the environment during the time course of cultivation. In batch culture, equations 1 to 4 will apply without adjustment to the value of μ only during the portion of the growth cycle in which the changes in the growth environment have no influence on population growth (i.e., during the exponential growth phase).

10.1. BATCH SYSTEMS

10.1.1. Principles

A batch-culture system is one in which nothing (with the frequent exception of the gas phase) is added to or removed from an environment after medium of appropriate composition is inoculated with living cells; this is called a **closed system**. It follows that a batch system can support cell multiplication for only a limited time and with progressive changes in the original medium and environment.

10.1.1.1. Normal growth cycle

Figure 1 shows an idealized normal growth cycle (curve) for a simple, homogeneous batch culture of bacteria. The normal growth cycle assumes constant morphology and asynchronous binary fission of the cells. Growth proceeds through a **lag phase,** during which cell mass increases but cell numbers do not, and into a **growth phase,** which usually is characterized by an exponential increase in numbers and follows the relationships of equations 1 to 4. Ultimately, changes in the chemical or physical environment result in a phase of no net increase in numbers, the **maximum stationary phase.** Cells in the stationary phase still require an energy source for the maintenance of viability. Since the availability of an energy source in a batch culture is limited, there ensues a **death phase,** which is often characterized by an exponential decrease in the number of living cells.

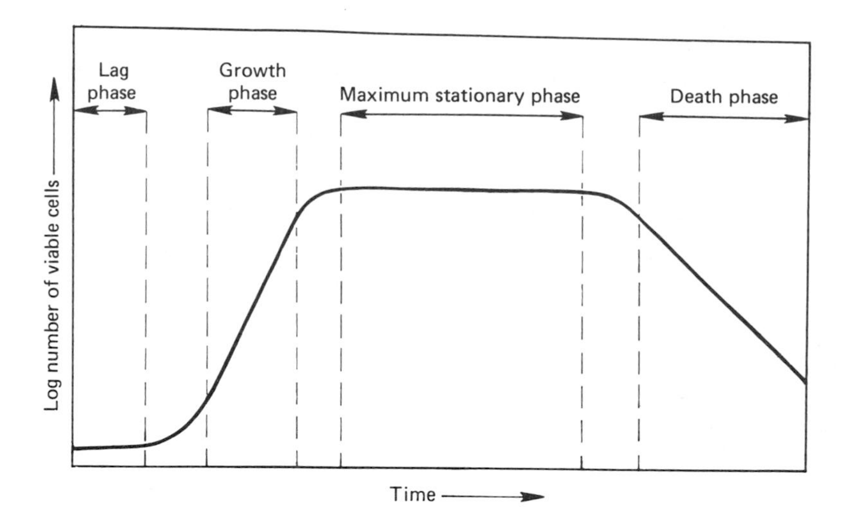

FIG. 1. Idealized normal growth cycle for a bacterial population in a batch culture system.

The lag phase may be brought about by the shock of rapid change in the culture environment. In fresh medium, the length of the lag phase depends on the size of the inoculum, the age of the inoculum, and the changes in nutrient composition and concentration experienced by the cells. A small volume of inoculum transferred to a large volume of fresh medium may result in outward diffusion of vitamins, cofactors, and ions which are required for many intracellular enzyme activities. If cells are inoculated from a rich medium to a minimum medium, the lag time may be affected by inoculum size as a result of carryover of trace nutrients from the original medium.

The age of the inoculum will influence the lag in a fresh medium as a result of toxic materials accumulated and essential nutrients depleted within the cells during their prior growth. Both a positive and a negative effect on the length of the lag phase in fresh medium can occur. In general, an increasing inoculum age lengthens the lag phase when cells are inoculated from a nutritionally simple medium into a richer one. A plot of the relationship between inoculum age and length of the lag phase for inoculation from rich to simple medium may show a definite minimum point because of the trade-off between nutrient loss and toxic-product buildup.

Finally, changes in nutrient composition and concentration between the inoculum culture and the fresh medium may trigger the control and regulation of enzyme activities within the cells or of morphological differentiation, such as spore formation. If the cells are transferred from a simple medium to a richer medium, both time and nutrients will be expended to allow an increase in the concentration of enzymes essential for metabolism. When cells are inoculated from a rich to a less rich medium, the cells may resume exponential growth immediately but at a lower rate.

The fact that constant exponential growth can occur for even a limited time in batch culture shows that the growth rate can be virtually unaffected by changes in substrate concentrations over wide ranges. Under these conditions the culture is said to be in **balanced growth,** whose rate can be described by a single numerical value, μ. Eventually, the culture will deviate from constant exponential growth and can no longer be described only by the value of μ, even though it is possible to calculate this value for the case of limitation by a single nutrient substrate. In a classic article (37), Monod described the relationship between substrate concentration and bacterial growth in simple systems at steady state as

$$\mu = \frac{\mu_{max} S}{(K_s + S)} \tag{5}$$

where μ is the specific growth rate, μ_{max} is the maximum value of μ obtained when $S >> K_s$, K_s is the saturation constant equivalent to a Michaelis-Menten constant, and S is the instantaneous or steady-state substrate concentration of a single limiting nutrient.

Several alternative models exist for growth response to substrate concentration (37). Equation 5 may be used to describe growth response to substrate limitation only under steady-state conditions with uncomplicated bacterial systems. When exponential growth ceases as a result of substrate limitation, conditions are no longer steady state for cell mass accumulation or substrate concentration, and equation 1 must be used with equations 5 and 6 to adequately model the response of the culture to diminishing substrate. Simultaneous solution of equations 1, 5, and 6 yields a somewhat bulky model for batch growth, which is at best a rough estimation for very simple systems.

The maximum population in a batch culture can be estimated from experimental data relating the increase in cell number or mass to the corresponding decrease in substrate concentration:

$$\frac{(X - X_0)}{(S_0 - S)} = Y \tag{6}$$

where X and S are the cell and substrate concentrations at time t, X_0 and S_0 are the cell and substrate concentrations at an earlier time, t_0, and Y is the overall yield coefficient. Equation 6 accurately describes the relationship between cell concentration and substrate concentration during exponential growth. If exponential growth continues unabated until the stationary phase is reached and substrate is completely consumed during exponential growth, the maximum cell number or concentration will be given by equation 6. This estimation assumes that the yield coefficient is constant throughout the growth cycle and neglects substrate consumption during the lag and stationary phases. In fact, the yield coefficient cannot be assumed to be constant for other than constant exponential growth conditions at a single specific growth rate. It follows that the prediction of a maximum population from equation 6 will lead to an overestimation. The concept of yield coefficient is discussed in greater detail later in this chapter (section 10.5). Prediction of the time required to attain maximum population density in batch culture requires simplifying assumptions (1, 5).

10.1.1.2. Aberrant growth cycles

Aberrations from the normal batch growth cycle are common. Morphological change in the culture (such as an increase in opacity, cellular refractive index, individual cell size, or cell aggregation) can lead to an apparent change in the growth cycle if growth is determined by optical measurement. For example, stationary bacterial cells are often more transparent than those growing in the exponential phase. As a result, a growth curve plotted as the logarithm of absorbance versus time may show an apparent decease in stationary-phase cell concentration compared with that attained at the end of exponential growth. In this case, the aberrant growth cycle represents a change in the morphological characteristics of the cell rather than a change in the number of cells.

Growth curves plotting the log of cell number versus time occasionally show an unusually rapid increase in cell number just after the lag phase, which then settles into a slower increase. The unusual burst of growth may in fact be an indication of partial or complete culture synchrony. This **synchronized growth** may result, for example, from culture acclimation to a new nutritional environment. Such culture synchrony degenerates rapidly, and asynchronous growth usually dominates within two generations. However, special techniques have been devised to synchronize cell divisions in a growing population as a way to mimic individual cell growth; these are discussed later in this chapter (section 10.3.2).

Arithmetic (linear) growth occurs when the supply of a critical nutrient is regulated by an arithmetic process (such as by dropwise addition or by diffusion) and this process becomes limiting. For example, limited diffusion of air through the cotton plug in a test tube or of nutrients through a membrane in a dialysis culture (section 10.3.5) may cause a shift from exponential to arithmetic growth. Although most bacterial cultures reproduce by binary fission as individual cells, some (e.g., actinomycetes) reproduce as filamentous extensions. If growth occurs primarily through extension of the tips of the hyphae, the increase in cell mass with respect to time will be arithmetic.

Filamentous bacteria often grow as pellets in liquid culture; when this occurs, biomass increases more slowly than the classical exponential rate and is proportional to the cube of time. Microbial growth in pellets may be severely affected by diffusion of nutrients into the pellet and by diffusion of metabolic products out from the pellet, a possibility that is not taken into account by equation 5.

Bacterial cells in complex environments often metabolize usable substrates in a sequential manner. That is, the presence of certain substrates may lead to repression of the enzymes for metabolism of other substrates. In this instance, only when the concentration of the repressing substrate has been reduced through bacterial consumption can the enzymes for metabolism of other substrates be elaborated. This regulation of bacterial physiology leads to an aberrant growth cycle that shows one or more intermediate but transient stationary phases. This response to a changing environment is termed **diauxic growth.** A classical example of diauxic growth is that of *Escherichia coli* growing in the presence of both glucose and lactose (Fig. 2). Rapid growth on glucose occurs first. At the point of glucose exhaustion, an inflection in the biomass curve occurs and there may even be a decline in biomass. A new enzyme system for metabolism of lactose is induced during this lag, and biomass accumulation at the expense of lactose then continues.

10.1.2. Culture Tubes

The lipless Pyrex-glass culture tube (usually 16 by 150 mm in size), plugged with nonabsorbent cotton or plastic foam, is the most convenient and widely used container for batch liquid culture of bacteria. However, for aerobic cultivation, test tubes usually provide only minimally effective conditions of oxygen supply. Fortunately, most bacteria are facultatively aerobic and most bacteriological uses of culture tubes do not require optimum growth conditions. To improve aerobiosis in a culture tube, increase the surface-to-volume ratio of the liquid medium by reducing the volume and slanting the tubes, or preferably, mount them on a rotary shaking machine to induce a vortex. Also increase the availability of air by using a small and loosely packed cotton plug, or a plastic or stainless-steel

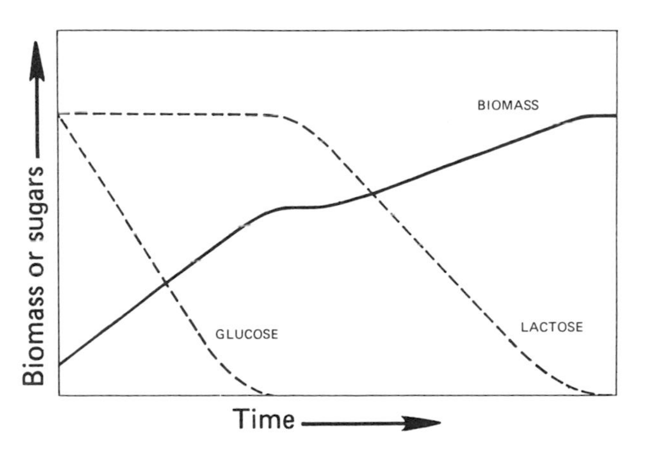

FIG. 2. Idealized idauxic growth of a bacterial population (biomass) in a batch culture system with two usable substrates (glucose and lactose).

Morton-type cap (for commercial sources, see section 10.6).

Conditions for the optimization of aerobiosis in laboratory culture are described more completely in the following sections on shake flasks (section 10.1.3), fermentors (section 10.1.6), and high-density culture strategies (section 10.3.7.1). Conditions that ensure strict anaerobic conditions for tube cultures are described elsewhere in this book (Chapter 6.6).

10.1.3. Shake Flasks for Aerobic Culture

An Erlenmeyer flask (capacity, 100 to 2,000 ml) is commonly used as a container for producing masses of cells in the laboratory or as the first stage in scale-up study of an industrial process involving batch liquid culture of bacteria. Consequently, maximally effective conditions of air (oxygen) supply are sought for obligately or facultatively aerobic types of bacteria.

Aerobic flask culture must be carried out with shaking of the flask to facilitate mass transfer of oxygen (as well as other gases and nutrients) at two levels: from the gas phase across the gas-liquid interface into the liquid phase, and from the liquid phase across the liquid-cell interface into the cell. Two major factors influence the ability of flask culture techniques to meet the oxygen requirements of the cells: gas exchange through the flask closure, and the liquid surface area available for oxygen transport from the gas phase.

Closure design for culture flasks must allow adequate gas exchange between the external environment and the flask interior yet must maintain asepsis. Although nonabsorbent cotton plugs wrapped in gauze or cheesecloth are often adequate for low-density aerobic cultivation of bacteria, they can lead to oxygen limitation in high-density bacterial cultures (55). Instead, prepare a cotton-and-gauze pad (see, e.g., Fig. 10) as follows.

1. Layer one thickness of *nonabsorbent* sheet cotton between two layers of surgical gauze.
2. Cut the cotton-and-gauze pad into squares sized to wrap over the tops of the flasks.
3. Place a square pad over the top of a flask, and secure the pad over the lip of the flask with a metal spring or rubber band. The metal springs can be prepared from heavy piano wire bent to form by local shop personnel or can be purchased from a laboratory supply house (section 10.6).

Morton-type caps also provide an adequate aseptic seal and good gas exchange, but they occasionally spin off a flask at high rotational speed. Foam plastic plugs also are effective aseptic closures for shake flask cultivation and provide moderate gas exchange, but the plugs must be sized to fit into the neck of the flask snugly yet without severe compression. Furthermore, foam plugs may contain plasticizers or other toxic chemicals, which must be thoroughly washed from the plugs prior to their use as closures for biological culture. Caps and plugs can be purchased from various laboratory supply houses (section 10.6).

The rate at which dissolved oxygen is consumed in a liquid culture of bacteria is determined by the cell density and growth rate. The demand for dissolved oxygen is created by cell consumption and is met by continuous diffusion of oxygen from the gas phase to the liquid phase and thence into the cell. The interfacial surface area between gas and liquid often controls the flux or volumetric rate of oxygen transfer. High interfacial surface area results in high volumetric rates of oxygen transfer and therefore allows rapid growth without oxygen limitation. The interfacial surface area can be maximized by maintaining low ratios of liquid volume to flask volume and by vigorously shaking the flasks. Oxygen limitation due to limited diffusion can usually be avoided by preparing flasks with liquid volumes of no more than 20% of flask volumes and by shaking the flasks at 200 to 350 rpm on a rotary shaker or at 150 to 250 strokes per min on a reciprocating shaker.

During incubation of the inoculated flasks at an appropriate temperature, rotary shaking should cause the liquid meniscus to rise to approximately two-thirds of the flask height. Reciprocal shaking, such as is found in most water bath shakers, should be sufficiently vigorous to cause significant breaking (turbulence) of the liquid wave as it moves from side to side in the flask but not so great as to wet the closure (19).

Oxygen transfer is further enhanced by baffling. Baffled flasks have indentations in the side and/or bottom surface, which act to break vortex formation in flasks on a rotary shaking machine and to induce turbulence. Baffled flasks with good design of the baffles are available commercially (e.g., Wheaton Science Products).

Optimum culture conditions for attaining the highest cell densities of aerobic bacteria involve the use of a flask specially designed with a widemouth opening, a barrel shape for a wide vortex and increased working capacity, effective baffling, and a closure consisting of a cotton-gauze pad or a silicone rubber membrane. Such a rubber (or polycarbonate) membrane allows good gas exchange but prevents the evaporation of water. Such special flasks are available commercially (BioQuip, Inc.).

10.1.4. Anaerobic Flask Culture

Erlenmeyer flasks, or flasks or bottles of any other convenient shape, are used to cultivate anaerobic bacteria on an intermediate size scale. Two main principles prevail: the nutrient medium must be prereduced to an E_h of -150 mV or lower, and air must be removed from the immediate environment of the culture. The special procedures for handling anaerobic bacteria are described more fully elsewhere in this manual (Chapter 6.6).

A few practical tips apply to mass cultivation of anaerobes. Fill the container nearly full to minimize the effect of the overlying gas phase. Use as large an inoculum as possible distributed from the bottom upward in a column of prereduced medium. Displace the gas phase with oxygen-free nitrogen or argon, plus 5 to 10% CO_2, and allow growth to begin with the medium quiescent. Once growth has progressed actively throughout the medium, provide agitation by means of a magnetic stirring bar or by purging nitrogen or natural gas through a tube with its outlet at the bottom of the container. Use a water trap to permit the escape of introduced and metabolic gases without the entrance of air. *CAUTION*: Some anaerobic bacteria produce hydrogen or methane gas, which may cause flammable or even explosive conditions, and most anaerobes produce CO_2

and other gases, which will increase the internal pressure dangerously if adequate venting is not provided.

10.1.5. Carboys

Carboys have traditionally been used to cultivate multiliter quantities of bacteria in liquid culture. However, several factors severely limit the usefulness of this technique. Handling and manipulating large glass carboys during medium preparation, sterilization, and inoculation are hazardous, and shaking of filled glass carboys for aerobic cultivation is risky, although the use of polypropylene or polycarbonate plastic carboys lessens these concerns. Gas dispersion in any carboy is poor, even with stirrers, so that adequate aeration is unlikely. Although preparation of media in plastic carboys is sometimes necessary for large-scale work, cultivation in carboys should be avoided.

10.1.6. Fermentors

The fermentor procedures described below are oriented primarily toward cultivation of aerobic bacteria but are applicable, with modification, to cultivation of anaerobes also. Fermentor design aspects discussed here were developed to minimize population heterogeneity. Other designs (such as plug-flow, closed-loop, and tubular fermentors) offer special research opportunities but are beyond the scope of this manual.

Fermentors (stirred tank reactors) for cultivation of bacteria are designed to provide a self-contained environment. The operational volume typically ranges from 500 ml to thousands of liters, with typical laboratory units having operating volumes of 1 to 20 liters. Since they are designed to minimize physical and chemical concentration gradients in the culture environment, fermentors are always agitated (stirred) in a manner that yields rapid and complete mixing of the liquid phase. Inadequate agitation will allow development of regions of poor mixing ("dead spots") and will result in uncontrolled local concentration gradients. The physiological result of poor mixing is a highly heterogeneous cell population. For example, inadequate agitation of aerobic cultures can result in insufficient contact between the gas phase and the liquid medium. As a result, some regions of the fermentor may be nearly anaerobic. If the organism is facultative, the cell population will have a wide range of respiratory characteristics and the cell yield will be reduced.

Although it is possible to build an adequate fermentor in a machine shop, the necessary attention to detail of design for mixing, aeration, sterilization, and maintenance of asepsis warrants the purchase of commercially available units (section 10.6). Such fermentors are capable of adequate aeration and agitation of relatively dense cultures and are usually equipped with a multiport head plate through which pH probes, dissolved-oxygen probes, and sampling or feed lines can be introduced into the fermentor.

10.1.6.1. Aeration and agitation

The agitation speed for adequate oxygen transfer in a fermentor is subject to change as the culture density (cell concentration) or operating conditions (volumetric air flow rate, temperature, and liquid volume) change. The operator should adjust the agitation speed and air flow rate so that the dissolved-oxygen concentration never falls below 30% of the initial saturated dissolved-oxygen concentration (48). Maintenance of dissolved-oxygen concentrations above 30% of the saturation value at ambient temperatures will allow growth of most bacterial cultures under conditions in which oxygen is the growth-limiting substrate and will keep aerobic cells growing exponentially. The fundamental theory behind this general rule for adequate agitation in sparged systems is based on classical Monod kinetics for microbial growth in response to rate-limiting substrates (37).

Some basic properties of oxygen supply and demand must be kept in mind when providing for adequate aeration in the cultivation of aerobic bacteria whether in a fermentor, a shake flask, or a test tube. Oxygen is quite insoluble; e.g., there are only about 9 ppm (0.0009%) in water at 20°C in equilibrium with air. As the temperature is raised, oxygen becomes even less soluble. Oxygen solubility is directly proportional to the partial pressure of oxygen in the gas phase but is substantially independent of the total pressure and the presence of other gases. Bacteria utilize only dissolved (not gaseous) oxygen. Oxygen is so insoluble that only a small reservoir of it exists in solution at any given time. Consequently the rate of dissolved-oxygen supply must at least equal the rate of oxygen demand by the culture. Fortunately, cell respiration proceeds at a rate that is independent of the dissolved-oxygen concentration as long as it remains above a critical concentration, which is considerably below the saturation value as noted in the foregoing paragraph. These basic properties of oxygen supply and demand are further developed and exemplified in a classic review on aeration and agitation in microbial culture by Finn (18) and in another excellent discussion in the monograph by Pirt (5).

Agitation requirements for adequate oxygen transfer are almost always well in excess of the agitation rquirements for mixing of highly soluble nutrients. This rule also applies to agitation conditions established for optimum dissolution of other sparingly soluble substrates, such as immiscible hydrocarbons and steroids. For oxygen transfer, both agitation speed and volumetric air flow rate can be varied to establish the desired oxygen concentration in the fermentation medium. Instrumentation for dissolved-oxygen monitoring and for control of air flow rate is useful for the cultivation of aerobic organisms and allows the operator to control dissolved-oxygen concentration as an independent fermentation parameter (14). Dissolved-oxygen control systems are available from all of the major fermentor manufacturers (section 10.6) and usually allow control of both agitation and air flow rate in a sequential manner.

10.1.6.2. Temperature control

Most commercial fermentors for laboratory-scale operation are equipped with integral heat exchangers. The temperature is controlled by a temperature control unit to maintain a constant temperature in the fermentation medium. In some cases, the fermentor is jacketed or submerged in a constant-temperature bath as a means of temperature control when accuracy is required. See Chapter 6.3 for principles of the effect of temperature on bacterial growth.

10.1.6.3. Air sterilization

The air or gas supply to sparged fermentors must be sterilized prior to injection into the fermentation vessel. This is most easily done by physical removal of airborne microorganisms with a fibrous-medium filter. Most commercially available laboratory fermentors contain fiberglass-packed pipe filters, although some contain membrane and sintered stainless-steel filters. In any case, the filtration medium must be changed regularly to ensure adequate performance. A supplementary or backup air sterilization filter may be needed for long-duration cultivation as a precaution against fouling of the duty filter.

An air sterlization filter can be constructed by using a 20-cm section of 2.5-cm (inside diameter) stainless-steel pipe threaded at both ends to mate with pipe caps. Drill and tap-thread the pipe caps, insert pipe nipples into the holes, and connect lengths of rubber tubing. Pack the thoroughly cleaned stainless-steel pipe body with approximately 90 to 150 kg of glass wool per m^3, and firmly screw the caps into place on each end of the filter. Apply Teflon pipe-sealing tape to the threads of the pipe body prior to attachment of the pipe caps to ensure a gas-tight seal.

Sterilize the air sterilization filter before its initial use by wrapping it in aluminum foil and heating at 170 to 190°C in an oven for 2 h. After the filter has cooled, make sure that the caps are securely in place and connect the filter to the air supply line and (aseptically) to the fermentor with the rubber tubing. The air filter can be used two or three times before repacking becomes necessary. This type of air filter is also useful for in-line sterilization of the oxygen-free gases supplied to anaerobic cultures. A more detailed description of fibrous air filter design is given in reference 6.

10.1.6.4. Foam control

Most well-aerated and agitated cultures will produce relatively stable foam at some point in the culture cycle. If foams are allowed to develop unchecked, they may wet the air filters and lead to back-contamination of the culture as well as to foam spillage. Foams may be controlled by mechanical foam breakers or by the addition of chemical antifoam agents. Many commercial fermentors are equipped with mechanical foam breakers as standard or optional equipment. Current designs for mechanical foam breakers are adequate for all but a very few culture conditions.

Chemical antifoam agents provide a less expensive means of foam control when intermittent or light use is expected (11). However, antifoam agents must be added to the fermentation medium and therefore contribute to the overall medium composition. Some antifoam agents such as vegetable (corn oil, cottonseed oil) and animal (lard oil) ones may be metabolized by the culture and therefore contribute to the carbon substrate pool. On the other hand, nonmetabolizable antifoam agents, such as the silicone antifoams (available from Dow Corning Co., General Electric Corp., and Union Carbide), may be toxic at high concentrations. However, their action as surfactant materials requires their use in only very low concentrations. Effectiveness and toxicity should be determined on an individual basis; Dow Corning antifoam B is suggested for initial trial. Antifoam agents can be added directly to the medium before sterilization or to the fermentation system through a feed line in the head plate (the antifoam agent, its reservoir, and the feed line must be sterilized prior to use). Automatic control units for antifoam addition are usually available commercially (section 10.6).

10.1.6.5. Sampling and inoculation

Preparation and operation of the fermentor require that all points of entry to or withdrawal from the system be designed for aseptic operation to prevent contamination of the fermentor from external sources. Most laboratory fermentors have a sample and inoculation line that penetrates the fermentor head plate and extends to within a few centimeters of the bottom of the fermentor vessel. Carefully follow the manufacturer's directions for sampling and inoculation. Note that positive pressure on the fermentor must always be maintained.

10.1.6.6. Anaerobic operation

The primary concern in system design for anaerobic cultivation in fermentors is the exclusion of oxygen. Fill the fermentor as full as practical, and sparge with oxygen-free gas during the early hours of cultivation to maintain a slight positive pressure. Use a large inoculum (10 to 20% of total operating volume) to allow rapid establishment of the culture and to reduce the sensitivity of the system to leaks of oxygen from the external environment. If pH is to be controlled or nutrients are to be added in a continuous or semicontinuous fashion, keep the flexible tubing connecting the reservoirs to the fermentor as short as possible, since oxygen will permeate natural or silicone rubber tubing; use butyl rubber instead. Tygon tubing is also relatively oxygen impermeable but softens during autoclaving, so it must be clamped or wired in place. Finally, fit the gas exit of the fermentor with a water-gas trap to prevent back-diffusion of oxygen into the fermentor head space.

A simple water trap can be constructed by placing a water-filled graduated cylinder or other glass container in an inverted position in a water bath. Place the gas exit line from the fermentor under the inverted container so that the escaping gas will collect within the vessel. The connecting tubing should be oxygen impermeable or as short as possible to minimize oxygen diffusion through the walls of the tubing. As gas collects within the container, the water will be displaced but will maintain an effective seal against atmospheric oxygen. The water bath container should be sufficiently large that the displaced water will not cause the bath to overflow. When one container is full, transfer a second water-filled container from its storage position in the water bath to the gas collection position. If graduated cylinders are used as the collection vessels, the volume of gas can be easily measured. A water seal is seldom necessary when active sparging of the fermentor by oxygen-free gas is in operation. However, once the anaerobic culture has become established in a growth pattern, the sparging of oxygen-free gas through the fermentor will probably not be necessary. If gas sparging is stopped, the water-gas trap must be installed immediately.

In all cases, prereduce the medium by heating within the fermentor, or charge prereduced medium into a fermentor that has been sparged with oxygen-free gas. Prepare and maintain acid, base, or nutrient solutions in an oxygen-deficient condition for addition to the fermentor during operation. Absolute exclusion of oxygen from acid, base, or other solutions to be added to the fermentor in small quantities is not necessary since the reducing characteristics of an active anaerobic culture will adequately cope with very slight additions of oxygen through the feed systems.

For futher precautions in cultivating anaerobes, see Chapter 6.6.

10.2. CONTINUOUS SYSTEMS

Continuous cultivation differs from batch cultivation in that a fresh supply of medium is added continuously at the same rate that culture is withdrawn (an **open system**). If the culture is well mixed, a sample that is representative of both the cell population and substrate concentration within the fermentor can be withdrawn. The technique of continuous cultivation theoretically allows continuous exponential growth of the culture in a system requiring constant addition of fresh medium and constant withdrawal of culture so that the culture volume and cell concentration remain constant with time. In its broadest sense, continuous cultivation does not require a constant cell concentration, but most literature accounts of quantitative study of continuous cultivation have dealt with such a situation and the principles developed below deal only with this situation.

The term "steady state" is often applied to continuous cultivation and means literally that no change in status occurs during the time span studied. In reality, this definition is too broad, since practical application of the continuous-cultivation theory often results in changes in some parameters while others remain constant. A continuous culture is therefore defined at steady state by an invariant biomass concentration with respect to the time span of observation. In contrast, a batch culture may have a steady-state dissolved-oxygen concentration maintained by constant replacement but is not a continuous culture because the biomass concentration changes with time.

Continuous cultivation at steady state is possible only when all factors contributing to the accumulation of biomass are exactly balanced by all factors contributing to the loss of biomass from the system. This is shown by the following general material balance on the bacterial cells:

cells added to the system	−	cells removed from the system	+	cells produced through growth	−	cells consumed through death	=	cells accumulated within the system

The balance is shown mathematically as follows:

$$\frac{FX_0}{V} - \frac{FX}{V} + \mu X - \alpha X = \frac{dX}{dt} \quad (7)$$

where F is the medium flow rate to and from the fermentor (liters per hour), V is the liquid volume within the fermentor (liters), X_0 and X are the cell masses (grams per liter) in the feed and fermentor, respectively, μ is the specific growth rate, α is the specific death rate (reciprocal hours), and dX/dt is the rate of change in cell mass (grams per liter-hour).

At steady state, $dX/dt = 0$. The volume of a true continuous culture is fixed in theory and undergoes negligible variation in practice. Therefore, the flow rates to and from the fermentor must be identical. Finally, the specific death rate is almost always much lower than the specific growth rate, so the death term (αX) may be ignored. (Exceptions to this rule may occur at very low growth rates, in the presence of toxic substances, or in an extreme biophysical environment.) Since the feed to the fermentor is usually sterile so that $X_0 = 0$, at steady state:

$$\mu = \frac{F}{V} = D \quad (8)$$

That is, *the specific rate of growth of the population within the fermentor is determined by the dilution rate,* D, where $D = F/V$.

Many types of continuous cultures are possible (1, 5). The following discussion is limited to the two major types of continuous culture. The **"chemostat"** achieves steady state by controlling the availability of a single growth-limiting substrate and the **"turbidostat"** achieves steady state by actually removing the cell mass and replacing it with fresh medium at the same rate as cell growth. Table 1 compares the chemostat and turbidostat with respect to various operating parameters. The chemostat mode allows precise control over the growth-limiting condition (nutrient limitation), in contrast to turbidostat operation. However, turbidostats are better suited to continuous-cultivation studies of growth at or near the maximum specific growth rate, μ_{max}.

Background reading on continuous culture should start with the classic papers in 1950 by Monod (38) and Novick and Szilard (43). Luedeking (31) in 1976 presented a very useful discussion of continuous-culture theory and practice. The article describes graphical design and analysis of continuous-culture systems, avoiding the requirement for an accurate model of growth response to changing environmental properties (such as the Monod model). An empirical approach to continuous cultivation based on batch data has its pitfalls, which were discussed by Luedeking; however, the approach is still useful for obtaining approximate design criteria. The very broad applications of continuous culture have been described in the proceedings of nine international symposia on the technique, the most recent in 1988 (27).

10.2.1. Chemostat

10.2.1.1. Theory

Deviations from simple chemostat theory abound, but most are the result of (i) product formation, which may require a sophisticated model of yield as a function of μ; (ii) imperfect mixing in the fermentor, which may

Table 1. Comparison of chemostat and turbidostat

Operating parameter	Chemostat	Turbidostat
Operation at or near maximum specific growth rate	Unstable	Stable, very nearly steady state
Operation at low specific growth rate	Stable steady states	Unstable, transient with pulsatile response
Dilution rate equals specific growth rate	Only at steady state	At all times
Cell concentration at constant specific growth rate	Depends on substrate concentration in the feed	Depends on substrate concentration in the feed
Dilution rate	Predetermined	Controlled as a function of cell mass
Substrate concentration for steady-state operation	Requires a single limiting substrate	All substrates may be present in excess

allow a stable dilution rate that is higher than the critical dilution rate; (iii) wall growth, which has an effect similar to imperfect mixing but more pronounced; and (iv) idiosyncracies of physiological response. Pirt (5) has discussed the first three of these deviations.

In its simplest form, chemostat continuous cultivation can be described as a collection of reaction steps in which the rate of growth of a culture is determined by the lowest rate of nutrient metabolism. Although a growing culture requires the metabolism of many different nutrients, the growth rate of an ideal culture at any given instant is determined by the rate of metabolism of a single limiting nutrient. Monod (37, 38) found that growth rate dependence on substrate concentration could be predicted by an equation whose form is essentially a Michaelis-Menten type function, as shown in equation 5. Monod's equation is most easily applied to experimental conditions of steady state. For usual purposes, a steady state exists when the substrate concentration does not fluctuate with time and the resulting cell population in the continuous-culture vessel is constant with time. When a chemostat is operated with a sterile feed ($X_0 = 0$) and without recycle, the specific growth rate is numerically equal to the dilution rate (equation 8). This identity is forced by controlling the availability of a limiting nutrient through the addition of fresh medium.

A limiting nutrient balance can be written for a chemostat:

input − output − consumed = accumulation

The mathematical expression for this, similar to equation 7 for cell mass, is as follows:

$$DS_0 - DS - \frac{\mu X}{Y_{x/s}} = \frac{dS}{dt} \qquad (9)$$

where D is the dilution rate ($D = F/V$); S_0 and S are the limiting substrate concentrations in the feed and fermentor, respectively; u is the specific growth rate of the culture in the fermentor; X is the dry cell mass in the fermentor; $Y_{x/s}$ is the overall yield coefficient (cells formed/substrate consumed); and dS/dt is the rate of change of substrate concentration in the fermentor. The overall yield term is a composite that includes contributions for both growth and maintenance (see section 10.5.). No products other than cells are assumed. If product formation does occur, an additional consumption term must be added. At steady state, equation 9 then becomes

$$D(S_0 - S) = \frac{\mu X}{Y_{x/s}} \qquad (10)$$

Substitution of equation 8 into equation 10 gives

$$X = Y_{x/s}(S_0 - S) \qquad (11)$$

Note that equation 11 was presented earlier without derivation for use with batch systems (equation 6). The overall growth yield is assumed to be dependent only on the limiting nutrient concentration and independent of specific growth rate. Exceptions to these assumptions are discussed in section 10.5. *Equations 8 and 11 are the steady-state equations for usual chemostat continuous cultivation.*

A model expressing the specific growth rate as a function of substrate concentration must be assumed before biomass concentration, substrate concentration, and specific growth rate can be related to define a stable set of operating conditions. Equation 5 provides an adequate model for many situations. Substitution of equation 8 into equation 5 yields

$$D = \frac{D_c S}{K_s + S} \qquad (12)$$

where D_c is the critical dilution rate corresponding to the maximum specific growth rate, μ_{max}. *Operation of the chemostat at dilution rates above D_c will result in complete washout of the culture.*

Equation 12 can be arranged to give

$$S = \frac{DK_s}{D_c - D} \qquad (13)$$

Substitution of equation 13 into equation 11 results in

$$X = Y_{x/s}\left(S_0 - \frac{DK_s}{D_c - D}\right) \qquad (14)$$

Equation 14 relates the steady-state biomass concentration to the dilution rate. Figure 3 depicts the generalized system response to changes in the dilution rate, as predicted by equations 13 and 14.

The Monod model (equation 5) is one of several models relating growth rate and substrate concentration. The assumptions implicit with this model are quite simple and do not adequately describe all sys-

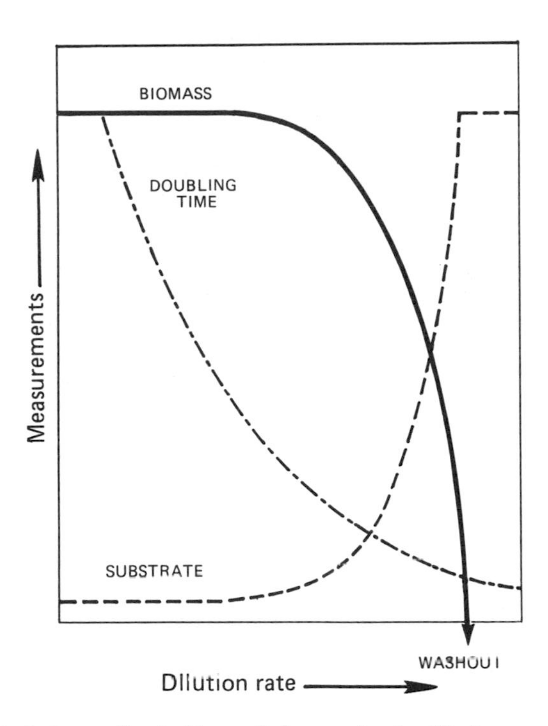

FIG. 3. Generalized effects of changes in the dilution rate on culture variables in a chemostat system for continuous cultivation, where the dilution rate approaches μ_{max} at washout.

tems. An excellent discussion of the fundamental theory of chemostat continuous culture is presented by Tempest (53).

10.2.1.2. Apparatus

Most of the equipment described above for batch cultivation of bacteria in stirred tanks (section 10.1.6) can be used directly for chemostat continuous cultivation, but some modification may be necessary. A dual-pump system (one for fresh medium addition and one for culture withdrawal) used with a level controller or weight scale can allow continuous culture operation by withdrawing culture through the sample port. Pump and level control systems for continuous culture are usually available from manufacturers of fermentation equipment (section 10.6).

Although the dual-pump and level control system for medium addition and culture withdrawal is the preferred method for continuous cultivation in multiliter fermentors, it may be more convenient to use gravity for overflow withdrawal of culture in subliter fermentors. Special culture vessels manufactured with a side arm for liquid withdrawal can be created by modifying standard glass or stainless-steel vessels. Special care must be taken to provide adequate surface mixing on a level with the withdrawal port. Adequate surface mixing can be accomplished by insertion of a liquid draft tube (15) or by installing a turbine paddle with a 5° forward pitch on the agitation shaft at the level of the withdrawal port. An all-glass continuous culture vessel with gravity overflow is available commercially (Bellco Glass, Inc.). This vessel is adequate for subliter cultivation of low-density bacterial cultures. Other manufacturers (section 10.6) market versatile continuous-culture units capable of meeting the aeration and mixing requirements of even the most dense bacterial cultures; however, these units also require withdrawal of culture by pumping.

10.2.1.3. Medium flow rate determination

Careful and precise control of medium flow rate into and out of the constant-volume fermentor is absolutely necessary for the establishment of steady-state conditions for continuous cultivation. The flow rate of fresh medium into the fermentor will determine the culture density and specific growth rate in chemostat-type continuous cultivation.

Figure 4 illustrates a simple device for the determination of medium flow rate. Carefully set and periodically check the medium flow rate during the fermentation. Fill the calibrated pipette (or burette) with a measured volume of fluid from the reservoir by briefly closing pinch clamp B, and then close pinch clamp A until ready for determination of the flow rate. Measure the flow rate by opening pinch clamp A while simultaneously closing pinch clamp C and measuring the time necessary for withdrawal of a measured volume from the pipette. Open pinch clamp C and close pinch clamp A to resume normal operation. Calculate the dilution rate (or specific growth rate for single culture vessels without recycle) by dividing the flow rate by the fluid volume in the fermentor.

10.2.1.4. Prevention of back-contamination

The aeration and agitation conditions of continuous culture result in the production of an aerosol containing large numbers of the cultured bacterium. As a result, the medium feed line is subject to back-contamination and must be fitted with a medium "break" tube, which acts as an aseptic seal between the culture vessel and the medium reservoir. Figure 5 illustrates the basic

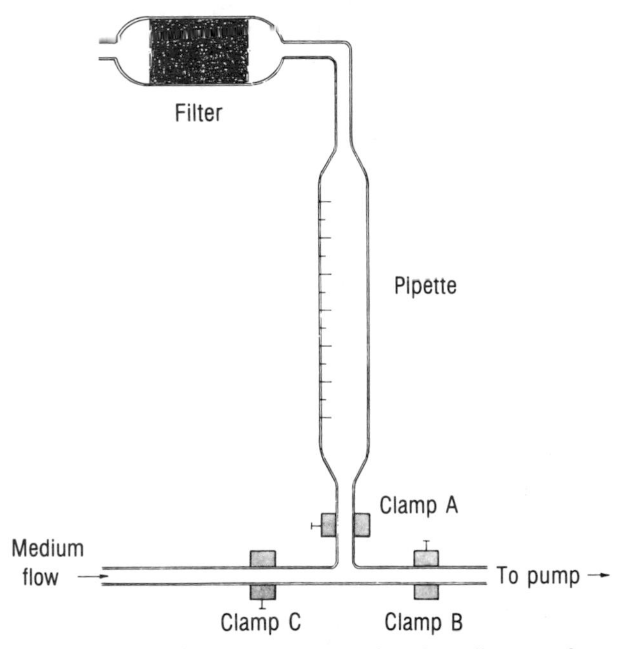

FIG. 4. Apparatus for determination of medium flow rate for a continuous-culture system.

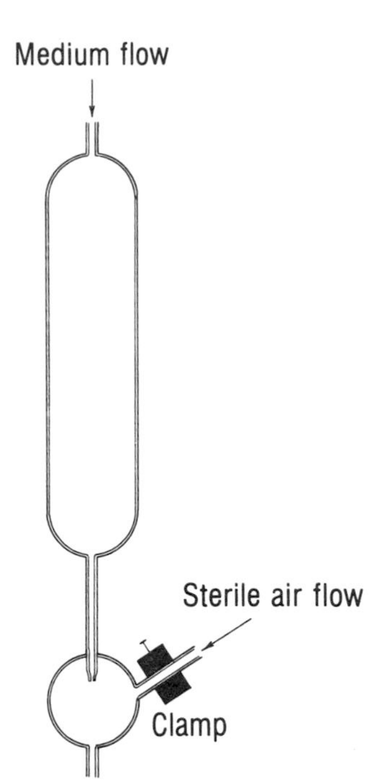

FIG. 5. Apparatus to prevent back-contamination of a medium reservoir for a continuous-culture or fed-batch culture system.

principle of such a tube (20). The lower section provides a slow steady flow of sterile air to prevent aerosol contamination of the medium feed line, and the upper chamber provides a secondary break in the liquid line.

10.2.1.5. Medium and product reservoirs

Continuous cultivation requires the use of rather large quantities of medium and produces equally large volumes of product culture. Table 2 lists the liquid volume of fresh medium necessary for 20 h of continuous operation at different rates of dilution. The corresponding mass doubling times are also listed. It is advisable to prepare enough medium for 20 h of operation to allow flexibility in the scheduling of reservoir transfer. Since most laboratory-scale operations require autoclaving of the medium as the means of sterilization, the required volumes of medium will probably be most conveniently prepared in carboys of appropriate size. (*CAUTION:* Exercise extreme care in handling large volumes of hot liquids. Use autoclavable plastic carboys rather than glass ones.) For example, a 14-liter fermentor with a 10-liter liquid working volume maintained so that the bacterial culture doubles every hour (dilution rate, 0.69 h^{-1}) would require 138 liters of fresh medium per 20-h interval. The same culturing conditions but with a working liquid volume of 1 liter would require 13.8 liters of medium. Small volumes of medium may be easily autoclaved and stored in standard 20-liter plastic carboys. The preparation of medium in small volumes (10 to 14 liters) will allow adequate heat sterilization without excessive deterioration.

Figure 6 illustrates a typical reservoir configuration for upright sterilization in an autoclave. The medium volume in culture reservoirs should not exceed 75% of the total reservoir volume, to minimize the chance of boil-over during autoclaving. Furthermore, it is essential that the gas space in the reservoir be vented during autoclaving. After sterilization, several reservoirs can be connected in series to increase the reservoir volume without changing the reservoir; this is accomplished by connecting line A of the first reservoir to line B of the adjacent reservoir aseptically.

Table 2. Medium requirements for continuous cultivation at various dilution rates[a]

Dilution rate[b] (h^{-1})	Corresponding mass doubling time		Medium vol per fermentor vol required for 20-h continuous operation (ml/ml)
	h	min	
2.77	0.25	15	55.4
1.39	0.50	30	27.8
0.92	0.75	45	18.4
0.69	1.00	60	13.8
0.46	1.50	90	9.2
0.35	2.00	120	7.0
0.17	4.00	240	3.4
0.09	8.00	480	1.8

[a]A bacterial culture maintained in 500 ml of liquid working volume at a mass doubling rate of 30 min (D = 1.39 h^{-1}) will require 13.9 liters of fresh medium per 20-h period.

[b]Dilution rate calculated for a single vessel without recycle.

10.2.1.6. Aseptic connections

Between one reservoir and another, and between the reservoirs and the fermentor, use the aseptic "quick connect" shown in Fig. 7. It consists of two sections of stainless-steel tubing sized so that one of the stainless-steel tubing sections will easily slip into the other and the rubber connecting lines fit securely over the free

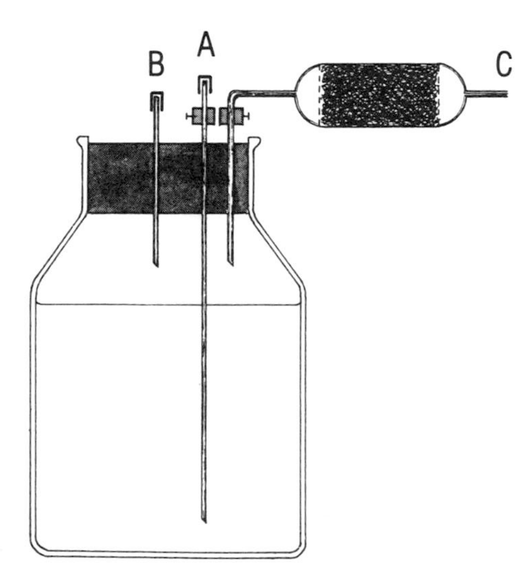

FIG. 6. Medium reservoir for upright sterilization in an autoclave. Lines A and B are fitted with gravity caps for aseptic closure, lines A and C are fitted with pinch clamps, and line C is fitted with an air filter. During use, line A is connected to the fermentor by a "quick connector" (see Fig. 7).

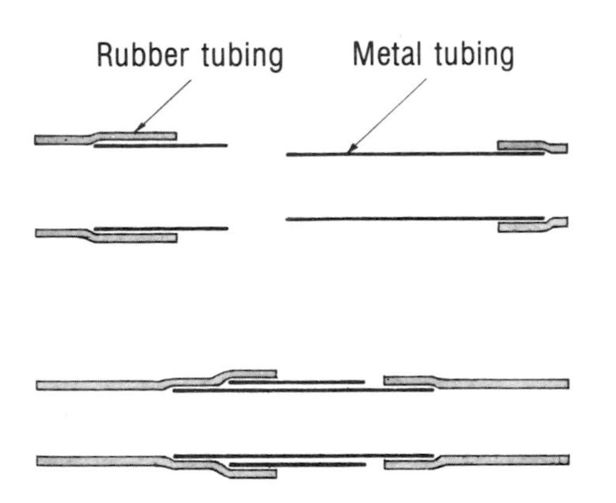

FIG. 7. Connector for quick and aseptic connections of tubing between reservoirs for continuous-culture systems. Top: disassembled connector parts. Bottom: assembled connector.

ends of both of the stainless-steel tubing sections. The individual tubing sections should be wrapped with gauze or aluminum foil prior to autoclaving. Aseptic connection is made by carefully removing the aluminum foil from the male and female sections to be joined, flaming each section, and then inserting the smaller-diameter tube into the larger until the smaller-diameter tube makes a secure fit with the tubing connected to the larger-diameter metal tube. The smaller metal tubing must be somewhat longer than the larger metal tubing section to allow complete penetration and seat in the rubber tubing line. The seal can be secured by wiring or using a small hose clamp. The quick connect should be held so that the larger female tube is placed with its open end pointing downward. The quick connect can also be rewrapped in sterile covering to prevent ingress of dust and airborne organisms between the tubes. These precautions will allow opening of the connection followed by flame sterilization and reconnection. The connector can be opened to allow changing of the reservoir by reversing the procedure, being sure to flame the smaller tubing prior to reconnection with the next sterile line.

10.2.1.7. Sterilization of medium

Standard procedures for the sterilization of liquid media are described elsewhere in this manual (see Chapter 7). Filter sterilization may be necessary to prevent destruction of heat-labile medium components. However, most simple media for cultivation of bacteria can be autoclaved prior to use, but larger volumes of medium require longer autoclaving times. Vessels containing approximately 10 liters of liquid should be autoclaved at 121°C for 30 to 90 min, depending on medium constituents. Media containing solids such as cornmeal flour or soy flour may require up to 90 min for complete sterilization of a 10-liter volume. However, media containing only dissolved components will usually be sterile after 30 min at 121°C. All vessels should be vented during autoclaving to allow equilibration of pressure.

For the cultivation of anaerobic bacteria, take special care to ensure establishment and maintenance of the chemically reduced state when autoclaving large volumes of prereduced medium. Connect vessels containing prereduced media to an oxygen-free gas supply immediately upon removal from the autoclave. As the steam in the gas head space of the vessel begins to condense, a partial vacuum will form. Sparging with oxygen-free gas will allow release of the vacuum formed during cooling without contamination of the medium by oxygen in the air.

10.2.1.8. Startup

Chemostat continuous cultivation is always preceded by transient batch cultivation during which time the cell mass accumulates at the expense of the substrate. Start the continuous culture while the batch cultivation is in the exponential phase of growth; this will minimize oscillations due to nutritional step-up and avoid inadvertent washout due to physiological lag. Start the dilution at a rate less than the desired operational dilution rate, and then increase to the operational dilution rate within one **residence time**, the average time that a cell remains in the vessel (i.e., $1/D$ or V/F in equation 8). This will minimize oscillations from toxic substrates. Chemostat response to toxic substrates is discussed in more detail by Pirt (5).

10.2.1.9. Sampling

Sampling is best accomplished with an independent sample line system. Alternatively, samples may be taken from the effluent line, although this procedure is not recommended when a small sample is needed, since the lumen of the effluent tube may become coated with adherent bacteria which may break away and result in nonrepresentative samples. Follow the fermentor manufacturer's instructions for sampling during continuous cultivation. If an external sampling system is used in place of sample collection from the effluent line, the sample size must be kept below 10% of the reactor working volume so that steady-state conditions will not be disturbed greatly. Finally, the sample line must be purged of its entire contents before samples are collected for analysis. Some fermentors are equipped with a sterile air purge which displaces the sample line contents back into the fermentor after a sample is taken. If the latter system is used, purging prior to sample collection is not necessary.

10.2.2. Turbidostat

The turbidostat is the simplest of the well-mixed continuous-cultivation systems (32, 41). Cell concentration (biomass) is monitored and maintained at a constant level by adjusting the feed rate of fresh nutrients. In contrast to the chemostat, a biomass concentration is chosen by the operator, and the dilution rate (and therefore the substrate concentration) is adjusted to maintain the predetermined biomass level. It follows that substrate need not be present in a limiting amount but, rather, will usually be present in excess. Operation is therefore most stable when the specific growth rate of the culture is near the maximum value, μ_{max}, for the particular medium in use. The system may then be operated over a wide range of biomass concentrations near the critical dilution rate as long as all medium components are in excess. These are precisely the conditions under which the chemostat is least stable. Also, unlike the chemostat, the dilution rate will always equal the specific growth rate, whether or not the system is at steady state, if biomass can be controlled with precision.

However, the precision with which the biomass can be controlled has historically been the weakest aspect of turbidostat cultivation. Most of the older methods of monitoring cell density rely on optical monitoring with some type of photoelectric sensor measuring scattered or transmitted light. These methods suffer from interference by bubbles and foam produced during aeration, wall growth on the fermentor and, even more critically, the optical device. Bubble interference can be eliminated by using an external flow-through optical cell. However, foaming remains a problem. Wall growth on optical surfaces can be partially remedied by wiping the surface, but the many designs for this have proved cumbersome and prone to failure. A recent article provides a good review of the turbidostat literature and describes a novel and effective turbidostat system in which cell density is monitored not optically but dielectrically (32).

The term "turbidostat" has come to include any technique that holds the cell concentration constant and includes monitoring techniques based on cellular metabolism. Since certain metabolic functions (such as oxygen uptake, carbon dioxide evolution, and, in some cases, pH change) are intimately linked to cell growth rate and ultimately to the specific growth rate, these parameters may be used as control variables for medium replacement.

Theoretically, any measurable metabolic parameter can be used as the control variable for turbidostat cultivation. Practically, only parameters that are tightly linked to cell growth (i.e., those that have small time delays in culture response) are useful control variables. The advent of sophisticated computer-sensor-fermentor control systems promises a powerful new level of use for turbidostat theory.

10.3. SPECIAL SYSTEMS

10.3.1. pH-Controlled Batch Culture

Controlled batch culture requires instrumentation that will monitor an environmental parameter and trigger an addition to the fermentor so that the indicator parameter is maintained at a steady value. An example of environmental parameter control in a fermentation system is the maintenance of pH at a constant value by the automatic addition of acid or base. The basic components of a pH control system are the pH electrode(s) with its shielded connector cable, the pH meter, an endpoint titrator, an acid or base reservoir with flexible tubing for connection to the fermentor, and a single peristaltic pump or solenoid valve. Although separate reference and pH electrodes can be used, it is usually more convenient to use a steam-sterilizable combination electrode to minimize space requirements in a small laboratory-scale fermentor. All components of the fermentor system, including pH and dissolved-oxygen electrodes, must be designed for steam sterilization or be compatible with other sterilization methods, such as ethylene oxide or formaldehyde-steam sparging.

The Ingold-type electrode (Ingold Electrodes, Inc.) is particularly reliable and can be repeatedly autoclaved. The pH meter and pH titrator must be interconnected and therefore should be purchased from the same manufacturer to ensure equipment match. Although a peristaltic pump can allow foolproof addition of acid or base against relatively high fermentor back-pressure, a simple tubing-pinch solenoid valve for control of gravity feed of acid or base to the fermentor is usually sufficient for laboratory-scale operation. Fermentor manufacturers (section 10.6) usually provide their own pH control systems. For more details about pH control, see Chapters 6.1 and 21.2.1 and reference 40.

10.3.2. Synchronous Batch Culture

Biochemical events associated with specific times in the individual cell cycle can be systematically studied in a population of cells by batch cultivation in a synchronous manner. Culture synchrony occurs when all of the cells divide at nearly the same instant, thus mimicking individual cell growth. In such cases, a plot of the logarithm of cell numbers versus time will resemble a stair step rather than a straight line. Although development of culture synchrony is relatively straightforward, maintenance of synchrony over long periods is a difficult task requiring precise control.

Culture synchrony can be achieved by periodically varying a critical environmental condition. The technique forces the synchronization of cell multiplication by interrupting, promoting, or retarding metabolic function in a cyclic manner. The population will gradually synchronize its response to these periodic disturbances of metabolic activity and will ultimately synchronize its growth pattern. The technique, however, requires severe disturbance of normal metabolic activity and is therefore of limited use in the study of growth cycle-linked bacterial physiology. One exception to this limitation is bacterial spore germination as a means of initiating culture synchrony. Step or pulse changes in the culture environment, or addition of a specific chemical germinant, can trigger spore germination and may closely model natural occurrences.

A generally useful method of synchronization is based on physical selection of a homogeneous fraction from a heterogeneous population of vegetative cells (16). This approach is often termed **selection synchrony** and avoids most of the problems of metabolic disturbance during synchronizaion by physically selecting cells that are in similar states of the cell growth cycle.

Kubitschek (26) and Poole (44) showed that the cell volume of *Escherichia coli* increases linearly during the cell cycle while its cell mass increases exponentially. The observation that cell volume is lowest just after cell division suggests that centrifugation in a density gradient might allow recovery of a population of new daughter cells from an asynchronous culture. However, the observation that the volume increase is linear while the mass increase is exponential means that cells which are just ready to divide or cells which have just divided will have the greatest cell density despite cell size differences. Density gradient centrifugation will result in cell fractions in which the most dense fraction contains both young (daughter) cells and mature (ready-to-divide) cells, while the least dense fraction will contain a homogeneous population of cells that have progressed through a common fraction of their cell cycles. See section 10.4.3. for details of centrifugal fractionation.

Selection synchrony through density gradient centrifugation cannot supply new daughter cells for direct study of cell cycle-linked physiology, because these fractions will always be contaminated with mature cells ready to divide. However, density gradient centrifugation can supply an adequate inoculum for synchronous culture growth. Cell cycle-linked physiology may then be studied by direct sampling of the synchronous culture. The procedure described below is presented as a general guideline for development of synchronous cultures and is based on the technique described by Mitchison and Vincent (36). This technique may be used as an initial guide for development of a synchronous cultivation technique specifically designed for the organism in use (30).

Procedure. Delay synchronization experiments until reproducible batch cultivation conditions can be established, including determination of asynchronous culture kinetics. When these prerequisites are met, establish batch cultivation conditions so that two to five cell mass doublings occur during logarithmic growth.

1. Prepare 500 ml of sterilized medium in several culture vessels (250-ml Erlenmeyer flasks are convenient). Inoculate half of the flasks, and incubate these under appropriate conditions. Store the remaining sterile flasks under identical conditions.

2. Harvest the cells from the batch culture at the mid-exponential growth phase. Rapidly cool the culture to 0 to 4°C by swirling the flasks in an ice bath. Harvest the cells by refrigerated centrifugation at 10,000 × *g* for 10 min.

3. Resuspend the sedimented cells in 2 ml of appropriate ice-cold buffer (0.1 M potassium phosphate [pH 7.0]) by vigorous agitation with a Vortex mixer. If the cells tend to aggregate, use mild sonication or mild homogenization with a blender or tissue grinder to prepare a suspension of discrete cells. For cultures that are particularly difficult to suspend, use 0.01% (wt/vol) Tween 80 in the suspending buffer before mechanical or sonic treatment.

4. Rapidly, but carefully, layer the suspended cells on a sterile, precooled density gradient prepared from Ficoll, sucrose, Percoll, or another appropriate medium as dictated by the cell system (Chapter 4.5). For example, to prepare an exponential sucrose gradient in a discontinuous manner, layer sterile ice-cold solutions of increasing sucrose concentration into presterilized, precooled (0°C) centrifuge tubes. Place 10 ml of 35% (wt/vol) sucrose in phosphate buffer into the bottom of a sterile centrifuge tube. Sequentially layer 10 ml each of sucrose-buffer solutions containing 26.5, 25.5, 24.5, 22,0, 19.0, and 15.0% (wt/vol), respectively, onto the 35% sucrose cushion (The buffers may all be prepared from the 35% sucrose stock solution.) Filter sterilize the sucrose solutions prior to use in forming the gradient. All solutions must be ice cold and preaerated (for aerobic cells).

5. Carefully centrifuge the cell-charged sealed centrifuge tubes in a precooled centrifuge at 2,500 × *g* for 15 to 20 min at 0°C. For more precise separations, adjust the time and speed of centrifugation so that the optically dense band of bacteria moves no more than two-thirds of the way down the centrifuge tube; speeds and times will vary somewhat depending on the culture being handled.

6. Inoculate prewarmed flasks of growth medium with 0.5 ml directly from the lightest density fraction of cells.

7. Carefully monitor the optical density of the newly inoculated culture flasks for a definite stepwise increase in optical density, an indication of synchronous growth.

Culture synchrony should be maintained for two to three cycles. Study over longer periods will require reestablishment of synchrony through the procedure described above. Figure 8 shows a typical growth curve for a selection-synchronized population of bacteria, with decay of synchrony after two cycles.

10.3.3. Immobilized-Cell Reactors

In most bacterial culture methods, maximum multiplication of cells is the objective and the medium nutrients are consumed mainly by growth metabolism. In an industrial fermentation process, however, maximum production of a metabolite product may be the objective, and the nutrients are consumed mainly for products rather than growth. A great deal of research has been directed toward the use of bacterial enzymes or cells that are "immobilized" within a reactor to maximize the continuous conversion of a substrate into a useful product via maintenance metabolism.

Cell immobilization is considered by Abbott (7) "as a physical confinement or localization of a microorga-

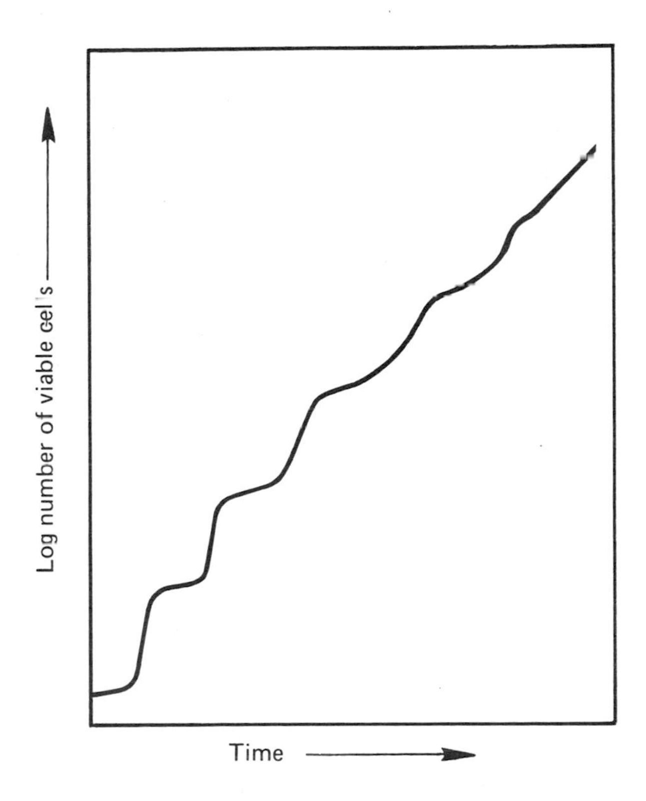

FIG. 8. Typical growth curve for a selection-synchronized population of bacteria.

nism that permits the economical reuse of the microorganisms." This operational definition can be replaced by a physiological one describing a continuous reactor that maximizes maintenance metabolism and minimizes growth metabolism of a microbial population.

The principle is exemplified by the old "quick vinegar" process in which acetic acid bacteria are immobilized as a colony film on a column of wood shavings, through which an ethanol solution is trickled downward while air passed upward, yielding an acetic acid solution. Once the bacterial film is established, little further growth occurs and bacterial maintenance energy is derived from the oxidation of ethanol to acetate. The process can be operated continuously over periods of weeks or months.

Physical adherence or chemical bonding to a solid carrier remains as one of the general methods for cell immobilization. Cell pellets or flocs are formed in other methods. Detailed description of methods for immobilized-cell reactors is beyond the scope of this book, and the reader is referred to reviews on the subject (3, 7, 8, 33, 57).

10.3.4. Fed-Batch Culture

In a batch culture, the transition from exponential growth to stationary phase may occur for a variety of reasons, including the depletion of an essential nutrient substrate or the buildup of a toxic metabolite product. When the transition results from nutrient depletion, growth will continue if fresh medium is added. The medium addition rate and the culture volume must be increased exponentially to maintain a constant rate of exponential growth. This technique for exponential growth maintenance (13, 28) is usually called fed-batch culture. Periodic removal of culture volume to allow additional feeding is called extended-batch culture or repeated fed-batch culture.

True fed-batch operation requires that the volume of the liquid medium in the fermentor increase during the fermentation. This requirement places an upper limit on the culture time based on feed rate and leads to changing conditions of aeration and agitation effectiveness. Furthermore, it is possible to back-contaminate the feed reservoir from the fermentor; for this reason, it is desirable to use a medium breaker (Fig. 5).

The fed-batch mode will allow substantial improvements in cell mass or product productivity over an ordinary batch operation (section 10.3.7). The technique of fed-batch culture may also be used to supply large quantities of a potentially toxic substrate while maintaining a low concentration of the substrate in the medium.

10.3.5. Dialysis Culture

Dialysis is a process for separation of solute molecules by means of their unequal diffusion through a semipermeable membrane because of a concentration gradient. The process is applied to the growth and maintenance of living cells by a technique called **dialysis** (or **diffusion** or **perfusion**) **culture**.

The history of dialysis culture can be traced back to 1896, when Metchnikoff inserted a collodion sac containing cholera bacteria into the peritoneum of a live guinea pig to learn whether the bacteria produced a diffusible toxin. The history, generalized design, mathematical theory, and practical applications of dialysis culture were brought together in a comprehensive review article by Schultz and Gerhardt in 1969 (46). Stieber (50) provided a comprehensive decade-updating review of the subject.

To use the technique of dialysis culture, a membrane is positioned between a culture chamber and a dialysate reservoir. For dialysis culture to be effective, the volume of the reservoir must be larger than that of the culture or else the reservoir must be replenishable. Further, the permeability and area of the membrane must be sufficient to permit useful diffusion in rate and amount. Such a system of dialysis culture can be operated in vitro or in vivo and batchwise, continuously, or a combination of these modes.

There are two basic principles underlying dialysis culture. First, it provides a means for achieving substrate-limited growth, i.e., fed-batch culture (section 10.3.4). Substrate in the dialysate reservoir diffuses through the membrane into the culture chamber, driven by a concentration gradient that results as the substrate is used for metabolism and growth. Second, dialysis culture provides a means for lowering the concentration of a diffusible metabolite product inhibitory to growth; the product in the culture chamber diffuses through the membrane and is diluted in the larger dialysate reservoir, thus relieving the feedback inhibition by the product that normally regulates its production. In the usual dialysis culture system, nutrient supply and product withdrawal occur concurrently; i.e., exchange dialysis occurs.

Three main types of membranes, differing in porosity, are applicable to dialysis culture. **Dialysis membranes** have a nominal pore size in the order of 10 nm in diameter so that they exclude cells and macromolecules but allow small molecules, such as the nutrients for bacterial growth, to pass. Dialysis membrane tubing in various sizes is manufactured from regenerated cellulose by the Visking process (Union Carbide Corp.). **Filter membranes** (or membrane filters) have a nominal porosity in the order of 100 nm, so they also exclude cells, but they allow macromolecules as well as small molecules to pass. Sheets of membrane are commercially available in a wide range of porosities and materials (section 10.6). Factors to consider for bacteriological use include autoclavability, pore size, inertness, and permeability (which is governed by porosity, void space, and thickness). **Solution transport membranes** have no pores and pass gases because of their solubility in the membrane material itself (e.g., oxygen solubility in silicone rubber or polycarbonate).

An interface between two physically different phases may also be used to separate the culture and dialysate (**interface dialysis culture**), such as in a liquid/solid biphasic system (Chapter 9.3).

Specific advantages and reasons for using dialysis for bacterial culture include (i) prolongation of the exponential growth phase in the batch cycle, which allows the attainment of very high densities of viable cells; (ii) extension of the maximum stationary phase in batch culture, thus permitting increased production of secondary metabolites associated with this phase; (iii) relief of product inhibition control by removal of metabolite products, thus enabling their greater production in batch or continuous culture; (iv) establishment of a steady-state population with mainly maintenance metabolism, thus immobilizing the cells for prolonged production of a metabolite product; (v) production of metabolites free from cells and, conversely, production of cells free from medium macromolecules; (vi) means to study and recover a cell population placed in an in situ or in vivo environment, such as in an ecological or animal system; and (vii) the capability for study of molecular interactions between separated populations of cells.

10.3.5.1. In vitro systems

Dialysis culture is often first attempted in the laboratory by suspending a membrane sac containing the culture in a tube, flask, or carboy of medium. The simplest such arrangement is to use a length of dialysis membrane tubing that is intussuscepted so as to form a double-walled tube containing culture in the annular space thus formed (Fig. 9).

The advantages of shake flasks (section 10.1.3) are combined with those of membrane dialysis culture in a unit assembled from flanged Pyrex glass pipe (Fig. 10). This and other flanged flask designs (50) enable the use of sheet membrane of any type. A liquid/solid biphasic shake flask system for interface dialysis culture is an effective and simple way to obtain the advantages of dialysis culture in the laboratory (Chapter 9.3).

Scale-up of dialysis culture to fermentors is possible by carrying out growth in a separate culture circuit that is connected with a separate dialysate circuit by means

FIG. 10. Dialysis culture flask, designed to be held on a carriage mounted on a rotary shaking machine. The bottom compartment is filled with sterile medium which is stirred by the rotating ball. The top compartment holds the culture, which is turbulently aerated by swirling and baffling. Reproduced from reference 21 with permission.

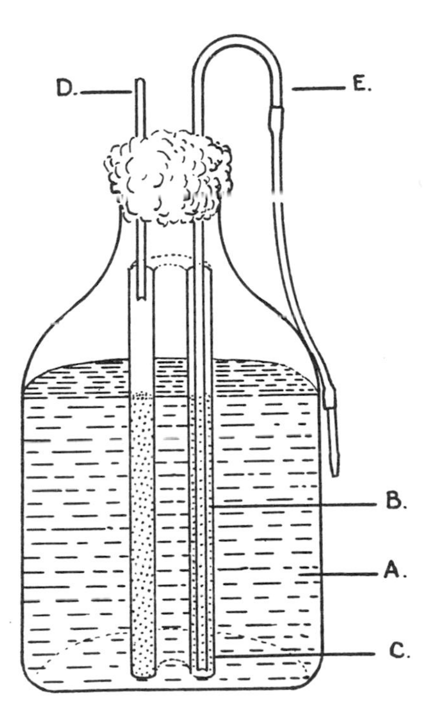

FIG. 9. Illustrative sac dialysis culture system. A tube of dialysis membrane is intussuscepted so as to form a double-walled sac (C), the annular space of which contains culture (B). The medium (A) is contained in a carboy. There also is a tube for sampling and inoculating (D) and a harvesting siphon (E). Reproduced from reference 4 with permission.

of an intermediate dialyzer (20). Such a dialyzer-dialysis culture system, depicted schematically in Fig. 11, can be operated batchwise in both circuits (Fig. 11a), continuously in both circuits (Fig. 11b), or in combinations of these modes (Fig. 11c and d). The key to effective operation of such systems is a suitably designed dialyzer. The best principle of design resembles a plate-and-frame filter press and contains molded silicone rubber separators that promote turbulent flow of the culture and dialysate on the opposing sides of the membranes (46). Various types of hollow-fiber dialyzers have also been used (45); a bundle of hollow fibers provides a large membrane surface area, but much of it is masked and turbulent flow is not usually attained.

The design of the system can be optimized for a given culture situation. For a situation in which growth is limited by a toxic metabolite product (e.g., lactic acid), a completely continuous system is operated most effectively as shown schematically in Fig. 12. The substrate (S_{f°) in relatively high concentration is fed directly into the fermentor, and the usual reservoir vessel for dialysate is eliminated. Instead, the dialysate circuit consists only of the tubing, pump, and dialysate side of the dialyzer and contains a relatively low volume (V_d). Only water is fed into the dialysate circuit at a relatively high flow rate (F_d). This system provides the greatest concentration gradient possible for dialysis. This operational mode has been used to maximize the conversion of concentrated whey lactose into lactic acid (52).

10.3.5.2. Mathematical modeling and computer simulation

The foregoing developments in system design and practical uses of dialysis continuous culture (fermenta-

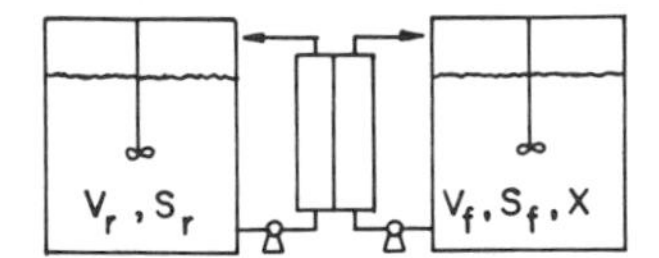

a) BATCH RESERVOIR AND FERMENTOR

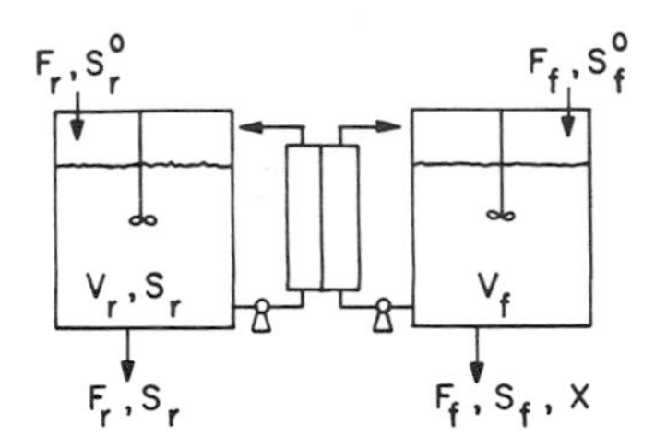

b) CONTINUOUS RESERVOIR AND FERMENTOR

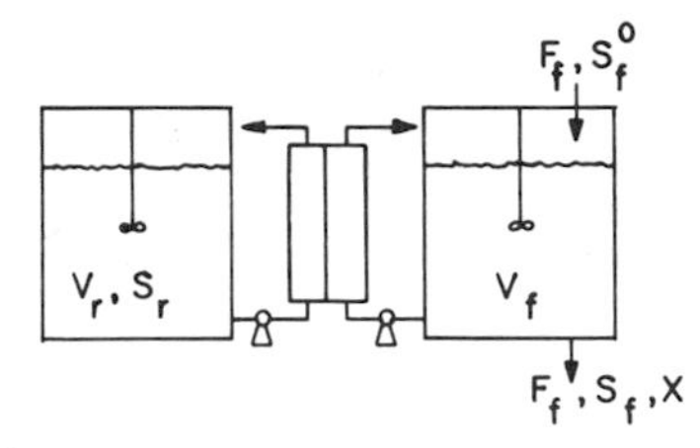

c) BATCH RESERVOIR, CONTINUOUS FERMENTOR

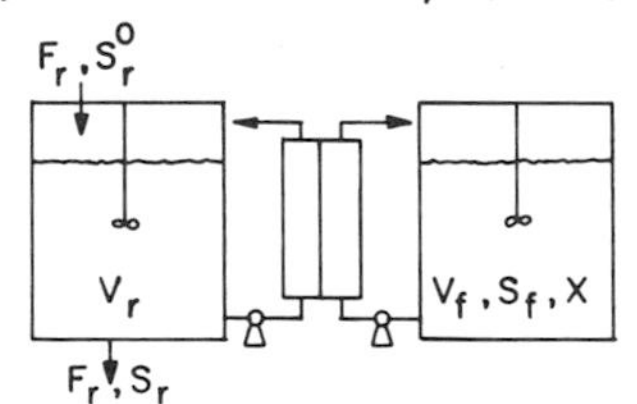

d) CONTINUOUS RESERVOIR, BATCH FERMENTOR

FIG. 11. Four modes of operating a dialysis culture. Although a dialyzer-dialysis system is diagrammed, the principles are intrinsically applicable to any design. The symbol *F* is the flow rate, *S* is the substrate concentration, *V* is the volume, and *X* is the cell concentration into, within, and out of the reservoir and fermentor vessels. Reproduced from reference 46 with permission.

tion) also exemplify the power of mathematical modeling and computer simulation for predicting the results to be expected in a bacterial growth process. Laboratory experiments need to be conducted only to validate the theoretical predictions by using a relatively limited number of changes in experimental conditions at preselected critical points. The experimental results in turn are used to establish growth constants and to indicate the need for additional terms in the equations. By this process of successive theoretical prediction and experimental validation, the mathematical model becomes increasingly accurate and useful for predicting optimization of a fermentation process.

An example of such a combined theoretical and experimental approach to a bacterial process is provided by the results (12, 51, 54) with the ammonium lactate

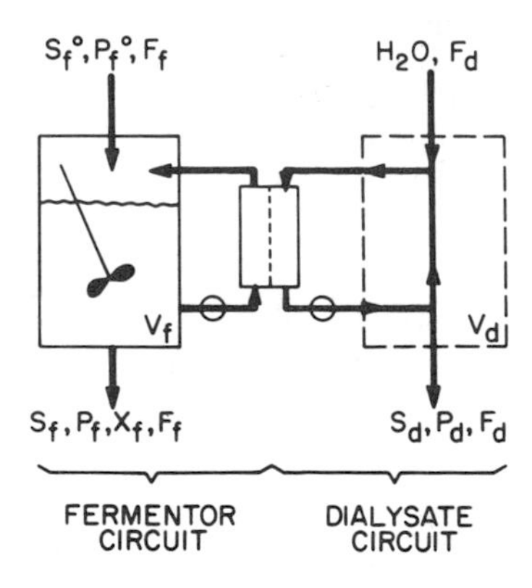

FIG. 12. Schematic of fermentor-fed dialysis continuous-culture system in which a stream of water is used in the dialysate circuit to maximize the withdrawal of a toxic metabolite product. Symbols are as in Fig. 11. Reproduced from reference 12 with permission.

fermentation. A set of material balance and rate relationship equations are developed for substrate, product, and cell mass in the fermentor and dialysate circuits. These equations are combined, the variables are defined in dimensionless parameters, and the time derivatives are set to zero to obtain a generalized solution for the steady state which consists of five quadratic equations. These are programmed on a digital computer by using selected real values for the various terms. A typical comparison of simulated predictions with experimental results (Fig. 13) shows the close correlation that can be attained.

Such a method of theoretical prediction and experimental validation is commonly used in the field of industrial fermentations but much less so in other fields of bacteriology. This powerful method could be much more widely used in medical bacteriology, e.g., to predict the outcome of infectious diseases.

10.3.5.3. In vivo systems

The term "diffusion chamber" is often used to describe a small dialysis culture unit that can be implanted within a living experimental animal (e.g., in the peritoneum or rumen or beneath the skin). Among a number of systems, the best are made with a nondegradable filter membrane sealed on both sides of an inert plastic cylinder (like a drum) through which sampling access can be provided. Bacteria introduced into diluent within the chamber grow entirely on diffusible nutrients from the host. Such systems have been used in bacteriology to study immune reactions and the growth of fastidious pathogens, e.g., *Mycobacterium leprae, Treponema pallidum,* and *Neisseria gonorrhoeae.*

10.3.5.4. Ex vivo systems

The carotid-jugular bloodstream of a large experimental animal (e.g., a goat) can be surgically externalized and circulated across one side of a membrane,

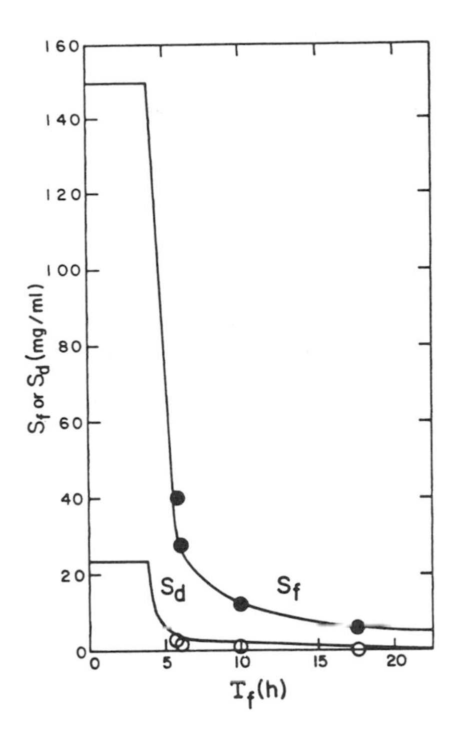

FIG. 13. Effects of cell retention time (T_f) on residual lactose in the fermentor circuit (S_f) and the dialysate circuit (S_d) during dialysis continuous fermentation. The curves were plotted by computer from the mathematical model, and the points were plotted from experimental data to demonstrate the close fit between the experimental results and the simulated predictions. Reproduced from reference 52 with permission.

on the other side of which is contained a bacterial culture. Such an ex vivo hemodialysis culture system enables the bacteria to grow entirely on the nutrients of the blood, yet separate from the macromolecular and cellular defense mechanisms, if a dialysis type of membrane of appropriate molecular exclusion is used. In one such system, the bacteria are contained in an ordinary laboratory fermentor and the culture is continuously circulated through the jacket of a hollow-fiber type of artificial-kidney hemodialyzer, which is connected by tubing to the bloodstream of the restrained animal (45). In another such system, the bacteria are contained in a small (3-ml) clear-plastic (polycarbonate) chamber over a membrane, on the opposite side of which is a passage for the bloodstream; the entire unit is mounted on the neck of the ambulatory animal. The latter system was used successfully with many facultatively aerobic (but no obligately anaerobic) bacteria and was designed to study host-parasite reactions of fastidious pathogenic bacteria (22).

10.3.6. Product Removal Culture Systems

10.3.6.1. Microfiltration and electrodialysis

The product removal and consequent cell-concentrating feature of dialysis culture can be accomplished much more efficiently by other systems involving membranes in which the driving force is greater than a concentration gradient, such as a pressure gradient for microfiltration and an electrical gradient for electrodialysis. However, these systems are more complicated in design, tend to foul the membrane more than dialysis does, and have specialized applicability beyond the scope of this book. An example of a continuous microfiltration culture system is given in reference 42, and an example of an electrodialysis system is given in reference 25.

10.3.6.2. Extraction, sorption, and evaporation

Product removal and relief of product feedback inhibition of bacterial growth can also be attained by a variety of other means. These include extraction by use of nonaqueous solvents, biphasic aqueous systems, and catalytic membranes; sorption by solvents; vacuum evaporation and membrane pervaporation; and precipitation. These extractive culture systems are reviewed in a publication edited by Mattiasson and Holst (34).

10.3.7. High-Density Batch Culture Strategies

Ordinary liquid-culture methods produce relatively low densities of bacterial cell mass (biomass), which is best expressed as grams of cell dry weight per liter for applicability to cells of various types and sizes. Usual maximum growth densities of aerobic or facultatively aerobic bacteria in batch systems are on the order of 0.1 in quiescent test tubes, 1 in shake flasks, and 10 in aerated and agitated fermentors (all expressed as grams of cell dry weight per liter). Cell counts and other direct measures of growth increase similarly by decades. Maximum densities of anaerobic bacteria are generally about a decade less than those of aerobes.

Further large increases to still higher cell densities can be attained by strategically using one or more of the critical limiting factors in batch growth discussed earlier in this chapter. High-density culture (concentrated culture) strategies for aerobic bacteria have resulted in concentrations as high as 150 g of dry cells/liter (20), but the maximum practical concentration is limited by oxygen transfer capability to about 70 g of dry cells/liter. Increasing the cell density makes more sense than making replicates or increasing the vessel volume. Optimization of a process for high cell density is desirable whether the process is for bacterial biomass production, metabolic-product formation, or substrate consumption.

The key principle underlying high-density culture is to relieve the feedback inhibition of bacterial growth by preventing the depletion of an essential nutrient and limiting the accumulation of toxic metabolic products. This is done by (i) supplying essential nutrients in increasing amounts to meet but not exceed growth needs, (ii) removing toxic products as they are formed, and (iii) controlling physicochemical growth factors at optimum levels. These are best accomplished in a fermentor system (section 10.1.6) and are usually applied to aerobic bacteria, but they are also applicable to anaerobes. A summary review of the theoretical and historical background of high-density culture was presented by Shiloach et al. (48).

10.3.7.1. Aerobes

High-density culture of aerobes is best exemplified with facultative bacteria such as *E. coli* and other enteric bacteria, which oxidize glucose and other carbon and energy sources completely to CO_2 via aerobic respiration in the presence of an adequate oxygen supply but incompletely to organic acids, particularly acetic acid, via anaerobic respiration in the presence of an inadequate oxygen supply. It is this accumulation of acetic acid by anaerobic respiration as the result of oxygen depletion that mainly inhibits growth.

Oxygen is almost always the limiting nutrient for aerobic bacterial growth because of its limited solubility in aqueous media. Oxygen affects both the rate and extent of growth (Chapter 6.5; section 10.1.6.1). To keep aerobic cells growing exponentially in a fermentor, the dissolved-oxygen supply should be increased as growth increases and kept above 30% saturation; for a culture of 10 g of dry cells/liter, this amounts to 300 mmol of oxygen per 30 min per liter (48). This enormous oxygen demand can only partially be met by maximizing the airflow and agitation rates in a fermentor. To further increase the dissolved-oxygen concentration, oxygen-enriched air must be provided by adding pure oxygen from a liquid-oxygen cylinder or by using oxygen-enriched (48%) air from a membrane separation system (17). Dissolved-oxygen concentration is monitored with a probe (14). The increases in the oxygen-enriched airflow and agitation rates corresponding to the increase in growth are best controlled by a computer system (17).

Increasing the addition of essential nutrients as exponential growth progresses (fed-batch culture [section 10.3.4]) is also usually essential and a primary consideration for high-density culture of aerobes. Glucose (or another carbon and energy source) must usually be maintained at a low level to eliminate toxic-product formation, e.g., acetic acid by *E. coli.* The system must be monitored with a glucose or acetate probe and controlled by a computer (to <2 g of glucose per liter or <1.5 g of acetate per liter for *E. coli*). This usual situation was contradicted in the publication of work in which a high initial glucose concentration (40 g/liter) was used without further glucose additions but with an extraordinarily high oxygen supply from oxygen-enriched air (48), which may account for the insusceptibility to the high glucose concentration.

Removing a toxic metabolic product as it is formed (sections 10.3.5 and 10.3.6) is another common strategy for attaining a high density of bacterial cells. Although essential for anaerobes (section 10.3.7.2), product removal alone should usually be a secondary strategy for aerobes if bacterial biomass production is the objective: it is better to limit product formation than to correct it. In dialysis culture (section 10.3.5), product removal is coupled with nutrient supply; this combination strategy is especially applicable for small-scale laboratory use by use of the biphasic shake flask system (Chapter 9.3).

Maintenance of pH at an optimum level (section 10.3.1) is necessary for high-density culture and has become commonplace in fermentor systems with provision for a pH probe and either manual or computer control by the addition of ammonium hydroxide, which also maintains the nitrogen supply necessary for high-density culture.

Control of temperature is also usual in fermentor systems (section 10.1.6.2). For high-density culture, it is crucial not to exceed the optimum temperature, beyond which growth rates fall off precipitously (Chapter 6.3). Indeed, a lower growth temperature (for *E. coli*, 22°C) has been used to lower the growth rate and so to extend the exponential phase (47), resulting in about a doubled yield but at the expense of a doubled duration. A practical compromise between these considerations is to use a moderately lower growth temperature, e.g., 30°C for *E. coli.*

10.3.7.2. Anaerobes

High-density culture of anaerobes (or of anaerobically growing facultative anaerobes) depends on relieving feedback inhibition of growth by limiting the accumulation of toxic metabolic products, just as for aerobes. The difference lies with the toxicity of oxygen for obligate anaerobes and with their complete reliance for energy on fermentative respiration and the inevitable accumulation of intermediary by-products to eventually toxic concentrations. Consequently, increased density of anaerobic cells must rely on the strategy of product removal by the most efficient means possible (sections 10.3.6.1 and 10.3.6.2).

10.4. HARVESTING AND PURIFICATION

The choice of a process for separation of bacterial cells from culture broth will be dictated by the ultimate disposition of the fractions. If cells are to be used for further study, cake filtration is not usually desirable owing to contamination of the cells by the filter aid material. Similarly, recovery of the culture broth by batch centrifugation may suffer from discarding a portion of the broth to avoid resuspension of the cell pellet during removal of the supernatant. For general reviews on separation processes, see references 9, 54, and 56.

10.4.1. Filtration Harvesting

10.4.1.1. Cake filtration

Two types of filtration are usually encountered in separating bacterial cells from culture broths: cake and membrane filtration. The former utilizes a filtration medium that is highly porous. Often, the filtration medium is a composite of a filter support (often filter paper cloth of relatively large pore diameter) with a relatively thick overlay of suspendible filter aid such as cellulose, diatomaceous earth, kieselguhr, or other material that can be freely dispersed in aqueous media to form a pad whose porosity is low enough to retain bacteria (54). The use of filter aid increases the number of bacteria that can be removed by a single filter. Filter aid may be added directly to the culture broth or prelayered on the filter support. The practice of adding filter aid to the culture broth can allow filtration of as much as 10 times the amount of culture broth that can be treated by prelayering the same quantity of filter aid on the filtration support.

To assemble a simple but effective cake filtration system, fit a Buchner funnel to a standard filtration flask with a side arm. Place a piece of highly porous filter paper, such as Whatman no. 1, on the funnel so that the paper covers the holes in the funnel plate. Cover the filter paper with a layer of diatomaceous earth 2 to 5 mm deep by suspending the appropriate amount in water and gently but uniformly pouring this suspension onto the filter paper while vacuum is applied. Gently scrape the cake surface during filtration to remove bacteria as they accumulate; this allows prolonged filtration without having to change the filter cake.

Cake filtration of bacterial cultures can remove most of the cells from the culture broth, leaving a highly clarified filtrate. However, this technique is not a reliable method of filter-sterilizing liquid media. Cells recovered from the filtration cake will usually be contaminated with large amounts of filter aid, which usually precludes further studies on the cells themselves.

10.4.1.2. Membrane filtration

Unlike cake filtration, membrane filtration does not rely on the depth or tortuosity of pores in the filtration medium for physical removal of particles from the fluid. Rather, the pore diameter at the surface of the membrane is controlled to within narrow tolerances. The technique of membrane filtration is based on exclusion of the bacteria at the surface of the filtration medium. Pressure drop accross the membrane is often high, and practical operation of a membrane filtration system requires pressurized feed. However, conventional through-flow membrane filtration soon results in an increased pressure drop and decreased flow as a result of the buildup of cells and their compressible nature.

This limitation can be largely overcome by application of **cross-flow (tangential-flow) membrane filtration.** In this method a bacterial suspension is fed continuously, under pressure, through a filtration module with the flow directed tangentially across the membrane surface at a rate high enough to induce turbulence. Some of the medium (filtrate) is continuously removed through the membrane, and the remaining thickened suspension (retentate) is continuously returned to the culture container. Since the retentate is continuously recirculated, there results an increasing concentration of cells in the culture container. Sludges containing 50% (wet wt/vol) cells can be attained, limited mainly by viscosity. For further purification of cells, the sludge is diluted with a suitable diluent and the cross-flow filtration is repeated as many times as necessary. The accumulation of cells on the filter is minimized, because its surface is continuously swept clean, so that filtration can be repeated or maintained over long periods. Eventually the membrane becomes fouled, mainly because of deposition of proteinaceous material on the membrane surface and within the pores, which necessitates cleaning or replacement of the membrane.

The choice of a specific membrane depends on the culture characteristics and whether the cell or filtrate fraction must be recovered. Since an average bacterium is roughly 1.5 by 0.5 μm, absolute removal of bacteria for the purpose of filter sterilization will require the use of a microporous membrane filter whose mean pore diameter is less than 0.5 μm; the use of membrane filtration for sterilization of media is described elsewhere in this book (Chapter 7.4.2). The use of microporous membrane filters with a mean pore diameter of 0.45 ± 0.04 μ m will generally allow rapid and efficient harvesting of bacterial cultures but will not ensure sterility of the filtrate.

Membrane filtration apparatus and membranes of various pore size are available commercially (section 10.6).

Additional reading on membrane filtration can be found in references 9, 10, 23, 35, 39, and 54.

10.4.2. Centrifugation Harvesting

Differential centrifugation (Chapter 4.3.1) usually results in less complete separation of the culture broth from the cells than can be achieved by filtration, but it is the method most often used for laboratory-scale harvesting of cells.

The properties which govern the rate of sedimentation of cells during centrifugation are the viscosity of the fluid through which sedimentation occurs, the size (diameter) of the particles, and the density difference between the suspending fluid and the cells. Although these factors interact to dictate specific centrifugation times for pellet formation, most bacteria can be removed from the usually nonviscous culture broths (such as nutrient broth) by centrifugation at $10{,}000 \times g$ for 10 to 15 min. Since the packing characteristics of bacteria are dependent on cell shape and cultivation conditions, it may be necessary to adjust the centrifugal force (rotational speed of the rotor) or centrifugation time so that the bacterial pellet is less subject to resuspension during removal of the centrifuge tubes and withdrawal of the supernatant.

Laboratory centrifuges are available for either batch or continuous operation, and often a single device can be equipped for both types of operation. Many rotor designs are available, but all operate on the same principle, i.e., controlled application of centrifugal force through rotation of the centrifuge head. Since the average diameter of bacterial cells is minute, their centrifugal recovery usually requires high-speed rotation. High-speed centrifuges inevitably generate large amounts of heat and therefore should be equipped with a self-contained refrigeration system or be amenable to operation in a cold room. Furthermore, since centrifuge heads are available in many sizes, the user should be aware that centrifugal force (usually expressed as multiples of the gravitational force, g) should be used rather than rotational speed, together with time, in designing and reporting centrifugation procedures. Specific details of rotor speed and equivalent centrifugal force are available from the manufacturers (section 10.6).

Large volumes of culture can be treated by continuous centrifugation. The DuPont-Sorvall refrigerated centrifuges can be equipped with a special rotor and delivery system to convert them to continuous operation, which is quite adequate for most bacterial cultures but does not adequately treat highly flocculated cultures that tend to plug the feed and distributor lines.

Sharples centrifuges have long, narrow, spindle-like rotors through which the culture is passed continuously in an upward flow. The high speed of spindle rotation causes the cells to settle on the walls of the centrifuge vessel. This type of continuous centrifuge offers very rapid processing capability and the potential for differential centrifugation of particles of different size or density. Models are available with either finite cell sludge capacity or continuous cell sludge discharge.

An example of a batch differential centrifugation procedure is illustrated below. The procedure allows aseptic preparation of washed cells. It may be simplified if asepsis is not required.

1. Prepare an appropriate diluent (see Chapter 11.3.1) for washing the cells recovered by centrifugation. Prepare enough diluent that four to six times the culture volume is available during the washing step. Be sure that the ionic strength and pH of the diluent closely approximate those of the culture medium. Sterilize the washing diluent, and refrigerate it prior to use.

2. Precool the appropriate centrifuge rotor by storage in a refrigerator. *CAUTION:* Do not place a room-temperature rotor onto the centrifuge spindle and then allow the centrifuge refrigeration system to cool the head. If this is done, the centrifuge rotor will contract and seize the spindle, making it very difficult to remove later.

3. Sterilize several capped centrifuge tubes (polypropylene and polycarbonate tubes are autoclavable), and cool them before use. Wrap aluminum foil around the cap and lip of each centrifuge tube prior to autoclaving to maintain sterility of the centrifuge tube lip during handling.

4. Determine the volume of the centrifuge tubes in use, and aseptically transfer two-thirds of that volume of bacterial culture into each centrifuge tube. Be sure that the pairs of centrifuge tubes weigh exactly the same. If the last centrifuge tube contains less than the others, balance it against a separate centrifuge tube containing enough water to allow the balance of tube weights. Be sure that the sterile caps are securely in place on each centrifuge tube.

5. Insert the centrifuge tubes in the rotor so that each tube is positioned opposite its balanced partner.

6. Carry out the centrifugation under appropriate conditions and using the procedures specified by the manufacturer. Centrifugation at $10,000 \times g$ for 10 to 15 min will suffice for most bacterial cultures. Some centrifuge systems have a special protective rotor head plate that must be secured in place prior to centrifugation. Removal of this head plate after the rotor has come to a gentle stop has a potential to jar the rotor, an action that can resuspend some of the bacterial pellet. Remove the head plate carefully while bracing the rotor to prevent rotor movement on the spindle.

7. Carefully remove each centrifuge tube, noting the position of the bacterial pellet at the bottom of the tube. Aseptically transfer the supernatant to a sterile reservoir by decanting. It will probably be impossible to prevent partial resuspension of the bacterial pellet during this procedure. Indeed, on the pellet surface there may be a fluffy layer containing dead cells and cell debris, which should be removed by gently loosening and washing it away with a stream of sterile diluent from a wash bottle.

8. Aseptically fill each centrifuge tube containing a bacterial pellet with two-thirds of the tube volume of precooled sterile washing diluent. Replace and secure the centrifuge caps, and resuspend the bacterial pellets by agitation on a Vortex mixer. Repeat steps 4 through 6 (substituting washing diluent for culture) for a minimum of four washings.

9. After the last wash diluent has been aseptically removed, resuspend the bacteria in a small amount of sterile diluent. This suspension may then be used for further study.

The number of washings required in either the filtration or centrifugation technique will depend on the culture conditions, volume of wash diluent, and volume of wet cell mass. If the wet cell mass occupies roughly 10% of the volume of the total culture suspension, for washings with volumes equal to the original culture suspension volume will result in approximately 99.99% removal of soluble contaminants. It is important to use multiple washes rather than a single wash whose volume equals that of the total multiple washes, since each sequential washing reduces the contaminant level by roughly a factor of 10 for the conditions described above.

10.4.3. Density Gradient Centrifugation Purification

Although the foregoing differential centrifugation harvesting procedure is satisfactory for most purposes, especially if the fluffy layer on the surface of the pellet is removed after each washing, a more rigorous procedure is required for the greatest purification of cells. This can best be accomplished by equilibrium density gradient centrifugation (Chapter 4.3.3), i.e., buoyant sedimentation in a density gradient of a medium of chemical density higher than the density of the cells. The gradient may be continuous or incrementally discontinuous. There results a suspended band of highly purified cells, although in very small amount. The procedure is exemplified by the separation of bacterial spore types and the determination of their buoyant densities (29).

A gradient medium of choice is Percoll (Pharmacia), which has minimal osmotic and water activity effects and does not penetrate the bacterial cell wall, unlike a medium such as sucrose. A newer gradient medium, Nycodenz (Accurate Chemical & Scientific Corp.; Robbins Scientific Corp.), has high density, is completely nonionic, and offers wide applicability but may penetrate the cell walls of gram-positive bacteria. See Chapter 4.3.5 for further information on gradient media.

10.5. GROWTH YIELD CALCULATIONS

The concept of growth yield is developed from a material balance of a stable cultivation system. Monod originally defined the yield constant, Y, in terms of mass units:

$$Y = \frac{\text{grams of dry cells formed}}{\text{grams of substrate consumed}} \qquad (15)$$

The yield from substrate is often indicated by $Y_{x/s}$, indicating a ratio of the respective concentrations. The convention of mass ratios for expressing the yield of cells on the basis of carbon substrate consumption is fairly well accepted. Yields can be easily calculated by determining the production of dry cell mass over a given period of the cultivation and dividing by the mass of carbon substrate consumed during the same period.

If the yield is to be expressed on the basis of oxygen consumed or ATP produced during growth, the cell yield can be defined as follows:

$$Y_{x/ATP} = \frac{\text{grams of dry cells formed}}{\text{moles of ATP formed}} \quad (16)$$

$$Y_{x/O_2} = \frac{\text{grams of dry cells formed}}{\text{moles of oxygen formed}} \quad (17)$$

Under many conditions, the yield with respect to ATP synthesis or oxygen consumption is relatively constant for many organisms. For example, the yield of cells per mole of ATP synthesized under conditions of energy substrate limitation and fairly high growth rate is approximately 10 ± 2 g of dry cells formed per mol of ATP.

However, the yield of cells per mole of ATP is not constant for all bacteria and is generally variable if the energy substrate does not limit growth or if the growth rate of the culture is significantly lower than the maximum rate. Also, carbon and energy sources may be consumed not only for cellular growth but also for product formation. Furthermore, energy substrates are involved in cell maintenance as well as in cell growth. Since cells can use energy substrates for endogenous respiration without growth, maintenance energy requirements will contribute to the consumption of energy substrates without a concomitant increase in dry cell weight. Although the influence of endogenous metabolism on the calculation of cell growth yield is minor when the culture growth rate is near its maximum value with the energy substrate as the limiting nutrient, consumption of energy substrates for maintenance can be quite significant at growth rates lower than the maximum or in an extreme growth environment.

Since meaningful calculation of yield values requires careful and precise control of the cultivation conditions, yield data are best obtained from continuous-culture systems. In this way, the growth-limiting substrate and the specific rate of growth, u, can be specified and controlled. Calculation of the overall growth yield from continuous-culture data assumes steady-state substrate and cell concentrations and therefore is most easily accomplished in a chemostat (section 10.2.1). The medium for chemostat operation must be designed so that a single nutrient limits the rate of cell growth. This can be easily accomplished by ensuring that all nutrients, other than the one on which yields will be based, are in excess. An insight into this medium design can be obtained from an elemental analysis of the bacteria to be grown in the chemostat. The medium must supply all of the components of the cell and can therefore be based on elemental analysis. The ultimate level of growth will be determined by the concentration of the limiting nutrient. Since cell concentration in a chemostat will automatically adjust to match the rate of limiting-nutrient addition, the ratios of medium components are more important than their absolute amounts. Media for the determination of overall growth yield should therefore be designed such that the ratios of elements supplied by the medium to the element chosen as the limiting nutrient are considerably higher than predicted by taking the ratio of elements from elemental analysis.

The calculations described below require the establishment of steady-state conditions in terms of cell concentration and substrate concentration. Make sure that this condition is met by monitoring them both over a long enough period to allow displacement of one reactor volume of medium. Whenever conditions are changed (such as increasing or decreasing the rate of medium addition to a chemostat), allow sufficient time for a return to steady state. This can be ensured by allowing at least four turnovers of fermentor liquid volume. Since a continuous reactor operates on a dilution principle, a step of pulse change in operating conditions would cause deviation from the preexistent steady state. This would approach a new steady-state value at a rate predicted by the dilution of the culture; that is, four mean residence times would be necessary for a close asymptotic approach to the new steady-state equilibrium value. For example, consider a 2-liter chemostat in which the rate of medium addition and withdrawal is 0.5 liter/h with a dilution rate of $F/V = D = 0.25\ h^{-1}$. The time necessary for four turnovers of medium volume is calculated by dividing 4 by the dilution rate. In this case, the reactor medium volume will have been completely replaced four times after 16 h (4/0.25 h^{-1}). That is, at least 16 h must elapse after changing the growth conditions of a continuous culture prior to the achievement of steady-state conditions.

In reality, the time necessary to obtain a close asymptotic approach to the new steady-state value may be dependent on the way in which operating conditions have been changed. If the change in operating conditions involves a nutritional step-down, for instance, four mean residence times should be adequate. If the change involved a nutritional step-up under conditions of rapid growth, the system may require more than four mean residence times to achieve steady state. In the latter case the rapidly growing cells may require elaboration of new enzymes for metabolism of the added nutrients.

Once steady-state continuous cultivation is achieved, the overall growth yield is determined as shown below:

$$Y_{x/s} = \frac{(X - X_0)}{(S_0 - S)} = \frac{\text{grams of dry cells produced}}{\text{grams of substrate used}} \quad (18)$$

where X is grams of dry cells per liter of effluent medium, X_0 is grams of dry cells per liter of influent medium (usually zero), S is grams of substrate per liter of effluent medium, and S_0 is grams of substrate per liter of influent medium. If the medium added to the fermentor is sterile, the value of X_0 is zero and the yield becomes $Y_{x/s} = X/(S_0 - S)$.

To determine the overall growth yield, first determine the dry cell weight by quantitatively filtering a precisely measured volume of culture effluent through a preweighed 0.2-μm membrane filter. Wash the collected cells with buffer or distilled water to remove excess medium. Dry the membrane filter and cells to constant weight in an oven whose temperature is not greater than 105°C. Lower oven temperatures may be required to prevent extensive browning and therefore weight loss of the sample. *CAUTION:* Many filters lose weight to a variable extent after washing and drying. Therefore, several samples should be filtered and weighed, and several filters should be washed, dried, and weighed as controls. It is more accurate to centrifuge and wash cells as described above (Section 10.4.2) from a larger volume of medium and then to resuspend in a small volume of water, dry, and weigh (Chapter 11.5.2).

Table 3. Summary of terms for growth-yield calculations

Symbol[a]	Definition
$Y_{x/s}$, Y_s, Y	Overall cellular yield coefficient, $Y_{x/s}$ = grams of cell formed per gram of nutrient consumed
$Y_{x/m}$	Molar yield coefficient, $Y_{x/m}$ = grams of cell formed per mole of nutrient consumed
$Y_{x/s(max)}$, Y_{max}	Theoretical or maximum yield coefficient, Y_{max} = grams of cell formed per gram of nutrient consumed
Y_G	Growth-specific yield coefficient, Y_G = grams of cell formed per gram of nutrient consumed for the production of cells alone
Y_{x/O_2}, Y_{O_2}	Overall cellular yield coefficient, Y_{x/O_2} = grams of cell formed per mole of oxygen (O_2) consumed
$Y_{x/ATP}$, Y_{ATP}	Overall cellular yield coefficient, $Y_{x/ATP}$ = grams of product chemical formed per mole of ATP formed
$Y_{p/s}$, Y_p	Overall product yield coefficient, $Y_{p/s}$ = grams of product chemical produced per gram of substrate consumed
$Y_{p/x}$	Specific product yield coefficient, $Y_{p/x}$ = grams of product chemical produced per gram of cell formed
$Y_{x/s(max)}$, $Y_{p(max)}$	Theoretical or maximum product conversion yield coefficient, $Y_{p/s(max)}$ = grams of product per gram of substrate converted to product
μ	Specific growth rate (grams of cells formed per grams of cells per hour)
m	Maintenance coefficient (grams of substrate consumed per gram per cell · hours)

[a]The symbol listed first is the preferred designation.

Overall growth yield is related to maintenance and growth requirements for limiting substrates that act as energy sources according to the following equation:

$$\frac{1}{Y_{x/s}} = \frac{m}{\mu} + \frac{1}{Y_G} \qquad (19)$$

where μ is specific growth rate (reciprocal hours), m is specific rate of substrate uptake for cellular maintenance (reciprocal hours), $Y_{x/s}$ is overall yield, and Y_G is growth specific yield. The values of m and Y_G can be estimated by plotting $1/Y_{x/s}$ versus $1/\mu$ on rectangular paper. (Remember that $\mu = D$ only for simple continuous culture without cell recycle.) If the data form a straight-line relationship, the intercept will be the value of $1/Y_G$ and the slope of the line will be the value of m.

Although growth yields are conventionally expressed in terms of mass of cells formed per mass or moles of substrate used, this mixture of terms is often confusing. Herbert (24) has suggested that yields should be expressed in terms of gram-atoms of cellular carbon formed per gram-atom of element consumed from the limiting substrate. Adoption of this convention would standardize reporting procedures and eliminate much of the confusion in interpreting yield data. Expression of growth in terms of gram-atoms of cellular carbon formed does not require a determination of the complete elementary composition of the cells. Rather, only the carbon content of the cells need be determined to allow use of this convention.

Table 3 presents a summary of terms related to growth yield calculations.

10.6. COMMERCIAL SOURCES

The equipment and other products mentioned in this chapter are mostly available commercially. Representative manufacturers were cited and tabulated, and their addresses were referenced, in the 1981 version of this chapter (4). Since then the number of commercial sources has increased greatly; e.g., there are almost 50 companies for fermentation equipment. Consequently, except for specific items with limited availability cited parenthetically in context, the reader is referred to the annual buyers' guide issues of journals such as *American Laboratory, Chemical Engineering*, and *Science* and to *Biotech Buyers' Guide* published annually by the American Chemical Society.

10.7. REFERENCES

10.7.1. General References

1. **Bailey, J. E., and D. F. Ollis.** 1986. *Biochemical Engineering Fundamentals*, 2nd ed. McGraw-Hill Book Co., New York.
2. **Bu'Lock, J.D., and B. Kristiansen (ed.).** 1987. *Basic Biotechnology.* Academic Press Ltd., London.
   This book deals with the basics of biotechnology for novice scientists and engineers. The first section covers fundamentals and principles of relevant microbiology and bioengineering. The second section covers examples of applications.
3. **Demain, A. L., and N. A. Solomon (ed.).** 1986. *Manual of Industrial Microbiology.* American Society for Microbiology, Washington, D.C.
   This manual includes several chapters about methods for liquid culture of bacteria, notably shake flasks, continuous fermentation, and cell immobilization.
4. **Drew, S. W.** 1981. Liquid culture, p. 151–178. *In* P. Gerhardt, R. G. E. Murray, R. N. Costilow, E. W. Nester, W. A. Wood, N. R. Krieg, and G. B. Phillips (ed.), *Manual of Methods for General Bacteriology.* American Society for Microbiology, Washington, D.C.
5. **Pirt, J. S.** 1975. *Principles of Microbe and Cell Cultivation.* John Wiley & Sons, Inc., New York.
   This classic monograph may well be the best work available on the theory of cell cultivation.
6. **Rehm, H.-J., and G. Reed (ed.).** 1981, 1985. *Biotechnology. A Comprehensive Treatise in 8 Volumes.* Vol. 1, *Microbial Fundamentals*, 1981. Vol. 2, *Fundamentals of Biochemical Engineering*, 1985. Verlagsgesselschaft mbH, Weinheim, Germany.
   These first two volumes in the series provide a thorough background of microbiological and bioengineering principles.

10.7.2. Specific References

7. **Abbott, B. J.** 1977. Immobilized cells. *Annu. Rep. Ferment. Proc.* **1**:205–233.
8. **Abbott, B. J.** 1978. Immobilized cells. *Annu. Rep. Ferment. Proc.* **2**:91–124.
9. **Asenjo, J. A. (ed.).** 1990. *Separation Processes in Biotechnology.* Marcel Dekker, Inc., New York.
10. **Brock, T. D.** 1983. *Membrane Filtration: A User's Guide and Reference Manual.* Science Tech, Inc., Madison, Wis.

11. **Bryant, J.** 1970. Anti-foam agents, p. 187–203. *In* J. R. Norris and D. W. Ribbons (ed.), *Methods in Microbiology*, vol. 2, Academic Press, Inc., New York.
12. **Coulman, G. A., R. W. Stieber, and P. Gerhardt.** 1977. Dialysis continuous process for ammonium-lactate fermentation of whey: mathematical model and computer simulation. *Appl. Environ. Microbiol.* **34:**725–732.
13. **Dunn, I. J., and J. R. Mor.** 1975. Variable-volume continuous culture. *Biotechnol. Bioeng.* **17:**1805–1822.
14. **Elsworth, R.** 1972. The value and use of dissolved oxygen measurement in deep culture. *Chem. Eng.* (London) **258:**63–71.
15. **Evans, C. G. T., D. Herbert, and D. W. Tempest.** 1970. The continuous cultivation of micro-organisms. 2. Construction of a chemostat, p. 277–327. *In* J. R. Norris and D. W. Ribbons (ed.), *Methods in Microbiology*, vol. 2. Academic Press, Inc., New York.
16. **Evans, J. B.** 1975. Preparation of synchronous cultures of *Escherichia coli* by continuous-flow size selection. *J. Gen. Microbiol.* **91:**188–190.
17. **Fass, R., T. R. Clem, and J. Shiloach.** 1989. Use of a novel air separation system in a fed-batch fermentative culture of *Escherichia coli. Appl. Environ. Microbiol.* **55:**1305–1307.
18. **Finn, R. K.** 1954. Agitation-aeration in the laboratory and in industry. *Bacteriol. Rev.* **18:**254–274.
19. **Freedman, D.** 1970. The shaker in bioengineering, p. 175–185. *In* J. R. Norris and D. W. Ribbons (ed.), *Methods in Microbiology*, vol. 2, Academic Press, Inc., New York.
20. **Gallup, D. M., and P. Gerhardt.** 1963. Dialysis fermentor systems for concentrated culture of microorganisms. *Appl. Microbiol.* **11:**506–512.
21. **Gerhardt, P., and D. M. Gallup.** 1963. Dialysis flask for concentrated culture of microorganisms. *J. Bacteriol.* **86:**919–929.
22. **Gerhardt, P., J. M. Quarles, T. C. Beaman, and R. C. Belding.** 1977. Ex vivo hemodialysis culture of microbial and mammalian cells. *J. Infect. Dis.* **135:**42–50.
23. **Henry, J. D., and R. Allred.** 1972. Concentration of bacterial cells by crossflow filtration. *Dev. Ind. Microbiol.* **13:**177–189.
24. **Herbert, D.** 1976. Stoichiometric aspects of microbial growth, p. 1–30. *In* A. C. R. Dean, D. C. Ellwood, C. G. T. Evans, and J. Melling (ed.), *Continuous Culture 6: Applications and New Fields.* Ellis Horwood Ltd., Chichester, England.
25. **Hongo, M., Y. Nomura, and M. Iwahara.** 1986. Novel method of lactic acid production by electrodialysis fermentation. *Appl. Environ. Microbiol.* **52:**314–319.
26. **Kubitscek, H. E.** 1987. Buoyant density variation during the cell cycle in microorganisms. *Crit. Rev. Microbiol.* **14:**73–97.
27. **Kyslik, P., E. A. Dawes, V. Krumphanzl, and M. Novak (ed.).** 1988. *Continuous Culture.* Academic Press, Inc., New York.
28. **Lim, H. C., B. J. Chen, and C. C. Creagan.** 1977. An analysis of extended and exponentially-fed-batch cultures. *Biotechnol. Bioeng.* **19:**425–433.
29. **Lindsay, J. A., T. C. Beaman, and P. Gerhardt.** 1985. Protoplast water content of bacterial spores determined by buoyant density sedimentation. *J. Bacteriol.* **163:**735–737.
30. **Lloyd, D. L., J. C. Edwards, and A. H. Chagla.** 1975. Synchronous cultures of micro-organisms: large-scale preparation by continuous-flow size selection. *J. Gen. Microbiol.* **88:**153–158.
31. **Luedeking, R.** 1976. Fermentation process kinetics, p. 181–243. *In* N. Blackebrough (ed.), *Biochemical and Biological Engineering Science*, vol. 1. Academic Press, Inc., New York.
32. **Markx, G. H., C. L. Davey, and D. B. Kell.** 1991. The permittistat: a novel type of turbidostat. *J. Gen. Microbiol.* **137:**735–743.
33. **Mattiasson, B. (ed.).** 1983. *Immobilized Cells and Organelles.* CRC Press, Inc., Boca Raton, Fla.
34. **Mattiasson, B., and O. Holst (ed.).** 1991. *Extractive Bioconversions.* Marcel Dekker, Inc., New York.
35. **McGregor, W. C.** 1986. *Membrane Separations in Biotechnology.* Marcel Dekker, Inc., New York.
36. **Mitchison, J. W., and W. S. Vincent.** 1965. Preparation of synchronous cell cultures by sedimentation. *Nature* (London) **205:**987–989.
37. **Monod, J.** 1949. The growth of bacterial cultures. *Annu. Rev. Microbiol.* **3:**371–394.
38. **Monod, J.** 1950. La technique de culture continue. Theorie et applications. *Ann. Inst. Pasteur Paris* **79:**390–410.
39. **Mulvany, J. G.** 1969. Membrane filter techniques in microbiology, p. 205–253. *In* J. R. Norris and D. W. Ribbons (ed.), *Methods in Microbiology*, vol. 1. Academic Press, Inc., New York.
40. **Munro, A. L. S.** 1970. Measurement and control of pH values, p. 39–89. *In* J. R. Norris and D. W. Ribbons (ed.), *Methods in Microbiology*, vol. 2. Academic Press, Inc., New York.
41. **Munson, R. J.** 1970. Turbidostats, p. 349–376. *In* J. R. Norris and D. W. Ribbons (ed.), *Methods in Microbiology*, vol. 2. Academic Press, Inc., New York.
42. **Nipkow, A., J. G. Zeikus, and P. Gerhardt.** 1989. Microfiltration cell-recycle pilot system for continuous thermoanaerobic production of exo-β-amylase. *Biotechnol. Bioeng.* **34:**1075–1084.
43. **Novick, A., and L. Szilard.** 1950. Experiments with the chemostat on spontaneous mutations of bacteria. *Proc. Natl. Acad. Sci. USA* **36:**708–714.
44. **Poole, R. K.** 1977. Fluctuations in buoyant density during the cell cycle of *Escherichia coli* K-12: significance for the preparation of synchronous cultures by age selection. *J. Gen. Microbiol.* **98:**177–186.
45. **Quarles, J. M., R. C. Belding, T. C. Beaman, and P. Gerhardt.** 1974. Hemodialysis culture of *Serratia marcescens* in a goat-artificial kidney-fermentor system. *Infect. Immun.* **9:**550–558.
46. **Schultz, J. S., and P. Gerhardt.** 1969. Dialysis culture of microorganisms: design, theory, and results. *Bacteriol. Rev.* **33:**1–47.
47. **Shiloach, J., and S. Bauer.** 1975. High-yield growth of *E. coli* at different temperatures in a bench scale fermentor. *Biotechnol. Bioeng.* **17:**227–239.
48. **Shiloach, J., M. Van de Walle, J. B. Kaufman, and R. Fass.** 1991. High density growth of microorganisms for protein production, p. 33–46. *In* M. D. White, S. Reuveny, and A. Shafferman (ed.), *Biologicals from Recombinant Microorganisms and Animal Cells.* VCH Publishers, New York.
49. **Sterne, M., and I. M. Wentzel.** 1950. A new method for the large-scale production of high-titre botulinum formal-toxoid types C and D. *J. Immunol.* **65:**175–183.
50. **Stieber, R. W.** 1979. Dialysis continuous processes for microbial fermentations: mathematical models, computer simulations, and experimental tests. Ph.D. thesis. Department of Microbiology and Public Health, Michigan State University, East Lansing.
51. **Stieber, R. W., G. A. Coulman, and P. Gerhardt.** 1977. Dialysis continuous process for ammonium-lactate fermentation of whey: experimental tests. *Appl. Environ. Microbiol.* **34:**733–739.
52. **Stieber, R. W., and P. Gerhardt.** 1979. Dialysis continuous process for ammonium-lactate fermentation: improved mathematical model and use of deproteinized whey. *Appl. Environ. Microbiol.* **37:**487–495.
53. **Tempest, D. W.** 1970. The continuous cultivation of microorganisms. 1. Theory of the chemostat, p. 259–276. *In* J. R. Norris and D. W. Ribbons (ed.), *Methods in Microbiology*, vol. 2. Academic Press, Inc., New York.
54. **Thomson, R. O., and W. H. Foster.** 1970. Harvesting and clarification of cultures—storage of harvest, p. 377–405. *In* J. R. Norris and D. W. Ribbons (ed.), *Methods in Microbiology*, vol. 2. Academic Press, Inc., New York.
55. **Van Suijdam, J. C., N. W. F. Kossen, and A. C. Joha.** 1978. Model for oxygen transfer in a shake flask. *Biotechnol. Bioeng.* **20:**1695–1709.
56. **Wang, D. I. C., and A. J. Sinskey.** 1970. Collection of microbiol cells. *Adv. Appl. Microbiol.* **12:**121–152.
57. **Woodward, J. (ed.).** 1985. *Immobilized Cells and Enzymes. A Practical Approach.* IRL Press, Oxford.
58. **Zeikus, J. G., and E. A. Johnson (ed.).** 1991. *Mixed Cultures in Biotechnology.* McGraw-Hill Book Co., New York.

Chapter 11

Growth Measurement

ARTHUR L. KOCH

11.1. PRINCIPLES

Principles of bacterial growth are also discussed in Chapter 10 of this volume and in references 1 through 7.

11.1.1. Definitions of Growth

To measure bacterial growth, precise definitions are needed. Probably the most basic definition of growth is based on the ability of individual cells to multiply, i.e., to initiate and complete cell division. This definition implies monitoring the increase in total number of discrete bacterial particles. There are two basic ways to do this: by microscopic enumeration of the particles or by electronic enumeration of the particles passing through an orifice. This definition would also include the fragmentation of nongrowing filamentous organisms as growth, although this kind of apparent growth cannot continue indefinitely.

A second definition of growth involves determining the increase in CFU. Since some cells may be dead or dying, this definition of growth may be different from the one based on the detection of discrete particles. In the long run, the increase in the number of organisms capable of indefinite growth is the only important consideration. This is the reason why colony counting and most-probable-number (MPN) methods of measurement are so important. Viable-counting methods, which seem so natural to a bacteriologist, are really quite special in that cultures are diluted so that individual organisms cannot interact. For example, these methods cannot in principle be applied to obligately sexually reproducing organisms requiring male-female interaction or to colonial organisms such as myxobacteria that under certain conditions need to be part of a large mass of organisms that produces sufficient exoenzyme. Even when applied to procaryotes, there are special restrictions and limitations; e.g., exogenous CO_2 must be available in sufficient concentration, although this need not be supplied if many organisms are present (see Chapter 7.1.4).

A third definition of growth is based on an increase in biomass. Macromolecular synthesis and increased capability for synthesis of cell components is the obvious basis for the measurement of growth by the bacterial physiologist, the biochemist, and the molecular biologist. From their point of view, cell division is an essential but minor process that seldom limits growth; what usually limits growth is the ability of enzymatic systems to rapidly utilize resources to form biomass.

A fourth definition of growth is based upon the action of the organisms in chemically changing their environment as a consequence of the increase in biomass.

11.1.2. Balanced Growth

The four definitions of growth mentioned above become synonymous under a single circumstance: balanced growth (17). An asynchronous culture can be said to be in balanced growth when all extensive properties increase regularly with time. "Extensive" is a term from physical chemistry and refers to the properties of the system that change when there are altered amounts of substances of various kinds in the system. Thus, biomass, DNA, RNA, and cell number are extensive properties of a culture in a fixed volume, but temperature or constituent ratios (such as DNA or RNA per cell) are intensive properties that do not change during balanced growth. The application of this physicochemical principle to bacteriology lies in the thought that, if a culture is grown for a long enough duration when growth is sparse enough not to alter the environment significantly, sooner or later the bacteria will come to achieve a particular growth state for any particular constant environment no matter what the condition of the cell initially. Once this balanced growth state has been achieved, if the conditions remain the same the culture will remain in balanced growth indefinitely (until a nutrient is exhausted or the culture is altered by mutation and selection). The criteria given above are too stringent to be fully met, but it is readily possible to study cultures that are substantially held in balanced growth by growing cultures maintained at low density by dilution and by using only one measure (such as biomass measured turbidimetrically) to monitor growth. If the doubling time remains constant over an extended period, it is a good assumption that growth is balanced.

Consider any extensive property (such as biomass, DNA, or RNA) of a culture in balanced growth and call this property X. The rate of formation of X will be proportional to the amount of biomass, m. Call the proportionality constant C_1. Then

$$dX/dt = C_1 m$$

Because growth is balanced, X/m is constant; call this constant C_2. Then

$$dX/dt = (C_1/C_2)X$$

where (C_1/C_2) is a proportionality constant (equal to the specific growth rate, most usually designated by μ). Through the laws of calculus, this equation can be integrated and a boundary condition ($X = X_0$ when $t = 0$) can be imposed, yielding

$$X = X_0 e^{\mu t}$$

That is, for every cellular substance or even an extracellular product of the cells, the increase of X after growth becomes balanced will be exponential in time. If the ratio of one substance to every other substance is to remain constant, the same proportionality constant must apply for every choice of X. This common value of μ is called the **specific growth rate or growth rate constant.** Not only will any substance chosen to be X increase exponentially but also any combination of substances or the rate of change of any substance will increase exponentially with the same specific growth rate, μ. Practically, this allows the rate of oxygen uptake or metabolite production to be used as an index of growth.

Intensive properties, such as μ and the ratio of the concentrations of different cell constituents, must necessarily remain constant under conditions of balanced growth. In addition, the distribution of cell sizes stays constant. An average cell will have a constant rate of carrying out every cellular process, and newly arisen daughter cells will have a constant probability of being able to form a colony (i.e., the percentage of nonviable cells will remain constant). Therefore, under these special conditions, no matter what measure of growth is used (whether it is particle counting, colony formation, or chemical determination of a cell substance or consumption or excretion of a substance), the same specific growth rate will be obtained and that rate will be constant through time.

Such favorable conditions in principle occur when bacteria are cultivated under long-term continuous culture. Balanced growth should arise during chemostat growth that is limited by the rate of a single nutrient addition and during turbidostat growth when the culture is mechanically diluted to maintain a constant amount of biomass in the growth vessel and μ is limited by the nature of nutrients and not their amount. Balanced cultures can be formed by repeated dilution under batch growth (45, 59), if carried out in such a way that the culture never goes into the lag or stationary phase.

11.1.3. Changes in Cell Composition at Different Growth Rates

Organisms respond to environmental conditions, both physical and chemical, by altering their own composition. These changes have been well documented for certain enteric bacteria (36, 45) but may occur with all procaryotes. In general, favorable growth conditions mean faster growth, which requires a higher concentration of ribosomes and associated proteins. In terms of gross composition, the most obvious change is an increase in RNA content. Also, under favorable growth conditions the cells can lay down reserve materials such as glycogen and poly-β-hydroxybutyric acid. These changes in composition lead to the possible pitfall of the bacteriologist falsely relating one measure of growth to others when comparing growth in different environments.

11.1.4. Unbalanced Growth

Although balanced growth conditions lead to reproducible cultures, much of physiology deals with the responses of organisms to changes in their environment which lead to progressive changes in the organism during the ensuing "unbalanced growth." When a stationary-phase culture is inoculated into fresh medium, the properties of organisms change drastically through the course of the batch culture cycle. Though only well documented for certain enteric bacteria (1, 14), similar changes probably apply with all procaryotes. The exact course of changes in composition and morphology depends on the medium and on the age and condition of the inoculum. Culture cycle phenomena have relevance to growth measurements. These phenomena are particularly important in ecological studies, where the conditions under which organisms grow are critical and, to a large degree, uncontrollable. The changes in characteristics are also involved in response to the fluctuations in natural conditions (36, 39).

A typical bacterial culture cycle progresses as follows. When a stationary-phase culture is diluted into rich medium, macromolecular synthesis accelerates. The components of the protein-synthesizing system (i.e., ribosomal proteins and RNA) are made first. Only after considerable macromolecular growth does cell division take place. During this lag phase, the average size of cells increases greatly. When the capacity of the medium to support balanced exponential rapid growth is exceeded, cellular processes slow down. Different processes slow differently in a way to make small, RNA-deficient cells. Finally, the cells die (20) and may eventually lyse.

Workers using the techniques from this book may employ intentional perturbations of growth. These perturbations can result from nutritional shifts or deficiencies, the use of growth inhibitors (particularly antibiotics), or the effects of radiation or other extreme physical conditions (e.g., high or low temperature and osmotic pressure). Their effect on growth can be quite complex, and the bacteriologist must be both cautious and critical.

11.1.5. Pitfalls in Growth Measurement

There are four classes of pitfalls in bacterial growth measurement. The first major pitfall is the general tendency of most organisms toward clumping or a filamentous habit of growth, which can occur even under mildly toxic conditions with bacteria that ordinarily divide regularly. The second pitfall is the differential viability of injured bacteria under different culture conditions. Repair processes may permit the recovery of viable cells under some but not other conditions (11). A third major pitfall is the possible development of resistant stages. Bacteria known to form resistant stages, such as spores, pose no problem since the controls and measurements to correct for such forms are well known; but when resistant stages are not suspected, error can arise. The fourth pitfall pertains to the way in which the inoculum is exposed to the new environment. Different results may be obtained if the concentration of an agent is raised gradually, if it is raised discontinuously, or if a high concentration is temporarily presented and then removed or lowered. The cell concentration at the time of challenge can be critical. Bacteria may have special ways, sometimes inducible and sometimes unknown, to protect themselves against toxic agents. The protection may be dependent on the number of organisms cooperating in detoxification.

The phenomena related to the fourth pitfall can be illuminated with a single example from work on the interaction of rifampin and *Escherichia coli* (41). Rifampin is a large molecule that can only very slowly enter the cell. Doses of rifampin can be detoxified by a sufficiently large number of cells, and the process appears to be inducible. Therefore, higher doses of drug can be tolerated when the cells have previously been gradually exposed to the antibiotic. Independent of this phenomenon, fast-growing bacteria are more resistant than slow-growing ones. This observation can be explained largely by the shorter time available for the drug to penetrate relative to the doubling time of the growing bacteria. For example, under conditions where the low dose by itself would not inhibit growth, a higher dose of antibiotic that is subsequently diluted to a low dose can block growth. Thus, when growth is slowed by the brief high dose, the inhibition remains permanent because now sufficient time is available to allow penetration of enough drug to maintain the blockage of growth.

11.1.6. Mycelial Growth

The evaluation of mycelial growth (the first pitfall) is both easier and much more difficult than is the evaluation of "well-behaved" organisms that engage in binary fission and then promptly separate. The problems and methods for mycelial growth have been discussed by Calam (16) and Koch (38) and will be briefly dealt with here. Filtering filamentous cells, with or without drying, is easier than filtering smaller nonfilamentous cells. In addition, the increase in the physical size of a mycelial colony of filamentous cells can be monitored. In the extreme case, a 1-m-long tube containing a layer of nutrient agar is inoculated at one end, and the mycelial mass grows along the tube and may even be continued successively into other tubes (56). Such colonial growth is one dimensional, whereas growth of mycelial colonies on agar surfaces is two dimensional. In liquid shake cultures (Chapter 10.1.3) mycelial growth is three dimensional. Under all three conditions the linear dimensions increase linearly (not exponentially) with time.

The rate of increase in size of the colony in one, two, or three dimensions depends on the rate of elongation of the terminal hyphae that happen to be growing perpendicular to the surface of the colony, but the mobilization of resources into the mycelial mass depends on the surface area of the colony. Therefore, with shake cultures particularly, the results depend on the nature and size of inoculated fragments; with large fragments, growth becomes limited at an earlier stage by diffusion of nutrients into the mycelial mass. Thus, the major pitfall with the mycelial habit of growth is that the growth quickly deviates from exponential growth and depends on the geometry of growth and the nature of the inoculum. In shake cultures, the apparent growth may depend on the shape of the vessel and on the shaking speed, its character (circular or reciprocal), and the distance moved, because all of the above can affect the tendency of the mycelia to break into smaller pieces.

11.1.7. Cell Differentiation

The change of enteric bacteria from large RNA-rich forms in the exponential phase to small RNA-poor forms in the stationary phase has many of the aspects of differentiation. Bacteriology, however, has much clearer examples of cell differentiation in the cases of transition of rod to coccus and of vegetative bacillus to endospore and in the formation of exospores, cysts, and buds. The tendency to form filaments in certain circumstances can also be considered a differentiation. These changes and their reversal pose potential pitfalls for all approaches to growth measurement.

11.1.8. Cell Adsorption to Surfaces

Many bacteria naturally adhere to certain surfaces or can adapt and mutate to achieve a high avidity for solid surfaces including glass. Many experiments with chemostat culture have failed to achieve their primary goal because the organisms adhered to the vessel walls. Therefore plastic or Teflon should be used whenever there is long-term contact of the organism with a culture vessel (36). Other approaches to minimizing the effects of growth on vessel walls include the use of large culture volumes in large containers, the use of violent agitation, and the frequent subculture of the bacteria into fresh glassware. In addition, the use of detergents, vegetable oils, silicone coatings, and a high-ionic-strength medium may to some degree alleviate this problem.

This problem is particularly pertinent during dilution of the culture for measurement by highly sensitive means, such as microscopic and plate counts. It is therefore given further consideration in the discussions of those methods (sections 11.2.1, 11.3.1, and 11.3.2).

11.1.9. Growth in Natural Environments

If the problem at hand is to measure growth under natural conditions, tremendous difficulties must be overcome (12). In nature, growth almost always involves mixed cultures with bacteria attached to each other or to solid particles. Six kinds of approaches have been used to deal with the measurement problem in these cases. First, $^{14}CO_2$ fixation has been used (68) to monitor biomass increase. Second, tritiated thymidine autoradiography has been used to identify cells engaged in DNA replication (13). Third, antibodies tagged with fluorescent labels have been used to scan soil samples and identify particular organisms (57). Fourth, nucleic acids within cells have been stained with DAPI (4′,6-diamidino-2-phenylindole) and acridine orange and then observed under the fluorescence microscope. Fifth, sophisticated mass spectrometry of the gases emanating from natural soil samples has been used to provide information about the growth of organisms within the sample (49). Sixth, agents that block cell division but not enlargement have been used. Cells that become bigger under such conditions can be considered to have been alive.

11.2. DIRECT COUNTS

11.2.1. Microscopic Enumeration

Microscopic enumeration in a counting chamber is a common technique that is quick and cheap and uses equipment readily available in the bacteriological laboratory. Microscopic direct counts can also be made by membrane filter sampling and staining (Chapter 2.1.4).

The counting chamber technique is subject to errors, but these can be overcome to a large degree by using improvements suggested by the work of Norris and Powell (51). The major difficulty in direct microscopic enumeration is the reproducibility of filling the counting chamber with fluid. In the technique recommended below, the thickness of the fluid filling is measured by focusing on bacteria attached to the top and bottom interfaces and measuring the distance between them with the micrometer scale on the microscope. The horizontal dimensions between the scribe marks in commercial chambers are quite accurate and cause no problem.

A second major difficulty is the adsorption of cells on the surfaces of glassware, including pipettes. The procedure described below avoids adsorption during the dilution process but desirably encourages it in the counting chamber. In the dilution process, it can be decreased by carrying out dilutions in high-ionic-strength medium (e.g., physiological saline or many minimal media without their carbon source) instead of water or in a solution of formaldehyde (any concentration between 0.5 and 5% is satisfactory) that has been neutralized with K_2HPO_4 together with a trace of anionic detergent such as sodium dodecyl sulfate (51). The formaldehyde stops growth and motility, and the K_2HPO_4-detergent combination prevents aggregation, but the detergent may lyse certain organisms even though it is used only in minute amounts. Alternatively, plastic containers and plastic pipette tips can be used for the dilution to decrease the loss by adsorption on surfaces. In the procedure described below, 0.1 N HCl is used as the final diluent to favor adsorption on glass surfaces of the chamber.

The Hawksley counting chamber (A70 Helber; Hawksley, Ltd., Lansing, England) is recommended in preference to the Petroff-Hausser counting chamber (Arthur H. Thomas Co., Philadelphia, Pa.). The prime advantage of the former is that its optical path with an ordinary coverslip is short enough that the chamber can be used under an oil-immersion objective at high power. Although most counts are done under a high-dry objective, it is sometimes necessary to use the oil-immersion objective, either because there are too many cells or because they tend to clump. Both chambers are 20 μm deep with scribe marks defining areas of 50 μm by 50 μm.

One major advantage of microscopic examination is that one can gain additional information about the size and morphology of the objects counted. Oil-immersion and high-power objectives make counting more tedious, but critical distinctions can be made. Alternative procedures are supplied by the manufacturers of chambers. Additional discussion has been presented by Meynell and Meynell (5) and Postgate (55). For a discussion of statistical considerations, see section 11.7.

Procedure

1. Clean the chamber and the coverslip with water containing a small amount of anionic detergent; rinse with water and then alcohol, blot, and let air dry.

2. Make a preliminary estimation of the concentration of cells. Proceed to the next step if the concentration is less than 3×10^8 cells per ml. Otherwise, make a primary dilution of the cell suspension in Norris-Powell diluent (51) prepared as follows. Add 5 ml of formalin (37% formaldehyde) to 1 liter of water. Adjust to pH 7.2 to 7.4 (indicator paper is sufficiently accurate) by adding solid K_2HPO_4; the amount of the phosphate needed will vary with the amount of formic acid in the formalin. Add a few milligrams of sodium dodecyl sulfate, and repeat until bubbles do not break immediately when air is passed through the solution with a Pasteur pipette.

3. Carry out a final single dilution of the sample in a ratio of at least 1:1 with 0.1 N HCl. This will kill the cells, had the formaldehyde not sufficed, and gives their surface a net positive charge so that they will not aggregate but will instead adsorb onto glass.

4. Immediately fill the Hawksley chamber with approximately 5 μl of the diluted sample, using a Pipetteman, Eppendorf, Centaur, or other plastic-tipped pipette. Let the chamber stand for 1 to 2 min.

5. Examine with a phase-contrast microscope under a high-dry or oil-immersion objective. Most of the cells will have attached to the bottom interface; a few cells will have attached to the top interface; and only a few cells will remain suspended and exhibit Brownian motion.

6. First, focus on the cells that have attached to the bottom interface, and read the markings on the dial of the focusing knob. On many microscopes of high quality, this dial reads directly in micrometers. Next, focus the cells on the top interface. Note the distance between the top and bottom, and augment the difference by one bacterial diameter. This total distance will quite accurately measure the filling, which is nominally 20 μm. When gaining familiarity with the technique, make depth measurements in several well-separated regions of the chamber to find the uniformity of the depth of filling.

7. Count the cells lying within small squares. Optimally, the number in each small square should be in the range of 5 to 15. Score the cells that lie on the boundaries of a square if they are on the upper or right side but not if they are on the lower or left side. A hand tally counter is convenient to count the cells within a square. A second hand tally is convenient to keep track of the number of squares. At least 600 total organisms should be counted for accurate work (see below), but this need not be done with a single filling. Some authors recommend multiple fillings of the chamber, thus averaging the variability of the fillings; however, this is not necessary with the method described here, because the thickness of the filling is measured each time. It is best to count squares chosen in a systematic fashion, such as the four corner squares and the major diagonal squares. This prevents counting the same square twice and averages a possible geometric gradient of cells in the chamber.

8. Calculate the number of cells per ml of undiluted culture by use of the following formula:

$$\frac{\text{total bacteria counted} \times \text{dilution factor} \times 4 \times 10^8}{\text{number of small squares counted} \times \text{filling depth (in micrometers)}}$$

If the coverslip is precisely positioned, the volume of the filling on top of a small square is $50 \times 20\ \mu m^3 = 5 \times 10^4\ \mu m^3 = 5 \times 10^{-8}$ ml. The reciprocal of this, 2×10^7/ml, is the usual factor in the formula quoted in the instructions supplied by the manufacturer; thus the two formulas are the same if the thickness of the filling is exactly 20 μm. For the procedure to be successful, only a few cells need be attached to each surface for the thickness measurement. It is most convenient, however, if most of the cells are on one surface for counting. Then the final act of counting a small square is the focusing through the suspension to count the cells not attached and the ones on the interface that has fewer cells. Then tally the square, refocus on the original surface, and move on to the next square.

11.2.2. Electronic Enumeration

The Coulter Counter (Coulter Electronics, Hialeah, Fla.), its commercial competitors, and particularly the laboratory-built versions have been important in the development of bacteriology over the last 40 years. Such instruments are used routinely in clinical hematology. They are also very useful in the enumeration of nonfilamentous yeasts and protozoa but not of mycelial or filamentous organisms. Although use of these counters has led to important concepts in bacteriology, the technique is difficult to apply in a valid way to estimate the volume of bacteria because of their small size and usually elongated shape. Attempts to improve and validate the technique for use in bacteriology have been made at the research level by people with backgrounds in physics or engineering. Evidence of the difficulties involved is the fact that many of the people who helped develop the technique for measuring distribution of cell volumes no longer use it. This article, therefore, only presents the principles, mentions the difficulties and the attempted solutions to these difficulties, and directs the reader to published literature. Then the reader will be able to consider the applicability of the technique to a specific bacteriological problem.

The principle of electronic enumeration is as follows. A fixed volume of diluted cell suspension is forced to flow through a very small orifice connecting two fluid compartments. Electrodes in each compartment are used to measure the electrical resistance of the system. Even though the medium conducts electricity readily, the orifice is so small that its electrical resistance is very high. Consequently, the electrical resistance of the rest of the electrical path is negligible by comparison. When a cell is carried through the orifice, the resistance further increases since the conductivity of the cell is lower than that of the medium. This change in resistance is sensed by a measuring circuit and converted into a voltage or current pulse. The pulses are counted by an electronics circuit similar to that used in counting radioactivity. Very small pulses are eliminated by a discriminator circuit. Very high pulses, which might be due to dirt or other irrelevant particles, are eliminated by an upper discriminator. In advanced models, the pulses may be analyzed by size and stored in a multichannel analyzer; later, the data are recovered and plotted in a histogram, and the numbers, mean size, and standard deviation are calculated. All the data may be collected and the discrimination against pulses that are too high or too low are carried out as the data are analyzed. The instrument needs some method of forcing an accurately known volume through the orifice during the counting period. This is usually done by displacing the fluid in contact with a mercury column past triggering electrodes that conduct effectively through the mercury but not through the diluent medium.

There are three major problem areas in electronic enumeration. First, some bacterial cells are very small (less than 0.4 μm^3), and the resistance pulses produced as they pass through the orifice are comparable to the noise generated by the turbulence that develops in the fluid flowing through the orifice. The discriminator dial on the instrument can be set to reject the turbulence noise, but then sample information, particularly about newly divided cells, is lost. Blanks can be run and their values can be subtracted, but blanks are particularly variable for small cell sizes. In addition, a pattern of turbulence can become established, remain for a while, and then be replaced with another pattern. Finally, the overall error increases when the blank has large statistical variation (section 11.7.3). There is little problem in the study of bacteria growing in rich medium in which even newly formed cells are larger than the pulses produced by the turbulence or if interest is restricted to the relatively few large bacteria such as *Azotobacter agilis* and *Lineola longa*.

The second major problem in electronic counting results from the failure of cells to separate promptly from each other after cell division. This, and the tendency to form filaments and aggregates, can be minimized by careful choice of the organism and the conditions. The choice of *E. coli* for physiological and genetic studies was fortuitous because of the relatively small extent to which it remains as pairs or chains or forms aggregates. Various physical techniques such as mild ultrasound treatment or vigorous blending in a Vortex mixer can be used to try to separate pairs, disperse aggregates, or break up filaments. There is a related problem of coincidence (section 11.7.5), i.e., the passage of more than one cell through the orifice in a short enough time that a single larger cell is registered by the electronics. This problem can be dealt with (see below) by increasing the dilution so that the probability of coincidence is altered. Coincidence is an especially vexing problem when cell size distributions are to be accurately measured.

The third major problem in electronic counting is clogging of the orifice. The resistance change is a smaller proportion of the total resistance across the orifice when the orifice diameter is larger. Consequently, for small rod-shaped bacteria such as *E. coli*, orifices with diameters in the range of 12 to 30 μm must be used. The exact choice is a trade-off between an increased signal and the increased noise and chance of becoming clogged. Clogging is best prevented by ultrafiltration of all reagents. Alternatively, the diluent can be prepared and allowed to settle for a long time (months) in a siphon bottle so that the particulate-free solution can be withdrawn. Choosing solvents that do not tend to generate particulate matter can also be of help. Kubitschek (43) recommended 0.1 N HCl for this reason; it is entirely volatile and does not leach materials out of the glass, which later may form precipitates. However, the problems in practice are severe and become worse in some of the modifications needed to size bacterial cells more accurately.

Electronic counting has influenced the study of bacterial growth more because of its presumed ability to measure the size distribution of bacterial cells than because of its ability to enumerate them. In fact, it is very difficult to measure cell size accurately because of the nature of the resistance pulse generated by a cell passing through an orifice. Attempts have been made to overcome this difficulty by using a relatively long pore (100 μm) (42), but the resulting slower flow and longer path increase the chance of clogging. A second approach involves special hydrodynamic focusing of solutions in such a way that the bacterial cells pass very nearly down the center of the pore surrounded by fluid containing no particles (60, 62, 71).

Additional information about this technique can be found in references 2, 22, and 43 and in instruction manuals for the instruments.

11.2.3. Flow Cytometry

Flow cytometry has become an extremely powerful method for the studies of many aspects of the biology of eucaryotes, but the methods are only now coming into their own in the study of the biology of procaryotes, simply because the latter are smaller. The instruments that are now common in hospitals and research laboratories operate by forming a small-diameter stream of the sample suspension. This is encased in a stream of fluid that is added in such a way that the stream of the sample is made still narrower in diameter. The flowing stream is examined with laser light of various frequencies and angles, and the output of the measurement circuits is used to detect when a particle passes through. The design permits the analysis of biomass by light-scattering methods and by staining of chemical components such as DNA. The electronic circuits allow cells to be counted very rapidly. Growth can be monitored by measuring the increase in counts in sample counted for a fixed duration corresponding to a fixed volume. The problems for the application to procaryotes are in sensitivity and background noise. General applications in microbiology are described in references 22a and 58. Commercially available equipment has been used to study a variety of problems in ecology (15). Other applications are described in references 8 and 50.

Another approach (63) depends on flow directly on a microscope slide on an inverted microscope.

11.3. COLONY COUNTS

Bacteriology really became an experimental science when Robert Koch listened to Fannie Hesse and developed the agar plate. This allowed not only the cloning of pure strains but also the enumeration of colonies arising from individual viable cells or colony-forming units (CFU). Various colony count methods have been used: (i) **pour plates**, in which an aliquot sample of diluted cells is pipetted into an empty sterile petri dish, molten but cool (45°C) agar medium is poured onto the sample, and the contents are mixed by swirling and then allowed to harden; (ii) **spread plates**, in which the sample is pipetted onto the surface of solidified agar medium in a petri dish and the cells are distributed with a wire, glass, or Teflon spreader; (iii) **thin-layer plates**, in which the sample is pipetted into a tube containing a small volume (2.5 to 3.5 ml) of molten but cool soft (0.6 to 0.75%) agar medium, the mixture is poured onto hardened agar medium in a petri dish, and this overlay is allowed to harden; (iv) **layered plates**, which are like the thin-layer plates except that an additional layer of agar medium is poured onto the newly congealed soft-agar medium containing the cells so that all colonies are subsurface; and (v) **membrane filter methods**, in which the diluted cells are filtered onto an appropriate membrane filter, which is then placed on an agar medium plate or onto blotter pads containing liquid medium. Sometimes it is necessary to carefully prewash the membranes and the pads with water or medium and then sterilize them.

The pour plate methods have variations in which the cells are grown in roller tubes or microtubes (53) and are examined with low-power microscopes when the colonies are small (54). There are many individual variations or techniques, sometimes resulting from historical accidents and sometimes resulting from special bacteriological circumstances. Automation of colony counting has put additional special restrictions on techniques but allows colonies on petri dishes to be counted rapidly without operator error.

Two colony count methods are detailed below, and another is described briefly. The first is the spread plate, in which all colonies are surface colonies. It is chosen for first presentation because it is reliable and because surface colonies are required to produce the proper color responses with many indicator agars. In many cases, different colors are given from subsurface colonies because the oxygenation is different; therefore, the acid production and reducing potential are different from those on the surface.

The second method is the layered plate. It is very useful because all colonies are subsurface and therefore much smaller and compact. They can be intensely colored. Many more colonies may be present, and yet the coincidence by fusion of colonies is small; this means that several thousand colonies per plate can be used to give meaningful results. This approach is especially recommended because the main difficulty with usual colony-counting methods is the lack of dynamic range. Rules have been issued that between 30 and 300 colonies are required. The lower limit is set by statistical accuracy, and the upper limit is set by coincidence limitations. This 10-fold range is inconvenient for many purposes, because in many cases one cannot predict the number within a factor of 10 when choosing the dilution factor. The extra care needed to prepare the overlay plates is justified because it allows one to count in the larger range of 30 to 2,000. A second major advantage is flexibility for nutrient supplementation. Minimal agar can be used to pour the basal layer in a large number of petri plates for indefinite storage. Stock supplies of the minimal soft agar can also be kept on hand. Then 10- to 50-fold excesses of needed special nutrients can be added to the aliquots used for the molten top agar. In some cases dyes and chromogenic substrates, to allow screening of the colonies, are added as needed to the soft agar. Toxic substances can be incorporated to measure frequencies of resistant mutants.

The third method is the pour plate. This method, although commonly employed, must be used with caution because some bacteria may be killed by the elevated temperature or the sudden change in temperature. The method also lacks the advantages of the foregoing two methods. Before routine use, test and compare pour plates with one of the other methods.

Several articles and books have been devoted to attempts to speed and automate growth measurement (2, 28). This is a field that is in so much flux that further discussion is not practical here.

Certain organisms are very sensitive to substances present in agar. Meynell and Meynell (5) presented an excellent discussion of these problems. Injured organisms may have additional special requirements, and the entire 26th symposium of the Society for General Microbiology (24) was devoted to these problems. Genetically defective organisms pose their own individualistic problems that can be research problems on their own, e.g., the ability of various repair mutants to form countable colonies (18).

The problem of quantitating the number of organisms in cultures of strict anaerobes is dealt with in references 19, 29, and 30 and in Chapter 6.

11.3.1. Diluents (with Philipp Gerhardt)

Bacterial cells often must be diluted from their original dense concentration to a sparse concentration suitable for observation in a microscope, measurement of cell numbers, analysis for genetic or metabolic properties, or washing preparatory to study. Whatever the use, the diluted cells must retain their original characteristics. The preservation of viability and metabolic activity is particularly important when measuring the CFU. Diluted cells are more likely to be harmed by an unfavorable environment than are cells in dense suspensions, in which the environment contains higher levels of materials leached from the cells. Consequently, care must be taken to use a suitable diluting solution.

Distilled or tap water alone should usually *not* be used, because it is osmotically hypotonic to all bacterial cells (except dormant spores) and unbuffered against pH change. Traces of heavy-metal contaminants and detergents in water supplies may also cause inactivation of cells in very dilute suspension. "Physiological saline" (0.85% NaCl) also is usually inadequate because it is isotonic only to mammalian cells and is unbuffered. Viability may be reduced by 50% or more in such diluents.

A common general-purpose diluent is phosphate-buffered saline with a protein, usually gelatin or peptone. The presence of the protein stabilizes fragile bacteria by binding metals and detergents, and the diluent is buffered at neutrality by the phosphate salts. Buffered saline with protein contains 8.5 g of NaCl, 0.3 g of anhydrous KH_2PO_4, 0.6 g of anhydrous Na_2HPO_4, and 0.1 of gelatin or peptone per liter of distilled or deionized water. Adjust to pH 7.0 if necessary.

For critical or unusual situations, pretest the diluent for adverse effects and use a diluent adapted to the particular conditions. For plate counts, do not transfer bacteria growing rapidly at an incubator temperature into cold diluent and do not suspend them in diluent for more than 30 min at room temperature, to minimize death or multiplication. The growth medium without the carbon source, to slow or arrest further growth, may be the best diluent for cultures. Magnesium chloride (0.2 g/liter, sterilized and cooled separately as a 4% aqueous stock solution) may be added to preserve membrane integrity, especially with gram-negative bacteria. With anaerobic bacteria, the diluent should contain a reducing agent and be freed from dissolved oxygen (see Chapter 6.6.3 and 6.6.4).

A detergent (e.g., 0.1% Tween 80) is sometimes suggested as a dispersing agent but is effective mainly for mycobacteria and other cells with hydrophobic surfaces and may be harmful to other cells. Polymerized organic salts of sulfonic acids of the alkyl-aryl type are effective dispersing agents and apparently are not harmful to representative bacteria (47).

Selection and standardization of diluents are especially important in microbiological tests for safeguarding public consumption of food and water, which are detailed in standard methods manuals published by the American Public Health Association. A particularly good review of the literature and methods for using diluents is contained in *Standard Methods for the Examination of Dairy Products,* 16th ed. (48).

Meynell and Meynell (5) have also reviewed the use of diluents, and an entire symposium has dwelt on the death and survival of vegetative microorganisms (24). Further details of diluents are provided in section 11.2.1 and Chapter 24.1.1.

11.3.2. Spread Plates

Prepare suitable agar medium in ordinary glass (old-fashioned but ecologically sound) or plastic (modern and convenient) 9-cm petri plates. Various procedures have been suggested to pour and dry the plates. The following procedure is recommended.

1. Place the covered container of molten agar, shortly after removal from the autoclave and after addition of mildly thermolabile substances (sugars, dyes, antibiotics, chromogenic substances), into a dishpan or other large vessel containing hot tap water (45 to 50°C). The agar cools quickly (temporarily warming the water further) but uniformly throughout the vessel and reaches a temperature plateau above its solidification point. For 1 liter of agar, 30 min is sufficient. With this procedure there is no gelling around the edges of the container.
2. Pour the agar into the petri plates. If the agar is sufficiently cooled, little condensation will form on the undersurface of the petri plate lid, and further condensation can be minimized by covering the plates with a sheet of newspaper. If bubbles form on the agar surface, direct the flame of a Bunsen burner momentarily downward on them until they burst (be careful not to melt plastic petri plates). With 9-cm petri plates, 15 to 20 ml of agar is satisfactory, corresponding to a thickness of 0.24 to 0.32 cm. If thicker, the less contrast there will be between the colonies and their background. If thinner plates are used, colonies may be small and present in reduced numbers because some nutrient is limiting or the plate becomes locally dried.
3. Dry the plates (to remove the water of condensation on the plate lids and water of syneresis on the agar) at room temperature overnight or for 24 h, depending on the relative humidity. Frequently the humidity is lower in incubation rooms or laminar flow hoods, so use them.
4. Store the plates indefinitely at 4°C in closed plastic containers. Be sure to allow them to warm to room temperature before use.
5. Prepare a suitable dilution of the cell suspension based on all the information and hunches at your disposal. Calculate to get 100 to 200 colonies, but use the results of plates containing between 30 and 300. If the organism forms only small colonies, up to 500 may be counted. Make the dilutions with a diluent solution that does not favor adsorption to glass, if ordinary glassware and pipettes are to be used. High ionic strength, pH between 4 and 5 (if not harmful), and the presence of small amounts of anionic detergents (if not toxic) are helpful in this regard (section 11.3.1). Alternatively, use presterilized plastic vessels and pipette tips for a Pipetteman-type device.
6. The actual dilutions can be carried out in many ways. Historically, large volumes of diluent (99 ml) and 1-ml samples were used. As pipette and volumetric ap-

paratuses have been improved, smaller volumes have been used, economizing on reagents and mixing time. Modern plastic-tipped semiautomatic pipettes allow very small samples of bacterial culture to be used, but then the problem is proper sterilization. This can be done in an autoclave or in a microwave oven. The microwaves are absorbed by bacteria or other biological material much more readily than by the plastic (because the bacteria have dipoles and ions), so that the plastic remains cool but the cells and even spores are heated to lethal temperatures. For many purposes, 0.1-ml serological pipettes may be used to deliver a full 0.1-ml volume of culture into 0.9 or 4.9 ml of diluent contained in tubes measuring 13 by 100 mm or into 9.9 ml of diluent contained in tubes measuring 16 by 150 mm. These dilution factors are about optimum for ease of making an accurate dilution and ease of mixing adequately (which is the most critical factor). Moreover, with this rule it is easy to check that the dilution was properly carried out and that no trivial error was introduced.

7. Pipette 0.1 ml of the final dilution onto the agar surface of the petri plate. Form two or three free-falling drops off center on the surface, and then blow the remaining fluid onto the surface. If using a pipetting device, push to the second stop of the pipettor. The cells may have a tendency to become immediately attached in situ, so do not delay in spreading the drops.

8. Sterilize a spreader (Fig. 1) by dipping it in 70% alcohol (do not use the autoclave), shaking off the excess alcohol, and igniting. A spreader prepared from a large paper clip is recommended. Bend it out, being careful to keep the longest straight part of the wire unbent. Then place a length of no. 16 shrinkable Teflon tubing (Small Parts, Inc., Miami, Fla.) over the straight part, and gently heat to shrink it onto the wire. The part serving as a handle can be bent to the user's convenience. This spreader has the advantage of low heat capacity, it cools quickly compared with solid glass rods, and it has much less affinity for bacterial cells. A spreader can also be made quickly from a glass Pasteur pipette by using a Bunsen burner with a wing tip.

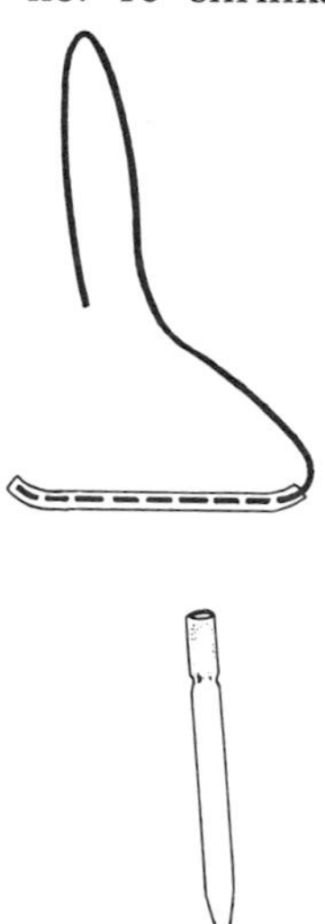

FIG. 1. Spreaders. The one at the top is bent from a paper clip and fitted into a length of Teflon tubing. The one at the bottom is made by fusing the end of a Pasteur pipette and then bending it sharply in two places by using a Bunsen burner with a wing tip to define the spreading region. The handles of both are bent to the convenience of the user. Both spreaders have low heat capacity. The spreader at the top absorbs fewer bacteria because of its Teflon construction.

9. Spread the plate. Try to achieve a uniform coverage as close to the edges as possible. The major difficulty with this method is learning the technique for uniform distribution of the cells. Therefore, after the plates have been incubated, examine the distribution of colonies on all of the plates to learn how uniform the spreading technique has become. Plates with a larger number of colonies are especially useful in this regard, even if they contain too many colonies to count. Also note whether the number of colonies is larger in the vicinity of the original droplets. If so, then the technique must be improved. The agar surface should be dry enough that the 0.1 ml delivered from the pipette is absorbed in 15 to 20 s by the agar. Some workers prefer to work with turntables, which speed the work of spreading the sample uniformly.

10. Incubate the inverted plates in a constant-temperature room or chamber with good temperature regulation (Chapter 6.3). Incubating in closed containers is advantageous, but do not overfill the containers, as this will increase the time required for the temperature of the plates to become equilibrated to the temperature of the incubator. This is especially important when using temperature-sensitive mutants. Storing in closed containers also avoids the effects of any noxious gases which may be present in general-purpose constant-temperature rooms (e.g., acetic acid fumes from gel destaining). Opaque closed containers protect against inactivation of colored drugs and dyes by keeping out light. Finally, in closed containers the plates do not dry out and so can be retained for slow- or late-developing colonies (the container can also contain filter paper or paper towels that are kept moist to maintain high humidity). Another point of concern has to do with CO_2. Even organisms not usually isolated in a high-CO_2 atmosphere may have a CO_2 requirement. This may be particularly evident when single cells are spread at high dilution and incubated in normal laboratory air. Some of these considerations may be of small importance in any particular case, but all should be kept in mind.

11. Observe the plates before the colonies have fully matured. It is often possible to see that too many colonies are developing. To obtain reliable information with less coincidence correction, it may be possible to make fresh dilutions of a sample that had been retained in the cold or to count colonies under a dissecting microscope while still small.

12. Count the colonies. Depending on the circumstance, various types of illumination are advantageous and may not be obtainable with commercial colony counters. Experiment with various types of magnifying glasses that can be worn or clipped onto one's own glasses. Also try various lamps that have a magnifying lens as an integral part. The colonies may be enumerated by marking the bottom of the petri plate with a marking pen, or they may be counted by hand with an electronic counter, hand tally, or television-based scanning equipment. One technique that clearly marks the

individual colonies is to stick the point of a colored pencil into the colony. Not all brands of colored pencils transfer color well; Eberhard Faber Mongol colored pencils do this well, but some colors work better than others (e.g., French Green no. 898). Several colors can be used for differential counting. If using an electronic scanning counter, pay careful attention to be sure that false-positive counts are not registered as a result of dirt and imperfections in the petri dishes (which would not cause difficulty to a human eye). Dyes or pigments can be incorporated into the agar to increase the contrast needed to avoid these errors during the electronic scan.

11.3.3. Layered Plates

1. Prepare the petri plates with a base of 1.5% agar medium approximately 0.4 cm thick. Pour the plates on a level surface. Check the work area with a level carefully; if it is not true, use any shim material (metal strips, wood strips, pieces of paper) to level a piece of plate glass, and pour the three layers of agar while the plate rests on this level surface.

Prepare small test tubes (13 by 100 mm) with 2.5 to 3 ml of soft 0.7% agar medium. It is convenient to prepare stock bottles with this strength of agar in the basal medium. Melt the agar in these bottles. (A microwave oven is the quick way, but experience is necessary to find the setting that melts but does not boil the agar explosively.) In any case, the agar must be thoroughly melted, or mixing cannot be uniform. Add nutrients, dyes, and inhibitors at this point, and then pipette aliquots into small test tubes (13 by 100 mm) previously placed in a water bath or heating block at 45°C. This pipetting can be done while the agar medium is still very hot, decreasing the chance of contamination. The agar will remain liquid at 45°C for several hours. It remains liquid for a longer time at 50°C, but this temperature is likely to cause some killing.

2. Dilute the sample as in the previous procedure (section 11.2.1).

3. Bring the base-medium petri plates from storage to at least 25°C (30°C is better).

4. Pipette 0.1 ml of the diluted sample near the inside lip of the tube. Do this on an identifiable side of the tube (e.g., the side with a trademark, or a side that has a frosted spot or an identifying mark made with a marking pen).

5. Immediately pour the soft agar out of the tube onto the base agar. Pour it over the side of the tube on which the sample has been pipetted. This will wash *all* of the organisms onto the base agar in such a way that they will be quantitatively transferred and uniformly mixed with the rest of the soft agar. They will not have a chance to become adsorbed to the glass of the test tube or to cluster in spots on the base agar.

6. Tilt and swirl the petri plate so that the melted agar covers the surface.

7. Place the petri plate on the level work area to congeal the agar.

8. Carry out the platings needed for the remainder of the experiment.

9. Pour or pipette 2.5 to 3 ml of additional soft agar onto the congealed agar surface, and distribute by rocking. Then let this third layer congeal on the level surface.

10. Incubate the plates either upright or upside down, since contamination and degree of dryness are much less critical with this technique than with spread plates.

11. Examine and count the colonies. The subsurface colonies will be compact and lens shaped with well-defined edges and will be smaller than the usual surface colonies. This makes subsurface colonies a little more difficult to count, but the magnifying glass of a colony counter helps. Magnifying glasses on a headband also help, and these glasses have built-in prisms that reduce eye strain. A dissecting microscope can also be used.

The extra trouble involved in the overlay technique is worthwhile when many colonies are present. The colonies should be uniformly distributed, and this can be checked by visual inspection. A fraction of a plate can be counted and the count can then be prorated. A low-power dissecting binocular microscope allows virtually all errors due to coincidence to be eliminated because the colonies can be visualized in three dimensions. It is laborious to count under such conditions, but it can save a very carefully performed experiment from being incomplete and inconclusive.

11.3.4. Pour Plates

1. Prepare a number of tubes with 20 to 30 ml of thoroughly molten agar.

2. Place in a 45°C water bath or heated aluminum block. Allow adequate time for temperature equilibration.

3. Mix the sample aliquot. Do this well but without causing bubbles to form.

4. Pour the contents into a sterile petri dish, and allow them to set.

5. Incubate, examine, count, and calculate.

11.4. MOST PROBABLE NUMBERS

The concentration of viable cells can be roughly estimated by the MPN method (also called the **fraction-negative** or **endpoint-quantal method**), which involves the mathematical inference of the viable count from the fraction of multiple cultures that fail to show growth in a series of dilution tubes containing a suitable medium.

This method consists of making a number of replicate dilutions in a growth medium and recording the fraction of tubes showing bacterial growth. The tubes exhibiting no growth presumably failed to receive even a single cell that was capable of growth. Since the distribution of such cells must follow a Poisson distribution (see below), the mean number plated at this dilution can be calculated from the formula $P_0 = e^{-m}$, where m is the mean number and P_0 is the ratio of the number of tubes with no growth to the total number of tubes. The mean number, estimated by $-\ln P_0$ (ln designates the natural logarithm), is then simply multiplied by the dilution factor and by the volume inoculated into the growth tube to yield the viable count of the original sample. The MPN method is a very inefficient method from the point of view of statistics, because each tube corresponds to a small fraction of the surface of a petri dish in a plate count. Consequently, very many tubes or wells in a titer plate must be used, or the worker must be prepared to settle for a very approximate answer.

When is the MPN method of advantage? First, it can be used if there is no way to culture the bacterium on solidified medium. Second, it is preferred if the kinetics of growth are highly variable. Suppose some cells grow immediately and rapidly and end up making a large colony on solid agar that spreads over and obscures colonies of the organism of interest that form later. The small-colony formers may be more numerous but unmeasurable on plates because of the fewer but highly motile or rapidly growing bacteria. Third, it is preferred if other organisms not of interest are present in the sample and no selective method is available. Thus, the method has utility when the bacterium of interest produces some detectable product (e.g., a colored material, specific virus, or antibiotic). Then, even though contaminating organisms may overgrow the culture, the numbers of the bacterium in question can be estimated by the fraction of the tubes that fail to produce the characteristic product. Fourth, if agar and other solidifying materials have some factors (such as heavy metals) that may alter the reliability of the count or interfere with the object of the experimental plan, the MPN method can be used to avoid these difficulties.

Modern developments in laboratory techniques can be used to speed the execution of the MPN method. Machines are available to fill the wells of plastic trays that have as many as 144 depressions. Scanning devices designed for other purposes can be used to aid in counting the number of wells with no growth. Similarly, automatic and semiautomatic pipettes can be used to fill small test tubes. Because these procedures make it possible to examine many more cultures, the classical tables of fixed numbers of tubes and fixed dilution series are obsolete and should be abandoned. A different approach and method of calculation is needed and is provided below.

The MPN method used at a single level of dilution when the mean number of bacterial cells capable of indefinite growth is 1.59 per tube is statistically the most accurate (23). This number would result in 20.8% of the tubes remaining sterile (Fig. 2). Consequently, if the expected number is known precisely and the assay is conducted for confirmation, all the available tubes should be seeded with a dilution that is expected to have the value 1.59. The accuracy falls off rapidly with deviations from this optimum, particularly outside the range of an average of 1 to 2.5 cells per tube. Thus, the dilutions 10-fold away from the optimum contribute almost no information about the mean number of cells. Consequently, if the viable count is known within a narrow factor, the optimum strategy is to use a narrow range of dilution factor. If the uncertainty is greater, a wider range should be used.

Usually, prior knowledge is not available and growth tubes at several dilutions are needed. The simplest approach is to separate the dilution levels 10-fold. For this approach, little is lost by discarding the data from dilutions at which the percentage of sterile tubes lies outside of 8 to 36% or outside of the range of 5 to 50% if a larger range of error is acceptable. The error from the dilution within the range can then be read or interpolated from Fig. 2. This method is simple but more wasteful than is necessary. Consequently, statistical methods have been developed to combine data from different dilution levels when a specified number of tubes is run at each level.

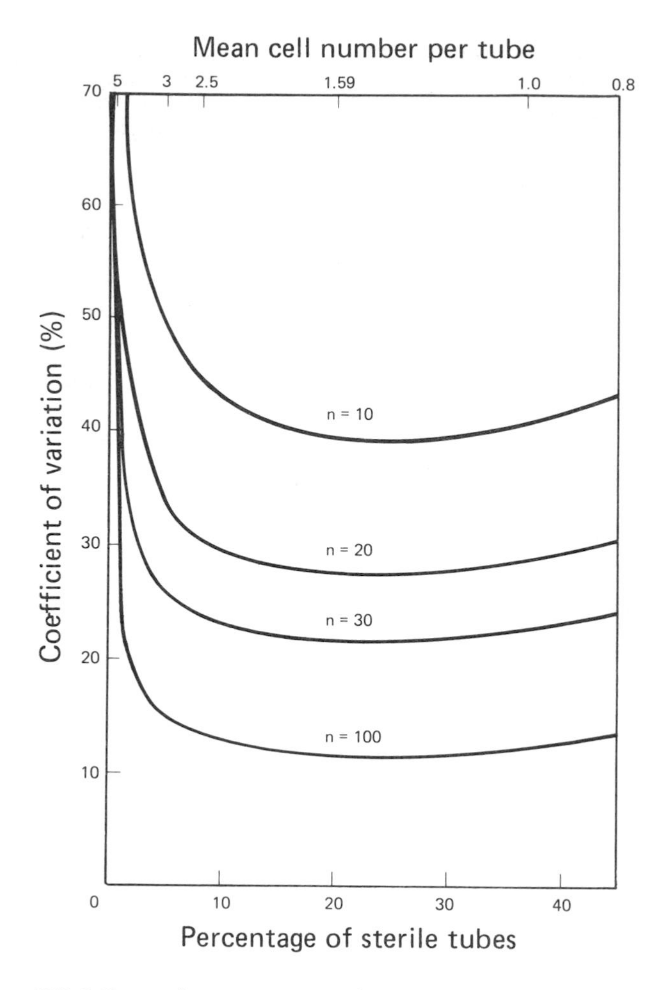

FIG. 2. Errors that arise in using the MPN method. The CV for a single level of dilution is shown as a function of the percentage of sterile tubes (bottom abscissa) and the mean number of cells per tube (top abscissa). The respective optima occur at 20.8% and 1.59 mean cells per tube for a varying number of tubes (*n*) prepared at a single dilution level.

Classically 10 cultures at each of three successive 10-fold dilutions have been used most often. Of the 1,000 possible outcomes of such tests, Halvorson and Ziegler (27) listed the 210 cases that could happen with $P >$ 0.01%. Finney (23) gave a good exposition of the statistical aspects and the approaches to calculate the MPN of replicates at each dilution. He pointed out that there can be a definite advantage in computation when carrying out error calculations if the dilution series extends over a range such that every tube shows growth in at least one dilution and every tube shows no growth at a higher dilution. Even then the proper statistical calculations are cumbersome and very wasteful of time and material.

With modern computer programs (see Table 1), it is easier, better, and more flexible to do away with tables and to calculate the MPN for an individually chosen and optimized arrangement of tubes and dilutions. Even for a standard series of dilutions (for example, the data shown in Table 2), a computer program is of value since sometimes outcomes (codes) not listed in the published tables are obtained.

Table 2 shows an example for the case of 10 tubes at each of three 10-fold dilutions. Their t_1, t_2, and t_3 are each 10. The data are $g_1 = 8$, $g_2 = 5$, and $g_3 = 1$. Although 30 tubes are used, the accuracy is lower than if all 30 tubes had been used at a single dilution (as given by the curve method $n = 30$ in Fig. 2) and would have been much poorer if wider range had been chosen.

When designing an MPN measurement, one must choose first how many dilution tubes at each level and how many dilution levels are to be used for the analysis and how far apart they will be spaced. To do this, one must consider the expected titer and the confidence that one has in that estimate. The less sure one is of the titer of the sample, the wider is the range of dilution levels that one should test; conversely, if the titer can be quite accurately estimated or inferred, one should choose fewer levels. This flexibility, compared with existing tables, is really the point of using the computer program, because it allows the investigator to make optimum use of the MPN technique.

BASIC computer program procedure to calculate MPN values

1. Install the program in a PC, MacIntosh, or microcomputer. The code can be copied from Table 1 in a few minutes and then stored. Because it is written in a very primitive form of BASIC, it should run with substantially any dialect of BASIC, such as Quick Basic or ZBasic. Alternatively to keying the code into the computer, the code can be obtained on disk both in BASIC and in a compiled stand-alone version. In any case, test the program first with the sample data given in Table 2, which also shows the relevant formula.

2. Run the program. The program first asks for the dilution factor. This is the fold dilution between the experimental culture or field sample and the solution actually pipetted into the MPN tubes. Key this number in, and then key **<ENTER>**. The computer then asks for the number of levels of dilution. Key this number in, and again key **<ENTER>**. For the example shown in Table 1, these numbers are 10-fold dilution and three levels.

3. Next, the program requests sets of three numbers about a given level used in the procedure: key the volume of the diluted test suspension placed in each tube, the number of tubes inoculated at that level, and the observed number of positive cultures. Key these numbers separated by commas, and then key **<ENTER>**. So for the first entry, enter 10, 10, 8; for the second, 1, 10, 5; and for the third, 0.1, 10, 1. The first two must be followed by commas and the last with the **<ENTER>** command. If the **<ENTER>** key is used instead of the comma, the computer can still use the data and will request the remaining data. If the same number of tubes is used at all levels, it is necessary to **<ENTER>** the number of tubes only at the first level and then it can be omitted later. However, be sure to put two commas between the volume of sample and the number of positive cultures.

4. After the data entry for the last level is completed, the computer uses an interpolation routine (the Illinois version of the regula falsi algorithm) to find the value that best fits the data according to the formula given in Table 2 and prints out the results. The reiterative process is shown on the screen as the computer finds suc-

Table 1. BASIC computer program for MPN determination

```
00010 'MPN, ARTHUR L. KOCH, 11/2/90
00020 DIM D(15,5):TV=0:DEFSNG A-Z:CLS
00030 INPUT "DILUTION FACTOR TO TEST SOLUTION ";DIL
00040 INPUT "NUMBER OF LEVELS TESTED ";I
00050 PRINT "VOLUME, # OF TUBES, # WITH GROWTH"
00060 FOR J=1 TO I:INPUT V,N,G
00070 IF J=1 THEN NN=N
00080 D(J,1)=J:D(J,2)=V:D(J,3)=N:D(J,4)=G
00090 IF D(J,3)=0 THEN D(J,3)=NN
00100 TV=TV+D(J,2)*D(J,3)
00110 NEXT J
00120 X1=.001:X2=.8
00130 X=X1
00140 GOSUB 340
00150 Y1=Z
00160 X=X2
00170 GOSUB 340
00180 Y2=Z
00190 X=X2-(X2-X1)*Y2/(Y2-Y1)
00200 GOSUB 340
00210 Y=Z:PRINT X;Y
00220 IF ABS(Z)<.0001 THEN 270
00230 IF Y*Y2>0 THEN 260
00240 X1=X2:Y1=Y2
00250 X2=X:Y2=Y:GOTO 190
00260 Y1=Y1/2:GOTO 250
00270 PRINT:PRINT "VOLUME, # OF TUBES, #WITH GROWTH"
00280 FOR J=1 TO I:PRINT D(J,2),D(J,3),D(J,4)
00290 NEXT J
00300 PRINT
00310 PRINT "MOST PROBABLE NUMBER =";X
00320 PRINT "ORIGINAL TITER =";X*DIL
00330 END
00340 SUMMATION SUBROUTINE
00350 ZZ=0
00360 FOR J=1 TO I
00370 ZZ=ZZ+D(J,2)*D(J,4)/(1-EXP(-X*D(J,2)))
00380 NEXT J
00390 Z=ZZ-TV
00400 RETURN
```

Table 2. Formula and sample data for MPN computation

General formula

$$\Sigma \, a_i g_i / (1 - e^{-a_i n_i}) = a_i t_i$$

where a_i is the volume tested, g_i is the number of tubes showing growth, and t_i is the total number of tubes in level i.

Sample data

Level (i)	Vol (a_i)	Total no. (t_i)	No. positive (g_i)
1	10	10	8
2	1	10	5
3	0.1	10	1

cessively better values of the MPN. For the example of Table 2, the MPN is 0.27.

The program as written prints the output only on the screen. It is expected that the user will wish to modify the program to obtain a printed output instead. Printer directions have been omitted because format statements vary widely among computer systems. If the same volumes or the same number of levels are always used, the program can be customized to fix these parameters and to delete some of the requests that are made by the supplied program.

A statistician might properly point out that the program does not compute the error statistics. Although that feature could have been included in the program, it was omitted to make the program briefer, simpler, and more readily understandable and adaptable. The program is short enough that the user, it is hoped, will undertake the one-time-only job of keying it in. However, there are two even better reasons for not calculating the error directly from the statistical theory for the MPN. First, the most reliable data come from the level in which 20% of the cultures show no growth. The error can be quickly estimated with adequate accuracy by neglecting the rest of the data and using the level with the closest proportion to 20% and values read from Fig. 2. Second, as mentioned in section 11.2, the errors in pipetting and sampling are not included in such an error computation. Consequently, it is strongly recommended that several independent samples be taken and processed independently. Consider an example in an ecological context in which the problem is to estimate the number of cells in a stream. It is far better to take samples from different locations within the stream or sample at a series of different times than to take one sample and analyze it with many tubes. With more original samples and fewer tubes in each individual MPN determination, the estimation of the mean and standard deviation (and standard error) of the independent MPN values will give an estimate of the combined error due to variation of the biological source, errors in the dilution procedure, and errors in the MPN determination itself.

11.5. BIOMASS MEASUREMENT

(with Philipp Gerhardt)

A measure of the mass of bacterial cell constituents frequently is used as the unit basis for measurement of a metabolic activity or of a morphological or chemical constituent; i.e., "biomass" and cell numbers provide the two basic parameters of bacterial growth.

The methods for measuring biomass seem obvious and straightforward, but in fact they are complicated if accuracy is sought. Furthermore, the results may be expressed in different ways, and, in some of these ways, the values may be more relative than absolute.

11.5.1. Wet Weight

A nominal wet weight of bacterial cells originally in liquid suspension is obtained by weighing a sample in a tared pan after separating and washing the cells by filtration or centrifugation. In either case, however, diluent is trapped in the interstitial (intercellular) space and contributes to the total weight of the mass. The amount of interstitial diluent may be substantial. A mass of close-packed, rigid spheres contains an interstice of 27% independent of sphere size, and a pellet of bacterial cells close packed by centrifugation may contain an interstice of 5 to 30%, depending on cell shape and amount of deformation.

A method for obtaining the actual wet weight of the cells themselves is to correct the nominal wet weight of a packed-cell pellet by subtracting the experimentally determined weight of diluent in the interstitial space. This is determined by use of a nonionic polymeric solute that permeates into the interstitial space but is too large to penetrate into bacterial cell walls, such as dextrans of very high average molecular weight. The method is described and discussed in Chapter 24.1.2.

11.5.2. Dry Weight

A nominal dry weight (solids content) of bacterial cells originally in a liquid suspension is obtained by drying a measured wet weight or volume in an oven at 105°C to constant weight. The cells must be washed with water (possibly extracting cell components), or (better) a correction must be made for medium or diluent constituents that are dried along with the cells. Separating the cells by filtration poses particular problems (Chapter 10.4.1.2). Volatile components of the cells may be lost by oven drying, and some degradation may occur, as evidenced by discoloration (particularly if a higher temperature is used). Some regain of moisture occurs during the transferring and weighing process in room atmosphere, so this should be done quickly within a fixed time for all replicate samples, especially if the relative humidity is high. It is best, of course, to use tared weighing vessels that can be sealed after drying.

Possibly more accurate determination can be made by drying the sample to constant weight in a desiccator vessel with P_2O_2 and under oil-pump vacuum at 80°C or by lyophilization (Chapter 12.2.1). Results with the three methods were indistinguishable within 1% when compared for bacterial spores (10). An excellent discussion of dry-weight procedures and errors is given by Mallette (46).

The dry weight of cells may be expressed on a wet-weight basis (grams of solids per gram of wet cells) or on a wet-volume basis (grams of solids per cubic centimeter of wet cells or of cell suspension).

11.5.3. Water Content

The amount of cell water in fully hydrated cells is equal to the difference between the wet weight and the dry weight of the cells themselves, determined as described above.

Water content also can be determined relative to the humidity of the atmosphere or to the water activity of the solution in which the cells occur (Chapter 6.2). Completely dried cells equilibrate with an atmosphere of controlled, known humidity in successively increasing increments up to saturation (67), resulting in a "sorption isotherm" curve (Fig. 3). The initial phase of water sorption, at very low humidities, represents tightly bound, monolayer-adsorbed water; the intermediate plateau phase represents loosely bound, multilayer-adsorbed water; and the terminal phase, at high humidities, represents bulk solution, "free" water. The total amount of cell water in fully hydrated cells is obtained at 100% humidity, the intercept for which occurs at a steep rise in the curve and consequently is difficult to determine precisely.

Another way to estimate the amount and distribution of cell water is to carry out isotope dilution measurements with a probe for water (e.g., ^3H_2O), a probe that will not pass the cytoplasmic membrane (e.g., [^{14}C]sucrose), and a probe excluded from the cell wall (e.g., [^{14}C]dextran). See Chapter 24.1.5 and reference 64.

The water content of cells may be expressed on a wet-weight basis (grams of water per gram of wet cells), a dry-weight basis (grams of water per gram of dry cells), or a wet-volume basis (grams of water per cubic centimeter of wet cells). The wet-weight basis is perhaps most commonly used by bacteriologists, but the dry-weight basis is less subject to experimental error. The two bases are correlated by the equation, $WC_{dry} = WC_{wet}/(1 - WC_{wet})$, where WC_{dry} and WC_{wet} are the dry and wet water content determinations.

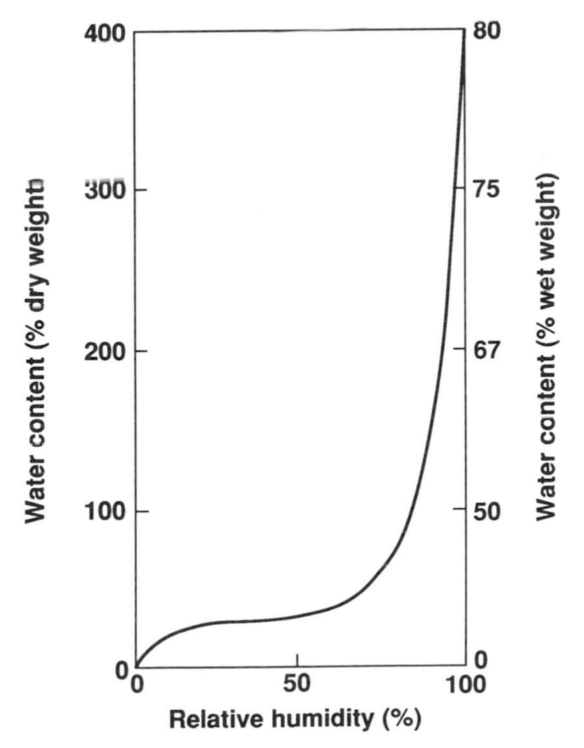

FIG. 3. Typical water sorption isotherm curve for bacterial cells.

11.5.4. Volume

The volume of a mass of wet cells (or the average size of a single cell) is best obtained by the procedure described in Chapter 24.1.4.

11.5.5. Wet and Dry Densities

The density (i.e., the specific gravity) of a bacterial cell may be obtained as either the wet density, based on the total of the solids and water contents, or the dry ("chemical") density, based on the solids content. Both densities are expressed in units of weight per unit of volume (grams per cubic centimeter). The wet density may be obtained simply by dividing the cell wet weight by the cell dry volume. Unfortunately, the dry volume is difficult to determine accurately. The cells first are completely dried, e.g., by lyophilization (Chapter 12.2.1). Occluded gas and residual water vapor are removed by holding a fairly large mass (>2 g) of the dried cells under a high vacuum until the pressure becomes constant (e.g., at 0.01 mmHg [ca 1.33 mPa]) The volume of an inert and nonadsorbing gas (e.g., helium or nitrogen) displaced by the degassed cells then is measured in a volumetric adsorption apparatus. This procedure was exemplified with bacterial spores by Berlin et al. (1). The equipment may be obtained commercially (Fekrumeter; Gallard-Schlesinger, Inc., Carle Place, New York).

Wet and dry densities can also be determined by equilibrium buoyant density gradient centrifugation (25, 44, 66). For determination of the wet density of hydrated cells by this technique, gradients of Percoll (Chapter 4.3.5.3) almost completely avoid variations in osmotic tonicity and water activity and have replaced earlier gradient media. Percoll gradients can be formed in the usual gradient maker or can be self-generated by prior centrifugation. Percoll is manufactured by the addition of polyvinylpyrrolidine to colloidal silica particles to create a stable suspension of dense particles that do not penetrate cell walls. Additional precision has been achieved by the use of colored beads of known densities to provide density markers (Chapter 4.3.5.3) and laser-based machines to locate the bands (9). A major finding with the Percoll gradients is that some organisms maintain a constant density through the cell cycle (44) and others do not (21, 44).

Other substances such as Nycodenz and Metrizamide can be very useful under certain circumstances (Chapter 4.3.5.4). They are derivatives of triiodobenzene and have a high density, and their contribution to the osmotic pressure can be significant; however, they afford a dense and transparent material for gradient construction.

11.6. LIGHT SCATTERING

Light-scattering methods are the techniques most generally used to monitor the growth of pure cultures. They can be powerful and useful, but they can lead to erroneous results. Their major advantages are that they can be performed quickly and nondestructively. However, they may give information about a quantity not of primary interest to the investigator. They give information mainly about macromolecular content (dry weight) and not about the number of cells. The physics and mathematics of light scattering are complex; still, without difficult physics, elementary considerations can give most of the needed answers.

The basic principle is that of Huygens and is the same one needed to understand the properties of lenses, such as those in microscopes. Electromagnetic radiation interacts with the electronic charges in all matter. When the light energy cannot be absorbed, a light quantum of the same energy (color) must be reradiated. This light can emerge in any direction. This means that all atoms in a physical body serve as secondary sources of light. Photons that happen to go in the direction of the original wave will stay in phase with the wave arriving directly from the light source, but light reirradiated from different points in the body that go in other directions will differ in phase at an observation point. The phase will differ depending on the distance that the emitting photon must travel throughout the object, because light going through matter is slowed; the degree of slowing is measured by the index of refraction.

It is the above circumstance that controls how light is bent and focused in large bodies such as prisms, lenses, raindrops, or a pane of glass. In the last case, all the light scattered in every direction except straight ahead cancels, leaving a beam of light going in the original direction. However, it is slightly retarded relative to a light ray not going through the pane; this is the basis leading to indices of refraction greater than unity.

11.6.1. Turbidimetry

Bacteria in suspension are in between the size limits of atoms and objects such as window panes. Consequently, most of the scattered light is deviated only slightly; i.e., it is directed almost but not quite in the same direction as the incident beam (Fig. 4). The light scattered from an atom or a very small particle depends inversely on the fourth power of the wavelength of light. Such particles therefore scatter blue light more strongly and transmit red light more efficiently. This is

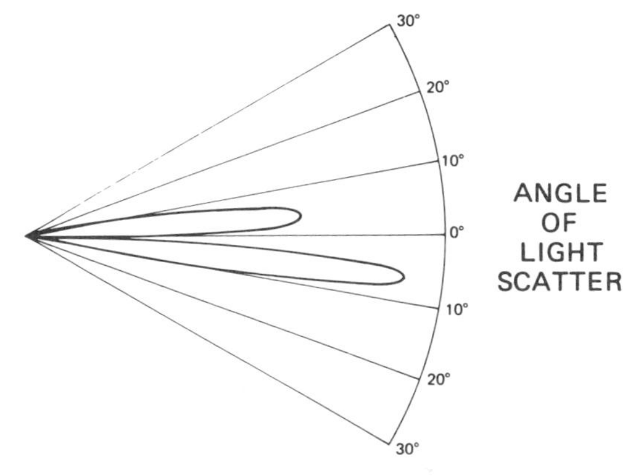

FIG. 4. Light-scattering patterns from randomly oriented bacteria showing the angular distribution of light scattered by a beam traversing from left to right through a suspension of cells with the size and physical properties of *Escherichia coli* grown in minimal medium. The upper pattern shows the distribution if the cells are ellipsoidal with an axial ratio of 4:1. The lower pattern shows the distribution for spherical cells.

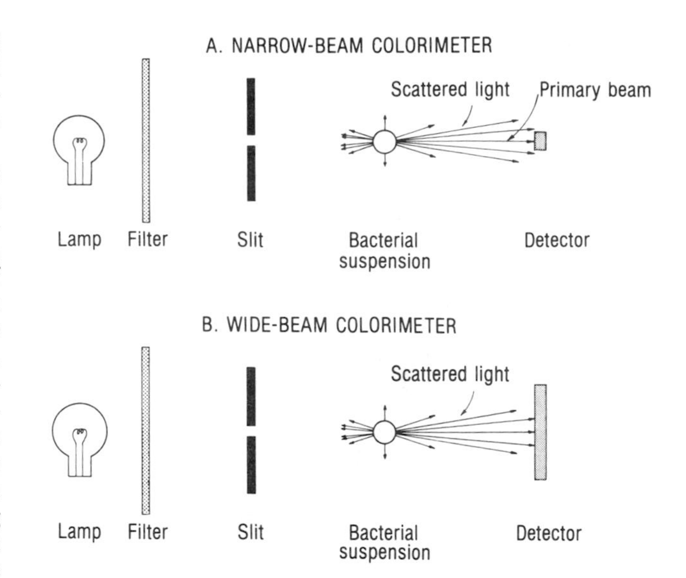

FIG. 5. Schematic designs for filter-type "colorimeter" instruments. (A) Narrow-beam instrument. Almost all the scattered light deviates enough to escape falling on the photodetector. (B) Wide-beam instrument. A variable fraction of the scattered light falls on the detector, and thus the sensitivity is lowered.

the reason that the sky appears blue, smoke and many virus suspensions are bluish, and sunsets are orange and red. For a pane of glass, there is no light left uncanceled in any but the original direction, so the light retains its original hue and the pane appears transparent. For bacteria, the light scattered away from the original direction most nearly approximates an inverse second power (33). Therefore, suspensions of bacteria (like a cloud in the sky) are intermediate and appear white and turbid. Approximations of the turbidity of bacterial suspensions may be made by visual comparisons with **McFarland turbidity standards** consisting of $BaSO_4$ suspensions (see Chapter 25.4.9).

The common practice in bacteriology is to use a colorimeter or spectrophotometer to measure turbidity directly. Ideally, such instruments measure only the primary beam of light that passes through the sample and reaches the photocell without deviation (Fig. 5A). Usually, the measurement is made of the light intensity relative to that which reaches the photocell when a sample of the suspending medium has replaced the cell suspension. From the above discussion, an ideal photometer must be designed with a narrow beam and a small detector so that only the light scattered in the forward direction reaches the photocell. That is, the instrument must have well-collimated optics. Such an ideal instrument gives larger apparent absorbance values than do simple instruments with poorly collimated optics, since in the latter a large percentage of the light scattered by the suspension is still intercepted by the phototube (Fig. 5B). Therefore, the measuring system responds as if there were less light scattered than is actually deviated from the forward direction.

The larger the number of bacteria in the light path, the lower the intensity of the light that goes through the sample without deviation. At low turbidities this is a simple geometric relationship, since the intensity of the unscattered light decreases exponentially as the number of bacteria increases. The geometric relationship between turbidity and bacterial numbers can be deduced by considering a suspension of bacteria that reduces the light intensity to 1/10 of the original intensity. Now consider two such identical suspensions in two cuvettes in a row such that the light beam goes through both suspensions. By how much would the light intensity be reduced after it goes through both tubes? The answer is 0.1 × 0.1, or 0.01, of the intensity of the incident beam. In the ideal case the same relationship holds when the concentration of bacteria is doubled and only one tube is used. When this is expressed mathematically, the intensity of the unscattered light (I) is equal to the intensity of the incident light (I_0) multiplied by $10^{-w/w_{10}}$ is that concentration of bacteria which gives a 10-fold decrease in the light intensity. That is:

$$I = I_0 \cdot 10^{-W/W_{10}}$$

Therefore, if $W = W_{10}$, I would be 0.1 of I_0. If this concentration of bacteria were doubled, I would be 0.01 of I_0. By taking base-10 logarithms (log) of both sides of the equation, this equation can be rewritten:

$$-\log(I/I_0) = \log(I_0/I) = W/W_{10}$$

A similar law, the Beer-Lambert law (Chapter 21.1.1), works for the absorption of light by colored samples. It could be derived in the same way. Therefore, most instruments have a scale that directly reads the quantity $\log I_0/I$. This is called **absorbance** (A) or, on older instruments, **optical density** (OD). Optical density is the more general term and can be used for turbid as well as light-absorbing solutions, but most workers prefer the symbol A. From the above equation, ideally, a plot of A of bacterial cultures versus cell numbers yields a straight line going through the origin. This could be written

$$A = K \cdot W$$

where K is the slope (and is $1/W_{10}$) and A has replaced $\log I_0/I$. As shown in Fig. 6, the actual absorbance becomes increasingly lower than predicted by the formula (see Chapter 2.1.1). More light gets through the turbid suspension than expected, for two reasons: first, because light scattered from one bacterium is rescattered by another bacterium so that the light is redirected back into the phototube; and second, because the bacteria interfere with the Brownian motion of one another so that they become more evenly distributed and scatter less light away from the direction of this beam (like the pane of glass considered above).

These considerations allow one to grasp some practical implications about bacterial turbidimetry. The results of both theoretical (33, 34) and experimental (26, 33, 35, 36) studies show that dilute suspensions of most bacteria, independently of cell size, have nearly the same absorbance per unit dry weight concentration. However, very different absorbances are found per particle or per CFU with different sizes of bacterial cells. An approximate rule has been proposed that the dry-weight concentration is directly proportional to the absorbance (33). This rule applies to both cocci and rods and is a first approximation to a more precise rule, i.e., that the light scattered out of the primary beam is proportional to the four-thirds power of the average vol-

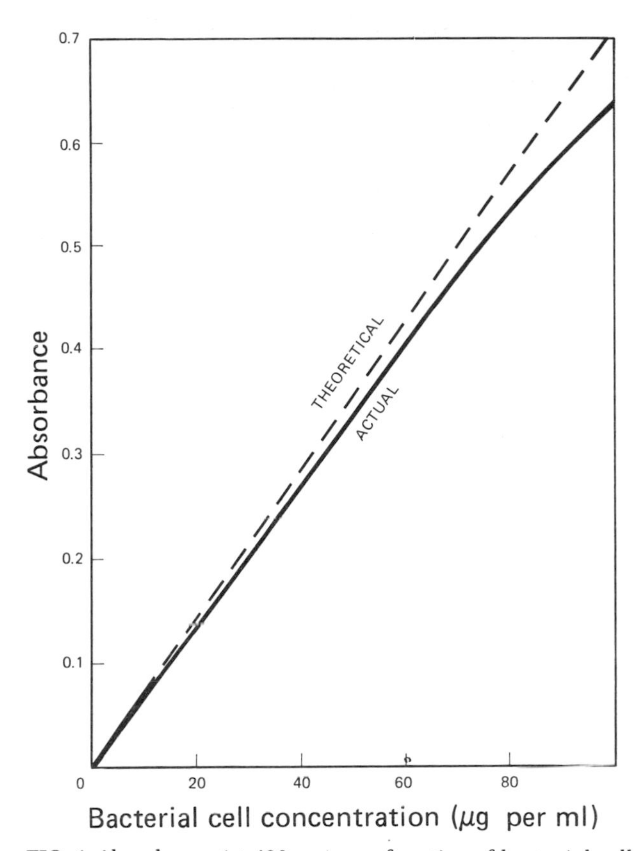

FIG. 6. Absorbance (at 420 nm) as a function of bacterial cell concentration. The dashed line shows the theoretical relationship. The solid line shows an actual result obtained with 15 dilutions of a bacterial culture, whose concentrations were measured by dry-weight determinations and whose plot was obtained by the best-fitting quadratic curve that was forced to pass through the origin.

ume of cells in the culture. Objects smaller than bacteria, such as suspensions of viruses, do not obey these rules. The rules also do not apply when the suspensions of small bacteria have a bluish cast. Larger objects (e.g., filaments and aggregates of bacteria) may still appear cloudy and not necessarily colored but may not obey the rule. When the rule applies, the proportionality constant relating the absorbance measurement to the dry-weight concentration is the same for any good, well-collimated photometer. In such instruments the phototube is at a considerable distance from the sample and/or the instruments are so well collimated that a narrow undeviated beam passes through and only this narrow beam is intercepted by the detector. The Zeiss instruments possess this design character. Other instruments can be assessed by examination of their optical design, or by comparison of the same suspension of cells within a short period when measured in an instrument from Carl Zeiss, Inc., Thornwood, N.Y., or other narrow-beam instrument and in the instrument under consideration.

If the instrument or the size of organism does not meet these criteria, valid measurements in many cases can be made by establishing a standard curve on either a relative or an absolute basis. Difficulty arises only as physical or biological conditions vary with the samples to be compared, since the experimenter cannot be sure that the same standard curve applies.

The above considerations apply only to dilute suspensions. Deviation from Beer's law must be applied after the absorbance exceeds about 0.3 when light of 500 nm or shorter is used (see the example in Fig. 6). A good discussion of this problem was given by Kavenagh (31). Failure to take this deviation into consideration is the most common difficulty encountered in the bacteriological literature that involves turbidimetry. One way to avoid this problem is to restrict culture densities and use only dilute cultures for measurement (i.e., to keep the absorbance below 0.3). However, the photometric measurements at this level are less accurate; an additional dilution may introduce significant pipetting error; for such small absorbance measurements, it is also critical that the cuvettes must be well matched and scrupulously clean; and the cuvette must be precisely placed in the spectrophotometer. It is usually better to make measurements at higher densities and correct for deviations from Beer's law.

11.6.1.1. Procedure

The important point about turbidimetric methods is that there is no set procedure. One uses the equipment at hand and seeks useful results. The most costly spectrophotometers increase the spectrophotometric precision and the stability and accuracy of wavelength settings. However, the main advantage of such instruments for the present purpose is that they have a larger dynamic range. Lower absorbances have validity when measured on good equipment because of electronic stability and increased reproducibility in positioning the cuvettes, and because the cuvettes have plane parallel faces and are of accurate dimension. Higher absorbances, for which very little light reaches the phototube, have validity when measured on such equipment because there is less stray light. Moreover, the electronics can be compensated by a dark-current (zero) adjustment. Finally, the premier instruments may be read more accurately.

An important distinction among various types of equipment for turbidity measurements is that between single- and double-beam photometers. The critical test of the instrument is to obscure the phototube with the shutter or to place an opaque object in the light path and monitor the kinetics of decreases in transmission. In cheaper instruments, there will be a precipitous fall which is followed by a much slower but significant fall that may continue for a very long time. Similarly, when the shutter is subsequently opened or the opaque object is removed from cheaper instruments, after the transmission has risen quickly it then rises further very slowly but never achieves a truly stable value. These drifts result from changes in phototube response because of memory to prior illumination conditions. Single-phototube, double-beam instruments avoid this because the same photodetector alternately measures the blank and the sample. However, simple single-beam instruments can be used with very little error if the phototube is alternately exposed to blanks, sample, and dark-current measurements in a routine standardized way with fixed timing.

There is a second reason for adopting a routine timing procedure for measuring cell turbidity, even with the best spectrophotometer. This is because turbidity fluctuates depending on the alignment of rod-shaped particles. The longer the rods, the more severe the alignment will be in recently stirred cultures. For organisms with moderate ratios of length to width, such as *E. coli*, one approach is to mix the culture in the cuvette, place it in the spectrophotometer, wait for a definite period (e.g., 45 s), and then record the reading. Waiting for a long time in hopes of constancy is not recommended. This is because the cuvette will heat up from the light source, creating convection currents that cause partial vertical alignment of rod-shaped organisms. Another approach for rod-shaped organisms is to make the measurements in a slowly flowing system, where rod-shaped organisms will be oriented by the laminar flow (31). The orientation is greater nearer the walls than the center of the cuvette because the gradient of velocity is greater there. Of course, the procedure must be calibrated under the same flowing conditions with organisms of the same axial ratio.

11.6.1.2. Calibration

The procedure adopted for a given culture and instrument must be calibrated in some way to allow for comparison with other work. It is not valid to report turbidity measurements in absorbance units at a specified wavelength, for reasons to be itemized below, although this is commonly done. On the other hand, if a narrow-beam spectrophotometer is used, if the cells have a size range from 0.4 to 2 μm_3, and if reporting the measurement in terms of dry-weight concentration is acceptable, the conversion factors and deviations from Beer's law given below and elsewhere (35, 37) can be used.

To calibrate any adopted procedure, prepare a concentrated suspension of the organism grown under the appropriate conditions. Carry out a series of dilutions of the well-mixed suspension, and make measurements as quickly as possible so that growth need not be stopped or interfered with. Alternatively, place the concentrated suspension in an ice bath and make dilutions and measurements and/or introduce an agent such as chloramphenicol to stop protein synthesis. These are the effective ways to minimize biomass changes while measurements are made. In any case, make a repeat of the first dilution from the stock after all other measurements have been made. The first and last measurement with the same dilution factor must check. Plot the data as they are obtained on a working graph similar to Fig. 6, but use an abscissa expressed as the percent concentration relative to the undiluted stock. A feeling for the shape of the curve is achieved with this working graph. Inspect the graph after each point is added; this will allow the detection of observational errors. Repeat particular dilutions when error is suspected. Make additional dilutions to define the parts of the curve where there is more curvature. Repeating a dilution can give an immediate estimation of error, but an additional intermediate dilution defines the curvature better. There need be no significant pipetting errors since the stock can be adjusted such that dilutions need not be high and relatively large volumes of solution can be pipetted into volumetric glassware. This working graph is sufficient to permit accurate estimates of growth of a culture, since only the relative change in biomass with time is needed for calculating specific growth rates and doubling times. However, for many other uses an absolute instead of a relative measure is needed. In addition, if biomass is measured on an absolute basis, the work is reportable in a way that allows accurate replication in other laboratories. Therefore, it is suggested that calibration be done on an absolute basis, as follows.

For absolute calibration it is necessary to measure accurately the stock suspension content by some other measure of bacterial biomass or numbers. This may be done by dry-weight determinations (46) (section 11.5.2) or by chemical determinations of a nucleic acid (Chapter 22.6), protein (Chapter 22.7.6 to 10), or total nitrogen (Chapter 22.7.11). This must be done with adequate replication and quality control. Use many independent dilutions for viable counts, and use many independent fillings and adequate counts in the counting chamber for microscopic counts. Accuracy at this stage is to be emphasized, since the value obtained will pervade future work.

Prepare a graph that includes all the data that relate meter readings on the photometric instruments to cell number or mass, such as shown in Fig. 6. If computer facilities are available, calculate the best linear, quadratic, or cubic fit to the data. Programs available at most computing centers and available statistical software programs for PCs and MacIntosh computers carry out an *F*-test to define which of the types of curves fit the data most efficiently. The best-fitting curve can then be calculated by simple programs with whatever computer facilities are available. Even programmable pocket calculators are adequate.

In work from this laboratory, we calibrated bacterial suspensions by using the narrow-beam Cary and Zeiss spectrophotometers. We also used and tested a range of commercial instruments with bacterial cells of various sizes (35). Over the size range from 0.4 to 2 μm^3, the results were very nearly the same when expressed on a dry-weight basis. Since the same curve applied to all the bacteria tested, the conversion formulas no doubt will apply to other unpigmented procaryotes whose mean size is within these limits. When the absorbance range is limited to the region below 1.1, a quadratic when passed through the origin fitted the data adequately: $A_\lambda = a_\lambda W - b_\lambda W^2$, where W is the dry weight concentration of bacteria and a_λ and b_λ are constants that apply at a given wavelength λ. The empirical equations for two convenient wavelengths are as follows:

$$A_{420} = 7.114 \times 10^{-3}\ W - 7.702 \times 10^{-6}\ W^2$$

$$A_{660} = 2.742 \times 10^{-3}\ W - 0.138 \times 10^{-6}\ W^2$$

The coefficient of the first power W (a_λ) decreases with increasing wavelength. The coefficient can be calculated at other wavelengths since it depends inversely on the wavelength raised to approximately the second power, actually the 2.11 power. Consequently, a_λ can be calculated at other intermediate wavelengths from

$$a_\lambda = 2{,}439/\lambda^{2.11}$$

There is no rule to estimate b_λ at other wavelengths, although Kavenagh (31) measured this quantity at a number of wavelengths. This constant depends more critically on cell shape, cell internal heterogeneity, and photometer design. In fact, the quadratic fit is only marginally satisfactory at higher turbidities. For this reason and for photometric reasons, it is best to work at absorbances less than 1. Increasing the wavelength decreases b_λ, decreases multiple scattering of light into the detector, and improves linearity.

For routine calculation, the equations given above can be solved to yield W in the units of micrograms of dry weight per milliliter:

$$W = 461.81(1 - \sqrt{1 - 0.60881\ A_{420}})$$

and

$$W = 9929(1 - \sqrt{1 - 0.07347\ A_{660}})$$

The latter expression can be expanded with no loss of accuracy to:

$$W = 364.74A_{660} + 6.7A_{660}$$

All three expressions can be used for optical densities up to 1.

In the absence of pigments, the choice of wavelengths should be based on the following considerations. The lower the wavelength, the larger the turbidity. Therefore, lower concentrations can be directly measured. Conversely, low sensitivity is desirable for some purposes because more dense suspensions are of interest and an extra step of dilution can then be avoided. An additional reason for using a higher wavelength is that many growth media contain substances that absorb near the blue end of the visible spectrum and therefore appear yellow or brown. Then 660 nm is a desirable wavelength because the medium blank is not much different from a water blank. Any higher wavelengths would be inconvenient because a different phototube would be required in many instruments. Chemical changes due to bacterial action, air oxidation, or variation in autoclaving alter the light transmission in complex media at shorter wavelengths. In any case, it is important when using a new medium that the supernatant of a culture of high density be compared with the original growth medium to see whether pigments are introduced or removed by bacterial growth.

Photometric methods can be used with colored organisms. It may be possible to choose a suitable wavelength to avoid absorption bands. Alternatively, choose a wavelength where the absorption by cellular pigments is maximum, and report the biomass using the pigment as an indicator. The latter procedure has the difficulty that in many cases the pigment content varies with culture conditions.

If a narrow-beam spectrophotometer is not available for routine use, the above formula for dry weight can be used if there is temporary access to a narrow-beam instrument. Simply carry out the calibration procedure, but instead of the other methods listed above, also measure the turbidity of a known dilution on a narrow-beam instrument and use the formulas given above.

With wide-beam instruments the measured turbidity is less than in the narrow-beam instruments when a cuvette of the same size and shape is used; therefore, the sensitivity is lower (see above). This is a minor factor in most work. It is more significant that the calibration may change every time the phototube or the light bulb is changed; in some instruments, simply removing and replacing the housing compartment for the measuring cuvette causes measurable changes. Frequent restandardization of a wide-beam instrument against a narrow-beam instrument is recommended. As an example, the reading on an instrument of the DU type (Beckman Instruments, Inc., Fullerton, Calif.) exhibits dramatic changes in sensitivity as thermospacers are inserted around the cuvette compartment (35). This is not unexpected, since the addition of thermospacers lengthens the working distance and hence the instrument more closely approaches the quality of a narrow-beam instrument with its greater sensitivity.

Special quality control must be used with spectrophotometers in which gratings are used to disperse the spectrum. As the grating replicas age, they become warped and an increased amount of stray light hits the phototube.

11.6.1.3. Klett-Summerson colorimeter

Of the many colorimeters and relatively inexpensive spectrophotometers, the Klett-Summerson colorimeter must be singled out for special comment. It was one of the first photoelectric instruments, it is rugged and stable, and it has been used by many workers for more than 50 years. It is still being used in many laboratories. It is stable because it is a double-beam instrument so that fluctuations in the light source are canceled out. It is not sensitive, since it has a simple CdS sensor which produces only enough current so that a sensitive galvanometer is needed. The detector is far enough from the cuvettes that less of the scattered light is trapped than in many other instruments.

The Klett colorimeter is calibrated on a different scale from that used for more modern instruments: an absorbance of 1 on other instruments corresponds to 500 on the Klett scale. It reads so sluggishly that the turbidity of rod-shaped bacteria becomes stable by the time the dial can be properly adjusted; by then, the cells have achieved random orientation. Take care in reading the instrument dial and adjusting the dial to achieve a null reading on the galvanometer. Stand in a reproducible position so that parallax error is avoided. The major fault of this instrument is that there is no dark-current (zero) adjustment. Light can pass around the contents of the cuvette by being scattered on instrument surfaces and reflected through the glass walls of the cuvette. Test this (and other instruments without dark-current adjustments) by filling a cuvette with a highly absorbing material having an essentially infinite optical density (0% transmittance). In the Klett colorimeter, the reading should be beyond the printed scale of the instrument and in fact should be right on the vertical line that forms the left-hand side of the letter R in COPYRIGHT. Frequently the Klett colorimeter is out of calibration, and then there is an adventitious instrument error which creates additional deviations to the Beer's law relationship. Special adjustment, repair, or replacement of the light source is needed. The problem can be partially alleviated, in some cases, by reblackening the internal surfaces.

11.6.2. Nephelometry

Although most of the light scattered by bacteria is nearly in the forward direction, instruments that measure the light scattered at 90° from the primary beam have also been used to measure bacterial concentrations. Instruments designed to make right-angle measurements of light scattering are called nephelometers (Greek *nephelos* = cloud). Bacterial concentration may be measured with this principle in a spectrofluorometer when the dials are set so that the excitation monochromator and the emission monochromator are at the same wavelength.

Nephelometric measurements can be ultrasensitive, so that very low concentrations of bacterial cells can be measured. There are two difficulties in making routine measurements. The first is caused by false signals from particulate impurities in the medium; this can be partially overcome by ultrafiltration of all reagents. The second is that it is necessary to standardize the instrument repeatedly (if the light intensity from the lamp varies) during a series of measurements. A different kind of standardization is needed with a nephelometer or spectrofluorometer from that with a spectrophotometer. With the former, the operator sets the dark-current (zero) adjustment so that the scale reads zero with a reagent blank. Then the gain is set to give a fixed response from a standard scattering material. That is, it is necessary to run a one-point standard curve routinely. Since a suspension of bacteria would not be stable, a secondary standard such as a piece of opal glass must be used. One can also use a suspension of glass beads, such as those used in rupturing bacteria and other biological material. Commercial preparations are also available. A tube can be prepared and sealed and used to calibrate the instrument. Then a standard curve can be prepared relating meter readings after standardization to bacterial concentration. The standard curves are not linear at higher bacterial densities.

Although various specialized instruments have been made and used over the years to measure the light scattered at different angles, these instruments are of limited value because most of the light that is scattered from particles in the size range of bacteria is nearly in the forward direction (Fig. 4). If this light is to be sensed, the instrument must be capable of orientation to within a very few degrees of the direction of the incident light without intercepting undeviated light from the primary beam. This means that very well collimated light is necessary. Today, this is readily achieved with a laser beam. Instruments that use a low-angle range between 2 and 12° from the beam are available, but none is commercially available at a reasonable price. Measurements at higher angles contain information about the amount of cell material, but they also include information about internal structure and distribution of material within the cell. If it is known which factors are important and whether elongated particles are oriented or randomly distributed in space, light-scattering measurement over a range of angles can give information about the state of aggregation of the protoplasm, such as whether the ribosomes are in polysomes or as monosomes within the living cell (6); the thickness of the cell envelope (32, 34, 40, 69); and the distribution of cell mass from the center of the cell (32, 69).

Wyatt (69, 70) has developed an apparatus to measure the angular dependency of the light-scattering signal from 30 to 150°. In this range, different organisms, different treatments of a culture, and cultures in different phases of growth give characteristic patterns. These may be of use as a fingerprint technique for the diagnosis and study of drug action. For the reasons stated above, the light-scattering signals in the high-angle range cannot be used reliably for growth measurement because they are sensitive to details of subcellular structure.

Light-scattering measurements made over a range of angles from 0 to 20° are relatively independent of these three factors but are more critically dependent on overall size. Measurement in this range could nondestructively yield the biomass concentrations, the number concentration, the average axial ratio, and a measure of average biomass distribution around the center of the cells for bacteria in balanced growth.

For additional information on nephelometry, see Chapter 21.1.2.

11.7. STATISTICS, CALCULATIONS, AND CURVE FITTING

Several aspects relevant to the measurement of bacterial growth are almost never included in courses in basic microbiology or statistics taken by undergraduate microbiology majors. For the former, this is because the needed material is a little beyond the scope of the courses, and for the latter, it is because the counts of colonies or particles is an all-or-none event and of lesser utility for most users of statistics. The following description is intended to make these mathematical tools readily available.

11.7.1. Population Distributions

The **binomial distribution** describes the chance of occurrence of two alternative events. For example, it provides the answer to the question, "What is the chance of having five boys in a family of nine, assuming that births of boys represent 0.56 of the total live births?" The numerical answer is $P_5 = 0.2600$. The relevant formula is as follows:

$$P_r = \frac{n!}{(n-r)!r!} p^r (1-p)^{n-r}$$

where p is the chance of a specified response on a single try, n is the total number of trials, and r is the number of specified responses and would vary from 0 to n. The symbol, !, means factorial; i.e., $n! = n(n-1)(n-2)\ldots 1$. The formula shows that in different families of nine children more families would have five boys than any other number. For example, $P_4 = 0.2044$, $P_5 = 0.2600$, and $P_6 = 0.2207$. The plot of P against r is an example of a distribution histogram. The binomial distribution provides a way to estimate the mean and standard deviation of p from experimental data. Assuming that there are data on only a single family with n children and that there are by chance r boys in that family, statistical theory shows that the best estimate of p is r/n. Numerically for the example where $r = 5$ and $n = 9$, p is $5/9 = 0.5556$ and its standard deviation is

$$\sqrt{p(1-p)/n} = \sqrt{(5/9)(4/9)/9} = 0.1656$$

This result is not very reliable because the coefficient of variation (CV) is nearly 30% (CV = 0.1656 × 100%/0.5556 = 29.81%). The precision in estimating p could be improved by looking for a family which had many more children and applying the same formula. It would be not only easier but better to pool the census data of a number of families. Suppose 50,000 boys are counted in 90,000 children in a large pool of families. Then the same formula yields $p = 0.5556$, but now CV will be 0.2981%. Not only is this more precise because large numbers are counted, but also many families, for genetic and sociological reasons, may have different values of p. *It may be desirable to estimate the value of p that applies generally to the entire population.*

The formula for the distribution is cumbersome to calculate when the numbers are large. An important contribution of Gauss was to rewrite the binomial distribution for this large-number case. Thus, the **Gaussian distribution** applies as a generalization of the binomial distribution for the case in which the numbers involved are so large that they can be treated as a continuous distribution instead of one with discrete variables. The variables of the Gaussian distribution replacing n and p are the population mean (m) and standard deviation (σ). The formula then becomes

$$P_x = \frac{1}{\sqrt{2\pi\sigma}} e^{-(x-m)^2/2\sigma^2}$$

where e is 2.71828.

In this formula the continuous quantity (x) replaces the positive integer variable (r) as the measurement of response. This distribution, like the binomial, can be mathematically manipulated so that one can go from data to estimations of these two parameters. The Gaussian distribution is also called the **normal distribution,** partly because it is symmetrical about the mean and partly because it is so frequently observed.

For data that follow a Gaussian distribution, the estimate of the mean is called m and is given by $m = \Sigma x/n$. The estimate of the standard deviation is called s and is given by:

$$s = \sqrt{\frac{\Sigma x^2 - (\Sigma x)^2/n}{n-1}}$$

The coefficient of variation, CV, is given by s/m.

The other limiting distribution of the binomial is the **Poisson distribution.** It applies for the case where n is very large and p is very small but the product np is finite. The best estimate of np is N, the observed number of total responses of a specified kind. This distribution would be useful if, for example, boys occurred very rarely (say, 1 in 10,000) but families were large (say, 100,000 children). Then, an average family would contain $N = 10$ boys and the standard deviation would be

$$n\sqrt{p(1-p)/n} = \sqrt{n(N/n)(1-N/n)} =$$

$$\sqrt{10{,}000 \cdot (10/10{,}000)(1-10/10{,}000)} = 3.1621$$

A keen simplification, due to Poisson, was to assume $N/n << 1$. Then the formula for the binomial distribution simplifies to a one-parameter distribution

$$p_r = \frac{e^{-m} \cdot m^r}{m!}$$

The best estimate of the mean is $m = N$, and the best estimate of the standard deviation is $\sqrt{N}$. Note that this is not much different from the binomial distribution, since $\sqrt{10} = 3.1623$ is not much different from 3.1621. Note that again n and p are replaced by different symbols, in this case by a single one, m. The important point is that *the count of the number of discrete objects provides not only the best estimate of the mean value (i.e., $N = x = m$) but also an estimate of the precision of the estimate ($\sqrt{N} = s = \sigma$).*

Consequently, in the enumeration of objects it does not matter how they are subdivided. Two replicate plates with 200 colonies on each are no better or worse from the point of view of Poisson statistics than is one plate with 400 colonies. In both cases the standard deviation, s, of the measurement is, $\sqrt{400} = 20$ and the CV is $\sqrt{400}/400 = 5\%$. Therefore, to get the best estimate from a group of plates from the same or different dilutions of the same sample, simply add the total counts on all the plates and divide by the total volume of the original solution. The standard deviation is the square root of the total count divided by the plated volume of solution.

As an example, imagine that duplicate plates were made at dilutions of both 10^{-5} and 10^{-6} with counts of 534 and 580 and of 32 and 60, respectively. The total count is 1,206. If 0.1 ml of these dilutions were plated, 2.2×10^{-6} ml would be the total volume of original culture used to make the four plates. Therefore, the best estimate of the concentration is $1{,}206/2.2 \times 10^{-6} = 5.46 \times 10^8$/ml. The standard deviation is $\sqrt{1{,}206}/2.2 \times 10^{-6} = 0.16 \times 10^8$/ml.

Justification for treating the results at the two different dilutions separately and not pooling them, as done above, depends on other kinds of errors being larger. Then, by comparing the results at different levels of dilution, some estimate is obtained of the variability due to the additional pipetting operation and to other sources of error that are not included in the calculation of the Poisson sampling error. Although this variability is of interest, sources of error could be more directly measured by carrying out independent dilutions from the same cell suspension. As an example, imagine that 0.1-ml aliquots were plated on single plates from each of 12 independent 10^{-5} fold dilutions and that the following set of colony counts were obtained: 534, 580, 760, 643, 565, 498, 573, 476, 555, 634, 514, 693. The sum is 7,026, and the Poisson standard deviation is $\sqrt{7{,}026} = 83.8$. The mean of the numbers is 585.5, and the standard deviation by the Gaussian formula is 81.2. Calculation of the bacterial count of the original suspension together with the two different estimates of error yields count $= 7{,}026/1.2 \times 10^{-5}$ ml $= 5.85 \times 10^8$/ml; Gaussian error $= \pm 81.2/0.1 \times 10^{-5} = \pm 0.81 \times 10^8$/ml; and Poisson error $= \pm 83.8/1.2 \times 10^{-5} = \pm 0.07 \times 10^8$/ml. The comparison of these two estimates of error suggests that considerable error is due to sources other than random sampling. An attempt should be made to find and reduce these sources of error. Until that is done, it is necessary to make many independent dilutions and use Gaussian statistics. *Until the source of error is found, the Poisson error is irrelevant.*

This same point can be made in another way from this example. Imagine that only one plate had been made, say the first one, in which case only the Poisson error would be available for consideration. The count then would be $5.34 \times 10^8 \pm 0.23 \times 10^8$, and the real error would be underestimated fourfold. It is therefore cautioned not to rely on Poisson statistics until their use has been justified for the conditions actually in use. Instead, make a comparison on at least several occasions with a Gaussian statistic measurement of error as indicated above.

11.7.2. Statistical Tests

Much of the statistics taught in elementary courses is concerned with whether a body of data is consistent with a hypothesis. Usually the **probability** (P) that the observed deviations from the hypothesis could occur by chance is computed. If P is small, but not very small, the hypothesis could still be false although improbable. If P is very small, the hypothesis can, with some confidence, be said to be true. These statistical tests are generally made on the assumption that the data follow a Gaussian distribution. In many cases in bacteriology, this assumption should be questioned and other appropriate statistical tools should be used.

The standard deviation has been defined above. This is frequently confused with another term, the **standard error,** also called **standard deviation of the mean.** The standard deviation measures the deviation of an individual measurement from the mean of many measurements. The standard error measures the mean of all the data observed from the mean of a hypothetical data base containing an infinite number of observations and is a measure of how close the average is to the "true" mean value.

The only statistical test mentioned in the text below is Student's ***t* test**. This applies to the difference in the means of two groups. The difference is divided by the standard error of the combined data. Thus the t value measures the difference in the means by using the standard error as its unit of measurement. This ratio is compared with values given in tables. Use of the tables generates P values but requires a knowledge of the number of measurements and whether potential deviations can occur on both sides or only one side of the mean. The bigger that t is, the smaller P becomes; if P is not very small, the hypothesis that the two populations were identical may not be rejected.

In recent years, the **analysis of variance** (ANOVA), which is a subbranch of statistics, has been elaborated so that it now can be applied to many problems and replaces many of the more specialized techniques used previously. The availability of packaged programs for various computers means that it requires work, but much less work than previously, to learn to use and apply statistical methods when they are appropriate in bacteriology.

11.7.3. Error Propagation

The accuracy of an estimate depends on the accuracy of its component measurements. The Poisson error of a colony count and the error of the dilution procedure both contribute to the error in the estimated concentration of organisms of the original undiluted suspension. Additional errors can only further blur the results or make them less precise. Even though errors in one part of the estimate may compensate for errors in another part, on the average random errors will make them larger. When errors in one measurement are independent of (uncorrelated with) errors in another measurement, the overall error can be calculated by two rules for **propagation of errors,** as follows.

1. If two quantities (x and y) are to be added or subtracted, the standard deviation (s) of the combined quantities is

$$s_{x+y} = s_{x-y} = \sqrt{s_x^2 + s_y^2}$$

2. If two quantities are to be multiplied or divided, the coefficient of variation (CV) of the combined quantities is

$$CV_{xy} = CV_{x/y} = \sqrt{CV_x^2 + CV_y^2}$$

As an example, apply the second rule to estimate the overall error in a single plate count containing colonies from the series of 10^5-fold dilutions. Assume that the dilutions were performed in five steps of 10-fold each and that the pipetting error of a single 10-fold dilution has a CV of 0.02. Then the overall error of the five dilution steps is $\sqrt{5} \times 0.02$. This result is obtained by the repeated use of the second rule. It then must be combined with the Poisson error. Since the best estimate of the Poisson error CV is $1/\sqrt{585.5}$, then the overall CV is as follows:

$$CV = \sqrt{1/585.5 + 5(0.02^2)}$$

$$= \sqrt{0.001709 + 0.00200} = 6.1\%$$

This 6.1% error is composed of a Poisson counting error of 4.1% = $100/\sqrt{585.5}$ and an error due to the cumulative pipetting errors of 2% × $\sqrt{5}$ = 4.47%. The rule to combine them gives a value smaller than their sum (4.1 + 4.47 = 8.57%) but larger than the largest contributor to the error.

Two important experimental considerations derive from this example. First, *there is no reason for increasing the accuracy of one part of an experiment unless other sources of error comparable to it are also decreased or eliminated.* Second, *if an operation is to be done many times, it is worthwhile to devise a way to do it accurately and then obviate the need to carry out elaborate statistical calculations.* In the previous section, the pipetting error was neglected because it was assumed that pipetting can be and was done accurately. This is a reasonable thing to do if the CV of this error is smaller than the Poisson counting error. As an example, imagine that each pipetting operation had been carried out with an accuracy of 1% instead of 2%. Then the overall pipetting error would have been 1% × $\sqrt{5}$ = 2.23% and the overall total CV consequently would have been $\sqrt{4.1\%^2 + 2.23\%^2}$ = 4.7%, only a little bit larger than the Poisson counting error by itself.

Similar logic follows for cases in which blank values and background values are to be subtracted (in which case the first rule for propagation of errors applies) or in which the instrument calibration factors are used to multiply observed values (and the second rule applies). In the measurement of controls used repeatedly in the calculation of data, errors should be reduced by repetitions or by more accurate measurement than for individual experimental values so that the control factors do not contribute significantly to the overall error of measurement.

11.7.4. Ratio Accuracy

There is a very powerful and general statistical method applicable to diverse experimental situations varying between large natural ecosystems at one extreme and a drop of culture on an electron microscope grid at the other. This method is to add a known number of reference particles, which may be bacteria, ferritin particles, polystyrene spheres, abortively transduced bacteria, plasmids, viruses, etc. After mixing takes place, samples are taken and the ratio of the number of objects of interest to the number of reference particles is determined by appropriate means. The method can be illustrated for microscopic smears containing a class of recognizable organism of unknown number and polystyrene beads of known concentration. The smear must be prepared from a known volume of cellular suspension and a known volume of suspension of beads of known concentration. The concentration of the beads in the final suspension is multiplied by the ratio of the counts of the unknown cells relative to those of the reference beads to calculate the concentration of unknown cells. The second rule for the propagation of error applies in this case. If the concentration of the reference particles is known without error in the original stock solution, the coefficient of variation of the unknown particles is given by

$$CV = \sqrt{\frac{1}{N_u} + \frac{1}{N_r}}$$

where N_u and N_r are the counts of the unknown and reference, respectively. To minimize the number of total counts following the first argument in the previous action, N_u should be about equal to N_r. Then the CV will be about $\sqrt{2}$ = 1.4-fold larger than if a very large number of reference cells (or unknown cells) had been counted.

11.7.5. Coincidence Correction

Coincidence corrections must be applied when too many colonies are on a plate or too many cells are on a square of a counting chamber or if too many radioactive decays are recorded by a radioactivity counter in a unit of time. For the case of a colony count, assume that, if two cells initially are closer together than a distance r, they will be counted as a single colony. Let N_t be the true count and N_a be the actual count, and assume that the radius of the petri dish is R. Consider a single cell; the chance that another cell on the plate is within a distance r is $N_t r^2/R^2$; thus, the count is decreased by $N_t r^2/R^2$. The number of colonies not counted will be $N_a N_t r^2/R^2$. Therefore,

$$N_t = N_a + N_a N_t r^2/R^2 = N_a(1 + cN_t)$$

where $c = r^2/R^2$. It is usually convenient to substitute N_a for N_t on the far-right-hand side when the correction is small. From this formula it is clear why a fourfold reduction in colony size reduces the coincidence correction at a given count by 16-fold. This is the basis of the use of layered plates (section 11.3.3), which makes the colonies smaller and thus reduces coincidence.

11.7.6. Exponential-Growth Calculations

Under constant conditions after a long enough time when cell-cell interaction is small, growth measured in any manner is expected to proceed according to $X = X_0 e^{\mu t}$. This can be written in any of the following equivalent ways:

$$\ln X = \ln X_0 + \mu t$$

$$\log X = \log X_0 + \mu t/2.303$$

$$X = X_0 2^{t/T_2}$$

In the last equation, T_2 is the **doubling time** and can be calculated from the following:

$$T_2 = (\ln 2)/\mu = 0.6931/\mu$$

Many symbols other than μ have been used for the specific growth rate including a, k, and λ. Knowing these other symbols is important because they are used without definition in many papers. Confusion arises with μ, which designates the specific growth rate in the literature on continuous culture (Chapter 10.2) and in microbial ecology. This is the usage employed here. Unfortunately, μ is also used to symbolize the number of doublings per hour in the literature on cell physiology. The latter usage differs from the former usage by a factor of ln 2 = 0.6931. Although any time unit could be used, reciprocal hours appears to be nearly standard. The doubling time (T_2) is reported in the literature in either minutes or hours.

11.7.7. Plotting and Fitting Exponential Data

There are several alternative ways to fit data to the exponential-growth model that are equally valid but differ in their precision and in the additional information yielded to the experimenter. The most simple conceptually is to look up the natural logarithms of the concentration of cells (or the dry weight of biomass, or other measurement of an extensive property of the bacteria) and plot them on ordinary arithmetic graph paper against the time that the measurements were made. Then draw a straight line through the datum points as close to the points as possible, and determine its slope by the rise-over-fall method. If the time scale is in hours, then the slope is in units of reciprocal hours. At this point, ask two questions: How appropriate is a straight line to the data? Is the line drawn a good summary of the data?

Common (base 10) logarithms may be used, in which case the slope must be multiplied by 2.303 = ln 10 to obtain μ. Base 2 logarithms also may be used, in which case multiply by 0.6931 = ln 2. There is an advantage in using base 2 logarithms: the reciprocal of the slope gives the doubling time directly. Whatever the base of logarithm used, plot both the characteristic and mantissa numbers on the arithmetic scale of the graph paper. Do not make the all-too-frequent blunder of mixing them (e.g., $\log_{10}$ of 4×10^8 cells = $8.6 \neq 8.4$)!

It usually is more convenient to use semilogarithmic graph paper. Paper that has six divisions between darker lines on the arithmetic scale and two cycles on the logarithmic scale is recommended. Define the abscissa (x axis) according to hours or days on the major lines so that an even number of minutes or hours is represented by the minor lines. Mark each of the three unit labels with the appropriate powers of 10 on the ordinate (y axis). The printed scales may be multiplied by a constant, but it is invalid to add or subtract a constant from the logarithmic scale. Find the point corresponding to the amount of biomass, cell numbers, or other measure of extent of growth on the y axis, and mark the point exactly. It is useful to make a small point mark for exactness and surround it with a larger circle for better visualization. It also is useful, when light-scattering methods are used, to plot each point as soon as it is obtained. This frequently shows when errors have been made, whether they be biological, instrumental, arithmetic, or human. If the error is detected immediately, one has the opportunity to restart a culture, remeasure the culture, or replot a point. Once all the data have been plotted, draw the best straight line to fit the points.

The mathematical procedure that does this best is called the **least-squares** fit, which minimizes the square of the vertical distance of all points to the proposed line. This can be approximated visually by mentally noting the distance from the few points that are farthest from the position of a proposed straight line by moving a transparent ruler. By readjusting the ruler and remembering that the actual distances, even though the graph paper is semilogarithmic, are to be squared, one can do a quite accurate job of drawing very nearly the line which would be generated by the mathematical least-squares procedure.

Figure 7 shows an actual growth curve with an "eyeballed line" which is essentially the same as the computer-fitted line. This example is drawn from a carefully executed experiment with accurate data over a period permitting a 30-fold increase in cell mass. Note that the datum points fit close to the line. This implies that if overt or gross errors were made during the experiment, they were discovered. This example was chosen to show how sensitive growth rate measurements are to the accuracy of the data and to show the way the line is drawn. The drawn line is almost as good a representation of the data as (and is indistinguishable from) the computed line over the range from 10 to 20 μg (dry weight)/ml but corresponds to a 2.7% difference in μ or T_2.

Note that there are many kinds of semilogarithmic graph paper classified by how many cycles (powers of 10) they span. The fewer the number of cycles, the more spread out the points are and the easier it is to define the line accurately. Note also that there is only one type of

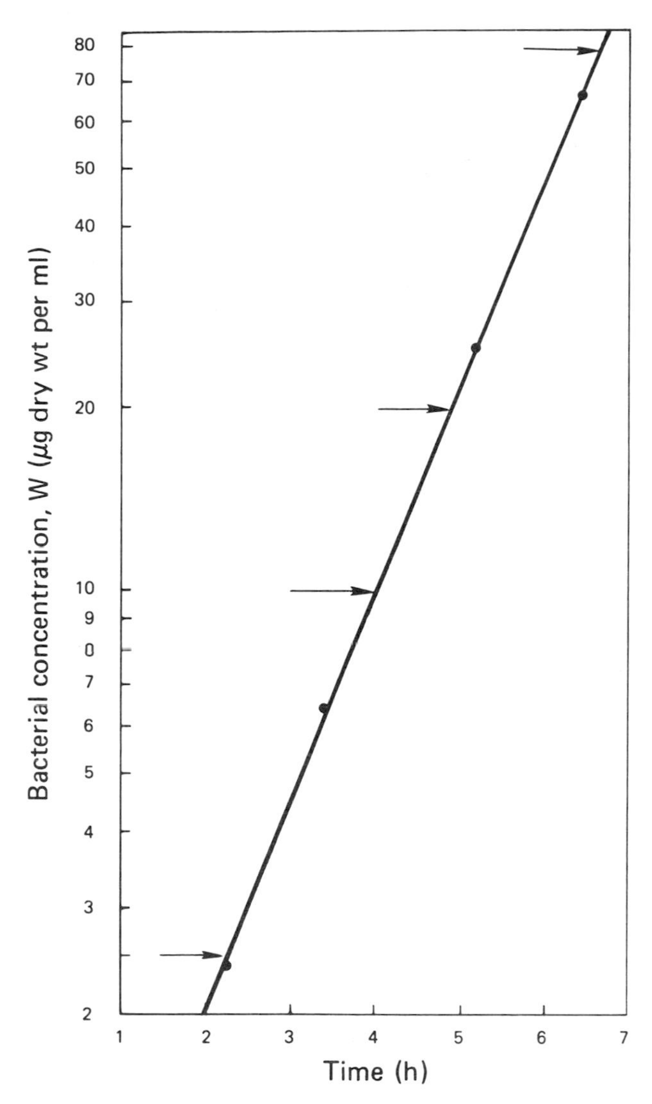

FIG. 7 Growth curve plotted on semilogarithmic paper. See the text for conditions. The horizontal arrows are drawn to the curve at values of bacterial concentrations that are convenient powers of 2 apart, which facilitates estimations of the doubling time.

semilogarithmic paper. The same paper is used whether the user chooses to work and think with natural, common, or base 2 logarithms.

Use the graph to find the doubling time, since this is the measure of growth rate that has the most intuitive appeal to bacteriologists. The doubling time is easily found by measuring the time for the bacterial concentration to change twofold. Put a tick mark on the graph where the line crosses some major division on the logarithmic scale of the paper; make another on the division two-fold higher or lower. Read both times off the abscissa and subtract them to obtain the doubling time. Alternatively, better precision can be obtained by using fourfold, eightfold, or 2^N multiples of mass and then dividing the time difference by 2, 3, or N to correct the resulting time difference to one doubling time. The specific growth rate or other measures of growth rate can then be calculated.

11.7.8. Least-Squares Fitting

11.7.8.1. By pocket calculator

The least-squares procedure is more precise than visual fitting of exponential growth data, and there is no reason not to make the calculation this way since pocket calculators with linear-regression capability are readily available at low cost. Table 3 shows the data for Fig. 7 treated on a pocket calculator with least-squares capability. By simply following the instruction manual, one can use any calculator with linear-regression functions.

The first step is to accumulate the sums, sums of squares, and sums of cross-products. Take care to calculate the natural logarithm (ln) of the cell count (or other measure of cell substance) and to use it as the Y variable. Key the time in for the X variable. Take care to convert the time to decimal units before executing the cumulation procedure ($\Sigma+$). Some calculators convert hours and minutes into decimal hours; thus, if 10.30 is keyed in and this is thought of as 10 h and 30 min, a function available on some calculators converts this to decimal time 10.50, which is then suitable for regression. Be careful if time goes past noon (or midnight) to add 12 or 24 h to the time. Alternatively, calculate the elapsed time from the start of the growth run or use an elapsed-time counter that reads in decimal time during the execution of the experiment. Many clocks available for laboratory use read in minutes and decimals of minutes. The clock can also be watched and measurements can be made at the hour, the half hour, or at any regular time interval.

It is important to use time as the X variable. This is because the formulas used in the linear regression have been derived by assuming that all error in the measurement is error in the Y variable and that there is none in the X variable. For growth measurements, the time can be precisely defined (if necessary, to the second), and the major error is in the measurement of the number of cells or other indices of biomass.

The calculator derives the best estimate of the intercept and slope by the formulas applicable to the linear equation $Y = a_0 + a_1X$, i.e.:

$$a_0 = \frac{(\Sigma X^2)(\Sigma Y) - (\Sigma X)(\Sigma XY)}{n(\Sigma X^2) - (\Sigma X)^2}$$

$$a_1 = \frac{n(\Sigma XY) - (\Sigma X)(\Sigma Y)}{n(\Sigma X^2) - (\Sigma X)^2}$$

where a_0 is the estimate of the value of Y when X is zero, a_1 is the slope of the regression, and n is the number of observations. For the case of exponential growth, the value of a_0 is the natural logarithm of the initial cell density. This is of lesser interest than the slope (a_1), which is the best estimate of the specific growth rate (μ).

A very real problem arises with the usual use of a linear regression in its unweighted form. To see this more clearly, imagine a hypothetical situation in which the problem does not arise. Imagine that growth is measured by viable counts and that dilutions are made at every time point in such a way that precisely 400 colonies developed on each plate. In this case the coefficient

Table 3. Sample data for exponential growth by graphical estimation or pocket calculator computation

Sample Data. These data were obtained with a subculture of *E. coli* ML308 growing in 0.2% glucose-M9 medium at 28°C

As recorded		As converted	
Time (h:min) (1)	A_{420} (2)	Time (min) (3)	W (μg/ml) (4)
2:13	0.017	133	2.39
3:23	0.045	203	6.36
3:57	0.071	237	10.09
5:10	0.174	310	25.14
6:23	0.439	383	66.50

Graphical estimation. From Fig. 6 (which is the plot of data in column 1 versus data in column 4), determine the time at any value of W and also at twice that value (e.g., at W = 10 and 20 μg/ml):

Time (h:min)	W (μg/ml)
4:00	10
4:52	20

The time difference is 52 min; T_2 = 52 min; and the ratio is 2.

Alternatively, determine the time at any chosen value of W_0 and the time that the culture reaches $2^N W_0$. For example, if W_0 = 2.5 μg/ml and N = 5, then $W_5 = 2.5 \times 2^5$ = 80 μg/ml. The time when the culture had W_0 = 2.5 μg/ml as read from the graph is 2:15 p.m., and the time when W_5 = 80 μg/ml as read from the graph is 6:38 p.m. The difference in time, 263 min, is the time for five generations of growth, and therefore T_2 = 263 min/5 = 52.6 min.

Pocket calculator computation. Obtain the regression of the logarithms of data in column 4 against data in column 3 by the following steps:

1. Clear memory
2. Key 2.39, ln, key 133, Σ+
3. Key 6.36, ln, key 203, Σ+
4. Key 10.09, ln, key 237, Σ+
5. Key 25.14, ln, key 310, Σ+
6. Key 66.50, ln, key 383, Σ+
7. Key linear regression. Read ln W_0 = −0.85456.
8. Key X ↔ Y. Read = 0.01321/min.
9. These results can be expressed as ln $W = -0.85456 + 0.013213t$, or $W = 0.4255\, e^{+0.013213t}$ when t is expressed in minutes.
10. To reexpress the specific growth rate on a per-hour basis, multiply by 60 to give = 0.793/h.
11. To reexpress in terms of doubling time, divide into ln 2 to give T_2 = 52.47 min.
12. For an error calculation (see text), T_2 = 52.47 ± 0.65 min and CV = 1.24%.

of variation is $\sqrt{400}/400$ = 5% and is constant throughout the growth period, even though growth may have increased the number of cells per unit volume of culture many orders of magnitude during the experiment. This means that each value of ln N is known with a standard deviation of ln (1.05) = 0.05 above and ln (0.95) = −0.05 below standard mean value and is thus constant throughout the experiment. Constancy of the error in the Y variable was another of the basic assumptions made in deriving the least-squares unweighted linear regression formula.

Now imagine that growth is monitored turbidimetrically without diluting samples for measurement. At low turbidities the coefficient of variation is high because of the difficulty of matching cuvettes and balancing the blank. At an apparent absorbance of 0.4, the relative error is at a minimum; at higher absorbances, the photometric precision falls and the scales cannot be read accurately. Consequently, the coefficient of variation rises. In addition, there are errors involved in correction at high absorbance for deviations from Beer's law and for problems resulting from stray light.

For these reasons, after logarithms have been taken, the standard deviation is not constant but falls to a minimum at intermediate absorbances and then becomes larger. Therefore, the data at intermediate absorbances have higher accuracy but do not determine the slope of the line set by the experimental data as well as is done by points at the low and high end that span a longer stretch of time. One can overcome this problem and use the data in a balanced way by carrying out a weighted regression. To do so it is necessary to estimate the error of each datum in one way or another. Ordinarily this involves a great deal of experimental and computational work and is not routinely done. Alternatively, make turbidimetric measurements by using techniques that are very precise so that the errors are very small. Additionally, take more datum points and exercise more care in regions at the low and high ends of the curves.

If the data are treated by pocket calculator (or by computer), no arithmetic mistakes are likely to be made. However, human copying error can easily invalidate the results. For this reason, it is strongly recommended that a graph be made so that the results of the computation can be compared and qualitatively checked with the graphical results. Alternatively, the error of the slope should be calculated (this is done automatically for the BASIC program of Table 4). The error of the slope of a linear regression is given by

$$s_{a_1} = s_\lambda = \sqrt{\frac{(\Sigma Y^2 - a_0\Sigma Y - a_1\Sigma XY)n}{(n-2)(\Sigma X)^2/n}}$$

Except for ΣY^2, most of the summations in this formula were also needed by the formula for calculating a_0 and a_1. Fortunately, ΣY^2 is needed for calculating correlation coefficients and for the standard deviation of the Y values. For this reason, most modern pocket calculators and statistical computer packages calculate everything needed for the error calculation when the data pair is entered (with the $\Sigma+$ key). With ΣY^2 accumulated, it is no great difficulty to program the programmable versions of the hand calculators to calculate the error of the slope. The error given in Table 3 was calculated this way. One could also calculate errors of the intercept, the means of X and of Y, and the confidence interval. However, the error of the specific growth rate is the crucial value and worth the effort to calculate. The standard error of the doubling time can be calculated to sufficient accuracy by assuming that μ and T_2 ($= \ln 2/\mu$) have the same coefficient of variation. Algebraic manipulation shows that the standard deviation of the doubling time is the standard deviation of the specific growth rate multiplied by ln 2 and divided by μ^2; i.e.,

$$s_{T_2} = s_\mu \ln 2/\mu^2$$

The error in the slope of the regression has some of the same properties as does the Poissonian $\sqrt{N}$ as an estimate of the error of a count. It tells about the internal accuracy of the measurement but does not assess external sources of variation.

The specific growth rate can be measured with precision within a run better than the reproducibility of the specific growth rate from run to run or of the reproducibility of growth rate of subcultures of bacteria from day to day. The sources of this variation are unknown at present, but this is an important field of research for the future. However, these run-to-run variations must be taken into consideration for the practical purposes of making and using growth rate measurement to study aspects of microbiology.

Most often, the reason for measuring growth rates is to measure the effect of some treatment. This requires comparison of growth with and without treatment. Ultimately, comparison requires replicate control cultures and replicate experimental cultures. Then, the standard t test of the specific growth rates of the two groups suffices to test the probability that the groups are not different. Other statistical procedures should be used when it is evident that there is a difference and the effect of the treatment is to be quantitated (see any statistics text, e.g., reference 61). The real value of the calculation of s_μ is in allowing the experimenter to replicate the treatments and the controls less often once the relationship of the internally measured error to the external measured error has been measured often enough to instill confidence. If they are always comparable when examined, single growth runs have more significance—when backed up with the error of the slope.

Another way to increase the precision of the measurement of the effect of a treatment when the variation from subculture to subculture is important is the "subject as its own control" method. Start by monitoring the growth of an untreated culture. When the growth rate has been determined, treat the culture and monitor growth to remeasure the specific growth rate. This procedure isolates the treatment as the sole variable between the two halves of the growth run. It is a smart move to run control experiments in which the treatment is only a sham treatment.

11.7.8.2. By BASIC computer program

In the first edition of this manual only the pocket calculator routine was provided to calculate the growth parameters. Now a decade later, a BASIC program written so as to be especially user friendly is given. The program can be keyed in from Table 4 or imported. It is arranged in such a way that it can be conveniently used during the execution of a physiological experiment without curtailing other uses of the computer. This facility is possible because the data, as entered, are stored into a file external to the working program. This means that the program can be terminated and the computer can be used for other purposes. Later, the growth program can be reactivated and new data can be recorded. A second user-friendly feature is that one can study and monitor many independent cultures simultaneously.

For use, assign a different single letter for each culture. Run the program. The program offers several choices. Initially, at this information request, key **S** (for start-up). The computer will then offer an opportunity to change two constants to compute biomass from the measurements of absorbance. If the **<ENTER>** key is struck twice, the program will use the default constants appropriate if 1.0-cm cuvettes are being used in a reasonably good spectrophotometer at 420 nm. If runs have been already started, instead of keying **S**, key **A** and then **<ENTER>** for added data. If just the growth rate parameters are desired, key **G** and then **<ENTER>** when growth rate estimates alone are needed.

For each datum point, key the letter designation of the culture, then press **Comma**, key the time in hours, **Comma**, the minutes during the hour, **Comma**, the optical density, and finally **<ENTER>**. These four numerical values must be separated by commas. If the time is available from a clock reading in minutes only, key a second **Comma** before keying in the elapsed time in minutes. If, on the other hand, a clock is used that gives the time in hours and decimals, key this time followed by **Comma, Comma.** If the growth runs continue past noon, use time expressed on a 24-h basis. If time goes past midnight or for several days it will be necessary to key in the hours as time passed from midnight of the day before the run started, unless the time the culture was started was taken as zero.

After the culture letter and the three datum pairs (hours, minutes, and absorbance) for a particular culture have been entered, the computer repeats the data about that particular chosen culture. After three time points have been entered for the culture, if **G** and **<ENTER>** are keyed and then the desired culture designation is keyed followed by **<ENTER>**, the com-

Table 4. BASIC computer program for determination of growth rate

```
00010 'GROWTH PROGRAM, A.L. KOCH, 11/6/90
00020 DEFSNG A-Z
00030 AL1=481.81:BL1=.60881:CLS
00040 INPUT "START-UP(S),ADD DATUM(A), OR GROWTH RATE(G) "; D$
00050 IF D$="G" OR D$="g" THEN 200
00060 IF D$="A" OR D$="a" THEN 110
00070 IF D$<>"S" AND D$<>"s" THEN 40
00080 INPUT "CONVERSION CONSTANTS, AL AND BL "; AL,BL
00090 IF AL=0 THEN AL=AL1:IF BL=0 THEN BL=BL1
00100 OPEN "O",1,"DATA":CLOSE
00110 OPEN "A",1,"DATA":CLS
00120 PRINT "CULTURE, HOUR, MIN, ADORB. ":INPUT C$,H,M,OD
00130 T=H*60+M:BM=AL*(1-SQR(1-BL*OD))
00140 PRINT #1,C$;",","H;",";M;",";T;",";OD;",";BM
00150 PRINT C$;H;":";M;T;OD;BM:CLOSE
00160 PRINT "MORE DATA (Y) OR QUIT (Q) OR GROWTH CALC. (G)  ": INPUT D$
00170 IF D$="G" OR D$="g" THEN 200
00180 IF D$="Y" OR D$="y" THEN 110
00190 IF D$="Q" OR D$="q" THEN END
00200 INPUT "FOR CULTURE DESIGNATED WITH WHICH LETTER  ";C$
00210 OPEN "I",1,"DATA"
00220 X1=0:X2=0:Y1=0:Y2=0:XY=0:N=0
00230 IF EOF(1) THEN 320
00240 INPUT #1,G$,H,M,T,OD,BM
00250 IF C$<>G$ THEN 230
00260 N=N+1:X1=X1+T:X2=X2+T*T:LBM=LOG(BM)
00270 Y1=Y1+LBM:Y2=Y2+LBM*LBM:XY=XY+T*LBM
00280 PRINT N;G$;H;":";M;T;OD;BM
00290 IF N<>1 THEN 230
00300 T0=X1:BI=BM
00310 GOTO 230
00320 PRINT:PRINT N;X1;X2;Y1;Y2;XY:IF N<4 THEN END
00330 DEN=N*X2-X1*X1:Y0=(Y1*X2-XY*X1)/DEN
00340 SLOPE=(N*XY-X1*Y1)/DEN:PRINT Y0,SLOPE:PRINT
00350 T2=.69315/SLOPE:MU=SLOPE*60:DT=2.3036/SLOPE
00360 PRINT "SPECIFIC GROWTH RATE =";MU;"/H"
00370 PRINT "DOUBLING TIME       = ";T2;"MIN"
00380 PRINT "D-TIME              = ";DT;"MIN"
00390 ES2=(Y2-Y0*Y1-SLOPE*XY)*N*N/((N-2)*DEN)
```

(Continued next page)

Table 4. BASIC computer program for determination of growth rate (*continued*)

```
00400 CV=SQR(ES2)/SLOPE
00410 PRINT "CV                    = ";CV"
00420 BE=EXP(Y0+SLOPE*T0):BF=EXP(Y0+SLOPE*T)
00430 PRINT:PRINT "                    BIOMASS OBSERVED     BIOMASS EXPECTED"
00440 PRINT "FIRST POINT          ";BI,"        ";BE
00450 PRINT "LAST POINT           ";BM,"        ";BF
00460 PRINT "MORE DATA (Y) OR QUIT (Q) OR GROWTH CALC. (G) OR INTERPOLATE (I)
": INPUT D$
00470 IF D$="Y" OR D$="y" THEN 110
00480 IF D$="Q" OR D$="q" THEN END
00490 IF D$="G" OR D$="g" THEN 200
00500 IF D$="I" OR D$="i" THEN PRINT
00510 INPUT "TIME IN HOURS, MIN "; H,M
00520 TX=H*60+M:BX=EXP(Y0+SLOPE*TX):PRINT "AT TIME ";TX"THE BIOMASS IS";BX
00530 GOTO 460
```

puter lists the data for that culture and then continues to calculate the specific growth rate, the doubling time, and the coefficient of variation of the estimates. The computer then requests whether it should add more data, quit, compute growth output, or extrapolate or interpolate the biomass to some other time. This is a feature that allows the experimenter to estimate when another phase of the experimental program should be initiated such as diluting the culture or adding a drug. If a desired time is not keyed in the computer, but instead <**ENTER**> is struck, it automatically also makes the same comparison for the first datum pair in the series. The computer gives the best estimate of the culture biomass at the time of the last datum point and matches that with the culture biomass actually observed at that time. Obviously there should be close agreement between the calculated and observed values. This is one way to assess whether the culture was actually growing exponentially. A second test for exponential growth is provided because the CV of the logarithmic slope is part of the output. The CV will be quite small if growth is balanced, enough data are collected, and experimental errors are small. If the CV is small, confidence is instilled in the growth rate estimates. If the CV is large, it would discourage confidence but should cause the experimenter to inspect the datum points to try to detect a typographical error. If one is found, the data file can be corrected or a datum pair can be deleted. Subsequent recomputation by the computer would then yield valid growth estimates.

There are many modifications that the user may wish to make. Almost certainly, changes will be needed for the output format and printing to output hard copy. It is suggested, however, that one test the example of Table 4 and then a sample of one's own experimental data before modifying the program.

11.7.9. Nonexponential Growth

Except under exponential batch culture or continuous culture, growth does not stay balanced (if this is ever achieved) for very long. Even under continuous-culture conditions there can be long-term changes that cause deviations from a constant state of balanced growth. So in all cases, the assumption of exponential growth is a limited one from the point of view of the biological growth process, independent of the difficulties and imprecision of making measurements. It becomes necessary, therefore, to limit the range of measurement. For batch cultures, this means excluding low- and high-density data or conducting the experimental measurements only over a limited range of growth density. Both of these approaches are attempts to obtain balanced growth by restricting the part of the total growth process considered. All such choices and judgments are difficult, and they are impossible to justify from a statistical point of view. It is possible to refine growth measurements to the point at which the constancy of growth can be assessed. This has been attempted by using a computer-linked double-beam spectrometer and a long-path aerating cuvette to achieve accuracy and high sensitivity (45). With this equipment and under optimum conditions, the measurements are so numerous and accurate that the standard deviation is less than 1% of the specific growth rate in a measurement period of only 200 s.

11.8. REFERENCES

11.8.1. General References

1. **Dawson, P. S. S. (ed.).** 1974. *Microbial Growth.* Halsted Press, New York.
   A collection of papers that earlier dominated the study of physiology of bacterial growth.
2. **Gall, L. S., and W. A. Curby.** 1979. *Instrumental Systems for Microbiological Analysis of Body Fluids.* CRC Press, Inc., Boca Raton, Fla.
3. **Kavenagh, F. (ed.).** 1963. *Analytical Microbiology,* vol. 1. Academic Press, Inc., New York.

3a. **Kavenagh, F. (ed.).** 1972. *Analytical Microbiology,* vol. 2. Academic Press, Inc., New York.

4. **Meadows, P., and S. J. Pirt (ed.).** 1969. Microbial growth. *Symp. Soc. Gen. Microbiol.* **19:**1–450.
5. **Meynell, G. G., and E. Meynell.** 1970. *Theory and Practice in Experimental Bacteriology,* 2nd ed. Cambridge Press, Cambridge.
   Excellent, critical discussion of techniques and methods.
6. **Miller, J. H.** 1972. *Experiments in Molecular Genetics.* Cold Spring Harbor Laboratory, Cold Spring Harbor, N.Y.
7. **Pirt, S. J.** 1975. *Principles of Microbe and Cell Cultivation.* Blackwell Scientific Publications, Oxford.

11.8.2. Specific References

8. **Amann, R. I., B. J. Binder, R. J. Olson, S. W. Chisholm, R. Devereux, and D. A. Stahl.** 1990. Combination of 16S rRNA-targeted oligonucleotide probes with flow cytometry for analyzing mixed microbial populations. *Appl. Environ. Microbiol.* **56:**1919–1925.
9. **Baldwin, W. W., and A. L. Koch.** Unpublished data.
10. **Black, S. H., and P. Gerhardt.** 1962. Permeability of bacterial spores. IV. Water content, uptake, and distribution. *J. Bacteriol.* **83:**960–987.
11. **Bridges, B. A.** 1976. Survival of bacteria following exposure to ultraviolet and ionizing radiation. *Symp. Soc. Gen. Microbiol.* **26:**183–208.
12. **Brock, T. D.** 1967. Bacterial growth in the sea; direct analysis by thymidine autoradiography. *Science* **155:**81–83.
13. **Brock, T. D.** 1971. Microbial growth rates in nature. *Bacteriol. Rev.* **35:**39–58.
14. **Buchanan, R. E.** 1918. Life phases in a bacterial culture. *J. Infect. Dis.* **23:**109–125.
    This classic has been reprinted in reference 1.
15. **Button, D. K., and B.R. Robertson.** 1989. Kinetics of bacterial processes in natural aquatic systems based on biomass determined by high-resolution flow cytometry. *Cytometry* **10:**558–563.
16. **Calam, C. T.** 1969. The evaluation of mycelial growth, p. 567–591. *In* J. R. Norris and D. W. Ribbons (ed.), *Methods in Microbiology,* vol. 1. Academic Press, Inc., New York.
17. **Campbell, A.** 1957. Synchronization of cell division. *Bacteriol. Rev.* **21:**263–272.
18. **Capaldo, F. N., and S. D. Barbour.** 1975. The role of the *rec* genes in the viability of *Escherichia coli* K12, p. 405–418. *In* P. C. Hanawalt and R. B. Setlow (ed.), *Molecular Mechanisms for Repair of DNA,* part A. Plenum Publishing Corp., New York.
19. **Cato, E. P., C. S. Cufflins, L. V. Holdeman, J. L. Johnson, W. E. C. Moore, R. M. Smibert, and K. D. S. Smith.** 1970. *Outline of Culture Methods in Anaerobic Bacteriology.* Virginia Polytechnic Institute and State University, Blacksburg.
20. **Clifton, C. E.** 1966. Aging of *Escherichia coli. J. Bacteriol.* **92:**905–912.
21. **Dicker, D. T., and M. L. Higgins.** 1987. Cell cycle changes in the buoyant density of exponential-phase cells of *Streptococcus faecium. J. Bacteriol.* **169:**1200–1204.
22. **Drake, J. F., and H. M. Tsuchiya.** 1973. Differential counting in mixed cultures with Coulter counters. *Appl. Microbiol.* **26:**9–13.

22a. **Edwards, C., J. Porter, J. R. Saunders, J. Diaper, J. A. W. Morgan, and R. W. Pickup.** 1992. Flow cytometry and microbiology. *SGM Q.* **19:**105–108.

23. **Finney, D. J.** 1964. *Statistical Method in Biological Assay,* 2nd ed., p. 570–586. Hafner Publishing Co., New York.
24. **Gray, T. R. G.** 1976. Survival of vegetative microbes in soil. *Symp. Soc. Gen. Microbiol.* **26:**327–364.
25. **Guerrero, R., C. Pedrós-Alió, T. M. Schmidt, and J. Mas.** 1985. A survey of buoyant density of microoganisms in pure cultures and natural samples. *Microbiologia* **1:**53–65.
26. **Gunther, H. H., and F. Berhgter.** 1971. Bestimmung der Trockenmasse von Zellsuspensionen durch Extinktionsmessungen. *Z. Allg. Mikrobiol.* **11:**191–197.
27. **Halvorson, H. O., and N. R. Ziegler.** 1933. Application of statistics to problems in bacteriology. I. A means of determining bacterial population by the dilution method. *J. Bacteriol.* **25:**101–121.
28. **Heden, C.-G., and T. Illeni.** 1975. *Automation in Microbiology and Immunology.* John Wiley & Sons, Inc., New York.
29. **Holdeman, L. V., E. P. Cato, and W. E. C. Moore (ed.).** 1977. *Anaerobe Laboratory Manual,* 4th ed. Virginia Polytechnic Institute and State University, Blacksburg.
30. **Hungate, R. E.** 1971. A roll tube method for cultivation of strict anaerobes, p. 117–132. *In* J. R. Norris and D. W. Ribbons (ed.), *Methods in Microbiology,* vol. 3a. Academic Press, Inc., New York.
31. **Kavenagh, F.** 1972. Photometric assaying, p. 43–121. *In* F. Kavenagh (ed.), *Analytical Microbiology.* Academic Press, Inc., New York.
32. **Kerker, M., D. D. Coke, H. Chew, and P. J. McNulty.** 1978. Light scattering by structural spheres. *J. Opt. Soc. Am.* **68:**592–601.
33. **Koch, A. L.** 1961. Some calculations on the turbidity of mitochondria and bacteria. *Biochim. Biophys. Acta* **51:**429–441.
34. **Koch, A. L.** 1968. Theory of angular dependence of light scattered by bacteria and similar-sized biological objects. *J. Theor. Biol.* **18:**133–156.
35. **Koch, A. L.** 1970. Turbidity measurements of bacterial cultures in some available commercial instruments. *Anal. Biochem.* **38:**252–259.
36. **Koch, A. L.** 1971. The adaptive responses of *Escherichia coli* to a feast and famine existence. *Adv. Microb. Physiol.* **6:**147–217.
37. **Koch, A. L.** 1975. Lag in adaptation to lactose as a probe to the timing of permease incorporation into the cell membrane. *J. Bacteriol.* **124:**435–444.
38. **Koch, A. L.** 1975. The kinetics of mycelial growth. *J. Gen. Microbiol.* **89:**209–216.
39. **Koch, A. L.** 1976. How bacteria face depression, recession, and derepression. *Perspect. Biol. Med.* **20:**44–63.
40. **Koch, A. L., and E. Ehrenfeld.** 1968. The size and shape of bacteria by light scattering measurements. *Biochim. Biophys. Acta* **165:**262–273.
41. **Koch, A. L., and G. H. Gross.** 1979. Growth conditions and rifampin susceptibility. *Antimicrob. Agents Chemother.* **15:**220–228.
42. **Kubitschek, H. E.** 1964. Apertures for Coulter counters. *Rev. Sci. Instrum.* **35:**1598–1599.
43. **Kubitschek, H. E.** 1969. Counting and sizing microorganisms with the Coulter counter, p. 593–610. *In* J. R. Norris and D. W. Ribbons (ed.), *Methods in Microbiology,* vol. 1. Academic Press, Inc., New York.
44. **Kubitschek, H. E.** 1987. Buoyant density variation during the cell cycle in microorganisms. *Crit. Rev. Microbiol.* **14:**73–97.
45. **Maaløe, O., and N. O. Kjeldgaard.** 1966. *Control of Macromolecular Synthesis.* W. A. Benjamin, Inc., New York.
46. **Mallette, M. F.** 1969. Evaluation of growth by physical and chemical means, p. 521–566. *In* J. R. Norris and D. W. Ribbons (ed.), *Methods in Microbiology,* vol. 1. Academic Press, Inc., New York.

47. **Marquis, R. E., and P. Gerhardt.** 1959. Polymerized organic salts of sulfonic acid used as dispersing agents in microbiology. *Appl. Microbiol.* **7:**105–108.
48. **Marshall, R. T.** 1993. Media, p. 89–101. *In* R. T. Marshall (ed.), *Standard Methods for the Examination of Dairy Products,* 16th ed. American Public Health Association, Washington, D. C.
49. **Mitruka, B. M.** 1976. *Methods of Detection and Identification of Bacteria.* CRC Press, Inc., Cleveland.
50. **Nir, R., Y. Yisraeli, R. Lamed, and E. Sajar.** 1990. Flow cytometry sorting of viable bacteria and yeasts according to β-galactosidase activity. *Appl. Environ. Microbiol.* **56:**3861–3866.
51. **Norris, K. P., and E. O. Powell.** 1961. Improvements in determining total counts of bacteria. *J. R. Micros. Soc.* **80:**107–119.
52. **Novick, A., and L. Szilard.** 1951. Experiments with the chemostat on the spontaneous mutations of bacteria. *Proc. Natl. Acad. Sci. USA* **36:**708–719.
53. **Perfilèv, B. V., and D. R. Gabe.** 1969. *Capillary Methods of Investigation of Microorganisms.* University of Toronto Press, Toronto.
54. **Postgate, J. R.** 1967. Viability measurements and the survival of microbes under minimal stress. *Adv. Microb. Physiol.* **1:**2–23.
55. **Postgate, J. R.** 1969. Viable counts and viability, p. 611–628. *In* J. R. Norris and D. W. Ribbons (ed.), *Methods in Microbiology,* vol. 1. Academic Press, Inc., New York.
56. **Ryan, F. J., G. W. Beadle, and E. L. Tatum.** 1943. The tube method of measuring the growth rate of neurospora. *Am. J. Bot.* **30:**784–799.
57. **Schmidt, E. L., R. O. Bantkole, and B. B. Bohlool.** 1968. Fluorescent-antibody approach to study of rhizobia in soil. *J. Bacteriol.* **95:**1987–1992.
58. **Shapiro, H. M.** 1988. *Practical Flow Cytometry,* 2nd ed. Alan R. Liss, New York.
59. **Shehata, T. E., and A. G. Marr.** 1971. Effect of nutrient concentration on the growth of *Escherichia coli. J. Bacteriol.* **107:**210–216.
60. **Shuler, M. L., R. Aris, and H. M. Tsuchiya.** 1972. Hydrodynamic focusing and electronic cell-sizing techniques. *Appl. Microbiol.* **24:**384–388.
61. **Sokal, R. R., and F. J. Rohlf.** 1969. *Biometry.* W. H. Freeman & Co., San Francisco.
62. **Spielman, L., and S. L. Goren.** 1968. Improving resolution in Coulter counting by hydrodynamic focusing. *J. Colloid Interface Sci.* **26:**175–182.
63. **Steen, H. B.** 1990. Flow cytometric studies of microorganisms, p. 605–622. *In* M. R. Melamed, T. Linmo, and M. L. Mendelsohn (ed.), *Flow Cytometry and Sorting,* 2nd ed. Wiley-Liss, New York.
64. **Stock, J. B., B. Rauch, and S. Roseman.** 1977. Periplasmic spaces in *Salmonella typhimurium* and *Escherichia coli. J. Biol. Chem.* **252:**7850–7861.
65. **Taney, T. A., J. R. Gerke, and J. F. Pagano.** 1963. Automation of microbiological assays, p. 219–247. *In* F. Kavenagh (ed.), *Analytical Microbiology.* Academic Press, Inc., New York.
66. **Tisa, L. S., T. Koshikawa, and P. Gerhardt.** 1982. Wet and dry bacterial spore densities determined by buoyant sedimentation. *Appl. Environ. Microbiol.* **43:**1307–1310.
67. **Troller, J. A., and J. H. B. Christian.** 1978. *Water Activity and Food.* Academic Press, Inc., New York.
68. **Vollenweider, R. A.** 1974. *A Manual on Methods for Measuring Primary Production in Aquatic Environments.* IBP handbook no. 12. Blackwell Scientific Publications, Oxford.
69. **Wyatt, P. J.** 1973. Differential light scattering techniques for microbiology, p. 183–263. *In* J. R. Norris and D. W. Ribbons (ed.), *Methods in Microbiology,* vol. 8. Academic Press, Inc., New York.
70. **Wyatt Technology.** 1988. *Literature on Model F and Model B Dawn Instruments.* Wyatt Technology, Santa Barbara.
71. **Zimmerman, U., J. Schultz, and G. Pilwat.** 1973. Transcellular ion flow in *Escherichia coli* B and electrical sizing of bacterias. *Biophys. J.* **13:**1005–1013.

Chapter 12

Culture Preservation

ROBERT L. GHERNA

Most bacteriology laboratories maintain stock cultures for educational, research, bioassay, industrial, or other purposes. The proper preservation of cultures is extremely important and should be (but often is not) given the same attention as the standardization of equipment and the selection of chemical compounds.

Numerous programs have been hampered by the loss or variation of a stock culture as a result of improper preservation.

The primary aim of culture preservation is to maintain the organism alive, uncontaminated, and without variation or mutation, that is, to preserve the culture in a condition that is as close as possible to the original isolate. Many methods have been used to preserve bacteria, but not all species respond in a similar manner to a given method. In fact, some strains of the same species give variable results with the same procedure. It should be emphasized, however, that the success or failure of any preservation method also depends on the use of the proper medium and cultivation procedure and on the age of the culture at the time of preservation. This is particularly true when working with bacteria that contain plasmids or recombinant DNA, or that exhibit growth phases such as morphogenesis or spore formation. The availability of equipment, storage space, and skilled labor often dictates the method employed.

Record keeping is an important and often neglected aspect of stock culture maintenance. Too often, isolates are not given strain designations or, if they are, the designations are not documented in strain data files. A strain designation (e.g., by number or letter) refers to a specific strain with distinct phenotypic characteristics which should be recorded. Culture collections, for example, use an identifying acronym and a unique numbering system (such as ATCC 12301, NCTC 34567, and NRRL 23456) to identify their cultures. Another method is to use the initials of the institution followed by the day, month, year, and isolate number (such as UCLA 090892:2).

It is also extremely important to obtain as much documentation as possible on the history of the strain. Data such as the identity of the culture, the individual who isolated it, the source, geographical information, and the date of isolation are important especially if the organism or its utility will be the subject of a patent application. If the culture is received from another investigator, the documentation must also include the name of the investigator, history if available, date of acquisition, and notation on whether a new strain designation is being used since the same number may be assigned to other strains by different workers. Some strain variations that have been attributed to inadequate preservation conditions have turned out to be due to poor strain documentation; e.g., the wrong number was noted on the data card or on the containers.

12.1. SHORT-TERM METHODS

12.1.1. Subculturing

The traditional method of preserving bacterial cultures is by periodic serial transfer to fresh medium. The interval between such transfers varies with the organism, the medium used, and the external conditions. Some bacteria must be transferred every other day, whereas others need be transferred only after several weeks or months. The major disadvantages of the serial-transfer technique are the risks of contamination, transposition of strain numbers or designations (mislabeling), selection of variants or mutants, and possible loss of culture, as well as the required storage space. Three conditions must be determined when using this method for the preservation of cultures: suitable maintenance medium, ideal storage temperature, and frequency between transfers. Subculturing is not recommended as a long-term preservation method.

12.1.1.1. Maintenance medium

Minimal media are preferred for subculturing because they lower the metabolic rate of the organism and thus prolong the period between transfers. However, some bacteria require a complex medium (21) for growth, or their retention of specific physiological properties necessitates the presence of complex compounds in the medium. When a complex medium is used, more frequent transfers may be necessary as a result of accelerated growth or metabolite accumulation. A variety of bacteria, including members of the *Archaeobacteria* (17, 31, 32), have been preserved by subculturing in biphasic (liquid-solid) medium (Chapter 9.3).

12.1.1.2. Storage

Storage in a refrigerator is the preferred method for subcultures; however, if refrigeration presents a problem, the simplest method is storage at room temperature in a test tube rack or in a specially constructed storage box containing shelves with holes for test tubes. Cultures stored in this fashion require constant care since they tend to dry quickly and, unless the laboratory has a controlled environment, are subject to temperature fluctuations.

To minimize the dehydration, use screw caps with rubber liners, wrap the tube tops in Parafilm (VWR Scientific, Inc.; see section 12.5 for addresses of suppliers), or place the tubes in a plastic bag. To reduce the metabolic rate of the organism, place the culture in a refrigerator at about 5 to 8°C. Most bacteria can be kept for 3 to 5 months between transfers if these precautions are used.

12.1.1.3. Transfer schedule

Determine the subculturing interval by experience. Keep subculturing to a minimum to avoid the selection of variants. Maintain duplicate tubes as a precaution against loss. Examine the cultures for purity after each transfer, and perform an abbreviated characterization check periodically to monitor any changes in phenotypic characteristics. *Do not select single colonies in transferring cultures,* since the chances of selecting a mutant are greater when this technique is used.

12.1.2. Immersing in Oil

Many bacterial species can be successfully preserved for months or years by the relatively simple and cheap method of immersion in sterile medicinal-grade mineral oil. Paraffin oil of specific gravity 0.865 to 0.890 is also satisfactory.

Sterilize 5 ml of oil per Pyrex test tube (18 by 150 mm), sealed with a metal or other heat-resistant closure, by dry heat at 180°C for 2 h; autoclaving is not recommended (36). Allow sufficient time for the hot-air oven and the contents of the test tubes to reach sterilization temperature before beginning the timing of the cycle. Use commercially available strips (section 12.5.1) impregnated with *Bacillus subtilis* spores to check the efficacy of the sterilization equipment and cycle. Do not exceed the recommended sterilization temperature, since the oil will smoke and may ignite. Contamination of cultures when this method is used is often due to improperly sterilized mineral oil. Each lot of sterilized oil must be tested for sterility by streaking on an appropriate medium such as nutrient agar or Trypticase soy agar.

Grow the culture in the appropriate medium as an agar slant or stab or as a broth culture. After growth reaches late logarithmic phase or sporulation occurs, add the sterile mineral oil aseptically to a depth of at least 2 cm (a slant must be entirely covered) to prevent dehydration and to reduce metabolic activity and growth of the culture (13, 34).

Store the oil-covered culture upright in a refrigerator. Perform viability tests periodically by removing a loopful of growth and touching it to a sterile filter paper strip to remove the oil. Streak the growth on agar plates containing a suitable growth medium such as nutrient agar or Trypticase soy agar. Incubate the plates at the appropriate growth temperature for the strain, and assess the resultant growth. Confluent growth represents good survival, but plates which show few colonies indicate a low viability, and new stocks must be prepared.

Use an inoculating needle to transfer the culture under oil to fresh medium, and overlay the subsequent growth with sterile oil. Take care to cool the needle after flaming, since otherwise the oil will splatter and contaminate the surrounding area and personnel. Keep the original culture for at least several weeks to enable the recovery of the culture in the event that the subculture is contaminated or shows aberrant characteristics. The disadvantages of this method are the same as those of ordinary subculturing (section 12.1.1). In addition, the oil is messy.

12.1.3. Ordinary Freezing

The preservation of bacteria in the freezing compartment of a refrigerator or in an ordinary freezer with a temperature range of 0 to −20°C produces variable results, and its success depends on the bacterial species. Some bacteria can be kept by ordinary freezing for 6 months to 2 years. In general, ordinary freezing is not recommended for preservation because the freezing process damages the cells.

When the temperature of a cell suspension drops below 0°C, the extracellular liquid water begins to freeze and form external ice crystals; the actual temperature at which this occurs depends on the supercooling properties of the suspension and can be depressed by the addition of a cryoprotective agent such as glycerol or dimethyl sulfoxide (DMSO) to the cell suspension. As ice crystals continue to form in the menstruum, the solute concentration increases and the internal cellular water begins to migrate out of the cell because of the difference in osmotic pressure. Removal of too much of the internal water results in cell damage. Eventually, all of the free water crystallizes as pure ice, leaving only the concentrated solute and its water of hydration. The concentrated solute eventually also freezes at what is called the eutectic temperature. Too rapid a drop in temperature below 0°C often results in the formation of internal ice crystals, which damages the cells. A balance between damage caused by solute concentration and internal ice formation is achieved by controlling the rate of cooling. A rate of 1 to 10°C/min is satisfactory for all bacteria. *CAUTION:* Most modern refrigerators that have a freezer compartment in the range of 0 to −20°C are frost free. The frost-free condition is achieved by an alternating warming and cooling cycle, which can result in further damage to cells with a concomitant loss of viability.

In addition, the concentration of solutes during freezing can alter medium components by the formation of complexes or the lowering of pH in poorly buffered media. The eutectic temperature of a solution will vary depending on its composition; for example, a sodium chloride solution has a eutectic temperature of −21.8°C (5, 24).

12.1.4. Deep Freezing

The deep-freezing preservation of bacteria at −70°C in a mechanical freezer is superior to ordinary freezing at −20°C. The deep-freezing procedure involves the freezing of a bacteria suspension in the presence of a cryoprotective agent such as glycerol or DMSO to prevent cellular damage during the freezing process. Many bacterial species have been successfully preserved for several years by using this method (Table 1).

Grow the cells in an appropriate medium. Harvest the bacteria as described in section 12.2.2 (culture preparation). Prepare the cell suspensions in broth containing 30 to 50% (vol/vol) glycerol. Dispense the cell into vials as described below (section 12.2.2), and place the vials in a mechanical freezer set at −70°C. Recover the cells by rapidly thawing the frozen cell suspensions in a 37°C water bath.

Freezing at −70°C in a mechanical freezer with glass beads has also been used to preserve a wide variety of bacteria with 15% glycerol as the cryoprotectant (10). Prepare the glass beads as described in section 12.1.5. Place approximately 30 beads in 2-ml glass screw-cap vials (Wheaton no. 224841), and sterilize for 15 min at 121°C. The vials must be labeled with ink that will not come off easily at ultralow temperatures. Prepare a cell suspension containing approximately 10^8 cells per ml in sterile 15% glycerol, and dispense 0.5 ml into each vial. Gently shake the vial to ensure that all of the beads are wetted with the bacterial suspension, and then remove the excess from the vial with a sterile Pasteur pipette. Place the labeled vials in a storage box or tray and then into a mechanical freezer at −70°C. Recover the bacteria by aseptically removing a bead from the vial with a sterile spatula or forceps and placing it in a suitable broth medium. *CAUTION:* Do not allow the contents of the vial to thaw; withdraw the vial from the freezer, and place it in a box containing crushed dry ice or in a frozen paraffin block containing holes for the vials.

12.1.5. Drying

Most cultures die if left to dry in laboratory media. However, some cultures, especially bacterial spores,

Table 1. Expected shelf life of 33 representative bacterial genera preserved by various methods

Genus	Shelf life of bacteria preserved by:				
	Subculturing[a] (mo)	Immersing in oil (yrs)	Deep freezing (yrs)	Freeze-drying (yrs)	Ultra freezing/ liquid nitrogen (yrs)
Acetobacter	1–2	1	1–3	>40	>40
Achromobacter	1	1–2	1–3	>40	>40
Acinetobacter	1 wk			>40	>40
Actinobacillus	1 wk	2–3		>40	>40
Actinomyces	1		2–3	>40	>40
Agrobacterium	1–2	1–2		>40	>40
Arthrobacter	1–2		1–2	>40	>40
Bacillus	2–12	1	2–3	>40	>40
Bacteroides	1 wk		1	>40	>40
Bifidobacterium	1 wk			>40	>40
Chromatium	1			>5	>40
Clostridium	6–12	1–2	2–3	>40	>40
Corynebacterium	1–2	1	1–2	>40	>40
Enterobacter	1–4	1–2		>40	>40
Erwinia	1–4	1–2		>40	>40
Escherichia	1–4	1–2		>40	>40
Flavobacterium	1	2		>40	>40
Gluconobacter	1			>40	>40
Haemophilus	1 wk	1 mo (37°C)		>40	>40
Klebsiella	1–4	1	1–2	>40	>40
Lactobacillus	1 wk			>40	>40
Methanobacterium[b]	1			>5	>40
Methanomonas[b]	1			>5	>40
Micromonospora	1		1	>40	>40
Neisseria	1			>40	>40
Nocardia	1–4	1	1–2	>30	>40
Proteus	1–2	1	1–2	>40	>40
Pseudomonas	1–3			>40	>40
Spirillum[c]	1 wk				>40
Staphylococcus	1–2		1	>40	>40
Streptococcus	1–2	1		>40	>40
Streptomyces	1–8	1–2	1–3	>40	>40
Xanthomonas			1–2	>40	>40

[a]The transfer schedule depends on the medium used. The times listed are approximations, and species variation occurs within a genus.
[b]Depends on species.
[c]*Spirillum volutans.*

can be preserved for years by being dried on a suitable menstruum (3, 44). Bacterial spores have been shown to survive most preservation procedures including drying. Indeed, spores are resistant mainly because of their dehydrated protoplast, and most spores become more resistant if they are dried further in a dried and more hygroscopic medium (such as soil). It is greatly advantageous to grow spore-forming bacteria under conditions which favor spore formation (Chapter 4.4.9).

12.1.5.1. Paper

A relatively simple and inexpensive method of preserving various bacteria involves drying them on sterile filter paper (Whatman no. 4; VWR Scientific) strips or discs. The technique is ideally suited for the maintenance of quality control cultures. Many discs containing the same culture can be stored in a single test tube or screw-cap vial. A single disc can be removed aseptically with sterile forceps as needed and inoculated into a suitable broth. Members of the *Enterobacteriaceae,* as well as a variety of other bacteria, are successfully preserved for several years in this manner.

The procedure for this method is as follows. Place the sterile filter paper strips or discs into a sterile petri plate, and add a bacterial suspension containing approximately 10^8 cells or more per ml (cell concentrations can be determined by direct plate counts of cell dilutions and correlated with optical densities [OD] or nephelometric readings; a faint turbidity is about 10^7 cells per ml; an OD of 1.0 is about 10^9 cells per ml; and an OD of 0.1 is about 10^8 bacteria per ml) dropwise onto each strip or disc until saturated, and dry in air or vacuum desiccator (dry under vacuum for greater survival). Store the strips or discs sealed in tubes in desiccators or between two layers of sterile clear plastic. Store the desiccators in a refrigerator, because this extends the shelf life of the culture.

12.1.5.2. Gelatin

Many heterotrophic bacteria can be preserved in dried gelatin drops or discs (2, 40). The storage temperature is important for successful preservation. In general, −20°C is superior to refrigerator or room temperature.

Prepare the gelatin cultures as follows. Grow the culture in the appropriate medium, and harvest by aseptic centrifugation. Resuspend the pellet in a small amount of broth, and inoculate a tube containing 2.0 to 5.0 ml of melted sterile nutrient gelatin held at 30°C to yield a density of 10^8 to 10^9 cells per ml. Using a sterile Pasteur pipette or syringe, place a drop of the bacterial suspension on the bottom of a sterile petri plate. Place the petri plate in a desiccator containing phosphorus pentoxide, and evacuate with a vacuum pump. After the gelatin drops have dried, aseptically transfer them to sterile screw-cap tubes and store in a refrigerator or, better, in a freezer. Propagate the cultures by aseptically transferring a gelatin drop into a tube containing a suitable medium.

12.1.5.3. Silica gel

A variety of bacteria have been successfully preserved by drying on sterile silica gel granules (12). The procedure is simple and inexpensive. Sterilize screw-cap test tubes (16 by 125 mm) half filled with silica gel (6- to 12-mesh, grade 40, desiccant activated; W. R. Grace and Co.) in an oven at 180°C for 2 h. Prepare spore or cell suspensions by suspending in 1 to 2 ml of sterile 10% (vol/vol) skim milk (no. 0032; Difco). Add the cell or spore suspension dropwise (0.5 ml) to prechilled silica test tubes (which must be kept in an ice bath to diminish the effect of heat generated by absorption of the culture to the anhydrous granules), and hold at 0°C for approximately 10 min. Place the tubes with cap loosened half a turn in a desiccator containing activated silica gel for 1 week at room temperature or dry at 25°C for 2 days. Seal the tubes, and store in a close container over indicating silica gel (grade 42; 6–16 mesh, "Tel-Tale" brand; Fisher Scientific Co.).

12.1.5.4. Porous glass and porcelain beads

A diverse group of heterotrophic bacteria have been successfully preserved by drying under vacuum on perforated glass beads (30) or porous porcelain beads (18). Only 13 of a total of 202 cultures were nonviable after 21 years of storage on glass beads.

Wash glass or porcelain beads (diameter, 2 to 5 mm) in distilled water containing a mild detergent, and then rinse thoroughly in distilled water. Soak the beads in 2% hydrochloric acid for 2 h to eliminate the alkalinity of the beads; then rinse the beads in distilled water until the pH is neutral, and dry them in an oven. Place the beads in screw-cap tubes, and sterilize in a dry-heat oven at 180°C for 2 h. Tighten the caps securely while hot to prevent entry of moist air into the tubes. Do not use rubber or plastic closures since these will melt in the oven; instead, use metal caps or other suitable closures. Sterilize test tubes half filled with blue indicator silica gel and a wad of glass wool on top of the gel in a dry-heat oven at 180°C for 2 h.

Prepare a cell or spore suspension containing approximately 10^8 cells per ml in sterile 10% (vol/vol in water) skim milk. Aseptically transfer 20 to 40 sterile glass beads into a sterile test tube, and add 1 ml of the cell suspension to the tube. Gently shake the tube to ensure that all of the beads are wetted with the suspension. Remove the excess suspension from the tube with a Pasteur pipette or syringe, and transfer the beads into sterile test tubes containing the indicator silica gel and glass wool plug. Tighten the caps securely to prevent entry of moist air. Store the test tubes at room temperature for 1 week and then in a refrigerator.

12.2. LONG-TERM METHODS

12.2.1. Freeze-Drying (Lyophilization)

Freeze-drying (lyophilization) is one of the most economical and effective methods for long-term preservation of bacteria and other microorganisms. Many physiologically diverse bacterial species and bacteriophages have been successfully preserved by this technique and have remained viable for over 50 years. The method enables large numbers of vials to be produced, and the small size of the vial facilitates storage. Although the freeze-drying procedure is relatively simple, the theoretical aspects are complex. The reader is referred to several excellent reviews for in-depth discussions (15, 16, 28).

Freeze-drying involves the removal of water from frozen bacterial suspensions by sublimation under reduced pressure; that is, the water is evaporated without going through a liquid phase. The dried cells can be stored for long periods if kept away from oxygen, moisture, and light. They can, at any time, be easily rehydrated and restored to their previous state.

Lyophilization can be performed in several ways, for which various types of apparatus have been devised. A simple lyophilization system can be assembled at a reasonable cost. The system consists of a vacuum pump capable of an ultimate pressure of less than 10 μmHg (a pump rated at 35 to 50 liter/min is usually adequate), a small stainless-steel condenser for cooling with dry ice, a thermocouple vacuum gauge to monitor the vacuum system, and heavy-walled vacuum/pressure tubing to connect the components (Fig. 1 illustrates component assembly). Details of other procedures and equipment are given by Haynes et al. (14).

Two of the most common methods used are centrifugal freeze-drying and prefreezing. In centrifugal freeze-drying, the cell suspension is first centrifuged to eliminate frothing due to the removal of dissolved gases. The bacterial suspension is then frozen by the loss of heat during evaporation of the water by vacuum. After primary freeze-drying, the vials are constricted with a narrow flame and then placed on a manifold for secondary freeze-drying. For a detailed description of centrifugal freeze-drying, see Rudge (38) and Lapage et al. (4).

The more frequently used prefreezing method is as follows.

12.2.1.1. Equipment

Freeze-drying equipment varies from a simple and relatively inexpensive condenser-vacuum pump system to a complex commercial freeze-dryer which costs thousands of dollars. Reproducibility and shelf life depend on the system used. However, excellent results can be obtained with a simple system consisting of a high-vacuum pump, a condenser, and a chamber or manifold (Fig. 1 and 2).

12.2.1.2. Vials

Although dimensions and shapes of vials are variable, there are two basic designs used for freeze-drying. When the cells are to be freeze-dried in a chamber (Fig. 1), double vials are recommended (Fig. 3). Such vials are easier to handle, less susceptible to contamination, and

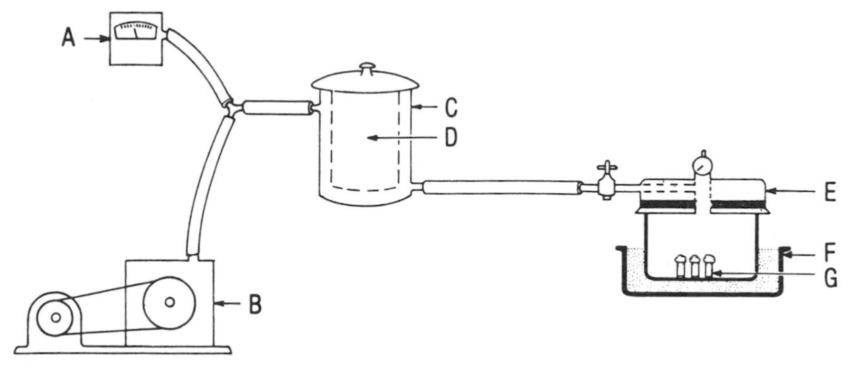

FIG. 1. Double-vial method of freeze-drying. (A) Vacuum gauge: (B) vacuum pump; (C) VirTis condenser; (D) reservoir filled with dry ice and ethylene glycol; (E) Acrylic top plate; (F) stainless-steel pan filled with crushed dry ice and ethylene glycol; (G) specimen vial.

safer for preserving pathogens than are single manifold vials (46), which are dried and sealed while directly attached to a manifold (Fig. 2).

For double vials, prepare the outer vials (14.25 by 85.0 mm; Glass Vials, Inc.) by adding a small quantity of indicator silica gel granules (grade 42, 6-16 mesh, "Tel-Tale" brand) into the vial to cover approximately half of the bottom, and then add a small wad of cotton on top of the granules to cushion the inner vial. Heat the vials in an oven at 100°C for 6 to 8 h, and then cool them in a dry box (< 10% relative humidity). The silica gel should be dark blue after heating and should remain blue during sealing and storage of the cultures.

For the single-vial manifold procedure, two styles of vials are used. A 1.0-ml bulb-shaped vial (outside diameter, 8 mm; Kontes Glass Co.) is used for freezing a shell of the cell suspension, and it is generally used when sucrose is the cryoprotective agent. A tubular type of vial (outside diameter 8 mm; Bellco Glass Co.) is used when a pellet is desired, and this is used when the culture is suspended in skim milk. Rinse the manifold vials by soaking them in distilled water, and autoclave them for 15 min at 121°C to ensure penetration of the water into areas where air is trapped. Dry the vials in an oven and then plug them lightly with non-oil absorbent cotton (USP grade). Sterilize the manifold vials for 30 min at 121°C.

Prepare the inner vials (11.5 by 35 mm, Glass Vials, Inc.) by rinsing them in distilled water and drying them in an oven. Plug the vials with non-oil absorbent cotton (USP grade) which extends about 4 cm into the vial. Allow a sufficient amount of cotton to extend above the vial to facilitate removal for dispensing the bacterial suspension. Sterilize the vials by autoclaving for 30 min at 121°C, with the vials on their sides to allow steam penetration.

Vials must be properly labeled with ink that does not come off easily. A labeling machine (model 135A; Markem Machine Co.) with specially formulated ink (no. 7224; Markem Machine Co.) is recommended. This ink will withstand ultra-low temperatures and brief exposure to solvents such as alcohol. The letters and numbers are assembled on a master plate and imprinted on a pad. The inner vials are rolled over the pad to pick up the print, and then the vials are heated at 160°C in an oven for 20 min to set the ink. After labeling, sterilize the inner vials again by autoclaving for 30 min at 121°C. Manifold vials are labeled in the same manner as the inner vials.

12.2.1.3. Culture preparation

Successful freeze-drying depends on using healthy cells grown under optimum conditions on the medium of choice for each strain, which ensures the retention of the desired features of the bacterium. Grow a sufficient number of cells to provide a suspension of at least 10^8 cells per ml (OD, 0.1). Harvest the culture at the time of

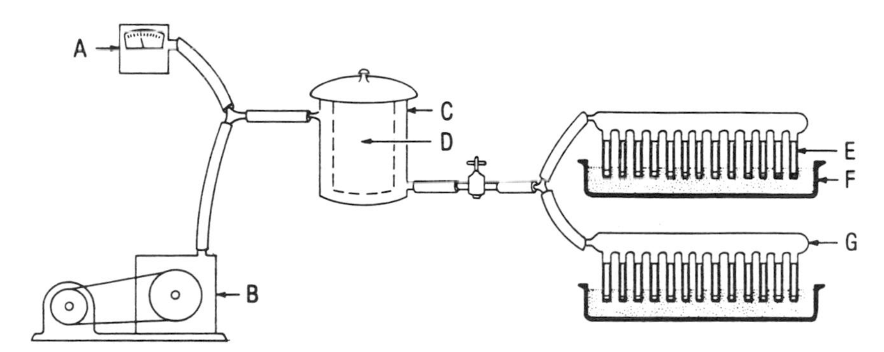

FIG. 2. Manifold method of freeze-drying. (A) Vacuum gauge; (B) vacuum pump; (C) VirTis condenser; (D) reservoir filled with dry ice and ethylene glycol; (E) specimen vial; (F) stainless-steel pan filled with crushed dry ice and ethylene glycol; (G) manifold.

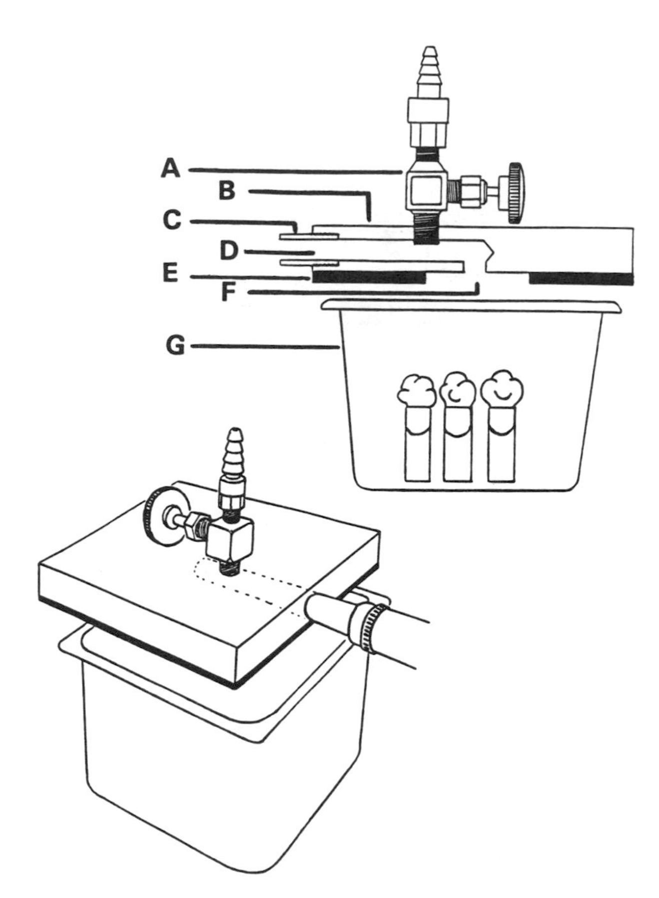

FIG. 3. Sectional view (top) of acrylic top plate, chamber double-vial method. (A) vacuum release valve; (B) Acrylic plastic top (7 in. wide by 7 in. long by 1 in. deep); (C) stainless steel inlet port (outer diameter, 11/16 in.; inner diameter, 1/2 in.); (D) hole (inner diameter, 1/2 in.) drilled in acrylic plastic to accept the stainless steel tube; (E) rubber gasket (7 in. wide by 7 in. long by 1/4 in. deep) with 2 3/8-in. hole drilled in the center; (F) center hole in bottom of acrylic plastic plate (diameter, 0.5 in.); (G) stainless steel specimen pan (5 in. long by 5 in. wide by 4 in. deep). Top view (bottom) of acrylic top plate, showing Tygon pressure tubing attached to the stainless steel tube and the pressure release valve on top.

maximum stability and viability, in the late logarithmic or early stationary phase.

12.2.1.4. Cryoprotective agents (also see section 12.2.2.3)

Cryoprotective agents are chemical compounds such as glycerol or DMSO that are added to a bacterial suspension to help reduce the damage caused during freezing. These compounds must possess certain properties such as being nontoxic, able to penetrate the cell membrane easily, and able to bind either the electrolytes that increase in concentration during freezing or the water molecules to delay freezing (9, 29). Prepare the cells for lyophilization by suspending them in a cryoprotective agent. The American Type Culture Collection (ATCC) has experienced considerable success in long-term preservation of physiologically diverse bacteria by using either 20% (wt/vol) skim milk for the double-vial method or a 24% (wt/vol) sucrose solution diluted equally with growth medium to yield a 12% (wt/vol) sucrose solution (final concentration) for the single-vial manifold procedure. Others have used 10% (wt/vol) dextran, horse serum, inositol, raffinose, trehalose, and other cryoprotective chemical agents (1, 5, 24, 37).

12.2.1.5. Double-vial chamber method

The following procedures are used for the double-vial method. Prepare a 20% (wt/vol) solution of skim milk (no. 0032; Difco), and sterilize it in small volumes (5 ml) at 116°C for 20 min. Avoid overheating, which can cause caramelization of the milk. For cultures grown on agar surfaces, harvest by aseptically washing the growth off with the 20% skim milk solution. Harvest broth cultures by aseptic centrifugation, and suspend the pellet with the sterile skim milk to yield a cell suspension containing at least 10^8 cells per ml. This can be standardized by resuspending the pellet in a small amount of sterile 0.01 M phosphate buffer (pH 7.0) and measuring the OD (an OD of 0.1 equals 10^8 cells per ml). Recentrifuge the cells aseptically, and resuspend the pellet with the same volume of sterile 20% skim milk to yield the desired cell concentration. *CAUTION:* Do not use skim milk for bacteria that are inhibited by milk; instead use sucrose and the manifold single-vial procedure for these bacteria.

As soon as the cell suspensions are prepared, dispense 0.2 ml into each vial, and trim the cotton plugs with scissors to the tops of the vials. The interval between dispensing and the freeze-drying process should be kept to a minimum to avoid possible alteration of the culture.

Figure 1 shows a typical chamber double-vial freeze-drying system. All vacuum lines are made of Tygon tubing (inner diameter, 3/8 in. [0.95 cm]; outer diameter, 7/8 in. [2.22 cm]; Norton Co.). The system is monitored by a thermistor vacuum gauge (model 10-324; The VirTis Co., Inc.) and should be evacuated below 30 μmHg. When the vacuum sensor is placed between the product and the condenser, it will show an increase in pressure as drying occurs. However, when the drying is complete, the pressure should return to below 30 μmHg.

Place the filled "inner" vials upright in a stainless steel pan. For efficient drying, use only a single layer of vials. Freeze the cells by placing the stainless steel pan containing the vials on the bottom of a mechanical freezer maintained at −60 to −70°C for 1 h.

Prepare a moisture trap by placing chunks of dry ice into a cylindrical condenser (The VirTis Co., Inc.), and add ethylene glycol (no. E-180; Fisher Scientific Co.). After the condenser has cooled for approximately 30 min, turn on the vacuum pump. Close the stopcock in the vacuum line between the acrylic plate and the condenser. The plastic plate has an inlet port and a vacuum release valve on the top surface and a rubber gasket on the bottom surface. These plates seal the stainless steel pans to enable a vacuum to be drawn and can be easily made by a machine shop (Fig. 3). Evacuate the system below 30 μmHg.

At the end of 1 h at −60 to −70°C, place the stainless steel pan containing the vials into a larger shallow pan containing sufficient crushed dry ice to cover the bottom and surround the sides of the stainless steel container. Let stand for 3 h. Take care to avoid getting dry ice into the pan containing the vials, since this will interfere with the evacuation of the system. Attach the acrylic plate to the pan immediately, and hold it firmly in place while opening the stopcock in the vacuum line. Monitor the stopcock to ensure that no pinhole leaks occur. Drying proceeds under a vacuum of 20 to 30 μmHg for 18 h. For convenience, a run can be started late in the afternoon and allowed to proceed overnight.

Remove the outer vials containing the silica gel from the oven, place them in a dry box (maintained below 10% relative humidity), and allow them to cool. Close the stopcock between the acrylic plate and the condenser. Connect one end of a rubber tube to the inlet port on the plate, which leads to a column of silica gel located inside the dry box. Open the stopcock on the inlet port to allow air through the silica gel column; the pressure in the stainless-steel pan will reach atmospheric. This procedure minimizes the introduction of moisture into the samples. Transfer the vials to the dry box, and insert each inner vial into a soft-glass outer vial. Tamp a 0.6-cm plug of glass-fiberpaper (no. 1821-915; Whatman, Inc.) above the cotton-plugged inner vial.

Remove the double vials from the dry box, and heat the outer vial just above the glass-fiberpaper with a compressed-air/gas (natural, mixed, propane, butane) torch. Rotate the vial and flame just above the glass-fiberpaper until the glass begins to constrict. Pull the vial slowly with forceps until the constriction is a narrow capillary.

Allow the vials to cool in a dry box, and then attach them to a vacuum manifold (Fig. 2) by using single-hole rubber stoppers which fit the open end of the vials. Evacuate them to less than 50μmHg, and seal them with a double-flame air/gas torch at the capillary constriction.

12.2.1.6. Single-vial manifold method

In the single-vial manifold method, dispense 0.2 ml of the bacterial suspension into each sterile vial. Push the cotton plug about 1.3 cm below the rim of the vial with a sterile probe, and flame the rim of the vial to eliminate protruding cotton fibers that would interfere with the integrity of the vacuum system.

Attach a 2.5-cm piece of sterile nonpowdered amber latex IV tubing (no. 17610-163; Baxter Diagnostics Inc., Scientific Products Division) to the rim of each filled vial by using a tube stretcher. When the tubular-type vials are used, freeze the cell suspension by direct immersion in a dry-ice–ethylene glycol bath and attach the vials to the manifold while they are still immersed in the dry-ice bath. When the bulb-type vials are used, immerse them in the dry-ice–ethylene glycol bath and rotate them to shell-freeze the suspension before placing the vials on the manifold.

Prepare a moisture trap as described for the double-vial freeze-drying method. The vials attached to the manifold remain in the dry-ice–ethylene glycol bath during the freeze-drying process. Dry with a vacuum below 30 μmHg for 18 h. During the freeze-drying cycle, the bath will warm to ambient temperature.

After completion of the drying process, seal the vials below the cotton plug by using a dual-tipped air/gas torch and continuously moving the flame up and down the vials within a 2.5-cm area. Allow the vials to cool.

12.2.1.7. Storage

Freeze-dried cultures, in either double or single vials, are stored at 2 to 8°C. Extended shelf life has been obtained when cultures are stored at −30 or −70°C in a mechanical freezer. Room temperature storage of freeze-dried cultures should be avoided. Since most modern refrigerators have frost-free freezer compartments, which use a heating and cooling cycle, vials must be stored in the refrigerator compartment only. Walk-in coolers or mechanical freezers set for a maximum temperature of 4°C can also be used for storage.

12.2.1.8. Recovery

Open the double vials by heating the tapered end of the outer container vigorously in a Bunsen burner flame, and quickly add a drop of water to crack the hot glass (Fig. 4). Remove the broken tip of the container carefully (preferably in a hood) with a sharp blow of a forceps, and then withdraw the glass-fiber plug and inner vial. Use sterile forceps to gently remove the cotton plug, and rehydrate the culture with 0.3 to 0.4 ml of appropriate broth medium.

Open the single vials by first scoring the vial with the edge of a file approximately 2.5 cm from the tip. Disinfect the vial with a piece of gauze dampened with 70% alcohol (70% alcohol will kill most vegetative cells, but it is not effective against mycobacteria; for additional information on disinfectants the reader is referred to reference 45. Wrap sterile gauze around the vial, and break at the scored area. Flame the opened end gently, prior to rehydration of the culture. Open the vials in a laminar-flow hood.

Rehydrate the culture by adding the broth with a sterile needle, and remove with the same needle. Open vials containing pathogenic bacteria only in a closed safety cabinet. Some investigators use a tungsten needle, heated in an oxygen gas flame to white heat, to poke a small hole in the tip of the ampoule; this allows air to enter the vial more slowly and sterilizes the area around the hole.

Rehydrate a freeze-dried culture immediately after opening by aseptically adding 0.3 to 0.4 ml of suitable sterile broth to the contents of the vial. Mix well so that the pellet dissolves completely. Reuse the syringe to transfer the cell suspension to a test tube containing 5 ml of the rehydrating broth. After thorough mixing, transfer 0.2 ml of the rehydrated culture to an agar slant or a semisolid medium of the same composition.

Pre- and postlyophilization purity checks must always be done. To accomplish this, serially dilute the prelyophilization cell suspension or the reconstituted culture and streak on solid medium. Incubate the tubes and plates at the optimum growth temperature, and subculture onto fresh medium as soon as growth appears and purity is ascertained. Freeze-dried cultures often exhibit a prolonged lag period and should be allowed to incubate for an extended period before the culture is considered dead.

12.2.1.9. Monitoring viability

Although freeze-drying has facilitated the long-term preservation of bacteria, viability checks must be done before and after freeze-drying to determine the effectiveness of the process. In addition, periodic viability

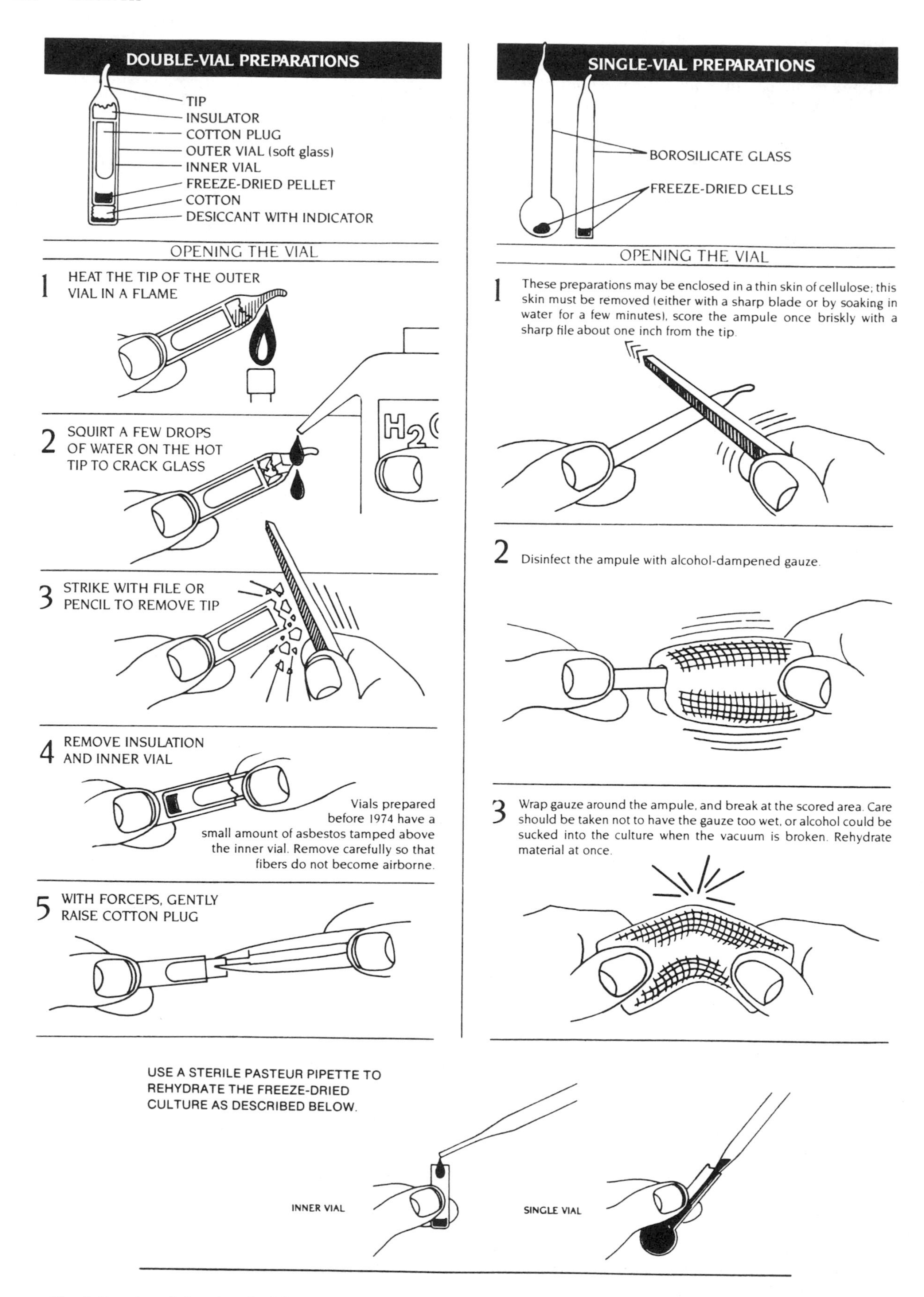

Fig. 4. Opening of chamber double vials and manifold single vials. Reproduced from reference 10a with permission.

tests must be done to ascertain the shelf life of the cultures. Characterization tests should also be performed on cultures that have been lyophilized to determine whether any changes have occurred as a result of the freeze-drying procedures or during storage.

12.2.2. Ultrafreezing

Long-term preservation of bacterial species not amenable to freeze-drying has been achieved through storage in the frozen state at the temperature of liquid nitrogen (−196°C) or above the liquid (vapor phase, −150°C, well-insulated tanks). The ATCC has successfully preserved many fastidious bacteria in liquid nitrogen for over 30 years without the loss of phenotypic properties.

Long-term preservation of bacterial cultures also can be achieved by using ultra-low-temperature mechanical freezers (e.g., from Revco, Inc.) with a temperature of −70°C. This method of preserving cultures is fairly successful for a large variety of bacteria. Precautions must be taken, however, against electrical shutdowns or compressor malfunction. Adequate back-up systems (such as alarms, back-up freezers, or an electrical generator) will help to prevent the loss of a valuable collection. Commercial freezers are now equipped with battery-operated alarm systems as standard equipment. However, additional alarm systems are recommended, such as the sound/off power-temperature monitor (Arthur H. Thomas Co.), which has a temperature range of −75 to +200°C (±0.3°C). The unit is battery powered and can be mounted on a wall. Divided storage of lots in different freezers is an added precaution.

12.2.2.1. Equipment

Liquid nitrogen has been considered an expensive method for the long-term preservation of microorganisms. However, the recent history of the successful storage of physiologically diverse bacteria by this method, along with a decreased need for handling, makes liquid nitrogen feasible to use, especially when one considers the cost of labor.

A variety of liquid-nitrogen refrigerators are now commercially available, with a wide assortment of features and storage capacity (e.g., MVE, Linde, Union Carbide Corp.). The sizes range from 10 to 1,000 liters, permitting storage of 300 to 40,000 vials.

12.2.2.2. Vials

Various kinds of liquid-nitrogen vials are available commercially. Types that have been used successfully are heavy-walled, prescored, and unscored borosilicate cryule vials and also special wide-mouth gold-band cell and tissue culture dryule vials (Wheaton). These vials have a capacity of 1.2 ml and an indentation groove to facilitate opening. The ATCC uses presterilized polypropylene tubes (no. 3-68632 and 3-75353; Nunc) with screw caps and silicone washers in place of the glass vials (*see caution in section 12.2.2.6*).

12.2.2.3. Cryoprotective agents (also see section 12.2.1.4)

Cryoprotective compounds fall into two types: agents such as glycerol and DMSO, which readily pass through the cell membrane and appear to provide both intracellular and extracellular protection against freezing; and agents such as dextran, glucose, lactose, mannitol, polyglycol, polyvinylpyrrolidone, sorbitol, and sucrose, which appear to exert their protective effect external to the cell membrane. The former type has proven to be more effective, and glycerol and DMSO appear to be equally effective in preserving a wide range of bacteria. The exact choice of cryoprotective agent, however, depends on the bacterial species. A tolerance test must be done when freezing new species to ascertain whether the cryoprotective agent is toxic or beneficial.

Glycerol and DMSO are routinely used at concentrations of 10% (vol/vol) and 5% (vol/vol), respectively, in an appropriate growth medium. Glycerol is usually prepared in double strength (20%, vol/vol) and then mixed with an equal amount of the cell suspension. Sterilize the glycerol by autoclaving at 121°C for 15 min, and store in 6-ml volumes at 2 to 8°C. DMSO (reagent grade) is sterilized with 0.22-μm pore-size Teflon (polytetrafluoroethylene [PTFE]) membrane filters (Gelman or Millipore) that have been prewashed with methanol and DMSO, collected in 10–15 ml quantities in sterile test tubes, and then stored in the frozen state at 5°C (DMSO freezes at 18°C) and protected from light. Because of the accumulation of oxidative breakdown products, an opened bottle of DMSO should not be used for more than 1 month.

12.2.2.4. Culture preparation

The physiological condition of the culture plays an important part in the survival of the bacterium to liquid-nitrogen freezing. In general, use actively growing cells at the mid- to late logarithmic phase of growth (41).

Grow the cells in an appropriate medium. For broth cultures, harvest by aseptic centrifugation and resuspend the pellet with sterile fresh medium containing either 10% (vol/vol) glycerol (prepared by adding sterile 20% glycerol to an equal volume of sterile broth) or 5% (vol/vol) DMSO (prepared by adding the appropriate amount of sterile 100% DMSO to the sterile broth). For agar cultures, wash the growth from the agar surface with sterile broth containing the suitable cryoprotective agent. Dispense 0.4 ml of the cell suspension containing at least 10^8 cells per ml into each sterile, prelabeled vial. Plastic presterilized screw-cap vials (Nunc) are recommended for bacteria. Since the screw caps can leak, store the vials in the vapor phase (*see caution in section 12.2.2.6*). For glass vials, precool the vials for a minimum of 30 min at 4°C before sealing. Then heat the vials above the prescored area by rotating in the torch flame for several seconds to remove moisture. This procedure avoids pressure inside the vial, which could result in bubbling during sealing. Hold the vial at an angle to prevent the culture suspension from contacting the hot glass, and partially withdraw the cotton plug, stopping just short of complete removal. Seal the vials, preferably with a semiautomatic sealer equipped with an oxygen gas torch (e.g., from Kahlenberg Globe Equipment Co.) which is designed to pull and seal, thus minimizing the formation of pinhole leaks.

Carefully remove the hot vial with a forceps, and place the vial in a test tube rack immersed in a cool-water bath deep enough to immerse the vial to a depth of 0.6 cm. Avoid getting water on the hot portion of the vial; otherwise, the glass will crack. When the ampoules have cooled, they are ready for freezing. If this process is performed satisfactorily, the temperature of the specimen should not rise above 25°C. If the semiautomatic sealer is unavailable, the vials can be sealed with a gas torch by hand, but caution must be used to prevent the presence of pinhole leaks. To detect the presence of leaks, allow the sealed glass ampoules to stand for approximately 30 min at 5°C in a 0.05% methylene blue solution. Improperly sealed vials are detected by penetration of the dye into the vial.

12.2.2.5. Freezing

Much has been written about the effect on survival exerted by the cooling rate during freezing (5, 6). In general, the best results in the preservation of bacterial cultures have been obtained with slow cooling (e.g., 1°C/min).

The procedures for this method are as follows. Place the filled vials onto aluminum canes (no. A545; Nasco), and insert the canes into boxes (open-end cartons with rectangular holes, 2.5 by 2.5 by 11 in. [6.4 by 6.4 by 27.9 cm]; Murray and Heister, Inc.). Place the boxes into the freezing chamber of a programmable freezer (Linde BF 3-2; Union Carbide Corp.) which has an adjustable cooling rate. Freeze the cells at a controlled rate of 1°C/min to −40°C and then at a more rapid drop of 10°C/min to −90°C. After this temperature is achieved, transfer the vials to a liquid-nitrogen tank and store them immersed in the liquid phase at −196°C or above in the vapor phase at −150°C.

If a programmable freezer is not available, slow cooling can be achieved by either of the following procedures. (i) Place filled glass or plastic vials in a stainless-steel pan on the bottom of a mechanical freezer at −60°C for 1 h, and then plunge them into a liquid-nitrogen bath for 5 min. The rate of cooling to −60°C with this method is approximately 1.5°C/min. (ii) Canale-Parola (7) recommended immersing the sealed vials (placed on a cane) in 95% ethanol contained in a graduated cylinder. The cylinder containing the vials is placed in a mechanical freezer set at −85°C and allowed to reach that temperature before the canes and ampoules are placed in liquid nitrogen.

12.2.2.6. Thawing

It has been reported that rapid thawing of frozen cultures results in the greatest recovery of bacteria (5, 24, 27). This has also been the experience with bacterial cells frozen at the ATCC.

To recover frozen cultures, rapidly thaw them with moderate agitation in a 37°C water bath until all the ice melts. This usually takes about 50 s for glass vials and about 90 s for polypropylene ones. Immediately after thawing, remove the vial from the water bath and wipe with 70% ethanol to disinfect. Open and flame the vial, and aseptically transfer the culture to fresh medium.

CAUTION: Because of its extremely low temperature, liquid nitrogen can be hazardous if not used correctly. Improperly sealed glass vials can explode when retrieved from liquid nitrogen because of the rapid expansion of the liquid nitrogen that enters the vials through microscopic holes. Wear protective gloves and a face shield when handling frozen glass vials. Plastic screw-cap vials that are not properly sealed can fill with liquid nitrogen if stored in the liquid phase. Retrieval of the vials to warmer temperatures can result in spraying of the contents as the liquid nitrogen expands. Cell suspensions in plastic vials should be stored in the vapor phase.

Check the viability of the culture to determine the effectiveness of the procedure for a given species. The expected shelf life of various bacteria is shown in Table 1.

12.3. PRESERVATION OF REPRESENTATIVE GENERA AND SPECIFIC GROUPS

12.3.1. Representative Genera

The reader should consult Table 1 and the various sections of *Bergey's Manual of Systematic Bacteriology* volumes 1 through 4 (22, 39, 43, 48) for information on the maintenance and preservation of representative bacterial genera. Certain groups of bacteria have special requirements, as follows.

12.3.2. Anaerobes

It is essential to maintain strictly anaerobic conditions during growth, harvesting, dispensing, and freezing of anaerobic bacteria. The cryoprotective agents must also be prereduced.

Grow anaerobes such as *Ruminococcus* species in prereduced broth in Hungate test tubes (no. 2047 and 2048; Bellco Glass, Inc.) with rubber stoppers. Harvest the cells in the late logarithmic phase by centrifugation. Conduct all subsequent manipulations under oxygen-free gas flow with a sterile cannula. Remove the supernatant with a sterile Pasteur pipette connected to a filtration flask. Flush all pipettes, syringes, and the freezing or freeze-drying vials with oxygen-free gas before use. Harvest anaerobes grown on slants by washing down the growth with sterile prereduced medium containing the appropriate cryoprotective agent. Pool the pellets or slants to form a cell suspension of about 10^8 cells per ml (OD, 1.0). For freezing the cryoprotectant is glycerol (final concentration, 10% [vol/vol]). For freeze-drying the ATCC uses cryoprotectant reagent 18, which has the following composition.

Trypticase soy broth	1.5 g
Sucrose	10.0 g
Bovine serum albumin fraction V	5.0 g
Distilled water	100.0 ml

Filter sterilize the solution through a 0.22-μm filter.

Flush the freezing or freeze-drying vials with oxygen-free gas, and dispense 0.2 ml of the cell suspension into each vial. Freeze immediately in a dry-ice–ethylene glycol bath prior to freeze-drying, or store in the liquid-nitrogen vapor phase if freeze-drying is delayed. Bacteria that will be frozen in liquid nitrogen are placed immediately in a rack that is partially immersed in liquid nitrogen and then transferred to a liquid-nitrogen tank.

For more details on the cultivation of anaerobic bacteria, see Chapter 6.6 and reference 23, and for their preservation, see references 20, 25, and 26.

12.3.3. Cyanobacteria

The most successful method of preservation of cyanobacteria is freezing and storage in liquid-nitrogen vapor. The cryoprotectant of choice for most strains is DMSO (final concentration, 5% [vol/vol]) in maintenance broth. Some cyanobacteria have been freeze-dried, notably the strains that produce heterocysts and/or akinetes. In these instances, the cells are harvested at the early stationary phase by centrifugation and the pellet is resuspended in sterile cryoprotectant reagent 20 mixed equally with fresh sterile growth broth. The formulation for reagent 20 is as follows.

Bovine serum albumin fraction V	10.0 g
Sucrose.............................	20.0 g
Distilled water	100.0 ml

Filter sterilize the solution through a 0.22-μm filter.

Reagent 20 is used instead of reagent 18 for cyanobacterial cultures because they may be adversely affected by Trypticase soy broth. Dispense 0.2 ml of the cell suspension into sterile glass inner vials (section 12.2.1), and process as described above (section 12.2.1). The freezing is done in a commercial freezer instead of by the chamber method.

12.3.4. Methanogens

The methanogens are a diverse group of the class *Archaeobacteria* that require stringent anaerobic conditions for growth. The cultures can be maintained by periodic subculturing in an appropriate medium and periodic replenishment of the appropriate gaseous atmosphere (17). Freezing and storage at low temperature is the most successful method of preservation of this group of bacteria (17).

Most methanogens grow on an H_2–CO_2 gas mixture, and special heavy glass-walled test tubes or serum bottles must be used to accommodate the pressurized atmosphere of the gas. Hungate test tubes (no. 2047-16125; Bellco Glass, Inc.), Balch serum tubes (no. 2048-18150; Bellco Glass, Inc.), and serum bottles are ideal containers for growing small volumes of the methanogenic bacteria.

A modified heavy-walled bottle with a neck that can be closed with a screw cap has been described by Hippe (17). This bottle facilitates the handling of methanogens because it fits in normal laboratory centrifuges and can be used for both the cultivation and harvesting of the cultures. Grow the cultures to an OD of 0.300 at 600 nm in 20 ml of the appropriate medium under H_2–CO_2 in using the Hippe container. Harvest the cells by centrifugation directly in the unopened bottle; after removal of the closure, insert a sterile cannula and flush the bottle with oxygen-free gas. Resuspend the pellet with 2 ml of sterile fresh culture medium containing 10% (vol/vol) glycerol or 5% (vol/vol) DMSO (final concentrations). This suspending medium must be prepared just prior to use and must be gassed to ensure anaerobiosis. Dispense 0.2 ml of the cell suspension into sterile glass cryules (no. 12523; Wheaton) while flushing the vials with oxygen-free gas. It is important to flush all pipettes and syringes with oxygen-free gas prior to use. *CAUTION:* Use an N_2–CO_2 gas mixture instead of H_2-containing mixtures for flushing the vials. After dispensing, carefully move the gassing cannula toward the top of the vial and continue gassing. Seal the middle of the vial by using an air-gas torch, and then place the vial immediately in a rack immersed (only the bottoms of the vials are immersed) in an ice bath to prevent heating of the suspension from the heated vial. Transfer the vials to aluminum canes (Shur-Bend Manufacturing Co.), and store them in the vapor phase of a liquid-nitrogen container. Other members of the *Archaeobacteria* can also be preserved by similar methods (43).

12.3.5. *Neisseria, Haemophilus, Campylobacter,* and *Helicobacter* Species

Neisseria, Haemophilus, Campylobacter, and *Helicobacter* species have been shown to be sensitive to the preservation process during different stages in the growth phase. Therefore, it is important to harvest these cells at the proper stage in the growth curve before subjecting them to the preservation process. Preservation experiments should be conducted with cells harvested at different stages of growth to obtain optimum conditions. For example, *Campylobacter* cells should be harvested after 48 h of growth on Brucella Albimi broth (MO8600; GIBCO, Grand Island, N.Y.) containing 2% agar under an atmosphere of 5% CO_2, 6% O_2, and 89% N_2. Prepare bacterial suspensions in sterile 10% (vol/vol) glycerol to yield approximately 10^9 cells per ml (OD, 1.0), and dispense into presterilized plastic screw-cap vials (33). Ultrafreeze the bacteria as described in section 12.2.2.

12.3.6. Plasmid-Containing Bacteria

Preservation of organisms containing plasmids has become an important part of biotechnology programs. These organisms are less stable genetically and are more likely to lose their foreign genetic material (19).

Successful preservation of plasmid-containing bacteria depends on the stability of their plasmids (35). Stable *Escherichia coli* plasmids can be freeze-dried or frozen by standard preservation methods with excellent recovery of both the host and plasmids. The recovery of unstable *E. coli* plasmids can be enhanced by growing the cells in the presence of the appropriate antibiotic, both before and after preservation. In general, bacteria containing unstable plasmids should be grown on antibiotic-containing liquid medium at 30°C, frozen with 10% (vol/vol) glycerol, and stored in the vapor over liquid nitrogen.

12.3.7. Sporeformers

Sporeformers have been successfully preserved for years by use of a mixture of sterile, air-dried soil (2, 8, 11, 44, 47). Sterilize 1 g of soil per screw-cap tube (16 by 125 mm) by autoclaving at 121°C for 3 h on three consecutive days. Inoculate 1 ml of spore suspension into each tube, and allow the tubes to stand at room temperature, or preferably under vacuum over Drierite or P_2O_5, with the cap loosened half a turn until visibly dry. Remove the cap aseptically, replace it with a sterile rubber stopper, and store the tube in the refrigerator.

12.4. CULTURE COLLECTIONS AND THEIR FUNCTIONS

Culture collections represent valuable resources of microbial germplasm. They range from those that serve specific research aims to those that house a wide diversity of microorganisms, cell lines, viruses, and genetic material. In general, there are three kinds of culture collections: specialized, reference, and national collections.

12.4.1. Specialized Collections

Specialized collections are usually of a personal nature and are not intended to be permanent. They may be narrow in the variety of microorganisms they hold but extensive in the number of strains of each species. These collections are indispensable as resources in studies of taxonomy, genetics, and strain variabilities. An example of a specialized collection was the bacterial plant pathogen collection of the late Mortimer Starr at the University of California at Davis.

12.4.2. Reference Collections

Reference collections are involved in collaborative efforts of scientists, industry, academe, and government agencies and are concerned mainly with one major group or function. These collections may be located in government agencies, universities, or industries associated with the production of goods, chemicals, and drugs. Examples of these collections are as follows.

The *Bacillus* Genetic Stock Center
Dr. D. H. Dean, Curator
Department of Microbiology
The Ohio State University
484 W. 12th Avenue, Columbus, OH 43210
Telephone: 614-422-5550

The *Escherichia coli* K-12 Stock Culture Collection
Dr. B. Bachmann, Curator
Department of Biology, Yale University
255 omi, P.O. Box 666, New Haven, CT 06511-7444
Telephone: 201-449-9395

Collection of Microorganisms
Dr. C. Kurtzman, Curator
National Center for Agricultural Utilization Research
1815 N. University St., Peoria, Il 61604-3999
Telephone: 309-685-4011

12.4.3. National Collections

The national collections, such as the ATCC in Rockville and the National Collection of Type Cultures in England, are intended to be permanent national resources. They are often called national service collections, and their main function is to acquire, authenticate, preserve, and distribute authentic reference and type cultures. These collections attempt to be comprehensive, they often are broad in the variety of species held, but they may be narrow in the number of strains maintained within a species. Many also serve as national and international patent depositories. Many provide a wide variety of services such as identification of unknown strains, proprietary safe deposit of microorganisms, and specialized training. They also represent an invaluable source of strain information including history, strain data, and special applications. These data bases are being made available through computer networking and provide information to the scientific community through a variety of computer networks. National service collections regularly publish catalogs of their holdings; e.g., see reference 10a.

The policy for acquisition of cultures varies among the service collections. A detailed description of their policies can be found in their respective catalogs. Individuals wishing to deposit strains should write to the collection to determine whether there is an interest in acquiring the strains.

The following is a partial list of national service collections. More comprehensive lists can be found in the *World Directory of Collections of Cultures of Microorganisms* (42) and in a directory of over 300 national culture collections available from the World Data Center for Collections of Cultures of Microorganisms at the Japan Collection of Microorganisms (address below).

American Type Culture Collection, Dr. R. L. Gherna, Head, Department of Bacteriology, 12301 Parklawn Drive, Rockville, MD 20852
Telephone: 301-231-5542
Fax: 301-770-1848

Center for General Microbiological Culture Collection, China Committee for Culture Collections of Microorganisms, Institute of Microbiology, Academia Sinica, Beijing, China

Coleccion Espanola De Cultivos Tipo, Dr. F. Uruburu, Curator, Departamento de Microbiologia, Facultad de Ciencias Biologicas, Universidad de Valencia, E-46100 Burjasot, Spain
Telephone 034-6-3864300

Collection Nationale de Cultures De Microorganismes, Mme Y. Cerisier, Institut Pasteur, 25 rue du Docteur Roux, F-75724 Paris Cedex 15, France
Telephone: 033-1-45688251
Fax: 033-1-45688236

Culture Collection of the Laboratorium Voor Microbiologie en Microbiele Genetica, Dr. K. Kersters, Curator, Faculteit der Wetenschappen, K. L. Ledeganckstraat 35, B-9000 Gent, Belgium
Telephone: 032-91-227821
Fax: 032-91-223149

Czechoslovak Collection of Microorganisms, Dr. I. Sedlácek, Masaryk University, Jostova 10, 662 43 Brno, Czechoslovakia
Telephone: 042-5-23407

DSM-Deutsche Sammlung von Mikro-Organismen und Zellkulturen GmbH, Dr. R. M. Kroppenstedt, Managing Director, Mascheroder Weg 1b, D-3300 Braunschweig, Germany
Telephone: 049-531-61870
Fax: 049-531-618718

Hungarian National Collection of Medical Bacteria, Dr. B. Lanyi, Curator, National Institute of Hygiene, Gyali ut 2-6, H-1097 Budapest, Hungary
Telephone: 036-1-142250

Japan Collection of Microorganisms, Dr. T. Nakase, RIKEN, Wako-shi, Saitama 351-01, Japan
Telephone: 0484-62-1111 ext 5100
Fax: 0484-64-5651

National Bank for Industrial Microorganisms and Cellular Cultures, Dr. I. Tepavitcharova, 25 'Lenin' Bd, B12, 1113 Sofia, Bulgaria
Telephone: 0359-2-720865

National Collection of Agricultural and Industrial Microorganisms, Dr. T. Deak, Department of Microbiology and Biotechnology, University of Horticulture and Food Industry, Somloi ut 14-16, H-1118 Budapest, Hungary
Telephone: 036-1-665411
Fax: 036-1-666220

National Collection of Food Bacteria, Dr. P. Jackman, AFRC Institute of Food Research, Reading Laboratory, Shinfield, Reading RG2 9AT, United Kingdom
Telephone: 044-734-883103
Fax: 044-734-884763

National Collections of Industrial and Marine Bacteria Ltd., Dr. I. Bousfield, 23 St. Machar Drive, Aberdeen AB2 1RY, United Kingdom
Telephone: 044-224-273332
Fax: 044-224-487658

National Collection of Plant Pathogenic Bacteria, Dr. E. Stead, Central Science Laboratory, Hatching Green, Harpenden, Hertfordshire AL5 2BD, United Kingdom
Telephone: 044-5827-5241
Fax: 044-5827-62178

National Collection of Type Cultures, Dr. L. R. Hill, PHLS Central Public Health Laboratory, 61 Colindale Avenue, London NW9 5HT, United Kingdom
Telephone: 044-81-2004400
Fax: 044-81-2007874

Russian Collection of Microorganisms, Dr. L. V. Kalakoutskii, Institute of Biochemistry and Physiology of Microorganisms, Russian Academy of Sciences, Puschino, Moscow Region 142297, Russia
Telephone: 07-095-231-6576

VTT Collection of Industrial Microorganisms, Dr. Maija-Liisa Suihko, VTT Biotechnical Laboratory afrc. frin. Finland, Tietotie, SF-02150 Espoo, Finland
Telephone: 0358-0-4565133
Fax: 0358-0-4552028

12.5. COMMERCIAL SOURCES

AMSCO Scientific, 2424 23rd St., Erie, PA 16514

Baxter Diagnostics Inc., Scientific Products Division, 1430 Waukegan Rd., McGaw Park, IL 60085

Bellco Glass, Inc., 340 Edrudo Rd., P.O. Box "B," Vineland, NJ 08360

Difco Laboratories, P.O. Box 1058A, Detroit, MI 48232

Fisher Scientific Co., 711 Forbes Ave., Pittsburgh, PA 15219

Gelman Sciences, Inc., 600 S. Wagner Rd., Ann Arbor, MI 48106

GIBCO Laboratories, Division of Life Technologies, Inc., 3175 Staley Rd., Grand Island, NY 14072

Glass Vials, Inc., 1352 James St., Baltimore, MD 21223

W. R. Grace and Co., Davison Chemical Division, 10 E Baltimore, Baltimore, MD 21202

Kahllenberg-Globe Equipment Co., Sarasota, FL 33577

Kontes, Spruce St., P. O. Box 739, Vineland, NJ 03860

Markem Machine Co., 150 Congress St., Box 480, Keene, NH 03431

Millipore Filter Corp., 800 Ashby Rd., Bedford, MA 01730

Minnesota Valley Engineering, Inc., New Prague, MN 56071

Murray & Heister, Inc., 10738 Tucker St., Beltsville, MD 20705

Norton Performance Plastics, P.O. Box 3660, Akron, OH 44309

Nunc, Inc., 2000 North Aurora Rd., Naperville, IL 60563

Revco, Inc., 1100 Memorial Drive, West Columbia, SC 29169

Shur-Bend Manufacturing Co., 4612 N. Chatsworth, St. Paul, MN 55126

Arthur H. Thomas Co., P.O. Box 779, Philadelphia, PA 19105

Union Carbide Corp., Linde Division, 270 Park Ave., New York, NY 10017

The VirTis Co., Inc., Gardiner, NY 12525

VWR Scientific, Inc., 3202 Race St., Philadelphia, PA 19104

Whatman, Inc., 9 Bridewell Place, Clifton, NJ 07014

Wheaton, 1501 N. 10th St., Millville, NJ 08332

12.6. REFERENCES

12.6.1. General References

1. **Cabasso, V. J., and R. H. Regamy (ed.).** 1977. International Symposium on Freeze-Drying of Biological Products. *Dev. Biol. Stand.* **36:**1–398.
A collection of papers on the theoretical aspects, available equipment, and research in freeze-drying.
2. **Heckly, R. J.** 1978. Preservation of microorganisms. *Adv. Appl. Microbiol.* **24:**1–53.
This review presents a detailed discussion of the preservation of bacteria by a variety of methods and includes a description of the lyophilization process.
3. **Kirsop, B. E., and A. Doyle.** 1991. *Maintenance of Microorganisms and Cultured Cells,* 2nd ed. Academic Press Ltd., London.
This manual of laboratory methods provides a complete update of preservation methods for bacteria, fungi (including yeasts), algae, and protozoa.
4. **Lapage, S. P., J. E. Shelton, T. G. Mitchell, and A. R. Mackenzie.** 1970. Culture collections and the preservation of bacteria, p. 135–228. *In* J. R. Norris and D. W. Ribbons (ed.), *Methods in Microbiology,* vol. 3A. Academic Press Ltd., London.
This paper discusses the activities of maintaining a large service culture collection and describes the use of centrifugal freeze-drying.
5. **Meryman, H. T.** 1966. Review of biological freezing, p. 1–114. *In* H. T. Meryman (ed.), *Cryobiology.* Academic Press, Inc., New York.
In addition to the review chapter cited, this book contains numerous reports dealing with the theoretical aspects of freezing and freeze-drying of bacteria.
6. **Rinfret, A. P., and B. LaSalle (ed.).** 1975. *Round Table Conference on the Cryogenic Preservation of Cell Cultures,* p. 1–78. National Academy of Sciences, Washington, D.C.
This report contains papers describing the preservation of bacteria, industrial starter cultures, algae, and protozoa by using liquid nitrogen storage.

12.6.2. Specific References

7. **Canale-Parola, E.** 1973. Isolation, growth and maintenance of anaerobic free-living spirochetes, p. 61–73. *In* J. R.

Norris and D. W. Ribbons (ed.), *Methods in Microbiology,* vol. 8. Academic Press, Inc., New York.

8. **Chang, L. T. and R. P. Elander.** 1986. Long-term preservation of industrially important microorganisms, p. 49–55. *In* A. L. Demain and N. A. Solomon (ed.), *Manual of Industrial Microbiology and Biotechnology.* American Society for Microbiology, Washington, D.C.
9. **Fahy, G. M.** 1986. The relevance of cryoprotectant "toxicity" to cryobiology. *Cryobiology* **23:**1–13.
10. **Feltham, R. K. A., A. K. Power, P. A. Pell, and P. H. A. Sneath.** 1978. A simple method for storage of bacteria at −76°C. *J. Appl. Bacteriol.* **44:**313–316.

10a. **Gherna, R. L., P. Pienta, and R. Cote (ed.).** 1992. *American Type Culture Collection Catalogue of Bacteria and Phages,* 18th ed. American Type Culture Collection, Rockville, Md.

11. **Gordon, R. E., and T. K. Rynearson.** 1963. Maintenance of strains of *Bacillus* species, p. 118–128. *In* S. M. Martin (ed.), *Culture Collections: Perspectives and Problems.* University of Toronto Press, Toronto.
12. **Grivell, A. R., and J. F. Jackson.** 1969. Microbial culture preservation with silica gel. *J. Gen. Microbiol.* **58:**423–425.
13. **Hartsell, S. E.** 1956. Maintenance of cultures under paraffin oil. *Appl. Microbiol.* **4:**350–355.
14. **Haynes, W. C., L. J. Wickerham, and C. W. Hesseltine.** 1955. Maintenance of cultures of industrially important microorganisms. *Appl. Microbiol.* **3:**361–368.
15. **Heckly, R. J.** 1961. Preservation of bacteria by lyophilization. *Adv. Appl. Microbiol.* **3:**1–76.
16. **Heckly, R. J.** 1985. Principles of preserving bacteria by freeze-drying. *Dev. Ind. Microbiol.* **26:**379–395.
17. **Hippe, H.** 1991. Maintenance of methanogenic bacteria, p. 101–113. *In* B. E. Kirsop and A. Doyle (ed.), *Maintenance of Microorganisms and Cultured Cells,* 2nd ed. Academic Press Ltd., London.
18. **Hunt, G. A., A. Gourevitch, and J. Lein.** 1958. Preservation of cultures by drying on porcelain beads. *J. Bacteriol.* **76:**453–454.
19. **Imanaka, T., and S. Aiba.** 1981. A perspective on the application of genetic engineering: stability of recombinant plasmids. *Ann. N.Y. Acad. Sci.* **36:**1–14.
20. **Imprey, C. S., and B. A. Phillips.** 1991. Maintenance of anaerobic bacteria, p. 71–80. *In* B. E. Kirsop and A. Doyle (ed.), *Maintenance of Microorganisms and Cultured Cells,* 2nd ed., Academic Press Ltd., London.
21. **Kauffman, F.** 1966. *The Bacteriology of Enterobacteriaceae,* p. 1–400. The Williams & Wilkins Co., Baltimore.
22. **Krieg, N. R., and J. G. Holt (ed.).** 1984. *Bergey's Manual of Systematic Bacteriology,* vol. 1. The Williams & Wilkins Co., Baltimore.
23. **Ljungdahl, L. G., and J. Wiegel.** 1986. Working with anaerobic bacteria, p. 84–96. *In* A. L. Demain and N. A. Solomon (ed.), *Manual of Industrial Microbiology and Biotechnology.* American Society for Microbiology, Washington, D.C.
24. **Mackenzie, A. P.** 1977. Comparative studies on the freeze-drying survival of various bacteria: gram type, suspending medium, and freezing rate, p. 263–277. *In* V. J. Cabasso and R.H. Regamey (ed.), *International Symposium on Freeze-Drying of Biological Products.* S. Kargel AG, Basel.
25. **Malik, K. A.** 1990. Use of activated charcoal for the preservation of anaerobic phototrophic and other sensitive bacteria by freeze-drying. *J. Microbiol. Methods* **12:**117–124.
26. **Malik, K. A.** 1990. A simplified liquid-drying method for the preservation of microorganisms sensitive to freezing and freeze-drying. *J. Microbiol. Methods* **12:**125–132.
27. **Mazur, P.** 1966. Physical and chemical basis of injury in single-celled microorganisms subjected to freezing and thawing, p. 213–315. *In* H. T. Meryman (ed.), *Cryobiology.* Academic Press, Inc., New York.
28. **Meryman, H. T.** 1966. Freeze-drying, p. 609–663. *In* H. T. Meryman (ed.), *Cryobiology.* Academic Press, Inc., New York.
29. **Meryman, H. T.** 1971. Cryoprotective agents. *Cryobiology* **8:**173–183.
30. **Miller, R. E., and L. A. Simmons.** 1962. Survival of bacteria after twenty-one years in the dried state. *J. Bacteriol.* **84:**1111–1113.
31. **Miller, T. L., and M. J. Wolin.** 1985. *Methanosphaea stadtmaniae* gen. nov., sp. nov.: a species that forms methane by reducing methanol with hydrogen. *Arch. Microbiol.* **141:**116–122.
32. **Miller, T. L., M. J. Wolin, and E. A. Kusel.** 1986. Isolation and characterization of methanogens from animals. *Syst. Appl. Microbiol.* **8:**234–238.
33. **Mills, C. K., and R. L. Gherna.** 1988. Cryopreservation studies of *Campylobacter. Cryobiology* **25:**148–152.
34. **Nadirova, I. M., and V. L. Zemliakov.** 1970. A study on the degree of stability of typical properties of bacteria preserved under mineral oil. *Mikrobiologiya* **39:**1106–1109.
35. **Nierman, W. C., and T. Feldblyum.** 1985. Cryopreservation of cultures that contain plasmids. *Dev. Ind. Microbiol.* **26:**423–434.
36. **Perkins, J. J.** 1976. *Principles and Methods of Sterilization in Health Science,* p. 286–292. Charles C Thomas, Publishers, Springfield, Ill.
37. **Redway, K. F., and S. P. Lapage.** 1974. Effect of carbohydrates and related compounds on the long-term preservation of freeze-dried bacteria. *Cryobiology* **11:**73–75.
38. **Rudge, R. H.** 1991. Maintenance of bacteria by freeze-drying, p. 31–44. *In* B. E. Kirsop and A. Doyle (ed.), *Maintenance of Microorganisms and Cultured Cells,* 2nd ed. Academic Press Ltd., London.
39. **Sneath, P. H. A., N. S. Mair, M. E. Sharpe, and J. G. Holt (ed.).** 1986. *Bergey's Manual of Systematic Bacteriology,* vol. 2. The Williams & Wilkins Co., Baltimore.
40. **Snell, J. J. S.** 1984. Maintenance of bacteria in gelatin discs, p. 41–45. *In* B. E. Kirsop and J. J. S. Snell (ed.), *Maintenance of Microorganisms.* Academic Press Ltd., London.
41. **Speck, M. L., and R. A. Cowan.** 1970. Preservation of microorganisms by drying from the liquid state, p. 241–250. *In* H. Iizuka and T. Hasegawa (ed.), *Proceedings of the First International Conference on Culture Collections.* University of Tokyo Press, Tokyo.
42. **Staines, J. E., V. F. McGowan, and V. B. D. Skerman.** 1986. *World Directory of Collections of Cultures of Microorganisms.* World Data Center, University of Queensland, Brisbane, Australia.
43. **Staley, J. T., M. P. Bryant, N. Pfennig and J. G. Holt (ed.).** 1989. *Bergey's Manual of Systematic Bacteriology,* vol. 3. The Williams & Wilkins Co., Baltimore.
44. **Vela, G. R.** 1974. Survival of *Azotobacter* in dry soil. *Appl. Microbiol.* **28:**77–79.
45. **Vesley, D., and J. Lauer.** 1986. Decontamination, sterilization, disinfection, and antisepsis in the microbiology laboratory, p. 182–198. *In* B. M. Miller, D. H. M. Grochel, J. H. Richardson, D. Vesley, J. R. Songer, R. D. Housewright, and W. E. Barkley (ed.), *Laboratory Safety: Principles and Practices.* American Society for Microbiology, Washington, D.C.
46. **Weiss, F. A.** 1957. Maintenance and preservation of cultures, p. 99–119. *In* H. J. Conn (ed.), *Manual of Microbiological Methods, Committee on Bacteriological Technic.* Society of American Bacteriologists. McGraw-Hill Book Co., New York.
47. **Williams, S. T., and T. Cross.** 1971. Actinomycetes, p. 295–334. *In* C. Booth (ed.), *Methods in Microbiology,* vol. 5. Academic Press Ltd., London.
48. **Williams, S. T., M. E. Sharpe, and J. G. Holt (ed.).** 1989. *Bergey's Manual of Systematic Bacteriology,* vol. 4. The Williams & Wilkins Co., Baltimore.

Section III

MOLECULAR GENETICS

Introduction to Molecular Genetics

PHILIPP GERHARDT

Bacterial genetics and molecular biology have become so melded that genetic phenomena in bacteria are now expressed almost entirely in molecular terms. Molecular genetics has permeated all aspects of bacteriology and so is contained in various degrees in most chapters of this book, from the molecular morphology of DNA in Chapter 3 to the molecular sequence similarities of DNA and RNA in Chapters 26 and 27. Molecular genetics has become a premier part of research in bacteriology.

The growth of knowledge about the molecular genetics of bacteria has accelerated like a chain reaction through a series of technological advances over the past 50 years. It may be said to have started with the demonstrations of DNA as the essential material of genetic information in bacteria by Avery, MacLeod, and McCarty in 1944 and of genetic recombination in *Escherichia coli* by Lederberg and Tatum in 1946. It continued with the demonstration of the molecular structure of DNA by Watson and Crick in 1953, the elucidation of the genetic code by several investigators in 1966, and the general development of recombinant DNA technologies in the next two decades. All of these milestone findings arose from new techniques, which continue to catalyze research in molecular genetics: witness the current impact of the polymerase chain reaction.

This explosion in the methodology of molecular genetics seemed so rapidly changing and specialized in 1981 that only limited and very basic coverage of the field was included in the prior edition of this book (5a), with chapters on Gene Mutation, Gene Transfer (in gram-negative bacteria), and Plasmids. In the present book these three technologies are retained and updated, but five important new ones are added: Gene Transfer in Gram-Positive Bacteria, Transposon Mutagenesis, Gene Cloning and Expression, Polymerase Chain Reaction, and Nucleic Acid Analysis. Furthermore, the chapter on Genetic Characterization in the section on Systematics in the 1981 version is expanded and updated in the present book in Chapters 25 to 27 on DNA Sequence Similarities, rRNA Hybridization and Gene Sequencing, and Nucleic Acid Probes. Taken together, these 11 chapters on molecular genetics provide a set of reliable, basic methods for practical use in the laboratory.

As stated in the Preface, the present book "is intended to serve as a first resource for undertaking a laboratory method and then for seeking further information or alternative methods in general and molecular bacteriology." For molecular genetics there is a plethora of journals and books for this second objective. Many of these are generalized for eucaryotic as well as procaryotic organisms, including the periodicals (1, 5, 10) and books (3, 4, 10, 11) recommended here. However, there are some equally excellent ones that relate specifically to bacteria (6–9). In addition to those listed in the references of this sectional introduction, there are a number of methodological books that deal with only one aspect of molecular genetics and are referenced in the appropriate chapter.

Molecular genetics has a language of its own, often expressed in acronyms and abbreviations and seldom user friendly for the neophyte. In the introduction to Chapter 13, the author provides the basics of genetic nomenclature. For genetic nomenclature in other chapters and for more details, the reader is referred to the basics of genetic nomenclature provided on pages 33 to 38 of the *ASM Style Manual for Journals and Books* (2).

Finally, I express appreciative acknowledgment to Simon D. Silver for his help in getting the section on molecular genetics initially organized and started. Later on, colleagues Frans J. de Bruijn and Michael Bagdasarian also helped substantially.

REFERENCES

1. **Abelson, J. N., and M. I. Simons (ed.).** *Methods in Enzymology.* Academic Press, Inc., San Diego, Calif.
Several volumes are published each year, including some on aspects of molecular genetics such as reference 8 (below).
2. **American Society for Microbiology.** 1991. *ASM Style Manual for Journals and Books.* American Society for Microbiology, Washington, D.C.
3. **Ausubel, F. M., R. Brent, R. E. Kingston, D. D. Moore, J. G. Seidman, J. A. Smith, and K. Struhl.** *Current Protocols in Molecular Biology.* Green Publishing Associates, Inc., and John Wiley & Sons, Inc., New York.
The core book is published in two thick (~1,600 total pages) ring-bound volumes. Supplements (each ~150 pages) are sent to subscribers every few months.
4. **Ausubel, F. M., R. Brent, R. E. Kingston, D. D. Moore, J. G. Seidman, J. A. Smith, and K. Struhl.** 1992. *Short Protocols in Molecular Biology,* 2nd ed. Greene Publishing Associates, Inc., and John Wiley & Sons, Inc., New York.
A concise (834 pages), less expensive edition of the above core book and its supplements.
5. **Eaton Publishing Co.** *BioTechniques.* Eaton Publishing Co., Natick, Mass.
This is essentially a product-advertising journal but contains useful original technical and research reports on techniques for molecular biology. It is available at no cost by request from individuals actively engaged in laboratory bioresearch and at reduced cost to such individuals in Canada.

5a. **Gerhardt, P., R. G. E. Murray, R. N. Costilow, E. W. Nester, W. A. Wood, N. R. Krieg, and G. B. Phillips (ed.).** 1981. *Manual of Methods for General Bacteriology.* American Society for Microbiology, Washington, D.C.

6. **Harwood, C. R., and S. M. Cutting.** 1990. *Molecular Biological Methods for Bacillus.* John Wiley & Sons, Inc., New York.
With so much of bacterial molecular genetics focused so much on Escherichia coli and other gram-negative bacteria, this book helps to broaden the perspective to include gram-positive bacteria. Among these, Bacillus subtilis 168 is the prototypic and main model.

7. **Maloy, S. R.** 1990. *Experimental Techniques in Bacterial Genetics.* Jones and Bartlett Publishers, Boston.
An excellent, short (180 pages) manual suitable for an undergraduate laboratory course.

8. **Miller, J. H. (ed.).** 1991. Bacterial genetics systems. *Methods Enzymol.* **204:**1–706.
A collection of genetic methods for a diverse collection of bacteria including E. coli, S. typhimurium, B. subtilis, Haemophilus influenzae, Neisseria spp., myxobacteria, Caulobacter crescentus, Agrobacterium spp., Rhizobium meliloti, cyanobacteria, Streptomyces spp., the family Rhodospirillaceae, Pseudomonas spp., Vibrio spp., mycobacteria, streptococci, and staphylococci.

9. **Miller, J. H.** 1992. *A Short Course in Bacterial Genetics. A Laboratory Manual and Handbook for Escherichia coli and Related Bacteria.* Cold Spring Harbor Laboratory, Cold Spring Harbor, N.Y.
This new publication in two volumes is intended for students with training in biochemistry and recombinant DNA techniques but without training in bacterial genetics techniques. The laboratory manual contains units with 34 experiments on working with bacterial strains, the lac system, mutagenesis, gene transfer, and transposable elements. The handbook is an extensive, detailed compilation of reference information on genetic maps, vectors, data bases, tables, charts, recipes, etc.

10. **Perbal, B. (ed.).** *Methods in Molecular and Cellular Biology.* John Wiley & Sons, Inc., New York.
This new international journal contains articles describing new and improved protocols related to broad aspects of molecular biology.

11. **Sambrook, J., E. F. Fritsch, and T. Maniatis.** 1989. *Molecular Cloning: A Laboratory Manual,* 2nd ed. Cold Spring Harbor Laboratory, Cold Spring Harbor, N.Y.
This "bible" of molecular biology practice is published in three thick, comb-bound, relatively inexpensive volumes. There are 18 chapters which start with plasmid vectors and end with detection and analysis of proteins expressed from cloned genes. Seven appendices start with bacterial media, antibiotics, and bacterial strains and end with a list of suppliers.

Chapter 13

Gene Mutation

ERIC EISENSTADT, BRUCE C. CARLTON, and BARBARA J. BROWN

Genetic and biochemical investigations in bacteriology are often initiated by the isolation of mutants. As examples, the identification of an *Escherichia coli* mutant that lacked DNA polymerase I (21) permitted the detection of new biochemical activities associated with DNA replication and led ultimately to the biochemical characterization of the DNA replication machine; a collection of motility and chemotaxis mutants helped to define the 50-odd cytoplasmic and membraneous components that make up the bacterial flagellum (44); and the discovery and characterization of mutants blocked at specific stages of spore development inaugurated modern investigations of the regulation and mechanisms of bacterial sporulation (53).

The power of mutational analysis derives from its ability to query an organism incisively. Two examples illustrate the insights and experimental handles offered by the isolation of bacterial mutants. Kato and Shinoura (33) asked whether there were *E. coli* genes whose products were essential for the production of mutations following UV irradiation. They developed a procedure for identifying nonmutable cells and thereby discovered a genetic locus whose products are required for efficient mutagenesis by UV and many chemicals. In another example, Tormo et al. (57) asked whether there were genes essential for the survival of *E. coli* in the stationary phase. They developed procedures for identifying mutants with diminished viability in the stationary phase. These mutants led to the discovery of genetic loci whose products are dispensable for growing cells yet are essential for the survival of stationary-phase cells. In each of these two examples, the isolation of mutants signaled the discovery of new genes, implied the existence of previously undescribed functions, and brought investigators to a new plateau from which to launch genetic and biochemical analyses of complex biological phenomena.

The identification of new genetic loci via the isolation of mutants underscores what is, perhaps, the most valuable attribute of a genetic approach to a biological problem: mutant isolation and characterization can operate independently of an investigator's preconceived biases about which genes, gene products, or processes are relevant to the biological phenomenon. Of course, mutants supply only raw substance for further genetic and biochemical refinement, but a thoughtfully obtained collection of mutants can yield enough fertile material to last a lifetime. For bacteria without established genetic systems, mutant isolation can be a critical early step toward developing genetic exchange methods.

This chapter presents basic information about the principles and modern practice of mutant isolation and analysis. The hope is that it will prove to be useful for investigators and students who have the urge—and the courage!—to explore unknown biological territory but who may not be aware that genetic analyses can provide lanterns to guide the way. Techniques that have proven to be useful for analyzing genetically tractable organisms such as *E. coli* and *Salmonella typhimurium* are emphasized in this chapter, but similar techniques have been applied to a variety of other bacteria, including archaeobacteria (7, 12, 19, 27, 30, 36, 37, 39, 46, 48, 52, 55a, 65).

Although the experimental details provided in this chapter may not address specific technical needs, the principles discussed here should provide guidance for developing procedures to isolate mutants of a novel organism. Novices to bacteriology might consider practicing with *E. coli* or *S. typhimurium* strains as a way of becoming familiar with mutagenesis and mutant isolation procedures. Exercising with these prototype organisms will bolster confidence and permit the calibration of results with novel organisms.

The previous version of this chapter by Carlton and Brown (1) contained many useful procedures and descriptions, and several of these have been retained with only slight modification. The general references (2–8) listed at the end of this chapter provide more background material on mutagenesis and bacterial genetics. Other chapters in this section of the book provide complementary information about genetic analysis.

A brief description of the bacterial genetic nomenclature used throughout this chapter is provided here as a guide to novices. Genetic nomenclature for bacteria has been standardized since 1966 (22) and is described in detail in the *ASM Style Manual for Journals and Books* (10). **Genotype** refers to the genetic constitution of an organism. **Phenotype** refers to the observable properties of the organism. Genes are designated by a three-letter code written in italic lowercase letters, e.g., *abc*. *abc* signifies the name of a locus on the bacterial chromosome. Phenotypes attributable to the *abc* locus are designated by a three-letter code written in roman letters whose first letter is capital, e.g., Abc. The genotypes of different isolates (occurrences) of a mutant with an Abc phenotype may be designated *abc-1*, *abc-2*, etc. If the mutations are allelic, these are designated *abc1*, *abc2*, etc. If the Abc phenotype is attributable to several different loci, these genes can be designated by *abcA*, *abcB*, etc. Their corresponding phenotypes would be AbcA,

AbcB, etc. Wild-type alleles of a gene are designated with a "+" superscript, as in *abc*$^+$. The ASM style does not use the "−" superscript to designate a mutant genotype. Thus, depending upon the context, *abc* designates a locus or a mutant allele. Wild-type and mutant phenotypes are designated with "+" and "−" superscripts, respectively (e.g., Abc$^+$ or Abc$^-$).

13.1. PRINCIPLES OF MUTANT ISOLATION

Even following exposure to mutagens, mutations in a particular gene might occur, at most, with an incidence of 10^{-3}. In the absence of mutagens, the incidence of spontaneous mutation might be only 10^{-10}. Thus, mutant isolation requires the use of selection and/or screening techniques that can reveal or identify the rare colonies arising from mutant cells.

13.1.1. Selection

Selection strategies use conditions that permit growth of only the desired mutant. Selection techniques are powerful because rare mutants can be isolated from a large background of nonmutants in a single step. When there are several different solutions to the selective challenge imposed on the microbial population, more than one kind of mutant can be isolated. For example, resistance to the lethal effects of an antibiotic can arise in a variety of ways including via mutations affecting antibiotic permeability, antibiotic degradation, or antibiotic binding by the target molecule. Thus, one should always be alert to the possibility that a bacterium has many ways to "solve" the selection problem. One always gets what is selected for whether or not the organism's solution has been anticipated!

13.1.2. Screening

Many mutations of interest cannot be selected directly. For such mutations one has to use screens that discriminate between wild-type and mutant phenotypes. Screening strategies use tests that are often applied directly to colonies to distinguish wild-type from mutant phenotypes. Nontoxic chemical dyes or chromophore-bearing synthetic derivatives of enzyme substrates, for example, can be incorporated into a growth medium solidified with agar, where they will register specific metabolic capabilities of individual colonies (e.g., carbohydrate metabolism, esterase activity). The sensitivity of such screens makes it possible to detect rare mutants among large numbers of organisms, typically at frequencies approaching 10^{-5}. Screens can be designed to be highly specific at the biochemical level and are especially useful when there is no obvious selection for the specific phenotype. The original *polA* mutant of *E. coli* (deficient in DNA polymerase I) was detected via an enzymatic screen and found on the 3,478th trial (21)!

Screening is often used in conjunction with replica plating, a technique that is based on the property that colonies of many bacterial species can be imprinted on filter paper or on a piled fabric such as velveteen (section 13.5.4). Replica plating facilitates transfer of thousands of colonies at a time from one surface to another while preserving their spatial arrangement and location. Replicate plates incubated under different physiological conditions can be used to identify conditional mutants such as temperature-sensitive mutants or auxotrophs. Filter replicas of the plates can be probed in a variety of ways (e.g., with radiolabeled, chromophore-labeled, or fluorophore-labeled nucleic acid probes or enzymatic substrates or with antibodies directed against specific gene products or cell structures) and used to identify a desired mutant. A splendid example of enzymatic screening is described in a paper by Raetz (51) on the isolation of *E. coli* mutants carrying specific defects in lipid biosynthesis.

13.1.3. Enrichment

The efficiency of screening can be enhanced considerably by using enrichment procedures that favor the survival of the desired mutant. Two or three cycles of an enrichment can improve the chances of finding a particular mutant by several orders of magnitude.

Enrichments typically involve exposures to antibiotics that preferentially kill growing bacteria. The classic enrichment procedure with penicillin was independently developed by Davis (20) and Lederberg and Zinder (42). The principle behind penicillin selection is straightforward: bacterial cultures treated with penicillin are enriched for the nongrowing cells in the population. For example, if one is seeking a collection of temperature-sensitive mutants, penicillin treatment of cultures grown at the elevated temperature will preferentially kill growing cells and enhance the proportion of temperature-sensitive mutants. Since penicillin does not kill all growing cells, one only *enriches* cultures for the nongrowing organisms. A single cycle of penicillin treatment can achieve a 10^3- to 10^4-fold enrichment. Repeated cycles of penicillin treatment can improve the enrichment, but, because the conditions also select for penicillin-resistant bacteria, there are diminishing returns after two to three cycles. Additional cycles of enrichment can be achieved if penicillin selections are combined with other treatments that preferentially kill growing cells (e.g., nalidixic acid or cycloserine).

For further enrichment strategies, see Chapter 8.

13.1.4. Selection and Screening Combined

Combinations of selection and screening can be used to identify pleiotropic mutants or to characterize conditional mutants. The strategy in both cases is to look for phenotype B among a set of mutants possessing phenotype A when one suspects that the B phenotype will be represented among those with the A phenotype. The following example illustrates the use of this approach. Sporulation-defective mutants (Spo$^-$) of *B. subtilis* were identified among a collection of mutants selected for resistance to the RNA polymerase antibiotic rifampin (56). The Spo$^-$ mutants had an altered *rpoB* gene encoding an RNA polymerase β subunit which was not only resistant to rifampin but was also defective in its interactions with a specific transcriptional cofactor that was essential for the expression of genes participating in the onset of sporulation. The isolation of a

Spo− *rpoB* mutant was a critical observation contributing to the notion that the development of spores is under transcriptional control.

Screens are commonly applied to collections of conditional-lethal mutants such as temperature-sensitive mutants. A set of temperature-sensitive mutants can be a rich source of mutants affected in many essential cell functions and biosynthetic pathways. Clever screening can exploit a collection of conditional mutants to yield scientific treasures!

13.2. PRINCIPLES OF MUTAGENESIS

Knowledge of the transfer of genetic information from DNA to RNA to protein has led to a rational classification of mutations based on changes in nucleotide sequence and the impact of a mutation on gene product synthesis. Much of the information on mutation types summarized below is covered in more detail in introductory genetics texts.

13.2.1. Base Pair Substitutions

Base substitution mutations are either transitions (A·T ↔ G·C, i.e., a substitution of one purine-pyrimidine pair by another) or transversions (A·T ↔ T·A or C·G; G·C ↔ C·G or T·A, i.e., a substitution of a purine·pyrimidine pair by a pyrimidine·purine pair). Base substitution can generate a missense mutation (a new amino acid codon) or a nonsense mutation (a stop codon). The nonsense codons (TAG, TAA, TGA) can arise from specific sense codons (Lys [AAA, AAG], Gln [CAA, CAG], Glu [GAA, GAG], Tyr [TAC, TAT], Ser [TCA, TCG], Leu [TTA, TTG], Trp [TGG], or Cys [TGC, TGT]) via each of the single-base-pair substitutions except the A·T-to-G·C transition.

The impact of a missense mutation will depend on how it affects the structure and, hence, the function of the mutant protein. Mutants with marginally altered phenotypes compared with the wild type often result from missense mutations that only modestly affect the function of the gene product. Nonsense mutations will usually have a devastating impact on protein structure and generate nonfunctional gene products.

13.2.2. Frameshift Mutations

Frameshift mutations are additions or deletions of any number of base pairs that are not multiples of three, which cause the translation reading frame to shift ± 1 or 2. Frameshift mutations radically alter the coding sequence and therefore the structure and function of the gene product.

13.2.3. Deletions and Insertions

Deletion and insertion mutations are losses or additions of any number of base pairs and can be quite large. These mutational events can affect many genes simultaneously, because large deletions can eliminate DNA from more than one gene and insertions in one gene can influence the expression of neighboring genes.

13.2.4. Origins of Mutations

Spontaneous mutations arise in the absence of deliberate exposure to mutagenic agents at frequencies that range from 10^{-10} to 10^{-6}. **Induced mutations** arise following treatment of cultures with mutagens and can occur at frequencies approaching 10^{-3}.

The advantage of working with spontaneous mutants is a consequence of their rarity: a given mutant is not likely to carry additional and possibly confounding mutations. The advantages of working with induced mutants are that they occur more frequently and provide a wider spectrum of mutational events and, consequently, a wider variety of phenotypes. The disadvantage of induced mutants is that a particular isolate can harbor additional mutations which complicate an analysis. Proper characterization of a mutant generated by exposure to mutagens requires genetic transfer of the mutation to an unmutagenized host.

13.2.4.1. Spontaneous mutations

DNA sequence changes that involve multiple bases in a gene seem to account for a major fraction of spontaneous mutations in *E. coli*. For example, 80% of spontaneous *lacI* mutants arise via deletions (ranging from 4 to 123 bases), additions (4 bases), or insertions of transposable insertion sequences (26). Approximately half of the deletions, including all those that showed multiple independent occurrences, occur between direct repeats (ranging from 5 to 8 bp in length) situated up to many hundreds of bases apart.

The pattern of spontaneous mutagenesis in *E. coli* is, however, probably not generally applicable to other bacteria. For example, insertion sequences do not appear to play as important a role in spontaneous mutagenesis in *S. typhimurium* as in *E. coli* (38). On the other hand, in the archaeobacterium *Halobacterium halobium* insertion sequences are major contributors to spontaneous mutagenesis (19). Attempting to draw definitive conclusions about bacterial mutagenesis by comparing mutational spectra among different bacterial species when different mutational targets and selection conditions have been used is a treacherous exercise. Nonetheless, the available data suggest that no meaningful generalization can be made about the pattern of spontaneous mutation in bacteria. Each new organism should be studied with this limitation (and research opportunity) in mind.

13.2.4.2. Induced mutations

Induced mutations can arise via either of two mechanisms. One mechanism is a direct consequence of base mispairing during replication of a template containing a modified base (e.g., 2-aminopurine or methylation of the O^6 position of guanidylate yields a base that mispairs with dTTP). These mutagenic events appear to have no special requirement for inducible gene products, presumably because the responsible lesions are inherently replicable.

The second mechanism seems to involve the processing of inherently nonreplicable DNA lesions such as UV photoproducts. Mutagenesis by such lesions is not inevitable; e.g., mutants of *E. coli* without appropriate inducible gene products and responses to DNA damage cannot be efficiently mutated by UV photoproducts or other nonreplicable DNA lesions. Mutagens in this category are often referred to as **SOS-dependent mutagens** (60). From a practical point of view it is important to know whether an organism is capable of SOS mutagenesis, because UV irradiation is convenient (no hazardous mutagens to handle or dispose of) and potent (high mutation frequencies and a wide variety of mutational events). Some bacteria that do not naturally display efficient SOS mutagenesis can be provided with plasmid-borne genes conveying or enhancing UV mutability (50, 60).

13.2.5. Independently Arising Mutations

If there is interest in capturing a wide variety of biologically interesting variants, it is desirable to assemble a collection of independently arising mutants. Several factors conspire to make it trickier than might have been imagined to get variety in a collection of mutants with mutations affecting a particular locus. One important factor is that not all sites in a gene are equally mutable. Therefore, some mutational events will be represented more often than others; e.g., 75% of spontaneously arising *E. coli lacI* mutants had mutations that mapped to a single site in the gene (26). Another factor that limits variety among collections of mutants with missense mutations is that the sites at which mutations yield phenotypes are not randomly distributed across a gene. This is simply a consequence of the fact that the effect of an amino acid substitution varies with its impact on protein structure and function; i.e., some substitutions are tolerated better than others. An additional factor that can limit the variety of mutants is that some desired mutations might have deleterious side effects on cell survival and/or growth.

Fortunately, the distribution of mutations varies with mutagenic treatment. Collections of mutants consisting of spontaneous and mutagen-induced isolates are likely to represent a wide spectrum of different mutational events at many different sites.

13.2.6. Phenotypic Lag

Mutagens do not always induce phenotypes that can be immediately selected for. The sequence of events leading to the appearance of new phenotypes is roughly as follows. Mutagens cause DNA damage; DNA damage is processed by the bacterial DNA repair and DNA replication machinery to yield altered DNA sequences; altered sequences are expressed; and new phenotypes emerge via the appearance of mutant gene products or the loss of wild-type products by dilution during cell division. An additional process contributing to a lag in the full expression of a mutant phenotype is the segregation into daughter cells of mutant chromosomes from wild-type chromosomes. Because of the time dependence of these events, one must delay the application of lethal selections (e.g., resistance to bacteriocidal agents such as antibiotics or bacteriophage) to allow for the expression of the new phenotype. Expression is easily achieved by permitting cultures to grow for several generations following treatment with a mutagen before applying the selective pressure.

13.2.7. Mutagens

To increase the mutation frequency, one must generate DNA damage and then allow endogenous cellular DNA repair and DNA replication processes to generate mutations. There are three categories of mutagenic agents: radiation, chemical, and biological.

13.2.7.1. Radiation

Radiation damage is conveniently generated by a UV light source such as a 15- or 25-W germicidal lamp. γ-Irradiation from a ^{60}Co source is also a convenient method for generating DNA damage. The virtues of radiation treatment are severalfold: the dosage is easily controlled by varying the exposure time and/or the intensity of the source; radiation generates a wide variety of lesions across a genetic target and, consequently, a broad spectrum of mutational events including base substitutions, frameshifts, and deletions; and there is no hazardous chemical to handle or dispose of. A drawback of radiation mutagenesis is that the lesions generated by UV or γ-irradiation are not efficiently fixed as mutations unless the cells are capable of carrying out SOS repair.

13.2.7.2. Chemicals

Chemically induced mutations generally arise either via structural damage to DNA, as when alkylating or base-pair-intercalating agents are used, or through the misincorporation of base analogs. As with radiation-induced damage, chemically induced DNA damage depends on endogenous DNA repair and DNA replication processes to operate on the damaged template and generate mutant sequences.

The chemical mutagens of choice for initial efforts to generate base substitution mutants are the alkylating agents 1-methyl-3-nitro-1-nitrosoguanidine (MNNG) and ethyl methanesulfonate (EMS). Both of these agents induce primarily transition mutations at G·C or A·T base pairs, and they have similar site specificities (16). Importantly, many of the lesions that these agents generate do not require processing by SOS repair. Thus, most organisms are likely to be efficiently mutagenized by either MNNG or EMS.

Another procedure for generating base pair substitution mutations is to grow bacteria in the presence of a base pair analog such as 2-aminopurine (2-AP). The incorporation of this base during DNA replication (following uptake and conversion to the nucleotide) generates both G·C→A·T and A·T→G·C substitutions. 2-AP and other base analogs, however, are weakly mutagenic in comparison with MNNG and EMS.

Frameshift mutations can be generated by growing bacteria in the presence of the acridine derivative ICR-191. Deletion mutations, on the other hand, are not specifically induced by mutagenic treatments. Nitrous acid has been shown to induce a high frequency of deletion mutations in *E. coli* and *S. typhimurium* (9, 54), but it is not known how general a phenomenon this is.

CAUTION: Associated with the use of chemical mutagens is the risk of adverse health or environmental effects as a consequence of exposure (section 13.8). Many chemical mutagens, including the ones described here, are potent animal carcinogens. Investigators who handle chemical mutagens should be sure to follow their institutional environmental and occupational health guidelines.

13.2.7.3. Transposons

Mutations may also be induced by the rearrangement of DNA sequences catalyzed by sequence specific recombination pathways that work on genetic elements known as **transposons** (see Chapter 17). When transposons are linked to genetic markers such as antibiotic resistance determinants, they form a robust genetic tool for inducing and analyzing mutants. Insertion mutations generated by transposons carrying a determinant for an antibiotic resistance are marked by two independent phenotypes: one is a consequence of the gene knockout, and the other is a property of the transposon. The advantage of transposon mutagenesis is that every isolate selected for its transposon-encoded phenotype (e.g., ampicillin resistance) is a mutant. For a recent illustration of the virtues of transposon mutagenesis, see the report by Matsunaga et al. (46) on the identification of genomic fragments required for magnetosome biosynthesis in magnetotactic bacteria.

Artificial transposons carrying readily visualized gene products such as those encoded by *lacZ*, *luxAB*, or *phoA* have also been constructed (see Chapter 17). These transposons permit novel approaches to the identification of genes, the study of gene regulation and function, and the characterization of coordinately regulated gene families.

13.3. MUTANT CHARACTERIZATION

Imagine that one has developed a cunning selection for isolating mutations affecting a new "subtle" (Sub$^+$) phenotype and has used it to isolate 100 independently arising *sub* mutants. The three steps below show how one might begin to characterize *sub* mutants.

13.3.1. Stability and Suppression

The genetic stability of the *sub* mutants can provide some crude but useful information about the nature of the mutation. For example, if reversion to the Sub$^+$ phenotype occurs spontaneously and is stimulated by several orders of magnitude following mutagen treatment with MNNG, the mutant is likely to carry a base substitution mutation, possibly an A·T to G·C. If the mutant reverts in response to treatment with a frameshift mutagen but not with a base substitution mutagen, it is likely to carry a frameshift mutation. If the mutant does not revert even after mutagenesis, it may carry either a deletion or insertion mutation. Knowledge of the stability not only provides information about the nature of the mutational event at the DNA level but is also of practical value as a guide to handling the mutants.

Reversion to the Sub$^+$ phenotype can occur, in principle, via three different pathways. Reversion via restoration of the wild-type (or near wild-type) sequence is the most obvious mechanism and is an expected occurrence for base pair substitution mutations or some frameshift mutations. A second pathway involves a mutational event at a different site within the mutant version of the gene which compensates or suppresses the original mutation and thereby restores the phenotype to wild type (or near wild type). Such suppressor mutations are referred to as intragenic suppressors. The identification of intragenic suppressors can fuel hypotheses concerning interactions between different sites on a polypeptide that contribute to the structure and function of the gene product. The third pathway involves a mutational event at a different locus (extragenic) which generates a phenotype that suppresses the Sub$^-$ phenotype. Two types of suppressors arising via second-site mutations are possible. One type is an informational suppressor, e.g., a tRNA allele that suppresses nonsense mutations by supplying an amino-acylated tRNA to translate stop codons. A particular informational suppressor can operate, on a given class of mutation (nonsense or, in some instances, frameshift), at many sites in many genes. The second and more significant type is an allele-specific suppressor. This type is specific for a given *sub* allele and identifies a gene whose product substitutes for, interacts with, or regulates the expression of the *sub* product. The isolation and characterization of suppressor mutants is a powerful and incisive genetic approach to dissecting the components of the Sub phenotype. Recent examples of studies involving the use of second-site mutations can be found in references 25 and 40.

Some mutants will display "leaky" phenotypes; i.e., they will not be fully mutant. Leaky phenotypes are most easily evaluated by plating the mutant cells at low density on the selective or nonpermissive medium and observing the extent of growth or other expression of that particular property. Leakiness can be a problem with mutants which are to be used for other physiological studies such as enzymatic assays or for complementation tests with other mutants. Leakiness frequently is a property of missense mutations that result in only modestly disruptive amino acid substitutions in the gene product. Nonsense, frameshift, or deletion mutations, which more dramatically affect genes and their products, are less likely to result in leaky phenotypes.

13.3.2. Complementation and Mapping

One of the reasons for isolating 100 independent Sub$^-$ mutants might have been to determine how many loci and gene products contribute to the Sub$^-$ phenotype. The classical way to proceed is to examine the Sub phenotype that results when combinations of two independently arising *sub* mutants are present in the same organism. This kind of analysis, known as **complementation analysis**, is performed by making cells heterozygous for the two mutations. If the two mutations reside in different genes coding for different proteins, one would expect the heterozygote to display a wild-type phenotype. If, on the other hand, the two mutations reside in the same gene, the heterozygote will retain the mutant phenotype. Such analyses can identify how many discrete functional units are contributing to the Sub phenotype.

Genetic mapping of the *sub* mutants will independently provide information about how many different loci are represented in the collection. Both complementation and mapping studies require that a genetic system be available for the organism.

13.3.3. Physiological Characterization

Physiological characterization of mutant phenotypes is a critical first step in interpreting a collection of mutants. The evaluation is usually guided by the specific biological questions that formed the rationale for the mutant hunt in the first place. Although there are no universal guidelines governing the physiological characterization of a mutant collection, one should bear in mind the observation referred to in section 13.1.1: an independently arising set of mutants displaying the desired phenotype can consist of a heterogeneous set of defects in related genes or in unrelated genes. Two examples of how one might initially pursue analysis of a collection of mutants are briefly discussed below.

For a collection of mutants with mutations affecting genes that specify related enzymes of a common biochemical pathway, the biochemical nature of the defect can often be elucidated by a combination of growth response and intermediate accumulation tests. In ideal situations, for any sequence of biochemical reactions, mutants blocked at any step in the sequence will grow on any diffusible intermediate located beyond the point of the block. The intermediate(s) just prior to the point of the block will accumulate in the growth medium. Thus, for the sequence of reactions $X \xrightarrow{1} A \xrightarrow{2} B \xrightarrow{3} C \xrightarrow{4}$ EP (end product) catalyzed by the enzymes 1, 2, 3, and 4, respectively, a mutant with a mutation in the gene which specifies enzyme 1 will accumulate intermediate X and will grow if supplied with A, B, C, or EP. A mutant defective in enzyme 2 will accumulate intermediate A (and perhaps X) and will grow on B, C, or EP.

In practice, the characterization of mutants by accumulation and growth response tests is limited by several factors. First, many intermediates cannot be taken up by (transported into) the cells; for example, normally phosphorylated intermediates cannot be taken up by cells and thus cannot be used to satisfy growth requirements. Second, some intermediates are chemically unstable and are rapidly degraded to other compounds or may be metabolized by other enzyme systems. Third, the utility of these tests is restricted by the availability of the appropriate intermediates and of simple tests for identifying them (e.g., colorimetric assays). Despite these limitations, accumulation and growth response tests can be extremely useful in distinguishing the different genetic and enzymatic functions in biosynthetic pathways for amino acids, nucleosides, and vitamins.

Mutants selected for a particular phenotype can also arise via mutations affecting unrelated genes that alter different physiological functions. For example, resistance to a given antibiotic might be achieved by defects either in a membrane transport function or in extracellular polysaccharide synthesis, either one of which might prevent efficient antibiotic uptake. Alternatively, a mutation affecting the binding of an antibiotic by its target molecule (e.g., an RNA polymerase ß subunit) could result in a resistant phenotype. One way to distinguish permeability mutants from target mutants is to test the sensitivity of the resistant mutants to other, unrelated bacteriocidal agents. Gross changes in permeability, such as might arise via a mutation that leads to overproduction of extracellular polysaccharide, often convey multiple resistance phenotypes to a cell. Such changes might also be accompanied by marked differences in colony appearance (e.g., mucoidy).

13.3.4. Molecular Characterization

Cloning and sequencing of genes and mutant alleles contribute to a mechanistic understanding of their function and regulation. Sequence information also provides experimental tools for further analysis. For example, the nucleotide sequences of a gene can be altered via site-specific mutagenesis to test specific hypotheses about the structure and function of a gene product. Cloned genes and the sequences derived from them are also useful for developing molecular probes to characterize mutants, to study gene expression, and to identify related genes in other organisms. Chapter 17 on transposon mutagenesis and Chapter 19 on the polymerase chain reaction describe procedures that can be adapted for the molecular characterization of mutations.

13.4. MUTAGEN DETECTION

Several genetic systems have been developed for detecting mutagenic agents and for monitoring specific mutational events. The simplest of these systems are based on selections for auxotrophic revertants (e.g., His^- to His^+) or carbohydrate utilization (e.g., Lac^- to Lac^+). The systems are valuable because of both their high sensitivity and their rapid diagnoses of mutagenic specificity. Yanofksy et al. isolated a set of *trpA* mutants to help establish the colinearity of genes and proteins (63) and to contribute to the solution of the coding problem (64). Several of these mutants subsequently became useful for analyzing the mutagenic specificity of base substitution mutagens (62).

A vast collection of His auxotrophs has been assembled (29) as part of an effort to study the genetics and regulation of histidine biosynthesis in *S. typhimurium*. Several of these *his* alleles have proven to be useful for detecting mutagenic agents. In fact, the most widely used system today is based on the analysis of His^+ revertants in *S. typhimurium*. It was developed by Ames and his colleagues (45) for the detection of mutagens in environmental samples and human body fluids. The use of such bacterial tests established a striking correlation between mutagenicity and carcinogenicity (23, 47) and provided evidence in support of the notion that DNA damage can be a critical step in chemical carcinogenesis.

Specific *his* alleles that have proven to be most useful for mutagenicity testing include *hisG46*, a missense mutation that reverts by all possible base substitution events at G·C base pairs (24, 49); *hisG428*, a nonsense mutation that has been used to monitor all possible base substitution events at A·T base pairs (43); and the *hisC3076* and *hisD3052* alleles, which are frameshift mutations and can be used to detect frameshift mutagens (45).

An efficient screen to rapidly identify the base change in His$^+$ revertants of several different *his* alleles was developed by Miller and Barnes (49) and extended by Cebula and Koch (14). The screen involves colony hybridizations with specific oligonucleotide probes directed against each of the possible base sequences that can arise among revertants. This approach makes it possible to characterize hundreds of revertants at a time and thereby to develop a statistically significant picture of mutagenic specificities at a particular site.

Alper and Ames (9) developed a positive selection for deletions spanning a specific region of the *S. typhimurium* chromosome, from the *gal* operon through *chlA*. Since no essential genes are known to map in this region, it is possible to detect deletions covering as much as 1 min (~40 kb) of the chromosome. The selection for *chlA* requires anaerobic conditions, thus precluding the possibility of studying deletions of this kind during aerobic growth. Agents that increase the frequency of these deletions include nitrous acid (50-fold induction), nitrogen mustard, mitomycin C, and fast neutrons (5- to 8-fold induction). UV irradiation was one of 28 mutagenic treatments that did not enhance the frequency of deletion formation under these conditions. This method was said to work with *E. coli* (J. McCann, personal communication, cited in reference 9).

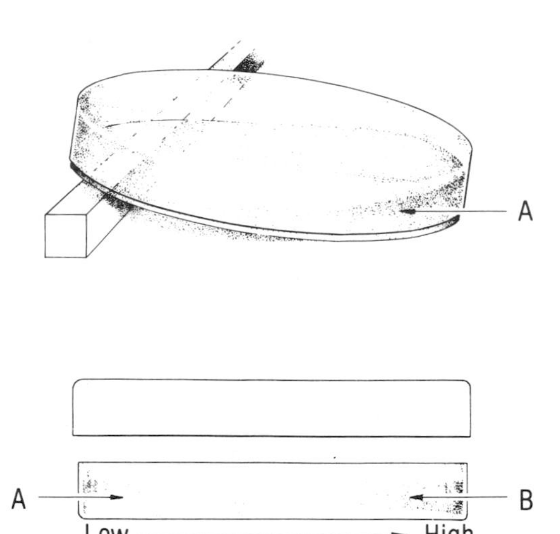

FIG. 1. Preparation of gradient plate for isolation of antibiotic-resistant mutants. Top, step 1; bottom, step 2. (A) Nutrient agar; (B) antibiotic agar. The resulting antibiotic gradient extends from a low concentration on the left to high on the right of the plate.

13.5. MUTANT ISOLATION

13.5.1. Antibiotic-Resistant Mutants

Antibiotic-resistant mutants can be easily isolated from most bacteria and even from archaeobacteria (27, 36). The mutations are useful as genetic markers because they provide a readily selectable phenotype. For gram-negative enteric bacteria, the antibiotics commonly used are rifampin (100 μg/ml), streptomycin (100 μg/ml), and naladixic acid (100 μg/ml). A compendium of antibiotics, antimetabolites, and other conditions that can be used to select directly for *S. typhimurium* or *E. coli* mutants is given by Vinopal (59).

13.5.1.1. Gradient plate technique

A major difficulty encountered with direct selection is choosing the optimal concentration of the selective agent. For example, when selecting mutants resistant to an antibiotic, one must choose antibiotic concentrations that prevent growth of all but the resistant mutants. Determining optimal antibiotic concentrations for selection of antibiotic-resistant mutants can be aided by the use of a simple gradient plate technique as follows.

1. Measure 10 ml of autoclaved nutrient agar (section 13.9.13) into sterile petri plates 100 mm in diameter, and allow to cool with one edge elevated so that the agar level just reaches the intersection of the bottom and side of the plate (Fig. 1).
2. After the agar has hardened, level the plates and add an additional 10 ml of antibiotic agar (section 13.9.14) that contains streptomycin sulfate (100 μg/ml), ampicillin (500 μg/ml), or rifampin (100 μg/ml).
3. Allow the plates to cool on a level table, producing a concentration gradient of the antibiotic from 0 μg/ml at one edge to the maximal concentration (100 to 500 μg/ml) at the opposite edge.
4. Mark the edges to show which way the gradient is oriented. The plates should be prepared at least 24 h before use so that the antibiotic gradients will become established.

13.5.1.2. Rifampin-resistant mutants

1. Grow the organism overnight in a rich broth to mid-logarithmic phase (2×10^8 cells per ml).
2. Spread 0.1-ml aliquots onto rifampin gradient plates prepared as described above, and incubate the plates for 24 to 36 h in the dark. Rifampin-resistant colonies will appear in areas where there is no distinguishable bacterial lawn.
3. Pick several colonies with a sterile toothpick or loop, and streak onto a second gradient plate in the direction of increasing rifampin concentration. This will allow estimation of the approximate level of resistance.
4. To more accurately determine the level of resistance to rifampin, streak the mutants on each of several plates containing different rifampin concentrations.
5. To evaluate the stability of these mutants, quantitatively grow them nonselectively in broth to mid-log phase and plate dilutions onto media with and without antibiotic. The number of CFU should be identical (within experimental error).
6. Freeze aliquots of these mutants as described in Chapter 12. These mutants may be used in the transposon mutagenesis procedure described below.

Equivalent procedures can be used to isolate mutants resistant to other antibiotics or metabolic inhibitors of choice. A compilation of antibiotics and inhibitors that have been useful for *E. coli* procedures can be found in reference 8.

13.5.2. UV-Induced Mutants

The procedure outlined below provides conditions that work well for generating mutations in *E. coli* or *S. typhimurium* by UV radiation. Both of these organisms are able to carry out SOS mutagenesis (section 13.2.4.2). Since this ability is not found in all bacterial genera, it is necessary to determine whether a given organism is mutable by UV. A simple way to do this is to select among UV-irradiated bacteria for an antibiotic- or inhibitor-resistant phenotype that arises spontaneously. If the incidence of resistant mutants is increased by UV irradiation, it is a good indication that UV mutagenesis has taken place. For this exercise the incidence of rifampin-resistant mutants will be used as a convenient assay for assessing the mutagenic effects of UV irradiation.

13.5.2.1. Preparation

Two factors that can interfere with efficient UV mutagenesis are the presence of UV-absorbing nutrients (aromatic compounds such as tryptophan or nucleic acid bases) and the shielding of cells by high cell densities. The protocol suggested below avoids these factors by irradiating bacteria suspended in a UV-transparent buffer at low densities. An additional interfering factor is photoreactivation. Many bacteria are able to repair some kinds of UV-induced DNA damage (pyrimidine dimers) in a visible-light-dependent reaction. To avoid photoreactivation, carry out UV irradiation and postirradiation treatment in dim light. The procedure is as follows.

1. Grow an overnight (16-h) culture of bacteria in a rich medium such as LB to early stationary phase ($\sim 2 \times 10^9$ to 5×10^9 cells per ml)
2. Dilute the overnight culture 100-fold into fresh broth, and grow to a cell density of approximately 2×10^8 cells per ml.
3. Wash the bacteria by filtration or centrifugation, and resuspend them in phosphate-buffered saline (PBS) (section 13.9.1) at 2×10^8 cells per ml.
4. Plate 0.1-ml aliquots of 10^{-5} dilutions in duplicate on solid media. This will provide a measure of the number of viable cells per milliliter prior to irradiation.

13.5.2.2. Irradiation conditions

Germicidal bulbs (e.g., no. 15T8; General Electric, Schnectady, N.Y.) are available from electric supply stores or laboratory supply houses. These bulbs fit in standard fluorescent light fixtures, which are also available at electric supply stores. Lamps of 15 W provide sufficient radiation fluxes for most organisms. A more expensive alternative to germicidal bulbs is a mineral light. If there is access to a mineral light, be sure to set it for short-UV wavelengths. The UV flux to the sample will vary with distance from the lamp; 25 to 30 cm is a standard distance that, for a 15-W lamp, will produce biological effects within a few minutes of exposure of bacteria whose UV sensitivity is comparable to that of *E. coli*. *CAUTION*: UV radiation is a potent mutagen and human carcinogen. To avoid exposure to either direct or reflected light, wear UV-opaque glasses and cover hands and forearms. A simple wooden frame can be constructed to hold the UV light source, over which a black curtain can be hung vertically to screen users from the UV light. The lamp can be turned on and off by using a light switch placed outside the curtain and connecting to the fixture inside the curtain.

The UV flux of a germicidal lamp can be calibrated with a flux meter. Meters can be quite expensive, however. A biological dosimeter can be developed by the use of *E. coli* phage such as T2 or T4. The UV sensitivity of these phages is well studied and reproducible and can be used to calibrate the output of a germicidal lamp. Thus, to achieve a 10% reduction in the survival of a population of phage T2 requires ~ 8 J/m^2 (28). For an organism that is not a host for these phage, the dosimetry on experimental samples can be determined by adding the phage to bacterial suspensions. The phage can subsequently be titered on appropriate hosts. A UV dose of 50 J/m^2 leaves approximately 40% of a population of *E. coli* bacteria viable.

The procedure is as follows.

1. Distribute 5- to 6-ml aliquots of bacteria suspended in PBS (section 13.9.1) at 2×10^8 cells per ml to 100-mm glass petri dishes. Place the dishes (be sure to remove the tops, since glass and plastic petri dishes are UV opaque!) on a shaker or rotator behind the curtain underneath the UV light source. Glass dishes are preferable to plastic because the liquid will not bead up on the surface. If plastic is used, a benign surfactant (e.g., gelatin, 0.1 to 1.0% [wt/vol]) can be added to the bacterial suspension to promote homogeneous dispersal of the bacterial suspension.
2. Irradiate the aliquots for various periods. The exposure time may have to be varied over a 10- to 100-fold time scale (e.g., from 15 s to 20 min) depending on the UV sensitivity of the organism.
3. For each UV dose, remove a 0.1-ml aliquot of cells, dilute serially into PBS, and spread aliquots onto nonselective plates to determine the surviving fraction of cells. Higher UV doses will result in more killing than lower doses will. Consequently, heavily irradiated samples will not have to be diluted as much as unirradiated or lightly irradiated samples. A UV dose that yields 0.1 to 1% survival of the bacterial population generates optimal yields of mutants for bacteria that perform SOS mutagenesis. Experiment to find the optimal doses for the organism.
4. For each UV dose, dilute a 0.5- to 1.0-ml aliquot into 10 ml of fresh growth medium and grow overnight to stationary phase. Spread 0.1-ml aliquots of undiluted as well as 10^1- and 10^2-fold dilutions of cultures onto rifampin-containing broth plates (section 13.9.4). Determine the viable-cell count by diluting the overnight cultures and planting aliquots on broth plates without antibiotic (e.g., 0.1-ml aliquots of 10^5-fold dilutions for bacterial cultures that reach a cell density of 2×10^9 cells per ml. These data should provide a quantitative estimate of the incidence of UV-induced mutations to rifampin resistance.

13.5.2.3. Postirradiation treatment

UV-irradiated bacteria should be grown under nonselective conditions to permit processing of DNA damage and expression of mutant phenotypes. Typically, the irradiated bacteria are diluted into rich growth medium (e.g., LB broth) and grown for several hours. Aliquots of these cultures can then be spread onto plates containing ingredients that will select for the desired phenotype (as in the example above, for an antibiotic resistance). If replica plating is to be performed (section 13.5.4), the irradiated bacteria can be spread immediately on nonselective plates and incubated until the colonies are visible (typically, an overnight incubation). At this point, the colonies can be replica plated onto selective plates or screening plates.

13.5.2.4. Enhancement of UV mutagenesis

The products of the *recA* and the *umuDC* operons are essential for efficient UV mutagenesis in *E. coli* and *S. typhimurium* (35, 60). In some bacteria, poor mutability by UV can be overcome by the introduction of a plasmid (such as pKM101) that carries a homolog of *umuDC* (50). pKM101 even enhances the UV mutability of bacteria that are well mutagenized by UV. The procedure outlined below describes a mating technique for moving pKM101 from *S. typhimurium* to *E. coli*. A similar procedure was used to move the R46 plasmid (the parent of pKM101) into *Proteus mirabilis* (30). If the organism is poorly mutable by UV and can be mated with *E. coli* or *S. typhimurium*, the introduction of a plasmid such as pKM101 may be of value.

1. Grow overnight cultures of the pKM101 donor strain *S. typhimurium* TA100 in LB containing 100 μg of ampicillin per ml. Resistance to ampicillin is encoded by pKM101 and provides a phenotype that can be used to monitor plasmid transfer to a recipient strain. Grow the recipient bacterium, an *E. coli* strain resistant to nalidixic acid, in plain LB broth. Nalidixic acid resistance provides the recipient with a phenotype that can be used to select against the growth of donor cells.
2. Prepare an LB plate containing ampicillin and nalidixic acid at 100 and 40 μg/ml, respectively. Growth on plates containing both antibiotics will occur only when nalidixic acid-resistant recipient cells receive and express the plasmid encoding ampicillin resistance.
3. Spot a loopful (~5 μl) of the recipient *E. coli* strain onto the plate, and let it dry. Then spot a loopful of the donor *S. typhimurium* on top of the *E. coli* spot. In addition, spot loopfuls of donor alone and recipient alone elsewhere on the plate. Incubate overnight at 37°C.
4. The plate should contain a healthy growth of bacteria only where donor and recipient bacteria were spotted together. Pick cells from this area, and streak for single colonies onto a fresh LB plate containing both antibiotics. Restreak isolated colonies to ensure their purity.
5. A UV mutagenesis experiment along the lines suggested above can now be performed with the pKM101-containing derivative. In comparison with the results obtained with the parental *E. coli* strain not carrying pKM101, a dramatic (10- to 100-fold) increase in mutation yield should be seen.

13.5.3. Fermentation Mutants

One way to screen for a desired mutant is to use a suitable chromogenic substrate in the plating medium. In this way, the mutant colonies can be distinguished from the wild-type ones by a different color in, or adjacent to, the bacterial colonies. One example of this application, presented below, involves the detection of lactose-nonfermenting mutants by incorporating a tetrazolium dye into the plating medium.

1. Grow an inoculum of an appropriate lactose-fermenting bacterium in LB broth (section 13.9.3) or nutrient broth (section 13.9.12).
2. Mutagenize the culture as described for the UV mutagenesis procedure above.
3. Plate appropriate dilutions on lactose-tetrazolium plates (section 13.9.7) so as to obtain 50 to 100 colonies per plate.
4. Incubate until colonies are 2 to 3 mm in diameter. The lactose-positive parents are detected as white colonies, whereas the lactose-negative variants are bright red. The basis for this differential staining is that lactose utilizers produce acid as a product of the fermentation, which in a weakly buffered medium lowers the pH and prevents reduction of the tetrazolium dye. The lactose-nonfermenting colonies have normal ability to reduce the tetrazolium to its red form.

It is important that the colonies not be overcrowded on the plates, because the release of acid by an excess of wild-type colonies will interfere with the reduction of the tetrazolium by the nonfermenting lactose-negative colonies. Dilutions should be adjusted so as to obtain no more than 100 colonies per plate. It is also important that high-quality, glucose-free lactose be used; otherwise, the results may be obscured by the leakiness often observed when lactose is contaminated with glucose.

This general type of screening assay can be used for any mutant for which there is a suitable chromogenic substrate available to distinguish mutant and wild-type strains. Some examples are *p*-nitrophenylphosphate for distinguishing phosphatase mutants, Giemsa or methyl green for detecting nuclease mutants, and diazotized derivatives of casein, collagen, or albumin for detecting proteolytic enzyme mutants. Although gelatin is not a chromogenic indicator per se, its solubilization from an opaque form to a transparent form on plates has long been used as a measure of proteolytic activity. Similarly, nuclease activity can be detected by the solubilization of RNA and DNA incorporated in an acid-insoluble form in agar plates.

13.5.4. Replica Plating for Mutant Selection

In 1952, Lederberg and Lederberg (41) devised a replica-plating technique for indirect selection of bacterial mutants. This technique, diagrammed in Fig. 2, permits the simultaneous transfer of a large number of colonies from one plating medium to another in a single operations by use of cotton velveteen cloth. Since the exact pattern of colonies on the master plate is reproduced on the replica plate, one can rapidly examine large numbers of individual colonies for mutant characteristics. A suitable dilution of mutagenized cells is first plated on the surface of a nonselective complete medium. After colonies form, they are replica plated to a selective or minimal medium. Colonies that grow on the supplemented but not on the minimal plate are then picked and tested for the desired mutation. By this technique one can readily examine thousands of colonies in a short time, without having to pick and transfer each one separately. It is a particularly useful technique for isolating auxotrophic mutants, which grow on a complete medium but not on a minimal medium. By appropriate

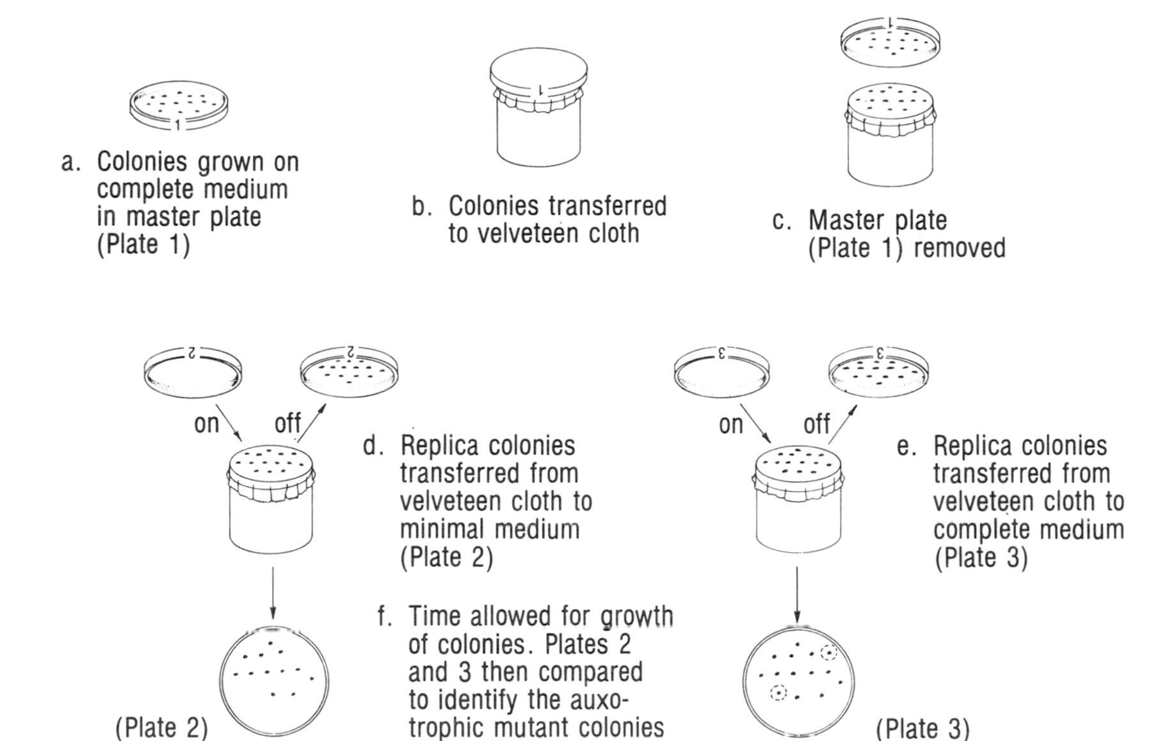

FIG. 2. Replica-plating method for screening auxotrophic mutants of a bacterium.

supplementation of the replica plate, one can identify auxotrophs of a specific type (e.g., a single amino acid, vitamin, or nucleic acid precursor). When used together with a highly efficient mutagen such as nitrosoguanidine, this technique can yield up to 20% total auxotrophs, or 1 to 2% of a particular mutant type. This approach is illustrated by the following procedure for isolating various types of auxotrophic mutants.

1. Cut ordinary cotton velveteen cloth into square pieces about 15 cm on a side. Wrap them in packets of 10 to 20 in heavy paper or foil. Autoclave for 10 to 15 min at 121°C on a fast exhaust and dry cycle. Replica blocks can be made from a no. 15 rubber stopper onto which has been glued (with paper cement) an 8- to 9-mm thick piece of Plexiglas whose outer diameter (8 cm) is flush with the narrow part of the rubber stopper. A metal ring (e.g., outer diameter 100 mm; inner diameter 90 mm; width ~ 14 mm) can be used to secure the velveteen cloth to the replica-plating blocks. Blocks can also be made from aluminum automobile pistons (diameter, 8 cm) or the bottoms of 400-ml beakers. Velveteen cloths can also be secured to the blocks with rubber bands instead of a metal ring.

2. Mutagenize the bacterium of interest, preferably with a mutagen having a wide spectrum of mutagenic effects (such as UV light).

3. Allow the mutagenized cells to grow out in nutrient broth (section 13.9.12) to express the mutations. If a number of independent mutations are desired, distribute the treated cell culture into a number of subcultures before the outgrowth period.

4. Plate appropriate dilutions on the surface of nutrient agar (section 13.9.13) in petri plates and incubate upside down at 37°C. Aim for 100 to 200 colonies per plate.

5. After colonies reach about 3 mm in diameter, replicate them to minimal agar (section 13.9.10) and nutrient agar (section 13.9.13) plates. Remember to mark the orientation of the plates on the replica block. After the colonies have grown, identify those which grew on the nutrient agar but not on the minimal agar plates.

The procedure to this point (and as depicted in Fig. 2) identifies only the total auxotrophs; it does not distinguish among those altered in different metabolic functions. To identify the various types of auxotrophs more specifically, carry out the following additional steps.

6. Pick the identified auxotrophic colonies with sterile toothpicks, streak ~ 0.5-cm-wide patches onto fresh plates of nutrient agar (section 13.9.13) and incubate at 37°C. Each plate should accommodate 40 to 50 such patches. Use a grid drawn on graph paper placed underneath the plate to organize the distribution of patches on the plate.

7. Replica plate the patched clones to supplemented minimal agar plates (section 13.9.10) containing the nine combinations of metabolic pools shown in Table 1, as well as to unsupplemented minimal agar plates.

8. Determine the group(s) of metabolic pools on which a given mutant will grow. From this information it should be possible to deduce the nature of the mutational defect. For example, a mutant that grows only on pools 3 and 6 is most probably a cysteine-requiring mutant; one that grows only on pools 5 and 9 is probably an arginine auxotroph. Test each clone further by streak-

Table 1. Combinations of supplements in minimal agar plates for diagnosis of auxotrophs[a]

	Composition of pool:				
Pool	1	2	3	4	5
6	Adenine	Guanine	Cysteine	Methionine	Thiamine
7	Histidine	Leucine	Isoleucine	Valine	Lysine
8	Phenylalanine	Tyrosine	Trytophan	Threonine	Proline
9	Glutamic acid	Serine	Alanine	Aspartic acid	Arginine

[a]The compositions of pools 1 to 5 are listed vertically and the compositions of pools 6 to 9 are listed horizontally. All supplements are at a final concentration of 20 μg/ml, except for thiamine, which is at 4 μg/ml. Pools may be made up at 10× concentration.

ing it out on a minimal agar plate supplemented with the suspected growth requirement to verify the preliminary conclusions.

Some auxotrophs may respond to only a single metabolic pool or to none at all. The former may be a consequence of several factors: (i) two independent mutations, leading to unrelated growth requirements which happen to be included in the same metabolic pool; (ii) a mutation at an early step in a complex pathway (e.g., a polyaromatic acid-requiring strain, which would grow only on pool 8); or (iii) inhibitory effects of other metabolites in the pool.

Mutants that respond to none of the pools may be (i) mutants requiring some other metabolite (e.g., a different vitamin) or (ii) double mutants requiring a combination of two metabolites not represented in these pools. These mutants can be tested on plates containing a complex vitamin mixture or on RNA hydrolysates or a complete mixture of amino acids (e.g., acid-hydrolyzed casein or Casamino Acids supplemented with L-Tryptophan). If auxotrophs that fail to respond to any of the nine suggested metabolic pools are isolated, try other compounds such as mixtures of vitamins, amino acids, or nucleic acid hydrolysates.

13.5.5. Temperature-Sensitive Mutants

Temperature-sensitive mutations have been exploited for many years by geneticists to help identify genes whose products are essential. The isolation and characterization of temperature-sensitive mutations was first described by Horowitz and Leupold (31). Even at the permissive growth temperature, many temperature-sensitive mutants display phenotypes that provide important clues about the cellular function of the gene and its product.

A procedure for isolating temperature-sensitive mutants employs MNNG, a potent mutagen that has been a staple for bacterial geneticists. MNNG is an alkylating agent that generates high levels of O^6-methyl guanidylate and O^4-methyl thymidylate. When these methylated bases serve as templates during DNA replication, they induce G·C → A·T and A·T → G·C substitutions, respectively. MNNG treatment of *E. coli* can yield mutants at frequencies approaching 1% of the survivors. A significant problem associated with the use of MNNG is its tendency to induce clusters of closely linked mutations (15). Therefore, mutants that have been isolated following MNNG treatment must be carefully analyzed genetically or by studying revertants to assess whether the phenotype is a consequence of a single mutation. To minimize the number of multiple mutations, aim for MNNG treatments that kill no more than 50% of the bacteria. The procedure outlined below is based on the Miller protocol (8).

1. Grow the bacterium to a density of 3×10^{85} to 5×10^8 cells per ml in LB.

2. Centrifuge 5 ml of the culture for 10 min at 8,000 x *g*. Resuspend the pellet in 5 ml of citrate buffer (section 13.9.2.), repeat the centrifugation step two more times, and then resuspend the pellet in 4 ml of citrate buffer. Add 0.2 ml of a 1 mg/ml solution of MNNG in citrate buffer. This will yield a final MNNG concentration of approximately 50 μg/ml. Stock solutions of MNNG in citrate buffer can be prepared in advance and frozen. They are stable for months.

3. Incubate the treated bacteria for 30 min at 37°C without shaking. The time of incubation required to achieve 50% killing may vary from strain to strain and certainly will vary from genus to genus. Therefore, bacterial survival as a function of incubation time must be determined before a large-scale experiment to isolate a desired mutant phenotype is performed.

4. Centrifuge the treated bacteria as in step 2, but now wash the pellet with phosphate buffer. Repeat this once. Resuspend the washed pellet in 10 ml of LB medium, and grow overnight. If screening for temperature-sensitive mutants, incubate the overnight cultures at the permissive temperature. To avoid collecting auxotrophs, grow the bacteria in a minimal medium.

5. Plate aliquots of diluted cells onto broth (or minimal plates) at the permissive temperature, and incubate until colonies are 1 to 2 mm in diameter. Aim for between 150 and 250 colonies per plate. These plates are the master plates.

6. Replicate the master plates onto each of two broth plates. Be sure to mark the master and replica plates on the side or bottom so that the orientation is known. The first replica should be incubated at the nonpermissive temperature, and the second replica, together with the master plate, should be incubated at the permissive temperature. To reduce residual growth of temperature-sensitive cells, prewarm the plates to be incubated at the nonpermissive temperature.

7. After an overnight incubation, remove the plates and "read" them to identify colonies that grow only at the permissive temperature. Mark and pick them and then immediately resuspend them in medium and freeze an aliquot (following the preservation procedures outlined in Chapter 12). These temperature-sensitive isolates are now ready for further characterization and screening.

Other temperature-sensitive mutants can be isolated by similar procedures. Cold-sensitive mutants often identify assembly mutants whose proteins are components of multiprotein "machines" such as ribosomes or the DNA replication apparatus (58).

New classes of conditional-lethal mutants were recently identified by a technique involving the use of heavy water (D_2O) (11). The rationale for this procedure was to identify mutant forms of proteins which do not fold and/or function properly in heavy water. Bartel and Varshavsky (11) developed a screen for yeast mutants that grew poorly or not at all on media containing 90% D_2O but were otherwise robust enough to grow on normal media. Following mutagenesis, the screen enabled them to identify a large number of conditional-lethal mutations affecting the yeast cell cycle. The Bartel and Varshavsky method may be useful for identifying conditional-lethal mutations when growth at different temperatures is not a tractable approach.

13.5.6. Ethyl Methanesulfonate-Induced Mutants

Other alkylating agents that are potent bacterial mutagens include ethyl methanesulfonate, diethyl sulfate, and methyl methanesulfonate (MMS). EMS ethylates guanine at the O-6 position, thereby inducing primarily G·C → A·T transition mutations (16). In *E. coli*, EMS mutagenesis is not dependent on specific DNA repair activities whereas MMS mutagenesis is largely SOS dependent (60). EMS may therefore be a more tractable mutagen for organisms not known to be capable of carrying out SOS mutagenesis.

A representative procedure involving EMS is as follows.

1. Grow a 5-ml bacterial culture to 2×10^8 to 5×10^8 cells per ml. Centrifuge for 10 min at 8,000 × *g*, and resuspend in only 2.5 ml of a minimal medium (section 13.9.9) buffered to pH 7.5 but containing no carbon source.
2. Add 2.5 ml of a freshly prepared EMS stock solution (0.1 ml of EMS in 2.4 ml of minimal medium buffered to pH 7.5) warmed to 37°C.
3. Aerate the mixture on a shaking incubator at 37°C for 1 to 2 h.
4. Wash the bacteria twice by centrifugation and resuspension in minimal medium.
5. Dilute the treated bacteria 10-fold into fresh broth or minimal medium (this time with a carbon source!), and allow them to grow out for several hours or overnight. Dilutions can be plated directly for mutants, enrichment, or in preparation for replica plating.

13.5.7. Frameshift-Induced Mutants

Frameshift mutagens induce the addition or deletion of bases (see, e.g., reference 13). The potency of the frameshift mutagen known as ICR-191 (developed at the eponymous Institute for Cancer Research in Philadelphia) is attributed to its ability to alkylate DNA bases and subsequently intercalate its acridine moiety between the stacked base pairs of DNA. High levels of mutation induction by ICR-191 and other frameshift mutagens appear to require growth for several generations in the presence of the mutagen. A useful procedure based on that of Miller (8) is as follows.

1. Prepare an overnight LB broth (section 13.9.3) culture. Dilute it 10^{-4} (to $\sim 10^5$ cells per ml), and distribute 0.1 ml to each of 10 tubes containing 3 ml of enriched A medium and the following concentrations of ICR-191: 0, 2, 4, 6, 8, 10, 12, 14, 16, and 20 μg/ml. An ICR-191 stock solution at 1 mg/ml in sterile distilled water should be prepared just before use. Culture tubes should be wrapped in foil, and all ICR-191 solutions should be kept in the dark (or in dim light).
2. Incubate the tubes at 37°C overnight on a shaking machine. Culture tubes with intermediate levels of growth can be selected for evaluation of mutation frequency either by direct plating or by screening. Growth is not expected in tubes with the highest ICR-191 concentrations. If the organism has a different sensitivity to ICR-191 from that of *E. coli*, adjust the range of ICR-191 concentrations appropriately.

13.5.8. Transposon Mutants via Phage Mu

As described above (section 13.2.7.3), transposon mutagenesis is a potent tool for genetic analyses. A method for carrying out transposon mutagenesis with Mu phage will be outlined here. Transposon mutagenesis is covered in more detail in Chapter 17.

Mu is a phage that lysogenizes by transposition. Mu derivatives carrying the *lacZ* and *lacY* portions of the *lac* operon can be used to randomly insert *lacZY* into chromosomal DNA. Since the *lacZ* promoter is missing in these constructs, synthesis of ß-galactosidase will occur only if the transposon integrates in the correct orientation downstream from a chromosomal promoter sequence. Expression of ß-galactosidase will then be under the control of that new promoter element.

A virtue of carrying out transposon mutagenesis with an operon fusion system is that the collection of random operon fusions can be screened not only for phenotypes resulting from the insertion of a transposon but also for gene expression (e.g., ß-galactosidase) under specific physiological or environmental circumstances. *lac* operon fusions have been used to identify families of genes whose expression is regulated by DNA damage (34, 55), phosphate deprivation (61), or developmental programs (65). The procedure outlined below uses materials and methods developed for *S. typhimurium* by Hughes and Roth (32). Figure 3 diagrammatically illustrates key features of the procedure.

13.5.8.1. Materials

1. *S. typhimurium* TT10289 (*hisD9953*::MudI1734 *hisA9944*::MudI) carries two insertions in the *his* operon. One insertion, in the *hisD* locus, consists of the MudI1734 element. MudI1734 is a defective Mu phage encoding a kanamycin resistance determinant and the *lacZ* and *lacY* genes without a promoter. The Mu sequences at the ends of MudI1734 are substrates for the Mu specific transposase. MudI1734 is itself defective for transposition. The second insertion, MudI, is located in the *hisA* locus, approximately 4 kb away from the first insertion. MudI is fully functional for transposition because it carries the functional transposase genes A and B. MudI is 37 kb in length, and MudI1734 is 11.3 kb in length.
2. Phage P22 is a temperate *S. typhimurium*-transducing phage that packages 44 kb of chromosomal DNA. Therefore, P22 cannot package enough chromosomal DNA to simultaneously transduce both Mu elements in strain TT10289 in their entirety (37 + 4 +

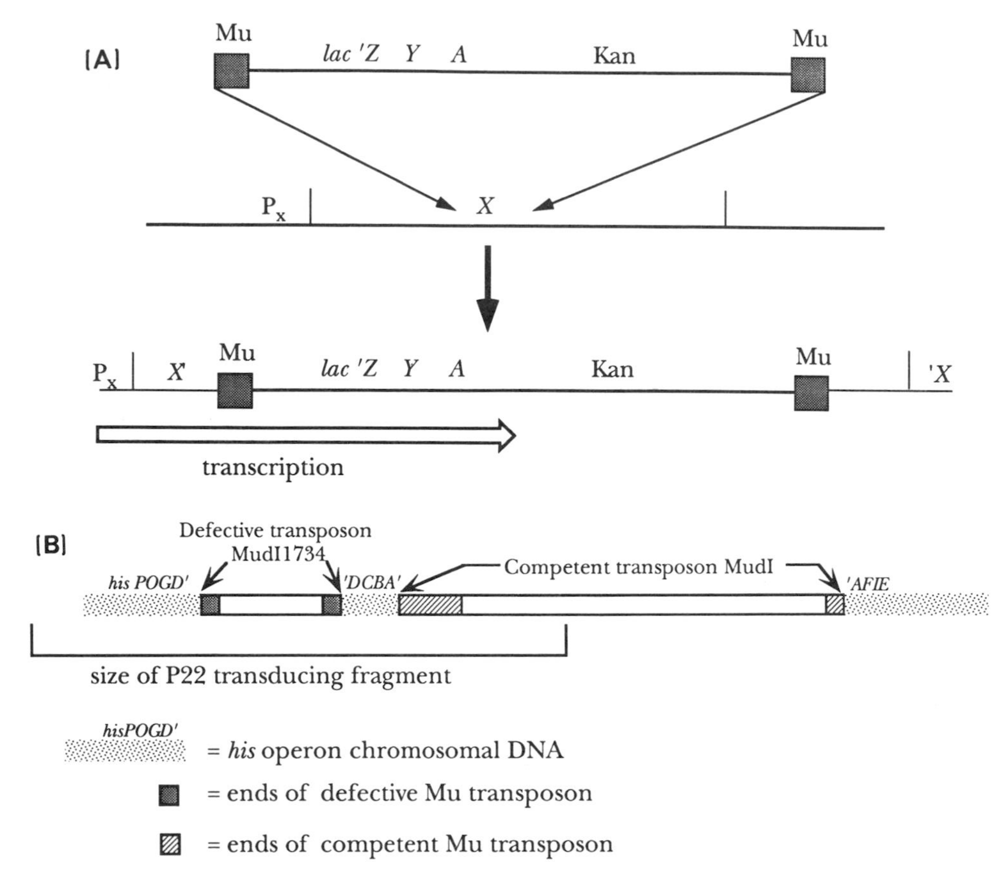

FIG. 3. Transposition mutagenesis with MudI1734. (A) Structure and transposition of MudI1734. The ends of Mu (shaded) flank DNA encoding a promoterless *lac* operon and kanamycin resistance. In the presence of Mu transposase, the transposon can insert itself randomly into a gene (e.g., *X*) on chromosomal DNA. If the MudI1734 transposes into gene *X* in the orientation shown above, transcription of *lac'ZYA* will fall under the control of the *X* promoter element P_X. The *X* gene will be inactivated by the insertion event, and cells will display an X^- phenotype. (B) Configuration of MudI1734 and MudI on the donor chromosome of *S. typhimurium* TT10289 (32). The two Mud insertions lie within the *his* operon in close proximity to each other. The left end of MudI codes for the transposition functions of Mu. If a P22 transducing fragment that includes all of MudI1734 plus the left end of MudI is introduced into a recipient cell, the transposition proteins are synthesized and act to promote transposition of the MudI1734 element into the recipient chromosome.

11.3 kb = 52.3 kb). However, P22 can package sufficient DNA to include all of the MudI1734 element plus the neighboring transposase genes, A and B, from the nearby MudI element. Such transducing fragments, when introduced into a recipient cell, will transiently express the transposase proteins and permit transposition of MudI1734 into the chromosome. Recipients that harbor a transposed MudI1734 can be selected by demanding growth on medium containing kanamycin. Kanamycin-resistant bacteria can then be screened for expression of ß-galactosidase under conditions of interest to the investigator. The P22 strain used in this procedure is a variant (P22 HT, *int*) that transduces at high frequencies and is integration defective (2).

13.5.8.2. Procedures

Because strain TT10289 carries a "live" transposon, one must screen for cultures in which the MudI1734 element has not transposed to another site. This is done by transducing a collection of independent cultures to His^+ and asking whether the transductants have simultaneously lost kanamycin resistance. Cultures that yield fewer than 1% kanamycin-resistant His^+ colonies have not accumulated Mu insertions elsewhere, and P22 transducing lysates prepared on these cultures can be used for the transposition procedure.

1. Streak *S. typhimurium* TT10289 onto a nutrient broth (section 13.9.12) plate containing both kanamycin (50 μg/ml) and ampicillin (30 μg/ml), and incubate overnight at 30°C.
2. Pick five to eight single colonies to start independent overnight cultures in 2 ml of nutrient broth at 30°C.
3. Remove a 1-ml aliquot from each culture, and use it as a recipient in a transduction experiment with P22 grown on a His^+ *S. typhimurium* strain. Use the rest for preparing a P22 lysate. The lysate, transduction, and transposition procedures are as follows.

Lysate preparation

1. Prepare a fresh P22 (HT, *int*) stock by suspending a single plaque in 4 ml of P22 broth (section 13.9.6). The suspended plaque will yield a suspension of about 10^6 phage per ml. P22 plaques can be generated by mixing 0.1 ml of various dilutions of a P22 stock with 0.1 ml of an overnight culture of a His$^+$ *S. typhimurium* in 2 ml of LB top agar (section 13.9.5) and pouring onto an LB plate (section 13.9.4). The plates should not be too dry (judge by observing how moist and unwrinkled the plate surface is).

2. Prepare a high-titer P22 stock by adding 1 ml of an overnight bacterial culture to 4 ml of the resuspended plaque and incubating from 5 to 18 h at 37°C. The culture may not show much lysis. Remove cells by centrifuging for 10 min at 8,000 × *g*, and store the supernatant over 0.5 ml of chloroform. The phage preparation should contain 10^{10} to 10^{11} phage per ml.

3. When the P22 lysate for the transposition experiment (see the transduction procedure below) is identified, determine its titer by mixing 0.1 ml of a series of dilutions (from 10^{-6} to 10^{-8}) with 0.1 ml of an overnight culture of *S. typhimurium* in 2 ml of LB top agar, pouring onto LB plates, and incubating overnight at 37°C.

Transduction procedure

1. Mix 0.1 ml each of undiluted and 10^{-1}, 10^{-2}, and 10^{-3}-fold-diluted P22 grown on a His$^+$ strain of *S. typhimurium* with 0.1 ml of the independent cultures of TT10289 in sterile tubes (10 by 75 mm). Incubate for 20 min at room temperature, add 20 μl of 1 M EGTA [ethylene glycol-bis (β-aminoethyl ether)-*N,N,N',N'*-tetraacetic acid] (to prevent further cycles of P22 infection), and mix.

2. Add 2 ml of minimal top agar (section 13.9.10), and pour immediately onto a minimal medium (section 13.9.9) plate. When the agar hardens, incubate plates upside down at 37°C for 16 to 24 h. Prepare control plates that have bacteria only (no phage) and phage only (no bacteria) to verify the procedure.

3. Grid 100 His$^+$ colonies from each transduction onto minimal plates. When these colonies have grown, replica plate them onto LB plates containing 50 μg of kanamycin per ml. The P22 lysates from cultures yielding 0 or 1 kanamycin-resistant colony can be used for the transposition procedure.

Transposition procedure

1. Grow the *S. typhimurium* to be mutagenized in 10 ml of LB broth at 37°C overnight.

2. Mix 3.75 ml of P22 broth (section 13.9.6), 1.5 ml of a P22 lysate (adjusted to a titer of 4 to 5 × 10^9 cells per ml) grown on TT10289, and 3.75 ml of the *S. typhimurium* overnight culture. The total volume of this mixture is 9 ml, and the multiplicity of infection (ratio of phage particles to bacteria) is ~0.5.

3. Incubate for 20 min at room temperature with occasional manual shaking.

4. Add 15 ml of P22 broth containing 15 mM EGTA (to inhibit further cycles of P22 infection).

5. Incubate for an additional 60 min at 37°C.

6. Centrifuge the bacteria (8,000 × *g* for 10 min at 4°C), and wash twice with 25 ml of M9 medium (section 13.9.9) with no carbon source and containing 10 mM EGTA.

7. Resuspend the final pellet in 5 ml of M9 medium containing 10 mM EGTA and 15% glycerol.

8. Divide 4.5 ml of suspended bacteria into a series of 1- and 0.5-ml portions, and freeze at −70°C. This will provide stocks of mutagenized bacteria which can be plated and screened for mutants and/or ß galactosidase expression.

9. Plate duplicate 0.1-ml aliquots of the suspended bacteria directly onto LB plates (section 13.9.4) containing 50 μg of kanamycin per ml. Plate duplicate 0.1-ml aliquots of 10^{-6} dilutions of the suspended bacteria onto LB plates with no antibiotic. The number of viable bacteria in the stocks should be approximately 3 × 10^9 cells per ml. The number of kanamycin-resistant bacteria should be approximately 6,000 cells per ml; 60% of these will be transposition mutants. The remainder will have arisen as a result of homologous recombination at the *his* locus. Thus, in 5 ml there will be over 15,000 mutants to screen for a favorite phenotype! (To avoid the latter class of kanamycin-resistant bacteria, plate on a M9 medium supplemented with a cocktail containing all amino acids except for histidine. If seeking auxotrophic mutants arising from transposition, supplement the medium with other essential nutrients and vitamins.)

10. Thaw a frozen aliquot of cells, and plate an appropriate volume of the bacterial suspension on LB-kanamycin plates to achieve a distribution of 200 to 300 colonies per plate.

11. Incubate the LB-kanamycin plates at 37°C overnight.

12. Replica plate to screen for the desired phenotype. If the chromogenic β-galactosidase substrate 5-bromo-4-chloro-3-indolyl-β-D-galactopyranoside (X-Gal) (34) is incorporated into the plates, the blue colonies will represent those expressing *lacZ*. By replica plating bacteria on plates with and without a DNA- alkylating agent (e.g., MNNG), one can identify insertion mutations that express *lacZ* only on plates containing an alkylating agent. A collection of such mutants could be the start of an investigation of genes that are specifically inducible by alkylating agents.

13.6 DETECTION OF AUXOTROPH REVERSION

The following methods are based on the detailed procedures developed by Ames and his colleagues for testing the mutagenicity of chemicals and environmental samples. The His$^-$ *S. typhimurium* strains used in the Ames test (45) carry defects in DNA repair and in the lipopolysaccharide coat as well as a mutation-enhancing plasmid, the combined effects of which are to increase the sensitivity of the organisms to mutagenic treatments. One can also incubate the tester strains in the presence of liver homogenates to metabolize chemicals to the highly reactive and carcinogenic species (usually epoxides) that are normally generated as a byproduct of xenobiotic metabolism. The procedure outlined below focuses simply on the essentials of monitoring reversion of His auxotrophs and can easily be adapted for detecting or characterizing reversion of any auxotrophy.

For details about the **Ames test for mutagenicity,** consult reference 45.

Preparation of materials

1. Pour 30 ml of E minimal agar (Section 13.9.11), containing 1.5% agar and 2% glucose, into petri plates (100 by 15 mm).

2. Prepare 100 ml of top agar (section 13.9.15) containing 0.65% agar and 0.5% NaCl. Add L-histidine·HCl to 0.05 mM final concentrations (10 ml of a stock solution which is 0.5 mM histidine added to 100 ml of top agar), and maintain at 45°C until used. Additional required nutrients can be added as needed to the top agar (or incorporated beforehand in the bottom agar).
3. Distribute 2 ml of top agar to sterile tubes (10 by 75 mm).

Assay procedure
1. For the standard pour plate assay, add 0.1 ml of an overnight culture of a His auxotroph grown in Oxoid broth (section 13.9.8) to 2 ml of top agar (section 13.9.15) at 45°C.
2. Add no more than 0.5 ml of the compound whose mutagenicity is to be tested. Compounds can be dissolved in water, DMSO, ethanol, or formamide, depending on their solubility. Experiment with a range of solvents and concentrations to arrive at mixtures that are not toxic to the bacteria. In addition to the test plates, control plates of bacteria alone and bacteria plus known mutagens should be prepared.
3. Mix the contents immediately on a Vortex mixer or by rotating the tube between the palms, and pour the contents over the surface of a minimal E glucose agar plate (section 13.9.11). To prevent immediate gelling of the top agar, the agar plates should be no cooler than 25°C.
4. Tilt the plate gently both back and forth and side to side to evenly distribute the top agar layer. The entire operation should be carried out in 20 s or less.
5. Allow the agar to harden in the dark for a few minutes; then incubate the plates upside down in the dark at 37°C.
6. After 2 days, score the histidine revertants (prototrophs). A slight background growth of bacteria should be visible on the plate as a result of the small amount of histidine added to the medium. The growth-limiting amounts of histidine permit growth of all bacteria in the population of five or six generations and allow processing of DNA damage as well as segregation and expression of the mutant phenotype.

Mutagens can also be tested by a spot test. In that case the mutagen is omitted from the top agar mixture and is applied after the top agar containing the bacterial inoculum has been poured and hardened. The mutagen can be added directly as a few crystals or onto a sterile 6-mm filter disk as an aqueous, alcoholic, or DMSO solution.

CAUTION: S. typhimurium is a human and mouse pathogen. It is one of the main agents of food poisoning and, if ingested in high doses, can cause severe diarrhea, vomiting, and gastroenteritis. Although most laboratory strains of *S. typhimurium* LT2 are attenuated, especially those lacking the external lipopolysaccharide capsule, they should still be handled with caution. All pipettes should be cotton plugged, and mouth pipetting of the organism should be avoided. Cultures should be autoclaved before disposal. Laboratory workers taking antibiotics should be especially cautious in working with strains carrying plasmids encoding antibiotic resistances. Also see the caveats in section 13.8.

Table 2. *E. coli* strains for monitoring specific mutational events[a]

E. coli strain	Reversion event to restore phenotype
CC101	A·T→C·G
CC102	G·C→A·T
CC103	G·C→C·G
CC104	G·C→T·A
CC105	A·T→T·A
CC106	A·T→G·C
CC107	+1G
CC108	−1G
CC109	−2 (-C-G-)
CC110	+1A
CC111	−1A

[a]For more details, see references 16 and 17.

13.7. Lac⁻ STRAINS FOR MUTAGENICITY TESTING

Cupples and Miller (17, 18) have described the construction and use of 11 *lacZ* alleles (Table 2) that permit the rapid detection of specific base substitution or frameshift mutations by selecting for reversion of a Lac^- to a Lac^+ strain. The assay procedure is as follows.
1. Grow Lac^- strains overnight in LB broth at 37°C.
2. Spread 0.1 ml of each overnight culture with 2 ml of top agar (section 13.9.15) onto Lac minimal plates (M9 medium [section 13.9.9] containing 0.5% lactose).

As in the His reversion procedure, up to 0.5 ml of a compound to be tested can be added to the soft agar. Alternatively, pure compound as a crystal or a concentrated solution can be added to the plates after the top agar has hardened. Experiment with plating conditions (e.g., adding growth-limiting amounts of broth or a carbon source such as glucose or glycerol) to enhance the mutagenic response.

For quantitative mutagenesis with these strains, follow the protocols such as those provided for UV (section 13.5.2) or MNNG (section 13.5.5) mutagenesis and determine survival on glucose minimal plates as well as reversion on lactose minimal plates.

13.8. CAVEATS FOR HANDLING MUTAGENS, CARCINOGENS, AND PATHOGENS

1. Never pipette mutagens by mouth!
2. Wear gloves and a laboratory coat.
3. Dispense solid chemicals or concentrated solutions in a certified fume hood.
4. To weigh mutagenic chemicals, use tared (preweighed) tubes. Solid chemicals can be dispensed with disposable wooden applicator sticks that have been split to leave flat surfaces.
5. To dispose of mutagens and carcinogens, consult with the local Environmental Safety Office.
6. Also see Chapter 29 on laboratory safety!

13.9. RECIPES

13.9.1. Phosphate-Buffered Saline (buffer for suspending bacteria)

NaCl.................................... 8.5 g

Add to 1 liter of 0.066 M $NaPO_4$ buffer adjusted to pH 7.0 with 2 N NaOH. Autoclave for 15 min at 15 lb/in^2 (121°C).

13.9.2. Citrate Buffer (for MNNG mutagenesis)

Citric acid .... 10.5 g
NaOH .... 4.4 g

Add to 500 ml of distilled water, and adjust to pH 5.5 with 2 N NaOH.

13.9.3 LB Broth (complex medium for rapid growth of *E. coli* and other members of the *Enterobacteriaceae*)

Tryptone .... 10 g
Yeast extract .... 5 g
NaCl .... 5 g
Distilled water .... 1 liter

Add 1 M NaOH to adjust pH to 7.0. Autoclave for 15 min at 121°C.

13.9.4. LB Agar (standard rich plating medium for many bacteria)

Add 15 g of agar per liter of LB broth. Pour 20 to 25 ml into petri plates (100 by 15 mm).

13.9.5. LB Top Agar (for titer determination of P22 lysates)

Use 8 g of agar per 100 ml of LB medium. Autoclave as for LB broth, and store at room temperature in screw-cap bottles. Top agar can be melted in boiling water or in a microwave oven. *CAUTION:* To avoid a messy and possibly dangerous explosion, loosen or, better, remove the bottle cap before turning on the microwave!

13.9.6. P22 Broth (for preparing and storing P22 transducing lysates)

LB medium .... 97 ml
50 × E salts (section 13.9.11) .... 2 ml
20% glucose .... 1 ml

13.9.7. Lactose-Tetrazolium Plates (for detecting lactose-nonfermenting mutants)

Difco antibiotic medium no. 2 .... 25.5 g
Distilled water .... 950 ml
2,3,5-Triphenyltetrazolium chloride (Sigma Chemical Co.) .... 50 mg

Dissolve the antibiotic medium no. 2 first; then add the tetrazolium chloride, and heat until it dissolves *completely*. Autoclave for 15 min at 121°C, and add 15 ml of filter-sterilized, glucose-free 20% lactose after autoclaving. Pour into petri plates (100 by 15 mm).

13.9.8. Oxoid Broth

Oxoid medium no. 2 .... 25 g
Distilled water .... 1 liter

Autoclave at 121°C for 15 min to sterilize.

13.9.9. M9 Minimal Medium (standard minimal growth medium for *E. coli* and other bacteria not having fastidious growth requirements)

Na_2HPO_4 .... 6 g
KH_2O_4 .... 3 g
NaCl .... 0.5 g
NH_4Cl .... 1 g

Add distilled water to 1 liter, and autoclave for 15 min at 121°C. After the medium has cooled to about 50°C aseptically add:

1 M $MgSO_4 \cdot 7H_2O$ .... 1 ml
0.01 M $CaCl_2$ .... 10 ml

Supplement with 10 ml of a sterile 20% glucose solution for use as a growth medium.

13.9.10. M9 Minimal Agar (basic plating medium)

Prepare 100-ml aliquots of 10× M9 minimal medium, and sterilize by autoclaving. Add 15 g of agar to 900 ml distilled water, and autoclave for 15 min at 121°C. After the medium has cooled to about 60°C, aseptically add 100 ml of sterile 10 × M9, 10 ml of 20% glucose, and amino acid and vitamin pools as required. Mix well, allow bubbles to dissipate, and pour into plates. If bubbles persist on plates, blow them off by quickly passing a Bunsen burner flame over the surface of the still-molten agar.

Agar is a biological product obtained from seaweed and can contain chemical species (metals, sugars, amino acids, etc.) in sufficient concentrations to confound a microbiological experiment. One should therefore be alert to the possibility that impurities in agar will interfere with some experiments. For example, to make minimal plates that are limiting in phosphate, high-quality agar that has been extensively washed must be used (Chapter 9.1.1).

13.9.11. E medium (standard minimal medium for *S. typhimurium*)

A stock solution of 50× E medium is prepared by dissolving each of the following salts in the order given.

Distilled water .... 670 ml
$MgSO_4 \cdot 7H_2O$ .... 10 g
Citric acid·$1H_2O$ .... 100 g
K_2HPO_4, anhydrous .... 500 g
$NaHNH_4PO_4 \cdot 4H_2O$ .... 175 g

Store the 50× stock solution over chloroform, and sterilize when diluted to 1×. Add sugars and other nutritional supplements as required. For E plates, use 15 g of agar per liter of 1× E medium.

13.9.12. Nutrient Broth (general-purpose medium)

Nutrient broth powder .... 8 g
Distilled water .... 1 liter

13.9.13. Nutrient Agar

Nutrient broth powder .... 8 g
Agar .... 15 g
Distilled water .... 1 liter

Autoclave at 121°C for 15 min. Pour 20 to 25 ml into petri plates (100 by 15 mm).

13.9.14. Antibiotic Agar (for selection of antibiotic-resistant mutants by the gradient plate technique and for plasmid curing)

After autoclaving nutrient agar (section 13.9.13), add one of the following antibiotic solutions, sterilized by membrane filtration.

Streptomycin sulfate (Sigma) .... 100 μg/ml
Penicillin G (Sigma or Squibb) .... 500 μg/ml
Rifampin (Sigma) .... 100 μg/ml

13.9.15. Top Agar (for reversion tests)

NaCl .... 5 g
Agar .... 6.5 g
Distilled water .... 1 liter

Dissolve by heating, and dispense 100-ml amounts into bottles. Autoclave for 15 min at 121°C, and cool to about 50°C. Add, after autoclaving, per 100 ml of top agar, 10 ml of a sterile stock mixture of

L-Histidine·HCl (0.5 mM) .... 10.5 mg/100 ml

In 2 ml of this top agar, introduce 100 nmol of histidine per plate. This is sufficient to permit five or six generations of growth from an initial inoculum of ~ 2 × 10^8 to ~ 1 × 10^{10} bacteria per plate. Other nutrients should be added to top agar or incorporated into bottom agar as required.

Top agar can be stored at room temperature for many weeks in glass bottles and melted as needed in a boiling-water bath or in a microwave oven. *CAUTION:* To avoid a messy and possibly dangerous explosion, loosen or, better, remove the bottle cap before turning on the microwave!

13.10. SOURCES OF STRAINS AND SPECIAL MATERIALS

13.10.1. Strains

Escherichia coli strains: E. coli Genetic Stock Center, Dr. Barbara Bachmann, Department of Human Genetics, Yale University School of Medicine, 333 Cedar St., New Haven, CT 06510

Lac⁻Escherichia coli tester strains: Dr. Jeffrey H. Miller, Department of Biology, UCLA, 405 Hilgard Avenue, Los Angeles, CA 90024

Bacillus subtilis strains: Curator, *Bacillus* Genetic Stock Center, Department of Microbiology, Ohio State University, Columbus, OH 43210

Salmonella typhimurium strains for mutagen testing: Dr. Bruce N. Ames, Department of Biochemistry, University of California, Berkeley, CA 94720

Other Salmonella typhimurium strains: Dr. Kenneth E. Sanderson, *Salmonella* Genetic Stock Center, Department of Biological Sciences, University of Calgary, Alberta, Canada T2N 1N4

13.10.2. Media, Chemicals, and Reagents

Most of these following materials can be obtained through scientific supply houses.

Growth media. Agar, antibiotic media, nutrient broth, dextrose, *etc.*: Difco Laboratories, Detroit, MI 48232; BBL Microbiology Systems, Cockeysville, MD 21030

Phosphate salts and other inorganic salts. Mallinckrodt Chemicals, St. Louis, MO 63147

Amino acids, 2,3,5-triphenyltetrazolium chloride, 2-aminopurine, ethyl methanesulfonate, nitrosoguanidine, rifampicin, ampicillin, streptomycin sulfate: Sigma Chemical Co., P.O. Box 14508, St. Louis, MO 63178

ICR-191. Polysciences, Inc., Paul Valley Industrial Park, Wamington, PA 18976

Lactose. Lactose CHR (glucose-free): U.S. Biochemical Corp., Cleveland, OH 44122

MMS, EMS. Eastman Organic Chemicals, Rochester, NY 14650

Casamino Acids. Acid-hydrolyzed casein or Difco vitamin-free Casamino Acids: Difco Laboratories, Detroit, MI 48232. Casein hydrolysate-acid: U.S. Biochemical Corp., Cleveland, OH 44122

DMSO (spectral grade). Schwarz/Mann, Orangeburg, NY 10962

Oxoid medium no. 2. K. C. Biological, P.O. Box 5441, Lenaxa, KS 66215

13.11. REFERENCES

13.11.1. General References

1. **Carlton, B. C., and B. J. Brown.** 1981. Gene mutation, p. 222–242. *In* P. Gerhardt, R. G. E. Murray, R.N. Costilow, E. W. Nester, W. A. Wood, N. R. Krieg, and G. B. Phillips (ed.), *Manual of Methods for General Bacteriology.* American Society for Microbiology, Washington, D.C.
The original version of this chapter.
2. **Davis, R. W., D. Botstein, and J. R. Roth.** 1980. *A Manual for Genetic Engineering: Advanced Bacterial Genetics.* Cold Spring Harbor Laboratory, Cold Spring Harbor, N.Y.
A practical guide to the genetics of E. coli and S. typhimurium with an emphasis on the use of bacteriophages and transposons. Full of many useful tables and tips for making media.
3. **Drake, J. W.** 1970. *The Molecular Basis of Mutation.* Holden-Day, San Francisco.
Highly worthwhile history for background reading on the subject of mutagenesis.
4. **Eisenstadt, E.** 1987. Analysis of mutagenesis, p. 1016–1033. *In* J. L. Ingraham, K. B. Low, B. Magasanik, F. C. Neidhardt, M. Schaechter, and H. E. Umbarger (ed.), *Escherichia coli and Salmonella typhimurium: Cellular and Molecular Biology.* American Society for Microbiology, Washington, D.C.
A review of mutagenesis and methods for analyzing mutational spectra in E. coli and S. typhimurium.
5. **Harwood, C. R., and S. M. Cutting.** 1990. *Molecular Biological Methods for Bacillus.* John Wiley & Sons, Chichester, England.
A modern manual devoted to manipulating Bacillus species.

6. **Hayes, W.** 1968. *The Genetics of Bacteria and Their Viruses: Studies in Basic Genetics and Molecular Biology.* John Wiley & Sons, Inc., New York.
An unparalleled account of the fundamentals of bacterial genetics.
7. **Miller, J. H. (ed.).** 1991. Bacterial genetic systems. *Methods Enzymol.* **204:**1–706.
A collection of genetic methods for a diverse collection of bacteria including E. coli, S. typhimurium, B. subtilis, Haemophilus influenzae, Neisseria spp., myxobacteria, Caulobacter crescentus, Agrobacterium spp., Rhizobium meliloti, cyanobacteria, Streptomyces spp., members of the family Rhodospirillaceae, Pseudomonas spp., Vibrio spp., mycobacteria, streptococci, and staphylococci. The volume includes a beautiful introductory chapter by Jon Beckwith on strategies for finding mutants.
8. **Miller, J. H.** 1992. *A Short Course in Bacterial Genetics. A Laboratory Manual and Handbook for Escherichia coli and Related Bacteria.* Cold Spring Harbor Laboratory, Cold Spring Harbor, N.Y.
This new publication in two volumes is intended for students with training in biochemistry and recombinant DNA techniques but without training in bacterial genetics techniques. The laboratory manual contains units with 34 experiments on working with bacterial strains, the lac system, mutagenesis, gene transfer, and transposable elements. The handbook is an extensive, detailed compilation of reference information on genetic maps, vectors, data bases, tables, charts, recipes, etc.

13.11.2. Specific References

9. **Alper, M. D., and B. N. Ames.** 1975. Positive selection of mutants with deletions of the *gal-chl* region of the *Salmonella* chromosome as a screening procedure for mutagens that cause deletions. *J. Bacteriol.* **121:**259–266.
10. **American Society for Microbiology.** 1991. *ASM Style Manual for Journals and Books*, p. 33–38. American Society for Microbiology, Washington, D.C.
11. **Bartel, B., and A. Varshavsky.** 1988. Hypersensitivity to heavy water: a new conditional phenotype. *Cell* **52:**935–941.
12. **Belas, R., D. Erskine, and D. Flaherty.** 1991. Transposon mutagenesis in *Proteus mirabilis. J. Bacteriol.* **173:**6289–6293.
13. **Calos, M. P., and J. H. Miller.** 1981. Genetic and sequence analysis of frameshift mutations induced by ICR-191. *J. Mol. Biol.* **153:**39–66.
14. **Cebula, T. A., and W. H. Koch.** 1990. Sequence analysis of *Salmonella typhimurium* revertants. *Clin. Biol. Res.* **340:**367–377.
15. **Cerdá-Olmedo, E., P. C. Hanawalt, and N. Guerola** 1968. Mutagenesis of the replication point by nitrosoguanidine: map and pattern of replication of the *Escherichia coli* chromosome. *J. Mol. Biol.* **33:**705–719.
16. **Coulondre, C., and J. H. Miller.** 1977. Genetic studies of the *lac* repressor. IV. Mutagenic specificity in the *lacI* gene of *Escherichia coli. J. Mol. Biol.* **117:**577–606.
17. **Cupples, C. G., and J. Miller.** 1989. A set of *lacZ* mutations in *Escherichia coli* that allow rapid detection of specific frameshift mutations. *Genetics* **125:**275– 280.
18. **Cupples, C. G., and J. Miller.** 1989. A set of *lacZ* mutations in *Escherichia coli* which allow rapid detection of each of the six base substitutions. *Proc. Natl. Acad. Sci. USA* **86:**5345–5349.
19. **Dassarma, S., J. T. Halladay, J. G. Jones, J. W. Donovan, P. J. Giannasca, and N. Tandeau de Marsac.** 1988. High-frequency mutations in a plasmid-encoded gas vesicle gene in *Halobacterium halobium. Proc. Natl. Acad. Sci. USA* **85:**6861–6865.
20. **Davis, B. D.** 1948. Isolation of biochemically deficient mutants of bacteria by penicillin. *J. Am. Chem. Soc.* **70:**4267.
21. **De Lucia, P., and J. Cairns.** 1969. Isolation of an *E. coli* strain with a mutation affecting DNA polymerase. *Nature* (London) **224:**1164–1166.
22. **Demerec, M., E. A. Adelberg, A. J. Clark, and P. E. Hartman.** 1966. A proposal for a uniform nomenclature in bacterial genetics. Genetics **54:**61–76.
23. **Dunkel, V. C., E. Zeiger, D. Brusick, E. McCoy, D. McGregor, K. Mortelmans, H. S. Rosencranz, and V. F. Simmon.** 1985. Reproducibility of microbial mutagenicity assays. II. Testing of carcinogens and noncarcinogens in *Salmonella typhimurium* and *Escherichia coli. Environ. Mutagen.* **7** (Suppl. 5):1–248.
24. **Eisenstadt, E., J. K. Miller, L.-S. Kahng, and W. M. Barnes.** 1989. Influence of *uvrB* and pKM101 on the spectrum of spontaneous, UV and γ-ray induced base substitutions that revert *hisG46* in *Salmonella typhimurium. Mutat. Res.* **210:**113–125.
25. **Engstrom, J., A. Wong, and R. Maurer.** 1986. Interaction of DNA polymerase III γ and β subunits *in vivo* in *Salmonella typhimurium. Genetics* **113:**499–516.
26. **Farabaugh, P., U. Schmeissner, M. Hofer, and J. H. Miller.** 1978. Genetic studies of the *lac* repressor. VII. On the molecular nature of spontaneous hotspots in the *lacI* gene of *Escherichia coli. J. Mol. Biol.* **126:**847–863.
27. **Grogan, D. W.** 1991. Selectable mutant phenotypes of the extremely thermophilic archaebacterium *Sulfolobus acidocaldarius. J. Bacteriol.* **173:**7725–7727.
28. **Harm, W.** 1980. *Biological Effects of Ultraviolet Radiation.* Cambridge University Press, Cambridge.
29. **Hartman, P. E., Z. Hartman, R. C. Stahl, and B. N. Ames.** 1971. Classification and mapping of spontaneous and induced mutations in the histidine operon of *Salmonella. Adv. Genet.* **16:**1–34.
30. **Hofemeister, J., H. Kohler, and V. D. Filippov.** 1979. DNA repair in *Proteus mirabilis.* VI. Plasmid (R46-)mediated recovery and UV mutagenesis. *Mol. Gen. Genet.* **176:**265–273.
31. **Horowitz, N. H., and U. Leupold.** 1951. Some recent studies bearing on the one gene-one enzyme hypothesis. *Cold Spring Harbor Symp. Quant. Biol.* **16:**65–74.
32. **Hughes, K. T., and J. R. Roth.** 1988. Transitory *cis* complementation: a method for providing transposition functions to defective transposons. *Genetics* **119:**9–12.
33. **Kato, T., and Y. Shinoura.** 1977. Isolation and characterization of mutants of *Escherichia coli* deficient in induction of mutations by ultraviolet light. *Mol. Gen. Genet.* **156:**121–131.
34. **Kenyon, C. J., and G. C. Walker.** 1980. DNA damaging agents stimulate gene expression at specific loci in *Escherichia coli. Proc. Natl. Acad. Sci. USA* **77:**2819–2823.
35. **Koch, W. H., T. A. Cebula, P. L. Foster, and E. Eisenstadt.** 1992. UV mutagenesis in *Salmonella typhimurium* is *umuDC* dependent despite the presence of *samAB. J. Bacteriol.* **174:**2809–2815.
36. **Kondo, S., A. Yamagishi, and T. Oshima.** 1991. Positive selection for uracil auxotrophs of the sulfur-dependent thermophilic archaebacterium *Sulfolobus acidocaldarius* by use of 5-fluoroorotic acid. *J. Bacteriol.* **173:**7698–7700.
37. **Kroos, L., and D. Kaiser.** 1984. Construction of Tn *lac*, a transposon that fuses *lacZ* expression to exogenous promoters, and its introduction into *Myxococcus xanthus. Proc. Natl. Acad. Sci. USA* **81:**5816–5820.
38. **Lam, S., and J. R. Roth.** 1983. IS200: a *Salmonella*-specific insertion sequence. *Cell* **34:**951–960.
39. **Lam, W. L., and W. F. Doolittle.** 1989. Shuttle vectors for the archaebacterium *Halobacterium volcanii. Proc. Natl. Acad. Sci. USA* **86:**5478–5482.
40. **Lancy, E. D., M. R. Lifsics, D. G. Kehres, and R. Maurer.** 1989. Isolation and characterization of mutants with deletions in *dnaQ*, the gene for the editing subunit of DNA polymerase III in *Salmonella typhimurium. J. Bacteriol.* **171:**5572–5280.
41. **Lederberg, J., and E. M. Lederberg.** 1952. Replica plating and indirect selection of bacterial mutants. *J. Bacteriol.* **63:**399–406.

42. **Lederberg, J., and N. D. Zinder.** 1948. Concentration of biochemical mutants of bacteria with penicillin. *J. Am. Chem. Soc.* **70:**4267–4268.
43. **Levin, D. E., and B. N. Ames.** 1986. Classifying mutagens as to their specificity in causing the six possible transitions and transversions: a simple analysis using the *Salmonella* mutagenicity assay. *Environ. Mutagen.* **8:**9–28.
44. **Macnab, R. M.** 1987. Flagella, p. 70–83. *In* J. L. Ingraham, K. B. Low, B. Magasanik, F. C. Neidhardt, M. Schaechter, and H. E. Umbarger (ed.), *Escherichia coli and Salmonella typhimurium: Cellular and Molecular Biology.* American Society for Microbiology, Washington, D.C.
45. **Maron, D. M. and B. N. Ames.** 1983. Revised methods for the *Salmonella* mutagenicity test. *Mutat. Res.* **113:**173–215.
46. **Matsunaga, T., C. Nakamura, J. G. Burgess, and K. Sode.** 1992. Gene transfer in magnetic bacteria: transposon mutagenesis and cloning of genomic DNA fragments required for magnetosome synthesis. *J. Bacteriol.* **174:**2748–2753.
47. **McCann, J., E. Choi, E. Yamasaki, and B. N. Ames.** 1975. Detection of carcinogens as mutagens in the Salmonella/microsome test: assay of 300 chemicals. *Proc. Natl. Acad. Sci. USA* **72:**5135–5139.
48. **Micheletti, P. A., K. Sment, and J. Konisky.** 1991. Isolation of a coenzyme M-auxotrophic mutant and transformation by electroporation in *Methanococcus voltae. J. Bacteriol.* **173:**3414–3418.
49. **Miller, J. K., and W. M. Barnes.** 1986. Colony probing as an alternative to standard sequencing as a means of direct analysis of chromosomal DNA to determine the spectrum of single-base changes in regions of known sequence. *Proc. Natl. Acad. Sci. USA* **83:**1026–1030.
50. **Perry, K. L., S. J. Elledge, B. B. Mitchell, L. Marsh, and G. C. Walker.** 1985. *umuDC* and *mucAB* operons whose products are required for UV light- and chemical-induced mutagenesis: UmuD, MucA, and LexA proteins share homology. *Proc. Natl. Acad. Sci. USA* **82:**4331–4335.
51. **Raetz, C. R. H.** 1975. Isolation of *Escherichia coli* mutants defective in enzymes of membrane lipid synthesis. *Proc. Natl. Acad. Sci. USA* **72:**2274–2278.
52. **Sandman, K., R. Losick, and P. Youngman.** 1987. Genetic analysis of *Bacillus subtilis spo* mutations generated by Tn*917*-mediated insertional mutagenesis. *Genetics* **117:**603–617.
53. **Schaeffer, P.** 1969. Sporulation and the production of antibiotics, exoenzymes, and exotoxins. *Bacteriol. Rev.* **33:**48–71.
54. **Schwartz, D. O., and J. R. Beckwith.** 1966. Mutagens which cause deletions in *Escherichia coli. Genetics* **61:**371–376.
55. **Smith, C. M., Z. Arany, C. Orrego, and E. Eisenstadt.** 1991. DNA damage-inducible loci in *Salmonella typhimurium. J. Bacteriol.* **173:**3587–3590.
55a. **Sonenshein, A. L., J. A. Hoch, and R. Losick (ed.).** 1993. *Bacillus subtilis and Other Gram-Positive Bacteria: Biochemistry, Physiology, and Molecular Genetics.* American Society for Microbiology, Washington, D.C.
56. **Sonenshein, A. L., and R. Losick.** 1970. RNA polymerase mutants blocked in sporulation. *Nature* (London) **227:**906–909.
57. **Tormo, A., M. Almirón, and R. Kolter.** 1990. *surA,* an *Escherichia coli* gene essential for survival in stationary phase. *J. Bacteriol.* **172:**4339–4347.
58. **Traub, P., and M. Nomura.** 1969. Structure and function of *E. coli* ribosomes. VI. Mechanism of assembly of 30S ribosomes studied *in vitro. J. Mol. Biol.* **40:**391–413.
59. **Vinopal, R. T.** 1987. Selectable phenotypes, p. 990–1015. *In* J.L. Ingraham, K. B. Low, B. Magasanik, F. C. Neidhardt, M. Schaechter, and H. E. Umbarger (ed.), *Escherichia coli and Salmonella typhimurium: Cellular and Molecular Biology.* American Society for Microbiology, Washington, D.C.
60. **Walker, G. C.** 1984. Mutagenesis and inducible responses to deoxyribonucleic acid damage in *Escherichia coli. Microbiol. Rev.* **48:**60–93.
61. **Wanner, B. L., and R. McSharry.** 1982. Phosphate-controlled gene expression in *Escherichia coli* K12 using Mu*d1*-directed *lacZ* fusions. *J. Mol. Biol.* **158:**347–363.
62. **Yanofsky, C.** 1971. Mutagenesis studies with *Escherichia coli* mutants with known amino acid (and base-pair) changes, p. 283–287. *In* A. Hollaender (ed.) *Chemical Mutagens: Principles and Methods for Their Detection,* vol. 1. Plenum Press, New York.
63. **Yanofsky, C., B. C. Carlton, J. R. Guest, D. R. Helinski, and U. Henning.** 1964. On the colinearity of gene structure and protein structure. *Proc. Natl. Acad. Sci. USA* **51:**266–272.
64. **Yanofsky, C., J. Ito, and V. Horn.** 1966. Amino acid replacements and the genetic code. *Cold Spring Harbor Symp. Quant. Biol.* **31:**151–162.
65. **Youngman, P., P. Zuber, J. B. Perkins, K. Sandman, M. Igo, and R. Losick.** 1985. New ways to study developmental genes from spore-forming bacteria. *Science* **228:**285–291.

Chapter 14

Gene Transfer in Gram-Negative Bacteria

DAVID L. PROVENCE and ROY CURTISS III

Three processes of gene transfer are known in bacteria: transformation, transduction, and conjugation. These means of gene transfer, along with mutation, are the driving force behind genetic diversity and hence the evolution of new bacterial species. In each transfer, genetic information from one bacterium, the donor, is acquired by another bacterium, the recipient. **Transformation** is a gene transfer process in which DNA from the donor is taken up by the recipient. A recipient cell that now expresses a donor genetic trait is called a transformant. **Transduction** is a gene transfer process in which a bacterial virus propagating in the donor picks up some of its genetic information and, on infection of the recipient, causes a heritable change in the recipient. A recipient cell that acquires a donor trait by this process is called a transductant. **Conjugation** is a gene transfer process in which the donor makes physical contact with and transfers genetic material to the recipient. A recipient that acquires donor genetic information is called a transconjugant. These three processes have been central in allowing the bacterial geneticist to analyze the genetic and biochemical bases of bacterial functions and to establish principles of gene structure, function, and regulation as well as of the more complex processes of macromolecular synthesis, growth, and cell division. In addition, continuing advances in the development of recombinant DNA technologies are due in large part to the ability to transfer recombinant molecules into bacteria, mainly *Escherichia coli*. Thus, gene transfer has allowed the bacterial geneticist to analyze the genetic and biochemical principles of bacterial function by using both classical genetic techniques and the techniques of molecular biology.

In this chapter, detailed methods for gene transfer are limited to three gram-negative bacteria: *E. coli*, *Salmonella typhimurium*, and *Acinetobacter calcoaceticus*. Methods describing gene transfer in other gram-negative bacteria have recently been described for *Haemophilus* spp. (7), *Neisseria* spp. (72), and *Pseudomonas* spp. (66); also see reference 2. The methods for gene transfer of gram-negative bacteria are not applicable to gram-positive bacteria, but Chapter 15 details methods for gene transfer in the latter.

The most essential requirement of any gene transfer procedure is the ability to select for and identify a successful recipient. In the gene transfer procedures described below, this is often accomplished by using a recipient that contains mutations causing inability to synthesize a metabolite or utilize a carbon source, or by using a recipient that possesses a wild-type sensitivity to an antibiotic or phage. The DNA to be transferred, on the other hand, possesses a readily selectable allele of that trait. Selection is most often accomplished by growing the recipients on media that will allow successful recipients to grow (form colonies) while killing, inhibiting, or not supporting the growth of unsuccessful recipients. Sometimes the inherited trait is not selectable, e.g., the inability to synthesize a metabolite. Procedures involving the transfer of a nonselectable trait usually require the growth of the population of successful and unsuccessful recipients on nonselective media and then screening of individual colonies for the inherited trait.

There are two basic factors that affect gene transfer efficiencies: factors affecting the efficiency of DNA entry into the recipient, and factors affecting the efficiency with which traits persist after entry and are stably inherited. In the former, these factors are often influenced by the growth conditions used to make the recipient receptive to the DNA to be transferred. Also, the recipient genotype often plays an important role in the efficiency of DNA uptake. The efficiency of stable inheritance of a trait is dependent on the recipient genotype. The stable inheritance of a plasmid requires host gene products to allow the plasmid to replicate. Often, a recipient that is capable of recombining DNA into the chromosome is required for stable inheritance. Thus, in *E. coli* and *S. typhimurium*, the wild-type *recA* allele is required for homologous recombination to occur. One recipient phenotype affecting the inheritance of traits is restriction. **Restriction** is a process in which a bacterium recognizes entering DNA as either foreign or like self. DNA is recognized as foreign if it is not properly methylated; this DNA is degraded. The use of restriction-deficient recipients containing a mutant *hsdR* allele can therefore increase the stable inheritance of traits.

Many of the procedures described below involve recipients that are in the exponential phase of growth. During exponential growth the number of chromosomes per cell is usually more than one and can increase further. This increasing number of chromosomes has an important influence on the kinetics of detection of stably inherited traits. This is especially true when the stable inheritance of a trait requires recombination of transferred DNA with the chromosome. In this case, immediately after DNA transfer only one copy of the chromosome will contain the transferred trait. The recipient cell must divide several times before the chromosomes that do not contain the newly inherited trait are segregated from those that do contain it. This delay in time before a transferred trait is present in all chromosomal copies of a daughter cell is called **genotypic** or **segregational lag.** If the transferred trait is dominant, its phenotype will be detectable immediately, before the transferred dominant trait has segregated from chromosomes containing the recessive trait. If the transferred trait is recessive, the phenotype of the inherited trait will not be detectable until the transferred recessive trait has segregated from the dominant trait to yield a progeny cell containing only the recessive trait on all chromosomal copies. Sometimes a recessive trait is not detectable until several further cell divisions have occurred. This delay in time before a newly inherited trait is detectable is called **phenotypic lag.**

Phenotypic lag is important in procedures in which the phenotype to be selected is resistance that is recessive to alleles encoding sensitivity. Examples of this include phage resistance and resistance to certain antibiotics, e.g., nalidixic acid, rifampin, and streptomycin. To produce a resistant cell, time must be allowed for cell division and also for the degradation or dilution of preexisting structures encoding sensitivity. Phenotypic lag is also displayed by some dominant phenotypes, most notably resistance determinants to bacteriocidal antibiotics, e.g., ampicillin and kanamycin. Exposure of recipients too soon after the stable inheritance of these genes results in cell death. For this reason, cells that have received DNA encoding resistance determinants to bacteriocidal antibiotics are incubated under nonselective conditions to allow for phenotypic expression and then exposed to the appropriate selective agent. This is not as crucial when the selective agent is a bacteriostatic antibiotic, e.g., chloramphenicol or tetracycline. The phenotypic expression of resistance to these antibiotics can sometimes be expressed slowly even in the presence of the antibiotics.

14.1. TRANSFORMATION

Transformation is known to occur and has been studied in a variety of bacterial genera. The occurrence of transformation as a natural process is dependent on the ability of a recipient strain to exhibit competence. **Competence** is defined as the ability of the recipient strain to transport DNA from the culture medium into the cell.

Successful development of a transformation system depends on the use of double-stranded DNA and the discovery of conditions that achieve maximum competence of the recipient-cell population. Some species, such as *A. calcoaceticus*, easily develop competence under various culture conditions (42). However, competence in other species, such as *E. coli*, is artificially induced by using carefully defined conditions (49). There are two general methods to transform bacteria that are not naturally able to take up DNA. One of these involves artificially inducing competence by using divalent cations and thermal shock. The other uses a high-strength electrical field (electroporation). In the former case, important conditions for inducing competence include the stage in the growth cycle of the recipient population, the nutritive value of the culture medium, the degree of aerobiosis or anaerobiosis, the pH, and the presence of appropriate concentrations of divalent cations. Ca^{2+} is the divalent cation used most frequently to induce competence in *E. coli* and *Salmonella*. Under optimal conditions, more than 5×10^8 transformants per μg of plasmid DNA can be produced for certain strains of *E. coli* (31). The genotypic properties of the recipient can also affect competence. The structure of the cell surface, the production of competence factors, the ability to undergo autolysis, and the presence or absence of prophages (genomes of temperate bacterial viruses replicating in synchrony with bacterial host genomes) are among the more important genotypic attributes influencing competence.

Electroporation is a relatively new and powerful technique that is similar to transformation in that DNA is taken up by competent cells. In this technique, bacteria are subjected to a high-strength electrical field to form pores in the cells. While the pores are present, DNA in solution can diffuse into the cells. Electroporation has proven to be useful in a variety of *E. coli* K-12 strains and, under optimal conditions, results in frequencies of 10^9 to 10^{10} transformants per μg of plasmid DNA.

One novel aspect of electroporation that sets it apart from transformation and other classical means of gene transfer is its applicability to establishing gene transfer in genera that do not have characterized gene transfer mechanisms. Electroporation has been used to introduce DNA into *Campylobacter jejuni*, a species in which no other means of transformation or transduction currently exists (52). It has also been used in a variety of other gram-positive and gram-negative microorganisms (32). The identification and optimization of electroporation conditions are different for different genera, and often for different species, and require systematic testing of several variables. These variables include but are not limited to growth phase, cell density, DNA concentration, field strength, and temperature. Some investigators have increased electroporation frequencies by modification of the cell wall, either by growth in penicillin G (62) or through the use of the cell wall-degrading enzymes isonicotinic acid hydrazine or lysozyme (35).

14.1.1. Isolation and Quantitation of Transforming DNA

Specialized methods for the isolation of plasmid DNA for transformation are described in Chapter 16. The isolation of chromosomal DNA for transformation is described in section 14.1.2.2.

Transforming DNA is usually suspended in a buffer that contains sodium EDTA to chelate divalent cations such as magnesium, which are essential cofactors for most DNases. The DNA preparations are usually stored in TE buffer (10 mM Tris, 1 mM EDTA [pH 8.0]).

The concentration of a DNA solution can be quantitated by measuring its A_{260} and A_{280}. The A_{260} in a cuvette with a 1-cm light path divided by 20 gives the DNA concentration in milligrams per milliliter. Thus, 50 μg of DNA per ml has an A_{260} of 1.0. The ratio of A_{260}/A_{280} is a measure of the purity of the DNA solution and should be 1.8 to 2.0.

An alternate method to quantitate DNA is to visually compare the DNA with other DNA standards of known concentrations. This is easily done by resolving the DNAs on a 0.7% agarose gel to which ethidium bromide has been added to 500 ng/ml. Ethidium bromide is a dye that intercalates into the DNA and will fluoresce when exposed to light of 254 or 302 nm. (*CAUTION:* Ethidium bromide is a powerful mutagen, and care should be exercised when working with solutions containing it.) The following procedure using DNA standards is based on that of Sambrook et al. (68).

1. Use a microwave oven or other heat source to dissolve a solution of 0.7% electrophoresis grade agarose in 0.5× electrophoresis buffer (TBE) (a 5× solution contains 54 g of Tris base, 27.5 g of boric acid, and 20 ml of 0.5 M EDTA [pH 8.0] per liter). After cooling to approximately 60°C, add ethidium bromide to 500 ng/ml, pour into a horizontal minigel apparatus, and allow to solidify.
2. Add enough 0.5× TBE to submerge the gel.
3. Add 4 μl of the unknown DNA to 0.8 μl of 6× gel-loading buffer (6× solution is 0.25% bromophenol blue and 40% [wt/vol] sucrose in water, stored at 4°C), and load this into a well in the gel.
4. Add 4 μl of each of the DNA standards to 0.8 μl of gel-loading buffer, and load each into a separate well. The standards should range from 2.5 to 50 μg of DNA per ml.
5. Electrophorese until the bromophenol blue has migrated at least 1 cm into the gel.
6. Photograph the gel under UV light of 254 or 302 nm. Compare the intensity of the unknown DNA to the standards to estimate the unknown DNA concentration.

14.1.2. Transformation of *Acinetobacter calcoaceticus*

A. calcoaceticus is a gram-negative, oxidase-negative, nonmotile, aerobic coccus which develops competence under various culture conditions and in which transformation is independent of the presence of divalent cations (42). Since competence in *A. calcoaceticus* is easily induced, it is useful in laboratory courses to demonstrate transformation. Sawula and Crawford (70) showed that the *trpA* and *trpB* genes, which encode the two subunits of tryptophan synthase, are closely linked and cotransformable. They developed a method to easily distinguish between recipients stably transformed with the wild-type *trpB* gene and those stably transformed with the mutant *trpA* linked to the wild-type *trpB* alleles. The method is based on the observation that *trpA* mutant strains form large colonies on agar media containing either 10 μg of L-tryptophan per ml or 40 μg of indole per ml and form small colonies on media containing only 5 μg of indole per ml. Thus, if a *trpB* mutant, which can grow only when supplied with 10 μg of L-tryptophan per ml, is transformed with DNA from a *trpA* mutant donor and plated on agar media containing 5 μg of indole per ml, both small and large colonies arise. The small colonies are due to cotransformation of the *trpA* mutant and wild-type *trpB* donor alleles, and the large colonies are transformants that have inherited only the donor *trpB* wild-type allele.

14.1.2.1. Strains and media

The donor and recipient strains were isolated from strain BD413, which is a microencapsulated mutant of the transformable wild-type *A. calcoaceticus* BD4 (42). The donor strain has the *trpA23* mutation, and the recipient strain has the *trpB18* mutation (70). These strains are grown at 30 or 35°C and are stored on L-agar slants at 4°C. Fresh transfers should be made every 2 months.

A. calcoaceticus can be cultivated by use of a variety of complex media such as L broth and L agar, Penassay broth and agar, or brain heart infusion broth and agar. Various minimal media can also be used, although VB minimal medium E is suggested. In using the *trpB* and *trpA* mutants, use an acid hydrolysate of casein such as Casamino Acids at 0.5% to stimulate more rapid growth of prototrophic recombinants while precluding growth of mutants, since acid hydrolysis destroys all tryptophan. See section 14.4 for methods of medium preparation.

14.1.2.2. DNA isolation

Transforming chromosomal DNA from the mutant *trpA* donor strain can be isolated as follows (51, 70).

1. Grow donor cells with aeration at 30 to 35°C to early stationary phase in 45 ml of L broth (section 14.4.1). Aeration can be achieved by using bubbler tubes (screw-cap tubes [25 by 200 mm] with a 4- to 5-mm-diameter glass tube encased in a rubber tube sleeve passed through the cap and with a cotton plug at the top of the glass tube) connected to an aquarium pump. Alternatively, use flasks agitated by a reciprocating or rotary shaker (see Chapter 10).
2. Sediment cells by centrifugation, and suspend in 5 ml of saline-EDTA buffer (0.15 M NaCl, 0.1 M EDTA [pH 8.0]).
3. Add 0.1 ml of 20% sodium dodecyl sulfate, and incubate the suspension at 37°C for 10 min to lyse the cells.
4. Clear the lysate by adding 0.2 ml of 2.5 M KCl, and vortex gently to mix.
5. Add 5 ml of chloroform-isoamyl alcohol (24:1), vortex for 10 s, allow to stand for 2 min, and vortex again for 10 s.
6. Pellet unlysed bacteria and cellular debris at 27,000 × *g* at 4°C for 15 min.
7. Carefully transfer the aqueous (top) layer to a 50-ml tube. Heat this solution of DNA at 65°C for 10 min to kill any surviving cells and to inactivate any nucleases present.
8. Test the sterility of the DNA preparation by streaking a small amount on an L agar plate, and incubate overnight at 30 to 35°C.
9. If desired, precipitate the DNA by adding twice the volume of cold ethanol, and wind it on a glass rod which has a small hook on one end. Dissolve the DNA in TE buffer.

10. Determine the DNA concentration by using one of the methods outlined in section 14.1.1.
11. Store the DNA preparation at 4°C.

14.1.2.3. Transformation procedure

1. Grow the *trpB* mutant recipient strain in L broth with aeration at 35°C overnight. (Actually, *A. calcoaceticus* can grow at any temperature from 30 to 37°C, although it grows best at 35°C.)
2. Dilute the overnight culture 1:10 into fresh prewarmed L broth, and aerate by either bubbling or shaking at 35°C for 2 h.
3. On each VB minimal agar plate (supplemented with 0.5% Casamino Acids, 0.5% glucose, and 5 μg of indole per ml), spot 0.1 ml of the competent recipient culture and a different amount of the donor DNA preparation (incremental amounts of 0.01 to 0.2 ml, or 20 to 500 μg).
4. Immediately mix the two preparations on the plate, and distribute them evenly over the surface of the agar by using a glass spreader which has been immersed in alcohol, flamed, and cooled on the agar surface.
5. Spot and spread 0.1 ml of the recipient-cell population on one selective medium plate as a control, to verify the absence of wild-type *trpB* revertants or other contaminants.
6. Spot and spread 0.1 ml of the donor DNA preparation on one selective medium plate as an additional control, to verify that it is free from contaminating cells.
7. Add 0.01 ml of 100 μg/ml DNase in 0.2 M $MgCl_2$ to 0.1 ml of the donor DNA preparation (which is in TE buffer) about 2 to 3 min before mixing on a selective plate with 0.1 ml of competent recipient cells, as a control to verify that destruction of DNA eliminates transformant colonies.
8. Invert the plates and incubate at 35°C overnight or until colonies are clearly discernible.
9. Examine the plates and count the total numbers of small and large colonies on each plate without destroying the colonies.
10. Purify some of the small-colony types to determine whether they have the properties expected of *trpA23* wild-type *trpB* cotransformants (i.e., they still require either tryptophan or indole and form large colonies on media with 40 μg of indole per ml) and also purify some of the large-colony types to determine whether their growth is independent of either indole or tryptophan. This can be done as follows.

(i) Pick 10 to 20 individual colonies of each type (with either sterile toothpicks or a sterile needle) into 1 to 2 ml of buffered saline with gelatin (BSG) in test tubes (13 by 100 mm), into a few drops of BSG contained in individuals wells of a microtiter plate, or into droplets of BSG contained on the inside of a plastic petri dish.

(ii) Streak these suspensions on selective agar plates in a manner to ensure obtaining isolated colonies. This is most routinely achieved by using a sterile needle if the colonies were suspended in very small volumes of BSG or by using a sterile loop with the excess fluid tapped off on the side of the test tube if the colonies were suspended in 1 to 2 ml of BSG. The selective agar plates can be marked into 4 to 10 equal-sized pie-shaped segments to allow purification of 4 to 10 isolates per plate.

(iii) Incubate these plates at 35°C for 1 to 2 days.

(iv) Pick individual isolated colonies from each original transformant into BSG, and streak these suspensions on minimal agar and minimal agar supplemented with either 5 μg of indole per ml, 40 μg of indole per ml, or 10 μg of tryptophan per ml. Incubate these plates at 35°C.

11. Analyze the results to determine whether the total number of transformants increases proportionally with increases in the amount of donor DNA used. This would be expected except at high concentrations of DNA, when saturation may occur.
12. Calculate the frequency of cotransformation of the *trpA23* allele with the wild-type *trpB* allele, and determine whether this value is constant independent of DNA concentration.

14.1.2.4. Procedural modifications

The above-described procedure can be made quantitative by adding known quantities of donor DNA (1 to 500 ng/ml) to broth cultures of competent recipient cells at 1×10^8 to 2×10^8 cells per ml. After 10 to 15 min at 35°C, add 0.01 ml of 0.5-mg/ml DNase, suspended in 1 M $MgCl_2$, per ml of donor DNA–recipient-cell mixture to stop the reaction. Plate appropriate dilutions on selective media, and plot the log of the absolute titers of transformants against the log of the DNA concentrations.

The state of maximum competence in the growth cycle of the recipient culture can be determined by a slight modification of the more quantitative method described in the preceding paragraph. Precision in the design of this procedure is facilitated by first plotting a growth curve for the recipient strain growing in L broth at 35°C with aeration. The general procedure for determining the effect of stage of growth on transformation frequency is as follows. Sediment the cells in a stationary aerated overnight culture by centrifugation, and suspend them at the same density in fresh L broth. Dilute the suspension 1:20 into a flask containing L broth, and then incubate at 35°C with shaking. Remove an aliquot immediately and every 15 min for mixing with saturating concentrations of donor DNA to assess the competence of cells. Also, take samples from the culture at the same times to determine the number of viable cells present. By using these titers and the number of transformed cells, the growth curve can be graphed and the frequency of transformants for each point (number of transformants/number of viable cells) can be plotted on the same graph. Peak levels of competence should be noted for late lag to early log phase and for late log to early stationary phase.

Procedures to determine the absolute fraction of the recipient population competent to take up DNA involve either the use of radioactively labeled donor DNA and radioautographic analysis or else a mathematical analysis of data on frequencies of transformation of two unlinked donor markers singly and together as a function of donor DNA concentration. These methods are discussed in general terms by Freifelder (1) and Notani and Setlow (59).

14.1.3. Transformation of *Escherichia coli*

Mandel and Higa (49) discovered that $CaCl_2$ treatment of *E. coli* cells, followed by incubation in the cold and then a heat shock, resulted in the uptake of DNA. This discovery led to a succession of refinements to artificially induce competence in various bacterial genera to allow transformation with chromosomal or plasmid DNA.

14.1.3.1. Strains and media

Table 1 lists strains that can be used as donors and recipients in the following transformation experiments, as well as those for use in transduction and conjugation experiments. Certain mutations can be used in strains to be transformed with DNA. These mutations can increase the efficiency of stable inheritance after transformation or aid in the analysis of transformed strains. Elimination of the exonuclease V and exonuclease I in *recB recC sbcB* (61) strains or elimination of the exonuclease V subunit D in *recD* (11) strains (strains JC7623 and DPB353, respectively) facilitates transformation of *E. coli* with linear DNA yet has no effect on transformation with plasmid DNA. *E. coli* K-12 contains a type 1 restriction enzyme, *Eco*K. If DNA is transformed into a strain containing this restriction system and the DNA is not appropriately methylated, the DNA will be restricted and greatly reduce the transformation frequency. Strains containing an *hsdR* mutation (e.g., strain LE392) will not restrict but will modify transformed DNA.

Some plasmid vectors, for instance pUC19, contain the regulatory region of the *lac* operon (*lacI*) and a short segment of DNA encoding the amino-terminal end of *lacZ*, the gene encoding ß-galactosidase. When this type of plasmid is present in an *E. coli* strain containing the appropriate *lacZ* mutation (*lacZ*ΔM15, present in strain DH5α), expression of the amino-terminal portion of *lacZ* encoded by the plasmid can be induced by isoprophy-ß-D-thiogalactopyranoside (IPTG) (38). This peptide combines with the defective ß-galactosidase produced by the *lacZ*ΔM15 mutation and results in a functional ß-galactosidase. This complementation is referred to as α-complementation (79), and the phenotype of the functional ß-galactosidase can be visualized as a blue color when the chromogenic substrate 5-bromo-4-chloro-3-indolyl-ß-D-galactopyranoside (X-Gal) is included in the medium. The plasmid usually contains a series of closely spaced restriction enzyme sites (often referred to as a multiple cloning region) to allow cloning of DNA fragments directly upstream of *lacZ*. When a DNA fragment is cloned into this multiple cloning region, the functional activity of the plasmid *lacZ* is usually disrupted. Transformation of this recombinant plasmid into a *lacZ*ΔM15 host will no longer result in a blue colony, thus allowing one to easily distinguish between transformants that contain a recombinant plasmid and those that contain a plasmid with no insert.

L broth and agar or Penassay broth and agar can be used as complex media. A diversity of minimal media can be used, including M9, VB medium E, and ML. Agar is added to 1.5% for solidified media. Amino acids, purines, pyrimidines, and vitamins are added to optimal concentrations as appropriate (Table 2) depending on the genotype of the recipient strain used. Carbon sources are generally added to a final concentration of 0.5% (Table 3), and antibiotics are added at concentrations that are dependent on the level of resistance conferred by the drug resistance mutation and/or gene (Table 4). When using α-complementation to screen for plasmids containing insert DNA, IPTG and X-Gal are also added to the media. SOC medium is used to suspend the cells immediately after electroporation. See section 14.4 for medium preparation methods.

14.1.3.2. $CaCl_2$ transformation procedure

The use of $CaCl_2$ treatment to make *E. coli* competent was originally reported by Mandel and Higa (49). A modification of this procedure, as published by Cohen et al. (17), is routinely used with a variety of *E. coli* K-12 strains and can be used for plasmid or chromosomal DNA. The procedure is as follows.

1. Grow a 3-ml overnight culture of the recipient strain in L broth as a standing culture at 37°C. Standing overnight cultures (which have not been aerated by either bubbling or shaking) exhibit a much shorter lag phase prior to initiation of exponential growth when subcultured the next day than do aerated overnight cultures, which have already reached the stationary phase of growth.
2. Dilute this culture 1:10 into 10 ml of prewarmed L broth, and incubate with aeration by either bubbling or shaking at 37°C until the cells reach early log phase (about 1×10^8 to 2×10^8 cells per ml). This is easily done by monitoring the optical density at 600 mm (OD_{600}) of the culture while it is growing. Since different strains have different optical densities at the same stage of growth, the OD_{600} that is indicative of early log phase must be determined for each strain to be used.
3. Transfer the early-log-phase culture to a chilled glass test tube (16 by 125 or 150 mm), and keep chilled with crushed ice for 10 min. Sediment the cells at 2,700 × *g* at 4°C for 10 min.
4. Remove all traces of the supernatant fluid, and gently suspend the pellet in 5 ml of ice-cold 10 mM Tris HCl (pH 7.6)–50 mM $CaCl_2$.
5. Hold this suspension with ice for 15 min, and then sediment the cells as described in step 3.
6. After removing all traces of supernatant fluid, gently suspend the cells in 670 μl of the same buffer as used in step 4.
7. Distribute 0.2-ml samples of cells to three chilled 1.5-ml microcentrifuge tubes. The competent bacteria can be used immediately or stored with ice. After 24 h the competence of the cells decreases rapidly.
8. Add plasmid or chromosomal DNA using no more than 0.01 ml. The DNA can be in water or TE. If a larger volume of DNA is to be added, dilute the DNA in the same buffer used for the competent cells. Two controls should be included. One 0.2-ml sample should receive a plasmid known to be capable of transforming the strain, and another 0.2-ml sample should not receive any DNA. Incubate each of these on ice for 30 min.
9. Heat shock each sample at 42°C for 90 s. The rapidity of the temperature shift is important. Use a heat block with wells filled with water equilibrated to 42°C to ensure a quick change in temperature from 0 to 42°C.
10. Dilute each sample 1:5 into growth medium, and incubate at 37°C for 60 to 90 min.
11. Plate (with or without dilution, depending on the type of DNA used and on whether dilution was already made to permit expression) on medium to select transformants. Since the treatment of cells with $CaCl_2$ causes them to become fragile, it is important to use

Table 1. *E. coli* K-12 and *S. typhimurium* strains, plasmids, and phages[a]

Strain	Mating type	Genotype[b]	Phenotype[c]	Reference
E. coli				
JC7623	F^-	(λ^-) *thr-1 ara-14 leu-6 proA2 lacY1 tsx-33 sup-37 galK2 sbcB15 his-4 recB21 recC22 rpsL31 xyl-5 mtl-1 argE3 thi-1*	Exonuclease I^-, exonuclease V^-	
DPB353	F^-	(λ^-) *galK2 pyrD34 trp-45 recD1901*::Tn*10 his-68 thyA25 malA1 rpsL118 xyl-7 mtl-2 thi-1*	Exonuclease V^-, Tc^r	11
LE392	F^-	(λ^-) *lacY1 galK2 galT22 glnV44 tyrT58 metB1 hsdR574 trp R55*	Suppresses amber mutations, modifies but will not restrict DNA	56
DH5α	F^-	(λ^-) (ϕ80d*lacZ*Δ*M15*) Δ(*lacZYA-argF*)*U169 glnV44 deoR gyrA96 recA1 relA91 endA1 thi-1 hsdR17*	$LacZ^-$ but can be complemented by LacZα	29
W3102	F^-	(λ^-) *galK2*	Gal^-	
χ289	F^-	(λ^-) *glnV42*	Gal^+	
χ226	F^+	(λ^-) *lacZ81 glnV42*	$LacZ^-$	
χ353	F^-	(λ^-) *proC63 glnV42 cycA1*	$LacZ^+$ $ProC^-$	
χ1301	F′	(λ^-) (F14 *lac*$^+$) Δ(*proB-lac*)*41 glnV42 cycA1*	Lac^+ present on F′	
χ1997	F^-	(λ^-) *rpoB309* Δ (*proB-lac*) *41 glnV42 his-53 gyrA95 xyl-14 cycB2 cycA1*	Lac^- Nal^r Tc^s Cm^s St^s Sp^s Su^s	
CC118		*araD139* Δ(*ara. leu*)*7697* Δ*lacX74* Δ*phoA20 galE galK recA1 rpsE argE*(Am)*rpsB thi*	$PhoA^-$, suppressor free	50
χ1784	R^+	(λ^-) (R100-1 *drd-1 tet*$^+$ *cat*$^+$ *aadA*$^+$ *sul*$^+$ IncFII) *tsx-63 glnV42 his-53 lysA32 xyl-14 arg-65*	Tc^r Cm^r Sm^r Sp^r Su^r	
SM10	Hfr	(λ^-) *thr-1 leuB6 tonA21 lacY1 glnV44 recA*::RP4-2-*tet*::Mu *aphA thi-1*	Encodes IncP transfer functions, Km^r	74
S17-1	Hfr	(λ^-) *thi pro recA hsdR* RP4-2-*tet*::Mu *aphA*::Tn*7* (location of F insertion unknown)	Encodes IncP transfer functions, Sm^r Sp^r Tp^r	74
SM10 λ*pir*	Hfr	(λ*pir*) *thi-1 thr-1 leuB6 lacY1 tonA21 glnV44 recA*::RP4-2-*tet*::Mu *aphA thi-1*	Encodes IncP transfer functions, Km^r π^+	54
χ433	HfrOR4	(λ^-) O-*thr leu proA lac ... pyrB* F	Prototrophic, Nal^s Str^s	
χ2634	F^-	(λ^-) *ara-14 leu-6 azi-6 tonA23 proA70 lacY1 tsx-67 purE42 glnV44 galK2 trpE38 gyrA94 rpsL109 xyl-5 mtl-1 thi-1*	Leu^- $ProA^-$ $PurE^-$ $TrpE^-$ Nal^r Str^r Thi^-	
χ585	F^-	(λ^-) *leu-50 glnV42 rpsL206*	Leu^- Str^r	
S. typhimurium				
LT-2			Prototrophic	69
SA2649		*proC131*	$ProC^-$	69
LB5010		*hsdL hsdSA hsdSB metA22 metE551 trpD2 leu val rpsL120 galE*	$GalE^-$	69
Plasmids[d]				
pUC19		*lacI lacZ*α	Encodes the amino-terminal portion of LacZ; will complement the *lacZ*Δ*M15* mutation of DH5α	84
pGP704		pBR322 *oriR6K mobRP4 bla*	Mobilizable, requires π for replication, Ap^r	54
pRT733		pBR322::Tn*phoA oriR6K mobRP4 bla aphA*	Mobilizable, requires π for replication, Ap^r Km^r	78
pAMH62		ϕ (*ompR′-lamB*) *bla* (derived from pBR322)	$LamB^+$ Ap^r	34
Phages				
P1 *kc*				45
P1 L4		*c*I		13
P22 HT *int*		*int*-201		4
λ::Tn*5 kan-1*		λ *b*515 *b*519 *ral*::Tn*5 c*I857 *Sam7*		9
λ::Tn*10* (λ561)		λ *b*221 *c*I857::Tn*10 Oam29 Pam80*		81

[a]The strains and plasmids have been selected because they possess the simplest constellation of suitable genotypic properties to illustrate the concepts and methodologies described in the text. These specific strains have not all been used or tested by the authors to conduct the procedures described in the text.

[b]Genetic nomenclature conforms to that of Demerec et al. (25) and uses notations and symbols described for chromosomal (6, 69) and plasmid (60) traits.

[c]Abbreviations: Ap, ampicillin; Cm, chloramphenicol; Km, kanamycin; Nal, nalidixic acid; Sm and Str, streptomycin; Sp, spectinomycin; Su, sulfanilamide; Tc, tetracycline; Tp, trimethoprim.

[d]Plasmids have no phenotype, so only strains that contain a plasmid display the phenotypes listed.

Table 2. Amino acid, purine, pyrimidine, and vitamin concentrations for stock solutions and for supplementation of minimal media[a]

Substance	Concn of stock solution (mg/ml)	Amt of stock solution (ml)/liter of minimal medium	Final concn in minimal medium (μg/ml)[b]
Amino acids[c]			
DL-Alanine[d]	50.0	2.0	100
L-Arginine HCl	11.0	2.0	22
L-Asparagine	10.0	10.0	100
L-Aspartic acid	10.0	10.0	100
L-Cysteine HCl[e]	11.0	2.0	22
Glycine	50.0	2.0	100
L-Glutamic acid	10.0	10.0	100
L-Glutamine	20.0	5.0	100
L-Histidine HCl	11.0	2.0	22
L-Isoleucine	10.0	2.0	20
L-Leucine	10.0	2.0	20
L-Lysine HCl	44.0	2.0	88
DL-Methionine	10.0	2.0	20
L-Phenylalanine	10.0[f]	2.0	20
L-Proline	15.0	2.0	30
DL-Serine	10.0	10.0	100
DL-Threonine	40.0	2.0	80
L-Tryptophan	4.0	5.0	20
L-Tyrosine	2.0[g]	10.0	20
DL-Valine	20.0	2.0	40
Purines and pyrimidines			
Adenine	2.0[h]	20.0	40
Thymidine	10.0	0.4 or 4.0	4 or 40[i]
Thymine	5.0	0.4 or 8.0	2 or 40[i]
Uracil	2.0	20.0	40
Vitamins			
Biotin	0.25	2.0	0.5
Niacin	0.5	2.0	1
Calcium pantothenate	0.5	2.0	1
Pyridoxine HCl	0.5	2.0	1
Thiamine HCl	0.5	2.0	1

[a]Prepare all stock solutions in distilled or deionized water and stored then at 4°C, unless otherwise indicated. Sterilize either by autoclaving at 121°C (15 min for volumes of 100 ml per bottle or less) or by membrane filtration.
[b]Concentrations give optimum growth of *E. coli* auxotrophs. Somewhat different concentrations might be needed for mutant strains of other bacterial species.
[c]The L isomers have been listed when their price per gram is no more than twice the price of the DL racemic mixture.
[d]Both the D and L isomers are usable by bacteria that possess alanine racemase.
[e]Prepare fresh (at least every week), since cysteine is rapidly oxidized to cystine (which is very insoluble). Alternatively, overlay the stock solution of cysteine with sterile mineral oil after autoclaving.
[f]Prepare in 0.001 N NaOH, and store at room temperature.
[g]Prepare in 0.01 N NaOH, and store at room temperature.
[h]Prepare in 0.03 N HCl.
[i]*thyA* mutants of enteric bacteria require relatively high concentrations of either thymine or thymidine. *thyA* strains with either *deoB* or *deoC* mutations (which eliminate functional deoxyribomutase and deoxyriboaldase, respectively) cannot degrade deoxyribose 1-phosphate and thus grow with lower concentrations of either thymine or thymidine.

selective agar medium that has been freshly prepared and is not too dry. It is also important not to spread the plates to dryness but, after distributing the diluted suspension over the surface of the agar, to allow it to dry into the agar by leaving the lid of the petri dish slightly ajar for a few minutes.

12. Incubate the plates at 37°C for 1 or 2 days as appropriate, depending on the selected marker and the growth rate of the recipient strain. The control sample that received no DNA will indicate whether the recipient strain is able to grow on the selective medium. This is especially important when the selective drug is ampicillin, since plating cells at high density on plates containing ampicillin can lead to false-positive colonies. The other sample that received the control plasmid DNA will indicate whether the strain was competent. This latter control is especially useful when the transforming DNA is either derived from in vitro construction or uncharacterized in its ability to replicate in the recipient or there is any reason to expect the transformation frequency to be low.

14.1.3.3. Electroporation procedure

The following procedure is a modification of the procedure recommended by Bio-Rad Laboratories. It is routinely used to transform *E. coli* and *S. typhimurium*. The apparatus is a Bio-Rad Gene Pulser with a Pulse Control Unit (Bio-Rad Laboratories, Richmond, Calif.).

1. Grow a 3-ml overnight culture of the recipient strain as a standing culture at 37°C.
2. Dilute the culture 1:10 into 25 ml of prewarmed L broth, and incubate with aeration until early log phase (1×10^8 to 2×10^8 cells per ml).

Table 3. Concentrations and preparations of sterile carbohydrate stock solutions

Carbohydrate	Concn of stock solution (%)[a]
L-Arabinose	20
D-Galactose	30
D-Glucose	40
D-Glycerol[b]	40
Lactose	20
Maltose	30
D-Mannitol	10
Sodium succinate	20
D-Xylose	30

[a]All concentrations are in weight per unit volume. Heating of distilled (or deionized) water facilitates solution of the carbohydrates, but the solutions should be allowed to cool to room temperature before they are brought to the correct final volume. All solutions should be sterilized by autoclaving at 121°C (20 min for volumes of 250 ml or less) prior to use and should be stored at 4°C.
[b]Since glycerol is a liquid, add 31.7 ml/68.3 ml of water to achieve a 40% (wt/vol) solution.

3. Transfer the cells to a chilled tube (16 by 125 or 150 mm), and incubate with ice for 15 min. Keep the cells cold through the rest of the procedure. Pellet the cells at 2,700 × *g* for 10 min at 4°C.

4. Suspend the pellet in 10 ml of cold 1 mM *N*-2-hydroxyethylpiperazine-*N'*-2-ethanesulfonic acid (HEPES; pH 7.0), and then pellet as in step 3 above. It is crucial to thoroughly suspend the pellet at this point and dilute the conductive salts present in L broth. If the salt levels are too high when the cells are electroporated, arcing will result. Arcing often causes a large decrease in the frequency of electroporation.

5. Suspend the pellet in 10 ml of cold 1 mM HEPES (pH 7.0), and pellet the cells as above.

6. Suspend the cells in 5 ml of cold 10% glycerol, and pellet as above.

7. Suspend the cells in 0.5 ml of cold 10% glycerol. This should yield approximately 3×10^{10} cells per ml. This cell suspension can be aliquoted in 0.1- to 0.2-ml amounts into microcentrifuge tubes and stored at −70°C. When needed for electroporation the frozen cells can be thawed on ice and used as described below.

8. Mix 0.1 to 0.2 ml of cells with 0.005 ml of DNA in a chilled microcentrifuge tube. The DNA should be in a low-ionic-strength TE buffer, or arcing may result. Transfer the cell-DNA mixture to a cold 0.2-cm cuvette, shake to the bottom, and place in the Gene Pulser cuvette holder. Apply one pulse with the gene pulser set at 2.5 kV and 25 μF and the pulse controller set at 200 Ω. If the volume of DNA to be added is greater than 5 μl, suspend the DNA in water to decrease the likelihood of arcing.

9. Immediately suspend the cells in 1 ml of SOC. It is important to add SOC immediately, since waiting will result in lower transformation frequencies. Transfer the cells to a tube, and incubate at 37°C for 1 h to allow recovery of transformants and expression of genes encoding antibiotic resistance.

10. Plate out the cells as described above for transformation.

It should be noted that electroporation of *E. coli* with ligated DNA results in low transformation frequencies unless the DNA is first transferred to a low-ionic-strength buffer (40, 83, 85). Since extremely high frequencies are possible (up to 10^{10} transformants per μg of plasmid for *E. coli*), electroporation is a more efficient means of transforming a recipient with plasmid libraries.

14.1.3.4. Applications

Transformation has become a central technique in the manipulation of DNA to form novel recombinant molecules. The ease with which *E. coli* can be transformed and the vast knowledge of *E. coli* genetics have made this the species of choice for propagating recombinant plasmids. Plasmids containing recombinant DNA from a wide variety of bacterial genera have been

Table 4. Antibiotic concentrations for stock solutions and medium supplementation

Antibiotic	Concn of stock solution (mg/ml)	Storage temp (°C)	Final concn (μg/ml) in medium for:	
			Plasmid-specified resistance	Chromosomal mutation
Ampicillin	50[a]	−20	50–100	25–50
Chloramphenicol	34[b]	−20	25–50	25–50
Kanamycin	40[c]	−20	30–50	30–50
Nalidixic acid	10[d]	4	—[e]	50[f]
Rifampin	2[g]	−20	—	50
Streptomycin	50[c]	−20	20–50	200–400
Tetracycline[h]	12.5[i]	−20	2–20[j]	12.5–20

[a]Suspend in sterile water. Correct for buffers or other ingredients to achieve the appropriate concentration of pure ampicillin.
[b]Suspend in 100% ethanol.
[c]Suspend in sterile water.
[d]Suspend in 0.1 N NaOH, which converts the acid to the sodium salt.
[e]—, not applicable.
[f]Add at 75 μg/ml to EMB agar.
[g]Dissolve at 20 mg/ml in 100% methanol, and then dilute to 2 mg/ml in 10% methanol with sterile water. Store in the dark, and prepare fresh each week.
[h]Tetracycline is incompatible with magnesium ions (68) and hence cannot be used in minimal medium.
[i]Suspend in 50% ethanol, and store in the dark. Correct for the presence of high concentrations of ascorbic acid in most tetracycline preparations.
[j]Unlike other antibiotics, an increase in the copy number of Tn*10* does not lead to increased levels of resistance to tetracyline (16). The Tn*10 tet* gene product is thought to be harmful when present at high levels (55).

transformed into and propagated in *E. coli*. *E. coli* transformation is used for mapping of genes on recombinant DNA fragments by subcloning, in vitro and in vivo mutagenesis, and complementation analysis. Genes on recombinant plasmids can also be mapped by in vivo mutagenesis with transposons known to function in *E. coli*.

Transformation is also used as a tool in strain construction. A chromosomal gene can be inactivated by transposon mutagenesis, with the transposon being introduced on a plasmid by transformation. Also, strains can be constructed by transformation of a temperature-sensitive replication-defective or suicide plasmid containing mutagenized genes to replace wild-type genes by allele replacement (30, 54).

The methods described above can also be used to transform strains of the appropriate genetic background with linear DNA. **Linear DNA** can be derived either from the bacterial chromosome or from a purified recombinant plasmid. Linear DNA of chromosomal origin consists of a heterogeneous population of DNA fragments, and only a small percentage of these will contain the gene to be transformed. Linear DNA produced by restriction endonuclease digestion of a recombinant plasmid will result in a homogeneous population of DNA fragments (68). Transformation with such a homogeneous population will result in a higher frequency of stable transformants since all of the transforming DNA fragments will contain the desired gene.

14.2. TRANSDUCTION

Bacterial viruses are classified as either virulent, in which case all infected cells die with liberation of new phage particles, or temperate, in which the infected bacteria either lyse with liberation of new phage progeny or become lysogenic (1). In the lysogenic state, the phage genome, now called a prophage, replicates in synchrony with the bacterial chromosome either by establishing a plasmid form or by integrating into the chromosome. Although virulent phages and virulent mutants of temperate phages are capable of transduction, temperate phages capable of establishing lysogeny probably account for most transduction-mediated gene flow among bacteria in nature and provide systems that are most easily studied.

Two types of transduction are known: specialized and generalized (1). In **specialized transduction,** only donor genetic markers adjacent to the integrated prophage are capable of being transduced. For this reason, specialized transduction is observed only when the phage lysate is obtained either by spontaneous or induced liberation of phage from a lysogenic donor and never when the phage is propagated by the lytic productive cycle. Specialized transducing phages contain segments of donor genetic information that have replaced phage genome segments. Thus, these phages are recombinants containing both phage and host genetic information. On infection and lysogenization of a suitable recipient strain, specialized transducing phages establish a state of partial diploidy such that both the donor genetic information contained in the transducing phage and the homologous genetic information in the recipient chromosome persist. In forming a specialized transducing phage, if the donor genetic information replaces phage genetic information that is vital for the propagation of the phage, the resulting transducing phage becomes defective and is unable to multiply and propagate in the absence of an intact helper phage genome. In certain instances, however, the specialized transducing phage does not lose essential functions and is able to form plaques and to exhibit a nearly normal productive lytic cycle.

In **generalized transduction,** any donor trait (whether it be on a chromosome, plasmid, or prophage) is capable of being transduced. With generalized transducing phages, donor genetic information can be picked up either after induction of a lysogenic donor culture or by lytic productive infection of the donor culture. In the systems most extensively studied, the formation of a generalized transducing phage involves the inadvertent packaging of donor host genetic material in lieu of phage genetic information. Thus, phage particles capable of transducing donor information are completely defective. A recipient cell infected with such a particle will acquire only the donor genetic material and no phage genetic information, provided that the recipient cell is not multiply infected with infective phage particles along with a transducing phage particle. On entry into the recipient cell, the donor genetic information (if it is chromosomal in origin) may recombine with and replace recipient genetic information or may persist without recombination or replication to give rise to what is termed an abortive transductant. In the latter case, the transduced donor genetic material does express the genes encoded by the DNA fragment. If these transduced genes complement a mutation in the recipient, growth will occur. However, the nonreplicating nature of the transduced DNA results in growth limited to form a microcolony. Generalized transducing phages can also package and transduce plasmid genetic information. Since plasmids are capable of autonomous replication, recombination is not required for their inheritance in transductants.

Transduction is usually used to modify the recipient genotype, but several advances have made transduction more powerful and useful. The limitation of specialized transduction was the ability to package and transduce only DNA adjacent to the site where the phage integrated into the chromosome. Specialized transducing phages such as phage lambda (λ) have been altered to form vectors that can carry any DNA clone. Now transduction can be used to introduce any DNA fragment into *E. coli*, *S. typhimurium*, and other gram-negative bacteria. λ is also used extensively as a vector for transposon mutagenesis. The limitation to generalized transduction was often the lack of an easily selectable marker near the gene to be transduced. Now, transposons encoding easily selectable antibiotic markers can be moved close to any gene. Since the gene of interest will be cotransduced with the transposon at a frequency that varies with the distance between the transposon insertion site and the gene to be transduced, selecting for the antibiotic resistance encoded by the transposon will enrich for recipients that were transduced with DNA containing the gene. This is especially useful when transducing traits that have no easily assayable or selectable function.

The preparation of specialized transducing lysates requires the use of a donor lysogenic for the transducing phage. The lysate can be prepared by inducing this lysogen with either UV or mitomycin C, depending on the system, or, even better, by using a lysogen carrying a heat-inducible prophage in which lysogeny is stably maintained at 30°C but the lytic response is induced at 37°C (64). In preparing generalized transducing lysates, one usually infects donor bacteria and obtains the phage lysate by growing the infected culture in a liquid medium or by plating with the soft agar overlay method to obtain a confluent plate lysate (53). In either case, one must grow the bacteria and infect them in a medium with the appropriate cation concentrations to maximize phage yields. Most transducing phage lysates are stored over chloroform, which also kills any uninduced or uninfected donor cells.

The conditions to ensure optimum infection of the recipient strain depend on the phage-host system being used, but they usually involve careful attention to the ionic environment and the means of growing the recipient. With most phage-host systems, phages are preadsorbed to the recipient cells prior to plating for transductants. Since most transducing lysates contain wild-type phages and are able to kill recipient cells, it is necessary to circumvent infection and killing of transductant cells. This can be accomplished by using a multiplicity of less than one phage particle per recipient cell to minimize multiple infections. Addition of either antiserum against the phage or a chelating agent, if the phage requires divalent cations for adsorption, after preadsorption can also decrease multiple infections. The last two methods generally prevent infection of potential transductants by phages released from recipient cells that were productively infected. When phage strains that are unable to replicate in the recipient can be used, multiple infections with the phages are not a concern.

14.2.1. Specialized Transduction in *E. coli* with Phage Lambda

14.2.1.1. Strains, phages, and media

All strains that are to be infected with phage λ must contain the wild-type *lamB* gene. *lamB* encodes a protein that transports maltose across the outer membrane and is also the protein used by λ to attach to *E. coli*. Since λ usually integrates into the *E. coli* chromosome at *att*λ between the *gal* and *bio* genes, it is these genetic traits that are normally transducible by λ (1). Transduction can be demonstrated by using a nonlysogenic recipient strain that possesses a *gal* mutation such as W3102 (Table 1) and a wild-type *gal* donor strain such as χ289 (Table 1).

When λ is used as a vehicle to deliver transposons, it usually contains several mutations that make it incapable of replication and integration in certain bacterial hosts. The presence of these mutations greatly facilitates the selection of transposable element-induced mutants because the λ vehicle is unable to persist in the host. Most λ transposon delivery vehicles contain amber mutations in essential genes, usually *O* or *P* and sometimes *S*. These amber mutations result in the stop codon UAG within the mRNA encoded by the gene and yield a nonfunctional gene product. The mutated gene can be suppressed by the appropriate suppressor mutation in the bacterial host. A suppressor mutation results in a tRNA that recognizes the UAG codon but inserts an amino acid instead of terminating translation of the mRNA and synthesis of the nascent peptide. For this reason, λ delivery vehicles containing *O*am29 or *P*am80 (or both) must be propagated on a strain containing the suppressor *glnV* (also referred to as *supE*) and those containing the amber mutation *S*am7 or *S*am100 must be grown on a *tyrT* (also referred to as *supF*) host. Bacterial strains that allow mutant phage to propagate are called permissive hosts. Other important mutations in λ delivery vehicles are deletions in the *b* region (*b*221, *b*515, *b*519) and *c*I857. Deletions in *b* disrupt the phage *attP* site and stop λ from lysogenizing by integrating into the host attachment site *att*λ. The *c*I857 mutation confers a temperature-sensitive phenotype to the repressor protein necessary to establish and maintain lysogeny. λ strains containing this mutation can stably lysogenize the bacterial host at 30°C but lyse the bacterial host at 37°C. Permissive bacterial strains used as hosts to propagate λ transposon delivery vehicles should also contain the *hsdR* mutation. A useful bacterial strain that is *glnV tyrT hsdR* is LE392 (Table 1).

The recipient strain must be suppressor free (strain CC118; Table 1). When λ :: Tn*5* containing amber mutations is injected into the recipient, it is incapable of DNA replication (*O*am *P*am) or cell lysis (*S*am), and the cells that are kanamycin resistant (Km^r) will be due to transposition of Tn*5* into the host genome.

Super broth containing 10 mM $MgCl_2$, lambda agar or soft agar, EMB agar containing 1% galactose (EMB + Gal) or containing no added carbohydrate (EMBO), VB minimal agar containing 0.5% galactose (MA+Gal) (and any other nutritional supplements required for growth of the recipient used), and 10 mM $MgCl_2$ solution in water are used in experiments to propagate λ, and TMGS is used to dilute and store λ. Strains containing transposon insertions are selected by plating on complex media containing the appropriate antibiotics. See section 14.4 for medium preparation methods.

14.2.1.2. Preparation of phage lysate

A transducing phage lysate is most easily prepared by first lysogenizing the donor strain with λ *c*I857 and then inducing the lysogenized donor cells to undergo lysis. The procedure is as follows.

1. Grow a few milliliters of the donor culture in super broth at 37°C to an approximate titer of 1×10^8 to 2×10^8 cells per ml. The titer of the growing culture can be determined by monitoring the optical density as described in 14.1.3.2.

2. Add approximately 0.2 ml of this culture to 2 ml of molten lambda soft agar held at 45°C, and immediately pour the contents onto the surface of a lambda agar plate. Rapidly tilt the plate back and forth to distribute the soft agar mixture uniformly over the plate, and then let it sit flat on the countertop at room temperature for several minutes until the soft agar layer solidifies. (The lambda agar plates should be equilibrated to room temperature or even prewarmed, since use of plates that are too cool will cause the soft agar to solidify before it can be evenly distributed on the surface of the plates.)

3. Place 1 or 2 drops of a λ *c*I857 lysate containing 10^9 to 10^{10} PFU/ml onto the center of the plate. (If the λ lysate has been stored over $CHCl_3$, the residual $CHCl_3$ must be removed prior to use by incubating a sample of

the lysate at 37°C with intermittent bubbling until $CHCl_3$ vapors are no longer detectable.) Leave the petri dish lid ajar until the spot has dried into the plate. Invert the plate, and incubate at 30°C overnight. The next day there will be a confluent lawn of bacteria over the surface of the plate with a large plaque in the center caused by lysis of some of the cells infected with λ *c*I857. The center of this plaque will be somewhat turbid, however, because of the survival and growth of donor cells that have become lysogenized with λ *c*I857.

4. Use a sterile toothpick, wood applicator stick, or loop to scrape off some of the turbid growth from the center of the plaque to inoculate 10 ml of super broth contained in a 125-ml sterile Erlenmeyer flask.

5. Incubate this culture at 30°C on a gyratory shaker to get good aeration. Monitor the optical density of the culture as described in 14.1.3.2. When the culture reaches approximately 2 × 10^8 cells per ml, shift the flask to 45°C for 1 to 2 min and then to 37°C. Continue aerated incubation for approximately 2 h, when lysis should be complete.

6. Add 1 drop of $CHCl_3$ to kill bacteria, shake vigorously to disperse, and continue shaking at 37°C for 10 to 15 min.

7. Remove bacterial debris from the lysate by centrifugation at 7,000 × *g* for 5 to 10 min. Pour the supernatant fluid into a sterile bottle or screw-cap tube, and add another drop of $CHCl_3$.

14.2.1.3. Determination of phage titer

The λ *c*I857 lysate should have a titer of about 10^{10} PFU/ml, and the exact titer can be determined as follows.

1. Grow a 10-ml culture of either the nonlysogenic donor or recipient strain in super broth with aeration by either bubbling or shaking at 37°C until 1 × 10^8 to 2 × 10^8 cells per ml is reached. The titer of the growing culture can be determined by monitoring the optical density as described in 14.1.3.2.

2. Sediment the cells in this culture by centrifugation at 7,000 × *g* for 5 to 10 min, and suspend the pellet in 10 mM $MgCl_2$.

3. Sediment the cells as a washing step, and suspend the pellet in 10 ml of 10 mM $MgCl_2$.

4. Starve this suspension for 60 min by shaking in a sterile Erlenmeyer flask at 37°C.

5. Distribute 0.9-ml amounts in sterile test tubes (13 by 100 mm), and place these at 37°C.

6. Dilute the λ lysate serially using 10 mM $MgCl_2$.

7. Add 0.1 ml of each of the 10^{-4}, 10^{-5}, 10^{-6}, and 10^{-7} dilutions to the 0.9-ml samples of bacterial cells. Allow 15 min for preadsorption of phage to the bacteria. Plate 0.2 ml from each tube by mixing it with 2 ml of molten lambda soft agar (equilibrated to 45°C) and immediately pouring the contents onto the surface of a lambda agar plate as described above.

8. After the agar has solidified, invert the plates and incubate at 37°C.

9. Count the plaques after overnight incubation.

It is often wise to plate each dilution in duplicate in case one of them smears because of excess moisture. The most reliable titer is determined from assay plates that contain between 50 and 400 plaques. The titer will be equal to the mean plaque count per plate times the reciprocal of the initial dilution that gave rise to the plaque count times 50. At the time of determining the phage lysate titer, it is wise to take some of the lysate and streak on a nonselective L agar plate to verify that the lysate is free from all viable donor bacteria.

14.2.1.4. Procedure for specialized transduction

1. Grow the mutant *gal* recipient in 10 ml of super broth at 37°C with aeration to 2 × 10^8 to 4 × 10^8 cells per ml.

2. Sediment, wash, and suspend the cells in 10 ml of 10 mM $MgCl_2$; then starve the cells with aeration at 37°C for 60 min prior to infection.

3. Distribute 0.9-ml amounts of this culture in sterile test tubes (13 by 100 mm), and incubate at 30°C.

4. Dilute the transducing phage lysate appropriately in 10 mM $MgCl_2$ so that addition of 0.1 ml to 0.9 ml of bacteria will result in multiplicities of infection of 0.1, 1.0, and 3.0 phage per bacterium. Be sure to remove residual $CHCl_3$ by bubbling air with a pipette if the lysate is not diluted more than 10^{-1}, to prevent killing of recipient cells.

5. Add these diluted amounts of lambda to individual 0.9-ml samples of recipient cells, and leave one 0.9-ml recipient culture uninfected to serve as a control to verify that the *gal* mutation in the recipient is stable and that the number of Gal^+ transductants is in excess of the number of Gal^+ revertants, should any appear.

6. Allow 15 min for preadsorption, and then plate 0.1-ml amounts from each mixture by spreading on the surface of each of four MA+Gal plates. Incubate two plates from each infection at 30°C and two at 37°C.

7. Incubate the plates for 2 days; then count the number of Gal^+ colonies.

This procedure can also be used to prepare *E. coli* for transduction with λ-derived vectors containing cloned DNA. If *E. coli* is being transduced with a λ-derived vector containing cloned DNA, the multiplicity of infection should be 0.1 phage per bacterium to ensure that no bacterial cell is infected more than once. Plate the bacteria-phage mixture on media containing the appropriate antibiotic at a density that will yield 50 to 200 colonies per plate.

14.2.1.5. Characterization of transductants

To be characterized, transductants must first be picked and purified to yield individual colonies. They can be purified on MA+Gal, EMB+Gal, or MacConkey+Gal plates which should, of course, be incubated at 30°C. As the last step, pick individual isolated colonies from each transductant into 2 ml of Super broth and grow to 1 × 10^8 to 2 × 10^8 cells per ml at 30°C. Gal^+ transductants should have several properties. The procedure is as follows.

1. The transductants should be lysogenic for defective or nondefective derivatives of λ *c*I857 and thus should be unable to grow at 37 or 42°C. Test this by spotting loopfuls of each culture on an EMB+Gal plate and incubating at 37 or 42°C. As a control, test for growth at 30°C.

2. The transductants should be immune to λ, and this is most easily determined by a cross-streak test. Distribute approximately 0.05 ml of the λ *c*I857 lysate evenly

down the center of an EMBO agar plate. After 5 to 10 min to allow the lysate to dry into the agar surface, streak full loopfuls of the individual Gal$^+$ transductant cultures across the phage streak. For this test, the donor and recipient strains should be included as controls. Incubate plates at 30°C. The growth to one side of the λ *c*I857 streak will serve as the uninfected control and will appear white to pink on the EMBO plates. Sensitivity to λ is revealed by the dark-red coloration of the surviving bacterial growth at the juncture of the λ *c*I857 streak with the streak of a λ-sensitive strain. The change in color is due to acid production and/or liberation as a consequence of lysis of some of the bacteria. EMBO is thus a very sensitive indicator of sensitivity of bacteria to temperate phage that do not lyse all infected cells.

3. The transductants should be partially diploid and heterozygous, containing both wild-type and mutant *gal* genetic information. This is most easily determined by diluting a culture sufficiently so that when 0.1 ml is spread on the surface of an EMB+Gal plate, 100 to 300 colonies will result after incubation at 30°C. If the transductants are partially diploid and heterozygous, approximately 1 to 3% of the colonies will give a fermentation-negative appearance and be white to pink and the remainder will give a fermentation-positive reaction and appear dark purple or black. The Gal$^-$ types arise as a result of recombination leading to homozygosity for the *gal* mutation originally in the recipient.

4. Most Gal$^+$ transductants arising from the low-multiplicity infections will harbor only the λ *gal* transducing phage genome. Test some of these to see whether infectious lambda particles are liberated at 37°C. To do this, (i) dilute the cultures of purified transductants 1:10 into 1 to 2 ml of super broth in test tubes (13 by 100 mm), (ii) incubate these subcultures at 37°C for 2 h, (iii) add a drop of $CHCl_3$ to kill surviving bacteria, (iv) spot loopfuls of each "lysate" onto either an EMBO or lambda agar plate that has been overlaid with 2 ml of lambda soft agar inoculated with 0.1 to 0.2 ml of a log-phase culture of a λ-sensitive strain, and (v) incubate the plates overnight at 37°C. These Gal$^+$ transductants will probably contain λ d*gal* particles and will not be able to yield plaque-forming particles to lyse the λ-sensitive indicator strain. This is because λ genes essential for the lytic response have been replaced by the wild-type *gal* gene.

5. Most Gal$^+$ transductants formed in the higher-multiplicity infections will be double lysogens having a wild-type λ *c*I857 prophage as well as a λ d*gal* prophage. Incubation of these cultures at 37°C will lead to spontaneous lysis with liberation of a more or less equal mixture of λ d*gal* particles and λ *c*I857 particles that will be able to form plaques. The plaque-forming ability of these lysates can be verified by the test outlined in step 4. Such lysates also have a high frequency of non-plaque-forming λ *gal* particles and can transduce 10 to 50% of the mutant *gal* recipients to Gal$^+$. This can be verified by repeating the transduction as outlined in section 14.2.1.4 above, with the modification that the infection mixture should be diluted (10^{-3} to 10^{-5}) prior to plating on the MA+Gal plates. These have been called high-frequency-transducing (HFT) lysates. This is in contrast to the low-frequency-transducing (LFT) lysate formed after the λ *c*I857 donor lysogen was initially heat induced at 37°C.

14.2.1.6. Phage plaque purification

When λ delivery vehicles are propagated in *E. coli*, mutations in λ that result in faster-growing phage will occur (5). These faster-growing phage will be enriched after propagation and may be the result of a mutation in a gene crucial to the experiment the phage was to be used in, e.g., loss of an amber mutation or loss of the transposon. For this reason, stocks of phage containing crucial genetic markers should be prepared from phage isolated from a single plaque. This method of producing a lysate allows minimal opportunity for selection of faster-growing λ mutants.

1. Grow a 10-ml culture of a *glnV tyrT hsdR* strain (e.g., LE392) in super broth with aeration at 37°C until a titer of 1 × 10^8 to 2 × 10^8 cells per ml is reached.

2. Sediment the cells in this culture by centrifugation at 2,700 × *g* for 10 min, and suspend the pellet in 10 ml of 10 mM $MgCl_2$.

3. Sediment the cells as a washing step, and suspend the pellet in 10 ml of 10 mM $MgCl_2$.

4. Starve this suspension for 60 min by shaking in a sterile 125-ml Erlenmeyer flask at 37°C.

5. Distribute 0.9-ml amounts in sterile glass test tubes (13 by 100 mm), and place these tubes at 37°C.

6. Dilute the λ lysate serially with 10 mM $MgCl_2$, or, if the titer is known to a reasonable approximation, dilute by a calculated amount to yield plates in the step below with about 100 plaques.

7. Add 0.1 ml each of the 10^{-4}, 10^{-5}, 10^{-6}, and 10^{-7} dilutions, or a calculated dilution to individual 0.9-ml samples of bacteria in the test tubes (13 by 100 mm) at 37°C. Allow 15 min for preadsorption, and take 0.2 ml from each tube and mix with 2 ml of molten lambda soft agar (equilibrated to 45°C). Immediately pour the contents onto the surface of a lambda agar plate.

8. After the soft agar has solidified, invert the plates and incubate at 37°C.

9. After incubation for 6 h, choose a plate that has well-separated plaques. The plates should not be incubated for excessive periods, since this will enrich for faster-growing phage variants (5). Using a sterile Pasteur pipette, harvest a plug of soft agar from the middle of a plaque. Transfer this to 0.2 ml of TMGS, and vortex to break up the plug and disperse the phage. This should yield a small lysate with a titer of approximately 1 × 10^7 to 1 × 10^8 PFU/ml.

14.2.1.7. Large-scale phage plaque purification

The large-scale procedure is essentially identical to the plaque purification procedure described above, but instead of harvesting phage from a single plaque the phage from an entire plate with plaques that have just become confluent is harvested.

1. Follow steps 1 to 8 in section 14.2.1.6. However, dilute the phage serially in TMGS to 10^{-5} instead of 10^{-7}, and add 0.1 ml of each of the 10^{-2} through 10^{-5} dilutions to 0.9-ml samples of bacterial cells.

2. At 6 h after placing the plates at 37°C, inspect the plates. Those inoculated with the 10^{-5} dilution of phage should have few isolated plaques and will serve as an indicator of bacterial growth. The plate inoculated with the 10^{-2} dilution will be clear as a result of complete lysis of all bacterial cells. One dilution in between 10^{-2} and 10^{-5} will result in a plate with plaques that are confluent. Add 3 ml of TMGS to this plate, and place at

4°C for several hours to overnight. The plates should not be incubated for excessive periods, since this will enrich for faster-growing phage variants (5).

3. Using a sterile spreader, break up the soft agar overlay and transfer to a sterile 15-ml polypropylene tube. Add 0.2 ml of chloroform to kill bacteria, and gently mix end over end to break up the soft agar and distribute the chloroform.

4. Pellet the agar and cellular debris at 7,000 × *g* for 10 min. Carefully pour the supernatant fluid into a sterile bottle or screw-cap tube, and add another drop of chloroform.

14.2.1.8. Mutagenesis of plasmid DNA

Since a plasmid is only a small fraction of the total DNA present in the cell, it is necessary to enrich for plasmids that contain the transposable element. This is easily accomplished by isolating plasmid DNA after Tn*5* mutagenesis and transforming or electroporating a kanamycin-sensitive (Km^s) strain. The total number of transposon insertions into the plasmid is a function of the transposition frequency and the size and copy number of the plasmid. The procedure described below will yield about 1,400 individual insertions in a plasmid the size and copy number of pBR322.

1. Grow the suppressor-free recipient strain containing the plasmid to be mutagenized in super broth with aeration to a titer of 1×10^8 to 2×10^8 cells per ml. Include the appropriate antibiotic to select for maintenance of the plasmid.

2. Sediment, wash, and suspend the cells in 10 ml of 10 mM $MgCl_2$; then starve the cells with aeration at 37°C for 60 min prior to infection.

3. Distribute 0.9 ml of this culture in a sterile test tube (13 by 100 mm), and incubate at 37°C.

4. Dilute the transducing phage lysate appropriately in 10 mM $MgCl_2$ so that addition of 0.1 ml to 0.9 ml of bacteria will result in a multiplicity of infection of 5 phage per bacterium. Be sure to remove residual $CHCl_3$ from the lysate by blowing a stream of air through the lysate.

5. Add 0.1 ml of diluted λ lysate to a 0.9-ml sample of recipient cells.

6. Allow 15 min for preadsorption. Add 1 ml of L broth to the bacteria, and incubate at 37°C for 2 h to allow expression of the transposon-encoded antibiotic resistance. Add kanamycin to the appropriate level, and continue to incubate at 37°C overnight. This period of growth will select for cells that successfully received the transposon from λ.

7. Purify the plasmid DNA from the overnight culture as described in Chapter 16, and use this plasmid DNA to transform the appropriate strain of Km^s bacteria as described in section 14.1.3. All the Km^r transformants contain a plasmid that has been mutagenized with Tn*5*.

14.2.1.9. Mutagenesis of chromosomal DNA

The aim of chromosomal DNA mutagenesis is to introduce Tn*5* elements throughout the *E. coli* chromosome to make a random Tn*5* insertion library. "Library" is used to mean a pool of bacterial cells each containing one Tn*5* element inserted into the chromosome. A sufficient number of insertions must occur to ensure that there is a high probability of inactivating any given gene on the bacterial chromosome. A certain fraction of inserts will be in genes specifying vital functions, and these insertions will be absent from the library since they are lethal. Genes vary in size, but the *E. coli* chromosome has approximately 4,700,000 bp (43, 76) and at least 1,800 genes (63). At least 10,000 (and preferably more) independent Tn*5* insertions must be obtained and screened to ensure that the insertion library includes at least one insertion in each gene in the genome. Since the insertion of Tn*5* could very probably impose a nutritional requirement, the Tn*5* insertion library is made in cells growing in a rich medium that contains an abundance of amino acids, purines, pyrimidines, vitamins, etc., and a variety of carbon sources.

1. Grow the suppressor-free recipient strain to be mutagenized in Super broth with aeration to a titer of 1×10^8 to 2×10^8 cells per ml.

2. Sediment, wash, and suspend the cells in 10 ml of 10 mM $MgCl_2$; then starve the cells with aeration at 37°C for 60 min prior to infection.

3. Distribute 0.9-ml amounts of this culture in sterile test tubes (13 by 100 mm), and incubate at 37°C.

4. Dilute the λ :: Tn*5* transducing phage lysate appropriately in 10 mM $MgCl_2$ so that addition of 0.1 to 0.9 ml of bacteria will result in a multiplicity of infection of 5 phage per bacterium.

5. Add 0.1 ml of diluted λ lysate to a 0.9-ml sample of recipient cells.

6. Allow 15 min for preadsorption. Add 1 ml of L broth to the bacteria, and incubate at 37°C for 2 h to allow expression of the transposon-encoded antibiotic resistance.

7. Approximately 90,000 Km^r colonies will result from each sample of culture that contains 1.8×10^8 cells infected with λ :: Tn*5* at a multiplicity of infection of 5. Make serial dilutions of the bacteria to 10^{-2}, and plate 0.1-ml volumes of the 10^{-2} dilution on L agar containing kanamycin to yield about 90 colonies per plate. The resulting Km^r bacteria each contain Tn*5* inserted into the genome. Each colony should be streaked on L agar containing kanamycin for individual colonies to ensure that each colony is descended from a single cell.

14.2.1.10. Mutagenesis of other gram-negative bacteria

λ uses the outer membrane protein LamB to attach to *E. coli* and inject its DNA. Since *E. coli* is essentially the only bacterium that naturally expresses a LamB that will act as a receptor for λ, nearly all experiments done with λ have used *E. coli*. Recently, investigators have constructed plasmids containing the *E. coli lamB* gene that is constitutive or fused to the promoter of another gene, *ompR*. These plasmids have been used to make *S. typhimurium* (26), *Klebsiella pneumoniae* (26), *Erwinia* spp. (67), and *Vibrio cholerae* (33) sensitive to infection by λ. The only strain that was tested and not made sensitive to λ was *Pseudomonas aeruginosa* (26). Incorporation of these plasmids into the above bacteria does not enable them to propagate λ, because of the requirement for other host factors present in *E. coli* but not in the other bacteria.

Effective adsorption of λ to *Salmonella* spp. containing *E. coli* LamB requires that the recipient be incapable of synthesizing a complete lipopolysaccharide (LPS) and instead produce an incomplete or rough LPS. Since the outer core of *Salmonella* LPS contains galactose, mutations that disrupt the production of galactose result in a rough LPS phenotype. Strains containing these mutations are easily selected by their ability to grow on minimal agar containing 2-deoxygalactose. Mutations in *galE* will also confer a galactose sensitivity on a strain containing the mutation. The following procedure uses the plasmid pAMH62 (34) to make *S. typhimurium* sensitive to λ and the method of Nnalue and Stocker (58) to produce an *S. typhimurium galE* mutant strain.

1. Grow a standing overnight culture of LT-2 or another *S. typhimurium* strain in L broth at 37°C.

2. Spread 0.1-ml samples of the overnight culture onto several minimal agar plates that contain 0.1% (wt/vol) 2-deoxygalactose and 0.5% galactose as carbon sources. Incubate at 37°C for 2 to 3 days until colonies appear. Purify the colonies by streaking for individual colonies on the same medium.

3. Pick a well-isolated colony from each streak, streak onto minimal agar containing 0.5% glycerol and onto minimal agar containing 0.5% glycerol and 0.5% (wt/vol) galactose, and incubate at 37°C for 4 days. Streak out the 2-deoxygalactose-sensitive parent as a control. Compare the size of the colonies on each plate. The *galE* mutation causes strains grown in galactose to accumulate UDP-galactose, a toxic intermediate. Therefore, strains containing the *galE* mutation will produce smaller colonies on the agar containing galatose than on the agar containing glycerol alone.

4. Verify that the 2-deoxygalactose-resistant galactose-sensitive mutants are *galE* mutants by cross-streaking with P22 HT *int*. *galE* mutant strains are insensitive to infection by P22 HT *int* because of the loss of LPS. The parental strain should be cross-streaked as a P22 HT *int*-sensitive control, and a *galE* mutant strain (LB5010; Table 1) should be cross-streaked as a P22 HT *int*-resistant control. Cross-streaking is described in detail in section 14.2.1.5, but be sure to use P22 HT *int* and not λ.

5. Transform or electroporate a *galE Salmonella* mutant with pAMH62 by using the methods described in section 14.1.3, and select for ampicillin-resistant transformants on L agar containing 100 μg of ampicillin per ml.

6. Transduce a transformant with λ::Tn5 as described in section 14.2.1.9. pAMH62 contains an *ompR'-lamB* fusion under control of the *ompR* promoter and does not require growth in media lacking glucose to increase the expression of *lamB*. All colonies that arise are *Salmonella* colonies containing Tn5.

14.2.1.11. Applications

Specialized transduction facilitates the genetic analyses of bacteria, including gene arrangement and regulation. *E. coli* has a number of λ-like phages that integrate into different locations of the chromosome and allow specialized transduction of adjacent chromosome markers. Even more important, by using a donor host that carries a deletion for the normal integration site of the λ prophage (*att*λ), λ will integrate at low frequency into abnormal sites scattered throughout the chromosome. Induction of these lysogens will lead to formation of transducing phage carrying almost any gene desired (73). This approach can be used with other specialized transducing phages of *E. coli* as well as with those that propagate on other bacterial species. Many of the methods to accomplish these objectives, as well as other uses of specialized transducing phages, are described by Miller (53).

Historically, λ has played a crucial role in the development of *E. coli* genetics, and it continues to be an important tool. λ is used as a vector to carry a variety of transposons, including those that can be used to generate fusions. A fusion is produced when a transposon inserts into a gene or an operon. All transposons capable of producing fusions carry a reporter gene that is transcribed and translated only when a productive fusion is made to sequences upstream of the transposon. The reporter genes frequently used are *lacZ* and, to a lesser extent, *cat*, *lux*, and *phoA*. Each of these genes encodes a protein that is easily assayed quantitatively by using the appropriate chromogenic or luminescent reagents. Transposons can be used to generate two types of fusions. Type I (also referred to as operon or transcriptional) fusions result from insertion of a transposon in the correct orientation into the transcriptional unit of an operon. The reporter genes used to generate this type of fusion are *lacZ*, *cat*, and *lux* sequences lacking a promotor but retaining a Shine-Dalgarno sequence. The reporter genes rely on the operon fusion for transcriptional signals but are then able to direct their own translation. Type II (or protein or translational) fusions result from the insertion of a transposon into a gene in the correct reading frame and orientation. *lacZ* and *phoA* sequences used to generate type II fusions lack both promotor and Shine-Dalgarno sequences. A type II fusion results in a hybrid protein with the targeted gene encoding the amino-terminal portion and the transposon-encoded gene (*lacZ* or *phoA*) encoding the carboxy-terminal end of the hybrid protein. The generation of fusions has facilitated the study of operon expression, gene expression, protein structure, and gene product placement in the bacterial cell. The large number of transposons that have been described (see reference 8 for a review) is indicative of how important these genetic elements have become in bacterial genetics.

In addition to being used as a vector for transposons, nonessential regions of the λ genome have been identified and deleted to form cloning vectors. Through deletion of these regions, λ has been used in the construction of cloning vectors that retain the wild-type self-replicating nature of λ. λ has also been used to construct cosmids, i.e., cloning vectors that can be packaged into phage λ particles and infect strains that express the LamB protein but then replicate as a plasmid. These cloning vectors have in turn made *E. coli* the preeminent host for propagation and analysis of cloned genes from other genera. Complementation of *E. coli* mutations by recombinant DNA has provided a great deal of information about genes in other bacterial genera, but not all foreign bacterial genes are expressed in *E. coli*, and some gene products, particularly those located in or passing through the membrane, are toxic. A potentially promising way to replace *E. coli* as the host for λ-derived cloning vectors in the future may be to introduce the *lamB* gene into other genera to allow λ to attach and inject its DNA.

14.2.2. Generalized Transduction in *E. coli* with Phage P1

14.2.2.1. Strains, phages, and media

Phage P1 is a generalized transducing phage with an extremely broad host range. It is easy to isolate mutant derivatives of wild-type P1 that will plate with high efficiency on approximately 70% of *E. coli* strains obtained from clinical specimens and also to isolate derivatives that will readily infect bacterial strains that are members of the genera *Shigella, Salmonella, Enterobacter, Klebsiella,* and *Citrobacter* (28). Because of this attribute and the fact that P1 can transduce fragments of donor strain DNA of 91 to 100 kb, it is a very useful transducing phage. On the other hand, compared with other frequently used coliphages, P1 adsorbs to bacteria more slowly, forms very small plaques, and yields lysates that are somewhat unstable unless completely purified from bacterial cell debris. P1 *kc* (45) is a P1 derivative that plates with high efficiency on *E. coli* K-12 and forms a relatively clear plaque that facilitates its assay. However, P1 *kc* is able to lysogenize bacterial hosts. P1 L4 (13) is a mutant derivative of a strain of P1 *kc* that possesses a mutation in the *c*I repressor gene and is thus unable to lysogenize.

Although a diversity of donor and recipient strains of *E. coli* K-12 can be used for P1-mediated transduction, it is convenient to illustrate the principles involved by using a system that allows a simple demonstration of cotransduction. Thus the donor strain (χ226; Table 1) should have a *lacZ* mutation conferring inability to synthesize ß-galactosidase and thus inability to utilize lactose as a carbon source. The recipient (χ353; Table 1) should have the phenotype LacZ$^+$ ProC$^-$, so that selection of ProC$^+$ transductants will sometimes allow inheritance of the donor *lacZ* mutation by cotransduction with wild-type *proC* allele. The two genes are approximately 30% cotransducible. Recently, Singer et al. (75) described 182 Tn*10*-induced mutations spaced at approximately 1-min intervals around the *E. coli* K-12 chromosome. This collection of strains can be used to prepare lysates and transduce Tn*10* selectively into any area of the chromosome of P1-sensitive strains and aid in strain construction.

Use L broth, L agar (1.2% agar), and L soft agar (0.65% agar) containing 2.5 mM $CaCl_2$ for growth of bacteria and P1 lysate preparation and assays; Penassay agar for bacterial assays; and VB medium E agar supplemented with 1% lactose, 0.4% disodium succinate, and 50 mg of triphenyltetrazolium chloride per ml for selection of transductants. On this last medium, LacZ$^+$ ProC$^+$ transductants will form white colonies and LacZ$^-$ ProC$^+$ transductants will be able to grow because of utilization of succinate as a carbon source and will appear as red colonies. The citrate in VB medium E will chelate the Ca^{2+} needed for P1 adsorption and will minimize P1 infection of bacteria on the selective medium. See section 14.4 for methods for preparation of media.

14.2.2.2. Determination of phage titer

1. Add 1 ml of a standing overnight culture of the P1-sensitive donor χ226 grown in L broth (without $CaCl_2$) to 9 ml of L broth containing 2.5 mM $CaCl_2$.

2. Aerate this culture at 37°C for 60 to 90 min until the culture reaches a titer of approximately 2 × 10^8 cells per ml. The titer of the growing culture can be determined by monitoring the optical density as described in 14.1.3.2.

3. While the culture is growing, dilute a P1 L4 (or P1 *kc*) lysate (which should have a titer of approximately 10^{10} PFU/ml) 10^{-5}, 10^{-6}, and 10^{-7} in L broth containing 2.5 mM $CaCl_2$.

4. Place 0.9-ml amounts of the grown P1-sensitive strain in test tubes (13 by 100 mm) at 37°C.

5. After a few minutes for temperature equilibration, add 0.1 ml of each P1 dilution and allow 20 or 30 min for adsorption of P1 to the bacterial cells.

6. Add 0.2 ml of each phage-bacterium mixture to 2 ml of L soft agar, and immediately distribute on the surface of an L agar–2.5 mM $CaCl_2$ plate. After 5 min, invert the plate and incubate at 37°C.

Plaques become visible after 6 h of incubation and are 0.5 to 1 mm in diameter. Plaque assay plates for P1 L4 can be incubated overnight prior to counting. Plaque assay plates for P1 *kc* should be read after about 6 h since continued growth of lysogenized cells in the center of P1 *kc* plaques causes them to become turbid and difficult to visualize if plates are incubated overnight. The phage titer is equal to the plaque count times the reciprocal of the dilution times 50.

14.2.2.3. Preparation of phage lysate

1. Grow the desired bacterial donor strain as described above (steps 1 and 2).

2. Dilute the P1 L4 (or P1 *kc*) lysate in L broth plus 2.5 mM $CaCl_2$ so that it contains between 6 × 10^7 and 8 × 10^7 PFU/ml, and add 0.1 ml to 0.9 ml of log-phase donor bacteria in test tubes (13 by 100 mm) at 37°C.

3. Allow 20 to 30 min for P1 adsorption.

4. Add 0.2 ml of the adsorption mixture per 2 ml of L soft agar containing 2.5 mM $CaCl_2$, and immediately distribute on L agar–2.5 mM $CaCl_2$ plates. Prepare five to eight plates in this way. (Each plate will be seeded with 6 × 10^5 to 8 × 10^5 P1 L4-infected [or P1 *kc*-infected] donor bacteria and an excess of uninfected donor bacteria.) After 5 min, invert the plates and incubate at 37°C for 6 h.

5. Add 3 to 5 ml of L broth containing 10 mM $MgSO_4$ (but no $CaCl_2$) per plate. Using a sterile glass spreader, break up the soft agar layer and scrape into a 40- to 50-ml polypropylene (or another $CHCl_3$-resistant material) centrifuge tube. Add approximately 0.1 ml of $CHCl_3$ for each 4 ml of lysate, cap the centrifuge tube tightly, and blend vigorously on a Vortex mixer for 1 to 2 min. Let this tube sit for 2 h at room temperature or overnight in a refrigerator to permit the P1 to diffuse into the aqueous phase.

6. Centrifuge the crude lysate at 7,000 × g for 10 to 15 min, and carefully decant the supernatant fluid into a bottle or screw-cap test tube. Add a few drops of $CHCl_3$ to this lysate, and shake vigorously. Store the lysate at 4°C.

The next day, determine the titer of the lysate as described above and test for sterility by streaking a small amount of lysate on the surface of an L or Penassay agar plate. These lysates are relatively stable but will drop 1 to 2 log units in titer over 1 year when stored at 4°C. Complete stability of P1 can be achieved by banding the phage in CsCl and removing CsCl by dialysis against L broth containing 10 mM $MgSO_4$ (13).

14.2.2.4. Construction of P1 *kc* lysogens

It is wise to use recipient strains that are lysogenic for P1 *kc* when the transduction experiment is performed to yield quantitative data on gene frequencies in donor cells, cotransduction, etc. This is neither necessary nor desirable when using transduction to construct strains. Lysogenic recipient strains are constructed as follows.

1. Grow the recipient strain χ353 as described in section 14.2.2.2, steps 1 and 2.

2. Add 0.9 ml of culture to a test tube (13 by 100 mm), and equilibrate at 30°C for about 5 min.

3. Add 0.1 ml of a P1 *kc* lysate to achieve a multiplicity of infection of about 5 PFU/bacterium. Be sure to remove traces of $CHCl_3$ by bubbling air with a pipette if the P1 *kc* lysate is not diluted more than 10^{-1}, to prevent killing of recipient cells.

4. Incubate at 30°C for 2 to 3 h. With P1 infection, the lysogenic response is favored at low temperatures (20 to 30°C) and the lytic response is favored at high temperatures (37 to 42°C).

5. Streak out growth on a Penassay or L agar plate to obtain single colonies. Alternatively, plate an appropriate dilution to achieve single colonies. Incubate these plates at 30°C.

6. Pick and purify 10 well-isolated colonies (nonmucoid and with regular smooth edges) on Penassay or L agar by the general methods described in section 14.1.2.3.

7. Pick a well-isolated colony from each of the 10 original isolates into 2 ml of L broth containing 1 mM citrate (no $CaCl_2$) in test tubes (13 by 100 mm). Number the cultures from 1 to 10, and incubate at 42°C for 4 to 6 h, to about 2 × 10^8 cells per ml. Growth at 42°C increases the rate of spontaneous induction of P1 lysogens and the absence of Ca^{2+} permits the P1 phage liberated as a consequence of spontaneous induction to accumulate.

8. Test whether the putative P1 lysogens liberated P1 during growth at 42°C. This is done as follows.

(i) Place 1 ml of each of the 10 cultures and 1 ml of a culture of the P1-sensitive starting strain (as a control) into numbered test tubes (13 by 100 mm) at 42°C.

(ii) Add a drop of $CHCl_3$ to each culture, shake, and continue incubating for 15 to 20 min.

(iii) Make serial 10-fold dilutions of a P1 *kc* lysate from 10^{-1} to 10^{-8}.

(iv) Add 0.2 ml of a log-phase P1-sensitive strain grown in L broth containing 2.5 mM $CaCl_2$ to 2 ml of molten L soft agar containing 2.5 mM $CaCl_2$, and pour onto the surface of an EMBO agar (containing 2.5 mM $CaCl_2$) plate. Let the soft agar layer solidify.

(v) Spot full loopfuls of each of the 11 chloroform-treated cultures and the 8 serial P1 dilutions onto the surface of the EMBO plate. By touching the side of the loop to the surface of the agar, the entire contents of the loop should be delivered. Leave the lid ajar until the spots have dried into the soft agar. Mark the bottom of the plate so that the locations of all spots are known.

(vi) It is best to leave the EMBO plate right side up (i.e., uninverted) for incubation at 37 or 42°C overnight.

Areas of lysis (confluence or plaques) should be discernible for each chloroform-treated extract made from a P1 lysogen provided that sufficient P1 was liberated during growth at 42°C. The spots made from the P1 dilutions should indicate the sensitivity of the method.

9. Test whether the putative P1 lysogens are immune to P1. This is done as follows.

(i) Distribute 0.05 to 0.1 ml of a P1 L4 lysate (ca. 10^{10} PFU/ml) down the center of an EMBO agar (containing 2.5 mM $CaCl_2$) plate. Leave the lid ajar until the streak is dry. Mark the plate bottom to indicate the orientation and placement of streaks.

(ii) Streak each of the 10 cultures of viable putative P1 lysogens and the culture of the viable P1-sensitive starting strain across the P1 L4 streak.

(iii) Invert the plate, and incubate overnight at 37°C.

A P1 L4 immune isolate that liberates P1 *kc* during growth at 42°C can be stocked as a P1 *kc* lysogen. P1 lysogens are somewhat unstable, presumably because the P1 prophage exists as a plasmid which is sometimes lost. Thus, P1 lysogens that have been subcultured for a period may require reisolation by the procedures and tests enumerated above. Growth of P1 lysogens in the presence of 2.5 mM $CaCl_2$ at 37°C also permits spontaneously liberated P1 to infect nonlysogenic segregants and either lyse or relysogenize them.

14.2.2.5. Generalized transduction procedure

When using a nonlysogenic recipient, it is customary to use a multiplicity of infection of 1 or slightly less so that cells infected with a transducing P1 particle will not also be infected with an infectious P1 particle and thereby be lost as a result of lysis. Since P1 does not adsorb very rapidly (only 30 to 50% of the phage will attach and inject within 20 to 30 min at 37°C), an initial multiplicity of 2 to 3 phage per bacterium is used to achieve an actual multiplicity of infection of 1. If a P1 lysogen is used as a recipient, double infections will not decrease the recovery of transductants, since the lysogenic cells are immune to superinfection with P1 L4. To illustrate these points, the P1 L4 lysate grown on the donor parent can be added to the nonlysogenic and lysogenic recipient strains at multiplicities of infection of 0.5, 2.5, and 12.5 per bacterium. The highest multiplicity may be difficult to achieve unless the P1 L4 lysate grown on the donor strain has a titer of 10^{10} PFU/ml or higher. The transductions are conducted as follows.

1. Grow the P1 *kc* lysogenic and nonlysogenic LacZ$^+$ ProC$^-$ recipient χ353 to 2 × 10^8 cells per ml as described in section 14.2.2.2.

2. While the recipient strains are growing, place some of the P1 L4 lysate grown on the LacZ$^-$ ProC$^+$ donor at 37°C and aerate to remove all residual $CHCl_3$. Also, calculate the amount of diluted or undiluted lysate (usually 0.1 to 0.5 ml) to add to a selected volume of recipient cells at about 2 × 10^8 ml (usually 0.5 to 1.9 ml) to achieve a multiplicity of infection of 12.5 PFU per bacterium. Then dilute the P1 L4 lysate in L broth containing 2.5 mM $CaCl_2$ so that the multiplicities of infection of 0.5, 2.5, and 12.5 PFU per bacterium can be achieved by the addition of a constant volume of phage lysate to a constant volume of recipient cells.

3. Add the calculated volume of bacteria of each recipient strain to each of four test tubes (13 by 100 mm), and place the eight cultures at 37°C. Determine the titers of the recipient cultures by plating appropriate dilutions on L or Penassay agar.

4. Add the calculated amounts of the donor P1 L4 lysate to three of the four cultures of each recipient

strain and an equal volume of L broth containing 2.5 mM $CaCl_2$ to the fourth culture, which will serve as a control. It is best to stagger the infections at about 3-min intervals.

5. After 30 min for preadsorption, add 0.1 ml of 0.2 M sodium citrate to each phage-bacterium mixture to chelate the Ca^{2+} and stop further infection of P1.

6. Spread 0.1-ml amounts of undiluted and 10^{-1}-diluted samples from each infected and uninfected culture per VB minimal agar plate (containing lactose, succinate, and triphenyltetrazolium chloride). Incubate these plates at 37°C for 2 days prior to counting.

7. Plate appropriate dilutions of the infected and uninfected cultures on L or Penassay agar to determine the viable titers. Incubate these plates overnight at 37°C.

8. Redetermine the titer of the P1 L4 lysate by the method described in section 14.2.2.2. This is especially important with newly made P1 lysates, since the titer sometimes increases two- to threefold during the first week because of disaggregation by the action of $CHCl_3$ of P1 particles stuck together in bacterial debris and agar remaining in the lysate.

9. Plate 0.1 ml of the P1 L4 lysate on the selective VB minimal agar to verify the absence of viable donor cells.

The number of transductants will be proportional to the titer of phage used when the P1 *kc* lysogenic recipient is used. When the nonlysogenic recipient is used, fewer transductants should be observed when multiplicities of infection are in excess of 1 than when they are 1 or less. In either case, the frequency of cotransduction of the *lacZ* mutation with the wild-type *proC* allele should be approximately the same within the limits of statistical variation. The frequencies of killing of the nonlysogenic recipient for each infection can be calculated and used in conjunction with the Poisson distribution to estimate the actual multiplicities of infection (1).

14.2.3. Generalized Transduction in *S. typhimurium* with Phage P22

14.2.3.1. Strains, phages, and media

Experiments with phage P22 provided the first evidence of generalized transduction (87). The bacterial receptor of P22 is the repeating polysaccharide of the O antigen 12 specificity, serogroups A, B, and D1, present on *S. typhimurium* and other related *Salmonella* species (86). P22 is capable of readily adsorbing onto sensitive strains. P22 is closely related to λ and, like λ, will integrate into the host genome at a specific site, *ataA*, between *proA* and *proC* in the *Salmonella* chromosome. Packaging of P22 DNA into phage heads is signaled by *pac* sites present on the phage DNA. Unlike λ and its host *E. coli*, however, the *Salmonella* chromosome contains sequences that are similar to the P22 *pac* sites, and these cause host DNA to be packaged in phage heads at a low frequency; approximately 39 to 42 kb of contiguous host DNA is packaged into 1 to 5% of the phage heads. The host DNA packaged by wild-type P22 starts at these sites, and this is reflected in the observed decrease in transduction frequency of chromosomal genes that are farther away from the sites than of genes near the sites (15).

Generalized transduction experiments have been greatly facilitated by P22 HT *int*. P22 HT is a mutant that does not require *pac* sites to package DNA and transduces genes located around the chromosome at a high frequency (71). The wild-type *int* gene of P22 is required for integration of P22 into the chromosome of the host. P22 containing an *int* mutation is thereby incapable of forming a true lysogen after infecting a nonlysogenized strain. Experiments with P22 in this laboratory are carried out with P22 HT *int*.

As is the case for *E. coli* and P1, a diversity of *S. typhimurium* strains can be used for P22-mediated transduction. The experiments described below illustrate simple transduction of a $ProC^-$ strain (SA2649; Table 1) to $ProC^+$ with LT-2 (Table 1) as a donor. Also, the transduction of a $ProC^+$ strain to $ProC^-$ is described. This transduction makes use of Tn*10* to tag the mutant *proC* allele and facilitate recovery of $ProC^-$ transductants.

Use LB broth and LB agar for growth of bacteria and P22 preparation and assays; minimal liquid and minimal agar media supplemented with proline and LB agar supplemented with tetracycline for selection and analysis of transductions; Green indicator plates to identify colonies undergoing active P22 infection; and EMBO agar to check strains for sensitivity to P22. Add 10 mM ethylene glycol-bis(ß-aminoethyl ether)-*N,N,N′,N′*-tetraacetic acid (EGTA; pH 8.0) to the appropriate plates when selecting and purifying transductants.

The methods described below are based on procedures described by Davis et al. (24).

14.2.3.2. Determination of phage titer

1. Grow the recipient strain as a standing overnight culture in 3 ml of LB broth at 37°C.

2. Dilute the overnight culture 1:10 into 9 ml of warm LB broth, and grow with aeration at 37°C until the mid-log phase.

3. Distribute 0.1-ml samples of bacteria to test tubes (13 by 100 mm), and place these at 37°C.

4. Dilute the P22 lysate serially in BSG or TMGS.

5. Add 0.1 ml of each of the 10^{-5}, 10^{-6}, 10^{-7}, 10^{-8}, 10^{-9}, and 10^{-10} dilutions to the 0.1-ml samples of bacterial cells, mix gently, and place at 37°C for 10 min to preadsorb the phage. Add 2 ml of molten soft agar (equilibrated to 45°C) to each bacterium-phage mixture, mix gently, and immediately pour the contents onto the surface of a LB agar plate that was poured 2 days before and left at room temperature.

6. After the agar has solidified, invert and incubate the plates at 37°C.

Plaques become visible after 4 to 6 h of incubation and should be turbid. The plaque size is dependent on the growth rate of the host strain and the quality of the LB agar plate and the LB soft agar. Plaque assay plates for P22 HT *int* can be incubated overnight before being counted. The phage titer is equal to the plaque count times the reciprocal of the dilution times 10.

14.2.3.3. Preparation of phage lysate

1. Grow the recipient strain as a standing overnight culture in 3 ml of LB broth at 37°C.

2. Dilute the overnight culture 1:10 into 9 ml of warm LB broth, and grow with aeration at 37°C until mid-log phase.

3. Transfer the bacteria to a 125-ml flask, and infect the donor with P22 HT *int* diluted to yield a multiplicity

of infection of 0.01 phage per bacterium. The low multiplicity is used to reduce the carryover of transducing particles containing DNA from the strain on which the lysate was produced.

4. Incubate this at 37°C with aeration until lysis occurs. If lysis is not complete after 18 h, continue with step 5. Complete lysis is not required to achieve high-titer lysates with P22 HT *int*.

5. Add 0.1 ml of chloroform, and continue to aerate for 30 min to lyse any unlysed cells. Transfer to a sterile 15-ml polypropylene tube, and sediment cellular debris at 7,000 × *g* for 10 min. Place the supernatant fluid in a sterile glass screw-cap tube, add a few drops of chloroform, and vortex to mix and sterilize the lysate. Store this at 4°C. Streak a loopful of the lysate on LB agar, and incubate at 37°C to verify that the lysate is sterile.

14.2.3.4. Generalized transduction procedure

The following procedure details the transduction of a ProC⁻ recipient to ProC⁺. Since P22 is capable of generalized transduction, this procedure can be used to transduce any chromosomal allele of a donor into a recipient.

1. Grow a standing overnight culture of the ProC⁻ recipient strain in LB broth at 37°C.

2. Dilute the strain 1:10 in 9 ml of prewarmed LB broth, and grow with aeration to 2×10^8 cells per ml.

3. Distribute 0.9-ml samples of bacteria into two tubes (13 by 100 mm), and place them at 37°C.

4. Dilute a transducing phage lysate produced on the ProC⁺ donor so that addition of 0.1 ml of phage will yield a multiplicity of infection of 0.1 phage per bacterium. The plating efficiency of P22 on different strains will change as a result of differences on the O antigen and the host restriction systems of the donor and recipient strains. For this reason it is appropriate to determine the titer of a transducing lysate on each recipient it will be used to transduce. Add phage to one tube, and keep the other tube of recipient cells as an uninfected control. Incubate these at 37°C 10 min to allow adsorption to occur.

5. Dilute the bacterium-phage mixture, and plate 0.1-ml samples of the 10^0, 10^{-1}, and 10^{-2} dilutions onto minimal medium containing 10 mM EGTA. Inoculate a single plate with 0.1 ml of cells from the uninfected cells. Incubate these plates at 37°C for 2 days. The plate inoculated with the uninfected cells will indicate the reversion frequency of the *proC* mutation in the recipient.

6. Pick approximately 20 transductants, and purify by streaking for isolated colonies on Green indicator medium. On this type of plate, colonies that contain P22 as a pseudolysogen are dark green and those colonies not infected will be light colored. Incubate at 37°C until the color of the colonies is easily distinguished.

7. Pick a well-isolated light-colored colony from each streak, and inoculate into 1 ml of LB broth. Incubate these cultures without aeration (i.e., statically) overnight at 37°C.

8. Verify that the ProC⁺ transductants are not lysogens by cross-streaking the ProC⁺ transductants against P22 HT *int* on EMBO agar that does not contain EGTA. Include the recipient as a control. Incubate this plate at 37°C overnight. The growth to one side of the P22 HT *int* streak will serve as the uninfected control and will appear as white to pink on the EMBO plates. Sensitivity to P22 HT *int* is revealed by the dark-red coloration of the surviving bacterial growth at the juncture of the P22 HT *int* streak with the streak of a P22 HT *int*-sensitive strain. The ProC⁺ transductants should be sensitive to P22 HT *int*.

14.2.3.5. Strain construction by transduction of tagged alleles

Generalized transduction of a recipient with an allele that can be directly selected for is a powerful technique. However, identification of recipients transduced with an unselectable allele can often involve a great deal of work. One method to select easily for transduced alleles with no selectable phenotype involves placing a transposon that encodes drug resistance close to or within the allele to be transduced. Since P22 heads are capable of packaging up to 42 kb of DNA (equivalent to approximately 1 min of the chromosome), alleles in close proximity to each other can be cotransduced into the recipient. When a transposon is located close to an allele that confers no selectable phenotype, plating the P22-infected recipient on media containing the drug to which the transposon encodes resistance will enrich for transductants which also received the nonselectable allele.

The procedure described below involves four major components: generation of a Tn*10* insertion library in a prototrophic strain, identification of transposon insertions that cotransduce with the wild-type *proC* allele, verification of linkage of the Tn*10* insertion to the wild-type *proC* allele and isolation of Tn*10* inserted near the mutant proC allele, and, finally, transduction of the recipient to ProC⁻ and tetracycline resistance (Tc^r). The method below uses λ :: Tn*10* (81), and other methods in which a specially constructed P22 :: Tn*10* is used have been described (14). It should be noted that a substantial number of transposable element-induced mutations have been produced and mapped in *S. typhimurium* (69). These can be used to transduce a strain to antibiotic resistance and thereby insert transposable elements selectively into a variety of locations throughout the *S. typhimurium* chromosome.

Transduction of *S. typhimurium* with λ::Tn*10*

The purpose of the following procedure is to construct a library of *S. typhimurium* containing Tn*10* inserted throughout the chromosome. This library will then be used to isolate a Tn*10* element that is linked to the *proC* allele.

1. Prepare the wild-type ProC⁺ strain (LT-2; Table 1) to be transduced with λ::Tn*10* by transforming it with pAMI162 as described in section 14.2.1.10.

2. Transduce an ampicillin-resistant transformant, isolated in step 1, with λ::Tn*10*, as described in section 14.2.1.10, step 6, but use λ::Tn*10* instead of λ::Tn*5*. Also, transduce a total of 11 to 12 samples of bacterial cells. Pellet each bacterium-λ mixture, suspend in 0.1 ml of TMGS, and plate each sample on an L agar plate containing tetracycline and 0.5% galactose instead of glucose. If the rate of Tn*10* transposition is 10^{-6} per transposable element per cell generation (81), each 0.9-ml sample that contains 1.8×10^8 cells will yield approximately 900 Tc^r cells. Incubate these plates overnight at 37°C.

3. Add 3 ml of LB broth that contains 0.5% galactose instead of glucose to each plate, and use a sterile

spreader to suspend the cells. Transfer the cells to a single 250-ml flask, dilute the cells to approximately 10^8 cells per ml, and add tetracycline to 12.5 μg/ml. Incubate at 37°C overnight with aeration. The titer of the cells removed from the plate can be determined by diluting the cells and measuring the OD_{600} as described in 14.1.3.2.

4. Remove 1 ml of the culture containing the Tn*10* mutagenized cells, and transfer to a test tube (16 by 125 or 150 mm). Infect this with P22 HT *int* at a multiplicity of infection of 0.1 phage per bacterium by adding 0.1 ml of a dilution of phage containing 10^9 phage per ml.

5. Incubate this at 37°C with aeration until lysis occurs. If lysis is not complete after 18 h, continue with step 6. Complete lysis is not required to achieve high-titer lysates with P22 HT *int*.

6. Add 0.1 ml of chloroform, and continue to aerate for 30 min to lyse any remaining cells. Transfer to a sterile 15-ml polypropylene tube, and sediment the cellular debris at 7,000 × *g* for 10 min. Place the supernatant fluid in a sterile glass screw-cap tube, add a few drops of chloroform, and vortex to mix and sterilize the lysate. Store this at 4°C. Streak a loopful of the lysate on nonselective LB agar, and incubate at 37°C to verify that the lysate is sterile.

Identification of transposon insertions that cotransduce with the wild-type *proC* allele

1. Grow a standing overnight culture of the strain containing the mutant *proC* allele (SA2649; Table 1) in LB broth at 37°C.

2. Dilute the strain 1:10 in warm LB broth, and grow with aeration to 2 × 10^8 cells per ml.

3. Distribute 0.9-ml samples of bacteria into two tubes (13 by 100 mm), and place at 37°C.

4. Dilute the transducing phage lysate prepared on the prototrophic strain in the above subsection so that addition of 0.1 ml of phage will yield a multiplicity of infection of 0.1 phage per bacterium. Add phage to one tube of cells, and keep the other as an uninfected control. Incubate these at 37°C for 10 min to allow adsorption to occur.

5. Plate 0.1-ml samples of the bacterium-phage mixture onto LB agar plates containing tetracycline and 10 mM EGTA to select for Tc^r cells. Inoculate a single plate with 0.1 ml of cells from the uninfected cells. Incubate these plates overnight at 37°C. Replica plate (44) these Tc^r transductants onto minimal medium. The resulting prototropic Tc^r colonies are most probably due to the cotransduction of the wild-type *proC* allele and a closely linked Tn*10*.

6. Pick several of the prototrophic Tc^r colonies, and purify by streaking for isolated colonies on Green indicator medium containing tetracycline. Incubate at 37°C until the color of the colonies is easily distinguished.

7. Pick a well-isolated light-colored colony from each streak, and inoculate into 1 ml of LB broth containing tetracycline. Incubate these cultures overnight at 37°C. Also patch each of these strains onto LB agar containing tetracycline. After overnight incubation at 37°C, store this plate at 4°C.

8. Verify that the transductants are not lysogens by cross-streaking loopfuls of each colony culture against P22 HT *int* on EMBO agar that does not contain EGTA. Cross-streaking is described in section 14.2.1.5.

Verification of linkage of the Tn*10* insertion to the *proC* allele and isolation of Tn*10* inserted near the mutant *proC* allele

1. Prepare a P22 HT *int* lysate on each prototrophic Tc^r transductant isolated as above by using the purified colonies set aside in step 7 above.

2. Infect separate cultures of the strain containing the mutant *proC* allele (SA2649; Table 1) with the individual lysates, and plate on LB agar containing tetracycline. Streak 100 resultant Tc^r colonies from each transduction onto minimal media and minimal media containing proline. This will indicate which colonies transduced with Tn*10* were cotransduced with the wild-type *proC* allele.

3. For each lysate, count the number of $ProC^+$ Tc^r colonies and divide by the total number of Tc^r colonies. This will give the frequency of cotransduction of the wild-type *proC* allele and the Tn*10* insertion. The lysate with the highest frequency of cotransduction will indicate which strain isolated above in step 7 above contains Tn*10* most closely linked to *proC*.

4. Isolate one $ProC^-$ Tc^r colony from step 2 that was derived by infecting strain SA2649 with the lysate yielding the highest frequency of cotransductants. This colony can then be used to produce a P22 HT *int* lysate and used to transduce strains to $ProC^-$ and Tc^r as described in the next step.

Transduction to $ProC^-$ and Tc^r. The following procedure is identical to that described above with one exception: the bacterium-phage mixture should be plated on LB agar containing EGTA and tetracycline. The Tc^r colonies can then be screened to identify those colonies that are $ProC^-$. The frequency of $ProC^-$ Tc^r colonies will be dependent on the cotransduction frequency calculated above.

14.2.3.6. Applications

Generalized transduction with P1 and P22 has been very useful in unambiguously determining the order of genes on the bacterial chromosome. This has been accomplished by using appropriately marked strains and monitoring inheritance of two or three nonselected alleles in conjunction with the selected marker and by performing the transductions in all possible reciprocal combinations for selected and unselected markers. As described above, generalized transduction is useful to construct strains of desired genotypes. A mutant allele of known properties can be cotransduced into a strain by selecting for inheritance of an adjacent wild-type marker or transposon.

Another useful method was developed by Hong and Ames (37) to induce mutations in unknown genes that are closely linked to known genes. By treating a transducing phage lysate with nitrous acid, hydroxylamine, or nitrosoguanidine and then selecting transductants for inheritance of a known wild-type allele at 30°C, it is possible to isolate a high frequency of mutations causing thermosensitive defects in some nutritional or vital function in genes adjacent to the selected allele.

P1- and P22-mediated transduction can also be used to determine whether a trait is chromosomally specified or plasmid specified. A P1 lysate propagated on the donor is irradiated with increasing doses of UV light, and these irradiated and unirradiated lysates are used to transduce a recipient strain with selection for inheritance of the donor trait in question. If the trait is specified by a chromosomal gene, small doses of UV will actually increase the total number of transductants since UV is recombinogenic, whereas if the trait is specified by a plasmid, the frequency of transductants inheriting the donor trait will decrease with increasing UV dose since UV-induced damage will preclude the replication of the plasmid after it is introduced into the recipient by transduction.

The availability of P1 derivatives, such as P1CM*clr100* (65), that possess translocatable drug resistance elements permits positive selection of rare lysogens and facilitates the development of P1 transducing systems among other members of the *Enterobacteriaceae* (28). One can also select rare derivatives of P1 that are better able to propagate on a given host strain as well as mutant derivatives of a host strain that are better able to propagate P1.

14.3. CONJUGATION

Conjugation in gram-negative bacteria is a cell-mediated type of gene transfer that requires that the donor parent possess a conjugative plasmid. This plasmid may either replicate in the cytoplasm in synchrony with the bacterial chromosome or be integrated into and replicate as a part of the bacterial chromosome. Conjugative plasmids possess *tra* genes that specify and/or control the synthesis of appendages termed donor pili that are obligatory, at least for most conjugative plasmids, to allow donor cells to make contact with recipient cells, and also control the substances that minimize the occurrence of donor-donor matings. In addition to *tra* genes, conjugative plasmids contain the locus *bom* (basis of mobility). This locus contains the transfer origin site (*oriT*), where conjugative transfer of plasmid or chromosomal DNA begins (82). Conjugative plasmids also possess genes to control (i) their vegetative replication, (ii) the number of copies of plasmid per chromosomal DNA equivalent, and (iii) incompatibility functions (described in Chapter 16). Conjugative plasmids may or may not possess genes for other phenotypic attributes such as antibiotic resistance, heavy-metal ion resistance, bacteriocin production, surface antigen production, and enterotoxin production.

The best-studied conjugative plasmid is the fertility factor F of *E. coli* K-12. This conjugative plasmid can exist either in an autonomous cytoplasmic state in an F^+ donor or in an integrated state in an Hfr donor. Recipient strains do not possess F and are termed F^-. Conjugative transfer of F begins at *oriT* and proceeds unidirectionally. For this reason, only sequences *cis* to and on one side of *oriT* can be transferred. When F is autonomous, only F is transferred. If F has integrated into the chromosome to form an Hfr, transfer also begins at *oriT* and proceeds undirectionally. Hfr donors can transfer all chromosomal markers to the recipient, but the start site, orientation, and frequency of transfer depend on the location and orientation of the integrated F. F^+ donors transfer the F plasmid to essentially all F^- cells with which they mate but are able to transfer chromosomal genetic information only at low frequency. All F^+ donor chromosomal markers are transferred at nearly equal frequency (ca. 10^{-5} for any donor marker per donor cell), and most of these transconjugants inherit F. The low frequency of transfer from an F^+ donor results from the low frequency of spontaneous integration of F into the chromosome. Genetic markers near the origin of transfer of the Hfr chromosome are transferred at highest frequencies, with frequencies of transfer and inheritance decreasing proportionately as markers become more distal from the origin. This behavior is observed because of spontaneous random interruption of chromosome transfer. Most transconjugants inheriting Hfr chromosomal material remain F^- unless the mating is of long enough duration to allow transfer of the entire Hfr chromosome, since parts of the integrated F plasmid are transferred as the terminal markers. The rate of Hfr chromosome transfer is constant at 37°C. By using a multiply mutant F^- strain and periodically interrupting mating, it is possible to determine the mating time necessary to transfer any marker to the recipient and the interval of time between the transfer of sequentially transferred markers. In using Hfr matings at 37°C, it has been possible to determine that transfer of the entire *E. coli* K-12 chromosome would take 100 min and that approximately 47 kb of DNA is transferred per min. The integration of F into the *E. coli* chromosome is often due to recombination between an insertion sequence (IS element) on the F plasmid and a similar IS element in the *E. coli* chromosome (23). Some 25 to 30 F integration sites have been identified by genetic techniques, giving rise to Hfr donors with different origins and orientations of chromosome transfer (20, 48).

Although conjugative plasmids contain both *tra* and *bom*, there are other plasmids that contain only *bom*. Since this latter class of plasmids cannot synthesize the donor pili and therefore cannot mediate interaction of donor and recipient cells required for conjugation to occur, these plasmids are nonconjugative. When a nonconjugative plasmid that contains the *bom* locus is present in a cell that also contains a conjugative plasmid and when conjugation is allowed to proceed, recipient cells are found to contain both the conjugative and nonconjugative plasmids. This results from the interaction of the nonconjugative plasmid with the *tra* gene products encoded by the conjugative plasmid. The *tra* genes are able to act in *trans* and mediate the conjugation of plasmids that contain *bom*. Plasmids that contain *bom* and can be mobilized by another conjugative plasmid are referred to as Mob^+. Not all Mob^+ plasmids can be mobilized by all conjugative plasmids. RP4 is an IncP plasmid, and conjugative IncP plasmids are noted for their ability to conjugate with a broad range of gram-negative bacteria.

In experiments in which the kinetics of conjugational transfer are being studied, it is important to be able to terminate the conjugational event at precise intervals of time and to preclude further mating on the selective medium. The latter can occur when high densities of donor and recipient cells are plated. The simplest way to accomplish these goals is to use a recipient strain that is resistant to nalidixic acid because of a mutation in the *gyrA* gene. Nalidixic acid inhibits conjugational DNA transfer, and its addition to a mating mixture with a nalidixic acid-sensitive (Nal^s) donor and a Nal^r recipient immediately terminates transfer of either plasmid or chromosomal DNA. Recipients resistant to rifampin as a result of a mutation in the *rpoB* gene also have been used, but rifampin does not cause cessation of conjugational DNA transfer once it is in progress. Rifampin does terminate transcription of donor genetic information, which is highly advantageous when attempting to detect low-frequency transfer of plasmids specifying resistance to drugs such as ampicillin, since continued production of ß-lactamase by the donor leads to inactivation of the ampicillin included in the selective medium and precludes detection of Ap^r transconjugants. When experiments are conducted in which precise termination of conjugation is not crucial, the chromosomal marker for streptomycin resistance, *rpsL*, can be used instead of *gyrA* or *rpoB* to select against the donor.

The donor phenotype of most conjugative plasmids isolated from strains of bacteria in nature is repressed such that only 1 of 1,000 to 10,000 cells harboring the conjugative plasmid is able to transfer it. Since, in some instances, transient derepression occurs on transfer to a recipient cell, the detection of conjugational plasmid transfer sometimes requires that matings be conducted over a period of several hours or longer and with periodic dilution of the mating mixture to maintain appropriate bacterial densities. Derepressed mutants of some repressed conjugative plasmids have been isolated such that essentially all donor cells harboring the conjugative plasmid are capable of acting as genetic donors. F is one of the best characterized of the derepressed conjugative plasmids.

Although some conjugative plasmids promote conjugational DNA transfer when matings are conducted in liquid media, others are more efficient at promoting transfer when the donor and recipient cells are either mated on agar plates or impinged on a membrane filter that is placed on a nutritive agar medium (12). In the latter case, the plates are incubated uninverted for a suitable period, after which the filter is moved to a selective agar plate that selects against the donor parent. Alternatively, the filter can be immersed in buffered saline and the cells can be removed by intermittent blending in a vortex mixer, after which suitable dilutions are plated on selective agar medium.

Another variable that must be considered in optimizing detection of conjugational transfer is temperature. Many conjugative plasmids found in members of the *Enterobacteriaceae* exhibit optimal transfer frequencies when matings are conducted at or near 37°C. With some of these, transfer is all but undetectable at 25°C, and leaving donor cultures at room temperature for a time prior to mixing with recipient cultures at 37°C can lead to a significant drop in the frequency of transfer. On the other hand, there are conjugative plasmids such as those in the IncH group that transfer optimally at 25°C and not at all at 37°C. In examining whether a trait might be transferred on a conjugative plasmid, it is wise to examine mating at 37 versus 25°C. As pointed out above, a restriction proficient recipient can decrease the stable inheritance of traits. This problem of restriction can frequently be eliminated by heating the recipient parent prior to mating for 5 to 10 min at 50°C or 15 to 20 min at 45°C.

14.3.1. Conjugative Plasmid Transfer in *E. coli* with F′ or R Plasmids

14.3.1.1. Strains, plasmids, and media

The study of conjugative plasmid transfer is facilitated if the plasmid specifies a readily selectable trait. The integrated F plasmid of an Hfr donor can be excised from the chromosome at a low frequency. Occasionally chromosomal DNA sequences that were adjacent to the integrated F plasmid are excised with F to form a plasmid containing both F and host genetic information. This plasmid is referred to as F′. A donor such as χ1301 (Table 1) which harbors an F′*lac*$^+$ plasmid can readily transfer the Lac$^+$ trait at high frequency to an F$^-$ recipient such as χ1997 (Table 1), and Lac$^+$ Nal^r transconjugants can be easily selected.

The transfer of conjugative R plasmids is also easily studied. The transfer of derepressed R-plasmid derivatives such as the R100 *drd-1* plasmid in χ1784 (Table 1) is easily detected since R100 *drd-1* confers resistance to 25 to 50 μg of tetracycline per ml, 50 μg of chloramphenicol per ml (Cm^r), 50 μg of streptomycin per ml (Sm^r), 50 μg of spectinomycin per ml (Sp^r), and (on synthetic media) 25 to 40 μg of sulfanilamide per ml (Su^r). The donor harboring R100 *drd-1* should be sensitive to nalidixic acid, and the recipient, such as χ1997 (Table 1), should not express resistance to any of the antibiotics for which R100 *drd-1* expresses resistance and should have a *gyrA* mutation.

Conjugational transfer of nonconjugative Mob$^+$ plasmids is usually done with either SM10 or S17-1 as the donor (Table 1). Each of these strains contains the conjugative plasmid RP4-2-*tet* :: Mu integrated into the chromosome. Integration of this plasmid into the chromosome results in an Hfr donor that can mediate the high-frequency transfer of Mob$^+$ plasmids in the cell by means of the *trans*-acting *tra* gene products. Since the conjugative plasmid is integrated into the chromosome, it is not transferred; however, chromosome transfer does occur.

The plasmid to be mobilized by the donor must contain *bom* and must belong to an incompatibility group on which the transfer functions of the integrated RP4 plasmid can act. Most conjugation experiments involve the transfer of plasmid or cosmid libraries to complement mutations in *E. coli* or *S. typhimurium*. Mob$^+$ plasmids containing transposable elements are used to mutagenize strains that can then be screened for a specific mutation. Also, in vitro-constructed mutations are mobilized into strains to replace the wild-type gene with the in vitro mutagenized gene by homologous recombination between the plasmid and the chromosome. Transposon mutagenesis and allele replacement experiments that rely on a Mob$^+$ plasmid as a vector are best accomplished when the plasmid is incapable of replication in the recipient. This type of plasmid is often referred to as a suicide plasmid, and Mob$^+$ suicide plasmids are available. Both pGP704 (54) and pRT733 (78) are plasmids with an origin of replication that requires the *pir* gene product, π, to successfully initiate replication. In the donor strain, SM10 λ*pir* (Table 1), the *pir* gene is present in the chromosome on a λ prophage. When the plasmid is transferred to a recipient lacking the *pir* gene the plasmid is incapable of replication. pGP704 is used for allele replacement, and pRT733 contains Tn*phoA* and can be used for transposon mutagenesis.

Either L or Penassay broth allows optimal growth and maximum expression of donor and recipient phenotypes. Penassay agar (with 8 g of NaCl per liter) can be used for titer determination of donor and recipient cultures and, when supplemented with tetracycline, chloramphenicol, or streptomycin and nalidixic acid (see above and section 14.4; Table 4), to select R-plasmid- and recombinant plasmid-containing transconjugants. L agar supplemented with kanamycin and 5-bromo-4-chloro-3-indolyl phosphate p-toluidine salt (XP) will facilitate selection and identification of *phoA* fusions. VB medium E agar supplemented with 0.5% lactose, proline, and 50 μg of nalidixic acid per ml can be used to select Lac$^+$ Nal^r transconjugants. See section 14.4 for methods for preparation of media.

14.3.1.2. Procedure for F′-plasmid transfer

1. Inoculate 5 ml of Penassay broth contained in test tubes (16 by 125 or 150 mm) with loopfuls of growth off slants of each parent, and incubate overnight at 37°C without aeration (i.e., as standing cultures).

2. Dilute these overnight cultures 1:100 or 1:200 into 10 to 20 ml of prewarmed Penassay broth. Aerate these cultures by shaking, and grow to a density of about 2 × 10^8 cells per ml. The titer of the growing culture can be determined by monitoring the optical density as described in 14.1.3.2. It is advantageous for optimum expression of the donor phenotype to cease shaking the donor culture 30 min before mating begins since nonaerated growth conditions favor synthesis of maximum numbers and lengths of donor pili specified by F. This, however, may not be the case for all conjugative plasmid types.

3. Place either flat-bottomed 125-ml micro-Fernbach flasks or 250-ml Erlenmeyer flasks in a 37°C water bath with lead collars around the necks of the flasks so that the water in the water bath surrounds the flasks to a level just below the caps. The lead collars can be cut in a doughnut shape out of 6- to 8-mm-thick lead sheets. In the absence of lead collars, clamp the flasks so they will not float when immersed in the water bath. The immersion of flasks is important if the temperature of the mating culture is to be maintained at 37°C.

4. Add 9 ml of recipient culture to the mating flask to equilibrate, and spread a 10^{-6} dilution on Penassay agar to determine its titer. Spread 0.1-ml amounts of the undiluted recipient culture on the selective agar medium to verify that the culture is free from contaminants and contains few, if any, revertants.

5. After 5 min, add 1 ml of donor culture to the mating flask and gently swirl to mix the donor and recipient cells. Determine the titer of the donor culture, and plate 0.1 ml of the undiluted donor culture on selective medium. Note that the total volume of 10 ml gives a 2-mm depth in the mating flask, which allows sufficient oxygen diffusion to the mating cells and obviates any need for agitation by shaking of the mating mixtures. Although it is customary to use an excess of recipient cells in matings, different ratios ranging from 1:20 to 1:1 can be used by varying the volumes and/or densities of recipient and donor cultures added to the mating flask.

6. At 5-, 10-, or 15-min intervals up to 1 h, dilute the mating mixture and plate 0.1-ml samples (0.05 ml per plate) on selective agar medium. Total dilutions of 10^{-2} to 10^{-3} should be suitable up to 30 min of mating, and total dilutions of 10^{-3} and 10^{-4} should be suitable for longer durations of mating. Invert plates, and incubate them at 37°C.

Donor and recipient cultures grown in rich complex media give higher frequencies of transconjugants than do those grown in synthetic media. However, the selection on synthetic media of transconjugants that are formed from broth-grown and mated parents constitutes a step-down growth condition that sometimes results in detection of fewer transconjugants than are actually formed. This potential difficulty can be circumvented by adding 10% (vol/vol) Penassay broth to the BSG diluent used to dilute the mating mixtures, since the small amount of broth added to the selective minimal agar medium will permit all Lac$^+$ Nal^r transconjugants to form colonies.

The frequency of transconjugants for each time of interruption is calculated by dividing the titer of transconjugants per milliliter of mating mixture by the titer of donor cells in the mating mixture at the commencement of mating.

14.3.1.3. Procedure for R-plasmid transfer

Transfer of R100-*drd-1* (χ1784; Table 1) is the same as steps 1 to 5 described above for F′ transfer (section 14.3.1.2). If step 6 is performed as described above, Tc^r Nal^r and Cm^r Nal^r transconjugants will be detected at higher frequencies than Sm^r Nal^r transconjugants. This will be especially apparent for the early times at which mating is interrupted, and the frequencies detected for all transconjugant classes will be lower than the frequencies actually formed in the mating. These observations are due to the different times of phenotypic lag displayed by resistance to bacteriostatic and bacteriocidal antibiotics. These problems in phenotypic expression are even more pronounced when working with R plasmids that express resistance to either ampicillin or kanamycin.

To circumvent the above problems and to achieve an accurate measure of the rate and total frequency of R-plasmid transfer at each time of interruption of mating, place 0.1 ml of the mating mixture into 0.9 ml of Penassay broth containing 50 μg of nalidixic acid per ml in a test tube (13 by 100 mm) at 37°C. Leave these tubes at 37°C for 20 to 30 min prior to further dilution and plating on selective media.

To verify that the various drug resistance genes are indeed transferred together on the same plasmid, pick 10 or so colonies of each transconjugant type with sterile toothpicks and streak onto the other types of selective media. For instance, if the drug resistance genes are transferred together on the same plasmid, Tc^r Nal^r colonies will also be Cm^r and Sm^r.

14.3.1.4. Mobilization of cosmid libraries

Cosmids are plasmid cloning vectors that are derived from λ phage and contain large segments of cloned DNA (18). These vectors can be packaged in λ phage particles in vitro by using commercially available λ packaging extracts. Packaged cosmid libraries can also be amplified in vivo (41) and then conveniently stored as phage lysates. Packaged cosmids can then be used to infect *E. coli* or other strains that express the LamB protein. However, as discussed in section 14.2.1, not all bacteria are easily transduced with λ.

Mobilization provides a convenient means of introducing cosmid libraries into other bacterial genera, as well as *E. coli*. Since cosmids in their plasmid form are often unstable, care should be taken to reduce the number of generations of growth of the donor strain before mobilizing the cosmid library into a recipient strain. This is easily accomplished by transducing the donor strain with the λ- packaged cosmid library immediately before beginning conjugation. As discussed above, only cosmids that contain *bom* are Mob$^+$. Plasmids that contain *bom* can also be mobilized by this procedure. However, mobilization of plasmids does not require the transduction described in steps 1 through 7.

1. Grow a 3-ml standing overnight culture of the donor strain in Super broth and the recipient strain in L broth. The recipient must contain a marker to allow for selection against the donor.
2. Dilute the donor (SM10 or S17-1) 1:10 into Super broth, and grow with aeration to a titer of 1×10^8 to 2×10^8 cells per ml.
3. Sediment, wash, and suspend the cells in 10 ml of 10 mM $MgCl_2$; then starve the cells with aeration at 37°C for 60 min prior to infection.
4. Distribute 0.9 ml of this culture to a sterile test tube (13 by 100 mm), and incubate at 37°C.
5. Dilute a cosmid library (which has been packaged in λ particles) appropriately in 10 mM $MgCl_2$ so that addition of 0.1 ml to 0.9 ml of bacteria will result in a multiplicity of infection of 0.1 phage per bacterium.
6. Add 0.1 ml of diluted cosmid library to the 0.9-ml samples of cells.
7. Allow 15 min for preadsorption. Add 1 ml of L broth to the bacteria, and incubate at 37°C. Keep these bacteria at 37°C until the recipient cells in the next step have been prepared.
8. Dilute the recipient 1:10 into warm L broth, and incubate until early to mid-log phase. The culture can be started while the donor is being prepared in steps 2 through 7 above. Place a sterile 0.4-μm polycarbonate 25-mm diameter membrane filter (Nuclepore, Pleasanton, Calif.) in a filtration device that has been kept at 37°C. A glass microanalysis funnel and base with fritted glass (Millipore Corp., Bedford, Mass.) can be used as a filtration device. Prewet the filter with prewarmed L broth, and then add both donor and recipient together to yield a donor-to-recipient ratio of 1:10. Gently apply vacuum to concentrate the bacteria onto the surface of the filter, and then immediately remove the filter and place cell side up on a prewarmed L agar plate. Incubate this at 37°C for 1 h.
9. Remove the filter from the plate with sterile forceps, and place it in a test tube (13 by 100 mm) containing 1 ml of BSG. Vortex this to suspend the cells, and plate on medium containing the appropriate antibiotics to select for the recipients containing the mobilized cosmid and to select against the donor.

14.3.1.5. Production of *phoA* fusions

The procedure described in this section uses Tn*phoA* to produce type II fusions (50). PhoA encodes the enzyme alkaline phosphatase and is active only when present outside the cytoplasm. When a fusion is produced between a cytoplasmic protein and the transposon-encoded PhoA, the resulting hybrid protein is expressed in the cytoplasm and there is no PhoA activity. However, when the fusion is to a protein exported out of the cytoplasm, the result is a hybrid protein that often has PhoA activity. This activity can be visualized by including the chromogenic substrate XP in the medium. The PhoA-catalyzed dephosphorylation of XP yields a blue color that is easily distinguished from white cells unable to catalyze the reaction. The requirements for the recipient phenotype are (i) to allow selection against the donor; (ii) to be Km^s, allowing selection of cells that contain Tn*phoA* in the genome; and (iii) to be white when grown on media containing XP. The last requirement is met by using an *E. coli* strain that contains a mutant *phoA* gene. Other bacterial genera may or may not form blue colonies on media containing XP.

This procedure uses an alternate method to conjugate cells. Instead of mixing donor and recipient cells on the surface of a filter, the two cell types are mixed on the surface of a nonselective plate. The procedure is based on that described by Taylor et al. (78).

1. Streak out SM10 λ*pir* containing pRT733 for isolated colonies on an L agar plate containing ampicillin, and do the same for the recipient but use a nonselective agar plate. Incubate overnight until colonies approximately 2 mm in diameter are present.
2. Using a sterile loop, remove a colony of the donor strain and transfer to a nonselective plate. Repeat this with a colony of the recipient strain, and place it next to the transferred donor colony. Mix the cells from the two colonies with a sterile loop, and spread over a 4- to 5-mm-diameter patch. Several patches can be spread on the same plate. Incubate this at 37°C overnight.
3. Remove the cells from the plate with a sterile loop, and streak onto an agar plate containing XP and kanamycin to isolate individual colonies. The type of agar plate, supplements, or antibiotics used will depend on how the donor will be selected against. Incubate the plates at 37°C until colonies are present. Storage of these plates at 4°C for 1 to 2 days will allow further development of color and is sometimes helpful in permitting one to distinguish between blue and white colonies. Whereas all the Km^r recipients contain Tn*phoA* in the genome, the blue colonies contain Tn*phoA* inserted into a gene that encodes an exported protein.

14.3.1.6. Applications

The above protocols can be used to study conjugational transfer of a diversity of plasmid-mediated traits. Of course, the detection of transfer of Col plasmids (3) and plasmids that specify traits such as antigen production and enterotoxins (27) requires special procedures since direct selection of the plasmid-mediated trait is not possible. Some conjugative plasmids are able to integrate into the bacterial chromosome to produce Hfr-type donors. On excision they may pick up a piece of chromosomal genetic information to generate R′, Col′, or F′ plasmids. Such plasmids can be used to determine dominance-recessiveness and complementation relationships between alleles and aspects of gene regulation. They also can be used to construct strains. These applications are described by Miller (53). Since these plasmids share homology with the chromosome, they can be used in lieu of Hfr-type donors to mobilize and transfer chromosomal genes in an oriented sequential manner. This can be particularly useful since the R′, Col′, and F′ plasmids can mobilize the chromosome of strains (or species) of bacteria in which Hfr donors have not been isolated and are closely related to the strain that gave rise to the R′, Col′, or F′ plasmid.

The above procedures detailing conjugation of recombinant plasmids and cosmids can be used to mobilize libraries into *E. coli*. If broad-host-range plasmids are used, recombinant libraries can also be mobilized into other genera. *E. coli* has been used to mobilize plasmids into *Rhizobium* spp. (22), *Pseudomonas* spp. (21, 22), *Proteus mirabilis* (21), *Agrobacterium tumefaciens* (36), and other gram-negative bacteria.

Conjugation is also useful for in vitro and in vivo strain construction. Mob$^+$ plasmids containing transposable elements can be used to mutagenize strains that can then be screened for a specific mutation. Also, in vitro-constructed mutations can be mobilized into strains to replace the wild-type gene with the in vitro-mutagenized gene by homologous recombination between the plasmid and the chromosome.

14.3.2. Chromosome Transfer in *E. coli* F$^-$ with *E. coli* Hfr

14.3.2.1. Bacterial strains and media

Numerous *E. coli* Hfr strains with differing origins and directions of chromosome transfer have been isolated (20, 48). The choice of a specific Hfr strain thus depends on the region of the bacterial chromosome to be studied with regard to the mapping of genetic loci. New Hfr donors can be readily isolated in different genetic backgrounds by using either a modified fluctuation test with an F$^-$ strain (39) or integrative suppression in an F$^+$ or R$^+$ strain containing a *dnaA*(Ts) mutation (57). The methods for conducting Hfr × F$^-$ crosses can be illustrated by using an Hfr strain such as the prototrophic χ433 (Table 1) which transfers its chromosome O-*thr leu proA lac...pyrB* F and is sensitive to nalidixic acid and streptomycin and an F$^-$ strain such as χ2634 (Table 1) that contains mutations conferring auxotrophy for leucine, proline, adenine, tryptophan, and thiamine and resistance to both nalidixic acid and streptomycin. χ2634 also contains other mutations, and inheritance of their wild-type alleles can be monitored as unselected markers. The F$^-$ χ585 strain (Table 1) has fewer mutations than does χ2634 and is more conveniently used to test the stability of the Hfr phenotype in χ433.

Either L or Penassay broth allows optimum growth and maximum expression of donor and recipient phenotypes. Penassay agar (with 8 g of NaCl per liter) can be used to titer donor and recipient cultures. VB minimal agar medium E, appropriately supplemented, can be used to select transconjugants. For example, Leu$^+$ Nal^r transconjugants in a cross between χ433 and χ2634 would be selected on a medium containing 0.5% glucose and proline, adenine, tryptophan, thiamine, and nalidixic acid (see Tables 2 through 4 for concentrations of supplements). BSG can be used as the diluent but should be supplemented with 10% (vol/vol) broth when mating mixtures, which are in a complex medium, are diluted prior to plating on a synthetic medium. This small amount of broth ensures nutrients for recombination and phenotypic expression of donor traits and gives rise to maximal recovery of transconjugants. See section 14.4 for methods for preparation of media.

14.3.2.2. Stability testing of Hfr phenotype

Since Hfr donors arise by integration of a conjugative plasmid into the bacterial chromosome, it follows that they can revert to a state in which the conjugative plasmid replicates freely in the cytoplasm. Thus, it is important to verify that the majority of cells in an Hfr population do exhibit the Hfr phenotype prior to use of the strain in conjugation experiments. This can be done by either a replica-plating or a picking and streaking technique.

Replica-plating technique

1. Dilute the Hfr culture and plate on Penassay agar to achieve about 100 well-isolated colonies per plate. Incubate at 37°C for 12 to 16 h.
2. Spread 0.05 to 0.1 ml of a broth-grown log-phase culture of the F− χ585 (Table 1) on a minimal agar plate selective for Leu$^+$ Str^r transconjugants.
3. Replica plate the colonies from the Hfr culture onto the minimal agar plate spread with the F$^-$ strain. Mark the bottoms of both plates so that the orientation and placement are known. Incubate the minimal agar plate at 37°C for 24 to 48 h.
4. Confluent patches of recombinant growth should be evident for each colony from the Hfr culture that displays the Hfr phenotype and only 0, 1, or 2 recombinant colonies for a colony from the Hfr culture that has reverted to the F$^+$ state.
5. If the culture contains an appreciable frequency of F$^+$ revertants, the master Penassay agar plate can be used to pick a colony that displays the Hfr phenotype.

Picking and streaking technique

1. Plate the Hfr culture to get isolated colonies on a Penassay agar plate.
2. Pick 10 to 20 isolated colonies into individual test tubes (13 by 100 mm) containing 2 ml of broth.
3. Incubate cultures at 37°C for 4 to 6 h to achieve densities of about 2 × 10^8 to 5 × 10^8 cells per ml. Also grow cultures of F$^-$ and F$^+$ strains such as χ289 and χ226 (Table 1) to be included as controls.
4. Place 0.05 to 0.1 ml of a broth-grown log-phase culture of the F$^-$ χ585 down the center of a minimal agar plate selective for Leu$^+$ Str^r transconjugants. Leave the lid ajar until the streak has dried. (Two streaks can be accommodated per plate if desired.)
5. Immediately streak full loopfuls of each "Hfr" culture (being careful not to touch sides of tubes with the loop, which will cause loss of fluid and yield variable results) and the F$^-$ and F$^+$ cultures across the F$^-$ streak. Let the streaks dry, and incubate the plate for 24 to 48 h at 37°C.
6. A culture giving the Hfr phenotype will give confluent growth of transconjugants at the juncture of the two streaks, whereas a culture that contains principally F$^+$ cells will give only 5 to 10 isolated transconjugant colonies. An F$^-$ × F$^-$ cross streak should, of course, give no transconjugants.

Both of the above techniques are based on the fact that, although the donor cells are ultimately killed by the streptomycin in the selective minimal agar, they are still able to transfer their chromosomes in the presence of streptomycin (albeit at a somewhat reduced frequency). Nalidixic acid cannot be used to select against the donor in such "plate" matings since it causes an immediate cessation of genetic transfer by Nal^s donors.

14.3.2.3. Procedure for interrupted mating

Use the reisolated Hfr, if necessary, and F$^-$ strains. Follow the procedure outlined in steps 1 through 5 in section 14.3.1.2. However, use minimal agar that is se-

Table 5. Final dilutions for plating to recover various transconjugant classes in Hfr × F^- mating[a]

Time of interruption (min)	Dilution for transconjugant class:			
	Leu^+ Nal^r	Pro^+ Nal^r	Ade^+ Nal^r	Trp^+ Nal^r
5	10^{-2}	10^{-2}	—[b]	—
10	10^{-2} 10^{-3}	10^{-2}	10^{-2}	—
15	10^{-3} 10^{-4}	10^{-2} 10^{-3}	10^{-2}	—
20	10^{-3} 10^{-4}	10^{-2} 10^{-3}	10^{-2} 10^{-3}	—
25	10^{-4} 10^{-5}	10^{-3} 10^{-4}	10^{-2} 10^{-3}	10^{-2}
30	10^{-4} 10^{-5}[c]	10^{-3} 10^{-4}	10^{-3} 10^{-4}	10^{-2}
35	10^{-4} 10^{-5}	10^{-3} 10^{-4}	10^{-3} 10^{-4}	10^{-2} 10^{-3}
40	10^{-4} 10^{-5}	10^{-4} 10^{-5}[c]	10^{-3} 10^{-4}	10^{-2} 10^{-3}
45	10^{-4} 10^{-5}	10^{-4} 10^{-5}	10^{-4}[c]	10^{-3} 10^{-4}
50	10^{-4} 10^{-5}	10^{-4} 10^{-5}	10^{-4}	10^{-3} 10^{-4}
55	10^{-4} 10^{-5}	10^{-4} 10^{-5}	10^{-4}	10^{-3} 10^{-4}
60	10^{-4} 10^{-5}	10^{-4} 10^{-5}	10^{-4}	10^{-3} 10^{-4}

[a]Based on the results expected when χ433 and χ2634 (Table 1) are grown and mated under the optimal conditions.
[b]Plating is not necessary, since no transconjugants are expected for these times of interruption.
[c]Transconjugant titer should be nearly at the maximum expected frequency. Platings can be omitted for some or all of subsequent times of mating interruption.

lective for Leu^+ Nal^r, Pro^+ Nal^r, Ade^+ Nal^r, and Trp^+ Nal^r transconjugants and the methods for dilution and plating that follow.

1. At appropriate intervals (every 2, 3, 5, or 10 min, but the more frequent sampling will require two people to do plating), take 0.1 ml of the mating mixture and add to 0.9 ml of BSG containing 50 μg of nalidixic acid per ml in test tubes (13 by 100 mm) at 37°C.

2. Incubate the 10^{-1} dilution tubes at 37°C for 30 to 40 min. Choose a time that is about 30 s after a mating mixture is to be sampled, since this will maximize the time left to make further dilutions and platings.

3. Plate either 0.05 ml on each of two selective agar plates or 0.1 ml on one selective agar plate of the appropriate dilution for each time of mating interruption. Table 5 gives the approximate final dilutions that should be plated on each selective agar medium for each time of mating interruption. These values are based on a donor-to-recipient cell ratio of 1:10, a total mating density of about 2×10^8 cells per ml, use of a pure Hfr culture, and use of conditions that optimize the Hfr phenotype and are optimal for Hfr chromosome transfer.

4. For titer determinations, incubate Penassay agar plates overnight at 37°C. If the Hfr was not reisolated prior to use, the Penassay plates can be used to determine the fraction of Hfr cells by one of the methods described in section 14.3.2.2.

5. Incubate minimal selective agar plates at 37°C for 2 days. Save plates, and do not count colonies by destruction since they can be used to test inheritance of unselected markers.

6. Calculate transconjugant frequencies for each time of interruption by dividing the titers of each type of transconjugant by the Hfr titer in the mating mixture at the commencement of mating. For each transconjugant type, plot the frequency of transconjugants versus the time of mating interruption. Extrapolation of the line through the points to the abscissa defines the time of entry for each Hfr gene into the recipient.

14.3.2.4. Analysis of transconjugants

It is sometimes either not possible to directly select for inheritance of some donor markers or not feasible because of the number of markers that are subject to inheritance in a given mating. In these cases, a more classical recombination analysis can be performed to evaluate the order of genetic loci. Customarily, this is done by picking 100 to 200 colonies of a given transconjugant type with a needle or toothpicks into BSG and streaking each of these small suspensions on the same selective medium to obtain purified isolated colonies. These isolated colonies can then be grown as small cultures and used to test the genotypes by streaking on appropriate media or patched onto a suitable agar medium for subsequent analysis by replica plating. When using the latter technique, the master plate should be selective for the transconjugant phenotype, and each test replica plate should evaluate one additional nutritional requirement or capability in addition to maintaining selection for the original transconjugant phenotype. This method precludes occasional ambiguity due to failure to completely purify the transconjugant type away from other transconjugant or parental types prior to patching on the master plate.

Since chromosome transfer by Hfr donors is spontaneously interrupted, the determination of an unambiguous gene order by recombinant analysis demands that recombinants inheriting a terminally transferred Hfr marker be tested for inheritance of proximally transferred markers. For the cross between χ433 and χ2634 (Table 5), 200 to 400 Trp^+ Nal^r transconjugants should be picked, purified, and tested for inheritance of the Hfr wild-type alleles *ara, leu, tonA, proA, lac, tsx, purE,* and *galK.* Testing for inheritance of these markers can be done as follows.

1. Patch 50 to 100 purified Trp^+ Nal^r transconjugants per minimal agar plate selective for Trp^+ Nal^r transconjugants. Incubate at 37°C overnight, and use to replica plate sequentially to minimal agar that is selective for Ara^+ Trp^+ Nal^r, Leu^+ Trp^+ Nal^r, Pro^+ Trp^+ Nal^r, Lac^+ Trp^+ Nal^r, Ade^+ Trp^+ Nal^r, Gal^+ Trp^+ Nal^r, and Trp^+ Nal^r transconjugants. The last plate is included as a control to verify that cells were replicated to each plate.

2. Testing for resistance or sensitivity to T5 (*tonA*) and T6 (*tsx*) is most accurately done by cross-streaking broth cultures of purified Trp^+ Nal^r transconjugants

against the respective phage on EMBO plates (see step 2 in section 14.2.1.5). This is laborious, however, and replica plating can sometimes give reasonably accurate results. For this purpose, the EMBO plates are spread with 0.05 to 0.1 ml of a T5 or T6 lysate (ca. 10^{10} PFU/ml) and a separate piece of velveteen is used to replica plate from the master plate to each of these since carryover of phage from one plate to another will obscure the results. Even though the *tonA* mutation also confers resistance to phage T1, this phage should not be used unless scrupulous care is taken and there is rigid adherence to all rules and aseptic technique. T1 is resistant to desiccation and has a very short latent period. It has been responsible for wiping out *E. coli* research in more than one laboratory.

After scoring the phenotype of all Trp^+ Nal^r transconjugants, the number in each recombinant class can be counted and used to calculate map distances between all markers and to determine the order of the markers on the Hfr chromosome.

14.3.2.5. Applications

By using a diversity of Hfr parents with different origins and directions of chromosomal transfer in conjugation with the cross-streak method of mating, it is easy initially to identify the general location of the wild-type allele for a mutant allele isolated in an F^- recipient (20). Interrupted matings with the appropriate Hfr strain can then be used to more precisely map this gene with respect to other loci. P1-mediated generalized transduction may be necessary, however, to unambiguously map this gene in relation to other closely linked loci since conjugation is often inadequate for this purpose. Mapping the location of a gene by P1-mediated generalized transduction and Hfr analysis can be done with genes that have not been cloned. When mapping a cloned *E. coli* gene, it is possible to use hybridization to a set of ordered λ clones that together contain the entire *E. coli* chromosome (43).

14.4. MEDIA AND STRAINS

The media commonly used in gene transfer procedures are ordinary and commercially available from laboratory supply companies. Only media requiring modification and various supplements are described below. Sources of bacterial and phage strains also are listed.

14.4.1. L Broth and L Agar (45)

L broth contains 10 g of tryptone, 5 g of yeast extract, 5 g of NaCl, and 1 g of glucose per liter of water. The pH is adjusted to 7.0 by adding approximately 1.7 ml of 1 N NaOH prior to autoclaving. L agar is L broth containing 1.5% agar and supplemented with $CaCl_2$. Before pouring, $CaCl_2$ is added to a final concentration of 2.5×10^{-3} M.

14.4.2. LB Broth and LB Agar (10)

LB broth contains 10 g of tryptone, 5 g of yeast extract, 10 g of NaCl, and 1 g of glucose per liter of water. The pH is adjusted to 7.0 by adding approximately 1.7 ml of 1 N NaOH prior to autoclaving. LB agar is LB broth containing 1.5% agar and supplemented with $CaCl_2$. Before pouring, $CaCl_2$ is added to a final concentration of 2.5×10^{-3} M.

14.4.3. Lambda Agar

Attachment and injection of λ require the presence of the maltose transport proteins on the *E. coli* cell surface, whose synthesis is inhibited by glucose as a result of catabolite repression. Thus, lambda agar cannot contain glucose. It contains 10 g of tryptone, 5 g of yeast extract, 5 g of NaCl, and 12 g of agar per liter of water. The pH is adjusted to 7.0 by adding 1.7 ml of 1 N NaOH prior to autoclave. Lambda agar plates should be poured 2 days in advance and then allowed to sit at room temperature until used. Lambda soft agar has the same ingredients as lambda agar but contains 0.65% agar and 10 mM Mg^{2+}.

14.4.4. Super Broth (47)

Preparation of high-titer lysates of λ and many other phages is best achieved by using a rich medium for growth of the host bacteria. Super broth (one such medium) contains 32 g of tryptone, 20 g of yeast extract, and 5 g of NaCl per liter of water. The pH is adjusted to 7.0 by adding about 5 ml of 1 N NaOH prior to autoclaving. For propagation of λ, add 10 mM $MgCl_2$ after autoclaving. The $MgCl_2$ can be prepared as a 1 M solution in water.

14.4.5. Penassay Agar

Commercial Penassay agar is satisfactory as a general complex plating medium, but adding 8 g of NaCl per liter very much improves its utility in many bacterial genetics experiments.

14.4.6. Brain Heart Infusion Medium

Brain heart infusion is a very rich medium appropriate for the culture of fastidious microorganisms.

14.4.7. EMB and MacConkey Agars

The EMB and MacConkey base agars commercially available are satisfactory fermentation indicator media for working with gram-negative enteric bacteria. However, since many of the *E. coli*, *Salmonella*, and other strains used in gene transfer experiments have mutations conferring auxotrophy for purines, pyrimidines, and/or vitamins, 5 g of yeast extract should be added per liter of medium. The addition of 5 g of NaCl per liter is also advisable if the media are to be used for testing strains of bacteria for sensitivity to phage. EMB and MacConkey base agars are usually supplemented with sugars to a final concentration of 1% and are designated EMB+Gal, Mac+Lac, etc. EMB agar without added sugar is used for testing bacteria for sensitivity to phage and is designated EMBO agar. Table 3 lists the concentrations for sterile stock solutions of the most commonly used sugars.

14.4.8. VB Minimal Medium E (80)

To prepare a 50× stock, dissolve successively in 670 ml of distilled or deionized water 10 g of $MgSO_4 \cdot 7H_2O$, 100 g of citric acid·H_2O, 500 g of anhydrous K_2HPO_4, and 175 g of $NaNH_4PO_4 \cdot 4H_2O$. The final volume will be 1 liter. Autoclave to sterilize. To prepare 1 × VB minimal liquid medium E, add 20 ml of sterile 50× salts to about 960 ml of sterile distilled or deionized water so that, on addition of sterile carbohydrate and other required supplements (see below), the final volume will be 1 liter. To prepare 1 × VB minimal agar medium E, add 15 g of agar to about 960 ml of distilled or deionized water, autoclave, and, when cooled to 60 to 70°C, add 20 ml of 50× salts and appropriate sterile carbohydrate and other supplements. When triphenyltetrazolium chloride is used as a fermentation indicator, it should be added to the agar prior to autoclaving.

14.4.9. ML Medium (19)

To prepare a 10× stock, dissolve successively in 670 ml of distilled or deionized water 50 g of NH_4Cl, 10 g of NH_4NO_3, 20 g of anhydrous Na_2SO_4, 90 g of K_2HPO_4, and 30 g of KH_2PO_4. The final volume must be adjusted to 1 liter and autoclaved to sterilize. To prepare 1 × ML, add 100 ml of 10× salts to sterile distilled or deionized water and add $MgSO_4 \cdot 7H_2O$ to 0.01%. Add sterile carbohydrates and other required supplements so that the final volume is 1 liter. 1× MA is prepared in the same way as ML with the single addition of 15 g/liter of agar to the distilled or deionized water before autoclaving. Add the $MgSO_4 \cdot 7H_2O$ after the mixture has cooled to 60 or 70°C. The 1× minimal liquid (ML) and minimal agar (MA) media have a pH of 7.

14.4.10. M9 Medium

To prepare a 10× stock, dissolve successively in distilled or deionized water 60 g of Na_2HPO_4, 30 g of KH_2PO_4, 5 g of NaCl, and 10 g of NH_4Cl. Adjust the final volume to 900 ml and autoclave to sterilize. After autoclaving, add 100 ml of a 0.01M $CaCl_2$ solution. To prepare 1 × M9 minimal liquid medium, add 10 ml of sterile 10× salts and 1 ml of a 1 M solution of $MgSO_4 \cdot 7H_2O$ to about 880 ml of sterile distilled or deionized water so that, on addition of sterile carbohydrate and other required supplements (see below), the final volume will be 1 liter. To prepare 1 × M9 minimal agar medium, add 15 g of agar to about 880 ml of distilled or deionized water, autoclave, and when cooled to 60 to 70°C, add 100 ml of 10× salts, 1 ml of a 1 M solution of $MgSO_4 \cdot 7H_2O$, and appropriate sterile carbohydrate and other supplements.

14.4.11. Green Indicator Plates (46, 77)

Green indicator medium is useful in distinguishing colonies infected with P22 from those not infected. Infected colonies are green, whereas uninfected colonies are light colored. Add 8 g of tryptone, 1 g of yeast extract, 15 g of NaCl, and 15 g of agar, and bring to 1 liter. Autoclave to sterilize, and add the following sterile solutions: 34 ml of 40% glucose, 25 ml of 2.5% alizarin yellow G,GG (EM Science, Gibbstown, N. J.), and 6.6 ml of 2% aniline blue (Sigma Chemical Co., St. Louis, Mo.). Alizarin yellow is insoluble at room temperature and must be heated just before addition.

14.4.12. SOC

SOC is used to suspend *E. coli* and *Salmonella* spp. after electroporation. It consists of 2% Bacto-Tryptone, 0.5% Bacto-Yeast extract, 10 mM NaCl, 2.5 mM KCl, 10 mM $MgCl_2$, 10 mM $MgSO_4 \cdot 7H_2O$, and 20 mM glucose.

14.4.13. Buffered Saline with Gelatin (19)

Buffered saline with gelatin (BSG) is a general-purpose diluent. The presence of gelatin stabilizes phages and fragile bacterial cells and improves the accuracy of dilutions. The medium contains 8.5 g of NaCl, 0.3 g of anhydrous KH_2PO_4, 0.6 g of anhydrous Na_2HPO_4, 10 ml of 1% gelatin, and 990 ml of water.

14.4.14. TMGS

TMGS is a diluent similar to BSG that contains $MgSO_4$. The addition of the $MgSO_4$ makes it appropriate for manipulations involving λ, including dilution of the phage for infection and long-term storage of λ prepared from plate lysates. It contains 1.2 g of Tris base, 2.46 g of $MgSO_4 \cdot 7H_2O$, 0.01% gelatin, and 8.5 g of NaCl per liter of solution. The pH is adjusted to 7.4 with HCl before autoclaving.

14.4.15. Supplements

Carbohydrate energy sources (Table 3) are customarily added to minimal media to a final concentration of 0.5%.

Casamino Acids or other acid hydrolysates of casein are often added to minimal media to a final concentration of 0.5%. Sterile 10% (wt/vol) stock solutions can be stored at 4°C. The addition of amino acids to minimal media stimulates more rapid growth of bacteria and is particularly useful in conducting genetic experiments with mutants requiring tryptophan (which is destroyed during acid hydrolysis of proteins), purines, pyrimidines, and vitamins. In the last case, use vitamin-free Casamino Acids.

The optimal concentrations for supplementing minimal media with amino acids, purines, pyrimidines, and vitamins to permit growth of strains with auxotrophic mutations must often be determined empirically. Important factors in deciding on concentrations to try include (i) the amount of the compound in cellular matter, (ii) the efficiency of transport of the compound, (iii) use of the compound for synthesis of other cellular constituents, and (iv) the ability of the bacterium to degrade the compound. Table 2 lists the concentrations of amino acids, purines, pyrimidines, and vitamins in sterile stock solutions and also the levels for supplementation of minimal media to achieve maximal growth of mutant strains of *E. coli* and other gram-negative bacteria.

Strains of bacteria that are resistant to antibiotics as a result of chromosomal mutations or the presence of R plasmids are frequently used in gene transfer experiments. Table 4 lists the concentrations of antibiotics for stock solutions and for supplementing media with antibiotics commonly used in gene transfer experiments with gram-negative enteric bacteria. Sterilization of antibiotic stock solutions is usually not necessary, provided that precautions are taken during their preparation. These include use of weighing paper that has been taken from a closed box or envelope container (paper products as packaged are generally sterile), a sterile container (bottle or tube), and sterile water or alcohol for suspension. Alternatively, stock solutions can be prepared, filter sterilized, and stored at −20°C. Stock solutions of nalidixic acid are stable when stored at 4°C. Ampicillin, rifampin, and tetracycline are relatively unstable; therefore, media containing these antibiotics should be used within several days. Storing media with tetracycline and rifampin in the dark prolongs their useful life.

Stock solutions of X-Gal and XP are prepared as 20 mg/ml in dimethylformamide. Each is sensitive to light and should be stored wrapped in aluminum foil at −20°C. A stock solution of IPTG is 200 mg/ml in H_2O. This solution is filter sterilized and stored in aliquots at −20°C. X-Gal is expensive and is therefore added to premade plates. X-Gal (0.04 ml) and IPTG (0.004 ml) are added to plates containing the appropriate antibiotics and then spread across the surface of the plates and allowed to dry before using. XP is added to agar media at a concentration of 40 mg/ml before pouring.

14.4.16. Preparation of Agar Media

Agar plates used to prepare or determine the titer of phage lysates must be poured on a level surface. Phage yield and plaque size are increased by using 35 to 45 ml of agar medium per plate. It is also important that plates for plaque assays be neither too dry nor too wet. A satisfactory procedure is to pour plates either 2 days prior to use and leave them at room temperature or the day before use and incubate them with the lids closed at 37°C for about 12 h. When preparing lysates by the confluent plate method, agar plates should be poured the day before use and not dried at 37°C.

Minimal agar plates, which must often be incubated for 2 days, should contain 25 to 30 ml of medium. Fermentation reactions on EMB and MacConkey agars are also improved if the plates contain at least 25 ml of medium.

14.4.17. Storage of Agar Media

Most agar media, except those containing unstable antibiotics (Table 4), can be placed in plastic bags and stored at 4°C for several months. It is wise to store media that contain dyes and other pigmented compounds in the dark. This is especially true for EMB agar which, because of photooxidation, becomes inhibitory to bacterial growth.

14.4.18. Sources of Bacterial and Phage Strains

E. coli strains: Coli Genetic Stock Center, Dr. Barbara J. Bachmann, Department of Biology, Yale University, New Haven, CT 06511-7444.

Salmonella strains: Salmonella Genetic Stock Centre, Dr. Kenneth E. Sanderson, Department of Biological Sciences, University of Calgary, Alberta, Canada T2N 1N4.

Other bacterial and phage strains: American Type Culture Collection, 12301 Parklawn Drive, Rockville, MD 20852.

14.5. REFERENCES

14.5.1. General References

1. **Friefelder, D. M.** 1987. *Microbial Genetics. The Jones and Bartlett Series in Biology.* Jones and Bartlett Publishers, Inc., Boston.
This textbook is a good source for descriptions of the concepts of classical microbial genetics.
2. **Miller, J. H. (ed.).** 1991. Bacterial genetic systems. *Methods Enzymol.* **204:**305–636.
This contains a comprehensive compilation of methods and protocols for experiments in microbial genetics with various gram-negative bacteria as well as other bacterial types.

14.5.2. Specific References

3. **Achtman, M.** 1973. Genetics of the F sex factor in enterobacteriaceae. *Curr. Top. Microbiol. Immunol.* **60:**79–124.
4. **Anderson, R. P., and J. R. Roth.** 1978. Tandem chromosomal duplications on *Salmonella typhimurium*: fusion of histidine genes to novel promoters. *J. Mol. Biol.* **119:**147–166.
5. **Arber, W., L. Enquist, B. Hohn, N. E. Murray, and K. Murray.** 1983. Experimental methods for use with lambda, p. 433–466. *In* R. W. Hendrix, J. W. Roberts, F. W. Stahl, and R. A. Weisberg (ed.), *Lambda II.* Cold Spring Harbor Laboratory, Cold Spring Harbor, N.Y.
6. **Bachmann, B. J.** 1990. Linkage map of *Escherichia coli* K-12, edition 8. *Microbiol. Rev.* **54:**130–197.
7. **Barcak, G. J., M. S. Chandler, R. J. Redfield, and J.-F. Tomb.** 1991. Genetic systems in *Haemophilus influenzae. Methods Enzymol.* **204:**321–341.
8. **Berg, C. M., D. E. Berg, and E. A. Groisman.** 1989. Transposable elements and the genetic engineering of bacteria, p. 879–926. *In* D. E. Berg and M. M. Howe (ed.), *Mobile DNA.* American Society for Microbiology, Washington, D. C.
9. **Berg, D. E., J. Davies, B. Allet, and J.-D. Rochaix.** 1975. Transposition of R factor genes to bacteriophage λ *Proc. Natl. Acad. Sci. USA* **72:**3628–3632.
10. **Bertani, G.** 1952. Studies on lysogenesis. I. The mode of phage liberation by lysogenic *Escherichia coli. J. Bacteriol.* **62:**293–300.
11. **Biek, D. P., and S. N. Cohen.** 1986. Identification and characterization of *recD*, a gene affecting plasmid maintenance and recombination in *Escherichia coli. J. Bacteriol.* **167:**594–603.
12. **Bradley, D. E., D. E. Taylor, and D. R. Cohen.** 1980. Specification of surface mating systems among conjugative drug resistance plasmids in *Escherichia coli* K-12. *J. Bacteriol.* **143:**1466–1470.
13. **Caro, L., and C. M. Berg.** 1971. P1 transduction. *Methods Enzymol.* **22D:**444–458.
14. **Chan, R. K., D. Botstein, T. Watanabe, and Y. Ogata.** 1972. Specialized transduction of tetracycline resistance by

phage P22 in *Salmonella typhimurium. Virology* **50:**883–898.

15. **Chelala, C. A., and P. Margolin.** 1974. Effects of deletions on cotransduction linkage in *Salmonella typhimurium*: evidence that bacterial chromosome deletions affect the formation of transducing DNA fragments. *Mol. Gen. Genet.* **131:**97–112.
16. **Chopra, I., S. W. Shales, J. W. Ward, and L. J. Wallace.** 1981. Reduced expression of Tn*10*-mediated tetracycline resistance in *Escherichia coli* containing more than one copy of the transposon. *J. Gen. Microbiol.* **126:**45–54.
17. **Cohen, S. N., A. C. Y. Change, and L. Hsu.** 1972. Nonchromosomal antibiotic resistance in bacteria: genetic transformation of *Escherichia coli* by R-factor DNA. *Proc. Natl. Acad. Sci. USA* **69:**2110–2114.
18. **Collins, J., and B. Hohn.** 1978. Cosmids: a type of plasmid gene-cloning vector that is packageable in vitro in bacteriophage λ heads. *Proc. Natl. Acad. Sci. USA* **75:**4242–4246.
19. **Curtiss, R., III.** 1965. Chromosomal aberrations associated with mutations to bacteriophage resistance in *Esherichia coli. J. Bacteriol.* **89:**28–40.
20. **Curtiss, R., III, F. L. Macrina, and J. O. Falkinham III.** 1974. *Escherichia coli*—an overview, p. 115–134. *In* R. C. King (ed.), *Handbook of Genetics.* Plenum Press, New York.
21. **Datta, N., and R. W. Hedges.** 1972. Host ranges of R factors. *J. Gen. Microbiol.* **70:**453–460.
22. **Datta, N., R. W. Hedges, E. J. Shaw, R. B. Sykes, and M. H. Richmond.** 1971. Properties of an R factor from *Pseudomonas aeruginosa. J. Bacteriol.* **108:**1244–1249.
23. **Davidson, N., R. C. Deonier, S. Hu, and E. Ohtsubo.** 1975. Electron microscope heteroduplex studies of sequence relations among plasmids of *Escherichia coli.* X. Deoxyribonucleic acid sequence organization of F and of F-primes and the sequences involved in Hfr formation, p. 56–65. *In* D. Schlessinger (ed.), *Microbiology—1974.* American Society for Microbiology, Washington, D.C.
24. **Davis, R. W., D. Botstein, and J. R. Roth.** 1980. *Advanced Bacterial Genetics.* Cold Spring Harbor Laboratory, Cold Spring Harbor, N.Y.
25. **Demerec, M., E. A. Adelberg, A. J. Clark, and P. E. Hartman.** 1966. A proposal for a uniform nomenclature in bacterial genetics. *Genetics* **54:**61–76.
26. **De Vries, G. E., C. K. Raymond, and R. A. Ludwig.** 1984. Extension of bacteriophage λ host range: selection, cloning, and characterization of a constitutive λ receptor gene. *Proc. Natl. Acad. Sci. USA* **81:**6080–6084.
27. **Falkow, S.** 1975. *Infectious Multiple Drug Resistance.* Pion Limited, London.
28. **Goldberg, R. B., R. A. Benber, and S. L. Steicher.** 1974. Direct selection for P1-sensitive mutants of enteric bacteria. *J. Bacteriol.* **118:**810–814.
29. **Grant, S. G. N., J. Jesse, F. R. Bloom, and D. Hanahan.** 1990. Differential plasmid rescue from transgenic mouse DNAs into *Escherichia coli* methylation-restriction mutants. *Proc. Natl. Acad. Sci. USA* **87:**4645–4649.
30. **Hamilton, C. M., M. Aldea, B. K. Washburn, P. Babitzke, and S. R. Kushner.** 1989. New method for generating deletions and gene replacements in *Escherichia coli. J. Bacteriol.* **171:**4617–4622.
31. **Hanahan, D.** 1983. Studies on transformation of *Escherichia coli* with plasmids. *J. Mol. Biol.* **166:**557–580.
32. **Hanahan, D., J. Jesse, and F. R. Bloom.** 1991. Plasmid transformation of *Escherichia coli* and other bacteria. *Methods Enzymol.* **204:**63–113.
33. **Harkki, A., T. R. Hirst, J. Holmgren, and E. T. Palva.** 1986. Expression of *Escherichia coli lamB* gene in *Vibrio cholerae. Micro. Pathog.* **1:**283.
34. **Harkki, A., and E. T. Palva.** 1985. A *lamB* expression plasmid for extending the host range of phage λ to other enterobacteria. *FEMS Microbiol. Lett.* **27:**183–187.
35. **Hermans, J., J. G. Boschloo, and J. A. M. de Bont.** 1990. Transformation of *Mycobacterium aurum* by electroporation: the use of glycine, lysozyme and isonicotinic acid hydrazide. *FEMS Microbiol. Lett.* **72:**221–224.
36. **Hernalsteens, J. P., R. Vilarroel-Mandiola, M. Van Montagu, and J. Schell.** 1977. Transposition of Tn*1* to a broad-host range drug resistance plasmid, p. 179–183. *In* A. I. Bukari, J. A. Shapiro, and S. L. Adhya (ed.), *DNA Insertion Elements, Plasmids, and Episomes.* Cold Spring Harbor Laboratory, Cold Spring Harbor, N.Y.
37. **Hong, J.-S., and B. N. Ames.** 1971. Localized mutagenesis of any specific small region of the bacterial chromosome. *Proc. Natl. Acad. Sci. USA* **12:**3158–3162.
38. **Horwitz, J. P., J. Chua, R. J. Curby, A. J. Tomson, M. A. DaRooge, B. E. Fisher, J. Mauricio, and I. Klundt.** 1964. Substrates for cytochemical demonstration of enzyme activity. I. Some substituted 3-indolyl-ß-D-glycopyranosides. *J. Med. Chem.* **7:**574–575.
39. **Jacob, F., and E. L. Wollman.** 1956. Recombinaison gététique et mutants de fertilité chez *Escherichia coli. C.R. Acad. Sci. Paris* **242:**303–306.
40. **Jacobs, M., S. Wnendt, and U. Stahl.** 1990. High-efficiency electro-transformation of *Escherichia coli* with DNA from ligation mixtures. *Nucleic Acids Res.* **18:**1653.
41. **Jacobs, W. R., J. F. Barrett, J. E. Clark-Curtiss, and R. Curtiss III.** 1986. In vivo repackaging of recombinant cosmid molecules for analyses of *Salmonella typhimurium, Streptococcus mutans,* and mycobacterial genomic libraries. *Infect. Immun.* **52:**101–109.
42. **Juni, E., and A. Janik.** 1969. Transformation of *Acinetobacter calcoacetis (Bacterium anitratum). J. Bacteriol.* **98:**281–288.
43. **Kohara, Y., K. Akiyama, and K. Isono.** 1987. The physical map of the whole *E. coli* chromosome: application of a new strategy for rapid analysis and sorting of a large genomic library. *Cell* **50:**495–508.
44. **Lederberg, J., and E. M. Lederberg.** 1952. Replica plating and indirect selection of bacterial mutants. *J. Bacteriol.* **63:**399–406.
45. **Lennox, E. S.** 1955. Transduction of linked genetic characters of the host by bacteriophage P1. *Virology* **1:**190–206.
46. **Levine, M.** 1957. Mutations in the temperate phage P22 and lysogeny in *Salmonella. Virology* **3:**22–41.
47. **Lodish, H. F.** 1970. Secondary structure of bacteriophage f2 ribonucleic acid and the initiation of in vitro protein biosynthesis. *J. Mol. Biol.* **50:**689–702.
48. **Low, K. B.** 1972. *Escherichia coli* K-12 F-prime factors, old and new. *Bacteriol. Rev.* **36:**587–607.
49. **Mandel, M., and A. Higa.** 1970. Calcium-dependent bacteriophage DNA infection. *J. Mol. Biol.* **53:**159–162.
50. **Manoil, C., and J. Beckwith.** 1985. Tn*phoA:* a transposon probe for protein export signals. *Proc. Natl. Acad. Sci. USA* **82:**8129–8133.
51. **Marmur, J.** 1961. A procedure for the isolation of deoxyribonucleic acid from micro-organisms. *J. Mol. Biol.* **3:**208–218.
52. **Miller, J. F., W. J. Dower, and L. S. Tompkins.** 1988. High-voltage electroporation of bacteria: genetic transformation of *Campylobacter jejuni* with plasmid DNA. *Proc. Natl. Acad. Sci. USA* **85:**856–860.
53. **Miller, J. H.** 1992. *A Short Course in Bacterial Genetics. A Laboratory Manual and Handbook for Escherichia coli and Related Bacteria.* Cold Spring Harbor Laboratory, Cold Spring Harbor, N.Y.
54. **Miller, V. L., and J. J. Mekalanos.** 1988. A novel suicide vector and its use in construction of insertion mutations: osmoregulation of outer membrane proteins and virulence determinants in *Vibrio cholerae* requires *toxR. J. Bacteriol.* **170:**2575–2583.
55. **Moyed, H. S., T. T. Nguyen, and K. P. Bertrand.** 1983. Multicopy Tn*10 tet* plasmids confer sensitivity to induction of *tet* gene expression. *J. Bacteriol.* **155:**549–556.
56. **Murray, N. E., W. J. Brammer, and K. Murray.** 1977. Lambdoid phages that simplify the recovery of in vitro recombinants. *Mol. Gen. Genet.* **150:**53–61.
57. **Nishimura, Y., L. Caro, C. M. Berg, and Y. Hirota.** 1971. Chromosome replication in *Escherichia coli.* IV. Control of chromosome replication and cell division by an integrated episome. *J. Mol. Biol.* **55:**441–456.

58. **Nnalue, N. A., and B. A. D. Stocker.** 1986. Some *galE* mutants of *Salmonella choleraesuis* retain virulence. *Infect. Immun.* **54:**635–640.
59. **Notani, N. K., and J. K. Setlow.** 1974. Mechanisms of bacterial transformation and transfection. *Prog. Nucleic Acid Res.* **14:**39–100.
60. **Novick, R. P., R. C. Clowes, S. N. Cohen, R. Curtiss III, N. Datta, and S. Falkow.** 1976. Uniform nomenclature for bacterial plasmids: a proposal. *Bacteriol. Rev.* **40:**168–189.
61. **Oishi, M., and S. D. Cosloy.** 1972. The genetic and biochemical basis of the transformability of *Escherichia coli* K-12. *Biochem. Biophys. Res. Commun.* **49:**1568–1572.
62. **Park, S. F., and G. S. A. B. Stewart.** 1990. High-efficiency transformation of *Listeria monocytogenes* by electroporation of penicillin-treated cells. *Gene* **94:**129–132.
63. **Phillips, T. A., V. Vaughn, P. L. Bloch, and F. C. Neidhardt.** 1987. Gene-protein index of *Escherichia coli* K-12, p. 919–966. *In* F. C. Neidhardt, J. L. Ingraham, K. B. Low, B. Magasanik, M. Schaechter, and H. E. Umbarger (ed.), *Escherichia coli and Salmonella typhimurium: Cellular and Molecular Biology,* vol. 2. American Society for Microbiology, Washington, D.C.
64. **Roberts, J., and R. Deveret.** 1983. Lysogenic induction, p. 123–144. *In* R. W. Hendrix, J. W. Roberts, F. W. Stahl, and R. A. Weisberg, (ed.), *Lambda II.* Cold Spring Harbor Laboratory, Cold Spring Harbor, N.Y.
65. **Rosner, J. L.** 1972. Formation, induction, and curing of bacteriophage P1 lysogens. *Virology* **49:**679–689.
66. **Rothmel, R. K., A. M. Chakrabarty, A. Berry, and A. Darzins.** 1991. Genetic systems in *Pseudomonas. Methods Enzymol.* **204:**485–514.
67. **Salmond, G. P. C., J. C. D. Hinton, D. R. Gill, and M. C. M. Pérombelon.** 1986. Transposon mutagenesis of *Erwinia* using phage λ vectors. *Mol. Gen. Genet.* **203:**524–528.
68. **Sambrook, J., E. F. Fritsch, and T. Maniatis.** 1989. *Molecular Cloning: A Laboratory Manual,* 2nd ed. Cold Spring Harbor Laboratory, Cold Spring Harbor, N.Y.
69. **Sanderson, K. E., and J. R. Roth.** 1988. *Linkage map of Salmonella typhimurium, edition VII. Microbiol. Rev.* **52:**485–532.
70. **Sawula, R. V., and I. P. Crawford.** 1972. Mapping of the tryptophan genes of *Acinetobacter calcoaceticus* by transformation. *J. Bacteriol.* **112:**797–805.
71. **Schmieger, H.** 1972. Phage P22-mutants with increased or decreased transduction abilities. *Mol. Gen. Genet.* **119:**75–88.
72. **Seifert, H. S., and M. So.** 1991. Genetic systems in pathogenic Neisseriae. *Methods Enzymol.* **204:**342–357.
73. **Shimada, K., R. A. Weisberg, and M. E. Gottesman.** 1972. Prophage lambda at unusual chromosomal locations. I. Location of the secondary attachment sites and the properties of the lysogens. *J. Mol. Biol.* **63:**483–503.
74. **Simon, R., U. Priefer, and A. Pühler.** 1983. A broad range mobilization system for in vivo genetic engineering: transposon mutagenesis in gram negative bacteria. *Bio/Technology* **1:**784–791.
75. **Singer, M., T. A. Baker, G. Schnitzler, S. M. Deischel, M. Goel, W. Dove, K. J. Jaacks, A. D. Grossman, J. W. Erickson, and C. A. Gross.** 1989. A collection of strains containing genetically linked alternating antibiotic resistance elements for genetic mapping of *Escherichia coli. Microbiol. Rev.* **53:**1–24.
76. **Smith, C. L., J. G. Econome, A. Schutt, S. Klco, and C. R. Cantor.** 1987. A physical map of the *Escherichia coli* K12 genome. *Science* **236:**1448–1453.
77. **Smith, H. O., and M. Levine.** 1967. A phage P22 gene controlling integration of prophage. *Virology* **31:**207–216.
78. **Taylor, R. K., C. Manoil, and J. J. Mekalanos.** 1989. Broad-host range vectors for delivery of Tn*phoA*: use in genetic analysis of secreted virulence determinants of *Vibrio cholerae. J. Bacteriol.* **171:**1870–1878.
79. **Ullman, A., F. Jacob, and J. Monod.** 1967. Characterization by in vitro complementation of a peptide corresponding to an operator-proximal segment of the ß-galactosidase structural gene of *Escherichia coli. J. Mol. Biol.* **24:**339–343.
80. **Vogel, H. J., and D. M. Bonner.** 1956. Acetylornithinase of *Escherichia coli*: partial purification and some properties. *J. Biol. Chem.* **218:**97–106.
81. **Way, J. C., M. A. Davis, D. Morisato, D. E. Roberts, and N. Klecker.** 1984. New Tn*10* derivatives for transposon mutagenesis and for construction of *lacZ* operon fusions by transposition. *Gene* **32:**369–379.
82. **Willets, N. S.** 1972. Location and origin of transfer of the sex factor F. *J. Bacteriol.* **112:**773–778.
83. **Wilson, T. A., and N. M. Gough.** 1988. High voltage *E. coli* electro-transformation with DNA following ligation. *Nucleic Acids. Res.* **16:**11820.
84. **Yanisch-Perron, C., J. Vieira, and J. Messing.** 1985. Improved M13 phage cloning vectors and host strains: nucleotide sequences of the M13mp18 and pUC19 vectors. *Gene* **33:**103–119.
85. **Zabarovsky, E. R., and G. Winberg.** 1990. High efficiency electroporation of ligated DNA into bacteria. *Nucleic Acids Res.* **18:**5912.
86. **Zinder, N. D.** 1958. Lysogenization and superinfection immunity in *Salmonella. Virology* **5:**291–326.
87. **Zinder, N. D., and J. Lederberg.** 1952. Genetic exchange in *Salmonella. J. Bacteriol.* **64:**679–699.

Chapter 15

Gene Transfer in Gram-Positive Bacteria

SIMON M. CUTTING and PHILIP YOUNGMAN

Methods of choice for transfer of DNA into and between strains of bacteria are dictated ultimately by the biological properties of those bacteria. Thus in *Escherichia coli* and its close relatives, the properties of naturally existing fertility factors, such as the F-factor, allow conjugation to be used to position mutations on the genetic map and the properties of the cell envelope permit Ca^{+}- conditioned heat shock to be used for efficient introduction of plasmid DNA into cells (see Chapter 14). Bacteria with different biological properties require different approaches. This is illustrated here by using *Bacillus subtilis* as the model organism. *B. subtilis* is the best characterized of the gram-positive bacteria and exhibits distinctive biological properties which dictate the use of gene transfer strategies quite different in many cases from those presented in Chapter 14. Since Hfr-like conjugation mechanisms are unknown in *B. subtilis* (and other gram-positive species), coarse-level genetic mapping is accomplished instead by using an extremely large generalized transducing phage, PBS1. On the other hand, *B. subtilis* exhibits certain gene transfer mechanisms unknown in *E. coli*. Of particular importance is the natural mechanism of DNA uptake and incorporation known as transformation competence, which allows for highly efficient introduction of DNA into *B. subtilis* under certain circumstances. Natural transformation competence in *B. subtilis* is the basis for fine-structure genetic mapping and allele replacement strategies not possible in *E. coli*. Some of the gene transfer techniques described in this chapter are applicable to gram-positive species other than *B. subtilis*. For example, nearly all of the generalizations made about the transformation of naturally competent *B. subtilis* strains apply equally well to naturally competent strains of streptococci. Nevertheless, it should be recognized that both *B. subtilis* and *E. coli* are unusually well endowed in terms of natural gene transfer mechanisms and naturally occurring extrachromosomal elements which are easily adapted for use as genetic tools. In most other species, less attractive alternatives must often be used. Specific examples will be noted where relevant.

The organization of this chapter is patterned after that of Chapter 14, in part to facilitate a comparison of methods or approaches available in *B. subtilis* with those used in *E. coli*. General gene transfer concepts, such as transformation, transduction, and conjugation, are defined in Chapter 14.

15.1. TRANSFORMATION

Three major options exist for the introduction of DNA into *B. subtilis* by transformation: (i) transformation of naturally competent bacteria, (ii) polyethylene glycol (PEG)-mediated transformation of protoplasts, and (iii) electroporation. The method of choice for a particular application will depend on the nature of the DNA molecules to be introduced into the cell and the genotype of the cell. When linear fragments of DNA are introduced with the intention of obtaining incorporation into chromosomal or plasmid-borne sequences by homologous recombination, transformation of naturally competent bacteria is the only realistic option. Linear fragments of DNA introduced into bacteria by protoplast transformation or electroporation enter the cell in a double-stranded form and are thus inefficiently incorporated by homologous recombination; they are instead lost to nucleolytic degradation. Transformation with covalently closed circular plasmid DNA capable of autonomous replication could in most cases be carried out by any of the three procedures mentioned above, although certain recipient genotypes, such as recombination-deficient (*rec*) mutations or certain sporulation-deficient (*spo*) mutations, might prohibit the use of naturally competent bacteria. These and other considerations that might influence the choice of methods for introducing plasmid DNA into *B. subtilis* or other gram-positive bacteria will be discussed below. Transformation with ligated DNA involves still other special considerations that will also be discussed.

15.1.1. Natural Transformation Competence

The genetics, biochemistry, and physiology of the competent state in *B. subtilis* have recently been reviewed in detail (16), and methods for genetic transformation are well established (5, 14, 17, 41). It should be noted that the term transformation competence has a very different meaning when applied to *B. subtilis* than when applied to *E. coli*. Competence for uptake of DNA is not a normal aspect of *E. coli* natural history and is induced in the laboratory by subjecting mid-log-phase bacteria to a brief heat shock after incubation in the cold in the presence of Ca^{+}. In response to this treatment, *E. coli* bacteria presented with double-stranded plasmid DNA can take up intact molecules with sufficient efficiency to typically produce on the order of 10^7 transformants per microgram of DNA. In contrast, competence for DNA uptake by *B. subtilis* bacteria is a natural aspect of its biological behavior. In response to nutritional stress, a significant fraction of a *B. subtilis* population can undergo a developmental transition to a special physiological condition referred to as the competent state (16). This developmental transition involves the expression of many new specialized genes, whose products participate in the binding, uptake, and recombinational incorporation of transforming DNA into genomic or plasmid DNA already present in the transformation recipient. Of particular importance for the understanding of many of the experimental approaches presented below is the fact that DNA taken up by *B. subtilis* bacteria in the natural state of transformation competence does not remain intact; instead, DNA bound to the surface of the cell suffers double-strand cleavage and, apparently coincident with actual entry into the cell, one of the two strands is stripped away.

Natural competence is achieved simply by growing cells in a defined minimal medium to the stationary phase of growth. As cells leave the exponential phase of growth, an increasing number of cells become competent and thus have the ability to bind and take up DNA. By use of the procedures outlined below, it is routinely possible to obtain cultures in which 5 to 10% of the cell population is in the competent state. Therefore, transformation of naturally competent bacteria can be highly efficient. As noted above, however, DNA uptake by competent cells involves nucleolytic fragmentation and apparently quantitative conversion of the transforming DNA to a single-stranded and highly recombinogenic form. Thus, transformation with linear fragments of DNA, resulting in marker replacement recombination, is extremely efficient whereas transformation with plasmid DNA may be quite inefficient unless sequences homologous to the plasmid DNA are already present in the transformation recipient (see section 15.1.1.7). The implications of this for the design of specific experiments will be considered below. It should be apparent from the obligatory role of recombination in the transformation of naturally competent bacteria that Rec^- mutants (strains deficient in homologous recombination) cannot be transformed with either chromosomal or plasmid DNA. Mutations in the *spo0A* regulatory locus also severely reduce transformation competence, since they affect the pathway of gene expression required for development of the competent state (16).

Although simpler procedures may result in the preparation of adequately competent bacteria (4), the method of choice for generating a large number of physiologically competent cells remains the classical two-step procedure of Dubnau and Davidoff-Abelson (17). This procedure is reliable and allows a large quantity of cells to be prepared and frozen indefinitely in the competent state. Before outlining the two-step procedure, we describe a procedure for the isolation of chromosomal transforming DNA, as follows. See section 15.4 for preparation of the media used below and in the following procedures.

15.1.1.1. Chromosomal transforming DNA isolation

1. Inoculate 25 ml of Luria-Bertani (LB) medium with a single, fresh colony. Incubate at 30°C overnight (or at 37°C until late log phase [optical density at 600 nm, OD_{600}, 1.5 to 2.0] is reached).
2. Pellet cells (8,000 × *g* for 10 min at 25°C) in a plastic (phenol- and chloroform-resistant) centrifuge tube, resuspend and repellet (wash) once in 10 ml of lysis buffer (50 mM EDTA, 0.1 M NaCl [pH 7.5]), and resuspend in 4 ml of lysis buffer.
3. Add approximately 4 mg of lysozyme (e.g., 0.2 ml of a freshly prepared solution at 20 mg/ml), vortex briefly, and incubate at 37°C for 10 min.
4. Add 0.3 ml of 20% (wt/vol) Sarkosyl (*N*-lauroylsarcosine [Sigma]), mix gently by inverting the tube several times (the cell suspension will visibly clarify), and continue incubation at 37°C for 5 min.
5. Add an equal volume of phenol. Vortex vigorously for about 20 s. Centrifuge (10,000 × *g* for 10 min at 25°C), and carefully remove the upper, aqueous phase with the reverse end of a glass 10-ml pipette (or Pasteur pipette); dispense into a new centrifuge tube.
6. Repeat the extraction procedure of step 5 twice, first with phenol-chloroform and then with chloroform alone.
7. To the aqueous, extracted DNA phase, add 0.1 volume of 3 M sodium acetate (pH 5.2) followed by 2 to 3 volumes of ice-cold ethanol (99%). Cap and gently invert the tube until the DNA precipitate has condensed into a small clump. The DNA can now be carefully "fished" out with a clean Pasteur pipette (the DNA rapidly adheres to glass), air dried, and dissolved in 1.0 ml of sterile TE buffer. (*Note:* complete desiccation will make it difficult to redissolve the DNA.) The DNA solution can be stored in aliquots at 4 or −70°C.

For the preparation of transforming DNA, this method works well and is both quick and reliable. Milligram quantities are typically obtained, and the DNA can be stored at 4 or −70°C for long periods. In addition, the extraction procedure works equally well on frozen cell pellets, and the method can readily be scaled up or down as necessary. The following precautions should be observed, however.

(i) When using phenol in step 5, it is necessary to vortex the cell-DNA suspension quite vigorously. High concentrations of chromosomal DNA render the solution very viscous, and it may prove impossible to separate the aqueous and phenolic phases after centrifugation unless some hydrodynamic shearing of the DNA has occurred. When the aqueous phase is being removed after phenol extraction, it should be noticeably viscous. If it is not, the DNA has probably been trapped in the "white" interface between the aqueous and phenolic phases. In this case it can be recovered by adding more TE buffer to the phenolic phase and reextracting.

(ii) The DNA will initially come out of solution as an extended coagulum, yet with continued inversion it will condense further into a very tight clump (the size of a pinhead). This tight clump may be difficult to remove from the extracted phase with a glass pipette but will dissolve completely in TE buffer if left overnight at 4°C, provided that the DNA has not been dried completely. It does not appear to matter if the DNA precipitate is slightly wet with ethanol. Alternatively, the DNA may be "spooled" from the aqueous phase onto a glass rod or Pasteur pipette. The simplified method of Juni (26) is an excellent alternative for rapid preparation of a small quantity of DNA for transformation of *B. subtilis* strains, provided that care is taken to eliminate spores from the DNA sample or to select against the donor strain.

15.1.1.2. Transformation procedure

Preparation of competent cells

1. Streak out the strain to be made competent on an LB agar plate as a large patch, and incubate overnight at 30°C.
2. The following morning, scrape the cell growth from the plate and resuspend it (by vortexing) in about 2.0 ml of SpC medium.
3. Use the cell suspension to inoculate 20 ml of fresh, prewarmed (37°C) SpC medium in a 250-ml Erlenmeyer flask to give a starting culture OD_{600} of about 0.5 (with practice this can be done by eye!). *Note:* SpC should be supplemented with all nutrients required by the strain for growth on minimal media!
4. Incubate the culture with moderate orbital shaking at 37°C. Take periodic OD_{600} readings (every 30 min at early times and every 15 min as the culture approaches the stationary phase).
5. Just as or slightly before the culture reaches stationary phase (i.e., less than a 5% increase in OD_{600} readings in a 15-min time span), use 20 ml of this culture to inoculate 200 ml of fresh, prewarmed, SpII medium in a 2-liter flask. Incubate at 37°C with moderate orbital shaking.
6. After 90 min of incubation, pellet the cells (8,000 × *g* for 5 min) at room temperature.
7. Carefully decant the supernatant, and save it in a separate container.
8. Gently resuspend the cell pellet in 17 ml of the saved supernatant, and add 2 ml of sterile 70% (vol/vol) glycerol; mix gently.
9. Use the competent cells immediately or aliquot the cells to be frozen (e.g., 0.5-ml samples) in screw-cap

tubes. Freeze rapidly in liquid nitrogen, and store at −70°C.

Transformation

1. Thaw a frozen tube of competent cells by immersing the tube in a 37°C water bath or by holding it under running tap water. After the tube contents are partially thawed (when a small piece of frozen culture remains in the tube), vortex until the remaining ice disappears.
2. Immediately, to 1 volume of competent cells, add an equal volume of SpII + EGTA at room temperature. Mix gently. Dispense the cells into test tubes appropriate for incubation in a roller drum or reciprocal shaker (200 to 500 μl per tube would be typical for most routine experiments).
3. To these test tubes, add DNA dissolved in a volume of TE less than half the volume of the competent-cell suspension (see below). Incubate in a roller drum or shaker at 37°C for 15 to 20 min.
4. Dilute the transformed cells as appropriate in T base, and plate on suitable selective media.

Comments

When this procedure is used, a robust strain of *B. subtilis* (e.g., 168 or PY79) will take approximately 4 to 5 h to reach a stationary phase cell density corresponding to an OD_{600} of about 2 to 3 in SpC at 37°C. Growth may be significantly slower at lower temperatures or under antibiotic selection. Strains with multiple auxotrophic requirements may also grow more slowly and may reach stationary phase at an OD_{600} of less than 2.0. When growing an unfamiliar strain to competence, careful attention to growth parameters may be required for optimal results. On the other hand, optimal competence is often not required for many experiments, and acceptable levels of competence may be achieved with only casual attention to growth behavior; visual inspection of the culture may be sufficient, with some experience and a familiar strain. Even if growth in SpC is very poor, it may still be worth proceeding, since it is nearly always possible to obtain a low yield of competent cells. Certain strains may also exhibit an extended lag phase when transferred to SpC after overnight growth on a rich solid medium. In some cases, this can be remedied by substituting tryptose blood agar base (TBAB) or a glucose-minimal medium such as TSS (TSSA medium minus the agar). If overnight growth on LB or TBAB is very poor, it should be noted that some auxotrophic supplements (e.g., thymine) are required even on complex, rich media. On the other hand, excess concentrations of some supplements (e.g., cysteine) may actually inhibit growth.

When growing temperature-sensitive strains (or strains containing temperature-sensitive plasmids, such as those used to select for Tn*917* transpositions) at 30°C, the time of incubation in SpII should be increased from 90 to 120 min. The most frequent mistake in this procedure is forgetting to save the supernatant in step 7. Using freshly prepared SpII medium at this point will not work. The frozen competent cells, if stored properly, will retain competence for at least 5 years. Nevertheless, if the highest transformation frequencies are required, unfrozen competent cells should be used. Use of freshly prepared growth media (SpC and SpII) is recommended to avoid contamination, although SpII + EGTA can be filter sterilized and frozen in small aliquots.

15.1.1.3. DNA concentrations

Chromosomal transforming DNA should be prepared by one of the methods described above. When chromosomal DNA is used in genetic-mapping experiments, low DNA concentrations are most desirable (a final concentration in the competent-cell mixture of no more than 50 ng/ml). When using chromosomal DNA to introduce genetic markers by congression (section 15.1.1.6), saturating levels of DNA should be used (final concentration, 1 to 5 μg/ml). When transforming with plasmid DNA, concentrations should generally be in the range of 1 to 5 μg/ml, unless sequences homologous to the plasmid are already present in the transformation recipient, in which case much lower concentrations should be sufficient.

15.1.1.4. Selection for prototrophy and drug resistance

In *B. subtilis*, detection of a transformation-mediated gene transfer event normally requires a direct selection. Typically this involves selecting for a recombinational event, e.g., conversion of an auxotrophic marker to prototrophy, or the introduction of a drug resistance gene. These selections are applied as follows.

Prototrophy selection

To select for conversion of an auxotrophic marker (e.g., the *pheA1* mutation) to prototrophy (i.e., Phe^+) the transformation mixture must be plated directly on minimal agar plates that lack the growth requirement (phenylalanine) being selected for. The following two minimal growth media are recommended for such prototrophic selections (formulations are given in section 15.4).

Tris-Spizizen salts agar (TSSA). TSSA is a general-purpose minimal growth medium for *B. subtilis*. It is formulated to maximize growth rate; for most strains, colonies can easily be seen after overnight growth at 37°C. Colonies may acquire a distinctive pink coloration on this medium, but they do not sporulate efficiently.

Lactate-glutamate minimal agar (LGMA). LGMA is formulated to allow not only selection for prototrophic recombinants but also analysis of their sporulation phenotype. Because of the presence of lactate as the primary carbon source in this medium, bacteria typically grow much more slowly than on TSSA; often, 2 days or more is required to form a colony. If bacteria are Spo^+, they will sporulate and form brown (Pig^+) colonies, whereas Spo^- recombinants will form white-translucent colonies. Therefore, LGMA is appropriate for the mapping of *spo* mutations relative to auxotrophic reference markers.

Growth requirements. Table 1 provides a list of amino acid, purine, and pyrimidine concentrations that should be used in TSSA or LGMA medium. The same final concentrations of these supplements should also be added to the SpC and SpII media used to induce transformation competence. For detailed information concerning particular mutant phenotypes and selections or screens used to distinguish them from the wild type, the original reference to the mu-

Table 1. Prototrophy medium supplements[a]: amino acids, purines, and pyrimidines

Compound	Stock solution concn (mg/ml)	Stock solution added to medium (ml/liter)	Final concn (μg/ml)[b]
Amino acids			
D-Alanine	10	5.0	50
L-Arginine HCl	10	5.0	50
L-Asparagine	10	5.0	50
L-Aspartic acid	50	10	500
L-Cysteine HCl[c]	4.0	10	40
L-Glutamic acid[d]	50	10	500
L-Glutamine	10	5.0	50
Glycine	10	5.0	50
L-Histidine HCl	10	5.0	50
L-Isoleucine	10	5.0	50
L-Leucine	10	5.0	50
L-Lysine	10	5.0	50
L-Methionine	10	5.0	50
L-Phenylalanine[e]	10	5.0	50
L-Proline	10	5.0	50
L-Serine	10	5.0	50
L-Threonine	10	5.0	50
L-Tryptophan	2.0	10	20
L-Tyrosine[f]	2.0	10	20
L-Valine	10	5.0	50
Purines and pyrimidines			
Adenine/adenosine[g]	2.0	10	20
Cytidine	5.0	10	50
Guanine/guanosine	2.0	10	20
Thymine/thymidine	5.0	10	50
Uracil[h]	4.0	10	40

[a]All stock solutions are prepared in distilled water unless otherwise indicated. Sterilization can be performed by autoclaving or filtration, with the exception of tryptophan which must be filter sterilized.
[b]These concentrations are sufficient for growth. Lower concentrations can be used in some cases, although use of greater concentrations may inhibit growth.
[c]Cysteine should be made fresh since it is quickly oxidized to cystine crystals. Alternatively, the stock solution can be overlaid with sterile mineral oil.
[d]pH to 7.0 with 1 M NaOH.
[e]Prepare in 0.001 M NaOH, and store at room temperature.
[f]Prepare in 0.01 M NaOH, and store at room temperature.
[g]Prepare in 0.03 M HCl.
[h]Prepare in 0.1 M KOH.

tant should be consulted. Alternatively, the *Bacillus* Genetic Stock Center provides a handbook listing detailed information on almost all *B. subtilis* genotypes; this handbook is also a convenient reference guide to the primary literature and includes a catalog of *Bacillus* strains, phages, and plasmids available for distribution to academic or industrial researchers (see section 15.4.1).

Selection for drug resistance

When selection is made for the introduction of a drug resistance gene following gene transfer, time must often be allowed for the gene to express itself. In a transformation or transduction experiment, this can be done in some cases by increasing the time of incubation of donor with recipient to a full 60 min. This approach often increases the number of transformants and transductants significantly, but it is not always necessary. In some cases, particularly when the drug resistance gene is carried on a multicopy plasmid, omitting the extra incubation period will make little difference. When using an unfamiliar selection, it is best to start with conservative assumptions.

Table 2 gives the concentrations of antibiotics used in plates for some of the more common drug-resistance genes used in *B. subtilis*. An additional consideration when selecting for drug resistance in *B. subtilis* is that expression of some of the most commonly used resistance genes is inducible. Two of the most important examples are the Tn*917*-associated *erm* gene (35), which confers inducible resistance to erythromycin and other MLS (macrolide-lincosamide-streptogramin B) antibiotics, and the *cat* gene, which is derived from the *Staphylococcus* plasmid pC194 (24) and confers inducible resistance to chloramphenicol. When selecting for the transfer of a single copy of Tn*917* (e.g., as an insertion into chromosomal DNA), induction of the *erm* gene can significantly enhance the efficiency with which transformant colonies are recovered. The *cat* gene is often used as a selectable marker in integrational vectors to insert recombinant gene constructions into the *B. subtilis* chromosome (see below). Failure to induce expression of the *cat* gene in such a situation can lead to the recovery of multiple integrated copies of the vector, producing misleading results. An effective method for inducing expression of the *erm* and *cat* genes during gene transfer selections is to use a double-overlay method in which transformant or transductant populations are first plated in a soft-agar overlay with a subselective level of the antibiotic, followed after a period

Table 2. Selective antibiotic concentrations

Antibiotic	Selective concn (μg/ml)	Solvent
Chloramphenicol[a]	5	95% ethanol
Erythromycin[a,b]	1	95% ethanol
Lincomycin[b]	25	50% ethanol
Neomycin	5	Distilled H_2O
Kanamycin	10	Distilled H_2O
Phleomycin[c,d]	0.1–0.5	Distilled H_2O
Tetracycline[c]	20	50% ethanol

[a]Cm^r and Em^r are frequently conferred in gram-positive bacteria by inducible antibiotic resistance genes. In these cases, the first overlay should contain a subselective concentration of the antibiotic (see section 15.1.1.4).

[b]When selecting for MLS^r both erythromycin and lincomycin are used.

[c]Best to select in double overlay: plate cells in LB soft-agar overlay with no drugs; incubate for 2 h at 37°C, and then add 2.5 ml of molten LB overlay agar containing the drug to give final concentration shown above; incubate overnight.

[d]Bleomycin is not recommended as an alternative to phleomycin, since it increases the mutation rate. Phleomycin can be obtained from CAYLA, Toulouse, France.

of incubation with a second overlay containing selective levels, as explained below.

Induction of MLS resistance genes

1. Add 0.1 ml of 10 μg/ml erythromycin solution to a small test tube containing 2.5 ml of LB soft agar (0.7% agar), kept molten at 45°C. Mix gently.
2. Add transformed/transduced bacteria, mix, and pour onto a drug-free LB agar plate (35 to 40 ml of agar).
3. Allow the overlay to set, and incubate at 37°C.
4. After 2 h of incubation, add a second overlay (2.5 ml) of LB soft agar containing 0.1 ml of a mixture of 0.4 mg/ml of erythromycin and 10 mg/ml of lincomycin. Allow to set, and incubate overnight at 37°C.
5. Colonies will grow out onto the surface of the agar plate.

Induction of chloramphenicol resistance genes

Use the method for MLS genes as described above but with 0.1 ml of a 50 μg/ml chloramphenicol solution in the first overlay and 0.1 ml of a 2 mg/ml solution in the second overlay.

Comments

When selecting for the induction of temperature-sensitive plasmids containing *erm* or *cat* genes, such as the commonly used vectors for generating Tn*917* insertions, induction and selection must be carried out at 32°C. In this case, the induction period should be 3 h.

15.1.1.5. Genetic mapping

Transformation is commonly used for establishing the genetic distance or linkage between a mutation and a reference marker by means of a two- or three-factor genetic cross. Linkage between two genetic markers is defined as the cotransformation linkage and is expressed as a percentage. Linkage can also be translated directly into arbitary **transformation map units** by using the equation (100 − percent cotransformation frequency)/100. As an example, for two mutations that are 60% cotransformed, the distance between the two markers is said to be 0.40 transformation map units. Map units, as the name implies, are used by convention to indicate linkage relationships on genetic maps.

Since the average size of a linear fragment of transforming DNA is about 30 kb, two genetic markers must be separated by no more than 30 kb of DNA for their genetic distances from each other to be determined by cotransformation. Moreover, linkage between two markers is considered to be linear only at distances of less than 10 kb; this is due in part to the phenomenon of "congression" (see below) and in part to the endonucleolytic fragmentation of DNA on uptake by competent cells (15). Thus, transformation is most appropriately viewed as a fine-structure mapping tool. To establish truly accurate estimates of genetic distance, especially when mapping weakly linked markers, it is necessary to use nonsaturating concentrations of transforming DNA (less than 50 ng/ml). Otherwise, an apparent linkage may be an artifact of congression, as explained below.

15.1.1.6. Congression and strain construction

Congression is a term used to describe transformation of the same competent cell by two genetically unlinked markers as the result of independent DNA uptake events. At "saturating" DNA concentrations (1 to 5 μg/ml), congression is observed to occur at a frequency of 2 to 3%. Thus, as indicated above, congression can be an artifactual source of spurious transformation linkage. On the other hand, congression can be an extremely useful strain construction tool, since it enables the use of any selectable marker to transfer any unlinked mutation into a new strain. For example, consider donor strain A, which otherwise is wild type but contains a *spoIIE* mutation (blocking sporulation at "stage II") and recipient strain B, which contains a *metB* mutation (conferring auxotrophy for methionine). Suppose one wishes to transfer the *spoIIE* mutation from strain A to strain B. This can readily be accomplished, even though the *spoIIE* locus is not genetically linked to the *metB* locus. All one need do is transform strain B to Met^+ (prototrophy for methionine) with chromosomal DNA from strain A at a high DNA concentration (>1.0 μg/ml) and then screen transformants for their sporulation phenotypes. Since the *spoIIE* mutation should be transferred to strain B by congression at a frequency of 2 to 3%, one need screen only 100 transformants to expect to find at least one Spo^- recombinant.

15.1.1.7. Plasmid DNA

The mechanism by which naturally competent cells take up DNA during transformation involves random nicking, fragmentation, and conversion to a single-stranded form. Thus, introduction of plasmid DNA by transformation into naturally competent bacteria generally requires multiple uptake events by a single cell or transformation with complex multimeric plasmid forms. In most cleared-lysate preparations from either gram-positive or gram-negative bacteria, plasmid dimers or higher-order multimers are abundant enough to yield transformants of *B. subtilis* strains at a level of 10^3 to 10^5/μg. However, plasmid DNA isolated from Rec^- bacteria contains a much lower abundance of multimers and may therefore yield transformants at levels 2 to 3 orders of magnitude lower. The efficiency of transformation with plasmid DNA can be dramatically enhanced when the transformation recipient contains sequences homologous to the plasmid, either inserted in the chromosome or on a plasmid already present in the recipient. In this case, transformants may be generated at levels greater than $10^6/\mu g$, even with monomeric forms of plasmid DNA.

15.1.1.8. Cloning systems

The manner in which transforming DNA is processed during uptake by naturally competent bacteria also has several important implications for recombinant DNA manipulations. Simple insertion of DNA fragments by ligation into plasmid momomers will not produce transformants. When vector-insert ligation mixtures do produce transformants, this is only because some multimeric ligation products are present. Moreover, treatment of vector preparations with alkaline phosphatase, which is often done in routine cloning experiments in which *E. coli* is the transformation host, will abolish the biological activity even of multimeric ligation products. Thus, a cloning experiment with *B. subtilis* modeled after procedures developed for *E. coli* can have disastrous consequences.

On the other hand, cloning strategies have been developed specifically to take advantage of the recombinogenic nature of DNA uptake by naturally competent gram-positive bacteria. Among the most elegant is a plasmid-based cloning system developed by Haima et al. (22), which exploits the concept of using homologous sequences in a resident plasmid to promote recombinational integration of vector sequences containing a cloned insert (20). In this **marker rescue cloning system**, a multiple-cloning site is embedded in a copy of the *lacZα* gene fragment and placed adjacent to a Cm^r marker (e.g., in pHPS9 [Fig. 1]). Transformation recipients contain plasmid pHPS9R, which lacks the *lacZα* gene fragment and a functional copy of the Cm^r marker. Restriction fragments to be cloned are simply ligated in the presence of similarly digested pHPS9 DNA at high DNA concentration, to form linear concatemers of vector and insert, and the ligation mixture is used to transform competent cells that already contain pHPS9R, selecting for Cm^r.

As shown in Fig. 1, a combination of ligation and homologous recombination results in the incorporation of a cloned insert fragment into the resident plasmid. The presence of a *lacZ*ΔM15 gene fragment in the chromosome of the transformation recipient permits the failure of *lac* α-complementation to be used to identify transformants with inserts. A conceptually similar approach referred to below (section 15.2.2.5) exploits the recombinogenic nature of DNA uptake by naturally competent cells for cloning into a temperate phage of *B. subtilis*.

15.1.2. Protoplast Transformation

15.1.2.1. *B. subtilis*

For *B. subtilis* strains not naturally competent for transformation, the most attractive alternative for introducing plasmid DNA is often PEG-mediated protoplast transformation. Protoplasts in this case are generated by treatment of the bacteria with lysozyme (to remove cell walls) in the presence of an osmotic stabilizer. As with many gram-positive species, protoplasts of *B. subtilis* become permeable to DNA in the presence of a high concentration of PEG. When plated under appropriate conditions, transformed protoplasts can regenerate cell walls and form colonies (section 15.4.4). See section 15.4.3 for media and reagents used in protoplast preparation and transformation. The method given below was developed by Chang and Cohen (11) and remains the most widely used for *B. subtilis*.

Preparation of protoplasts

1. Inoculate 10 ml of PAB medium (in a 100-ml flask) with a single colony. Grow overnight at 37°C with orbital shaking (200 rpm).
2. Add 0.2 ml of overnight culture to 10 ml of fresh, warm PAB medium (in a 100-ml flask). Incubate at 37°C with orbital shaking (200 rpm).
3. When the culture has reached an OD_{600} of 0.5 (this takes about 2 h), pellet the cells in a capped plastic centrifuge tube (8,000 × *g* for 10 min at 25°C) and resuspend in 5 ml of warm SMMP.
4. Add ca. 5 mg of lysozyme, and mix by gently inverting the capped tube several times. Incubate at 37°C in a roller drum or reciprocal shaker.
5. After 2 h of incubation, approximately 99% of the cells should be converted to protoplasts (this must be checked microscopically, best by phase-contrast microscopy; protoplasts have a spherical or globular appearance).
6. Harvest the protoplasts by gentle centrifugation (5,000 × *g* [no centrifugal brake] for 10 min at 25°C).
7. Gently resuspend the pellet in 5 ml of SMMP.
8. The protoplasts are now ready for use, although aliquots (e.g., 0.5 ml) can be frozen at −80°C for future use. (*Note:* Glycerol addition is not required for protoplast viability!)

Transformation

1. Transfer 450 μl of PEG to a microcentrifuge tube.
2. Mix 1 to 10 μl of DNA (ca. 0.2 μg) with 150 μl of protoplasts, and add the mixture to the microcentrifuge tube containing PEG. Mix immediately and very gently.
3. Leave at 25°C for 2 min, and then add 1.5 ml of SMMP and mix.
4. Harvest the protoplasts by centrifugation (8,000 × *g* for 7 min), and pour off the supernatant. Remove the remaining droplets with a tissue.
5. Add 300 μl of SMMP (do not vortex), and incubate for 60 to 90 min at 37°C in a water bath orbital shaker (ca. 100 rpm) to allow for expression of antibiotic resistance markers (gentle shaking will also resuspend the pelleted protoplasts).
6. Make appropriate dilutions in 1 × SSM and plate 0.1-ml aliquots on DM3 plates.

Comments

The most common reason for failing to obtain protoplasts is dirty glassware. It is imperative to have completely detergent-free glassware. Use disposable plasticware whenever possible. When this is not possible, use acid-washed glassware rinsed extensively with distilled H_2O. Protoplasts of *B. subtilis* cells are extremely fragile and will lyse if subjected to either mechanical

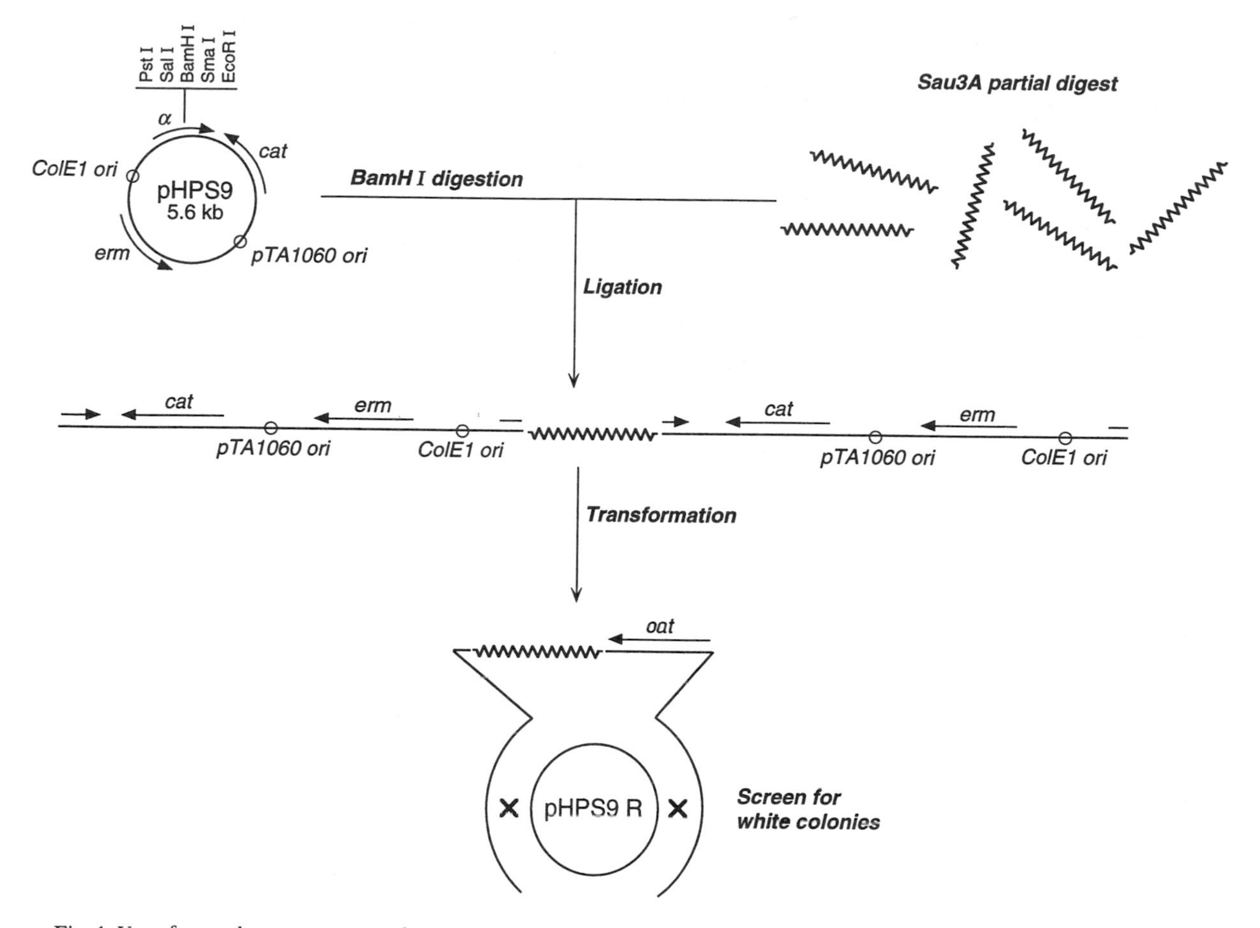

Fig. 1. Use of a marker rescue transformation system to clone DNA by ligation/recombination. Plasmid pHPS9 (22) is a chimeric replicon, containing replication functions which operate in both *E. coli* (ColE1 ori, from pUC9 [30]) and *B. subtilis* (pTA1060 ori, from a stable, low-copy *Bacillus* plasmid [9]). Arrows mark the positions and orientations of relevant plasmid-encoded genes: *erm*, erythromycin resistance (originally from pE194 [24]); *cat*, chloramphenicol resistance (originally from pC194 [24]) α, *lacα* complementation fragment from pUC9 (30). Jagged lines represent fragments generated by *Sau*3A digestion of genomic DNA from a cloning source; X represents recombination between a ligation product taken up by competent cells and the resident plasmid.

stress or trace quantities of surfactants. Dryness of DM3 agar used for plating a protoplast suspension (step 6) can significantly influence the regeneration rate and overall transformation efficiency. Use plates poured within 24 to 48 h, whenever possible. Alternatively, plate transformed protoplasts in soft-agar overlays (DM3 with 0.7% Bacto Agar). To obtain efficient and reproducible protoplast transformation, great care should be taken with the preparation of all media and reagents (section 15.4.3). Under optimal conditions, transformants generated with small, multicopy plasmids can be expected to arise at a frequency of 10^6 to 10^7/μg of DNA.

15.1.2.2. Other gram-positive species

Protoplast transformation has been used successfully to introduce plasmid DNA into a wide variety of gram-positive bacteria. However, as noted above, protoplasts of *Bacillus* species and members of at least some other gram-positive species are unusually fragile, require long regeneration times, and exhibit quite variable levels of regeneration (21, 25, 29). Although careful attention to special biological features of particular species can lead to modifications of standard procedures that significantly improve the efficiency and/or reproducibility of protoplast transformation (40), electroporation has become the method of choice for introduction of plasmid DNA into many species (see below). On the other hand, protoplasts of certain other gram-positive species, notably members of the genus *Streptomyces*, are more easily handled and exhibit reproducibly high levels of regeneration (6, 7, 31). When one is confronted with the need to develop plasmid transformation in a gram-positive species for which good established procedures are not available, both protoplast transformation and electroporation should be evaluated.

15.1.3. Electroporation of *B. thuringiensis*

Electroporation or "electrotransformation" refers to the introduction of plasmid DNA mediated by a pulse of high-voltage electric shock and is a method applicable to both procaryotic and eucaryotic systems (32). Although extremely high efficiencies have been reported for some bacteria (>$10^{10}/\mu g$) (10), electrotransformation of *B. subtilis* and other gram-positive species has been generally much less impressive, apparently because of low or variable viability of electroshocked cells (38). The principal advantages of electroporation over protoplast transformation for most gram-positive species are in the simplicity of the manipulations and the elimination of the regeneration time (2 to 6 days) required for protoplasts. It should also be noted that some gram-positive species are not sensitive to lysozyme or other commercially available cell wall hydrolases, which prohibits the use of protoplast transformation. The method given below was developed by Schurter et al. (34) for electroporation of *B. thuringiensis*.

1. Inoculate 10 ml of LB in a 100-ml flask with a single colony. Incubate overnight at 27°C with orbital shaking at 200 rpm.
2. Add 0.4 ml to 40 ml of LB in a 250-ml flask, and incubate at 30°C with orbital shaking at 250 rpm until the culture has reached an OD_{600} of 0.2.
3. Pellet cells by centrifugation (8,000 × *g* for 10 min at 4°C), and resuspend in 1 ml of ice-cold electroporation buffer (400 mM sucrose, 1 mM $MgCl_2$, 7 mM potassium phosphate [pH 6.0]).
4. Repeat centrifugation, and resuspend again in the ice-cold electroporation buffer (above).
5. Transfer 800 μl of cell suspension to a Bio-Rad Genepulser cuvette chilled to 4°C.
6. Add plasmid DNA, and maintain at 4°C for 10 min.
7. In a Bio-Rad Genepulser device, subject to discharge at 1.3 kV and 25 μF.
8. Within 10 min, dilute into 1.2 ml of LB and incubate for 2 h at 30°C in a roller drum or reciprocal shaker.
9. Dilute, and plate on LB agar containing selective levels of appropriate antibiotics.

To adapt this method for use with other types of electroporation devices or with other species of gram-positive bacteria, set the capacitance at the level recommended by the manufacturer for electrotransformation of *E. coli* and determine the optimal field strength empirically for the organism at hand. Other parameters that might be varied for use with different strains or species of bacteria include sucrose concentration and pH of electroporation buffer. The growth phase of the culture at the time of harvesting for electroporation is typically an extremely critical determinant of overall transformation efficiency. Thus growth parameters, including temperature, growth medium, aeration, and OD_{600} at the time of harvest might be varied as well. It should be noted that when gene transfer techniques such as electroporation are adapted to relatively uncharacterized species, restriction of foreign DNA can be a significant barrier. Even in the absence of restriction, however, electrotransformation of *Bacillus* strains is typically relatively inefficient (<$10^4/\mu g$).

15.2. TRANSDUCTION

Many virulent or temperate phages of *B. subtilis* are known, and several are useful for either generalized or specialized transduction. As noted above, Hfr-like conjugation-mediated DNA transfer has not been observed in the gram-positive bacteria, and in *B. subtilis* transformation of naturally competent bacteria is a highly effective tool for fine-structure genetic mapping. Thus, generalized transduction is used mainly in this species as a tool for coarse-level positioning of markers on the genetic map, and the phage of choice for this purpose is an extremely large (>200-kb) pseudolysogenic phage, PBS1. This phage is referred to as pseudolysogenic because it makes very turbid plaques on a *B. subtilis* host yet does not enter a true lysogenic state. Among the most useful tools for specialized transduction in *B. subtilis* is the large (ca. 130-kb) temperate phage SPβ. Basic techniques for preparing lysates of these phages and examples of standard genetic applications for which these phages are well suited are given below.

15.2.1. Generalized Transduction in *B. subtilis* with Phage PBS1

The bacteriophage PBS1 infects cells of *B. subtilis* by adsorption to flagella; therefore, successful gene transfer requires the preparation of highly motile cells (27, 33). The procedure for PBS1 transductions is first to prepare a PBS1 lysate from the donor strain, and next to use this donor lysate to transduce a motile recipient strain. Problems in PBS1 transductions nearly always result from either poor transducing lysates or the failure to prepare highly motile cells. Preparing a good donor lysate itself requires infection of the donor strain with a PBS1 lysate capable of producing a high titer of transducing particles. Such a lysate is referred to for simplicity as the producer lysate. The producer lysate may be any PBS1 lysate; that is, the genotype of the bacterial strain used to raise the lysate is of no importance, as long as it contains an adequate titer of PBS1 phage. It should be noted, however, that the titer of PFU in a PBS1 lysate has no consistent correlation with the titer of transducing particles or the potential to generate a high titer of transducing particles. Moreover, the potential for a lysate to generate a high titer of transducing particles on subsequent infection decays with storage of a lysate. Because factors responsible for maintaining a high transducing potential are poorly understood, it is worth passaging a producer lysate every 6 months or so and monitoring its capacity for generalized transduction.

15.2.1.1. Preparation of phage donor lysate

1. Inoculate 10 ml of brain heart infusion broth (BHIB) (in a 250-ml flask) with a single, very fresh colony of the appropriate strain. Incubate overnight at 30°C with slow orbital shaking.
2. The following morning, check the culture by microscopic examination for cell motility. If it is motile, proceed to step 3; if not, remove from the shaker and leave on the bench at ambient temperature for 15 min and check again for motility.
3. Inoculate a test tube containing 10 ml of BHIB with 0.1 ml of motile culture. Incubate at 37°C on a roller drum shaker.
4. After 5 min inoculate the tube with 0.1 ml of a PBS1 producer lysate (see above), and return the tube to the shaker.
5. After a total of 8 to 9 h of incubation, remove the tube from the shaker and leave on the bench at ambient temperature overnight.

6. In the morning, decant the contents of the tube into a centrifuge tube and spin hard to pellet any unlysed cells (8,000 × *g* for 15 min at 25°C).
7. Carefully decant the lysate supernatant into a sterile glass tube. Sterilize the lysate by either filter sterilization or chloroform treatment (see below). Store at 4°C.

The most common problem with this procedure is in not getting highly motile cells. There is a tremendous variation among strains, and some strains will never become motile. Normally, up to 80% of the population will be seen to be swimming vigorously. Leaving the flask on the bench causes anaerobic conditions and induces motility rapidly, which may help (if in doubt, it's best to go ahead regardless!).

After overnight incubation at room temperature, PBS1-containing cells will lyse and release phage particles. It is possible to get completely clear lysates the following day, although more often there will be some degree of turbidity due to regrowth or the presence of PBS1-resistant cells. It is unlikely that there will be no phage present. Ideally, the centrifuged lysate should be sterilized by adding 0.5 ml of chloroform, vortexing, and leaving for at least 24 h at 4°C before checking the sterility. If in a hurry, the lysate can be filter sterilized, although this will reduce the active titer as a result of phage adsorbing to the nitrocellulose membrane.

The origins and properties of the producer lysate often cause confusion. This reflects the fact that factors controlling the abundance of transducing particles in a PBS1 lysate are poorly understood (section 15.2.1). The following guidelines are recommended. Obtain a source of PBS1 phage as a lysate (only 0.1 ml is necessary); it is not necessary to worry about its titer, just to ensure that it is sterile. Use this lysate to infect a wild-type strain of *B. subtilis* and generate a PBS1-liquid lysate as described above. Repeat this procedure with the new PBS1 lysate. Two rounds of infection and lysis should provide a PBS1 lysate of adequate titer ($>10^7$/ml). (The titer of PBS1 can be determined by using *B. pumilus* 8A1 as an indicator strain, which can be obtained from the *Bacillus* Genetic Stock Center; see section 15.4.1.) Alternatively, the titer of the producer lysate can be enriched by successive rounds of infection and lysis. Normally, the phage titer will increase after each round of infection and lysis. A PBS1 producer lysate stored at 4°C is normally viable for many years, although its transducing titer will drop. To minimize the problems encountered with using PBS1, it is best to regenerate the producer lysate about every 6 months and for mapping experiments to use the PBS1-donor lysate within a month of preparation.

15.2.1.2. Generalized transduction procedure

1. Inoculate 10 ml of BHIB (in a 250-ml flask) with a very fresh colony of the recipient strain. Incubate overnight at 30°C with slow orbital shaking.
2. The following morning, check cell motility. If it is motile, proceed to step 3; if not, remove from the shaker and leave on the bench for 15 min before checking again for motility.
3. Use 2 ml of motile culture to inoculate 10 ml of warm BHIB (in a 250-ml flask), and incubate with slow orbital shaking at 37°C.
4. After 3 to 4 h, check cell motility. If it is motile, proceed to step 5; if not, remove from the shaker and leave static on the bench for a further 15 min.
5. Add 1 ml each of motile culture to two capped centrifuge tubes. To one tube add 0.1 ml of PBS1-transducing lysate. The other tube will serve as the uninfected control.
6. Incubate both tubes at 37°C on a roller drum shaker.
7. After 30 min, centrifuge both tubes to pellet cells (8,000 × *g* for 10 min at 25°C). Wash once with 10 ml of T base.
8. Resuspend cell pellets in 1 ml of T base, and plate 0.1-ml aliquots on selective media.

The same selection procedures are followed as described in section 15.1.1.4 for transformation selections. Normally the yield of transductants is much lower than with transformation crosses. Indeed, for most crosses it is worth plating 0.1-ml aliquots of the original transduction mixture directly on a plate. On a good day 200 to 500 transductants can be expected, although less than 100 is the norm! If drug resistance is to be selected for, it is not necessary to wash the cells with T base.

15.2.1.3. Gene mapping

The primary use of PBS1 is in the initial and coarse-level mapping of new mutations. PBS1 is ideal because in a single cross up to 10% of the *B. subtilis* chromosome can be transferred, which is approximately 10 times that transferred in a transformation cross. Therefore, mutations are first mapped on the chromosome by PBS1 transduction. After an approximate position has been determined by two-factor genetic crosses with auxotrophic reference markers, DNA-mediated transformation is used to position the mutation more accurately. Traditionally, mapping with PBS1 is done with a kit of mapping strains. The mapping kit consists of nine isogenic strains, each containing the *trpC2* marker and one or more auxotrophic reference markers. These strains are shown in Table 3 and were constructed by Dedonder et al. at the University of Paris (15); they can be obtained from the *Bacillus* Genetic Stock Center (see section 15.4.1). Together, these strains contain 22 genetic markers spaced at relatively even intervals around the chromosome. To map an unknown mutation, a PBS1 transducing lysate is made from the mutant strain (which should also contain the wild-type allele of most or all of the auxotrophic reference markers carried by the kit strains) and used to infect each kit strain; this is followed by selection for Aux^+ (loss of particular auxotrophic requirements) on selective media and screening for the mutant phenotype. In principle, at least one reference marker should be linked to the mutation, allowing an approximate chromosomal position of the mutation to be determined.

An alternative technique is to use the extensive collection of "silent" chromosomal insertions of transposon Tn*917* (37). These strains are obtainable from the *Bacillus* Genetic Stock Center and are easier to use than the mapping-kit strains, since a transducing lysate is made from the Tn*917* insertion strain and used to infect the mutant strain, and this is followed by selection for MLS^r. Drug-resistant transductants are then screened for correction of the mutant phenotype. In this way,

Table 3. *B. subtilis* mapping kit

Strain	BGSC designation[a]	Genotype[b]
QB944	1A3	*cysA14* (11) *purA26* (348) *trpC2*
QB928	1A4	*aro1906* (26) *dal-1* (38) *purB33* (55) *trpC2*
QB934	1A5	*glyB133* (74) *metC3* (115) *tre-12* (62) *trpC2*
QB943	1A6	*ilvA1* (194) *pyrD1* (139) *thyA1* (168) *thyB1* (195) *trpC2*
QB922	1A7	*gltA292* (180) *trpC2*
QB935	1A8	*aroD120* (226) *lys-1* (210) *trpC2*
QB936	1A9	*ald-1* (280) *aroG932* (264) *leuA8* (247) *trpC2*
QB917	1A10	*hisA1* (298) *thr-5* (284) *trpC2*
QB123	1A11	*ctrA1* (324) *sacA321* (330) *trpC2*

[a]BGSC, *Bacillus* Genetic Stock Center.
[b]Numbers in parentheses are chromosomal map positions.

lysates of each Tn*917* strain can be made and used repeatedly for mapping different mutations, with the need for only one selection procedure.

15.2.2. Specialized Transduction in *B. subtilis* with Phage SPβ

SPβ is a large temperate phage similar in many of its general characteristics to phage lambda of *E. coli* (46). Therefore, it is capable of generating defective or nondefective specialized transducing derivatives by the same mechanisms as lambda (45). The main difference between the two is size: SPβ is roughly three times larger than lambda. The large size of SPβ creates special opportunities in the context of its use as a cloning vehicle, as discussed below. SPβ has also proved highly useful as a portable platform for the expression of regulated gene fusions (i.e., a mobile form of *lac* fusion which can easily be transferred to different genetic backgrounds by specialized transduction). Given below are methods for preparing lysates and determining the titer of the phage and a description of basic techniques in which SPβ is widely used.

15.2.2.1. Preparation of phage lysate

As with most temperate phages of *E. coli* and *B. subtilis*, SPβ prophages can be induced to enter the lytic cycle by exposure of a lysogen to mitomycin C or other DNA-damaging agents. A more convenient alternative, however, is to take advantage of the *c2* mutation, which renders the phage repressor heat labile. Lysogens of *c2* mutants are quite stable at 37°C but are efficiently induced at 50 to 52°C. The phage itself is unable to replicate efficiently at such high temperatures, however. Thus, strategies for preparing a heat-induced lysate involve a brief exposure of lysogens to 50 to 52°C, followed by rapid cooling to 37°C and further incubation until lysis has occurred, as described below.

1. Inoculate 2 ml of Penassay broth (PAB) or LB medium in a 15-mm glass test tube with a single fresh colony of the SPβ *c2* lysogen.
2. Incubate at 37°C with aeration by a roller drum or reciprocal shaker until light turbidity is apparent (OD_{600}, ca. 0.2 to 0.4). If the culture is highly turbid at the time of induction, lysis will be inefficient.
3. Remove the tube, and transfer it to a shaking or stationary water bath at 50 to 52°C. Incubate for 6 min. Longer incubation will significantly reduce the phage titer.
4. Immediately return the tube to 37°C, and continue incubation with aeration for an additional 80 min. Clearing of the culture as a result of lysis should be obvious by visual inspection.
5. Centrifuge the culture (8,000 × *g* for 10 min at 25°C), and filter sterilize the supernatant. Store the lysate at 4°C.

Because of the need for a brief exposure to high temperature and rapid return to 37°C, scaling up heat induction can be tricky. Up to 40 ml can readily be handled in a rotary shaking water bath. For larger-scale heat induction (e.g., for DNA isolation), the following method is recommended.

1. Inoculate 1,000 ml of LB broth at 37°C in a 4-liter Erlenmeyer flask with a log phase culture of an SPβ *c2* lysogen, prepared as described above, to an initial OD_{600} of less than 0.05.
2. Incubate with shaking at 37°C until the culture reaches an OD_{600} of 0.3.
3. Heat rapidly to about 51°C by holding the flask directly over the flame of a Fisher burner (a high-temperature blast burner) while swirling. Monitor the temperature with a thermometer held in the open mouth of the flask (suspended by a wire, if necessary, to reach into the broth). *Note:* A sterile technique is unnecessary, since introduction of minor contaminants is of no consequence at this stage.
4. After 6 min at 51°C, cool rapidly to 37°C by immersion in a slurry of ice-water, being careful to keep the culture mixed by swirling at frequent intervals, monitoring the temperature continuously.
5. When the culture has reached 37°C, incubate with shaking for an additional 80 min, removing samples to check the OD_{600} for lysis.
6. Harvest the lysate by centrifugation, as described above.

When using most strains of SPβ, lysis of the culture should be quite obvious 80 min following heat induction. This may not be true for derivatives of the phage carrying large cloned inserts or situations in which defective phages are induced in the presence of helpers. If obvious lysis is not apparent, proceed with the harvesting anyway; the lysate may yield a reasonably good titer despite the lack of obvious clearing. In no case should the induced culture be allowed to incubate for more than 90 min following heat induction; released phage particles will absorb to unlysed bacteria, drastically reducing the titer.

15.2.2.2. Determination of phage titer

Most common laboratory strains of *B. subtilis* (the standard "168 derivatives" are not lysogenic for SPβ. Nevertheless (for unknown reasons), none of these common laboratory strains are suitable for SPβ plaque formation; even clear-plaque mutants of the phage make extremely small plaques on the common nonlysogens. To determine the titer of SPβ, use strain CU1050, obtainable from S. A. Zahler at Cornell University, or its equivalent, BGSC strain IA459, as plating bacteria.

15.2.2.3. Specialized transduction procedure

1. Inoculate 5 ml of PAB or LB medium with a single colony of the recipient strain, and incubate at 37°C in a roller drum or reciprocal shaker.
2. When cells have reached the mid-log phase of growth (OD_{600}, 0.5), remove two 1-ml aliquots and place them in two separate test tubes.
3. To one tube add 0.1 ml of SPβ lysate prepared as described above, and then incubate both tubes at 37°C for 30 to 60 min. The other tube will serve as the uninfected negative control.
4. Plate out 0.1-ml aliquots for selection.

Many SPβ derivatives in common use carry an insertion of Tn*917* into nonessential phage DNA (44). This makes it possible to select for lysogen formation (as in specialized transduction) simply by selecting for MLS^r. Under some circumstances, it may be desirable to select directly for acquisition of phage immunity. This can be done by plating transduction mixtures in LB soft agar (LB with 0.7% agar) with 0.1 ml of a lysate of a clear plaque-forming mutant (e.g., SPβ*c1*, which can be obtained from the *Bacillus* Genetic Stock Center; see section 15.4.1). Infected cells that become lysogens acquire immunity to phage infection and are therefore able to form a colony even in the presence of virulent phages. Colony-purified lysogens may be distinguished from nonlysogens either by testing for sensitivity to a clear-plaque mutant or by scoring for betacin resistance. Betacin is a bacteriocin produced by lysogens to which only nonlysogens are sensitive. Therefore, colonies to be tested for harboring an SPβ prophage can simply be picked into lawns of tester strains known to be lysogens or nonlysogens. Colonies or patches of growth made by strains harboring a prophage will develop a clear halo on the nonlysogen lawn as a result of the action of betacin.

15.2.2.4. Transduction of *lacZ* fusions

Because SPβ is a relatively large phage, it can tolerate up to 10 kb of extra DNA without exceeding its packaging capacity. To exploit this, methods have been developed to allow the integration of foreign DNA into the SPβ prophage by homologous recombination during the transformation of naturally competent cells (43). Among these is a method for integrating transcriptional *lacZ* fusions into the phage, as illustrated in Fig. 2. First, a promoter-bearing DNA fragment is subcloned into *E. coli* plasmid pTK*lac* (or some similar vector) in an orientation that places the expression of the promoterless *lacZ* fragment under control of the promoter to be studied. The resulting construction (which is incapable of autonomous replication in *B. subtilis*) is then used to transform competent cells of an SPβ lysogen containing a special SPβ prophage engineered to contain extensive sequence similarity to pTK*lac*, but without the *cat* gene present in pTK*lac*. Thus selection for Cm^r can be used to force recombinational transfer of the *lacZ* fusion into the prophage, as shown in Fig. 2. This approach has several important advantages over introducing fusions onto plasmids or into ordinary genomic DNA. First, unlike plasmid-borne fusions, fusions inserted into a prophage can be studied in single copy, a situation more likely to reproduce an authentic pattern of regulated expression. Second, unlike fusions inserted into ordinary genomic DNA, phage-borne fusions are readily cured from strains and transferred to other strains.

15.2.2.5. Construction of recombinant phages

Two different temperate phages have been developed for use as cloning vectors in *B. subtilis*: ϕ105, a phage approximately the same size as coliphage lambda, and SPβ, a much larger phage (discussed above). Both systems were devised to allow the construction of recombinant specialized transducing phages for the purpose of cloning *B. subtilis* genes by complementation (19). For the case of the SPβ system, the cloning strategy involves the insertion of recombinant sequences into prophage DNA by recombination during transformation of competent cells, an approach conceptually similar to that described in section 15.1.1.8. As illustrated in Fig. 3, plasmid vectors such as pCV2 are constructed in which cloning sites are associated with a plasmid replication origin that functions in *E. coli* (ColE1), a drug resistance gene selectable in *B. subtilis* (*cat*) and flanked by DNA homologous to a Tn*917* insertion in SPβ. Genomic restriction fragments ligated to pCV2 can thus be forced into an SPβ::Tn*917 c2* prophage by recombination, selecting for Cm^r (Fig. 3). To identify recombinants containing genes that complement specific mutations in *B. subtilis*, Cm^r transformants are pooled. A lysate is prepared by heat induction (section 15.2.2.1), used to infect a strain containing the mutation (section 15.2.2.3), and plated with selection or screening for complementation of the mutant allele (e.g., Aux$^+$). The configuration of sequences flanking the cloned DNA in specialized transducing phages (Fig. 3) allows the recovery of cloned DNA in *E. coli* by digesting a crude preparation of phage DNA with the restriction enzyme *Not*I, followed by ligation and transformation of an *E. coli* strain with a selection for Ap^r.

15.3. CONJUGATION

Conjugative plasmids are very common among gram-positive species and include several different varieties, some of which are strikingly different from the gram-negative paradigms. Particularly interesting are the conjugative plasmids of *Enterococcus faecalis*, which have coevolved with their bacterial hosts a complex system of **pheromone-mediated plasmid exchange** (18). In these systems, plasmid-free strains (potential recipients) excrete short oligopeptide pheromones (clump-inducing agents) to which plasmid-containing strains (potential donors) respond by synthesizing cell surface proteins (aggregation substance) which promote the

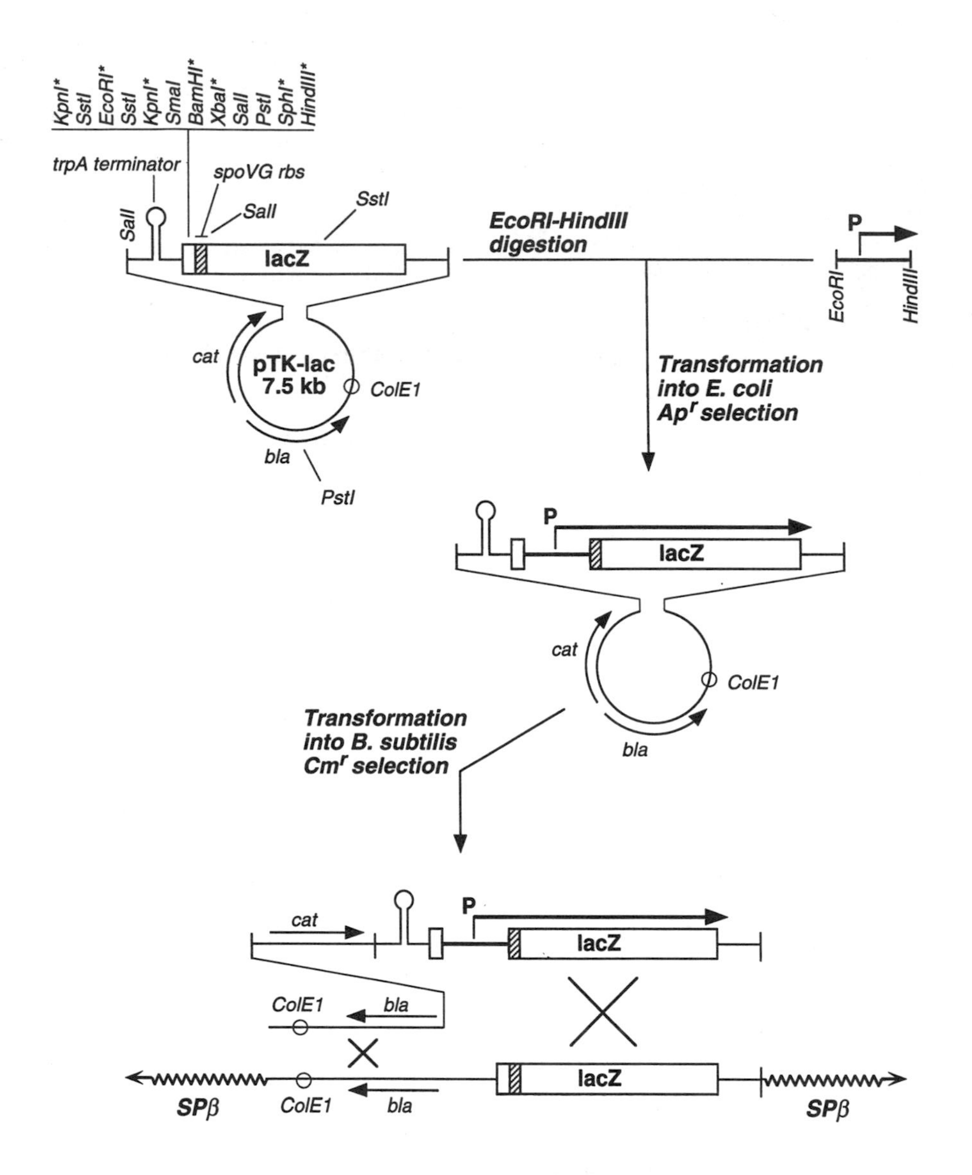

Fig. 2. Use of recombination during transformation of naturally competent cells to integrate a *lacZ* fusion into an SPβ prophage. Plasmid pTK-*lac* (28) is an *E. coli* replicon containing a promoterless copy of *lacZ* modified by replacing its natural ribosome-binding site with one expected to function more efficiently in *B. subtilis* (47). Upstream of the *lacZ* coding sequence is a multiple cloning site, containing several restriction sites unique to the plasmid (asterisks). Upstream of the cloning site is a *rho*-independent transcription terminator, originally from the *trpA* gene of *E. coli* (12). Also associated with this plasmid is a *cat* gene from pC194 (24), which confers resistance to chloramphenicol in *B. subtilis* when present in a single copy, and a *bla* gene from pBR322 (8), which confers resistance to ampicillin in *E. coli*. As illustrated, when a restriction fragment containing a promoter is inserted into the multiple cloning site of pTK-*lac* in the appropriate orientation, a transcriptional *lacZ* fusion results. When the promoter-containing plasmid is then used to transform a strain of *B. subtilis* containing a special derivative of SPβ, ZB307 (47), selecting for Cm^r, the *lacZ* fusion is transferred into the prophage by homologous recombination.

close physical association of donors and recipients. Subsequent transfer of the conjugative plasmid from donor to recipient can be extremely efficient (13). Less efficient but more promiscuous plasmid transfer systems have been identified in many other gram-positive genera, some of which are known to promote transfer of conjugative plasmids across genus barriers. As noted above, however, none of these conjugation systems are capable of Hfr-like transfer of chromosomal DNA and thus are not useful agents for genetic exchange of chromosomal markers. Only in the genus *Streptomyces* are conjugative plasmids useful for mapping chromosomal mutations (23).

Although conjugation is not generally an important means of mapping chromosomal genetic loci in gram-positive bacteria, it has become an extremely valuable alternative to protoplast transformation and electroporation for introducing plasmid DNA into gram-posi-

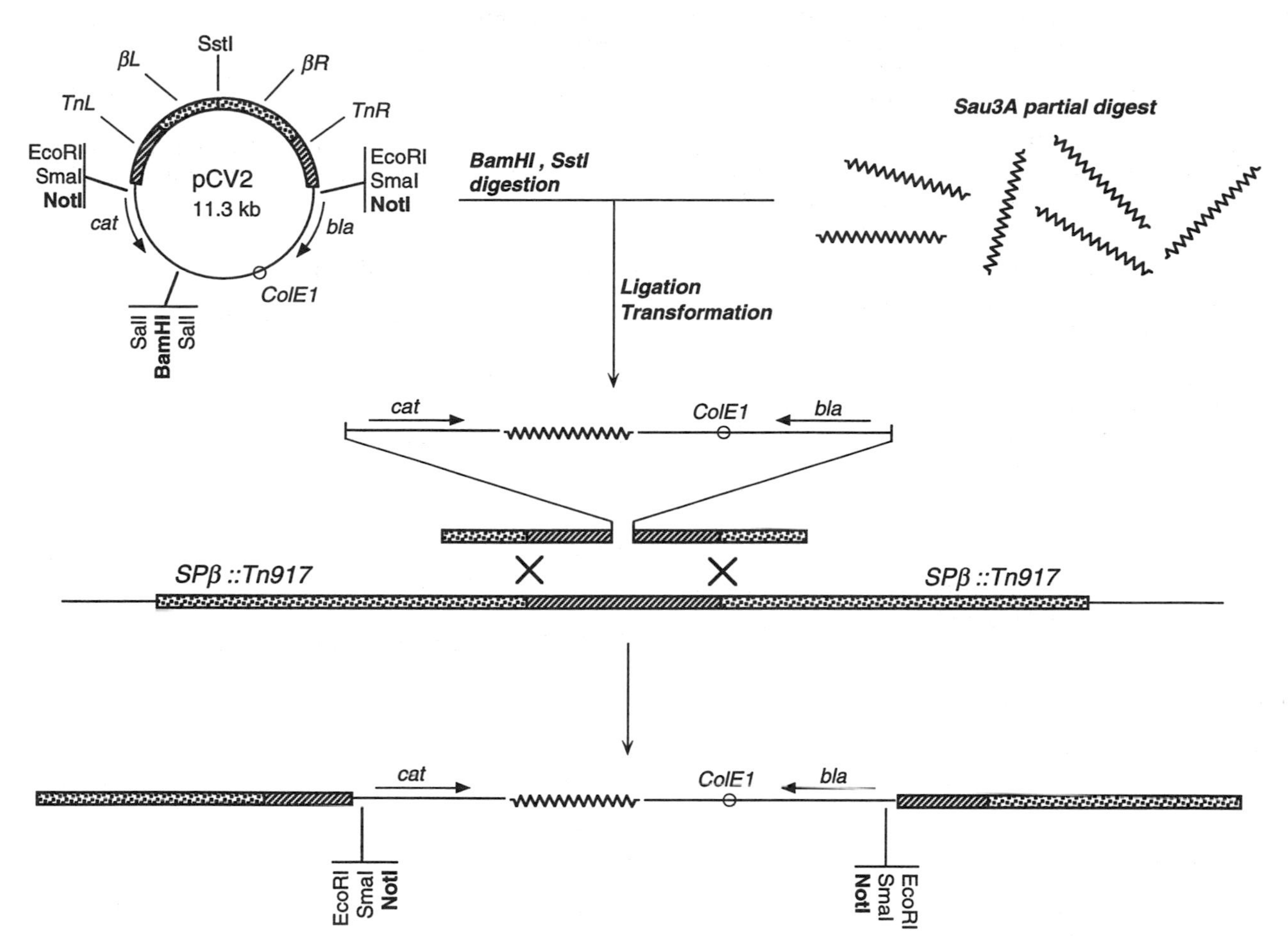

Fig. 3. Cloning DNA into the temperate phage SPβ by homologous recombination. Plasmid pCV2 (42) is an *E. coli* vector containing sequences homologous to both Tn*917* (hatched area) and SPβ (stippled area) in the vicinity of an SPβ :: Tn*917* insertion; TnL, DNA homologous to the "left" arm of Tn*917* (*erm* proximal); TnR, DNA homologous to the "right" arm (*erm* distal); βL, DNA homologous to SPβ sequences immediately adjacent to the "left" (*erm*-proximal) insertion junction of SPβ :: Tn*917*; βR, DNA homologous to the "right" (*erm*-distal) junction. Other features of pCV2 are as indicated for pTK-*lac* in the legend to Fig. 2. After digestion with both *Sst*I and *Bam*HI, vector fragments of pCV2 are ligated to DNA fragments to be cloned (here, a partial *Sau*3A digest of genomic DNA). When the ligation mixture is used to transform naturally competent cells of a strain containing an SPβ :: Tn*917* prophage, a selection for Cm^r forces the insert into the prophage by homologous recombination. Cloned DNA fragments can subsequently be rescued back into *E. coli* by digestion of phage DNA with *Not*I, circularization by ligation, and selection for Ap^r. This would not be applicable for the rare instance in which the cloned insert also contains a *Not*I site.

tive species that have proved refractory to those techniques. The basic strategy is that outlined by Trieu-Cuot et al. (36). This involves the use of shuttle plasmids containing a ColE1 origin of replication, replication functions derived from the gram-positive plasmid pAMβ1, and an origin-of-transfer sequence (*oriT*) from the gram-negative broad-host-range conjugative plasmid RK2. Such shuttle plasmids can be mobilized in "filter matings" by RK2 or other IncP-type plasmids or by other plasmid-borne or chromosomally inserted DNA fragments carrying IncP *tra* functions. In filter matings, donor and recipient strains are grown separately in broth cultures and then combined at high cell density on filter membranes which are incubated on solid media for several hours (usually overnight) to allow mating to occur. By using such procedures, *oriT*-containing shuttle vectors can be transferred with high efficiency to a wide variety of gram-positive species, including members of the genus *Clostridium*, which have generally proved refractory to more conventional plasmid introduction strategies (39). Thus the approach may represent an important breakthrough in the adaptation of recombinant DNA techniques for use with more exotic varieties of bacteria. The following method may be used to mobilize *oriT*-pAMβ1 shuttle vectors from *E. coli* into *B. subtilis*.

1. Inoculate 10 ml of BHIB with single fresh colonies of *E. coli* donor and *B. subtilis* recipient strains. Add a selective antibiotic to the donor culture appropriate for maintaining the particular shuttle vector used.
2. Incubate with good aeration at 37°C overnight.
3. The following morning, mix 2 ml of donor culture with 0.2 ml of recipient culture and harvest the bacteria on a 2.5-cm membrane filter (0.2-μm porosity).

4. Place the filter, bacteria side up, on BHIB agar, and incubate at 37°C for between 6 and 24 h.
5. Using sterile forceps, transfer the filter to a large test tube and suspend the bacteria in 1 ml of BHIB by vortexing.
6. Plate out suitably diluted samples on appropriate selective media to enable the number of transconjugant, donor, and recipient bacteria to be determined. For matings between *E. coli* donors and *B. subtilis* recipient strains, it is convenient to select transconjugants on BHIB containing the appropriate antibiotic at 48°C (*E. coli* will not grow at this temperature). A viable count of recipient bacteria can be obtained simply by plating the mating mixture on BHIB without antibiotic at 48°C; donor bacteria (which should be in significant excess) can be counted by plating on BHIB at 37°C. Although both *E. coli* and *B. subtilis* will grow at 37°C, the two are readily distinguished by colony morphology. Alternatively, MacConkey agar (Difco) can be used to exclude the growth of most gram-positive species.

Alternative media such as LB can be used in place of BHIB. In step 2, overnight growth is not required; transfer may also be effected by using exponentially growing cells. In step 3, the ratio of donor to recipient bacteria should be at least 10:1 to optimize the probability that a given recipient cell will interact with a donor cell; increasing the ratio even further will increase this probability slightly but will reduce the total number of exconjugants. For some recipients it may be worthwhile to resuspend donor bacteria in broth lacking antibiotic before mixing. In step 4, the composition of the filter itself is unimportant; nitrocellulose, cellulose acetate, or mixed cellulose ester membrane filters (pore size, 0.45 or 0.2 μm) from a variety of commercial sources can be used. In step 6, BHIB can be used as diluent; other procedures based on appropriate auxotrophic or resistance markers may be used for counterselecting donor or recipient bacteria.

15.4. STRAINS AND MEDIA

15.4.1. Strains

Bacillus strains can be obtained from the *Bacillus* Genetic Stock Center, The Ohio State University, Department of Biochemistry, 484 West 12th Avenue, Columbus, Ohio 43210-1292 (telephone: 614-292-5550; fax: 614-292-1538).

15.4.2. Media

15.4.2.1. Brain heart infusion broth (BHIB)

Prepare brain heart infusion broth (BHIB) (Difco) as recommended by the manufacturer but containing 0.5% Bacto Yeast Extract. Sterilize by autoclaving (15 lb/in^2 for 30 min).

15.4.2.2. Luria-Bertani (LB) medium/agar

General purpose growth medium containing, per liter

Bacto-tryptone .... 10.0 g
Bacto-yeast extract .... 5.0 g
NaCl.. 10.0 g
1M NaOH .... 1.0 ml

Autoclave (15 lb/in^2 for 30 min). For LB agar, solidify with 1.2% (wt/vol) agar; for LB soft overlay, use 0.7% agar.

15.4.2.3. Penassay broth (PAB)

PAB is antibiotic medium no. 3 (Difco). Prepare as recommended by the manufacturer. Sterilize by autoclaving (15 lb/in^2 for 30 min).

15.4.2.4. Tryptose blood agar base (TBAB)

TBAB is Tryptose blood agar base (Difco). Note that this medium already contains agar. Prepare as recommended by the manufacturer. Sterilize by autoclaving (15 lb/in^2 for 30 min).

15.4.2.5. Lactate glutamate minimal agar (LGMA)

$MgSO_4 \cdot 7H_2O$.......................... 9.9 g
Sporulation salts (see below).......... 900 ml
Agar (Lab M)* .... 15 g
*Lab M agar (Davis) is recommended but not essential.

Autoclave (15 lb/in^2 for 30 min), cool to 50 to 55°C, and add the following sterile solutions.

5% (wt/vol) glutamate* .... 40 ml
35% (vol/vol) sodium lactate......... 8 ml
0.1 M $CaCl_2$ .... 5 ml
L-Alanine (20 mg/ml) .... 5 ml
Growth requirements .... see Table 1
*Adjust pH of glutamate solution to 7.0 with NaOH.

15.4.2.6. Sporulation salts

Sporulation salts is made from two stock solutions.

Distilled H_2O .... 989 ml
Solution A .... 1 ml
Solution B .... 10 ml
Aliquot and autoclave (15 lb/in^2 for 30 min).

Solution A
$FeCl_3 \cdot 6H_2O$........................ 0.089 g
$MgCl_2 \cdot 6H_2O$ .... 0.830 g
$MnCl_2 \cdot 4H_2O$ .... 1.979 g
Distilled H_2O........................ 100 ml
Do not autoclave. Store at 4°C.

Solution B
NH_4Cl .... 53.5 g
Na_2SO_4 .... 10.6 g
KH_2PO_4................................. 6.8 g
NH_4NO_3 .... 9.7 g
Dissolve in 800 ml of distilled H_2O, adjust pH to 7.0 with 2 M NaOH, and bring the volume to 1 liter. The solution can be autoclaved. Store at 4°C.

15.4.2.7. Tris Spizizen salts agar (TSSA)

10× TSS salts (see below)	100 ml
Distilled H_2O	870 ml
Agar	15 g

Autoclave (15 lb/in² for 30 min), cool to 50 to 55°C, and add the following sterile solutions.

50% (wt/vol) glucose	10 ml
0.1M $MgSO_4 \cdot 7H_2O$	10 ml
0.4% $FeCl_3$–0.4% trisodium citrate·$2H_2O$	10 ml
Growth requirements	see Table 1

10× TSS salts

NH_4Cl	20 g
$K_2HPO_4 \cdot 3H_2O$	3.5 g
Tris base	60 g

Dissolve in distilled H_2O, adjust pH to 7.4 with HCl, and bring to 1 liter. Autoclave (15 lb/in² for 30 min).

15.4.2.8. T base

$(NH_4)_2SO_4$	2 g
$K_2HPO_4 \cdot 3H_2O$	18.3 g
KH_2PO_4	6 g
Trisodium citrate·$2H_2O$	1 g

Autoclave (15 lb/in² for 30 min). This solution can be made as a 10× stock.

15.4.2.9. SpC medium

Make fresh on day of use from the following sterile solutions.

T base (see above)	20 ml
50% (wt/vol) glucose	0.2 ml
1.2% (wt/vol) $MgSO_4 \cdot 7H_2O$	0.3 ml
10% (wt/vol) Bacto yeast extract	0.4 ml
1% (wt/vol) Bacto Casamino Acids	0.5 ml
Growth requirements	see Table 1

15.4.2.10. SpII medium

Make fresh on day of use from the following sterile solutions.

T base (section 15.4.2.8)	200 ml
50% (wt/vol) glucose	2 ml
1.2% (wt/vol) $MgSO_4 \cdot 7H_2O$	14 ml
10% (wt/vol) Bacto Yeast Extract	2 ml
1% (wt/vol) Bacto Casamino Acids	2 ml
0.1 M $CaCl_2$	1 ml
Growth requirements	see Table 1

15.4.2.11. SpII + EGTA medium

SpII + EGTA is SpII (200 ml) without $CaCl_2$ but containing 4 ml of 0.1 M ethylene glycol-bis(β-aminoethyl ether)-*N,N,N′,N′*-tetraacetic acid (EGTA; pH 8.0). SpII + EGTA medium can be prepared and frozen in aliquots at −20°C.

15.4.3. Media and Reagents for Protoplast Preparation and Transformation

2× SMM. Sucrose (342 g), sodium maleate (4.72 g), $MgCl_2 \cdot 6H_2O$ (8.12g); add enough distilled H_2O to dissolve; adjust pH to 6.5 with concentrated NaOH; adjust volume to 1 liter with distilled H_2O; autoclave 50-ml aliquots (5 lb/in² for 10 min).

4× PAB. 7 g of Difco Antibiotic Medium no. 3 per 100 ml; autoclave (15 lb/in² for 30 min).

SMMP. Mix equal volumes of 2× SMM and 4× PAB.

PEG. 10 g of polyethylene glycol 8000 (Sigma) in 25 ml of 1× SMM; autoclave (5 lb/in² for 10 min).

15.4.4. DM3 Medium for Protoplast Plating and Regeneration

DM3 regeneration medium should be made up fresh on the day of use from the following sterile solutions, each of which should be maintained at 60°C in a water bath.

1 M sodium succinate (adjusted to pH 7.3 with concentrated NaOH)	250 ml
5% (wt/vol) Bacto Casamino Acids	50 ml
10% (wt/vol) Bacto Yeast Extract	25 ml
Phosphate buffer (3.5 g of K_2HPO_4 and 1.5 g of KH_2PO_4 per 100 ml)	50 ml
20% (wt/vol) glucose	15 ml
1 M $MgCl_2$	10 ml
2% (wt/vol) bovine serum albumin (fraction V [Sigma]; fractionation method not critical); filter sterilize	2.5 ml
4% (wt/vol) molten agar	100 ml
Antibiotics if required	see Table 2

15.5. ACKNOWLEDGMENTS

Preparation of this chapter was supported by Public Health Services grant GM35495 (P.Y.) from the National Institutes of Health.

15.6. REFERENCES

15.6.1. General References

1. **Dean, D. H., and D. R. Ziegler (ed.).** 1990. *Bacillus Genetic Stock Center Catalog of Strains*, 4th ed. The Ohio State University, Columbus.
2. **Harwood, C. R., and S. M. Cutting (ed.).** 1990. *Molecular Biological Methods for Bacillus.* John Wiley & Sons, Ltd., Chichester, England.
3. **Sonenshein, A. L., J. A. Hoch, and R. Losick (ed.).** 1993. *Bacillus subtilis and Other Gram-Positive Bacteria: Biochemistry, Physiology, and Molecular Genetics.* American Society for Microbiology, Washington, D.C.

15.6.2. Specific References

4. **Albano, M., J. Hahn, and D. Dubnau.** 1987. Expression of competence genes in *Bacillus subtilis. J. Bacteriol.* **169:**3110–3117.
5. **Anagnostopoulos, C., and J. Spizizen.** 1961. Requirements for transformation in *Bacillus subtilis. J. Bacteriol.* **81:**741–746.
6. **Baltz, R. H., and P. Matsushima.** 1983. Advances in protoplast fusion and transformation in *Streptomyces*, p. 143–148. *In* I. Potrykus (ed.), *Protoplasts 1983.* Birkhäser Verlag, Basel.

7. **Bibb, M. J., J. M. Ward, and D. A. Hopwood.** 1978. Transformation of plasmid DNA into *Streptomyces* protoplasts at high frequency. *Nature* (London) **274:**398–400.
8. **Bolivar, F., R. L. Rodriguez, M. C. Betlach, and H. W. Boyer.** 1977. Construction and characterization of new cloning vehicles. I. Ampicillin-resistant derivatives of the plasmid pMB9. *Gene* **2:**75–93.
9. **Bron, S., P. Bosma, M. van Belkum, and E. Luxen.** 1987. Stability function in the *Bacillus subtilis* plasmid pTA1060. *Plasmid* **18:**8–15.
10. **Calvin, N. M., and P. C. Hanawalt.** 1988. High-efficiency transformation of bacterial cells by electroporation. *J. Bacteriol.* **170:**2796–2801.
11. **Chang, S., and S. N. Cohen.** 1979. High frequency transformation of *Bacillus subtilis* protoplasts by plasmid DNA. *Mol. Gen. Genet.* **168:**111–115.
12. **Christie, G. E., P. J. Farnham, and T. Platt.** 1981. Synthetic sites for transcription termination and a functional comparison with tryptophan operator terminator sites in vitro. *Proc. Natl. Acad. Sci. USA* **78:**4180–4184.
13. **Clewell, D. B.** 1990. Movable genetic elements and antibiotic resistance in enterococci. *Eur. J. Clin. Microbiol. Infect. Dis.* **9:**90–102.
14. **Cutting, S. M., and P. B. Vander Horn.** 1990. Genetic analysis, p. 27–74. *In* C. R. Harwood and S. M. Cutting (ed.), *Molecular Genetic Methods for Bacillus.* John Wiley & Sons, Ltd., Chichester, England.
15. **Dedonder, R. A., J. A. Lepesant, J. Lepesant-Kejzlarova, A. Billault, M. Steinmetz, and F. Kunst.** 1977. Construction of a kit of reference strains for rapid genetic mapping in *Bacillus subtilis. Appl. Environ. Microbiol.* **33:**989–993.
16. **Dubnau, D.** 1991. Genetic competence in *Bacillus subtilis. Microbiol. Rev.* **55:**395–424.
17. **Dubnau, D., and R. Davidoff-Abelson.** 1972. Fate of transforming DNA following uptake by competent *Bacillus subtilis.* I. Formation and properties of the donor-recipient complex. *J. Mol. Biol.* **56:**209–221.
18. **Dunny, G. M., B. Brown, and D. B. Clewell.** 1978. Induced cell aggregation and mating in *Streptococcus faecalis:* evidence for a bacterial sex pheromone. *Proc. Natl. Acad. Sci. USA* **75:**3479–3483.
19. **Errington, J.** 1990. Gene cloning techniques, p. 175–220. *In* C. R. Harwood and S. M. Cutting (ed.), *Molecular Biological Methods for Bacillus.* John Wiley & Sons, Ltd., Chichester, England.
20. **Gryczan, T., S. Contente, and D. Dubnau.** 1980. Molecular cloning of heterologous chromosomal DNA by recombination between a plasmid vector and a homologous resident plasmid in *Bacillus subtilis. Mol. Gen. Genet.* **177:**459–467.
21. **Haima, P., S. Bron, and G. Venema.** 1988. A quantitative analysis of shotgun cloning in *Bacillus subtilis* protoplasts. *Mol. Gen. Genet.* **213:**364–369.
22. **Haima, P., D. van Sinderen, S. Bron, and G. Venema.** 1990. An improved beta-galactosidase alpha-complementation system for molecular cloning in Bacillus subtilis. *Gene* **93:**41–47.
23. **Hopwood, D. A.** 1972. Genetic analysis in microorganisms, p. 29–158. *In* J. R. Norris and D. W. Ribbons (ed.), *Methods in Microbiology*, vol. 7B. Academic Press Ltd., London.
24. **Iordanescu, S.** 1976. Three distinct plasmids originating in the same *Staphylococcus aureus* strain. *Arch. Roum. Pathol. Exp. Microbiol.* **35:**111–118.
25. **Jensen, K. K., and M. F. Hulett.** 1989. Protoplast transformation of *Bacillus licheniformis* MC14. *J. Gen. Microbiol.* **135:**2283–2287.
26. **Juni, E.** 1972. Interspecies transformation of *Acinetobacter*: genetic evidence for a ubiquitous genus. *J. Bacteriol.* **112:**917–931.
27. **Karamata, D., and J. D. Gross.** 1970. Isolation and genetic analysis of temperature-sensitive mutants of *Bacillus subtilis* defective in DNA synthesis. *Mol. Gen. Genet.* **108:**277–287.
28. **Kenney, T. J., and C. P. Moran, Jr.** 1991. Genetic evidence for interaction of σ^A with two promoters in *Bacillus subtilis. J. Bacteriol.* **173:**3282–3290.
29. **Mann, S. P., C. G. Orpin, and G. P. Hazlewood.** 1986. Transformation of *Bacillus* spp.: an examination of the transformation of *Bacillus* protoplasts by the plasmids pUB110 and pHV33. *Curr. Microbiol.* **13:**191–195.
30. **Messing, J.** 1983. New M13 vectors for cloning. *Methods Enzymol.* **101:**20–78.
31. **Okanishi, M., K. Suzuki, and H. Umezawa.** 1974. Formation and reversion of streptomycete protoplasts: cultural conditions and morphological study. *J. Gen. Microbiol.* **80:**389–400.
32. **Potter, H.** 1988. Electroporation in biology: methods, applications and instrumentation. *Anal. Biochem.* **174:**361–373.
33. **Raimondo, L. M., N. Lundh, and R. J. Martinez.** 1968. Primary adsorption site of phage PBS1: the flagellum of *Bacillus subtilis. J. Virol.* **2:**256–264.
34. **Schurter, W., M. Geiser, and D. Mathé.** 1989. Efficient transformation of *Bacillus thuringiensis* and *B. cereus* via electroporation: transformation of acrystalliferous strains with a cloned delta-endotoxin gene. *Mol. Gen. Genet.* **218:**177–181.
35. **Tomich, P. K., F. Y. An, and D. B. Clewell.** 1980. Properties of erythromycin-inducible transposon Tn*917* in *Streptococcus faecalis. J. Bacteriol.* **141:**1366–1374.
36. **Trieu-Cuot, P., C. Carlier, P. Martin, and P. Courvalin.** 1987. Plasmid transfer from *Escherichia coli* to Gram-positive bacteria. *FEMS Microbiol. Lett.* **48:**289–294.
37. **Vandeyar, M. A., and S. A. Zahler.** 1986. Chromosomal insertions of Tn*917* in *Bacillus subtilis. J. Bacteriol.* **167:**530–534.
38. **Vehmaanperä, J.** 1989. Transformation of *Bacillus amyloliquifaciens* by electroporation. *FEMS Microbiol. Lett.* **61:**165–170.
39. **Williams, D. R., D. I. Young, and M. Young.** 1990. Conjugative plasmid transfer from *Escherichia coli* to *Clostridium acetobutylicum. J. Gen. Microbiol.* **136:**819–826.
40. **Wu, L., and N. E. Welker.** 1989. Protoplast transformation of *Bacillus stearothermophilus* NUB36 by plasmid DNA. *J. Gen. Microbiol.* **135:**1315–1324.
41. **Young, F. E., and J. Spizizen.** 1961. Physiological and genetic factors affecting transformation of *Bacillus subtilis. J. Bacteriol.* **81:**823–829.
42. **Youngman, P.** 1987. Plasmid vectors for recovering and exploiting Tn*917* transpositions in *Bacillus* and other gram-positives, p. 79–103. *In* K. Hardy (ed.), *Plasmids: A Practical Approach.* IRL Press, Oxford.
43. **Youngman, P.** 1990. Use of transposons and integrational vectors for mutagenesis and construction of gene fusions in *Bacillus* species, p. 221–266. *In* C. R. Harwood and S. M. Cutting (ed.), *Molecular Biological Methods for Bacillus.* John Wiley & Sons Ltd., Chichester, England.
44. **Youngman, P. J., J. B. Perkins, and R. Losick.** 1983. Genetic transposition and insertional mutagenesis in *Bacillus subtilis* with *Streptococcus faecalis* transposon Tn*917. Proc. Natl. Acad. Sci. USA* **80:**2305–2309.
45. **Zahler, S. A.** 1982. Specialized transduction in *Bacillus subtilis*, p. 269–305. *In* D. Dubnau (ed.), *The Molecular Biology of the Bacilli.* Academic Press, Inc., New York.
46. **Zahler, S. A.** 1988. Temperate bacteriophages of *Bacillus subtilis*, p. 559–592. *In* R. Calendar (ed.), *The Bacteriophages.* Plenum Press, New York.
47. **Zuber, P., and R. Losick.** 1987. Role of *abrB* in *spo0A*- and *spo0B*-dependent utilization of a sporulation promoter in *Bacillus subtilis. J. Bacteriol.* **169:**2223–2230.

Chapter 16

Plasmids

JORGE H. CROSA, MARCELO E. TOLMASKY, LUIS A. ACTIS, and STANLEY FALKOW

The term "plasmid" was originally used by Lederberg (58) to describe all extrachromosomal hereditary determinants. Currently, the term is restricted to autonomously replicating extrachromosomal DNA. Plasmid sizes range from 1 to more than 200 kbp (22), and even larger megaplasmids were detected in *Rhizobium* spp. (14, 17). Although they replicate autonomously, plasmids rely on host-encoded factors for their replication. Although not essential for the survival of bacteria, plasmids may encode a wide variety of genetic determinants which permit their bacterial hosts to survive better in an adverse environment or to compete better with other microorganisms occupying the same ecological niche. Plasmids are found in a wide variety of bacteria, and it is as difficult to generalize about plasmids as it is to generalize about the bacteria that harbor them. The medical importance of plasmids that encode antibiotic resistance and specific virulence traits has been well documented. Plasmids have also been shown to influence significantly several properties contributing to the usefulness of bacteria in agriculture and industry (8, 16, 25, 27, 42, 69, 70, 72, 99). The types of genes that encode these traits are frequently located in transposable elements (12), producing great variation and flexibility in the constitution of plasmids. These extrachromosomal elements are of equal importance, however, for the study of the structure and function of DNA. Plasmids have taken on paramount importance in recombinant DNA technology, particularly as cloning vehicles for foreign genes.

Most bacterial plasmids are usually covalently closed circular (CCC) DNA; however, linear DNA plasmids, as well as single-stranded plasmids which are intermediates in replication, have also been described (10, 36, 44, 53, 74). Plasmids have also been found in eucaryotic microorganisms (42).

This chapter deals with the isolation, purification, and characterization of bacterial plasmids. Many of the techniques currently used for the isolation of plasmid DNA from bacteria are based on its supercoiled CCC configuration. All of the techniques require some means of gently lysing bacterial cells so that the plasmid DNA is preserved intact and can be physically separated from the more massive chromosomal DNA. This separation is more easily achieved with smaller plasmids, and the degree of difficulty increases as the size of the plasmid increases, particularly with some megaplasmids. This complication is often compounded by the fact that very large plasmids are normally present in low copy numbers. Even more specialized techniques are needed to isolate linear plasmids.

The initial characterization of a bacterial plasmid usually is at the genetic level. If a bacterial trait is suspected of being plasmid mediated, gene transfer experiments often document transmissibility of plasmid determinants independently of chromosomal determinants. Moreover, the elimination (curing) of a genetic trait by treating the bacterial population with chemical or physical curing agents such as acridine dyes, ethidium bromide, sodium dodecyl sulfate (SDS), antibiotics, or high temperature (24, 46, 78, 91) indicates that the expression of that genetic trait is linked to the presence of a plasmid. Curing causes loss of the plasmid and therefore is irreversible, as opposed to mutations. Most curing agents are also mutagenic agents. Electroporation (67) has been used successfully to cure plasmids (41). In most cases, however, it is essential to document that a plasmid is present and unequivocally associated with the genetic trait in question. Clinically, the plasmid content of a cell can be a useful epidemiological marker (40, 92).

If possible, it is best to transfer a plasmid by physical or genetic means into a bacterial host that is known to be devoid of plasmids. The advantage, of course, is that any single plasmid transferred to and subsequently isolated from such a host can be analyzed without fear of contamination by a preexisting plasmid.

In some cases, genetic methods are not available, and so one must directly examine a bacterium to determine its plasmid content. Such an analysis usually can be performed to determine simply whether a plasmid is not present. Subsequently, one can examine a strain "cured" of a trait by chemical or physical means to determine whether a plasmid is lost concomitantly with a particular host cell function.

The compositions of the commonly used reagents mentioned throughout this chapter are indicated in Table 1.

Additional information on bacterial plasmids can be found in general references 1 to 6.

16.1. ISOLATION

In this section are described methods of isolating plasmid DNA that do not require the use of commercially available materials such as prepared chromatography columns or reagents. Some characteristics are explained before each procedure. When dealing with a new isolated bacterium, a trial-and-error strategy in selecting the appropriate method is recommended.

16.1.1. Ordinary Plasmids by Large-Scale Methods

Large-scale methods permit the preparation of ordinary plasmid DNA in amounts such that repetitive analysis (restriction endonuclease digestions, preparative isolation of specific DNA fragments, cloning and subcloning, etc.) can be carried out. Generally, these methods are used in conjunction with a subsequent purification step, such as equilibrium density gradient centrifugation in CsCl-ethidium bromide gradients or precipitation with polyethylene glycol. Most of the methods described below are generally used for isolating ordinary plasmid DNA with molecular weight no greater than 100×10^6.

16.1.1.1. Cold Triton X-100

The cold Triton X-100 procedure (20, 55) works well for isolating plasmid DNA from *Escherichia coli* K-12 strains, with the cells lysed by the nonionic detergent Triton X-100.

1. Grow the cells at 37°C either in a rich medium or in a suitable minimal medium (under selection, if possible), with gentle aeration achieved by shaking the flask at a rate just sufficient to keep the surface of the medium in motion. The details that follow are for a 100-ml culture in a 250-ml Erlenmeyer flask, but the method can be scaled up or down proportionally. To ensure

Table 1. Composition of commonly used reagents

Reagent	Composition
Acetate solution	60 ml of 5 M potassium acetate, 11.5 ml of glacial acetic acid, 28.5 ml of water
Boiled RNase	5 mg of RNase per ml in 0.15 M NaCl, heated at 100°C for 20 min to destroy DNase
50x Denhardt's solution	1% Ficoll, 1% polyvinylpyrrolidone, 1% bovine serum albumin
Gel-loading buffer	0.25% bromophenol blue, 0.25% xylene cyanol FF, 30% glycerol
Phenol	Equilibrated with 0.1 M Tris-HCl (pH 8.0) and 0.2% β-mercaptoethanol; oxidation is prevented by the addition of 0.1% 8-hydroxyquinoline
20x salt-sodium citrate (SSC)	175.3 g of NaCl and 88.2 g of sodium citrate per liter adjusted to pH 7.8
20x salt-sodium phosphate-EDTA (SSPE)	3.6 M NaCl, 0.2 M sodium phosphate (pH 7.7), 0.02 M EDTA
Sucrose-Triton-EDTA-Tris (STET)	8% sucrose, 0.5% Triton X-100, 0.05 M EDTA, 0.05 M Tris-HCl (pH 8.0)
Tris-EDTA (TE) buffer	0.01 M Tris-HCl, 0.001 M EDTA (pH 8.0)
Tris-EDTA-salt (TES) buffer	0.05 M Tris-HCl, 0.005 M EDTA, 0.05 M NaCl (pH 8.0)
Tris-acetate (electrophoresis buffer)	0.04 M Tris, 0.001 M EDTA (pH 8.0), 0.02 M glacial acetic acid
Tris-borate (electrophoresis buffer)	0.089 M Tris, 0.0025 M EDTA (pH 8.0), 0.089 M boric acid
Triton X-100 lytic mixture	1 ml of 10% Triton X-100 in 0.01 M Tris-HCl (pH 8.0), 25 ml of 0.25 M EDTA (pH 8.0), 5 ml of 1 M Tris-HCl, (pH 8.0), and 69 ml of water

optimum lysis, harvest the cells in the mid-logarithmic phase of growth.

2. Harvest the cells by centrifugation in a 250-ml bottle (e.g., for 10 min at 12,000 × *g* at 5°C), resuspend the pellet with 15 ml of TES buffer (Table 1), transfer to a 40-ml tube, and centrifuge for 10 min at 12,000 × *g* at 5°C. Resuspend the pellet with 1 ml of ice-cold 25% sucrose solution (in 0.05 M Tris–0.001 M EDTA [pH 8.0]), and place the tube in crushed ice for 30 min.

3. Add 0.2 ml of a freshly prepared lysozyme solution (5 mg/ml in 0.25 M Tris-HCl [pH 8.0]). Mix the contents by swirling the tube several times, and then place the tube in ice for 10 min.

4. Add 0.4 ml of 0.25 M EDTA (pH 8.0). Swirl the tube, and then place it in ice for another 10 min.

5. Add 1.6 ml of Triton X-100 lytic mixture (1 ml of 10% Triton X-100 in 0.01 M Tris-HCl [pH 8.0], 25 ml of 0.25 M EDTA [pH 8.0], 5 ml of 1 M Tris-HCl [pH 8.0], 69 ml of water). After very gentle swirling to mix the contents thoroughly, place the tube in ice for 20 min.

6. Centrifuge at 35,000 × *g* at 5°C for 20 min, decant the supernatant, and discard the pellet.

About 95% of the plasmid is separated from the pellet containing the chromosomal DNA and cellular debris and can be used for further analysis. The plasmid-enriched supernatant fraction (3.2 ml) can also be purified and concentrated by centrifugation in a CsCl-ethidium bromide density gradient as described in section 16.2.

16.1.1.2. Hot Triton X-100

The hot Triton X-100 method is a modification of the foregoing procedure which improves the sharpness of the DNA bands in agarose gels for lysates obtained from certain bacteria, e.g., *Pseudomonas aeruginosa, Serratia marcescens, Proteus rettgeri,* and *Klebsiella pneumoniae.* This procedure was successfully used in the analysis of plasmid profiles of clinical isolates of *Acinetobacter calcoaceticus* subsp. *anitratus* (new species name *baumanii*) (40). This method is useful for isolation of plasmids of sizes ranging from 2 to 100 kbp.

1. Grow the culture in 40 ml of brain heart infusion broth or any other appropriate rich medium in a 100-ml flask on a shaker, as in step 1 above.

2. Harvest the cells by centrifugation at 3,000 × *g* for 10 min.

3. Suspend the cell pellet in 5 ml of TES buffer as in step 2, above.

4. Centrifuge the cells as in step 2, above.

5. Suspend the cell pellet in 2 ml of a 25% sucrose solution (in 0.001 M EDTA–and 0.05 M Tris-HCl [pH 8.0]). Place the tube in ice for 20 min.

6. Add 0.4 ml of a lysozyme solution (10 mg/ml in 0.25 M Tris HCl [pH 8.0]) to the suspension. Place the tube in ice for 20 min.

7. Add 0.8 ml of 0.5 M EDTA (pH 8.0) to the cell suspension.

8. Lyse the cells with 4.4 ml of Triton lytic mixture as in step 5, above. Mix gently.

9. Heat the tube at 65°C for 20 min.

10. Sediment the cellular debris by centrifugation at 27,200 × *g* for 40 min.

11. Decant, and then adjust the supernatant solution to 0.5 M NaCl and 10% polyethylene glycol by use of stock solutions of 5 M NaCl and 40% polyethylene glycol (molecular weight, 1,000 to 6,000).

12. Store the tube at 4°C overnight.

13. Sediment the resulting precipitate by centrifugation at 3,000 × *g* for 10 min, and resuspend the pellet in 1

to 2 ml of 0.25 M NaCl containing 0.001 M EDTA and 0.01 M Tris-HCl (pH 8.0) at 4°C.

14. Precipitate the DNA by adding 2 volumes of 95% ethanol at −20°C, and let the tube stand overnight at −20°C. Alternatively, the DNA can be precipitated by letting the tube stand for 30 min at −70°C.

15. Collect the plasmid DNA pellet by centrifugation at 12,000 × *g* at 4°C. Resuspend it in TES.

16.1.1.3. Alkaline pH

The alkaline pH method described below is a shorter version of the one described by Birnboim and Doly (13). It is used routinely to prepare plasmid DNA from *E. coli, Vibrio anguillarum, K. pneumoniae,* and *Enterobacter cloacae* and has been used by others for many other genera. The pH should not be allowed to go above 12.5, to avoid plasmid DNA alteration.

1. Harvest cells from a 50- to 200-ml (depending on the plasmid copy number) culture by centrifugation at 12,000 × *g* at 4°C. Resuspend the pellet with 5 ml of a solution containing 0.05 M Tris-HCl, 0.010 M EDTA, and 0.050 M glucose (pH 8.0).
2. Add 10 ml of a solution containing 0.2 N NaOH and 1% SDS. Mix gently. The solution should clear and turn viscous.
3. Add 7.5 ml of acetate solution (make the solution by adding 60 ml of 5 M potassium acetate, 11.5 ml of glacial acetic acid, and 28.5 ml of water). Mix thoroughly by shaking. A white precipitate should form.
4. Centrifuge at 30,000 × *g* for 20 min at 4°C.
5. Discard the pellet. If the supernatant contains pellet particles or is milky, filter it to obtain a clear solution. Add 12.5 ml of isopropanol, and let stand 15 min at room temperature.
6. Sediment the plasmid DNA by centrifugation at 12,000 × *g* at room temperature.
7. Resuspend the DNA in TE buffer (Table 1).

The original technique recommends that, after addition of the alkali and acetate solutions in steps 2 and 3, the sample stand in ice for a certain time. However, skipping these waiting steps results in the same yield and quality of plasmid DNA.

16.1.1.4. Sodium dodecyl sulfate

The SDS method below is a modification (37) of that described by Hirt (47).

1. Sediment the cells by centrifugation at 12,000 × *g* for 10 min at 4°C, and then resuspend them in 1.5 ml of 25% sucrose containing 0.05 M Tris-HCl and 0.001 M EDTA (pH 8.0).
2. Add 0.2 ml of lysozyme solution (10 mg of lysozyme per ml in 0.25 M Tris-HCl [pH 8.0]) to the cell suspension, and mix gently. Place the tube in ice for 15 min.
3. Add 0.1 ml of 0.25 M EDTA at pH 8.0, mix gently by slow inversion of the tube, and replace the tube in ice for 10 min.
4. Add 0.1 ml of 20% SDS solution. Mix the suspension gently, and keep in ice for 10 min. During this time, the cells lyse and the solution becomes viscous.
5. Precipitate chromosomal DNA by adding 0.9 ml of 3 M NaCl (to bring the final concentration to 1 M NaCl). Place the tube in ice for at least 2 h to allow complete precipitation.
6. Centrifuge the precipitated chromosomal DNA and any remaining cell debris at 35,000 × *g* for 30 min at 4°C, and decant the supernatant (enriched for plasmid DNA) into a 15-ml Corex tube.
7. Remove the RNA by adding 1 volume of distilled water and 4 μl of a 5 mg/ml boiled RNase solution in 0.15 M NaCl (the enzyme must be heated at 100°C for 20 min before use to destroy DNases). Incubate the tube at 37°C for an additional 1 h.
8. Add an equal volume of Tris-saturated phenol to deproteinize the mixture (as described in section 16.2.1). Shake the mixture vigorously, and then centrifuge at 3,000 × *g* for 20 min at 20°C. Remove the aqueous (upper) phase containing the plasmid DNA. Repeat this step once. Extract the aqueous phase once with 1 volume of chloroform-isoamyl alcohol (24:1, vol/vol) to remove residual phenol.
9. Transfer the aqueous phase to a 30-ml Corex tube, and add 0.6 ml of 3 M sodium acetate to make the final concentration 0.3 M.
10. Add 2 volumes of cold (−20°C) 95% ethanol. Mix the solution well, and place at −20°C overnight.
11. Sediment the precipitated DNA at 12,000 × *g* for 20 min at −10°C, decant the ethanol thoroughly, and resuspend the plasmid DNA in 0.2 ml of 0.006 M Tris-HCl (pH 7.5):

16.1.1.5. Boiling

The boiling method described below is a modification by Reddy et al. (77) of the original method of Holmes and Quigley (48), which is commonly used as a small-scale method. A related modification was also recently reported by Lev (59) and Lev and Seveg (60). The following method has been used with *E. coli* cells, but it is possible that it could be used with other bacteria.

1. Collect cells from a 1-liter culture by sedimentation at 16,000 × *g* for 10 min.
2. Resuspend the cells in 10 ml of a solution containing 0.050 M Tris-HCl (pH 7.5), 0.062 M EDTA, 0.4% Triton X-100, and 2.5 M LiCl. Add 0.5 ml of 20% lysozyme in water. Mix gently, and incubate at 25°C for 10 min.
3. Transfer the solution to a plastic bag (freezer bag; Seal-a-Meal) (8 by 8 cm), and seal. Immerse the bag in a boiling-water bath for 1 min.
4. Transfer the contents to a centrifuge tube, let it stand in ice for 5 min, and centrifuge at 27,200 × *g* at 4°C for 20 min.
5. Transfer the supernatant to a fresh tube, add 2 volumes of ethanol, mix, and keep the tube in dry ice for 10 min.
6. Sediment the DNA by centrifugation as in step 4, wash with 70% ethanol, and let dry.
7. Resuspend the DNA in 0.5 ml of TE buffer (Table 1), and add 10 μl of a boiled RNase solution (10 mg/ml dissolved in 0.15 M NaCl). Incubate at 25°C for 20 min.
8. Extract the plasmid DNA-containing solution with phenol as described in section 16.2.1, and precipitate with $^{1}/_{9}$ volume of 5 M potassium acetate (pH 4.8) and 2.5 volumes of ethanol.
9. Collect the plasmid DNA by centrifugation, and resuspend in TE buffer.

16.1.1.6. Lysostaphin

The following procedure has been used successfully to obtain plasmid DNA from gram-positive bacteria, as well as from gram-negative bacteria that are resistant to lysis by Triton X-100 or SDS.

1. Pellet the cells from a 30-ml overnight culture by centrifugation at 16,000 × *g* for 10 min, and suspend them in 1.5 ml of 0.0075 M NaCl–0.050 M EDTA (pH 7.0).

2. Add lysostaphin to a final concentration of 15 μg/ml. (Double the enzyme concentration for *Staphylococcus epidermidis.*) Incubate the suspension at 37°C for 15 min with gentle agitation, and then place the flask in ice.

3. Add 1.5 volumes of a mixture containing 0.4% deoxycholate, 1% Brij 58, and 0.3 M EDTA (pH 8.0) to achieve cell lysis. Mix the viscous contents of the tube gently, and put the tube in ice for 15 min.

4. Pellet the cellular debris by centrifugation at 23,000 × *g* for 20 min at 4°C, and decant the supernatant fluid into a 15-ml Corex tube.

5. Add 1 volume of distilled water and 4 μl of boiled RNase solution (1 mg/ml) to the supernatant fluid. Incubate the tube at 37°C for 1 h. Proceed with plasmid DNA purification as described for SDS lysis, step 8.

16.1.1.7. Lysostaphin and lysozyme

The lysostaphin and lysozyme method was adapted from that of Lyon et al. (61).

1. Collect the cells from a 20-ml culture, wash them, and resuspend them with 1 ml of TES buffer (Table 1). Add lysostaphin and lysozyme to a final concentration of 75 μg/ml for each enzyme. Incubate for 1 h at 37°C.

2. Add 1 ml of a solution containing 2.5% Sarkosyl, 0.4 M EDTA (pH 8), and 100 μg proteinase K. Mix by inversion, and incubate at 65°C for 30 min.

3. Add 2 ml of a solution containing 3 M sodium acetate and 0.005 M EDTA (pH 7.0), and keep in ice for 30 min.

4. Centrifuge at 40,000 × *g* at 4°C for 30 min. Transfer the supernatant to a fresh tube, and add 1 volume of isopropanol. Let stand at room temperature for 10 min.

5. Collect the DNA pellet by centrifugation at 27,000 × *g* for 10 min.

6. Dissolve the DNA pellet in 0.04 ml of water. Add boiled RNase to a final concentration of 50 μg/ml. Incubate for 30 min at 37°C. This DNA can be phenol extracted and ethanol precipitated.

16.1.1.8. Sucrose

The sucrose method can be used with gram-negative as well as gram-positive bacteria. Bacterial lysis is performed by applying an osmotic shock and then treating with detergent (79).

1. Harvest cells from 50 ml of culture, and wash them with 0.7 ml of 0.05 M Tris-HCl (pH 8.0) containing 25% sucrose.

2. Resuspend the cells in 0.7 ml of 0.05 M Tris-HCl (pH 8.0) containing 100% sucrose. Mix by pipetting, which takes time since 100% sucrose solution is very thick. Incubate for 3 h at 37°C.

3. Add 3 ml of a solution made by mixing 0.5 parts of 0.25 M EDTA (pH 8.1) with 2.3 parts of 1% Brij 58–0.4% sodium deoxycholate–0.05 M Tris-HCl (pH 8.0)–0.06 M EDTA–2 M NaCl. Mix gently, and incubate at room temperature for 1 h.

4. Pellet the cellular debris for 30 min at 10,000 × *g* at 4°C.

5. Add 0.6 volume of isopropanol to the supernatant, mix, and keep for 15 min.

6. Centrifuge at 10,000 × *g* at 4°C, and resuspend the pellet in 0.7 ml of TE.

7. This preparation can be phenol extracted and ethanol precipitated.

16.1.1.9. Penicillin

The following procedure can be used for obtaining plasmid DNA from gram-positive and gram-negative organisms that are resistant to Triton and SDS but sensitive to penicillin. If the bacterium under study produces a β-lactamase, however, a penicillinase-resistant antibiotic (e.g., a cephalosporin) may be effective. It is not certain whether other antimicrobial agents active against cell wall biosynthesis would prove successful, although it may be useful to investigate this possibility.

1. Grow bacteria to the mid-logarithmic phase, add 1 mg of penicillin G per ml, and incubate the culture at the optimum growth temperature for an additional 2 h. (*Note:* It is convenient to perform a preliminary experiment to determine the time and antibiotic concentration that give optimum protoplast formation; the goal is to achieve this maximum without massive lysis. The addition of 1 M sucrose to the growth medium may be beneficial in achieving this goal.)

2. Harvest the penicillin-treated cells by centrifugation, and wash them twice with an equal volume of 0.01 M Tris-HCl (pH 8.2) of the original culture.

3. Suspend the cell pellet in 0.25 volume of 0.02 M Tris-HCl (pH 8.2)–0.5 volume of 1 M sucrose–0.25 volume of lysozyme (4 mg/ml) in 0.02 M Tris-HCl (pH 8.2).

4. Incubate the cell suspension at 37°C for 1 h with shaking. Centrifuge at 27,000 × *g* for 15 min.

5. Gently suspend the cells in 0.02 M Tris-HCl (pH 8.2) and 0.01 M EDTA, taking care to ensure that the cells are homogeneously dispersed.

6. Add 0.1 volume of 10% SDS to the cells. Lysis should be complete within 10 min. Proceed from here as described for the SDS lysis (section 16.1.1.4, step 8).

Among gram-positive bacteria such as streptococci, lactococci, and lactobacilli, addition of 0.01 M L-threonine to the growth medium weakens the cell wall and leads to easier lysis after lysozyme treatment followed by detergent lysis (18, 63). Glycine added at the early to mid-log phase can be just as effective as penicillin but, like the latter and unlike threonine, will stop growth of the culture (63).

16.1.2. Ordinary Plasmids by Small-Scale Methods

Small-scale methods permit the rapid preparation of ordinary plasmid DNA in small amounts, normally enough for a few assays such as restriction endonuclease digestion for screening or preparation of probes. Some of them produce a DNA of high enough quality to perform a few sequencing reactions, and some produce a DNA of inferior quality. These methods do not require a subsequent purification step for the stated purposes.

16.1.2.1. Cold alkaline pH

The cold alkaline pH method is a scaled-down version of the one described in large scale (section 16.1.1.3). Use it to prepare *E. coli* and *V. anguillarum* plasmid DNA that is not intended to be used for sequencing.

1. Harvest cells from 1.5 ml of culture by centrifugation in a microcentrifuge (16,000 × *g*) for 10 s. Resuspend the pellet with 100 μl of a solution containing 0.05 M Tris-HCl, 0.010 M EDTA, and 0.050 M glucose (pH 8.0).
2. Add 200 μl of a solution containing 0.2 N NaOH and 1% SDS. Mix gently. The solution should clear and turn viscous.
3. Add 150 μl of acetate solution (make the solution with 60 ml of 5 M potassium acetate, 11.5 ml of glacial acetic acid, and 28.5 ml of water). Mix thoroughly by shaking. A white precipitate should form.
4. Centrifuge in a microcentrifuge (16,000 × *g*) for 5 min.
5. Transfer 400 μl of the supernatant to a fresh Eppendorf tube, and add 1 ml of cold ethanol.
6. Recover the plasmid DNA by centrifugation for 5 min, pour off the supernatant, wash once with 70% cold ethanol, and dry.
7. Resuspend the DNA in 20 to 50 μl of TE buffer.

This DNA is suitable for restriction endonuclease analysis and Southern blot hybridization (see section 16.3.4). After RNase treatment and either phenol extraction or purification with silica powder (e.g., Geneclean [Bio101, La Jolla, Calif.]), the DNA can be used for probe preparation and cloning and subcloning experiments. *CAUTION:* (i) If the plasmid DNA is to be used as a probe against genomic DNA from an organism that is homologous to the host from which the plasmid is isolated, hybridization might occur between the blotted DNA and the residual chromosomal DNA from the host. This situation can be avoided either by purifying the plasmid DNA by dye buoyant-density gradient centrifugation or by preparing plasmid DNA by the acid-phenol method described below. (ii) Sometimes the plasmid DNA is not clean enough at this point for restriction endonuclease analysis. In this case, a phenol extraction usually renders the preparation suitable for this purpose.

16.1.2.2. Hot alkaline pH

The hot alkaline pH method is a modification (51) of the alkaline lysis method described above (section 16.1.2.1). This rapid miniscreen method has been used to analyze large and small plasmids in many types of gram-positive as well as gram-negative bacteria (50a). Figure 1 shows plasmid DNA prepared by this method.

1. Harvest cells from 2 ml of culture by centrifugation, and resuspend the cell pellet in 1 ml of 0.002 M EDTA–0.04 M Tris-acetate (pH 8.0).
2. Add 2 ml of a solution containing 0.05 M Tris and 3% SDS (pH 12.6) (adjust the pH with 2 N NaOH). Mix uniformly, and incubate at 60°C for 30 min.
3. Add 6 ml of phenol-chloroform (1:1), and mix until complete emulsification occurs. Separate the aqueous phase by centrifugation at 10,000 × *g* for 15 min at 4°C.
4. Transfer the aqueous phase to a fresh tube containing 1 volume of chloroform. Mix, and again separate the phases.
5. Recover the aqueous phase containing the plasmid DNA.

16.1.2.3. Acid phenol

The acid phenol method (96) is based on the addition of an acid phenol extraction to the cold alkaline lysis

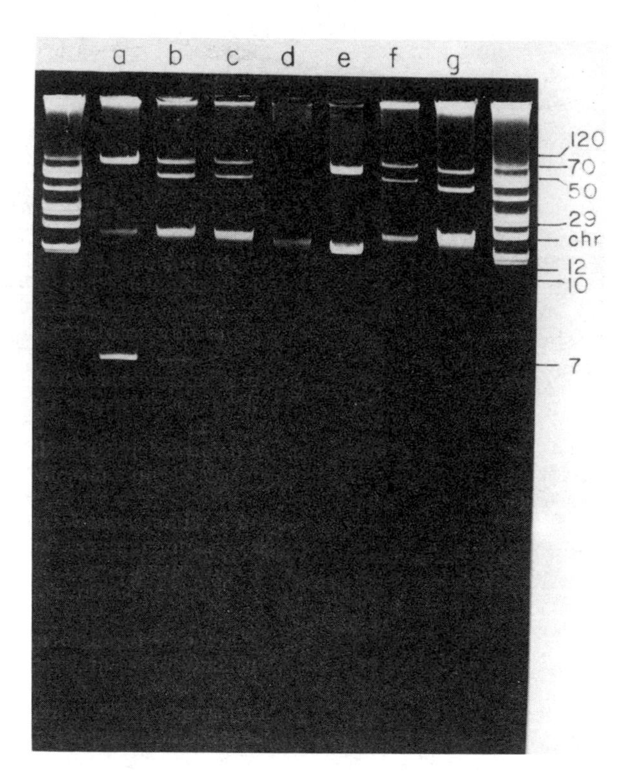

FIG. 1. Agarose gel electrophoresis of plasmid DNA obtained by the method of Kado and Liu (51). Plasmids in lanes a through g are from *Erwinia herbicola* 2D3, 25D31, 25D32, 25D33, 25D34, 25D35, and 25D36, respectively. Flanking lanes are plasmids from *E. stewartii* SW2 used as molecular weight standards. (Kindly supplied by C. Kado.)

preparation of DNA (section 16.1.2.1) and has been used very successfully. The additional use of phenol produces a preparation of similar quality to that obtained after separation in a CsCl-ethidium bromide gradient. This DNA can be used for sequencing and other procedures such as exonuclease III-generated nested deletions. Acid phenol removes nicked and linear DNA, leaving only the supercoiled molecules, by a mechanism still not well understood (100).

1. Proceed through step 4 as described for the alkaline lysis method (section 16.1.2.1). Transfer 400 μl of the supernatant to a fresh Eppendorf tube, and add 400 μl of acid phenol (made by equilibration of distilled phenol with 0.05 M sodium acetate [pH 4.0]). Vortex and separate the layers in a microcentrifuge (16,000 × *g*).
2. Transfer the upper phase to another Eppendorf tube, and extract with 1 volume of chloroform-isoamyl alcohol (24:1, vol/vol).
3. Transfer the upper phase to another Eppendorf tube, and add 2.5 volumes of cold ethanol.
4. Recover the plasmid DNA by centrifugation for 5 min, pour off the supernatant, wash once with 70% cold ethanol, and dry. Resuspend the pellet in 50 μl of TE buffer (Table 1) containing 50 μg of boiled RNase per ml.

16.1.2.4. Phenol and chloroform

The phenol-chloroform method involves the lysis of bacterial cells by phenol-chloroform, which renders a DNA preparation suitable for restriction endonuclease treatment as well as transformation (7, 76, 82). Broth cultures or pooled bacterial growth from solid surfaces can be used for this method.

1. Transfer 1.5 ml of culture or suspended scraped cells from a plate to an Eppendorf tube, and centrifuge for 3 min.

2. Resuspend the pellet in 0.05 ml of a solution containing 0.01 M Tris-HCl (pH 8.0), 0.1 M NaCl, and 0.001 M EDTA. Add 0.05 ml of phenol-chloroform solution (made by mixing phenol, chloroform, and isoamyl alcohol [25:24:1, vol/vol/vol]), and vortex.

3. Centrifuge for 5 min to yield a clear supernatant. Transfer the upper phase to another tube, add ammonium acetate to a final concentration of 2 M, and add 2 volumes of cold ethanol. Let stand in ice for 15 min.

4. Centrifuge for 10 min to collect the DNA. Wash the pellet with 70% cold ethanol, dry it, and resuspend it in 0.02 ml of TE buffer.

16.1.2.5. Boiling

The boiling method described here is the original method of Holmes and Quigley (48), used effectively for screening of high-copy-number plasmids.

1. Transfer 1.5 ml of culture to an Eppendorf tube, and pellet the cells.

2. Resuspend in 85 μl of STET buffer (8% sucrose, 0.5% Triton X-100, 0.05 M EDTA, 0.05 M Tris-HCl [pH 8.0]). Add 10 μl of 10 mg/ml lysozyme dissolved in water.

3. Place in a boiling-water bath for 90 s. The optimal time should be determined for different strains, since effectiveness seems to be strain dependent.

4. Immediately centrifuge for 10 min.

5. Remove the viscous pellet with a pipette tip, and add 85 μl of cold isopropanol. Let stand at −20°C for 20 min.

6. Collect the DNA by centrifugation in a microcentrifuge (16,000 × *g*), wash the pellet once with 70% cold ethanol, and dry it. Resuspend the pellet in 20 μl of TE buffer.

Another method involving lysis by boiling is reported in reference 101.

16.1.3. Large Plasmids

Large plasmids have been found in several bacterial systems, including the tumor-inducing and root-inducing plasmids of *Agrobacterium tumefaciens* and *A. rhizogenes* (93, 98), the sym plasmids of *Rhizobium* spp. (9), the H incompatibility group of antibiotic resistance plasmids (22), the camphor-degradative plasmid of *Pseudomonas putida* (16, 72), and the various F′ factors (64, 83, 86). Several methods have been developed which permit the isolation of small as well as large plasmids (8, 28, 38, 50, 94). In addition, several techniques have been developed which permit the use of very small working volumes or even material from a single colony (31). A method similar to those mentioned above has been applied to lactic acid bacteria and several other groups of gram-positive organisms (8).

16.1.3.1. Lysis in solution

The following method has the advantage of using small working volumes, shares the best features of the previously published methods, and permits the isolation of small as well as large plasmids. It takes advantage of the resistance of plasmid DNA molecules to strand separation by alkali, which may account for the improved yields in plasmid DNA of very high molecular weight (up to 350×10^6).

This method has been used for the isolation of plasmids from gram-negative bacteria (e.g., *A. tumefaciens, Yersinia enterocolitica, Salmonella typhi, Klebsiella* spp., *E. coli*) and also gram-positive bacteria (e.g., *Staphylococcus* and *Streptococcus* spp.). Consequently, it is recommended for the routine screening of all types of bacteria for all types of plasmids.

Plasmid DNA obtained by this method from even 2 ml of culture is relatively free from chromosomal DNA and, after a few additional steps, can be used directly for restriction endonuclease analyses (section 16.3.1.2). The general procedure is as follows.

1. Grow 2 ml of culture overnight in an appropriate medium (e.g., brain heart infusion broth) in 120-mm screw-cap tubes on a shaker. The cells may be used directly or, better, can be used to obtain logarithmically growing cells by diluting the overnight culture 1:20 into 2 ml of fresh medium and incubating the subculture for 2 to 3 h. Harvest the cells by centrifugation at about 2,500 × *g* for 10 min.

2. Resuspend the cells in 2 ml of TE buffer (Table 1), recentrifuge, and resuspend the cell pellet thoroughly in 40 μl of TE. For marine vibrios and other halophilic bacteria, it is necessary to carry out this and subsequent washing steps in the presence of high salt concentrations (up to 1 M NaCl).

3. Prepare lysis buffer (4% SDS in TE [pH 12.4] daily. It is important to determine the pH accurately (use a high-pH electrode), since values higher than 12.5 will irreversibly denature the plasmid DNA. Add 0.6 ml of the lysis buffer to a 1.5-ml centrifuge tube, and, with a Pasteur pipette, transfer 40 μl of the cell suspension into the lysis buffer. Mix the suspension well, but avoid vigorous agitation or the use of mechanical mixing devices. (The culture could be grown directly in a 1.5-ml centrifuge tube to avoid the need for transfer in this step.)

4. Incubate the suspension at 37°C for 20 min to achieve full lysis of the cells.

5. Neutralize the solution by adding 30 μl of 2.0 M Tris-HCl (pH 7.0). Slowly invert the tube until a change in viscosity is noted.

6. Precipitate chromosomal DNA by adding 0.24 ml of 5 M NaCl. This step should be done immediately after step 5. For complete removal of the chromosomal DNA, put the tube in ice for 4 h. If a large number of cultures are being screened and maximum plasmid purity is not important, the time in ice can be shortened to 1 h.

7. Centrifuge in a microcentrifuge (16,000 × *g*) for 10 min to sediment debris, and then pour the supernatant fluid into another 1.5-ml centrifuge tube (do not attempt to obtain the small amount of fluid remaining in the bottom of the tube).

8. Add 0.55 ml of isopropanol to the supernatant fluid to precipitate the DNA. After mixing, place the tube at −20°C for 30 min.

9. Centrifuge for 3 min in a microcentrifuge (16,000 × *g*), pour off the supernatant fluid, and invert the tube on a paper towel. Then dry the tube under vacuum.

10. Resuspend the precipitate in 30 μl of TES. Allow

the DNA to dissolve overnight at 4°C. Ordinarily, 10 μl of this solution will show readily visible DNA bands after gel electrophoresis.

Plasmid DNA obtained by this method should be further purified for restriction endonuclease analysis by the following additional steps.

11. Resuspend the pellet obtained in step 9 in 100 μl of 0.010 M Tris (pH 8.0). Add 100 μl of phenol equilibrated with 0.010 M Tris-HCl (pH 8.0). Mix well. Add 100 μl of chloroform.

12. Centrifuge for 30 s in a microcentrifuge (16,000 × g) to separate the aqueous phase from the phenol-chloroform phase. Remove the upper, aqueous phase.

13. Precipitate the plasmid DNA with 2 volumes of ice-cold 95% ethanol, and proceed as in step 9. Resuspend the precipitate in a suitable buffer for the restriction endonuclease reactions.

For lactic streptococci, another lysis protocol is described in reference 8.

16.1.3.2. Lysis in agarose gel

The following method is based on bacterial cell lysis directly in the wells of a vertical agarose gel (31). Since most of the chromosomal DNA remains intact after the lysis, it remains in the well, and therefore less than 0.5% of the total chromosomal DNA appears in the gel as a band of linear DNA. Megaplasmids have been successfully detected by this method.

1. Pick cells from a colony with a flat toothpick, and resuspend them in the well of a gel containing 15 μl of a solution consisting of 7,500 U of lysozyme per ml, 0.3 U of boiled RNase per ml, 20% Ficoll 400,000, and 0.05% bromophenol blue in electrophoresis buffer (0.09 M Tris base, 0.012 M EDTA, 0.09 M boric acid) (for gram-negative bacteria) or 75,000 U of lysozyme per ml, 0.3 U of boiled RNase per ml, 20% Ficoll 400,000, 0.05 M EDTA (pH 8.0), 0.1 M NaCl, and 0.05% bromophenol blue in electrophoresis buffer (for gram-positive bacteria). The gel dimensions recommended are 110 by 140 by 2.5 mm, and the well sizes should be 6.5 by 15 by 2.5 mm. Gel concentrations may vary from 0.75 to 1.2% agarose.

2. On top of the bacterium-lysozyme mixture add 30 μl of a solution containing 0.2% SDS and 10% Ficoll 400,000 in electrophoresis buffer for gram-negative bacteria or 30 μl of the same solution but also containing 2% SDS for gram-positive bacteria.

3. Gently mix the two layers with a toothpick by moving it from side to side (not more than twice for gram-negative bacteria). Complete mixing should be avoided. The two layers must still be distinguishable.

4. On top add 0.1 ml of a solution containing 0.2% SDS and 5% Ficoll 400,000 in electrophoresis buffer without disturbing the viscous DNA lysate.

5. Seal the wells with hot agarose, fill both electrophoretic chambers with electrophoresis buffer, and proceed with electrophoresis for 60 min at 2 mA and then for 60 to 150 min at 40 mA.

16.1.4. Linear Plasmids

Although most plasmids are composed of circular DNA, linear plasmids have also been described. They were found in lankamicin- and lankacidin-producing *Streptomyces rochei* as well as in *Borrelia burgdorferi* and *B. coriaceae*, the ethiological agents of Lyme disease and epizootic bovine abortion, respectively (10, 44, 74, 80).

The protocols to isolate linear plasmids involve cell lysis, in some cases followed by CsCl-ethidium chloride gradient purification, and the separation of the linear plasmid from the chromosomal DNA by either sucrose gradient or agarose gel electrophoresis (10, 44, 74). A procedure designed to isolate linear plasmids of *B. coriaceae* is as follows.

1. Pellet cells from a 20-ml culture, and wash them with phosphate-buffered saline containing 0.005 M $MgCl_2$.

2. Resuspend the pellet in 0.4 ml of a solution containing 0.05 M Tris-HCl (pH 8.0) and 0.05 M EDTA.

3. Add 0.1 ml of 10% SDS and 25 μl of 0.02 M proteinase K. Incubate at 37°C for 1 h.

4. Load this mixture onto a 10 to 40% sucrose gradient prepared in 0.05 M Tris-HCl (pH 8.0)–0.05 M EDTA. Run overnight at 100,000 × g in an SW41 rotor (Beckman Instruments).

5. Collect 3-drop fractions from the bottom of the gradient.

6. Run the fractions in 0.3% agarose gel electrophoresis, and use UV transillumination to locate those containing the linear plasmid DNA.

7. Pool these fractions, and precipitate the DNA by addition of 2.5 volumes of cold ethanol. Wash with 70% cold ethanol, and dry. Resuspend the DNA in TE buffer.

16.1.5. Single-Stranded Plasmids

Several plasmids of gram-positive bacteria may be present as both single- and double-stranded DNA. All these plasmids replicate via a rolling-circle mechanism, and so the single-stranded DNA is a replicative intermediate (36). Single-stranded DNA plasmids have been isolated from gram-positive bacteria: lysozyme or lysostaphin in combination with Sarkosyl were used with *Bacillus subtilis* and *Staphylococcus aureus* (88), and electroporation was recently applied to *Clostridium acetobutylicum* for the same purposes (54). The method developed by te Riele et al. (88) is as follows.

1. Collect cells from a culture of optical density at 650 mm (OD_{650}) = 1, and wash with a solution containing 0.1 M EDTA (pH 8) and 0.15 M NaCl.

2. Resuspend the cells in a solution containing 0.01 M EDTA (pH 8.0) and 0.15 M NaCl.

3. Add 10 mg of lysozyme per ml (for *B. subtilis*) or 50 μg of lysostaphin per ml (for *S. aureus*), and incubate at 37°C.

4. Add Sarkosyl to a final concentration of 1%, and incubate at 65°C for 20 min.

5. Extract the cell lysate with phenol and chloroform.

6. Add boiled RNase to 50 μg/ml, and incubate at 37°C for 20 min.

The single-stranded plasmid can be recognized by performing Southern blot hybridization (as described in section 16.3.4) but skipping the alkali treatment of the gel to impede the binding of the double-stranded DNA to the membrane.

Further purification can be achieved by hydroxyapatite column chromatography. Dialyze the extract obtained in step 6 against 0.075 M sodium phosphate buffer (pH 6.8). Apply this sample to a hydroxyapatite column equilibrated with the same buffer. After washing, elute the single-stranded DNA with a linear gradient consisting of 0.075 to 0.250 M phosphate buffer.

16.1.6. Single-Stranded Phagemids

Several techniques have been developed for working with a cloned DNA fragment. Several of these techniques (sequencing, generation of DNA probes labeled in only one strand, site-directed mutagenesis, etc.) require that the cloned DNA fragment be in the form of a single-stranded template. Filamentous bacteriophages (34), such as M13, f1, or fd, can be used for this purpose. However, the recently developed phagemids are more convenient. Dotto et al. (30) demonstrated that a plasmid carrying the major intergenic region of f1, in the presence of a helper phage, enters the f1 replication pathway by generating a single-stranded DNA that is packaged and exported in phage-like particles. The major intergenic region of f1 contains all of the sequences required in *cis* for initiation and termination of viral DNA synthesis and for morphogenesis of bacteriophage particles. Phagemids are plasmid vectors that have the ColE1 origin of replication, an antibiotic resistance gene for selection, a multiple cloning site, and the origin of replication of one filamentous phage. Most of these phagemids also have the T3 and/or T7 promoter to allow the synthesis of specific RNA by using the clone as template. The cloned DNA can be propagated as a regular recombinant plasmid. Single-stranded DNA is obtained in the culture supernatant after the *E. coli* cells harboring the phagemid are coinfected with a helper phage. Different phagemids are available from several commercial sources (Stratagene, Bio-Rad, and others). In the following paragraphs a general protocol is described for the generation of single-stranded phagemid DNA.

1. Inoculate 5 ml of 2× YT broth with a single colony of *E. coli* harboring a phagemid, and add 2×10^7 PFU of helper phage suspension (e.g., phage M13KO7 [Bio-Rad], which has a kanamycin resistance gene) per ml.
2. Incubate at 37°C for 30 min with strong agitation (250 to 300 cycles/min), and then add kanamycin to a final concentration of 70 μg/ml. This addition prevents growth of cells not infected with the helper phage.
3. Incubate in a rotary incubator (250 to 300 cycles/min) at 37°C for 18 h.
4. Transfer 1.5 ml of the culture to an Eppendorf tube, and pellet the cells in a microcentrifuge (16,000 × *g*) for 5 min at 4°C.
5. Transfer 1.2 ml of the supernatant to a new tube, add 7 μg of boiled RNase, and incubate for 30 min at room temperature.
6. Add 240 μl of a solution containing 20% polyethylene glycol 8000 and 15% NaCl (a solution containing 3.5 M sodium acetate instead of NaCl may also be used) to precipitate the phage particles. Let the mixture stand in ice for 30 min.
7. Collect the phage by centrifugation in a microcentrifuge (16,000 × *g*) for 15 min. Remove the supernatant. Resuspend the phage pellet in 200 μl of a solution containing 0.3 M NaCl, 0.1 M Tris-HCl (pH 8.0), and 0.001 M EDTA. Transfer to another tube, and let stand in ice for 30 min. Centrifuge again, and transfer the supernatant to a fresh tube (this step removes insoluble products). This phage suspension can be stored at 4°C for 1 week.
8. To isolate the phagemid DNA, extract the phage suspension with 1 volume of phenol (prepared as described in section 16.2.1). Transfer the aqueous phase to a fresh tube.
9. Extract once with 1 volume of phenol-chloroform-isoamyl alcohol (25:24:1), recover the aqueous phase, and reextract several times with chloroform-isoamyl alcohol (24:1) until no interphase is visible.
10. To the aqueous phase add 1/10 volume of 7.8 M ammonium acetate and 2.5 volumes of cold ethanol. Keep at −70°C for 30 min.
11. Recover the phagemid DNA by centrifugation, wash the pellet with 70% cold ethanol, dry, and resuspend in 20 μl of TE buffer.

16.2. PURIFICATION

The subsequent analysis of plasmid DNA determines the degree of purity required. For example, to analyze and compare plasmid profiles for epidemiological purposes, further purification is normally not required. Conversely, other purposes such as restriction endonuclease analysis, cloning, and probe preparation require further purification. Furthermore, a higher degree of purification is required for nucleotide sequencing in which all linear or nicked DNA molecules as well as RNA must be eliminated. After purification, the plasmid DNA concentration can be determined by either classical methods such as spectrophotometric determination or ethidium bromide fluorescence or commercial methods such as colorimetric quantitation with DipStick (Invitrogen, San Diego, Calif.). This is a rapid method that consists of spotting a small amount of the nucleic acid solution on the DipStick, dipping the stick in a solution provided with the kit, and determining the concentration by matching the color intensity of the sample to a standard chart. A detailed description of DNA quantitation is described in Chapter 22.6.

16.2.1. Extraction with Organic Solvents

Extraction with organic solvents is used to remove proteins that are normally present after plasmid DNA isolation or after boiled RNase or restriction endonuclease treatment.

1. Mix the plasmid DNA solution with 1 volume of phenol (Table 1) or phenol-chloroform-isoamyl alcohol (25:24:1, vol/vol/vol) until emulsified.
2. Separate the phases by centrifugation in a microcentrifuge (16,000 × *g*) for 10 s.
3. Transfer the aqueous upper phase to a fresh tube, add 1 volume of either chloroform-isoamyl alcohol (24:1) or water-saturated ether, and mix until the aqueous phase becomes translucent.
4. Separate the phases by centrifugation as in step 2, and recover the aqueous phase containing the plasmid DNA. With chloroform extraction, the aqueous phase is on top; with ether extraction, the aqueous phase is below.
5. Add 1/9 volume of 5 M potassium acetate (pH 4.8) and 2.5 volumes of cold ethanol. Precipitation can also

be achieved by addition of 0.5 volume of 7.5 M ammonium acetate and 3 volumes of ethanol.

6. Pellet the DNA by centrifugation in a microcentrifuge (16,000 × *g*) for 5 min, wash the pellet with 70% cold ethanol, and dry.

7. Resuspend the DNA in TE buffer.

To obtain a higher degree of purification, perform an extraction with phenol followed by another extraction with phenol-chloroform-isoamyl alcohol (25:24:1).

16.2.2. Equilibrium Gradient Centrifugation

Equilibrium gradient centrifugation is the most common method to purify plasmid DNA from large preparations. Ethidium bromide is one in a series of phenathridinium dyes that bind to DNA and RNA and inhibit nucleic acid functions (21, 43). The DNA is sedimented to equilibrium by high-speed centrifugation. Ethidium bromide intercalates into the DNA, thus reducing its density (19, 75). A linear DNA molecule or an open-circular (OC) (nicked) molecule of plasmid DNA does not have the same physical constraints that are imposed on a CCC molecule of plasmid DNA. Consequently, the former types of DNA can bind significantly more ethidium bromide molecules and are rendered less dense than the latter type of DNA. This difference in binding permits the separation of the different forms of plasmid DNA. OC and linear plasmid DNA that may form as a result of handling during the preparation, as well as linear chromosomal DNA (also a form due to handling), will all appear in the same band unless the G+C contents of plasmid and chromosome are drastically different. Only CCC plasmid DNA and some replicative intermediates will be separated from the linear and OC DNAs. In gradients of cesium chloride, these forms can be visualized as discrete bands when the gradient is illuminated with UV light, as a result of the fluorescence of the ethidium bromide intercalated in the DNA.

1. To a polyallomer ultracentrifuge tube (5/8 by 3 in. [1.6 by 7.6 cm]), add about 5 g of cesium chloride, 2 ml of TES buffer, and 3.2 ml of the DNA solution.

2. Mix the contents of the tube until the cesium chloride is in solution, and add 0.2 ml of ethidium bromide solution (10 mg/ml in TES buffer). Add the ethidium bromide after the CsCl, since the relaxation protein present in some plasmids (which leads to nicking of the DNA) can be activated by ethidium bromide (56). The high CsCl concentration (7 M) inactivates the relaxation complex of most plasmids (56), and consequently higher yields of DNA can be obtained. Even at a high salt concentration, visible light (in the presence of ethidium bromide) can induce nicking of plasmid DNA. Therefore, all operations must be conducted under indirect illumination.

3. Adjust the refractive index of the solution to 1.3925 ± 0.001 g/ml by adding TES buffer or solid cesium chloride. The refractive index is a suggested value and may vary for each instrument; it must be standardized by doing trial experiments. If necessary, fill the tubes with mineral oil. *Rule of thumb:* To the dried DNA pellet resuspended in 4.0 ml of TE buffer, add 4.5 g of CsCl and 0.1 ml of 10 mg/ml ethidium bromide. Under these conditions, good separation of the bands is usually obtained.

4. Prepare the tubes for ultracentrifugation as specified by the centrifuge manufacturer, and place them in the appropriate rotor. Fixed-angle or vertical rotors can be used. However, in vertical rotors the equilibrium is reached faster.

5. Centrifuge the tubes at 180,000 × *g* at 15°C for 6 to 8 h in a vertical rotor or 12 to 16 h in a fixed-angle rotor. After centrifugation, examine the gradients by illuminating the tubes with UV light and looking for the presence of fluorescent bands due to the DNA-ethidium bromide complex. Two bands should be observed, the upper one corresponding to chromosomal and linear or nicked plasmid DNA, and the lower band corresponding to CCC plasmid DNA. In some cases only the lower band is observed. Collect the plasmid DNA by puncturing the bottom, or just below the plasmid DNA, band with a needle.

6. Free the DNA of ethidium bromide by extracting the preparation at least three times with cesium chloride-saturated isopropanol.

7. Dialyze the DNA against TE buffer for 2 to 3 h to eliminate the cesium chloride. Alternatively, the cesium chloride can be eliminated by twofold dilution of the DNA solution with TE buffer and ethanol precipitation.

16.2.3. Precipitation with Polyethylene Glycol

1. Add 1 volume of 5 M LiCl (to precipitate high-molecular-weight RNA) to the DNA solution. Mix, and centrifuge at 12,000 × *g* at 4°C for 10 min.

2. Transfer the supernatant to a fresh tube, and add 1 volume of isopropanol. Mix and centrifuge as in step 1 but at room temperature.

3. Discard the supernatant, wash the pellet with cold 70% ethanol, and allow it to dry.

4. Dissolve the DNA pellet in 0.5 ml of TE buffer containing boiled RNase (20 μg/ml). Let stand for 30 min at room temperature.

5. Add 0.5 ml of a solution containing 1.6 M NaCl and 13% polyethylene glycol 6000. Mix.

6. Collect the plasmid DNA by centrifugation in a microcentrifuge (16,000 × *g*) for 5 min, and discard the supernatant.

7. Dissolve the DNA pellet in 20 μl of TE buffer, extract with phenol, and precipitate the DNA as described above.

16.2.4. Chromatography

Several authors have described the application of high-pressure liquid chromatography and fast protein liquid chromatography to the purification of plasmid DNA (65, 97). These methods can be fast and convenient, although they require special equipment.

16.2.5. Commercial Methods

Many companies have introduced products for nucleic acid purification that are rapid and nontoxic alternatives to organic-solvent extraction. These products can be used to purify and concentrate DNA preparations, to extract-purify DNA fragments from agarose gels, and to separate labeled DNA from unincorporated nucleotides. The methods consist basically of the adsorption of the DNA to glass or silica powder or of minicolumns for adsorption or ion-exchange chromatography.

The principle of glass or silica matrix methods is the selective binding of DNA to these matrixes in the presence of sodium iodide. The contaminants (RNA, proteins, salts, agarose, and polysaccharides normally present in agarose) are eliminated by washing steps. Finally the DNA is recovered from the matrix by elution in low-ionic strength buffer. Examples of these commercial products are Geneclean II (Bio 101, La Jolla, Calif.), which is used to purify DNA fragments 200 bp or larger; and Mermaid (Bio 101), which is used to purify DNA fragments smaller than 200 bp. An alternative method is based on the use of a hydroxylated silica resin that has high affinity for proteins but low affinity for nucleic acid at neutral pH, such as StrataClean Resin (Stratagene, La Jolla, Calif.). Quick minicolumn chromatography products are based on ion exchange or adsorption. Some of these columns are prepacked in small devices, such as pipette tips or small columns that can be centrifuged for easier elution, and are components of DNA purification kits. Commercial kits such as Magic Minipreps DNA purification columns (Promega, Madison, Wis.) and Plasmid Quick push column (Stratagene) are appropriate for plasmid DNA purification.

More recently, plasmid DNA isolation and purification have become almost routine in most clinical and research laboratories. Therefore, several companies have developed products to make nucleic acid isolation and purification easier and faster. Most of these products are basically combinations of alkaline or boiling lysis methods followed by a purification step such as those described above. Some examples of these kits are Prep-A-Gene (Bio-Rad, Richmond, Calif.), Magic DNA purification system (Promega), PlasmidQuick (Stratagene), QIAGEN (QIAGEN Inc., Chatsworth, Calif.), and Miniprep Kit Plus (Pharmacia, Piscataway, N.J.). Some of these kits work with both gram-positive and gram-negative organisms.

16.3. CHARACTERIZATION

Physical characterization of plasmid DNA is covered in sections 16.3.1 through 16.3.4. For details of genetic characterization, see sections 16.3.5 through 16.3.7.

16.3.1. Size

16.3.1.1. Gel electrophoresis of native plasmid DNA

Simple adaptations of gel electrophoresis methods (Chapter 21.5.1) (31, 68) are suitable for the detection and preliminary size characterization of plasmid DNA present in clinical or environmental isolates and in laboratory strains of gram-negative and gram-positive bacteria.

When plasmid DNA is subjected to electrophoresis in gel, the migration of the different DNA species is inversely related to their molecular weights; i.e., the higher the molecular weight, the slower the migration. Procedures involve either a vertical or a horizontal slab of agarose gel. Even partially purified lysates (generally 10 to 20 μl of lysate is sufficient) can be subjected to electrophoresis. This technique has been used for purposes ranging from the simple visualization of a plasmid in a bacterial cell to epidemiological studies made by analysis of plasmid profiles (40).

For electrophoresis, use standard vertical or horizontal gels. Gel concentrations vary from 0.25 to 1.5% agarose in a standard electrophoresis buffer (e.g., 0.04 M Tris-acetate and 0.001 M EDTA [pH 8.0], prepared from a 50× concentrated solution that is made by addition of 242 g of Tris base, 57.1 ml of glacial acetic acid, and 100 ml of 0.5 M EDTA [pH 8.0] for 1 liter). Tris-acetate rather than Tris-borate is used because it is recommended when DNA is to be extracted from the gel by using Geneclean (see section 16.1.1.2). Also with this buffer system, precipitation of salts in the electrophoresis tanks is avoided. However, it is advisable to try two different buffer systems when examining plasmid DNA for the first time. Plasmids in the 10- to 15-kbp range migrate with the chromosomal DNA band in Tris-borate gels, whereas plasmids in the 20- to 23-kbp range migrate with the chromosomal band in Tris-acetate gels. Therefore, if a plasmid is masked by residual chromosomal DNA in one buffer, it will be revealed by running the same sample in the second buffer. If speed is important, plasmids in Tris-borate migrate much faster than those in Tris-acetate.

The sample (in about 20 μl) is mixed with 2 μl of gel-loading buffer (0.25% bromophenol blue, 0.25% xylene cyanol FF, and 30% glycerol in water). Electrophoresis is ordinarily carried out for 2 h at room temperature, 60 mA, and 120 V or until the bromophenol blue band (the dark blue band) reaches the end of the gel. The gel is then placed in a solution of ethidium bromide (0.4 to 1 μg/ml) and stained for 15 min. The DNA then can be visualized in a UV light transilluminator.

Plasmids of known molecular weight can be used as standards to determine the molecular weight of an unknown plasmid by agarose gel electrophoresis. A good source of plasmids of different molecular weight is *E. coli* V517 (62). By plotting the $\log_{10}$ of the distance migrated from the origin by the plasmid DNA versus the $\log_{10}$ of the molecular mass of the plasmid DNA, a straight-line relationship is obtained from about 1 to 100 MDa. Interpolation permits the determination of molecular weights of the various plasmids.

16.3.1.2. Gel electrophoresis of digested plasmid DNA

After restriction endonuclease treatment, plasmid DNAs (and all forms of DNA, for that matter) have characteristic fragment patterns which depend on the number and spacing of the specific recognition sites within the genome. These patterns can be analyzed by electrophoresis on either agarose or polyacrylamide gels. In general, larger DNA fragments will be better resolved in agarose than polyacrylamide gels, whereas smaller DNA fragments will be better separated in polyacrylamide gels. The separation of the DNA fragments also depends on the concentration of agarose or polyacrylamide used to cast the gels. For example, 0.7% agarose gels are suitable to resolve fragments within 0.5 to 10 kbp, wherease 1.5% agarose gels will resolve fragments from 0.2 to 3 kbp. Normally no more than 2% agarose gels are used for DNA electrophoresis. Instead, for example, 15% polyacrylamide gels are good to separate fragments ranging from 0.025 to 0.250 kbp and 20% polyacrylamide gels are suitable to resolve fragments smaller than 0.1 kbp. Once electrophoresed, the DNA fragments can be transferred from the agarose gel to nitrocellulose or nylon membranes. After hybridization with a radioactive DNA "probe," the fragments containing sequences homologous to those present in the probe can be detected as sharp bands by autoradiography.

Agarose gels are prepared with the same buffer described above, but for digested DNA submarine horizontal gels (the gel is completely submerged in the buffer) are used. These type of gels can also be used for the analysis of undigested plasmid DNAs. The size of these gels, as well as the agarose concentration, varies according to the size of the fragments to be separated. Polyacrylamide gels are commonly used in a vertical apparatus; however, in our laboratories we use submarine horizontal minigels. To prepare these gels, mix 3 ml of solution A (18.1% Tris base, 0.24% *N,N,N',N'*-tetramethylethylenediamine [TEMED], 0.25 N HCl), 4.2 ml of solution B (28% acrylamide, 0.75% bisacrylamide), and 4.8 ml of 0.14% ammonium persulfate. This makes a 12% polyacrylamide gel of 5 by 6 by 0.3 cm, which is electrophoresed by using 0.06% Tris–0.28% glycine as the electrophoresis buffer. Under these conditions the migration of the bromophenol blue dye corresponds to 20-bp oligonucleotides and that of the xylene cyanol corresponds to 70-bp oligonucleotides. Depending on the sizes of the DNA fragments to be analyzed, the porosity of the gel can be adjusted by varying the concentration of acrylamide. The samples for both kinds of gels are prepared as mentioned in the previous paragraph.

Traditional agarose gel electrophoresis techniques such as those described above use an electric field of constant strength and direction. These methods are good for separating linear DNA up to approximately 50 kbp. Schwarz and Cantor (81) introduced a new concept in agarose gel electrophoresis, called **pulsed-field gel electrophoresis**. It involves the use of pulsed alternating orthogonal electric fields to separate large DNA fragments. With this technique, fragments up to 10,000 kbp have been resolved. Applications of this technique include separation of not only very large DNA fragments but also fragments between 20 and 50 kbp, commonly present in cosmid cloning. Resolution of restriction fragments in the analysis of restriction fragment length polymorphisms is also achieved by this technique. Simske and Scherer (85) described an application of this method for the analysis of CCC and OC DNA. Many apparatuses for pulsed-field gel electrophoresis currently are on the market (Hoefer, San Francisco, Calif.), and a profuse literature describing the different designs and protocols is available (15, 39, 57, 81).

In some circumstances, the ability to extract fragments that have been separated on agarose or polyacrylamide gels is highly desirable (33). These fragments can be useful for physical and chemical studies, as well as for fine-structure restriction analysis, superhelical density determination, electron microscopy, DNA sequencing, heteroduplex analysis, subcloning, and hybridization by the S1 endonuclease method (23) or by the method of Southern (87) (section 16.3.4).

Below are described techniques that permit the extraction of a variety of physically intact, biologically active DNAs in high yield directly from agarose gel bands. These techniques include electroelution, low-melting-point agarose, silica powder, and treatment with GELase (Epicentre Technologies, Madison, Wis.) for agarose gels; and "crush and soak" for polyacrylamide gels. The newer silica powder and GELase methods are the easiest and fastest and provide good results. However, some shearing can occur when eluting fragments larger than 15 kbp with silica powder. Two different kits are offered by Bio 101: Geneclean for DNA fragments between 0.5 and 15 kbp and Mermaid for DNA fragments from 10 to 500 bp. Electroelution has the advantage that no commercial products are required.

a. Electroelution

1. Visualize DNA fragments separated on an agarose gel by illumination with a UV lamp, and slice out the fragments with a razor blade. Locate the DNA band with a hand-held, long-wavelength UV light to minimize DNA damage.
2. Fill a dialysis bag with agarose electrophoresis buffer, transfer the gel slice containing the DNA to be eluted, and close the bag, avoiding air bubbles. The amount of buffer should be the minimum required to keep the gel in contact with it.
3. Immerse the bag in the electrophoresis buffer in the electrophoresis tank. Apply 4 to 5 V/cm during 2 to 3 h.
4. Reverse the polarity of the current for 1 min.
5. Recover the buffer containing the extracted DNA in solution.
6. The DNA can be purified further by extraction with organic solvents or by the use of glass or silica powder as described in section 16.2.2.

Commercial options are also available for electroelution of DNA. Analytical Unidirectional Electroeluter (IBI, New Haven, Conn.) and the SixPac electroelution unit (Hoefer) are examples of these devices.

b. Low-melting-point agarose

1. Once a regular agarose gel is prepared, cut out a transversal strip with respect to the lane in which the fragment of interest is going to run. Fill this space with low-melting-point agarose, and let it solidify in a cold room.
2. Load and run the DNA sample in a cold room, monitoring the migration of the band of interest, preferably with a hand-held, long-wavelength UV light to minimize DNA damage.
3. Once this band enters the portion of low-melting-point agarose, stop the agarose electrophoresis.
4. Cut out a slice of low-melting-point agarose containing the band, and transfer it to a plastic tube.
5. Add 5 volumes of TE buffer, and incubate for 5 min at 65°C. Then treat it as a regular DNA solution; i.e., purify the DNA by extraction with organic solvents.

c. Silica powder

Here the protocol for extraction of DNA with Geneclean is described. Protocols involving other similar products are similar and are provided by the supplier.

1. After electrophoresis, cut out a slice of agarose containing the fragment of interest. Weigh it, and add 3 volumes of 5 M sodium iodide.

2. Incubate at 55°C for about 5 min, shaking occasionally until the agarose melts.
3. Add about 5 μl of the suspension of silica powder, vortex, and let the solution remain in contact with the powder for 5 min at room temperature, mixing occasionally to prevent sedimentation of the powder.
4. Recover the powder by centrifugation in a microcentrifuge (16,000 × g) for 5 s, and wash three times with the washing solution provided by the supplier.
5. Extract the DNA by incubation at 55°C in the presence of TE buffer. Recover the DNA solution by centrifugation.

d. GELase

GELase is an enzyme that degrades low-melting-point agarose into small oligosaccharides that are soluble in water and alcohol. The product of this reaction is a clear liquid that does not become viscous even in an ice bath. Two protocols have been developed by the supplier.

Fast protocol for GELase

1. Cut out the low-melting-point agarose portion containing the DNA band of interest, and determine the weight of the gel slice.
2. Add 50× GELase buffer (1 μl/50 mg of gel), and completely melt the gel by incubation at 70°C.
3. Equilibrate the temperature of the molten gel at 45°C, add the proper amount of GELase (1 U/300 mg of molten 1% low-melting-point agarose when the agarose electrophoresis buffer was Tris-acetate or morpholinepropanesulfonic acid [MOPS], or 1 U/80 mg when Tris-borate buffer was used). Incubate at 45°C for 1 h.

High-activity protocol for GELase

1. Cut out the low-melting-point agarose portion containing the DNA band of interest, and determine the weight of the gel slice.
2. Replace the electrophoresis buffer by soaking the gel slice in 10 volumes of 1× GELase buffer at room temperature for 1 h.
3. Melt the agarose slice at 70°C for 20 min, and then equilibrate the tube at 45°C.
4. Digest the molten agarose by adding 1 U of GELase per 600 mg of low-melting-point agarose for 1 h at 45°C.

The alcohol precipitation steps that follow are common for both GELase protocols.

1. Add 1 volume of 5 M ammonium acetate and 2 volumes of ethanol, and mix gently and thoroughly.
2. Recover the DNA by centrifugation in a microcentrifuge (16,000 × g) for 5 min.
3. Wash the pellet with 70% ethanol, and dry.
4. Dissolve the DNA in the appropriate buffer.

e. "Crush and soak"

1. Locate the band of interest in the polyacrylamide gel, and cut it out.
2. Transfer to a plastic tube, crush the agarose polyacrylamide with a pipette tip, and add 1 to 2 volumes of a solution containing 0.5 M ammonium acetate, 0.01 M magnesium acetate, 0.001 M EDTA (pH 8.0), and 0.1% SDS.
3. Incubate at 37°C on a rotary shaker for 3 to 24 h depending on the size of the DNA fragment (fragments larger than 0.5 kbp will take between 12 and 24 h).
4. Centrifuge the sample in a microcentrifuge (16,000 × g) for 1 min at 4°C, and transfer the supernatant to a fresh tube.
5. Reextract the polyacrylamide pellet with 0.5 volume of the same solution.
6. Combine the two supernatants, and precipitate the DNA by addition of 2 volumes of cold ethanol. Recover the DNA by centrifugation.
7. Redissolve the DNA in 0.2 ml of TE buffer. Reprecipitate the DNA by the addition of 1/9 volume of 5 M potassium acetate (pH 4.8) and 2.5 volumes of cold ethanol.
8. Collect the DNA by centrifugation, wash the pellet with cold 70% ethanol, dry, and resuspend in TE buffer.

16.3.2. Copy Number

The number of copies of a plasmid present per chromosome equivalent is a parameter that characterizes a plasmid and also gives information about the nature of its replication. There are several ways to express copy number, e.g., per chromosome equivalent, per cell volume, or per mass. Plasmid copy number can be estimated by centrifugation of a [^{3}H]thymine-labeled total-cell lysate in an ethidium bromide-cesium chloride gradient or by agarose gel electrophoresis (26). A method based on in situ colony lysis with known cell volumes was developed by Shields et al. (84) for mini-F, a low-copy-number plasmid. A specific probe for mini-F was used for hybridization with DNA immobilized on nitrocellulose. Although this method offers the advantage of simplicity as well as sensitivity and low cost, a potential drawback is the requirement for a specific probe that recognizes sequences only in the plasmid DNA.

16.3.2.1. CsCl-ethidium bromide gradient centrifugation

If the bacterial strain requires thymine, use an adequate concentration of thymine for growth plus [^{3}H]thymine (7 μCi/ml). For a strain that does not require thymine, use 1 μg of thymine, 250 μg of deoxyadenosine, and 7 μCi of [^{3}H]thymine per ml. For some gram-positive bacteria, [^{3}H]thymidine is all that need be added to the cultures in rich media to label the DNA.

The procedure for gram-negative bacteria is as follows.

1. Grow a culture (1 ml) of a plasmid-containing strain at 37°C in the presence of 7 μCi of [^{3}H]thymine per ml to a density of 2×10^8 cells per ml.
2. Centrifuge the cells, wash them in TE buffer, and resuspend them in 400 μl of 0.5 M sucrose (ultrapure, RNase-free sucrose in 0.5 M Tris-HCl [pH 8.0]).
3. Add 0.2 ml of 0.2 M EDTA [pH 7.8] to the cell suspension, and mix well.
4. Add 0.2 ml of a fresh solution containing 1% lysozyme (in 0.25 M Tris-HCl [pH 8.0]) to the cell suspension, and mix well. Shake gently, and let stand in ice for 15 min.
5. Add 1.2 ml of 1.2% Sarkosyl to the suspension, and then immediately add 0.1 ml of pronase (1 mg/ml) in 0.01

M Tris-HCl (pH 8.1) and 0.02 M EDTA. Pronase should be first self-digested for 2 h at 37°C and then heated for 2 min at 80°C before use. Mix gently, and incubate at 37°C until the suspension clears completely.

6. Draw the DNA solution rapidly in and out of a 5-ml pipette (to shear the chromosomal DNA) until the solution is no longer viscous.

7. Centrifuge a sample containing about 500,000 cpm of radioactive DNA in a 5-ml ethidium bromide-cesium chloride gradient (the refractive index must be 0.0025 lower than the refractive index used for Triton X-100 lysate CsCl-ethidium bromide gradient). Also centrifuge a sample in a 5 to 20% linear sucrose gradient. Sucrose gradients can be prepared in a gradient mixing chamber or, alternatively, by making a sucrose solution of an average concentration between the two desired extreme concentrations. Freezing of this solution at −20°C and slow thawing at 4°C (overnight) permits the development of a gradient suitable for analytical purposes. For example, a frozen 12.5% sucrose solution after thawing will render a gradient that is approximately 5 to 20%.

8. Collect the gradients on Whatman 3 MM filter discs.

9. Precipitate radioactive DNA with cold 5% trichloroacetic acid containing 50 μg of thymine per ml, wash with 95% ethanol, and dry. Determine radioactivity in a scintillation spectrometer.

Figure 2 shows a cesium chloride-ethidium bromide gradient with the profile obtained for ^{3}H-labeled total-cell DNA from a plasmid-containing strain of *E. coli*. The ratio between the plasmid and chromosome peaks can be used for the calculation of copy numbers per chromosome equivalent, as follows.

$$\text{Copy number} = \frac{\text{cpm of plasmid peak}}{\text{cpm of chromosome peak}} \times \frac{MW_c}{MW_p}$$

where MW_c and MW_p are the molecular weights of the chromosome and the plasmid, respectively.

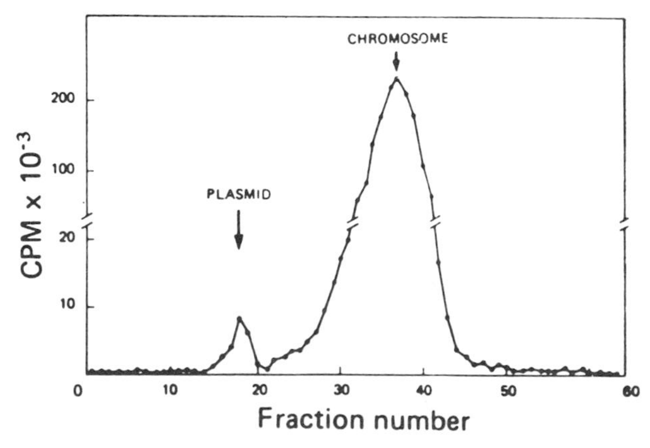

FIG. 2. Determination of plasmid copy number. A plasmid-containing bacterial strain was grown in a minimal salts medium plus Casamino Acids and glucose. Labeling and lysis were carried out, and the lysate was centrifuged in the CsCl-ethidium bromide gradient, as described in section 16.3.2.1. Seven-drop fractions were collected on microtiter trays and assayed for radioactivity.

Comments about determination of plasmid copy number

Plasmid forms other than CCC DNA are detected in the same cesium chloride-ethidium bromide gradient (or in sucrose gradients) by their banding characteristics. For example, dimers of two CCC DNA molecules will coband with CCC DNA in a CsCl-ethidium bromide gradient but will sediment ahead of monomeric CCC DNA in a sucrose gradient. Dimers of one CCC and one OC DNA molecule will band at a position intermediate between CCC DNA and OC DNA in the CsCl-ethidium bromide gradient. Dimers of two OC DNA molecules will coband with OC DNA in a CsCl-ethidium bromide gradient but will band ahead of OC DNA in a sucrose gradient.

16.3.2.2. Agarose gel electrophoresis

A faster method to estimate copy number by using agarose gel electrophoresis (95) is as follows.

1. Grow cells until the OD_{600} reaches 0.8. Pellet the cells by centrifugation of 1.5 ml of the culture.

2. Wash the cells twice with 0.85% NaCl. Freeze the cells at −20°C for 1 h.

3. Thaw the cells, resuspend in 30 μl of a solution containing 0.01 M EDTA, 10 μg of boiled RNase per ml, and 2 mg of lysozyme per ml, and incubate for 15 min at 37°C.

4. Add 30 μl of 2% SDS, mix, and incubate for 10 min at 60°C.

5. Add 10 μl of loading solution (40% sucrose, 0.25% bromophenol blue).

6. Load in the well of a horizontal agarose gel (section 16.3.1), and after all air bubbles are removed (this can be achieved by holding a flamed loop over the well), fill the well with agarose to prevent the viscous lysate from floating out of the well. Run the electrophoresis.

7. Stain the gel with ethidium bromide, photograph, and scan with a densitometer.

8. Cut out the peak areas from the densitometry plots, and weigh them. Some densitometers have an integrator; if so, the integrated values can be used to calculate the copy number. In the formula shown above, counts per minute can be replaced by the weight or the integrated values of the peaks.

This method can also be performed by using a coresident plasmid of known copy number as an internal control, as follows.

1. Transform *E. coli* cells already carrying a control plasmid (such as the low-copy-number plasmid pSC101) with the plasmid of unknown copy number.

2. Prepare plasmid DNA from this *E. coli* strain by any of the methods described above.

3. Subject the plasmid DNA to agarose gel electrophoresis as described in section 16.3.1.1. Stain the gel with ethidium bromide, photograph, and scan the bands corresponding to the control and problem plasmids with a densitometer.

4. Cut out the peak areas from the densitometry plots, and weigh them. Some densitometers have an integrator; if so, the integrated values can be used to calculate copy number by using the copy number and molecular weight of the reference plasmid (rp) in the following formula.

$$\text{Copy number} = \frac{\text{weight of plasmid peak}}{\text{weight of rp peak/rp copy number}} \times \frac{MW_{rp}}{MW_p}$$

If both the control and problem plasmids contain unique restriction sites, the following protocol can be applied.

1. Transform *E. coli* cells already carrying a control plasmid with the plasmid of unknown copy number.
2. Prepare plasmid DNA from this *E. coli* strain by using any of the methods described above. Digest the plasmids with a restriction endonuclease that cuts each plasmid once.
3. End label the plasmid DNAs by mixing 5 μl of 10× kinase buffer (0.5 M Tris-HCl [pH 9.5], 0.1 M $MgCl_2$, 0.05 M dithiothreitol, 50% glycerol), 10 U of T4 polynucleotide kinase, 150 μCi of [γ-^{32}P]ATP, 9.8 μl of DNA (0.1 ng/μl in TE buffer), and 19.2 μl of distilled water. Incubate the reaction for 60 min at 30°C.
4. Precipitate the end-labeled plasmid with 0.5 volume of 6 M ammonium acetate and 2 to 3 volumes of ethanol. Let stand at −20°C for 1 h. Collect the DNA by centrifugation in a microcentrifuge (16,000 × *g*) for 5 min. Wash the pellet three times with 70% ethanol, and dry under vacuum.
5. Resuspend the DNA in 20 μl of water, add 2 μl of gel-loading buffer (see section 16.3.1), mix, and separate the DNAs by agarose gel electrophoresis. Localize the bands by UV transillumination, and excise the gel fragments containing each plasmid. Place each gel fragment in a scintillation vial, add scintillation cocktail, and determine radioactivity in a liquid scintillation counter. The copy number of the plasmid is determined by normalizing to the reference plasmid. In this case the determination is independent of the molecular weights of the plasmids.

16.3.3. Base Composition

Another means of characterizing plasmid DNA is by its base composition. The percentage of guanosine and cytosine (G+C content) in the total plasmid DNA content is a characteristic of a particular molecule of DNA. For example, by comparing the G+C content of plasmids or chromosomes, it can be determined whether the plasmids are related to each other or to the chromosome of the bacterial host. Methods to determine the G+C content of DNA are described in Chapter 26.4.

16.3.4. DNA "Homology" (see Chapter 26 Introduction)

The dissociation of double-stranded plasmid DNA followed by specific hybridization (reassociation, reannealing) with single-stranded DNA from a homologous or heterologous source is the basis of all homology studies. The following methods use single-strand-specific endonuclease and immobilized DNA for hybridization. The technique based on the use of single-strand-specific endonuclease is more sensitive and is thus the preferred method of quantitatively determining the homology among DNA molecules. Hybridization with immobilized DNA on nitrocellulose or nylon membranes is easier and faster, but the relationship between the degree of homology and the signal is not linear. Therefore, this last technique is convenient when DNA fragments with a high level of homology (usually more than 70%) must be detected.

DNA homology experiments can be carried out to study the characteristics of a certain DNA, in which case the method that uses single-strand-specific endonuclease is recommended, or as a tool to detect or identify DNA fragments, in which case the membrane-immobilized DNA method is chosen.

16.3.4.1. Single-strand-specific endonuclease

Plasmid homo- and heteroduplexes can be analyzed with the single-strand specific endonuclease, S1, of *Aspergillus oryzae*. The following procedure (23) permits an accurate and rapid determination of polynucleotide sequence relationships. It is particularly useful for surveys and for other investigations which require a large number of DNA-DNA hybridization assays.

Purified, radiolabeled plasmid DNA is used as the reference probe and is hybridized with purified total-cell DNA from plasmid-containing strains. DNA from a plasmid-free strain serves as a control. DNA-DNA reassociation reactions assayed by the S1 endonuclease method are performed by incubating approximately 5,000 to 10,000 cpm (typically less than 0.001 μg of DNA) of sheared, denatured, purified plasmid DNA with 150 μg of unfractionated, total, sheared, denatured bacterial DNA preparations from plasmid-containing and plasmid-free bacteria in a total volume of 1 ml of 0.42 M NaCl. The DNA mixtures are incubated at a temperature which depends on the G+C content of the plasmid DNA. The time of reassociation is such that essentially complete reassociation for the homologous reaction is achieved. The time required for a given plasmid to reassociate completely is a function of its molecular mass and the ionic concentration. The time of reassociation can be determined by calculating the $C_0t_{1/2}$. C_0 is the initial concentration of DNA; $t_{1/2}$ is the time taken for 50% reassociation of the DNA (at C_0 concentration) at a given temperature, salt concentration, and DNA fragment size. The empirical relationship (14) is as follows.

$$C_0t_{1/2} = \frac{\text{molecular weight of plasmid}}{3 \times 10^7}$$

The approximate time of incubation to get complete reassociation is equivalent to about 8 $C_0t_{1/2}$ in 0.42 M NaCl. Thus,

$$8\ C_0t_{1/2} = C_0t_x$$

where C_0 is the initial concentration of plasmid DNA (which can be calculated or estimated from the plasmid copy number) and t_x is the time of reassociation.

Under optimum conditions, less than 10% of the labeled plasmid DNA incubated alone should reassociate with itself, while more than 85% of the labeled plasmid DNA should reassociate with its homologous unlabeled DNA. The reassociation of labeled plasmid DNA with DNA of a plasmid-free bacterial strain is also included in each experiment, and this value (about 10% or less) is subtracted from the values of all reactions.

a. Hybridization step

1. Prepare the reassociation mixture to contain 150 μg of unlabeled sonicated total-cell DNA, 5,000 to

10,000 cpm of labeled sonicated plasmid DNA, and sufficient NaCl to bring the solution to 0.42 M in a total volume of 1.0 ml.
2. Denature the DNA by boiling the solution for 10 min. Then place the DNA solution in ice.
3. Incubate the reassociation mixture at 55 to 70°C depending on the G+C content of the plasmid DNA.

[^{3}H]thymine-labeled plasmid DNA is prepared by isolating CCC DNA by the Triton X-100 lysis technique (section 16.1.1.1), or it can be labeled in vitro. The specific activity of the [^{3}H]thymine plasmid DNA obtained by this method is usually about 10^6 cpm/μg. The plasmid DNA is subjected to sonic treatment to obtain fragments of an approximate molecular weight of 2.5×10^5.

Total-cell DNA is prepared as described previously (49). This DNA is also degraded by sonic treatment to an approximate molecular weight of 2.5×10^5.

b. S1 endonuclease reaction step

1. Prepare stock solution of S1 reaction mixture (0.125×10^{-3} M $ZnSO_4$, 0.087 M NaCl, 0.038 M sodium acetate buffer [pH 4.5], 25 μg of calf thymus DNA per ml). Store frozen at −20°C in 0.8-ml samples. The reaction mixture will then contain 0.8 ml of the S1 reaction mixture and 0.2 ml of the reassociation mixture. The final reagent concentrations are 0.1×10^{-3} M $ZnSO_4$, 0.150 M NaCl (this includes the NaCl added with 0.2 ml of the reassociation mixture), 0.03 M sodium acetate buffer (pH 4.5), and 20 μg of calf thymus DNA per ml. For each reassociation reaction, prepare two S1-treated and two untreated samples.
2. Add 187.5 U of S1 endonuclease to start the reaction. Incubate at 50°C for 20 min.
3. Stop the reaction by placing the tubes in an ice bath; add 50 μg of calf thymus or salmon sperm DNA per ml as carrier, and add 0.3 ml of cold 20% trichloroacetic acid. Collect the trichloroacetic acid precipitate by vacuum filtration on a membrane filter (type HA; Millipore Corp.). Dry the filters at 70°C, and determine the radioactivity by counting in a liquid scintillation spectrometer. Controls should include native-labeled plasmid DNA, which will detect any double-stranded nuclease activity in the S1 enzyme preparation; denatured and reassociated labeled DNA alone, which is a control for the amount of self-reassociation of the labeled DNA under the conditions of reassociation; and denatured labeled DNA, which is a control for the efficiency of the single-stranded activity of the S1 endonuclease preparations. Table 2 shows how to transform the raw radioactivity results into percent DNA homology. The results are expressed as the percentage of the untreated controls normalized to the values obtained for the homologous reaction.

16.3.4.2. DNA immobilization and hybridization

DNA can be immobilized on nitrocellulose or nylon membranes without any isolation or purification steps (colony hybridization) or after agarose gel electrophoresis (Southern blot hybridization).

a. Colony transfer

1. Transfer cells from each colony to be tested with a sterile toothpick to two plates, one a regular solid medium (master plate) and the other on whose surface a membrane has been laid. Incubate.
2. After incubation, keep the master plate at 4°C and lyse the cells in situ by placing the membrane, colony side up, on top of a filter paper soaked with 0.5 N sodium hydroxide. Let stand for 10 min at room temperature.
3. Transfer the membrane to a filter paper soaked with 0.5 M Tris-HCl (pH 7.5) for 5 min.
4. Transfer the membrane to a filter paper soaked with a solution containing 0.5 M Tris-HCl (pH 7.5) and 1.5 M NaCl for 5 min.
5. Transfer the membrane to a filter paper soaked with 2× SSC solution (20× SSC solution is 175.3 g of NaCl and 88.2 g of sodium citrate per liter adjusted to pH 7.8).
6. Immobilize the DNA by either baking under vacuum for 2 h at 80°C or UV cross-linking. UV cross-linking is performed with commercial equipment (Stratalinker; Stratagene). The entire cross-linking process takes only 30 to 60 s, in contrast to the 2 h needed for baking.
7. Keep the membrane in a sealed plastic bag until hybridization. Any commercial bag sealer will be suitable for this purpose. Also, many laboratories store blots between tissue wiper sheets (Kimwipes) in a drawer for long periods.

This technique can be applied to *B. subtilis* and other *Bacillus* species without modification. However, other gram-positive bacteria such as *S. aureus* and streptococci are more difficult to lyse. In these cases it is recommended to soak the nitrocellulose paper in a lysostaphin solution prior to its use (74a).

b. Southern blot transfer

1. Soak the agarose gel containing the DNA in 0.25 N HCl, and gently rock for 15 min. Make sure that the gel is totally covered. Rinse twice with distilled water quickly.
2. Denature the DNA by soaking the gel in a solution containing 1.5 M NaCl and 0.5 M NaOH with gentle rocking for 20 min. Rinse twice with distilled water quickly.
3. Neutralize by soaking the gel in a solution containing 1.5 M NaCl and 1 M Tris-HCl (pH 8.0) with gentle rocking for 20 min. Rinse twice with water.
4. Place the gel on top of a stack of glass plates covered with a strip of filter paper soaked with 6× SSC. The width of the filter paper should be such that it can accommodate the gel and its length should overhang the plates. Soak a nitrocellulose or nylon membrane the size of the gel with 6× SSC, and place it on top of the gel, avoiding air bubbles between the gel and the membrane (Fig. 3).
5. Place this device in a pan containing 6× SSC, and allow the overhanging filter to touch the solution.
6. Place on top of the membrane a piece of filter paper the size of the membrane. Avoid air bubbles between the membrane and the filter paper.
7. Cut a stack of paper towels just smaller than the filter, and place on the filter paper. Add a glass plate and a weight of about 500 g on top of the paper towels.
8. Allow transfer of the DNA for about 10 h.

Table 2. Determination of homology between heterologous plasmids by use of single-strand specific endonuclease

Sample	cpm after S1 nuclease[a]		Raw %	Avg %	Corrected and normalized homology[b] (%)
	Plus nuclease	No nuclease			
1. Heat-denatured ^{3}H-labeled plasmid DNA	20	821	2.4	2.3	
	18	832	2.2		
2. Heat-denatured and reassociated ^{3}H-labeled plasmid DNA	50	850	6.0	6.0	
	53	890	6.0		
3. Heat-denatured reassociated mixture of ^{3}H-labeled plasmid DNA and total-cell DNA containing the same plasmid	790	850	92.9	94.8	100
	798	825	96.7		
4. Heat-denatured and reassociated mixture of ^{3}H-labeled plasmid DNA and total-cell DNA containing a heterologous plasmid	45	820	5.5	6.3	<0.3
	63	881	7.1		
5. Heat-denatured and reassociated mixture of ^{3}H-labeled plasmid DNA and total-cell DNA from a plasmidless derivative	60	860	7.0	7.0	<1
	62	880	7.0		

[a]The count-per-minute values were obtained by subtracting a background of 20 cpm.
[b]Values in this column were obtained by subtracting the value in line 2 from the respective values in lines 3, 4, and 5 and then normalizing to the corrected value in line 3.

9. Remove the membrane, immobilize the DNA by baking or UV cross-linking, and keep the membrane as described for colony blot (section 16.3.4.2.a).

c. Probe labeling

Two methods are available to radioactively label DNA, nick translation and random priming. The latter method provides probes of higher specific activity. Presently, commercial kits to label DNA by either of these two methods are available from many companies. A protocol for the random-priming method is as follows.

1. In an Eppendorf tube containing 10 μl of oligolabeling buffer (OLB), mix 2 μl of 10 mg/ml gelatin or the same concentration of nuclease-free bovine serum albumin, 2 U of Klenow fragment of DNA polymerase I, and 5 μl of [α-^{32}P]dATP. To make OLB, mix solutions A (1.25 M Tris-HCl [pH 8.0], 0.125 M $MgCl_2$, 0.25 M 2-mercaptoethanol, 0.0005 M deoxynucleoside triphosphates except dATP), B (2 M *N*-2-hydroxyethylpiperazine-*N'*-2-ethanesulfonic acid [HEPES; pH 6.6]), and C (90 OD_{260} units/ml random hexanucleotides dissolved in 0.001 M Tris-HCl, [pH 7.0]–0.001 M EDTA) in a proportion of 100:250:150. Keep the solutions at −20°C. The hexanucleotides are commercially available from different companies, such as Boehringer Mannheim, Indianapolis, Ind., and Pharmacia LKB Biotechnology Inc., Piscataway, N.J.
2. Add 8 μl of the DNA to a separate Eppendorf tube, and denature by placing it in a heating block at 100°C for 5 min. Transfer the tube to an ice-water bath.
3. Transfer the DNA to the tube containing the reaction mixture, and incubate at room temperature for 2 h.
4. To determine the incorporation, spot 1 μl of the mixture on a filter paper. Dry, and determine the total counts per minute. Precipitate the labeled DNA by submerging the filter in cold 5% trichloracetic acid, and let it stand for 10 min in ice. Pour off the liquid, and wash once with cold 5% trichloracetic acid and once with cold ethanol. Dry, and determine the incorporated counts per minute. The yield is calculated by dividing the incorporated counts per minute by the total counts per minute. Normally about 90% of labeled nucleotide is incorporated.

Other systems of labeling probes and detection of the

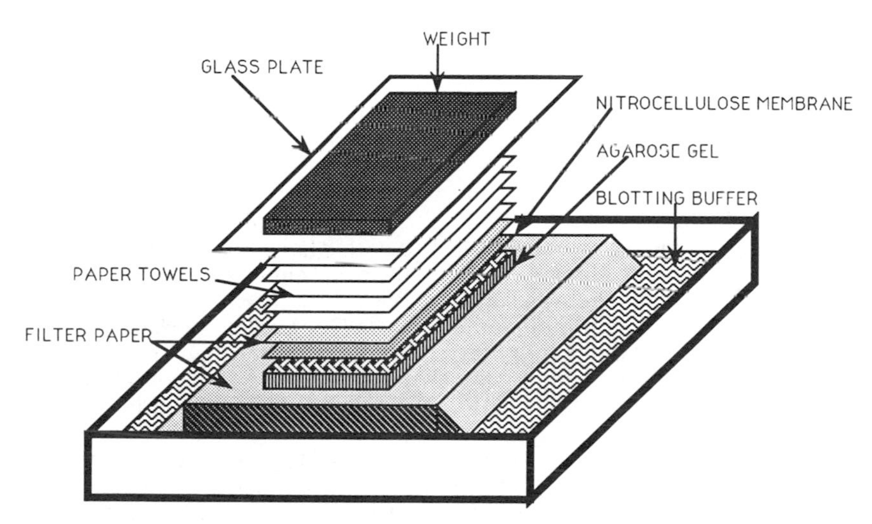

FIG. 3. Diagram of a Southern blot assembly.

hybrid products by different strategies that do not involve radioactivity have been described and can be obtained as kits from several companies. They are based on modified nucleotides that can be recognized by specific ligands such as the complex streptavidin-biotin or antigen-antibody. These complexes can be detected either by enzyme (phosphatase or peroxidase)-specific chromogenic substrates that produce a visible signal or by chemiluminescence. The reagent 4-methoxy-4(3-phosphatephenyl)spiro[1,2-dioxetane-3,2′-adamantane] (LUMI-PHOS 530; Lumigen, Inc., Detroit, Mich.) has been used for this purpose and has a sensitivity comparable to that obtained using radiolabeled probes. For further details on nonisotopic probes, see Chapter 28.4.2.

d. Hybridization

1. To the membrane contained in a plastic bag, add enough prehybridization solution to keep it wet. Prehybridize for 2 h at 37°C in prehybridization solution (6× SSC, 0.5% SDS, 50% formamide, 5× Denhardt's solution). 50× Denhardt's solution is 1% Ficoll, 1% polyvinylpirrolidone, and 1% bovine serum albumin. The stringency can be modified by changing the concentration of formamide. Each increase of 1% in the formamide concentration in the hybridization buffer lowers the melting temperature of duplex DNA by 0.7°C (66).
2. Add 10^6 cpm of the radiolabeled probe to a tube containing 200 μl of 100 μg/ml sonicated salmon sperm DNA. Denature by heating at 100°C for 5 min.
3. Hybridize at 37°C for about 3 h.
4. Remove the membrane from the bag, and wash it with a solution containing 2× SSC and 0.1% SDS at 65°C for 1 h, changing the washing solution three times, once every 20 min. Depending on the degree of homology between the DNAs, different temperatures ranging from 45 to 70°C should be tested.
5. Wash the membrane briefly with 2× SSC.
6. Dry the membrane, and expose it to an X-ray film. For nylon membranes, drying should be avoided if reuse with a different probe is intended. To remove the radioactive bound probe from used nylon membranes, wash them at 65°C for 30 min in a solution containing 50% formamide and 6× SSPE (20× SSPE is 3.6 M NaCl, 0.2 M sodium phosphate [pH 7.7], and 0.02 M EDTA) and rinse with 2× SSPE.

16.3.5. Replication and Maintenance Functions

All plasmids possess essential determinants that constitute the minimal replicon. These determinants, including self-perpetuation (replication and stable inheritance) in dividing bacteria, are known as replication and maintenance functions. Maintenance of a plasmid includes different functions, the main one being a partition system. Others include site-specific recombination systems, anticlumping systems, and killing of cells that have lost the plasmid (89). Plasmids with high copy number under conditions where they can freely diffuse within the cell may not require a partition system, because the probability that one cell gets all the copies and the other gets none during cell division is very low. It should be said that most plasmids are stably maintained in a limited number of bacterial species. Some, such as RK2, have the characteristic of being stable in many species and are called wide-host-range plasmids (32). Others, such as ColE1, P15A, and pSC101, can exist in only a few species and are designated narrow-host-range plasmids (11). The host range of these three plasmids is limited to *E. coli* and closely related enterobacterial species. Besides these determinants, several large low-copy-number plasmids also encode conjugation functions that promote their transfer to bacterial cells of either the same or different species. In addition to inheritance functions, most plasmids encode a wide range of specialized functions, such as resistance to antimicrobial agents, acquisition of essential nutrients, degradation of organic compounds, and virulence factors. However, there are cryptic plasmids for which none of these latter functions have been assigned.

Replication and maintenance functions are studied by introducing the plasmid into the bacterial cell. This can be achieved by using transformation or transduction for naked and packaged DNA, respectively. Protocols to carry out these procedures are described in Chapters 14 and 15. Replication and maintenance can be monitored by taking advantage of one of the specialized functions. Isolation of cells expressing the specialized function, for example antibiotic resistance, is an indication that the plasmid replicates in that strain. However, the possibility of a genetic rearrangement must be discounted by isolating the original plasmid from transformants.

Maintenance can be studied as follows (73).

1. Inoculate 1 ml of the appropriate liquid medium without antibiotic with one isolated colony grown on selective agar medium.
2. Take samples of the culture at different times until it reaches stationary phase. The times vary according to the bacterial species used; for *E. coli*, take samples every 3 h.
3. Dilute each sample to allow colony counting, and spread on plates with and without antibiotic.
4. After incubation, count the colonies grown under both conditions and estimate the frequency of plasmid loss by plotting the percentage of plasmid-containing cells versus the number of generations. The number of generations is calculated by dividing the time at which the sample was taken by the doubling time of the bacterial strain. A stable plasmid should display about the same percentage of plasmid-containing cells throughout the growth of the liquid culture. It is convenient to confirm the plasmid presence by screening some of the colonies growing in the presence of the antibiotic.

16.3.6. Incompatibility

Incompatibility is the inability of two plasmid species that have been introduced or transferred to a single bacterial clone to be stably maintained in the same cell line in the absence of selective pressure for both plasmids (22, 35, 71). Incompatibility is a consequence of relatedness between the plasmids; i.e., incompatible plasmids share some or all functions involved in replication or maintenance. Plasmids have been classified by being assigned to incompatibility groups (plasmids in the same group are incompatible).

A simple method to determine incompatibility between two plasmids, provided they have different markers, is as follows.

1. Transform a strain carrying plasmid A (resident) with plasmid B (incoming). This transfer can also be accomplished by conjugation. Select transformants or transconjugants with antibiotics selective for both plasmids. It is convenient to use a *recA* bacterial host to prevent recombination between the two plasmids.
2. Inoculate broth containing both antibiotics with the strain carrying both plasmids, and allow growth up to the stationary phase.
3. Dilute the culture 10^6-fold into antibiotic-free broth, and incubate until the stationary phase. At this point, plate a suitable dilution on antibiotic-free agar medium to obtain isolated colonies.
4. Toothpick the colonies to two agar plates, each containing one antibiotic, to determine the frequency of cells carrying one or the other of the two plasmids. Growth of colonies in both plates indicates that both plasmids are compatible.

Incompatibility testing presents some potential problems: the lack of resistance markers in one or both plasmids or the presence of the same marker in the plasmids, the inhibition of entry of the incoming plasmid (entry or surface exclusion), and the presence of more than one incompatibility function on either plasmid. To overcome these difficulties, a series of miniplasmids (representing six major plasmid incompatibility groups) containing incompatibility functions of reference plasmids and the *gal* operon was designed by Davey et al. (29). The miniplasmids are constructed by cloning plasmid DNA fragments that do not include surface exclusion functions. The use of galactose utilization permits one to determine incompatibility by monitoring the loss of the miniplasmid, revealed by the appearance of Gal^- segregants on indicator plates.

Gal^- indicator plates are prepared by adding 1% galactose and 25 μg of 2,3,5-triphenyltetrazolium chloride per ml to sterile L-agar medium (90a). Colonies that ferment galactose are white, and those that do not ferment the sugar are red.

Despite these improvements, incompatibility tests can still give misleading results because of the presence of more than one replicon in a single plasmid or the presence of some replicons that, as a result of their mechanism of control of replication by using antisense RNAs, either can be detected as compatible even though they differ in just a few bases or can show cross-incompatibility (22).

To address these problems, a classification scheme based on replicon typing was developed for plasmids present in gram-negative bacteria; it involved the use of specific DNA probes to detect the presence of a known basic replicon by DNA-DNA hybridization (22). Before performing replicon typing, the plasmid profile of the strain to be tested should be determined. If only a single plasmid replicon is present, typing can be carried out directly by colony hybridization. If two or more plasmids are present, they must be separated by gel electrophoresis and transferred to nitrocellulose or nylon membrane before the hybridization reaction. The protocol can be summarized as follows.

1. Immobilize the DNA in nitrocellulose or nylon, obtained by colony lysis or plasmid purification and gel electrophoresis as described in section 16.2.
2. Digest the hybrid plasmid carrying the probe with the appropriate restriction enzyme. Separate the fragments by agarose gel electrophoresis, extract the DNA band to be used as probe, and label the DNA as described in section 16.3.1.4.
3. Hybridize the immobilized DNA with the labeled probe, wash, and detect the radioactive hybrids using the conditions described in section 16.3.1.4.

Plasmids sharing highly related replicons will show a positive signal on X-ray films.

16.3.7. Curing

Elimination ("curing") of a plasmid from a bacterial culture is the best test to substantiate the relationship between a genetic trait and carriage of that specific plasmid by the culture. A phenotype linked to the presence of the plasmid will not be expressed in cured cultures. Reintroduction of the plasmid or a recombinant clone derivative carrying the specific genetic trait must be accompanied by regaining of the phenotype.

Curing of plasmids was first reported more than 25 years ago (52). The artificial elimination of F was reported by Hirota more than 35 years ago, before the nature of the "F factor" was even understood (45). Since then, various protocols involving chemical and physical agents have been developed to cure plasmids. Depending on the nature of the bacterial host and/or the plasmid, some methods work better than others. Protocols for curing plasmids consist basically of exposure of a culture to different concentrations of the chemical agent or to higher than normal temperature, followed by plating and analysis of the presence of a trait carried by the plasmid. Colonies that lose the trait are then analyzed for their plasmid content to confirm loss of the plasmid. Some plasmid curing protocols are described below.

16.3.7.1. Elevated incubation temperature

The elevated incubation temperature protocol was used to cure *V. anguillarum* (24, 90), which grows normally at 30°C.

1. Inoculate 1 ml of Trypticase soy broth supplemented with 1% NaCl with 0.01 ml of an overnight culture of *V. anguillarum*. Incubate at 37°C for 16 h.
2. Use 0.01 ml of this culture as an inoculum for 1 ml of the same broth, and incubate at 37°C for another 16 h. This procedure can be repeated several times.
3. Spread the cells on Trypticase soy agar plates supplemented with 1% NaCl, and incubate at 30°C.
4. Analyze colonies for the presence of the plasmid. With this protocol, about 3% of the colonies are cured.

Basically the same protocol was used to cure a plasmid from *Acinetobacter baumanii*, but L medium and incubation at 42°C were used.

16.3.7.2. Intercalating dyes

1. Inoculate 2 ml of L broth in tubes with 20 μl of an overnight inoculum. Add to each tube increasing concentrations of an intercalating dye (e.g., ethidium bromide or acridine orange) ranging from 10 to 300 μg/ml.
2. Incubate overnight, and select the culture with the highest concentration of the acridine dye at which growth occurred.

3. Dilute the culture, plate on L agar, and incubate to obtain isolated colonies.
4. Analyze the colonies for plasmid content.

16.3.7.3. Sodium dodecyl sulfate

1. Inoculate 2 ml of broth containing 10% SDS with 20 μl of an overnight inoculum.
2. Incubate overnight at 37°C.
3. Plate after dilutions to obtain isolated colonies. Incubate the plates at 37°C overnight.
4. Analyze the colonies for the presence of the plasmid.

16.3.7.4. Electroporation

Electroporation was used originally for the transformation of bacteria with DNA (67) (Chapter 14.1.3.3). More recently, this method was used to cure a pBR322 derivative from *E. coli* DH1 (41), as follows.

1. Incubate the cells in 10 ml of L broth.
2. Pellet the cells, and wash three times with distilled water.
3. Let the cells stand in ice for 5 min.
4. Subject the cells to one electrical pulse (Gene Pulser [Bio-Rad]; pulse at 2.5 kV and 25 μF, time constant ca. 4.5 ms).
5. Add 1 ml of SOC (2% tryptone, 0.5% yeast extract, 0.01 M NaCl, 0.0025 M KCl, 0.01 M $MgCl_2$, 0.01 M $MgSO_4$, 0.02 M glucose) immediately after the pulse, and incubate for 20 min at 37°C.
6. Make dilutions in SOC to plate the cells on L agar to obtain isolated colonies.
7. Cells can be analyzed by toothpicking to plates with and without ampicillin (100 μg/ml). Colonies that are sensitive to ampicillin are cured. In this particular case, related to the presence of a plasmid encoding resistance to ampicillin, the frequency of curing was 80 to 90%.

16.4. REFERENCES

16.4.1. General References

1. **Berger, A., and A. Kimmel (ed.).** 1987. Guide to molecular cloning techniques. *Methods Enzymol.* **152**.
2. **Grinsted, J., and P. M. Bennet (ed.).** 1988. Plasmid technology. *In Methods in Microbiology,* vol. 21. Academic Press, Inc., San Diego, Calif.
3. **Hardy, K. (ed).** 1987. *Plasmids—A Practical Approach.* IRL Press, Washington, D.C.
4. **Old, R., and S. Primrose.** 1985. *Principles of Gene Manipulation—An Introduction to Genetic Engineering.* Blackwell Scientific Publications, Oxford.
5. **Sambrook, J., E. F. Fritsch, and T. Maniatis.** 1989. *Molecular Cloning: A Laboratory Manual,* 2nd ed. Cold Spring Harbor Laboratory, Cold Spring Harbor, NY.
6. **Winnacker, E. (ed.).** 1987. *From Genes to Clones—An Introduction to Gene Technology.* VCH, Weinheim, Germany.

16.4.2. Specific References

7. **Alter, D., and K. Subramanian.** 1989. A one step, quick step, miniprep. *BioTechniques* **7:**456–457.
8. **Anderson, D., and L. McKay.** 1983. Simple and rapid method for isolating large plasmid DNA from lactic streptococci. *Appl. Environ. Microbiol.* **46:**549–552.
9. **Banfalvi, Z., E. Kondorosi, and A. Kondorosi.** 1985. *Rhizobium meliloti* carries two megaplasmids. *Plasmid* **13:**129–138.
10. **Barbour, A., and C. Garon.** 1987. Linear plasmids of the bacterium *Borrelia burgdorferi* have covalently closed ends. Science **237:**409–411.
11. **Barth, P., L. Tobin, and G. Sharpe.** 1981. Development of broad host-range plasmid vectors, p. 439–448. *In* S. Levy, R. Clowes, and E. Koenig (ed.), *Molecular Biology, Pathogenicity, and Ecology of Bacterial Plasmids.* Plenum Press, New York.
12. **Berg, D. E., and M. M. Howe (ed.).** 1989. *Mobile DNA.* American Society for Microbiology, Washington, D.C.
13. **Birnboim, H., and J. Doly.** 1979. A rapid extraction procedure for screening recombinant plasmid DNA. *Nucleic Acids Res.* **7:**1513–1525.
14. **Burkardt, B., D. Schillik, and A. Puhler.** 1987. Physical characterization of *Rhizobium meliloti* megaplasmids. *Plasmid* **17:**13–25.
15. **Carle, G., and M. Olson.** 1985. An electrophoretic karyotype for yeast. *Proc. Natl. Acad. Sci. USA* **82:**3756–3760.
16. **Chakrabarty, A. M.** 1976. Plasmids in *Pseudomonas. Annu. Rev. Genet.* **10:**7–30.
17. **Charles, T., and T. Finan.** 1990. Genetic map of *Rhizobium meliloti* megaplasmid pRmeSU47b. *J. Bacteriol.* **172:**2469–2476.
18. **Chassy, B., and A. Giuffrida.** 1980. Method for the lysis of gram-positive, asporogenous bacteria with lysozyme. *Appl. Environ. Microbiol.* **39:**153–158.
19. **Clayton, D. A., R. W. Davis, and J. Vinograd.** 1970. Homology and structural relationships between the dimeric and monomeric circular forms of mitochondrial DNA from human leukemic leukocytes. *J. Mol. Biol.* **47:**137–153.
20. **Clewell, D. B., and D. R. Helinski.** 1969. Supercoiled circular DNA-protein complex in *Escherichia coli:* purification and induced conversion to an open circular form. *Proc. Natl. Acad. Sci. USA* **62:**1159–1166.
21. **Clowes, R. C.** 1972. Molecular structure of bacterial plasmids. *Bacteriol. Rev.* **36:**361–405.
22. **Couturier, M., F. Bex, P. Bergquist, and W. Maas.** 1988. Identification and classification of bacterial plasmids. *Microbiol. Rev.* **52:**375–395.
23. **Crosa, J., D. Brenner, and S. Falkow.** 1973. Use of a single-strand specific nuclease for analysis of bacteria and plasmid deoxyribonucleic acid homo- and heteroduplexes. *J. Bacteriol.* **115:**904–911.
24. **Crosa, J., L. Hedges, and M. Schiewe.** 1980. Curing of a plasmid is correlated with an attenuation of virulence in the marine fish pathogen *Vibrio anguillarum. Infect. Immun.* **27:**897–902.
25. **Crosa, J. H.** 1989. Genetics and molecular biology of siderophore-mediated iron transport in bacteria. *Microbiol. Rev.* **53:**517–530.
26. **Crosa, J. H., L. K. Luttropp, F. Heffron, and S. Falkow.** 1975. Two replication initiation sites on R-plasmid DNA. *Mol. Gen. Genet.* **140:**39–50.
27. **Crosa, J. H., J. Olarte, L. J. Mata, L. K. Luttropp, and M. E. Penoranda.** 1977. Characterization of a R-plasmid associated with ampicillin resistance in *Shigella dysenteriae* type 1 isolated from epidemics. *Antimicrob. Agents Chemother.* **11:**553–558.
28. **Currier, T. L., and E. W. Nester.** 1976. Isolation of covalently closed circular DNA of high molecular weight from bacteria. *Anal. Biochem.* **76:**431–441.
29. **Davey, R., P. Bird, S. Nikoletti, J. Praszkier, and J. Pittard.** 1984. The use of mini-gal plasmids for rapid incompatibility grouping of conjugative R-plasmids. *Plasmids* **11:**234–242.
30. **Dotto, G., V. Enea, and N. Zinder.** 1981. Functional analysis of bacteriophage f1 intergenic region. *Virology* **114:**463–473.
31. **Eckhardt, T.** 1978. A rapid method for the identification of plasmid deoxyribonucleic acid in bacteria. *Plasmid* **1:**584–588.

32. **Figurski, D., and D. Helinski.** 1979. Replication of an origin-containing derivative of plasmid RK2 dependent on a plasmid function provided in trans. *Proc. Natl. Acad. Sci. USA* **76:**1648–1652.
33. **Finkelstein, M., and R. H. Rownd.** 1978. A rapid method for extracting DNA from agarose gels. *Plasmid* **1:**557–562.
34. **Fulford, W., M. Russel, and P. Model.** 1986. Aspects of the growth and regulation of the filamentous phages. *Progr. Nucleic Acid Res.* **33:**141–168.
35. **Grindley, N. D. F., G. O. Humphreys, and E. S. Anderson.** 1973. Molecular studies of R-factor compatibility groups. *J. Bacteriol.* **115:**387–398.
36. **Gruss, A., and S. Ehrlich.** 1989. The family of highly interrelated single-stranded deoxyribonucleic acid plasmids. *Microbiol. Rev.* **53:**231–241.
37. **Guerry, P., D. J. LeBlanc, and F. Falkow.** 1973. General method for the isolation of plasmid deoxyribonucleic acid. *J. Bacteriol.* **116:**1064–1066.
38. **Hansen, J. B., and R. H. Olsen.** 1978. Isolation of large bacterial plasmids and characterization of the P2 incompatibility group plasmids PMG1 and PMG5. *J. Bacteriol.* **135:**227–238.
39. **Hardy, D., J. Bell, E. Long, T. Lindsten, and H. McDevitt.** 1986. Mapping of the class II region of the human major histocompatibility complex by pulsed-field electrophoresis. *Nature* (London) **323:**453–455.
40. **Hartstein, A., A. Rashad, J. Liebler, L. Actis, J. Freeman, J. Rourke, T. Stibolt, M. Tolmasky, G. Ellis, and J. Crosa.** 1988. Multiple intensive care unit outbreak of *Acinetobacter calcoaceticus* subspecies *anitratus* respiratory infection and colonization associated with contaminated, reusable ventilator circuits and resuscitation bags. *Am. J. Med.* **85:**624–631.
41. **Heery, D., R. Powell, F. Gannon, and K. Dunican.** 1989. Curing of a plasmid from *E. coli* using high-voltage electroporation. *Nucleic Acids Res.* **17:**10131.
42. **Helinski, D., S. Cohen, D. Clewell, D. Jackson, and A. Hollaender (ed.).** 1985. *Plasmids in Bacteria.* Plenum Press, New York.
43. **Helinski, D. R.** 1973. Plasmid determined resistance to antibiotics: molecular properties of R-factor. *Annu. Rev. Microbiol.* **27:**437–470.
44. **Hirochika, H., and K. Sakaguchi.** 1982. Analysis of linear plasmids isolated from *Streptomyces:* association of protein with the ends of the plasmid DNA. *Plasmid* **7:**59–65.
45. **Hirota, Y.** 1956. Artificial elimination of the F factor in *Bact. coli* K-12. *Nature* (London) **178:**92.
46. **Hirota, Y.** 1960. The effect of acridine dyes on mating type factors in *Escherichia coli. Proc. Natl. Acad. Sci. USA* **46:**57–64.
47. **Hirt, B.** 1967. Selective extraction of polyoma DNA from infected mouse cell cultures. *J. Mol. Biol.* **26:**365–369.
48. **Holmes, D., and M. Quigley.** 1981. A rapid boiling method for the preparation of bacterial plasmids. *Anal. Biochem.* **114:**193–197.
49. **Hull, R., R. Gill, P. Hsu, B. Minshew, and S. Falkow.** 1981. Construction and expression of recombinant plasmids encoding type 1 or D-mannose-resistant pili from a urinary tract infection *Escherichia coli* isolate. *Infect. Immun.* **33:**933–938.
50. **Humphreys, G. O., G. A. Willshaw, and E. S. Anderson.** 1975. A simple method for the preparation of large quantities of pure plasmid DNA. *Biochim. Biophys. Acta* **383:**457–463.
50a. **Kado, C.** Personal communication.
51. **Kado, C., and S. Liu.** 1981. Rapid procedure for detection and isolation of large and small plasmids. *J. Bacteriol.* **145:**1365–1373.
52. **Kawakami, M., and O. Landman.** 1965. Experiments concerning the curing and intracellular site of episomes. *Biochem. Biophys. Res. Commun.* **18:**716–724.
53. **Keen, C., S. Mendelovitz, G. Cohen, Y. Aharonowitz, and K. Roy.** 1988. Isolation and characterization of a linear DNA plasmid from *Streptomyces clavuligerus. Mol. Gen. Genet.* **212:**172–176.
54. **Kim, A., A. Vertes, and H. Blaschek.** 1990. Isolation of a single-stranded plasmid from *Clostridium acetobutylicum* NCIB 6444. *Appl. Environ. Microbiol.* **56:**1725–1728.
55. **Kupersztoch, R. M., and D. R. Helinski.** 1973. A catenated DNA molecule as an intermediate in the replication of the resistance transfer factor R6K in *Escherichia coli. Biochem. Biophys. Res. Commun.* **54:**1451–1459.
56. **Kupersztoch-Portnoy, Y. M., M. A. Lovett, and D. R. Helinski.** 1974. Strand and site specificity of the relaxation complex of the antibiotic resistance plasmid R6K. *Biochemistry* **13:**5484–5490.
57. **Lai, E., B. Birren, S. Clark, M. Simon, and L. Hood.** 1989. Pulsed field gel electrophoresis. *BioTechniques* **7:**34.
58. **Lederberg, J.** 1952. Cell genetics and hereditary symbiosis. *Physiol. Rev.* **32:**403–430.
59. **Lev, Z.** 1987. A procedure for large-scale isolation of RNA-free plasmid and phage DNA without the use of RNase. *Anal. Biochem.* **160:**332–336.
60. **Lev, Z., and O. Seveg.** 1986. The RNA transcripts of *Drosophila melanogaster src* are differentially regulated during development. *Biochim. Biophys. Acta* **867:**144–151.
61. **Lyon, B., J. May, and R. Skurray.** 1983. Analysis of plasmids in nosocomial strains of multiple-antibiotic-resistance *Staphylococcus aureus. Antimicrob. Agents Chemother.* **23:**817–826.
62. **Macrina, F., D. Kopecko, K. Jones, D. Ayers, and S. McCowen.** 1978. A multiple plasmid-containing *Escherichia coli* strain: convenient source of size reference plasmid molecules. *Plasmid* **1:**417–420.
63. **Macrina, F., J. Reider, S. Virgill, and D. Kopecko.** 1977. Survey of the extrachromosomal gene pool of *Streptococcus mutans. Infect. Immun.* **17:**215–226.
64. **Manis, J. J., and H. J. Whitfield.** 1977. Physical characterization of a plasmid cointegrate containing an *F′ his gnd* element and the *Salmonella typhimurium* LT2 cryptic plasmid. *J. Bacteriol.* **129:**1601–1606.
65. **McClung, J., and R. Gonzalez.** 1989. Purification of plasmid DNA by fast protein liquid chromatography on Superose-6 preparative gradient. *Anal. Biochem.* **177:**378–382.
66. **McConaughy, B., C. Laird, and J. McCarthy.** 1969. Nucleic acid reassociation in formamide. *Biochemistry* **8:**3289–3295.
67. **Miller, J., W. Dower, and L. Tompkins.** 1988. High-voltage electroporation of bacteria: transformation of *Campylobacter jejuni* with plasmid DNA. *Proc. Natl. Acad. Sci. USA* **85:**856–860.
68. **Myers, J. A., D. Sanchez, L. P. Elwell, and S. Falkow.** 1976. A simple agarose gel electrophoretic method for the identification and characterization of plasmid deoxyribonucleic acid. *J. Bacteriol.* **127:**1529–1537.
69. **Nester, E., and T. Kosuge.** 1981. Plasmids specifying plant hyperplasias. *Annu. Rev. Microbiol.* **35:**531–565.
70. **Nies, D., M. Mergeay, B. Friedrich, and H. Schlegel.** 1987. Cloning of plasmid genes encoding resistance to cadmium, zinc, and cobalt in *Alcaligenes eutrophus* CH34. *J. Bacteriol.* **169:**4865–4868.
71. **Novick, R.** 1987. Plasmid incompatibility. *Microbiol. Rev.* **51:**381–395.
72. **Palchaudhuri, S.** 1977. Molecular characterization of hydrocarbon degradative plasmids in *Pseudomonas putida. Biochem. Biophys. Res. Commun.* **77:**518–525.
73. **Perez-Casal, J., and J. Crosa.** 1987. Novel incompatibility and partition loci for the REPI replication region of plasmid ColV-K30. *J. Bacteriol.* **169:**5078–5086.
74. **Perng, G., and R. Lefebre.** 1990. Expression of antigens from chromosomal and linear plasmid DNA of *Borrelia coriaceae. Infect. Immun.* **58:**1744–1748.
74a. **Projan, S.** Personal communication.
75. **Radloff, R., W. Bauer, and J. Vinograd.** 1967. A dye buoyant density method for the detection and isolation of closed circular duplex DNA: the closed circular DNA in HeLa cells. *Proc. Natl. Acad. Sci. USA* **57:**1514–1522.

76. **Reddy, B., P. Billingsley, and L. Etkin.** 1990. A quick protocol for plasmid DNA analysis. *BioTechniques* **9:**717–718.
77. **Reddy, S., K. Hamsabhushanam, and P. Jagadeeswaran.** 1989. A rapid method for large-scale isolation of plasmid DNA by boiling in a plastic bag. *BioTechniques* **7:**820–822.
78. **Rusansky, S., R. Avigad, S. Michelis, and D. Gutnik.** 1987. Involvement of a plasmid in growth on and dispersion of crude oil by *Acinetobacter calcoaceticus* RA57. *Appl. Environ. Microbiol.* **53:**1918–1923.
79. **Saha, B., D. Saha, S. Nigoyi, and M. Bal.** 1989. A new method of plasmid DNA preparation by sucrose-mediated detergent lysis from *Escherichia coli* (Gram-negative) and *Staphylococcus aureus* (Gram-positive). *Anal. Biochem.* **176:**344–349.
80. **Sakaguchi, K., H. Hirochika, and N. Gunge.** 1985. Linear plasmids with terminal inverted repeats obtained from *Streptomyces rochei* and *Kluyveromyces lactis*. *Basic Life Sci.* **30:**433–451.
81. **Schwarz, D., and C. Cantor.** 1984. Separation of yeast chromosome-sized DNAs by pulsed field electrophoresis. *Cell* **37:**67–75.
82. **Serghini, M., C. Rotzenthaler, and L. Pinck.** 1989. A rapid and efficient "miniprep" for isolation of plasmid DNA. *Nucleic Acids Res.* **17:**3604.
83. **Sharp, P. A., M. T. Hsu, E. Ohtsubo, and N. Davidson.** 1972. Electron microscope heteroduplex studies of sequence relations among plasmids of *Escherichia coli*. 1. Structure of R-prime factors. *J. Mol. Biol.* **71:**471–497.
84. **Shields, M., B. Kline, and J. Tam.** 1986. A rapid method for the quantitative measurement of gene dosage: mini-F plasmid concentration as a function of cell growth rate. *J. Microbiol. Methods* **6:**33–46.
85. **Simske, J., and S. Scherer.** 1989. Pulsed-field gel electrophoresis of circular DNA. *Nucleic Acids Res.* **17:**4359–4365.
86. **Skurray, R. A., H. Nagaishi, and A. J. Clark.** 1976. Molecular cloning of DNA from F sex factor of *Escherichia coli* K-12. *Proc. Natl. Acad. Sci. USA* **73:**64–68.
87. **Southern, E. M.** 1975. Detection of specific sequences among DNA fragments separated by gel electrophoresis. *J. Mol. Biol.* **98:**503–517.
88. **te Riele, H., B. Michel, and S. Ehrlich.** 1986. Single-stranded plasmid DNA in *Bacillus subtilis* and *Staphylococcus aureus*. *Proc. Natl. Acad. Sci. USA* **83:**2541–2545.
89. **Tolmasky, M. E., L. A. Actis, and J. H. Crosa.** 1992. Plasmids, p. 431–442. *In* J. Lederberg (ed.), *Encyclopedia of Microbiology*. Academic Press, Inc., San Diego, Calif.
90. **Tolmasky, M., L. Actis, A. Toranzo, J. Barja, and J. Crosa.** 1985. Plasmids mediating iron uptake in *Vibrio anguillarum* strains isolated from turbot in Spain. *J. Gen. Microbiol.* **131:**1987–1997.

90a. **Tolmasky, M. E., R. J. Staneloni, and L. F. Leloir.** 1982. Lipid-bound saccharides in *Rhizobium meliloti*. *J. Biol. Chem.* **257:**6751–6757.

91. **Tomoeda, M., M. Inuzuki, N. Kubo, and S. Nokamura.** 1968. Effective elimination of drug resistance and sex factors in *Escherichia coli* by sodium dodecyl sulfate. *J. Bacteriol.* **95:**1078–1089.
92. **Tompkins, L., N. Troup, T. Woods, W. Bibb, and R. McKinney.** 1987. Molecular epidemiology of *Legionella* species by restriction endonuclease and alloenzyme analysis. *J. Clin. Microbiol.* **25:**1875–1880.
93. **Van Larebekke, N., G. Engler, M. Holsters, S. Van den Elsacker, I. Zaenen, R. Schilperoort, and J. Schell.** 1974. Large plasmid in *Agrobacterium tumefaciens* essential for crown-gall inducing ability. *Nature* (London) **252:**169–170.
94. **Watson, B., T. C. Currier, M. P. Gordon, M.-D. Chilton, and E. W. Nester.** 1975. Plasmid required for virulence of *Agrobacterium tumefaciens*. *J. Bacteriol.* **123:**255–264.
95. **Wegrzyn, G., P. Neubauer, S. Krueger, M. Hecker, and K. Taylor.** 1991. Stringent control of replication of plasmids derived from coliphage λ. *Mol. Gen. Genet.* **225:**94–98.
96. **Weickert, M., and G. Chambliss.** 1989. Acid-phenol minipreps make excellent sequencing templates. *USB Editorial Comments* **16:**5–6.
97. **Whisenant, S., B. Rasheed, and Y. Bhatnagar.** 1988. Plasmid purification using high-performance gel filtration chromatography. *Nucleic Acids Res.* **16:**5202.
98. **White, F., and E. Nester.** 1980. Hairy root: plasmid encodes virulent traits in *Agrobacterium rhizogenes*. *J. Bacteriol.* **141:**1134–1141.
99. **Woloj, M., M. Tolmasky, M. Roberts, and J. Crosa.** 1986. Plasmid-encoded amikacin resistance in multiresistant strains of *Klebsiella pneumoniae* isolated from neonates with meningitis. *Antimicrob. Agents Chemother.* **29:**315–319.
100. **Zasloff, M., G. Ginder, and G. Felsenfeld.** 1978. A new method for the purification and identification of covalently closed circular DNA molecules. *Nucleic Acids Res.* **5:**1139–1152.
101. **Ziai, M., C. Hamby, R. Reddy, K. Hayashibe, and S. Ferrone.** 1989. Rapid purification of plasmid DNA following acid precipitation of bacterial proteins. *BioTechniques* **7:**147.

Chapter 17

Transposon Mutagenesis

FRANS J. DE BRUIJN and SILVIA ROSSBACH

Transposable elements are distinct DNA segments which have the unique capacity to move (transpose) to new sites within the genome of their host organisms. The transposition process is independent of the classical homologous recombination (*rec*) system of the organism. Moreover, the insertion of a transposable element into a new genomic site does not require extensive DNA homology between the ends of the element and its target site. Transposable elements have been found in a wide variety of procaryotic and eucaryotic organisms, where they can cause null mutations, chromosome rearrangements, and novel patterns of gene expression on insertion in the coding region or regulatory sequences of resident genes and operons (for a general review, see reference 4).

Procaryotic transposable elements can be roughly divided into three different classes. Class I consists of simple elements such as insertion sequences (IS elements), which are approximately 800 to 1,500 bp in length. IS elements normally consist of a gene encoding an enzyme required for transposition (transposase), flanked by terminally repeated DNA sequences which serve as substrate for the transposase. IS elements were initially identified in the lactose (*lac*) and galactose (*gal*) utilization operons of enteric bacteria, where they were found to cause often unstable, polar mutations on insertion (117).

Class II consists of composite transposable elements; members of this class have also been referred to as transposons or Tn elements (23). Transposons in procaryotes have been defined as a class of complex transposable elements, often containing simple IS elements (or parts thereof) as direct or inverted repeats at their termini, behaving formally like IS elements but carrying additional genes unrelated to transposition functions, such as antibiotic resistance, heavy-metal resistance or pathogenicity determinant genes (23, 24). Transposons carrying antibiotic resistance genes were first identified in the mid-1970s after transposition from drug resistance transfer plasmids into other replicons in the cell, such as bacteriophage λ (15, 45, 47, 59; see reference 4 for a review). A simple italic number is assigned to each independent isolate from nature, e.g., Tn*5* (23, 24). An abbreviated list of the most extensively studied transposons is shown in Table 1; for a more complete list, see reference 1. The insertion of a transposon into a particular genetic locus or replicon (phage) is designated by using a double colon, e.g. *lacZ*::Tn*5* or λ::Tn*5* (23, 24). These designations will also be used in this chapter.

Class III includes "transposable" bacteriophages, such as Mu and its relatives. Phage Mu is, in fact, both a virus and a transposon and has been extremely useful in elucidating the mechanism of transposition of transposable elements (85). It was first discovered in the early 1960s as a novel phage that, on lysogenization, could integrate at multiple sites in the host chromosome, thereby frequently causing mutations (mutator phage [85, 122]).

It was recognized soon after the discovery of transposons that these elements could be used as extremely efficient tools in bacterial genetics (7, 22) because of the following characteristics.

(i) The transposition or insertion of a transposon into a gene generally leads to its inactivation, and the resulting null mutations are relatively stable since the frequency of precise excision of the transposon is very low.

(ii) If a transposon integrates into an operon, it usually exerts a strong polar effect on genes located downstream of its insertion site, allowing transposon mutagenesis to be used to determine operon structure.

(iii) On integration, transposons introduce new genetic and physical markers into the target locus, such as antibiotic resistance genes, new restriction endonuclease cleavage sites, and unique DNA sequences which can be identified by genetic means, DNA-DNA hybridization, or electron-microscopic heteroduplex analysis. The genetic markers can be used to rapidly map and transduce the mutated loci.

(iv) Transposons can generate a variety of genomic rearrangements, such as deletions, inversions, translocations, or duplications, and can be used to introduce specific genes into the genome of a target bacterium.

These characteristics have made transposon mutagenesis a valuable addition to the more classical mutagenesis techniques described in Chapter 13. Moreover, recent advances in molecular biology have paved the way for sophisticated applications of transposon mutagenesis to clone and sequence genes, to construct reporter gene fusions, to construct correlated physical and genetic maps of cloned DNA segments, to map entire bacterial genomes by pulsed-field electrophoresis, to construct conditional mutations with portable promoters, and to introduce desired origins of conjugal transfer or replication into chromosomes and plasmids of interest.

In this chapter a subset of these applications and the corresponding experimental protocols are described. The focus is on one of the transposons, Tn*5*, which has been extensively used in bacterial molecular genetics, but the principles and most of the experimental strategies described are applicable to a variety of other transposons. For a treatise on the biology of transposons and their transposition mechanisms, the interested reader is referred to the recent publication by Berg and Howe (4). For a detailed description of the application of transposon mutagenesis to more classical bacterial genetics, the reader is referred to the still very pertinent article by Kleckner et al. (7). For an in-depth overview of the use of different transposable elements in the

Table 1. Common bacterial transposable elements and their characteristics

Designation	Antibiotic resistance	Length (kb)	Transposition frequency (per cell generation)	Insertion specificity	Reference
Tn*3*	Ampicillin (Ap^r)	4.957	10^{-7}–10^{-5}	Prefers A+T-rich sequences, attracted to sites having some similarity to transposon ends	107
Tn*5*	Kanamycin (Km^r)/neomycin (Nm^r), bleomycin (Bm^r)[a] streptomycin (Sm^r)[a]	5.818	10^{-3}–10^{-2}	Nearly none, hot spot in pBR322	3
Tn*7*	Trimethoprim (Tp^r), streptomycin (Sm^r), spectinomycin (Sp^r)	14	High frequency if target site *att*Tn*7* is present, low frequencies to secondary sites	Very high to target site *att*Tn*7*	28
Tn*9*	chloramphenicol (Cm^r)	2.6	2×10^{-5}	Many sites	45
Tn*10*	Tetracycline (Tc^r)	9.3	10^{-7}	Insertion hot spots (symmetrical 6-bp sequence), prefers nontranscribed regions	58
Phage Mu		37	High frequency to chromosome	Essentially random	85

[a]Expressed in certain nonenteric bacteria.

genetic engineering of bacteria, as well as comprehensive lists of transposable elements and the map positions of known insertions in *Escherichia coli*, the reader is referred to articles by Berg and Berg (1), Berg et al. (2), Sherratt (107), Craig (28), and Kleckner (58). For a description of a novel transposable element (m$\gamma\delta$), not yet described in these review articles but particularly useful for plasmid mutagenesis, the reader is referred to references 11 and 12.

Because of the great diversity of bacterial strains, transposons, delivery vehicles, and gene transfer methods available today, it is not possible to present a complete set of transposon mutagenesis protocols in this chapter. Nevertheless, the methods, examples, and references included here should provide some basic tools.

17.1. Tn*5* AS A MODEL TRANSPOSON

Transposon Tn*5* is a 5,818-bp composite transposon, consisting of two inverted repeats of 1,533 bp (IS*50*L and IS*50*R), flanking a unique region which carries three genes that confer on certain bacterial hosts resistance to the antibiotics kanamycin/neomycin (*kan* or neomycin phosphotransferase *nptII* gene), bleomycin (*ble* gene) and/ or streptomycin (*str* gene) (3) (Fig. 1). The *nptII, ble,* and *str* genes are organized in an operon and are transcribed from a promoter located at the inside end of IS*50*L (3) ("p" in Fig. 1). IS*50*R encodes the transposase responsible for Tn*5* transposition (*tnp* gene), as well as a transposition inhibitor (*inh* gene) (3) (Fig. 1). The first 19 bp of both IS*50*L and IS*50*R have been found to be essential for efficient transposition of Tn*5* (3).

Tn*5* is capable of transposing at a very high frequency in *E. coli* (10^{-2} to 10^{-3}) by using an apparently conservative "cut-and-paste" mechanism, and it generates a 9-bp target site duplication on insertion (3). Tn*5* has a low insertional specificity and therefore can insert into a large number of locations in bacterial genomes, as well as into multiple positions within single genes, although occasional hot spots for insertion have been reported (3, 5, 6, 14). Tn*5*-induced insertion mutations are very stable and only occasionally revert to wild type via a process known as precise excision, although the reversion frequency is somewhat dependent on the precise location of Tn*5* in the genetic locus, the nature of the replicon, and the host genotype (3). These parameters are important for the experiments described below and should be examined when Tn*5* mutagenesis experiments are performed in new organisms or when problems are encountered with the protocols presented here. For in-depth discussions of these parameters, see reviews by Berg and colleagues (2, 3, 14) and de Bruijn and Lupski (6).

Tn*5* mutagenesis experiments can be divided into two distinct categories: random mutagenesis and region-directed mutagenesis (5). The terms "site-directed" and "site-specific" transposon mutagenesis have also been used in the literature to describe the latter category (5). The term "region-directed" mutagenesis is used here to distinguish this category of Tn mutagenesis from base-pair-specific oligonucleotide and chemical mutagenesis methods, which are usually referred to as "site specific" (see Chapters 13 and 19). **Random mutagenesis** involves the introduction of Tn*5* into the bacterial species of interest via transformation, transduction, or conjugation, using "suicide" plasmid or phage vectors, followed by selection for the antibiotic resistance marker(s) carried by Tn*5*. Transposition and insertion of Tn*5* into the genome of the recipient bacterium are detected after the vector has been lost by segregation, and the phenotype of the Tn*5*-induced mutations can be analyzed subsequently. **Region-directed mutagenesis** usually involves the isolation of Tn*5* insertion mutations in genes cloned into (multicopy) plasmids in *E. coli*, followed by characterization of the mutant phenotype in this (heterologous) host or in the organism of origin of the cloned genes after reintroduction of the Tn*5*-mutated loci by gene replacement (5). Experimental strategies for these two types of mutageneses are presented in the following sections and have been reviewed previously (2, 5, 6).

17.2. RANDOM Tn*5* MUTAGENESIS

To generate Tn insertion mutations in the genome of bacteria, the transposon must be introduced into its

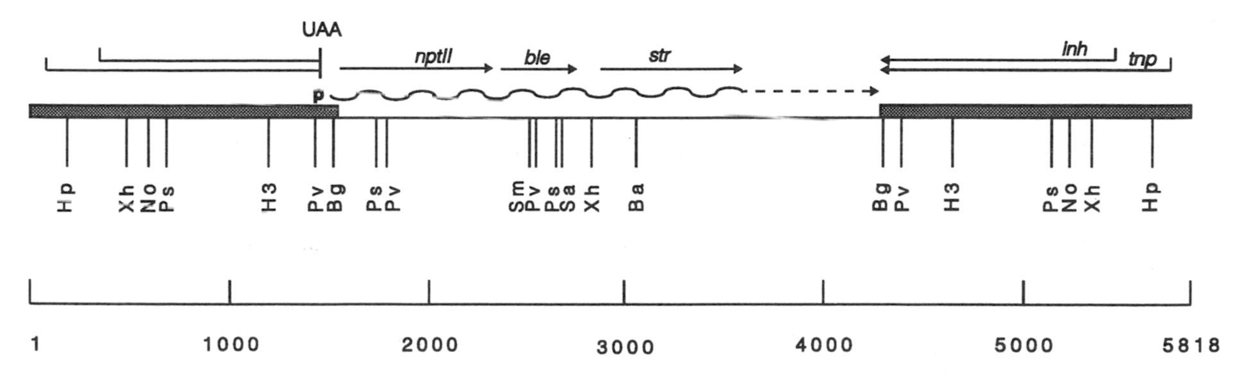

FIG. 1. Structure of the transposable element Tn*5*. Stippled bars represent the insertion sequences IS*50*L (left) and IS*50*R (right), present as terminal inverted repeats in Tn*5*. Gene symbols: *tnp*, coding region for the transposase; *inh*, coding region for the inhibitor; UAA, stop codon responsible for the truncated versions of the transposase and the inhibitor in IS*50*L; p, promoter for the transcript of the operon coding for the three antibiotic resistances (*nptII*, neomycin/kanamycin resistance; *ble*, bleomycin resistance; *str*, streptomycin resistance). Abbreviations for restriction enzymes: Ba, *Bam*HI; Bg, *Bgl*II; H3, *Hin*dIII; Hp, *Hpa*I; No, *Not*I; Ps, *Pst*I; Pv, *Pvu*II; Sa, *Sal*I; Sm, *Sma*I; Xh, *Xho*I.

new host and cells in which transposition has occurred must be identified. These objectives are achieved by the use of so-called suicide vectors, which can be of phage or plasmid origin. In both cases, after introduction of the suicide vector::Tn*5* DNA molecule into the cell, the vector must be incapable of stable replication or integration, so that selection for one or more of the antibiotic resistance markers on the Tn element leads to the identification of cells in which the element has transposed from the vector into the genome.

A commonly used phage vector for the introduction of transposons into bacterial cells is a modified derivative of bacteriophage λ, which can be used for the mutagenesis of all bacteria that are naturally susceptible to λ adsorption and infection or have been genetically modified to be λ sensitive. The protocol for generating the latter category of engineered organism has been described previously (67, 84) and is not presented here. For bacteria which are not λ sensitive or cannot be engineered to become λ sensitive, the transposon can be introduced into the cells by using plasmids, which can be conjugally mobilized (see Chapter 14.3) or electroporated (38) (see Chapter 14.1.3.3) into recipient cells but lack the appropriate (wide host range) origins of replication for stable maintenance.

Phage P1 has also been used to introduce transposons such as Tn*5* into different bacterial species, and examples of the use of this vector system have been described previously (9, 64).

The choice of suicide vector and Tn element to be used depends on the purpose of the random mutagenesis experiment and a number of criteria, including phage sensitivity and intrinsic antibiotic resistance of the bacterium to be mutagenized, as well as the availability of an *E. coli* to target organism conjugal transfer or direct transformation system. A short list of transposons and their relevant properties is shown in Table 1, but for a discussion of the above parameters and an extensive list of transposons available for specific purposes, the reader is referred to previous reviews (1, 2, 6).

This chapter focuses on the use of bacteriophage λ and conjugatable narrow-host-range or conditionally replicating plasmid vectors to deliver Tn*5* and its derivatives and describes experimental protocols that are commonly used to carry out random mutagenesis of (predominantly gram-negative) organisms. For a discussion of chemical transformation and high-voltage electroporation techniques (38) to introduce plasmids carrying Tn*5* into cells, the reader is referred to Chapter 14.1.3. For a discussion of transposons in gram-positive organisms and their use for mutagenesis experiments, the reader is referred to articles by Murphy (80) and Youngman (129).

17.2.1. Phage λ Suicide Vectors

For λ::Tn*5*-mediated random mutagenesis experiments, a derivative of phage λ (λ467) is routinely used; it has the genotype λ *b*221 *rex*::Tn*5* *c*I857 *O*am29 *P*am80 (6, 7). The *b*221 mutation is a deletion in the λ genome that removes the phage attachment (*att*) site. This prevents lysogenization of the target strain. The *rex* gene is a nonessential λ gene and carries the Tn*5* insertion. Both the *O* and *P* genes are involved in phage replication, and therefore amber mutations in these genes will prevent phage replication in a suppressor-negative (Su^0) background. The λ467-type phage can be propagated in a Su^+ *E. coli* strain (see below), will adsorb to a λ-sensitive strain, and will inject its DNA; however, this DNA will be unable to replicate and will be lost by segregation. Selection for Km^r Nm^r (and, in selected cases, Bm^r or Sm^r) survivors results in the identification of cells in which Tn*5* has transposed and integrated into the genome. This type of Tn*5* mutagenesis has been successfully used to isolate mutations in many genes of *E. coli* (1, 69, 106).

17.2.1.1. Phage lysate preparation

To prepare λ::Tn*5* lysates, the Su^+ *E. coli* LE392 (F^- *hsdR514 supE44 supF58 lacY1 galK2 galT22 metB1 trpR55* λ^-) and the following protocol are used (modified from reference 6).

1. Start an overnight culture of *E. coli* LE392 at 37°C in 10 ml of LB medium (see section 17.5) plus 0.2% maltose.
2. Pellet the cells by centrifugation, wash once with 5 ml of 10 mM $MgSO_4$, and pellet again.
3. Resuspend the cells in 4 ml of 10 mM $MgSO_4$.
4. Prepare five sterile glass tubes with 0.1 ml of washed LE392 cells, and add 10 μl of λ::Tn*5* lysate or the phages from a single plaque to four of the tubes. Leave one tube uninfected as a control.
5. Mix gently, and let the phages adsorb for 2 h at room temperature without agitating the tubes.
6. Mix the infected cells with 3 ml of soft agar (45°C), and pour the mixture evenly on λ agar plates (see section 17.5).
7. Incubate the plates for 6 to 8 h (or overnight) at 37°C, and scrape off the clear soft agar into a fresh tube.
8. Mix the agar-phage-cell suspension well but gently with 2.5 ml of SM buffer (see section 17.5) and 0.5 ml of chloroform in a sterile (Corex) centrifuge tube. Incubate the mixture for 5 min on ice, and centrifuge it for 15 min at 1,000 × *g* (e.g., at 3,000 rpm in an SS34 fixed-angle rotor) at 4°C.
9. Transfer the supernatant solution into a small glass tube, add 20 drops of chloroform, and store at 4°C.
10. Determine the titer of the lysate by infecting LE392 cells (prepared as described in steps 1 to 3), with several dilutions of the lysate (from 10^{-2}, 10^{-4}, to 10^{-12}) and counting the resulting single plaques on the indicator plates.

Controls and comments. The plaques on the titer plates should be small and well defined (pinpoint). No plaques should be observed on the plates with bacterial cell suspensions to which no phage was added. The titer of the lysate should be in excess of 5×10^{10} for successful use in subsequent Tn*5* mutagenesis experiments.

17.2.1.2. Infection and mutant selection

1. Start an overnight culture of the λ^s Su^0 bacterial strain to be mutagenized in 5 ml of complete medium for the respective strain, supplemented with 0.2% maltose.
2. Inoculate 100 μl (1:100) of the saturated culture into 10 ml of complete medium + 0.2% maltose, and incubate the culture for 2 to 4 h.
3. Pellet the cells, wash them with a sterile solution of 10 mM $MgSO_4$, and pellet them again. Resuspend the pelleted bacteria in 4 ml of 10 mM $MgSO_4$.

4. Infect 1-ml aliquots of the washed cells with 100 μl of λ:Tn*5* lysate (10^8 to 10^9 CFU/ml), and keep 1 ml of uninfected cells as a control.

5. Mix the phage and the cells gently, and allow the phage to adsorb for 2 h at room temperature without agitating the tubes.

6. Plate 200-μl aliquots of the infected cells on complete-medium plates containing kanamycin (or neomycin, bleomycin, and streptomycin) at the desired concentration, and incubate the plates at the appropriate temperature until clearly defined colonies appear on the selection plates (but not on the control plates).

7. Purify the resistant colonies on complete-medium plates with the desired antibiotics, and analyze them as described in sections 17.2.3 and 17.2.4.

Controls and comments. No Km^r colonies should be observed on the plates onto which the uninfected cells were plated.

17.2.2. Plasmid Suicide Vectors

The first class of suicide plasmids to be used for random Tn*5* mutagenesis of λ-insensitive bacterial strains was based on an IncP-type plasmid carrying a copy of Mu in addition to Tn*5* (pJB4JI [18]). The presence of Mu in this replicon was found to prevent its stable maintenance in different gram-negative bacteria, and Tn*5* transpositions could be identified after antibiotic resistance selection (18). However, for several bacterial species it was found that a large percentage of the insertion mutants carried Mu sequences at the insertion site, in addition to Tn*5* (74), and pJB4JI was observed to be relatively stable in other species (112), leading to a reduced use of this class of vectors.

A second class of Tn*5*-containing suicide plasmids is based on replicons carrying a temperature-sensitive origin of replication, allowing the selection for Tn*5* transposition at elevated temperatures (65). More recently, derivatives of the broad-host-conjugative plasmids R388 (pCHR71) and pRK290 (pRK340), carrying temperature-sensitive mutations in plasmid replication functions, have been described and their utility in Tn*5* mutagenesis of, for example, *Shigella* and *Legionella* species has been demonstrated (56, 102, 103).

A third class of suicide vectors for Tn*5* mutagenesis, based on narrow-host-range IncW- and IncN-type plasmids, have been described and used for mutagenesis of different bacterial species, such as *Rhizobium meliloti* (pGS9) (104), *Bradyrhizobium japonicum* (pGS9) (96), *Azorhizobium caulinodans* (pGS9) (49), *Pseudomonas solanacearum* (pWI281) (79), and *Azotobacter vinelandii* (57) to mention a few examples.

A fourth class of suicide vectors is derived from the R6K replicon and is based on the principle of *trans*-complementation-dependent maintenance of the R6K replication origin (61). In this system, a plasmid carrying (a fragment of) the R6K replication origin (e.g., pRK703) can be stably maintained only if a *trans*-acting factor (the π protein encoded by the *pir* gene) is provided in the bacterial cell, for example, by a λ*pir* prophage (61). Miller and Mekalanos (78) inserted into plasmid pRK703 an IncP-type origin of conjugal transfer (*mob*, *oriT* of RP4), to allow high-frequency transfer to a variety of gram-negative bacteria with the help of RP4 mobilization functions provided in *trans* (see below), and a polylinker containing several cloning sites (pGP704). Transposon Tn*5* derivatives (Mini-Tn*5*s), consisting of 19 bp of the Tn*5* inverted repeats required for transposition coupled to antibiotic resistance or other selectable marker genes (see section 17.4.4), have been inserted into this vector, in addition to the Tn*5* transposase (*tnp*) gene, to construct an elegant suicide Tn mutagenesis system. The suicide plasmid::Mini-Tn*5* is conjugally mobilized (see sections 17.2.2.1 and 17.2.2.2) from an *E. coli* strain harboring λ*pir* to the gram-negative bacterium to be mutagenized, and selection for one of the mini-Tn*5*-borne marker genes is applied. This results in the selection of transconjugants in which the mini-Tn*5* has transposed from the vector, which is no longer capable of replicating in the absence of the *pir* gene, into the resident genome. Once inserted into the genome, mini-Tn*5* is incapable of further transposition, since it is lacking the suicide vector-borne *tnp* gene, resulting in a very stable integration event (48). The latter is especially relevant when Tn*5*-carried genes must be stably maintained in the absence of antibiotic selection in natural environments (48). For a detailed description of the use of these vectors, the reader is referred to the articles by Herrero et al. (48) and de Lorenzo et al. (33).

The most universally employed class of suicide vectors for Tn*5* mutagenesis in gram-negative bacteria other than *E. coli* is based on replicons derived from plasmid RP4, carrying the same broad-host-range conjugal transfer or mobilization functions and sites as used for the above-described vectors but a narrow-host-range origin of replication (111, 112). These vectors can be mobilized at a high frequency from *E. coli* to other gram-negative bacteria but cannot be stably maintained in the recipient species.

These vectors contain, in addition to Tn*5*, the IncP-type mobilization (*mob*; *oriT*) site and are based on commonly used *E. coli* cloning vectors, such as pACYC184 (25) and pBR325 (20), which cannot replicate in nonenteric bacteria (pSUP series (110–112). pSUP-type plasmids can be mobilized in diparental mating experiments by providing the transfer functions in *trans* from a chromosomally integrated copy of the IncP plasmid RP4 in the donor strain itself (e.g., strain S17-1 [RP4-2-Tc::Mu, Km::Tn*7* Tp^r Sm^r *thi pro hsdR*] [111; see section 17.2.2.1]) or in triparental mating experiments by providing the transfer functions from plasmid pRK2013, harbored by a nondonor, nonrecipient helper strain of *E. coli* (*E. coli*/pRK 2013 [35, 36; see section 17.2.2.2]). The use of this vector system will be described here; for a detailed review of the conjugal transfer of pSUP vectors, the reader is referred to other recent reviews (110, 112). As an alternative to conjugation methods to transfer suicide vectors, high-voltage electroporation can be used (38) (Chapter 14.1.3.3).

17.2.2.1. Diparental conjugation and mutant selection

1. Start a 10-ml culture of the recipient bacterial strain in complete medium, and grow it until it has reached saturation at the appropriate temperature. Reinoculate the cells 1:10 in fresh medium, and incubate the cultures for another 2 to 5 h (depending on the doubling time of the bacterial strain used).

2. Start a 10-ml overnight culture of the donor strain *E. coli* S17-1, harboring the Tn*5* delivery plasmid (110,

111), in LB medium supplemented with the appropriate antibiotics (e.g., usually kanamycin for Tn*5* and tetracycline, ampicillin, or chloramphenicol for the pSUP vector), and grow the culture until saturation at 37°C. Reinoculate the cells 1:10 in fresh medium, and grow for 1 to 2 h at 37°C.

3. Pellet the logarithmically growing recipient and donor cultures (1,000 × *g* for 10 min), wash the bacteria with 0.9% NaCl to remove the antibiotics, and pellet them again.

4. Resuspend the pelleted bacteria in 1 ml of 0.9% NaCl.

5. Spot 100 μl of the donor and recipient cell cultures separately and 100 μl of a 1:1 mixture of both on conjugation plates containing the complete medium suitable for the recipient strain, in the absence of antibiotics. Allow the spots to dry, and incubate the plates at the appropriate temperature for the recipient strain for 12 to 36 h.

6. Scrape the mating mixtures and the single spots of donor and recipient cells off the plate with a sterile spatula, resuspend the bacterial cells in 1 ml of 0.9% NaCl (or H_2O for osmotically stable recipient bacteria, such as rhizobia), and plate out the suspensions (or appropriate dilutions thereof) on selective plates containing antibiotics to select for the recipient strain (if the recipient strain harbors a gene conferring resistance to an antibiotic to which the donor strain is sensitive) and for the presence of the Tn*5* (kanamycin, neomycin, bleomycin, or streptomycin).

7. Purify the single colonies appearing on the selective plates once on the same selective medium to avoid contamination with donor bacteria, or (when screening large numbers of colonies for mutant phenotypes) characterize them directly.

Km^r derivatives of the recipient strain should be the result of the introduction of the suicide vector, subsequent (random) transposition of the Tn*5* into the resident genome, and loss of the vector. These colonies can now be subjected to further studies (see sections 17.2.3 and 17.2.4).

Controls and comments. No colonies should appear on the selection plates onto which the donor and recipient cells were plated separately. If colonies do appear repeatedly on the control plates, alternative counterselection protocols should be considered (see below). Care should be taken to treat the cells gently during pelleting and washing and to keep them at the growth temperature of the recipient strain.

Instead of or in addition to the antibiotic selection, a specific minimal medium suitable for the recipient bacteria but lacking the appropriate carbon source for *E. coli* (e.g., containing sucrose) or lacking the amino acid/vitamin requirements of the *E. coli* donor strain (e.g., thiamine and proline for strain S17-1 [111]) can be used to counterselect the *E. coli* donor bacteria. The latter method can of course be used only if one is not trying to isolate Tn*5*-induced auxotrophic mutations in the target bacterial strain (see section 17.2.2.3).

17.2.2.2. Triparental conjugation and mutant selection

1. Prepare a culture of logarithmically growing recipient bacteria as described above. Similarly, prepare a logarithmically growing culture of a donor *E. coli* strain harboring the RP4-derived Tn*5* mutagenesis vector.

In this case, the donor strain does not contain a chromosomally integrated copy of RP4 to provide transfer functions but can be any *E. coli* strain, preferably carrying one or more auxotrophic markers for counterselection (see above).

In addition, prepare a logarithmically growing culture of a third *E. coli* (helper) strain, harboring a plasmid (pRK2013) (35, 36), providing the RP4 transfer functions in *trans* and allowing the conjugal transfer of the Tn*5* mutagenesis vector from the donor to the recipient bacteria in this type of triparental mating experiment. Since the pRK2013 plasmid carries a Km^r (Nm^r) gene, kanamycin or neomycin should be added to the growth medium to select for the presence of the plasmid.

2. Pellet the cultures (1000 × *g* for 10 min), wash the bacteria with 0.9% NaCl, and pellet them again. Resuspend the pelleted bacteria in 1 ml of 0.9% NaCl.

3. Spot 100 μl of the donor, helper, and recipient cell cultures separately, in pairwise combinations (1:1), and all three together (1:1:1) on conjugation plates containing the complete medium suitable for the recipient strain in the absence of antibiotics. Incubate the plates at the appropriate temperature for the recipient strain for 12 to 36 h.

4. Follow steps 6 and 7 of the diparental mating protocol.

Controls and comments. No colonies should appear on the selection plates onto which the donor, helper, or recipient strains were plated separately or in pairwise combinations. The use of an unsupplemented minimal medium and auxotrophic donor and helper *E. coli* strains should help reduce background on the selection plates. If colonies do form consistently on the control plates, the individual strains should be checked or other (or modified) counterselection conditions should be explored (e.g., using a different minimal medium, using an alternative antibiotic, or raising the concentration of the antibiotic).

17.2.2.3. Mutagenesis protocols and mutant characterization

The protocols outlined above normally result in approximately 10^{-3} to 10^{-6} Tn*5*-containing colonies per recipient bacterial cell. If the total number of Tn*5*-containing strains is too small for the planned mutant-screening program, the mutagenesis protocol should be optimized. For the conjugation protocols, the total number of ratio of donor and recipient cells should be varied (e.g., 2- to 20-fold excess of donor to recipient cells) and the conjugation time should be extended. In addition, instead of spotting the conjugation mixtures directly, the cell mixtures should be combined onto 0.2-μm-pore-size nitrocellulose filters placed on (prewarmed) conjugation plates, incubated for the desired time, and removed from the filters by vortexing in 0.9 M NaCl (or H_2O) before plating on selection plates. The concentration of the antibiotic(s) used for (counter)selection should also be varied.

When first carrying out random Tn*5* mutagenesis experiments on a particular bacterial strain, it is useful to examine the frequency of Tn*5*-mediated gene inactivation. This is most conveniently done by determining the percentage of auxotrophic mutants resulting from a given Tn*5* mutagenesis experiment. This requires the availability of a defined minimal medium for the strain under study. Five hundred to several thousand Km^r colonies are picked and streaked or replica plated (66) (see Chapter 13.5.4) onto complete and minimal medium, and the frequency of colonies unable to grow on minimal medium is examined. Approximately 0.1 to 3.0% of the colonies should fall in the latter category. The occurrence of different auxotrophic mutants can be examined by using the Holliday test (52). The actual percentage of different auxotrophic mutants observed represents an important piece of data that can be used to determine how many random Tn*5*-containing colonies must be screened to find an insertion in a particular gene of interest, since it reflects the insertional specificity of Tn*5* in the bacterial strain under study.

Once it has been established that Tn*5* mutagenesis can indeed lead to the identification of different mutations in the bacterial strain of interest, it should be determined whether the Km^r (mutant) colonies isolated carry a simple insertion of a single Tn*5* into a particular genomic locus. These events should be distinguished from multiple Tn*5* insertions or aberrant cointegration events involving the suicide vector. The latter is of particular importance, since in some bacterial species predominantly Tn*5*-mediated cointegrate formation, resulting in the integration of the entire Tn*5*-carrying suicide vector into the genome, is the cause of the observed mutant phenotype (for examples, see references 37 and 49). For a detailed description of this phenomenon, the reader is referred to reference 3.

The cointegration of the entire suicide vector can usually be ruled out conveniently by examining the Km^r colonies, resulting from a random Tn*5* mutagenesis experiment, for resistance to one or more of the antibiotic resistance genes carried by the non-Tn*5* part of the vector (e.g., Tc^r, Ap^r, or Cm^r for the pSUP vectors [110, 111]). Single versus multiple Tn*5* insertions and cointegration of parts of the suicide vector should also be examined by DNA-DNA hybridization (Southern blotting [see section 17.2.3.2]).

17.2.3. Physical Mapping of Insertions

To physically characterize Km^r colonies resulting from the random Tn*5* mutagenesis protocols outlined in sections 17.2.1 and 17.2.2, total genomic DNA is isolated from cultures derived from the respective colonies, digested with restriction enzymes, separated by agarose gel electrophoresis, transferred onto nitrocellulose or nylon membranes, and hybridized with Tn*5*- and suicide vector-derived DNA probes.

The method for isolating total genomic DNA from bacterial cultures depends strictly on the genus and species being analyzed, and therefore no universal protocol is available. However, the following method (modified from references 5 and 74) is applicable to a variety of gram-negative enteric and soil bacteria.

17.2.3.1. Isolation of genomic DNA

1. Generate a 1-ml culture of logarithmically growing cells.

2. Pellet the cells (e.g., in a 1.5-ml microcentrifuge tube), and wash them twice with 1 M NaCl to remove (most) extracellular polysaccharides. Resuspend the pelleted cells in 1 ml of TE (10 mM Tris, 1 mM EDTA [pH 8.0]).

3. Incubate the cells for 20 min at 37°C in the presence of 0.1 ml of a freshly prepared 2 mg/ml lysozyme solution in TE, with gentle mixing.

4. Add 0.125 ml of a 10% Sarkosyl–5 mg/ml pronase solution in TE (which has been predigested for 1 h at 37°C), and incubate the mixture for 1 h at 37°C.

5. Extract the viscous mixture of lysed cells with TE (pH 8.0)-saturated phenol two or three times, and extract the final aqueous phase with chloroform.

6. Add ammonium acetate to a final concentration of 0.3 M, and precipitate the nucleic acids by the addition of 2.5 volumes of ice-cold ethanol. "Spool out" the DNA strands with a glass rod, or pellet the nucleic acids by centrifugation. Resuspend the spooled out DNA or the pellet in 0.1 to 0.5 ml of TE by gentle mixing. Leave the DNA solution at 4°C overnight, and/or mix gently to achieve optimal resuspension.

7. Determine the concentration of the DNA preparations, and load 0.1 to 0.5 μg of undigested DNA onto an agarose gel to verify its high molecular weight and to check for the presence of nucleases in the preparations.

17.2.3.2. Southern blotting and hybridization

The total genomic DNA from the Km^r colonies (1 to 2 μg) can now be digested with different restriction endonucleases. For the first analysis, an enzyme which does not cut Tn*5* should be used (e.g., *Eco*RI, *Kpn*I, or *Cla*I). For modified Tn*5* derivatives carrying additional sequences, the restriction maps should be examined to verify that these enzymes do not cut; if necessary, other enzymes should be selected. Insertion of Tn*5* in a specific *Eco*RI or *Cla*I fragment (X) should give rise to a new, larger *Eco*RI or *Cla*I fragment of a larger size (X + Tn*5*; Tn*5* ≈ 5.8 kb). In the second level of analysis, an enzyme should be used which cuts the transposon twice (in the inverted repeats; *Hin*dIII, *Bgl*II [Fig. 1]), thereby generating distinct Tn*5* target junction fragments, as well as constant internal Tn*5* fragments (5). To visualize the Tn*5*-containing (junction) fragments, the digested DNA should be separated by agarose gel electrophoresis and transferred to membranes by Southern blotting (for details see Chapter 16.3.4.2.b and reference 70). To determine the orientation of the Tn*5* insertion relative to known restriction sites in the mutated locus, enzymes which cut the transposon once (asymmetrically, e.g., *Sal*I, *Sma*I [Fig. 1]) should be used.

The membranes should be hybridized with labeled Tn*5* sequences (e.g., the *Hpa*I internal fragment of Tn*5* [Fig. 1]), and the hybridizing fragments should be visualized by autoradiography or other appropriate visualization technique (see Chapter 26).

For the enzymes not cutting Tn*5*, single hybridizing fragments of different sizes (all larger than 5.8 kb) should be observed in the DNA of different Km^r colonies, suggesting that they are the result of single Tn*5* transpositions into the genome. For the enzymes cutting Tn*5* once, two distinct hybridizing fragments should be observed in the DNA of each Km^r colony, representing defined transposon-host junction sequences. For the enzymes cutting Tn*5* twice, in the inverted repeats, two distinct junctions and one constant internal Tn*5* fragment should hybridize.

The Southern blots carrying the digested genomic DNA of the Km^r colonies should also be probed with labeled suicide vector sequences (without Tn*5*). If simple Tn*5* transpositions occurred, fragments hybridizing with vector sequences should not be observed.

The Km^r colonies have now been shown to be the result of the simple insertion of a single copy of Tn*5* and can be characterized further, both genetically (see section 17.2.4) and physically, by more extensive restriction mapping, cloning, and DNA sequencing (see sections 17.2.5, 17.4.5, and 17.4.6 and Chapters 18 and 20).

17.2.4. Genetic Mapping of Insertions and Tn*5*-Induced Mutations

Once a Tn*5* insertion, a Tn*5*-induced mutation, or a collection thereof have been isolated and characterized, they can be used for genetic mapping experiments to determine their relative position with reference to previously identified mutations or to create a genetic map of an organism. This procedure is facilitated by the presence of the antibiotic resistance marker on the transposon, which permits selection for inheritance of the integrated transposon and selection of or screening for the mutations caused by its insertion (7). Transposon-induced insertion mutations behave essentially as standard genetic markers and therefore can be used in two-point crosses to determine genetic distances and in three-point crosses to establish gene order (7). By using standard F′ or R′ formation methods (see Chapter 14.3.1), these crosses can be carried out and genetic map positions can be determined. This is further facilitated by the use of Tn*5* derivatives carrying an origin of conjugal transfer (see section 17.4.3).

Detailed strategies for carrying out mapping experiments with Tn insertions are described by Kleckner et al. (7) and exceed the scope of this chapter. Examples of the mapping of specific Tn*5*-induced insertion mutations have been described previously (46, 106), and a list of Tn*5*-induced insertion mutations which have been mapped in *E. coli* and *Salmonella typhimurium* has been compiled (1). Examples of the use of Tn*5* mutagenesis to create a genetic map have also been published (8, 91). For an example of a fine-structure analysis of Tn*5* insertions within a single operon, see Miller et al. (77).

Once the genetic map of a region (operon, regulon) has been established with the help of Tn*5*-induced insertion mutations and a recombinant plasmid corresponding to the wild-type operon (regulon) has been obtained, a correlated physical and genetic map can be prepared by using restriction mapping and Southern blotting. For a description of the strategy used and an example of this type of analysis, see Riedel et al. (95).

17.2.5. Cloning of Tn*5*-Mutated Genes

On the basis of hybridization data (see section 17.2.3), a fragment is usually chosen that is generated by a restriction enzyme that does not cut in the Tn*5* sequence. This enables the cloning of regions upstream as well as downstream of the Tn*5* insertion. The fragment should be large enough to carry, besides Tn*5*, the sequences coding for the gene(s) of interest, at least with a high probability. Any suitable vector can be used, but of course it will facilitate the cloning procedure if the antibiotic resistance marker of the vector is not kanamycin resistance.

A great advantage of this step of the random mutagenesis is the use of a Tn*5* that carries an origin of replication (see section 17.4.3) in addition to the Tn*5* antibiotic resistance genes. In this case, the chromosomal DNA can be isolated, digested with a restriction enzyme that does not cut in the Tn*5* derivative, self-ligated, and transformed into *E. coli*. Only fragments containing the Tn*5* derivative with the origin will replicate and propagate in *E. coli*.

As soon as the mutated gene region is cloned, it can be used to conduct a gene replacement experiment (see section 17.3.4) to verify that the Tn*5* insertion indeed caused the observed phenotype.

Finally, the mutated gene region can be used as a probe to clone the equivalent wild-type region from the original bacterial strain. Different strategies are available for this purpose; they are summarized in the following paragraphs (for detailed protocols, see recent molecular biology manuals [87, 98]).

1. After the plasmid carrying the Tn*5*-mutated region is mapped with restriction enzymes, prepare a DNA probe from the cloned and mutagenized bacterial locus. This probe could consist of a restriction fragment carrying DNA sequences immediately adjacent to the Tn*5* insertion or a fragment carrying Tn*5* plus target sequences. Alternatively, if the DNA sequence of (part of) the mutagenized locus has been determined (see also section 17.4.5), a DNA oligonucleotide probe can be synthesized or a specific polymerase chain reaction (PCR)-generated probe can be prepared (see Chapter 19) for hybridization experiments.
2. Hybridize a Southern blot, containing genomic DNA from both the Tn*5*-mutated and wild-type strains digested with enzymes which do not cut the transposon (see section 17.2.3.2), with the DNA probe prepared in step 1. This analysis should lead to the identification of the same single hybridizing fragments found to hybridize with the Tn*5* probe (see section 17.2.3.2) in the Tn*5*-containing mutant strains and corresponding smaller single fragments (minus the length of Tn*5*) in the wild-type strain.
3. Create a complete clone bank of DNA from the wild-type bacterial strain, which was used to generate the Tn*5* mutants, in a plasmid or phage cloning vector (see Chapter 18). Alternatively, a partial clone bank can be constructed by purifying DNA fragments of the expected size (see step 2) from a gel and cloning them in a plasmid vector.
4. Screen the library (step 3) with the probe generated in step 1, and purify hybridizing plaques or colonies (see Chapter 18).
5. Hybridize a Southern blot containing genomic DNA of the strain and DNA of the recombinant phages/plasmids, digested with the same enzymes as above (step 2), with the probe (step 1) to verify that the fragment of interest has been cloned.
6. Further analyze the cloned wild-type locus by restriction mapping, complementation studies, DNA se-

quencing (see Chapter 20), and expression studies (see Chapter 18). In addition, the cloned region can be subjected to region-directed Tn*5* mutagenesis to create a precise, correlated physical and genetic map (see section 17.3.3) or subjected to transposon mutagenesis to create gene fusions for regulation studies (see section 17.4.1).

In selected cases, the wild-type equivalent of a Tn*5*-mutated locus can also be cloned by direct complementation. A readily screenable or selectable phenotype of the complemented strain would have to be available. In this case, the clone bank of the wild-type strain (step 3) would be introduced into the mutant strain via transformation, transduction, or conjugation and individual transformants, tranductants, or transconjugants would be examined for restoration of the wild-type phenotype.

17.3. REGION-DIRECTED Tn*5* MUTAGENESIS

When a wild-type locus has been cloned, region-directed (Tn*5*) mutagenesis can be used to delimit the genes and operons carried by the cloned region. The whole procedure can be carried out most efficiently if the presence of the cloned locus in *E. coli* results in an observable phenotype. The latter includes situations in which the cloned locus of interest is able to complement an auxotrophic mutation or growth defect of a special *E. coli* strain, endows such strains with the ability to catabolize or grow in the presence of novel compounds, or produces protein products which can be detected enzymatically or immunologically. If the cloned locus does not carry genes whose presence results in an observable phenotype, region-directed Tn*5* mutagenesis can still be performed as described below. The insertions are mapped, and a subset is chosen for gene replacement experiments (see section 17.3.4) to analyze the effect of the different Tn*5* insertions when introduced into the genomic locus of the original bacterial strain.

17.3.1. Vectors

In principle, every plasmid cloning vector carrying a cloned region but not encoding a kanamycin resistance gene can be used as a target for region-directed Tn*5* mutagenesis. The length of the vector relative to the length of the insert should be considered. Assuming a random insertional specificity of Tn*5* (3, 6), the longer the vector sequence the more insertions will be in the vector and not in the cloned locus of interest. In addition, the copy number of the cloning vector is important. High-copy-number vectors will be easier to mutagenize, and it will be more convenient to isolate DNA from strains harboring them. In addition, high-copy-number plasmids can be used for Tn*5* mutagenesis experiments with a chromosomal copy of Tn*5* as the donor (16, 109). This protocol will not be presented here but is detailed in reference 5.

Some commonly used cloning vectors, such as pBR322 (21), carry hot regions for Tn*5* insertion (2) and may therefore be less useful. However, successful Tn*5* mutagenesis of cloned genes has been performed not only with small, high-copy-number vectors such as pACYC184 (25) but also with low-copy broad-host-range vectors such as pLAFRI and pRK290 (36, 40).

The most commonly used Tn*5* delivery (mutagenesis) vector for region-directed mutagenesis is λ::Tn*5* (see section 17.2.1), and a protocol (modified from references 5 and 6) for this type of experiment follows.

17.3.1.1. Phage λ::Tn*5*

1. Transform the plasmid to be mutagenized into any λ-sensitive *E. coli* host (see Chapter 14.1.3). The plasmid-selectable marker should not be kanamycin or neomycin resistance.
2. Purify a single colony, make a small-scale plasmid DNA preparation (see Chapter 16.1.2), and verify the monomeric structure of the recombinant plasmid to be mutagenized. (Analyze the size of the undigested plasmid by agarose gel electrophoresis.)
3. Start an overnight culture of the *E. coli* strain with the target plasmid in 5 ml of LB supplemented with the appropriate antibiotic (plasmid vector marker gene) and 0.2% maltose at 37°C.
4. Inoculate 100 μl (1:100) of the saturated culture into 10 ml of LB–antibiotic–0.2% maltose, and let it grow for 2 h at 37°C.
5. Spin down the cells, wash them with 10 mM $MgSO_4$, and pellet them again. Resuspend the pelleted bacteria in 4 ml of 10 mM $MgSO_4$.
6. Infect 1-ml aliquots of cells with 100 μl of λ::Tn*5* lysate, prepared as described in section 17.2.1.1 (10^8 to 10^9 CFU/ml), and keep 1 ml of cells uninfected as a control.
7. Mix gently, and let the phage adsorb for 2 h at room temperature without agitating.
8. Plate 200-μl aliquots of infected cells on LB plates containing kanamycin (25 μl/ml) and the antibiotic selecting for the presence of the plasmid. Incubate the plates for 2 days at 37°C.
9. Wash the colonies off the plates with 1 to 2 ml of LB medium by using a sterile spreading rod, and pellet the cells by centrifugation. Resuspend the pellet in 2 to 5 ml of LB medium. Isolate plasmid DNA from the cell suspension by using a (scaled-up) miniprep plasmid DNA preparation protocol (see Chapter 16.1.2).
10. Transform competent *E. coli* cells (see Chapter 14.1.3) with the plasmid DNA isolated in step 9, and select for kanamycin- and plasmid marker-resistant transformants. Store the desired number of transformants on master plates for further analysis. These colonies should harbor plasmids carrying (generally) a single copy of Tn*5* inserted in a distinct plasmid location (pXX::Tn*5*), which can be determined by restriction mapping (see section 17.3.2).

17.3.1.2. Mutagenesis protocol and mutant characterization

Generally, step 8 of the above protocol should yield plates with at least 500 to 1,000 Km^r colonies, from which plasmid DNA is to be isolated. If fewer than 500 colonies appear, the titer of the λ :: Tn*5* lysate should be checked. Steps 9 and 10 will select colonies in which the Tn*5* has inserted into the plasmid and not into the chromosome of the original *E. coli*/pXX strain infected with λ :: Tn*5*. The effect of the Tn*5* insertion on the expression of the gene(s) or operon(s) carried on the cloned DNA of plasmid pXX can now be directly examined, if there is a measurable phenotype.

1. If the cloned DNA is capable of complementing an auxotrophy or growth requirement of the *E. coli* strain used in step 10, plate the *E. coli*/pXX :: Tn*5* transformants on the appropriate minimal medium and select pXX :: Tn*5* derivatives no longer capable of complementing. These plasmids are expected to carry Tn*5* insertions in the locus of interest, either in the structural gene(s) or on their control region(s). For examples of this type of analysis, see references 30 and 31. A similar strategy would apply when analyzing a cloned DNA region conferring on the host *E. coli* the ability to catabolize certain compounds. For an example of this type of analysis, see reference 42.

2. If the detectable "phenotype" of the cloned region is solely the production of a polypeptide of a known molecular weight, introduce the pXX :: Tn*5* plasmids and the parental plasmid into minicells (92) or maxicells (99) via transformation and analyze the plasmid-encoded polypeptides by sodium dodecyl sulfate-polyacrylamide gel electrophoresis (SDS-PAGE) (see Chapter 18.3.3). Insertion of Tn*5* in the locus would result in the disappearance of the polypeptide of interest. In many cases, because of the presence of transcriptional as well as translational stop signals on Tn*5*, truncated polypeptides will appear. The size of these truncated polypeptides is indicative of the position of the Tn*5* in the gene (or the operon) relative to the amino-terminal end, and this information can be used to deduce the direction of transcription of the gene (operon), once combined with the physical mapping data (section 17.3.2). Examples of this type of analysis have been described previously (31, 32).

3. If the cloned region encodes an enzymatic activity detectable in *E. coli* or produces a protein for which an antibody has been isolated, screen *E. coli*/pXX :: Tn*5* transformants appropriately for loss of function. For antibody-screening methods, see Chapter 18.3.4, and for an example of the latter type of Tn*5* mutant screening, see reference 68.

Clearly, there are many variations not listed here on the theme of how to screen for the desired phenotype. For an extensive list of transposon-induced insertion mutants in distinct genes of *E. coli* and *S. typhimurium* (and screens used to identify them), the reader is referred to reference 1. For further discussion, see references 3 and 6.

17.3.2. Physical Mapping of Plasmid-Borne Insertions

1. Isolate plasmid DNA from the desired *E. coli*/pXX :: Tn*5* strains by using a rapid, miniprep isolation protocol (see Chapter 16.1.2).

2. Digest the DNA with several restriction enzymes. Several convenient enzymes are available for the physical mapping of Tn*5* insertions on pXX :: Tn*5* plasmids, some of which do not cut Tn*5* (e.g., *Eco*RI, *Kpn*I, and *Cla*I), or cut Tn*5* once (e.g., *Sal*I, *Sma*I, and *Bam*HI) or twice (e.g., *Hin*dIII, *Bgl*II, *Not*I, and *Hpa*I) (Fig. 1). Depending on the restriction sites available in the pXX plasmid, different single and double digests can be carried out to map the position and relative orientation of the Tn*5* insertion. As discussed above for the mapping of chromosomal Tn*5* insertions (section 17.2.3.2), first use an enzyme which does not cut Tn*5* and subsequently use an enzyme which cuts (twice) in the inverted repeats, in order to generate specific Tn*5*-pXX junction fragments. By measuring the length of the junction fragments and subtracting the number of base pairs contributed by Tn*5* (Fig. 1), the approximate distance of the Tn*5* insertion to the nearest cleavage site for the enzyme used can be deduced. To help determine the relative orientation of the transposon within the cloned DNA, an enzyme cutting Tn*5* asymmetrically (e.g., *Sma*I) can be used. For a detailed discussion of mapping strategies, see references 5 and 6.

17.3.3. Correlated Physical and Genetic Maps

The physical mapping data (section 17.3.2) can now be combined with the data obtained by examining the phenotype of the different insertions (section 17.3.1.2) to create a correlated physical and genetic map. The Tn*5* insertions that disturb the expression of the cloned locus under analysis would be expected to cluster and to be flanked by Tn*5* insertions that do not affect its expression. Thus the extent of the loci can be mapped to an accuracy of approximately 100 bp, depending on the extent of the saturation mutagenesis experiment and the lack of insertional specificity of Tn*5* in the particular cloned region under investigation. For examples of correlated maps thus created, see references 31 and 32, and for discussions of insertional specificity of Tn*5*, see references 3, 6, and 14.

The information thus obtained on the precise location of a gene of interest in a large fragment of cloned DNA can now be used to characterize this gene further, for example by directed DNA sequencing with Tn*5*-derived primers (see section 17.4.5), characterization of its regulatory regions and expression (see section 17.4.1), and the creation, by directed gene replacement, of well-characterized Tn*5* insertion mutants of the bacterial species from which the gene was cloned (section 17.3.4).

Since Tn*5* insertion mutations are generally polar on genes located downstream of the mutated gene in the same operon (3), this type of mutagenesis can also be used to determine the structure of an operon carried on a cloned segment of DNA, especially when combined with an analysis of the polypeptides encoded by the (mutagenized) recombinant plasmids (see section 17.3.1.2). However, Tn*5*-mediated polarity is not always observed (17), especially when analyzing Tn*5*-mutated operons carried by multicopy plasmids (32), and therefore one should exercise caution when interpreting the results obtained with region-directed Tn*5* mutagenesis of cloned DNA segments (for a discussion, see references 3 and 6).

17.3.4. Gene Replacement Vectors and Methods

A powerful extension of the Tn*5* mutagenesis protocol is the use of gene replacement techniques to substitute the wild-type gene in the original bacterial strain with its well-characterized Tn*5*-mutated analog carried by a plasmid in *E. coli*. This allows the determination of the phenotype of specific Tn*5*-induced insertion mutations in the original organism. The general principle is to introduce the Tn*5*-mutated gene region back into the original organism and force a double-crossover event between the wild-type gene sequences in the bacterial genome and the equivalent sequences flanking the Tn*5*, followed by the loss of the wild-type gene.

Two methods widely applicable to gram-negative bacteria other than *E. coli* have been described previously (5). The first method was developed by Ruvkun and Ausubel (97). In this method, the broad-host-range vector pRK290 (36) is used to introduce the Tn*5*-mutated gene into the original bacterial strain by triparental conjugation (see section 17.2.2.2) with the helper plasmid pRK2013 (36). This is followed by the introduction of a plasmid incompatible with pRK290, e.g., pPH1JI (50), into the same strain. Selection for the antibiotic resistance marker of Tn*5* (kanamycin) and of the newly introduced plasmid pPH1JI (gentamicin) forces a double homologous crossover event and the subsequent loss of the pRK290 replicon carrying the wild-type sequences.

The second method uses narrow-host-range plasmids carrying the ColE1 mobilization (*bom*) site (such as pBR322). These plasmids are able to replicate only in *E. coli*, but they are transferable to a wide range of other gram-negative bacteria if the ColE1 Mob and Tra functions (see Chapter 14) are provided in *trans* (see section 17.2.2). Van Haute et al. (123) described the use of *E. coli* GJ23 harboring the plasmids pGJ28 (Mob$^+$) and R64*drd*11 (Tra$^+$) to transfer pBR322 derivatives carrying a Tn*5*-mutated gene effectively into *Agrobacterium tumefaciens*. The ColE1 replicon cannot be stably maintained, and therefore on selection for the resistance marker of Tn*5*, the mutated region will integrate into the genome via homologous recombination between cloned plasmid-borne and resident sequences. In most cases a cointegration of the vector will occur via a single crossover, but at a certain frequency (depending on the host and the mutated gene region) a double crossover will occur, leading to a bona fide gene exchange. Alternatively, the above-described vectors of the pSUP series (see section 17.2.2) can be used for the same purpose, since they carry the mob region of RP4 cloned into the narrow-host-range plasmids such as pACYC184 and pBR325 (111). Therefore, they can be mobilized into other gram-negative bacteria by providing the Tra functions in *trans*. This is normally achieved by using strain S17-1, which carries a chromosomally integrated copy of RP4 (Tra$^+$).

Gene replacement techniques can also be used to replace already existing Tn*5* insertions with other Tn*5* derivatives, carrying alternative selectable marker genes, reporter gene fusions, portable promoters, DNA primers of restriction sites, and origins of transfer or replication (see section 17.4 and references 17, 34, and 90). This allows a large degree of versatility. The only limitation in these types of replacement experiments is the length of homologous sequences needed for the double-recombination event. This is particularly relevant for transposons used for the construction of gene fusions and those carrying portable promoters, in which very limited Tn*5* sequences may be present between the reporter gene/promoter and the end of the transposon.

17.3.4.1. Broad-host-range and incompatible plasmids

1. Clone the Tn*5*-mutated region in the broad-host-range vector pRK290.
2. Perform a triparental mating, as outlined in section 17.2.2.2, with the recipient strain (from which the cloned region was derived), an *E. coli* strain harboring the Tn*5*-mutated gene region cloned in pRK290, and an *E. coli* strain harboring the helper plasmid pRK2013. Select for the recipient, Tn*5* (Km^r), and pRK290 (Tc^r).
3. Purify the obtained exconjugant once on the same selective plates.
4. Perform a diparental mating (section 17.2.2.1) with the recipient strain harboring the Tn*5*-mutated gene on the pRK290 replicon and an *E. coli* strain harboring pPH1JI. Select for the recipient strain, Tn*5* (Km^r), and pPH1JI (Gm^r).
5. Pick 100 to 1,000 exconjugants.
6. Screen for the loss of the pRK290 replicon by streaking the exconjugants on master plates with and without tetracycline.
7. Purify about 10 different tetracycline-sensitive exconjugants once or twice on selective plates.
8. Isolate total chromosomal DNA of these tranconjugants, digest it with one or two restriction enzymes, separate it by agarose gel electrophoresis, blot it onto a filter, and hybridize with the cloned wild-type gene region. The wild-type region should be replaced by the Tn*5*-mutated region. (In a digest with a restriction enzyme that does not cut Tn*5*, such as *Eco*RI, the wild-type fragment X should be replaced by a fragment X + 5.8 kb in size [see section 17.2.3.2 and reference 5].)

17.3.4.2. Narrow-host-range plasmids

1. Clone the Tn*5*-mutated region in a narrow-host-range vector, such as pBR322 or pSUP202.
2. Transform the clone obtained in step 1 in an *E. coli* strain providing the Mob and Tra functions. (Use strain GJ23 for pBR322 derivatives and strain S17-1 for pSUP derivatives.)
3. Perform a diparental mating (section 17.2.2.1) with the recipient strain (from which the region was cloned) and the *E. coli* strain harboring the narrow-host-range vector with the cloned Tn*5*-mutated region. Select for the recipient strain and Tn*5* (Km^r).
4. Pick 100 to 1,000 transconjugants.
5. Screen for loss of the vector sequences by streaking the exconjugants on master plates with and without the antibiotic resistance marker of the vector used but containing kanamycin for the selection of Tn*5*.
6. Purify, once or twice on selective plates, about 10 colonies that do not exhibit resistance to the antibiotic encoded by the vector.
7. Isolate total chromosomal DNA of these transconjugants, digest it with one or two restriction enzymes, separate it by agarose gel electrophoresis, blot it onto a filter, and hybridize it with the wild-type gene region. The wild-type region should have been replaced by the Tn*5*-mutated region (see section 17.3.4.1, step 8).

17.3.4.3. Screening for double crossovers

In some bacterial systems, depending on the host or the special gene region mutagenized, it may be difficult to find double-crossover events. Sometimes more than 1,000 colonies must be screened to find exconjugants that have undergone the double homologous crossover event (86). Recently, a conditionally lethal gene has been used to improve the gene replacement protocol. This gene allows selection and screening for the bona fide gene exchange event in several bacterial species. The *sacB* gene of *Bacillus subtilis* (coding for levansucrase) confers sensitivity to 5% sucrose; i.e., gram-negative organisms expressing the *sacB* gene cannot survive on solid medium containing 5% sucrose (44). If the vector used for the gene replacement experiment carries the *sacB* gene, all single-crossover events leading to cointegration of the vector will be lethal. Therefore, only exconjugants in which double-crossover events have occurred will be obtained in the presence of 5% sucrose. Before trying this approach, the sensitivity of the strain of interest, containing the cloned *sacB* gene, to the presence of sucrose in the growth medium should be tested.

17.4. USES OF Tn*5* DERIVATIVES

17.4.1. Reporter Gene Fusions

The utility of transposon Tn*5* mutagenesis has recently been extended to the generation of gene fusions for regulation studies. An abbreviated list of Tn*5*-based vehicles for the construction of chimeric genes is shown in Table 2. For a more complete list, see reference 2.

Transposons for the generation of both transcriptional and translational **reporter gene** fusions are now available. In the former case, the reporter gene carries its own start codon and DNA sequences recognized by the translational machinery. In the latter case, the reporter gene lacks its own start codon and ribosome-binding site, and its expression is dependent not only on transcription originating from the promoter of the mutagenized gene but also on the simultaneous creation of an in-frame fusion of the open reading frame of the mutagenized gene with the truncated reporter gene.

The reporter genes which are commonly used to create gene fusions include the β-galactosidase [β-Gal; *lacZ*(*YA*)], β-glucuronidase (Gus; *uidA*), alkaline phosphatase (AP; *phoA*), and luciferase (Lux; *lux*) genes. Convenient dyes to detect their expression on solid growth media, enzymatic assays, and histochemical staining methods for the detection of these reporter gene products are available, making them very useful to monitor the quantitative and in situ expression of chimeric genes. β-Gal, Gus, and Lux activities are cytoplasmic, whereas while AP activity can be measured only if the protein is periplasmic or excreted. Therefore, saturation mutagenesis of a gene with, for example, Tn*5*-*lac* and Tn*5*-*phoA* can reveal sequences that are translated into cytoplasmic protein domains or transmembrane domains, respectively. Moreover, Tn*5*-*phoA* fusions can be used to identify protein signal sequences required for excretion (71, 72).

For detailed descriptions of the use of the *lac* (108), *phoA* (73), *uidA* (43, 54), and *lux* (75, 119) reporter gene systems, the reader is referred to the respective references. For the detection of β-Gal enzyme activity in vivo on plates or in cell extracts, see Chapter 23 and reference 76. For the detection of AP activity, see Chapter 23 and reference 101. For the detection of Gus activity, see references 43, 54, and 105. For the detection of Lux activity, see references 75, 119, and 126.

In addition to reporter genes lacking transcriptional and/or translational start signals, a number of other useful DNA sequences and genes have been cloned into Tn*5* to facilitate their random or region-directed introduction into bacterial chromosomes or recombinant plasmids. A number of examples of useful Tn*5* derivatives are discussed in the following sections. For a more comprehensive list of such derivatives and examples of their use, see reference 2. In most cases, the protocols for the generation and characterization of Tn*5*-containing replicons are the same as those described above (sections 17.2 and 17.3).

The actual mutagenesis protocols for the generation of Tn*5*-*lac*, Tn*5*-*phoA*, Tn*5*-*uidA*, and Tn*5*-*lux* insertions in genomic loci or cloned DNA fragments carried by recombinant plasmids are the same as described above (sections 17.2 and 17.3). For details of generating gene fusions with the Mini-Tn*5*-reporter gene transposons, see de Lorenzo et al. (33).

Table 2. Tn*5* derivatives used for the creation of gene fusions

Reporter gene	Designation	Antibiotic resistance	Length (kb)	Type of fusion	Reference
lacZ	Tn*5*-*lac*	Km^r, Nm^r	12.0	Transcriptional	63
lacZ	Tn5ORF*lac*	Tc^r	~12	Translational	62
lacZ	Tn*5*-B20	Km^r, Nm^r	8.3	Transcriptional	113
lacZ	Tn*5*-B21	Tc^r	8.4	Transcriptional	113
lacZ	Tn*5*-B22	Gm^r	9.4	Transcriptional	113
lacZ	Mini-Tn*5lacZ1*	Km^r, Nm^r	5.4	Transcriptional	33
lacZ	Mini-Tn*5lacZ2*	km^r, Nm^r	5.3	Translational	33
phoA	Tn*5*-*phoA*	Km^r, Nm^r	7.7	Translational	71
phoA	Mini-Tn*5phoA*	Km^r, Nm^r	3.5	Translational	33
nptII	Tn*5*-VB32	Tc^r	~5	Transcriptional	10
nptII	Tn*5*-B30	Tc^r	6.2	Transcriptional	113
uidA[a]	Tn*5*-*gusA1*	Km^r, Nm^r, Tc^r	8.3	Transcriptional	105
uidA[a]	Tn*5*-*gusA2*	Km^r, Nm^r, Tc^r	8.3	Translational	105
uidA[a]	Tn*5*-*gusA9*	Sm^r, Sp^r	5.0	Transcriptional	105
uidA[a]	Tn*5*-*gusA10*	Sm^r, Sp^r	5.0	Translational	105
luxAB	Tn*5*-1063	Km^r, Nm^r, Bm^r, Sm^r	7.8	Transcriptional	126
luxAB	mini-Tn*5luxAB*	Tc^r	5.2	Transcriptional	33
xylE	mini-Tn*5xylE*	Km^r, Nm^r	3.7	Transcriptional	33

[a]Designated as *gusA* by Sharma and Signer (105).

The only difference from conventional Tn*5* mutagenesis is that, at the stage of selecting Tn*5*-reporter gene-containing strains or plasmids, the appropriate indicator compounds (e.g., 5-bromo-4-chloro-3-indolyl-β-D-galactoside [X-Gal] for β-Gal activity, 5-bromo-3-chloro-indolyl-phosphate [XP] for AP activity, and 5-bromo-4-chloro-3-indolyl-β-D-glucuronide [X-Gluc] for Gus activity) are added to the solid media and "positive" colonies containing the active transcriptional or translational fusions are identified under the proper physiological (inducing/repressing) conditions for the gene of interest. For the Tn*5*-*lux* transposons, colonies containing proper gene fusions can be identified by their light emission, using a photonic detection system (126). The map position and orientation of the Tn*5*-reporter gene insertions are mapped in the genome or on plasmids as described in sections 17.2.3 and 17.3.2.

Once an interesting Tn*5*-reporter gene-mediated gene fusion on a recombinant plasmid has been generated and physically characterized, it can be integrated into the genome of the host organism by gene replacement (section 17.3.4) or cointegrate formation. In the latter case, the protocols described in sections 17.3.4.1 and 17.3.4.2 can be used, but instead of screening for transconjugants which have lost the antibiotic marker(s) of the plasmid vector carrying the fusion, colonies which retain the vector marker are selected and the presence of the vector is verified by Southern blotting (see section 17.2.3.2).

To generate mutations in *trans*-acting regulatory genes which control the expression of the gene fusions of interest, strains carrying these fusions can be remutagenized with a Tn*5* derivative containing a selectable antibiotic resistance gene different from the marker gene on the original Tn*5*-reporter gene transposon (see section 17.4.4) (Table 3) by using the methods described in section 17.2.

17.4.2. Portable Promoters

Tn*5* derivatives carrying strong, regulatable, outward-facing promoters have been constructed in vitro. The synthetic, inducible *tac* promoter and the constitutive promoter for the neomycin phosphotransferase (*nptII*) gene have been introduced into Tn*5*, near one of the ends, to generate Tn*5*tac1 and Tn*5*-B50, respectively (27, 113; see Table 3 for additional derivatives). On insertion of these Tn elements upstream of the 5′ end of the coding sequence of a certain gene, the expression of this gene is rendered inducible by isopropyl β-D-thiogalactoside (for Tn*5*tac1) or constitutive (for Tn*5*-B50). This also applies to genes positioned distally in an operon. When Tn*5* derivatives such as Tn*5*tac1 or Tn*5*-B50 insert into an operon, the downstream genes come under the control of the Tn*5*-borne promoters. By using random or region-directed Tn*5* mutagenesis protocols, cryptic or normally down-regulated genes can be activated in this fashion. This can be very helpful for the overproduction of proteins.

17.4.3. Transfer and Replication Origins

Tn*5* derivatives have been constructed in which the central region of Tn*5* has been replaced with the IncP-specific mobilization target site (Tn*5*-Mob or Tn*5oriT* [109, 128]) (see Table 3 for additional derivatives). Insertion of these Tn elements allows the target replicon to be efficiently mobilized via the broad-host-range mobilization functions of RP4 or RK2. If the insertion is in the chromosome, the insertion site becomes the origin of Hfr-like oriented transfer (see Chapter 14.3.2). This technique is therefore very useful for the construction of genetic maps (see references 26, 39, 83, and 128 for examples; also see section 17.2.4). If the insertion is in a (recombinant) plasmid, it becomes F′-like (see Chapter 14.3.1) and allows the plasmid to be conjugally mobilized at high frequency in di- or triparental mating experiments (see sections 17.2.2.1 and 17.2.2.2); for reviews, see references 112 and 113).

Tn*5* derivatives carrying a narrow-host-range origin of replication have also been constructed. Furuichi et al. (41) have cloned the origin of replication of the ColE1-type plasmid pSC101 into Tn*5* to construct Tn*5*-V (Table 3). Tn*5*-V-mutated chromosomal loci generated by random Tn*5* mutagenesis (see section 17.2) can be readily cloned by using the antibiotic resistance marker and the replication origin carried on the transposon, by digestion of the total DNA with an enzyme which does not cut Tn*5*-V, religation, and transformation (see section 17.2.5). Wolk et al. (126) have also constructed a Tn*5* derivative which carries a P15A (pACYC184) origin of replication in addition to a *lux* reporter gene (Tn*5*-1063; see section 17.4.1) (Table 2), which is extremely useful for recloning Tn*5*-1063-mutagenized chromosomal loci with *lux* gene fusions.

17.4.4. Alternative Marker Genes

Many derivatives of the original Tn*5* transposon, containing a variety of additional antibiotic resistance and other marker genes, have been constructed (Table 3) (2). Therefore, not only can the original kanamycin and neomycin resistance markers (which are expressed in all organisms tested thus far) or the streptomycin and bleomycin resistance markers (which are expressed in several nonenteric bacteria) be used to monitor the presence of Tn*5* derivatives, but also genes encoding resistance to tetracycline, ampicillin, chloramphenicol, trimethoprim and gentamicin can be used (Table 3). These derivatives are especially useful if the target bacterial strain has an intrinsic resistance to the antibiotics normally used to select for Tn*5* (kanamycin, neomycin, bleomycin, and streptomycin) or if a secondary mutagenesis of a strain already harboring a Tn*5* derivative (e.g., a Tn*5*-reporter gene fusion; see 17.4.1) is to be carried out. They can also be used to replace already existing genomic Tn*5* insertions via homologous recombination (gene replacement) to exchange antibiotic resistance marker genes.

Mini-Tn*5* derivatives, consisting of the terminal 19 bp of IS*50*L and IS*50*R flanking the genes encoding resistance to the herbicide bialaphos (*bar*), mercury and organomercurial compound resistance (*mer*), or arsenite resistance (*ars*), have been constructed by Herrero et al. (48). The transposase required for transposition of these derivatives is provided by a (modified) *tnp* gene, cloned immediately adjacent to the mini-Tn*5* on the suicide vector pGP704 (48, 49) (see 17.2.1).

Table 3. Tn*5* derivatives carrying portable promoters, alternative marker genes, and useful sites

Use	Designation in original reference	Antibiotic resistance and novel sequences	Length (kb)	Reference
Outward reading promoters	Tn*5*-tac1	Km^r, *ptac*	4.6	27
	Tn*5*-B50	Tc^r, *pnptII*	5.0	113
	Tn*5*-B60	Km^r, *ptac*	5.0	113
	Tn*5*-B61	Gm^r, *ptac*	5.5	113
Replication/mobilization origins	Tn*V*	Km^r, *ori* pSC101	~6	41
	Tn*5*-*oriT*	Km^r, *oriT* (RK2)	6.5	128
	Tn*5*-Mob	Km^r, *oriT* (RP4)	7.8	109
	Tn*5*-B11	Gm^r, *oriT* (RP4)	8.4	113
	Tn*5*-B12	Gm^r, Sp^r, *oriT* (RP4)	10.2	113
	Tn*5*-B13	Tc^r, *oriT* (RP4)	9.4	113
Sequencing primer sites	Tn*5*seq1	Km^r, pT7, pSP6	3.2	81
	Tn*5supF*	*supF*	0.3	89
Novel antibiotic resistance genes	Tn*5*-Tc2	Tc^r	5.1	103
	Tn*5*-Cm	Cm^r	3.9	103
	Tn*5*-Gm	Gm^r	7.5	103
	Tn*5*-Tp	Tp^r	5.2	103
	Tn*5*-Sm	Sm^r	5.2	103
	Tn*5*-Ap	Ap^r	5.0	103
	Tn*5*.7	Tp^r, Sm^r, Sp^r	7.6	130
	Tn*5*-233	Gm^r/Km^r, Sm^r/Sp^r	6.6	34
	Tn*5*-235	Km^r, *'lacZY*	10.0	34
	Tn*5*-GmSpSm	Gm^r, Sp^r, Sm^r	6.8	51
	Tn*5*-*751*	Km^r, Tp^r	9.0	93
	Mini-Tn*5* Sm/Sp	Sm^r, Sp^r	2.1	33
	Mini-Tn*5* Tc	Tc^r	2.2	33
	Mini-Tn*5* Cm	Cm^r	3.5	33
	Mini-Tn*5* Km	Km^r	2.3	33
Other marker genes	Tn*5* *Bgl*+[a]	Km^r, *bgl*	11.3	125
	Tn*5* *Amy*+[b]	Km^r, *amy*	8.7	125
	Tn*5*-*tox*[c]	Km^r, *tox*	10.5	82
	Tn5-Lux[d]	Km^r, *luxAB*	6.6	19
	Mini-Tn*5* *bar*[e]	*bar*	2.5	48
	Mini-Tn*5* *mer*[f]	*mer*	4.5	48
	Mini-Tn*5* *ars*[g]	*ars*	5.0	48
Restriction mapping	Tn*5cos*	Km^r, Nm^r	6.2	131
Plasmid curing	Tn*5*-*rpsL*	Km^r, Nm^r	7.6	120
	Tn*5*-B12S	Km^r, Nm^r, *sacBR*	9.9	53
	Tn*5*-B13S	Km^r, Nm^r, Tc^r, *sacBR*	13.2	53
Multiple unique restriction sites	Tn*5*-K20	Km^r, Nm^r, Tc^r	9.5	55
	Tn*5*-K28	Sm^r, Tc^r	7.4	55

[a] *Bgl*+, ability to ferment cellobiose.
[b] *Amy*+, ability to degrade starch.
[c] *tox*, delta endotoxin gene from *Bacillus thuringiensis*.
[d] Lux, bioluminescence.
[e] *bar*, resistance to the herbicide bialaphos; vector designated as pUT/PTT by Herrero et al. (48).
[f] *mer*, resistance to mercuric salts and organomercurial compounds; vector designated as pUT/Hg by Herrero et al. (48).
[g] *ars*, resistance to arsenite; vector designated as pUT/Ars by Herrero et al. (48).

Several additional non-antibiotic-resistance genes have been inserted in Tn*5* to facilitate their transfer to and maintenance in a range of gram-negative bacteria. For example, genes encoding β-glucosidase and amylase, tryptophan biosynthesis, catechol-2,3-dioxygenase, and delta endotoxin production have been inserted into Tn*5* (Table 3), and expression of these genes can be used to monitor the (stable) presence of the modified transposons in different bacterial species. To facilitate cloning of other genes into Tn*5* for stable integration into the genome of bacteria, Tn*5* derivatives carrying many unique restriction sites have been constructed, such as Tn*5*-K20 and K28 (55).

A derivative of Tn*5* carrying the λ *cos* site has recently been constructed; it allows the rapid restriction mapping of large plasmids by using λ terminase (131). After transposition of Tn*5cos* into the desired plasmid, the DNA is linearized at the *cos* site; the left and right termini thus generated are labeled selectively in vitro. A partial digest of the linearized DNA is carried out with the restriction enzyme of choice, and the resulting (partial) fragments are visualized by means of the (radioactive or nonradioactive) label. This method should also be applicable to the establishment of a restriction map of a specific region of a bacterial chromosome after Tn*5cos* mutagenesis (131).

A Tn*5* derivative that can be used for the direct selection for curing and deletion of large indigenous plasmids in bacteria has been constructed by inserting the *Bacillus subtilis sacBR* genes and the RP4 *mob* site into

Tn*5* (Tn*5*-B12S or Tn*5*-B13S [Table 3]) (53). When this modified transposon inserts into a large cryptic plasmid, the resulting tagged plasmid can be mobilized to other bacterial strains. However, when the gram-negative bacteria containing the Tn*5*-B12/13S-tagged plasmid are grown in the presence of sucrose, the expression of the *sac* genes is lethal (see also section 17.3.4.3) and therefore curing of the plasmid or deletion formation events are selected (53).

An additional derivative of Tn*5* useful in plasmid-curing experiments has recently been described (120). Tn*5*-*rpsL* carries the *E. coli rpsL* gene, which confers streptomycin sensitivity to strains carrying the transposon. After an Sm^r derivative of the desired bacterial strain containing a plasmid to be cured has been isolated, it is mutagenized with Tn*5*-*rpsL* and insertions in the plasmid are isolated. The resulting strains are now Sm^s again. By selecting for Sm^r derivatives of the plasmid :: Tn*5*-*rpsL*-containing strain, loss of this plasmid is selected for, resulting in plasmid curing and deletion formation.

A Tn*5* deletion derivative has also been reported which has the ability to act as a recombinational switch, activating distal gene expression when inserted in one orientation (Tn*5*-112) (13). Already existing Tn*5* insertions at any site can be replaced with this transposon, via homologous recombination, and can be inserted to activate (or silence) downstream genes.

17.4.5. Primers for DNA Sequencing and PCR

Tn*5* derivatives have been developed that facilitate directed DNA sequencing by the chain termination method (100) (also see Chapter 20.1). To use this method, it is important to bring the DNA to be sequenced into the immediate vicinity of unique sequences that can be used as oligonucleotide-priming sites. Because of the nature of its inverted repeats, Tn*5* does not contain unique primer sites at its ends. Detailed mutational analyses have shown that only the outermost 19 bp is needed for efficient transposition of Tn*5* (88). Nag et al. (81) have constructed Tn*5*seq1, which contains unique primer sites derived from the promoters of the phages SP6 and T7 adjacent to these essential sequences at the subterminal ends of Tn*5*. These unique sequences, normally not present in the bacterial genome or plasmids, can be used as primer hybridization sites for commercially available oligonucleotides. Because Tn*5* transposes in vivo into many different sites, its ends can serve as portable priming sites for DNA sequencing, avoiding subcloning or deletion experiments (see Chapter 20).

Another mini-Tn*5*-derived tool for DNA sequencing of phage λ-borne DNA segments, Tn*5supF*, has been constructed by Phadnis et al. (89). This transposon carries a synthetic *E. coli supF* gene cloned between the 19-bp termini of Tn*5* and, when inserted after transposition, allows amber mutant phage λ derivatives (such as the cloning vector Charon 4A) to form plaques on a Sup^0 strain. By using unique primers hybridizing to the ends of the transposon, DNA sequencing in both directions reading outwards from the Tn*5supF* insertion can be carried out (89). This transposon is also very suitable for gene replacement experiments in *E. coli* (90; also see section 17.3.4).

With the help of PCR (see Chapter 19), Rich and Willis (94) have developed a method of isolating DNA sequences adjacent to the ends of a chromosomal Tn*5* insertion. They have used a single oligonucleotide complementary to and extending outward from the inverted repeats of Tn*5* to amplify the target sequences. By digesting the chromosomal DNA with a restriction enzyme not cutting in Tn*5* (*Eco*RI) and self-ligating the chromosomal DNA at a low concentration to favor intramolecular ligation, the circular DNA molecule containing the Tn*5* and the adjacent target sequences will be the only plasmid containing the two priming sites for the PCR. The resulting PCR product can then be used as a probe for cloning of the wild-type region.

A combination of the PCR and the Tn*5supF* mutagenesis protocol (89; see above) has been used to further facilitate Tn*5supF*-based DNA sequencing of λ phage clones (60).

Avoiding any cloning or ligation steps, Subramanian et al. (121) have used primers based on the ends of Tn*5* and of the so-called REP sequences to map chromosomal Tn*5*-induced mutations. The REP sequence (repetitive extragenic palindrome) is a highly conserved 35-bp long palindromic sequence, first identified in a number of intercistronic regions of numerous operons in *E. coli* and *S. typhimurium* (118), but recently also found in a variety of other eubacteria (29, 124). It is estimated that its copy number is greater than 500; therefore, these sequences make up to 1% of the whole bacterial genome and are very useful for genomic fingerprinting of bacteria (29, 124). For the PCR, one primer homologous to the end of Tn*5* and another degenerate primer homologous to the genomically anchored REP sequence is used. Thus, a unique segment of chromosomal DNA will be amplified, which can be used as a probe to identify the corresponding region in a genomic library of the bacterial species of interest.

17.4.6. Genome Mapping

Tn*5* and its derivatives are also useful for genomic mapping. Pulsed-field gel electrophoresis is a novel tool for separating large DNA fragments (114) (see Chapter 16.3.1.2). To obtain large DNA fragments, several new restriction endonucleases that cut only at rare sites in the genome are available. For example, in A+T-rich organisms, the restriction enzyme *Not*I, which recognizes an octameric sequence consisting only of G and C residues, is used to generate large fragments. In G+C-rich organisms, *Pac*I and *Swa*I, which recognize octameric sequences consisting of A and T residues only, can be used. For *E. coli*, a genomic map based on the 22 different *Not*I fragments has been established (115). *Not*I also cuts in the inverted repeats (IS*50*) of Tn*5* (Fig. 1); therefore, insertion of Tn*5* in the *E. coli* genome will generate two additional *Not*I sites, which will result in the appearance of two new fragments (plus a small internal Tn*5* fragment). By measuring the size of the two new (junction) fragments, the position of the Tn*5* on the *Not*I restriction map can be determined. Smith and Kolodner (116) have used this method to assign Tn*5* insertions to specific *Not*I fragments of the *E. coli* genome. A similar strategy has been used by Wolk et al. (126) to map Tn*5lux*-induced mutations in *Anabaena* mutants.

Mini-Tn*5* derivatives (33) have been constructed which carry rare restriction sites such as M.*Xba*I/*Dpn*I,

*Not*I, *Swa*I, *Pac*I, and *Spe*I (127). These derivatives have been used to construct a map of *S. typhimurium* via pulsed-field gel electrophoresis and should be extremely useful for genomic mapping of chromosomes and large replicons of other bacteria, since they introduce a variety of rare cutting sites on (random) insertion. In addition, already existing Tn*5* insertions will be able to be converted into derivatives carrying these rare sites via gene replacement, further extending the utility of this method for large-scale mapping.

17.5. RECIPES FOR MEDIA

The media listed here are used to prepare the λ :: Tn*5* lysate for region-directed Tn*5* mutagenesis.

LB medium. Tryptone (10 g/liter), yeast extract (5 g/liter), NaCl (5 g/liter).
SM buffer. 0.1 M NaCl, 0.02 M Tris (pH 7.5), 0.01 M $MgSO_4$, 0.01% gelatin.
λ agar medium. Tryptone (10 g/liter), NaCl (2.5 g/liter), Bacto Agar (Difco; 11 g/liter).
λ soft medium. Tryptone (10 g/liter), NaCl (2.5 g/liter), Bacto Agar (6 g/liter).

For all other experiments, the appropriate growth medium and antibiotic concentration for the target organism should be used. For a general discussion about nutrition and a list of basic media, see Chapter 7.

17.6. ACKNOWLEDGMENTS

We thank Uwe Rossbach for help in preparing the manuscript and figures, Peter Wolk for helpful comments on the manuscript, and Jim Lupski for many stimulating discussions about the use of transposon mutagenesis in bacterial molecular genetics.

17.7. REFERENCES

17.7.1. General References

1. **Berg, C. M., and D. E. Berg.** 1987. Uses of transposable elements and maps of known insertion, p. 1071–1109. *In* F. C. Neidhardt, T. L. Ingraham, K. B. Low, B. Magasanik, M. Schaechter, and H. E. Umbarger (ed.), *Escherichia coli* and *Salmonella typhimurium: Cellular and Molecular Biology.* American Society for Microbiology, Washington, D.C.
A general review of different parameters of transposon mutagenesis, including a list of useful transposable elements and comprehensive lists of all known transposon-induced mutations in E. coli and S. typhimurium.
2. **Berg, C. M., D. E. Berg, and E. A. Groisman.** 1989. Transposable elements and the genetic engineering of bacteria, p. 879–926. *In* D. E. Berg and M. M. Howe (ed.), *Mobile DNA.* American Society for Microbiology, Washington, D.C.
A general review of methods used for transposon mutagenesis, including a complete list of useful Tn elements and a comprehensive list of bacterial species in which transposons have been used.
3. **Berg, D. E.** 1989. Transposon Tn*5*, p. 163–184. *In* D. E. Berg and M. M. Howe (ed.), *Mobile DNA.* American Society for Microbiology, Washington, D.C.
*A comprehensive review of the characteristics of Tn*5*, including structure, transposition parameters, insertional specificity, etc.*
4. **Berg, D. E. and M. M. Howe.** 1989. *Mobile DNA.* American Society for Microbiology, Washington, D.C.
A comprehensive monograph describing transposable elements in procaryotes and eucaryotes.
5. **de Bruijn, F. J.** 1987. Transposon Tn*5* mutagenesis to map genes. *Methods Enzymol.* **154:**175–196.
*A review about transposon Tn*5* mutagenesis and its application to the construction of Tn*5* mutants. Includes a detailed description of general and region-directed mutagenesis protocols and gene replacement methods.*
6. **de Bruijn, F. J., and J. R. Lupski.** 1984. The use of transposon Tn*5* mutagenesis in the rapid generation of correlated physical and genetic maps of DNA segments cloned into multicopy plasmids—a review. *Gene* **27:**131–149.
*A review of the different parameters of Tn*5* mutagenesis and its application to the construction of correlated physical and genetic maps. Includes detailed protocols for region-directed mutagenesis and mapping of cloned genes.*
7. **Kleckner, N., J. Roth, and D. Botstein.** 1977. Genetic engineering *in vivo* using translocatable drug-resistant elements: new methods in bacterial genetics. *J. Mol. Biol.* **116:**125–159.
A still relevant review about Tn elements and their use in bacterial genetics. Includes detailed strategies for genetic mapping of transposon-induced mutations, etc.

17.7.2. Specific References

8. **Barrett, J. T., R. H. Croft, D. M. Ferber, C. J. Gerardot, P. V. Schoenlein, and B. Ely.** 1982. Genetic mapping with Tn*5*-derived auxotrophs of *Caulobacter crescentus. J. Bacteriol.* **151:**888–898.
9. **Belas, R., A. Mileham, M. Simon, and M. Silverman.** 1984. Transposon mutagenesis of marine *Vibrio* spp. *J. Bacteriol.* **158:**890–896.
10. **Bellofatto, V., L. Shapiro, and D. A. Hodgson.** 1984. Generation of a Tn*5* promoter probe and its use in the study of gene expression in *Caulobacter crescentus. Proc. Natl. Acad. Sci. USA* **81:**1035–1039.
11. **Berg, C. M., N. B. Vartak, G. Wang, X. Xu, L. Liu, D. J. MacNeil, K. M. Gewain, L. A. Wiater, D. E. Berg.** 1992. The mγδ-1 element, a small γδ (Tn*1000*) derivative useful for plasmid mutagenesis, allele replacement and DNA sequencing. *Gene* **113:**9–16.
12. **Berg, C. M., G. Wang, L. D. Strausbaugh, and D. E. Berg.** 1993. Transposon-facilitated sequencing of DNAs cloned in plasmids. *Methods Enzymol.,* **218:**279–306.
13. **Berg, D. E.** 1980. Control of gene expression by a mobile recombinational switch. *Proc. Natl. Acad. Sci. USA* **77:**4880–4884.
14. **Berg, D. E., and C. M. Berg.** 1983. The prokaryotic transposable element Tn*5*. *Bio/Technology* **1:**417–435.
15. **Berg, D. E., J. Davies, B. Allet, and J. D. Rochaix.** 1975. Transposition of R factor genes to bacteriophage lambda. *Proc. Natl. Acad. Sci. USA* **72:**3628–3632.
16. **Berg, D. E., M. A. Schmandt, and J. B. Lowe.** 1983. Specificity of Transposon Tn*5* insertion. *Genetics* **105:**813–828.
17. **Berg, D. E., A. Weiss, and L. Crossland.** 1980. Polarity of Tn*5* insertion mutations in *Escherichia coli. J. Bacteriol.* **142:**439–446.
18. **Beringer, J. E., J. L. Beynon, A. V. Buchanan-Wollaston, and A. W. B. Johnston.** 1978. Transfer of the drug-resistance transposon Tn*5* to *Rhizobium. Nature* (London) **276:**633–634.
19. **Boivin, R., F. P. Chalifour, and P. Dion.** 1988. Construction of a Tn*5* derivative encoding bioluminescence and its introduction in *Pseudomonas, Agrobacterium* and *Rhizobium. Mol. Gen. Genet.* **213:**50–55.
20. **Bolivar, F.** 1978. Construction and characterization of new cloning vehicles. III. Derivatives of plasmid pBR322

carrying unique *Eco*RI sites for selection of *Eco*RI generated recombinant molecules. *Gene* **4:**121–136.

21. **Bolivar, F., R. L. Rodriguez, P. J. Greene, M. C. Betlach, H. L. Heyneker, H. W. Boyer, and J. H. Crosa.** 1977. Construction and characterization of new cloning vehicles. II. A multipurpose cloning system. *Gene* **2:**95–113.
22. **Bukhari, A. I., J. A. Shapiro, and S. L. Adhya (ed.).** 1977. *DNA Insertion Elements, Plasmids, and Episomes.* Cold Spring Harbor Laboratory, Cold Spring Harbor, N.Y.
23. **Campbell, A., D. E. Berg, D. Botstein, E. M. Lederberg, R. P. Novick, P. Starlinger, and W. Szybalski.** 1979. Nomenclature of transposable elements in prokaryotes. *Gene* **5:**197–206.
24. **Campbell, A., D. E. Berg, D. Botstein, R. Novick, and P. Starlinger.** 1977. Nomenclature of transposable elements in prokaryotes, p. 15–22. *In* A. I. Bukhari, J. A. Shapiro, and S. L. Adhya (ed.), *Insertion Elements, Plasmids, and Episomes.* Cold Spring Harbor Laboratory, Cold Spring Harbor, N.Y.
25. **Chang, A. C. Y., and S. N. Cohen.** 1978. Construction and characterization of amplifiable multicopy DNA cloning vehicles derived from the P15A cryptic miniplasmid. *J. Bacteriol.* **134:**1141–1156.
26. **Charles, T. C., and T. M. Finan.** 1990. Genetic map of *Rhizobium meliloti* megaplasmid pRmeSU47b. *J. Bacteriol.* **172:**2469–2476.
27. **Chow, W.-Y. and D. E. Berg.** 1988. Tn*5*tac1, a derivative of transposon Tn*5* that generates conditional mutations. *Proc. Natl. Acad. Sci. USA* **85:**6468–6472.
28. **Craig, N. L.** 1989. Transposon Tn*7*, p. 211–225. *In* D. E. Berg and M. M. Howe (ed.), *Mobile DNA.* American Society for Microbiology, Washington, D.C.
29. **de Bruijn, F. J.** 1992. Use of repetitive (repetitive extragenic palindromic and enterobacterial repetitive intergeneric consensus) sequences and the polymerase chain reaction to fingerprint the genomes of *Rhizobium meliloti* isolates and other soil bacteria. *Appl. Environ. Microbiol.* **58:**2180–2187.
30. **de Bruijn, F. J., and F. M. Ausubel.** 1983. The cloning and characterization of the *glnF* (*ntrA*) gene of *Klebsiella pneumoniae*: role of *glnF* (*ntrA*) in the regulation of nitrogen fixation (*nif*) and other nitrogen assimilation genes. *Mol. Gen. Genet.* **192:**342–353.
31. **de Bruijn, F. J., S. Rossbach, M. Schneider, P. Ratet, W. W. Szeto, F. M. Ausubel, and J. Schell.** 1989. *Rhizobium meliloti* 1021 has three differentially regulated loci involved in glutamine biosynthesis, none of which is essential for symbiotic nitrogen fixation. *J. Bacteriol.* **171:**1673–1682.
32. **de Bruijn, F. J., I. L. Stroke, D. J. Marvel, and F. M. Ausubel.** 1983. Construction of a correlated physical and genetic map of the *Klebsiella pneumoniae hisDGO* region using transposon Tn*5* mutagenesis. *EMBO J.* **2:**1831–1838.
33. **de Lorenzo, V., M. Herrero, U. Jakubzik, and K. N. Timmis.** 1990. Mini-Tn*5* transposon derivatives for insertion mutagenesis, promoter probing, and chromosomal insertion of cloned DNA in gram-negative eubacteria. *J. Bacteriol.* **172:**6568–6572.
34. **De Vos, G. F., G. C. Walker, and E. R. Signer.** 1986. Genetic manipulations in *Rhizobium meliloti* utilizing two new transposon Tn*5* derivatives. *Mol. Gen. Genet.* **204:**485–491.
35. **Ditta, G.** 1986. Tn*5* mapping of *Rhizobium* nitrogen fixation genes. *Methods Enzymol.* **118:**519–528.
36. **Ditta, G., S. Stanfield, D. Corbin, and D. R. Helinski.** 1980. Broad host range DNA cloning system for gram-negative bacteria. Construction of a gene bank of *Rhizobium meliloti. Proc. Natl. Acad. Sci. USA* **77:**7347–7351.
37. **Donald, R. G. K., C. K. Raymond, and R. A. Ludwig.** 1985. Vector insertion mutagenesis of *Rhizobium* sp. strain ORS571: direct cloning of mutagenized DNA sequences. *J. Bacteriol.* **162:**317–323.
38. **Dower, W. J., J. F. Miller, and C. W. Ragsdale.** 1988. High efficiency transformation of *E. coli* by high voltage electroporation. *Nucleic Acids Res.* **16:**6127–6145.
39. **Finan, T. M., I. Oresnik, and A. Bottacin.** 1988. Mutants of *Rhizobium meliloti* defective in succinate metabolism. *J. Bacteriol.* **170:**3396–3403.
40. **Friedman, A. M., S. R. Long, S. E. Brown, W. J. Buikema, and F. M. Ausubel.** 1982. Construction of a broad host range cosmid cloning vector and its use in the genetic analysis of *Rhizobium* mutants. *Gene* **18:**289–296.
41. **Furuichi, T., M. Inouye, and S. Inouye.** 1985. Novel one-step cloning vector with a transposable element: application to the *Myxococcus xanthus* genome. *J. Bacteriol.* **164:**270–275.
42. **Furukawa, K., S. Hayashida, and K. Taira.** 1991. Gene-specific transposon mutagenesis of the biphenyl/polychlorinated biphenyl-degradation-controlling *bph* operon in soil bacteria. *Gene* **98:**21–28.
43. **Gallagher, S. R. (ed.).** 1992. *GUS protocols: Using the GUS Gene as a Reporter of Gene Expression.* Academic Press, Inc., San Diego, Calif.
44. **Gay, P., D. Le Coq, M. Steinmetz, T. Berkelman, and C. I. Kado.** 1985. Positive selection procedure for entrapment of insertion sequence elements in gram-negative bacteria. *J. Bacteriol.* **164:**918–921.
45. **Gottesman, M. M., and J. L. Rosner.** 1975. Acquisition of a determinant for chloramphenicol resistance by coliphage lambda. *Proc. Natl. Acad. Sci. USA* **72:**5041–5045.
46. **Harayama, S., E. T. Palva, and G. L. Hazelbauer.** 1979. Transposon-insertion mutants of *Escherichia coli* K12 defective in a component common to galactose and ribose chemotaxis. *Mol. Gen. Genet.* **171:**193–203.
47. **Hedges, R. W., and A. E. Jakob.** 1974. Transposition of ampicillin resistance from RP4 to other replicons. *Mol. Gen. Genet.* **132:**31–40.
48. **Herrero, M., V. de Lorenzo, and K. N. Timmis.** 1990. Transposon vectors containing non-antibiotic resistance selection markers for cloning and stable chromosomal insertion of foreign genes in gram-negative bacteria. *J. Bacteriol.* **172:**6557–6567.
49. **Hilgert, U., J. Schell, and F. J. de Bruijn.** 1987. Isolation and characterization of Tn*5*-induced NADPH-glutamate synthase (GOGAT$^-$) mutants of *Azorhizobium sesbaniae* ORS571 and cloning of the corresponding *glt* locus. *Mol. Gen. Genet.* **210:**195–202.
50. **Hirsch, P. R., and J. E. Beringer.** 1984. A physical map of pPH1JI and pJB4JI. *Plasmid* **12:**139–141.
51. **Hirsch, P. R., C. L. Wang, and M. J. Woodward.** 1986. Construction of a Tn*5* derivative determining resistance to gentamicin and spectinomycin using a fragment cloned from R1033. *Gene* **48:**203–209.
52. **Holliday, R.** 1956. A new method for identification of auxotrophic mutants in micro-organisms. *Nature* (London) **178:**987.
53. **Hynes, M., J. Quandt, M. P. O'Connell, and A. Puehler.** 1989. Direct selection for curing and deletion of *Rhizobium* plasmids using transposons carrying the *Bacillus subtilis sacB* gene. *Gene* **78:**111–120.
54. **Jefferson, R. A., S. M. Burgess, and D. Hirsh.** 1986. β-Glucuronidase from *Escherichia coli* as a gene-fusion marker. *Proc. Natl. Acad. Sci. USA* **83:**8447–8451.
55. **Kaniga, K., and J. Davison.** 1991. Transposon vectors for stable chromosomal integration of cloned genes in rhizosphere bacteria. *Gene* **100:**201–205.
56. **Keen, M. G., E. D. Street, and P. S. Hoffman.** 1985. Broad-host-range plasmid pRK340 delivers Tn*5* into the *Legionella pneumophila* chromosome. *J. Bacteriol.* **162:**1332–1335.
57. **Kennedy, C., R. Gamal, R. Humphrey, I. Ramos, K. Brigle, and D. Dean.** 1986. The *nifH, nifM* and *nifN* genes of *Azotobacter vinelandii*: characterization by Tn*5* mutagenesis and isolation from pLAFR1 gene banks. *Mol. Gen. Genet.* **205:**318–325.
58. **Kleckner, N.** 1989. Transposon Tn*10*, p. 227–268. *In* D. E. Berg and M. M. Howe (ed.), *Mobile DNA.* American Society for Microbiology, Washington, D.C.
59. **Kleckner, N., R. K. Chan, B.-K. Tye, and D. Botstein.** 1975. Mutagenesis by insertion of a drug-resistance element carrying an inverted repetition. *J. Mol. Biol.* **95:**561–575.

60. **Krishnan, B. R., D. Kersulyte, I. Brikun, C. M. Berg, and D. E. Berg.** 1991. Direct and crossover PCR amplification to facilitate Tn*5supF*-based sequencing of λ phage clones. *Nucleic Acids Res.* **19:**6177–6182.
61. **Kolter, R., M. Inuzuka, and D. R. Helinski.** 1978. Trans-complementation-dependent replication of a low molecular weight origin fragment from plasmid R6K. *Cell* **15:**1199–1208.
62. **Krebs, M. P., and W. S. Reznikoff.** 1988. Use of a Tn*5* derivative that creates *lacZ* translational fusions to obtain a transposition mutant. *Gene* **63:**277–285.
63. **Kroos, L., and D. Kaiser.** 1984. Construction of Tn*5 lac*, a transposon that fuses *lacZ* expression to exogenous promoters, and its introduction into *Myxococcus xanthus. Proc. Natl. Acad. Sci. USA* **81:**5816–5820.
64. **Kuner, J. M., and D. Kaiser.** 1981. Introduction of transposon Tn*5* into *Myxococcus* for analysis of developmental and other nonselectable mutants. *Proc. Natl. Acad. Sci. USA* **78:**425–429.
65. **Laird, A. J., and I. G. Young.** 1980. Tn*5* mutagenesis of the enterocholin gene cluster of *Escherichia coli. Gene* **11:**359–366.
66. **Lederberg, J., and E. M. Lederberg.** 1952. Replica plating and indirect selection of bacterial mutants. *J. Bacteriol.* **63:**399.
67. **Ludwig, R. A.** 1987. Gene tandem-mediated selection of coliphage λ-receptive *Agrobacterium, Pseudomonas,* and *Rhizobium* strains. *Proc. Natl. Acad. Sci. USA* **84:**3334–3338.
68. **Lupski, J. R., L. S. Ozaki, J. Ellis, and G. N. Godson.** 1983. Localization of the *Plasmodium* surface antigen epitope by Tn*5* mutagenesis mapping of a recombinant cDNA clone. *Science* **220:**1285–1288.
69. **Lupski, J. R., Y. H. Zhang, M. Rieger, M. Minter, B. Hsu, T. Koeuth, and E. R. B. McCabe.** 1990. Mutational analysis of the *Escherichia coli glpFK* region with Tn*5* mutagenesis and the polymerase chain reaction. *J. Bacteriol.* **172:**6129–6134.
70. **Maniatis, T., E. F. Fritsch, and J. Sambrook.** 1982. *Molecular Cloning: A Laboratory Manual,* 2nd ed. Cold Spring Harbor Laboratory, Cold Spring Harbor, N.Y.
71. **Manoil, C., and J. Beckwith.** 1985. Tn*phoA*: a transposon probe for protein export signals. *Proc. Natl. Acad. Sci. USA* **82:**8129–8133.
72. **Manoil, C., and J. Beckwith.** 1986. A genetic approach to analyzing membrane protein topology. *Science* **233:**1403–1408.
73. **Manoil, C., J. J. Mekalanos, and J. Beckwith.** 1990. Alkaline phosphatase fusions: sensors of subcellular location. *J. Bacteriol.* **172:**515–518.
74. **Meade, H. M., S. R. Long, G. B. Ruvkun, S. E. Brown, and F. M. Ausubel.** 1982. Physical and genetic characterization of symbiotic and auxotrophic mutants of *Rhizobium meliloti* induced by transposon Tn*5* mutagenesis. *J. Bacteriol.* **149:**114–122.
75. **Meighen, E. A.** 1991. Molecular biology of bacterial bioluminescence. *Microbiol. Rev.* **55:**123–142.
76. **Miller, J. H.** 1972. *Experiments in Molecular Genetics.* Cold Spring Harbor Laboratory, Cold Spring Harbor, N.Y.
77. **Miller, J. H., M. P. Calos, M. Hofer, D. Buechel, and B. Mueller-Hill.** 1980. Genetic analysis of transpositions in the *lac* region of *Escherichia coli. J. Mol. Biol.* **144:**1–18.
78. **Miller, V. L., and J. J. Mekalanos.** 1988. A novel suicide vector and its use in construction of insertion mutations: osmoregulation of outer membrane proteins and virulence determinants in *Vibrio cholerae* requires *toxR. J. Bacteriol.* **170:**2575–2583.
79. **Morales, V. M., and L. Sequeira.** 1985. Suicide vector for transposon mutagenesis in *Pseudomonas solanacearum. J. Bacteriol.* **163:**1263–1264.
80. **Murphy, E.** 1989. Transposable elements in gram-positive bacteria, p. 269–288. *In* D. E. Berg and M. M. Howe (ed.), *Mobile DNA.* American Society for Microbiology, Washington, D.C.
81. **Nag, D. K., H. V. Huang, and D. E. Berg.** 1988. Bidirectional chain termination nucleotide sequencing: transposon Tn*5*-seq1 as a mobile source of primer sites. *Gene* **64:**135–145.
82. **Obukowicz, M. G., F. J. Perlak, K. Kusano-Kretzmer, E. J. Mayer, and L. S. Watrud.** 1986. Integration of the delta endotoxin gene of *Bacillus thuringiensis* into the chromosome of root-colonizing strains of pseudomonads using Tn*5. Gene* **45:**327–331.
83. **Osteras, M., J. Stanley, W. J. Broughton, and D. N. Dowling.** 1989. A chromosomal genetic map of *Rhizobium* sp. NGR234 generated with Tn*5*-Mob. *Mol. Gen. Genet.* **220:**157–160.
84. **Palva, E. T., P. Liljestroem, and S. Harayama.** 1981. Cosmid cloning and transposon mutagenesis in *Salmonella typhimurium* using phage λ vehicles. *Mol. Gen. Genet.* **181:**153–157.
85. **Pato, M. L.** 1989. Bacteriophage Mu, p. 23–52. *In* D. E. Berg and M. M. Howe (ed.), *Mobile DNA.* American Society for Microbiology, Washington, D.C.
86. **Pawlowski, K., P. Ratet, J. Schell, and F. J. de Bruijn.** 1987. Cloning and characterization of *nifA* and *ntrC* genes of the stem nodulating bacterium ORS571, the nitrogen fixing symbiont of *Sesbania rostrata*: regulation of nitrogen fixation (*nif*) genes in the free living versus symbiotic state. *Mol. Gen. Genet.* **206:**207–219.
87. **Perbal, B.** 1988. *A Practical Guide to Molecular Cloning.* John Wiley & Sons, Inc., New York.
88. **Phadnis, S. H., and D. E. Berg.** 1987. Identification of base pairs in the outside end of insertion sequence IS*50* that are needed for IS*50* and Tn*5* transposition. *Proc. Natl. Acad. Sci. USA* **84:**9118–9122.
89. **Phadnis, S. H., H. V. Huang, and D. E. Berg.** 1989. Tn*5supF*, a 264-base-pair transposon derived from Tn*5* for insertion mutagenesis and sequencing DNAs cloned in phage λ. *Proc. Natl. Acad. Sci. USA* **86:**5908–5912.
90. **Phadnis, S. H., S. Kulakauskas, B. R. Krishnan, J. Hiemstra, and D. E. Berg.** 1991. Transposon Tn*5supF*-based reverse genetic method for mutational analysis of *Escherichia coli* with DNAs cloned in λ phage. *J. Bacteriol.* **173:**896–899.
91. **Pischl, D. L., and S. K. Farrand.** 1984. Characterization of transposon Tn*5*-facilitated donor strains and development of a chromosomal linkage map for *Agrobacterium tumefaciens. J. Bacteriol.* **159:**1–8.
92. **Reeve, J.** 1979. Use of minicells for bacteriophage directed polypeptide synthesis. *Methods Enzymol.* **68:**493–503.
93. **Rella, M., A. Mercenier, and D. Haas.** 1985. Transposon insertion mutagenesis of *Pseudomonas aeruginosa* with a Tn*5* derivative: application to physical mapping of the *arc* gene cluster. *Gene* **33:**293–303.
94. **Rich, J. J., and D. K. Willis.** 1990. A single oligonucleotide can be used to rapidly isolate DNA sequences flanking a transposon Tn*5* insertion by the polymerase chain reaction. *Nucleic Acids Res.* **18:**6673–6676.
95. **Riedel, G. E., F. M. Ausubel, and F. C. Cannon.** 1979. Physical map of chromosomal nitrogen fixation (*nif*) genes of *Klebsiella pneumoniae. Proc. Natl. Acad. Sci. USA* **76:**2866–2870.
96. **Rostas, K., P. Sista, J. Stanley, and D. P. S. Verma.** 1984. Transposon mutagenesis of *Rhizobium japonicum. Mol. Gen. Genet.* **197:**230–235.
97. **Ruvkun, G. B., and F. M. Ausubel.** 1981. A general method for site-directed mutagenesis in prokaryotes. *Nature* (London) **289:**85–88.
98. **Sambrook, J., E. F. Fritsch, and T. Maniatis.** 1989. *Molecular Cloning: A Laboratory Manual,* 2nd ed. Cold Spring Harbor Laboratory, Cold Spring Harbor, N.Y.
99. **Sancar, A., A. M. Hack, and W. D. Rupp.** 1979. Simple method for identification of plasmid-coded proteins. *J. Bacteriol.* **137:**692–693.
100. **Sanger, F., S. Nicklen, and A. R. Coulson.** 1977. DNA sequencing with chain-terminating inhibitors. *Proc. Natl. Acad. Sci. USA* **74:**5463–5467.

101. **Sarthy, A., S. Michaelis, and J. Beckwith.** 1981. Deletion map of the *Escherichia coli* structural gene for alkaline phosphatase. *J. Bacteriol.* **145:**288–292.
102. **Sasakawa, C., K. Kamata, T. Sakai, S. Makino, M. Yamada, and N. Okada.** 1988. Virulence-associated genetic regions comprising 31 kilobases of the 230-kilobase plasmid in *Shigella flexneri* 2a. *J. Bacteriol.* **170:**2480–2484.
103. **Sasakawa, C., and M. Yoshikawa.** 1987. A series of Tn*5* variants with various drug-resistance markers and suicide vector for transposon mutagenesis. *Gene* **56:**283–288.
104. **Selvaraj, G., and V. N. Iyer.** 1983. Suicide plasmid vehicles for insertion mutagenesis in *Rhizobium meliloti* and related bacteria. *J. Bacteriol.* **156:**1292–1300.
105. **Sharma, S. B., and E. R. Signer.** 1990. Temporal and spatial regulation of the symbiotic genes of *Rhizobium meliloti* in planta revealed by transposon Tn*5-gusA*. *Genes & Dev.* **4:**344–356.
106. **Shaw, K. J., and C. M. Berg.** 1979. *Escherichia coli* K-12 auxotrophs induced by insertion of the transposable element Tn*5*. *Genetics* **92:**741–747.
107. **Sheratt, D.** 1989. Tn*3* and related transposable elements, p. 163–184. *In* D. E. Berg and M. M. Howe (ed.), *Mobile DNA*. American Society for Microbiology, Washington, D.C.
108. **Silhavy, T. J., and J. R. Beckwith.** 1985. Uses of *lac* fusions for the study of biological problems. *Microbiol. Rev.* **49:**398–418.
109. **Simon, R.** 1984. High frequency mobilization system for *in vivo* constructed Tn*5*-Mob transposon. *Mol. Gen. Genet.* **196:**413–420.
110. **Simon, R., M. O'Connell, M. Labes, and A. Puehler.** 1986. Plasmid vectors for the genetic analysis and manipulation of rhizobia and other Gram-negative bacteria. *Methods Enzymol.* **118:**640–659.
111. **Simon, R., U. Priefer, and A. Puehler.** 1983. A broad host range mobilization system for *in vivo* genetic engineering: transposon mutagenesis in gram negative bacteria. *Bio/Technology* **1:**784–791.
112. **Simon, R., and U. B. Priefer.** 1990. Vector technology of relevance to nitrogen fixation research, p. 13–49. *In* P. M Gresshoff (ed.), *Molecular Biology of Symbiotic Nitrogen Fixation*. CRC Press, Inc., Boca Raton, Fla.
113. **Simon, R., J. Quandt, and W. Klipp.** 1989. New derivatives of transposon Tn*5* suitable for mobilization of replicons, generation of operon fusions and induction of genes in Gram-negative bacteria. *Gene* **80:**161–169.
114. **Smith, C. L., and C. R. Cantor.** 1987. Purification, specific fragmentation, and separation of large DNA molecules. *Methods Enzymol.* **151:**449–467.
115. **Smith, C. L., J. G. Econome, A. Schutt, S. Klco, and C. R. Cantor.** 1987. A physical map of the Escherichia coli K12 genome. *Science* **236:**1448–1453.
116. **Smith, C. L., and R. D. Kolodner.** 1988. Mapping of *Escherichia coli* chromosomal Tn*5* and F insertions by pulsed field gel electrophoresis. *Genetics* **119:**227–236.
117. **Starlinger, P., and H. Saedler.** 1976. IS-elements in microorganisms. *Curr. Top. Microbiol. Immunol.* **75:**111–152.
118. **Stern, M. J., G. Ferro-Luzzi Ames, N. H. Smith, E. C. Robinson, and C. F. Higgins.** 1984. Repetitive extragenic palindromic sequences: a major component of the bacterial genome. Cell **37:**1015–1026.
119. **Stewart, G. S. A. B., and P. Williams.** 1992. *lux*-genes and the applications of bacterial bioluminescence. *J. Gen. Microbiol.* **138:**1289–1300.
120. **Stojiljkovic, I., Z. Trgovcevic, and E. Salaj-Smic.** 1991. Tn*5-rpsL*: a new derivative of transposon Tn*5* useful in plasmid curing. *Gene* **99:**101–104.
121. **Subramanian, P. S., J. Versalovic, E. R. B. McCabe, and J. R. Lupski.** 1992. Rapid mapping of *Escherichia coli*::Tn*5* insertion mutations by REP-Tn*5* PCR. *PCR Methods Applic.* **1:**187–194.
122. **Taylor, A. L.** 1963. Bacteriophage-induced mutation in *Escherichia coli*. *Proc. Natl. Acad. Sci. USA* **50:**1043–1051.
123. **Van Haute, E., H. Joos, M. Maes, G. Warren, M. Van Montagu, and J. Schell.** 1983. Intergenic transfer and exchange recombination of restriction fragments cloned in pBR322: a novel strategy for the reversed genetics of the Ti-plasmids of *Agrobacterium tumefaciens*. *EMBO J.* **2:**411–417.
124. **Versalovic, J., T. Koeuth, and J. R. Lupski.** 1991. Distribution of repetitive DNA sequences in eubacteria and application to fingerprinting of bacterial genomes. *Nucleic Acids Res.* **19:**6823–6831.
125. **Walker, M. J., and J. M. Pemberton.** 1988. Construction of transposons encoding genes for beta-glucosidase, amylase and polygalacturonate *trans*-eliminase from *Klebsiella oxytoca* and their expression in a range of gram-negative bacteria. *Curr. Microbiol.* **17:**69–75.
126. **Wolk, C. P., Y. Cai, and J.-M. Panoff.** 1991. Use of a transposon with luciferase as a reporter to identify environmentally responsive genes in a cyanobacterium. *Proc. Natl. Acad. Sci. USA* **88:**5355–5359.
127. **Wong, K. K., and M. McClelland.** 1992. Dissection of the *Salmonella typhimurium* genome by use of a Tn*5* derivative carrying rare restriction sites. *J. Bacteriol.* **174:**3807–3811.
128. **Yakobson, E. A., and D. G. Guiney, Jr.** 1984. Conjugal transfer of bacterial chromosomes mediated by the RK2 plasmid transfer origin cloned into transposon Tn*5*. *J. Bacteriol.* **160:**451–453.
129. **Youngman, P.** 1990. Use of transposons and integrational vectors for mutagenesis and construction of gene fusions in *Bacillus* species, p. 221–266. *In* C. R. Harwood and S. M Cutting (ed.), *Molecular Biological Methods for Bacillus*. John Wiley and Sons, Chichester, England.
130. **Zsebo, K. M., F. Wu, and J. E. Hearst.** 1984. Tn*5*.7 construction and physical mapping of pRPS404 containing photosynthetic genes from *Rhodopseudomonas capsulata*. *Plasmid* **11:**182–184.
131. **Zuber, U., and W. Schumann.** 1991. Tn*5cos*: a transposon for restriction mapping of large plasmids using phage lambda terminase. *Gene* **103:**69–72.

Chapter 18

Gene Cloning and Expression

MICHAEL BAGDASARIAN and MIRA M. BAGDASARIAN

The isolation of genes, their manipulation in vitro, and their expression in different organisms are the most powerful and versatile research strategies in modern biology. Often referred to as gene cloning or genetic engineering, this set of methods allows directed biochemical alteration of genes and directed alteration of the host genomes. These methods are of particular value for studying and engineering poorly characterized genetic systems, such as bacteria isolated from natural environments, because they are independent of the relatedness of the organisms under investigation to the well-characterized laboratory strains. The methods of gene cloning do not require the availability of detailed information on the structure and organization of the donor or host genomes, nor do they depend on the existence of direct in vivo transfer of genetic information from the donor to the host organism.

In the past two decades of an explosive development of gene cloning, many excellent articles and laboratory manuals on in vitro gene manipulation have appeared. More detailed information and many variations of the basic techniques may be found in references 1a to 11.

Gene cloning and expression involve four essential steps.

(i) Generation of DNA fragments from the donor organism.

(ii) Biochemical linking of these donor DNA fragments to the DNA of autonomously replicating genetic elements called vectors.

(iii) Insertion of the recombinant DNA molecules thus obtained into a suitable host bacterium.

(iv) Detection of the clones containing the recombinant DNA molecules and/or their gene products in the new host.

A scheme illustrating these steps is presented in Fig. 1.

Selection of a strategy most appropriate for the particular cloning experiment may have a decisive effect on the success of the experiment and may save considerable time and effort. The following concerns are important: (i) the source and method of preparation of donor DNA; (ii) the restriction endonucleases; (iii) the type of cloning vector; and (iv) the host bacteria.

The procedures described in this chapter apply primarily to the use of gram-negative bacteria, such as *Escherichia coli* and *Pseudomonas* spp., as host bacteria. Comparable procedures for gram-positive bacteria, particularly *Bacillus* spp., are described in reference 4.

A simple and convenient method for the preparation of total DNA from many different bacterial species is described below (section 18.2.2.4). If the gene of interest is present on a plasmid or bacteriophage or if it has already been inserted into such a vector, the DNA of the appropriate extrachromosomal element should be purified (see Chapter 16.2). This will increase the abundance of the gene to be isolated by 2 to 4 orders of magnitude in comparison with the total DNA.

The choice of a restriction endonuclease will depend largely on the available information about the DNA fragment to be cloned. If there is a probe for hybridization to the gene of interest, digests with several endonucleases are obtained on an agarose gel, transferred to

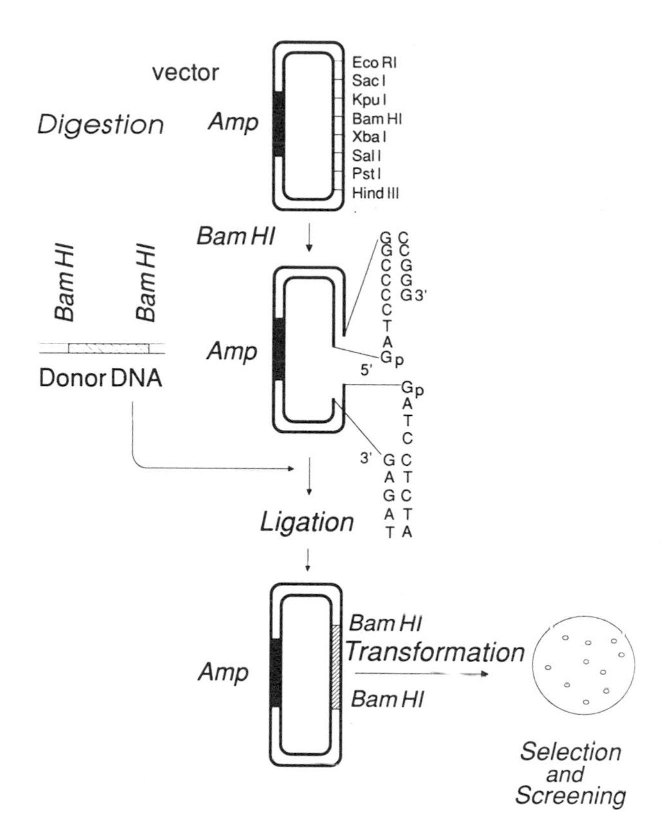

FIG. 1. The basic steps involved in gene cloning. *bla*, the β-lactamase gene, is a selectable marker.

a membrane filter, and hybridized to the probe to reveal the most convenient size of the fragment (see Chapter 26.6.3). The fragments of this size can be subsequently isolated and used for ligation, thus achieving a considerable degree of enrichment for the desired gene.

Selection of an appropriate vector is of utmost importance for the success of cloning. The following properties should be considered: (i) copy number and size; (ii) host range and selective markers; (iii) the method of selection or screening for the vectors carrying the insertion; and (iv) expression of the cloned genes.

18.1. GENERATION OF DONOR DNA FRAGMENTS

In most cases the donor DNA fragments are generated by treatment of a sample of total chromosomal DNA, plasmid DNA or bacteriophage DNA with a restriction endonuclease (1a–11). A large selection of restriction endonucleases, with different specificity in the sequence and the number of nucleotides they recognize or the type of ends they generate, are available from commercial sources, and new ones are continuously being introduced into the market. It is usually possible, therefore, to find a suitable endonuclease, or a combination of the enzymes, that will enable one to excise a convenient fragment of DNA carrying a gene or a site from the donor, if some information already exists on the genetic site in question. The most important information on the individual restriction enzymes, their specificity, optimal conditions for use, and other important properties are found in the catalogs from the manufacturers (e.g., New England BioLabs, Beverly, MA 01915-9931; Bethesda Research Laboratories, P.O. Box 6009, Gaithersburg, MD 20877; Boehringer GmbH, P.O. Box 31 01 20, 6800 Mannheim 31, Germany).

*Sau*3A or *Mbo*I		*Bam*HI	
5′..NNN	**GATC**NNN..3′	5′..NNNG	**GATC**CNNN..3′
3′..NNN**CTAG**	NNN..5′	3′..NNNC**CTAG**	GNNN..5′

FIG. 2. Cohesive ends generated by different restriction endonucleases.

It may not be known in advance which endonuclease-sensitive sites exist within the gene to be cloned. In such cases, one of the endonucleases that cuts DNA at multiple sites should be used, but only a partial digest should be obtained. This procedure generates a mixture of fragments that resemble those obtained by random fragmentation methods, such as mechanical shearing. Convenient endonucleases to use in such cases are *Sau*3A and *Mbo*I. The cohesive ends generated by these enzymes are complementary to the protruding ends produced by *Bam*HI digestion. As shown in Fig. 2, the fragments thus obtained may be inserted into the *Bam*HI sensitive site present in many cloning vectors (see also Fig. 1 and 4).

Many manufacturers of restriction endonucleases will supply an appropriate buffer with the purchase of an enzyme. It is convenient to have a general restriction buffer (GRB) kept as 10× concentrated stock. GRB works with most of the commonly used endonucleases (except *Sma*I) provided that NaCl is added (from a 1.0 M stock solution) when required. GRB may also be used as ligation buffer (see below).

The compositions of frequently used reagents are described in Table 1. The nomenclature for restriction endonucleases, bacteriophages, plasmids, and other genetic elements is described in reference 1.

a. Restriction endonuclease digestion

1. Mix

10× GRB	2.0 μl
DNA	0.2–2.0 μg
Endonuclease (e.g., *Bam*HI)	5–10 U in 1.0 μl
1 M NaCl	2.0 μl
H_2O	to 20 μl

2. Incubate at 37°C for 1 h. Check for completeness of digestion by running an aliquot (0.2 to 0.5 μg) on agarose gel electrophoresis (see Chapter 16.3.1.2).
3. Shake with an equal volume of the phenol-chloroform mixture (1:1).
4. Centrifuge to break the emulsion, and save the aqueous (upper) phase.
5. Extract the aqueous phase once with chloroform.
6. Add 0.1 volume of 3 M sodium acetate and 2 volumes of ethanol or 0.7 volume of isopropanol.
7. Chill at 0°C for 10 min, and centrifuge at 12,000 rpm (17,000 × g) for 10 min at 4°C.
8. Wash the pellet with 70% ethanol, and dry under vacuum for 1 min.
9. Dissolve in 1 × TE buffer for storage or in 1 × GRB if used promptly for ligation or further restriction.
10. Check an aliquot on agarose gel electrophoresis.

b. Ligation

1. Mix

10× GRB	2.0 μl

Table 1. Composition of reagents frequently used in cloning

Reagent	Composition	Comments
10× TE Buffer	0.1 M Tris·HCl (pH 8.0), 1.0 mM EDTA	Autoclave and keep sterile
10× GRB	0.2 M Tris·HCl (pH 8.0), 0.1 M $MgCl_2$	Autoclave and keep sterile in aliquots at ambient temperature
10× *Sma*I buffer	0.15 M Tris·HCl (pH 8.5), 0.15 M KCl, 0.06 M $MgCl_2$	Autoclave and keep sterile
STE buffer	20% sucrose, 50 mM EDTA, 0.1 M Tris·HCl (pH 8.0)	Autoclave and keep sterile
10× TBS buffer	NaCl, 160 g; KCl, 4 g; Tris, 60 g; adjust pH to 8.0 with HCl and add H_2O to 1,500 ml	Store at room temperature
Colony lysis buffer	100 mM Tris·HCl (pH 7.8), 150 mM NaCl, 5 mM $MgCl_2$, 1.5% BSA, 1 μg of pancreatic DNase I per ml, 40 μg of lysozyme per ml	Store at −20°C
AP buffer	100 mM Tris·HCl (pH 9.5), 100 mM NaCl, 5 mM $MgCl_2$	Store at 4°C
Phenol-chloroform	Liquid phenol extracted with 1.0 M Tris·HCl until pH is >7.6, then equilibrated with 0.1 M Tris·Cl (pH 8.0); add an equal volume of chloroform and 0.1% 8-hydroxyquinoline	Store at −20°C in small aliquots; may be stored at 4°C for up to 1 month
BSA	1.0% bovine serum albumin in 1× TBS, 0.05% Tween-20	Store at −20°C
NBT	250 mg of Nitro Blue Tetrazolium dissolved in 5 ml of 70% dimethylformamide	Store at −20°C
BCIP	50 mg of 5-bromo-4-chloro-3-indolylphosphate per ml of 100% dimethylformamide	Store at −20°C

Digested vector DNA .... 0.05–0.2 μg
Digested insert DNA............... 0.2–0.5 μg
10 mM ATP .... 2.0 μl
100 mM dithiothreitol (DTT) .... 2.0 μl
T4 DNA ligase .... 2.0–10.0 U
H_2O .... to 20.0 μl

2. Incubate for 2 h at room temperature or overnight at 4 to 14 °C.
3. Introduce aliquots by transformation into a suitable host (Chapter 14.1).

For ligation of blunt ends, use half of the above concentration of ATP, since concentrations of ATP higher than 0.08 mM inhibit blunt-end ligation. Higher concentrations of ligase (10.0 to 100.0 U) and of DNA (200 to 300 μg/ml) are also recommended.

18.2. VECTORS

Vectors are well-characterized molecules of DNA to which a fragment of DNA, called an insert, may be ligated. Vectors allow the introduction into, and the stable maintenance of the resulting recombinant molecules within, suitable host cells. They are constructed from plasmid or bacteriophage replicons by the addition of DNA sequences that provide them with specific properties required for their use. The basic functional parts of a vector are (i) a **replicon**, consisting of an origin of replication and genes that allow the replication of this molecule as an extrachromosomal element; (ii) **selection markers**, usually genes encoding resistances to toxic substances such as antibiotics, or genes that complements auxotrophies of the host bacterium; and (iii) a **cloning site**, one or several restriction endonuclease sensitive sites that are unique in the molecule.

Many vectors have additional features that facilitate screening for the recombinant molecules, allow controlled expression of the cloned genes, allow introduction into a wide range of bacterial hosts, or allow selection for specific functions such as promoters or terminators. Figure 3 represents a physical and genetic map of a vector which, in addition to the basic elements of a cloning vector, encodes a built-in system for controlled expression and whose replicon has a very wide host range. Properties of numerous cloning vectors are listed in Table 2. This list will facilitate the selection of vectors most convenient for the type of experiments planned.

18.2.1. General-Purpose Plasmid Vectors

Plasmids derived from pBR322 (16–19) or from pUC replicons (56, 57) are the most frequently used vectors for general-purpose cloning in *E. coli* or closely related bacteria such as *Salmonella typhimurium*. In the case of pBR322 and other similar vectors, the fragments are inserted into unique sites located in either the Ap^r or Tc^r gene. Very often this allows the expression of the inserted genes from the promoters of the Ap^r or Tc^r genes of the vector. Vectors of the pUC series carry unique cloning sites for as many as 13 different restriction enzymes. These sites are located within the gene encoding a fragment of β-galactosidase. Insertions into any of these sites allows the expression of cloned genes from the *lac* promoter of the vector. In addition, the insertion of foreign DNA fragments into the *lac* gene facilitates screening for clones that carry recombinant plasmids (see section 18.3). These pUC vectors are present in *E. coli* in multiple copies of 40 to 70 molecules per cell. Other vectors are able to autoamplify at the late logarithmic phase (29). This usually results in the synthesis of large amounts of gene products from the cloned genes. In some cases this may be desirable. Often, however, a high concentration of foreign gene products is inhibitory or even toxic to the cell. In such cases it may be more convenient to use low-copy-number vectors such as pACYC177, pACYC184 (24, 44), or the deriv-

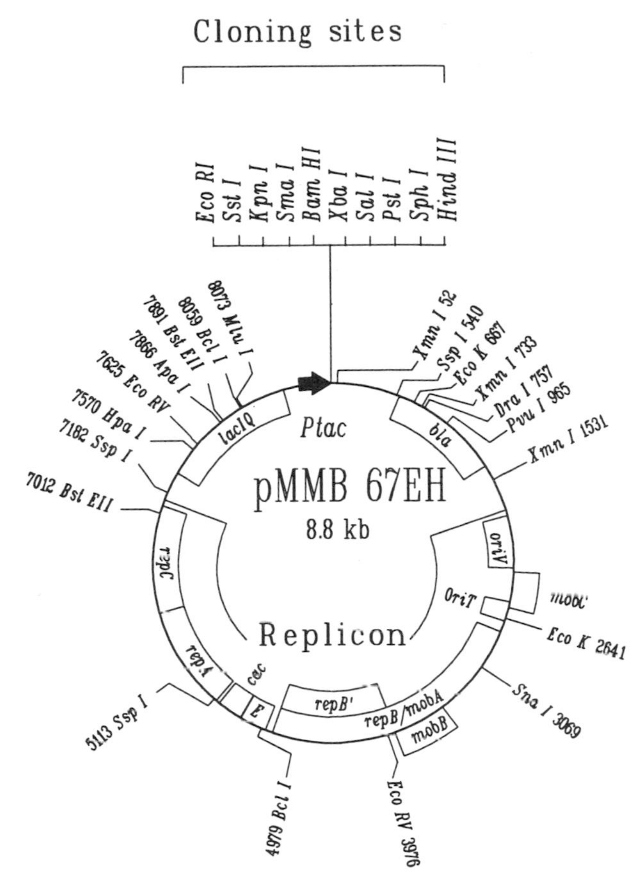

FIG. 3. Physical and genetic map of a broad-host-range, controlled expression vector. *oriV*, origin of vegetative replication; *oriT*, origin of transfer replication; *rep*, genes essential for replication; *mob*, genes essential for mobilization; P_{tac}, hybrid *trp-lac* promoter; *lacI*^{Q}, gene overexpressing the LacI repressor; *bla*, gene for β-lactamase (resistance to ampicillin).

atives of plasmid RSF1010 (27, 28, 30, 37) described below.

Generally, the frequency of obtaining viable recombinant molecules decreases with the size of the inserted fragment of DNA, particularly for high-copy-number vectors. Although fragments as long as 20 to 30 kb have successfully been inserted into and maintained in general-purpose plasmid vectors, the size of an insert more easily maintained is of the order of a few kilobases.

18.2.2. Special-Purpose Vectors

18.2.2.1. Bacteriophage lambda

Vectors derived from bacteriophage λ are useful for cloning large inserts (up to 30 to 40 kb) for generation of gene libraries and detection of genes by hybridization to an existing DNA or RNA probe (35, 51). Vectors, such as λgt11, may be also used for the detection of gene products by immunological techniques (33). Another advantage is a very efficient transfer of recombinant DNA, packaged in bacteriophage particles, into host bacteria. These efficiencies are orders of magnitude higher than those obtained by transformation.

The disadvantage of using λ-derived vectors is their very narrow host range. They may be used only in *E. coli*. Attempts to extend their host range to other bacterial species by introduction of the λ receptors, cloned from *E. coli*, have not yet given satisfactory results for practical use. The inconvenience of having to prepare in vitro bacteriophage-packaging mixtures has been overcome by the commercial availability of ready-to-use packaging mixtures that are highly efficient and are delivered with detailed instructions for use. They are supplied by most companies that sell restriction enzymes (see section 18.1).

18.2.2.2. Bacteriophage M13

The principal use of vectors constructed from M13 (50, 56, 57) is in the generation of single-stranded DNA for nucleotide sequence determination (Chapter 20.1) and for oligonucleotide-directed mutagenesis of cloned genes (Chapter 13.2).

18.2.2.3. Phagemids

Phagemids are a special class of bacteriophage-derived vectors. These vectors contain a replicon from a plasmid and an origin from a filamentous phage. They can replicate as plasmids but, on infection with helper phage, produce single-stranded DNA and package it into bacteriophage particles. The main advantage of phagemids is the elimination of the need to subclone isolated DNA fragments into M13 bacteriophages for sequence determination or site-directed mutagenesis. In addition, they are usually more stable than the M13 vectors if large DNA fragments are inserted. The properties of two phagemid vectors are included in Table 2 (see also references 5 and 50).

18.2.2.4. Cosmids

Cosmid vectors are plasmid vectors containing one or two *cos* sites from bacteriophage λ. They have the advantage of small size, the replication mode of plasmids, and the ability to be packaged in vitro into bacteriophage λ heads. Their main use is in the generation of gene libraries, since they allow cloning of large fragments (30 to 40 kb), thus reducing the number of recombinant clones representing the entire library. A scheme describing the steps used in cosmid cloning is presented in Fig. 4. DNA preparations used for cosmid cloning must consist predominantly of long strands. Otherwise, digestion of this DNA with restriction endonucleases will generate molecules considerably smaller than 30 kb, which will not be packaged into bacteriophage λ heads.

a. Preparation of total DNA from bacterial cells

All steps are carried out in the cold unless indicated otherwise.

1. Prepare a 200-ml overnight culture in LB medium, and centrifuge at 7,000 rpm (6,000 × *g*) for 10 min.
2. Wash once in 50 ml of 20% sucrose in 50 mM Tris (pH 8.0).
3. Suspend the cells in 25 ml of the same sucrose solution per 3.0 g of cells.
4. Add lysozyme (10 mg per 1.0 g of cells).

Table 2. Properties of cloning vectors

Vector	Size (kb)	Detection markers	Cloning sites	Insertional alteration	Reference
General-purpose vectors					
pBR322	4.4	Tc^r	*Sca*I, *Pvu*I, *Pst*I, *Ase*I	$Ap^r \rightarrow Ap^s$	16, 17, 19
			*Eco*RI, *Cla*I, *Ava*I	None	
		Ap^r	*Hin*dIII, *Eco*RV, *Bam*HI, *Sph*I, *Eco*NI, *Nru*I, *Bsp*MI	$Tc^r \rightarrow Tc^s$	
pBR325	5.7	Tc^r, Cm^r	*Pst*I	$Ap^r \rightarrow Ap^s$	18, 19
			*Ava*I	None	
		Ap^r, Cm^r	*Hin*dIII, *Bam*HI, *Sal*I	$Tc^r \rightarrow Tc^s$	
		Ap^r, Tc^r	*Eco*RI	$Cm^r \rightarrow Cm^s$	
pUC18	2.7	Ap^r, *lacZα*	*Hin*dIII, *Sph*I, *Pst*I, *Sal*I, *Acc*I, *Hin*cII, *Xba*I, *Bam*HI, *Sma*I, *Kpn*I, *Sac*I, *Eco*RI	*lacZα*⁺ → *lacZα*	57
pUC19	2.7	Ap^r, *lacZα*	Same as pUC18, reverse orientation	*lacZα*⁺ → *lacZα*	57
pACYC177	3.5	Km^r	*Pst*I	$Ap^r \rightarrow Ap^s$	24, 44
		Ap^r	*Bam*HI, *Hin*dIII, *Sma*I, *Xho*I	$Km^r \rightarrow Km^s$	
pACYC184	3.9	Tc^r	*Eco*RI	$Cm^r \rightarrow Cm^s$	24, 44
		Cm^r	*Bam*HI, *Sal*I, *Hin*dIII	$Tc^r \rightarrow Tc^s$	
Bacteriophage-derived vectors					
EMBL 3	44 (14)[a]	Plaque	*Sal*I, *Bam*HI, *Eco*RI	*spi*⁺ → *spi*	35
λDASH	44 (14)[a]	Plaque	*Xba*I, *Sal*I, *Eco*RI, *Bam*HI, *Hin*dIII, *Sac*I, *Xho*I	*spi*⁺ → *spi*	51
λgt11	43.7	Plaque	*Eco*RI	*lacZ*⁺ → *lacZ*	33
M13mp18	7.3	Plaque	*Eco*RI, *Sac*I, *Kpn*I, *Sma*I, *Bam*HI, *Xba*I, *Sal*I, *Pst*I, *Sph*I, *Hin*dIII	*lacZ*⁺ → *lacZα*	4, 50, 57
M13mp19	7.3	Plaque	Same as M13mp18 but reverse orientation	*lacZ*⁺ → *lacZα*	4, 50, 57
Phagemid vectors					
pSL1180	3.4	Ap^r, *lacZα*	64 palindromic hexamer recognition sequences	*lacZα*⁺ → *lacZα*	22
pSL1190	3.4	Ap^r, *lacZα*	64 palindromic hexamer reverse orientation to that in pSL1180	*lacZα* → *lacZα*	22
Cosmid vectors					
pHC79	6.4	Tc^r	*Pst*I	$Ap^r \rightarrow Ap^s$	32
		Ap^r	*Eco*RV, *Cla*I, *Bam*HI, *Sal*I	$Tc^r \rightarrow Tc^s$	
			*Eco*RI,*Hin*dIII, *Bst*EII	None	
pHSG274	3.4	Km^r	*Bam*HI, *Hin*dIII, *Sal*I	None	20
pLAFR5	21.5	Tc^r, *mob*⁺	*Eco*RI, *Sma*I, BamHI, SalI, *Pst*I, *Hin*dIII	None	36
p*cos*2EMBL	6.1	Tc^r, Km^r	*Bam*HI, *Sal*I	$Tc^r \rightarrow Tc^s$	41
			*Xho*I	None	
pMMB33	13.8	Km^r, *mob*⁺, broad host range	*Bam*HI, *Eco*RI, *Sac*I	None	28
Promoter detection vectors					
pMC1403	9.9	Ap^r	*Bam*HI, *Eco*RI, *Sma*I	lac → lac⁺	23
pKK231-1	5.1	Ap^r	*Sma*I, *Bam*HI, *Sal*I, *Hin*dII	$Cm^s \rightarrow Cm^r$	21
pKT240	12.9	Km^r, Ap^r, broad host range	*Eco*RI, *Hpa*I	$Sm^s \rightarrow Sm^r$	15
Broad-host-range vectors					
pMMB66EH	8.8	Ap^r, p_{tac}, *lac*I^Q, *mob*⁺	*Eco*RI, *Sma*I, *Bam*HI, *Sal*I, *Pst*I, *Hin*dIII	None	30, 37
pMMB66HE[b]	8.8	Ap^r, p_{tac}, *lac*I^Q, *mob*⁺	*Hin*dIII, *Pst*I, *Sal*I, *Bam*HI, *Sma*I, *Eco*RI	None	30, 37
pMMB67EH	8.9	Ap^r, p_{tac}, *lac*I^Q, *mob*⁺	*Eco*RI, *Sst*I, *Kpn*I, *Sma*I, *Bam*HI, *Xba*I, *Sal*I, *Pst*I, *Hin*dIII	None	30, 37
pMMB67HE[b]	8.9	Ap^r, p_{tac}, *lac*I^Q, *mob*⁺	*Hin*dIII, *Sph*I, *Pst*I, *Sal*I, *Xba*I, *Bam*HI, *Sma*I, *Kpn*I, *Sst*I, *Eco*RI	None	30, 37
pMMB207	9.0	Cm^r, p_{tac}, *lac*I^Q, *mob*⁺	*Eco*RI, *Sac*I, *Kpn*I, *Sma*I, *Bam*HI, *Xba*I, *Sal*I, *Pst*I, *Sph*I, *Hin*dIII	None	38, 39
pRK2501	11.1	Tc^r	*Hin*dII, *Xho*I	$Km^r \rightarrow Km^s$	34
		Km^r	*Sal*I	$Tc^r \rightarrow Tc^s$	
			*Eco*RI, *Bgl*II	None	
pRK310	20.4	Tc^r	*Bam*HI, *Hin*dIII, *Pst*I	*lacZα*⁺ → *lacZα*	26
			*Bgl*II	None	

(Continued)

Table 2. Continued

Vector	Size (kb)	Detection markers	Cloning sites	Insertional alteration	Reference
Controlled-expression vectors					
pKK223-3	4.6	Ap[r], P_{tac}	*Eco*RI, *Sma*I, *Bam*HI, *Sal*I, *Acc*I, *Hinc*II, *Pst*I, *Hind*III	None	21
pJF118	5.3	Ap[r], P_{tac}	*Eco*RI, *Sma*I, *Bam*HI, *Sal*I, *Pst*I, *Hind*III	None	30
pT7-1	2.5	Ap[r], $P_{T7\phi10}$	*Nde*I, *Eco*RI, *Sac*I, *Sma*I, *Bam*HI, *Xba*I, *Sal*I, *Pst*I, *Hind*III, *Cla*I	None	53
pMMB66, pMMB67, pMMB207		See broad-host-range vectors			37, 38

[a]The size of the "stuffer" fragment is indicated in parentheses.
[b]Reverse orientation of the cloning cassette with respect to P_{tac}.

5. Incubate at 37°C for 30 min (or until the cells start to lyse).
6. Add pronase (or proteinase K) at 5.0 mg per 3.0 g of cells.
7. Incubate at 37°C with gentle shaking for 15 min.
8. Add 25% SDS, 0.15 ml per 1.0 ml of cell suspension.
9. Incubate at 60°C if necessary to complete lysis (the suspension should become very viscous and almost clear).
10. Add an additional portion (5.0 mg per 3.0 g of cells) of pronase; incubate at 37°C for 30 min.
11. Add an equal volume of phenol saturated with 50 mM Tris (pH 8.0), and shake to form the emulsion.
12. Centrifuge at 10,000 rpm (12,000 × *g*) for 15 min to separate the phases.
13. Transfer the upper phase into a small beaker.
14. Add 1/10 volume of 3 M sodium acetate, and mix.
15. Add 0.6 volume of isopropanol (cold) by overlaying.
16. Mix the phases with a glass rod, and spool the DNA threads onto the rod.
17. Wash the DNA by immersion in 70% ethanol.
18. Dissolve the DNA in 10 ml of TE buffer by gentle shaking in the cold room overnight.
19. Add 1/100 volume of 1.0 M $MgCl_2$.

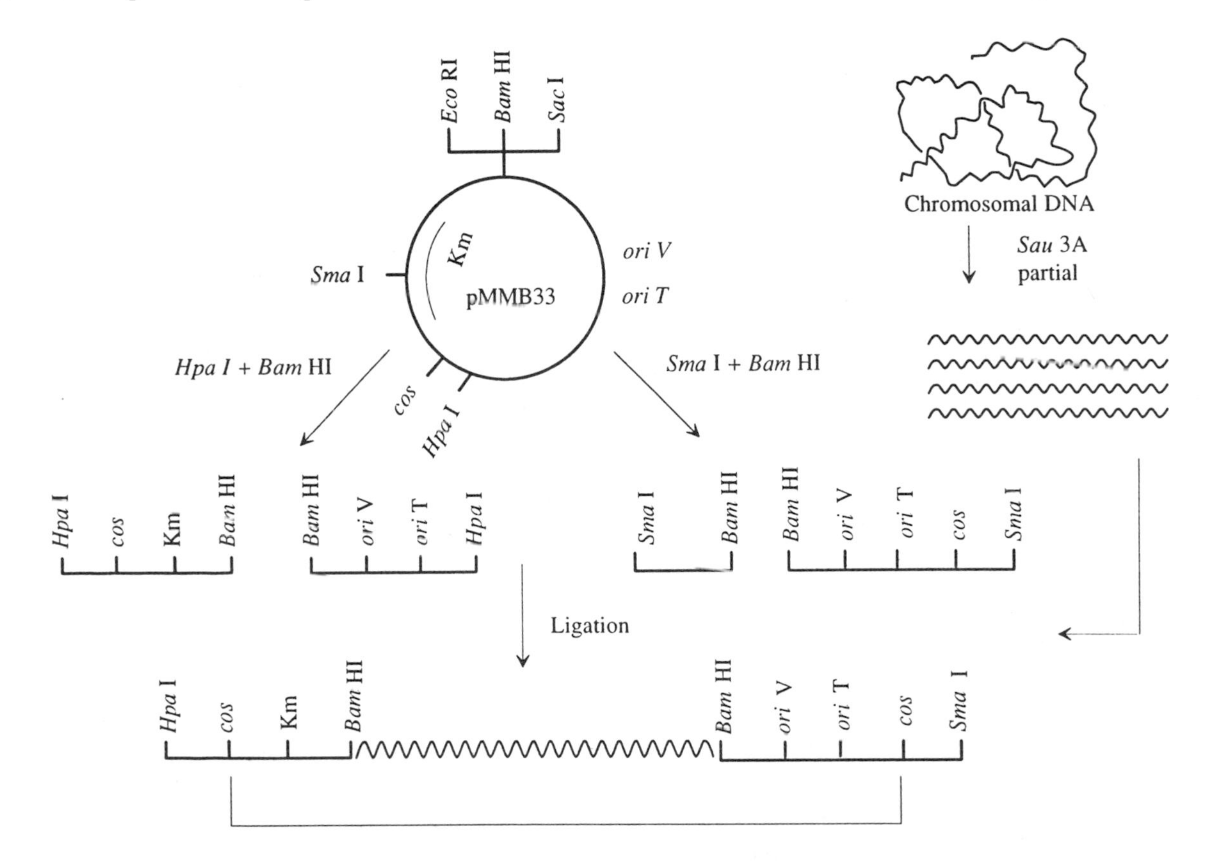

FIG. 4. Steps involved in cosmid cloning.

20. Add RNase (DNase free!) to a final concentration of 50 μg/ml.
21. Incubate at 37°C for 30 min.
22. Extract with phenol as in steps 10 to 13.
23. Repeat the phenol extraction until no interphase remains after phase separation.
24. Precipitate with isopropanol as in steps 14 to 17.
25. Dissolve DNA in 5 ml of TE buffer. If the solutions are very viscous, the volumes should be increased.
26. Keep frozen at −20°C or, better, at −80°C in small aliquots.

b. Generation of partial DNA digest with *Sau*3A

The activity of restriction endonucleases is inhibited by intercalative dyes such as ethidium bromide. A good way to obtain partial digestion of DNA with restriction enzymes is to check the effect of different concentrations of ethidium bromide on the distribution of the DNA fragments obtained and to select the conditions that best suit the particular requirement.

1. Mix

Chromosomal DNA (ca 5 μg)	20 μl
Ethidium bromide	5 μl
10× GRB	5 μl
1.0 M NaCl	5 μl
H_2O	14 μl
*Sau*3A (ca. 20 U)	1 μl

2. Use several different concentrations of ethidium bromide (for example, four solutions with concentrations between 10 and 50 μg/ml).
3. Incubate the mixtures at 37°C for 1 h, and then inactivate *Sau*3A by incubation at 65°C for 15 min.
4. Electrophoretically separate aliquots of each digest on a low concentration (0.5 to 0.7%) of agarose gel along with suitable size markers to determine the optimal conditions for generation of suitably sized fragments. If more digested DNA is required, run multiple samples of the same digest rather than scaling up a digest by changing concentrations and volumes.
5. Separate selected digests on a preparative agarose gel, cut out the appropriate portion of the gel, and isolate the fragments by electroelution in a dialysis sac.
6. Purify the eluted fragments by phenol extraction followed by ethanol precipitation or, better, by passing through Elutip columns (Schleicher & Schuell, Keene, NH 03431).

c. Cloning with cosmid vectors

The cosmid vector cloning procedure described below is applicable to cosmid vectors that contain unique restriction sites on either side of the *cos* region. This strategy minimizes ligation and packaging of DNA fragments consisting of vector DNA without an insert. The following example is described for the broad-host-range cosmid pMMB33 (27, 28).

1. Digest two separate aliquots of the pMMB33 vector DNA (2 to 5 μg of DNA each), one with *Hpa*I and *Bam*HI, the other with *Sma*I and *Bam*HI, by using the protocol for digestion described in section 18.1.1.
2. Extract the digests with phenol, and precipitate the DNA with ethanol (see section 18.1).
3. Mix with the purified DNA fragments from the partial *Sau*3A digest with a molar excess of insert fragments (section 18.1).
4. Ligate as suggested in section 18.1.1.
5. Use the ligated mixture for packaging into bacteriophage λ heads. The best results are obtained with commercially available packaging mixes.

Cosmid vectors pMMB33 and pMMB34 (Table 2) may be conveniently mobilized from *E. coli* into many different bacterial species. This property greatly facilitates the transfer of the recombinant cosmids, which are difficult to move by transformation because of their size. It also enables the transfer of recombinant cosmid plasmids into species for which transformation or electroporation conditions have not been worked out (14, 28).

To mobilize the derivatives of pMMB33, introduce the packaged cosmids into strain S17-1 [*recA pro hsr hsm*$^+$ RP4-2(Tc::Mu Km::Tn7) Tp Sm] (49). This strain will transfer RSF1010 derivatives at high frequency but will not transfer the RP4 plasmid.

d. Mobilization of RSF1010-based plasmids into different bacterial species

Mobilization of RSF1010-based plasmids into bacteria has been used successfully by the authors to transfer broad-host-range vectors, based on the RSF1010 replicon (Table 2), from *E. coli* into different bacterial species.

1. Grow donor and recipient cells in LB medium to late logarithmic phase.
2. Spot 10 μl of the donor strain (S17-1 containing the RSF1010-derivative vector to be transferred) and 30 to 40 μl of the recipient strain onto sterile Millipore filters (such as HAWP 013 00) placed on the surface of an LB agar plate.
3. Place control filters containing the donor only and the recipient only on the same plate.
4. Incubate for the desired time (overnight incubation may be required if the frequency of mobilization is low) at a convenient temperature.
5. Place the filters into 1.0 ml of sterile saline solution, and vortex for 1 min.
6. Plate appropriate dilutions on selective plates.

The main problem encountered in using cosmid cloning is the instability of large fragments of foreign DNA in the new host. Frequently, large deletions in the inserted fragments are found, and the cells harboring the recombinant cosmids grow more slowly than do plasmid-free cells. The cosmid vectors described above are present in a relatively small number of copies (13 copies per cell) in *E. coli* and *Pseudomonas* spp. This helps to circumvent some of these difficulties. It should be stressed that use of cosmids has given excellent results in numerous cloning experiments. If, however, the problems described above are encountered, it is recommended that low-copy-number cosmid vectors, such as the pLAFR series (36) or λ replacement vectors, be used.

18.2.2.5. Promoter detection plasmids

Sequences that regulate gene expression can be identified and analyzed by using specialized cloning vectors. Promoter detection plasmid vectors are constructed with cloning sites placed immediately upstream of a promoterless reporter gene. Inserting a promoter into these cloning sites increases reporter gene expression, which can be monitored by selection, screening, or quantitative determination (40). The *lacZ, galK, cat, str,* and *lux* genes have been used as reporters in these vectors. A diagram explaining the use of promoterless *str* genes, present in the vector pKT240 (15), is shown in Fig. 5. By application of the same principles, vectors for detection of terminator sequences are constructed by placing cloning sites between a promoter and a reporter gene.

18.2.2.6. Controlled-expression plasmids

The ability to control the expression of cloned genes is particularly helpful in the following circumstances: (i) if the product of the cloned gene is not tolerated well by the host organism and the gene must be maintained under repression until its induction is desired; (ii) if a gene product is needed in large quantities for isolation and purification; and (iii) if the effects of a particular gene function on cellular metabolism or regulation are being studied.

Different regulated promoters have been used to construct controlled-expression plasmid vectors. Among the most widely used systems are the *trp* promoter and the *trpR* regulatory genes (54); the *lac* promoter and the *lacI* or *lacI*^{Q} regulatory gene (56, 57); and the hybrid *trp-lac* (*tac*) promoter and the *lacI*^{Q} regulatory gene (12, 15, 25, 30). The vector pMMB67, presented in Fig. 2, is an example of a p_{tac}-based vector carrying the regulatory gene *lacI*^{Q} (30, 37). Of the other regulated systems, the promoters of bacteriophage λ are engineered into vectors together with their thermosensitive repressor, the *c*I857 gene (42, 43).

Very high expression of genes has been obtained by the use of the bacteriophage T7 promoter, ϕ10. This promoter is not recognized by the *E. coli* RNA polymerase, and no transcripts are initiated from it unless T7 polymerase is present in the cells. This expression system has been constructed such that the gene for T7 RNA polymerase is present on a separate plasmid, where it is under control of the λ p_L promoter and the *c*I857 gene or under control of the *tac* promoter and *lacI*^{Q} gene. Alternatively, the polymerase may be supplied by infection of the cells with a bacteriophage carrying the T7 RNA polymerase gene (52, 53).

The repression of genes inserted under the control of these expression systems is not complete. Small quantities of products are still synthesized in the host cells. The addition of isopropyl-β-D-thiogalactopyranoside (IPTG) usually increases the amount of product 50- to 100-fold (15).

To obtain expression with these promoters, grow the culture in LB or suitably supplemented M9 medium to mid-logarithmic phase and induce expression by shifting to inducing conditions (add 0.5 mM IPTG for the *lacI*- or *lacI*^{Q}-repressed system; shift to 43°C for 30 min and continue to grow at 37°C thereafter for the *c*I857-repressed systems). Continue growth 2 to 3 h after the increase in optical density stops. For obtaining optimal yields, proteins of interest should be monitored by sodium dodecyl sulfate-polyacrylamide gel electrophoresis (SDS-PAGE) or another suitable detection method (activity determination, immunoassay). In many cases, gels stained with Coomassie blue will give satisfactory results. The best results are obtained by monitoring the induced protein synthesis by labeling in "maxicells" (see section 18.3.3).

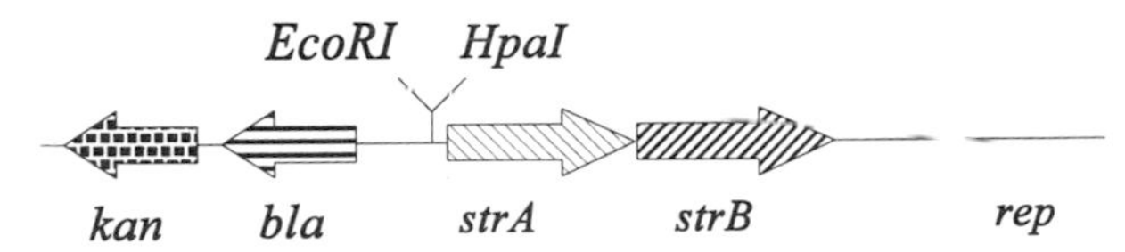

FIG. 5. Physical and genetic map of the promoter-detection vector pKT240. The map is not drawn to scale. *strA* and *strB* represent two promoterless aminoglycoside phosphotransferase genes from the plasmid RSF1010 (47, 48). Insertion of promoter sequences at *Eco*RI or *Hpa*I sites activates the expression of these genes and confers strong resistance to streptomycin (15). *kan,* kanamycin resistance gene; *bla,* β-lactamase gene; *rep,* replication genes; arrows indicate the direction of transcription.

18.2.2.7. Broad-host-range plasmids

The system of *E. coli* host bacteria and of *E. coli* plasmid- and bacteriophage-derived vectors is without doubt the most sophisticated and most highly developed cloning system. However, the use of the *E. coli* system has certain limitations. (i) The first is the narrow host range of the vectors and, consequently, the inability of narrow-host-range *E. coli* vectors to be transferred to and maintained in other species (13, 14). Most of the cloning vectors discussed above are specific for *E. coli* or for very closely related species. The narrowest host range is exhibited by the bacteriophage-based vectors, since, in addition to the host specificity of their replicon, they are strictly specific for the receptors on the host membranes during the adsorption step. (ii) The second is poor recognition by *E. coli* of transcriptional and translational signals from many other species; as a consequence, heterologous genes may be expressed poorly in this host. (iii) The third is the impossibility of being able to study the function of genes that are parts of metabolic or regulatory pathways not present in *E. coli* (e.g., hydrocarbon degradation, dechlorination, pathogenicity, secretion, and nitrogen fixation). (iv) The fourth is the toxicity of heterologous gene products for *E. coli* (e.g., membrane components, secreted proteins).

Plasmids of the incompatibility groups C, N, P, Q, and W exhibit a host range much wider than those used as replicons for *E. coli*-specific vectors. The replicons of the plasmid RP1/RP4/RK2 (IncP1 group) and RSF1010 (IncQ group) are the most extensively studied replicons to date (27, 46–48). They have the widest host range of all known plasmids among gram-negative bacteria. Vectors based on the RP1/RP4/RK2 and RSF1010 replicons are among the most versatile and user-friendly vectors available. For some of them, such as pMMB66 and pMMB207, the entire nucleotide sequence has been determined (37, 38).

18.3. DETECTION OF RECOMBINANT GENES

18.3.1. Insertional Inactivation

Most vectors are constructed such that the donor DNA fragment is inserted into a gene whose function is easily scored. Thus, in the vector pBR322, *Pst*I, *Pvu*I, and *Sca*I sites are located in Ap^r, the gene encoding β-lactamase, whereas *Hin*dIII, *Bam*HI, and several other sites are located in the gene encoding resistance to tetracycline. Insertions of DNA fragments in these sites will inactivate the corresponding genes, and the colonies containing recombinant plasmids may be easily identified by screening for the loss of appropriate antibiotic resistance. This phenomenon of insertional inactivation was the first to be used for the identification of clones containing recombinant DNA (31, 55). In the newer and more advanced vectors, such as the pUC series or the M13mp series, an even more convenient screening method is used. Insertion of a DNA fragment into the cloning site of these plasmids inactivates the gene encoding a fragment of the β-galactosidase, the so-called α-fragment. In appropriate host strains such as the JM strains of *E. coli* (57) carrying a deletion of *lac* genes on the chromosome and an F plasmid with the Δ(*lacZ*)M15 mutation, able to express the so-called Ω fragment of β-galactosidase, colonies carrying recombinant plasmids may be detected as white (*lacZ*) on agar plates containing 40 μg of X-Gal (5-bromo-4-chloro-3-indolyl-β-galactopyranoside) per ml. Colonies carrying the vector without an insert are blue ($lacZ^+$). Similarly, plaques resulting from a recombinant M13mp bacteriophage are white on such plates, whereas plaques resulting from the uncleaved or religated bacteriophage M13mp vectors are blue.

18.3.2. Colony and Plaque Hybridization

If a fragment of a gene to be detected is available, colonies harboring recombinant plasmids or plaques resulting from recombinant bacteriophages may be screened by DNA-DNA hybridization. If at least a partial sequence of the gene is known (for example, by obtaining a partial amino acid sequence of the protein encoded by this gene), a synthetic oligonucleotide may be used as a probe for such hybridization. A protocol used in colony hybridization is given below.

1. Pick colonies onto two replicate plates with a grid pattern: LB covered with a membrane filter (nitrocellulose or nylon) and LB. Incubate until colonies appear.
2. Place the filter with colonies on Whatman 3MM paper soaked in 0.5 M NaOH (colony side up) for 2 to 3 min.
3. Move the filter to another 3MM paper soaked in 0.5 M NaOH for 2 to 3 min.
4. Move the filter to 3 MM paper soaked in 1 M Tris·HCl (pH 7.4) for 2 to 3 min.
5. Move the filter to another 3MM paper soaked in 1 M Tris·HCl (pH 7.4) for 2 to 3 min.
6. Move the filter to 3MM paper soaked in 1.5 M NaCl in 0.5 M Tris·HCl (pH 7.4).
7. Air dry the filter (30 min), place in a container with 1.5 M NaCl in Tris·HCl (pH 7.4), and thoroughly remove all bacterial debris with Kleenex tissue while holding the filter under a layer of buffer.
8. Dip the filters in 0.3 M NaCl, and air dry.
9. Cross-link in a UV cross-linking chamber (Bio-Rad) at 125 mJ, or bake at 65°C overnight or 80°C for 3 to 4 h in a vacuum oven.

A protocol for plaque hybridization is as follows.

1. Mix an aliquot of phage particles to be screened containing an appropriate number of PFU (with skill, up to 50,000 plaques may be screened on one 150-mm petri dish) with 0.3 ml of indicator bacteria. Incubate the mixture for 10 min at 37°C.
2. Add 6.5 ml of molten (47°C) agarose (0.7%), and pour onto a 150-mm LB agar plate. The bottom agar has to be dry, otherwise the top agarose layer may peel off with the filter in step 6. Agarose is recommended rather than agar for the same reason.
3. Incubate the plates at 37°C until the plaques are about 1.0 to 1.5 mm in diameter.
4. Chill the plates at 4°C.
5. Place dry, circular filters (nitrocellulose or nylon) on the surface of the plates. Avoid trapping air bubbles beneath the filters, and wear gloves to prevent getting fingerprints on the filters. Filters may be marked with a soft pencil or by making notches with scissors. The petri dish should be marked appropriately with a marker pen.
6. After 30 to 60 s, remove the filter from the plate and place it, DNA side up, on Whatman 3MM paper soaked with 0.5 N NaOH–1.5 M NaCl for 5 min.
7. Transfer the filter to a neutralizing solution (0.5 M Tris·Cl [pH 7.4], 1.5 M NaCl) for 5 min.
8. Rinse the filter in 2× SSC (1× SSC is 0.15 M NaCl plus 0.015 M sodium citrate), and air dry; then cross-link as in the colony hybridization procedure above.

Prepare labeled probes by end labeling with polynucleotide kinase or by nick translation using oligonucleotide-primed translation. Then hybridize, wash, and expose the filters in the same way as for Southern hybridization described in Chapter 26.6.3.

18.3.3. Detection of Gene Products in "Maxicells"

The so-called maxicell strain is a recombination- and repair-defective strain of *E. coli*. After irradiation with UV light, its chromosome is irreversibly damaged and chromosomal genes are not expressed. Since plasmid vectors are smaller than 1% of the chromosome, the probability of the damage of their DNA is much lower. The plasmids are able to replicate and express their genes for some time after chromosomal genes are inactivated. If such cells are labeled with radioactive amino acids at this stage, the proteins encoded by the plasmid genes are predominantly labeled (45).

1. Grow the strain CSR603 (45), harboring the plasmid under investigation, overnight in LB with appropriate antibiotics or in M9 medium with necessary supplements. Inoculate 0.25 ml into 25 ml of the same medium and grow to an optical density at 650 nm of 0.2.
2. Irradiate 20 ml of culture in a standard petri dish under a UV lamp. The UV dose must be calibrated for each UV lamp. Determine the time and distance from the UV source required to give 1% survivors. Remember that the energy is inversely proportional to the square of the distance from the lamp.
3. Incubate 10 ml of irradiated culture with shaking at 37°C for 1 h.
4. Add D-cycloserine to a final concentration of 200 μg/ml (from a freshly prepared solution).

5. Incubate with shaking at 37°C for 16 h.
6. Collect the cells from 2 ml of culture by centrifugation at 5,000 *g* for 10 min.
7. Wash the cells three times with 1 ml of M9 medium.
8. Suspend the cells in 1 ml of Methionine Assay Medium (Sigma) or M9 medium with all amino acids except methionine (50 μg/ml), and incubate at 37°C for 1 h. Add 0.5 μl of [^{35}S]methionine (1 to 2 mCi), and incubate at 37°C for 1 h.
9. Add 300 μl of nonradioactive L-methionine (1%).
10. Collect the cells, and suspend in 50 μl of sample buffer for SDS-PAGE.
11. Boil the sample for 2 to 4 min, and load half of this sample onto an SDS-polyacrylamide gel.
12. After electrophoresis, dry the gel and expose to X-ray film.

Control cells, containing only the vector, should be processed in parallel with the assay sample. The comparison of bands visualized by autoradiography will determine which bands are present in the sample containing the recombinant plasmid but absent in the sample containing only the vector. This comparison is greatly facilitated if a controlled-expression vector, such as that depicted in Fig. 3, is used. In this case two parallel cultures are processed. One contains IPTG added to a final concentration of 0.05 to 1.0 mM before the addition of labeled methionine. The other, serving as the control sample, does not receive IPTG. Incubate both cultures at 37°C for 1 h. Comparisons of radiographs will reveal which bands were induced in the culture containing IPTG.

18.3.4. Detection of Gene Products by Antibodies

Protein products of genes that are expressed from plasmid or bacteriophage vectors may be detected by specific antibodies. The best results are obtained with polyclonal antibodies, since they react with many different epitopes and will generally give strong signals even with denatured proteins. On the other hand, such antibodies may also recognize a fragment of the protein being sought. Another problem encountered with polyclonal antisera is the fact that they often contain antibodies directed against different components of the *E. coli* cells. These antibodies must be removed before the antisera can be used for screening.

1. Colonies to be screened may be grown directly on a nitrocellulose filter either by spreading the appropriate number of cells on the filter placed on a surface of a LB agar plate or by transferring individual colonies onto the filter with a toothpick and allowing them to grow.
2. A replica of the plate with the filter is made either by performing a parallel transfer of colonies to an LB agar plate or by taking a second filter, wetted by placing it first on an LB agar, and then gently pressing it onto the master filter with the grown colonies. The replica filter is then placed on an LB agar plate, and the colonies are allowed to grow. They are preserved at 4°C until the results of immunodetection are obtained.
3. If controlled-expression vectors are used for cloning, the master filters are placed under induction conditions for 2 to 4 h. Use an LB agar plate containing 0.5 mM IPTG for vectors carrying *lac* or *tac* promoters and an LB agar plate incubated at 42°C for the systems containing thermosensitive repressor (such as the $\lambda\ p_R$ promoter and *c*I857 repressor).
4. Transfer the filter onto a damp paper towel, and cover it with a glass or plastic hood. Place an open glass petri dish containing chloroform under the hood. Alternatively, hang the filter by two paper clips in a beaker with chloroform at the bottom and cover the beaker with tinfoil. Expose the colonies to chloroform for 15 min.
5. Transfer the filter to a dish containing the colony lysis buffer (Table 1), and gently agitate overnight (12 to 16 h).
6. Transfer the filters to a dish containing 1× TBS (Table 1) and agitate gently for 30 min. Repeat this step once with fresh TBS.
7. Transfer the filter to a dish with fresh TBS, and thoroughly remove the residues of colonies from the filter by using Kimwipes.
8. Transfer the filter for at least 1 h to a dish containing BSA (some protocols recommend this "blocking" step to be carried out overnight).
9. Wash the filter in TBS containing 0.05% Tween-20. Repeat this step once.
10. Transfer the filters to TBS containing 0.05% Tween-20, 0.1% BSA, and antibody. Incubate for 2 to 4 h at room temperature. The dilution of antibody should be determined experimentally. The highest dilution that still gives a satisfactory signal with 50 pg of antigen (spotted onto an identical membrane filter), while giving a low background, should be used.
11. Wash the filters in three successive changes of TBS containing 0.05% Tween-20; allow 10 min for each wash.
12. Transfer the filter to the secondary antibody solution. This is an enzyme-coupled antibody that reacts with species-specific determinants of the primary antibody used. Alkaline phosphatase-coupled anti-rabbit antibody or similar reagents are available commercially. Typically, 5 μl of a 0.1 mg/ml solution is diluted into 7.5 ml of TBS containing 0.05% Tween-20 and incubated with gentle shaking for 30 min.
13. Wash the filter in two successive portions of TBS–0.05% Tween-20.
14. Transfer the filter to 10 ml of AP buffer containing 66 μl of NBT and 33 μl of BCIP (Table 1), and shake at room temperature until the reaction is easily visible and the background is still low.
15. Stop the reaction by washing the filter with water three or four times.

18.5. REFERENCES

18.5.1. General References

1. **American Society for Microbiology.** 1991. *ASM Style Manual for Journals and Books,* p. 35–38. American Society for Microbiology, Washington, D.C.

1a. **Ausubel, F. M., R. Brent, R. E. Kingston, D. D. Moore, J. G. Seidman, J. A. Smith, and K. Struhl (ed.).** 1992. *Short Protocols in Molecular Biology.* Greene Publishing Associates, Inc., and John Wiley & Sons, Inc., New York.

2. **Brown, T. A.** 1990. *Gene Cloning, An Introduction,* 2nd ed. Chapman & Hall, London.

3. **Goeddel, D. V. (ed.).** 1990. Gene expression technology. *Methods Enzymol.* **185.**

4. **Harwood, C. R., and S. M. Cutting.** 1990. *Molecular Biological Methods for Bacillus.* John Wiley & Sons, Inc., New York.
5. **Sambrook, J., E. F. Fritsch, and T. Maniatis.** 1989. *Molecular Cloning: A Laboratory Manual,* 2nd ed. Cold Spring Harbor Laboratory, Cold Spring Harbor, N.Y.
6. **Wu, R. (ed.).** 1987. Recombinant DNA, part F. *Methods Enzymol.* **155.**
7. **Wu, R., and L. Grossman (ed.).** 1987. Recombinant DNA, part D. *Methods Enzymol.* **153.**
8. **Wu, R., and L. Grossman (ed.).** 1987. Recombinant DNA, part E. *Methods Enzymol.* **154.**
9. **Wu, R., L. Grossman, and K. Moldave (ed.).** 1983. Recombinant DNA, part B. *Methods Enzymol.* **100.**
10. **Wu, R., L. Grossman, and K. Moldave (ed.).** 1983. Recombinant DNA, part C. *Methods Enzymol.* **101.**
11. **Zyskind, J. W., and I. S. Bernstein.** 1989. *Recombinant DNA Laboratory Manual.* Academic Press, Inc., New York.

18.5.2. Specific References

12. **Amann, E., J. Brosius, and M. Ptashne.** 1983. Vectors bearing a hybrid *trp-lac* promoter useful for regulated expression of cloned genes in *Escherichia coli. Gene* **25:**167–178.
13. **Bagdasarian, M., R. Lurz, B. Rückert, F. C. H. Franklin, M. M. Bagdasarian, and K. N. Timmis.** 1981. Specific purpose plasmid cloning vectors, II. Broad host range, high copy number, RSF1010-derived vector system for gene cloning in *Pseudomonas. Gene* **16:**237–247.
14. **Bagdasarian, M., and K. N. Timmis.** 1982. Host: vector systems for gene cloning in *Pseudomonas Curr. Top. Microbiol. Immunol.* **96:**47–67.
15. **Bagdasarian, M. M., E. Aman, R. Lurz, B. Rückert, and M. Bagdasarian.** 1983. Activity of the hybrid *trp-lac* (*tac*) promoter of *Escherichia coli* in *Pseudomonas putida.* Construction of broad-host-range, controlled-expression vectors. *Gene* **26:**273–282.
16. **Balbàs, P., X. Soberòn, F. Bolivar, and R. L. Rodriguez.** 1988. The plasmid, pBR322, p. 5–41. *In* R. L. Rodriguez and D. T. Denhardt (ed.), *Vectors.* Butterworths, Boston, Mass.
17. **Balbàs, P., X. Soberòn, E. Merino, M. Zurita, H. Lomeli, F. Valle, N. Flores, and F. Bolivar.** 1986. Plasmid vector pBR322 and its special-purpose derivatives. *Gene* **50:**3–40.
18. **Bolivar, F.** 1978. Construction and characterization of new cloning vehicles. 3. Derivatives of plasmid pBR322 carrying unique EcoRI sites for selection. *Gene* **4:**121–136.
19. **Bolivar, F., L. R. Rodriguez, P. J. Greene, M. C. Betlach, H. L. Heyneker, H. W. Boyer, J. H. Crosa, and S. Falkow.** 1977. Construction and characterization for new cloning vehicles. II. A multicopy cloning system. *Gene* **2:**95–113.
20. **Brady, G., H. M. Jantzen, H. U. Bernard, R. Brown, G. Schütz, and T. Hashimoto-Gotoh.** 1984. New cosmid vectors developed for eukaryotic DNA cloning. *Gene* **27:**23–232.
21. **Brosius, J.** 1984. Plasmid vectors for the selection of promoters. *Gene* **24:**151–160.
22. **Brosius, J.** 1989. Superpolylinkers in cloning and expression vectors. *DNA* **8:**10.
23. **Casadaban, M. J., J. Chou, and S. N. Cohen.** 1980. In vitro gene fusions that join an enzymatically active β-galactosidase segment to amino-terminal fragments of exogenous proteins. *Escherichia coli* plasmid vectors for the detection and cloning of translational initiation signals. *J. Bacteriol.* **143:**971–980.
24. **Chang, A. C. Y., and S. N. Cohen.** 1978. Construction and characterization of amplifiable multicopy DNA cloning vehicles derived from the P15A cryptic miniplasmid. *J. Bacteriol.* **134:**1141–1156.
25. **de Boer, H. A., L. J. Comstock, and M. Vasser.** 1983. The *tac* promoter: a functional hybrid derived from the *trp* and *lac* promoters. *Proc. Natl. Acad. Sci. USA* **80:**21–25.
26. **Ditta, G., T. Schmidhauser, E. Yacobson, P. Lu, X.-W. Liang, D. R. Finlay, D. Guiney, and D. R. Helinski.** 1985. Plasmids related to the broad host range vector, pRK290, useful for gene cloning and for monitoring gene expression. *Plasmid* **13:**149–153.
27. **Frey, J., and M. Bagdasarian.** 1989. The molecular biology of IncQ plasmids, p. 79–94. *In* C. Thomas and F. C. H. Franklin (ed.), *Promiscuous Plasmids of Gram Negative Bacteria.* Academic Press Ltd., London.
28. **Frey, J., M. Bagdasarian, D. Feiss, F. C. H. Franklin, and J. Deshusses.** 1983. Stable cosmid vectors that enable the introduction of cloned fragments into a wide range of Gram negative bacteria. *Gene* **24:**299–308.
29. **Frey, J., M. Bagdasarian, and K. N. Timmis.** 1984. Escherichia coli cloning vectors that auto-amplify in the stationary phase of bacterial growth, p. 377–383. *In* S. Mitsuhashi and V. Krcmery (ed.), *Plasmids and Gene Manipulation.* Avenicum Praha and Springer, Heidelberg, Germany.
30. **Fürste, J. P., W. Pansegrau, R. Frank, H. Blocker, P. Scholz, M. Bagdasarian, and E. Lanka.** 1986. Molecular cloning of the plasmid RP4 primase region in a multi-host-range *tacP* expression vector. *Gene* **48:**119–131.
31. **Hershfield, V., H. W. Boyer, C. Yanofsky, M. A. Lovett, and D. R. Helinski.** 1974. Plasmid colE1 as a molecular vehicle for cloning and amplification of DNA. *Proc. Natl. Acad. Sci. USA* **71:**3455–3459.
32. **Hohn, B., and J. Collins.** 1980. A small cosmid for efficient cloning of large DNA fragments. *Gene* **11:**291–298.
33. **Huynh, T. V., R. A. Young, and R. W. Davis.** 1985. Constructing and screening cDNA libraries in λgt10 and λgt11, p. 49–78. *In* D. Glover (ed.), *DNA Cloning. A Practical Approach,* vol. 1. IRL Press, Oxford.
34. **Kahn, M., R. Kolter, C. Thomas, D. Figurski, R. Meyer, E. Remaut, and D. R. Helinski.** 1978. Plasmid cloning vehicles derived from plasmid ColE1, F, R6K and RK2. *Methods Enzymol.* **68:**268–280.
35. **Kaiser, K., and N. E. Murray.** 1985. The use of phage lambda replacement vectors in the construction of representative genomic DNA libraries, p. 1–47. *In* D. Glover (ed.), *DNA Cloning: A Practical Approach,* vol. 1. IRL Press, Oxford.
36. **Keen, N. T., S. Tamaki, D. Kobayachi, and D. Trollinger.** 1988. Improved broad-host-range plasmids for DNA cloning in gram-negative bacteria. *Gene* **70:**191–198.
37. **Morales, V., M. M. Bagdasarian, and M. Bagdasarian.** 1990. Promiscuous plasmids of the IncQ group: mode of replication and use for gene cloning in gram-negative bacteria, p. 229–241. *In* S. Silver, A. Chakrabarty, B. Iglewski, S. Kaplan, and I. C. Gunsalus (ed.), *Pseudomonas: Biotransformations, Pathogenesis, and Evolving Biotechnology.* American Society for Microbiology, Washington, D.C.
38. **Morales, V. M., A. Bäckman, and M. Bagdasarian.** 1991. A series of wide-host-range low-copy-number vectors that allow direct screening for recombinants. *Gene* **97:**39–47.
39. **Morales, V. M., and M. Bagdasarian.** 1991. Cloning in bacteria: vectors and techniques, p. 3–28. *In* A. Prokop, R. Bajpal, and C. Ho (ed.), *Recombinant DNA. Technology and Applications.* McGraw-Hill Book Co., New York.
40. **Nierman, W. C.** 1988. Vectors for cloning promoters and terminators, p. 153–170. *In* R. L. Rodriguez and D. T. Denhardt (ed.), *Vectors.* Butterworths, Boston, Mass.
41. **Poustka, A., H. R. Rackwitz, A. M. Frischauf, B. Hohn, and H. Lehrach.** 1984. Selective isolation of cosmid clones by homologous recombination in *Escherichia coli. Proc. Natl. Acad. Sci. USA* **81:**4129–4133.
42. **Remaut, E., P. Stanssens, and W. Fiers.** 1981. Plasmid vectors for high-efficiency expression controlled by the pL promoter of coliphage lambda. *Gene* **15:**81–93.
43. **Remaut, E., P. Stanssens, and W. Fiers.** 1983. Inducible high level synthesis of mature human fibroblast interferon in *Escherichia coli.* Nucleic Acids Res. **11:**4677–4688.
44. **Rose, R. E.** 1988. The nucleotide sequence of pACYC177. *Nucleic Acids Res.* **16:**355–356.
45. **Sancar, A., A. M. Hack, and W. D. Rupp.** 1979. Simple method for identification of plasmid-coded proteins. *J. Bacteriol.* **137:**692–693.

46. **Scherzinger, E., M. M. Bagdasarian, P. Scholz, R. Lurz, B. Rückert, and M. Bagdasarian.** 1984. Replication of the broad host range plasmid RSF1010: requirement for three plasmid-encoded proteins. *Proc. Natl. Acad. Sci. USA* **81:**654–658.
47. **Scholz, P., V. Haring, E. Scherzinger, R. Lurz, M. M. Bagdasarian, H. Schuster, and M. Bagdasarian.** 1985. Replication determinants of the broad host-range plasmid RSF1010, p. 243–259. *In* D. Helinski, S. Cohen, D. Clewell, D. A. Jackson, and A. Hollander (ed.), *Plasmids in Bacteria.* Plenum Press, New York.
48. **Scholz, P., V. Haring, B. Wittmann-Liebold, K. Ashman, M. Bagdasarian, and E. Scherzinger.** 1989. Complete nucleotide sequence and gene organization of the broad-host-range plasmid RSF1010. *Gene* **75:**271–288.
49. **Simon, R., U. Priefer, and A. Pühler.** 1983. A broad host range mobilization system for *in vivo* genetic enegineering: transposon mutagenesis in gram negative bacteria. *Bio/Technology* **1:**784–790.
50. **Smith, G. P.** 1988. Filamentous phages as cloning vectors. p. 61–83. *In* R. L. Rodriguez and D. T. Denhardt (ed.), *Vectors.* Buttersworths Boston, Mass.
51. **Sorge, J. A.** 1988. Bacteriophage lambda cloning vectors, p. 43 60. *In* R. L. Rodriguez and D. T. Denhardt (ed.), *Vectors.* Butterworths, Boston, Mass.
52. **Studier, F. W., A. H. Rosenberg, J. J. Dunn, and J. W. Dubendorff.** 1990. Gene expression technology. *Methods Enzymol.* **185:**60–88.
53. **Tabor, S., and C. C. Richardson.** 1985. A bacteriophage T7 RNA polymerase/promoter system for controlled exclusive expression of specific genes. *Proc. Natl. Acad. Sci. USA* **82:**1074–1078.
54. **Tacon, W. C. A., W. A. Bonass, B. Jenkins, and J. S. Emtage.** 1983. Expression plasmid vectors containing *Escherichia coli* tryptophan promoter transcriptional units lacking the attenuator. *Gene* **23:**255–266.
55. **Timmis, K. N., F. Cabello, and S. N. Cohen.** 1974. Utilization of two distinct modes of replication by a hybrid plasmid constructed *in vitro* from separate replicons. *Proc. Natl. Acad. Sci. USA* **71:**4556–4560.
56. **Vieira, J., and J. Messing.** 1982. The pUC plasmids, an M13mp7-derived system for insertion mutagenesis and sequencing with synthetic universal primers. *Gene* **19:**259–268.
57. **Yanisch-Perron, C., J. Vieira, and J. Messing.** 1985. Improved M13 phage cloning vectors and host strains: nucleotide sequences of the M13mp18 and pUC19 vectors. *Gene* **33:**103–119.

Chapter 19

Polymerase Chain Reaction

RONALD M. ATLAS and ASIM K. BEJ

The polymerase chain reaction (PCR) is an in vitro method for replicating a defined (target) DNA sequence so that its amount is increased exponentially. Whereas previously only minute amounts of a specific gene could be obtained from a cell, now even a single gene copy can be amplified to a million copies within a few hours by PCR.

PCR consists of repetitive cycles of DNA denaturation through melting at elevated temperature to convert double-stranded DNA to single-stranded DNA, annealing of oligonucleotide primers to the target DNA, and extension of the DNA by nucleotide addition from the primers by the action of DNA polymerase (1, 2, 4, 5, 7–10, 83) (Fig. 1). The target region for amplification is defined by unique oligonucleotide primers that flank a DNA segment. The oligonucleotide primers are designed to hybridize to regions of DNA that flank a desired target gene sequence, annealing to complementary strands of the target sequence. The primers are then extended across the target sequence by using a heat-stable DNA polymerase (frequently *Taq* DNA polymerase, a thermostable and thermoactive enzyme from *Thermus aquaticus*) in the presence of free deoxynucleoside triphosphates (dNTPs), resulting in a doubled replication of the starting target material. By repeating the three-stage process many times, a nearly exponential increase in the amount of target DNA results (Fig. 2). The product of each PCR cycle is comple-

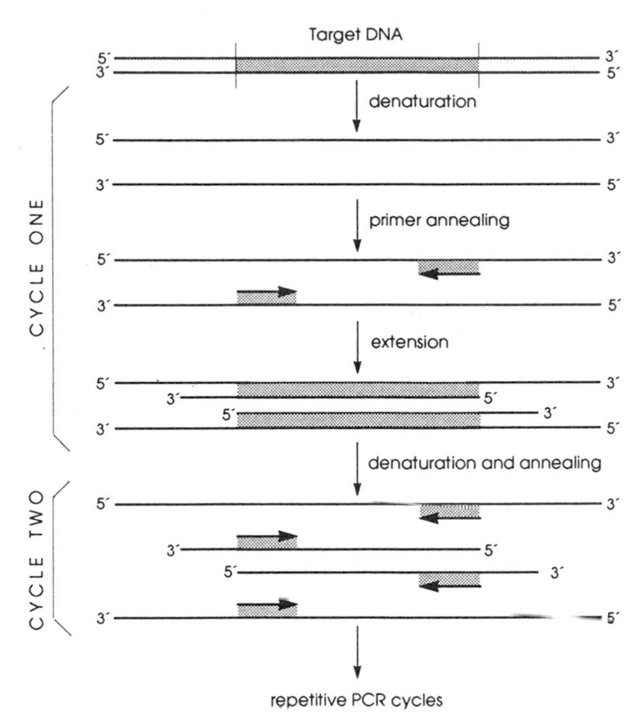

FIG. 1. Diagram of PCR showing the amplification of a target region. The three stages of each cycle are DNA denaturation, primer annealing, and primer extension. The primers define the region that is to be amplified. The cycles are repeated to replicate the target DNA.

mentary to and capable of binding the primers, and so the amount of DNA is potentially doubled in each successive cycle (9).

19.1. REAGENT PREPARATION

The essential reagent components for PCR are a thermostable DNA polymerase, oligonucleotide primers, deoxynucleotides (dNTPs), template (target) DNA, and magnesium ions. Generally the total reaction volume is 50 to 100 μl.

The reagent preparation steps for a typical PCR protocol (see section 19.2) are as follows.

1. Prepare template (target) DNA (use appropriate procedures for sample processing and preparation of purified DNA as described in other chapters).
2. Prepare reaction buffer as a 10× stock solution of 500 mM KCl, 100 mM Tris·HCl, and 15 mM $MgCl_2$ (pH 8.4) or as a 10× stock solution of 500 mM KCl, 500 mM Tris·HCl, and 25 mM $MgCl_2$ (pH 8.9) (see section 19.1.6).
3. Prepare dNTPs as a 10× stock solution of 2 mM dATP, 2 mM dCTP, 2 mM dGTP, and 2 mM dTTP (pH 7.0).
4. Prepare primers as a 10× solution of 2.4 μM of primer 1 and 2.5 μM of primer 2 in TE buffer (10 mM Tris·HCl, 0.1 mM EDTA [pH 8.0]).
5. Combine 10 μl each of the 10× reaction buffer, 10× dNTPs, and 10× primer solutions and 20 μl of distilled water to form a 2× reaction mixture.
6. Add 50 μL of 2× reaction mixture to 50 μl (0.5 to 1.0 μg) of target DNA in a 0.5-ml microcentrifuge tube. (Reduce volume to 50 μl total by using 25 μl of 2× reaction mixture and 24.5 μl [0.5 to 1.0 μg] of target DNA for some thermocyclers.)

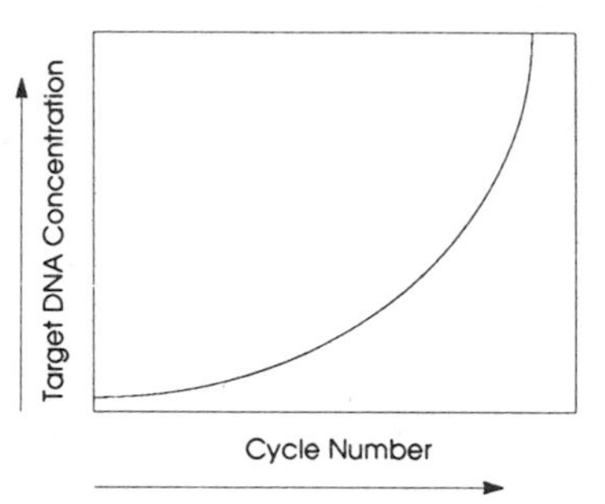

FIG. 2. Diagram showing exponential increase of target DNA concentration with increasing number of PCR cycles. Each cycle results in a potential doubling of the target DNA sequence.

7. Add 2.5 U (0.5 μl) of *Taq* DNA polymerase.
8. Overlay with 80 to 100 μl of mineral oil or Ampliwax (optional).
9. Maintain at 5°C until starting the first PCR cycle.

19.1.1. Target DNA

For PCR there must be at least one intact copy of the target gene. A greater number of target copies enhances the probability of successful DNA amplification. Any damage, such as a nick, in the target DNA will block PCR amplification. The target gene can be from <0.1 to a few kilobases. The total amount of DNA typically used for PCR is 0.05 to 1.0 μg; this permits detection of single copies of target genes. A sample need not be highly purified and may be prepared by lysing the cells, by boiling with Chelex or in a hypotonic solution, or by freeze-thaw cycling. However, some impurities—such as formalin, heme, heparin, humic acids, Mg^{2+}-chelating agents, detergents, and heavy metals (especially Fe^{2+} and Fe^{3+})—must be eliminated or diluted so as not to inhibit PCR amplification of DNA.

19.1.2. Oligonucleotide Primers

Pairs of primers are selected that flank the DNA region to be amplified. Primers are oligonucleotides which are short single-stranded segments of DNA, complementary to the 5′ ends of the strands of target DNA to be amplified. For primer annealing to occur specifically at the sites flanking the DNA region to be amplified, there must be nearly complete identity between the target DNA and the primer nucleotide sequence. To obtain specific amplification of the target DNA, the primers must not match other, nontarget sequences of the DNA in the reaction. Typically the primers are 15 to 30 nucleotides long, preferably with no complementary overlap at their 3′ ends (101). Several computer programs are available that aid in the selection and design of optimal primers for the amplification of specific gene sequences. These programs help avoid mispriming at nontarget sites. They also aid in selection of primers with similar G+C contents so that an efficient annealing temperature can be used. Additionally, they assist in avoiding the selection of primers and target sequences without secondary structures that would interfere with efficient DNA amplification.

It is important that the 5 or 6 nucleotides at the 3′ ends of the primers exhibit precise base pairing with the target DNA. An exact base pair match is generally required at the 3′ end for effective PCR amplification of target DNA. The stringency of complete matching can be overcome by using degenerate primers, which are primers that allow some mismatches (wobble) with the target sequence, either by incorporating inosines at specific positions or by having a mixture of multiple nucleotides at specified positions.

Although the primer should match the target, complementarity between the paired primers at the 3′ ends should be avoided because they may produce primer artifacts (100, 101). A primer artifact is a double-stranded fragment of DNA that is formed when one primer is extended by the polymerase over the other primer and has a length close to the sum of the two primers. An amplification artifact can become the predominant product in the PCR.

All primers for a target amplification should have nearly the same melting temperature (T_m), or, more specifically, temperature of annealing (T_a) to the template, and an average G+C content of 40 to 60% with no stretches of polypurines or polypyrimidines to enhance specificity. The T_a of the primers should be kept as close to 72°C as possible for specific priming and removal of unnecessary "ghost" amplified DNA bands (12, 101). The T_a can be calculated either from the melting temperature (T_m) of the primers by using the equation $T_a = T_m - 5°C = 2(A+T) + 4(G+C) - 5°C$ or by using a computer-aided program such as OLIGO (99).

Significant secondary structures (measured by the negative value of ΔG) internal to the primers or immediately downstream (for the left-hand primer) or immediately upstream (for the right-hand primer) of the template should be avoided (100, 101). Also, three or more C's or G's at the 3′ ends of primers may promote mispriming at G+C-rich sequences and should be avoided. Palindromic sequences within primers likewise must be avoided. Primer concentrations of 0.1 to 0.5 μM are recommended. Higher primer concentrations may promote nonspecific-product formation and may increase the generation of a primer artifact. Lower primer concentrations may allow complementary-strand annealing to compete with primer annealing, reducing product yield.

For optimum results, primers should be purified after synthesis by high-pressure liquid chromatography (HPLC) or gel purification. The primers should be stored at −20°C when not in use; the shelf life of oligonucleotide primers is at least 6 months when stored in liquid and 12 to 24 months when stored dry after lyophilization.

19.1.3. DNA Polymerase

The key to the development and wide use of PCR was the discovery of a DNA polymerase that is both thermostable and thermoactive, that is, one that can withstand repetitive exposure to temperatures required for DNA denaturation and that is active at temperatures at which annealing of primers to target DNA will occur with high stringency. The DNA polymerase from *T. aquaticus* (*Taq* DNA polymerase) meets the requirements (72). A recombinant *Escherichia coli* was constructed to produce this polymerase. The enzyme called *AmpliTaq* is available commercially (Roche Molecular Corp., Alameda, Calif.).

The recommended concentration range of *Taq* DNA polymerase for PCR is 1 to 5 U per 100 μl. If the enzyme concentration is too high, nonspecific background products often form. If the enzyme concentration is too low, an insufficient amount of desired product is made. The *Taq* DNA polymerase has an optimum activity around 70°C and is not inactivated by short incubations at temperatures at which PCR-generated fragments will denature (usually 90 to 95°C). *Taq* DNA polymerase has a half-life of approximately 45 min at 95°C. Exposure for this duration to temperatures above 96°C causes faster enzyme inactivation and should be avoided.

Although *Taq* DNA polymerase and its recombinant equivalent (*AmpliTaq* DNA polymerase) are most commonly used for PCR, other thermostable enzymes can also be used. A fragment of *Taq* DNA polymerase (produced by a recombinant *E. coli*) that has 289 amino acids deleted from the N-terminal region, called a Stoffel fragment, is approximately twofold more thermostable, exhibits optimal activity over a broader range of magnesium ion concentrations (2 to 10 mM), and lacks 5′-to-3′ exonuclease activity compared with the complete *Taq* DNA polymerase. The denaturation temperature can be raised when using the Stoffel fragment, which is useful for templates that are G+C rich or that contain complex secondary structure. The thermostable DNA polymerase from *Thermus thermophilus* (*Tth*) is useful for detecting gene expression because in the presence of $MnCl_2$ at high temperature it has significant reverse transcriptase activity (3). *Tth* DNA polymerase also is relatively resistant to blood components that potentially inhibit *Taq* DNA polymerase, making it useful for detection of pathogens in blood. A thermostable DNA polymerase from *Thermococcus litoralis* (vent DNA polymerase) has 3′-to-5′ nuclease (proofreading) activity, so that using this enzyme may lower the misincorporation rate, which is especially important when PCR is used for DNA sequencing (3). Vent DNA polymerase retains its activity for 2 h at 100°C, with a primer extension capacity up to 13 kb. A thermostable DNA polymerase from *Pyrococcus furiosus* (*Pfu*) has both 5′-to-3′ and 3′-to-5′ exonuclease activities; it has excellent thermostability at 95°C and has a 12-fold lower misincorporation frequency than *Taq* DNA polymerase does (79).

19.1.4. Deoxynucleotides

Free dNTPs are required for DNA synthesis. The dNTP concentrations for PCR should be 20 to 200 μM for each deoxynucleotide to give optimal specificity and fidelity at high incorporation rates. The four dNTPs (dATP, dTTP, dGTP, and dCTP) should be used at equivalent concentrations to minimize misincorporation errors. A final concentration of more than 50 mM total dNTP in the PCR inhibits *Taq* DNA polymerase activity (57). The lowest dNTP concentration appropriate for the length and composition of the target sequence should be used to minimize mispriming at nontarget sites and reduce the likelihood of extending misincorporated nucleotides; 20 μM each dNTP in a 100-μl reaction mixture is theoretically sufficient to synthesize 2.6 μg of a 400-bp sequence (57).

19.1.5. Magnesium Ions

PCR requires divalent magnesium ions. The magnesium ion concentration affects primer annealing, DNA melting temperatures, and enzyme activity. PCRs should contain 0.5 to 2.5 mM magnesium over the total dNTP concentration. Various concentrations of $MgCl_2$, usually in the range of 1.5 to 4 mM, can be used for enhanced specificity and higher yield of the amplified products (90). The $MgCl_2$ concentration can be increased up to 10 mM when the Stoffel fragment of *AmpliTaq* DNA polymerase is used. The presence of EDTA or other chelating agents in the reaction may interfere with the Mg^{2+} concentration. The optimal Mg^{2+} concentration for each primer set should be determined by running replicate PCRs with different Mg^{2+} concentrations.

19.1.6. Reaction Buffer

In addition to the reagents directly involved in the reaction, PCR requires a suitable buffer. The recommended reaction buffer is 10 to 50 mM Tris·HCl (pH 8.3 to 8.9). Changing the buffering capacity of the PCR sometimes increases the yield of the amplified DNA products, for example by increasing the concentrations of the Tris·HCl up to 50 mM or by increasing the pH up to 8.9. Up to 50 mM KCl can be included in the reaction mixture to facilitate primer annealing. KCl greater than 50 mM inhibits *Taq* DNA polymerase activity (57). Gelatin or bovine serum albumin (100 μg/ml) and nonionic detergents such as Tween 20 can be used to stabilize the *Taq* DNA polymerase when diluting the enzyme below 1 U/100 μl.

19.1.7. Cosolvents

Since some template DNA for PCR amplification may not denature completely because of the presence of a high G+C content and primers may not anneal at regions of the template DNA with secondary structure, cosolvents in the PCR such as 1 to 10% dimethyl sulfoxide or 5 to 20% glycerol may greatly increase the amount of PCR-amplified DNA (113, 131). Use of 15 to 20% glycerol as a cosolvent in PCR helps to amplify larger DNA fragments (of about 2.5 kb).

19.1.8. Oil or Wax Overlay

An overlay of 80 to 100 μl of light mineral oil on top of the reaction mix often is used to prevent evaporation of the liquid and to speed the approach to the denaturation temperature. The overlay may maintain heat stability and limits evaporation, so that salt concentrations are maintained. The use of light mineral oil can increase the yield of the amplified product approximately fivefold (82). Alternatively, a hot wax (Ampliwax) can be used to maintain precise temperatures within the reaction mixture, thereby enhancing the specificity of PCR. With some thermocyclers, such as the Perkin-Elmer DNA thermocycler 9600, in which the reaction volumes are low, such overlays are not used.

19.2. REACTION CONDITIONS

PCR typically involves repetitive cycling between a high temperature to separate the strands of the DNA, a relatively low temperature to allow the primers to hybridize (anneal) with the complementary region of the target DNA, and an intermediate temperature for primer extension (Fig. 3). Temperature cycling can be achieved by using an automated thermal cycler. The most critical temperatures are the denaturation and annealing temperatures. Inadequate heating leads to failure of DNA melting, and therefore no DNA amplification takes place. The annealing step determines the specificity of PCR: too low a temperature results in mispriming and amplification of nontarget sequences, and too high a temperature causes a lack of primer annealing and hence no DNA amplification.

The reaction conditions for a typical PCR protocol are as follows.

1. Place the reaction mixture (prepared as described in section 19.1) into the thermal cycler.
2. Melt the DNA by heating to 94°C for 3 min.
3. Anneal primers by cooling to 50°C for 1 min (or appropriate temperature based on T_m of primers).
4. Raise temperature to 72°C for 1 min to extend primers.
5. Raise the temperature to 94°C for 1 min to remelt the DNA
6. Repeat steps 3 to 5 20 to 30 times.
7. End the last cycle at step 4 by maintaining 72°C for 3 min.
8. Cool to 5°C, and remove tube for analysis. (*Note*: If uracil *N'*-glycosylase [UNG] is used for contamination control as described in section 19.2.6, the samples should be held at −72°C and then transferred to a −20°C freezer.)

19.2.1. Thermal-Cycler Instrumentation

The temperatures for PCR are critical, so it is important that thermal cyclers, which permit the automated cycling of temperatures according to a temperature program, have accurate temperature regulation. Since the introduction of the first DNA thermal cycler by Perkin-Elmer Cetus Corp., a dozen more companies have undertaken the manufacture of thermal cyclers with different designs. Some thermal cyclers use a series of heated water baths, whereas others use solid metal heating blocks. Typically 24 to 96 reactions can be run simultaneously.

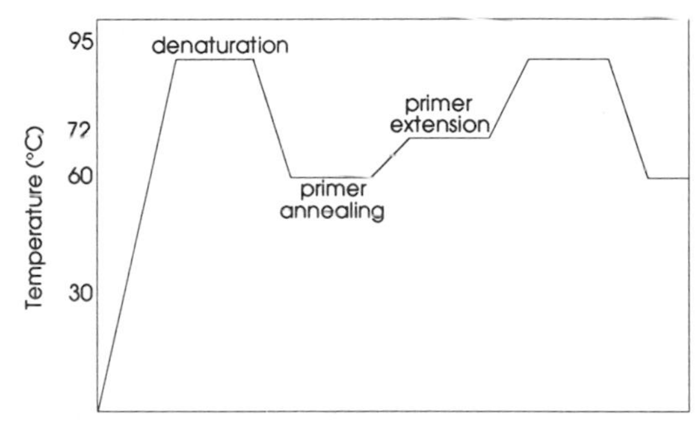

FIG. 3. Diagram of temperature changes during a PCR cycle. There are three critical temperatures, one for separation of DNA strands (denaturation), one for primer annealing that depends upon the specific annealing temperatures of the primers, and one for primer extension at which *Taq* DNA polymerase has optimal activity.

It is essential that a thermal cycler give uniform heating and cooling, consistently give the same results in all reaction wells, and produce a profile in the sample tube that is essentially the same as the cycle that had been programmed. The temperature in each reaction well must be precise so as to provide the required stringency for primer annealing and to provide accurate conditions for denaturation and DNA extension. The temperature ramping between temperature settings must be rapid to permit accurate and short cycles. New thermal-cycler models have increased the rate of heating and cooling and the rate of efficient heat transfer to specialized reaction vessels (42). A thermal cycler with a metal block heated by a heating element pad or a cell cooled by a Peltier pump gives the best results. Newer thermal cyclers do not require addition of mineral oil overlay because of the specialized tubes used in the system (42, 132). PCR amplification in capillary tubes allows rapid and efficient thermal cycling and has reduced a complete cycle of denaturation, primer annealing, and primer extension to 20 s, but sample handling can be tricky.

19.2.2. Primer Annealing Temperature

The appropriate temperature for primer annealing depends on the base composition, length, and concentration of the amplification primers (56). The ideal annealing temperature generally is 5°C below the true T_m of the primers. Annealing temperatures in the range of 55 to 72°C generally yield the best results. At typical primer concentrations of 0.2 μM, annealing requires only a few seconds. Increasing the annealing temperature enhances discrimination against incorrectly annealed primers and reduces addition of incorrect nucleotides at the 3′ end of the primers. Stringent annealing temperatures, especially during the first several cycles, help to increase specificity. If the temperature is lower than the optimum, the reaction products frequently include additional nontarget DNA fragments.

19.2.3. Primer Extension Temperature

Extension time depends on the length and concentration of the target sequence and on the temperature used for the extension reaction. The activity of *Taq* DNA polymerase varies by 2 orders of magnitude between 20 and 85°C (36). Primer extensions typically are performed at 72°C, which is optimal for *Taq* DNA polymerase.

Although the use of low extension temperatures, particularly with high dNTP concentrations, favors extension of misincorporated nucleotides, a rapid and convenient two-step PCR amplification can be performed when amplifying a short target sequence ($\leq$100 to 300 bases) (1–3). In two-step PCR, the primer extension step is set at the same temperature as the reannealing temperature. Using only two temperatures (e.g., 55 to 75°C for annealing and extension, and 94°C for DNA melting) can yield better PCR results.

For amplification of greater than 1 kb of target DNA, the primer extension step can be between 1 and 7 min according to the length of the target DNA to be amplified (59). In combination with a longer prime extension time, agents such as gelatin or bovine serum albumin are required in the reaction buffer for longer activity and stability of the *Taq* DNA polymerase.

19.2.4. Denaturation Temperature

To ensure complete denaturation of target DNA, a temperature of 94 to 95°C for 3 min typically is used prior to the first cycle. In subsequent cycles, typical denaturation conditions are 94 to 95°C for 30 to 60 s (56). The selection of temperature and denaturation time depends on the length and G+C content of the target DNA sequences. Using too low a temperature may result in incomplete denaturation of the target template and/or the PCR product and therefore in failure of the PCR. In contrast, DNA denaturation steps that are too high and/or too long cause enzyme denaturation and loss of essential enzyme activity.

19.2.5. Number of Cycles

The number of cycles used in PCR depends on the degree of amplification required and the need to amplify selectively the target DNA sequence. Too many cycles can increase the amount and complexity of nonspecific background products. Too few cycles give low product yield. The initial amount of target DNA and the sensitivity of the detection method are critical in determining the number of cycles. Twenty cycles theoretically could produce 1 million copies of a target sequence ($2^{20} = 10^6$), but because of reaction inefficiencies 30 to 40 cycles are usually used, which should produce at least a 10^6-fold increase in the concentration of the target DNA sequence.

19.2.6. Specificity Enhancement

To improve specificity, at least one critical reactant (for example, *Taq* DNA polymerase) can be omitted until after the reaction is heated to 94 to 95°C (3). This "hot-start" method is based on the fact that nonspecific priming and subsequent production of unwanted amplified DNA bands are generally produced as a result of the retention of considerable enzymatic activity at temperatures below the optimum for DNA synthesis. Therefore, in the initial heating step of the PCR, primers that anneal nonspecifically to a partially single-stranded template region can be extended and stabilized before the reaction reaches 72°C for extension of specifically annealed primers. If the DNA polymerase is activated only after the reaction has reached high (>70°C) temperatures, nontarget amplification can be minimized. This can be done by manual addition of an essential reagent (e.g., DNA polymerase) to the reaction tube at elevated temperatures. This approach improves specificity and minimizes the formation of primer dimers.

19.2.7. Contamination Control

Because of the amplification power of PCR, it is critical to avoid even a trace of contamination with DNA containing the target sequence. Contamination of the

PCR amplification with products of a previous PCR, cross-contamination between samples, contamination with exogenous nucleic acids from the laboratory environment, and contamination even from the skin of the laboratory personnel can create a false-positive result. As general preventive measures, precautions such as prealiquoting reagents, using pipettes exclusively for specific steps of the reaction, using positive-displacement pipettes, using tips with barriers to prevent contamination of the pipette barrel, and physically separating the amplification reaction preparation from the area of amplified-product analysis can minimize the possibility of such contamination (70, 89). Addition of multiple negative control reactions without any target DNA is essential as quality control with every set of PCRs to reveal and monitor possible contamination.

Several methods have been developed to prevent PCR amplification of contaminants (35, 58, 60, 74). The reaction mixture can be exposed to UV irradiation (300 or, preferably, 254 nm) for 5 to 20 min prior to addition of actual target DNA and the *Taq* DNA polymerase. This causes a 10^5- to 10^6-fold reduction in contaminating DNA that could be amplified (106).

Contaminating DNA can be photochemically modified with psoralen or isopsoralen (58, 60). The reaction mixture is incubated prior to the addition of the template or primer DNA with a psoralen (e.g., 8-methoxypsoralen) at a final concentration of 460 μM in the dark for 30 min to 12 h; then the reaction mixture is exposed to UV light (365 nm) for 1 h.

Contaminating DNA can also be reduced by treating reaction mixtures with 0.5 to 1.0 U of DNase I or 10 to 20 U of the restriction enzymes that recognize short (e.g., 4-bp) sequences. The template DNA and the *Taq* DNA polymerase are added after appropriate inactivation of the DNase I or restriction endonucleases.

Another approach that can be used is to make contaminating PCR products susceptible to degradation (Fig. 4). This method involves substituting dUTP for dTTP in the PCR mix and treating subsequent PCR amplifications with uracil *N'*-glycosylase (UNG) (74). From 1 to 3 U of UNG is added to the reaction, and the sample is incubated at room temperature (18 to 22°C) for 10 to 15 min, decontaminating dU-containing PCR products from the previous amplification. The sample is then incubated at 96°C for 10 min to inactivate the UNG, to cleave the dU-containing contaminating PCR products, and to denature the templated DNA for PCR amplification. When using this method it is important that the primer-annealing temperature be above 55°C because UNG is active below 55°C; this avoids the risk of degradation of the newly synthesized dU-containing PCR products by residual UNG. At the end of the PCR amplification, the samples should be held at 70 to 72°C, rather than at 4°C, until they are removed from the thermal cycler; they then should be transferred immediately to −20°C or an equal volume of chloroform should be added to prevent degradation of dU-containing PCR products by residual UNG.

19.3. PROCEDURAL VARIATIONS

In addition to the amplification of a target DNA sequence by the typical PCR procedures already described, several specialized types of PCR have been developed for specific applications.

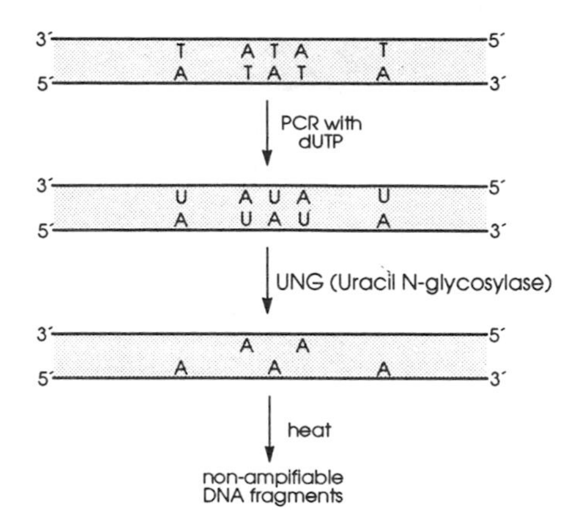

FIG. 4. Diagram showing the method for preventing carryover from previous PCRs so that actual target DNA in a sample will be amplified but contaminating DNA from a previous PCR will not. During PCR, dUTP is included in the reaction mixture so that uracil instead of the normal thymine is incorporated into the PCR-produced DNA. Prior to another PCR, the reaction mixture is treated with UNG and heated to digest any DNA containing uracil. The resulting single-stranded DNA fragments are not amplifiable with the primers for the target DNA. Target DNA not produced by PCR, which does not contain uracil, is not digested and is amplified by the PCR reaction. dUPT is incorporated again during that PCR, so that carryover contamination can again be eliminated.

19.3.1. Repetitive PCR

Nested sets of primers can be used to improve PCR yield of the target DNA sequence (Fig. 5). PCR with nested primers is performed for 15 to 30 cycles with one primer set and then for an additional 15 to 30 cycles with a second primer set for an internal region of the first amplified DNA product (44). The first primer set is removed by centrifugation through Centricon 100 molecular filters or by using limiting amounts of the initial primers, before the second primer set is used. The second round of PCR may be performed with the same primer set, or with "nested" primers designed to recognize regions within the initial amplified region (7, 44). This provides an additional level of specificity and increases amplification efficiency by minimizing nonspecific primer annealing.

One method for increasing the success of specific amplification of environmental DNA is to perform a number of PCR cycles, to dilute the sample (1:10 to 1:200), and then to perform an additional 10 to 30 PCR cycles (7, 93). This process may effectively dilute potential inhibitors to an acceptable level and allow successful amplification. The diluted DNA will also contain relatively high ratios of target sequences to total, nontarget, background DNA.

A diphasic amplification strategy called booster PCR can be used to improve the efficiency of amplifying target sequences that are present in small numbers (98). Initial PCR cycles are performed with a low concentration of primers (2.5 to 83 pM); after 20 PCR cycles, primer concentrations are increased to 0.1 μM and up to 50 more PCR cycles are performed. This procedure minimizes the formation of primer dimers, which may

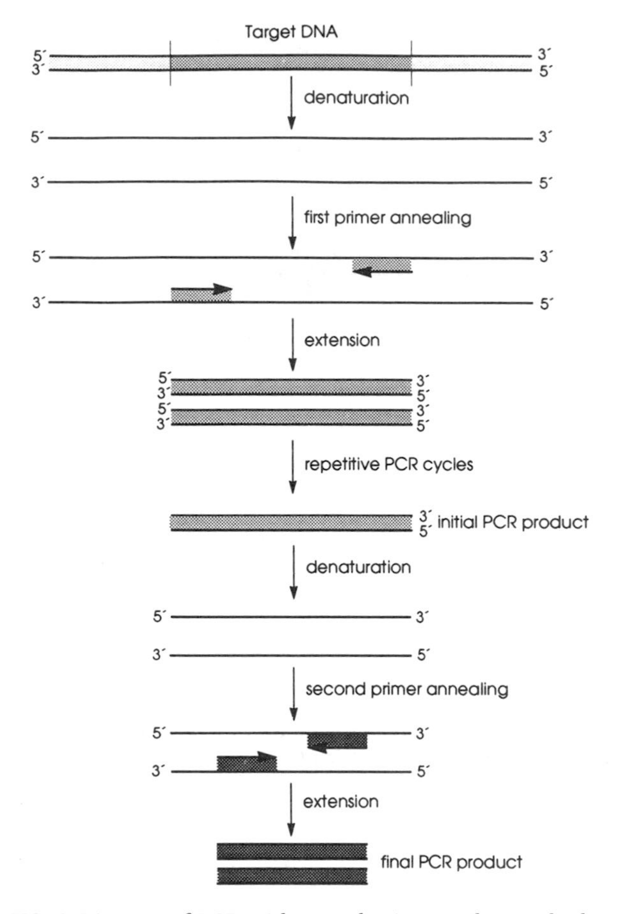

FIG. 5. Diagram of PCR with nested primers. This method is used to increase the fidelity of the reaction. The first PCR is run with a set of primers that produce a product larger than the eventual end product. The stringency of the amplification conditions used in this first PCR generally is low; this is achieved by using a relatively low annealing temperature. A second PCR is run with a set of primers that anneal to an internal region of the first PCR product. The second PCR is run with high stringency so that the fidelity of producing the final PCR product, which is smaller than the initial PCR product, is high.

result from primer excess (101) in the early PCR cycles and compete for polymerase molecules.

19.3.2. Multiplex PCR

Several DNA segments can be simultaneously amplified by using multiple pairs of primers (22). For such "multiplex" PCR amplification, primers are chosen with similar T_a values. A difference of $\pm 10°C$ in the T_a values of the two sets of primers may lead to differential amounts of amplified products, and often there is no visible amplification for one or the other target. The lengths of the target DNAs should be similar; large differences in the lengths of target DNAs will favor the amplification of the shorter target over the longer one, and this will result in differential yields of amplified products. In cases requiring different T_a values, two sets of primers, and/or variable target DNA lengths, staggered DNA amplification and variable amounts of primers can be used to achieve equal amounts of amplified products (16).

19.3.3. RNA Target PCR

To use PCR for the amplification of a target RNA sequence, a DNA copy (cDNA) must be synthesized by using reverse transcriptase before the PCR is begun. Avian myeloblastosis virus (AMV) and Moloney murine leukemia virus polymerases often are used to make cDNA from the target RNA. *Taq* polymerase also has reverse transcriptase activity and can be used to make cDNA (61), and *Tth* has reverse transcriptase activity, particularly in the presence of $MnCl_2$ at high temperatures (3, 80).

The RACE (rapid amplification of cDNA ends) protocol generates cDNAs by PCR amplification of the region between a single point in the transcript and the 3′ or 5′ end (33) (Fig. 6). For the 3′ end, a "hybrid" primer containing 17 oligo(dT) residues linked to a unique 17-mer primer (adapter) is used for reverse transcription. The adapter primer, together with another unique primer, is used for subsequent PCR amplification. For the 5′ end, cDNA is synthesized by using a gene-specific primer. The first-strand reaction product has a poly(A) tail. Then PCR amplification is performed with the hybrid primer and a gene-specific primer. One of the problems encountered with this method is nonspecific amplification. This can be minimized by raising the annealing temperature and choosing primers with similar melting temperatures.

A typical protocol for RACE amplification of cDNA 3′ ends is as follows.

1. Mix 2 μl of 10× reverse transcriptase buffer, 10 U of RNasin, 0.5 μg of dT_{17}-adapter primer, and 10 U of AMV reverse transcriptase in a total volume of 3.5 μl to form the reverse transcription mixture.
2. Heat 16.5 μl of target RNA at 65°C for 3 min, and then cool to 5°C.
3. Mix target RNA and reverse transcription mixture, and incubate for 1 h at 42°C and then for 30 min at 52°C.
4. Dilute with 1 ml of TE buffer, and maintain at 4°C.
5. Perform PCR as described in sections 19.1 and 19.2 with the dT_{17}-adapter primer and a second gene-specific primer.

A typical protocol for RACE for amplification of cDNA 5′ ends is as follows.

1. Mix 2 μl of 10× reverse transcriptase buffer, 10 U of RNasin, 0.5 μg of gene-specific primer, and 10 U of AMV reverse transcriptase in a total volume of 3.5 μl to form the reverse transcription mixture.
2. Heat 16.5 μl of target RNA at 65°C for 3 min, and then cool to 5°C.
3. Mix target RNA and reverse transcription mixture, and incubate for 1 h at 42°C and then for 30 min at 52°C.
4. Remove excess primer by passage through a Centricon 100 spin filter.
5. Dilute with 2 μl of 0.1× TE buffer, and maintain at 4°C.
6. Concentrate to 10 μl.
7. Add 4 μl of 5× tailing buffer (Bethesda Research Laboratories), 4 μl of 1 mM dATP, and 10 U of terminal deoxyribonucleotidyl transferase (TdT).
8. Incubate at 37°C for 5 min and then at 65°C for 5 min.
9. Dilute to 0.5 μl with TE buffer.
10. Perform PCR as described in sections 19.1 and 19.2 with the dT_{17}-adapter primer and the gene-specific primer.

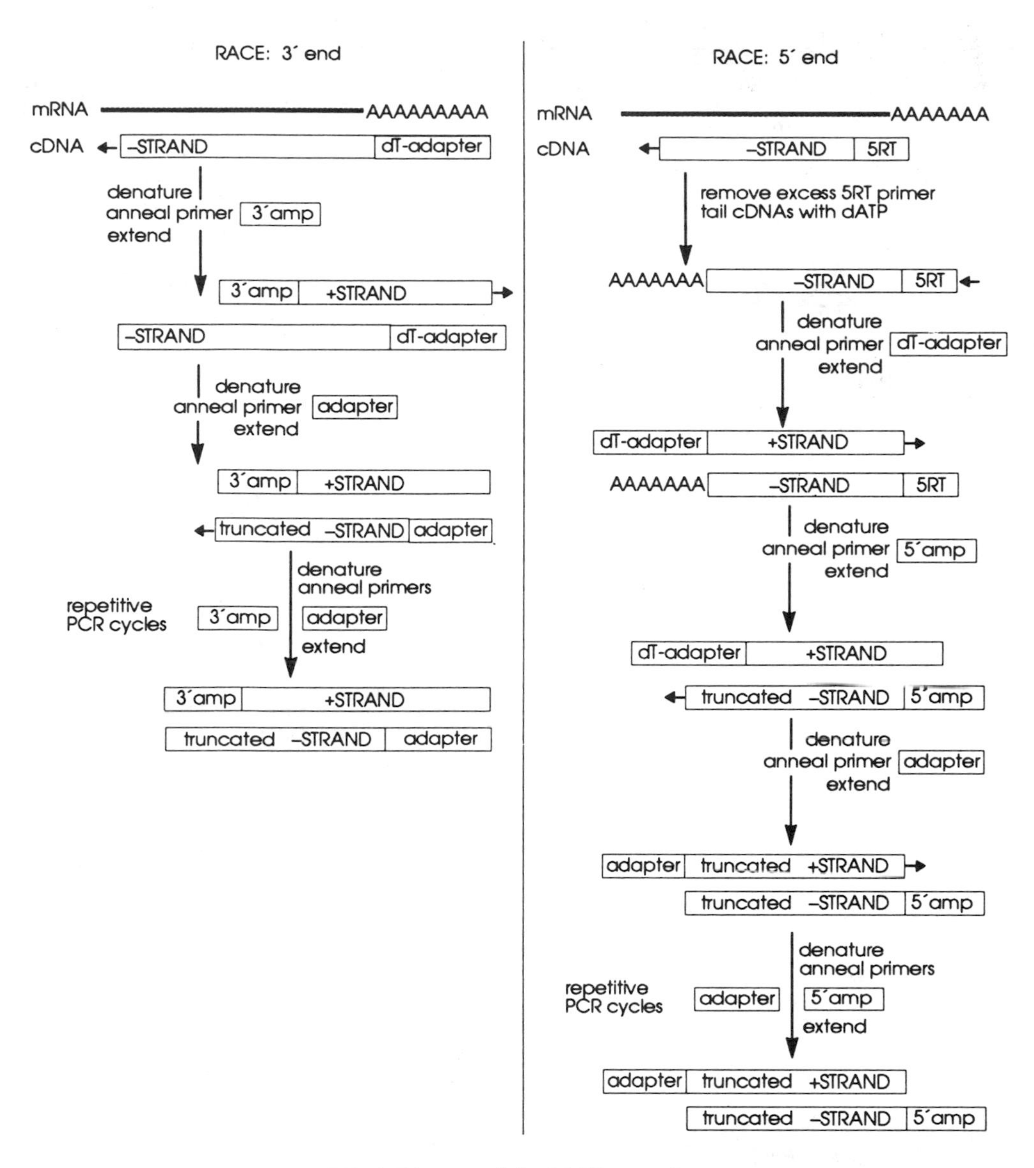

FIG. 6. Diagram of the RACE procedure.

19.3.4. Inverse PCR

Inverse PCR permits the amplification of DNA flanking a region of known sequence (86). In this reaction the DNA is synthesized outward from the primer pairs rather than inward between the two primers (Fig. 7). The source DNA is cleaved with restriction enzyme and circularized with DNA ligase before amplification. The primers used for inverse PCR are synthesized in the opposite orientations to those used in normal PCR. Upstream and/or downstream segments are thus amplified.

19.3.5. Membrane-Bound PCR

Binding the target DNA to a membrane is useful when there are limited amounts of template DNA or when the DNA is contaminated with large amounts of nontarget DNA or impurities that inhibit PCR (68). The DNA can be purified by electrophoresis and then blotted, or it can be blotted and then washed prior to PCR. The efficiency of membrane-bound PCR is somewhat lower than that of solution PCR.

19.3.6. Expression Cassette PCR

In expression cassette PCR (ECPCR), a contiguous coding sequence is inserted between sequences that direct high-level protein biosynthesis in *E. coli* (75) (Fig. 8). The gene expression cassettes obtained by ECPCR are inserted in a regulated overexpression plasmid, which is then transferred into competent *E. coli* cells by transformation. ECPCR permits the generation of proteins with N- and/or C-terminal truncations by modifying the 5′ or 3′ end of a coding sequence.

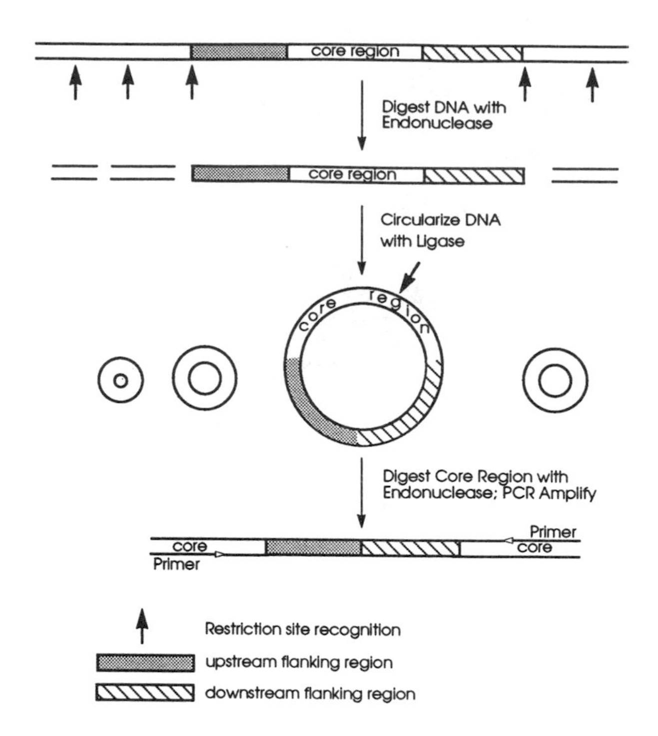

FIG. 7. Diagram of inverse PCR. DNA is digested with endonuclease to form a DNA fragment with a core region and both upstream and downstream flanking regions. The linear DNA is circularized with a ligase. The circular DNA is then linearized by cutting with an endonuclease that has a unique recognition site within the core region; the endonuclease used here is different from the original endonuclease used to digest the DNA. The core and flanking DNA regions are then amplified by PCR with primers that anneal to the core regions that are now present at the ends of the DNA segment.

19.3.7. Quantitative PCR

Quantitative estimates of a target DNA based on PCR products are difficult because small differences in any of the parameters that affect the efficiency of amplification can dramatically affect the outcome of the reaction. The target DNA can be quantified, however, by coamplifying target DNA in the presence of known quantities of a defined standard DNA (37, 38, 124) that serves as an internal standard. Because the primers are the same for the target DNA and the internal standard DNA, the two DNAs should be amplified with the same efficiency, provided that the amplification of the competitive DNA is not affected by the presence of the intervening sequences. The ratio of the known amount of internal standard added to the amount after PCR amplification gives the efficiency of the PCR and permits calculation of the starting concentration of the target DNA sequence from the measured amount of target DNA after PCR. This can be expressed as follows.

$$\frac{\text{[Target DNA before PCR]}}{\text{[Target DNA after PCR]}} = \frac{\text{[Internal standard added]}}{\text{[Internal standard after PCR]}}$$

19.3.8. Arbitrarily Primed PCR

Fingerprints of genomes can be generated by using a variant of PCR called arbitrarily primed PCR (126). The resulting pattern of PCR products can be used to differentiate specific bacterial strains. Arbitrarily primed PCR uses primers chosen without regard to the sequences of the genome to be fingerprinted and involves several cycles (typically two) of low stringency followed by PCR at higher stringency. The low stringency is achieved by using a low annealing temperature, such as 30 to 40°C, and the high stringency is achieved by using annealing temperatures close to 60°C.

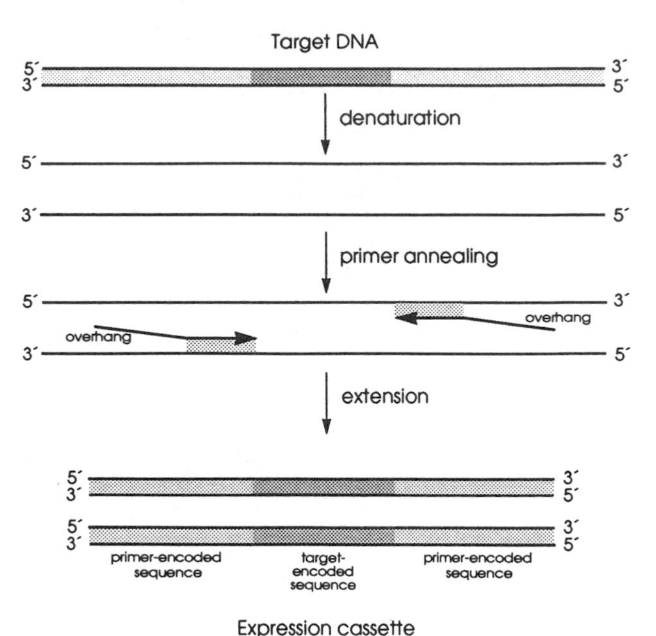

FIG. 8. Diagram of expression cassette PCR. The primers used in this procedure differ from those used in normal PCR in having extra (overhang) sections. The extra sections contain specific endonuclease recognition sites so that the amplified DNA segment can be cut as a cassette for cloning. Additionally, sequences are incorporated via the primers that enhance gene expression, for example, strong regulated promoters and efficient transcription terminators. In this manner, when the PCR-amplified DNA is cloned, it results in overexpression of encoded genes.

19.4. APPLICATIONS

19.4.1. Gene Cloning

PCR provides a relatively simple method for cloning genes (34, 102, 107, 135). By amplifying gene sequences, PCR produces sufficient amounts of a target gene to increase the probability of successful cloning of the target gene. Besides simply making sufficient amounts of a gene for cloning, PCR can be used to make new gene sequences (53, 134), to add expression sequences (e.g., promoters) (75, 129), and to insert or delete sequences into DNA for gene cloning (50, 65).

PCR can be used for gene cloning even when only limited or virtually no specific sequence information is known (64, 97, 116, 120, 127). Degenerate primers can be used so that every combination of nucleotides that could code for a given amino acid sequence can be generated and used to amplify the target DNA (28, 39, 73,

128, 135). Deoxyinosine can be inserted into the primers to establish degenerate primers (67, 92).

Several parameters are important for DNA amplification with degenerate primers (27, 73, 113). Generally, short oligomeric primers (≤20 nucleotides) with a limited degeneracy (<64-fold) appear to provide the most satisfactory amplification results, but longer primers (>30 nucleotides) with greater degeneracies (>500-fold) also provide adequate amplification in some instances (27, 114). When short degenerate primers are used, annealing temperatures higher than the calculated T_m of the primers may produce sufficient amplification (76). With primers of 17 to 20 nucleotides, great degrees of degeneracy may be acceptable provided the first three nucleotides on the 3′-end are a perfect match (114). The first few cycles of PCR can be performed by using lower annealing temperature to ensure priming, and later cycles can then be performed at a higher annealing temperature. Alternately, the ramping rate from annealing temperature to extension temperature can be decreased to allow for adequate priming (27).

19.4.2. DNA Sequencing

PCR products either can be cloned prior to sequencing or can be sequenced directly. Because *Taq* DNA polymerase has no 3′-to-5′ exonuclease ("proofreading") activity but has 5′-to-3′ exonuclease activity, misincorporation sometimes occurs during polymerization. The misincorporation rate has been estimated at about 10^{-4} per cycle; it can be reduced to 10^{-5} per cycle by lowering the dNTP and $MgCl_2$ concentrations, by raising the annealing temperature, and by using shorter extension times. Because *Taq* polymerase incorporates one incorrect nucleotide per 10^{-4} to 10^{-5} base additions, several independent isolates may have to be analyzed to determine the correct sequence.

By using different amounts of primers, it is possible to generate single-stranded DNA by PCR (130). This method is called asymmetric PCR and is likely to revolutionize the DNA-sequencing method by saving time, accuracy, and effort (40, 81). Asymmetric PCR, however, is less efficient than conventional PCR, and more cycles must be run to achieve maximum yield.

Two examples of PCR-based approaches for sequencing are genomic amplification with transcript sequencing (GAWTS) (118) (Fig. 9) and RNA amplification with transcript sequencing (RAWTS) (104, 105) (Fig. 10). The four steps involved in the procedures are (i) cDNA synthesis for RAWTS (genomic DNA is used in GAWTS), (ii) PCR in which either or both primers contain a phage promoter, (iii) transcription with a T7 phage polymerase, and (iv) dideoxy sequencing with reverse transcriptase. The attachment of a T7 phage promoter onto at least one of the PCR primers is required in these procedures. The transcription step involving the T7 phage polymerase generates an abundance of single-stranded template for reverse transcriptase-mediated dideoxy sequencing. An end-labeled reverse transcriptase primer complementary to the desired sequence generates the additional specificity needed to generate unambiguous sequence data.

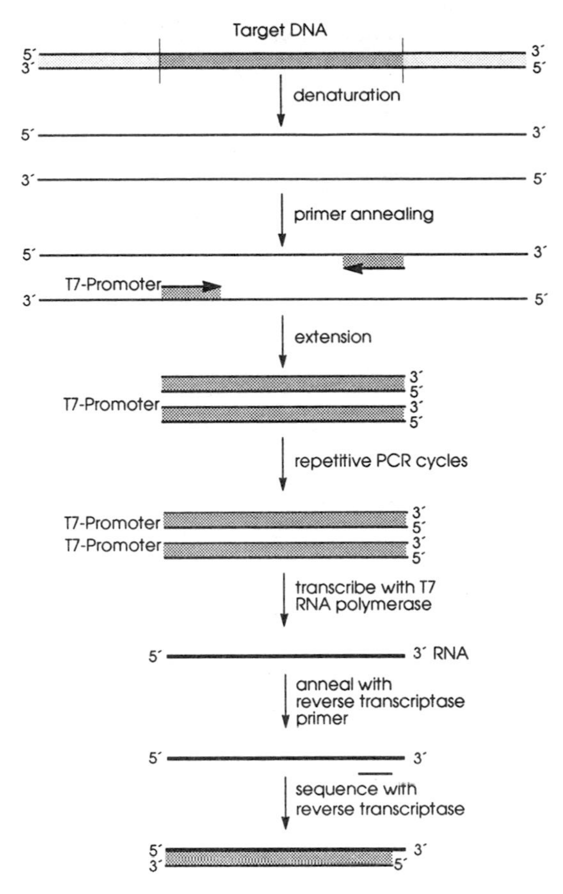

FIG. 9. Diagram of GAWTS procedure, using PCR for gene sequencing starting with target DNA. The primer includes a terminal T7 phage promoter so that, after the PCR amplification of the target DNA, an RNA polymerase can be used to transcribe and to make single-stranded RNA transcripts. Reverse transcriptase is then used to make single-stranded DNA for sequencing.

PCR can be used for direct sequencing of DNA (69). The general procedure for direct sequencing of PCR-amplified DNA is as follows.

1. Amplify the target DNA and generate single-stranded template DNA by asymmetric PCR or by the affinity capture method.
2. Purify the amplified DNA, if required.
3. Distribute 5 μl of appropriate termination mixture in a microtiter well or in microcentrifuge tubes.
4. Combine 10 μl of PCR-amplified DNA, 2 μl of 5× sequencing reaction buffer, 1 μl of radiolabeled primer (0.5 to 1.0 pmol), 2 μl of *Taq* DNA polymerase (diluted to 1 U/μl), and distilled water to 22 μl.
5. Mix well by gentle vortexing, and centrifuge briefly to collect the liquid, if any, from the wall of the tube.
6. Transfer 5 μl of the reaction mix to the termination mix, and then incubate at 72°C for 5 min.
7. Cool the reaction mixture to room temperature, and add 4 μl of stop solution with electrophoresis dye. Store at −20°C until used.
8. Denature the samples by heating at 65 to 70°C before loading in sequencing gel.

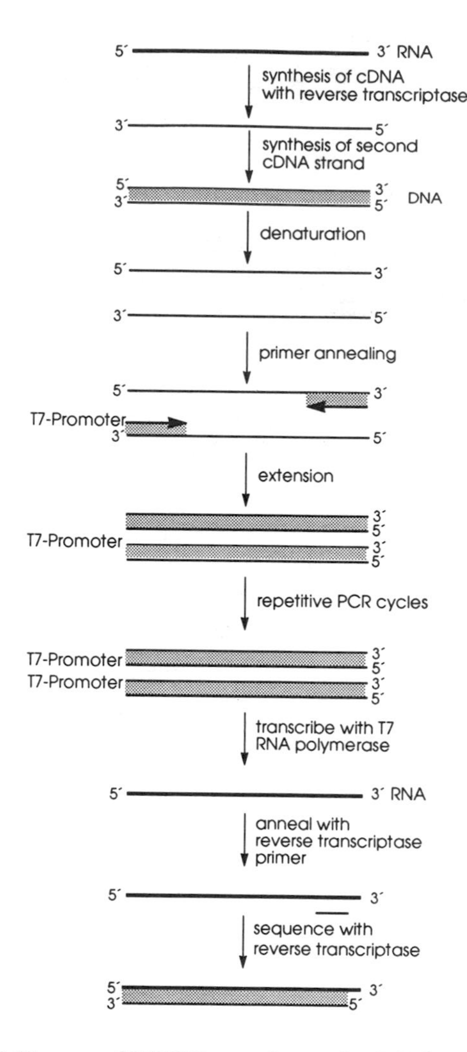

FIG. 10. Diagram of RAWTS procedure, using PCR for sequencing a target RNA. This procedure is identical to GAWTS except that the original target is RNA and the cDNA for PCR amplification is formed by using reverse transcriptase.

The use of PCR to generate single-stranded DNA and the rapid amplification of a segment of a specific gene segment has facilitated obtaining nucleotide sequence information. Single-stranded template DNA generation is required for applying the chain termination method of DNA sequencing (103). DNA sequencing templates are readily generated by asymmetric PCR (41), in which one strand is preferentially amplified by limiting the concentration of one of the two primers. Primer ratios of 50:1 to 100:1 are most commonly used to generate single-stranded products. The primer which is complementary to the template DNA to be sequenced is kept in larger quantity. During the first 15 to 20 cycles, double-stranded DNA accumulates in an exponential fashion until limited by depletion of one of the primers. The single-stranded DNAs continue to act as templates for the unexhausted primer, resulting in a total production of single-stranded DNA template.

Purification of the asymmetric-PCR products from unincorporated deoxynucleotides and other unwanted materials may be required before sequence analysis is performed. This step is generally performed by using a commercially available microconcentrator (Centricon 100, Amicon, or Millipore Spin Column) or selective alcohol precipitation as follows.

1. Add an equal volume of 4 M ammonium acetate to the amplified DNA (e.g., in 100 μl of PCR sample, add 100 μl of 4 M ammonium acetate).
2. Add an equal volume (200 μl) of isopropanol, mix well by inverting the tube several times, and incubate at room temperature for 10 min.
3. Centrifuge in a microcentrifuge for 10 min at 12,000 $\times$ g.
4. Carefully pour off the supernatant without disturbing the pellet.
5. Wash the pellet with 70% alcohol, centrifuge, discard the alcohol, and dry the pellet in vacuo.
6. Dissolve the dry DNA pellet in TE (pH 8.0) buffer, and use it for the DNA sequence reaction.

The single-stranded DNA templates may also be purified by affinity capture if they are labeled with biotin at their 5′ end. Following amplification in which one of the primers is biotinylated, the double-stranded amplified products are passed through a column containing a streptavidin ligand binder to which the biotinylated primer-derived DNA strands will bind (Fig. 11). The nonbiotinylated complementary strands are denatured by NaOH treatment and eluted with elution buffer. This

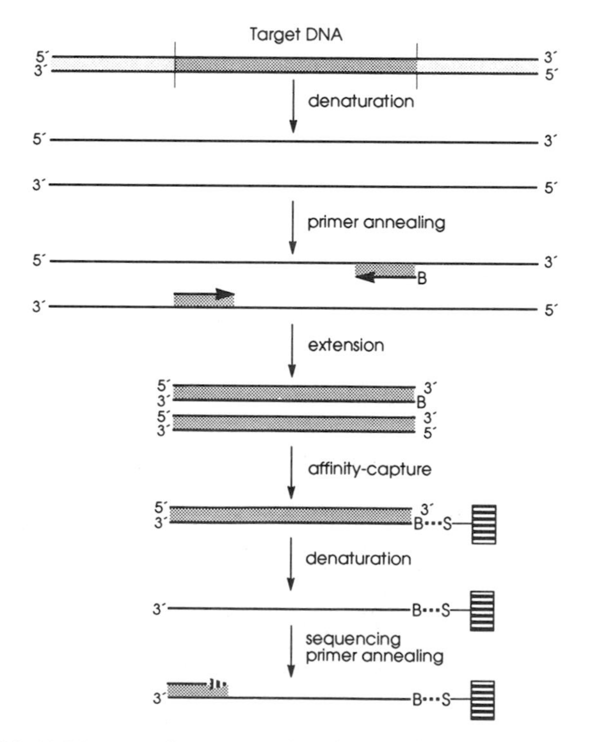

FIG. 11. Diagram of PCR procedure for direct DNA sequencing. One of the primers is biotinylated (B). After PCR amplification, the DNA strand with the incorporated biotin is captured by streptavidin (S) bound to a solid support matrix. This procedure of affinity chromatography is useful for the recovery of purified single-stranded DNA for sequencing. A DNA-sequencing primer is then used along with the chain termination reaction to determine the base sequence of the target DNA segment.

eluted non-biotin-labeled DNA can be used as template for DNA sequence analysis. Alternatively, the biotinylated affinity matrix-bound DNA strands can be eluted and used for DNA sequence analysis.

The disadvantage of this method is that the amplified DNA must be the specific target gene or the DNA segment of interest, since nonspecific amplified DNA products will be labeled with biotin and a mixture of heterologous DNAs will result. These problems can be minimized or eliminated by designing primers with high T_m values, optimizing amplification specificity, and/or using nested primer amplification.

Sequencing of the double-stranded amplified DNA of the desired gene or the DNA segment can also be performed. The advantage of this approach is that both strands can be sequenced simultaneously in two separate sequencing reactions (19, 62). Unused primers, dNTPs, and other materials are removed from the PCR sample by using a Centricon 100, selective alcohol precipitation, gel purification, enzymatic digestion of the gel matrix with agarase, or HPLC (19).

PCR amplification conditions for the removal of "ghost" DNA, which may interfere with the sequencing reaction, and for an increase in sequencing band resolution are as follows (19) (i) The first temperature cycle consists of an initial denaturation step for 5 min at 94°C followed by 30 s of annealing at 55°C and 30 s of extension at 72°C. (ii) Each of the additional 24 cycles is performed with 30 s at 94°C, 30 s at 55°C, and 1 min at 72°C followed by 5 min for the last extension. (iii) After purification, the standard chain termination method (103) is used with PCR primers or nested primers for sequence analysis of the amplified DNA segment.

DNA sequencing can also be performed from asymmetrically amplified DNA directly from a bacterial colony, without purification of the DNA (51). Bacteria are transferred from a single colony into 40 μl of PCR mixture containing 1× PCR buffer (10 mM Tris·HCl [pH 9.0], 50 mM KCl, 1.5 mM $MgCl_2$, 0.01% gelatin, 0.1% Triton X−100), 20 pmol of upstream primer, and 0.2 pmol downstream primer (the limiting primer for asymmetric PCR). The mixture is then heated to 95°C for 15 min and cooled quickly in ice. Thermostable DNA polymerase is then added to the sample, and PCR amplification is performed with appropriate PCR parameters. The DNA sequencing is then performed with the limiting primer (here, the downstream primer) radiolabeled at the 5′ ends. In this method the amplified DNA need not be purified from unused dNTPs, primers, etc. However, purification of the amplified DNA may be required if expected results are obtained. This approach is fast and produces the same resolution as with the purified DNA.

19.4.3. Genome Mapping

Physical mapping of the genomes of some bacteria can be achieved by repetitive extragenic palindrome (REP) PCR methods, which take advantage of the repetitive sequences present in the genome of the bacteria (119). This method allows one to map insertion mutations rapidly by using the insertion element as one primer-binding site and the REP sequence as the other genomically anchored primer-binding site for PCR amplification. The unique sequence spanned by the amplification product can be used as a probe to identify genomic regions corresponding to the insertion site (119, 122). The REP element has been predicted to make up to 1% of the genome of *E. coli*. This approximately 35-bp palindromic sequence has been identified in the intercistronic regions of numerous bacterial operons, and its putative stem-loop structure has been proposed to decrease expression of downstream genes while stabilizing transcription of upstream sequences. When an outward-facing primer corresponding to either half of the REP palindrome and a Tn*5*-specific primer are used with chromosomal DNA from the Tn*5* insertion mutant strain as a template, PCR amplification products contain unique sequences that lie between the transposon and adjacent REP site(s) as well as between REP sequences themselves.

By using PCR conditions to favor amplification from Tn*5* to REP and by comparing with an amplification involving REP primers alone, the resulting Tn*5*-REP amplification products can be identified readily. These DNA fragments are then used as probes for the mutated locus by screening an ordered genomic library of *E. coli* or the organisms of interest. This REP-Tn*5* PCR-based technique can be applied for genome mapping and characterization of insertion mutation studies in *E. coli*, *Salmonella* spp., and other organisms for which repeat elements such as REP and ERIC have been defined. REP- and ERIC-like sequences have been found in a diverse group of eubacterial species. As new repeat sequences are discovered in diverse bacterial species, the utility of this method is likely to increase.

19.4.4. Mutagenesis

PCR can be used for site-directed mutagenesis (84, 85). Primers with one or several nucleotide base differences from the wild-type sequence can be incorporated into the PCR product during amplification. The altered DNA sequences are then cleaved with restriction enzymes and ligated in place of the similarly digested wild-type sequences (48, 63).

Inverse PCR can also be used to create site-specific mutation in a target DNA (47). In this method, an entire circular plasmid can be amplified and the mutation can be generated at any specific place in the plasmid. Two primers are located back-to-back on the opposing DNA strands of the template. One primer contains one or more mismatches which will generate the desired site-specific mutation in the target DNA.

By using four different primers, site-specific mutations can be performed with linear double-stranded DNA (85) (Fig. 12). One primer in each primer set contains a single or multiple base mismatches to direct mutagenesis at the desired locations on the target DNA. The other primers are designed for selective amplification of the mutated sequence. Two PCR products from different sections of the same template or from two different templates are made such that the resultant fragments overlap in sequence. These products are mixed, denatured, and allowed to reanneal. One of the heteroduplex forms containing the DNA strands that overlap at the 3′ ends is extended to form a complete double-stranded DNA. By using primers specific at their extended 5′ ends, amplifications of the complete DNA molecules are made with the desired mutation at the overlapping joining point.

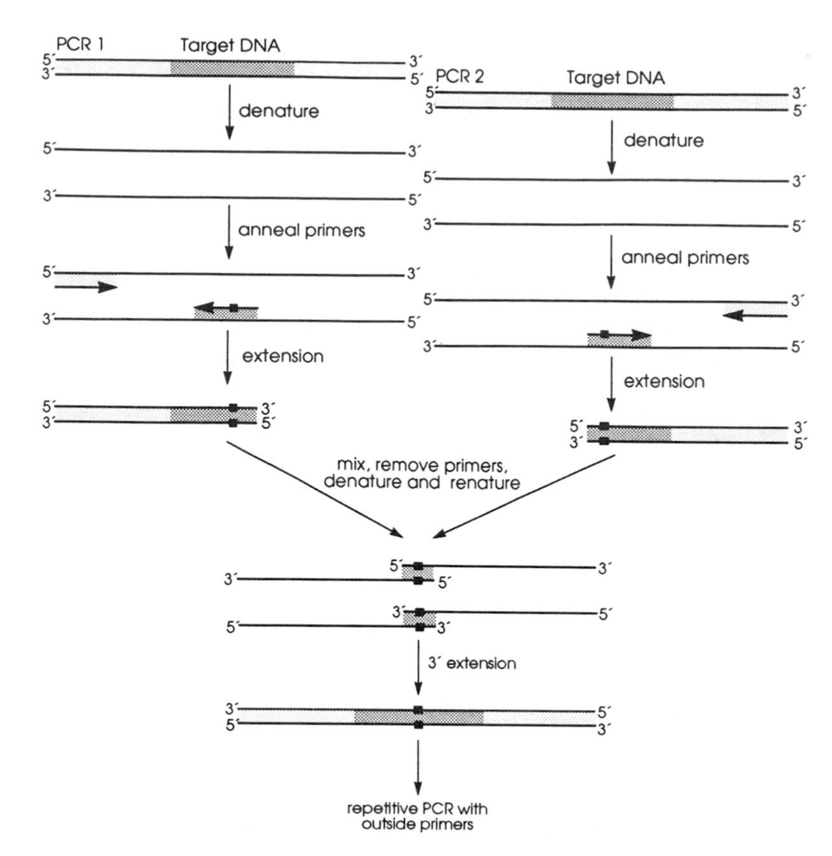

FIG. 12. Diagram of PCR procedure for generating a site-specific mutation. The procedure uses two sets of primers. One of the primers in each set has a specific mismatch with the target DNA, and this results in a change at a specific site in the PCR-amplified DNA product. This is shown as a black square in the diagram.

An analogous procedure can be used to insert DNA sequences (Fig. 13). In this method the primers contain additional noncomplementary regions. Insertional mutagenesis can also be performed by using two sequential PCR steps (65). In the first step, a sequence from the template is amplified with a downstream primer consisting of the sequence to be inserted flanked by short sequences complementary to the template. In the second step, 5 μl (approximately 300 ng) of each of the amplified DNAs from the first PCR amplification can be used for priming. Since the overlap necessary to effect the combination need not exist in the natural template but can be made as add-on sequences to the primers, PCR can be used to create specific site substitutions, deletions, and insertions at nearly any position in a target DNA sequence.

19.4.5. Diagnostic Bacteriology

PCR is useful for the rapid detection of pathogens, especially those whose in vitro cultivation is difficult or not possible (6). The list of pathogens that can be diagnosed by PCR is increasing rapidly (Table 1). There are several advantages to using PCR for their diagnosis. PCR can be applied to fixed tissues (frozen or formalin fixed) (109), reducing the potential dangers involved in transport and handling of specimens with live virulent pathogens. Diagnosis of diseases can be made in hours rather than days, as required by some other methods. Specimens can be examined directly without culturing pathogens, so that diagnoses can be accomplished rapidly and safely. Although blood can interfere with PCR, the sensitivity of PCR permits dilution of samples so that this interference is eliminated.

PCR is so sensitive that the presence of pathogens can be detected very early in an infection. Only a few cells of the pathogen in a specimen are necessary to yield a positive diagnosis. PCR also has great specificity. By using PCR, pathogens with specific virulence factors can be specifically diagnosed. Both invasive factors and toxins can be detected. Multiplex PCR permits simultaneous screening for multiple pathogens that might be causing a disease condition. In cases of multiple-agent infections, all the pathogens causing the disease can be detected in a single reaction.

19.4.6. Environmental Monitoring

PCR can usefully be applied to environmental samples (13, 117). Water samples can be filtered, and the filters can be placed directly into the PCR mixture. For this purpose, Teflon filters are used; if other filters are used, the cells must be washed from the filters and the filters must be removed from the reaction (15). Soils and sediments contain humic acids that interfere with PCR. Differential centrifugation to remove cells from soil and/or sediment particles or cleanup procedures such as cesium chloride centrifugation and hydrox-

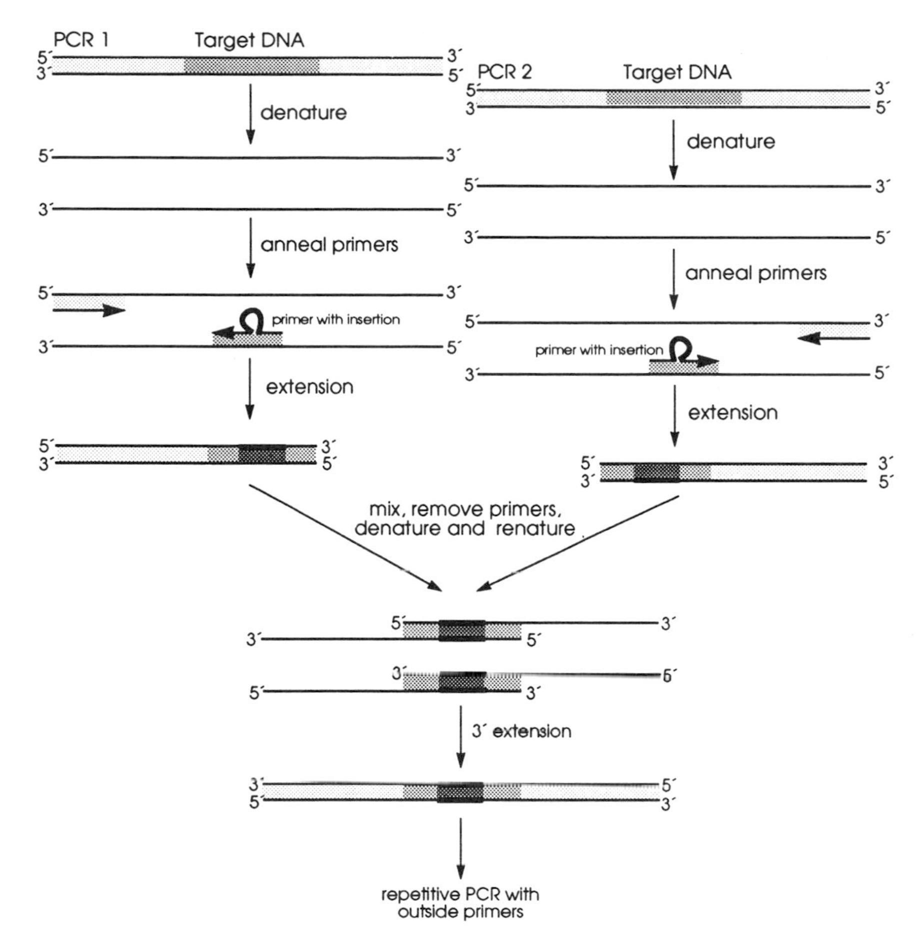

FIG. 13. Diagram of PCR procedure for generating insertional mutation. The procedure uses two sets of primers. One of the primers in each set has a specific additional segment of DNA. The additional segment is shown as a darkened line in the DNA. The DNA that is amplified by PCR contains the additional sequence coded for by the additional segments of the primers.

yapatite chromatography may be required; the use of polyvinylpyrrolidone is useful in removing interfering humics (116).

PCR can be used to monitor engineered microorganisms in soil and water. Amplification of a 1.3-kb probe-specific region of DNA from the recombinant 2,4,5-T herbicide-degrading bacterium *Pseudomonas cepacia* AC1100 can detect a concentration of 1 target organism per g of soil against a background of 10^9 diverse nontarget organisms (116). PCR can also be used to detect genetically engineered *E. coli* containing a marker gene sequence (23).

PCR is useful for the identification of pathogens and indicator species for environmental surveillance (11, 17). PCR amplification and gene probe detection of regions of *lacZ, lamB,* and *uidA* genes can be used to detect the coliform bacterium *E. coli* as well as the enteric pathogens *Shigella* and *Salmonella* species in environmental waters (12, 18, 26). PCR provides the necessary specificity and sensitivity for monitoring the bacteriological quality of water to ensure the safety of water supplies.

Legionella species, particularly the human pathogen *Legionella pneumophila,* can be detected in environmental water sources by using PCR and gene probes (77, 115). All *Legionella* species, including all 15 serogroups of *L. pneumophila,* are detectable by PCR. Both viable culturable and viable nonculturable cells of *L. pneumophila,* resulting from exposure to hypochlorite, show positive PCR amplification, whereas nonviable cells do not (14). To confirm the presence of live *L. pneumophila,* brief treatment with chloramphenicol can be used to inhibit protein synthesis; the mRNA for *L. pneumophila* macrophage infectivity potentiator (*mip*) can be copied to cDNA followed by PCR amplification, and, since *mip* mRNA has a half-life less than 1 min, the presence of live *L. pneumophila* can be confirmed (78).

In environmental monitoring it is often important to detect the activities rather than the simple presence of organisms. At the genetic level this can be accomplished by measuring gene expression. Amplification of mRNA rather than DNA templates allows the analysis of gene expression. This approach requires the conversion of the mRNA template to cDNA with reverse transcriptase and oligonucleotide primers. *Taq* DNA polymerase has some reverse transcriptase activity and can be used for the production of cDNA and subsequent DNA amplification. *Tth* DNA polymerase, in particular, can be used for reverse transcription and the cDNA can be used for PCR in a single reaction tube.

Table 1. Some bacterial pathogens that can be diagnosed by PCR

Pathogen	Disease	Reference(s)
Aeromonas hydrophila	Gastroenteritis	55, 96
Bordetella pertussis	Whooping cough	54
Borrelia burgdorferi	Lyme disease	94
Chlamydia trachomatis	Trachoma, urethritis	52
Chlamydia psittaci	Parrot fever	52
Chlamydia pneumoniae	Pneumonia	52
Clostridium difficile	Pseudomembranous enterocolitis	66, 133
Escherichia coli (enterotoxigenic)	Traveler's diarrhea	32, 87, 88, 95, 123
Helicobacter pylori	Peptic ulcers, gastritis	24, 25, 31, 43, 49, 121
Legionella pneumophila	Legionnaires' disease	77, 114
Mycobacterium leprae	Leprosy	45
Mycobacterium tuberculosis	Tuberculosis	20, 29, 30, 108, 111
Mycoplasma genitalium	Urethritis	112
Rickettsia typhi	Typhus	125
Treponema pallidum	Syphilis	21, 46
Shigella flexneri	Dysentery	32, 71
Vibrio cholerae	Cholera	91, 110

19.5. REFERENCES

19.5.1. General References

1. **Bej, A. K., M. H. Mahbubani, and R. M. Atlas.** 1991. Amplification of nucleic acids by polymerase chain reaction (PCR) and other methods and their applications. *Crit. Rev. Mol. Biol.* **26:**301–334.
A comprehensive review of studies conducted with PCR and other methods of nucleic acid amplification; includes over 200 references.
2. **Erlich, H. A. (ed.).** 1989. *PCR Technology: Principles and Applications for DNA Amplification.* Stockton Press, New York.
An excellent collection of articles on the various applications of PCR and its underlying principles.
3. **Erlich, H. A., D. Gelfand, and J. J. Sninsky.** 1991. Recent advances in the polymerase chain reaction. *Science* **252:**1643–1651.
A timely update discussing several advances in PCR technology that overcome previous problems associated with PCR, such as excessive primer dimer formation, poor thermal cycler performance, and contamination due to PCR product carryover.
4. **Gibbs, R. A.** 1990. DNA amplification by the polymerase chain reaction. *Anal. Chem.* **62:**1202–1214.
A thorough review of PCR.
5. **Innis, M. A., D. H. Gelfand, J. J. Sninsky, and T. J. White (ed.).** 1990. *PCR Protocols: A Guide to Methods and Applications.* Academic Press, Inc., San Diego, Calif.
A comprehensive collection of articles on PCR, the articles provide protocols for various uses of PCR.
6. **Lynch, J. R., and J. M. Brown.** 1990. The polymerase chain reaction: current and future clinical applications. *J. Med. Genet.* **27:**2–7.
A review of the applications of PCR for clinical diagnoses, including diseases caused by bacterial pathogens.
7. **Mullis, K. B., and F. A. Faloona.** 1987. Specific synthesis of DNA *in vitro* via a polymerase-catalyzed chain reaction. *Methods Enzymol.* **155:**335–351.
A detailed description of the enzymatic basis of PCR.
8. **Mullis, K. B.** 1990. The unusual origin of the polymerase chain reaction. *Sci. Am.* **262**(4):56–65.
A well-illustrated and very readable overview of PCR by its inventor.
9. **Saiki, R. K., D. H. Gelfand, S. Stoffel, S. J. Scharf, R. Higuchi, G. T. Horn, K. B. Mullis, and H. A. Erlich.** 1988. Primer-directed enzymatic amplification of DNA with a thermostable DNA polymerase. *Science* **239:**487–494.
The classic scientific report of PCR.
10. **White, T. J., N. Arnheim, and H. A. Erlich.** 1989. The polymerase chain reaction. *Trends Genet.* **5:**185–189.
A review of PCR and its applications in genetics.

19.5.2. Specific References

11. **Atlas, R. M., and A. K. Bej.** 1990. Detecting bacterial pathogens in environmental water samples by using PCR and gene probes, p. 399–407. *In* M. A. Innis, D. H. Gelfand, J. J. Sninsky, and T. J. White (ed.), *PCR Protocols: A Guide to Methods and Applications.* Academic Press, Inc., San Diego, Calif.
12. **Bej, A. K., J. L. DiCesare, L. Haff, and R. M. Atlas.** 1991. Detection of *Escherichia coli* and *Shigella* spp. in water by using the polymerase chain reaction and gene probes for *uid. Appl. Environ. Microbiol.* **57:**1013–1017.
13. **Bej, A. K., and M. H. Mahbubani.** 1992. Applications of the polymerase chain reaction in environmental microbiology. *PCR Methods Applic.* **1:**151–159.
14. **Bej, A. K., M. H. Mahbubani, and R. M. Atlas.** 1991. Detection of viable *Legionella pneumophila* in water by polymerase chain reaction and gene probe methods. *Appl. Environ. Microbiol.* **57:**597–600.
15. **Bej, A. K., M. H. Mahbubani, J. DiCesare, and R. M. Atlas.** 1991. PCR-gene probe detection of microorganisms using filter-concentrated samples. *Appl. Environ. Microbiol.* **57:**3529–3534.
16. **Bej, A. K., M. H. Mahbubani, R. Miller, J. L. DiCesare, L. Haff, and R. M. Atlas.** 1990. Multiplex PCR amplification and immobilized capture probes for detection of bacterial pathogens and indicators in water. *Mol. Cell. Probes* **4:**353–365.
17. **Bej, A. K., S. C. McCarty, and R. M. Atlas.** 1991. Detection of coliform bacteria and *Escherichia coli* by multiplex polymerase chain reaction: comparison with defined substrate and plating methods for water quality monitoring. *Appl. Environ. Microbiol.* **57:**2429–2432.
18. **Bej, A. K., R. J. Steffan, J. DiCesare, L. Haff, and R. M. Atlas.** 1990. Detection of coliform bacteria in water by polymerase chain reaction and gene probes. *Appl. Environ. Microbiol.* **56:**307–314.
19. **Bevan, I. S., R. Rapley, and M. R. Walker.** 1992. Sequencing of PCR amplified DNA. *PCR Methods Applic.* **1:**222–228.

20. **Brisson-Noel, A., D. Lecossier, X. Nassif, B. Gecquel, V. Levy-Frebault, and A. J. Hance.** 1989. Rapid diagnosis of tuberculosis by amplification of mycobacterial DNA in clinical samples. *Lancet* **ii:**1069–1071.
21. **Burstain, J. M., E. Grimprel, S. A. Lukehart, M. V. Norgard, and J. D. Radolf.** 1991. Sensitive detection of *Treponema pallidum* by using the polymerase chain reaction. *J. Clin. Microbiol.* **29:**62–69.
22. **Chamberlain, J. S., R. A. Gibbs, J. E. Ranier, P. N. Nguyen, and C. T. Cashey.** 1988. Deletion screening of the Duchenne muscular dystrophy locus via multiplex DNA amplification. *Nucleic Acids Res.* **16:**11141–11156.
23. **Chaudhry, G. R., G. A. Toranzos, and A. R. Bhatti.** 1989. Novel method for monitoring genetically engineered microorganisms in the environment. *Appl. Environ. Microbiol.* **55:**1301–1304.
24. **Clayton, C. L., H. Kleanthous, P. J. Coates, D. D. Morgan, and S. Tabaqchali.** 1992. Sensitive detection of *Helicobacter pylori* by using polymerase chain reaction. *J. Clin. Microbiol.* **30:**192–200.
25. **Clayton, C. L., M. J. Pallen, H. Kleanthous, B. W. Wren, and S. Tabaqchali.** 1991. Molecular cloning and sequencing of *Helicobacter pylori* urease genes and detection of *H. pylori* using polymerase chain reaction technique, p. 74–78. *In* P. Malfertheiner and H. Ditschuneit (ed.), *Helicobacter pylori Gastritis and Peptic Ulcer.* Springer-Verlag, New York.
26. **Cleuziat, P., and J. Robert-Baudouy.** 1990. Specific detection of *Escherichia coli* and *Shigella* species using fragments of genes coding for β-glucuronidase. *FEMS Microbiol. Lett.* **72:**315–322.
27. **Compton, T.** 1990. Degenerate primers for DNA amplification, p. 39–45. *In* M. A. Innis, D. H. Gelfand, J. J. Sninsky, and T. J. White (ed.), *PCR Protocols: A Guide to Methods and Applications.* Academic Press, Inc., San Diego, Calif.
28. **Cooper, D. L., and N. Isola.** 1990. Full-length cDNA cloning utilizing the polymerase chain reaction, a degenerate oligonucleotide sequence and a universal mRNA primer. *BioTechniques* **9:**60–64.
29. **De Wit, D., L. Steyn, S. Shoemaker, and M. Sogin.** 1990. Direct detection of *Mycobacterium tuberculosis* in clinical specimens by DNA amplification. *J. Clin. Microbiol.* **28:**2437–2441.
30. **Eisenbach, K. D., J. T. Crawford, and J. H. Bates.** 1989. Repetitive DNA sequences as probes for *Mycobacterium tuberculosis. J. Clin. Microbiol.* **26:**2240–2246.
31. **Foxall, P. A., L. Hu, and H. L. T. Mobley.** 1992. Use of polymerase chain reaction-amplified *Helicobacter pylori* urease structural genes for differentiation of isolates. *J. Clin. Microbiol.* **30:**739–741.
32. **Frankel, G., J. A. Giron, J. Vallmassoi, and G. K. Schoolnik.** 1989. Multi-gene amplification: simultaneous detection of three virulence genes in diarrhoeal stool. *Mol. Microbiol.* **3:**1729–1734.
33. **Frohman, M. A., M. K. Dush, and G. R. Martin.** 1988. Rapid production of full-length cDNAs from rare transcripts: amplification using a single gene-specific oligonucleotide primer. *Proc. Natl. Acad. Sci. USA* **85:**8998–9002.
34. **Fuqua, S. A. W., S. D. Fitzgerald, and L. M. McGuire.** 1990. A simple polymerase chain reaction method for detection and cloning of low-abundance transcripts. *BioTechniques* **9:**206–211.
35. **Furrer, B., U. Candrian, P. Wieland, and J. Luthy.** 1990. Improving PCR efficiency. *Nature* (London) **346:**324.
36. **Gelfand, D. H., and T. J. White.** 1990. Thermostable DNA polymerases, p. 129–141. *In* M. A. Innis, D. H. Gelfand, J. J. Sninsky, and T. J. White (ed.), *PCR Protocols: A Guide to Methods and Applications.* Academic Press, Inc., San Diego, Calif.
37. **Gilliland, G., S. Perrin, K. Blanchard, and H. F. Bunn.** 1990. Analysis of cytokine mRNA and DNA: detection and quantitation by competitive polymerase chain reaction. *Proc. Natl. Acad. Sci. USA* **87:**2725–2729.
38. **Gilliland, G., S. Perrin, and H. F. Bunn.** 1990. Competitive PCR for quantitation of mRNA, p. 60–69. *In* M. A. Innis, D. H. Gelfand, J. J. Sninsky, and T. J. White (ed.), *PCR Protocols: A Guide to Methods and Applications.* Academic Press, Inc., San Diego, Calif.
39. **Girgis, S. I., M. Alevizaki, P. Denny, G. J. M. Ferrier, and S. Legon.** 1988. Generation of DNA probes for peptides with highly degenerate codons using mixed primer PCR. *Nucleic Acids Res.* **16:**10371.
40. **Gyllensten, U.** 1989. Direct sequencing of in vitro amplified DNA, p. 45–60. *In* H. A. Erlich (ed.), *PCR Technology.* Stockton Press, New York.
41. **Gyllensten, U. B., and H. A. Ehrlich.** 1988. Generation of single-stranded DNA by polymerase chain reaction and its application to direct sequencing of the HLA-DQA locus. *Proc. Natl. Acad. Sci. USA* **85:**7652–7656.
42. **Haff, L., J. G. Atwood, J. DiCesare, E. Katz, E. Picozza, J. F. Williams, and T. Woudenberg.** 1991. A high-performance system for automation of the polymerase chain reaction. *BioTechniques* **10:**102–112.
43. **Hammar, M., T. Tyszkiewicz, T. Wadstrom, and P. W. O'Toole.** 1992. Rapid detection of *Helicobacter pylori* in gastric biopsy material by polymerase chain reaction. *J. Clin. Microbiol.* **30:**54–58.
44. **Haqqi, T. M., G. Sarkar, C. S. David, and S. S. Sommer.** 1988. Specific amplification with PCR of a refractory segment of genomic DNA. *Nucleic Acids Res.* **16:**11844.
45. **Hartskeerl, R. A., M. Y. L. De Wit, and P. R. Klatser.** 1989. Polymerase chain reaction for the detection of *Mycobacterium leprae. J. Gen. Microbiol.* **135:**2357–2364.
46. **Hay, P. E., J. R. Clark, R. A. Strugnell, D. Taylor-Robinson, and D. Goldmeier.** 1990. Use of the polymerase chain reaction to detect DNA sequences specific to pathogenic treponemes in cerebrospinal fluid. *FEMS Microbiol. Lett.* **68:**233–238.
47. **Hemsley, A., N. Arnheim, M. D. Toney, G. Cortopassi, and D. J. Galas.** 1989. A simple method for site-directed mutagenesis using the polymerase chain reaction. *Nucleic Acids Res.* **17:**6545–6551.
48. **Higuchi, R., B. Krummel, and R. K. Siaki.** 1988. A general method of in vitro preparation and specific mutagenesis of DNA fragments: study of protein and DNA interactions. *Nucleic Acids Res.* **16:**7351–7367.
49. **Ho, S., J. A. Hoyle, F. A. Lewis, et al.** 1991. Direct polymerase chain reaction test for detection of *Helicobacter pylori* in humans and animals. *J. Clin. Microbiol.* **29:**2543–2549.
50. **Ho, S. N., H. D. Hunt, R. M. Horton, J. K. Pullen, and L. R. Pease.** 1989. Site-directed mutagenesis by overlap extension using the polymerase chain reaction. *Gene* **77:**51–57.
51. **Hoffman, M. A., and D. A. Brian.** 1991. Sequencing PCR DNA amplified directly from a bacterial colony. *BioTechniques* **11:**30–31.
52. **Holland, S. M., C. A. Gaydos, and T. C. Quinn.** 1990. Detection and differentiation of *Chlamydia trachomatis, Chlamydia psittaci,* and *Chlamydia pneumoniae* by DNA amplification. *J. Infect. Dis.* **162:**984–987.
53. **Horton, R. M., H. D. Hunt, S. N. Ho, J. K. Pullen, and L. R. Pease.** 1989. Engineering hybrid genes without the use of restriction enzymes: gene splicing by overlap extension. *Gene* **77:**61–68.
54. **Houard, S., C. Hackel, A. Herzog, and A. Bollen.** 1989. Specific identification of *Bordetella pertussis* by the polymerase chain reaction. *Res. Microbiol.* **140:**477–487.
55. **Howard, S. P., W. J. Garland, M. J. Green, and J. T. Buckley.** 1987. Nucleotide sequence of the gene for the hole-forming toxin aerolysin of *Aeromonas hydrophila. J. Bacteriol.* **169:**2869–2871.
56. **Innis, M. A., and D. H. Gelfand.** 1990. Optimization of PCRs, p. 3–12. *In* M. A. Innis, D. H. Gelfand, J. J. Sninsky, and T. J. White (ed.), *PCR Protocols: A Guide to Methods and Applications.* Academic Press, Inc., San Diego, Calif.
57. **Innis, M. A., K. B. Myambo, D. H. Gelfand, and D. Brow.** 1988. DNA sequencing with *Thermus aquaticus* DNA

polymerase and direct sequencing of polymerase chain reaction-amplified DNA. *Proc. Natl. Acad. Sci. USA* **85**:9436–9440.

58. **Isaacs, S. T., J. W. Tessman, K. C. Metchette, J. E. Hearst, and G. D. Cimino.** 1991. Post-PCR sterilization: development and application to an HIV-1 diagnostic assay. *Nucleic Acids Res.* **19**:109–116.
59. **Jeffreys, A. J., V. Wilson, R. Neumann, and J. Keyte.** 1988. Amplification of human minisatellites by the polymerase chain reaction: towards DNA fingerprinting of single cells. *Nucleic Acids Res.* **16**:10953.
60. **Jinno, Y., K. Yoshiiura, and N. Niikawa.** 1990. Use of psoralen as extinguisher of contaminated DNA in PCR. *Nucleic Acids Res.* **18**:6739.
61. **Jones, M. D., and N. S. Foulkes.** 1989. Reverse transcription of mRNA by *Thermus aquaticus* DNA polymerase. *Nucleic Acids Res.* **17**:8387–8388.
62. **Jung, M., A. Dritscholi, and U. Kasid.** 1992. Reliable and efficient direct sequencing of PCR amplified double-stranded genomic DNA template. *PCR Methods Applic.* **1**:171–174.
63. **Kadowaki, H., T. Kadowaki, F. E. Wondisford, and S. I. Taylor.** 1989. Use of the polymerase chain reaction catalyzed by *Taq* DNA polymerase for site-specific mutagenesis. *Gene* **76**:161–166.
64. **Kalman, M., E. T. Kalman, and M. Cashel.** 1990. Polymerase chain reaction (PCR) with a single specific primer. *Biochem. Biophys. Res. Commun.* **167**:504–506.
65. **Kammann, M., J. Laufs, J. Schell, and B. Gronenborn.** 1989. Rapid insertional mutagenesis of DNA by polymerase chain reaction (PCR). *Nucleic Acids Res.* **17**:5404.
66. **Kato, N., C.-Y. Ou, H. Dato, S. Bartley, V. K. Brown, V. R. Dowell, and K. Ueno.** 1991. Identification of toxigenic *Clostridium difficile* by the polymerase chain reaction. *J. Clin. Microbiol.* **29**:33–37.
67. **Knoth, K., S. Roberts, C. Poteet, and M. Tamkun.** 1988. Highly degenerate, inosine-containing primers specifically amplify rare cDNA using the polymerase chain reaction. *Nucleic Acids Res.* **16**:10932.
68. **Kodokami, Y., and R. V. Lewis.** 1990. Membrane bound PCR. *Nucleic Acids Res.* **18**:3082.
69. **Kusukawa, N., T. Uemori, and I. Kato.** 1990. Rapid and reliable protocol for direct sequencing of material amplified by the polymerase chain reaction. *BioTechniques* **9**:66–72.
70. **Kwok, S., and R. Higuchi.** 1989. Avoiding false positives with PCR. *Nature* (London) **339**:237–238.
71. **Lampel, K. A., J. A. Jagow, M. Trucksess, and E. E. Hill.** 1990. Polymerase chain reaction for detection of invasive *Shigella flexneri* in food. *Appl. Environ. Microbiol.* **56**:1536–1540.
72. **Lawyer, F. C., S. Stoffel, R. K. Saiki, K. Myambo, R. Drummond, and D. H. Gelfand.** 1989. Isolation, characterization and expression in *Escherichia coli* of the DNA polymerase gene from *Thermus aquaticus*. *J. Biol. Chem.* **264**:6427–6437.
73. **Lee, C. C., and C. T. Caskey.** 1990. cDNA cloning using degenerate primers, p. 47–53. *In* M. A. Innis, D. H. Gelfand, J. J. Sninsky, and T. J. White (ed.), *PCR Protocols: A Guide to Methods and Applications.* Academic Press, Inc., San Diego, Calif.
74. **Longo, M. C., M. S. Berninger, and J. L. Hartley.** 1990. Use of uracil DNA glycosylase to control carry-over contamination in polymerase chain reaction. *Gene* **93**:125–128.
75. **MacFerrin, K. D., M. P. Terranova, S. L. Schreiber, and G. L. Verdine.** 1990. Overproduction and dissection of proteins by the expression-cassette polymerase chain reaction. *Proc. Natl. Acad. Sci. USA* **87**:1937–1941.
76. **Mack, D. H., and J. J. Sninsky.** 1988. A sensitive method for the identification of uncharacterized viruses related to known virus groups: hepadvirus model system. *Proc. Natl. Acad. Sci. USA* **85**:6977– 6981.
77. **Mahbubani, M. H., A. K. Bej, R. Miller, L. Haff, J. DiCesare, and R. M. Atlas.** 1990. Detection of *Legionella* with polymerase chain reaction and gene probe methods. *Mol. Cell. Probes* **4**:175–187.
78. **Mahbubani, M. H., A. J. Bej, R. D. Miller, R. M. Atlas, J. L. DiCesare, and L. A. Haff.** 1991. Detection of bacterial mRNA using polymerase chain reaction. *BioTechniques* **10**:48–49.
79. **Mathur, E. J., K. S. Lunberg, D. D. Shoemaker, K. B. Nielson, S. A. Mathur, B. R. Scott, and J. A. Sorge.** 1991. Characterization of thermostable DNA polymerase with high fidelity. *Genome Sequence Conference III.* Rockville, Md. American Chemical Society, Washington, D.C.
80. **Mayers, T. W., and D. H. Gelfand.** 1991. Reverse transcription and DNA amplification by a Thermus thermophilus DNA polymerase. *Biochemistry* **30**:7661–7666.
81. **McCabe, P. C.** 1990. Production of single-stranded DNA by asymmetric PCR, p. 76–83. *In* M. A. Innis, D. H. Gelfand, J. J. Sninsky, and T. J. White (ed.), *PCR Protocols: A Guide to Methods and Applications.* Academic Press, Inc., San Diego, Calif.
82. **Mezei, L. M.** 1990. Effect of oil overlay on PCR amplification. *Amplifications* **4**:11.
83. **Mullis, K., F. Faloona, S. Scharf, R. Saiki, G. R. Horn, and H. Ehrlich.** 1986. Specific enzymatic amplification of DNA *in vitro:* the polymerase chain reaction. *Cold Spring Harbor Symp. Quant. Biol.* **51**:263–273.
84. **Nassal, M., and A. Rieger.** 1990. PCR-based site-directed mutagenesis using primers with mismatched 3′-ends. *Nucleic Acids Res.* **18**:3077–3078.
85. **Nelson, R. M., and G. L. Long.** 1989. A general method of site-specific mutagenesis using a modification of the *Thermus aquaticus* polymerase chain reaction. *Anal. Biochem.* **180**:147–151.
86. **Ochman, H., J. W. Ajioka, D. Garza, and D. L. Hartl.** 1989. Inverse polymerase chain reaction, p. 105–111. *In* H. A. Erlich (ed.), *PCR Technology.* Stockton Press, New York.
87. **Olive, D. M.** 1989. Detection of enterotoxigenic *Escherichia coli* after polymerase chain reaction amplification with a thermostable DNA polymerase. *J. Clin. Microbiol.* **27**:261–265.
88. **Olive, D. M., A. I. Atta, and S. K. Setti.** 1988. Detection of toxigenic *Escherichia coli* using biotin-labelled DNA probes following enzymatic amplification of the heat labile toxin gene. *Mol. Cell. Probes* **2**:47–57.
89. **Orrego, C.** 1990. Organizing a laboratory for PCR work, p. 447–454. *In* M. A. Innis, D. H. Gelfand, J. J. Sninsky, and T. J. White (ed.), *PCR Protocols: A Guide to Methods and Applications.* Academic Press, Inc., San Diego, Calif.
90. **Oste, C.** 1989. Optimization of magnesium concentration in the PCR reaction. *Amplifications* **1**:10.
91. **Palmer, L. M., and R. R. Colwell.** 1991. Detection of luciferase gene sequence in nonluminescent *Vibrio cholerae* by colony hybridization and polymerase chain reaction. *Appl. Environ. Microbiol.* **57**:1286–1293.
92. **Patil, R. V., and E. E. Dekker.** 1990. PCR amplification of an *Escherichia coli* gene using mixed primers containing deoxyinosine at ambiguous positions in degenerate amino acid codons. *Nucleic Acids Res.* **18**:3080.
93. **Paul, J. H., L. Cazares, and J. Thurmond.** 1990. Amplification of the *rbcL* gene from dissolved and particulate DNA from aquatic environments. *Appl. Environ. Microbiol.* **56**:1963–1966.
94. **Pershing, D. H., S. R. Telford, A. Speilman, and S. W. Barthold.** 1990. Detection of *Borrelia burgdorferi* infection in *Ixodes dammini* ticks with the polymerase chain reaction. *J. Clin. Microbiol.* **28**:566–572.
95. **Pollard, D. R., W. M. Johnson, H. Lior, D. Tyler, and K. R. Rozee.** 1990. Rapid and specific detection of verotoxin genes in *Escherichia coli* by the polymerase chain reaction. *J. Clin. Microbiol.* **28**:540–545.
96. **Pollard, D. R., W. M. Johnson, H. Lior, D. Tyler, and K. R. Rozee.** 1990. Detection of the aerolysin gene in *Aeromonas hydrophila* by the polymerase chain reaction. *J. Clin. Microbiol.* **28**:2477–2481.
97. **Roux, K. H., and P. Dhanarajan.** 1990. A strategy for single site PCR amplification of dsDNA: priming digested

cloned or genomic DNA from an anchor-modified restriction site and a short internal sequence. *BioTechniques* **8:**48–57.

98. **Ruano, G., W. Fenton, and K. K. Kidd.** 1989. Biphasic amplification of very dilute DNA samples via "booster" PCR. *Nucleic Acids Res.* **17:**5407.
99. **Rychilk, W., and R. E. Rhoads.** 1989. A computer program for choosing optimal oligonucleotides for filter hybridization, sequencing and *in vitro* amplification of DNA. *Nucleic Acids Res.* **17:**8543–8551.
100. **Saiki, R.** 1990. Amplification of genomic DNA, p. 13–20. *In* M. A. Innis, D. H. Gelfand, J. J. Sninsky, and T. J. White (ed.), *PCR Protocols: A Guide to Methods and Applications.* Academic Press, Inc., San Diego, Calif.
101. **Saiki, R. K.** 1989. The design and optimization of the polymerase chain reaction, p. 7–22. *In* H. A. Erlich (ed.), *PCR Technology.* Stockton Press, New York.
102. **Saiki, R. K., P. S. Walsh, C. H. Leverson, and H. A. Erlich.** 1989. Genetic analysis of amplified DNA with immobilized sequence-specific oligonucleotide probes. *Proc. Natl. Acad. Sci. USA* **86:**6230–6234.
103. **Sanger, F., F. Nicklen, and A. R. Coulson.** 1977. DNA sequencing with chain-terminating inhibitors. *Proc. Natl. Acad. Sci. USA* **74:**5463–5467.
104. **Sarkar, G., and S. S. Sommer.** 1988. RNA amplification with transcript sequencing (RAWTS). *Nucleic Acids Res.* **16:**5197.
105. **Sarkar, G., and S. S. Sommer.** 1989. Access to a messenger RNA sequence or its protein product is not limited by tissue or species specificity. *Science* **244:**331–334.
106. **Sarkar, G., and S. S. Sommer.** 1990. More light on PCR contamination. *Nature* (London) **347:**340–341.
107. **Scharf, S. J., G. T. Horn, and H. A. Erlich.** 1986. Direct cloning and sequence analysis of enzymatically amplified genomic sequences. *Science* **233:**1076–1078.
108. **Shibata, D., W. J. Martin, M. D. Appleman, D. M. Causey, J. M. Leedom, and N. Arnheim.** 1988. Detection of cytomegalovirus DNA in peripheral blood of patients infected with human immunodeficiency virus. *J. Infect. Dis.* **158:**1185–1192.
109. **Shirai, H., M. Nishibuchi, K. T. Ramamurthy, S. K. Bhattacharya, S. C. Pal, and Y. Takeda.** 1991. Polymerase chain reaction for detection of the cholera enterotoxin operon of *Vibrio cholerae. J. Clin. Microbiol.* **29:**2517–2521.
110. **Sjobring, U., M. Mecklenburg, A. B. Andersen, and H. Miorner.** 1990. Polymerase chain reaction for detection of *Mycobacterium tuberculosis. J. Clin. Microbiol.* **28:**2200–2204.
111. **Skov Jensen, J., S. A. Uldum, J. Sondergard-Andersen, J. Vuust, and K. Lind.** 1991. Polymerase chain reaction for detection of *Mycoplasma genitalium* in clinical samples. *J. Clin. Microbiol.* **29:**46–50.
112. **Smith, K. T., C. M. Long, B. Bowman, and M. M. Manos.** 1990. Using cosolvents to enhance PCR amplification. *Amplifications* **5:**16.
113. **Sommer, R., and D. Tautz.** 1989. Minimal homology requirements for PCR primers. *Nucleic Acids Res.* **17:**6749.
114. **Starnbach, M. N., S. Falkow, and L. S. Tompkins.** 1989. Species specific detection of *Legionella pneumophila* in water by DNA amplification and hybridization. *J. Clin. Microbiol.* **27:**1257–1261.
115. **Steffan, R. J., and R. M. Atlas.** 1988. DNA amplification to enhance the detection of genetically engineered bacteria in environmental samples. *Appl. Environ. Microbiol.* **54:**2185–2191.
116. **Steffan, R. J., and R. M. Atlas.** 1991. Polymerase chain reaction: applications in environmental microbiology. *Annu. Rev. Microbiol.* **45:**137–161.
117. **Steigerwald, S. D., G. P. Pfeifer, and A. D. Riggs.** 1990. Ligation-mediated PCR improves the sensitivity of methylation analysis by restriction enzymes and detection of specific DNA strand breaks. *Nucleic Acids Res.* **18:**1435–1439.
118. **Stoflet, E. S., D. D. Koeberl, G. Sakar, and S. S. Sommer.** 1988. Genomic amplification with transcript sequencing. *Science* **239:**491–494.
119. **Subramanian, P. S., J. Versalovic, E. R. B. McCabe, and J. R. Lupski.** 1991. Rapid mapping of *Escherichia coli*:Tn5 insertion mutations by REP-Tn5 PCR. *PCR Methods Applic.* **1:**187–194.
120. **Timblin, C., J. Battey, and W. M. Kuehl.** 1990. Application for PCR technology to subtractive cDNA cloning: identification of genes expressed specifically in murine plasmacytoma cells. *Nucleic Acids Res.* **18:**1587–1593.
121. **Valentine, J. L., R. R. Arthur, H. L. T. Mobley, and J. D. Dick.** 1991. Detection of *Helicobacter pylori* by using the polymerase chain reaction. *J. Clin. Microbiol.* **29:**689–695.
122. **Versalovic, J., T. Koeuth, and J. R. Lupski.** 1991. Distribution of repetitive sequences in eubacteria and application of fingerprinting of bacterial genomes. *Nucleic Acids Res.* **19:**6823–6831.
123. **Victor, T., R. DuToit, J. VanZyl, A. J. Bester, and P. D. Van Helden.** 1991. Improved method for the routine identification of toxigenic *Escherichia coli* by DNA amplification of a conserved region of the heat-labile toxin A subunit. *J. Clin. Microbiol.* **29:**158–161.
124. **Wang, A. M., and D. F. Mark.** 1990. Quantitative PCR, p. 70–75. *In* M. A. Innis, D. H. Gelfand, J. J. Sninsky, and T. J. White (ed.), *PCR Protocols: A Guide to Methods and Applications.* Academic Press, Inc., San Diego, Calif.
125. **Webb, L., M. Carl, D. C. Malloy, G. A. Dasch, and A. F. Azad.** 1990. Detection of murine typhus infection in fleas by using the polymerase chain reaction. *J. Clin. Microbiol.* **28:**530–534.
126. **Welsh, J., and M. McClelland.** 1990. Fingerprinting genomes using PCR with arbitrary primers. *Nucleic Acids Res.* **18:**7213–7218.
127. **Wieland, I., G. Bolger, G. Asouline, and M. Wigler.** 1990. A method for difference cloning: gene amplification following subtractive hybridization. *Proc. Natl. Acad. Sci. USA* **87:**2720–2724.
128. **Wilks, A. F., R. R. Kurban, C. M. Hovens, and S. R. Ralph.** 1989. The application of the polymerase chain reaction to the cloning of the protein tyrosine kinase family. *Gene* **85:**67–74.
129. **Wilson, E. M., C. A. Franke, M. E. Black, and D. E. Hruby.** 1989. Expression vector pT7:TKII for the synthesis of authentic biologically active RNA encoding vaccinia virus thymidine kinase. *Gene* **77:**69–78.
130. **Wilson, R. K., C. Chen, and L. Hood.** 1990. Optimization of asymmetric polymerase chain reaction for rapid fluorescent DNA sequencing. *BioTechniques* **8:**184–189.
131. **Winship, P. R.** 1989. An improved method for directly sequencing PCR amplified material using dimethyl sulphoxide. *Nucleic Acids Res.* **17:**1266.
132. **Wittwer, C. T., G. C. Fillmore, and D. J. Garling.** 1990. Minimizing the time required for PCR amplification by efficient heat transfer to small samples. *Anal. Biochem.* **186:**328–331.
133. **Wren, B., C. Clayton, and S. Tabaqchali.** 1990. Rapid identification of toxigenic *Clostridium difficile* by polymerase chain reaction. *Lancet* **i:**423.
134. **Yon, J., and M. Fried.** 1989. Precise gene fusion by PCR. *Nucleic Acids Res.* **17:**4895.
135. **Zehr, J. P., and L. A. McReynolds.** 1989. Use of degenerate oligonucleotides for the amplification of the *nifH* gene from the marine cyanobacterium *Trichodesmium thiebautii. Appl. Environ. Microbiol.* **55:**2522–2526.

Chapter 20

Nucleic Acid Analysis

WILLIAM G. HENDRICKSON and TAPAN K. MISRA

This chapter describes several of the most common nucleic acid analyses performed in vitro to characterize a cloned DNA segment coding for a gene or genes. However, Southern hybridization, which is done to locate a cloned DNA segment within a physical map, is described in Chapter 16.3.4.2.b and will not be discussed here. DNA sequencing, described in section 20.1, is essential for many other methods and is often one of the first procedures done with a cloned DNA. In section 20.2 are described gel mobility shift, DNase protection and chemical-protection ("footprinting") assays, used to locate DNA-binding sites for regulatory proteins at promoter and control regions of genes. The precise 5′ end of the mRNA can be located by S1 nuclease protection or by primer extension methods, as described in section 20.3. The size of an mRNA is determined by doing a Northern blot hybridization, also described in section 20.3. Both the S1 protection and Northern (RNA) blot hybridization methods provide information on the abundance of mRNA for studies of gene regulation.

20.1. DNA SEQUENCING

Two basic methods are in use to determine the sequence of nucleotides in a DNA fragment. One of the methods is based on the ability of certain chemicals to cleave DNA at specific base positions (36, 37). End-labeled DNA is partially cleaved by using these chemicals to generate sequencing "ladders" (oligonucleotides of variable sizes). Another method (47) is based on the generation of sequencing ladders by using a DNA oligomer as a primer, the single-stranded form of the target DNA (ssDNA) as the template, a modified form of the *Escherichia coli* DNA polymerase I (**Klenow fragment**), deoxynucleoside triphosphates (dNTPs), and 2′,3′-dideoxynucleoside triphosphates (ddNTPs). The polymerase I extends the primer from the 5′ to the 3′ direction. The ddNTPs are analogs of dNTPs, but they lack a hydroxyl residue at the 3′ position of deoxyribose. When a ddNTP residue is incorporated into the growing chain of the newly synthesized DNA strand, it cannot form a phosphodiester bond with another dNTP, and consequently the DNA chain terminates. By using one specific ddNTP in a reaction mixture, the DNA chain is terminated at specific base positions, generating sequence (oligonucleotide) ladders of random sizes with a fixed 5′ terminus. Separate reactions are run for each of the four nucleotides. The DNA oligomers are separated by denaturing polyacrylamide gel electrophoresis. A maximum of 300 to 400 bases can be read accurately from one gel.

The enzymatic method of nucleotide sequencing is simpler than the chemical method of sequencing and therefore is included in this chapter. The enzymatic method is widely used for the determination of unknown sequences of large target DNA. During the early stages of the application of the enzymatic method, randomly generated small DNA fragments were used to generate a library of sequences. Overlapping sequences were determined and ordered by computer analysis. Methods of ordered ("nested") deletions from the 5′ end of the target DNA have since been developed and are now used for the determination of overlapping nonrandom sequences. Two different methods of generating nested deletions, using *Bal* 31 and exonuclease III enzymes, will be described in this chapter.

There is a tremendous potential for the application of DNA sequence information in biomedical science (basic and applied). Automated techniques of DNA sequencing that allow large-scale sequencing projects have been developed. All these techniques use the dideoxy-chain termination principle. Although automated procedures are still being refined, it is expected that they will be exclusively used for sequencing projects with large DNA in the near future. Nevertheless, simple procedures of nucleotide sequencing will be used in the laboratory for routine molecular biology, for example, promoter mapping by primer extension, determination of mutation in a small DNA fragment, and footprinting studies.

20.1.1. Dideoxy-Chain Termination Inhibitors

20.1.1.1. DNA polymerases

The large fragment of the *E. coli* DNA polymerase (the Klenow fragment), originally used by Sanger et al. (47) for DNA sequencing, has a low 3′-to-5′ exonuclease activity and an intermediate level of 5′-to-3′ polymerase activity. The exonuclease can hydrolyze 3′-terminal deoxy or dideoxy monophosphates, thus interfering with the analysis of polymerase reactions. Other polymerases that are used for DNA sequencing include avian myeloblastosis virus reverse transcriptase, modified T7 phage-derived DNA polymerase (54, 55), and a thermostable DNA polymerase that was originally isolated from the thermophilic bacterium *Thermus aquaticus* (12) and is now available in genetically engineered form (*AmpliTaq*; Perkin-Elmer, Emeryville, Calif.). The modified T7 phage-derived polymerase is marketed under the trade name Sequenase [United States Biochemicals, Cleveland, Ohio; telephone: (800) 321-9322].

It is useful to know the important characteristics of the DNA polymerases that are obtained from different sources and used for sequencing reactions. The Klenow fragment polymerase has low processivity (dissociates from and reassociates with the template before 10 nucleoside triphosphates [NTPs] are incorporated) and incorporates ddNTPs at a 10^3-fold lower rate than dNTPs. These properties result in differential band intensities, net low incorporation of labeled substrate used, and limits in the extension of the primer chain to approximately 350 nucleotides in a typical reaction. The AMV reverse transcriptase has approximately 10-fold higher processivity, lacks exonuclease activity, and does not discriminate much between dNTPs and ddNTPs for the polymerase reaction, but the polymerization is very slow (about 4 nucleotides per s). Sequenase (version 2.0) lacks exonuclease activity and is highly processive (for several thousand nucleotides), and the rate of polymerization reaction is very high (about 70-fold faster than for the AMV reverse transcriptase). *Taq* DNA polymerase lacks 3′-to-5′ exonuclease activity, has intermediate rates of processivity and polymerization reaction, but has 5′-to-3′ exonuclease activity and works best at 70 to 80°C (30). This enzyme is particularly useful for sequencing ssDNA templates that form localized secondary structures, which are melted at high temperature (65 to 75°C), allowing polymerase to read through these regions. It is also useful for sequencing double-stranded DNA (dsDNA), especially when double-stranded plasmid DNA from "mini" preparations, double-stranded phage, or polymerase chain reaction (PCR) products are used as templates. There is a high degree of selectivity of priming, since the sequencing reaction is carried out at high temperature. A genetically modified version of the *Taq* DNA polymerase (Delta *Taq* version 2.0 DNA polymerase; United States Biochemicals) lacks any exonuclease activity and works efficiently at 90 to 95°C. DNA polymerases from other thermophilic microorganisms are also now in use for DNA sequencing. These polymerases have specific characteristics that are suitable for special needs. For a comprehensive update of these enzymes and their use in DNA sequencing, see reference 2.

Most companies that market the DNA polymerases and dNTPs or ddNTPs provide excellent literature support for DNA sequencing free of charge on request. The protocols are constantly being modified for improved reactions. Basic protocols for sequencing and for using Klenow fragment and *Taq* DNA polymerase will be described below. These protocols have been tested, but they should not be considered superior to any other procedures. Choosing among the available enzymes is dictated not only by the properties of the enzymes but also by personal convenience, cost, and experience. Beginners are advised to start with a commercially available sequencing kit and to try the protocol that is available from the vendor supplying the sequencing reagents. United States Biochemicals, the marketer of Sequenase (version 2.0) and Delta *Taq* (version 2.0) DNA polymerases, provide the protocols free of charge, and hence these protocols are not included in this chapter. The Klenow fragment is least expensive and is available from a large number of sources. It is still considered appropriate for sequencing DNA by the dideoxy-chain termination method and is recommended for use by novices. For large sequencing projects it might be useful to start with Sequenase and use other enzymes (for example, *Taq* DNA polymerase) to resolve ambiguous sequences.

20.1.1.2. DNA templates

Both dsDNA and ssDNA are used as templates for sequencing. dsDNA is obtained mostly from plasmids, and methods for preparation of plasmid DNA are described in Chapter 16. Plasmid DNA preparations should be free from RNA. The best result is obtained if CsCl-gradient-purified DNA is used. Plasmid DNA prepared by "minipreparations" must be treated with DNase-free RNase to completely digest RNA.

The original sequencing method introduced by Sanger et al. (47) relies on the use of single-stranded M13 bacteriophage-derived DNA. Relatively pure M13 DNA is easy to isolate. Phage M13 does not kill the host bacterium and synthesizes double-stranded replicative-form (RF) DNA and ssDNA. The phage that contains ssDNA is released into the medium, while RF DNA remains within the cell, like plasmid DNA. The RF DNA can be isolated and purified by standard methods for the preparation of plasmid DNA (see Chapter 16). The ssDNA is purified from the culture supernatant by precipitation with polyethylene glycol and then removing phage proteins by extraction with phenol and chloroform. Chimeric vectors containing the phage M13 origin of replication in plasmids have been used to prepare ssDNA in the presence of the helper phage F1 (59, 66). The chimeric vectors can replicate as double-stranded plasmid DNA in the absence of phage F1 and can stably maintain inserted foreign DNA (in contrast to single-stranded M13, which is less stable). DNA that will be used as the template for the sequencing reaction, whether prepared from bacteriophage M13 or from plasmid derivatives, should be free from small fragmented DNA and DNA nucleases. The best result is obtained by using *E. coli* DH5αF′ (for phage M13) or DH5α (for plasmids), which lack a functional *endA1* gene (encodes DNA endonuclease). Other commercially available *E. coli* strains, for example, JM101-109 (65), have been used successfully.

The following procedure works well for the preparation of M13 ssDNA (38, 39).

1. Grow an *E. coli* host strain in a rich liquid medium (for example, 2× NY consisting of 25 g of NZ-amine [also sold as casein hydrolysate, enzymatic digest], 10 g of yeast extract, and 2.5 g of NaCl per liter of H_2O) to late log phase, no longer than 24 h.
2. To a test tube containing 5 ml of 2× NY, transfer 20 to 25 μl of the freshly grown bacterial suspension and 20 to 25 μl of phage stock. Phage stock can be prepared by using a toothpick to transfer phage from a plaque into a test tube containing 5 ml of 2× NY medium, incubating at 37°C for 8 to 12 h with shaking, centrifuging the culture in a microcentrifuge for 4 to 5 min, transferring the upper three-quarters of the supernatant fluid to a microcentrifuge tube, and heating at 70°C for *exactly* 10 min. Phage stock prepared in this way can be stored at 4°C for up to 6 months. Alternatively, a tiny agar block containing a single plaque can be removed from a plate by using a Pasteur pipette and substituted for the phage stock. Incubate the mixture at 37°C (no less than 36°C) for 8 to 16 h with agitation.
3. Transfer 1.4 ml of the phage-infected culture into duplicate microcentrifuge tubes. Centrifuge for 2 min at room temperature (RT).
4. Transfer 0.9 ml of the clear supernatant fluid into duplicate microcentrifuge tubes. Add 0.25 ml of a solution of 15% polyethylene glycol and 2.5 M NaCl (Table 1) to each tube. Mix, and leave the mixture on ice for 30 min or longer. Centrifuge for 5 to 6 min, and carefully decant the supernatant fluid without disturbing the pellet at the bottom of the tube. Gently wash the pellet once with 0.5 ml of TES buffer (Table 1).
5. Resuspend the pellet in 0.15 ml of TES buffer by vigorously shaking at RT, and combine the contents of the duplicate tubes into one tube. Remove proteins from the samples by extracting with an equal volume of phenol-chloroform (1:1), and precipitate DNA with twice the aqueous volume of ethanol or an equal volume of isopropanol in the cold (see Chapter 16).
6. Pellet DNA by centrifugation in a microcentrifuge for 5 to 6 min, remove the solvent, wash the pellet once with cold (4°C) 70% ethanol, and dry the pellet under vacuum (see Chapter 16).
7. Resuspend the DNA in 30 to 40 μl of TE buffer (Table 1) at RT over a period of 30 min with occasional shaking. DNA is ready for use as the sequencing template. Store the DNA sample frozen at −20°C or below for future use. The average yield of DNA should be about 20 μg. Determine the approximate amount of DNA by running samples and standard ssDNA solutions on an agarose gel and comparing the intensities of ethidium bromide-stained bands in the gel (for agarose gel electrophoresis and staining procedure, see Chapter 16).

20.1.1.3. Deoxynucleoside and dideoxynucleoside triphosphates

DNA polymerase uses dNTPs as substrates for the extension of the primer that hybridizes with the ssDNA template. ddNTPs are substrate analogs, whose incorporation results in the termination of the polymerization reaction. Polymerases from different sources have different affinities for dNTPs and ddNTPs. It is essential to adjust the ratio of dNTPs and ddNTPs in the sequencing mixtures for reactions with DNA polymerases obtained from different sources. A lower dNTP/ddNTP ratio in the reaction mixture results in better incorporation of the labeled dNTP into the newly synthesized DNA oligomers (sequencing ladders) and shorter extensions to assist reading of the sequence proximal to the primer, in contrast to a higher dNTP/ddNTP ratio, which is suitable for reading sequences distal to the primer. If dGTP is used in the sequencing

Table 1. Preparation of buffers and solutions

Buffer or solution	Ingredient	Amt	
Alkaline phosphatase buffer (10×)	0.5 M Tris HCl (pH 8.4)	200	μl
	1.0 M $MgCl_2$	10	μl
	0.1 M $ZnCl_2$	100	μl
	Water	690	μl
Bal 31 buffer (5×)	2 M Tris HCl (pH 8.0)	62.5	μl
	1 M $MgCl_2$	60.0	μl
	1 M $CaCl_2$	60.0	μl
	5 M NaCl	600.0	μl
	0.2 M disodium EDTA (pH 8.0)	25.0	μl
	Water	192.5	μl
Exonuclease III buffer	2 M Tris HCl (pH 8.0)	330	μl
	1 M $MgCl_2$	660	μl
	Water	9.01	ml
Formamide dye mixture	Xylene cyanol FF	10	mg
	Bromophenol blue	10	mg
	0.5 M disodium EDTA (pH 8.0)	200	μl
	95% deionized formamide	To 10	ml
Ligase buffer (10×)	2 M Tris HCl (pH 8.0)	150	μl
	1 M $MgCl_2$	40	μl
	0.2 M disodium EDTA (pH 8.0)	50	μl
	10 mM ATP	40	μl
	0.5 M DTT	200	μl
	Water	520	μl
NaTMM buffer (10×)	2 M Tris HCl (pH 8.0)	100	μl
	2 M NaCl	50	μl
	1 M $MgCl_2$	200	μl
	14.2 M 2-mercaptoethanol	14	μl
	Water	1,636	μl
PEG solution	Polyethylene glycol 6000 or 8000 (Carbowax)	15	g
	NaCl	14.6	g
	Water	To 100	ml
S1 buffer (10×)	3 M Potasium acetate (pH 4.6)	1.1	ml
	5 M NaCl	5	ml
	Glycerol	5	ml
	$ZnSO_4$	30	mg
Taq sequencing buffer (5×)	1 M Tris HCl (pH 9.0)	0.5	ml
	1 M $MgCl_2$	0.1	ml
	Water	1.4	ml
TBE buffer (5×)	Tris base	60.55	g
	Boric acid	25.8	g
	Disodium EDTA	1.85	g
TE buffer	5 M NaCl	0.2	ml
	2 M Tris HCl (pH 8.0)	0.5	ml
	0.2 M disodium EDTA (pH 8.0)	50	μl
	Water	To 100	ml
TES buffer	5 M NaCl	6	ml
	2 M Tris HCl (pH 8.0)	5	ml
	0.2 M disodium EDTA (pH 8.0)	0.5	ml
	Water	To 100	ml
TMM buffer (10×)	2 M Tris HCl (pH 8.0)	100	μl
	1 M $MgCl_2$	200	μl
	14.2 M 2-mercaptoethanol	14	μl
	Water	1,700	μl

reaction, electrophoresis occasionally results in anomalous separations of oligonucleotides because of compression of G-C stretches present in the DNA, which forms localized secondary structure. This problem can be alleviated by substituting dGTP with dITP, at four times the concentration of dGTP used, or with 7-deaza-2′-deoxyguanosine 5′-triphosphate (c^7dGTP), at the same concentration as dGTP, in the polymerization reaction. dNTPs and ddNTPs can be obtained commercially as powders or in solutions (100 mM). The concentrations of dNTPs and ddNTPs are critical for the sequencing reaction; therefore these solutions must be prepared accurately. Each reaction mixture contains all four dNTPs and only one ddNTP. The reaction mixture containing the specific ddNTP is labeled accordingly: the mixture containing ddATP is labeled A, the mixture containing ddTTP is labeled T, etc. Since the elongation of the DNA chain terminates with the incorporation of one molecule of ddNTP, the oligonucleotides in the sequencing ladder will end with the specific ddNTP used in the reaction. By using four different ddNTPs in separate reactions and then separating the oligonucleotides extended from the primer in parallel in a polyacrylamide gel, the sequence of bases is read (Fig. 1).

dNTP and ddNTP solutions are stored frozen at −70°C or below as 10 or 100 mM stock solutions in H_2O or in a buffer solution containing 10 mM Tris-HCl (pH 7.5 to 8.0) and 0.1 mM EDTA. The solutions should be stored in small aliquots since repeated freezing and thawing of the working solutions (more than 20 times) is not recommended. Preparation of dNTP-ddNTP mixtures for use with the Klenow fragment is described in Table 2. The concentrations of the dNTPs and ddNTPs in the sequencing reactions vary with different enzyme sources. The reader is advised to follow the directions of the vendor supplying the enzyme and reagents for the sequencing reactions.

The radiolabeled dNTP ([α^{35}S]dNTP or [α^{32}P]dNTP) should be obtained fresh. It is best to use [^{35}S]dATP. ^{35}S-labeled oligomers give sharp bands on polyacrylamide gel electrophoresis and exposure of the gel to an X-ray

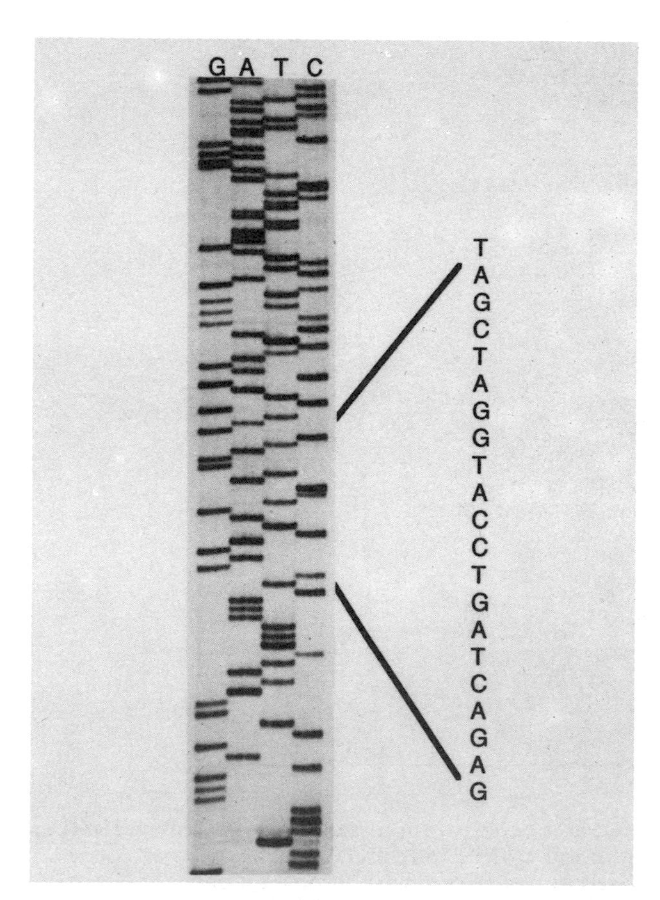

FIG. 1. Autoradiogram of a sequencing gel. A central portion of the gel is shown, approximately 100 bases from the primer. Lanes are identified at the top, and a portion of the sequence is indicated at the side.

Table 2. Preparation of dNTP-ddNTP mixtures for sequencing reactions with the Klenow fragment polymerase[a]

Stock solution (10 mM)	Amt of stock solution (μl) needed to prepare 1 ml of nucleotide mixture for determining:				
	T	A	G	C	I[b]
dGTP	40	40	5.6	40	
dCTP	40	40	40	15	40
dTTP	3.4	40	40	40	40
dITP					20
ddTTP	100				
ddATP		35			
ddGTP			100		35
ddCTP				100	
Water	816.6	845	814.4	805	865

[a] Reference 4.
[b] I is substituted for G (see text).

film. [^{35}S]dNTP can be used over a longer period (the half-life of ^{35}S is 87.4 days) than that with ^{32}P-labeled dNTP (half-life of ^{32}P is 14.3 days). ^{35}S has a shorter path length and requires about 10 times longer exposure time as compared with an equivalent amount of ^{32}P. Store radiolabeled compounds at −20°C.

For some applications, such as sequencing with templates from DS phage or PCR products, it may be preferable to end label the primers by using T4 polynucleotide kinase and [γ-^{32}P]NTP or [γ-^{35}S]NTP, as described in section 20.3. There is no need for radiolabeled dNTP, if end-labeled primer is used. The appearance of background ghost bands on the autoradiograph (see section 20.1.1.8, step 15) of the sequencing gel is minimized if labeled primer is used for the sequencing reaction, since reactions primed from contaminating DNA or RNA are not labeled and therefore are not seen.

20.1.1.4. Sequencing reactions with Klenow fragment polymerase and ssDNA template

1. The following materials and reagents are needed: template DNA (1 to 3 μg), DNA oligonucleotide primer (2.5 ng/μl, ≥15 mer), dNTP-ddNTP mixes, TMM buffer, four dNTP solutions (concentrations, 10 mM or higher), Klenow fragment polymerase (usually 5 U/μl), 0.02 mM dATP, and [α^{35}S]dATP (>600 Ci/mmol) or [α^{32}P]dATP (>400 Ci/mmol).
2. Prepare a water bath by warming water in a beaker or a pan, and adjust the temperature to 70°C.
3. While warming water, start thawing all frozen solutions. On thawing, shake solutions for uniform mixing and then place on ice.
4. Label four microcentrifuge tubes (preferably 0.5-ml capacity) A, T, G, and C. In a separate microcentrifuge tube, add 5 μl of 10× TMM buffer, 1 to 3 μg of template DNA, 1 μl of the oligonucleotide primer solution, and water to a final volume of 20 μl. Mix the contents of the tube, incubate at 70°C, and allow the temperature to drop gradually to near RT over a period of 10 to 30 min (the time will depend on the volume of the water and the shape of the container).
5. Mix 8 to 10 μCi of [^{35}S]dATP and 4 μl of nonradioactive 0.02 mM dATP, 2.5 U of Klenow fragment, and water to a total volume of 30 μl.
6. Briefly (5 s) centrifuge the tube containing the annealed DNA (step 4). Mix the annealed DNA with 25 μl of the solution containing radioactive dATP.

7. Distribute 8 μl of the above mixture into each of the four microcentrifuge tubes marked A, T, G, and C. Add 1 μl of dNTP-ddNTP mixes (Table 2) in the respective tubes (the tube marked A should contain ddATP and no other ddNTP, the tube marked T should contain ddTTP and no other ddNTP, etc.). Incubate the tubes at 45°C for 20 min.

8. While the tubes are being incubated, prepare a mixture of all four dNTPs and Klenow fragment (1 mM each dNTP and 0.2 U of Klenow fragment in a total volume of 6 μl).

9. After 20 min of incubation of the tubes at 45°C, add 1 μl of the dNTP and Klenow fragment mixture to each tube. Mix the contents of the tubes, and incubate them at 30°C for additional 15 min.

10. Add 6 μl of formamide-dye mixture (Table 1). The samples are then ready for electrophoresis. If the samples are to be stored, place at −70°C, and do not add the formamide-dye mixture prior to freezing. The reaction mixture with added formamide-dye solution should be placed in a hot-water bath (95 to 100°C) for about 2 min prior to loading on the gel.

20.1.1.5. Sequencing reactions with Klenow fragment polymerase and dsDNA template

1. The following materials and reagents are needed: template DNA (2 to 4 μg for a 3- to 6-kb plasmid), DNA oligonucleotide primer (17- to 30-deoxynucleotide oligomer, 10 ng/μl), dNTP-ddNTP mixes, TMM buffer, 2 M NaOH containing 2 mM EDTA, four dNTP solutions (10 mM or higher), 2 M ammonium acetate (pH 4.5, adjusted with acetic acid), Klenow fragment (usually 5 U/μl), 0.02 mM dATP, and [α-^{35}S]dATP (>600 Ci/mmol) or [α-^{32}P]dATP (>400 Ci/mmol).

2. Thaw all the frozen solutions. Then shake the solutions for uniform mixing, and place them on ice.

3. Mix, in a 0.5-ml microcentrifuge tube, 2 to 4 μg of the template DNA, 10 to 15 ng of the primer, 2 μl of 2 M NaOH containing 2 mM EDTA, and water to a total volume of 20 μl. Incubate the mixture for 5 min at room temperature (23 to 25°C) for denaturation (strand separation), and then add 3 μl of 2 M ammonium acetate (pH 4.5) to neutralize the solution. Precipitate DNA in the cold by adding twice the aqueous volume of ethanol, as described in section 20.1.1.2. Dry and resuspend the DNA in 10 μl of water at room temperature over a period of 30 min. Add 5 μl of 10× NaTMM buffer and water to a final volume of 20 μl. Mix the contents of the tube, incubate at 70°C, and allow the temperature to drop gradually to near RT over a period of 10 to 30 min (the time will depend on the volume of the water and the shape of the container). Then follow steps 5 to 10 of section 20.1.1.4.

20.1.1.6. Sequencing reactions with *Taq* polymerase and ssDNA template

Sequencing protocols with *Taq* DNA polymerase are adapted from *Protocols and Applications Guide* (42a). The reactions are done in two basic steps: extension/labeling, and termination by the ddNTPs. Primers of 20- to 30-nucleotide oligomers are used. The annealing and initial primer labeling/extension are done at 37°C followed by complete extension at 70°C. Although *Taq* DNA polymerase works best at 65 to 75°C, reaction at 37°C slows down the incorporation rate of dNTPs and thus allows reading of the sequence closer to the primer. This may also be facilitated by shortening the duration of labeling/extension reaction.

1. The following materials and reagents are needed: template DNA (2 to 4 μg), DNA oligonucleotide primer (8 ng/μl), dNTP-ddNTP mixes (Table 3), *Taq* sequencing buffer, chase solution (1 mM each dNTP), extension/labeling mix (7.5 μM each dGTP, dTTP, and dCTP), *Taq* DNA polymerase (usually 5 U/μl), [α-^{35}S]dATP (>600 Ci/mmol) or [α-^{32}P]dATP (>400 Ci/mmol), and formamide-dye solution (Table 1).

2. Thaw all frozen solutions. Then shake solutions for uniform mixing, and place on ice.

3. In a 0.5-ml microcentrifuge tube, add 1 to 2 μl of template DNA (2 to 4 μg), 1 to 1.5 μl of primer, 5× *Taq* sequencing buffer, 2 μl of extension/labeling mix, and water to 25 μl. Mix the contents of the tube, and incubate at 37°C for 10 min.

4. To the above mixture, add 8 to 10 μCi of [α-^{35}S]dATP or [α-^{32}P]dATP (approximately 10 mmol; add non-radioactive dATP, if necessary), 4 U of *Taq* DNA polymerase, and mix by pipetting up and down gently. Incubate at 37°C for 5 min.

5. Mark four tubes A, T, G, and C, add 1 μl of the appropriate dNTP-ddNTP mixture (Table 3) to each tube, and place the tubes on ice while incubating the reaction mixture in step 4.

6. Briefly (5 s) centrifuge the tube from step 4. Add 6 μl of the mixture to each of the tubes marked A, T, G, and C. Centrifuge briefly to get down all the liquid at the bottom of the tube. Incubate at 70°C for 15 min.

7. Add 4 μl of formamide-dye mixture (Table 1), and the samples are ready for electrophoresis. If the samples are to be stored, store at −70°C and do not add the formamide-dye mixture prior to freezing. The reaction mixture with added formamide-dye solution should be placed in a hot-water bath (95 to 100°C) for about 2 min prior to loading on the gel.

20.1.1.7. Sequencing reactions with *Taq* DNA polymerase and dsDNA template

1. The following materials and reagents are needed: template DNA (2 to 4 μg), DNA oligonucleotide primer (8 ng/μl), dNTP-ddNTP mixes (Table 3), *Taq* sequencing buffer, chase solution (1 mM each dNTP), extension/labeling mix (7.5 μM each dGTP, dTTP, and dCTP), *Taq*

Table 3. Preparation of dNTP-ddNTP mixtures for sequencing reactions with *Taq* DNA polymerase

Stock solution (10 mM)	Amt of stock solution (μl) needed to prepare 1 ml of nucleotide mixture for determining:			
	T	A	G	C
dATP	25	2.5	25	25
dGTP	25	25	2.5	25
dCTP	25	25	25	2.5
dTTP	2.5	25	25	25
ddTTP	30			
ddATP		35		
ddGTP			2.5	
ddCTP				16
Water	892.5	887.5	920.0	906.5

DNA polymerase (usually 5 U/μl), 2 M NaOH containing 2 mM EDTA, 2 M ammonium acetate (pH 4.5, adjusted with acetic acid), [α-^{35}S]dATP (>600 Ci/mmol) or [α-^{32}P]dATP (>400 Ci/mmol), and formamide-dye solution (Table 1).

2. Thaw all frozen solutions. Then shake solutions for uniform mixing and place on ice.

3. Mix, in a 0.5-ml microcentrifuge tube, 2 to 4 μg of the template DNA, 10 to 15 ng of the primer, 2 μl of 2 M NaOH containing 2 mM EDTA, and water to a total volume of 20 μl. Incubate the above mixture for 5 min at RT (23 to 25°C) for denaturation (strand separation), and then add 3 μl of 2 M ammonium acetate (pH 4.5) to neutralize the solution. Precipitate DNA in the cold by adding twice the aqueous volume of ethanol, as described above. Dry, and resuspend DNA in 10 μl water at room temperature over a period of 30 min. Then follow steps 3 to 7 of section 20.1.1.6.

20.1.1.8. Gel electrophoresis (also see Chapter 21.5.1)

Gel apparatus of various lengths and widths are commercially available. Larger or wider gels are difficult to handle. Apparatus from Bio-Rad, Bethesda Research Laboratories, and International Biotechnologies Inc. all work well. Gels of 40 by 20 cm (0.3 to 0.4 cm thick) can be handled conveniently. Plastic spacers placed between glass plates determine the thickness of the gel. To read the maximum number of nucleotides from a gel, wedge-shaped spacers (tapered, 0.3 to 0.8 cm) may be used (1, 41); however, they are more difficult to dry. Polyacrylamide gradient gels (4 to 18%) or buffer gradient gels have also been used (6, 28, 51). Buffer gradient gels with different ionic concentrations of the electrode buffer are easiest to handle among all different kinds of gradient gels. Two different kinds of combs are used for forming wells in gels: conventional square or rectangular-shaped wells that are spaced a few millimeters apart from each other, or wells separated from each other by a minimum spacing formed by using a "shark's-tooth" comb. To form conventional wells, the comb is inserted into the gel mold before the gel solution polymerizes. In contrast, the flat end of the shark's-tooth comb is inserted into the gel mold prior to polymerization of the gel solution. The comb is removed after the gel is polymerized, the comb is rotated 180°, and wells are formed by slightly inserting the sharp teeth of the comb into the gel. Loading of samples on the shark's-tooth gel is more difficult, and leaking of samples from one well to another is a common problem. On the other hand, reading of sequences from such a gel is improved since the lanes run very close together.

For the preparation of sequencing gels, use a solution containing 38% (wt/vol) acrylamide and 2% (wt/vol) *N,N'*-methylenebisacrylamide as stock. Acrylamide is a neurotoxin, so wear gloves while handling it and use a mask while handling powdered acrylamide. Make the solution in warm water (40 to 50°C), deionize by stirring with 5 g of amberlite MB-1 resin (Mallinckrodt or Sigma Chemical Co.) per liter or an equivalent monobed resin for a period of 30 min or longer and then filter through Whatman no. 1 filter paper (preferably fluted). The stock solution can be stored in the dark at 4°C for up to 12 months. Ammonium persulfate and TEMED (*N,N,N',N'*-tetramethylethylenediamine) are used as catalysts for the polymerization reaction. Stock solution of ammonium persulfate (10% [wt/vol] in water) can be stored up to a week at 4°C.

The volume of the solutions required for a gel depends on the size of the glass plates and the thickness of the spacers used. For casting gels, follow the recommendation of the supplier of the gel apparatus. Polymerized gels can be stored clamped with large binder clamps and wrapped with plastic wrap at RT to prevent dehydration from the area where the comb is inserted. Gels should not be stored for more than 24 h under these conditions. Use clean glass plates to prevent trapping of bubbles in the gel sandwiched between the glass plates. The glass plates are best cleaned with warm water, without detergents or acids, immediately following electrophoresis. Spray ethanol on the washed glass plates, and wipe with a few pieces of Kimwipe. Applying a layer of silicone to one side of one of the plates (with Sigmacoat [Sigma Chemical Co.] or an equivalent chemical) facilitates separation of the gel that is sandwiched between the glass plates. A protocol for the preparation of a 7% gel (20 by 40 cm, 0.3 cm thick) is given below.

1. The following materials and solutions are required: glass plates, spacers, combs, large binder clamps, 40% acrylamide–bisacrylamide solution (38:2), 10% ammonium persulfate, TEMED, ultrapure urea, 5× TBE (Table 1).

2. Two glass plates (one notched) are put together with spacers in between the plates on the longitudinal edges and at the bottom (optional). The spacers must be placed on the siliconized surface of one of the glass plates. Use clamps to hold the plates and spacers together. The bottom can be sealed with a tape (3M yellow electrical or its equivalent) instead of using the bottom spacer. The bottom can also be sealed by inserting a piece of a filter paper between the plates and soaking the filter paper with a mixture of 12 to 18% acrylamide solution and high concentrations of catalysts (60).

3. To cast two 7% gels, mix the following in a beaker: urea, 33 g; 40% acrylamide solution, 13.1 ml; 10% ammonium persulfate, 0.38 ml; 5× TBE, 15 ml; water to 75 ml. Addition of warm water (about 50°C) will help in dissolving urea. After the solutions are thoroughly mixed, filter the solution through a fluted Whatman no. 1 filter paper if any suspended particles are visible (some batches of urea contain insoluble particles). Cool the solution on ice, add 38 μl of TEMED, and mix by swirling. Cold solution will polymerize slowly in the presence of TEMED, allowing sufficient time to pour the gel solution within the gel mold.

4. Draw the above solution into a 25-ml pipette or a hypodermic syringe. Hold the glass plates at an angle (45°), and pour gel solution through one corner of the gel mold at a uniform rate. If any air bubbles are trapped, the plates are not clean. However, air bubbles usually can be removed by tapping on the upper plate (the wooden handle of a large spatula works well for tapping). When the mold is filled, lay the plates on the top of a rubber stopper at near-horizontal position. Insert the comb. Place clamps holding the two glass plates together in the area where the comb is inserted. Allow the gel to polymerize for 1 to 2 h at RT before electrophoresis. Efficient polymerization can be achieved in 30 min at 37°C.

5. When ready to use the gel for electrophoresis, remove the clamps, the bottom spacer (or the piece of

sealing tape), and the comb. Wash the top part of the gel with 1× TBE. Remove any polymerized gel particles trapped in the wells by forcing 1× TBE through a Pasteur pipette (one end drawn to a capillary shape). If a shark's-tooth comb is used, insert the comb with the teeth just penetrating the acrylamide. Make sure all the teeth have penetrated, but do not insert the comb so far that the surface of the well is bowed.

6. Pour electrode buffer (1 × TBE or a different buffer for buffer gradient gel) in the lower buffer reservoir. Place the plates vertically (unnotched plate facing outside), and secure them to the gel apparatus with clamps or screws. Air bubbles between the lower edge of the gel and the lower electrode buffer should be avoided (forcing of buffer through a Pasteur pipette may help remove any trapped air bubble). Fill the upper buffer reservoir with 1× TBE, making a connection between the upper part of the gel and the upper electrode buffer.

7. Prerun the gel for 30 min to warm up to 40 to 45°C (set the power as recommended by the supplier of the gel apparatus).

8. Disconnect power, and wash urea from the wells by squirting 1× TBE with a Pasteur pipette or syringe.

9. Load 1 to 3 μl of samples in each well with a pipettor and capillary tips (commercially available as "flat" tips or "gel-loading" tips). Load in the order G, A, T, and C for each template. Rinse the pipette tip by pipetting buffer in and out at the lower buffer reservoir in between loading of each sample. When loading samples with reactions from many templates, load samples for one to three templates and then electrophorese for 3 min before loading another set of samples. Remove urea from the wells, as above, between each loading.

10. Turn on the power, and run the gel for about 20 min after the bromophenol blue dye runs out of the bottom of the gel. Previously loaded samples can be loaded again in adjacent wells at this time, and electrophoresis is continued for an equal time to obtain maximum sequence information for each template from a single gel. The choice of using multiple gels with different times of runs or staggered loading of one gel depends on the number of templates and the width of the gel (number of wells) used.

11. After electrophoresis is complete, disassemble the glass plates from the electrophoresis apparatus, wash the plates with tap water, and lay the plates flat on a sheet of an absorbent paper (for safe handling of radioactive materials). Pry the plates apart from one corner. Observe carefully, and lay down the plate to which the gel remains attached. If desired, cut about the top 3 cm of the gel with a pizza cutter (works best) or a scalpel and discard the top portion of the gel. This part of the gel contains the maximum amount of radioactivity, and sequences cannot be read in this area. Note that the buffer in the lower reservoir is radioactive and should be disposed of properly.

12. Place the gel with the support of the attached glass plate in 1 liter (or more) of 5% acetic acid bath (a plastic tray can be used for making the acetic acid bath). Soak the gel in acetic acid for 30 min at RT without shaking. Soaking with acetic acid helps fix the oligonucleotide bands in the gel and removes urea.

13. Carefully remove the gel with the support of the glass plate from the acetic acid. Soak excess acetic acid from the top of the gel by placing two or three layers of tissue paper (Kimwipes) over the gel and slowly rolling the tissue papers off the gel from one corner. Place a piece of a Whatman 3 MM paper on the gel, and press the paper slowly to adhere to the gel surface.

14. Invert the plate with hand palm support on the Whatman paper, and slowly peel the plate away from the gel. Note the orientation of the gel to ascertain the proper order of the samples loaded. Place a piece of plastic wrap on the gel, and dry the gel under vacuum at 70 to 80°C. Gel-drying apparatuses are commercially available.

15. Remove the plastic wrap (leave the wrap on if [^{32}P]dATP was used), and expose the dried gel to an X-ray film (Kodak XAR-5 or its equivalent) for autoradiography overnight.

16. After the X-ray film is developed, sequences can be read from the film. Mark the lanes in the appropriate order of loading. Start reading from the bottom of the film in ascending order of bands present in the lanes (G, A, T, C). Note that the sequence read is complementary to the sequence of the template DNA strand. Commercially available sequence readers help by entering data automatically while the sequence is read. For manual reading of the film, it is better if one person reads and another person writes the sequence. In any case, the sequence should be recorded at least twice, compared, and edited. Characteristics of the autoradiogram (when Klenow fragment is used for the sequencing reactions) include the following: (i) in a run of consecutive C bands, the fastest-moving band is the most weakly labeled, and (ii) within a stretch of A bands, the topmost band is the least intense.

20.1.2. Nested Deletions with *Bal* 31 Exonuclease

Bal 31 is an exonuclease that cleaves ssDNA or dsDNA from both the 3′ and 5′ ends. The target DNA is first cloned into the polylinker cloning region of a vector plasmid or phage M13. There must be at least two separate unique restriction enzyme sites adjacent to the ends of the cloned target DNA. dsDNA prepared from the hybrid clone (target DNA cloned into the vector) is cut with a unique restriction enzyme, whose site is adjacent to one end of the target DNA. The linearized DNA is digested with *Bal* 31 for various lengths of time. *Bal* 31-digested DNA is then cleaved with another unique restriction enzyme whose site is located adjacent to the covalently closed end of the target DNA. The digested target DNA fragments are separated in an agarose gel and recloned into phage M13 vector or another vector for sequencing (42). However, by modification of this procedure (38, 39), one can ligate the mixture of the *Bal* 31-digested target DNA and the M13 vector DNA fragments with intact linear M13 vector RF DNA and transform competent M13-sensitive *E. coli* host cells for selection of hybrid clones of intact M13 DNA and partially deleted target DNA. Partially deleted M13 DNA fragments are also ligated with the intact linear M13 vector DNA. The hybrid DNA will contain M13-derived DNA fragments in the reverse orientation of the intact M13 vector DNA, and no plaque will be formed on transformation.

An outline of the following procedure is diagrammed in Fig. 2.

1. Clone target DNA fragment into the M13mp18/M13mp19 vector in two different orientations by using standard procedures for cloning into plasmid DNA vectors (see Chapter 18).

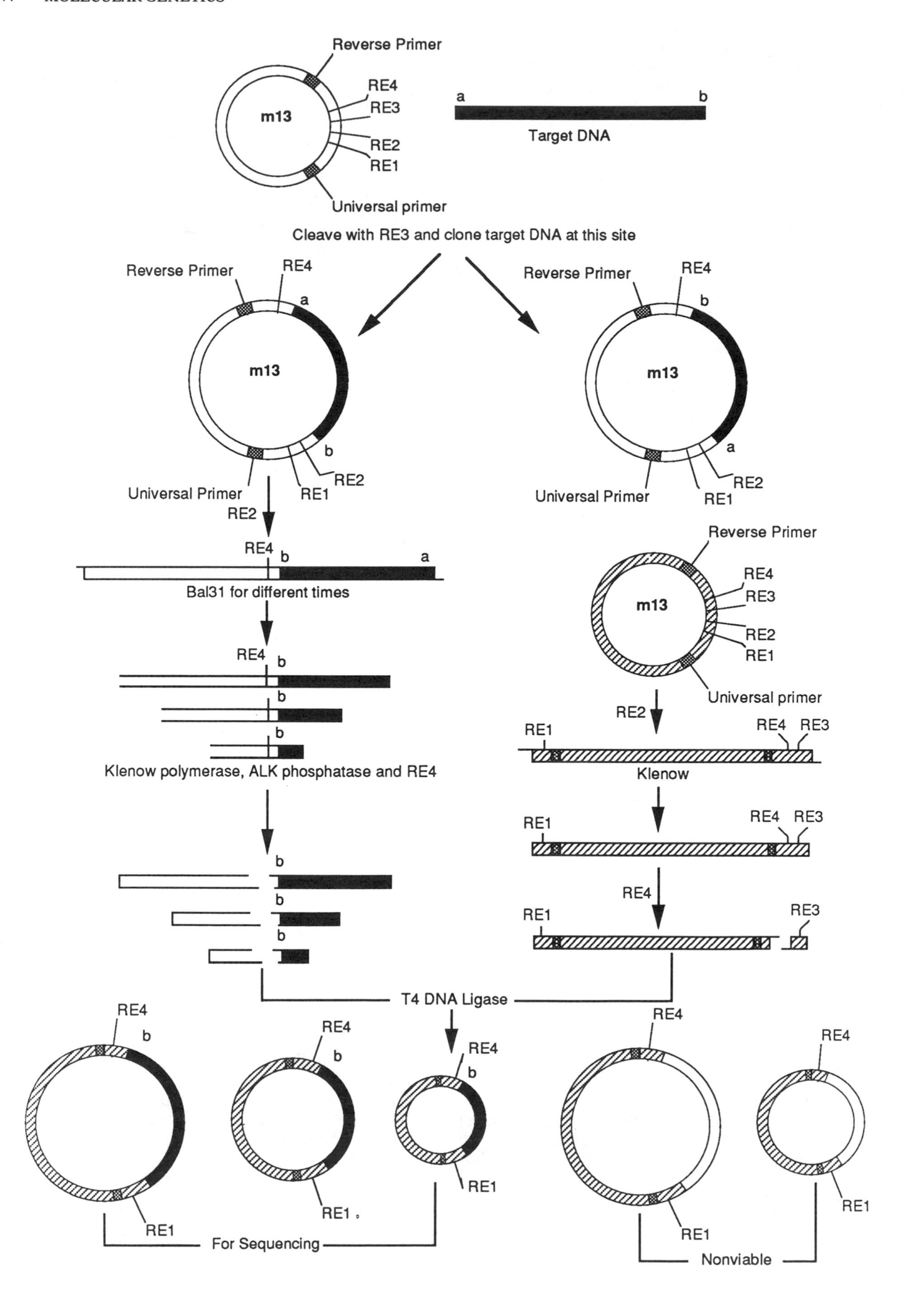

Reverse Primer
RE4
RE3
RE2
RE1
m13
Universal primer
a
b
Target DNA
Cleave with RE3 and clone target DNA at this site
Reverse Primer
RE4
a
m13
b
RE2
RE1
Universal Primer
Reverse Primer
RE4
b
m13
a
RE2
RE1
Universal Primer
RE2
RE4
b
a
Bal31 for different times
RE4
b
b
b
Klenow polymerase, ALK phosphatase and RE4
b
b
b
Reverse Primer
RE4
RE3
m13
RE2
RE1
Universal primer
RE2
RE1
RE4 RE3
Klenow
RE1
RE4 RE3
RE4
RE1
RE3
T4 DNA Ligase
RE4
b
RE1
RE4
b
RE1
RE4
b
RE1
For Sequencing
RE4
RE1
RE4
RE1
Nonviable

2. Prepare RF DNA (by any of the methods of preparing purified plasmid DNA) from the above clones. Proceed to generate deletions from each DNA by following the steps described below.

3. Digest 5 μg of the RF DNA with a unique restriction enzyme (e.g., RE2, as shown in Fig. 2) in a 50-μl volume. Make sure that the digestion is complete by analyzing small aliquots of digested and undigested DNA samples on an agarose gel. Store the RE2-cleaved DNA at −20°C while testing for complete digestion.

4. Thaw the DNA sample from above, add 50 μl of 5× *Bal* 31 buffer and water to a final volume of 250 μl, mix, and incubate at 37°C in a water bath. Transfer 50 μl of 0.2 M disodium EDTA (pH 8.0) in a separate microcentrifuge tube. After 5 to 10 min of incubation at 37°C, add 1.25 U of *Bal* 31 enzyme to the tube containing the DNA sample, mix, and continue incubation.

5. Withdraw 10 μl of *Bal* 31-digested samples at 1-min intervals starting from the first minute of incubation, and collect into the tube containing disodium EDTA solution to stop further digestion by *Bal* 31. Disodium EDTA chelates Mg^{2+}, rendering *Bal* 31 enzyme inactive. Approximately 200 bp is removed per min by *Bal* 31 from each terminus of the linear DNA. Adjust the ratio of DNA to *Bal* 31, if desired results are not achieved. *Note: Bal* 31-digested aliquots can be collected in separate tubes containing 2.5 μl of 0.2 M disodium EDTA (pH 8.0) to analyze deletions at every time point of the termination of the *Bal* 31 digestion. This may facilitate deletion analysis, but it will involve ligation and subsequent transformation of multiple samples.

6. Remove proteins with phenol-chloroform, and precipitate DNA with ethanol or isopropanol. Resuspend dried DNA in 35 μl of water–5 μl of 10× NaTMM buffer (Table 1). Add 5 μl of chase solution (all four dNTPs, 1 mM each) and 2 U of Klenow fragment, and incubate at 37°C for 15 min to make the DNA termini blunt.

7. Remove proteins, precipitate DNA, and resuspend DNA in the reaction buffer for cleavage with a restriction enzyme whose site is located adjacent to the undigested end of the target DNA (e.g., RE4, as shown in Fig. 2). Completely digest DNA with RE4, deproteinize, precipitate, and resuspend in 50 μl of alkaline phosphatase buffer (Table 1).

8. Add 1 to 1.5 U of calf intestinal alkaline phosphatase, mix, and incubate at 37°C for 15 min and then at 55°C for 45 min. Incubate the alkaline phosphatase-treated sample at 65°C for 1 h to inactivate the phosphatase activity. Deproteinize, precipitate, and finally resuspend DNA in 20 to 25 μl of TE buffer. *Note:* Alkaline phosphatase removes 5′-terminal phosphates and prevents self-ligation of DNA fragments or ligation of two DNA fragments from which 5′-terminal phosphates have been removed.

9. In parallel, digest 5 μg of M13mp18 or M13mp19 (use the same vector that was used to clone the target DNA [described in step 3]) with RE1, RE2, or any other unique restriction enzyme whose site is located downstream from RE4 in the polylinker cloning site. On cleavage with the restriction enzyme, if 3′ or 5′ recessed ends are produced, make the ends blunt with Klenow fragment polymerase or T4 DNA polymerase and dNTPs.

10. Remove proteins, precipitate, and then completely digest DNA with the restriction enzyme RE4 (Fig. 2).

11. Remove proteins, precipitate, and resuspend DNA in 20 to 25 μl of TE buffer.

12. Ligate DNA fragments from steps 8 and 9 with T4 DNA ligase. Mix 5 μl of DNA solutions, 1.5 μl of 10× ligase buffer, 0.2 U of T4 DNA ligase, and water in a total volume of 15 μl. Incubate at 16°C for 5 to 6 h. Under these conditions, sticky ends are ligated efficiently.

13. Add 0.5 μl of 10× ligase buffer, 2 μl of 10 mM hexamine cobalt chloride, 2 μl of 10 mM ATP, and 5 U of T4 DNA ligase to the sample from step 12. Incubate at 4°C for 12 to 16 h. The blunt-ended termini are ligated efficiently under these conditions. *Note: The definition of units for ligase varies among companies. The enzyme from New England BioLabs is suggested, although preparations from other companies work as well.*

14. Use 3 and 0.3 μl, in two sets, of the ligated DNA to transform competent M13-sensitive *E. coli* host cells (see Chapter 18 for details of the transformation procedure).

15. Select transformants on LB agar plates. Overlay a mixture of cells from the transformation reaction, 2.5 ml of 0.7% agar (first melted by boiling and then equilibrated at 45 to 46°C), 20 μl of a stock solution of 40 mg of X-Gal (5-bromo-4-chloro-3-indolyl-β-D-galactopyranoside) per ml of dimethylformamide, and 10 μl of a stock solution of 80 mg of IPTG (isopropyl-β-D-galactopyranoside) per ml of water on each LB agar plate. (X-Gal and IPTG stock solutions can be stored at −20°C in the dark.) Uniform overlaying can easily be achieved if LB agar plates are prewarmed at 37°C. Pick "white" plaques (those not colored blue). Grow hybrid M13 phage, with inserts of deleted fragments of target DNA, in 2× NY medium. Any blue plaque should have no inserted target DNA fragment.

16. Determine the relative size of the inserted DNA by agarose gel analysis of the lysed culture and by using RF DNAs from the M13 vector and the hybrid M13 (with an insert of the intact target DNA) as reference markers. Order the hybrid DNAs according to their relative sizes. The procedure reported as the "toothpick assay" (3, 4) works best for such an analysis by agarose gel electrophoresis. An ethidium bromide-stained gel will show bands corresponding to the host chromosomal DNA, nicked RF phage DNA, intact RF DNA, phage ssDNA, and tRNAs in descending order from the top to the bottom of the gels. M13-infected cells are easily lysed by mixing 200 μl of the culture suspension with 25 μl of a solution containing 5% sodium dodecyl sulfate (SDS), 0.1 M disodium EDTA, 50% (vol/vol) glycerol, and 0.1%

FIG. 2. Nested deletions with *Bal* 31. Black areas are DNA to be cloned; open areas are vector; small shaded areas are regions that hybridize to commercially available oligonucleotide primers. RE1 to RE4 indicate restriction enzyme cleavage sites within the polycloning region of the vector. The target DNA ends are identified by a and b to follow orientation at different steps. Not drawn to scale. Modified from reference 39.

bromophenol blue (as an indicator dye) and then incubating the mixture at 70°C for 10 min. The heated mixture is viscous and can be readily used for the toothpick assay. Alternatively, RF DNA (prepared by the method described for plasmid DNA preparation [see Chapter 16]) can be linearized with the restriction enzyme (RE4 in Fig. 2) and the linear DNA can be analyzed on agarose gels for size determination.

20.1.3. Nested Deletions with Exonuclease III

The method of nested deletions with exonuclease III was first reported by Henikoff (23, 24) and relies on the ability of exonuclease III to digest linear blunt-ended DNA or DNA with a 5′ protruding end, but not with a 3′ protruding end, from the 3′ to the 5′ direction. The single-stranded portion of the DNA remaining after exonuclease III digestion is removed by S1 nuclease treatment. The DNA is recircularized after the ends were blunted with Klenow fragment in the presence of dNTPs and transformed into the appropriate host cells to obtain desired unidirectional deletions. The presence of nicked DNA interferes with the deletion analysis because of random distribution of nicking sites on the DNA and the ability of exonuclease III to digest the nicked DNA strand from the 3′ to the 5′ direction. A description of the procedure is given below. All manipulations are done in microcentrifuge tubes (capacity, 0.5 or 1.5 μl).

1. Clone target DNA in the polylinker cloning region in two different orientations into a plasmid, phage M13, or a phagemid vector. Note that there should be at least two unique restriction enzyme sites in the polylinker region of the vector, one producing a 3′ protruding end (e.g., RE1 as shown in Fig. 2) and the other producing a blunt end or a 5′ protruding end (e.g., RE2 as shown in Fig. 2) on cleavage of the DNA with the respective enzymes.

2. Digest 10 μg of plasmid or RF DNA from step 1 with RE1 (Fig. 2) for complete cleavage. Make sure that the digestion is complete by analyzing small aliquots of digested and undigested DNA samples on an agarose gel. Store the RE1 cleaved DNA at −20°C, while testing for complete digestion.

3. Thaw the DNA sample from step 2. Adjust the buffer of the reaction mixture to the optimum for cleavage of DNA by RE2. Alternatively, precipitate DNA from step 2, dry, and resuspend in RE2 reaction buffer.

4. Remove proteins by extracting with phenol-chloroform, precipitate and dry the DNA from step 3, and resuspend it in 60 μl of exonuclease III buffer (Table 1). In the meantime, prepare the S1 buffer-enzyme mixture by mixing 172 μl of water, 27 μl of 10× S1 buffer, and 60 U of S1 nuclease (1 U produces 1 μg of acid-soluble nucleotide per min at 37°C). Add 7.5 μl of S1 buffer-enzyme mixture in a series of 10 to 20 tubes, and place them on ice.

5. Incubate the tube containing DNA from step 4 at 37°C in a water bath for 3 min. Add 500 U (up to 12 μl) of exonuclease III, and continue incubation at 37°C.

6. Withdraw 2.5 μl of exonuclease III-digested samples at 30-s intervals, and collect into the tubes containing S1 buffer-enzyme mixture. Alternatively, all samples can be collected in one tube containing 7.5 n μl of S1 buffer, where n is the number of samples to be withdrawn. Approximately 175 to 200 bases are deleted by exonuclease III under these conditions.

7. Remove proteins, and precipitate and dry the DNA. Resuspend dried DNA in 17.5 μl of water–2.5 μl of 10× NaTMM buffer (Table 1). Add 2.5 μl of chase solution (all four dNTPs, 1 mM each) and 0.5 to 1 U of Klenow fragment. Incubate at 37°C for 15 min to make the DNA termini blunt.

8. Recircularize DNA from step 7 by ligation with T4 DNA ligase. Remove proteins, precipitate, and dry the DNA. Resuspend the dried DNA in 14 μl of water and 2 μl of 10× ligase buffer. Add 2 μl of 10 mM hexamine cobalt chloride, 2 μl of 10 mM ATP, and 2 U of T4 DNA ligase. Mix, and incubate at 4°C for 12 to 16 h for efficient ligation (see step 13 in section 20.1.2).

9. Use the DNA from step 8 (above) to transform appropriate competent cells. Screen and order deletion as described in section 20.1.2, steps 14 to 16.

20.2. DNA-PROTEIN BINDING

A variety of techniques are available to study the factors that influence regulation of a gene. If sequence-specific DNA-binding proteins are thought to be involved, it is of interest to determine the locations of the binding sites and the biochemical properties of the protein-DNA interactions. In bacteria these sites can be close to the transcription start or more than 1 kb away. Analysis in vivo of deletion and point mutants can be used to obtain information on the location of the sites, but ultimately biochemical assays must be used. To study in vitro the factors that influence transcription, several DNA-binding assays can be used. If the location of the suspected binding site is known within a few hundred base pairs, nuclease or chemical protection ("footprinting") assays can be used (see section 20.2.2). If the location of the site is not suspected, or if the protein is of low abundance and purity, it is better to use filter-binding or gel mobility shift assays. The filter-binding assay was the first routine binding method developed (44). The amount of labeled DNA bound to protein is detected by filtering the complexes through nitrocellulose filters; DNA is retained only if bound to protein. This method is still useful for highly accurate measurements (see reference 2 for details), but it has been largely superseded by the more versatile gel mobility shift and footprinting assays. The basic aspects of the gel mobility shift assay with purified protein are described first, and then the cell lysate assay is discussed. Finally, protection assays are considered. Further discussion of all the assays may be found in reviews (11, 25, 43) and methods manuals (2, 9).

20.2.1. Gel Mobility Shift Assay

The gel mobility shift assay (also called the gel retardation assay) is based on the differences in electrophoretic mobility between nucleic acid fragments and nucleic acid-protein complexes (14, 19). Although the assay has been used successfully to study protein binding to RNA (61), this discussion will be limited to the most common use of protein binding to dsDNA. A solution of protein is mixed with DNA fragments in an appropriate binding buffer. After allowing the protein to bind, the mixture is applied to a nondenaturing gel and, following a period of electrophoresis, the positions of the DNA fragments are determined. If the protein binds the DNA stably, the increased mass and alteration in charge of the complex will generally cause it to migrate more slowly than the DNA alone. The relative amounts of "free" DNA and complex bands can be analyzed quantitatively to obtain information about the concentrations of the reactants, the equilibrium and kinetic binding constants, the stoichiometry, and some aspects of the structure of the complex (13–17, 20, 35, 49, 53, 63).

The greatest advantage of the gel shift assay over other methods is its versatility. Specific protein-DNA interactions can be detected by using a mixture of DNA fragments, such as a digest of a whole plasmid. This is valuable in quickly locating binding sites within a large segment of DNA. Interactions with multiple proteins usually can be resolved by using a single DNA fragment, since different proteins or different stoichiometries of binding yield characteristic shift positions. Thus, binding of multiple activator or repressor proteins can be distinguished from binding of one species, which in turn can be distinguished from binding of a large molecule such as RNA polymerase (29, 53). This allows functional assays in studies of protein interactions at promoters. Even in a crude cell lysate, specific binding of one protein can be observed if appropriate controls are performed to demonstrate specificity. This is perhaps the most powerful aspect of the assay, since it provides a rapid and convenient method to quantify protein from new constructs or to assay fractions from purification schemes. The method also can be used preparatively to obtain the bound and free fractions of the DNA, to isolate bound proteins from a crude mixture, or to obtain the protein-DNA complexes after treatments such as UV cross-linking or chemical modification (25, 43).

All these applications rely on the basic finding that protein-DNA complexes are more stable in the gel than in solution. This observation has been explained on the basis of the "caging" effect of the gel; i.e., the complexes are not more stable, but dissociated protein reassociates at a greatly increased rate, so that the reactants do not have a chance to migrate apart. Since the electrostatic component of the binding interaction is stronger at lower salt concentrations, additional stability of complexes may be obtained by using low-salt gels.

20.2.1.1. DNA fragments

The most common DNA templates used in mobility shift assays are restriction fragments, synthetic oligodeoxynucleotides, and PCR products (see Chapter 19). The best results (easily observable shifts) are obtained with fragments of less than 1 kbp, although assays with larger fragments and even whole plasmids have been successful. It should be kept in mind that the larger the DNA fragment, the smaller the shift in mobility of the complexes and the greater the amount of nonspecific protein binding. If the fragment of interest can be cut by restriction enzymes, leaving only one or two large vector fragments, the mixture may be used in the assay; otherwise the fragment should be isolated by gel electrophoresis. Synthetic oligodeoxynucleotides and fragments produced by PCR also should be purified to eliminate any contaminants that may interfere with binding. Mixtures containing fragments that may comigrate with the protein-DNA complexes should be avoided. Unlabeled DNA can be detected by soaking the gel in 1 μg of ethidium bromide per ml after electrophoresis. Labeling the fragments with ^{32}P greatly reduces the necessary amount of reactants and is the method of choice. One of the standard end-labeling or internal-labeling protocols may be used; however, protocols such as nick translation and random-primer labeling should be used with caution since they may leave partially filled-in DNA, which will significantly alter quantitative measurements and produce smeared bands. For most applications, labeling to high specific activity is not necessary since as little as 200 cpm is sufficient to detect bands after overnight exposure of the X-ray film with an intensifying screen. After labeling, the unincorporated nucleotides may be removed by chromatography on a 1-ml Sephadex G-50 column or by similar methods. Purification or removal of unincorporated nucleotides by using DNA adsorption to silica (e.g., Magic minipreps and Magic DNA clean up kits; Promega, Madison, Wis.) may be used, but frequently a small portion of the DNA remains bound with silica and is observed as a band that fails to migrate into the gel. Gel purification can be used to isolate the fragment from vector fragments as well as free nucleotides (see Chapters 16 and 22). The DNA should be stored in a buffer such as 10 mM Tris (pH 7.5)–50 mM NaCl–1 mM EDTA. Storage of small fragments in low-salt buffers such as TE may allow some denaturation of the DNA, resulting in the appearance of a band migrating more slowly than the intact dsDNA, exactly where a protein-DNA complex is anticipated.

20.2.1.2. Gels and buffers

Polyacrylamide and agarose gels have both been used successfully for mobility shift assays. Polyacrylamide gels may be 4 to 10% with 0.33 to 0.7% bisacrylamide, depending on the size of the DNA and the protein. For most regulatory proteins and DNA fragments of less than 600 bp, 4 to 6% acrylamide is a good concentration for pilot experiments. NuSieve agarose (FMC Corp.) at 3 to 4% may be substituted in most cases since it has sieving properties similar to those of polyacrylamide. Regular agarose is generally useful only when the protein is large or multiprotein complexes are bound.

Many different gel buffers have been used. In the first gel mobility shift experiments, very low salt TE (10 mM Tris, 1 mM EDTA [pH 7.5]) buffer was used to maximize protein binding during electrophoresis (14, 19). The low buffering capacity requires that a circulator be used during electrophoresis; in as little as 30 min without circulation, the pH of the gel can drop to ≤ 5. Usually 0.5× TBE (45 mM Tris-borate, 1 mM EDTA [pH 8.3]) and TAE (40 mM Tris-acetate, 15 mM Sodium acetate, 1 mM EDTA [pH 8.0]) may be substituted; these buffers have the advantage that no circulation is necessary. Care should be taken with TBE, however, since borate may significantly affect the DNA-binding properties of protein. The gel buffer need not be the same as the buffer used for the binding reaction. Some cofactors required for stable binding of protein to DNA also may be necessary in the gel, e.g., metal ions such as Mg^{2+} or allosteric effectors, such as cyclic AMP (cAMP) for *E. coli* cAMP receptor protein.

20.2.1.3. DNA labeling

The following protocol is for filling in the ends of restriction fragments with 5′ overhangs. End labeling with T4 polynucleotide kinase (section 20.4) may be used for 5′ overhangs or blunt-end fragments. For fragments with 3′ overhangs, T4 polymerase can be used to digest the ends back and then to fill in with labeled nucleotides (46). Incorporation of radioactivity may be measured by placing a portion of the sample in the scintillation counter without using scintillation fluid and by using the discriminator setting for tritium or an open setting. This method of counting by the "Cerenkov effect" is only about 50% as efficient as counting ^{32}P with a scintillation fluid, but the method is quick and does not destroy the counted sample.

1. Use an appropriate restriction enzyme in a volume of 40 μl to digest 5 to 10 μg of plasmid DNA containing the region of interest (see Chapter 16 for plasmid purification and handling). Inactivate the enzyme by heating to 70°C for 10 min. For heat-resistant enzyme, remove by phenol extraction and ethanol precipitation or with a silica cleanup system (Magic DNA clean up). Tables for the heat inactivation of restriction enzymes may be found in the catalogs supplied by the manufacturers, such as New England BioLabs.

2. If a low- or moderate-salt buffer or a 1× concentration of a "universal" buffer (50 mM potassium acetate or glutamate) is used for the restriction digestion, proceed to step 3. If a high-salt buffer is used or the sample was treated with phenol, remove the buffer by ethanol precipitation with 1/10 volume of 3 M sodium acetate (pH 5)–2.5 volumes of ethanol at −20°C overnight or −70°C for 30 min. Alternatively, isolate the DNA by using a silica method. If a silica method is used, follow the manufacturer's instructions but repeat the final elution step and carefully draw off the supernatant to reduce the amount of silica carried over in the sample. Resuspend the DNA in 40 μl of TE (Table 1).

3. Add 50 μCi of [α-^{32}P]dNTP and 1 μl of a 10 mM stock of each of the other dNTPs as necessary to fill in the fragment ends. The nucleotides used will depend on the restriction enzyme, but it is best that the labeled nucleotide is not the last incorporated. If the DNA is in TE, add 5 μl of 10× polymerase buffer (500 mM Tris HCl [pH 8.0], 100 mM $MgCl_2$, 10 mM dithiothreitol, 500 mg of bovine serum albumin [BSA] per ml). Add 2 U of Klenow DNA polymerase fragment, and incubate at room temperature for 15 min.

4. Add 2 μl of a solution of all four nucleotides (5 mM each), and continue the incubation for 5 min.

5. Ethanol precipitate as described in step 2, and dry. Centrifuge for 10 min in a microcentrifuge, remove the supernatant, add 70% ethanol to the pellet without disturbing it, centrifuge for 2 min, remove the supernatant, and dry the pellet in air or in a vacuum concentrator. Resuspend the DNA in 20 μl of TE, and add 2 μl of 10× loading buffer (50% glycerol, 0.05% bromophenol blue).

6. Purify the fragment of interest by electrophoresis on agarose or NuSieve agarose, depending on size. NuSieve agarose (3% NuSieve, 1% agarose) will separate fragments well in the low range of 50 to 500 bp. After electrophoresis in the presence of 0.6 μg of ethidium bromide per ml, visualize the band with UV light and make an incision in the gel just ahead of the band. Place a small piece of NA 45 membrane (Schleicher & Schuell, Keene, N.H.) in the incision down to the gel tray. Electrophorese the DNA onto the membrane at twice the voltage used for the gel run; take care not to run additional bands onto the membrane. Remove the membrane, wash in 1 ml of low-salt buffer (10 mM Tris [pH 7.5], 100 mM NaCl), and elute for 45 min in 300 μl of high-salt buffer (10 mM Tris [pH 7.5], 1 M NaCl) at 70°C as specified by the manufacturer. Remove the membrane, and precipitate the DNA by adding 1.5 μl of 1 M $MgCl_2$ and 2.5 volumes of ethanol, hold at −70°C for 30 min, centrifuge for 10 min, and wash the pellet with 70% ethanol. Resuspend in 100 μl of TES (Table 1), and store frozen. This method is simple and produces quite clean DNA with a specific activity of $>5 \times 10^6$ cpm per μg of fragment.

Silica methods also work well (Qiaex gel extraction kit, Qiagen Inc., Chatsworth, CA) to isolate the DNA from NuSieve or standard agarose gels, but the silica and irreversibly bound DNA must be removed by a repeated centrifugation step. For fragments less than 100 bp, such as synthetic oligonucleotides, use the Mermaid kit (Bio 101, La Jolla, CA). The gel purification may be done before labeling, especially if several fragments are produced by the restriction digestion. In this case, remove the free nucleotides by column chromatography after labeling.

20.2.1.4. Gel preparation

The binding assay can be done most easily on horizontal, "submarine" gels (26); however, since horizontal acrylamide gel apparatuses are not used widely, the following procedure is for a standard vertical system. It is best if the vertical gel apparatus has fittings for buffer circulation, but this is not essential if the running time is limited.

1. Cast a 1.5-mm-thick gel with standard size plates (usually 14 to 16 cm long) and a comb to produce wells 5 to 10 mm wide. Use 6% polyacrylamide–0.1% bisacrylamide in TAE buffer, which is filtered and degassed. A 20% polyacrylamide–0.33% bisacrylamide stock can be stored in the dark at 4°C for up to several months; store TAE as a 10× stock.

2. Polymerize with 20 μl of TEMED and 200 μl of fresh 10% ammonium persulfate for 50 ml of gel.

3. If available, set up a pump to circulate the gel buffer. The gel should be run in the cold for pilot experiments until the stability of the complex is known.

4. Soak the gel in "running buffer" (the same as the gel buffer) for several hours (preferably overnight) to allow the TEMED and persulfate to diffuse out. Add the ligands necessary for the protein (such as cAMP for the cAMP receptor protein) at this time. For horizontal gels, soak for 1 h.

5. Prerun the gel at 5 to 10 V/cm for 30 min to 1 h. If the buffer is not circulated, mix the buffer from the upper and lower tanks or replace the buffer before loading the samples.

20.2.1.5. Binding reactions and electrophoresis

1. Make an appropriate binding buffer. Usually some salt is added to reduce nonspecific binding, components are added to stabilize the protein, and glycerol is added to allow layering on the gel. A typical buffer consists of 10 mM Tris, 50 mM KCl, 1 mM EDTA, 5% glycerol, 50 μg of BSA per ml, and 1 mM dithiothreitol (DTT). Nonidet P-40 (0.05%) may be added if the protein is

unstable and/or aggregates. An easy method is to make a 5×-concentrated stock with the first four items (store them frozen) and to add the BSA, DTT, water, any additional factors, and DNA to a single tube just before use. Make up 10% more mix than required, to allow for pipetting error.

2. Add enough DNA to make 500 cpm (Cerenkov) for each sample (assuming a single labeled DNA fragment). Pipette 20 μl of the complete binding mix into 1.5-ml microcentrifuge tubes for each binding reaction.

3. Make up the binding mix without DNA for diluting the protein. Make 10-fold serial dilutions of the protein, and maintain on ice for a limited time. *Never vortex solutions containing protein;* mix gently with the pipette tip or by flicking the tube with a finger. For a pilot experiment, alternate adding 1 and 3 μl of each dilution to reaction tubes. For a purified protein, the lowest target concentration can be estimated from the amount of DNA in each sample (generally less than 1 ng) and assuming 100% active protein. A protein at 1 mg/ml should require a dilution of 10^{-4} to 10^{-5}. One tube should have no protein added. Thus, the full range of concentrations can be assayed in 12 reactions.

4. After adding protein, incubate the reactions for 30 min at room temperature. Many proteins require only 5 to 10 min.

5. Gently add 1 μl of 0.05% bromophenol blue (pH 7.5) (the dye should be neutralized with Tris or other buffer before use). Turn off the power (and circulator if used), and load the samples carefully but quickly onto the gel. An important consideration of the assay is that the binding characteristics may be changed when the sample is mixed with gel running buffer. To minimize this, samples are loaded with a minimum of mixing, and frequently the samples are loaded with the current turned on. (*CAUTION:* Take great care to avoid electric shock.) It takes about 1 to 3 min for the complexes to enter the gel. After the dye has entered the gel, the circulator may be turned on.

6. Electrophorese at 5 to 10 V/cm (usually 100 to 150 V) until the dye is one-third to one-half of the way down the gel (1.5 to 2 h). The dye should run with a DNA fragment of about 50 bp. The voltage should be adjusted so that minimal heating occurs.

7. Gently pry the plates apart with a spatula, and pull the top plate away, leaving the gel on one plate. Gently press a piece of filter paper on the gel, and pull it off the plate. Cover with a piece of plastic wrap, and dry in a vacuum dryer (1 h at 80°C).

8. Expose the dried gel to X-ray film. Expose overnight with an enhancing screen (such as DuPont Cronex) for 300 to 1,000 cpm (Cerenkov) per band. Place the screen in the film holder, then add the film, and finally add the gel (facing down). For shorter exposures, or if intensifying screens are not available, use 2,000 to 5,000 cpm of DNA per band.

An example of a protein titration by using the gel mobility shift assay is given in Fig. 3. The relative mobility of the shifted band can vary widely with different combinations of protein and DNA. If only a slight shift is seen, the gel concentration can be increased. Running conditions may also greatly affect the mobility. For *trp* repressor-operator complexes, reducing the pH is necessary to noticeably lower the mobility of the complexes (10). Thus, running a sample of protein alone (detected by silver staining the gel) may show whether the protein runs far into the gel at the pH selected, and conditions can be altered to maximize the difference in mobility of the protein and free DNA. If the shifted band is near the top of the gel, repeat the assay with an agarose gel. Sometimes the free-DNA band disappears with increasing protein, but no complex band appears. In this case the protein is probably dissociating during the run. A lower-salt gel, different binding conditions, or different running buffer may allow more stable complexes.

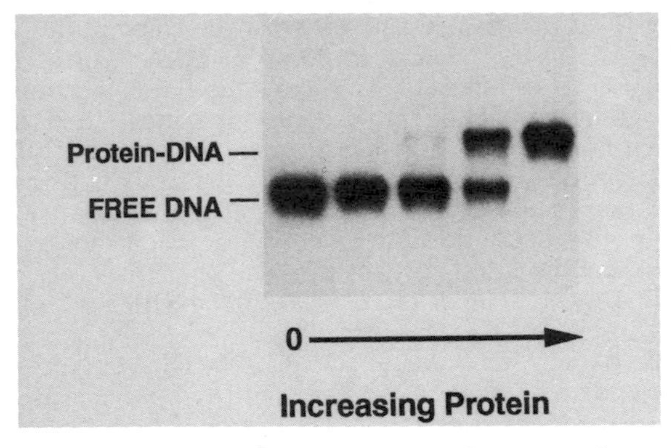

FIG. 3. Gel mobility shift assay. An autoradiogram is shown of an assay with 500 cpm of a 200-bp DNA fragment per sample run on a 6% polyacrylamide gel for 1 h at 8 V/cm. The second through fifth lanes contain samples with threefold increases in protein (purified *E. coli* AraC protein). The film was exposed overnight at −70°C with an intensifying screen.

A common observation is that a single shifted band is observed at one protein concentration and multiple higher bands are seen at higher protein concentrations until all the DNA is near the gel origin at the highest concentrations. Since weak protein-DNA interactions are stabilized in the gel, nonspecific binding is readily detected. Once one molecule of protein is bound at its specific site, additional protein added to the reaction results in moderately stable binding of multiple proteins. Even if no specific binding occurs, nonspecifically bound protein-DNA bands may be observed. Therefore, it is important to carefully titrate the DNA and to characterize the binding with proper controls. One simple control is to add an equal amount of a labeled, irrelevant DNA fragment to the binding mix along with the specific DNA fragment. The specific fragment should shift in position while the control fragment remains unchanged. This is easiest to observe if the control fragment is somewhat smaller than the specific fragment.

The gel mobility shift assay can be used to obtain reasonably accurate quantitative information about the binding reaction. Several methods can be used to determine the equilibrium constant for binding. For any of these methods, the concentration of protein active in specific DNA binding must be determined accurately. The simplest method to accomplish this is to accurately determine the concentration of the DNA fragment containing a single, specific binding site and then titrate the protein against the DNA.

20.2.1.6. Protein concentration

1. Carefully measure the concentration of plasmid DNA containing the binding site of interest by measur-

ing the A_{260} against a blank with the same buffer. An A_{260} of 1 is equivalent to 50 μg of DNA per ml. The A_{260}/A_{280} ratio should be approximately 1.8. It is important that the DNA be free of contaminating RNA. RNase treatment is not sufficient, since mono- and oligonucleotides will still be present even after CsCl purification of the DNA. Use DNA that has been purified twice by CsCl-ethidium bromide gradients or some other method that ensures removal of RNA.

2. Digest 20 μg of the plasmid DNA with restriction enzyme(s).

3. Label a small portion (1 μg) of the DNA by filling in the ends with Klenow fragment of DNA polymerase as described in section 20.2.1.3.

4. Inactivate the enzymes in the labeled sample by heating to 70°C for 20 min or by phenol extraction. Remove the unincorporated label by purification on a 1-ml Sephadex G-50 column equilibrated with TE buffer or with a silica method.

5. Inactivate the restriction enzyme in the unlabeled sample by heating to 70°C. Do not use procedures, such as ethanol precipitation, that could cause loss of DNA during handling.

6. Mix the labeled and unlabeled samples. The molar concentration of the binding site is calculated from the molecular weight of the plasmid. Since the labeled sample is only 5% of the total, even a significant loss during handling will not affect the calculation. If the affinity of the protein for the specific site is thought to be weak or the nonspecific binding is strong, protein binding to the vector sequences will alter the results. In such cases, at least 1 μg of the fragment must be purified after restriction cutting and the fragment concentration must be determined spectrophotometrically. Do not add carriers such as tRNA or BSA, since these will interfere with the readings.

7. Perform the binding assay as described in section 20.2.1.5, using less than a twofold increase in protein content for each sample. The DNA concentration should be high so that the measurements are above the dissociation binding constant of the protein. For most regulatory proteins, a concentration above 10 nM is sufficient, and 100 nM would be sufficient for even very low affinity specific binding reactions. The binding can be quantitatively measured by densitometry of the gel as long as data are within the linear range of the film. Generally the film is preflashed (cover a camera flash attachment with two layers of 3-mm paper, and flash from a distance of about 2 ft [61 cm]) so that the A_{450} is about 0.1. The film is then tested by exposing a range of labeled standards. Alternatively, the bands may be cut from the gel and measured directly in a scintillation counter. As long as the excised gel fragments are nearly equal in size, quenching of radiation for ^{32}P-labeled samples is not a problem.

8. For the most accurate calculation, plot (1 − fraction DNA not bound) versus protein. The fraction of free DNA in each sample should be calculated by dividing the counts in the sample band by the counts in the band from a control lane with DNA but with no protein. Since some of the protein-DNA complex may be lost during the gel run, the fraction of DNA in the complex sometimes is not an accurate reflection of the amount of DNA bound in solution. The plot should be linear to well over 50% DNA bound. At the point at which 50% of the DNA is bound, the protein is equal to one-half of the amount of DNA, and the active-protein concentration can be calculated. It is quite common for highly purified proteins to be 10 to 50% active in specific binding. This method is also useful for estimating the amount of active protein from partially or unpurified samples (see below) as long as the impurities do not strongly interfere with the binding reaction.

20.2.1.7. Equilibrium binding constant

The simplest method to estimate the equilibrium binding constant is to perform a titration as described in section 20.2.1.5; however, in this assay the DNA concentration is held at a lower level than the equilibrium binding concentration. If the midpoint for binding is obtained under conditions where the protein concentration is in vast excess over the DNA concentration, the equilibrium dissociation constant is identical to the protein concentration. This follows from the simple binding relationship $P_f + D_f \rightleftharpoons P \cdot D$, where P_f is free protein, D_f is free DNA, and PD is the complex. The equilibrium dissociation constant, K_d, is equal to $[P_f][D_f]/[PD]$. At the point at which 50% of the DNA is bound, $[PD] = [D_f]$. Substituting $[D_f]$ for $[PD]$ in the equation gives $K_d = [P_{f50\%}]$.

If total-protein concentration $[P_t] \gg [D_t]$, the amount of protein in the complex is negligible and $[P_t] \approx [P_f]$. Thus, $K_d \approx [P_t]$. If the DNA concentration is 10-fold lower than the protein concentration, $[P_f]$ is only about 5% lower than $[P_t]$ and the value of K_d obtained will be well within the experimental variation of the assay. The equilibrium constant may also be expressed as the equilibrium association constant, K_a, which is simply the inverse of the dissociation constant.

Equilibrium dissociation constants of 10^{-7} to 10^{-13} M have been obtained for regulatory proteins in *E. coli*, with most falling in the 10^{-9} to 10^{-12} M range (8–10, 14, 16, 17). The constant may be affected strongly by the binding conditions. The electrostatic contribution to the binding free energy is highly dependent on the concentration and type of salt in the buffer. Many proteins bind with much higher affinity at lower salt concentrations and with anions such as acetate or glutamate than with chloride. All sequence-specific DNA-binding proteins also bind nonspecifically. The nonspecific binding constant is mostly electrostatic and therefore more salt sensitive. To maximize specific over nonspecific binding, an optimum salt concentration and type can be determined. A good starting point is to use 50 mM KCl and test increasing concentrations. A buffer with potassium acetate may also be beneficial.

1. Use DNA labeled to high specific activity. Since only 100 cpm is necessary per reaction, DNA concentrations in the range of 10^{-11} M are easily obtained in a 20-μl reaction. For lower concentrations, alternative labeling schemes are needed. T4 polymerase can be used to digest and then fill in one strand of DNA with labeled nucleosides (46), or nick translation may be used if the procedure is optimized so that intact fragments are obtained.

2. Perform a gel mobility shift assay as described in section 20.2.1.5. Allow plenty of time for the reaction to come to equilibrium. The approach to equilibrium is proportional to the product of the concentrations of the reactants and the association and dissociation rate constants, so at very low concentrations the time needed is

greatly extended. Nevertheless, most systems will be at equilibrium within 30 min. Use a wide range of protein concentrations starting with the estimated concentration of the DNA. An easy method is to use 1 and 3 μl of each of a series of 10-fold serial dilutions. In this way a 10^4-fold range in protein concentration can be tested in just eight samples. Very carefully load the gel while it is running to minimize the mixing of the sample with running buffer. Since protein binding is salt sensitive, the effects of exposure to a low-salt running buffer are a major concern; however, experiments with a number of well-characterized proteins suggest that data from carefully performed gel mobility shift assays with low-salt running buffer closely match binding data from other methods such as DNase footprinting. At 8 V/cm, a sample in 50 mM KCl will fully enter the gel in about 1 min. Run the gel for a sufficient time to resolve the complexes, dry it, and expose it to X-ray film. At the lowest levels of label, exposures of several days may be necessary without an enhancing screen.

3. Determine the fraction of free DNA as described in section 20.2.1.6, and plot the fraction free (or fraction bound) against the active-protein concentration. If necessary, correct for any DNA that is not bound even at very high protein concentrations. This is sometimes a factor when using hybridized, synthetic oligonucleotides. If densitometry of autoradiograms is performed, the linearity of the response must be checked under the conditions of the binding assay and film exposure. Accurate quantitation also may be achieved by direct counting of the gel in a beta-scanning device or excising the bands and placing them in a scintillation counter. Mincing the gel slices and placing them in an aqueous counting cocktail is sufficient for good counting efficiency of ^{32}P-labeled samples. An alternative, simple method to obtain a reasonable estimate ($\pm$ a factor of 2) of the binding affinity is to visually estimate the binding half-point. For initial characterization of a protein, such accuracy is usually sufficient.

For the simple, bimolecular reaction, the binding curve should follow the relation $K_d = [P_t](1 - f)/f$, where f is the fraction of DNA bound. If the binding data do not fit a theoretical curve assuming this relation, cooperativity, multiple binding sites, interference by partially active protein, or some other factor may be involved (8, 9, 29). The gel mobility shift assay is especially useful in determining the quantitative parameters of cooperative binding, and detailed discussions of such measurements have been published (8, 9, 29). It is important to validate such measurements by an independent method, and DNase I protection assays (see section 20.2.2.1) are frequently useful.

20.2.1.8. Kinetics

Rate measurements are often used to confirm the equilibrium measurements, as well as to compare differences in related binding sites such as those obtained from mutants. Dissociation rates for many regulatory proteins, with reaction half times of minutes, are easily measured with great precision. Since the dissociation reaction is first order, it is independent of the concentration of the reactants. The dissociation assay is as follows.

1. Perform a gel mobility shift assay (section 20.2.1.5). Add just sufficient protein to bind most of the DNA. Do not add excess. Allow the binding reaction to come to equilibrium.

2. Add a 20- to 100-fold molar excess of an unlabeled DNA containing the binding site. This DNA can be synthetic oligonucleotides, a purified restriction fragment, or a whole plasmid. The excess unlabeled site sequesters the protein as it dissociates; therefore, the most precise measurements are made with the maximum competitor. Experiments should also be performed with different levels of competitor to ensure that there are no concentration-dependent effects.

3. At intervals, load samples on the gel while it is running. These samples can be removed from a single reaction tube or a series of parallel tubes. The dissociation is stopped by loading on the gel. In moderate-salt buffer (50 mM KCl), permeation into the gel adds about 1 min to the effective dissociation time. If samples are removed from a single tube, each time point will run on the gel for a different length of time. For quick dissociation reactions this can work well, but for reactions spanning more than 1 h it can be problematic. If separate tubes are used, the starting points for each sample can be adjusted so that all the reactions are loaded on the gel within a short period (30 s for each sample).

4. Quantitate the amount of DNA in the lanes as described in section 20.2.1.6. The dissociation rate constant is equal to the slope of a plot of the log fraction bound against time. Differences in dissociation rates of less than twofold are easily detected by this assay. Adjustments in the salt concentration or type usually will alter the rate so that it falls within a range that can be measured; however, proteins with very weak binding affinities may dissociate substantially during the time required for the complexes to enter the gel.

Association rates are more difficult to measure and will not be treated in detail here (50). Since the association rate is proportional to the product of the free DNA and protein concentrations, both must be determined. The reaction is stopped by adding an unlabeled competitor or by loading on the gel. Unlabeled competitor can be used only if the dissociation rate is insignificant compared with the time required to process the sample, and quenching by loading on the gel is useful only for relatively slow reactions that are much longer than 1 min.

20.2.1.9. Stoichiometry

Accurate determination of the stoichiometry of the protein-DNA complex can be difficult. Double-label experiments are performed with ^{32}P-labeled DNA and ^{3}H-, ^{125}I-, or ^{35}S-labeled protein. The relative amounts of label in the complexes are determined by cutting out and counting bands after a mobility shift experiment. To calculate the molar ratios of protein to DNA, specific activities of both must be known. One method is given below; detailed treatments of several methods have been published (15, 20, 27, 64).

1. Label and calculate the specific activity of the DNA as in section 20.2.1.6. High levels of DNA in the mobility shift assay are required to produce sufficient complexes to measure labeled protein; however, the specific activity must be low to minimize overlap with the labeled protein during scintillation counting. Adjust the DNA specific activity to approximately 100 cpm/ng of fragment.

2. Label the protein by growing cells in medium lacking methionine with 1 mCi of [^{35}S]methionine added per 5 ml of culture. Purify the protein. The protein also can be labeled after purification (15). If the extinction coefficient is known, the specific activity can be calculated directly and a correction can be applied for the percentage of inactive protein. The protein activity is measured by running a mobility shift assay with excess, unlabeled DNA and measuring the ratio of labeled protein in specific complexes against input protein. This can be done only if the complexes are stable during the gel run. If the extinction coefficient is not known, the protein specific activity is calculated from the specific activity of the methionine in the growth medium and the number of labeled amino acids in the protein (10, 27). Alternatively, the protein concentration can be deduced from quantitative amino acid analysis (64). Labeling before purification is feasible only if the protein is overproduced. Protein overproduced with the T7 polymerase-promoter system (57) can be specifically labeled by adding rifampin after induction of the promoter. This eliminates the background of labeled contaminants and may allow measurements with partially purified proteins.

3. Perform the gel mobility shift assay (section 20.2.1.5) with 20 ng of DNA per sample and sufficient protein to bind 90% of the DNA.

4. Expose the wet gel to film or stain the gel for 30 min in ethidium bromide. Use the developed film or UV-illuminate a stained gel to locate bands containing complexes. Cut out the bands, and place in scintillation vials.

5. Dissolve the gel slice by adding 1 ml of 21% H_2O_2 and 17% perchloric acid and incubating overnight in a tightly capped scintillation vial at 60°C. Add 19 ml of aqueous scintillation fluid, and store the samples overnight in the dark before counting. Excise blank regions of the gel, and treat similarly to determine backgrounds. It is especially important to count sections of the gel where complexes migrate by using a sample with protein alone to determine background in the region. Calculate quenching by adding known quantities of DNA or protein to blank gel fragments. Calculate the overlap of ^{32}P into the other channel of the scintillation counter by using identically treated standards.

6. Convert the counts to moles of DNA and moles of protein by applying the quenching and overlap corrections to the calculated specific activities. Ratios of protein monomer to DNA are then calculated.

20.2.1.10. Assay with crude protein

One of the most useful aspects of the gel mobility shift assay is the capability it provides to identify and monitor the purification of sequence-specific DNA-binding proteins. Proteins from mutants also can be screened rapidly for qualitative or quantitative changes in DNA binding. The assay is performed as usual, with the exception that binding of irrelevant proteins to the target DNA must be inhibited. In addition, the specificity of binding must be conclusively demonstrated by competition assays. The simplest assay is to mix cell extract with two labeled fragments simultaneously; one carries the specific protein-binding site, and the other is a nonspecific control. Differential shifting of the specific DNA indicates the level of protein necessary to achieve specific binding. Proteins present in low concentrations or with very low binding affinity may not be detected above the nonspecific background. In such cases the protein must be overproduced or initial purification steps must be performed prior to the assay. Small amounts of the protein may be purified by a preparative mobility shift procedure; the yield is governed by the level of DNA in the binding reaction.

The following protocol is for detecting binding proteins from *E. coli* extracts. For other bacteria the lysis procedure may be altered and additional protease inhibitors may be added as needed.

1. Label a DNA fragment containing the specific binding site. Small fragments will have fewer sites for nonspecific binding and are preferred for lysate assays. Use 2,000 cpm per reaction so that as little as 5% binding can be detected. Higher levels of DNA can be used, but nonspecific binding becomes more of a problem as the amount of protein needed to bind the DNA is increased. Also label a second DNA fragment, such as another fragment from the plasmid vector, to use as a nonspecific control. The second fragment should be easily distinguished from the target fragment when run on the gel. It is easier to observe binding if the target is larger than the nonspecific fragment so that the protein-DNA complexes are not obscured on the gel.

2. Culture 10 ml of bacteria to late log phase (maximum density) in rich medium. Centrifuge, resuspend in 1 ml of TES buffer (Table 1), and transfer to a 1.5-ml centrifuge tube. Centrifuge again, and resuspend in 1 ml of lysate buffer (50 mM Tris HCl [pH 7.5], 100 mM KCl, 10% glycerol, 1 mM EDTA, 1 mM DTT, 160 μg of phenylmethylsulfonyl fluoride per ml [16-mg/ml stock in 95% ethanol made up fresh]).

3. Sonicate with a microprobe. Set the sonicator on the maximum setting appropriate for the probe, and sonicate the sample on ice for two to four 10-s pulses, allowing 30 s between pulses. It is important to keep the sample cool and to avoid creating foam, which can occur if the probe is too close to the meniscus. If foaming is a problem, double the volumes in step 2. If necessary, the energy level of the probe may be reduced. Check the cell lysis by viewing a few microliters of the sample under a microscope. Use the minimum sonication necessary to lyse 90% of the cells. Monitoring the reduction in A_{550} with a spectrophotometer will also provide an estimate of cell lysis; look for about 80% reduction.

4. Centrifuge at $\geq 12{,}000 \times g$ for 20 min at 4°C. Centrifugation at $40{,}000 \times g$ can be used to obtain a slightly more pure preparation. Transfer the supernatant to a new tube, and add one-fifth volume of cold glycerol. Mix gently with a pipette. Remove a portion of the sample for assay, and store aliquots of the remainder at −70°C. Many proteins will remain active for days at −20°C under these conditions. Some activity (up to 50%) is lost on freezing, but the remaining activity is usually stable for months at −70°C. Thaw aliquots by placing them at −20°C, and gently mix after thawing.

5. Make serial dilutions, and perform a gel mobility shift assay (section 20.2.1.5). Add equal counts per minute of the specific and nonspecific DNA to all samples. If the protein is highly overproduced ($\geq$1% of the total protein), use the normal binding buffer. The chromosomal DNA in the lysate should block binding of nonspecific proteins. For lysates with lower concentrations of the specific protein, add a competitor to the reaction. Both the concentration and type of competi-

tor must be determined empirically. Start with 0.1 to 1 μg of poly(dI-dC) per reaction. This competitor has the lowest probability of inhibiting specific binding. Other competitors include calf thymus, salmon sperm, or *E. coli* DNA. A large number of samples are run to test different concentrations of competitor with various levels of protein. The goal is to find the lowest levels of competitor and protein that produce a shift in the specific DNA fragment without shifting the nonspecific fragment. With crude lysates, only a portion of the specific DNA may shift at protein levels below those at which nonspecific binding interferes. Further purification of the protein or its overproduction will increase the signal.

6. Small amounts of column buffers do not strongly affect the assay, so fractions from purification steps can be assayed directly. A total monovalent salt concentration above 200 mM may cause the bands in the gel to smear badly, as will >20 mM divalent cations and >100 mM phosphate. Since binding highly labeled DNA requires less than 1 ng of active protein, dilutions of 100- to 1,000-fold are usually required for proteins from lysates of overproducer strains or from fractions obtained during purifications.

7. To confirm the specificity of the binding reaction, perform additional competition assays. In a series of tubes, add the labeled specific DNA and increasing levels of *unlabeled* specific DNA competitor. Then add sufficient protein to bind the labeled DNA alone; do not add excess. The competitor concentration should vary from 0- to 100-fold excess over the labeled DNA. A similar set of reactions is done with unlabeled nonspecific competitor. The nonspecific competitor should be as similar as possible to the specific competitor; do not use calf thymus DNA. If the assay is detecting specific binding, the level of specific competitor required to block binding should be significantly lower than the level of nonspecific competitor.

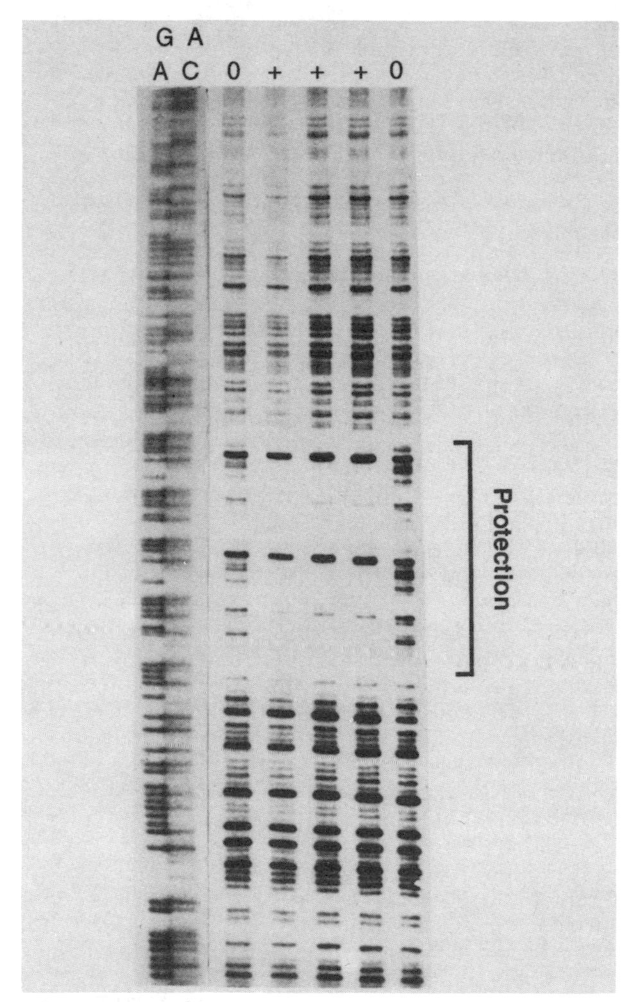

FIG. 4. DNase I footprint. An autoradiogram is shown of a footprint assay of AraC protein binding to a synthetic binding site cloned into the *araFGH* promoter. G+A and A+C chemical sequencing reactions (37) were run in the first two lanes. 0, reactions with no protein; +, reactions with a twofold excess of protein to DNA. A 5,000-cpm portion of each sample was loaded onto an 8% sequencing gel. The protected region is indicated by a bracket. Two bands of strongly enhanced cleavage appear within the protected region.

20.2.2. DNA Protection (Footprinting) Assays

An alternative DNA-protein-binding assay is to cleave or modify DNA bound by a protein such that, on average, each DNA molecule is cleaved once. On a sequencing gel, a "ladder" of cleavage products is produced, much like a sequencing reaction. The DNA region covered by protein will be protected from cleavage or modification, creating a footprint in the cleavage pattern as compared with an unbound control (Fig. 4). Sequencing reactions of the same DNA are run in parallel to determine the position of the binding site. Both enzymatic and chemical modification techniques have been developed (reviewed in reference 56). An assay involving cleavage by DNase I is capable of localizing the protein-binding site to within a few base pairs (18). An assay relying on the modification of DNA by dimethyl sulfate (DMS) provides additional information beyond the general region covered by the protein (52). This method also identifies the close association of protein with guanine residues in the DNA major groove and adenine residues in the DNA minor groove. Both methods can also provide quantitative data on binding affinity, kinetics, cooperativity, and stoichiometry. Both also have been adapted to obtain protection data with supercoiled DNA and whole cells (21, 31, 40, 48). For an in vivo technique, DNA is modified by adding DMS (see below) to cells. After the modified DNA has been isolated, primer extension reactions with a labeled oligonucleotide (section 20.3.2) are performed to create a set of fragments ending at the modified sites. When run on sequencing gels, the data appear similar to those obtained by the direct method with labeled restriction fragments.

Other methods can provide additional information. The recently developed hydroxyl radical footprinting technique provides high-resolution data on the phosphates of the DNA backbone that are protected by bound protein (57, 58), and the hydroxyl radical interference assay indicates specific base interactions with protein (22). The hydroxyl radical is produced by the reaction of hydrogen peroxide with iron(II)-EDTA. The advantage of this method is that all bases are modified, whereas cleavage by DNase I is not random and gaps may appear in the regions sampled. Since the DNase I molecule is large compared with the hydroxyl radical, the enzymatic approach is also much lower in resolution. Chemical agents such as copper phenanthroline and potassium permanganate have been used to study changes in the DNA structure in the complex (7, 33). Both have been used to detect DNA unwinding at promoters. Two of the simplest footprinting methods are described here; for the other methods, see references 2, 56, and 58.

20.2.2.1. DNase I protection assay

1. DNA must be labeled on one end only. Do this by filling in the end with Klenow fragment polymerase (section 20.2.1.3.) or by adding a labeled phosphate with T4 polynucleotide kinase (section 20.3.1.3). The easiest method is to cut a plasmid containing the insert to be studied with one restriction enzyme, label the DNA end, and extract with phenol to inactivate the enzyme. Then cut on the opposite end of the DNA insert with a second enzyme, and purify the single end-labeled fragment by electrophoresis onto NA45 membranes or by a silica protocol (section 20.2.1.3). The overall length of the DNA fragment is not critical, but the labeled end must be within 300 bp of the binding site. Beyond this distance, the separation on a sequencing gel is insufficient to obtain a convincing footprint.

2. Perform sequencing reactions on the DNA as described in section 20.1. If the chain termination method (section 20.1.1) is used, the 5′ end of the sequencing primer must exactly match the 5′ end of the labeled DNA fragment used for footprinting.

3. Add 10,000 to 20,000 cpm (Cerenkov) of DNA to each binding reaction. Various buffer conditions are acceptable (0 to 200 mM salt, pH 6 to 8), but all reactions must contain at least 3 mM Mg^{2+} for the DNase I. $CaCl_2$ is usually added to optimize the enzyme also. A common buffer contains 20 mM Tris HCl (pH 7.5), 3 mM $MgCl_2$, 1 mM $CaCl_2$, 0.1 mM EDTA, 50 mM KCl, 1 mM DTT, and 50 μg of BSA per ml in a volume of 100 μl per reaction. Calf thymus or salmon sperm DNA also can be added (50 ng per reaction) to avoid variations in the activity of the DNase I caused by differences among the samples in the availability of DNA substrate.

4. Make serial dilutions of the protein in the binding buffer, and keep on ice. Make a 1-mg/ml solution of DNase I in 20 mM Tris HCl (pH 7.5)–3 mM $MgCl_2$–1 mM $CaCl_2$–0.1 mM EDTA–50 mM KCl–50% glycerol. This solution can be stored for months at −20°C. Just before use, dilute the DNase I to 2 μg/ml in binding buffer and keep on ice.

5. Set up reaction tubes with no protein and with increasing concentrations of protein, and incubate at room temperature for 30 min. Add 3 μl of DNase I (final concentration, 0.06 μg/ml) to the first tube, mix gently but quickly, and incubate for 30 s at room temperature. Stop the reaction by adding 50 μl of 4 M ammonium acetate and 5 μg of yeast tRNA (or any other carrier such as DNA or glycogen), vortex, and place the sample on ice. Repeat the procedure for each sample. To a second set of reactions, repeat the procedure with 10 μl of DNase I added. If carrier DNA was added, or if the protein is not purified, it may be necessary to add higher concentrations of DNase I (up to a final concentration of 1 μg/ml).

6. Add 400 μl of 95% ethanol to each sample, vortex, and incubate at −20°C for several hours or at −70°C for 30 min. Centrifuge for 10 min (preferably at 4°C), carefully discard the supernatant, rinse the pellet with 70% ethanol, and dry. Scintillation count the samples (Cerenkov) to determine recoveries, and resuspend in 80% formamide–dye mix, adjusting the volumes so that equal counts are loaded on the gel. Also check to make sure that the DNA is resuspended in the formamide by removing the sample and counting the empty tube. If much of the DNA remains in the tube, samples can be resuspended in 5 μl of TE and then mixed with 5 μl of formamide-loading buffer. If the problem persists, it may be necessary to extract the protein in the samples with phenol before the precipitation step.

7. Denature the DNA by heating the sample tubes to 90°C for 3 min, and load no more than 5 μl on a sequencing gel as described in section 20.1.1.8. Also load the sequencing reactions. The gel running time will depend on the location of the binding site with respect to the labeled end of the DNA. A second loading of samples after 1.5 h can be used to maximize the readable sequence.

8. Expose the gel to X-ray film. If the samples loaded on the gel contain less than 5,000 cpm, an enhancing screen (e.g., DuPont Cronex) can be used to increase the sensitivity by fivefold; however, some resolution will be lost since the bands will be more diffuse. Expose the film overnight at room temperature (−70°C if an enhancing screen is used), and develop. The location of the protein-binding site is determined by comparing the bands in the no-protein sample with the bands in the other samples. Missing bands indicate a footprint. Within the footprint there also may be one or more darker bands indicating enhanced cleavage caused by protein binding.

The two concentrations of DNase should provide maximum intensity of bands over a broad stretch of DNA. If the bands are too light and most of the DNA is uncut at the top of the gel, increase the DNase concentration and/or increase the incubation time to 1 min. If there is little or no uncut DNA, decrease the DNase concentration. Purified proteins are usually added in such small amounts that the protein will not interfere with the sequencing gel. If the bands are smeared, it may be necessary to extract the DNA with phenol before performing the ethanol precipitation in step 6. It is advisable to perform the phenol extraction for crude lysates or partially purified proteins. At high concentrations of protein or in samples with significant protein contaminants, the DNA may be coated with protein, causing changes in DNA bands over a large stretch of DNA. The apparent activity of the DNase also may be reduced. If necessary, add a competitor such as poly(dI-dC) to reduce background binding and add additional DNAse.

20.2.2.2. Dimethyl sulfate protection assay

The DMS assay is performed much like the DNase protection assay, except that the sites modified by the DMS must be cleaved in a subsequent reaction (52). The procedure is as follows.

1. End label and purify the DNA as described in 20.3.1.3.

2. For each sample, add 20,000 cpm of DNA to binding buffer (20 mM Tris HCl [pH 7.5], 50 mM KCl, 1 mM

EDTA, 50 μg of BSA per ml) in a volume of 100 μl in a 1.5-ml centrifuge tube. Set up sufficient tubes for a no-protein standard and the necessary protein dilutions.

3. Just before use, make protein dilutions in binding buffer. Add protein to sample tubes, and incubate at RT for 30 min.

4. To the first sample, add 1 μl of DMS and incubate at RT for 1 min. (*CAUTION:* DMS is a carcinogen! Use it in a hood, and take proper precautions.)

5. Stop the reaction by adding 25 μl of DMS stop solution (1 M Tris HCl [pH 7.5], 1.5 M sodium acetate, 1 M β-mercaptoethanol, and 50 mM magnesium acetate, stored at 4°C), 5 μg of yeast tRNA, and 400 μl of 95% ethanol. Place on ice, and repeat steps 4 and 5 with the rest of the samples.

6. Precipitate for several hours at −20°C or 30 min at −70°C. Centrifuge for 10 minutes, discard the supernatant, wash the pellet with 70% ethanol, dry, and resuspend in 100 μl of 1 M piperidine. The piperidine is highly volatile and should be diluted to 1 M (the stock is 10 M) on the day it is used. Place the samples in tightly capped 0.5 ml centrifuge tubes, and heat in a 90°C water bath for 30 min to cleave the DNA at the modified sites. It may be necessary to place a weight on the caps to prevent them from popping open. This reaction will cleave guanines methylated at the N-7 position in the major groove of the DNA and will partially cleave adenines modified at the N-3 position in the minor groove. If increased cleavage at adenines is desired, resuspend the pellet in 85 μl of 20 mM sodium acetate instead of piperidine, add 15 μl of 1 M NaOH, vortex, and incubate at 90°C for 30 min.

7. After the incubation, neutralize samples containing NaOH with 15 μl of 1 M HCl and add 5 μg of carrier tRNA. Add 250 μl of 95% ethanol to all samples, precipitate, centrifuge, wash with 70% ethanol, and dry.

8. Measure the counts per minute by Cerenkov counts. Resuspend the pellet in 80% formamide loading buffer, adjusting the volumes so that equal counts are loaded for each sample. Heat the samples to 90°C, and load no more than 5 μl on a sequencing gel along with the sequence standard. Process the gel as described (section 20.1.1.8), and expose the film overnight. Reduced and enhanced cleavages are seen at the position of the protein-binding site, as with the DNase protection assay. For overnight exposure of less than 5,000 cpm, an enhancing screen should be used.

20.3. RNA TRANSCRIPT MAPPING

Once a gene or promoter has been cloned, it is of interest to determine the start site and size of the RNA transcript. There are two common methods for determining the start site; each has advantages and disadvantages. For the S1 nuclease protection assay, a restriction fragment that spans the region near the beginning of the transcript is end labeled and hybridized to cellular RNA. The portion of DNA that is not hybridized is removed by S1 endonuclease digestion, and the position of the end of the RNA is then inferred from the size of the remaining DNA fragment. In the primer extension assay, a labeled oligonucleotide primer is hybridized to the RNA and reverse transcriptase extends the DNA to the 5′ end of the RNA. Again, the start site is inferred from the size of the resulting fragment. Both procedures require that the position (sequence) of the 5′ end of the probe be known accurately, and both require only that the 5′-proximal portion of the RNA be intact. The S1 method also can be used to map the 3′ end of the RNA. A simple procedure for determining the overall size of the transcript is to perform a Northern blot hybridization on cellular RNA. Total cellular RNA is separated by size in a denaturing gel and transferred to a membrane, and the specific RNA of interest is detected by hybridization of a labeled probe.

20.3.1. S1 Nuclease Protection Assay

Make sure that all buffers, glassware, and other materials are RNase free. Liquids should be decontaminated of RNase by adding diethylpyrocarbonate (DEPC) (final concentration, 0.2%; Sigma Chemical Co.). Shake the mixture vigorously to dissolve as much DEPC as possible, incubate overnight, and then autoclave to decompose the DEPC to CO_2. Do not treat Tris solutions with DEPC; instead, make them up with DEPC-treated and autoclaved water. Handle all materials only with clean gloves. Autoclaving generally does not destroy all RNase activity. RNases can be readily removed from plasticware by washing with a strong detergent such as Absolve (DuPont) or by soaking in DEPC solution; however, disposables such as microtubes and tips are not contaminated in the manufacturing process, so careful handling should eliminate the need for cleaning. Bake all glassware and spatulas at ≥150°C for ≥2 h. Then proceed with RNA isolation and the assay as follows.

20.3.1.1. Crude-RNA isolation

A simple method for isolating total crude RNA from gram-negative bacteria (45) is as follows.

1. Grow a 50-ml cell culture in a rich medium to late log phase (an A_{550} of 0.7 to 0.8 for *E. coli*).

2. Harvest the cells by pouring the culture into a centrifuge bottle containing 50 ml of ice, 5 mg of chloramphenicol (50-mg/ml stock in 95% ethanol), and 5 mM NaN_3. Pellet the cells by centrifugation at 8,000 × *g* for 10 min. Discard the supernatant, and resuspend the cells in 2 ml of 10 mM Tris HCl (pH 7.4)–10 mM KCl–5 mM $MgCl_2$.

3. Add 0.7 ml of lysozyme from a freshly prepared stock of 10 mg/ml in 10 mM Tris (pH 7.4). Mix well, incubate on ice for 15 min, and freeze in a bath of dry ice-ethanol.

4. Add 0.3 ml of 10% SDS, and thaw in a 65°C water bath. Incubate until the sample clears (<5 min). For a cleaner preparation, proteinase K can be added (0.1 ml of 5-mg/ml stock) and the sample can be incubated at 37°C for 20 min.

5. Add 0.1 ml of 3 M sodium acetate (pH 5.2) followed by 1 volume of TE-saturated phenol. Extract for 5 min by placing the tube on a water bath shaker at 65°C, and set on moderate shaking (best mixing is obtained by capping the tube and laying it on its side). Centrifuge the sample briefly to separate the phases, remove the aqueous phase to a new tube, and repeat the phenol extraction. After recovering the aqueous phase again, measure the volume (estimate by filling a similar tube to the same level).

6. Add sufficient 5 M NaCl to bring the concentration to 2 M NaCl, and then add 1.5 volumes of ethanol and

precipitate at −20°C for 60 min. Centrifuge for 10 min at ≥12,000 × *g*, discard the supernatant, rinse the pellet three times with 70% ethanol, and resuspend it in 1 ml of 10 mM Tris (pH 7.4)–10 mM NaCl. Clarify by centrifugation for 5 min in a microcentrifuge, and remove the sample to a new tube.

7. Read the A_{260} in a spectrophotometer. Most of the absorbance is due to RNA, most of which is rRNA and tRNA. The concentration of RNA is approximately 1 A_{260} unit = 40 μg/ml. A highly expressed mRNA species will represent no more than 1% of the total RNA.

20.3.1.2. Pure-RNA isolation

An alternative method for isolating total RNA is adapted from a method for eukaryotic cells described by Sambrook et al. (46). This method removes DNA and gives a much cleaner intact RNA preparation, especially from more difficult cells containing large amounts of RNase and including gram-positive as well as gram-negative bacteria.

1. Pour a late log-phase culture (50 to 200 ml) over an equal volume of ice-cold ethanol-acetone (1:1), and pellet the cells by centrifugation at 7,000 rpm (Du Pont Sorvall GSA rotor) for 10 min.

2. Lyse the cells. For gram-positive bacteria such as *S. aureus*, wash (resuspend and pellet again) once with SMM (1 M sucrose, 40 mM maleic acid, 40 mM $MgCl_2$ [pH 7.6]) and suspend the cells in 9 ml of SMM. Prepare protoplasts by adding 1 ml of lysostaphin (1 mg/ml in SMM), and incubate in ice for 30 min. Pellet the cells by centrifugation, and resuspend them in 2 ml of 4 M guanidinium thiocyanate (in 0.1 M Tris-HCl [pH 7.5]–1% β-mercaptoethanol). Add sodium lauroyl sarcosinate (Sarkosyl) to a final concentration of 0.5%. For *E. coli* cultures, chill the culture in ethanol-acetone, pellet and wash the cells with 0.8% NaCl, and add 2 ml of the 4 M guanidinium thiocyanate–β-mercaptoethanol solution. The guanidinium thiocyanate should be filtered through Whatman no. 1 filter paper and stored at room temperature, and the β-mercaptoethanol should be added just before use.

3. For all cultures, freeze-thaw twice by alternately plunging into dry ice-ethanol and 45°C baths. Centrifuge at 6,000 × *g* for 10 min, and recover the supernatant.

4. Pellet the RNA. Prepare a solution containing 5.7 M CsCl and 10 mM disodium EDTA (pH 7.5), and add 3 ml to a centrifuge tube for an SW50.1 or similar swinging-bucket rotor. Layer the cleared lysate on top of the CsCl solution, and centrifuge at 160,000 × *g* at 20°C for at least 24 h. With a pipette held at the meniscus so that some air is drawn up along with the liquid, carefully draw off the supernatant and discard it. Rinse the RNA pellet with 75% ethanol to remove the CsCl, allow it to dry, and dissolve it in 200 μl of 10 mM Tris (pH 7.5). Remove to a microcentrifuge tube, and precipitate with 1/10 volume of sodium acetate (pH 5.2)–2.5 volumes of ethanol at −20°C for 1 h. Centrifuge, discard the supernatant, rinse the pellet with 75% ethanol, dry it, and dissolve it in 100 μl of water. If the RNA is to be stored, add 3 volumes of ethanol and keep at −70°C. Add sodium acetate, and precipitate aliquots as needed.

20.3.1.3. Assay procedure

1. Label the DNA probe. The DNA fragment should have one 5′ overhang that is labeled with [^{32}P]ATP and T4 polynucleotide kinase. Use 0.2 to 1 μg of appropriately cut DNA fragment. For quantitative results, the amount of probe must be in at least 10-fold molar excess to the mRNA being detected. The labeled end of the fragment should be within the region that hybridizes to the RNA (Fig. 5). A DNA restriction fragment is chosen so that one 5′ end that is 50 to 400 bases downstream from the transcription start can be labeled. After hybridization and digestion of the nonhybridized, single-stranded portion of the DNA fragment with S1 nuclease, the 3′ end of the DNA strand will end at the beginning of the transcript. If this DNA is run on a sequencing gel with similar DNA that has undergone sequencing reactions (the 5′ end of the sequencing primer must correspond to the 5′ end of the labeled probe), the position of the band will correspond to the transcription start (5, 62). The expected direction of transcription also must be considered to determine which strand will hybridize. Separate probes labeled at opposite ends can be used to determine the direction of transcription.

To label a restriction fragment, remove phosphates from the ends of the DNA. Digest with a restriction enzyme, and isolate a DNA fragment if necessary. Do not add tRNA as a carrier for precipitation, since it will interfere with subsequent reactions. If a carrier is necessary, use 10 mg of glycogen (Boehringer Mannheim) per ml. Add DNA and 5 μl of 10× phosphatase buffer (10× = 0.2 M Tris [pH 8.5], 10 mM $MgCl_2$, 10 mM $ZnCl_2$, 1 mM EDTA) plus water to a final volume of 50 μl. If a low- or moderate-salt buffer is used for the restriction enzyme, precipitation is not necessary; just omit the 10× buffer and add 1 μl of 1 M Tris HCl (pH 9.5)–50 mM $ZnCl_2$. Add 0.5 U of calf intestine alkaline phosphatase, and incubate at 37°C for 60 min. Heat to 65°C for 15 min. Extract the protein by adding 50 μl of 0.6 M sodium acetate and 100 μl of TE-saturated phenol. Vortex for 1 min, centrifuge for 1 min, and remove the aqueous (top) phase to a new tube. Repeat the extraction with phenol-chloroform (see Chapter 16.2), and precipitate the DNA as usual.

Resuspend the DNA in 21 μl of TE, and add 3 μl of 10× kinase buffer (0.5 M Tris HCl [pH 7.6], 0.1 M $MgCl_2$, 50 mM DDT, 1 mM spermidine HCl, 1 mM EDTA) and 5 μl of [γ-^{32}P]ATP (3,000 Ci/mmol; 10 mCi/ml). Add 10 U of T4 polynucleotide kinase, and incubate at 37°C for 45 min. Heat to 65°C for 10 min, and remove the unincorporated nucleotides by chromatography or silica beads (section 20.2.1.3).

Digest the probe with a restriction enzyme to remove one labeled end, and isolate the appropriate fragment by gel electrophoresis.

2. Hybridize the probe. Add 2×10^4 to 1×10^5 cpm (Cerenkov) probe to 25 μg of RNA in 1/10 volume of 3 M sodium acetate, and then add 2.5 volumes of 95% ethanol and precipitate at −70°C for 30 min. For low-abundance mRNA, the amount of RNA can be increased by up to eightfold. Centrifuge, remove the supernatant, and wash and air-dry the pellet. Do not overdry, or the pellet will be difficult to resuspend. Resuspend the pellet in 20 to 30 μl of hybridization buffer [80% deionized formamide, 40 mM piperazine-*N*,*N*′-bis(2-ethanesulfonic acid) adjusted with HCl (PIPES HCl) (pH 6.4), 400 mM NaCl, 1 mM EDTA]. It may be necessary to pipette the sample up and down and incubate it for several hours to completely dissolve it. Heat to 80°C for

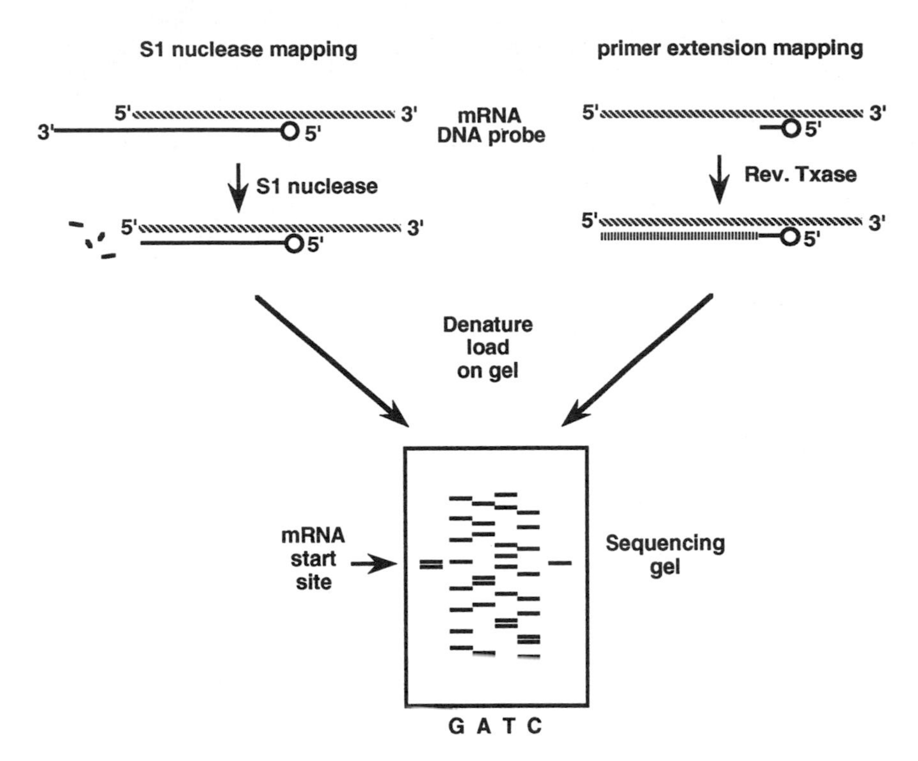

FIG. 5. S1 and primer extension assays. Cross-hatched line, mRNA; solid line, DNA; circle, labeled 5′ end of DNA. Cartoon of sequencing gel shows sequencing reactions in the center, S1 reaction to the left, and primer extension to the right.

10 min, and then quickly place in a water bath at the hybridization temperature. The hybridization temperature depends on the length of the hybrid and its G+C content. The temperature is determined empirically and usually falls within 42 to 60°C. Start by incubating samples at 55°C for hybrids expected to be greater than 100 bases and 50% G+C. Additional samples can be hybridized at 5°C above or below this temperature to ensure proper hybridization. Since RNA-DNA hybrids will form at about 10°C higher than DNA-DNA hybrids, incubation at the optimal temperature will maximize the desired hybridization and give quantitative results. Allow to hybridize for 6 to 12 h.

3. Dilute the samples with 400 μl of cold S1 buffer (30 mM sodium acetate [pH 4.5], 1 mM $ZnSO_4$, 250 mM NaCl, 5% glycerol). Add 30 U of S1 nuclease, and incubate at 37°C for 30 min. Digestion conditions may be varied to obtain maximal digestion. The temperature can be increased to 45°C, although this may result in some cleavage of A+T-rich regions. The concentration of S1 also may be increased to 500 U. Stop the reaction with 100 μl of 2.5 M ammonium acetate–50 mM EDTA.

4. Add 500 μl of isopropanol, precipitate for 30 min at 4°C, centrifuge, discard the supernatant, and resuspend the pellet in 100 μl of 0.3 M sodium acetate. Precipitate with 2.5 volumes of ethanol, rinse with 70% ethanol, and dry.

5. Resuspend in 5 μl of formamide loading buffer (as in section 20.1). Heat to 90°C for 3 min, and load on a sequencing gel along with sequence standards. Run the gel, and expose to X-ray film as with sequencing samples (section 20.1).

6. A single band or, more commonly, a cluster of a few bands in the lanes containing the S1-treated samples indicates the start site of transcription. The position is determined by comparison with the sequencing lanes. The S1 may "nibble" the DNA end, so multiple bands do not necessarily indicate more than one start position. S1 may also cleave the DNA at sites of high A+T content, giving a false start position. Because of these artifacts, it is often useful to confirm the start site by another means such as the primer extension assay.

20.3.2. Primer Extension Assay

Isolate RNA by one of the above methods (section 20.3.1.1 or 20.3.1.2). The 5′ end of an mRNA can be mapped by hybridizing a primer to the RNA and synthesizing a DNA strand from the 3′ end of the primer extending to the 5′ end of the RNA. Reverse transcriptase is used for the synthesis. The products of the reaction are loaded onto a sequencing gel along with a set of sequencing reactions as described for S1 mapping. Generally, the yield of DNA is lower than that obtained by the S1 procedure, but some of the artifacts of S1, such as bands appearing at A+T-rich regions, are avoided. Restriction fragments labeled at one end can be used as primers, but synthetic oligonucleotides of 20 to 30 bases are preferable. The oligonucleotide is labeled to high specific activity, and there is no competition with a second DNA strand during hybridization with the RNA. Choose a sequence that is 100 to 400 bases from the expected position of the 5′ end of the RNA, and make sure that it is the appropriate strand and polarity.

1. Label 100 ng of the oligonucleotide with [γ-^{32}P]ATP by using T4 polynucleotide kinase as described in section 20.3.1.3. A 5′-phosphate is not present on the oli-

gonucleotide, so treatment with phosphatase is unnecessary. Separate the labeled DNA from unincorporated label by ethanol precipitation in the presence of 2 M ammonium acetate or by column chromatography.

2. Mix 10^4 to 10^5 cpm of the probe with 10 to 100 μg of RNA prepared as described for S1 mapping. If the extensions do not work, a higher-purity RNA preparation may help (2). A large excess of probe can increase the number of premature bands during the extension reaction, so it is advisable to try several levels of probe. Add 1/10 volume of 3M sodium acetate (pH 5.2) and 2.5 volumes of ethanol, precipitate at −20°C for 60 min, and recover the nucleic acids. Air-dry the pellet after washing it with 70% ethanol.

3. Dissolve the pellet in 30 μl of hybridization buffer, and hybridize as described in section 20.3.1.3, step 2. Heat to 80°C for 10 min, and transfer to a water bath at hybridization temperature. For synthetic oligonucleotides, the optimal temperature is usually between 30 and 42°C; try 32°C first. Incubate for 6 to 12 h. Add 170 μl of 0.3 M sodium acetate and 2.5 volumes of ethanol, and precipitate at −20°C for 60 min. Centrifuge, rinse with 70% ethanol, and carefully remove all ethanol. Allow to air dry.

4. Dissolve the sample in 25 μl of reverse transcriptase buffer (50 mM Tris HCl [pH 7.6], 60 mM KCl, 10 mM $MgCl_2$, 1 mM DTT, 50 μg of actinomycin D per ml, 25 U of RNasin, 0.4 mM each dNTP). Add 40 U of AMV reverse transcriptase, and incubate at 42°C for 90 min. (The actinomycin D destabilizes DNA-DNA hybrids, and the RNasin inhibits contaminating RNases.) Heat at 70°C for 10 min.

5. Add 1 μl of 0.5 M EDTA and 1 μl of 1-mg/ml RNase A, and incubate for 30 min at 37°C.

6. Add 100 μl of 2.5 M ammonium acetate, extract by vortexing for 1 min with an equal volume of phenol-chloroform, centrifuge for 1 min, and remove the aqueous phase to a new tube. Precipitate the DNA with 300 μl of ethanol at −20°C for 60 min, centrifuge, and rinse and dry the pellet.

7. Resuspend the pellet in 3 μl of TE buffer, add 3 μl of formamide loading buffer, and heat at 80°C for 3 min. Load onto a sequencing gel with DNA sequence reactions.

20.3.3. Northern Blot Hybridization Assay

Northern blot hybridization is a method of determining the size of an RNA species (34). RNA from a cellular extract is run on a denaturing gel and then transferred to nitrocellulose paper, much like a Southern blot procedure (see Chapter 16.3.4.2.b). The specific RNA is detected by hybridization of a labeled probe and autoradiography. The most difficult step in the procedure is to obtain relatively intact RNA without degradation by RNases. This is especially important for mRNA, which has a short half-life. All buffers, water, glass, and plasticware must be RNase free.

20.3.3.1. Cell lysate RNA preparation

The following procedure is for RNA in cell lysates (32), *not* for purified RNA. It has the advantage that minimal handling will result in the least degradation of the RNA, and quantitative recoveries are possible. Therefore, the assay can be used to determine relative levels of transcription during induction of a gene. This procedure has worked well for *S. aureus*, *E. coli*, and *Bacillus subtilis*.

1. Grow a 10-ml culture of bacteria in rich medium at 37°C to late log phase.

2. Rapidly cool in ice-water, and immediately collect the cells by centrifugation at 4°C for 3 min in a microcentrifuge. Remove the supernatant, and resuspend the pellet in 100 μl of cold lysis buffer (20% sucrose, 20 mM Tris HCl [pH 7.6], 10 mM disodium EDTA, 50 mM NaCl, 100 μg of lysostaphin per ml [for *S. aureus*] or 1 mg of lysozyme per ml [for *B. subtilis* and *E. coli*]). Incubate on ice until the suspension clears (10 to 20 min).

3. Complete the cell lysis by adding 100 μl of 2% SDS and 50 μl of proteinase K (5-mg/ml solution). Shake at room temperature for 15 min.

4. Freeze and thaw the lysate twice in a dry ice-ethanol bath and a 45°C water bath. Add 50 μl of loading dye (50% glycerol, 200 mM EDTA, 0.1% bromophenol blue, 0.1% xylene cyanol), and mix well.

5. Check the integrity of the RNA by electrophoresis of a 6-μl aliquot on a 1% agarose gel. Run the gel until the blue dye is two-thirds the distance from the origin. The rRNA should form two distinct bands after the gel is stained with ethidium bromide.

20.3.3.2. Assay procedure

1. Prepare a 1.2% denaturing agarose gel. Mix 1.2 g of agarose with 73 ml of distilled water, and heat until boiling. After the agarose has dissolved, cool to 65°C and add 10 ml of 10× running buffer (0.2 M morpholinepropanesulfonic acid [MOPS; pH 7.0], 50 mM sodium acetate, 1 mM EDTA), and 16.2 ml of 37% formaldehyde. Stir briefly, and pour into the gel apparatus. After the agarose has hardened, cover the gel with running buffer–2.2 M formaldehyde and 1 μg of ethidium bromide per ml.

2. Mix 10 μl of lysate prepared as above with 20 μl of deionized formamide (50%) and 4 μl of 10× running buffer. Denature the RNA by heating at 65°C for 10 min. After chilling on ice, load the entire 34-μl sample on the gel. Also load RNA markers: 6 μl of RNA markers (Bethesda Research Laboratories) mixed with 4 μl of loading dye that has been denatured and chilled.

3. Carry out electrophoresis at room temperature in a hood at a constant voltage of 100 V. Circulate the buffer from the cathode to the anode. A peristaltic pump is sufficient. Run the gel until the blue dye is two-thirds the distance from the origin. The running time and voltage can be adjusted depending on the size of the RNA of interest and the particular apparatus used.

4. After electrophoresis, cut off the lanes of the gel containing the marker RNA and stain in a 1-mg/ml ethidium bromide solution for 20 min followed by a 15-min destain in water. View the gel with UV light, and measure the migration of the markers from the origin.

5. Transfer the RNA to a nitrocellulose filter. This procedure is quite similar to that described in Chapter 16.3.1, in which buffer formulas and transfer details are given. The same blotting setup is used (Fig. 3 in Chapter 16.3). No pretreatment (denaturation or nicking) is necessary; however, the formaldehyde may be removed by soaking the gel in two changes of 500 ml of water, 15 min for each. Blot the RNA from the gel onto

the filter (GeneScreen Plus; Du Pont) overnight with 20× SSC as the transfer buffer. The origins of the gel lanes can be marked on the filter with a soft pencil to aid in molecular weight calculations. The transfer should be done in a fume hood if the formaldehyde was not removed.

6. After transfer, place the filter between two pieces of Whatman 3 mm paper and bake at 80°C for 2 h in a vacuum oven.

7. Prehybridize the membrane by soaking it with agitation in 10 ml (for a 12- by 12-cm filter) of 2× Denhardt's solution (Chapter 16.3.1) (0.04% each BSA, Ficoll, and polyvinylpyrrolidone) plus 10 mM EDTA, 0.2% SDS, and 5× SSC for 3 h at −60°C in a sealed plastic bag.

8. Open a corner of the bag, and add 0.2 ml of denatured sonicated salmon sperm DNA (5-mg/ml stock) and 1 × 10^7 cpm of denatured probe. The probe can be prepared by random-primed synthesis as described in Chapter 16.3.1.

9. Reseal the bag, and incubate it with gentle agitation for 18 h at 60°C.

10. Remove the membrane from the bag, and wash it twice in 100 ml of 2× SSC–0.1% SDS for 5 min at room temperature with constant agitation. Then wash it twice with 200 ml of 2× SSC–1% SDS at 60°C for 30 min and twice with 100 ml of 0.1× SSC at RT for 30 min.

11. Place the membrane on a sheet of filter paper, and allow it to dry. Place the filter paper on top of a piece of X-ray film. Mark the origins of the gel lanes on the film with a pencil or mark the filter with an autoradiography pen (a fluorescent dye that will expose the film). Expose overnight. An intensifying screen (DuPont Lightning Plus) can be used if the signal is weak; expose overnight at −70°C or below with the screen behind the film. Use the distances measured for the marker RNAs to construct a log-linear graph of RNA size (molecular weight or number of bases) versus migration distance. A linear relationship should be obtained, which can be used to interpolate the size(s) of the RNA of interest from the developed autoradiogram.

20.4. REFERENCES

1. **Ansorge, W., and S. Labeit.** 1984. Field gradients improve resolution on DNA sequencing gels. *J. Biochem. Biophys. Methods* **10:**237–243.
2. **Ausubel, F. M., R. E. Kingston, R. Brent, D. D. Moore, J. G. Seidman, J. A. Smith, and K. Struhl.** 1991. *Current Protocols in Molecular Biology.* Greene Publishing Associates & Wiley-Interscience, New York.
3. **Barnes, W. M.** 1977. Plasmid detection and sizing in single colony lysates. *Science* **195:**393–394.
4. **Barnes, W. M., M. Bevan, and P. H. Son.** 1983. Kilo-sequencing: creation of an ordered nest of asymmetric deletions across a large target sequence carried on phage M13. *Methods Enzymol.* **101:**98–122.
5. **Berk, A. J., and P. A. Sharp.** 1977. Sizing and mapping of early adenovirus mRNAs by gel electrophoresis of S1 endonuclease-digested hybrids. *Cell* **12:**721–732.
6. **Biggin, M. D., T. J. Gibson, and G. F. Hong.** 1983. Buffer gradient gels and ^{35}S label as an aid to rapid DNA sequence determination. *Proc. Natl. Acad. Sci. USA* **80:**3963–3965.
7. **Borowiec, J. A., L. Zhang, S. Sasse-Dwight, and J. D. Gralla.** 1987. DNA supercoiling promotes formation of a bent repression loop in lac DNA. *J. Mol. Biol.* **196:**101–111.
8. **Brenowitz, M., E. Jamison, A. Majumdar, and S. Adhya.** 1990. Interaction of the Escherichia coli gal repressor protein with its DNA operators in vitro. *Biochemistry* **29:**3374–3383.
9. **Brenowitz, M., D. F. Senear, M. A. Shea, and G. K. Ackers.** 1986. Quantitative DNase I footprint titration: a method for studying protein-DNA interactions. *Methods Enzymol.* **130:**132–181.
10. **Carey, J.** 1988. Gel retardation at low pH resolves trp repressor-DNA complexes for quantitative study. *Proc. Natl. Acad. Sci. USA* **85:**975–979.
11. **Carey, J.** 1991. Gel retardation. *Methods Enzymol.* **208:**103–117.
12. **Chien, A., D. B. Edgar, and J. M. Trela.** 1976. Deoxyribonucleic polymerase from the extreme thermophile *Thermus aquaticus. J. Bacteriol.* **127:**1550–1557.
13. **Crothers, D. M., M. R. Gartenberg, and T. E. Shrader.** 1991. DNA bending in protein-DNA complexes. *Methods Enzymol.* **208:**118–146.
14. **Fried, M., and D. M. Crothers.** 1981. Equilibria and kinetics of lac repressor-operator interactions by polyacrylamide gel electrophoresis. *Nucleic Acids Res.* **9:**6505–6525.
15. **Fried, M., and D. M. Crothers.** 1983. CAP and RNA polymerase interactions with the lac promoter: binding stoichiometry and long range effects. *Nucleic Acids Res.* **11:**141–158.
16. **Fried, M. G., and D. M. Crothers.** 1984. Equilibrium studies of the cyclic AMP receptor protein-DNA interaction. *J. Mol. Biol.* **172:**241–262.
17. **Fried, M. G., and D. M. Crothers.** 1984. Kinetics and mechanism in the reaction of gene regulatory proteins with DNA. *J. Mol. Biol.* **172:**263–282.
18. **Galas, D. J., and A. Schmitz.** 1978. DNase footprinting: a simple method for the detection of protein-DNA binding specificity. *Nucleic Acids Res.* **5:**3157–3170.
19. **Garner, M. M., and A. Revzin.** 1981. A gel electrophoresis method for quantifying the binding of proteins to specific DNA regions: application to components of the Escherichia coli lactose operon regulatory system. *Nucleic Acids Res.* **9:**3047–3060.
20. **Garner, M. M., and A. Revzin.** 1982. Stoichiometry of catabolite activator protein/adenosine cyclic 3′, 5′-monophosphate interactions at the lac promoter of Escherichia coli. *Biochemistry* **21:**6032–6036.
21. **Gralla, J. D.** 1985. Rapid "footprinting" on supercoiled DNA. *Proc. Natl. Acad. Sci. USA* **82:**3078–3081.
22. **Hayes, J. J., and T. D. Tullius.** 1989. The missing nucleoside experiment: a new technique to study recognition of DNA by protein. *Biochemistry* **28:**9521–9527.
23. **Henikoff, S.** 1984. Unidirectional digestion with exonuclease III creates targeted breakpoints for DNA sequencing. *Gene* **28:**351–359.
24. **Henikoff, S.** 1987. Unidirectional digestion with exonuclease III in DNA sequence analysis. *Methods Enzymol.* **155:**156–165.
25. **Hendrickson, W.** 1985. Protein-DNA interactions studied by the gel electrophoresis-DNA binding assay. *BioTechniques* **4:**198–207.
26. **Hendrickson, W., and R. Schleif.** 1984. Regulation of the Escherichia coli L-arabinose operon studied by gel electrophoresis DNA binding assay. *J. Mol. Biol.* **178:**611–628.
27. **Hendrickson, W., and R. Schleif.** 1985. A dimer of AraC protein contacts three adjacent major groove regions of the araI DNA site. *Proc. Natl. Acad. Sci. USA* **82:**3129–3133.
28. **Hong, G. F.** 1987. The use of DNase I, buffer gradient gel, and ^{35}S label for DNA sequencing. *Methods Enzymol.* **155:**93–110.
29. **Hudson, J. M., and M. G. Fried.** 1990. Co-operative interactions between the catabolite gene activator protein and the lac repressor at the lactose promoter. *J. Mol. Biol.* **214:**381–396.
30. **Innis, M. A., K. B. Myambo, D. H. Gelfand, and M. A. D. Brow.** 1988. DNA sequencing with *Thermus aquaticus* DNA polymerase and direct sequencing of polymerase chain reaction-amplified DNA. *Proc. Natl. Acad. Sci. USA* **85:**9436–9440.

31. **Khoury, A. M., H. S. Nick, and P. Lu.** 1991. *In vivo* interaction of *Escherichia coli lac* repressor N-terminal fragments with the *lac* operator. *J. Mol. Biol.* **219:**623–634.
32. **Kornblum, J. S., S. J. Projan, S. L. Moghazeh, and R. P. Novick.** 1988. A rapid method to quantitate non-labeled RNA species in bacterial cells. *Gene* **63:**75–85.
33. **Kuwabara, M., and D. S. Sigman.** 1987. Footprinting DNA-protein complexes in situ following gel retardation assays using 1,10-phenanthroline-copper ion: *Escherichia coli* RNA polymerase-*lac* promoter complexes. *Biochemistry* **26:**7234–7238.
34. **Lehrach, H., D. Diamon, J. M. Wozney, and H. Boedtker.** 1977. RNA molecular weight determinations by gel electrophoresis under denaturing conditions: a critical examination. *Biochemistry* **16:**4743–4749.
35. **Liu-Johnson, H. N., M. R. Gartenberg, and D.M. Crothers.** 1986. The DNA binding domain and bending angle of E. coli CAP protein. *Cell* **47:**995–1005.
36. **Maxam, A. M., and W. Gilbert.** 1977. A new method for sequencing DNA. *Proc. Natl. Acad. Sci. USA* **74:**560–564.
37. **Maxam, A. M., and W. Gilbert.** 1980. Sequencing end-labeled DNA with base-specific chemical cleavages. *Methods Enzymol.* **65:**499–560.
38. **Misra, T. K.** 1985. A new strategy to create ordered deletions for rapid nucleotide sequencing. *Gene* **34:**263–268.
39. **Misra, T. K.** 1987. DNA sequencing: a new strategy to create ordered deletions, modified M13 vector, and improved reaction conditions for sequencing by dideoxy chain termination method. *Methods Enzymol.* **155:**119–139.
40. **Nick, H., and W. Gilbert.** 1985. Detection in vivo of protein-DNA interactions within the lac operon of Escherichia coli. *Nature* (London) **313:**795–797.
41. **Olson, A., T. Moks, and A. B. Gaal.** 1984. Uniformly spaced banding pattern in DNA sequencing gels by use of a field strength gradient. *J. Biochem. Biophys. Methods* **10:**83–90.
42. **Poncz, M., D. Solowiejczyk, M. Ballantine, and S. Surey.** 1982. "Nonrandom" DNA sequence analysis in bacteriophage M13 by the dideoxy chain-termination method. *Proc. Natl. Acad. Sci. USA* **79:**4298–4302.

42a. **Promega Biologicals.** 1990. *Protocols and Applications Guide.* Promega Biologicals, Madison, Wis.

43. **Revzin, A.** 1989. Gel electrophoresis assays for DNA-protein interactions. *BioTechniques* **7:**346–355.
44. **Riggs, A. D., H. Suzuki, and S. Bourgeois.** 1970. lac repressor-operator interaction. I. Equilibrium studies. *J. Mol. Biol.* **48:**67–83.
45. **Salzer, W., R. F. Gestland, and A. Bolle.** 1967. In vitro synthesis of bacteriophage lysozyme. *Nature* (London) **215:**588–591.
46. **Sambrook, J., E. F. Fritsch, and T. Maniatis.** 1989. *Molecular Cloning: A Laboratory Manual,* 2nd ed. Cold Spring Harbor Laboratory, Cold Spring Harbor, N.Y.
47. **Sanger, F., S. Nicklen, and A. R. Coulson.** 1977. DNA sequencing with chain-terminating inhibitors. *Proc. Natl. Acad. Sci. USA* **74:**5463–5467.
48. **Sasse-Dwight, S., and J. D. Gralla.** 1988. Probing co-operative DNA-binding in vivo: the *lac* O_1:O_3 interaction. *J. Mol. Biol.* **202:**107–109.
49. **Shanblatt, S. H., and A. Revzin.** 1983. Two catabolite activator protein molecules bind to the galactose promoter region of Escherichia coli in the presence of RNA polymerase. *Proc. Natl. Acad. Sci. USA* **80:**1594–1598.
50. **Shanblatt, S. H., and A. Revzin.** 1984. Kinetics of RNA polymerase-promoter complex formation: effect of nonspecific DNA-protein interactions. *Nucleic Acids Res.* **12:**5287–5306.
51. **Sheen, J.-Y., and S. Brian.** 1988. Electrolyte gradient gels for DNA sequencing. *BioTechniques* **6:**942–944.
52. **Siebenlist, U., and W. Gilbert.** 1980. Contacts between E. coli RNA polymerase and an early promoter of phage T7. *Proc. Natl. Acad. Sci. USA* **77:**122–126.
53. **Straney, D. C., S. B. Straney, and D. M. Crothers.** 1989. Synergy between Escherichia coli CAP protein and RNA polymerase in the lac promoter open complex. *J. Mol. Biol.* **206:**41–57.
54. **Tabor, S., and C. C. Richardson.** 1985. A bacteriophage T7 RNA polymerase/promoter system for controlled exclusive expression of specific genes. *Proc. Natl. Acad. Sci. USA* **82:**1074–1078.
55. **Tabor, S., and C. C. Richardson.** 1987. DNA sequence analysis with a modified bacteriophage T7 DNA polymerase. *Proc. Natl. Acad. Sci. USA* **84:**4767–4771.
56. **Tullius, T. D.** 1989. Physical studies of protein-DNA complexes by footprinting. *Annu. Rev. Biophys. Biophys. Chem.* **18:**213–237.
57. **Tullius, T. D., and B. A. Dombroski.** 1986. Hydroxyl radical "footprinting": high-resolution information about DNA-protein contacts and application to lambda repressor and cro protein. *Proc. Natl. Acad. Sci. USA* **83:**5469–5473.
58. **Tullius, T. D., B. A. Dombroski, M. E. A. Churchill, and L. Kam.** 1987. Hydroxyl radical footprinting: a high resolution method for mapping protein-DNA contacts. *Methods Enzymol.* **155:**537–558.
59. **Vieira, J., and J. Messing.** 1987. Production of single stranded plasmid DNA. *Methods Enzymol.* **153:**3–11.
60. **Wahls, W. P., and M. Kingzette.** 1988. No runs, no drips, no errors: a new technique for sealing polyacrylamide gel electrophoresis apparatus. *BioTechniques* **6:**308–309.
61. **Walden, W. E., M. M. Patino, and L. Gaffield.** 1989. Purification of a specific repressor of ferritin mRNA translation from rabbit liver. *J. Biol. Chem.* **264:**13765–13769.
62. **Weaver, R. F., and C. Weissman.** 1979. Mapping of RNA by a modification of the Berk-Sharp procedure: the 5′ termini of 15S β-globin mRNA precursor and mature 10S β-globin mRNA have identical map coordinates. *Nucleic Acids Res.* **7:**1175–1193.
63. **Wu, H. M., and D. M. Crothers.** 1984. The locus of sequence-directed and protein-induced DNA bending. *Nature* (London) **308:**509–513.
64. **Yang, C.-C., and H. A. Nash.** 1989. The interaction of E. coli IHF protein with its specific binding sites. *Cell* **57:**869–880.
65. **Yanisch-Perron, C., J. Vierira, and J. Messing.** 1985. Improved M13 phage cloning vectors and host strains: nucleotide sequences of the M13amp18 and pUC19 vectors. *Gene* **33:**103–119.
66. **Zagursky, R., and K. Baumeister.** 1987. Construction and use of pBR322 plasmids that yield single-stranded DNA for sequencing. *Methods Enzymol.* **155:**139–155.

Section IV

METABOLISM

Introduction to Metabolism

W. A. WOOD

The metabolic activities of bacterial cells constitute the organized and regulated biosynthetic and biodegradative enzymatic sequences that are responsible for maintenance, growth, motion, and reproduction. These functions proceed at the expense of energy sources and other nutrients of extracellular and intracellular origin. The metabolic systems that support these functions are studied through a variety of biological, physical, chemical, and enzymatic methods. The biological methods prominently include various aspects of growth as well as nutrition and molecular genetics. These aspects, as well as the use of microscopy, are described in other sections of this manual.

This section deals with the physical, chemical, and enzymatic methods used in the study of bacterial metabolism. Included in the following chapters are physical and instrumental techniques, fractionation and analysis of chemical components, preparation and assay of enzymes, and measurement of permeability and transport processes. In each of these areas, the state of the art has been so highly developed that books dedicated to a single methodology and its applications are available. Here, however, the authors have restricted the scope to describe only reliable and easy methods which are widely applicable to basic studies of bacterial metabolism by a neophyte user.

Chapter 21

Physical Analysis

W. A. WOOD and J. R. PATEREK

This chapter deals with the basic physical analytical methods and separation procedures of photometry, ion electrodes, chromatography, radioactivity, electrophoresis, and manometry. Their use in bacteriology has in many cases been major and almost characteristic of the field. The approach in this manual is to give a general description along with strengths and weaknesses of a method, specific techniques and preparations, and examples of application to investigations with bacteria. The purification of enzyme proteins is particularly exemplified.

In this chapter, especially, it has been necessary to confine the scope to the simple procedures which are appropriate for laboratory classes or are of value in initiating a research program. In-depth treatment of the six basic methods can be found in books specifically dedicated to one technique, some of which are listed in section 21.8. In addition, catalogs and commercial literature from manufacturers and suppliers contain a wealth of information that is highly valuable to the investigator. Some specialized techniques such as thin-layer chromatography, analysis of macromolecules by sedimentation procedures, and unusual types of electrophoresis have been mentioned only briefly. On the other hand, some specialized techniques which are useful but seldom encountered by bacteriologists are presented.

21.1. PHOTOMETRY

A wide variety of photometric methods have found important uses in bacteriological research and applications. Absorption photometry, fluorimetry (1, 2), and nephelometry have been particularly useful. These have facilitated identification of compounds, determination of structure, estimation of concentrations, and measurement of reaction rates. For uses of photometry in measuring bacterial growth, see Chapter 11.6.

21.1.1. Absorption Photometry

UV-visible spectrophotometers measure the ability of solutes to absorb light at specified wavelengths. A plot of the light absorbed versus the wavelength gives an absorption spectrum, which aids in the identification of compounds and gives information about the structure of the chromophore. The light absorbed at a given wavelength can yield concentration information, and plots of the energy absorbed versus time can yield reaction rates.

The monochromator part of the instrument selects and transmits a narrow range of wavelengths from a radiant source through the sample to a photometer, which quantitates the energy received and expresses it as the ratio of the transmitted light (I_T) to the incident light (I_O). When I_O is set at 100, the instrument response to light passing through the sample is percent transmittance (%T). This can be related to the molar concentration (C) of an absorbing solute through the Beer-Lambert law:

$$-\log \%T = \epsilon l C = A$$

where ϵ is the molar extinction coefficient of the absorbing species at a specified wavelength and l is the length of the light path in centimeters. Since both ϵ and l are constant in practice, C is proportional to the negative logarithm of %T or directly proportional to absorbance (A). For this reason, newer spectrophotometers reading linearly in A are more convenient and more accurate for samples of moderate to high absorbance.

21.1.1.1. Performance factors

For laboratory spectrophotometers, quality is determined by the purity of the light presented to the sample, the intensity and area of the beam (numerical aperture), the accuracy of wavelength calibration, the accuracy of photometer calibration, the linearity of response, noise, and short- and long-term drift. Manufacturers of some instruments go to great lengths and expense to achieve high quality. However, for most biological experiments it is preferable to know the limitations of an ordinary instrument rather than to feel compelled to acquire or seek out a high-quality unit.

Effect of spectral bandwidth. Spectral purity of light from a monochromator, or spectral bandwidth, is the width in nanometers at half the height of the emission energy peak at a specified nominal wavelength (Fig. 1B). This bandwidth is specified for the instrument and is not readily determined in the laboratory unless a very narrow bandpass (ca. 1-nm) interference filter is available. Bandwidth is reasonably constant across the wavelength range in monochromators in which a grating is the dispersing element but varies with wavelength for prism monochromators. It varies with slit width in both cases. The accuracy of a measured absorbance, assuming no stray light (see below), depends on the ratio of the spectral bandwidth (a property of the instrument) to the natural bandwidth (the width in nanometers at half the peak height) of the peak of the solute in the light beam (a property of the chromophore in solution) (Fig. 1A). Therefore, in an inexpensive spectrophotometer with a spectral bandwidth at 340 nm of 8 nm, a peak with an 80-nm natural bandwidth (giving a ratio of 0.1) allows absorbance measurements with 99.5% accuracy (4). Similarly, with an 8-nm bandwidth, 99% accuracy is achieved with a chromophore with a 50-nm natural bandwidth. The most commonly measured material at 340 nm, NADH, has a natural bandwidth of 58 nm at that wavelength. Therefore, it is readily apparent that NADH can be measured at better than 99% accuracy in this inexpensive spectrophotometer (4).

Effect of stray light. Another consequence of wide spectral bandwidth, or low spectral purity, is an increased content of stray light (i.e., energies at wavelengths removed from the nominal value and not part of the spectral bandwidth wavelengths). Since for prism monochromators spectral bandwidth is related to both slit width and wavelength, so also is stray-light content. Stray light derives from incomplete removal of unwanted wavelengths owing to faulty design or dirty optics. The stray-light content of the exit beam diminishes the linear range of response to chromophore concentration, i.e., the range where the Beer-Lambert law applies. The operator should determine the useful linear range in the wavelength region at which determinations are to be made.

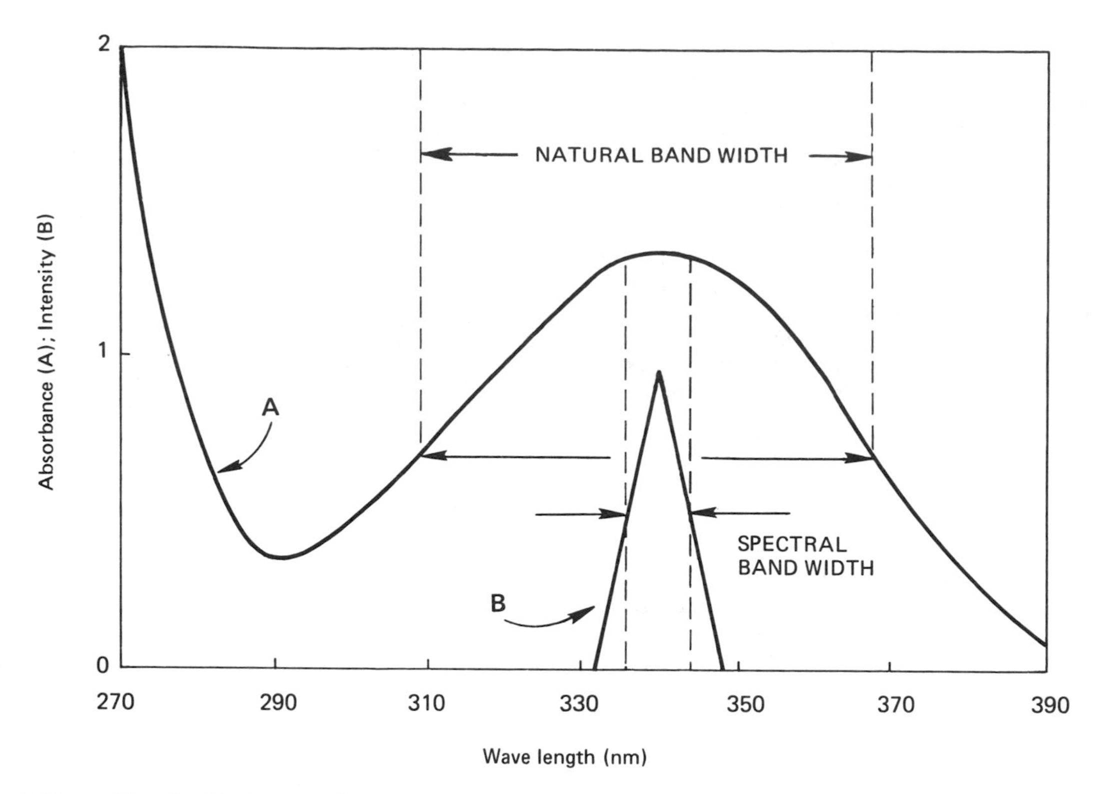

FIG. 1. (A) Natural bandwidth diagram of NADH; the bandwidth is 58 nm. (B) Spectral bandwidth diagram (schematic) of spectrophotometer exit beam; the spectral bandwidth is 8 nm.

Since the stray-light content increases as the wavelength decreases, a conservative and convenient approach to determine the stray-light general performance of the instrument is to use common solvents as sharp-cutoff filters (5). Figure 2A shows absorbance-versus-wavelength plots for a series of solvents. The sharp departure from the rapid upward path of the measured absorbance by a common laboratory spectrophotometer indicates that stray-light energy not absorbed predominates over that of the selected wavelength. The result is an inability to measure the absorbance of solutes in the absorbance region at which the stray-light content becomes appreciable. The effect of percent stray light on adherence to the Beer-Lambert law is shown in Fig. 2B (4). Percent stray light is the percent transmittance measured at a wavelength at which a sharp-cutoff filter should be opaque.

Newer spectrophotometers in wide use can make accurate absorbance measurements in the 2.0- to 3.0-Å (0.2- to 0.3-nm) region, but this ability can be lost when optical components become dirty, the sample is fluorescent, or there is a light leak.

Photometric accuracy and linearity. For certain spectrophotometers, electronic adjustment of photometric calibration is not available and linearity of response is assumed. However, for at least one line of spectrophotometers (Gilford Instrument Laboratories, Oberlin, Ohio), adjustment of calibration with supplied absorbance standards is a basic feature. In either case, both absorbance accuracy and linearity of absorbance measurement should be established and an error table should be developed if necessary. For instruments for which absorbance calibration is possible, neutral density filter standards of known absorbance at specified wavelengths are supplied and are used to adjust the photometer response to give the same absorbance value as the standards at one or a few absorbance values. With linearity between these points assumed, the total absorbance range is calibrated. The National Bureau of Standards and some instrument companies supply glass filter standards. In addition, the procedure for preparation of liquid absorbance standards has been published (3, 7).

Linearity can be ascertained at fixed wavelength by various versions of the following procedure, using either a standard of known absorbance of about 0.4 A (A_T) or a neutral-density filter or solution in that absorbance range (A_N) as follows. (i) Install the filter in a cuvette carrier so that either the air (or water) blank or the filter can be placed in the light beam, and null the photometer on the blank. (ii) Place the filter in the beam and read the absorbance (A_{o1}). (iii) Return to the air blank and close the slit so that the blank has an arbitrary absorbance reading (A_X) of, for example, 0.1. (iv) Return to the filter, and read the absorbance (A_{o2}). The true absorbance of this reading is A_T (or A_N) + A_X. The observed absorbance is $A_{o1} + A_X$. This process can then be repeated cyclically by closing the slit with the blank in the beam to give the absorbance reading of A_Y and then reading the filter (A_{o2}). The result is a set of true and observed absorbance values increasing from A_T at intervals across the absorbance range. A lower value for A increases the number of points in the determination of linearity. Figure 3A shows a plot of observed versus the true absorbance values for a particular instrument.

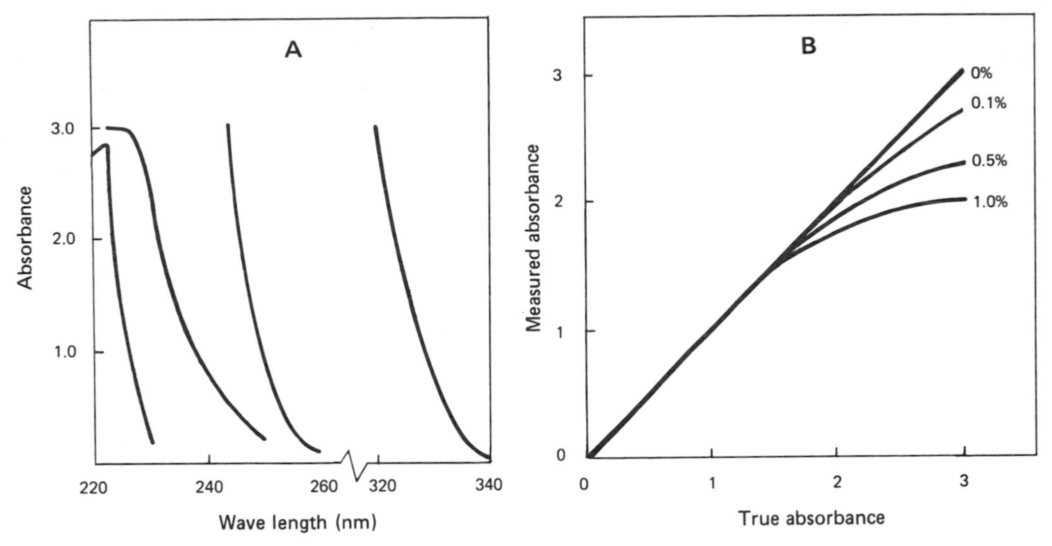

FIG. 2. (A) Absorbance measurements on solvents (left to right: 0.1 M sodium bromide, chlorethanol, dichlorethane, and acetone) for determination of stray light behavior of a monochromater. (B) Effect of various percentages of stray light on the Beer-Lambert behavior of a chromophore.

When the increment is constant, the photometric response is linear. Figure 3B shows a plot of the error ($A_o - A_T$) in observed absorbance (A_o) versus true absorbance. This plot is especially useful in showing departure from linearity and the magnitude of the error.

Effect of slit width. For prism monochromators working at fixed slit setting, there is a very marked nonlinear dependency of band pass on wavelength. Operating manuals from spectrophotometer manufacturers and books on spectrophotometry contain graphs of this function, and an example for one instrument is shown in Fig. 4. Thus, to make measurements at constant band pass with prism optics, take the slit setting at the wavelength of measurement from the graph. To be rigorous, determine the solute concentration at the slit width setting used for determination of its extinction coefficient. Conversely, when reporting an extinction coefficient, state the band pass used. For prism instruments this involves both the slit width and the graph of band pass versus wavelength for that instrument.

Similarly, spectra should be acquired at constant band pass, not constant slit width. The absorbance values at fixed wavelength or the absorbance spectra determined at two different slit widths seldom are identical. The error is magnified when the absorbance peaks of the solutes are sharp and may be negligible when peaks are broad. Therefore, it is prudent to use the minimum slit width.

Slit width also is a function of the sensitivity of the photometer, the spectral characteristics of the photo-

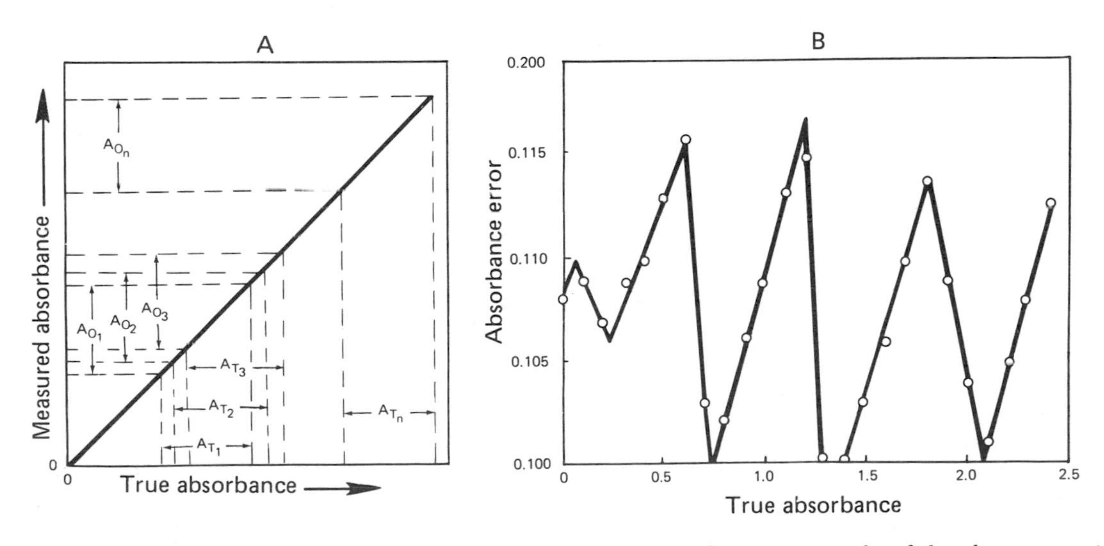

FIG. 3. (A) Plot of measured (A_o) versus true (A_T) absorbance for a spectrophotometer. (B) Plot of absorbance error ($\Delta A_o/\Delta A_T$) versus true absorbance.

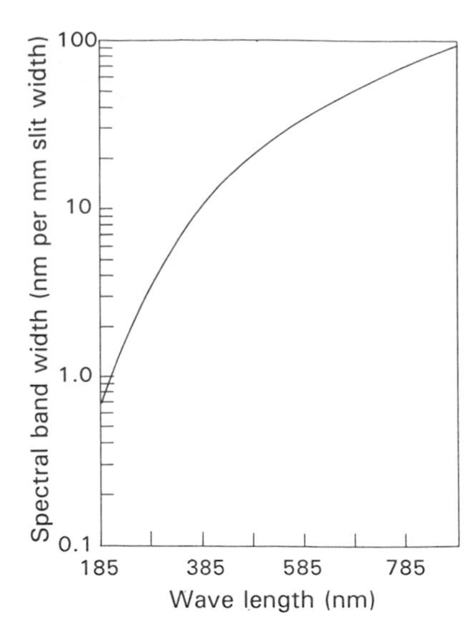

FIG. 4. Effect of wavelength on spectral bandwidth for a prism monochromator exemplified by data for a Gilford model 2400 spectrophotometer.

detector, and the source. More importantly, intensity of the lamp, cleanliness of optical surfaces, and critical focusing of the source on the slit greatly affect slit width. As a rule of thumb, the average spectrophotometer should be capable of nulling by opening the slit to 2 mm at 210 nm or lower. If this is not possible, one or more of the above factors may require attention.

21.1.1.2. Single-beam versus double-beam spectrophotometers

Double-beam instruments are constructed so that there are two light beams giving a simultaneous or parallel comparison between I_o and I_T. Single-beam units make these two measurements serially, whereas double-beam spectrophotometers split the light beam so that part of the beam passes through the solution used to set the instrument at $I_o = \%T = 100$ or $A_o = 0$, while the second part passes through the sample to measure $\%T$ or A_1. For this approach to be valid, the parallel optical systems and their associated electronics must perform identically. Considerable attention to design is necessary to achieve identity, and this has been accomplished for many available instruments. In single-beam units, the same end is achieved by first setting $\%T = 100$ or $A_o = 0$ with the blank and then reading $\%T$ or A_1 when the sample is introduced into the light beam.

There is a widespread belief that double-beam instruments are superior to single-beam units, the implication often being that single-beam units are less than adequate. Only in narrowly restricted situations is this the case, i.e., when rapid acquisition of spectra is required or when there is rapid drift or instability in the light source of a single-beam unit. However, double-beam units are highly convenient for the direct acquisition of difference spectra. Conversely, there is no reason to use a double-beam unit for individual determinations of absorbance, as in colorimetric determinations and other routine spectrophotometric measurements.

21.1.1.3. Preparation of spectra

Spectra are most conveniently obtained in a double-beam unit with an associated recorder. Use one cuvette for the reagent (buffer) blank and the second for the sample in the same reagent. Keep the slits as narrow as possible, and use a slow scan rate to obtain increased resolution. For critical work, ascertain that the concentration of chromophore is within the linear response region, that fluorescence is not a problem at high concentrations (an indication can be obtained with a fluorimeter), and that the unit is in wavelength calibration.

For determination of spectra with a single-beam unit, manually set the wavelength at acceptable increments. Zero the instrument at each wavelength setting with the reagent blank, and then record the absorbance of the sample. Construct a plot of wavelength (nanometers) on the x axis versus A on the y axis. This process will be more laborious but will suffice very well for most purposes.

21.1.1.4. Determination of solute concentration

With a properly calibrated spectrophotometer and standard 1-cm light path cuvettes, read the absorbance at the wavelength specified for the extinction coefficient (ϵ). Calculate A/ϵ as moles per liter or millimoles per milliliter. However, a high-quality spectrophotometer and knowledge of the extinction coefficient are not required if pure material is available to standardize even a filter colorimeter. In that case, prepare a standard regression line which relates absorbance to concentration for a set of concentration standards which may be measured directly or have been run through a procedure for color development. The plot will show the linear region and the variability of individual determinations. Standard statistical considerations apply, i.e., replication, least-square fits of the data to straight lines, determination of standard deviation from the mean, and standard error. Although linearity is highly desirable, it is also possible to use a standard curve (if the line is reproducible) by taking values directly from the graph.

21.1.1.5. Determination of particle concentration (6)

Spectrophotometers and colorimeters can also be used to determine the concentration of particles. The method depends on attenuation of the transmitted beam as a result of light scattering and hence is relatively insensitive to cell concentration. The photometric response does not follow the Beer-Lambert law; however, a calibration curve relating instrument response to bacterial cell concentration, for instance, gives satisfactory results (see Chapter 11.6). For more sensitive determinations of particle concentrations, see section 21.1.2 below.

21.1.1.6. Determination of reaction rates

When either the product or the reactant of a reaction is a chromophore, the rate of chromophore production or utilization can be determined at fixed wavelength by reading the absorbance at time intervals or by monitoring the absorbance change versus time with a strip chart recorder. A plot of absorbance versus time permits determination of velocity from the slope of the line. Specialized spectrophotometers and attachments to standard spectrophotometers have been developed for this purpose, Some of these, especially designed for determination of enzyme-catalyzed rates, present absorbance values to a chart recorder and can monitor several reactions simultaneously by incorporating a timed automatic cuvette changer.

21.1.1.7. Automated spectrophotometers

Manual spectrophotometers apply an analog voltage output from the photodetector to some type of comparator circuit such as a Wheatstone bridge or potentiometric recorder. With the advent of analog to digital conversion and computer technology, the output from the photodetector can be digitized and manipulated. Accordingly, modern spectrophotometers are now equipped with computers which can perform numerous transformations and displays of the photometer output. In addition, incorporation of a clock and computer-controlled stepping motors for the wavelength drive and slit adjustment has facilitated the acquisition of data with respect to time as well as with continuous wavelength scanning. Such instruments afford much greater convenience of operation, especially for double-beam spectrophotometers. Until recently, however, the basic optical system has been unchanged.

Invention of the photodiode array detector has made possible a different configuration for a spectrophotometer. This detector is a linear assemblage of extremely thin individual photodiodes. A parallel beam of light with a wide range of wavelengths (white light) is presented to the sample (and blank), whereupon absorption of light at various wavelengths characteristic of the chromophores present takes place. The exiting beam is then dispersed by a fixed prism or diffraction grating to produce a spectrum of energies in the visible and UV range that are "radially fanned out" as in any monochromator. These energies impinge on the photodiode array, which then puts out a very large number of individual voltages that are proportional to the light intensities at very narrow wave bands (nominally wavelengths). Thus the full information content of the exiting beam is acquired in a very short time interval and is made available to the computer. Since progressive scanning of wavelengths is not involved, automated wavelength drive and coupled slit adjustment mechanisms are not needed. Spectrophotometers incorporating all of these changes are widely available. However, since a beam of all wavelengths passes through the sample, it is important to confine measurements to the absorbance range specified and to consider photodecomposition and fluorescence of the sample.

21.1.2. Nephelometry

Photometric determination of scattered light from a particle suspension at 45 to 150° gives a very sensitive response to particle concentration. This is due to the ability to measure a weak photometer response against the nearly zero response of the solvent alone. The Beer-Lambert law is not followed in nephelometers, but at very low concentrations of particles the instrument response approaches linearity (see Chapter 11.6.2).

21.1.3. Fluorimetry

Fluorescence is the emission of light by a chromophore (fluorophore) in response to absorption of light at a lower wavelength. Since the measurement involves the appearance of radiant energy against the very low level of background light from the blank, the signal-to-noise ratio is very high, and hence fluorimetric methods are inherently more sensitive than photometric methods that are based on absorption of light. For instance, the fluorescent method for determination of NADH is roughly 100 times more sensitive than the spectrophotometric method. Fluorescence intensity depends on the concentration of fluorophore and, unlike absorption, also depends on the intensity of the excitation beam. Since different wavelengths are involved in excitation and fluorescent emission, the instrument must be designed to prevent excitation energy from reaching the photodetector. This is usually accomplished by placing the detector at 90° to the excitation beam. Stray light arising from scattering both in the solution and from surfaces can be a major problem, but this can be minimized by placing before the photodetector a sharp-cutoff filter, which passes light at the wavelength of fluorescent emission but blocks excitation energy.

Since fluorescence intensity is dependent on excitation intensity, lamp emission characteristics and stability are of major importance. Some instruments monitor beam intensity through a second photodetector and display the ratio of fluorescence to excitation intensities.

At low fluorescent-emission intensities, photodetector output voltage is nearly linear with intensity; hence, concentration is proportional to the equivalent of transmittance (not absorbance) as in absorption photometry. The concentration of a fluorophore can be determined within limits by measuring relative fluorescence. The process basically involves preparation of a solvent blank and a set of standards, construction of a standard curve, and comparison with unknowns. Since the sensitivity is greater, it is more likely that measurements will be affected by artifacts than in absorption photometry.

21.1.3.1. Factors affecting accuracy

Instrument variation. Care should be taken to correct for or prevent fluctuations in excitation intensity as a result of variations in lamp intensity. It may be necessary to introduce the solvent blank and standards frequently.

Absorbance of solutions. High absorbance of solutions produces two kinds of errors: (i) reabsorption of fluorescent energy to give an apparently lower fluorescence yield and (ii) nonuniform intensity across the face (emitting surface) for a right-angle instrument. This results from variations due to high absorbance at the excitation wavelength. For this reason, absorbance should not exceed 0.2 Å (0.02 nm), and independent absorbance measurements on standards and unknowns should be made at the same excitation and emission wavelengths.

Fluorescence of impurities. The limiting sensitivity is determined by background fluorescence. This is minimized by use of pure reagents. In addition, errors can result from nonuniform contamination of samples and glassware by a variety of fluorescent materials, including rubber, grease, fingerprints, and extracts of filter paper or dialysis tubing. Turbidity contributes to error through scattering, which, unless an excitation energy-blocking filter is present, is measured as fluorescence. Therefore, maintenance of scrupulously clean glassware for measurement and storage of solutions is necessary. Many of the above effects can be minimized by the use of blocking filters at the photomultiplier.

Deterioration and other artifacts. Samples may deteriorate before or during measurement. Photochemical decomposition and adsorption to glass surfaces, especially for macromolecules, are quite common. The former can be detected as a loss of fluorescence with time, and the latter can be suspected if there is an apparent lag in fluorescence with increased concentration of sample or fluorophore.

Effect of temperature. Since fluorescent phenomena are temperature dependent, all measurements should be made at constant temperature, preferably in a thermostated cell.

21.2. ION ELECTRODES (11)

The Nernst equation

$$E = E_0 + \frac{RT}{nF} \log \frac{[\text{ox}]}{[\text{red}]}$$

can be used to measure the concentration (activity), or the ratio of concentrations, of certain molecules or ions. E is the measured potential, E_0 is the standard electrode potential, R is 8.315 J/°C, T is the absolute temperature, n is the number of electrons, F (Faraday) is 96,500 coulombs, and [ox] and [red] are the concentrations of oxidized and reduced members of the half-cell pair, respectively.

This basic behavior has made possible the determination of a number of biologically important ions, e.g., hydrogen, oxygen, ammonium, chloride. Incorporation of these principles into a usable instrument involves (i) the availability of a measurement electrode capable of sensing the potential generated by the internal oxidant-reductant system, (ii) a standard half-cell of known potential, and (iii) an electrometer or potentiometer which measures the voltage between the two half-cells. Solution of the Nernst equation for the measuring half-cell results from the relationship

$$E_{(\text{measuring half-cell})} = E + E_0$$

The ability to measure a specific ion is dependent on the construction of the measuring electrode. For instance, a glass electrode develops a potential proportional to hydrogen ion concentration, and several ion-selective electrodes develop potentials for specific ions. An ion-specific oxidation-reduction occurs in the coating material that isolates the electrode from the medium. There are, in addition, many other strategies for determination of concentration of various ions on the basis of electrode behavior. The oxygen electrode device, for instance, is a form of polarograph, because there must be a polarizing voltage to ionize the oxygen.

21.2.1. Hydrogen Ion Electrode (also see Chapter 6.1)

The glass electrode is the only practical means of measuring hydrogen-ion activity (pH). A thin membrane of special glass isolates a silver-silver chloride half-cell from the measured solution (12). The electrical potential of the glass electrode varies linearly with pH over a wide range, pH 1 to 11. In the region of pH 11 and above, there is nonlinearity unless a special alkali-resistant glass is used. The reference is a standard calomel (Hg/Hg_2Cl_2) half-cell, which is connected to the measured solution by a salt bridge of KCl. Both the glass electrode and the standard electrode are immersed in the sample or assembled into a single, combination electrode. The latter can be constructed to enter long test tubes or to accommodate very small samples.

In addition to the static determination of the pH of samples, the pH meter or an automated version can be used to monitor over time the reactions in which a hydrogen ion is evolved or consumed. In this application, alkali or acid is added to maintain constant pH; a plot of acid or alkali consumption versus time expresses the reaction progress. For instance, the oxidation of glucose and glucose 6-phosphate either enzymatically or chemically and the phosphorylation of substrates with adenosine triphosphate by kinases are reactions that can be tracked with the glass electrode.

21.2.1.1. Calibration of the pH meter

The system must be calibrated by using buffers of known pH. A popular but expensive approach to calibration is to buy prepackaged liquid or solid buffers of specified pH, but this is by no means necessary or superior. Table 1 lists some standard buffers (10) which have dependable pH values. With linear electrode behavior, a single standard buffer should serve to calibrate across the useful range. However, electrode behavior is seldom ideal; hence, it is wise to use two standard buffers (for instance, pH 4 and 6.88) and to bracket the pH region of the unknown.

21.2.1.2. Problems (13)

The major problems in measurement of pH come from nonlinearity in the alkaline region, absorption of CO_2 in neutral and alkaline samples and in standards, and slow evaporation of standards. Additional problems of unsatisfactory instrument behavior commonly arise from poor maintenance of electrodes. The glass electrode needs an equilibrium-hydrated silica gel layer for accurate and constant pH measurement. Electrode manufacturers recommend that a new or cleaned

Table 1. pH of standard buffers at 20°C[a]

pH	Buffer
1.10	0.1 M HCl
2.16	0.01 M potassium tetroxalate
4.00	0.05 M potassium hydrogen phthalate
6.88	0.025 M KH_2PO_4 + 0.025 M Na_2HPO_4
9.22	0.01 M borax
10.02	0.025 M $NaHCO_3$ + 0.025 M Na_2CO_3

[a] Data from reference 10.

electrode be soaked in water, or in buffer followed by water, for up to several days to establish the gel layer. They also recommend that the electrodes be maintained in distilled water between measurements. Electrodes must be cleaned after use by washing and careful wiping to remove adhering materials.

Unfortunately, incomplete cleaning plus storage in water can result in a buildup of contaminants and fouling of the glass surface. Lipid and protein contribute to the nonlinearity of electrode behavior and produce a slow response. In addition, bacterial growth may occur, with both fouling of the glass surface and plugging of the asbestos fiber which functions as a salt bridge between the standard electrode and the solution being measured. The KCl level in the calomel electrode should be maintained, and the cap should be removed during measurement so that KCl drainage through the fiber will prevent impurities from lodging in the fiber.

Once the glass electrode is fouled, the temptation is to replace it with a new one. However, it can be cleaned effectively by soaking alternately in 0.1 N HCl and 0.1 N NaOH at 50°C for several cycles, followed by reestablishment of the gel layer by soaking. In difficult cases, immersion in 10% HF for a few seconds followed by 5 N HCl and then thorough soaking in distilled water is recommended.

21.2.1.3. Effect of salt

A high ionic strength of the solution being measured results in errors in pH measurement by as much as 0.5 pH unit. This problem is commonly encountered in preparing high-molarity ammonium sulfate solutions in the neutral and alkaline pH region. It is common practice to dilute such samples 1:20 before pH measurement. Dilution of solutions of completely dissociated salts has little effect on pH because the ratio of ions remains nearly constant. The magnitude of the salt effect for specific circumstances is ascertained by making up standards in the salt solutions involved.

21.2.2. Other Ion-Specific Electrodes

Ion-specific electrodes are available for electrometric determination of a large number of ions and gases (8, 11). Although most of these have applications in industrial processes, several determine biologically important ions with sufficient sensitivity and selectivity to be of use in bacteriological investigations. Ion-specific membrane probes fall into four categories: glass, solid or precipitate, gas-sensing, and liquid electrodes (Table 2). Electrode response is, broadly speaking, the result of an ion-exchange process, and potentials follow the Nernst equation or one of its expanded forms.

Since there is no system for absolute calibration of probes as there is for pH with the hydrogen electrode, it has been common practice to use dilutions of stock solution standards made up of a completely dissociating salt (17).

With respect to sensitivity, measurements can be made over the range of 10^{-6}–10^{-7} M to 1 M for the favored ion. However, at lower concentrations some time is required to establish equilibrium with the electrode. Seldom is the sensor highly specific. In fact, interference with other ions is quite common. For instance, a calcium electrode gives 10% response to equivalent amounts of Pb^{2+}, Zn^{2+}, Cu^{2+}, and Fe^{2+} in the 10^{-5} to 10^{-6} M range. Chloride electrodes respond similarly to iodide and bromide, i.e., 10% response at 10^{-5} to 10^{-6} M. On the other hand, the K^+ electrode is quite insensitive to Na^+, Li^+, Cs^+, and H^+. The specific details of sensitivity and selectivity depend on the particular electrode being used. Refer to the manufacturer's literature for detailed information.

Table 2. Some ion-selective electrodes with potential use in bacteriology[a]

Type of electrode	Ions
Liquid membrane	H^+, monovalent cations
Solid	Halides, HS^-, SO_4^{2-}
Gas sensing	CO_2, NH_3, SO_2, NO_2, H_2S, HCN
Liquid membrane	Br^-, I^-, NO_3, CO_3^{2-}, SO_4^{2-}, K^+, Ca^{2+}, PO_4^{2-}, organics

[a] Data from reference 8.

21.2.2.1. Ammonia electrode

The ammonia electrode uses a gas-permeable membrane (15) through which ammonia diffuses to react with the filling solution:

$$NH_3 + H_2O = NH_4^+ + OH^-$$

The hydroxide level of the filling solution is then measured against a reference electrode. The electrode is specific for ammonia; the ammonium ion concentration is also measured after addition of NaOH to pH 11. Only volatile amines interfere, in the range from 1×10^{-7} to 5×10^{-7} M. Nitrate and nitrite can also be measured after reduction to ammonia, and Kjeldahl nitrogen can be measured after digestion and addition of NaOH (see Chapter 22.7). Urea-N can also be measured in a special adaptation of the ammonia electrode in which a second membrane encloses a urease-containing solution. Alternatively, the samples may be pretreated with urease and the normal ammonia electrode may be used for measurement (16).

21.2.2.2. Iodide glucose-sensing electrode

Ion-specific electrodes have been adapted to measure activities of enzymes and concentrations of compounds which are substrates for certain enzymes (17). For instance, the Hg_2S/Hg_2I_2 crystal membrane electrode which is designed to monitor iodide ion can be used for the determination of both glucose oxidase and glucose according to the equations

$$\text{Glucose} + H_2O + O_2 \xrightarrow[\text{glucose oxidase}]{} \text{gluconic acid} + H_2O_2$$

$$H_2O_2 + 2I^- + 2H^+ \xrightarrow[\text{Mo (IV)}]{} I_2 + 2H_2O$$

In addition, a flowthrough iodide-sensing electrode has been developed so that up to 70 samples per hour can be analyzed with a commercial autoanalyzer. With excess glucose oxidase and limiting glucose in the sample, the potentiometric response measures glucose in the sample. With excess glucose and limiting glucose oxidase, the rate of potential change can be used to determine the amount of oxidase present.

21.2.3. Oxygen Polarograph (9, 14; also see Chapter 6.5)

A platinum electrode at −0.5 to −0.8 V consumes oxygen according to the equation

$$O_2 + 2H_2O + 4e^- \rightarrow 4OH^-$$

The equation shows that the process occurs with the flow of electrons or current toward the electrode. This current flow through a resistor appears as a voltage across the resistor. The "potential" of such a half-cell, when used with a standard Hg/Hg_2Cl_2 or Ag/AgCl (silver wire in chloride-containing solution) half-cell, can be measured if the two are connected by a salt bridge. Alternatively, the current may be measured with a galvanometer or the voltage can be measured after amplification with a potentiometer, and these do not require a reference electrode.

The platinum electrode may be stationary, rotated, or oscillated to minimize diffusion gradients. However, such electrodes tend to be poisoned in biological measurements. A popular solution to this problem is the Clark oxygen electrode (14), in which the platinum electrode is covered with a gas-permeable membrane. Whereas the platinum electrode measures the number of oxygen molecules present at its surface as governed mostly by collision theory, the Clark electrode response is proportional to the rate of diffusion across the membrane isolating the electrode. In this arrangement a gradient across the membrane results from the zero oxygen concentration at the electrode surface where oxygen consumption occurs. Fortunately, the rate of diffusion is linearly dependent on oxygen concentration or partial pressure. The current output is also dependent on the platinum cathode area.

A plot of current versus polarizing voltage shows a flat or nearly constant response in the region of −0.5 to −0.8 V. Therefore there should be a polarizing voltage adjusted to the center of the constant-current region. The behavior in response to polarizing voltage is the major distinction of the oxygen electrode.

21.2.3.1. Effects of temperature

The sensor current is quite sensitive to temperature because the membrane material permeability has a high temperature coefficient (about 3 to 5%/°C). In addition, oxygen solubility varies with temperature. For this reason, good temperature control and adequate equilibration are required.

21.2.3.2. Calibration and calculations (14)

Calibration is based on the known oxygen solubility of a solution under defined conditions. Zero oxygen level or "bleed current" can be defined as the response in the presence of added sodium dithionite. Full response is set at 100% with an oxygen-saturated solution. The oxygen concentration at the operating temperature is obtained from oxygen solubility tables in the *Handbook of Chemistry and Physics* (17a) or similar reference publication. The oxygen content of solutions to be used as working standards can then be measured. For most work the actual oxygen concentration of standards is not needed; rather, a reproducible concentration is necessary to set the instrument. The measurements needed are the relative differences between experimental samples. For example, if Ringer's solution contains 5 μl of O_2/ml when air saturated, 3 ml of solution giving an instrument response of 92% saturation contains 13.8 μl of O_2 ($3 \times 5.0 \times 0.92$). Rates of oxygen consumption or evolution are either calculated by taking measurements at two times or obtained from the slope of plots of percent full scale versus time involving several readings. For example, with an instrument response at 0 and 5 min of 92 and 70%, respectively, the calculation is as follows: 92 − 70 = 22% consumed/5 min and 5.0 × 3 = 15 μl of O_2 × 0.22 = 3.3 μl/5 min = the reaction rate.

Unfortunately, it is unlikely that reaction mixtures under study have the same composition as the oxygen standard, and hence there will be some error. However, the error can be minimized by maintaining a low total solute concentration. When solutions of greater oxygen concentration are needed, increased partial pressures can be generated by saturation of standards and working solutions with oxygen or oxygen-nitrogen mixtures.

The oxygen content of standards and reagents is influenced by temperature and altitude. Barometric pressure fluctuations seldom change oxygen concentration by more than 3%. Correction of the oxygen solubility coefficient for barometric change involves the relationship

$$\frac{\text{observed barometric pressure}}{760} \times \text{solubility at 760 mmHg (101.3 kPa)}$$

21.2.3.3. Operating information

At the beginning of an experiment, the probe should be tested for (i) plateau setting of polarizing voltage, (ii) response time for a step change in polarizing voltage, (iii) noise, and (iv) drift. The lack of horizontal plateau indicates a dirty electrode; however, some slope can be tolerated in many instances. Response time for a step change in polarizing voltage should be 1 to 1.5 min. Noise may result from poor connections and grounding, a damaged membrane (folds, holes, KCl crystals, or drying), or poor contact with the silver electrode, in which case cleaning with ammonia is recommended.

A properly functioning probe should have less than 2% drift per h. Drift may be caused by atmospheric changes or by contamination with organic material. Thus, proper care and cleaning of the probe are essential. In one study, reproducible results were obtained when the probe and receptacle were washed with reagents used in the test.

21.2.3.4. Assembly of electrode membrane

The demanding aspect of the Clark electrode is attaching the membrane. Directions and spare membranes are furnished by manufacturers of the probe. Care must be taken in assembly to ensure that the membrane is smoothly stretched over the probe, that there is a true seal at all points, and that no air bubbles are trapped on the electrode side of the membrane. To obtain good wetting in assembly, add 3 or 4 drops of Kodak Photoflow per eyedropper bottle of KCl used in assembly.

21.3. CHROMATOGRAPHY

Chromatography and related separation techniques depend upon many kinds of distribution processes, which take place repeatedly in a small local environment (18–23). Therefore, the operations of ion-exchange, solvent partition (including reversed-phase), adsorption-desorption, hydrophobic interaction, ion pair and hydrogen bonding, and affinity (association-dissociation) chromatography, as well as the operationally related gel permeation (size exclusion) chromatography based on a sieving principle, can all be carried out with the same laboratory equipment to make separations based on widely different properties of the solute molecules.

21.3.1. Ion-Exchange Chromatography (25)

Sorption of solutes is based on ionic charge, although other types of bonding may have some influence on separation. Charged groups on a permeable matrix are able to exchange sorbed ions with those in solution

$$-(X)-SO_3Na^+ + K^+Cl^- \rightarrow -(X)-SO_3^-K^+ + Na^+Cl^-$$

Matrix materials include cross-linked, substituted, or derivatized resins such as polystyrene or polymethacrylate, as well as derivatized carbohydrate-containing natural polymers such as cellulose, dextran, or agarose.

Exchangers are available with a great variety of functional groups, support materials, "purity," porosity, capacity, size, shape, selectivity, and stability. It is therefore necessary to consult technical literature from manufacturers for the specific information needed, for instance, Dow Chemical Co. for Dowex resins, Rohm and Haas Co. for Amberlites, Bio-Rad Laboratories for a wide variety of exchangers, Pharmacia Fine Chemicals for Sephadexes and Sepharoses, and Reeve Angel Co. and Whatman Co. for cellulosic exchangers.

21.3.1.1. Resin exchangers

Many synthetic resins are available with strong and weak acidic and basic exchange groups as well as with different degrees of cross-linking. Lower cross-linking results in higher permeability and water uptake and an ability to interact with larger molecules. For instance, for Bio-Rad resins of the AG 1 series, 2% cross-linking produces an exclusion limit of 2,700 Da whereas 8% cross-linked resin excludes 1,000-Da molecules. Thus, high cross-linking favors sorption of substrates and products of enzyme reactions while discriminating against sorption of the enzyme involved. Ion-exchange wet-volume capacity and selectivity increase with higher cross-linking and also increase as the particle size decreases. Thus, fine-mesh exchangers have increased resolution per bed length simply because of the increased number of exchange cycles taking place. However, the increased packing of small particles decreases the flow rate, which in turn requires increased operating pressures.

Strong anion, cation, and chelating resin beads have been cast into membranes of pore size 0.45 μm. Such membranes offer the advantages of speed and ease of use for certain applications.

Sorption, purification, and fractionation are major applications of exchangers. In sorption, a desired ion is selectively removed from extraneous material by ion exchange. In purification, the ion desired is selectively sorbed and eluted while other ions are not sorbed or are retained. Fractionation is the process of separating several similarly charged ions by introduction of solvent-solute systems which progressively elute sorbed ions of increasing charge. The charge state is often a function of the pK of dissociating groups and the pH of the environment.

As a rule of thumb, resolution is related to bed length, whereas for a given resolution, capacity is related to bed cross-section. Since binding of soluble ions is stoichiometric and reversible, the amount of resin needed can be calculated from its published capacity. In general, strength of binding of inorganic ions is directly related to valence (number of charges) and inversely related to atomic number. For weak acids and bases, strength of binding is related to pK of the ionizing group if the pH of operation is in the dissociation range.

Exchange of counterions and deionization. Since ion exchangers are charged with a counterion (for instance, for Dowex-1 this might be OH^-, Cl^-, or formate), it is possible to exchange virtually completely an ion in the solvent with the charged counterion. For instance, chloride can be exchanged for OH^- with Dowex-1 and H^+ can be exchanged for Na^+ with Dowex-50. Therefore, a major application of exchangers is the deionization of uncharged solutes such as urea and polyols. This can be accomplished by serial passages of the sample through cationic and anionic exchangers or through a mixed bed. However, the mixed bed is more difficult to regenerate because the cationic and anionic exchangers must first be separated on the basis of density difference. Alternatively, the spent exchanger is discarded.

In deionization of carbohydrates, the noncharged sugars may behave as weak acids and be sorbed by a strong anionic exchanger; for this reason, a weak basic exchanger should be used.

In exchange of ions, there is a hierarchy of affinities of ions. For this reason it is difficult or inefficient to substitute directly a low-affinity ion for one of high affinity. For example, it is difficult to convert Dowex-1–Cl to Dowex-1–formate simply by washing with formic acid or sodium formate. The more efficient route involves conversion of chloride to hydroxide and from hydroxide to formate. The order of affinities can be obtained from the manufacturer's literature or from reference books.

Selective removal of divalent and trivalent cations is accomplished by specialized resins which have EDTA, iminodiacetate, or related groups to chelate the metal ion. Chelex (Bio-Rad), Chelite (Serdolite), and other products perform this function.

Concentration. As a consequence of the high affinity for ions plus the fact that fresh resin is encountered with movement of the sample through the column, it is possible to bind ions from a large volume of dilute solution. The sorbed ions can then be recovered in a small volume of eluant containing an ion of higher affinity or of higher concentration. In this way, any number of common and trace inorganic elements as well as radionuclides and trace organic compounds have been concentrated for analysis.

Purification and fractionation. If two ions have different affinities for an exchanger at a given pH, they can be separated by finding conditions which sorb and/or elute one ion but not the other. In such extreme cases, there is a "frontal" elution of each ion. In cases when the affinity is similar, separation is still possible by finding conditions which selectively elute the ion. Development of the column with eluants then leads to a gradual separation. If the column length is sufficient, separation can be attained. However, diffusion of ions limits the ability of increased column length to resolve the components. Changing pH or ionic strength elutes sorbed ions of different pK values or of increasing affinities.

Gradient elution. Elution of sorbed molecules by continuous change of eluant composition is a useful way to maximize the separation potential of a chromatographic system. Rather than making step changes in concentration by application of different eluants, the change is made to be continuous within given limits. Devices and procedures range from the very complicated, as in the nine-chamber device (72) for separation of amino acids on Dowex-50 $(Na)^+$, which produces complex gradients, to the relatively simple linear concentration gradient involving mixing of two components. Gradient-making pumps are also available. In recent years the simple linear gradient has been used most frequently (71). For this purpose, two reservoirs are connected so that the outlet of one is fed into the other (Fig. 5). Another outlet of the second reservoir feeds the column. A mixer in the second vessel keeps the eluant composition homogeneous. When both vessels are open to the atmosphere and each vessel has the same cross-sectional area, the eluant composition will change linearly from the limit in the second (mixing) reservoir to the limit in the first. The steepness of the gradient is determined by the volume in the two reservoirs. (See Chapters 4.3 and 11.5.5 for the use of gradients for cell fractionation and determination of cell density.)

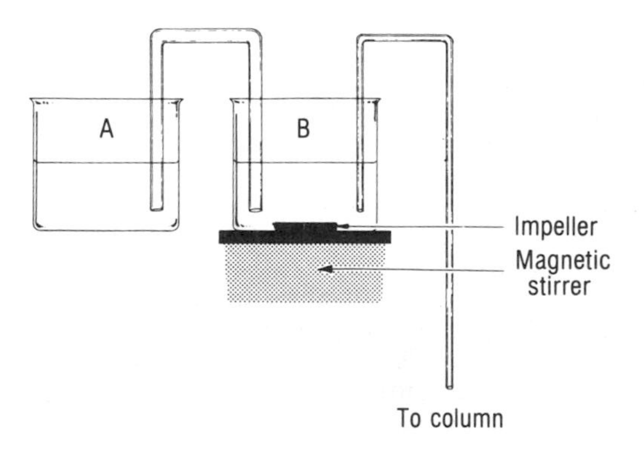

FIG. 5. Gradient maker for linear gradients. The surface areas of beakers A and B are equal.

Suppose it is desired to construct a gradient such that 25 10-ml fractions can be collected in which the salt concentration increases linearly from 0.01 M potassium phosphate buffer (pH 7) to 0.01 M phosphate (pH 7)–0.1 M KCl. The total volume needed for the gradient is 250 ml. Now arrange two 250-ml beakers level so that they can be connected with a loop of glass tubing (Fig. 5), or connect beakers modified to have outlets at the bottom, one outlet for the first beaker and two for the second (mixing) beaker. Place the second beaker on a magnetic stirrer, and connect the second outlet to the column. Add 125 ml of phosphate buffer plus KCl to the first beaker and about 125 ml (should be equal weight) of phosphate buffer without KCl to the second. Independently run some 0.01 M phosphate buffer down the column to equilibrate the system with the starting eluant. Then start the eluant from the second beaker. The elution continues with stirring until the contents of both beakers have been transferred to the column. Independent determinations of chloride (chloride electrode, above) or of salt concentration (by refractive index measurement) in the fractions will verify the linearity of the gradient. It is also necessary to prevent intermixing of the two reservoirs before elution is started. Therefore, use pinch clamps in the siphon tube and at the outlet of the second beaker.

When the ratio of cross-section of beaker 1 to beaker 2 is greater than 1, the gradient will be convex, and when the ratio is less than 1, the gradient will be concave, as dictated by the equation

$$C_2 = C_1 - D(1 - v)A_1/A_2$$

where C_2 is the concentration in the second vessel (lower limit), C_1 is the concentration in the first vessel (upper limit), D is the difference in concentrations, v is the fraction of initial volume withdrawn up to the point in question, and A_1 and A_2 are the cross-sectional areas of the vessel containing the higher concentration and the mixing vessel, respectively.

A common method of making gradients, which involves dropping eluant of higher concentration from a separatory funnel into a closed mixing reservoir of constant volume, produces convex gradients, which are less useful.

More complex gradients can be constructed by placing more beakers in series with mixing in all but the farthest removed. The eluant in each beaker can be different to effect a compound gradient. A nine-chamber gradient device has been developed for this purpose (72).

Ion exchange and dialysis. Solutions of proteins and other sensitive macromolecules can be successfully treated with ion exchangers for the purposes of deionization or ion replacement. However, steps should be taken to isolate macromolecules from exchangers in the H^+ or OH^- form. This is accomplished by placing the solution to be treated in a dialysis bag and suspending the bag in a stirred suspension of the exchanger.

Nucleic acids. Nucleotides, nucleosides, and purine and pyrimidine bases are separated on anionic and cationic exchangers of low cross-linkage. The methods involve control of ionic strength, pH, and nonionic binding. With Dowex-50W-X4 (Dow Chemical Co.) or AG 50W-X4 (Bio-Rad Laboratories), make up samples in 0.1 to 1 ml of elution buffer, and use a flow rate of 1 ml/min. Separate the 5′-nucleotides on a 40-cm column with 0.1 M ammonium formate adjusted to pH 3.2 with formic acid. For 2′- and 3′-nucleotides, use a column length of 85 cm and elution buffer of 0.25 M ammonium formate adjusted to pH 4.1 with formic acid (38).

Another system in which a salt gradient and ethanol are used to control nonionic binding separates ribo- and deoxyribo-oligonucleotides by base composition, chain length, and 3′- or 5′-terminal phosphate group (33). For one of the procedures, use AG 1-X2 (400 mesh; Bio-Rad Laboratories) in a 60- by 0.3-cm column. Elute at 12 ml/h with 400 ml of 20% ethanol containing a linear gradient of 0 to 1 M NH_4Cl (pH 10). For a second procedure, use Dowex-1-X4 (400 mesh) in a 100- by 0.2-cm column and program the sample addition and elution as follows: (i) 0.4 ml of 1 M NaOH, (ii) 1 ml of 20% ethanol, (iii) 1 ml water and sample, (iv) 1 ml of 20% ethanol, and (v) 200 ml of 20% ethanol containing a linear gradient of 0 to 0.5M NH_4Cl (pH 10). Set the flow rate to 6 to 10 ml/h.

For cyclic nucleotides in a homogenized, deproteinized, and neutralized sample, carry out the separation in a 0.5- by 4-cm column of AG 1-X8-formate (200 to 400 mesh) equilibrated in 0.1 N formic acid and elute with 10 ml each of 8 M formic acid containing 20 mM, 100 mM, and 1 M sodium formate (60).

Citric acid cycle intermediates (61). Adjust the sample to pH 6, and add 0.5 ml of sample to a 0.9- by 14-cm column of AG 1-X10-formate. Elute with a 150-ml gradient formed by adding 3 M formic acid to a 250-ml closed mixing flask filled with water. Follow this with 2 N ammonium formate at 2 ml/min. β-Hydroxybutyrate, acetoacetate, succinate, malate, pyruvate, citrate-isocitrate, α-ketoglutarate, and fumarate elute as separate peaks in that order.

21.3.1.2. Cellulose, dextran, and agarose exchangers (71, 75)

For separation of macromolecules, natural polymers have been modified to contain weak acid or base functional groups. These matrices also have capacities and other properties generally more suitable than resins for fractionation of biomolecules. With dextrans and agaroses, cross-linking and particle size distribution can be controlled, and this controls shrinkage and swelling and produces exchangers with high capacity for macromolecules as well as high and uniform flow rates. Generally, agaroses better maintain bed volume when ionic strength increases and when the pH is outside the neutral region.

Since proteins are amphoteric and have a large number of charges, often in clusters, it is possible to treat them as cations or anions by choosing a pH for ion-exchange separations that is at least 1 unit above or below the isoionic point or the pI value. Cycles of binding and elution make use of a pH change which can cause disappearance of ionic groups, e.g., carboxyl or amino groups on the protein or on the exchanger. Alternatively, elution is effected by displacement with another ion.

The DEAE derivative of cellulose, dextran, or agarose is a weak anion exchanger (pK of about 9.5). Hence, in the neutral- or lower-pH region, the amino group is protonated and binds proteins and peptides if the pH is near or above their pI values.

The carboxymethyl (CM) derivatives of the same supports are cationic exchangers with a pK of about 3.5. Separations are generally carried out above pH 4. When the pH of the separation is below the isoelectric point of the protein and above pH 4, sorption is efficient.

ECTEOLA-cellulose and a similar derivative of dextran have a pK of 7.5. This exchanger has been used in chromatography of nucleic acids and nucleoproteins.

Cellulose and dextran derivatives used in ion-exchange chromatography are listed in Table 3.

Practical considerations. Successful use of exchangers rests on the judicious choice of a number of parameters. For the exchanger these include the type of charged group, matrix, particle size and shape, and porosity; for the eluant they include the buffer type and pH; for the system they include the column dimensions, flow rate, and elution program. In addition, attention must be paid to exchanger packing, sample addition, maintenance of low sample viscosity, and column design, since these also determine the degree of resolution. Treatment of these aspects in adequate detail is beyond the space available here, and hence only general guides for each will be given. Reference 38 and the publications by Whatman, Inc. (29), for exchanger celluloses and by Pharmacia, Inc. (26), for the corresponding Sephadexes are recommended for the details.

In general, uniform spherical dextran beads (Sephadex) give higher flow rates, although the cellulose derivatives now available also perform well. Smaller bead or particle sizes give lower flow rates and increased resolution. Greater porosity increases the capacity for a given molecule and is particularly important for work with macromolecules. The choice of exchanger is based on the ionic forms to be separated as described above. The pH chosen should be at least 1 pH unit removed from the isoelectric point or the pK of the dissociating group. When possible, cationic buffers (alkylamines, ammonia, barbital, imidazole, and Tris) should be used with anionic exchangers; anionic buffers (acetate, barbital, citrate, glycine, and phosphate) are used with cationic exchangers. The recommended pH range for DEAE-Sephadex is 2 to 9, and that for CM-Sephadex is 6 to 10. Ionic strength influences the binding and therefore the capacity for ions. A relatively high ionic strength of ca. 0.1 M is recommended, but the concentration should be somewhat lower than is needed to elute the ions desired.

Column height determines the resolution achieved. When the zone of absorbed ions is 1 to 2 cm, a 20-cm bed height is recommended as a starting point. At fixed height, the column capacity is linearly related to the cross-sectional area.

Elution is effected either by changing the pH to cause the ionic charge to diminish or disappear or by increasing the ionic strength to the point at which the buffer ion displaces the ion of interest. These can be done by either stepwise or gradient change. For anion exchangers the pH is decreased or the ionic strength is increased; for cation exchangers the elution is effected by increasing the pH or the ionic strength.

Table 3. Derivatives of cellulose and dextran used in ion-exchange chromatography

Derivative	Symbol	pK
Sulfopropyl	SP	Strong acid
Phosphocellulose	P	1.5, 6
Carboxymethyl	CM	4
Mixed amine	ECTEOLA	7.5
Benzoylated diethylaminoethyl	BD	9
Diethylaminoethyl	DEAE	9.5
Triethylaminoethyl	TEAE	9.5
Diethyl-2 hydroxypropyl aminoethyl	QAE	Strong base

Either the sample total ionic content or column size must be determined first, because available capacity is the property of a specified volume of exchanger for each exchanger type. The ionic composition of the sample ideally should be that of the starting buffer. When necessary, the ionic composition is changed to that desired by means of gel filtration on Sephadex G-25, dialysis, or dilution. The sample volume is of minor importance for all elution programs except when using only the starting buffer for development. The sample must be added without disturbing the bed and without becoming distributed in the oncoming developing buffer.

As with other types of chromatography, it is essential to use glassware designed for ease of packing, addition of eluant and sample, application of pressure (if needed), regulation of flow rate, prevention from running dry, and low holdup volume above and below the exchanger bed. It is also necessary to establish criteria for achieving the objective in mind and to analyze the effluent to determine performance of the column.

21.3.2. Adsorption Chromatography

Adsorption involves van der Waals interactions as well as electrostatic, hydrophobic, and steric factors. Adsorption of a single component is typically described by Langmuir's adsorption isotherm, a rectangular hyperbola in a plot of solute concentration versus amount adsorbed. The same curve describes the saturation behavior of substrate for an enzyme and the "law of diminishing returns" among others. Normal phase surface sorption and elution has been successful with silicic acid and magnesium silicate (Florosil), kieselguhr (diatomaceous earth), aluminas, and charcoal. Proteins and nucleic acids have been purified by either batch treatment or column chromatography with alumina $C\gamma$ or various forms of calcium phosphate including hydroxyapatite, brushite, and calcium phosphate gel (62).

21.3.2.1. Types of adsorbant

Silicic acid has been used as adsorbant for purification of simple and complex lipids, fatty acids, sterols, phenols, and hydrocarbons. Elution is effected by changing the solvent composition from partially nonpolar toward polar character by use of a concentration gradient. Neutral alumina (pH 6.9 to 7.1) separates steroids, alkaloids, hydrocarbons, esters, organic acids and bases, and other organic compounds in nonaqueous solution. Basic alumina (pH 10 to 10.5) has similar properties in adsorbing compounds from organic solvents. However, with aqueous media it acts as a strong cation exchanger, sorbing basic substances such as amino acids and amines. Acid-washed alumina (pH 3.5 to 4.5) acts as an anion exchanger in aqueous medium.

Adsorbents for purification of macromolecules in aqueous solution are alumina $C\gamma$, calcium phosphate gel, and hydroxyapatite. Some of these do not lend themselves well to column chromatography and hence are used with batch methods. Column methods for hydroxyapatite are well known (49).

21.3.2.2. Separation of proteins

Batch separations of proteins are made by proper manipulation of adsorption and elution phases. The conditions for their use must be established by pilot tests with each batch of adsorbent and usually with each preparation of protein (enzyme or antibody) to be separated. Figure 6 schematically represents the way in which adsorption and elution cycles are used; these kinds of data are derived from pilot tests. In Fig. 6A are depicted three kinds of adsorption behavior influenced by the nature of the material being purified, the contaminants, and other conditions, e.g., pH and ionic strength. With increasing amounts of adsorbent, the protein of interest may be adsorbed either better than (left), the same as (center), or poorer than (right) the contaminating protein. If the protein of interest is adsorbed better, add a sufficient amount of adsorbent to remove most of the desired protein from solution while leaving most of the contaminants behind (positive adsorption). If the protein is adsorbed less well, add and remove adsorbent until the desired protein is slightly (i.e., 10%) adsorbed. In this way contaminating proteins are removed first (negative adsorption), after which the desired and partially purified material may now be adsorbed, or it may be subjected to another type of purification step. If both contaminant and desired proteins are adsorbed at equal rates, adsorption can provide little purification. Nevertheless, the protein being purified can be adsorbed and subjected to a useful elution cycle.

Similarly, Fig. 6B shows three types of elution behavior of proteins adsorbed in Fig. 6A. With increasing elution variable (for instance, ionic strength, pH change), the desired material may elute before (left), with (center), or after (right) the protein impurities. If the target material elutes preferentially, treat the adsorbent with only enough eluant or eluant cycles to remove most of the desired protein, leaving behind the impurities. If the other proteins are eluted preferentially, treat the adsorbent with eluant of increasing ionic strength with or without pH change until significant amounts of desired material appear in the eluate. Discard this volume of eluate. Then change the elution conditions just enough to elute the desired material. It

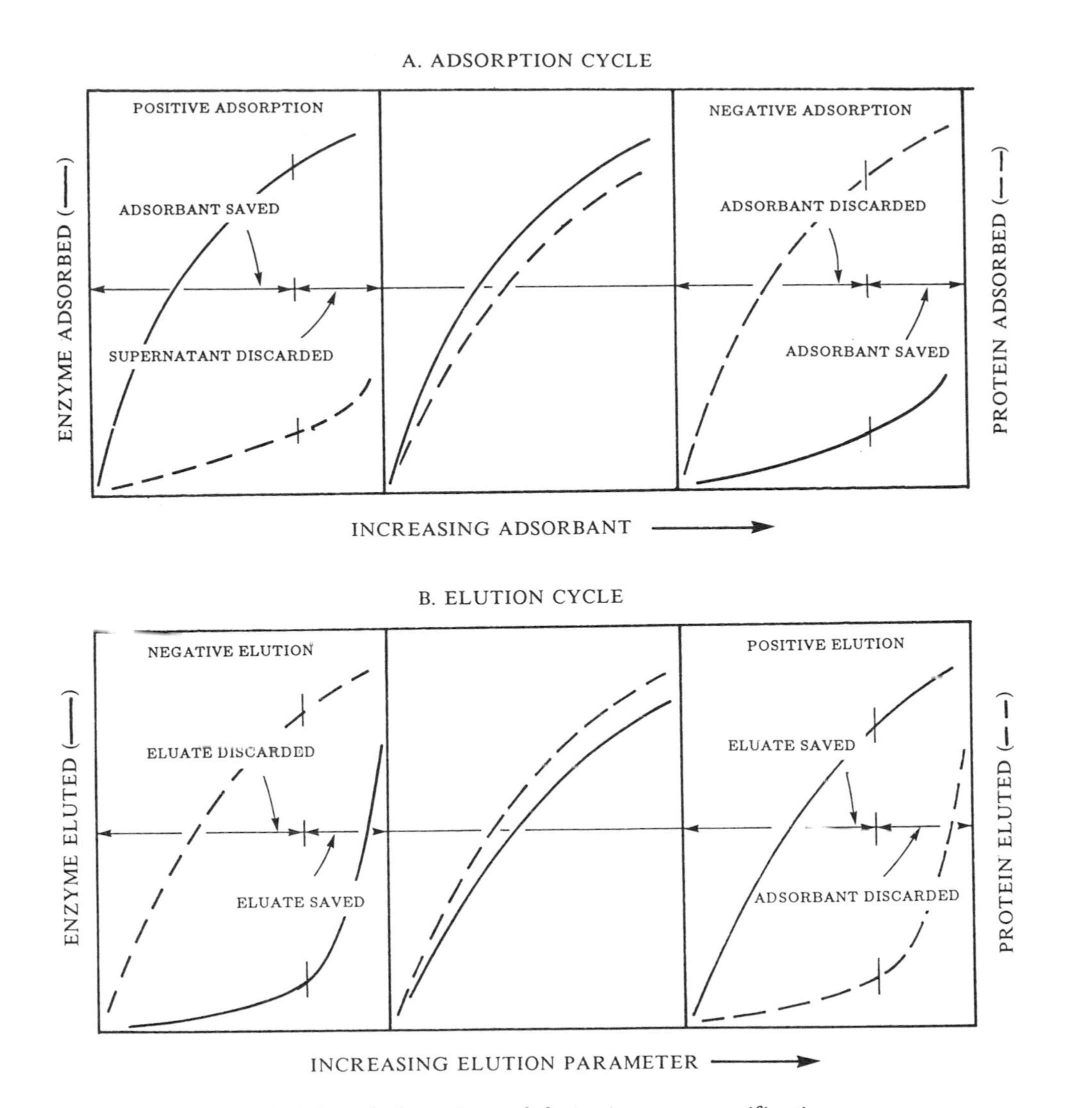

FIG. 6. Use of adsorption and elution in enzyme purification.

is obvious that if there is no differential behavior between the desired material and the impurities on either adsorption or elution cycles, this procedure should be discarded, new conditions should be found, or a new purification strategy should be developed.

There is a great deal of empiricism in this kind of work, but construction of a diagram from small pilot runs showing the behavior of desired material versus impurities greatly aids in designing the total procedure. In general, increasing the ionic strength and changing the pH toward the isoelectric point of a protein enhance elution. Further, substrates, effectors, and other ligands for enzymes and proteins often act as specific eluting agents effecting substantial purification.

The above approach applies in principle to all kinds of purification procedures.

21.3.3. Reversed-Phase and Hydrophobic Chromatography

Both reversed-phase and hydrophobic interaction chromatography separate molecules on the basis of hydrophobicity (28). For reversed-phase work, an alkyl or phenyl arm is bonded to silica and organic solvents are used as eluting agents. This method is useful for solutes that are not adversely affected by an organic solvent, usually small molecules. However, some enzymes and bioactive peptides have been purified by elution from the matrix with a water-acetonitrile gradient containing trifluoroacetic acid.

In hydrophobic chromatography, a phenyl or alkyl group is attached to a polar matrix such as Sepharose. Hydrophobic interaction between the matrix ligands and hydrophobic regions of a protein is enhanced by raising the hydrophilicity of the aqueous mobile phase with salt. Decreasing the salt concentration, often followed by increasing the hydrophobicity of the eluant with an alcohol or glycol, elutes the sorbed proteins. This procedure is especially useful following an ammonium sulfate precipitation in which the dissolved precipitate contains a relatively high concentration of ammonium sulfate. The salt is useful in enhancing binding of proteins, but it must also be removed before adsorption or ion-exchange chromatography can be used as a next purification step.

21.3.4. Liquid-Liquid Partition Chromatography

When a two-phase solvent system is set up, a large number of solutes, partially soluble in each phase, reach an equilibrium distribution between the two phases which reflects the relative solubility in each phase. This has led to useful solvent extraction procedures, both batch and continuous. The commercial Craig and Podbilniak countercurrent extractors are examples of these techniques at their best. However, the ability to separate similar materials is limited by the number of distribution cycles one can make. Fortunately, when one phase of a two-phase system is immobilized as a coating on a small particle, the distribution process can occur repeatedly in a column or paper chromatographic arrangement. Thus, paper and cellulose column chromatography separate by partition between an aqueous phase coating on the cellulose fiber and a less polar phase which flows through the interstitial spaces. Silicic acid (Celite) is another support which has been widely used in partition chromatography, both in column and in thin layers.

Gas-liquid chromatography and some forms of high-performance liquid chromatography (HPLC) also separate by solute partition. Both gas-liquid chromatography and HPLC have been very highly developed in terms of equipment, instrumentation, column packing matrices, and procedures, originally for analytical purposes, for which high resolution is desired. These specialized forms of partition chromatography warrant separate treatment (see sections 21.3.5 and 21.3.6, respectively).

21.3.5. Gas-Liquid Chromatography

Gas-liquid chromatography (gas chromatography [GC]) has a long history in the separation, identification, and quantitation of compounds that are volatile or can be made volatile at temperatures up to 450°C. Many compounds of biological interest exhibit these characteristics. One of the first reports by James and Martin in 1952 (55) involved volatile fatty acids. Separation was improved with the introduction of capillary columns in 1957. GC is widely used because of its ability to (i) separate diverse mixtures, (ii) separate chemically and physically similar components, (iii) analyze mixtures with a wide variety of components, and (iv) achieve rapid analysis.

21.3.5.1. Principles

Separation by GC is defined as the partitioning of a solute between a mobile phase of gas and a stationary phase of polar or nonpolar material that is liquid at the operating temperature of the system. The material is coated on inert particles packed into a column or bound to the interior surface of a capillary column. The components have various affinities for the liquid phase and hence move or are retained to various degrees. This retention of each component or analyte is a function of the chemical nature of the analyte, column-packing material (liquid phase), temperature, and flow rate of the mobile phase. Chemical reaction between the analyte and carrier is minimal and is usually disregarded. Each solute moves through the column as a band with a specific speed, and the bands exit the column and enter the detector distributed as symmetrical peaks.

Four parameters, resolution, efficiency, selectivity, and capacity, are considered in the effective separation of the sample of interest. Under specific operating conditions, each substance has a net retention time, $t_{R'}$, in the stationary phase. The total retention time, $t_{R'}$, is the sum of the net retention time and time spent in the mobile phase, (t_0); thus,

$$t_R = t_0 + t_{R'}$$

All analytes should have equal t_0 volumes under the same operating conditions. The term t_0 is also known as dead time. Its value is obtained by injecting an inert gas such as ethane and recording the elution time under operating conditions. Because the flow rate of the mobile phase is a factor, retention volume, V_R, is often another term calculated. The liquid phase is the major factor in the behavior of the analytes being analyzed, so choosing the proper column for a specific sample is critical.

Because of the separation technique involved, chromatography is readily described as analytical distillation, and performance of the column with the particular sample can be measured by using the concept of theoretical plates. Separation or resolution (R) of the components in the sample is calculated by the following equation:

$$R = \frac{n^{1/2} (\alpha - 1)\kappa}{4\alpha (\kappa + 1)}$$

where n is the number of theoretical plates, α is the relative retention time, and κ is the capacity (or partition) ratio. The number of theoretical plates represents efficiency and is determined by

$$n = 5.545\ t_R{}^2/W_h$$

where t_R is the retention time of the peak of interest and W_h is the peak width at half-height of peak of interest.

The relative retention time, α, is the ratio of retention times for two solutes for which separation is desirable. The capacity or partition ratio, κ, is derived from the ratio of the corrected retention time $t_{R'}$ of the solute and the column dead time.

Efficiency is a measure of the broadening of the solute plug or front as it moves through the column and generates a peak at the detector. A number of methods are available to calculate the efficiency of a column separation. A common method is to assume a Gaussian peak shape and to calculate efficiency in terms of a theoretical plate number, n (above). Other efficiency methods that are beyond the scope of this manual include relating efficiency to the average linear velocity of the mobile or gas phase by using van Deemter's equation and trennzahl or separation numbers when using capillary columns. These topics are examined in depth in specialized texts on chromatography in general and GC in particular (41, 48).

Selectivity is the third criterion of the separation, and this refers to the interactions between the analyte(s) and the chromatographic system. In GC this factor is controlled by the liquid component of the stationary phase. Components controlling selectivity include a mixture of physicochemical interactions that may be dispersive and dipolar. The degree of polarity of the stationary phase is given by the **McReynolds constant.** This constant is defined by using a set of reference compounds and comparing the retention of the reference compounds on a "standard" liquid phase with that in the liquid phase in question. The relative retention of an analyte pair expressed as α (above), is a common expression of selectivity. If $\alpha > 1$, separation is achievable. Selectivity is a major consideration in analyses involving packed columns which are usually rated at low efficiency. The final parameter of separation is capacity, and this is affected by the partitioning of the solute between the mobile and stationary phases. The liquid phase and the supporting solid phase or carrier both contribute to the separation of the analyte and to the total sample-carrying capacity of the system. Manipulation of the liquid- and solid-phase polarity can be used to optimize analysis. Absorption to the solid phase can be used to enhance detection of trace components in gas analysis (56). This is a dynamic equilibrium defined by the distribution coefficient, K_D. A complete discussion of this factor is available in texts on chromatography (37, 41, 48).

21.3.5.2. Instrumentation

There is a great variety of instruments available, but all consist of a few major components: injectors and inlets, column ovens for temperature control and programming, detector, data collection and handling, and a source of gas for the mobile phase. The choice of column is the first step in setting up a GC system.

A survey of published methods shows that most bacteriologist/chromatographers use commercial columns rather than preparing their own. Choosing the proper column from the vast array of those available is very difficult. The two types of columns used differ in the attachment sites of the liquid phase. Packed columns contain an inert carrier, such as silica, which is coated with the liquid-phase material. Capillary and large-bore (megabore) columns lack a carrier and have the liquid phase sheathing the interior walls of the column.

The advantages of packed columns are lower expense, the ability to accept larger injection volumes (1 to 10 μl), the capacity to analyze samples with high concentrations of analytes, the ability to perform in the presence of interfering compounds such as salts, and the variety of liquid phases and inert carriers available. The shortcomings of packed columns include lower selectivity, especially for trace analysis, difficulty in separating complex mixtures, and requirements for higher volumes of the carrier phase (gas).

Capillary and megabore coated columns limit or eliminate these problems. The megabore columns have a greater internal diameter than the capillary columns but perform in a similar manner. Capillary columns are used with complex mixtures at low concentrations and when similar compounds are present (65a). The greatest disadvantage is their expense.

The selection of the liquid phase in the column is not controlled by any specific guidelines, but a common rule of thumb is to separate analytes with liquid phases that are chemically similar. For instance, use a polar liquid phase for polar analytes. The suppliers of packed and capillary columns furnish a variety of systems for most compounds of interest to bacteriologists.

The most commonly used detector system is flame ionization detection (FID). The sample enters a hydrogen-air flame and is burned to its component ions. All the ions are collected and measured at a collector electrode that surrounds the flame. The collector has an applied potential of -150 V or greater to ensure efficient ion collection. This detector responds to virtually all organic compounds. It does not respond to the rare gases, oxygen, nitrogen, carbon monoxide, carbon dioxide, hydrogen, hydrogen sulfide, halogen, hydrogen halides, ammonia, or oxides of nitrogen. Sensitivity depends on the gas flow rates, for both the flame gases of air and hydrogen and the carrier gas. The detector is destructive and mass sensitive (its response depends on the sample component entry rate). This allows a broad range of sensitivities. An example is hydrocarbons with a detection limit of 10^{-11} g of carbon per s. The concentration range in a sample can be 0.1 ppm to 100% molar (41).

The thermal conductivity detector (TCD) is one of the oldest and simplest detection systems available. The sample exits the column and passes over an electrically heated element in a thermostat-controlled metal block. The changes in the thermal conductivity of the gas stream as a result of the sample components cause changes in the heat loss of the element, and these changes are recorded as signal (peaks). Since the signal results from a minute change in temperature, dual elements are used to compare changes in signal as a result of environmental temperature, changes in block temperature, and changes in the temperature, pressure, and flow rate of the carrier gas. This detector has a broad sensitivity range because the component in the carrier gas changes its thermal conductivity. The detection limits are not as low as for FID; for example, hydrocarbons can be determined in concentrations ranging from approximately 50 ppm to 100% molar. Analysis of permanent gases is a common use of TCD.

The electron capture detector (ECD) reports changes in a standing current generated by a β-emitting source such as ^{63}Ni in an ionization chamber creating a standing current.

$$\beta + N_2 \rightarrow N_2^+ + e^-$$

The analytes capture the electrons to form stable ions, and some ions can further react with charged carrier gas.

$$\text{Solute (X)} + e^- \rightleftharpoons X^-$$
$$X^- + N_2^+ \rightarrow \text{uncharged products}$$

The changes in the standing current so created generate the signals. Electron capture detectors are now available without radionuclides to generate electrons. Polyatomic compounds such as alkyl halides and conjugated carbonyls capture electrons. Many compounds of interest to microbiologists, such as hydrocarbons, alcohols, and aldehydes, do not capture electrons and hence are not detected.

Mass-selective detectors (MSD) and dedicated mass spectrometers have also been coupled to the gas chromatograph (52, 73). The MSD can identify and quantify analytes on the basis of mass, charge, and abundance of the analyte ion and its fragmentation ions. The most common of the "small" (benchtop) MSDs is the negative-ion quadrupole system (model 5970; Hewlett Packard Co.). This system generates a stream of ions derived from the analytes, with the mass analysis being accomplished at a detector while changing the voltage applied to the quadropoles. The masses of the ions in the gas stream are scanned, and signal or peaks are generated on the basis of the abundance of each mass level (in atomic mass units). Because of the availability of computer data handling systems and pattern recognition software, the compounds in the sample can be identified by comparison of the mass spectrum of the analyte(s) and its fragments to scans of compounds in a computer library. The major limitation to the systems is the requirement for a strong vacuum in the detector, so only capillary columns with a lower carrier gas flow rate can be used.

21.3.5.3. Sample preparation and derivatization

Some samples, e.g., culture supernatants, cell extracts, and biological fluids, may contain interfering substances. These include highly polar and/or highly reactive components. Preliminary separations or preparations of the samples may be required. The compounds of interest can be extracted with a nonpolar solvent, or solid-phase extraction (SPE) systems can be used to clean up the sample. An internal standard should be added to the sample prior to any treatment, to permit calculation of the recovery efficiency of the pretreatment. The internal standard consists of a compound chemically similar to the analytes of interest but not present in the sample. This material is added at a known concentration estimated to be near that of the analytes. The recovery percentage of this internal standard can be used to correct for any sample component loss. The procedures for the pretreatment or extraction can be found in general references and from the vendors of the extraction material, as for the SPE.

Some analytes are of insufficient volatility and hence must be converted to a more volatile derivative, e.g., methyl esters of fatty acids, trimethylsilyl ethers of sugars, and methyl esters and N-acyl forms of amino acids and peptides. These methods and others are available from references (60a, 68, 72a) and from vendors supplying the derivatization compounds.

21.3.5.4. Metabolic products

Bacteria generate end products during their growth on various substrates. Aerobic bacteria generate only carbon dioxide and water as products, but fermenting anaerobic bacteria produce end products including alcohols (C_1 to C_5), volatile and nonvolatile acids (C_2 to C_6), aldehydes, and ketones.

The following procedure (76) allows the determination of end products to assess the extent and nature of fermentation and aids in the identification of anaerobic bacteria involved.

Acidify filtrates from pure cultures of an anaerobic bacterium, fungus, or sheep rumen liquor by the addition of 50 μl of formic acid (98% [wt/wt]) and 385 μl of periodic acid (17.5% [wt/vol]) per ml of sample. Mix the samples, let stand for 10 min on ice, centrifuge in an Eppendorf desktop centrifuge for 5 min, and store at 4 to 10°C for up to 48 h prior to analysis. Use an analytical system consisting of a 1.8 m glass column packed with 80/100 Chromosorb WAW with a liquid phase of GP 10% SP-1200–1% H_3PO_4 (Supelco, Inc., Bellefonte, Pa.) maintained at 110°C. Use nitrogen saturated with concentrated formic acid as the carrier gas, and detect the effluent components by flame ionization. Before each sample application, prepare the column by preinjecting five 5-μl samples of 4.4% formic acid. Use acetone, alcohols, and volatile fatty acids (VFAs) as external standards. If preliminary samples show no isobutyric acid, add this as an internal standard.

This method allows the separation and analysis of alcohols, VFAs, lactic acid, and acetone without tailing or ghosting and leads to stable column performance after over 5,000 injections.

21.3.5.5. Gas production and consumption

Especially during anaerobic growth, bacteria can consume or produce various gases. The presence and concentration of these gases can be determined readily by GC. One method involves the use of the thermal conductivity detector and a column packed with a solid-phase carrier material, such as 100/120 Carbosieve S-II (Supelco Inc.). Because there is no liquid phase, the chromatography is a form of adsorption separation not partitioning. Pack the material into a 6- to 10-ft. (183- to 305-cm) stainless-steel column (diameter, 1/8 in. [0.32 cm]). Hold the detector at 250°C, and program the column oven to hold at 35°C for 7 min and then to increase the temperature to 225°C at a rate of 32°C/min. When analyzing gases other than hydrogen, use helium at a flow rate of 30 ml/min as the carrier gas. For determination of hydrogen, the carrier is nitrogen at the same settings. Collect and inject 500-μl samples by using a Pressure-Lok gas syringe. This protocol is based on the authors' experience in measuring production of hydrogen, methane, and carbon monoxide by pure-culture and mixed-culture fermentations from marine and hypersaline ecosystems, as well as the consumption of hydrogen, methane, ethylene, carbon monoxide, and ethane in other experiments.

21.3.5.6. Membrane lipid analysis for identification of bacteria (also see Chapter 25.2.4)

The membranes of bacteria contain lipids as a major component. The fatty acid composition of these lipids is characteristic of the different taxa. Specific fatty acids of these lipids can be used as biomarkers, and the quantitative recovery and analysis of these fatty acids can be used to define the members of a microbial community (68).

A number of methods for the analysis of whole-cell lipid fatty acids have been developed (39, 44, 45, 65a, 68, 78, 79). The lipids are extracted from the cells with organic solvent, and specific classes of lipids such as the phospholipids are isolated by silicic acid chromatography (78). Following saponification of the phospholipids, the fatty acids are made volatile by an esterification by methanolysis. The methyl esters of the fatty acids are separated and detected by capillary GC with FID. The sensitivity of this method, based on palmitic acid, is approximately 2×10^{-12} mol (68).

Samples with low biomass, such as biofilms, ground water, and subsurface soils, usually require greater sensitivity. Picomolar concentrations of phosopholipid fatty acids can be measured by using GC and chemical ionization mass spectrometry (78). Lyophilize and extract samples of bacteria with 5.0 ml of a mixture consisting of chloroform, methanol, and water (1:2:0.8 by volume). Sonicate samples for 1 min, and extract for 2 h. Centrifuge the samples, and wash the residue twice with 1 ml each of the above mixture. Separate the combined extracts into two phases by addition of 1.8 ml of chloroform and 1.8 ml of water. After a 2-min vortexing, centrifuge the mixture at 4,000 × *g* and discard the aqueous layer. Concentrate the organic phase by evaporation under N_2, and place the total volume on a 100-mg column of silicic acid. Fractionate the neutral, glyco-, and phospholipid by using 1 ml of chloroform, 4 ml of acetone, and 1 ml of methanol, respectively. Recover the methanol fraction containing phospholipid, and dry it under N_2. Hydrolyze the dried fraction in 1 ml of 15% KOH in methanol-water (1:1, vol/vol) at 80°C for 30 min. Extract the unsaponified lipids by adding 0.5 ml of H_2O and 1.5 ml of hexane to the cooled samples. Extract and analyze the free fatty acids. Purify the phospholipids by extraction with 1 M phosphate buffer (pH 7) and by chromatography on a short silicic acid column. Analyze by GC-mass spectrometry with a fused silica column coated with a nonpolar cross-linked methyl silicone phase (HP-1; Hewlett-Packard). Maintain the oven temperature at 80°C for 1 min, and then increase it to 150°C at 10°C/min, then to 240°C at 3°C per min, and finally at 5°C/min to 300°C. Use helium at 30 cm/min as the carrier gas. Use isobutane as the reagent gas for detection by chemical ionization. Set the source temperature of the mass-selective detector at 150°C. Characterize the bacterial fatty acids by selective ion monitoring. The sensitivity of this method for a 16:0 fatty acid is 0.01 pmol; 125 pmol (mean) of phospholipid occurs in deep subsurface sediments, which, expressed by weight, represents a concentration of 5.8 pmol of phospholipids per g of dried sediment.

A combined hardware (Hewlett-Packard Co.) and software (Microbial ID, Inc., Newark, Del.) system identified as the Microbial Identification System (MIS) is commercially available. The MIS analysis is based on fatty acids of 9 to 20 carbons that have straight and branched chains, saturated or unsaturated, and contain hydroxy and cyclopropane groups. MIS uses computerized high-resolution GC to determine the fatty acid composition of isolates of interest and then searches against libraries of known composition for species identification. Libraries include those available from Microbial ID and those generated by the user or other researchers (44). The software included by Microbial ID allows the generation of dendrograms and two-dimensional association plots from the fatty acid data from the unknown organism compared with the library of known compositions of selected bacterial groups.

Whole-cell fatty acid analysis has been applied to the identification of nonfermenting gram-negative rods in clinical specimens (79), benthic microbial communities (39), and members of the Archaeobacteria (65a).

21.3.5.7. Biomineralization of xenobiotic organic compounds

Microorganisms are able to mineralize most if not all known naturally occurring organic compounds (the doctrine of microbial metabolic infallibility), and this sweeping ability is being put to use by biotechnology. Manmade and naturally occurring toxic organic compounds contaminate soil, sediments, and waters at many locations. The use of microorganisms to convert these polluting xenobiotics to harmless end products is the goal of a number of microbiology laboratories. GC can be used to monitor the concentration of these contaminating compounds (35).

Phenol is on the Environmental Protection Agency priority pollutant list and is the basic aromatic structure of many synthetic organic compounds. The concentration of phenol and possible degradation intermediates are detectable by GC with a 5% phenylmethyl-silicone capillary column (25 m by 0.2 mm [inner diameter]) (35). For this determination, use nitrogen as the carrier gas and detect components by FID. Hold the temperature of the column at 50°C, inject 2 ml of sample, and increase the temperature to 230°C at 11°C/min. Set the injector and detector temperatures at 250 and 300°C, respectively. Prepare the sample (1.0 ml) for injection by addition of 0.2 ml of 9 M H_2SO_4. Then add 0.4 g of NaCl and 1 ml of ethyl ether. After agitation and centrifugation, remove the ether layer. Perform extraction for the possible intermediates, cyclohexanol and cyclohexanone, at pH 7.0. The ether phase (0.5 ml) is evaporated under nitrogen and derivatized with *N,O*-bis-(trimethylsilyl)trifluoroacetamide in acetonitrile (1:4). Add an internal standard, *m*-cresol, at a concentration of 150 mg/liter, and incubate the mixture at 70°C prior to injection. Run standards for tentative identification of peaks by retention time and for quantification. Verify the identification of peaks by GC-mass spectrometry. In this study, GC is also used to monitor production of methane gas, an end product of the consortium growing on the phenol (see above for gas analysis).

GC has also been used to monitor numerous compounds that contaminate ecosystems such as aromatic and saturated compounds in crude oil (47), chlorinated and brominated phenols and biphenyls (70), and organic sulfur compounds such as methanethiol, dimethyl sulfide, and dimethyl disulfide (40).

21.3.6. High-Performance Liquid Chromatography

HPLC (also known as high-pressure liquid chromatography and high-precision liquid chromatography) is a chromatographic separation in which the mobile phase is not a gas (as in GC) but a liquid. Most of the theory described in the previous section on GC applies to HPLC. The compounds of biological interest that can be analyzed by HPLC are more varied than those amenable to GC investigation. This advantage over GC is due to the removal of the volatility requirement. The lower-molecular-mass compounds (<80 to 100 Da) must compete with the mobile solvent phase that is present in relative excess for sorption sites on the stationary phase. The larger molecules (>2,000 to 3,000 Da) are affected by exclusion mechanisms, so they are best analyzed by gel permeation chromatography (46) (see also section 21.3.8).

Examples of compounds that can be resolved well by HPLC include VFAs such as formic, acetic, propionic, and butyric acids; nonvolatile fatty acids such as pyruvic, lactic, fumaric, and oxalic acids; carbohydrates such as cellobiose, maltose, lactose, and glucose; amino acids such as glycine, alanine, leucine, and lysine (64); nucleotides and nucleosides (57, 66, 67); secondary products of metabolism (77); and pigments such as carotenoids, chlorophyll, and bacteriochlorophylls (59, 69, 81).

Since the general introduction of modern liquid column chromatography in 1969, the method has undergone spectacular growth (46). The development of sensitive detectors, the transfer of knowledge and techniques from GC, the advances in pump and metering technology, and the availability of narrow columns with small particles (50 μm or less) have led to this rapid increase in use. High-efficiency columns are available (Supelco, Bellefonte, Pa.) with particles of 3 to 5 μm, which deliver >120,000 to 80,000 theoretical plates per meter, respectively. To maintain a reasonable flow rate of 0.1 to 5.0 cm/s or greater, the mobile liquid phase is operated at elevated pressures from 1,000 to 40,500 kPa (10 to 400 atm).

21.3.6.1. Instrumentation

HPLC systems are available either as a complete unit such as model 1098 (Hewlett-Packard) or in a modular form in which various components are interconnected. Both systems have the same major components. These include high-pressure liquid pumps (one pump for isocratic analysis, or two or more pumps for gradients of two or three solvents), pressure transducer and recorder, high-pressure sample addition valve, column and oven, and detector system. Columns and detector systems are the components that are variable and are the areas of the fastest growth and technical improvements.

The column form dictates the mode of chromatography. An uncoated normal phase packing consists of silica particles with pore diameter ranging from 5 to 20 pm and is used for adsorption chromatography. Bonded phase columns have particles with a bonded liquid phase that incorporate one of the following groups: cyanopropylmethylsilyl, aminopropylsilyl, and 3-glycerylpropylsilyl for normal phase or trimethylsilyl (C_1), butyldimethylsilyl (C_4), octyldimethylsilyl (C_8), and octadecyldimethylsilyl (C_{18}) for reverse-phase separations. A number of resin-based columns are available that separate analytes by affinity, hydrophobic interaction, gel filtration, and ion-exchange chromatography. Microbore columns with a diameter of 1 mm are also commercially available. These columns have the advantage of lower sample volume requirement and enhanced sensitivity. The decreased diameter of microbore columns decreases the dilution within the column (at the same solvent velocity), thus increasing sensitivity.

No universal liquid chromatography detector exists, so most instruments require two or more detector types for broad applicability. A good approach is to have a "specific" detector such as a UV or fluorescence photometer, conductivity, or polarographic detector, as well as a "differential" detector, such as a refractometer. The introduction of the microbore column with decreased total flow has facilitated the direct connection of a mass spectrometry detector, which is a very powerful analytical tool.

The UV detector has a low sensitivity to temperature and flow rate fluctuations. It can use single, adjustable, or multiple-wavelength monitoring, the last by means of photodiode array detectors (see section 21.1.1). Spectrophotometers can be used with simple HPLC systems in which a fraction collector receives the column effluent. The major disadvantage of UV detectors lies in their specificity, but this limitation can be minimized with photodiode array monitoring of multiple wavelengths.

The fluorescence detector is useful because many compounds of biological interest are fluorescent, or can be derivatized to a fluorescent form. Amino acids, alkaloids, and some amines can be converted to dansyl derivatives (treatment with 1-dimethylaminonaphthalene-8-sulfonic acid) for ready separation and detection at low concentrations. Methods are available in the literature and in the catalogs of reagent suppliers to create dansyl and other fluorescent derivatives of compounds that may be of interest to bacteriologists.

21.3.6.2. Separation of organic acids produced by anaerobes

VFAs and nonvolatile fatty acids are produced by some anaerobic bacteria (e.g., *Clostridium, Bacteroides, Veillonella,* and *Propionibacterium* spp.) and can be used to differentiate between closely related bacteria. Use an Aminex HPX-87H column (30 cm by 7.8 mm [inner diameter]) with a mobile phase of 3.5 mM sulfuric acid–10.8% acetonitrile for separation of these acids. Set the flow rate at 0.4 ml/min and the UV detector at 210 nm to detect all carboxylic acids (64).

21.3.6.3. Guanine-plus-cytosine content of DNA (also see Chapter 26.4)

The G+C content of the bacterial chromosome (expressed as the percent G+C of the total base content) is a commonly used taxonomic marker in characterization of an organism. A number of indirect physical methods and a direct chemical method are used to determine this value.

The DNA thermal denaturation temperature, or melting temperature (T_m), is related to the G+C content. This value is obtained by using isolated and purified DNA (65b). Alternatively, it can be obtained by using intact cells, in which case intracellular melting of DNA is observed by differential scanning calorimetry (65). The content has also been related to the buoyant density of DNA as determined by ultracentrifugation, usually in a cesium chloride gradient (72b). All of these methods require calibration with DNAs or cells of known G+C content.

Direct HPLC measurement of the bases following hydrolysis of DNA is a precise and rapid alternative to the indirect methods above. Nuclease P1 can be used to hydrolyze DNA isolated from the microorganism to 5′-deoxynucleoside monophosphates (dNMP) which can readily be identified and quantitated by HPLC (57). Isolate and dissolve the DNA in 0.15 M NaCl–15 mM trisodium citrate (pH 7.0) at about 2 mg/ml. Treat the solution with 10 μg of proteinase K per ml for 1 h at

37°C, extract twice with phenol, and keep the aqueous layer. Precipitate the nucleic acids with cold ethanol, and recover the precipitate by centrifugation. Two nuclease P1 treatments are used; in the first, to digest single-stranded DNA and RNA, add 2 ng of nuclease P1 and 1 mg of DNA in 2 ml of 10 mM acetate buffer (pH 5.3), and incubate for 30 min at 40°C. Dialyze overnight against two changes of the above buffer, then heat the mixture to 100°C for 10 min and rapidly cool in an ice bath. For the second treatment, add 2.0 μg of nuclease P1 to the mixture and incubate at 40°C for 15 min. Heat the mixture to 100°C for 10 min, and centrifuge at 15,000 × *g* for 25 min. Analyze 2 to 5 μl of supernatant (10 to 15 μg of dNMP) on an anion-exchange column of Zorbax NH_2 (25 cm by 4.6 mm [inner diameter]) (Shimadzu-Dupont Co., Kyoto, Japan). Use a mobile phase of 20 mM KH_2PO_4 (pH 2.8) at a flow rate of 1 ml/min. Set the detector at 258 nm. Determine the concentration of each deoxynucleotide by using standards.

An alternate HPLC procedure for determination of G+C content has been reported (66, 67); in this method, the DNA is digested by the same proteinase and nuclease as described above but with different conditions and extraction protocols. To quantify the dNMPs, use a C_{18} reversed-phase column (Econosphere; Alltech Associates, Inc., Deerfield, Ill.), a mobile phase of 12% methanol–20 mM triethylamine phosphate (pH 5.1), and a flow rate of 1.0 ml/min. Detect the hydrolysis products by measuring the A_{254}. Also, see references 66 and 67 for a procedure for isolation and analysis of DNA from single colonies.

Two shortcomings of the method have been noted in practice. One is the difficulty of obtaining pure DNA by the published methods. For example, the traditional lysis and isolation procedures may not perform as expected for bacteria isolated from an extreme environment such as brine. In that event, the intact-cell method (65) should be tried, or a new DNA isolation procedure must be developed. The second shortcoming is the appearance of many more peaks in the column effluent than expected, as a result of incomplete digestion or the presence in the DNA of methylated bases such as 5-methylcytosine. Incomplete digestion can be identified and overcome by using less DNA, by adding more of the hydrolytic enzymes, or by using longer incubation times. The presence of methylated bases can be ascertained with standards. Following quantitation, the content of methylated base(s) is added to the content of the related nonmethylated base.

21.3.6.4. Separation and quantitation of bacterial photosynthetic pigments

Anaerobic photosynthetic (phototrophic) bacteria can be studied in pure culture, or in natural samples, by analysis of the bacteriochlorophyll present. Bacteriochlorophylls *a*, *b*, *c*, and *e*, as well as plant chlorophyll *a*, can be separated by HPLC. Pigment analysis can be used as a signature to aid in the taxonomic characterization of pure cultures of phototrophic bacteria and in the identification of species in natural samples. Yacobi et al. (81) described a method for comparing pigment extracts from isolated and identified species with those from lake water samples as follows. Filter water samples and culture suspensions of phototropic bacteria onto Whatman GF/C filters. Use 2 ml of pure culture or various volumes of lake water. Extract the pigments from the filters in 3 ml of HPLC grade methanol for 30 min in the dark at 2 to 4°C. Filter the extract through Whatman GF/F filters, and analyze by reversed-phase chromatography with a Spherisorb C_{18} column (25 cm by 4.6 mm [inner diameter]) (Supelco). For the mobile phase, use a gradient-producing pump to make a gradient by mixing 30% 1 M ammonium acetate in double-distilled water–70% HPLC grade methanol (component 1) with 30% ethyl acetate–70% methanol (component 2). Start the gradient with 80% component 1 and 20% component 2. Program the pump to produce a linear increase from 20 to 60% component 2 in 7 min, a hold at 60% for 5 min, a linear increase from 60 to 100% in 12 to 20 min, and a final hold at 100% component 2 for 20 min. Detect and scan eluting components with a photodiode array spectrophotometer that covers the range from 300 to 800 nm. This method is effective in developing pigment profiles that are detectable and quantifiable in the raw lake water samples.

21.3.7. Ion Chromatography

HPLC with an ion-exchange column provides excellent separations of anions and cations, both inorganic and organic, on the basis of their pK values and relative affinities for the exchanger. In practice, however, application to most ions has suffered from the lack of chromophoric behavior in the visible or UV spectral region above 220 nm. Therefore, it has been necessary either to resort to derivatization by a chromophoric reagent or to monitor the column effluent by conductivity. To decrease the background conductivity of the eluant buffer and thereby to increase sensitivity, postcolumn conductivity suppression is required. Although cumbersome, conductivity monitoring has proven useful, especially because of improvements in suppression methods (63).

Of more use to many investigators is column effluent monitoring by indirect photometric chromatography as developed by Small and Miller (74), in which a chromophoric ion buffer is used as the eluant. Proper choice of eluant buffer and its concentration, matched to the appropriate column packing, allows detection and quantitation of a large number of ions of interest to the microbiologist using HPLC equipped with a photometer designed for chromatographic work.

This strategy makes use of low-capacity and low-affinity ion exchangers, which in turn require low concentrations of eluant to displace the sorbed ions. The wavelength for detection is selected to give the best signal-to-noise ratio (53). The broad absorbance peaks of the eluant ions allow use of broad bandwidth conditions (see section 21.1.1.1), which allows low concentrations of eluant and increased sensitivity of the method (53).

The light-absorbing eluant ion passes continuously through the photometer, giving a high and constant absorbance. When an ion elutes, a corresponding amount of displacing eluant is bound to the exchanger, resulting in a decrease in absorbance. The area of the negative "peak" is proportional to the concentration of ion being eluted. Thus, conditions are adjusted such that the concentration of eluant ion both is able to displace from the exchanger the ions of interest in a reasonable amount of time and is only moderately higher than that of the eluting ions.

21.3.7.1. Instrumentation

The standard components of HPLC, including the photometer plus columns specific to ion chromatography, are all that is required. A variable-wavelength photometer either with substantial electronic zero baseline offset or with double-beam optics and reference cell, although not absolutely required, is highly desirable. Such photometers have very low volume flow cells designed for column work.

Both low-capacity silica- and resin-based exchangers have been used successfully (36, 50). Silica exchangers are said to be more efficient, whereas resin exchangers have more alkali resistance. Vyodac columns (Separations Group, Hesperia, Pa.) of the 300 series for anions and 400 series for cations have performed well.

21.3.7.2. Ions determined

The following selected inorganic and organic anions and cations can be quantitated by ion chromatography: halogen, carbonate, nitrate, nitrite, sulfate, sulfite, P_i, PP_i, acetate, citrate, glycolate, propionate, succinate, sodium, ammonium, amines, and potassium. Joint determination of anions and cations is possible under some circumstances (74).

21.3.7.3. Eluants

Anions are most commonly determined by this method, primarily because of a wider variety of chromophoric anion buffers. The order of eluting power for anions is 1,2-sulfobenzoate > 1,3,5-benzenetricarboxylate > phthalate > iodide. The most common eluants for cations are copper sulfate and copper nitrate.

21.3.7.4. Sensitivity

Detection limits are well below 50 ppb for several common anions (51) when careful attention is given to (i) controlling the temperatures of both the column and the eluant, (ii) eliminating pump noise with a pump-damping accessory, (iii) minimizing the dead volume of the entire system, and (iv) using the maximum sensitivity of the photometer-recorder system (0.005 to 0.01 absorbance units). Less stringent observance of these requirements still leads to determinations of less than 1 ppm.

21.3.7.5. Anion analysis

For analysis of carbonate, nitrite, nitrate, phosphate, and sulfate, equip the HPLC apparatus with a Vydac 300IC column (The Separation Group) and equilibrate the system with the eluant buffer 1.5 mM potassium acid phthalate adjusted to pH 8.9 with trimethylamine. Inject 100 μl of sample or standard mixture containing 0.5 to 3 ppm of the anions. Develop the column at a flow rate of 4 ml/min, and monitor the effluent at 260 nm. Set the sensitivity of the photometer-recorder system to 0.05 absorbance units full scale. Adjust the sensitivity of the photometer if needed. The order of elution is solvent front, carbonate, nitrite, nitrate, *o*-phosphate, and sulfate over a 3- to 4-min period. Minimum detectable levels are in the 50- to 300-ppb range (50).

21.3.8. Affinity Chromatography

Affinity chromatography makes use of highly specific binding equilibria in which at least one member is a complex macromolecule to effect separations that usually cannot be accomplished by other means (20, 22, 24, 27, 30). The principle simply involves immobilization of one component of a reversible binding system on a solid support followed by mixing with the other component(s) of the equilibrium in solution. Following binding between members of the complementary pair, separation of the solid phase leaves the contaminants in solution. Thus, the basis of separation is the specificity of the binding system conferred by the three-dimensional location of the complementary binding groups rather than simple differences in physical and chemical properties.

Affinity chromatography techniques have been highly developed so that they make use of (i) small molecule-biopolymer interactions such as hormone-binding protein and enzyme-ligand interactions, (ii) the group-specific interactions among proteins and nucleic acids, and (iii) the exceptionally high specificity of interaction between proteins such as the binding between antibody and antigen. The major factors involved are the nature and chemistry of the matrix, the chemistry of ligand attachment, the requirement for a spacer arm, and conditions for adsorption and elution of the mobile binding member of the system. General information is available in references 20 and 22 and in manufacturers' literature.

Many types of affinity materials are commercially available, with choices of (i) the type of matrix or support, either activated but without spacer arm; (ii) the presence and kind of spacer arm and terminating functional groups; and (iii) bonding to certain biopolymers which act as group-specific binding agents.

21.3.8.1. Coupling of ligand to matrix

Although small ligands can attach to an activated matrix directly, large ligands such as proteins are often hindered sterically, and the reaction is slow and incomplete. When proteins do react, it is likely that several covalent bonds will be formed with the matrix. For this reason, the specific ligand is often attached to the matrix through a spacer arm. This arm generally is a linear saturated aliphatic chain of 6 to 12 carbon atoms containing reactive groups at each end to facilitate bonding to both the matrix and the ligand conferring specificity. The matrix material is first activated by derivatization, for example with *N*-hydroxysuccinimide or cyanogen bromide or by conversion to the hydrazide. The first two activated supports react with amino groups of spacer arms or ligands, and the last activated matrix reacts with carbohydrate moieties such as that of the Fc region of immunoglobulin, IgG. The latter special case provides oriented coupling to IgG as opposed to random coupling through amino groups as occurs with the other activated supports. Spacer arms can terminate in amino, carboxyl, thiol, or other functional groups. These can act as affinity ligands as such or, alternatively, can undergo further coupling through these functional groups to biospecific ligands such as proteins and nucleic acids.

Matrix material can be derivatized by the investigator (42) or purchased with an arm which is terminated with a reactive group such as succinyl-*N* succinimide ester. This group reacts with alkyl and aryl amines. With CNBr-activated Sepharose, the most common arms are produced by reaction with 1,6-diaminohexane and 6-aminohexanoic acid. Ligand bonding then involves coupling with a water-soluble carbodiimide to form amides. For this reaction, 1-ethyl-3-(3-dimethylaminopropyl)carbodiimide hydrochloride or 1-cyclohexyl-3-(2-morpholinoethyl)carbodiimide metho-*p*-toluene sulfonate gives good yields in aqueous and nonaqueous media at pH 4.5 to 6.0.

Activated dextrans such as Sepharose 4B (Pharmacia Fine Chemicals) and agaroses such as Bio-Gel A (Bio-Rad Laboratories) have the necessary stabilities, flow rates, and pore sizes to be useful matrices.

Use of cyanogen bromide-activated Sepharose. Activated Sepharose can be purchased as a freeze-dried powder (1 g, ca. 3.5 ml of swollen gel) or prepared by the method of Axen et al. (34). The dry material is washed and reswollen on a sintered-glass filter (Corning type G-3) with several wash cycles of 200 ml of 0.1 M HCl, the supernatant being removed by suction. It is then used immediately.

Procedure for bonding ligand or spacer arm. Dissolve the protein- or amino group-containing ligand in 0.1 M $NaHCO_3$ (pH 8.3) containing 0.5 M NaCl (for proteins only). Use 5 to 10 mg of protein per ml of gel, and remember that amino buffers cannot be used. Mix the gel and ligand slowly for 2 h at room temperature or overnight at 4°C; do not use a magnetic stirrer. Wash the excess protein away with coupling buffer, and remove the excess coupling groups by treatment with 1 M ethanolamine (pH 9) for 2 h. Other amines, including Tris buffer, can be used to neutralize coupling groups. Wash the final product four or five times with high- and low-pH buffers such as 0.1 M acetate (pH 4) and 0.1 M borate (pH 9) containing 1 M NaCl. Changes in pH are usually necessary to remove adsorbed protein. Store suspensions in a refrigerator (without freezing), and add a bacteriostat if long storage is planned. Of course, the above procedure should be modified to meet the stability requirements of any particular protein. As a guide, it should be noted that CNBr-Sepharose is stable (not very reactive) at low pH and reactive at high pH.

With aminohexyl-Sepharose or 6-aminohexanoic acid-Sepharose, coupling is carried out by carbodiimide methods. Weigh out the dried material, and swell as above in 0.5 M NaCl (1 g = 4 ml of gel). Dissolve the ligand (protein or small molecule) to be coupled in water or water-solvent to give 10 mg of swollen gel per ml, and adjust the pH to 4.5. Add 1-ethyl-3-(3-dimethylaminopropyl)carbodiimide-HCl in water. Mix the gel and carbodiimide overnight at room temperature; do not use a magnetic stirrer. Remove excess reagents by washing with the same water-solvent mixture to be used in the ensuing ligand-coupling step. This conjugate can be stored at 4 to 8°C until used. Dissolve the ligand to be coupled in water or a water-solvent mixture for which the pH of the water phase has been adjusted to 4.5; do not use a buffer. The pH, which decreases mostly in the early part of the reaction, is maintained at 4.5 by the addition of NaOH. Pour the ligand-derivatized gel into a column, and wash thoroughly with coupling solution; blocking of excess groups is unnecessary. Wash the gel in distilled water and finally in the buffer to be used in affinity chromatography. The material can be stored at neutral pH below 8°C without freezing.

Derivatization of agarose-Affi-Gel 10. The affinity material Agarose-Affi-Gel 10 is available from Bio-Rad Laboratories as a wet slurry (1 g [dry weight]/25 ml). Attach the spacer arm and succinimide ester group through a succinylated aminoalkyl arm to a Bio-Gel A matrix. On a filter, wash the gel with three bed volumes of water. Mix the gel with buffer, e.g., 0.1 M acetate, morpholinopropanesulfonic acid (MOPS), *N*-tris (hydroxymethyl)methyl-2-aminomethanesulfonic acid (TES), or HCO_3^- (pH 5 to 8.5). Alternatively, add the ligand in nonaqueous solvent directly to the moist gel pellet or to the gel washed and suspended in solvent.

21.3.8.2. Ligand affinity agents (22)

Calculate the amount of material from the furnished or determined capacity, making some assumptions about the experimental conditions to be used and the binding affinity of the material to be adsorbed. Usually, excess agent is used and the adsorbed material forms a band at the top of the bed; therefore, small columns are used. Pack columns as for gel filtration to achieve homogeneity and free flow.

For adsorption, choose the starting buffer to maximize binding and minimize protein-protein interactions. Therefore, consider the pH and presence of specific ions or factors, and use a buffer of high ionic strength, e.g., 0.5 M NaCl. Add the sample in the starting buffer, and, if necessary, exchange the buffer ion in the sample either by dialysis or by membrane or gel filtration.

Elution may involve one or a combination of pH shift, change in ionic strength, or displacement by a solute related or identical to the immobilized ligand. For the first two methods, the specificity characteristic of affinity chromatography lies only in the adsorption phase. In elution by specific displacement of immobilized ligand, specificity is exhibited in both the adsorption and elution phases.

21.3.8.3. Other aspects

In instances when a highly reactive support material is to be coupled to macromolecules that are highly sensitive to excessive multipoint coupling, it may be necessary to reduce the number of coupling groups in the matrix. This is accomplished for CNBr-Sepharose and some other materials by controlled hydrolysis of the reactive group and for other activated matrices by controlled pretreatment with a small ligand. Even without such sensitivity, it is likely that some distortion of macromolecular structure will occur. For enzymes, this is displayed by a lower specific activity and other changes in catalytic properties.

After completion of the procedure for binding a ligand to a matrix, any remaining reactive sites on the matrix should be destroyed by adding an excess of small molecule containing the same reactive group; e.g., for CNBr-activated Sepharose, add ethanolamine.

21.3.9. Gel Permeation Chromatography (19, 26)

The availability of spherical gel beads of relatively homogeneous pore size distribution makes possible a number of kinds of molecular separations which are based to a great degree on molecular size. When a sample is layered on a gel filtration column, molecules much larger than the pore size of the beads proceed through the column with the displacing solvent without penetrating the beads and exit as a peak after the void or interstitial volume, V_o, has passed through. Small molecules, which freely equilibrate with the internal space of the beads, V_i, theoretically proceed down the column by eluant displacement as though beads were not present and exit after passage of the total column volume minus the volume of the matrix. Thus, the total column volume, V_t, is the sum of the excluded volume, V_o, the internal volume, V_i, and the matrix or bead material volume, V_g.

$$V_t = V_o + V_i + V_g$$

Molecules intermediate between these extremes "partition" between interstitial and internal spaces largely on the basis of molecular size. Since a wide variety of pore sizes, and hence fractionation ranges, of gel filtration media are available, it is possible to separate molecules over a molecular mass range of 1×10^2 to 4×10^7 Da. Although many molecules behave "ideally" as described above, many others are retarded more than expected. This is due to adsorption, ion exchange with a small number of carboxyl groups, or interaction between aromatic rings and the matrix. The behavior of molecules as they pass through the column is usually deduced by analyzing the column effluent. However, a procedure for computer-controlled scanning of the intact gel column has been developed (43).

21.3.9.1. Filtration materials

At least four kinds of gel beads are available: cross-linked dextrans and derivatives, polyacrylamides, agaroses and cross-linked agaroses, and styrene-divinylbenzene copolymer (see Chapter 9.1 for further information about the nature of these gels). Gel filtration media are available in many bead sizes and porosities; from only two manufacturers, 180 kinds are available. It is therefore essential to consult the manufacturers' literature for details of their properties.

21.3.9.2. Multicomponent analysis

By proper choice of bead porosity and carefully worked-out conditions, it is possible to separate closely related molecules or, conversely, to separate classes of widely different molecules. Analysis of mixtures places the greatest demand on separation, but under these conditions a high flow rate is not as important. Since quantitative yields are usually obtained on small columns, separation of components into nonoverlapping bands followed by quantitative analysis by nonspecific methods is possible. In this way a series of simple sugars such as raffinose, maltose, and glucose can be separated and also desalted on Sephadex G-15 (27). Usually, such separations are based on molecular weight. However, such separations have also been made because particular molecules are retarded for other reasons, such as adsorption of a polar solute from a nonpolar solvent.

21.3.9.3. Determination of protein-ligand binding constant

The amount of a ligand reversibly or irreversibly bound to an unknown amount of macromolecule can be used to calculate the equilibrium constant or the stoichiometry of binding, respectively (31, 54, 58) (Fig. 7). These can be determined by incubation of the macromolecule and ligand, followed by gel filtration on Sephadex G-50 which has been equilibrated with and developed in the same concentration of ligand as used in the binding experiment. During column development, the concentration of ligand in the protein band is higher than the baseline level used in the elution buffer, the excess being due to binding to the protein which is in the eluate at that time. In the region of the total volume or "salt" volume, there will be a trough. This region contains the rest of the original incubation mixture, which is depleted by the amount of ligand bound to the protein that has been removed. The area of the trough equals the area of the peak. Conversion of these areas to the amount of ligand by integrating the area and by having areas from standards applied to the same column allows calculation of the equilibrium constant or stoichiometry of binding if the binding is very tight. Chromatography on a coarse or fine Sephadex G-50 column (40 by 2 cm) is effective. The conditions used in Fig. 7 (54) are for a column of coarse Sephadex G-25 (0.4 by 10 cm), a sample size of 0.25 ml, and an elution rate of 0.3 ml/min. V_o and $V_i + V_o$ should be determined with blue dextran and ferricyanide, respectively, to obtain the volume that separates the two peaks. The sample size can be increased so that the first peak occupies the region between the two peaks. However, because peak broadening occurs, no more than 66% of the maximum sample volume should be used. For instance, on a 5- by 50.8-cm column, KCl and albumin are separated in a 30-ml sample with fivefold dilution and a large elution volume between the peaks. The same degree of separation could also be obtained with a 300-ml sample with a 1.5-fold dilution; the eluting agent can be water or buffer. The same procedure is used to remove small molecules after chemical treatment of proteins (i.e., soluble fluorescence reagents) and to exchange buffers. For the latter, a "desalting" column is used, but it is equilibrated and developed with the new buffer.

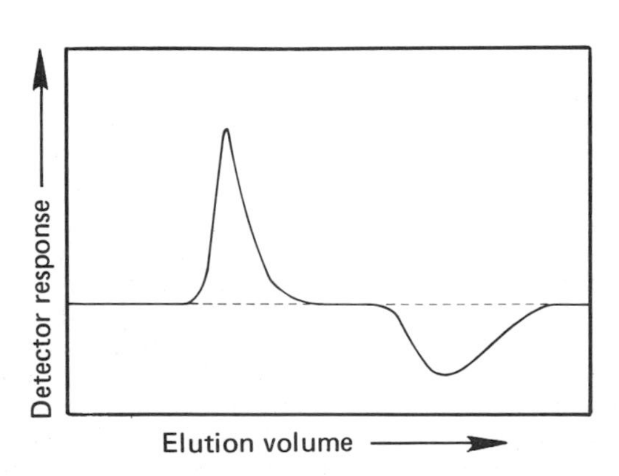

FIG. 7. Elution diagram for determination of equilibrium binding of a ligand to a protein.

21.3.9.4. Fractionation of proteins and determination of molecular weight

For gel filtration on a preparative scale such as in the purification of enzymes and antibodies, it is often desirable to compromise between resolution and flow rate. Thus, some resolution is sacrificed by using a larger gel particle size to accommodate the volume involved. The porosity is chosen (often empirically) to resolve as many components as possible and to place the desired component intermediate in the elution diagram between V_o and $V_o + V_i$. Successive gel filtrations or a gel filtration step in combination with other separation procedures has been used effectively. The molecular weight of soluble proteins can be estimated by gel filtration on Sephadex G-75 and G-100 (32, 80). A regression line of the elution position V_x/V_o, where V_x is the volume to the peak of the unknown and V_o is the void volume, versus the log of molecular weight is established with a group of enzymes and proteins of known molecular weight. The molecular weight of the unknown is taken from the graph. This method, although very useful, has largely been replaced by the more rapid and convenient polyacrylamide gel electrophoresis method (see section 21.5.1).

21.3.9.5. Solute concentration

Macromolecules can be concentrated very simply without change in pH or ionic strength by addition of dry Sephadex G-25, coarse grade. After 10 min or more, the excluded volume is recovered by centrifugation in a basket centrifuge for large volumes or with filter adapters for centrifuge tubes for small volumes.

21.3.9.6. Practical considerations

Choice of gel filtration medium. It is necessary to choose a medium with a fractionation range to suit the application in mind. For total exclusion, use a gel with a molecular weight fractionation range well below that of the material, i.e., Sephadex G-25 (fractionation range, 100 to 5,000 molecular weight) for exclusion of most proteins or Sephadex G-10 (fractionation range, 0 to 1,500 molecular weight) for molecules above 3,000 molecular weight. For total equilibration with the gel interior, V_i, and elution with the "salt" volume, $V_o + V_i$, select a gel with a fractionation range whose lower limit is above the molecular weight of the molecule of interest. For fractionation, the gel should have a molecular weight fractionation range such that the molecule(s) of interest will be partially retarded, i.e., Sephadex G-200 (exclusion limit, 200,000) for fractionation of serum proteins or 0.5 M Bio-Gel A (exclusion limit, 500,000) for molecular weight determination.

The efficiency of resolution is a function primarily of the flow rate, which in turn is related to bead size and uniformity. Thus, very small beads and slow flow give the highest resolution. For Bio-Gels P and A, flow rates of 2 to 10 ml/h/cm^2 of bed cross-section give the highest resolution, and this is much slower than the maximum gravity flow rate for some bead sizes and porosities. With the most porous gels, for instance, Bio-Gels A 50 and 150 M (200 to 400 mesh), increased hydrostatic head is needed. With Bio-Gels P-2, P-4, and P-6, elevated pressures of several atmospheres may be used with 400-mesh beads capable of high-resolution separation of small molecules. When pressure is used, measures must be taken to ensure use of appropriate vessels and columns. The maximum recommended pressure should be ascertained from the manufacturer's literature. With some highly porous materials (Sephadex 200), pressures above that recommended by the manufacturer lead to collapse of the bead structure and very low flow rates.

Preparation of biological materials rarely involves chemical conditions that would threaten the integrity of the gel filtration medium. However, the general stability characteristics are included here as a guide. Dextrans such as Sephadex are stable in salt solution, organic solvents, alkaline solutions, and weakly acidic solvents but are hydrolyzed in strong acid. Dextrans are stable to 30 min at 121°C, wet or dry, at neutral pH. Agaroses such as Sepharoses do not contain cross-links and hence should be maintained between pH 4 and 9 and between 2 and 30°C. They are stable to 1 M NaCl and 2 M urea. Cross-linked agarose, i.e., Sepharose CL, is stable between pH 3 and 9, but oxidizing conditions should be avoided. It is highly stable in 3 M potassium thiocyanate, a strong chaotropic agent. It can be transferred from water to a number of organic solvents with little change in pore size. It can be sterilized repeatedly at neutral pH at 110 to 120°C. Polyacrylamides such as Bio-Gel P are reasonably stable in dilute organic acids, 8 M urea, 6 M guanidine HCl, chaotropic agents, and detergents. Buffers with 0.5 M ionic strength are recommended for protein separations. Miscible solvents, e.g., alcohol up to 20%, do not materially change the pore size. Bio-Gel P is stable between pH 2 and 10 at room temperature and to autoclaving (121°C) at pH 6 to 8.

Column design and size. Many gel filtration applications such as desalting do not require high resolution and hence can be carried out on homemade or disposable columns of ordinary design. However, success in critical applications requires properly designed columns and associated equipment. "Mixing reservoirs" above and below the gel bed must be avoided; otherwise, the gradient shape or the step change in eluant composition, as well as eluted peaks, is distorted by passing through the reservoirs. Commercial columns are available which have virtually no eluant retention below the gel bed. Others have fittings both above and below the gel bed to add eluant and collect effluent in the absence of reservoirs. Such devices are also essential for upward flow work. Sintered-glass supports are to be avoided, particularly for low-cross-linked and high-porosity gels.

The column length affects resolution, and the column cross-sectional area is proportional to capacity. On the other hand, finer particle size increases resolution and requires shorter gel beds. Since the sample size also dictates the initial bandwidth, the length and diameter become functions of the degree of separation needed for a given sample volume, longer for greater separation and wider for a narrower starting band. A 51-cm bed height will easily desalt a protein in a small volume, whereas very long columns (300 cm) are needed to partially separate a set of peptides derived from a protein by treatment with cyanogen bromide. An internal diameter of 2.5 cm may be a good starting point, although a somewhat narrower column can also be used. With very narrow columns, there are wall effects caused by a greater proportion of laminar flow, which broadens the bands. With excessive diameter, there is unnecessary dilution and difficulty in maintaining horizontal zones.

Column packing. Calculate the amount of dry gel needed from the swollen volume furnished in the manufacturers' literature (for preswollen gels, this step is unnecessary). Swell the gel in twice the expected bed volume of eluant. The length of time varies with volume regain. Highly cross-linked and lower-regain gels such as Sephadexes G-10, G-15, G-25, G-50, and LH-50 and Bio-Gel P require 3 to 4 h at room temperature, whereas Sephadexes G-75 to G-200 require 1 to 3 days. At 90 to 100°C, these times are reduced to 1 to 5 h. Swelling in boiling water aids in removing bubbles, which at room temperature must be removed by vacuum deaeration. Dilute preswollen gels enough to allow bubbles to escape, but avoid stirring. Allow the swollen gel to stand, and remove excess eluant to give a thick slurry of a consistency which will still allow air bubbles to escape. Dilute the preswollen gels if necessary to give a slurry of the same consistency.

Mount and align the column vertically. Fill the column with eluant, preferably from the bottom, until the dead space below the gel bed is occupied and all air bubbles are removed. With the outlet closed, pour the gel slurry smoothly down the column wall or down a rod all at one time. Add eluant to fill the column completely, connect an eluant reservoir, and remove all bubbles through an air vent in the column top piece. If the total gel suspension cannot be added at one time, use a column extension. Start the flow immediately. Bed formation from a thin suspension, addition of a slurry to a column containing eluate, or packing in stages are not recommended. Two or three column volumes of eluant are passed through at about the flow rate to be used. The column must not be allowed to go dry. This can be prevented by making a loop in the outlet tube which is higher than the gel surface.

Sample addition. Before sample addition, place a disc of filter paper or perforated plastic, or a sample applicator cup, on the gel surface. The cup is a plastic cylinder with a nylon net bottom. Determine the homogeneity of packing, as well as the void volume (V_o) and fluid volume ($V_o + V_i$) by passage of a mixture of blue dextran (2 mg/ml) and potassium ferricyanide or other colored small molecule through the column.

Remove excess eluant, layer the sample on the gel surface, and open the outlet until the sample has moved into the bed. Wash the column wall in the region of the surface with a small amount of eluant. Carefully fill the column with eluant, and connect the column to a reservoir of eluant to eliminate any bubbles in the column or tubing.

It is also possible to layer the sample on the gel bed and beneath the eluant. The sample must be more dense than the eluant; if necessary, sucrose or salt is added. Layering is done with a pipette or capillary tubing attached to a syringe. Pharmacia Inc. has developed flow adaptors for its columns. The adaptors consist of pluglike devices attached to threaded tubes which pass through the column ends and terminate in capillary tubing. Two of the adaptors with tight fits to the internal walls can be adjusted via the threaded rods to fit snugly at the bottom and top of the gel bed. With this arrangement, sample followed by eluate is added directly to the gel bed and eluate is removed from the bed bottom.

Elution. The flow rate can be regulated by hydrostatic pressure, i.e., by the difference between eluant surface and outlet, as long as the maximum pressure permitted for a particular gel is not exceeded. The effluent reservoir (Mariotte flask), any other devices, and the outlet tube end are then placed so that the difference in fluid level in the adjustable vent tube of the Mariotte flask and the column outlet end does not exceed the permissible difference in height (hydrostatic pressure). The Mariotte flask is a vessel with an outlet at the bottom to connect to the column via tubing and closed at the top with a stopper or plug through which passes an adjustable-height glass tube; the tube extends below the eluate surface and maintains a constant head. A pump can also be used under the same limitations. For more highly cross-linked gels, the maximum gravity flow rates greatly exceed the flow rate for optimal separation or peak sharpness. With less highly cross-linked gels, the flow rate may be limited by the permissible hydrostatic pressure.

Factors affecting peak sharpness. Gel permeation is a diffusion-controlled process; hence, peak sharpness and resolution require a very low flow rate and bead size uniformity. In other words, the highest resolution is obtained when the void and internal volumes of the beads are in equilibrium. In analytical work, resolution is primary; in preparative-scale work, resolution is usually secondary to the speed of separation. For Sepharoses, use 2 to 8 ml/cm^2/min; for desalting on Sephadex G-15, use 12 ml/cm^2/h; and for Sephadex G-200, adjust the flow to 5 to 20 ml/cm^2/h.

Elution patterns, i.e., peak shapes, are highly sensitive to the viscosity of the sample. Therefore, for maintenance of peak shape, samples containing sucrose or very high protein concentrations must be diluted so that the relative viscosity is not more than twice that of the eluant.

Storage. If the column is to be inactive for more than 1 week at a time, it is necessary to prevent microbial growth by adding either sodium azide (0.02%), Cloretone (0.05%), Merthiolate (0.005%), or chlorhexidine (0.002%). Chloroform, butanol, and toluene should not be used. Before the column is returned to use, the above agents can be removed by passing eluant through the column.

21.4. RADIOACTIVITY (82–84)

In most bacteriological studies, radioisotopes can be used to advantage in the experimental program. The choice of radionuclide depends on the half-life, the kind of decay process and its energy, and a number of other factors peculiar to the experiment. The half-life, short or long, determines the practicality of a given strategy as well as the necessity to correct for decay during the experiment. The type of decay and its energy determine the kind of counting instrument to be used as well as the resolution on photographic emulsions. Table 4 gives information on the isotopes most likely to be used in

Table 4. Physical properties of some radionuclides

Radionuclide	Half-life	Type	Energy (MeV)	Sp act: Theoretical maximum (mCi/matom[a])	Sp act: Common values for compounds (mCi/mmol)
Tritium (^3H)	12.26 yr	β	0.018	2.92×10^4	10^2–10^4
Carbon-14 (^{14}C)	5,730 yr	β	0.159	62.4	1–10^2
Sulfur-35 (^{35}S)	87.2 days	β	0.167	1.5×10^6	1–10^2
Chlorine-36 (^{36}Cl)	3×10^5 yr	β	0.714	1.2	?
Phosphorus-32 (^{32}P)	14.3 days	β	1.71	9.2×10^6	10–10^3
Iodine-131 (^{131}I)	8.06 days	β	0.81	1.62×10^7	10^2–10^4
		γ	0.08–0.72		
Iodine-125 (^{125}I)	60 days	γ	0.035	2.18×10^6	10^2–10^4

[a]A milliatom is the atomic weight of the element in milligrams.

bacteriological work. Unfortunately, there are no useful radioactive isotopes of nitrogen or oxygen.

Most measurements of radioactivity in biology are of particles (electrons) and X-rays. β radiation, for instance, from tritium and ^{14}C, can be measured in a gas counting chamber attached to an electrometer, by Geiger counter, by proportional methods (usually as barium carbonate), or by scintillation techniques. In recent years the liquid scintillation spectrometer has replaced all other methods and is the only system described here. γ counters are used primarily to measure ^{131}I and ^{125}I.

21.4.1. Safety

The conduct of experiments involving radioisotopes can involve risks to human health. Therefore, it is essential that safety practices and procedures be strictly followed. Such procedures not only minimize the risk to investigators but also improve the conduct of the experiment by requiring a range of good laboratory practices.

The use of radioisotopes for any purpose is strictly regulated in many countries. In the United States, regulation is carried out by the Nuclear Regulatory Commission of the Federal Government as well as by regulatory agencies in many states. Licensing and periodic inspection are usually required for the parent institution under which the research is conducted. In turn, the institution, through a radiation safety officer, ensures that safe practices and records are maintained, and it supplies working rules for the laboratory.

1. Conduct work in a laboratory area that is appropriate for the kind of radioisotope compound involved. Often, this is a separate area with a hood if volatile radioactive materials (e.g., ^3H_2O) are involved. Control the amounts used and their disposition. Work in a setup that has specific waste and cleanup capabilities available and ready for use.
2. Do not eat, drink, or smoke in this area.
3. Do not pipette by mouth.
4. Wear a laboratory coat, gloves, and safety glasses.
5. Label all containers with specific radioisotope labels, and give specific data.
6. Use disposable laboratory items when possible, and deposit them in designated solid-waste containers.
7. Monitor and decontaminate all laboratory glassware before returning it to general use.
8. Maintain an isotope inventory and record of isotope use.
9. Monitor the work area, as well as clothes and hands, before leaving the laboratory.
10. Report spills both to others in the laboratory and to the radiation safety officer.

A thin-window Geiger counter can monitor all isotopes except tritium. For tritium, working areas are sampled with absorbent wipes which are then evaluated by scintillation counting. Film badges are required for persons using γ-emitters in quantities above 100 μCi and for persons using ^{32}P and other strong β-emitters. Persons using ^{14}C, ^3H, and ^{35}S do not need film badges. However, for persons using millicurie levels of these isotopes, urinalyses should be performed. This service is often provided, along with a film badge service and waste disposal, by the radiation safety officer.

Protect counting equipment against contamination. This means having a separate location and paying special attention to cleanliness of the *outside* of counting vials, using leak-free closures tightly sealed, and maintaining the sample changer to prevent mishaps with vials. Distinguish between and separate high-level and low-level work. Clean up. Do not leave contaminated glassware and equipment where others can handle it.

Tritium and ^{14}C in the quantities used in most experiments present no radiation danger. However, ^3H as in ^3H_2O is volatile, and $^3H_2$ gas as used in compound labeling is of very high specific activity; these require special precautions. ^{14}C, although a weak β-emitter, has a very long half-life (5,000 years). Hence, its incorporation into biological material (DNA and bone) presents a long-term hazard. ^{35}S and ^{32}P are strong β-emitters with relatively short half-lives (87.1 and 14.3 days, respectively). High levels of these should be handled behind a radiation shield (lead bricks), and the radiation level at the worker location should be monitored.

21.4.2. Purity of Isotopically Labeled Materials

Isotopic materials supplied by commercial sources are generally of high quality. However, one should keep in mind that both radiochemical purity and chemical purity may be critical in the proposed experiment. At least, it should be known that impurities, if present, will have no effect. A compound that is radiochemically pure contains one kind of radioactive molecule, but determinations of radiochemical purity give no indication of nonradioactive contamination. Radiochemical purity may be determined chromatographically and by reverse isotope dilution (see below), that is, by addition of the pure nonradioactive compound (carrier) followed by isolation and determination of specific activity. If the compound is radiochemically pure, the new specific activity is that expected by simple dilution with nonradioactive compound. If radioactive impurities are present, in reisolation with carrier these would be expected to be lost and the specific activity would be lower than expected. If the impurity is not radioactive, the specific activity after reisolation is higher than expected after dilution with carrier. If the impurity is known, its concentration can be determined by isotope dilution, i.e., by addition of a pure nonradioactive contaminant, isolation, and determination of specific activity. Chemical purity is analyzed in the usual ways, e.g., spectrophotometry, optical rotation, or GC.

Substantial changes in the radiochemical and chemical purities of a compound may occur during storage, primarily as a result of radiation damage from high specific activities. This process can be minimized by adherence to the specific conditions recommended by the supplier or more generally to one of the following: (i) storing at the lowest practical specific activity; (ii) subdividing into smaller lots; (iii) keeping dry if solid; (iv) storing in vacuo or under inert gas; (v) storing in pure benzene at 5 to 10°C; (vi) for water-soluble, benzene-insoluble compounds, adding 2 to 10% ethyl alcohol; and (vii) storing at the lowest possible temperature. Aqueous solutions of ^{3}H-labeled compounds should not be frozen. Storage of solids often does not require a vacuum or inert gas, and hence solids can be stored as dry as possible in screw-cap vials. Volatile materials and liquids should be resealed in ampoules under vacuum or inert gas.

21.4.3. Isotope Dilution Techniques

Radioactive compounds can be used to determine the amount of the same unlabeled compound in a mixture. It is necessary to know the specific activity and weight of the radioactive compound added and to isolate some of the pure compound for determination of its specific activity. The amount of unknown is derived from the equation

$$\text{weight of unknown (g)} = \text{weight of added radioactive compound (g)} \times \frac{SA_{\text{initial}}}{SA_{\text{final}}} - 1$$

where SA is specific activity. Purity is attained when the specific activity after several reisolations (or recrystallizations) is constant.

A reverse approach can measure the total amount of a radioactive compound in a mixture. In this method, pure nonradioactive compound is added and the compound is reisolated. The initial specific activity of the compound and the amount of nonradioactive carrier must be known, and the final specific activity (after reisolation) is determined. The weight of isotopic material is calculated from

$$\text{weight of unknown (radioactive) (g)} = \text{total weight (carrier plus unknown) (g)} \times \frac{SA_{\text{final}}}{SA_{\text{initial}}}$$

A double-isotope dilution technique is valuable for isolation of very small quantities of, say, a sterol from a homogenate by a procedure involving many steps and partial recoveries. For instance, a ^{14}C-labeled sterol of known total activity can be added. The total ^{14}C radioactivity of the pure isolated compound is determined by measuring the recovery of the sterol in the procedure. In another procedure, the sterol in the sample is derivatized either in the crude extract if practical or at the earliest possible purification stage. For instance, [^{3}H]acetic anhydride can be used to form the acetyl ester of the sterol. Then determination of ^{3}H in the isolated material gives the amount of sterol originally present. For this, only the specific activity of the acetic anhydride is needed; i.e.,

$$\text{Sterol recovery} = \frac{\text{cpm recovered}}{\text{cpm added}}$$

$$\text{Amount of sterol } (\mu\text{mol}) = \frac{\mu\text{mol of } ^3\text{H-compound}}{\text{recovery}}$$

$$= \frac{\dfrac{\text{cpm of } ^3\text{H/counting efficiency}}{\text{SA of acetic anhydride } (\mu\text{Ci}/\mu\text{mol})}}{\text{recovery}}$$

where 1 μCi = 2.22 × 10^6 dpm.

21.4.4. Calculation of Amount of Isotope Required

In general, the total amount of activity should be at least threefold above the minimum needed to make the final or lowest activity measurement. In this way, radiation effects and laboratory contamination are avoided and radiation safety is maximized. The easiest method, of course, is to use activities already reported for similar experiments, making adjustments for changes in volumes and instrument characteristics. When starting from scratch, first decide on the minimum count rate which will be satisfactory (section 21.4.8). Then calculate back to the start of the experiment to obtain the total amount (disintegrations per minute) that is required. To do this, it is necessary to bring into the calculations many of the following factors.

Half-life of radioactive compound
Duration of experiment
Initial sample size
Dilution factor
Counting efficiency
Molar specific activity
Size of sample counted
Biological half-life
Losses in isolation
Purity and stability of labeled compound
Metabolism of labeled compound

As an example, assume that it is desired to isolate on a column a tritium-labeled metabolite derived from the incubation of glucose in ^3H_2O. The material will be distributed in five 10-ml fractions from the column and will contain 1 atom of ^{3}H per molecule of metabolite. The yield from glucose added after the column chromatography is 5%, taking into account all factors of metabolite pools, side reactions, and losses in isolation. One-half of the total incubation mixture is used for this isolation.

In this experiment, ^{3}H from water is incorporated into the material isolated. In the incorporation process, protons compete very effectively with $^3H^+$ both because there are many more and because the mass and the chemical characteristics of $^3H^+$ create an isotope discrimination effect, which effect, for this example, is assumed to be 4:1 against $^3H^+$.

The counting solution to be used functions well with a maximum of 0.5 ml of aqueous sample; 0.2 ml is used in this case. The counting efficiency for ^3H is 43%, and there is 20% quenching. Sometimes these two together are called counting efficiency. For ^{14}C and ^3H, no calculations are needed to correct for decay of radioactivity.

From the standpoint of counting-error statistics (see below), 5,000 count events are desired. Since the scintillometer cannot be tied up for more than 1 to 2 h, this count should be obtained in 10 min per sample. Therefore, a 0.2-ml sample should contain radioactivity giving 500 cpm for *the lowest-activity fraction.* The radioactivity is estimated to be distributed into five fractions as 5, 20, 50, 20, and 5% of the total; hence, the count rate in each 0.2-ml sample should be 500, 2,000, 5,000, 2,000, and 500 cpm, or 8,000 cpm for the five samples counted out of a total of 50 ml of eluate. To continue: 8,000 cpm × 50 (total sample) × 2 (total reaction mixture) × 4 (isotope effect)/0.43 (counting efficiency)/0.8 (quench)/0.05 (yield) = 1.86×10^8 total dpm of ^3H_2O to be added to the reaction mixture. The ^3H_2O purchased contains 25 mCi/ml or 25 mCi × 2.22×10^9 (dpm/mCi) = 5.55×10^{10} dpm/ml in stock solution. Thus $1.86 \times 10^8/5.55 \times 10^{10}$ = 0.00335 μl, or, better, 3.35 ml of 1:1,000 dilution of ^3H_2O solution.

Total activity calculations thus provide information about quantities to be used. However, it is necessary to make specific activity calculations as well, i.e., disintegrations per minute per mole or disintegrations per minute per micromole, to determine how many atoms of tritium have exchanged or whether other processes equilibrating with hydrogen atoms are involved. In the above example, it was calculated that 5.55×10^{10} dpm was present in 1 ml or 55.56 mmol of water or 111 meq of hydrogen per ml. The specific is $5.55 \times 10^{10}/1.11 \times 10^2$ or 5×10^8 dpm/meq of H.

In the 10-ml incubation, the dilution of specific activity is 10 ml/0.00335 ml of ^3H = 2,985-fold, giving a specific activity of $5 \times 10^8/2{,}985 = 1.67 \times 10^5$ dpm/meq of H. If the metabolite isolated had the same disintegrations per minute per millimole, there was 1 eq of $^3H^+$ incorporated with the assumed isotope effect of 4. If the specific activity is higher or lower than this value, the magnitude of the isotope effect may be different, or there may be exchanges with hydrogen atoms not derived from ^3H_2O (lower specific activity), or more than one proton from ^3H_2O has been incorporated (higher specific activity). The isotope effect is significant with tritium but is small enough with ^{14}C, ^{32}P, and ^{35}S to be ignored. It is often necessary, as in this example, to make an independent determination of the isotope effect by measuring the rate with H_2O or ^2H_2O and the rate with ^3H_2O.

21.4.5. Enzyme Assays with Isotopes (also see Chapter 23)

A wide range of enzyme assays which depend on radioisotopes have been developed. In many cases there are no other convenient methods. Further, these have intrinsically high sensitivity, which is demanded in many kinds of work. The amount of enzyme is proportional to the rate of substrate utilization or product appearance. Hence, determination of the concentration of either substrate or product with time is needed. For an assay to be valid, the initial rate must be constant for a short period and must also be proportional to the level of enzyme in the assay. In some instruments such as spectrophotometers and pH meters, rates can be monitored continuously on a single assay mixture, often with the progress of the reaction being displayed as a strip chart recording. In other cases, however, such as with radioisotope enzyme assays, the rate must be established by measuring radioactivities on discrete aliquots of a reaction mixture taken at intervals.

One of the major problems characteristic of such assays is the necessity of separating the radioactive substrate from the radioactive products. Usually, the labeled substrate is present in an amount required to saturate the enzyme and produce pseudo-zero-order kinetics in which the rate of reaction is constant with time for an appreciable period. In sensitive and valid assays, the amount of product formed may be only 1% of the substrate present. It is therefore necessary to separate a small amount of radioactive product from a very large amount of radioactive substrate. Hence, the success of the method depends on finding an efficient separation that will give 0.1% or less cross-contamination. These procedures are often specialized and hence are outside the scope of this manual. The general types will be summarized. For details, consult the general reference by Oldham (84).

The separation methods include (i) precipitation of macromolecules synthesized or utilized, i.e., polynucleotides, polypeptides, and polysaccharides; (ii) release or uptake of a volatile radionuclide which can be separated by distillation or an equivalent process, i.e., ^3H_2O, $^{14}CO_2$, or VFAs; (iii) solvent extraction; (iv) ion exchange; (v) paper and thin-layer chromatography and electrophoresis; (vi) adsorption and elution; (vii) isolation of derivatives; (viii) conversion of unlabeled product to labeled derivative; (ix) dialysis and gel filtration; and (x) reverse isotope dilution.

The degree of cross-contamination of substrate into product in the isolation procedure limits the sensitivity of the method. For instance, assume that there is a 0.1% cross-contamination of substrate into product in the separation used. Further assume that the zero-time (i.e., with no product formed) count rate due to cross-contamination is 100 cpm. It is desirable to have the count rate due to product be at least equal to the rate at zero time or 100 cpm for 0.5% conversion of substrates. In this case the signal-to-noise ratio is 1:1, or the minimum sensitivity without very long count rates (see section 21.4.8). To attain these minimum count rates, 100,000 cpm in the substrate is the minimum to be used if the total sample is counted. The use of larger amounts of radioactivity will not improve the sensitivity, because the zero-time value increases correspondingly. Of course, larger amounts of radioactivity in the product above the zero-time count rate decrease the error of the assay. However, a very substantial amount of product formation causes loss of the pseudo-zero-order condition needed for a valid assay.

From the example, it is clear that, although it is necessary to add enough substrate to saturate the enzyme, its radioactivity need not be high. On the other hand, if separation of substrate and product is complete, the sensitivity is proportional to the specific activity of the substrate and is limited only by cost, practicality, and availability. Under these circumstances, investigators may prefer to use ^3H- rather than ^{14}C-labeled substrates because of the availability of 100-fold-higher specific activities at similar costs (see section 21.4.6 for a discussion of problems with ^3H substrates).

Dilution of radioactive substrate with unlabeled substrate affords a number of advantages in addition to facilitating saturation of the enzyme with substrate. It reduces (i) cost by using less isotope, (ii) effects of contaminants and radiation degradation products in radioactive substrates, (iii) the effect of an unwanted stereoisomer in radioactive substrate containing a mixture of isomers by adding carrier substrate of the desired stereo configuration, and (iv) radiation decomposition of the substrate. Such dilution also increases the accuracy of determinations.

Since enzyme activity is expressed in international units (IU), where 1 IU is 1 μmol of substrate converted per min under standard conditions, it is necessary to convert count rates to micromoles per minute by using the specific activity of the substrate (which will be converted to product presumably of the same specific activity). This value is subject to errors in measurement of both radioactivity and concentration. Although the specific activity of the radionuclide as applied is usually known to $\pm 5\%$, it is seldom used as such. Rather, the specific activity is diluted with large amounts of the nonradioactive form, with attendant errors of measurement and purity. It is wise, therefore, to carefully determine the specific activity of the substrate as used in the assay. It may also be necessary to demonstrate radiochemical purity of the substrate or to purify it prior to using it to remove decomposition products and other impurities.

21.4.6. Special Problems with Tritiated Substrates

Tritium shows isotope effects relative to hydrogen because of the large change in mass involved. The effect is most pronounced when making or breaking the bond between tritium or hydrogen and another atom is a rate-limiting step in the reaction. This effect may amount to virtual exclusion of reaction with the tritiated species, as has been observed in the enzymatic reductive carboxylation of L-ribulose-5-phosphate by [^3H]NADH-phosphate and CO_2. The isotope effect may be displayed as a change in V_{max} of the enzyme when using a tritiated substrate or as a change in K_m. If the K_m is lower, the effect would not be observed when the enzyme is saturated with substrate. If the K_m is higher, the rate would decrease if the substrate concentration is no longer saturating.

When tritium is used as a label only to monitor the conversion of substrate to product, it must be remembered that (unlike ^{14}C) ^3H often is released as $^3H^+$ into the medium in unrelated spontaneous reactions, giving substrate and products a lower specific activity and hence erroneously low enzyme activities. In some cases the side reaction is enzyme catalyzed, not necessarily accompanied by the complete reaction. That is, there is a proton-triton exchange reaction which is independent of, and often more rapid than, the total reaction. In addition, it is characteristic of tritiated compounds to "leak" $^3H^+$ by chemical reaction. This varies with the location of ^3H in the molecule. For example, α-^3H-amino acids lose ^3H in many enzyme-catalyzed conversions, whereas ^3H in other positions, especially if remote from functional groups, is relatively stable. ^3H may also be lost by spontaneous exchange reactions during isolation of the product.

Some enzyme assays are based on release of ^3H from a specific position in the substrate. Errors result if ^3H in the substrate is not exclusively located in the position assumed and/or in the correct stereo location, because it has been demonstrated many times that enzymes selectively or specifically labilize tritium in one of two or more bonding positions occupied by ^3H or H.

The low energy of ^3H radiation causes low counting efficiency and losses due to quenching and absorption by solid materials in the counting vial. These in turn lead to greater inaccuracies or even lower apparent enzyme velocities. Good efficiency and quench corrections made on a sample-by-sample basis often are required.

21.4.7. Liquid Scintillation Counting

Liquid scintillation counting (LSC) has now supplanted Geiger, proportional, and gas counting methods for ^3H, ^{14}C, ^{35}S, and ^{32}P, even though the instruments are quite expensive. This preference is due to several advantages of liquid scintillation counters such as elimination of self-absorption, more favorable counting efficiency, and greater convenience in sample preparation.

In LSC, the radiation event excites the emission of a light pulse from a fluorescent material, which is then "seen" by a pair of photomultiplier tubes and registered as a radiation event or count. The sophistication of this instrument lies in the fact that both photomultiplier tubes must see the light flash before a "count" is registered. This eliminates a large number of spurious, electronically derived pulses. In addition, the photomultiplier produces voltage pulses whose amplitude or height is proportional to the energy of the radiation event. By use of pulse height analyzers, with adjustable discriminator settings, the instrument can be set to count only pulses appearing in the "window" corresponding to desired minimum and maximum radiation energies. Therefore the instrument can be set to (i) eliminate many low background pulses, (ii) selectively count one isotope (one class of pulse heights), or (iii) count all isotopes (all pulse heights); it usually does these simultaneously via several parallel analyzer channels.

The radiation event is converted into a scintillation via solvent molecules that become activated by the decay event. The activated solvent then transfers the excitation energy to a compound which fluoresces (fluor). The excited fluor either emits a photon of light at a characteristic wavelength or transfers the excitation energy to a secondary fluor, which in turn emits a photon of light at a more advantageous wavelength with respect to the spectral sensitivity of the photomultiplier. 2,5-Diphenyloxazole (PPO) and 1,4-bis-2-(5-phenyloxazolyl)benzene (POPOP) and its dimethyl derivative are the most popular primary and secondary fluors. These are highly nonpolar substances which are dissolved in a nonpolar solvent such as toluene to make up the scintillation solution. The principal drawback in scintillation counting is the difficulty of introducing aqueous samples containing polar solutes into the nonpolar counting solution. Further, water and solvents of increasing polarity tend to diminish the counting efficiency; that is, they, along with colored compounds, quench the normal chain of events leading to light emission.

Because of the problems described above, a variety of scintillation mixtures have been developed for polar and nonpolar materials in solution, for solid materials, and for other specific purposes (see below). In addition, high background counting rates can occur as a result of fluorescent light or sunlight-induced phosphorescence, which may require hours or days to decay. Also, chemiluminescence of samples in counting medium gives very large counts which decay slowly. Glass vials may contain ^{40}K, which raises the background level. On the other hand, a number of materials quench the count rate; trichloroacetic acid, $HClO_4$, water, pyridine, and colored compounds are major problems.

21.4.7.1. Solid materials

Small pieces of paper or removed sections of a thin-layer plate may be counted in scintillation medium. The main problems are associated with self-absorption by the particles, uniformity of orientation, uniformity of solubilization of radioactive material, and difficulties in using the external standard method for determining counting efficiency (see section 21.4.3.4). Since there is no direct way of determining self-absorption, only relative radioactivity of a series of samples counted under the same conditions can be obtained. Variability is greatly decreased if the material to be counted is either completely dissolved or is completely insoluble in the scintillant. For instance, hyamine hydroxide 10-X in methanol has been used to remove amino acids and sugars from paper, and then a toluene scintillant is added. When all of the material cannot be eluted, it may be necessary to burn the paper, trap the CO_2 in an organic base or as sodium or barium carbonate, and count these materials. When papers containing insoluble radioactivity are counted, the most reproducible orientation is to place the paper in the bottom of the vial.

Glass filter discs and strips are now used to advantage because of higher counting efficiencies with tritium, i.e., 10% at best with $\pm 5\%$ reproducibility compared with 2 to 8% on cellulose paper. With stronger β-emitters such as ^{14}C, reproducibility and efficiency are higher.

21.4.7.2. Polyacrylamide gels

The main problem with polyacrylamide gels is to solubilize them rapidly. This is best accomplished by using a cross-linking reagent that is readily cleaved to yield linear polyacrylamide chains that are soluble in scintillation mixtures. *N,N′*-Diallyltartardiamide has been used as a cross-linking agent to make up the gel because it is readily cleaved by periodic acid (86). A series of gels of 7% (wt/vol) acrylamide and 0.27% (wt/vol) cross-linking agent are treated with 0.5 ml of 2% periodic acid for 2 h. The scintillant is added, vigorously mixed on a Vortex mixer, and used as such; alternatively, a portion can be transferred to a scintillation vial. The scintillation cocktail is designed to maintain solubility (85).

21.4.7.3. Carbon dioxide

Measurement of ^{14}C radioactivity as CO_2 is of major importance in metabolic studies and in the large number of applications that require combustion or digestion of carbonaceous material. For scintillation counting, the CO_2 is trapped in an organic base, for instance, hyamine hydroxide 10-X, Primene 81-R, ethanolamine-ethylenediamine, or phenylethylamine, or in barium hydroxide or NaOH. With all of these, the CO_2 must be released with acid and swept through a train of traps containing the base. The base with trapped CO_2 is then added to a scintillant.

For information on combustion methods, consult reference 89.

21.4.7.4. Instrument operation

Single-isotope counting. The goal in single-isotope counting is to set the scintillation counter for the highest efficiency and the lowest background. Two types of adjustment are provided for each channel: (i) high and low discriminator settings, and (ii) photomultiplier voltage and amplifier gain. Higher settings of the latter pair increase the pulse height. The high and low discriminator settings determine the pulse height range that will be accepted. The pulse height or energy spectrum of β-emission is determined by setting the upper and lower discriminator with a fixed differential and then counting a radioisotope at increasing positions of the discriminator band, e.g., 0 to 10, 10 to 20, 20 to 30. A plot of the count versus discriminator range midpoint gives the spectrum. Because β-emitters have different energies of emission, each isotope has a characteristic peak height distribution, and hence the discriminators in various channels may be set to include or exclude various parts of the isotope energy spectrum. Once this has been determined, the lower discriminator is set to exclude most of the background (i.e., set at 10) and the upper discriminator is set to just include the maximum pulse height of the isotope being measured. In this way, many of the low-energy and high-energy background events are eliminated.

Determination of the peak height spectrum at various photomultiplier voltage settings also gives a family of curves (Fig. 8) which are especially useful in separating the radiation of each isotope in dual-labeled samples. Note that lower voltage settings narrow the discriminator settings necessary to produce the window.

In all cases, the background count rate must be determined with a vial of scintillation solution with no radioisotope and subtracted from each sample count. It is assumed that quench of the background measured as above is nonexistent when high sample count rates are measured. However, at very low count rates it may be necessary to determine the effect of sample quench as described below.

Dual-label counting. Dual-label counting can be accomplished when the energies of the isotope pair are

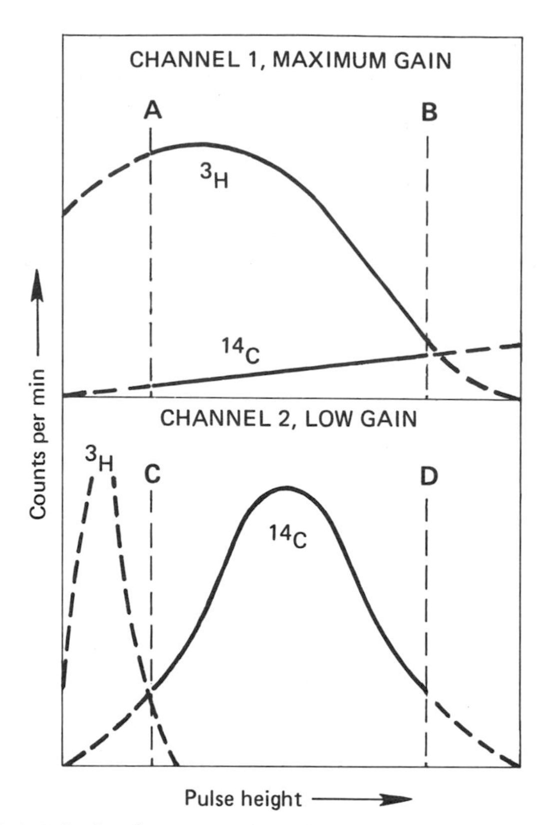

FIG. 8. Pulse height spectrum as a function of photomultiplier voltage. A, B, C, and D represent the settings of four different discriminators.

sufficiently different that there is resolution of the two pulse height spectra. This is practical for the ^{3}H and ^{14}C, ^{3}H and ^{35}S, ^{3}H and ^{32}P, and ^{14}C and ^{32}P pairs. The more energetic isotope is counted exclusively in channel 1. However, the count for the weaker isotope in channel 2 will always have counts of the more energetic member of the pair. It is necessary to determine counting efficiency (instrument efficiency plus quench) for each isotope separately, preferably by internal or external standard methods, and then to calculate the count for the second channel from simultaneous equations:

$$\text{dpm of } ^{14}\text{C} = \frac{\text{net count in channel 1}}{\text{efficiency of } ^{14}\text{C in channel 1}}$$

$$\text{dpm of } ^{3}\text{H} = \frac{(\text{net counts in channel 2}) - (\text{dpm of } ^{14}\text{C} \times \text{counting efficiency of } ^{14}\text{C in channel 2})}{\text{tritium efficiency in channel 2}}$$

The counting efficiency of channel 2 for the lower-energy isotope (^{3}H in this case) can be optimized first by adjustment of discriminator settings with a tritium-containing standard. Then the lower discriminator of channel 1 is set to exclude counts from ^{3}H and the upper discriminator is set to give maximum efficiency for ^{14}C. The above procedure applies if the quench and efficiency of samples correspond to those of standards used. When quench is encountered, more of the higher-energy counts appear in the low-energy channel. The original relationship of channels can be reestablished by changing the discriminator settings if the needed information is available from similarly quenched standards. Alternatively, the gain of the photomultiplier or amplifier can be advanced to reestablish the pulse height pattern of the original window settings. Dual-label counting is fraught with difficulties and errors when low count rates are involved.

Quench correction and counting efficiency. Under ideal conditions, not all radiation events are detected even in the best scintillation counters, although efficiencies of above 90% are obtained with intermediate and strong β-emitters, e.g., ^{14}C and ^{32}P. Instrument efficiency can be determined by several means (described below) with a nonpolar counting medium which has neither chemical nor physical quenchers. More commonly, however, a combined instrument efficiency and sample quench correction factor is produced by one of the three methods, each of which has advantages and disadvantages.

Internal-standard method. The original method used, the internal-standard method, is relatively simple and easy to understand. It is highly reliable and potentially very accurate for counting single isotopes. The correction includes both chemical and color quenching. It consists of first counting a sample (C) and then adding a small volume containing a known number of disintegrations per minute (D_s) of the same isotope: e.g., for ^{14}C counting, [^{14}C]benzoic acid, and for ^{3}H counting, [^{3}H]toluene. Then the sample plus standard (C_t) is recounted. From this, the overall efficiency is $(C_t - C)/D_s$, and the radioactivity of the sample is

$$\text{Sample dpm} = C \times \frac{D_s}{C_t - C}$$

The main disadvantages are that the vials must be removed from the counter to add the standard and then be recounted. The sample alone cannot be recounted as an afterthought. Addition of the standard to the vial must be accurate. For large numbers of vials, this is now accomplished with improved accuracy by using various kinds of automatic pipettes or syringes.

Channel ratio method. The channel ratio method is based on the observation that the spectrum of pulse heights in a given scintillant-isotope system shifts with the efficiency. In an instrument with two counting channels, the discriminators in each channel are set differently and the count in each channel is determined simultaneously and expressed as a ratio. This method can also be used in a single-channel instrument; however, it is necessary to make two counts, each at a different discriminator setting.

Using a set of standards prepared to have different efficiencies, a standard curve of efficiency versus channel ratio is constructed and used to correct the unknowns. In both channels the lower discriminator is set above the pulse height of background noise. One discriminator setting, either upper or lower, may be the same in both channels, with the other different. Alternatively, both upper and lower discriminator settings may be different for the two channels. Since quenching may be due to colorless chemicals such as water and CCl_4 and also to colored compounds, it may be difficult to prepare standards of the same quencher content as the unknowns. Therefore, there will be some error if colored quenchers are present in the unknowns, especially at high concentrations. However, this method applies over a wide range of conditions (87). Dual-channel systems require less time unless low counts are involved. In that case, much more counting time is required to obtain a ratio of given precision than for the internal-standard method. In this method, sample counting can be repeated.

External-standard method. Newer instruments are equipped with a γ-source of known disintegrations per minute, which, on call by the instrument program, is positioned directly below the counting vial. Counts are made with and without the γ-source. A standard curve of external standard counts versus counting efficiency (produced by introducing quenchers) is constructed and used to correct counts in unknowns. The method is susceptible to variations in volume of scintillant, thickness of the glass vial, and different scintillants. In addition, inhomogeneity of solutions or counting of paper strips produces errors. Long counts are not needed at low radioactivity levels as for the channel ratio method. It is possible to improve the external-standard method described above by making a channel ratio approach with and without the γ-source. In this case the external standard is counted in two channels with discriminators set above the pulse height of the isotope being counted to obtain counting efficiency. This method requires the availability of four channels or two counting runs at different discriminator settings.

21.4.7.5. Counting solutions

1. **Nonpolar applications** (minimum temperature, −30°C)
   2,5-Diphenyloxazole (PPO), 5 g/liter
   1,4-Bis-2-(5-phenyloxazolyl)benzene (POPOP), 0.3 g/liter
   Toluene
   Very high efficiency (90 to 95%). Can be used to count filter discs (Millipore Corp.) if dry. Discs must be translucent when in toluene; if white, drying is not complete.

2. **Homogeneous solutions or emulsions** (depending on amount of water; 4 to 14% water produces a solution, 20 to 30% produces an emulsion) (91, 92).
   Toluene (667 ml) containing 0.4% PPO (2.67 g) and 0.1% POPOP (0.07 g)
   Triton X-100 (Rohm and Haas Co., Philadelphia, Pa.), 333 ml.
   Counting efficiency for ^{14}C, 52 to 68%; for ^{3}H, 7 to 10%, but varies with batch of Triton X-100.

3. **Stable gel with high water content** (forms emulsions in about 35% water and a stable gel at low temperature [to −5°C] which resolubilizes when warmed) (91).
   Toluene (538 ml) containing 0.4% PPO (2.1 g) and 0.1% POPOP (0.05 g)
   Triton X-100, 462 ml
   Counting efficiency for ^{14}C, 43 to 55% at 43% water. Use with tritium not recommended unless a gel is needed. ^{3}H efficiency about 5%. Useful with tissue suspensions as well as powders such as barium carbonate and other white materials.

4. **Homogeneous solution with up to 35% water** (88)
   Triton X-100, 257 ml
   Ethylene glycol, 37 ml
   Ethanol, 106 ml
   Xylene, ca. 500 ml (to make 1 liter)
   PPO, 3 g (POPOP not used)
   Counting efficiencies unquenched are 87% for ^{14}C and 47% for ^{3}H. With maximum water, the values are 82% for ^{14}C and 35% for ^{3}H. The relationship between efficiency and external standard ratio (see below) is linear.

5. **Solubilized polyacrylamides**
   Triton X-114 (Rohm and Haas Co.), 250 ml
   Xylene, 750 ml
   PPO, 3 g (POPOP is not needed)
   Maximum counting efficiency is 93% for ^{14}C and 47% for ^{3}H. Up to 30% water may be used with a decrease in efficiency to 50 to 60% for ^{14}C and 10% for ^{3}H. A stable emulsion is formed at intermediate water concentrations.

6. **CO_2 counting** (90)
   Ethanolamine, 1 part
   Ethylene glycol monomethyl ether, 8 parts
   Toluene, 10 parts
   PPO, 5 g/liter

21.4.8. Statistics of Counting (82, 83)

Determinations of radioactivity from count and time measurements under ideal conditions contain random errors which cause the measured count rate to depart from the true value. For this reason, one should have at hand statistical tools which aid in estimating error. The intelligent use of these methods requires that one be informed of the assumptions and terminology involved if not how the final useful expressions are derived. It is not within the scope of this manual to present the application of these tools with the appropriate background, because various tables and graphs are needed to perform the operations. Rather, it is recommended that general references 82 and 83 be consulted.

Nevertheless, several important statistical tools are listed and their equations are given below. These include average of several counts, standard deviation estimates of single and multiple counts, percent standard error (%SE) of counts and count rates, and time required for a desired %SE of count rates of samples with and without background for equal and unequal times for counting background. Additional expressions and tables are available for determination of times required for counting background and for total count with a given standard error, limits of detection of radioactivity, and the correct way to eliminate a count value which varies widely from the mean.

There are a number of different kinds of error: probable error, for which the true value is likely to exceed the error expressed 50% of the time; standard error, 68% of the time; and 99/100 error, 99% of the time. Since standard error has a coefficient of 1 in many statistical derivations, and since this expression is most commonly used, only "standard" forms of equations will be given here. This means that, among members of a set of values, one value will not vary from the mean by ± the standard deviation more than 32% of the time. Since count errors are independent of time, one can determine the number of counts required for a given %SE by using the expressions below. For instance, for 1% SE, 10,000 counts are required; for 2.5% SE, 1,600; for 5% SE, 400; and for 10% SE, 100. From the corresponding expression below, the times required to determine both background and total count rates at 1% SE when both rates are equal (see calculations of the amount of radionuclide to use) are 804 min for background and 1,140 min for sample plus background; for 10% SE, 8 min and 11 min are required, respectively.

Equations for counting statistics are as follows.

1. **Average of several counts**

$$\overline{X} = \frac{\Sigma \text{ individual counts } (X)}{\text{number of determinations } (n)}$$

2. **Standard deviation from a number of counting cycles**

$$\text{SD} = \left(\frac{\Sigma \text{ (deviations of } X \text{ from } \overline{X})^2}{n-1}\right)^{1/2}$$

3. **SE (%) in count rate**

$$\text{SE of rate (\%)} = \frac{100}{\text{rate } (R) \times \text{time } (t)^{1/2}}$$

4. **Time required to give the desired SE (%)**

$$t = \frac{10^4}{R(\%\text{SE})^2}$$

5. **%SE of net sample count when the times of counting sample and background are equal**

$$\%\text{SE} = \frac{100\,(X_{\text{total}} - X_{\text{background}})}{X_{\text{total}} - X_{\text{background}}}$$

6. **%SE of sample count without background when there are unequal counting times for background and sample**

$$\%\text{SE} = \frac{200\,(R_t/t + R_b/t_b)^{1/2}}{R_t - R_b}$$

where R_t and R_b are the total and background rates, respectively, and t_t and t_b are the times for R_t and R_b.

21.5. ELECTROPHORESIS

A number of separation methods make use of the electrophoretic force that propels charged molecules through various media. The original methods involving free flow and a moving boundary through a column of liquid have largely been supplanted by methods which separate molecules by a combination of electrophoretic mobility plus another method of molecular discrimination.

21.5.1. Gel Electrophoresis

Separation of macromolecules by electrophoresis through a sieving gel has become a very important tool in biological and medical research as well as in clinical diagnosis and monitoring of industrial processes. Since the inception of the discontinuous disc gel method in 1964 by Ornstein (99) and Davis (97), for separation of blood proteins, two general strategies have emerged: continuous zone and discontinuous/multizone electrophoresis. In the first, the ionic content of the gel is continuous throughout. This strategy is used extensively in molecular biology, particularly for Western, Southern, and Northern blots. In the second, two different ions are present, one with greater mobility and the other with lesser mobility than the mobilities of the macroions of the biological material to be separated. In each case, conditions are adjusted so that the sample macroions are sandwiched between the fast- and slow-moving buffer ions. When electrophoresis starts, free movement in the stacking gel concentrates the macroions into a series of very narrow bands at the top of the separation gel. When this is accomplished, the two strategies give comparable resolutions. Applications have expanded beyond visualization of the number of components in a sample to include determination of apparent molecular weights of protein monomers and oligomers, estimation of isoelectric point (pI) by isoelectric focusing, and visualization and sizing of intact and fragmented DNA molecules. Intensive investigations have led to a large body of theory, detailed information on conditions, and many variants of the buffer and gel system (93).

It is the intent here to describe only the simpler discontinuous system. Refer to reference 93 for more comprehensive treatment of current general- and special-purpose methods, which include buffers for operation at constant and higher pH as well as a conductivity shift to aid in concentrating the sample. For pulsed-field gel electrophoresis, see Chapter 16.3.1.2.

21.5.1.1. Continuous-zone electrophoresis

The essential step in continuous-zone electrophoresis is to reduce the conductance of the sample to 20 to 50% the ionic strength of the buffer in the separating gel by dialysis and/or dilution (93, 108). In homogeneous gels, separation is based on charge and size. Gel concentration gradients have been used to make separations based on size alone (93).

21.5.1.2. Discontinuous-zone/multizone electrophoresis

Discontinuous-zone/multizone electrophoresis is carried out either in cylindrical gel columns (in glass tubes) or in a gel slab about 10 cm by 10 cm by <0.1 to 4 mm thick. Three layers of gel are involved: a separation gel at the bottom, a spacer gel in the middle, and a sample gel at the top. Each is composed of polyacrylamide of specified percent cross-linking. The gel is formed in a container (tube or flat plate) by mixing acrylamide (monomer), *N,N′*-methylenebisacrylamide (BIS) with buffer, and polymerizing agent. The polymerization is catalyzed either chemically with ammonium persulfate or photochemically by light, which produces free radicals when oxygen is present.

Separation of proteins in the discontinuous system is the result of two major factors: differences in migration as a result of the sieving effect of the cross-linked gel causing separations based on size, and differences due to charge. In addition, unusually sharp bands exist after electrophoreses as a result of concentration of the proteins into very thin bands or discs before they enter the separation gel. Thus, molecular sieving is controlled by the degree of cross-linking, which is in turn dictated by the concentration of cross-linking acrylamide. Concentrating proteins into a narrow band before they enter the separation gel is characteristic of the discontinuous system and results from an ingenious manipulation of discontinuities of pH, buffer ion composition, and gel pore size of the sieving gel.

Concentration is obtained by choosing a pair of buffer ions and pH such that the mobility multiplied by the degree of dissociation for one ion (Cl^-) is greater than the same value for protein, which in turn is greater than the value for the second ion (glycine) (see p. 23 of reference 93). To accomplish this, a zone of low cross-linked stacking gel is layered over the longer, smaller-pore separation gel. Above this, the sample in less highly cross-linked gel is layered over the stacking gel (Fig. 9). The separation gel is more alkaline (pH 8.9), whereas the stacking gel and sample are in a different buffer at pH 8.3. When a potential is applied, the proteins quickly traverse the spacer gel and form a very thin band at the top of the separation gel; in fact, the proteins are sandwiched between Cl^- and glycinate. The proteins then slowly penetrate the separation gel. Migration rate in the separation gel is then a result of charge (dictated by pH) and molecular size (dictated by pore size or degree of cross-linking). Since the bands were originally very thin, resolution is exceptionally high. For instance, blood serum gives 15 to 20 bands. With an optimized discontinuous buffer system, pore gradient gels (12 to 19% cross-linking), and increased gel length (28 cm), the separation of 150 components has been reported (101).

Following electrophoresis, the gels can be analyzed in place by UV scanning or can be extruded and analyzed by UV scanning, staining, or sectioning. The sections can be soaked to elute components, or they can be solubilized and radioactivity can be determined. Gels or longitudinal sections can be dried for autoradiography.

Column method (93, 99). The column system requires very simple equipment, either commercial or home built. It involves a watertight tray with holes in the bottom to hold 1 to *n* rubber stoppers holding glass tubes, 5 mm (inner diameter) by 65 to 70 mm long. The construction is such that buffer plus electrode (cathode) is in the tray above and in contact with the top of the gel and the same buffer plus anode is in contact with the bottom of the gel (Fig. 10). The edges of the tubes must be planar, and the diameter must be uniform. However, these tubes can be made in the laboratory or glass-blowing shop. The tubes must be scrupulously clean and may be dipped in a nonionic detergent (e.g., Kodak Photo-Flo) before being dried.

The original sequence of filling tubes with acrylamide (97, 99) started by polymerizing the separation

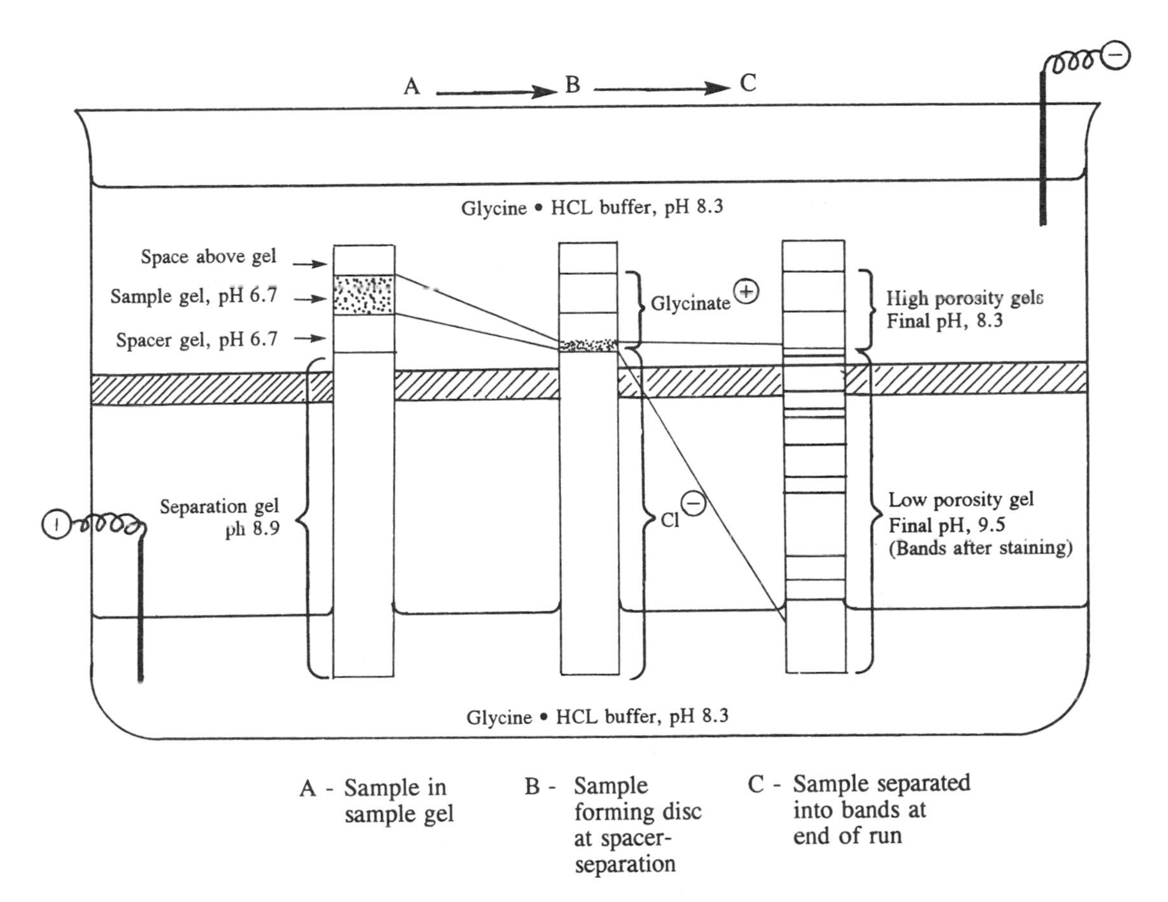

FIG. 9. Gel layers and ionic behavior in discontinuous polyacrylamide electrophoresis.

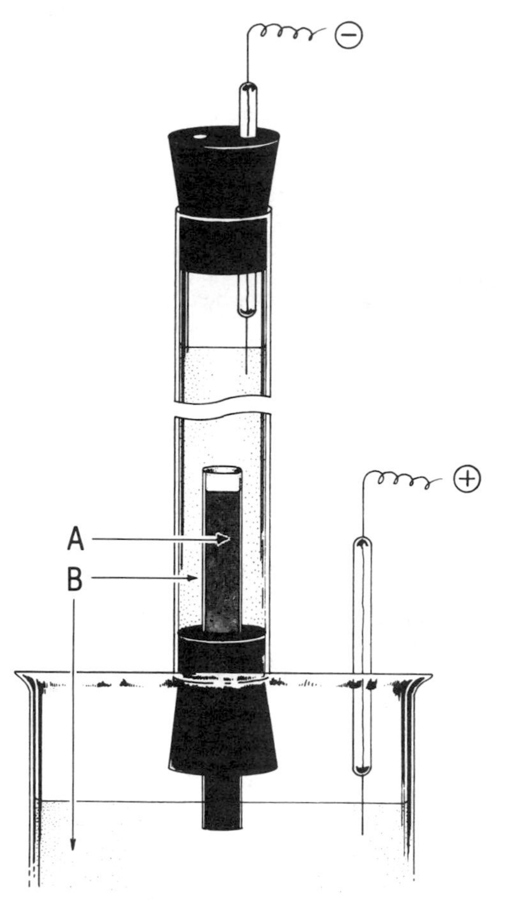

FIG. 10. Simple arrangement for column electrophoresis. A, agar-solidified buffer; B, buffer solution.

gel in the bottom section. A reverse procedure has since been recommended (93) as producing a more uniform interface between the spacer and separation gels, as follows. Transfer 0.2 ml of 40% sucrose to a rubber cap which will fit tightly when the gel tube is firmly pressed to the bottom of the cap; this prevents entrapment of air bubbles. Align the tube vertically; then prepare about 0.15 ml of spacer gel mixture (see below), take it up in a small syringe equipped with an 8-cm needle (inner diameter, 0.6 to 0.8 mm), and layer it on the sucrose solution. Layer water very carefully over the spacer gel layer so that a sharp boundary is formed with no mixing. High performance results largely from a sharp uniform boundary between the spacer and separation gels. Polymerize the gel by irradiation with a fluorescent lamp for 20 to 30 min at a distance of 2 to 3 cm. Carefully and completely remove the water layer by using a syringe and then a wick of lint-free material.

Next, prepare the separation gel. Rinse the space over the spacer gel quickly twice with separation gel mixture, and fill the tube with separation gel mixture by syringe to give a convex meniscus, taking care not to entrap air bubbles. Then cover the surface tightly with plastic film. This must be done within 5 min after completion of spacer gel preparation. After 30 to 40 min of polymerization time, carefully remove the cap at the bottom, invert the tube, carefully place the cap on the new bottom, remove the sucrose solution from the new top, and then rinse the top of the spacer gel with spacer gel mixture. Apply the sample in spacer gel mixture, and carefully cover with electrode buffer to the upper rim, taking care that the two layers do not mix. Then photolyze the sample layer as above. This procedure can be managed easily for many tubes if suitable carriers are prepared in advance and the fluorescent light is of sufficient size.

Place the tubes in the apparatus, and carefully add precooled electrode buffer above and below the tubes to at least 4 cm for the lower buffer. Add bromothymol blue to the upper buffer to track the migrating boundary. Then connect the power supply so that the anode (positive terminal) is at the bottom, and adjust the current of the power supply to 1 mA per gel tube for 2 min and then to 2 to 5 mA per tube until the tracking-dye disc has migrated to about 1 cm from the bottom of the tube.

The voltage drop across the gel produces heat that is proportional to current flow. Excessive heat causes uneven bands. A 40-min run starting at 20°C will increase the temperature to 30 to 40°C. The temperature rise can be diminished by lower current flow, i.e., 1 mA per gel with an increase in electrophoresis time to 140 min. However, there will be some band broadening due to diffusion. Alternatively, electrophoresis can be run at 4°C with stirring of the electrode buffers.

Separation gels can be stored for several days, but after sample and spacer gels have been added, electrophoresis should be performed within 1 h. After electrophoresis, the electrode buffers may be removed and saved for future runs, but anode and cathode buffers must be kept separate. The buffers should be discarded after several runs or when the pH changes. The gels are then analyzed either before or after removal from the tubes. Since the conductivity of the gel column changes during electrophoresis, changes in voltage and current can be expected to occur during the run. A current-regulated power supply should be used so that current can be maintained constant throughout the run. When n gels are run simultaneously, the current required is n times the value chosen for one tube and is divided equally among the tubes only if the cross-section and composition are identical.

Slab method. The same electrophoretic separations can be made in slabs of polyacrylamide of 100 by 100 by <0.1 to 3.5 mm. Special trays are used to form the slab with slots for each sample, and special equipment is needed to perform the electrophoresis. The apparatus provides a way of holding the slab vertically, with the slots of the upper edge in contact with the upper buffer and the opposite edge in the lower buffer. Provision must be made to circulate coolant across the face of the slab on both sides. The main advantage of the slab system is the ability to run multiple samples under identical conditions so that comparisons among samples can be made easily. In addition, very efficient cooling of thin slabs can be arranged easily. It is claimed that slab gels are easier to read in a densitometer, but a special apparatus is needed. Slabs are also easier to cut into strips and analyze for (i) different components with different stains, (ii) radioactivity, and (iii) drying and storage. However, more acrylamide is needed, especially when only a few samples are run. As with tube gels, a wide variety of continuous and discontinuous buffer systems have been developed. The methods of preparing gels, running, and analyzing are similar to those for tube gels.

Preparation of gels. Details for preparation of one of the more common discontinuous gel systems are presented here. Many others are given in references 93 and 94. *CAUTION:* Acrylamide must not be inhaled or allowed to contact the skin.

The spacer sample gel is prepared as follows.

1 part A (25.6 ml of H_3PO_4, and 5.7 g of Tris per 100 ml)
2 parts B (10 g of acrylamide and 2.5 g of BIS per 100 ml)
1 part C (4 mg of riboflavin per 100 ml)
4 parts D (40% sucrose)

The separation gel is prepared as follows.

1 part E (48 ml of 1 N HCl, 36.6 g of Tris, and 0.23 ml of TEMED per 100 ml)
2 parts F (30 g of acrylamide and 0.8 g of BIS per 100 ml)
1 part water
4 parts G (0.14 g of ammonium persulfate per 100 ml)

Polymerization with persulfate yields gels of variable porosity and electrophoretic behavior. Therefore, it is important to establish a standard set of polymerization conditions. Persulfate also may cause artifacts. These may be eliminated by substitution of riboflavin in the gel mixture and photolytic polymerization, by use of reducing agents, e.g., thioglycolate (0.01 to 0.001 part per 1 part of glycine in buffers), or by addition of 5 mM mercaptoethanol to the spacer gel, sample, and upper buffers. Preelectrophoresis has also been used to remove persulfate and other impurities, but the discontinuous nature of the system is destroyed, resulting in a continuous electrophoretic system, which may or may not have the required resolution.

A number of protein-dissociating agents can be incorporated into the gel formulations, e.g., 0.5% sodium dodecyl sulfate (see below), deoxycholate, Triton X-100, and urea.

Electrode buffer. Electrode buffer consists of 0.6 g of Tris, 2.88 g of glycine, and 2 ml of 0.001% filtered bromothymol blue (upper buffer only) (pH 8.3). A large number of general and special systems have been reported, which may be useful (see reference 93 for a procedure and table showing the composition required to obtain any desired level of cross-linking).

The purity of acrylamide and BIS are important for obtaining good results, especially in UV scanning for proteins or nucleic acids (see below).

Sample. As nearly as possible, the sample should have the same buffer composition, pH, and ionic strength as the spacer gel. High ionic strength must be avoided. For a small number of bands, 5 to 50 μg of protein per gel is quite sufficient; a single band containing 1 μg of protein is normally visible after staining with Coomassie blue. Care must be taken that some of the sample is not lost to the electrode solution.

Analysis of gels

Direct UV scanning of gels. For both proteins and nucleic acids, the possibility of direct UV scanning in a spectrophotometric device is often overlooked. This method is rapid, is amenable to repetitive scans (for tubes) at intervals during electrophoresis to obtain electrophoretic mobility, and is free of damage that may result from expulsion of gels from tubes. The disadvantages are two- to threefold decreased sensitivity (for proteins measured at 280 nm) compared with Coomassie blue staining and a requirement for UV-transmitting tubes, pure reagents, and precise alignment in an optical system designed to work with a circular cross-section. Inexpensive UV-transmitting tubes are easily made by cutting ordinary Vycor tubing (Corning Glass Works) of inner diameter 5 mm to the required length with a diamond saw and candling and selecting tubes for uniform optical characteristics. The problem of optical alignment of cylindrical gels is basically the same for both stained and unstained gels. Some scanners are designed to function optimally with such gels.

Analysis after removal of gel. Removed gels can be optically scanned as above, with or without staining. Stained gels can be scanned or photographed (see Chapter 30 for details of photographing gels). Unstained gels also can be sectioned and material can be eluted for analysis, or sections can be solubilized for analysis, including determination of radioactivity. Remove the gel immediately, and stain to avoid diffusion of the bands. The removal process is difficult and may require practice to prevent scratching the gel. In one simple procedure, slowly insert a syringe needle about 8 cm long while simultaneously moving the needle around the inner circumference and discharging water or 50% glycerol between the wall and the gel. Advance the needle until the gel is free or can be extruded with pressure from a medicine dropper bulb filled with water. Perform the procedure over or in a tray of water.

Staining choices. Most analyses are done with prior staining. Visualization procedures are available for proteins, glycoproteins, nucleic acids, and carbohydrate polymers (106), and many formulations are given in reference 93. Amido black 10B, Procion brilliant blue RS, and Coomassie blue R250 are dyes commonly used for proteins, with Coomassie blue being the most sensitive (93). Recently, silver staining has become a prominent visualization method, with at least 100-fold greater sensitivity as well as the ability to produce patterns of color which can give further information.

To remove up to 0.1% sodium dodecyl sulfate (SDS), rinse the gels several times in 10% (wt/vol) trichloroacetic acid–33% (vol/vol) methanol in water before staining.

Staining with Coomassie blue (93). Prepare Coomassie blue stain reagent by dissolving 4 g of Coomassie brilliant blue R250 in 950 ml of 95% ethyl alcohol. Add 850 ml of distilled water and 200 ml of glacial acetic acid, and filter. Place the gel in the stain solution at 50 to 60°C for 1 min for 125-μm-thick gels, 10 min for 1-mm-thick gels, or 30 to 45 min for 3-mm-thick gels. Destain at 50 to 60°C in a solution made up of 210 ml of 95% ethanol, 80 ml of glacial acetic acid, and 510 ml distilled water. Replace the destain solution when blue to obtain a clear background. Enhance the color by placing the gel in 10% acetic acid at 55°C for several minutes. The band intensity is lost by storage in acids. The method gives linear densitometer readings between 0.1 and 10 μg of protein (serum albumin).

Silver-staining choices. Methods have been developed for staining proteins with silver from silver nitrate or as the silver diamine complex augmented with an aldehyde (93, 111). Many variations of these general strategies exist, but the basic chemistry of the process has not been established. Fixation of the proteins is required to capitalize on the sensitivity of the method, particularly in thin gel sections. Immersion in 20% trichloracetic acid for 10 min or more is often used.

Ionic silver staining. Treat the gel with 20% trichloracetic acid for 20 min. Wash it three times for 10 min each in 10% ethanol–5% acetic acid. Place it in an oxidizing solution of 0.0034 M potassium dichromate and 0.0032 M nitric acid for 5 min. Treat it with 0.012 M silver nitrate for 20 min and then with reducing solution composed of 0.28 M sodium carbonate plus 0.5 ml of 37% formalin; 15 to 30 min is required for a 1-mm-thick gel and longer for thicker gels. Change the solution at least twice to prevent buildup of silver on the surface of the gel.

Diamine silver staining. Fix the gel in 10% glutaraldehyde or ethanol–15% acetic acid for 45 min. Rinse it for 45 min in double-distilled water with agitation. Prepare a staining solution by combining 2.8 ml of concentrated NH_4OH, 42 ml of 0.36% NaOH, and 8 ml of 20% $AgNO_3$ and diluting to 200 ml. Add the stain to the gel and agitate for 1 min and then replace the stain and agitate for 15 min. Place the gel in a new container with distilled water for 2 min and then in another new container with 200 ml of fresh developing solution (0.05 g of citric acid–0.5 ml 37% formaldehyde per liter) until background appears. Wash for 1 h with distilled water. Destain if necessary in 25% Kodak Rapid Fixer A, and use 25% Kodak Hypo Clearing Agent as the destain stop.

Location of enzymes. An enzyme in the gel can be located by immersing the gel or gel sections in a catalytic reaction mixture in which conversion of the substrate leads to a colored product. After incubation for a few minutes, color develops in a disc where the diffusing reactants contact the enzyme. Alternatively, when reagents do not diffuse into the gel, the reaction produces a ring around the gel where the band of enzyme is exposed.

Recording results and quantitation. As a minimum, sketches of stained gels can be recorded in a notebook. More quantitative measurements are obtained from (i) photographs of stained gels followed by a densitometer scan of the negative (see Chapter 30); (ii) optical scanning of stained or unstained gels (at 280 nm) to give strip chart recordings of absorbance along the gel length; and (iii) sectioning, elution, and analysis of small (1-mm) cross-sections or liquefaction into a stream and segmentation into fractions.

Molecular weight determination of native proteins. The method of Hedrick and Smith (107) is widely used to determine molecular weights of oligomeric proteins because SDS-gel electrophoresis as described below would cause dissociation into randomly coiled monomers. The method for native proteins requires preparation of two plots. The first is a plot of the log of R_m (the log of the ratio of migration of a protein to that of the dye front) versus gel concentration for protein standards. Each standard is run at several gel concentrations, i.e., at different degrees of gel cross-linking (Fig. 11A). The second is a plot of the slopes from the first plot versus the molecular weight of each standard (Fig. 11B). Both plots are linear. Standards are chosen such that the plot covers the molecular weight range expected for

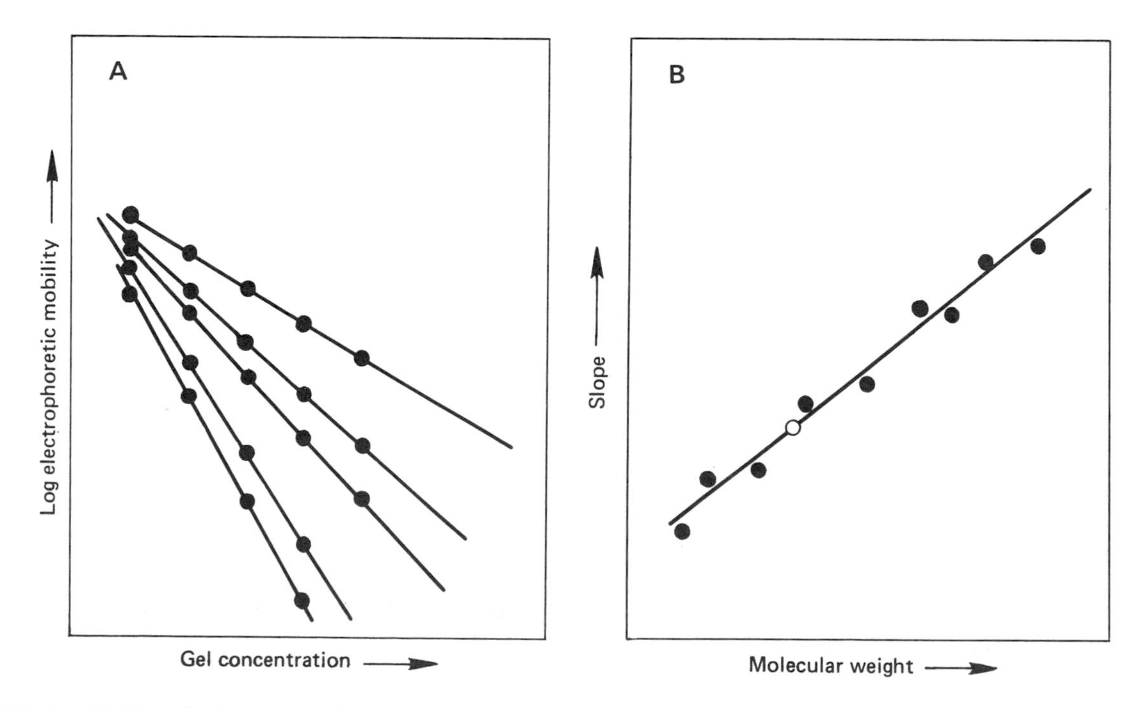

FIG. 11. (A) Plot of relative mobility (R_m) versus degree of cross-linking (gel concentration) for several reference proteins. (B) Plot of the slopes of lines in panel A versus the molecular weight of the reference proteins. The open circle represents the unknown protein.

the unknown. The slope of the unknown protein derived by plotting log R_m versus gel concentration can then be applied to the molecular weight plot to find the molecular weight of protein under study.

Procedure. Prepare and run the gels as described above for the general method with the following exception. Combine equal volumes of sample and diluent (below), and carefully layer the diluted sample between the spacer gel and buffer with a syringe. Following electrophoresis, mark the dye front by insertion of a 32-gauge copper wire and stain the gels as above. Calculate the R_m for each gel, and prepare a plot the log of R_m versus gel concentration (at least four concentrations) and a plot of the slopes of these plots for each protein and standard against the molecular weight of each (Fig. 11B). The molecular weight of the unknown is obtained by locating its log R_m value on the line.

Separation gels (pH 7.9)
- 5 parts A (Tris-HCl, 0.48 M based on Cl^-, and 0.46 ml of TEMED per 100 ml [pH 7.9])
- 10 parts B (24 g of acrylamide and 0.8 g of BIS per 100 ml) (gives 6% gel)
- 5 parts water

The gel concentrations needed for this work vary between 3 and 15%. Prepare these gels by diluting concentrated reagent B to obtain the desired concentration, i.e., the 6% level above. For example, a 3% gel contains 12 g of acrylamide and 0.4 g of BIS per 100 ml.

Spacer gel (pH 5.7)
- 2 parts C (imidazole-HCl, 0.48 M based on Cl^- [pH 5.7])
- 4 parts D (10 g of acrylamide and 2.5 g of BIS per 100 ml)
- 2 parts E (4 mg of riboflavin per 100 ml)
- 8 parts water

Diluent, pH 5.7
Imidazole-HCl, 0.06 M based on Cl^-, in 50% glycerol

Reservoir buffer
0.034 M asparagine plus 5×10^{-5}% bromothymol blue neutralized to pH 7.3 with Tris

Molecular weight of denatured monomeric proteins or subunits of oligomeric proteins. An effective way to minimize charge effects in electrophoresis of proteins so that migration is related only to size is to carry out the electrophoresis in SDS. SDS disrupts tertiary and secondary structures. The procedure of Weber and Osborn (115) is widely used with purified proteins as well as with membranes and ribosomes, in which case SDS also releases and solubilizes otherwise insoluble components. In this method, only one gel concentration is needed and a plot of relative mobility (distance of dye front/distance of protein band) versus the log of the molecular weight is linear over a fairly wide range of molecular weights. Therefore, a plot constructed with proteins of known molecular weight is used to find the molecular weight of the unknown. For a given gel concentration, e.g., 10%, the plot is linear over a range of molecular weights, e.g., 10,000 to 100,000 (104). With other gel concentrations, the range is either higher or lower.

Procedure. The separation gel containing SDS is poured and layered with water containing SDS, etc., as above for discontinuous gel electrophoresis. However, no spacer gel is used, and the sample is added directly to the gel surface.

Incubate protein solutions (0.2 to 0.6 mg/ml) at 37°C for 2 h in 0.01 M sodium phosphate buffer (pH 7.0)–1% SDS–1% β-mercaptoethanol. To obtain complete unfolding and dissociation, it may be necessary to boil the solution, add urea to 8 M, and/or reduce and carboxymethylate the thiol groups (103). In many cases the solution can be used directly, but if salts interfere, dialyze the sample for several hours at room temperature against 500 ml of 0.01 M sodium phosphate buffer (pH 7.0) containing 0.1% SDS and 0.1% β-mercaptoethanol.

For sample preparation, mix 3 μl of tracking dye (0.05% bromothymol blue), 1 drop of glycerol, 5 μl of mercaptoethanol, and 50 μl of reagent A (below) in a small tube. Add 10 to 50 μl of protein solution, mix, and apply to the surface of the gel. Carefully layer reagent A diluted 1:1 over the sample to fill the tubes. Fill the electrode compartments with the buffer of reagent A diluted 1:1. Use tubes measuring 10 cm by 6 mm (115), and perform the electrophoresis with a constant current of 8 mA per tube. The time required is 4 h. Remove, stain, and measure the gels as usual, with a 32-gauge copper wire to mark the dye front. Usually, 10 μg of protein is used, but less can be used if the stain is sensitive enough to detect the band. With 10 μg, the band has a sharp leading edge and a diffuse trailing edge. However, measurements of the leading edge give good results.

10% Separation gel
- 15 parts A (0.78 g of $NaH_2PO_4 \cdot H_2O$, 3.86 g of $Na_2HPO_4 \cdot 7H_2O$, and 0.2 g of SDS per 100 ml), deaerated
- 13.5 parts B (22.2 g of acrylamide and 0.6 g of BIS per 100 ml), filtered through Whatman no. 1 paper. Keep at 4°C in a dark bottle. For increased or decreased cross-linking, vary the amount of BIS accordingly. Deaerate after mixing A and B.
- 1.5 parts C (1.5 g of ammonium persulfate per 100 ml), freshly prepared
- 0.045 part TEMED

Molecular weight of nucleic acids. The R_m (electrophoretic mobility) of low-molecular-weight RNA (110) and rRNA (109) is inversely related to the sedimentation coefficient. Hence, the log molecular weight is inversely linear with mobility (113). The range of this proportionality depends on gel cross-linking.

Fractionation of low-molecular-weight oligodeoxynucleotides has been accomplished (105), but molecular weight measurement on high-molecular-weight DNAs did not become routine until the development of **pulsed-field gel electrophoresis.** Repeated reorienting of the direction of the applied electric field means that high-molecular-weight DNAs traverse the gel in compact bands whose migration distance can be related to those of standards of known molecular weight. See reference 93a for a practical guide and also Chapter 16.3.1.2.

Micromethod (capillary gel electrophoresis). Microgram to picogram amounts of protein can be separated on gel columns in capillary tubes of various sizes by the discontinuous-electrophoresis method. The procedure requires an apparatus adapted to small-scale filling, electrophoresing, and analyzing operations.

Preparative procedures. The above methods cannot be converted to preparative procedures of similar attributes simply by expanding the dimensions of the gel column. For increased capacity, a large series of small gels can be run, but this is laborious. With large gel columns, provision must be made for recovery of bands, dissipation of heat, and support of a large gel column. Commercial units in one way or another attempt to diminish the effect of these problems. In addition, retrieval of the desired component may require cutting out bands and elution from the gel by diffusion or electrophoresis. Alternatively, the bands are allowed to exit from the gel, where they are swept away to a fraction collector by a continuous flow of buffer. Such units require that the gel be maintained in the column by hydrostatic equilibrium between upper and lower buffers, that the temperature rise be controlled as much as possible by a circulating system, and that the volume of the elution chamber be small.

21.5.2. Immunoelectrophoresis

See reference 98 and Chapter 5.5.3.3 and 5.5.3.4.

21.5.3. Isoelectric Focusing (93, 96, 100)

Isoelectric focusing is a special kind of electrophoresis that involves migration of charged molecules in an electric field across a liquid column with a stable pH gradient. Amphoteric molecules such as proteins change charge as they traverse the pH gradient and come to a region where the pH is the same as the isoelectric point (pI) of the protein. At this point of equal plus and minus charges, the protein no longer migrates and the band is maintained in an equilibrium between the forces of diffusion and electrophoretic movement. Thus, separation is based on different pI values. It is necessary to stabilize the column against convection by use of either a sucrose density gradient, Sephadex, or polyacrylamide. The column is charged with a commercially produced material (Ampholine, Pharmalyte) that is a mixture of ampholytes of different pI values over a specified range. In an electric field, the various ampholyte molecules arrange themselves according to pI to produce a stable pH gradient of high conductivity and buffering capacity. A mixture of proteins then introduced separates into zones at the pI values of the components of the mixture. The procedure can separate components that differ by 0.005 pH unit or less. Ampholytes are available over wide and narrow pH ranges. Both microanalytical (102, 114) and preparative (93, 100) methods have been developed.

It is necessary to provide efficient cooling to prevent the accumulation of "Joule heat" produced by the voltage gradient. This has led to the universal use of slab gels and has resulted in the development of methods for preparing and running of ultrathin gels as well as the development of improved power supplies, special cooling devices, and complex protocols for voltage application. For preparative isoelectric focusing, a flat plate apparatus for polyacrylamide gel electrophoresis is preferred for heat dissipation and granulated beads of Bio-Gel P-60 or Sephedex G-200 superfine are used as an anticonvectant.

Specialized aspects are beyond the scope of general methods covered by this volume. References 93 and 100 are excellent sources concerning these specialized methods.

21.5.4. Two-Dimensional Electrophoresis

A single band on disc gel electrophoresis can contain more than one protein, and incomplete separation of complex mixtures can occur. The most effective approach to achieve separation is to base the separation on two strategies, size and pI. Thus, a sample can first be subjected to isoelectric focusing in polyacrylamide. Then a narrow slice with partially separated components is excised from the slab and laid along the top edge of a second slab gel in the position where the normal slots would be and is held in place by molten agarose. The second stage, SDS-gel electrophoresis, is carried out, and the slab is stained. A particularly effective example of the original method is described by O'Farrell (112), who separated 1,000 proteins from an *Escherichia coli* extract. Many improvements have been made since the mid-1970s, and the procedure has of necessity become very complex. However, two-dimensional methods offer the best possibility for separation and visualization of the protein complement of cells. Completely instrumented and computer-driven data acquisition systems are now commercially available. For more complete information, consult references 93, 95, and 101.

21.6. MANOMETRY

Volumetric measurement of gas exchanges has historically been a major tool in metabolic research. For many applications it still is the method of choice, and at times it is the only method available. However, the complexity of the system, together with the advent of other methods such as ion-specific electrodes and GC, has greatly diminished the popularity of the manometric method, until it now threatens to become a lost art. Certainly, it is difficult to teach manometry in formal laboratories with the insufficient time available. Therefore, the investigator becomes proficient with manometry only in a research setting.

Because manometric methods can measure oxygen, hydrogen, CO_2, methane, and hydrogen ion production or utilization (all essentially continuously), the method is directly applicable to a wide spectrum of uses with tissue homogenates, tissue cells, bacterial suspensions, and enzyme preparations. The spectrum of uses is further broadened by indirect methods, such as assays for kinases based on hydrogen ion production.

The main source of information in this field, and indeed a pioneering reference on many aspects of metabolism, is *Manometric Techniques* by Umbreit et al. (116).

21.7. ENZYME PURIFICATION

Investigation of the function of enzymes and other biologically active proteins often requires purification to remove the influence of extraneous factors and facilitate measurements. The purification process involves the application of a series of treatments which differentially separate components to purify and concentrate the target protein. Usually, preservation of biological activity is required, in which case an acceptable proportion of the original activity (units) must be recovered and the specific activity of the purified fraction (units per milligram of protein) should be greatly increased as well.

A detailed presentation of many aspects of this field is beyond the scope of this volume. The investigator is referred to specialized books (117, 119), as well as to the series *Methods in Enzymology* (118), for comprehensive treatment of and procedures for specific enzymes.

In most cases, one or more purification procedures will have been published for a protein of interest. Therefore, it should be a simple matter to follow the procedure. The general experience has been, however, that initial attempts to follow a procedure, even by experienced workers, often give disappointing results. There are many reasons for this, including an inability to closely follow the directions, the omission of details not recognized as important by the author(s), and variation in chemicals, apparatus, and starting material. Further, performing the procedure "blind," that is, without monitoring for biological activity and protein content, almost always results in failure. Nevertheless, careful work accompanied by appropriate assays will usually yield the desired material.

The design and testing of a new purification procedure is a challenging but rewarding venture. There are numerous individual operations that can be applied to purification of the target protein; some of these are described in this and other chapters. A primary but often unrecognized purification strategy can be effected before the extract is prepared, by simply enhancing the activity present in the cells. Understanding growth conditions for maximizing enzyme production, including enzyme induction and genetic manipulation for increased expression and secretion, has the effect of raising the starting specific activity and hence amounts to a purification step. A second strategy involves assembling a series of fractionation steps that individually improve purity and preserve activity and that work well together.

To produce an efficient procedure, the operational parameters (pH, temperature, salt concentration, etc.) of each step must be optimized, usually on the basis of a series of pilot tests, as follows. Select a separation process, and apply it to a series of small aliquots of the preparation to be purified while varying the level of one parameter to be optimized (e.g., the quantity of adsorbant). Following separation (often into a series of fractions), assay for activity and protein. Construct graphs of activity recovered and specific activity versus levels of the parameter being studied to establish the optimum for that parameter. Repeat this procedure for each parameter under study. When this is finished, run a second pilot test using the optimal value established for each parameter. If there is purification (i.e., a satisfactory increase in specific activity) and good recovery of the total units treated, process a larger amount of the preparation by using the best conditions found for that step to obtain material for pilot tests performed by a different separation method. Proceed in this fashion with a multiplicity of steps until a satisfactory level of purification is achieved and there is sufficient activity remaining for the experiments planned. This exercise consumes considerable material and may require starting over several times. In the repetitious process, iteration on a better procedure takes place.

It is useful to develop an accounting spreadsheet in which several vertical columns record important information as to volumes, assay results, units per milliliter, total units, protein content, specific activity, and purification factor. Horizontal rows are assigned to different steps in the process. From this assembly of data, it is easy to monitor success or failure as the procedure is applied.

Steps must be taken to protect the activity of enzymes, especially in crude extracts. Losses are caused by proteolytic and other enzyme activities, by oxidation of sulfhydryl groups, and by dissociation of coenzyme and metal cofactors. At a minimum, the preparation of crude extracts and manipulation of various fractions should be performed in an ice bath. In addition, proteinase inhibitors, sulfhydryl-reducing agents, and coenzyme/cofactor and other treatments may help stabilize activity.

In developing a procedure, a twofold improvement or less may be considered sufficient when starting with a crude extract, because of the large volume involved and the large amount of protein and other solubles present. However, a greater purification must be achieved in subsequent steps, otherwise too many steps (with unacceptably high losses in activity) would be needed to reach reasonable purity. For many steps a 5- to 10-fold purification or more is to be expected. Similarly, up to a 50% loss in total activity may be acceptable in the first step but not in the following steps.

This section describes a few of the popular and traditional procedures that were not covered earlier in this chapter. Many types of chromatography, treated above in a general way, adequately describe application to enzyme purification and will only be referenced here. Thus, ion exchange (21.3.1), adsorption (21.3.2), reversed-phase and hydrophobic (21.3.3), high-performance (21.3.6), affinity (21.3.8), and gel permeation (21.3.9) chromatographic methods have been successfully used in enzyme purification. The following additional techniques are specific to enzyme purification.

21.7.1. Preparation of Cell Extract

Preparation of a cell extract, which must precede the fractionation operations, is essentially a purification step. The goal is to achieve complete breakage with minimal loss in activity. Following centrifugation to recover the extract, considerable inactive material is removed. For a discussion of the various breakage methods, see Chapter 23.2.3.

21.7.2. Removal of Nucleic Acids

In general, bacterial extracts differ from eucaryotic extracts in the high content of nucleic acids (mostly DNA), which interfere with the differential separation usually obtained with many methods of purification.

Therefore, the first step is to dissociate nucleic acid-protein complexes and precipitate nucleic acids. This situation holds true whether the DNA is highly polymerized, fragmented, or hydrolyzed.

Two different strategies are used, depending on whether there is substantial polysaccharide in the crude extract. In the absence of polysaccharides, add solid ammonium sulfate to a final concentration of 0.2 M and then add protamine sulfate (2%, pH 5) to 20% of the crude extract volume, and centrifuge. When polysaccharide is present, totally precipitate the protein with ammonium sulfate (2.8 to 3.2 M) or with about 1.3 volumes of acetone (section 21.7.3). Recover the precipitate by centrifugation, and dissolve the protein in a buffer (pH 6 to 7). Add protamine sulfate as above, and centrifuge to remove the protamine-DNA complex.

A rough evaluation of the success of this step is made by determining the ratio of A_{280}/A_{260}. The ratio of nucleic acid to protein can then be determined (124). The ratio for a crude bacterial extract is usually about 0.5. Following treatment with protamine, the value usually increases to an unimpressive 0.6 to 0.8, and hence the success of the step is in doubt. However, the general experience is that in subsequent steps, the ratio continues to rise and approaches the value for pure protein (ca. 1.75).

21.7.3. Fractionation by Precipitation

Proteins are precipitated either by high salt concentrations, by solvents, or by approaching the isoelectric point of the target protein. At best, fractionation by precipitation gives a modest purification increase as a result of coprecipitation of other proteins. However, such a step early in the procedure may be advantageous as a means of handling a large volume and removing a large amount of nonproteinaceous material. Pilot tests are used to determine the cuts to make. Each fraction is recovered by centrifugation and dissolved in buffer at a higher protein concentration.

Ammonium sulfate is commonly used for salt fractionation. Enzyme grade salt (crystallized from EDTA) is powdered in a mortar and added slowly with stirring to the desired concentration. Alternatively, a saturated solution (with or without pH adjustment) may be used. In the past "percent saturation" has been used to express the concentration of salt involved. This has proven to be a crude and unreliable term, largely because the concentration at saturation varies, especially with temperature. This term has been largely supplanted by the use of molarity. Consult reference 125 for amounts and volumes to use to achieve desired ammonium sulfate concentrations. The protein precipitates are removed by centrifugation and dissolved in buffer.

Additional precautions are necessary when using acetone or ethanol for purification. Close temperature control is required because there is a substantial temperature rise in diluting acetone or ethanol. In addition, proteins may denature unless the temperature is maintained near 0°C or below.

A general solvent precipitation procedure is as follows. Dilute the acetone or ethanol to 90%, and chill to −20°C. Set up a stirrer, a thermometer, and a separatory funnel to serve the contents of a beaker in a salt-ice bath. Arrange the funnel so that the solvent is introduced to the wall of the beaker. Add the preparation to the beaker, and start rapid stirring. As the temperature approaches 0°C, start the slow addition of cold solvent. The temperature should continue to drop as solvent is added, but the solution should not freeze. When solvent addition is complete, rapidly centrifuge the preparation in prechilled vessels at −10 to −15°C. Rapidly dissolve the precipitate in buffer, and, if necessary, remove insoluble material by centrifugation. Some solvent will remain in the recovered fraction, and this could lead to inactivation with time. If a subsequent fractionation step can follow immediately, then proceed. Otherwise, completely precipitate the protein with ammonium sulfate and dissolve the precipitate in buffer.

21.7.4. Fractionation by Denaturation

Some enzymes are remarkably stable under conditions that denature and precipitate most other proteins. If such a situation can be defined by pilot tests, the target protein may be substantially purified merely by removing the denatured protein. For this purpose, heat, acid, and solvents are commonly used.

Heat denaturation is pH dependent; therefore, run the pilot tests over a temperature range at several pH values. Prepare graphs to visualize the profiles obtained. Take care in scaling up from pilot studies. The rate of temperature rise and fall will be quite difficult for the larger volume. Therefore, place the preparation in a stainless-steel beaker, and immerse it in a water bath with a temperature substantially above the desired holding temperature. When the desired temperature is reached, transfer the beaker to a water bath at that temperature and hold for the specified period. Then rapidly cool the beaker in an ice bath with stirring. The time/temperature relationship should be held as closely as possible to that of the selected pilot condition.

For acid denaturation, the pI and the pH of maximal activity usually are not in the region of maximal stability. Hence, pilot tests are required to find the pH region for maximal stability. Either HCl or acetic acid is added, depending on the pH to be achieved.

A dramatic example of acid denaturation is seen in the purification of 2-keto-3-deoxy-6-phosphogluconate aldolase from *Pseudomonas putida* (121). This is carried out as follows. Add 4 N HCl to the crude extract at 1 to 4°C to give a final concentration of 0.2 N HCl. After 15 min, remove the copious precipitate by centrifugation and, without prior neutralization, perform an ammonium sulfate fractionation on the supernatant. A total of 70% of the activity is recovered, and a 10-fold purification is achieved in the acid step.

Solvent denaturation takes place at a higher temperature than solvent fractionation. Pilot studies should establish the pH, salt concentration, temperature, and solvent type and concentration for optimal results.

21.7.5. Purification by Chromatography

Separations by chromatography and gel filtration make use of charge/pH, hydrophobic/hydrophilic, ligand-binding, and size properties of proteins. Specialized supports and matrices which optimize the binding of macromolecules have been developed. Thus, ion-exchange purification of enzymes uses derivatized celluloses, dextrans, and agaroses which have better operating characteristics than do the corresponding resins. See section 21.3.1.2 for details. Similarly, media for adsorption and elution (section 21.3.2) in both batch and column mode and hydrophobic (section 21.3.3) and affinity (section 21.3.8) strategies are well developed and are especially suited to protein purification in standard laboratory equipment. HPLC (section 21.3.6), while developed as an analytical tool, has been adapted to larger-scale separations by using specifically developed equipment. Gel permeation chromatography (section 21.3.9.4), using media with highly controlled pore sizes, also provides useful separations.

21.7.6. Affinity Elution with Substrate

In a few but striking instances, sorbed enzyme can be specifically eluted from an ion exchanger with substrate or competitive inhibitors (and presumably also with effectors). Because of the specific binding of these ligands to the enzyme, this process is a form of affinity elution chromatography. In many cases the binding to substrate imparts an elution behavior best explained by the cancellation of charges on the enzyme by the substrate. In others, a change in teritiary structure in response to substrate binding may be involved. Therefore, the adsorption of positively charged enzymes to CM-cellulose (negatively charged) can be overcome by binding the substrate (negatively charged) to the enzyme, which effectively displaces the CM-cellulose and elutes the enzyme. See reference 120 for detailed information.

The following elution of fructose-1,6-bisphosphatase is a useful example (122). This phosphatase, with a pI of 8, is positively charged at pH 6. CM-cellulose, with a pK of 4, is negatively charged at pH 6. Treat CM-cellulose with 5 mM malonate buffer (pH 6), and add NaOH to maintain the pH at 6. Pour the exchanger into a column, and saturate the column with the phosphatase by addition of enzyme until 5 to 10% of the activity breaks through. The amounts of exchanger and enzyme preparation to use are determined by pilot tests (described above). Wash the column with malonate buffer until no more protein appears in the effluent. Elute the phosphatase by addition of 0.05 mM sodium fructose 1,6-bisphosphate in malonate buffer. A purification of 50- to 300-fold is attained depending on the purity of the starting material.

21.7.7. Dye Ligand Chromatography

A number of dyes bind to proteins, especially enzymes. When the dye is bound to a porous support such as dextran or agarose, the enzyme is selectively picked up from solution. Following a wash to remove free material, the enzyme is eluted, with usually substantial purification.

Originally, Cibachron Blue F3GA was found to bind a number of enzymes, especially those that had nucleotide phosphate coenzyme- or substrate-binding sites. Recently, this method has been highly developed (117, 120). A large number of dyes are available, and procedures for attachment to solid supports, selection of dye ligand for adsorption, and conditions for elution have been published.

The preferred strategy is to select a pair of dye-liganded supports such that one binds a substantial fraction of contaminating proteins but not the target enzyme and the second binds the target enzyme preferentially. A series of dye-supports are screened to find the best pair for the purpose. A comprehensive dye ligand kit is available from the Centre for Protein and Enzyme Technology, La Trobe University, Bundoora, Australia. Dye-liganded supports are also available from Amicon Co., Danvers, Mass. Elution may be nonspecifically effected by high salt (1 M NaCl), high pH, incorporation of solvent (ethylene glycol), or chaotropic agents. In addition, affinity elution, usually with better purification, can be effected with substrate or competitive inhibitor.

The following illustrates purification of 6-phosphogluconic dehydratase from *Zymomonas mobilis* (123). Prepare an extract of cells in 20 mM K-MES buffer (pH 6.5) containing 30 mM NaCl, 5 mM $MnCl_2$, 0.5 mM freshly prepared ferrous ammonium sulfate, and 10 mM β-mercaptoethanol. Pass the extract from 30 g of cells (150 ml, 8 to 10 mg of protein per ml) serially through a column (cross-section, 8 cm^2) containing 50 ml of Scarlet MX-G (Reactive Red 8) and an identical column containing Blue HE-G (Reactive Blue 187). Follow with 100 ml of the above buffer. The second column picks up the dehydratase. Next, pass buffer containing 20 mM Na_2SO_4 through the second column only until the protein content of the effluent is near zero. Carry out affinity elution of the dehydrase with MES buffer containing 20 mM α-glycerophosphate, which is a strong competitive inhibitor. Half the activity is recovered, with 43-fold purification.

21.7.8. Crystallization

Crystallization of enzymes can be a useful means of purification. However, crystallization from impure preparations is difficult, and hence this step is usually the last in a purification procedure. Considerable experimentation can be performed quite satisfactorily with noncrystalline but substantially purified material. In addition, crystallization has proven to be a dubious criterion of high purity of the protein. However, chemical and physical analyses of enzyme composition and structure require a purity level obtained by crystallization, and, of course, the crystals themselves are needed for X-ray crystallography.

Crystallization occurs when a gradual change in conditions causes a transition from the solution phase to the solid phase. When this occurs rapidly, disordered precipitation results, yielding amorphous material; when it occurs slowly, ordered deposition-yielding crystals are obtained. Dialysis is a preferred means of developing a time-dependent decrease or increase in salt or solvent (polyethylene glycol) concentration. In some cases, however, gradual addition of a salting-out agent is effective. Microdialysis/diffusion devices which use 100 μl or less of material and which provide for regulation of the rate of concentration increase or decrease have been developed (124).

Removal of water by diffusion from the protein solution to a concentrated solute (polyethylene glycol or sucrose) in closed cells is effective. A large number of tests can be run in readily available microtiter plates. Place the concentrated polyethylene glycol or sucrose solution in the bottom of the wells. Place a drop of purified enzyme solution on a coverslip. Invert and center the drop over the well. Seal the coverslip with Vaspar or Vaseline, and hold at constant temperature. Observe crystal formation through the coverslip by using a magnifying glass or microscope.

Protein concentration, pH, temperature, salt or solvent concentration, presence or absence of coenzymes, metal cofactors, substrates, substrate analogs, and inhibitors are parameters that can be varied to effect crystallization.

21.8. REFERENCES

21.8.1. Photometry

21.8.1.1. General references

1. **Bauman, R. P.** 1962. *Absorption Spectrophotometry.* John Wiley & Sons, Inc., New. York.
2. **Brewer, J. M., A. J. Pesce, and R. B. Ashworth.** 1974. *Experimental Techniques in Biochemistry.* Prentice-Hall, Inc., Englewood Cliffs, N.J.

21.8.1.2. Specific references

3. **Burke, R. W., and R. Mavrodineonu.** 1977. *Certification and use of Acidic Potassium Dichromate Solutions as an Ultraviolet Absorbance Standard.* National Bureau of Standards Special Publication 260-54. National Bureau of Standards, Washington, D.C.
4. **Lucas, D. H., and R. E. Blank.** 1977. Spectrophotometric standards in the clinical laboratory. *Am. Lab.* **9(11):**77–89.
5. **Slavin, W.** 1963. Stray light in ultraviolet, visible and near infrared spectrophotometry. *Anal. Chem.* **35:**561–566.
6. **Vanous, R. D.** 1978. Understanding nephelometric instrumentation. *Am. Lab.* **10(7):**67–79.
7. **West, M. A., and D. R. Kemp.** 1977. Practical standards for UV absorption and fluorescence spectrophotometry. *Am. Lab.* **9(3):**37–49.

21.8.2. Ion Electrodes

21.8.2.1. General references

8. **Buck, R. P.** 1978. Ion selective electrodes. *Anal. Chem.* **50:**17R–29R.
9. **Hitchman, M. L.** 1978. Measurement of dissolved oxygen. *Chem. Anal. Ser. Monogr. Anal. Chem.* **49:**1–255.
10. **Long, C.** 1961. *Biochemists Handbook,* p. 21. Van Nostrand Co., Princeton, N.J.
11. **Orion Research.** 1982. *Handbook of Electrode Technology.* Orion Research, Inc., Cambridge, Mass.
12. **Westcott, C. C.** 1978. Biomedical pH measurements. *Am. Lab.* **10(3):**131–145.
13. **Westcott, C. C.** 1978. Selection and care of pH electrodes. *Am. Lab.* **10(8):**71–73.

2.1.8.2.2. Specific references

14. **Estabrook, R. W.** 1967. Mitochondrial respiratory control and the polarographic measurement of ADP:O ratios. *Methods Enzymol.* **10:**41–47.
15. **LeBlanc, P. J., and J. F. Sliwinski.** 1973. Specific ion electrode analysis of ammonium nitrogen in wastewaters. *Am. Lab.* **5(7):**51–54.
16. **Rechnitz, G. A.** 1974. Clinical analysis with ion- and gas-sensing membrane electrodes. *Am. Lab.* **6(2):**13–21.
17. **Toth, K., and E. Pungor.** 1976. The problem of standards especially in the field of ion-selective and voltamprometric analysis. *Am. Lab.* **8(6):**9–18.

17a. **Weast, R. C. (ed.).** 1980. *CRC Handbook of Chemistry and Physics,* 60th ed. CRC Press, Inc., Boca Raton, Fla.

21.8.3. Chromatography

21.8.3.1. General references

18. **Brewer, J. M., A. J. Pesce, and R. B. Ashworth.** 1974. *Experimental Techniques in Biochemistry.* Prentice-Hall, Inc., Englewood Cliffs, N.J.
19. **Determann, H.** 1968. *Gel Chromatography.* Springer-Verlag, New York.
20. **Dunlap, R. B. (ed.).** 1974. *Immobilized Biochemical and Affinity Chromatography.* Plenum Publishing Corp., New York.
21. **Fischer, L.** 1980. *Gel Filtration Chromatography.* Elsevier/North Holland Biomedical Press, New York.
22. **Gelotte, B., and J. Porath.** 1967. *In* E. Heftman (ed.), *Chromatography,* 2nd ed., p. 343–369. Reinhold Publishing Corp., New York.
23. **Heftman, E.** 1975. *Chromatography.* Van Nostrand-Reinhold, New York.
24. **Jakoby, W. B., and M. Welchek (ed.).** 1974. Literature on affinity chromatography. *Methods Enzymol.* **34:**3–171.
25. **Khym, J. X.** 1974. *Ion Exchange Procedures in Chemistry and Biology: Theory, Equipment, Technology.* Prentice-Hall, Inc., Englewood Cliffs, N.J.
26. **Pharmacia Fine Chemicals.** 1976. *Sephadex. Gel Filtration in Theory and Practice.* Pharmacia Fine Chemicals AB, Uppsala, Sweden.
27. **Pharmacia Fine Chemicals.** 1976. *Affinity Chromatography: Principles and Methods.* Pharmacia Fine Chemicals AB, Uppsala, Sweden.
28. **Shalteil, S.** 1984. Hydrophobic chromatography. *Methods Enzymol.* **104:**69–96.
29. **Whatman Inc.** 1977. *Liquid Chromatography Guide.* Bulletin no. 123. Whatman Inc., Clifton, N.J.
30. **Wilchek, M., T. Miron, and J. Kohn.** 1984. Affinity chromatography. *Methods Enzymol.* **104:**3–55.

21.8.3.2. Specific references

31. **Andrea, J. M.** 1985. Measurement of protein-ligand interactions by gel chromatography. *Methods Enzymol.* **117:**346–354.
32. **Andrews, P. B.** 1964. Estimation of molecular weights of proteins by Sephadex gel filtration. *Biochem. J.* **91:**222–233.
33. **Asteriadis, G. T., M. A. Armbruster, and P. T. Gilham.** 1976. Separation of oligonucleotides, nucleotides and nucleosides on columns of polystyrene anion-exchangers with solvent systems containing ethanol. *Anal. Biochem.* **70:**64–74.
34. **Axen, R., J. Porath, and S. Ernback.** 1967. Chemical coupling of peptides and proteins to polysaccharides by means of cyanogen halides. *Nature* (London) **214:**1302–1304.
35. **Bechard, G., J.-G. Bisaillon, and R. Beaudet.** 1990. Degradation of phenol by a bacterial consortium under methanogenic conditions. *Can. J. Microbiol.* **36:**573–578.
36. **Benning, M.** 1988. Single ion chromatography. *Am Lab.* **20:**74–79.
37. **Betsch, W.** 1988. *Practical Capillary Chromatography.* Hewlett-Packard Co., Palo Alto, Calif.

38. **Blattner, F. R., and H. P. Erickson.** 1967. Rapid nucleotide separation by chromatography on cation-exchange columns. *Anal. Biochem.* **18:**220–227.
39. **Bobbie, R. J., and D. C. White.** 1980. Characterization of benthic microbial community structure by high resolution gas chromatography of fatty acid methyl esters. *Appl. Environ. Microbiol.* **39:**1212–1222.
40. **Cho, K.-S., L. Zhang, M. Harai, and M. Shoda.** 1991. Removal characteristics of hydrogen sulfide, and methanethiol by *Thiobacillus sp.* isolated from peat in biological deodorization. *J. Ferment. Bioeng.* **71:**44–49.
41. **Cowper, C. J., and A. J. DeRose.** 1983. *The Analysis of Gases by Chromatography.* Pergamon Press, Oxford.
42. **Cuatrecasas, P., and I. Parikh.** 1972. Adsorbants for affinity chromatography. Use of N-hydroxysuccinimide esters of agarose. *Biochemistry* **11:**2291–2299.
43. **Davis, L. C., T. J. Sokolovsky, and G. A. Radke.** 1985. Computer-controlled scanning gel chromatography. *Methods Enzymol.* **117:**116–142.
44. **DeTurck, J., D. Brooks, M. Pease, and D. Bobey.** 1987. Critical evaluation of computerized gas-liquid chromatography system for bacterial identification, p. 333. *Abstr. 87th Annu. Meet. Am. Soc. Microbiol.* 1987. American Society for Microbiology, Washington, D.C.
45. **Dworzanski, J. P., L. Berwald, and H. L. C. Meuzelaar.** 1990. Pyrolytic methylation-gas chromatography of whole bacterial cells for rapid profiling of cellular fatty acids. *Appl. Environ. Microbiol.* **56:**1717–1724.
46. **Engelhardt, H. B.** 1979. *High Performance Liquid Chromatography.* Springer-Verlag KG, Berlin.
47. **Fedorak, P. M., and D. W. S. Westlake.** 1981. Microbial degradation of aromatics and saturates in Prudhoe Bay crude oil as determined by glass capillary gas chromatography. *Can. J. Microbiol.* **27:**432–443.
48. **Freeman, R. R. (ed).** 1981. *High Resolution Gas Chromatography.* Hewlett-Packard Co., Palo Alto, Calif.
49. **Gorbunoff, M. J.** 1985. Protein chromatography on hydroxyapatite columns. *Methods Enzymol.* **117:**370–380.
50. **Harrison, K., W. C. Beckham, Jr., T. Yates, and C. D. Carr.** 1985. Rapid, cost effective ion chromatography. *Am. Lab.* **17:**114–121.
51. **Heckenberg, A. L., and P. R. Haddad.** 1984. Determination of inorganic ions at parts per billion levels using single column ion chromatography without preconcentration. *J. Chromatogr.* **299:**301–305.
52. **Hewlett Packard Co.** 1990. *GC/MS for the Chromatographer.* Hewlett Packard Co. Palo Alto. Calif.
53. **Hordijk, C. A., C. P. C. M. Hagenaars, and T. E. Cappenberg.** 1984. Analysis of sulfate at the mud-water interface of freshwater lake sediments using indirect photometric chromatography. *J. Microbiol. Methods* **2:**49–56.
54. **Hummel, J. P., and W. J. Dreyer.** 1962. Measurement of protein binding phenomena by gel filtration. *Biochim. Biophys. Acta* **63:**530–532.
55. **James, A. T., and A. J. P. Martin.** 1952. Gas-liquid partition chromatography: the separation and microestimation of volatile fatty acids from formic to dodecanoic acid. *Biochem. J.* **50:**679–690.
56. **Janak, J.** 1990. The role of adsorption in gas-liquid systems and its use for enrichment of trace amounts. *Chromatographia* **30:**489–492.
57. **Katayama-Fujimura, Y., Y. Komatsu, H. Kuraishi, and T. Kanerko.** 1984. Estimation of DNA base composition by high performance liquid chromatography of its nuclease P1 hydrolysate. *Agric. Biol. Chem.* **48:**3169–3172.
58. **Kido, H., A. Vita, and B. L. Horecker.** 1985. Ligand binding to proteins by equilibrium gel penetration. *Methods Enzymol.* **117:**342–346.
59. **Korthals, H. J., and C. L. M. Steenbergen.** 1985. Separation and quantification of pigments from natural phototrophic microbial populations. *FEMS Microbiol. Ecol.* **31:**177–185.
60. **Krishnan, N., and G. Krishna.** 1976. A simple and sensitive assay for guanylate cyclase. *Anal. Biochem.* **70:**18–31.

60a. **Laine, R. A., W. J. Esselman, and C. F. Sweeley.** 1972. Gas-liquid chromatography of carbohydrates. *Methods Enzymol.* **28:**159–178.
61. **LaNoue, K., W. J. Nicklas, and J. R. Williamson.** 1970. Control of citric acid cycle activity in rat heart mitochondria. *J. Biol. Chem.* **245:**102–111.
62. **Levin, O.** 1962. Column chromatography of proteins: calcium phosphate. *Methods Enzymol.* **5:**27–33.
63. **Linhardt, R. J., K. N. Gu, D. Loganathan, and S. R. Carter.** 1989. Analysis of glycosaminoglycan-derived oligosaccharides using reversed-phase ion-pairing and ion-exchange chromatography with suppressed conductivity detection. *Anal. Biochem.* **181:**288–296.
64. **Lucarelli, C., L. Radin, R. Corio, and C. Eftimiadi.** 1990. Applications of high-performance chromatography in bacteriology. *J. Chromatog.* **515:**415–434.
65. **Mackey, B. M., S. E. Parsons, C. A. Miles, and R. J. Owen.** 1988. The relationship between the base composition of bacterial DNA and its intracellular melting temperature as determined by differential scanning calorimetry. *J. Gen. Microbiol.* **134:**1185–1195.

65a. **Mancuso, C. A., P. D. Nichols, and D. C. White.** 1986. A method for the separation and characterization of archaebacterial signature ether lipids. *J. Lipid Res.* **27:**49–56.

65b. **Marmur, J., and P. Doty.** 1962. Determination of the base composition of deoxyribonucleic acid from its thermal denaturation temperature. *J. Mol. Biol.* **5:**109–118.
66. **Mesbah, M., U. Premachandran, and W. B. Whitman.** 1989. Precise measurement of the G+C content of deoxyribonucleic acid by high-performance liquid chromatography. *Int. J. Sys. Bacteriol.* **39:**159–167.
67. **Mesbah, M., and W. B. Whitman.** 1989. Measurement of deoxyguanosine/thymidine ratios in complex mixtures by high-performance liquid chromatography for determination of the mole percentage guanine + cytosine of DNA. *J. Chromatogr.* **479:**297–306.
68. **Miller, L., and T. Berger.** 1985. *Bacterial Identification by Gas Chromatography of Whole Cell Fatty Acids.* Application note 228-41. Hewlett-Packard Co., Palo Alto, Calif.
69. **Nelis, H. J. and A. P. De Leenheer.** 1989. Profiling and quantitation of bacterial carotenoids by liquid chromatography and photodiode array detection. *Appl. Environ. Microbiol.* **55:**3065–3071.
70. **Paterek, J. R.** 1991. Reduction dehalogenation by marine anaerobic bacteria associated with decaying red algae, p. 282. *Abstr. 91st Meet. Am. Soc. Microbiol.* 1991. American Society for Microbiology, Washington, D.C.
71. **Peterson, E. A., and H. A. Sober.** 1962. Column chromatography of proteins: substituted celluloses. *Methods Enzymol.* **5:**3–27.
72. **Peterson, E. A., and H. A. Sober.** 1959. Variable gradient device for chromatography. *Anal. Chem.* **31:**857–862.

72a. **Pisano, J. J.** 1972. Gas-liquid chromatography (GLC) of amino acid derivatives. *Methods Enzymol.* **25:**27–44.

72b. **Preston, J. F., and D. R. Boone.** 1973. Analytical determination of the buoyant density of DNA in acrylamide gels after preparative CsCl gradient centrifugation. *FEBS Lett.* **37:**321–324.
73. **Schmid, E. R.** 1990. Chromatography and mass spectrometry—an overview. *Chromatographia* **30:**573–576.
74. **Small, H., and T. E. Miller.** 1982. Indirect photometric chromatography. *Anal. Chem.* **54:**262–269.
75. **Sober, H. A., F. J. Gutter, M. M. Wykoff, and E. A. Peterson.** 1956. Fractionation of proteins. II. Fractionation of serum proteins on anion exchange cellulose. *J. Am. Chem. Soc.* **78:**756–763.
76. **Teunissen, M. J., S. A. E. Marras, H. J. M. Op den Camp, and G. D. Vogels.** 1989. Improved method for simultaneous determination of alcohols, volatile fatty acids, lactic acid or 2,3-butanediol in biological samples. *J. Microbiol. Methods.* **10:**247–254.
77. **Thrane, U.** 1990. Grouping Fusarium section Discolor isolates by statistical analysis of quantitative high perfor-

mance liquid chromatographic data on secondary metabolite production. *J. Microbiol. Methods* **12**:23–39.

78. **Tunlid, A., D. Ringelberg, T. J. Phelps, C. Low, and D. C. White.** 1989. Measurement of phospholipid fatty acids at picomolar concentrations in biofilms and deep subsurface sediments using gas chromatography and chemical ionization mass spectrometry. *J. Microbiol. Methods* **10**:139–153.
79. **Veys, A., W. Callewaert, E. Waelkens, and K. Van Den Abbeele.** 1989. Application of gas-liquid chromatography to the routine identification of nonfermenting gram-negative bacteria in clinical specimens. *J. Clin. Microbiol.* **27**:1538–1542.
80. **Whitaker, J. R.** 1963. Determination of molecular weights of proteins by gel filtration on Sephadex. *Anal. Chem.* **35**:1950–1953.
81. **Yacobi, Y. Z., W. Eckert, H. G. Truper, and T. Berman.** 1990. High performance liquid chromatography detection of phototrophic bacterial pigments in aquatic environments. *Microb. Ecol.* **19**:127–136.

21.8.4. Radioactivity

21.8.4.1. General references

82. **Brewer, J. M., A. J. Pesce, and R. B. Ainsworth.** 1974. *Experimental Techniques in Biochemistry.* Prentice-Hall, Inc., Englewood Cliffs, N.J.
Chapter 8 gives a good theoretical treatment of measurement of radioactivity.
83. **Horrocks, D. D., and C.-T. Peng.** 1971. Organic *Scintillators and Liquid Scintillation Counting.* Academic Press, Inc., New York.
84. **Oldham, K. G.** 1976. *Radiochemical Methods in Enzyme Assay.* Amersham Searle Co., Arlington Heights, Ill.

21.8.4.2. Specific references

85. **Anderson, L. E., and W. O. McClure.** 1973. An improved scintillation cocktail of high solubilizing power. *Anal. Biochem.* **51**:173–179.
86. **Anker, H. S.** 1970. Solubilization of acrylamide gel for electrophoresis. *FEBS Lett.* **7**:293.
87. **Bush, E. T.** 1963. General applicability of the channels ratio. *Anal. Chem.* **35**:1024–1029.
88. **Fricke, U.** 1975. Tritosol: a new scintillation cocktail based on Triton X-100. *Anal. Biochem.* **63**:555–558.
89. **Jeffay, H.** 1962. Oxidation techniques for preparation of liquid scintillation samples, p. 1–8. *Packard Technical Bulletin no. 10.* Packard Instrument Co., Inc., Rockville, Md.
90. **Jeffay, H., and J. Alvarez.** 1961. Liquid scintillation counting of carbon-14. *Anal. Chem.* **33**:612–615.
91. **Patterson, M. S., and R. C. Green.** 1965. Measurement of low energy β-emitters in aqueous solution by liquid scintillation counting of emulsions. *Anal. Chem.* **37**:854–857.
92. **Turner, J. C.** 1967. *Sample Preparation for Liquid Scintillation Counting.* The Radiochemical Centre, Amersham, England.

21.8.5. Gel Electrophoresis

21.8.5.1. General references

93. **Allen, R. C., C. A. Saravis, and H. R. Maurer.** 1984. *Gel Electrophoresis and Isoelectric Focusing of Proteins: Selected Techniques.* Walter de Gruyter, New York.
93a. **Birren, B., and E. Lai.** 1993. *Pulsed Field Gel Electrophoresis, a Practical Guide.* Academic Press, Inc., New York.
94. **Blackshear, P. J.** 1984. Systems for polyacrylamide gel electrophoresis. *Methods Enzymol.* **104**:237–255.
95. **Bravo, R.** 1984. *Two-Dimensional Gel Electrophoresis of Proteins.* Academic Press, Inc., New York.
96. **Catsimpoolas, N.** 1975. Isoelectric focusing: fundamental aspects. *Sep. Sci.* **10(1)**:55–76.
97. **Davis, M. B. J.** 1964. Disc electrophoresis. II. Method and application to human serum proteins. *Ann. N. Y. Acad. Sci.* **121**:404–427.
98. **Kochwa, S.** 1980. Electrophoretic and immunoelectrophoretic characterization of immunoglobulins, p. 121–134. *In* N. R. Rose and H. Friedman (ed.), *Manual of Clinical Immunology,* 2nd ed. American Society for Microbiology, Washington, D.C.
99. **Ornstein, L.** 1964. Disc electrophoresis. I. Background and theory. *Ann. N.Y. Acad. Sci.* **121**:321–403.
100. **Richetti, P.G.** 1984. *Isoelectric Focusing Theory, Methodology and Applications.* Elsevier Biochemical Press, New York.

21.8.5.2. Specific references

101. **Anderson, D., and C. Peterson.** 1981. High resolution electrophoresis of proteins in SDS polyacrylamide gels, p. 41. *In* R. C. Allen and P. Arnaud (ed.) *Electrophoresis.* Walter deGruyter, Berlin.
102. **Catsimpoolas, N.** 1968. Microisoelectric focusing in polyacrylamide columns. *Anal. Biochem.* **26**:480–482.
103. **Crestfield, A. M., S. Moore, and W. H. Stein.** 1973. Preparation and enzymatic hydrolysis of reduced and S-carboxymethylated protein. *J. Biol. Chem.* **238**:622–627.
104. **Dunker, A. K., and R. R. Rueckert.** 1969. Observations on molecular weight determinations on polyacrylamide gel. *J. Biol. Chem.* **244**:5074–5080.
105. **Elson, E., and T. M. Jovin.** 1969. Fractionation of oligonucleotides by polyacrylamide gel electrophoresis. *Anal. Biochem.* **27**:193–204.
106. **Grässlin, D., H. Weicker, and D. Z. Barwich.** 1970. Über die Glycoproteine, Plasmaproteine und Glykosaminoglykane in Normalurin. *Z. Klin. Chem. Klin. Biochem.* **8**:288–291.
107. **Hedrick, J. L., and A. J. Smith.** 1968. Size and charge isomer separation and estimation of molecular weights of proteins by disc gel electrophoresis. *Arch. Biochem. Biophys.* **126**:155–164.
108. **Hjerten, S., S. Jerstedt, and A. Tselius.** 1965. Some aspects of the use of "continuous" and "discontinuous" buffer systems in polyacrylamide gel electrophoresis. *Anal. Biochem.* **11**:219–223.
109. **Loening, U. E.** 1969. The determination of the molecular weights of ribonucleic acid by polyacrylamide gel electrophoresis. *Biochem. J.* **113**:131–138.
110. **McPhie, P., J. Hounsell, and W. B. Gratzer.** 1966. The specific cleavage of yeast ribosomal ribonucleic acid with nucleases. *Biochemistry* **5**:988–993.
111. **Merril, C. D., D. Goldman, S. A. Sedman, and M. H. Ebert.** 1981. Ultrasensitive stain for proteins in polyacrylamide gels shows regional variation in cerebralspinal fluid proteins. *Science* **211**:1437.
112. **O'Farrell, P.** 1975. HIgh resolution two-dimensional elecrophoresis of proteins. *J. Biol. Chem.* **250**:4007–4021.
113. **Peacock, A. C., and C. W. Dingman.** 1968. Molecular weight estimation and separation of ribonucleic acid by electrophoresis in agarose-acrylamide composite gels. *Biochemistry* **7**:668–674.
114. **Riley, R. F., and M. K. Coleman.** 1978. Isoelectric fractionation of proteins on a microscale in polyacrylamide and agarose matrices. *J. Lab. Clin. Med.* **72**:714–720.
115. **Weber, K., and M. Osborn.** 1969. Reliability of molecular weight determinations by dodecyl sulfate-polyacrylamide gel electrophoresis. *J. Biol. Chem.* **244**:4406–4412.

21.8.6. Manometry

21.8.6.1. General reference

116. **Umbreit, W. W., R. H. Burris, and J. F. Stauffer.** 1957. *Manometric Techniques.* Burgess Publishing Co., Minneapolis.

21.8.7. Enzyme Purification

21.8.7.1. General references

117. **Amicon Corp.** 1990. *Dye-Ligand Chromatography.* Amicon Corp, Danvers, Mass.
118. **Colowick, S. P., and N. O. Kaplan (ed.), and J. N. Abelson and M. I. Simon (ed.).** *Methods in Enzymology.* Academic Press, Inc., San Diego, Calif.
119. **Deutscher, M. P. (ed.).** 1990. Guide to protein purification. *Methods Enzymol.* **182:**1–894.
120. **Scopes, R. P.** 1987. *Protein Purification: Principles and Practice.* Springer-Verlag, New York.

21.8.7.2. Specific references

121. **Hammerstedt, R. H., H. Möhler, K. A. Decker, and W. A. Wood.** 1971. Structure of 2-keto-3-deoxy-6-phosphogluconate aldolase. I. Evidence for a three-subunit Molecule. *J. Biol. Chem.* **246:**2069–2074.
122. **Pogell, B. M.** 1966. Enzyme purification by specific elution procedures with substrate. *Methods Enzymol.* **9:**9–15.
123. **Scopes, R. K., and K. Griffith-Smith.** 1984. Use of differential dye-ligand chromatography with affinity elution for enzyme purification: 6-phosphogluconate dehydratase from *Zymomonas mobilis. Anal. Biochem.* **136:**530–534.
124. **Warburg, O., and W. Christian.** 1942. Isolierung und Kristallisation des Gärungsferments Enolase. *Biochem. Z.* **310:**384–421.
125. **Wood, W. I.** 1976. Tables for the preparation of ammonium sulfate solutions. *Anal. Biochem.* **73:**250–257.

Chapter 22

Chemical Analysis

LACY DANIELS, RICHARD S. HANSON, and JANE A. PHILLIPS

The methods selected for this chapter are most commonly used in research, but most are also suited to advanced undergraduate and graduate teaching laboratories. Many other methods and more detailed descriptions of those selected are presented in publications listed at the end of the chapter. Techniques which require costly instrumentation or significant training are not described in detail. Specialized methods (which vary with the nature of the samples, equipment, and expertise) are treated by providing references to other sources of information rather than by describing experimental details.

22.1. FRACTIONATION OF MAIN CHEMICAL COMPONENTS

The incorporation of radioactively labeled compounds into bacterial cells is used as a means of estimating the rates and amounts of synthesis as well as the distribution of the main small-molecule and macromolecule fractions of the cells (2). After the cells have been incubated with the labeled substrate and the various molecular components have been fractionated, the radioactivity in each fraction can be determined.

Macromolecules are precipitated by cold dilute solutions of trichloroacetic acid. The (denatured) macromolecules are separated from the small molecules by centrifugation. The supernatant fraction of small molecules contains the pooled metabolites of sugars, sugar derivatives, organic acids, amino acids, nucleotides, and coenzymes; oligonucleotides and basic proteins with fewer than 20 nucleotide or amino acid residues are also contained in this fraction (2). (Note that some coenzymes will be destroyed by acid extraction; see section 22.2 for alternate methods of coenzyme measurements.) The precipitate fraction of macromolecules is sequentially extracted with organic solvents for lipids, alkali for RNAs, and hot trichloroacetic acid for DNAs; the precipitate remaining after these extractions contains the cell proteins and peptidoglycan. A flow diagram of this fractionation procedure is shown in Fig. 1.

No single procedure gives complete separation of all of the chemical components and total recovery of a given type of macromolecule in a single fraction. The procedures described by Kennell (1) are presented below. They are modifications of the methods of Roberts et al. (2).

22.1.1. Reagents and Materials

Saline solution: 0.85% (wt/vol) NaCl in distilled water.
Trichloroacetic acid solutions (*CAUTION*: See section 22.10.29): 50, 20, 10, and 5% (wt/vol) trichloroacetic acid in distilled water.
Ethanol (*CAUTION*: See section 22.10.8): 70% (vol/vol) in distilled water.
Diethyl ether (*CAUTION*: See section 22.10.7):
Sodium hydroxide (0.5 N) (*CAUTION*: See section 22.10.26). Dissolve 2.0 g in 100 ml of distilled water.
Bovine serum albumin. Dissolve 0.1 g in 10 ml of distilled water. Store in a refrigerator.
Scintillation fluid (available from a variety of vendors).
Glassware: 20-ml Pyrex glass beakers, 10-ml Erlenmeyer flasks, 25-ml volumetric flasks, capped 15- and 30-ml centrifuge tubes.
Filters: fiberglass or synthetic membrane filters (pore size, 0.45 μm), usually 25 mm in diameter. The filters must be acid and solvent resistant. Pyrex filter holders with fritted-glass filter supports.
Vacuum filtering flask: 125-ml flask and test tubes cut to fit below the filter holder inside the flask. The test tubes are used to collect filtrates.
Heating lamp: 250-W infrared heat lamp, and a suitable reflector for sample drying; alternatively, an oven heated at 50 to 80°C is suitable but should not be used with significant amounts (>1 ml) of ether.
Ice-water bath.
Water baths, adjustable to 80°C.
Refrigerated centrifuge, capable of achieving 10,000 × *g* with a centrifuge head that holds 30-ml centrifuge tubes.
Scintillation counter.
Vortex mixer.

22.1.2. Radioactive Labeling

(*CAUTION*: See section 22.10.22.) The use of radioactive isotopes is strictly regulated. All personnel involved in their use must be familiar with the regulations (see also Chapter 21.4.1).

The culture volume to be used will vary with the organism and growth conditions. For example, 30 ml of a mid-exponential-phase culture of *Escherichia coli* is adequate to provide the 10 mg of cells required.

If the purpose of the experiment is to determine cell composition, a uniformly labeled substrate such as [^{14}C]glucose (5 μCi/30 ml of a culture containing 1% glucose as the only carbon source) is suitable (1 μCi = 37,000 Bq). The incorporation of an amino acid into protein can be measured by adding 2 to 5 μCi of a labeled amino acid to 30 ml of a culture growing on a mineral salts medium plus glucose. Ideally, an auxotrophic mutant that is unable to synthesize the amino acid should be used when incorporation into protein is measured. In this case, 10 μg of the unlabeled amino acid per ml and 2 to 5 μCi of the labeled amino acid should be added to the medium. It is important to use a nonmetabolizable amino acid as a labeled protein precursor. Leucine, lysine, phenylalanine, and tyrosine are not significantly metabolized by most bacteria. When labeled thymidine is used as a precursor of DNA, and when uracil or uridine is used as a precursor of RNA, auxotrophs should also be used to obtain incorporation of a larger fraction of the label added to the medium.

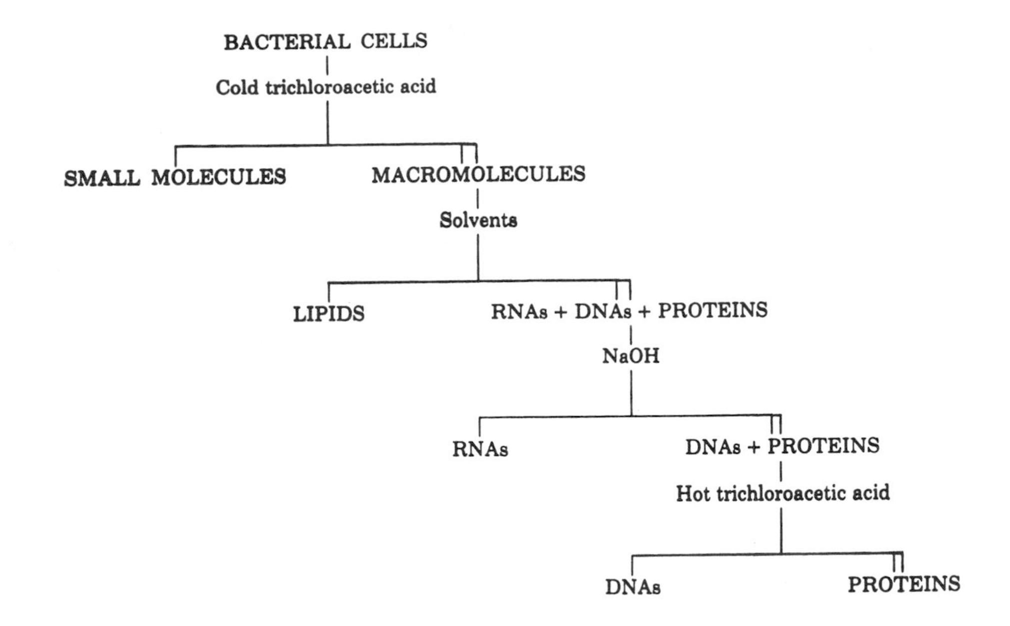

FIG. 1. Flow diagram for fractionation of bacterial cells into their main chemical components. Soluble material is removed by centrifugation or filtration in the fractions on the left, and precipitates are removed on the right.

The cultures (inoculated with healthy, mid-exponential-phase cultures) should be harvested during exponential growth when cell composition studies are performed. The cell composition changes during the lag and stationary phases of growth.

22.1.3. Small Molecules

1. Cool the culture rapidly by pouring it over ice. Centrifuge the radioactively labeled cells (approximately 10 mg [dry weight]) from the growth medium (5,000 × g for 10 min at 5°C), and wash the sedimented cells twice by centrifugation with ice-cold (5°C) 0.85% saline (one-fifth of the culture volume).

2. Suspend the cells with 10 ml of distilled water in the centrifuge tube. Place the suspension in an ice bath until cold, and add 10 ml of ice-cold 20% trichloroacetic acid. Cap the centrifuge tube, and mix the contents by inversion. Place the tube in the ice bath for 5 min to allow complete precipitation.

3. Centrifuge (5,000 to 10,000 × g for 15 min) at 0 to 4°C. Carefully decant the supernatant fraction, without disturbing the precipitate, into a 25-ml volumetric flask.

4. Add 4 ml of ice-cold 10% trichloroacetic acid to the precipitate, suspend it, and centrifuge again. Decant the supernatant fraction into that obtained after the first centrifugation. Adjust the volume of the combined supernatant fractions to 25 ml with distilled water.

5. Determine the radioactivity of a sample of the combined supernatant fraction containing the small molecules (metabolite pools) directly in a solution used for determining radioactivity in aqueous samples, as described in the preceding chapter (see Chapter 21.4). Internal or external standards should be used to correct for quenching by trichloroacetic acid. Alternatively, trichloroacetic acid can be removed by three extractions with equal volumes of ether. However, ether extraction will remove organic acids from the small-molecule fraction. Samples of the ether-extracted portion can be counted separately in a counting solution used for aqueous solvents or can be dried in vials with the assistance of a heat lamp; large volumes (>1 ml) of ether solutions should not be dried in an oven because of the explosion hazard, but it is possible to evaporate most of the ether at room temperature, preferably in a fume hood, and then place the container in the oven. After drying, the residue can be counted in a suitable scintillation cocktail (see Chapter 21.4 and section 22.2).

22.1.4. Lipids

1. Separate lipids from the precipitated fraction of macromolecules in the centrifuge tube (see above). Suspend the sediment in 10 ml of ice-cold 10% trichloroacetic acid with the aid of a vortex mixer. (*CAUTION*: Cap the tube to avoid aerosols of trichloroacetic acid and radioisotopes.)

2. Filter 1 ml of the homogeneous suspension, and wash the filter twice with 2 ml of ice-cold 10% trichloroacetic acid. Wash the filter with 3 ml of ice-cold 70% ethanol to remove residual trichloroacetic acid, which may cause solubilization of some protein in the next step. Discard the filtrate.

3. Place the filter holder (with the filter and suction flask) and a test tube for collecting the filtrate in an incubator at 45°C, and allow the assembly to warm to temperature. Then remove the filter flask from the incubator, quickly add 10 ml of 70% ethanol (warmed in a water bath to 45°C), and allow the solvent to pass slowly through the filter. Collect the filtrate in a test tube. The filtration should take approximately 10 min to effect the extraction.

4. Wash the filter in a hood with 2- to 5-ml volumes of ethanol-diethyl ether (1:1, vol/vol) warmed to 45°C in a

water bath by allowing the solvents to pass slowly through the filter (10 min). Collect the filtrate in a test tube, and combine it with the 45°C ethanol wash. Retain the filter containing the residual precipitated fraction of macromolecules for further fractionation.

5. Adjust the volume of the combined filtrates to 25 ml, and dry a sample in a scintillation vial under a stream of filtered air or nitrogen at 40°C. Gas filtration (to remove oil and particulates) can be accomplished by passing compressed gas through glass wool or an in-line 0.45-μm bacterial filter.

6. Determine the radioactivity of the dried lipid sample by use of a suitable scintillation cocktail (see Chapter 21.4).

22.1.5. RNA

1. Separate RNA from the precipitated fraction of macromolecules, with the lipids now removed. Remove the filter containing the precipitate from the filter holder (step 4 above), and place upside down in a 20-ml beaker. Allow the filter to air dry in a hood (about 10 min).

2. Add 2 ml of 0.5 N NaOH (warmed to 37°C) to the beaker, and incubate with occasional shaking at 37°C for precisely 40 min. Make sure the filter is completely covered with the NaOH solution. Longer incubation causes hydrolysis of some proteins (1, 3).

3. Transfer the beaker to an ice bath, and cool rapidly. Add 0.6 ml of ice-cold 50% trichloroacetic acid, mix, and leave the beaker in the ice bath for about 30 min.

4. Transfer the liquid in the beaker to a new filter held in a filter holder, apply a vacuum, and collect the acid-soluble products in a test tube. Rinse the beaker and original filter with 3 ml of ice-cold 5% trichloroacetic acid, transfer the liquid to the same filter, and collect the filtrate in the same test tube. Adjust the volume of the combined filtrates to 15 ml in a volumetric flask with distilled water. The filtrate contains about 95% of the ribonucleotides in RNA.

5. Determine the radioactivity of the hydrolyzed RNA fraction in the same way as for the small-molecule fraction (see above and Chapter 21.4).

22.1.6. DNA

1. After removing lipids and RNA from the macromolecular fraction, separate the DNA from the precipitated fraction of macromolecules by treatment with hot trichloroacetic acid. Use a duplicate sample from the suspended sediment in cold trichloroacetic acid, as described in the first steps of the lipid fractionation (section 22.1.4). Repeat steps 1 through 4 for removal of lipids. After washing the filter with ethanol and ethanol-ether to remove lipids, place the filter in a beaker, add 2 ml of 0.5 N NaOH, mix, and incubate for 90 min at 37°C. The incubation time with NaOH is increased to give total hydrolysis of RNA. DNA is present in smaller amounts than RNA, and traces of unhydrolyzed RNA can significantly increase estimates of radioactivity in DNA. The solubilization of some proteins will not affect the estimations of DNA.

2. Cool the sample on ice, add 0.6 ml of cold 50% (wt/vol) trichloroacetic acid, and keep it on ice for 30 min; then pour the liquid in the beaker onto a new filter that has been presoaked in cold 5% trichloroacetic acid. Apply a vacuum to facilitate filtration. Wash the original filter twice with 3 ml of ice-cold 5% trichloroacetic acid, and discard the filtrate.

3. Place the two filters in a 10-ml Erlenmeyer flask, add 3 ml of 5% trichloroacetic acid, and heat for 30 min in a water bath at 80°C; this step hydrolyzes purines from the DNA. Make sure the filters are entirely immersed in the liquid.

4. Add 0.1 ml of a solution (1.0 mg/ml) of bovine serum albumin, mix rapidly, cool the flask for 30 min in an ice bath, and filter the entire contents of the Erlenmeyer flask through a new filter. The bovine serum albumin ensures precipitation of minute amounts of macromolecules that may otherwise remain in suspension. Use a slight vacuum to increase the filtration rate. Collect the filtrate in a test tube placed inside the vacuum flask. Wash the Erlenmeyer flask and filters twice with 3 ml of cold 5% trichloroacetic acid. Collect the filtrates in the same test tube. Adjust the volume of the combined filtrates to 15 ml with water in a volumetric flask. Discard the filters.

5. Determine the radioactivity of the hydrolyzed DNA fraction in the same way as for the small-molecule fraction (see above and Chapter 21.4).

22.1.7. Proteins

1. Separate proteins from the precipitated fraction of macromolecules. Use a duplicate sample from the suspended sediment in cold trichloroacetic acid, as described in the first step of the lipid fractionation (section 22.1.4). Transfer 1 ml into a test tube (15 by 100 mm). Dilute with 1 ml of water to give a trichloroacetic acid concentration of 5% (wt/vol).

2. Heat the diluted sample for 30 min at 80°C to hydrolyze both RNA and DNA.

3. Cool the hydrolyzed sample on ice for 30 min to precipitate proteins. Collect the precipitate on a filter, and discard the filtrate. Wash the test tube twice with 2 ml of ice-cold 10% trichloroacetic acid, and apply both washes to a filter; wash the filter once with 3 ml of ice-cold 70% ethanol to remove residual trichloroacetic acid that would otherwise solubilize some protein in the next step.

4. Wash the filter twice with 5 ml of 70% ethanol (45°C), twice with 5.0 ml of ethanol-diethyl ether (1:1) at 45°C, and once with 5 ml of diethyl ether.

5. Air dry the filter, place it in a scintillation vial, and dry it under a heat lamp.

6. Determine the radioactivity of the dried protein by use of a suitable scintillation cocktail (see Chapter 21.4).

22.1.8. Applications

The techniques described above are usually applied to the fractionation of bacterial cells labeled with a medium component that is nonspecifically incorporated into all of the cell molecules. When bacterial cells are incubated in the presence of a specific precursor of a macromolecule to measure its rate of synthesis, it is important to determine that the macromolecule of interest contains all of the radioactivity.

This can be determined by fractionating the cells and estimating the radioactivity of each fraction by the techniques described above. If all the radioactivity occurs in the fraction of interest, subsequent experiments can be simplified. For example, if radioactive thymidine is used to measure the amount of DNA synthesis and preliminary experiments indicate that it is not found in other fractions, the cells in their culture medium can be cooled and treated with ice-cold 10% trichloroacetic acid. The precipitate containing the DNA can then be collected on filters, washed with cold ethanol, and dried and the radioactivity can be counted. This procedure is suitable for measuring the amount of DNA synthesis. The same technique can be applied to measure the amount of protein or RNA synthesis if the precursor appears only in the one fraction. Other factors must be considered in calculating the rates of synthesis. A primary concern is the dilution of exogenously added substrates by unlabeled precursors in the cell pools. In kinetic experiments it is necessary to determine the rate of saturation of the pool by radioactive precursors.

Although the methods described above provide a means of obtaining an estimate of the distribution of carbon from a ^{14}C-labeled compound in bacteria, several cell components are not accounted for. Products of hydrolysis of teichoic acids and polyphosphates will appear in the RNA fraction, and peptidoglycan will be precipitated with protein.

Poly-β-hydroxybutyrate, if present, can be extracted with chloroform-methanol (2:1, vol/vol) at 60°C after the ether-ethanol washes (see also section 22.4.9). If this solvent is used, care should be taken to use filters that are stable in methanol and chloroform. (*CAUTION*: The extraction should be performed in a hood [see sections 22.10.6 and 22.10.13].)

Polysaccharides, if present, are partially extracted by cold trichloroacetic acid and can significantly affect estimates of the small molecules. When present in large amounts, polysaccharides present a variety of troublesome problems if nonspecific precursors of macromolecules such as radioactive glucose are used. When these problems occur, these procedures must be modified (3).

Some bacterial cells leak part of their pools of small molecules when chilled during centrifugation and washing. Some cells have high activities of nucleases or proteases that may be activated during harvesting and washing.

Significant losses of macromolecules occur when smaller amounts of material than indicated are used in the first steps. It is possible to add unlabeled bacterial cells as a carrier to facilitate the precipitation of macromolecules. Precipitation of DNA and protein can also be facilitated by addition of serum albumin to the solutions.

Each organism potentially presents a unique problem in fractionation. Although the procedures described above have been widely used, the data obtained should be interpreted with proper regard for the problems of obtaining total separation of the small molecules and macromolecules (3).

Sequential methods for treating a single sample have been described (2, 3). However, they are likely to result in greater cross-contamination of fractions than the procedure described here.

22.2. SMALL-MOLECULE EXTRACTION AND ANALYSIS

In some cases the small molecules of interest are unstable to acid treatment or require further fractionation or analysis for quantitation. Also, coenzyme analysis may be needed without the need for any of the other analyses described above for macromolecules. For example, coenzymes could be extracted under quite different conditions from those described above, from either labeled or unlabeled cells, and analyzed further. In some cases, anaerobic methods are required to maintain the stability of oxygen-sensitive components, e.g., by conducting extraction under nitrogen or argon gas as is done for tetrahydromethanopterin from methanogens (5). In other cases, some coenzymes (e.g., B_{12}) are sensitive to light, and procedures should be conducted under low-light conditions.

One type of more gentle extraction is the use of 70% aqueous ethanol or 50% aqueous acetone at 4°C to resuspend the original cell pellet (6, 7). (*CAUTION*: See sections 22.10.4 and 22.10.8.) The suspension is stirred at 4°C for 16 h and centrifuged, and the supernatant is saved; the pellet is then extracted one more time with fresh solvent. The combined supernatants contain virtually all the water-soluble, free coenzymes. In some cases this extract (diluted or evaporated to remove possibly inhibitory solvents in some cases) can be examined by using bioassay techniques, as described in Chapter 7.3.5. However, more commonly the supernatant can then be fractionated by chromatography on suitable columns, e.g., DEAE-Sephadex, or C-18 reverse-phase high-pressure liquid chromatography (HPLC). Coenzyme elution can be monitored by absorbance at an appropriate UV or visible wavelength; in some cases, if the material is sufficiently pure after one chromatographic step, the coenzyme can be quantitated by measuring the absorbance or the fluorescence of the pooled column fractions. Alternatively, the area under the HPLC chromatographic profile can be measured; care should be taken to ensure that the proper extinction coefficient is used for the pH and solvent in which the material is dissolved as it passes through the detector, since organic solvents and pH can greatly affect the spectra of some small molecules (see Chapter 21.1.1.4 and 21.1.3.6).

A good example of where either of these analysis methods can work is with the 5-deazaflavin known as F_{420}, found in methanogenic bacteria (6); acid exposure and heating cause loss of the coenzyme. A variety of other coenzymes can be quantitated by HPLC methods if retention times are known and if there are no interfering peaks; the use of a rapid spectral scan of HPLC peaks while the material passes through the detector is an excellent tool to examine the purity of peaks and thus confirm the validity of the quantitation and the purity of the material. Examples of HPLC methods for quantitation of ascorbic acid, thiamine, biotin, NADH, flavins, deazaflavins, folates, and methanopterin species have been described (4).

If labeled material has been used to feed the cells prior to coenzyme analysis, the specific radioactivity of the pure coenzymes can be determined by comparing the molar quantity of coenzyme and the radioactivity measured by scintillation counting. This approach may be useful in the study of precursors of coenzyme production.

In addition to coenzymes, other small molecules of interest, including amino acids and sugars, can be extracted by this general method and analyzed as described below (sections 22.3 and 22.7).

22.3. CARBOHYDRATES

Bacteria contain a large number of different carbohydrates: free sugars, sugar derivatives, simple polysaccharides (polymers of a single sugar or sugar derivative, such as glycogen), and complex polysaccharides (polymers of more than one different sugar, amino sugars, uronic acids, etc.). Some macromolecules contain both carbohydrate and noncarbohydrate components (lipopolysaccharides, peptidoglycans, teichoic acids, lipoteichoic acids, teichuronic acids, nucleic acids, glycoproteins, etc.).

The colorimetric methods based on the **Molisch test** for carbohydrates (8, 15, 16, 27) are useful for estimating the amount of simple sugars and their polymers. Specific assays for pentoses in nucleic acids (ribose and deoxyribose) are described below in the sections that deal with nucleic acids (sections 22.6.1 and 22.6.2). Specific enzymatic assay examples are given for glucose, galactose, lactose, and sucrose. More complex descriptions of analytical methods for the analysis of carbohydrates have been presented by Work (11), Herbert et al. (10), Colowick and Kaplan (8), and Ginsberg (9). Details of automated enzymatic methods can be obtained from the suppliers of enzyme kits or from instrument manufacturers.

22.3.1. Total Carbohydrates by Anthrone Reaction

The most convenient assays for total carbohydrates involve heating of media, cells, or isolated carbohydrates with sulfuric acid to hydrolyze the polysaccharides and dehydrate the monosaccharides to form furfural from pentoses and hydroxymethylfurfural from hexoses. The solutions of furfural and hydroxymethylfurfural are then treated with a reagent (an aromatic compound or phenol) to produce a colored compound, which is measured in a spectrophotometer. Many variations of this Molisch test have been described (8, 15, 16). Pentoses, hexoses, heptoses, and their derivatives (except the amino sugars) yield colored products, but trioses and tetroses do not.

None of the methods accurately measures all the carbohydrate in bacteria, because different sugars give different color intensities at any selected wavelength (10). The anthrone and the phenol methods (described below) are presented because of their simplicity, relative insensitivity to interference by other cell components, and similar response to most common hexoses in cells, thus providing a reliable index of "total carbohydrates."

Either the anthrone or the phenol procedure is suitable for estimation of total carbohydrate in bacteria that contain 10% or more hexose polymers. The pentoses in nucleic acid interfere with these essays when the hexose content is low. Two notable disadvantages of these methods are that individual sugars are not identified and that working with either concentrated sulfuric acid or phenol is dangerous.

a. Reagents and equipment

Saline solution: 0.85% (wt/vol) NaCl in distilled water.

Stock glucose solutions. Dissolve 50 mg of glucose in 50 ml of 0.15% (wt/vol) benzoic acid, which is added as a preservative. Store at 5°C. This solution is stable for several months. Dilute 1:10 in distilled water just before use to give a solution containing 100 μg/ml. Prepare standard solutions (10 to 100 μg/ml) from the diluted solution.

Sulfuric acid solution (75%, vol/vol). Add 150 ml of reagent-grade concentrated sulfuric acid to 50 ml of distilled water (*CAUTION*: Sulfuric acid is very dangerous [see section 22.10.28]). Also, avoid contact of paper with either of the sulfuric acid-containing reagents, since cellulose will be hydrolyzed to sugars, resulting in a high background; one way in which this contamination can occur is to wipe acid off a pipette tip with a paper towel and then to continue using the pipette. Also, the concentrated sulfuric acid should be of good quality, and free of excess water; excess water in the assay causes turbidity.

Anthrone reagent. Add 100 mg of anthrone to 2.5 ml of absolute ethanol, and make up to 50 ml with 75% H_2SO_4 (*CAUTION*: See sections 22.10.8 and 22.10.28). Stir until dissolved. Prepare fresh daily, and store in a refrigerator.

Thick-walled Pyrex boiling tubes.

Glass marbles to fit boiling tubes (to avoid water dropping in, and to avoid evaporation; a drop of water can cause turbidity).

Spectrophotometer or colorimeter.

Glass cuvettes or tubes.

Goggles.

b. Procedure

Wash samples of bacterial cells free from medium components by centrifugation at 5,000 × *g* for 10 min. Resuspend in saline solution, and centrifuge again. Resuspend the sedimented cells in distilled water, and pipette samples (0.06 to 0.6 ml) of the cell suspension into the boiling tubes. Adjust the volume of all samples to 0.6 ml with distilled water. Prepare a reagent blank of distilled water (0.6 ml) and a standard curve by adding 0.6 ml of standard solutions containing 10 to 100 μg of glucose per ml. Chill the tubes and the anthrone reagent in an ice-water bath until cold. Add 3.0 ml of cold anthrone reagent, mix rapidly by swirling the tubes in ice water, and continue mixing in the ice-water bath for 5 min. Then transfer the tubes to a bath of boiling water for precisely 10 min. Return the tubes to the ice-water. Measure the absorbance of each solution, A_{625} (section 22.1.1.4). Determine the concentration of sugars in the samples from a standard curve prepared by plotting the absorbances of the standards versus the concentration of glucose.

22.3.2. Total Carbohydrates by Phenol Reaction

a. Reagents and equipment

Phenol reagent. Dissolve 5 g of reagent-grade phenol in 100 ml of distilled water (*CAUTION*: See section 22.10.19).

Reagent-grade concentrated sulfuric acid (*CAUTION*: See section 22.10.28).
Glucose standard (see anthrone procedure above).
Thick-walled Pyrex boiling tubes (15 by 2.5 cm).
Spectrophotometer or colorimeter.
Glass cuvettes or tubes.
Goggles.

b. Procedure

Pipette the samples to be analyzed into thick-walled Pyrex tubes. Adjust the volume of all samples to 0.5 ml with distilled water. Prepare a reagent blank by adding 0.5 ml of distilled water, and prepare a standard curve by adding 0.5 ml of solutions containing 10 to 100 μg of glucose per ml. Add 0.5 ml of the phenol reagent; mix rapidly and thoroughly. Add 2.5 ml of concentrated sulfuric acid, mix rapidly, and let stand for 10 min. (*CAUTION*: Addition of sulfuric acid to aqueous solutions causes rapid heating and occasionally boiling; see also section 22.10.28.) Place the tubes in a water bath at 25°C for 15 min. Read the A_{488} of each tube against the blank prepared without glucose (see Chapter 21.1.1). Determine the concentration of sugars in the sample from a standard curve prepared by plotting the absorbances of standards versus the concentration of glucose. (See also section 22.3.1, on the anthrone reaction, for a discussion of the disadvantages of these two assays, and see below for more specific assays.)

22.3.3. D-Glucose, Galactose, Lactose, and Sucrose

A large number of monosaccharides are liberated by the hydrolysis of simple and complex polysaccharides in bacterial cells. Because the polymers exist as mixtures, except in homopolymers such as glycogen, the estimation of specific sugars sometimes presents difficult analytical problems that can be best solved with HPLC, ion chromatography, or gas-liquid chromatography procedures (9, 12, 32, 34, 40). However, some sugars can be estimated by specific colorimetric reactions or enzymatic methods (8, 18, 28, 36, 37); two good examples are the hexoses glucose and galactose, for which a variety of enzymatic methods are available. A specific example is the use of glucose oxidase to assay glucose, but a very similar system works for galactose.

Glucose oxidase catalyzes the following reaction:

$$\beta\text{-D-glucose} + O_2 \rightarrow \text{gluconic acid} + H_2O_2$$

The hydrogen peroxide, in the presence of peroxidase and a reduced chromogen, causes the production of a brown or blue oxidized chromogen that can be measured spectrophotometrically. Different chromogenic compounds are used in commercially available kits, which also contain the needed enzymes and sometimes glucose standards. The glucose oxidase assay is specific, sensitive, and easily performed. Only glucose and 2-deoxyglucose are known to produce the colored products.

Galactose oxidase assay kits are also available from commercial sources. The kits contain galactose oxidase, chromogen, and standards. Since galactose oxidase oxidizes the *C*-6-hydroxymethyl group of galactose to produce galacturonic acid, the enzyme will also oxidize some galactosides.

Commercial kits and/or the constitutent enzymes are available from many sources, but a few examples are Sigma Chemical Co., St. Louis, Mo.; Worthington Biochemical Co., Freehold, N.J.; and Boehringer Mannheim Biochemicals, Indianapolis, Ind.

A variation on these two assays has been developed for use in an automated system, in which hydrogen peroxide is detected electrochemically (25, 29, 30, 41, 44, 45). This method is particularly attractive in classroom settings, where a large number of students have to do assays in a limited period and with a higher degree of safety than is possible with the anthrone and phenol assays; it is also cost effective in an industrial setting, where many assays are done routinely.

Several other sugars can also be assayed by a system very similar to those above with an enzyme that generates hydrogen peroxide. Lactose and sucrose are two examples of additional sugars that can be determined with the automated analyzer from Yellow Springs Instrument Co. (YSI), Yellow Springs, Ohio (25, 44, 45). The same system can also estimate ethanol and lactate by using an enzymatic approach.

An example of the use of glucose oxidase in a manual spectroscopic system is as follows. Assay of galactose would be very similar.

a. Reagents

Glucose stock solution. Dissolve 300 mg of D-glucose in 100 ml of 0.15% (wt/vol) benzoic acid. Store in a refrigerator. Dilute 1:10 with distilled water, and adjust to pH 7.0 before use.
Commercial glucose oxidase-peroxidase preparation. Prepare as specified by the manufacturer.
4 M HCl (*CAUTION*: See section 22.10.10).
Chromogen: available as part of the commercial kits. Prepare as specified by the manufacturer.

b. Procedure

Dilute samples with distilled water so that they contain 50 to 300 μg of glucose per ml. Adjust the pH to between 6.8 and 7.2 with 0.1 M KOH or 0.1 M HCl. A pH indicator paper is adequate. Prepare a blank and a set of standards containing 50 to 300 μg of glucose in 1.0 ml distilled water. Add the enzyme and chromogen, and incubate the mixture as specified by the manufacturer. Add 1 drop of 4 M HCl with a Pasteur pipette, and read the A_{400} (section 22.1.1.4). Determine the concentration of glucose in the samples from a standard curve prepared by plotting the absorbances of the standards versus the concentration of glucose.

c. Applications

The oxidase tests for the sugars glucose, galactose, lactose, and sucrose are suitable for estimating levels of each in growth media, in neutralized hydrolysates of polysaccharides, and in enzyme reaction mixtures. For example, the enzymatic hydrolysis of cellulose can be monitored by measurement of glucose released, and the metabolism of sucrose in a fermentor containing molasses can be monitored by measurement of sucrose disappearance. A particularly attractive monitoring system is found in the YSI analyzer equipped for simultaneous measurement of both glucose consumption and either ethanol or lactate production due to metabolizing microorganisms, e.g., yeasts or lactic acid bacteria, respectively (44, 45).

One of the major attractions of enzymatic methods for sugar analyses are their relative specificity; in general, there are many fewer interferences than with chemical methods. However, caution is always a good practice in analysis, and appropriate controls should always be used; it is wise to add a standard to each type of unknown solution to check for interference by salt and other possible inhibitors of oxidases.

22.3.4. Glycogen

It is difficult to quantify any simple bacterially produced polysaccharide except glycogen by unsophisticated methods. The method for isolation of glycogen is presented because this polymer functions as a reserve material in many bacteria and can account for a significant fraction of the biomass in cultures. Methods for isolation of other commonly occurring complex polysaccharides (capsules, lipopolysaccharides, and peptidoglycan) are presented in separate sections. Glycogen, like other polysaccharides, is resistant to hydrolysis to alkali, but it is readily soluble in water and insoluble in ethanol. The quantitative isolation of the polymer takes advantage of these properties. After isolation, and after hydrolysis for 6 h in 4 N HCl at 100°C in a sealed ampoule, the amount of glucose can be estimated by the anthrone or phenol reaction or by enzymatic assay.

a. Reagents and equipment

Absolute ethanol (*CAUTION*: See section 22.10.8).

Potassium hydroxide (30% wt/vol) in distilled water (*CAUTION*: See section 22.10.21).

Refrigerated centrifuge, capable of achieving 10,000 × *g*.

30-ml glass centrifuge tubes.

Vacuum desiccator or lyophilizer.

b. Procedure

Harvest bacterial cells by centrifugation at 5,000 to 10,000 × *g* for 15 min, resuspend the cells in an equal volume of 0.85% (wt/vol) NaCl, and centrifuge again. Wash the cells by centrifugation once more, and then lyophilize (see Chapter 12.2.1). Treat 100 mg of dry cells in a Pyrex screw-cap test tube with a Teflon-lined cap with 1 ml of 30% KOH at 100°C in a steamer or a boiling-water bath for 3 h. If lyophilization is not practical, suspend 500 mg of packed wet cells in 0.6 ml of 50% (wt/vol) KOH and heat for 3 h at 100°C to solubilize the glycogen. Cool without opening the tube, and then add 3 ml of distilled water and 8 ml of ethanol to precipitate the glycogen. Centrifuge at 10,000 × *g* for 15 min, and wash the precipitate with 8 ml of 60% (vol/vol in distilled water) ice-cold ethanol by centrifugation. Dry the precipitate in a vacuum desiccator or lyophilize. The washed precipitate can be analyzed for glucose by the anthrone or phenol method (see section 22.3.1 or 22.3.2), or by enzymatic methods (see section 22.3.3).

Mechanical disruption of gram-positive cells is sometimes necessary to obtain complete extraction of glycogen (17). If cells have a polysaccharide capsule, the capsular material will contaminate the glycogen preparation since, in most cases, it has similar solubility properties in ethanol.

22.3.5. Starch

Although starch is not a common component of microbial cells, it is used as a substrate for a variety of bacteria and yeasts; it is therefore sometimes of value to know the level at which starch is present in a microbial culture. Soluble starch can be obtained free of cells by use of selective centrifugation, and some organic solvents can be of value in its precipitation and purification. It is easily hydrolyzed by acid and heat, or by using enzymes (e.g., amylglucosidase), and the resultant glucose can be estimated by any of the methods above in sections 22.3.1, 22.3.2, and 22.3.3. An automated enzymatic method is also available in the YSI analyzer discussed above (45). Insoluble crude starch is more problematic, because of the difficulty of purification separate from the growing cells.

22.3.6. Hexosamines in Purified Peptidoglycans

Hexosamines occur in lipopolysaccharides, peptidoglycans, and teichoic acids of bacteria (10, 11, 16, 17, 19, 21, 23). The most common hexosamines are glucosamine, galactosamine, and muramic acid. All N-acetylated hexosamines react with *p*-dimethylaminobenzaldehyde to produce a red product. The test is very sensitive and specific for hexosamines but does not distinguish among glucosamine, muramic acid, and galactosamine. Muramic acid can be estimated separately by other procedures (section 22.3.8).

The following procedure modified by Ghuysen et al. (21) is more precise and convenient than the Morgan-Elson procedure as originally described and is applicable to estimation of total hexosamines in purified peptidoglycans. However, estimation of hexosamines in complex polymers or bacterial cells requires separation of hexosamines from neutral sugars by ion-exchange chromatography, as described in the following method.

a. Reagents, equipment, and supplies

Morgan-Elson reagents. Dissolve 16 g of *p*-dimethylaminobenzaldehyde in sufficient acetic acid to give a volume of 95 ml. Add 5 ml of concentrated HCl. Dilute 1:5 with acetic acid to prepare the color reagent. (*CAUTION*: See sections 22.10.2 and 22.10.10.)

3 N HCl (*CAUTION*: See section 22.10.10).

3 N NaOH (*CAUTION*: See section 22.10.26).

Saturated solution of sodium bicarbonate: Add ca. 15 g of $NaHCO_3$ to 100 ml of distilled water, and stir at room temperature for 20 min.

5% (vol/vol) acetic anhydride in distilled water. Prepare just before use, and keep on ice until use. (*CAUTION*: See section 22.10.3).

5% Potassium tetraborate in distilled water.

Glucosamine standard. Dissolve 17.92 mg of glucosamine in 10 ml of 3 N HCl; store this stock solution in the refrigerator (for no longer than 1 week). This standard contains 0.2 μmol in 20 μl. Dilute with 3 N HCl to obtain standards containing 0.01 to 0.2 μmol/20 μl.

Ampoules: 1-ml lyophilization ampoules or 1-ml screw-cap test tubes with Teflon-lined caps.

Pyrex test tubes, 3 ml.

Spectrophotometer or colorimeter.

Glass cuvettes or tubes for spectrophotometry, 1 ml.
Lyophilizer or vacuum desiccator.

b. Procedure

Lyophilize, or dry in a vacuum desiccator, samples containing between 0.01 and 0.1 μmol of hexosamine. Add 20 μl of 3 N HCl to each sample. Prepare standards of glucosamine (0.01 to 0.2 μmol/20 μl). Heat the samples and standards at 95°C for 4 h in a sealed ampoule or tube. Cool and neutralize the hydrolysates with 20 μl of 3 M NaOH. Transfer a 30-μl sample of each hydrolysate and standard to separate 3-ml test tubes. Add 10 μl of a saturated solution of $NaHCO_3$ and 10 μl of freshly prepared 5% acetic anhydride. Allow the tubes to stand at room temperature for 10 min, during which N-acetylation of the hexosamines occurs; then immerse them in boiling water for precisely 3 min to destroy the unreacted acetic anhydride. Add 50 μl of 5% $K_2B_4O_7$, mix, and heat in a boiling water bath for 7 min. Cool, and add 0.7 ml of the color reagent (a 1:5 dilution of the Morgan-Elson reagent). Mix again, and incubate at 37°C for 20 min. Read the A_{585} of each tube.

Determine the amount of total hexosamine in each sample from a standard curve prepared by plotting the A_{585} versus the amount of hexosamine in each standard (section 22.1.14). Reaction products of glucosamine, galactosamine, and muramic acids have the same extinction coefficients.

c. Applications

As mentioned above, this procedure is not suited for analyses of hexosamines in complex mixtures of sugars and sugar derivatives. Hexosamine derivatives such as *N*-acetylmuramic acid substituted with peptides produced by the action of lysozyme on cell walls yield the same colored products.

The precise neutralization of the HCl after hydrolysis of samples is critical to the reaction. Therefore, it is important to prepare the HCl and NaOH carefully, and to be sure that the ampoules or tubes are tightly sealed during hydrolysis.

Glucosamine, galactosamine, and muramic acid can be separately estimated by a procedure described by Stewart-Tull (39). An enzymatic assay for galactosamine has been described previously (37). Other assays for muramic acid are described in section 22.3.8.

22.3.7. Hexosamines in Complex Solutions

Hexosamines in complex solutions and those produced by the hydrolysis of mixed polymers can be separated from neutral sugars that interfere with the Morgan-Elson reaction by ion-exchange chromatography. The purified hexosamines can then be quantified by the modified reaction described above.

a. Reagents, equipment, and supplies

4 N HCl (*CAUTION*: See section 22.10.10).
1 N NaOH (*CAUTION*: See section 22.10.26).
Lyophilization ampoules or Pyrex screw-cap test tubes with Teflon-lined caps.
Dowex-50 (H^+) columns:

Preparation (precycling) procedure: Add Dowex-50 resin (1 g for each sample to be analyzed) to 2 N NaOH (20 ml/g of resin). Mix for ca. 8 min, and filter onto a Buchner funnel. Wash the resin with distilled water (50 ml/g), and slowly stir the resin with 3 N HCl (20 ml/g). Remove the HCl by filtration on a Buchner funnel, and wash the resin on the funnel with ca. 2 liters of distilled water to remove the acid. Remove the water by applying suction, and suspend the resin in distilled water (1 g/5 ml). This preparation procedure is important for removal of impurities in the Dowex resin.

Use: Transfer 5 ml of the suspended resin to a Pasteur pipette or a 3-ml disposable syringe with a ca. 18-gauge needle, plugged at the bottom with glass wool, or to a commercially prepared minicolumn. Attach rubber or silicon tubing and a clamp at the bottom to control effluent flow. Do not allow the liquid surface to pass below the top of the resin bed.

Vacuum desiccator containing NaOH pellets.
Nitrogen gas supply (*CAUTION*: See section 22.10.16).
Sintered-glass funnel, fine porosity.
Buchner funnel.

b. Procedure

Mix the sample to be assayed with 4 N HCl in a small lyophilization ampoule. The amount of material and volume of acid will vary with the material used. Samples should contain 50 to 100 mg (dry weight) of bacterial cells and lesser amounts of purified material in 3 ml of 4 N HCl. Each sample should contain 0.15 to 0.50 mg of hexosamine in 3 ml. Hydrolyze the preparation for 6 h at 100°C in a vial which has been flushed with nitrogen gas and sealed.

Hydrolysis times should be optimized by using several subsamples treated for different times. The amount of hexosamine will increase to a maximum and then decrease with time of hydrolysis because of hexosamine destruction. Corrections for losses can be made by adding a known amount of hexosamine to one sample.

Dry the hydrolysate in vacuo over NaOH to remove HCl. Drying can be accomplished in a vacuum desiccator containing NaOH pellets; add water, and redry. Dissolve the hydrolysate in water, and filter to remove insoluble material. Add the filtered hydrolysate to a small column containing 1 g of Dowex-50 resin. Wash the column with 15 ml of water, and discard the eluate. Elute the hexosamines with 10 ml of 2 N HCl, and dry the eluate in a vacuum desiccator containing NaOH pellets. Add 2 ml of water, and redry.

Drying of the hydrolysates and column eluates causes considerable destruction of muramic acid as a result of concentration of HCl. The losses of this compound are variable; this procedure tends to underestimate total hexosamines.

Dissolve the dried hexosamine fraction in 0.1 ml of water, and proceed with the estimation of hexosamines as described above (section 22.3.6). Because hydrolysis has been completed, treatment with 3 N HCl and subsequent neutralization can be eliminated.

c. Applications

This procedure for purification of hexosamines is necessary when complex polymers other than peptidoglycan (e.g., lipopolysaccharides) are examined. It can be applied to whole bacterial cells with the knowledge that muramic acid recoveries will be variable. It is also applicable to estimation of hexosamines in culture media. For galactosamine analysis, an alternative to separation of the components of complex polymers is possible when galactose oxidase is used in an enzymatic assay (37).

22.3.8. Muramic Acid

Muramic acid occurs only as a component of the cell wall of eubacteria (i.e., "typical bacteria") and is not found in any of the archaeobacteria (17, 19, 21, 35, 88). The amount of muramic acid in a bacterial cell suspension is a measure of the presence and the amount of peptidoglycan.

The most convenient and specific assays for muramic acid involve treatment of acid hydrolysates of cell walls with alkali to remove D-lactate from the 3-position of muramic acid. The D-lactate is then estimated by an enzymatic method (42) involving lactate dehydrogenase or, as mentioned above (section 22.3.3), in an automatic enzymatic method, or by the colorimetric method of Hadzija (23) (see below). (Some archaeobacteria contain a peptidoglycan cell wall material called pseudomurein, but although it contains hexosamines, it does not contain the D-lactate or muramic acid and thus would be negative in the assay described here [88].)

a. Reagents, equipment, and glassware

Copper sulfate solution. Dissolve 4 g of $CuSO_4$ in 100 ml of distilled water.

p-Hydroxydiphenyl reagent. Dissolve 1.5 g of *p*-hydroxydiphenyl in 100 ml of 95% (vol/vol) ethanol (*CAUTION*: See section 22.10.8).

Muramic acid stock solution. Dissolve muramic acid (Sigma Chemical Co.) in water to give 0.25 mg/ml. Muramic acid contains 30% by weight of O-ether-linked lactic acid.

Screw-cap test tubes, 20 ml with Teflon-lined screw caps.

Spectrophotometer and glass cuvettes or tubes.

All glassware must be carefully cleaned, rinsed with sulfuric acid, and then rinsed very well with high-quality distilled water. After cleaning, contamination with dust and fingerprints must be avoided.

Concentrated sulfuric acid (*CAUTION*: See section 22.10.28).

b. Procedure

Neutralize samples of hydrolyzed peptidoglycan (in 3 N HCl [see section 22.3.6]) containing 5 to 50 μg of muramic acid in screw-cap test tubes with Teflon-lined caps by adding an equal volume of 3 N NaOH. Prepare standards containing 1 to 100 μg of lactic acid from the muramic acid stock solution; also prepare a reagent blank. Adjust the volume of the samples, standards, and the reagent blank to 1.0 ml with distilled water. Add an additional 0.5 ml of 1.0 N NaOH, and incubate the solution for 30 min at 37°C. Add 10 ml of concentrated sulfuric acid, seal the tubes, and heat them in boiling water for 10 min. After cooling, add 0.1 ml of the copper sulfate reagent and 0.2 ml of the *p*-hydroxydiphenyl reagent and mix rapidly. Incubate the tubes at 30°C for 30 min in a water bath, and read the A_{560}. Determine the concentration of muramic acid in the samples from a standard curve prepared by plotting the absorbance of each standard versus the concentration of muramic acid in the standard.

c. Applications

The method described is a simple, rapid, inexpensive procedure for detection of muramic acid when metabolic lactic acid is not present. Other components of the muramic acid do not interfere, nor do uronic acids and neutral sugars up to 250 μg per sample.

22.3.9. 2-Keto-3-Deoxyoctanoic Acid

The 2-keto-3-deoxyoctanoic acid (KDO) in lipopolysaccharide can be determined by the method of Weisbach and Hurwitz (43) as modified by Osborn (33). When oxidized with periodate, KDO (as well as other 2-keto-3-deoxy carbohydrates) yields formylpyruvic acid, which reacts with thiobarbituric acid to yield a chromogen with an absorption maximum at 550 nm. Deoxy sugars other than 2-keto-3-deoxy sugar acids produce different products, which form chromogens that absorb light maximally outside of the 545- to 550-nm spectral range.

a. Reagents

Periodate reagent. Dissolve HIO_4 in 0.125 N H_2SO_4 to give a concentration of 0.025 N. (*CAUTION:* Iodic acid and periodate are dangerous acids and should be handled in the hood, using precautions as described for other volatile acids in section 22.10.)

Dilute sulfuric acid, 0.02 N (*CAUTION:* See section 22.10.28).

Sodium arsenite reagent: 0.2 N sodium arsenite dissolved in 0.5 N HCl (*CAUTION:* See section 22.10.23).

Thiobarbituric acid reagent: 0.3% thiobarbituric acid in high-quality distilled water; adjust to pH 2.

b. Procedure

Dissolve lyophilized polysaccharide containing 0.01 to 0.25 μmol of KDO in 0.1 ml of 0.02 N H_2SO_4, and heat at 100°C for 20 min to release KDO from the polymer.

Adjust samples of the hydrolysate to 0.2 ml with 0.02 N H_2SO_4. Add 0.25 ml of the periodate reagent, mix, and let stand for 20 min at room temperature. Add 0.5 ml of 0.2 N sodium arsenite solution. Incubate for 2 min at room temperature, and add 2 ml of the thiobarbituric acid reagent. Mix, and heat at 100°C for 20 min. Cool, and read the A_{548}. One micromole of KDO per ml gives an A_{548} of 19.0 in a 1-cm cuvette.

c. Applications

The procedure described above is suitable for the determination of KDO in purified lipopolysaccaride or cell wall preparations.

22.3.10. Heptoses

The product formed when a heptose is heated with sulfuric acid reacts with cysteine hydrochloride to produce a chromogen with an intense orange color. This chromogen changes to a pink compound which absorbs at 505 nm (33).

a. Reagents, supplies, and equipment

Sulfuric acid reagent. Add 6 volumes of concentrated H_2SO_4 to 1 volume of water (*CAUTION:* See section 22.10.28).

Cysteine hydrochloride reagent: 0.3 g of cysteine hydrochloride in 10 ml of distilled water.
Water bath at 20°C.
Ice-water bath.
Boiling-water bath.
Spectrophotometer.
Cuvettes or tubes for spectrophotometer.

b. Procedure

Adjust sample volume to 0.5 ml. Place in an ice bath. Add 2.25 ml of the sulfuric acid reagent to samples. Mix the tubes in the ice-water bath for 3 min. Transfer to the 20°C water bath for 3 min, and then heat in boiling water for 10 min. Cool, add 0.05 ml of the cysteine reagent, and mix well. Read the A_{505} and A_{545} 2 h after adding cysteine reagent. Use a blank containing the same reagents described above, but lacking the cysteine, to zero the spectrophotometer at each wavelength.

Subtract the A_{545} from the A_{505}. After this subtraction, 1 μmol of L-glycero-D-mannoheptose in 1 ml gives an absorbance of 1.07 in a 1-cm cuvette.

22.3.11. Polysaccharides and Complex Carbohydrates

In some cases, experiments require analysis of multicomponent carbohydrates containing a variety of possible sugars and sometimes organic acids, proteins, and lipids. Analysis of this type of material is considerably more complex than that of simple polymers of monosaccharides, and often the structural information required may involve the use of nuclear magnetic resonance (NMR) and mass spectral data, both of which are beyond the scope of this chapter. However, a variety of capsule or slime polysaccharides provide examples of the utility of techniques available in a large number of well-equipped laboratories and departments; in some cases, noncarbohydrate components, if they are limited in number, can be analyzed by chromatography systems identical to or not too dissimilar from those already being used for sugar analysis.

Hydrolysis needs will vary somewhat, depending on the material, and thus it is advisable to conduct hydrolysis under several conditions (in some cases with internal sugar controls to estimate sugar destruction), and to examine the products by relevant and informative methods, e.g., HPLC (see section 22.3.6) or ion chromatography (see section 22.3.7). Both of these chromatographic techniques allow the separation of a number of compounds and the use of variable detection methods, e.g., refractive index, pulsed amperometric detection, or absorbance at various wavelengths. Evaluation of polysaccharides and complex carbohydrates will be greatly strengthened by additional confirmatory work, e.g., by enzymatic assays for glucose or galactose, so that chromatographic information can be verified when multiple and sometimes overlapping peaks are present in elution profiles. Likewise, thin-layer chromatography can play a valuable confirmatory role in identification.

Although exocellular polysaccharides vary considerably in their sugar content, most can be harvested and partly purified by solvent precipitation, and the component sugars can be determined by HPLC or ion chromatography. Described below is an example of a procedure that would be suitable for recovery and analysis of exopolysaccharides from *Acinetobacter calcoaceticus* (12, 24), *Klebsiella* spp. (12), *Rhodotorula glutinis* (20), and (possibly with a slight modification to use ethanol or isopropanol instead of acetone) *Xanthomonas campestris* (13), *Pseudomonas mendocina* (22), and *Azotobacter vinelandii* (14). The general problem of microbial polysaccharide recovery is also discussed by Smith and Pace (38).

a. Reagents and materials

Sodium chloride.
Acetone (*CAUTION:* See section 22.10.4).
Sodium azide (*CAUTION:* See section 22.10.24).
Incubator at 4°C.
Dialysis tubing.
Screw-cap tubes with Teflon-lined caps.
Trifluoroacetic acid (8 N aqueous solution) (*CAUTION:* See section 22.10.30).
Boiling-water bath.
Vacuum desiccator or rotary evaporator.
Suitable analysis instrument (e.g., HPLC or ion chromatograph) and supplies for its use.
Centrifuge, refrigerated.
Magnetic stirrer.
Lyophilizer.

b. Procedure

Grow cells in 50-ml volumes in a 250-ml flask with an appropriate medium for polysaccharide production (e.g., see above references). Remove cells by centrifugation at 5,000 × *g* for 20 min. To the supernatant of ca. 45 ml add 0.05 g of NaCl, and mix. Pour this mixture into 200 ml of acetone, and stir well (acetone can be replaced with 95% ethanol or neat isopropanol in most cases). Allow this to sit without stirring overnight at 4°C, and the following day centrifuge the material at 10,000 × *g* for 20 min (at 4°C) and decant most of the supernatant, leaving the precipitate undisturbed. Resuspend the precipitate in a minimum of the remaining solution, and place the slurry in a dialysis bag. (If necessary, the polysaccharide could be washed one or more times with ethanol or another solvent, as described above, before starting dialysis.) Dialyze overnight at 4°C in distilled water containing 0.01% sodium azide; replace the water with pure distilled water, and dialyze for an additional 3 to 4 h. Lyophilize the remaining slurry to make a dry polysaccharide powder.

Place 4 mg of polysaccharide powder in a Teflon-lined screw-top glass tube, add 1 ml of 8 N trifluoroacetic acid, and tightly cap the tube. Heat the tube at 100°C for 2 h. Cool the tube to room temperature, uncap, and dry under reduced pressure (e.g., in a vacuum desiccator containing NaOH pellets, under continuous vacuum) for ca. 48 h until only a residue remains (12, 32).

The residue can then be examined for sugar content by any of a variety of methods including HPLC (9; also see informational literature from HPLC suppliers), gas-liquid chromatography (31, 32) (see Chapter 21.3.5), or ion chromatography (12, 26, 34) (see Chapter 21.3.7; e.g., with a Dionex Q1C analysis system equipped with a HPIC-AS6 anion-exchange column). The enzymatic methods described above (section 22.3) are also excellent and specific methods for some possible component sugars. See also Chapter 21.3 for a description of types of chromatographic analysis.

22.4. LIPIDS

Lipids are a chemically diverse group of biological molecules that are insoluble in water but soluble in organic solvents and that commonly contain long-chain hydrocarbon groups. This description includes long-chain hydrocarbons, alcohols, aldehydes, fatty acids, and their derivatives including glycerides, phospholipids, glycolipids, and sulfolipids. Sterols, fatty acid esters of sterols, vitamins A, D, E, and K, carotenoids, and other polyisoprenoids often are considered in this class because they, like lipids, are associated with membranes and can be extracted by lipid solvents. A survey of the types and composition of microbial lipids has been compiled by O'Leary (48), and a more recent reference (60) describes archaeobacterial lipids in greater detail.

The cytoplasmic membrane of all living cells contains lipids. Bacterial membranes are much less complex than those of eucaryotic cells. However, bacteria are known to contain a wide variety of lipids in addition to phospholipids. These include sphingolipids, neutral lipids, glycolipids, and a variety of unusual lipids associated with organisms of specific genera such as *Corynebacterium*, *Nocardia*, and *Mycobacterium* (48). Procaryotes were believed not to contain sterols; however, it is now known that some bacteria can contain squalene and a variety of sterols (49, 64). Gram-negative bacteria contain unique lipid constituents in lipopolysaccharide complexes as part of their outer membrane (48). Poly-β-hydroxybutyrate is accumulated in large amounts in the cytoplasm of several bacteria, and serves as a carbon and energy reserve.

Archaeobacterial lipids are quite different from those found in eubacteria (57–60, 68, 69). Archaeobacterial lipids have an ether bond linking the hydrophobic polyisoprenoid group to the hydrophilic glycerol, whereas eubacterial lipids have an ester bond between the glycerol and an alkane chain. A diversity of novel variations on this ether-linkage and polyisoprenoid structural motif can be found among the archaeobacteria. Since no archaeobacterium has yet been demonstrated to have an outer membrane, the major location of these lipids is the cytoplasmic membrane. In general, total archaeobacterial lipids account for about 2 to 5% of the cell dry weight.

Because of the complexity of bacterial lipids, a description of procedures that would provide a total analysis of each class is impractical here. For variations of the procedures described below, and for detailed techniques required for a more complete analysis of bacterial lipids, the reader should consult one of the specialized articles on the subject (46, 47, 50, 54–60, 63–71).

22.4.1. Crude Total Lipids

The procedure for extraction of total crude lipids described by Bligh and Dyer (50) is less time-consuming and requires less sample manipulation than other methods. The procedure has been widely used with bacteria. The bacterial cells are mixed with chloroform and methanol in amounts that yield a monophasic solution when combined with water in the tissue (53). Upon dilution with water or chloroform, a biphasic solution results. The lipids are retained in the chloroform phase, and nonlipid materials are retained in the methanol-water phase. The chloroform layer is isolated, and the lipids can be purified and separated into classes as described below. The recovery of total crude lipids after separation from the nonlipid contaminants should exceed 95%.

Hara and Radin (54) have described an alternative to this procedure which does not involve chloroform use and which has the advantage that centrifugation can be used instead of the filtration steps in the Bligh and Dyer procedure. A hexane-isopropanol mixture is used to extract lipids, and nonlipid contaminants are removed by washing the extract with aqueous sodium sulfate. The lipid extract can be fractionated on silica gel columns by use of hexane, isopropanol, and water mixtures.

a. Reagents, materials, and equipment

Methanol (*CAUTION:* See section 22.10.13).
Chloroform (*CAUTION:* See section 22.10.6).
Filter paper, Whatman no. 1.
Porcelain Büchner funnel, ca. 43 mm diameter.
Suction flask: glass, 125 ml.
Fume hood.
Nitrogen gas supply (*CAUTION:* See section 22.10.16).

b. Procedure

Mix a thick cell paste prepared from 5 g of wet packed cells, or 1 g of lyophilized cells in 4 ml of distilled water, in a glass-stoppered tube with 5 ml of chloroform and 10 ml of methanol. Shake the mixture for 5 min, and leave it at room temperature with intermittent shaking for 3 to 4 h. Add another 5 ml of chloroform and 5 ml of distilled water, and shake the mixture at room temperature for another 30 min.

Rapidly filter the extract through Whatman no. 1 filter paper with slight suction. Transfer the filtrate to a graduated glass cylinder to allow separation of the layers. Completely remove the methanol-water layer by aspiration.

Reextract the residue of cell material on the filter paper by shaking with 5 ml of chloroform. Filter the mixture once again, and rinse the filter with 12.5 ml of chloroform. Add the combined filtrates to the chloroform layer from the first extraction.

Add the chloroform extracts to a tared glass or porcelain container, and evaporate to dryness under a stream of nitrogen in a water bath at 35 to 40°C to give a dry crude total-lipid preparation. Work in the fume hood. Measure the amount of crude total lipid with an analytical balance.

(Note that the procedure of Hara and Radin [54] can be used as an alternative, to avoid the use of the dangerous chloroform [see section 22.10.6]. This is particularly attractive for larger-scale preparations, in which large volumes of chloroform must be handled.)

c. Applications

This procedure is the starting point for most types of lipid analysis. It can be used to obtain the total dry weight of lipids in cells, as the starting point for the purification of many types of specific lipids, and to determine the structures of specific lipids.

22.4.2. Purified Total Lipids (71)

a. Reagents and materials

Solvents. Mix 200 parts of redistilled or fresh chloroform and 100 parts of methanol with 75 parts by volume of water. Shake vigorously in a separatory funnel, and allow the mixture to separate into an upper phase (solvent I) and a lower phase (solvent II). Collect the phases separately, and store them at room temperature. (*CAUTION:* See sections 22.10.6 and 22.10.13.)

Sephadex G-25, fine bead form (Pharmacia Fine Chemicals, Inc., Piscataway, N.J.).

Chromatography tube: ca. 1 by 15 cm with sintered-glass bottom and Teflon stopcock or other flow control device.

Paraffin shavings.

b. Procedure (see Chapter 21.3.9 for principles of gel filtration)

Soak 25 g of Sephadex G-25 beads overnight in 100 ml of solvent I. Allow the beads to settle, decant the solvent, and rinse the beads four times with solvent I. Avoid rough treatment of beads; e.g., do not use stirring bars to mix the solution.

Prepare a thick slurry of beads (total volume, 50 ml) in solvent I. Degas the slurry by placing it in a vacuum flask, and draw a vacuum with an aspirator for ca. 12 min. Swirl occasionally. Pour the slurry of beads into a column containing a few milliliters of solvent I. Continue adding slurry until a bed height of 10 cm is obtained. Apply a pressure of ca. 0.1 atm (1 atm = 1.0129×10^5 Pa) from a nitrogen tank to settle the resin bed, and place a fiberglass filter disc on top of the resin bed. Rinse the column with 10 ml of solvent I and 10 ml of solvent II.

Dissolve ca. 175 mg of the dry crude total lipids (see the preceding procedure) in ca. 4 ml of solvent II. Filter the dissolved lipid fraction through a medium-porosity sintered-glass funnel directly onto the Sephadex column. Rinse the funnel with 2 ml of solvent II. Apply a pressure of 0.15 to 0.2 atm with a nitrogen tank to obtain a flow rate of 0.5 to 1.0 ml/min. Add solvent II, and elute the column until 25 ml of solvent is collected.

Concentrate the eluate under reduced pressure at room temperature until a cloudy aqueous emulsion forms. Lyophilize this emulsion to dryness. Dissolve the dried lipid in chloroform-methanol (2:1, vol/vol), transfer it to tared glass-stoppered vials, and evaporate the solvents under a stream of nitrogen at ca. 37°C. Complete the drying in an evacuated vacuum desiccator containing paraffin shavings. After drying, weigh the vial containing the dried lipid extract.

Recovery of lipids can be checked by using known amounts of purified total lipids in the extraction and purification steps.

22.4.3. Lipid Classes

Lipids can be bound to a solid adsorbent material in a chromatography tube; the adsorbent materials most often used are silicic acid, alumina, and magnesium oxide (see also Chapter 21.3). The lipids are bound by polar, ionic, and van der Waals forces. Separation is effected by using solvents of increasing polarities, which separates lipids into classes determined largely by the number and types of polar groups in the molecules of each class (46, 47).

In the method described below, a chromatography grade of silicic acid is used. The solvents used to elute the adsorbed lipids are chloroform, a chloroform-acetone mixture, and methanol. These have been selected because of the relative ease and reproducibility of the technique and because all of the phospholipids are eluted in a single fraction.

a. Reagents and materials

Silicic acid for chromatography: 100 g of Bio-Sil A (100-200 mesh, activated at 125°C for 4 h; Bio-Rad Laboratories, Richmond, Calif.).

Chloroform (*CAUTION:* See section 22.10.6).

Acetone (*CAUTION:* See section 22.10.4).

Methanol (*CAUTION:* See section 22.10.13).

Glass chromatographic tube approximately 1 by 30 cm, fitted with a solvent reservoir (separatory funnel with a Teflon stopcock) coupled to the top of the column by a 24/40 or standard taper greaseless glass joint. The column should be fitted with a fritted-glass disc or a glass-wool plug and greaseless Teflon stopcock at the bottom.

b. Procedure

Make a slurry of 12 g of Bio-Sil A in redistilled chloroform. Place the slurry in a vacuum flask or desiccator, and apply a vacuum for 10 min to degas it. Pour the degassed slurry into the chromatography tube. Allow the chloroform to drain through the column until the meniscus is about 1 mm above the resin.

Dissolve 150 to 200 mg of the purified total lipids (see the preceding procedure) in 3 ml of chloroform. Add the solution to the column, and drain the column until the chloroform just enters the resin layer. Add an additional 5 ml of chloroform, and drain the column again until all the lipid just enters the resin bed. Carefully add 50 ml of chloroform to the column and reservoir without disturbing the resin bed surface. Allow the solvent to flow through the column into a glass collection vessel at a rate not exceeding 1 ml/min. When the chloroform just enters the resin bed, change collection vessels and add 60 ml of chloroform-acetone to the column and reservoir. Collect the effluent at a rate not exceeding 1 ml/min in the second collection flask. When this solvent mixture just enters the resin bed, add 60 ml of acetone and collect the eluate in the same vessel; then add 60 ml of methanol to the column and collect the eluate in the third vessel.

The first collection vessel will contain neutral lipids (hydrocarbons, carotenoids, chlorophyll, sterols and sterol esters, glycerides, waxes, long-chain alcohols, and aldehydes) and fatty acids. The second reservoir will contain glycolipids, sulfolipids, and occasionally some phosphatides. The third reservoir will contain phospholipids. These can be detected and quantified by assaying for organic phosphate (see section 22.4.5).

22.4.4. Phospholipid Fractionation

There are many different solvent systems and choices of developing chambers for thin-layer chromatography plates (46, 47, 66).

a. Reagents and materials

Silica gel glass plates, 20 by 20 cm, 250 μm thick. Prepared plates are available from several manufacturers, or they can be prepared by coating silica gel onto glass plates (47). Develop the plates to the top in a chromatography chamber with redistilled acetone. Air dry in a fume hood, and heat for 30 min at 110°C. Cool in a desiccator, and use on the same day.

Glass chromatography tanks. Thin-layer chromatography tanks or sandwich developing tanks can be purchased from several sources. Add freshly mixed solvent (60 ml) to multiplate tanks 2 h before developing the chromatograms. Line the tanks with sheets of filter paper saturated with solvent, and seal the tanks to equilibrate the atmosphere with the solvent.

Solvents (*CAUTION:* See sections 22.10.5, 22.10.6, and 22.10.13).

Solvent I: Chloroform-methanol-water, 65:25:4 (vol/vol), prepared from fresh solvents.

Solvent II: Chloroform–methanol–7 N ammonium hydroxide, 60:35:5 (vol/vol/vol).

Nitrogen gas: in a compressed gas cylinder; use carefully (*CAUTION:* See section 22.10.16).

Micropipettes or microsyringes, 50 μl, graduated at intervals.

Iodine crystals (*CAUTION:* See section 22.10.11).

b. Procedure

Dissolve the dry phospholipid fraction from the Bio-Sil column in chloroform-methanol (1:1, vol/vol [about 100 mg/ml]). Record the volume of the solution. Apply a measured amount of each solution as a spot, <10 mm in diameter, 2.5 cm from the bottom and 2.5 cm from the right side of a separate chromatography plate. Use a graduated 50-μl disposable pipette, micropipettor, or microsyringe. Dry between applications with a stream of nitrogen gas (an alternative is to use a hair dryer). Apply up to 0.1 ml of a solution to a plate. Dry the spot, and place in a developing tank containing solvent I. Seal the tank, and allow the chromatogram to develop until the solvent front reaches the top of the plate (1 to 2 h). Remove and air dry the plate. Place the plate, rotated 90° clockwise, relative to the first solvent flow direction in another tank containing solvent II, and develop in the second direction. Air dry the developed plate.

Stain the developed chromatogram with iodine vapors as described below to visualize the phospholipid spots. Wet the chromatogram slightly with a water mist, and scrape each spot off with a stainless-steel spatula or a razor blade, collecting the powder on weighing paper. Transfer all the silica gel from each spot to a separate Pasteur pipette plugged with glass wool in the constricted end (or into a commercially available minicolumn). Tap the Pasteur pipette until the silica gel forms a layer on the glass wool. Rinse the Pasteur pipette and silica gel with 2 ml of chloroform-methanol (1:1, vol/vol) followed by 2 ml of methanol to elute the phospholipids. Collect the eluates in glass-stoppered tubes. Distribute each phospholipid into three equal portions in two glass-stoppered 15-ml test tubes and a lyophile ampoule for further analysis. Evaporate the solvents at 37°C under a stream of nitrogen. One tube containing each of the separated phospholipids is to be assayed for total organic phosphorus to determine the amount of each phospholipid present. Other samples of each phospholipid can be subjected to acid and alkaline hydrolysis followed by identification of the products to establish their probable structures.

Iodine stain for detection of lipids on chromatograms (46,47). Working in a hood, add crystals of iodine to a chromatographic tank, cover the tank and warm it in a water bath at 60°C (or with a hair dryer) to vaporize the iodine (*CAUTION:* Iodine vapors are dangerous; see section 22.10.11). Place the chromatograms in the tank on glass supports to hold them 2.5 cm off the bottom. Cover the tank for 1 min, remove the chromatogram, and view it in daylight and with a UV lamp at 365 nm (*CAUTION:* UV light is dangerous; see section 22.10.31). The lipids will appear as brown spots in daylight and as very dark spots under UV light. Mark the outline of the spots with a pencil before the color fades.

22.4.5. Phosphate in Phospholipids

The amount of phosphate in each phospholipid fraction obtained by thin-layer chromatography can be estimated by the procedure of Bartlett, as described by Kates (47) and as follows.

a. Reagents and materials

Clear Pyrex test tubes, 13 by 100 mm. Clean in hot 1 N nitric acid for 1 h, and rinse with distilled water. All glassware must be carefully cleaned to remove traces of detergent and other sources of phosphorus contamination. (*CAUTION:* Nitric acid is dangerous; see section 22.10.15.)

Acid-washed glass beads (ca. 1 to 2 mm in diameter) and marbles (sized to fit top of the Pyrex tubes above). Clean with the test tubes above.

Perchloric acid (72%, vol/vol) in distilled water (*CAUTION:* Perchloric acid is dangerous; see section 22.10.17).

Ammonium molybdate solution (5%, wt/vol) in distilled water.

Amidol reagent. Add 0.5 g of 2,4-diaminophenol dihydrochloride to 50 ml of 20% (wt/vol) sodium bisulfite solution. Filter.

Phosphorus standard (10 μg of phosphorus per ml). Dissolve 1.097 g of KH_2PO_4 in 250 ml of distilled water. Dilute 1:10 with distilled water to give 10 μg/ml.

Spectrophotometer to read at 830 nm (near-IR).

b. Procedure

Place a measured volume of lipid sample in test tubes, and evaporate the solvent under a stream of filtered air or nitrogen. Dry the phosphate standards containing 1 to 10 μg of phosphorus per tube (0.1 to 1.0 ml of standard) in similar fashion. Add 0.4 ml of 72% perchloric acid and an acid-washed glass bead. Heat at 100°C until the solution is clear and colorless. Work in a fume hood that has been approved for use with perchloric acid. Add 4.2 ml of distilled water, 0.2 ml of ammonium molybdate solution, and 0.2 ml of the amidol reagent. Mix, and cover the tubes with acid-washed glass marbles. Heat in a boiling-water bath for 7 min. Cool rapidly, and read the A_{830} after 15 min (see Chapter 21.1.1). A blank without phosphate is used to zero the spectrophotometer. Determine the concentration of phosphate in each sample from a standard curve prepared by plotting the amount of phosphorus in each standard tube versus the absorbance of the standard. The absorbance is linear up to 10 μg of phosphorus per ml. The total phospholipid in each fraction can be converted into micromoles of each phospholipid by assuming that each phospholipid contains one phosphate group.

22.4.6. Phospholipid Identification (46, 47, 66)

Phospholipids can be deacylated by mild alkaline hydrolysis to yield water-soluble glycerol phosphate esters and fatty acids, which are soluble in organic solvents. The glycerol phosphate esters can be identified by specific staining reactions and their R_f values as determined by paper chromatography (48) or thin-layer chromatography. The fatty acids can best be identified by gas-liquid chromatography of methyl esters (55, 67, 70a) (see Chapter 21.3.5).

The identity of the phospholipids can be further confirmed by analysis of the free bases liberated by acid hydrolysis. Free bases are identified by R_f values and color reactions after thin-layer or paper chromatography (46, 66).

a. Reagents and materials

Methanolic sodium hydroxide (0.2 N). Dissolve 0.4 g of NaOH in 50 ml of methanol. Prepare fresh (*CAUTION:* See sections 22.10.13 and 22.10.26).

Methanolic ammonium hydroxide (1.5 N). Dilute 10 ml of concentrated ammonium hydroxide to 100 ml with methanol (*CAUTION:* Ammonium hydroxide is dangerous; see sections 22.10.5 and 22.10.13).

Sodium hydroxide (1.25 N) (*CAUTION:* See section 22.10.26).

Chloroform (*CAUTION:* See section 22.10.6).

Methanol (*CAUTION:* See section 22.10.13).

Cation-exchange resin: Bio-Rad AG 50W × 8, Dowex-50 (H^+), Amberlite IR-100 (H^+), or a similar resin.

Hydrochloric acid, 1 N (*CAUTION:* See section 22.10.10).

Water-saturated phenol. Place equal volumes of fresh, high-quality liquid phenol and water in a well-capped bottle, shake well, and remove the phenol layer. Store in a brown bottle at 4°C. Prepare fresh for each use from redistilled phenol stored at −20°C. (*CAUTION:* Phenol is dangerous; see section 22.10.19.)

Methanol-water (10:9, vol/vol) (*CAUTION:* See section 22.10.13).

Ethanol, reagent grade, absolute (*CAUTION:* See section 22.10.8).

Acetic acid, reagent grade, absolute (*CAUTION:* See section 22.10.2).

Hexane, spectroscopic grade (*CAUTION:* See section 22.10.9).

Glass-stoppered 15-ml test tubes.

Whatman no. 1 chromatography paper. The size of the paper depends on the developing tank available; alternatively, use cellulose thin-layer chromatography plates, available from many commercial sources.

Tank for descending chromatography, or for thin-layer chromatography.

Lyophile ampoules, 10 ml.

Lyophilizer or flash evaporator.

Phospholipid standards (available from commercial sources).

Phosphatidylcholine, distearoyl ester.

Phosphatidylserine from bovine brain.

Phosphatidylethanolamine from *E. coli.*

Phosphatidyl-*N,N*-dimethylethanolamine, dipalmitoyl ester.

Diphosphatidylglycerol (cardiolipin) from bovine heart.

Dragendorf spray reagent. Dissolve 1.7 g of bismuth nitrate in 100 ml of 20% acetic acid. Dissolve 10 g of potassium iodide in 25 ml of water. Just before use, mix 5 ml of the potassium iodide solution, 20 ml of the bismuth nitrate solution, and 70 ml of water to use as a spray reagent.

Ninhydrin reagent. Dissolve 0.25 g of reagent-grade ninhydrin in 100 ml of acetone-lutidine (2,6-dimethyl pyridine) (9:1, vol/vol). Prepare fresh for each use. (*CAUTION:* Some individuals are allergic to ninhydrin.)

b. Procedures

Acid hydrolysis of phospholipids and identification of free bases

Add 2 ml of 1 N HCl to the lyophile ampoule containing the purified individual phospholipids from the thin-layer chromatographic plates (section 22.3.4). Seal the ampoule, and heat to 100°C for 4 h. Cool, open the ampoule, transfer the contents to a 10-ml screw-cap tube, and add 2 ml of redistilled hexane. Mix on a Vortex mixer, and let stand to allow the phases to separate. Remove the aqueous phase, and lyophilize to dryness or dry in a rotary evaporator.

Dissolve the residue in 0.1 ml of distilled water. Spot three samples of each aqueous phase 10 mm apart on Whatman no. 1 chromatographic paper, and develop the chromatogram by descending chromatography by using the solvent system phenol-ethanol-acetic acid (50:5:6, vol/vol/vol). Known phospholipids treated identically, or choline, ethanolamine, serine, and *N,N*-dimethylethanolamine, can be used as standards for chromatography and identification of the free bases (46). With minor modifications, thin-layer chromatography can be used for this separation.

After the chromatogram is developed, dry it over night in a fume hood and cut it into strips. Spray one strip, containing a standard or an unknown, with ninhydrin reagent, and spray another strip, with the same unknown or standard, with the Dragendorf reagent. Serine and ethanolamine will give blue spots with R_f values of approximately 0.27 and 0.51, respectively, after treatment with ninhydrin. Choline and *N,N*-dimethylethanolamine do not react with the ninhydrin spray. Choline gives an orange spot and dimethylethanolamine gives a weak orange spot with the Dragendorf spray reagent. The R_f values for choline and *N,N*-dimethylethanolamine are approximately 0.93 and 0.86, respectively.

Alkaline hydrolysis of phospholipids and identification of glycerol phosphate esters (46, 47)

Add 0.2 ml of chloroform, 0.3 ml of methanol, and 0.5 ml of methanolic sodium hydroxide to 1 to 5 mg of phospholipid prepared by chromatography on Bio-Sil A (section 22.4.3). Mix in a glass-stoppered test tube on a Vortex mixer, and leave at room temperature for 15 min.

Add 1.9 ml of chloroform–methanol–1.25 N NaOH (1:4:4.5, vol/vol/vol). Transfer to a conical centrifuge tube, mix on a Vortex mixer, and centrifuge for 1 min at 600 × *g*. Remove the upper methanol-water phase with a Pasteur pipette. Add cation-exchange resin to the methanol-water phase with vigorous mixing until the pH, measured with indicator paper, is slightly acidic (pH 5 to 6). Centrifuge, and remove the liquid from the resin with a Pasteur pipette. Add 1 or 2 drops of methanolic ammonium hydroxide to make the supernatant slightly alkaline (pH 7.5 to 9.0). Check the pH with indicator paper. Concentrate to dryness under a stream of nitrogen at 30°C, and dissolve the residue in 0.1 ml of methanol-water (10:9, vol/vol).

The dissolved glycerol phosphate esters can be identified by cochromatography with products of known phospholipids treated in an identical fashion (46). Three 25-μl aliquots of each sample should be spotted 10 cm apart on Whatman no. 1 paper and developed by descending chromatography with a solvent system of phenol-ethanol-acetic acid (50:5:6). After the chromatogram is developed, dry it overnight in a fume hood and cut into strips. Spray one of the strips containing the glycerol-phosphate ester sample with the Dragendorf spray reagent, one with the ninhydrin reagent, and another with the salicylsulfonic acid-ferric chloride spray reagent of Vorbeck and Marinetti (70) (see below). Phosphatidylethanolamine (R_f = 0.66) and phosphatidylserine (R_f = 0.28) give mauve spots when sprayed with the ninhydrin spray and left for a few hours at room temperature. Phosphatidylcholine (R_f = 0.86) and phosphatidyldimethylethanolamine (R_f = 0.80) give orange spots and weak orange spots, respectively, with the Dragendorf reagent and no color with the ninhydrin spray reagent. All react with the salicylsulfonic acid-ferric chloride reagent described below.

Detection of glycerol phosphate esters on chromatograms by use of the salicylsulfonic acid-ferric chloride reagent (46)

Dissolve 1.5 g of $FeCl_3 \cdot 6H_2O$ in 30 ml of 0.3 N HCl. Dilute with 90 ml of acetone. Dip paper strips from chromatograms in this reagent, dry in a fume hood, and dip in a solution prepared by dissolving 1.5 g of salicylsulfonic acid in 100 ml of acetone. Glycerol phosphate esters appear as white spots on a mauve background.

Note that thin-layer chromatography can be used as a substitute for the paper chromatography in all of the above procedures, with appropriate modifications.

22.4.7. Archaeobacterial Lipids

The lipids of archaebacteria are structurally distinct from lipids of eubacteria: the glycerol moiety is bound to the hydrophobic portion by an ether bond, not an ester bond; the hydrophobic portion is composed of a polyisoprenoid chain, not an alkane; in some cases a tetraether structure occurs, where two glycerol moieties are bound to an intervening double-length polyisoprenoid (57–60, 68, 69). Despite these major differences, these ether lipids can be extracted and examined by using methods very similar to those used for eubacteria such as *E. coli* or *Bacillus* species. Since variation in lipid content between genera or species may be of taxonomic value, lipid analysis of archaeobacteria is in some cases useful for identification, and classification of new isolates. Provided here is a general overview of how archaeolipids can be extracted, separated, and analyzed.

22.4.7.1. Extraction

The Bligh and Dyer extraction method (50) described above (section 22.4.1) is appropriate for most archaeobacteria, but modified procedures have also been described (57, 68). The nonchloroform procedure described by Hara and Radin (54) should work, but its use has not yet been described. One gram of dry cell weight should yield 20 to 50 mg of total lipids.

22.4.7.2. Separation

a. Reagents and materials

Silicic acid for chromatography: 10 g of Bio-Sil A, 100 to 200 mesh, activated at 125°C for 4 h (Bio-Rad).
Glass chromatography column (ca. 1 by 30 cm or 2 by 10 cm).
Chloroform (*CAUTION*: See section 22.10.6).
Acetone (*CAUTION*: See section 22.10.4).
Methanol (*CAUTION*: See section 22.10.13).

b. Procedures

Make a slurry of 10 g of Bio-Sil A in chloroform. Place the slurry in a vacuum flask or desiccator, and apply a vacuum for 10 min to degas it. Pour the degassed slurry into the chromatography column. Drain the column until the meniscus is about 1 mm above the resin. (*CAUTION*: chloroform is hepatotoxic and a carcinogen. Avoid exposure to either liquid or vapor, and use a fume hood. See section 22.10.6.)

Dissolve 50 to 200 mg of the total lipid in 3 ml of chloroform. (Alternatively, lipid treated to yield "purified total lipid" as described in section 22.4.2 can be used.) Add the solution to the column, drain until the liquid is about 1 mm above the resin surface, carefully add another 5 ml of chloroform, and drain in the same way. Carefully add 5 ml of chloroform, and elute with another 45 ml of chloroform; collect this fraction 1 in a glass vessel (contains neutral lipids). Elute the column with 50 ml of acetone by using a similar technique; this fraction 2 should contain principally glycolipids and phosphatidyl glycerol. Elute the column with 65 ml of chloroform-methanol (3:2, vol/vol); this fraction 3 should contain principally phosphoglycolipids. Elute the column with 200 ml of methanol; fraction 4 should contain principally phosphoglycolipids.

22.4.7.3. Analysis

The polar lipids (fractions 2, 3, and 4) and neutral lipids (fraction 1) can be analyzed by a variety of methods. The general scheme for certain identification of the structures of lipids is to further purify the polar lipids by thin-layer chromatography, with and without methanolic-HCl hydrolysis, to degrade the lipid products by cleavage and reduction (e.g., with HI and $LiAlH_4$), and to analyze the resultant products by gas chromatography coupled with mass spectrometry. Structural identity is further strengthened with infrared and NMR data, and information on hydrophilic groups is obtained by sugar assays, HPLC, and NMR. These methods and strategies are described in a variety of references (47, 55–60, 68, 69) and are beyond the scope of this chapter. However, we do provide here a method by which the distribution of diethers and tetraethers in polar lipids can be determined by thin-layer chromatography.

a. Reagents and materials

Silica gel glass thin-layer chromatography plates, and chromatography tank (as described in section 22.4.4).

Thin-layer chromatography solvent: *n*-hexane, diethyl ether, and acetic acid (80:20:1, vol/vol/vol) (*CAUTION*: See sections 22.10.2, 22.10.7, and 22.10.9).

Hydrolysis reagent: 2.5% anhydrous methanolic HCl (made by bubbling anhydrous methanol with gaseous HCl). (*CAUTION*: This procedure is extremely dangerous and requires the use of a fume hood for the containment of a gaseous HCl tank and apparatus for making the reagent. Gaseous HCl is a very corrosive and toxic gas. If unfamiliar with this type of procedure, consultation with staff familiar with the techniques is essential.)

I_2 crystals (*CAUTION*: See section 22.10.11).

Petroleum ether (*CAUTION*: See section 22.10.18).

b. Procedure

Hydrolyze each polar lipid fraction by addition to anhydrous 2.5% methanolic HCl, and reflux for 1 to 2 h, using a $CaCl_2$ tube on the reflux exit. Extract the products with petroleum ether. Apply the extracts to TLC plates, and develop until the solvent front reaches the top of the plates. Remove the plates, and dry with a hair dryer. Use the iodine stain (see section 22.4.4) to detect the lipids. Diphytanyl glycerol diether will have an R_f value of ca. 0.12, and dibiphytanyl diglycerol tetraether will have an R_f of ca. 0.46.

A variety of methanogens, including *Methanobacterium* and *Methanospirillum* species, have both diethers and tetraethers, whereas *Methanococcus vannielii* and *Methanosarcina barkeri* have only the diether (68). This technique is valuable for characterization and identification of newly isolated archaeobacteria of any group and may also be valuable in monitoring lipid changes in response to growth conditions.

22.4.8. Fatty Acids

One of the most convenient methods for identification of fatty acids in bacterial cells is gas-liquid chromatography of their methyl esters prepared from phospholipids, total lipids, or other lipid fractions. Methanolysis of lipids is accomplished by heating the lipids in sulfuric acid and methanol. The methyl esters can be extracted into hexane, and quantitative measurements of each fatty acid can be made with a gas chromatograph equipped with appropriate columns (see Chapter 21.3.5).

Several procedures are available for preparation of methyl esters of fatty acids. One method selected for its simplicity and convenience is present below. Alternative methods are described in references 55, 67, and 70a.

a. Reagents and materials

Gas chromatograph, equipped with a flame ionization detector and suitable recorder. Several choices for columns and temperature programs are available. Investigators should consult one of the references to select a gas-liquid chromatography procedure suitable for the instrument available and the material to be analyzed (47, 55, 63, 65, 70a). In most cases, catalogs available from commercial gas chromatography suppliers are very helpful in selection of the proper column, especially when samples of chromatographic elution profiles are provided.

n-Hexane, spectroscopic grade (*CAUTION*: See section 22.10.9).

Sulfuric acid-methanol (5%). Add 5 ml of reagent-grade concentrated sulfuric acid to 95 ml of water-free methanol. Prepare fresh before each use. (*CAUTION*: See sections 22.10.13 and 22.10.28).

Standards: methyl esters of straight- and branched-chain C-13 to C-20 fatty acids (available from a variety of gas chromatography supply companies).

Glass-stoppered 15-ml Pyrex test tubes.

b. Procedure

Transfer 0.5 to 1 mg of the purified total lipids (section 22.4.2) to glass-stoppered vials, and evaporate the solvent under a stream of nitrogen at 30 to 37°C. Add 4 ml of 5% (wt/vol) sulfuric acid in methanol to the dried lipid, and heat for 2 h at 70°C. After cooling, add 4 ml of *n*-hexane and shake the tube for 10 min. Remove and save the hexane phase. Reextract the methanol phase with 4 ml of hexane. Use the pooled hexane phases for gas-chromatographic analysis (see Chapter 21.3.5). The fatty acid methyl esters can be identified by comparing retention times with those of known fatty acids. Quantitation is achieved by comparing peak areas with those for an internal standard added to each sample (55).

22.4.9. Poly-β-Hydroxybutyrate

Poly-β-hydroxybutyrate (PHB), although a lipid, is not extracted from bacterial cells by the usual solvents used for lipid extraction (61). In the procedure described by Law and Slepecky (61), PHB is extracted into hot chloroform from cell material that is obtained by first treating bacterial cells with hypochlorite to remove interfering substances. The extract is hydrolyzed to yield β-hydroxybutyrate, which undergoes dehydration to crotonic acid in concentrated sulfuric acid. Crotonic acid absorbs UV light with an absorption maximum at 235 nm. (Although butyrate-based polymers are the most common naturally occurring polyesters, in many cases polymers containing valerate and other longer chains may be present, as discussed below in section 22.4.10.)

a. Reagents and materials

Sodium hypochlorite solution: a commercial bleach solution such as Clorox or other bleach (*CAUTION*: See section 22.10.27).

Cuvettes: 1 cm, made of quartz or other material transparent to UV light.

Spectrophotometer, capable of UV absorbance measurements.

Polypropylene or glass centrifuge tubes, 12 by 100 mm, or 30 ml. The plastic centrifuge tubes should be washed with hot chloroform and ethanol to remove plasticizers. (*CAUTION*: See sections 22.10.6 and 22.10.8).

Centrifuge: a refrigerated centrifuge capable of achieving 10,000 × *g*.

Pyrex glass boiling tubes.
Glass marbles to fit boiling tubes.
Volumetric flask, 10 ml.
Boiling-water bath.
Chloroform (*CAUTION*: See section 22.10.6).
Acetone (*CAUTION*: See section 22.10.4).
Reagent-grade sulfuric acid. The quality of the sulfuric acid is important. Low grades contain materials that absorb UV light and cause high absorbance in the blanks used to zero the spectrophotometer. (*CAUTION*: See section 22.10.28.)
Standard. Dissolve 3-hydroxybutyric acid (0.5 mg/ml) in ethanol. Dilute 1:10 in ethanol to obtain a solution of 50 μg/ml. (*CAUTION*: See section 22.10.8.)

b. Procedure

Centrifuge bacterial cells from a volume of culture containing 10 to 20 mg (dry weight) for 15 min at 10,000 × *g* and 4°C. Determine the dry weight of cells from an equal volume of culture. Add a volume of sodium hypochlorite equal to the original culture volume, and transfer the suspension to a centrifuge tube. Incubate the suspension at 37°C for 1 h. Centrifuge the suspension at 10,000 × *g* for 15 min to sediment the lipid granules. The temperature during centrifugation is not important. Suspend the sediment in water, and centrifuge at 10,000 × *g* for 15 min. Discard the supernatant fraction, and wash lipid granules adhering to the centrifuge tube walls with 5 ml of acetone followed by 5 ml of ethanol. The washings which remove water are accomplished by suspending the granules in 5 ml of acetone with a Vortex mixer followed by centrifugation at 10,000 × *g* for 15 min. Suspend the sediment in 5 ml ethanol by using a Vortex mixer, and centrifuge again at 10,000 × *g* for 15 min.

Dissolve the washed granules with boiling chloroform by heating with 3 ml of chloroform in a boiling-water bath for 2 min. Cool the solution, and centrifuge it at 10,000 × *g* for 15 min at room temperature. Save the supernatant fraction in a 10-ml volumetric flask, and reextract the sediment twice with 3 ml of chloroform at 100°C for 2 min. If the solution is not clear, filter while hot through Whatman no. 1 filter paper (previously washed with hot chloroform). The chloroform extracts are pooled and made to 10 ml with chloroform. Work in a fume hood while extracting with chloroform. (*CAUTION*: See section 22.10.6 for precautions involved in the use of chloroform.)

Add samples of the chloroform extract containing 5 to 50 μg of the polymer and 5 to 10 standards containing 5 to 50 μg of 3-hydroxybutyric acid in ethanol to glass boiling tubes. Immerse the tubes in a boiling-water bath until all the solvents are evaporated. Add 10 ml of sulfuric acid, cap the tubes with a glass marble, and heat them in a boiling-water bath for 10 min. Cool the solutions, and mix them thoroughly.

Read the A_{235} of the solutions against a sulfuric acid blank (see section 22.1.1.4). Calculate the amount of β-hydroxybutyrate from the standard curve. The amount of PHB per gram (dry weight) of bacterial cells can be determined from the dry-weight measurements of the cells (see Chapter 11.5.2).

c. Applications

The technique described above is suitable for measuring the amount of PHB in freeze-dried cells or cell pastes obtained by centrifugation. Other β-hydroxy acids and some sugars interfere because they are converted to products that absorb at 235 nm. Sugars are removed by hypochlorite treatment followed by extraction of PHB into chloroform. Other substances in cells occasionally produce products that absorb UV light. The absorption spectrum of the product from bacterial cells should be compared with that of the standard to ensure that the absorbance is due to crotonic acid. (See section 22.1.1.3 for determining absorption spectra.)

22.4.10. Poly-β-Hydroxyalkanoate Polymers

A wide variety of organisms can make poly-β-hydroxyalkanoate (PHA) polymers that contain repeating units with chain lengths longer than butyrate in the polymer (51, 52, 62). The makeup of the major carbon source can, in some cases, determine the chain length of the monomers; e.g., with *Pseudomonas oleovorans*, alkanes with chain lengths of 6 to 10 result in monomers of different lengths, generally with most monomers of the same length as the alkanoic substrates (51). In some cases, the polyester makes up more than 50% of the dry weight of the culture, and these polymers are being examined as the basis for renewable and biodegradable plastic production.

Analysis by an assay that is effective at detecting β-hydroxybutyrate only by its conversion to crotonic acid would obviously be inaccurate for some polymers but may be acceptable for many natural samples, because of the predominance of butyrate in the naturally occurring polymers. If nonbutyrate polymers are to be examined, a gas-chromatographic method is suitable. Described here is a method by which monomers containing 4 to 12 carbons can be estimated.

a. Reagents and materials

Sodium hypochlorite solution: a commercial bleach solution such as Clorox or other bleach (*CAUTION*: See section 22.10.27).
Polypropylene or glass centrifuge tubes: 12 by 100 mm, or 30 ml. The plastic centrifuge tubes should be washed with hot chloroform and ethanol to remove plasticizers (*CAUTION*: See section 22.10.6).
Centrifuge: a refrigerated centrifuge capable of achieving 10,000 × *g*.
Pyrex glass boiling tubes with glass marbles to fit tubes, or screw-cap Pyrex tubes with Teflon-coated caps.
Volumetric flask, 10 ml.
Boiling-water bath.
Chloroform (*CAUTION*: See section 22.10.6).
Methanol (*CAUTION*: See section 22.10.13).
Sulfuric acid (*CAUTION*: See section 22.10.28).
Gas chromatograph with flame ionization detector; column of Durabond Carbowax M15 or its equivalent; injector, 230°C; detector, 275°C; column, starting at 80°C for 4 min, then with a temperature ramp of 8°C/min to reach 160°C, then 160°C for 6 min.

b. Procedure

The dry weight of the PHA polymers can be determined by using the extraction method described above for PHB, with sufficient cells to produce an accurate measurement by weighing the extracted polymer. After the extraction, place the solution in tared weighing boats and dry it to a constant weight in an oven at ca. 100°C. (Alternatively, a Soxhlet extraction as described by Brandl et al. [51] can be used to extract the PHA.) This weight can be compared with the dry cell weight, which is the sum of the weights of the cell components and PHA.

After the PHA is extracted by either method, hydrolyze and methylate the polymer and perform gas-chromatographic analysis. (Alternatively, whole dried cells can be treated in the same way; if PHA makes up a substantial percentage of the dry cell weight, this is a good method, but if not, cellular lipids may contribute to a background of peaks which reduce the accuracy of the analysis.)

For hydrolysis, derivatization, and analysis, use the following procedure (51). Place the PHA fraction (ca. 2 mg) or dried cells (ca. 4 mg) in a small screw-cap test tube, and react for 140 min at 100°C with a solution of 1.0 ml of chloroform, 0.85 ml of methanol, and 0.15 ml of sulfuric acid. This degrades the PHA to its constituent β-hydroxyalkanoic methyl esters. After the reaction, add 0.5 ml of distilled water and shake the tube vigorously for 1 min. The solution then resolves into two phases; transfer the bottom, organic phase to another vial, and either store it frozen or analyze it. Inject a 2-μl volume into the gas chromatograph system as described above, using temperature programming. Use standards of C-4 to C-12 β-hydroxyalkanoic methyl esters to determine the retention times of the compounds being separated; peak areas of different concentrations of the compounds can be used to establish a standard curve.

22.5. CELL WALL POLYMERS

The term peptidoglycan refers to a large structural component, composed of both sugars and peptides, found in the cell wall structure of most bacteria; it most commonly refers to the murein found in eubacteria but also as a general term refers to the pseudomurein found in archaeobacteria (88). The specific type of peptidoglycan called murein (containing *N*-acetylmuramic acid) may make up 40 to 90% of the weight of isolated cell walls of gram-positive eubacterial cells (15 to 20% of the cell dry weight), usually less than 10% of the gram-negative bacterial cell walls, and as little as 1.2% of the cell wall of some pseudomonads (72–75, 81–83, 98). Murein controls the shape of procaryotes except mycoplasmas and some species of procaryotes belonging to the taxonomic grouping of archaeobacteria (86–88); the archaeobacteria have a variety of structural alternatives to murein, some of which are described in section 22.5.4.

The peptidoglycan (murein) of most bacteria has a backbone containing alternate residues of *N*-acetylglucosamine and *N*-acetylmuramic acid linked by β(1-4)-glycosidic bonds. Minor exceptions to this backbone structure are found in a few *Streptomyces* species (75) in which *N*-glycolylmuramic acid exists, and a more modified peptidoglycan containing a muramyl lactam is found in the cortex of bacterial spores (95).

A tetrapeptide is attached to the carboxyl group of the 3-*O*-D-lactic acid linked to many of the *N*-acetylmuramic acid residues. The tetrapeptide most frequently contains amino acids in the order L-alanine, D-glutamic acid, a diamino acid, and a terminal D-alanine. The diamino acids found in the tetrapeptide include *meso*- and LL-diaminopimelic acid, L-lysine, L-ornithine, and L-diaminobutyric acid. Descriptions of these and several variations of the tetrapeptide sequences can be found in reviews by Ghuysen (77), Rogers et al. (82), and Cummins (75).

Many of the tetrapeptide chains are cross-linked to one another. The extent of cross-linking and the nature of the linkages vary from organism to organism. The linkage may be a direct peptide bond from a dibasic amino acid on one tetrapeptide to a D-alanine on another tetrapeptide, or it may involve one or several amino acids as bridges between one tetrapeptide and another (73, 75, 82, 85).

The unique occurrence of *N*-acetylmuramic acid in murein provides a means of determining the presence of peptidoglycan in bacterial cells and can be used to estimate eubacterial biomass in soil and water samples. Diaminopimelic acid is found in macromolecules only as a component of peptidoglycan, and positive assays for this compound indicate the presence of peptidoglycan. However, not all peptidoglycan polymers contain this diamino acid (75, 88).

Cell walls of gram-positive eubacteria often also contain teichoic acids, teichuronic acids, proteins, and lipoteichoic acids. Teichoic acids are complex polymers of polyols (ribitol and glycerol) linked together in a backbone by phosphodiester bonds. Several variations in the backbone structure have been described previously (75, 84). These polymers, which make up 50% of the cell wall of some gram-positive bacteria, are good antigens, and their serological reactivity has been used to define antigenic groups of several genera (84). Cell wall teichoic acids are covalently linked to peptidoglycan. Teichoic acids of the glycerol phosphate polymer type also occur as lipoteichoic acids which contain covalently linked glycolipid. They are found in the plasma membrane of cells. The most common antigenic determinants are sugars or amino sugars attached through the hydroxyl groups of the polyols of the backbone. Alanine is often found attached to the polyols as well.

Teichuronic acids are acidic polysaccharides which contain uronic acids but no phosphate. Polymers of glucuronic acid with *N*-acetylglucosamine and aminomannuronic acid and glucose occur in *Bacillus licheniformis* and *Micrococcus lysodeikticus*, respectively. Other uronic acid polymers have been observed as cell wall components of other gram-positive bacteria (75).

Cell wall polysaccharides and lipoteichoic acids of gram-positive eubacteria, particularly the streptococci, are responsible for antigenic differences which have enabled the separation of these bacteria into serological groups (75). Cell wall polysaccharides of other gram-positive bacteria form a heterogeneous group of macromolecules that usually contain neutral sugars and occasionally contain amino sugars. These polysaccharides are of value in distinguishing between strains and species, especially when unusual or uncommon sugars are present.

Proteins occur widely as surface components of gram-positive bacteria. Surface antigens of streptococci include numerous acid-soluble proteins (75).

The cell walls of gram-negative eubacteria are more complex than those of gram-positive cells. They have a comparatively low peptidoglycan content, and in enteric bacteria the peptidoglycan is covalently bonded to lipoproteins, which probably form a bond between the peptidoglycan and the outer membranous layer of the cell wall (92). The outer membranous layer is rich in lipopolysaccharide.

Lipopolysaccharides are extremely complex molecules with molecular weights higher than 10,000 (72, 73, 76, 78, 81, 89, 91, 92, 94, 97). The lipopolysaccharide from *Salmonella* spp. is made up of three different parts: the lipid moiety, a core polysaccharide, and the *O*-polysaccharide. The lipid moiety, termed lipid A, contains a phosphorylated glucosamine disaccharide heavily esterified through hydroxyl and amino groups with fatty acids that are 10 to 22 carbon atoms in length. A major fatty acid component of the lipid A fraction is β-hydroxymyristic acid, which is unique to the lipid A portion of the lipopolysaccharide molecule.

The core polysaccharide serves as a link between lipid A and the *O*-polysaccharide. It is composed of four basal sugars, KDO, phosphate, and ethanolamine. KDO is unique to this part of the lipopolysaccharide molecule.

In *S. typhimurium* the sugars of the core polysaccharide are glucose, galactose, heptose, and glucosamine (72, 97). The core polysaccharide is covalently bonded through C-2 of KDO to C-2 of a glucosamine residue in lipid A. The core polysaccharide is also covalently linked to the variable *O*-polysaccharide chains. Several lipopolysaccharide subunits may be bonded together by pyrophosphate bonds in the lipid A portion of the molecule. The lipid A portion is embedded in the outer membrane with the *O*-polysaccharides extending outward (72, 73). The *O*-polysaccharides are the major antigenic determinants (the *O* antigens) of the surface of gram-negative bacteria. They often serve as receptors of bacteriophage attachment.

The fine structure and composition of the outer membrane of gram-negative bacteria suggest that it is a bilayer containing phospholipids and proteins with variable amounts of lipopolysaccharide. The outer membrane encloses the periplasmic space, an enzyme-containing compartment bounded by the plasma membrane on the inside (72, 73). Several gram-negative cells possess proteinaceous outer coats or thick carbohydrate layers outside the lipopolysaccharide zone extending from the outer membrane.

It is impractical to present here all methods used for isolating and characterizing structural components of cell walls. The methods used depend on the organism, the facilities available, and the objective of the investigation. For some investigations, intact highly purified polymers are not required; for others, macromolecular integrity and purity are required. Persons who wish to thoroughly characterize one of these polymers from a bacterium should first consult the literature describing the different isolation techniques and the advantages and limitations of each (72, 73, 85–87, 97, 98). The simplest methods for isolation and preliminary characterization of murein peptidoglycans and lipopolysaccharides from the more ordinary laboratory strains of bacteria are presented here. As mentioned below, these methods are generally applicable to isolation of the pseudomureins found in some archaeobacteria, but their characterization requires the examination of different components (section 22.5.4).

22.5.1. Peptidoglycan

a. Reagents and equipment

Saline solution: 0.9% (wt/vol) NaCl in distilled water.

Trypsin: crystalline enzyme preparation.

Tris buffer. Dissolve 1.21 g of Tris in 90 ml of distilled water. Adjust the pH to 7.0 with 5 N HCl, and add additional water to 100 ml. (*CAUTION:* HCl is dangerous; see section 22.10.10.)

DNase. Dissolve 5 mg of dry powdered enzyme preparation in 5.0 ml of Tris buffer.

RNase. Dissolve 5 mg of dry powdered enzyme preparation in 1.0 ml of Tris buffer.

Refrigerated centrifuge, capable of achieving 22,000 × g.

Heavy-walled glass or Pyrex 30-ml centrifuge tubes, centrifuge bottles (100 to 250 ml), and rotors to accommodate each. Plastic centrifuge tubes and bottles can be used if they are stable to chloroform. (*CAUTION:* Thin-walled Pyrex glass tubes will not withstand 22,000 × *g*.)

Sonicator, French pressure cell, or colloid mill.

b. Procedures

The following procedure is appropriate for ca. 200 to 1,000 ml of culture (corresponding to ca. 0.4 to 4 g of dry-cell weight) and can be scaled up as needed. Cool the bacterial cells in their growth medium to 4°C with ice, and harvest by centrifugation in 100- to 250-ml centrifuge bottles (5,000 × *g* for 15 min). Resuspend the cells in 20 ml of cold saline solution, and centrifuge again. Use the packed cells immediately, or lyophilize them and store them in a freezer to prevent autolysis. Suspend the cells homogeneously in 20 ml of cold water. If the cells have been lyophilized, suspend them in stages by first wetting the cells with cold water to form a paste and subsequently adding liquid slowly while stirring the suspension. Mix the cells mechanically or with a hand-operated homogenizer to obtain complete dispersal of the cells. Resuspension by use of a Pasteur pipette is adequate if mechanical homogenizers are unavailable. Keep the suspension cold to prevent autolysis of cell walls during all manipulations. Disintegrate the cells by using a sonicator, mechanical grinding device, or French pressure cell.

As soon as possible after disruption of the cells, heat the suspension to 75°C; keep the suspension at this temperature for 15 min to inactivate autolytic enzymes. (For mesophiles, this is effective. However, thermophiles may have to be heated at higher temperatures, ca. 30°C above their maximal growth temperature; alternatively, autolytic activity may best be controlled by keeping the cells at 4°C. Experience with a given organism is needed to choose the best approach.) Centrifuge at a low speed (1,000 × *g* for 10 min) to remove unbroken cells. Discard the sediment, and centrifuge the supernatant fraction in 30-ml centrifuge tubes at 22,000 × *g* for 15 min. A small opaque layer of unbroken cells under a translucent layer of cell walls should be obtained. Remove the upper layer carefully by gently washing the sediment with a stream of ice-cold 0.1 M Tris buffer (pH 7) delivered from a Pasteur pipette. When the upper layer is totally suspended in 15 ml of buffer, centrifuge again at 1,000 × *g* for 10 min, discard the sediment, and centrifuge the supernatant fraction at 22,000 × *g* for 15 min. Resuspend the upper layer in the Tris buffer. Repeat these steps until a homogeneous cell wall fraction, free from intact cells, is obtained.

Suspend the cell wall fraction in 100 ml of Tris buffer that has been saturated with chloroform. Add DNase (0.5 ml of a 1-mg/ml solution) and RNase (1.0 ml of a 5-mg/ml solution), and incubate the suspension at 37°C for 30 min. Add 100 μg of crystalline trypsin, and continue the incubation for 6 h at 37°C. Trypsin is added to remove denatured proteins. This enzyme causes lysis of the peptidoglycan of some bacteria. If this occurs, centrifugation will not yield a precipitate, and the trypsin treatment must be omitted. Sediment the cell walls by centrifugation, and wash them as described above. Continue the sedimentation and washing until a homogeneous layer is obtained.

Purification of peptidoglycan from most gram-negative cells involves the use of much larger amounts of cell material (20 to 30 g) because of the small amount of peptidoglycan per unit weight of cells and the greater susceptibility of the peptidoglycan of these bacteria to autolysis. A procedure for preparation of 150 mg of peptidoglycan from *E. coli* has been described by Weidel et al. (96).

The amounts of hexosamine and muramic acid can be estimated after acid hydrolysis of the purified cell walls (see sections 22.3.7 and 22.3.8). The amino acid content can be determined with an amino acid analyzer, by HPLC, or by thin-layer chromatography (98). The presence of many amino acids indicates contamination by membrane, cell wall, or proteins.

Good examples of approaches for a more complete characterization of cell walls involving a variety of chemical assays to determine reducing sugars, unsubstituted amino groups of amino acids, etc., have been provided by Kottell et al. (90), Ghuysen et al. (77, 85), and Kandler et al. (86–88).

22.5.2. Total Lipopolysaccharides (79)

a. Reagents, materials, and equipment

Saline solution: 0.9% (wt/vol) NaCl in distilled water.

Phenol solution. Using a fume hood, add 10 ml of distilled water to 90 ml of warm (60°C) high-quality phenol. (*CAUTION:* Phenol is dangerous; see section 22.10.19.)

Centrifuges, rotors, and tubes: a centrifuge capable of achieving 10,000 × *g* and an ultracentrifuge capable of achieving 100,000 × *g*, centrifuge bottles (100 to 250 ml) and a rotor for them to be used for harvesting cells from the medium, 30-ml centrifuge tubes and rotors to hold them for both centrifuges.

Dialysis tubing.

Rotary evaporator.

Lyophilizer or vacuum desiccator and a vacuum source.

b. Procedures

Centrifuge the bacteria from ca. 3 liters of growth medium at 10,000 × *g* for 15 min. Suspend the sedimented cells in 300 ml of ice-cold saline solution, and centrifuge again. Repeat the resuspension and centrifugation to remove medium constituents. Lyophilize the bacteria in the final sediment, or store the packed cells in a freezer until used.

Thoroughly suspend 20 g (dry weight) of bacteria in 350 ml of water at 65 to 68°C. Heat 350 ml of phenol solution to 65 to 68°C, and add it to the bacterial cell suspension (*CAUTION:* Phenol is dangerous; see section 22.10.19). Stir vigorously, and keep the mixture at 65°C for 15 min; then cool it to 10°C in an ice bath. Centrifuge the emulsion at 10,000 × *g* for 30 min. Three layers result: a water layer, a phenol layer, and a sediment. Sometimes a layer of denatured protein is formed at the phenol-water interphase. Remove the upper, aqueous layer, and save it. Remove and discard the material at the interface. Add another 350 ml of water to the phenol layer and sediment, and mix thoroughly while the temperature is held at 65 to 68°C for 15 min. Centrifuge the mixture again, and combine the upper, water layer with that obtained from the first centrifugation step. Dialyze the aqueous suspension for 3 to 4 days at 4°C against distilled water to remove the phenol and low-molecular-weight contaminants. Concentrate the dialyzed solution to 5 ml at 37°C under a vacuum. A rotary evaporator is useful for this purpose if one is available. Centrifuge the concentrate at 3,000 × *g* for 10 min to remove insoluble material. Discard the sediment. Half the suspended organic material in the supernatant fraction should be lipopolysaccharide. The major contaminant is RNA. Centrifuge the supernatant fraction at 100,000 × *g* for 1 h. Thoroughly resuspend the sediment in saline solution, and centrifuge again at 100,000 × *g* for 1 h. Repeat the centrifugation and resuspension of the sediment five times to wash it free from contaminating material. Finally, dissolve the sediment in distilled water, and lyophilize the material until dry. If a lyophilizer is unavailable, the suspension can be dried in a vacuum desiccator. A yield of 100 to 500 mg of lipopolysaccharide should be obtained from 20 g (dry weight) of species of the family *Enterobacteriaceae*. This lipopolysaccharide is a potent endotoxin (72, 76, 92) and contains firmly bound lipid A; it should be handled carefully to avoid contact that could lead to ingestion or inhalation.

22.5.3. Lipid A and Polysaccharide from Lipopolysaccharides

The lipid A moiety can be obtained as a water-insoluble product after mild acid hydrolysis of purified lipopolysaccharide. The polysaccharide portion of the polymer is soluble in water.

a. Reagents, materials, and equipment

Acetic acid solution (1%, vol/vol). Add 1 ml of acetic acid to 99 ml of distilled water. (*CAUTION:* See section 22.10.2.)

Lyophilized lipopolysaccharide preparation.

Lyophilization ampoules, 15 ml.

Dialysis tubing.

Tank of compressed nitrogen gas (*CAUTION:* See section 22.10.16).

Vacuum desiccator or lyophilizer.

Centrifuge capable of achieving 5,000 × *g*, 15-ml centrifuge tubes, and an appropriate rotor. Temperature control is not important.

b. Procedure

Suspend 0.1 g of the dried lipopolysaccharide, obtained as described above, in 10 ml of a 1% acetic acid solution that has been gassed with nitrogen for 10 min.

Seal the suspension in an ampoule under N_2, and heat it in a steamer for 4 h. Cool and open the vial, transfer the contents to a centrifuge tube, and centrifuge at 5,000 × *g* for 10 min. Carefully remove the supernatant from the insoluble lipid A. Save the supernatant fraction which contains the polysaccharide portion of the lipopolysaccharide.

Resuspend the sediment in distilled water, and centrifuge again. Repeat the resuspension and centrifugation five times. Discard the supernatant fractions. Dry the sediment in a vacuum desiccator or by lyophilization.

Dialyze the polysaccharide remaining in solution after acid hydrolysis against 200 volumes of water, and dry the contents of the bag by lyophilization or in a vacuum desiccator. Between 40 and 100 mg of lipid-free polysaccharide and 30 to 150 mg of lipid A should be obtained from 100 to 500 mg of lipopolysaccharide.

Assays for KDO and heptose can be used to estimate the amount of lipopolysaccharide or lipid-free polysaccharide (section 22.3.9). Hexosamine assays (section 23.3.6) can be used to estimate the amount of lipid A in purified preparations.

22.5.4. Archaeobacterial Cell Wall Components

The archaeobacteria as a group are different in several ways from eubacteria; these differences include the lack of murein-containing peptidoglycan, D-alanine, muramic acid, and ester-linked alkane lipids (60, 88). Most archaeobacteria do have a cell wall and can be characterized by the color of staining as gram positive or gram negative; however, no outer membrane or lipopolysaccharide is present in the gram-negative types and no common cell wall component is known within the archaeobacteria. Instead, a variety of cell wall structures are present. The most common gram-positive cell walls include the pseudomurein found in *Methanobacterium* and *Methanobrevibacter* species and the heteropolysaccharides found in *Methanosarcina* and *Halococcus* species. Several types of protein and glycoprotein cell walls have been found in the gram-negative archaeobacteria.

Both pseudomurein and the heteropolysaccharide can be isolated by methods similar to those used with eubacteria, as described above in section 22.5.1, i.e., physical disruption and trypsin degradation (88). Amino acid components can be determined after acid hydrolysis by using methods similar to those in section 22.7.13, and carbohydrates and aminosugars can be analyzed by methods similar to those in section 22.3. The pseudomurein is characterized by the presence of *N*-acetylglucosamine and of *N*-acetyltalosaminuronic acids as a replacement for *N*-acetylmuramic acid. No known archaeobacterial cell walls contain D-alanine or diaminopimelic acid. The heteropolysaccharide cell walls contain principally glucose, galactosamine, and uronic acids.

22.6. NUCLEIC ACIDS (also see Chapter 26.3)

The most widely used method for estimation and identification of deoxyribose and ribose nucleic acids (DNA and RNA) in cell extracts involves extraction and hydrolysis with a hot solution of trichloroacetic (or perchloric) acid followed by specific colorimetric assays for pentose components of nucleic acids (100). A more convenient means of measuring the amounts of DNA and RNA in pure solutions takes advantage of their A_{260}. Rapid and reliable quantitation of nanogram to microgram amounts of DNA can be achieved by fluorescence techniques (101, 103, 104, 107). These techniques are much more sensitive than are either of the spectrophotometric methods, and one fluorescent method using bisbenzimide is less sensitive to interference by small amounts of RNA, protein, and other macromolecules. Other procedures less applicable to the purposes of this manual are described in other reference books (99, 105, 108).

The estimation of nucleic acids in bacterial cultures by the techniques described below requires that the cells be present in sufficient numbers in a noninterfering medium or that they be able to be harvested from the growth medium, washed free from interfering medium components, and resuspended at an appropriate concentration without lysis of the cells or hydrolysis of the nucleic acids by nucleases. Total extraction of nucleic acids from the cells is essential for accurate determination of their nucleic acid content.

22.6.1. Total DNA by Diphenylamine Reaction

The reaction of DNA with diphenylamine in a mixture of acetic and sulfuric acids is the most widely used of the colorimetric procedures for the determination and identification of this polymer in complex mixtures. The chemistry of the reaction and the nature of the chromophore formed remain unknown. The Burton method (102), described below, has the advantages that it is sensitive and that few substances in bacterial cells interfere with its accurate determination of DNA.

a. Reagents, materials, and equipment

Perchloric acid: 0.5, 1.0, and 2.5 N solutions (*CAUTION:* Perchloric acid is very dangerous; See section 22.10.17).

Diphenylamine reagent. Dissolve 1.5 g of purified high-quality diphenylamine (white crystals; melting point, 52.9°C) in 100 ml of acetic acid, and add 1.5 ml of reagent-grade concentrated sulfuric acid (*CAUTION:* Acetic and sulfuric acids are dangerous; see sections 22.10.2 and 22.10.28). On the day it is used, add 0.1 ml of a 16-mg/ml aqueous solution of acetaldehyde (1 ml of acetaldehyde in 50 ml of water) to each 20 ml of the diphenylamine reagent (*CAUTION:* Acetaldehyde is very volatile and toxic and boils at 21°C; see section 22.10.1).

Saline-citrate solution (SSC): 0.15 M NaCl plus 0.015 N trisodium citrate (pH 7.0).

Standards: Dissolve crystalline DNA (0.4 mg/ml), from one of several commercial sources, in 5 mM NaOH. Prepare standards every 2 weeks by mixing measured volumes of the stock DNA solution with an equal volume of 1 N perchloric acid, and heat at 70°C for 15 min. Prepare dilutions in 0.5 N perchloric acid. (*CAUTION:* See section 22.10.17.)

Most DNA preparations contain some salts and water, and it is difficult to obtain or prepare DNA totally free from water to use in the preparation of standards. One should be aware of the inaccuracies resulting from the use of DNA as a standard and should account for the salts and an estimate of the water, both of which can be obtained from the supplier. One alternative is to calibrate the DNA by using the UV method described below in section 22.6.4. As an alternative, a DNA standard can be hydrolyzed, and its phosphate content per milligram of DNA can be determined (106). The results can then be reported as micrograms (or milligrams) of DNA-phosphorus rather than micrograms of DNA.

Deoxyribose prepared as an aqueous solution (1.0 mM) is also a convenient standard. Solutions are stable for 3 months in a refrigerator. Dilutions made in 0.5 N perchloric acid can be used to construct a standard curve for the diphenylamine assay. Results are reported as deoxyribose equivalents of DNA if this standard is used.

Spectrophotometer or colorimeter and glass cuvettes.

Centrifuge, capable of achieving 10,000 × g, which can be refrigerated or placed in a cold room.

Centrifuge tubes, 15 or 30 ml.

Water baths at 70°C and an incubator or water bath at 30°C.

Test tubes, Pyrex glass, 15 by 125 mm.

b. Procedure

A suspension of 5 to 10 mg (dry weight) of bacterial cells per ml (about 5×10^{10} to 1×10^{11} cells per ml for bacteria the size of *E. coli*) is required. Harvest the cells from the medium if the culture does not contain sufficient cells or if the medium contains interfering compounds. Complex media contain nucleic acid components and sugars that must be eliminated by harvesting the cells from the media and washing them by resuspension and centrifugation. Most bacteria can be harvested by centrifugation at 5,000 × g for 15 min at 4 to 5°C. Suspend the sedimented cells in 1/10 of the culture volume of SSC. Centrifuge the cells at 5,000 × g for 15 min at 5°C, and resuspend them in ice-cold SSC. If the cells are grown in mineral salts medium and the cell concentration is sufficient, samples of the culture can be used directly for extraction of DNA.

Acidify each sample of bacterial culture or suspension of bacteria in SSC with sufficient 2.5 N perchloric acid to give a final concentration of 0.25 N perchlorate. Cool the acidified sample on ice for 30 min, and centrifuge at 10,000 × g for 10 min. Using a glass rod, suspend the precipitate containing the nucleic acids in 0.5 ml of 0.5 N perchloric acid. Add an additional 3.5 ml of 0.5 N perchloric acid, and heat the suspension at 70°C for 15 min. It is important to stir or mix the suspension occasionally during this step. Centrifuge the heated suspension at room temperature for 10 min at 5,000 × g and carefully decant the supernatant fraction into a 10-ml graduated test tube. Reextract the precipitate in an identical fashion. Combine the two extracts, and mix. Record the volume of the combined extracts.

For the diphenylamine reaction, mix samples of various volumes from the combined extracts with sufficient 0.5 N perchloric acid to bring the final volume to 2 ml. It is usually wise to use a 10-fold range of sample sizes from each extract (0.2, 1.0, and 2.0 ml) to ensure that the amount of color developed in one sample falls on the standard curve. After the volumes of each sample have been adjusted to 2.0 ml, add 4.0 ml of the diphenylamine reagent to each tube, and mix thoroughly. Adjust the volume of the liquid in tubes containing dilutions of the standard (5 to 100 μg/ml) to 2.0 ml with 0.5 N perchloric acid, and mix the contents with 4.0 ml of the diphenylamine reagent. Prepare a blank for the spectrophotometer by adding 4 ml of the diphenylamine reagent to 2.0 ml of 0.5 N perchloric acid. Incubate the tubes at 30°C for 16 to 20 h (or overnight). If the contents of the tubes are cloudy, centrifuge the tubes at 10,000 × g for 15 min to remove unhydrolyzed protein. Determine the A_{600} of the supernatant fractions against the blank. Estimate the concentration of DNA in each tube from the absorbance values by using a standard curve prepared by plotting the absorbance of standard solutions against the DNA, deoxyribose, or phosphate concentrations of the standards. (If small [1-ml] cuvettes are available, the assay volumes can be appropriately reduced.)

It is not important to control the temperature carefully. If all tubes, including the standards, are incubated at the same temperature, the results should not be affected. A standard curve should be prepared with each set of determinations.

22.6.2. Total DNA by Fluorescence Spectroscopy

Two dyes that bind to DNA, bisbenzimide and ethidium bromide, have commonly been used for fluorescence measurements (101, 103, 107). Bisbenzimide (also called Hoechst 33258 dye) has little affinity for RNA and preferentially binds to regions of DNA rich in adenosine and thymidine residues. The fluorescence of the dye is enhanced when bound to DNA, and the excitation maximum shifts from 356 nm for the unbound dye to 365 nm when the dye is bound to DNA. The emission spectrum of the free dye peaks at 492 nm, and the emission maximum in the presence of DNA is 458 nm.

Ethidium bromide intercalates into the DNA double helix, and its fluorescence is enhanced. This reaction is often used to detect DNA in agarose gels. Detection of DNA with ethidium bromide is not as sensitive as with Hoechst 33258. However, the binding of ethidium bromide to DNA is not affected by the base composition. DNA samples with high guanosine and cytosine contents give less fluorescence than do DNA samples rich in adenosine and thymidine when Hoechst 33258 is used. Ethidium bromide also binds to RNA and thus should not be used to quantitate DNA if significant amounts of RNA are also present.

22.6.2.1. Bisbenzimide (Hoechst 33258 dye) binding

a. Reagents and equipment

Concentrated dye solution: 10 mg of Hoechst 33258 dissolved in 10 ml of distilled water. The solution should be stored refrigerated and protected from light. It is stable for 6 months.

10× TNE buffer: 100 mM Tris buffer, 1 M NaCl, 100 mM EDTA (pH 7.4).

Working dye solution A: suitable for measuring solution containing 10 to 400 μg of DNA per ml. Add 10 ml of 10× TNE buffer to 90 ml of distilled water; mix and filter through a nitrocellulose filter (pore size, 0.45 μm). Add 10 μl of the concentrated dye solution to 100 ml of the filtered 1× TNE buffer.

Working dye solution B: suitable for measuring DNA in solution at concentrations between 100 ng/ml and 15 μg/μl. Prepare and filter 1X TNE buffer as described for working dye solution A. Add 100 μl of the concentrated dye solution to 99.9 ml of the 1X TNE buffer.

DNA standard solutions: A DNA with approximately the same base composition as the DNA sample should be used as a standard. Supercoiled plasmid DNA has different dye-binding characteristics from those of linear DNA. Therefore, a plasmid DNA such as pBR322 should be used as a standard when the sample to be assayed is plasmid DNA. Prepare solutions of DNA from 25 to 250 mg/ml in 1 × TNE buffer. The range of standards to be used will depend on the concentration of DNA in the samples to be assayed. A 10-fold range of DNA concentrations in the standard should be used to establish a standard curve.

Fluorimeter. Dedicated instruments with preset activation and emission wavelengths are relatively inexpensive (e.g., Hoeffer model TKO 100). A scanning spectrofluorimeter is also suitable.

Fluorometric cuvettes: glass cuvettes with 1-cm sides.

Liquid measurement devices to accurately measure 2.0 ml and 2.0 μl.

b. Procedure

Add 2.0 ml of the working dye solution A or B to a cuvette, depending on the amount of DNA in the standard or sample solutions. Set the excitation wavelength at 365 nm and the emission wavelength at 458 nm if a scanning spectrofluorimeter is used. Set the fluorimeter to read zero with the working dye solution as a blank. Add 2 μl of a standard DNA solution and mix with a Teflon stirring rod or by capping and inverting the cuvette. Record the emission in relative fluorescence units. Repeat with fresh working dye solutions and the other standards, and plot a standard curve of relative fluorescence versus DNA concentrations. If a concentration readout is available, the readout can be set to equal that of a DNA standard. Add 2.0 μl of the sample solutions, and read the relative fluorescence. Estimate the concentration of DNA in the samples by comparison with the standard curve or by using the concentration readout.

22.6.2.2. Ethidium bromide binding

a. Reagents and equipment

10 × TNE buffer: see above.

Ethidium bromide assay solution. Add 10 ml of 10 × TNE buffer to 89.5 ml of distilled water. Filter through a 0.45-μm nitrocellulose filter, and add 0.5 ml of a solution containing 1 mg of ethidium bromide per ml in distilled water.

Fluorimetric cuvettes: glass cuvettes with 1-cm sides.

Scanning spectrofluorimeter.

Liquid measurement devices to accurately measure 2.0 ml and 2.0 μl.

b. Procedure

Add 2.0 ml of the ethidium bromide assay solution into a cuvette. Set the excitation wavelength at 546 nm and the emission wavelength at 590 nm. The remainder of the procedure is identical to that described for assaying DNA samples by using Hoechst 33258 dye.

22.6.3. Total RNA by Orcinol Reaction

The reaction of aldopentoses with acidified orcinol to produce a green chromogen is a classical reaction of sugar chemistry. This reaction detects ribose moieties present in RNA. Deoxyribose gives 20% of the color of ribose. The method is suitable for the measurement of RNA or ribonucleotides in solution as well as in tissue or cell extracts (100, 106).

a. Reagents and equipment

Ferric chloride reagent. Dissolve 100 mg of $FeCl_3 \cdot 6H_2O$ in 100 ml of reagent-grade concentrated HCl. (*CAUTION:* HCl is a strong and volatile acid. Use in a fume hood. See section 22.10.10.)

Ethanolic orcinol solution. Dissolve 6.0 g of orcinol in 100 ml of ethanol (*CAUTION:* See section 22.10.8). Store in a dark bottle. If the crystals are yellow or brown, use a new bottle or seek advice from a chemist on recrystallization.

Orcinol reagent: On the day the reagent is to be used, mix equal volumes of the ferric chloride reagent with the ethanolic orcinol solution. The complete reagent solution should be bright yellow when prepared (*CAUTION:* This is a strong HCl solution; see section 22.10.10).

Perchloric acid (*CAUTION:* This is a dangerous acid; see section 22.10.17).

Standard solutions for RNA analysis. Heat a solution containing 200 μg of RNA per ml with 0.25 N perchloric acid at 70°C for 30 min. Add an equal volume of 0.1 N HCl to give a final RNA concentration of 100 μg/ml.

AMP is a preferred standard because its concentration can be accurately measured. Prepare a solution containing 100 μg/ml by heating 200 μg/ml in 0.25 N perchloric acid. Dilute the solution with an equal volume of 0.1 N HCl.

Spectrophotometer or colorimeter and cuvettes or tubes.

Water bath, set at 90°C.

Test tubes, Pyrex glass, 18 by 125 mm.

Clear glass marbles, approximately 20 mm in diameter.

b. Procedure

Prepare hydrolysates from suspensions of bacterial cells by treatment with perchloric acid exactly as described for the extraction of DNA (section 22.6.1). Dilute the cell hydrolysates with 0.1 N HCl so that they contain 10 to 100 μg of RNA per ml. (Cell hydrolysates from 0.8 to 1 mg [dry weight] of cells contain ca. 100 μg of RNA.) Mix the diluted perchlorate hydrolysates of the cells or standards with 2 volumes of the orcinol reagent, cap the tubes with marbles, and heat them at 90°C for 30 min. (The volumes used can vary depending on the spectrophotometer or colorimeter available. For most spectrophotometers and colorimeters, mix 1.0 ml of the extracts or standards with 2 ml of the orcinol reagent.) Cool the tubes under running tap water, and determine the A_{665} of each tube. Determine the concentration of RNA or nucleotides from a standard curve prepared with known amounts of AMP or RNA. The standard curve should be repeated each time a sample or series of samples is assayed. The assay can reliably measure 15 to 100 μg of RNA in an extract of bacterial cells. When AMP is used as a standard, the results should be reported as purine riboside equivalents of RNA.

22.6.4. DNA and RNA by Other Methods

Pure DNA or RNA solutions, free from contaminating protein and carbohydrate, can be measured by UV spectroscopy. A solution containing 1 mg of pure nucleic acid per ml has an A_{260} of ca. 22 to 23.

A rapid and sensitive alternative to the colorimetric analysis of DNA in cells and tissues has been described by Gauchel and Zahn (105). This procedure is based on the assay of thymine in hydrolysates of cell samples by gas chromatography. This procedure has the disadvantage that access to an appropriate gas chromatograph is required, but if DNA assays are routinely performed, the use of a gas chromatograph for this purpose can be justified because of the savings in time.

a. Applications

Perchloric acid extracts approximately 95% of the nucleic acid from bacterial cells and endospores. Alternative extraction procedures can be used if incomplete extraction is indicated by the presence of DNA in the soluble fraction after a third extraction of the precipitate from the second hot perchloric acid treatment. Cells can also be extracted conveniently with 5% (wt/vol in water) trichloroacetic acid by heating at 90°C for 15 min. The extract can be diluted to the desired volume with 5% trichloroacetic acid and assayed for ribose by the orcinol procedure described above (22.6.3).

When small amounts of DNA are available, the fluorescence assays described in section 22.6.3 are useful because very small amounts of sample are consumed and the assays are much more sensitive than other methods. These methods are particularly useful for measuring polymerase chain reaction products, plasmids, and other DNA preparations prior to ligation reactions and transformation of cells during cloning of recombinant DNA molecules. These reactions are not subject to interference by small amounts of protein, and they have been used to measure DNA levels in cellular homogenates (101).

If an estimate of the concentration of purified DNA in solution is to be made, the sample of the solution can be treated once in 0.5 N $HClO_4$ at 70°C for 15 min. Portions of the treated DNA solution can be assayed by the diphenylamine reaction described above. The precipitation of the nucleic acid with cold perchloric acid causes losses when small amounts of DNA or other macromolecules are present. Therefore, this step should be omitted when pure DNA solutions are assayed. If it is necessary to precipitate the DNA from dilute solutions to remove interfering compounds such as nucleotides, the addition to each tube of 100 μg of a protein such as bovine serum albumin facilitates complete precipitation of the DNA (except when UV methods are to be used, since protein also absorbs in the UV).

The colorimetric and UV absorption assays are applicable to any class of RNA. All RNA classes have the same extinction coefficient when the colorimetric assay is used. Proteins, polysaccharides, and many other UV-absorbing compounds (including phenol, which is often used in the preparation of nucleic acids) compromise the accuracy of the UV absorption assay. Deviations from a normal spectrum indicate contamination with protein, carbohydrate, or other compounds.

The orcinol procedure for measuring RNA is subject to interference by hexoses, which produce a brown color. Therefore, the UV absorbance assay is preferred for the measurement of RNA in sucrose gradients or solutions containing significant levels of hexoses or hexose polymers.

The UV absorbance assays are nondestructive, a matter of importance when small amounts of purified nucleic acids are available, and they are more convenient to perform.

It is possible to calculate the DNA content per cell from the DNA content of a suspension if the number of cells per unit volume is known. The nucleotide content per cell can be determined, and these data can be used to calculate the genome size of an organism (99).

22.7. NITROGEN COMPOUNDS

Choice of the most suitable method for analysis of nitrogen in a compound depends on the form of nitrogen analyzed, the concentration of nitrogen, and the presence of interfering compounds. For nitrite analysis, the most widely accepted method involves modification of the diazotization and coupling reactions. For nitrate nitrogen, ammonia nitrogen, and organic nitrogen, many other methods have been used. A comparison of the sensitivities and uses of some of these analyses is given in Tables 1 and 2. For purposes of this manual, the simplest widely applicable method for each nitrogen compound will be described. References to other methods are given in Tables 1 and 2.

Because of the diversity of substances that may be analyzed for nitrogen (such as milk, fertilizer, soil, and plant and animal materials), many of the methods in Table 1 must be modified or the sample must be pretreated to fit the method. Many variations of the procedures listed or described here have been published, and the reader should refer to the literature for procedures directly applicable to a particular kind of sample.

22.7.1. Nitrite

The analysis of nitrite nitrogen is based on the reaction of NO_2^- with sulfanilamide in an acid solution to form a diazo compound. This compound reacts with *N*-(1-naphthyl)ethylenediamine to form a colored dye whose A_{543} is measured in a spectrophotometer (119, 127). Sulfanilic acid and α-naphthylamine can be substituted for the two analytical chemicals; however, α-naphthylamine has been identified as a carcinogen and should be used only under carefully controlled conditions.

This reaction can detect as little as 1 μg of NO_2^- per liter when cuvettes with a 10-cm light path are used. The absorbance of the colored compound follows Beer's law (see Chapter 21.1.1) up to 180 μg of NO_2^- per liter (114). Ions which interfere by causing precipitation are listed in Table 1. The presence of any of these may require the use of an alternate method of NO_2^- detection, such as reduction and subsequent analysis as NH_4^+. As

Table 1. Comparison of analytical techniques available for analysis of nonprotein nitrogen compounds

Compound	Method	Sensitivity (μg of N/liter in sample)	Interference	Advantages	Disadvantages	References
NO_2^-	Diazotization	1	Cl^-, colored samples, particulates; Sb, Fe(III), Pb, Hg, Ag, Cu(III) ions	Simple	Cl^- and Fe^{3+} interfere	109, 114, 119, 127, 142
	Reduction to NH_4^+, further analysis	>2,000	NO_3^-, NH_4^+	Good for high nitrite	Hot alkali, distillation, nonspecific	114, 116
NO_3^-	Reduction to NO_2^-, further analysis[a]					114, 119, 128, 142
	Measurement of A_{220}	40	Many organics, NO_2^-, surfactants, chromate, colored samples	Simple	Requires UV spectrophotometer, interference by organics	114, 121, 122
	Brucine	100	Colored samples, NO_2^-, oxidants, reductants, high organics; Cl, Fe, Mn ions.	None	Brucine toxicity, low sensitivity	114
	Chromotropic acid	100	Colored samples; Ba, Pb, I, Se, Cr ions.	Color stable for 24 h	Low sensitivity	114
	Nitrophenoldisulfonic	20	Organics, Cl^-	Simple	Requires drying	109, 110, 144, 156
	Reduction to NH_4^+, further analysis	>2,000	NO_2^-, NH_4^+	Good for high nitrate	Hot alkali, distillation, nonspecific	109, 110, 114, 131
NH_4^+	Oxidation to NO_2^-, further analysis	Varies	Amino acids, atmospheric NH_3	No distillation	Some amino acids yield nitrite	119
	Indophenol blue (phenate)	10	Particulates, colored samples, excess acid or base, phenol, small N compounds, Cl^-	Simple; preferred Kjeldahl method	Phenol; may need distillation	109, 114, 153
	Nessler reagent	400	Particulates, colored samples; Mg, Fe, Ca, S, Mn, Cl ions; small organics	Good for pure water samples	Hg; may need distillation	109, 114
	Acidimetric	2,000	Cl^-	Good for high ammonium; after distillation, simple; preferred for Kjeldahl method	Distillation	109, 114, 116
	Pyridine-pyrazole	50	Fe, Zn, Ag, Cu ions; cyanate	Specific for NH_4^+	Pyridine	119, 137
Total N	Kjeldahl digestion, reduction to NH_4^+	Variable, insensitive	Few	Quantitative	Distillation; sulfuric acid	109–111, 114, 116, 117, 126, 131

[a]For futher information and sensitivity, see analysis listed under appropriate nitrogen class.

Table 2. Comparison of analytical techniques available for protein assay

Method	Sensitivity[a] (µg/ml in assay solution)	Interference	Advantages	Disadvantages	References
Lowry	ca. 8	Ammonium sulfate, amino acid and peptide buffers, mercaptans	Very sensitive	Variation with amino acid content; nonlinear with Beer's Law; multistep and timed	123a, 129, 139, 140
Biuret	ca. 200	Amino acids and peptide buffers	Rapid and simple	Not highly accurate or sensitive; variation with amino acid content	112, 116, 138
Bradford dye binding	1–4	Some detergents	More uniform response than Lowry to proteins; rapid, simple, very sensitive	Use of phosphoric acid; nonlinear with Beer's law	125, 149, 154
Bicinchoninic acid/copper	1–5	Thiols and mercaptans, Uric acid, Others[b]	Very sensitive	Timed reaction	133, 148, 155
UV absorbance	ca. 0.5	Nucleic acids, any aromatic or phenolic compound, many coenzymes, some buffers	Very sensitive, simple and rapid	Requires UV spectrophotometer and quartz cuvette; serious interference	136, 138, 158
Fluorometric	ca. 0.0003	Amino acid or peptide buffers, some coenzymes	Very sensitive, simple and rapid	Variable protein response; requires fluorometer and special cuvette; dioxane use	124, 147

[a]Sensitivity is given in micrograms per milliliter of solution used in the assay; this can be converted easily to micrograms per milliliter of original sample by accounting for all the dilution factors. Where a range between two numbers is given, both the extrasensitive microassay and the more routine, less sensitive assays are described. Interfering substances may reduce this sensitivity but still allow accurate assays in an appropriate range.
[b]Glucose, Tris buffer, and ammonium sulfate can cause interference but can be corrected with proper blanks or with proper pH adjustment.

with any colorimetric procedure, colored or particulate substances in the sample may interfere; this is particularly important with 10-cm cuvettes. Interference by colored substances may be lessened by dilution prior to the reaction if the NO_2^- concentration is high enough to permit this. Alternatively, treatment with activated charcoal often is effective for removal of interfering compounds. Particulate substances (such as bacterial cells) can be removed by centrifugation or filtration through a small-pore (0.45-µm) filter.

An ion-specific electrode is also available for estimation of nitrite (see Chapter 21.2.2).

a. Sampling

Samples should be assayed immediately or frozen at −20°C until the assay can be run. Samples for NO_2^- assays should not be preserved with acid. Neutralization of all samples to pH 7.0 is suggested before analysis.

b. Reagents

NO_2^--free water. Most sources of distilled or demineralized water are free from NO_2^-. If not, water may be prepared as follows. Add 1 ml of concentrated H_2SO_4 (*CAUTION:* This acid is corrosive and dangerous; see section 22.10.28) and 0.2 ml of $MnSO_4$ solution (36.4 g of $MnSO_4 \cdot H_2O$ per 100 ml of distilled water) to each liter and make pink with 1 to 3 ml of $KMnO_4$ solution (400 mg of $KMnO_4$ per liter).

Sulfanilamide reagent. Add 5 g of sulfanilamide to 300 ml of distilled water containing 50 ml of concentrated HCl (*CAUTION:* Concentrated HCl is corrosive and dangerous; see section 22.10.10). Adjust the volume to 500 ml with distilled water. The solution is stable at room temperature for many months.

Standard NO_2^- solutions (114). Prepare a stock solution by dissolving 0.13 g of high-purity $NaNO_2$ in 500 ml of distilled water. Use a new bottle of $NaNO_2$ since NO_2^- is readily oxidized in the presence of water. Keep the stock bottle tightly stoppered. One milliliter contains ca. 50 µg of NO_2^--N. Dilute the stock to prepare standards for the standard curve.

Spectrophotometer and cuvettes.

c. Procedure

Add 0.1 ml of the sulfanilamide solution to 5.0 ml of the sample. Allow it to react for a minimum of 2 min but not longer than 8 min. Undesirable side reactions and decomposition are significant after 10 min (119). Add 0.1 ml of *N*-(1-naphthyl)ethylenediamine solution, and mix immediately. Allow color development for at least 10 min, and then measure the A_{543} (see Chapter 21.1.1.4).

Prepare a set of standards from the 50-µg/ml NO_2^- stock solution to give 0 to 25 µg of NO_2^--N per ml. Plot $[NO_2^-]$ versus A_{543} (see Chapter 21.1.1). Using the standard curve, determine the concentration of NO_2^--N in the sample by reading the concentration from the curve or by determining the extinction coefficient (ϵ) (the slope of the line) from linear regression analysis. This extinction coefficient can be used in successive determinations with the same reagents. A new standard curve should be prepared when new reagents are prepared.

22.7.2. Nitrate

The method for analysis of NO_3^- outlined here was chosen because of its relative simplicity, and the reagents prepared for the NO_2^- determinations are also used in this assay. The basis for this method is the reduction of NO_3^- to NO_2^-, followed by the analysis of NO_2^-. Two procedures have been commonly used: reduction by zinc (128, 142) and reduction by cadmium (114); both are described here.

The reduction of NO_3^- to NO_2^- with Zn is a simple procedure involving the reaction of the sample with Zn dust for 30 min, followed by filtration. The disadvantage of this procedure is the manipulation time required for each sample.

The reduction of NO_3^- with Cd amalgamated with Cu is performed by passing the sample treated with NH_4Cl (to prevent deactivation of the column) through a previously prepared column of Cd-Cu^{2+} filings. This procedure requires little effort at the time of analysis but requires prior preparation of the reduction column.

After either of the reduction procedures described here, the analysis for NO_2^- is performed as outlined above in the method for analysis of nitrite. Since the analysis is for total nitrite, part of the sample must be analyzed for NO_2^- before reduction, and this amount must be subtracted from the total NO_2^- to determine the NO_3^- concentration.

An ion-specific electrode is also available for measurement of nitrate (see Chapter 21.2.2).

22.7.2.1. Sampling

Samples should be treated as described above for nitrite.

22.7.2.2. Zinc reduction procedure

a. Reagents

Use of NO_2^--free distilled water (section 22.6.1) is recommended in the preparation of all reagents.

Buffer. Mix 50 ml of 0.2 M KCl with 34 ml of 0.2 M HCl, and dilute to 400 ml. Adjust the pH to 1.5 with HCl. (*CAUTION:* See section 22.10.10.)

NH_4Cl (2%, wt/vol).

Zinc powder.

b. Procedure

Add 1.0 ml of 2% NH_4Cl and 2 ml of buffer to 25 ml of each sample in 50-ml stoppered flasks, and bring to 20 to 25°C. Add 0.2 g of zinc powder, and let stand for 30 min with occasional shaking. Filter out the zinc particles by using a low-porosity filter paper.

22.7.2.3. Cadmium reduction procedure

a. Reagents

Copper sulfate reagent: 2% (wt/vol) $CuSO_4 \cdot 5H_2O$.

Concentrated NH_4Cl. Add 175 g of NH_4Cl to 500 ml of distilled water.

Dilute NH_4Cl. Dilute 1 volume of concentrated NH_4Cl with 40 volumes of NO_2^--free water to give 4.38 g of NH_4Cl in 500 ml of water.

Reagent grade cadmium sticks.

b. Preparation of columns

Columns should be ca. 1 by 30 cm with an appropriate opening (e.g., thistle tube or funnel) at the top so that the total sample can be placed on the column at once. The bottom of the column should have a burette-type stopcock or be attached to rubber tubing with a pinch clamp to allow the flow to be stopped before the column runs dry. The bottom of the column should be packed with glass wool or fine Cu "wool" turnings. A small amount of glass wool or copper filings is often added to the top of the column to prevent disturbance of the amalgamated cadmium-copper filings when adding the sample.

Cadmium filings can be prepared by filing sticks of reagent-grade cadmium metal with a coarse metal hand file (about second cut). Collect the fraction that passes a 2-mm sieve but is retained on a 0.5-mm sieve. Stir approximately 50 g of filings with 250 ml of 2% $CuSO_4 \cdot 5H_2O$ until the blue color disappears from solution and semicolloidal Cu particles begin to enter the liquid above the sediment. Allow the suspension to settle for 10 min. Fill the column with dilute NH_4Cl solution or with the supernatant fraction from the Cd-Cu amalgamation; then slowly pour in the Cd-Cu filings, and let them settle until a column height of about 30 cm is achieved. Wash the column three times with 100 ml of the dilute NH_4Cl solution. The flow rate should be adjusted to approximately 10 ml/min.

c. Procedure

Add 2 ml of concentrated NH_4Cl reagent to 100 ml of the sample. Add 5 ml of this solution to the column, and allow the liquid to drain to the top of the column. This small addition ensures that when the bulk of the sample is added to the column, no error results from mixing with the previously run sample.

Add the remainder of the sample NH_4Cl solution. Collect and discard the first 25 to 30 ml of the eluting liquid. Then collect 50 ml for the analysis. Discard the remainder of the eluting liquid. A column can be reused up to 100 times and usually need not be rinsed between samples. However, if the column will not be used for a few hours, rinse it with dilute ammonium chloride.

The 50 ml of eluate collected for analysis is assayed for NO_2^- (section 22.6.1) as soon as possible after reduction.

22.7.3. Ammonium by Indophenol Blue Reaction

Many methods are available to determine the concentration of NH_4^+ in samples (Table 1). An easy determination of NH_4^+ is based on the reaction of ammonia, hypochlorite, and phenol to form indophenol blue. The reaction is catalyzed by Mn^{2+}. The color is measured in a colorimeter or spectrophotometer at 630 nm (109, 114, 153, 156). The major disadvantages of this method are the use of phenol and the instability of the reagents.

Another easy determination of NH_4^+ is Nessler's method, in which the NH_4^+ reacts with mercury and iodide in an alkaline solution to yield a colored complex which is measured in a colorimeter or spectrophotometer (109, 114). The major disadvantage of this method is the use of mercury compounds. Nessler's method is given in section 22.7.4.

An alternative, and almost as simple, method, which can use the reagents prepared for NO_2^- determination, involves the oxidation of NH_4^+ to NO_2^- (119) with subsequent analysis of NO_2^-. However, the oxidation requires several reagents, as well as 3.5 h for the oxidation; therefore, the indophenol blue method or Nessler's method is recommended. If a technique very specific for NH_4^+ is required, the pyridine-pyrazole method (119, 137) may be more satisfactory than the other techniques. Also see Chapter 21.2.2.1 for information on commercially available NH_3 electrodes. Some applications may benefit from the ammonium ion-specific electrodes, but they are generally not as accurate or as sensitive as the methods described here; however, they can provide a continuous readout, which can be very useful for some applications, e.g., monitoring the contents of a fermentor.

a. Samples

Samples should be treated as described in section 22.7.1. Samples high in interfering compounds (Table 1) may require distillation before the analysis is performed (114).

b. Reagents

Ammonium-free water. Add 0.1 ml of concentrated H_2SO_4 (*CAUTION:* This is a dangerous acid; see section 22.10.28) to each liter of distilled water, and redistill. Prepare fresh for each sample group, since NH_4^+ fumes from the laboratory will rapidly contaminate the water. Alternatively, an ion-exchange column may be used. Resins should be selected for maximum binding of NH_4^+ and other interfering organic compounds.

0.003 M $MnSO_4$. Dissolve 50 mg of $MnSO_4 \cdot H_2O$ in 100 ml of distilled NH_4^+-free water.

Hypochlorite reagent. Add 10 ml of commercial bleach (5% sodium hypochlorite solution) to 40 ml of NH_4^+-free distilled water. Adjust the pH to 6.5 to 7.0. Prepare fresh weekly. (*CAUTION:* See section 22.10.27.)

Phenate reagent: Dissolve 2.5 g of NaOH and 10 g of phenol in 100 ml of NH_4^+-free distilled water. Prepare fresh weekly. (*CAUTION:* Phenol is dangerous [see section 22.10.19], and NaOH is a dangerous base [see section 22.10.26].)

Standard NH_4^+ solution: Prepare a stock NH_4^+ solution by dissolving 381.9 mg of anhydrous NH_4Cl (dried at 100°C) in NH_4^+-free water, and adjust the volume to 1 liter. One ml contains 100 μg of N and 122 μg of NH_3. To prepare standards, dilute the stock solution to obtain a range of 0.1 to 5 μg of NH_4^+-N per 10 ml.

c. Procedure

Place 10 ml of each sample and 0.05 ml (1 drop) of the $MnSO_4$ solution in a small beaker or flask. With constant and vigorous stirring, preferably with a magnetic stirrer, add 0.5 ml of the hypochlorite reagent. Immediately add 0.6 ml of phenate reagent (alkaline phenol) dropwise.

The reaction is complete in 10 min, and the color is stable for 24 h. Measure the color in a colorimeter or spectrophotometer at 630 nm. Prepare a blank and a standard curve each day, since the color intensity may change with the age of reagents.

22.7.4. Ammonium by Nessler Reaction

a. Samples

Samples should be treated as described in section 22.7.1. Those high in interfering compounds (Table 1) may require distillation before analysis (114).

b. Reagents

Ammonium-free water. Prepare as described in section 22.7.3.

Reagents for removing large amounts of interfering cations. Prepare $ZnSO_4$ solution by dissolving 100 g of $ZnSO_4 \cdot 7H_2O$ in ammonium-free water to a final volume of 1 liter. Prepare 6 N NaOH by dissolving 240 g of NaOH in ammonium-free water to a final volume of 1 liter. (*CAUTION:* NaOH is a dangerous base. See section 22.10.26.)

Reagent for removing trace amounts of interfering cations. Dissolve 50 g of disodium EDTA in 60 ml of ammonium-free water containing 10 g of NaOH. (Heat may be required to dissolve the chemicals.) Adjust the volume of the cooled solution to 100 ml with ammonium-free water.

Nessler's reagent. Dissolve 100 g of HgI_2 and 70 g of KI in a minimal amount of water (100 to 200 ml). Add this slowly and with stirring to 500 ml of 8 N NaOH at 20 to 25°C. Adjust the volume of this mixture to 1 liter. Store in a rubber-stoppered glass vessel in the dark. The reagent is stable for up to 1 year. (*CAUTION:* This reagent is toxic; see sections 22.10.12 and 22.10.26.) Check the reagent monthly by comparing a standard curve of NH_4^+-N using the reagent with the standard curve prepared when the reagent was new. When the curves deviate significantly, discard the reagent.

Standard NH_4^+ solution. Prepare a stock NH_4^+ solution as outlined in section 22.7.3. To prepare the standards, dilute the stock solution to obtain a range of 20 to 500 μg NH_4^+-N per 50 ml.

c. Procedure

To remove interfering ions of Ca, Mg, S, and Fe, two methods are used. For large quantities of interfering ions, add 1 ml of the $ZnSO_4$ reagent to 100 ml of sample. Adjust the pH to 10.5 with 0.4 to 0.5 ml of 6 N NaOH by using a pH meter. The interfering ions precipitate out of solution within 15 min. The precipitate usually sediments other suspended matter as well. Remove the precipitate by filtration (use NH_4^+-free filter paper) or by centrifugation. If experience with the samples indicates that only trace amounts of the interfering ions are present, add 0.05 ml (1 drop) of the EDTA solution, with thorough mixing, to 50 ml of sample. Samples pretreated with $ZnSO_4$ and alkali are often also treated with EDTA. Samples with no interfering cations may be treated with Nessler's reagent without pretreatment.

The Nessler reaction is performed by adding 1 ml of Nessler's reagent to 50 ml of sample. If EDTA has been added to the sample, 2 ml of Nessler's reagent should be used. Mix the sample-reagent solution well, and let it stand for at least 10 min at room temperature. (For low NH_4^+ levels, 30 min is preferred.) Keep the temperature and reaction time constant for all samples, standards, and blanks. Read the absorbance of the resulting colored compound at a single wavelength between 400 and 425 nm for samples and standards with low NH_4^+ (20 to 250 μg of NH_4^+-N per 50 ml) and at a wavelength between 450 and 500 nm for those with high NH_4^+ (250 to 500 μg/50 ml). This reaction can be used with proportionally smaller volumes of sample, standards, and reagents as required by the investigator. The sensitivity of the method can be increased if a cuvette with a 5-cm light path is available. Calculate the amount of ammonia in each sample from a standard curve prepared by plotting the absorbance of each standard versus the concentration of ammonia in the standard solution.

22.7.5. Protein Analysis Problems

A wide variety of methods are available for determination of protein in biological samples, five of which are described below. Most sensitivities vary from ca. 1 to 1,000 μg/ml, and this factor may determine which assay is preferable. Kjeldahl analysis of crude protein is also possible (152), but is seldom used for bacteria. Several problems prevent any one method from being the best for all purposes. First, a variety of substances interfere to different degrees with the different assays, e.g., commonly used detergents, reducing agents, or buffers (Table 2). Second, different proteins often give different responses, including the proteins used for standardization. Third, some assays are more time-consuming to perform, or require careful timing for proper quantitation. From ca. 1960 till 1980, the Folin reaction (Lowry assay) (139) was the most commonly used, but in the past decade, the Coomassie blue G method (Bradford assay) has become increasingly popular (125, 154). The newest method gaining popularity is the bicinchoninic acid-copper (BCA) assay (155).

When conducting protein assays, the practitioner is well advised to examine the literature for known interferences and to examine blanks experimentally, using the buffer and other components minus protein, for possible problematic reactions. In most cases, a high level of precision is not required, e.g., when measuring protein eluting from a chromatography column or even when measuring specific activities during a purification procedure. When a fairly precise quantification is needed, the use of buffer controls minus protein is mandatory, and all assays should be performed in duplicate; the use of two types of assays is recommended to counter the variation in the response of individual proteins; two such good pairs are the Bradford and BCA or the Bradford and Lowry assays. If samples of a specific protein of interest are available, it would be most appropriate to use this protein as a standard.

Protein assay is also used in the development of gel electrophoresis experiments. The Coomassie blue R dye method is commonly used, but the silver stain method is gaining popularity, especially because of its high sensitivity. However, the gel-staining methods are not described in this section; see Chapter 21.5.1.

22.7.6. Protein by Folin Reaction (Lowry Assay)

From ca. 1960 till 1980, the most widely used colorimetric method for estimation of proteins was that of Lowry et al. (139), using the Folin reaction. The blue color appearing in the assay is due to the reaction of protein with copper ion in alkaline solution (the biuret reaction) and the reduction of the phosphomolybdate-phosphotungstic acid in the Folin reagent by the aromatic amino acids in the treated protein. The following procedure is useful for proteins that are already in solution or that are soluble in dilute alkali.

a. Reagents

Reagent A. Dissolve 20 g of Na_2CO_3 in 1 liter of 0.1 N NaOH (*CAUTION:* see section 22.10.26).

Reagent B. Dissolve 0.5 g of $CuSO_4 \cdot 5H_2O$ in 100 ml of a 1% (wt/vol) aqueous solution of sodium tartrate.

Reagent C. Just before use, mix 50 ml of reagent A and 1 ml of reagent B. Discard after 1 day.

Reagent D. Diluted Folin reagent. Concentrated Folin-Ciocalteau phenol reagent can be purchased commercially from many suppliers and is diluted as specified by the manufacturer; normally, it is provided as a 2 N acid solution by suppliers and should be diluted to 1 N with distilled water on the day of use. This reagent is usually purchased premade and is not made in individual laboratories. (*CAUTION:* This reagent contains a strong volatile acid [HCl] and phosphoric acid; see also sections 22.10.10 and 22.10.20.)

Standard protein solution. Any one of several commercially available proteins or standards is suitable. Crystalline bovine serum albumin or prepared solutions of bovine serum albumin that can be purchased from biological supply houses produce suitable linear standard curves. The stock standard solution should contain 0.5 mg of protein per ml in distilled water. Different proteins give different standard curves. The protein standard used should be specified when results are reported.

b. Procedure

Add up to 0.5 ml of a sample containing 25 to 125 μg of protein to a clean 10-ml test tube. Adjust the volumes of each sample to 0.5 ml with distilled water. Add 2.5 ml of reagent C, and mix well. Allow this mixture to stand for at least 10 min at room temperature. Add 0.25 ml of reagent D, and mix immediately. The reactivity of this reagent lasts only seconds after addition to the alkaline, protein-containing solution. After 30 min or longer at room temperature, measure the A_{500} in a spectrophotometer or colorimeter. Greater sensitivity can be achieved by measuring the A_{750} if an appropriate spectrophotometer is available.

If 1-ml cuvettes and a suitable instrument are available, all volumes can be reduced by a factor of 2.5 and reagents and sample can be conserved. The samples and standards can be adjusted to 0.2 ml, and 1.0 ml of reagent C and 0.1 ml of reagent D are then added at the time intervals indicated above.

When protein is to be measured in bacterial-cell suspensions or when proteins are insoluble in dilute alkali, it is necessary to heat the suspension at 90°C in 1 N NaOH for 10 min to obtain complete solubilization; the suspension should generally be centrifuged to remove particulates prior to assay, and it may be advisable to wash the pellet with 1 N NaOH to obtain good recovery of the proteins. The standard should be treated in the same way, because this treatment reduces the intensity of the color. When this procedure is used to dissolve or extract proteins, a reagent containing 20 g of Na_2CO_3 in 1 liter of water (no NaOH) is substituted for reagent A. The assay procedure is otherwise identical.

Ammonium sulfate in amounts greater than 0.15% reduces color development. Investigators should be aware of this when this assay is used for proteins purified by fractionation procedures based on differential solubilities in ammonium sulfate solutions. High concentrations (1 to 5 M) of many other salts, some sugars, glycine, Tris buffer, some chelating agents, and compounds with sulfhydryl groups also interfere with the assay. Aromatic amino acids produce color with the phenol reagent, and glycine interferes with color development.

A modification of the Lowry assay that permits its use with samples containing sucrose or EDTA and with membrane and lipoprotein preparations without prior solubilization of the proteins has been described by Markwell et al. (140). The procedure employs a detergent (sodium dodecyl sulfate) in the alkaline carbonate reagent to improve solubilization of samples containing lipoidal material and an increase in the amount of the copper tartrate reagent to overcome interference by sucrose and EDTA. This procedure has several applications, including assays of proteins in fractions from sucrose gradients.

A commercially available modified Lowry kit has recently been described (123a). Its advantages include a more convenient Lowry assay and one which is insensitive to detergent interference.

22.7.7. Protein by Biuret Reaction

Peptide bonds of proteins form blue chelates with copper ions in alkaline solutions (114, 130). This reaction is much less sensitive than is the Lowry procedure but has the advantage that it is less time-consuming and can be used in the presence of ammonium salts and other materials that interfere with the Lowry procedure. There is also less variability among proteins in the amount of color formed per unit weight of protein.

The modification of the biuret procedure (130) described below involves the use of tartrate to form a copper complex that is soluble in NaOH. The protein displaces the copper ion from the complex to form a copper-protein complex with a different color and absorption intensity.

a. Reagents

Dissolve 1.5 g of cupric sulfate$\cdot 5_2$O and 6 g of sodium potassium tartrate$\cdot 4H_2O$ in 500 ml of water. Add, with stirring, 300 ml of a 10% (wt/vol) solution of sodium hydroxide in water (*CAUTION:* NaOH is a dangerous base; see section 22.10.26). Adjust the volume of the mixture to 1 liter with distilled water. Discard if a reddish or black precipitate is evident.

b. Procedure

Add 2.0 ml of the above biuret reagent to a neutralized sample (0.5 ml) containing 0.5 to 5 mg of protein. Mix thoroughly, and incubate at room temperature for 30 min. The time and temperature for all samples, standards, and blanks should be the same. Measure the A_{500}. Prepare a blank with 0.5 ml of water or aqueous solution to dissolve proteins and 2.0 ml of biuret reagent. A standard curve can be prepared by using any completely soluble protein such as crystalline bovine serum albumin. The response is not linear with protein concentration because of competition between tartrate and protein for copper ions.

c. Applications

A modification of this procedure described by Stickland (157) is suitable for the analysis of total protein in bacterial cells. In this procedure the cellular protein is dissolved in 1.0 N NaOH, and then $CuSO_4$ without tartrate is added. The insoluble cellular material and insoluble $Cu(OH)_2$ are removed by centrifugation, leaving the colored Cu-protein complex in solution.

22.7.8. Protein by Coomassie Blue Reaction (Bradford Assay)

Proteins bind several dyes, causing a shift in the absorption spectrum of the dye. In the procedure described below, Coomassie brilliant blue G is used as the dye. This dye exists in two forms: the red anionic form is converted to a blue form when the dye binds to amino groups of proteins. This procedure, described in detail by Bradford, is often more sensitive and less subject to interference by many compounds that restrict the use of other assays (125, 154). The dye assay can be performed more rapidly than those described above. The sensitivity of the dye assay is a little higher than that of the Lowry procedure (Table 2). The response to different proteins is less variable than the color reactions in the other assays described above.

a. Reagent

Dissolve 100 mg of Coomassie brilliant blue G250 in 50 ml of 95% ethanol (*CAUTION:* See section 22.10.8). Add 100 ml of 85% (wt/vol) phosphoric acid (*CAUTION:* This is a strong acid; see section 22.10.20). Adjust the volume to 1 liter with distilled water. (Commercially prepared reagent kits are also available, e.g., from Bio-Rad Laboratories or Pierce Chemical Co.)

b. Procedure

Add a protein solution containing 5 to 50 μg of protein in 0.05 ml to a test tube. Add 2.5 ml of the dye reagent, and thoroughly mix the contents (or follow instructions for the commercially prepared reagents). Measure the A_{595} after 2 min and before 1 h against a blank containing 0.05 ml of the buffer or salt solution and 2.5 ml of the dye reagent. Relatively inexpensive plastic cuvettes can be used multiple times for assays; moderately expensive glass cuvettes can also be used; however, quartz cuvettes bind to the dye more firmly and, if used, must be rinsed with ethanol or acetone at appropriate intervals.

The protein content of the sample is determined from a standard curve obtained by plotting the absorbance of standard solutions containing 10 to 60 μg of protein in the same 2.55-ml assay volume as the samples, versus the concentration of protein in each standard solution. The same buffer used to dissolve the unknown samples is used with the standards.

c. Applications

The Coomassie blue procedure is applicable to the assay of many soluble proteins. Detergents interfere with color development; therefore, glassware should be rinsed well. The assay is linear from 25 to 75 μg of protein per sample. Other limitations of the assay are described by Pierce and Suelter (149). A microassay procedure (125) and another variation that has high reproducibility and can detect 1.0 μg of protein in an assay volume of 1 ml have also been described (154).

22.7.9. Protein by Bicinchoninic Acid-Copper Reaction

In an alkaline solution, proteins react with Cu^{2+} to produce Cu^{+}, as in the biuret assay; bicinchoninic acid

(BCA) reacts with the Cu^+, forming a purple, water-soluble product which can be measured by its A_{562} (155). The sensitivity of the BCA method is about equal to that of the Bradford method and a little higher than the Lowry method (Table 2). It is compatible with most detergents used in the laboratory and with many buffer components; it may be the preferred assay for laboratories in which detergent is commonly used in protein solutions (155). The assay is interfered with by high levels of ammonium sulfate, EDTA, and sulfhydryl reagents; the sulfhydryl types of interference can be remedied by their modification with iodoacetamide (133). The assay requires mixing two or three reagents on the day of the assay, but the individual solutions are stable for many days. The response variation between different proteins is approximately similar to that of the Lowry and Bradford methods and can vary a maximum of 20 to 40%, depending on the protein standard used. The assay time is slightly shorter than required for the Lowry method and involves incubation for a set time and at a specific temperature; the assay is less complex than the Lowry method but more so than the Bradford method.

a. Reagents

A commercially prepared reagent is available from Pierce Chemical Co.; it is essentially the same as described here and as originally described by Smith et al. (148, 155). Both a regular assay and a more sensitive micro modification (148) are possible; the regular assay is described here.

The regular working reagent is prepared by mixing two stock solutions (reagent A and reagent B). The stock solutions can be stored indefinitely at room temperature, but the mixture of the two can be stored for only 1 day.

Reagent A. Dissolve 10 g of sodium bicinchoninate in a 1-liter aqueous solution containing 20 g of $Na_2CO_3 \cdot H_2O$, 1.6 g of sodium tartrate, 4.0 g of NaOH, and 9.5 g of $NaHCO_3$; the pH may be adjusted, if necessary, to pH 11.25 by the addition of concentrated NaOH or solid $NaHCO_3$. (*CAUTION:* This reagent is a strong base and contains both sodium hydroxide and sodium carbonate; see sections 22.10.25 and 22.10.26.)

Reagent B. Dissolve 4 g of $CuSO_4 \cdot 5H_2O$ in 100 ml of deionized or distilled water.

b. Procedure

A working reagent is made by mixing 100 volumes of reagent A with 2 volumes of reagent B, resulting in an apple-green solution (148, 155). Add 1 volume of protein sample or buffer blank to 20 volumes of the working reagent in an appropriate tube (e.g., 0.1 ml of sample plus 2.0 ml of reagent), and mix gently. Incubate the tubes at 37°C for 30 min, and then cool them to room temperature; the protein-containing tubes will turn purple. (Alternatively, assays can be incubated at room temperature for 2 h or at 60°C for 30 min; the latter method enhances the sensitivity ca. fourfold.) Read the A_{562}, using a water blank. Subtract the absorbance of the buffer blank from all sample absorbances, and compare with a standard curve of protein concentration versus absorbances; bovine serum albumin is a suitable standard protein and should be diluted in the same buffer as used for the samples to be assayed and at concentrations similar to those of the protein being assayed, generally in the range of 10 to 120 μg/ml of assay working reagent.

c. Applications

The BCA method is a good method for the determination of proteins for a variety of purposes, but it is particularly appropriate for solutions containing most detergents, since most of those used as solubilizing agents do not interfere with the assay. Although thiol-containing reagents interfere with this method, suitable methods involving iodoacetamide have been developed to circumvent some of these problems (133); trichloroacetic acid precipitation or other methods may also be of value (143). Alternative procedures are available for enhancement of sensitivity (the micro BCA method [148]) by about fivefold; this variation is typically incubated at 60°C and is more subject to interference problems.

22.7.10. Protein by UV Absorption

The absorption maximum of proteins is 280 nm and is due to the presence of the aromatic amino acids tyrosine and tryptophan. These two amino acids are present in nearly all proteins, and their proportions relative to other amino acids usually vary over a narrow range. The absorption of UV light is a suitable means of estimating proteins in solution if the protein solution does not contain more than 20% by weight of other UV light-absorbing compounds, such as nucleic acids or phenols, and if the solution is not turbid. Correction for the absorbance of smaller amounts of nucleic acids in protein can be made by the method of Warburg and Christian (158) or of Kalckar (136).

a. Equipment

UV spectrophotometer, calibrated with an absorbance standard (see Chapter 21.1.1.4).

Cuvettes, made from quartz or silica, which do not prevent transmittance of UV light.

b. Procedure

The A_{280} and A_{260} are measured with the protein dissolved in a suitable buffer. A solvent blank containing the buffer is used to zero the instrument at each wavelength. If the ratio of A_{280}/A_{260} is not greater than 1.70, except with solutions known to contain pure protein, the following equation (136) should be used to calculate the concentration of protein, or an appropriate table (138) should be consulted. The equation is used to subtract the contribution of nucleic acids to the UV absorbance of the solution.

$$\text{protein concentration (mg/ml)} = 1.45A_{280} - 0.74A_{260}$$

c. Applications

When preservation of a protein sample is important because of its limited availability, this nondestructive assay procedure is the method of choice, provided that the protein is sufficiently pure to permit its application. Different proteins can yield different extinction coefficients, and a standard prepared from an identical protein (if available) should be used if precise measurements are important.

The use of absorbance to estimate protein is common during protein purification involving column chromatography, but the correction for A_{260} is seldom used. This method should be used with caution, since coenzymes present in early fractions can contribute significant nonprotein A_{280} values.

22.7.11. Total Nitrogen (Kjeldahl Digestion)

A sample may be analyzed for total nitrogen by adding the results of analyses for the separate nitrogenous components (NO_2^-, NO_3^-, NH_4^+, and organic nitrogen). However, if one is interested in only the total nitrogen, performing all these analyses can be time-consuming. The most commonly used analysis for total nitrogen is the Kjeldahl digestion of organic nitrogen compounds to yield NH_4^+, which is then assayed by one of the techniques for NH_4^+ analysis. NO_3^- and NO_2^- can be included in this analysis when desired by reduction to NH_4^+ with hot alkali (114) or salicylic acid and zinc (110). An alternative method, which allows use of reagents already prepared for the NO_2^- analysis, is based on the oxidation of the NH_4^+ from digestion to nitrite and subsequent analysis of nitrite.

The procedure used to digest the organic material depends on the state of the sample (solid or liquid). Because the procedures for the Kjeldahl digestions are variable, complex, and easily available in various texts, the reader is referred to the literature that treats the specific procedure. Kjeldahl digestion techniques developed for liquid samples include those for seawater (119), lake water (109, 114), and wastewater (109, 114). Techniques for solids include those for soil (109–111,117), sediment (114), plant materials (109, 116), fertilizers (116), animal materials (116), and microorganisms (114).

22.7.12. Molecular Nitrogen

The need to detect molecular nitrogen transformation in a biological system is usually limited to nitrogen fixation studies in which nitrogen (dinitrogen) is reduced by an enzyme system (called nitrogenase) to ammonia, which can then be assimilated by the cell. Nitrogen fixation can be estimated or monitored in two very different ways: by direct measurement of N_2 assimilated, or by assay for the enzyme nitrogenase (132).

Direct measurement could be made of the disappearance of N_2 (determined by gas chromatography or other physical analysis), but the abundance of N_2 in the atmosphere makes such analysis difficult. Direct measurement of ammonia formation is also possible, using analysis described in the above sections (sections 22.7.3 and 22.7.4).

Neither of these direct measurements are made in practice. Instead, the stable isotope, $^{15}N_2$, is used. This isotope is provided to cells or plants, enriched significantly above the natural abundance of 0.4%. In one variation of this approach, after time has been allowed for dinitrogen metabolism, the total organic nitrogen, as described in the section above, is obtained and then analyzed by mass spectrometry (120, 150). A more contemporary alternative is the use of ^{15}N-NMR, in which cells are allowed to metabolize the isotope and either extracts or whole cells are examined for ^{15}N enrichment in the proteins (123). Use of either the mass-spectrometric or NMR approach may be the best way to monitor N_2 metabolism, as well as metabolism of other nitrogenous substrates (e.g., nitrate, nitrite, and ammonium), and even some specific intermediates may be identified by the NMR approach (123). However, these approaches require the use of facilities dedicated to these types of analyses and equipment operated by well-trained staff. Despite this, most major universities and industrial companies in the United States have such facilities, and these methods are not particularly expensive by modern research standards.

However, the nitrogen fixation capability of cells is commonly measured by the ability of nitrogenase to reduce acetylene to ethylene. This requires less-expensive equipment than other methods and can be conducted in most microbiology and biochemistry laboratories. Acetylene and ethylene can be detected and quantified by use of a gas chromatograph (see Chapter 21.3.5). For qualitative determination of nitrogenase in systematic bacteriology, and for further discussion, see Chapter 25.2.7.

22.7.13. Specific Nitrogenous Compounds

A variety of specific nitrogen-containing compounds may be of interest to the investigator. These compounds include amino acids and nitrogen-containing cellular components not discussed above. Many of these components can be most conveniently measured by amino acid analyzers (sometimes with alternative detection methods) normally present in central facilities, but they can also be measured by the individual scientist using a variety of methods. Broadly applicable analysis methods include ion-exchange chromatography or HPLC, coupled with ninhydrin assay of eluates or coupled with spectrophotometric detectors of several sorts. Gas chromatography and mass spectrometry may also be of value (115, 118, 120). Discussed below are three specific examples of nitrogenous-compound analysis particularly applicable to the bacterial physiologist. For analysis of nucleotides, nucleosides, purines, and pyrimidines, see Chapter 21.3.1.1.

22.7.14. Total Amino Acids

It is often necessary to determine the amino acid composition of proteins, either from whole cells or from those specifically purified. Most research institutions have specific facilities for this analytical task; because of the large number of amino acids normally found in proteins (twenty) and a smaller number of unusual or atypical amino acids, such analysis is not feasible in most individual laboratories. However, HPLC methods coupled with sophisticated detectors can be used by some laboratories to supplant the more traditional "amino acid analyzer"; information on the variety of equipment for HPLC amino acid analysis can be obtained from a variety of suppliers (e.g., Waters, Beckman, Bio-Rad, Ranin) and from institutional facilities. In general, for work needed by most bacteriologists, protein hydrolysis and submission to a central facility is the best route; this is also likely to be the fastest and least expensive.

For analysis by central facilities, a wide array of methods are available for sample hydrolysis to convert the protein to individual amino acids. In general, treatment involves heating solutions of the protein in a strong acid in a sealed glass tube; most methods create difficulties for the stability and quantitation of at least one amino acid, and thus two parallel methods must often be used for measurement of all the desired amino acids. The resulting amino acid mixture can also be separated by a variety of methods, usually either ion-exchange or C-18 reverse-phase chromatography. Detection can be carried out by a postcolumn assay (e.g., the ninhydrin reaction) or by spectrophotometric or fluorometric methods involving either pre- or post-column derivitization (e.g., with dimethylaminonaphthalene sulfonyl chloride, phenylisothiocyanate, or orthophthalaldehyde). These methods are very sensitive and require very small amounts of protein; a corollary to this is that the material must be very pure, and contamination must be avoided during sample preparation (e.g., amino acids left by a fingerprint can be significant).

22.7.15. *meso*-Diaminopimelic Acid

meso-Diaminopimelic acid (DAP) is found in most gram-negative peptidoglycans and also serves as the precursor of dipicolinic acid in spores (113, 145, 146, 159). DAP is not a normal component of proteins, but its measurement in cell walls or spores is sometimes of interest. One analytical strategy is to submit a hydrolyzed sample for amino acid analysis as described above, because this can provide a good quantitative method. Another approach is to use HPLC methods, as indicated above for amino acids.

Described here is an alternative colorimetric procedure for the estimation of DAP, in which the ninhydrin reaction is used as described by Work (160). In this method, the reaction is carried out under acidic conditions, which makes it more selective for diamino acids.

a. Reagents, supplies, and equipment

Acetic acid (*CAUTION*: See section 22.10.2).
Ninhydrin reagent: 250 mg of ninhydrin dissolved in 6 ml of acetic acid and 4 ml of aqueous 0.6 M phosphoric acid (*CAUTION*: See section 22.10.20).
Spectrophotometer.
Glass or quartz cuvettes or spectrophotometer tubes (plastic can also be used if it is stable to concentrated acetic acid).
Glass test tubes capable of holding 10 ml.
DAP standard.
Samples, properly prepared for analysis.
Water bath at 37°C.

b. Procedures

Samples containing DAP can be prepared in any of a variety of ways. A processed portion of cells (e.g., peptidoglycan as described above in section 22.5.1) can be treated as described for protein hydrolysis in section 22.7.14.1, or alternative hydrolysis methods can be used. In most cases samples will contain low levels of a variety of amino acids, but contamination with significant quantities of protein, and the accompanying amino acids, creates difficulties with the assay unless some purification step is performed. Prior to hydrolysis, many protein-containing solutions can be deproteinized by addition of 1 volume of acetic acid to 1 volume of sample, mixing, and centrifugation to remove the protein as a pellet. The supernatant is used as the sample.

The following procedure can be conducted with samples containing lysine, ornithine, proline, cysteine, or tryptophan at levels similar to the DAP. Mix a 0.25-ml sample with 2.0 ml of acetic acid, and add 0.25 ml of the ninhydrin reagent. Heat the solution at 37°C for 1.5 h, and then read the A_{440} of the standard and sample by using a blank containing water instead of the sample added to the acetic acid and ninhydrin reagent. Construct a standard curve by using sample solutions containing ca. 10 to 450 μg of DAP per ml. (Hydrolysis of the original sample should not be done with HCl.)

This assay would also be appropriate for examination of fractions eluted during column chromatography of samples; e.g., acid-hydrolyzed samples could be applied to an AG 50W-X8 (H^+ form) column and eluted with a gradient of acid, with or without additional salts. The resulting fractions could be assayed with this procedure. (See also Chapter 21.3.1 for information on ion-exchange chromatography.)

22.7.16. Dipicolinic Acid

Dipicolinic acid (DPA) is found almost entirely in bacterial endospores (spores made by *Bacillus* and *Clostridium*), where it is a major and unique component, contributing 5 to 14% of the spore dry weight (145, 146). Following its extraction from spores, it can be estimated spectrophotometrically.

a. Reagents, supplies, and equipment

Culture of spore-forming bacterium grown on a medium appropriate for sporulation (alternatively, a pure spore preparation).
Methylene chloride (*CAUTION*: See section 22.10.14).
Toluene (*CAUTION*: Toluene is flammable and toxic; use a hood, and keep away from flames).
Centrifuge and centrifuge tubes.
Refrigerator or cold room at ca. 4°C.
Boiling-water bath.
Sulfuric acid (0.1 and 1.0 M solutions) (*CAUTION*: See section 22.10.28).
KOH (1.0 M solution) (*CAUTION*: See section 22.10.21).
Diethyl ether (*CAUTION*: See section 22.10.7).
pH meter.
DPA standard.
Spectrophotometer capable of use in UV and visible ranges.
Quartz cuvettes (or glass cuvettes for alternative assay).
$Fe(NH_4)_2(SO_4)_2 \cdot 6H_2O$.
Ascorbic acid.
Acetic acid (*CAUTION*: See section 22.10.2).

b. Procedures

Preparation of spores

Grow any *Bacillus* or *Clostridium* species under spore-forming conditions (see Chapter 4.4.9). Harvest cells and spores by centrifugation (ca. 5,000 × *g* for 15 min), and wash them once with water. In many cases, virtually all cells have undergone sporulation, but cell debris and vegetative cells will contaminate the spores, so some cleanup of the preparation is needed for some purposes; for other purposes, further cleanup is not needed. Either of two general approaches can be used to purify spores: degradation of nonspore material, or density gradient centrifugation (see Chapter 4.3.3).

In the degradation approach, residual vegetative cells are removed by treatment of a water suspension of cells plus spores with a protease (1 mg of papain per ml) at 4°C for 5 to 20 days in the presence of 0.1% of a 1:1 methylene chloride-toluene mixture (adapted from reference 161 with modifications). After this treatment, spores should be washed six times with a 20-fold volume excess of high-purity distilled water and either used directly, stored frozen, or dried at 80°C overnight and then stored dry at 4°C in a dry closed container.

Extraction of DPA

Since chelating metals should be avoided for analysis, all water should be high-quality distilled water. Resuspend spores (ca. 0.05 g dry weight or 0.15 g frozen wet) in 10 ml of 0.1 M H_2SO_4, heat in boiling water for 5 min, and then neutralize with 1.0 M KOH. Extract the suspension with 30 ml of alkali-washed diethyl ether, and, after phase separation, discard the ether and adjust the aqueous phase to pH 1.0 with 1.0 M H_2SO_4. Extract this solution with 30 ml of alkali-washed diethyl ether, and discard the aqueous phase. Add 5.0 ml of water to the ether, and evaporate the ether. This aqueous extract will contain the DPA. (Adopted from reference 141 with modifications.)

Estimation of DPA

Dilute the aqueous extract and prepared DPA standards in water as needed, to yield concentrations in the range of 30 to 150 μg/ml. Add 0.2 ml of these solutions to 1.8 ml of 0.1 M H_2SO_4 in a quartz spectrophotometer cuvette, and read against a water blank at 273 nm. Compare the samples with a DPA standard curve; a DPA concentration in the assay cuvette of 10 μg/ml will give an A_{273} of ca. 0.4. (Adopted from references 141 and 159 with modifications.)

An alternative assay method that is not sensitive to contamination by UV-absorbing material is available (134) and may not require ether extraction prior to assay. The sample from the ether extraction above (or in some cases from a direct dilution of the spore hydrolysis solution) is diluted in water to the same DPA concentrations as above. To 4.0 ml of this dilution is added 1.0 ml of the freshly prepared reagent containing 1.0% $Fe(NH_4)_2(SO_4)_2 \cdot 6H_2O$ and 1.0% ascorbic acid, both dissolved in 0.5 M potassium acetate buffer adjusted to pH 5.5 after their addition. After the sample and reagent are mixed, the A_{440} is measured within 2 h, and the DPA concentration is calculated from a standard curve. This assay has the advantage that quartz cuvettes are not required and a spectrophotometer capable of use below 320 nm is not required; as well, the assay is more specific, and with proper caution it can be used without ether extraction to purify the DPA prior to analysis.

22.8. ORGANIC ACIDS AND ALCOHOLS

A number of short-chain organic acids and alcohols are formed in bacterial cells as intermediates or end products of the citric acid, glycolytic, and other metabolic pathways. These compounds may be accumulated in the cytoplasm or excreted into the medium. In bacterial cell fractions, these compounds occur in the small-molecule, supernatant fluid fraction after precipitation of the macromolecules with cold trichloroacetic acid (Fig. 1). The small-molecule fraction may include acetate, acetoacetate, citrate, α-ketoglutarate, formate, lactate, succinate, pyruvate, malate, fumarate, ethanol, and many other compounds.

These small molecules may be analyzed by ion-exchange chromatography (Chapter 21.3.1), absorption chromatography (Chapter 21.3.2), HPLC with a variety of column materials, or gas-liquid partition chromatography (Chapter 21.3.5). Lactate, ethanol, acetate, and succinate may be also analyzed by enzymatic or colorimetric methods (see also section 22.2.8).

22.9. MINERALS

The amounts of different elements in bacterial cells, and the biological roles of trace elements, are described in references 162 and 164; references 162 and 166 describe sample preparation and storage. Metal analysis in samples of interest to bacteriologists can be accomplished by colorimetric methods, atomic absorption, photometry, spectrophotometry, flame emission photometry, nuclear activation analysis, the use of ion-specific electrodes, and other means; these and other methods for determining metals are described in references 161 to 164.

A rapid, accurate, and convenient titrimetric procedure that requires no specialized equipment can be applied to the analysis of some elements, such as calcium. Interfering elements and compounds present in small amounts can sometimes be removed by treatment of the samples. These procedures are described in references 164 and 165.

Flame emission photometry (162, 163) is ideally suited to the analysis of alkali and alkaline earth metals. **Atomic absorption analysis** (161, 163) is also an accurate and relatively convenient method for the analysis of several metals. Both of these methods require experienced personnel and a specialized spectrophotometer. The inaccessibility of proper advice or of a suitable instrument may limit the use of the technique. Several elements and compounds interfere with atomic absorption analysis of an element, but proper treatment of samples can reduce or eliminate interferences.

Nuclear activation analysis (163) involves the bombardment of a sample to make one or more elements radioactive. The radioactive species are identified and measured quantitatively. This method of elemental analysis has the advantages of sensitivity, and many elements can be assayed in a single small sample. The use of this technique requires access to a facility that has a high-energy neutron source (a nuclear reactor or accelerator), sophisticated isotope detection equipment, a suitable computer, and expert personnel to operate the facility. When available, this method is relatively inexpensive in comparison with other methods if time and other costs are taken into account and if several elements are determined simultaneously in a single sample. This is particularly true when the availability of sample material for analysis is limited.

X-ray emission spectroscopy (163) is a nondestructive method of analysis applicable to the estimation and identification of some elements that are present in sufficient abundance in biological material.

Electron probe X-ray microanalysis can be applied to the estimation and location of some elements in situ. The resolution of the technique is suitable for use in individual bacterial cells or spores (168), and even in procaryotic cells some estimation of the location of the elements within the cell can be made. This technique also requires sophisticated equipment and expertise. Thus, its application is generally limited to specialized research laboratories, but many universities and industrial companies have access to such instrumentation via their electron microscopy central facilities.

Methods for colorimetric and turbidimetric analysis of metals and inorganic compounds, including automated analysis, have been described by Snell and Snell (164) and in publications from the U.S. Environmental Protection Agency (165).

Commercially available selective ion electrodes also provide a very convenient and inexpensive means of measuring many elements. Their application is limited to a few types of samples and elements, including sodium, chloride, bromide, fluoride, copper, silver, and calcium (see Chapter 21.2.2).

22.10. HAZARDS AND PRECAUTIONS

The methods described above almost always require the use of chemicals, supplies, or equipment that are hazardous when handled or applied improperly. Provided here is a listing of chemical precautions needed to safely carry out most of the methods in this chapter; however, these warnings and precautions should be used in an intelligent way by suitably trained personnel and are meant to supplement their awareness of laboratory hazards. A more complete coverage of these and related topics is found in Chapter 29. In addition, any unfamiliar laboratory work should be attempted only after consultation with more experienced personnel. The laboratories to be used should meet minimum standards for safety, including adequate fire safety, availability of safety devices, plans for spills, and availability of fume hoods. Common sense should be used in all laboratory work.

If any serious exposure or injury occurs, seek medical attention promptly, after immediately initiating appropriate decontamination or first aid procedures.

Hazards of using chemicals are often dependent on the scale of chemical use. In many cases an experiment can be adjusted in size to require less material and thus to create less hazard in the conduct of the work, as well as to generate a smaller amount of dangerous waste to process. Conducting experiments at the smallest scale that will provide good data will be both safer and cheaper.

Several general rules apply to all chemicals as follows.

(i) Never mouth pipette anything in the laboratory, but use a pipetting device instead.

(ii) Avoid contact with chemicals by using proper laboratory practices.

(iii) Learn about the chemicals used before an accident occurs with them.

Listed here in alphabetical order are a variety of chemical and physical hazards that will be encountered in the work described above. Each of these is denoted by "*CAUTION:*" in the relevant section, and the reader is referred to this section by a specific section number.

22.10.1. Acetaldehyde

Vapors and liquid are toxic. Flammable. Chill bottle before opening, since the boiling point is 21°C. Use only in a fume hood.

22.10.2. Acetic Acid

A strong and volatile acid. Avoid exposure to both liquid and vapors. Use in a fume hood. Wear gloves and eye protection.

22.10.3. Acetic Anhydride

Flammable; avoid contact with flames or sparks. Irritates tissue; avoid contact with skin and eyes. Use in a fume hood.

22.10.4. Acetone

Highly flammable. Avoid contact of liquid or vapors with flame or sparks. Skin and respiratory tract irritant. Use in a fume hood when large quantities are handled.

22.10.5. Ammonium Hydroxide

Fumes and liquid are both caustic bases. Avoid contact with liquid and vapor. Use in a fume hood. Wear eye protection.

22.10.6. Chloroform

Chloroform is hepatotoxic, is a carcinogen, and acts as an anesthetic. Contact with liquid or vapor should be strictly avoided. Use in a fume hood.

22.10.7. Diethyl Ether

Ethers are very flammable and may explode when vapors contact flames or electrical sparks. Keep the minimum amount necessary in a laboratory. Warm all solutions in water baths rather than incubators. Avoid inhalation of vapors by use of a fume hood; ethers cause irritation of mucous tissues and can cause fainting. Ethyl ether can form explosive peroxides when evaporated to dryness; use only fresh containers, and properly dispose of older containers. Recycling of ethers should be done with great caution; if one container is used for several evaporations without being cleaned to remove peroxides, a spontaneous and life-threatening explosion may occur.

22.10.8. Ethanol

Flammable. Avoid using near open flame and electrical appliances that cause sparks.

22.10.9. Hexane

Very flammable. Avoid use near open flames and electrical appliances that cause sparks. Respiratory irritant. Use in a fume hood.

22.10.10. Hydrochloric Acid

A strong and volatile acid. Avoid exposure to both liquid and vapors. Use in a fume hood. Wear gloves and eye protection.

22.10.11. Iodine

Iodine vapors are a corrosive and dangerous eye, respiratory, and skin irritant; avoid vapors by using in a fume hood.

22.10.12. Mercuric Iodide

Poisonous. Wash hands after use. Avoid dust. Dispose of properly.

22.10.13. Methanol

Very flammable; avoid contact of liquid or vapor with flames or sparks. Toxic when swallowed or inhaled, and causes blindness. Use in a fume hood when dealing with large quantities.

22.10.14. Methylene Chloride

Narcotic in high concentrations.

22.10.15. Nitric Acid

Strong acid and oxidizing agent. Avoid contact with liquid or vapor. Use in a fume hood. Wear eye protection and gloves. Concentrated nitric acid can react violently with organics, including some alcohols. If skin or eyes are exposed, wash with abundant quantities of tap water.

22.10.16. Nitrogen Gas Cylinder

Nitrogen gas is nontoxic and not flammable. However, compressed gas cylinders are always dangerous. Use a proper regulator. Secure the cylinder carefully to prevent it from falling. Make sure that the pressure used does not exceed the strength of any hosing attached to carry gas from the tank. Use a proper cart to move the cylinder, and never move a cylinder that does not have a cap. If the top of a gas cylinder is damaged by a fall, the entire container can become an extremely dangerous missile and is capable of penetrating walls.

22.10.17. Perchloric Acid

A very strong acid and oxidizer. Liquid and vapors are corrosive and cause severe burns. Mixture with water is exothermic, and exposure of the concentrated reagent to organic compounds may cause a fire or explosion. Use only in a fume hood. Wear eye protection. If skin is exposed, wash with abundant quantities of cold water. Purchase and use the minimum quantity needed.

22.10.18. Petroleum Ether

Highly flammable; use only in absence of flames and sparks. Vapors may explode if lighted. Use in fume hood.

22.10.19. Phenol

Poisonous and caustic. Exposure of skin, eyes, or the respiratory tract to phenol or its vapors can cause tissue damage. Wear gloves and eye protection. Use a fume hood if the phenol is hot. If skin or eyes are exposed, wash with abundant quantities of tap water.

22.10.20. Phosphoric Acid

Strong acid. Avoid contact with skin and eyes. Wear eye protection. If skin or eyes are exposed, wash with abundant quantities of tap water.

22.10.21. Potassium Hydroxide

A strong base. Avoid contact with skin and eyes; wash with abundant quantities of water if contact occurs. Wear eye protection. Be aware when making strong solutions that the dissolution of KOH in water or alcohol is exothermic, and the solution can get very hot.

22.10.22. Radioactive Chemicals

Use of radioactive material is strictly regulated, and materials may be purchased only by licensed laboratories and must be used under careful supervision. The hazards of different materials vary considerably. Consult appropriate references and local supervising offices for details.

22.10.23. Sodium Arsenite

Very poisonous; do not breathe dust. Wash hands well after use.

22.10.24. Sodium Azide

Poisonous. Avoid contact. Wash hands well after use.

22.10.25. Sodium Carbonate

Strong alkali. Avoid contact with aqueous solution. If skin or eye exposure occurs, wash with abundant quantities of tap water. Wear eye protection.

22.10.26. Sodium Hydroxide

A strong base. Avoid contact with skin, wash with abundant quantities of tap water if skin is exposed to

NaOH solutions. Wear eye protection. Be aware when making strong solutions that the dissolution of NaOH in water is exothermic and may make the solution very hot.

22.10.27. Sodium Hypochlorite

Oxidant. Ingestion of the liquid is dangerous and will cause severe tissue destruction. Avoid vapors and contact with skin.

22.10.28. Sulfuric Acid

A very strong acid. Avoid contact with skin, and wash with abundant quantities of cold tap water if skin or eyes are exposed to this acid. Wear eye protection. Be aware that the dissolution of sulfuric acid in water is very exothermic and can make the solution very hot; the heat generated may be sufficient to burn skin and to break non-Pyrex containers. Do not mix by adding water to the acid, since it can spatter; instead, add acid to the water. For preparation of fairly concentrated solutions, cooling the mixing container with cold water or ice water may be appropriate, but do not place an already very hot glass container in cold water, since it may break. Be aware of where the nearest safety shower is when handling bottles of concentrated acid; purchase the smallest appropriate bottle.

22.10.29. Trichloroacetic Acid

Very corrosive. Causes skin irritation and is dissolved in lipids of the skin. Avoid contact by wearing protective gloves, appropriate laboratory clothing, and eye protection. If skin contact occurs, wash well with abundant quantities of tap water.

22.10.30. Trifluoracetic Acid

Strong acid. Avoid contact with liquid and vapor. Wear eye protection. Use in a fume hood. If exposure to skin occurs, wash with abundant quantities of tap water.

22.10.31. UV Light

UV light can cause severe retinal burns. Chromatograms and other material should be observed in UV light only when plastic goggles are used to cover the eyes, although normal glass eyeglasses decrease the UV somewhat. A variety of suppliers will provide special goggles that reduce the UV exposure even more.

22.11. REFERENCES

22.11.1. Fractionation and Radioactivity

1. **Kennell, D.** 1967. Use of filters to separate radioactivity in RNA, DNA, and protein. *Methods Enzymol: Nucleic acids* **12A:**686–692.
2. **Roberts, R. B., D. B. Cowie, E. T. Bolton, P. H. Abelson, and R. J. Britten.** 1955. Studies of biosynthesis in *Escherichia coli*. Publ. 607. Carnegie Institute of Washington.
3. **Sutherland, I. W., and J. F. Wilkinson.** 1971. Chemical extraction methods of microbial cells, p. 345–383. *In* J. R. Norris and D. W. Ribbons (ed.), *Methods in Microbiology*, vol. 5B. Academic Press, Inc., New York.

22.11.2. Small Molecules

4. **Chytil, F., and D. B. McCormick (ed.).** 1986. Vitamins and coenzymes. *Methods Enzymol.* **122:**3–425.
5. **Escalente-Semerena, J. C., K. L. Rinehart, Jr., and R. S. Wolfe.** 1984. Tetrahydromethanopterin, a carbon-carrier in methanogenesis. *J. Biol. Chem.* **259:**9447–9455.
6. **Purwantini, E.** 1991. Coenzyme F_{420}: factors affecting its purification from *Methanobacterium thermoautotrophicum*. M.S. Thesis. University of Iowa, Iowa City.
7. **Schonheit, P., H. Keweloh, and R. K. Thauer.** 1981. Factor F_{420} degradation in *Methanobacterium thermoautotrophicum* during exposure to oxygen. *FEMS Microbiol. Lett.* **12:**347–349.

22.11.3. Carbohydrates

22.11.3.1. General references

8. **Colowick, J., and N. Kaplan (ed.).** 1966. *Methods Enzymol: Complex carbohydrate.* **8:**1–759.
9. **Ginsburg, V. (ed.)** 1989. Complex carbohydrates. *Methods Enzymol.* **179:**3–287.
10. **Herbert, D., P. J. Phipps, and R. E. Strange.** 1971. Chemical analysis of microbial cells, p. 209–344. *In* J. R. Norris and D. W. Ribbons (ed.), *Methods in Microbiology*, vol. 5B. Academic Press, Inc., New York.
11. **Work, E.** 1971. Cell walls, p. 361–418. *In* J. R. Norris and D. W. Ribbons (ed.), *Methods in Microbiology*, vol. 5A. Academic Press, Inc., New York.

22.11.3.2. Specific references

12. **Bryan, B. A., R. J. Linhardt, and L. Daniels.** 1986. Variation in composition and yield of exopolysaccharides produced by *Klebsiella* sp. strain K32 and *Acinetobacter calcoaceticus* BD4. *Appl. Environ. Microbiol.* **51:**1304–1308.
13. **Cadmus, M. C., C. A. Knutson, A. A. Lagoda, J. E. Pittsley, and K. A. Burton.** 1978. Synthetic media for production of quality xanthan gum in 20 liter fermentors. *Biotechnol. Bioeng.* **20:**1003–1014.
14. **Chen, W.-P., J.-Y. Chen, S.-C. Chang, and C.-L. Su.** 1985. Bacterial alginate produced by a mutant of *Azotobacter vinelandii. Appl. Environ. Microbiol.* **49:**543–546.
15. **Dische, Z.** 1953. Qualitative and quantitative colorimetric determination of heptoses. *J. Biol. Chem.* **204:**983–997.
16. **Dische, Z.** 1962. Color reactions of carbohydrates. *Methods Carbohydr. Chem.* **1:**477–514.
17. **Elson, L. A., and W. T. Morgan.** 1933. A colorimetric method for the determination of glucosamine and chondrosamine. *Biochem. J.* **27:**1824–1828.
18. **Ford, J. D., and J. C. Haworth.** 1964. The estimation of galactose in plasma using galactose oxidase. *Clin. Chem.* **10:**1002–1006.
19. **Fox, G. E., L. J. Magrum, W. E. Balch, R. S. Wolfe, and C. R. Woese.** 1977. Classification of methanogenic bacteria by 16S ribosomal RNA. *Proc. Natl. Acad. Sci. USA* **74:**4537–4541.
20. **Fukagawa, K., H. Yamaguchi, D. Yonezawa, and S. Murao.** 1974. Isolation and characterization of polysaccharide produced by *Rhodotorula glutinis* K-24. *Agr. Biol. Chem.* **38:**29–35.
21. **Ghuysen, J.-M., D. J. Tipper, and J. L. Strominger.** 1966. The preparation and properties of ^{15}N-deuterated DNA

from bacteria. *Methods Enzymol.: Nucleic Acids.* **12A:**695–699.
22. **Hacking, A. J., I. W. F. Taylor, T. R. Jarman, and J. R. W. Govan.** 1983. Alginate biosynthesis by *Pseudomonas mendocina. J. Gen. Microbiol.* **129:**3473–3480.
23. **Hadzija, O.** 1974. A simple method for the quantitative determination of muramic acid. *Anal. Biochem.* **60:**512–517.
24. **Kaplan, N., E. Rosenberg, B. Jann, and K. Jann.** 1985. Structural studies of the capsular polysaccharide of *Acinetobacter calcoaceticus* BD4. *Eur. J. Biochem.* **152:**453–458.
25. **Kosinski, E. D.** 1981. The indirect determination of lactose using a glucose analyzer with particular reference to solutions containing low levels of lactose in high levels of glucose and galactose. *J. Soc. Dairy Technol.* **43:**28–31.
26. **Lee, Y. C.** 1990. High-performance anion-exchange chromatography for carbohydrate analysis. *Anal. Biochem.* **189:**151–162.
27. **Linhardt, R. J., R. Bakhit, L. Daniels, F. Mayerl, and W. Pickenhagen.** 1989. Microbially produced rhamnolipid as a source of rhamnose. *Biotechnol. Bioeng.* **33:**365–368.
28. **Lott, J. A., and K. Turner.** 1975. Evaluation of Trinder's glucose oxidase method for measuring glucose in serum and urine. *Clin. Chem.* **21:**1754–1760.
29. **Mason, M.** 1982. A new method for ethanol measurement utilizing an immobilized enzyme. *J. Am. Soc. Brew. Chem.* **40:**78–79.
30. **Mason, M.** 1983. Determination of glucose, sucrose, lactose, and ethanol in foods and beverages, using immobilized enzyme electrodes. *J. Assoc. Off. Anal. Chem.* **66:**981–984.
31. **Mukumoto, T., and H. Yamaguchi.** 1977. The chemical structure of a mannofucogalactan from the fruit bodies of *Flammulina velutipes* (Fr.) Sing. *Carbohydr. Res.* **59:**614–621.
32. **Nesser, J.-R., and T. F. Schweizer.** 1984. A quantitative determination by gas-liquid chromatography of neutral and amino sugars (as O-methyloxime acetates), and a study on hydrolytic conditions for glycoproteins and polysaccharides in order to increase sugar recoveries. *Anal. Biochem.* **142:**58–67.
33. **Osborn, M. J.** 1963. Studies on the Gram-negative cell wall. I. Evidence for the role of 2-keto-3-deoxyoctanoate in the lipopolysaccharide of *Salmonella typhimurium. Proc. Natl. Acad. Sci. USA* **50:**499–506.
34. **Rocklin, R. D., and C. A. Pohl.** 1983. Determination of carbohydrates by anion exchange chromatography with pulsed amperometric detection. *J. Liquid Chromatogr.* **6:**1577–1590.
35. **Rondle, C. J. M., and W. T. J. Morgan.** 1955. The determination of glucosamine and galactosamine. *Biochem. J.* **61:**586–590.
36. **Roth, H., S. Segal, and D. Bertoli.** 1965. The quantitative determination of galactose—an enzymic method using galactose oxidase, with applications to blood and other biological fluids. *Anal. Biochem.* **10:**32–52.
37. **Sempre, J. M., C. Gancedo, and C. Asensio.** 1965. Determination of galactosamine and N-acetylgalactosamine in the presence of other hexosamines with galactose oxidase. *Anal. Biochem.* **12:**509–515.
38. **Smith, I. H., and G. W. Pace.** 1982. Recovery of microbial polysaccharides. *J. Chem. Technol. Biotechnol.* **32:**119–129.
39. **Stewart-Tull, D. E. S.** 1968. Determination of amino sugars in mixtures containing glucosamine, galactosamine, and muramic acid. *Biochem. J.* **109:**13–18.
40. **Sweeley, C. C., W. W. Wells, and R. Bentley.** 1966. Gas chromatography of carbohydrates. *Methods Enzymol.: Complex Carbohydr.* **8:**95–107.
41. **Taylor, P. J., E. Kmetec, and J. M. Johnson.** Design, construction, and applications of a galactose selective electrode. *Anal. Chem.* **49:**789–794.
42. **Tipper, D. J.** 1968. Alkali-catalyzed elimination of D-lactic acid from muramic acid and its derivatives, and the determination of lactic acid. *Biochemistry* **7:**1441–1449.
43. **Weisbach, A., and J. Hurwitz.** 1959. The formation of 2-keto-3-deoxyheptanoic acid in extracts of *E. coli* B. *J. Biol. Chem.* **234:**705–712.
44. **Yellow Springs Instrument Co.** 1978. *Application Note 109: Dextrose and/or Sucrose Measured with the Model 27 Industrial Analyzer.* Yellow Springs Instrument Co., Yellow Springs, Ohio.
45. **Yellow Springs Instrument Co.** 1980. *The YSI Model 27 Industrial Analyzer Now Measures Starch, as well as Dextrose, Sucrose, and Lactose.* Yellow Springs Instrument Co., Yellow Springs, Ohio.

22.11.4. Lipids

22.11.4.1. General references

46. **Dittmer, J. C., and M. A. Wells.** 1969. Quantitative and qualitative analysis of lipids and lipid components. *Methods Enzymol.: Lipids* **14:**482–530.
47. **Kates, M.** 1972. Techniques of lipidology isolation, analysis, and identification of lipids, p. 275–327. *In* T. S. Work and E. Work (ed.), *Laboratory Techniques in Biochemistry and Molecular Biology*, vol. 3. American Elsevier Publishing Co., Inc., New York.
48. **O'Leary, W.** 1974. Chemical and physical characterization of fatty acids, p. 275–327. *In* A. I. Laskin and H. A. Lechevalier (ed.), *Handbook of Microbiology,* vol. 2. Chemical Rubber Co., Cleveland.

22.11.4.2. Specific references

49. **Bird, C. W., J. M. Lynch, S. J. Pirt, W. W. Reid, C. J. Brooks, and B. S. Middleditch.** 1971. Steroids and squalene in *Methylococcus capsulatus* grown on methane. *Nature* (London) **230:**473–474.
50. **Bligh, E. G., and W. J. Dyer.** 1959. A rapid method of total lipid extraction and purification. *Can. J. Biochem. Physiol.* **37:**911–917.
51. **Brandl, H., R. A. Gross, R. W. Lenz, and R. C. Fuller.** 1988. *Pseudomonas oleovorans* as a source of poly-β-hydroxyalkanoates for potential applications as biodegradable polyesters. *Appl. Environ. Microbiol.* **54:**1977–1982.
52. **Brandl, H., R. A. Gross, R. W. Lenz, and R. C. Fuller.** 1990. Plastics from bacteria and for bacteria: poly-β-hydroxyalkanoates as natural, biocompatible, and biodegradable polyesters. *Adv. Biochem. Eng. Biotechnol.* **41:**77–93.
53. **Folch, J.** 1957. A simple method for the isolation and purification of total lipids from animal tissue. *J. Biol. Chem.* **226:**497–509.
54. **Hara, A., and N. S. Radin.** 1978. Lipid extraction with a low-toxicity solvent. *Anal. Biochem.* **90:**420–426.
55. **Johnson, A. R., and R. B. Stocks.** 1971. Gas chromatography of lipids, p. 195–218. *In* A. R. Johnson and J. B. Davenport (ed.), *Biochemistry and Methodology of Lipids.* Wiley-Interscience, New York.
56. **Kates, M.** 1964. Simplified procedure for hydrolysis or methanolysis of lipids. *J. Lipid Res.* **5:**132–135.
57. **Kates, M., S. C. Kushwaha, and D. G. Sprott.** 1982. Lipids of purple membrane from extreme halophiles and of methanogenic bacteria. *Methods Enzymol.* **88:**98–111.
58. **Kates, M., L. S. Yengoyan, and P. S. Sastry.** 1965. A diether analog of phosphytidyl glycerophosphate in *Halobacterium cutirubrum. Biochim. Biophys. Acta* **98:**252–268.
59. **Kushwaha, S. C., M. Kates, G. D. Sprott, and I. C. P. Smith.** 1981. Novel complex polar lipids from the methanogenic archaebacterium *Methanospirillum hungatei. Science* **211:**1163–1164.
60. **Langworthy, T. A.** 1985. Lipids of archaebacteria, p. 459–497. *In* C. R. Woese and R. S. Wolfe (ed.), *The Bacteria*, vol. 8. Academic Press, Orlando, Fla.
61. **Law, J. H., and R. A. Slepecky.** 1961. Assay of poly-β-hydroxybutyric acid. *J. Bacteriol.* **82:**33–36.

62. **Lenz, R. W., B. W. Kim, H. W. Ulmer, K. Fritzche, E. Knee, and R. C. Fuller.** 1990. Functionalized poly-β-hydroxyalkanoates produced by bacteria, p. 23–35. *In* E. A. Dawes (ed.), *Novel Biodegradable Microbial Polymers.* Kluwer Academic Publishers, Dordrecht, The Netherlands.
63. **Paton, J. C., E. J. McMurchie, B. K. May, and W. H. Elliot.** 1978. Effect of growth temperature on membrane fatty acid composition and susceptibility to cold shock of *Bacillus amyloliquefaciens. J. Bacteriol.* **135:**754–759.
64. **Patt, T. E., and R. S. Hanson.** 1978. Intracytoplasmic membrane, phospholipid, and sterol content of *Methylobacterium organophilum* cells grown under different conditions. *J. Bacteriol.* **134:**636–644.
65. **Rilfors, L., A. Wieslander, and S. Stahl.** 1978. Lipid and protein composition of membranes of *Bacillus megaterium* variants in the temperature range of 5 to 70°C. *J. Bacteriol.* **135:**1043–1052.
66. **Skipski, V. P., and M. Barlday.** 1969. Detection of lipids on thin-layer chromatograms. *Methods Enzymol.: Lipids* **14:**541–548.
67. **Stein, R. A., V. Slawson, and J. F. Mead.** 1976. Gas-liquid chromatography of fatty acids and derivatives, p. 857–896. *In* G. V. Marinetti (ed.), *Lipid Chromatographic Analysis,* 2nd ed. Marcel Dekker, Inc., New York.
68. **Tornabene, T. G., and T. A. Langworthy.** 1979. Diphytanyl and dibiphytantanyl glycerol ether lipids of methanogenic archaebacteria. *Science* **203:**51–53.
69. **Tornabene, T. G., R. S. Wolfe, W. E. Balch, G. Holzer, G. E. Fox, and J. Oro.** 1978. Phytanyl-glycerol ethers and squalenes in the archaebacterium *Methanobacterium thermoautotrophicum. J. Mol. Evol.* **11:**259–266.
70. **Vorbeck, M. L., and G. V. Marinetti.** 1965. Intracellular distribution and characterization of the lipids of *Streptococcus faecalis. Biochemistry* **4:**296–305.
70a. **Welch, D. F.** 1991. Applications of fatty acid analysis. *Clin. Microbiol. Rev.* **4:**422–438.
71. **Wuthier, R. E.** 1966. Purification of lipids from non-lipid contaminants on Sephadex bead columns. *J. Lipid Res.* **7:**558–561.

22.11.5. Cell Wall Polymers

22.11.5.1. General references

72. **Braun, V.** 1978. Structure-function relationship of the Gram-negative bacterial cell envelope. *Symp. Soc. Gen. Microbiol.* **28:**111–138.
73. **Braun, V., and K. Hantke.** 1974. Biochemistry of bacterial cell envelopes. *Annu. Rev. Biochem.* **43:**89–121.
74. **Costerton, J. W., J. M. Ingram, and K. J. Cheng.** 1974. Structure and function of the cell envelope of gram-negative bacteria. *Bacteriol. Rev.* **38:**87–110.
75. **Cummins, C. S.** 1974. Bacterial cell wall structures, p. 251–284. *In* A. I. Laskin and H. A. Lechevalier (ed.), *Handbook of Microbiology,* vol. 2. *Microbial Composition.* Chemical Rubber Co., Cleveland.
76. **Elin, R. J., and S. M. Wolff.** 1974. Bacterial endotoxin, p. 674–731. *In* A. I. Laskin and H. A. Lechevalier (ed.), *Handbook of Microbiology,* vol. 2. *Microbial Composition.* Chemical Rubber Co., Cleveland.
77. **Ghuysen, J. M.** 1968. Use of bacteriolytic enzymes in determination of wall structure and their role in cell metabolism. *Bacteriol. Rev.* **32:**425–464.
78. **Leive, L. (ed.).** 1973. *Bacterial Membranes and Walls.* Marcel Dekker, Inc., New York.
79. **Luderitz, O., A. M. Staub, and O. Westphal.** 1966. Immunochemistry of O and R antigens of *Salmonella* and related *Enterobacteriacae. Bacteriol. Rev.* **30:**192–255.
80. **Milner, K. C., J. A. Rudbach, and E. Ribi.** 1971. General characteristics, p. 1–65. *In* G. Weinbaum, S. Kadis, and S. Ajl (ed.), *Microbial Toxins: Bacterial Endotoxins,* vol. 4. Academic Press, Inc., New York.
81. **Reaveley, D. A., and R. E. Burge.** 1972. Walls and membranes in bacteria. *Adv. Microb. Physiol.* **7:**1–81.
82. **Rogers, H. J., J. B. Ward, and I. D. J. Burdett.** 1978. Structure and growth of the walls of Gram-positive bacteria. *Symp. Soc. Gen. Microbiol.* **28:**139–176.
83. **Schleifer, K. H., and O. Kandler.** 1972. Peptidoglycan types of bacterial cell walls and their taxonomic implications. *Bacteriol. Rev.* **36:**407–477.

22.11.5.2. Specific references

84. **Davidson, A. L., and J. Badiley.** 1964. Glycerol teichoic acids in walls of *Staphylococcus epidermidis. Nature* (London) **202:**874.
85. **Ghuysen, J. M., D. Tipper, and J. L. Strominger.** 1965. Structure of the cell wall of *Staphylococcus epidermidis* strain Copenhagen. IV. The soluble glycopeptide and its sequential degradation by peptidase. *Biochemistry* **4:**2245–2256.
86. **Kandler, O., and H. Hippe.** 1977. Lack of peptidoglycan in the cell walls of *Methanosarcina barkeri. Arch. Microbiol.* **113:**57–60.
87. **Kandler, O., and H. Konig.** 1978. Chemical composition of the peptidoglycan-free cell walls of methanogenic bacteria. *Arch. Microbiol.* **118:**141–152.
88. **Kandler, O., and H. Konig.** 1985. Cell envelopes of archaebacteria, p. 413–457. *In* C. R. Woese and R. S. Wolfe (ed.), *The Bacteria,* vol. 8. Academic Press, Orlando, Fla.
89. **Knox, K. W.** 1966. The relation of 3-deoxy-2-oxo-octanoate to the serological and physical properties of a lipopolysaccharide from a rough strain of *Escherichia coli. Biochem. J.* **100:**73–78.
90. **Kottel, R. H., K. Bacon, D. Clutter, and D. White.** 1975. Coats from *Myxococcus xanthus*: characterization and synthesis during myxospore differentiation. *J. Bacteriol.* **124:**550–557.
91. **Osborn, M. J.** 1963. Studies on the Gram-negative cell wall. *Proc. Natl. Acad. Sci. USA* **50:**499–514.
92. **Raetz, C. R. H.** 1990. Biochemistry of endotoxins. *Annu. Rev. Biochem.* **59:**129–170.
93. **Strominger, J. L., and D. J. Tipper.** 1974. Structure of bacterial cell walls: the lysozyme substrate, p. 169–184. *In* E. F. Osserman, R. E. Canfield, and S. Beychok (ed.), *Lysozyme.* Academic Press, Inc., New York.
94. **Taylor, A., K. W. Knox, and E. Work.** 1966. Chemical and biological properties of an extracellular liposaccharide from *Escherichia coli* grown under lysine-limiting conditions. *Biochem. J.* **99:**53–61.
95. **Warth, A. W., and J. L. Strominger.** 1969. Structure of the peptidoglycan of bacterial spores: occurrence of the lactam of muramic acid. *Proc. Natl. Acad. Sci. USA* **64:**528–535.
96. **Weidel, W., H. Frank, and H. H. Martin.** 1960. The rigid layer of the cell wall of *Escherichia coli* strain B. *J. Gen. Microbiol.* **22:**158–166.
97. **Westphal, O., and K. Jan.** 1965. Bacterial lipopolysaccharide extraction with phenol-water and further applications of the procedure. *Methods Carbohydr. Chem.* **5:**83–91.
98. **Work, E.** 1971. Cell walls, p. 361–418. *In* J. R. Norris and D. W. Ribbons (ed.), *Methods in Microbiology,* vol. 5A. Academic Press, Inc., New York.

22.11.6. Nucleic Acids

22.11.6.1. General references

99. **DeLey, J.** 1971. The determination of the molecular weight of DNA per bacterial nucleoid, p. 301–311. *In* J. R. Norris and D. W. Ribbons (ed.), *Methods in Microbiology,* vol. 5A. Academic Press, Inc., New York.
100. **Parish, J. H.** 1972. *Principles and Practice of Experiments with Nucleic Acids.* Longman Group, Ltd., London.

22.11.6.2. Specific references

101. **Brunk, C. F., K. C. Jones, and T. W. James.** 1979. Assay for nanogram quantities of DNA in cellular homogenates. *Anal. Biochem.* **92:**497–500.
102. **Burton, K.** 1957. A study of the conditions and mechanism of the diphenylamine reaction for the colorimetric estimation of deoxyribonucleic acid. *Biochem. J.* **62:**315–323.
103. **Cesarome, C., C. Bolognesi, and L. Santi.** 1979. Improved microfluorometric DNA determination in biological material using 33258 Hoechst. *Anal. Biochem.* **179:**401–403.
104. **Gallagher, S.** 1989. Quantitation of DNA and RNA with absorption and fluorescence spectroscopy, p. A.3.9–A.3.15. *In* F. A. Ausubel, R. Brent, R. E. Kingston, D. D. Moore, J. M. Seidman, J. A. Smith, and K. Struhl (ed.), *Current Protocols in Molecular Biology*, supplement 8. Greene Publishing Associates and John Wiley & Sons, Inc., New York.
105. **Gauchel, F. D., and R. Z. Zahn.** 1970. Micro-determination of DNA in biological materials by gas-chromatographic and isotope dilution analysis of thymine content. *FEBS Lett.* **6:**141.
106. **Griswold, B. L., F. L. Humoller, and A. R. McIntyre.** 1951. Inorganic phosphates and phosphate esters in tissue extracts. *Anal. Chem.* **23:**192–194.
107. **Labarca, C., and K. Paigen.** 1980. A simple, rapid, and sensitive DNA assay procedure. *Anal. Biochem.* **102:**344–352.
108. **Tilzer, L., S. Thomas, and R. F. Moreno.** 1989. Use of silica gel polymer for DNA extraction with organic solvents. *Anal. Biochem.* **183:**13–15.

22.11.7. Nitrogen Components

22.11.7.1. General references

109. **Allen, S. E., H. M. Grimshaw, J. A. Parkinson, and C. Quarmby.** 1974. Inorganic constituents: nitrogen, p. 184–206. *In* S. E. Allen (ed.), Chemical analysis of ecological materials. Blackwell Scientific Publications, London.
110. **Black, C. A. (ed.).** 1965. *Methods of Soil Analysis*, part 2. *Chemical and Microbiological Properties.* American Society of Agronomy, Inc., Madison, Wis.
111. **Bremner, J. M., and D. R. Keeney.** 1965. Steam distillation methods for determination of ammonium, nitrate, and nitrite. *Anal. Chim. Acta* **32:**485–495.
112. **Cooper, T. G.** 1977. *The Tools of Biochemistry.* John Wiley & Sons, Inc., New York.
113. **Davis, B. D., R. Dulbecco, H. N. Eisen, and H. S. Ginsberg.** 1990. *Microbiology*, p. 45–49. J. B. Lippincott, Philadelphia.
114. **Franson, M. A. (ed.).** 1976. *Standard Methods for the Examination of Water and Wastewater*, 14th ed. American Public Health Association, Washington, D.C.
115. **Gudinowicz, B. J., M. J. Gudinowicz, and H. F. Martin.** 1976. *Fundamentals of Integrated GC-MS.* Part II. *Mass Spectrometry.* Marcel Dekker, Inc., New York.
116. **Horowitz, W. (ed.).** 1975. *Official Methods of Analyses of the Association of Official Analytical Chemists*, 12th ed. Association of Official Analytical Chemists, Washington, D.C.
117. **Jackson, M. L.** 1958. *Soil Chemical Analysis.* Prentice-Hall, Inc., Englewood Cliffs, N.J.
118. **Safe, S., and O. Hutzinger.** 1973. *Mass Spectrometry of Pesticides and Pollutants.* CRC Press, Inc., Cleveland.
119. **Strickland, J. D. H., and T. R. Parsons.** 1960. *A Practical Handbook of Seawater Analysis.* Fisheries Research Board of Canada, Ottawa.
120. **Waller, G. R. (ed.).** 1972. *Biochemical Applications of Mass Spectrometry.* John Wiley & Sons, Inc., New York.

22.11.7.2. Specific references

121. **Armstrong, F. A. J.** 1963. Determination of nitrate in water by ultraviolet spectrophotometry. *Anal. Chem.* **35:**1292.
122. **Baca, P., and H. Freiser.** 1977. Determination of trace levels of nitrates by an extraction-photometric method. *Anal. Chem.* **49:**2249–2250.
123. **Belay, N., R. Sparling, B.-S. Choi, M. Roberts, J. E. Roberts, and L. Daniels.** 1988. Physiological and ^{15}N-NMR analysis of molecular nitrogen fixation by *Methanococcus thermolithotrophicus, Methanobacterium bryantii*, and *Methanospirillum hungatei. Biochim. Biophys. Acta* **971:**233–245.

123a. **Bio-Rad Laboratories.** 1992. *Bio-Rad DC Protein Assay.* Bio-Rad Laboratories, Richmond, Calif.

124. **Boehlen, P., S. Stein, W. Dairman, and S. Udenfriend.** 1973. Fluorometric assay of proteins in the nanogram range. *Arch. Biochem. Biophys.* **155:**213–220.
125. **Bradford, M. M.** 1976. A rapid and sensitive method for the quantitation of microgram quantities of protein utilizing the principle of protein-dye binding. *Anal. Biochem.* **72:**248–254.
126. **Bremner, J. M.** 1960. Determination of nitrogen in soil by the Kjeldahl method. *J. Agric. Sci.* **55:**11–33.
127. **Canney, P. J., D. E. Armstrong, and J. H. Wiersma.** 1974. *Determination of Nitrite and Nitrate Ions in Natural Waters Using Aromatics or Diamines as Reagents.* Technical Report, University of Wisconsin Water Resources Center, Madison.
128. **Chow, T. J., and M. S. Johnstone.** 1962. Determination of nitrate in sea water. *Anal. Chim. Acta* **27:**441–446.
129. **Coakley, W. T., and C. J. James.** 1978. A simple linear transformation for the Folin-Lowry protein calibration curve to 1.0 mg/ml. *Anal. Biochem.* **85:**90–97.
130. **Gornall, A. G., C. S. Bardawill, and M. M. David.** 1949. Determination of serum protein by means of the Biuret reaction. *J. Biol. Chem.* **177:**751–756.
131. **Guiraud, G., J. C. Fardeau, G. Llimous, and M. A. Barral.** 1977. Determination of nitrates in soils and plants by the Kjeldahl method. *Ann. Agron.* **28:**329–333.
132. **Hardy, R. W. F., and A. H. Gibson (ed.).** 1977. *A Treatise on Dinitrogen Fixation.* Section IV. *Agronomy and Ecology.* John Wiley & Sons, Inc., New York.
133. **Hill, H. D., and J. G. Straka.** 1988. Protein determination using bicinchoninic acid in the presence of sulfhydryl reagents. *Anal. Biochem.* **170:**203–208.
134. **Janssen, F. W., A. J. Lund, and L. E. Anderson.** 1957. Colorimetric assay for dipicolinic acid in bacterial spores. *Science* **127:**26–27.
135. **Kalb, V. F., Jr., and R. W. Bernlohr.** 1977. A new spectrophotometric assay for protein in cell extracts. *Anal. Biochem.* **82:**362–371.
136. **Kalckar, H. M.** 1947. Differential spectrophotometry of purine compounds by means of specific enzymes. I. Determination of hydroxypurine compounds. *J. Biol. Chem.* **167:**429–475.
137. **Kruse, J. M., and M. G. Mellon.** 1953. Colorimetric determination of ammonia and cyanate. *Anal. Chem.* **25:**1188–1192.
138. **Layne, E.** 1957. Spectrophotometric and turbidometric methods for measuring proteins. *Methods Enzymol.* **3:**447–454.
139. **Lowry, O. H., N. J. Rosebrough, A. L. Farr, and R. J. Randall.** 1951. Protein measurement with the Folin phenol reagent. *J. Biol. Chem.* **193:**265–275.
140. **Markwell, M. A., S. M. Haas, L. L. Bieber, and N. E. Tolbert.** 1978. A modification of the Lowry procedure to simplify protein determination in membrane and lipoprotein samples. *Anal. Biochem.* **87:**206–210.
141. **Martin, H. H., and J. W. Foster.** 1958. On the chromatographic behavior of dipicolinic acid. *Arch. Mikrobiol.* **31:**171–178.
142. **Matsunaga, K., and M. Nishimura.** 1969. Determination of nitrate in sea water. *Anal. Chim. Acta* **45:**350–353.
143. **McClard, R. W.** 1981. Removal of sulfhydryl groups with 1,3,4,6-tetrachloro-3a-6a-diphenylglycoluril: application to the assay of protein in the presence of thiol reagents. *Anal. Biochem.* **112:**278–281.

144. **Mubarak, A., R. A. Howald, and R. Woodriff.** 1977. Elimination of chloride interferences with mercuric ions in the determination of nitrates by the phenol disulfonic acid method. *Anal. Chem.* **49:**857–860.
145. **Murrell, W. G.** 1969. Chemical composition, p. 249–251. *In* G. W. Gould and A. Hurst (ed.), *The Bacterial Spore.* Academic Press, Inc., New York.
146. **Murrell, W. G., and A. D. Warth.** 1965. Composition and heat resistance of bacterial spores, p. 1–24. *In* L. L. Campbell and H. O. Halvorson (ed.), *Spores III.* American Society for Microbiology, Washington, D.C.
147. **Nakamura, H., and J. J. Pisano.** 1976. Sensitive fluorometric assay for proteins. Use of fluorescamine and membrane filters. *Arch. Biochem. Biophys.* **172:**102–105.
148. **Pierce Chemical Co.** 1989. *BCA Protein Assay Reagent: Instructions.* Pierce Chemical Co., Rockford, Il.
149. **Pierce, J., and C. H. Suelter.** 1977. An evaluation of the Coomassie brilliant blue G-250 dye-binding method for quantitative protein determination. *Anal. Biochem.* **81:**478–480.
150. **Postgate, J. R. (ed.).** 1971. *The Chemistry and Biochemistry of Nitrogen Fixation.* Plenum Press, New York.
151. **Raganowicz, E., and A. Niewiadomy.** 1976. Colorimetric determination of the nitrites content in water by means of 2-sulfanilamidothiazole method. *Pol. Arch. Hydrobiol.* **23:**1–4.
152. **Rexroad, P. R., and R. D. Cathey.** 1976. Pollution-reduced Kjeldahl method for a crude protein. *J. Assoc. Off. Anal. Chem.* **59:**1213–1217.
153. **Russel, J. A.** 1944. The colorimetric estimation of small amounts of ammonia by the phenol-hypochlorite reaction. *J. Biol. Chem.* **156:**457–461.
154. **Sedmak, J. J., and S. E. Grossberg.** 1977. A rapid, sensitive, and versatile assay for protein using Coomassie brilliant blue G-250. *Anal. Biochem.* **79:**544–552.
155. **Smith, P. K., R. I. Krohn, G. T. Hermanson, A. K. Mallia, F. H. Gartner, M. D. Provenzano, E. K. Fujimoto, N. M. Goeke, B. J. Olson, and D. C. Klenk.** 1985. Measurement of protein using bicinchoninic acid. *Anal. Biochem.* **150:**76–85.
156. **Snell, F. D., and C. T. Snell.** 1945. *Colorimetric Methods of Analysis.* D. Van Nostrand Co., Inc., New York.
157. **Stickland, H. L.** 1951. The determination of small quantities of bacteria by means of the Biuret reaction. *J. Gen. Microbiol.* **5:**698–703.
158. **Warburg, O., and W. Christian.** 1942. Isolierung und kristallisation des garungsferments enolase. *Biochem. Z.* **310:**384–421.
159. **Warth, A. D., D. F. Ohye, and W. G. Murrell.** 1963. The composition and structure of bacterial spores. *J. Cell Biol.* **16:**579–592.
160. **Work, E.** 1963. α,ε-Diaminopimelic acid. *Methods Enzymol.* **6:**624–634.

22.11.8. Minerals

22.11.8.1. General references

161. **Christian, G. D., and F. J. Feldman.** 1970. *Atomic Absorption Spectroscopy. Applications in Agriculture, Biology and Medicine.* Wiley-Interscience, New York.
162. **Heirman, R., and C. T. J. Allkemade.** 1963. *Chemical Analysis by Flame Photometry,* 2nd ed. Translated by P. T. Gilbert, Jr., Interscience Publishers, Inc., New York.
163. **Morrison, G. H. (ed.).** 1956. *Trace Analysis.* Interscience Publishers, Inc., New York.
164. **Snell, D. S., and C. T. Snell.** 1963. *Colorimetric Methods of Analysis,* vol. 2, 3rd ed. D. Van Nostrand Co., Inc., Princeton, N.J.
165. **U.S. Environmental Protection Agency.** 1974. *Methods for Chemical Analysis of Water and Wastes.* Office of Technology Transfer, Washington, D.C.
166. **Weinberg, E. D. (ed.).** 1977. *Microorganisms and Minerals.* Marcel Dekker, Inc. New York.

22.11.8.2. Specific references

167. **Stewart, M., A. P. Somlyo, A. V. Somlyo, H. Shuman, J. A. Lindsay, and W. G. Murrell.** 1980. Distribution of calcium and other elements in cryosectioned *Bacillus cereus* T spores, determined by high-resolution scanning electron probe X-ray microanalysis. *J. Bacteriol.* **143:**481–491.

Chapter 23

Enzymatic Activity

ALLEN T. PHILLIPS

The study of enzymes and enzyme-catalyzed reactions has contributed greatly to our understanding of microbial metabolism, just as current advances in protein structure promise new insights into the detailed workings of enzymes at the molecular level. Nevertheless, there remain enormous gaps in our knowledge regarding the biochemistry and physiology of all but the best-studied bacteria, and thus a thorough analysis of enzymes, their action, and their properties continues to be an important effort for bacteriologists. This chapter is intended to provide a brief description of the most important principles involved in enzyme activity measurements, for the purpose of helping investigators perform meaningful and accurate evaluations of catalytic and regulatory properties. The discussion focuses almost exclusively on quantitative methods and the principles needed for reliable enzyme measurement. Because the intent is to present information useful to the bacteriologist who seeks to assay enzymes or use them as reagents, descriptions are keyed mainly toward small-scale and simple operations that should be possible in almost any laboratory. One or two specific assays are detailed to exemplify each of the six enzyme classes.

23.1. PRINCIPLES

23.1.1. Units and Specific Activity

Although enzymes are proteins (except perhaps catalytic RNA species) and can be quantified by methods specific for proteins, it is almost invariably useful to express the quantity of an enzyme present in terms of its catalytic activity rather than on the content of protein. This has the advantage of relative specificity, assuming no interfering activities are present, and measurements can usually be done in impure systems containing numerous other protein species as well as on purified preparations. The most common definition of one **unit (U) of enzyme activity** is the amount of enzyme which catalyzes the conversion of 1 μmol of substrate to product in 1 min under a prescribed set of assay conditions, i.e., at a stated pH, temperature, buffer type and concentration, and substrate and cofactor concentrations. This definition, although somewhat dated (12), is widely used and is applicable to most situations. Where the resulting numbers are inconveniently large or small, multipliers such as kilounits or milliunits are preferable to defining a nonstandard unit.

Two exceptions to this general practice exist. The first is the need for a different definition when the molecular weight of a substrate is unknown and thus a micromole quantity cannot be specified; in this instance it is acceptable to use some convenient mass of substrate converted to product per minute as the basis for the unit definition. The second is where one chooses to employ the International Union of Biochemistry's currently recommended unit, the **katal**. One katal is the amount of activity that converts 1 mol of substrate to product in 1 s in the defined assay (31). The katal conforms to International System units, but despite being officially sanctioned, it is not widely used. One enzyme unit as defined earlier is equal to 16.67 nkat.

Although an enzyme unit is a measure of quantity, expressed in catalytic terms, the relative purity of an enzyme is specified by its **specific activity**. Specific activity is the number of enzyme units present per milligram of protein. For a given enzyme, this quantity increases during selective biosynthesis or during purification up to a limit value at which all protein present is the active enzyme. Since estimates of protein content are much influenced by the method used for the determination, values for specific activity should always be accompanied by information about how the protein was quantified (both the method and the reference protein used) as well as how an enzyme unit was defined. Specific activity expressed in IUB-recommended terms is katals per kilogram of protein or suitable multiples thereof (e.g., microkatals per milligram).

23.1.2. Reaction Classification and Nomenclature

The current system of enzyme nomenclature recognizes six classes of enzymes and defines the basis for a systematic name for each enzyme. A listing of enzymes along with their recommended names and numerical **enzyme codes (EC numbers)** is periodically updated by the IUB and should be consulted when preparing results of enzyme studies for publication (32). The EC number (e.g., 3.2.1.23) is made up of four parts; the first decimal indicates the enzyme's class; the next two decimals classify the enzyme according to type of bonds broken, coenzymes used, and other important characteristics; and the final decimal is a serial number used to distinguish among enzymes otherwise classified together. Because EC numbers provide generally unambiguous identification of enzymes, they are of increasing importance in assisting the cataloging of sequence data for computer searches. A brief explanation of the six enzyme classes is as follows.

Class 1: Oxidoreductases. Enzymes in class 1 catalyze oxidation-reduction reactions and are assigned systematic names based on the form **hydrogen donor:acceptor oxidoreductase**. The name **dehydrogenase** is recommended for normal use, but **reductase** is acceptable where this more correctly describes the usual reaction direction; **oxygenase** or **oxidase** is used when O_2 is the acceptor species. Thus, the NAD^+-dependent isocitrate dehydrogenase of the Krebs cycle has the assigned systematic name *threo*-D_s-isocitrate:NAD^+ oxidoreductase (decarboxylating), EC 1.1.1.41.

Class 2: Transferases. Class 2 enzymes carry out group transfer from a donor to an acceptor. Accordingly, the systematic designation for the enzyme catalyzing the ATP-dependent phosphorylation of adenosine to yield AMP + ADP is ATP:adenosine 5′-phosphotransferase, EC 2.7.1.20, more commonly known as adenosine kinase. Enzymes catalyzing transamination (aminotransferases) and hexose transfer (e.g., glycogen synthase, with its systematic name UDPglucose:glycogen 4-α-D-glucosyltransferase, EC 2.4.1.11) are also included in this class.

Class 3: Hydrolases. Class 3 is a broad category, which covers enzymes that catalyze the hydrolytic cleavage of a substrate. In cases such as proteolytic enzymes (carboxypeptidase or collagenase) or phosphatases, specificity is usually directed toward a family of related substrates containing common structural features rather than specific compounds. Although their systematic names always include the term **hydrolase**, as in urea amidohydrolase (EC 3.5.1.5), members of this class are frequently referred to by the name of their substrate with the suffix -ase (e.g., urease, penicillinase, and β-galactosidase). Also found in this class are proteinases (proteases), many with nonstandard but still widely used names, such as trypsin and papain.

Class 4: Lyases. Class 4 enzymes catalyze cleavage of C—C, C—O, and C—N bonds (plus a few others) with elimination of a group to leave an unsaturated residue; obviously the converse addition of groups to double bonds is included as well. Most lyase reactions are referred to as uni-bi type, meaning that they involve conversion of a single substrate to two products. Their systematic names are based on the form **substrate group-lyase**, as in *S*-adenosylmethionine carboxy-lyase, EC 4.1.1.50 (also known as a **decarboxylase**), and L-malate hydro-lyase, commonly called fumarase or fumarate hydratase. Several enzymes involving coenzyme A (CoA)-dependent cleavages, as in citrate synthase, or addition of another element concomitant with substrate cleavage (e.g., anthranilate synthase) are less obvious members of the lyase family.

Class 5: Isomerases. Class 5 enzymes catalyze geometric or structural changes within one molecule. Class 5 includes enzymes variously known as **racemases, epimerases, isomerases, tautomerases**, and **mutases**. There is no single form for the systematic name, and thus one finds alanine racemase (interconverting D- and L-alanine, EC 5.1.1.1) as well as D-ribulose-5-phosphate 3-epimerase (EC 5.1.3.1) and D-glyceraldehyde-3-phosphate ketol-isomerase (EC 5.3.1.1) (better recognized as triose-phosphate isomerase).

Class 6: Ligases (Synthetases). Ligases catalyze the joining of two molecules, coupled with the hydrolysis of a pyrophosphate bond from ATP or similar triphosphate. Their systematic names are based on the pattern substrate 1:substrate 2 ligase (ADP-forming). The recommended class name is **synthetase**, not to be confused with **synthase**, which simply describes an enzyme that joins parts of two molecules without involvement of pyrophosphate bond cleavage. Examples of synthetases are pyruvate carboxylase [pyruvate:carbon-dioxide ligase (ADP-forming), EC 6.4.1.1], glutamine synthetase [L-glutamate:ammonia ligase (ADP-forming), EC 6.3.1.2], and leucyl-tRNA synthetase [L-leucine:$tRNA^{Leu}$ ligase (AMP-forming), EC 6.1.1.4].

23.1.3. Cofactors

Many enzyme reactions require substances other than the enzyme protein and substrate for optimal activity. These accessory materials, termed cofactors, are in most cases unchanged at the end of a catalytic cycle and are regarded as important (perhaps essential) parts of the reaction mechanism. Cofactors are usually divided into two classes. **Coenzymes** are organic molecules which have a precisely defined role in the reaction. **Activators** enhance catalysis or stabilize enzyme structure; these include inorganic ions as well as a variety of organic substances and often assist in the binding of substrates. A **prosthetic group** is a special category of coenzyme which remains firmly attached to the same enzyme molecule throughout catalysis. In many cases a prosthetic group is covalently bonded to the enzyme protein (e.g., biotin, lipoic acid). When an enzyme reaction is considered outside of the cell environment, a coenzyme such as NAD^+, which cannot be regenerated without the participation of another reaction that consumes NADH, is in fact indistinguishable from a substrate. On the other hand, the in vivo coenzymatic functions of ATP, NAD^+, and CoA are easily accepted when one recognizes that these materials migrate from one enzyme molecule to another within the cell and thus are regenerated for continued use, unlike a traditional substrate.

23.2. ENZYME PREPARATIONS

23.2.1. Native Intact Cells

Assays of individual enzymes other than for semiquantitative assessment are infrequently conducted on native intact cells, largely because of the inability to make substrates available to the enzyme in known and nonlimiting quantities. Nevertheless, most investigators recognize that enzymes ought to be studied under conditions in which protein-protein or protein-membrane interactions are those normally found in the cell. This is especially important now that there is growing evidence for association of enzymes within some pathways, leading to a channeling of metabolites from one enzyme to another without release into the cellular milieu (39). Nuclear magnetic resonance (NMR) analysis is one of the best methods for obtaining in vivo activity evaluations; and because this yields information on the concentration of substrate(s) available to the enzyme as well as the amounts and nature of the products formed, it can be applied to the assay of bacterial enzymes in growing cells as well as in the resting state. The low sensitivity of NMR necessitates high cell densities in order to have sufficient concentrations of metabolic intermediates; for many analyses, 10^{10} cells per ml are required, and this presents some difficulty if oxygenation is important.

Despite these problems, a number of successful studies have dealt with intact cells of *Saccharomyces cerevisiae* and *Escherichia coli*. For example, ^{31}P-NMR and ^{13}C-NMR have enabled whole-cell measurements of the rate of phosphofructokinase action in acetate-grown yeast cells maintained under aerobic and anaerobic conditions and have identified several compounds which serve to regulate its activity (4). Applications to other organisms are now appearing. *Pseudomonas putida* and *Staphylococcus aureus* were shown to have a formaldehyde dismutase which generated methanol and formate from formaldehyde, the enzymatic equivalent of the base-catalyzed Cannizzaro reaction involving formaldehyde (23). This whole cell study not only used deuterium NMR analysis to confirm the existence of the enzyme but also provided considerable mechanistic detail, including absolute stereochemistry for a cross-dismutation involving formaldehyde and benzaldehyde. The properties of the enzyme revealed through the in vivo measurements were in good agreement with those found by studying isolated preparations.

On a simpler level, gas exchange techniques have long been popular for measurement of enzyme activity in resting cells, but these may suffer from questions about actual concentrations of substrates available to the enzyme under assay. The Clark-type oxygen electrode is suitable for whole-cell measurements of an enzyme reaction in which O_2 is a substrate; likewise, manometry can be used for monitoring gas uptake or evolution and is easily adapted for use with O_2, CO_2, or H_2. Additional information pertaining to these techniques is found in Chapter 21. Both methods are useful with whole resting cells, provided all substrates can be taken up and endogenous energy reserves can be depleted prior to the assay. For oxygenase assays in which the oxidized product is rapidly metabolized further, inhibitors of electron transport (e.g., sodium azide) may be needed to prevent additional O_2 uptake coupled to NADH reoxidation. Decarboxylases can be conveniently assayed by manometry as an alternative to the use of $^{14}CO_2$ production from labeled substrate. A simple means of reducing endogenous activity for manometric or oxygen electrode experiments is to aerate a suspension of cells at high density (at least 10^9 cells per ml) in medium lacking an energy source or in a nonmetabolizable buffer. Typically, 2 h of this treatment at the usual growth temperature will burn up sufficient metabolic reserves to permit enzyme assays.

23.2.2. Permeabilized Intact Cells

Cells rendered permeable to substrates by chemical or enzymatic treatment are quite suitable for enzyme assay, provided the effectiveness of the permeabilization agents can be standardized and cell lysis can be avoided or minimized. When the "intactness" of the cells is questionable, it might be preferable to use truly broken preparations, which can be more reproducibly assayed and for which one clearly recognizes the in vitro nature of the results. Nevertheless, permeabilization methods are popular because they are rapid and are readily adapted to large numbers of individual samples of cells. Keep in mind, however, that the presence of intact cells can have considerable effect on direct spectrophotometric analysis as a result of light scattering, and therefore such applications are probably best avoided unless measurements are performed on a spectrophotometer that easily permits corrections for turbidity. This need not be a problem if the cells are removed by centrifugation or precipitation prior to product estimation.

Among the most gentle permeabilization methods is Tris-EDTA treatment. Solutions of these (0.1 M Tris-HCl [pH 7.5], 1 mM EDTA) alter the permeability of the lipopolysaccharide-containing outer membrane of gram-negative bacteria by a combination of chelating Mg^{2+} ions and some incompletely understood effects of Tris buffer on the integrity of the outer membrane (13). In using this method, however, keep cell densities moderate (1 mg [dry weight]/ml or less); in addition, run assays both with and without the Tris-EDTA treatment to ensure that the permeabilization is effective for the substrate being used.

Toluene is a widely used permeabilization agent, which probably disrupts membrane lipid regions, permitting a wide variety of low-molecular-weight materials access to the cell interior (but also allowing many metabolites to diffuse out!). Normally 1 drop (50 μl) in a 1-ml assay volume is sufficient to render *E. coli* cells permeable to various amino acids and carbohydrates.

Another organic solvent treatment that has been found of some use is chloroform plus a detergent; this has been used for enteric organisms and is said to be as effective as sonic disruption (36). To use this, harvest cells in mid-log phase, centrifuge, and resuspend to the same density in cold 25 mM phosphate buffer (pH 7.2); add 15 μl of chloroform and 50 μl of a 1% (wt/vol) sodium deoxycholate solution per ml of cell suspension. Incubate on ice for 5 min, and then conduct the assay.

Some detergents have found application as permeabilization agents, although in many instances it is not clear whether the cells remain intact during the subsequent assay. The cationic detergent cetyltrimethylammonium bromide (CTAB) is representative of these compounds. In a typical case, CTAB is placed in the assay mixture at a final concentration of 0.05 to 0.1 mg/ml, the cell suspension is added, and the reaction is started by substrate addition after a 5-min preincubation period.

In all applications with permeabilized cells, there may be some variation in results as a function of the concentration of cells used, and standardization of cell density in the assay may be desirable for maximum reproducibility. Include controls with no permeabilizing agent to ascertain that the treatment is effective.

23.2.3. Cell Extracts

The great majority of enzyme assays are conducted on cell-free preparations in order to have known substrate and cofactor concentrations, to use coupled assays where desired, and to provide good control over reaction conditions such as pH and ionic strength. There are numerous physical methods for the release of enzymes from cells; this discussion is limited to the three most widely used techniques: hydrodynamic shear, ultrasonic disruption, and homogenization with glass beads. In addition, enzymatic lysis techniques are mentioned because these are especially convenient when many individual samples must be handled. Further information on cell disruption techniques is found in Chapter 4.1.

23.2.3.1. Hydrodynamic shear

The **French press** (SLM Instruments., Inc., Urbana, IL 61801) is popular for its high efficiency, good recovery of activity, and simplicity of operation. Most units are restricted to volumes of less than 50 ml per treatment. The disruption chamber consists of a stainless-steel cylinder fitted with a plunger that is inserted from one end. The other end of the cylinder contains a removable base, above which is located a small orifice in the cylinder connecting the inside chamber with the outside; a needle valve controls access through this opening. Upon pressure drop as a cell suspension passes through the orifice, the cells are disrupted by hydrodynamic shear.

The French press is generally effective for all bacteria but considerably less so for gram-positive cocci such as *S. aureus* than for others. It is relatively efficient for mycobacteria, although several passes may be needed. In operation, the steel cylinder with its plunger inserted partway is precooled to 0°C and a thick suspension of the microorganism is added through the open (base) end. The base is then inserted, and the plunger is adjusted upward to expel all air inside the cylinder through the needle valve (*CAUTION*: Failure to remove all air can result in violent discharge of the last sample portion). When all air is displaced and the cell suspension just begins to emerge, close the needle valve and place the unit base-down in a hydraulic press capable of achieving a pressure of 30,000 lb/in^2 (210 MPa). Once the unit is pressurized to the extent desired, open the needle valve slightly to allow the sample to exit through the orifice into a chilled receiver. The internal pressure must be maintained near the desired level as the sample emerges through the needle valve; this will necessitate operating the hydraulic press at a rate sufficient to counter the pressure drop. With adequate pumping speed, pressure can be maintained such that the entire sample can be processed within 2 to 3 min for a 40-ml suspension. Difficult cell types should be reprocessed as necessary. Centrifuge the homogenate for at least 20 min at 30,000 × *g* to clarify the sample and remove unbroken cells, large membrane fragments, and general cell debris. Higher speeds, including ultracentrifugation at 100,000 × *g* for 1 to 2 h, are appropriate when the sample is desired free of ribosomes or other particulate materials.

Among the advantages offered by the French press are its lack of adverse effects on most enzymes, applicability over a wide range of organisms, and speed of operation. Extrusion from pressures above 10,000 lb/in^2 generally leads to almost complete disruption of an *E. coli* suspension as viewed microscopically and produces fairly uniform-sized cell wall and membrane fragments. Pressures above 20,000 lb/in^2 may be necessary to give maximum disruption of *Bacillus* cells. Within the limits of pressure normally encountered, most enzymes are unaffected by the pressure treatment, and the high protein content of the extract, coupled with the rapidity of the process, leads to good recovery of activity for all but the most labile enzymes. It is inadvisable, however, to keep the suspension at pressure above 20,000 lb/in^2 for periods exceeding 5 min, since some enzymes may undergo denaturation under these conditions.

For processing very large volumes (minimum, 0.5 liter), the **Manton-Gaulin homogenizer** (Gaulin Corp., Everett, MA 02149) is preferred over the standard French press. Like the French press, this unit disrupts mainly by hydrodynamic shear upon pressure drop and is designed to operate continuously at pressures of 6,000 to 10,000 lb/in^2. The high velocity and pressure operation leads to considerable heating of the sample. A temperature rise of 20 to 30°C for the exiting sample is not uncommon, but the use of a well-chilled sample at the outset and rapid cooling of the emerging homogenate generally minimize loss of enzyme activity. A typical minimum-sized suspension would contain 250 g of wet packed cells dispersed in 500 ml of a suitable ice-cold buffer. Such a concentrated sample enhances enzyme recovery but must be free of obvious lumps of cells or other debris; if necessary, filter the suspension through cheesecloth before homogenization. Two processing passes are often needed for complete disruption. Centrifuge the final product in the cold at 20,000 × *g* for 30 min to obtain a cell extract.

23.2.3.2. Ultrasonic cavitation

Numerous ultrasound-producing instruments designed for cell breakage are available commercially; most involve a piezoelectric principle in which a crystal transducer resonates at a frequency near 20 kHz, producing movement in a titanium metal tip. When the tip comes into contact with a liquid, this movement sets up sonic waves, which, when of sufficient amplitude, produce cavitation. The cavitation process is the formation and subsequent violent collapse of minute bubbles of about 150 μm in diameter, with a resulting localized high pressure and temperature change. Fragile cells such as *E. coli* can be disrupted simply by the shock waves resulting from bubble vibration, whereas the most difficult cell types (*S. aureus*, *Mycobacterium phlei*) probably require true cavitation for significant breakage. Glass beads can also be present during the sonic treatment for additional disruption.

For disruption, select a probe (tip) based on the volume of material to be treated. Microtips which can process volumes as small as 0.25 ml are available, and larger probes (diameter, 0.5 or 0.75 in. [1.27 or 1.91 cm]) are used for volumes of 50 to 500 ml. Placement of the tip in the solution is important; too shallow or too deep positioning can result in partial or complete displacement of the liquid from around the tip, especially at higher power settings, and interferes with breakage as well as leading to probe damage. Optimum positioning is at a depth no less than one-third and no more than two-thirds of the solution height. Prepare the cell suspension in a suitable buffer such that the resulting mixture is not excessively viscous. Cavitation is suppressed by high viscosity, and this limits the quantity of cells that can be processed at one time. A reasonable upper limit is a 20% (wt/vol) suspension (packed wet weight cells).

With the sample cooled in an ice-water bath, apply power initially by increasing the power setting over several seconds toward the limit prescribed for the probe used. Optimum power is that which results in an irritating sizzling or hissing sound, with visible cavitation streams noted (in a clear solution) and a gentle rocking or vibration of the liquid surface. Sharp squeals, foaming, large-bubble formation, or violent turbulence are indicators of improper probe placement or excessive power and should be promptly corrected. The time of treatment is dependent on the volume and power setting and must be worked out empirically, but it is advisable to use numerous short bursts (0.5 to 2 min each) rather than continuous operation, to minimize temperature rise. The use of metal or glass containers for rapid heat conduction and an ice-NaCl bath can help to counter heating tendencies. Newer instruments have a variable duty cycle, which automatically permits brief cooling during the nonpower periods.

Disruption efficiency is affected by many variables; among the most important are viscosity, sonic power intensity, probe depth, and vessel shape. High viscosity, either through the presence of viscous agents (e.g., glycerol) or resulting from very thick cell suspensions, adversely affects disruption. Probe intensity is determined by the power limits of the unit and the probe tip diameter, with units commonly delivering between 125 and 450 W/cm². When this power output is channeled into a microtip, greatly increased turbulence can result and necessitates reduced power operation compared with the larger tips. Because cavitation can generate minor amounts of free radicals, it is useful to include a small amount of dithiothreitol (0.1 to 1.0 mM) or other SH-containing compound in the disruption buffer. Alternatively, the solution can be degassed before breakage or the sample can be processed in a helium atmosphere.

Attachments for many sonic disruptors which permit continuous-flow operation or operation in a sealed chamber (for breakage of pathogens) are available. Some units have a cup horn so that the sample does not come in direct contact with the probe. This accessory is of low efficiency and is satisfactory only with easily disrupted cells; its advantages are that the sample does not become contaminated with tip metal, the sample can be in a sealed container during breakage, and cooling of the sample during prolonged operation is quite simple.

23.2.3.3. Grinding with glass beads

Grinding with glass beads is a simple technique that can be performed in any laboratory on almost any scale. Although older procedures involved grinding in a mortar with an abrasive such as alumina or sand, the use of glass minibeads (diameter, 0.1 to 0.5 mm) in an apparatus that provides for mechanical agitation is more satisfactory. One such device adaptable to volumes from 5 to 200 ml is the Bead Beater (Biospec Products, Bartlesville, Okla.). This unit can be obtained with several chamber sizes. Heating due to vigorous mixing is a major concern, and this limits treatment times to between 30 s and 1 min each, with 1 to 2 min of cooling between agitation cycles. Although virtually any appropriate solution can be used for suspending the cells, high cell densities are preferred (usually equal weights of packed cells and suspending solution). Add to the cold cell suspension an equal volume of minibeads that have been washed well with the suspending solution and prechilled. Pour this mixture into the proper Bead Beater chamber, and agitate for several minutes total elapsed time; it is advisable that the chamber be filled near capacity to improve heat transfer to the outer cooling chamber. After agitation, allow the beads to settle and then wash them twice with small amounts of the suspending solution, finally combining all supernatant solutions. Following centrifugation to remove unbroken cells and other debris, the extract is ready for assay.

23.2.3.4. Enzymatic lysis

The cell wall of many gram-positive cells is attacked by lysozyme, and treatment with Tris-EDTA can render gram-negative cells susceptible to this hydrolytic enzyme. A simple yet highly effective method for lysis of many cell types thus uses lysozyme in combination with Tris-EDTA and osmotic shock. Harvest and rapidly wash the cells once with 0.1 M NaCl. Resuspend the cell pellet in twice its weight of lysis solution; if the quantity of cells is less than 25 mg (wet weight), use 50 μl of solution in a microcentrifuge tube. The lysis solution contains 100 mM Tris-HCl (pH 7.5), 5% (vol/vol) glycerol, 1 mM dithiothreitol, 2 mM EDTA, and 1 mM phenylmethylsulfonyl fluoride. (*CAUTION:* This is a moderately toxic substance, which should be handled with gloves, particularly when preparing and transferring stock solutions, normally 0.2 M in isopropyl alcohol; the material is not stable in aqueous solutions, so prepare only as much lysis solution as can be used immediately.) Add a freshly prepared solution of chicken egg-white lysozyme (15 mg/ml) to give a concentration of 300 μg/ml. Mix, and place the tube on ice for 15 to 60 min. Examine the suspension by microscopy to ascertain that breakage has occurred. If not, quick-freeze the samples in ethanol-dry ice, and then thaw them rapidly in a 35°C water bath; this procedure should be repeated three times or until breakage is evident. Centrifuge at high speed in the cold, and remove the supernatant solution to another tube to await assay.

Variations of this procedure include incubation for 15 min at 4°C with 10 mM $MgCl_2$ and DNase I (20 μg/ml) following the lysozyme incubation, or the addition of 0.2% Triton X-100 to the lysis buffer to facilitate release of some enzymes. Some investigators prefer to conduct the lysozyme digestion in the presence of 10% sucrose and then lyse the resulting spheroplasts by a freeze-thaw procedure (47). On the other hand, once spheroplasts have been formed, periplasmic enzymes are usually released and can be assayed without further treatment of the cells. Lysostaphin from "*S. staphylolyticus*" can be used in place of lysozyme when disrupting *S. aureus*.

Several problems may be encountered in the use of enzymatic lysis methods. First, it may be difficult to evaluate the degree of lysis. This can be estimated by removing test portions of the lysate at various times of treatment and determining the protein content of the supernatant following centrifugation to remove cells. After correcting for the lytic enzyme added, a mostly disrupted sample should result in a protein content of roughly 25 mg/ml when a 1:2 weight ratio of cells and lysis buffer was used. If cells contain large amounts of nonprotein components (e.g., poly-β-hydroxybutyrate or polysaccharide materials), the weight of cells used should be adjusted accordingly. More important, perhaps, is the reproducibility when multiple samples are being treated, and this is usually very good. Of course, if the enzyme of interest is particulate, enzymatic procedures, like most other methods, may have to be followed by a mild detergent treatment to assist in solubilization of the enzyme.

23.2.4. Crude versus Purified Enzymes

An investigator should always be able to indicate the reason(s) for conducting an enzyme assay. Serious kinetic studies and documentation of most enzyme properties are best conducted with purified preparations, but there are a few situations in which quantitative assays of enzymes in unpurified extracts are desirable and even preferable; this is particularly true when some important properties are likely to be affected by purification, as, for example, sensitivity to certain allosteric effectors. More commonly, assays are performed with crude extracts simply because only comparative activities are sought among samples in a group and conditions need not be optimum for the desired comparisons. This might be the case when conducting a physiological study, such as the effect of a growth condition on enzyme production, or when assaying mutants.

The principal limitations on the use of crude extracts for assays are the presence of competing activities, endogenous inhibitors, and proteolytic enzymes which might reduce the assayable activity; problems can also be encountered from endogenous substrate, as discussed in section 23.3.6. Frequently, dialysis or gel filtration can be used to remove contaminating low-molecular-weight components. Furthermore, there is a reasonable selection of protease inhibitors (metal chelators for metalloproteases, serine protease inhibitors such as phenylmethylsulfonyl fluoride, peptides such as leupeptin), although these may be unnecessary because of the high protein concentration of most crude extracts unless extracts are to be stored prior to assay, an inadvisable option. The problem of competing activities, e.g., contaminating ATPase activity in assays which measure kinase-dependent production of ADP from ATP, is clearly a major limitation on conduct of assays and one for which there is no general solution. Fortunately, innovative assay methods or a modified design can help in some instances, but when that is not possible, partial or complete purification is the only recourse.

23.3. KINETIC CONSIDERATIONS

23.3.1. Enzyme Concentration and Reaction Time

Fundamental to any enzyme assay is establishing that the observed reaction velocity (rate) is a function of the concentration of enzyme present. The quantitative basis for the relationship between velocity observed and the amount of enzyme used is found in the Michaelis-Menten equation for enzyme-catalyzed reactions, shown here for a single substrate:

$$v = \frac{V_{max}[S]}{K_m + [S]} \text{ or } \frac{k_{cat}[E]_T[S]}{K_m + [S]} \qquad (1)$$

In this equation, observed velocity (v) is seen to be a function of substrate concentration [S], plus two other parameters, V_{max} and K_m. K_m, the Michaelis constant for a substrate, is described in the next section. V_{max} is defined as the velocity produced by the action of a given concentration of enzyme operating in a vast excess of substrate and is the product of a rate constant, k_{cat}, and $[E]_t$, the total concentration of enzyme present. V_{max} can be best understood by considering a single-substrate mechanism of the type

$$E + S \underset{k_{-1}}{\overset{k_1}{\rightleftharpoons}} ES \xrightarrow{k_2} P + E \qquad (2)$$

where free enzyme (E) combines reversibly with substrate (S) to produce an intermediate complex (ES) with rate constant k_1; this complex decomposes either with first-order rate constant k_2 to produce product (P) and release free enzyme or k_{-1} to regenerate the starting materials. Reversibility from P to ES can be ignored if only initial velocities in the forward direction are considered and no P has accumulated or is present.

The observed forward enzyme velocity is always proportional to k_2[ES]; and when [S] is sufficiently large to tie up all enzyme as the ES intermediate (i.e., essentially no free enzyme exists and thus $[ES] = [E]_T$), the reaction velocity is at its maximum value and therefore $v = V_{max} = k_2[E]_T$. In more complex reaction schemes, k_2 may not be the only rate constant relating velocity to the amount of ES complex present. The inability to identify which constants are involved in the limiting rate step often prompts the use of a less committal designation k_{cat}, which is a composite of any rate constants that limit reaction velocity. k_{cat} is equal to the **molecular activity** of the enzyme (formerly referred to as its turnover number), which describes the molecules of substrate consumed per minute per molecule of enzyme. It has units of reciprocal minutes for a single substrate enzyme when V_{max} is given as micromoles of substrate consumed per milliliter per minute (i.e., millimolar per minute or units per milliliter), and $[E]_T$ is in micromoles per milliliter (i.e., millimolar). Molecular activity can also be calculated from the specific activity at saturating substrate if the molecular weight of the enzyme is known. The reciprocal of k_{cat} is the time required for one catalytic cycle. For example, alanine racemase of *P. putida* has a k_{cat} of 2.4×10^5 min^{-1} and thus completes each cycle in 250 μs (1).

An accurate and reproducible assay is essential for good enzyme studies. Reaction velocity must be constant over the time of measurement; this is of particular importance for fixed-time (endpoint) assays. The Michaelis-Menten equation (equation 1) reveals how velocity varies with substrate concentration. Note, however, that when [S] is very high relative to K_m, the observed velocity, v, approaches V_{max}. Under such a saturating substrate condition, the reaction rate is influenced by [E] but not by [S]. This zero-order kinetics situation provides the most straightforward analysis of enzyme activity and usually gives the highest velocity per amount of enzyme, provided there is no inhibition by excess substrate.

Regardless of whether [S] is large or small, equation 1 shows that velocity is a function of $[E]_T$. Thus, provided [S] is essentially unchanged during the assay, measurements of reaction velocity at any [S] will be proportional to the amount of enzyme present. For normal assessment of enzyme activity and the related kinetic constants (described in the next section), initial velocity conditions (i.e., a negligible decrease in the starting substrate concentration and no product accumulation) are essential. In practice, substrate consumption should not exceed 5% over the course of the measured period to maintain the initial velocity.

One of the first tests of an enzyme assay must be to demonstrate a linear dependence of velocity on the amount of enzyme added. Set up a series of identical assays, each with a different amount of enzyme added. For reactions in which the rate of product formation (or substrate disappearance) can be monitored continuously, identify a time span over which the velocity ($-d[S]/dt$ or $d[P]/dt$) is constant; this period must be early in the reaction course to minimize substrate depletion and product accumulation. The results should resemble Fig. 1 when plotted as shown, with a region of direct proportionality between measured velocity and enzyme supplied, usually followed by a region of declining increments in velocity with further enzyme in

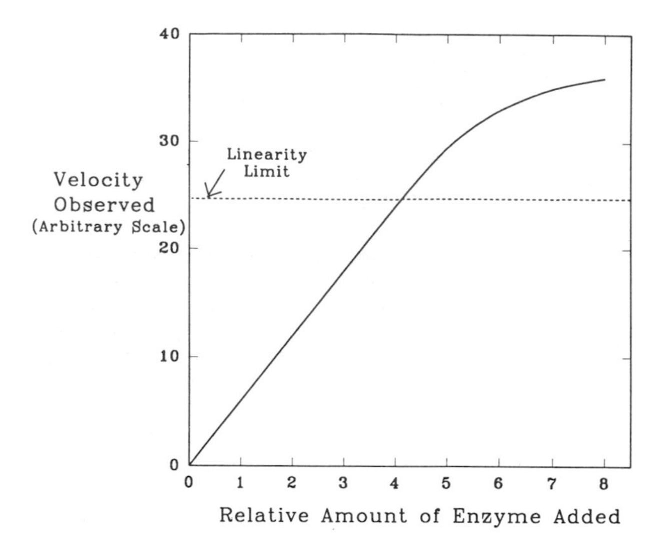

FIG. 1. Proportionality between reaction velocity and amount of enzyme added. The deviation from a linear relationship marks the highest velocity measurable with assurance that it is directly proportional to the quantity of enzyme present.

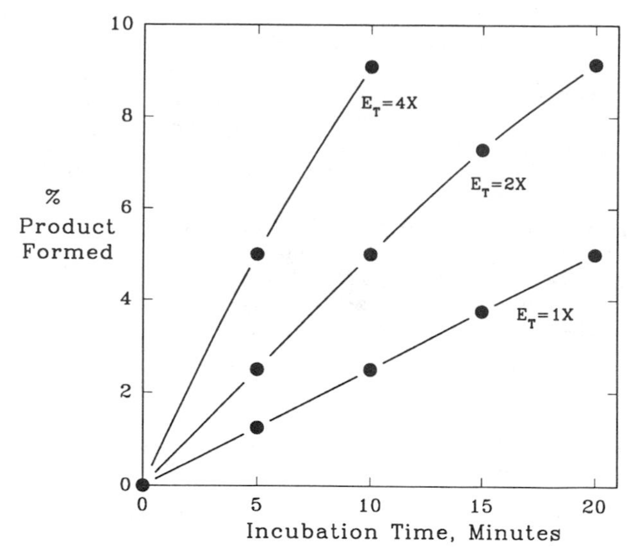

FIG. 2. Illustration of linear formation of product with time and its relationship to the quantity of enzyme added as well as to the extent of the reaction. For this example, enzyme concentration (E_T) was varied over a fourfold range and product formation was assumed to become nonlinear with time as the reaction extent exceeded 5%. Nonlinear behavior above 5% could be due to approaching equilibrium, a decreased substrate concentration, or product inhibition.

creases. From such data, estimate the velocity (in any convenient terms) above which linearity breaks off. In routine use, accept no assays in which the experimentally observed velocity exceeds this linear rate limit. Instead, dilute the enzyme sample and reassay to establish that the measured velocity is below the maximum linear rate value. Some frequently encountered reasons for a departure from linearity at too high a velocity include significant substrate depletion, reversibility or inhibition due to product accumulation, and pH changes in inadequately buffered reactions which produce or consume protons.

The previous test assumed that reaction velocity, determined as either substrate consumption or product formation with time, is monitored continuously. However, for assays in which a single fixed-time incubation is conducted, it is necessary to establish that reaction velocity is constant throughout the incubation period; this must be known even before checking for proportionality between velocity and enzyme added. To evaluate the constancy of velocity, set up several identical assays (with the same enzyme concentration throughout) and terminate each at a different time; determine the amount of product formed at each time, and plot the results as shown in Fig. 2. Repeat the experiment, if desired, with a different enzyme concentration, and plot those results also.

Using the data presented in Fig. 2 to illustrate the point, when an incubation time of 10 min is selected, product formation is linear throughout the 10-min period provided enzyme concentration is equal to or less than $2X$, where X is some activity value in units per milliliter; however, a 20-min assay period would not have a constant rate of product formation over the entire period if $[E]_T = 2X$ because of excessive product formation (substrate consumption). On the other hand, if a 5-min incubation were used, a constant velocity could be maintained for that period even at relative enzyme amounts of $4X$. Since one normally does not know the actual amount of enzyme activity being added, the choice of incubation period is arbitrary, but, once it is selected, it is important to know that the velocity is constant throughout the chosen period. Only then can the range of velocities that is a direct function of $[E]_T$ be established with confidence. Too short an incubation often leads to hurried assays or timing errors; too long an incubation risks enzyme inactivation.

23.3.2. Michaelis Constant Determination for Single-Substrate Reactions

The Michaelis constant for a particular combination of substrate and enzyme is a type of equilibrium constant made up of rate constants appropriate for a specific mechanism. K_m is defined as $(k_{-1} + k_2)/k_1$ for the single-substrate reaction scheme described above in equation 2. It is common, albeit not always justified, to assume that k_2 is small relative to k_1 and k_{-1} and thus to consider that K_m approximates the dissociation constant for the dissociation of the ES complex to S + E, i.e., k_{-1}/k_1. Although this assumption fails in some instances, K_m values usually reflect the affinity of the enzyme for different substrates and thus are frequently used for comparing relative binding strengths between an enzyme and its various substrates.

Knowledge of K_m is also important in establishing a rational level of substrate to be used in an assay. Since the Michaelis-Menten equation (equation 1) illustrates that K_m is numerically equal to the substrate concentration giving a velocity of 0.5 V_{max}, one can use K_m values to estimate reasonable substrate concentrations for assays, above which the rate becomes increasingly insensitive to further changes in substrate concentration. For instance, when $[S] = 5\ K_m$, $v = 0.83\ V_{max}$, whereas when $[S] = 10\ K_m$, $v = 0.91\ V_{max}$, and when $[S] = 100\ K_m$, $v = 0.99\ V_{max}$. Unless a substrate is very inexpensive and can be shown to be noninhibitory in excess, substrate levels over 10 K_m are seldom needed for simple assays. Estimates of K_m are also important for use with coupled enzyme assays, as described in section 23.3.5.

Under limiting substrate conditions when $[S] \ll K_m$, equation 1 describes a first-order-type reaction in which velocity is clearly a function of [S] and the kinetic constants V_{max} and K_m. This situation is likely to be encountered under conditions of substrate scarcity, high cost, insolubility, or inhibition; it is also common when a high-specific-radioactivity substrate is being used in an enzyme assay. It is important to realize that reliable assays for enzyme activity can be readily achieved even under these low-substrate conditions, provided the initial substrate concentration is constant from assay to assay and is not sufficiently changed during the course of an assay so as to affect the observed velocity.

Estimates of K_m are generally obtained from velocity measurements made in the range between true first-order kinetics (where $[S] \ll K_m$) and zero-order kinetics (where [S] is not rate limiting). In determining K_m for a single substrate, an initial rough estimate is helpful to permit selection of the appropriate range of substrate concentrations to be tested, normally severalfold above and below the estimated K_m value. Obtain this approximation by assaying an enzyme sample with excess substrate to approach V_{max}; then reassay at lower substrate levels but with the same enzyme concentration until a rate roughly equal to 0.5 V_{max} is determined; this [S] is the crude K_m value. Having established the approximate value and assuming that a double-reciprocal plot (i.e., Lineweaver-Burk plot) will be used for K_m assessment, choose substrate concentrations that give evenly spaced reciprocals, with the reciprocal of the estimated K_m being used as the mid-value for the 1/[S] range. Generally it is satisfactory to limit the range of substrate concentrations to 0.3 to 3.0 times the K_m estimate. Because assays must be conducted at constant enzyme concentration, it is critical that the enzyme sample used for the assays be stable over the time required to carry out all assays. The addition of glycerol, a protectant protein (e.g., bovine serum albumin), or reducing agents may be needed to stabilize the enzyme for several hours. All conditions normally controlled during regular enzyme assays, such as pH, buffer type, ionic strength, and cofactors, must be standardized for this analysis. Moreover, because substrate concentrations are being varied and are rate limiting, it is important that initial velocities be measured; otherwise, the substrate concentration is not constant during the assay and the initial concentration used is not representative of the average concentration seen by the enzyme. Failure to limit substrate depletion to less than 5% at each substrate concentration tested can seriously jeopardize the validity of the results.

The Lineweaver-Burk double-reciprocal plot is possibly the most common form for graphing kinetic data to estimate K_m and V_{max}. It follows from the Michaelis-Menten equation (equation 1) that

$$\frac{1}{v} = \left(\frac{K_m}{V_{max}} \frac{1}{[S]}\right) + \frac{1}{V_{max}} \tag{3}$$

Thus a plot of $1/v$ versus 1/[S] gives a line of slope K_m/V_{max} and a y-axis intercept of $1/V_{max}$. Also, setting $1/v$ equal to 0, equation 3 reveals that the x-axis intercept will be $-1/K_m$. These relationships are illustrated in Fig. 3.

Section 23.3.4 describes several other tools for estimating these important kinetic values. Although graphical methods such as the Lineweaver-Burk double-reciprocal plot are convenient and readily understood by most investigators, linearized plotting procedures in

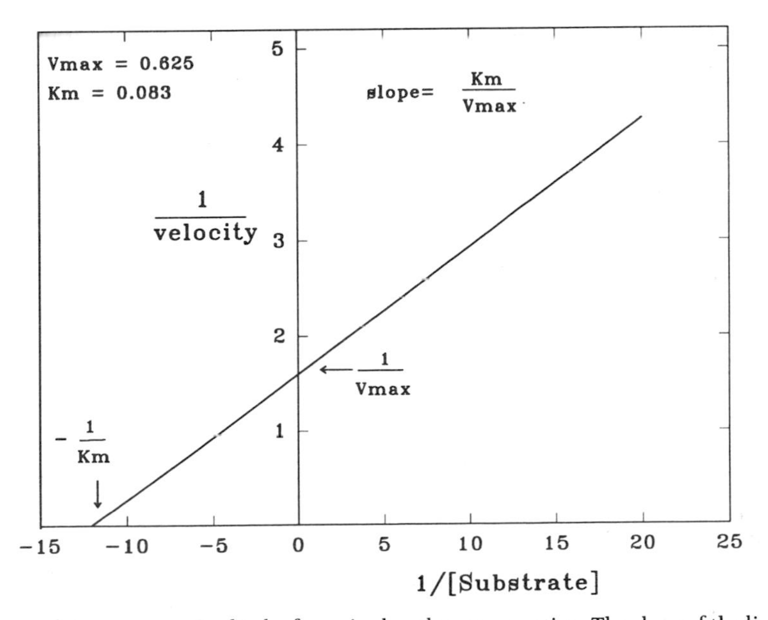

FIG. 3. Double-reciprocal (Lineweaver-Burk) plot for a single-substrate reaction. The slope of the line defined by the plot of $1/v$ versus 1/[S] is K_m/V_{max}. Extrapolation of this line to the y axis intersects at $1/V_{max}$, whereas extrapolation to the negative x axis yields the value $-1/K_m$.

general are being replaced by computer programs which fit values of velocity and [S] directly to the Michaelis-Menten equation and derive solutions for K_m and V_{max} by curve-fitting methods. Subsequent discussions will make use largely of the Lineweaver-Burk plot to illustrate patterns and show how various kinetic constants may be estimated, but investigators are urged to use computerized analysis for final calculation of results prior to publication.

23.3.3. Michaelis Constant Determination for Bisubstrate Reactions

Although enzyme kinetics are frequently explained by analyzing the single-substrate situation illustrated in equation 2, most of the important biological reactions involve two (or more) substrates. In a situation involving multiple substrates, K_m determinations require several additional considerations. First, it must be recognized that K_m for a particular substrate will be influenced by the concentrations of all other substrates, with the "true" K_m being estimated only when all cosubstrates are present at saturating concentrations. Values for K_m determined under subsaturating conditions of cosubstrates are only "apparent" K_m estimates and change as the concentration of these vary; apparent K_m values are of little use and generally should not be reported. Likewise, a true V_{max} value is measured only when all substrates are at saturating concentrations.

If the aim is only to obtain K_m estimates for the substrates, one can determine K_m for one substrate by varying its concentration while maintaining the other(s) at a kinetically saturating level. Analyze each set of v-versus-[S] data as described for the single-substrate case. This simple approach cannot be used in many instances, however, since a very large excess of the cosubstrate is required to achieve a condition in which a change in its concentration would have no effect on the overall reaction velocity. Frequently a substrate cannot be used in an assay at a saturating level because of expense or availability or because its physical solubility limit is reached before achieving the concentration which represents kinetic saturation. Also, excess concentrations of a substrate can be inhibitory, thereby preventing its inclusion in a reaction mixture at a high level. These situations mandate a more thorough and general approach but one that also can provide additional information about the enzyme and its action. In this approach, both substrates are varied and the condition of kinetic saturation with each is achieved by extrapolation to infinite concentration.

Obviously, bisubstrate reactions are potentially much more complicated than the unisubstrate type. Questions concerning the order of substrate binding and product release, whether new intermediates form before all substrates are bound, and how inhibitors might affect overall kinetics while interacting with only one form of the enzyme are all of some interest and can be approached through kinetic analysis. Three examples of different types of bisubstrate reactions will suffice to illustrate the approach required and information gained. Experimentally all situations are treated the same, in that velocity is measured as a function of varied concentrations of both substrates. The results are then plotted in a linearized format to establish the order of substrate binding, where possible, and estimate kinetic constants. How these are extracted, however, depends on the patterns revealed, which, in turn, provide some mechanistic detail.

23.3.3.1. Random binding

If a reaction is of the type where either of its two substrates can bind to the enzyme, binding of one may alter the dissociation constant for the other. To analyze this situation, it is necessary to make a simplifying assumption regarding the kinetic characteristics of the reaction, namely that an equilibrium exists among all steps prior to the rate-limiting decomposition of the enzyme-substrate complex to products. This "rapid-equilibrium" treatment is less comprehensive than the "steady-state" treatment, which assumes only that the rates of formation of enzyme-containing intermediates are balanced by their rate of breakdown, but the rapid equilibrium treatment is required in this particular instance to eliminate terms in the velocity equation which involve $[A]^2$ or $[B]^2$. The rapid-equilibrium assumption is generally valid if the kinetics follow a Michaelis-Menten model and yield linear double-reciprocal plots.

The following scheme illustrates the rapid equilibrium situation involving a random addition of two substrates to an enzyme. Reversibility is ignored in this example, since we are considering only initial velocity conditions.

$$\begin{array}{ccccc} & \mathrm{A} \overset{K_A}{\rightleftharpoons} \mathrm{EA} + & \mathrm{B} & & \\ + & & \Updownarrow \alpha K_B & & \\ \mathrm{E} & & \mathrm{EAB} & \xrightarrow[k_{cat}]{} & \mathrm{E} + \mathrm{Product(s)} \\ + & K_B & \Updownarrow \alpha K_A & & \\ & \mathrm{B} \rightleftharpoons \mathrm{EB} + & \mathrm{A} & & \end{array} \quad (4)$$

K_A and K_B are dissociation constants for the EA and EB complexes, respectively, but, as shown shortly, these are not necessarily the same as the Michaelis constants. When α, termed the binding interaction factor, is equal to 1, the binding of either substrate to the enzyme is independent of the other. If, however, $\alpha > 1$, the binding of one substrate decreases the affinity of the enzyme for the other substrate, whereas if $\alpha < 1$, the binding of one substrate is enhanced by the presence of the other substrate. Information pertaining to α is obviously important for interpreting kinetic data and fortunately can be obtained along with other constants.

The velocity equation for such a reaction is the following:

$$v = \frac{V_{max}[A][B]}{\alpha K_A K_B + \alpha K_B [A] + \alpha K_A [B] + [A][B]} \quad (5)$$

This can be rearranged to resemble the Michaelis-Menten expression if written so that the initial velocity is a function of the varied substrate.

$$v = \frac{V_{max}[A]}{\alpha K_A(1 + K_B/[B]) + [A](1 + K_B/[B])} \quad (6)$$

A similar type of equation can be written for B as the varied substrate. From equation 6 it is clear that if [B] is very high and $\alpha = 1$, the situation essentially collapses to that for a single substrate. Equation 6 also reveals that αK_A is the Michaelis constant for A since it is equal to [S] when v equals 0.5 V_{max} and [B] is infinite. Only when $\alpha = 1$ is K_A, the EA dissociation constant, equal to the Michaelis constant for A. Likewise, αK_B is the Michaelis constant for B.

If reaction velocities are measured as a function of both [A] and [B], this permits all terms in the velocity expression to be evaluated. Depending on the complexity of the assay, expense involved, and time required, between four and eight concentrations of each substrate should be used, centered around its preliminary K_m estimate and ranging three- to fourfold above and below that. If a reciprocal plot method is to be used for graphical analysis, plan for evenly spaced reciprocals of substrate concentrations to avoid clustering of points. The total number of assays should be between 16 (four concentrations of the varied substrate at each of four concentrations of the unvaried substrate) and 64 (eight concentrations of each substrate). It is advisable to conduct an initial experiment of only 16 assays in case a poor choice of substrate concentrations was made and the velocities do not span a satisfactory range. As in the single-substrate analysis, enzyme concentration should be the same for all assays, so special attention must be given to stabilization of the enzyme sample if all assays are not done concurrently.

Analysis of data for velocity at different concentrations of the two substrates is conducted as follows. The reciprocal version of equation 6 is

$$\frac{1}{v} = \frac{\alpha K_A}{V_{max}}\left(1 + \frac{K_B}{[B]}\right)\frac{1}{[A]} + \frac{1}{V_{max}}\left(1 + \frac{\alpha K_B}{[B]}\right) \quad (7)$$

The analogous equation expressed for B as the varied substrate is symmetrical to equation 7 and is

$$\frac{1}{v} = \frac{\alpha K_B}{V_{max}}\left(1 + \frac{K_A}{[A]}\right)\frac{1}{[B]} + \frac{1}{V_{max}}\left(1 + \frac{\alpha K_A}{[A]}\right) \quad (8)$$

To illustrate their use, a plot of $1/v$ versus 1/[A] will have both its slope and its y intercept varying as a function of [B]; a similar situation prevails for plots of $1/v$ versus 1/[B]. Figure 4 reveals the type of result expected for a plot of equation 7. The lines corresponding to the different fixed levels of B always intersect at a common point whose x-axis value is $-1/K_A$. The intersection occurs above the x axis if $\alpha < 1$, below the x axis if $\alpha > 1$, and on the x axis if $\alpha = 1$. Although in Fig. 4 and several other instances a line is illustrated for infinite concentration of cosubstrate, such a condition cannot be truly achieved in practice and is presented here only to provide the limiting case. It is for this reason that it will probably not be possible to read off values of V_{max} and the Michaelis constants in multisubstrate reactions as was possible in the single-substrate case.

Although values for K_A, K_B, αK_A, αK_B, and V_{max} can be obtained from the y-intercept values and slope results of double-reciprocal plots (and is simplified if a really saturating level of the fixed substrate can be used), it is quite easy to extract these values from secondary plots of $1/V_{max}$ (apparent) versus 1/[B] or plots of slope values (from the 1/[A] lines) versus 1/[B]. These "replots" make use of all the data, and their basis is as follows. Figure 4 and equation 7 show that each line's intersection on the y axis (i.e., $1/V_{max_{app}}$) for the plot of $1/v$ versus 1/[A] can be rearranged in a form that is easily seen to be a linear function of 1/[B]; that form is

$$\frac{1}{V_{max_{app}}} = \frac{\alpha K_B}{V_{max}}\frac{1}{[B]} + \frac{1}{V_{max}} \quad (9)$$

Thus, making a plot of $1/V_{max_{app}}$ versus 1/[B] results in a straight line whose slope is $\alpha K_B/V_{max}$, y intercept is $1/V_{max}$, and $-x$ intercept is $-1/\alpha K_B$. Similarly, Fig. 4 and equation 7 show that the slope for each line in a plot of $1/v$ versus 1/[A] is a linear function of 1/[B]:

$$\text{Slope of 1/[A] plot} = \frac{\alpha K_A K_B}{V_{max}}\frac{1}{[B]} + \frac{\alpha K_B}{V_{max}} \quad (10)$$

In this case, constructing a secondary plot of the slope (for each 1/[A] line) versus 1/[B] gives a straight line of slope $\alpha K_A K_B/V_{max}$, a y intercept equal to $\alpha K_B/V_{max}$, and a $-x$ intercept of $-1/K_A$. Replot analysis provides simple graphical estimates of all the desired kinetic constants for this mechanism. Because the reaction scheme is symmetrical, plots of $1/v$ versus 1/[B] will resemble those illustrated in Fig. 4 and permit direct determination of K_B; replot analysis of these data will provide comparable estimates of αK_A, K_B, αK_B, and V_{max}. Obviously the different methods for calculation of the kinetic constants should provide similar results. Note that a single V_{max} value should be observed, regardless of the varied substrate, as long as E_T is constant.

The presence of a third substrate, a relatively common situation, causes more complexity; however, if it also binds randomly, the resulting velocity equation is surprisingly similar to that shown in equation 5 but with new terms for [C], K_c, and two additional interaction factors that reflect possible changes in dissociation constants for a substrate when other substrates are bound (37). A random terreactant reaction can be analyzed similar to a random bisubstrate reaction if one of the substrates is set at a saturating value while another is varied and a third is fixed but unsaturating. A complete analysis will require six such sets of data, each with as few as 16 assays.

23.3.3.2. Ordered binding

The previous case involving random binding used the simplifying assumption that all steps leading to the formation of the EAB complex were rapid and in equilibrium, with the breakdown of EAB being rate limiting (equation 4). The assumption of rapid equilibrium conditions for an ordered addition of substrates is unnecessary (and may be inappropriate) since in this instance one can analyze the less restrictive steady-state situation without undue difficulty.

A scheme that depicts the details of this type of reaction is shown in equation 11.

$$\begin{array}{ccccccc} E + A & \underset{k_{-1}}{\overset{k_1}{\rightleftharpoons}} & EA & & EQ & \underset{k_{-4}}{\overset{k_4}{\rightleftharpoons}} & Q + E \\ & & + & & + & & \\ & & B & & P & & \\ & & k_2 \upharpoonleft\downharpoonright k_{-2} & & k_{-3} \upharpoonleft\downharpoonright k_3 & & \\ & & (EAB & \underset{k_{-c}}{\overset{k_c}{\rightleftharpoons}} & EPQ) & & \end{array} \quad (11)$$

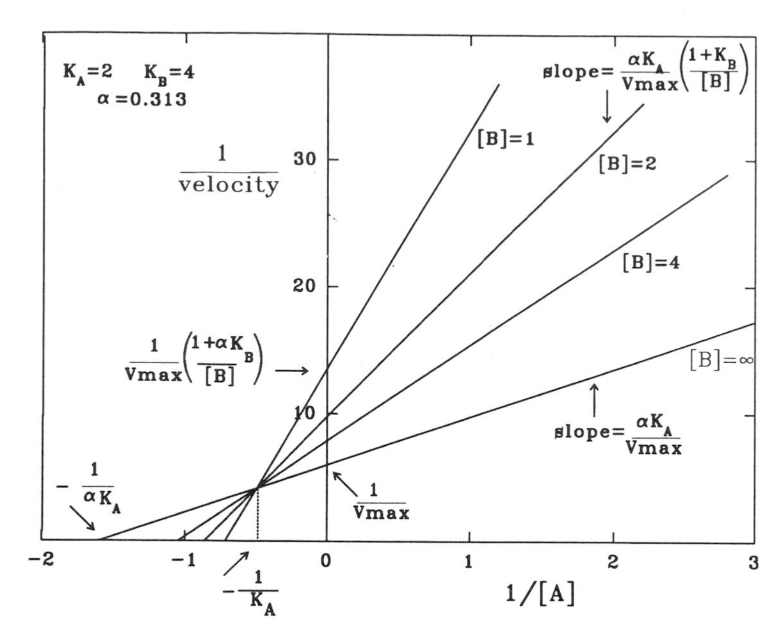

FIG. 4. Double-reciprocal plots for a bisubstrate reaction with a random but sequential addition of substrates and assuming a rapid equilibrium among all species prior to the catalytic step. In this example, A is the varied substrate in each plot at a fixed level of B, as shown. Plots in which B is the varied substrate at fixed levels of A have a generally similar appearance.

There is but one path leading to the EAB complex, and that involves binding of A before B. For the analysis, one does not have to know which substrate binds first, but this knowledge is useful and can be easily obtained from product inhibition studies. Similarly, products leave the enzyme-product complex in an ordered fashion. The velocity equation in the forward direction when no products are present is given in equation 12.

$$v = \frac{V_{\max}\,[A][B]}{K_i^A\,K_m^B + K_m^B\,[A] + K_m^A\,[B] + [A][B]} \qquad (12)$$

The similarity of this equation to that for the random bisubstrate case (equation 5) should be carefully noted. The major changes are in the definitions of the Michaelis constants and in the introduction of a new constant, K_i^A, called an inhibition constant for reasons that need not be described here. In the random rapid equilibrium condition, the Michaelis constants are considered as dissociation constants for the EAB complex to EB + A (i.e., αK_A) or to EA + B (i.e., αK_B). Under the steady-state assumption, Michaelis constants are simply defined as certain groups of rate constants and herein are termed K_m^A and K_m^B without detailing which rate constants make up each K_m term. In either situation, the value of a substrate's Michaelis constant is equal to the concentration that gives a half-maximal velocity when all other substrates are saturating and no products are present. Thus K_m^A is analogous to αK_A, whereas K_m^B corresponds to αK_B. The term K_i^A is likewise defined as a ratio of rate constants and specifically in the ordered binding situation, K_i^A is equal to k_{-1}/k_1 and thus is the dissociation constant for the EA complex.

Differences between the random rapid equilibrium and ordered steady-state situations begin to be apparent when equation 12 is modified to illustrate the velocity dependence of changing [A] while holding [B] fixed, and vice versa. Whereas in the random case these were symmetrical equations, the ordered equations are different for the two substrates. To illustrate varied [A], equation 12 can be rearranged as follows:

$$v = \frac{V_{\max}\,[A]}{K_m^A\left(1 + \dfrac{K_i^A\,K_m^B}{K_m^A\,[B]}\right) + [A]\left(1 + \dfrac{K_m^B}{[B]}\right)} \qquad (13)$$

For varied [B], rearrangement of equation 12 yields

$$v = \frac{V_{\max}\,[B]}{K_m^B\left(1 + \dfrac{K_i^A}{[A]}\right) + [B]\left(1 + \dfrac{K_m^A}{[A]}\right)} \qquad (14)$$

Since equations 13 and 14 are not symmetrical (note the differences in the left denominator terms), the kinetic behavior of A will be different from that of B.

Conduct of the experiment to enable calculation of the desired kinetic constants is identical to that described in the previous random bisubstrate case. Again it should be emphasized that the nature of the mechanism (e.g., ordered or random) need not be known and substrates do not have to be saturating. For the bisubstrate cases, it is only necessary to vary one substrate while holding the second at a fixed level and then repeat the analysis at another fixed level of the second substrate and so on until each substrate has been examined at four to eight concentrations while the other is held at four to eight different levels.

Reciprocal plots of data for the ordered bisubstrate situation are illustrated in Fig. 5. Perhaps surprisingly, they appear very similar to those seen above in Fig. 4 for

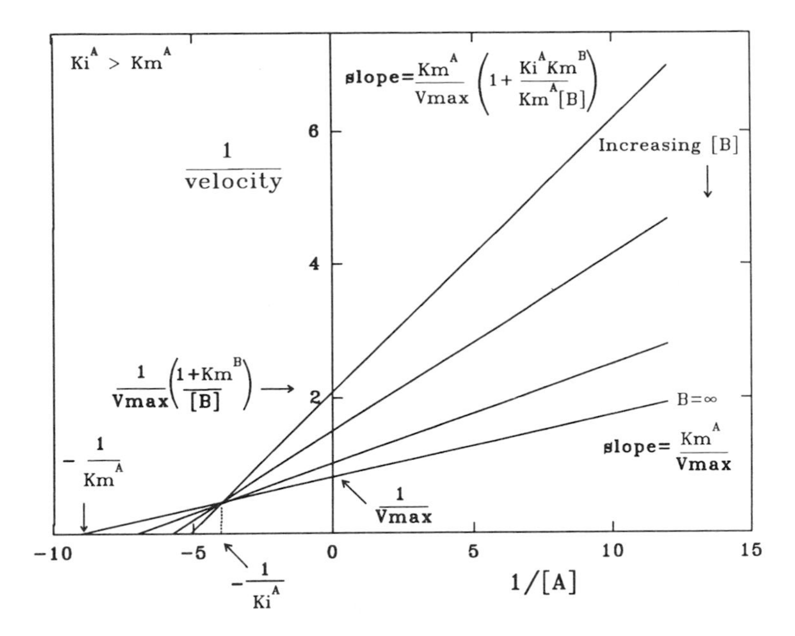

FIG. 5. Double reciprocal plots for a bisubstrate steady-state reaction in which the reactants bind in an ordered fashion. Substrate A is the first to bind, followed by B. Each plot of $1/v$ versus 1/[A] corresponds to a fixed level of B. The intersection point of the extrapolated lines is above the negative x axis when $K_i^A > K_m^A$ but will be below that axis if $K_i^A < K_m^A$ and will be on the axis if $K_i^A = K_m^A$.

the random-mechanism case, and thus it is not possible to distinguish ordered from random binding simply by reciprocal plots. For each substrate analyzed at several fixed levels of the other substrate, a family of lines will result that intersect at a common point to the left of the y axis. The intersection may be above, on, or below the $-x$ axis, depending on whether K_i^A is greater than, equal to, or less than K_m^A, respectively. For a plot of $1/v$ versus 1/[A] at fixed [B], the y intercept and slope of each line can be determined, and taking the reciprocal of equation 13 indicates the terms included in each:

$$\frac{1}{v} = \frac{K_m^A}{V_{\max}}\left(1 + \frac{K_i^A K_m^B}{K_m^A[\mathrm{B}]}\right)\frac{1}{[\mathrm{A}]} + \frac{1}{V_{\max}}\left(1 + \frac{K_m^B}{[\mathrm{B}]}\right) \quad (15)$$

Although the x-axis value corresponding to the point where the various lines intersect is indicated as $-(1/K_i^A)$ in Fig. 5, this is correct only for the first substrate (A), not the second (B), and thus the interpretation of this point requires knowing which substrate actually binds first. Replot analysis of the various y-intercept values (i.e., $1/V_{\max_{app}}$) against 1/[B] gives a line whose slope is $K_m^B/V_{\max}$, y intercept is $1/V_{\max}$, and $-x$ intercept is $-1/K_m^B$.

Comparing the reciprocal equation for $1/v$ versus 1/[B] at fixed [A] (taken from equation 14), we find that

$$\frac{1}{v} = \frac{K_m^B}{V_{\max}}\left(1 + \frac{K_i^A}{[\mathrm{A}]}\right)\frac{1}{[\mathrm{B}]} + \frac{1}{V_{\max}}\left(1 + \frac{K_m^A}{[\mathrm{A}]}\right) \quad (16)$$

Note that the y-intercept term is similar to that in equation 15, and so a replot of $1/V_{\max_{app}}$ versus 1/[A] gives a line of slope $K_m^A/V_{\max}$, a y intercept of $1/V_{\max}$, and a $-x$ intercept of $-1/K_m^A$.

Although considerable differences exist in the slope terms for equations 15 and 16, the two plots corresponding to these equations are similar in form. It is therefore not possible to interpret the slope terms until A, the first substrate, can be distinguished from B. Once that is determined, however, equation 16 reveals that K_i^A can be estimated from the slopes for lines from $1/v$ versus 1/[B]. A replot of slope value (from the 1/[B] plot) versus 1/[A] has its slope equal to $K_m^B K_i^A/V_{\max}$, its y intercept equal to $K_m^B/V_{\max}$, and its $-x$-axis intercept equal to $-1/K_i^A$. K_i^A can also be evaluated from the intersection point of the lines in a plot of $1/v$ versus 1/[A], as shown in Fig. 5, but only when the identity of A is known; there is no analogous constant for B in this mechanism.

Since the most common need from these kinetic experiments is to obtain values for Michaelis constants for each substrate and $V_{\max}$ for the forward reaction, standard analyses provide information on K_m^A (or αK_A) and K_m^B (or αK_B) plus $V_{\max}$. To go beyond this and estimate K_A, K_B, and α (for a random reaction mechanism) or K_i^A for an ordered reaction requires that the reaction types be distinguished and, if ordered, which substrate is A. This can most readily be done by product inhibition studies. For a thorough explanation of these, including the equations for product inhibition effects on initial velocity, consult reference 37; a brief summary of possible results is provided in Table 1. As shown there, inhibition of the forward reaction by individual products gives different types of inhibition for the two mechanisms. Definitions of these inhibition patterns are given in section 23.5.1, but the various types (competitive, noncompetitive, and uncompetitive) can all be easily distinguished by graphical methods.

One example will suffice to illustrate the power of this analysis. Consider the reaction catalyzed by lactate dehydrogenase. For this discussion, first assume that the reaction is ordered, with NADH binding before py-

Table 1. Product inhibition patterns for some bisubstrate reactions[a]

Mechanism	Substrate		Product	Pattern[b]
	Variable	Fixed (conc)[c]		
Ordered	A	B (U or S)	Q	C
Ordered	B	A (U)	Q	NC[d]
Ordered	B	A (S)	Q	None
Ordered	A	B (U)	P	NC[d]
Ordered	A	B (S)	P	UC
Ordered	B	A (U or S)	P	NC[d]
Random[e]	A	B (U)	Q	C
Random[e]	A	B (S)	Q	None
Random[e]	A	B (U)	P	C
Random[e]	A	B (S)	P	None

[a]Patterns are described for double-reciprocal plots of 1/[variable substrate] versus 1/[product] at the stated level of fixed substrate.
[b]C, competitive; NC, noncompetitive; UC, uncompetitive; None, not inhibitory.
[c]Fixed-substrate concentration range: U, unsaturated; S, saturated.
[d]Convergence point is not necessarily on the horizontal axis but is always to the left of the vertical ($1/v$) axis.
[e]In this mechanism, assignments of A and B are arbitrary, as are P and Q. The patterns stated apply only if there are no dead-end complexes, such as EAP or EBQ.

ruvate; products are thus released in the order lactate before NAD^+. According to Table 1, for this ordered reaction where NADH is substrate A, pyruvate is substrate B, lactate is product P, and NAD^+ is product Q, NAD^+ should be a competitive product inhibitor toward NADH, regardless of whether the amount of pyruvate is rate limiting or saturating. (Adding one compound to serve as a product inhibitor does not violate initial velocity conditions, because the reverse reaction can only proceed if both products are present.) Moreover, Table 1 further indicates that the second substrate (pyruvate) should be noncompetitively inhibited by its counterpart product (lactate). Thus, if the reaction is truly ordered, examination of the type of inhibition resulting from the presence of the counterpart product can help to identify which substrate binds first. Several other patterns of inhibition are indicated in Table 1 for an ordered reaction, and because of the relative simplicity of these studies, all should be examined before accepting the identity of A versus B. Note, however, that if the mechanism is random rather than ordered, a different set of patterns would be found. Analyzing the forms of inhibition which result from individually testing the effect of each product can therefore permit one to distinguish between mechanisms as well as identify the binding order where appropriate.

23.3.3.3. Ping-Pong binding

Whereas the previous two examples involved the addition of substrates to the enzyme before any products were released (usually termed **sequential** mechanisms), many transferase-type enzymes operate by a process wherein one product is released prior to the addition of the second substrate. Transaminases (aminotransferases) are good examples of a Ping-Pong mechanism in that one amino acid is bound to the enzyme and its keto-acid analog is released, followed by the binding of another keto acid and its conversion to a second amino acid. This scheme usually involves some sort of substituted or modified enzyme as an intermediate and is illustrated in equations 17 and 18.

$$E + A \rightleftharpoons (EA \rightleftharpoons E'P) \rightleftharpoons E' + P \qquad (17)$$

$$E' + B \rightleftharpoons (E'B \rightleftharpoons EQ) \rightleftharpoons E + Q \qquad (18)$$

Although the reactions seem similar, A and B (as well as P and Q) are easily distinguished on the basis of the chemical changes involved and because the reaction begins with E rather than E′. Thus in a transaminase reaction, A and Q correspond to the two amino acids, whereas B and P are the two keto acids; moreover, A converts to P upon transamination, just as B changes to Q. The enzyme form E contains pyridoxal-P and is converted transiently to the pyridoxamine-P intermediate form, E′.

The steady-state velocity equation for this type of reaction in the absence of significant P and Q is

$$v = \frac{V_{max}[A][B]}{K_m^B[A] + K_m^A[B] + [A][B]} \qquad (19)$$

Rewriting this to illustrate A as the varied substrate gives

$$v = \frac{V_{max}[A]}{K_m^A + [A]\left(1 + \frac{K_m^B}{[B]}\right)} \qquad (20)$$

Likewise,

$$v = \frac{V_{max}[B]}{K_m^B + [B]\left(1 + \frac{K_m^A}{[A]}\right)} \qquad (20)$$

Determination of K_m for a Ping-Pong bisubstrate reaction is like the others discussed. Although the velocity equations are symmetrical, the identities of A and B are often sufficiently obvious that it is simple to decide which substrate precedes the other. Evidence for a Ping-Pong mechanism is also easily discerned from the nature of the plots which result. The reciprocal form of equation 20 is

$$\frac{1}{v} = \frac{K_m^A}{V_{max}}\frac{1}{[A]} + \frac{1}{V_{max}}\left(1 + \frac{K_m^B}{[B]}\right) \qquad (22)$$

and a symmetrical version can be written for B as the varied substrate. For equation 22, the plots of $1/v$ versus $1/[A]$ at fixed levels of B are parallel (i.e., all have the same slope), as is illustrated in Fig. 6. This figure reveals the parallel nature indicative of a Ping-Pong mechanism; similar relationships are seen with fixed [A] and varying [B]. In Fig. 6 the x- and y-intercept values increase (apparent K_m^A and apparent V_{max} values decrease) as the concentration of the fixed substrate decreases. Replots of $1/V_{max_{app}}$ versus $1/[B]$ give a line with y intercept of $1/V_{max}$ and x intercept of $-1/K_m^B$.

23.3.4. Kinetic Constants

Direct plots of velocity observed as a function of substrate concentration(s), in contrast to double-reciprocal plots, have not been widely used in the past for estimation of kinetic constants. This situation is changing,

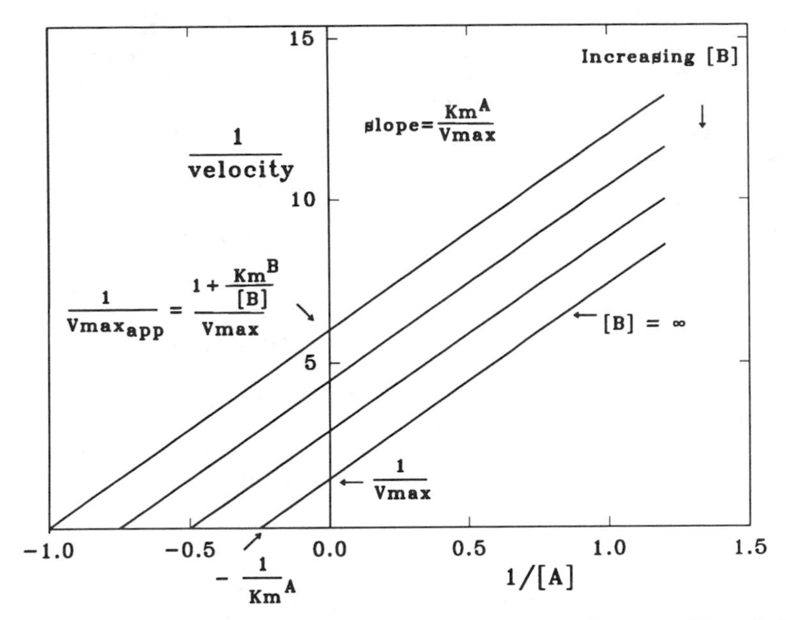

FIG. 6. Double-reciprocal plots for a bisubstrate reaction with a Ping Pong mechanism. Although the plots with varied [B] and fixed [A] are generally similar to those shown here for 1/[A] and fixed [B], A is the leading substrate and binds to a form of enzyme different from that to which B binds.

however, because there are computationally sophisticated nonlinear regression programs that produce best fits to the limb of a rectangular hyperbola which describes the relationship of v to [S]. Several commercial enzyme kinetics programs are marketed for microcomputers, and these not only lead to the desired kinetic constants but also offer various statistical weighting routines to avoid biasing the results with lesser reliable data. These programs are available from several companies, in particular Biosoft, Cambridge, United Kingdom, and On-line Instrument Systems Inc., Jefferson, GA 30549. The use of enzyme kinetics software is mandatory for the most accurate results and will increasingly become the standard approach for kinetic data analysis. Despite this, graphical methods for estimating kinetic constants based on linearized velocity equations are likely to remain popular because of their simplicity. Among these, the double-reciprocal (Lineweaver-Burk) plot is the best known. Its popularity is due in part to the fact that the dependent variable ($1/v$) and the independent variable (1/[S]) are plotted on opposing axes, whereas most of the other linearized versions involve both variables on one of the axes (e.g., the Eadie-Hofstee plot of v versus v/[S]). The major criticism of the double-reciprocal plot is the tendency for low-substrate-concentration data to exert an overimportant role. This can be largely offset in the experimental design by selection of evenly spaced values for 1/[S] rather than an even progression of [S].

The most important fact to keep in mind is that no method for analysis is better than the data obtained. Most lines are satisfactorily drawn by a visual fit to the linearized data. Use of linear regression techniques on a double-reciprocal or other linearized equation can compound errors unless the rearranged data are properly weighted. Moreover, K_m values are frequently stated with excessive precision. It is almost never appropriate to report values such as 2.87 mM for a K_m, since repetition of the experiment and replotting by the same or another method is likely to give a result which differs by as much as ± 1 mM. Therefore, use only one or two significant figures when reporting K_m values and always indicate the units as concentration. Also keep in mind that V_{max} is *not* a true kinetic constant, since it includes a term for enzyme concentration. Reporting a number for V_{max} is useful only when making relative comparisons or when the actual enzyme (protein) concentration is known, in which case it really becomes a specific activity at saturating substrate concentration(s).

Two other linearized plotting methods are sometimes encountered. One is the Hanes transformation of the Michaelis-Menten equation. In this method, equation 1 is rewritten as

$$\frac{[S]}{v} = \frac{1}{V_{max}}[S] + \frac{K_m}{V_{max}} \tag{23}$$

Then if [S]/v is plotted versus [S], a line of slope $1/V_{max}$ is obtained and the y intercept will be equal to K_m/V_{max}. Extending the line to its $-x$ intercept (i.e., [S]/v = 0) gives the value $-K_m$.

The second method, generally preferred by those who dislike the double-reciprocal method, is the Eadie-Hofstee plot. This is based on the equation

$$v = -K_m\frac{v}{[S]} + V_{max} \tag{24}$$

A plot of v versus v/[S] will normally have a negative slope (i.e., $-K_m$) and a y intercept of V_{max}. Extrapolation of the line to the x intercept (where $v = 0$) will give V_{max}/K_m. This plot has the obvious advantage that in most instances K_m and V_{max} can be determined graphically without resorting to the use of a negative axis. The Eadie-Hofstee plot is also more sensitive to deviations from Michaelis-Menten behavior than are the other plot forms, and thus a nonlinear plot of v versus v/[S] could be indicative of cooperative substrate binding, perhaps of the allosteric type. Be cautious in drawing such a conclusion, however, since nonlinearity is more often due to poor data than to non-Michaelis behavior.

Although there are some reasonable arguments which favor either the Hanes or Eadie-Hofstee transformations over the Lineweaver-Burk form, the fact that most persons are familiar with the latter and need less help in understanding how to interpret its results continues to keep the double-reciprocal method widely used. Also, if the graphical shortcomings mentioned earlier are recognized and experiments are properly designed to circumvent them, the reciprocal form presentation is adequate for less rigorous applications. For a more thorough description of plotting methods and regression techniques, consult reference 18.

23.3.5. Coupled Assays

Measurement of reaction products is often made easier by adding an auxiliary enzyme which converts the primary product to a new material concomitant with providing a useful detection system such as a convenient absorbance change. An example would be the coupled assay of phosphoglucomutase with purified glucose-6-phosphate dehydrogenase (glucose-6-P dehydrogenase), wherein the phosphoglucomutase-catalyzed formation of glucose-6-P is monitored as a change in A_{340} due to the reduction of $NADP^+$ to NADPH. Several conditions must be met, however, for this coupling of reactions to provide valid measurements of the rate for the first enzyme. These are as follows: (i) the first reaction must be zero order; (ii) the second reaction must be first order; and (iii) neither reaction should be significantly reversible under the assay conditions. In the example cited above, this would mean that excess glucose-1-P would have to be supplied for the phosphoglucomutase assay and that the coupling reaction of glucose-6-P with $NADP^+$ to produce 6-P-gluconolactone and NADPH must be conducted with excess $NADP^+$ and glucose-6-P dehydrogenase so that the velocity of this step is a direct function only of the glucose-6-P concentration being produced by phosphoglucomutase. Functional irreversibility of the two reactions is achieved by having the dehydrogenase remove glucose-6-P as it is formed and by monitoring the overall velocity before the final products (here 6-P-gluconolactone and NADPH) have accumulated appreciably. The requirement that the coupling enzyme be in excess can be met by a trial-and-error process in which its concentration is increased until further additions have no effect on the measured velocity of the overall reaction.

Alternatively, equation 25 can be used to calculate the amount of second enzyme required:

$$V_{max}{}^{II} = -K_m{}^{II}\, 2.3 \log (1 - F_s)\cdot(t_L)^{-1} \qquad (25)$$

where $V_{max}{}^{II}$ is the desired activity of E^{II} (the coupling enzyme), $K_m{}^{II}$ is the Michaelis constant for its limiting substrate, t_L is the lag time permitted before the steady-state concentration of the limiting substrate is attained and thus the overall coupled rate is constant, and F_s is the fraction of the steady-state concentration required by time t_L. For most coupled reactions, satisfactory values for F_s and t_L are 0.98 (98% of the true steady-state value) and 6 s (0.1 min), respectively. In the example mentioned above, we must know the K_m of glucose-6-P dehydrogenase for glucose-6-P at the pH, temperature, and ionic strength to be used in the assay of phosphoglucomutase in order to calculate the amount of glucose-6-P dehydrogenase needed. Assuming this K_m value to be 0.2 mM, the calculation yields the result $V_{max}{}^{II} = -0.2\ \text{mM} \cdot 2.3 \log (1 - 0.98) \cdot 10\ \text{min}^{-1} = 7.8$ mM min^{-1}. This result indicates that a concentration of glucose-6-P dehydrogenase equal to 7.8 U/ml is needed in the phosphoglucomutase assay to maintain a steady-state level of glucose-6-P sufficiently low that a first-order condition can prevail and be achieved within 0.1 min after reaction initiation.

When a second coupling reaction is part of the assay, rough estimates of V_{max}/K_m must be made for both coupling enzymes and it must be ascertained which of the enzymes has the lower value. Then if a large excess of the other is present (at least 5- to 10-fold), equation 25 can be used to calculate the t_L that will be due to the slowest of the coupling enzymes. A detailed discussion of the kinetic consequences of multiply coupled enzymes is found in references 25 and 41.

23.3.6. Impure and Endogenous Substrates

Assays in which the substrate is impure seldom give erroneous activity measurements unless the impurities themselves exert some effect on catalytic action. In the case where a contaminating analog is present in the substrate, this compound might serve as an inhibitor through its catalytically unproductive binding to the enzyme. Even though this binding might be competitive with the true substrate, its effect cannot be overcome, because increasing substrate concentration also increases contaminant concentration in the assay. Not infrequently, a substrate can contain inhibitory cations (e.g., Ba^{2+}, Hg^{2+}) or anions (SO_4^{2-}, F^-) whose action may not be recognized until other substrate preparations which lack these ions are examined. Because these impure compounds give altered activity levels in regular assays, it is clear that kinetic determinations of K_m and/or V_{max} will also be affected, the actual result depending on the nature of the inhibition type. Overcoming this problem requires further purification of the substrate.

The most common impurity present in most substrates is water, and its content in the form of water of hydration or simply as moisture in an insufficiently dried product is often variable or unrecognized. Since water is not inhibitory, its only direct effect is in reducing the actual amount of substrate present in the assay. When substrate levels can be set high, this contamination is of no concern. When substrate concentrations must be known and are not saturating, as in K_m determinations, the presence of contaminating water leads to erroneously low values for substrate concentration and thus to incorrect K_m values, although V_{max} estimates may be valid. This situation can be easily corrected by moisture determinations on the substrate.

Contamination of a crude extract with endogenous substrate can be a serious source of error in enzyme assays, especially when the enzyme activity is moderately low and the enzyme cannot be diluted greatly. The problem is encountered when assays are done at nonsaturating substrate levels, in which case increasing enzyme concentration also leads to an increase in substrate concentration, thereby causing a nonlinear relationship between velocity observed and enzyme added. The problem disappears at saturating substrate levels. Activators present in the enzyme extract also can give rise to nonlinear plots of velocity versus $[E]_T$, and depending on the effect (i.e., if the apparent V_{max} is increased rather than the apparent K_m being decreased), this may even occur at saturating substrate levels.

As expected, the presence of endogenous substrate in the enzyme sample gives some velocity value in the absence of added substrate, thereby interfering with analyses of velocity as a function of [S] for determination of V_{max} and K_m. Attempts to correct for this by subtracting the observed velocity at [S] = 0 from each of the observed velocities at other [S] values can give data which result in a linear double-reciprocal plot but the calculated V_{max} and K_m results are incorrect. For this reason, when rates are observed in the absence of added substrate, serious kinetic analyses should not be attempted. However, in many instances, gel filtration or dialysis of the enzyme sample removes the contaminating material and permits valid results. One additional caution may be needed. Monitoring the conversion of a radioactive substrate to product as the basis of an assay can be in error if significant amounts of unlabeled substrate are unknowingly present in the enzyme. This contamination not only lowers the actual specific radioactivity of the substrate but also does so as a function of enzyme added.

23.4. MEASUREMENT OF ACTIVITY

23.4.1. Optimization

Whether newly developed or simply adapted from a method used for the same enzyme from another source, an assay should be optimized for the specific enzyme being studied if the intent is to report properties of that enzyme. Such characteristics as K_m, specific activity in the homogeneous state, and pH optimum can be greatly influenced by assay conditions (e.g., temperature, buffer type, and ionic strength), and it is therefore desirable to determine these properties under conditions which seem most favorable for catalytic activity. Although it can be argued in some instances that these are not "physiological" conditions, in the absence of sufficient information to describe the intracellular environment it seems reasonable to study the enzyme under otherwise optimal conditions.

Most of the steps for accomplishing optimization are relatively obvious. The pH, buffer type, ionic strength, concentrations of reactants and cofactors, and assay temperature are among the most important parameters to consider optimizing. Some of the results may be interactive, however. For example, it is possible that the pH optimum varies with buffer type; in such an instance, select the buffer and pH condition which gives the highest relative catalytic activity. On the other hand, avoid impractical situations even if they appear to be optimum, such as temperatures or pH conditions that give high activity but under which the enzyme is detectably unstable. It is sometimes necessary to adopt an assay condition which is suboptimal in order to improve day-to-day reproducibility. Similarly, if using kinetically saturating substrate or cofactors is not feasible or appropriate, simply be aware of the limitations imposed by assaying at less than optimal reactant concentrations.

Other useful pieces of information to gather are (i) the best solution for performing enzyme dilutions, (ii) the stabilizers that might be added to the dilution buffer (e.g., bovine serum albumin, glycerol, dithiothreitol), (iii) the length of time a given dilution can be held before activity loss is observed, and (iv) whether the enzyme is possibly cold labile.

Finally, the assay procedure itself can influence observed properties, particularly when different catalytic features of the enzyme are being exploited. One example of this is the assay of glutamine synthetase from *E. coli*. There are several methods available for monitoring the conversion of glutamate, ammonia, and ATP to glutamine plus ADP, but the so-called γ-glutamyl transferase activity of the enzyme is also frequently measured. The transferase assay uses a colorimetric procedure to detect the formation of γ-glutamylhydroxamate from the reaction of glutamine with hydroxylamine, catalyzed by the enzyme in the presence of ADP and arsenate (see reference 40 for more details on this assay). Unlike in the reaction measuring glutamine formation, in which only the deadenylylated form of the enzyme is active, both adenylylated and deadenylylated enzyme forms respond in the transferase assay. The point here is that properties of the enzyme determined by the transferase assay are not necessarily identical to those observed when glutamine formation is being monitored.

23.4.2. Standardization and Calibration

The importance of establishing the dependence of velocity on the amount of enzyme added has been amply discussed. Along with this, however, is the necessity for knowing that the assay measures product formation or substrate disappearance in absolute, not just relative, terms. Although not all applications require accurate knowledge of micromoles of product formed per minute, results can be exchanged between investigators only if assays are calibrated with standards of known purity. In an assay which monitors an absorbance change, determination of an extinction coefficient for either product or substrate provides the basis for quick and simple estimation of quantity formed or consumed. If an extinction coefficient is not available from the literature for the conditions to be used in the assay, a value must be calculated from absorbance measurements on pure material. More commonly, it is necessary to standardize colorimetric procedures by performing mock assays which contain all ingredients but one (e.g., no substrate or no enzyme) and then adding known amounts of the material to be analyzed. For instance, in the assay of a phosphatase which hydrolyzes phosphate ester substrates to produce P_i, the procedure should be standardized by incorporating various known amounts of P_i in reaction mixtures lacking enzyme before performing a quantitative evaluation for P_i content by one of the established methods.

In the following sections are presented specific examples of a variety of assays, most of which can be calibrated such that a precise estimate can be made of the amount of product formed per unit time. These are intended only to illustrate how one or more member enzymes of each class can be assayed. For a much more complete description of specific assays, consult such multivolume collective works as *Methods in Enzymology* and *Methods of Enzymatic Analysis.*

23.4.3. Oxidoreductases

23.4.3.1. Glucose-6-phosphate dehydrogenase (EC 1.1.1.49)

Oxidoreductases are among the easiest enzymes to assay because of the variety of spectral characteristics exhibited during their catalytic process. Those which utilize the pyridine nucleotide coenzymes NAD^+ or $NADP^+$ as electron acceptors exhibit very cleanly measured increases in A_{340} as a result of the production of NAD(P)H. The assay of glucose-6-P dehydrogenase from *P. fluorescens* is described here (19, 24); with minor variations, this procedure can be followed for almost any dehydrogenase with $NAD(P)^+$ as coenzyme. The reaction measured is

$$\beta\text{-D-Glucose-6-P} + NAD(P)^+ \rightleftharpoons \text{D-gluconolactone-6-P} + NAD(P)H + H^+$$

When assayed in the forward [$NAD(P)^+$ reduction] direction, assays are usually run at relatively high pH to consume protons and shift the equilibrium to the right; the oxidation of NAD(P)H is normally run at a lower pH since H^+ is a substrate in that reaction. This adjustment of the assay pH with reaction direction is particularly necessary when assays reveal a very short reaction period of linear product formation, indicative of an unfavorable equilibrium. For situations when the assayed direction is sufficiently favorable, the choice of pH can be dictated by the optimum pH value for the enzyme. The assay procedure is as follows.

In a 1-ml quartz spectrophotometer cell with a 1-cm light path, place 0.7 ml of 50 mM 2-amino-2-methyl-1,3-propanediol buffer (pH 8.9) followed by 0.1 ml of 30 mM NAD^+ or $NADP^+$ (prepared fresh daily in the same buffer or obtained from a frozen stock solution neutralized to pH 7) and 0.1 ml of 200 mM glucose-6-P, also dissolved in buffer. Allow the cuvette contents to come to 30°C in the spectrophotometer, and then add 0.1 ml of appropriately diluted enzyme sample to initiate the reaction. When operating the spectrophotometer at ambient temperature, be sure to note the actual temperature of the cell housing. Monitor the change in A_{340}. If the spectrophotometer is not equipped to monitor absorbance continuously as a function of time, manually record the A_{340} every 20 s for at least 4 min to ascertain that the rate is constant. Include a control in which buffer replaces the glucose-6-P. On the basis of the millimolar extinction coefficient of NADH or NADPH at 340 nm (6.22 mM^{-1} cm^{-1}), a change of 0.10 optical density unit (OD unit)/min corresponds to the production of 0.016 μmol of NAD(P)H per min. Because some preparations of NAD^+ are quite acidic, check the pH of the assay mixture after all components are added whenever a new batch is used. Do not store solutions of NAD^+ or $NADP^+$ at pH values over 7, since they are unstable under this condition.

Unlike most organisms, *P. fluorescens* contains two glucose-6-P dehydrogenases, one involved in the hexose monophosphate pathway (HMP) and another concerned with the Entner-Doudoroff pathway (EDP). Their individual activities can be measured in extracts based on differences in the ratio of reaction rates for NAD^+ to those for $NADP^+$. The EDP enzyme exhibits a ratio of 1.29 for $NAD^+/NADP^+$, whereas the corresponding ratio for the HMP enzyme is 0.55. The equations needed to calculate each activity in mixtures are described in reference 24. Methylotrophic pseudomonads contain a glucose-6-P dehydrogenase that is involved in the assimilation of formaldehyde via the ribulose monophosphate pathway (27).

23.4.3.2. Protocatechuate dioxygenase (EC 1.13.11.3)

Another significant group of oxidoreductases are those which catalyze oxygen uptake, and these offer the opportunity to measure activity by manometric or polarographic procedures. An example of this type is the assay of protocatechuate-3,4-dioxygenase by using a Clark-type oxygen electrode (46), although in this particular instance one could also use a spectrophotometric assay at 290 nm, monitoring protocatechuate disappearance. The product of the dioxygenase action is β-carboxy-*cis,cis*-muconate, resulting from the Fe-dependent intradiol cleavage of protocatechuate. The crystal structure of the enzyme from *P. aeruginosa* has been determined (33), and the dioxygenase from *Brevibacterium fuscus* has been studied extensively for its mechanistic features (45).

Prepare 50 mM 3-(*N*-morpholino)propanesulfonic acid (MOPS) –NaOH buffer (pH 7.0). Dissolve 77 mg of protocatechuic acid in MOPS buffer, adjust it to pH 7.0 with 5 M NaOH (approximately 0.1 ml), and dilute it to 5 ml with MOPS buffer to give a 0.1 M protocatechuate solution. Place this in a serum vial, and seal it with a septum stopper. Evacuate the vial and replace with a nitrogen or argon atmosphere four to six times. The nitrogen should be freed of oxygen by passage through a commercial gas purifier (e.g., OMI-1 purifier; Supelco Corp., Bellefonte, Pa.).

Fill the sample chamber of a polarograph (containing a Clark electrode) with MOPS buffer which has been well equilibrated with air, and allow for temperature stabilization at 25°C. The final volume must be known in order to calculate the protocatechuate concentration. Add protocatechuate solution to 2 mM, removing an equal volume of MOPS buffer from the chamber if necessary. Record O_2 uptake until a stable baseline reading is obtained, and then add a precise volume of dioxygenase to be assayed, diluted if necessary in MOPS buffer. Monitor the initial rate of oxygen uptake, ascertaining that the rate is constant for several minutes.

Rates (normally measured as recorder units per minute) are converted to micromoles of O_2 consumed per minute by conducting a calibration reaction in a similar fashion but with a limiting and known amount of protocatechuate (e.g., 0.1 μmol [total]) in the presence of excess enzyme; a commercial preparation (Sigma) can be used for the standardization. The recorder displacement corresponding to 0.1 μmol of protocatechuate (and oxygen) consumption can then be used to convert recorder units to micromoles of O_2 taken up. Calibrations should be conducted daily, if possible, and whenever the electrode membrane is changed.

The enzyme from *B. fuscus* has a K_m of 120 μM for protocatechuate at saturating O_2 levels and pH 7.0. The K_m of oxygen is about 800 μM, and thus its concentration in the assay, about 260 μM, is not in excess.

23.4.4. Transferases

23.4.4.1. Phosphofructokinase (EC 2.7.1.11)

Many kinases catalyze important control steps in metabolism and thus are frequently studied in detail to reveal both their regulatory and catalytic features. Among the most important is phosphofructokinase, which catalyzes the ATP-dependent phosphorylation of fructose-6-P to produce fructose-1,6-bis-P and ADP. The assay described here is a spectrophotometric one in which fructose-1,6-bis-P formation is coupled to NADH oxidation via the linking enzymes fructose bisphosphate aldolase, triose-phosphate isomerase, and α-glycerol-P dehydrogenase. The conditions were optimized for the phosphofructokinase from *Bacillus licheniformis* (22).

Prepare the coupling enzyme system by mixing 1.5 mg of bovine serum albumin with 40 U (4 mg) of fructose bisphosphate aldolase from rabbit muscle, 200 U (0.05 mg) of triose-phosphate isomerase from rabbit muscle, and 50 U (0.3 mg) of α-glycerol-P dehydrogenase from rabbit muscle; dilute with 1 ml of water to dissolve all materials. Dialyze this mixture at 4°C against 1 mM glycylglycine–NaOH buffer (pH 8.2) containing 1 mM mercaptoethanol to remove sulfate ions, and then dilute to 2.5 ml with the same buffer. Store the mixture in ice until needed for the assay.

Each assay should contain, as final concentration in a 1 ml cuvette, 20 mM imidazole-HCl (pH 6.8), 0.1 M KCl, 3 mM $MgCl_2$, 1 mM mercaptoethanol, 1 mM ATP (20 mM stock solution adjusted to pH 7), 1 mM fructose-6-P, 0.25 mM NADH, and 0.1 ml of dialyzed coupling enzymes; add sufficient water to make the final assay to 1.0 ml after the enzyme is added. When the cuvette temperature has stabilized (30°C), add enzyme extract to initiate the reaction and monitor the decrease in A_{340} in a recording spectrophotometer. The molar absorbance coefficient of NADH is 6.22×10^3 cm^{-1} M^{-1}. Calculation of units of phosphofructokinase present in the assay must take into account that for each mole of hexose-P consumed, 2 mol of triosephosphate is formed in the assay and thus 2 mol of NADH is oxidized.

Many phosphofructokinases are subject to inhibition by high ATP levels, especially when the fructose-6-P concentration is low, and in some cases the loss of activity is time dependent and irreversible. Ascertain that velocity is constant over the assay period examined, and, if necessary, adjust the ATP concentration accordingly. The K_m for ATP in the phosphofructokinase from *B. licheniformis* is 0.07 mM, whereas that for fructose-6-P is 0.05 mM. Under the prescribed assay conditions, these compounds are in sufficient excess to approach a zero-order kinetic condition, yet the ATP level used is not inhibitory, provided fructose-6-P is also present.

Although the particular coupling system used here is usually satisfactory even in relatively crude extracts, another system has been widely used for kinases when the ATPase activity is low. In this case, the reactions are

$$\text{Fructose-6-P} + \text{ATP} \rightleftharpoons \text{Fructose-1,6-bisP} + \text{ADP}$$
(phosphofructokinase)
$$\text{ADP} + \text{phosphoenolpyruvate} \rightleftharpoons \text{pyruvate} + \text{ATP}$$
(pyruvate kinase)
$$\text{pyruvate} + \text{NADH} + \text{H}^+ \rightleftharpoons \text{lactate} + \text{NAD}^+$$
(lactate dehydrogenase)

Monitor the decrease in A_{340} as an indication of phosphofructokinase activity. Obviously, contamination with ATPase or NADH oxidase interferes with this spectrophotometric analysis.

23.4.4.2. Chloramphenicol acetyltransferase (EC 2.3.1.28)

Chloramphenicol acetyltransferase (CAT) catalyzes the acetylation of chloramphenicol by acetyl-CoA to form chloramphenicol-3-acetate plus CoA-SH; it is found in many bacteria, where it forms the basis for their resistance to the antibiotic chloramphenicol since the acetoxy derivative has no antibiotic activity. Its structural gene is often extrachromosomal, residing on plasmids where expression is usually constitutive but regulated to some degree by catabolite repression. In a few instances (e.g., *Agrobacterium tumifaciens* and *Streptococcus faecalis*), CAT activity has been found to be inducible, with the resulting competition between chloramphenicol (serving as the CAT inducer) and its antibiotic action in blocking protein synthesis. The CAT gene also has been inserted into phages P1 and lambda.

Several methods are used for the CAT assay. Probably the most common is the thin-layer silica gel chromatographic separation of [^{14}C]chloramphenicol from the labeled acetylated product and subsequent autoradiographic analysis. This assay is quite satisfactory when one is studying on/off regulation of a promoterless CAT gene to detect function of a suspected promoter sequence, and it can be used to detect CAT activity in growing cultures by chromatographic analysis of the culture supernatant after extraction of the labeled compounds into ethyl acetate. If a densitometer is available, it can be used to quantify results by scanning the autoradiogram. It is also possible to scrape the appropriate band of chloramphenicol acetate from the thin-layer chromatography plate and quantitatively estimate the product by scintillation counting. The following assays are preferred for more rapid quantitative determinations.

The most sensitive nonchromatographic method for CAT assay uses [^{14}C]acetyl-CoA and chloramphenicol; the labeled chloramphenicol acetate is then extracted into an organic solvent under conditions where [^{14}C]acetyl-CoA or [^{14}C]acetate is not. A cycling system composed of acetate kinase, phosphotransacetylase, and pyruvate kinase is used to maintain an excess of [^{14}C]acetyl-CoA throughout the reaction (38).

In a 15-ml conical glass centrifuge tube, add 50 μl of 0.5 M Tris-HCl (pH 7.8); 15 μl of 100 mM $MgCl_2$; 10 μl of 10 mM dithiothreitol; 20 μl of 50 mM ATP; 50 μl of 100 mM phosphoenolpyruvate; 20 μl of 5 mM CoA (reduced); 20 μl of 5 mM [^{14}C]acetate (specific radioactivity, 5 mCi/

mmol); 1 U each of acetate kinase, phosphotransacetylase, and pyruvate kinase; CAT preparation; and water to a final volume of 240 μl. Mix the contents, and bring to 37°C in a water bath. Start the reaction by adding 10 μl of 5 mM D-*threo*-chloramphenicol. Cover the tubes with Parafilm to reduce evaporation, and incubate for 20 min. Terminate the reaction by adding 25 μl of 1 M sodium arsenate. Blanks in which water replaces CAT or chloramphenicol should be included for each set of incubations.

Add 2 ml of toluene to each tube, vortex to mix the phases well, and then centrifuge in a clinical centrifuge to speed up phase separation. Transfer most of the upper, organic layer with a Pasteur pipette to a scintillation vial, and repeat the extraction of the reaction mixture with another 2 ml of toluene, combining the toluene phases for each sample. Add 6 ml of scintillation fluid to the vial, and count.

This assay is an example of one that uses coupling enzymes to keep the concentration of a possibly unstable substrate (acetyl-CoA) high in the presence of crude extracts. Pyruvate kinase and phosphoenolpyruvate are used to keep the level of ATP high for maximum activity of acetate kinase. This enzyme, in turn, uses [^{14}C] acetate and ATP to produce [^{14}C]acetylphosphate, which is then converted to [^{14}C]acetyl-CoA by phosphotransacetylase. Because these enzymes must all be in excess, it is best to test each of the coupling enzymes separately by adding it in a 5- to 10-fold larger quantity than specified here to see whether any increase in rate is observed for a given amount of CAT activity. Since most CAT enzymes exhibit K_m values near 50 μM for acetyl-CoA and 20 μM for chloramphenicol, the concentrations recommended are near saturating levels, provided CAT activity does not produce more than 10 nmol of acetylated product during the incubation (i.e., is not over 0.5 mU per assay).

With less sensitivity, CAT can be assayed spectrophotometrically by monitoring the reaction of 5,5′-dithio-*bis*(2-nitrobenzoate) (DTNB, Ellman's reagent) with CoA-SH released from acetyl-CoA by the enzyme. This assay is subject to serious interferences by other thiol compounds and by contaminating thiolesterases, and therefore it is recommended only for partially purified extracts or when CAT levels are moderately high.

23.4.5. Hydrolases

23.4.5.1. β-Galactosidase (EC 3.2.1.23)

The simple sugar hydrolase β-galactosidase has proven to be one of the most frequently assayed enzymes, albeit often qualitatively, because of the many *lacZ* gene fusions that have been made. However, it is also easily assayed by a quite precise spectrophotometric procedure as outlined here, using the chromogenic substrate *o*-nitrophenyl-β-D-galactoside (ONPG) which is converted by the enzyme to galactose and *o*-nitrophenol. The latter compound is faintly yellow, and at high pH, at which the nitrophenolate ion dominates, a more intense yellow color is produced, which can be measured at 420 nm. The procedure given here is that of Miller (28).

Prepare a solution of ONPG (4 mg/ml) in 0.1 M phosphate buffer (pH 7.0), and equilibrate it to 28°C. In a tube (12 by 75 mm), place 0.9 ml of Z buffer (Z buffer contains 0.1 M sodium phosphate [pH 7.0], 10 mM KCl, 1 mM $MgSO_4$, and 50 mM mercaptoethanol). Add 0.1 ml of diluted enzyme sample (or cell suspension as described below). After this solution is allowed to come to 28°C in a water bath, add 0.2 ml of the ONPG solution and mix well; the final volume should be 1.2 ml. The reaction can be monitored visually and terminated once a yellow color is observed, noting the time elapsed since reaction initiation; alternatively, a fixed period such as 30 min can be used. Stop the reaction by the addition of 0.5 ml of 1 M Na_2CO_3, which takes the pH to around 11 and inactivates β-galactosidase. Read the color at 420 nm; the assay is usually linear at least up to 1 OD unit. The millimolar extinction coefficient for *o*-nitrophenol at alkaline pH is 4.5 at 420 nm. Include blanks which contain no ONPG (to correct for color due to the sample) and no sample (to correct for spontaneous hydrolysis of ONPG).

It is possible to adapt this assay for whole-cell measurements, as follows. Combine an appropriate amount of cell suspension (usually 0.1 to 0.5 ml containing 1×10^8 to 2×10^8 cells that have been induced for β-galactosidase) with Z buffer to give a volume of 1.0 ml. Add 2 drops of $CHCl_3$ and 1 drop of 0.1% sodium dodecyl sulfate solution, vortex to mix, and then proceed with the assay by adding ONPG as above. After the addition of sodium carbonate, centrifuge the sample to pellet the cell debris and then remove a portion of the supernatant fraction for spectrophotometric analysis.

23.4.5.2. *Eco*RI restriction endonuclease (EC 3.1.21.4)

Aside from the usefulness of type II restriction endonucleases as tools for preparation and analysis of recombinant DNA, this group of enzymes is of obvious importance in the destruction of foreign DNA on entry into an organism. The study of these enzymes has supplied useful insights into the restriction process and DNA-protein interaction, as well as the practical aspects of site specificity. The *Eco*RI endonuclease catalyzes the hydrolytic cleavage of double-stranded DNA containing the hexanucleotide sequence

deoxy-5′. . .pG↓pApA*pTpTpCp. . . .
deoxy-3′. . .pCpTpTp*ApAp↓Gp. . . .

provided the adenine residues marked * are not methylated. The sites of cleavage in the two strands are indicated by arrows, leaving a staggered cut near the 5′ ends of the twofold-symmetrical hexanucleotide sequence.

The usual assay for *Eco*RI endonuclease activity, as well as that for most restriction endonucleases, involves measuring the amount of enzyme need to digest 1 μg of lambda DNA in 1 h, with digestion completeness being determined by agarose gel electrophoresis; the sizes of the five resulting *Eco*RI-generated fragments are 4.878, 5.643, 5.804, 7.421, and 24.756 kbp. Details of this type of assay are available from most of the commercial restriction enzyme suppliers. A more rapid assay has been described by Modrich and Zabel (29) and is based on the conversion of radioactively labeled circular colicin E1 (ColE1) DNA containing one *Eco*RI site to a linear form, which then is hydrolyzed by exonuclease V (*recBC* DNase) to acid-soluble nucleotides.

Prepare ^{3}H-labeled covalently closed circular ColE1 DNA by growing *E. coli* JC411 *thy* (ColE1) in the presence of [^{3}H]thymidine and then isolating the plasmid (29). Dissolve the DNA in 20 mM Tris-HCl (pH 7.6) containing 50 mM NaCl and 1 mM EDTA to give a concentration of 1 mM with respect to nucleotide content (A_{260} = 6.7); the specific radioactivity should be roughly 10 μCi/μmol of nucleotide. A suitable partially purified preparation of exonuclease V can be obtained from *E. coli* by the procedure of Eichler and Lehman (7) through the phosphocellulose chromatography step; adjust to a concentration of approximately 1,000 U/ml. Alternatively, commercial preparations of exonuclease V are available (e.g., from Sigma), although the unit definitions are not the same in all cases and corrections for the amounts to be added should be made.

In a 1.5-ml conical plastic microcentrifuge tube, place 20 μl of 1 M Tris-HCl [pH 7.6] (determine the pH at 25°C on a 1:10 dilution of the stock), 10 μl of 1.0 M NaCl, 5 μl of 10 mM disodium EDTA (pH 8.0), 10 μl of 10 mM dithiothreitol, 10 μl of 2 mM trisodium ATP, 10 μl of the labeled ColE1 DNA (approximately 80,000 cpm), and 7 U of exonuclease V; adjust the final volume to 200 μl with water, allowing for the amount of *Eco*RI endonuclease to be added. When all materials have been mixed, add the enzyme sample and incubate the tube at 37°C for 10 min. Enzyme should be diluted as needed in 20 mM phosphate buffer (pH 7.4) containing 0.2 M NaCl, 0.5 mM dithiothreitol, 0.2 mM EDTA, 0.2 mg of nuclease-free bovine serum albumin per ml, and 10% (wt/vol) glycerol. Terminate the reaction by adding 50 μl of 3 mM salmon sperm DNA (dissolved in 20 mM Tris-HCl [pH 7.6] – 1 mM EDTA) immediately followed by 50 μl of 30% (wt/vol) trichloroacetic acid. Vortex the tube gently to mix, and then place it at 0°C for 10 min to allow complete precipitation. Centrifuge the tube at full speed in a microcentrifuge for 10 min, and carefully remove 200 μl of the supernatant solution for counting by scintillation spectrometry. One unit of *Eco*RI endonuclease activity in this assay has usually been based on the conversion of 1 pmol of ColE1 circular DNA to an exonuclease V-sensitive form per min under the stated conditions.

By using this assay, it has been found that the K_m for ColE1 DNA is 8 nM and that the enzyme exhibits Michaelis-type kinetics. Presumably, other related substrates (e.g., pBR322) could be substituted for ColE1 DNA if desired. Considerable insight into the mechanism of DNA cleavage by the *Eco*RI endonuclease has been obtained from studies in which this assay was used (35).

23.4.6. Lyases

23.4.6.1. Ribulose-1,5-bisphosphate carboxylase (EC 4.1.1.39)

The central importance of ribulose-1,5-bisphosphate carboxylase in carbon fixation in autotrophs is well recognized, and its comparative properties in a variety of species have been studied. Although the carboxylation of ribulose bisphosphate leads to the formation of two 3-P-glycerates whose appearance can be monitored spectrophotometrically by a coupled reaction involving 3-P-glycerate kinase in conjunction with glyceraldehyde-3-P dehydrogenase, the method described here is a $^{14}CO_2$ fixation procedure that is simple and moderately rapid. It is particularly suited to assay of crude materials, conditions under which spectrophotometric analysis may suffer some interferences. This description is based on assays designed for hydrogen bacteria and *Rhodospirillum rubrum* (26), with some minor improvements suggested by Lorimer et al. (20).

Prepare 0.11 M Tris-HCl containing 5.5 mM $MgCl_2$ and adjusted to pH 8.0 at 25°C. Dissolve tetrasodium ribulose bisphosphate in 10 mM Tris-HCl (pH 8.0) to give a final concentration of 10 mM. Although frozen solutions are stable, an inhibitor forms over time, and thus freshly prepared solutions are preferred. Finally, make a 0.5 M solution of $NaH^{14}CO_3$ containing 50 μCi of ^{14}C per ml; the specific activity of this solution must be accurately known.

The reactions are conducted directly in scintillation vials, either the 20-ml size or 7-ml minivials, fitted with appropriately sized serum caps. Place the vials in a constant-temperature bath at 30°C in a well-ventilated hood. For the larger vials, pipette 900 μl of Tris-Mg buffer into each, followed by 50 μl of bicarbonate solution and 50 μl of ribulose bisphosphate solution (or an equivalent amount of 10 mM Tris-HCl [pH 8.0] in no-substrate controls). Seal the vials, and allow temperature equilibration for 5 min. Inject through the stopper 10 μl of enzyme preparation, and swirl to mix thoroughly. Allow the reaction to proceed for a precisely known time (1 to 2 min is usually an appropriate range), and then terminate it with 100 μl of 6 M acetic acid. Carefully remove the vial caps in the hood, and allow the vials to stand open for 30 min to remove most of the unreacted bicarbonate as $^{14}CO_2$. Place the samples in a forced-draft oven at 90°C until dry, and then, when cooled, add 5 ml of scintillation fluid and count. The volumes used in this procedure can be scaled down by half when using minivials.

23.4.6.2. Histidine ammonia-lyase (EC 4.3.1.3)

As the first enzyme in the degradation of histidine, this histidine ammonia-lyase (usually called histidase) has been studied widely because its formation is often carefully regulated according to the nitrogen and carbon status of the cell. Histidase catalyzes the nonoxidative deamination of L-histidine to form urocanate (imidazole acrylate) and ammonia. A variety of assays have been suggested, but those based on monitoring ammonia formation via colorimetric procedures or monitoring urocanate production by measuring the increase in A_{277} are the most common. The latter procedure, described here, is based on the method of Rechler and Tabor (34) but is optimized for histidase from *P. putida*. Most of the other common ammonia-lyases can be assayed by a similar procedure, since their unsaturated products absorb at wavelengths at which the substrates do not, e.g., fumarate, the product of aspartate ammonia-lyase, and cinnamate, the product of phenylalanine ammonia-lyase.

Prepare the following stock solutions: 0.3 M L-histidine, adjusted to pH 9 with KOH and stored frozen; 2 mM $MnCl_2$; 26 mM 2-mercaptoethanol, mixed fresh daily; and 0.5 M diethanolamine-HCl buffer (pH 9.0) stored at 4°C and replaced if discolored.

In a 1-ml (working volume) quartz cuvette with a 1.0-cm path length, place 0.2 ml of diethanolamine buffer, 0.05 ml of $MnCl_2$, 0.05 ml of mercaptoethanol solution,

0.2 ml of histidine, water, and diluted enzyme sample (usually 0.02 to 0.05 ml) to produce a final volume of 1.0 ml. Mix by gentle stirring or inversion, and monitor the change in A_{277} at 25°C. Blanks containing no enzyme or no histidine should also be included. A reaction rate of 0.05 to 0.1 OD unit/min is best, but the assay is generally linear over a wide range of initial rates. The calculation of units is based on the millimolar extinction coefficient for urocanate of 18.8 $mM^{-1} cm^{-1}$ at 277 nm; under the assay conditions described here, an OD change of 0.10/min corresponds to a urocanate formation rate of 5.3 nmol/min.

Some histidase preparations exhibit a rather pronounced lag before attaining a linear rate, presumably as a result of partial oxidation of the enzyme during preparation or storage. If a lag is observed, preincubate the enzyme as follows. Place 0.05 ml of enzyme sample (undiluted), 0.25 ml of 26 mM mercaptoethanol, 0.05 ml of diethanolamine buffer, and 0.15 ml of water in a 1.5-ml microcentrifuge tube. Incubate the solution on ice for 30 min, and then perform the assay with this or a further dilution as the enzyme source.

23.4.7. Isomerases (Ribosephosphate Isomerase) (EC 5.3.1.6)

Ribosephosphate isomerase has been found in a wide variety of heterotrophic and autotrophic organisms. It is a key member of the pentose phosphate pathway as well as a component of the Calvin carbon reduction cycle; its function can vary depending on the organism and its physiological state. The assay procedure described here (21) involves the detection of ribulose-5-P formed in a fixed time incubation by its reaction with cysteine-carbazole reagent to produce a reddish purple color.

In a tube (18 by 150 mm), place 0.3 ml of 0.1 Tris-HCl (pH 7.4) containing 1 mM EDTA and 1 mM mercaptoethanol; this mixture should be prepared daily to avoid oxidation of mercaptoethanol. Add 0.1 ml of 0.03 M ribose-5-P and then 0.1 ml of appropriately diluted enzyme sample (or water for the blank) to give a final volume of 0.5 ml. Incubate at 25 or 37°C for 15 min.

Stop the reaction by the addition of 0.5 ml of 0.03 M cysteine-HCl (this must be freshly prepared), and then immediately pipette in 5 ml of 70% (vol/vol) H_2SO_4 (prepare each day if possible) followed by 0.2 ml of 0.1% carbazole (recrystallized from xylene) in absolute ethanol; alcoholic carbazole is quite stable. Mix carefully by vortexing the solution, and let the color develop for 30 min at 37°C. Promptly read the A_{540} in a 1-cm cuvette. An A_{540} of 0.14 corresponds to the formation of 0.1 μmol of ribulose-5-P (2). Blank readings should be <0.05 OD unit. Linearity of the assay is limited to approximately 0.5 OD units or roughly 0.3 μmol of ribulose-5-P formed. K_{eq} for ribulose-5-P formation is 0.32 at 37°C.

23.4.8. Ligases (Glutaminyl-tRNA Synthetase) (EC 6.1.1.18)

The aminoacyl-tRNA synthetases which activate amino acids for protein synthesis are among the most important members of the relatively small class of enzymes called ligases. Although the use of pyrophosphate bond energy is the common feature of the group, this fact seldom is of direct use in their assay. For the amino acyl-tRNA synthetases, a radiochemical procedure is the most common assay performed. In this assay a labeled amino acid becomes bound to a specific tRNA species, which is then precipitated and counted. The description here is for the assay of glutaminyl-$tRNA^{Gln}$ synthetase (GlnRS) from *E. coli* (11).

In a small conical plastic centrifuge tube, place 10 μl of 0.5 M sodium cacodylate (pH 7.0), 10 μl of 100 mM magnesium acetate, 10 μl of 20 mM ATP (pH 7.0), 2 A_{260} units of unfractionated tRNA from *E. coli*, 10 μl of 10 mM sodium glutamate, 10 μl of 40 μM L-[^{14}C]glutamine (specific radioactivity, 250 mCi/mmol), and diluted enzyme extract; adjust the final volume to 100 μl with water. Enzyme should be diluted in a solution consisting of 50 mM sodium cacodylate (pH 7.0), 20 mM mercaptoethanol, 1 mg of bovine serum albumin per ml, and 10% glycerol. Start the reaction by the addition of diluted enzyme, and incubate at 37°C. At various times, remove an aliquot to a 24-mm-diameter disc of Whatman 3MM filter paper, and immerse the disc in a 5% trichloroacetic acid solution kept at 4°C. The filters can then be transferred to a vacuum filtration manifold (to allow washing of multiple filters under reduced pressure) and washed four times with 4 ml of cold 5% trichloroacetic acid, followed by once with 5 ml of ethanol and once with diethyl ether. Place the dried discs in scintillation vials, add counting fluid, and determine the radioactivity.

Unlabeled glutamate is included in the assay to block the formation of appreciable [^{14}C]glutamyl-tRNA as a result of the action of glutaminase on [^{14}C]glutamine to produce [^{14}C]glutamate. By analyzing several aliquots of each reaction at different times, linearity with time can be ascertained directly. K_m values are 110 μM for both glutamine and ATP. Because this assay uses saturating ATP but limiting glutamine, careful attention must be given to avoiding variations in substrate added and its excessive utilization. At an initial concentration of 4 μM glutamine, observed velocity is directly proportional to its content (i.e., pseudo-first-order kinetics are followed). In some instances, the same concern may apply to the acceptor tRNA, in which case it is important to use accurately prepared solutions of tRNA. Recalibrate the assay when changing batches of tRNA.

23.5. REGULATION OF ACTIVITY

23.5.1. Enzyme Inhibitors

The use of enzyme inhibitors as tools for control of enzyme activity, both in vivo and in vitro, is widespread, but often there is insufficient appreciation for the factors which influence the effectiveness of inhibition and for what inhibitors can tell us regarding the properties or mechanism of an enzyme. At the molecular level, inhibitor action ranges from blocking a specific step in the catalytic cycle of an enzyme to covalent modification of individual amino acid residues on an enzyme. A detailed analysis of this action generally requires some amount of kinetic characterization to provide useful information. The current discussion describes a few basic aspects of this kinetic analysis and what potential pitfalls might be encountered.

Enzyme inhibitors can act in either a reversible or an irreversible fashion. **Reversible enzyme inhibitors** enter into some noncovalent interaction with an enzyme, whereas **irreversible enzyme inhibitors** form covalent unions with a targeted group(s) on the enzyme, even though these might be reversed under some conditions. For example, Ellman's reagent (DTNB) reacts stably with -SH groups in many enzymes but can be removed by added thiol; it is nevertheless classified as an irreversible inhibitor. Most reversible inhibitors are quite specific for a particular enzyme since they must be recognized and bound by the enzyme at some site; a substrate analog which interferes with normal substrate binding or reaction progress but which can be readily reversed by dilution or dialysis would be a typical reversible-type inhibitor. Only this class can be analyzed by steady-state or rapid equilibrium kinetic treatments, and it is critical that the inhibitor type be established prior to undertaking extended kinetic analysis.

The analysis of inhibitor type can be done in several ways. First, determine whether the degree of inhibition exhibited is time dependent. Irreversible inhibitors react with enzyme in a time- and concentration-dependent fashion until the reaction is complete; therefore, if the degree of inhibition increases with time, the inhibitor is probably acting in an irreversible manner. Conversely, if inhibition is greater with increasing concentration of inhibitor but each new inhibited level is attained rapidly and then without further change with time, the inhibitor is probably of the reversible type. When monitoring reaction velocity in an assay containing inhibitor, one can see that the irreversible inhibitor generally shows a steadily decreasing velocity throughout the assay period whereas the reversible inhibitor shows a reduced but constant velocity compared with a no-inhibitor control. Also, incubation of enzyme with inhibitor, then dilution into an assay without any inhibitor, should reveal no obvious inhibitory effect if the compound is acting as a reversible inhibitor, provided there was no carryover of inhibitor, regardless of the incubation time.

Besides the usually obvious time dependence of an irreversible inhibitor, another technique for distinguishing these two situations is to measure V_{max} as a function of total enzyme units in the assay in the presence and absence of the inhibitor. Irreversible inhibitors always result in a decrease in V_{max}, whereas only some types of reversible inhibitors exhibit this property. Plotting observed V_{max} versus enzyme units in the absence of an inhibitor yields a straight line which goes through the origin with a value of zero V_{max} at zero enzyme concentration. For a reversible inhibitor that affects V_{max}, this type of plot will go through the origin but will have a lower slope; the irreversible inhibitor plot should produce the same slope as the control (no inhibitor) plot but will intersect the enzyme unit axis at some positive value equivalent to the amount of enzyme that has been irreversibly inactivated. Of course, V_{max} must be estimated at a fixed time of assay because the degree of inhibition changes with time for an irreversible inhibitor. If the concentration of substrates can be set very high (at least 10 times K_m), this single condition may suffice for approximating V_{max} and construction of the plot therefore requires only enough different assays varying $[E]_T$ to establish a linear V_{max} versus $[E]_T$ relationship.

23.5.1.1. Reversible inhibitors

A reversible inhibitor is usually classified as **competitive, noncompetitive,** or **uncompetitive,** depending on how it interacts with the enzyme. Furthermore, these inhibitors may be of a **complete** or **partial** type; in partial inhibition, an inhibitor-enzyme-substrate complex can form, but, in contrast to complete-type inhibitors, this complex leads to some product release. The present discussion deals mainly with complete-type reversible inhibitors acting in a single substrate reaction (or a bisubstrate reaction in which one substrate is saturating, i.e., a pseudo-first-order condition).

At the mechanistic level, a competitive inhibitor (I) interacts reversibly only with free enzyme to form an EI complex; there can be no ESI complex since substrate and inhibitor are mutually exclusive. A noncompetitive inhibitor, on the other hand, can interact with either free enzyme or enzyme-substrate complex to form EI or ESI, respectively; substrate and inhibitor do not exclude one another, and the ESI complex does not proceed to form product (except when I is a partial-type inhibitor). An uncompetitive inhibitor in the case of a single-substrate reaction interacts only with ES and not E to form an ESI complex. This type of inhibition is seldom encountered for a single-substrate situation but is rather common in multisubstrate reactions.

These inhibitors can be easily distinguished by differences which are apparent from double-reciprocal plots of $1/v$ versus $1/[S]$ at varying concentrations of the inhibitor. Sample plots illustrating the most fundamental of these situations are shown in Fig. 7. The inhibition described in each instance is of the complete (not partial) type, and thus the degree of inhibition is proportional to inhibitor concentration.

In the double-reciprocal plot for a competitive inhibitor (Fig. 7A), only the slope of each line increases directly as a function of the inhibitor level. Because the slope of a plot of $1/v$ versus $1/[S]$ is a function of the ratio K_m/V_{max}, and since V_{max} is unchanged (as shown by the convergence of all lines on the y axis), we can conclude that a competitive inhibitor must alter only the apparent affinity of the enzyme for the substrate (i.e., it affects the apparent K_m value). The plots shown in Fig. 7A are described explicitly by equation 26.

$$\frac{1}{v} = \frac{K_m}{V_{max}}\left(1 + \frac{[I]}{K_i}\right)\left(\frac{1}{[S]}\right) + \frac{1}{V_{max}} \qquad (26)$$

The slope in the presence of inhibitor differs from that in the absence of inhibitor by the term $(1 + [I]/K_i)$, where [I] is the inhibitor concentration and K_i is an **inhibitor constant**, in simple cases being the dissociation constant for the EI complex. Reciprocal plots of $1/v$ versus $1/[S]$ as a function of [I] allow estimation of the inhibitor constant if K_m and V_{max} are known or are evaluated under the condition of no inhibitor present. K_i can be calculated either from the slope at some [I] or from any of the intercepts on the negative x axis, which will be $-1/(K_m)(1 + [I]/K_i)$, but these methods suffer from the disadvantage that they do not necessarily use all the data available. When data have been obtained at a number of inhibitor concentrations, as is always preferable, construct a secondary plot (replot) of the slope from each reciprocal plot versus [I]. This should produce a

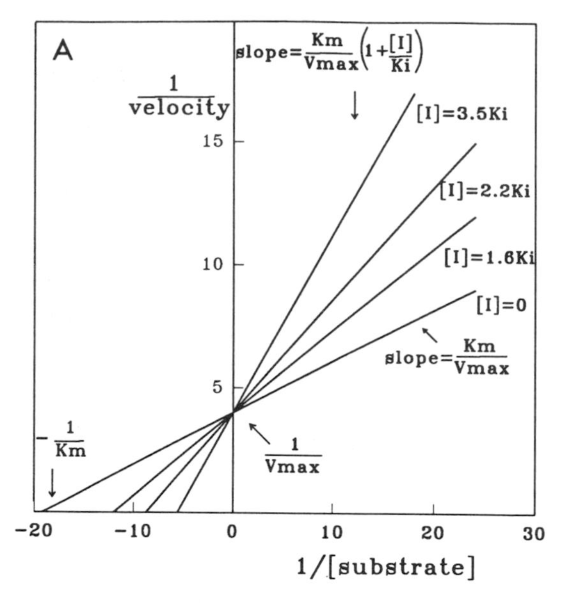

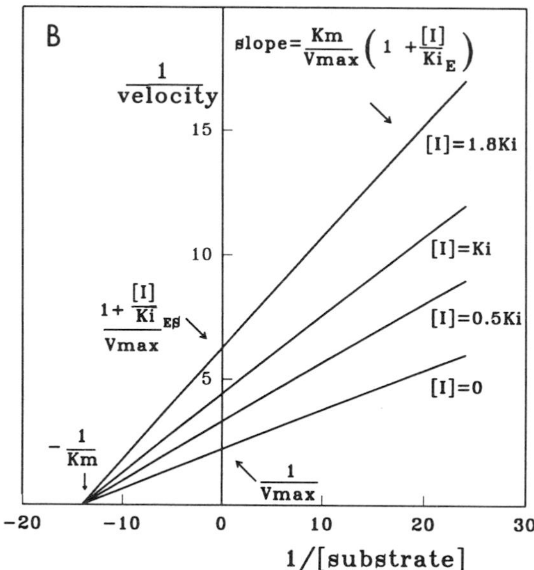

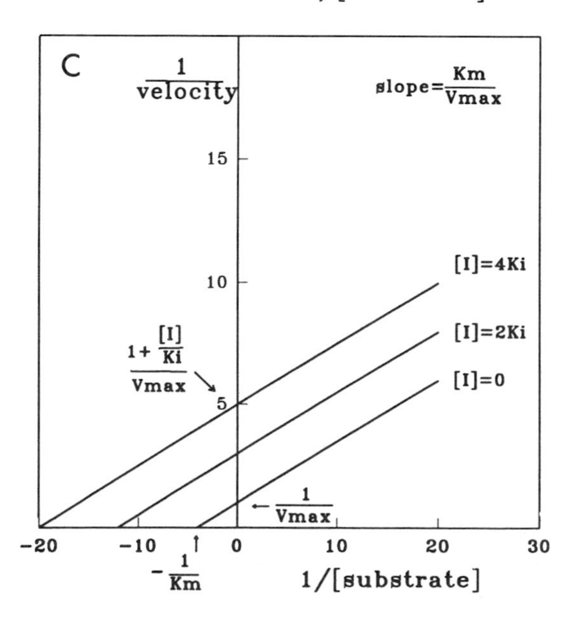

straight line whose intercept on the extended x axis is $-K_i$.

A reversible noncompetitive inhibitor reduces the amount of active enzyme in the sample (i.e., affects V_{max}) and may or may not have an effect on K_m. This is illustrated by Equation 27:

$$\frac{1}{v} = \frac{K_m}{V_{max}}\left(1 + \frac{[I]}{K_{i_E}}\right)\frac{1}{[S]} + \frac{1}{V_{max}}\left(1 + \frac{[I]}{K_{i_{ES}}}\right) \qquad (27)$$

For a reversible noncompetitive inhibitor, a plot of $1/v$ versus $1/[S]$ as a function of [I] gives a series of lines which intersect somewhere to the left of the $1/v$ axis and above, below, or on the negative x axis. These various patterns arise because the inhibitor may bind differently to the enzyme depending on whether substrate is present. This results in a dissociation constant, K_{i_E} for the enzyme-inhibitor complex, and another, $K_{i_{ES}}$, for the enzyme-substrate-inhibitor complex. When $K_{i_E} = K_{i_{ES}}$, meaning that there is no effect of substrate on inhibitor binding and vice versa, the intersection will occur on the x axis, as shown in Fig. 7B; this situation is termed **simple noncompetitive inhibition.** Values of $K_{i_{ES}} > K_{i_E}$ result in an intersection above the x-axis, whereas the converse produces intersection below that axis. Nonidentity of the K_i terms is sometimes referred to as **mixed-type inhibition,** but it can be considered just a form of noncompetitive inhibition.

For simple noncompetitive inhibition, estimate the single K_i value from either the slope or y intercept at any nonzero concentration of I (Fig. 7B), and repeat the calculation for all values of [I] to arrive at an average. When the inhibition is mixed-type, estimate and report both constants, K_{i_E} from the slope term and $K_{i_{ES}}$ from the y-intercept values. A value for $K_{i_{ES}}$ can also be obtained from a replot of the different y-intercepts versus [I], with $K_{i_{ES}}$ being the intercept on the negative x axis. A similar replot of slopes versus [I] will provide an estimate for K_{i_E} at the intersection of the replot with the negative x axis.

An uncompetitive inhibitor is readily identified from the characteristic parallel lines which result from plots of $1/v$ versus $1/[S]$ as a function of [I], illustrated in Fig.

FIG. 7. Double-reciprocal plots illustrating the effects of different types of inhibitors on reactions involving a single substrate (or when a second substrate is present at a saturating level). In each instance the inhibitor is capable of full inhibition at infinite concentration, as opposed to a partial inhibition. (A) Inhibitor (I) is competitive toward the varied substrate. As [I] approaches infinity, the double-reciprocal plot approaches the vertical axis; all plots intersect at a common point on that axis. (B) Inhibitor is noncompetitive with respect to the varied substrate. The special case shown, in which all lines converge at a common point on the negative x axis, is seen when the inhibitor binds with equal affinity to free enzyme and to the enzyme-substrate complex. When this is not the case, convergence will be either above or below the extended x axis. These latter situations, although simply variations of the noncompetitive case, are sometimes referred to as "mixed-type" inhibitions. (C) Inhibitor is uncompetitive toward the varied substrate. This condition is seen in single-substrate reactions when I binds only to the enzyme-substrate complex. Uncompetitive inhibition is encountered more frequently and its interpretation is more complex in a multisubstrate enzyme reaction.

7C. Equation 28 reveals that an uncompetitive inhibitor decreases both V_{max} and K_m by the term $(1 + [I]/K_i)$, with a corresponding effect on only the y intercept of a reciprocal plot.

$$\frac{1}{v} = \frac{K_m}{V_{max}} \frac{1}{[S]} + \frac{1}{V_{max}} \left(1 + \frac{[I]}{K_i}\right) \quad (28)$$

There are several specialized plots sometimes used for estimating K_i from inhibition experiments; the Dixon plot of $1/v$ versus [I] is perhaps the most commonly used for this purpose, but it is used mainly for competitive and simple noncompetitive inhibition. Most of the enzyme kinetics software available for K_m and V_{max} determinations also have routines for handling inhibitor data, and these can provide more precise values for inhibition constants than is usually possible from graphical solutions.

Partial inhibitors are detected by their inability to inhibit fully even when [I] is in vast excess. Calculation of K_i for these inhibitors is more complex, as is K_i determination for inhibitors which bind at multiple sites on an enzyme; consult references 6 and 37 for procedures to estimate K_i in these instances. Finally, there are conditions under which noncompetitive inhibitors can give rise to nonlinear reciprocal plots that might be mistaken for allosteric effects but that result instead from interconversions between EI and EIS which affect catalytic rate. Be alert to this possibility; although it cannot be easily treated kinetically, it should be distinguished from allosteric inhibition, which, by definition, is due to conformational changes on enzyme subunits (see section 23.5.4).

23.5.1.2. Irreversible inhibitors

Inhibitors of the irreversible type which lead to enzyme inactivation are among the most common and useful of inhibitors. They frequently provide information about functional groups or cofactors involved in the enzyme and are of practical benefit when the goal is destruction of enzyme activity. However, these must not be treated in the same fashion as for reversible ones. Instead, a time course should be constructed for the inhibition (measuring enzyme activity remaining as a function of time of inhibitor treatment) at a variety of inhibitor concentrations. These time course incubations are best conducted with sufficient enzyme that a sample can be taken and diluted to a negligible inhibitor concentration for assay of activity remaining. When the concentration of the inhibitor is much greater than that of the enzyme (usually true except for extremely tight-binding inhibitors), the inhibitor concentration will not significantly change during inactivation and the process will be pseudo-first order. The slope of a plot of log (activity remaining) versus time gives an observed first-order rate constant (in units of reciprocal time [$time^{-1}$]) for the concentration of inhibitor used. If the reaction is of the simple type $E + I = EI^*$, where EI^* is inactive, a plot of observed rate constant versus [I] will be linear with a slope reflecting the influence of [I] on the inactivation rate; this result has the units of a second-order rate constant, $molar^{-1}$ $time^{-1}$. Report this as a measure of the effectiveness of the inhibitor, rather than terms such as the inhibitor concentration necessary to give 50% inhibition in a given time. When the inactivation proceeds in two steps, initially forming an EI intermediate which then is converted to EI^*, the plot of observed rate constant versus [I] will be hyperbolic rather than linear. Linearization by plotting $1/k_{obs}$ versus $1/[I]$ and computing the inverse of the slope provides a second-order constant that reflects the inhibitor potency.

23.5.2. pH

Enzymes are affected by pH through alterations in protein structure which influence stability as well as catalytic reactivity, or pH may simply modify the kinetic properties of the enzyme. In practice it often suffices to establish how pH affects enzyme stability and then to gather data on the pH-versus-activity profile.

Determining the effect of pH on stability is quite straightforward, although the results obtained may vary with the conditions under which the study is conducted. Prepare a concentrated stock of enzyme by ultrafiltration or other appropriate treatment such as $(NH_4)_2SO_4$ precipitation followed by dialysis. Prepare a selection of buffers adjusted to the desired pH values at the temperature to be used, and with the same ionic strength, μ, defined as $\frac{1}{2} \Sigma c_i z_i^2$ where c is the concentration of each ionic species (i) and z is the charge on each ion. Keep in mind that because buffers usually change ionic strength as a function of pH, buffer concentrations may have to be varied to hold ionic strength at a fixed value as pH changes. Alternatively, an innocuous salt such as KCl can be added to adjust μ without affecting the pH. Dilute the enzyme into the various buffers, and incubate at a set temperature for a predetermined time. If possible, remeasure the pH to establish that dilution of the enzyme into the various buffers did not alter their pH. If small changes result, the actual pH values can be used. Then remove a portion of each sample for subsequent assay under normal assay conditions. Make certain that the addition of buffered enzyme to the assay does not change the assay pH. From the activity results, construct a plot of enzyme remaining (units) versus pH for the chosen time and temperature of preincubation.

Depending on the relative stability of the enzyme under these different pH conditions, one usually finds a pH region where activity recovery (stability) is highest. It may be necessary to increase the time and temperature of incubation to induce moderate instability for proteins which are inherently more stable in order to see a true optimum pH for stability. The resulting pH should be an appropriate one for purification studies or for storage of enzyme.

The effect of pH on catalytic properties is much more complicated to analyze, although the information return is potentially greater. Normally the pH dependence of an enzyme is studied simply to establish an optimum assay pH. When the purpose is to design a reliable assay, it is best if the assay pH selected can vary ± 0.1 pH unit without appreciable activity change, to prevent small variations in buffer preparation from leading to large changes in activity.

A situation often encountered is that of an effect of buffer type as well as pH on reaction velocity. Since most pH studies will involve at least two buffers, conduct several assays in which the different buffers have one or more overlapping pH values to assess any differences exerted by buffer type on velocity at a single common pH. If different, normalize to the higher value in constructing an activity-versus-pH plot, but advise others of the buffer effect and size of the correction. Also take care to ensure that ionic strength is constant across all buffers and pH values tested. One possible solution to buffer effects is to use a mixed buffer so that a broader pH range can be tested without addition of a new buffer species. Two mixed buffers sometimes used are Tris-maleate (pH 5.5 to 8.5) and citrate-phosphate (pH 3.5 to 7.5), although these are not ideal in terms of overlapping pK_a values for each buffer pair. When testing any buffer with a single pK_a in the range of interest, avoid extending the pH more than ± 1pH unit from the pK_a of the buffer; instead, select another buffer with a $pK_a \leq 2$ pH units from the pK_a for the first buffer. Avoid buffers with strong metal-chelating tendencies, especially when dealing with metalloenzymes; many of the zwitterionic buffers such as *N*-2-hydroxyethylpiperazine-*N'*-2-ethanesulfonic acid (HEPES), MOPS, and morpholineethanesulfonic acid (MES) are recommended in these instances, whereas citrate and PP_i are potent metal chelators.

If a large excess of substrate(s) is maintained at all pH values, a true measure of the effect of pH on V_{max} can be obtained. Otherwise, there may be some influence of pH on substrate binding, as a result of either effects on the charge of a substrate or protonation and deprotonation of groups on the enzyme. Many pH curves found in the literature are really composites of the effects of pH on V_{max} and on K_m, because substrate concentrations were not sufficiently high at all pH values tested to saturate the enzyme. The result almost always is a curve which shows an optimum in activity as a function of pH, and this may be sharp or broad, but few conclusions can be drawn about the reasons for the changes. On the other hand, measurement of V_{max} as a function of pH gives information regarding the influence of pH on ionization of groups in the ES complex that affect activity, since all enzyme is tied up in the ES form and changes in the ionization state of free enzyme or free substrate have no effect on the results. In such instances in which detailed information is desired regarding the pK_a values of groups ionizing on the enzyme, the substrate, or the ES complex, a full examination of the effects on pH on K_m and V_{max} is needed. A discussion of the Michaelis pH functions which describe the quantitative relationships necessary to derive this information is beyond the scope of this chapter but can be found in several references included herein (6, 37).

Despite a relatively solid theoretical basis for the effects of pH on kinetic constants, such studies are seldom informative with respect to identification of specific groups involved in catalysis or substrate binding. Not only are most studies of this sort done inadequately, but also there are frequently anomalous pK_a values observed for important groups in proteins, and this often leads to incorrect identification of the residue(s) involved. In the absence of prior definitive information concerning groups associated with these functions, one should avoid drawing conclusions, from pH studies, about the nature of amino acid side chains involved in catalysis and binding.

23.5.3. Temperature

Because of the relatively limited temperature range over which most enzymes operate efficiently, this parameter is studied primarily to permit assay optimization and to provide measurement of the **Arrhenius energy of activation,** E_a, the threshold energy needed for a substrate to undergo reaction. Enzymes can be denatured and lose catalytic activity when exposed to temperatures much above the physiological range. As a consequence, observable increases in reaction rate as a function of temperature are usually noted only up to some "optimum" value of temperature. Above this, the rate of denaturation and the associated activity loss are greater than any rate enhancement due to increased frequency of molecular collisions, and thus a plot of apparent velocity versus temperature of the assay will show an increase with temperature up to a maximum value followed by a decrease above the temperature optimum.

This "optimum" temperature is not an inherent property of the enzyme, however, and is greatly influenced by incubation conditions (ionic strength, pH, cofactor, and substrate concentrations) as well as by the length of the assay. For example, a fixed-time enzyme assay conducted at 60°C for 30 min might give less total product formation than one done at 40°C for the same period, thereby indicating that 60°C was above the temperature optimum. However, conducting the same assays for only 10 min could reveal greater activity at 60°C than at 40°C, since there may be fewer effects of denaturation in the shorter assay. It is this type of variability that has led some to define the optimum temperature for an enzyme as the highest temperature that permits a constant level of activity over the entire assay time. This optimum value is determined by incubating a fixed amount of enzyme at different temperatures in the usual assay mixture minus substrate for the period of the assay and then assaying the enzyme which remains, using a lower temperature at which the enzyme is known to be stable. Such a result is the highest temperature for full stability over the time needed for a normal assay; in most cases an assay then conducted at that temperature will exhibit the highest rate consistent with full activity retention during the course of the assay. However, note that this temperature optimum is often different from one determined by just conducting assays at increasingly higher temperatures. Because of this, it is necessary to indicate which method was used when reporting the temperature optimum of an enzyme; the method with full activity retention throughout the assay is clearly preferable.

The Arrhenius activation energy is a term found in the empirical relationship describing the effect of temperature on a rate constant, namely

$$d(\ln k)/dT = E_a/RT^2 \qquad (29)$$

where E_a is the activation energy. From transition-state theory it can be determined that E_a is equal to $\Delta H^\ddagger + RT$; $\Delta H^\ddagger$ is the enthalpy of formation of the transition state, and it reflects for the rate-limiting step (assuming that this is the rate constant being used in equation 29) the amount of energy needed to assemble the transition-state complex. Although there is considerable mechanistic information to be learned from evaluation of $\Delta H^{\ddagger}$ and the associated entropy of formation of the transition state, $\Delta S^{\ddagger}$, these are seldom of interest where the main intent is to characterize an enzyme rather than the reaction it catalyzes. On the other hand, estimation of E_a for an enzymatic reaction is easily done and provides an indication of how the reaction rate will increase as a function of temperature.

Integration of equation 29 provides a form that is more easily used for estimation of E_a from rate measurements at different temperatures. The result is

$$\ln k = -E_a/RT + \ln A \qquad (30)$$

in which $\ln A$ is a constant of integration. A plot of $\ln k$ versus $1/T$ should yield a linear relationship whose slope is $-E_a/R$ (or $-E_a/2.303\,R$ if $\log k$ is used). Because the slope is desired, it is inadvisable to use semilog paper for constructing the plot; also, T must be in kelvin and R must be evaluated as 1.98 cal mol^{-1} K^{-1} for E_a to have units of calories/mol.

For most enzymes, standard assays conducted at saturating substrate and a fixed enzyme level are performed at different temperatures and the initial velocity values are taken as proportional to k_{cat}. It is usually desirable to use excess substrate because K_m may vary with temperature and this would introduce a rate change unrelated to the direct effect of temperature on k_{cat}.

There are many examples in which plots of log V_{max} versus $1/T$ have a biphasic nature, leading to two slopes and thus two values of E_a. This break in the plot may indicate a change in the rate-limiting step with temperature or can be due to a conformational alteration in the enzyme. A dramatically downward shift in log V_{max} with increasing temperature usually points to enzyme denaturation during the assay.

23.5.4. Allosteric Effectors

Metabolic regulation at the enzyme level effectively complements long term control mechanisms exerted on enzyme biosynthesis (i.e., induction and repression) and protein turnover as means to control flux through pathways. Enzyme-level regulation operates rapidly, is usually specific and reversible, and is brought about by cellular metabolites whose identity and action can be understood from an analysis of the function and properties of the regulatory enzyme. Most examples of enzyme-level regulation are found in either allosteric enzymes which are controlled by noncovalent interactions or enzymes which are covalently modified, sometimes as part of a multicomponent enzyme system. The latter type is discussed briefly in section 23.5.5.

Although models exist to explain many of the detailed properties of allosteric enzymes, from a practical viewpoint it is usually sufficient to recognize the existence of allosteric properties and identify the effector(s) which influence catalytic activity. Non-Michaelis kinetic characteristics are often the most obvious indications of an allosteric enzyme, and it therefore becomes important to recognize their presence and offer a valid interpretation of their basis. A recent volume (10) summarizing the kinetic and structural properties of the best-studied allosteric enzymes provides specific examples to supplement the more general discussion here.

Plots of velocity versus initial substrate concentration for an allosteric enzyme can deviate considerably from the expected hyperbolic behavior of a nonallosteric enzyme. In many instances there is a pronounced sigmoidal nature to the curve, making recognition of an allosteric condition simple, but in some cases there is little or no sigmoidicity to the velocity-versus-[S] curve. Fortunately the existence of non-Michaelis kinetics is easily discerned by looking for deviations from a hyperbolic relationship. For a hyperbolic curve, there is a predictable ratio of substrate concentrations which correspond to velocities of 0.9 V_{max} and 0.1 V_{max}. This ratio of $[S]_{0.9}/[S]_{0.1}$ is 81 for a hyperbolic condition, as shown in Fig. 8. Values significantly less than 81 (usually below 40) are characteristic of conditions showing **positive cooperativity** whereas values above 81 (often 250 or higher) indicate **negative cooperativity.** Where a single ligand, namely the substrate, is involved, these are termed positive or negative **homotropic cooperative responses.** Figure 8 illustrates the velocity-versus-[S] relationship for an enzyme exhibiting Michaelis-Menten hyperbolic kinetics, with a V_{max} of 100 and a K_m of 2 (expressed in any concentration terms). Also shown is an enzyme exhibiting slight positive cooperativity. The sigmoidal nature is evident in the latter, even though the Hill coefficient, n_H, is only 1.4 and $[S]_{0.5}$ is 3 (see below for an explanation of these terms); this situation produces an $[S]_{0.9}/[S]_{0.1}$ ratio of 23.

Positive homotropic cooperativity in a multisite allosteric enzyme results from an increase in the catalytic efficiency of one active site as others become saturated with substrate. This is most commonly due to an increased affinity of unoccupied sites for substrate as a

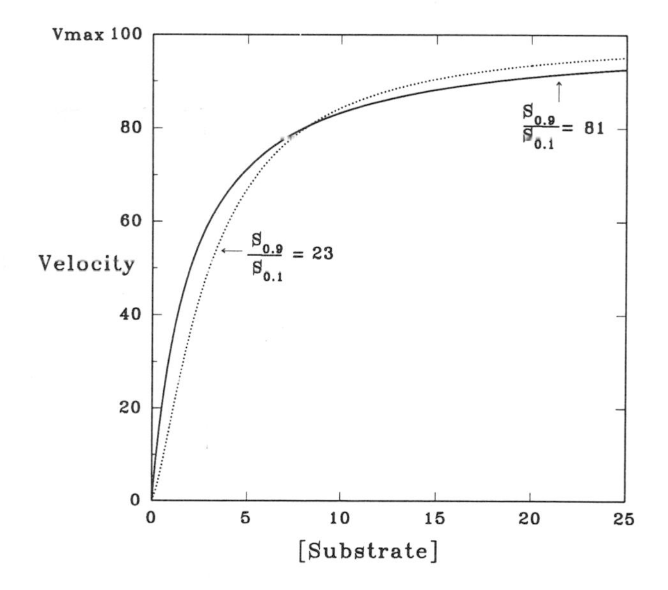

FIG. 8. Comparison of velocity-versus-substrate concentration curves for enzymes following Michaelis-Menten hyperbolic kinetics and enzymes exhibiting slight positive cooperativity. The ratio $[S]_{0.9}/[S]_{0.1}$ is always 81 for the hyperbolic situation, but it is lower for an enzyme with positive cooperativity, shown here with a ratio of 23. $[S]_{0.9}$ is the substrate concentration corresponding to a velocity equal to 0.9 V_{max}, whereas $[S]_{0.1}$ is the substrate concentration when velocity equals 0.1 V_{max}.

consequence of their interaction with occupied sites, but it may also reflect a change in k_{cat} as a function of the number of sites filled. By analogy to the Hill equation for the binding of oxygen to hemoglobin, an expression can be written which describes the saturation function for many allosteric enzymes based on initial velocity measurements:

$$v = \frac{V_{max}\,[S]^{n_H}}{[S]_{0.5}^{n_H} + [S]^{n_H}} \qquad (31)$$

In this equation, n_H is the Hill coefficient, an empirical parameter without physical significance except that it is always less than or equal to the number of interacting sites on the enzyme, and $[S]_{0.5}$ is the substrate concentration when $v = \frac{1}{2} V_{max}$. In the situation where $n_H = 1.0$, equation 31 becomes equivalent to the Michaelis-Menten relationship for a single-substrate enzyme with independently saturating sites (i.e., no cooperativity). If V_{max} is known, n_H and $[S]_{0.5}$ can be calculated from the relationship

$$\log\,[v/(V_{max} - v)] = n_H \log\,[S] - n_H \log\,[S]_{0.5} \qquad (32)$$

A plot of $\log[v/(V_{max} - v)]$ versus $\log[S]$ yields a line of slope n_H. $[S]_{0.5}$ is equal to [S] when $\log\,[v/V_{max} - v)] = 0$. Equation 32 must be used with care in the calculation of n_H, however, since the slope at very low and very high substrate concentrations approaches 1.0 even for interacting multisite allosteric enzymes. In practice, one should attempt to estimate n_H only from data over an intermediate substrate concentration range. Figure 9 depicts the use of equation 32 for n_H determination. Frequently n_H is determined as the slope at $v = \frac{1}{2} V_{max}$ to avoid uncertainty about where the nonlinear regions might be encountered. Values of n_H in excess of 1.0 are indicative of positive cooperativity, although errors in its estimation are such that one should be cautious about interpretation in the range between 1.0 and 1.5. Values less than 1.0 suggest negative cooperativity.

The non-Michaelis nature of most allosteric enzymes renders standard kinetic treatment inappropriate. When a velocity-versus-[S] curve is not truly hyperbolic, the corresponding double-reciprocal plot will not be linear, although how pronounced this deviation is from a linear relationship depends on the extent of sigmoidicity. Plots of v versus v/[S] (Eadie-Hofstee plots) are much more sensitive to nonlinearity and should be conducted if there is ambiguity about whether a true hyperbolic pattern exists. Also estimate the $[S]_{0.9}/[S]_{0.1}$ ratio for velocity-versus-[S] data.

Negative homotropic cooperativity results from a decrease in binding affinity of the remaining unfilled sites as substrate (ligand) is bound. Because the velocity-versus-[S] curve is neither sigmoidal nor hyperbolic, negative cooperativity may be difficult to detect. Calculation of the $[S]_{0.9}/[S]_{0.1}$ ratio, constructing a v-versus-v/[S] plot, or performing a Hill plot should reveal whether negative cooperativity is exhibited by an enzyme.

A **heterotropic effector,** often structurally unrelated to the substrate for the enzyme which it affects, influences catalytic activity by altering in some manner the interaction of an enzyme with substrate. These effectors are nearly always related to the metabolic function of the enzyme, such as being the end product in a pathway for which the enzyme serves as a major control point. Their action may be stimulatory to catalysis (activation) or inhibitory. Just as cooperativity may exist among substrate-binding sites, cooperative binding may be noted for these effectors. Moreover, heterotropic effectors can also exaggerate (or reduce) a sigmoidal response of velocity to substrate concentration.

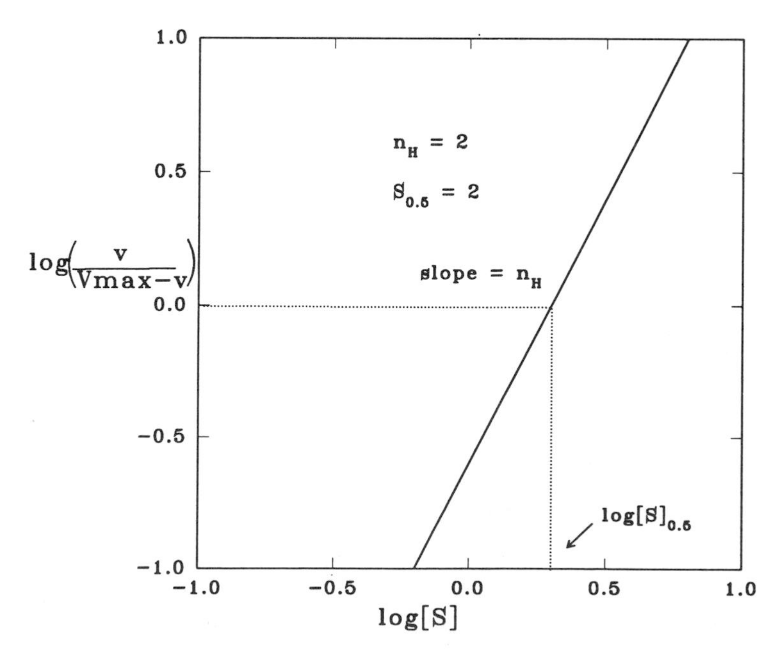

FIG. 9. Hill plot for the determination of n_H in a multisite allosteric enzyme. In this example, the line shown corresponds to an enzyme having at least two interacting sites. $[S]_{0.5}$ is the concentration of substrate when velocity is 0.5 V_{max}.

When a sigmoidal response is noted in a plot of velocity versus effector concentration (at a fixed substrate concentration), the relationship can be analyzed much as that for cooperative substrate binding. For an effector serving as an allosteric inhibitor, the velocity (v_i) measured at some concentration of effector, $[I]_0$, is compared with the uninhibited velocity, v_0. If the inhibition is not complete when [I] is in great excess, the quantity v_{lim} should also be determined; v_{lim} is the rate observed when $[I]_0$ is infinite. Equation 33 then provides a means for calculating n_H, the Hill coefficient for I, and $[I]_{0.5}$, the concentration at which $(v_0 - v_i)$ equals $\frac{1}{2}(v_0 - v_{lim})$. A Hill-type plot of $\log[(v_0 - v_i)/(v_i - v_{lim})]$ versus $\log[I]_0$ has a slope of n_H.

$$\log[(v_0 - v_i)/(v_i - v_{lim})] = n_H \log[I]_0 - n_H \log[I]_{0.5} \quad (33)$$

Kurganov (15) cautions against assuming that v_{lim} is zero when actually it is not and cites examples where this error has occurred. If the effector is an activator rather than an inhibitor, a related equation can be used to calculate n_H (15).

The positive cooperativity which results in sigmoidal kinetics renders the enzyme's rate more sensitive to change in the concentration of its substrate or effector compared with the expected behavior for hyperbolic binding; in the case of negative cooperativity, the response is less sensitive. It is important to keep in mind, however, that a sigmoidal response does not necessarily indicate a regulatory function or imply the presence of multiple interacting sites. Monomeric enzymes with single binding sites can exhibit sigmoidal velocity-versus-[S] plots as a consequence of mechanistic features or interaction with cofactors. Thus, it is unwise to draw a conclusion about a regulatory role on the basis of kinetic analysis alone. Studies of subunit structure and direct-binding data can augment kinetic information and provide stronger evidence for allosteric interactions when these exist.

Before assigning allosteric properties to an enzyme, establish that it contains multiple subunits (and thus has the potential for multiple interacting sites). Also, where possible, examine the mechanistic features of the reaction sufficiently to establish that such situations as a random steady-state bisubstrate mechanism do not apply; this mechanism is one which leads to nonhyperbolic kinetics under some conditions and can allow an erroneous conclusion about the allosteric nature of the enzyme. Finally, insist on a metabolic logic for the enzyme to be subject to kinetic regulation. If the enzyme is not situated at some branch point in a pathway or if the effector is not related to the substrate on a precursor-end product or similar basis, it is rather unlikely that the enzyme exhibits a true allosteric nature.

23.5.5. Covalent Modifications

Many instances of reversible covalent changes in bacterial proteins now exist, and these doubtless are of regulatory significance. Although the best-documented instance may be the adenylylation of glutamine synthetase of *E. coli*, numerous proteins are known to undergo a variety of changes, including phosphorylation (e.g., the *E. coli dnaK* protein), methylation (e.g., chemotaxis transducer proteins) and ADP-ribosylation (nitrogenase reductase of *Rhodospirillum rubrum*) as a means of regulating activity. Because the changes are not intended to be permanent, enzymes exist for carrying out the modifications (i.e., adenylyl transferase, specific protein kinases, or methyltransferases) and for their removal (i.e., phosphatases, methylesterases, adenylyl-removing enzyme). The key to understanding these systems is first to detect the existence of a covalent modification and determine its nature, then to establish the catalytic consequences of the modification, and finally to identify effectors or conditions which influence the relative activities of the modification and demodification enzymes.

One example of this process is covalent control of isocitrate dehydrogenase (ICDH) of *E. coli* by reversible phosphorylation. Several reviews have recently discussed this or other phosphorylated protein systems and should be consulted for further information (3, 5, 30). ICDH serves as a control point for isocitrate metabolism, in which it can permit the oxidation of isocitrate to α-ketoglutarate for either glutamate formation or further oxidation in the Krebs cycle, or, when inoperative, shunting isocitrate toward glyoxylate and succinate (via isocitrate lyase) in the glyoxylate bypass. Therefore it is not surprising to find that the level of activity of ICDH is very sensitive to carbon source. Growth of *E. coli* on acetate results in much less ICDH activity than is seen during growth on glucose or glycerol. This low activity can be rapidly increased by the addition of glucose to a culture growing on acetate, and, on exhaustion of the glucose, enzyme activity again declines as acetate utilization resumes. The experimental procedures based on those of Garnak and Reeves (9) illustrate how phosphorylation and its effect on activity can be observed in this instance.

Grow two 25-ml cultures of *E. coli* K-12 in sidearm flasks on a low-phosphate medium consisting of 0.05 M Bis-Tris buffer (pH 7.5), 1 mM $MgSO_4$, 15 mM $(NH_4)_2SO_4$, 40 μM $Ca(NO_3)_2$, 2 μM $FeSO_4$, 30 mM KCl, 0.5 mM Na_2HPO_4, 30 μg of thiamine per liter, and 0.1% glycerol. Monitor aerobic growth at 37°C by turbidimetric measurements until cells approach the stationary phase. Add rifampin to both cultures to give a final concentration of 200 μg/ml. After 10 min, add 250 μCi of [^{32}P]orthophosphoric acid, carrier free, to each culture and continue shaking at 37°C. At 5 min after the isotope addition, add to one culture 0.25 ml of 2.5 M sodium acetate to give a final concentration of 25 mM; add nothing to the control culture. Immediately remove 5 ml from each flask to chilled centrifuge tubes, and spin down the cells at 4°C and 10,000 × *g* for 10 min; carefully remove the radioactive supernatant fluid, and hold the cell pellet on ice for the duration of the experiment. Continue incubation of the original cultures and sample in an identical fashion at 30 and 60 min following the addition of acetate; these cells can be used for subsequent determination of fractional phosphorylation levels as described below. After the removal of the 60-min samples, harvest the remaining 10 ml of each culture by centrifugation. In this instance, however, resuspend the cell pellets each in 100 μl of a solution containing 1% sodium dodecyl sulfate (SDS), 0.05 M Tris-HCl (pH 7.5), and 0.1 M mercaptoethanol, and heat at 90°C for 2 min. Subject these samples to SDS-polyacrylamide gel electrophoresis, stain them, and perform autoradiography as described by Garnak and

Reeves (9). To reduce background from unincorporated $^{32}P_i$, it may be desirable to electroblot the proteins onto nitrocellulose before autoradiography. Electrophoresis of the sample from the acetate-treated culture should reveal a pronounced ^{32}P-containing protein species of ca. 46,000 molecular weight, whereas the control culture should show no significant phosphorylated species of similar size.

Phosphorylation and dephosphorylation of ICDH in *E. coli* are conducted by a single protein, ICDH kinase/phosphatase, the product of the *aceK* gene (16). ICDH itself is a dimer of identical subunits, and phosphorylation occurs on a single residue (Ser-113) in each subunit. The diphosphorylated enzyme is inactive, and progressive dephosphorylation leads to proportional activity appearance. As a result of this situation, ICDH phosphatase can be used to estimate the relative amounts of phosphorylated and nonphosphorylated ICDH present in a culture at any point during growth. This is described here for the cells obtained earlier following acetate addition to a glycerol-grown culture and the procedure of LaPorte et al. for treatment with ICDH phosphatase (17).

Remove the cell pellets from ice, and resuspend each in 0.1 ml of Tris-EDTA solution plus lysozyme (see section 23.2.3). Disrupt by three freeze-thaw cycles, and then centrifuge at high speed to provide a crude extract. Use this for dephosphorylation as follows. Incubate cell extract in a mixture containing 25 mM MOPS buffer (pH 7.5), 100 mM NaCl, 5 mM $MgCl_2$, 1 mM dithiothreitol, 2 mM ADP, 5 mM 3-phosphoglycerate, 1 mM isocitrate, and 0.1 U of ICDH phosphatase in a total volume of 0.15 ml. Another reaction with the same amount of cell extract but without phosphatase should also be prepared. Withdraw samples at 15-min intervals, and assay for ICDH activity by the procedure given below. The purification and assay of ICDH phosphatase are described by Laporte and Koshland (16).

For assay of ICDH, place the following solutions in a 1-cm-path-length spectrophotometer cell (volume, 1 ml): 0.1 ml of 250 mM MOPS (pH 7.5), 0.1 ml of 50 mM $MgCl_2$, 0.25 ml of 1 mM $NADP^+$, 0.25 ml of 2 mM DL-isocitrate, and water to a final volume of 0.97 ml. Bring the contents to 37°C in the spectrophotometer, and then add 0.03 ml of dehydrogenase sample to be assayed. Monitor the change in A_{340} as a measure of the amount of active ICDH present. The activity seen in samples not treated with phosphatase represents nonphosphorylated ICDH, whereas that seen in phosphatase-treated samples is total ICDH activity; the difference represents the amount of phosphorylated ICDH present. Analyze multiple time points for the phosphatase-treated samples to ascertain when dephosphorylation is complete.

23.6. ANALYSIS OF METABOLIC PATHWAYS

Most investigations of intermediary metabolism in microbial systems have focused on elucidating the detailed aspects of metabolic pathways, with emphasis on identification of intermediates, the reactions involved, and the enzymes responsible for the interconversions. Although the pace of discovery of new routes has slowed in recent years, there remain many organisms for which little metabolic detail is available, and extension of metabolic studies to these less well characterized species continues. Meanwhile, interest has quickened in the evaluation of flux through known pathways and particularly in estimating comparative values for competing routes plus obtaining *in vivo* evidence for regulatory sites. It is nevertheless essential for these quantitative-type goals that the intermediates and enzymes that make up a pathway be clearly identified and characterized; therefore, techniques which facilitate this continue to be of importance.

Some of the following are among the most widely used general approaches to establish the nature of intermediates. Metabolic inhibitors are frequently used to block pathways and force accumulation of intermediates prior to an affected step; the inhibition of IMP synthesis by sulfonamides with the resulting accumulation of 5′-phosphoribosyl-4-carboxamide-5-aminoimidazole and compounds leading to its formation is an example of this method, and current related applications would probably use a labeled known precursor along with chromatographic separations and identification, preferably by gas chromatography-mass spectrometry or high pressure liquid chromatography (HPLC) but even with less sophisticated chromatographic methods (e.g., thin-layer chromatography). Where some details of the pathway and its enzymes are known, irreversible enzyme inhibitors or substrate analogs can be used in a similar fashion to produce a block in the pathway, as in the use of phosphonoacetylornithine, which inhibits ornithine carbamoyltransferase needed in arginine biosynthesis. Surely the most common technique for interrupting the flow of metabolites through a pathway is the use of mutants affected in various steps. With the increasing power of genetic techniques and their application to less well studied organisms, it is likely that mutagenesis procedures, especially transposon insertional mutagenesis, will provide the most insight into a newly uncovered pathway or an incompletely detailed scheme.

On the more quantitative level, radiotracer studies and NMR analyses are essential tools for establishing the rate of movement through a pathway. Although in some instances this information can be estimated by using the kinetic properties of the enzymes in conjunction with available information on the intracellular levels of metabolites, it is generally more satisfactory to use in vivo measurements when possible. In one such study, Walsh and Koshland (43) evaluated the fluxes through the Krebs cycle and the glyoxylate bypass for *E. coli* growing on acetate. Their approach used radioactive acetate incorporation into various cell constituents (e.g., saponifiable lipids) as well as $^{14}CO_2$ production and labeling of intercellular metabolites (separated by HPLC). These labeling data were supplemented with ^{13}C-NMR results from metabolites formed during growth on [^{13}C]acetate, most notably the labeling pattern in glutamate, and by kinetic values determined for isocitrate dehydrogenase and isocitrate lyase. The fluxes estimated from the in vivo labeling measurements were in general agreement with fluxes calculated from kinetic data on the enzymes, a satisfying situation but one that may not hold in other instances.

This and similar studies conducted over the past 10 years illustrate how the techniques of HPLC and, to a lesser extent, gas chromatography have facilitated obtaining high-quality data on the labeling of pathway intermediates. High-performance ion-exchange or reverse-phase liquid-chromatographic systems have largely supplanted thin-layer or paper chromatography as the methods of choice for separation and quantification of metabolites. When used with rapid sampling of growing cultures by membrane filtration and a rapid quench (usually with acid) to halt metabolic changes, HPLC analysis of labeled intermediates provides a very attractive means for establishing precursor-product relationships. Although not nearly so frequently used, the technique of radiorespirometry, in which there is a continuous sampling of $^{14}CO_2$ evolved from cultures metabolizing specifically labeled substrates, can provide reasonably quantitative information on pathway involvement. Although the equipment needed for this method is less sophisticated than the chromatographic instruments, one is frequently restricted to studying pathways in which several specifically labeled substrates can be synthesized or otherwise obtained (44).

The use of in vivo NMR analysis is likewise increasing, both for dissection of pathways and for obtaining information on their rates of operation. The most favored nuclei for these purposes are ^{13}C, ^{15}N, and ^{31}P, used singly or in combination. A number of applications of this noninvasive technique have been described previously (8). Among these is a study on the pathway for glyphosate metabolism in a *Pseudomonas* species by using solid-state ^{13}C- and ^{15}N-NMR on lyophilized intact-cell samples harvested during growth on ^{13}C- and/or ^{15}N-labeled glyphosate (*N*-phosphonomethylglycine) (14). This investigation involved the relatively new technique of cross-polarization magic-angle spinning NMR to demonstrate conclusively that glyphosate in this organism is metabolized directly to glycine by cleavage of the phosphonomethyl carbon-nitrogen bond. It was possible to trace the glycine released from glyphosate into other compounds, specifically protein (as glycine or serine) and into purines. The phosphonomethyl carbon appeared in certain positions of purines as well as in the methyl groups of thymine and methionine, indicating its entry into the one-carbon pool mobilized by tetrahydrofolate. Solid-state NMR is particularly recommended for analysis of pathways in which steady-state levels of intermediates are low but the rate of formation of polymeric material (proteins, nucleic acids, etc.) linked to the intermediates is high.

Studies with ^{31}P-NMR have been conducted primarily to monitor in vivo pH gradients and energy metabolism. This method can distinguish internal from external P_i, and many of the phosphorylated glycolytic intermediates can be identified and quantified; in addition, sugar nucleotides (e.g., UDPglucose), NAD, and nucleoside di- and triphosphates can be measured. Whereas ^{13}C-substrates (or ^{15}N-substrates) must be prepared with a high level of isotopic enrichment to have adequate sensitivity, ^{31}P is the naturally occurring stable isotopic form, and thus perturbations in the levels of phosphorylated intermediates are relatively easy to track. For example, the intracellular levels of fructose 1,6-bisphosphate, 2- and 3-phosphoglycerate, and phosphoenolpyruvate could be monitored in *Streptococcus lactis* during glycolysis and starvation by obtaining ^{31}P-NMR spectra and independently by ^{14}C-fluorography on cells supplied [^{14}C] glucose (42). These measurements provided in vivo evidence that the activity of pyruvate kinase in this organism was influenced by the relative levels of fructose 1,6-bisphosphate and inorganic phosphate.

Several problems sometimes observed in NMR studies with whole cells were also encountered in this study of pyruvate kinase in *S. lactis*. Broadening of line widths, possibly as a result of accumulation of paramagnetic ions, can present a problem in making quantitative determinations. Moreover, because of the concentrated nature of the cell suspensions needed for NMR analysis, metabolism continuing within the NMR tube can result in a significant change in pH for organisms which produce acidic products, and this may have an effect on the chemical shifts observed. These effects detract somewhat from the usefulness of NMR but also suggest ways in which they may be overcome. Obviously not all cell types produce acidic products, and a flowthrough system involving trapped or immobilized cells can circumvent the accumulation of products as well as provide fresh nutrients. Furthermore, as improvements in sensitivity continue, the requirement for highly concentrated cell suspensions should decrease, and this also will allow for a better control of the pH and nutrient state of cells being analyzed.

23.7. REFERENCES

1. **Adams, E.** 1976. Catalytic aspects of enzymatic racemization. *Adv. Enzymol. Relat. Areas Mol. Biol.* **44:**69–138.
2. **Ashwell, G., and J. Hickman.** 1957. Enzymatic formation of xylulose 5-phosphate from ribose 5-phosphate in spleen. *J. Biol. Chem.* **226:**65–76.
3. **Bourret, R. B., K. A. Borkovich, and M. I. Simon.** 1991. Signal transduction pathways involving protein phosphorylation in prokaryotes. *Annu. Rev. Biochem.* **60:**401–441.
4. **Campbell-Burk, S. L., and R. G. Shulman.** 1987. High-resolution NMR studies of *Saccharomyces cerevisiae*. *Annu. Rev. Microbiol.* **41:**595–616.
5. **Cozzone, A. J.** 1988. Protein phosphorylation in prokaryotes. *Annu. Rev. Microbiol.* **42:**97–125.
6. **Dixon, M., and E. C. Webb.** 1979. *Enzymes,* 3rd ed. Academic Press, Inc., New York.
7. **Eichler, D. C., and I. R. Lehman.** 1977. On the role of ATP in phosphodiester bond hydrolysis catalyzed by the *recBC* deoxyribonuclease of *Escherichia coli*. *J. Biol. Chem.* **252:**499–503.
8. **Fernandez, E. J., and D. S. Clark.** 1987. N.m.r. spectroscopy: a noninvasive tool for studying intracellular processes. *Enzyme Microb. Technol.* **9:**259–271.
9. **Garnak, M., and H. C. Reeves.** 1979. Phosphorylation of isocitrate dehydrogenase of *Escherichia coli*. *Science* **203:**1111–1112.
10. **Herve, G.** 1989. *Allosteric Enzymes.* CRC Press, Inc., Boca Raton, Fla.
11. **Hoben, P., N. Royal, A. Cheung, F. Yamao, K. Biemann, and D. Soll.** 1982. *Escherichia coli* glutaminyl-tRNA synthetase. II. Characterization of the *glnS* gene product. *J. Biol. Chem.* **257:**11644–11650.
12. **International Union of Biochemistry.** 1961. *Report of the Commission on Enzymes.* Pergamon Press, Oxford.
13. **Irvin, R. T., T. J. MacAlister, and J. W. Costerton.** 1981. Tris (hydroxymethyl) aminomethane buffer modification of *Escherichia coli* outer membrane permeability. *J. Bacteriol.* **145:**1397–1403.
14. **Jacob, G. S., J. Schaefer, E. O. Stejskal, and R. A. McKay.** 1985. Solid-state NMR determination of glyphosate metabolism in a *Pseudomonas* sp. *J. Biol. Chem.* **260:**5899–5905.

15. **Kurganov, B. I.** 1982. *Allosteric Enzymes. Kinetic Behavior,* p. 56–64. John Wiley & Sons, Inc., New York.
16. **LaPorte, D. C., and D. E. Koshland, Jr.** 1982. A protein with kinase and phosphatase activities involved in regulation of tricarboxylate acid cycle. *Nature* (London) **300:**458–460.
17. **LaPorte, D. C., P. E. Thorsness, and D. E. Koshland, Jr.** 1985. Compensatory phosphorylation of isocitrate dehydrogenase: a mechanism for adaptation to the intracellular environment. *J. Biol. Chem.* **260:**10563–10568.
18. **Leatherbarrow, R. J.** 1990. Using linear and non-linear regression to fit biochemical data. *Trends Biochem. Sci.* **15:**455–458.
19. **Lessmann, D., K.-L. Schimz, and G. Kunz.** 1975. D-Glucose-6-phosphate dehydrogenase (Entner-Doudoroff enzyme) from *Pseudomonas fluorescens.* Purification, properties and regulation. *Eur. J. Biochem.* **59:**545–559.
20. **Lorimer, G. H., M. R. Badger, and T. J. Andrews.** 1977. D-Ribulose-1,5-bisphosphate carboxylase-oxygenase. Improved methods for the activation and assay of catalytic activities. *Anal. Biochem.* **78:**66–75.
21. **MacElroy, R. D., and C. R. Middaugh.** 1982. Bacterial ribosephosphate isomerase. *Methods Enzymol.* **89:**571–579.
22. **Marschke, C. K., and R. W. Bernlohr.** 1973. Purification and characterization of phosphofructokinase of *Bacillus licheniformis. Arch. Biochem. Biophys.* **156:**1–16.
23. **Mason, R. P., and J. K. M. Sanders.** 1989. In vivo enzymology: a deuterium NMR study of formaldehyde dismutase in *Pseudomonas putida* F16a and *Staphylococcus aureus. Biochemistry* **28:**2160–2168.
24. **Maurer, P., D. Lessmann, and G. Kurz.** 1982. D-Glucose-6-phosphate dehydrogenases from *Pseudomonas fluorescens. Methods Enzymol.* **89:**261–270.
25. **McClure, W. R.** 1969. A kinetic analysis of coupled enzyme assays. *Biochemistry* **8:**2782–2786.
26. **McFadden, B. A., F. R. Tabita, and G. D. Kuehn.** 1975. Ribulose-diphosphate carboxylase from the hydrogen bacteria and *Rhodospirillum rubrum. Methods. Enzymol.* **42:**461–472.
27. **Miethe, D., and W. Babel.** 1990. Glucose-6-phosphate dehydrogenase from *Pseudomonas* W6. *Methods Enzymol.* **188:**346–350.
28. **Miller, J. H.** 1972. *Experiments in Molecular Genetics,* p. 352–355. Cold Spring Harbor Laboratory, Cold Spring Harbor, N.Y.
29. **Modrich, P., and D. Zabel.** 1976. EcoRI endonuclease. Physical and catalytic properties of the homogeneous enzyme. *J. Biol. Chem.* **251:**5866–5874.
30. **Nimmo, H. G.** 1987. The tricarboxylic acid cycle and anaplerotic reactions, p. 156–169. *In* F. C. Neidhardt, J. L. Ingraham, K. B. Low, B. Magasanik, M. Schaecter, and H. E. Umbarger (ed.), *Escherichia coli and Salmonella typhimurium: Cellular and Molecular Biology,* vol. 1. American Society for Microbiology, Washington, D.C.
31. **Nomenclature Committee of the International Union of Biochemistry.** 1979. Units of enzyme activity—recommendations 1978. *Eur. J. Biochem.* **97:**319–320.
32. **Nomenclature Committee of the International Union of Biochemistry.** 1984. *Enzyme Nomenclature, Recommendations 1984.* Academic Press, Inc., New York.
33. **Ohlendorf, D. H., J. D. Lipscomb, and P. C. Weber.** 1988. Structure and assembly of protocatechuate 3,4-dioxygenase. *Nature* (London) **336:**403–405.
34. **Rechler, M. M., and H. Tabor.** 1971. Histidine ammonialyase *(Pseudomonas). Methods Enzymol.* **17B:**63–69.
35. **Rubin, R. A., and P. Modrich.** 1978. Substrate dependence of the mechanism of EcoRI endonuclease. *Nucleic Acids Res.* **5:**2991–2997.
36. **Searles, L. L., and J. M. Calvo.** 1988. Permeabilized cell and radiochemical assays for ß-isopropylmalate dehydrogenase. *Methods Enzymol.* **166:**225–229.
37. **Segel, I. H.** 1975. *Enzyme Kinetics.* John Wiley & Sons, Inc., New York.
38. **Shaw, W. V.** 1975. Chloramphenicol acetyltransferase from chloramphenicol-resistant bacteria. *Methods Enzymol.* **43:**737–755.
39. **Srere, P. A.** 1987. Complexes of sequential metabolic enzymes. *Annu. Rev. Biochem.* **56:**89–124.
40. **Stadtman, E. R., P. Z. Smyrniotis, J. N. Davis, and M. E. Wittenberger.** 1979. Enzymic procedures for determining the average state of adenylylation of *Escherichia coli* glutamine synthetase. *Anal. Biochem.* **95:**275–285.
41. **Storer, A. C., and A. Cornish-Bowden.** 1974. The kinetics of coupled enzyme reactions. Applications to the assay of glucokinase, with glucose-6-phosphate dehydrogenase as coupling enzyme. *Biochem. J.* **141:**205–209.
42. **Thompson, J., and D. A. Torchia.** 1984. Use of ^{31}P nuclear magnetic resonance spectroscopy and ^{14}C fluorography in studies of glycolysis and regulation of pyruvate kinase in *Streptococcus lactis. J. Bacteriol.* **158:**791–800.
43. **Walsh, K., and D. E. Koshland, Jr.** 1984. Determination of flux through the branch point of two metabolic cycles: the tricarboxylic acid cycle and the glyoxylate shunt. *J. Biol. Chem.* **259:**9646–9654.
44. **Wang, C. H.** 1972. Radiorespirometric methods, p. 185–230. *In* J. R. Norris and D. W. Ribbons (ed.), *Methods in Microbiology,* vol. 6B. Academic Press, Inc., New York.
45. **Whittaker, J. W., J. D. Lipscomb, T. A. Kent, and E. Munck.** 1984. *Brevibacterium fuscum* protocatechuate 3,4-dioxygenase. Purification, crystallization and characterization. *J. Biol. Chem.* **259:**4466–4475.
46. **Whittaker, J. W., A. M. Orville, and J. D. Lipscomb.** 1990. Protocatechuate 3,4-dioxygenase from *Brevibacterium fuscum. Methods Enzymol.* **188:**82–88.
47. **Wood, E. R., and S. W. Matson.** 1987. Purification and characterization of a new DNA-dependent ATPase with helicase activity from *Escherichia coli. J. Biol. Chem.* **263:**15269–15276.

Chapter 24

Permeability and Transport

ROBERT E. MARQUIS

Two general types of processes for entry and exit of solutes across cell membranes can be distinguished, as follows.

Permeation denotes passive movement of a solute into (uptake) or out of (efflux) a cell by diffusion, which may be facilitated by chemical interaction between the solute and some cell structure, e.g., a membrane. A cell may be permeable or impermeable to a particular solute. **Permeability** is a property of the cell and not of the solute. The following equation applies to permeation of a solute through a membrane of a cell:

$$ds/dt = PA\,(\Delta C)$$

where ds/dt is the change in amount of internal solute per unit of time, P is the permeability coefficient (which has units of distance divided by time), A is the surface area of the membrane or other structure across which diffusion occurs, and ΔC is the difference in solute concentration between the interior and the exterior of the cell. If the internal solute concentration (C_{in}) is plotted as a function of time, the resulting curve is parabolic with a plateau at $C_{in} = C_{out}$, i.e., when the internal concentration equals the external concentration or, more exactly, when the internal electrochemical activity of the solute equals that of the exterior. For such a curve, C_{in} = (maximum C_{in}) $(1 - e^{-kt})$ where k is a constant and t is the time.

Transport denotes active movement of a solute into or out of a cell by specific transport systems which may be coupled to metabolism and may involve energized carriers. The two most thoroughly studied classes of transport systems in bacteria are phosphotransferase systems (51) and so-called **permease** systems (6). The phosphotransferase system (PTS) serves in facultatively anaerobic bacteria for transport of sugars, whereas essentially all bacteria have permease systems for transport of amino acids, inorganic ions, nucleic acid precursors, and sugars. The general regulatory role of the PTS has only recently become more fully appreciated, and in at least some organisms its major function is regulatory rather than for uptake of sugars (55).

A curve relating solute uptake to time for a specific transport process does not differ qualitatively from one for a general permeation process. However, active transport does not stop when $C_{in} = C_{out}$ but continues until a substantial concentration gradient has developed and the process becomes one of **concentrative transport.** Maximum uptake is then determined by the rates of the inward and outward processes.

As indicated, transport may result in movement of solutes into or out of cells. For many solutes, a transport process of so-called **exchange diffusion** occurs in which there is a one-for-one exchange of internal and external solutes across the membrane. This exchange is generally assessed by use of radioactively labeled solutes; details of the technique are presented by Maloney et al. (7). The exchange may also involve antiporters and heterologous exchange. For example, Driessen et al. (20) have characterized the arginine/ornithine antiporter of *Lactococcus lactis,* which catalyzes exchange of ornithine, the product of the arginine deiminase system, for arginine, the substrate of the system, without need for ATP or net energy transfer. Permease proteins such as the *lacY* gene product may catalyze various types of transport processes including active transport energized by ΔpH or $\Delta\psi$, facilitated diffusion, counterflow, exchange, or solute efflux from cells.

Concentrative solute uptake does not necessarily require a membrane and transport catalysts. Ion-exchange resins, for example, can take up and concentrate ions. Most of the organic components of a cell are not in aqueous solution but are in solid phases in the cell wall, membranes, ribosomes, the nuclear body, etc. Thus, solutes may be taken up and retained in the same manner as those in a resin. This aspect of solute transport and permeation has been reviewed by Damadian (18).

24.1. PERMEABILITY

24.1.1. Assay Procedure

The most commonly used technique for assessing solute permeation into bacterial cells or cell structures is the **space** (or thick-suspension) **technique** described by Conway and Downey (17) and modified by Mitchell and Moyle (49), MacDonald and Gerhardt (41), and others. Large masses of cells are commonly used for the assay—generally 3 g or more (wet weight) per assay tube. Obtaining this quantity of cells can be a problem but is not a great one when common laboratory bacteria are used. Techniques that require fewer cells and are based on use of membrane filters or a minicentrifuge are described in the section on transport (section 24.2). In any case, careful attention should be paid to details of growth and harvest since the permeability of bacteria may change during the culture cycle in batch cultures or as a result of washing procedures used to rid the cells of medium constituents. There is often an advantage in using cells obtained from continuous cultures. It is generally best to take large harvests from the main culture vessel rather than to use so-called runoff cells, which may have altered permeability because of storage during collection. Changes in growth medium, growth temperature, pH, etc., can also be expected to alter permeability.

The space technique is designed specifically to give an estimate of the fractional space (volume) of cells that can be penetrated by the solute being tested. The technique is based simply on determination of the degree of dilution of a solution containing the solute as a result of mixing the solution with a pellet of centrifuged cells. The following information is needed: the initial volume and concentration of the test solution, the volume of the cell pellet, and the final concentration of the solute in the extracellular phase of the mixture of pellet and solution. Cell pellets contain not only cells but also interstitial (intercellular) space. The fractional volume of the interstitial space is generally estimated by use of high-molecular-weight polymeric solutes which can penetrate the interstitium but not the cell.

The stepwise procedure for the space technique is as follows.

24.1.1.1. Washing of cells

After harvest, it is generally necessary to wash the cells before they can be used for permeability assays. Usually a single wash with a large volume of fluid is sufficient to nearly rid the cells of medium constituents but not to deplete them excessively of internal solutes. Often two washings are done, but there is little justification for a third wash. Many bacteria, especially gram-positive ones, are hardy and can be washed with water or dilute buffers without great harm, especially if the wash fluids are chilled. However, other bacteria, particularly gram-negative ones, are much more sensitive and can be damaged in major ways by being washed with water. Moreover, with gram-negative and some gram-positive bacteria, it is often best to use warm wash fluids instead of chilled ones. Commonly used wash solutions for gram-negative bacteria contain magnesium ions (usually 50 mM), a buffer (e.g., 50 mM phosphate buffer or Tris buffer), and an osmotic stabilizer (e.g., 0.4 M NaCl or 0.5 M sucrose). Maloney et al. (7) recommend using growth medium 63 for washing *Escherichia coli* cells, but with deletion of the carbon source. With any previously untested bacterium, it is worthwhile to try a variety of wash solutions to find the best one. The criterion for efficacy is that the cells should lose medium constituents but retain internal pools of potassium, amino acids, or P_i. No wash solution is perfect, and generally some compromise must be accepted. When dense cell suspensions are used for permeability assays, the composition of the wash (and suspension) fluid is not as critical as when dilute suspensions are used for transport assays.

24.1.1.2. Centrifugation of cells

Once the cells have been washed, they must be centrifuged to obtain a workably tight pellet. The centrifugation schedule varies depending on the particular organism. A packing curve of the type shown in Fig. 1 is generally prepared by plotting pellet weight against

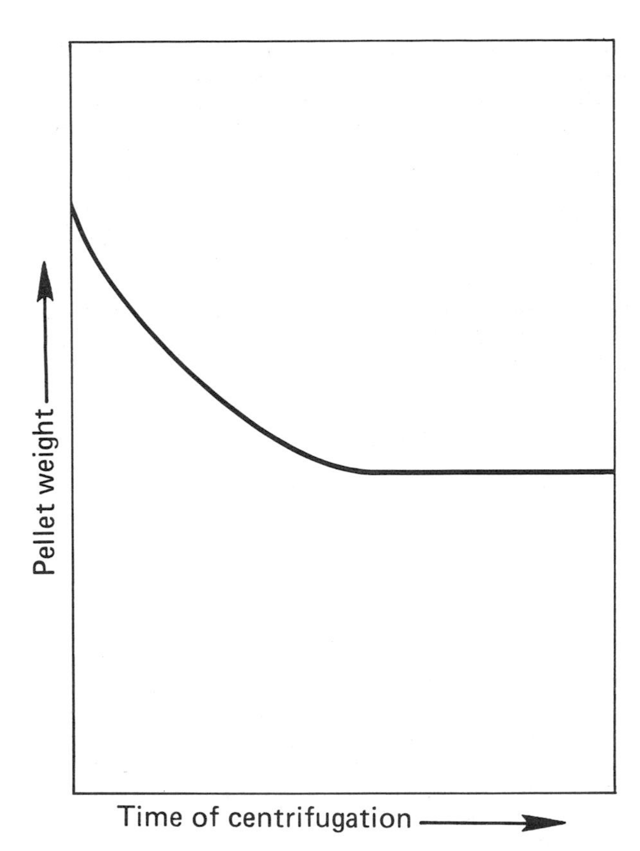

FIG. 1. Typical packing curve for a cell pellet.

centrifugation time, using a series of cell suspensions to obtain the curve. The centrifugation time chosen for assays should be well out into the plateau region so that minor variations in centrifugation time do not affect the pellet weight. The final centrifugation before solute addition is generally done with a tared centrifuge tube so that the pellet weight can be obtained by difference. Tubes with nominal 50-ml capacity are best for pellets of about 10 g (wet weight), and 15-ml tubes are best for pellets of about 3 g. Weighing to within 0.01 g is sufficiently accurate for routine assays. Prior to weighing, it is best to decant the supernatant fluid carefully, to invert the tubes for maximum and uniform draining, and to wipe excess moisture from the inside of each tube with absorbent tissue. However, attempts to blot surface water from the pellet are best avoided because cells may be removed with the water.

24.1.1.3. Measurement of pellet weight and volume

Pellet wet weight is usually an adequate indicator of pellet volume and may be used instead of volume in calculations, since the wet densities of bacterial vegetative cells are generally about 1.02 to 1.04 g/ml. However, bacterial spores have significantly higher wet densities, as high as 1.22 g/ml for *Bacillus stearothermophilus* spores, and the pellet weight of these cells is not a good substitute for pellet volume. Certain inclusion granules (e.g., poly-β-hydroxybutyrate) of vegetative cells also have relatively high wet densities and may significantly increase the difference between pellet weight and volume.

Pellet volumes can be estimated readily by use of a standard pycnometer or a volumetric flask. For example, suppose one has a cell pellet that weighs about 3 g and wishes to know its volume. The pellet can be mixed with water and transferred to a 50-ml, tared pycnometer vessel or volumetric flask, and the mixture is made up to volume with water. Suppose that the weight of the suspension determined with an analytical balance is found to be 50.110 g (weight of mixture plus pycnometer minus weight of pycnometer) and the pycnometer has a calibrated volume of exactly 50.000 ml at 25°C, the temperature of weighing. Suppose also that the pellet has been found with the analytical balance to weigh exactly 3.000 g. Then the weight of water added to the pellet must be 50.110 − 3.000, or 47.110 g. At 25°C, this weight of water would occupy a volume of 47.249 ml, based on the water density at that temperature obtained from tables in a standard chemical handbook. Therefore, the pellet volume must be 50.000 − 47.249, or 2.752 ml. The density of the pellet is then 3.000 g/2.752 ml, or 1.090 g/ml. Once the density is known, and if standard centrifugation conditions are used, the pellet volume can be readily calculated from the pellet weight.

The pellet volume can also be measured directly in a tared centrifuge tube closed with a tared microscopy coverglass. The open ends of glass tubes must be ground flat, but polycarbonate tubes are manufactured with flat ends. The total volume in the tube is obtained by filling it completely with distilled water and closing it with the coverglass, taking care to exclude air bubbles and to wipe excess water from the external surfaces. The weight of water in the tube is measured and converted to volume with the density of water at the ambient temperature. Pellet volumes also have been estimated directly by use of calibrated centrifuge tubes which yield cytocrit data equivalent to the hematocrit data obtained routinely in hematology; this cytocrit method can give useful volume estimates, but it has difficulties in practice.

24.1.1.4. Mixing and incubation of cells with solute

The cell pellet to be used for a permeability assay is brought to the desired experimental temperature and mixed with a solution containing the test solute by use of glass or Teflon-coated stirring rods, which do not absorb fluid. The volume of solution used should be about equal to the pellet volume. Basically, one wants to have sufficient dilution of the solution that concentration differences are readily determined. After mixing of pellet and solution, the suspension should be incubated for a set time at the experimental temperature to allow for diffusional equilibration, generally 15 min to 1 h. However, it is best to determine that equilibrium has in fact been reached during the incubation period by use of multiple pellets and incubation times.

24.1.1.5. Recentrifugation

After each mixture of pellet and test solution has been incubated, it should be centrifuged sufficiently to obtain a cell-free supernatant fluid for assay. The supernatant fluid may require an additional, clarifying centrifugation.

24.1.1.6. Determination of solute concentrations

The concentrations of solute must be determined both in the initial solution that was mixed with the cell pellet and in the final solution that was obtained after the recentrifugation. These measurements generally can be made on small samples by means of gravimetric or refractometric techniques. However, these techniques are not specific, and other substances purposely added to or leached from the cells may interfere. If so, a specific chemical or radioactive determination must be used for the test solute.

24.1.2. Determination of Interstitial Space

It is necessary to know the interstitial volume of the cell pellet in order to be able to calculate the volume of the cells themselves available to the test solute. Consequently, one pellet in each set should be mixed with a solution of a large nonionic polymeric solute in which all of the component molecules are too large to penetrate through cell walls, moving only into the interstitium. A dextran preparation with an average molecular weight of 2,000,000 (Dextran T2000; Pharmacia Fine Chemicals, Inc., Piscataway, N.J.) serves this purpose. Dextrans are polydisperse, but even the smallest molecules in the distribution of this large a dextran preparation do not penetrate even relatively porous bacterial cell walls (56).

Dextran concentrations generally can be determined gravimetrically, by means of refractometry, or chemically by the phenol-sulfuric acid method (22). However, these assays are not specific for dextrans, and other substances purposely added to or leached from the cells may interfere. It would be convenient to use radioactively labeled dextrans, but the largest ones commercially available (e.g., from New England Nuclear Corp.) have average molecular weights of only 70,000. The smallest molecules in this dextran preparation are sufficiently large to be excluded by the walls of some bacteria but not others. Therefore, radioactive dextrans may or may not be useful, and before they can be used for a particular bacterium, it is necessary to determine the dextran exclusion threshold. Of course, it would be possible to label larger dextran molecules with [^{14}C]cyanide by conventional chemical techniques to obtain carboxyl-dextran. Blue dextrans (Pharmacia) have been used for colorimetric assays, but bacteria can become stained after contact with blue dextran because of transfer of the dye.

A number of other polymeric molecules have been used instead of dextrans. Proteins (such as serum albumin) have an advantage in being essentially monodisperse and specifically assayable. However, they have a disadvantage in that they may bind to cells, and this binding would lead to anomalously high uptake values. Inulin also has been used, but it is capable of penetrating the cell walls of many bacteria, especially gram-positive ones (56). Percoll (see Chapter 4.3.5.3) should be suitable but has not yet been tested.

Yet another problem in determining interstitial volumes arises from the swelling-shrinking responses of cells to many test solutes, especially to salts. It may be necessary to assess the interstitial volume in the presence of such a test solute in addition to the dextran. If the additional solute is of small molecular size (say, NaCl or KCl), it is possible to dialyze the supernatant fluid with retention of the large dextran molecules within the dialysis sac and loss of the small solutes. Then all of the dextran can be recovered from the dialysis sac for assay.

The interstitial-volume fraction may represent an appreciable part of the total volume. With close-packed nondeformable spheres such as glass beads, regardless of their size, the interstitium represents 27% of the total volume. With bacterial cells packed by centrifugation to essentially constant volume, the interstitium typically is 10 to 15% because of the cylindrical shape, deformation, and irregular size of the cells.

24.1.3. Expression of Results

A commonly used permeability index is the space value or ***S* value,** which can be calculated by use of the formula

$$S = [V_s/V_p]/[(C_o/C_f) - 1]$$

where V_s is the volume of the solution added to the pellet, V_p is the volume of the pellet (commonly, the wet weight of the pellet, W_p, is used in place of V_p), C_o is the initial concentration of solute, and C_f is the final concentration. *S* values indicate the fraction of the cell pellet that is penetrated by the solute.

However, the desired information is the fraction of the actual cell volume penetrated rather than that of the pellet volume; this cell volume fraction is called the ***R* value.** It is calculated by use of the formula

$$R = (S_{sol} - S_{dex})/(1 - S_{dex})$$

where S_{sol} is the space value for the solute under study and S_{dex} is the dextran or interstitial space. A zero value for *R* indicates that the cells are impermeable to the solute. Values from zero to one indicate various degrees of penetration. Values greater than one indicate that the cells have concentrated the solute.

When pellet weight is used as a measure of pellet volume, a superscript *w* is often used with the symbols *S* and *R*. Also, it is often desirable to express results with units of cell dry weight or cell water as the base, e.g., the sucrose-permeable volume per gram of cell dry weight or per milliliter of cell water.

24.1.4. Cell Volume Determination

Useful cytological information can be gained from space values. For example, the technique offers by far the most accurate means available for determining the average cell size of a bacterium. The pellet volume minus the interstitial volume is equal to the volume of the cells. If the pellet is then resuspended in water to some known volume (say, 50 ml) in a volumetric flask, the number of cells can be assessed by use of a Petroff-Hausser counter, and the average volume of a cell can be calculated. The technique also can be used to determine the volume of the protoplast in a cell (section 24.3.2).

Electronic counters used to obtain accurate estimates of bacterial numbers can also yield volume estimates. However, the estimates are based on differences in electrical conductivity between the cell and its suspending medium, and so only relative estimates of volume are obtained, especially with bacterial cells that may have thick cell walls with relatively high conductivities as a result of mobile counterions associated with charged groups in wall polymers.

24.1.5. Cell Water Content Determination

The resuspended cells can also be used for a dry-weight determination from which one can estimate the dry weight per cell. The amount of cell water is equal to the difference between the wet weight of the cell and the dry weight. Since the weight or volume of the interstitial water is known, it is possible to estimate cell water from information about the amount of pellet water.

Sometimes it is convenient to estimate the amount of cell water directly from the *R* value for solutes such as ^3H_2O and [^{14}C]urea or glycerol, which readily permeate but are not concentrated by resting cells. One should be aware that the tritium of ^3H_2O may exchange rapidly with hydrogen in various cell polymers and monomers, especially with ionizable components, but this exchange is generally not a problem in short-term incubations.

24.1.6. General Considerations

The major advantages of the space technique include its simplicity and reliability. However, it is not very useful for kinetic measurements because of the relatively long periods required to mix the cells, test solutions, and then separate them. Thus, the technique is good for equilibrium but not rate determinations. Because of the very thick suspensions used, there are generally no problems due to excessive leaching of physiologically important solutes (such as potassium ions) from cells. However, the thickness of the suspensions makes it difficult to control pH or to provide adequate aeration for aerobic organisms. Thus, the technique is most useful for resting cells and much less useful for actively metabolizing ones. As mentioned above, there is often a problem in obtaining the large masses of cells required. However, the accuracy of the technique depends on having large quantities, especially since solute uptake is estimated from determinations of the change in concentration of the suspending medium.

24.2. TRANSPORT

24.2.1. Assay Procedure

24.2.1.1. Preparation of cells

For studies of the rates of solute transport, it is necessary to have some means of rapidly separating cells from their suspending medium. This separation is generally done by filtration through membrane filters; therefore it is necessary to use rather dilute suspensions, generally with less than about 200 μg (dry weight)/ml if 1-ml samples are to be filtered through filters 25 mm in diameter.

As with permeation assays, it is necessary with transport assays to pay close attention to controlling the conditions of cell growth and harvest. Wash and suspension media tend to be more critical in preparing adequate cell suspensions for transport assays, because cells in dilute suspensions are more likely to be harmed by unfavorable environments than are cells in dense suspensions, in which the environment contains high levels of materials leached from the cells.

For transport studies, it may be necessary to pretreat the cells. For example, they may be starved to deplete endogenous pools of amino acids. This starvation may be of particular advantage when radioactively labeled solutes are added exogenously, since it reduces changes in specific activity resulting from mixing with endogenous unlabeled solute. However, starvation may also deplete cells of energy reserves and usually diminishes the proton motive force across the membrane. Therefore, some fuel for the transport process may have to be added along with the test solute, although even starving cells generally maintain at least low-level pools of ATP and phosphoenolpyruvate.

24.2.1.2. Mixing and incubating cells with solute

The cell suspension and the test solution are brought to the experimental temperature separately and then mixed at zero time. Rapid mixing is generally required because of the rapidity of most transport reactions, and the incubation of the reaction mixture is generally of short duration (e.g., less than 10 min).

24.2.1.3. Sampling and separation

At short intervals (e.g., 1 min), samples of the reaction mixture are withdrawn. The cells and the test solution are rapidly separated, generally by filtration through a membrane filter with 0.45-μm-diameter pores, which should be prewetted with wash solution or test solution. Lower-porosity filters may be used, but they offer more resistance to fluid flow and slow down the filtration process. The 0.45-μm filters are adequate for filtration of most bacterial cells. Generally, filters of cellulose acetate and cellulose nitrate may be used, especially if scintillation counting is planned. Vacuum is developed within a filter flask by use of line vacuum in the laboratory or, if that is not adequate, with a vacuum pump. The common clip-on funnel attachment for the filter unit may be replaced with a custom-made heavy brass ring to hold down the filter disk. The ring can then be rapidly placed and removed to allow for repeated use of a single filter unit. Also, it is possible to purchase or fabricate manifold filter units for very rapid sampling.

There may be an advantage to mixing ingredients in a syringe with a Swinney filter attachment, from which samples of cell-free medium can be expressed. The obvious disadvantage of this method is that one must depend on determination of differences in solute concentration in the suspending medium when estimating uptake. Moreover, the removal of samples results in progressively increased cell concentration in the remaining mixture.

Often, centrifugation can be used to separate the cells, even when rate data are required. Micro-centrifuges, which accommodate small tubes of 250- to 1,500-μl capacity, can attain speeds of 12,000 × g in only a few seconds.

Also, it is often possible to stop transport reactions by adding samples directly to ice or chilled buffer. Centrifugation is then carried out in the cold. For this procedure to be effective, chilling must stop influx but not induce efflux. Satisfactory results are more likely to be obtained with gram-positive than with gram-negative bacteria.

In some cases, it is not necessary to separate the cells from the test solution. For example, ion-specific electrodes allow for continuous monitoring of pH, partial pressure of oxygen (pO_2), potassium, or other ions. Moreover, in studies of salt uptake, it is often necessary to monitor changes only in conductivity of the suspending medium with a conventional conductivity cell.

24.2.1.4. Washing of cells

Cells retained on filters must generally be washed to remove adherent and interstitial medium. For gram-negative bacteria especially, the wash liquid should be osmotically buffered because the transport systems are osmotically sensitive. Concentrated sugar solutions (e.g., 0.5 M sucrose solution) are often used for washing. Magnesium ions (usually 50 mM) may also be added as the chloride or sulfate salt. An ideal wash solution should remove test solute from the cell pad and filter primarily in the first wash but not in subsequent washes. Some compromise must generally be tolerated (see section 24.1.1 for additional information).

It is possible sometimes to avoid the problems of washing cells by centrifuging them through a layer of liquid silicones, as described by Hurwitz et al. (26). The cells pass through the silicone to form a pellet that does not contain interstitial water. Some adherent water may accompany the cells, and the cell wall water will remain undisturbed.

24.2.1.5. Removal and assay of cells for solute content

The cells may be eluted from the filter, resuspended to a known volume, and assayed for the test solute by chemical or radiometric methods.

When radioactive solutes are used, the filters can be placed directly into scintillation fluid for counting. Often, the filters are dried initially, especially if the fluid to be used is not compatible with aqueous samples. It is also possible to estimate the amount of solute incorporated into polymeric materials by treating the cells on the filter with cold trichloroacetic acid solution and then washing them with water before counting. Commercial brochures or articles on scintillation counting should be consulted for advice.

Membrane filters may bind test solutes, especially ionized ones, and it may be necessary to correct for this binding. An estimate of the extent of binding can be obtained by passing some of the test solution without cells through a filter, washing in the same way that cells are washed, and then assaying the filter for test solute.

24.2.2. Expression of Results

Transport uptake or release is generally expressed in terms of base units of cell weight, cell volume, or cell water. Transport results may also be expressed on a per-cell basis, and such values are useful in comparing transport capacities of various cells. The most commonly used base for expressing the results is cell water. It is also of value to calculate concentration ratios, i.e., the concentration of the test solute within the cell divided by its concentration in the suspending medium, so that assessment can be made of whether or not concentrative transport has occurred. When using radioactively labeled solutes, it is desirable to convert data from counts per minute to molar units.

Transport processes are generally considered to be a subclass of enzyme-catalyzed reactions, and the principles and techniques of enzyme kinetics (see Chapter 23.3) can be applied to transport studies. For example, Lineweaver-Burk double-reciprocal plots of the inverse of the transport rate versus the inverse of the solute concentration commonly yield straight lines from which one can calculate values for K_m (Michaelis constant) and V_{max} (maximum velocity). For a specific example, see reference 11. Alternatively, one can prepare Eadie-Hofstee plots of the velocity of transport versus the velocity divided by substrate concentration. The complications of kinetics that arise in the study of enzyme kinetics arise also in the study of transport.

It is not uncommon for cells to have more than one transport system for a particular solute. For example, many bacteria have a high-affinity transport system that is highly specific for an individual aromatic amino acid as well as a low-affinity system that is much less specific. Methods for assessing these multiple systems are described by Ames (11).

In addition, many solutes can enter cells passively by diffusion (permeation) as well as actively by transport. The diffusion component generally does not show saturation at high substrate concentrations, and its extent can be estimated from plots such as that shown in Fig. 2. With intact bacteria, the diffusion component may include diffusion of solute into the cell wall water (45), which can account for some 50% of the cell volume in organisms such as *Micrococcus luteus*.

24.2.3. General Considerations

A major advantage to the use of dilute suspensions of cells is the ease with which kinetic data can be obtained. Also, with dilute suspensions, it is possible more readily to control pO_2, pH, ionic strength, etc., and to maintain nearly constant solute concentration in the suspending medium.

The disadvantages include problems associated with solute leaching from cells during initial suspension or washing, the general need to assess solute uptake or loss from changes in intracellular rather than extracellular concentration, and the inherent vagaries in the use of dilute suspensions in which the cells may be highly sensitive to changes in environmental conditions.

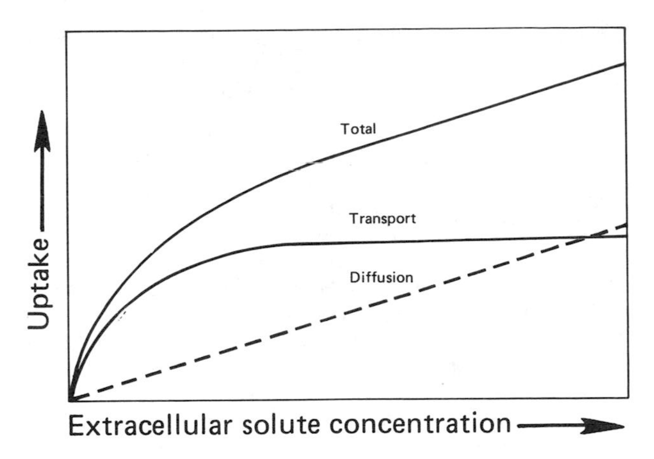

FIG. 2. Graphic estimation of the diffusion component in the total uptake of a solute by bacterial cells. The transport uptake is the total uptake minus the passive uptake. The uptake parameter may be either the rate or the extent of uptake.

24.3. SPECIAL METHODS

24.3.1. Proton Motive Force, Intracellular pH, and Proton Permeation

The **proton motive force** is considered to be a major means for energy conversion and transfer in biological systems. It is defined as

$$\Delta p = \Delta\psi - 2.3\frac{RT}{F}(\Delta \text{pH})$$

where Δp is the proton motive force, $\Delta\psi$ is the membrane potential, R is the gas constant (8.318 V-C/K charge equivalent), T is the temperature (kelvin), F is the faraday (96,500 C/charge-equivalent), and ΔpH is the difference in pH between the interior and the exterior of the cell. At about 25°C, the value of $2.3RT/F$ (which is often given the symbol z) is approximately 59 mV. Thus, to estimate Δp, it is necessary to estimate $\Delta\psi$ and ΔpH separately. Techniques for estimating $\Delta\psi$ and ΔpH have been reviewed critically by Kashket (5).

$\Delta\psi$ for giant cells of *E. coli* has been assessed directly by means of microelectrodes (23), but the method is difficult and is not applicable to normal bacterial cells. Generally $\Delta\psi$ is assessed indirectly from the distribution of permeant cations between the interior and exterior of cells. Harold and Altendorf (24) have indicated that a useful indicator for $\Delta\psi$ must "diffuse rapidly across the membrane . . . be fully dissociated by physiological pH, metabolically innocuous and not subject to translocation by a biological transport system." These criteria can be met by K^+ or Rb^+ in cells that have been treated with 1 to 10 μM valinomycin to render the membrane permeable to K^+. Since the membrane potential in bacterial cells is generally about 180 mV with the interior negative, the potassium concentration in the cytoplasm of valinomycin-treated cells is some 20 times the concentration in the suspending medium. Potassium concentrations can be determined by means of atomic absorption spectrophotometry or flame photometry or by use of an ion-specific electrode. The uptake of K^+ is determined by the space technique, or a variation of it, and the concentration is calculated in terms of moles of K^+ taken up per unit of cell water. Radioactive $^{86}Rb^+$ can be used for estimations of Rb^+ concentrations.

$\Delta\psi$ can be assessed also by determining the distribution of permeant, lipophilic organic cations or by use of fluorescent dyes. Kashket (5) recommends tetraphenylphosphonium (TPP^+) for assessing $\Delta\psi$ and describes basic procedures for its use. There is some worry with the use of TPP^+ with some organisms in that Midgley (47) has found that *E. coli* has an efflux system with broad specificity for complex organic cations, including TPP^+. This efflux would upset $\Delta\psi$ determinations, and so there may be value in the use of the K^+ or Rb^+/valinomycin technique.

For determination of ΔpH, a weak acid or base to which the cell is permeable is used. The compound chosen must dissociate at physiological pH values so that its distribution between the cell and its environment will depend on the pH difference between the two phases. The membrane of the cell should then be permeable to the undissociated form but impermeable to the ionized form. Compounds that are commonly used for ΔpH estimates include acetate, benzoate, 5,5-dimethyl-2,4-oxazolidinedione, and salicylate when the cytoplasm is alkaline relative to the exterior pH value. When the cytoplasm is more acidic than the environment, weak bases such as methylamine, hexylamine, or benzylamine are used. As Kashket (5) points out, the ΔpH calculation generally is valid only when the pK of the probe is one pH unit or more different than the exterior pH value.

Considering the Henderson-Hasselbalch equation for a weak acid, let A stand for the anionic form, let HA stand for the undissociated form, and use the subscripts i and o to refer to concentrations inside and outside of the cell. Therefore,

$$\text{pH}_i = \text{pK}_i + \log[\text{A}_i]/[\text{HA}_i]$$

$$\text{pH}_o = \text{pK}_o + \log[\text{A}_o]/[\text{HA}_o]$$

It is generally assumed that $\text{pK}_i = \text{pK}_o$ and, since the membrane is permeable to the undissociated form, $[\text{HA}_i] = [\text{HA}_o]$. Therefore,

$$\text{pH}_i - \text{pH}_o = \log[\text{A}_i]/[\text{HA}_o] - \log[\text{A}_o]/[\text{HA}_o]$$

$$= \log[\text{A}_i]/[\text{HA}_i] + \log[\text{HA}_o]/[\text{A}_o]$$

$$= \log([\text{A}_i]/[\text{HA}_i] \times [\text{HA}_o]/[\text{A}_o])$$

$$\Delta\text{pH} = \log[\text{A}_i]/[\text{A}_o]$$

The ΔpH value can then be assessed from estimates of the distribution of, say, benzoate between cells and suspending medium by the methods described above for permeability assays. From the external benzoate concentration and pH, one can calculate the external concentration of HA, which is equal to the internal concentration. The internal concentration of A is equal to the total internal benzoate concentration minus the HA concentration. From the ratio $[\text{A}_i]/[\text{HA}_i]$ and a knowledge of pK_i, an estimate of the internal pH can be derived with the Henderson-Hasselbalch equation.

Again, the space technique can be used to estimate the uptake of the pH-probing compounds. It is possible also to use techniques that do not require large quantities of cells if one uses some nonpermeant compound to estimate the noncytoplasmic water concentration. For example, on page 29 of reference 7 an example is presented of an experiment in which [^{3}H]sorbitol is used to estimate the noncytoplasmic water in a suspension of *Lactococcus lactis* cells and [^{14}C]DMO is used as a pH probe. One should realize that ΔpH and $\Delta\psi$ determinations are only estimates and that there are problems associated with specific and nonspecific binding of charged species to biopolymers. However, one is often most interested in changes in ΔpH or $\Delta\psi$ rather than in their absolute magnitudes.

A flow dialysis method that does not require separation of the cells and the suspending medium has been developed and described in detail by Ramos et al. (52, 53). It is particularly useful for work with membrane vesicles.

Methods have been described by Avison et al. (14) for estimation of the **intracellular pH** of bacterial cells by use of nuclear magnetic resonance techniques. pH_i has also been assessed by using fluorescent dyes such as 9-aminoacridine (5). If pH_i, is known, pH_o is readily determined with a glass electrode, and ΔpH can be calculated.

Internal pH values of cells can also be determined by a so-called null point method. For this method, cells are suspended in thick suspensions in a set of dilute buffers at various pH values. Butanol is added to the cells to disrupt the cell membrane, and changes in pH values of the suspensions as a result of butanol addition are recorded. The changes are then plotted against the pH values of the suspensions before butanol addition. The initial pH value at which there is no change in pH upon addition of butanol can then be taken as the internal pH value of the cells. Disruptive agents other than butanol can be used (see Chapter 23.2.2). The method is labor intensive and requires large quantities of cells. Therefore, its use is limited.

Assessments of Δp, $\Delta\psi$, and ΔpH are subject to many inaccuracies, as reviewed by Kashket (5). Moreover, any manipulation of cells involved in preparing them for measurements will affect the measured values. Often a better view of proton currents across membranes can be obtained by assessing the permeabilities of cells to protons. Methods for assessing **proton permeation** are described by Bender et al. (15) and are based on modifications of procedures described earlier by Scholes and Mitchell (57) and Maloney (42). Cells are often considered to be impermeable to protons, but in fact, half times for pH equilibration across cell membranes after an acid pulse generally take minutes. Thus, to maintain more than a minimal ΔpH across the membrane, a cell must actively extrude protons. Otherwise the ΔpH quickly dissipates.

Buffer capacities of cells can be estimated by methods described by Krulwich et al. (36) for assessing external buffering (B_o), total buffering (B_t), and internal buffering (B_i) on the basis of acid-base titrations of thick suspensions of intact or permeabilized cells and the difference between the two. As expected, buffering capacities vary with pH and tend to be dominated by carboxyl, phosphate and amino groups. Capacity is generally expressed in units of nanomoles of H^+ per pH unit per unit of cell weight or cell protein.

24.3.2. Osmotically Sensitive Cells

When a solute is taken up by a cell, the process results in a lowering of the water activity in the interior and a consequent influx of water, with resultant swelling of the cell. Biological membranes are highly permeable to water, and the swelling reaction is extremely rapid. For example, Matts and Knowles (46) used a stopped-flow spectrophotometer to show that the osmotic movement of water from *E. coli* cells transferred into a 0.4 M $MgCl_2$ solution is complete within 50 ms at 37°C. Since water movement is so rapid, the net rate of swelling that accompanies solute uptake can be related directly to transport. However, some caution must be applied, especially for bacterial cells with walls sufficiently elastic to resist swelling. This elasticity of the wall results in buildup of hydrostatic pressure within the cell, called **turgor pressure.** The turgor pressure ($P^{in} - P^{out}$) within the cell is expressed as

$$P^{in} - P^{out} = (2.30\, RT/V)(\log a^{out}/a^{in})$$

where R is the gas constant, T is the Kelvin temperature, V is the molar volume of water, a^{out} is the water activity outside of the cell, and a^{in} is the water activity within the cell. At a temperature of 25°C, turgor pressure is equal to 311 MPa $\times$ log a^{out}/a^{in}. The symbol π is often used for turgor pressure.

The swelling and shrinking of an ideal osmometer can be predicted by use of the van't Hoff-Boyle equation,

$$V - b = a/\pi$$

where V is the total volume of the cell, b is the so-called osmotically dead space (which generally is approximately equal to the calculated volume of the dry matter of the cell), a is a constant (the number of osmoles in the cell), and π is the osmolality of the suspending medium. For an ideal osmometer, a plot of V against $1/\pi$ yields a straight line with intercept b and slope a, whereas the actual behavior of bacterial cells yields a curved line (Fig. 3). Clearly, bacteria are not truly ideal osmometers.

Osmotic responsiveness is markedly reduced in media of high osmolality, and one can obtain an experimentally determined value for b only by extrapolation. It is not known just why the cell behaves anomalously in concentrated media. The behavior could have to do with bound water and gelling of the cytoplasm or with decreased water permeability.

When placed in dilute media, bacterial cells are resistant to osmotic swelling primarily because they have an elastic cell wall. When placed in concentrated media, they become plasmolyzed; that is, the protoplast shrinks so that the membrane is pulled away from the wall and plasmolysis vacuoles are formed between the wall and the protoplast membrane. These vacuoles can often be seen microscopically, especially in gram-negative bacteria. However, even if they cannot be seen, their presence can be detected by permeability determinations. Thus, one can assess the total cell volume in

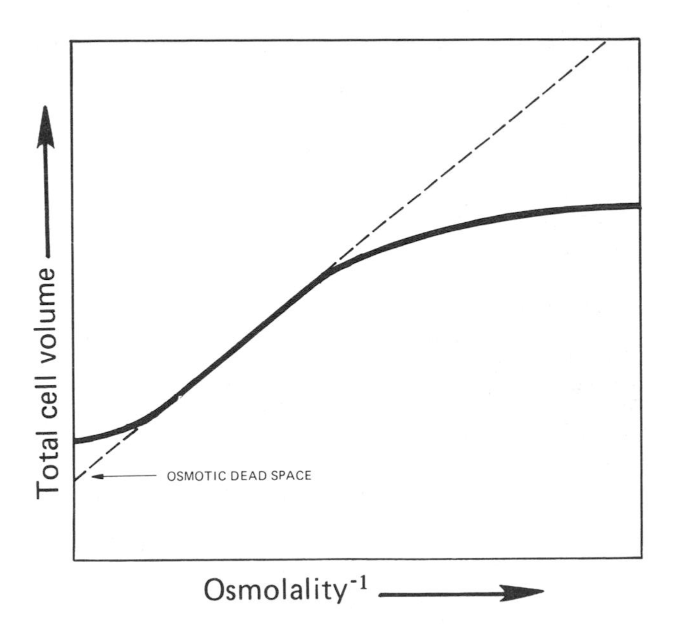

FIG. 3. Osmotic volume changes for bacterial cells. The dashed line indicates the ideal behavior predicted by the van't Hoff-Boyle equation. The solid curve indicates the actual behavior.

terms of the dextran-impermeable space. The protoplast volume is essentially equal to the raffinose-impermeable volume under conditions in which raffinose does not pass through the protoplast membrane. The aqueous volume in the cell wall and the plasmolysis vacuoles together is then equal to the dextran-impermeable volume minus the raffinose-impermeable volume. A reasonable estimate of the aqueous wall volume alone can be made in terms of the raffinose-permeable space of turgid nonplasmolyzed cells.

If the plasmolyzed cells are now returned to more dilute media, water flows into the protoplast across the water-permeable membrane, the protoplast swells to occupy the volume previously occupied by the plasmolysis vacuoles, and the optical density of the suspension increases. There is a direct relationship between the protoplast volume and the inverse of the optical density (44). Initially, when the protoplast swells, there is little resistance to swelling and the process is nearly ideal; i.e., it follows the van't Hoff-Boyle relationship. However, when the protoplast abuts against the cell wall, the elasticity of the wall resists further swelling, and wide deviation from the ideal situation occurs. Now water uptake involves an increase not only in the volume of the protoplast but also in that of the whole cell, with stretching of the wall. A break in the curve for V versus $1/\pi$ occurs at what is called the point of incipient plasmolysis. Further reductions in medium osmolality result in the buildup of turgor pressure. It is considered that the medium osmolality at the point of incipient plasmolysis is approximately equal to the internal osmolality of the cell, as discussed primarily for plant cells by Stadelmann (60). It is possible also to obtain another estimate of internal osmolality from the value of the constant a (the number of ideal molar equivalents within the cell) and a knowledge of the protoplast volume. Generally, estimates from the two sources agree reasonably well.

The elasticity of the cell wall does not affect the swelling of isolated protoplasts, which have been used often for transport studies, and is thought to contribute little to the swelling of spheroplasts, which retain wall remnants. However, if protoplasts are made to swell slowly so that their membranes do not undergo brittle fracture, the membrane can become taut and the magnitude of swelling is not predictable from the van't Hoff-Boyle equation (44).

As mentioned, the inverse of optical density can be used as an indicator of cell volume and therefore of solute uptake, although of course one should always be aware that it is actually water movement that is being monitored. It is important also to know that the optical density depends on the difference in refractive index between the cell and the suspending medium and that corrections may have to be made for changes in medium refractive index as a result of increased or decreased solute concentration. Ideally, one would like to have suspending media of different solute concentrations but constant refractive index. In many studies of solute uptake, the external concentration does not change greatly, and so no corrections are needed. However, in studies of osmotic volume changes, concentrations generally do vary widely. If salt solutions are used, the changes in refractive index due to changes in solute concentration are smaller than when sugar solutions are used. However, in either case, turbidity values may have to be corrected for changes in refractive index. It is useful to prepare matching solutions of polymeric dextrans that have high refractive increments but low osmolalities. Thus, for example, it would be possible to prepare a concentrated sucrose solution of high refractive index and high osmolality and a solution of dextran T2000 of the same high refractive index but nearly zero osmolality. The two solutions could be mixed in various proportions to give solutions of various osmolalities but all with the same refractive index. The refractive increment (a) for carbohydrates is approximately 0.00143. Therefore, the refractive index of a carbohydrate solution can be calculated by means of the formula $n = n_s + aC$, where n is the refractive index, n_s is the refractive index of water (1.33252 at 25°C), and C is the carbohydrate concentration in grams per 100 ml of solution. Of course, it is possible also to determine the refractive indices of the solution with a refractometer.

In a study of regulation of turgor pressure in *Bacillus subtilis*, Whatmore and Reed (64) used a Coulter counter (Coulter Electronics, Inc., Hialeah, Fla.) to assess changes in cell volume, and these changes were used to construct a van't Hoff-Boyle plot. These authors also found that a spectrophotometric analysis of volume changes gave comparable but somewhat lower estimates for turgor. For bacteria with internal gas vesicles, changes in light scattering as a result of collapse of the vesicles under pressure can be used to estimate cell turgor, as described by Reed and Walsby (54).

Bacterial cells require turgor for growth, cell division, and certain other functions. Therefore, they have developed means to maintain turgor when placed in hypertonic solutions through pooling of solutes, including K^+, glycine betaine, proline, and so-called compatible solutes (25). The buildup of tension in the cell wall associated with turgor appears to be required for normal cell division, as proposed in the surface stress hypothesis (33). The mechanical properties of cell walls have been assessed by use of bacterial threads (61).

24.3.3. Nonmetabolized Analogs for Separation of Transport from Metabolism

For many studies, there is advantage in dissociating transport of the test solute from its subsequent metabolism, especially if one wants to demonstrate concentrative uptake. With phosphotransferase systems, complete dissociation is not possible because the solute is phosphorylated as part of the transport process. There are three major ways of making a separation: by the use of nonmetabolizable analogs that can be transported, by the use of mutants blocked in metabolism but not transport, and by the use of membrane vesicles that lack cytoplasmic enzymes.

The use of analogs to study uptake of β-galactosides by *E. coli* has been well documented (31). Unfortunately, analogs may interact differently with different organisms. For example, thiomethylgalactoside is transported by means of Δp-driven systems in *E. coli* (9) and *S. lactis* (30), but *S. mutans* also shows PTS activity for the analog (19). Thus, a significant amount of background work may be required before a specific analog can be used knowledgeably with a specific organism.

Toxic analogs may be useful also for the isolation of transport-negative mutants. A good example was described by Jacobson et al. (27). They used the antibiotic streptozotocin, which is an analog of *N*-acetylglucosamine, to isolate both pleiotropic and specific carbohydrate-negative mutants of *S. mutans*. Wild-type cells were able to take up the antibiotic and were killed by it, whereas transport-negative mutants were spared. 2-Deoxyglucose has been used in a similar way (40, 62). Toxic analogs of amino acids also may be useful for mutant selection. However, the mutants obtained may simply be hyperproducers of the natural amino acid.

24.3.4. Metabolism-Blocked Mutants

The selection of *Salmonella typhimurium* mutants for histidine transport studies is described in detail by Ames (12). Mutants of *E. coli*, which lack β-galactosidase but retain the transport system for β-galactosides, have been used extensively and can be obtained from genetic stock culture collections (see Chapter 13.10).

24.3.5. Membrane Vesicles and Liposomes

Membrane vesicles have become increasingly useful for physiologic studies since they were introduced by Kaback and Stadtman (29).

Generally, the first step in preparing vesicles is to convert the test bacteria into osmotically sensitive forms, which are then lysed under conditions that allow for membrane resealing. Spheroplasts of gram-negative bacteria such as *E. coli* can be prepared by the lysozyme-EDTA method (28), which leaves the outer membrane of the cell wall associated with the spheroplast. Protoplasts of gram-positive bacteria can often be completely freed from cell wall structures by treating the cells with a muralytic enzyme. Lysozyme has been used for *B. megaterium, B. subtilis, S. faecalis*, and *M. luteus;* lysostaphin has been used for *Staphylococcus aureus* (34).

Kobayashi et al. (32) isolated vesicles from *Enterococcus hirae* ATCC 9790, and their procedure is illustrative. They first suspended cells in an osmotic stabilizing solution of 500 mM glycylglycine plus 2 mM $MgSO_4$ (pH 7.2), added 0.5 mg of lysozyme per ml, and incubated the mixture at 37°C for 1 h (the protoplasts can also be stabilized with sucrose instead of glycylglycine). The protoplast suspension was centrifuged; resuspended in a hypotonic solution of 50 mM Tris-maleate (pH 7.4), 250 mM sucrose, and 5 mM $MgSO_4$; and then treated with DNase I, which requires magnesium ions for activity. The suspension was chilled, homogenized with a tissue homogenizer, centrifuged at 5,000 × *g* for 10 min to remove debris, and then centrifuged at 27,000 × *g* for 10 min to pellet the membranes. The membranes were resuspended in the buffer solution of Tris-maleate plus sucrose plus $MgSO_4$ and then passed through a French press (10,000 lb/in^2 or 68 MPa). The resulting suspension was twice centrifuged, once at 27,000 × *g* to remove large particles and once at 48,000 × *g* for 90 min to pellet the vesicles.

There are many possible variants of this procedure, and some preliminary experimentation is needed to determine optimum procedures for a particular organism. A shortened, "one-step" procedure for use with gram-positive bacteria, notably *B. subtilis*, was described by Konings et al. (35).

Vesicle preparations tend to be heterogeneous in size, and the preparation of Kobayashi et al. (32) contained both right-side-out and wrong-side-out vesicles, as indicated by their appearance in electron micrographs of freeze-fracture preparations. With *E. coli* it is possible to prepare populations of vesicles of which all are right side out. For example, Kaback (28) described a procedure involving osmotic lysis without passage through a French pressure cell. *E. coli* vesicles prepared in the French pressure cell may contain mainly wrong-side-out forms which accumulate Ca^{2+} instead of extruding it and which cannot take up solutes such as proline (10).

The membrane continuity of vesicles is best tested by assessing the extent of swelling or shrinking in response to changes in the osmolality of the suspending medium. The procedures outlined in section 24.3.2, including determinations of changes in light scattering, can be used. The functionality of the vesicles is tested by determining their abilities to transport solutes such as proline.

The workings of individual transport components such as permease proteins, antiporters, symporters, or ion-translocating ATPases can often best be studied by isolating the components from the cell membrane of the producing organism and incorporating them into liposomes or proteoliposomes. The net result is a simpler system than that of membrane vesicles, with only a single transport system and more complete control of internal and external compartments on either side of the membrane. A thorough description of preparation and use of proteoliposomes containing *E. coli* phospholipids and *lac* permease has been presented by Viitanen et al. (8).

A limitation to the use of proteoliposomes is that Δp

can be developed only passively, for example, by rapidly lowering the pH value of the proteoliposome suspension so as to create a transient ΔpH. Driessen et al. (21) have made use of a model system composed of membrane vesicles from streptococci fused with liposomes containing beef heart cytochrome *c* oxidase. When cytochrome *c* is added to suspensions of these structures, a $\Delta\psi$ (inside negative) and ΔpH (inside alkaline) are generated, and this combination mimics the common Δp in metabolizing cells.

Preparation of proteoliposomes is often difficult, especially if vesicles free of lamellae and with not too great a size range are desired. Therefore a significant amount of experimentation is required in a particular laboratory with the use of a particular system. However, there are major advantages in using the structures; i.e., the results obtained with them can generally be more directly interpreted than can results obtained with cells or membrane vesicles containing multiple systems of primary and secondary transporters.

24.3.6. Isolation of Transport Components

With development of detailed knowledge of the genetics and molecular biology of various transport systems, there has been increased need to isolate and characterize individual components. Many of the components of transport systems are not integral parts of cell membranes or walls, and so their isolation follows standard procedures for cytoplasmic enzyme isolation. Also, periplasmic substrate-binding proteins can be isolated and purified from so-called shockate fluids derived from osmotically shocked cells or, alternatively, from disrupted cells, for example as described by Kustu et al. (38) for histidine-binding protein J and binding proteins for glutamine and arginine from *Salmonella typhimurium*. A more recently developed, simpler procedure involves release of the proteins in a chloroform extraction technique (13).

Enzyme 1 and HPr proteins of the phosphotransferase system are cytoplasmic and can be isolated from cell lysates, for example, as described by Kukuruzinska et al. (37) and Beneski et al. (16) for gram-negative *Salmonella typhimurium* cells or by Mimura et al. (48) and Vadeboncoeur et al. (63) for gram-positive oral streptococci.

When components of transport systems are integral parts of membranes, their isolation commonly involves initial isolation of the appropriate membranes. For example, isolation of porin proteins of the outer, cell wall membrane of gram-negative bacteria may require separation of inner (lower buoyant density) and outer (higher density) membranes. Procedures for isolation of specific porins and other outer membrane components have been described by Nikaido and Vaara (50).

Ion-transporting ATPases are structurally integrated into cytoplasmic membranes or organelle membranes. In fact, the working forms of the enzymes are commonly proteolipids, including membrane lipids. Their isolation begins with membrane isolation and is followed by further fractionation of the isolated membranes, commonly with detergents. Perhaps the best example for consideration here is the isolation of the proton-translocating F_1F_0-ATPase. Leimgruber et al. (39) described procedures for isolation of the enzyme from *Enterococcus hirae* (*Streptococcus faecium*) ATCC 9790. After growth and harvest, cells of the organism were converted to protoplasts by treatment with lysozyme in the presence of an osmotic stabilizer, in this instance 0.4 M glycylglycine. The protoplasts were then sedimented and lysed by osmotic shock involving resuspension in a solution containing 10 mM $MgCl_2$ and 1 μg of DNase per ml but no osmotic stabilizer. The membranes were then sedimented at 78,000 × *g* for 20 min and washed with 10 mM $MgCl_2$ solution. There is advantage in working with lysozyme-sensitive organisms that can be converted to protoplasts or spheroplasts. Bender et al. have used mutanolysin (15) for preparing osmotically sensitive forms of oral streptococci previously grown in medium supplemented with 20 mM DL-threonine. With gram-negative organisms, the inner membrane containing the ATPase can be separated from the outer membrane by means of density gradient centrifugation.

Isolated membranes contain the F_1 or hydrolytic part of the enzyme attached to the F_0 or proton-translocating part, which is embedded in the membrane. ATP hydrolysis is then obligatorily coupled to proton movements across the membrane, and the enzyme is sensitive to the inhibitor dicyclohexylcarbodiimide. The F_1 part of the enzyme can be washed free of the membrane and of the F_0 part with low-ionic-strength buffer by the procedure described by Senior et al. (58). The recommended buffer contains glycerol as a so-called osmolyte (43) to help stabilize the active form of this enzyme, which is composed of five types of subunits, usually with a stoichiometry of $\alpha_3\beta_3\gamma\delta\epsilon$. Also, once cells have been lysed, there may be advantage in adding protease inhibitors to avoid unwanted degradation. There are many recipes for inhibitor cocktails, for example, that described by Solioz and Furst (59) for use with *E. hirae*. After the F_1 enzyme has been washed from the membrane, it can be further purified by means of $(NH_4)_2SO_4$ precipitation and column or gel chromatography. The component proteins can be separated by polyacrylamide gel electrophoresis.

Leimgruber et al. (39) separately isolated F_1 by low-ionic-strength washing and F_0 by detergent extraction of the stripped membranes before recombining them to yield F_1F_0 complete enzyme. However, it is often possible directly to separate F_1F_0 enzyme from membranes by means of solubilization with detergents such as deoxycholate, Triton X-100, or octylglucoside. Solioz and Furst (59) described the isolation from *E. hirae* of an E_1E_2 or P-type enzyme which is involved in ion transport. The procedure is similar to that for isolation of F_1F_2 enzymes.

A good example of a procedure for isolation of a permease protein is presented by Viitanen et al. (8) for the *lac* permease from *E. coli*. This protein could then be incorporated in a fully functional form into proteoliposomes and could carry out all of its normal functions including $\Delta\psi$ or ΔpH driven active transport, facilitated diffusion, counterflow, exchange, or efflux.

24.3.7. Permeabilization of Cells

Methods to render cells permeable to solutes by chemical or enzymatic treatment are described in Chapter 23.2.2.

24.4. REFERENCES

24.4.1. General References

1. **Fleischer, S., and B. Fleischer (ed.)** 1986. Biomembranes. Part M. Transport in bacteria, mitochondria and chloroplasts: general approaches and transport systems. *Methods Enzymol.* **125:**1–788.
This volume contains multiple chapters (19 to 45) on bacterial transport systems with orientation to methodology. Genetic techniques useful for study of transport systems are described also in the earlier chapters on membrane characteristics and chemistry.
2. **Fleischer, S., and B. Fleischer (ed.).** 1988. Biomembranes. Part Q. ATP-driven pumps and related transport: calcium, proton, and potassium pumps. *Methods Enzymol.* **157:**1–731.
This volume has very useful chapters (38 to 53) on methods for study of ATP-driven proton and K^+ pumps in various procaryotic and eucaryotic systems.
3. **Fleischer, S., and B. Fleischer (ed.).** 1989. Biomembranes. Part R. Transport theory: cells and model systems. *Methods Enzymol.* **171:**1–926.
This volume contains chapters on production and characteristics of model membrane systems, including liposomes and planar lipid-protein membranes.
4. **Harold, F. M.** 1986. *The Vital Force: A Study of Bioenergetics.* W. H. Freeman and Co., New York.
This book does not have primary orientation to techniques but provides an excellent background for understanding the bases on which standard techniques were developed for bioenergetics, especially the energetics of transmembrane ion currents.
5. **Kashket, E. R.** 1985. The proton motive force in bacteria: a critical assessment of methods. *Annu. Rev. Microbiol.* **39:**219–242.
As its title indicates, this article reviews critically the common methods for assessing $\Delta\psi$, ΔpH and Δp with bacterial cells. It can be considered required reading for anyone wanting to determine these parameters.
6. **Krulwich, T. A. (ed.)** 1990. *The Bacteria,* vol. 12. *Bacterial Energetics.* Academic Press, Inc., San Diego, Calif.
This book contains review chapters on many bacterial transport systems with an orientation to energetics.
7. **Maloney, P. C., E. R. Kashket, and T. H. Wilson.** 1975. Methods for studying transport in bacteria. *Methods Membr. Biol.* **5:**1–49.
This well-known article reviews basic techniques for studying bacterial transport systems, presents sample data, and discusses many important technical details. It focuses on lactose transport, but the techniques described are generally useful ones.
8. **Viitanen, P., M. J. Newman, D. L. Foster, T. H. Wilson, and H. R. Kaback.** 1986. Purification, reconstitution, and characterization of the *lac* permease of *Escherichia coli. Methods Enzymol.* **125:**429–452.
This article is an excellent supplement to the article described above in extending consideration of the lactose transport system to molecular aspects with the use of proteoliposomes to study function.

24.4.2. Specific References

9. **Ahmed, S., and I. R. Booth.** 1981. Quantitative measurements of the proton-motive force and its relation to steady state lactose accumulation in *Escherichia coli. Biochem. J.* **200:**573–581.
10. **Altendorf, K. H., and L. A. Staehelin.** 1974. Orientation of membrane vesicles from *Escherichia coli* as detected by freeze-cleave electron microscopy. *J. Bacteriol.* **117:**888–899.
11. **Ames, G. F.-L.** 1974. Two methods for the assay of amino acid transport in bacteria. *Methods Enzymol.* **32:**843–849.
12. **Ames, G. F.-L.** 1974. Isolation of transport mutants in bacteria. *Methods Enzymol.* **32:**849–856.
13. **Ames, G. F.-L., C. Prody, and S. Kustu.** 1984. Simple, rapid, and quantitative release of periplasmic proteins by chloroform. *J. Bacteriol.* **160:**1181–1183.
14. **Avison, M. J., H. P. Hetherington, and R. G. Shulman.** 1984. Applications of NMR to studies of tissue metabolism. *Annu. Rev. Biophys. Biophys. Chem.* **15:**377–402.
15. **Bender, G. R., S. V. W. Sutton, and R. E. Marquis.** 1986. Acid tolerance, proton permeabilities, and membrane ATPases of oral streptococci. *Infect. Immun.* **53:**331–338.
16. **Beneski, D. A., A. Nakazawa, N. Weigel, P. E. Hartman, and S. Roseman.** 1982. Sugar transport by the bacterial phosphotransferase system: isolation and characterization of a phosphocarrier protein HPr from wild type and mutants of *Salmonella typhimurium. J. Biol. Chem.* **257:**14492–14498.
17. **Conway, E. J., and M. Downey.** 1950. An outer metabolic region of the yeast cell. *Biochem. J.* **47:**347–355.
18. **Damadian, R. V.** 1973. Cation transport in bacteria. *Crit. Rev. Microbiol.* **2:**377–422.
19. **Dashper, S. G., and E. C. Reynolds.** 1990. Characterization of transmembrane movement of glucose and glucose analogs in *Streptococcus mutans* Ingbritt. *J. Bacteriol.* **172:**556–563.
20. **Driessen, A. J. M., D. Molenaar, and W. N. Konings.** 1989. Kinetic mechanism and specificity of the arginine-ornithine antiporter of *Lactococcus lactis. J. Biol. Chem.* **264:**10361–10370.
21. **Driessen, A. J. M., C. van Leeuwen, and W. N. Konings.** 1989. Transport of basic amino acids by membrane vesicles of *Lactococcus lactis. J. Bacteriol.* **171:**1453–1458.
22. **Dubois, M., K. A. Gilles, J. K. Hamilton, P. A. Rebers, and F. Smith.** 1956. Colorimetric method for determination of sugars and related substances. *Anal. Chem.* **28:**350–356.
23. **Felle, H., J. S. Porter, C. L. Slayman, and H. R. Kaback.** 1980. Quantitative measurements of membrane potential in *Escherichia coli. Biochemistry* **19:**3585–3590.
24. **Harold, F. M., and K. Altendorf.** 1974. Cation transport in bacteria: K^+, Na^+, and H^+. *Curr. Top. Membr. Transp.* **5:**1–50.
25. **Higgins, C. F., J. Cairney, D. A. Stirling, L. Sutherland, and I. R. Booth.** 1987. Osmotic regulation of gene expression: ionic strength as an intracellular signal. *Trends Biochem. Sci.* **12:**339–344.
26. **Hurwitz, C., C. B. Braun, and R. A. Peabody.** 1965. Washing bacteria by centrifugation through a water-immiscible layer of silicone. *J. Bacteriol.* **90:**1692–1695.
27. **Jacobson, G. R., F. Poy, and J. W. Lengeler.** 1990. Inhibition of *Streptococcus mutans* by the antibiotic streptozotocin: mechanisms of uptake and the selection of carbohydrate-negative mutants. *Infect. Immun.* **58:**543–549.
28. **Kaback, H. R.** 1971. Bacterial membranes. *Methods Enzymol.* **22:**99–120.
29. **Kaback, H. R., and E. R. Stadtman.** 1966. Proline uptake by an isolated cytoplasmic membrane preparation of *Escherichia coli. Proc. Natl. Acad. Sci. USA* **55:**920–927.
30. **Kashket, E. R., and T. H. Wilson.** 1974. Protonmotive force in fermenting *Streptococcus lactis* 7962 in relation to sugar accumulation. *Biochem. Biophys. Res. Commun.* **59:**879–884.
31. **Kepes, A.** 1970. Galactoside permease of *Escherichia coli. Curr. Top. Membr. Transp.* **1:**101–134.
32. **Kobayashi, H., J. Van Brunt, and F. M. Harold.** 1978. ATP-linked calcium transport in cells and membrane vesicles of *Streptococcus faecalis. J. Biol. Chem.* **253:**2085–2092.
33. **Koch, A. L.** 1990. Growth and form of the bacterial cell wall. *Am. Sci.* **78:**327–341.
34. **Konings, W. N.** 1977. Active transport of solutes in bacterial membrane vesicles. *Adv. Microb. Physiol.* **15:**175–251.
35. **Konings, W. N., A. Bisschop, M. Veenhuis, and C. A. Vermeulen.** 1973. New procedure for the isolation of mem-

brane vesicles of *Bacillus subtilis* and an electron microscopic study of their ultrastructure. *J. Bacteriol.* **116:**1456–1465.

36. **Krulwich, T. A., R. Agus, M. Schneier, and A. A. Guffanti.** 1985. Buffering capacity of bacilli that grow at different pH ranges. *J. Bacteriol.* **162:**768–772.
37. **Kukuruzinska, M. A., W. F. Harrington, and S. Roseman.** 1982. Sugar transport by the bacterial phosphotransferase system. Studies on the molecular weight and association of Enzyme I. *J. Biol. Chem.* **257:**14470–14476.
38. **Kustu, S. G., N. C. McFarland, S. P. Hui, B. Esmon, and G. F.-L. Ames.** 1979. Nitrogen control in *Salmonella typhimurium:* coregulation of synthesis of glutamine synthetase and amino acid transport systems. *J. Bacteriol.* **138:**218–234.
39. **Leimgruber, R. M., C. Jensen, and A. Abrams.** 1981. Purification and characterization of the membrane adenosine triphosphate complex from the wild-type and *N,N'*-dicyclohexylcarbodiimide-resistant strains of *Streptococcus faecalis. J. Bacteriol.* **147:**363–372.
40. **Liberman, E. S., and A. S. Bleiweis.** 1984. Role of the phosphoenolpyruvate-dependent glucose phosphotransferase system of *Streptococcus mutans* GS5 in the regulation of lactose uptake. *Infect. Immun.* **43:**536–542.
41. **MacDonald, R. E., and P. Gerhardt.** 1958. Bacterial permeability: the uptake and oxidation of citrate by *Escherichia coli. Can. J. Microbiol.* **4:**109–124.
42. **Maloney, P. C.** 1979. Membrane H^+ conductance of *Streptococcus lactis. J. Bacteriol.* **140:**197–205.
43. **Maloney, P. C., and S. V. Ambudkar.** 1989. Functional reconstitution of prokaryote and eukaryote membrane proteins. *Arch. Biochem. Biophys.* **269:**1–10.
44. **Marquis, R. E.** 1967. Osmotic sensitivity of bacterial protoplasts and the response of their limiting membrane to stretching. *Arch. Biochem. Biophys.* **118:**323–331.
45. **Marquis, R. E., and P. Gerhardt.** 1964. Respiration-coupled and passive uptake of α-aminoisobutyric acid, a metabolically inert transport analogue, by *Bacillus megaterium. J. Biol. Chem.* **239:**3361–3371.
46. **Matts, T. C., and C. J. Knowles.** 1971. Stopped-flow studies of salt-induced turbidity changes of *Escherichia coli. Biochim. Biophys. Acta* **249:**583–587.
47. **Midgley, M.** 1987. An efflux system for cationic dyes and related compounds in *Escherichia coli. Microbiol. Sci.* **4:**125–127.
48. **Mimura, C. S., L. B. Eisenberg, and G. R. Jacobson.** 1984. Resolution of the phosphotransferase enzymes of *Streptococcus mutans:* purification and preliminary characterization of a heat-stable phosphocarrier protein. *Infect. Immun.* **44:**708–715.
49. **Mitchell, P., and J. Moyle.** 1959. Permeability of the envelopes of *Staphylococcus aureus* to some salts, amino acids, and non-electrolytes. *J. Gen. Microbiol.* **20:**434–441.
50. **Nikaido, H., and M. Vaara.** 1987. Outer membrane, p. 7–22. *In* F. C. Neidhardt, J. L. Ingraham, K. B. Low, B. Magasanik, M. Schaechter, and H. E. Umbarger (ed.) *Escherichia coli and Salmonella typhimurium: Cellular and Molecular Biology.* American Society for Microbiology, Washington, D.C.
51. **Postma, P. W., and J. W. Lengeler.** 1985. Phosphoenolpyruvate: carbohydrate phosphotransferase system of bacteria. *Microbiol. Rev.* **49:**232–269.
52. **Ramos, S., and H. R. Kaback.** 1977. The electrochemical proton gradient in *Escherichia coli* membrane vesicles. *Biochemistry* **16:**848–854.
53. **Ramos, S., S. Schuldiner, and H. R. Kaback.** 1976. The electrochemical gradient of protons and its relationship to active transport in *Escherichia coli* membrane vesicles. *Proc. Natl. Acad. Sci. USA* **73:**1892–1896.
54. **Reed, R. H., and A. E. Walsby.** 1985. Changes in turgor pressure in response to increases in external NaCl concentration in the gas-vacuolate cyanobacterium *Microcystis* sp. *Arch. Microbiol.* **143:**290–296.
55. **Reizer, J., A. Peterkofsky, and A. H. Romano.** 1988. Evidence for the presence of heat-stable protein (HPr) and ATP-dependent HPr kinase in heterofermentative lactobacilli lacking phosphoenolpyruvate:glycose phosphotransferase activity. *Proc. Natl. Acad. Sci. USA* **85:**2041–2045.
56. **Scherrer, R., and P. Gerhardt.** 1971. Molecular sieving by the *Bacillus megaterium* cell wall and protoplast. *J. Bacteriol.* **107:**718–735.
57. **Scholes, P., and P. Mitchell.** 1970. Acid-base titration across the plasma membrane of *Micrococcus denitrificans:* factors affecting the effective proton conductance and the respiratory rate. *J. Bioenerg.* **1:**61–72.
58. **Senior, A. E., J. A. Downie, G. B. Cox, F. Gibson, L. Langman, and D. R. H. Fayle.** 1979. The uncA gene codes for the α-subunit of the adenosine triphosphate of *Escherichia coli. Biochem. J.* **180:**103–109.
59. **Solioz, M., and P. Furst.** 1988. Purification of the ATP-ase of *Streptococcus faecalis. Methods Enzymol.* **157:**680–689.
60. **Stadelmann, E. J.** 1966. Evaluation of turgidity, plasmolysis, and diasmolysis of plant cells. *Methods Cell Physiol.* **2:**144–216.
61. **Thwaites, J. J., and N. H. Mendelson.** 1991. Mechanical behaviour of bacterial cell walls. *Adv. Microb. Physiol.* **32:**173–222.
62. **Vadeboncoeur, C.** 1984. Structure and properties of phosphoenolpyruvate:glucose phosphotransferase system of oral streptococci. *Can. J. Microbiol.* **30:**495–502.
63. **Vadeboncoeur, C., M. Proulx, and L. Trahan.** 1983. Purification of proteins similar to HPr and Enzyme I from the oral bacterium *Streptococcus salivarius.* Biochemical and immunochemical properties. *Can. J. Microbiol.* **29:**1694–1705.
64. **Whatmore, A. M., and R. H. Reed.** 1990. Determination of turgor pressure in *Bacillus subtilis:* a possible role for K^+ in turgor regulation. *J. Gen. Microbiol.* **136:**2521–2526.

Section V

SYSTEMATICS

Introduction to Systematics

NOEL R. KRIEG

Systematics in bacteriology is concerned with three interrelated topics: (1) **classification,** the arranging of bacteria with similar phenotypic or genetic characteristics into groups; (ii) **nomenclature,** the naming of bacteria according to internationally agreed-upon principles, rules, and recommendations which allow precise communications between bacteriologists concerning the various kinds of bacteria (12); and (iii) **identification,** the comparing of unknown organisms with bacteria that have already been classified, for the purpose of determining the identities or names of the unknown organisms. Systematics requires first learning as much as possible about the characteristics of bacteria and then deciding how the bacteria should be arranged or grouped. The resulting groups are then given names. Only after bacteria have been characterized and classified can reliable schemes be constructed for determining the identity of unknown organisms. In this way, systematics seeks to bring order from chaos and to organize the many kinds of bacteria into a coherent, understandable, and useful system.

Several levels or ranks are used in bacterial classification, but the basic unit is the **species** (16). In bacteriology, a species consists of a type strain together with all other strains that are deemed to be sufficiently similar to the type strain as to be included with it in the species. The **type strain** is a strain that has been designated as the permanent example of what is meant by the species, although it may not necessarily be the most typical or representative strain in the species. It is also the name bearer (nomenclatural type) of the species. All other strains being considered for inclusion in a species must be compared with the type strain of that species. Cultures of type strains can be purchased from various reference collections of bacteria (3). In the above definition of a bacterial species, it is the phrase "sufficiently similar" that causes most of the difficulties in the classification of bacteria, because what is deemed sufficiently similar by one person may not necessarily be considered so by another person. However, the more that is known about a group of bacteria, the more likely it is that various investigators can agree on a suitable classification scheme.

The historic way to characterize bacteria is to describe qualitatively as many phenotypic properties as possible, such as morphology, structure, cultivation, nutrition, biochemistry, metabolism, pathogenicity, antigenic properties, and ecology. Routine and special tests for determining many of the phenotypic characteristics of bacteria are described in Chapter 25. Many new and sophisticated methods have been developed for classifying bacteria on the basis of the chemical composition and molecular architecture of bacterial cells, i.e., the methods of **chemotaxonomy** (5). **Numerical taxonomy** methods are useful for quantifying similarities on the basis of phenotypic characteristics and for grouping strains into species; they can achieve the objectivity that is sometimes lacking when the bacteria are grouped according to intuitive judgments (2, 11).

Phenotypic similarities do not necessarily indicate **phylogenetic relationships** (relationships based on the ancestry of organisms). For instance, the members of the genus *Aquaspirillum*—aerobic freshwater spirilla—are similar phenotypically but diverse phylogenetically (10). In practical terms, the great advantage of a phylogenetic classification is that it provides stability. Stability in bacterial classification is highly desirable, considering the drastic rearrangements that bacterial classifications have undergone in the past. With a phylogenetic classification, new discoveries about the phenotypic characteristics of an organism would not affect its relatedness to other organisms.

Phylogenetic relationships among bacteria can be deduced through analysis of bacterial DNA. The initial use of DNA in bacterial classification was limited to determination of the overall nucleotide base composition (moles percent G+C) for various strains. This helped to resolve many taxonomic problems on the basis of the principle that strains having significantly different DNA base compositions do not belong in the same species. However, DNA base composition was only a very crude measure of bacterial relatedness. Two bacterial strains might have exactly the same G+C content and yet be unrelated. What was really needed was a much more precise means of comparing the DNA from one strain with that from other strains.

DNA hybridization techniques were able to provide this precision. These techniques allowed the entire genome of one bacterial strain to be compared with that of other strains in terms of nucleotide base sequence. These methods are described in Chapter 26. DNA hybridization is most useful at the species level of classification. Its practical value lies in the fact that bacterial strains usually tend to be either very closely related or not. This eliminates many of the problems posed by the continua that often occur between species that are defined only by their phenotypic characteristics. In fact, DNA-DNA duplexes do not even form if the base pair mismatches between two DNA molecules exceed 10 to 20%. A species that is based on DNA hybridization can usually be readily defined in phenotypic terms, because the strains in the species tend to be very similar to each other not only in genotype but also in phenotype. The species is presently the only bacteriological taxonomic unit that can be defined in phylogenetic terms (17). DNA hybridization is much less useful at the genus level or higher than it is at the species level of classification. This is because DNA hybridization values between different species are usually too low to provide much information about the relatedness among the species within a genus.

Analysis of the nucleotide base sequence in certain cistrons of bacterial DNA makes it possible to deduce broader phylogenetic relationships among bacteria. These cistrons have changed more slowly during evolution than the bulk of the genome, thereby allowing them to be used as molecular/evolutionary chronometers. In particular, 16S rRNA cistrons, present in all bacteria, have been widely used for this purpose. The various techniques for comparing the 16S rRNA cistrons of bacteria are described in Chapter 27.

Despite the extensive use of these techniques, there is still no general agreement on the definition of a genus or higher taxon in phylogenetic terms. Part of the difficulty is that phylogenetic parameters do not have the same meaning when applied to different bacterial groups (14). Another factor is that the traditional phenotypic features of bacteria such as cell shape, motility, and type of metabolism may not correlate well with phylogenetic groups. In other words, a bacterial group that is phylogenetically coherent may not be phenotypically coherent (see reference 6 for examples). In some instances the seemingly dissimilar features may be the result of not using the same characterization methods for all the members of the phylogenetic group. In other instances the features may result from preconceived notions about the characteristics of the organisms. For example, rRNA analyses of campylobacters and their relatives suggest that four species of anaerobic bacteria should be grouped with the campylobacters, which are microaerophilic (9). These results seemed puzzling until further phenotypic characterization studies showed that the four anaerobes were really microaerophiles capable of respiring with oxygen (4, 8). Although this resolved one important aspect of the systematics of this particular bacterial group, other difficulties remain. For instance, some of the species in the group are straight rods whereas others are vibrioid in shape. This illustrates that some of the traditional characteristics used in the past for classifying bacteria, such as cell shape, are not necessarily good indicators of phylogenetic relationships. Taxonomists may be able to find other, less traditional characteristics to define some phylogenetic groups in phenotypic terms, and serious attempts should be made to find such features. If this is not possible, practical considerations may have to take precedence over phylogeny in some cases (6). However, an integration of phenotypic and phylogenetic data is the most satisfactory overall approach to bacterial classification (14) and should be done whenever possible.

Bacterial classification is not the only beneficiary of molecular biology techniques: bacterial identification has also benefited, through the development of **nucleic acid probes.** As applied to bacterial identification, a probe is a piece of single-stranded DNA or RNA whose nucleotide base sequence is complementary to a sequence that has been determined to be unique to a particular bacterial species or group. The probe is labeled with either a radioisotope or a nonradioactive "reporter" molecule and mixed with nucleic acid suspected of containing the unique sequence. If the sequence is present in the suspect nucleic acid, the probe will hybridize with it to form a labeled (and thus detectable) heteroduplex. Probes can be used not only to identify a particular bacterial isolate definitively but also to detect the presence of the organism in natural samples such as water or soil. Because they can identify a bacterium definitively, probes may eventually be used extensively in bacteriology laboratories, especially if the methodology can be further simplified or even automated. The methodology and uses of probes in bacteriology are described in Chapter 28.

Where can one find information about the characteristics and relationships of a particular group of bacteria? One major source is *Bergey's Manual of Systematic Bacteriology* (7, 13, 15, 18). This is a comprehensive work that describes thousands of genera and species of bacteria and includes identification tables and keys. The taxonomic descriptions were written by hundreds of contributors, each of whom had extensive experience with a specific group of bacteria. Preparation of *Bergey's Manual* was begun in 1980, when knowledge of the phylogenetic relationships among bacteria was still very fragmentary. Consequently, the editors chose to rely on a largely practical phenotypic arrangement of bacteria at the genus level and higher, although phylogenetic information was provided whenever possible. This arrangement was intended to be a purely interim one until a comprehensive classification based on phylogeny could be prepared.

Another major source of information about specific bacterial groups is the second edition of *The Prokaryotes*, published in 1992 (1). In this work, bacterial groups are arranged along phylogenetic lines. Like *Bergey's Manual, The Prokaryotes* depends on a large number of contributors and includes many tables and keys useful for the identification of bacteria. Because the phylogenetic arrangement often resulted in phenotypically dissimilar organisms being placed next to one another, several initial chapters that describe various bacterial groups from a physiological or ecological viewpoint are included.

The methods to be described in this section of the book have been selected on the basis of their interest and usefulness for persons who have had little experience with bacterial systematics but who wish to cope with this area of bacteriology. Emphasis is given to the practical performance of the various methods rather than to the principles and theoretical considerations involved. However, general and specific references are provided at the end of each chapter for those who wish to pursue any topic further.

REFERENCES

1. **Balows, A., H. G. Trüper, M. Dworkin, W. Harder, and K.-H. Schleifer** (ed.). 1992. *The Prokaryotes. A Handbook on the Biology of Bacteria: Ecophysiology, Isolation, Identification, Applications,* 2nd ed. Springer-Verlag, New York.
2. **Colwell, R. R., and B. Austin.** 1981. Numerical taxonomy, p. 444–449. *In* P. Gerhardt, R. G. E. Murray, R. N. Costilow, E. W. Nester, W. A. Wood, N. R. Krieg, and G. B. Phillips (ed.), *Manual of Methods for General Bacteriology.* American Society for Microbiology, Washington, D.C.
3. **Gibbons, N. E., P. H. A. Sneath, and S. P. Lapage.** 1989. Reference collections of bacteria—the need and requirement for type strains, p. 2325–2328. *In* S.T. Williams, M. E. Sharpe, and J. G. Holt (ed.), *Bergey's Manual of Systematic*

Bacteriology, vol. 4. The Williams & Wilkins Co., Baltimore.

4. **Han, Y.-H., R. M. Smibert, and N. R. Krieg.** 1991. *Wolinella recta, Wolinella curva, Bacteroides ureolyticus,* and *Bacteroides gracilis* are microaerophile, not anaerobes. *Int. J. Syst. Bacteriol.* **41:**218–222.
5. **Jones, D., and N. R. Krieg.** 1989. Serology and chemotaxonomy, p. 2313–2316. *In* S. T. Williams, M. E. Sharpe, and J. G. Holt (ed.), *Bergey's Manual of Systematic Bacteriology,* vol. 4. The Williams & Wilkins Co., Baltimore.
6. **Krieg, N. R.** 1988. Bacterial classification: an overview. *Can. J. Microbiol.* **34:**536–540.
7. **Krieg, N. R., and J. G. Holt (ed.).** 1984. *Bergey's Manual of Systematic Bacteriology,* vol 1. The Williams & Wilkins Co., Baltimore.
8. **Ohta, H., and J. C. Gottschal.** 1988. Microaerophilic growth of *Wolinella recta* ATCC 33238. *FEMS Microbiol. Ecol.* **53:**79–86.
9. **Paster, B. J., and F. M. Dewhurst.** 1988. Phylogeny of campylobacters, wolinellas, *Bacteroides gracilis,* and *Bacteroides ureolyticus* by 16S ribosomal ribonucleic acid sequencing. *Int. J. Syst. Bacteriol.* **38:**56–62.
10. **Pot, B., M. Gillis, and J. De Ley.** 1992. The genus *Aquaspirillum,* p. 2569–2582. *In* A. Balows, H. G. Trüper, M. Dworkin, W. Harder, and K.-H. Schleifer (ed.), *The Prokaryotes. A Handbook on the Biology of Bacteria: Ecophysiology, Isolation, Identification, Applications,* 2nd ed. Springer-Verlag, New York.
11. **Sneath, P. H. A.** 1989. Numerical taxonomy, p. 2303–2305. *In* S. T. Williams, M. E. Sharpe, and J. G. Holt (ed.), *Bergey's Manual of Systematic Bacteriology,* vol. 4. The Williams & Wilkins Co., Baltimore.
12. **Sneath, P. H. A.** 1989. Bacterial nomenclature, p. 2317–2321. *In* S. T. Williams, M. E. Sharpe, and J. G. Holt (ed.), *Bergey's Manual of Systematic Bacteriology,* vol 4. The Williams & Wilkins Co., Baltimore.
13. **Sneath, P. H. A., N. S. Mair, M. E. Sharpe, and J. G. Holt (ed.).** 1986. *Bergey's Manual of Systematic Bacteriology,* vol 2. The Williams & Wilkins Co., Baltimore.
14. **Stackebrandt, E.** 1992. Unifying phylogeny and phenotypic diversity, p. 19–47, *In* A. Balows, H. G. Trüper, M. Dworkin, W. Harder, and K.-H. Schleifer (ed.), *The Prokaryotes. A Handbook on the Biology of Bacteria: Ecophysiology, Isolation, Identification, Applications,* 2nd ed. Springer-Verlag, New York.
15. **Staley, J. T., M. P. Bryant, N. Pfenning, and J. G. Holt (ed.)** 1989. *Bergey's Manual of Systematic Bacteriology,* vol 3. The Williams & Wilkins Co., Baltimore.
16. **Staley, J. T., and N. R. Krieg.** 1989. Bacterial classification of procaryotic organisms: an overview, p. 2299–2302. *In* S. T. Williams, M. E. Sharpe, and J. G. Holt (ed.), *Bergey's Manual of Systematic Bacteriology,* vol 4. The Williams & Wilkins Co., Baltimore.
17. **Wayne, L. G., D. J. Brenner, R. R. Colwell, P. A. D. Grimont, O. Kandler, M. I. Krichevsky, L. H. Moore, W. E. C. Moore, R. G. E. Murray, E. Stackebrandt, M. P. Starr, and H. G. Trüper.** 1987. Report of the ad hoc committee on reconciliation of approaches to bacterial systematics. *Int. J. Syst. Bacteriol.* **37:**463–464.
18. **Williams, S. T., M. E. Sharpe, and J. G. Holt (ed.).** 1989. *Bergey's Manual of Systematic Bacteriology,* vol 4. The Williams & Wilkins Co., Baltimore.

Chapter 25

Phenotypic Characterization

ROBERT M. SMIBERT and NOEL R. KRIEG

During a long-term study of a group of strains, stock cultures may be subject to genetic variation. Therefore, the cultures of an organism that are used near the end of a study may not contain quite the same organism as that used earlier. If cultures are properly preserved at the beginning of the study, one can always have a source of the organism with its original characteristics. See Chapter 12 for various preservation methods.

In identifying bacteria, certain general characteristics are of primary importance for determining the major group to which the new isolate is most likely to belong. The investigator should determine whether the organism is phototrophic, chemoautotrophic, or chemoheterotrophic and whether it is aerobic, anaerobic, microaerophilic, or facultative. Similarly, certain morphological features should be determined: the Gram reaction and the cell shape (e.g., rod, coccus, vibrioid, helical, or other). The presence of special morphological features (e.g., endospores, exospores, true branching, acid fastness, sheaths, cysts, stalks or other appendages, fruiting bodies, gliding motility, or budding division) is valuable. The arrangement of the cells, e.g., cocci occurring in chains or in irregular clusters, may be helpful. Three physiological tests are of primary importance: the oxidase test, the catalase test, and the determination of whether sugars are oxidized or fermented. The presence of special physiological characteristics (e.g., nitrogen-fixing ability, degradation of cellulose, bioluminescence, requirement for seawater for growth, production of pigments, or a heme requirement) may narrow the field of possibilities considerably.

It is desirable to use an orderly approach to identification, based on common sense and on the use only of tests that are pertinent. Avoid a "shotgun" approach, in which all sorts of tests are performed in the desperate hope that some of them may be helpful. The table of contents and the major keys given in *Bergey's Manual of Systematic Bacteriology* (6, 10, 11, 13) will be of great assistance, as will the general descriptions of the various genera and species of bacteria. After the initial possibilities have become more restricted, consult the detailed keys and tables from various sources (many of which are listed in references 1 to 13) for the identification of genera and species. In the final steps of identification, compare the new strain with either the type strain of the suspected species or with some other named strain whose inclusion in the species has been firmly established.

In many cases, especially with pathogenic bacteria, one may be able to make a rapid presumptive identification of a genus or species by the use of antisera. However, such tests are often not absolutely specific, and additional physiological tests are usually required for confirmation of the identification. For example, agglutination of gram-negative enteric rods by polyvalent *Salmonella* serum is presumptive evidence for a *Salmonella* strain, but this must always be confirmed by a number of biochemical tests.

In performing phenotypic characterization tests, the organisms used for inoculation of test media should be from fresh transfers and in good physiological condition. Old stock cultures, which may contain mostly dead or dying cells, are not satisfactory. Incubate the inoculated test media under conditions that are optimum for the organisms or for the characteristic being tested, e.g., optimum temperature, pH, gaseous conditions, and ionic conditions. Furthermore, apply characterization methods not only to the organism to be tested but also to strains known to be positive and negative for the characteristic, as a check on the reliability of the reagents and media.

At the end of this chapter is listed a selection of general references (1–13) that provide identification schemes for various groups of bacteria, further information on methodology, and additional characterization methods.

25.1. ROUTINE TESTS

25.1.1. Acetamide Hydrolysis

From a dilute suspension, streak the surface of a slant of an acetamide agar slant (section 25.3.2). Incubate for up to 7 days. A red or magenta color signals a positive test; no color change signals a negative test.

Examples: positive, *Pseudomonas aeruginosa;* negative, *Pseudomonas fluorescens.*

25.1.2. Acid-Fast Reaction

See Chapter 2.3.6 for two staining methods. A positive reaction indicates the presence of a member of the genus *Mycobacterium* (which is strongly acid fast), *Nocardia,* or *Rhodococcus* (the last two genera are partially acid fast) (10).

25.1.3. Acidification of Carbohydrates

25.1.3.1. Method 1

Incorporate a pH indicator into the medium during its preparation. For a list of the most commonly used indicators and their properties, see Table 8. In cases when an indicator may be toxic to the bacteria or may be reduced during their growth, add drops of indicator (0.02 to 0.04% alcoholic solution) to the culture after it has grown to maximum turbidity.

Comments. A relatively high concentration of sugars, e.g., 1 to 2% (wt/vol), is usually added to the basal medium. On the other hand, the peptone concentration in the basal medium should be kept as low as possible while still allowing good growth, to minimize buffering and alkali formation. This is particularly important when characterizing aerobic organisms, which produce comparatively little acidity when oxidizing sugars. For instance, in determining the acidification of sugar media by various *Aquaspirillum* species, the peptone level in the medium should not exceed 0.2% and a pH indicator, such as phenol red, that will change color with only slight acidity should be used (59).

In general, sugar-containing media to be used for testing acidification can be sterilized by autoclaving at 115 to 118°C for 10 to 15 min. However, heating is known to alter some sugars, especially in alkaline solution, or to convert one sugar to another (77). Consequently, when using xylose, lactose, sucrose, arabinose, trehalose, rhamnose, or salicin, it is prudent to sterilize the complete sugar-containing medium by filtration or to sterilize a 5 or 10% stock solution of the sugar separately by filtration and then add appropriate amounts aseptically to portions of the autoclaved, cooled basal medium (2).

Some organisms, such as *Neisseria* spp., show acidity best when grown in a semisolid medium.

Care should be taken to ensure that the basal medium to which the carbohydrates are added does not contain fermentable or oxidizable sugars, and control media without any added carbohydrates should always be included during testing of cultures. Moreover, basal media should not contain the oxidizable salts of organic acids, such as malate, pyruvate, or succinate, because oxidation of these substances by the test organism will cause an alkaline reaction that may obscure acidification due to carbohydrate oxidation or fermentation.

With cultures grown in defined media with an ammonium salt or nitrate as the major or sole nitrogen source, utilization of the ammonium salt will cause a decrease in pH and utilization of the nitrate will cause an increase in pH (76). This may lead to incorrect conclusions about acidification of sugar media. For instance, almost all of the acidity produced in cultures of the aerobe *Azospirillum lipoferum* grown in a defined fructose-containing medium is due to utilization of ammonium sulfate and not to production of organic acids from the fructose (45).

25.1.3.2. Method 2

To provide a more quantitative measurement of acidification, use a pH meter to measure the pH of the culture after it has grown to maximum turbidity. Use a long, thin, combination pH electrode which can be inserted directly into a culture tube. Affix a flat rubber washer with a diameter larger than that of the culture tube to the upper portion of the electrode to prevent the electrode tip from striking the bottom of the culture tube when the electrode is inserted. When determining the pH of a number of cultures of pathogenic bacteria, dip the electrode into a beaker of distilled water between cultures. This water should later be autoclaved. When finished with the electrode, rinse it with a suitable disinfectant (e.g., 3% hydrogen peroxide or 70% ethanol).

For anaerobes cultured in prereduced PY broth (section 25.3.30) containing carbohydrates, the oxygen-free carbon dioxide used to purge the tube during their inoculation will usually lower the pH of the medium to 6.2 to 6.4. Consequently, pH values for PY carbohydrate broth cultures are usually interpreted as acidification when they are 5.5 to 6.0 (weak acid) or below 5.5 (strong acid). For PY broth cultures containing arabinose, ribose, or xylose, a pH of 5.7 or below usually indicates acidification, because the sterile media purged with carbon dioxide may have a pH of 5.9 after 1 to 2 days of incubation (4). In any case, the pH of cultures in PY broth lacking any carbohydrate should be determined as a control, because acids may be formed from peptones in certain cases.

25.1.4. Agar Corrosion and Digestion

Colonies of "agar-corroding" or "agar-pitting" bacteria appear as if they are in a shallow pit in the surface of the agar medium. The term "corrosion" is used rather than "digestion" because the agar seems not to be softened or liquefied. Examples are *Eikenella corrodens, Bacteroides ureolyticus, Kingella kingae,* and various *Moraxella* species.

Several species of gliding bacteria are able to digest agar by means of agarolytic enzymes. For example, the colonies of *Nannocystis exedens* may deeply corrode the surface of the medium and may even penetrate the medium, producing holes or tunnels in the agar. *Chondromyces* and *Polyangium* species can pit, erode, and penetrate agar media. *Vibrio alginolyticus* and some *Cytophaga* species can soften or totally liquefy agar gels, whereas others, such as *Cytophaga fermentans,* merely produce gelase fields and craters on agar plates.

25.1.5. Ammonia from Arginine

Inoculate arginine broth (see sections 25.3.3 and 25.3.4). Also inoculate a control lacking arginine. After incubation for 2 to 3 days, test samples of the culture in a spot plate with Nessler's solution (section 25.4.11). A positive test is indicated by a yellow or orange color compared with the control.

Examples: positive, *Enterococcus faecalis;* negative, *Streptococcus salivarius.*

25.1.6. Arginine Dihydrolase Activity

25.1.6.1. Method 1

The following method is widely used to distinguish among members of the family *Enterobacteriaceae.* Inoculate Møller's broth base (section 25.3.25) supplemented with 1% L-arginine monohydrochloride (or 2% of the DL form). Also inoculate a control lacking arginine. After inoculation, overlay the broth with 10 mm of sterile mineral (paraffin) oil. Examine daily for 4 days. A positive test is indicated by a violet or reddish-violet color due to increase in pH. Weak reactions are bluish gray.

Examples: positive, *Enterobacter cloacae;* negative, *Proteus vulgaris.*

25.1.6.2. Method 2

The following method is suitable for a wide variety of facultatively anaerobic bacteria. Inoculate Thornley's semisolid medium containing arginine (see section 25.3.35) and also a control lacking arginine. After stab inoculation, seal the medium with a layer of sterile melted petrolatum and incubate. A positive test is indicated by a change from yellow-orange to red within 7 days due to increase in pH.

25.1.6.3. Method 3 (115)

The following method is suitable for aerobic and facultative bacteria and is based on the direct measurement of the disappearance of arginine. Make a dense suspension of the bacteria in 0.033 M phosphate buffer (pH 6.8) (see section 25.4.18). Purge 4 ml of the suspension by bubbling nitrogen through the suspension for several minutes, and add 1 ml of 0.001 M L-arginine monohydrochloride. After purging again, stopper the tubes, incubate them for 2 h, and heat them at 100°C for 15 min. After removing the cells by centrifugation, determine the concentration of arginine in the supernatant by the method of Rosenberg et al. (102) as follows. Mix 1 ml of the sample with 1 ml of 3 N NaOH, 2 ml of developing solution (see section 25.4.4), and 6 ml of water; read the tubes at 30 min against a blank prepared without arginine by using a colorimeter equipped with a green filter (540 nm); and compare the readings with those obtained with an uninoculated control containing arginine. A positive test is indicated by the disappearance of some or all of the arginine.

25.1.7. Aromatic Ring Cleavage (96)

Grow the culture in chemically defined medium containing an aromatic substrate such as 0.1% sodium *p*-hydroxybenzoate as the carbon source. (Also grow it on a yeast extract agar without the aromatic substrate to determine whether the enzymes are constitutive or inducible.) Scrape growth from the agar, and suspend it in 2 ml of 0.02 M Tris buffer (pH 8.0). Shake the tubes with 0.5 ml of toluene, and add 0.2 ml of either 0.1 M catechol or 0.1 M sodium protocatechuate. A bright yellow color appearing in a few minutes indicates *meta* cleavage. If no color appears, shake the tubes for 1 h at 30°C. Add solid $(NH_4)_2SO_4$ to saturation, adjust the pH to approximately 10 by adding 2 drops of 5 N ammonium hydroxide, and then add 1 drop of freshly prepared 25% (wt/vol) sodium nitroprusside (nitroferricyanide). A deep purple color indicates *ortho* cleavage.

Examples: meta cleavage, *Pseudomonas acidovorans; ortho* cleavage, *Pseudomonas fluorescens;* negative, *Escherichia coli.*

25.1.8. Arylsulfatase Activity

Aseptically add a sufficient amount of a filter-sterilized 0.08 M solution of tripotassium phenolphthalein disulfate to a sterile liquid medium to give a final concentration of 0.001 M. Media containing methionine as the sole source of sulfur are best for synthesis of arylsulfatase; sulfur sources such as sulfate, sulfite, thiosulfate, or cysteine may repress synthesis (14, 78). Inoculate the medium, incubate it for 7 days, and add 1 N NaOH or 1 M Na_2CO_3 drop by drop. A positive test is indicated by a faint pink to light red color. An uninoculated control similarly treated should remain colorless.

Examples: positive, *Proteus rettgeri;* negative, *Chromobacterium violaceum.*

25.1.9. Bacitracin Sensitivity

Use sterile commercially available differentiation (not sensitivity) discs (Difco Laboratories, Detroit, Mich.; BBL Microbiology Systems, Cockeysville, Md.) or sterile filter paper discs impregnated with 0.04 U of bacitracin (Sigma Chemical Co., St. Louis, MO.). Place a disc on an inoculated blood agar plate (see section 25.3.8), and incubate for 24 h. A positive test is indicated by a zone of growth inhibition around the disc.

Examples: positive, *Streptococcus pyogenes;* negative, other beta-hemolytic streptococci.

25.1.10. Bile Solubility

Centrifuge a 24-h culture grown in 10 ml of Todd-Hewitt broth (see section 25.3.36), and discard the supernatant into a flask of disinfectant. Suspend the cells in 0.5 ml of 0.067 M phosphate buffer (pH 7.0) (see section 25.4.18). Add 0.5 ml of 10% sodium deoxycholate, and incubate the mixture at 37°C for 15 to 30 min. If the suspension becomes clear, the test is positive.

Examples: positive, *Streptococcus pneumoniae;* negative, other alpha-hemolytic streptococci.

25.1.11. Bile Tolerance

25.1.11.1. Method 1

Streak a plate of bile agar (section 25.3.7). Compare the growth on the plates with that occurring in the absence of the oxgall.

Examples: positive (10 and 40% bile), *Enterococcus faecalis;* positive (10 but not 40% bile), *Streptococcus salivarius;* negative (neither 10 nor 40% bile), *Streptococcus dysgalactiae.*

25.1.11.2. Method 2

Use method 2 for anaerobes (4). Inoculate PY broth (section 25.3.30) containing 2% oxgall and 1% glucose. Inoculate tubes with a Pasteur pipette while flushing them with oxygen-free carbon dioxide. Stopper the tubes, and incubate. Observe the growth response, and compare it with that in the absence of the oxgall.

Examples: growth, *Prevotella oralis;* no growth, *Prevotella melaninogenica.*

25.1.12. CAMP Test (β-Hemolysis Accentuation)

The CAMP test (the acronym derives from the originating authors' names) has been used mainly in the identification of *Streptococcus agalactiae* and of *Listeria monocytogenes* and *Listeria seeligeri,* but it might be useful for other bacteria as well.

Make a single streak of a strain of β-hemolysin-producing strain of *Staphylococcus aureus* (e.g., ATCC 25923) across the middle of a plate of blood agar (containing sheep erythrocytes that have been washed to remove natural antibodies to hemolysins). Make a single streak of each test organism perpendicular to but not quite touching the *S. aureus* streak (about 3 to 4 mm away). For a positive control, make a similar streak of a reference strain of *S. agalaciae, L. monocytogenes,* or *L. seeligeri;* for a negative control, make a streak of a reference strain of *Streptococcus pyogenes* or *Listeria innocua* (18). Incubate the plate aerobically overnight. A positive test is indicated by the appearance of a crescent- or arrowhead-shaped zone of enhanced hemolysis at the juncture of the test organism and the *S. aureus* streak.

The test has also been used in the identification of *Listeria ivanovii;* in this case, however, a strain of *Rhodococcus equi* (e.g., NCTC 1621) is substituted for the *S. aureus. L. ivanovii* gives a positive reaction, whereas *L. monocytogenes* and *L. seeligeri* give a negative reaction (18).

25.1.13. Carbon Dioxide Requirement

Some aerobic, microaerophilic, or facultatively anaerobic chemoheterotrophs are characterized by a requirement for 3 to 15% CO_2 (vol/vol) in the gaseous atmosphere in the culture vessel. Such bacteria are called **capnophilic bacteria.** The CO_2 is required to initiate growth; after the organisms have begun to grow in a liquid or semisolid medium, they usually can provide themselves with CO_2 from their own metabolism. However, when growing in direct contact with the gaseous atmosphere, as on the surface of a solid medium, they may continue to require a supply of exogenous CO_2.

Some examples of capnophilic bacteria are as follows:

- *Capnocytophaga ochracea*—facultative anaerobe that requires 5% CO_2 (55).
- *Campylobacter jejuni* and other *Campylobacter* species—microaerophiles that require 3 to 5% CO_2 (108).
- *Neisseria gonorrhoeae*—aerobe that requires 3 to 10% CO_2 (123).
- *Brucella ovis* and some strains of *Brucella abortus*—aerobes that require 5 to 10% CO_2 (28).
- Some strains of *Streptococcus pneumoniae, Streptococcus mutans,* and *Streptococcus milleri*—facultative anaerobes that require 5% CO_2 (49).

There are some bacteria that do not have an absolute requirement for exogenous CO_2 but whose growth is nevertheless markedly stimulated by CO_2. For instance, *Neisseria meningitidis* grows much better when 5 to 8% CO_2 is provided (123). Similarly, many streptococci grow better in the presence of an enriched CO_2 atmosphere (49).

25.1.14. Casein Hydrolysis

Combine sterile (autoclaved) skim milk at 50°C with an equal volume of double-strength nutrient agar (see section 25.3.27) or other carbohydrate-free agar medium at 50 to 55°C. Incubate streaked plates for up to 14 days, and look for clear zones surrounding the growth. Confirm by flooding the plates with 10% HCl. *Note:* Acid production from the lactose in the milk may in rare instances inhibit the casein hydrolysis and necessitate prior dialysis of the skim milk.

Examples: positive, *Bacillus polymyxa;* negative, *Bacillus macerans.*

25.1.15. Catalase Activity

25.1.15.1. Method 1

Inoculate a nutrient agar slant (see section 25.3.27) or other medium lacking blood. After incubation, trickle 1 ml of 3% hydrogen peroxide down the slant. Examine immediately and after 5 min for the evolution of bubbles, which indicates a positive test. Alternatively, add a few drops of 3% peroxide to colonies on a plate or to the heavy growth that may occur at or near the surface of semisolid media. For broth cultures, add 0.5 ml of 3% hydrogen peroxide to 0.5 ml of culture and observe for continuous bubbling. *Note:* Media containing blood must not be used for catalase tests, because blood contains catalase activity unless the blood has been heated (see section 25.1.15.3). Also note that some bacteria (e.g., certain lactic acid organisms) make a nonheme "pseudocatalase" in media containing little or no glucose (127). Pseudocatalase can be prevented by incorporating 1% glucose into the medium. When anaerobes are tested for catalase activity, it is important to expose the cultures to air for 30 min before adding peroxide (4).

Examples: positive, *Staphylococcus epidermidis;* negative, *Lactococcus lactis.*

25.1.15.2. Method 2

The following method eliminates problems of penetration that might occur when peroxide is added to colonies on a plate or to growth on a slant. Scrape the growth from a slant or plate with a nonmetallic instrument, and suspend it in a drop of 3% hydrogen peroxide on a slide. Examine immediately and at 5 min for bubbles, either macroscopically or with a low-power microscope. If the catalase activity is weak, a coverslip placed over the wet mount can help to capture the bubbles.

25.1.15.3. Method 3

Use the following method for certain bacteria that can make catalase only if grown on a heme-containing medium, e.g., certain lactic acid bacteria (127). To sterile blood agar base (section 25.3.34) containing 1% glucose to inhibit pseudocatalase formation, add 5% (vol/vol) of a 1:1 mixture of defibrinated blood (section 25.3.8) and sterile water. Heat the medium at 100°C for 15 min to inactivate blood catalase, cool to 45 to 50°C, and dispense into plates. Test growth on the plates directly with 3% hydrogen peroxide, or use method 2.

25.1.16. Cellulase Activity

25.1.16.1. Method 1 (57)

The following method of preparing cellulose gives the best form of native cellulose for testing cellulolytic activity; for other methods, see references 8 and 9. Incorporate finely divided cellulose into appropriate carbohydrate-free agar media. To prepare the cellulose, wet-grind 3% (wt/vol) Whatman no. 1 filter paper in a pebble mill as follows. Place 30 g of the paper (torn into small pieces) into a porcelain jar (ca. 4-liter capacity) with 1 liter of water; add enough flint pebbles (porcelain balls are not as satisfactory) that the liquid just covers them; roll the jars for 24 h at 74 rpm or until a very fine state of suspension has been achieved and the suspension has become viscous; stop before the viscosity decreases and copper-reducing substances appear. (To test for the latter, remove 1 ml of the suspension and test as described in section 25.1.39 with 1 ml of Benedict's reagent [section 25.4.1].) A positive test is indicated by the appearance of clear zones around the colonies on the cellulose agar. Long periods of incubation may be required.

Examples: positive, *Cellulomonas* species; negative, *Escherichia coli.*

25.1.16.2. Method 2

Method 2 is less sensitive than Method 1. Place a strip of Whatman no. 1 filter paper in tubes of carbohydrate-free broth before autoclaving. A portion of the strip should extend above the level of the broth. A positive test is indicated by partial or complete disintegration of the paper strip during growth of the culture.

25.1.17. Citrate Utilization

25.1.17.1. Method 1

Prepare a dilute suspension of the organisms in sterile water or saline. Make a single streak up a slant of Simmons' citrate agar (section 25.3.33). Incubate for up to 7 days. A blue color indicates the utilization of citrate as the sole carbon source (i.e., a positive test).

Examples: positive, *Enterobacter aerogenes;* negative, *Escherichia coli.*

25.1.17.2. Method 2

From a dilute suspension, streak the entire surface of a slant of Christensen citrate agar (section 25.3.12). Incubate for up to 7 days. A positive test is indicated by a red or magenta color. *Note:* A positive reaction indicates that citrate is used but not necessarily as the sole carbon source; i.e., an organism could give a positive reaction on Christensen agar and a negative reaction on Simmons' citrate agar.

25.1.18. Coagulase Activity

Mix one loopful of growth from an agar slant, 0.1 ml of broth culture, or a single colony from an agar plate with 0.5 ml of undiluted rabbit plasma or plasma diluted 1:4 with saline. Incubate at 37°C, and examine at 4 and 24 h. A positive test is indicated by solid clot or a loose clot suspended in the plasma. Granular or ropy formations are inconclusive.

Examples: positive, *Staphylococcus aureus;* negative, *Staphylococcus epidermidis.*

25.1.19. Coccoid Bodies

Use phase-contrast microscopy to examine cultures that are held static (not shaken during growth) for up to

4 weeks. Coccoid bodies may occur as early as 2 or 3 days. A positive test is indicated by a predominance of round refractile forms which have a cell diameter greater than that of the original cells and which lack a thickened cell wall. Many of the forms have a discrete, dark peripheral region of cytoplasm in an otherwise empty-appearing cell. Coccoid bodies occur in certain spirilla, vibrios, and campylobacters.

Examples: positive, *Aquaspirillum itersonii;* negative, *Aquaspirillum serpens.*

25.1.20. Colonies

Measure the colony diameter in millimeters, describe the pigmentation, and describe the form, elevation, and margin as indicated in Fig. 1. Low-power microscopy may be necessary for observation of the margin. Also indicate whether the colonies are smooth (shiny glistening surface), rough (dull, bumpy, granular, or matte surface), or mucoid (slimy or gummy appearance). Record the opacity of the colonies (transparent, translucent, or opaque) and their texture when tested with a needle: butyrous (butter-like texture), viscous (gummy), or dry (brittle or powdery). Describe the colonies from both young and old cultures. The medium, age of the culture, gaseous conditions, exposure to illumination, and other culture conditions may affect the colony characteristics.

25.1.21. Cysts and Microcysts

Examine cultures daily by phase-contrast microscopy. Initially rod-shaped cells become spherical or ovoid with thickened walls in older cultures. These forms may be optically dense or may be refractile. They do not have the heat resistance of endospores (section 25.1.71), except for the cystlike forms made by certain *Nocardia* species, but they are extremely resistant to desiccation. In *Azotobacter* strains the forms are termed cysts. In the order *Myxobacteriales* they are termed microcysts; here, the term cyst refers to the sporangium, if any, which contains the microcysts. In *Nocardia* strains the forms are termed microcysts or chlamydospores.

The following methods can be used to compare the desiccation resistance of cyst-enriched cultures of an organism to cultures lacking cysts. See also Chapter 2.3.8 for methods of staining cysts.

25.1.21.1. Method 1 (67)

Spread suspensions of the cells onto plates of media containing 2% agar, and incubate until the agar has dried completely to a thin film. Cut the agar film into pieces with sterile scissors, and transfer each piece aseptically to a 30-ml screw-cap vial half full of silica gel desiccant. Each vial should contain a sterile paper disc or funnel to separate the agar film from the desiccant. Store vials in the dark at room temperature. At various periods, remove a piece of agar film from a vial, place it upon fresh agar culture medium, and incubate for up to 7 days to see whether growth occurs.

25.1.21.2. Method 2 (103)

Prepare a suspension of the cells with sterile distilled water, and then prepare a series of 10-fold dilutions of this suspension. Transfer 100-μl portions of each dilution to sterile 1.5-ml Eppendorf microcentrifuge tubes. Place the open tubes in a sterile petri dish, and incubate at 30°C until dry. Immediately add 100 μl of sterile distilled water to some of the tubes, then agitate vigorously to suspend the cells, and plate onto an appropriate agar medium to determine the initial colony count. After various storage periods, repeat the hydration and plating procedure with the remaining tubes to estimate the percent survival.

25.1.22. Dye Tolerance

The ability to grow in the presence of low concentrations of various dyes added to culture media has been used to characterize and differentiate bacteria, especially those that have few other readily determinable

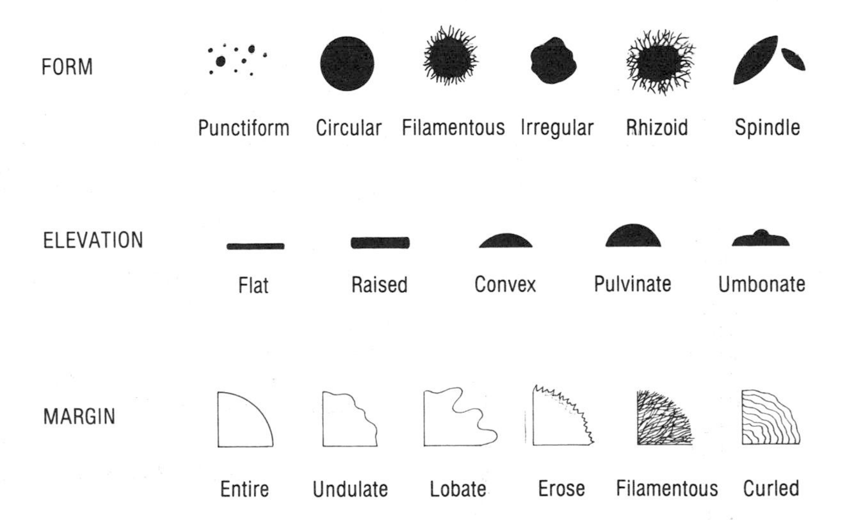

FIG. 1. Diagram illustrating the various forms, elevations, and margins of bacterial colonies. (Adapted by permission of the McGraw-Hill Book Co. from reference 97.)

phenotypic characteristics. For example, *Wolinella curva* can be differentiated from *Wolinella recta* by its ability to grow in the presence of Janus green, basic fuchsin, methyl orange, indulin scarlet, and safranin (117, 118).

Some dyes that have been used in dye tolerance tests for characterizing bacteria are listed in Table 1. In some instances, dye tolerance has been determined by soaking paper strips in a dye solution, placing the strips on inoculated agar plates, and noting whether growth occurs around the strips after incubation. For instance, taxonomic studies of *Bordetella* species have used paper strips soaked in basic fuchsin (0.0005%), methyl violet (0.00025%), pyronin G (0.000125%), safranin O (0.000125%), and thionine (0.000125%) (60).

25.1.23. Esculin Hydrolysis

Grow cultures in broth or agar media supplemented with 0.01% esculin and 0.05% ferric citrate. A positive test occurs when the medium becomes brownish black as a result of a reaction between 6,7-dihydroxycoumarin and Fe^{3+}.

Examples: positive, *Enterococcus faecalis;* negative, *Streptococcus mitis.*

25.1.24. Esterase Activity (see also Lipase Activity [section 25.1.44])

The hydrolysis of ester linkages between fatty acids and 4-methylumbelliferone results in liberation of the latter compound, which is fluorescent when exposed to UV light at 362 to 365 nm. Commercially available substrates include 4-methylumbelliferyl (4-MU) acetate, 4-MU propionate, 4-MU butyrate, 4-MU heptanoate, 4-MU nonanoate, 4-MU oleate, 4-MU elaidate, and 4-MU palmitate.

25.1.24.1. Method 1 (122)

A rapid spot test for detection of butyrate esterase can be done as follows. Dissolve 100 mg of 4-MU butyrate (Sigma) in 10 ml of dimethyl sulfoxide (DMSO) and 100 μl of Triton X-100 detergent (Sigma). Dilute this stock solution 1:10 in citrate buffer (0.1 M citric acid, 0.05 M $Na_2HPO_4 \cdot 12H_2O$[pH 5.0]). This diluted reagent can be stored in 250-μl portions at −70°C for at least 1 month. To test a culture, place a few drops of the reagent onto a piece of filter paper. Streak two or three colonies from a culture of the test organism onto the paper. After 30 s, expose the paper to a UV source (362 to 365 nm, or the long wavelength of a Wood's lamp). A positive test is indicated by light blue fluorescence. A negative test has no fluorescence.

Examples: positive, *Branhamella catarrhalis, Staphylococcus aureus;* negative, *Neisseria sicca, N. lactamica.*

25.1.24.2. Method 2 (46)

The following quantitative method for esterases has been applied to mycobacteria but can be adapted for use with other bacteria. Prepare 0.04 M stock solutions of 4-MU acetate, 4-MU propionate, 4-MU butyrate, and 4-MU heptanoate in DMSO. Prepare 0.02 M stock solutions of 4-MU oleate, 4-MU elaidate, and 4-MU palmitate in DMSO. Store at −20°C until needed. Immediately before use, add 0.2 to 9.8 ml of 0.2 M phosphate buffer. The pH of the buffer depends on the esterase and the species. For mycobacterial esterases, the optimum pH is 8.0 to 8.5; however, spontaneous hydrolysis of the substrates is reported to occur at this pH. Therefore, buffer with a pH of 7.3 has been used instead (46). Reaction mixtures contain 0.1 ml of a suspension of the bacteria to be tested, 0.9 ml of phosphate buffer, and 1.0 ml of the buffered 4-MU ester. (Also prepare a blank containing 0.1 ml of distilled water instead of bacteria.) Incubate the mixtures at 37°C for the desired period, and then stop the reaction by adding 1.0 ml of glycine buffer (0.2 M solution of glycine, adjusted to pH 10.5 with 1.0 N NaOH). Measure the fluorescence in quartz cuvettes at 37°C with a spectrofluorimeter, using an excitation wavelength of 362 nm and an emission wavelength of 450 nm.

Table 1. Some dyes that have been used in dye tolerance tests for characterizing bacteria

Dye	% Concn (wt/vol)	Reference
Alizarin red	0.032	118
Azure II	0.0025	117
Basic fuchsin	0.005	117
	0.002	28
Brilliant green	0.001	35
Crystal violet	0.0005	117
	0.0001	35
Indulin scarlet	0.05	117
Janus green	0.01	117
Malachite green	0.001	35
Methyl orange	0.032	117
Methylene blue	0.1	84
	0.3	85
Safranin	0.05	117
Thionine	0.002	28

25.1.25. Flagella

Use flagella staining with light microscopy (see Chapter 2.3.11) or electron microscopy (see Chapters 3.2.2 and 3.2.6.1). Agitate the bacteria as little as possible during their preparation, because flagella may be easily broken off the cells. Flagellar stains with light microscopy are occasionally misleading when applied to polar-flagellated bacteria, because a fascicle of several flagella sometimes seems to be a single flagellum (130). Culture conditions may be important; e.g., some bacteria, such as *Vibrio parahaemolyticus,* form a single polar flagellum when grown in liquid media but also form lateral flagella of shorter wavelength when grown on solid media (105). Some bacteria, such as *Listeria monocytogenes,* may not form flagella at certain temperatures but do form them at lower temperatures (63). In some instances glucose-containing media inhibit the formation of flagella (15).

25.1.26. Fluorescent Pigment

Make a single streak across a plate of King medium B (see section 25.3.6). Examine the plates with the covers removed at 24, 48, and 72 h under UV light (below 260 nm). A shortwave lamp used for examination of mineral specimens is suitable. Plates should not be reincubated after examination, because the UV illumination may be bactericidal. A positive test is indicated by a fluorescent zone in the agar surrounding the growth. Fluorescent pseudomonads produce a yellow-green pigment; certain *Azotobacter*, *Azomonas*, and *Beijerinckia* species may form green or white fluorescent pigments.

Examples: positive, *Pseudomonas fluorescens;* negative, *Escherichia coli.*

25.1.27. Formate-Fumarate Requirement

Use the formate-fumarate requirement test to distinguish certain bacteria that, when grown anaerobically, oxidize formate and concomitantly reduce fumarate to succinate. Prepare formate-fumarate (F-F) stock solution (section 25.4.6), and add 1 drop of the solution per ml of prereduced PY broth (section 25.3.30) when inoculating the medium under oxygen-free carbon dioxide. Compare the growth response with that occurring in medium lacking the F-F. If the growth is greatly stimulated by the F-F, the test is positive. In some cases, growth may not occur at all unless the supplement is added.

Examples: positive, *Bacteroides ureolyticus*; negative, *Prevotella oralis.*

25.1.28. Gas Production from Sugars

25.1.28.1. Method 1

Before autoclaving the sugar broth, place a small vial (ca. 10 by 75 mm) in an inverted position in the tube. After autoclaving, the vial will become completely filled with medium. For thermolabile sugars add filter-sterilized stock solutions aseptically after the base medium has been autoclaved and cooled; allow the tubes to stand for a day or so to allow diffusion of the sugar into the inverted vials. Inoculate the medium and incubate. A positive test is indicated by accumulation of gas within the vial. *Note:* If media are stored in a refrigerator before use and then are inoculated and incubated, dissolved gases in the medium may be liberated and accumulate in the vial, giving a false-positive reaction. Sterile controls should be used to detect this occurrence. Also, if the organism forms only small amounts of carbon dioxide, detection may be difficult because of the great solubility of carbon dioxide and because of rapid diffusion into the air. Methods 2 and 3 are designed to eliminate this difficulty.

25.1.28.2. Method 2

Method 2 is the same as method 1 but with the following change. After inoculation add ca. 1 ml of sterile mineral (paraffin) oil to prevent diffusion of carbon dioxide into the air. Alternatively, use sugar-containing agar media cooled to 45°C after autoclaving (or, for thermolabile sugars, add a filter-sterilized stock solution of the sugar to the base agar medium, which has been melted and cooled to 45°C). Inoculate the tubes, and allow them to solidify. Overlay with ca. 6 mm of sterile nonnutrient 2% agar (i.e., agar prepared only with water) and then a thin layer of sterile mineral oil. Incubate the cultures. A positive test is indicated by cracks in the medium and upward movement of the agar overlay.

25.1.28.3. Method 3 (4)

Use method 3 only for anaerobes. Cool melted prereduced PY agar (section 25.3.29) containing 1% glucose to 45°C, and inoculate it with 2 to 3 drops of culture. Cover the tubes with sterile aluminum foil (do not stopper them), allow the agar to solidify, and incubate the tubes. A positive test is indicated by formation of gas bubbles in the medium or breaks in the agar.

25.1.29. Gelatin Hydrolysis

25.1.29.1. Method 1

Use broth supplemented with 12% gelatin.

(A) Incubate at 20 to 22°C for up to 6 weeks. A positive test is indicated by liquefaction of at least a portion of the medium. Evaporation should be prevented during prolonged incubation.

Examples: positive, *Proteus vulgaris;* negative, *Escherichia coli.*

(B) Incubate at the optimum growth temperature for the organism if greater than 22°C; then chill in a refrigerator along with an uninoculated control. If the test culture fails to solidify, the organism has hydrolyzed the gelatin. If the test culture does solidify, remove the culture and the control, and incubate in a tilted position at room temperature or in a 30°C incubator for ca. 30 min. If the culture liquefies before the control, a weakly positive reaction is indicated. However, if the test culture takes as long as the control to liquefy, this indicates a negative reaction.

(C) Method 1C is the same as method 1B but with 4% gelatin. This is a more sensitive method than method 1B.

25.1.29.2. Method 2

Method 2 is more sensitive than method 1A or 1B. Streak plates of agar medium supplemented with 0.4% gelatin. Incubate them at the optimum temperature for the organism. Flood the plates with gelatin-precipitating reagent (15% $HgCl_2$ in 20% [vol/vol] concentrated HCl). A positive test is indicated by clear zones around colonies.

25.1.29.3. Method 3

Method 3 is sensitive and provides a more readily observable reaction. Culture the organism in a tube of broth containing a charcoal-gelatin disc (section 25.4.3). If the charcoal granules become dispersed in the medium, the test is positive.

25.1.30. Glycosidase Activity

25.1.30.1. Method 1

The following method depends on the use of chromogenic substrates—in particular, glycosides in which the sugar moiety is linked by a glycosidic bond to *o*-nitrophenol or *p*-nitrophenol. For example, the substrate for β-galactosidase is *o*-nitrophenyl β-D-galactopyranoside (ONPG). Numerous other substrates are commercially available; some examples may be found in Table 2. For some enzymes, either *para*- or *ortho*-nitrophenyl substrates are available.

Dissolve 60 mg of the chromogenic substrate in 10 ml of 0.1 M sodium phosphate buffer (pH 7.0), and sterilize by filtration. Aseptically add 1 volume of this solution to 4 volumes of sterile peptone broth (0.1% peptone, 0.05% NaCl), and dispense 0.5-portions of the mixture into small sterile capped tubes. The mixture can be stored for up to 1 month in a refrigerator or up to 6 months at −20°C. It must not show a yellow coloration before use.

Culture the bacterium to be tested on an agar-solidified medium, preferably in the presence of the appropriate glycoside substrate (e.g., lactose for β-D-galactosidase; maltose for α-D-glucosidase), because some glycosidases are inducible. Remove some of the growth from the surface of the agar, and suspend the cells to a dense concentration in a tube of the peptone broth-chromogenic substrate. Incubate the tube in a water bath at 37°C, and examine it at 1 h and at intervals up to 24 h. A control lacking the cell suspension should also be prepared and incubated. A positive test is indicated by a yellow color.

Note: Merely because cells hydrolyze a chromogenic substrate does not necessarily mean that the cells contain the glycosidase: the enzyme may instead be a phosphoglycosidase, which acts only on the phosphorylated form of the glycoside. For instance, *Staphylococcus aureus* cells produce *o*-nitrophenol when incubated with ONPG; however, they do this by transporting the ONPG into the cell by means of a membrane-bound phosphoenolpyruvate phosphotransferase system which phosphorylates the ONPG; the phospho-ONPG is then hydrolyzed by a cytoplasmic phospho-β-galactosidase, yielding galactose-6-phosphate and *o*-nitrophenol. Therefore *S. aureus* is ONPG positive but β-galactosidase negative. Also note that the fact that a cell suspension hydrolyzes a chromogenic substrate (e.g., ONPG) does not necessarily mean that the cells can ferment or grow on the natural substrate (e.g., lactose). This is because some bacterial strains are cryptic: they possess the hydrolytic enzyme but lack an effective transport system for the substrate. For example, many strains of *Salmonella arizonae* can produce detectable amounts of yellow *o*-nitrophenol from ONPG but are unable to ferment or grow on lactose.

Examples (for β-D-galactosidase): positive, *Escherichia coli;* negative, *Proteus vulgaris.*

25.1.30.2. Method 2 (73)

Method 2 is a qualitative spot method that depends on the use of fluorogenic substrates—in particular, glycosides in which the sugar moiety is linked by a glycosidic bond to 4-methylumbelliferone. For example, the substrate for β-galactosidase is 4-MU-β-D-galactopyranoside. Numerous other substrates are commercially available; a few examples are found in Table 3.

Dissolve 15 μmol of each substrate (use 5 μmol for 4-MU-β-D-galactopyranoside) in 0.2 ml of DMSO. Bring the volume of each solution up to 10 ml of Dulbecco phosphate buffered saline (0.8% NaCL, 0.02% KCl, 0.115% Na_2HPO_4, 0.02% KH_2PO_4 [pH 7.3]). Remove a sample of the colony to be tested with a glass rod, and rub it vigorously onto a Whatman no. 1 qualitative filter paper. Then cover the bacterial smear with 2 drops (0.067 ml) of substrate. As controls, (i) prepare another bacterial smear nearby on the filter paper and cover it with 2 drops of the DMSO in which the substrate is dissolved, and (ii) place 2 drops of substrate nearby on the filter paper (no bacteria). Incubate for 30 min. Cover the spots on the filter paper with 2 drops of 0.1 N NaOH to increase the intensity of fluorescence. Examine in a dark room under UV light (362 to 365 nm, or the long wavelength of a Wood's lamp). A positive test is indicated by light blue fluorescence. A negative test has no fluorescence.

Examples (for β-D-galactosidase): positive, *Escherichia coli;* negative, *Proteus vulgaris.*

Table 2. Chromogenic substrates for glycosidase activity

Enzyme	Substrate
α-D-Acetylgalactosaminidase	*o*-Nitrophenyl *N*-acetyl-α-D-galactoasaminide
β-D-Acetylgalactosaminidase	*p*-Nitrophenyl *N*-acetyl-β-D-galactoasaminide
α-D-Acetylglucosaminidase	*o*-Nitrophenyl *N*-acetyl-α-D-glucosaminide
β-D-Acetylglucosaminidase	*p*-Nitrophenyl *N*-acetyl-β-D-glucosaminide
α-L-Arabinofuranosidase	*p*-Nitrophenyl α-L-arabinofuranoside
α-L-Arabinopyranosidase	*p*-Nitrophenyl α-L-arabinopyranoside
α-L-Fucosidase	*p*-Nitrophenyl α-L-fucopyranoside
β-D-Fucosidase	*p*-Nitrophenyl β-D-fucopyranoside
β-L-Fucosidase	*p*-Nitrophenyl β-L-fucopyranoside
α-D-Galactosidase	*p*-Nitrophenyl α-D-galactopyranoside
β-D-Galactosidase	*o*-Nitrophenyl β-D-galactopyranoside
β-D-Galacturonidase	*p*-Nitrophenyl β-D-galacturonide
α-D-Glucosidase	*p*-Nitrophenyl α-D-glucopyranoside
β-D-Glucosidase	*p*-Nitrophenyl β-D-glucopyranoside
α-D-Mannosidase	*p*-Nitrophenyl α-D-mannopyranoside
β-D-Mannosidase	*p*-Nitrophenyl β-D-mannopyranoside
α-L-Rhamnosidase	*p*-Nitrophenyl α-L-rhamnopyranoside
β-D-Xylosidase	*o*-Nitrophenyl β-D-xylopyranoside

Table 3. Fluorogenic substrates for glycosidase activity

Enzyme	Substrate
β-D-Glucosidase	4-MU-β-D-glucopyranoside
β-D-Galactosidase	4-MU-β-D-galactopyranoside
β-D-Glucuronidase	4-MU-β-D-glucuronide
N-Acetyl-β-D-glucosaminidase	4-MU-2-acetamido-2-deoxy-β-D-glucopyranoside
N-Acetyl-β-D-galactosaminidase	4-MU-2-acetamido-2-deoxy-β-D-galactopyranoside
α-D-Mannosidase	4-MU-α-D-mannopyranoside
α-L-Arabinosidase	4-MU-α-L-arabinopyranoside

25.1.30.3. Method 3

The following quantitative method for glycosidases has been applied to mycobacteria but can be adapted for use with other bacteria (46). Prepare 0.04 M stock solutions of 4-MU glycosides in DMSO. Store at −20°C until needed. Immediately before using, add 0.2 ml to 0.8 ml of 0.2 M phosphate buffer. The pH of the buffer depends on the glycosidase and the bacterial species; for mycobacterial glycosidases, optimal pH values occur within the range 5.5 to 7.0. Reaction mixtures contain 0.1 ml of a suspension of the bacteria to be tested, 0.1 ml of the buffered 4-MU substrate, and 0.8 ml of 0.2 M phosphate buffer. (Also prepare a blank containing 0.1 ml distilled water instead of bacteria.) Incubate the mixtures at 37°C for the desired period, and then stop the reaction by adding 1.0 ml of sodium glycinate buffer (0.2 M solution of glycine, adjusted to pH 10.5 with 1.0 N NaOH). Measure fluorescence in quartz cuvettes at 37°C with a spectrofluorimeter, using an excitation wavelength of 362 nm and an emission wavelength of 450 nm.

25.1.31. Gram Reaction

See Chapter 2.3.5 for two Gram staining methods.

25.1.32. Hemolysis

25.1.32.1. Method 1

Streak a plate of blood agar (section 25.3.8). Incubate it at 37°C for 24 to 48 h. If beta-hemolysis is occurring, a clear, colorless zone surrounds the colonies and the erythrocytes in the zone are completely lysed. If alpha-hemolysis is occurring, an indistinct zone of partial destruction of the erythrocytes surrounds the colonies, often accompanied by a greenish to brownish discoloration of the medium. If Alpha prime hemolysis is occurring, a small halo of intact or partially lysed erythrocytes is present adjacent to the colony, with a zone of complete hemolysis extending farther out into the medium; the colony can be confused with a beta-hemolytic colony when examined only macroscopically. *Note:* Some bacteria, such as certain staphylococci, produce beta-hemolysis only if the inoculated plates are first incubated for 48 h at 37°C and then chilled in a refrigerator for 1 h ("hot-cold hemolysis").

Sharper zones of hemolysis may be obtained by the following modification. Prepare a plate of blood agar base (without added blood), and allow it to solidify. Pour an overlay of 10 ml of blood agar onto the blood agar base, and allow it to solidify. Streak the surface of the blood agar overlay with the test culture.

25.1.32.2. Method 2

Streak a plate of blood agar base (without added blood). Pour an overlay of 10 ml of blood agar (at 45 to 50°C) onto the streaked plate. Hemolysis (alpha or beta) may be more pronounced than in method 1 because the colonies are submerged, offering some protection for hemolytic enzymes (hemolysins) that may be oxygen labile (such as streptolysin O).

25.1.33. Hippurate Hydrolysis

25.1.33.1. Method 1

Prepare a broth medium supplemented with 1.0% sodium hippurate. Mark the level in each tube after autoclaving. Inoculate the medium, and incubate it along with a sterile control tube for 4 days. If evaporation occurs, restore the volumes in the tubes to their original levels with distilled water at the end of the incubation period. Place 1.0 ml of the sterile control broth in a small tube, and add 0.10-ml increments of ferric chloride solution (12 g of $FeCl_3 \cdot 6H_2O$ in 100 ml of 2% HCl). Agitate the tube after each addition. At first a precipitate of iron hippurate will form, but on further addition of ferric chloride this will dissolve, leaving a clear solution. Determine the smallest total amount of ferric chloride that will allow the initial precipitate to dissolve. Centrifuge the broth culture, add 1.0 ml of the supernatant to another small tube, and then add an amount of ferric chloride solution equivalent to that determined previously for the control. A positive test is indicated by formation of a heavy permanent precipitate of iron benzoate.

Examples: positive, *Streptococcus agalactiae;* negative, *Streptococcus pyogenes.*

25.1.33.2. Method 2 (58)

Use the following method for a more rapid test. Prepare a 1% solution of sodium hippurate in distilled water, and dispense 0.4-ml portions into tubes. Store the corked tubes in a freezer at −20°C until needed. To test an organism, thaw a tube and mix a large loopful of the growth from the surface of a solid medium into the hippurate solution to form a very cloudy suspension. Incubate the tube in a heating block or water bath at 37°C for 2 h. After incubation, add ca. 0.2 ml of ninhydrin reagent (3.5 g of ninhydrin in 100 ml of a 1:1 [vol/vol] mixture of acetone and butanol). Do not shake the tube (40). Continue incubation at 37°C for 10 min. A positive test is indicated by formation of a deep purple color as a result of the glycine which is formed from hippurate hydrolysis. A negative test shows no color reaction or only a faint tinge of purple color. Controls containing known positive and negative organisms should be included.

25.1.34. Hydrogen Sulfide Production from Thiosulfate

25.1.34.1. Method 1

Use a deep tube of a solid or semisolid agar medium supplemented with sodium thiosulfate (0.008 to 0.03%) and 0.025% ferric ammonium citrate or 0.015% ferrous sulfate. Sulfide production usually occurs best under anaerobic conditions; therefore the medium should be inoculated by stabbing. A positive test is indicated by blackening of the medium during growth, as a result of formation of iron sulfide (see also section 25.1.73).

Examples: positive, *Proteus vulgaris;* negative, *Escherichia coli.*

25.1.34.2. Method 2 (101)

The following method has been used with microaerophiles that do not produce H_2S anaerobically. Use triple-sugar iron agar slants (section 25.3.37) that have been freshly prepared so that there is water of syneresis at the junction of the slant with the butt. (If the water of syneresis is not present, a small amount of sterile water can be added to the slant.) Inoculate the entire surface of the slant, and incubate it under microaerobic conditions (e.g., 6% O_2, 5% CO_2, 89% N_2) for up to 7 days. A positive test is indicated by blackening of the liquid at the slant-butt junction. A negative test shows no blackening.

Examples: positive, *Campylobacter coli;* negative, *Campylobacter jejuni.*

25.1.35. Hydrogen Sulfide Production from Cysteine

Cut filter paper into strips 5 by 20 mm, and soak them in a solution of 5% lead acetate. Autoclave the strips in cotton-stoppered test tubes, and dry them in a drying oven. Inoculate a tube of liquid medium containing 0.05% cysteine. Before incubation, suspend a lead acetate strip in the tube so that the end of the strip is approximately 1.3 cm above the level of the medium; the top of the strip can be held by the cotton plug or other tube closure. The strip must not be allowed to come into contact with the medium. Prepare uninoculated controls in a similar manner. A positive test is indicated by blackening (as a result of formation of lead sulfide) of the lower portion of the strip after incubation for several days.

Examples: positive, *Aeromonas hydrophila;* negative, *Aeromonas caviae.*

25.1.36. Indole Production

25.1.36.1. Method 1

Inoculate a liquid medium that is free of carbohydrate, nitrate, and nitrite and contains a source of tryptophan (e.g., 1% tryptone broth or medium supplemented with 0.1% L-tryptophan). Test cultures of different ages by the following method. Add 1 ml of xylene (or water-saturated toluene), agitate the culture vigorously to extract the indole, and then allow the xylene to form a layer at the top of the broth. With the tube held in a slanted position, trickle 0.5 ml of Ehrlich's reagent (section 25.4.5) down the wall of the tube to form a layer between the xylene and the broth. Allow the tube to stand for 5 min. A position test is indicated by the formation of a red ring just below the xylene layer. A tube of uninoculated medium should also be tested as a control. *Note:* If the medium in which the organisms are growing contains a carbohydrate, synthesis of the enzyme tryptophanase, responsible for indole formation, may be repressed (20). Also, nitrite at levels of 0.75 mg/ml or higher may block detection of indole (111); therefore, the presence of nitrate in the medium may lead to a false-negative test if the organism can reduce the nitrate to nitrite.

Examples: positive, *Escherichia coli;* negative, *Enterobacter aerogenes.*

25.1.36.2. Method 2

Method 2 is not as sensitive as method 1; however, method 2 eliminates the necessity for using flammable and toxic solvents. Use Kovacs' reagent (section 25.4.7) rather than Ehrlich's reagent, and omit extraction with xylene. Add 0.5 ml of the reagent to the broth culture, and shake the tube gently. A positive test is indicated by a red color.

25.1.36.3. Method 3

Cut filter paper into strips, and soak them in a warm saturated solution of oxalic acid. Crystals of the acid will form on the strips upon cooling. Dry the strips thoroughly (sterilization by heat seems unnecessary), and insert them aseptically into the inoculated culture tube before incubating the culture. The strip should not come into contact with the medium but rather should be bent so that it presses against the wall of the tube and remains near the mouth when the cotton stopper or screw cap is replaced. If the test is positive, the paper strip becomes pink or red during growth of the culture.

25.1.37. Indoxyl Acetate Hydrolysis (80)

Add 50 μl of a freshly prepared 10% (wt/vol) solution of indoxyl acetate in acetone to a blank paper disc (0.25-in. [ca. 0.64-cm]-diameter concentration discs [Difco]). Allow the disc to air dry. Store dried discs in a tightly capped amber or opaque bottle at 4°C in the presence of silica gel desiccant; the shelf life of the discs is at least 1 year.

Place a colony or a large loopful of growth from an agar medium onto a test disc, and immediately add 1 drop of sterile distilled water. Incubate the mixture for 5 to 10 min. A positive test is indicated by development of a dark blue color. For a negative test, the blue color is absent.

Examples: positive, *Campylobacter jejuni;* negative, *Campylobacter lari.*

25.1.38. Iron-Porphyrin Compounds (33)

The following test detects the presence of iron-porphyrin compounds, such as cytochromes, catalase, and nitrate reductase. Dissolve 1.0 g of benzidine · 2HCl in 20 ml of glacial acetic acid, add 30 ml of distilled water, heat the solution gently, cool, and add 50 ml of 95% ethanol. Prepare a 5% solution of hydrogen peroxide by diluting a 30% stock solution. To perform the test, flood the bacterial growth on an agar plate with the benzidine solution and then immediately add an approximately equal volume of the 5% hydrogen peroxide solution. The benzidine solution must come into contact with all of the microbial growth before the hydrogen peroxide is added. A positive test is indicated by prompt development of a blue-green to deep blue coloration.

Examples: positive, *Staphylococcus aureus;* negative, *Lactococcus lactis.*

CAUTION: Benzidine is a carcinogen and is rapidly absorbed through the skin. On ingestion it may produce nausea, vomiting, and liver and kidney damage. Consequently, all precautions pertinent to the use, handling, and subsequent disposal of carcinogens should be observed for this compound.

25.1.39. 2-Ketogluconate from Glucose Oxidation

Grow the culture in a medium such as Haynes' broth (section 25.3.16) containing 4% potassium gluconate (sterilized by filtration and added aseptically). After various periods of incubation, remove 1-ml samples and add 1 ml of Benedict's reagent (section 25.4.1). Heat the mixture in a boiling-water bath for 10 min, and cool rapidly. A positive test is indicated by a yellowish-brown precipitate of cuprous oxide.

Examples: positive, *Pseudomonas aeruginosa;* negative, *Escherichia coli.*

25.1.40. 3-Ketolactose from Lactose Oxidation (17)

Grow the organism initially on slants of medium containing 1% yeast extract, 2% glucose, 2% $CaCO_3$, and 1.5 to 2.0% agar. With a loop needle, scrape growth from the slant and deposit it as a small heap, approximately 5 mm in diameter, on a plate of lactose agar (1% lactose, 0.1% yeast extract, 2% agar). Four to six different strains may be tested on the same plate. After incubation for 1 to 2 days, flood the plate with a shallow layer of Benedict's reagent (section 25.4.1). A positive test is indicated by formation of a yellow ring of cuprous oxide around the growth, reaching a maximum in 2 h (approximately 2 to 3 cm).

Examples: positive, *Agrobacterium tumefaciens* and *A. radiobacter;* all other bacteria so far studied are negative.

25.1.41. Lactic Acid Optical Rotation (4, 24)

To 10 ml of the culture, and also to 10 ml of uninoculated medium, add 0.2 ml of 50% (vol/vol) H_2SO_4. Use 1 ml of the acidified culture to determine the concentration of lactic acid (milliequivalents per 100 ml) by gas chromatography (see section 25.2.1). This concentration must be at least 1 meq/100 ml to determine optical rotation by the method below.

See section 25.4.8 for preparation of perchloric acid solution, buffer, lactic dehydrogenase solution, and NAD solution.

Prepare the L-(+)-lactic acid working standard as follows. Add 0.0296 g of L-(+)-lactic acid (calcium salt) to 5.0 ml of acidified uninoculated medium. Add 5.0 ml of distilled water. Add 1.0 ml of this solution to 1.0 ml of perchloric acid solution. Centrifuge and save the supernatant, which is the working standard [= 1.0 μmol of L-(+)-lactic acid per 0.1 ml].

To 1.0 ml of acidified culture, add 1.0 ml of distilled water and 2.0 ml of perchloric acid solution. Centrifuge, and save the supernatant, which is the sample to be tested for optical rotation.

To 1.0 ml of acidified uninoculated medium, add 1.0 ml of distilled water and 2.0 ml of perchloric acid solution. Centrifuge, and save the supernatant, which is the blank stock solution.

Use seven glass tubes, 12 by 75 mm. Add 0.1 ml of the blank stock solution to tube 1. Add 0.02, 0.04, 0.06, 0.08, and 0.10 ml of the working lactic acid standard to tubes 2 through 6, respectively. Make up the volume in each tube to 0.10 ml by adding 0.08, 0.06, 0.04, 0.02, and 0.00 ml of blank stock solution, respectively. To tube 7, add 0.10 ml of the sample to be tested. Add 3.0 ml of buffer to each of the seven tubes, followed by 0.03 ml of lactic dehydrogenase solution. Then add 0.1 ml of NAD solution, cover each tube with Parafilm, and invert gently several times to mix (do not shake the tubes). Incubate at 30°C for 60 min, being careful to start timing from the time the NAD was added to the first tube.

Read the absorbance of the contents of each tube. Use a wavelength of 340 nm if possible. If the instrument is not capable of reading at this wavelength, use a wavelength as near to 340 nm as possible; for example, if a Spectronic 20 spectrophotometer (Bausch & Lomb, Inc., Rochester, N.Y.) is used, set the wavelength to 366 nm. Use tube 1 as the blank to set the instrument to 0.00 absorbance. Read the absorbance values, being careful to use the same sequence as that used for addition of the NAD. Use the same cuvette for all readings; pour out the contents after making a reading, rinse out the cuvette, and fill it with the next solution to be read.

Plot micromoles of L-(+)-lactic acid (0.0, 0.2, 0.4, 0.6, 0.8, and 1.0 μmol, respectively, for tubes 1 through 6) versus absorbance on linear graph paper. Draw the line of best fit. By comparison with this standard curve, determine the micromoles of L-(+)-lactic acid in the test sample (tube 7). Multiply this value by 4 to obtain milliequivalent of L-(+)-lactic acid per 100 ml of culture. Calculate the percent L-(+)-lactic acid by the following formula:

$$\frac{\text{meq of L-(+)-lactic acid (from enzyme reactions) per 100 ml}}{\text{meq of total lactic acid (from gas chromatography) per 100 ml}} \times 100$$

25.1.42. β-Lactamase Activity (91)

The following method uses the chromogenic cephalosporin substrate nitrocefin (obtainable from Glaxo Laboratories, Greenford, England, or from BBL). Dissolve 10 mg of nitrocefin powder in 1 ml of DMSO. Dilute the solution 1:20 with 0.1 M phosphate buffer (pH 7.0) to a concentration of 500 μg/ml. The solution is yellow to light orange and is stable for many weeks at 4 to 10°C. For use, add some of the solution to the well of a microtiter plate or small test tube. Prepare a heavy suspension of the organism to be tested, and mix it with the nitrocefin solution. A positive test is indicated by development of a red color, usually within 10 to 30 min, although incubation for as long as 6 h is needed in some instances.

25.1.43. Lecithinase Activity

Streak a plate of egg yolk agar (section 25.3.15) to obtain well-isolated colonies. A positive test is indicated by a cloudy (opaque) zone within the medium around the colony (see also section 25.1.44).

Examples: positive, *Bacillus cereus, Clostridium perfringens;* negative, *Bacillus subtilis, Clostridium butyricum.*

25.1.44. Lipase Activity

25.1.44.1. Method 1 (4)

Streak a plate of egg yolk agar (section 25.3.15). After incubation, remove the petri dish cover and examine the growth closely under oblique illumination. A positive test is indicated by an oily, iridescent sheen or pearly layer over the colony and on the surface of the surrounding agar.

Examples: positive, *Clostridium sporogenes;* negative, *Clostridium butyricum.*

25.1.44.2. Method 2

Use the following method for palmitic, stearic, and oleic acid esters (106). Use an agar medium supplemented with 0.01% $CaCl_2 \cdot H_2O$. Sterilize Tween 40 (palmitic acid ester), Tween 60 (stearic acid ester), or Tween 80 (oleic acid ester) by autoclaving for 20 min at 121°C. Add sterile Tween to the molten agar medium at 45 to 50°C to give a final concentration of 1% (vol/vol). Shake the medium until the Tween is completely dissolved. Dispense into petri dishes, allow to solidify, and streak the plates. If the test is positive, an opaque halo forms around the colonies. Microscopically, the halo consists of crystals of calcium soaps.

Examples: positive (Tween 80), *Pseudomonas aeruginosa;* negative, *Pseudomonas putida.*

25.1.44.3. Method 3

Use the following method for fats of various types (8). Prepare fat of the desired type stained with Nile blue sulfate (section 25.4.12). Add 1 ml of melted fat emulsion to 20 ml of an agar growth medium which has been melted and cooled to 45 to 50°C. Mix well, and dispense into petri dishes. Allow to solidify, and streak the plates. A positive test is indicated by formation of a blue halo around the colonies.

25.1.45. Lysine Decarboxylase Activity

See arginine dihydrolase method 1 (section 25.1.6). Substitute 1% L-lysine dihydrochloride (or 2% of the DL form) for the arginine.

Examples: positive, *Serratia marcescens;* negative, *Proteus vulgaris.*

25.1.46. Lysozyme Resistance (2, 75)

The following test has been used to differentiate various aerobic actinomycetes. Inoculate several bits of a colony of the isolate to be tested into a tube of basal medium without lysozyme (control) and a tube of basal medium containing lysozyme (section 25.3.22). Incubate for up to 7 days at room temperature. A positive test (i.e., resistance to lysozyme) is indicated by good growth in the presence and absence of lysozyme. A negative test (i.e., susceptibility to lysozyme) is indicated by good growth only in the medium lacking lysozyme.

Examples: positive, *Nocardia asteroides;* negative, *Streptomyces griseus.*

25.1.47. Malonate Utilization

Inoculate a tube of malonate broth (section 25.3.23) from a young agar slant or broth culture. Incubate it at 37°C for 48 h. A positive test is indicated by a change from green to deep blue.

Examples: positive, *Klebsiella pneumoniae;* negative, *Escherichia coli.*

25.1.48. Meat Digestion by Anaerobes (4)

Inoculate prereduced chopped-meat broth (section 25.3.11) while purging the tube with oxygen-free carbon dioxide. Incubate for up to 21 days. If the test is positive, meat particles disintegrate, leaving a fluffy powder at the bottom of the tube.

Examples: positive, *Clostridium sporogenes;* negative, *Clostridium butyricum.*

25.1.49. Methyl Red Reaction

Inoculate a tube of MRVP broth (section 25.3.26). Incubate at 30°C for 5 days (in many cases incubation at 37°C for 48 h may also be satisfactory). Add 5 to 6 drops of methyl red solution (dissolve 0.1 g of methyl red in 300 ml of 95% ethanol, and make up to 500 ml with distilled water). A positive test is indicated by a bright-red color, indicating a pH of 4.2 or less. A negative test gives a yellow or orange color, and a weakly positive test gives a red-orange color.

Examples: positive, *Escherichia coli;* negative, *Enterobacter aerogenes.*

25.1.50. Milk Reactions

25.1.50.1. Method 1

Inoculate a tube of litmus milk (section 25.3.21). Reactions of cultures are as follows. Acid reaction, pink; alkaline reaction, blue; reduction of litmus, milk appears white due to discoloration of the litmus (usually occurs in the bottom portion of tube); acid curd, a firm pink clot which does not retract and is soluble in alkali;

rennet curd, a soft curd which retracts and expresses a clear grayish fluid (whey) and is insoluble in alkali; Peptonization (proteolysis), the milk becomes translucent as a result of hydrolysis of the casein, which often begins at the top of the medium; also, the medium frequently becomes alkaline.

25.1.50.2. Method 2 (4)

Use the following method for anaerobes. Inoculate a tube of prereduced milk medium (section 25.3.24) while the tube is being purged with oxygen-free carbon dioxide. Incubate the tube for up to 3 weeks. Check for curd and/or peptonization.

25.1.51. Motility, Swimming

Prepare wet mounts from both cultures that are young (e.g., still in the exponential phase), which are grown at two temperatures (e.g., 37 and 20°C). In some instances glucose-containing media may inhibit the formation of flagella (15). For methods of light-microscopic observation of motility, see Chapter 2.2.1.1. Examine wet mounts immediately after they are prepared. For anaerobes, the central portion of the mount is best for examination. For aerobes, areas near air bubbles or near the edge of the coverslip are best. Not all the cells in the mount need be motile. For a bacterial strain to be considered motile, at least one cell should be seen to alter its location in relation to at least two other cells (8). *Note:* Motility must be distinguished from Brownian motion (random "jiggling about" of cells) and from the passive motion of cells being carried about by currents under the coverslip. Anaerobes may also be tested by drawing up some of the broth culture into a capillary tube and immediately sealing both ends of the capillary in a narrow oxygen flame (29). Examine the capillary with a high-power dry objective. Some bacteria, such as the members of the orders *Myxobacteriales* and *Cytophagales,* lack flagella but exhibit "gliding motility" when in contact with a solid surface. For methods of detection and observation of gliding motility, see Chapter 8.1.5.4 and 8.1.5.5.

25.1.52. Motility, Twitching (53)

Twitching motility is a type of motility in which cells on an agar surface exhibit small, intermittent jerks covering only short distances and often changing the direction of movement, which is not regularly related to the long axis of the cell. The cells move predominantly singly, although slower-moving aggregates occur. Sometimes a small number of cells lying together may be seen to make a jump, in any direction, in a body. The maximum speed of cells is about 1 to 5 μm/min. Twitching motility is independent of flagella but has been associated with the presence of pili. The phenomenon has been reported in relatively few bacteria, e.g., species of *Neisseria, Moraxella, Acinetobacter, Kingella,* and *Eikenella*; these species do not exhibit either swimming or gliding motility. Most authors study twitching motility by microscopy of cultures growing on thin agar plates poor in nutrients. Use of a coverslip may prevent cell movement because of adherence of the cells to the glass. The presence of a thin film of water at the agar surface, as with incubation in a humidified atmosphere, facilitates twitching motility.

When streaked in a line across a plate of agar, cultures of a twitching strain often exhibit spreading zones that vary in width, appearance, and distribution along the sides of the central streak of growth. The spreading may occur along the entire streak or as a varying number of fan-shaped zones that may be unevenly distributed.

Twitching motility also has been reported in at least one species of gliding bacteria, i.e., *Agitococcus lubricus,* which glides across agar surfaces and exhibits an intense twitching motility when colonies are covered with a drop of water and a coverslip (41).

25.1.53. Nalidixic Acid Susceptibility (4)

Susceptibility or resistance to nalidixic acid is widely used for differentiation of *Campylobacter* species. Streak a plate of an appropriate medium (e.g., Brucella agar), and on the portion of the plate with the heaviest inoculum place a sterile paper disc impregnated with 30 μg of nalidixic acid. Such discs can be obtained commercially (e.g., NA30 Sensi-Disc [BBL]) in a convenient dispenser. After incubation, organisms producing any size zone of growth inhibition around the disk are considered to be susceptible.

Examples: susceptible, *Campylobacter jejuni;* resistant, *Campylobacter fetus.*

25.1.54. Nitrate Reduction and Denitrification

See section 25.4.13 for preparation of reagents and caution about their carcinogenic properties.

25.1.54.1. Method 1 (87)

The following method is quantitative for nitrite production from nitrate. Inoculate a suitable broth medium supplemented with 0.1% KNO_3 and 0.17% agar (the latter to give a semisolid consistency and promote semianaerobic conditions). Gently mix the inoculum with the medium to distribute it throughout the tube. Examine the cultures periodically during incubation for gas bubble formation, which is presumptive evidence for denitrification. Control cultures grown without nitrate should not exhibit gas and should not give a positive test for nitrite for the procedure given below. To detect the ability of the organisms to reduce nitrate to nitrate, remove 0.1 ml samples from cultures of various ages, and add 2 ml of test reagent (section 25.4.13). Add water to make the final volume up to 4 ml, and allow the color to develop for 15 min. Estimate the intensity of the purple-red color visually on a scale from 0 to 5. The intensity of the color can also be estimated by using a spectrophotometer at 540 nm. Accumulation of nitrite with increasing age of the culture indicates the reduction of nitrate to nitrite, whereas initial formation of nitrite followed by a decrease in its level indicates a further ability to reduce the nitrite. If no red color develops, add up to 5 mg of zinc powder per ml of the mixture. If the nitrate in the medium has not been reduced, the zinc will reduce it chemically to nitrite and a red color will appear. If no color appears, the organism has reduced all of the nitrite and denitrification has probably occurred. In this case an earlier examination of the culture might give a positive reaction for nitrite. *Note:* Zinc powder may become inactive with long storage, and its ability to reduce nitrate to nitrite should be checked periodically with uninoculated medium.

Examples: denitrification, *Pseudomonas aeruginosa;* nitrate to nitrite, *Escherichia coli;* negative, *Bacillus sphaericus.*

25.1.54.2. Method 2

The following method is qualitative for nitrite. Grow the organisms as described for method 1. Examine the cultures periodically for gas bubbles, which are presumptive evidence for denitrification. To cultures of different ages, add 1 ml of solution A and 1 ml of solution B (section 25.4.13). A pink or red color indicates the presence of nitrite. If no red color occurs, test the culture with zinc powder as described in method 1 to see whether nitrate is present. If neither nitrite nor nitrate can be demonstrated, denitrification has probably occurred. The controls described for method 1 also apply to method 2.

An alternative way to demonstrate gas production is to grow the organism in nitrate broth (no agar) containing a small, inverted vial. This allows the visible accumulation of gas during denitrification.

25.1.54.3. Method 3 (87)

The following method allows the detection of nitrous oxide formed during denitrification and provides stronger evidence for denitrification than does method 1 or 2. Inoculate tubes or flasks of semisolid nitrate medium. Seal the vessels with serum bottle stoppers. Inject acetylene gas, generated as described in section 25.2.7, through the stopper to give a final concentration of 1% (vol/vol). The acetylene will prevent N_2O formed during denitrification from being further reduced to N_2 (16). After the culture has grown, remove gas samples with a syringe, and determine the presence of N_2O by gas chromatography with a Porapak Q (80- to 100-mesh) column (274 cm by 4.8 mm) at 35°C with helium carrier gas and a thermal conductivity detector.

25.1.55. Nitrite Reduction

See section 25.4.13 for preparation of reagents and caution about their carcinogenic properties.

Inoculate a suitable broth medium supplemented with 0.01 to 0.10% $NaNO_3$ or KNO_3 and 0.17% agar (the latter to give a semisolid consistency and promote semianaerobic conditions). Gently mix the inoculum with the medium to distribute it throughout the tube. To cultures of different ages, add 1 ml of solution A and 1 ml of solution B (section 25.4.13). A positive test is indicated by lack of a pink or red color. A negative test is indicated by pink or red color.

Comments: Nitrite at a concentration of 0.1% is toxic to some bacteria, so it may be necessary to use 0.01% or an even lower concentration. For instance, *Neisseria gonorrhoeae* gives positive results with 0.01% nitrate but negative results with 0.1% nitrite (65).

The type of medium used may also affect the results. For instance, *Kingella kingae* and *Kingella indologenes* reduce nitrite when a rich basal medium such as Mueller-Hinton broth with yeast extract is used, but when less rich media are used the two species are negative (113).

25.1.56. Nucleic Acid Hydrolysis

25.1.56.1. Method 1

Use the following method to test for both DNase and RNase. Dissolve DNA or RNA (Sigma) in distilled water at a concentration sufficient to give 0.2% when added to an agar growth medium. DNA is readily soluble in water. To dissolve RNA in water, slowly add 1 N NaOH but do not allow the pH to become greater than 5.0. Add the nucleic acid solution to the agar medium just prior to autoclaving (121°C for 15 min), and pour plates when the medium has cooled to 50°C. Make a single streak of the organism across the solidified medium. After incubation, flood the plate with 1 N HCl. In a positive test, a clear zone around the growth indicates DNase or RNase activity.

Examples: positive (DNase), *Serratia marcescens;* negative (DNase), *Enterobacter aerogenes.*

25.1.56.2. Method 2

Use the following method to test for only DNase. Add 100 mg of toluidine blue O (Allied Chemical Corp., New York, N.Y.) per liter of DNA-containing test medium (see method 1) prior to autoclaving (104). Alternatively, add methyl green solution (section 25.4.10) to give a final concentration of 0.005% (110). For the toluidine blue medium, make a single streak of the test organism across the plate; for the methyl green medium, streak the plate to obtain isolated colonies. For a positive test with the toluidine blue plates, a bright rose-pink zone surrounds the growth; for a positive test with the methyl green plates, the colonies are surrounded by a colorless zone on an otherwise green plate. These methods provide a more readily observable reaction for DNase than method 1. *Note:* The two dye methods should not be used with bacteria grown under anaerobic conditions, because the indicator system may be nullified (98).

25.1.56.3. Method 3 (37)

Use the following method as a more rapid test for DNase, although it may not detect very weak DNase activity. Prepare a test medium by combining 31.5 mg of DNA (Difco), 5.0 ml of 0.1 M $CaCl_2$, and 5.0 ml of 0.1 M $MgCl_2$ in 100 ml of Tris buffer (pH 7.4). Make a heavy (milky) suspension of the organism in 0.5 ml of this medium, and incubate it for 2 h at 37°C. Add 1.6 ml of methyl green solution (section 25.4.10) to 100 ml of distilled water; then add 2 drops of this solution to the bacterial suspension. Mix the contents of the tube, and incubate for an additional 2 h at 37°C. A positive reaction is indicated by a colorless tube. A negative test has a blue-green tube.

25.1.57. 0/129 Antibacterial Susceptibility

Place one or two crystals of 2,4-diamino-6,7-diisopropylpteridine phosphate (Sigma) on an inoculated agar plate. Alternatively, soak sterile paper discs (6 to 8 mm in diameter) in a solution of the reagent (0.1 g in 100 ml of acetone), dry the discs, and apply them to an inoculated agar plate. A positive test is indicated by a zone of growth inhibition around the crystals or discs.

Examples: positive, *Vibrio parahaemolyticus;* negative, *Escherichia coli.*

25.1.58. Optochin Susceptibility

Place a sterile, commercially available (Difco; BBL) optochin differentiation disc or a sterile paper disc impregnated with 0.02 ml of a 0.025% solution of optochin (ethylhydrocupreine hydrochloride [Sigma]) on an inoculated agar plate. A positive test is indicated by a zone of growth inhibition around the disc.

Examples: positive, *Streptococcus pneumoniae;* negative, other alpha-hemolytic streptococci.

25.1.59. Ornithine Decarboxylase Activity

See method 1 for arginine dihydrolase (section 25.1.6). Substitute 1% L-ornithine dihydrochloride (or 2% of the DL form) for the arginine.

Examples: positive, *Enterobacter aerogenes;* negative, *Proteus vulgaris.*

25.1.60. Oxidase Activity (119)

Place ashless filter paper (Whatman no. 40, quantitative grade) into a petri dish, and wet it with 0.5 ml of 1% solution of tetramethyl-*p*-phenylenediamine (no. 16020-2 [not HCl salt]; Aldrich Chemical Co., Milwaukee, Wis.) in certified-grade DMSO. (*CAUTION:* Tetramethyl-*p*-phenylenediamine is a respiratory and skin irritant and should be handled in a fume hood.) Pick up one large isolated colony with a cotton-tipped swab, and allow the inoculum to dry on the swab for about 5 s. Then tamp the swab lightly 10 times onto the wet filter paper. A positive test is indicated by development of a blue-purple color on the tip of the swab within 15 s. The reagent solution is stable under refrigeration for at least 1 month.

Examples: positive, *Pseudomonas aeruginosa;* negative, *Escherichia coli.*

25.1.61. Oxidation/Fermentation Catabolism

A test carbohydrate (usually glucose) should be prepared as a stock solution, sterilized by filtration, and added aseptically to an autoclaved semisolid basal medium at 45 to 50°C to give a final concentration of 0.5 to 1.0%. The most widely used basal medium is that of Hugh and Leifson (see section 25.3.18). If the organism cannot grow in this medium because of complex nutritional requirements, the further addition of 0.1% yeast extract or 2% blood serum may permit growth. In cases in which acid production may be masked by the production of alkaline substances from the peptone, a medium such as that devised by Board and Holding (see section 25.3.9) may be suitable. In cases in which the bromothymol blue indicator present in the medium is inhibitory to the organism being tested, substitution of another indicator such as phenol red or bromocresol purple may be necessary. For marine bacteria, the modified O/F (MOF) medium of Leifson (see section 25.3.20) is useful.

To perform the test, place the tubes of the semisolid carbohydrate medium in a boiling-water bath for a few minutes to remove most of the dissolved oxygen; cool the tubes quickly, and inoculate them by stabbing with a straight needle. After inoculation, overlay the medium in only one of the tubes with 10 mm of sterile melted petrolatum; this tube is the "anaerobic" tube. No overlay is used for the second tube (the "aerobic" tube). Prepare the following controls to detect nonspecific reactions: (i) inoculate similar media lacking the carbohydrate, and (ii) incubate sterile media containing the carbohydrate.

Interpret the results as follows. (i) If only the aerobic tube is acidified, the organism catabolizes the carbohydrate by oxidation (an O reaction). Growth should be evident in the aerobic tube. (ii) If both the aerobic and the anaerobic tubes are acidified, the organism is capable of fermentation (an F reaction). Growth should be evident in both tubes. (iii) If neither tube becomes acidified, the organism is unable to catabolize the carbohydrate. If no growth is evident, the medium may be lacking some nutrient required for the organism being tested.

Examples: F reaction with glucose, *Staphylococcus epidermidis;* O reaction with glucose, *Micrococcus luteus;* negative reaction with glucose, *Pseudomonas lemoignei.*

25.1.62. Oxygen Relationships

One of the most fundamental phenotypic characteristics of a bacterium is its relationship to O_2. Although various categories have been devised to describe the oxygen relationships of bacteria, the following four major categories are useful.

(i) Aerobes. Aerobes can use O_2 as an electron acceptor at the terminus of an electron transport chain; tolerate a level of O_2 equivalent to or higher than that present in an air atmosphere (21%); have a strictly respiratory type of metabolism. **Respiration** is an energy-yielding process that generates a proton motive force for oxidative phosphorylation by means of an electron transport chain, for which an exogenous terminal electron acceptor is required. It differs from **fermentation** which is an energy-yielding process that does not involve an electron transport chain and an exogenous terminal electron acceptor but instead relies on substrate-level phosphorylation and an endogenously generated electron acceptor (e.g., pyruvate from glycolysis, which can be reduced to lactate).

Note that some aerobes may be capable of growing anaerobically with electron acceptors other than O_2. For example, *Pseudomonas aeruginosa* is an aerobe even thought it can grow anaerobically by respiring with nitrate.

(ii) Anaerobes. Anaerobes are incapable of O_2-dependent growth (although many anaerobes take up O_2, they do not use it for respiration but instead make toxic derivatives from it, such as H_2O_2 or $O_2^{\cdot -}$), by autooxidation of oxygen-labile metabolites, by reactions catalyzed by various oxidases, or by other means and cannot grow in the presence of an oxygen concentration equivalent to that present in an air atmosphere (21% oxygen).

Some anaerobes, such as *Clostridium perfringens*, have a strictly fermentative type of metabolism. Other anaerobes may carry out anaerobic respiration in which an exogenous terminal electron acceptor other than O_2 is required. For example, *Desulfobacter postgatei* has a strictly respiratory type of metabolism in which sulfate or other oxidized sulfur compounds serve as terminal electron acceptors, being reduced to H_2S. *Desulfococcus multivorans* can respire in a similar manner with sulfate or other oxidized sulfur compounds, but in the absence of such electron acceptors it can grow by fermenting lactate or pyruvate.

(iii) Facultative anaerobes. Facultative anaerobes grow both in the absence of oxygen and in the presence of a level of oxygen equivalent to that in an air atmosphere; they also grow under anaerobic conditions by fermentation, not by respiration.

Some facultative anaerobes, such as *Escherichia coli* or *Staphylococcus aureus*, grow aerobically by respiring with oxygen and grow anaerobically by fermentation. Others, such as *Streptococcus pyogenes* and *Lactobacillus plantarum*, have a strictly fermentative type of metabolism under both aerobic and anaerobic conditions.

(iv) Microaerophiles. Microaerophiles are capable of oxygen-dependent growth and cannot grow in the presence of a level of oxygen equivalent to that present in an air atmosphere (21% oxygen), although oxygen-dependent growth does occur at lower oxygen levels.

Although most microaerophiles are capable of respiring with O_2, *Borrelia* species seem to have a strictly fermentative type of metabolism. The reason why they require a low level of O_2 for growth is not known.

Some microaerophiles are capable not only of respiring with O_2 but also of respiring anaerobically with electron acceptors other than O_2. For example, *Wolinella recta* is an H_2- or formate-requiring organism that was originally thought to be an anaerobe that respires with fumarate or nitrate as terminal electron acceptors. It was later recognized as a microaerophile when the discovery was made that it could use O_2 as an electron acceptor if this molecule was present at very low concentrations (92). Similar results have been reported for other putative anaerobes, i.e., *Wolinella curva*, *Bacteroides ureolyticus*, and *Bacteroides gracilis* (48).

Some microaerophiles, such as *Campylobacter jejuni* and *Spirillum volutans*, use only O_2 as a terminal electron acceptor and are not able to respire anaerobically.

A clue to the oxygen relationships of an isolate may be obtained by the following procedure. Melt a narrow (e.g, 16- by 150-mm) tube filled to three-fifths of its capacity with an appropriate agar medium. Cool the medium to 45°C, and inoculate it with the isolate. Gently mix the inoculum with the medium to distribute it uniformly, and then allow the agar to solidify. On incubation, the following growth responses may occur.

(a) Growth occurs only at the surface of the medium. Growth only on the surface suggests that the organism is aerobic. (However, a fermentable substrate should be present in the medium in case the organism is a facultative anaerobe that not only respires with O_2 but also grows anaerobically by a fermentative type of metabolism.)

(b) Growth occurs only in the bottom regions of the tube and not near the surface. Growth only on the bottom suggests that the organism is anaerobic. (However, anaerobes that are extremely intolerant of O_2 may not be able to grow even in the bottom-most portion of the tube, because of the presence of small amounts of dissolved O_2.)

(c) Growth occurs throughout the tube. Growth throughout the tube suggests that the organism is a facultative anaerobe. (The medium should contain a fermentable substrate, but no terminal electron acceptors other than O_2 should be present, since some aerobes can respire anaerobically with terminal electron acceptors other than O_2.)

(d) Growth occurs only in the form of a disc a short distance below the surface. Growth in a small area a short distance below the surface suggests that the organism is a microaerophile.

For motile bacteria, a semisolid medium (0.2% agar) may be preferable to a solid medium, because it can be inoculated by stabbing with an inoculating needle after the agar has gelled. This makes it unnecessary to mix the medium with the inoculum and thus avoids incorporating some dissolved O_2 in the medium. Motile organisms usually exhibit negative or positive aerotaxis, which results in their translocation to a zone in the tube where the oxygen level is optimal for growth.

More quantitative information about the oxygen relationships of an isolate can be obtained by incubating cultures (preferably inoculated onto the surface of agar media, where they are most susceptible to O_2 in the gas phase) under a series of atmospheres containing various O_2 levels. The most convenient way to do this is to use sealable vessels which can be connected by a manifold to a vacuum pump and various gas cylinders. Use polycarbonate jars such as those that are used routinely for cultivation of anaerobes, provided they have a vent to allow evacuation and refilling with gases. Alternatively, adapt ordinary aluminum pressure cookers, obtainable at a department store, by removing the rubber gasket from the lid, cleaning all traces of grease from the gasket and the lid, and then gluing the gasket back into place with a silicone rubber adhesive. After the glue has cured thoroughly (which may take several days), the vessel should be connected to a vacuum pump to confirm that it can maintain a vacuum without leaking.

To obtain a particular oxygen level, merely evacuate the appropriate amount of air from the jar and replace it with N_2. This is accomplished most conveniently by measuring the decrease in pressure in the vessel by means of a vacuum gauge or mercury manometer. For instance, suppose that, at a particular geographical location, the uncompensated barometric pressure is 29 in. of mercury (737 mmHg). To obtain an atmosphere of 7% O_2 i.e., 7/21 of the 21% O_2 in an air atmosphere, pump air out of the jar until the manometer or gauge indicates a vacuum of 19.3 in. (14/21 of 29 in.) or 491 mmHg (14/21 of 737 mmHg.) Then add N_2 to the vessel until the manometer or gauge reads 0 in. of vacuum (i.e., atmospheric pressure), and seal the jar. Further examples are as presented in Table 4.

Table 4. Examples of evacuation requirements to obtain various oxygen levels

O_2 level desired, %	Evacuate air from vessel until vacuum gauge reads:
0	See footnote[a]
1	27.6-in. (702-mm) vacuum
3	24.9-in. (632-mm) vacuum
5	22.1-in. (562-mm) vacuum
7	19.3-in. (491-mm) vacuum
9	16.6-in. (421-mm) vacuum
12	12.4-in. (316-mm) vacuum
15	8.3-in. (211-mm) vacuum
18	4.1-in. (105-mm) vacuum
21	No evacuation of vessel

[a]To achieve strictly anaerobic conditions, merely evacuating the vessel and refilling it with N_2 (even if repeated several times) may not suffice. This is because culture media usually contain dissolved O_2, especially after they have been stored for several hours. This dissolved O_2 bleeds off slowly into the gas phase within the culture vessel; meanwhile, some bacteria that are extremely efficient at scavenging O_2 may be able to respire with it and grow, or some bacteria that are highly susceptible to O_2 may be killed or inhibited. One way to minimize dissolved O_2 in agar media is to use only media that are freshly prepared and that are inoculated immediately after they have solidified. An alternative method is to add to the vessel a palladium catalyst (of the type used in commercial anaerobic jar systems), evacuate the jar, and refill it with N_2 several times and finally with H_2; the H_2, with catalyst, will then remove any residual O_2. Even this method will not work with anaerobes that are extremely intolerant of O_2, and more stringent methods, such as using prereduced media, may be necessary (see Chapter 6.6.4).

When testing an isolate that requires CO_2, an appropriate amount of CO_2 should be added from a tank of pure, compressed CO_2 before the final filling with N_2. When testing an isolate that requires H_2 as an electron donor, such as *Campylobacter mucosalis*, H_2 can be added instead of N_2 during the final filling of the culture vessel. (*CAUTION:* Gas mixtures containing O_2 and H_2 are highly explosive. Do not use H_2 near open flames or any source of sparks, and when evacuating vessels and adding H_2, enclose the vessels in wire netting or a metal mesh cage. Wear safety goggles when using H_2. When incubating culture vessels, make sure that the incubators have thermoregulators that are isolated from the incubation chamber since some thermoregulators can generate sparks. When opening the culture vessels after incubation, first make sure that there are no flames or sparks in the area that could ignite the escaping gas mixture.)

25.1.63. Peptidase Activity

The test for peptidase activity depends on the use of amino acid naphthylamides, in which an amino acid moiety is linked by a peptide bond to β-naphthylamine. For example, the substrate for glycine aminopeptidase is glycine β-naphthylamide. Quantitative estimation of enzyme activity is made by using standardized bacterial suspensions or cell extracts and measuring the rate of formation of β-naphthylamine by using a fluorimeter (124, 125). For qualitative purposes, the following methods may be useful. However, as noted by Westley et al. (125), the culture medium and the age of the cells, unless carefully standardized for a particular group of organisms, can influence the reproducibility of the results.

25.1.63.1. Method 1

The following method was originally used to differentiate mutants of *Salmonella typhimurium* that were unable to hydrolyze L-alanine β-naphthylamide (79). Stain colonies grown on agar media by pouring over them a mixture containing 5 ml of 0.2 M Tris-HCl buffer (pH 7.5), 0.2 ml of a solution of amino acid-naphthylamide substrate (10 mg/ml in *N,N*-dimethylformamide), and 10 mg of Fast Garnet GBC (Sigma). Colonies hydrolyzing the amino acid-naphthylamide to β-naphthylamine develop a dark red color, usually in 1 to 2 min.

25.1.63.2. Method 2 (31)

Dissolve amino acid-naphthylamide substrates in 0.1 M Tris-phosphate buffer (pH 8.0 for most substrates, pH 7.2 for hydroxyproline β-naphthylamide, and pH 7.6 for γ-glutamyl β-naphthylamide) to a concentration of 10^{-3} M. Solubilize the crystalline solids initially in a small volume of 50% ethanol or *N,N*-dimethylformamide, and then add the appropriate amount of buffer to achieve the final concentration. Dispense 40-μl volumes of the substrate solutions into small wells or microtubes. Dry the wells or microtubes under a vacuum below 50°C, and store at 4°C until used.

Suspend cells from a culture on an agar medium in physiological saline (0.85% NaCl) to a turbidity equivalent to a McFarland no. 3 standard (section 25.4.9), and add 20 to 40 μl of cell suspension to each well or microtube. Incubate for 1 to 4 h in a humid chamber, and then add 20 to 40 μl of detector reagent A and 20 μl of detector reagent B (section 25.4.14). Controls are prepared in the same manner as the test wells but without any substrate. Incubate for 15 min. A positive test is indicated by an orange color. Negative tests show no color change compared with the controls. (To eliminate any residual yellow color in negative reactions after reagent B is added, expose the wells or microtubes to direct sunlight for 2 to 4 min.)

Examples (with γ-glutamic acid β-naphthylamide as the substrate): positive, *Neisseria meningitidis, Moraxella urethralis;* negative, *Neisseria gonorrhoeae, Neisseria sicca, Moraxella lacunata.*

Some substrates that are available commercially from biochemical supply houses are listed in Table 5.

CAUTION: The β-naphthylamine that results from hydrolysis of these substrates is a carcinogen that causes bladder tumors. Consequently, appropriate precautions should be taken during the use, handling, and subsequent disposal of all materials containing this substance.

25.1.64. Phenylalanine Deaminase Activity

Inoculate heavily a phenylalanine agar slant (section 25.3.28). After incubation, allow 4 to 5 drops of 10% ferric chloride solution to run down the slant. A positive test is indicated by a green color, which is due to formation of phenylpyruvic acid.

Examples: positive, *Proteus vulgaris;* negative, *Escherichia coli.*

Table 5. Some substrates available commercially from biochemical supply houses[a]

L-Alanine β-naphthylamide
L-Arginine β-naphthylamide
L-Asparagine β-naphthylamide
L-Aspartic acid α-(β-naphthylamide)
L-Cystine di-β-naphthylamide
L-Glutamic acid γ-(α-naphthylamide)
L-Glutamic acid γ-(β-naphthylamide)
Glycine β-naphthylamide
L-Histidine β-naphthylamide
trans-4-Hydroxy-L-proline β-naphthylamide
L-Isoleucine β-naphthylamide
L-Leucine β-naphthylamide
L-Leucine 4-methoxy-β-naphthylamide
L-Lysine β-naphthylamide
DL-Methionine β-naphthylamide
L-Phenylalanine β-naphthylamide
L-Proline β-naphthylamide
L-Pyroglutamic β-naphthylamide
L-Serine β-naphthylamide
L-Tryptophan β-naphthylamide
L-Tyrosine-β-naphthylamide
L-Valine β-naphthylamide

[a]*CAUTION:* The β-naphthylamine that results from hydrolysis of these subtrates is a carcinogen that causes bladder tumors. Consequently, all precautions pertinent to the use, handling, and subsequent disposal of all materials containing it should be observed.

25.1.65. Phosphatase Activity

25.1.65.1. Method 1

Use the following method to test for both acidic and alkaline phosphatases. To an agar growth medium such as nutrient agar (section 25.3.27), melted and cooled to 45 to 50°C, aseptically add a sufficient amount of a filter-sterilized 1% solution of the sodium salt of phenolphthalein diphosphate (Sigma) to give a final concentration of 0.01%. Incubate streaked plates for 2 to 5 days. Place 1 drop of ammonia solution (specific gravity, 0.880 [28 to 30%]) in the lid of the inverted plate, and replace the culture over it to allow the ammonia fumes to reach the colonies. If the test is positive, colonies become red, as a result of the presence of free phenolphthalein.

Examples: positive, *Staphylococcus aureus;* negative, *Micrococcus luteus.*

25.1.65.2. Method 2 (21)

Use method 2 to test only for acid phosphatase. Prepare a very dense suspension of cells in saline (0.85% NaCl). Add 0.3 ml of the suspension to 0.3 ml of a substrate solution consisting of 0.01 M citrate buffer (pH 4.8) containing 0.01 M disodium *p*-nitrophenyl phosphate (Sigma). Incubate the mixture for up to 6 h at 37°C. Then add 0.3 ml of 0.04 M glycine buffer which has been adjusted to pH 10.5 with NaOH. Also use a control lacking the bacteria. A positive test is indicated by a yellow color due to acid phosphatase activity.

25.1.65.3. Method 3

Use method 3 to test for only alkaline phosphatase. This method is similar to method 2, but the *p*-nitrophenyl phosphate is dissolved in the glycine buffer instead of the citrate buffer. No further addition of glycine buffer is necessary to develop the yellow color.

25.1.66. Poly-β-Hydroxybutyrate Presence (68, 94, 116)

Grow a culture of the organism under conditions conducive to the formation of poly-β-hydroxybutyrate (PHB). For aerobic bacteria these conditions usually are nitrogen and also oxygen limitations of exponential growth in the presence of excess carbon and energy source. PHB occurs as refractile intracellular granules.

25.1.66.1. Staining procedures

Staining may be used as presumptive evidence for PHB granules; however, chemical analysis is required for a definitive test. The most specific staining procedure is performed as follows. Prepare and filter a 1% aqueous solution of Nile Blue A (Matheson, Coleman & Bell stock NX0395) just before use. Stain a heat-fixed smear of the bacteria with this solution for 10 min at 55°C in a coplin staining jar. Wash the slide briefly with tap water to remove excess stain, and place it in 8% aqueous acetic acid solution for 1 min. Wash with tap water and blot dry. Remoisten with tap water, and place a coverslip over the smear. Place immersion oil on the coverslip, and examine the preparation under an epifluorescence microscope, using an exciter filter that provides an excitation wavelength of approximately 460 nm. If the test is positive, PHB granules exhibit a bright orange fluorescence. Other cell inclusion bodies such as polyphosphate do not fluoresce in this manner.

25.1.66.2. Chemical analysis

Place 1 ml of the culture into a conical glass centrifuge tube, add 9 ml of alkaline hypochlorite reagent (see section 25.4.20), and incubate at room temperature for 24 h (116). Alternatively, add 8 ml of commercial 5% hypochlorite (Clorox bleach) to 2 ml of culture and incubate (93). The hypochlorite will digest most cell components but will not digest the PHB granules. Centrifuge the mixture to collect the granules, and discard the clear supernatant fluid. Wash the sediment by suspending it in 10 ml of distilled water, centrifuging, and discarding the supernatant fluid. Wash the sediment once more with water, twice with 10-ml portions of acetone, and finally twice with 10-ml portions of diethyl ether. Dry the sediment, and add 2 ml of concentrated sulfuric acid. Place the tube in a boiling-water bath for 10 min. Cool it to room temperature. Using a UV spectrophotometer and far-UV quartz cuvettes, determine the absorption spectrum of the sample at 5-nm increments from 215 to 255 nm against a blank of plain concentrated sulfuric acid. (The sample may have to be diluted 1:10 or more with sulfuric acid if its A_{235} is greater than 1.0.) A positive test is indicated by the occurrence of an absorption peak at 235 nm as a result of crotonic acid formed from PHB by the action of the sulfuric acid.

Examples: positive, *Pseudomonas lemoignei;* negative, *Pseudomonas aeruginosa.*

25.1.67. Poly-β-Hydroxybutyrate Hydrolysis (115)

Hydrolysis of PHB is very useful for distinguishing among the members of the genus *Pseudomonas;* however, the PHB granule suspension required for the test cannot be purchased commercially and is tedious to prepare. Details of its preparation are given in section 25.4.17.

Prepare a plate of mineral agar, such as that described in section 25.2.2, lacking carbon sources. To another portion of the medium that has been melted and cooled to 45 to 50°C, add a sufficient amount of the sterile concentrated PHB granule suspension to give a final concentration of 0.25% (wt/vol) in the medium. Pour the granule-agar suspension over the surface of the plates of solidified medium to form a thin overlay.

After the overlay solidifies, inoculate a washed suspension of the organism to be tested onto a small area of the plate. (Several strains can be tested on a single plate.) Incubate the plate for several days. A positive test is indicated by growth of the organism at the site of inoculation and clearing of the surrounding medium as a result of depolymerization of the PHB granules.

Example: positive, *Pseudomonas lemoignei;* negative, *Pseudomonas acidovorans.*

25.1.68. Proteinase Activity

Various chromogenic substrates for detection and quantitative assays of proteinases are available commercially from biochemical supply houses. These substrates consist of a dye that is chemically conjugated to a particular protein; hydrolysis of the protein results in liberation of the dye. Examples of substrates are azoalbumin, azocasein, and Azocoll (a dye-impregnated cowhide powder). Assay methods involve incubating a solution of the substrate (or a suspension, in the case of insoluble substrates) with the supernatant fluid from a bacterial culture (most bacterial proteinases of interest are exocellular), removing the residual unhydrolyzed substrate, and measuring the amount of soluble dye that has been liberated.

A typical protocol with azoalbumin is as follows. To a test tube add 0.3 ml of test culture supernatant, 1.7 ml of distilled water, and then 0.1 ml of azoalbumin solution (5 mg/ml in 0.1 M Tris buffer [pH 7.5]). Also prepare a blank containing 2.0 ml of distilled water and 1.0 ml of azoalbumin solution. Incubate the test tubes for 1 h in a water bath at 30°C. Then add 2.0 ml of 8% trichloroacetic acid to stop the reaction and to precipitate the unhydrolyzed azoalbumin. Pour the mixture into a polypropylene centrifuge tube, and centrifuge at 10,000 × *g* or at a speed sufficient to sediment the precipitated protein. Using a pipette and suction bulb, carefully transfer 2.0 ml of the clear supernatant fluid to a tube. Add 2.0 ml of 0.5 M NaOH to intensify the color. Read the A_{400} of the solution. Although the units of enzyme activity are arbitrary, it is often convenient to let each 0.001 increase in absorbance beyond the reading of the blank equal 1 U of activity.

Azocoll, an insoluble substrate, is used in the form of a suspension (e.g., 50 mg suspended in 5 ml of 100 mM phosphate buffer [pH 7.0] [Calbiochem, catalog no. 194932 or 194933]). Instead of centrifuging the reaction mixture after incubation, remove the residual unhydrolyzed protein by filtration and measure the A_{520}.

See also sections 25.1.14, 25.1.29, 25.1.48, and 25.1.50.

25.1.69. Pyocyanin and Other Phenazine Pigments

Inoculate a slant of King's medium A (section 25.3.1). Examine the slant at 24, 48, and 72 h. A positive test is indicated by pigments that diffuse into the agar and stain it. In some cases the pigments are only slightly water soluble and precipitate as crystals in the medium.

Examples: pyocyanin, blue color which stains the medium (produced by *Pseudomonas aeruginosa);* phenazine-1-carboxylic acid, orange-yellow and may crystallize in the colonies and surrounding medium (produced by *Pseudomonas aureofaciens);* oxychloraphin, stains the medium pale yellow (produced by *Pseudomonas chlororaphis);* chlororaphin, only very sparingly soluble and accumulates as isolated green crystals in the growth at the butt of the slant (produced by *P. chlororaphis).*

25.1.70. Pyrazinamidase Activity

25.1.70.1. Method 1 (as applied to *Corynebacterium* spp.) (114)

Inoculate a tube of the test medium (section 25.3.31) with bacteria from an overnight growth on chocolate agar slants (ca. 5 × 10^9 cells). Incubate at 37°C overnight. To each butt add 1.0 ml of freshly prepared 1% (wt/vol) aqueous ferrous ammonium sulfate. Refrigerate the tube for 1 to 4 h. A positive test is indicated by the appearance of a pink band in the agar, indicating the formation of pyrazinoic acid. A negative test shows no pink band.

For a rapid test, omit the Dubos broth base from the culture medium and decrease the agar concentration to 0.2%. Prepare a physiological saline (0.85% NaCl) suspension of the bacteria to be tested from growth on chocolate agar slants, and, with a pipette, mix the suspension into the top one-half of the tube of test medium. Incubate for 2 h in a 37°C water bath. Add 1 ml of ferrous ammonium sulfate solution (see above), and incubate for 1 to 5 min. A positive test is indicated by a pink color. A negative test shows no pink color.

Examples: positive, *Corynebacterium xerosis;* negative, *Corynebacterium pseudotuberculosis.*

25.1.70.2. Method 2 (as applied to *Yersinia* spp.) (61)

Inoculate slants of the test medium (section 25.3.31) with bacteria grown overnight on soybean-casein digest agar (section 25.3.34), and incubate for 48 h at 25 to 30°C. Flood the slant with 1 ml of ferrous ammonium sulfate solution (see method 1), and incubate for 15 min. A positive test is indicated by a pink color. A negative test shows no pink color.

Examples: positive, *Yersinia intermedia;* negative, *Yersinia pseudotuberculosis.*

25.1.71. Spores (Endospores)

For staining of spores, see Chapter 2.3.7. The media used for growth of cultures should contain adequate levels of Ca^{2+} and Mn^{2+}. Media with and without carbohydrates should be inoculated, and cultures of different ages should be examined. *Note:* Large inclusion granules (e.g., PHB) may be mistaken for endospores.

If sporelike structures are observed, a heat test should be performed for confirmation. Inoculate a tube of broth with growth from solid or liquid media in which sporulation is suspected to occur. Take care not to touch the wall of the tube with the inoculum. Place the tube in a water bath at 80°C along with an uninoculated tube of broth containing a thermometer. Make sure the level of the water in the bath is higher than the level of the broth. Heat for 10 min; begin timing when the thermometer reaches 80°C. Cool the tube, and incubate at the optimum growth temperature to see if growth occurs. *Note:* Some endospores, such as occur with certain strains of *Clostridium botulinum*, may be killed at 80°C, necessitating the use of lower temperatures for testing heat resistance (e.g., 75 or 70°C). Also, apply the "popping" test (Chapter 2.3.7.1). Also, certain *Nocardia* species form chlamydospores or microcysts, which can survive 80°C for several hours and yet are not true endospores.

25.1.72. Starch Hydrolysis

Prepare an agar medium on which the organism can grow well, and add 0.2% soluble starch before boiling. The medium should preferably contain no glucose since this may diminish starch hydrolysis (126). Sterilize the medium at 115°C for 10 min. Make a single streak of the organism across the center of a plate of the starch agar. After incubation, flood the plate with iodine solution (the iodine normally used for the Gram stain is suitable). A positive test is indicated by a colorless area around the growth.

Examples: positive, *Bacillus subtilis;* negative, *Escherichia coli.*

25.1.73. Triple-Sugar Iron Agar Reactions

Inoculate a slant of triple-sugar iron agar (section 25.3.37) by using a straight needle. First, stab the butt down to the bottom, withdraw the needle, and then streak the surface of the slant. Use a cotton stopper or loosely fitting closure to permit access of air. Read results after incubation at 37°C for 18 to 24 h. Three kinds of data may be obtained from the reactions.

(i) Sugar fermentations

Acid butt, alkaline slant (yellow butt, red slant): glucose has been fermented but not sucrose or lactose.

Acid butt, acid slant (yellow butt, yellow slant): lactose and/or sucrose has been fermented.

Alkaline butt, alkaline slant (red butt, red slant): neither glucose, lactose, nor sucrose has been fermented.

(ii) Gas production. Indicated by bubbles in the butt. With large amounts of gas, the agar may be broken or pushed upward.

(iii) Hydrogen sulfide production from thiosulfate. Indicated by a blackening of the butt as a result of reaction of H_2S with the ferrous ammonium sulfate to form black ferrous sulfide.

25.1.74. Urease Activity

25.1.74.1. Method 1

Inoculate the surface of a Christensen urea agar slant (section 25.3.13) and incubate. A positive test is indicated by a red-violet color.

Examples: positive, *Proteus vulgaris;* negative, *Escherichia coli.*

25.1.74.2. Method 2 (59)

The following method is useful for organisms that cannot grow on Christensen urea agar. Prepare the following solution: 0.1065% BES buffer [*N,N*-bis-(2-hydroxyethyl)-2-aminoethanesulfonic acid (ICN Pharmaceuticals, Cleveland, Ohio)]–2.0% urea–0.001% phenol red (pH 7.0). Sterilize by filtration, and aseptically dispense 2.0-ml volumes into small tubes. Centrifuge a broth culture of the organism, and suspend the cells in sterile distilled water or saline to a dense concentration. Add 0.5 ml of the cell suspension to a tube of the urea medium. A control medium lacking urea should be inoculated similarly. Incubate the tubes for 24 h. A positive test is indicated by a red-violet color.

25.1.75. Voges-Proskauer Reaction

Inoculate two tubes of MRVP broth (section 25.3.26). Incubate one tube at 37°C and the other at 25°C. After 48 h, remove 1 ml of culture to another tube, add 0.6 ml of 5% (wt/vol) α-naphthol dissolved in absolute ethanol, and mix thoroughly. Then add 0.2 ml of 40% aqueous KOH. Mix well, and incubate the tube in a slanted position to increase the surface area of the medium (the reaction is dependent on oxygen). Examine after 15 and 60 min. A positive test is indicated by a strong red color that begins at the surface of the medium.

Examples: positive, *Enterobacter aerogenes;* negative, *Escherichia coli.*

25.1.76. X and V Factor Requirements

Inoculate a plate of soybean-casein digest agar (section 25.3.34) as follows. Moisten a sterile cotton swab in the suspension, squeeze out excess fluid by pressing and rolling the swab against the wall of the tube, and roll the swab across the entire surface of the plate (do not "scour" the surface of the agar). Use sterile paper strips impregnated with the X factor (heme) or the V factor (NAD; β-nicotinamide adenine dinucleotide). The strips are obtainable commercially (Difco, BBL). With a flamed forceps place an X strip on the surface of the seeded agar, flame the forceps, and then place a V strip on the agar approximately 2 cm from the X strip. Incubate the plates for 1 to 2 days, and interpret the results as follows. A requirement for either X or V factor, but not both, is indicated by growth only around the appropriate strip. A requirement for both factors is indicated by occurrence of growth only between the two strips. If growth occurs over the entire plate, neither factor is required.

Examples: positive for both X and V, *Haemophilus influenzae;* positive for V only, *Haemophilus parainfluenzae;* negative both X and V, *Escherichia coli.*

25.2. SPECIAL TESTS

25.2.1. Fermentation Products

25.2.1.1. Organic acids and alcohols by gas chromatography (4) (also see Chapter 21.3.5)

The following procedures were designed for identification of anaerobes, but they can also be used for aerobic and facultative organisms. Acidify 1 ml of culture medium containing glucose or other fermentable carbohydrate in a 12- by 75-mm tube by adding 0.2 ml of 50% (vol/vol) H_2SO_4; samples thus acidified can be stored in a refrigerator for future analysis. Add 0.4 g of NaCl and 1 ml of diethyl ether. Stopper the tube with a cork stopper. Mix the contents of the tube by gently inverting the tube about 20 times to extract the free fatty acids into the ether phase. Centrifuge the mixture for a few minutes to break the emulsion, and remove the ether (upper) layer from the aqueous layer. It is advisable not to remove too much of the ether layer in order to avoid contaminating the ether with the water. Place the ether layer into a new 12- by 75-mm tube, and add anhydrous $CaCl_2$ (4 to 20 mesh) to about one-fourth the volume of the ether, to remove traces of dissolved water. Inject 14 μl into the gas chromatograph to detect short-chain volatile fatty acids and alcohols. Uninoculated medium should be acidified, extracted, and examined as a control.

Pyruvic, lactic, fumaric, and succinic acids are nonvolatile and must first be converted to their methyl esters for detection by gas chromatography. Place 1 ml of the culture in a 12- by 75-mm tube, and add 2 ml of methanol and 0.4 ml of 50% H_2SO_4 (2 ml of boron trifluoride methanol [obtainable from any chromatographic supply house] may be added in place of plain methanol). Heat the mixture at 60°C for 30 min. Add 1 ml of water and 0.5 ml of chloroform to the tube, stopper the tube tightly, and gently invert it about 20 times. Remove the chloroform (bottom) layer, and inject 14 μl into the gas chromatograph.

A gas chromatograph equipped with a thermal conductivity (TC) detector is recommended. A flame ionization detector may be used; however, it will not detect formic acid. Use a stainless-steel, aluminum, or glass column measuring 6 ft (1.8 m) by 0.25 in. (6 mm). Pack the column with Supelco SP-1000, 100 to 200 mesh (Supelco Inc., Bellefonte, Pa.), or with 5% FFAP on Chromosorb-G. Both volatile and methylated fatty acids are detected with the same column. Set the column oven temperature at 135 to 145°C. The carrier gas is helium at a flow rate of 120 cm^3/min. If the instrument is equipped with separate injector and detector oven heaters, set these 20 to 30°C higher than the column temperature. Use a 1-mV recorder with a chart speed of 0.5 in. (1.3 cm)/min. Prepare standards for volatile acids, alcohols, and nonvolatile acids (section 25.4.19), and extract them in the same manner as the culture samples. Such standards may also be purchased from chromatography supply houses.

The presence or absence, as well as the approximate amounts, of the various metabolic end products constitute patterns that are very helpful in identifying bacteria. More quantitative information can be obtained by use of careful technique and by use of standard curves based on known amounts of each acid or alcohol. The order of elution of alcohols and volatile fatty acids from the gas chromatograph column is as follows: ethanol, *n*-propanol, isobutanol, *n*-butanol, isopentanol, n-pentanol, acetic acid, formic acid, propionic acid, isobutyric acid, *n*-butyric acid, isovaleric acid, *n*-valeric acid, isocaproic acid, *n*-caproic acid, and heptanoic acid. For methyl esters of the nonvolatile acids, the order of elution is as follows: pyruvic acid, lactic acid, oxalacetic acid, oxalic acid, methylmalonic acid, malonic acid, fumaric acid, and succinic acid.

25.2.1.2. Organic acids and alcohols by HPLC (109) (also see Chapter 21.3.6)

Add 3 to 5 drops of 50% sulfuric acid to a culture grown in a broth medium with glucose or other fermentable carbohydrate. Pipette 1 ml of the culture into a Clin-Elute, CE-1001 column (Analytechem International Inc., Harbor City, Calif.). Elute the organic acids from the column by adding 9 ml of methyl *tert*-butyl ether (in 3-ml portions). Add 1 ml of 0.2 M NaOH to the 9 ml of ether extract, and shake. Remove the aqueous layer containing the sodium salts of the fatty acids and wash it once with methyl *tert*-butyl ether. Inject 20 μl of the aqueous extract into a high-pressure liquid chromatography (HPLC) analyzer equipped with a UV detector set at 214 nm to detect the double bond of the carboxyl group of the fatty acids. The column is an HPX-87H organic acid column (Bio-Rad Laboratories, Richmond, Calif.); other manufacturers also make organic acid columns. Elute organic acids from the column with 5% acetonitrile in 0.013 N sulfuric acid at a rate of 0.8 ml/min; the column temperature is 60°C. Record peaks on an integrator/recorder.

Alcohols can be separated on the organic acid column; they are eluted with distilled water and detected with a refractive index monitor. Organic acid and alcohol standards are commercially available.

25.2.1.3. Hydrogen and methane by gas chromatography

Analyze cultures grown in a medium containing glucose or other fermentable carbohydrates for the production of hydrogen and methane. Stopper the cultures with a rubber stopper, and incubate them. After the cultures have grown, remove gas samples from the culture tubes by pushing the needle of a syringe through the rubber stopper and into the gas phase of the culture. Use a 1- or 5-ml disposable plastic syringe with a 1.5-in. (3.8-cm) needle, and start the needle into the culture tube at a point where the rubber stopper and glass wall meet; push the needle through the stopper at an angle that will permit the tip of the needle to enter the gas phase of the culture. Withdraw 0.7 to 1.0 ml of the gas phase, and inject it into the gas chromatograph.

The gas chromatograph must be equipped with a TC detector. Use either nitrogen or argon as the carrier gas, with a flow rate of 25 to 30 cm^3/min. Use an aluminum or stainless-steel column measuring 6 ft by 0.25 in., packed with silica gel (grade 12, 80 to 100 mesh; Supelco Inc.). Set the column oven temperature to 25 to 30°C; if the instrument is equipped with separate injector and detector ovens, set these at the same temperature as the column oven or a few degrees higher. Standards for gas analysis can be purchased from any chromatography supply house; natural gas can be used to determine the elution point of methane. For mixtures of hydrogen and methane, hydrogen will be eluted from the column before methane.

25.2.2. Substrate Utilization Patterns

25.2.2.1. Auxanographic nutritional method

Seed a molten basal agar medium (45 to 50°C) lacking carbon sources with a washed suspension of the organism to be tested. After the plates have solidified, dip the edges of sterile filter paper discs into sterile solutions of various carbon sources, allowing the fluid to be absorbed by capillary action. Such discs may be applied immediately to the surface of the seeded agar plates or may be dried and stored in a sterile container until needed. Place each disc near the periphery of the seeded plate, using three to four discs per plate, or, alternatively, place the discs at suitable distances on larger sheets of seeded agar (e.g., in a Pyrex baking dish). After incubation, determine whether a carbon source has been used by looking for a halo of turbid growth around the disc. In some cases growth occurs only as a line between two discs, indicating that the organism requires both carbon sources for growth. If growth occurs over the entire plate, the inoculum was probably too dense; use a more dilute inoculum to seed the agar.

25.2.2.2. Multipoint inoculators

The widespread commercial availability of sterile, plastic microtiter plates (plastic dishes with 96 miniature wells, each having a volume of ca. 0.2 ml) has led to development of miniaturized procedures for determination of acid production from carbohydrates, gas production, and other biochemical reactions. One can make a multiple inoculator by simply inserting the pointed ends of 27-mm stainless-steel pins into wax-filled wells or a microtiter plate (38). One can also make an inoculator by inserting pins or brads into a wood, plastic, or metal template so that they are in the exact pattern of the wells of a microtiter plate (43). An inoculator made entirely of metal has the advantage of being autoclavable, but one can effectively "sterilize" the other inoculators by dipping the pinheads into alcohol and then flaming them for 1 to 2 s with the pinheads pointing upwards (42). Charge an inoculator by lowering the device onto a master microtiter plate in which each well contains a different strain of bacteria; in this way each pinhead becomes charged with ca. 0.006 ml of culture. Then lower the inoculator onto a microtiter plate with its wells filled with carbohydrate broth, thereby causing each pinhead to inoculate its particular strain of bacteria into a separate well of broth. Return the inoculator to the master plate for recharging, and use it to inoculate a second microtiter plate that has its wells filled with a different carbohydrate broth, etc. To determine gas production from carbohydrates, cover each inoculated well with a sterile overlay of a sealant such as 1 part of Amojell (American Oil Co.) and 1 part of mineral oil (44). Cover the inoculated plates with their lids, and incubate them. Acid production is determined by a change in the color of the pH indicator in the medium. Gas production is indicated by the accumulation of bubbles under the overlay.

Filling the wells of a microtiter plate with sterile media can be greatly facilitated if one uses a multichannel automatic pipette (Cooke Engineering Co., Alexandria, Va.) or similar apparatus. One can construct a homemade device such as that described by Wilkins et al. (129) as follows. Remove the tips and hubs from eight 18-gauge hypodermic needles, and braze the needles into holes in a 13.5-cm-long piece of 7-mm stainless-steel tubing so that the needle centers are the same distance apart as the microtiter wells (Fig. 2). Close off one end of the tubing, and braze a Luer-Lok connector onto the other end. After sterilizing the assembly, aseptically connect the Luer-Lok to a sterile automatic pipetting syringe (Cornwall type; Becton, Dickinson & Co., Rutherford, N.J.) equipped with a sterile filling hose (Fig. 2). With the end of the filling hose immersed in a flask of sterile medium, depress the plunger of the adjustable syringe several times to fill the syringe. Then place the eight-needle manifold into a row of eight wells on a microtiter plate, and depress the plunger of the syringe to deliver 0.2 ml of medium simultaneously to each of the eight wells. Continue to fill rows of wells in this manner until all of the wells on the plate have been filled. Rinse the assembly by flushing it with sterile water (or boiling water, if an agar medium has been dispensed), and then flush it several times with the next medium to be dispensed.

Biochemical tests other than acid and gas production can be performed with microtiter plates. For example, perform the indole test by adding 2 drops of Kovács' reagent to the well cultures, or for the Voges-Proskauer test add one loop of α-naphthol followed by one loop of KOH (43). Some tests may not be suitable for microtiter plates; e.g., in testing for urease, the ammonia produced by a urease-positive organism may diffuse to adjacent wells to give false-positive reactions. Pinhead inoculators may be used to inoculate the surface of diagnostic agar media contained in conventional petri dishes (42, 71). Also, inoculators with the pointed ends of brads or pins projecting from the block may be used to pierce the agar surface (86). However, when one deposits various strains on a single petri dish, problems of overgrowth, spreading, or mutual interference may occur. These problems can be eliminated if one uses a petri dish divided into compartments (Dyos Plastics Ltd., Surbiton, England) (112), mini-ice-cube trays (42), or microtiter plates (43, 128). Tiny slants, such as triple-sugar iron agar slants, may be prepared in microtiter plates and inoculated by a piercing inoculator (43).

For multipoint inoculation of anaerobes, dispense sterile semisolid media into microtiter plate wells, and

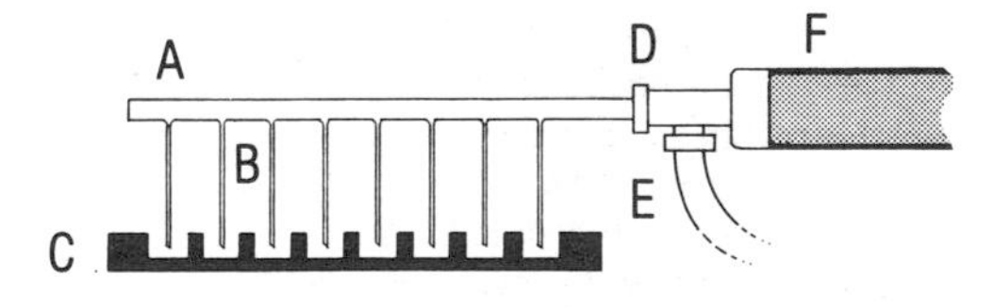

FIG 2. Diagram of apparatus for filling eight microtiter plate wells simultaneously. (A) Stainless-steel tubing; (B) 18-gauge hypodermic needles brazed to tubing; (C) microtiter plate with wells; (D) Luer-Lok connector brazed to end of tubing; (E) filling hose attached to Cornwall syringe; (F) barrel and plunger of Cornwall syringe. By depressing the plunger, the eight wells can be filled simultaneously with sterile medium.

incubate the plates with their lids in position overnight aerobically at 37°C to check for contamination and also to dry the plates slightly (128). Place the plates in an anaerobic glove box, and allow them to reduce for 2 days prior to inoculation. Inoculate the plates with a piercing inoculator within the glove box. Although each plate contains 96 wells, inoculate only 48 of them (in a checkerboard pattern) and leave 48 uninoculated to serve as controls to detect nonbiological pH changes or possible migration of motile bacteria from one well to another. After inoculation, cover each plate with a sheet of sterile plastic tape (nontoxic plate sealer; Cooke Engineering Co.) by placing the tape with adhesive side up on a rubber mat and pressing the inverted plate onto it. Roll the plates with the tape roller to ensure a firm seal on each well. To allow for the escape of gases during growth, punch a pinhole into the tape over each well by means of the tape perforator. Incubate the plate within the anaerobic glove box for 3 days. Determine acid production by means of a narrow pH electrode (such as an S-30070-10 electrode [Sargent-Welch Co., Skokie, Ill.]) inserted into the medium of each well. Do not rinse the electrode between wells, because each insertion pushes the previous agar medium upward away from the electrode tip (128).

25.2.2.3. Velveteen for multiple inoculation

The multiple-inoculation method using velveteen is designed to determine the nutritional characteristics of a large number of strains. The procedure described below, in which the ability of 267 strains of pseudomonads to use 146 different organic compounds as carbon and energy sources are determined (115), exemplifies the method.

Prepare a mineral agar base with the following composition: 40 ml of 10 M Na_2HPO_4–KH_2PO_4 buffer (pH 6.8), 20 ml of Hutner's vitamin-free mineral base (section 25.3.19), 1.0 g of $(NH_4)_2SO_4$, 20 g of agar, and 940 ml of distilled water. Supplement the medium with 0.1 to 0.2% of the carbon source to be tested, and dispense the medium into petri dishes. Dry the sterile plates by incubating them at 37°C for 2 to 3 days, and store them at room temperature in plastic bags until used (up to 4 or 5 weeks of storage). Prepare a master plate by "patching" (depositing inoculum in a small area) up to 23 different strains on a plate of mineral agar base containing 0.5% yeast extract. After growth occurs, prepare submaster plates by the velveteen replication method originally devised by Lederberg and Lederberg (69); also see Chapter 13.5.4. In this method, a piece of sterilized velveteen cloth is affixed to a piston having a diameter slightly smaller than that of the petri dish. By gentle pressing of the velveteen onto the master plate and then onto a series of sterile plates, the plates become inoculated with the various strains in exactly the same pattern as on the master plate. Up to nine submaster plates can be prepared from a single master plate. After growth occurs on the submaster plates, use each plate to print test plates for nine different carbon sources, followed by a terminal yeast extract agar plate to verify that transfer of all the strains has occurred. Read the plates at 48 and 96 h, and score the results, as follows: 0, growth no greater than on a control plate lacking any carbon source; +, good growth; ±, scanty growth but significantly greater than on the control plate; and +M, late growth of a few colonies in the area of the patch, presumed to arise from mutants in the inoculum. When strains of markedly different nutritional characters are being tested together on a plate, cross-feeding (diffusion of nutrients produced by one strain across to adjacent strains, resulting in a false-positive growth response of the latter strains) or interference between strains may occur. Consequently, follow up each initial random screening by a second screening with a different arrangement or pattern of strains on the master plate.

25.2.2.4. Commercial system based on nutritional patterns

The Biolog Nutritional System (Biolog, Inc., Hayward, Calif.) is based on the utilization of nutrients and other chemicals by bacteria. A 96-well microplate with a different dehydrated nutrient in each well along with a tetrazolium redox indicator is used. A cell suspension is pipetted into the wells, and the plate is incubated for a few hours or overnight. Reduction of the colorless tetrazolium indicator to a purple formazan shows which substrates have been oxidized. Sterile pipettes and saline for the cell suspension must be used. The Biolog GN plate is available for gram-negative aerobic bacteria, and the GP plate is used for gram-positive bacteria. Another plate, the MT plate, contains only the tetrazolium redox dye and a buffered medium; the user adds various substrates to the wells according to personal choice. The system allows computerized identification of many bacterial species. Manual data entry software that accesses all of the system's data bases is available; alternatively, an automated plate reader can be used. There is also a computer program that allows users to develop their own data bases.

25.2.3. PAGE Patterns (also see Chapter 21.5.1)

Polyacrylamide gel electrophoresis (PAGE) patterns of water-soluble proteins can be useful for bacterial characterization and identification. Strains of a given species will have similar patterns, especially of the major proteins. Although the patterns are similar, they are not necessarily identical; for instance, some protein bands may be seen as dark bands in one strain and light ones in another strain. Some strains will yield distinctive patterns that help to identify the organism readily, whereas other organisms may show patterns that are not readily distinguishable from those of other species, and some additional tests may be needed to verify identification of the bacteria. In an electrophoretic study of clostridia, Cato et al. (23) found that strains with greater than 80% DNA homology usually produced identical patterns; strains with ca. 70% DNA homology produced patterns with some minor heterogeneity but an overall similarity; and strains that were unrelated by DNA homology produced patterns with major differences. Other examples of the usefulness of PAGE can be found in references 88, 90, and 100. Highly reproducible patterns can be produced under rigorously standardized conditions and can be analyzed by rapid, computerized, numerical analysis (62).

Both native (nondenaturing) gels and sodium dodecyl sulfate (SDS)-gels can be used. Although an SDS gel yields more bands, this seems to offer no advantages and, in fact, makes visual comparison of the patterns more difficult. The following method has been reported to be useful for the rapid identification of various gram-positive and gram-negative isolates from human gingival crevice floras (82).

Resolving gel (8.5%)
Add 11.3 g of acrylamide and 0.3 g of *N,N'*-methylenebisacrylamide to 50 ml of 0.375 M Tris-chloride buffer (pH 8.8). (The buffer can be prepared by dissolving 23.85 g of Tris preset pH crystals [pH 8.8 at 25°C; Sigma] in 500 ml of distilled water.) Heat the mixture to dissolve the ingredients, and make up to a volume of 133 ml with pH 8.8 buffer. Filter the solution through rapid-flow filter paper, and add 0.06 ml of TEMED (*N,N,N',N'*-tetramethylethylenediamine). Degas the solution by passing 32 ml through a 50-ml syringe. Add a 0.3-ml portion of a freshly prepared aqueous 10% solution (wt/vol) of ammonium persulfate, pour the gels, and allow them to polymerize for 10 to 20 min. Rinse the top of the gels with distilled water to remove any unpolymerized acrylamide.

Stacking gel (4.7%)
Add 2.5 g of acrylamide and 0.07 g of *N,N'*-methylenebisacrylamide to 15 ml of 0.15 M Tris-chloride buffer (pH 7.0). (The buffer can be prepared by dissolving 5.82 g of Tris preset pH crystals [pH 7.0 at 25°C; Sigma] in 250 ml of distilled water.) Store the solution in 10-ml amounts for no more than 1 week. Before use, add 7 drops of tracking dye (0.025% aqueous bromophenol blue) and 0.01 ml of TEMED to 10 ml of the solution. Degas, and add 0.1 ml of freshly prepared 10% aqueous ammonium persulfate. Add a comb to the resolving-gel plate, and pour the stacking gel. After the stacking gel has solidified (20 min), remove the comb and rinse the wells.

Electrode buffer (0.025 M Tris–0.25 M glycine buffer)
Dissolve 5.4 g of Tris and 27.0 g of glycine in distilled water, and make up to a final volume of 1,000 ml.

Sample preparation
Harvest the bacterial cells from a 5-ml broth culture by centrifuging at 8,000 × *g* for 10 min. To the unwashed cell pellet, add approximately 0.15 g of glass beads (74 to 110 μm in diameter, class IV, type C [Cataphore Division of Ferro Corp., Jackson, Miss.]) and 0.15 ml of Tris-chloride buffer (pH 7.0). Disrupt the cells by holding the tube against a Vortex mixer for 4 min. (Alternatively, the bacterial cells can be broken using other methods.) Add one-third volume of powdered sucrose to the sample. For gram-negative bacterial species, first centrifuge the extracts at 8,000 × *g* and heat at 55°C for 5 min before adding the sucrose, to reduce background in the gels.

Electrophoresis
Add 30 μl of sample to a well, and operate the electrophoresis apparatus at room temperature at 150 V (constant) and 33 mA initial current. Stop electrophoresis when the tracking dye reaches the bottom of the gel (about 2.5 h).

Gel staining
Place the gels in 12% trichloroacetic acid for 20 min. Then incubate them for 10 min in 0.08% Coomassie blue stain (4.0 g of Brilliant Blue R [60% dye content], 900 ml of methanol, 400 ml of glacial acetic acid, 1,800 ml of distilled water). Destain the gels in 10% (vol/vol) glacial acetic acid. Gels can be saved by being placed into plastic Ziploc bags.

25.2.4. Fatty Acid Profiles (also see Chapter 21.3.5.6)

Long-chain fatty acids (9 to 20 carbons) found in the membranes of bacteria can be used to characterize and identify these organisms. The kinds and relative amounts of the fatty acids in each bacterial species yield a profile that is unique to the species. By using a computer program and statistical analysis of the individual cellular fatty acids, the most probable identity of the isolate can be made when the results are compared with data bases in a library of known species. The accuracy of the system depends on using the recommended medium, incubation temperature, and time of incubation.

25.2.4.1. Gas chromatography of cellular long-chain fatty acids (83)

Suspend the growth from an agar medium in 0.5 ml of sterile distilled water, and place the suspension in a 13- by 100-mm tube with a Teflon-lined screw-cap. Saponify the cells by adding 1 ml of NAOH-methanol reagent (15% NaOH in 50% aqueous methanol), and heat at 100°C for 30 min. Cool to room temperature, and add 1.5 ml of HCl-methanol reagent (25% HCl in methanol). Heat the mixture to 100°C for 15 min, and then cool it to room temperature. Extract the methyl esters of the fatty acids by adding 1.5 ml of ether-hexane (1:1, vol/vol). Shake, and allow to stand for 1 to 2 min. Remove the aqueous (lower) layer with a Pasteur pipette, and discard it. Add 1 ml of phosphate buffer (pH 11.0) to wash the organic phase, shake, and allow to stand for 2 to 3 min. Remove and save ca. two-thirds of the organic (top) layer. Inject samples of this into a gas chromatograph equipped with a flame ionization detector. Use a fused silica capillary column (25 m by 0.2 mm [inner diameter]) containing cross-linked methylphenyl silicone (SE 54) as the stationary phase. Set the detector temperature to 300°C and the injector temperature to 250°C, and program the column oven temperature to increase from 170 to 300°C at a rate of 5°C/min and then maintain 300°C for 1 min. Peaks are recorded on an integrating recorder, and the retention times are compared with those of peaks obtained by using a standard mixture of methyl esters of fatty acids.

25.2.4.2. MIDI Microbial Identification System of bacteria by gas chromatography of cellular fatty acids

The MIDI Microbial Identification System (Midi, Newark, Del.) is an automated system for bacterial identification. Most organisms in the system's computer libraries are clinical bacteria; however, the system can be used for bacteria from any source if users create their own library of data from known organisms. To identify an isolate, harvest cells from colonies, free the fatty acids from the cellular lipids by saponification, and then methylate the fatty acids to form the volatile methyl esters. These are extracted and automatically injected into a gas chromatograph for separation, identification, and quantification. A computer compares the fatty acid profile with other profiles in a data base library of known organisms and selects the most likely name for the isolate. The system includes a gas chromatograph with automatic injection, an integrator, and a computer complete with data libraries for facultative, aerobic, and anaerobic bacteria. A custom library is also available for user-generated data on organisms not included in the available data libraries.

25.2.5. Isoprenoid Quinones (32, 83)

Isoprenoid quinones can be useful in characterizing bacteria. Some genera, such as *Campylobacter, Wolinella,* and *Helicobacter,* have been differentiated on the basis of their patterns of quinones.

Harvest the bacteria from the surfaces of five agar plates. Add 0.2 ml of 50% aqueous potassium hydroxide and 3 ml of 1% pyrogallol in methanol. Heat at 100°C for 10 min, and cool to room temperature. Add 1 ml of saturated aqueous NaCl solution and 5 ml of acetone-hexane (1:4, vol/vol), and shake for 5 min on a wrist action shaker. Let the phases separate, and remove the organic (upper) layer. Extract the aqueous layer two or three additional times, and combine all of the organic layers. Evaporate to dryness under a stream of nitrogen gas. Dissolve in 0.5 ml of methanol. Separate the quinones by reverse-phase HPLC. Use a 300- by 3.9-mm, 10-μm-particle size, C_{18} reverse-phase column (μBondapak; Waters Associates, Inc., Milford, Mass.). Elute the quinones with methanol-isopropanol-water (75:25:5, vol/vol/vol) at a flow rate of 1 ml/min. Monitor the quinones with a variable-wavelength detector at 248 and 275 nm. To identify the quinones, compare their retention times with those of authentic standards (such as those supplied by Hoffman-La Roche Co., Basel, Switzerland).

25.2.6. Multitest Systems

Various systems that allow a large number of biochemical characterization tests to be performed conveniently and rapidly are commercially available. The number of such test systems has grown in recent years. Complete information about the method can be obtained from the manufacturers. Charts, tables, coding systems, or characterization profiles designed to be used with the systems are also available from the manufacturers. Many of the systems now have computer programs that are used for bacterial identification. Table 6 provides a listing of some of these systems; although the list is not complete, it serves to indicate the diversity of multitest identification systems presently available. The reader should be aware that the purpose of these systems is the identification of clinical isolates and that they may not be reliable for environmental bacteria or isolates from nonclinical sources. In that case, investigators may have to develop their own charts and identification profiles by using species from culture collections as reference strains. The reader should see references 1 and 2 for additional information and information on the accuracy and reliability of these systems in clinical microbiology. It should be noted that no identification system is 100% accurate, although some systems are more reliable and accurate than others.

25.2.7. Nitrogenase Activity and Nitrogen Fixation

The development of the acetylene reduction method has made the testing of bacterial cultures for nitrogenase activity relatively easy in comparison with the use of $^{15}N_2$ incorporation methods, although the latter remain the most definitive. The principle of the acetylene reduction method is that the nitrogenase complex of enzymes is capable of reducing acetylene to ethylene. The organism to be tested is cultured under conditions conducive to nitrogenase formation. Acetylene gas is then added to the culture vessel, and the formation of ethylene is determined after a period of incubation. Controls involving media that contain ammonium ions, which repress nitrogenase synthesis (except for *Rhizobium* spp.), should exhibit little or no ethylene production in comparison. Comprehensive reviews of the acetylene reduction method and its application are available (50, 51, 99).

It is important to culture the organisms under conditions that are favorable for nitrogenase synthesis. This involves the use of media free from fixed nitrogen but containing a suitable carbon and energy source and also a source of molybdenum. Also, an appropriate gaseous atmosphere should be supplied. For anaerobes such as *Clostridium pasteurianum,* anaerobic conditions are essential. Anaerobic conditions are also the most favorable for nitrogenase synthesis by facultative anaerobes such as *Klebsiella* species. Many nitrogen fixers require oxygen for growth and energy production, but the oxygen tolerance for growth under nitrogen-fixing conditions varies considerably among such organisms. For example, *Azospirillum* cells grow best under nitrogen-fixing conditions at ca. 1% oxygen and do not grow at all under an air atmosphere (21% oxygen), whereas *Azotobacter* cells can tolerate an air atmosphere. Even aerobic nitrogen fixers, however, fix nitrogen best when grown at oxygen levels lower than that of air.

25.2.7.1. Aerobic nitrogen fixers

Grow the organisms in a mineral salts medium (liquid or solid) in cotton-stoppered tubes or serum bottles. An example of such a medium is the Burk medium for *Azotobacter* spp. (section 25.3.10). It is sometimes desirable to add a low level of a source of fixed nitrogen (e.g., 0.0001 to 0.005% yeast extract) to initiate growth. At the end of the incubation period, replace the cotton stopper with a sterile serum bottle stopper. Generate the acetylene gas (*CAUTION:* Flammable!) in a fume hood by the following procedure (99). Add a lump (ca. 1 g) of calcium carbide to a tube half filled with 15 ml of water; stopper the tube immediately with a one-hole stopper fitted to a piece of rubber tubing, and immerse the distal end of the tubing in a beaker of water. After air has been flushed out of the tubing by the acetylene but while the acetylene is still being formed, use a syringe equipped with a 26-gauge hypodermic needle to withdraw acetylene from the tubing. Inject the acetylene into the culture vessel through the serum bottle stopper

Table 6. Some commercial multitest systems for identification of bacteria

Manufacturer	System	Used for identification of:	Description
Abbott Laboratories Diagnostic Division, Abbott Park, Ill.	Quantum II	*Enterobacteriaceae*, gram-negative nonfermenters, and other gram-negative bacteria	Cartridge of multicompartment wells containing dehydrated substrates for 20 characterization reactions. Incubation, 4–5 h.
Analytab Products, Plainview, N.Y.	An-Ident	Anaerobes	Plastic strip of microtubes containing dehydrated substrates for 21 characterization reactions. Incubation (aerobic conditions), 4 h.
	API 20E	*Enterobacteriaceae*, gram-negative nonfermenters, and other gram-negative bacteria	Plastic strip of microtubes containing various dehydrated substrates for 20 characterization reactions. Incubation, 24–48 h.
	API 20S	Streptococci	Plastic strip of microtubes containing various dehydrated substrates for 20 characterization reactions. Incubation, 4 h.
	API 20A	Anaerobes	Plastic strip of microtubes containing various dehydrated substrates for 21 characterization reactions. Incubation (anaerobic conditions), 24 h.
	Quad FERM+	*Neisseria* spp. and *Branhamella catarrhalis*	Plastic strip of microtubes containing dehydrated substrates for 4 carbohydrate reactions and β-lactamase. Incubation, 2 h.
	Rapid E	*Enterobacteriaceae*	Similar to API 20E but with unbuffered media and smaller microtubes; 21 characterization reactions. Incubation, 4 h.
	Rapid STREP	Streptococci	Plastic strip of microtubes containing dehydrated substrates for 20 characterization reactions. Incubation, 4–24 h.
	Staph-Ident	Staphylococci	Plastic strip of microtubes containing various dehydrated substrates for 10 characterization reactions. Incubation, 5 h.
	Staph-TRAC	Human and veterinary staphylococci and micrococci	Plastic strip of microtubes containing dehydrated substrates for 20 characterization reactions. Incubation, overnight.
	UniScept 20PG	Staphylococci and streptococci	Plastic strip of of microtubes containing various dehydrated substrates for 20 characterization reactions. Incubation, 18–24 h. Automated instrumentation is available for inoculation, incubation, video image reading, addition of reagents, and interpretation.
	UniScept 20E	*Enterobacteriaceae*, gram-negative nonfermenters, and other gram-negative bacteria	Similar to UniScept 20PG; 23 characterization reactions. Incubation, 24–48 h.
	Rapid NFT	Gram-negative nonfermenters and some fermentative bacteria not belonging to the family *Enterobacteriaceae*	Plastic strip of microtubes containing various dehydrated substrates for 20 characterization reactions. Incubation, 24 h.
Austin Biological Laboratories, Austin, Tex.	RIM-N	*Neisseria* spp., *Branhamella catarrhalis*	Kit containing tubes with substrate solutions for four carbohydrate utilization tests. Incubation, 30–60 min.
Baxter Diagnostics, MicroScan Division, West Sacramento, Calif.	MicroScan	Gram-negative fermenters and nonfermenters, gram-positive cocci, *Listeria* spp., *Neisseria* spp., and anaerobes	Trays of various dehydrated substrates. Number of tests and incubation period vary with the bacterial group. Fluorogenic substrates yield fluorescence reactions that can be read with an automated reader, used in conjunction with a data base.
BBL Microbiology Systems, Cockeysville, Md.	Minitek II	*Enterobacteriaceae*, gram-negative nonfermenters, gram-positive cocci, or *Neisseria* spp., depending on the particular regimen of discs used	Plastic trays with wells to which are added an assortment of various paper discs impregnated with dehydrated substrates for four characterization reactions (for *Neisseria* spp.), 20 reactions (for gram-positive cocci), or 21 reactions (for *Enterobacteriaceae* and nonfermenters). Incubation period varies depending on the organisms being tested.
	Minitek Anaerobe II	Anaerobes	Similar to Minitek II; 20 characterization reactions. Incubation (anaerobic conditions), 48 h.

(Continued)

Table 6. *Continued*

Manufacturer	System	Used for identification of:	Description
Difco/Pasco Laboratories, Wheat Ridge, Colo.	Tri Panel	Gram-negative and gram-positive bacteria	Microplate containing hydrated substrates for 30 characterization tests. Plate stored at −20°C. Incubation, 16–20 h or 40–44 h, depending on organisms.
Innovative Diagnostic Systems, Atlanta, Ga.	RapID ANA II	Anaerobes	Plastic tray with cavities containing dehydrated substrates for 19 characterization reactions. Incubation (aerobic conditions), 4–6 h.
	RapID NH	*Neisseria, Haemophilus,* and *Moraxella* spp.	Plastic strip with cavities containing dehydrated substrates for 13 characterization tests. Incubation, 1–4 h.
	RapID SS/U	Urinary organisms: *E. coli, Klebsiella* spp., *Enterobacter* spp., *Proteus* spp., *Serratia* spp., *Citrobacter* spp., *Providenica* spp., *Morganella morganii*, enterococci, *Pseudomonmas* spp., staphylococci	Plastic panel containing dehydrated substrates for 11 characterization reactions. Incubation, 2 h.
	RapID STR	Group A, B, C, G streptococci, staphylococci, and some gram-positive rods	Plastic strips with cavities containing dehydrated substrates for 14 characterization tests. Incubation, 4 h.
Micro-Media Systems, San Jose, Calif.	Quad Enteric Panel	*Enterobacteriaceae*	Plastic microplate containing substrates for 20 characterization tests. Plate stored at −20°C. Incubation, 18–24 h.
Organon Teknika Corp., Durham, N.C.	Micro-ID	*Enterobacteriaceae*	Molded styrene tray containing 15 reaction chambers with various test substrates incorporated into paper discs; 15 characterization reactions. Incubation, 4 h.
Remel, Lenexa, Kans.	Uni-N/F-Tek	Gram-negative nonfermenters	Plastic plate with 11 pie-shaped wells and one center well, containing hydrated agar media; also provided are two screening tubes of agar media (N/F screen); 18 characterization reactions. Incubation, 18–24 and 48 h.
Roche Diagnostic Systems, Nutley, N.J.	Enterotube II	*Enterobacteriaceae*	A colony is picked onto the end of a long skewer that is drawn through a large tube containing a series of compartments, each having a hydrated agar medium with a different substrate; 15 characterization reactions. Incubation, 18–24 h.
	Oxi-Ferm	Gram-negative nonfermenters	Similar to Enterotube II; nine characterization reactions. Incubation, 24 and 48 h.
Sensititre, Salem, N.H.	Sensititre	*Enterobacteriaceae* and nonfermenters	Plastic microtiter plate containing dehydrated, fluorogenic substrates for 32 characterization tests. Incubation, 5–18 h. Fluorometer required.
Vitek Systems, Hazelwood, Mo.	ANI	Anaerobes	Cards with wells containing dehydrated substrates for 28 characterization tests. Incubation (aerobic conditions), 4 h.
	Bacillus Biochemical Card	*Bacillus* spp.	Cards with wells containing dehydrated substrates for 29 characterization tests. Incubation, 15 h.
	EPS (Enteric Pathogens Screen Card)	*Edwardsiella tarda, Salmonella* spp., *Salmonella typhi, Shigella* spp., *Yersinia enterocolitica*	Cards with wells containing dehydrated substrates for 10 characterization tests. Incubation, 4–8 h.
	GNI (Gram Negative Identification Card)	*Enterobacteriaceae* and other gram-negative rods	Cards with wells containing dehydrated substrates for 29 characterization tests. Incubation, 4–13 h.
	GPI (Gram Positive Identification Card)	Streptococci, staphylococci, and a few gram-positive rods	Cards with wells containing dehydrated substrates for 29 characterization tests. Incubation, 4–15 h.

(Continued)

Table 6. *Continued*

Manufacturer	System	Used for identification of:	Description
	NHI	*Neisseria* spp., *Haemophilus* spp.	Cards with wells containing dehydrated substrates for 15 characterization tests. Incubation, 4 h.
	Urine Identification 3 Card	*Citrobacter freundii*, *Serratia* spp., *Klebsiella* spp., *Enterobacter* spp., *Proteus* spp., *E. coli*, *Pseudomonas aeruginosa*, group D streptococci, staphylococci	Cards with wells containing dehydrated substrates for nine characterization tests. Incubation, 1–13 h.

to give 10% acetylene (vol/vol). After various periods of incubation, remove 1-ml gas samples from the culture vessel with a syringe, and test for the presence of ethylene by gas chromatography (see below).

25.2.7.2. Facultative and anaerobic nitrogen fixers

Bubble oxygen-free filtered nitrogen gas through sterile culture media in tubes, and inoculate the media while the tubes are being flushed with nitrogen. Seal the tubes with sterile serum bottle stoppers. Strictly anaerobic conditions may not be required for facultative bacteria such as *Klebsiella* spp. because the organisms themselves may use up small residual amounts of oxygen and create highly anaerobic conditions. For examples of media, see section 25.3.38 for *Klebsiella* medium, section 25.3.14 for *Clostridium pasteurianum* medium, and section 25.3.17 for *Bacillus* medium.

Alternatively, use 40-ml Pankhurst tubes (Astell Laboratory Service Co., Ltd., London, England) (22). The Pankhurst tube (Fig. 3) consists of two test tubes, one larger (tube A) and one smaller (tube C), with a horizontal connecting tube (tube B) filled with nonabsorbent cotton as a gas filter. Dispense 5- to 10-ml volumes of culture medium into the larger tube, and temporarily stopper the tube with cotton. Sterilize the medium by autoclaving. Inoculate the cooled medium, and then aseptically seal both the larger and the smaller tubes with sterile serum bottle stoppers. Flush the assembly with nitrogen gas (this step may not be required [99]) by inserting hypodermic needles into both stoppers, one to serve as an inlet for the flushing gas and the other to serve as an exit. After flushing, remove the needles and inject 0.75 ml of a saturated solution of pyrogallol into the smaller tube, followed by 0.75 ml of a solution containing 10% (wt/vol) NaOH and 15% (wt/vol) K_2CO_3. These additions are soaked up by a little absorbent cotton previously placed in the bottom of the small tube as shown in Fig. 3. The reagents react with residual oxygen and use it up, creating anaerobic conditions. If the Pankhurst tube has not been flushed with nitrogen, 2 to 3 h is required to establish anaerobic conditions, and ca. 12 ml of nitrogen should be added at this time to replace the oxygen that has been removed (99).

After the culture has grown, inject acetylene through the serum bottle stopper to give a final concentration of 10% (vol/vol). Incubate the culture for various periods, and use a syringe to withdraw 1-ml gas samples for gas chromatography to determine the presence of ethylene.

25.2.7.3. Microaerophilic nitrogen fixers

Semisolid nitrogen-free media have been successfully used for demonstrating acetylene reduction by nitrogen-fixing spirilla; e.g., see section 25.3.5 for a medium for *Azospirillum* spp. Semisolid media can be incubated in an air atmosphere because they provide a stratified oxygen gradient from the surface of the medium downward. This allows microaerophilic bacteria to initiate growth at whatever oxygen level is most suitable. Growth begins as a thin band or pellicle several millimeters below the surface of the medium; as cell numbers increase, the pellicle becomes denser and migrates closer to the surface. The tube should not be agitated during growth or during subsequent addition of acetylene, since any disturbance of the pellicle may allow air to reach the organisms and inactivate the nitrogenase.

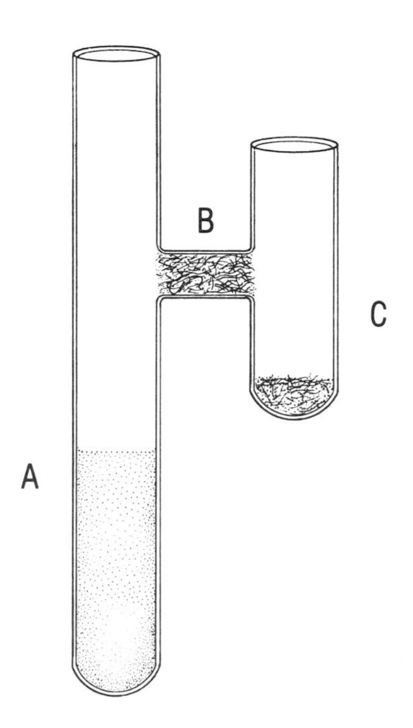

FIG. 3. Pankhurst tube. (A) Larger test tube (1.9 by 15.7 cm) containing culture; (B) connecting tube (0.8 by 1.8 cm) with nonabsorbent cotton as gas filter; (C) smaller test tube (1.9 by 6.0 cm) with absorbent cotton to hold alkaline pyrogallol.

After growth has occurred, replace the cotton stopper of the tube or vial with a serum bottle stopper, and inject acetylene to give a final concentration of 10% (vol/vol). After incubation for various periods, remove 1-ml gas samples for gas chromatography (see below).

25.2.7.4. Gas chromatography (also see Chapter 21.3.5)

A gas chromatograph equipped with a hydrogen flame detector should be used. The column is usually ca. 5 ft (1.5 m) by 2 mm (inside diameter), filled with Porapak R or N packing (Waters Associates). Keep the column oven temperature at 50°C, although any constant temperature between room temperature and 80°C will be satisfactory. Use nitrogen as the carrier gas, with a flow rate of ca. 50 cm³/min. Calibrate the chromatograph with dilutions of commercial ethylene (99% purity) in nitrogen.

Use plastic disposable syringes for injection of gas samples and ethylene standards. Because of retention of some ethylene on the plastic surfaces, discard any syringe that has previously contained ethylene and replace it with a fresh syringe. Rubber serum bottle stoppers may also become contaminated with ethylene and should not be used again.

Hypodermic needles may punch out a hole in a serum bottle stopper or in the rubber septum of the gas chromatograph injector port. To prevent this, bend the tip of the needle with a small pliers as shown in Fig. 4.

During chromatography with gas samples containing both acetylene and ethylene, the acetylene will take longer to pass through the Porapak column than the ethylene. Do not inject new samples into the gas chromatograph until the acetylene from the previous sample has passed from the column.

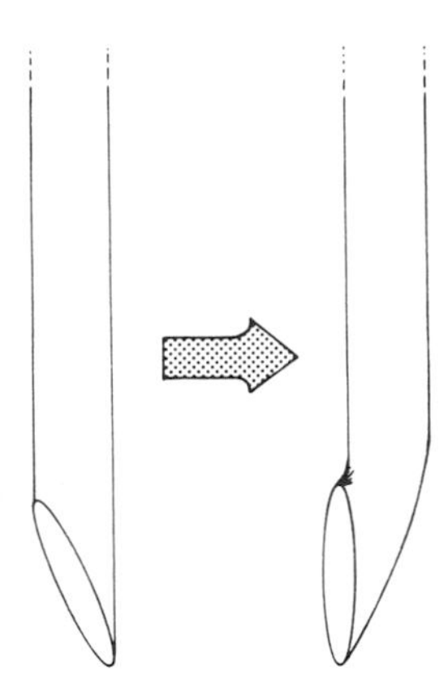

FIG. 4. Diagram showing how the end of a hypodermic needle can be bent with a needle-nose pliers so that the needle will not punch out a hole in a serum bottle stopper or in the rubber septum of a gas chromatograph injector port.

25.2.8. Antigen-Antibody Reactions

For a more detailed discussion of the principles on which these tests are based, see Chapter 5 and reference 121.

25.2.8.1. Direct bacterial agglutination

Slides with ceramic rings ca. 14 mm in diameter are preferred. However, glass plates marked off into 2.5-cm squares with a wax glass-marking pencil will suffice. Rehydrate the lyophilized antiserum as specified by the manufacturer. Place a drop of the antiserum into one of the rings, and place a drop of saline (0.85% NaCl) or normal rabbit serum in another ring. Mix a loopful of the growth from the surface of an agar medium into each drop, so that a homogeneous suspension of cells without lumps is obtained. The suspension should be visibly turbid but not extremely dense. Rock the slide in a circular fashion for 1 min to further mix the cells and liquid, being careful not to spill over the boundary of the ring. Examine the slide by eye; for best visibility hold the slide near a bright light and view it against a dark background. If the test is positive, the cells mixed with the antiserum show clumping (i.e., the suspension appears to be granular or, with strong reactions, even to "curdle"). Clumps are best seen by tilting the slide slightly so that fluid drains down toward the lower boundary of the ring. The saline or normal serum control should not show clumping. If it does, this usually indicates a rough (R) strain instead of a smooth (S) strain. *Note:* The bacteria-antiserum mixture must not be allowed to dry, since no indication of agglutination can be obtained if drying occurs. If drying is a problem, repeat the test using a larger drop of antiserum.

CAUTION: After completion, discard the slide into a container of disinfectant (e.g., 1% Lysol). Wash hands and bench surface with disinfectant. This is important because during the mixing of cells with antiserum, small droplets may be scattered about on the slide and bench surface. The antiserum will not kill the bacteria.

25.2.8.2. Coagglutination

This test is based on the use of *Staphylococcus aureus* cells that have an outer coating of protein A, which can combine directly with the Fc portion of immunoglobulin G (IgG) molecules from almost all mammals. When killed staphylococci are incubated with antibodies of the IgG class, the Fc portion of the antibodies binds to the protein A on the surface of the staphylococci. This leaves the Fab portions of the antibodies available for binding to a specific antigen. The antibody-coated staphylococci are suspended in saline and stored in a refrigerator until needed.

When the staphylococci are mixed with a saline suspension of bacterial cells for which the IgG antibodies are specific, the Fab portions of the IgG bind to the surface of the cells, causing visible clumping of the cells. A positive test is indicated by a granular appearance, whereas a negative test is indicated by a homogenous, milky appearance.

An example of a commercially available coagglutination system is the Phadebact system for the identification of streptococci (Karo Bio Diagnostics AB, Huddinge, Sweden; distributed in the United States by Remel, Lenexa, Kans.). In this system, stained staphylococcal cells are coated with antibodies against various Lancefield streptococcal grouping antigens (for example, the group A polysaccharide antigen characteristics of *Streptococcus pyogenes*). The coagglutination reaction produces readily visible clumps on a plastic-coated card, and no magnification is required. The procedure is simpler and much more rapid than that for the classical precipitin tests for Lancefield antigens (see below), but the manufacturer's directions should be strictly followed.

25.2.8.3. Latex agglutination

In latex agglutination, tiny beads of polystyrene latex serve as an inert carrier of antibodies or antigens. The Fc portion of IgG molecules can bind spontaneously to latex; once the Fc portion of an IgG molecule attaches hydrophobically to the polystyrene surface, the attachment is virtually irreversible. Various soluble protein antigens, such as the proteins from a bacterial sonicate, may also bind spontaneously to latex. Latex beads can be obtained in various sizes but usually have an average diameter of 0.81 or 1.0 μm. They can be obtained with amino, carboxyl, and other functional chemical groups so that antigens can be covalently coupled to the beads. Colored, fluorescent, and magnetic latex beads are also available. Latex beads and coupling reagents can be obtained from Difco and from Polyscience, Inc., Warrington, Pa.

To prepare antibody-coated beads, add the inert latex particles (usually obtained as a 2.5% suspension) to antibodies of the IgG class and incubate the latex-antibody mixture for several hours (the time should be determined by the investigator). Then wash the particles, and suspend them in saline. Place a drop of the latex-antibody suspension on a slide (commercially available slides that have ceramic rings are convenient), and mix a drop of an antigen (either a soluble antigen or a whole-cell antigen) with the antibody-coated beads. A positive test is indicated by agglutination or clumping of the particles; a negative test shows no change in the milky suspension.

An example of a latex slide agglutination test is the Campyslide Test (BBL), which is used to confirm the identification of various *Campylobacter* species at the genus level. The test is based on the use of antibody-coated latex beads for detection of certain cell wall antigens that are shared by the species (*C. jejuni, C. coli, C. fetus, C. lari*).

25.2.8.4. Hemagglutination

In hemagglutination, erythrocytes are used as carriers instead of bacterial cells or latex beads. Many polysaccharide and protein antigens can bind directly to the surface of erythrocytes. Treating erythrocytes with tannic acid stabilizes the cell membrane and gives the erythrocytes a longer shelf life.

To prepare tanned erythrocytes, pipette 3 ml of sheep blood into a 15-ml centrifuge tube, add 10 ml of physiological saline, mix, and centrifuge for 5 min at 500 × *g*. Wash the pellet three times in saline. After the final centrifugation, suspend the cells to a density of 0.5 ml of packed erythrocytes per 20 ml of saline. Add 2 ml of this cell suspension to 10 ml of distilled water. Read the optical density at 520 nm, and adjust it to 0.4 and 0.6. Pipette 3 ml of the cell suspension into a centrifuge tube, and add 3 ml of 0.005% tannic acid (prepared in saline). Mix, and incubate at 37°C in a water bath for 10 min. Centrifuge at 500 × *g* for 5 min. Suspend the cells in buffered saline (pH 7.2), wash them once, and suspend them in 3 ml of buffered saline.

To coat the tanned erythrocytes with a soluble antigen, add 4 ml of buffered saline to a centrifuge tube, and add 1 ml of antigen solution and 1 ml of tanned erythrocytes. Incubate the mixture at room temperature for 10 min. Centrifuge it at 500 × *g* for 5 min. Suspend the cells in 2 ml of a diluent consisting of a 1:100 dilution of normal rabbit serum in saline.

A hemagglutination test for antibodies, antigens, or other substances that react with coated erythrocytes can be done by using 13- by 100-mm tubes or a plastic microplate with round-bottom wells. A positive hemagglutination reaction is indicated by an even layer of erythrocytes coating the bottom of the tube; a negative test is indicated by a discrete button of cells at the bottom of the tube.

An interesting example of a hemagglutination test is the Staphyloslide test for identification of *S. aureus* (BBL). In this test, fibrinogen-coated sheep erythrocytes are used to detect a cell wall polypeptide called clumping factor, which is produced by most strains of *S. aureus*. After the staphylococcal cells have been mixed with the coated erythrocytes, the fibrinogen on the erythrocytes reacts with the clumping factor and is converted to insoluble fibrin; this in turn causes the erythrocytes to agglutinate.

25.2.8.5. Quellung reaction

Spread a loopful of broth culture on a clean slide, and allow it to air dry without heating. (If a similar smear stained with crystal violet shows more than 15 to 25 organisms per oil immersion field, dilute the culture, since too high a concentration of cells may obscure the results.) Rehydrate the antiserum, and place a large loopful of it on a coverslip. Mix a loopful of 1% aqueous methylene blue into the antiserum, and then place the coverslip, fluid side down, onto the dried smear on the slide. Examine by oil immersion. If the test is positive, the capsules around the blue-stained cells will appear distinctly outlined. A comparison should always be made with a control slide prepared with normal rabbit serum instead of antiserum. If the results of the test are negative, reexamine after 1 h; prevent the preparation from drying by incubating it in a moist chamber or by sealing the edges with melted Vaspar (a mixture of equal volumes of petrolatum and mineral oil).

25.2.8.6. Fluorescent-antibody tests

Rehydrate the fluorescent antiserum (termed the conjugate) and fluorescent normal serum. Prepare a series of twofold dilutions of the fluorescent antiserum (e.g., 1:5, 1:10, 1:20, 1:40, and 1:80) in phosphate-buffered saline (PBS; see section 25.4.16). Using the staining procedure given below with a known culture of the organism, and using a suitable fluorescence microscope system (see Chapter 1.6), determine the degree of fluorescence of the bacterial cells as obtained with each antiserum dilution. Rate the fluorescence on a scale of 0, 1+, 2+, 3+, and 4+ (maximum). The working dilution of the antiserum is one-half of the highest dilution that gives 4+ fluorescence. For example, if a 1:20 dilution of the serum is the highest dilution that gives 4+ fluorescence, the working dilution would be 1:10. An equivalent dilution of the fluorescent normal serum should give no fluorescence.

Centrifuge a young broth culture of the organism to be tested, and decant the supernatant fluid. Suspend the cells in PBS, and recentrifuge. Suspend the cells to a dense concentration in a small amount of PBS. With a diamond or tungsten carbide pencil, circumscribe a circle about the diameter of a quarter (ca. 20 to 25 mm in diameter) on a glass slide. Clean the slide thoroughly so that it is free from grease, and smear a loopful of bacterial suspension within the circle. Allow the smear to air dry without heat. Place the slide in a jar of 95% ethanol for 1 min, remove, and allow it to air dry.

Add several drops of the working dilution of fluorescent antiserum to the smear. Distribute the antiserum over the entire smear with an applicator stick, but do not touch the smear itself with the stick. Place the slide in a petri dish in which a moistened disc of filter paper has been affixed to the lid; this serves as a humid chamber to prevent evaporation of the antiserum. Incubate at 37°C for 15 to 20 min.

Drain off the excess antiserum by tilting the slide onto a piece of absorbent paper. Place the slide in a jar of PBS for 10 min, occasionally moving the slide up and down. Repeat this two more times, using a fresh jar of PBS each time. After the third washing in PBS, dip the slide only once into a jar of distilled water to remove the PBS and then immediately blot the slide (without rubbing) with absorbent paper. Allow the slide to dry. Place a small drop of buffered glycerol mounting fluid (section 25.4.2) on the smear, and apply a coverslip. Examine the smear under the high-power dry lens or oil-immersion lens, using a fluorescence microscope (see Chapter 1.5). When using oil immersion, be sure to use only nonfluorescing oil.

At the same time as the test organism is being stained, prepare two controls by a similar procedure. Make one control with the working dilution of the fluorescent antiserum and a known culture of the organism. Make the second control with the organism being tested, but use a dilution of fluorescent normal serum. The first control should give 4+ fluorescence, and the second should give no fluorescence.

Note: When examining smears under the fluorescence microscope, be sure to move the field often, because the fluorescence fades quickly (in ca. 15 s) when any particular field is being observed. Fluorescence is always brightest when a fresh field is moved into position.

Note: The working dilution of the fluorescent antiserum should be prepared fresh daily. The undiluted stock antiserum is stable when stored in a refrigerator below 8°C.

25.2.8.7. Precipitin test (for identification of streptococci)

Culture the organism in 40 ml of Todd-Hewitt broth (section 25.3.36) contained in 50-ml plastic (polycarbonate or polypropylene) screw-cap centrifuge tubes. After incubation at 37°C for 18 to 24 h, centrifuge the culture at top speed in a clinical-type centrifuge with a "swinging-bucket" type of rotor. Decant the supernatant fluid carefully into an Erlenmeyer flask containing disinfectant, and wipe the lip of the centrifuge tube with a towel moistened with disinfectant. Be sure to decant all of the supernatant fluid without suspending the sediment (cells) at the bottom of the tube. Add 0.5 ml of saline (0.85% NaCl) to the sediment, and suspend the organisms by gently agitating the tube from side to side. Replace the screw cap tightly, and autoclave the tube for 15 min at 121°C. Centrifuge again at top speed to sediment the autoclaved cells. Taking care not to disturb the packed cells, withdraw most of the supernatant fluid with a Pasteur pipette into a small tube. The fluid must be perfectly clear; if it is cloudy, it must be recentrifuged. The supernatant fluid contains the soluble cell wall antigens which have been extracted from the streptococcal cell walls.

Rehydrate the antiserum as specified by the manufacturer. Dip a capillary tube (75 to 90 mm by 0.7 to 1.0 mm [inside diameter]) in a tilted position into the antiserum, so that 2 to 3 cm of serum rises into the capillary. Place the forefinger over the opposite end of the capillary, remove the capillary from the serum bottle, and wipe off excess serum from the outside of the capillary. Dip the end of the capillary tube into the antigen extract, and remove the forefinger to allow a rise of 2 to 3 cm of antigen. There must be no airspace between the antiserum and the extract in the capillary as the extract rises. Replace the forefinger, and withdraw the capillary. Tilt the capillary to a nearly horizontal position, and with the finger removed, allow the column of fluid to move to the center part of the capillary. Leave an airspace of at least 1 cm at each end. Plunge the end of the capillary into a small lump of modeling clay. Wipe the surface of the capillary free from fingerprints, and incubate it for 10 to 15 min. Examine by holding the capillary near a strong light and viewing it against a dark background with a hand lens. A positive test is indicated by development of a milky haze near the center of the capillary at the junction of the antiserum and extract. With further incubation this haze will become a precipitate that settles to the bottom of the column of fluid.

25.3. MEDIA

25.3.1. "A" Medium of King et al. (64)

Peptone	20.0 g
Glycerol	10.0 ml
K_2SO_4 (anhydrous)	10.0 g
$MgCl_2$ (anhydrous)	1.4 g
Agar (dried)	13.6 g
Distilled water	1,000 ml

Combine ingredients, and adjust the pH to 7.2. Boil to dissolve the agar, and sterilize by autoclaving at 121°C for 15 min. Cool in a slanted position.

25.3.2. Acetamide Agar (47)

Acetamide	10.0 g
NaCl	5.0 g
K_2HPO_4	1.39 g
KH_2PO_4	0.73 g
$MgSO_4 \cdot 7H_2O$	0.5 g
Phenol red	0.012 g
Agar	15.0 g
Distilled water	1,000 ml

Combine ingredients, and adjust the pH to 6.9 to 7.2.

Boil to dissolve the agar, and sterilize by autoclaving at 121°C for 15 min. Cool in a slanted position.

25.3.3. Arginine Broth of Niven et al. (89)

Yeast extract	5.0	g
Tryptone	5.0	g
K_2HPO_4	2.0	g
Glucose (dextrose)	0.5	g
L-Arginine · HCl	3.0	g
Distilled water	1,000	ml

Note: Although the original formula designated the D form of arginine, the L form is satisfactory (29).

Combine the ingredients, and dissolve with heating. Adjust the pH to 7.0. Boil, filter, and sterilize by autoclaving.

25.3.4. Arginine Broth of Evans and Niven (38)

Tryptone	10.0	g
Yeast extract	5.0	g
NaCl	5.0	g
L-Arginine · HCl	3.0	g
Distilled water	1,000	ml

Combine the ingredients, and dissolve with heating. Adjust the pH to 7.0. Boil, filter, and sterilize by autoclaving.

25.3.5. Azospirillum Semisolid Nitrogen-Free Malate Medium (36)

KH_2PO_4	0.4	g
K_2HPO_4	0.1	g
$MgSO_4 \cdot 7H_2O$	0.2	g
NaCl	0.1	g
$CaCl_2 \cdot 2H_2O$	0.026	g
$FeCl_3 \cdot 6H_2O$	0.017	g
$Na_2MoO_4 \cdot 2H_2O$	0.002	g
L-Malic acid	3.58	g
(or sodium malate)	(5.00	g)
Bromothymol blue, 0.5% alcoholic solution	5.0	ml
Agar	1.75	g
Distilled water	1,000	ml

Combine the ingredients, and adjust the pH to 6.8 with KOH. Sterilize by autoclaving. *Note:* Some *Azospirillum* strains require biotin. Add 10 mg of the vitamin to 100 ml of distilled water, heat until dissolved, and add 1 ml of this stock solution per liter of nitrogen-free malate medium.

25.3.6. "B" Medium of King et al. (64)

Pancreatic digest of casein USP	10.0	g
Peptic digest of animal tissue USP	10.0	g
K_2HPO_4	1.5	g
$MgSO_4 \cdot 7H_2O$	1.5	g
Agar (dried)	14.0	g
Distilled water	1,000	ml

Combine the ingredients, and adjust the pH to 7.2. Boil to dissolve the agar, and sterilize by autoclaving at 121°C for 15 min. Cool in a slanted position.

25.3.7. Bile Agar (29)

Beef extract	10.0	g
Peptone	10.0	g
NaCl	5.0	g
Distilled water	1,000	ml

Combine the ingredients, and dissolve with heating. Adjust the pH to 8.0 to 8.4 with 10 N NaOH. Boil for 10 min. Filter, adjust the pH to 7.2 to 7.4, and add 10.0 g of dehydrated oxgall. Readjust the pH if necessary. Add 15.0 g of agar, boil, and sterilize by autoclaving. Cool to 55°C, and aseptically add 50 ml of sterile blood serum. Mix, and dispense into plates. *Note:* 1% dehydrated oxgall = 10% bile (vol/vol); to make "40% bile agar," use 4% dehydrated oxgall.

25.3.8. Blood Agar

To sterile blood agar base (see section 25.3.34) which has been melted and cooled to 45 to 50°C, add 5% (vol/vol) sterile defibrinated blood, mix well, and dispense into plates. Avoid getting plates with bubbles or froth on the surface. Defibrinated sheep or rabbit blood is preferred. Horse blood may give incorrect hemolytic reactions. Human blood-bank blood contains citrate and glucose; the citrate may be inhibitory to some bacteria, and the glucose may cause false greening rather than true alpha-hemolysis. Antibiotics or chemotherapeutic agents should not be present in blood used for making blood agar.

Prepare defibrinated blood by drawing whole blood aseptically with a syringe equipped with an 18-gauge needle (to prevent hemolysis caused by mechanical damage) and dispensing the blood immediately into a sterile flask containing a layer of sterile glass beads (ca. 3 mm). Shake the flask from side to side for 10 min. The fibrin formed during clotting will be deposited on the beads. Decant the supernatant containing blood cells and serum into a sterile container, and store it in a refrigerator.

25.3.9. Board and Holding Medium (19)

$(NH_4)H_2PO_4$	0.5	g
K_2HPO_4	0.5	g
Yeast extract	0.5	g
Bromothymol blue	0.03	g
Mineral solution (see section 25.3.19)	20	ml
Agar	5.0	g
Distilled water	880	ml

Dissolve the first three ingredients in the distilled water; add the indicator and the mineral solution. Adjust the pH to 7.2. Add the agar, and boil to dissolve. Disperse 9.0-ml volumes into tubes, and sterilize by holding momentarily at 22 lb/in². Cool to 45 to 50°C. A precipitate may form during autoclaving but should redissolve as the solution cools. To each tube, add 1.0 ml of a 5% solution of carbohydrate (usually glucose) which has previously been sterilized by filtration. Mix, and allow the medium to cool.

25.3.10. Burk Medium, Modified (95)

Solution A

$MgSO_4 \cdot 7H_2O$	0.2	g
$CaSO_4 \cdot 2H_2O$	0.1	g
$FeSO_4 \cdot 7H_2O$	0.005	g
$Na_2MoO_4 \cdot 2H_2O$	0.00024	g
Glucose	10.0	g
Distilled water	500	ml

Solution B

Potassium phosphate buffer, 0.005 M (pH 7.1)	500	ml

Autoclave solutions A and B separately, and combine equal volumes when cool.

25.3.11. Chopped-Meat (CM) Broth, Prereduced (4)

Ground beef, fat-free; use lean beef (cutter-canner grade is suitable) or horsemeat	500	g
Distilled water	1,000	ml
NaOH, 1 N	25	ml

Remove fat and connective tissue before grinding meat. Mix meat, water, and NaOH; bring to a boil while stirring. Cool to room temperature, skim fat off the surface, and filter, retaining both meat particles and filtrate. To the filtrate, add sufficient distilled water to restore the original 1-liter volume. To this filtrate add:

Trypticase	30.0	g
Yeast extract	5.0	g
K_2HPO_4	5.0	g
Resazurin solution, 0.025%	4.0	ml

Boil the mixture in a flask equipped with a chimney (to prevent boiling over). For a 1-liter flask, use 750 ml of the broth; for a 750-ml flask, use 500 ml of broth. Boil until the resazurin changes from pink to colorless. Remove the flask from the heat, and replace the chimney with a two-hole stopper that has a gas cannula in one hole. Bubble oxygen-free CO_2 through the medium while cooling it to room temperature in an ice bath. Add the following ingredients (per liter):

Cysteine	0.5	g
Hemin solution (see section 25.3.29)	10	ml
Vitamin K_1 solution (see section 25.3.29)	0.2	ml

Adjust the pH to 7.2 with 8 N NaOH. Continue bubbling CO_2 until the pH is lowered to 7.0. Bubble oxygen-free nitrogen through the medium while dispensing it into 18- by 142-mm tubes (Bellco Glass, Inc., Vineland, N.J.) that are being flushed with oxygen-free nitrogen. The tubes should already contain the meat particles prepared above so as to give 1 part of meat particles per 4 to 5 parts of added broth. Stopper the tubes with black rubber stoppers (size no. 1) as the gas cannula is being removed from the tubes. Autoclave the tubes in a press (to prevent the stoppers from being blown off) for 30 min. Cool. Discard any tubes that develop a light pink color on standing (indicating oxidation of the medium).

25.3.12. Christensen Citrate Agar (26)

Sodium citrate	3.0	g
Glucose	0.2	g
Yeast extract	0.5	g
Cysteine hydrochloride	0.1	g
Ferric ammonium citrate	0.4	g
KH_2PO_4	1.0	g
NaCl	5.0	g
Sodium thiosulfate	0.08	g
Phenol red	0.012	g
Agar	15.0	g
Distilled water	1,000	ml

Combine ingredients, and adjust the pH to 6.7. Sterilize by autoclaving. Cool in a slanted position for a 2.5-cm butt and 3.8-cm slant.

25.3.13. Christensen Urea Agar (25)

Peptone	1.0	g
Glucose	1.0	g
NaCl	5.0	g
KH_2PO_4	2.0	g
Phenol red	0.012	g
Agar	20.0	g
Distilled water	1,000	ml
Yeast extract (optional)	(0.1)	g

Combine ingredients, and adjust the pH to 6.8 to 6.9. Boil to dissolve the agar, and sterilize by autoclaving. Cool to 50°C. Aseptically add sufficient 20% urea solution (sterilized by filtration) to give a final concentration of 2% urea. Mix, and aseptically dispense 2- to 3-ml volumes into sterile small tubes. Cool in a slanted position to give a 1.3-cm butt and a 2.5-cm slant.

25.3.14. *Clostridium pasteurianum* Nitrogen-Free Medium (30)

$MgSO_4 \cdot 7H_2O$	0.0493	g
$FeCl_3 \cdot 6H_2O$	0.0541	g
$MnSO_4 \cdot H_2O$	0.0034	g
$Na_2MoO_4 \cdot 2H_2O$	0.0048	g
$ZnSO_4 \cdot 7H_2O$	0.00058	g
$CuSO_4 \cdot 5H_2O$	0.00050	g
$CoCl_2 \cdot 6H_2O$	0.00048	g
K_2HPO_4	1.132	g
Biotin	Trace	
Sucrose	20.0	g
$CaCO_3$	3.0	g
Distilled water	1,000	ml

Note: The amount of $CaCO_3$ can be increased to 30 g for increased buffering.

Combine ingredients, and autoclave. The medium must be made anaerobic before inoculation by bubbling oxygen-free nitrogen through it.

25.3.15. Egg Yolk Agar (4)

Peptone	20.0 g
Na_2HPO_4	2.5 g
NaCl	1.0 g
$MgSO_4$, 0.5% (wt/vol) solution	0.1 ml
Glucose	1.0 g
Agar	12.5 g
Distilled water	500 ml

Combine ingredients, and adjust the pH to 7.3 to 7.4. Boil to dissolve the agar, and sterilize by autoclaving. Cool to 60°C in a water bath. Disinfect the surface of an egg with alcohol, and allow the egg to dry. Crack the egg, and separate the yolk from the white. Add the yolk aseptically to the melted agar, and mix to obtain a homogeneous suspension. Dispense into plates, and allow to solidify. For anaerobes, use plates within 4 h or store them in an anaerobe jar until needed.

25.3.16. Haynes' Broth (52)

Tryptone	1.5 g
Yeast extract	1.0 g
K_2HPO_4	1.0 g
Potassium gluconate	40.0 g

Combine ingredients, and adjust the pH to 7.0. Sterilize by filtration.

25.3.17. Hino and Wilson Nitrogen-Free Medium for *Bacillus* Species (54)

Solution A

Sucrose	20.0 g
$MgSO_4 \cdot 7H_2O$	0.5 g
NaCl	0.01 g
$FeSO_4 \cdot 7H_2O$	0.015 g
$Na_2MoO_4 \cdot 2H_2O$	0.005 g
$CaCO_3$	10.0 g
Distilled water	500 ml

Solution B

p-Aminobenzoic acid	10 μg
Biotin	5 μg
K_2HPO_4-KH_2PO_4 buffer, 0.1 M (pH 7.7)	500 ml

Autoclave solutions A and B separately. When cool, combine equal volumes aseptically.

25.3.18. Hugh and Leifson O/F Medium (56)

Peptone (pancreatic digest of casein)	2.0 g
NaCl	5.0 g
K_2HPO_4	0.3 g
Bromothymol blue	0.03 g
Agar	3.0 g
Distilled water	1,000 ml

Combine ingredients, and adjust the pH to 7.1. Boil to dissolve the agar, and dispense 3- to 4-ml volumes into 13- by 100-mm tubes. Sterilize by autoclaving. Cool to 45 to 50°C. Add sufficient 10% carbohydrate solution (usually glucose) sterilized by filtration to each tube to give a final concentration of 1.0%. Mix, and allow the tubes to cool.

25.3.19. Hutner's Mineral Base (27)

Mineral solution

Nitrilotriacetic acid	10.0 g
$MgSO_4$	14.45 g
$CaCl_2 \cdot 2H_2O$	3.335 g
$(NH_4)_6Mo_7O_{24} \cdot 4H_2O$	0.00925 g
$FeSO_4 \cdot 7H_2O$	0.099 g
Stock salts solution (see below)	50 ml
Distilled water	950 ml

Stock salts solution

EDTA	2.5 g
$ZnSO_4 \cdot 7H_2O$	10.95 g
$FeSO_4 \cdot 7H_2O$	5.0 g
$MnSO_4 \cdot H_2O$	1.54 g
$CuSO_4 \cdot 5H_2O$	0.392 g
$Co(NO_3)_2 \cdot 6H_2O$	0.248 g
$Na_2B_4O_7 \cdot 10H_2O$	0.177 g
Distilled water	1,000 ml

Add a few drops of H_2SO_4, to the stock salts solution to decrease precipitation. To prepare the mineral solution, dissolve the nitrilotriacetic acid in the distilled water, and neutralize with approximately 7.3 g of KOH. Add the remaining ingredients, and adjust the pH to 6.8.

25.3.20. Leifson Modified O/F Medium (70)

Casitone (Difco)	1.0 g
Yeast extract	0.1 g
$(NH_4)_2SO_4$	0.5 g
Tris buffer	0.5 g
Phenol red	0.01 g
Agar	3.0 g
Artificial seawater (see section 25.3.32)	500 ml
Distilled water	500 ml

Dissolve ingredients in the distilled water, adjust the pH to 7.5, and sterilize by autoclaving. Autoclave the seawater separately. After cooling to 45 to 50°C, mix the two solutions aseptically. Add a sufficient volume of a 10 or 20% solution of the desired carbohydrate (usually glucose) sterilized by filtration to give a final concentration of 0.5 or 1.0%. Dispense 2.5- to 3.0-ml volumes of the medium aseptically into 13- by 100-mm tubes, and allow them to cool.

25.3.21. Litmus Milk (8)

Alcoholic litmus solution. Grind 50 g of litmus in a mortar with 150 ml of 40% ethanol. Transfer to a flask, and boil gently on a steam bath for 1 min. Decant the fluid, and save it. Add another 150 ml of 40% ethanol to the residue, and boil again for 1 min. Decant, and combine the two supernatants. Allow to settle overnight, and dilute up to 300 ml with 40% ethanol. Add 1 N HCl drop by drop to adjust the pH to 7.0.

Preparation of medium. Add sufficient litmus solution to skim milk to give a bluish-purple color (usually 40 ml/liter). Adjust the pH to 7.0 with 1 N NaOH. Sterilize by autoclaving at 115°C for 10 min. Overheating will result in caramelization.

25.3.22. Lysozyme Resistance Test Medium (2)

Peptone (Difco)	5.0	g
Beef extract (Difco)	3.0	g
Glycerol	70	ml
Distilled water	1,000	ml

Dispense 500 ml of this medium into 16- by 125-mm screw-cap glass test tubes, 5 ml per tube. Dispense the remaining medium into screw-cap dilution bottles, 95 ml per bottle. Autoclave the tubes and bottles for 15 min at 121°C. After cooling, tighten the screw caps and store at 4°C for a maximum of 2 months. To prepare lysozyme-containing medium, add 5 ml of sterile lysozyme solution (see below) to a 95-ml portion of the sterile medium and then dispense 5.0-ml portions aseptically into sterile 16- by 125-mm screw-cap test tubes. The lysozyme-containing medium can be stored at 4°C for a maximum of 2 weeks.

Lysozyme solution

Lysozyme (Sigma)	100	mg
HCl, 0.01 N	100	ml

Sterilize by filtration through a membrane filter (pore size, 0.45 μm).

25.3.23. Malonate Broth (39)

Yeast extract	1.0	g
$(NH_4)_2SO_4$	2.0	g
K_2HPO_4	0.6	g
KH_2PO_4	0.4	g
NaCl	2.0	g
Sodium malonate	3.0	g
Glucose	0.25	g
Bromothymol blue	0.025	g
Distilled water	1,000	ml

Combine ingredients, and adjust the pH to 7.0. Sterilize by autoclaving.

25.3.24. Milk Medium, Prereduced (4)

Fresh skim milk	100	ml
Resazurin solution, 0.025%	0.4	ml
Hemin solution (section 25.3.30)	1.0	ml
Vitamin K_1 solution (section 25.3.30)	0.02	ml

Add the milk and resazurin to a flask, and prepare as described for PY broth (section 25.3.30). Add the hemin and vitamin K_1 after boiling and cooling. The pH after bubbling with CO_2 should be 7.1. Dispense the medium into tubes being purged with nitrogen, stopper the tubes, and autoclave them in a press for 12 min at 121°C.

25.3.25. Møller's Broth Base (81)

Peptide digest of animal tissue USP, (see *Note*)	5.0	g
Beef extract	5.0	g
Bromocresol purple solution, 1.6%	0.625	ml
Cresol red solution, 0.2%	2.5	ml
Glucose	0.5	g
Pyridoxal	0.005	g
Distilled water	1,000	ml

Note: The original formula designates Orthana special peptone; however, Thiotone (BBL; peptic digest of animal tissue) is satisfactory (10).

Combine ingredients, and adjust the pH to 6.0 or 6.5. Dispense into 13- by 100-mm tubes, and sterilize by autoclaving.

25.3.26. MRVP (Methyl Red–Voges-Proskauer) Broth

Polypeptone or buffered peptone	7.0	g
K_2HPO_4 (see *Note*)	5.0	g
Glucose	5.0	g
Distilled water	1,000	ml

Note: For testing the Voges-Proskauer reaction of members of the genus *Bacillus*, replace the phosphate with 5.0 g of NaCl (8).

Combine ingredients, and adjust the pH so that after autoclaving and cooling the pH is 6.9 (6.5 when used for *Bacillus* species). Dispense in 5-ml volumes, and sterilize at 121°C for 10 min.

25.3.27. Nutrient Agar

Beef extract	3.0	g
Peptone	5.0	g
Agar	15.0	g
Distilled water	1,000	ml

Combine ingredients, and adjust the pH to 6.8. Sterilize by autoclaving.

25.3.28. Phenylalanine Agar

Yeast extract	3.0	g
L-Phenylalanine	1.0	g
Na_2HPO_4	1.0	g
NaCl	5.0	g
Agar	12.0	g
Distilled water	1,000	ml

Combine ingredients, and adjust the pH to 7.3. Boil to dissolve the agar, dispense into tubes, and sterilize by autoclaving. Cool the tubes in a slanted position.

25.3.29. PY Agar, Prereduced (4)

Prepare PY broth (section 25.3.30), and dispense 10-ml volumes into tubes containing 0.2 g of agar while purging the tubes with oxygen-free nitrogen. Stopper the tubes while removing the gas cannula, and autoclave them in a press for 15 min, using fast exhaust. After autoclaving, mix by inverting the tubes in the press times.

25.3.30. PY Broth, Prereduced (4)

Peptone	5.0	g
Trypticase	5.0	g
Yeast extract	10.0	g
Resazurin solution, 0.025%	4.0	ml
Salts solution (see below)	40.0	ml
Hemin solution (see below)	10.0	ml
Vitamin K_1 solution (see below)	0.2	ml
Cysteine · HCl	0.5	g
Distilled water	1,000	ml

Salts solution
$CaCl_2$ (anhydrous), 0.2 g; $MgSO_4 \cdot 7H_2O$, 0.48 g; K_2HPO_4, 1.0 g; KH_2PO_4, 1.0 g; $NaHCO_3$, 10.0 g; NaCl, 2.0 g. Mix the $CaCl_2$ and $MgSO_4$ in 300 ml of distilled water until dissolved. Add 500 ml of distilled water, and, while swirling, add the remaining salts. After the salts are dissolved, add 200 ml of distilled water.

Hemin solution
Dissolve 0.05 g of hemin in 1 ml of 1 N NaOH. Dilute to 100 ml with distilled water. Autoclave at 121°C for 15 min.

Vitamin K_1 solution
Dissolve 0.15 ml of vitamin K_1 in 30 ml of 95% ethanol.

To prepare PY broth, place dry ingredients (except cysteine) into a flask equipped with a chimney to prevent boiling over. For a 1-liter flask, use ingredients sufficient to prepare 750 ml of medium; for a 750-ml flask, prepare 500 ml of medium. Add water, salts solution, and resazurin. Boil until the resazurin changes from pink to colorless. Remove the flask from the heat, and replace the chimney with a two-hole stopper that has a gas cannula in one hole. Bubble oxygen-free CO_2 through the medium while cooling to room temperature in an ice bath. Add hemin solution, vitamin K_1 solution, and cysteine. Adjust the pH to 7.1 with 8 N NaOH. Continue bubbling CO_2 until the pH is lowered to 6.9. Then bubble oxygen-free nitrogen through the medium while dispensing into 18- by 142-mm tubes (Bellco Glass, Inc.) that are being flushed with nitrogen. Stopper with black rubber stoppers (size no. 1) as the gas cannula is being removed from the tubes. Autoclave the tubes in a press (to keep the stoppers from being blown off) for 15 min, using fast exhaust. Cool. Discard any tubes that develop a light pink color (indicating oxidation) on standing.

25.3.31. Pyrazinamide Medium

For method 1

Dubos broth base, dehydrated (Difco)	6.5	g
Pyrazinamide	0.1	g
Sodium pyruvate	2.0	g
Agar	15.0	g
Distilled water	1,000	ml

Combine ingredients, boil to dissolve the agar, and dispense 5-ml volumes into 16- by 125-mm screw-cap tubes. Sterilize by autoclaving. Allow the medium to solidify upright to form a butt.

For method 2

Tryptic soy agar, dehydrated (Difco)	30.0	g
Yeast extract	3.0	g
Pyrazinamide	1.0	g
Tris-maleate (Sigma)	47.4	g
NaOH solution, 1 N	(ca. 120	ml)
Distilled water	to 1,000	ml

Add the solid ingredients to ca. 800 ml of distilled water, and dissolve. Adjust the pH to 6.0 with the 1 N NaOH solution, and add water to bring the final volume to 1,000 ml. Boil to dissolve the agar. Dispense 5-ml volumes into 16- by 160-mm screw-cap tubes, and sterilize by autoclaving. Slant the tubes before the medium solidifies.

25.3.32. Seawater, Artificial

Formula 1 (97)

NaCl	27.5	g
$MgCl_2$	5.0	g
$MgSO_4$	2.0	g
$CaCl_2$	0.5	g
KCl	1.0	g
$FeSO_4$	0.001	g
Distilled water	1,000	ml

Formula 2 (133)

NH_4NO_3	0.002	g
H_3BO_3	0.027	g
$CaCl_2$	1.140	g
$FePO_4$	0.001	g
$MgCl_2$	5.143	g
KBr	0.100	g
KCl	0.690	g
$NaHCO_3$	0.200	g
NaCl	24.320	g
NaF	0.003	g
Na_2SiO_3	0.002	g
Na_2SO_4	4.060	g
SrCl	0.026	g
Distilled water	1,000	ml

25.3.33. Simmons' Citrate Agar (107)

Sodium citrate	2.0	g
NaCl	5.0	g
$MgSO_4$	0.2	g
$(NH_4)H_2PO_4$	1.0	g
K_2HPO_4	1.0	g
Bromothymol blue	0.08	g
Agar	15.0	g
Distilled water	1,000	ml

Combine the ingredients, and adjust the pH to 6.9. Boil to dissolve the agar, dispense into tubes, and sterilize by autoclaving. Cool in slanted position.

25.3.34. Soybean-Casein Digest Agar (see *Note*)

Pancreatic digest of casein USP	15.0	g
Papaic digest of soy meal USP	5.0	g
NaCl	5.0	g
Agar	15.0	g
Distilled water	1,000	ml

Combine ingredients, and adjust the pH to 7.3. Boil to dissolve the agar, and sterilize by autoclaving. *Note:* This medium is also known as tryptic soy agar or Trypticase soy agar; it can be used as a blood agar base.

25.3.35. Thornley's Semisolid Arginine Medium (120)

Peptone	1.0	g
NaCl	5.0	g
K_2HPO_4	0.3	g
Phenol red	0.01	g
L-Arginine · HCl	10.0	g
Agar	3.0	g
Distilled water	1,000	ml

Combine ingredients, and adjust the pH to 7.2. Boil to dissolve the agar, and sterilize by autoclaving.

25.3.36. Todd-Hewitt Broth, Modified (40)

Beef heart, infusion from	1,000	ml
Neopeptone	20.0	g

Combine, and adjust the pH to 7.0 with 1 N NaOH. Then add the following:

NaCl	2.0	g
$NaHCO_3$	2.0	g
Na_2HPO_4	0.4	g
Glucose	2.0	g

Adjust the pH to 7.8. Boil for 15 min, filter through paper, and dispense into tubes. Sterilize by autoclaving at 121°C for 10 min.

25.3.37. Triple-Sugar Iron Agar

Pancreatic digest of casein USP (see *Note*)	10.0	g
Peptic digest of animal tissue USP (see *Note*)	10.0	g
Glucose	1.0	g
Lactose	10.0	g
Sucrose	10.0	g
Ferrous sulfate or ferrous ammonium sulfate	0.2	g
NaCl	5.0	g
Sodium thiosulfate	0.3	g
Phenol red	0.024	g
Agar	13.0	g
Distilled water	1,000	ml

Note: The following combination of ingredients can substitute for the first two components listed: beef extract, 3.0 g; yeast extract, 3.0 g; and peptone, 20.0 g.

Combine ingredients, and adjust the pH to 7.3. Boil to dissolve the agar, and dispense into tubes. Sterilize by autoclaving at 121°C for 15 min. Cool in a slanted position to give a 2.5-cm butt and a 3.8-cm slant.

25.3.38. Yoch and Pengra Nitrogen-Free Medium for *Klebsiella* Species (132)

Solution A

Na_2HPO_4	6.25	g
KH_2PO_4	0.75	g
Distilled water	500	ml

Solution B

$MgSO_4 \cdot 7H_2O$	6.25	g
$FeSO_4 \cdot 7H_2O$	0.04	g
Na_2MoO_4 (anhydrous)	0.005	g
Sucrose	20.0	g
NaCl	8.5	g
Distilled water	500	ml

Autoclave solutions A and B separately, and aseptically combine equal volumes when cool.

25.4. REAGENTS

25.4.1. Benedict's Reagent

Sodium citrate	17.3	g
Na_2CO_3	10.0	g
$CuSO_4 \cdot 5H_2O$	1.73	g

Dissolve the first two ingredients in 80 ml of water by heating. Filter. Dilute to 85 ml. Dissolve the $CuSO_4$ in 10 ml of water, and add to the carbonate-citrate solution with stirring. Dilute to 100 ml with distilled water.

25.4.2. Buffered Glycerol Mounting Fluid

Glycerol	90	ml
PBS (section 25.4.16)	10	ml

Combine ingredients, and adjust the pH to 8.0 with NaOH. The solution should be tightly stoppered to prevent absorption of carbon dioxide. Fresh solutions should be prepared monthly.

25.4.3. Charcoal-Gelatin Discs (66)

Mix 15 g of dehydrated nutrient gelatin (Difco) with 100 ml of tap water. Heat to dissolve. Add 3 to 5 g of finely powdered charcoal, shake thoroughly, and pour into petri dishes to form a layer about 3 mm thick. Pour the mixture when quite cool so that it sets quickly before the charcoal can sediment. It is advisable to smear the bottom of the dish very thinly with petrolatum before pouring the gelatin. After the mixture has set, lift off the whole sheet from the dish and place it in 10% Formalin for 24 h. Then punch the formalinized sheet into discs (about 1 cm in diameter) or cut it into strips (about 5 to 8 mm by 20 mm). Wrap the pieces in gauze, and place them in a basin under running tap water for 24 h. Then place the pieces into screw-cap bottles, and cover them with water. Sterilize by steaming for 30 min or by repeated heating in a water bath at 90 to 100°C for 20 min each time. Do not autoclave. After sterilization, decant the water and add the pieces aseptically to sterile broth (one piece per tube). Incubate the tubes containing the broth to test for sterility. The prepared media are then ready for inoculation.

25.4.4. Developing Solution for Arginine (102)

α-Naphthol, 25% (wt/vol) solution in *n*-propanol .... 20 ml
Diacetyl, 1% (wt/vol) in *n*-propanol ..... 2.5 ml

Combine the ingredients, and dilute to 100 ml with *n*-propanol.

25.4.5. Ehrlich's Reagent

p-Dimethylaminobenzaldehyde .... 1.0 g
Ethanol, 95% .... 95 ml
HCl, concentrated .... 20 ml

Dissolve the aldehyde in the ethanol, and then add the acid. Protect from the light, and store in a refrigerator.

25.4.6. Formate-Fumarate (F-F) Solution (4)

Sodium formate .... 3.0 g
Fumaric acid .... 3.0 g
Distilled water .... 50 ml

Combine ingredients. Add 20 pellets of NaOH with stirring until the pellets are dissolved and the fumaric acid is in solution. Adjust the pH to 7.0 with about 15 drops of 4 N NaOH. Sterilize by filtration.

25.4.7. Kovács' Reagent

p-Dimethylaminobenzaldehyde .... 3.0 g
Pentanol or butanol (amyl or butyl alcohol) .... 75 ml
HCl, concentrated .... 25 ml

Dissolve the aldehyde in the alcohol at 50 to 55°C. Cool, and add the acid. Protect from the light, and store in a refrigerator.

25.4.8. Lactic Acid Reagents for Optical Rotation (4)

Perchloric acid solution
Perchloric acid, 70% .... 5.8 ml

Make up to 100 ml with distilled water.

Table 7. Preparation of McFarland turbidity standards

McFarland scale no.	Amt of 1% $BaCl_2$ (ml)	Amt of 1% H_2SO_4 (ml)
0.5	0.05	9.95
1	0.1	9.9
2	0.2	9.8
3	0.3	9.7
4	0.4	9.6
5	0.5	9.5
6	0.6	9.4
7	0.7	9.3
8	0.8	9.2
9	0.9	9.1
10	1.0	9.0

Buffer solution
Glycine .... 3.75 g
Hydrazine sulfate .... 5.2 g

Add ingredients to 60 ml of distilled water. Adjust the pH to 9.0 with 8 N NaOH, and make the volume up to 100 ml with distilled water.

Lactic dehydrogenase solution
Lactic dehydrogenase, rabbit muscle (no. L-2375; Sigma) .... 0.8 ml
Distilled water .... 0.8 ml

Make up the solution in an ice bath, and store in ice. The solution should be prepared on the same day it is used.

NAD solution
NAD (no. N-7004; Sigma) .... 0.160 g
Distilled water .... 8.0 ml

Dissolve the NAD in the water. Store in a refrigerator. The solution should be freshly prepared.

25.4.9. McFarland Turbidity Standards (74)

Prepare a 1% solution of anhydrous $BaCl_2$ and a 1% solution of H_2SO_4. The McFarland standards consist of mixtures of the two solutions in various proportions (Table 7) to form a turbid suspension of $BaSO_4$.

Tightly seal the tubes containing these mixtures, and store them at room temperature in the dark. They will remain stable for at least 6 months. Before comparing the turbidity of a bacterial suspension with that of the standards, be sure to invert the standards several times to suspend the $BaSO_4$ precipitate evenly. Comparison of a bacterial suspension with the standards is best done by placing a white card with thick, horizontal, black lines behind the tube of bacterial suspension and the tube of standard and viewing the black lines through the tubes.

25.4.10. Methyl Green for DNase Test (110)

Methyl green (no. M295, Certified Biological Stain, C.I. no. 43590; Fisher Scientific Co., Pittsburgh, Pa.) ...... 0.5 g
Distilled water .... 100 ml

Dissolve the dye in the water. Extract the solution with approximately equal volumes of chloroform until the chloroform is colorless (usually six to eight extractions). Store the resulting aqueous solution of methyl green at 4°C. The dye may be added to DNase medium before autoclaving, or it may be sterilized by filtration and added aseptically to the autoclaved medium at 45 to 50°C.

25.4.11. Nessler's Solution

KI	5.0 g
Distilled water, ammonia-free	5.0 ml

Dissolve the KI in the water, and add a saturated solution of $HgCl_2$ (about 2.0 g in 35 ml) until an excess is indicated by the formation of a precipitate. Then add 20 ml of 5 N NaOH, and dilute to 100 ml. Let any precipitate settle out, and draw off the clear supernatant.

25.4.12. Nile Blue Sulfate Fat (8)

Nile blue sulfate solution. Prepare a saturated aqueous solution of Nile blue sulfate. Add 1 N NaOH drop by drop until precipitation is complete. Filter, and wash the precipitate with distilled water at pH 7.5. Dry, and store for use.

Staining of fat. Prepare a saturated solution of the Nile blue sulfate oxazine base, using the prepared dried precipitate prepared above. Mix 1 ml of the solution with 10 ml of the fat (tripropionine, tributyrin, tricaproine, tricaprylin, triolein, beef tallow, butterfat, coconut oil, corn oil, cottonseed oil, lard, linseed oil, or olive oil, as desired). If necessary, work in a heated water bath to liquefy the fat and maintain it in a liquid state throughout the subsequent washing procedure.

Washing procedure. With fats that are liquid at room temperature, add 2 volumes of diethyl ether to the dye-fat mixture in a separatory funnel; separate the red ether-soluble layer from the water layer, and wash it several times with water. (*CAUTION:* Ether is flammable! For all operations involving ether, use a fume hood and avoid any flames or sparks that could cause an explosion.) Finally, separate the ether-fat layer and evaporate off the ether. Separate the fat from the residual water, and sterilize by autoclaving. Store in a refrigerator. For use, add 1 ml of the dyed fat to 10 ml of sterile melted basal medium, mix well, and dispense into plates.

With fats that are solid at room temperature, wash the dyed fat with several changes of hot water, and then disperse the fat in a melted, neutral 0.5% agar solution, using 10 ml of fat per 90 ml of agar. Sterilize by autoclaving. For use, melt the fat emulsion and add 1 ml of fat emulsion to 20 ml of the sterile melted basal medium. Mix well, and dispense into plates.

25.4.13. Nitrite Test Reagents

For method 1

Solution A (see *CAUTION*):

N-(1-Naphthyl)ethylenediamine dihydrochloride (Sigma)	0.02 g
HCl, 1.5 N solution	100 ml

Dissolve by gentle heating in a fume hood.

Solution B

Sulfanilic acid	1.0 g
HCl, 1.5 N solution	100 ml

Dissolve by gentle heating.

Mix equal volumes of solutions A and B just before use to make the test reagent.

For method 2

Solution A (see *CAUTION*):

N,N-Dimethyl-1-naphthylamine dihydrochloride (Sigma)	0.6 g
(or use α-naphthylamine)	(0.5 g)
Acetic acid, 5 N solution	100 ml

Dissolve by gentle heating in a fume hood.

Solution B

Sulfanilic acid	0.8 g
Acetic acid, 5 N solution	100 ml

Dissolve by gentle heating in a fume hood.

CAUTION: α-Naphthylamine is a carcinogen. Although *N*-(1-naphthyl)ethylenediamine and *N,N*-dimethyl-1-naphthylamine have not been listed as carcinogenic, one would suspect on the basis of structural similarity that they might also be dangerous. All precautions pertinent to the use, handling, and subsequent disposal of carcinogens should be observed for these compounds.

25.4.14. Peptidase Developing Reagents (31)

Reagent A

Tris	250 g
HCl, 37%	110 ml
SDS	100 g
Distilled water	to 1,000 ml

The final pH should be 7.6.

Reagent B

Fast blue BB	3.0 g
2-Methoxyethanol	1,000 ml

25.4.15. pH Indicators for Culture Media

Some common pH indicators for use in culture media are listed in Table 8. Also see Chapter 6.1.1.

25.4.16. Phosphate-Buffered Saline (PBS)

10× stock solution

Na_2HPO_4, anhydrous, reagent grade	12.36 g
$NaH_2PO_4 \cdot H_2O$, reagent grade	1.80 g
NaCl, reagent grade	85.00 g

Dissolve ingredients in distilled water to a final volume of 1,000 ml.

Working solution (0.01 M phosphate, pH 7.6)

Stock solution	100 ml
Distilled water	900 ml

Table 8. pH indicators for culture media

Name	pK′	pH range and colors[a]	Concn usually used in media (g/liter)[b]	Amt (ml) of 1 N NaOH needed to dissolve 0.1 g
Phenol red	7.8	6.9(Y)–8.5(R)	0.010–0.030	28.2
Bromothymol blue	7.1	6.1(Y)–7.7(B)	0.010–0.032	16.0
Bromocresol purple	6.2	5.4(Y)–7.0(P)	0.010–0.032	18.5
Chlorophenol red	6.0	5.1(Y)–6.7(R)	0.015	23.6
Bromocresol green	4.7	3.8(Y)–5.4(B)	0.020	14.3

[a]Abbreviations: Y, yellow; B, blue; P, purple; R, red.
[b]For addition to culture media, dissolve in alcohol or prepare an aqueous solution with 0.01 N NaOH.

25.4.17. Poly-β-Hydroxybutyrate Granule Suspension (34)

Bacillus megaterium KM (ATCC 13632) was originally used as the source of the granules, but *Pseudomonas* strains may be used alternatively and may give higher yields of PHB (96a). Culture the bacterial strain under conditions conducive to PHB production. For *B. megaterium* these conditions are described by Macrae and Wilkinson (72); for pseudomonads, see reference 115. When the level of PHB is sufficiently high, as judged by staining with Nile Blue A (see section 25.1.66), harvest the cells by centrifugation and suspend them to a concentration of 15% (wt/vol) in ice-cold 0.02 M phosphate buffer, pH 7.2 (see section 25.4.18). Disrupt the cells by sonication (see Chapter 4.1.1.2), and centrifuge at high speed. Suspend the pellet in the alkaline hypochlorite reagent of Williamson and Wilkinson (see section 25.4.20) to a density of ca. 8 mg (dry weight)/ml or less (131). Alternatively, commercial 5% hypochlorite (Clorox bleach) can be used (96a). Incubate at room temperature for 2 days with stirring. Either allow the suspension to settle and carefully remove and discard the supernatant fluid, or collect the granules by centrifugation. Suspend the granules in water, and dialyze against running tap water until free from chloride (use $AgNO_3$ on a small portion; there should be no precipitate of AgCl formed). Dialyze further against distilled water. In a series of centrifugations, wash the granules several times with acetone. Place the granules in a Soxhlet apparatus, and continuously extract them with acetone-ether (2:1, vol/vol) for 3 days to remove non-PHB lipids. Further extract the granules with hot diethyl ether, pulverize them, and dry them under a vacuum. Determine the dry weight of the granules. After drying, disperse the granules in 0.02 M phosphate buffer by means of a tissue grinder and sonic oscillator to form a stable, concentrated suspension. Sterilize the suspension by autoclaving, and store it in a refrigerator. Calculate how much of the concentrated suspension to add to molten mineral agar to give a final concentration of 0.25% (wt/vol) in the overlay described in section 25.1.67.

25.4.18. Potassium Phosphate Buffer

Prepare 0.2 M solutions of K_2HPO_4 and KH_2PO_4. Mix the solutions in the ratios shown in Table 9 to obtain the appropriate pH value. The mixture (0.2 M phosphate buffer) can be further diluted to give the desired molarity. Also see Chapter 6.1.2.

25.4.19. Standard Solutions for Gas Chromatography (4)

Standard mixture of volatile fatty acids (ca. 1 meq/100 ml)

Formic acid	0.037 ml
Acetic acid, glacial	0.057 ml
Propionic acid	0.075 ml
Isobutyric acid	0.092 ml
Butyric acid	0.091 ml
Isovaleric acid	0.109 ml
Valeric acid	0.109 ml
Isocaproic acid	0.126 ml
Caproic acid	0.126 ml
Heptanoic acid	0.142 ml
Distilled water	100 ml

When using the standard mixture, acidify, and extract 1 ml of the solution with ether for gas chromatography.

Standard mixture of alcohols
The following list shows the amount of each component to be added to 100 ml of distilled water to obtain the final concentration shown in parentheses.

Ethanol	0.1 ml	(1.7 mM)
Propanol	0.035 ml	(0.5 mM)
Isobutanol	0.005 ml	(0.05 mM)
Butanol	0.01 ml	(0.1 mM)
Isopentanol	0.005 ml	(0.05 mM)
Pentanol	0.005 ml	(0.05 mM)

Extract 1 ml of the mixture with ether for gas chromatography.

Table 9. Number of parts of K_2HPO_4 solution and KH_2PO_4 solution required to obtain potassium phosphate buffer of various pH levels

Relative amts of components		
K_2HPO_4	KH_2PO_4	pH
49	51	6.8
55	45	6.9
61	39	7.0
67	33	7.1
72	28	7.2
77	23	7.3

Standard mixture of nonvolatile acids (ca. 1 meq/100 ml)

Pyruvic acid	0.068 ml
Lactic acid, 85% syrup	0.084 ml
Oxalacetic acid	0.06 g
Oxalic acid	0.06 g
Methylmalonic acid	0.06 g
Malonic acid	0.06 g
Fumaric acid	0.06 g
Succinic acid	0.06 g
Distilled water	100 ml

Methylate 1 ml of the mixture and extract with chloroform.

25.4.20. Williamson and Wilkinson Hypochlorite Reagent (131)

Triturate 200 g of fresh bleaching powder (chlorinated lime) with a little distilled water, and make the volume up to 1 liter. With stirring, add 1 liter of 30% (wt/vol) Na_2CO_3. Allow the mixture to stand for 2 to 3 h with shaking at intervals. Filter through paper. Adjust the pH of the filtrate to 9.8 with concentrated HCl. Warm the solution to 37°C, and remove the precipitate by filtration. Store the clear filtrate in a stoppered bottle in a refrigerator. The solution is stable for several months.

25.5. REFERENCES

25.5.1. General References

1. **Balows, A., W. J. Hausler, Jr., K. L. Herrmann, H. D. Isenberg, and H. J. Shadomy.** 1991. *Manual of Clinical Microbiology,* 5th ed. American Society for Microbiology, Washington, D.C.
   Contains microscopic, cultural, biochemical and serological methods for the identification of pathogenic microorganisms.
2. **Baron, E. J., and S. M. Finegold.** 1990. *Bailey & Scott's Diagnostic Microbiology,* 8th ed. The C. V. Mosby, Co., St. Louis.
   Contains a wealth of information about clinical characterization methods as well as recipes for reagents and culture media.
3. **Bascomb, S.** 1987. Enzyme tests in bacterial identification, p. 105–160. *In* R. R. Colwell and R. Grigorova (ed.), *Methods in Microbiology,* vol. 19. Academic Press, Inc., New York.
   A comprehensive review of the various substrates and testing methods for detecting activity of enzymes such as esterases, glycosidases, oxidases, nucleases, peptidases, and phosphatases.
4. **Holdeman, L. V., E. P. Cato, and W. E. C. Moore (ed.).** 1977. *Anaerobe Laboratory Manual,* 4th ed. Virginia Polytechnic Institute and State University, Blacksburg.
   Identification charts, characterization methods, and media for anaerobic bacteria; describes in detail the preparation and use of prereduced media and culture techniques.
5. **Holding, A. J., and J. G. Colee.** 1971. Routine biochemical tests, p. 2–32. *In* J. R. Norris and D. W. Ribbons (ed.), *Methods in Microbiology,* vol. 6A. Academic Press, Inc., New York.
   Principles and procedures for a wide variety of routine biochemical characterization tests.
6. **Krieg, N. R., and J. G. Holt (ed.).** 1984. *Bergey's Manual of Systematic Bacteriology,* vol. 1. The Williams & Wilkins Co., Baltimore.
   Describes gram-negative chemoheterotrophic eubacteria that have clinical, industrial, or agricultural importance. Includes many identification tables and keys.
7. **Lányi, B.** 1987. Classical and rapid identification methods for medically important bacteria, p. 1–67. *In* R. R. Colwell and R. Grigorova (ed.), *Methods in Microbiology,* vol. 19. Academic Press, Inc., New York.
   An extensive compilation of morphological, cultural, and biochemical methods that are commonly used for identifying bacteria in the clinical microbiology laboratory.
8. **Skerman, V. B. D.** 1967. *A Guide to the Identification of the Genera of Bacteria.* The Williams & Wilkins Co., Baltimore.
   Based on the 7th edition of Bergey's Manual of Determinative Bacteriology. The methods section is a goldmine of useful procedures and media for many kinds of bacteria.
9. **Skerman, V. B. D.** 1969. *Abstracts of Microbiological Methods.* Wiley-Interscience, New York.
   A wide variety of routine physiological and biochemical characterization methods, abstracted from original articles.
10. **Sneath, P. H. A., N. S. Mair, M. E. Sharpe, and J. G. Holt (ed.).** 1986. *Bergey's Manual of Systematic Bacteriology,* vol. 2. The Williams & Wilkins Co., Baltimore.
    Describes gram-positive chemoheterotrophic eubacteria that have clinical, industrial, or agricultural importance. Includes many identification tables and keys.
11. **Staley, J. T., M. P. Bryant, N. Pfennig, and J. G. Holt (ed.).** 1989. *Bergey's Manual of Systematic Bacteriology,* vol. 3. The Williams & Wilkins Co., Baltimore.
    Describes phototrophs, autotrophs, sheathed and budding bacteria, myxobacteria, archaeobacteria, and other aquatic and soil bacteria. Includes many identification tables and keys.
12. **Tilton, R. C. (ed.).** 1982. *Rapid Methods and Automation in Microbiology. Proceedings of the Third International Symposium on Rapid Methods and Automation in Microbiology.* American Society for Microbiology, Washington, D.C.
    Rapid methods, automation, computers, and miniaturized techniques relating to characterization and identification of bacteria.
13. **Williams, S. T., M. E. Sharpe, and J. G. Holt (ed.).** 1989. *Bergey's Manual of Systematic Bacteriology,* vol. 4. The Williams & Wilkins Co., Baltimore.
    Describes the genera and species of actinomycetes. Includes many identification tables and keys.

25.5.2. Specific References

14. **Adachi, T., Y. Murooka, and T. Harada.** 1973. Depression of arylsulfatase synthesis in *Aerobacter aerogenes* by tyramine. *J. Bacteriol.* **116:**19–24.
15. **Adler, J.** 1966. Chemotaxis in bacteria. *Science* **153:**708–716.
16. **Balderston, W. L., B. Sherr, and W. J. Payne.** 1976. Blockage by acetylene of nitrous oxide reduction in *Pseudomonas perfectomarinus. Appl. Environ. Microbiol.* **31:**504–508.
17. **Bernaerts, M. J., and J. De Ley.** 1963. A biochemical test for crown gall bacteria. *Nature* (London) **187:**406–407.
18. **Bille, J., and M. P. Doyle.** 1991. *Listeria* and *Erysipelothrix,* p. 287–295. *In* A. Balows. W. J. Hausler, Jr., K. L. Herrmann, H. D. Isenberg, and H. J. Shadomy (eds.), *Manual of Clinical Microbiology,* 5th ed. American Society for Microbiology, Washington, D.C.
19. **Board, R. G., and A. J. Holding.** 1960. The utilization of glucose by aerobic gram-negative bacteria. *J. Appl. Bacteriol.* **23:**xi.
20. **Botsford, J. L., and R. D. DeMoss.** 1971. Catabolic repression of tryptophanase in *Escherichia coli. J. Bacteriol.* **105:**303–312.

21. **Bürger, H.** 1967. Biochemische Leistungen nicht-proliferierender Mikroorganismen. II. Nachweis von Glycosid-Hydrolasen, Phosphatasen, Esterasen und Lipasen. *Zentralbl. Bakteriol. Parasitenkd. Infektionskr. Hyg. Abt. I Orig.* **202:**97–109.
22. **Campbell, N. E. R., and H. J. Evans.** 1969. Use of Pankhurst tubes to assay acetylene reduction by facultative and anaerobic nitrogen-fixing bacteria. *Can. J. Microbiol.* **15:**1342–1343.
23. **Cato, E. P., D. E. Hash, L. V. Holdeman, and W. E. C. Moore.** 1982. Electrophoretic study of *Clostridium* species. *J. Clin. Microbiol.* **15:**688–702.
24. **Cato, E. P., and W. E. C. Moore.** 1965. A routine determination of the optically active isomers of lactic acid for bacterial identification. *Can. J. Microbiol.* **11:**319–324.
25. **Christensen, W. B.** 1946. Urea decomposition as a means of differentiating *Proteus* and paracolon cultures from each other and from *Salmonella* and *Shigella* types. *J. Bacteriol.* **52:**461–466.
26. **Christensen, W. B.** 1949. *Hydrogen Sulfide Production and Citrate Utilization in the Differentiation of Enteric Pathogens and Coliform Bacteria.* Research Bulletin no. 1. Weld County Health Department, Greeley, Colo.
27. **Cohen-Bazire, G., W. R. Sistrom, and R. Y. Stanier.** 1957. Kinetic studies of pigment synthesis by nonsulfur purple bacteria. *J. Cell. Comp. Physiol.* **49:**25–68.
28. **Corbel, M. J., and W. J. Brinley-Morgan.** 1984. Genus *Brucella*, p. 377–388. *In* N. R. Krieg and J. G. Holt (ed.), *Bergey's Manual of Systematic Bacteriology*, vol. 1. The Williams & Wilkins Co., Baltimore.
29. **Cowan, S. T.** 1974. *Cowan and Steel's Manual for the Identification of Medical Bacteria*, 2nd ed. Cambridge University Press, London.
30. **Daesch, G., and L. E. Mortensen.** 1972. Effect of ammonia on the synthesis and function of the N_2-fixing enzyme system in *Clostridium pasteurianum*, *J. Bacteriol.* **110:**103–109.
31. **D'Amato, R. F., L. A. Enriquez, K. M. Tomfohrde, and E. Singerman.** 1978. Rapid identification of *Neisseria gonorrhoeae* and *Neisseria meningitidis* by using enzymatic profiles. *J. Clin. Microbiol.* **7:**77–81.
32. **Dees, S. B., C. W. Moss, D. G. Hollis, and R. E. Weaver.** 1986. Chemical characterization of *Flavobacterium odoratum, Flavobacterium breve,* and *Flavobacterium*-like groups IIe, IIh, and IIf. *J. Clin. Microbiol.* **23:**267–273.
33. **Deibel, R. H., and J. B. Evans.** 1960. Modified benzidine test for the detection of cytochrome-containing respiratory systems in microorganisms. *J. Bacteriol.* **79:**356–360.
34. **Delafield, F. P., M. Doudoroff, N. J. Palleroni, C. J. Lusty, and R. Contopoulos.** 1965. Decomposition of poly-β-hydroxybutyrate by pseudomonads. *J. Bacteriol.* **90:**1455–1466.
35. **De Ley, J., and J. Swings.** 1984. Genus *Gluconobacter*, p. 275–278. *In* N. R. Krieg and J. G. Holt (ed.), *Bergey's Manual of Systematic Bacteriology*, vol. 1. The Williams & Wilkins Co., Baltimore.
36. **Döbereiner, J., and J. M. Day.** 1976. Associative symbioses in tropical grasses: characterization of microorganisms and dinitrogen fixing sites, p. 518–538. *In* W. E. Newton and C. J. Nymans (ed.), *Symposium on Nitrogen Fixation.* Washington State University Press, Pullman.
37. **Elder, B. L., I. Trujillo, and D. J. Blasevic.** 1977. Rapid deoxyribonuclease test with menthyl green. *J. Clin. Microbiol.* **6:**312–313.
38. **Evans, J. B., and C. F. Niven.** 1950. A comparative study of known food-poisoning staphylococci and related varieties. *J. Bacteriol.* **59:**545–550.
39. **Ewing, W. H., B. R. Davis, and R. W. Reavis.** 1957. Phenylalanine and malonate media and their use in enteric bacteriology. *Public Health Lab.* **15:**153.
40. **Finegold, S. M., W. J. Martin, and E. G. Scott.** 1978. *Bailey and Scott's Diagnostic Microbiology*, 5th ed. The C. V. Mosby Co., St. Louis.
41. **Franzmann, P. D., and V. B. D. Skerman.** 1989. Genus *Agitococcus*, p. 2133–2135. *In* J. T. Staley, M. P. Bryant, N. Pfenning, and J. G. Holt (ed.), *Bergey's Manual of Systematic Bacteriology*, vol. 3. The Williams & Wilkins Co., Baltimore.
42. **Fung, D. Y. C., and P. A. Hartman.** 1972. Rapid characterization of bacteria, with emphasis on *Staphylococcus aureus. Can. J. Microbiol.* **18:**1623–1627.
43. **Fung, D. Y. C., and P. A. Hartman.** 1975. Miniaturized microbiological techniques for rapid characterization of bacteria, p. 347–370. *In* C. G. Hedén and T. Illéni (ed.), *New Approaches to the Identification of Microorganisms.* John Wiley & Sons, Inc., New York.
44. **Fung, D. Y. C., and R. D. Miller.** 1970. Rapid procedure for the detection of acid and gas production by bacterial cultures. *Appl. Microbiol.* **20:**527–528.
45. **Goebel, E. M., and N. R. Krieg.** 1984. Fructose catabolism in *Azospirillum brasilense* and *Azospirillum lipoferum. J. Bacteriol.* **159:**86–92.
46. **Grange, J. M.** 1978. Fluorimetric assay of mycobacterial group-specific hydrolase enzymes. *J. Clin. Pathol.* **31:**378–381.
47. **Greenberg, A. E., R. R. Trussel, and L. C. Clesceri (ed.).** 1985. *Standard Methods for the Examination of Water and Wastewater*, 16th ed. American Public Health Association, Washington, D.C.
48. **Han, Y.-H., R. M. Smibert, and N. R. Krieg.** 1991. *Wolinella recta, Wolinella curva, Bacteroides ureolyticus,* and *Bacteroides gracilis* are microaerophiles, not anaerobes. *Int. J. Syst. Bacteriol.* **41:**218–222.
49. **Hardie, J. M.** 1986. Genus *Streptococcus*, p. 1043–1047. *In* P. H. A. Sneath, N. S. Mair, M. E. Sharpe, and J. G. Holt (eds.), *Bergey's Manual of Systematic Bacteriology*, vol. 2. The Williams & Wilkins Co., Baltimore.
50. **Hardy, R. W. F., R. C. Burns, and R. D. Holsten.** 1973. Applications of the acetylene-ethylene assay for measurement of nitrogen fixation. *Soil Biol. Biochem.* **5:**47–81.
51. **Hardy, R. W. F., R. D. Holsten, E. K. Jackson, and R. C. Burns.** 1968. The acetylene-ethylene assay for N_2 fixation: laboratory and field evaluation. *Plant Physiol.* **43:**1185–1207.
52. **Haynes, W. C.** 1951. *Pseudomonas aeruginosa*—its characterization and identification. *J. Gen. Microbiol.* **5:**939–950.
53. **Henrichsen, J.** 1972. Bacterial surface translocation: a survey and a classification. *Bacteriol. Rev.* **36:**478–503.
54. **Hino, S., and P. W. Wilson.** 1958. Nitrogen fixation by a facultative bacillus. *J. Bacteriol.* **75:**403–408.
55. **Holt, S. C., and S. A. Kinder.** 1989. Genus *Capnocytophaga*, p. 2050–2058. *In* J. T. Staley, M. P. Bryant, N. Pfenning, and J. G. Holt (ed.), *Bergey's Manual of Systematic Bacteriology*, vol. 3. The Williams & Wilkins Co., Baltimore.
56. **Hugh, R., and E. Leifson.** 1953. The taxonomic significance of fermentative versus oxidative metabolism of carbohydrates by various gram-negative bacteria. *J. Bacteriol.* **66:**22–26.
57. **Hungate, R. E.** 1966. *The Rumen and Its Microbes.* Academic Press Inc., New York.
58. **Hwang, M., and G. M. Ederer.** 1975. Rapid hippurate hydrolysis method for presumptive identification of group B streptococci. *J. Clin. Microbiol.* **1:**114–115.
59. **Hylemon, P. B., J. S. Wells, Jr., N. R. Krieg, and H. W. Jannasch.** 1973. The genus *Spirillum:* a taxonomic study. *Int. J. Syst. Bacteriol.* **23:**340–380.
60. **Johnson, R., and P. H. A. Sneath.** 1973. Taxonomy of *Bordetella* and related organisms of the families *Achromobacteraceae, Brucellaceae,* and *Neisseriaceae. Int. J. Syst. Bacteriol.* **23:**381–404.
61. **Kandolo, K., and G. Wauters.** 1985. Pyrazinamidase activity in *Yersinia enterocolitica* and related organisms. *J. Clin. Microbiol.* **21:**980–982.
62. **Kersters, K., and J. De Ley.** 1975. Identification and grouping of bacteria by numerical analysis of their electrophoretic patterns. *J. Gen. Microbiol.* **87:**333–342.
63. **Killinger, A. H.** 1974. *Listeria monocytogenes*, p. 135–139. *In* E. H. Lennette, E. H. Spaulding, and J. P. Truant (ed.),

Manual of Clinical Microbiology, 2nd ed. American Society for Microbiology, Washington, D.C.

64. **King, E. O., M. K. Ward, and D. E. Raney.** 1954. Two simple media for the demonstration of pyocyanin and fluorescein. *J. Lab. Clin. Med.* **44:**301–307.
65. **Knapp. J. S.** 1988. Historical perspectives and identification of *Neisseria* and related species. *Clin. Microbiol. Rev.* **1:**415–431.
66. **Kohn, J.** 1953. A preliminary report of a new gelatin liquefaction method. *J. Clin. Microbiol.* **6:**249.
67. **Lamm, R. B., and C. A. Neyra.** 1981. Characterization and cyst production of azospirilla isolated from selected grasses growing in New Jersey and New York. *Can. J. Microbiol.* **27:**1320–1325.
68. **Law, J. H., and R. A. Slepecky.** 1961. Assay of poly-β-hydroxybutyric acid. *J. Bacteriol.* **82:**33–36.
69. **Lederberg, J., and E. M. Lederberg.** 1962. Replica plating and indirect selection of bacterial mutants. *J. Bacteriol.* **63:**399–406.
70. **Leifson, E.** 1963. Determination of carbohydrate metabolism of marine bacteria. *J. Bacteriol.* **85:**1183–1184.
71. **Lovelace, T. E., and R. R. Colwell.** 1968. A multipoint inoculator for petri dishes. *Appl. Microbiol.* **16:**944–945.
72. **Macrae, R. M., and J. F. Wilkinson.** 1958. Poly-β-hydroxybutyrate metabolism in washed suspensions of *Bacillus cereus* and *Bacillus megaterium*. *J. Gen. Microbiol.* **19:**210–222.
73. **Maddocks, J. L., and M. J. Greenan.** 1975. Technical method. A rapid method for identifying bacterial enzymes. *J. Clin. Pathol.* **28:**686–687.
74. **McFarland, J.** 1907. Nephelometer: an instrument for estimating the number of bacteria in suspensions used for calculating the opsonic index and for vaccines. *JAMA* **14:**1176–1178.
75. **McGinnis, M. R., R. F. D'Amato, and G. A. Land.** 1982. *Pictorial Handbook of Medically Important Fungi and Aerobic Actinomycetes.* Praeger Publishers, New York.
76. **Meynell, G. G., and E. Meynell.** 1965. *Theory and Practice in Experimental Bacteriology.* Cambridge University Press, London.
77. **Meynell, G. G., and E. Meynell.** 1970. *Theory and Practice in Experimental Bacteriology*, 2nd ed. Cambridge University Press, London.
78. **Milazzo, F. H., and J. W. Fitzgerald.** 1967. The effect of some culture conditions on the arylsulfatase of *Proteus rettgeri*. *Can. J. Microbiol.* **13:**659–664.
79. **Miller, C. G., and K. MacKinnon.** 1974. Peptidase mutants of *Salmonella typhimurium*. *J. Bacteriol.* **120:**355–363.
80. **Mills, C. K., and R. L. Cherna.** 1987. Hydrolysis of indoxyl acetate by *Campylobacter* species. *J. Clin. Microbiol.* **25:**1560–1561.
81. **Møller, V.** 1955. Simplified test for some amino acid decarboxylases in *Enterobacteriaceae*. *Acta Pathol. Microbiol. Scand.* **36:**158–172.
82. **Moore, W. E. C., D. E. Hash, L. V. Holdeman, and E. P. Cato.** 1980. Polyacrylamide slab electrophoresis of soluble proteins for studies of bacterial floras. *Appl. Environ. Microbiol.* **39:**900–907.
83. **Moss, C. W., P. L. Wallace, D. G. Hollis, and R. E. Weaver.** 1988. Cultural and chemical characterization of CDC groups EO-2, M-5, and M-6, *Moraxella* (*Moraxella*) species, *Oligella urethralis*, *Acinetobacter* species, and *Psychrobacter immobilis*. *J. Clin. Microbiol.* **26:**484–492.
84. **Mundt, J. O.** 1986. Enterococci, p. 1063–1065. *In* P. H. A. Sneath, N. S. Mair, M. E. Sharpe, and J. G. Holt (ed.), *Bergey's Manual of Systematic Bacteriology*, vol. 2. The Williams & Wilkins Co., Baltimore.
85. **Mundt, J. O.** 1986. Lactic acid streptococci. p. 1065–1066. *In* P. H. A. Sneath, N. S. Mair, M. E. Sharpe, and J. G. Holt (ed.), *Bergey's Manual of Systematic Bacteriology*, vol. 2. The Williams & Wilkins Co., Baltimore.
86. **Neal, J. L., Jr., K. C. Lu, W. B. Bollen, and J. M. Trappe.** 1966. Apparatus for rapid replica plating in rhizosphere studies (soil surrounding and influenced by roots of plants). *Appl. Microbiol.* **14:**695–696.
87. **Neyra, C. A., J. Döbereiner, R. LaLande, and R. Knowles.** 1977. Denitrification by N_2 fixing *Spirillum lipoferum*. *Can. J. Microbiol.* **23:**300–305.
88. **Nicolet, J., P. Paroz, and M. Krawinkler.** 1980. Polyacrylamide gel electrophoresis of whole-cell proteins of porcine strains of *Haemophilus*. *Int. J. Syst. Bacteriol.* **30:**69–76.
89. **Niven, C. F., Jr., K. L. Smiley, and J. M. Sherman.** 1942. The hydrolysis of arginine by streptococci. *J. Bacteriol.* **43:**651–660.
90. **Noel, K. D., and W. J. Brill.** 1980. Diversity and dynamics of indigenous *Rhizobium japonicum* populations. *Appl. Environ. Microbiol.* **40:** 931–938.
91. **O'Callaghan, C. H., A. Morris, S. M. Kirby, and A. H. Shingler.** 1972. Novel method for detection of β-lactamases by using a chromogenic cephalosporin substrate. *Antimicrob. Agents Chemother.* **1:**283–288.
92. **Ohta, H., and J. C. Gottschal.** 1988. Microaerophilic growth of *Wolinella recta*. *FEMS Microbiol. Ecol.* **53:**79–86.
93. **Okon, Y., S. L. Albrecht, and R. H. Burris.** 1976. Carbon and ammonia metabolism of *Spirillum lipoferum*. *J. Bacteriol.* **128:**592–597.
94. **Ostle, A. G., and J. G. Holt.** 1982. Nile Blue A as a fluorescent stain for poly-β-hydroxybutyrate. *Appl. Environ. Microbiol.* **44:**238–241.
95. **Page, W. J., and H. L. Sadoff.** 1976. Physiological factors affecting transformation of *Azotobacter vinelandii*. *J. Bacteriol.* **125:**1080–1087.
96. **Palleroni, N. J.** 1984. Genus *Pseudomonas*, p. 141–199. *In* N. R. Krieg and J. G. Holt (ed.), *Bergey's Manual of Systematic Bacteriology*, vol. 1. The Williams & Wilkins Co., Baltimore.

96a. **Palleroni, N. J.** Personal communication.

97. **Pelczar, M. J., Jr. (ed.).** 1957. *Manual of Microbiological Methods.* McGraw-Hill Book Co., New York.
98. **Porschen, R. K., and S. Sonntag.** 1974. Extracellular deoxyribonuclease production by anaerobic bacteria. *Appl. Microbiol.* **27:**1031–1033.
99. **Postgate, J. R.** 1972. The acetylene reduction test for nitrogen fixation, p. 343–356. *In* J. R. Norris and D. W. Ribbons (ed.), *Methods in Microbiology*, vol. 6B. Academic Press, Inc., New York.
100. **Razin, S., and S. Rottem.** 1967. Identification of *Mycoplasma* and other microorganisms by polyacrylamide gel electrophoresis of cell proteins. *J. Bacteriol.* **94:**1807–1810.
101. **Roop, R. M., II, R. M. Smibert, J. L. Johnson, and N. R. Krieg.** 1984. Differential characteristics of catalase-positive campylobacters correlated with DNA homology groups. *Can. J. Microbiol.* **30:**938–951.
102. **Rosenberg, H., A. H. Ennor, and V. F. Morrison.** 1956. The estimation of arginine. *Biochem. J.* **63:**153–159.
103. **Sadasivan, L., and C. A. Neyra.** 1987. Cyst production and brown pigment formation in aging cultures of *Azospirillum brasilense* ATCC 29145. *J. Bacteriol.* **169:**1670–1677.
104. **Schreier, J. B.** 1969. Modification of deoxyribonuclease test medium for rapid identification of *Serratia marcescens*. *Am. J. Clin. Pathol.* **51:**711–716.
105. **Shinoda, S., and K. Okamoto.** 1977. Formation and function of *Vibrio parahaemolyticus* lateral flagella. *J. Bacteriol.* **129:**1266–1271.
106. **Sierra, G.** 1957. A simple method for the detection of lipolytic activity of microorganisms and some observations on the influence of the contact between cells and fatty substrates. *Antonie van Leeuwenhoek J. Microbiol. Serol.* **23:**15–22.
107. **Simmons, J. S.** 1926. A culture medium for differentiating organisms of the typhoid-colon aerogenes groups and for isolation of certain fungi. *J. Infec. Dis.* **39:**201–214.
108. **Smibert, R. M.** 1984. Genus *Campylobacter*, p. 111–118. *In* N. R. Krieg and J. G. Holt (ed.), *Bergey's Manual of Systematic Bacteriology*, vol. 1. The Williams & Wilkins Co., Baltimore.

109. **Smibert, R. M., J. L. Johnson, and R. R. Ranney.** 1984. *Treponema socranskii* sp. nov., *Treponema socranskii* subsp. *socranskii* subsp. nov., *Treponema socranskii* subsp. *buccale* subsp. nov., and *Treponema socranskii* subsp. *paredis* subsp. nov. isolated from the human periodontia. *Int. J. Syst. Bacteriol.* **34:**457–462.
110. **Smith, P. B., G. A. Hancock, and D. L. Rhoden.** 1969. Improved medium for detecting deoxyribonuclease-producing bacteria. *Appl. Microbiol.* **18:**991–993.
111. **Smith, R. F., R. R. Rogers, and C. L. Bettge.** 1972. Inhibition of the indole test reaction by sodium nitrite. *Appl. Microbiol.* **23:**423–424.
112. **Sneath, P. H. A., and M. Stevens.** 1967. A divided petri dish for use with multipoint inoculators. *J. Appl. Bacteriol.* **30:**495–497.
113. **Snell, J. J. S.** 1984. Genus *Kingella*, p. 307–309. *In* N. R. Krieg and J. G. Holt (ed.), *Bergey's Manual of Systematic Bacteriology*, vol. 1. The Williams & Wilkins Co., Baltimore.
114. **Sulea, I. T., M. C. Pollice, and L. Barksdale.** 1980. Pyrazine carboxylamidase activity in *Corynebacterium*. *Int. J. Syst. Bacteriol.* **30:**466–472.
115. **Stanier, R. Y., N. J. Palleroni, and M. Doudoroff.** 1966. The aerobic pseudomonads: a taxonomic study. *J. Gen. Microbiol.* **43:**159–271.
116. **Stockdale, H., D. W. Ribbons, and E. A. Dawes.** 1968. Occurrence of poly-β-hydroxybutyrate in the *Azotobacteraceae*. *J. Bacteriol.* **95:**1798–1803.
117. **Tanner, A. C. R., M. A. Listgarten, and J. L. Ebersole.** 1984. *Wolinella curva* sp. nov.: "*Vibrio succinogenes*" of human origin. *Int. J. Syst. Bacteriol.* **34:**275–282.
118. **Tanner, A. C. R., and S. S. Socransky.** 1984. Genus *Wolinella*, p. 646–650. *In* N. R. Krieg and J. G. Holt (ed.), *Bergey's Manual of Systematic Bacteriology*, vol. 1. The Williams & Wilkins Co., Baltimore.
119. **Tarrand, J. J., and D. H. M. Gröschel.** 1982. Rapid, modified oxidase test for oxidase-variable bacterial isolates. *J. Clin. Microbiol.* **16:**772–774.
120. **Thornley, M. J.** 1960. The differentiation of *Pseudomonas* from other gram-negative bacteria on the basis of arginine metabolism. *J. Appl. Bacteriol.* **23:**37–52.
121. **Tinghitella, T. J., and S. C. Edberg.** 1991. Agglutination tests and *Limulus* assay for the diagnosis of infectious disease, p. 61–72. *In* A. Balows, W. J. Hausler, Jr., K. L. Herrmann, H. D. Isenberg, and H. J. Shadomy (ed.), *Manual of Clinical Microbiology*, 5th ed. American Society for Microbiology, Washington, D.C.
122. **Vaneechoutte, M., G. Verschraegen, G. Claeys, and P. Flamen.** 1988. Rapid identification of *Branhamella catarrhalis* with 4-methylumbelliferyl butyrate. *J. Clin. Microbiol.* **26:**1227–1228.
123. **Vedros, N. A.** 1984. Genus *Neisseria*, p. 290–296. *In* N. R. Krieg and J. G. Holt (ed.), *Bergey's Manual of Systematic Bacteriology*, vol. 1. The Williams & Wilkins Co., Baltimore.
124. **Watson, R. R.** 1976. Substrate specificities of aminopeptidases: a specific method of microbial differentiation, p. 1–14. *In* J. R. Norris (ed.), *Methods in Microbiology*, vol. 9, Academic Press Ltd., London.
125. **Westley, J. W., P. J. Anderson, V. A. Close, B. Halpern, and E. M. Lederberg.** 1967. Aminopeptidase profiles of various bacteria. *Appl. Microbiol.* **15:**822–825.
126. **Wheater, D. M.** 1955. The characteristics of *Lactobacillus acidophilus* and *Lactobacillus bulgaricus*. *J. Gen. Microbiol.* **12:**123–132.
127. **Whittenbury, R.** 1964. Hydrogen peroxide formation and catalase activity in the lactic acid bacteria. *J. Gen. Microbiol.* **35:**13–26.
128. **Wilkins, T. D., and C. B. Walker.** 1975. Development of a micromethod for identification of anaerobic bacteria. *Appl. Microbiol.* **30:**825–830.
129. **Wilkins, T. D., C. B. Walker, and W. E. C. Moore.** 1975. Micromethod for identification of anaerobic bacteria: design and operation of apparatus. *Appl. Microbiol.* **30:**831–837.
130. **Williams, M. A.** 1960. Flagellation in six species of *Spirillum*—a correction. *Int. Bull. Bacteriol. Nomencl. Taxon.* **10:**193–196.
131. **Williamson, D. H., and J. F. Wilkinson.** 1958. The isolation and estimation of poly-β-hydroxybutyrate inclusions in *Bacillus* species. *J. Gen. Microbiol.* **19:**198–209.
132. **Yoch, D. C., and R. M. Pengra.** 1966. Effect of amino acids on the nitrogenase system of *Klebsiella pneumoniae*. *J. Bacteriol.* **92:**618–622.
133. **Zobell, C. E.** 1946. *Marine Microbiology*. Chronica Botanica Co., Waltham, Mass.

Chapter 26

Similarity Analysis of DNAs

JOHN L. JOHNSON

Elucidation of the structure and the physical properties of DNA and RNAs has enabled investigators to determine phylogenetic and taxonomic relationships among bacteria by comparing their genomes. A preliminary comparison between two organisms can be made with respect to the overall nucleotide base composition of DNA. This is usually expressed as the moles percent guanine plus cytosine, which is considered a constant feature of that organism. A large difference in the moles percent G+C values from DNAs of two organisms indicates the lack of a close genetic relatedness; however, a similarity in the moles percent G+C does not necessarily mean that the two organisms are closely related with regard to the sequence of nucleotides in their DNAs. In such cases, methods more precise than DNA base composition are required for comparing the genomes, namely **DNA-DNA reassociation** (the formation of duplexes between single-stranded DNA fragments by complementary base pairing), **rRNA-DNA hybridization** (the formation of duplexes between single-stranded DNA and rRNA), and the nucleotide sequencing of specific genes and rRNAs. Methods involving DNA reassociation will be covered in this chapter, and those involving rRNA hybridization and nucleotide sequencing will be dealt with in the following chapter.

The reassociation of denatured DNA fragments or the hybridization between denatured DNA and rRNA have commonly been referred to as "homologies," with the results reported as relative "percent homology values." In retrospect, the word "homology" was not the best choice to describe these experiments, and it will not be used in this chapter because of its long use in a different sense among classical evolutionalists: **homology** (or homologous) refers to phylogenetic development, in which, for example, the arm and the wing are viewed as homologous appendages, with degrees of similarity between the two. Also, in the literature on DNA reassociation, the relative amounts of DNA from one organism forming duplexes with DNA from the same organism or from another organism have been referred to in the literature as **percent similarity, percent identity,** or **percent reassociation.** For consistency, only the term "percent similarity" will be used in this chapter.

DNA reassociation studies allow comparison of organisms with respect to the linear arrangements of the nucleotides along their entire DNA strands. Percent similarity values are average measurements of nucleotide sequence similarity which use the entire genome of each organism being compared. As a result, these values are used to express similarities between closely related bacteria, i.e., closely related species, subspecies, and in some cases strains of a species.

In theory and in practice, DNA reassociation experiments are relatively easy to do. However, there are many technical details or problems associated with the experiments. These include the initial isolation of high-purity DNA, experimental conditions, and the type of experiment. Some problems may be unique for a given group of bacteria, and therefore there are no universal solutions. The methods described in this chapter are, for the most part, those that have been successfully employed in the author's laboratory, and familiarity has contributed to the selection of the specific methods. In some cases specialized instruments are required, such as those found in core facilities (e.g., determination of moles percent G+C in intact cells by flow cytometry or differential scanning colorimetry).

Additional background information on DNA reassociation methods, as well as on DNA base composition methods, can be found in the general references at the end of this chapter (1–6).

26.1. CELL GROWTH AND DISRUPTION

The rapid growth of many bacteria, relative to other organisms, is advantageous for the quick generation of cell crops from which to isolate DNA. The growth rates and the extent of growth, however, vary greatly and will dictate how long cultures must be grown and what volumes are produced.

26.1.1. Growth Conditions (also see Chapters 6 to 10)

The group of organisms being investigated will determine the culture medium requirements. A peptone-yeast extract-based medium supplemented with a carbohydrate energy source works well for cultivating many chemoheterotrophs. If high levels of acidic by-products are produced, inclusion of a buffer (e.g., 50 mM potassium phosphate [pH 7.0]) may result in larger cell numbers.

The consideration of oxygen requirements is very important. Although reducing agents are usually added to anaerobic media, many anaerobic bacteria grow well in a freshly prepared medium with a nitrogen-filled head-space (24, 38) and a 5% (vol/vol) inoculum of an actively growing culture. For anaerobic bacteria requiring CO_2, sterile sodium bicarbonate can be added at the time of inoculation. Gentle agitation (stirring bar or shaking) may also be important for organisms that tend to settle under static growth conditions. Facultatively anaerobic organisms are best cultured under aerobic conditions, because smaller amounts of acidic end products are produced. Actively growing aerobic cultures rapidly deplete oxygen in the medium; therefore, the rate at which oxygen is dissolved from the atmosphere limits the growth rate. Aeration is optimized by using small volumes of medium per flask (e.g., 250 ml per 2-liter flask), flasks with fluted walls, and rapid shaking (250 to 450 rpm). The oxygen requirements of microaerophilic organisms are the most difficult to fulfill. The medium is first equilibrated by sparging with a mixture of air and CO_2, and the sparging is continued during growth (89).

The preferred time for harvesting a culture is at the early stationary phase, when the amount of growth is near maximum and the death rate is still minimal. The amount of attention that must be given to the growth curve will depend on the organism. Many organisms have a long stationary phase with little cell death, so that the exact harvest time is not very critical. Others may start dying quite soon after reaching maximal growth. With sporulating organisms, vegetative cell DNA can be converted into sporal DNA during the 1- to 3-h sporulation period of the stationary phase.

26.1.2. Cell Disruption (also see Chapter 4.1)

The first step in DNA isolation is the disruption of the bacterial cells. Depending on the nature of the cell wall, bacterial cells can be disrupted (i) with a detergent alone, (ii) with a combination of detergents and hydrolytic enzymes, or (iii) by physical methods such as sonic oscillation, shearing release from a pressure cell, or shaking in the presence of glass beads.

The general protocols are for culture volumes of 250 to 350 ml, and, after centrifugation, the cell pellets are suspended in 25 ml of suspending buffer (see below). These volumes can be scaled up or down depending on the amount of growth or the amount of DNA needed.

26.1.2.1. Gram-negative bacteria

Detergents alone can cause many gram-negative bacteria to lyse, but some species may lyse only partially or not at all. An initial incubation of the cells in the presence of hen egg white lysozyme will often provide a more uniform lysis on subsequent treatment with the detergent. (Lysozyme is a muramidase, a hydrolytic enzyme that acts on the peptidoglycan of the cell wall. It is inhibited by high salt concentrations and low pH.)

Suspend the cells in a buffer consisting of 10 mM Tris-HCl (pH 8.0), 1 mM sodium EDTA, and 0.35 M sucrose. The sucrose stabilizes the resulting spheroplasts until the addition of the lysing solution. Add lysozyme (0.5 to 1.0 mg/ml), and incubate at room temperature for 5 min; then add the lysing solution (see section 26.2.1). If the cells are not readily lysed by the lysing solution, use a temperature of 37°C for the initial lysozyme digestion.

26.1.2.2. Gram-positive bacteria

In contrast to gram-negative bacteria, nearly all gram-positive bacteria must be digested with a lytic enzyme before they can be lysed by a detergent. In addition to lysozyme, which is the enzyme most commonly used, several other enzymes are available. These include *N*-acetylmuramidase, which is isolated from *Streptomyces globisporus* and also cleaves the muramic acid backbone; lysostaphin, an endopeptidase that is isolated from *Staphylococcus* sp. strain K-6-WI and is specific for the cross-linking peptides of other staphylococci; and achromopeptidase, a peptidase that is isolated from *Achromobacter lyticus* and is also active on the cross-linking peptides.

Lysozyme. Use the same suspending buffer described previously for gram-negative organisms. Add 1 to 3 mg of lysozyme per ml, and incubate at 37°C until the cells are susceptible to lysis by a detergent.

***N*-Acetylmuramidase.** First isolated by Yokogawa et al. (99), *N*-acetylmuramidase has the same substrate specificity as lysozyme, but the range of organisms that can be lysed by it differs (98). The recommended buffer for this enzyme is 20 mM Tris-HCl (pH 7.0). Add approximately 6 μg of the enzyme per ml to the cell suspension and incubate at 50°C.

Lysostaphin. Lysostaphin is specific for the cross-linking peptides in the cell walls of staphylococci (76). The recommended buffer for the enzyme is 50 mM Tris-HCl–0.145 M NaCl (pH 7.5). Add the enzyme to a concentration of 25 μg/ml, and incubate at 37°C.

Achromopeptidase. The *Achromobacter lyticus* strain that produces achromopeptidase was first isolated at Takeda Chemical Industries, Ltd., Japan. The enzyme has been found useful for plasmid isolation (39) and is also useful for routine DNA isolations. The recommended buffer is 10 mM Tris-HCl (pH 8.2). Add the enzyme to a concentration of 50 μg/ml, and incubate at 50°C until the cells become sensitive to lysis.

26.1.2.3. Recalcitrant bacteria

There are many bacteria whose cell walls are not susceptible to enzymic digestion and detergent lysis. Although this group may include some gram-negative bacteria, most are typical gram-positive bacteria (such as *Actinomyces, Streptococcus, Peptococcus,* and *Propionibacterium* species) and archaeobacteria (many of which have cell walls containing pseudomurein). There are two approaches available for disrupting these recalcitrant bacteria. The first involves making the cells susceptible to one or more lytic enzymes by growing them in the presence of wall-component analogs or antibiotics, and the second involves the use of any of several physical methods for cell disruption.

One can make cells more susceptible to lysozyme by growing them in the presence of glycine, threonine, or lysine (20, 97). However, a problem with using high concentrations of these amino acids is that they may inhibit growth. Another way of making cells susceptible to lysozyme is by adding penicillin at late log phase to inhibit wall synthesis. Here one has to be concerned that the cells do not lyse before or during harvesting. Required penicillin levels may vary from strain to strain. Lysozyme-induced lysis of some gram-negative bacteria can be improved by washing the cells with a mild detergent (0.1% Sarkosyl) and then subjecting them to a mild osmotic shock (77).

Many physical methods have been used to disrupt bacterial cells. The following are some methods that have been routinely used for DNA isolation.

Sonication. The major problems with sonication are that it may not disrupt cells having thick cell walls or a small cell size and that it causes substantial fragmentation of the DNA. The amount of fragmentation increases with time of sonication. DNA preparations from cells disrupted in this manner are not useful for restriction endonuclease digestion and cloning, but if the sonication times can be standardized the DNA preparations can be used in reassociation experiments.

French pressure cell. Disruption by passage through a French pressure cell has the same limitations as sonication. The advantage of the French pressure cell is that one obtains more evenly sized DNA fragments.

Glass bead disruption. When suspensions of organisms are shaken with glass beads at high speed (4,000 oscillations per min), the cells rupture by being crushed between the colliding beads. The bead size is very important; for bacteria, small beads of about 100 μm in diameter are used. There are several instruments available commercially for shaking cells with glass beads. One that works well is the Braun Cell Homogenizer (B. Braun Instruments, Melsungen, Germany). The colliding beads produce so much kinetic energy that the sample vessel must be cooled with liquid CO_2. Although it can disrupt all cell types, the homogenizer does not fragment DNA to the same extent as does sonication or passage through a French pressure cell.

Alkaline hydrolysis. Sodium hydroxide at a concentration of 0.03 N has been used for the mild disruption of gram-negative bacilli and subsequent isolation of native DNA (9) (see section 26.2.1). It has also been used at a concentration of 0.4 N for not only disrupting bacteria but also obtaining denatured (single-stranded) DNA, as in the following procedure (66).

Suspend the bacterial cells in water, and add NaOH from a 10 N stock solution to give a final concentration of 0.4 N. (*CAUTION:* Add the stock solution slowly, and be sure to wear safety goggles.) Incubate in a water bath at 80°C for 30 min. Under these conditions the RNA, proteins, and many of the polysaccharides will be hydrolyzed but the denatured DNA will remain intact except for some depurination and strand scission (90). The alkaline lysate can be used directly to immobilize DNA on nylon membranes; alternatively, after neutralization, the denatured DNA can be isolated by using hydroxylapatite (see section 26.2.3).

26.2. DNA ISOLATION

The pioneering work of Kirby et al. (51, 52) and Marmur (63) has provided the basis for many nucleic acid isolation protocols. Procedures involving cetyltrimethylammonium bromide (CTAB) precipitation have evolved from the work of Jones (48). Britten et al. (15) were the first to use hydroxylapatite (HA) adsorption for DNA purification.

26.2.1. Marmur Procedure

The Marmur procedure (63) is probably the most widely used for DNA isolation. The bacterial cells are lysed with sodium dodecyl sulfate (SDS) in the presence of sodium chloride. Na-EDTA, and sodium perchlorate. Sodium perchlorate is a chaotropic salt that helps dissociate proteins and polysaccharides from nucleic acids. The proteins are then extracted with chloroform-isopentanol and discarded, and the DNA is precipitated with ethanol. After treatment with RNase and more chloroform-isopentanol extractions, the DNA is again precipitated. The isolated DNA has reasonably large fragments such that, in addition to all of the standard DNA reassociation and RNA hybridization experiments, it can be used in endonuclease restriction digestions and partial digestions for fragment size estimations, fragment size polymorphism studies, and construction of genomic libraries. The major problem with the procedure, depending on the organism, is the coisolation of high-molecular-weight polysaccharides, which have ethanol precipitation properties very much like those of DNA. The current variation of the Marmur procedure that is used in the author's laboratory includes the addition of the reducing agent 2-mercaptoethanol (which was initially used to inactivate nucleases in eucaryotic tissue but also seems to work with bacterial cells) and proteinase K to degrade proteins. The procedure is described for culture volumes of 250 to 500 ml but is amenable to scaling up or down. The protocol below is based on the use of lysozyme to render the cells susceptible to SDS disruption, but the suspending buffer can be modified for use with other lytic enzymes.

Reagents

Cell-suspending buffer: 10 mM Tris-HCl (pH 8.0), 1 mM sodium EDTA, and 0.35 M sucrose

Na-EDTA, 0.5 M solution (pH 8.0)

Sodium dodecyl sulfate (SDS), 20% (wt/vol) solution in water

Proteinase K, 20 mg/ml in Tris-EDTA (TE) buffer. The latter consists of 10 mM Tris-HCl (pH 8.0) containing 1 mM sodium EDTA (pH 8.0).

Lysing solution. The lysing solution consists of 100 mM Tris-HCl buffer (pH 8.0), 0.3 M NaCl, 20 mM EDTA, 2% (wt/vol) SDS, 2% (vol/vol) 2-mercaptoethanol, and 100 μg/ml proteinase K. Add the proteinase K and 2-mercaptoethanol at the time of use to the amount of lysing solution needed; do not keep the mixture longer than a day. Also, just before use, add 0.5 vol of 5 M sodium perchlorate to the volume of lysing solution needed. A precipitate will form but will quickly dissolve if the solution is warmed in a 50 to 60°C water bath. Add 1.5 vol of this mixture to each volume of cell suspension.

Sodium acetate, 3 M (pH 6.0)

Standard saline-citrate (SSC): 0.15 M NaCl, 0.015 M trisodium citrate (pH 7.0). Other concentrations of SSC are indicated by numbers; for example, 20× SSC = 20 fold concentrate, 0.1× SSC = one-tenth the standard concentration.

Chloroform-isopentanol: chloroform containing 3% (vol/vol) isopentanol

Phenol-$CHCl_3$: water-saturated phenol mixed with an equal volume of $CHCl_3$ that contains 3% (vol/vol) isopentanol. Add 0.1% (wt/vol) hydroxyquinoline to the mixture.

Ethanol, 80%: 80 parts of 95% ethanol and 20 parts distilled water. Store and use at −20°C.

RNase A: bovine pancreatic RNase A (Sigma Chemical Co., St. Louis, Mo.), 1 mg/ml dissolved in 0.15 M NaCl (pH 5.0). Heat the solution at 80°C for 10 min to inactivate any trace of DNase. Store in tubes containing small portions of the solution at −20°C.

RNase T_1. Dissolve RNase T_1 (Sigma) in 0.02 M Tris-HCl buffer (pH 6) to a concentration of 200 U/ml.

Procedure

1. Harvest the cells by centrifugation. After decanting the spent medium, remove any residual medium with a Pasteur pipette. Suspend the cells in 20 ml of suspending buffer, and place them into a ground-glass-stoppered 250-ml Erlenmeyer flask.
2. Add dry lysozyme to the 20-ml cell suspension with a measuring spoon (one-eighth teaspoon gives about 2.5 mg/ml). Incubate the mixture at room temperature or at 37°C until the cells become susceptible to lysis by the addition of the next set of reagents. For gram-negative organisms only about 5 min may be needed; gram-positive organisms may also become susceptible in a few minutes but sometimes require several hours. Remove small samples (0.1 to 0.2 ml) to a small test tube, and mix with 1.5 vol of lysing solution to test for lysis. On lysis, the cell suspension will change from turbid to opalescent and become very viscous.
3. Lyse the cells by adding 30 ml of the warm lysing solution. Quickly swirl to uniformly mix the compo-

nents and obtain a uniform lysis of the cells. Incubate at 50 to 60°C for 1 to 4 h.

4. Add 12 ml of the phenol-$CHCl_3$ mixture, and swirl and shake to obtain a homogeneous mixture. (Shaking may have to be vigorous for a highly viscous lysate.) Shake on a wrist-action shaker for 20 min at a setting that will provide vigorous mixing; however, if the phenol-$CHCl_3$ is not first mixed in well by hand, the extraction will be incomplete.

5. Place the extracted lysate into centrifuge tubes (dividing the lysate between two 50-ml straight-wall polypropylene tubes works well), and centrifuge for 10 min at 17,000 × *g*. Carefully decant and/or pipette the upper aqueous layer from the centrifuge tubes back into a clean Erlenmeyer flask. An inverted 5-ml pipette placed in a pipetting device or pump works well for removing the last of the viscous lysate. Discard the phenol-$CHCl_3$ bottom layer into an organic-waste container.

6. Add 10 ml of phenol-$CHCl_3$ mixture to the lysate, and again shake for 20 min on a wrist-action shaker. Centrifuge, and collect the upper layer as in step 5. Repeat this step until there is no protein layer at the aqueous-$CHCl_3$ interface. Each time discard the phenol-$CHCl_3$ bottom layer into an organic-waste container.

7. Place the aqueous phase into a 125-ml ground-glass-stoppered Erlenmeyer flask, and add 0.6 volume of isopropanol. Swirl the flask so that the precipitated DNA will form a clot. Decant the lysate, leaving the DNA clot in the flask. Add about 20 ml of cold 80% ethanol, swirl, and decant after about 5 min. Repeat this ethanol wash step twice. After the final ethanol wash, allow the clot to stick to the bottom of the flask (this can usually be done by slowly tilting the flask) and then invert the flask on a paper towel to drain. Air dry for 15 to 30 min in a 37°C incubator.

8. Dissolve the DNA by adding 20 ml of 0.1× SSC to the flask. After the DNA is completely dissolved, add 0.25 ml of RNase A solution and 2.5 ml of RNase T_1 solution. Incubate at 37°C for 30 min.

9. Add 5 ml of $CHCl_3$-isopentanol to the DNA solution, and shake on a wrist-action shaker for 20 min. Centrifuge at 17,000 × *g* for 10 min. Remove the aqueous phase with a pipette, and save it. If there is a substantial protein layer, repeat the $CHCl_3$-isopentanol step one more time. Finally, transfer the aqueous phase to an 80- or 100-ml beaker.

10. Again precipitate the DNA, as follows. Add 2.0 ml of 3 M sodium acetate to the DNA solution, swirl to mix, and then overlay the solution with 2 volumes of 95% ethanol. "Spool" the DNA onto a glass rod (20 cm by 7 mm) by slowly stirring and spinning the rod in the mixture (Fig. 1). Continue the spooling until the two phases are totally mixed. Gently press and turn the rod on the wall of the beaker to squeeze out the remaining ethanol. Wash the rod with 10 to 20 ml of cold 80% ethanol. Invert and stand the rod in a test tube rack to air dry. Dissolve the DNA by placing the rod in screw-cap test tube containing 3 to 5 ml of 0.1× SSC or TE buffer. Usually after about 30 min the DNA can be slipped off the glass rod by gently shaking the rod up and down (be careful not to hit the bottom of the test tube at this point; the rod can easily punch through the bottom). Alternatively, place the tube and rod overnight in a refrigerator to dissolve the DNA.

FIG. 1. Spooling DNA onto a glass rod during precipitation by ethanol. Photo by D. Arbour.

11. Finally dissolve the DNA in 3 to 5 ml of 0.1× SSC or TE, and store in a freezer.

Comments

There are many DNA isolation procedures, but the Marmur procedure is the best overall method. The following are some important variations from the procedure as originally described. (i) The low-ionic-strength suspending buffer gives lysozyme the best opportunity to weaken the cell walls. (ii) The single lysing solution allows for an even distribution of all of the components in what will become a very viscous solution. (iii) The proteinase K digestion helps free DNA from tenacious proteins. For some organisms, the binding of protein to the DNA can cause the DNA to be extracted with the proteins. (iv) Initial protein extraction with phenol-$CHCl_3$ seems to be more effective than the use of $CHCl_3$-isopentanol alone. (v) The use of 0.6 volume of isopropanol for the initial DNA precipitation helps to leave many polysaccharides in solution. In the original procedure, this selective precipitation step was near the end. (vi) Letting the DNA form a clot permits efficient washing with 80% ethanol and rapid redissolving in buffer. (vii) In many cases, the use of a single RNase (RNase A) leaves RNA fragments that are still large enough to coprecipitate with the DNA. Using both RNase A and RNase T_1 reduces this problem.

There have been other variations of the Marmur procedure reported in the literature, such as dialyzing DNA preparations to remove small RNA fragments, replacing sodium perchlorate with NaCl (30), and performing an initial alkaline lysis of the cells (8). The major features of the procedure of Beji et al. (9) are as follows.

Harvest late-log-phase cells (500 to 600 ml of culture), and resuspend them in 50 ml of saline-EDTA (150 mM NaCl, 100 mM EDTA [free acid] [pH 8.0]). Centrifuge again, and determine the wet weight of the cells. Lyse the cells with 0.03 M NaOH (10 ml/g [wet weight] of cells); leave the cells in contact with NaOH for a maximum of 2 min at room temperature. Then add (per gram [wet weight] of cells) 1.5 ml of 25% (wt/vol) SDS and 35 ml of saline-EDTA (pH 7). Mix well, and centrifuge at 17,000 × *g* for 5 min at 4°C. After treatment with 2.5 mg of RNase at 60°C for 30 min and 0.6 mg of proteinase K at 37°C for 15 min, add 5 M NaCl to increase the NaCl concentration to 1 M. Precipitate the DNA with 1 vol of 95% ethanol, and redissolve in 20 ml of 0.1× SSC. Add 0.45 M NaCl, and again precipitate the DNA by adding an equal volume of isopropanol. Dissolve the final product in 0.1× SSC, and dialyze it against 0.1× SSC.

26.2.2. Cetyltrimethylammonium Bromide Lysis

The CTAB procedure has been used extensively for fungal materials, which contain polysaccharide-like contaminants that have plagued DNA isolation procedures (68, 100). The procedure described here works well for actinomycetes and propionibacteria also.

The cetyltrimethylammonium ion is a cationic detergent and cannot be used in conjunction with SDS or phenol, because these mixtures will form very insoluble complexes. DNA is soluble in the presence of CTAB if there is also a high concentration of monovalent cations such as Na^+ or NH_4^+.

The protocol given here (47) is for recalcitrant organisms that must be physically disrupted. This disruption results in fragmentation of the DNA, and the DNA works well for reassociation experiments involving only fragmented DNA, i.e., the optical, hydroxylapatite, and S1 nuclease procedures. However, it may not work well as membrane-bound DNA in DNA reassociation experiments or in rRNA hybridization (see Chapter 27), and it would not be adequate for restriction endonuclease digestions. The procedure can be used for the isolation of high-molecular-weight DNA if lysozyme-digested cells (or cells digested by some other lytic enzyme) are lysed by the CTAB. The procedure may be useful when polysaccharide contamination of DNA preparations is a problem.

Reagents

Cell-suspending buffer (TE buffer): 10 mM Tris-HCl buffer (pH 8.0) containing 1 mM Na-EDTA (pH 8.0)
CTAB lysing solution: 50 mM Tris-HCl buffer (pH 8.0), 0.7 M NaCl, 10 mM EDTA, 1% CTAB (wt/vol), 1% (vol/vol) 2-mercaptoethanol. 2× and 5× CTAB lysing solutions can be prepared by using two and five times the concentrations of all of the components, respectively.
Proteinase K: see section 26.2.1
Chloroform-isopentanol: see section 26.2.1
CTAB dilution-precipitation buffer: 50 mM Tris-HCl, 10 mM EDTA, 1% CTAB (wt/vol)
CTAB wash solution: 0.4 M NaCl
CTAB-DNA salt dissolving solution: 2.5 M ammonium acetate
SSC: see section 26.2.1
RNase A: see section 26.2.1
RNase T_1: see section 26.2.1
Phenol-$CHCl_3$: see section 26.2.1
Sodium acetate, 3 M (pH 6.0)
80% ethanol: see section 26.2.1

Procedure

1. Harvest the cells by centrifugation. After decanting the spent medium, remove any residual medium with a Pasteur pipette. Transfer the cell pellet to a tared Braun Disintegrator shaking bottle, and make up to 10 g with TE buffer. Add 10 ml of 2× CTAB lysing buffer, 0.2 ml of 2-mercaptoethanol, and 20 ml of glass beads (diameter range, ca. 100 μm). Place the bottles in the Braun Disintegrator, and shake them for 5 min with CO_2 cooling. The exact shaking time required depends on the particular organisms; check for cell breakage by microscopic observation (phase contrast or Gram staining).
2. Separate the lysate from the glass beads by filtration through a sintered-glass filter (coarse pore) into a vacuum flask. Filtration is facilitated by a vacuum (a water aspirator can produce a sufficient vacuum). Use an additional 20 ml of 1× lysing buffer to wash the remaining lysate away from the glass beads. Add small amounts of buffer at a time, and use a rubber-tipped stirring rod to mix it with the glass beads. Metal or glass stirring rods will damage the surface of the sintered-glass filter.
3. To the pooled lysate add proteinase K to a final concentration of 50 μg/ml, and incubate the mixture for 1 to 2 h at 60°C.
4. Add 15 ml of $CHCl_3$-isopentanol to the lysate, shake on a wrist-action shaker for 20 min, and centrifuge at 17,000 × *g* for 10 min.
5. Transfer the lysate (upper layer) into a 250-ml centrifuge bottle, and add an equal volume of CTAB dilution-precipitation buffer. Mix well, and pellet the CTAB-nucleic acid salts by centrifuging at 8,000 × *g* for 10 min. Wash the precipitate two or three times with 40-ml volumes of cold 0.4 M NaCl wash solution to remove free CTAB. A glass homogenizer works well to disperse the pellet for more efficient washing.
6. Dissolve the CTAB-nucleic acid pellet in 8 ml of 2.5 M ammonium acetate. Use a 50-ml screw-cap polypropylene centrifuge tube, and warm the material at 50°C to facilitate dissolution. After the nucleic acids are in solution, place the tube on ice for 15 min to precipitate the large RNAs. Remove the RNA by centrifuging at 17,000 × *g* at 4°C for 10 min. Transfer the supernatant to another centrifuge tube, and precipitate the remaining nucleic acids with 2 volumes of 95% ethanol. Collect the precipitate by centrifuging at 17,000 × *g* for 10 min.
7. Dissolve the pellet in 10 ml of 0.1× SSC, and add 0.25 ml of RNase A and 2.5 ml of RNase T_1. Incubate at 37°C for 1 h.
8. Extract the digest once with 5 ml of phenol-$CHCl_3$. After centrifugation, add 1 ml of 3 M sodium acetate to the extracted digest, and precipitate by adding 2 volumes of 95% ethanol. Collect the precipitate by centrifugation, and wash the pellet with 10 ml of 80% ethanol.
9. Finally dissolve the DNA in 3 to 5 ml of 0.1× SSC or TE and store in a freezer.

Comments

The procedure described above can be used for organisms that are difficult to lyse and have high levels of nuclease activity. A technical difficulty is that the detergent causes foaming and tends to reduce the velocity of the beads, thereby reducing their kinetic energy. Depending on the organism, this may result in inefficient cell breakage. If nuclease activity is not a problem, suspend and disrupt the cells in TE buffer (it may be helpful to increase the EDTA concentration to 10 mM). Also add the RNase at this time. Then add 0.25 volume of 5× CTAB lysing solution after removing the glass beads by filtration. Then proceed to step 3 of the procedure.

If the main reason for using this procedure is to eliminate contaminating polysaccharides in DNA preparations for organisms that otherwise lyse with a lytic enzyme plus detergent, one can start as if doing the Marmur procedure but then lyse cells with 2× CTAB lysing solution.

26.2.3. Hydroxylapatite Adsorption

Britten et al. (15) were the first to isolate DNA by adsorbing it to HA. Since then, several modifications of the method have been described (2, 44, 61). The following adaptation is useful.

Reagents

Cell-suspending buffer (TE buffer): see section 26.2.2
SDS: see section 26.2.1
RNase A: see section 26.2.1
RNase T_1: see section 26.2.1
Phenol: water-saturated phenol containing 0.1% (wt/vol) 8-hydroxyquinoline
HA, DNA grade Bio-Gel HTP (Bio-Rad Laboratories, Richmond, Calif.)
Phosphate buffer (PB), 1.0 M (pH 6.8) (prepare by mixing equal volumes of 1 M Na_2HPO_4 and 1 M NaH_2PO_4). This is used for preparing the lower concentrations as indicated in the text.

Procedure

1. Suspend the centrifuged cells in 25 ml of TE buffer, add lysozyme or other lytic enzyme, and incubate until the cells are susceptible to detergent lysis (see section 26.2.1, step 2).
2. Add 0.25 ml of RNase A and 2.5 ml of RNase T_1. Swirl the flask, and warm in a 50 to 60°C water bath until lysis is complete. For bacteria not disrupted by detergent alone, see Comments, section 26.2.2, for alternate methods.
3. Reduce the viscosity of the lysate by briefly subjecting it to sonic oscillation or by use of a mechanical tissue homogenizer. Proteinase K may be added at this time (50 μg/ml). For some organisms proteinase digestion is not needed, but for others it is essential. If the proteinase is added, incubate it with the lysate at 50°C for 1 h.
4. Add 7 ml of water-saturated phenol, shake by hand to get the two phases well mixed, and then shake the flask on a wrist-action shaker for 20 min.
5. Centrifuge at 17,000 × g in a polypropylene centrifuge tube, using a refrigerated centrifuge at 0 to 4°C. Carefully draw off the upper (aqueous) layer (see step 5 of section 26.2.1), and return it to the flask.
6. Repeat steps 4 and 5.
7. After again returning the aqueous phase to the flask, add 2.0 ml of 1.0 M PB. Then add 2 g (1 measuring teaspoon is convenient and sufficiently accurate) of dry HA. Suspend well, and gently shake on a rotary or reciprocal shaker for 1 h at a speed sufficient to keep the HA from settling out.
8. Transfer the suspension to a 50-ml polypropylene centrifuge tube, and centrifuge for 2 to 3 min at 5,000 × *g* at room temperature. Return the supernatant layer (lysate) to the flask (this can be used for a second DNA adsorption cycle), and use the sedimented HA (to which DNA is now adsorbed) in step 9.
9. To the sedimented HA, add 8 ml of 0.10 M PB. Suspend the HA with the aid of a Vortex mixer. Immediately add an additional 24 ml of the PB (an automatic pipettor works well for these additions). The PB should be added with some force, so that the HA will become evenly mixed for maximum dilution of nucleotides and phenol. Allow the HA to settle for 1 to 2 min, and then centrifuge for 2 to 3 min at 5,000 × *g*. Discard the supernatant.
10. Repeat step 9 six or seven times or until the A_{270} of the supernatant is less than 0.05 (the wavelength of maximum absorption by phenol).
11. Suspend the HA in 5.0 ml of 0.5 M PB to desorb the DNA. Centrifuge as before, but this time save the DNA-containing supernatant.
12. If most of the DNA has not been removed from the lysate in step 8, wash the HA from step 11 once with distilled water, and then add it to the lysate for a second adsorption cycle, Alternatively, fresh HA may be used. Add the DNA obtained from a second adsorption cycle to that from the first cycle.
13. Filter the DNA preparation through a fiberglass filter (2.4-cm diameter, type 934AH; Reeve Angel Inc., Clifton, N.J.) used in a syringe-type filter holder (Gelman Instrument Co., Ann Arbor, Mich.) to remove any remaining HA particles.
14. Dialyze the DNA preparation against TE buffer (pH 8.0). Cut 13- to 15-cm lengths of dialysis tubing (Fisher Scientific Co., Pittsburgh, Pa.), and wash out UV-absorbing materials by boiling the tubing in a 2 to 5% solution of sodium carbonate for a few minutes, thoroughly rinsing the lengths of tubing under running tap water, and then rinsing them under distilled water. The tubing can be stored in a refrigerator for 2 to 3 days in distilled water. Cellulolytic organisms may hydrolyze the tubing if it is stored for a longer time. Tie a knot at one end of the tubing length, pour in the DNA solution, and either tie another knot at the other end or tie it off with a string. Wrap a piece of tape around the string for a convenient label. Dialyze in a 400- to 500-volume excess of the buffer for about 3 h, change the buffer, and continue dialyzing overnight. Store the DNA preparation in a freezer, or add a few drops of chloroform and store in a refrigerator.

Comments

If the lysate is digested with proteinase K, a single phenol extraction will usually be sufficient.

An important feature of the HA procedure is that RNA is degraded to such an extent that it will not compete with DNA for adsorption sites on the HA. For some groups of organisms, the RNase is inhibited when added in step 2. In such cases, dialyzing the lysate against saline-EDTA buffer for several hours will remove the inhibitory effect.

The major contaminant of the DNA preparations, other than RNA, is polysaccharide. Many polysaccharides do not adsorb to HA and therefore are easily separated from DNA by this procedure. Others, however, can bind to HA and inhibit DNA adsorption.

A more dilute cell suspension of some organisms may be needed for this method to be efficient. The dilution appears to affect the completeness of lysis and/or RNA degradation. Scale up the volumes four or five times, and follow the above steps until the HA adsorption. Here, after adding the HA to part of the lysate, pour the HA suspension into a 60-ml sintered-glass filter. The HA will quickly settle, and the lysate will flow through the HA as through a column, resulting in more efficient adsorption. Then transfer the HA to a centrifuge tube, and complete the procedure as described above.

26.3. DNA MEASUREMENT (also see Chapter 22.6)

26.3.1. UV Spectrophotometry

The molar extinction coefficient of native DNA is in the range 6,650 to 6,700 (46, 59). Conversion of the DNA concentration from molarity to milligrams per milliliter yields the following formula, for cuvettes with a 1-cm light path. For native (double-stranded) DNA, the formula is

$$\text{Concentration (mg/ml)} = A_{260}/20 \quad (1)$$

For denatured (single-stranded) DNA and for RNA, the formula is

$$\text{Concentration (mg/ml)} = A_{260}/23 \quad (2)$$

In practice, most stock DNA preparations will range from 0.5 to 2 mg of DNA per ml. It is convenient to dilute them 20-fold (0.1 ml of DNA solution added to 1.9 ml of buffer) because the resulting A_{260} can be read directly as milligrams of DNA per milliliter of the stock preparation (see equation 1).

UV spectrophotometry has also been used for detecting contaminating materials (46). (i) Proteins absorb UV light most strongly at 280 nm. The A_{260}/A_{280} ratio of a nucleic acid solution has been used as an indicator of protein contamination. The extinction coefficient for protein, however, is very low compared with that for nucleic acids; consequently, the A_{260}/A_{280} ratio is not a very sensitive indicator. (ii) The absorption spectrum for nucleic acids has maxima at approximately 208 and 260 nm and a minimum at 234 nm. The peak at 208 nm is not specific for nucleic acids, because most compounds will absorb light of that wavelength. If contaminating material is present in a nucleic acid preparation and contributes to a large peak at 208 nm, this peak will overlap the 234-nm minimum for the nucleic acid. Therefore, the A_{260}/A_{234} ratio is a more sensitive indicator of contaminating material in a DNA preparation than the A_{260}/A_{280} ratio. (iii) The absorption maximum of phenol is 270 nm; consequently, the A_{260}/A_{270} ratio is quite sensitive for detecting phenol contamination in a nucleic acid preparation.

26.3.2. Hyperchromic Shift

The major contaminant of DNA preparations is usually RNA, and, unfortunately, the two nucleic acids cannot be differentiated directly by their UV absorbance. However, the hyperchromic shift (absorbance change) that occurs when DNA is thermally denatured (see section 26.4.1) can be used to estimate the fraction of the preparation that is DNA. A preparation of pure native DNA will increase in A_{260} by approximately 40% (Fig. 2) when denatured. If the secondary structure of the contaminating RNA has been destroyed by RNase, the RNA will have no hyperchromic shift. If the secondary structure is intact, the RNA will have a hyperchromic shift of approximately 30%. The hyperchromic shift for RNA occurs over a lower and wider temperature range than for DNA, so the profiles for RNA and DNA can be easily distinguished. Therefore, to estimate the A_{260} of a nucleic acid preparation that is due only to the DNA, divide the hyperchromic shift due to DNA by 0.4. For example, assume that the A_{260} of a sample (or a dilution of it) is 1.25 and that the change in A_{260} as a result of denaturation is 0.4. The calculated DNA A_{260} would be 0.4/0.4, or 1.00 (i.e., 1/1.25 or 80% of the A_{260} due to DNA).

26.3.3. Colorimetric Method (Diphenylamine Reaction)

Many colorimetric assays have been developed and modified over the years for the measurement of DNA (8, 16, 17, 34, 81, 96). Although all can be used for measuring purified DNA, these procedures were often designed for measuring DNA in tissue material. Although

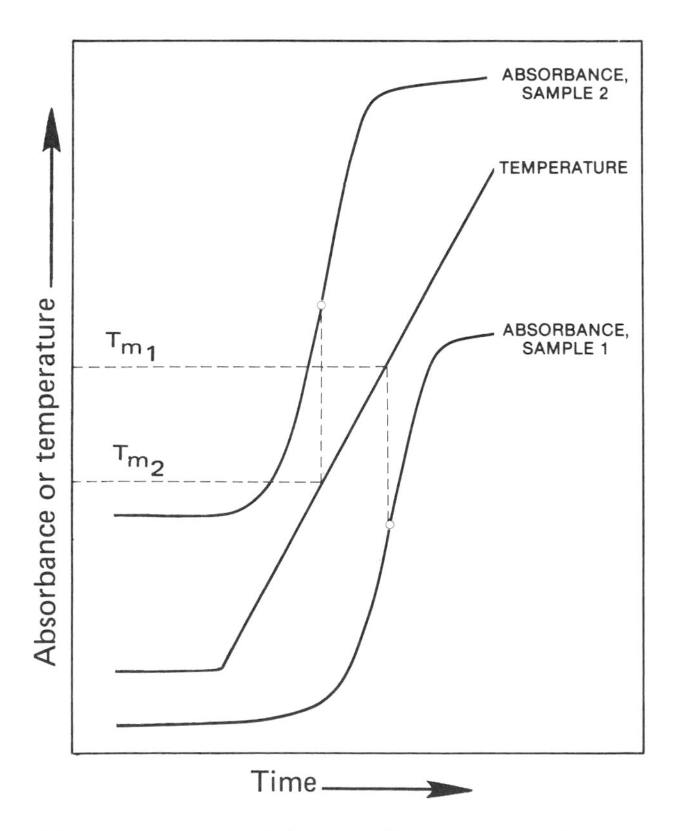

FIG. 2. Hyperchromic shift tracings for two samples of DNA in an automatic recording spectrophotometer. The points at which any vertical line crosses the absorbance and temperature curves give values for these two parameters at a given time. The inflection points are estimated at one-half of the hyperchromic shift, and the temperature at that time is the melting temperature (T_m).

they all can be made specific for measuring DNA, they have the disadvantage of using high concentrations of mineral acids and of requiring long periods for color development.

The diphenylamine reaction is specific for the deoxyribose in DNA. Although the Burton procedure (16) has often been used, the following variation of the more sensitive Giles and Meyers modification (32) is preferable.

1. Place 1.0 ml of the DNA sample (5 to 50 μg) in a 16- by 125-mm screw-cap tube and add 1.0 ml of 20% (wt/vol) perchloric acid.

2. Add 2.0 ml of glacial acetic acid containing 4% (wt/vol) diphenylamine.

3. Add 0.2 ml of 0.16% (wt/vol) solution of acetaldehyde.

4. Mix, and incubate overnight at 30°C.

5. Read the A_{595} and A_{700}, and subtract the A_{700} from the A_{595}. Construct a standard curve, using known amounts (5 to 50 μg) of a pure standard DNA (calf thymus or salmon sperm DNA; Sigma Chemical Co.), and plot the microgram amount of DNA against the ($A_{595} - A_{700}$) for each DNA level. Determine the concentration of DNA in the sample from the standard curve.

The diphenylamine procedure can also be used for quantifying the DNA bound to nitrocellulose or nylon membranes in hybridization studies. It is particularly useful when one cannot readily measure the amount of DNA being bound to a filter (for example, see section 26.6.3). After the amounts of probe (radioactive or nonradioactive) bound to the membranes have been measured (use a nonaqueous counting solution so that the membranes will not be dissolved or the DNA eluted), dry the membranes and place each membrane in 2.0 ml of 10% perchloric acid (duplicate membranes can be placed together in the same tube). Heat in a boiling-water bath for 5 min. After cooling to room temperature, proceed with step 2 above.

26.3.4. Fluorimetric Dye-Binding Methods

Fluorescent compounds, usually dyes, have been used for characterizing and quantitating DNA. After interacting with the DNA, the resulting complex causes an enhancement of the fluorescence. The complexes are very specific, in that molecules other than DNA (such as RNA, proteins, and carbohydrates) seldom interfere with assays. A major use of these procedures is the measurement of DNA in tissue homogenates. Certain dyes may complex specifically with high-A+T or high-G+C regions of the DNA, whereas the binding of others is independent of the base composition. As a result, the moles percent G+C content of the DNA may have to be taken into account when constructing standard curves, and in some cases differential binding of two dyes can be used for estimating this value. The following are some of the more commonly used fluorimetric methods for measuring DNA.

26.3.4.1. Mithramycin

The antibiotic mithramycin binds to double-stranded DNA, and the amount of fluorescence is directly proportional to the amount of DNA (37). The antibiotic binds to a lesser extent to denatured DNA and does not bind to DNase-treated DNA or to RNA.

Reagent

Prepare a stock solution containing 200 μg of mithramycin per ml in 300 mM $MgCl_2$.

Procedure

1. Dilute a sample of a DNA standard (a preparation containing no RNA, with the concentration determined by UV absorbance) with distilled water to give amounts ranging from 0.4 to 32 μg in a final volume of 1.9 ml.

2. Dilute the test DNA preparations so that they are within the same range.

3. Add 0.1 ml of the mithramycin stock solution to the standard DNA dilutions and test DNA dilutions, mix, and determine the fluorescence (emission) at 540 nm with an excitation wavelength of 440 nm.

4. Plot a standard curve from the values obtained for the DNA standards, and use it to estimate the concentrations of the test DNA samples.

26.3.4.2. 4′,6′-Diamidino-2-phenylindole·2HCl

4′,6′-Diamidino-2-phenylindole·2HCl (DAPI) forms a specific fluorescent complex with DNA, but the manner by which it binds (i.e., by intercalation, or by a nonintercalating mode) is uncertain (58). DAPI does exhibit a preference for several continuous A·T base pairs (49). The dye is very stable and is specific for DNA. It has been used to measure DNA directly in fixed cells of *Escherichia coli* and might be similarly useful with other bacteria.

Reagents

DAPI stock solution, 3 μg/ml in distilled water. This stock solution is stable at 4°C for months.

Dilution buffer: 10 mM NaCl, 6.6 mM Na_2SO_4, 5 mM *N*′-2-hydroxyethylpiperazine-*N*′-2-ethanesulfonic acid (HEPES; pH 7.0)

Toluene-fixed cells. Mix samples of bacteria with toluene (1% vol/vol), and shake vigorously. Store at 4°C, at which they are stable for months.

Procedure

1. Dilute a sample of a DNA standard in the dilution buffer to give concentrations ranging from 5 ng/ml to 1.0 μg/ml in a final volume of 3.0 ml.

2. Similarly dilute the fixed bacterial cells in the dilution buffer to a final turbidity of 0.008 to 0.08 at 600 nm (turbidity values may differ from one spectrophotometer to another; the only important requirement is that the fluorescence values for DAPI [see step 3] be linear within the turbidity range used).

3. Add 100 μl of DAPI stock solution to each tube, mix, and measure the fluorescence intensity with excitation at 350 nm and emission at 450 nm. (The mixtures are stable at 4°C for several weeks.) Plot a standard curve from the values obtained for the DNA standard, and use it to estimate the concentrations of the test DNA samples.

26.3.4.3. Bisbenzimide (Hoechst 33258)

The Hoechst 33258 dye is probably the most widely used of DNA-binding dyes (22, 27, 53, 70, 71, 85, 86). The dye binds specifically to regions rich in A·T base pairs in the large groove of the DNA helix (85). Consequently, the DNA used for the standard curve should have a moles percent G+C content similar to that of the test DNAs.

Reagents

20× phosphate-buffered SSC (PB-SSC): 20× SSC (see section 26.2.1) prepared in 200 mM sodium phosphate buffer (pH 7.0)

Hoechst 33258 stock solution: 2 μg/ml in 1× PB-SSC. (This is equivalent to a concentration of 3.7×10^{-6} M; the molecular weight of the dye is 533.89.) The dye concentration can be measured with a spectrophotometer at 338 nm; the molar extinction coefficient is 4.2×10^{4} M^{-1} (88).

Procedure

1. Dilute samples of a DNA standard (a preparation containing no RNA, with the concentration determined by UV absorbance) in 1× PB-SSC to give amounts ranging from 0.01 to 10 μg in final volumes of 2.0 ml.
2. Dilute the test DNA preparations so that they are within the same range.
3. Add 2 ml of dye solution to each tube of DNA, mix, and measure the fluoresence intensity with excitation at 360 nm and emission at 450 nm. Plot a standard curve from the values obtained for the DNA standard, and use it to estimate the concentrations of the test DNA samples.

26.3.4.4. Ethidium bromide

The intercalating dye ethidium bromide binds to double-stranded DNA with little dependence on the base composition. Although it binds to a lesser extent to intact RNA, this can be eliminated by treating the DNA preparation with RNase (50, 58, 70, 71).

Reagents

20× phosphate-buffered SSC (PB-SSC): see section 26.3.4.3

Ethidium bromide. Prepare a working solution of 2 μg/ml in 2× PB-SSC.

Procedure

1. Dilute samples of a DNA standard (a preparation containing no RNA and the concentration determined by UV absorbance) in 2× PB-SSC to give amounts ranging from 0.01 to 10 μg in final volumes of 2.0 ml.
2. Dilute the test DNA preparations so that they are within the same range.
3. Add 2 ml of dye solution to each tube, mix, and measure the fluorescence intensity with excitation at 305 nm and emission at 615 nm. Plot a standard curve from the values obtained for the DNA standard, and use it to estimate the concentrations of the test DNA samples.

26.4. GUANINE-PLUS-CYTOSINE CONTENT OF DNA

The moles percent G+C content of DNA can be estimated by several different methods; see reference 1 for a review. The methods described in detail here use thermal denaturation, high-performance liquid chromatography, or dye-binding fluorimetry. Although buoyant-density centrifugation has also been used for many moles percent G+C estimations reported in the literature, there are few if any analytical ultracentrifuges still functioning. All of these methods require the prior isolation and purification of DNA.

Estimation of the moles percent G+C directly in intact whole cells is often desirable, especially in characterizing newly isolated bacteria in ecological and systematics studies or bacteria whose DNA is difficult to extract or purify. One of the dye-binding fluorimetric methods described below (section 26.4.3.2) can be used with formalin-fixed whole cells. A calorimetric method involving native intact cells is also described (section 26.4.4); this method requires expensive specialized instrumentation but can be performed quickly.

26.4.1. Thermal Melting Method

The midpoint temperatures of the thermal melting profiles (T_m) were first correlated with the base composition of DNA preparations by Marmur and Doty (64) (Fig. 2). The method has been reexamined by several investigators, and several equations correlating the T_m with the base composition have been proposed. With the exception of the equation proposed by Mandel et al. (62), which appears to have a different slope (2), all are similar to the one proposed by Marmur and Doty. The T_m values are greatly affected by the ionic strength of the buffer used, and this effect is the same on all DNA preparations. Consequently, one must either know the ionic strength of the buffer that is used or include a reference DNA preparation at the same ionic strength as that used for the unknown samples. The latter is easy to do: one merely dialyzes all of the DNA preparations (including the reference DNA) together in a single batch of buffer. Alternatively, if the DNA preparations are all dissolved in a common buffer, they can be diluted with the same buffer and tested directly.

Procedure

The following procedure is reliable and may be adapted to other buffers or to buffer concentrations other than those given.

1. Prepare 2 to 4 liters of 0.5× SSC (section 26.2.1) at pH 7.0. Save some of the buffer for diluting the DNA preparations and for rinsing out cuvettes just prior to putting the DNA samples into them. Divide the rest into two equal parts so that the buffer can be changed once during dialysis (see step 3).
2. Prepare 2- to 5-ml samples (depending on the size of the cuvettes to be used) of DNA at 50 μg/ml, using 0.5× SSC to dilute the stock DNA preparations. Prepare a 5- to 10-ml volume of the reference DNA (i.e., the DNA of *E. coli* b has a G+C content of 51 mol% and a melting point [T_m] of 90.5°C in 1× SSC), because it will have to be included in each instrument run.
3. Dialyze all of the preparations together overnight in the 0.5× SSC. Change the buffer once after the first few hours of dialysis. After dialysis, return the DNA preparations to screw-cap test tubes.
4. Determine the melting profile with an automatic recording spectrophotometer having a cuvette holder that is heated electronically. Start at 60°C, and increase the temperature linearly at 0.5 or 1.0°C per min over the entire range.
5. Determine the T_m values as in Fig. 2.

6. Calculate the moles percent G+C of the sample DNA by the following equation:

$$\text{Mol\% G+C} = \frac{\{T_m + [90.5 - T_{m(\text{ref})}]\} - 69.3}{0.41}$$

This is the Marmur and Doty equation, modified to normalize the T_m values to those in SSC.

Comments

The thermal stability of double-stranded DNA is highly dependent on the ionic strength of the buffer in which it is dissolved. If the DNA preparations have been spooled out of a sodium acetate-ethanol mixture and the spooled DNA has been rinsed with 80% ethanol (i.e., there is little or no residual salt in the DNA preparations), and if the DNA preparations have been dissolved in 0.1 × SSC, they can be diluted in a single batch of 0.1 × SSC and not dialyzed. The rinsing of the cuvettes can be a cause of error. If they are going to be reused immediately after being washed, give them a final rinse with the buffer being used in the T_m determination, so that the samples do not become diluted with variable amounts of water.

26.4.2. High-Performance Liquid Chromatography (also see Chapter 21.3.6)

The only direct approach for measuring the base composition of DNA is to hydrolyze it and quantify the individual nucleosides or nucleotides. Originally this was done by using paper chromatography to separate the nucleosides or nucleotides, which were then eluted and measured with a spectrophotometer. Many of the other methods have been based on correlations with these early results. However, because of the difficulty in quantifying these chromatographic analyses, few have been done since the early 1960s. Recent advances in high-performance liquid chromatography (HPLC), have enabled investigators to determine the base composition of DNA more rapidly and accurately (see reference 2 for a review). The procedure described here is that of Mesbah et al. (67), which includes degrading denatured DNA with the single-strand-specific nuclease P1 from *Penicillium citrinum*. The resulting 5′ nucleotides are then dephosphorylated with alkaline phosphatase, and the resulting nucleosides are separated on a reverse-phase column.

Instrumentation requirements

The HPLC setup includes a pump that will produce pulse-free isocratic separations of the nucleosides, an injector, a fixed-wavelength (254 nm) UV detector, and an integrator-recorder. The column is a Econosphere C18 reverse-phase column, with 5-μm particle size and column dimensions of 250 by 4.6 mm (Alltech Associates, Inc., Deerfield, Ill.).

Reagents

Stock triethylamine phosphate buffer, 0.5 M. Prepare by adjusting the pH of a solution of 0.5 M triethylamine to 5.1 with 85% phosphoric acid.

Chromatographic buffer: 20 mM triethylamine phosphate buffer (pH 5.1) containing 12% (vol/vol) HPLC grade methanol. Before using, filter the buffer through a cellulose triacetate membrane filter with a pore size of 0.45 μm (GA-6 Metricel; Gelman Sciences, Inc., Ann Arbor, Mich.).

P1 nuclease: 20 to 40 U/μl in a solution consisting of 10 mM sodium acetate (pH 6.0) and 50% (vol/vol) glycerol (GIBCO BRL, Gaithersburg, Md.). Before using, dilute to 1 U/μl with P1 buffer (see below).

Calf intestinal alkaline phosphatase: 20 to 30 U/μl dissolved in a solution consisting of 30 mM triethanolamine, 3 M NaCl, 1 mM $MgCl_2$, and 0.1 mM $ZnCl_2$ (pH 7.6) (GIBCO BRL). Before using, dilute to 0.2 U/μl with alkaline phosphatase buffer (see below).

P1 buffer: 30 mM sodium acetate buffer (pH 5.3), 2 mM $ZnCl_2$

TE buffer: see section 26.2.2

Alkaline phosphatase buffer: 0.1 M glycine · HC1 (pH 10.4)

DNA standard: lambda phage DNA (Sigma Chemical Co.). This has been sequenced so that the exact G+C content is known to be 49.858 mol%.

Procedure

1. Precipitate 10 to 25 μg of DNA in a 1.5-ml microcentrifuge tube by adding 0.5 volume of 7.5 M ammonium acetate and 2 volumes of 95% ethanol. Wash the pellet with 80% ethanol, and air dry it in a 37°C incubator. Dissolve the pellet in 5 μl of TE buffer and 20 μl of sterile distilled water. Denature the DNA by heating in a boiling-water bath for 2 min, and then rapidly cool it by placing it in an ice bath.
2. Add 50 μl of P1 buffer and 1 U of P1 nuclease. Incubate the mixture at 37°C for 2 h.
3. Add 5 μl of alkaline phosphatase buffer and 5 μl of diluted (0.2-U/μl) alkaline phosphatase, incubate at 37°C for 6 h.
4. Centrifuge the sample in a microcentrifuge for 5 min, transfer to a clean tube, and store at −20°C until chromatographed.
5. Determine the relative integrator values of the various nucleosides using the lambda DNA. The best nuclosides to use in the G+C calculations are deoxyguanosine and thymidine because they are not methylated and RNA degradation products do not coelute with them. Using the mole fractions of deoxyguanosine (G) and thymidine (T), calculate the moles % G+C by the following equation:

$$\text{mol\% G+C} = \frac{G}{G+T} \times 100$$

26.4.3. Fluorimetric Dye-Binding Methods

Although fluorimetric procedures have not been used extensively, they are rapid and sensitive, and some can be done directly with fixed cells. They rely on differential binding of fluorescent dyes to G+C and A+T regions of DNA.

26.4.3.1. Purified DNA

This procedure is designed for use with purified preparations of DNA (71). One sample, from a diluted DNA preparation, is assayed with ethidium bromide, which binds independently of the G+C content, whereas another sample is assayed with Hoechst 33258, which binds preferentially to regions of the DNA that are rich in A · T base pairs. There is about a 10-fold range in the relative response ratios (Hoechst 33258/ ethidium bromide) between poly(dA-dT) and poly(dG-dC). Use DNA samples that contain about 1 μg/ml in 1× PB-SSC, which is in the range that is linear for both assays.

Reagents

PB-SSC: prepare 1× and 2× PB-SSC from 20× PB-SSC (see section 26.3.4.3)
Hoechst 33258 solution: 2μg/ml in 2× PB-SSC
Ethidium bromide solution: 2 μg/ml in 2× PB-SSC

Procedure

1. Mix 1.0 ml of DNA solution (1 μg/ml in 1× PB-SSC), 1.0 ml of 2× PB-SSC, and 2.0 ml of Hoechst 33258 solution. Let stand for at least 30 s before measuring the fluorescence (excitation at 365 nm and emission at 458 nm).
2. Repeat step 1, substituting ethidium bromide solution for the Hoechst 33258 dye. Measure the fluorescence at an excitation setting of 305 nm and an emission setting of 615 nm.
3. Plot the ratio of fluorescence intensity of Hoechst 33258/ethidium bromide for each known DNA (i.e., known moles percent G+C content) solution, and calculate the moles percent G+C content of the test DNA samples from the graph.

26.4.3.2. Intact-cell DNA

This procedure uses a combination of two dyes, Hoechst 33258 and chromomycin A_3. The rationale for using this dye combination is that the Hoechst 33258 binds selectively in regions of the DNA that are rich in A+T, whereas chromomycin A_3 selectively binds to regions rich in G+C; thus the ratio of binding for the two dyes is a function of the moles percent G+C content. The procedures discussed here are for formalin-fixed whole cells analyzed in a conventional fluorimeter (60) or in a dual-beam flow cytometer (70, 75, 92). The protocol given here is a variation of the method of Lutz and Yayanos (60) with a conventional fluorimeter.

Reagents

Hoechst 33258 solution: see section 26.3.4.3. Prepare a 15 μM stock solution in distilled water. The stock solution can be stored in plastic tubes wrapped in aluminum foil at −20°C for several weeks. Prepare the working dye solution by diluting to 1.5 μM in 1× PB-SSC.
Chromomycin A_3 solution. Prepare a 15 μM stock solution in distilled water. Chromomycin A_3 has a molecular weight of 1183.3 and a molar extinction coefficient of 1.86×10^4 M^{-1} cm^{-1} at 281 nm when the dye is dissolved in ethanol. Storage of the stock solution and preparation of the working dye solution are as described for Hoechst 33258.
Formalin

Procedure

1. Grow cultures to 10^7 to 10^8 cells per ml. Fix the cells by adding 2 to 4% (vol/vol) formalin, mix, and incubate for 20 min. Centrifuge the cells, and suspend them in a similar volume of 1× PB-SSC. The fixed cells are stable for months when stored at 4°C.
2. To 2.0 ml of cell suspension, add 1.0 ml of each dye solution, mix vigorously, and incubate in the dark for no more than 10 min. Measure the fluoresence for each dye: for Hoechst 33258, use emission and excitation wavelengths of 450 and 360 nm, respectively; for chromomycin A_3, use 378 and 325 nm, respectively.

26.4.4. Differential Scanning Calorimetry of Intact-Cell DNA

The melting temperature of intracellular DNA, determined by differential scanning calorimetry (DSC), has been correlated with literature values of the moles percent G+C for 58 species of bacteria by Mackey et al. (60b). The method is simple and quick, obviating the need to isolate and purify DNA but requiring costly specialized instrumentation. The method has not been tested by the author, but its usefulness has been confirmed by investigators of Michigan State University (48a).

A small sample of cells from a centrifuged pellet or an agar colony (50 μl, 10 to 20 mg [dry weight]) is crimp-sealed in a gasket-containing stainless-steel sample pan and heat scanned in a Perkin-Elmer model DSC7 instrument at a rate of 10°C/min from 25 to 130°C, with an empty sealed pan used as reference. The first scan yields a thermogram of endothermic transitions which reflect the denaturation of membranes, ribosomes, cell walls, and DNA (60a). The sample pan is then rapidly cooled (200°C/min) to the initial temperature and immediately heat rescanned. The second scan yields only a single major endothermic transition, which reflects DNA reassociated (renatured) during the cooling step. The temperature at which the DNA transition peak is maximum (T_{max}) is characteristic for the melting point of the intracellular DNA of a particular organism. The moles percent G+C content is linearly correlated with T_{max} by the following equation: $T_{max} = 73.8 + 41.0(\text{mol\% G+C})$. The entire procedure can be completed within 1 h.

26.5. DNA LABELING

There are many options for preparing labeled probe DNA, either radioactive or nonradioactive. In addition, commercial kits with detailed instructions are available for performing many of the labeling procedures, and the procedures will be discussed here only in general terms. Methods for using nucleic acid probes are described in Chapter 28.

In vivo labeling of DNA is limited to radioactive precursors and is dependent on the medium required for culturing the organism and also on the nucleic acid precursors which the organism will take up. The ^{32}P-, ^{33}P-, ^{3}H- or ^{14}C-labeled nucleic acid precursors most often taken up by bacteria are adenine, thymidine, and deoxyadenosine. For details on the in vivo labeling of DNA, see reference 45.

In vitro labeling can be achieved by direct chemical reactions with the DNA or by DNA polymerase incorporation of nucleotides, using modified nucleoside triphosphates. Examples of direct chemical reactions include iodination, photoreaction (photobiotin), and glutaraldehyde coupling of horseradish peroxidase. DNA polymerases are used in the nick translation and random-primer procedures.

26.5.1. Iodination

Iodine can be bound to cytosine residues in the presence of thallium chloride (pH 4.8) (19, 21, 46, 69, 80, 88). The following is a protocol for labeling 2 to 10 μg of DNA.

Reagents and components

Buffer-salt solution: 7.2 M sodium perchlorate, 0.25 mM potassium iodide, 100 mM sodium acetate buffer (pH 4.8). Store in refrigerator, and keep on ice when labeling.

$TlCl_3$ solution: 1 mg/ml in 50 mM sodium acetate buffer (pH 4.8). Store in refrigerator, and keep on ice when labeling.

^{125}I: sodium [^{125}I]iodide (pH 10). Usually supplied at 100 μCi/μl. Some examples of sources are no. IMS.30, Amersham Corp., Arlington Heights, Ill., no. NEZ-033A, Dupont New England Nuclear, Boston, Mass.

PB-mercaptoethanol: 0.5 M Na_2HPO_4-NaH_2PO_4 buffer (pH 6.8) containing 0.1 M 2-mercaptoethanol.

DNA: denatured DNA fragments (i.e., as prepared for S1 nuclease or HA reassociation experiments; see section 26.5.2.1).

Reaction tubes. Autoinjection tubes (no. 063201, Chemical Research Supplies, Addison, Ill.) work well. They can be centrifuged directly in Brinkman horizontal 0.5-ml tube holders for collecting the denatured DNA, and they have serum caps that crimp on.

Gel filtration columns: Pharmacia NAP 25 columns. These disposable G-25 gel-filtration columns work well. Columns from other companies will also work well, but be sure to check the void and elution volumes.

HA columns: Bio-Rad DNA grade hydroxylapatite (HA)

HA buffer: 0.14 M Na_2HPO_4-NaH_2PO_4 (pH 6.8), 0.5% (wt/vol) SDS

Radiation safety: Check with your radiation safety office to make sure that your labeling facility is in compliance with safety requirements. In general, do all manipulations in a fume hood containing a safety shield, and equipped with an auxiliary hood having a charcoal filter. A survey meter that detects ^{125}I is also needed.

Procedure

1. Place 2 to 5 μg of denatured DNA fragments in a reaction tube. Add 3 M sodium acetate (pH 6) to give a final concentration of 0.3 M. Mix, add 2 volumes of 95% ethanol, mix again, and centrifuge for 10 min. Draw off the supernatant, wash the pellet with 100 μl of 80% ethanol, and then dry the pellet in an incubator.

2. Dissolve the DNA in 10 μl of the buffer-salt solution. Deposit 1 or 2 μl of sodium [^{125}I]iodide on one inside wall of the reaction tube, and then deposit 2.0 μl of the $TlCl_3$ solution on the opposite side of the tube. Crimp the serum cap onto the tube, mix the components, and centrifuge the tube briefly at low speed.

3. Place the reaction tube in a 70°C water bath for 20 min. Remove it, let it cool for 1 or 2 min, and inject approximately 0.1 ml of PB-mercaptoethanol buffer into the tube via a disposable tuberculin syringe. Return the tubes to the water bath for an additional 20 min. Remove, and cool.

4. Using a 1-ml disposable syringe with a 1.5-in. (3.8-cm) 22-gauge needle, inject approximately 0.15 ml of HA buffer into the reaction tube and then draw the contents of the tube up into the syringe. Load this material onto a NAP 25 gel filtration column that has been equilibrated with HA buffer. Add HA buffer to the column to give a total volume of 2.3 to 2.4 ml (i.e., sample plus buffer; the void volume is 2.5 ml), and discard the eluate. Place the collection tube containing 20 μg of carrier DNA (e.g., sheared salmon sperm DNA) under the column, and add an additional 2.0 ml of HA buffer to the column to collect the labeled DNA. Cap the column, and dispose of it in the laboratory's radioactive-waste container. Holding the tubes up to the survey meter will give you an immediate indication as to how well the iodination reaction worked.

5. Prepare a small HA column in a Pasteur pipette. Cut the tip off so that the pipette will just protrude from the top of a screw-cap culture tube (16 by 125 mm). Place glass wool in the pipette to retain the HA, wash with HA buffer, and add enough HA to make a column that is 0.75 to 1 in. (2 to 2.5 cm) high. Place the tube in a 70°C water bath, and wash the column with heated (70°C) HA buffer. Heat the labeled preparation in a boiling water bath for 3–5 min and then load onto the washed HA column. Transfer the column to a clean tube when the radioactivity begins to elute, and, after all of the sample has entered the column, add an additional 0.5 ml of HA buffer to elute the rest of the DNA from the column. Dispose of the HA column as follows. Break the tip by pressing it against the inside wall of an empty culture tube; this shortens the column so that the tube can now be capped. Place the capped tube in the laboratory's radioactive-waste receptacle.

6. Place the labeled DNA (approximately 1.8 to 2.2 ml) on another NAP 25 column, which has been equilibrated with 0.1× SSC. Add 2.5 ml (the void volume) of 0.1× SSC, place a collection tube under the column, and add an additional 3.5 ml of 0.1× SSC to collect the labeled product. Add SDS to the preparation to a final concentration of approximately 0.2%. Store at −20°C.

Comments

This labeling procedure is fast and gives consistent results. Usually the labeled DNA preparations are about 95% digested with S1 nuclease (single-stranded-specific nuclease). Occasionally a DNA preparation will contain contaminating material that will become iodinated, and this will raise the background level of S1-resistant material in the reassociation experiments. There are several advantages to using ^{125}I labeling. It does not affect the DNA fragment lengths during labeling, because the low energy release during decay does not cause fragmentation of the DNA. If one has a gamma scintillation counter, there is no need for scintillation fluids, and the 60-day half-life allows one to use the labeled DNA for up to 6 months without any change in relative reassociations. One disadvantage of using ^{125}I is that during the iodination step, when the pH is low, the resulting iodine is volatile and can be a safety hazard if the reaction tubes are not properly cleansed or if proper venting is not used.

26.5.2. Photobiotination

Photobiotin, a photoactive analog of biotin, was developed by Forster et al. (31) and is available in commercial kits. An example is no. 8186SA (GIBCO BRL). The labeling procedure does not alter the fragment lengths of the DNA preparation, and the labeled product is stable for at least several months.

26.5.3. Glutaraldehyde Procedure

The glutaraldehyde procedure involves treating horseradish peroxidase to give it a positive charge and then mixing it and glutaraldehyde with the denatured DNA (72). This results in covalent cross-linking between the DNA and the enzyme. The reagents are commercially available as a kit. An example is no. RPN.3000 (Amersham Corp., Arlington Heights, Ill.).

26.5.4. Nick Translation Procedure

The nick translation procedure is used for the labeling of high-molecular-weight DNA. It involves the generation of single-strand breaks (nicks) followed by nucleotide removal and replacements by *E. coli* polymerase I (74). Both radioactive analogs and other analogs (biotin or digoxigenin) can be introduced by this procedure. Kits incorporating this technology are available from several companies; some examples are no. 976776 (Boehringer Mannheim, Indianapolis, Ind.), no. U1001 (Promega, Madison, Wis.), and no. IB19080 (International Biotechnologies, Inc., New Haven, Conn.). The lengths of the labeled fragments are dependent on the reaction time: the longer the time, the shorter will be the fragments. Nicks within the synthesized regions will remove the introduced analogs.

26.5.5. Random Primer Procedure

The random-primer procedure was developed by Feinberg and Vogelstein (28, 29) for radioactively labeling DNA fragments to high specific activity. The procedure can also be used with nucleotide analogs containing biotin or digoxigenin. Labeling kits are available from a number of companies; examples include no. 80605A (GIBCO BRL), no. K1013-1 (Clontech, Palo Alto, Calif.); no. 300385 (Stratagene, LaJolla, Calif.), and no. 1550 (New England BioLabs, Beverly, Mass.).

26.6. DNA-DNA REASSOCIATION

Similarities in the linear arrangement of nucleotides along DNA strands from different organisms are determined by reassociation of labeled DNA with an excess of unlabeled DNA either in solution or immobilized on a membrane or surface. The method is used to measure the fraction of the genomes that can form duplexes under specific conditions of ionic strength and temperature. The results are then reported as percent similarity. In addition, the thermal stability of the duplexes can be measured and is used to estimate the degree of base mispairing in heterologous duplexes (hybrids).

There are two general methods for measuring DNA similarity. The first method, often referred to as the free-solution method, involves the reassociation of the single-stranded nucleic acids in solution. Reassociation of the DNA may be monitored optically by using UV spectrophotometry, or it may be measured by incubating a small amount of denatured DNA having a high specific radioactivity with a large excess of unlabeled, denatured DNA. In the latter instance, the ability of the labeled fragments to form duplexes with unlabeled DNA is assayed by adsorption to HA or by resistance to hydrolysis by S1 nuclease.

The second method, often referred to as the membrane method, involves immobilizing denatured (single-stranded) unlabeled DNA onto a nitrocellulose or nylon membrane filter, treating the filter so that additional DNA will not adsorb to it, and then incubating the filter in the presence of labeled denatured DNA fragments (or RNA, for RNA-DNA hybridization experiments). Duplexes of the labeled DNA that form by reassociation with the DNA fragments bound to the membrane are stable to the subsequent washing of the membrane, and the amount of duplex formation can be estimated by measuring the radioactivity of the membrane (or, for nonradioactive probes, some other unique activity).

26.6.1. Special Apparatus

Some of the specially constructed equipment that is useful for these measurements is shown in Fig. 3. The reaction vials can be constructed from 6-mm (outer diameter) Pyrex tubing cut into 7/8-in (2.2-cm) lengths and sealed at one end with a flame. The stoppers for these vials are serum vial stoppers which have had the outside flange cut off. Small polyethylene microcentrifuge tubes (no. 1049-00-0; Robbins Scientific) also work well and have the advantage of being disposable. When components are being added to the reaction vials, the vials are placed in a Plexiglas holder that has double rows of holes to help the investigator keep track of the additions. After being filled, the vials are stoppered and placed in a stainless-steel holder for incubation in a water bath. This water bath must be large enough to contain the submerged holder and should have a circulatory pump for maintaining an even temperature distribution. If the holder is not totally submerged or the bath not tightly covered, there is danger of distillation and collection of water drops on the vial sides and in the cap, which will greatly alter the ionic strengths and nucleic acid concentrations.

For membrane filter hybridization procedures, denatured DNA can be immobilized on a large filter by using dot blot or slot blot devices; alternatively, the DNA can be immobilized on the entire surface of a large membrane by using a filtration device. In the latter case, small membranes are punched out of the larger membrane for use in individual reaction vials. An oblong paper punch was found to be useful (45) but is no longer available; however, various other paper punches are available, although a reaction vial may have to be selected to fit the particular size of the punched-out membrane. Commercially available filtering devices can be used for immobilizing DNA on larger membranes, or custom-made ones can be con-

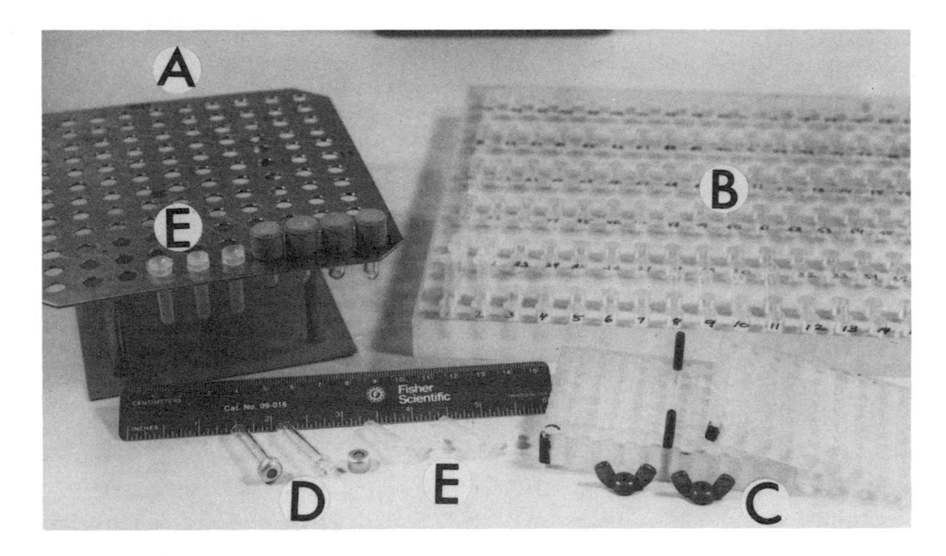

FIG. 3. Specialized equipment for DNA reassociation experiments. (A) Stainless-steel holder for vials, used during incubation of vials in a water bath. (B) Plexiglas rack for setting up experiments. (C) Plexiglas washing chamber for membrane filters. (D) Autoinjection tubes used for iodinating nucleic acids. (E) Polyethylene tubes (0.2 ml) for use in reassociation experiments.

structed (Fig. 4). A useful one can be constructed from a Plexiglas cylinder and a porous polyethylene filtration surface. The base of the apparatus is a short Plexiglas cylinder with an outside diameter the same as that of the membrane. The porous polyethylene is mounted on the inside and flush with the top. The base has a single exit tube, whose flow rate can be controlled with an adjustable pinch clamp. The top part of the device consists of a Plexiglas cylinder with a neoprene gasket (1/8 in. [3-mm] thick) glued to the bottom rim. The top part is placed on top of the membrane and is held on by spring tension. All of the Plexiglas apparatuses can be made by a local machine shop. Two different designs are illustrated in Fig. 4.

A high surface-to-volume ratio is needed for efficient annealing of probe DNA with the membrane-immobilized DNA. Therefore, the specific reaction vial will depend on the paper punch that is used to punch out the membrane. Another device that is useful for working with small membranes is a washing chamber to remove unreacted probe from the membrane surface. For oval membranes, one can use the Plexiglas washing chamber in Fig. 3. This chamber contains 120 compartments and consists of a top and bottom piece; the membrane filters are placed in the bottom piece, after which the top piece is bolted on. The assembly shown is small enough to fit into a 400-ml beaker.

The thermal stability of membrane-bound duplexes can be determined with the aid of the devices depicted in Fig. 5 and 6. The membrane skewer makes it easy to transfer the membranes from one tube to the next, while the tube rack fits into a circulating water bath

FIG. 4. Plexiglas filtration devices for immobilizing DNA on membranes.

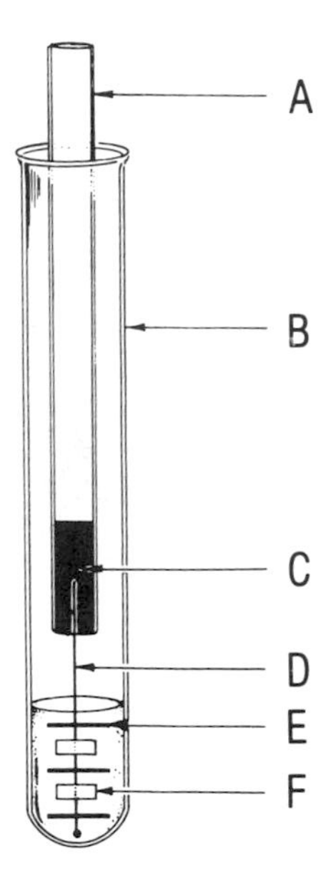

FIG. 5. Diagram of apparatus for testing the thermostability of hybrid duplexes. A piece of 7-mm-diameter glass tubing (A) is sealed with a rubber plug (C), which is made from a rubber stopper by use of a cork borer. Membrane filters (E) and Teflon washers (F) are impaled on an insect pin (D) close to the head of the pin. The pointed end of the pin is inserted into the rubber plug. This assembly is placed into a 13- by 10-mm glass test tube (B) containing 1.2 ml of 0.5 × SSC. The test tube is then placed in a heated water bath.

opening and reduces the exposed water surface, which will result in more efficient heating at the higher temperatures.

26.6.2. Free-Solution Methods

Free-solution reassociation of denatured DNA fragments can be measured optically by use of a recording spectrophotometer; if radioactive DNA is included, it can be measured by adsorption to HA or by resistance to S1 nuclease. A general scheme for performing these procedures is given in Fig. 7. Both require high-specific-activity labeled DNA (a very small amount) to be incubated with a large excess of unlabeled DNA fragments. The specific activity must be in the range of 5×10^4 to 2.5×10^5 cpm/μg, and the DNA should be at a concentration of less than 5 μg/ml.

26.6.2.1. Preliminary procedures

DNA fragmentation

For all of the free-solution reassociation methods, the DNA preparations must be fragmented to similar fragment sizes. The two common methods for fragmenting DNA are passage through a French pressure cell and sonication.

French pressure cell

1. Dilute the stock DNA to 0.4 mg/ml.
2. Pass the DNA preparation through the French pressure cell three times at a pressure of 16,000 lb/in^2. The flow rate should be a fast drip, with a constant pressure on the hydraulic pump being maintained. This method results in a Gaussian distribution of fragment sizes, with a mean size of 600 to 700 nucleotides. Check the uniformity of the fragment sizes in the DNA preparations by agarose gel electrophoresis, with 5-μg amounts of DNA.

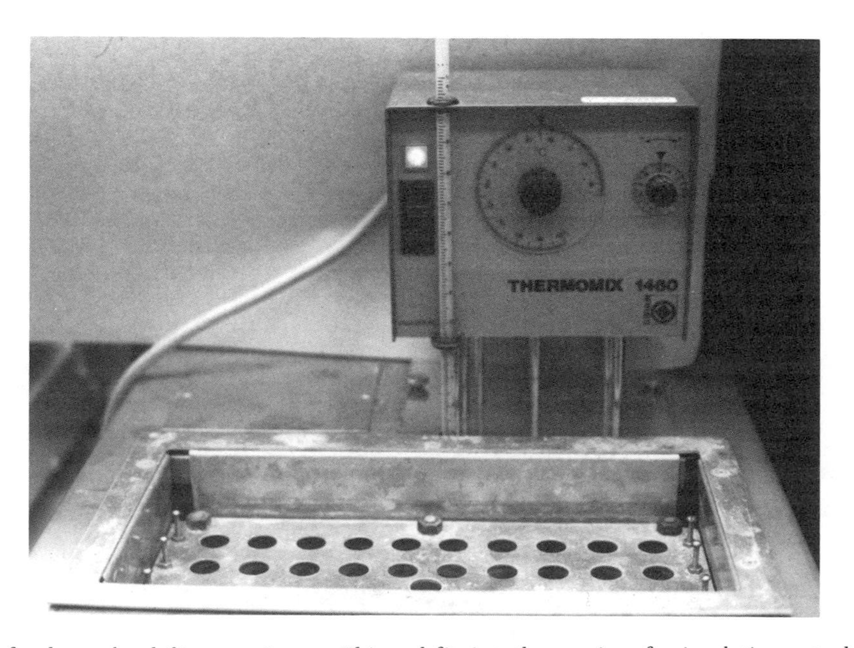

FIG. 6. Tube rack for thermal stability experiments. This rack fits into the opening of a circulating water bath. The small amount of exposed surface area cuts down on evaporation cooling.

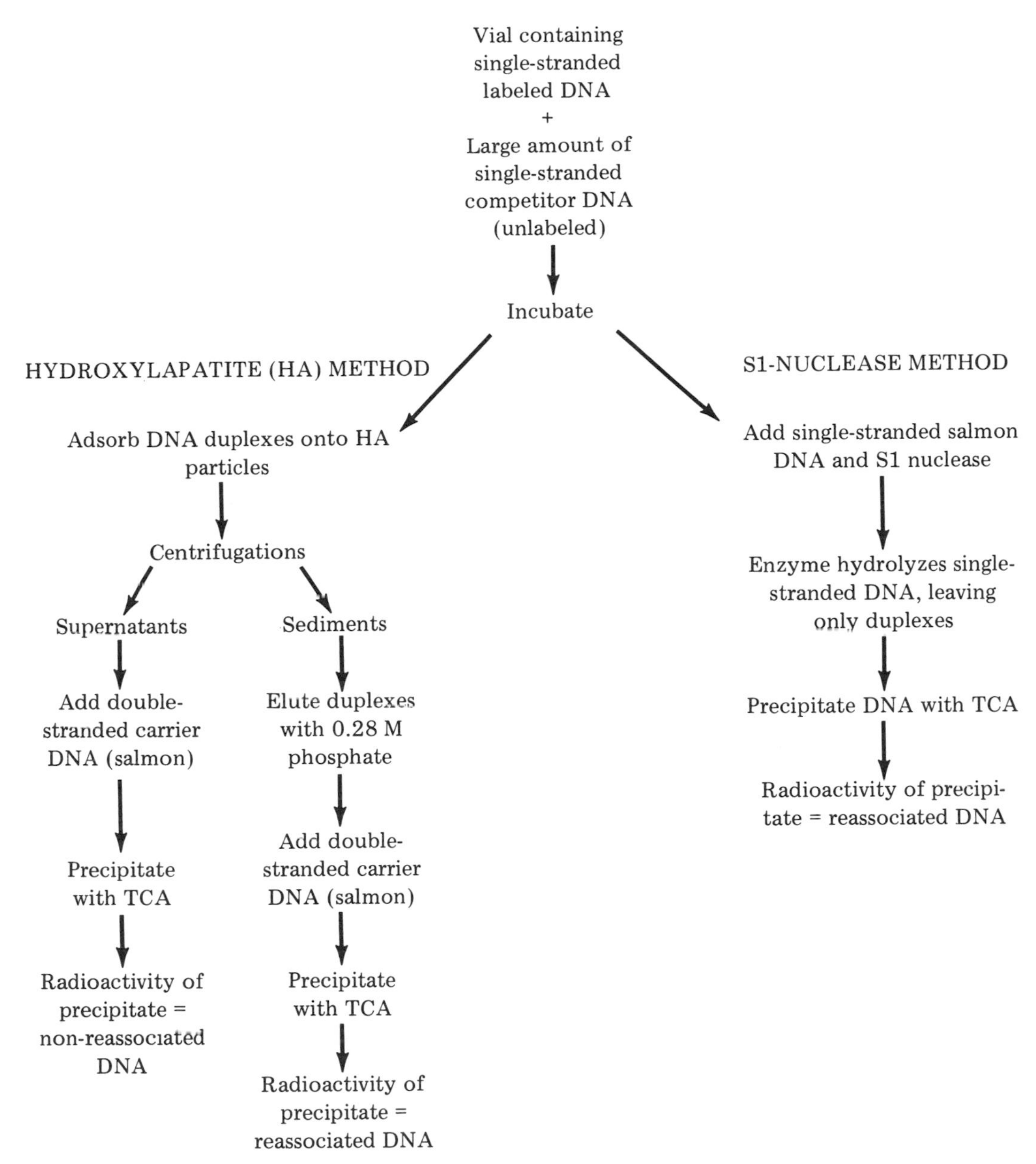

FIG. 7. Scheme for performing free-solution DNA reassociation experiments, assayed by either the hydroxylapatite (HA) or S1 nuclease procedure TCA, trichloroacetic acid.

Sonication

1. Dilute the stock DNA to 0.4 mg/ml.
2. Sonicate each preparation for two 30-s periods (e.g., in a Biosonic III sonicator with a 0.375-in. [9.5-mm] probe and an energy setting of 60). If there is a lot of heat buildup, it may be desirable to cool the preparations on ice between sonications. Check for fragment size uniformity by agarose gel electrophoresis as described above.

For optical reassociation experiments, centrifuge each fragmented DNA preparation at 17,000× g for 10 min to remove any particulate debris.

DNA denaturation

If the fragmented DNA preparations are to be assayed in reassociation experiments using S1 nuclease or HA, first denature the preparations by heating the tubes in a boiling-water bath for 5 min, cool on ice, and then centrifuge at 17,000 × g for 10 min. Store the DNA preparations at −20°C.

Preliminary denaturation of DNA is not done in the optical reassociation method; the denaturation occurs when the DNA is heated in the spectrophotometer cuvettes.

DNA labeling

Both the HA and S1 nuclease procedures require a

small amount of denatured labeled DNA with a high specific radioactivity to be incubated with a large excess of denatured unlabeled DNA fragments. In the case of iodination, use French pressure cell-fragmented, denatured DNA preparations for labeling; thus the labeled DNA fragments will have the same size distribution as those in the unlabeled DNA preparations. Dilute the labeled DNA preparations to 3×10^6 cpm/ml.

26.6.2.2. Reassociation measured by photometry

The protocol presented here is based on those of De Ley et al. (25) and Huss et al. (41). The major variables are the type of UV spectrophotometer and the type of computer interface that may be available. The spectrophotometer must be equipped with a temperature-controlled cuvette chamber or cuvette holding block and must have an automatic cuvette positioning mechanism. The reassociation rates can be calculated manually or by a microprocessor-controlled spectrophotometer, or the data can be uploaded into a computer system. Procedures presented here will be limited to sample preparation and the initial reassociation reactions.

Sample preparation

Adjust each fragmented DNA sample to a concentration of 40 μg/ml in 2× SSC–1 mM HEPES (pH 7.0) buffer by the following procedure. Prepare a large batch of 2× SSC–1 mM HEPES buffer and a smaller batch containing 4× SSC–2 mM HEPES. Dilute the fragmented DNA by adding 1 volume of the sample to an equal volume of 4× SSC; then add sufficient 2× SSC–1 mM HEPES to obtain the desired final concentration. Alternatively, each diluted DNA preparation can be dialyzed against 2× SSC–1 mM HEPES buffer. Pass each preparation through a 0.45-μm-pore-size filter to remove any particulate matter.

Procedure

1. Prepare mixtures (M) of each pair of DNA preparations (equal volumes) to be compared. When comparing two organisms (i.e., a DNA preparation from organism A and a DNA preparation from organism B) by using a spectrophotometer that has a cuvette holder for four cuvettes, fill one cuvette with 2× SSC–1 mM HEPES buffer, one with DNA from organism A, one with DNA from organism B, and the last with the equal-volume mixture of the two DNA preparations. Close the cuvettes tightly so that there is no evaporation during the denaturation and reassociation reaction; be sure to leave an air headspace in each cuvette to allow for the thermal expansion of the solutions. Teflon-stoppered cuvettes seal very well, and they can be filled to about 80 or 90% capacity.

2. Place the cuvettes in the spectrophotometer, heat the samples to 100°C, and hold at that temperature for 10 min. Then decrease the temperature to the optimal renaturation temperature (T_{OR}), and allow the reassociation to proceed for about 40 min. The optimal renaturation temperature can be calculated by the following formula (25).

$$T_{OR} = 0.51(\text{mol\% G+C}) + 47.0$$

3. Express reassociation rates (v') as the decrease in A_{260} per min, and calculate the percent similarity by the following formula (25). (The percent similarity was designated $\%D$ [degree of binding] in this formula by the original authors.)

$$\%D = \frac{100(4v'_M - v'_A - v'_B)}{2(v'_A \times v'_B)^{1/2}}$$

26.6.2.3. Reassociation measured with S1 nuclease

When radioactive probe DNA is reassociated with an excess of unlabeled DNA, the resulting heteroduplexes can be measured by resistance to S1 nuclease or by adsorption to HA. For the procedures described here, the initial reassociation mixtures are the same for both measurements.

Reassociation mixtures

When starting an experiment, thaw the DNA preparations and allow them to warm to room temperature. In addition to the test DNA preparations, prepare an unrelated DNA preparation (e.g., 0.4 mg of sheared salmon sperm DNA per ml), for measuring background reassociation. Prepare the reaction mixtures in 6- by 22-mm vials (or 250-μl microcentrifuge tubes) as follows.

Labeled DNA (3×10^6 cpm/ml)	10 μl
13.2× SSC–1 mM M HEPES buffer (pH 7.0)	50 μl
Unlabeled DNA, 0.4 mg/ml	50 μl
Total volume..........................	110 μl

Place sheared native salmon sperm DNA (which has no sequence similarity with bacterial DNA) into four control vials to measure the amount of self-renaturation of the labeled fragments. Use four vials for homologous reassociation and two vials for each of the heterologous DNA preparations. Incubate the vials at 25°C below the T_m of the reference DNA (as determined in 6× SSC) for 20 to 24 h. After incubation, quantitatively transfer the reassociation mixture to a large tube for determining the extent of reassociation by either the S1 nuclease or the hydroxylapatite procedure (see 26.6.2.4).

Comments

The accuracy of the results obtained from these methods is dependent on the reproducibility of the addition of the 10-μl amounts of labeled DNA to the vials. A multiple-dispensing micropipette (Eppendorf repeater pipette, no. 22-26-000-6; Brinkmann Instruments Inc., Westbury, N.Y.) works well for adding constant volumes to a large number of vials. The 110-μl volumes allow for accurate addition of components but are small enough that vast amounts of DNA are not required for the experiments. If the DNA preparations being compared have a high moles percent G+C value, formamide can be included to lower the T_m (46, 65). For instance, substituting the 50 μl of 13.2× SSC–1 mM HEPES with 25 μl of 26.4× SSC–2 mM HEPES and 25 μl of deionized formamide works well (this will lower the T_m by about 13.6°C).

Procedure

Under carefully controlled conditions, S1 nuclease will have little effect on double-stranded DNA but will hydrolyze single-stranded DNA (23). Therefore, the extent of duplex formation between radioactively labeled DNA fragments and an excess of unlabeled DNA fragments can be determined by measuring the amount of S1-resistant (i.e., acid-precipitable) radioactivity.

1. Quantitatively transfer the contents of each reaction vial to a 12- by 75-mm polypropylene test tube containing 1 ml of acetate-zinc buffer (0.05 M sodium acetate, 0.3 M NaCl, 0.5 mM $ZnCl_2$ [pH 4.6]) and 50 μl of denatured sheared salmon sperm DNA (0.4 mg/ml). Add 10 μl (100 U) of diluted S1 nuclease (no. 27-0920-01; Pharmacia), mix the tube contents on a Vortex mixer, and incubate the tubes in a 50°C water bath for 1 h.

2. Remove the tubes from the water bath, cool to room temperature, add 50 μl of sheared salmon sperm DNA (1.2 mg/ml) and 0.5 ml of HCl-PP_i solution (1 N HCl, 1% $Na_4P_2O_7 \cdot 10H_2O$, 1% NaH_2PO_4 [74]) to each tube, mix, and incubate at 4°C for 1 h.

3. Collect the precipitates on nitrocellulose membrane filters or on fiberglass filters (GF/F type; Whatman). Dry the filters, place them into scintillation vials, and count the radioactivity. Only the S1-resistant fragments (duplexes) are detected.

Calculate percent similarity values by dividing the counts per minute of the heterologous S1-resistant DNA by the counts per minute of the homologous S1-resistant DNA and multiplying by 100. A hypothetical example of the calculation is given in Table 1.

26.6.2.4. Reassociation measured with hydroxylapatite

The HA procedure separates double-stranded DNA from single-stranded DNA by selective adsorption to HA (Bio-Rad). Double-stranded DNA will adsorb to HA in 0.14 M phosphate buffer, whereas single-stranded DNA will not (11, 13, 14). There is variability in the adsorption capacity of different lots of HA, so it is a good idea to determine the binding capacity of each lot. The reassociation mixture in this procedure is the same as that described for the S1 nuclease procedure (section 26.6.2.3).

1. Place 0.5 g of dry HA into a 13- by 100-mm test tube, and add 3.0 ml of 0.14 M phosphate buffer (PB; equimolar NaH_2PO_4 and Na_2HPO_4) containing 0.4% SDS. Suspend the HA with a Vortex mixer, and warm the tube in a 60°C water bath. Centrifuge the tube at 3,000 × g for 3 to 4 min, decant the buffer, and resuspend the HA in 1 ml of PB.

2. Quantitatively transfer the reassociation mixture to the HA-containing tube, using PB for rinsing the reassociation vial. Mix on the Vortex mixer and return to the water bath for 5 min. Add an additional 2 ml of PB, mix on the Vortex mixer, and allow the HA to settle for 1 to 2 min.

3. Centrifuge the tube at 3,000 × g for 3 to 4 min, and decant the buffer into a collection tube. Add another 3 ml of 0.14 M PB to the HA. Repeat the mixing, warming, and centrifugation three times, and decant each supernatant into the collection tube.

4. Add 3 ml of 0.28 M PB to the HA, mix, and centrifuge. Decant the supernatant into a second collection tube. Add another 3 ml of buffer to the HA, mix, centrifuge, and decant the supernatant; decant it again into a collection tube. *Note:* When using an ^{125}I-labeled probe DNA, collect the supernatants in a tube that will fit into a gamma counter and measure the radioactivity directly. If the label is a beta emitter, the fragments must be precipitated and collected on filters before the radioactivity is measured. For this, pool the 0.14 M PB washes for the unbound fragments and the 0.28 M PB elutions before precipitating.

5. To each of the two collection tubes add 60 μg of carrier DNA (50 μl of sheared salmon sperm DNA, 1.2 mg/ml). Mix, and add 0.25 volume of HCl-PP_i solution, and again mix well. Cool in a refrigerator for 1 h.

6. Collect the precipitates on nitrocellulose or fiberglass (GF/F type; Whatman) filters. Dry the filters, place them into scintillation vials, and measure the radioactivity.

The sum of the counts in the two scintillation vials represents the total number of counts per minute in the reassociation vial. The radioactivity in the first scintillation vial represents the nonreassociated DNA and the radioactivity in the second scintillation vial represents reassociated DNA.

Calculate the percent similarity values by dividing the amount of heterologous reassociation by the amount of homologous reassociation and multiplying by 100. A hypothetical example of the calculation is given in Table 1.

Brenner et al. (11) have described an effective HA procedure in which larger amounts of DNA are used. Lachance (54) has described the use of microcolumns for separating the nonreassociated fragments from the reassociated fragments, and Sibley and Ahlquist (82) have devised an automated instrument for measuring thermal stabilities of duplexes.

Table 1. Hypothetical examples of the calculations of percent similarity values in the free-solution method by the HA and S1 nuclease procedures

Reaction vial	cpm S1 resistant	Net cpm S1 resistant	cpm not adsorbed	cpm adsorbed	% Adsorbed	Net % adsorbed	% Similarity
S1 nuclease procedure							
Labeled DNA fragments only	100						
Homologous reassociation	900	800					100
Heterologous reassociation	600	500					62
HA procedure							
Labeled DNA fragments only			900	100	10		
Homologous reassociation			100	900	90	80	100
Heterologous reassociation			400	600	60	50	62

26.6.2.5. Thermal-stability measurement

The extent of base pair mismatching can be estimated by comparing the thermal stability of heterologous duplexes with the stability of homologous duplexes. Thermal-stability measurements can be determined by using either the S1 nuclease or HA assay procedure and can also be measured by using membranes (see section 26.6.3.4). When using the S1 nuclease procedure, the decrease of S1-resistant radioactively labeled DNA is measured as the reassociated DNA is heated stepwise (usually in 2.5 or 5.0°C increments) in a circulating-water bath until denaturing temperatures are reached. When using HA, the nonreassociated DNA fragments are first washed from the HA with 0.14 M PB. Then, with the same 0.14 M PB, the HA is heated stepwise in temperature increments (again, 2.5 or 5.0°C increments) until all of the labeled DNA has denatured and eluted from the HA.

The reassociation mixtures and reassociation conditions are the same as when determining similarity values, except that the probe DNA has double or triple the specific activity (i.e., 10^7 cpm/ml) and the volume is doubled:

Labeled DNA (1 × 10^7 cpm/ml)	20 μl
13.2 × SSC–1 mM HEPES buffer (pH 7.0)	100 μl
Unlabeled DNA, 0.4 mg/ml	100 μl
Total volume	220 μl

These can be assayed by either the S1 nuclease or HA procedure.

S1 nuclease procedure

The S1 nuclease procedure for determining the thermostability of duplexes can be done as follows. After reassociation, dilute the mixture 12-fold by adding 2,420 μl of water to decrease the SSC concentration to 0.5× SSC. Then add 2.5 ml of 0.5× SSC–1 mM HEPES (pH 7.0) buffer. Place 500-μl amounts into 10 tubes (14.5- by 82-mm polypropylene tubes work well).

1. Start the thermal-stability profile by using a circulating-water bath set at 50°C. Maintain the first set of tubes (i.e., one from each reassociation mixture) at that temperature for 5 min, and then remove the tubes and cool them on ice or in a refrigerator. Increase the temperature of the water bath to 55°C, and place the next set of tubes in for 5 min. Continue the cycles through 90°C. Retain one tube that is not heated. *Note:* For low-G+C DNAs, the melting profile may have to begin at 45 or 50°C; for higher-G+C DNAs, it may have to begin at 60 or 65°C.

2. Add 2.0 ml of S1 nuclease buffer, 50 μl of 0.4-mg/ml denatured sheared salmon sperm DNA, and 10 μl (100 U) of diluted S1 nuclease to each tube. Mix the tube contents on a Vortex mixer, and incubate the tubes in a water bath at 50°C for 1 h.

3. Remove the tubes from the water bath, and cool them to room temperature. Add 50 μl of 1.2-mg/ml sheared salmon sperm DNA and 0.5 ml of HCl-PP_i solution to each tube, mix, and incubate at 4°C for 1 h.

4. Collect the precipitates on nitrocellulose membrane filters or fiberglass filters (GF/F type, Whatman). Dry the filters, place them into scintillation vials, and count the radioactivity.

Plot the S1-resistant counts per minute versus the temperatures. The temperature at which 50% of duplexes become S1 sensitive (i.e., have dissociated) is called the $T_{m(i)}$ ("*i*" meaning "irreversible separation of strands"). The difference between the homologous and heterologous $T_{m(i)}$ values is the $DT_{m(i)}$ value. Estimates for the amount of base mispairing for each 1°C $T_{m(i)}$ range from 1 to 2.2% (91).

Hydroxylapatite procedure

The hydroxylapatite procedure for determining the thermal stability of duplexes can be done by the following protocol.

1. Place 0.5 g of dry HA into a 13- by 100-mm test tube, and add 3.0 ml of 0.14 M PB (equimolar NaH_2PO_4 and Na_2HPO_4) containing 0.4% SDS. Suspend the HA with a vortex mixer, and warm the tube in a 50°C water bath. Centrifuge the tube at 3,000 × *g* for 3 to 4 min, decant the buffer, and resuspend the HA in 1 ml of PB.

2. Add 220 μl of 0.28 M PB buffer, and quantitatively transfer the reassociation mixture to the HA tube, using 0.14 M PB for rinsing the reassociation vial. Mix on the Vortex mixer, and return to the water bath for 5 min. Add an additional 2 ml of PB, mix on the Vortex mixer, and allow the HA to settle for 1 to 2 min.

3. Centrifuge the tube at 3,000 × *g* for 3 to 4 min, and decant the buffer into a collection tube. Add another 3 ml of 0.14 M PB to the HA. Repeat the mixing, warming, mixing, settling, and centrifugation a total of three times, decanting each supernatant into a collection tube.

4. Add 3 ml of 0.14 M PB to the HA, mix, and return to the water bath. Increase the temperature of the water bath by 5°C, allow the bath to reach the new temperature, and incubate the tubes for 5 min at that temperature. Mix the tubes on a Vortex mixer, allow the HA to settle, and centrifuge as above. Decant the supernatant into another collection tube. Add another 3 ml of buffer to the HA, and heat the water bath to the next temperature. Continue until the bath reaches 90 or 95°C. *Note:* When using ^{125}I-labeled probe DNA, collect the supernatants in tubes that will fit into a gamma counter, so that the radioactivity can be measured directly. If the label is a beta emitter, the eluate can be mixed with scintillation fluid that is miscible with aqueous samples or, if the volume is too great, the fragments can be precipitated and collected on filters before measuring the radioactivity.

5. For precipitating the eluted fragments, add 60 μg of carrier DNA (50 μl of sheared salmon sperm DNA, 1.2 mg/ml) to each collection tube. Mix, and add 0.25 volume of HCl PP_i solution. Mix well, and cool in a refrigerator for 1 h.

6. Collect the precipitates on nitrocellulose membrane filters or fiberglass filters (GF/F type, Whatman). Dry the filters, place them into scintillation vials, and measure the radioactivity.

The radioactivity from the three washes represents nonreassociated DNA fragments, and the radioactivity at each temperature represents the amount of denaturation that occurred at that temperature. Hypothetical data for a thermal stability profile are given in Table 2, and the plotting of the data is illustrated in Fig. 8.

Table 2. Calculations for hypothetical data from HA thermal stability profile

Temp (°C)	Homologous duplexes		Heterologous duplexes	
	cpm eluted	Sum cpm eluted	cpm eluted	Sum cpm eluted
50				
1st wash	350	350	2,100	2,100
2nd wash	125	475	750	2,850
3rd wash	25	500	150	3,000
55	300	800	1,800	4,800
60	400	1,200	1,400	6,200
65	1,000	2,200	1,200	7,400
70	2,300	4,500	1,400	8,800
75	2,500	7,000	800	9,600
80	2,200	8,200	200	9,800
85	600	9,800	100	9,900
90	200	10,000	100	10,000

26.6.2.6. Determination of C_0t values

Reassociation kinetics are easiest studied in solution; consequently, the optical, S1 nuclease, and HA procedures can all be used in these types of experiments. The following is a brief discussion of the equations that have been used for these determinations.

The initial reassociation of DNA in solution follows second-order reaction kinetics (3, 6, 14, 94, 95). The reassociation rate is a function of the genome size and has been used to estimate genome sizes. The usual equation for a second-order reaction (equation 3) is rearranged (equation 4) for plotting the so-called C_0t curves (14)

$$1/C - 1/C_0 = kt \tag{3}$$

$$C/C_0 = 1/(1 + kC_0t) \tag{4}$$

where t is the time in seconds, C is the concentration of single-stranded DNA at time t in moles of nucleotides per liter, C_0 is the concentration of single-stranded DNA at zero time in moles of nucleotides per liter, and k is the reassociation rate constant.

Calculate the moles of nucleotides per liter by dividing the milligrams of DNA per milliliter by 331 mg of DNA per millimole of nucleotide. The units (millimoles per milliliter) will be equal to moles per liter (M). Equation 4 is for a hyperbolic curve, for which the general formula is

$$y = 1/(1 + ax) \tag{5}$$

When $y = 1/2$ in equation 5, $x = 1/a = x_{1/2}$, and $a = 1/x_{1/2}$. If the logs of the x values are plotted, when $y = 1/2$, $\log a = \log 1/x_{1/2} = -\log x_{1/2}$. A general log C_0t plot is shown in Fig. 9. There is an almost linear region on the curve that extends for about 2 log units. This is because any significant change in C/C_0 occurs between $1/(1 + 0.1)$ and $1/(1 + 10)$, i.e., when kC_0t is between 0.1 and 10. The rate constant $k = 1/C_0t_{1/2}$. The units for the rate constant are reciprocal moles reciprocal hours.

Denatured DNA in free solution may be reassociated for kC_0t values well beyond 10, and it is not critical to know either the exact time of incubation or the exact concentrations of the DNA fragments. When determining $C_0t_{1/2}$ values, however, one must know these parameters. This is because a series of points between $kC_0t = 0.1$ and $kC_0t = 10$ must be determined. In practice, for bacterial DNA preparations, these will be C_0t values from about 0.1 to 50. A particular C_0t value can be reached by altering the time of incubation, the concentration of the unlabeled DNA, or both. Listed in Table 3 are C_0t values for four DNA concentrations and eight reassociation times.

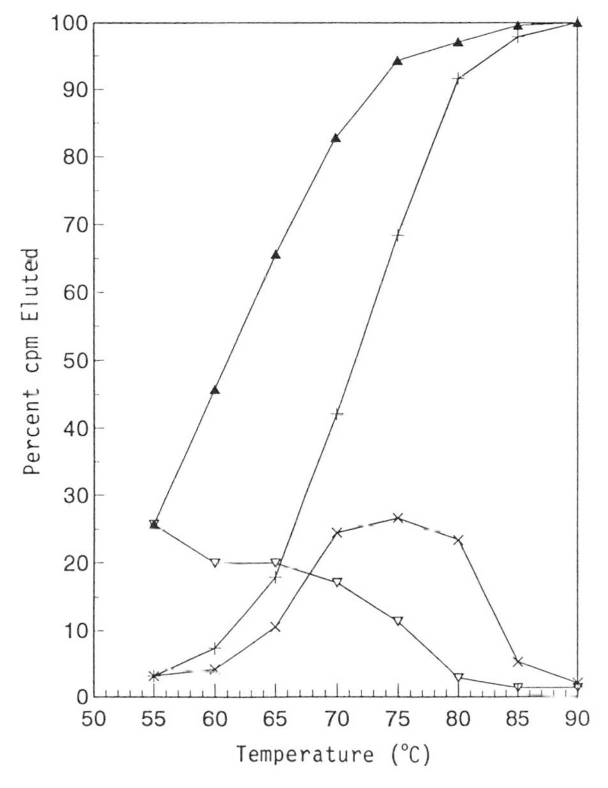

FIG. 8. Thermal stability profiles obtained with the hypothetical hydroxylapatite procedure data in Table 2. Symbols: ▽, percent cpm eluted at each temperature for homologous duplexes; ▲, sum of percent cpm eluted at each temperature for homologous duplexes; ×, percent cpm eluted at each temperature for heterologous duplexes; +, sum of percent cpm eluted at each temperature for heterologous duplexes.

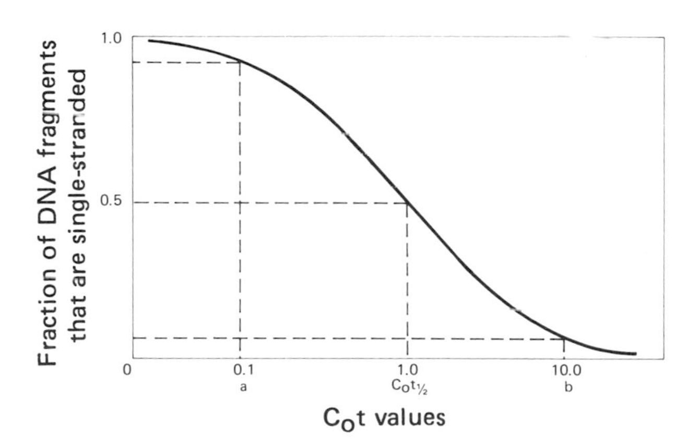

FIG. 9. Generalized C_0t plot. At C_0t values less than or equal to a, $1/(1 + C_0t)$ values are 0.9 or greater. At C_0t values equal to or greater than b, $1/(1 + C_0t)$ values are less than 0.1.

Table 3. C_0t values for various DNA concentrations and reassociation times

Unlabeled DNA (μg/110 μl)[a]	Time		C_0t
	h	s	
1	0.5	1.8 × 10³	0.049
5	0.5		0.247
20	0.5		0.99
30	0.5		1.48
1	1.0	3.6 × 10³	0.097
5	1.0		0.49
20	1.0		1.98
30	1.0		2.97
1	2.0	7.2 × 10³	0.194
5	2.0		0.986
20	2.0		3.95
30	2.0		5.93
1	4.0	1.44 × 10⁴	0.388
5	4.0		1.97
20	4.0		7.90
30	4.0		11.9
20	8.0	2.88 × 10⁴	15.8
30	8.0		23.7
20	16	5.76 × 10⁴	31.6
30	16		47.5
20	20	7.2 × 10⁴	39.5
30	20		59.3
20	24	8.64 × 10⁴	47.4
30	24		71.2

[a] Equivalents:
1 μg/110 μl = 9.1 μg/ml or 0.27 × 10^{-4} mol/liter.
5 μg/110 μl = 45 μg/ml or 1.37 × 10^{-4} mol/liter.
20 μg/110 μl = 182 μg/ml or 5.49 × 10^{-4} mol/liter.
30 μg/110 μl = 272 μg/ml or 8.24 × 10^{-4} mol/liter.

Set up reassociation mixtures in the usual manner (see section 26.6.2) with duplicate mixtures for each DNA concentration at each temperature. After incubation is completed for a given set of reaction vials, place them into a freezer until all have been incubated. Assay for the extent of duplex formation by either the HA or S1 nuclease procedure.

Plot the fractions of DNA fragments not bound to HA or the fractions which are S1 nuclease sensitive on the y axis and C_0t values on the x axis of four- to five-cycle semilog graph paper. The rate constant k can be calculated by using equation 4 when the duplexes are measured by the HA procedure. The results obtained by the S1 nuclease procedure, however, do not reflect second-order kinetics, because the ends of duplexed fragments are hydrolyzed. The relationship between the two procedures is

$$S/C_0 = [1/(1 + kC_0t)]^{0.45} \quad (6)$$

where S is the concentration of S1 nuclease-sensitive DNA (83). Equation 6 can be rearranged as

$$[S/C_0]^{2.222} = 1/(1 + kC_0t) \quad (7)$$

If one plots the S1 nuclease results according to equation 7, one will generate a C_0t curve comparable to that obtained by the HA procedure. Also important to keep in mind is that the reassociation rates are greatly affected by the ionic strengths of the reassociation mixtures, and therefore comparisons should be made by using a single ionic strength.

For additional information about reassociation kinetics and C_0t curves (including optical reassociation), see references 6, 13, 14, 79, 94, and 95.

26.6.3. Membrane Methods

The ability to immobilize DNA on a solid support has been one of the more important developments for the measurement of DNA reassociation. DNA was first successfully immobilized in an agar matrix (40), which was soon replaced by nitrocellulose membranes (33, 93). The immobilization of denatured DNA on nitrocellulose membranes involves slow filtration of a high-salt (e.g., 6× SSC)-denatured DNA solution through the membrane, followed by air drying and baking in an oven. The chaotropic salt NaI has also been used for immobilizing DNA to nitrocellulose membranes (12). Other membranes developed for immobilizing DNA include diazobenzyloxymethyl paper and membranes made from positively charge-modified nylon 66 (7, 18, 78). All of these products are available from scientific supply companies, together with detailed protocols for their use.

The membrane reassociation methods differ from the free-solution methods in that the unlabeled DNA fragments are unable to reassociate with each other. As a result, the rate of probe DNA reassociation to the immobilized DNA is linear for commonly used reassociation times and the rate of reassociation is dependent on the amount of DNA on the membrane. The advantages of membrane methods are as follows. (i) It is relatively easy to wash nonreassociated DNA—and nonspecifically labeled contaminants—away from the reassociated DNA. (ii) In addition to radioactively labeled probes, nonradioactive probes (i.e., biotin, digoxigenin, and direct enzyme cross-linking) can be used. (iii) Membranes can be probed more than once. The major disadvantage of membrane procedures is that the results are not as quantitative as the solution methods. An indirect membrane procedure (competition method), which is more quantitative (45), requires such large amounts of DNA that it is no longer used. Procedures given here will be for the direct binding of radioactively labeled DNA probes to membrane-immobilized DNA and for the determination of the thermal stabilities of the duplexes.

A general scheme for performing the membrane method is given in Fig. 10. The reassociation mixtures may include either a single small membrane in a vial or a larger membrane (dot blot or slot blot formats) containing immobilized DNA from a number of sources.

26.6.3.1. DNA immobilization on membranes

The following adaption of the Gillespie and Spiegelman (33) procedure is for a 15-cm-diameter BA 85 nitrocellulose membrane filter (Schleicher & Schuell Co., Keene, N.H.) having an effective filtering surface of about 173 cm². The immobilization of 4.35 mg of DNA will result in 25 μg/cm². For smaller membranes, reduce the amount of DNA and the buffer volumes accordingly.

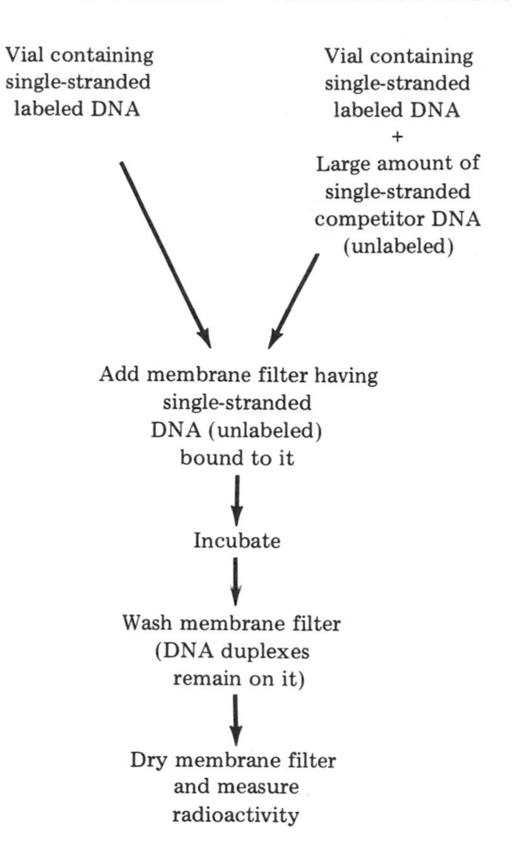

FIG. 10. Scheme for performing membrane DNA reassociation experiments by using either the direct-binding or competition procedure.

1. Dilute 4.35 mg of DNA to 50 μg/ml with 0.1× SSC (approximately 90 ml).

2. Place the solution in a 125-ml Erlenmeyer flask, and denature the DNA by heating in a boiling-water bath for 10 min. Cool the solution quickly by pouring it into 800 ml of ice-cold 6× SSC (to give about 5 μg of DNA per ml).

3. Float the 15-cm nitrocellulose membrane filter on distilled water so that the pores will be filled with water (if the membrane is submersed immediately, air pockets will form and cause uneven filtration).

4. Place the wet membrane filter on the filtration device (section 26.6.1), and wash the membrane with 500 ml of cold 6× SSC, using a flow rate of approximately 30 ml/min. Then pass the denatured DNA through the filter, and wash it again with 500 ml of 6× SSC.

5. Let the membrane dry initially at room temperature and then overnight at 60°C.

6. Label the membrane on the edge with a pencil, place in an envelope, and store at room temperature over desiccant.

Handle the membrane by the outside edge, and avoid touching the surface on which the DNA is bound. When cutting small membranes out of the large one, avoid getting too close to the edge.

26.6.3.2. Preincubation of DNA-containing membranes

The Denhardt preincubation reagent (26), which is used to block additional DNA-binding sites on the DNA-containing membrane, contains the following ingredients dissolved in 6× SSC.

Bovine serum albumin (fraction V; Sigma Chemical Co.) 0.02%
Polyvinylpyrrolidone (Calbiochem-Behring, La Jolla, Calif.) 0.02%
Ficoll 400 (Pharmacia Fine Chemicals, Uppsala, Sweden) 0.02%

1. Cut small membranes (about 5 μg of DNA per membrane) from the large one with a paper punch.

2. Preincubate the membranes in the Denhardt mixture at the incubation temperature of the experiment for 0.5 to 2 h. Swirl the membranes several times during the preincubation period.

3. Remove the membranes, place on a paper towel to blot off excess liquid, and then place into the reaction vials.

26.6.3.3. Reassociation measured by direct binding

In the direct-binding procedure, the amount of a labeled probe DNA that forms duplexes with DNA preparations immobilized on membrane filters is measured. Reaction mixtures contain the following.

Labeled DNA (1×10^6 to 3×10^6 cpm/ml) .... 10 μl
6.6× SSC .... 100 μl

1. Preincubate the required number of DNA membrane filters. Keep each kind separate.

2. Make up the reaction mixtures, and mix them on a Vortex mixer.

3. Place the membranes in vials, stopper the vials, and incubate them at 25°C below the T_m (as determined in SSC) of the labeled DNA. Incubate for 12 to 15 h.

4. Remove the membranes from the vials, and place them in the washing chamber (Fig. 3). Wash them in 2× SSC at the incubation temperature. Use two 300-ml volumes of washing buffer, and wash for 5 min in each. Move the washing chamber back and forth in the beaker several times during the washing period to ensure buffer flow through each compartment. Shake the washing chamber free from buffer. Remove the membrane filters, and dry them on a paper towel under a heat lamp.

5. Skewer the membranes on insect pins (if there are duplicates, place both membranes on the same pin, as shown in Fig. 11).

6. Place into scintillation vials (Fig. 11), and measure the radioactive counts.

The sequence similarity calculations are similar to those used for the S1 nuclease procedure (Table 1) in that the counts bound on the heterologous DNA membrane are divided by the counts bound on the homologous DNA membrane. This ratio is then multiplied by 100 to give the percent similarity.

Comments

The size of the membranes will dictate the size of the reaction vial and the volume of the reaction mixture.

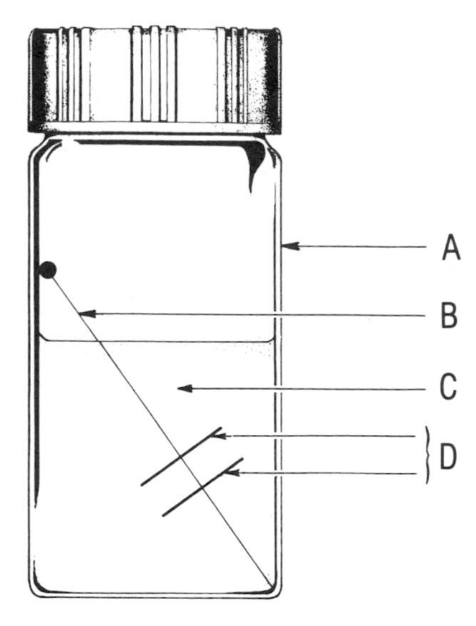

FIG. 11. Apparatus for counting radioactivity on membrane filters. A, scintillation vial; B, insect pin; C, toluene-based cocktail; D, duplicate filters impaled on pin.

For higher-G+C DNAs, include 20 to 30% deionized formamide so that the reassociation temperature can be lowered (0.6°C/1% formamide).

Although 1× Denhardt reagent (as given in section 26.6.3.2) works well as a blocking agent of nonspecific binding, other hybridization protocols (see Chapter 16.3.4.2) call for 5× Denhardt reagent (4) or the milk reagent called BLOTTO (Bovine Lacto Transfer Technique Optimizer) (42). The exact concentrations of blocking agents probably are not critical; their main function is to ensure that there is a low background of probe DNA on membranes containing either no DNA or an unrelated DNA, without causing any interference with the reassociation reaction.

The membrane-binding experiments are also amenable to the use of nonradioactive probes (31, 55–57, 72, 73, 87). The major advantage of these probes is that they are not subject to the safety precautions and regulations required when using radioactive substances; their major disadvantage is that an additional reaction is introduced into the experiment. This will increase the assay time; moreover, the experiment may not be as quantitative as when radioactive probes are used. Nonradioactive probes have their major impact in diagnostic bacteriology for identification of isolates.

26.6.3.4. Thermal-stability measurement

1. Follow the direct-binding procedure through the washing step (step 4 above).

2. Remove the membrane filters from the washing chamber, skewer them on an insect pin near the head of the pin, insert the pin into the end of a rubber-stoppered tube (Fig. 5), and place into the first elution tube. When using more than one filter, separate them with Teflon discs.

3. Use 13- by 100-mm elution tubes containing 1.2 ml of 0.5× SSC. Place the tubes into a water bath, and perform stepwise elution of the duplexes by increasing the bath temperature 5°C at 10-min intervals. At the end of each period, transfer the filters to the next elution tube and increase the temperature by 5°C. Start the elution profile about 10°C below the incubation temperature used for the direct-binding procedure. Use nine elution temperatures, and count the residual radioactivity on the filters (i.e., 10 scintillation vials per stability profile).

4. Empty the contents of each elution tube into a scintillation vial, and rinse the tube with 0.5 ml of 0.5× SSC, also adding that to the vial.

5. Add 10 ml of a scintillation fluid that is miscible with aqueous samples.

The integral elution profile curves are obtained by summing the radioactivities at each temperature and dividing by the total radioactivity; this is essentially the same as illustrated for HA data (Table 2; Fig, 8) except that one cannot readily measure the unbound DNA—i.e., the washes.

In obtaining the profile curves, place the 1.2 ml of 0.5× SSC in all of the 13- by 100-mm tubes at the same time by using an automatic pipette. Keep one elution tube ahead in the water bath so that each is preheated. Have an extra tube containing a thermometer, and record the exact tube temperature at the middle and the end of each temperature interval.

26.7. RELIABILITY AND COMPARISON OF RESULTS

One does not have to search the DNA reassociation literature very extensively to realize that there are variations in the published results, many of which are the result of experimental errors (84). Bacteriologists have been criticized for being a bit sloppy in doing and reporting DNA reassociation experiments, compared with those working with eucaryotes. Although some of this criticism is justified, there are additional reasons for greater variability in bacterial DNA experiments. First, the sheer number of bacteriology laboratories involved (often temporarily) in performing DNA reassociation experiments has contributed to the variability of results. Secondly, the HA procedure has been used for most eucaryotic studies, whereas all of the procedures discussed in this chapter have been used extensively by bacteriologists. Probably the major factor causing experimental variations is the quality of the DNA. Although there are few differences in the isolation of DNA from one mammal to another or one bird to another, the ease of isolation and the quality (i.e., degree of fragmentation and various contaminants) of the DNA from microorganisms are much more variable. Therefore, one cannot overemphasize the importance of pure DNA preparations and the size homogeneity of fragmented DNA preparations. Another variable that is more pronounced with bacteria is the range of moles percent G+C in their DNA; experimental conditions must be adjusted to take this into account. All of these characteristics of bacterial DNA must be considered in order to reduce the amount of experimental error in the experiments.

The choice of reference organisms for use in DNA reassociation experiments can also have a profound effect on the apparent relationships in cluster analysis (36). The ideal situation is one in which every strain is

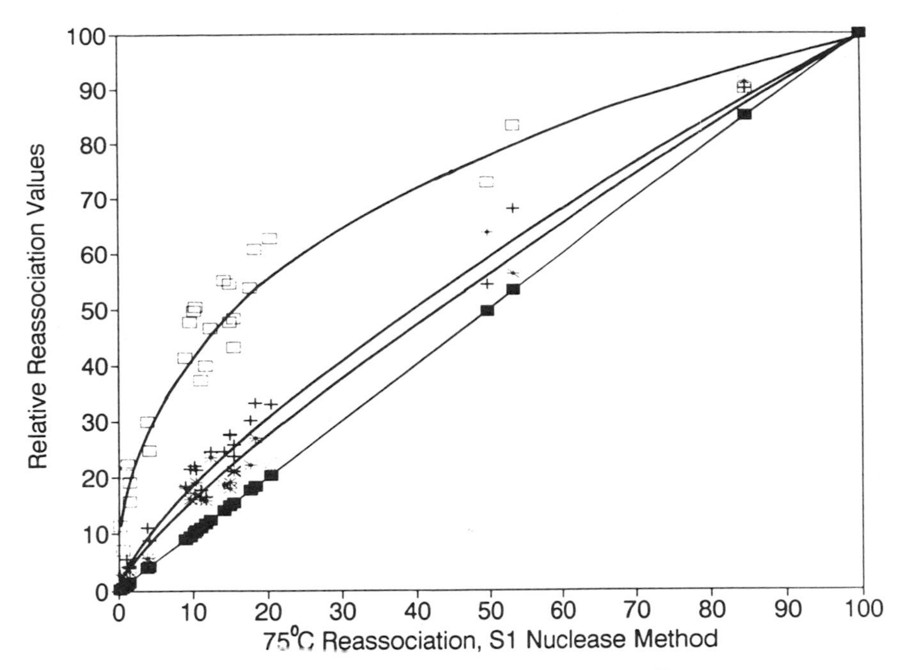

FIG. 12. Comparison of S1 and hydroxylapatite (HA) reassociation methods at different temperatures. Symbols; ■, S1 method, 75°C; *, HA method, 75°C; +, S1 method, 60°C; □, HA method, 60°C.

used as a reference organism. If such studies involve more than 10 to 20 organisms, they would soon become quite unwieldy; consequently, fewer reference strains must be carefully chosen. When investigating a group of unidentified organisms, a rule of thumb is that one must first select reference strains rather arbitrarily from each phenotypic cluster of organisms. Organisms whose DNA has 80 to 100% similarity with a given reference need not be used as references, whereas additional reference DNAs must be selected from each of the clusters that have lower percent similarity values. This process is repeated until the DNA from each organism used in the study either belongs to a cluster (80 to 100% similarity) or is unique when used as a reference organism. This type of data can be used for cluster analysis and dendrogram generation. It is also good to have more than one reference DNA preparation within a large cluster for better average similarity values. When named species are included in a study, use the type strains. If one wants only to demonstrate that an unidentified cluster of organisms has no measurable sequence similarity with a number of species, one does not have to use the DNAs of the type strains of these species as references, but the data are not suitable for cluster analysis.

26.7.1. DNA Similarity Values

For the discussion and evaluation of data, it is important to know how one method of determining DNA similarity values compares with another. Similar values are obtained when the photometric, membrane, and HA reassociation procedures are used. All are affected similarly by the reassociation temperature, with greater specificity at higher (T_m − 10) versus lower (T_m − 25) temperature. The optical method has higher background values, such that values up to 30% are not considered significant (41). The reassociation temperature has less effect on the S1 nuclease results, because the localized mismatched regions and overhanging ends are degraded. Correlations between procedures and reassociation temperatures (10, 35) are illustrated in Fig. 12.

26.7.2. DNA Thermal Stability Values

Theoretically, the $T_{m(i)}$ values are probably as good a measurement of similarity as any. These values correlate with the degree of base pair mismatches and are for the most part independent of the G+C content of the DNAs (43, 91) and the ionic strength of the elution buffer. The major disadvantage is that the experiments are laborious to perform.

26.8. REFERENCES

26.8.1. General References

1. **Hills, D. M., and C. Moritz (ed.).** 1990. *Molecular Systematics.* Sinauer Associates, Inc., Sunderland, Mass.
Although this useful book emphasizes the systematics of eucaryotic organisms, many of the techniques are directly applicable to bacteria.
2. **Johnson, J. L.** 1985. Determination of DNA base composition, p. 1–31. *In* G. Gottschalk (ed.), *Methods in Microbiology,* vol. 18. Academic Press Ltd., London.
Review the determination of moles percent G+C of DNA preparations.
3. **Johnson, J. L.** 1989. Nucleic acid hybridization: principles and techniques, p. 3–31. *In* B. Swaminathan and G. Prakash (ed.), *Nucleic Acids and Monoclonal Antibody Probes.* Marcel Dekker, Inc., New York.
Reviews reassociation and hybridization kinetics, effects of experimental variables, duplex and hybrid specificities, and techniques.

4. **Sambrook, J., E. F. Fritsch, and T. Maniatis.** 1989. *Molecular Cloning: A Laboratory Manual,* 2nd ed. Cold Spring Harbor Laboratory Press, Cold Spring Harbor, N.Y.
This large manual (it is published as three sequential books) includes protocols for nearly all molecular biology techniques.
5. **Stackbrandt, E., and M. Goodfellow (ed.).** 1991. *Nucleic Acid Techniques in Bacterial Systematics.* John Wiley & Sons Ltd., Chichester, England.
Provides an extensive range of molecular methods for studying bacterial classification by the techniques of molecular biology.
6. **Wetmur, J. G.** 1976. Hybridization and renaturation kinetics of nucleic acids. *Annu. Rev. Biophys. Bioeng.* **5:**337–361.
This review deals primarily with free-solution reassociation kinetics.

26.8.2. Specific References

7. **Alwine, J. C., D. J. Kemp, and G. R. Stark.** 1977. Method for detection of specific RNAs in agarose gels by transfer to diazobenzyloxymethyl-paper and hybridization with DNA probes. *Proc. Natl. Acad. Sci. USA* **74:**5350–5354.
8. **Bacon, M. F.** 1965. Analysis of DNA preparations by a variation of the cysteine-sulfuric acid test. *Anal. Biochem.* **13:**223–228.
9. **Beji, A., D. Izard, F. Gavini, H. Leclere, M. Leseine-Delstanche, and J. Krembel.** 1987. A rapid chemical procedure for isolation and purification of chromosomal DNA from Gram-negative bacilli. *Anal. Biochem.* **162:**18–23.
10. **Bouvet, P. J. M., and P. A. D. Grimont.** 1986. Taxonomy of the genus *Acinetobacter* with the recognition of *Acinetobacter baumannii* sp. nov., *Acinetobacter haemolyticus* sp. nov., *Acinetobacter johnsonii* sp. nov., and *Acinetobacter junii* sp. nov. and emended descriptions of *Acinetobacter calcoaceticus* and *Acinetobacter lwoffi. Int. J. Syst. Bacteriol.* **36:**228–240.
11. **Brenner, D. J., G. R. Fanning, A. V. Rake, and K. E. Johnson.** 1969. Batch procedure for thermal elution of DNA from hydroxyapatite. *Anal. Biochem.* **28:**447–459.
12. **Bresser, J., J. Doering, and D. Gillespie.** 1983. Quick-blot: selective mRNA or DNA immobilization from whole cells. *DNA* **2:**243–253.
13. **Britten, R. J., D. E. Graham, and B. R. Neufeld.** 1974. Analysis of repeated DNA sequences by reassociation. *Methods Enzymol.* **24E:**363–406.
14. **Britten, R. J., and D. E. Kohne.** 1968. Repeated sequences in DNA. *Science* **161:**529–540.
15. **Britten, R. J., M. Pavich, and J. Smith.** 1969. A new method for DNA purification. *Carnegie Inst. Wash. Year Book* **68:**400–402.
16. **Burton, K.** 1968. Determination of DNA concentration with diphenylamine. *Methods Enzymol.* **12B:**163–166.
17. **Byvoet, P.** 1966. Determination of nucleic acids with concentrated H_2SO_4. II. Simultaneous determination of ribo- and deoxyribonucleic acid. *Anal. Biochem.* **15:**31–39.
18. **Cannon, G., S. Heinhorst, and A. Weissbach.** 1985. Quantitative molecular hybridization on nylon membranes. *Anal. Biochem.* **149:**229–237.
19. **Chan, H. C., W. T. Ruyechan, and J. G. Wetmur.** 1976. *In vitro* iodination of low complexity nucleic acids without chain scission. *Biochemistry* **15:**5487–5490.
20. **Chassy, B. M., and A. Giuffrida.** 1980. A method for improved lysis of some gram-positive asporogenous bacteria with lysozyme. *Appl. Environ. Microbiol.* **39:**153–158.
21. **Commerford, S. L.** 1971. Iodination of nucleic acids *in vitro. Biochemistry* **10:**1993–1999.
22. **Cesarone, C. F., C. Bolognesi, and L. Santi.** 1979. Improved microfluorometric DNA determination in biological material using 33258 Hoechst. *Anal. Biochem.* **100:**188–197.
23. **Crosa, J. H., D. J. Brenner, and S. Falkow.** 1973. Use of single-strand-specific nuclease for analysis of bacterial and plasmid deoxyribonucleic acid homo- and heteroduplexes. *J. Bacteriol.* **115:**904–911.
24. **Cummins, C. S., and J. L. Johnson.** 1971. Taxonomy of the clostridia: wall composition and DNA homologies in *Clostridium butyricum* and other butyric acid-producing clostridia. *J. Gen. Microbiol.* **67:**33–46.
25. **DeLey, J., H. Cattoir, and A. Reynaerts.** 1970. The quantitative measurement of DNA hybridization from renaturation rates. *Eur. J. Biochem.* **12:**133–142.
26. **Denhardt, D. T.** 1966. A membrane-filter technique for the detection of complementary DNA. *Biochem. Biophys. Res. Commun.* **23:**641–646.
27. **Downs, T. R., and W. W. Wilfinger.** 1983. Fluorometric quantification of DNA in cells and tissue. *Anal. Biochem.* **131:**538–547.
28. **Feinberg, A. P., and B. Vogelstein.** 1983. A technique for radiolabeling DNA restriction endonuclease fragments to high specific activity. *Anal. Biochem.* **132:**6–13.
29. **Feinberg, A. P., and B. Vogelstein.** 1983. A technique for radiolabeling DNA restriction endonuclease fragments to high specific activity. *Anal. Biochem.* **137:**266–267. (Addendum.)
30. **Ferragut, C., and H. Leclerc.** 1976. Étude comparative des methodes de détermination du Tm de L'ADN Bactérien. *Ann. Microbiol. (Inst. Pasteur)* **127A:**223–235.
31. **Forster, A. C., J. L. McInnes, D. C. Skingle, and R. H. Symons.** 1985. Non-radioactive hybridization probes prepared by the chemical labeling of DNA and RNA with a novel reagent, photobiotin. *Nucleic Acids Res.* **13:**745–762.
32. **Giles, K. W., and A. Meyers.** 1965. An improved diphenylamine method for estimation of deoxyribonucleic acid. *Nature* (London) **206:**93.
33. **Gillespie, D., and S. Spiegelman.** 1966. A quantitative assay for DNA hybrids with DNA immobilized on a membrane. *J. Mol. Biol.* **12:**829–842.
34. **Gold, D. V., and D. Shochat.** 1980. A rapid colorimetric assay for the estimation of microgram quantities of DNA. *Anal. Biochem.* **105:**121–125.
35. **Grimont, P. A. D., M. Y. Popoff, F. Grimont, C. Coyault, and M. Lemelin.** 1980. Reproducibility and correlation study of three deoxyribonucleic acid hybridization procedures. *Curr. Microbiol.* **4:**325–330.
36. **Hartford, T., and P. H. A. Sneath.** 1988. Distortion of taxonomic structure from DNA relationships due to different choice of reference strains. *Syst. Appl. Microbiol.* **10:**241–250.
37. **Hill, B. T., and S. Whatley.** 1975. A simple, rapid microassay for DNA. *FEBS Lett.* **56:**20–23.
38. **Holdeman, L. V., E. P. Cato, and W. E. C. Moore.** 1977. *Anaerobe Laboratory Manual,* 4th ed. Virginia Polytechnic Institute and State University, Blacksburg.
39. **Horinouchi, S., T. Uozumi, T. Beppu, and K. Arima.** 1977. A new isolation method of plasmid deoxyribonucleic acid from *Staphylococcus aureus* using a lytic enzyme of *Achromobacter lyticus. Agric. Biol. Chem.* **41:**2487–2489.
40. **Hoyer, B. H., B. J. McCarthy, and E. T. Bolton.** 1964. A molecular approach in the systematics of higher organisms. *Science* **144:**959–967.
41. **Huss, V. A. R., H. Festl, and K. H. Schleifer.** 1983. Studies on the spectrophotometric determination of DNA hybridization from renaturation rates. *Syst. Appl. Microbiol.* **4:**184–192.
42. **Johnson, D. A., J. W. Gautsch, J. R. Sportsman, and J. H. Elder.** 1984. Improved technique utilizing nonfat dry milk for analysis of proteins and nucleic acids transferred to nitrocellulose. *Gene Anal. Techn.* **1:**3–8.
43. **Johnson, J. L.** 1973. Use of nucleic acid homologies in the taxonomy of anaerobic bacteria. *Int. J. Syst. Bacteriol.* **23:**308–315.
44. **Johnson, J. L.** 1978. Taxonomy of the bacteroides. I. Deoxyribonucleic acid homologies among *Bacteroides frag-*

ilis and other saccharolytic *Bacteroides* species. *Int. J. Syst. Bacteriol.* **28:**245–256.

45. **Johnson, J. L.** 1981. Genetic characterization, p. 450–472. *In* P. Gerhardt, R. G. E. Murray, R. N. Costilow, E. W. Nester, W. A. Wood, N. R. Krieg, and G. B. Phillips (ed.), *Manual of Methods for General Bacteriology.* American Society for Microbiology, Washington, D.C.
46. **Johnson, J. L.** 1985. DNA reassociation and RNA hybridization of bacterial nucleic acids, p. 33–74. *In* G. Gottschalk (ed.), *Methods in Microbiology,* vol. 18. Academic Press Ltd., London.
47. **Johnson, J. L., L. V. H. Moore, B. Kaneko, and W. E. C. Moore.** 1990. *Actinomyces georgiae* sp. nov., *Actinomyces gerencseriae* sp. nov., designation of two genospecies of *Actinomyces naeslundii,* and inclusion of *A. naeslundii* serotypes II and III and *Actinomyces viscosus* serotype II in *A. naeslundii* genospecies 2. *Int. J. Syst. Bacteriol.* **40:**273–286.
48. **Jones, A. S.** 1953. The isolation of bacterial nucleic acids using cetyltrimethylammonium bromide (CETAVLON). *Biochim. Biophys. Acta* **10:**607–612.

48a. **Kane, M. D., A. Brauman, and J. A. Breznak.** 1991. *Clostridium mayombei* sp. nov., an H_2/CO_2 acetogenic bacterium from the gut of the African soil-feeding termite, *Cubitermes speciosus. Arch. Microbiol.* **156:**99–104.

49. **Kapuciski, J., and B. Skoczylas.** 1977. Simple and rapid fluorometric method for DNA microassay. *Anal. Biochem.* **83:**252–257.
50. **Karsten, U., and A. Wollenberger.** 1977. Improvements in the ethidium bromide method for direct fluorometric estimation of DNA and RNA in cell and tissue homogenates. *Anal. Biochem.* **77:**464–470.
51. **Kirby, K. S.** 1957. A new method for the isolation of deoxyribonucleic acids: evidence on the nature of bonds between deoxyribonucleic acid and protein. *Biochem. J.* **66:**495–504.
52. **Kirby, K. S., E. Fox-Carter, and M. Guest.** 1967. Isolation of deoxyribonucleic acid and ribosomal ribonucleic acid from bacteria. *Biochem. J.* **104:**258–262.
53. **Labarca, C., and K. Paigen.** 1980. A simple, rapid, and sensitive DNA assay procedure. *Anal. Biochem.* **102:**344–352.
54. **Lachance, M. A.** 1980. A simple method for determination of deoxyribonucleic acid relatedness by thermal elution in hydroxyapatite microcolumns. *Int. J. Syst. Bacteriol.* **30:**433–436.
55. **Langer, P. R., A. A. Waldrop, and D. C. Ward.** 1981. Enzymatic synthesis of biotin-labeled polynucleotides: Novel nucleic acid affinity probes. *Proc. Natl. Acad. Sci. USA* **78:**6633–6637.
56. **Leary, J. J., D. J. Brigati, and D. C. Ward.** 1983. Rapid and sensitive colorimetric method for visualizing biotin-labeled DNA probes hybridized to DNA or RNA immobilized on nitrocellulose: bio-blots. *Proc. Natl. Acad. Sci. USA* **80:**4045–4049.
57. **Leary, J. J., and J. L. Ruth.** 1989. Nonradioactive labeling of nucleic acid probes, p. 33–57. *In* B. Swaminathan and G. Prakash (ed.), *Nucleic Acids and Monoclonal Antibody Probes.* Marcel Dekker, Inc., New York.
58. **Legros, M., and A. Kepes.** 1985. One-step fluorometric microassay of DNA in procaryotes. *Anal. Biochem.* **147:**497–502.
59. **Le Pecq, J.-B., and C. Paoletti.** 1966. A new fluorometric method for RNA and DNA determination. *Anal. Biochem.* **17:**100–107.
60. **Lutz, L. H., and A. A. Yayanos.** 1985. Spectrofluorometric determination of bacterial DNA base composition. *Anal. Biochem.* **144:**1–5.

60a. **Mackey, B. M., C. A. Miles, S. E. Parsons, and D. A. Seymour.** 1991. Thermal denaturation of whole cells and cell components of *Escherichia coli* examined by differential scanning calorimetry. *J. Gen. Microbiol.* **137:**2361–2374.

60b. **Mackey, B. M., S. E. Parsons, C. A. Miles, and R. J. Owen.** 1988. The relationship between the base composition of bacterial DNA and its intracellular melting temperature as determined by differential scanning calorimetry. *J. Gen. Microbiol.* **134:**1185–1195.

61. **Markov, G. G., and I. G. Ivanov.** 1974. Hydroxyapatite column chromatography in procedures for isolation of purified DNA. *Anal. Biochem.* **59:**555–563.
62. **Mandel, M., L. Igambi, J. Bergendahl, M. L. Dodson, Jr., and E. Schelgen.** 1970. Correlation of melting temperature and cesium chloride buoyant density of bacterial deoxyribonucleic acid. *J. Bacteriol.* **101:**333–338.
63. **Marmur, J.** 1961. A procedure for the isolation of deoxyribonucleic acid from microorganisms. *J. Mol. Biol.* **3:**208–218.
64. **Marmur, J., and P. Doty.** 1962. Determination of the base composition of deoxyribonucleic acid from its thermal denaturation temperature. *J. Mol. Biol.* **5:**109–118.
65. **McConaughy, B. L., C. D. Laird, and B. J. McCarthy.** 1969. Nucleic acid reassociation in formamide. *Biochemistry* **8:**3289–3294.
66. **McIntyre, P., and G. R. Stark.** 1988. A quantitative method for analyzing specific DNA sequences directly from whole cells. *Anal. Biochem.* **174:**209–214.
67. **Mesbah, M., U. Premachandran, and W. B. Whitman.** 1989. Precise measurement of the G+C content of deoxyribonucleic acid by high-performance liquid chromatography. *Int. J. Syst. Bacteriol.* **39:**159–167.
68. **Murray, M. G., and W. F. Thompson.** 1980. Rapid isolation of high molecular weight plant DNA. *Nucleic Acids Res.* **8:**4321–4325.
69. **Orosz, J. M., and J. G. Wetmur.** 1974. *In vitro* iodination of DNA. Maximizing iodination while minimizing degradation; use of buoyant density shifts for DNA-DNA hybrid isolation. *Biochemistry* **13:**5467–5473.
70. **Paul, J. H., and B. Myers.** 1982. Fluorometric determination of DNA in aquatic microorganisms by use of Hoechst 33258. *Appl. Environ. Microbiol.* **43:**1393–1399.
71. **Perkin-Elmer Corporation.** 1986. *Fluorometric Determination of DNA Concentration.* Biotechnology technical report L-913A (September). Perkin-Elmer Corp., Norwalk, Conn.
72. **Pollard-Knight, D., C. A. Read, M. J. Downs, L. A. Howard, M. R. Leadbetter, S. A. Pheby, E. McNaughton, A. Syms, and M. A. W. Brady.** 1990. Nonradioactive nucleic acid detection by enhanced chemiluminescence using probes directly labeled with horseradish peroxidase. *Anal. Biochem.* **185:**84–89.
73. **Renz, M., and C. Kurz.** 1984. A colorimetric method for DNA hybridization. *Nucleic Acids Res.* **12:**3435–3444.
74. **Rigby, P. W., J. M. Dieckmann, C. Rhodes, and P. Berg.** 1977. Labeling deoxyribonucleic acid to high specific activity in vitro by nick translation with DNA polymerase I. *J. Mol. Biol.* **113:**237–251.
75. **Sanders, C. A., D. M. Yajko, W. Hyun, R. G. Langlois, P. S. Nassos, M. J. Fulwyler, and W. K. Hadley.** 1990. Determination of guanine-plus-cytosine content of bacterial DNA by dual-laser flow cytometry. *J. Gen. Microbiol.* **136:**359–365.
76. **Schindler, C. A., and V. T. Schuhardt.** 1964. Lysostaphin: a new bacteriolytic agent for the *Staphylococcus. Proc. Natl. Acad. Sci. USA* **51:**414–421.
77. **Schwinghammer, E. A.** 1980. A method for improved lysis of some Gram-negative bacteria. *FEMS Microbiol. Lett.* **7:**157–162.
78. **Seed, B.** 1982. Attachment of nucleic acids to nitrocellulose and diazonium-substituted supports. *Genet. Eng.* **4:**91–102.
79. **Seidler, R. J., and M. Mandel.** 1971. Quantitative aspects of deoxyribonucleic acid renaturation: base composition, state of chromosome replication, and polynucleotide homologies. *J. Bacteriol.* **106:**608–614.
80. **Selin, Y. M., B. Harich, and J. L. Johnson.** 1983. Preparation of labeled nucleic acids (nick translation and iodination) for DNA homology and rRNA hybridization experiments. *Curr. Microbiol.* **8:**127–132.

81. **Setaro, F., and C. D. G. Morley.** 1977. A rapid colorimetric assay for DNA. *Anal. Biochem.* **81:**467–471.
82. **Sibley, C. G., and J. E. Ahlquist.** 1981. The phylogeny and relationships of the ratite birds as indicated by DNA-DNA hybridization, p. 301–335. *In* G. G. E. Scudder and J. L. Reveal (ed.), *Evolution Today. Proceedings of the Second International Congress of Systematic and Evolutionary Biology.* Carnegie-Mellon University, Pittsburgh.
83. **Smith, M. J., R. J. Britten, and E. H. Davidson.** 1975. Studies on nucleic acid reassociation kinetics: reactivity of single-stranded tails in DNA-DNA renaturation. *Proc. Natl. Acad. Sci. USA* **72:**4805–4809.
84. **Sneath, P. H. A.** 1989. Analysis and interpretation of sequence data for bacterial systematics: the view of a numerical taxonomist. *Syst. Appl. Microbiol.* **12:**15–31.
85. **Steiner, R. F., and H. Sternberg.** 1979. The interaction of Hoechst 33258 with natural and biosynthetic nucleic acids. *Arch. Biochem. Biophys.* **197:**580–588.
86. **Sterzel, W., P. Bedford, and G. Eisenbrand.** 1985. Automated determination of DNA using the fluorochrome Hoechst 33258. *Anal. Biochem.* **147:**462–467.
87. **Takahashi, T., T. Mitsuda, and K. Okuda.** 1989. An alternative nonradioactive method for labeling DNA using biotin. *Anal. Biochem.* **179:**77–85.
88. **Tereba, A., and B. J. McCarthy.** 1973. Hybridization of ^{125}I-labeled ribonucleic acid. *Biochemistry* **12:**4675–4679.
89. **Thompson, L. M., III, R. M. Smibert, J. L. Johnson, and N. R. Krieg.** 1988. Phylogenetic study of the genus *Campylobacter. Int. J. Syst. Bacteriol.* **38:**190–200.
90. **Ullman, J. S., and B. J. McCarthy.** 1973. Alkali deamination of cytosine residues in DNA. *Biochim. Biophys. Acta* **294:**396–404.
91. **Ullman, J. S., and B. J. McCarthy.** 1973. The relationship between mismatched base pairs and the thermal stability of DNA duplexes. *Biochim. Biophys. Acta* **294:**416–424.
92. **Van Dilla, M. A., R. G. Langlois, D. Pinkel, D. Yajko, and W. K. Hadley.** 1983. Bacterial characterization by flow cytometry. *Science* **220:**620–622.
93. **Warnaar, S. O., and J. A. Cohen.** 1966. A quantitative assay for DNA-DNA hybrids using membrane filters. *Biochem. Biophys. Res. Commun.* **24:**554–563.
94. **Werman, S. D., M. S. Springer, and R. J. Britten.** 1990. Nucleic acids I. DNA-DNA hybridization, p. 204–249. *In* D. M. Hillis and C. Moritz (ed.), *Molecular Systematics.* Sinauer Associates, Inc., Sunderland, Mass.
95. **Wetmur, J. G., and N. Davidson.** 1968. Kinetics of renaturation of DNA. *J. Mol. Biol.* **31:**349–370.
96. **Wiener, S. L., M. Urievetzky, S. Lendval, S. Shafer, and E. Meilman.** 1976. The indole method for determination of DNA: conditions for maximal sensitivity. *Anal. Biochem.* **71:**579–582.
97. **Yamada, K., and K. Komagata.** 1970. Taxonomic studies of coryneform bacteria. III. DNA base composition of coryneform bacteria. *J. Gen. Appl. Microbiol.* **16:**215–224.
98. **Yokogawa, K., S. Kawata, and Y. Yoshimura.** 1972. Bacteriolytic activity of enzymes derived from *Streptomyces* species. *Agric. Biol. Chem.* **36:**2055–2065.
99. **Yokogawa, K., S. Kawata, and Y. Yoshimura.** 1975. Purification and properties of lytic enzymes from *Streptomyces globisporus* 1829. *Agric. Biol. Chem.* **39:**1533–1543.
100. **Zolan, M. E., and P. J. Pukkila.** 1986. Inheritance of DNA methylation in *Coprinus cinereus. Mol. Cell. Biol.* **6:**195–200.

Chapter 27

Similarity Analysis of rRNAs

JOHN L. JOHNSON

Several types of RNA are synthesized by bacterial cells. When these RNAs are compared by RNA-DNA hybridization, the messenger RNA (mRNA) hybridization values are similar to those obtained by DNA-DNA reassociation experiments because a large portion of the genome is used for transcribing the mRNA molecules. In contrast, ribosomal RNA (rRNA) and transfer RNA (tRNA) are coded for by only small fractions of the genome; therefore, in hybridization experiments with these two types of RNA, only those small regions of the genome are being compared. In all groups of bacteria that have been systematically investigated, the arrangement of nucleotides in the rRNA and tRNA cistrons of the DNA appears to have evolved less rapidly than in the bulk of the cistrons. Therefore, rRNA-DNA hybridization methods are used to measure similarities between distantly related organisms, whereas DNA-DNA reassociation methods (see Chapter 26) are used to measure similarities between closely related organisms.

The 16S and 23S rRNA species were the first to be used for systematic studies, through hybridization to immobilized DNA (20). Only the 16S rRNA has been used in oligonucleotide similarities of T_1 RNase digests (16, 54, 55, 72). Because of its small size, 5S rRNA was used in early sequencing studies (14, 48, 59). Currently, most RNA sequence investigations involve the use of 16S rRNA or 16S rRNA genes (10, 41). However, the use of 23S rRNA or 23S rRNA genes is increasing (40).

Additional background information on rRNA-DNA hybridization, 16S and 23S rRNA sequencing, data analysis, and the use of rRNA sequence similarities for studying bacterial evolution can be found in the general references at the end of this chapter (1–4).

The protocols presented in this chapter will concentrate on rRNA-DNA hybridization and on the sequencing of 16S rRNA and of rRNA genes.

27.1. RNA ISOLATION AND MEASUREMENT

The bulk of the nucleic acid within a bacterial cell is RNA (mostly rRNA, with lesser amounts of tRNA and mRNA). This makes it easy to isolate RNA in large quantities.

The guidelines for growing and disrupting cells for DNA isolation (see Chapter 26.1) are also applicable for the isolation of RNA. One can nearly always isolate intact RNA from late-log-phase cultures. The cells usually can be disrupted by passage through a French pressure cell or by shaking with glass beads. Consequently, cell breakage is not a major problem for the isolation of rRNA.

27.1.1. RNase-Free Glassware and Reagents

The major problem in working with RNA is maintaining RNase-free conditions. One must first have an RNase-free preparation of RNA and then avoid introducing RNase during any of the experimental steps. The enzyme is not only ubiquitous but also very heat stable. Render glassware RNase free by baking it in an ashing oven or a dry-heat sterilizing oven. Disposable plasticware items are RNase free; for nondisposable plasticware, soak the items in water containing 0.1% diethylpyrocarbonate (DEPC) for at least 2 h, remove them from the solution, and then autoclave them at 121°C for 15 min. Treat water, aqueous buffers, and solutions by adding DEPC to a final concentration of 0.1%, allow to stand for at least 2 h, and then autoclave. Wear plastic gloves while isolating RNA, and do not touch the inside of tubes, beakers, or flasks.

27.1.2. Kirby Isolation Procedure

Kirby (37) has described in detail various procedures for isolating RNA. The following is a variation that works well for isolating bacterial RNA that is essentially free from DNA. The procedure involves the rapid disruption of the cells, in the presence of DEPC, by a French pressure cell directly into a deproteinizing-detergent solution to minimize RNase activity. After deproteinization and ethanol precipitation, DNA and small RNAs are eliminated from the preparation by selectively precipitating the large RNAs with ammonium acetate.

Reagents

Cell suspension buffer 10 mM Tris-HCl (pH 8.0), 1 mM Na-EDTA

Phenol-$CHCl_3$: water-saturated phenol and chloroform-isopentanol (chloroform containing 3% [vol/vol] isopentanol) mixed in equal volumes and containing 0.1% (wt/vol) 8-hydroxyquinoline

Ammonium acetate, 7.5 M (578 g/liter of distilled water)

Sodium chloride, 0.25 M (14.5 g/liter of distilled water)

Standard saline-citrate (SSC): 0.15 M NaCl, 0.015 M trisodium citrate (pH 7.0). Add 0.1% DEPC, and autoclave. Dilute with DEPC-treated water to obtain more dilute SSC buffer solutions such as 0.1× SSC.

Sodium dodecyl sulfate (SDS). Prepare 20% (wt/vol) solution in DEPC-treated water. Use electrophoretic-grade SDS for the solution that will be added to RNA preparations.

Na-EDTA, 0.5 M solution (pH 8.0)

1× SSC–1 mM HEPES (*N*-2-hydroxyethylpiperazine-*N*′-2-ethanesulfonic acid sodium salt [pH 7.0]) buffer

Procedure

1. Harvest the cells from 100 to 250 ml of broth culture by centrifuging at 7,000 to 8,000 × *g* for 20 min. Decant the supernatant and remove any remaining medium with a Pasteur pipette.

2. Suspend the cells in 15 ml of cell suspension buffer, and place them on ice.

3. Into a 125-ml ground-glass–stoppered Erlenmeyer flask, add 10 ml of phenol-$CHCl_3$, 8 ml of 0.25 M NaCl, 1.5 ml of 20% SDS, 0.25 ml of 0.5 M Na-EDTA, and 25 μl of DEPC.

4. Add 25 μl of DEPC to the cell suspension, mix, and place into a French pressure cell. Disrupt the cells at 16,000 lb/in^2, collecting them in the flask prepared in step 3. Keep the disrupted cells continually mixed with the phenol-$CHCl_3$ by gently swirling the flask.

5. Shake the flask for 20 min on a wrist-action shaker. Place the material in a 50-ml centrifuge tube, and centrifuge at 20,000 × *g* for 10 min. During centrifugation, rinse the flask with hot tap water and let it drain upside down on a paper towel.

6. Transfer the upper (aqueous) layer back to the Erlenmeyer flask, add an additional 10 ml of phenol-$CHCl_3$, and repeat the extraction and centrifugation. If a substantial protein layer is still present at the interface, do a third extraction and centrifugation.

7. Transfer the upper (aqueous) layer to a 250-ml sterile centrifuge bottle. Add 2 volumes of 95% ethanol, balance the bottles, place them in a freezer for 30 to 60 min, and centrifuge them at 8,000 × *g* for 15 min. Decant the ethanol, place the centrifuge bottles back into the centrifuge (make sure that the pellet is in the same orientation as it was after the first centrifugation), and briefly centrifuge at low speed. Then remove the rest of the ethanol with a Pasteur pipette.

8. Dissolve the pellet in 20 ml of 1× SSC–1 mM HEPES buffer. Transfer this RNA solution to a 50-ml screw cap centrifuge tube, add 10 ml of 7.5 M ammonium acetate, mix well, and place on ice for 20 min.

9. Centrifuge at 20,000 × *g* for 10 min. This will pellet the large rRNAs.

10. Dissolve the pellet in 10 ml of 1× SSC–1 mM HEPES, add 20 ml of 95% ethanol, and mix. Place on ice for 10 to 20 min, and centrifuge at 20,000 × *g* for 10 min.

11. Dissolve the pellet in 5 to 10 ml of 1× SSC–1 mM HEPES. Add electrophoretic grade 20% SDS to a final concentration of 0.5%, and store at −20°C. Alternatively, dissolve the RNA in 1.0 ml of DEPC-treated sterile distilled water, and add 4 ml of 95% ethanol before storage at −20°C. The RNA is stable under both conditions.

To remove SDS, place the portion of thawed RNA solution needed in a centrifuge tube, cool it on ice to precipitate the SDS, and remove the SDS by centrifugation in the cold.

For RNA stored in water-ethanol, place the portion needed in a centrifuge tube and add 0.1 volume of 3 M sodium acetate to precipitate the RNA. Mix, and then centrifuge to pellet the RNA.

27.1.3. Guanidine Isothiocyanate Isolation Procedure

The other major approach to isolating RNA has been to use a strong chaotropic agent such as guanidine hydrochloride or guanidine isothiocyanate (11). The following protocol is a variation of a guanidine isothiocyanate procedure described by Puissant and Houdebine (50) for the isolation of total RNA from eucaryotic tissue. The guanidine isothiocyanate-based lysing solution inhibits RNase activity, and the lowering of the pH with sodium acetate (pH 4.0) causes genomic DNA to precipitate with the proteins during extraction with phenol-$CHCl_3$.

Reagents

Suspending buffer. This consists of 10 mM Tris-HCl (pH 8.0), 1 mM Na-EDTA, and 0.35 M sucrose. Store in a refrigerator. Just before use, remove the required amount and heat it in a boiling-water bath for a few minutes to drive off most of the oxygen. Cool, and then add either dithiothreitol or 2-mercaptoethanol to a concentration of 5 mM (0.077 g/100 ml or 37 μl/100 ml, respectively).

Lysozyme: 5 mg/ml in Tris-EDTA (TE) buffer. The latter consists of 10 mM Tris-HCl (pH 8.0) containing 1 mM Na-EDTA.

Lysing solution: 4 M guanidine isothiocyanate, 25 mM sodium citrate (pH 7.0), 0.5% Sarkosyl, 0.1 M 2-mercaptoethanol

Sodium acetate, 2 M (pH 4.0)

Sodium acetate, 2 M (pH 5.0)

Phenol, water saturated, containing 0.1% 8-hydroxyquinoline

Chloroform containing no isoamyl alcohol

Ammonium acetate, 7.5 M (578 g/liter of distilled water)

SDS-TE: TE buffer (pH 7.5) containing 0.5% SDS

1× SSC–1 mM HEPES

Procedure

1. Harvest the cells from 40 to 100 ml of broth culture by centrifuging at 7,000 to 8,000 × *g* for 10 min. Decant the supernatant, then remove any remaining medium with a Pasteur pipette.

2. Suspend the cells in 2.0 ml of suspending buffer (a 50-ml polypropylene screw-cap centrifuge tube is a suitable container). Add 50 μl of lysozyme solution and digest at room temperature or at 37°C until the cells can be lysed by the lysing solution.

3. Add 10 ml of the lysing solution, and use a vortex mixer to mix thoroughly. Add the following reagents in succession, and mix thoroughly after each addition: 1.0 ml of 2 M sodium acetate (pH 4.0), 10 ml of phenol, and 2 ml of chloroform.

4. Centrifuge the tube at 20,000 × *g* for 10 min. Transfer the upper (aqueous) layer to a clean tube, and add 10 ml of isopropanol to precipitate the RNA. Pellet the RNA precipitate by centrifuging at 3,000 × *g* for 10 min. Drain the tube thoroughly.

5. Dissolve the RNA in 10 ml of TE buffer (containing no SDS). Add 5 ml of 7.5 M ammonium acetate solution. Mix well, and place on ice for 10 min. Then centrifuge at 20,000 × *g* for 10 min, and save the pellet.

6. Dissolve the pellet in 2.0 ml of TE-SDS, add 2 ml of chloroform, and mix on a vortex mixer. Save the upper (aqueous) phase, add 200 μl of 3 M sodium acetate (pH 6.0), and precipitate with an equal volume of isopropanol (2 ml). Pellet the RNA by centrifuging at 3,000 × *g* for 10 min. Wash the pellet with 70% ethanol.

7. Dissolve the pellet in 2 to 5 ml of 1× SSC–1 mM HEPES or in 0.5 ml of DEPC-treated sterile distilled water; then add 2 ml of 95% ethanol. Store at −20°C.

27.1.4. Direct Isolation from the Environment

Although too specialized to be included in this chapter, the characterization of rRNA or rRNA genes, the use of rRNA molecules for identifying organisms collected from the environment, and rRNA-specific probe hybridizations for estimating numbers of an organism in mixed populations are related areas of emerging importance (46). Unique nucleic acid isolation procedures are needed for the organisms collected from water (19, 60), soil (23, 27, 62), and bacterial mats of hot springs (70, 71). After treatment of bacteria with antibiotics, rRNA-DNA hybridization signals may or may not reflect viable cells (5).

27.1.5. RNA Measurement

The most common method for measuring RNA concentration is to measure the A_{260}, for which the specific extinction coefficient is about 23. Using water or low-ionic-strength buffer as a diluent, prepare 50- to 100-fold dilutions of the stock RNA preparations. Measure the A_{260}, and calculate the stock RNA concentrations from the following formula.

$$\text{mg of RNA per ml stock RNA} = (A_{260}/23) \times \text{dilution}$$

27.2. RNA FRACTIONATION

The various rRNA species can be separated by centrifugation through sucrose gradients (also see Chapter 4.3) or by electrophoresis through polyacrylamide or agarose gels (also see Chapter 21.5).

27.2.1. Sucrose Gradients

The rate of RNA sedimentation in sucrose gradients is a function of the rotor speed and the subunit sizes (42). Prepare 30-ml gradients (5 to 20% [wt/vol] sucrose) in 1- by 3-in. (2.54- by 7.62-cm) tubes or 35-ml gradients in 1- by 3.5-in. (2.54- by 8.89-cm) tubes. When covered with Parafilm, these can be stored in a refrigerator overnight or at −70°C for several months. Layer 1.0 to 1.5 mg of RNA (in 1 ml of 1× SSC) on top of a gradient, and centrifuge at 25,000 rpm (about 83,000 × *g*) for 15 h at 5 to 10°C. These swinging-bucket rotors can be run at 27,000 to 30,000 rpm (83,000 to 97,000 × *g*) and will require about 15 h of centrifugation to separate the subunits. For smaller, higher-speed rotors, the separations can be done in shorter times. The 23S rRNA component will sediment the most rapidly. The resolving of the components is best when the 23S band is near the bottom of the tube. However, since this is a sedimentation velocity centrifugation, the RNA components will pellet if the centrifugation time is too long. After centrifugation, pierce the tube at the bottom, collect the fractions sequentially with a fraction collector and measure the A_{260} of each fraction. Alternatively, after piercing the centrifuge tube, connect the piercing needle to a flowthrough cuvette placed in a spectrophotometer. This allows continuous monitoring of the A_{260} and permits collection of only the pertinent fractions. After collecting the fractions, precipitate the RNA with 2 volumes of 95% ethanol.

Under ideal conditions, there should be about twice the amount of 23S as 16S rRNA. If there has been RNase activity, degradation products of the 23S rRNA will end up in the 16S fraction.

27.2.2. Polyacrylamide Gel Electrophoresis

The following procedure has been described by Lane (40). Prepare a 14- by 15-cm vertical gel, 3 mm thick (model PROTEAN II xi electrophoresis unit; Bio-Rad, Richmond, Calif.), as follows, Pour 10 ml of solution containing 8% acrylamide, 0.4% bisacrylamide, 8 M urea, and 1× Tris-borate buffer (134 mM Tris, 45 mM H_3BO_3, 2.5 mM disodium EDTA) (40) into the bottom of the gel-casting form to generate a firm plug. After the plug has polymerized, continue filling with a solution containing 2.8% acrylamide, 0.14% bisacrylamide, 8 M urea, and 1× Tris-borate buffer. Sample wells that are 0.8 cm wide work well. Prepare the RNA samples (2 mg/ml in TE buffer), and add 1 mg of solid urea per μl of sample. Heat at 50 to 60°C for 5 min, and then load 50-μl amounts (about 50 μg) directly into the wells, after first washing residual urea from the wells with electrophoresis buffer. Load RNA standards (5 μg of 16S rRNA or 10 μg of total rRNA) on the outside lanes for visualization by UV shadowing (24). Electrophorese the gel at about 100 V for 8 h. Cut the 16S RNA band from the gel, and electroelute the RNA.

27.2.3. Agarose Gel Electrophoresis

The following protocol is primarily for evaluating the quality of an RNA preparation (i.e., the intactness of the 16S and 23S subunits, the presence of contaminating DNA, and the amount of low-molecular-weight RNAs). Prepare a 1.5% (wt/vol) agarose solution (70 ml) in Tris-acetate-EDTA electrophoresis buffer (40 mM Tris-acetate, 1 mM disodium EDTA [pH 8.0]) (2). After cooling to 50°C, add 0.5 μg of ethidium bromide per ml and 1% 1-mercaptoethanol, and pour into a 11- by 14-cm gel casting tray (Horizon 11-14 horizontal gel electrophoresis apparatus; GIBCO-BRL, Gaithersburg, Md.). Load 10 to 20 μl of sample (diluted in TE to 15 μl plus 3 to 4 μl of loading dye [15% Ficoll, 0.25% bromophenol blue]). Electrophorese at 70 to 80 v for 2 to 3 h. Rinse the gel with distilled water, and observe and photograph on a transilluminator. The intensity of the 16S rRNA bands should be about half that of the 23S rRNA bands; the intensities of the small rRNA and tRNA RNAs will be proportional to their amounts in the preparation. Since DNA stains more intensely with ethidium bromide, this is a rather sensitive method for detecting DNA contamination.

27.3. rRNA-DNA HYBRIDIZATION

Many bacterial taxonomists have determined phylogenetic relationships by using variations of the rRNA-DNA hybridization procedure introduced by Gillespie and Spiegelman (20). However, this procedure requires experimental measurements that must be repeated each time a new organism is included in the group to be analyzed. In contrast, determining the nucleotide base sequence of a rRNA or rRNA gene (section 27.4) has the advantage of being able to add the sequence to a continually expanding data base, where a sequence can be compared with that of any other rRNA or rRNA gene that has been sequenced or that will be sequenced in the future. For this reason, it is likely that rRNA hybridization experiments will be used in the future mainly as survey tools for assigning species to, or selecting species that belong to, specific rRNA similarity groups.

27.3.1. RNA Labeling

Radioactive labeling of intact RNA can be done either in vivo by including radioactive precursors in the growth medium or in vitro by using iodination. Labeling of RNA fragments and of rRNA genes is also described.

27.3.1.1. In vivo labeling

In vivo labeling of RNA is usually accomplished by including [^{3}H]uracil, [^{14}C]uracil, [^{3}H]uridine, or [^{14}C]uridine in the growth medium. A dilute peptone-yeast extract medium usually works best for incorporation of RNA precursors. The medium must be complete enough to support growth of the organism, but if excess yeast extract is present it will tend to dilute the labeled precursor. Minimal media often do not work well, because the organism that is synthesizing nucleotides via biosynthetic pathways may not readily take up a single preformed radioactive base or nucleoside.

The long half-lives of ^{3}H and ^{14}C result in rather low specific activities for the RNAs. Higher activities can be obtained using ^{32}P; however, the high energy of the beta radiation from this isotope makes it dangerous to work with; moreover, the beta particles will rapidly fragment the RNA preparation within a matter of days.

Isolate intact RNA from radioactively labeled cultures by using the same procedures that are described for unlabeled RNA (section 27.1).

27.3.1.2. In vitro random labeling by iodination

RNA molecules can be randomly labeled along their entire lengths by iodination. The procedure for labeling RNA by iodination is very similar to that for DNA (see Chapter 26.5.1), except that sodium perchlorate is not included in the labeling mixture and hydroxylapatite columns are not used. Iodine is directly linked to the cytosine residues in the presence of thallium chloride at a low pH.

Reagents

Reaction buffer: 50 mM sodium acetate, 25 μM potassium iodide (pH 4.8)

Other reagents: see Chapter 26.5.1

Procedure

1. Place 10 to 100 μg of RNA in a reaction tube (see Chapter 26.5.1). Add sufficient 3 M sodium acetate (pH 6) to give a final concentration of 0.3 M. Mix, add 2 volumes of 95% ethanol, mix again, and centrifuge for 10 min. Draw off the supernatant, wash the pellet with 80% ethanol, and then dry it in a 37°C incubator.
2. Dissolve the RNA in 35 μl of reaction buffer, and deposit 1 to 2 μl of $Na^{125}I$ (100 μCi/μl) on the inside wall of the reaction tube and 12 to 13 μl of $TlCl_3$ solution on the opposite inside wall. Cap the reaction tube, mix the contents, and incubate the tube in a 70°C water bath for 20 min.
3. Remove the tube from the water bath, and let it cool for 1 or 2 min. Using a disposable tuberculin syringe, inject approximately 0.2 ml of PB-mercaptoethanol buffer into the tube. Return the tube to the water bath, and incubate for an additional 20 min.
4. Equilibrate a NAP 25 gel filtration column with 1 × SSC–0.1% SDS (SSC-SDS) buffer. Using a 1-ml disposable syringe with a 1.5-in. (3.8-cm) 22-gauge needle, inject approximately 0.2 ml of the SSC-SDS buffer into the reaction tube. With the same syringe, remove all of the material from the tube and place it on the gel filtration column. Add SSC-SDS buffer to a volume of 2.5 ml (the void volume of the column), place a collection tube under the column, and add an additional 2.0 ml of SSC-SDS to the column.
5. Place the collected material in the 70°C water bath for an additional 20 min, and again pass the sample through another NAP 25 gel filtration column as described in step 4. This time, because of the larger sample volume, collect 3.5 ml of eluate. Store at −20°C.

27.3.1.3. In vitro end labeling of fragments

First the RNA is randomly fragmented (15, 58) and then labeled either at the 5′ end with bacteriophage T4 polynucleotide kinase or at the 3′ end with poly(A) polymerase (15). Randomly fragment an RNA preparation by incubating it in a borate buffer (2.5 M H_3BO_3, 1.2 M NaOH [pH 9.4]) at 70°C for 30 min (58). Precipitate the fragments with ethanol, and add a single label to the 5′ by using a commercial kit, such as DNA 5′-end-labeling system (no. U2010; Promega, Madison, Wis.), or add about 10 poly(A) residues to the 3′ end by using poly(A) polymerase (15).

27.3.1.4. In vitro random labeling of fragments

RNA fragments have been randomly labeled by using polythymidine primers to hybridize with polyadenylated RNA fragments and then synthesizing labeled RNA-complementary DNA (cDNA) with reverse transcriptase (7). It seems likely that cDNA could be synthesized directly by using a mixture of rRNA-sequencing primers (see Table 5) and reverse transcriptase from either avian myeloblastosis virus (no. M5101; Promega) or Moloney murine leukemia virus (no. M5301; Promega).

27.3.1.5. Labeling of rRNA genes

See Chapter 26.5.2 to 26.5.5.

27.3.2. DNA Immobilization on Membranes

For a general procedure for immobilizing DNA on nitrocellulose membranes, see Chapter 26.6.3.1. The procedure is also applicable to other membranes such as nylon, although the filtration buffers will differ. The membrane manufacturers provide the specific information.

The DNA-containing membranes can be preincubated by the following procedure.

Reagent

Denhardt reagent (13): 0.02% bovine serum albumin (fraction V; Sigma Chemical Co., St. Louis, Mo.), 0.02% polyvinylpyrrolidone (Calbiochem-Behring, La Jolla, Calif.), 0.02% Ficoll 400 (Pharmacia Fine Chemicals, Uppsala, Sweden) in water. This can be prepared as a 50× concentrate and stored at −20°C.

Procedure

1. First fix DNA to a large membrane; a large number of small membranes, with a reasonably even distribution of DNA, can then be punched out from the single large membrane. Punch out the required number of small membranes, and place them in a test tube. Dilute the concentrated Denhardt reagent to 1 × with 2 × SSC–1 mM HEPES (pH 7.0), and mix with the membranes (about 0.1 ml of reagent per membrane).
2. Preincubate the membranes at 50°C for 30 min to 2 h. Swirl the membranes several times during the preincubation period.
3. Remove the membranes, place them on a paper towel to blot off excess liquid, and then place them into the reaction vials. Do not allow the membranes to dry.

Comments

The amount of hybridization that will occur with the immobilized DNA (see below) is directly proportional to (i) the amount of DNA fixed on the membrane, (ii) the genome size of the organism from which the DNA is isolated, and (iii) the rRNA gene copy number per genome. The amount of DNA per membrane can be determined after measuring its radioactivity (30), but this is laborious and one must be careful in the handling of the residual radioactivity. The next best approach is to carefully measure the amount of DNA that is being immobilized. Most DNA preparations will be contaminated with various amounts of RNA degradation products, so that A_{260} measurements will include the degradation products. Using colorimetric or fluorometric methods that are specific for DNA (see Chapter 26.3.3 and 26.3.4) will reduce the effect of this variable. Also, do not use saturating amounts of DNA on the membrane (25 μg/cm² approaches saturation for nitrocellulose membranes). This is because minor contaminants in the DNA preparation may compete for binding sites on the membrane and thus cause variable amounts of DNA to be immobilized.

27.3.3. Hybridization Measurement on Membranes

The following reagents are used for the direct-binding (saturation hybridization) procedure, hybrid thermostability measurements, and competition hybridization experiments.

Reagents

^{125}I-labeled rRNA. This may be 16S rRNA, 23S rRNA, or a mixture of the two. The specific activity should be in the range 2×10^5 to 6×10^5 cpm μg^{-1} and at a concentration of about 30 μg/ml of 1× SSC–1 mM HEPES (pH 7.0) containing 0.1% SDS.

Immobilized DNA: nitrocellulose membrane containing 25 μg of DNA per cm²

Concentrated SSC. Prepare 17.6× SSC–5 mM HEPES (pH 7.0) buffer. The use of 25 μl/110 μl of reaction volume will result in a final concentration of 4× SSC.

RNase mixture: mixture of 10 μg of RNase A and 0.25 U of RNase T_1 per ml of 2× SSC

Competitor RNA: 2 mg of total RNA per ml of 1× SSC–1 mM HEPES–0.5% SDS. Heat each preparation in a boiling-water bath for 5 min at the time of preparation to disrupt the secondary structure of the RNA. Store at −20°C.

Deionized formamide. Slightly yellow-colored formamide indicates that there has been some decomposition. The decomposition products can be removed with a mixed-bed resin. Add 10 to 20 g of Dowex XG8 mixed-bed resin (Bio-Rex RG 501-X8; Bio-Rad) per liter of formamide. After mixing for 1 h, the resin can either be filtered out through Whatman no. 1 paper or left at the bottom of the bottle to exchange with future degradation products.

27.3.3.1. Direct binding procedure

Current methods for direct binding (12, 31, 36) are based on procedures described by Gillespie and Spiegelman (20), Denhardt (13), Fry and Artman (18), and Gillespie and Gillespie (21). A useful composite procedure is as follows.

1. Reaction mixtures consist of 10 μl of labeled rRNA (0.3 μg), 25 μl of 1× SSC–1 mM HEPES, 25 μl of 17.6× SSC, and 50 μl of deionized formamide. Mix the contents of each vial on a vortex mixer; a Thermolyne Maxi-Mix I mixer (Sybron/Thermolyne, Dubuque, Iowa) works well.

2. Place the DNA-containing membranes into the vials, and incubate at 50°C for 16 to 20 h.

3. After completion of the hybridization reaction, place the membranes into a washing chamber (Fig. 3 in Chapter 26) and wash them twice in 300-ml volumes of 2× SSC (in a 400-ml beaker) warmed to 50°C. Shake excess buffer from the chamber, place the chamber into a dish containing RNase mixture (the minimal volume that will cover the membranes), and incubate at 37°C for 1 h. Then wash the membranes in a final 300-ml volume of 2× SSC at room temperature for 5 min.

4. Dry the membranes, and measure their radioactivity in a gamma scintillation counter.

Comments

The extent of rRNA hybridization in part will be a function of the stringency of the hybridization, i.e., the difference between the dissociation temperature of the rRNA-DNA hybrids and the hybridization temperature. At low stringency (30 to 35°C below the dissociation temperature), most rRNA can be detected, whereas at medium stringency (20 to 25°C below) and high stringency (10 to 15°C below), more base pairing is required for hybrid formation. The hybridization conditions given above represent a temperature that is about 20°C below the denaturation temperature, so that the stringency is reasonably high but still allows for a range of similarity values. A disadvantage of direct-binding experiments is that DNA must be isolated from each organism in the study and fixed on membranes. Other problems associated with quantitation of direct-binding experiments are standardizing of the amount of DNA loaded per membrane, the genome size of the organisms, and the gene copy number per genome. Some of these have been addressed by Klink et al. (38).

27.3.3.2. Thermal stability profile

To generate an rRNA-DNA hybrid thermal elution profile, follow the direct-binding procedure as outlined above, but do not dry the membranes after the final wash. Then proceed as follows.

1. Remove the membranes from the washing chamber, skewer the replicate membranes onto a stainless-steel insect pin near the head end, and insert the pin into a rubber stopper core that is plugging a piece of glass tubing (see Fig. 5 in Chapter 26). Put spacers between each membrane. (Spacers can be made by cutting 5-mm-diameter Teflon discs from a Teflon weighing boat and drilling a small hole in the center of each disc.) Place the skewered membranes into the first elution tube.

2. For elution tubes, use appropriate-size polypropylene tubes that can fit directly into a gamma scintillation counter. Into each elution tube place 1.0 ml of 4× SSC–1 mM HEPES (pH 7.0) buffer–50% formamide. Start the elution profiles at 35°C, and increase the bath temperature at 5°C increments, with incubation of the membranes at each temperature for 5 min. Then transfer the membranes to the next tube and continue the stepwise temperature increases up to 80 or 85°C.

3. Measure the radioactivity eluted at each temperature and the residual radioactivity on the membranes. Generate the integral elution curves (Fig. 1) by plotting the sum of the counts eluted up to and including each temperature divided by the total counts eluted versus the elution temperature (Table 1).

Comments

When ^{3}H- or ^{14}C-labeled RNAs are used, the fragments eluted at each temperature must be mixed with a scintillation fluid designed for aqueous samples. The elution volumes can be decreased to 0.5 ml to minimize the amount of scintillation fluid needed, although this

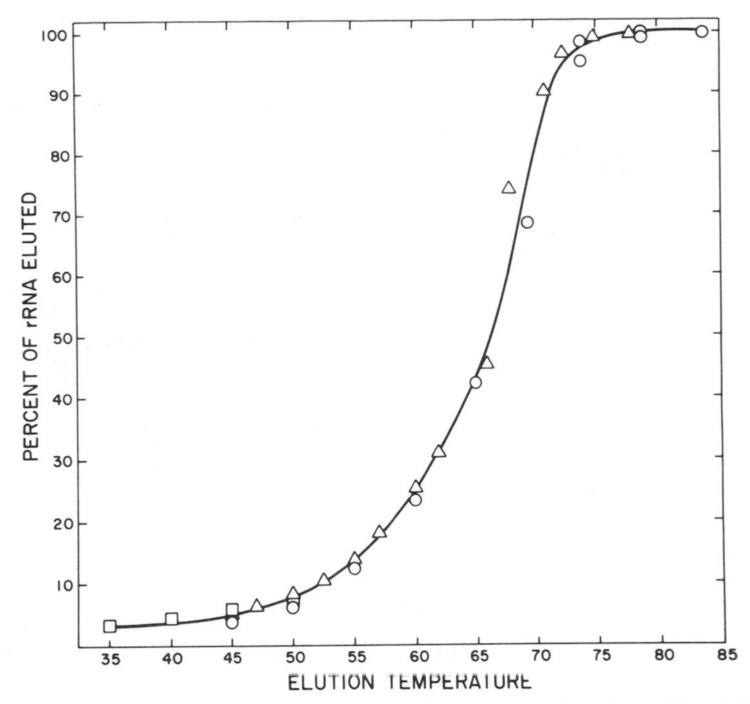

FIG. 1. rRNA-DNA hybrid thermal elution profile, based on data from Table 1 (31).

will increase the amount of carryover from one temperature to the next.

One of the technical difficulties is the cooling effect of evaporation from the water bath at higher temperatures. For even heat distribution, it is important to use a circulating water bath. Select one with a reasonably small water volume and large (1,500-W) heater. Minimize the open water surface by constructing a tube rack that will fit into the bath opening and have holes just large enough for the tubes (Fig. 2, Chapter 26). Use a thermometer in a tube (with the same volume of elution buffer) for measuring the elution temperature. Have only two sets of elution tubes in the water bath at a time, i.e., the set with the membranes in and the set for the next temperature. This allows enough time to equilibrate to the first temperature before increasing to the next while being rapid enough that the elution buffer is not altered by evaporation. Another option for generating the temperature profiles, is to use a thermocycler of the type used for a polymerase chain reaction (PCR). It can be manually set for any desired temperature, although one would have to modify the protocol to use 0.5-ml microcentrifuge tubes.

The midpoint values of the thermal stability profiles are reported either directly in °C (12) or as differences in °C with respect to a reference organism (31).

Table 1. Hypothetical example of thermal stability profile calculations

Elution temp (°C)	cpm eluted[a,b]	Percent eluted	Sum cpm eluted[b]	Percent of total cpm eluted
35	300	3.0	300	3.0
40	100	1.0	400	4.0
45	100	1.0	500	5.0
50	200	2.0	700	7.0
55	600	6.0	1,300	13.0
60	1,200	12.0	2,500	25.0
65	1,800	18.0	4,300	43.0
70	2,900	29.0	7,200	72.0
75	1,900	19.0	9,100	91.0
80	700	7.0	9,800	98.0
85	100	1.0	9,900	99.0

[a]Residual cpm (cpm not eluted) = 100.
[b]Sum total cpm = 9,900 + 100 = 10,000.

27.3.3.3. Competition hybridization

The major advantage of competition hybridization is that a large number of RNA preparations can be compared with reference organisms (30, 31). All that has to be prepared from the test organisms is total RNA. The procedure is similar to that for direct binding, except that the membrane-immobilized DNA and labeled RNA are from the same organism. An excess of unlabeled RNA is added to compete with the hybridization of the labeled RNA (the maximum competition occurs when the amount of unlabeled RNA is the same as the labeled RNA). Measurements include the amount of labeled rRNA hybridized in (i) the absence of unlabeled RNA (no competition), (ii) the presence of unlabeled RNA from the same organism (maximum competition), and (iii) the presence of the test RNA preparations (various levels of competition).

Procedure

1. Reaction mixtures consist of 10 μl of labeled rRNA (0.3 μg), 25 μl of 1× SSC–1 mM HEPES (or competitor RNA), 25 μl of 17.6× SSC, and 50 μl of deionized formamide. Mix the contents of each vial by vortexing.
2. Place the DNA-containing membranes into the vials, and incubate at 50°C for 16 to 20 h.

3. After completion of the hybridization reaction, place the membranes into a washing chamber (Fig. 3 in Chapter 26) and wash them twice in 300-ml volumes of 2× SSC (in a 400-ml beaker) warmed to 50°C. Shake excess buffer from the chamber, place the chamber into a dish containing RNase mixture (use a minimal volume that will cover the membranes), and incubate it at 37°C for 1 h. Then wash the membranes in 300 ml of 2× SSC at room temperature for 5 min.

4. Dry the membranes and measure their radioactivity in a gamma scintillation counter. The percent similarity calculations are illustrated in Table 2.

Comments

Competition experiments represent a reasonably quantitative measurement of sequence similarity, because the extent of competition, rather than absolute hybridization values, is being measured. Hence the results are not greatly affected by the amount of DNA on the membrane or other factors that may affect hybridization efficiency. The percent similarity values from reciprocal experiments are usually quite close, and they correlate well with 16S rRNA oligonucleotide similarity values (31, 32). It is likely that they will also correlate well with sequence similarity values.

27.4. DIRECT SEQUENCING OF rRNA AND OF rRNA GENES

The dideoxynucleotide-chain-terminating DNA sequencing method of Sanger et al. (55) introduced the era of rapid DNA sequencing. Evolving from these protocols was the introduction of the use of reverse transcriptase for the sequencing of RNA. In 1985, Lane et al. (41) described protocols for the direct sequencing of 16S rRNA. Although rRNA exhibits intramolecular base pairing, which causes premature termination during DNA synthesis by reverse transcriptase, the procedure of Lane et al. was fast, and its wide use led to the rapid accumulation of a vast 16S rRNA sequence data base. More recently, the development of PCR by Mullis and Faloona (44) has enabled investigators to amplify the 16S rRNA genes so that they can be cloned and sequenced, or sequenced directly via DNA sequencing protocols.

The protocols described in this section are, for the most part, those recently described in detail by Lane (40) for the sequencing of 16S rRNA.

27.4.1. 16S rRNA Sequencing

Reagents

Nucleotides. Deoxynucleoside triphosphates (dNTP) and dideoxynucleoside triphosphates (ddNTP) are available either in the dry form or as stock solutions from several commercial sources (e.g., Boehringer-Mannheim, Pharmacia). The purity and integrity of the compounds can be readily checked by thin-layer chromatography (40). If the dry form of an NTP is used, dissolve the powder in 10 mM Tris-HCl (pH 8.3) buffer to a concentration of approximately 10 mg/ml. Then measure the exact concentration of this stock solution as follows; (i) dilute 1 or 2 μl of each solution 1:500 with Tris-HCl buffer, (ii) measure the absorbance at the appropriate absorbance maximum with a UV spectrophotometer (Table 3), and (iii) calculate the concentration by the following formula.

$$\frac{\text{Absorbance} \times \text{dilution factor}}{\text{Millimolar extinction coefficient}} = \text{mM NTP}$$

Adjust the dNTP stock solutions to 10 mM and the ddNTP stock solutions to 1.0 mM (Table 3).

Sequencing nucleotide mixtures: prepare each sequencing nucleotide mixture as illustrated in Table 4 (40). Test the mixtures out on a known template rRNA to make sure that the dNTP/ddNTP ratios are correct. If the concentration of a particular ddNTP is too low, the intensity of the bands in that sequencing lane will be faint, with most of the radioactivity in the unresolved large-fragment region. If the concentration is too high, the shorter fragments will be dark but the intensity of the larger fragments will fade. Adjustments can be made by adding more ddNTP or making a smaller batch with no ddNTP and mixing the two together.

rRNA template: total rRNA that has been enriched for 16S and 23S species by salt precipitation (see sections 27.2.2 and 27.2.3); 2.0 μg/μl in 10 mM Tris-HCl (pH 8.0) buffer

Oligonucleotide primers. The reverse primers that are commonly used for 16S rRNA sequencing are listed in Table 5. Prepare oligonucleotides at a concentration of 100 ng/μl in TE buffer. Higher concentrations may be needed for some of the degenerate primers.

Avian myeloblastosis virus reverse transcriptase. This enzyme can be obtained from a number of suppliers

Table 2. Hypothetical example for the calculation of percent similarity in an rRNA competition experiment

Expt	cpm bound	cpm depression	% Similarity
Direct binding, no competitor reference organism	1,000		
Competitor RNA, reference organism	20	980	100
Competitor RNA, test organism A	60	940	96
Competitor RNA, test organism B	800	200	20

Table 3. Molecular weights, absorbance maxima, and millimolar extinction coefficients of dNTPs and ddNTPs at pH 7

Compound	Formula wt	Absorbance maximum (nm)	mM extinction coefficient
dATP	491.2	259	15.2
dCTP	467.2	271	9.1
dGTP	507.2	253	13.7
dTTP	482.2	267	9.6
dITP		249	12.2
[α-^{35}S]dATP		259	14.8
ddATP		259	15.2
ddCTP		272	9.1
ddGTP		253	13.7
ddTTP		267	9.6

Table 4. Sequencing nucleotide mixtures

Stock solution	Concnt	ddGTP	ddCTP	ddTTP	ddATP	No-dd	Chase
dGTP	10.0 mM	10	10	10	10	10	140
dCTP	10.0 mM	10	10	10	10	10	140
dTTP	10.0 mM	10	10	10	10	10	140
[α-^{35}S]dATP	10.0 mM	5	5	5	5	5	
dATP	10.0 mM						140
ddGTP	1.0 mM	9.5					
ddCTP	1.0 mM		15				
ddTTP	1.0 mM			15			
ddATP	1.0 mM				6.25		
Tris-HCl (pH 8.3)	10.0 mM	351	346	346	355	361	833
Dithiothreitol	0.1 M	4	4	4	4	4	
$MgCl_2$	1.0 M						7

such as Seikagaku America Inc., St. Petersburg, Fla. Promega; and GIBCO-BRL. It is usually supplied at a concentration of approximately 20 U/μl. Prepare a working solution of 1 U/μl by dilution with a buffer consisting of 50 mM Tris-HCl (pH 8.3), 2 mM dithiothreitol, and 50% (wt/vol) glycerol. The diluted enzyme is stable for about 1 week at −20°C.

Terminal deoxynucleotidyltransferase (TdT). TdT can be obtained from Supertechs, Inc., Bethesda, Md. It is supplied at a concentration of 15 to 30 U/μl. Dilute the amount needed to 5 U/μl just before use; use 100 mM HEPES (pH 7.0) buffer for the dilution. Discard any unused diluted enzyme.

Labeled [α-^{35}S]dATP (specific activity, 1,500 Ci/mmol). This is available from several suppliers, such as Amersham, Chicago, Ill., and New England Nuclear, Boston, Mass.

5× hybridization buffer (low stringency): 250 mM Tris-HCl (pH 8.5), 500 mM KCl

5× hybridization buffer (high stringency): 250 mM Tris-HCl (pH 8.5), 100 mM KCl

5× reverse transcription buffer (low stringency): 250 mM Tris-HCl (pH 8.5), 50 mM dithiothreitol, 50 mM $MgCl_2$, 250 mM KCl

5× reverse transcription buffer (high stringency): 250 mM Tris-HCl (pH 8.5), 50 mM dithiothreitol, 50 mM $MgCl_2$

Table 5. PCR and sequencing primers for 16S rRNA and rRNA genes

Name	Sequence[a]
PCR primers	
27f	5′-AGA GTT TGA TCC TGG CTC AG>
27fB	5′-Biotin-AGA GTT TGA TCC TGG CTC AG>
1522r	5′-AAG GAG GTG ATC CA(AG) CCG CA>
1522rB	5′-Biotin-AAG GAG GTG ATC CA(AG) CCG CA>
1522rN	5′-CAT GCG GCC GCA AGG AGG TGA TCC A(AG)C CGC A>
*Not*I-T7-PC5[b] (1492r)	5′-GGG CGG CCG CAA TTT AAT ACG ACT CAC TAT AGG GAT ACC TTG TTA CGA CTT>
*Not*I-T7-P3mod[b] (787f)	5′-GGG CGG CCG CGC GCA ATT AAC CCT CAC TAA AGG GAA ATT AGA TAC CCT (ATG)GT AGT CC>
Forward-sequencing primers	
27f	5′-AGA GTT TGA TCC TGG CTC AG>
123f	5′-ACG GGT GAG TAA (CT)G(CT) GT>
357f	5′-CTC CTA CGG GAG GCA GCA G>
530f	5′-GTG CCA GCA GCC GCG G>
704f	5′-GTA GCG GTG AAA TGC GTA GA>
926f	5′-AAA CTC AAA GGA ATT GAC GG>
1114f	5′-GCA ACG AGC GCA ACC C>
1242f	5′-CAC ACG TGC TAC AAT GG>
1406f	5′-TGT ACA CAC CTC CCG TG>
Reverse-sequencing primers	
109r	5′-AC(AG) C(AG)T TAC TCA CCC GT>
321r	5′ AGT CTG GAC CGT GTC TCA GT>
519r	5′-G(AT)A TTA CCG CGG C(GT)G CTG>
685r	5′-TCT ACG CAT TTC ACC GCT AC>
907r	5′-CCG TCA ATT CCT TTG AGT TT>
1069r	5′-CCA ACA T(TC)T CAC A(AG)C ACG AG>
1100r	5′-GGG TTG CGC TCG TTG>
1220r	5′-TTG TAG CAC GTG TGT AGC CC>
1392r	5′-ACG GGC GGT GTG T(AG)C>
1522r	5′-AAG GAG GTG ATC CA(AG) CCG CA>

[a]Mixtures, e.g., (AG), are equimolar.

[b]From reference 17. Underlined portions represent 16S rRNA gene sequences.

Stop solution-loading dye: 86% formamide, 10 mM sodium EDTA (pH 7.4), 0.08% xylene cyanol, 0.08% bromophenol blue

Procedure

1. For each reaction set, place the following into a 0.5-ml microcentrifuge tube: 2.8 μl of 5× hybridization buffer (low stringency for routine sequencing), 8 μl of 2-μg/μl template rRNA (16S-23S mixture), and 3 μl of an oligonucleotide primer (Table 5). Heat in a 90°C heating block for 2 min, cool at room temperature for 10 min, and then place on ice until the next additions.
2. In another 0.5-ml microcentrifuge tube, dry 60 μCi of [α-^{35}S]dATP with a SpeedVac vacuum centrifuge (Savant, Farmingdale, N.Y.). Then add 12 μl of 5× reverse transcription buffer (low stringency for routine sequencing), 12 μl of the hybridization mix (step 1), and 12 μl of reverse transcriptase (1.0 U/μl). Gently mix the components by stirring with a pipette tip.
3. Transfer 6.0-μl amounts of the mixture prepared in step 2 to each of five 0.5-μl microcentrifuge tubes labeled C, A, T, G, and N. Add 4.0 μl of ddCTP, ddATP, ddTTP, ddGTP, and no-ddNTP mixes (Table 4) to the respective tubes. Incubate for 5 min at room temperature, and transfer to 37°C (low stringency) or 55°C (high stringency) for 30 min.
4. Heat the tubes in a 90°C heating block for 2 min, and centrifuge briefly in a microcentrifuge to force any condensate to the bottom of the tube. Into another set of tubes (for TdT chase reactions) transfer 5.0 μl from each reaction tube.
5. Add 1.5 μl of TdT-chase mixture (1 volume of dNTP chase mix [Table 4] mixed with 2 volumes of TdT solution [5 U/μl]) to each TdT-chase reaction tube. Incubate at 37°C for 30 min.
6. Add 6.0 μl of stop solution-loading dye to each of the 10 reaction tubes. Store at −20°C. Heat the tubes in a 90°C heating block for 1 min immediately before loading 1 to 2 μl onto a sequencing gel.

27.4.2. rRNA Gene Sequencing

DNA sequencing has been described in Chapter 20.1 and thus will not be discussed further in this chapter. However, ways of preparing template DNA for DNA sequencing are described in section 27.6.

27.4.3. Sequencing Gels

Details of the setting assembly and running of polyacrylamide sequencing gels can be found in Sambrook et al. (2). The buffer gradient gels (9) are probably the most commonly used, although wedge-shaped spacers are available for generating wedge gels (8, 51). The purpose of these gels is to reduce the migration rates of the shorter fragments so that more sequence can be read from a single gel. For further discussion of sequencing gels, see reference 40.

27.5. GENERATION OF rRNA GENE TEMPLATES BY PCR

The development of PCR for amplifying specific regions of DNA (44, 53) has allowed investigators to generate rRNA genes from the DNA of most organisms. The following protocol works well for many rRNA genes.

27.5.1. Symmetrical Amplification

Symmetrical amplification, which occurs when there is a large excess of both amplification primers, is the most efficient type of amplification. It should be used for testing DNA templates and when large amounts of double-stranded DNA are wanted, as, for example, when doing direct double-stranded DNA sequencing, using a biotin-labeled primer, or planning to generate single-stranded sequencing template by T7 gene 6 nuclease digestion.

Reagents

10× polymerase buffer: 500 mM KCl, 100 mM Tris-HCl (pH 8.8), 15 mM $MgCl_2$, 1.0% Triton X-100. The buffer is usually supplied with the DNA polymerase.

dNTPs. Prepare 10 mM stock solutions of dATP, dCTP, dGTP, and dTPP (Boehringer-Mannheim) in sterile water. Adjust the pH of each solution to 8.0 with NaOH (approximately 2 mole equivalents). Mix equal volumes of the four stock solutions for a working stock solution. Add 8.0 μl per 100 μl of reaction mixture to obtain a final concentration of 200 μM each dNTP.

Primers. Prepare a 40 μM stock solution of each primer in TE buffer. For primers 20 to 30 nucleotides long, the amount will range from 0.2 to 0.4 μg/μl. See Table 5 for amplification primers.

Taq polymerase, purified from the thermophilic bacterium *Thermus aquaticus* (no. N801-0046, Perkin-Elmer Cetus, Norwalk, Conn., 5 U/μl).

Template DNA: native high-molecular-weight DNA (see Chapter 26.2.1) at 0.1 μg/ml in TE buffer.

Procedure

The procedure described here is for 0.5-ml microcentrifuge tubes and for reaction volumes of 25 to 100 μl. It is easiest to prepare a large batch of the reaction mixture with everything except the template DNA, which is added after aliquots have been dispensed to the amplification tubes. If the amplified DNA is going to be sequenced directly, prepare 500 to 600 μl of the complete reaction mixture (Table 6) and dispense it in 100-μl portions into the amplification tubes.

Overlay 25- and 50-μl reaction volumes with 25 μl of mineral oil and 100-μl reaction volumes with 50 μl of the oil. Place about 25 μl of the oil in each well of the thermocycler. Position the reaction tubes in the thermocycler,

Table 6. Composition of complete PCR mixture

Component	Amt (μl/100 μl of reaction mixture)
Water	77
10× buffer	10
dNTP mixture	8
Primer 1	1.25
Primer 2	1.25
Taq polymerase	0.5
Template DNA (100 ng/μl)	2.0

pushing each down firmly so that good contact is made with the aluminum block. Place a piece of 0.25-in. (0.64-cm)-thick neoprene on top of the tubes, and then a piece of plastic and a weight (ca. 1-2 kg) on top of that. This tends to keep the tubes from working up during the thermal cycles and losing contact with the heating block. For the first thermal cycle, heat to 96°C for 1.5 min, cool to 48°C for 0.5 min for annealing, and then heat to 76°C for 4.0 min for primer elongation. For the remaining cycles, reduce the denaturation temperature to 95°C for periods of 0.5 min and use annealing times of 15 s. Perform a total of 35 to 40 cycles.

The amount of DNA amplification can be quickly determined by electrophoresing a 5- to 10-μl sample of the reaction mixture on a 1% agarose gel (section 27.3.1.3). For cleaning up the reaction product (removing proteins, nucleotides, and salts), extract the sample once with phenol-$CHCl_3$ and precipitate the DNA with 0.6 volume of isopropanol. Wash the DNA pellet with 80% ethanol, dry it, and redissolve it in TE buffer.

Comments

The amplification conditions described above have worked well for a wide range of bacterial DNAs. There are several important parameters to keep in mind. (i) Keeping the template DNA (200 ng/100 μl of PCR reaction mixture) and primer (0.5 μM) concentrations rather low reduces the amount of nonspecific annealing during the initial PCR cycles. (ii) The advantage of using native high-molecular-weight DNA as the template is that there is essentially no nonspecific annealing and elongation while the reaction mixture is being prepared. Single-stranded template is not available until after the first denaturation step, and after that, only specific annealing conditions occur during the annealing step of each temperature cycle. (iii) The annealing temperature should be kept high enough to maintain annealing specificity. If a DNA preparation is not amplified, the first variable to change would be to lower the annealing temperature by 3 to 5°C.

27.5.2. Asymmetrical Amplification

The principle in asymmetrical amplification is that a lower concentration of one of the PCR primers results in less amplification of the corresponding DNA strand (6, 22, 26, 29). Experimental conditions are similar to those for the symmetrical amplification above, except that the concentrations of excess primer and the limiting primer are 1.0 and 0.02 μM, respectively. It is not as easy to evaluate the asymmetrical amplification, because the single-stranded DNA does not stain as intensely with ethidium bromide. As a result, one just has to test the asymmetrical amplification product as a sequencing template.

An alternative approach is to do a two-step amplification (39). First do a symmetrical amplification with a 0.25 μM concentration of each primer (35 to 40 cycles), and then transfer part of that reaction mixture (2 to 10 μl per 100 μl) to a second PCR (25 cycles) in which only one of the primers is included. The advantage of this is that the first amplification is improved and one can adjust the amount of material from the first PCR to add to the second PCR to get a consistent amount of single-stranded DNA product.

As a variation of the two-step PCR, the product from the symmetrical amplification can be electrophoresed on an agarose gel (39, 75). The product of interest can then be excised from the agarose and melted, and some of the melted agarose can be added to a standard asymmetrical reaction mixture. Load approximately 25 μl of the symmetrically amplified material per lane of low-melting-point agarose. After electrophoresis, excise the nearly full length rRNA band, place it into a microcentrifuge tube, and heat the tube in a 65°C water bath for 10 to 15 min. Add 10 μl of the melted agarose per 100 μl of asymmetrical amplification reaction. Thirty-five amplification cycles give good results.

27.5.3. Secondary Transcription of RNA

The use of transcript sequencing was introduced by Stoflet et al. (63). Frothingham et al. (17) incorporated a T3 RNA polymerase promoter in one PCR primer and a T7 RNA polymerase promoter in the other (Table 5). The PCR product was then used as a template for transcribing RNA from either one strand or the other. The resulting RNA preparations were used as sequencing templates.

27.6. PREPARATION OF SEQUENCING TEMPLATES FROM PCR PRODUCTS

The PCR products can either be sequenced directly or subcloned into M13 phage or a plasmid for use in the standard sequencing protocols (Chapter 20.1).

27.6.1. Purification and Concentration of PCR Products

The pooled product from three to six PCRs (300 to 600 μl) is usually sufficient for 15 to 20 direct DNA sequencing reaction sets.

Reagents and materials

TE buffer: 10 mM Tris-HCl, 1 mM sodium EDTA (pH 8.0)
Phenol-$CHCl_3$: water-saturated phenol mixed with an equal volume of $CHCl_3$ that contains 3% (vol/vol) isopentanol. Add 0.1% (wt/vol) hydroxyquinoline to the mixture.
Sodium acetate, 3 M (pH 6.0)
Microconcentrator: Centricon-100 (Amicon Division, W. R. Grace and Co., Beverly, Mass.)

Procedure

1. Pool the PCR products into a 1.5-ml microcentrifuge tube, add 200 μl of phenol-$CHCl_3$, mix with a vortex mixer, and centrifuge for 10 min.
2. Transfer the supernatant to a clean tube, and precipitate the DNA with 2 volumes of 95% ethanol. Wash the pellet with 500 μl of cold 80% ethanol. Dry the pellet, and then dissolve the DNA in 200 μl of TE buffer.
3. Following the manufacturer's instructions about centrifugation time and speed, wash a Centricon-100 unit with 1.0 ml of TE and then add the DNA plus 1.0 ml of TE. Centrifuge the unit until most of the sample has passed through the filter (usually about 30 to 40 min). Then add another 1.0 ml of TE, and repeat the centrifu-

gation. Repeat the 1.0-ml TE addition and centrifugation steps twice more. Place the collection cap on the upper portion of the unit, invert it, and centrifuge it briefly to collect the sample in the cap (usually about 100 μl). The DNA is now virtually free of PCR primers.

4. Transfer the purified DNA to a clean microcentrifuge tube, rinse the filtration unit by adding amounts of TE, and pool up to a volume of approximately 360 μl. Add 40 μl of 3 M sodium acetate (pH 6.0) and 800 μl of 95% ethanol, mix, cool on ice or in a refrigerator for 10 min, and centrifuge. Wash the pellet with 70% ethanol, air dry, and dissolve the DNA in 75 μl of TE buffer. Store at −20°C. A sample can be electrophoresed on an agarose gel to estimate the DNA concentration.

Comments

There are various commercial products designed for separating the amplified DNA from the primers, such as NACS Prepac Cartridge (no. 1526NP, GIBCO-BRL) and CHROMA SPIN-400 (no. K1303, Clontech Labs, Palo Alto, Calif.). Most of these products will work perfectly well for this purification step. Being a filtration device, the Centricon-100 does not have DNA capacity limits, which may be a disadvantage of some of the column devices. However, the primers are removed by simple dilution, and the volume of residual solution at the end of each centrifugation step will affect the efficiency of primer removal. Standard ethanol precipitation of the PCR product also removes a significant amount of primers, so that a combination of ethanol precipitation and another method (filtration, spin columns, or adsorption columns) is quite effective.

If the equipment is available, high-performance liquid chromatography (HPLC) is also an effective method for removing the PCR primers (33, 70). At step 2 in the procedure above, dissolve the amplified DNA in 50 to 100 μl of TE for loading directly on an HPLC column.

27.6.2. Direct Sequencing of PCR Products

The DNA sequencing template preparations described in this section are for use with Sequenase (United States Biochemical Corp.) DNA polymerase. The following sequencing protocols call for a 12-μl volume of primer-annealed template. This volume can be altered for other protocols. For a given protocol, it is prudent to optimize the amounts of both the template and primer.

27.6.2.1. Double-stranded DNA sequencing

The major problem with sequencing a double-stranded DNA fragment is the rapid reannealing of the strands during the sequencing primer annealing step. The following protocol for generating sequencing primer-annealed template reduces this problem (56, 73).

Reagents

Sequenase 5×sequencing buffer: 200 mM Tris-HCl (pH 7.5), 50 mM $MgCl_2$, 250 mM NaCl

Sequencing primers (Table 5). Prepare working solutions of 5 or 15 ng/μl of TE buffer.

Procedure

1. To a 1.5-ml microcentrifuge tube, add 1.0 μg of symmetrically amplified and purified DNA, 45 ng of sequencing primer, 2.4 μl of 5× Sequenase sequencing buffer, and sufficient sterile water to give a final volume of 12 μl.

2. Heat the tube in a boiling-water bath for 5 min, and then place the tube immediately into a dry ice-ethanol bath. Let the tube warm up to room temperature, maintain it at room temperature for 5 min (centrifuge for a few seconds to bring down any condensate), and then proceed with the sequencing reaction.

Comments

We have had average success with this protocol. The length and accuracy of sequencing are not as good as with, for example, the T7 gene 6 nuclease method (section 27.7.1.3). The use of heat-stable polymerases is another approach for direct sequencing of PCR product (43, 68).

27.6.2.2. Sequencing of asymmetrical PCR product

The amount of single-stranded DNA is difficult to estimate by standard ethidium bromide staining in agarose gels, because of a smaller amount of dye intercalation than with double-stranded DNA. Consequently, one must assume that there will be sufficient template and must arbitrarily determine the volume of template needed. A 300-μl portion of single-stranded PCR volume (section 27.6.2) that has been purified (section 27.7.1) and dissolved in 75 μl of TE buffer will provide sufficient template for 10 to 20 DNA sequencing reactions. Try 5 μl of template using a single sequencing primer. From this, one can arbitrarily determine whether more or less template should be used in subsequent sequencing reactions.

1. To a 1.5-ml microcentrifuge tube, add 5.9 μl of asymmetrically amplified and purified DNA, 15 ng of sequencing primer, 2.4 μl of 5× Sequenase sequencing buffer, and sufficient sterile water to give a final volume of 12 μl.

2. Heat the tube in a boiling-water bath for 5 min. Remove the tube, and place it in a 37°C water bath for 10 min. Then proceed with the sequencing reaction.

27.6.2.3. Digestion by bacteriophage exonuclease

Bacteriophage T7 gene 6 exonuclease is specific for double-stranded DNA. The enzyme degrades from the 5′ end, resulting in two half-length single strands (34, 35) as illustrated in Fig. 2. The exonuclease is not a processive enzyme: it dissociates frequently from the substrate during the course of digestion (66). The enzyme

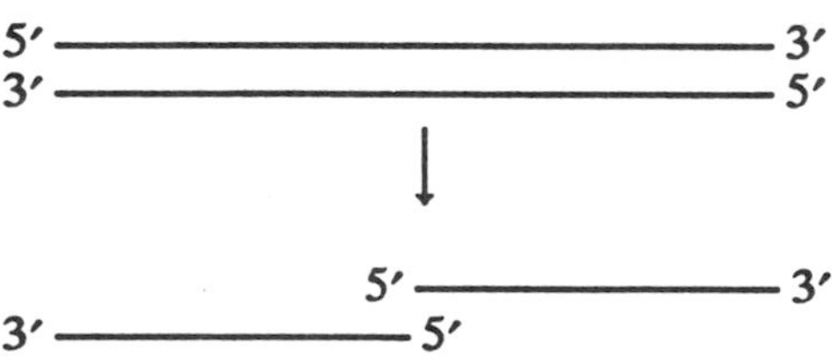

FIG. 2. Digestion by bacteriophage T7 gene 6 exonuclease. This enzyme degrades from the 5′ end, resulting in two half-length single strands.

has been used to generate single-stranded sequencing templates from plasmids (57), bacteriophage λ, and PCR product (52).

Reagents

T7 gene 6 exonuclease, 80 U/μl (United States Biochemical Corp.)
PCR product, prepared by symmetrical amplification and purified as described in section 27.7.1
10× gene 6 buffer: 500 mM Tris-HCl (pH 8.0), 200 mM KCl, 50 mM $MgCl_2$, 10 mM 2-mercaptoethanol
5× Sequenase buffer (United States Biochemical Corp.)
EDTA, 0.5 M (pH 8.0)
Glycogen: 20 μg/μl (no. 901 393; Boehringer Mannheim). This is an effective coprecipitant for precipitating small amounts of nucleic acid (67).
Sodium acetate, 3 M (pH 6.0)

Procedure

The protocol is written for 10 to 12 μg of symmetrically amplified and purified 16S rRNA gene DNA. There will be enough digested DNA for 10 to 12 DNA sequencing reactions.

1. To a 1.5-ml microcentrifuge tube, add 10 to 12 μg of DNA, 10 μl of 10× gene 6 buffer, water to 99 μl (total volume), and 1 μl of T7 gene 6 nuclease. Incubate at 37°C for 15 min. Add 8.0 μl of 0.5 M EDTA, mix, and heat in a 70°C water bath for 15 min.

2. Add 10 μl of 3.0 M sodium acetate and 1.0 μl of glycogen. Mix. Add 220 μl of 95% ethanol, mix, let stand on ice for 5 min, then centrifuge at high speed in a microcentrifuge for 15 min. Wash the pellet with 250 μl of 80% ethanol. Dissolve the single-stranded DNA in 20 μl of TE buffer, and use 1.0 to 3.0 μl per DNA sequencing reaction.

The following T7 gene 6 exonuclease protocol can be used for a single sequencing reaction which does not require precipitation of the digested DNA.

1. Use approximately 1.0 μg of PCR product for each sequencing reaction. To a 1.5-ml microcentrifuge tube, add 1.0 μg of DNA, 1.8 μl of 5× Sequenase buffer, and sufficient sterile water to give a final volume of 8.0 μl. Then add 1.0 μl of T7 gene 6 exonuclease (diluted to 5 U/μl), mix, and incubate at 37°C for 15 min. Stop the reaction by heating at 75°C for 15 min.

2. Briefly centrifuge the tube to collect any condensate, add 3 μl of sequence primer (5 ng/μl), and anneal at 37°C for 10 min. Continue with the DNA sequencing reaction.

Comments

T7 gene 6 exonuclease digestion results in an efficient single-stranded DNA template. The disadvantages of the procedure are that one is limited to sequencing the 5′ end of the 16S rRNA gene with the forward-sequencing primers and the 3′ end with reverse primers, whereas the middle of the DNA fragment cannot be sequenced well with either forward or reverse primers (Table 5). One can use either double-stranded sequencing (section 27.7.1.1) or asymmetrically amplified template (section 27.7.1.2) for sequencing the middle region of the fragment. Another approach would be to use one of the forward or reverse primers near the center of the 16S gene as PCR primers. This would move the center of the 16S gene to one end or the other of a smaller fragment, and again this would allow the use of the gene 6 nuclease.

Another 5′→3′ exonuclease that has been used for preparing single-stranded DNA sequencing templates has been obtained from bacteriophage lambda (25). This exonuclease initiates degradation much more efficiently when there is a 5′-phosphate group on the DNA substrate. As a result, one can obtain selective degradation of one strand or the other by using a 5′-phosphorylated PCR primer along with one that is not phosphorylated.

27.6.2.4. Magnetic beads for separating DNA strands

The sequencing of single-stranded DNA works better than that of double-stranded DNA, particularly when the double-stranded DNA is a fragment and not a covalently closed plasmid. Reassociation of the template DNA competes with annealing of the primer. A recent approach for separating the strands of PCR-generated DNA has been to attach biotin at the 5′ end of one of the PCR primers and then react the PCR-generated DNA with magnetic beads containing covalently coupled streptavidin (28, 61). The biotin will bind irreversibly with the streptavidin on the beads, and the complementary strand is then denatured with alkali and washed away from the beads. DNA sequencing can be done directly on the DNA that is bound to the beads. The complementary DNA strand, after being precipitated, can also be used as a sequencing template.

Reagents

Magnetic beads: Dynabeads M280-Streptavidin (Dynal AS, Oslo, Norway)
TE buffer: 10 mM Tris-HCl containing 1 mM Na-EDTA (pH 7.5)
Wash buffer: TE buffer containing 100 mM NaCl and 0.17% Triton X-100
NaOH, 0.15 M solution

Procedure

1. Perform the standard PCR amplification as indicated in section 27.6.1; however, it may be helpful to decrease the concentration of the biotin labeled primer to 0.25 μM. Also, the PCR mixture need not be extracted with phenol-$CHCl_3$.

2. Resuspend the Dynabeads. Then, for each sequencing reaction, place 15 μl into a 1.5-ml microcentrifuge tube. Wash the beads three times with 2 volumes of the wash buffer. Hold a magnet next to the tube to hold the beads against the wall while removing the wash buffers. Suspend the beads in wash buffer to their original volume.

3. Transfer 20 μl of the PCR product and 15 μl of the washed Dynabeads to a 1.5-ml microcentrifuge tube, gently mix, and incubate at 37°C for 30 min. Collect the beads at the side of the tube with a magnet, and remove the supernatant. Wash the beads with 100 μl of wash buffer, and then remove the buffer.

4. Resuspend the beads in 16 μl of distilled water, and add 4 μl of 0.5 M NaOH–2 mM Na-EDTA. Incubate at room temperature for 5 min.

5. Collect the beads at the side of the tube with a magnet, and transfer the NaOH supernatant, which contains the nonbiotinated strand, to another tube. Add 100 μl of wash buffer to the beads. Remove the buffer, and add the sequencing components.

6. Precipitate the nonbiotinated strand by adding 2 μl of 3 M sodium acetate (pH 6.0) and 22 μl of cold isopropanol. Centrifuge for 10 min. Remove the supernatant, wash the pellet with 80% ethanol, and dry the pellet for 5 min. Add the sequencing components.

27.6.3. Cloning into Bacteriophage M13

Another approach to generating DNA sequencing template is to clone the PCR fragments into bacteriophage M13 (74). The PCR primers (Table 5) may include 27f + 1522r for blunt-ended ligation and cloning or 27f + 1522rN or 27fN + 1522r for restriction endonuclease *Not*I-directed cloning. The following is a protocol for either blunt-ended or directed cloning.

Reagents and materials

Bacteriophage M13 with *Not*I site. A *Not*I site can be placed into M13mp18 or M13mp19 by ligating in a *Not*I linker or by cloning the *Eco*RI-*Hin*dIII multiple cloning region from plasmid pSPORT 1 (GIBCO-BRL) into *Eco*RI-*Hin*dIII-digested M13mp18 or M13mp19.

Klenow fragment of *E. coli* DNA polymerase I, 3 to 9 U/μl (GIBCO BRL).

Procedure

1. After a standard symmetrical amplification (50 or 100 μl), remove the reaction mixture from under the oil with a micropipette and place it into a clean tube. Add 1.0 U of the Klenow fragment of *E. coli* DNA polymerase I to the reaction mixture, and incubate it at 37°C for 30 min.

2. Extract the reaction mixture with phenol-$CHCl_3$, add 0.5 volume of 7.5 M ammonium acetate, and precipitate with 2 volumes of 95% ethanol. After centrifugation, wash the pellet with 80% ethanol, dry it, and then dissolve the DNA in 25 μl of TE buffer for blunt ligation or in 10 μl of TE for *Not*I digestion.

3. For *Not*I digestion, add 2.5 μl of 10× REact 7 buffer (GIBCO BRL), 11.5 μl of sterile water, and 1 μl (approximately 10 U) of enzyme. Incubate at 37°C for 2 to 4 h. Inactivate the enzyme by heating the tube at 65°C for 15 min.

4. For blunt-ended ligation, use an M13 vector digested with *Hin*cII (do not treat with alkaline phosphatase) and a two- to five-fold molar excess of the PCR fragment. Use standard M13 transformation procedures (2), and select clear plaques. The two fragment orientations can be determined by complement hybridization of the phage DNAs (2).

5. For directed cloning, use the modified M13 vector which contains a *Not*I site. First, digest two samples of the replicative form DNA, one with *Eco*RI and the other with *Hin*dIII, and then fill in the recessed 3′ termini with the Klenow fragment of *E. coli* DNA polymerase I and the four dNTPs (2). Digest each replicative-form DNA sample with *Not*I restriction enzyme. In one vector digest the rRNA gene will be cloned in one orientation; in the other digest it will be cloned in the opposite orientation.

Comments

Since there are no 5′ phosphates on the PCR primers, one cannot use phosphatase-treated vector for the blunt-ended ligations. This will result in a high background of plaques containing religated vector (blue plaques), with a smaller fraction of plaques containing the rRNA DNA insert (colorless plaques). One may have to test a range of insert DNA to vector DNA ratios in the ligation reaction to optimize insert incorporation.

With the modified M13 vector, there is no plaque color assay for identifying transformants containing the insert. However, if the digestions were complete, the number of transformants resulting from the vector-only ligation control should be quite small and the efficiency of the insert ligation high, so most of the plaques should represent insert-containing phage.

If there is a high background for the double-digested M13 vector control, this indicates that one of the restriction endonuclease digestions was not complete or that the small fragment released by the double digestion is being religated into the vector. Undigested vector DNA and the small fragment can be separated from the correctly digested DNA by electrophoresis of the final vector preparation on a low-melting-point agarose gel and excising the DNA band representing double-digested DNA.

The major disadvantage of this sequencing approach is that single amplified fragments are being selected for sequencing and thus an amplification reading error may be selected. In plasmid cloning, this problem has been circumvented by using the tranformation mixture as a broth culture inoculum (47). This results in a plasmid population containing a random cross-section of PCR-generated fragments. The PCR primers or internal primers can be used for sequencing. The same approach can be used for the directed cloning into M13, where part of the transformation mixture is grown in a broth culture. One would lose the advantage of single-stranded DNA sequencing using the nondirected ligation and transformation reactions.

27.7. SIMILARITY ANALYSIS OF rRNA GENES

Methods for measuring the sequence similarities of rRNAs and rRNA genes have been outlined in this chapter. An important consideration is how the results from the different procedures correlate. Although nucleotide sequencing results in a single pattern, results of the hybridization procedures will vary depending on the procedure and the reaction conditions under which it was carried out.

27.7.1. Base Sequence Analysis

Conceptually, comparing sequence A with sequence B is quite simple: the number of matching bases divided by the number of bases compared is the fraction that are similar. In practice, however, there are complications. The major problem is determining when a base mismatch is the result of a base change and not the result of a base insertion or a base deletion. To minimize the deletion/insertion effect, the conserved regions of the sequences have to be aligned. The test sequences are usually aligned with respect to the sequence of *Escherichia coli* 16S rRNA. Many sequence analysis program packages contain programs for aligning sequences, for example, the GAP and LINEUP programs in the Genetics Computer Group Sequence Analysis

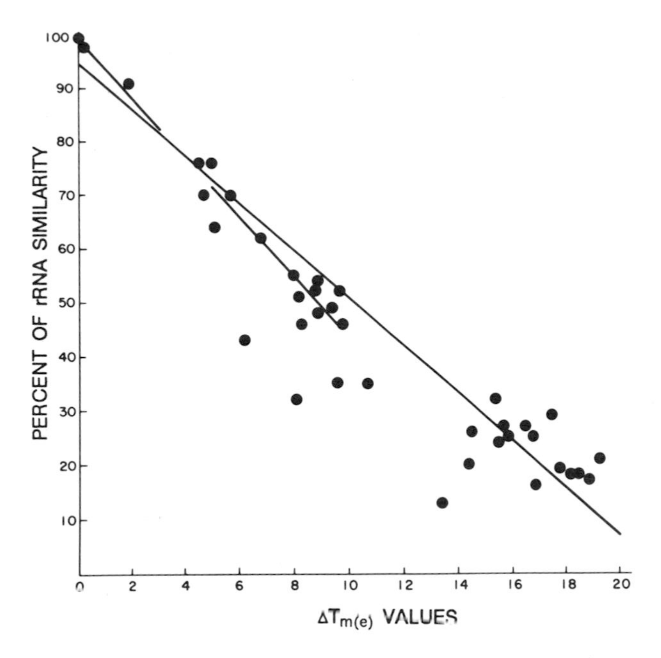

FIG. 3. Correlations between $\Delta T_{m(e)}$ and percent similarity values (31). The linear regression lines are for 50 to 100% similarity (short line) and 20 to 100% similarity values, respectively.

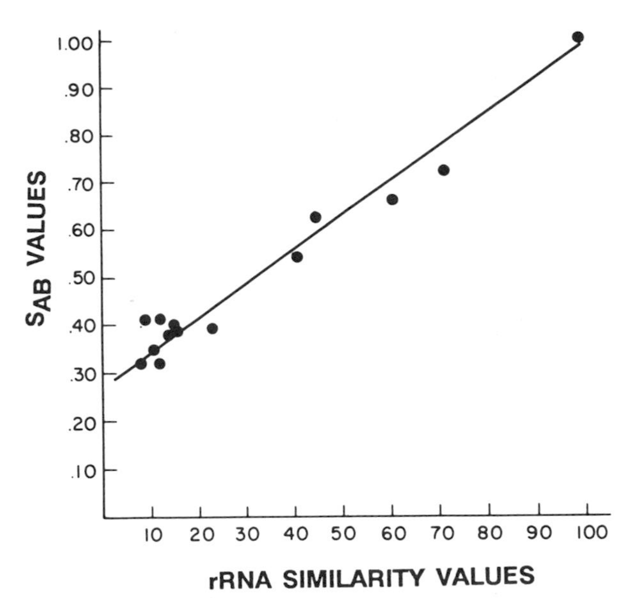

FIG. 4. Correlation between percent similarity values and S_{AB} values (31).

Software Package from the University of Wisconsin Biotechnology Center. C. R. Woese and G. J. Olsen, Principal Investigators for the Ribosomal Database Project at the University of Illinois, have a sequence editor program that has been used extensively for aligning rRNA sequences.

27.7.2. Correlations between Methods

The rRNA hybridization procedures tend to correlate quite well; results are very similar whether 16S, 23S, or a mixture of 16S and 23S rRNA is used in the experiments (31). Correlations between percent similarity values and thermal stability differences are illustrated in Fig. 3. Also, there is high correlation between hybridization and thermal stability experiments (31). The correlation between sequence similarity values and oligonucleotide similarity (S_{AB}) values is illustrated in Fig. 4 (31, 65). The correlation is reasonably good when the S_{AB} similarity values are above 0.4. At S_{AB} values less than 0.4, the hybridization similarity values tend to cluster in the 10 to 20% range. Cluster analysis of the two data sets results in similar clustering patterns (Fig. 5). Correlations between S_{AB} values and nucleotide sequence similarity values for 16S rRNA are illustrated in Fig. 6 (4). It is interesting that S_{AB} values of about 0.4 correspond to about 85% nucleotide sequence similarity values—values below which little hybridization would occur in the classical hybridization experiments. However, the highly conserved regions will always result in some hybridization.

27.7.3. Phylogenetic Relationships

The goal in most of the rRNA sequence similarity studies is to delineate phylogenetic relationships among the organisms being investigated. Since data analysis at this level is beyond the scope of this chapter, the reader is referred to several recent publications (45, 46, 49, 64) covering phylogenetic analysis. Some of the available software packages have been listed by Swofford and Olsen (64).

The major center for 16S rRNA sequence analysis

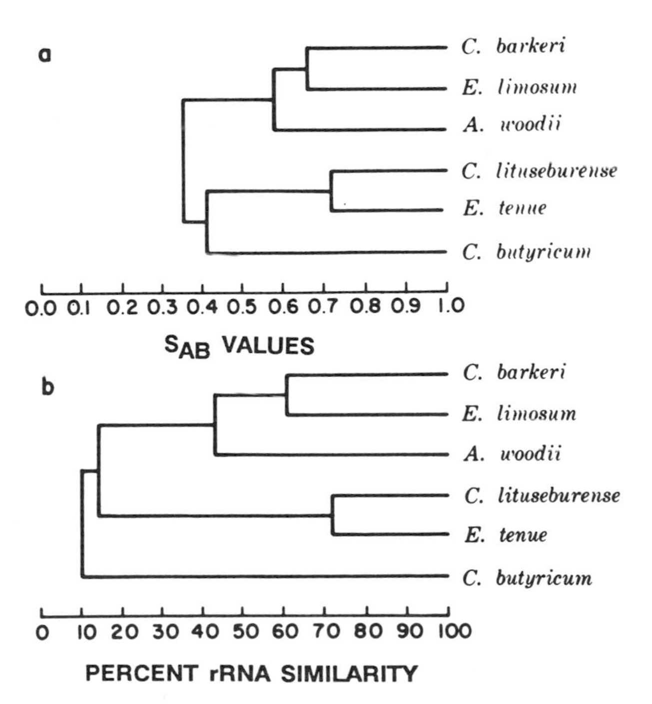

FIG. 5. Cluster analysis comparisons using percent rRNA similarity values and S_{AB} values (31).

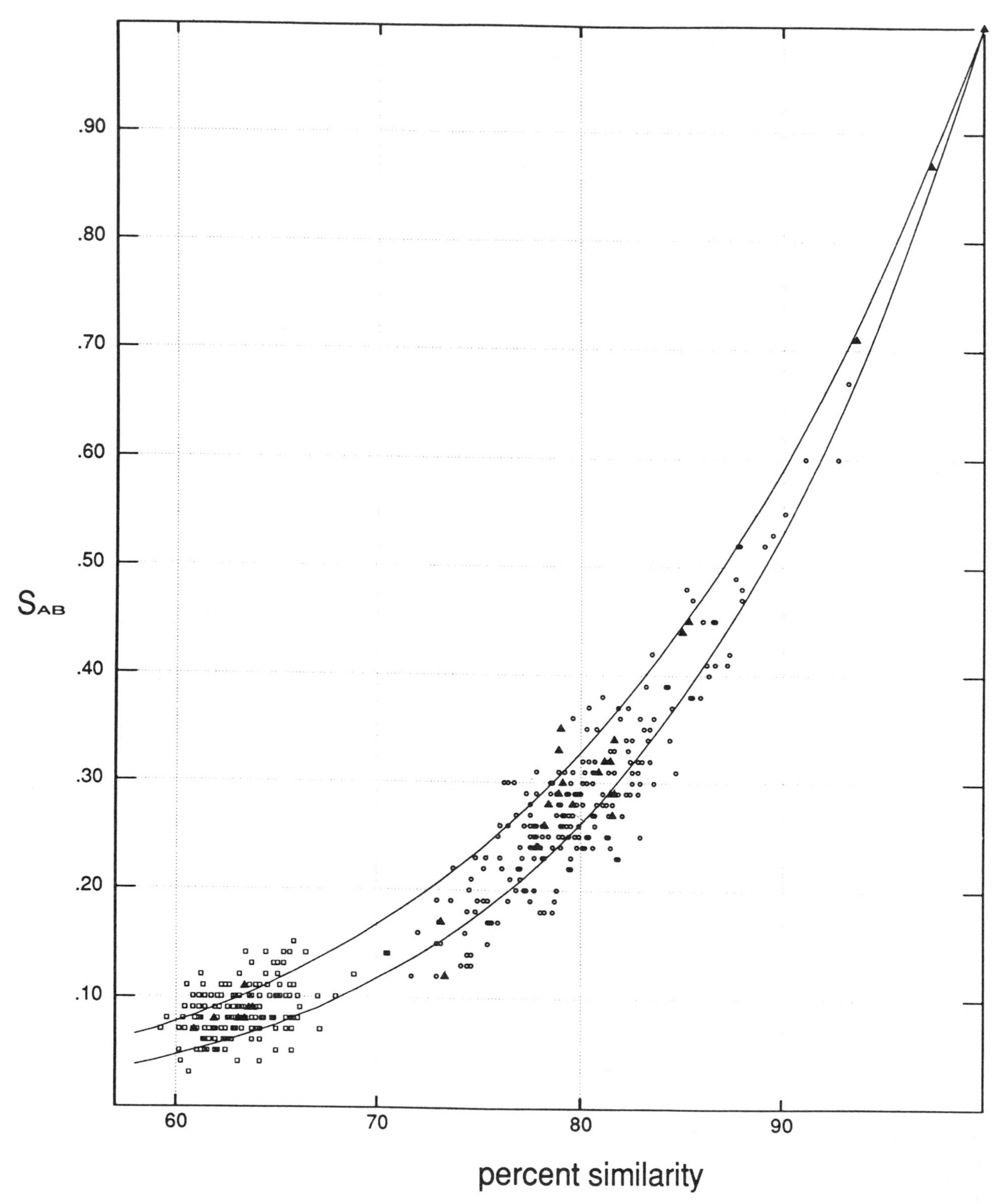

FIG. 6. Correlation between percent similarity values and S_{AB} values (4).

has been at the University of Illinois, where C. R. Woese and his colleagues have developed many of the procedures and collected many sequences. Through the Ribosomal Database Project, over 400 sequences are currently available in an aligned format. Other services that Woese and his colleagues may provide through the project are alignment and phylogenetic placement of 16S rRNA sequences.

27.8. REFERENCES

27.8.1. General References

1. **Johnson, J. L.** 1989. Nucleic acid hybridization: principles and techniques, p. 3–31. *In* B. Swaminathan and G. Prakash (ed.), *Nucleic Acids and Monoclonal Antibody Probes.* Marcel Dekker, Inc., New York.
Reviews reassociation and hybridization kinetics, effects of experimental variable, and duplex and hybrid specificities.
2. **Sambrook, J., E. F. Fritsch, and T. Maniatis.** 1989. *Molecular Cloning: A Laboratory Manual,* 2nd ed. Cold Spring Harbor Laboratory Press, Cold Spring Harbor, N.Y.
This large manual (it is published as three sequential volumes) includes protocols for nearly all molecular biology techniques.
3. **Stackebrandt, E., and M. Goodfellow (ed.).** 1991. *Nucleic Acid Techniques in Bacterial Systematics.* John Wiley & Sons Ltd., Chichester, England.
Provides an extensive range of molecular methods for studying bacterial classification by the techniques of molecular biology.
4. **Woese, C. R.** 1987. Bacterial evolution. *Microbiol. Rev.* **51:**221–271.
An extensive review on the use of 16S rRNA as a chronometer for measuring bacterial evolution. Outlines major groups delineated by this approach.

27.8.2. Specific References

5. **Adam, T., U. Göbel, and W. Bredt.** 1990. Effects of growth phase and antibiotics on quantitative DNA/RNA hybridization. *Mol. Cell. Probes* **4:**375–383.
6. **Aguilar-Cordova, E., and M. W. Lieberman.** 1991. Direct directional sequencing of PCR-amplified genomic DNA. *BioTechniques* **11:**62–65.
7. **Baharaeen, S., U. Melcher, and H. S. Vishniac.** 1983. Complementary DNA-25S ribosomal RNA hybridization: an improved method for phylogenetic studies. *Can. J. Microbiol.* **29:**546–551.
8. **Bankier, A. T., and B. G. Barrell.** 1983. Shotgun DNA sequencing, p. 1–34. *In* R. A. Flavel (ed.), *Techniques in the Life Sciences: Techniques in Nucleic Acid Biochemistry,* vol. B5. Elsevier, Limerick, Ireland.
9. **Biggin, M. D., T. J. Gibson, and G. F. Hong.** 1983. Buffer gradient gels and ^{35}S label as an aid to rapid DNA sequence determination. *Proc. Natl. Acad. Sci. USA* **80:**3963–3965.
10. **Brownlee, G. G., and E. M. Cartwright.** 1977. Rapid gel sequencing of RNA by primed synthesis with reverse transcriptase. *J. Mol. Biol.* **114:**93–117.
11. **Chirgwin, J. M., A. E. Przybyla, R. J. MacDonald, and W. J. Rutter.** 1979. Isolation of biologically active ribonucleic acid from sources enriched in ribonuclease. *Biochemistry* **18:**5294–5299.
12. **De Ley, J., and J. De Smedt.** 1975. Improvements of the membrane filter method for DNA:rRNA hybridization. *Antonie Leeuwenhoek J. Microbiol. Serol.* **41:**287–307.
13. **Denhardt, D. T.** 1966. A membrane-filter technique for the detection of complementary DNA. *Biochem. Biophys. Res. Commun.* **23:**641–646.
14. **Donis-Keller, H., A. M. Maxam, and W. Gilbert.** 1977. Mapping adenine, guanines, and pyrimidines in RNA. *Nucleic Acids Res.* **4:**2527–2537.
15. **Engel, J. D., and N. Davidson.** 1978. Addition of poly(adenylic acid) to RNA using polynucleotide phosphorylase: an improved method for electron microscopic visualization of RNA-DNA hybrids. *Biochemistry* **17:**3883–3888.
16. **Fox, G. E., L. J. Magrum, W. E. Balch, R. S. Wolfe, and C. R. Woese.** 1977. Classification of methanogenic bacteria by 16s ribosomal RNA characterization. *Proc. Natl. Acad. Sci. USA* **74:**4537–4541.
17. **Frothingham, R., R. L. Allen, and K. H. Wilson.** 1991. Rapid 16S ribosomal DNA sequencing from a single colony without DNA extraction or purification. *BioTechniques* **11:**40–44.
18. **Fry, M., and M. Artman.** 1969. Deoxyribonucleic acid-ribonucleic acid hybridization. *Biochem. J.* **115:**287–294.
19. **Fuhrman, J. A., D. E. Comeau, A. Hagstrom, and A. M. Chan.** 1988. Extraction from natural planctonic microorganisms of DNA suitable for molecular biological studies. *Appl. Environ. Microbiol.* **54:**1426–1429.
20. **Gillespie, D., and S. Spiegelman.** 1966. A quantitative assay for DNA-RNA hybrids with DNA immobilized on a membrane. *J. Mol. Biol.* **12:**829–842.
21. **Gillespie, S., and D. Gillespie.** 1971. Ribonucleic acid-deoxyribonucleic acid hybridizations in aqueous solutions and in solutions containing formamide. *Biochem. J.* **125:**481–487.
22. **Gyllensten, U. B., and H. A. Erlich.** 1988. Generation of single stranded DNA by the polymerase chain reaction and its application to direct sequencing of the HLA-DQalpha locus. *Proc. Natl. Acad. Sci. USA* **85:**7652–7656.
23. **Hahn, D., R. Kester, M. J. C. Starrenburg, and A. D. L. Akkermans.** 1990. Extraction of ribosomal RNA from soil for detection of *Frankia* with oligonucleotide probes. *Arch. Microbiol.* **154:**329–335.
24. **Hassur, S. M., and H. W. Whitlock.** 1974. UV shadowing—a new and convenient method for location of UV absorbing species in polyacrylamide gels. *Anal. Biochem.* **59:**162–164.
25. **Higuchi, R. G., and H. Ochman.** 1989. Production of single-stranded DNA templates by exonuclease digestion following the ploymerase chain reaction. *Nucleic Acids Res.* **17:**5865.
26. **Hofmann, M. A., and D. A. Brian.** 1991. Sequencing PCR DNA amplified directly from a bacterial colony. *BioTechniques* **11:**30–31.
27. **Holben, W. E., J. K. Jansson, B. K. Chelm, and J. M. Tiedje.** 1988. DNA probe method for the detection of specific microorganisms in the soil bacterial community. *Appl. Environ. Microbiol.* **54:**703–711.
28. **Hultman, T., S. Ståhl, E. Hornes, and M. Uhlén.** 1989. Direct solid phase sequencing of genomic and plasmid DNA using magnetic beads as solid support. *Nucleic Acids Res.* **17:**4937–4946.
29. **Innis, M. A., K. B. Myambo, D. H. Gelfand, and M. A. D. Brow.** 1988. DNA sequencing with *Thermus aquaticus* DNA polymerase and direct sequencing of polymerase chain reaction-amplified DNA. *Proc. Natl. Acad. Sci. USA* **85:**9436–9440.
30. **Johnson, J. L., and B. S. Francis.** 1975. Taxonomy of the clostridia: ribosomal ribonucleic acid homologies among the species. *J. Gen. Microbiol.* **88:**229–244.
31. **Johnson, J. L., and B. Harich.** 1983. Comparisons of procedures for determining ribosomal ribonucleic acid similarities. *Curr. Microbiol.* **9:**111–120.
32. **Johnson, J. L., and B. Harich.** 1986. Ribosomal ribonucleic acid homology among species of the genes *Bacteroides. Int. J. Syst. Bacteriol.* **36:**71–79.
33. **Kalnoski, M. H., M. F. McCoy-Haman, and G. F. Hollis.** 1991. A rapid method for purifying PCR products for direct sequence analysis. *BioTechniques* **11:**246–249.
34. **Kerr, C., and P. D. Sadowski.** 1972. Gene 6 exonuclease of bacteriophage T7: purification and properties of the enzyme. *J. Biol. Chem.* **247:**305–310.
35. **Kerr, C., and P. D. Sadowski.** 1972. Gene 6 exonuclease of bacteriophage T7. II. Mechanism of the reaction. *J. Biol. Chem.* **247:**311–318.
36. **Kilpper-Bälz, R.** 1991. DNA-rRNA hybridization, p. 45–68. *In* E. Stackebrandt and M. Goodfellow (ed.), *Nucleic Acid Techniques in Bacterial Systematics.* John Wiley & Sons Ltd., Chichester, England.
37. **Kirby, K. S.** 1965. Isolation and characterization of ribosomal ribonucleic acid. *Biochem. J.* **96:**266–269.
38. **Klenk, H. P., B. Haas, V. Schwass, and W. Zillig.** 1986. Hybridization homology: a new parameter for the analysis of

phylogenetic relations, demonstrated with the urkingdom of the Archaebacteria. *J. Mol. Evol.* **24:**167–173.
39. **Kocher, T. D., W. K. Thomas, A. Meyer, S. V. Edwards, and S. Pääbo.** 1990. Dynamics of mitochondrial DNA in animals: amplication and sequencing with conserved primers. *Proc. Natl. Acad. Sci. USA* **86:**6196–6200.
40. **Lane, D. J.** 1991. 16S/23S rRNA sequencing, p. 115–175. *In* E. Stackebrandt and M. Goodfellow (ed.) *Nucleic Acid Techniques in Bacterial Systematics.* John Wiley & Sons Ltd., Chichester, England.
41. **Lane, D. J., B. Pace, G. J. Olsen, D. A. Stahl, M. L. Sogin, and N. R. Pace.** 1985. Rapid determination of 16S ribosomal RNA sequences for phylogenetic analyses. *Proc. Natl. Acad. Sci. USA* **82:**6955–6959.
42. **McConkey, E. H.** 1967. The fractionation of RNA's by sucrose gradient centrifugation. *Methods Enzymol.* **12A:**620–634.
43. **Mead, D. A., J. A. McClary, J. A. Luckey, A. J. Kostichka, F. R. Witney, and L. M. Smith.** 1991. *Bst* DNA polymerase permits rapid sequence analysis from nanogram amounts of template. *BioTechniques* **11:**76–87.
44. **Mullis, K. B., and F. Faloona.** 1987. Specific synthesis of DNA *in vitro* via a polymerase catalyzed chain reaction. *Methods Enzymol.* **155:**335–350.
45. **Olsen, G. J.** 1988. Phylogenetic analysis using ribosomal RNA. *Methods Enzymol.* **164:**793–812.
46. **Olsen, G. J., D. J. Lane, S. J. Giovannoni, N. R. Pace, and D. A. Stahl.** 1986. Microbial ecology and evolution: a ribosomal RNA approach. *Annu. Rev. Microbiol.* **40:**337–365.
47. **Padgett, K. A., J. E. Mullersman, and B. K. Saha.** 1991. A novel approach to sequencing DNA amplified by the polymerase chain reaction. *BioTechniques* **11:**56–58.
48. **Peattie, D. A.** 1979. Direct chemical method for sequencing RNA. *Proc. Natl. Acad. Sci. USA* **76:**1760–1764.
49. **Penny, D.** 1991. From macromolecules to trees, p. 281–324. *In* E. Stackebrandt and M. Goodfellow (ed.), *Nucleic Acid Techniques in Bacterial Systematics.* John Wiley & Sons Ltd., Chichester, England.
50. **Puissant, C., and L.-M. Houdebine.** 1990. An improvement of the single-step method of RNA isolation by acid guanidinium thiocyanate-phenol-chloroform extraction. *BioTechniques* **8:**148–149.
51. **Reed, A. P., T. A. Kost, and J. T. Miller.** 1986. Simple improvements in ^{35}S dideoxy sequencing. *BioTechniques* **4:**306.
52. **Ruan, C. C., and C. W. Fuller.** 1991. Using T7 gene 6 exonuclease to prepare single-stranded templates for sequencing. *Editorial Comments,* vol. 18, p. 1–8. United States Biochemical Corp., Cleveland.
53. **Saiki, R. K., D. H. Gelfand, S. Stoffel, S. J. Scharf, R. Higuchi, G. T. Horn, K. B. Mullis, and H. A. Erlich.** 1988. Primer-directed enzymatic amplification of DNA with a thermostable DNA polymerase. *Science* **239:**487–491.
54. **Sanger, F., G. G. Brownlee, and B. G. Barrell.** 1965. A two-dimensional fractionation procedure for radioactive nucleotides. *J. Mol. Biol.* **13:**373–398.
55. **Sanger, F., S. Nicklen, and A. R. Coulson.** 1977. DNA sequencing with chain-terminating inhibitors. *Proc. Natl. Acad. Sci. USA* **74:**5463–5467.
56. **Seetharam, S., and I. B. Dicker.** 1991. A rapid and complete 4-step protocol for obtaining nucleotide sequence from *E. coli* genomic DNA from overnight cultures. *BioTechniques* **11:**32–34.
57. **Shon, M., J. Germino, and D. Bastia.** 1982. The nucleotide sequence of the replication origin of the plasmid R6K. *J. Biol. Chem.* **257:**13823–13827.
58. **Siberklang, H., A. M. Gillum, and U. L. RajBhandary.** 1979. Use of *in vitro* ^{32}P labeling in the sequence analysis of nonradioactive rRNAs. *Methods Enzymol.* **54G:**58–109.
59. **Simoncsits, A., G. G. Brownlee, R. S. Brown, J. R. Rubin, and H. Guilley.** 1977. A rapid gel sequencing method for RNA. *Nature* (London) **269:**833–836.
60. **Sommerville, C. C., I. T. Knight, W. L. Straube, and R. R. Colwell.** 1989. Simple, rapid method for direct isolation of nucleic acids from aquatic environments. *Appl. Environ. Microbiol.* **55:**548–554.
61. **Stahl, S., T. Hultman, A. Olsson, T. Moks, and M. Uhlén.** 1988. Solid phase DNA sequencing using the biotin-avidin system. *Nucleic Acids Res.* **16:**3025–3038.
62. **Steffan, R. J., J. Goksoyr, A. K. Bej, and R. M. Atlas.** 1988. Recovery of DNA from soils and sediments. *Appl. Environ. Microbiol.* **54:**2908–2915.
63. **Stoflet, E. S., D. D. Koeberl, G. Sarkar, and S. S. Sommer.** 1988. Genomic amplification with transcript sequencing. *Science* **239:**491–494.
64. **Swofford, D. L., and G. J. Olsen.** 1990. Phylogeny reconstruction, p. 411–501. *In* D. M. Hillis and C. Moritz (ed.), *Molecular Systematics.* Sinauer Associates, Inc., Sunderland, Mass.
65. **Tanner, R. S., E. Stackebrandt, G. E. Fox, and C. R. Woese.** 1981. A phylogenetic analysis of *Acetobacterium woodii, Clostridium barkeri, Clostridium butyricum, Clostridium lituseburense, Eubacterium limosum,* and *Eubacterium tenue. Curr. Microbiol.* **5:**35–38.
66. **Thomas, K. R., and B. M. Olivera.** 1978. Processivity of DNA exonucleases. *J. Biol. Chem.* **253:**424–429.
67. **Tracy, S.** 1981. Improved rapid methodology for the isolation of nucleic acids from agarose gels. *Prep. Biochem.* **11:**251–268.
68. **Tracy, T. E., and L. S. Mulcahy.** 1991. A simple method for direct automated sequencing of PCR fragments. *BioTechniques* **11:**68–75.
69. **Warren, W., T. Wheat, and P. Knudsen.** 1991. Rapid analysis and quantitation of PCR products by high-performance liquid chromatography. *BioTechniques* **11:**250–255.
70. **Weller, R., and D. M. Ward.** 1989. Selective recovery of 16S ribosomal RNA sequences from natural microbial communities in the form of cDNA. *Appl. Environ. Microbiol.* **55:**1818–1822.
71. **Weller, R., J. W. Weller, and D. M. Ward.** 1991. 16S rRNA sequences of uncultivated hot spring cyanobacterial mat inhabitants retrieved by randomly primed cDNA. *Appl. Environ. Microbiol.* **57:**1146–1151.
72. **Woese, C. R., and G. E. Fox.** 1977. Phylogenetic structure of the prokaryotic domain: the primary kingdoms. *Proc. Natl. Acad. Sci. USA* **74:**5088–5090.
73. **Wrischnik, L. A., R. G. Higuchi, M. Stoneking, H. A. Erlich, N. Arnhein, and A. C. Wilson.** 1987. Length mutations in human genomic DNA: direct sequencing of enzymatically amplified DNA. *Nucleic Acids Res.* **15:**529–542.
74. **Yanisch-Perron, C., J. Vieira, and J. Messing.** 1985. Improved M13 phage cloning vectors and host strains: nucleotide sequences of the M13mp18 and pUc19 vectors. *Gene* **34:**103–119.
75. **Zintz, C. B., and D. C. Beebe.** 1991. Rapid re-amplification of PCR products purified in low melting point agarose gels. *BioTechniques* **11:**158–162.

Chapter 28

Nucleic Acid Probes

JOSEPH O. FALKINHAM III

The foregoing chapters on nucleotide base similarities of DNA (see Chapter 26) and of rRNA (Chapter 27) provide the basis for understanding the development and rationale of the use of nucleic acid probes for the detection and identification of bacteria and of bacterial nucleic acid sequences in a wide variety of samples. In addition, several excellent compilations of probe technology (2, 4) and nucleic acid "hybridization" (1) are available. In Chapter 19 are described those uses of polymerase chain reaction (PCR) technology in which nucleic acid probes can be used.

Nucleic acid probes are either DNA or RNA single-stranded base sequences which can be used for detection of specific bacteria by forming "hybrids" with a complementary sequence in a target nucleic acid in a sample. The terms "hybridization" and "hybrid" are used in this chapter to describe the formation of nucleic acid duplexes from both DNA-DNA reassociation and RNA-DNA hybridization, as distinguished in the introduction to Chapter 26.

Detection is based on the formation of a double-stranded heteroduplex of the single-stranded probe and the target. The design, rationale, and advantages of rRNA probes have been convincingly presented (26a, 32a). Probes most commonly are labeled for detection by using either radioisotopes or nonradioactive reporter molecules. Because the stringency of duplex formation can be adjusted, exact sequence complementarity between the probe and the target nucleic acid can be expected, resulting in unambiguous detection. The sensitivity of detection can be measured by using serial dilutions of a single-stranded target nucleic acid sequence. Probes can be used in conjunction with amplification technologies (e.g., PCR) to achieve greater sensitivity. Further, the influence of a great excess of unrelated nucleic acid (e.g., human chromosomal DNA) on sensitivity can be measured. The specificity of detection by probes can be measured by using a collection of nucleic acid samples recovered from related and unrelated bacteria. Because the extent of hybridization under conditions of probe excess is proportional to the concentration of target sequences, the number of target sequences (and hence the number of bacteria) in a sample can be established. Because nucleic acid probes can be used to enumerate bacteria in samples, it is possible to use culture techniques to compare detection by probes.

Kits or processes that involve nucleic acid probes developed for the detection of individual bacteria or groups of bacteria are available commercially. Further, such probes, kits, or processes are patentable subject matter (2).

There is need for more rapid methods for the diagnosis of infections in humans, animals, and plants which are as sensitive and specific as culturing the etiologic agent. This is especially important for etiologic agents that are unculturable (e.g., *Mycobacterium leprae*), slow growing (e.g., *Mycobacterium tuberculosis*), difficult to grow (e.g., *Chlamydia* spp. and *Helicobacter pylori*), or highly contagious. Clearly, probes will find their most rapid application when the usual laboratory culturing is either impossible or time-consuming.

Recent publications have highlighted various other uses of nucleic acid probes. One such use is in the identification of potential pathogens in diseases for which the etiologic agent is unknown. For example, nucleic acid probes have been used to determine whether tissues from patients with sarcoidosis contain mycobacterial DNA (30, 36). It had long been suggested that sarcoidosis has a mycobacterial etiology, and this hypothesis has now been supported by the finding, by two independent groups, of mycobacterial nucleic acids in tissues from patients with sarcoidosis (30, 36). Probes can be used in epidemiological studies of bacterial infections. Recently a probe was used to identify the source of an outbreak of infection by multidrug-resistant *Mycobacterium tuberculosis* (13, 33). There is also a need to develop tools to monitor the deliberate or accidental release of genetically engineered organisms in the environment. In addition to monitoring the fate of whole organisms, nucleic acid probes can monitor the movement of unique nucleic acid sequences (e.g., cloned genes) or transcription of unique genes. Nucleic acid probes can be used to monitor contaminants of tissue culture (e.g., *Mycoplasma* spp.) used in either research or production of pharmaceutical agents. Finally, probes are expected to have legal utility. Not only can probes and probe technologies be patented (2) but also genetically engineered organisms, which are parts of a patented invention, can be identified. In this application, possible infringing products can be identified and their misuse can be discovered.

28.1. PROBE CRITERIA

Before identifying and isolating a nucleic acid probe, it is necessary to identify the criteria which the nucleic acid probe must meet and the conditions under which hybridization will be performed. An edited collection of articles, each dealing specifically with a specific genus or group of bacteria, is available describing the isolation, characteristics, and uses of a great variety of nucleic acid probes for bacteria (4). The process of identifying and isolating a nucleic acid probe entails not only defining the target organism but also taking into account other organisms which would be expected to be present in samples. For example, if a probe were to be specific for *Escherichia coli*, it would be necessary to test whether any hybridization occurred between the probe and any other fecal bacteria such as *Salmonella* spp., *Klebsiella* spp., and *Enterobacter* spp. Thus, it is necessary to obtain representative cultures of both the target organisms and others which might possibly compete and give a false-positive hybridization.

In addition to defining the target and potential interfering populations, the conditions of hybridization and detection of hybridization must be defined. Clearly, the conditions of hybridization directly influence the sensitivity and specificity of hybrid formation. Further, the conditions of hybridization may limit the choices for detection.

Finally, thought must be given to the characteristics of the probe. Self-complementary regions should be absent. The probe should be designed with an understanding of the type of hybridization (i.e., in solution or immobilized target nucleic acid) and the conditions of hybridization. Although small probes may be more specific and are capable of rapid diffusion to membrane-bound nucleic acids, they cannot carry as high a signal concentration and thus sensitivity may be lower unless such small probes are end labeled.

28.1.1. Isolation of Probes

There are a number of different methods for identifying and isolating nucleic acid probes. The different methods fall into three categories. (i) Clone a genomic library and screen for clones that meet the desired criteria. (ii) Enrich the nucleic acid by a method desired to select fragments that meet the criteria, and then clone from the enriched population. (iii) Identify unique sequences that meet the criteria, and synthesize one or more of these sequences.

One approach to the isolation of nucleic acid probes appears not to work. Although whole genomic DNA probes have been isolated (34, 38), they appear to lack the specificity required for successful use if the organism is a member of a closely related cluster (38). If the organism is unique, there is no problem with specificity and a whole genome probe will be 1,000 times more sensitive than a probe composed of a fragment of the genome.

28.1.1.1. Cloning and screening

A variety of nucleic acid probes have been developed by screening genomic libraries. For example, genomic libraries of *Salmonella typhimurium* (19), *Nocardia asteroides* (10), *Mycobacterium paratuberculosis* (6, 50), and *Mycobacterium tuberculosis* (14) have been screened and species-specific probes have been identified. The collection (4) of nucleic acid probes provides even more examples. The cloning and screening approach is limited only by the ability to isolate and clone nucleic acids from the organism chosen as the probe parent and the collection of strains used to demonstrate the specificity of the probe.

The probability of success when using this approach for isolation of nucleic acid probes is demonstrated by the frequency of group-specific clones among total clones screened. In *Salmonella typhimurium*, 8 *Salmonella*-specific clones were identified among 200 screened (19). In *Nocardia asteroides*, 2 *Nocardia*-specific clones were found among 50 screened (10). Only 1 *Mycobacterium paratuberculosis* clone among 650 screened was found that hybridized to *M. paratuberculosis* and not to the closely related *Mycobacterium avium* (6). Eisenach et al. (14) identified 3 *M. tuberculosis*-specific clones among 553 screened. Clearly, as demonstrated for *M. paratuberculosis*, the frequency of finding group-specific probes is dependent on the criteria chosen to define the probe. The frequency of group-specific probes falls as the specificity of the probe is narrowed. In the case of whole genomic probes, an *M. tuberculosis* whole genomic probe did not hybridize with other nonmycobacterial, respiratory pathogens; however, it did hybridize with *M. avium*, which is commonly found in respiratory specimens (38).

If the genomic library is cloned in a vector that is capable of carrying a large amount of DNA (e.g., lambda *gt*11), fewer clones must be screened. However, large fragments may lack the hybridization specificity expected of smaller fragments. Of the fragments cloned from *M. tuberculosis*, the three species-specific probes were fragments of a repeated (i.e., 10 to 16 times) sequence (14).

28.1.1.2. Hybrid selection

A U.S. patent (2) describes the isolation of group-specific rRNA sequences by hybrid selection (51). rRNA or DNA probes offer the advantage that cells contain a large number of copies of rRNA (e.g., 10^4 ribosomes per *Escherichia coli* cell), and thus detection of rRNA is expected to be more sensitive than detection of a single chromosomal gene (2). In the example given in the patent, cDNA was synthesized from *Mycoplasma hominis* rRNA by using reverse transcriptase. Selection of cDNA complementary to *M. hominis* rRNA was performed by hybridization in the presence of an excess of human rRNA, and the hybridized fraction was removed by adsorption with hydroxylapatite (2).

Hybrid selection was also used (22) to isolate an *M. paratuberculosis*-specific probe. An *M. paratuberculosis*-specific clone is difficult to isolate (6) because the organism shares almost 100% identity with *M. avium* (44). As expected, an *M. paratuberculosis* genomic library of 350 clones failed to yield any *M. paratuberculosis*-specific clones (22). Therefore, an enrichment was made for *M. paratuberculosis* sequences among 20,000 genomic clones by hybridization with *M. avium* genomic DNA bound to nitrocellulose (22). Unhybridized DNA was collected and used to generate additional clones. Of 584 selected clones tested, 12 hybridized only to *M. paratuberculosis* DNA and not to *M. avium* (22).

28.1.1.3. Synthesis from sequence data

It is logical to assume that if sufficient sequence data accumulate it will be possible to identify group-specific (e.g., species- or genus-specific) sequences. These could be synthesized and used as probes. Hance et al. (20) compared the sequences of the 65-kDa heat shock protein of mycobacteria to identify sequences and probes for the PCR product that could distinguish the *M. tuberculosis-M. bovis* pair from the *M. avium-M. paratuberculosis* pair and from *M. fortuitum*. Sufficient sequence data exist for rRNA sequences, and species- and genus-specific sequences are known (e.g., *Mycobacterium* [8]).

28.1.2. Commercially Available Probes

A limited number of nucleic acid probes have become commercially available. *M. tuberculosis* complex, *M. avium* complex (i.e., *M. avium-intracellulare*), *M. kansasii*, *M. gordonae*, *Mycoplasma*, *Legionella*, and *Chlamydia trachomatis* probes are available from Gen-Probe, Inc., San Diego, Calif. The kits can be purchased for either isotopic (only *Legionella* spp.) or nonisotopic detection. The *M. avium* probe does not distinguish between *M. avium* and *M. paratuberculosis* (45), which is expected because *M. paratuberculosis* and *M. avium* are subspecies of *M. avium* (44). A nucleic acid probe (i.e., IS*900*) which specifically detects *M. paratuberculosis* (50) is available through IDEXX Corp., Portland, Oreg. In addition to detection of *M. paratuberculosis* in cattle (i.e., Johne's disease), the probe may be useful in the diagnosis of Crohn's disease (29) and sarcoidosis (30, 36) in humans.

28.2. TARGET NUCLEIC ACID ISOLATION

28.2.1. Bacterial Cultures

A great variety of methods exist for the isolation of nucleic acid from cells in liquid culture or in colonies. Those for DNA isolation from gram-positive and gram-negative bacteria are outlined in Chapter 26.2, and those for RNA isolation are described in Chapter 27.1. Techniques for isolation of target DNA for PCR amplification (Chapter 19.1.1) may also be used. Buck et al. (11) compared methods for isolation of DNA suitable for amplification by PCR that may be suitable for detection by nucleic acid probe alone. Because of the need to analyze many samples it is necessary to develop simplified methods for isolation of nucleic acid preparations suitable for analysis by nucleic acid probes.

28.2.2. Clinical Patient Samples

There are two approaches to isolation of DNA from clinical patient samples. One approach involves the separation of bacterial cells from the sample (e.g., blood, serum, tissue, sputum, or feces), followed by lysis and isolation of nucleic acids from the cells. A method for the isolation of DNA from human serum and urine has been described which uses guanidinium thiocynate to lyse cells and inactivate nucleases and size-fractionated silica to isolate DNA (8a). Separation of bacterial cells from blood or tissue cells may require lysis of the host cells, especially for intracellular pathogens. In some instances, the extract is incubated to allow for growth of the target bacteria.

The alternative approach is a direct recovery method, which involves lysis and isolation of bacterial nucleic acids without the separate isolation of cells. Because of the need to reduce the number of manipulations and the difficulty in recovering every cell from a sample, the single-step direct-recovery method may prove to result in more sensitive detection. In addition to separation or extraction of cells from host tissue, experimental manipulations may be required to remove compounds which interfere with the activity of restriction endonucleases, hybridization, or detection hybridization. This may be a major problem in plant tissue, where polysaccharides contaminate nucleic acid preparations (12, 16).

A direct method for detection of *M. paratuberculosis*, the causative agent of Johne's disease in cattle, from bovine fecal samples has been described (22). The samples (30 g) are shaken with glass beads in 20 ml of Tris-EDTA buffer (pH 8), and large particulate material is removed by low-speed centrifugation (100 x *g* for 5 min). The remaining suspended material is pelleted by centrifugation (8,000 x *g* for 30 min), and cells are disrupted in either a pressure cell or bead beater (Mini-Beadbeater; Biospec Products, Bartlesville, Okla.). The nucleic acids in the crude lysate are purified by the Marmur procedure (Chapter 26.2.1). With this method a *M. paratuberculosis*-specific probe was able to detect 10^5 cells. Of 111 fecal samples from cattle with symptoms consistent with a diagnosis of Johne's disease, 21 culture-negative samples were identified as positive with the probe. Only one culture-positive sample was judged negative with the probe (22).

28.2.3. Water Samples

Isolation of nucleic acids from water samples is less complicated than isolation from clinical patient, soil, or sediment samples because most cells in water are not bound to particles that are removed by low-speed centrifugation or by filtration. Only in samples heavily contaminated with sediment will isolation of cells become a problem. A method for direct isolation of nucleic acids from water samples has been published (39). Water samples were filtered with Sterivex-GS filters (Millipore Corp., Bedford, Mass.), and nucleic acids were extracted from the material on the filter and within the housing by the Marmur procedure (Chapter 26.2.1). There was a correlation between the amount of DNA recovered and the number of cells in each sample if contaminating material was precipitated with ammonium acetate and DNA was precipitated with ethanol (39). There was no correlation between the cell number and the DNA recovered by using only ethanol precipitation, perhaps because of the absence of salts. Recovery was reported to be 1 ng of DNA from 10^6 cells added to a water sample. By contrast, 10 ng of DNA per 10^6 cells was recovered from a pure culture of *Vibrio cholerae* (39). Using that method, Knight et al. (25) was able to detect *Salmonella* spp. in estuarine water samples. Of 16 samples, 12 were *Salmonella* probe positive while only 5 were culture positive (25), strong evidence of the power of nucleic acid probe techniques for detection of bacteria in water samples.

28.2.4. Soil and Sediment Samples

As with techniques for nucleic acid isolation from clinical patient samples, isolation from environmental samples (e.g., soil and sediments) can be performed in two ways. One method starts with extraction or separation of bacterial cells from the sample, followed by cell lysis and isolation of nucleic acids. Alternatively, direct lysis and isolation of bacterial nucleic acids without the isolation of cells can be performed. Because of the need to analyze many samples and reduce the number of time-consuming steps, as well as the difficulty in extracting or separating cells from the soil or sediment particles, the single-step direct-recovery method is preferred. The procedures must allow for the fact that soils and sediments contain substances, such as humic acid, that may interfere with methods of either hybridization or detection (21, 43).

A procedure for isolation of DNA from soil, in which polyvinylpolypyrrolidone (PVPP) was used in the first step to remove humic acid from soil extracts, has been devised (21). Following humic acid extraction, soil bacteria were collected by centrifugation and lysed with a combination of Sarkosyl (CIBA-GEIGY Corp., Summit, N.J.), Na_2EDTA, and lysozyme in the presence of a proteinase (21). The method was able to detect 4.3×10^4 *Bradyrhizobium japonicum* CFU/g of soil.

Two methods, (i) cell separation followed by lysis and DNA recovery and (ii) direct cell lysis and alkaline DNA extraction and recovery, were compared for isolation of DNA from soils and sediments (43). Cells were labeled with tritiated thymidine following their addition to soil, and PVPP was used to remove humic acids. Cell separation was similar to that described by Holben et al. (21). In the direct-lysis procedure, cells were exposed to sodium dodecyl sulfate (SDS) at 70°C for 1 h and broken in a bead beater (Mini Beadbeater). Released DNA was coprecipitated with polyethylene glycol and further purified by phenol and potassium acetate extractions, ethanol precipitation, CsCl-ethidium bromide density gradient centrifugation, and adsorption with hydroxylapatite (43). Recovery was judged by measurement of total DNA and radioactivity. Although more DNA was recovered by the direct-lysis method (22 and 90% from soil and sediment, respectively), the ratio of labeled thymidine to total DNA was lower than that in the cell extraction method, suggesting that the direct-lysis method recovered DNA from sources other than cells actively synthesizing DNA (43).

A rapid method for extraction of DNA from small soil and sediment samples has been published (46). Soil slurry samples (1 g) were incubated in an EDTA-containing (pH 8) solution plus lysozyme for 2 h at 37°C. Cells were lysed in the presence of SDS by three freeze-thaw cycles. Released DNA was purified by phenol extractions and precipitated with isopropanol. Following removal of RNA by digestion with RNase, the DNA was purified with an Elutip-d column (Schleicher & Schuell, Keene, N.H.). Approximately 90% of the DNA from 8×10^7 or 4×10^6 CFU of *Pseudomonas* spp. per g of sediment or soil, respectively, was recovered (46).

A procedure for DNA isolation by electroelution of encapsulated soil or sediment samples following lysis

(35) was shown to be sensitive enough to detect 10^4 CFU of *Pseudomonas* spp. per g. Although humic acid was present in nucleic acids prepared by the last two methods, PVPP was not used to remove humic acid.

To examine in situ gene expression, a method for the extraction of mRNA from seeded soils was developed (47). Lysis of cells, RNA fixation, and DNA digestion were performed simultaneously by shaking a soil sample in 4 M guanidine thiocyanate containing 25 mM sodium citrate, 0.5% Sarkosyl, and 0.1 M 2-mercaptoethanol. After adjustment of the pH to 4 with 2 M sodium acetate, mRNA was extracted with phenol and chloroform-isoamyl alcohol and again with chloroform and precipitated with isopropanol. From 1 g of seeded soil containing 8.0×10^8 CFU of *Pseudomonas aeruginosa*, 17 μg of RNA (which contained mRNA) was isolated (47).

Although recovery values are high for the methods listed above, the values were obtained after bacteria were seeded in the soil and sediment samples. Recovery varies greatly from one bacterium to another. Quite possibly such seeded bacteria were more easily extracted from soil than indigenous bacteria would be. For example, only 5% of *M. avium* cells added to soil could be extracted (9). Thus, it is probably more valuable to attempt to isolate nucleic acids by a direct-lysis method rather than to attempt first to isolate bacterial cells. However, a recently published method uses a cation-exchange resin to extract bacteria from soil (23). Unfortunately, in all the studies reported there was no comparison between detection based on probe versus culture, as was done with water (see section 28.2.3).

28.3. PROBE-TARGET HYBRIDIZATION

Hybrid formation between the labeled nucleic acid probe and target can be performed in solution or on some type of support (e.g., nitrocellulose) to which the target nucleic acid is immobilized (1). The same method and criteria (i.e., conditions for hybridization) used for identification, isolation, and characterization of the probe must be maintained for detection of target nucleic acid in samples to ensure that the sensitivity and specificity of the probe are maintained.

28.3.1. Free-Solution Hybridization Methods

Liquid or free-solution methods for hybridization are presented in Chapter 26.6.3 and by Young and Anderson (52). Solution hybridization is described in a U.S. patent (2) and used in the kits supplied by Gen-Probe, San Diego, Calif.

28.3.2. Filter Hybridization Methods

Filter support hybridization methods are presented in Chapter 26, by Anderson and Young (7), and by Kricka (26). The technology is applicable to phage, plaque, and colony lifts and to dot blot and slot blot methodologies. Generally, the target nucleic acid is first denatured and then collected on a filter membrane to reduce the opportunity for the single-stranded nucleic acids to reanneal. Nylon membranes are generally preferred to nitrocellulose membranes because they tear less frequently, are easier to handle, and can be reused in different hybridizations without breakage. Further, nucleic acids can be covalently attached to nylon membranes by cross-linking thymine with the nylon amine groups by UV irradiation (254 nm) or baking (80°C for 2 h under vacuum). Nylon membranes were shown to have higher sensitivity than nitrocellulose membranes, and this sensitivity could be further increased by mildly denaturing the target nucleic acid before attachment to the membrane, rather than after (48).

After collection, the filter membrane with its bound nucleic acid is incubated with the probe. The specificity of hybridization can be regulated during either the hybridization reaction or the washing steps. It is important to point out that the parameters affecting the rate and extent of hybridization of a probe to membrane-bound target nucleic acids are not as well understood as those for hybridization in solution, and thus the calculations are more complicated and difficult. This may be because the rate and extent of hybridization are dependent on both the rate of hybridization and the diffusion of the probe to the filter (7). Rates of membrane hybridization are lower than rates of solution hybridization, in part because only a fraction of the membrane-bound nucleic acid is accessible (7). Diffusion can be increased by (i) using a small probe, (ii) using a small reaction volume, (iii) using a high incubation temperature, and/or (iv) shaking the reaction vessel. Note that temperature directly influences the stringency of the reaction (7).

28.3.3. Factors Affecting Hybridization

A variety of factors influence the rate and extent of hybridization whether it is carried out in solution or on membranes (1, 7, 52). (i) The more complex the probe, the lower the rate of hybridization. (ii) The temperature influences both the rate of hybridization and the stability of duplexes formed. (iii) The composition of the hybridization solution influences both the rate of hybridization and duplex stability. For example, a high salt concentration stabilizes duplexes. Thus, low ionic strength is preferred to increased probe specificity, even though low ionic strength reduces the rate of hybridization. Further, formamide can be included in the hybridization solution to decrease the temperature at which hybrids form, thereby permitting the reaction to be carried out at a lower and more convenient temperature (30 to 45°C).

28.3.4. Probe and Target Nucleic Acid Concentrations

The rate of hybrid formation is proportional to the concentration of probe and target DNA. Further, the ratio of probe to target nucleic acid directly influences the character of the hybridization reaction. Two conditions have been identified: (i) probe excess and (ii) target excess. Under single-stranded probe excess, the amount of probe hybridized is a direct measure of the amount of probe-specific sequences in the target DNA (52). Therefore, with a single assay point, both detection and quantitation can be accomplished (2). By contrast, under conditions of excess target sequences, the time required to reach reaction completion is not predictable and detection and quantitation cannot be accomplished with a single assay point (52).

28.4. PROBE LABELING

In any method in which nucleic acid probes are used for detection or quantitation of target DNA in a sample, the assay requires detection and/or measurement of the formation of a double-stranded hybrid. Further, the conditions of hybrid formation must be identical to those used to characterize the probe. Finally, the method for detection of hybrid formation is influenced by (i) the hybridization method, (ii) the composition of the hybridization mixture, and (iii) the nature of the label on the probe. The design of rRNA probes, their hybridization efficiency, and detection by reversible target capture and amplification have been reviewed (26a). Methods for labeling probes are as follows.

28.4.1. Isotopic Methods

A variety of methods for radioisotopic labeling of DNA (e.g., iodination, nick translation, and random primer) and RNA (e.g., iodination, end labeling, and RNA polymerase labeling) probes are presented in Chapters 26.5 and 27.3.1, respectively.

28.4.2. Nonisotopic Methods

As an alternative to labeling with radioisotopes, probes can be labeled with nonisotopic reporter molecules. Nonisotopic methods of DNA and RNA probe labeling are presented in Chapters 26.5 and 27.3.1, respectively, in chapters by Kessler (24) and Kricka (26), and in a collection of articles on nonisotopic labeling methods (3). Because of the possible widespread use of nucleic acid probes, nonisotopically labeled probes will become increasingly popular, and the methods deserve a detailed description.

There are a number of advantages to nonisotopic probes (26–28). (i) The nonisotopically labeled probe can be detected more rapidly. (ii) The sensitivity of detection of nonisotopically labeled probes can be equal to that of isotopically labeled probes. (iii) Nonisotopically labeled probes are more stable because the label does not decay over time. (iv) There is no radioactive waste, and personnel are not exposed to the hazards of radioisotope use.

28.4.2.1. Colorimetric methods

A variety of systems exist for colorimetric detection of labeled probes. Colorimetric methods can be used to produce a soluble colored product which can be detected by spectrophotometry. Alternatively, insoluble colored products can be produced to locate hybrids on solid supports.

One method (Genius; Boehringer Mannheim, Indianapolis, Ind.) involves incorporation of digoxigenin-dUTP into the probe during synthesis. Following the hybridization reaction, the labeled probe can be detected by using anti-digoxigenin antibodies which have been conjugated to alkaline phosphatase. In the presence of the alkaline phosphatase substrate 5-bromo-4-chloro-3-indoyl phosphate (BCIP) and nitroblue tetrazolium (NBT), an insoluble purple-blue deposit is produced at the site of the probe. This method is used with immobilized probes.

Biotin-labeled probes synthesized either by incorporation of biotinylated deoxynucleoside triphosphate or by photochemically linking biotin to DNA (Photobiotin; Bethesda Research Laboratories, Gaithersburg, Md.) can be detected colorimetrically with a streptavidin-alkaline phosphatase conjugate and the BCIP-NBT substrate (COOL-PROBE; Sigma Chemical Co., St. Louis, Mo.).

A third method relies on incorporation of sulfonated cytosines into the probe, binding of probe with a sulfonated-DNA monoclonal mouse antibody, and detection with an alkaline phosphatase-linked anti-mouse antibody (SulfoPROBE; Sigma).

One drawback of all these approaches is that the colorimetric reaction product can fade over time and no permanent record of the experimental results can be retained unless they are photographed.

28.4.2.2. Fluorescence

Fluorescence-labeled probes have been used widely for detecting evidence of in situ hybrid formation in tissue, and they are also useful in liquid hybridization. Although very low concentrations of fluorescein can be detected (e.g., as a fluorescein-labeled probe), background fluorescence may be high, especially in methods involving direct detection of hybridization in patient or environmental samples, and the reaction product can fade over time.

Fluorescent labels can be directly incorporated into a probe during synthesis, or can be conjugated with a molecule that reacts with a probe labeled with another agent. Fluorescein-12-dUTP (yellow-green), resourfin-10-dUTP (red), or hydroxy-coumarin-6- dUTP (blue) can be incorporated directly into DNA probes during synthesis. DNA probes can be end labeled with fluorescein-12-ddUTP. RNA probes can be labeled with fluorescein-12-UTP. Probes labeled with digoxigenin or biotin can be detected with anti-digoxigenin or anti-biotin antibodies, respectively, coupled with fluorescent molecules.

28.4.2.3. Chemiluminescence

Chemiluminescence methods offer the advantage of a detection system that provides a permanent record of the results. In addition, some of these techniques are as sensitive as isotopic methods. The probe is detected by light emission and exposure of X-ray film. The nucleic acid probe is labeled by incorporation of modified bases and detected with a label-specific molecule to which has been conjugated a reactive group required in the luminescence reaction. The latter reaction requires the use of additional reagents.

One chemiluminescence method is a variant of that used for colorimetric detection of a probe (Genius). A probe is labeled by incorporation of digoxigenin-dUTP during synthesis. Following the hybridization reaction, the hybridized labeled probe is reacted with anti-digoxigenin antibodies conjugated to alkaline phosphatase. In chemiluminescence, the probe is detected by reaction of alkaline phosphatase with 4-methoxy-4 (3-phosphatephenyl) spiro (1,2-dioxetane-3,2′-adamantane) disodium salt. Dephosphorylation results in formation of an unstable intermediate, which decomposes, releasing light (Lumi-Phos 530; Boehringer Mannheim) which can be used to expose X-ray film.

A similar approach involves the use of adamantyl-1,2-dioxetane arylphosphate (AMPPD), which is dephosphorylated to yield an unstable, light-emitting product (SOUTHERN-LIGHT; Tropix, Inc., Bedford, Mass.).

Another approach involves labeling the probe with biotin (e.g., biotin-12-dUTP for DNA or biotin-12-UTP for RNA) followed by detection of probe with alkaline phosphatase-conjugated streptavidin in the presence of an alkaline phosphatase substrate that yields an unstable, chemiluminescent product.

Acridinium ester-labeled DNA probes (ACCUPROBE; Gen-Probe) can be detected by light emission coincident with hydrolysis with H_2O_2 and NaOH. Unfortunately, the light is emitted in a flash; consequently, acridinium-labeled probes are not suitable for detection of hybridization on filter membranes (3).

28.4.2.4. Bioluminescence

Bioluminescence involves the reaction of luciferase with D-luciferin, resulting in light emission (18). Two methods for labeling probes to generate bioluminescence have been described. A biotin-labeled probe can be bound by a streptavidin-coupled alkaline phosphatase (18). Alkaline phosphatase will hydrolyze D-luciferin-*O*-phosphate to luciferin, which will yield light in the presence of luciferase (18). Alternative techniques for detection involve conjugating β-galactosidase or carboxypeptidase with streptavidin and using D-luciferin-*O*-β-galactoside or D-luciferin-*N*-arginine as substrates (18). These techniques can be used for either solution or filter hybridization with a suitable light detector.

A second bioluminescent labeling method uses glucose-6-phosphate dehydrogenase (G6PDH) coupled to streptavidin (32). Detection is carried out by using a bioluminescent chain reaction catalyzed by flavin mononucleotide (FMN) oxidoreductase and luciferase in the presence of decanal (32).

28.5. PROBES AND POLYMERASE CHAIN REACTION

PCR technology and related technologies are described in Chapter 19. Nucleic acid probes can be used with this powerful technique to unambiguously identify amplified products. In addition, PCR can be used to make the probes themselves.

Legionella pneumophila suspended in water was detected by PCR amplification followed by hybridization with a *Legionella*-specific probe (41). PCR was used with a probe to detect a herbicide-degrading strain of *Pseudomonas cepacia* in sediments (42). As few as 100 cells per 100 g of sediment could be detected—a 1,000-fold greater sensitivity than that of nonamplified controls (42).

Detection of *Mycobacterium tuberculosis* in either cultures (40) or sputum samples (15, 40) was successful when using PCR amplification and hybridization of products with a species-specific probe. Of 48 sputum samples which were smear or culture positive or both for *M. tuberculosis,* 47 were PCR and probe positive, and of 42 smear- and culture-negative sputum samples, 2 were PCR and probe positive (15). Further, of 29 cultures that were smear or culture positive for other mycobacteria, none were PCR or probe positive (15). In another study (40), 96 sputum samples gave the following results: 74 were positive by both PCR and culture, 8 were negative by both methods, and 14 were negative by culture but positive by PCR. None were positive by culture but negative by PCR. A 98% sensitivity and 100% specificity when using a DNA probe in combination with PCR on sputum samples has been reported (5). These data demonstrate that PCR and probe detection of *M. tuberculosis* infection in clinical samples is at least as sensitive and specific as is the combination of smear and culture.

The combination of PCR with a nucleic acid probe results in high sensitivity. For instance, as little as 10 fg (equivalent to two genomes) of *M. paratuberculosis* was detected by PCR amplification and probe hybridization (31).

28.6. COMPARISON OF PROBES WITH CULTURE TECHNIQUES

The sensitivity and specificity of nucleic acid probes in comparison with culture techniques has been cited above with the examples of clinical patient samples (section 28.2.2), water samples (section 28.2.3), soil and sediment samples (section 28.2.4), and probes and PCR (section 28.5). Nucleic acid probe technology faces the same problems encountered by serologic methods: the techniques may lack sensitivity or specificity. Low sensitivity is particularly a problem when direct detection in a clinical or environmental sample is required without any period of growth of target cells. Enzymatic target amplification (e.g., PCR to replace bacterial growth) may increase sensitivity. Low specificity would result if a probe reacted with related bacteria. This would be a particular problem in clinical and environmental samples that contain related bacteria.

Primers for PCR amplification of the 65-kDa heat shock protein gene of *M. tuberculosis* amplified the homologous gene from *M. intracellulare* (17). Although the sequence of the *M. intracellulare* product differed by only 2 of 20 bases from that of *M. tuberculosis,* the former hybridized to the *M. tuberculosis* 65-kDa heat shock protein gene probe, resulting in a false-positive signal (17). Probes will clearly find use most rapidly for bacterial infections for which laboratory diagnosis is either currently impossible or time consuming.

The limits of detection of *M. paratuberculosis* in feces were compared with those of a commercial nucleic acid probe test (IDEXX Corp.), PCR with a 16S rRNA probe, PCR with the IS*900* probe, and culture. The commercial test detected 10^4 to 10^5 organisms per g of feces, PCR coupled with a 16S rRNA probe detected 10^7 organisms per g of feces, PCR coupled with the IS*900* probe detected 10^4 to 10^5 organisms per g of feces, and culture detected 10 organisms per g of feces (49). Clearly in this instance, the culture technique was 1,000 times more sensitive than any of the PCR and probe tests.

In many instances, hybridization results are not expected to be simple yes-no answers. Whether detection of hybridization is performed by isotopic or nonisotopic methods, some true-negative samples will yield some signal (i.e., false positive) and some true-positive samples will fail to yield a strong signal (false negative). A cutoff range has been proposed (37) to exclude such false-positive and false-negative samples.

28.7. USES OF PROBES

Probes will probably find widespread use in all aspects of bacteriology. These uses are not limited to medical bacteriology. The real hope of nucleic acid probe technology will be that it will prove to be sensitive and specific for detection of bacteria in any sample directly without the necessity of culturing. The utility of nucleic acid probe technology will be dependent on the ability of the technique to detect and identify a target bacterium in the presence of related bacteria. Therefore the probes must be sufficiently specific to reduce cross-reactivity because of the likely presence of related bacteria. In addition, utility requires that the techniques be sufficiently sensitive. It is likely that sensitivity will be increased best by using a probe in conjunction with PCR.

28.7.1. Medical Diagnosis

Nucleic acid probes must be able to detect and identify bacteria in (i) pure culture, (ii) mixed cultures, (iii) bacterium-containing or -enriched extracts of clinical patient samples, (iv) naturally sterile clinical patient samples (e.g., blood), and (v) nonsterile clinical patient samples (listed in order of increasing difficulty). Although a variety of different probes have been shown to meet goals (i), (ii), and (iv), only time will tell whether they can be used successfully in the detection and identification of bacteria in clinical patient samples directly.

Probes are already finding routine use in clinical bacteriology as tools for the identification of bacteria in pure or mixed culture. Perhaps the widest use of nucleic acid probes involves identification of *M. avium* and *M. intracellulare (M. avium* complex). The availability of nucleic acid probe kits, the slow growth of these pathogens of patients with AIDS, and the lack of methods to distinguish between species have contributed to the rapid acceptance of nucleic acid probe technology.

Nucleic acid probes are also being used for detection of the difficult-to-grow pathogens *Neisseria gonorrhoeae* and *Chlamydia trachomatis* in urogenital swabs by using the Gen-Probe PACE assay.

Nucleic acid probes have been used to demonstrate that the slow-growing organisms which can be recovered from tissue samples from patients with Crohn's disease are identical to *M. paratuberculosis* (29). Recently, nucleic acid probes for mycobacteria have also been used to show that tissue from patients with sarcoidosis can harbor mycobacterial nucleic acid (30, 36). Although not all patients with sarcoidosis had tissue which contained probe-reactive nucleic acid, the approach demonstrates the power of nucleic acid probe technology to identify the etiologic agent of diseases whose causative agent is unknown.

28.7.2. Epidemiology

Nucleic acid probes are useful in epidemiological studies. In such studies, the nucleic acid probe used is complementary to sequences which are repeated in the genome of a bacterium. For instance, a transposable genetic element of *M. tuberculosis*, IS*6110*, has been found to be located in multiple copies and on different restriction endonuclease-generated fragments of genomic DNA. Recently, restriction fragment length polymorphism (RFLP) analysis of *M. tuberculosis* isolates recovered from a hospital outbreak of multidrug-resistant *M. tuberculosis* infection showed that 14 of 15 isolates shared identical RFLP patterns when IS*6110* was used as the probe (13). Therefore it was likely that all isolates came from a single individual (13). A similar study has also been reported (33).

28.7.3. Environmental Monitoring

Monitoring the movement and behavior of genetically engineered bacteria that have been released into the environment may be required in the future. With nucleic acid probes, not only can the movement and behavior of whole organisms be monitored but also the movement and behavior of specific DNA sequences can be checked. In the latter case, the movement of a genetically engineered sequence between related and unrelated bacteria (or other organisms) can be monitored.

Nucleic acid probes could be used in environmental monitoring for identification of bacteria in (i) pure cultures of bacteria recovered from the environment, (ii) mixed cultures recovered from the environment, (iii) bacterium-containing or -enriched extracts of environmental samples, (iv) sterile or sterilized environmental samples, and (v) nonsterile environmental samples. These uses are similar to those listed above for use in medicine. Although preliminary evidence suggests the utility of nucleic acid probe technology in environmental monitoring, no judgment can be made because too few examples are available.

Nucleic acid probe technology may prove to be of future value in environmental monitoring because it does not require the recovery and culture of a released, genetically engineered organism. Moreover, although viable but unculturable bacteria present in the environment cannot be recovered and enumerated by culture, their presence could be detected by a nucleic acid probe. The rationale and advantages of using rRNA probes in environmental studies (32a) and their use for analysis of a marine picoplankton community (36a) and uncultivated natural habitats (36b) have been presented.

28.7.4. Tissue Culture Contamination

Because of the widespread use of tissue culture in research and industry and the probability of contamination by *Mycoplasma* spp., there is a need for regular monitoring for the presence of these organisms. The difficulty in their culture makes probes particularly useful. In industry, the need for the incorporation of periodic and regular *Mycoplasma* monitoring as part of a quality assurance program will be a strong incentive for the use of probes.

28.7.5. Patents

Despite the novelty of nucleic acid probe technology, there are a large number of patents covering different aspects of this technology. These include (i) amplification of the target of a probe, (ii) assay format, (iii) devices, (iv) labeling, (v) label detection systems (with the largest number of entries), (vi) sample preparation, (vii) solid supports, (viii) separation procedures, and (ix) specific probes (3). Although it is too soon to judge, nucleic acid probe technology may prove to be the "gold standard" for detection and identification of bacteria. This would depend on demonstrating that the approach has sufficient sensitivity and specificity.

Nucleic acid probe technology may find utility in protecting patent rights from infringement. Nucleic acid probes, specific for a patented, individual, genetically engineered strain could be used to determine whether a competing product or process contains or uses that bacterium. Nucleic acid probe technology could also be used to detect patented, unique DNA sequences in novel hosts.

28.8. REFERENCES

28.8.1. General References

1. **Hames, B. D., and S. J. Higgins.** 1985. *Nucleic acid hybridization.* IRL Press, Oxford.
2. **Kohne, D. E.** July, 1989. Method for detection, identification and quantitation of non-viral organisms. U.S. patent 4,851,330.
3. **Kricka, L. J.** 1992. *Nonisotopic DNA Probe Techniques.* Academic Press, Inc., San Diego, Calif.
4. **Macario, A. J. L., and E. C. deMacario (ed.).** 1990. *Gene Probes for Bacteria.* Academic Press, New York.

28.8.2. Specific References

5. **Altamirano, M., M. T. Kelly, A. Wong, E. T. Bessuille, W. A. Black, and J. A. Smith.** 1992. Characterization of a DNA probe for detection of *Mycobacterium tuberculosis* complex in clinical samples by polymerase chain reaction. *J. Clin. Microbiol.* **30:**2173–2176.
6. **Ambrosio, R. E., Y. Harris, and H. F. A. K. Huchzermeyer.** 1991. A DNA probe for the detection of *Mycobacterium paratuberculosis. Vet. Microbiol.* **26:**87–93.
7. **Anderson, M. L. M., and B. D. Young.** 1985. Quantitative filter hybridization, p. 73–111. *In* B. D. Hames and S. J. Higgins (ed.), *Nucleic Acid Hybridization.* IRL Press, Oxford.
8. **Böddinghaus, B., T. Rogall, R. Flohr, H. Blöcker, and E. C. Böttger.** 1990. Detection and identification of mycobacteria by amplification of rRNA. *J. Clin. Microbiol.* **28:**1751–1759.

8a. **Boom, R., C. J. A. Sol, M. M. M. Salimans, C. L. Jansen, P. M. E. Wertheim-van Dillen and J. van der Noordaa.** 1990. Rapid and simple method for purification of nucleic acids. *J. Clin. Microbiol.* **28:**495–503.

9. **Brooks, R. W., K. L. George, B. C. Parker, J. O. Falkinham III, and H. Gruft.** 1984. Recovery and survival of nontuberculous mycobacteria under various growth and decontamination conditions. *Can. J. Microbiol.* **30:**1112–1117.
10. **Brownell, G. H., and K. E. Belcher.** 1990. DNA probes for the identification of *Nocardia asteroides. J. Clin. Microbiol.* **28:**2082–2086.
11. **Buck, G. E., L. C. O'Hara, and J. T. Summersgill.** 1992. Rapid, simple method for treating clinical specimens containing *Mycobacterium tuberculosis* to remove DNA for polymerase chain reaction. *J. Clin. Microbiol.* **30:**1331–1334.
12. **Do, N., and R. P. Adams.** 1991. A simple technique for removing plant polysaccharide contaminants from DNA. *BioTechniques* **10:**162–166.
13. **Edlin, B. R., J. I. Tokars, M. H. Grieco, J. T. Crawford, J. Williams, E. M. Sordillo, K. R. Ong, J. O. Kilburn, S. W. Dooley, K. G. Castro, W. R. Jarvis, and S. D. Holmberg.** 1992. An outbreak of multidrug-resistant tuberculosis among hospitalized patients with the acquired immunodeficiency syndrome. *N. Engl. J. Med.* **326:**1514–1521.
14. **Eisenach, K. D., J. T. Crawford, and J. H. Bates.** 1988. Repetitive DNA sequences as probes for *Mycobacterium tuberculosis. J. Clin. Microbiol.* **26:**2240–2245.
15. **Eisenach, K. D., M. D. Sifford, M. D. Cave, J. H. Bates, and J. T. Crawford.** 1991. Detection of *Mycobacterium tuberculosis* in sputum samples using a polymerase chain reaction. *Am. Rev. Respir. Dis.* **144:**1160–1163.
16. **Fang, G., S. Hammar, and R. Grumet.** 1992. A quick and inexpensive method for removing polysaccharides from plant genomic DNA. *BioTechniques* **13:**52–56.
17. **Fiss, E. H., F. F. Chehab, and G. F. Brooks.** 1992. DNA amplification and reverse dot blot hybridization for detection and identification of mycobacteria to the species level in the clinical laboratory. *J. Clin. Microbiol.* **30:**1220–1224.
18. **Geiger, R. E.** 1992. Detection of alkaline phosphatase by bioluminescence, p. 113–126. *In* L. J. Kricka (ed.), *Nonisotopic DNA Probe Techniques.* Academic Press, Inc., San Diego, Calif.
19. **Gopo, J. M., R. Melis, E. Filipska, R. Memeveri, and J. Filipski.** 1988. Development of a *Salmonella*-specific biotinylated DNA probe for rapid routine identification of *Salmonella. Mol. Cell. Probes* **2:**271–279.
20. **Hance, A. J., B. Grandchamp, V. Levy-Frébault, D. Lecossier, J. Rauzier, D. Bocart, and B. Gicquel.** 1989. Detection and identification of mycobacteria by amplification of mycobacterial DNA. *Mol. Microbiol.* **3:**843–849.
21. **Holben, W. E., J. K. Jansson, B. K. Chelm, and J. M. Tiedje.** 1988. DNA probe method for the detection of specific microorganisms in the soil bacterial community. *Appl. Environ. Microbiol.* **54:**703–711.
22. **Hurley, S. S., G. A. Splitter, and R. A. Welch.** 1989. Development of a diagnostic test for Johne's disease using a DNA hybridization probe. *J. Clin. Microbiol.* **27:**1582–1587.
23. **Jacobsen, C. S., and O. F. Rasmussen.** 1992. Development and application of a new method to extract bacterial DNA from soil based on separation of bacteria from soil with cation-exchange resin. *Appl. Environ. Microbiol.* **58:**2458–2462.
24. **Kessler, C.** 1992. Nonradioactive labeling methods for nucleic acids, p. 29–92. *In* L. J. Kricka (ed.), *Nonisotopic DNA Probe Techniques.* Academic Press, Inc., San Diego, Calif.
25. **Knight, I. T., S. Shults, C. W. Kaspar, and R. R. Colwell.** 1990. Direct detection of *Salmonella* spp. in estuaries by using a DNA probe. *Appl. Environ. Microbiol.* **56:**1059–1066.
26. **Kricka, L. J.** 1992. Nucleic acid hybridization test formats: strategies and applications, p. 3–28. *In* L. J. Kricka (ed.), *Nonisotopic DNA Probe Techniques.* Academic Press, Inc., San Diego, Calif.

26a. **Lane, D. J., and M. L. Collins.** 1991. Current methods for detection of DNA/ribosomal RNA hybrids, p. 54–75. *In* A. Vaheri, R. C. Tilton, and A. Balows (ed.), *Rapid Methods and Automation in Microbiology and Immunology.* Springer-Verlag, New York.

27. **Leary, J. J., D. J. Brigati, and D. C. Ward.** 1983. Rapid and sensitive colorimetric method for visualizing biotin-la-

beled DNA probes hybridized to DNA or RNA immobilized on nitrocellulose: bioblots. *Proc. Natl. Acad. Sci. USA* **80**:4045–4049.

28. **Lichter, P., and D. C. Ward.** 1990. Is non-isotopic *in situ* hybridization finally coming of age? *Nature* **345**:93–94.
29. **McFadden, J. J., P. D. Butcher, R. Chiodini, and J. Hermon-Taylor.** 1987. Crohn's disease-isolated mycobacteria are identical to *Mycobacterium paratuberculosis*, as determined by DNA probes that distinguish between mycobacterial species. *J. Clin. Microbiol.* **25**:796–801.
30. **Mitchell, I. C., J. L. Turk, and D. N. Mitchell.** 1992. Detection of mycobacterial rRNA in sarcoidosis with liquid-phase hybridization. *Lancet* **339**:1015–1017.
31. **Moss, M. T., E. P. Green, M. L. Tizard, Z. P. Malik, and J. Hermon-Taylor.** 1991. Specific detection of *Mycobacterium paratuberculosis* by DNA hybridization with a fragment of the insertion element IS900. *Gut* **32**:395–398.
32. **Nicolas, J.-C., P. Balaguer, B. Térouanne, M. A. Villebrun, and A.-M. Boussioux.** 1992. Detection of glucose-6-phosphate dehydrogenase by bioluminescence, p. 203–225. *In* L. J. Kricka (ed.), *Nonisotopic DNA Probe Techniques*, Academic Press, Inc., San Diego, Calif.

32a. **Olsen, G. J., D. J. Lane, S. J. Giovannoni, N. R. Pace, and D. A. Stahl.** 1986. Microbial ecology and evolution: a ribosomal RNA approach. *Annu. Rev. Microbiol.* **40**:337–365.

33. **Pearson, M. L., J. A. Jereb, T. R. Frieden, J. T. Crawford, B. J. Davis, S. W. Dooley, and W. R. Jarvis.** 1992. Nosocomial transmission of multidrug-resistant *Mycobacterium tuberculosis*. *Ann. Intern. Med.* **117**:191–196.
34. **Roberts, M. C., C. McMillan, and M. B. Coyle.** 1987. Whole chromosomal DNA probes for rapid identification of *Mycobacterium tuberculosis* and *Mycobacterium avium* complex. *J. Clin. Microbiol.* **25**:1239–1243.
35. **Rochelle, P. A., and B. H. Olson.** 1991. A simple technique for electroelution of DNA from environmental samples. *BioTechniques* **11**:724–728.
36. **Saboor, S. A., N. M. Johnson, and J. J. McFadden.** 1992. Detection of mycobacterial DNA in sarcoidosis and tuberculosis with polymerase chain reaction. *Lancet* **339**:1012–1015.

36a. **Schmidt, T. M., E. F. DeLong, and N. R. Pace.** 1991. Analysis of a marine picoplankton community by 16S rRNA gene cloning and sequencing. *J. Bacteriol.* **173**:4371–4378.

36b. **Schmidt, T. M., E. F. DeLong, and N. R. Pace.** 1991. Phylogenetic identification of uncultivated microorganisms in natural habitats, p. 37–46. *In* A. Vaheri, R. C. Tilton and A. Balows (ed.), *Rapid Methods and Automation in Microbiology and Immunology*, Springer-Verlag, New York.

37. **Sherman, I., N. Harrington, A. Rothrock, and H. George.** 1989. Use of a cutoff range in identifying mycobacteria by the Gen-Probe rapid diagnostic system. *J. Clin. Microbiol.* **27**:241–244.
38. **Shoemaker, S. A., J. H. Fisher, and C. H. Scoggin.** 1985. Techniques of DNA hybridization detect small numbers of mycobacteria with no cross-hybridization with non-mycobacterial respiratory organisms. *Am. Rev. Respir. Dis.* **131**:760–763.
39. **Somerville, C. C., I. T. Knight, W. L. Straube, and R. R. Colwell.** 1989. Simple, rapid method for direct isolation of nucleic acids from aquatic environments. *Appl. Environ. Microbiol.* **55**:548–554.
40. **Sritharan, V., and R. H. Barker, Jr.** 1991. A simple method for diagnosing *M. tuberculosis* infection in clinical samples using PCR. *Mol. Cell. Probes* **5**:385–395.
41. **Starnbach, M. N., S. Falkow, and L. S. Tompkins.** 1989. Species-specific detection of *Legionella pneumophila* in water by DNA amplification and hybridization. *J. Clin. Microbiol.* **27**:1257–1261.
42. **Steffan, R. J., and R. M. Atlas.** 1988. DNA amplification to enhance detection of genetically engineered bacteria in environmental samples. *Appl. Environ. Microbiol.* **54**:2185–2191.
43. **Steffan, R. J., J. Goksoyr, A. K. Bej, and R. M. Atlas.** 1988. Recovery of DNA from soils and sediments. *Appl. Environ. Microbiol.* **54**:2908–2915.
44. **Thorel, M.-F., M. Krichevsky, and V. V. Levy-Frébault.** 1990. Numerical taxonomy of mycobactin-dependent mycobacteria: emended description of *Mycobacterium avium* subsp. *avium* subsp. nov., *Mycobacterium avium* subsp. *paratuberculosis* subsp. nov., and *Mycobacterium avium* subsp. *silvaticum* subsp. nov. *Int. J. Syst. Bacteriol.* **40**:254–260.
45. **Thoresen, O. F., and F. Saxegaard.** 1991. Gen-Probe rapid diagnostic system for the *Mycobacterium avium* complex does not distinguish between *Mycobacterium avium* and *Mycobacterium paratuberculosis*. *J. Clin. Microbiol.* **29**:625–626.
46. **Tsai, Y.-I., and B. H. Olson.** 1991. Rapid method for direct extraction of DNA from soil and sediments. *Appl. Environ. Microbiol.* **57**:1070–1074.
47. **Tsai, Y.-I., M. J. Park, and B. H. Olson.** 1991. Rapid method for direct extraction of mRNA from seeded soils. *Appl. Environ. Microbiol.* **57**:765–768.
48. **Twomey, T. A., and S. A. Krawetz.** 1990. Parameters affecting hybridization of nucleic acids blotted onto nylon or nitrocellulose membranes. *BioTechniques* **8**:478–481.
49. **van der Giessen, J. W. B., R. M. Haring, E. Vauclare, A. Eger, J. Haagsma, and B. A. M. van der Zeijst.** 1992. Evaluation of the abilities of three diagnostic tests based on the polymerase chain reaction to detect *Mycobacterium paratuberculosis* in cattle: application in a control program. *J. Clin. Microbiol.* **30**:1216–1219.
50. **Vary, P. H., P. R. Andersen, E. Green, J. Hermon-Taylor, and J. J. McFadden.** 1990. Use of highly specific DNA probes and the polymerase chain reaction to detect *Mycobacterium paratuberculosis* in Johne's disease. *J. Clin. Microbiol.* **28**:933–937.
51. **Welcher, A. A., A. R. Torres, and D. C. Ward.** 1986. Selective enrichment of specific DNA, cDNA and RNA sequences using biotinylated probes, avidin and copper-chelate agarose. *Nucleic Acids Res.* **14**:10027–10044.
52. **Young, B. D., and M. L. M. Anderson.** 1985. Quantitative analysis of solution hybridization, p. 47–71. *In* B. D. Hames and S. J. Higgins (ed.), *Nucleic Acid Hybridization*, IRL Press, Oxford.

Section VI

GENERAL METHODS

Introduction to General Methods

PHILIPP GERHARDT

This section contains three chapters which describe methods that are generally applicable to those in all of the prior chapters but are not related to each other. Each of these general methods is important to successful research.

Laboratory biosafety (Chapter 29) is important in part to protect the experiment (e.g., to prevent contamination of other cultures) but especially to protect the experimenter (e.g., to prevent infection). As emphasized in the first version of this book, the generalist may consider that the bacteria usually handled are residents of the natural environment and therefore harmless to humans. In fact there is no distinction between nonpathogenic and pathogenic bacteria; rather, among bacteria there is a broad spectrum in degrees of virulence relative to human susceptibility. Yesterday's saprophyte becomes today's parasite and tomorrow's pathogen as exemplified by the histories of *Escherichia coli, Proteus vulgaris, Pseudomonas aeruginosa, Serratia marcescens,* and *Bacillus cereus.* Pathogenicity often becomes a matter of the infective dose (the greater the number of bacteria, the greater the likelihood of infection) and of the infectious route (inhalation of aerosols usually being the most dangerous). The lesson is precaution: *Treat ALL bacteria as potential hazards to human health.* Through this book, *CAUTION* is noted in context wherever a method may imperil the experimenter, physically and chemically as well as biologically. Radioactivity safety is dealt with in Chapter 21.4.1, carcinogen safety in Chapter 13.8, and chemical safety in Chapter 22.10; safety during photography procedures is covered in Chapter 30.2.3.1.

Chapter 30 on photography is new to this edition. The author of the chapter is a professional photographer who is experienced mainly in microbiological photography (and electron microscopy). Increasingly, researchers find it necessary or desirable to do their own photography. This chapter provides detailed advice on how to select photographic equipment and films, process the results, and deal with the particular requirements of macrophotography, photomicrography, electron micrography, and radioautography.

The final chapter, also newly added, deals with aspects of research that are too often considered something apart: recording and reporting of laboratory findings. In fact, these are integral and necessary to completion of the research process. However, in this chapter only, there seemed no need to provide details of the subject because there are inexpensive, authoritative books which the experimenter should have at hand or readily available. Consequently, only a few specific suggestions are provided in Chapter 31. The main purpose is to guide readers, particularly those in allied disciplines and around the world, with references and contents of these useful books which otherwise might not be known to them.

Chapter 29

Laboratory Safety

W. EMMETT BARKLEY and JOHN H. RICHARDSON

Laboratory safety requires an awareness of the possible risks associated with the handling of hazardous materials, knowledge of mechanisms by which exposures may occur, use of safeguards and techniques that reduce the potential for exposures, and vigilance against compromise and error. Microorganisms, radioisotopes, and hazardous chemicals are invariably found in bacteriology laboratories. The greatest risk of occupational infection in these laboratories is associated with the use of pathogenic microorganisms or the handling of materials contaminated with them. Biological safety and the prevention of laboratory-acquired infection is the principal subject addressed in this chapter. When pathogenic microorganisms are not present in the laboratory, the risk of physical injury and accidental exposure to hazardous chemicals and radioactive materials should be the primary concern of the laboratory worker, and a brief discussion of these hazards is also included.

Laboratory safety is also the purview of federal health and safety regulations and guidelines. The use of radioisotopes in laboratories has been regulated since 1946 by the Nuclear Regulatory Commission (41). The National Institutes of Health (NIH) established federal guidelines for handling recombinant DNA molecules in 1976 (37). These guidelines are mandatory for laboratory activities that are supported totally or in part by federal funds. Voluntary guidelines for the protection of laboratory workers who handle pathogenic microorganisms were published by Centers for Disease Control (CDC) and NIH in 1984 (39). The hazard communication standard, which was revised by the Occupational Safety and Health Administration (OSHA) in 1987, requires employers to provide laboratory workers with information and training about the use of hazardous chemicals used in the laboratory environment (42). The use of hazardous chemicals in laboratories was regulated by OSHA in 1990 (43). Occupational exposure to blood-borne pathogens was regulated by a standard issued by OSHA in 1991 (44).

Biological safety has always been an important concern for people who work in bacteriology laboratories. It is natural to want to protect oneself from harm, and thoughtful people would never knowingly do something that would increase the risk for a colleague to contract an occupational illness. The ability to prevent laboratory-acquired infection, however, requires certain skills and knowledge which can best be acquired through training and careful guidance provided by experienced colleagues. A study of the literature is not sufficient to prepare a laboratory worker for safely handling pathogenic microorganisms. It is necessary that proficiency in microbiological technique be acquired through practice with nonpathogenic microorganisms before higher-risk microorganisms are introduced into the laboratory routine.

Several books provide comprehensive coverage of the practice of biological safety (18, 21, 39). Consensus guidelines published by the CDC and the NIH (38) should be consulted before pathogenic bacteria are introduced into the laboratory. Authoritative books on chemical safety (20), radiation safety (33), physical safety (16), and disinfection and sterilization (3, 29a) are available. Technical handbooks (6, 31) on protocols in molecular biology are an excellent source of laboratory safety information. An excellent reference is also available for developing safety programs that can meet compliance requirements of pertinent regulatory standards (10). These references provide detailed information, which is beyond the basic scope of this chapter.

29.1. PRINCIPLES

29.1.1. Directors' Responsibilities

Laboratory directors have the primary responsibility for setting the expectations and means for protecting the health and safety of all laboratory workers under their supervision. This involves ensuring that (i) hazards associated with laboratory activities are recognized, (ii) routine and special practices that are helpful in controlling or eliminating those hazards are selected for use in the laboratory, and (iii) laboratory workers are provided with adequate instruction and training so that they understand the relative significance of the hazards that relate to them personally and acquire the proficiency necessary to protect themselves from these hazards. The laboratory director is also responsible for deciding who should have access to the laboratory and for establishing additional requirements aimed at reducing potential exposure of persons who may be at increased risk of infection. The preparation of a safety manual for the laboratory is helpful in clearly defining safety expectations and setting a positive attitude that will sustain a high level of safety awareness. The references cited above emphasize the importance and value of safety manuals that are relevant to the actual work activities of the laboratory and provide excellent guidance for developing such manuals.

The laboratory director has a legal obligation to prepare and use a safety manual or plan if the laboratory is involved in the handling of hazardous chemicals, human blood, or blood-borne pathogens. Hazardous chemicals, as defined by the OSHA standard on occupational exposure to hazardous chemicals in laboratories (43) are found in virtually every general or molecular bacteriology laboratory, and therefore the standard will apply to them all. Among the provisions of the standard is the requirement to develop and follow a written Chemical Hygiene Plan. The Plan must describe the specific measures that are to be taken to protect employees from health hazards associated with hazardous chemicals in the laboratory. Laboratories that handle human blood and blood-borne pathogens are required by the OSHA blood-borne pathogens standard (44) to establish a similar document called an Exposure Control Plan. The purpose of this Plan is to describe the emergency controls, laboratory practices, personnel protective equipment, and housekeeping practices that will eliminate or minimize employee exposure to human blood and blood-borne pathogens.

A healthful and safe laboratory environment cannot be sustained without every laboratory worker sharing in the responsibility for safety. All persons in the laboratory must be vigilant toward their own safety and the safety of their colleagues and neighbors. They must know and follow safety guidelines, regulations, and procedures that apply to their work. They should actively seek information and advice to improve their safety performance. They should always report accidents, injuries, and unsafe conditions to their supervisor or safety officer. It is imperative that all laboratory employees know how to respond to emergencies. Ideally, the methods and attitudes that enhance laboratory safety will become a conscious part of all work activity in the laboratory.

29.1.2. Biosafety Principles

Awareness of the potential risks, infectious and otherwise, associated with the procedure or activity being conducted is the safety cornerstone in clinical and research laboratories. To be aware of these risks, the individual worker must be knowledgeable about the agents, reagents, chemicals, equipment, and facilities being used. Knowledge implies appropriate training in risk recognition and prevention, in the proper execution of procedures, and in the use of equipment.

Universal precaution is a prudent standard (38). It is based on the realization that all clinical specimens, body fluids, tissues, and cultures are potentially infectious and should be handled with common sense practices and barrier precautions. Standard and special microbiological practices, containment equipment, and facilities recommended for biosafety level 2 (see section 29.4) constitute universal precaution for the laboratory and should be adopted as a *minimal* standard for work with any potentially infectious materials.

Containment of infectious agents within laboratory and research animal work areas is provided by appropriate combinations of laboratory practices, safety equipment, and facility design features. Primary containment, i.e., protection of the worker and the immediate work area from infection or contamination, is provided by consistent use of accepted good laboratory practices and appropriate use of safety equipment (e.g., a biosafety cabinet) for activities involving known or potentially infectious materials. The use of safe and efficacious vaccines may provide additional reductions in infection risks of workers.

Secondary containment, i.e., the protection of the environment external to the laboratory, is provided by a combination of facility design features (e.g., directional air flow, sealed penetrations, double doors), management of wastes and other infectious materials, and access control.

29.1.3. Facility Features

Safety features of laboratory facilities provide architectural and mechanical barriers which serve to segregate activities performed within a laboratory module from nonlaboratory areas such as administrative offices and cafeterias. The ventilation system provides for correct operations of safety cabinets and, in containment laboratories, can prevent the migration of airborne contaminants between occupied spaces. The design of entry areas to laboratories can be helpful in maintaining proper access control. It is also important that an autoclave for decontaminating infectious laboratory wastes be available within the facility.

The basic safety features of individual laboratory modules within a biosafety level 2 facility are floor and wall surfaces that are easily cleaned, a handwashing sink, and bench tops that are impervious to water and resistant to acids, alkalis, organic solvents, and moderate heat. Spaces between and beneath casework and equipment should be accessible for cleaning. A safety shower and eye wash fountain should also be easily accessible to all laboratory occupants.

Additional safety features are required for biosafety level 3 facilities. The laboratory is physically separated from nonlaboratory areas by entry corridors that require passage through two sets of doors. Floor, wall, and ceiling surfaces are water resistant. Penetrations through these surfaces should be sealed to allow space decontamination. Foot-operated, elbow-operated, or automatically operated handwashing sinks are located in each laboratory near the exit door. The ventilation system creates directional airflow by drawing air into the laboratory through the entry area. Exhaust air is discharged directly to the outside and dispersed away from occupied areas and air intakes of buildings.

29.2. HISTORICAL PERSPECTIVE

Occupational infections of laboratory workers have been documented for a century. Although infections with specific agents have varied by time and place, occupational infections have typically involved agents prevalent in the community and, consequently, likely to be present in materials received and handled in laboratories involved in diagnostic, production, reference, and research activities.

In 1941, the first survey of occupational infections in laboratory workers in the United States was published by Meyer and Eddy (17). In 1949, Sulkin and Pike published the first in a series of surveys of laboratory-associated infections listing 222 viral infections, 21 of which were fatal (34). A second survey, involving 5,000 laboratories and published in 1951, summarized 1,342 cases, only one-third of which had previously been reported (35). Brucellosis was the most common occupational infection; together, brucellosis, tuberculosis, tularemia, typhoid, and streptococcal infections accounted for 72% of all bacterial infections and 31% of the total number of cases caused by all categories of infectious agents. The overall case fatality rate was 3%.

The survey data were updated in 1965 and again in 1976, summarizing a cumulative total of 3,921 reported occupational infections worldwide (26, 27). Brucellosis was again the most commonly reported infection; brucellosis, typhoid fever, tularemia, tuberculosis, hepatitis, and Venezuelan equine encephalitis were the most frequently recorded occupational infections at that time.

Of 703 cases in which the source of infection was identified, 80% were due to contact with infectious materials from spills and splashes, accidents involving the use of needles and syringes, mouth pipetting, and injury from contaminated broken glass or other sharp objects. Bites and scratches by experimentally or naturally infected animals or ectoparasites was the fifth major cause of known accidents resulting in laboratory-acquired infection. The most frequently reported cases were 18 cases of rat-bite fever *(Streptobacillus moniliformis)*, 13 cases of leptospirosis *(Leptospira interrogans)*, and 9 cases of B-virus (herpesvirus simiae) infections. Awareness of occupational risks, good practices, and modest barrier precautions might have effectively prevented these occupational infections.

From these cited reports, as well as others, a consistent pattern and trend characterizing occupational infections in laboratory workers have emerged. Pertinent issues include the following.

1. Occupational infections are underreported. The Sulkin and Pike studies (34, 35) indicated that, historically, only one-third of the occupational infections that were brought to their attention were reported in the scientific literature. Furthermore, there is no national focus for the reporting or active surveillance of occupational infections.

2. Fewer than 20% of occupational infections are associated with an identified exposure, accident, or incident. The most plausible explanation is that diseases such as brucellosis, tuberculosis, tularemia, and those caused by certain arboviruses occur as a result of exposure to respirable airborne infectious particles (droplet nuclei). Diseases such as typhoid, cholera, and viral hepatitis are most likely to be associated with poor laboratory practices, poor personal hygiene, and failure to use simple barrier precautions.

3. The majority of cases in which the source of infection is identified could have been prevented by regular use of good laboratory safety practices.

In the years since Pike's 1976 summary (26), major changes have occurred in the diseases associated with occupational infections. In a joint advisory notice, the Department of Labor and the Department of Health and Human Services identified hepatitis B virus (HBV) as the primary occupational infectious hazard among the nation's 5 million health care workers, including an estimated 500,000 laboratory workers (40). These agencies estimated that 12,000 cases of HBV infection occur in health care workers each year, that an estimated 10% of these patients become chronic HBV carriers, and that 200 to 300 fatalities occur as a result of HBV-related disease. Other blood-borne pathogens (such as human immunodeficiency virus and the causative agents of syphilis, malaria, and hepatitis C) have similar transmission patterns and may pose an occupational hazard for health care workers.

These hazards prompted OSHA to promulgate regulations to protect health care workers, including laboratory personnel, from exposure to human blood and blood-borne pathogens. The OSHA blood-borne pathogens standard was published on 6 December 1991 and became effective on 6 March 1992 (44).

29.3. BIOHAZARD LEVELS

Although the number of laboratories in the United States is unknown, it is reasonable to assume that clinical laboratories (including physicians' office laboratories) constitute the most common type of laboratory operation. Clinical laboratory activities typically involve the handling, manipulation, and, sometimes, collection of blood, urine, feces, and other body fluids, excretions, or secretions. These clinical materials are all potentially infectious and may contain a variety of bacterial, fungal, viral, and protozoal agents. The concept of universal blood and body fluid precautions (universal precaution) should apply to all clinical specimens since it is not realistically possible to rule out all infectious agents on the basis of medical history and signs and symptoms at the time of medical examination or clinical diagnosis (38). It is also difficult to determine, with reasonable confidence, whether certain body fluids typically not associated with the transmission of blood-borne diseases may contain visible blood. Consequently, it is prudent to handle all body fluids, tissues, and cultures as potentially infectious. The estimated prevalence of HBV infection in the United States (approximately 300,000 new cases annually) would suggest that this viral disease is the occupational infection hazard most likely to be present in clinical materials submitted for laboratory examination.

HBV shares a common transmission pattern in the laboratory occupational setting with most bacterial pathogens that are regularly present in a community. Such agents may be transmitted parenterally, by ingestion, via broken skin, and following exposure of the mucous membranes of the mouth, eyes, or anterior nose. Characteristically, these pathogens are not transmitted in the laboratory via infectious aerosols.

Consequently, it is reasonable to use laboratory practices and barrier precautions that prevent exposure of laboratory workers from the infectious agent most likely to be present in specimens handled on a regular basis. It is especially important that clinical chemists, geneticists, and other scientists without specific training or experience in working with potentially infectious materials are made aware of such hazards and informed about appropriate preventive measures. These practices and precautions are succinctly summarized in the description of standard and special microbiological practices recommended for biosafety level 2 (39) and are discussed in detail in section 29.4.

A higher level of occupational hazard involves the handling of primary pathogens which may be transmitted not only by the routes listed above but also via infectious aerosols. Such infections include brucellosis, tularemia, tuberculosis, anthrax, rickettsial infections, and those caused by certain indigenous arboviruses. All of these except tuberculosis currently are diseases of low incidence and sporadic distribution. The standard and special practices and containment equipment described for biosafety level 3 are recommended for the handling of pathogenic organisms presenting an aerosol risk of infection (39); these are discussed further in section 29.5.

Tuberculosis, historically important as an occupational infection, continues to represent an especially significant infection hazard to laboratory workers, who may encounter this agent in sputum samples, biopsy materials, and in cultures. *Mycobacterium bovis* BCG vaccine, which is licensed and readily available, is not customarily used in the United States, so susceptibility to tuberculosis infection is essentially universal in this country. The adult infectious dose is estimated to be 10 or fewer organisms (46). Although the incidence of occupationally acquired tuberculosis is presently unknown, reports over several years attest to the continuing risk of infection with this classical and still important pathogen (15). A survey in 1976 indicated that the incidence of tuberculosis among medical laboratory workers in England was five times greater than that for the general community (11). Tuberculosis is an appropriate representative of occupational infections that may be transmitted via infectious aerosols.

Level 1 biosafety is not appropriate for work with pathogenic bacteria. This level was described in guidelines developed by CDC and NIH (39) to indicate basic practices appropriate for undergraduate and secondary educational training and teaching laboratories for which introductory microbiological protocols would involve only defined and characterized strains of viable microorganisms not known to cause disease in healthy adult humans. Emphasis is placed on the use of mechanical pipetting aids; handwashing; not eating, drinking, or smoking in the work area; and daily decontamination of work surfaces. This minimal level of containment has been adopted by molecular biology laboratories involved in research activities for which the host-vector systems in use have been exempted from the NIH guidelines for handling recombinant DNA molecules (37).

Level 2 biosafety is described below, and Level 3 is covered in section 29.5.

29.4. LEVEL 2 BIOSAFETY PRACTICES

The consistent use of good practices for the manipulation of potentially infectious materials is the single most important element in preventing occupational infections. As emphasized in the introduction to Section VI, *all* bacteria should be treated as potential hazards to human health. Consequently, level 2 safety practices should be adopted even in general and molecular bacteriology laboratories.

Level 2 biosafety practices are obligatory for most laboratory activities involving known pathogens and for experiments involving either the introduction of recombinant DNA into pathogenic bacteria or the introduction of DNA from pathogenic bacteria into nonpathogenic procaryotes or lower eucaryotes. *Brucella* spp., *Francisella tularensis, Mycobacterium bovis,* and *Mycobacterium tuberculosis* require level 3 containment because of the serious hazard of aerosol infection. However, DNA from these bacteria can be cloned in nonpathogenic procaryotic or lower-eucaryotic host-vector systems under level 2 containment (37). The deliberate formation of recombinant DNAs containing genes for the biosynthesis of toxic molecules lethal for vertebrates may require NIH approval before experiments are initiated.

Use the following level 2 biosafety practices routinely when handling even potentially pathogenic bacteria. The bacteria capable of causing serious disease when inhaled in infectious aerosols require level 3 containment.

1. Keep laboratory doors closed when work is in progress.
2. Post a hazard warning sign incorporating the universal biohazard symbol (Fig. 1) on the laboratory door when special provisions for entry are required. The OSHA blood-borne pathogens standard requires the sign to be orange or orange-red with symbol and lettering in a contrasting color (44). Also display the biohazard symbol on freezers, refrigerators, and other units used to store infectious materials.
3. Decontaminate work surfaces daily and immediately after spills of viable material.
4. Decontaminate infectious liquid or solid wastes before disposal. If the wastes have to be moved to a site away from the laboratory, place them in a durable leakproof container and close it before removal from the laboratory.
5. Always use a mechanical pipetting device when pipetting.
6. Do not permit eating, drinking, smoking, storage or preparation of food for human consumption, or cosmetics application in the laboratory.
7. Wear protective gloves when contact with infectious material may be unavoidable. Remove gloves carefully when they become soiled with contaminated material, and wash hands immediately. Put on a new pair of gloves before continuing the procedure. Do not wash or reuse soiled gloves.
8. Wash hands as soon as a procedure is completed and always before leaving the laboratory.
9. Avoid the use of syringes, needles, and other sharp instruments whenever possible. When they are required, handle them with caution to avoid accidental punctures or cuts. Use only needle-locking syringes or disposable syringe-needle units for injection or aspiration of infectious fluids.
10. Discard used needles and disposable cutting instruments into a puncture-resistant container which has a lid.
11. Wear laboratory coats, gowns, smocks, or uniforms while in the laboratory.
12. Exercise care in all procedures and manipulations to minimize aerosol formation. Use biological safety cabinets or physical containment devices such as centrifuge safety cups to contain operations that will generate aerosols capable of causing significant contamination or hazardous exposure.

Several routine operations in the laboratory can easily become the source of laboratory-acquired infections. They require vigilance to guard against compromise and error. The following sections emphasize the hazards associated with these operations and underscore the precautions necessary to conduct them safely.

29.4.1. Personal Hygiene

Appropriate personal-hygiene practices for laboratory workers have the simple objective of reducing personal risks of self-infection by avoiding ingestion, skin, or mucous-membrane exposure to infectious materials. Wash hands with soap and water following the completion of bench activities, the removal of protective gloves, and the overt contamination of skin and when leaving the laboratory. Handwashing is among the most important actions that workers can take in reducing the risk of infection by agents commonly handled in biomedical laboratories. A 10-s handwash with soap and water reduces the number of contaminating transient microorganisms to levels below the infective threshold. Use powdered soap from a dispenser rather than bar or liquid soap, which may harbor bacteria (see section 29.7.7). Follow handwashing by drying with disposable towels. Handwashing may also have the added benefit of reducing the level of cross-contamination between specimens or activities.

Wash hands promptly after removing protective gloves. Tests show that it is not unusual for bacteria to be present despite the use of gloves, as a consequence of unrecognized small holes, abrasions, or tears, entry at the wrist, or solvent penetration. A disinfectant wash or dip may be desirable, but its use must not be carried to the point of causing roughening, desiccation, or sensitization of the skin.

Work with bacteria should not be performed by anyone with a new or old cut, an abrasion, a lesion of the skin, or any open wound, including that resulting from a tooth extraction.

Food, candy, gum, and beverages for human consumption should not be stored or consumed in the laboratory. Smoking should not be permitted. A beard is undesirable because it retains particulate contamination and because a clean-shaven face is essential to the

ADMITTANCE TO AUTHORIZED PERSONNEL ONLY

Hazard identity: ______________________________

Responsible Investigator: ______________________

In case of emergency call:

Daytime phone ____________ Home phone ____________

Authorization for entrance must be obtained from
the Responsible Investigator named above.

FIG. 1. Biohazard warning sign for biomedical facilities.

adequate fit of a face mask or respirator. Keep hands away from the mouth, nose, eyes, face, and hair to prevent self-inoculation. Keep personal items such as coats, hats, storm rubbers or overshoes, umbrellas, and purses out of the laboratory.

29.4.2. Pipetting

Mouth pipetting and suctioning have been identified as practices that are significant and consistent causes of occupational infections (19, 21). Accidental aspiration of cultures or other infectious fluids into the mouth may result in infection from ingestion or oral exposure. Transfer of infectious agents from contaminated surfaces to the pipette mouthpiece by bare or gloved hands may also result in infection. A third main hazard of both mouth and manual pipetting is the creation of infectious aerosols when the residual contents of pipettes are forcibly expelled in the breathing zone of the worker.

There is also the danger of inhaling aerosols created in the handling of liquid suspensions when using unplugged pipettes, even if no liquid is drawn into the mouth. Additional hazards of exposure to aerosols are created by dropping liquid from a pipette to a work surface, by mixing cultures with alternate suction and blowing, by forcefully ejecting an inoculum onto a culture dish, or by blowing out the last drop.

Develop pipetting techniques that reduce the potential for creating aerosols. Plug pipettes with cotton. Avoid rapid mixing of liquids by alternate suction and expulsion. No material should be forcibly expelled from a pipette, and air should not be bubbled through liquids with a pipette. Pipettes that do not require expulsion of the last drop are preferable. Take care to avoid dropping cultures from the pipette. When pipetting materials containing pathogenic bacteria, place a disinfectant-soaked towel on the work surface. Discharge liquids from pipettes as close as possible to the fluid or agar level of the receiving vessel, or allow the contents to run down the wall of the receiving vessel. Avoid dropping the contents from a height into vessels. Place contaminated pipettes into a container with enough suitable disinfectant for complete immersion.

A variety of pipetting aids are available, ranging from simple bulb- and piston-actuated devices to sophisticated devices which contain their own vacuum pumps (18). In selecting a pipetting aid, consider the kind of procedure that is to be performed, the ease of operation, and the accuracy needed. Use several pipetting aids; none is available that is appropriate for all procedures.

29.4.3. Hypodermic Syringes and Needles

Accidents involving the use of syringes and needles are among the most common cause of occupational infections in laboratories and health care facilities. They account for an estimated one-fourth of laboratory-acquired infections that are caused by accidents (28). Needle "sticks" (punctures) account for an estimated one-third of work-related accidents in health care workers (12). In some health care facilities, annual injury rates from being accidentally stuck by a needle may exceed 100 per 1,000 employees, and these injuries may be underreported by 40 to 70% (1, 12). Accidental parenteral inoculation of infectious fluids during phlebotomy (bloodletting) procedures is the most obvious and common means of occupational infection associated with needles. Infections have also been associated with sprays or aerosols of infectious materials when needles separate from syringes and from direct or indirect skin contact with infectious fluids from leaking syringes or needles.

Of seven categories of phlebotomy devices used in a university hospital, disposable syringes are the devices most commonly associated with occupational needle sticks (12). However, when the data comparing the number of injuries with the number of devices purchased and presumably used were reviewed, disposable syringes had the lowest rate of associated injuries among all categories of devices used. Injuries associated with the use of disposable syringes most commonly occurred during the recapping of used needles.

In contrast to health care activities, when needles and syringes or other phlebotomy devices are commonly used to collect blood and other fluids from patients, there are relatively few indications for the use of such devices in most ordinary laboratory activities. Restrict the use of needles and syringes to the collection of blood or fluids from patients or experimental animals, the injection of materials into patients or experimental animals, and the aspiration of reagents or other materials from containers with diaphragm stoppers. Conduct all other manipulations of potentially infectious or hazardous fluids with mechanical pipettes or devices other than needles and syringes.

In response to the continuing problem of needle sticks, manufacturers of phlebotomy devices have developed a variety of "safety syringes" (e.g., a retractable needle). The use of these devices may provide additional safeguards during phlebotomy procedures. The wearing of gloves may also further reduce the risk of occupational exposures of skin to body fluids during phlebotomy and other procedures involving the collection or handling of potentially infectious body fluids.

The following practices are recommended for hypodermic needles and syringes when used for parenteral injections.

1. Avoid quick and unnecessary movements of the hand holding the syringe.
2. Examine glass syringes for chips and cracks, and examine needles for barbs and plugs. Do this before sterilization and before use.
3. Use needle-locking (Luer-Lok type) syringes only, and be sure that the needle is locked securely.
4. Wear surgical or other rubber gloves.
5. Fill syringes carefully to minimize air bubbles and frothing.
6. Expel excess air, liquid, and bubbles vertically into a cotton pledget moistened with a suitable disinfectant or into a small bottle of sterile cotton.
7. Do not use a syringe to forcefully expel infectious fluid into an open vial or tube for the purpose of mixing. Mixing with a syringe is appropriate only if the tip of the needle is held below the surface of the fluid in the tube.
8. If syringes are filled from test tubes, take care not to contaminate the hub of the needle, since this may result in transfer of infectious material to the fingers.
9. When removing a syringe and needle from a rubber-stoppered bottle, wrap the needle and stopper in a cotton pledget moistened with a suitable disinfectant. If there is danger that the disinfectant may contaminate sensitive experimental materials, a sterile dry pledget may be used and discarded immediately into a disinfectant solution.
10. Inoculate animals with the hand positioned behind the needle to avoid punctures.
11. Be sure that the animal is properly restrained prior to inoculation, and be alert for any unexpected movements.
12. Before and especially after injecting an animal, swab the injection site with a disinfectant.
13. Do not bend, shear, recap, or remove the needle from the syringe by hand. Place used needle-syringe units directly into a puncture-resistant container, and decontaminate before disassembly, reuse, or disposal.

29.4.4. Opening Containers

The opening of vials, flasks, petri dishes, culture tubes, embryonated eggs, and other containers of potentially infectious materials poses potential but subtle risks of creating droplets, aerosols, or contamination of the skin or the immediate work area.

Because the level of energy applied in such activities is relatively low, the efficiency for creating aerosols is also relatively low and the droplet is typically large. The potential for creating respirable droplets (in the 1- to 5-μm range) is relatively low. However, the potential for creating larger droplets, which may contaminate hands and immediate work surfaces, is relatively high.

The most common opening activity in most health care laboratories is the removal of stoppers from containers of clinical materials (e.g., blood, urine, spinal fluid, feces) submitted for examination. Seventeen percent of blood samples received in a clinical laboratory have blood on the container label, 6% of stool specimens have feces on the outside of the container, and 4 to 5% of laboratory specimen forms are visibly blood-stained (5). Specimens should be received and opened only by personnel who are knowledgeable about occupational infection risks and of the concept of Universal Precautions. Open clinical specimen containers in well-lighted designated areas with impervious and easily cleanable work surfaces. Wear a laboratory coat and suitable gloves. The use of plastic-backed absorbent paper towels on work surfaces facilitates cleanup and minimizes the creation of aerosols in the event of overt or unobserved spills of infectious materials. Some laboratories have found it to be advantageous to open all incoming specimens in the work chamber of a class I or class II biological safety cabinet, which is especially useful if incoming specimens are broken or leaking.

Appropriate techniques for opening sealed ampoules, dry powders, sealed containers under relatively negative pressure, embryonated eggs, and the like are described in detail elsewhere (19). The use of appropriate personal protection (coats, gloves), primary containment (biological safety cabinet), and physical barriers (e.g., acrylic "beta" shields) generally addresses the ubiquitous problems of exposure of personnel and contamination of work surfaces during the opening of containers.

Tubes containing bacterial suspensions should be manipulated with care. Simple procedures such as removing a tube cap or transferring an inoculum can create aerosols. Clearly mark tubes and racks of tubes containing biohazardous material. Use safety test tube trays in place of conventional test tube racks to minimize spillage from broken tubes. A safety test tube tray is one that has a solid bottom and sides deep enough to hold the liquid should the test tube break.

Vigorous shaking of liquid cultures creates a heavy aerosol. A swirling action will generally create a homogeneous suspension with a minimum of aerosol. When resuspending a liquid culture, allow a few minutes to elapse before opening the container, to reduce the aerosol.

The insertion of a sterile hot loop or needle into a liquid or slant culture can cause spattering and release of aerosol. To minimize aerosols, allow the loop to cool in the air or be cooled by touching to the inside of the container or to the agar surface where no growth is evident prior to contact with the culture or colony. After use of the inoculating loop or needle, sterilize it in an electric or gas incinerator specifically designed for this purpose rather than heating it in an open flame. These small incinerators have a shield to contain any material that may spatter from the loop. Disposable inoculating loops are available commercially; rather than decontaminating them with heat, discard them into a disinfectant.

Streaking an inoculum on rough agar results in aerosol production, created by the vibrating loop or needle. This generally does not occur if the operation is performed on smooth agar. Discard all rough agar poured plates intended for streaking with a wire loop.

Water of syneresis in petri dish cultures may contain viable bacterial and can form a film between the rim and lid of the inverted plate. Aerosols result when this film is broken by opening the plate. Vented plastic petri dishes, whose lid touches the rim at only three points, are less likely to create this hazard. Filter papers fitted into the lids reduce but do not prevent aerosols. If plates are obviously wet, open them in a safety cabinet.

Less obvious is the release of aerosols when screw-cap bottles or plugged tubes are opened. This happens when a film of contaminated liquid between the rim and the liner is broken during opening. Removing cotton plugs or other closures from flasks, bottles, and centrifuge tubes immediately after shaking or centrifugation can release aerosols. Removal of wet closures, which can occur if the flask or centrifuge tube is not held upright, is also hazardous. In addition, when using the centrifuge, there may be a small amount of foaming and the closure may become slightly moistened. Because of these possibilities, open all liquid cultures of infectious material in a safety cabinet and wear gloves and a long-sleeved laboratory garment while doing so.

When a sealed ampoule containing a lyophilized or liquid culture is opened, an aerosol may be created. Open ampoules in a safety cabinet. When recovering the contents of an ampoule, take care not to cut the gloves or hands or to disperse broken glass into the eyes, face, or environment. In addition, the material itself should not be contaminated with foreign organisms or with disinfectants. To accomplish this, work in a safety cabinet and wear gloves. Nick the ampoule with a file near the neck. Wrap the ampoule in disinfectant-wetted cotton. Snap the ampoule open at the nick, being sure to hold it upright. Alternatively, apply a hot wire or rod at the file mark to develop a crack. Then wrap the ampoule in disinfectant-wetted cotton, and snap it open. Discard the cotton and ampoule tip into disinfectant. Reconstitute the contents of the ampoule by slowly adding fluid to avoid aerosolizing the dried material. Mix the contents without bubbling, and withdraw them into a fresh container. Commercially available ampoules prescored for easy opening are available. However, prescoring may weaken the ampoule and cause it to break during handling and storage. Open ampoules of liquid cultures in a similar way.

Harvesting cultures from embryonated eggs is a hazardous procedure and leads to heavy contamination of the egg trays and shells, the environment, and the hands of the operator. Conduct operations of this type in a safety cabinet. A suitable disinfectant should be at hand and used frequently.

29.4.5. Centrifuging

Centrifugation is a common procedure in laboratories where infectious materials are handled. The demonstrated hazards are, however, relatively low and are typically confined to diseases which have low infective doses and which are efficiently transmitted by means of aerosols (e.g., tuberculosis, tularemia, brucellosis, Q fever, and certain arbovirus infections). Historically, infections associated with centrifugation typically have occurred as sporadic clusters of cases and are often associated with broken or leaking specimens in open, low-speed, clinical centrifuges (19, 45).

Blood-borne pathogens (e.g., *Treponema pallidum*, hepatitis B virus) are not transmitted by infectious aerosols in the community and have not been documented to be transmitted by infectious aerosols in the laboratory. The viscosity of blood and other body fluids and the use of low-speed centrifugation may further reduce the production of infectious droplet nuclei and minimize occupational infection risks.

The use of centrifuge safety cups and enclosed or safety centrifuges (45) for low-speed centrifugation and rotors with O-ring seals for high-speed centrifugation may effectively reduce the hazards associated with these centrifugation procedures. Mechanical failure of centrifuges can be minimized by appropriate servicing and maintenance and careful attention to good technique. Infectious or sensitizing materials, concentrated by centrifugation, should be handled by using appropriate primary containment and personal protection devices.

Use safety centrifuge trunnion cups when centrifuging infectious bacterial suspensions. Fill and open centrifuge tubes and trunnion cups in a class I safety cabinet. Decontaminate the outside surface of the cups before removing them from the cabinet. Since some disinfectants are corrosive, rinse with clean water after an appropriate contact time has elapsed.

Before centrifuging, eliminate tubes with cracks and chipped rims, inspect the inside of the trunnion cups for rough walls caused by erosion or adhering matter, and carefully remove bits of glass and other debris from the rubber cushion.

Add a disinfectant between the tube and trunnion cup to disinfect the materials in case of accidental breakage. This also provides an excellent cushion against shocks that might break the tube. Take care not to contaminate the culture material with the disinfectant. Also, if the tube breaks, the disinfectant may not completely inactivate the infectious material because of the dilution of the disinfectant and because of the high concentration of packing of the cells.

Avoid pouring supernatant fluid from centrifuge tubes. If this must be done, wipe off the outer rim with a disinfectant afterwards; otherwise, in a subsequent step, infectious fluid may be spun off as droplets and form an aerosol. Use of vacuum system with appropriate in-line safety reservoirs and filters is preferable to pouring from centrifuge tubes or bottles.

Avoid filling the centrifuge tube to the point that the rim, cap, or cotton plug becomes wet with culture. Screw caps or caps that fit over the rim are safer than plug-in closures. Some fluid usually collects between a plug-in closure and the rim of the tube. Even screw-cap bottles are not without risk; if the rim is soiled and sealed imperfectly, some fluid will escape down the outside of the tube.

Do not use aluminum foil to form centrifuge tube caps, because they often become detached or rupture during handling and centrifuging.

The balancing of buckets and trunnion cups is often improperly performed. Take care to ensure that matched sets of trunnions, buckets, and plastic inserts do not become mixed. If the components are not inscribed with their weights by the manufacturer, apply colored stains for identification.

High-speed rotor heads are prone to metal fatigue, and if there is a chance that they may be used on more than one machine, each rotor should be accompanied by its own log book indicating the number of hours run at top or derated speeds. Failure to observe this precaution can result in dangerous and expensive disintegration. Frequent inspection, cleaning, and drying are important to ensure the absence of corrosion or other damage that may lead to the development of cracks. If the rotor is treated with a disinfectant, rinse it with clean water and dry it as soon as the disinfectant has adequately decontaminated the rotor. Examine rubber O rings and tube closures for deterioration, and keep them lubricated as recommended by the manufacturer. When tubes of different materials (e.g., celluloid, polypropylene, stainless steel) are provided, take care that tube closures designed specifically for the type of tube are used. Caps are often similar in appearance but are prone to leak if applied to tubes of the wrong material. Properly designed tubes and rotors, when well maintained and handled, should never leak.

29.4.6. Mixing and Disruption

Manual or mechanical mixing, grinding, or disruption of infectious materials has a high potential for producing infectious aerosols. The more vigorous the manipulation, the more likely that respirable (1- to 5-μm) droplet nuclei may be created and liberated into the laboratory. Droplet nuclei, containing infectious agents, may remain suspended indefinitely and pose infection risks long after completion of the activity. Infectious droplet nuclei may also be transported to other laboratory and nonlaboratory areas by personnel or via the building ventilation system. Larger droplets tend to settle rapidly on horizontal surfaces and may serve as direct and indirect sources of infection or contamination.

All manipulations of known or potentially infectious materials should be conducted in such a manner that the contamination of hands, clothing, and work surfaces and the production of aerosols are avoided. These occupational hazards are best addressed by the routine use of leakproof equipment (safety blenders) in combination with protective clothing and primary containment devices (biological safety cabinets).

It is important that the selection of devices and containment used for grinding, mixing, or disruption be based on the assessed infection risks or the consequences of infection of the agent being handled. For example, although the recommended barrier and containment precautions for grinding brain tissue infected with rabies virus or yolk sacs infected with Q fever rickettsia are the same, the demonstrated infection risks are remarkably different. As few as 10 viable rickettsia can produce infection via the respiratory route. On the other hand, occupational infection with rabies virus via the respiratory route has not been demonstrated in personnel working with clinical materials and conducting diagnostic activities.

When using blenders, mixers, ultrasonic disintegrators, colloid mills, jet mills, grinders, and mortars and pestles with pathogenic bacteria, do the following.

1. Use a biological safety cabinet always.
2. Use safety blenders designed to prevent leakage from the rotor bearing at the bottom of the bowl. In the absence of a leakproof rotor, inspect the rotor bearing for leakage prior to operation. Test it in a preliminary run with sterile water, saline, or methylene blue solution.
3. Use a towel moistened with disinfectant over the top of the blender. Sterilize the device and residual contents promptly after use.
4. Do not use glass blender bowls with infectious materials.
5. Blender bowls sometimes require cooling to prevent destruction of the bearings and to minimize thermal effects on the contents.
6. Before opening the blender bowl, wait for at least 1 min to allow the aerosol to settle.

29.4.7. Handling Animals

Exposure to experimental animals with natural or induced infections is among the most frequently identified sources of occupational infections in laboratory workers. Of 703 occupational infections for which the source of infection was identified, almost 14% were associated with bites or scratches by experimental animals or exposure to their ectoparasites (39). The current use of research animals is, however, largely restricted to facilities involved in research and other specialty activities, rarely in clinical and diagnostic laboratories.

The practices, safety equipment, and facilities recommended for animal biosafety level 2 constitute the basic and minimal conditions for working with naturally or experimentally infected vertebrate animals (31).

Airflow in experimental-animal rooms should be inward and nonrecirculating. Ideally, only one species of animal and one infectious agent should be used for experiments within the same room. Access should be restricted to those working with the animals or providing services to the area. All personnel entering the animal room should wear appropriate protective clothing (e.g., coats and gloves) and meet the requirements for entry, including immunization(s). Physical and chemical restraints appropriate to the species involved in the study should be routinely used. Appropriate physical barriers and primary containment caging should be provided to minimize cross-contamination and exposure of personnel working with agents which may be transmitted by infectious aerosols. Cages for animals used for infectious-disease studies should be autoclaved or chemically decontaminated before cleaning and reuse. A biohazard warning sign indicating the agent(s) in use and conditions for entry should be posted on or near the animal room door.

These combined safety practices supplement the applicable requirements of the Department of Agriculture Animal Welfare Act and the recommendations of the Public Health Service *Guide for the Care and Use of Laboratory Animals* (36).

Animals are capable of shedding pathogens in saliva, urine, or feces. In the absence of information to the contrary, all animals should be regarded as shedders. Careful handling procedures should be used to minimize the dissemination of dust from animal and cage refuse, and used cages should be sterilized by autoclaving. Refuse, bowls, and watering devices should remain in the cage during sterilization.

Wear heavy gloves when feeding, watering, handling, or removing infected animals. Do not insert bare hands in the cage to move any object.

When injecting bacteria into animals, wear protective gloves. Make sure that animals are properly restrained (e.g., use a squeeze cage for primates) or tranquilized to avoid accidents that might disseminate infectious material or cause injury to the animal or personnel.

The oversize canine teeth of large monkeys present a biting hazard and are important in the transmission of naturally occurring, very dangerous monkey virus infections. Such teeth should be blunted or surgically removed by a veterinarian. Many zoonotic diseases, including infectious hepatitis and tuberculosis, can be transmitted from nonhuman primates to humans. Newly imported animals may be naturally infected with these or other diseases, and persons in close contact with such animals may become infected. Personal protective equipment or cage systems designed to contain infectious material should be used.

Keep doors to animal rooms closed at all times, except for necessary entrance and exit. Do not permit unauthorized persons in animal rooms.

Keep a container of disinfectant, prepared fresh each day, in each animal room for disinfecting gloves and hands and for general decontamination, even though no infectious animals are present. Wash hands, floors, walls, and cage racks with an approved disinfectant at the recommended strength at regular and frequent intervals.

Fill floor drains in animal rooms with water or disinfectant periodically to prevent backup of sewer gases. Do not wash shavings and other refuse on floors down the floor drain, because such refuse clogs sewer lines.

Maintain an insect and rodent control program in all animal rooms and in animal food storage areas. Take special care to prevent live animals, especially mice, from finding their way into disposable trash.

Carry out necropsy of infected animals in a safety cabinet. Wear surgeon's gowns over laboratory clothing, and wear rubber gloves. Wet the fur of animals with a suitable disinfectant.

On completion of a necropsy, place all potentially infectious materials in suitable containers and sterilize immediately. Place contaminated instruments in a horizontal tray containing a suitable disinfectant. Disinfect the inside of the safety cabinet as well as other potentially contaminated surfaces. Clean grossly contaminated rubber gloves in disinfectant before removal from the hands, preparatory to sterilization. Place dead animals in proper leakproof containers, autoclave, and tag properly before removing for incineration.

29.4.8. Vacuum Systems

The use of vacuum suctioning devices for aspiration and collection of biological fluids (e.g., tissue culture media, suspension fluids) for discard is a convenient and common laboratory practice. Vacuum systems are of two general types: small individual pumps serving a single laboratory or activity; and centralized "house" systems serving a building or multiple users. There are obvious safety advantages to the former system. Small systems, with short distances between the point of suction and the point of discard, can be effectively protected against aspiration of infectious aerosols or fluids into the vacuum system.

Protect the collection flask for discarded liquids by using a downstream overflow flask. The use of antifoaming agents and disinfectants is recommended for both of these containers. Distal to the collection and overflow containers, use an in-line cartridge-type filter to protect the vacuum pump (19). Empty the collection flask and overflow flask periodically or when the contents constitute approximately two-thirds of the container's capacity. Clamp off the flexible vacuum tubing during transport to the autoclave. Then remove the clamp and autoclave the entire apparatus. Following autoclaving, discard the treated fluids into the sanitary sewer. Use heavy-walled, autoclavable, nonglass (Nalgene) sidearm flasks for the primary and overflow containers. Autoclave the in-line filter, and replace it periodically or whenever it becomes overly contaminated with fluids from the overflow flask.

29.4.9. Miscellaneous Operations

Water baths used to inactivate, incubate, or test bacteria should contain a disinfectant. For cold-water baths, 70% propylene glycol is recommended. *CAUTION:* Sodium azide creates an explosive hazard and should not be used as a disinfectant.

Freezers, liquid nitrogen, dry-ice chests, and refrigerators should be checked, cleaned, and decontaminated periodically. Use rubber gloves and respiratory protection during cleaning. Label all infectious material stored in refrigerators or freezers.

Hazardous fluid cultures or viable powdered infectious materials in glass vessels should be transported, incubated, and stored in easily handled, nonbreakable, leakproof containers large enough to contain all the fluid or powder in case of leakage or breakage of the glass vessel.

Inoculated petri plates or other inoculated solid media should be transported and incubated in leakproof pans or leakproof containers.

Exercise care when using membrane filters to obtain sterile filtrates of infectious materials. Because of the fragility of the membrane and other factors, consider the filtrates to be infectious until culture tests have proved their sterility.

Examine shaking machines carefully for potential breakage of flasks or other containers being shaken. Use durable plastic or heavy-walled screw-cap glass flasks, and securely fasten them to the shaker platform. As an additional precaution, enclose each flask in a plastic bag with or without an absorbent material.

Never work alone on a hazardous operation.

29.5. LEVEL 3 BIOSAFETY PRACTICES

Practices that provide a higher level of containment than that afforded by biosafety level 2 must be used when handling microorganisms whose potential for causing serious or lethal infection by the inhalation route is real. Two principal practices separate biosafety level 3 from biosafety level 2. At biosafety level 3, the laboratory director should restrict entry into the laboratory to persons whose access is required for program or support purposes. Second, all activity involving infectious materials should be conducted in biological safety cabinets or other physical containment devices. This is important because of the risk of airborne transmission. In addition, the laboratory facility in which biosafety level 3 practices are used meet higher standards than those normally associated with conventional laboratories, as was indicated in section 29.1.3.

29.6. BIOSAFETY CABINETS

Biological safety cabinets are used to reduce the spread of contamination and to eliminate exposures to microbial aerosols. Activities involving pathogens that cause infection through inhalation should be conducted in safety cabinets or other containment devices since most laboratory techniques produce inadvertent aerosols. Class I and II biological safety cabinets are recommended for use at biosafety levels 2 and 3, respectively. Protection against aerosols generated by work activities is provided by the movement of room air into the front opening of the cabinet. Safety can be compromised if this inward air flow is disrupted by other room air currents. For this reason, biological safety cabinets should be located in laboratory areas away from doorways, supply air diffusers, and spaces of high activity. Generally, the best location for the cabinet is on a side wall at the position farthest from the door.

Cabinet users should be encouraged to wear long-sleeved gowns with knit cuffs and gloves to protect the hands and arms from contact contamination. The work surface of the cabinet should be decontaminated before equipment and material are placed into the cabinet. Ideally, everything needed for the procedure should be placed in the cabinet before work is started. In class II cabinets, nothing should be placed over the exhaust grilles at the front and rear of the work surface.

The arrangement of equipment and material on the work surface is an important consideration. Contaminated items should be segregated from clean items and located so that they are not passed over clean items. Discard trays should be located at the rear. The use of aseptic techniques is important to prevent contact contamination. All infectious items should be contained or disinfected before removal from the cabinet. Trays of discarded pipettes and glassware should be covered before they are removed. After all items have been removed from the cabinet, the work surface should be decontaminated.

Biosafety cabinets may be obtained from various manufacturers (e.g., LABCONCO, Kansas City, Mo.; Nu-Aire, Inc., Minneapolis, Minn.; Baker Co., Inc., Sanford, Maine).

29.6.1. Class I Cabinets

The class I cabinet (Fig. 2) may be used in one of three modes: with a full-width open front, with an installed front panel without gloves, or with the panel equipped with arm-length rubber gloves. Materials may be introduced and removed through the panel opening and, if provided, through a hinged front-view panel or a side airlock. Room air entering the cabinet prevents the escape of airborne contaminants. The air flows across the work space, over and under a back wall baffle, and out through a high-efficiency particulate air (HEPA) filter and blower in an overhead duct to the building air exhaust system or outdoors. When operated with the front open, a minimum inward face velocity of 75 ft (22.9 m)/min is needed.

The protection afforded by a class I cabinet can be compromised by sudden withdrawal of the hands, rapid opening and closing of the room door, or rapid movements past the front of the cabinet. Aerosols created forcefully and in large quantities may escape despite the inward flow of air. Also, the cabinet does not protect the hands and arms from contact with hazardous materials. Such protection is dependent on technique and on the use of gloves and other protective clothing.

29.6.2. Class II Cabinets

Class II cabinets (Fig. 3) utilize nonturbulent (laminar) airflow and have front openings for access to the work space and for introduction and removal of materials. Airborne contaminants in the cabinet are prevented from escaping by a curtain of air formed by unfiltered air flowing from the room into the cabinet and by HEPA-filtered air supplied from an overhead grille. The air curtain also prevents airborne contaminants in the room air from entering the work space of the cabinet. The air curtain is drawn through a grille at the forward edge of the work surface into a plenum, where it is HEPA filtered and recirculated through the overhead grille down into the cabinet. A portion of the air is used to maintain the air curtain, and the remainder passes down onto the work surface and is drawn out through a grille at the back edge of the work surface. The filtered air from the overhead grille flows in a uniform downward movement to minimize turbulence. This air provides and maintains a filtered air work environment. A percentage of air drawn through the front and back grilles of the work surface, which is equal to the flow of room air into the cabinet, is also filtered and exhausted from the cabinet. There are several configurations of class II cabinets; they vary primarily in the percentage of filtered air flow that is recirculated within the cabinet.

Consensus standards have been developed by the National Sanitation Foundation for the design, construction, and performance of class II biological safety cabinets (22). These standards should be strictly adhered to for the purchase and on-site certification of class II cabinets. It is generally recommended that class II cabinets be recertified annually or after they have been moved to another location.

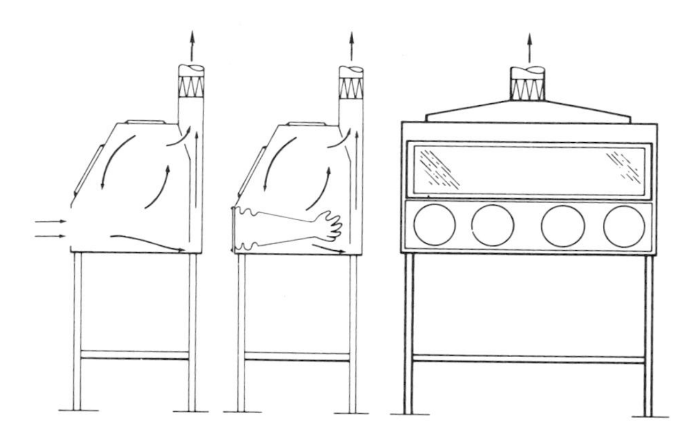

FIG. 2. Class I biological safety cabinet.

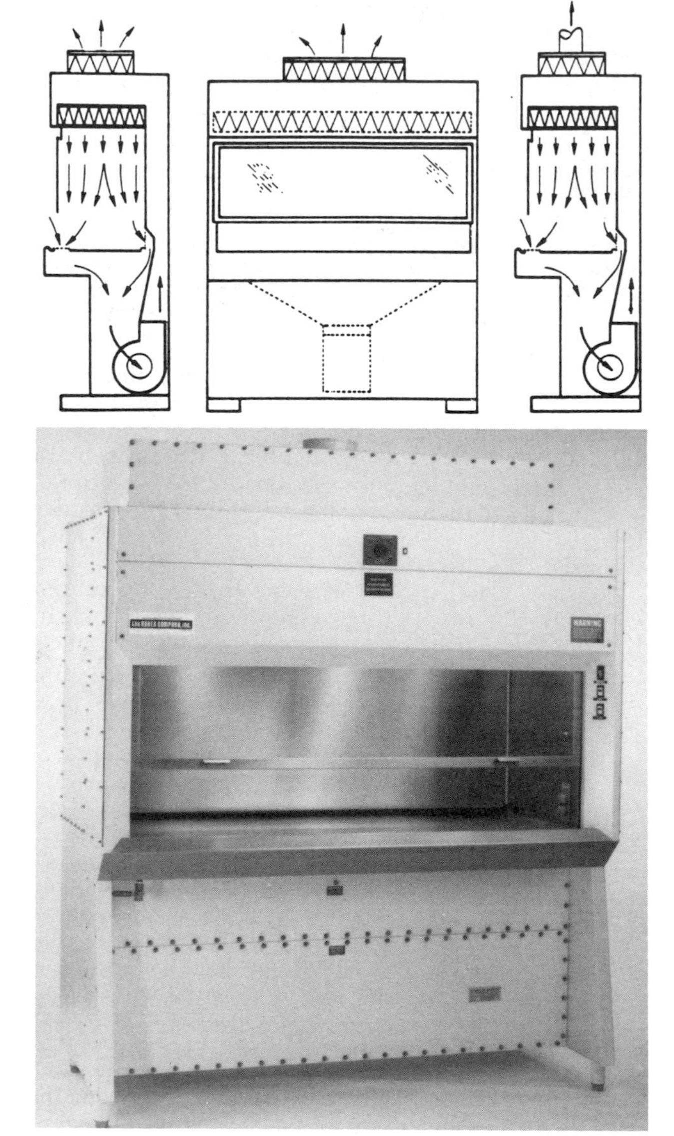

FIG. 3. Class II biological safety cabinet.

29.7. CHEMICAL DISINFECTION

Disinfection is the removal or destruction of pathogenic microorganisms from inanimate objects or surfaces, usually by use of a chemical agent, and is needed to protect the integrity of bacteriological test results or to prevent the occurrence and spread of disease. Chemical disinfection is necessary because the use of pressurized steam, the most reliable method of sterilization, and other physical methods are not normally feasible for disinfection of large spaces, surfaces, and stationary equipment. Moreover, high temperatures and moisture often damage delicate instruments, particularly those with complex optical and electronic components.

Recommendations for the selection and use of chemical disinfectants in health care facilities have been comprehensively and recently compiled by Rutala for the Association for Practitioners of Infection Control (APIC) (30). While focused primarily on patient care items and environments, these recommendations have generic application in bacteriological laboratories, including general ones. Recommendations for high-, intermediate-, and low-level disinfection methods described in *Guidelines for Prevention of Transmission of HIV and HBV in Healthcare and Public-Safety Workers* (38) may be interpolated and applied from the health care to the laboratory setting. For further general information, see references 3 and 29a.

29.7.1. General Guide

The information summarized in Table 1 provides a general guide for the selection and use of chemical disinfectants in bacteriological laboratories based on experience gained in health care facilities. The APIC recommendations (30) and manufacturer claims should be carefully reviewed in the selection and use of chemical disinfectants in laboratory workplaces.

As a general principle and to ensure the maximum efficacy of the chemical decontamination process selected, the potentially contaminated item or surface must first be cleaned of visible soil. Appropriate precautions, including the wearing of protective gloves, should be taken to prevent exposure of personnel during this predisinfection process. The chemical disinfectant selected should be of demonstrated efficacy against target organisms and prepared and used in accordance with the written instructions of the manufacturers.

The registration of a chemical disinfectant with the Environmental Protection Agency (EPA) as a *tuberculocidal hospital disinfectant* is based on information provided to this agency by the manufacturer. Verification testing of manufacturer claims of antimicrobial or antiviral efficacy is not conducted by the EPA. The effectiveness of chemical disinfectants in meeting label claims may be significantly influenced by the presence of organic materials on items or surfaces subjected to disinfection, the age of the use solution, the concentration of the chemical used, and the time of exposure or contact.

The combining of two or more chemical disinfectants or disinfectant chemicals and cleaning chemicals may reduce rather than enhance the effectiveness of the cleaning or decontamination effort. Additionally, the intentional or inadvertent mixing of certain compounds (e.g., household bleach and mineral or organic acids or ammonia compounds) may liberate reaction products into the work environment which are toxic, corrosive, or caustic to personnel.

29.7.2. Disinfectant Types

There are many disinfectants available under a wide variety of trade names. In general, disinfectants are classified as acids or alkalis, halogens, heavy metals, quaternary ammonium compounds, phenolic compounds, aldehydes, ketones, alcohols, amines, and peroxides. Unfortunately, the more active the disinfectant, the more likely it is to possess undesirable characteristics, such as toxic or corrosive properties. No disinfectant is equally useful or effective under all conditions. Resistance to the action of chemical disinfectants can be substantially altered by such factors as concentra-

Table 1. Some considerations in the selection and use of chemical disinfectants in biomedical laboratories

Disinfection	Effect	Method	Use	Examples
High-level disinfection	Destroys all microbial life forms except bacterial spores present in large numbers	Exposure of contaminated items to an EPA-registered "sterilant" chemical in accordance with manufacturer use recommendations	Reusable instruments and devices that cannot be subjected to autoclave temperatures	Chlorine dioxide; glutaraldehyde; solutions containing 1,000 ppm of free available chlorine (1:50 dilution of household bleach); stabilized hydrogen peroxide
Intermediate-level disinfection	Destroys *M. tuberculosis*, other vegetative bacteria, and most viruses and fungi but not bacterial spores	EPA-registered "hospital disinfectants" (label claim for tuberculocidal activity)	Items or surfaces that are visibly contaminated with blood, body fluids, or other potentially infectious materials; surfaces must be pre-cleaned before application of chemical disinfectant	Solutions containing 500 ppm of free available chlorine (1:100 dilution of household bleach); iodophor; phenolics
Low-level disinfection	Destroys most bacteria, some viruses and fungi, but not *M. tuberculosis* or bacterial spores	EPA-registered "hospital disinfectants" (no label claim for tuberculocidal activity)	For routine housekeeping or removal of soils from environmental or work surfaces (e.g., bench work, biological safety cabinets) in the absence of visible blood contamination	Ethyl or isopropyl alcohol; solutions containing 100 ppm of free dilution of household bleach); quaternary ammonium, iodophor, or phenolic germicidal detergent solutions prepared and used as specified on the product label

tion of active ingredient, duration of contact, pH, temperature, humidity, and presence of organic matter.

29.7.2.1. Alcohols

Ethyl or isopropyl alcohol in a concentration of 50 to 70% is often used to disinfect surfaces. However, these agents are slow in their germicidal action (several minutes being required) and ineffective against spores.

29.7.2.2. Formaldehyde

For use as a disinfectant, formaldehyde is usually marketed as a 37% concentration of the gas in water solution (formalin) or as a solid polymerized compound (paraformaldehyde). Formaldehyde in a concentration of 5% is an effective liquid disinfectant and is an effective space disinfectant for sterilizing rooms or buildings. However, formaldehyde loses considerable disinfectant activity at refrigeration temperatures, and its pungently irritating odor requires that care be taken when using it in solution in the laboratory. The use of formaldehyde is also regulated by OSHA, and strict practices for protection of human health are mandatory.

29.7.2.3. Phenolic compounds

Phenol is not often used by itself as a disinfectant. The odor is unpleasant, and a gummy residue remains on treated surfaces. Phenolic compounds, however, are basic to a number of popular disinfectants which are effective against vegetative bacteria at high dilutions and are essentially odorless. They are not effective in ordinary usage against bacterial spores.

29.7.2.4. Quaternary ammonium compounds

Quaternary ammonium compounds are strongly surface active, which makes them good surface cleaners. They attach to proteins, and so in dilute solutions they lose effectiveness. They clump bacteria and are neutralized by anionic detergents, such as soap. They are bactericidal at medium concentrations, but they are not tuberculocidal or sporicidal even at high concentrations. They have the advantages of being odorless, nonstaining, noncorrosive to metals, stable, inexpensive, and relatively nontoxic.

29.7.2.5. Chlorine

Chlorine is a universal disinfectant which is active against all bacteria, including bacterial spores, and is effective over a wide range of temperatures. It combines readily with protein, so that an excess of chlorine must be used if proteins are present. Free, available chlorine is the active element. It is a strong oxidizing agent and is corrosive to metals. Chlorine solutions gradually lose strength, so that fresh solutions must be prepared frequently. Sodium hypochlorite is usually used as a base for chlorine disinfectants. An excellent disinfectant can be prepared from household or laundry bleaches, which usually contain 5.25% (52,500 ppm) available chlorine. If one dilutes them 1:100, the solution will contain 525 ppm of available chlorine, and if 0.7% of a nonionic detergent is added, a very good disinfectant is created. This is an effective disinfectant against blood-borne pathogens and *M. tuberculosis.*

29.7.2.6. Iodophors

Iodophors, organic iodine compounds, constitute one of the most popular groups of disinfectants used in the laboratory. Recommended use concentrations for commercial products provide 75 ppm of free iodine. This small amount can be rapidly taken up by extraneous proteins present. Clean surfaces or clear water can be effectively treated by 75 ppm of available iodine, but difficulties may be experienced if an appreciable amount of protein is present. A concentration of 1,600 ppm of available iodine is required to achieve effective sporicidal properties. An alcoholic solution (tincture) of iodine is effective as a disinfectant but is irritating.

29.7.2.7. Heavy metals

Heavy metals are not recommended. Mercury and other heavy-metal preparations are toxic, and they are more bacteriostatic than bactericidal.

29.7.3. Disinfectant Characteristics

Although the ideal disinfectant does not exist, it is useful to consider what characteristics it should have, as follows.

High activity: effectiveness at high dilutions in the presence of organic matter.

Broad spectrum of antimicrobial activity: effectiveness against gram-positive, gram-negative, and acid-fast bacteria and against spores, viruses, and fungi.

Stability: retention of potency after storage for prolonged periods.

Homogeneity: no settling out of active ingredients.

Adequate solubility: solubility in water, fats, and oils for good penetration into microorganisms.

Low surface tension: penetration into cracks and crevices.

Minimum toxicity: lack of acute and chronic toxicity, mutagenicity, carcinogenicity, teratogenicity, allergenicity, irritability, and photosensitization.

Detergent activity: ability to solubilize and remove dirt and debris.

Minimum material effects: low or acceptable effects on metals, wood, plastics, and paint.

Odor control: pleasant odor, odorless, or having deodorizing properties.

Cost: inexpensive in relation to efficiency.

The effectiveness of a disinfection process can be maximized by attention to the following factors.

1. Select a disinfectant appropriate against the microorganism to be inactivated. If its identity is unknown, select an agent with as broad a spectrum of activity as possible.
2. Reduce the bioburden and the organic content to a minimum on surfaces or objects to be disinfected.
3. Take into account the fact that most chemical disinfectants have limited effectiveness against bacterial spores, and allow additional exposure time for the effective ones.
4. Use the disinfectant in the proper concentration. Inadequate concentrations may result in lack of disin-

fection, and excessively concentrated solutions may pose problems of toxicity and effects on materials. The concentration may determine whether the action is static or cidal.

5. Carefully consider the exposure or contact time necessary for disinfection.

6. Be sure that the chemical concentration and contact time are compatible with the temperature of disinfection. In general, lower temperatures require longer contact times and higher temperatures increase efficiency two- to threefold per 10°C rise in temperature.

7. Make sure that the water used for preparing use dilutions of disinfectants is of the proper quality. With certain chemicals, water hardness (e.g., calcium ions) in excess of 300 to 400 ppm destroys disinfecting ability.

8. Consider the amount of organic matter present on the object being treated. Organic matter reacts with disinfecting chemicals and, in effect, removes the active ingredient of the solution.

9. No single chemical disinfectant or method will be effective or practical for all situations. When selecting a chemical disinfectant, the following questions should be considered.

(i) What is the target organism(s)?

(ii) What disinfectants in what form are known to, or can be expected to, inactivate the target organism(s)?

(iii) What degree of inactivation is required?

(iv) In what menstruum is the organism suspended (i.e., simple or complex, on solid or porous surfaces, or airborne)?

(v) What is the highest concentration of cells anticipated to be encountered?

(vi) Can the disinfectant (as either a liquid, a vapor, or a gas) be expected to contact the organisms, and can effective duration of contact be maintained?

(vii) What restrictions apply with respect to compatibility of materials?

(viii) What is the stability of the disinfectant in use concentrations, and does the anticipated use situation require immediate availability of the disinfectant or will sufficient time be available for preparation of the working concentration shortly before its anticipated use?

10. Actual concentrations and contact times used in the laboratory may differ from the recommendations of the manufacturers. The efficacy of the selected disinfectant procedure should be validated by the individual user. This is particularly important for laboratory activities that involve serious pathogens, such as *M. tuberculosis*.

29.7.4. Environmental Surfaces

Routine disinfection of floor surfaces is often a function of the housekeeping department. For laboratories and other areas where infectious material is handled, clean floors daily with a suitable disinfectant-detergent at the specified use dilution. Use a broad-spectrum disinfectant-detergent which is active against bacteria, fungi, viruses, and acid-fast organisms and is not materially affected by hard water or organic matter.

Disinfect bench tops, tables, and large equipment by using a disinfectant-detergent at the use dilution with a clean cloth, sponge, or disposable towel. Allow surfaces to remain damp, moist, or wet. If 0.5% hypochlorite solutions are considered, it must be remembered that they are easily neutralized by excess organic matter, may discolor surfaces, and are highly corrosive to metals.

29.7.5. Bacterial Spills

A spill that is confined to the interior of the biological safety cabinet should present little or no hazard to personnel in the area. However, initiate chemical disinfection procedures at once, while the cabinet ventilation system continues to operate, to prevent the escape of contaminants. Spray or wipe walls, work surfaces, and equipment with a disinfectant. A disinfectant with a detergent will help clean the surfaces by removing both dirt and bacteria. A suitable disinfectant is a 3% solution of an iodophor or a 1:100 dilution of household bleach with 0.7% nonionizing detergent. Wear gloves during this procedure. Use sufficient disinfectant solution to ensure that the drain pans and catch basins below the work surface contain the disinfectant. Lift the front exhaust grille and tray, and wipe all surfaces. Wipe the catch basin, and drain the disinfectant into a container. Discard the disinfectant, gloves, wiping cloth, and sponges into a pan, and autoclave them. This procedure will not disinfect the filters, blower, air ducts, or other interior parts of the cabinet.

If the entire interior of the cabinet is to be disinfected, formaldehyde gas should be used. The procedure is potentially hazardous and requires special knowledge and equipment. It should be conducted only by experienced personnel. For this procedure, most institutions hire professional companies that certify biological safety cabinets.

If an organism that can transmit disease by infectious aerosols is spilled outside a biological safety cabinet, laboratory occupants can be placed at considerable risk. The first step in response to such an occurrence is to avoid inhaling airborne material by holding the breath and leaving the laboratory. Warn others in the area, and go directly to a washroom or changeroom area. If clothing is known or suspected to be contaminated, remove it with care and fold the contaminated area inward. Discard clothing into a bag, or place it directly into an autoclave. Wash all potentially contaminated areas of the body as well as the arms, face, and hands. Delay reentry into the laboratory for a period of 30 min to allow reduction of the aerosols generated by the spill. When entering the laboratory to clean the spill area, wear protective clothing (rubber gloves, autoclavable footwear, an outer garment, and a respirator). If the spill was on the floor, do not wear a surgical gown that may trail on the floor when bending down. Place a discard container near the spill, and (using forceps) transfer large fragments of material into it; replace the cover. Cover the spill area with paper towels or other absorbent materials. Using a freshly prepared 1:10 dilution of household bleach or other appropriate disinfectant, carefully pour the disinfectant around and into the visible spill. Avoid splashing. Allow a contact time of 20 min. Use paper or cloth towel to wipe up the disinfectant and spill, working toward the center of the spill. Discard towels into a discard container as they are used. Wipe the outside of the discard container, especially the bottom, with a towel soaked in disinfectant. Place the discard container and other materials in an autoclave, and sterilize them. Remove shoes, outer clothing, respirator, and gloves, and sterilize by autoclaving. Wash hands, arms, and face; if possible, shower. The above decontamination procedure can be begun immediately for spills of pathogens that do not present inhalation risks.

29.7.6. Instruments

Contaminated syringes, instruments, pipettes, thermometers, and other glassware should be decontaminated before being reprocessed. Whenever possible, carry out final processing with steam for heat-stable items or with ethylene oxide for heat-labile, moisture-sensitive materials. When decontamination is essential before handling, use containers of liquid disinfectant. Completely immerse the items in the solution, preferably for as long as 20 to 30 min. This allows a safety factor if the item is not extremely clean or if the solution has been used for some period previously. Before reuse, rinse the items carefully with distilled water and, if they are to be used in or on the body, with sterile distilled water.

Decontaminate laboratory pipettes by placing them in a vertical or horizontal container with disinfectant. Glassware should be immersed completely. Before glassware is handled and reprocessed, steam sterilization is recommended.

29.7.7. Hands

Vigorous washing with soap and water is the most practical means of removing organisms from the hands. When working with pathogenic bacteria, the hands should be washed after the removal of gloves and before exiting the laboratory.

Bar soap is discouraged, not only because of the inherent sloppiness of the soap dish but also because some organisms survive on the soap surface. Liquid soaps, unless they contain a preservative, may gradually develop large populations of organisms in the reservoir, which should be cleaned out routinely and new soap added. Powdered soaps and leaf soaps have the advantage of not being contaminated or allowing growth of organisms.

29.8. STERILIZATION (adapted from reference 16)

The objective of a sterilization process is to destroy or remove all living organisms in or on an item. Generally in microbiological laboratories, steam is used for the sterilization of culture media, equipment, and glassware; dry heat is used for glassware and metallic equipment; gas is used for instruments; and filtration is used for solutions. Sterilization is an evolving technology, with constant improvements in equipment design and control. For example, hydrogen peroxide is a promising sterilant for many laboratory, industrial, and hospital applications.

Sterilization is a probability function. Bacterial cells usually die at an exponential (logarithmic) rate similar to that of a first-order chemical reaction. Sterilization efficiency is expressed as the statistical probability that the process will yield a survivor (24). Terminal processes (moist heat, dry heat, ionizing radiation, and ethylene oxide) should have a probability of 10^{-6} or less of a survivor being present, based on extrapolation of the rate determined from measurable survivors.

With the exception of filtration, sterilization methods can be characterized by the rate of the killing process. A ***D*-value** is used to characterize sterilization and, for heat, represents the time required at a given temperature to cause a decimal (90%) reduction in a bacterial population; the temperature is often indicated as a subscript (e.g., D_{100}). The *D*-value can be derived graphically by plotting the logarithm of the number of survivors against time of exposure and is measured as the time on the abscissa corresponding to a decade of reduction in numbers on the ordinate (Fig. 4).

Deviations from exponential killing occur frequently, resulting in patterns different from the idealized rectilinear plot shown in Fig. 4. A common deviation is for a short lag or even an increase in initial cell numbers ("shoulder") to occur before the onset of an exponential regression. Another common deviation is for a long, concave, curvilinear decrease in cell numbers ("tailing") to occur after the rectilinear exponential regression. Other deviations, such as rectilinear biphasic or sigmoid ones, occur less frequently. Clearly, a *D*-value has meaning only in the primary rectilinear portion of the plot, and extrapolation from this portion is invalid if tailing occurs. These deviations seem especially prevalent in the killing of bacterial spores, which may be affected by population heterogeneity, clumping, activation, germination, adaptation, and other factors (29).

The efficacy of the sterilization process used in the laboratory should be validated by using biological indicators. These are bacterial spores which exhibit a predictable death rate when exposed to a defined treat-

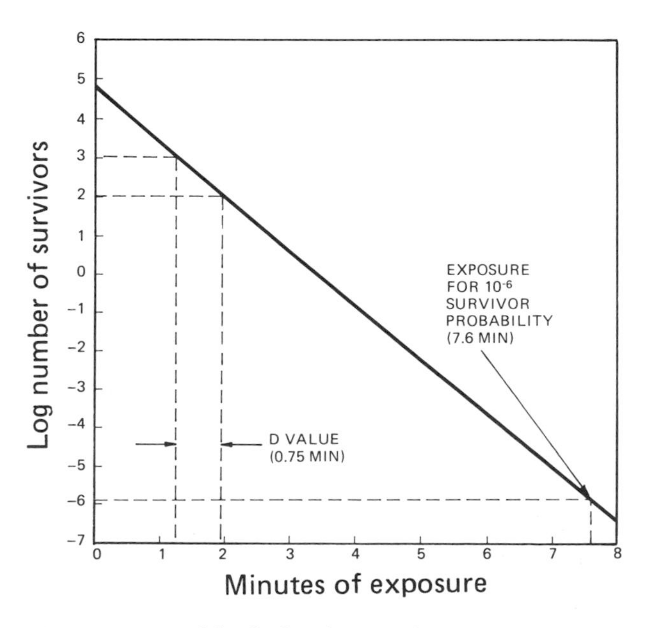

FIG. 4. Exponential death plot, showing the graphic determination of a *D*-value and a survivor probability.

ment. Highly resistant species of bacterial spores such as *Bacillus subtilis*, *B. stearothermophilus*, and *B. pumilus* are normally used in validation studies. Spore suspensions should be evaluated periodically to ensure that their resistance properties persist. Validation studies should include the use of at least three replicate units for each specific exposure time and condition that is to be evaluated. The accuracy of the result is a function of the number of replicate samples used. The equipment should be thoroughly evaluated for performance to ensure consistent sterilization results.

The authoritative compendium of Block (3) should be consulted for guidance on specific sterilization processes and validation methods, and Quesnel (28) provides basic methods for sterilization in biotechnology. For further general information, see references 3 and 29a.

29.8.1. Moist Heat

Steam sterilization is accomplished by using an autoclave with saturated steam under pressure for 15 min at a temperature of 121°C. Other time and temperature relationships can also result in effective sterilization.

Most laboratory autoclaves are gravity or downward-displacement devices which depend on the difference in density between air and steam. Problems occur with moisture penetration, superheating, entrainment or removal of air, and heat and/or moisture damage. Correct preparation of materials and their proper loading is important. The manufacturer of the autoclave can provide directions for installing thermocouples in materials to be sterilized to determine loading patterns within the autoclave.

Automatic autoclaves, equipped with cycle timers and automatic temperature controls, are most suitable for laboratory use. Several other moist-heat systems are available which reduce trapped air by high vacuum, steam and vacuum pulsations, and hot-water cascade or immersion. Air causes the chamber to heat more slowly and results in lower temperatures. Table 2 shows the lower temperature of several air-steam mixtures compared with the temperatures of pure steam. Laboratory autoclaves should be controlled by temperature readings (which are dependent on saturated steam pressures).

Interestingly, and contrary to commonly held view, carefully controlled studies have shown that air (oxygen) at 100% relative humidity significantly enhances the lethality of moist heat on bacterial spores, despite the practical effect of slower heating when air is present in an autoclave (32).

It is important to know the effects on steam penetrability of loading patterns, viscosity of solutions, size of containers, and densities. Also, the chemistry of the solution may influence microbial killing. Cool zones within loads of materials must be identified. Packages should be loosely arranged to facilitate steam penetration. Articles should be arranged to permit the downward displacement of the heavier air.

Table 3 illustrates the effect of the liquid volume and the number of containers within a load on the time required for sterilization. The time is shorter for small volumes. Thus, sterilization cycle times must be adjusted for the sizes or number of containers.

Precautions must be taken when sterilizing solutions in glass containers. Use a controlled rate of cooling and slow release of pressure to prevent glass shock when the cycle is completed. Some sterilizers use controlled water cool-down systems to avoid glass breakage. Take care to ensure that an autoclave has returned to ambient pressure before opening it. Most modern autoclaves prevent this problem through automatic control. When sterilizing flasks or bottles, loosen container closure fittings to permit air venting and prevent container breakage. Also place the containers in shallow metal or autoclavable plastic pans to capture liquid in the event of breakage. When the container is removed from the autoclave, a vacuum may result and cause an influx of air into the container, which can contaminate the solution unless the closure is made tight or other preventive measures are taken.

29.8.2. Dry Heat

Materials that cannot be sterilized by moist heat (e.g., greases, lubricants, mineral oil, waxes, and powders) are often sterilized by dry heat. Dry heat is also frequently used to sterilize laboratory glassware. Hot forced air and infrared energy have been used for dry-heat sterilization. There are a variety of commercial gas and electric dry-heat sterilizers, including batch ovens and continuous-heating tunnels.

Dry heat is less efficient than moist heat and requires higher temperatures for longer durations. The principal advantages of dry heat are that it is not corrosive for metals and instruments, does not affect glass surfaces, and is suitable for sterilizing powders and nonaqueous, nonvolatile viscous substances. Disadvantages include slow heat penetration and longer sterilization times. In

Table 2. Influence of incomplete air discharge on autoclave temperatures[a]

Gauge pressure (lb/in²)	Temp (°C)			
	Saturated steam, complete air discharge	Two-thirds air discharge	One-third air discharge	No air discharge
5	109	100	90	72
10	115	109	100	90
15	121	115	109	100
20	126	121	115	109
25	130	126	121	115
30	135	130	126	121

[a]Data from reference 23.

Table 3. Effect of liquid volume and number of containers on time required for liquid to reach 121°C in an autoclave[a]

Liquid vol/container (liters)	No. of containers/ load	Liquid temp at initiation of cycle (°C)	Time for liquid to reach 121°C (min)	Total time of cycle (min)
0.5	30	29	19	29
1.0	20	26	34	44
2.0	10	27	37	47
3.0	8	26	43	53
4.0	5	26	52	62
5.0	5	26	60	70
6.0	4	26	62	72

[a]Data from reference 26.

dry-heat sterilization, the higher temperatures may adversely affect some materials, and temperature stratification can occur unless the air is circulated.

29.8.3. Gases

Gases are used in sterilization because some materials and supplies cannot be sterilized by other means without damage or destruction. Gaseous processes are more difficult to control than other modes of sterilization because of the various process parameters. Factors such as gas stratification, temperature stratification, availability of moisture, chemically reactive barriers, physical diffusion barriers (such as packaging materials), devaporization, and polymerization must be considered when selecting a gaseous process (7–9). Sterilant gases are extremely hazardous to human health, and their use should be limited to persons who have been carefully trained in carrying out the sterilization process and are knowledgeable about the intrinsic hazards.

Safety and health regulations have been established for most gases that are commonly used as sterilants. These standards require that employers limit occupational uses to established concentration limits.

29.8.3.1. Ethylene oxide

Ethylene oxide (EtO) has been widely used in the sterilization of devices because of its wide compatibility with various materials. It is used in specially designed chambers or vessels. EtO is highly flammable and explosive by itself. Detailed instructions by the manufacturer of EtO sterilization equipment must be closely followed.

Mixtures of dichlorodifluoromethane and EtO are frequently used to minimize flammability. Although 100% EtO processes are used in the pharmaceutical industries, they are rarely used for laboratory purposes.

The use of EtO as a sterilization agent requires an understanding of the parameters affecting its activity (3). The diffusion of EtO, moisture, and heat into materials can be a limiting factor., Prehumidification overcomes diffusion barriers in packaging materials and permits better penetration of gas. Humidification prior to sterilization also ensures hydration and thus killing of bacterial spores.

29.8.3.2. Formaldehyde

Gaseous formaldehyde is a good space and surface sterilant and is the agent of choice for sterilizing biological safety cabinets. It is effective against bacteria and bacterial spores, fungi, viruses, and rickettsiae, as well as insects and other animal life (25).

Formaldehyde gas for sterilization purposes can be generated by heating paraformaldehyde, which is in a mixture of polyoxymethylene glycols containing 90 to 99% formaldehyde. The chemical composition of paraformaldehyde is $HO(CH_2O)_n$-H, where n represents 8 to 100 formaldehyde molecules. Paraformaldehyde depolymerizes to release gaseous formaldehyde when heated. Sterilization by formaldehyde gas generated from paraformaldehyde is more effective than sterilization by vaporized formaldehyde solution.

29.8.3.3. Hydrogen peroxide

Vapor-phase hydrogen peroxide is a promising surface sterilant that does not possess the toxic and carcinogenic properties of EtO or formaldehyde (13). The process is being adopted for the decontamination of biological safety cabinets.

29.8.4. Filtration

Filtration is one of the oldest methods used to sterilize solutions and is frequently used to remove microorganisms or particulate matter, or both, from solutions and gases. As a terminal process, it is less desirable than the use of moist heat because there is a higher probability that a microorganism will pass through a filter.

Filters function by entrapping microorganisms within the porous structure of the filter matrix. Vacuum or pressure is required to move solutions through the filter. There are two basic types: depth filters and membrane filters. Depth filters consist of fibrous or granular materials that are pressed, wound, fired, or bonded into a maze of flow channels. In these, retention of particulates is a matter of a combination of absorption and mechanical entrapment in the filter matrix. Membrane filters have a continuous structure, and entrapment occurs mainly on the basis of particle size.

Filtration is particularly applicable for oil emulsions or solutions that are heat labile. Membrane filtration is widely used to sterilize oils, ointments, ophthalmic solutions, intravenous solutions, diagnostic drugs, radiopharmaceuticals, tissue culture medium, and vitamin and antibiotic solutions.

For further information the reader is referred to the excellent monograph by Brock (4).

29.9. NONBIOLOGICAL SAFETY CONSIDERATIONS

Numerous physical and chemical hazards are present in general and molecular bacteriology laboratories. **Lacerations** are common laboratory injuries; they are most frequently caused by accidents involving the handling of razor blades, scalpels, scissors and other cutting instruments, glassware, and Pasteur pipettes. The **microwave oven** has become a common piece of laboratory equipment, but its use has introduced a significant new hazard to the laboratory. Careless operating procedures have resulted in the violent release of superheated fluids and explosions caused by rapid pressure buildup in accidentally sealed containers. Serious injury is always a potential consequence of the misuse of this equipment.

Although most commercially available **electrophoresis units** are designed with safeguards to interrupt the power source when electrodes are exposed or the fluid evaporates, older units and those fabricated locally may not include these safety devices. Such units present serious electrical shock and fire hazards. The considerable use of **flammable solvents,** in particular ethanol, in molecular bacteriology laboratories creates potential fire hazards associated with storage, use, and disposal practices. The storage of samples in **liquid nitrogen** creates the potential for injury from accidental skin contact and from explosion when improperly sealed vials are removed from storage.

Sulfur-35, phosphorus-32, iodine-125, carbon-14, and tritium are common isotopes in the molecular bacteriology laboratory. Good personal-hygiene habits are essential to prevent internal contamination, since the effect of internal **radioactivity** is generally more serious than that of external exposure. Plan experiments carefully so that exposure to the isotopes will be limited to the shortest time possible. To reduce exposure, use forceps and tongs when handling high activity, and always use appropriate shielding for storage locations, source containers, and waste containers. Use a low-density material such as plastic to shield phosphorus-32 and other high-energy beta emitters to minimize production of bremsstrahlung. It is important that the properties of isotopes used in the laboratory be known and that the radiation safety practices required under the laboratory's license be rigorously followed. See Chapter 21.4.1 for more detailed information on radioactivity safety.

The use of **hazardous chemicals** in the laboratory can cause serious injury to workers if they are not instructed and trained in how to handle them safely. Overt exposure can result in serious tissue damage and adverse acute effects. A major hazard is the potential for splashing of chemicals into the eyes. Corrosive chemicals such as phenol, for example, can have devastating effects. Safety goggles should be worn when there is a potential hazard of chemical splashes. Special eye protection is also important when working with lasers and UV light sources, glassware under reduced or elevated pressure, and other activities that could cause objects to strike the eye. See Chapter 13.8 for detailed information on chemical carcinogenicity, Chapter 22.10 for general chemical hazards, and Chapter 30.2.3.1 for hazards during photography procedures.

The brief discussion above is intended only to describe an array of hazards that are constantly present in the general or molecular bacteriology laboratory. Awareness of these hazards and biohazards that may be present is the first step toward establishing a safe laboratory environment. The attitudes and skill necessary for controlling these hazards and the continual reinforcement of good laboratory practice are essential to sustaining a safe and healthful general and molecular bacteriology laboratory.

29.10. REFERENCES

1. **Ausubel, F. M., R. Brent, R. E. Kingston, D. D. Moore, J. G. Seidman, J. A. Smith, and K. Struhl.** 1987. *Current Protocols in Molecular Biology.* Greene Publishing Associates, Inc., and John Wiley & Sons, Inc., New York.
2. **Billet, L. S., C. R. Parker, P. C. Tanley, and C. H. Wallace.** 1991. Needlestick injury rate reduction during phlebotomy—a comparative study of two safety devices. *Lab. Med.* **22:**120–124.
3. **Block, S. S. (ed.).** 1991. *Disinfection, Sterilization and Preservation,* 4th ed. Lea & Febiger, Philadelphia.
4. **Brock, T. D.** 1983. *Membrane Filtration: A User's Guide and Reference Manual.* Science Tech, Inc., Madison, Wis.
5. **Centers for Disease Control.** 1987. Recommendations for prevention of HIV transmission in health-care settings. *Morbid. Mortal. Weekly Rep.* **36(28):**1S–17S.
6. **Collins, C. H.** 1983. *Laboratory-Acquired Infections—History, Incidence, Causes and Prevention.* Butterworths, London.
7. **Ernst, R. R.** 1969. *Federal Regulations and Practical Control Microbiology for Disinfectants, Drugs, and Cosmetics.* Special publication no. 4, p. 55–60. Society for Industrial Microbiology, Linden, N.J.
8. **Ernst, R. R., and J. E. Doyle.** 1968. Sterilization with gaseous ethylene oxide—a review of chemical and physical factors. *Biotechnol. Bioeng.* **10:**1–31.
9. **Ernst, R. R., and J. E. Doyle.** 1968. Limiting factors in ethylene gaseous oxide sterilization. *Dev. Ind. Microbiol.* **9:**293–296.
10. **Gershey, E. L., E. Party, and A. Wilkerson.** 1991. *Laboratory Safety in Practice: A Comprehensive Compliance Program and Safety Manual for Industrial, Diagnostic and Clinical, Research and Development, Academic (School, College, University), Quality Control, Analytical and Testing Laboratories.* Van Nostrand Reinhold, New York.
11. **Harrington, J. M., and H. S. Shannon.** 1976. Incidence of tuberculosis, hepatitis, brucellosis and shigellosis in British medical laboratory workers. *Br. Med. J.* **1:**759–762.
12. **Jagger, J., E. H. Hunt, J. Brand-Elnagger, et al.** 1988. Rates of needlestick injury caused by various devices in a university hospital. *N. Engl. J. Med.* **319:**284–288.
13. **Klapes, N. A., and Vesley, D.** 1990. Vapor-phase hydrogen peroxide as a surface decontaminant and sterilant. *Appl. Environ. Microbiol.* **56:**503–506.
14. **Korczynski, M. S.** 1981. Sterilization, p. 476–486. *In* P. Gerhardt, R. G. E. Murray, R. N. Costilow, E. W. Nester, W. A. Wood, N. R. Krieg, and G. B. Phillips (ed.), *Manual of Methods for General Bacteriology.* American Society for Microbiology, Washington, D.C.
15. **Kubica, G. P.** 1991. Personal communication.
16. **Mahn, W. J.** 1991. *Fundamentals of Laboratory Safety: Physical Hazards in the Academic Laboratory.* Van Nostrand Reinhold, New York.
17. **Meyer, K. F., and B. Eddy.** 1941. Laboratory infections due to *Brucella. J. Infect. Dis.* **68:**24–32.
18. **Miller, B. M., D. H. M. Gröschel, J. H. Richardson, D. Vesley, J. R. Songer, R. D. Housewright, and W. E. Barkley (ed.).** 1986. *Laboratory Safety: Principles and Practices.* American Society for Microbiology, Washington, D.C.

19. **National Institutes of Health.** 1978. *Laboratory Safety Monograph.* Office of Research Safety, National Cancer Institute, and the Special Committee of Safety and Health Experts. Bethesda, Md.
20. **National Research Council.** 1981. *Prudent Practices for Handling Hazardous Chemicals in Laboratories.* Committee on Hazardous Substances in the Laboratory, Assembly of Mathematical and Physical Sciences. National Academy Press, Washington, D.C.
21. **National Research Council.** 1989. *Biosafety in the Laboratory: Prudent Practices for the Handling and Disposal of Infectious Materials.* Committee on Hazardous Biological Substances in the Laboratory, Board on Chemical Sciences and Technology, Commission on Physical Sciences, Mathematics, and Resources. National Academy Press, Washington, D.C.
22. **National Sanitation Foundation.** 1983. *Class II (Laminar Flow) Biohazard Cabinetry.* Standard 49. National Sanitation Foundation, Ann Arbor, Mich.
23. **Perkins, J. J.** 1976. *Principles and Methods of Sterilization in Health Sciences,* 2nd ed. Charles C Thomas, Publisher, Springfield, Ill.
24. **Pflug, I. J., and J. Bearman.** 1974. *Treatment of Sterilization Process Microbial Survivor Data in Environmental Microbiology as Related to Planetary Quarantine.* Progress report 9, NASA grant NGL 24-005-160. University of Minnesota Department of Food Science and Nutrition and School of Public Health. University of Minnesota, Minneapolis.
25. **Phillips, G. B., and W. S. Miller.** 1975. *Remington's Pharmaceutical Sciences,* 15th ed., p. 1389–1404. Mack Publishing, Easton, Pa.
26. **Pike, R. M.** 1976. Laboratory-associated infections—summary and analysis of 3921 cases. *Health Lab. Sci.* **13:**105–114.
27. **Pike, R. M., S. E. Sulkin, and M. L. Schultz.** 1965. Continuing importance of laboratory-acquired infections. *Am. J. Public Health* **55:**190–199.
28. **Quesnel, L. B.** 1987. Sterilization and sterility, p. 197–215. *In* J. Bu'lock and B. Kristiansen (ed.), *Basic Biotechnology.* Academic Press Ltd., London.
29. **Russell, A. D.** 1982. *The Destruction of Bacterial Spores.* Academic Press Ltd., London.

29a. **Russell, A. D., W. B. Hugo, and G. A. J. Ayliffe.** 1992. *Principles and Practice of Disinfection, Preservation and Sterilization,* 2nd ed. Blackwell Scientific Publications, Oxford.

30. **Rutala, W. A.** 1990. APIC guidelines for selection and use of disinfectants. *Am. J. Infect. Control* **18:**99–117.
31. **Sambrook, J., E. F. Fritsch, and T. Maniatis (eds).** 1989. *Molecular Cloning: A Laboratory Manual,* 2nd ed. Cold Spring Harbor Laboratory, Cold Spring Harbor, N.Y.
32. **Scruton, M. W.** 1989. The effect of air on the moist-heat resistance of *Bacillus stearothermophilus* spores. *J. Hosp. Infect.* **14:**339–350.
33. **Shapiro, J.** 1990. *Radiation Protection: A Guide for Scientists and Physicians,* 3rd ed. Harvard University Press, Cambridge, Mass.
34. **Sulkin, S. E., and R. M. Pike.** 1949. Viral infections contracted in the laboratory. *N. Engl. J. Med.* **241:**205–213.
35. **Sulkin, S. E., and R. M. Pike.** 1951. Survey of laboratory-acquired infections. *Am. J. Public Health* **41:**769–781.
36. **U.S. Department of Health and Human Services.** 1985. *Guide for the Care and Use of Laboratory Animals.* NIH publication no. 86-23, Public Health Service, National Institutes of Health. U.S. Government Printing Office, Washington, D.C.
37. **U.S. Department of Health and Human Services.** 1986. Guidelines for research involving recombinant DNA molecules. National Institutes of Health. *Fed. Regist.* **51:** 16958–16985.
38. **U.S. Department of Health and Human Services.** 1989. Guidelines for prevention of transmission of human immunodeficiency virus and hepatitis B virus to healthcare and public-safety workers. U.S. Public Health Service, Centers for Disease Control, National Institute for Occupational Safety and Health. *Morbid. Mortal. Weekly Rep.* **38**(5–6):1–36.
39. **U.S. Department of Health and Human Services.** 1992. *Biosafety in Micro-Biological and Biomedical Laboratories,* 3rd ed. HHS publication no. (CDC) 92-8395, U.S. Public Health Service, Centers for Disease Control and National Institutes of Health. U.S. Government Printing Office, Washington, D.C.
40. **U.S. Department of Labor, Department of Health and Human Resources.** 1987. Joint Advisory Notice: Protection against occupational exposure to hepatitis B virus (HBV) and human immunodeficiency virus (HIV). *Fed. Regist.* **52**(210):41818–41824.
41. **U.S. Nuclear Regulatory Commission.** 1946. *U.S. Code of Federal Regulations, Title 10, U.S. Nuclear Regulatory Commission, Part 20, (10 CFR 20).* U.S. Government Printing Office, Washington, D.C.
42. **U.S. Occupational Safety and Health Administration.** 1987. *U.S. Code of Federal Regulations, Title 29, Occupational Safety and Health Administration, Part 1910.1200 (29 CFR 1910.1200).* U.S. Government Printing Office, Washington, D.C.
43. **U.S. Occupational Safety and Health Administration.** 1990. *U.S. Code of Federal Regulations, Title 29, Occupational Safety and Health Administration, Part 1910.1450 (29 CFR 1910.1450).* U.S. Government Printing Office, Washington, D.C.
44. **U.S. Occupational Safety and Health Administration.** 1991. *U.S. Code of Federal Regulations, Title 29, Occupational Safety and Health Administration Part 1910.1030 (29 CFR 1910.1030).* U.S. Government Printing Office, Washington, D.C.
45. **Wedum, A. G.** 1973. Microbiological centrifuging hazards, p. 5–16. *In Centrifuge Biohazards: Proceedings of a Cancer Research Safety Symposium.* DHEW Publication no. (NIH) 78-373. Public Health Service, National Institutes of Health. U.S. Government Printing Office, Washington, D.C.
46. **Wedum, A. G., W. E. Barkley, and A. Hellman.** 1972. Handling of infectious agents. *J. Am. Vet. Med. Assoc.* **161:**1557–1567.

Chapter 30

Photography

TERRY J. POPKIN

Photography is an indispensable tool to the microbiologist. Within this "science" lies the ability to gather and record data quickly and easily. Mastery of the basic techniques described here will free you to make better use of your time spent in planning and performing experiments in the laboratory while noticeably enhancing your visual presentations.

In many laboratories it is commonplace to observe scientists scurrying about frantically seeking someone who might assist them with photographing their gels, copying their autoradiographs, and preparing slides for a presentation or publication. With a little knowledge, planning, and equipment, the scientists can easily and successfully perform these tasks themselves. After all, it is the scientist who wants to convey a specific message with photographs.

In my laboratory, I am actively engaged in the utilization of various photographic techniques including macrophotography, photomicrography, and electron microscopy. Frequently I need slides for presentation, copy negatives for reproduction, and prints for publication. I am reluctant to release my valuable, often irreplaceable "originals" to a laboratory or contract service for duplication. This necessitates an in-house photographic facility and the ability to efficiently produce high-quality photographs. I have found that a 35-mm single-lens reflex (SLR) camera, and related accessories, is generally satisfactory and usually produces results comparable to those from larger formats for most work. The 35-mm optics, if chosen properly, are superb and frequently outperform larger-format lenses. The 35-mm film choices are exceptional and are constantly being improved by the manufacturers. The physical space required for the system is minimal, and the cost is reasonable. Consequently, 35 mm systems are emphasized in this chapter. Camera formats up to 4 × 5 in. can provide exceptional results.

Although modern digital electronic imaging systems are evolving rapidly, they are very expensive and have not yet matched the resolving power of conventional film. Therefore the techniques and methods presented here require acquisition of a conventional film/camera system, preferably one with the finest optics. If the camera is to be shared, and often it is (in order to justify capital expenses in the laboratory), it is important that all users have a thorough knowledge of its operation. Indeed, all concerned might benefit from a formal course of instruction, but that is often not available or practical in terms of finances and time.

Good preparations, careful insight into what you must convey, and close attention to the various charts, tables, and information included in this chapter should provide you with the ability to produce excellent photographs. That is precisely what this chapter is all about—how to do it yourself, and better!

30.1. PRINCIPLES

30.1.1. The Nature of Light

Light is a form of electromagnetic energy and, as such, is commonly associated with wave forms. Although its wave characteristics usually predominate, light has the attributes of both waves and particles. These two characteristics stem initially from the absorption of energy by atoms, resulting in activation of shell electrons to higher energy levels. When these electrons ultimately fall back into their lower-energy orbits, photons are released and result in light that is characteristic of the associated atom. The photons travel in waves, yet they are able to strike objects (film) as if they were particles. They activate or disrupt the molecules within or on that object (silver compounds), subsequently changing the molecular structure and resulting in a change that is referred to as an "exposure."

Electromagnetic wavelengths have an extreme range from 10^{-9} mm (10^{-3} nm) for X rays and gamma rays to 6 miles (10^{13} nm) for radio waves (42). Photographic film can be sensitized to wavelengths ranging from X-rays up to the infrared (IR; 1,200 nm), although the wave forms referred to as visible light are fairly well confined to a small band (400 to 700 nm) within the recordable spectrum. Light waves can be modified with regard to their direction of travel (refraction, diffraction, and reflection), their plane of oscillation (polarization), their amplitude (brightness), and their wavelength (selective filtration). The ability of certain materials to alter light waves is utilized effectively in specialized equipment to achieve improvements in optical performance. For example, the unique chemical-physical properties of some of the rare earth elements are used in optical lens coatings because they can modify light wave reflections and thus decrease lens flare or reduce chromatic aberration.

Reflections can be thought of as light waves bounced from smooth surfaces that maintain an organized pattern and travel at specific uniform angles. **Refraction** is the change in direction of radiation caused by its passage obliquely from one medium into another in which its velocity of propagation is changed. The medium in which the radiation travels more slowly is said to have a higher refractive index, or density, in relation to the radiation. Different wavelengths are refracted through different angles, hence the separation of white light into the colors of the spectrum on emergence from glass to air. **Polarization** and **filtration** are modifications of light that allow you to limit the reflected or transmitted waves to specific angles or planes of oscillation or absorb them entirely. The characteristics of the modifiers are determined by their molecular structure and can be accurately defined. For example, a red filter can pass or transmit red light waves and absorb blue and green waves. Utilizing the aforementioned principles, you can take advantage of the properties of light in controlling and recording the images that you require.

30.1.2. Photographic Process

The process of creating an image on a piece of film involves physical-chemical reactions between the photons of light and the chemicals (usually silver compounds) dispersed in the emulsion (usually gelatin based) coated on the film matrix. It involves the formation of a latent image by photons striking silver ions and microscopically changing them physically and chemically. Subsequent chemical development enhances the change and permits a visible image to form. Although the process is very complex, it is totally reproducible and predictable. It is of prime importance in the laboratory, and particularly in microscopy, because it is the means of producing accurate and retrievable records of visual observations. Its proper execution is indispensable. A graphic illustration of this process is presented in the Time-Life Books series (42). As discussed in Chapter 3 of this volume, the actual reference material for the microscopist is the micrograph itself, and the results and interpretation of this image are dependent on accuracy and reproducibility in its creation. These should be the quintessential basis for all good science. Remember that if you are "sloppy" in making a photographic image, it is recorded for all to see.

Most laboratory work involves producing monochrome (black-and-white) images. The reason, in part, is the high cost of reproducing color images in journals and books. In addition, it is much simpler and requires less space to process black-and-white film. However, color film has particular advantages for some purposes and is widely used. Its processing today is far faster and less complicated than even in the recent past, and, except for certain highly specialized processing (Kodachrome), it can be done efficiently in the laboratory at a reasonable cost. If strict attention to detail is maintained, excellent results can be obtained by the average worker.

30.1.3. Films (1, 2, 3, 13, 21, 42)

Color film is basically a trilayered black-and-white emulsion with each layer sensitive to one of the three primary colors. After exposure, development produces three black-and-white negatives on the film (each corresponding to the exposure from the three primary colors), which then associate with specific colored dyes in proportion to the reduced silver. If color print film is being used, a first developer is used that reacts with the exposed silver compounds, which then become oxidized and chemically couple with dyes to produce a color negative. If color transparency film is being used, the unexposed silver compounds are used in the chemical process through the action of a second developer. The altered compounds then react with color couplers to form a permanent positive dye image on the film. The residual silver images are bleached away in a final chemical bath.

Each color film is formulated for balance at a **specific color temperature**, and the variations in color are measured on the Kelvin scale. Some films will record colors that are imperceptible to the human eye, but most are formulated to provide subtle distinctions between light waves in the visible and near-visible spectrum.

As points of reference, examples of measurements in the visible range that vary from warmer to cooler may be found in Fig. 1. Any "warmer" light (relatively rich in red rays) or "cooler" light (relatively rich in blue rays), corresponding to lower or higher color temperature, respectively, will have a pronounced effect on the color of the images produced on an inappropriate film. Color temperature errors on color negative films may go unnoticed to a person viewing a print because the color shift can be subjectively corrected (with filtration) when printing the images on paper. However, the same errors on transparency films appear bizarre. For example, if an object is photographed on a daylight-balanced film with a discontinuous light source (such as fluorescent bulbs), the images produced will appear blue-green. The same film will record objects lighted with incandescent bulbs as orange-yellow. These disparities can be visually disconcerting and frequently cannot be corrected. When scientific data are recorded on transparencies that are judged by colorimetric standards, the results of improper lighting and film balance can be disastrous.

Black-and-white film can also be balanced, filtered, or processed for specific spectral responses. Some films are panchromatic (sensitive to all visible wavelengths) and may be designed to have extended sensitivity in specific regions (usually red), whereas others are orthochromatic (usually insensitive to red). Exposure of the latter to an inappropriate light source (tungsten bulbs) can result in skewed tonal ranges or no exposure at all, so care must be taken. All special-purpose films have specific lighting and filtration data included in the instruction sheets, which should be followed closely.

The film packaging usually includes nomenclature which, when deciphered, will clearly indicate the conditions for which the film was intended. Films labeled type L are tungsten balanced emulsions designed for *long* exposures (0.1 to 100 s). Type S films are balanced for daylight or electronic flash and formulated for *short* exposures. Additional designations include type A for 3,400 K light, the suffix "color" (e.g., Kodacolor, Fujicolor) denoting a negative film, and the suffix "chrome" (e.g., Kodachrome, Agfachrome) indicating a positive or transparency film. A film should match the light source and conditions for which it was intended. If they are incompatible, specific methods (see section 30.2.1.4) can be used to achieve a correction.

There are confusing terms describing film exposure used throughout this chapter. (i) **Exposure value (EV),** a somewhat antiquated term, is used to describe the combination of lens aperture and shutter speed for a correct exposure on a particular film with lighting of a given intensity. A variation of 1 EV unit corresponds to a one f-stop change in either aperture or shutter speed or both. (ii) **Film speed (ISO)** is the standard exposure index assigned by the manufacturer. (iii) **Exposure index (EI)** is the "altered ISO" assigned by the user to account for variations in film speed as a result of special conditions in the applied photographic procedure.

Determining the best film to use for a particular task can be troublesome. The choices are many (55), and most comparable modern films are excellent. "Comparable" refers to grain structure (usually related to the sensitivity or the film speed) and color balance (daylight versus tungsten with color materials or orthochromatic versus panchromatic with black-and-white films). The strength of world market competition has resulted in better products—a distinct advantage to users. There are, however, some films which stand above all others, and these are decidedly advantageous to a scientist. These are the ultra-fine-grained and ultra-high-resolution films (e.g., Kodak's Kodachrome 25, Ektar 25, and Technical Pan films) which enable very large reproductions to be made while maintaining incredible sharpness. They are low-speed (ISO 25) films capable of recording images with a quality that enables a 35-mm camera user to produce an image which may rival a photograph made with a 4×5-in. camera. Of course, when these films are used in a medium- or larger-(negative) format camera and enlarged to the same degree, the quality is inspiring. However, for

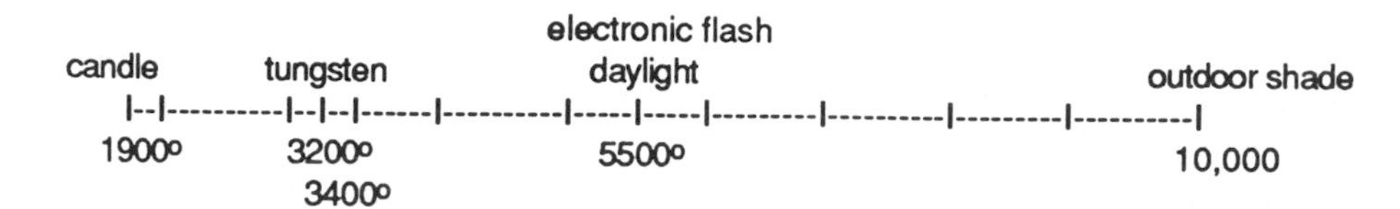

FIG. 1. Examples of the color temperature measurements of light in the visible range of the spectrum.

some technical uses these films can be impractical. An example is the recording of immunofluorescence (section 30.3.3.1), which usually requires specialized high-speed films because light-quenching occurs during the focusing, viewing, and exposure process, thereby diminishing the ability of a low-sensitivity film to faithfully record the microscope field. Specifics will be discussed under each topic, and appropriate film choices will be suggested. Keep in mind that these are only suggestions and guidelines. The protocols described in this chapter *work* and have withstood the test of time. However, once you are experienced and comfortable with any procedure, it is often useful to experiment.

30.2. IMAGE CAPTURE

30.2.1. Equipment

This discussion will center around the photographic applications of 35-mm, medium-format, and 4×5-in. photography. Selection and use of the camera and adjunct equipment is followed by selection and use of films for producing slides or negatives and then by processing them. The quality of these ultimate results, which relies on accuracy in all the intermediate steps, is the basis for judgment of the photographer's expertise.

30.2.1.1. Cameras and lenses (2, 13, 42, 48, 57, 59)

A professional 35-mm SLR camera (Canon, Nikon, Minolta, and others manufacture excellent camera bodies) provides stellar performance, but cameras having larger negative formats, such as Hasselblad or Rolleiflex in medium format or Sinar, Toyo, and Polaroid in 4×5 in., can be advantageous. The importance of excellent optics is a good application of the old adage: "a chain is only as strong as its weakest link." Always choose a manufacturer's professional series of lenses—those that include aspherical surfaces, low-dispersion glass, or synthetic crystal (fluorite) elements or those with designations indicating their design attributes. Examples are Canon's L lenses for their 35-mm cameras, Zeiss Makro-Planar and UV-Sonnar lenses for various medium-format cameras (Hasselblad), and Schneider's HM (high-modulation) lenses for 4×5 in. cameras. These esoteric lenses are manufactured in specific focal lengths, each of which is appropriate for a given task. The value of these fine optics lies in their ability to resolve the most minute detail at most lens apertures (always avoid the extremes when possible) while maintaining high contrast. There are circumstances when excessive contrast can be a problem. However, it is relatively simple to reduce resolution and contrast, if necessary (by using filtration, lighting, and processing modifications), but these characteristics must be present to begin with in order to get the most information on the film. To do that prudently, use the finest-grained film that is practical for the task at hand, use the best lenses available, and take appropriate steps, either in camera or in the darkroom, to improve the image and thus provide the photograph desired.

The following **camera specifications** are desirable.

1. Focusing screen interchangeability and a choice of viewfinder accessories. For example, specialized copy and reproduction work is facilitated by using focusing screens with grid lines etched on them. Viewfinders containing adjustable eyepieces, with diopter correction for the user, are indispensable for critical photography, since not all eyes and eyeglasses are the same.
2. Off-the-film-plane (OTF) or through-the-lens (TTL) light-metering capability. This is desirable, but excellent results are attainable with a small hand meter.
3. An excellent "macro" lens designed for the reproduction range of 1:10 or smaller and optimized for flat-field photography (copying two-dimensional objects). The lens diaphragms must have calibrated adjustments in 1/2 or 1/3 f-stops, because some specialized films are remarkably sensitive to small exposure increments and there is little latitude in this regard.
4. Lens shade. This will prevent bounced or stray light from hitting the front lens elements, which may cause a subsequent reduction in the contrast of the image. Many macro lenses have recessed front elements, and a shade may be unnecessary on these optics.
5. An extension tube or bellows attachment. This will allow focusing closer than possible with the lens internal helicoid focusing mechanism.
6. Manual-focus capability. On modern cameras, autofocus is practically unavoidable but can be a detriment for critical work. Be certain that if an autofocus camera system is purchased, it is one which allows for a manual-focus override control.
7. Motor-drive capability. This helps when producing slides and doing copy work. An automatic film advance mechanism prevents accidental twisting of the camera body on the copy stand and subsequent misalignment.
8. Multiple-exposure provision. This is required for some specialized work (section 30.3.1).

30.2.1.2. Light meters (6, 58)

Modern light meters are sophisticated minicomputers which quantitate light either fully or separately into the three primary colors. They can be designed to determine the color or intensity of light, compute lighting ratios, and indicate balances between ambient light and flash exposures. The meter design may incorporate single or multiple sensors to detect light. Most manufacturers use silicon cells or some rare earth elements in the meter sensor to enable measurements to be made over a broad spectral band. Older cadmium sulfide meter cells lacked red sensitivity, and even older meters, using selenium cells, had very little sensitivity at all. Many of the modern light meters are constructed to measure both incident light (the amount of light falling on a surface) and reflected light (the light bounced from the surface of a subject). Determinations of reflected light can be modified by incorporating light receptors having a focusing element which limits the field of view and permits "spot" readings. In-camera meters are always reflective and measure the light passing TTL or OTF. Some specialized meters can determine the wavelengths of light being measured (color meters). Others, both hand held and in camera, can differentiate between ambient light (that which exists continuously in the environment) and flash exposures (short pulses of high-intensity light provided by an electronic flash). All, however, have one thing in common: they are calibrated to indicate an exposure of what they read as a standard (18%) gray.

A major difficulty in producing good photographs is accurate light metering. Improper metering may result in unusable images. Transparencies that are overexposed will appear "washed out," and negatives that are underexposed will be "thin"; neither of these will contain critical detail that is useful. However, underexposed slides and overexposed negatives are generally salvageable because the image is recorded on the film and can be modified and improved (by copy techniques) if necessary. Nevertheless, there is no substitute for accurate exposures, and considering the selective nature of scientific work, a modern meter is recommended for critical work. An internal meter is an advantage when doing close-up work, especially with macro lenses or lens extensions used to give increased magnification. As a lens is extended from the film plane, the effective aperture is diminished, causing a loss of light. Because of its location, a TTL or OTF meter will sense this light loss and automatically compensate. By contrast, hand-held meters (either reflective or incident) cannot compensate for extension-caused light losses. Their use in each situation requires knowing the image magnification to derive the effective lens aperture, which is always smaller than that indicated by the meter and so requires either a compensatory decrease in shutter speed or some equivalent combination of lens opening and shutter speed (an EV change). The image magnification can be calculated by a formula involving the focal length of the lens and the length of the extension (2, 60) or found by reading extension factors or magnifications marked on the macro lens or bellows. The magnification is then used in another formula or, more simply, by using a handy formula-based calculator (Hasselblad makes an excellent one) to determine the corrected exposure.

Because a meter indicates an exposure based on the assumption that it has read an 18% gray equivalent, there are instances when the indicated exposure requires some thought or adjustment. For example, if either a white or a black area predominates in the field being read, a reflective meter will include and interpret it as 18% gray. This skews the reading to indicate a lesser or greater exposure, respectively, than that desired or appropriate for a subject. Learn to compensate for these skewed readings by adjusting either the aperture or the shutter speed of the camera. If the troublesome areas do not cover the entire field of view, a spot meter with a narrow acceptance angle may help in determining the proper exposure by averaging multiple readings. The simplest alternative is to place an 18% gray card (obtainable from Kodak) in the field to be photographed and meter it as a standard. Another solution would be the use of an incident meter because it is uninfluenced by tonal values of any object in the field it is reading. If there is any question about exposure accuracy, it is advisable to bracket (vary) the exposure around a theoretical optimum determination.

30.2.1.3. Optical filters (11, 13, 24, 36, 52)

Optical filters consist of colorants (organic dyes) dissolved in a substrate (gelatin or cellulose acetate) and coated on glass. The dry coat may be stripped from the glass and coated with lacquer or simply left in place and protected by a glass faceplate. By incorporating carefully mixed dyes to produce a specific color, the resulting filter can correct even the smallest deviations in the proper light-film relationship. If a colored object is to be recorded on black-and-white film, filters can be used to render the colors as tones of gray, each representing a specific color in the specimen. Some filters are designed to selectively absorb wavelengths of light that are out of the visible spectrum but record on film. Heat-absorbing filters (present in light microscopes and enlarging systems) are specifically designed to reduce IR emissions from xenon arcs and high-wattage tungsten lamps. Although IR is of little concern unless IR-sensitive film is used, it is a source of heat in an optical system and can be destructive to the system or a specimen. Ironically, additional color-balancing filters (magenta) are often required to offset the color introduced by a heat-absorbing filter in an optical system. The use of dichroic interference filters is an alternative means of eliminating IR by reflection rather than absorption (13). UV filters (Kodak Wratten 2B) will allow all visible wavelengths to pass but will efficiently absorb UV, which may record as blue on a UV-sensitive emulsion (24).

In color photographic processes, filters are used to add or subtract color to or from the image in much the same way that light is mixed optically to create color. The additive method assumes that the three primary colors of light are red, green, and blue (RGB). Mixing all three primaries (RGB) optically, in equal amounts, produces what is perceived to be white light. In the subtractive method of creating color, the primary colors are those that subtract or absorb RGB. These primaries are known as cyan (aqua-blue), magenta (pink-violet), and yellow, respectively. If mixed, they produce black by absorbing all the primary additive wavelengths. Dyes from both additive or subtractive color kits are available commercially and, when properly mixed, can produce any of the colors or hues imaginable, which can then be incorporated into filters.

The color temperature of the light source can be important and must be determined with a meter or from reference material (52). If a color temperature meter is used, it will indicate which subtractive or additive light-balancing filter(s) to use, either on the light source or in the optical path, to correct the color of the light recorded so that it is perceived as "true" by the film being used. In the absence of a color meter, the filtration instructions supplied with the film can be used, and results are usually good. Additional trial and error can further improve the results but at the expense of time rather than money for a color meter.

30.2.1.4. Enlargers

An enlarger consists of a chassis and a light source. The chassis contains all of the structural components necessary for the negative stage, the focusing mechanism, and the lens-mounting system. There are three distinct types of light sources available for enlargers: diffuser, condenser, and point source. For a given application, one type may be more desirable than another. For color printing, a diffusion source is preferred since it is least likely to show any defects in the three distinct emulsion layers on the films. In addition, colored light is provided by either three primary-color light sources or a single source that is filtered through a series of dichroic filters. For electron microscopy (EM), the diffusion enlarger is the least desirable since it is incapable of producing critical sharpness on a print. By contrast, the point source enlarger (a modified, specially aligned condenser system) provides supercritical sharpness but, at the same time, reveals every dust spot or scratch on either side of the film. All things considered, the conventional condenser enlarger is economical, readily available, and certainly the most widely used.

There are a few condenser enlarger features that are desirable, as follows.

1. The enlarger should have variable (or changeable) condensers that are adjustable for use with lenses of different focal lengths.
2. Filters should be accommodated internally and above the negative stage, to allow for the insertion of heat-absorbing glass and variable contrast or color filters without creating any aberrations in the optical path.
3. The negative stage and lens mount should be adjustable in all axes to allow for proper alignment. *Note:* This is a critical feature that is often overlooked. The use of a good enlarger alignment device (e.g., Salthill) to adjust for edge-to-edge negative sharpness in printing should be as routine a procedure as it is to align a light or electron microscope on a regular basis.

Once the enlarger is aligned and equipped with the finest lenses (Schneider, Rodenstock, and Nikon all make excellent apochromatic enlarging lenses), printing can begin.

The use of an automated photosensor exposure device (e.g., Lektra PTM-4), once it has been calibrated for each paper used, will allow perfect reproducibility.

30.2.2. Films and Slides

The scientist has to present findings to peers, supervisors, and review boards. The need for excellent prints or slides from data is constant, and its value is underrated; never forget that you may be judged by your presentation. This starts with selection of the best films and slides.

30.2.2.1. Continuous-tone black-and-white negatives (1, 3, 18)

Continuous-tone black-and-white negatives are best produced on films which are fine grained and have high resolution and accutance. Examples available in nearly all formats are Agfapan 25, Kodak Technical Pan, and Kodak T-Max 100. Polaroid products, such as type 665 positive/negative black-and-white film, can be used with medium- and large-format cameras. The negatives produced with this material are excellent and have reciprocal density curves with the positives, which permit accurate evaluation of the exposure by viewing the print (1). Polaroid's type 55 film (for 4×5-in. cameras only), although excellent, does not provide this convenience, since the negative/positive curves are asymmetrical, resulting in the production of a light print when an ideal negative is produced. Critical reproductions of micrographs or other original photographs are best when copied on Kodak's Professional Copy Film 4125, since this film was designed with a tonal gradation and contrast range expressly for this purpose.

30.2.2.2. High-contrast black-and-white negatives

High-contrast black-and-white negatives are required for printing or projection, and Kodak's Kodalith or Ektagraphic HC are certainly the best choices. These two films are identical, but Kodalith is available only in 100-ft. rolls whereas Ektagraphic HC is sold in 35-mm cassettes. The emulsion was designed for making line and halftone negatives and has a unique spectral sensitivity and contrast curve. The film speed varies between EI 3 under tungsten illumination and EI 8 with a daylight source. Although this is a black-and-white film, the color temperature of the illumination source is important, as seen by the relative EI. Kodalith (Ortho) has a lack of red sensitivity; although this is an advantage in allowing the processing of this film by inspection (a red safelight is usable), it is a disadvantage with regard to metering properly for exposure determination. (A considerable portion of the metered light is contributed by long-wavelength components of the spectrum, which will not record on this film.) Therefore, it is necessary to determine the film speed or EI (the ISO must be adjusted downward if tungsten light is used). This trait is not unique, and various manufacturers rate films differently for use with various illumination sources. This film demonstrates very little exposure latitude, and it is important to carefully examine the negatives in order to choose the perfect exposure (it is here that 0.5 f-stop detents on the lenses are critical). When properly exposed, the images can be extraordinary, and large poster displays can be printed from 35-mm Kodalith negatives. These negatives can also be used directly for projection as slides. Specific areas can be hand colored, or colored gels can be stacked or sandwiched with the negative for visual emphasis.

Kodak's Technical Pan film is usable as a higher film speed alternative. Although this film does not have the extreme density of Kodalith, it is quite good and is presently the only high-contrast film available in the United States for medium-format roll film users. Technical Pan has extended red sensitivity, the opposite of Kodalith, and this makes its response and metering predictable. Should other films be required, 4×5-in. films that are not available in smaller sizes can be trimmed and used in a cut film back designed for medium-format cameras (Table 1).

Table 1. Black-and-white films (negatives)

Black-and-white negative	ISO	35 mm	120/220	4 × 5	Comments[a]	Process[b]
Agfapan APX 25	25	•	•		a,d,g	O
Agfapan APX 100	100	•	•	•	d	O
Agfapan APX 400	400	•	•	•	d	O
Dupont Cronex	NA[c]				o,s,u,x	P
Fuji Neopan 125	125	•			d	F,G
Fuji Neopan 400	400	•			d	F,G
Fuji Neopan 1600	1,600	•			d	F,G
Ilford Pan F	50	•	•		a,d,g	A,G,L
Ilford FP4 Plus	125	•	•	•	d,g	G,L
Ilford HP5 Plus	400	•	•	•	d	G,L,N
Ilford XP2	400	•	•	•	c,d	CC
Ilford Delta 400	400	•			d,g	F,G
Kodak Technical Pan	25–200	•	•	•	a,d,e,g,s,v	C,I,M
Kodak T-Max 100	100	•	•	•	a,d,g,r,v	E,F,J
Kodak T-Max 400	400	•	•	•	d,g	J
Kodak T-Max 3200	3,200	•			b,d	J
Kodak Plus-X	125	•	•	•	d	E,F,G
Kodak Tri-X	400	•	•	•	d	A,B,E,F,H
Kodak Kodalith	8	•		•	a,h,o,s	K
Kodak Ektagraphic HC	8	•			a,h,o,s	K
Kodak Professional Copy	25			•	a,i,l,o,s	E
Kodak Commercial Copy	25			•	g,i,o,v	E
Kodak Hi-Speed Infrared	12–25	•			l,s,w	C,E,F
Kodak EM Film 4489	NA	•		•	e,o,y	C
Kodak SB	NA				o,s,u,x	P
Kodak X-Omat AR	NA				o,s,u,x	P
Konica Infrared 750	16–32		•		s,w	F,G
Polaroid Type 54	100			•	d,n−,q	U
Polaroid Type 55	50			•	d,n,q	U
Polaroid Type 57	3,000			•	d,n−,q	U
Polaroid Type 664	100			•	d,m,n−,q,y	U
Polaroid Type 665	80			•	d,m,n,q,y	U
Polaroid Type 667	3,000			•	d,m,n−,q,y	U

[c]NA, not applicable.

[a]Comment codes

a, Highest resolution
b, Lowest resolution
c, Chromogenic
d, Daylight source
e, Electron microscopy
f, Use with fluorescent lights
g, Low-grain structure
h, High-contrast
i, Internegative material
k, Kodak commercial processing
l, Low contrast
m, MP-4/pack film (3.25 × 4 in.)
n, Neg/positive (n−), no negative, only print
o, Orthochromatic
p, Designed to be "pushable to higher film speeds"
q, Rapid/self development
r, Reversal development
s, Special purpose
t, Tungsten source
u, UV sensitive
v, Variable contrast with development
w, Infrared sensitive
x, X-ray
y, Size variation (3.25 × 4 in., etc.)

[b]Process codes (italic letters refer to suppliers in section 30.5.2)

A, Acufine (*c, ee*)
B, Diafine (*c, ee*)
C, D-19 (*i*)
D, Dektol (*i*)
E, HC-110 (*i*)
F, D-76 (*i*)
G, ID-11 + (*o*)
H, Microdol-X (*i*)
I, Technidol (*i*)
J, T-Max (regular and reverse) (*i*)
K, Kodalith A + B (*i*)
L, Ethol (UFG and TEC) (*c, ee*)
M, Edwal FG7 (*c, ee*)
N, Perceptol (*c, ee*)
O, Rodinal (*h*)
P, X-ray (*i*)
Q, S-2 activator (*i*)
R, Royal print activator (*i*)
S, Royal print fixer (*i*)
T, Rapid fixer (*i*)
U, Self (with manual or power processor (*x*)
AA, Commercial process only
BB, E-6 (*h, i, p*)
CC, C-41 (*h, i, p*)
DD, EP-2 (*i*)
EE, RA-4 (*i*)
FF, R-3000 (*i*)
GG, Cibachrome P-30 (*o*)

30.2.2.3. Color negatives (11, 20)

There has been little demand for color negatives in the microbiology laboratory. This is unfortunate because the films are exceedingly good and the results are quite striking. The criteria for choosing color negatives are similar to those utilized for black-and-white: grain structure, resolution, and color balance (Table 2).

30.2.2.4. Black-and-white slides

Most facilities provide only for 35-mm projection equipment because of space limitations and cost factors. Medium-format photography has become more popular now that the manufacturers are providing a wide range of films for these cameras and an assortment of projectors are now being marketed in this format. The use of Polaroid's MP-4 copy camera in many laboratories has popularized 4×5-in. films, and the images produced in this size are useful for overhead projection.

For making 35-mm black-and-white slides, the Polaroid process is simple and rapid. There are two Polaroid versions, the high-contrast PolaGraph and the continuous-tone PolaPan films. Choose the latter if photographs or micrographs are being incorporated into slides, because they will reproduce better. There are alternatives which can provide superior quality but at the cost of more time. Kodak's Rapid Process Copy film, although very slow (EI 0.7), is an excellent film for direct positives and can be processed by manual techniques in tanks or in an X-ray processor by taping an old X-ray to the film as a leader and feeding it through the processor. Kodak's T-Max 100 film can also be processed as a direct positive by using a special reversal-processing kit produced by Kodak. T-Max 100 has a normal ISO of 100, but the film speed is dependent on the contrast level desired and the processing used (refer to the instructions included with the processing chemicals). For use as a normal contrast-positive material, the ISO must be adjusted to EI 50 and the reversal development must be used.

Kodak's LPD4 Precision Line Film is certainly the best choice for high-contrast positive line copy. It has a spectral sensitivity and contrast curve similar to that of Kodalith, and the only drawback lies in its film speed: EI 0.75 under tungsten illumination and EI 1.5 with a daylight source. Because of the very low EI, light metering may be difficult with some meters (see section 30.3.1).

Table 2. Color films (negatives)

Color negative	ISO	35 mm	120/220	4 × 5	Comments[a]	Process[a]
Agfacolor XRS 100	100	•	•		a,d	CC
Agfacolor XRS 200	200	•	•		d	CC
Agfacolor XRS 400	400	•	•		d	CC
Agfacolor XRS 1000	1,000	•	•		d	CC
Agfacolor Ultra 50	50	•	•		a,d,g	CC
Agfacolor Optima 125	125	•	•		d,g	CC
Agfacolor Portrait 160	160	•	•		d,l	CC
Fujicolor 100 Super HG	100	•	•		d,g	CC
Fujicolor 200 Super HG	200	•			d	CC
Fujicolor 400 Super HG	400	•	•		d	CC
Fujicolor 1600 Super HG	1,600	•			d	CC
Fujicolor Reala 100	100	•	•		a,d,f,g	CC
Fujicolor 160 NLP	160		•	•	t	CC
Fujicolor 160 NSP	160	•	•	•	d,g,l	CC
Fujicolor Internegative (IT-N)				•	i,l,s,t	CC
Fujicolor NHG	400	•	•		d	CC
Kodak Ektapress 100	100	•			d,g,p	CC
Kodak Ektapress 400	400	•			d,p	CC
Kodak Ektapress 1600	1,600	•			d,p	CC
Kodak Ektar 25	25	•	•		a,d,g,h	CC
Kodak Ektar 100	100	•			g,d	CC
Kodak Ektar 1000	1,000	•			d	
Kodak Kodacolor 100	100	•	•		a,d,g	CC
Kodak Kodacolor 200	200	•			d	CC
Kodak Kodacolor 400	400	•			d	CC
Kodak Kodacolor 1600	1,600	•			d	CC
Kodak Pro 400	400	•	•		a,d,g	CC
Kodak Vericolor VPL	100		•	•	t	CC
Kodak Vericolor VHC	100		•	•	d,g,h	CC
Kodak Vericolor VPS	160	•	•	•	d,g,l	CC
Kodak Vericolor VPH	400	•	•	•	d	CC
Kodak Vericolor 4112	12			•	i,l,s,t	CC
Polacolor 64T	64			•	d,m,n−,q,y	U
Polacolor 100	100			•	m,n−,q,t,y	U
Polaroid Type 59	80			•	d,n−,q	U
Polaroid Type 669	80			•	d,m,n−,q,y	U

[a]For comment and process codes, see Table 1 footnotes *a* and *b*.

Table 3. Black-and-white films (transparencies)

Black-and-white transparency (slide)	ISO	35 mm	120/220	4 × 5	Comments[a]	Process[a]
Kodak Rapid Process Copy	0.7	•			a,g,l,o,s	C,P
Kodak Fine Grain Release Positive	1.0	•			a,g,o,s	C,D
Kodak LPD4	1.5	•		•	a,g,h,o,s	K
Kodak T-Max 100	160	•	•	•	a,d,g,r,v	J
Polaroid PolaPan CT	125	•			d,g,q	U
Polaroid PolaGraph HC	400	•			d,g,h,q	U
Polaroid Type 55	50			•	d,m,n,q	U
Polaroid Type 665	80			•	d,n,q,y	U

[a]For comment and process codes, see Table 1 footnotes *a* and *b*.

For production of medium- and large-format black-and-white transparencies, T-Max 100, processed as a direct positive, is the best choice (Table 3).

30.2.2.5. Color slides

For 35-mm color slides, Kodak's Kodachrome 25 Professional and Fuji's Velvia transparency films provide the finest-grained, sharpest images possible. These are daylight-balanced films, but there are tungsten-balanced versions available, albeit somewhat more difficult to obtain (Table 4). *Note:* Always attempt to buy the professional versions of Kodak's films, since they are more accurately balanced for color and exposure than the amateur versions.

If a sacrifice in ultimate quality is acceptable, a plethora of medium- and high-speed 35-mm transparency films are available. The high-speed films should be used only when some means of camera support is unavailable or when they are required for special applications. A "quick" 35-mm slide is often needed, and, in that event, the acquisition of Polaroid's PolaChrome High Contrast or PolaBlue slide films and the necessary processor is indispensable.

For medium-format color slides, Kodak manufactures Kodachrome as an ISO 64 film, Fuji provides most of its films, including Velvia, in the 120 and 220 sizes (short and long rolls), and other manufacturers commonly produce ISO 100 speed films in these sizes (section 30.5).

Large-format color transparencies are useful for overhead projection, and the film choices are varied, commonly being of the ISO 50 to 100 varieties. Polaroid makes an instant type 691 3.25- by 4-in. color transparency material for this purpose, which can be used in a pack film holder designed for medium- or large-format cameras.

30.2.2.6. Slide mounting (9, 40)

The techniques involved in slide mounting can vary from simple manual methods to the use of complex automated machines. For work in the laboratory, the basic techniques will usually suffice, but if considerable production is required, not spending a few more dollars for minimal automation can be "penny-wise and dollar-foolish."

For making 35-mm slides, the Wess VR-FTU slide mounter is recommended. With this instrument an entire roll of film can be rapidly registered, cut, and mounted with a simple series of manual techniques, without the film's ever being physically handled. The alternatives are to use Polaroid's slide-mounting system or to manually cut the film and individually insert each frame into a slide mount on a light table. Be certain to "dot" completed slides so that a projectionist can properly insert them into a tray for viewing. The proper technique is to place a dot in the lower left-hand corner of the mount while viewing the slide correctly. The slide is then placed in a projection tray with that dot in the upper right-hand corner. Result: no inverted or reversed slides during presentations.

30.2.3. Processing and Printing (3, 4, 11, 18, 20)

The proper processing and printing of all exposed film is essential for obtaining ideal images. The development, fixation or stabilization, and washing all affect the permanence and quality of the final image. Most film manufacturers supply processing instructions with their products, but these generally provide the user only the information necessary to process their film in their chemicals. Expanded information is available from the manufacturers on request, and you should obtain all data sheets relevant to your work. These sheets usually include density (H and D) curves (3, 19) and reciprocity data for the films processed in a variety of available developers. Matching film density curves with exposure and development for the production of a perfect negative or slide is known as **sensitometry.** It is this exacting discipline which has established the standards for processing all color and black-and-white films. The chemical and physical properties of the processing solutions are inseparable from the sensitometry, and these concerns are the subject of entire volumes (6, 58). In this chapter, specific processing techniques for specified films are indicated in Tables 1 to 4 (see also Tables 6 and 7), which can assist in choosing a film-processing combination for a particular application. An excellent source of additional exposure and processing information is the *Photo Lab Index* (11), published quarterly to keep up with the rapid advances in this field.

30.2.3.1. Hazards (22, 23, 25–27, 56)

Hazards are often overlooked in photography. A good scientist will normally take all the precautions necessary to prevent biological and/or chemical contamination during a laboratory experiment (heed the notes in this and other chapters), but when the actual photographic procedures begin, all too frequently there is

Table 4. Color films (transparencies)

Color transparency (slides)	ISO	35 mm	120/220	4 × 5	Comments[a]	Process[a]
Agfachrome RS 50	50	•	•		d,g	BB
Agfachrome RS 100	100	•	•		d	BB
Agfachrome RS 200	200	•	•		d	BB
Agfachrome RS 1000	1,000	•	•		b,d	BB
Fujichrome Velvia	50	•	•	•	a,d,g,h	BB
Fujichrome 50-D	50	•	•	•	d	BB
Fujichrome 64-T	64	•	•	•	a,g,t	BB
Fujichrome 100-D	100	•	•	•	d	BB
Fujichrome 400-D	400	•	•		d	BB
Fujichrome P-1600	400	•			d,h,p	BB
Kodak Ektachrome 2483	16	•			a,h,s,t,u	BB
Kodak Ektachrome Lumiere 50	50	•	•		a,d,g,h	BB
Kodak Ektachrome Lumiere 50X	50	•	•		a,d,g,h	BB
Kodak Ektachrome 64	64	•	•	•	d	BB
Kodak Ektachrome 64X	64	•	•		d	BB
Kodak Ektachrome 64T	64	•	•	•	g,t	BB
Kodak Ektachrome 100	100	•	•	•	d	BB
Kodak Ektachrome Lumiere 100	100	•	•		a,d,g,h	BB
Kodak Ektachrome Lumiere 100X	100	•	•		a,d,g,h	BB
Kodak Ektachrome 100 Plus	100	•	•	•	d,g,h	BB
Kodak Ektachrome 160T	160	•	•		t	BB
Kodak Ektachrome 200	200	•	•	•	d	BB
Kodak Ektachrome 320T	320	•			g,t	BB
Kodak Ektachrome 400	400	•	•		d	BB
Kodak Ektachrome 400X	400	•	•		d,g	BB
Kodak Ektachrome P1600	400	•			d,h,p,s	BB
Kodak Ektachrome Slide Dupe 5071	NA[b]	•		•	a,d,i,s,u	BB
Kodak Ektachrome Infrared 2236	16	•			s,w	BB
Kodak Kodachrome 25	25	•			a,d,g,k	AA
Kodak Kodachrome 40	40	•			a,g,k,t	AA
Kodak Kodachrome 64	64	•	•		d,g,k	AA
Kodak Kodachrome 200	200	•			d,k	AA
Polaroid PolaBlue	16	•			d,h,q,s	U
Polaroid PolaChrome CS	40	•			d,q	U
Polaroid PolaChrome HC	40	•			d,h,q	U
Polaroid Type 691	80			•	d,m,q,y	U
Polaroid Presentation Chrome	100	•		•	d	BB

[a]For comment and process codes, see Table 1 footnotes *a* and *b*.
[b]NA, not applicable.

reckless abandon. Photography involves serious chemical hazards and risks which cannot be ignored. The consequences include contact dermatitis, chronic respiratory disease, immune-mediated organ disease, central and peripheral nervous system disorders, reproductive organ damage, and general toxicity. On a simpler note, protection against contact lens destruction is a major concern. Anyone involved with photographic chemicals on a routine basis should refer to the definitive book on photography safety by Shaw (56).

In this context, it is important that you be concerned with the environment as well as yourself. It would be prudent to investigate the disposal of photographic waste products in your laboratory and determine which potential contaminants, if any, are present. In addition, in larger installations the use of a simple cartridge or electrolytic system for silver recovery would be beneficial, economically as well as environmentally (23).

30.2.3.2. Black-and-white processing (3, 11, 18, 20)

Black-and-white processing is generally simpler to perform than color processing because of less stringent demands on time, temperature control, and chemistry preparation. Certain color films, such as Kodachrome, leave no options; they must be processed commercially. Others, such as Ektachrome or any of the E-6 compatible films, are relatively straightforward to process in a laboratory darkroom, especially with the availability of automated equipment.

For manual processing of black-and-white films, use the following guide.

1. Perform processing in a darkroom with a suitable safelight installed. If the film is panchromatic, no safelight is used (see section 30.2.2.10).
2. Choose a developing tank and reels sufficiently large to accommodate the film to be processed. There are many commercial sources for these accessories, but Jobo and other manufacturers make film reels that are simple to load (instructions are included) and tanks that are convenient to use.
3. Select the developer that is appropriate for the film used, and, after mixing, pour the required volume of solution into the developing tank.
4. Accurately measure the temperature of the solution in the tank, and determine the developing time from the data provided in Table 5, with the film, or in the *Kodak Darkroom Dataguide* (18).

Table 5. Quick reference for frequently used black-and-white negatives

Comment Key

Numbers = Time (minutes:seconds)
HFS = High Film Speed
LFS = Low Film Speed
Bold = Highly Recommended
(A) = Agfa
(I) = Ilford
(K) = Kodak

NOTE: Developing times may be varied to suit enlarging system or process temperature. All times are for 68°F except as noted.

		Reusable Developers (Replenished)								One Shot Developers (Dilute & Discard)								
Speed	Film	T-Max (Repl) (70°F)	Ethol UFG	Ethol TEC	Acufine	Diafine	D-19	D76 ID11 Plus	Microdol-X Perceptol	Rodinal (1:25)	Rodinal (1:50)	Rodinal (1:75)	FG-7/Sulfite* (1:15)	Technidol	Ethol Tec	D76 (1:1) ID11	HC-110 (Dil. B.)	T-Max (70°F)
Low Speed	(K) Tech-Pan (25-160)						**4:00** HFS						**3:00** LFS	**9:00** LFS			**8:00** (Dil "D")	
Low Speed	(A) Agfapan 25	5:45								**4:00**	5:00							6:00
Low Speed	(I) Pan-F (32-64)		**3:15** HFS	2:30	3:15 HFS			9:00	10:00	**4:00**			12:00		6:30 HFS	9:00	4:00	
Medium Speed	(I) FP 4 (80-250)		4:30 HFS		3:30			**8:30**	7:00 LFS	4:30			10:00		10:00 HFS	9:00	5:00	
Medium Speed	(K) Plus-X (100-200)	4:30	4:00 HFS	2:00	4:00 HFS	2:30		**5:30**	7:00	7:00	10:00		10:00		11:00 LFS	7:00	5:00	
Medium Speed	(A) Agfapan 100							9:00	13:30	**6:00**	14.00							
Medium Speed	(K) T-Max 100 (50-200)	5:45						9:00	13:30 LFS					26:00 LFS		12:30	7:00	**6:00**
High Speed	(K) T-Max 400 (200-800)	5:15						8:00	10:30 LFS							12:30	6:00	**5:30**
High Speed	(K) Tri-X (320-800)	5:15		4:00 HFS	**6:00** HFS	3:30 HFS		**8:00**	**11:00** LFS	8:00		13:00 LFS	9:00 LFS			**10:00**	**7:30**	**5:30**
High Speed	(I) HP-5 Plus (320-800)		**6:00** HFS		6:00 HFS	3:00 HFS		**7:00**	12:00 LFS	6:00		17:00 LFS				12:00	8:00 HFS	5:30
High Speed	(I) Delta 400 (400)							**6:00**	11:00							**9:00**	6:00	5:30
High Speed	(F) Neopan 400 (400)							7:30				13:00	13:00			9:30	4:00	5:30
High Speed	(A) Agfapan 400 (400-800)									**5:00**	9:00					15:00		
Super High Speed	(K) T-Max 3200 (800) (1600) (3200) (6400) (12,000)	7:00 7:30 **8:00** 11:00 13:00						9:30 10:00 10:30 13:30 16:00								9:30 10:00 10:30 13:30 16:00	6:30 7:00 7:30 10:00 12:00	7:00 7:30 **8:00** 11:00 13:00

*FG7 Sulfite: 30 ml (one film can) of sodium sulfite plus 30 ml (one film can) of FG7 conc. in 450 ml total volume (in H_2O).

5. Load the film carefully onto the reel(s), and immerse in the tank filled with the developer.

CAUTION: Never pour developer into a tank containing film since the random contact of the solution with the film may induce uneven development.

6. Replace the tank lid and, if possible, invert the tank 180° (first left, then right) quickly for five or six inversions within the first 30 s of development. Gently tap the can on the sink base to dislodge any air bubbles trapped on the film. Every 30 s to 1 min, invert the tank another three times and then tap the base again.

7. Repeat this process until 10 s before the allotted development time, and (while maintaining light-tight conditions) begin to pour the solution out of the tank until it is fully drained. Be sure to retain any reusable solutions in tightly sealed dark-colored bottles.

8. Fill the tank with water or a weak stop bath solution which is at the same temperature as the developer, agitate gently for about 30 s, and discard. Water might be better than a conventional acid stop bath since some developers contain carbonate ions, which are rapidly released when the pH is lowered and cause the emulsion to bubble and break.

9. Fill the tank with tempered fixing solution (e.g., Kodak Rapid Fixer). Use the same tank inversion techniques described for the development, but use 1-min intervals. Total fixing time should be about 3 min for most films, although a safe fixing time is generally twice the time it takes the film to clear in the fixer (determined by trial with each film).

10. Turn on the room lights, remove the tank lid, and place the reel(s) in a film washer or circulating water bath for 15 to 30 min. A plastic beaker, larger than the film reels (1 liter), with small holes punched in the bottom serves as an excellent washer. If the beaker is hung from a faucet by attaching a wire strap through the top, the water running from the tap will displace the higher-density but still solubilized fixer very quickly through the bottom. Be sure that the water level is sufficient to cover the reels.

11. The wash time may be reduced by using a commercial product which chemically neutralizes the fixer, such as Kodak's Hypo Clearing Agent or Heico's Permawash.

12. A final wash in water and then an immersion of the film into Kodak's PhotoFlo (surface active agent) will ensure clean, streak-free images. Unroll the film, and hang it from a clothesline with clamps or place in a film dryer until dry.

30.2.3.3. Black-and-white printing

Use monochromatic photographic paper and standard enlarging techniques for making black-and-white prints. If necessary, cut the negatives into strips (lengths) accommodated by the negative carriers for the enlarger. Be certain to choose an enlarging lens and a condenser setting (if adjustable) which are optimal for the size of the negative. (Every manufacturer publishes specific data indicating the recommended field coverage for its lenses.)

Place the negative into the enlarger, and expose portions of the photographic paper for various times to produce a test strip. A more scientific approach would be to use either a standard test negative (sold by Kodak) or light metering as a guide. The latter can be accomplished with the same light meter used for film exposure if a sensor receptor, specifically designed for enlarging, is available for it (Minolta makes a good one). A Lektra PTM4 Densitimer permits perfect reproducibility by determining reflected-light readings from the paper surface and computing the proper paper exposure time. This maximizes sharpness in the image, since an optimum f-stop on the enlarging lens (determined by testing) can be chosen.

If an exposure analyzing instrument is used, the darkroom safelight must be extinguished or its effect will be integrated into the exposure reading for the print. For this reason, use a safelight which can be turned on and off quickly. A filtered incandescent bulb or a fluorescent tube variety (Aristo) is excellent, but a sodium vapor light requires a warm-up and does not lend itself well to rapid on-off use.

Note: Be certain to test any safelight for "safety." A simple method is to lay a sample piece of photographic paper on a countertop under the safelight and cover an area with a coin. Allow the paper to expose for 20 times the normal exposure time. Process the paper, and determine whether an image of the coin (negative shadow) appears. If an image appears, the safelight is "unsafe." This can usually be corrected by changing the color of the bulb or by lowering the lamp intensity (covering its surface with a heat-resistant diffusion material, moving its location, or lowering the lamp wattage).

In the laboratory, not all photographs or micrographs are of even quality, and rescue of the image may be needed by "dodging" or "burning" in the printing process. The choice of black-and-white papers available should be reasonably wide because of these unavoidable variations among specimens and negatives and the emphasis desired. Variable-contrast papers, with an effective contrast range of grades 1 to 5, can be used for most printing. They incorporate multiple emulsions that respond individually to different colors. A green-sensitive layer responds to magenta filtration and produces higher-contrast images, whereas a blue-sensitive layer is affected by yellow filtration to produce the lower-contrast range. Additional flexibility can be provided by graded bromide papers and/or a choice of paper surface, weight, and image tone characteristics.

The exposed prints can be processed in trays or in machines. For black-and-white prints, tray processing (4) is a time-proven standard and the results are excellent and controllable (by inspection and comparison with a standard print having a full range of tonalities). In fact, if conventional fiber-based papers are used, it is the only method available. However, many of the modern papers incorporate an impervious plastic or resin coating on the paper base side and some have developing chemicals incorporated into the emulsion (Table 6). This permits chemical activation and rapid (10-s) development in a high-pH (13+) solution such as Kodak's Royal Print or S2 Activators. Activation may be done in standard trays, but this necessitates handling the solutions, and they are extremely caustic (*CAUTION:* Take care to protect your clothing and your skin). Machine processing is the best and most reproducible method. Agfa manufactures a roller transport processor (DD3700) which is well suited for black-and-white processing since it enables the production of a dry-to-dry print (exposed paper in/dry print out) in 1 min.

Table 6. Black-and-white printing materials

Black-and-white printing material	Contrast		Image tone			Base		Processing		Process[a]
	Graded	Multi	Warm	Neutral	Cool	Fiber	Resin	Tray	Machine[b]	
Agfa Brovira	•				•	•		•		D,T
Agfa Brovira Speed PE	•				•		•		•	Q,R,S
Agfa Insignia	•		•			•		•		D,T
Agfa Multicontrast (MC)		•			•		•		•	Q,R,S
Agfa Portriga-Rapid	•		•			•		•		D,T
Agfa Portriga-Speed	•		•				•		•	Q,R,S
Ilford Ilfobrome FB	•				•	•		•		D,T
Ilford Galerie FB	•				•	•		•		D,T
Ilford Multigrade FB		•	•	•		•		•		D,T
Ilford Multigrade III Deluxe & Express		•			•		•	•		D,T
Ilford Multigrade III Rapid		•			•		•		•	Q,R,S
Ilford Ilfospeed Delux	•		•				•		•	Q,R,S
Kodak Azo	•			•		•		•		D,T
Kodak Ektalure	•		•			•		•		D,T
Kodak Ektamatic SC		•		•		•			•	Q,R,S
Kodak Ektamax	•			•			•		•	EE
Kodak Elite	•			•	•	•		•		D,T
Kodak Kodabromide II RC	•			•			•		•	Q,R,S
Kodak Kodabromide	•			•		•		•		D,T
Kodak P-Max Art RC	•			•			•	•		D,T
Kodak Panalure Select	•		•	•			•		•	Q,R,S
Kodak Polycontrast Rapid III		•		•			•		•	Q,R,S
Kodak Polyfiber		•		•		•		•		D,T
Kodak Polymax RC		–[c]		•			•	•		D,T
Kodak Polyprint RC		•		•			•	•		D,T
Oriental Seagull Portrait	•		•	•	•	•	•	•		D,T
Oriental Seagull G	•				•	•		•		D,T
Oriental Seagull Select VC		•			•	•		•		D,T
Oriental Seagull Select VC-Rapid		•			•		•		•	Q,R,S
Oriental Seagull RP	•				•		•	•		D,T
Oriental Center	•		•			•		•		D,T
Oriental Panchromatic F	•		•				•	•		D,T

[a]For process codes, see Table 1, footnote *b*.
[b]All machine-processable papers may be conventionally processed in trays.
[c]Extended contrast range when used with special filters.

30.2.3.4. Color printing (30, 39, 46)

Color printing can be done directly from either negatives (C-prints) or positives (R-prints). The basic concepts are similar to black-and-white printing except that both density and color corrections must be made. Colored light is provided by an enlarger, and the paper is exposed and processed to produce a print. The contrast range of color paper is more limited than with black-and-white printing. Kodak has introduced a series of color papers (Ektacolor-RA) with three distinct contrast grades.

One of the most troublesome problems in color printing relates to the use of color filtration when changing the color balance of a print. If one prints both R-prints and C-prints only occasionally, it may be difficult to remember which colors to add or subtract from the color pack in the enlarger.

Note: When printing C-prints, to shift a print toward a particular color, subtract that color from the filter pack or add its complementary color. (Remember, this is a negative-to-positive process.) When printing R-prints, to shift toward a given color, either add the color directly or subtract its complementary color from the filter pack in the enlarger.

Another difficulty with color printing is that variations in the exposure time or electrical line voltage fluctuations may cause color shifts as a result of bulb color temperature variations. This concept is often overlooked but is quite significant. When an enlarger bulb is turned on, there is an initial warming followed by the remaining exposure time, and finally the bulb cools. The warming and cooling phases of the bulb activation have a direct influence on the color of the image. If the total exposure time is short, the relative amount of red light added is significant and will shift the color in that direction. In the same context, if the exposure time is lengthened, there will be proportionally less red light due to these bulb phases and the image color will shift toward cyan. This problem can be largely circumvented in color printing by using constant exposure times with a voltage-stabilized power source and/or an enlarger containing an additive electronic flashtube light source (Beseler/Minolta 45A). With this device, three flashtubes, each filtered with a primary additive color (RGB), are controlled precisely by a computer. The exact amount of light required in each primary wavelength is determined when the internal computer is directed to analyze the negative and program the light for the proper settings. The paper exposure consists of multiple short bursts of light, at precisely the proper wavelength from each tube, until the exact color and density are achieved.

Color prints can be processed in a special print-processing drum in a sink. It is preferable to have an inexpensive roller/agitator motor (made for use with the print drum) to provide consistent distribution of the chemistry over the print surface while it is being processed. Temperature control is critical in color processing and must be maintained within 0.25°C. Without a controlled processor, this is difficult but not impossible. A simple technique follows.

1. Determine the recommended temperature for the process being performed. Usually in color printing, the first developer is the only critical step in this regard, but there are exceptions.
2. Adjust the water temperature in the sink until the desired temperature is reached, and fill a large tray or bath with the tempered water.
3. Place the chemicals to be used in glass bottles, place the bottles in the water bath, and allow their temperatures to equilibrate.
4. *Accurately* (to 0.1°C) measure the water temperature in the bath and pour a volume of this water, equal to the volume required for the first developer in the process, into the tank. Roll the tank, either in the sink or on the roller motor, for the time recommended for the development. Now pour out the water and measure its temperature *accurately* once again. The second reading is usually lower.
5. Calculate the mean temperature for the two readings. The temperature will have drifted downward during the processing, and the actual temperature for the process can be estimated by the mean temperature.
6. From the manufacturer's processing guide, determine what the adjusted time for the process (determined by the temperature) will be.

The processing of all films and prints is time-temperature dependent, but color films and prints are infinitely more sensitive to variations than are black-and-white materials because both density and color changes take place.

A simple but expensive alternative is to use Jobo's ATL2-Plus rotary processor for color (or black-and-white) print and/or film processing. It is compact, accurate, and very reliable. All steps, except for drying, are programmable, and times and temperatures are controlled automatically.

Color-printing materials and processes are listed in Table 7.

30.3. APPLICATIONS

The photographic needs of the laboratory are many and varied. From the standpoint of subject matter, they can be viewed as involving increasing levels of magnification that range from the conventional (equipment or animals, copy work, computer graphics) through macrophotography (reproduction ratios from 1:10 to 20:1), photomicrography (imaging from 20:1 into light microscopy levels of 1,000:1), and, finally, electron microscopy at magnifications of 2,000:1 or greater.

30.3.1. Copy Work (31, 42, 59, 62)

30.3.1.1. Charts, graphs, photographs, and radiographs

Much of the photographic work in the laboratory involves copying charts, graphs, computer graphics, radiographs, and micrographs (or other photographs). Most of this is straightforward, and an excellent reference on the subject is provided by Kodak (62).

Once a camera system is in place, an appropriate and controlled light source is needed. For most copy work the use of an Aristo DA light box is recommended. This box incorporates a large (10- by 12-in. or 14- by 17-in.) rectangular daylight-corrected (5,500 K), fluorescent light source, which provides evenly distributed direct light on the copy subject placed therein and, when inverted, allows for transillumination through an opalescent plastic base panel. It emits very little heat, and the copy will lie flat. Even though it is daylight corrected (for viewing only), the light source will record on film as predominantly blue-green. This is ideal for most black-and-white emulsions but must be corrected for color film through magenta filtration. When measured with an incident-light meter, the smaller box has an EV of 11 if used with a film having a speed of ISO 100. As alternative light sources, use standard photofloods (bulbs with the designation BBA), commercial halogen bulbs, or small electronic flash units mounted on the copy stand light brackets. Be sure that any light source used has a known color temperature and, for the reasons discussed in the section on enlarging (section 30.2.2.4), that its voltage is regulated for consistency.

Countless difficulties may be encountered in copy work. These include problems relating to reflections, tonal-value reproduction, and contrast control. One potential problem is that the usable film speeds (ISO or EI) of some specialized but uniquely useful monochromatic films (e.g., Kodak LPD4, Kodalith, or Kodak Rapid Process Copy, with EIs from 0.75 to 8) are below the lowest range of most light meters. In such instances, a reading within the meter's range can be used by applying the following simple mathematical rule: for every halving of the film speed set on the meter that is required to attain the speed of the film in use, open the lens aperture one stop or decrease (slow) the shutter speed by a factor of 2. For example, the meter is set at its lowest setting (ISO 12), but the EI of the film in use is 1.5.

EI: (12—6—3—1.5)
f-stops: (0 1 2 3)

Therefore, read the meter at EI 12, and open the lens aperture three f-stops.

Another frequent problem is the increase in contrast in a second- or third-generation copy print. Unfortunately, this contrast gain is not entirely avoidable, but there are techniques to minimize it. Prefogging or presensitization of the film used in the copy camera is the method of choice. The following protocol should provide significant improvement in reducing the contrast of copy negatives in either color or black and white.

1. Set up the camera, frame the subject (X-ray, photography, etc.) in the viewfinder, and carefully focus the lens.
2. Take a light meter reading with the appropriate techniques (incident or reflective). Set the camera for the proper exposure. Be sure to include the exposure factor calculations for lens extension or filtration if an external light meter is used.
3. Place a nonabrasive "spacer" (a small block of wood or roll of tape) over the copy work, and place a smooth white mat board on top of the spacer so that it

Table 7. Color printing materials

Color printing material	Printed from:		Contrast			Process[a]
	Negative	Slide	Low	Medium	High	
Agfacolor Type 8	•			•		DD
Fujichrome Type 34		•		•		FF
Fujichrome Type 35		•		•		FF
Fujicolor Type 03	•			•		DD
Fujicolor Super FA	•			•		EE
Ilford Cibachrome A II (CPSA)		•			•	GG
Ilford Cibachrome A II (CF)		•		•		GG
Kodak Ektachrome Radiance		•		•		FF
Kodak Ektachrome Radiance Select		•		•		FF
Kodak Ektacolor Pro	•			•		DD
Kodak Ektacolor Plus	•				•	DD
Kodak Ektacolor Portra	•		•			EE
Kodak Ektacolor Supra	•			•		EE
Kodak Ektacolor Ultra	•				•	EE

[a]For process codes, see Table 1, footnote *b*.

completely covers the material being copied. (The spacer serves to raise the mat board from the copy work and defocus its surface.)

4. Install a 1.5 neutral-density (ND) gelatin filter over the lens. *Note:* ND filters are specified by their transmission density and are manufactured in 1/3 f-stop increments or multiples thereof. Each 1/3 f-stop is denoted by a value of 0.1. Therefore, a 1.5 ND filter corresponds to an exposure decrease of 5 f-stops (very dense).
5. Photograph the white board (now out of focus) by using the normal exposure determined by the light meter reading of the copy. Remember, the 1.5 ND filter is in place.
6. Remove the 1.5 ND filter, the white board, and the spacer material. This leaves only the original copy.
7. Reset the shutter without changing the f-stop or advancing or moving the film (this is where a camera with multiple exposure provision is essential).
8. Make a second, identical exposure.

This procedure involves photographing the subject on a piece of film that was prefogged by a previous exposure. The prefog level is determined by the amount of light necessary to sensitize the film so that any additional exposure will be recorded. It is simply a matter of adjusting the toe area of the characteristic curve (3, 19) to provide detail in the shadow areas of the copy. Experimentation can be performed by varying only the ND filtration. Higher ND values (less fog) will provide more contrast in the final image, whereas reduced ND values (more prefog) will yield diminished contrast in the copy.

As with any work in the laboratory, it is important to perform controls when doing photography. There must be some standardization between film exposure and development. An excellent means of accomplishing this is to photograph a standard gray scale or step wedge (Kodak's "step tablet #3" or "gray scale and color separation guides") and bracket (vary both up and down) the exposures by 1/2 f-stop over a 3 f-stop range. After processing, evaluation of the resulting images will allow close comparison with the original standards and indicate modifications for exposure and development.

Slide or negative duplication is frequently necessary in the laboratory. If only an occasional duplicate is needed, it is certainly simpler to have it made commercially. The cost of buying a slide duplicator is high, and it would not be cost-effective unless production was high as well. However, the techniques used are straightforward and are discussed fully in the instruction manual provided with any instrument. Both Kodak and Fuji manufacture low-contrast films specifically designed for this purpose (see section 30.5), and the type of light source used in the particular slide copier you have will dictate the choice of film. If there is a need to duplicate a film on an emergency basis, this can be done by photographing it transilluminated on the Aristo light box. Assuming the negative format is the same as the original slide (35 mm to 35 mm), a macro lens, extended to its 1:1 position, can be used to copy the slide directly. Remember, if this is done in color, it is essential to balance the film to the light source.

30.3.1.2. Computer graphics (33)

The production of graphics from computers is an art that is changing almost daily. The technology available today allows for the production of a video image on an inexpensive still video camera and, with the use of an electronic capture board, translation of that image onto a computer for modification. The generation of presentation slides on a computer (especially a Macintosh), with the superb graphics packages available, is simple and fast.

Computer graphics can be recorded on film in a variety of ways. Directly photographing the display screen, using a tripod-mounted camera, is the simplest method but also the least desirable. If this method is chosen, be certain to use a shutter speed of 1/15 or slower to ensure an even screen scan. Since the screen is a high-quality cathode ray tube (CRT), the images photographed in this way will be acceptable, but the resolution of the screen is lower than that which the computer is capable of producing or even that of a printed image from a laser printer.

The second choice is to use the aforementioned copy techniques (section 30.3.1) to photograph a laser printer hard copy (color or monochrome) with a copy camera. It is particularly helpful if the printer output is dimensionally formatted to the image aspect ratio for the camera being used (2:3 for a 35-mm camera) and is centered on the page. Finally, some facilities might have a computer-driven slide-making device which will download a digitized image of a graphic to a computer buffer or spooler and send this information directly to an internal 35-mm camera system and expose the film. The film choices are usually limited to those having an ISO compatible with the standards for the machine (generally ISO 100 and slower). Once the film is exposed, all that remains is to process the film and produce prints or mount slides.

30.3.2. Macrophotography (5, 7, 31, 50, 54)

Although there is some dispute with regard to the definition of macrophotography, the discussions here are concerned with images having magnifications from 1:10 (1/10 life size) to 20:1 (20 times that of the object). All are photographed (without the use of a microscope) with conventional cameras and with lenses formulated for this magnification range.

30.3.2.1. Density gradients (54)

Density gradient photography requires a tripod-mounted camera and, depending on the color or prominence of the bands being photographed, specific films and filters. As an example, a centrifuge tube may be photographed to show bands in a gradient. If the bands are faint, a continuous-tone, fine-grained, ultrasharp negative film is advised (see section 30.5) since these films usually exhibit high resolution and good exposure latitude. If slides are required, they can always be produced by copying the negatives (a negative of a negative is a positive) as described in section 30.3.1.1.

1. Once the image of the centrifuge tube is framed in the camera viewfinder, place a piece of matte-black material (cloth or board) behind the tube.
2. Direct a focusable light source (a microscope light works well) through either the bottom or top of the tube until the band(s) you wish to delineate appears defined. Use care: do not allow light to stray onto the black background, or the contrast will be diminished.
3. Metering this type of dark-field illumination can cause some difficulties especially when high magnification factors or filtration are used. If a large- or medium-format camera is being used, a Polaroid print can be taken to evaluate the correct exposure and/or be used for incorporation into a laboratory notebook.

30.3.2.2. Electrophoresis gels (5, 7, 54)

a. UV/Fluorography of DNA stained with ethidium bromide

When photography of gels is a routine procedure, the use of a Polaroid MP-4 camera with either a Schneider 135-mm Componon-S or 120-mm Makro-Symmar HM lens mounted in a Prontor Professional shutter is recommended. This system combines simplicity of operation, ease of viewing, and superior lens resolving power. The optimum light source is a transilluminator containing an efficient (>2,500 μW/cm^2) 302-nm UV tube beneath a compatible diffuser (54).

CAUTION: Protect eyes and skin with plastic goggles or a full-face shield that are designed to filter UV light, or serious burns or even blindness may result. Since ethidium bromide is a mutagen and is toxic, wear gloves.

The use of a deep yellow or orange filter (Kodak Wratten no. 15 or 22) is necessary to attenuate blue light and enhance fluorescence on black-and-white panchromatic films. These filters have an exposure factor of 1.5 f-stops (requiring a threefold increase in exposure time or a reduction of the EI to one-third the rated ISO).

1. With most gels, use Polaroid type 667 or type 57 (ISO 3000) films with an exposure of 1 s at f5.6 (as a starting point). Although no negatives are produced, these films are extremely sensitive (requiring short exposure times) and fluorescence quenching is not a problem as it is with slower films. For instance, when compared with Polaroid type 665 (ISO 80), there is a relative exposure differential of 5-1/2 f-stops (50-fold). The film speed difference would necessitate an increase of exposure time from 1 s to 1 min, the latter being less practical when quenching is a problem. It is for this very same reason that an excellent lens, with high resolution at near maximum aperture, should be chosen, i.e., to keep the time to a minimum.
2. For very faint gels, it may be necessary to use conventionally processed films in a 4×5-in. film holder on the MP-4 to enhance the contrast through a combination of exposure modification, film development, and printing. An excellent film choice, providing a good compromise between film speed, resolving power, and fluorescence quenching, is T-Max 400 push-processed to an EI of 800 to 1,200.

Once the photography of ethidium bromide-stained gels is standardized, it is possible to actually estimate the concentration of DNA in an unknown sample by directly comparing the photographs with those of a sample with known standards. The use of agarose is far superior to the use of polyacrylamide in preparing these gels, since the latter quenches the fluorescence and limits the resolution to 10 ng of DNA (7, 54).

b. Protein staining with Coomassie brilliant blue or silver salts

Protein bands can be visually detected in polyacrylamide gels by using either the nonspecific binding of Coomassie brilliant blue R or the specific binding (sulfhydryl or carboxyl groups) of silver salts. Coomassie blue does not bind to the gel material, and the bands appear as distinct blue lines that are easily photographed. The use of ammonium or nitrate silver salts to detect proteins is a significantly (100-fold) more sensitive technique but is more difficult to photograph because of the yellow-brown color of the bands (5, 7, 54). *Note:* Wear gloves to prevent staining artifacts from fingerprints, and be careful about excess pressure or drying to prevent the gels from cracking.

For photography of protein bands, a transilluminating daylight-corrected fluorescent light box (Aristo DA) should be used. If an MP-4 camera is used, Polaroid type 665 or type 55 film will provide both an excellent negative and a print for use in a laboratory notebook. However, any color or black-and-white film is usable, and for 35-mm or medium-format photography, the fine-grained, high-resolution films are best (see section 30.5).

Black-and-white photography of Coomassie blue-stained bands requires the use of a yellow filter (yellow is the color complementary of blue) to enhance the blue bands. Kodak's Wratten no. 15 or 22 (exposure factors of 1.5 f-stops) are excellent for this purpose and are the same filters used to photograph fluorescence with the ethidium bromide detection of DNA.

For silver salts, the use of a green filter, Kodak Wratten no. 58, is beneficial. This filter has an exposure factor of three f-stops (reduce the EI to one-eighth the rated ISO or increase the exposure time eightfold that without the filter).

Initial exposures are exemplified for Polaroid 665 film with an Aristo light box. Gels are variable in size, and the examples here are for 3- by 4-in. gels photographed on 3.25- by 4-in. film (i.e., a 1:1 reproduction ratio). For Coomassie blue-stained gels, use a no. 15 filter and expose for 0.25 s at f11. For silver-stained gels, use a no. 58 filter and expose for 1.0 s at f11. *Note:* The use of color film circumvents the need for colored filters and the related exposure factors, but be certain to balance the color of the light source to the film.

30.3.2.3. Immunodiffusion

The use of the **Ouchterlony technique** for the demonstration of antigen-antibody microprecipitation and migration is now somewhat outdated. Even so, there may be a need to use it for some basic initial experiments or for demonstration purposes.

Generally, this technique involves the coating of microscope slides with agar and cutting wells into the agar surface. Various antigen and antibody dilutions are placed in the wells, and the slide is incubated. Precipitin bands form as arcs between the wells containing homologous antigen and antibody. The bands can be easily photographed by dark-field techniques, as follows.

1. Support the ends of the slide above a dark cloth, and direct light onto the surface from above. The Aristo light box can be used for this purpose by removing the bottom diffuser.
2. Meter the exposure using either reflected or incident readings.

A direct-printing technique for these tests was developed in the author's laboratory, eliminating the use of a camera and utilizing the Ouchterlony plate as a negative, as follows.

1. Stain the Ouchterlony plate with neutral red for 30 min, and wash twice.
2. In the darkroom, with an appropriate safelight and orthochromatic film, make a contact negative of the slide by using the enlarger as a light source. *Note:* Orthochromatic film will not see the red precipitin band but will record only the surrounding enlarger light.
3. As an alternative, make a cardboard enlarger carrier for the microscope slide and insert it into the enlarger as if it were a negative. If a high-contrast graded paper is used (not polycontrast paper, since it is sensitive to a variety of colors), a direct print is easily made.

30.3.2.4. Mammalian cell migration (51)

The macrophage migration inhibitory factor assay permits quantitation of the interaction between lymphocytes or macrophages and a specific antigen. The cell suspensions are centrifuged and placed into small capillary tubes. The tubes are then placed in glass chambers containing 1 ml of cell nutrient and sealed with a coverslip. The normally motile macrophages migrate from the tube and adhere to the coverslip, forming a fan-like tuft of cells. In the presence of antigen, lymphocytes elicit a chemotactic factor that inhibits cell migration and reduces or eliminates the tuft. Planimetric measurements on the photographs of the tuft area can serve as a quantitative index of migration and/or antigen concentration.

There are some technical difficulties in photographing the tufts. First, the tuft size falls somewhere between the microscopic and the macrophotographic range. Second, the tufts are very fragile and lack contrast because of their monolayer configuration and translucency. The use of a table-top projector (such as a microfiche reader) with a ground-glass screen can enhance the magnification and markedly increase the contrast. The screen of the projector can be photographed, much like a computer screen, by using a tripod-mounted 35-mm camera and a macro lens. By using spot-metering and any fine-grained color or black-and-white film, excellent images can be produced which can then be measured on the planimeter.

30.3.2.5. Mammalian and bacterial cell clearing (plaques)

The key to successful black-and-white photography of vital-stained mammalian or bacterial cell plaques caused by viral lysing is to understand the concept of additive and subtractive colors. To enhance the contrast of an object, the use of a complementary color filter is necessary. These reciprocal colors are: red-cyan, green-magenta, and blue-yellow. However, care must be taken not to maximize the contrast in this manner, since fine details may be lost in the process. The choices for the photographer become simple once this is understood. Plaque assays incorporating neutral red will be photographically enhanced with a green filter, whereas those using trypan blue or crystal violet stains will benefit from the use of a light yellow filter. Similarly, blue plaques can be enhanced with a yellow filter. Naturally, photographing these in color is straightforward.

If plaques are faint, it is beneficial to place a translucent ruler over the transilluminated culture flasks or plates to provide a point of focus. Care must be taken to provide sufficient depth of field (using a small lens aperture) that the entire cell flask is in focus. Using an Aristo DA light (EV11) and a copy camera containing a film with an ISO of about 100 (i.e., MP-4 camera with Polaroid type 665 film), bracket the exposures around a camera setting of 1 s at f22. Do not neglect any filter factors.

30.3.2.6. Bacterial colonies

Bacterial colonies grown on agar plates can vary from translucent to brilliantly colored. As with plaque assays, vital stains can be incorporated into the growth medium and the resultant colonies will be depicted as either positively colored structures or clarified areas in the medium. If stained colonies are to be photographed, use the techniques described for plaques (section 30.3.2.5). Unstained colonies present a photographic

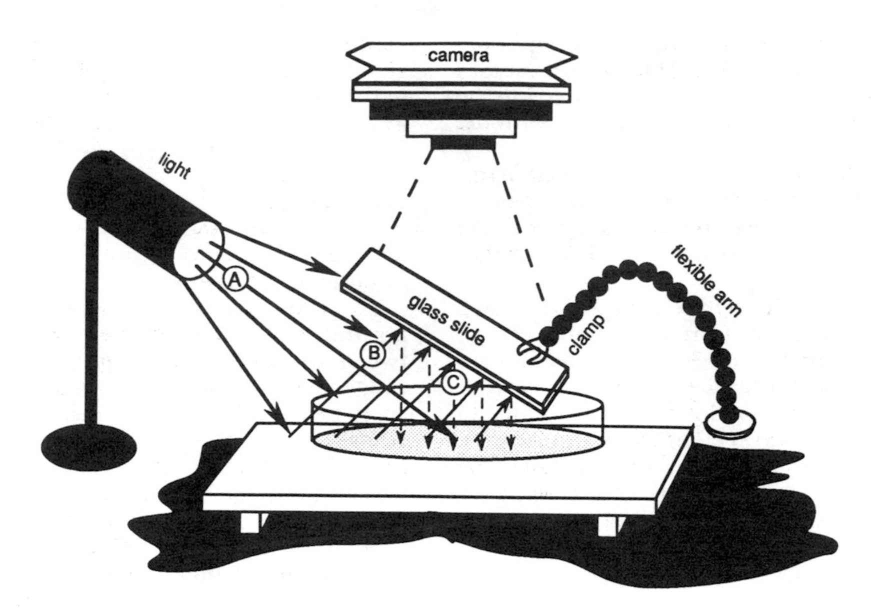

FIG. 2. A light beam-splitting technique for dark-field photography of bacterial colonies. (A) Incident light; (B) primary reflected light; (C) secondary reflected light.

challenge. Dark-field photography will enhance such colonies and, with the proper lighting, provide texture and form to the colony surface (Fig. 2).

1. Place the flask or plate over a dark background (cloth or paper). (*Note:* It is better if the plate is supported above the cloth by ca. 25 mm to permit the background to fall out of the plane of focus.)
2. Light the surface of the colonies with a diffused light source.
3. Insert an optically clear glass beam-splitter, supported by a flexible arm and clamp, into the light path between the colony surface and the lens.
4. Adjust the beam-splitter to redirect the desired amount of reflected light back onto the agar colony surface.
5. Compose, meter, and photograph what you see in the viewfinder.

A particularly useful method for the photography of small bacterial colonies is by means of the stereomicroscope (52). An SLR camera (without the lens) should be attached to the microscope with a specially designed tube fitted to the eyepiece assembly or camera port. Focusing should be performed through the camera viewfinder by using the microscope adjustments. Most stereomicroscopes have provisions for focusable lamps, which can be used both for epi-illumination and for substage transillumination. A judicious combination of the illumination sources will reveal both surface textures and internal detail of the colonies.

30.3.3. Photomicrography (12, 14, 28, 45, 52) (also see Chapter 1.8)

A thorough description of the mechanical adjustments in the light microscope and of optical techniques (such as dark-field, Kohler illumination, and phase-contrast microscopy) is given in Chapter 1 of this book and described in detail by Delly (12). What is pertinent here is to describe which films to use for a given task and to identify some definitive problems. Unfortunately, books or manuals published by manufacturers describe only their products,when, in fact, there may be a better material available. The following recommendations for specific films are based on the author's experience or the suggestions of experts in this field.

As mentioned above, it is important to optimize any image-forming system in order to obtain high-resolution and high-contrast photographs. Laboratories usually spend large sums of money on modern, high-quality microscopes and seldom compromise these instruments with low-quality lenses. It therefore appears that the difficulty in obtaining high-quality photomicrographs stems from poor photographic technique(s) resulting in the degradation of an otherwise excellent microscopic image.

The following precautions may assist in the elimination of some of the causes of lowered image quality.

1. Camera vibration, a primary concern, can usually be circumvented by mounting the microscope on a shock-absorbing table and/or turning off surrounding equipment which may cause a problem. Use of an internal camera with an integrated light meter is preferable to one mounted externally on tubes; but if there is no choice, interface the mounted camera through a bellows rather than a solid connection. When possible, always choose a leaf shutter camera over a focal plane shutter body when choosing a microscope camera and always use a cable release to actuate the shutter.
2. Check the mechanical adjustments on the microscope to see that they are secure. Loose or worn parts may cause slippage and result in out-of-focus images. Misalignment can cause uneven illumination and subsequently poor images. Be certain that the optical finder is adjusted for the observer's eyes and that the eyepiece reticle is clearly in focus.

3. Clean all optics before and after each use (see Chapter 1.1.6). All too often this simple precaution is overlooked.

4. Troubleshooting is made easy in Chapter 1.1.7.

30.3.3.1. Fluorescence (8, 37, 38, 43, 47, 53, 61)

All of the information regarding fluorescence photomicrography pertains to the 35-mm format. The use of larger-image formats is counterproductive because of substantial light losses inherent in the magnification process. Various fluorochromes may be employed, as follows.

a. Fluorescein

When fluorescein isothiocyanate is used as a fluorochrome, the secondary fluorescence is in the green portion of the visible spectrum at about 525 nm. Naturally, a film that renders green optimally will be best. Fujichrome P-1600 is unequaled in this regard. The greens recorded by this film are dramatic and allow a clear distinction between the cells of interest and any background autofluorescence. Other films, such as Ektachrome EES and P1600, are acceptable, but they truly lack the brilliance of the "Fuji greens."

Metering with this technique is difficult because of the dark backgrounds produced by the barrier filters (they prevent the passage of visible light other than that emitted by the fluorochrome). As discussed previously, dark backgrounds will indicate to an averaging meter the need for more exposure. If that occurs, the fluorescing cells (the main interest) will be rendered too light on the film. Unless the microscope is equipped with a spot meter having a very narrow field of view, it is best to rate Fujichrome P-1600 (or Ektachrome P1600) at an ISO of 1,600 on the microscope; however, it is important to note that the film must be processed as if it were exposed at ISO 800 (designate a one-stop push). *Note:* Fujichrome P-1600 and Ektachrome P1600 have a nominal ISO of 400 but are designed to be push-processed up to EI 1,600 with little image degradation.

The films suggested here are high-speed varieties which permit recording the image before fluorescence is lost by quenching. The rapid fading of fluorescein conjugates can be exploited by the labeling of two different antibodies with different fluorescent conjugates, particularly if the second fluorochrome resists quenching.

b. Rhodamine

Rhodamine is used in either of two configurations (53): the tetramethylrhodamine isothiocyanate or the sulfonyl chloride derivative of rhodamine B (RB 200). Both produce secondary fluorescence in the red-orange portion of the visible spectrum at about 590 nm. Note that the emitted secondary color is opposite that of the blue induced by stray UV light. An efficient UV barrier filter (Kodak Wratten 2E) is essential in preventing the neutralization of the true color of the rhodamine fluorescence on film. Any of the high-speed color transparency films are excellent with this technique (Fujichrome P-1600 or Ektachrome P1600). However, because rhodamine resists quenching more than fluorescein isothiocyanate (note that there is a shift in color from red to orange) (53), slower finer-grained films can be used. The same metering principles apply that were described for fluorescein.

c. Texas Red

The chromaticity (distinct red color) produced by Texas red is unmistakable. The brilliance is visually striking yet very difficult to meter accurately. Most photomicroscope meters do not respond well to long-wavelength emissions (615 nm) and may give erroneous readings. It would be wise to bracket exposures around an EI of twice the rated film speed (e.g., EI 800 on Fujichrome 400 or Ektachrome 400) but process the film as if it were exposed at the normal film speed.

d. DAPI

DAPI (4′,6-diamidino-2-phenylindole) was first used to complex with mitochondrial DNA in cesium chloride density gradients and effect a lowering of the buoyant density (61). Once it was found that the resulting compounds were fluorescent, DAPI became a widely used, highly sensitive DNA probe in cytological studies. The secondary emission is in the near-UV (420 nm) blue region of the visible spectrum. Recording the short-wavelength radiation is aided by the use of fluorite or synthetic fluorite objectives such as the Zeiss Neofluars or Plan-Neofluars. The film of choice is Kodak's Ektachrome 160 Professional (EPT). This is a specific example of the ability to exploit the color balance of a film in order to exaggerate the color of the image. EPT is a tungsten-balanced film (3,200 K) and therefore biased at the long-wavelength end of the spectrum. It will therefore record short-wavelength radiation with an exaggerated but representative blue color. With DAPI, any film used should be rated with an EI of twice the ISO but should be processed normally (e.g., EPT 160 rated at EI 320).

30.3.3.2. UV and infrared light (12, 13)

UV imaging is more frequently used in conventional forensic and industrial applications than in the biological sciences. Zeiss manufactures a UV-Sonnar lens for use on the Hasselblad camera. This is specifically designed for photography in the short-wavelength UV, through the visible, and into the near-IR part of the spectrum (215 to 700 nm). Normal optical glass absorbs UV radiation, so this lens is made entirely of quartz and fluorite elements. For use in conventional photography, a simple UV glass filter suffices. However, Zeiss also makes Ultrafluars (quartz elements) and an Acrostigmat 40 UV lens for its photomicroscopes. The latter is optimized for the 340-nm wavelength and is frequently used with UV irradiation for determining calcium ratios in bone studies.

Most images recorded under UV light are inherently low in contrast. Therefore, black-and-white films designed specifically for this purpose are higher in contrast than are normal panchromatic emulsions. If conventional films are used, take steps to enhance the contrast during development (use extended times or selected special chemicals). Color films should be carefully selected because most have a UV protective coat applied during manufacture, which will make them unsuitable for this purpose.

IR photomicrography is valuable in some areas of biology as well as in the forensic and material sciences. Some heavily pigmented or stained structures, found in insects or in bone, are opaque to visible light but translucent to IR emissions. IR film is manufactured by Fuji,

Konica, and Kodak. The first two are difficult to obtain in the United States, but are manufactured in varied formats. Kodak makes both black-and-white and color emulsions, but they are available only in 35-mm format. Accurate metering of IR is nearly impossible because the emissions are beyond the sensitivity range of most meters (750 nm). Trial and error with regard to exposure and filtration are essential.

Note: Carefully examine the "darkroom" in which IR film is to be processed. *The exclusion of visible light is not an indicator of safety!!* Windows, which have been sealed and painted black and which have proven themselves "safe" for normal films, can be extreme emitters of heat (IR) and will expose IR film readily. All areas in the darkroom must be insulated and proper barriers must be installed. Another difficulty with some IR films has to do with the type of base onto which the emulsion is coated. For example, Estar (Kodak)-based films can act as light pipes and actually transmit IR into the cartridge and expose the film as if it were a fiber-optic device.

30.3.3.3. Biological stains (12, 14)

The most common biological stains for photomicrography are tissue stains (e.g., hematoxylin and eosin, methylene blue and eosin [Wright's stain], and toluidine blue) and bacterial stains (e.g., Gram stain, Giemsa stain, and crystal violet). The suitability of any film for a given task is determined by its color sensitivity, resolution, contrast, granularity, speed, and exposure latitude (14). For this work, fine-grained high-resolution films are recommended for either black-and-white or color photography.

For black-and-white photography the use of complementary filtration (section 30.3.2.5) will heighten the contrast of the desired preparations. For example, the filters most often used for eosin are the Kodak Wratten 58 (green) or Wratten 11 (yellow-green), depending on the intensity of the staining (see Chapter 1.2.5).

For color photography, the use of selected films and filtration is necessary. The color of the illumination source may be affected by every component of the optical system (from the internal diffusers to the heat-absorbing glass). Light-balancing color correction filters (series 81 [warming] or, more probably, series 82 [cooling]) are usually required. Also, be aware that all stains are recorded differently on different dye sets used in the manufacture of color films and that some are more (or less) sensitive than others. The eosin reds may not be rendered properly on most color films, but Kodak specifically designed Photomicrography Color Film 2483 (recently discontinued) to faithfully reproduce hematoxylin and eosin without filtration. Some of the dye-film incompatibilities may be corrected with the use of special filters incorporating rare earth elements having specific spectral responses. An example is the didymium filter manufactured by Schott Glass (made by combining neodymium and praseodymium coatings), which enhances eosin in photographs. Careful experimentation will provide good results.

30.3.3.4. Phase and interference contrast microscopy

In the event an unstained preparation (mammalian or bacterial cells) needs to be photographed "for the record," it should be contrast enhanced before being photographed. Since color is of little or no concern, black-and-white films are most commonly used. A green filter (Kodak Wratten 58) is used in the microscope because the objectives are optimized for the green portion of the spectrum, as are most panchromatic films. The use of low-speed, high-resolution film is best unless motile cells are being photographed or light levels are very low, in which cases either a high-speed film or an electronic flash system can be used.

30.3.3.5. Dark-field microscopy

Dark-field microscopy is a technique used occasionally in bacteriological research. Cell inclusions or structures are easily visualized as bright objects contrasted against dark backgrounds. Just as with phase photomicrography, black-and-white films are most commonly used, and, depending on the subject and its motility, the slowest possible film should be employed.

30.3.3.6. Rheinberg differential coloration

Rheinberg differential coloration can be photographed in either color or black-and-white format. The use of various colored filters in the central and peripheral stops on the microscope permits the objects to be rendered any specific color against a background of any other color. This can be exploited in a black-and-white format by using the complementary filtration techniques (see section 30.3.2.5). The use of highly saturated color film (Fuji Velvia or Ektachrome 50 Lumiere) will provide very effective slides for a presentation.

30.3.3.7. Flow cytometry

The laser illumination imaging cytometer is a useful tool in cell biology. An argon laser is used to illuminate the cells being studied. It is focused to a spot size of about 1 μm through a microscope objective and, by using selective single-wavelength illumination, can excite fluorophors and produce secondary fluorescence. In bacteriology, flow cytometry is used for direct counts (see Chapter 11.2.3). In mammalian cell biology, the applications include membrane studies, nucleic acid measurements, cell viability studies, immunofluorescent intracellular antigens (41), and binding and uptake of glycoproteins (43). The imaging involves videography, since 5- to 20-mm cells are flowing through a 70-mm orifice at a rate of 10 m/s. Still photography is currently done by capturing a still video image but should be possible with a still camera and high-speed film pushed to its limits (e.g., Fuji P1600 rated at EI 6,400 to 12,800 and push-processed).

30.3.4. Electron Micrography (10, 23, 29, 32) (also see Chapter 3.3)

30.3.4.1. Practical considerations

The working and reference material of the electron microscopist consists of electron micrographs, and the proper taking and processing of the images is essential. Modern microscopes are designed for sheet films, although they can be adapted for glass plates or roll films (frequently used on older instruments). The cost of the new electron microscopes is so enormous that it is still commonplace to find aging units in use which utilize the older film systems. It would be wise to modify or update the film adapters to take the new sheet films. Glass plates, although still manufactured, are easily broken and require expensive and rigid storage facilities. The new sheet films are more easily filed and resist surface damage, but are a bit less dimensionally stable than glass. This stability is of no concern in the microscope, because the film base is completely supported within its carrier. The difficulty arises in a condenser enlarger, where heat from a tungsten bulb is absorbed by the film, which then buckles and changes its position in the optical system, resulting in a loss of focus. The use of alternative enlarger light sources to circumvent this problem was discussed in section 30.2.2.4. However, negative "popping" is largely avoidable if heat-absorbing glass and a well-designed negative carrier are used with a condenser system. Carlwen carriers are excellent, and their unique design combines negative support and handling convenience as well as adaptability to any enlarging system.

30.3.4.2. Films

Conventional factors influencing the choice of a film certainly have some bearing on the selection for use in transmission electron microscopy (TEM) (17). These factors include speed, resolution, contrast, and granularity. However, these considerations must be modified when considering a TEM film. Electrons interact with a photographic emulsion in a very different way from interactions of photons with films. Electrons strike the emulsion with much higher energy (50 to 125 kV) than photons (2 to 3 V) and therefore track randomly through the emulsion rather than being absorbed by a single silver grain. This electron tracking creates an effect visually similar to granularity, but it is, in fact, very different. Instead of a single silver grain being activated by many photons, one high-energy electron can activate many silver ions and create background graininess or "noise." It is the signal-to-noise ratio (S/N) which is unique to films exposed to electrons. Fortunately, excessive noise can be controlled by exposure and development modifications. Increasing the exposure and decreasing the development time improves the S/N because when more electrons are absorbed by the emulsion, the density and contrast rise linearly, while the random background noise increases far more slowly (10, 29). In this regard, it is advantageous to maximize the exposure time within the limits of specimen stability.

Kodak manufactures a variety of films for TEM (17). Electron Image glass plates and SO-163 Estar base films (SO refers to special order) have high speeds relative to 80-kV electrons and are easily modified with regard to S/N. Electron Microscope Film (no. 4489 in sheets and no. SO-281 in 35-mm and medium-format rolls) is the emulsion used for most routine work. This film has a micro-fine-grain structure and a thick emulsion designed to stop electrons, and exhibits intermediate speed and S/N when properly processed.

Some films, designed for other purposes, are suitable for TEM. Kodak recommends Eastman Fine Grain Release Positive Film 5302 for use as a direct positive film in a roll-film-adapted electron microscope. This material was designed for the motion picture industry for making contact copies with minimal quality loss. The film is very slow (the electron image speed is one-eighth that of 4489 EM film [three f-stops slower]) and exhibits very fine grain. Should a collaborator require a copy of an original electron micrograph negative, this film will enable the making of a suitable duplicate without risking the loss of a valuable, often irreplaceable, original.

Kodak Technical Pan Film, available in nearly all formats, is recommended for use with the unique fiber-optic exposure system found on the Zeiss EM-109 TEM. The superior resolution, ultrafine grain, and remarkable flexibility in contrast control make this emulsion a stellar performer for many specialized uses.

Scanning electron microscopy (SEM) involves internally recording a backscattered electron image on a scintillator plate, amplifying it through a video signal, and displaying it on a CRT. Polaroid films (type 52 and type 55) are usually the films of choice and a 90-s scan rate is the most common.

Note: Since film speed (ISO) is designated for pictorial use, it can be erroneous for CRT recording. Remember, just as with electron image films, the speed values for all uses must be determined by the specific application.

Kodak 4127 Commercial Film is excellent for this purpose since it is orthochromatic (i.e., it is sensitive to blue light and can be handled under a red safelight) and exhibits exceptionally fine grain. Kodak Technical Pan Film, with all of its inherent image contrast control, resolution, and nearly grainless structure, should be quite useful for SEM.

30.3.4.3. Exposure, processing, and printing (Tables 1 through 7)

For TEM the exposure conditions for electron image films must be carefully chosen by the microscopist. The appearance of the final image will be strongly affected by the S/N and the specimen stability and contrast. In addition, the exposure must be correlated to the relative energy of the electron source. Most published data for exposure recommendations assume an average energy for TEM to be 50 to 100 kV (17). If the microscopist is using a significantly different energy level, adjustments should be made to the exposure to account for the increased or decreased noise resulting from the higher or lower voltage. As with other films, the speed of the EM emulsion is measured by a characteristic density curve (19, 29). In fact, relative electron speeds were originally calculated from the slope of the low-density portion of the H and D curve. Since the curve is theoretically linear, the electron speed was rated as the reciprocal of the exposure in electrons per square millimeter required to produce a density of 1.0 above the base fog level. However, this determination did not adequately account for the S/N (17).

Currently, the relative electron speeds are determined by calculating the **detective quantum efficiency (DQE).** Kodak (17) defines the DQE as the ratio of the square of the S/N of the output to the square of the S/N of the input, as follows.

$$\mathrm{DQE} = \frac{(\mathrm{S/N})^2 \text{ of recorded image}}{(\mathrm{S/N})^2 \text{ electron beam}}$$

A perfect recording material with perfect processing would have a DQE of 1.0. Kodak has chosen an arbitrary number (100) as the standard relative speed for electron image films having a DQE of 0.5 at 80 kV and a density of 1.0 above minimum base fog level (10, 29). Increases in exposure and development can raise the electron speed rating and the DQE toward 1.0. Decreases can, similarly, lower the values.

Since contrast is always difficult to optimize with TEM, Kodak's D-19 is usually the developer of choice for electron image films. This is a high-contrast, high-energy, long-life solution originally designed for X-ray processing. For routine work, D-19 is diluted 1:2 and the electron image films are processed for 4 min at 68°F (20°C).

Note: Kodak frequently notates photographic dilutions differently from those in conventional chemistry. *The colon (:) actually refers to a plus (+).* Example: 1:2 indicates 1 part developer **plus** 2 parts water (which in normal scientific terms might be regarded as a 1:2 total dilution with 1 part developer and 1 part water).

Technical Pan Film is excellent when developed in HC-110, using various dilutions to take advantage of the remarkable ability of the film to vary contrast with controlled exposure and development. Using D-19 with Technical Pan will raise contrast levels to the extreme, approaching that of Kodalith.

In the SEM, exposure times are fixed by the scanning rate, usually 90 s. Most instruments have a setting for a higher scan rate to accommodate higher-speed films such as Polaroid type 53 (ISO 3000). Most SEM photography is done with Polaroid films, and the self-processing is uncomplicated. However, it is important to know the ambient room temperature to be certain that it is within the suggested range for processing. It is never detrimental to slightly overprocess Polaroid films, but underprocessing (normal time at a lower temperature) usually results in a solarized print and/or negative (1). The Sabatier effect, also known as solarization, is a phenomenon that occurs at transitions between developed and undeveloped silver grains and results in aberrant tonalities or image reversals between highlight and shadow areas. Kodak Technical Pan Film can be processed in various dilutions of HC-110 to provide precisely the contrast required for any application. Commercial Film 4127 should be processed in either HC-110 or D-19. The latter is ubiquitous in an electron microscopy laboratory.

30.3.4.4. Latent-image considerations

It should be a routine practice to process exposed electron microscopy films promptly (within a week). The latent image characteristics of 4489 EM Film tend to boost the image contrast excessively on prolonged storage after exposure unless the exposed plates are stabilized at −70°C.

30.3.4.5. Image enhancement (34, 35, 44, 49)

Specialized electron microscopes and computer-enhanced imaging are discussed in Chapter 3.1.4 and 3.1.5. The classical optical superposition techniques described by Horne and Markham (35) include rotational filtering and Markham averaging. Keep in mind that these techniques are relatively simple but usually inferior to numerical methods. Although it is possible to produce a filtered image in all cases, there has been some concern regarding the significance of the resulting structures. Statistics combined with computer analyses can assist in determining true significance (34, 44, 49).

a. Rotational filtering

The rotational-filtering technique involves superimposing the image of a particle n times and rotating it $360°/n$ with an exposure time of T/n with each rotation increment. The technique simply extracts structural information, making it more visible, but cannot add detail that was not resolved in the original micrograph (35).

b. Markham averaging

Markham averaging involves the superimposition of identical patterns to enhance structural detail. The choices of the images to be used must be selected from optical diffraction patterns, and the images must be physically aligned along the axis of a periodic structure. Multiple exposures, as with rotational filtering, will enhance any existing periodic structure.

30.3.5. Autoradiography (10, 15, 16)

Autoradiography entails the exposure of a photographic emulsion to radioisotopes and the subsequent imaging of their distribution. It has evolved from the optical level to the electron-microscopic level and is applicable to a wide variety of studies. At the electron-microscopic level, autoradiography permits monitoring of the fate or partition of substances labeled with radioactive isotopes. In molecular biology, it enables the delineation and localization of radioactive probes.

In the micro-autoradiographic range, liquid emulsion (Kodak nuclear track type NTB 3) is placed in contact with labeled sections of material on a microscope slide. The electron trails or grains are traced when the emulsion is stripped from the slide, developed, and observed at the microscopic level (optical or electron microscopy).

In the visible or macro-autoradiographic range, radioactively labeled specimens are placed in contact with X-ray films. The choice of film depends on the isotope used and the importance of exposure times. DuPont (Cronex) and Kodak (X-Omat) both manufacture radiographic films. Kodak makes an assortment of films with different characteristics, but the two which are most commonly used are SB Film and X-Omat AR. SB Film is a single-coated emulsion demonstrating medium speed, wide latitude, high resolution, and low background noise. This film is not compatible with rapid processors, and unless a long-cycle machine is available, manual processing is necessary. It is the film of choice for low-energy isotopes such as ^{3}H (0.018 MeV), ^{14}C (0.156 MeV), and ^{35}S (0.167 MeV), which have insufficient energy to completely penetrate any film incorporating an emulsion on both sides. X-Omat AR Film is a double-coated X-ray film compatible with rapid processing. Although the images on AR film do not have the resolution of those on SB, they provide increased sensitivity, especially with higher-energy β-emitters such as ^{45}Ca (0.256 MeV) and ^{32}P (1.710 MeV) as well as Γ-emitters such as ^{125}I (0.035 MeV). The additional emulsion layer on the second side of the film increases the efficiency (reduces the exposure times) and provides a basis for use with intensifying screens.

The photographic procedures for autoradiography are extensive and vary with each application. Kodak has published a selected bibliography on this subject (15), which is of immense value. The processing of X-ray films is generally automated in a rapid processor but can be performed manually by using the Kodak GBX series of chemicals or with D-19 developer and Rapid Fixer.

Intensifying screens provide a means of efficient image enhancement of short-half-life, high-energy β-emitters such as ^{32}P. These screens are phosphor-coated reflectors which convert absorbed energy and emit a spectral distribution of light to which the X-ray film is sensitive. When X-Omat Film is used, a screen can provide a second exposure to the two emulsion surfaces of the film: the first from the primary β-emission and the second from the blue or UV light emitted from the phosphor screen. Most intensifying screens utilize calcium tungstate and emit a relatively low-intensity light, which peaks at ca. 400 nm. A newer high-efficiency screen, DuPont Quanta III, is made with a rare earth phosphor, lanthanum oxybromide, and emits significantly (300%) more intense light in the 375- to 475-nm portion of the spectrum.

Fluorography is the chemical enhancement of autoradiographic images by impregnation with, or surface application of, fluorescent solutions to the medium containing the radioisotope. This technique is particularly useful with the weak β-emitters (tritium) and can increase sensitivity by as much as 10-fold. The most common scintillant is 1 M sodium salicylate (pH 6.0), but 2,5-diphenyloxazole and commercial preparations (e.g., New England Nuclear En³Hance) are also suitable. If fluorography is used, a film that responds to the blue and UV areas of the spectrum (Kodak SB) should be chosen.

Low-temperature exposure (−70°C) is suggested for use with intensifying screens and fluorography. The required exposure time is reduced significantly (especially with short half-life isotopes) because the lowered temperatures improve the stability of the atoms forming the latent image and also reduce the effects of low-intensity reciprocity failure (16).

Hypersensitization involves heating the film at 65°C in an 8% forming-gas environment (8% H_2, 92% N_2). The heating (baking) time must be determined empirically for each batch of film, but 8 to 16 h is usually sufficient. This technique effectively eliminates low-intensity reciprocity failure and increases sensitivity two- to fourfold.

Preflashing the film can enhance its responsiveness another twofold. This technique is analogous to the copy procedure described in section 30.3.1.1. By presensitizing the film to a brief, high-intensity, uniform, subdevelopable level of light (weak electronic flash), the silver grains in the emulsion will respond to a minimal additional exposure and become developable. The electronic flash should incorporate a diffuser (to allow for even illumination), neutral-density filtration (to lower the intensity), and a Kodak Wratten 22 filter (to reduce any incident blue light to which the film is sensitive). The amount of preflashing necessary will depend on the film used, the power of the flash unit, the amount of filtration used, and the distance of the flash to the film. Testing will provide the proper balance. Although this procedure produces some background fog, the autoradiographic image is enhanced disproportionately more. Kodak specifically notes that hypersensitization and preflashing effects are independent and additive since their mechanisms are different (26).

30.4. GLOSSARY

Accutance: a term used to describe the apparent sharpness of an image.

Acrostigmat: trademark for a specialized microscope manufactured by Zeiss.

Additive colors: see Primary colors.

Ambient: with reference to light, the amount existing in the environment.

Analyzer: an instrument used to measure, differentiate, or interpret. A color analyzer can measure the intensity and variation of each primary color relative to an established standard.

Aperture: an opening. The effective lens aperture describes the size of a lens diaphragm opening.

Apo: abbreviation for apochromatic. A lens able to focus the three primary wavelengths at one point.

Aspect ratio: the relative dimensions of a picture, document, or negative; 35-mm film measures 24 by 36 mm and has an aspect ratio of 2:3.

Aspheric (lens): a lens incorporating one or more elements that have been ground or cast in an aspherical configuration. The design is expensive but can correct spherical aberration or distortion.

Burning: additional exposure given to a selected area of a photographic material for the purpose of enhancement.

C-41: chemical process formulated by Eastman Kodak for the development of color negatives.

C-Print: color print made from a color negative.

Cathode ray tube (CRT): graphic display which produces an image by projecting electrons on a phosphor-coated surface in a vacuum tube. A monochrome or color monitor.

CdS: cadmium sulfide. A material used in vintage light meters. The material used senses light by absorbing specific wavelengths of light and emitting photons with a characteristic energy level.

Characteristic curve: see H and D curve.

Chemotaxis: cell migration induced by chemical stimulation.

Chromaticity: dominant color purity. Relates to the saturation of a wavelength (hue) without regard to brightness.

Chrome: referring to a slide or transparency, usually in color. Frequently used as a suffix indicating a transparency (for example, Ektachrome, Fujichrome).

Chromogenic: a black-and-white emulsion formed with dyes rather than silver ions and processed in standard C-41 color chemistry.

Cibachrome: trademark for a high-quality color R-print material manufactured by Ilford.

Color: a visual perception of the quality of an object resulting from the light reflected from it and measured by hue, saturation, and brightness. Frequently used as a suffix to denote the use of a color negative film (for example, Kodacolor, Agfacolor).

Color temperature: defines the color of light of a specific source relative to the visual appearance of a black body heated to incandescence. Measured with the Kelvin scale and specific for each light source.

Complementary colors: opposite colors. Can be either additive or subtractive (see Primary colors). Each color is a complement of the mixture of all the other colors.

Componon: trademark of Schneider Corporation for a series of high-quality enlarger lenses.

Coomassie brilliant blue: a stain used to detect proteins on polyacrylamide gels.

Copy work: to make copies by photographing two-dimensional objects (photos, graphs, etc.)

Cronex: DuPont trademark for radiographic (X-ray) films.

CRT: see cathode ray tube.

DAPI: 4′,6-Diamidino-2-phenylindole. A photofluor used in immunofluorescence techniques.

Dark-field illumination: a technique used in both photography and microscopy that renders the background darker than the object.

Depth of field: range of distances on the near and far side of the object photographed in which acceptable sharpness of the image is produced.

Developer: chemical used to enhance or complete the change produced on film through the formation of a latent image by light or electrons.

Dichroic filters: filters which are wavelength selective and designed for use with light at an incident angle of 45°. Cut-off filters which, when used in combination at various angles of incidence, can reflect unwanted wavelengths and transmit light of any hue and intensity.

Dodging: the reduced exposure given to a selected portion of a photographic material.

Download: the transmittal of information from one computer source to another.

DQE: detective quantum efficiency. A calculation used to determine the effective film speed of a material used to record electrons (see section 30.3.4.3).

E-6: a six-step chemical process formulated and marketed by Eastman Kodak Co. for the processing of Ektachrome (and similar) color transparencies. This process can be performed in any laboratory.

EES: The product designation for Kodak ultra-high-speed (ISO 400 to 1,600) Ektachrome film designed for push-processing.

EI: exposure index. A film speed value assigned by the photographer to adjust for variations in lighting or exposure based on a particular application. An "altered ISO."

Ektagraphic: trademark for a high-resolution copy film manufactured by Eastman Kodak.

Ektar: trademark for an ultrafine-grain high-resolution color negative film manufactured by Eastman Kodak.

Electronic flash: a high-intensity, short-duration burst of light generated by releasing a charge stored in an electronic capacitor through a pulsed xenon or similar tube. Usually balanced for a color temperature of 5,500 K.

Enlarger: a mechanical device used for the making of optically enlarged reproductions from films.

EPT: product designation for Kodak high speed (ISO 160) Ektachrome professional film which is balanced for tungsten light (3,200 K).

Estar: trademark for a film base material manufactured by Eastman Kodak.

EV: an antiquated term still used to describe the combination of lens aperture and shutter speed for a correct exposure on a particular film with lighting of a given intensity. A variation of 1 EV unit corresponds to a one f-stop change in either aperture or shutter speed or both.

Exposure (photographic): the subjecting of a film to a source of radiant energy that results in a chemical change in the emulsion and the formation of a latent image of the source.

Extension (lens): an increase in the distance of a lens from the film plane, resulting in an increase in image magnification and a loss of light falling on the film (requiring an exposure increase).

F-stop: ratio of the focal length of the lens to the size of the aperture of the lens diaphragm, irrespective of light transmission efficiency.

Film: coated and sensitized material (usually with silver) used to record images formed by electromagnetic waves.

Filter: a barrier. A photographic filter is designed to restrict or modify electromagnetic waves of a given wavelength. These are usually rated with regard to transmission efficiency. Interference or dichroic filters reflect unwanted wavelengths while those made from dyes absorb them.

Fixer: chemical used to stabilize or stop the development of a latent image.

Flare: the uncontrolled scattering of light from a lens surface resulting in decreased image contrast.

Flat-field: a designation for a lens design used to overcome field curvature in the image of the object being photographed. Most macro and enlarger lenses are flat-field designs to allow for the best quality image reproduction from two-dimensional objects.

Fluorescein: a fluorochrome (dye), fluorescein isothiocyanate, used to produce fluorescence in photomicrography.

Fluorescence: the ability of a material to absorb a quantum of radiation of one or a series of wavelengths and emit visible light at a secondary longer wavelength.

Fluorite: synthetic crystals of calcium fluoride. For apo-lenses, they are mechanically ground into optical elements having an index of refraction not obtainable in glass.

Fluorography: chemical enhancement of autoradiographs by using secondary fluorescence.

Format: negative size.

H and D curve: the characteristic curve used to describe the effect of light on light-sensitive materials. On the plot for the curve, the horizontal axis is represented by log E (exposure units) and the vertical axis is represented by D (density). The science of sensitometry was developed by Hurter and Driffield in the late 1890s, hence the "H and D curve."

H and E: hematoxylin and eosin double-staining technique used in microscopy.

Hue: basic attribute of color. The dominant wavelength of radiant energy in the visible portion of the spectrum.

Hypersensitization: a technique used for enhancing radiographs.

Hypo-Clearing Agent: trademark of Eastman Kodak for a product used to neutralize residual fixer left on films or photographic paper.

Image aspect ratio: the relative size (ratio) of the image height to width.
Incident light: light falling on an object. The color or brightness of an object is a perception resulting from the incident light reflected from or absorbed by it.
Intensifying screen: phosphor-coated reflector used to enhance radiographs.
Interference (filters): made of various coatings which create multiple and varied reflections of electromagnetic waves. If the reflected waves are optically coherent with one another, they can interfere and cause extinction of light.
Internegative: negative made from a transparency as an intermediate step in the production of a copy print.
IR: infrared. The portion of the electromagnetic spectrum beyond the visible red, usually 700 to 900 nm.
ISO: film speed (formerly ASA) assigned by the manufacturer in accordance with international standards.
K-14: a commercial process formulated by Eastman Kodak Co. to process Kodachrome film. This film is essentially three black-and-white emulsions which are exposed selectively to each of the three primary colors. In this very complicated 14-step process, stable color dyes are added to the film.
Kodalith: trademark for line copy materials (film and chemistry) manufactured by Eastman Kodak.
L-Lens: a trademark of Canon USA. Refers to the "luxury" lenses manufactured by Canon that incorporate low-dispersion glass, fluorite, or aspherical elements.
Lens: an optical or electromagnetic device, either simple or complex, used to focus electromagnetic waves.
Light meter: device used to measure the intensity of light and/or its color.
LPD-4: designation for a high-contrast line copy positive film manufactured by Eastman Kodak.
Lumiere: trademark for a series of high-quality low grain films manufactured by Eastman Kodak.
Macrophotography: term generally used to describe close-up or high-magnification photography without the use of a microscope. In this chapter the term is limited to image magnifications of 1:10 to 20:1.
Makro-Symmar: trademark of a flat-field lens manufactured by Schneider Corp.
Meter: see Light meter.
Methylene blue: a biological stain.
Microphotography: see Photomicrography.
Monochromatic: of or pertaining to a single wavelength or narrow band of wavelengths. A term frequently assigned to describe black-and-white film.
MP-4: trademark for a 4×5-in. copy camera manufactured by Polaroid Corp.
Negative: an image, on film, containing tonal values and/or hues that are the opposite of those that exist in nature. Used for the production of photographic prints.
Neofluar: trademark for a series of fluorite optics manufactured by Zeiss.
Noise: background exposure of silver ions creating graininess on film. Particularly troublesome with exposure to electrons that track randomly through an emulsion.
Orthochromatic: black-and-white photographic emulsions having a primary sensitivity to green light, although there is usually some sensitivity to blue and UV wavelengths but almost none to red.
OTF: off-the-film. Generally refers to a light meter, spot or averaging, that reads light reflected from the film plane.
Ouchterlony technique: a microprecipitation immunological technique.
Panchromatic: a film emulsion sensitive to all visible (and frequently beyond) electromagnetic radiation.
Photo-Flo: trademark by Eastman Kodak for a surface-active agent used to treat film surfaces and promote streak-free drying.
Photomicrography: photography through the microscope.
Polarization: technique of modifying light so as to transmit only that light oscillating in a specific plane. Used to eliminate stray reflections that tend to reduce contrast and to saturate color in photography.
Prefog (preflash): the technique of exposing film to a controlled intensity of light before exposing to the object being photographed. Reduces contrast and increases sensitivity.
Primary colors: A set of colors from which all other colors can be derived. The additive primaries are red, green, and blue (RGB), and the subtractive primaries are cyan, magenta, and yellow (CMY).
Print: an image on paper, a photographic print.
Push-process: term used to describe extended development times which are usually necessary to provide increased density when a film has been underexposed.
Radiograph: a negative image made on photographic film when exposed to X rays.
Reciprocity: the law of photographic reciprocity states that the exposure equals the illuminance on the film multiplied by the exposure time. Failure occurs because emulsions change sensitivity with extremes in time and/or brightness levels.
Reel: spiral device onto which film may be wound to permit uniform processing of all surfaces, generally designed for a specific film tank to ensure proper fit.
Reflection: light bounced from an object. A reflective meter measures the intensity of light reflected from an object irrespective of how much light is present.
Refraction: the change in direction of propagation of electromagnetic radiation caused on passage obliquely from one medium into a second medium in which the velocity of propagation is different from the first. Can be visualized as a color change.
Rheinberg differential coloration: a technique used in photomicrography to render objects a specific color against a background of a different color.
Rhodamine: a fluorochrome dye used for fluorescence in photomicrography.
R-print: a color print produced directly from a color transparency without the use of an internegative.
Sabatier effect: a phenomenon that occurs at transitions between developed and undeveloped silver grains and results in aberrant tonalities or image reversals between highlight and shadow areas. Sometimes erroneously referred to as solarization.
Safelight: a light used to illuminate an area where photographic emulsions are handled and to which the film is not sensitive.
SEM: scanning electron microscope.
Sensitometry: the science of measuring the responses of photographic emulsions to light. Established in the 1890s by Hurter and Driffield.

Shutter speed: the time that the shutter of a camera remains open to expose film.
Slide: a transparency made for projection.
Sonnar: trademark of a lens design manufactured by Zeiss.
Spotmeter: a light meter with a narrow angle field of view. Used to selectively evaluate reflective light values within the camera's viewing area.
Stop bath: a chemical used to rapidly stop the action of a developer and preserve the working life of a fixing bath.
Subtractive color: see Primary colors.
Tank: a receptacle designed to hold film reels or racks to facilitate processing.
Technical Pan: a trademark for a high-resolution, ultra-fine-grained black-and-white film manufactured by Eastman Kodak.
TEM: transmission electron microscopy.
Texas red: a fluorochrome used for fluorescence in photomicrography.
T-Max: trademark for a series of black-and-white films and developers manufactured by Eastman Kodak.
Toluidine blue: a biological stain.
Transparency: a photographic positive image in color or monochrome and viewed by transillumination.
TTL: through-the-lens. Generally refers to the location of a meter in a camera.
Tungsten light: light emitted from a bulb with a tungsten filament and having a specific color temperature.
Ultrafluar: trademark for a microscope objective made of quartz elements by Zeiss.
UV: ultraviolet light. Electromagnetic radiation below the visible range (300 to 400 nm).
Velvia: trademark of Fuji for an E-6-compatible high-quality transparency film.
Wratten: trademark used by Kodak to describe their color filters. C. Mees, the first director of Kodak's research laboratory, worked at Wratten and Wainwright Ltd. in England, hence the name.
Wright's stain: a biological stain consisting of methylene blue and eosin.
X-Omat: trademark of Eastman Kodak for radiographic emulsion.
X ray: extremely short wavelength electromagnetic radiation. Commonly used as a term to describe a radiograph.
Zone system: methodical approach to contrast and tone control in black-and-white photography. Utilizes selective exposure and development to render detail in highlights and shadows by altering the characteristic curve of the film to fit the conditions existing during exposure.

30.5. COMMERCIAL SOURCES

30.5.1. Materials

The letters following each of the photographic materials refer to the companies in the next section (30.5.2).

Analyzer (enlarging) (*h, r, s, aa*)
Cable release (*c, e, f, n, v, y, bb, gg*)
Camera (and lenses) (*f, s, v, y, gg, jj*)
Chemistry (processing) (*h, i, l, o, p, y, bb*)
Color chart (*i, p*)
Close-up calculator (*gg*)
Computer (for graphics) (*b*)
Copy stand (*e, y, bb*)
Dryer (film or print) (*c, n, p, w*)
Easel (enlarger) (*n, bb*)
Enlarger (*h, n, w, bb*)
Enlarger alignment device (*aa*)
Enlarger lenses (*n, v, cc*)
Filters (*e, f, h, i, m, n, bb, cc, dd, ee, ff, gg*)
Film (*a, i, k, o*)
Film clips (*n, p, w*)
Grain focusing device (*n, w, bb*)
Gray scale (step-tablet) (*i*)
Laser printer (*a, b, f, i*)
Light meter (*e, n, s, ff*)
Lights (photofloods) (*t*); (safelights) (*n, bb*); (UV) (*j*)
Negative carrier (*g, h, n, w, bb*)
Printing paper (*a, i, k, o, x*)
Processor (film and paper) (*a, i, p*)
Reels (developing) (*c, n, p*)
Slide projector (*i, n, cc*)
Slide-mounting device (*n, ii*)
Tank (processing) (*c, i, n, p*)
Thermometer (*i, n, bb*)
Timer (*h, n, w, aa, bb*)
Trays (development) (*n, bb*)
Tripod (*e, n, q, bb*)

30.5.2. Manufacturers

a. Agfa Corporation, 100 Challenger Road, Ridgefield Park, NJ 07660
b. Apple Computer Inc., 20525 Mariani Avenue, Cupertino, CA 95014
c. Argraph Corp., 111 Asia Place, Carlstadt, NJ 07072
d. Aristo Grid Lamps, 35 Lumber Road, Roslyn, NY 11576
e. Bogen Photo, 565 E. Crescent Avenue, Ramsey, NJ 07446
f. Canon USA, Inc., 1 Canon Plaza, Lake Success, Long Island, NY 11042
g. Carlwen Industries, P.O. Box 1748, Rockville, MD 20850
h. Charles Beseler Co., 1600 Lower Road, Linden, NJ 07036
i. Eastman Kodak Company, Rochester, NY 14650
j. Fotodyne, Inc., 16700 West Victor Road, New Berlin, WI 53151
k. Fuji Photofilm USA, Inc., 800 Central Blvd., Carlstadt, NJ 07072
l. Heico Marketing Corp., Route 611, Delaware Water Gap, PA 18327
m. Hoya Inc., 2180 Sand Hill Road, Menlo Park, CA 94025
n. HP Marketing Corp., 16 Chapin Road, Pine Brook, NJ 07058
o. Ilford Photo Corp., West 70 Century Road, P.O. Box 288, Paramus, NJ 07653
p. Jobo Fototechnic Inc., P.O. Box 3721, Ann Arbor, MI 48106
q. Karl Heitz Inc., 34-11 62nd Street, Woodside, NY 11377
r. Lektra SVC, 129-07 18th Avenue, College Point, NY 11356

s. Minolta Camera Company, 101 Williams Drive, Ramsey, NJ 07446
t. Nationwide Photolamp Supply Co., P.O. Box 206, Norfolk, VA 23501
u. New England Nuclear, 549 Albany Street, Boston, MA 02118
v. Nikon Inc., 623 Stewart Avenue, Garden City, NY 11530
w. Omega/Arkay, 1900 Hanover Pike, Hampstead, MD 21074
x. Oriental Photo Distributing Co., 3701 West Moore Avenue, Santa Ana, CA 92704
y. Polaroid Corporation, 575 Technology Square-9P, Cambridge, MA 02139
z. Research Organics, 4353 E. 49th St., Cleveland, OH 44125
aa. Salthill Photographic Instruments, Inc., Wildcat Road, Chappaqua, NY 10514
bb. Saunders Group, 21 Jet View Drive, Rochester, NY 14624
cc. Schneider Corporation, 400 Crossways Park Dr., Woodbury, NY 11797
dd. Schott Glass Technologies Inc., 400 York Avenue, Duryea, PA 18642
ee. Tiffen Mfg., 90 Oser Avenue, Hauppauge, NY 11788
ff. Uniphot Corp., 61-10 34th Avenue, Woodside, NY 11377
gg. Victor Hasselblad Inc., 10 Madison Road, Fairfield, NJ 07006
hh. Vivitar Corp., 9350 DeSoto Avenue, P.O. Box 2193, Chatsworth, CA 91313
ii. Wess Plastics, Inc., 70 Commerce Drive, Hauppauge, NY 11788
jj. Carl Zeiss Inc., 1 Zeiss Drive, Thornwood, NY 10594

The following two sources are Washington, D.C., metro area stores which will ship the above products nationwide:

Penn Camera Exchange, 915 E Street N.W., Washington, DC 20004 Fax (202) 347-6292

Photopro, 10630 Connecticut Avenue, Kensington, MD 20895 Fax (301) 949-8485

30.6. REFERENCES

1. **Adams, A.** 1978. *Polaroid Land Photography.* New York Graphic Society, Little, Brown & Co., Boston.
2. **Adams, A.** 1980. *The Camera.* New York Graphic Society, Little, Brown & Co., Boston.
3. **Adams, A.** 1981. *The Negative.* New York Graphic Society, Little, Brown & Co., Boston.
4. **Adams, A.** 1983. *The Print.* New York Graphic Society. Little, Brown & Co., Boston.
   The Ansel Adams series are classics in the photographic literature and are extremely valuable for the technically oriented photographer.
5. **Ausubel, F. M., R. Brent, R. E. Kingston, D. Moore, J. G. Seidman, J. A. Smith, and K. Struhl (ed.).** 1990. *Current Protocols in Molecular Biology,* vol. 1 and 2. John Wiley & Sons, Inc., New York.
6. **Baines, H., and E. S. Bomback (ed.).** 1971. *The Science of Photography.* John Wiley & Sons, Inc., New York.
7. **Berger, S. L., and A. R. Kimmel.** 1987. *Methods in Enzymology. Guide to Molecular Cloning Techniques.* Academic Press, Inc., San Diego, Calif.
8. **Beutner, E. H., R. J. Nisengard, and B. Albini (ed.).** 1983. *Defined Immunofluorescence and Related Cytochemical Methods. Ann. N.Y. Acad. Sci.* **420**.
9. **Biere, J.** 1990. *The Guide to Multi-Image.* Verlag Photographie AG, Schaffhausen, Switzerland.
   An excellent volume with valuable information.
10. **Clark Jones, R.** 1958. On the quantum efficiency of photographic negatives. *Photogr. Sci. Eng.* **2**:57.
11. **DeCock, L. (ed.).** 1991. *Photo Lab Index. The Cumulative Formulary of Standard Recommended Photographic Procedures.* Morgan and Morgan, Inc., Dobbs Ferry, N.Y.
12. **Delly, J. G.** 1980. *Photography through the Microscope.* P-2. Eastman Kodak Co., Rochester, N.Y.
    An excellent and descriptive reference volume.
13. **Driscoll, W. G., and W. Vaughan (ed.).** 1978. *Handbook of Optics.* McGraw Hill Book Co., New York.
14. **Eastman Kodak.** 1968. *Photography of Stained Specimens with Black-and-White Films.* N-8. Eastman Kodak Co., Rochester, N.Y.
15. **Eastman Kodak.** 1986. *Photographic Procedures for Light and Electron Microscope Autoradiography: A Selected Bibliography of Books and Papers.* M6-110. Eastman Kodak Co., Rochester, N.Y.
16. **Eastman Kodak.** 1987. *Autoradiography of Macroscopic Specimens.* M3-508. Eastman Kodak Co., Rochester, N.Y.
17. **Eastman Kodak.** 1987. Materials for TEM. *Tech Bits* **P-3**(3):4. Eastman Kodak Co., Rochester, N.Y.
18. **Eastman Kodak.** 1988. *Kodak Black-and-White Darkroom Data Guide.* Eastman Kodak Co., Rochester, N.Y.
19. **Eastman Kodak.** 1988. The Characteristic Curve. *Tech Bits* **P-3**(2):12. Eastman Kodak Co., Rochester, N.Y.
20. **Eastman Kodak.** 1989. *Kodak Color Darkroom Dataguide.* Eastman Kodak Co., Rochester, N.Y.
21. **Eastman Kodak.** 1989. *Pocket Photoguide.* Eastman Kodak Co., Rochester, N.Y.
22. **Eastman Kodak.** 1989. *The Prevention of Contact Dermatitis in Photographic Work.* J-4S. Eastman Kodak Co., Rochester, N.Y.
23. **Eastman Kodak.** 1989. *Choosing the Right Silver Recovery Method for Your Needs.* J-21. Eastman Kodak Co., Rochester, N.Y.
24. **Eastman Kodak.** 1990. *Handbook of Kodak Photographic Filters.* B-3. Professional Photography Division, Eastman Kodak Co., Rochester, N.Y.
25. **Eastman Kodak.** 1990. *Regulations Affecting the Disposal of Photographic Processing Solutions.* J-102. Eastman Kodak Co., Rochester, N.Y.
26. **Eastman Kodak.** 1990. *Disposal and Treatment of Photographic Effluent—In Support of Clean Water.* J-55. Eastman Kodak Co., Rochester, N.Y.
27. **Eastman Kodak.** 1990. *Safe Handling of Photographic Chemicals.* J-4. Eastman Kodak Co., Rochester, N.Y.
28. **Engel, C. E. (ed.).** 1968. *Photography for the Scientist.* Academic Press Ltd., London.
29. **Farnell, G. C., and R. B. Flint.** 1975. Photographic aspects of electron microscopy, p. 19. *In* M. A. Hayat (ed.), *Principles and Techniques of Electron Microscopy. Biological Applications,* vol. 5. Van Nostrand Reinhold Co., New York.
30. **Gassan, A.** 1981. *The Color Print Book.* Light Impressions Corp., Rochester, N.Y.
31. **Gibson, H. L.** 1970. *Close-up Photography and Photomacrography.* N-16. Eastman Kodak Co., Rochester, N.Y.
32. **Glauert, A. M. (ed.).** 1978. *Practical Methods in Electron Microscopy.* North-Holland, Amsterdam.
33. **Green, D., A. Marcus, A. H. Schmidt, and V. Gorter.** 1982. *The Computer Image: Applications in Computer Graphics.* Addison-Wesley Publishing Co., Inc., Reading, Mass.
34. **Hawkes, P. W.** 1978. Computer processing of electron micrographs, p. 262. *In* M. A. Hayat (ed.), *Principles and Techniques of Electron Microscopy. Biological Applications,* vol. 8. Van Nostrand Reinhold Co., New York.
35. **Horne, R. W., and R. Markham.** 1972. Applications of optical diffraction and image reconstruction techniques to electron micrographs, p. 327. *In* A. M. Glauert (ed.), *Practical Methods in Electron Microscopy,* vol. 1. North-Holland, Amsterdam.
36. **Hoya Corp.** (Yearly.) *Hoya Color Filter Glass.* Catalogue. Hoya Corp., Japan.

37. **Kawamura, A. (ed.).** 1977. *Fluorescent Antibody Techniques and Their Applications.* University Park Press, Baltimore.
38. **Kedersha, N. L.** 1991. Photomicrography of immunofluorescently labeled cells. *Am. Lab.* **23-6:**28.
39. **Krause, P., and H. Shull.** 1980. *The Complete Guide to Cibachrome Printing.* Ziff-Davis, New York.
40. **Lefkowitz, L.** 1985. *Polaroid 35mm Instant Slide System. A Users Manual.* Focal Press, Boston.
41. **Levitt, D., and M. King.** 1987. Methanol fixation permits flow cytometric analysis of immunofluorescent stained intracellular antigens. *J. Immunol. Methods* **96:**233.
42. **Mason, R. G. (ed.).** 1971. *Life Library of Photography.* Time-Life Books, New York.
    A useful series of 16 volumes on general and applied photography with superb graphics and concise ideas.
43. **Midoux, P., A. C. Roche, and M. Monsigny.** 1987. Quantitation of the binding, uptake and degradation of fluoresceinylated neoglycoproteins by flow cytometry. *Cytometry* **8:**327.
44. **Misell, D. L.** 1978. Image analysis, enhancement and interpretation. *In* A. M. Glauert (ed.), *Practical Methods in Electron Microscopy,* vol. 7. North-Holland, Amsterdam.
45. **Molbring, F. K.** *Microscopy from the Very Beginning. A Manual.* Carl Zeiss, Oberkichen, Germany.
46. **Nadler, B.** 1978. *The Color Printing Manual.* Amphoto, Garden City, N.Y.
47. **Nairn, R. C.** 1976. *Fluorescent Protein Tracing.* Churchill Livingstone, Edinburgh.
48. **Neblette, C. B.** 1973. *Photographic Lenses.* Morgan and Morgan, New York.
49. **Peters, K.-R.** 1977. Image reconstruction of electron micrographs by using equidensity integration analysis, p. 118. *In* M. A. Hayat (ed.), *Principles and Techniques of Electron Microscopy. Biological Applications,* vol. 7. Van Nostrand Reinhold Co., New York.
50. **Polaroid Corp.** 1990. *Photomacrography with the MP-4.* Polaroid, Cambridge, MA.
51. **Popkin, T. J., and E. Reiss.** 1974. Macrophotography of cell migration in an immune assay. *Photogr. Appl. Sci. Technol. Med.* **9:**28.
52. **Quesnel, L. B.** 1972. Photomicrography and macrophotography, p. 277. *Methods in Microbiology* vol. 7B.
    A good general reference. Film choices included may be outdated.
53. **Research Organics.** 1990. *Catalogue.* Research Organics, Cleveland, Ohio.
54. **Sambrook, J., E. F. Fritsch, and T. Maniatis.** 1989. *Molecular Cloning: A Laboratory Manual,* 2nd ed. Cold Spring Harbor Laboratory Press, Cold Spring Harbor, N.Y.
55. **Schaub, G.** 1991. *The Amphoto Book of Film* (1992 ed.). Amphoto, New York.
56. **Shaw, S.** 1991. *Overexposure: Health Hazards in Photography.* The Friends of Photography, Carmel, Calif.
    A valuable resource. Should be read and reviewed frequently.
57. **Sturge, J. M. (ed.).** 1977. *Neblette's Handbook of Photography and Reprography.* Van Nostrand Reinhold Co., New York.
58. **Todd, H. N., and R. D. Zakia.** 1969. *Photographic Sensitometry.* Morgan and Morgan, New York.
59. **Wildi, E.** 1987. *Medium Format Photography.* Focal Press, Stoneham, Mass.
60. **Wildi, E.** 1980. *The Hasselblad Manual.* Focal Press, London.
61. **Williamson, D. H., and D. J. Fennell.** 1975. The use of fluorescent DNA-binding agent for detecting and separating yeast mitochondrial DNA. *Methods Cell Biol.* **12:**335.
62. **Young, W. A., T. A. Benson, and G. T. Eaton.** 1985. *Copying and Duplicating in Black-and-White and Color.* M-1. Eastman Kodak Co., Rochester, N.Y.
    A superb reference volume.

Chapter 31

Records and Reports

PHILIPP GERHARDT

Scientific research is not complete until the results have been published. Therefore, a scientific paper is an essential part of the research process. Therefore, the writing of an accurate, understandable paper is just as important as the research itself. Therefore, the words in the paper should be weighed as carefully as the reagents in the laboratory. Therefore, the scientist must know how to use words. Therefore, the education of a scientist is not complete until the ability to publish has been established.

—ROBERT A. DAY (3)

The foregoing chapters provide detailed instructions on how to conduct laboratory methods; here limited suggestions are given on the recording and reporting of laboratory data. There are many literature resources about the writing of English prose, but here are suggested a few useful and inexpensive books specifically for writing about microbiology, plus a few general reference books. These should be at hand in the laboratory or easily accessible from a library for use, and they should be relied on for detailed guidance in recording and reporting laboratory findings.

31.1. LABORATORY RECORDS

Recording data in a laboratory notebook is an important but often slighted step between conducting experiments and reporting the results. Careful, proper, permanent records become the more important because of the possibility of patent applications and ethical concerns and for protection against fire and theft.

A generally useful and excellent little book, *Writing the Laboratory Notebook*, is published by the American Chemical Society (7). Here are a few tips from it and from my own experience.

Use an 8.5- by 11-in. case-bound notebook with tear-out, numbered pages that give carbon copies of everything written in them. Write with a black, fine ballpoint pen, legibly and grammatically so that you or another person can understand the meaning months or years later. Use the active voice in the first person. For each experiment have a descriptive heading, an introduction that includes the purpose, the procedure, the results, and a discussion that includes conclusions. Complete the entry immediately after the work is performed, and then date and sign it. Finally, remove the carbon copies and store them separately in a safe place.

The disclosure of an invention for a possible patent claim imposes especially stringent requirements. The following rules for record keeping are adopted from the publication *Patents and the University Inventor* (9).

1. Use bound notebooks for records, making entries on a daily basis. This "diary" format provides a day-to-day chronology.
2. Use the notebook to record a conception (a complete description of a means to accomplish a particular purpose or result), laboratory data, and drawings. Each entry should be headed with a title and continued on successive pages.
3. Make entries in ink, and do not erase. Draw a line through text or drawings to be deleted, and enter the material in corrected form. Draw a diagonal line through blank spaces on the page.
4. If separate sheets and photographs are pasted to notebook pages, refer to them in an entry. Material that cannot be incorporated in the notebook should be keyed to an entry.
5. Sign and date all entries at the time they are made, and have them witnessed by someone who has read the material and is capable of understanding it but had nothing to do with producing it. Secure additional witnesses when something important or highly unusual is discovered. Remember that an inventor and coinvestigators cannot serve as their own witnesses. Arrange to have two or more colleagues serve as witnesses on a regular basis, and offer reciprocal services.

31.2. WRITTEN RESEARCH REPORTS

Microbiology is fortunate in having a premier book on scientific writing, *How To Write and Publish a Scientific Paper*, authored by a former managing editor of the American Society for Microbiology (ASM), Robert A.

Day (3). The book is in its third edition; it is full of microbiological examples, is presented in "how to" cookbook style, and is humorously enjoyable to read. Consequently, go directly to Day's book for details.

31.2.1. Preparation and Writing

Published research reports in microbiology are organized into a highly structured format of parts including an Introduction, Methods, Results, and Discussion. This "IMRAD" format (originated by Pasteur) is supplemented with a title, authorship, abstract, acknowledgments, references, tables, and illustrations. How to prepare and write each of these component parts is the subject of a separate chapter in Day's book (3). Reading (or rereading) each of these chapters before starting a research report is highly recommended, even with advice from a mentor or colleague or with experience in report writing. Written directions are likely to be more complete and give time for reflection about what is to be done and why it is needed.

In my own experience, drafting the Methods first helps to get started because this part is the easiest to write. Drafting the title and abstract is suggested next, since these have to convey the essentials of a report with minimal but clear wording.

31.2.2. Publication

Publication of a primary research report involves properly preparing a manuscript, submitting it to an appropriate journal, dealing with external referees' and editor's review comments, dealing with printed galley proofs, and ordering and using reprints. Here, too, if inexperienced, consult Day's chapters (3) on each subject as it arises.

Each journal has its own publication format. Before starting a manuscript, carefully read the intended journal's Instructions to Authors (usually published in the journal's first issue each year) and examine a recent issue of that journal. In addition, consult a manual such as the *ASM Style Manual for Journals and Books* (1), which is used by copyeditors at ASM but is equally valuable to authors. For example, it lists and illustrates the marks that a copyeditor uses to direct the printer's typesetting; shows how to deal with numbers and measurements; and describes scientific nomenclature for chemistry, organisms, genetics, and immunology. Another excellent and more extensive resource is the *CBE Style Manual* published for the Council of Biology Editors (2), which is used by most biological journals.

31.2.3. Other Written Reports

Quite different formats apply to written reports other than original research papers, e.g., scientific reviews, conference reports, book reviews, and academic theses. Each of these has a special purpose and audience. Day's book (3) has separate chapters on how to write each of these.

31.3. ORAL PRESENTATIONS

Telling an audience about research involves much the same organization as writing a report, except that the presentation of the methods and citation of the literature should be reduced to bare essentials. The audience is usually more diverse, so the presentation should be more general and the introduction especially should be directed at the least informed listeners.

The problems in an oral presentation are usually those of ineffective communication. A paper that is read from script is rarely effective; learn to speak extemporaneously with the help of cue cards and projection slides. Illegible slides are the most common failure at a scientific meeting, and slides usually can be tested only after they are made. Some years ago Marvin Johnson at the University of Wisconsin invented a quick, reliable, three-finger method for evaluating the legibility of slide material before the slides are prepared. Here's how to do it.

1. Prepare a mockup of the table or figure on paper tacked onto a wall or on the screen of a computer.
2. Move back from the wall or computer with an arm extended until three fingers held upright and together just block out one-eyed vision of the width of the table or figure.
3. Lower the arm, open both eyes, and observe whether the mockup is easily legible at that distance. If so, proceed to make the slide with confidence that it will project well in a typical meeting room with a grateful audience. If not so, redesign and return to step 1.

Poor speaking is the next most common failure at a scientific meeting; it arises from insufficient attention to the requirements and from insufficient practice. Face the audience, and make eye contact for three full seconds with a sequence of individual listeners in different parts of the room. Speak measuredly, neither too slowly nor too rapidly, and loudly enough to be heard. Aim to tell a story, not to preach a sermon or give a speech. Use your face, hands, and body to support what you are saying. Use complete sentences, the shorter the better. Consciously learn to avoid "ah" and "you know." Whenever possible beforehand, practice an oral presentation with critical colleagues, and record it on tape for self-criticism.

Unfortunately, I know of no really useful book on scientific speaking. Day's book (3) has a short chapter on the subject. My best advice on how to learn to speak effectively is to listen, observe, and analyze how expert communicators perform.

31.4. POSTER PRESENTATIONS

Poster sessions are becoming increasingly popular at scientific meetings. Posters not only enable scientists to inform others about their research (as do oral presentations) but also provide the opportunity to meet personally with others who share interest in the field. Novice scientists especially are more comfortable with posters than with oral presentations.

As in the above section on oral presentations, resource books about poster presentations are in short supply, and Day's chapter is brief (3). The best tips I've seen were in a news article in *The Scientist* (5), adapted as follows.

1. Prepare a banner in very large type containing a descriptive title, the authors, and their affiliations. This banner should be situated high up on the

poster so it can be seen above people's heads from a distance of 15 to 20 ft.

2. Bracket the poster with an introduction at the left and a list of conclusions at the right. Place them at eye level. Remember that many people will read only these two parts of your poster. Use large type that will be easily readable from several feet away. Use lowercase (not all capitals) lettering for the text.
3. Make the flow of information in a poster explicit by numbering each statement. Organize the information in columns running down the poster, not in rows running across it.
4. The poster should be self-explanatory, so that its main points will be communicated even if you are not there. But don't load it down with large amounts of methodological detail or lists of references. Curious observers can ask you about these things directly.
5. Each illustration should have a prominent headline containing its take-home message in just a few words. The text below the illustrations should be in smaller type and should contain far more information than the typical figure legend. Only the most interested readers will spend time with this text.
6. Prepare oral comments of no more than five minutes (preferably two to four minutes) to guide interested parties quickly through your poster.
7. Make the poster well in advance, and practice the oral comments with your colleagues, much as you would practice an oral presentation.
8. Taking into account Murphy's Law, bring extra push pins (not thumbtacks) or Velcro tabs with you to the meeting. Also consider making up two complete copies of the poster. Mail one copy ahead or send it with a friend.
9. At the poster session, let people peruse your poster for 30 seconds or a minute before approaching them to ask if you may guide them through it. But don't be shy about introducing yourself, since the opportunity to meet people is one of the major advantages of poster sessions.
10. If you have a preprint of a publishable article already prepared, consider having a supply ready at the poster session to hand out to, or to be picked up by, people who are especially interested. If not, take down names and addresses and offer to send the preprint when it is ready.

31.5. ENGLISH USAGE

Scientific reporting ultimately depends on clarity, simplicity, and logic in use of the English language, which has become the universal language of science. The last chapters of Day's book (3) are directed at the use and misuse of English, avoidance of jargon, and the use of abbreviations. Now Day has produced a second and complementary book, *Scientific English: A Guide for Scientists and Other Professionals* (4), which I also recommend unreservedly for microbiologists, as did another reviewer (6). This second book is essentially a reference source on writing and grammar. Here you can find out whether to use "which" or "that," when to split an infinitive, where and how to punctuate, what words to avoid, etc.

There are a few other reference books I recommend for ready access. Certainly every scientific writer needs and probably has at hand an ordinary dictionary. Less common but equally useful is a thesaurus (a dictionary of synonyms, antonyms, related and contrasted words, and equivalent phrases). Roget's classic thesaurus has been superseded by *Webster's Collegiate Thesaurus* (8). A department library should afford Singleton and Sainsbury's *Dictionary of Microbiology and Molecular Biology* (10).

31.6. REFERENCES

1. **American Society for Microbiology.** 1991. *ASM Style Manual for Journals and Books.* American Society for Microbiology, Washington, D.C.
This excellent book is directed specifically toward copyediting of manuscripts for ASM books and journals but is useful generally. It contains the following chapters: Numbers and Measurement; Scientific Nomenclature; English; Sources for Materials; Abbreviations; References; Illustrations; Tables; Proofreading; Books; and Words, Abbreviations, and Designations.
2. **Council of Biology Editors Style Manual Committee.** 1983. *CBE Style Manual,* 5th ed. Council of Biology Editors, Inc., Bethesda, Md.
This outstanding book has long served as the standard reference for manuscript preparation and copyediting by the majority of biological journals. It contains the following chapters: Ethical Conduct in Authorship and Publications; Planning the Communication; Writing the Article; Prose Style for Scientific Writing; References; Illustrative Material; Editorial Review of Manuscripts; Application of Copyright Law; Manuscript into Print; Proof Correction; Indexing; General Style Conventions; Style in Specialized Fields (including microbiology); Abbreviations and Symbols; Word Usage; Secondary Services for Literature Searching; and Useful References with Annotations.
3. **Day, R. A.** 1988. *How To Write and Publish a Scientific Paper,* 3rd ed. Oryx Press, Phoenix, Ariz.
There are 30 chapters, which may be grouped into main topics: 2 chapters on the nature and origins of scientific writing; 1 chapter on the definition and organization of a scientific paper; 12 chapters on how to write or prepare each of the component parts of a paper; 4 chapters on the publishing process; 6 chapters on special kinds of reports, 1 chapter on ethics; 3 chapters on language use; and 1 chapter in summary. There also are three appendices on abbreviations, two appendices on language misuse, and a glossary of technical terms. Available paperbound.
4. **Day, R. A.** 1992. *Scientific English: A Guide for Scientists and Other Professionals.* Oryx Press, Phoenix.
This is a useful, readable, concise book on how scientists should write the English language simply, logically, and clearly. There are 19 chapters on using words, phrases, clauses, sentences, and paragraphs. Three appendices cover Punctuation Presented Plainly, Problem Words, and Words To Avoid. There are also lots of illustrative quotations and a few sprightly cartoons. Available paperbound.
5. **Finn, R.** 1993. Poster sessions can lead to networking opportunities. *The Scientist,* January 25, p. 20–21.
6. **Isenberg, H. D.** 1992. Book review of *Scientific English: A Guide for Scientists and Other Professionals. ASM News* **58:**570.
7. **Kanare, H. M.** 1985. *Writing the Laboratory Notebook.* American Chemical Society, Washington, D.C.
An excellent little handbook. Many examples are shown. Included are chapters on legal and ethical aspects, patents and invention protection, and electronic notekeeping.
8. **Merriam-Webster.** 1976. *Webster's Collegiate Thesaurus.* Merriam-Webster Inc., Springfield, Mass.
Compared with Roget's classic thesaurus, this is a totally

reorganized and user-friendly dictionary of synonyms, antonyms, idiomatic phrases, and related and contrasted words. Available paperbound.

9. **Research Corporation.** 1984. *Patents and the University Inventor.* Research Corporation, New York.
This brochure contains five precise and concise articles about patents for academics, as follows: Publish or Patent—or Both? Records Can Win or Lose Patents. How to Describe an Invention. Will the Real Investor(s) Stand Up? How Good Is a Patent?

10. **Singleton, P., and D. Sainsbury.** 1988. *Dictionary of Microbiology and Molecular Biology,* 2nd ed. John Wiley & Sons, Inc., New York.
A useful encyclopedia of terms, tests, techniques, taxa, and other topics. Entries range from short definitions to concise reviews. Expensive.

INDEX